Math Study Skills

Your overall success in mastering the material in your course depends on you. You must be ***committed*** to doing your best in this course. This commitment means dedicating the time needed to study math and to do your homework.

In order to succeed in math, you must know how to study it. The goal is to study math so that you understand and not just memorize it. The following tips and strategies will help you develop good study habits.

General Tips

ATTEND EVERY CLASS Be on time. If you must miss class, be sure to talk with your instructor or a classmate about what was covered.

MANAGE YOUR TIME School, work, family, and other commitments place a lot of demand on your time. To be successful, you must be able to devote time to study math every day. Writing out a weekly schedule that lists your class schedule, work schedule, and all other commitments with times that are not flexible will help you to determine when you can study. Use the companion resources that accompany the book, such as MyMathLab, so you can study online with tutoring help whenever you have the time.

Instructor Contact Information

Name:

Office Hours:

Office Location:

Phone Number:

E-mail Address:

Campus Tutoring Center

Location:

Hours:

DO NOT WAIT TO GET HELP If you are having difficulty, get help immediately. Since the material presented in class usually builds on previous material, it is very easy to fall behind. Ask your instructor if he or she is available during office hours or get help at the tutoring center on campus.

Tests

Tests are a source of anxiety for many students. Being well prepared to take a test can help ease anxiety.

BEFORE A TEST

- ❒ Review your notes and the sections of the textbook that will be covered on the test.
- ❒ Read through the Key Concepts and Skills in the textbook and your own summary from your notes.
- ❒ Do additional practice problems. Select problems from your homework to rework. The textbook also contains Mixed Practice and Review Exercises that provide opportunities to strengthen your skills. You can also review your understanding of concepts by completing the Say Why problems in the text and using the Study Plan in MyMathLab.
- ❒ Use the Posttest at the end of the chapter as your practice test. While taking the practice test, do not refer to your notes or the textbook for help. Keep track of how long it takes to complete the test. Check your answers with the Chapter Test Prep Videos on YouTube or in MyMathLab for any problems you did not answer correctly. If you cannot complete the practice test within the time you are allotted for the real test, you need additional practice in the tutoring center or in MyMathLab to speed up.

DURING A TEST

- ❒ Read through the test before starting.
- ❒ If you find yourself panicking, relax, take a few slow breaths, and try to do some of the problems that seem easy.
- ❒ Do the problems you know how to do first, and then go back to the ones that are more difficult.
- ❒ Watch your time. Do not spend too much time on any one problem. If you get stuck while working on a problem, skip it and move on to the next problem.
- ❒ Check your work, if there is time. Correct any errors you find.

AFTER A TEST

- ❒ When you get your test back, look through all of the problems.
- ❒ On a separate sheet of paper, do any problems that you missed. Use your notes and textbook, if necessary.
- ❒ Get help from your instructor or a tutor if you cannot figure out how to do a problem, or set up a meeting with your study group to go over the test together. Make sure you understand your errors.
- ❒ Attach the corrections to the test and place it in your notebook.

Best wishes in your studies!
— Geoffrey Akst   Sadie Bragg

FORM A STUDY GROUP A study group provides an opportunity to discuss class material and homework problems. Find at least two other people in your class who are committed to being successful. Exchange contact information and plan to meet or work together regularly throughout the semester either in person or via e-mail, MyMathLab, or phone.

USE YOUR BOOK'S STUDY RESOURCES There are additional resources and support materials available with this book to help you succeed. See the list below and in the preface.

Notebook and Note Taking

Taking good notes and keeping a neat, well-organized notebook are important factors in being successful. If you do your homework online through MyMathLab, you should still keep a notebook to stay organized.

YOUR NOTEBOOK Use a loose-leaf binder divided into four sections: (1) notes, (2) homework, (3) graded tests (and quizzes), and (4) handouts. Or combine the resources in MyMathLab with the MyWorkBook with Chapter Summaries.

TAKING NOTES

- ❒ Copy all important information. Also, write all points that are not clear to you so that you can discuss them with your instructor, a tutor, or your study group.
- ❒ Write explanations of what you are doing in your own words next to each step of a practice problem.
- ❒ Listen carefully to what your instructor emphasizes and make note of it.

The following resources are available in MyMathLab, through your college bookstore, and at **www.pearsonhighered.com:**

- Student's Solutions Manual
- Video Resources with Chapter Test Prep Videos
- MyMathLab
- MyWorkBook with Chapter Summaries

Full descriptions are available in the preface.

Class Time

BEFORE CLASS

- ❒ Review your notes from the previous class session.
- ❒ Read the section(s) of your textbook that will be covered in class to get familiar with the material. Read these sections carefully. Skimming may result in your not understanding some of the material and your inability to do the homework. If you do not understand something in the text, reread it more thoroughly or seek assistance.

DURING CLASS

- ❒ Pay attention and try to understand every question your instructor asks.
- ❒ Take good notes.
- ❒ Ask questions during class if you do not understand something. Chances are that someone else has the same question, but is not comfortable asking it. If you feel that way also, then write your question in your notebook and ask your instructor after class or see the instructor during office hours.

AFTER CLASS

- ❒ Review your notes as soon as possible after class. Insert additional steps and comments to help clarify the material.
- ❒ Reread the section(s) of your textbook, focusing on the Example and Practice side by sides and the Tips. After reading through an example, cover it up and try to do it on your own. Do the practice problem that is paired with the example. (The answers to the practice problems are given in the back of the textbook.) The Tips will help you avoid common errors and provide other suggestions to foster understanding.
- ❒ Work in MyMathLab and use the personalized study plan.

Homework

The best way to learn math is by doing it. Homework is designed to help you learn and apply concepts and to master certain skills. Some tips for doing homework are

- ❒ Do your homework the same day that you have class. Keeping up with the class requires you to do homework regularly rather than cramming right before tests.
- ❒ Review the section of the textbook that corresponds to the homework.
- ❒ Review your notes.
- ❒ If you get stuck on a problem, look for a similar example in your textbook or notes or use Help Me Solve This in MyMathLab.
- ❒ Write a question mark next to any problems that you just cannot figure out. Get help from your instructor or the tutoring center, or call someone from your study group.
- ❒ Check your answer after each problem. The answers to the odd-numbered problems are in the back of the textbook. If you are assigned even-numbered problems, try the odd-numbered problem first and check the answer. If your answer is correct, then you should be able to do the even-numbered problem correctly. If doing your homework in MyMathLab, keep your work and repeat problems until you can work through them successfully.

Developmental Mathematics

through Applications

Basic College Mathematics and Algebra

GEOFFREY AKST • SADIE BRAGG

Borough of Manhattan Community College, The City University of New York

PEARSON

Boston Columbus Indianapolis New York San Francisco Upper Saddle River

Amsterdam Cape Town Dubai London Madrid Milan Munich Paris Montréal Toronto

Delhi Mexico City São Paulo Sydney Hong Kong Seoul Singapore Taipei Tokyo

Editorial Director: Christine Hoag
Editor in Chief: Maureen O'Connor
Executive Content Editor: Kari Heen
Content Editors: Katie DePasquale, Katherine Minton
Assistant Editor: Rachel Haskell
Senior Managing Editor: Karen Wernholm
Senior Production Supervisor: Ron Hampton
Associate Director of Design: Andrea Nix
Senior Cover Designer: Barbara T. Atkinson
Senior Technical Art Specialist: Joseph K. Vetere
Image Manager: Rachel Youdelman
Text Design: Leslie Haimes
Composition: PreMediaGlobal
Media Producer: Aimee Thorne
Software Development: TestGen: Mary Durnwald; MathXL: Jozef Kubit
Executive Marketing Manager: Michelle Renda
Marketing Manager: Rachel Ross
Procurement Manager/Boston: Evelyn Beaton
Procurement Specialist: Debbie Rossi
Media Procurement Specialist: Ginny Michaud
Cover Photo: Bamboo on white © Subotina Anna/Shutterstock

Library of Congress Cataloging-in-Publication Data
Akst, Geoffrey.
Developmental mathematics through applications / Geoffrey Akst, Sadie Bragg.—1st ed.
p. cm
Includes index.
ISBN-13: 978-0-321-82604-6 ISBN-10: 0-321-82604-3 (student ed. : alk. paper)
1. Mathematics—Textbooks. I. Bragg, Sadie II. Title.
QA152.3.A468 2014
512.9—dc23 2012016617

4 2020

pearsonhighered.com

Photo Credits

1, Yakov Lapitsky; 11, left, Lightwise/123RF; 11, right, Gary Buss/Getty Images; 11, bottom, Al Behrman/AP Images; 12, top left, Shutterstock; 12, bottom left, Steve Herrmann/Shutterstock; 12, bottom right, Karin Hildebrand Lau/Shutterstock; 20, Keren Su/AP Images; 29, Jack Smith/AP Imges; 29, bottom, Angus Oborn/Dorling Kindersley, Ltd.; 41, Jochen Sand/Getty Images; 63, NASA; 70, bottom left, Stephen Frink/Getty Images; 70, bottom right, Library of Congress Prints and Photographs Division [LC-USZC4-13287]; 71, AP Images; 77, NASA Headquarters; 78, Richard Nowitz/Getty Images; 81, Archivs du 7e Art/Columbia Pictures/DR/The Image Works; 107, top right, Mark J. Terrill/AP Images; 107 bottom right, Claudiofichera/Shutterstock; 157, Laurence Gough/Shutterstock; 167, Ryan McVay/Getty Images; 169, Getty Images; 169, bottom right, Shutterstock; 170, Library of Congress; 171, Getty Images; 172, Pearson Education, Inc.; 186, Tom Brakefield/Getty Images; 203, Jeffrey M. Frank/Shutterstock; 203, right, Petr Masek/Shutterstock; 209, Rolf Nussbaumer/Alamy; 212, NASA; 213, Tatiana Morozova/Shutterstock; 222, Marek Szumlas/Shutterstock; 230, bottom left, Dreamstime; 230, bottom right, Stacy Gold/Getty Images; 236, Shutterstock; 240, bottom left, Stockbyte/Getty Images; 240, bottom right, Jake Schoellkopf/AP Images; 245, left, Jack Hollingsworth/Photodisc/Getty Images; 245, right, National Weather Service; 247, AP Images; 248, Purestock/Getty Images; 29, Jean-Pierre De Mann/Robery Harding Picture Library Ltd./Alamy; 268 Stockbyte/Getty Images; 255, Mark McClare/Shutterstock; 267, left, Don Farral/Getty Images; 267, right, Wavebreak Media Ltd/123RF; 272, right, AF Archive/Alamy; 274, left, PhotoLink/Getty Images; 274, right, Jon Helgason/Alamy; 275, Evan El-Amin/Shutterstock; 279, Jason Straziuso/AP Images; 283, Steve Allen/Stockbyte/Getty images; 289, Keith Wheatley/Shutterstock; 295, Dewitt/Shutterstock; 299, Dallas Events Inc/Shutterstock; 303, Steve Reed/Shutterstock; 312, Mypokcik/Shutterstock; 316, Jose Gil/Shutterstock; 321, Prasit Rodphan/123RF; 327, Ken Durden/Shutterstock; 332, DDCoral/Shutterstock; 348, Rod Catanach/KRT/Newscom; 350, Alvaro Pantoja/Shutterstock; 356, Mike Dotta/Shutterstock; 364, Rubberball/Getty images; 400, Detlev van Ravenswaay/Photo Researchers, Inc.; 407, left, Portrait of Lisa Gherardini ("Mona Lisa") (1503–1506), Leonardo da Vinci. Oil on poplar wood, 77 × 53 cm. Collection of the Louvre Museum; Purchased by François I in 1518 [Inv. 779]. Photograph by Stuart Dee/Photographer's Choice/Getty Images; 407, right, Horst Schafer/AP Images; 409, Library of Congress Prints and Photographs Division [LC-DIG-ppmsca-19926]; 410, Interfoto/Personalities/Alamy; 421, Stockbyte/Getty Images; 427, Shutterstock; 431, Juice Images170/Alamy; 433, Ragne Kabanova/Shutterstock; 457, NASA; 476, Jim DeLillo; 483, Harris Shiffman/Shutterstock; 509, Nucleus Medical Art, Inc./Alamy; 530, Timothy A. Clary/AFP/Getty Images; 550, Alexander Raths/Shutterstock; 554, left, Mark J. Terrill/AP Images; 578, Mauriehill/Dreamstime; 579, Stephen Coburn/Shutterstock; 593, Tselichtchev/Shutterstock; 642, Richard Drew/AP Images; 658, Pearson Education, Inc.; 659, Galina Barskaya/Shutterstock; 749, Dora Modly-Paris/Shutterstock; 789, StockLite/Shutterstock; 792, R. Nagy/Shutterstock; 794, Randall Fung/Corbis RF/Alamy; 801, left, Devi/Shutterstock; 801, right, Irina Tischenko/Shutterstock; 806, Mary Evans Picture Library/Alamy; 818, Pearson Education, Inc.; 825, Gordana Sermek/Shutterstock; 831, top right, Scott Camazine/Alamy; 831, bottom left, Sebastian Kaulitzki/Shutterstock; 842, Fotomilan011/Dreamstime; 843, DB Travel/DB Images/Alamy; 848, Joyce/Dreamstime; 851, left, Egomezta/Dreamstime; 851, right, U.S. Navy photo by Mass Communication Specialist 1st Class Chad Runge; 866, Pearson Education, Inc.; 868, Iorboaz Dreamstime; 887, left, Renata Sedmakova/Shutterstock; 887, right, Siamionau Pavel/Shutterstock; 891, Frederick News Post/Marny Malin/AP Images; 913, left, Ricknoll/Fotolia; 913, right, David Lee/Shutterstock; 932, NASA; 936, Gene Krebs/iStockphoto; 940, top left, Sean Pavone Photo/Shutterstock; 940, bottom left, IntraClique LLC/Shutterstock; 945, Jupiterimages/BananaStock/Thinkstock; 948, Triff/Shutterstock; 949, Brandon Bourdages/Shutterstock; 957, Bernhard Classen/Alamy; 985, Essxboy/iStockphoto; 991, Jose Gil/Shutterstock; 993, left, Jerry Horbert/Shutterstock; 993, right, Dennis Hallinan/Alamy; 1001, Sean Pavone Photo/Shutterstock; 1006, Grenland/Shutterstock; 1019, Roy Johnson/Alamy; 1020, Leshik/Shutterstock; 1030, Copestello/Shutterstock; 1047, Bill Fehr/Shutterstock; 1055, Vitalii Nesterchuk/Shutterstock; 1055, bottom, Oleg Golovnex/Shutterstock; 1066, Photos12/Alamy; 1067, Michaeljung/Shutterstock; 1078, Photo Researchers/Science History Images/Alamy Stock Photo; 1086, Sherri R. Camp/Shutterstock; 1095, Radu Razvan/Shutterstock

Contents

19. Quadratic Equations 1067

Appendixes

Preface

FROM THE AUTHORS

This text is intended for college students in a multiterm course sequence that covers basic mathematics and algebra. Alternatively, it may be used in an accelerated one-term course.

Our goal in writing *Developmental Mathematics through Applications: Basic College Mathematics and Algebra* was to create a text that would help students establish a strong foundation in basic mathematics and algebra for success in their following mathematics and mathematics-related courses.

Throughout, we emphasize an applied approach, which has two advantages. First of all, it can help students prepare to meet their future mathematical demands—in subsequent coursework, in everyday life, and on the job. Secondly, this approach can be motivating, convincing students that mathematics is worth learning and more than just a school subject.

We have attempted to make the text readable, with explanations that help students by fostering understanding and with sufficient exercises for honing skills. We have also put together a set of easy-to-grasp features, consistent across sections and chapters.

In an effort to address many of the issues raised by national professional organizations, such as AMATYC, NCTM, and NADE, we have been careful to stress applications that model real-world situations across many disciplines, the appropriate use of technology, and the integration of geometric visualization. We have included exercises in student writing, collaborative-learning exercises to encourage student interaction, and quantitative reasoning exercises involving the interpretation of tables and graphs.

Above all, we have tried to develop a flexible text that can meet the needs of students in both traditional and redesigned developmental courses, and that reflects our belief that mathematics is understandable and useful.

This text is part of the *through Applications* series that includes the following:

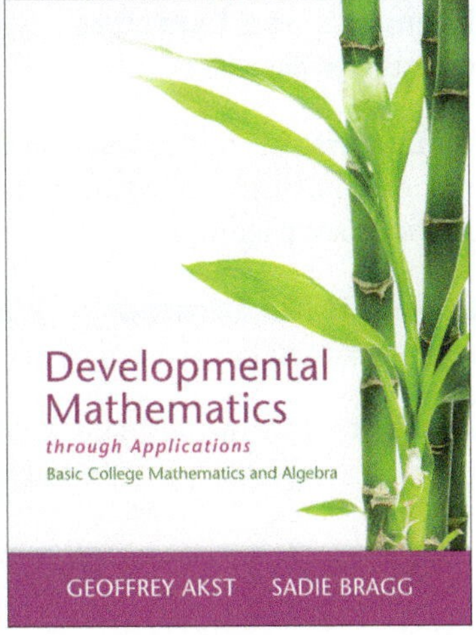

KEY FEATURES

Math Study Skills Foldout A full-color foldout with tips on organization, test preparation, time management, and more (see inside front cover).

Chapter Openers Extended real-world applications at the beginning of each chapter to motivate student interest and demonstrate how mathematics is used (see page 364).

Pretests and Posttests Chapter tests, which are particularly useful in a self-paced, lab, or digital environment (see pages 82 and 155).

Section Objectives with Paired Exercises Clearly stated learning objectives at the beginning of each section to identify topics to be covered. End-of-section odd/even paired exercises are closely aligned with the learning objectives in order to encourage and facilitate review (see pages 646 and 652–653).

Side-by-Side Example/Practice Format Distinctive side-by-side format that pairs each example with a corresponding practice exercise and gets students actively involved from the start (see page 404).

Examples and Exercises Based on Real Data Varied applied problems that are useful, realistic, and authentic (see page 814).

Tips Helpful suggestions and cautions for avoiding mistakes (see page 353).

Photos and Graphics Carefully selected photos to make problems seem more realistic, and relevant graphics to better meet the needs of visual learners (see page 11).

Journal Entries Writing assignments in response to probing questions interspersed throughout the text (see page 452).

Calculator Inserts Optional calculator and computer software instruction to solve section problems (see page 589).

Cultural Notes Glimpses of how mathematics has evolved across cultures and throughout history (see page 589).

For Extra Help Boxes at the beginning of every section's exercise set that direct students to helpful resources (see page 762).

Mathematically Speaking Exercises Vocabulary exercises in each section to help students understand and use standard mathematical terminology (see page 631).

Mixed Practice Exercises Problems in synthesizing section material (see page 169).

Application Examples and Exercises End-of-section problems to apply the topic at hand in a wide range of contexts (see page 252).

Mindstretcher Exercises Nonstandard section problems in critical thinking, mathematical reasoning, pattern recognition, historical connections, writing, and group work to deepen understanding and provide enrichment (see page 129).

Key Concepts and Skills Summary With a focus on descriptions and examples, the main points of the chapter organized into a practical and comprehensive chart (see pages 205–206).

Say Why Exercises Fill-in-the-blank problems, located at the beginning of each chapter review, providing practice in reasoning and communicating mathematical ideas (see page 737).

Chapter Review Exercises Problems for reviewing chapter content, arranged by section (see page 271).

Chapter Mixed Application Exercises Practice in applying topics across the chapter (see pages 886–887).

Cumulative Review Exercises Problems to maintain and build on the mathematical content covered in previous chapters (see pages 804–805).

Highlighting of Quantitative Literacy Skills Additional exercises that provide practice in number sense, proportional reasoning, and the interpretation of tables and graphs (see pages 386–388).

Online Bonus Material The following appendixes are available within the MyMathLab course for this book:

E. Solving Compound Inequalities
F. Solving Absolute Value Equations and Inequalities
G. Introduction to Functions
H. Solving Systems of Linear Inequalities
I. Variation
J. Radical Expressions and Rational Exponents
K. The Distance and Midpoint Formulas; the Circle

Easy-to-Locate Features Color borders added for back-of-book answer, glossary, and index pages.

Geometric Formulas A reference on the inside back cover of the text displaying standard formulas for perimeter, circumference, area, and volume.

U.S. and Metric Unit Tables Reference tables located opposite the inside back cover.

Coherent Development Texts with consistent content and style across the developmental math curriculum.

Robust MyMathLab Coverage! One of *every* problem type is assignable in MyMathLab.

Two MyMathLab Course Options

1. **Standard MyMathLab** courses allow instructors to build their course their way, offering maximum flexibility and complete control over all aspects of assignment creation.
2. **Ready-to-Go MyMathLab** courses provide students with all the same great MyMathLab features, but make it easier for instructors to get started.

WHAT SUPPLEMENTS ARE AVAILABLE?

For a complete list of the supplements and study aids that accompany *Developmental Mathematics through Applications: Basic College Mathematics and Algebra,* see p. xiii.

ACKNOWLEDGMENTS

We are grateful to everyone who has helped to shape this textbook by responding to questionnaires, participating in telephone surveys and focus groups, reviewing the manuscript, and using the text in their classes. We wish to thank Palma Benko, *Passaic County Community College*; Jennifer Caldwell, *Mesa Community College*; Edythe Carter, *Amarillo College*; Kristin Chatas, *Washtenaw Community College*; Bryan Cockerham, *Front Range Community College*; Vincent Conklin, *Bauder College*; Jonathan Cornick, *Queensborough Community College/CUNY*; Addie L. Davis, Ph.D., *Olive-Harvey College*; Mary Deas, *Johnson County Community College*; James Dressler, *Seattle Central Community College*; Karen Ernst, *Hawkeye Community College*; Mary Ellen Gallegos, *Santa Fe Community College*; Amadou Hama, *Kennedy King College*; Mary Beth Headlee, *State College of Florida*; Dr. Albert Hemenway, *Los Angeles Mission College*; Max Hibbs, *Blinn College*; Matthew Hudock, *St. Philip's College*; Sharon Jackson, *Brookhaven College*; Jennifer Johnson, *Delgado Community College*; Nancy R. Johnson, Ph.D., *State College of Florida*; Judy Kasabian, *El Camino College*; Dan Klienfelter, *College of the Desert*; Theodore Lai, *Hudson County Community College*; Thang Le, *College of the Desert*; Carol Lerch, *Daniel Webster College*; Mickey Levendusky, *Pima Community College, DTC*; Marcel Maupin, *Oklahoma State University–Oklahoma City*; Dena S. Messer-Herrera, *Rio Salado College*; Kathleen Offenholley; *Borough of Manhattan Community College/CUNY*; Ferdinand O. Orock, *Hudson County Community College*; Margaret Patin, *Vernon College*; Barbara Pearl, *Bucks County Community College*; Carol Perezluha, *Seminole State College of Florida*; Sara R. Pries, *Sierra Community College*; Sharonda Ragland, *ECPI College of Technology*; Sylvester Roebuck, *Olive-Harvey College*; Patricia C. Rome, *Delgado Community College*; Andrew Russell, *Queensborough Community College/CUNY*; Dr. Yojana Sharma, *Stark State College*; May Shaw, *Northcentral Technical College*; Mary Pat Sheppard, *Malcolm X College*; Lisa Winch, *Kalamazoo Valley Community College*; Michael D. Yarborough, *Cosumnes River College*; and Jeff Young, *Delaware Valley College*. In addition, we would like to extend our gratitude to the accuracy checker, Patricia Nelson, who helped us perfect the content in many ways.

Writing a textbook requires the contributions of many individuals. Special thanks go to Greg Tobin, President, English, Mathematics, and Student Success, Pearson Education, for encouraging and supporting us throughout the entire process. We thank Kari Heen, Katie DePasquale, and Katherine Minton for their patience and tact; Michelle Renda, Rachel Ross, and Maureen O'Connor for keeping us abreast of market trends; Rachel Haskell for attending to the endless details connected with the project; Ron Hampton, Elka Block, Lauren Foster, Andrea Stefanowicz, and Rachel Youdelman for their support throughout the production process; Barbara Atkinson for the cover design; and the entire Pearson developmental mathematics team for helping to make this text one of which we are very proud.

Geoffrey Akst   Sadie Bragg

Student Supplements

Student's Solutions Manual

By Deana Richmond

- Provides detailed solutions to the odd-numbered exercises in each exercise set and solutions to all chapter pretests and post-tests, practice exercises, review exercises, and cumulative review exercises

ISBN-10: 0-321-87281-9 ISBN-13: 978-0-321-87281-4

Video Resources on DVD with Chapter Test Prep Videos

- Complete set of digitized videos on DVD (also available in MyMathLab) for students to use at home or on campus
- Includes a full lecture for each section of the text
- Covers examples, practice problems, and exercises from the textbook that are marked with the icon
- Optional captioning in English is available
- Step-by-step video solutions for each chapter test
- Chapter Test Prep Videos are also available on YouTube (search by using author name and book title) and in MyMathLab

ISBN-10: 0-321-84765-2 ISBN-13: 978-0-321-84765-2

MyWorkBook with Chapter Summaries

By Carrie Green

- Provides one worksheet for each section of the text, organized by section objective, along with the end-of-chapter summaries from the textbook
- Each worksheet lists the associated objectives from the text, provides fill-in-the-blank vocabulary practice, and exercises for each objective

ISBN-10: 0-321-84777-6 ISBN-13: 978-0-321-84777-5

Instructor Supplements

Annotated Instructor's Edition

- Provides answers to all text exercises in color next to the corresponding problems
- Includes teaching tips

ISBN-10: 0-321-85682-1 ISBN-13: 978-0-321-85682-1

Instructor's Solutions Manual (download only)

By Deana Richmond

- Provides complete solutions to even-numbered section exercises
- Contains answers to all Mindstretcher problems

ISBN-10: 0-321-85681-3 ISBN-13: 978-0-321-85681-4

Instructor's Resource Manual with Tests and Mini-Lectures (download only)

By Elka Block

- Contains three free-response and one multiple-choice test form per chapter, and two final exams
- Includes resources designed to help both new and adjunct faculty with course preparation and classroom management, including sample syllabi, tips for using supplements and technology, and useful external resources
- Offers helpful teaching tips correlated to the sections of the text

ISBN-10: 0-321-84789-X ISBN-13: 978-0-321-84789-8

PowerPoint Lecture Slides (available online)

- Present key concepts and definitions from the text

TestGen® (available for download from the Instructor's Resource Center)

AVAILABLE FOR STUDENTS AND INSTRUCTORS

MyMathLab® Online Course (access code required)

MyMathLab delivers **proven results** in helping individual students succeed. It provides **engaging experiences** that personalize, stimulate, and measure learning for each student, and it comes from a **trusted partner** with educational expertise and an eye on the future. To learn more about MyMathLab, visit **www.mymathlab.com** or contact your Pearson representative.

MyMathLab® Ready-to-Go Course (access code required)

These new Ready-to-Go courses provide students with all the same great MyMathLab features but make it easier for instructors to get started. Each course includes preassigned homework and quizzes to make creating a course even simpler. Ask your Pearson representative about the details for this particular course or to see a copy of this course.

MathXL® Online Course (access code required)

MathXL® is the homework and assessment engine that runs MyMathLab. (MyMathLab is MathXL plus a learning management system.)

With MathXL, instructors can

- Create, edit, and assign online homework and tests using algorithmically generated exercises correlated at the objective level to the textbook.
- Create and assign their own online exercises and import TestGen tests for added flexibility.
- Maintain records of all student work tracked in MathXL's online gradebook.

With MathXL, students can

- Take chapter tests in MathXL and receive personalized study plans and/or personalized homework assignments based on their test results.
- Use the study plan and/or the homework to link directly to tutorial exercises for the objectives they need to study.
- Access supplemental animations and video clips directly from selected exercises.

MathXL is available to qualified adopters. For more information, visit our website at www.mathxl.com or contact your Pearson representative.

Index of Applications

Agriculture

Astronomy

Automotive

Biology

Business

●*Found in the online Appendix within MyMathLab.*

Chemistry

Construction

Consumer

INDEX OF APPLICATIONS

Economics

Education

Energy

Engineering

Entertainment

Environment

Finance

General Interest

INDEX OF APPLICATIONS

Geology

Geometry

Government

Health and Medicine

Labor

Physics

INDEX OF APPLICATIONS

Social Issues

Sports

Statistics/Demographics

Technology

Transportation

Weather

CHAPTER 1

Whole Numbers

Whole Numbers and YouTube

YouTube is a website where users can upload and view videos. These include movie clips, TV clips, music videos, and amateur content. This site made it feasible for anyone with an Internet connection to publish a video that could be seen by a worldwide audience within a few minutes.

In February 2005, the company was set up in a garage by several work colleagues. The first video posted on YouTube was *Me at the Zoo*, in which founder Jawed Karim is seen at the San Diego Zoo.

The usage of the site grew at an astonishing rate. By July 2006, more than 65,000 new videos were being uploaded every day, with about 10,000,000 visitors and 100,000,000 video views per day. Barely a year after its founding, the company was bought by Google for approximately $1,650,000,000.

YouTube has made sharing online video such an important part of Internet culture that it's been said "if it's not on YouTube, it's like it never happened."

(*Sources:* telegraph.co.uk, comscore.com, wikipedia.org, and cleancutmedia.com)

CHAPTER 1 PRETEST

To see if you have already mastered the topics in this chapter, take this test.

1. Insert commas as needed in the number 2 0 5 0 0 7. Then write the number in words.

2. Write the number one million, two hundred thirty-five thousand in standard form.

3. What place does the digit 8 occupy in 805,674?

4. Round 8,143 to the nearest hundred.

5. Add: $38 + 903 + 7{,}285$

6. Subtract 286 from 5,000.

7. Subtract: $734 - 549$

8. Find the product of 809 and 36.

9. Find the quotient: $27\overline{)7{,}020}$

10. Divide: $13{,}558 \div 44$

11. Write $2 \cdot 2 \cdot 2$, using exponents.

12. Evaluate: 6^2

Simplify.

13. $26 - 7 \cdot 3$

14. $3 + 2^3 \cdot (8 - 3)$

Solve and check.

15. The mathematician Benjamin Banneker was born in 1731 and died in 1806. About how old was he when he died? (*Source: The New Encyclopedia Britannica*)

16. At a certain college, students pay \$105 for each college credit. If a student takes 9 credits and pays with a \$1,000 voucher, how much change will he receive?

17. Phil Mickelson had scores of 67, 71, 67, and 67 for his four rounds at the 2010 Masters Tournament. What was his average score for a round of golf?

18. The Epson PictureMate Show Compact Photo Printer can print a 4-inch by 6-inch photo in 37 seconds, and the Epson Artisan 810 All-in-One Printer can print the same size photo in 10 seconds. How much longer would it take the Epson PictureMate Show to print twelve 4-inch by 6-inch photos? (*Source:* epson.com)

19. An insurance company offers an installment plan for paying auto insurance premiums. For a \$540 policy, the plan requires a down payment of \$81. The balance is paid in nine equal installments of \$55, which includes a service charge. How much money would be saved by paying for this policy without using the installment plan?

20. Which of the rooms pictured has the largest area? (feet = ft)

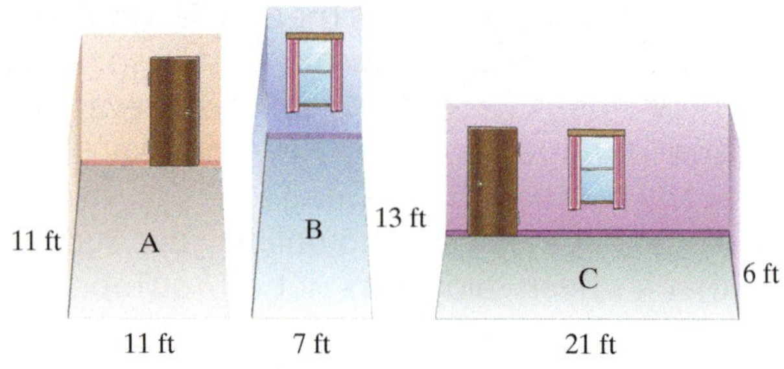

• Check your answers on page A-1.

1.1 Introduction to Whole Numbers

OBJECTIVES

- **A** To read or write whole numbers
- **B** To write whole numbers in expanded form
- **C** To round whole numbers
- **D** To solve applied problems involving reading, writing, or rounding whole numbers

What the Whole Numbers Are and Why They Are Important

We use whole numbers for counting, whether it is the number of *e*'s on this page, the number of stars in the sky, or the number of runs, hits, and errors in a baseball game.

The whole numbers are 0, 1, 2, 3, 4, 5, 6, 7, 8, 9, 10, 11, 12, 13, ... An important property of whole numbers is that there is always a next whole number. This property means that they go on without end, as the three dots above indicate.

Every whole number is either *even* or *odd*. The even whole numbers are 0, 2, 4, 6, 8, 10, 12, ... The odd whole numbers are 1, 3, 5, 7, 9, 11, 13, ...

We can represent the whole numbers on a number line. Similar to a ruler, the number line starts with 0 and extends without end to the right, as the arrow indicates.

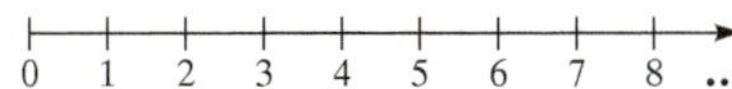

Reading and Writing Whole Numbers

Generally speaking, we *read* whole numbers in words, but we use the **digits** 0, 1, 2, 3, 4, 5, 6, 7, 8, and 9 to *write* them. For instance, we read the whole number *fifty-one* but write it *51*, which we call **standard form**.

Each of the digits in a whole number in standard form has a **place value**. Our place value system is very important because it underlies both the way we write and the way we compute with numbers.

The following chart shows the place values in whole numbers up to 15 digits long. For instance, in the number 1,234,056 the digit 2 occupies the hundred thousands place. Study the place values in the chart now.

TRILLIONS			BILLIONS			MILLIONS			THOUSANDS			ONES		
Hundred trillions	Ten trillions	Trillions	Hundred billions	Ten billions	Billions	Hundred millions	Ten millions	Millions	Hundred thousands	Ten thousands	Thousands	Hundreds	Tens	Ones
								1	2	3	4	0	5	6
					8	1	6	8	9	3	1	0	4	7

← Period
← Place value

TIP We read whole numbers from left to right, but it is easier in the place value chart to learn the names of the places *from right to left*.

When we write a large whole number in standard form, we insert *commas* to separate its digits into groups of three, called **periods**. For instance, the number 8,168,931,047 has four periods: *ones, thousands, millions*, and *billions*.

EXAMPLE 1

In each number, identify the place that the digit 7 occupies.

a. 207 **b.** 7,654,000 **c.** 5,700,000,001

Solution

a. The ones place

b. The millions place

c. The hundred millions place

PRACTICE 1

What place does the digit 8 occupy in each number?

a. 278,056

b. 803,746

c. 3,080,700,059

The following rule provides a shortcut for *reading a whole number*:

To Read a Whole Number

Working from left to right,

- read the number in each period and then
- name the period in place of the comma.

For instance, 1,234,056 is read "one million, two hundred thirty-four thousand, fifty-six." Note that the ones period is not read.

EXAMPLE 2

How do you read the number 422,000,085?

Solution Beginning at the left in the millions period, we read this number as "four hundred twenty-two million, eighty-five." Note that because there are all zeros in the thousands period, we do not read "thousands."

PRACTICE 2

Write 8,000,376,052 in words.

EXAMPLE 3

The display on a calculator shows the answer 3578002105. Insert commas in this answer and then read it.

Solution The number with commas is 3,578,002,105. It is read "three billion, five hundred seventy-eight million, two thousand, one hundred five."

PRACTICE 3

A company is worth $7372050. After inserting commas, read this amount.

Until now, we have discussed how to *read* whole numbers in standard form. Now, let's turn to the question of how they are *written* in standard form. We simply reverse the process just described. For instance, the number eight billion, one hundred sixty-eight million, nine hundred thirty-one thousand, forty-seven in standard form is 8,168,931,047. Here, we use the 0 as a **placeholder** in the hundreds place because there are no hundreds.

To Write a Whole Number

Working from left to right,

- write the number named in each period and
- replace each period name with a comma.

When writing large whole numbers in standard form, we must remember that the number of commas is always one less than the number of periods. For instance, the number one million, two hundred thirty-four thousand, fifty-six—1,234,056—has three periods and two commas. Similarly, the number 8,168,931,047 has four periods and three commas.

EXAMPLE 4

Write the number eight billion, seven in standard form.

Solution This number involves billions, so there are four periods—billions, millions, thousands, and ones—and three commas. Writing the number named in each period and replacing each period name with a comma, we get 8,000,000,007. Note that we write three 0's when no number is named in a period.

PRACTICE 4

Use digits and commas to write the amount ninety-five million, three dollars.

EXAMPLE 5

The treasurer of a company writes a check in the amount of four hundred thousand seven hundred dollars. Using digits, how would she write this amount on the check?

Solution This quantity is written with one comma, because its largest period is thousands. So the treasurer writes $400,700, as shown on the check below.

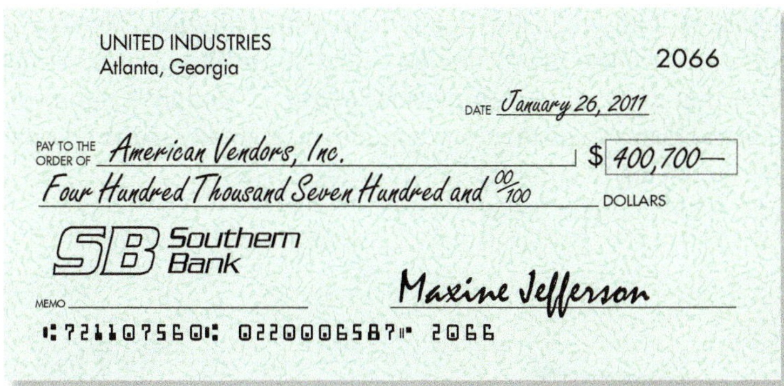
UNITED INDUSTRIES
Atlanta, Georgia
2066
DATE January 26, 2011
PAY TO THE ORDER OF American Vendors, Inc. $ 400,700—
Four Hundred Thousand Seven Hundred and 00/100 DOLLARS
SB Southern Bank
MEMO
Maxine Jefferson
⑆721107560⑆ 0220006587⑈ 2066

PRACTICE 5

A rich alumna donates three hundred seventy-five thousand dollars to her college's scholarship fund.

HARRIET YOUNG
3560 Ramstead St.
Reston, VA 22090
1434
DATE March 17, 2011
PAY TO THE ORDER OF Borough of Manhattan Community College $
Three Hundred Seventy-Five Thousand and 00/100 DOLLARS
MDBC
Bank USA Reston, VA 22090
MEMO
Harriet Young
⑆034005089⑆ 560240361⑈ 1434

Using digits, how would she write this amount on the check?

When writing checks, we write the amount in both digits and words. Why do we do this?

Writing Whole Numbers in Expanded Form

We have just described how to write whole numbers in standard form. Now, let's turn to how we write these numbers in **expanded form**.

Let's consider the whole number 4,025 and examine the place value of its digits.

$$4{,}025 = 4 \text{ thousands} + 0 \text{ hundreds} + 2 \text{ tens} + 5 \text{ ones}$$

This last expression is called the expanded form of the number, and it can be written as follows

$$4{,}000 + 0 + 20 + 5, \quad \text{or} \quad 4{,}000 + 20 + 5$$

The expanded form of a number spells out its value in terms of place value, helping us understand what the number really means. For instance, think of the numbers 92 and 29. By representing them in *expanded* form, can you explain why they differ in value even though their *standard* form consists of the same digits?

EXAMPLE 6

Write in expanded form:

a. 906 **b.** 3,203,000

Solution

a. The 6 is in the ones place, the 0 is in the tens place, and the 9 is in the hundreds place.

ONES		
Hundreds	**Tens**	**Ones**
9	0	6

So 906 is 9 hundreds + 0 tens + 6 ones = 900 + 0 + 6, or 900 + 6 in expanded form.

b. Using the place value chart, we see that

$$\begin{aligned} 3{,}203{,}000 &= 3 \text{ millions} + 2 \text{ hundred thousands} + 3 \text{ thousands} \\ &= 3{,}000{,}000 + 200{,}000 + 3{,}000. \end{aligned}$$

PRACTICE 6

Express in expanded form:

a. 27,013

b. 1,270,093

Rounding Whole Numbers

Most people equate mathematics with precision, but some problems require sacrificing precision for simplicity. In this case, we use the technique called **rounding** to approximate the exact answer with a number that ends in a given number of zeros. Rounded numbers have special advantages: They seem clearer to us than other numbers, and they make computation easier—especially when we are trying to compute in our heads.

Of these two headlines, which do you prefer? Why?

Daily Planet

7 MILLION PEOPLE UNEMPLOYED

By Clip Arttikil

Happy days are not here again. The unemploy… through to the end. Some believe that a down… far outweighs the benefits. The latest statistics… experts agree. These trends have shown no sign…

★ Late Edition City New…

7,183,208 Out of Work

By Lee Whay
City News Staff Writer

The numbers are in and the outlook is not rosy accor… across the nation report a down-turn. Leading the lis… but few find comfort in these trends. Many of the to… and have continued to decline…

Study the following chart to see the connection between place value and rounding.

Rounding to the nearest	Means that the rounded number ends in at least
10	One 0
100	Two 0's
1,000	Three 0's
10,000	Four 0's
100,000	Five 0's
1,000,000	Six 0's

Note in the chart that the place value tells us how many 0's the rounded number must have at the end. Having more 0's than indicated is possible. Can you think of an example?

When rounding, we use an underlined digit to indicate the place to which we are rounding.

Now, let's consider the following rule for rounding whole numbers:

To Round a Whole Number

- Underline the place to which you are rounding.
- The digit to the right of the underlined digit is called the *critical digit.* Look at the critical digit—if it is 5 or more, add 1 to the underlined digit; if it is less than 5, leave the underlined digit unchanged.
- Replace all the digits to the right of the underlined digit with zeros.

EXAMPLE 7

Round 79,630 to

a. the nearest thousand

b. the nearest hundred

Solution

a. $79{,}630 = 7\underline{9}{,}630$ ← Underline the digit in the thousands place.

$= 7\underline{9}{,}630$ ← The critical digit 6 is greater than 5; add 1 to the underlined digit.

$\approx 80{,}000$ ← Change the digits to the right of the underlined digit to 0's.

↑ This symbol means "is approximately equal to."

Note that adding 1 to the underlined digit gave us 10. As a result we regroup, that is, write 0, carry 1 to the next column, and change the 7 to 8.

b. First, we underline the 6 because that digit occupies the hundreds place: $79{,}\underline{6}30$. The critical digit is 3: $79{,}\underline{6}30$. Since 3 is less than 5, we leave the underlined digit unchanged. Then, we replace all digits to the right with 0's, getting 79,600. We write $79{,}630 \approx 79{,}600$, meaning that 79,630 when rounded to the nearest hundred is 79,600.

PRACTICE 7

Round 51,760 to

a. the nearest thousand

b. the nearest ten thousand

For Example 7a, consider this number line.

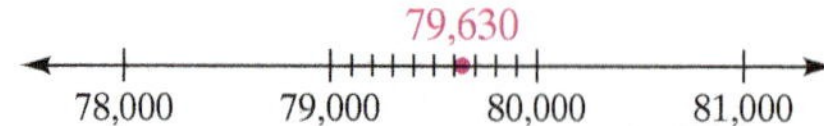

The number line shows that 79,630 lies between 79,000 and 80,000 and that it is closer to 80,000, as the rule indicates.

EXAMPLE 8

In an anatomy and physiology class, a student learned that the adult human skeleton contains 206 bones. How many bones is this to the nearest hundred bones?

Solution We first write 2̲06. The critical digit 0 is less than 5, so we do *not* add 1 to the underlined digit. However, we do change both the digits to the right of the 2 to 0's. So 2̲06 ≈ 200, and there are approximately 200 bones in the human body.

PRACTICE 8

Based on current population data, the U.S. Bureau of the Census projects that the U.S. resident population will be 419,845,000 in the year 2050. What is the projected population to the nearest million?

EXAMPLE 9

The following table lists five of the highest-grossing films of all time and the amount of money they took in.

Film	Year	World Total (in U.S. dollars)
Titanic	1997	$1,835,300,000
The Lord of the Rings: The Return of the King	2003	$1,129,219,252
Pirates of the Caribbean: Dead Man's Chest	2006	$1,060,332,628
The Dark Knight	2008	$1,001,921,825
Avatar	2009	$2,690,408,054

(*Source:* imdb.com)

a. Write in words the amount of money taken in by the film with the largest world total.

b. Round to the nearest ten million dollars the world total for *Titanic*.

Solution

a. *Avatar* has the largest world total. This total is read "two billion, six hundred ninety million, four hundred eight thousand, fifty-four dollars."

b. The world total for *Titanic* is $1,835,300,000. To round, we underline the digit in the ten millions place: 1,83̲5,300,000. Since the critical digit is 5, we add 1 to the underlined digit, and change the digits to the right to 0's. So the rounded total is $1,840,000,000.

PRACTICE 9

This chart gives the number of U.S. postsecondary teachers in the year 2008 as well as the projected number of postsecondary teachers for the year 2018.

Year	Number of Postsecondary Teachers
2008	1,699,200
2018	1,956,100

(*Source:* bls.gov)

a. Write in words the number of postsecondary teachers in the year 2008.

b. What is the number of projected postsecondary teachers in the year 2018 rounded to the nearest ten thousand?

1.1 Exercises

FOR EXTRA HELP MyMathLab  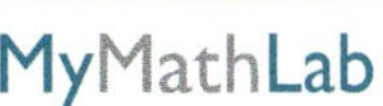    

Mathematically Speaking

Fill in each blank with the most appropriate term or phrase from the given list.

calculated	rounded	periods	odd
even	digits	whole numbers	standard form
placeholder	place value	expanded form	

1. The ____________ are 0, 1, 2, 3, 4, 5, ...

2. The numbers 0, 2, 4, 6, 8, 10, . . . are __________.

3. The numbers 1, 3, 5, 7, 9, . . . are __________.

4. The whole numbers are written with the __________ 0, 1, 2, 3, 4, 5, 6, 7, 8, and 9.

5. The number thirty-seven, when written as 37, is said to be in ____________.

6. In the number 528, the __________ of the 5 is hundreds.

7. In the number 206, the 0 is used as a __________ in the tens place.

8. Commas separate the digits in a large whole number into groups of three called __________.

9. When the number 973 is written as 9 hundreds + 7 tens + 3 ones, it is said to be in ____________.

10. The number 545 __________ to the nearest hundred is 500.

(A) *Underline the digit that occupies the given place.*

11. 4,867 Thousands place

12. 9,752 Thousands place

13. 316 Tens place

14. 728 Tens place

15. 28,461,013 Millions place

16. 73,762,800 Millions place

Identify the place occupied by the underlined digit.

17. $\underline{6}91{,}400$

18. $7\underline{2}{,}109$

19. $7{,}\underline{3}80$

20. $3\underline{5}1$

21. $\underline{8}{,}450{,}000{,}000$

22. $\underline{3}5{,}832{,}775$

Insert commas as needed, and then write the number in words.

23. 4 8 7 5 0 0

24. 5 2 8 0 5 0

25. 2 3 5 0 0 0 0

26. 1 3 5 0 1 3 2

27. 9 7 5 1 3 5 0 0 0

28. 4 2 1 0 0 0 1 3 2

29. 2 0 0 0 0 0 0 3 5 2

30. 4 1 0 0 0 0 0 0 0 7

31. 1 0 0 0 0 0 0 0 0 0

32. 3 7 9 0 5 2 0 0 0

Write each number in standard form.

33. Ten thousand, one hundred twenty

34. Twelve thousand, two hundred thirty

35. One hundred fifty thousand, eight hundred fifty-six

36. Two hundred forty thousand, seven hundred seventy-two

37. Six million, fifty-five

38. Two million, one hundred twenty-two

39. Fifty million, six hundred thousand, one hundred ninety-five

40. Thirty million, five hundred thousand, four hundred eighty-four

41. Four hundred thousand, seventy-two

42. Three hundred thousand, sixty-one

B *Write each number in expanded form.*

43. 3

44. 6,300

45. 858

46. 9,000,000

47. 2,500,004

48. 7,251,380

C *Round to the indicated place.*

49. 671 to the nearest ten

50. 838 to the nearest ten

51. 7,103 to the nearest hundred

52. 8,204 to the nearest hundred

53. 28,241 to the nearest ten thousand

54. 32,323 to the nearest ten thousand

55. 705,418 to its largest place

56. 806,329 to the largest place

57. 31,972 to its largest place

58. 52,891 to the largest place

Round each number as indicated.

59.

To the nearest	135,842	2,816,533
Hundred		
Thousand		
Ten thousand		
Hundred thousand		

60.

To the nearest	972,055	3,189,602
Thousand		
Ten thousand		
Hundred thousand		
Million		

Mixed Practice

Solve.

61. Write 12,051 in expanded form.

62. Identify the place occupied by the underlined digit in 26,543,009.

63. Underline the digit that occupies the ten thousands place in 40,059.

64. Write five hundred forty-two thousand, sixty-seven in standard form.

65. Insert commas as needed, and then write 1 0 5 6 1 0 0 in words.

66. Round 26,255 to the nearest thousand.

Applications

D *Write each whole number in words.*

67. Biologists have classified more than 900,000 species of insects. (*Source:* Smithsonian Institution)

68. In a recent year, there were 71,988 WiFi hotspots in the United States. (*Source:* jiwire.com).

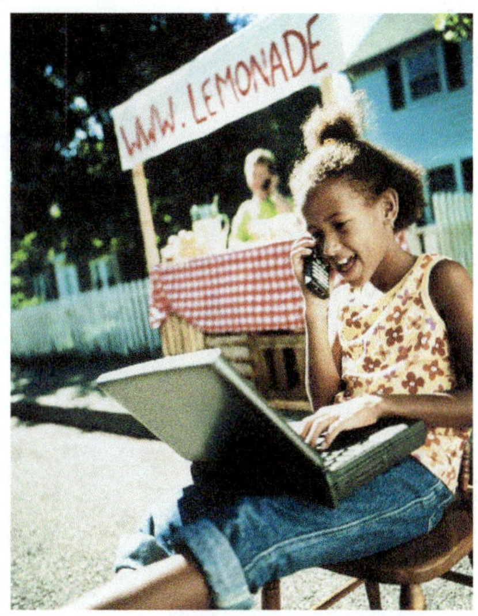

69. The land area of the Dominican Republic, located in the West Indies, is 48,381 square kilometers. (*Source*: infoplease.com)

70. Mercury is the closest planet to the Sun, a distance of approximately 36,000,000 miles.

71. Each pair of human lungs contains some 300,000,000 tiny air sacs. (*Source:* Sylvia S. Mader, *Inquiry Into Life*)

72. The Pyramid of Khufu in Egypt has a base of approximately 2,315,000 blocks. (*Source: The New Encyclopedia Britannica*)

Write each whole number in standard form.

73. Some one hundred billion nerve cells are part of the human brain.

74. Son of Beast, a roller coaster at Paramount's Kings Island in Ohio, has a track length of seven thousand thirty-two feet. (*Source:* American Coasters Network)

75. One of the largest giant sequoias in the United States is three thousand, two hundred eighty-eight inches tall. (*Source:* U.S. National Park Service)

76. The total land area of the United States is nine million, six hundred thirty-one thousand, four hundred eighteen square kilometers. (*Source: The World Factbook*, 2006)

77. In 1990, the U.S. public debt (in dollars) was three trillion, two hundred thirty-three billion, three hundred million. (*Source: The World Almanac 2010*)

78. The light-year is a unit of length used to measure distances to stars and other distances on an astronomical scale. A light-year is equal to about five trillion, eight hundred seventy-eight billion, six hundred thirty million miles. (*Source:* wikipedia.org)

Round to the indicated place.

79. The Statue of Liberty is 152 feet high. What is its height to the nearest 10 feet?

80. The Nile, with a length of 4,180 miles, is the longest river in the world. Find this length to the nearest thousand miles.

81. In 1949, Air Force Captain James Gallagher led the first team to make an around-the-world flight. The team flew 23,452 miles. What is this distance to the nearest ten thousand miles? (*Source:* Taylor and Mondey, *Milestones of Flight*)

82. The element copper changes from a liquid to a gas at the temperature 2,567 degrees Celsius (°C). Find this temperature to the nearest hundred degrees Celsius.

83. The South American country of Colombia is home to 1,897 bird species—more than any other country. How many species is this to the nearest hundred? (*Source: Avibase—the World Bird Database,* avibase.bsc-eoc.org)

84. The Rose Bowl stadium has a seating capacity of 92,542. Round this number to the nearest ten thousand. (*Source:* Rose Bowl Operating Company)

85. This chart displays the area of seven most heavily populated countries in the world.

Country	Area (in square miles)
China	3,600,930
India	1,147,950
United States	3,537,421
Indonesia	705,189
Brazil	3,265,061
Pakistan	300,664
Bangladesh	51,703

(*Source:* census.gov)

a. Write in words the area of China.

b. Round, to the nearest thousand, the area of Pakistan.

86. This chart displays the number of degrees awarded in the United States during a recent year.

Degree	Number Awarded
Associate	750,164
Bachelor's	1,563,069
Master's	625,023
Doctorate	63,712
Professional	91,309

(*Source:* nces.ed.gov)

a. Write in words the number of bachelor's degrees awarded.

b. Round, to the nearest hundred thousand, the number of associate degrees awarded.

• Check your answers on page A-1.

MINDStretchers

Mathematical Reasoning

1. I am thinking of a certain whole number. My number, rounded to the nearest hundred, is 700. When it is rounded to the nearest ten, it is 750. What numbers could I be thinking of?

Writing

2. How does the number 10 play a special role in the way that we write whole numbers? Would it be possible to have the number 2 play this role? Explain.

Groupwork

3. Here are three ways of writing the number seven: 7 VII 𝍸||

Working with a partner, express each of the numbers 1, 2, . . . , 9 in these three ways.

1.2 Adding and Subtracting Whole Numbers

OBJECTIVES

- **A** To add or subtract whole numbers
- **B** To solve applied problems involving the addition or subtraction of whole numbers

The Meaning and Properties of Addition and Subtraction

Addition is perhaps the most fundamental of all operations. One way to think about this operation is as *combining sets.* For example, suppose that we have two distinct sets of pens, with 5 pens in one set and 3 in the other. If we put the two sets together, we get a single set that has 8 pens.

5	+	3	=	8
Addend		Addend		Sum

So we can say that 3 added to 5 is 8, or here, 5 pens plus 3 pens equals 8 pens. Numbers being added are called *addends*. The result is called the *sum*, or *total.*

In the above example, note that we are adding quantities of the same thing, or *like quantities.*

Another good way to think about the addition of whole numbers is as *moving to the right on a number line.* In this way, we start at the point on the line corresponding to the first number, 5. Then to add 3, we move 3 units to the right, ending on the point that corresponds to the answer, 8.

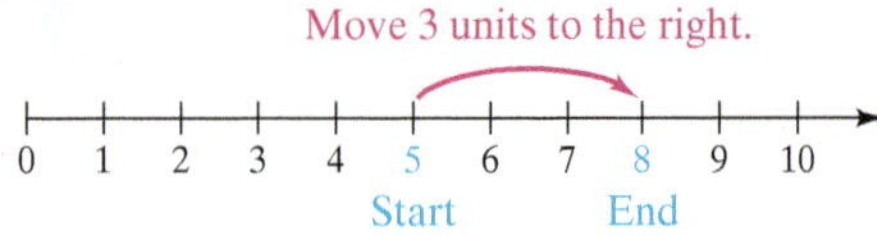

Now, let's look at subtraction. One way to look at this operation is as *taking away.* For instance, when we subtract 5 pens from 8 pens, we take 5 pens away from 8 pens, leaving 3 pens.

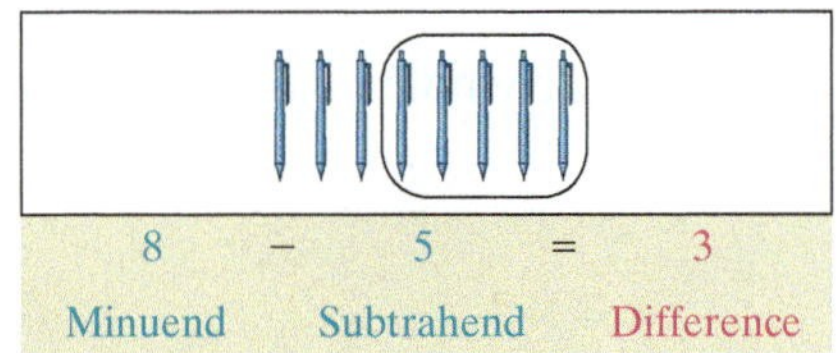

In a subtraction problem, the number from which we subtract is called the *minuend,* the number being subtracted is called the *subtrahend*, and the result is called the *difference*. In other words, the difference between two numbers is the first number take away the second number.

As in the preceding example, we can only subtract *like quantities*: we cannot subtract 5 pens from 8 scissors.

We can also think of subtraction as the *opposite of addition.*

$$\underbrace{8 - 5 = 3}_{\textbf{Subtraction}} \qquad \text{because} \qquad \underbrace{5 + 3 = 8}_{\textbf{Related addition}}$$

Note in this example that, if we add the 5 pens to the 3 pens, we get 8 pens.

Addition and subtraction problems can be written either horizontally or vertically.

$$5 + 3 = 8 \qquad 8 - 5 = 3$$

Horizontal

$$\begin{array}{r} 5 \\ +3 \\ \hline 8 \end{array} \qquad \begin{array}{r} 8 \\ -5 \\ \hline 3 \end{array}$$

Vertical

Either format gives the correct answer. But it is generally easier to figure out the sum and difference of large numbers if the problems are written vertically.

Now, let's briefly consider several special properties of addition that we use frequently. Examples appear to the right of each property.

The Identity Property of Addition

The sum of a number and zero is the original number.

$3 + 0 = 3$
$0 + 5 = 5$

The Commutative Property of Addition

Changing the order in which two numbers are added does not affect their sum.

$$\underbrace{3 + 2}_{5} = \underbrace{2 + 3}_{5}$$

The Associative Property of Addition

When adding three numbers, regrouping addends gives the same sum. Note that the parentheses tell us which numbers to add first.

We add inside the parentheses first

$$(4 + 7) + 2 = 4 + (7 + 2)$$
$$11 + 2 = 4 + 9$$
$$13 \qquad 13$$

Adding Whole Numbers

We add whole numbers by arranging the numbers vertically, keeping the digits with the same place value in the same column. Then, we add the digits in each column.

Consider the sum 32 + 65. In the vertical format at the right, the sum of the digits in each column is 9 or less. The sum is 97. When the sum of the digits in a column is greater than 9, we must **regroup (carry)** because only a single digit can occupy a single place. Example 1 illustrates this process.

$$\begin{array}{r} 32 \\ +65 \\ \hline 97 \end{array}$$

EXAMPLE 1

Add 47 and 28.

Solution First, we write the addends in expanded form. Then, we add down the ones column.

$$\begin{array}{rcrcr} 47 & = & 4 \text{ tens} + 7 \text{ ones} & = & \overset{1 \text{ ten}}{4 \text{ tens}} + 7 \text{ ones} \\ \underline{+28} & = & \underline{2 \text{ tens} + 8 \text{ ones}} & = & \underline{2 \text{ tens} + 8 \text{ ones}} \\ & & 15 \text{ ones} & & 5 \text{ ones} \end{array}$$

By regrouping, we express 15 ones as 1 ten + 5 ones. Then we carry the 1 ten to the tens place.

Next, we add down the tens column.

$$\begin{array}{l} \text{1 ten} \\ 4 \text{ tens} + 7 \text{ ones} \\ \underline{2 \text{ tens} + 8 \text{ ones}} \\ 7 \text{ tens} + 5 \text{ ones} = 75 \end{array}$$

PRACTICE 1

Add: 178 + 207

The following rule tells how to add whole numbers without using expanded form:

To Add Whole Numbers

- Write the addends vertically, lining up the place values.
- Add the digits in the ones column, writing the rightmost digit of the sum on the bottom. If the sum has two digits, carry the left digit to the top of the next column on the left.
- Add the digits in the tens column, as in the preceding step.
- Repeat this process until you reach the last column on the left, writing the entire sum of that column on the bottom.

EXAMPLE 2

Add: 9,824 + 356 + 2,976

Solution We write the problem vertically, with the addends lined up on the right.

$$\begin{array}{r} {}^{1} \\ 9,824 \\ 356 \\ +2,976 \\ \hline 6 \end{array}$$

The sum of the ones digits is 16 ones. We write the 6 and carry the 1 to the tens column.

$$\begin{array}{r} {}^{1\,1} \\ 9,824 \\ 356 \\ +2,976 \\ \hline 56 \end{array}$$

The sum of the tens digits is 15 tens. We write the 5 and carry the 1 to the hundreds column.

$$\begin{array}{r} {}^{2\,1\,1} \\ 9,824 \\ 356 \\ +2,976 \\ \hline 156 \end{array}$$

The sum of the hundreds digits is 21 hundreds. We write the 1 and carry the 2 to the thousands column.

$$\begin{array}{r} {}^{2\,1\,1} \\ 9,824 \\ 356 \\ +2,976 \\ \hline 13,156 \end{array}$$

The sum of the digits in the thousands column is 13, which we write completely—no need to regroup here.

The sum is 13,156.

PRACTICE 2

Find the total: 838 + 96 + 9,502

In Example 3, let's apply the operation of addition to finding the geometric perimeter of a figure. The **perimeter** is the distance around a figure, which we can find by adding the lengths of its sides.

EXAMPLE 3

What is the perimeter of the region marked off for the construction of a swimming pool and an adjacent pool cabana?

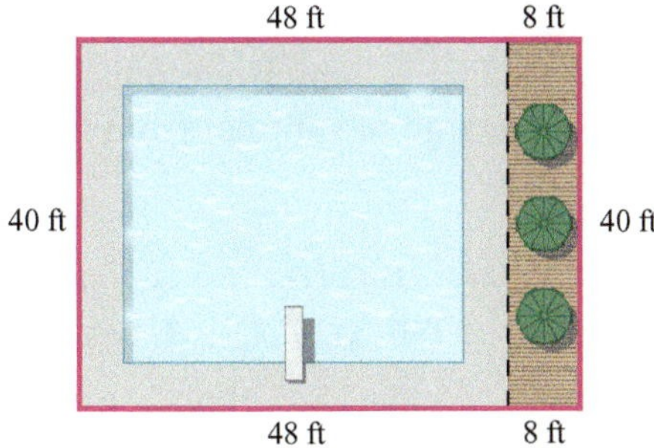

Solution This figure consists of two rectangles placed side by side. We note that the opposite sides of each rectangle are equal in length.

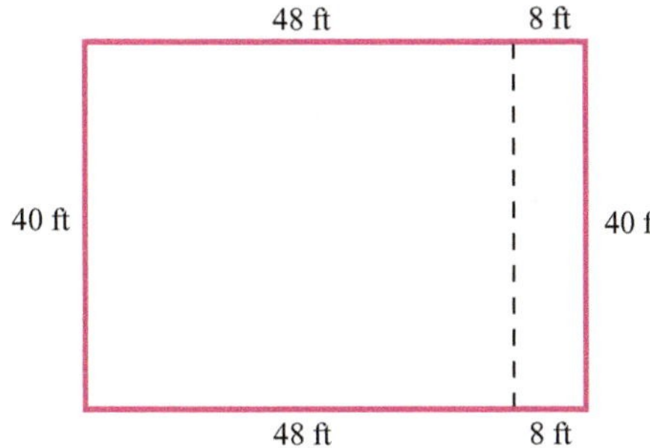

To compute the figure's perimeter, we need to add the lengths of all its sides.

$$\begin{array}{r} 40 \\ 48 \\ 8 \\ 40 \\ 8 \\ +\ 48 \\ \hline 192 \end{array}$$

The figure's perimeter is 192 feet.

PRACTICE 3

How long a fence is needed to enclose the piece of land sketched?

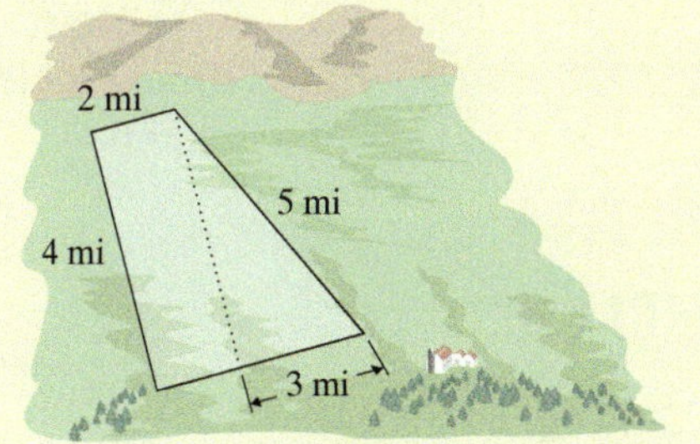

Subtracting Whole Numbers

Consider the subtraction problem 59 − 36, written vertically at the right. We write the whole numbers underneath one another, lined up on the right, so each column contains digits with the same place value. Subtracting the digits within each column, the bottom digit from the top, the result is a difference of 23.

$$\begin{array}{r} 59 \\ -36 \\ \hline 23 \end{array}$$

Keep in mind two useful properties of subtraction.

- When we subtract a number from itself, the result is 0: $6 - 6 = 0$
- When we subtract 0 from a number, the result is the original number: $25 - 0 = 25$

TIP When writing a subtraction problem vertically, be sure that

- the minuend—the number from which we are subtracting—goes on the top and that
- the subtrahend—the number being taken away—goes on the bottom.

Now, we consider subtraction problems in which we must regroup (*borrow*). In these problems a digit on the bottom is too large to subtract from the corresponding digit on top.

EXAMPLE 4

Subtract: $329 - 87$

Solution We first write these numbers vertically in expanded form.

$$\begin{array}{rcr} 329 & = & 3 \text{ hundreds} + 2 \text{ tens} + 9 \text{ ones} \\ -\ 87 & = & -\quad 8 \text{ tens} + 7 \text{ ones} \\ \hline \end{array}$$

We then subtract the digits in the ones column: 7 ones from 9 ones gives 2 ones.

$$\begin{array}{r} 3 \text{ hundreds} + 2 \text{ tens} + 9 \text{ ones} \\ -\quad 8 \text{ tens} + 7 \text{ ones} \\ \hline 2 \text{ ones} \end{array}$$

10 tens + 2 tens = 12 tens

We next go to the tens column. We cannot take 8 tens from 2 tens. But we can *borrow* 1 hundred from the 3 hundreds, leaving 2 in the hundreds place. We *exchange* this hundred for 10 tens (1 hundred = 10 tens). Then combining the 10 tens with the 2 tens gives 12 tens.

$$\begin{array}{r} \overset{2}{\not3} \text{ hundreds} + \overset{1}{}2 \text{ tens} + 9 \text{ ones} \\ -\quad 8 \text{ tens} + 7 \text{ ones} \\ \hline 2 \text{ ones} \end{array}$$

We next take 8 from 12 in the tens column, giving 4 tens. Finally, we bring down the 2 hundreds. The difference is 242 in standard form.

$$\begin{array}{r} \overset{2}{\not3} \text{ hundreds} + \overset{1}{}2 \text{ tens} + 9 \text{ ones} \\ -\quad 8 \text{ tens} + 7 \text{ ones} \\ \hline 2 \text{ hundreds} + 4 \text{ tens} + 2 \text{ ones} = 242 \end{array}$$

PRACTICE 4

Subtract: $748 - 97$

Although we can always rewrite whole numbers in expanded form so as to subtract them, the following rule provides a shortcut:

To Subtract Whole Numbers

- On top, write the number *from which* we are subtracting. On the bottom, write the number that is being taken *away*, lining up the place values. Subtract in each column separately.
- Start with the ones column.
  a. If the digit on top is *larger* than or *equal* to the digit on the bottom, subtract and write the difference below the bottom digit.
  b. If the digit on top is *smaller* than the digit on the bottom, borrow from the digit to the left on top. Then subtract and write the difference below the bottom digit.
- Repeat this process until the last column on the left is finished.

Recall that for every subtraction problem there is a related addition problem. So we can use addition to check subtraction, as in the following example.

EXAMPLE 5

Find the difference between 500 and 293.

Solution We rewrite the problem vertically.

$$\begin{array}{r} 500 \\ -293 \\ \hline \end{array}$$

We cannot subtract 3 ones from 0 ones, and we cannot borrow from 0 tens. So we borrow from the 5 hundreds.

$$\begin{array}{r} {}^{4}\not{5}\,{}^{1}0\,0 \\ -2\,9\,3 \\ \hline \end{array} \quad \longleftarrow \textbf{5 hundreds} = \textbf{4 hundreds} + \textbf{10 tens}$$

We now borrow from the tens column.

$$\begin{array}{r} {}^{4}\not{5}\ {}^{\not{10}\ 9}\not{0}\ {}^{1}0 \\ -2\ 9\ 3 \\ \hline 2\ 0\ 7 \end{array} \quad \longleftarrow \textbf{10 tens} = \textbf{9 tens} + \textbf{10 ones}$$

Check We check the difference by adding it to the subtrahend. The sum turns out to be the original minuend, so our answer is correct.

$$\begin{array}{r} 207 \\ +293 \\ \hline 500 \end{array}$$

PRACTICE 5

Subtract 3,253 from 8,000.

EXAMPLE 6

There are a total of 118 chemical elements. Ninety-four of these occur naturally on Earth and the rest are synthetic. How many chemical elements are synthetic? (*Source:* wikipedia.org)

Solution The total number of chemical elements equals the number of natural chemical elements plus the number of synthetic chemical elements.

Natural Chemical Elements (94)	Synthetic Chemical Elements (?)

All Chemical Elements (118)

To compute the number of synthetic chemical elements, we subtract the number of natural chemical elements from the total number of chemical elements.

Total number of chemical elements	118
Number of natural chemical elements	− 94
Number of synthetic chemical elements	24

So 24 chemical elements are synthetic.

PRACTICE 6

Of the 1,324 endangered and threatened species (plants and animals) in the United States, 574 are animal species. How many plant species are endangered and threatened? (*Source:* ecos.fws.gov)

EXAMPLE 7

The following graph shows the number of overseas visitors from various countries who came to the United States during a recent year.

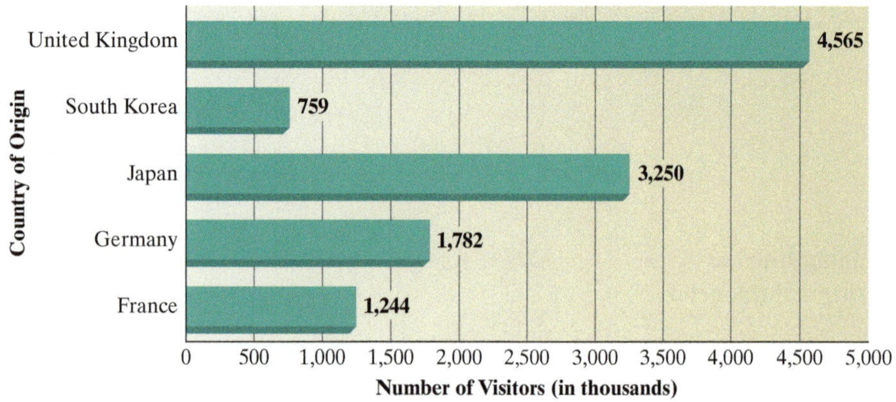

(*Source:* tinet.ita.doc.gov)

a. From which of these countries did the greatest number of visitors come?

b. How many visitors came from either the United Kingdom or Germany?

c. Was the number of German visitors greater or less than the total number of visitors who came from either South Korea or France?

Solution

a. More visitors came to the United States from the United Kingdom than from any other country.

b. To find how many visitors came to the United States from either the United Kingdom or Germany, we add the number of visitors from each country:

$$\begin{array}{rr} \text{United Kingdom} \rightarrow & 4{,}565{,}000 \\ \text{Germany} \rightarrow & \underline{+1{,}782{,}000} \\ \text{Sum} \rightarrow & 6{,}347{,}000 \end{array}$$

So 6,347,000 visitors came to the United States from either the United Kingdom or Germany.

c. The number of German visitors was 1,782,000. The total number of visitors who came from either South Korea or France is the sum of the number of visitors from each country:

$$\begin{array}{rr} \text{South Korea} \rightarrow & 759{,}000 \\ \text{France} \rightarrow & \underline{+1{,}244{,}000} \\ \text{Sum} \rightarrow & 2{,}003{,}000 \end{array}$$

Since 2,003,000 is greater than 1,782,000, the number of German visitors was less than the total number of visitors who came from either South Korea or France.

PRACTICE 7

The following chart shows the projected employment change in the United States from 2008 to 2018 for some of the fastest-growing occupations.

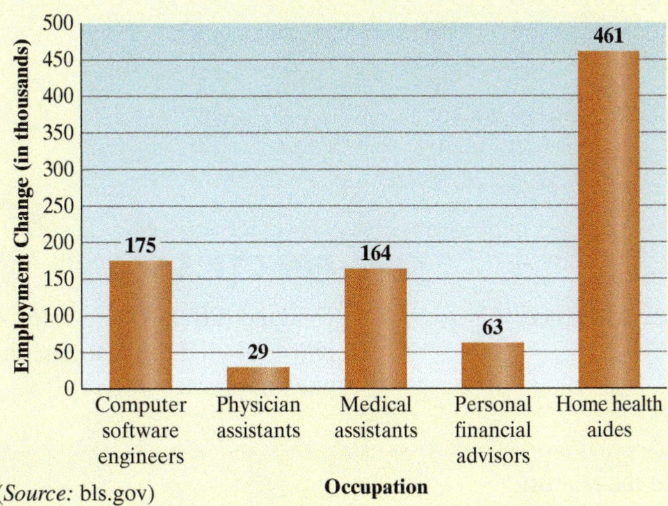

(*Source:* bls.gov)

a. How much greater is the projected employment growth for home health aides than that for computer software engineers?

b. What is the combined increase in employment projected for physician assistants and medical assistants?

c. Is the combined change in employment projected for computer software engineers and personal financial advisors greater or less than the projected change for home health aides?

Estimating Sums and Differences

Because everyone occasionally makes a mistake, we need to know how to check an answer so that we can correct it if it is wrong.

One method of checking addition and subtraction is by *estimation.* In this approach, we first compute and then estimate the answer. Then, we compare the estimate and our "exact answer" to see if they are close. If they are, we can be confident that our answer is reasonable. If they are not close, we should redo the computation.

We can get different estimates for an answer, depending on how we round. For addition, one good way involves rounding each addend to the largest place value, as shown in Example 8. Similarly, for subtraction, we can round the minuend and subtrahend to the largest place value, as Example 9 illustrates.

EXAMPLE 8

Compute the sum of 1,923 + 898 + 754 + 2,873. Check by estimation.

Solution

$$\begin{array}{r} 1{,}923 \\ 898 \\ 754 \\ +2{,}873 \\ \hline 6{,}448 \end{array} \leftarrow \text{Exact sum}$$

Check We round each addend to the largest place value.

$$\begin{array}{rcr} 1{,}923 & \approx & 2{,}000 \\ 898 & \approx & 900 \\ 754 & \approx & 800 \\ +2{,}873 & \approx & +3{,}000 \\ \hline & & 6{,}700 \end{array} \leftarrow \text{Estimated sum}$$

Our exact answer (6,448) is reasonably close to the estimate, so we are done.

PRACTICE 8

Add: 3,945 + 849 + 4,001 + 682. Check by estimating the sum.

EXAMPLE 9

Subtract 1,994 from 8,253. Check by estimating the difference.

Solution

$$\begin{array}{r} 8{,}253 \\ -1{,}994 \\ \hline 6{,}259 \end{array} \leftarrow \text{Exact difference}$$

Check We round the minuend and subtrahend to the nearest thousand.

$$\begin{array}{rcr} 8{,}253 & \approx & 8{,}000 \\ -1{,}994 & \approx & -2{,}000 \\ \hline & & 6{,}000 \end{array} \leftarrow \text{Estimated difference}$$

Our exact answer (6,259) and the estimated difference (6,000) are fairly close.

With practice, we can mentally estimate and check differences and sums quickly and easily.

PRACTICE 9

Find the difference between 17,836 and 15,045. Then estimate this difference to check.

We have already seen how estimation helps us to check an exact answer that is computed. But sometimes an approximate answer is good enough.

EXAMPLE 10

The top two languages used online are English and Chinese, with 478,582,273 and 383,212,617 Internet users, respectively. Estimate how many more English users there are than Chinese. (*Source:* internetworldstats.com)

Solution

$$\begin{array}{rrr} \text{English users} \rightarrow 478{,}582{,}273 & \approx & 500{,}000{,}000 \\ \text{Chinese users} \rightarrow 383{,}212{,}617 & \approx & \underline{-400{,}000{,}000} \\ & & 100{,}000{,}000 \end{array}$$

So there are about 100,000,000 more English users than there are Chinese users.

PRACTICE 10

Mount Everest is the highest mountain in the world, with an altitude of 29,028 feet above sea level. By contrast, Mauna Kea in Hawaii is considered to be the world's tallest mountain when its height is measured from the nearby ocean floor. Mauna Kea has an altitude of only 13,796 feet above sea level. Estimate the difference in altitude above sea level of the two peaks. (*Source:* geology.com)

Adding and Subtracting Whole Numbers on a Calculator

Calculators are handy and powerful tools for carrying out complex computations. But it is easy to press a wrong key, so be sure to estimate the answer and compare this estimate to the displayed answer to see if it is reasonable.

EXAMPLE 11

On a calculator, compute the sum of 3,125 and 9,391.

Solution

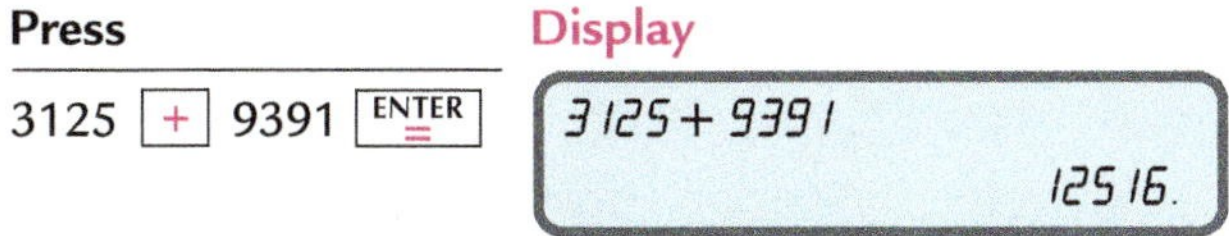

To check this answer, we mentally round the addends and then add.

$$\begin{array}{rrr} 3125 & \approx & 3{,}000 \\ 9391 & \approx & \underline{9{,}000} \\ & & 12{,}000 \end{array}$$

This estimate is reasonably close to our exact answer 12,516.

PRACTICE 11

Use a calculator to add: 39,822 + 9,710

Pressing the clear key cancels the number in the display. Press this key after completing a computation to be sure that no number remains to affect the next problem. Note that calculator models vary as to how they work, so it may be necessary to consult the manual for a particular model.

EXAMPLE 12

Calculate: $39 + 48 + 277$

Solution

Press	Display
39 [+] 48 [+] 277 [ENTER]	39+48+277 364.

A reasonable estimate is the sum of 40, 50, and 300, or 390—close to our calculated answer 364.

PRACTICE 12

Find the sum on a calculator:
$23{,}801 + 7{,}116 + 982$

When using a calculator to subtract,

- enter the numbers in the correct order—first enter the number **from which** we are subtracting and then the number **being** subtracted; and
- do not confuse the *negative sign key* [(−)] that some calculators have with the *subtraction key* [−].

EXAMPLE 13

Subtract on a calculator: $3{,}000 - 973$

Solution

A good estimate is $3{,}000 - 1{,}000$, or 2,000, which is close to 2,027.

PRACTICE 13

Use a calculator to find the difference between 5,280 feet and 2,781 feet.

1.2 Exercises

FOR EXTRA HELP MyMathLab MathXL PRACTICE WATCH READ REVIEW

Mathematically Speaking

Fill in each blank with the most appropriate term or phrase from the given list.

subtrahend	addends	left	estimates
commutative property of addition	right	identity property of addition	associative property of addition
	difference	sum	
	minuend		

1. The operation of addition can be thought of as moving to the ________ on a number line.

2. The __________________________ states that the sum of a number and zero is the original number.

3. The result of addition is called the ________.

4. The __________________________ states that changing the order in which two numbers are added does not affect the sum.

5. The __________________________ states that when adding three numbers, regrouping addends gives the same sum.

6. In an addition problem, the numbers being added are called ______________.

7. In a subtraction problem, the number being subtracted is called the ____________.

8. The result of subtraction is called the ___________.

A *Add and check by estimation.*

9. 100,250 + 77,528

10. 200,325 + 67,629

11. 8,132 + 6,578

12. 6,725 + 5,386

13. 7,481 + 702 + 5,819

14. 8,721 + 306 + 6,627

15. 49,002 + 1,999 + 5,187

16. 55,998 + 2,988 + 3,126

17. 1,903 + 5,075

18. 7,406 + 2,381

19. 800 + 20 + 4,000

20. 3,000 + 800 + 60

21. 31 + 93 + 277 + 12

22. 418 + 47 + 365 + 95

23. 3,911 + 2,947 + 8,007

24. 5,374 + 4,055 + 2,073

25. 6,482 meters + 9,027 meters

26. 17,812 miles + 4,283 miles

27. 35 hours + 47 hours

28. 25 square feet + 96 square feet

29. \$92,258 + \$7,447 + \$5,126

30. \$55,709 + \$2,822 + \$30,819

31. \$1,863 + \$1,089 + \$9,772

32. \$5,009 + \$7,993 + \$1,026

33. 8,300 tons + 22,900 tons

34. 7,400 tons + 32,600 tons

35. 3,088,281 + 5,658,137 + 4,550,239

36. 2,008,490 + 8,948,227 + 11,956,174

37. 638,719 + 40,003 + 984,035

38. 938,722 + 25,411 + 517,827

In each addition table, fill in the empty spaces. Check that the sum in the shaded empty space is the same working both downward and across.

39.

+	400	200	1,200	300	Total
300					
800					
Total					

40.

+	4,000	300	3,000	2,000	Total
100					
900					
Total					

41.

+	389	172	1,155	324	Total
255					
799					
Total					

42.

+	3,749	279	2,880	1,998	Total
134					
896					
Total					

In each group of three sums, one is **wrong**. *Use estimation to identify which sum is incorrect.*

43. a. $\begin{array}{r} 814 \\ 9{,}106 \\ +2{,}811 \\ \hline 15{,}731 \end{array}$ **b.** $\begin{array}{r} 30{,}812 \\ 47{,}045 \\ +\ 9{,}338 \\ \hline 87{,}195 \end{array}$ **c.** $\begin{array}{r} 183{,}066 \\ 78{,}911 \\ +\ 96{,}527 \\ \hline 358{,}504 \end{array}$

44. a. $\begin{array}{r} 1{,}035 \\ 5{,}210 \\ +7{,}992 \\ \hline 14{,}237 \end{array}$ **b.** $\begin{array}{r} 5{,}801 \\ 3{,}882 \\ +12{,}644 \\ \hline 32{,}327 \end{array}$ **c.** $\begin{array}{r} 801{,}716 \\ 78{,}001 \\ +5{,}009{,}635 \\ \hline 5{,}889{,}352 \end{array}$

45. a. $\begin{array}{r} \$711{,}488 \\ 102{,}663 \\ +\ 95{,}003 \\ \hline \$809{,}154 \end{array}$ **b.** $\begin{array}{r} \$62{,}933 \\ 51{,}858 \\ +\ 49{,}612 \\ \hline \$164{,}403 \end{array}$ **c.** $\begin{array}{r} \$106{,}729 \\ 99{,}821 \\ +\ 103{,}277 \\ \hline \$309{,}827 \end{array}$

46. a. $\begin{array}{r} \$9{,}512{,}622 \\ 8{,}038{,}517 \\ +\ 2{,}615{,}334 \\ \hline \$20{,}166{,}473 \end{array}$ **b.** $\begin{array}{r} \$4{,}277{,}020 \\ 915{,}611 \\ +\ 3{,}688{,}402 \\ \hline \$8{,}881{,}033 \end{array}$ **c.** $\begin{array}{r} \$200{,}312 \\ 102{,}683 \\ +\ 504{,}113 \\ \hline \$707{,}108 \end{array}$

Subtract and check.

47. $\begin{array}{r} 379 \\ -162 \\ \hline \end{array}$ **48.** $\begin{array}{r} 362 \\ -120 \\ \hline \end{array}$ **49.** $\begin{array}{r} 200 \\ -110 \\ \hline \end{array}$ **50.** $\begin{array}{r} 210 \\ -100 \\ \hline \end{array}$

51. $\begin{array}{r} 401 \\ -\ 39 \\ \hline \end{array}$ **52.** $\begin{array}{r} 728 \\ -\ 99 \\ \hline \end{array}$ **53.** $\begin{array}{r} 70{,}000 \\ -\ 1{,}759 \\ \hline \end{array}$ **54.** $\begin{array}{r} 80{,}000 \\ -\ 1{,}691 \\ \hline \end{array}$

55. $\begin{array}{r} 5{,}062 \\ -2{,}777 \\ \hline \end{array}$ **56.** $\begin{array}{r} 3{,}005 \\ -1{,}666 \\ \hline \end{array}$ **57.** $\begin{array}{r} 72{,}000 \\ -19{,}001 \\ \hline \end{array}$ **58.** $\begin{array}{r} 64{,}000 \\ -21{,}005 \\ \hline \end{array}$

59. $\begin{array}{r} 3{,}000 \\ -\ \ 57 \\ \hline \end{array}$ **60.** $\begin{array}{r} 7{,}000 \\ -\ \ 32 \\ \hline \end{array}$ **61.** $\begin{array}{r} 261{,}406 \\ -\ 57{,}941 \\ \hline \end{array}$ **62.** $\begin{array}{r} 729{,}888 \\ -\ 92{,}889 \\ \hline \end{array}$

Find the difference and check.

63. 550 − 182

64. 962 − 448

65. 6,000 − 1,004

66. 8,000 − 2,007

67. 3,570 − 2,588

68. 4,620 − 1,756

69. 5,000 miles − 3,005 miles

70. 4,000 miles − 1,008 miles

71. $800 − $131

72. $622 − $137

73. $4,812 − $1,203

74. $5,923 − $2,304

75. 500 books − 227 books

76. 537 pens − 196 pens

77. 527 meters − 318 meters

78. 642 meters − 214 meters

79. 30,000,000 − 27,999,000

80. 40,000,000 − 18,988,000

81. 13,402,331 − 12,588,902

82. 14,500,007 − 13,972,008

In each group of three differences, one is **wrong.** *Use estimation to identify which difference is incorrect.*

83. **a.** 817,770 − 502,966 = 314,804 **b.** 11,172,055 − 7,892,106 = 3,279,949 **c.** 71,384,612 − 32,016,594 = 29,368,018

84. **a.** 67,812 − 12,180 = 55,632 **b.** 3,997,401 − 1,125,166 = 1,872,235 **c.** 316,134 − 89,164 = 226,970

85. **a.** $381,882 − 173,552 = $108,330 **b.** $479,116 − 102,663 = $376,453 **c.** $200,072,639 − 150,038,270 = $50,034,369

86. **a.** $3,810,662 − 299,137 = $3,511,525 **b.** $4,718,287 − 1,002,875 = $5,721,162 **c.** $381,975 − 117,263 = $264,712

Mixed Practice

Perform the indicated operation.

87. 7,415 − 350

88. 90,316 + 10,882 + 5,281

89. 281 + 758 + 104 + 533

90. $5,233 + $481 + $82

91. 8,286 − 3,100

92. 410,700 miles − 280,900 miles

Applications

B *Solve and check.*

93. In 1900, the population of the United States was approximately 76,000,000. During the next 100 years, the population grew by about 205,000,000 people. What was the population in 2000? (*Source:* U.S. Bureau of the Census)

94. It is recommended that an adult over 50 consume 1,200 milligrams of calcium each day. Will an adult who has 645 milligrams of calcium at breakfast and 455 milligrams at lunch meet the recommended daily amount? (*Source:* niams.nih.gov)

95. Of the 6,000,000 square miles of tropical rainforest that existed on Earth before deforestation, 2,600,000 square miles remain. How many square miles of rainforest were destroyed? (*Source:* nature.org)

96. The Great Blue Norther of 1911 was the largest cold snap in U.S. history. On the day of the Norther, the temperature in Oklahoma City dropped from a record high of 83°F to a record low of 17°F in a 24-hour period. By how much did the temperature drop that day? (*Source:* National Weather Service)

97. The chart shows the 2010 Winter Olympic medal counts of selected countries.

Country	Gold	Silver	Bronze
Austria	4	6	6
Canada	14	7	5
Germany	10	13	7
Norway	9	8	6
United States	9	15	13

(*Source:* nbcolympics.com)

a. Calculate the total number of medals won by each country.

b. Which country won the most medals?

98. Consider the deposit slip shown.

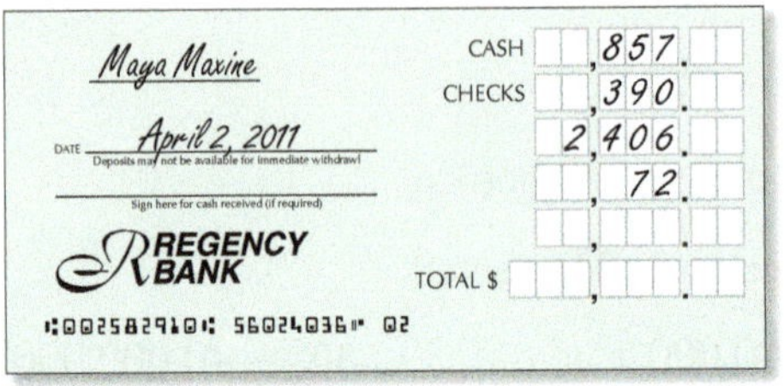

a. Estimate how much money is being deposited.

b. Fill in the exact total.

99. Blues singer Bessie Smith was born in 1894 and died in 1937. About how old was she when she died? (*Source: Encyclopedia of World Biography*)

100. The United States entered the First World War in 1917 and the Second World War in 1941. Approximately how many years apart were these two events?

101. A sign in an elevator reads: MAXIMUM CAPACITY: 1,000 POUNDS. The passengers in the elevator weigh 187 pounds, 147 pounds, 213 pounds, 162 pounds, 103 pounds, and 151 pounds. Will the elevator be overloaded?

102. According to a magazine article on the recent annual income of celebrities, the rapper 50 Cent earned \$150,000,000 whereas the talk-show host Oprah Winfrey earned \$125,000,000 more. If J.K. Rowling, the author of the *Harry Potter* series, earned the equivalent of \$300,000,000, who earned more—Rowling or Winfrey? (*Source: Forbes*)

103. The thermometer at the right shows the boiling point and the freezing point of water in degrees Fahrenheit (°F). What is the difference between these two temperatures?

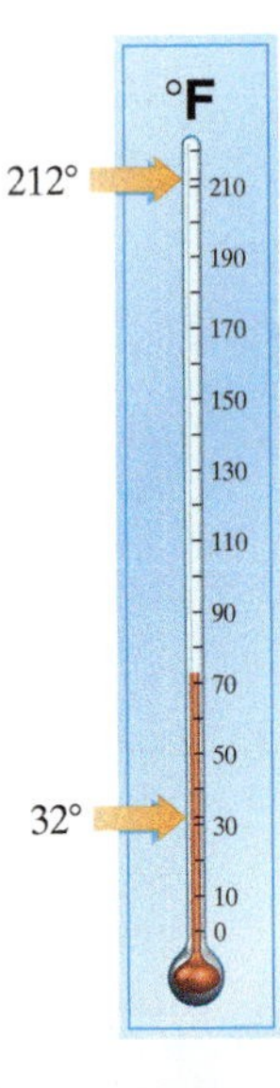

104. The following ad for a hybrid car was listed in a local newspaper. How much below the MSRP (manufacturer's suggested retail price) is the selling price?

105. Some friends drive from town A to town B, to town C, to town D, and then back to A, as shown below. How far did they drive in all?

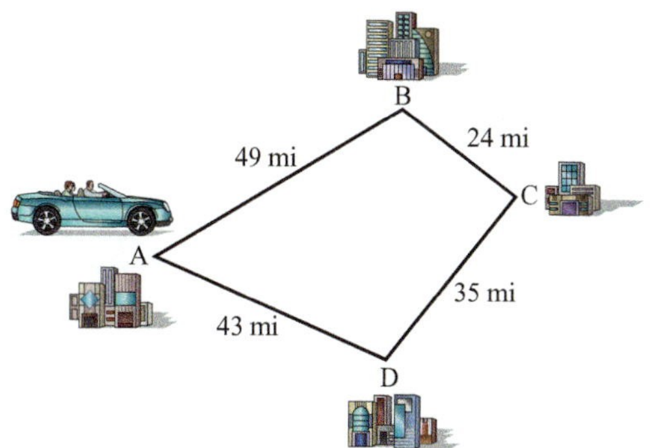

106. What is the length of the molding along the perimeter of the room pictured (yards = yd)?

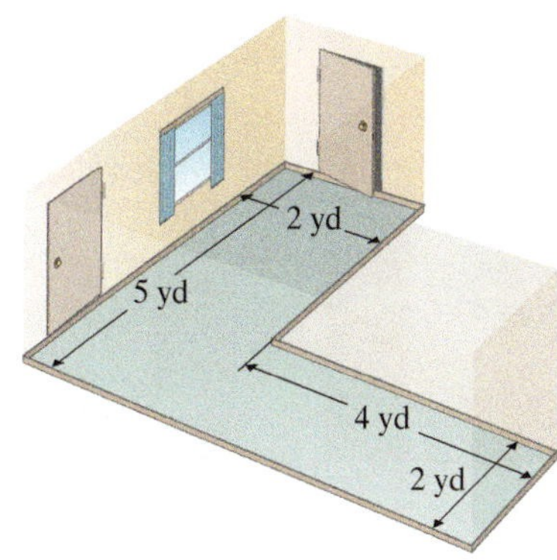

107. In a particular month, 163,038,000 home and work Internet users visited a site or launched an application owned by Google whereas 143,635,000 used a Microsoft-owned site or application. How many more users did Google have than Microsoft? (*Source:* nielsen.com)

108. In a recent year, the United States imported $21,615,426 in fish products from Japan. In the same year, the United States exported $19,579,138 in fish products to Japan. What is the difference between these two trade amounts? (*Source:* st.nmfs.noaa.gov)

109. The following table lists the number of deaths in New York City for a particular week.

Age Group	Number of Deaths
65 and over	1,049
45–64	303
25–44	67
1–24	21
Less than 1	14

(*Source:* cdc.gov)

How many deaths were there in New York City during that week?

110. An oil tanker broke apart at sea. It spilled 150,000 gallons (gal) of crude oil the first day, 400,000 gal the second day, and 1,000,000 gal the third day. How much oil was spilled in all?

111. The large auditorium at Carnegie Hall in New York City has been the site of many great concerts since the hall opened in 1891. The auditorium has five sections with the following seating capacity: Parquet—1,021, First Tier Boxes—264, Second Tier Boxes—238, Dress Circle—444, and Balcony—837. (*Source:* carnegiehall.org)

a. Is the total seating capacity of the auditorium more or less than 3,000?

b. For some concerts, 128 stage seats are also placed in the auditorium. What is the total seating capacity for these concerts?

112. A student has a credit card with a credit line of $3,000. On this card, there is a balance due of $1,369.

a. If the student uses this credit card to purchase an iPad computer for $499, how much credit is still available on the card?

b. If, instead of purchasing the computer, the student wishes to pay a tuition bill of $1,575, will there be enough credit available to pay the bill?

113. The following graph gives estimates of the number of species for various kinds of insects.

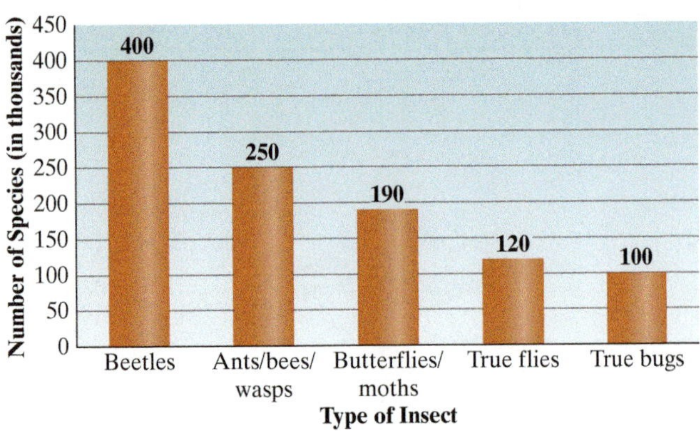

(*Source: Top Ten of Everything 2010*)

a. How many more beetle species are there than true-fly species?

b. Find the total number of species for all the insect types shown.

c. How many fewer beetle species are there than species of the other insect types shown?

114. The following graph shows the number of cases of measles for selected years in the United States as reported by the Centers for Disease Control (CDC).

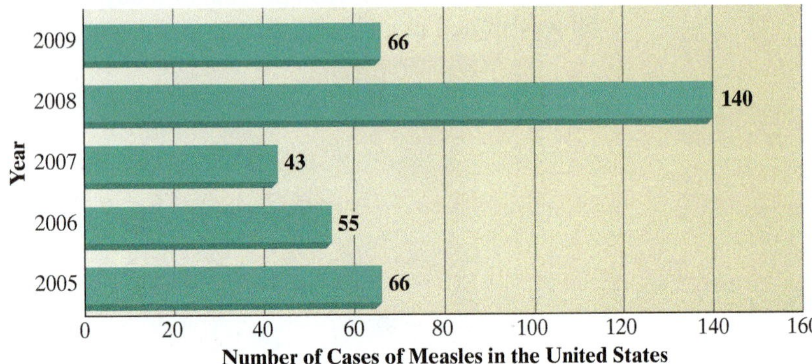

(*Source:* wonder.cdc.gov)

a. How many more cases of measles were reported in 2009 than in 2007?

b. What is the total number of cases reported for the years 2005–2009?

c. How many fewer cases were reported in 2008 than the other years combined?

115. A salesman for a car dealership works on commission. The following table gives his sales commission for each of the first six months of the year.

Month	Sales Commission
January	\$5,416
February	\$3,791
March	\$5,072
April	\$3,959
May	\$6,283
June	\$4,055

What is his total commission for the first half of the year?

116. At its first eight games this season, a professional baseball team had the following paid attendance:

Game	Attendance
1	11,862
2	18,722
3	14,072
4	9,713
5	25,913
6	28,699
7	19,302
8	18,780

What was the combined attendance for these games?

• Check your answers on page A-1.

MIND*Stretchers*

Writing

1. There are many different ways of putting numerical expressions into words.

   a. For example, 3 + 2 can be expressed as

   the sum of 3 and 2, 2 more than 3, or 3 increased by 2

   What are some other ways of reading this expression?

   b. For example, 5 − 2 can be expressed as

   the difference between 5 and 2, 5 take away 2, or 5 decreased by 2

   Write two other ways.

Critical Thinking

2. In a **magic square**, the sum of every row, column, and diagonal is the same number. Using the given information, complete the square at the right, which contains the whole numbers from 1 to 16. (*Hint:* The sum of every row, column, and diagonal is 34.)

16	3	2	
	10	11	
	6	7	

Groupwork

3. Two methods for borrowing in a subtraction problem are illustrated as follows. In method (a)—the method that we have already discussed—we borrow by taking 1 from the top, and in method (b) by adding 1 to the bottom.

a.
$$\begin{array}{r} \overset{7}{\cancel{8}}\ {}^{1}5\ \ 9 \\ -3\ \ 7\ \ 6 \\ \hline 4\ \ 8\ \ 3 \end{array}$$

b.
$$\begin{array}{r} 8\ \ {}^{1}5\ \ 9 \\ -\overset{4}{\cancel{3}}\ \ 7\ \ 6 \\ \hline 4\ \ 8\ \ 3 \end{array}$$

Note that we get the same answer with both methods. Working with a partner, discuss the advantages of each method.

1.3 Multiplying Whole Numbers

OBJECTIVES

- **A** To multiply whole numbers
- **B** To solve applied problems involving the multiplication of whole numbers

The Meaning and Properties of Multiplication

What does it mean to multiply whole numbers? A good answer to this question is *repeated addition.*

For instance, suppose that you buy 4 packages of pens and each package contains 3 pens. How many pens are there altogether?

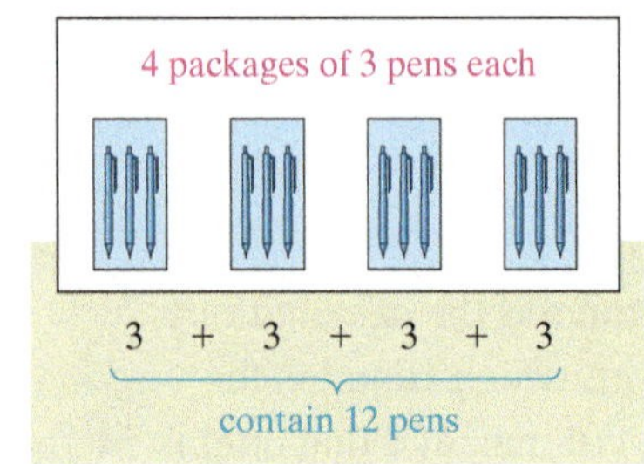

That is, $4 \times 3 = 3 + 3 + 3 + 3 = 12$. Generally, *multiplication means adding the same number repeatedly.*

We can also picture multiplication in terms of a rectangular figure, like this one, that represents 4×3.

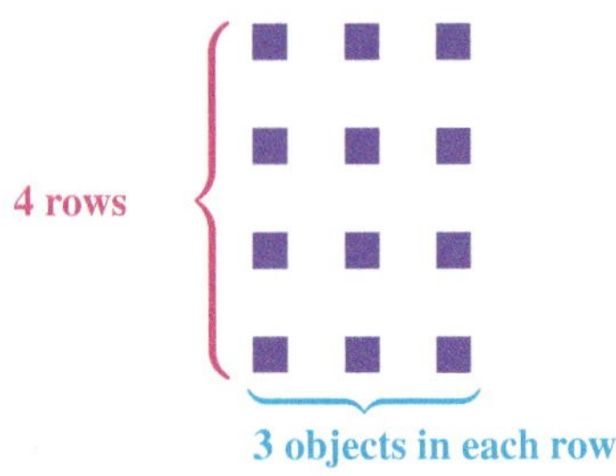

In a multiplication problem, the numbers being multiplied are called *factors*. The result is the *product*.

There are several ways to write a multiplication problem.

		Factor ↓		Factor ↓		Product ↓
×	the times sign	4	×	3	=	12
·	a multiplication dot	4	·	3	=	12
( )( )	parentheses	(4)(3)			=	12

Like addition and subtraction, multiplication problems can be written either horizontally or vertically.

$8 \times 5 = 40$
Horizontal

$$\begin{array}{r} 8 \\ \times\ 5 \\ \hline 40 \end{array}$$
Vertical

The operation of multiplication has several important properties that we use frequently.

The Identity Property of Multiplication

The product of any number and 1 is that number.

$$1 \times 12 = 12$$
$$5 \times 1 = 5$$

The Multiplication Property of 0

The product of any number and 0 is 0.

$$49 \times 0 = 0$$
$$0 \times 8 = 0$$

The Commutative Property of Multiplication

Changing the order in which two numbers are multiplied does not affect their product.

$$2 \times 9 = 9 \times 2$$
$$\downarrow \qquad \downarrow$$
$$18 = 18$$

The Associative Property of Multiplication

When multiplying three numbers, regrouping the factors gives the same product.

We multiply inside the parentheses first.

$$(3 \times 4) \times 5 = 3 \times (4 \times 5)$$
$$\downarrow \qquad\qquad \downarrow$$
$$12 \times 5 = 3 \times 20$$
$$\downarrow \qquad \downarrow$$
$$60 = 60$$

The next—and last—property of multiplication also involves addition.

The Distributive Property

Multiplying a factor by the sum of two numbers gives the same result as multiplying the factor by each of the two numbers and then adding.

$$2 \times (5 + 3) = (2 \times 5) + (2 \times 3)$$
$$2 \times 8 = 10 + 6$$
$$\downarrow \qquad \downarrow$$
$$16 = 16$$

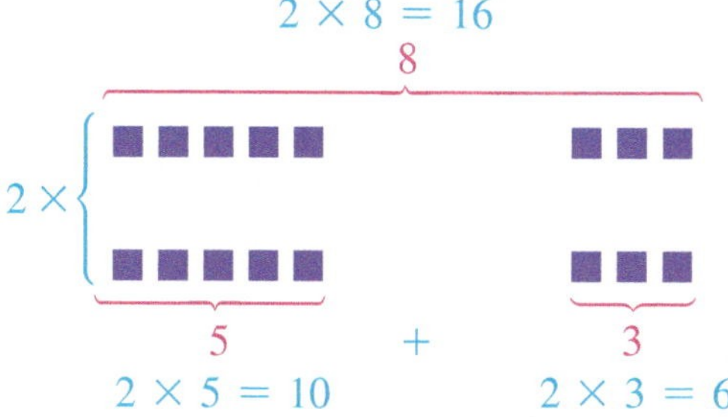

Before going on to the next section, study these properties of multiplication.

Multiplying Whole Numbers

Now, let's consider problems in which we multiply any whole number by a single-digit whole number.

Note that, to multiply whole numbers with reasonable speed, you must commit to memory the products of all single-digit whole numbers.

EXAMPLE 1

Multiply: $98 \cdot 4$

Solution We recall that the dot means multiplication. We first write the problem vertically.

$$\begin{array}{r} {}^{3} \\ 98 \\ \times\ 4 \\ \hline 2 \end{array}$$

← The product of 4 and 8 ones is 32 ones. We write the 2 and carry the 3 to the tens column.

We recall that the 9 in 98 means 9 tens.

$$\begin{array}{r} {}^{3} \\ 98 \\ \times\ \ 4 \\ \hline 392 \end{array}$$

← The product of 4 and 9 tens is 36 tens. We add the 3 tens to the 36 tens to get 39 tens.

So the product of 98 and 4 is 392.

PRACTICE 1

Find the product of 76 and 8.

EXAMPLE 2

Calculate: (806) (7)

Solution We recall that parentheses side-by-side mean to multiply. We write this problem vertically.

$$\begin{array}{r} {}^{4} \\ 806 \\ \times\ \ \ 7 \\ \hline 5{,}642 \end{array}$$

Here, 7×0 tens $= 0$ tens. Add the 4 tens to the 0 tens to get 4 tens.

The product of 806 and 7 is 5,642.

PRACTICE 2

Find the product: (705)(6)

Now, let's look at multiplying any two whole numbers.
Consider multiplying 32 by 48. We can write 32×48 as follows.

$$32 \times 48 = 32 \times (40 + 8)$$

We then use the distributive property to get the answer.

$$\begin{aligned} 32 \times (40 + 8) &= (32 \times 40) + (32 \times 8) \\ &= 1{,}280 + 256 \\ &= 1{,}536 \end{aligned}$$

Generally, we solve this problem vertically.

$$\begin{array}{r} {}^{1} \\ 32 \\ \times\ 48 \\ \hline 256 \\ 1280 \\ \hline 1{,}536 \end{array}$$

- 256 ← Partial product (8×32)
- 1280 ← Partial product (40×32)
- 1,536 ← Add the partial products.

Shortcut

$$\begin{array}{r} 32 \\ \times\ 48 \\ \hline 256 \\ 128 \\ \hline 1{,}536 \end{array}$$

- 256 ← (8×32)
- 128 ← (4×32)

If we use just the tens digit 4, we must write the product 128 leftward, starting at the tens column.

Example 2 suggests the following rule for multiplying whole numbers:

To Multiply Whole Numbers

- Multiply the top factor by the ones digit in the bottom factor, and write down this product.
- Multiply the top factor by the tens digit in the bottom factor, and write this product leftward, beginning with the tens column.
- Repeat this process until all the digits in the bottom factor are used.
- Add the partial products, writing down this sum.

EXAMPLE 3

Multiply: 300×50

Solution

```
    300
  × 50
    000 ← 0 × 300 = 0
  15 00  ← 5 × 300 = 1,500
  15,000
```

In Example 3, note that the number of zeros in the product equals the total number of zeros in the factors. This result suggests a shortcut for multiplying factors that end in zeros.

```
     300 ← 2 zeros
  ×   50 ← 1 zero
  15,000 ← 2 + 1 = 3 zeros
```

PRACTICE 3

Find the product of 1,200 and 400.

TIP When multiplying two whole numbers that end in zeros, multiply the nonzero parts of the factors and then attach the total number of zeros to the product.

EXAMPLE 4

Simplify: $739 \cdot 305$

Solution

```
     739
  ×  305
    3 695 ← 5 × 739
    0 00  ← 0 × 739 = 0
  221 7   ← 3 × 739
  225,395
```

We don't have to write the row 000. Here is a shortcut.

```
     739
  ×  305
    3 695
  221 70  ← This one 0 represents the product of the tens digit 0
  225,395   and 739. This 0 lines up the products correctly.
```

PRACTICE 4

Find the product of 987 and 208.

Now, let's apply the operation of multiplication to geometric area. Area means the number of square units that a figure contains.

In the rectangle at the right, each small square represents 1 square inch (sq in.). Finding the rectangle's area means finding the number of sq-in. units that it contains. A good strategy here is to find the number of units in each row and then multiply that number by the number of rows.

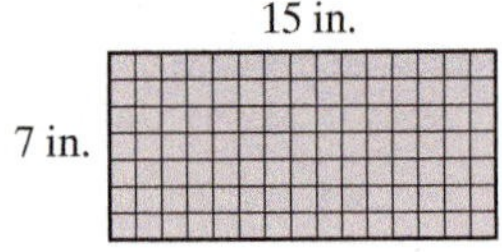

There are two ways to find that there are 15 squares in a row—either by directly counting the squares or by noting that the length of the figure is 15 in. Similarly, we find that the figure contains 7 rows. Therefore the area of the figure is 15 × 7, or 105 sq in.

In general, we can compute the *area of a rectangle* by finding the product of its length and its width.

EXAMPLE 5

Calculate the area of the home office shown in the diagram.

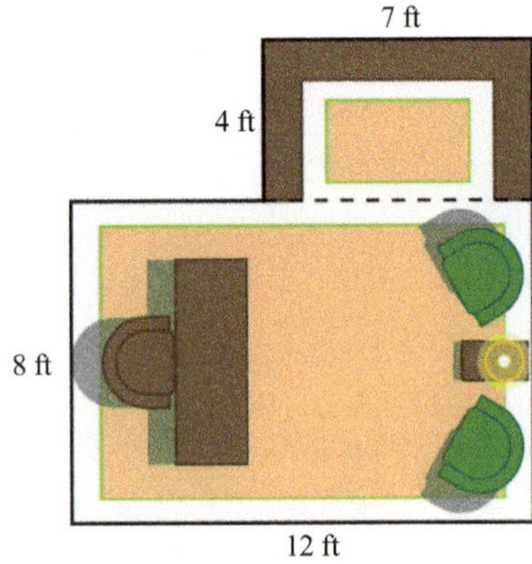

Solution The dashed line separates the office into two connected rectangles. The top rectangle measures 7 feet by 4 feet, and so its area is 7 × 4, or 28 square feet. The bottom rectangle measures 12 feet by 8 feet, and its area is 12 × 8, or 96 square feet. The entire area of the office is the sum of two smaller areas: 28 + 96, or 124 square feet. So the area of the home office is 124 square feet.

PRACTICE 5

Find the area of the room pictured.

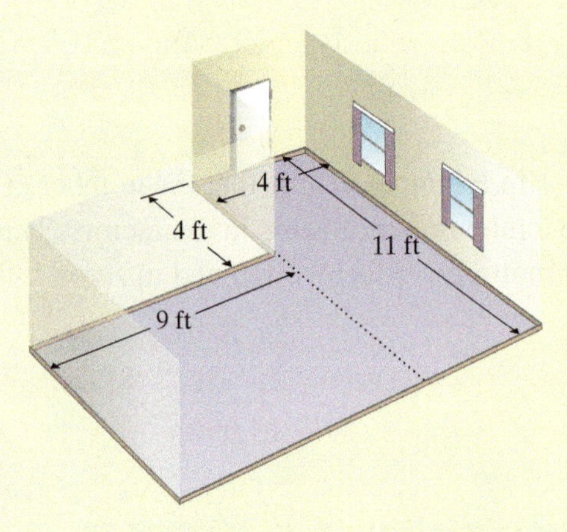

Estimating Products

As mentioned before, estimation is a valuable technique for checking an exact answer. When checking a product by estimation, round each factor to its largest place.

EXAMPLE 6

Multiply 328 by 179. Check the answer by estimation.

Solution

```
    328
  × 179
  -----
  2 952
 22 96
 32 8
 ------
 58,712 ← Exact product
```

PRACTICE 6

Find the product of 455 and 248. Use estimation to check your answer.

Check

$$\begin{array}{rcrl} 328 & \approx & 300 & \leftarrow \text{The largest place is hundreds.} \\ \times\ 179 & \approx & \times\ 200 & \leftarrow \text{The largest place is hundreds.} \\ \hline 58{,}712 & & 60{,}000 & \leftarrow \text{Estimated product} \end{array}$$

Our exact product (58,712) and the estimated product (60,000) are fairly close.

When solving some multiplication problems, we are willing to settle for—or even prefer—an approximate answer.

EXAMPLE 7

The director of a preschool budgeted \$900 to purchase supplies for an upcoming art project. She found handprint keepsake craft kits online for \$12 each for the 56 children in the preschool. By estimating, decide if the director set aside enough money to purchase craft kits for all the children.

Solution The total cost of the craft kits is the product of \$12 and 56. To estimate this product, we first round each factor to its largest place value so that every digit after the first digit is 0.

$$\begin{array}{rcrl} 12 & \approx & 10 & \leftarrow \text{The largest place is tens.} \\ \times\ 56 & \approx & 60 & \leftarrow \text{The largest place is tens.} \end{array}$$

Then, we multiply the rounded factors.

$$10 \times 60 = 600$$

Since the craft kits will cost about \$600 and \$900 is greater than \$600, we conclude that the director set aside enough money for the craft kits.

PRACTICE 7

Producing flyers for your college's registration requires 25,000 sheets of paper. If the college buys 38 reams of paper and there are 500 sheets in a ream, estimate to decide if there is enough paper to produce the flyers.

Multiplying Whole Numbers on a Calculator

Now, let's use a calculator to find a product. When you are using a calculator to multiply large whole numbers, the answer may be too big to fit in the display. When this occurs the answer may be displayed in scientific notation (see the Appendix).

EXAMPLE 8

Use a calculator to multiply: 3,192 × 41

Solution

A reasonable estimate for this product is 3,000 × 40, or 120,000, which supports our answer, 130,872.

PRACTICE 8

Find the product: 2,811 × 365

EXAMPLE 9

Calculate: 61 · 24 · 19

Solution

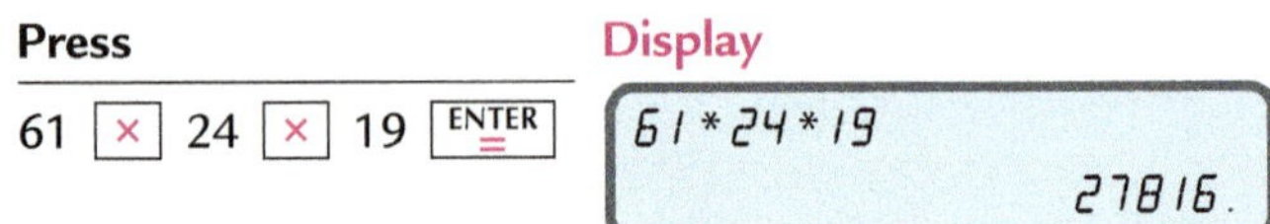

A good estimate is 60 · 20 · 20, or 24,000—in the ballpark of 27,816.

PRACTICE 9

Multiply: 2,133 · 18 · 9

1.3 Exercises

FOR EXTRA HELP MyMathLab    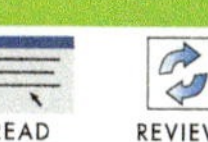

Mathematically Speaking

Fill in each blank with the most appropriate term or phrase from the given list.

associative property of multiplication	identity property of multiplication	multiplication property of 0	sum	subtraction
distributive property	perimeter	area	product	addition

1. The result of multiplying two factors is called their __________.

2. The ________________ is illustrated by $3 \times (7 + 2) = (3 \times 7) + (3 \times 2)$.

3. The __________________________ states that the product of any number and 1 is that number.

4. The ______________________ states that the product of any number and 0 is 0.

5. The multiplication of whole numbers can be thought of as repeated __________.

6. The __________ of a figure is the number of square units that it contains.

A *Compute.*

7. 4×100
8. 5×100
9. 710×200
10. 270×500
11. $8{,}500 \times 20$
12. $6{,}800 \times 30$
13. $10{,}000 \times 700$
14. $10{,}000 \times 800$

Multiply and check by estimation.

15. $6{,}350 \times 2$
16. $8{,}864 \times 7$
17. 209×2
18. 703×9
19. $812{,}000 \times 4$
20. $19{,}250 \times 8$
21. 882×74
22. 881×28
23. $43 \cdot 19$
24. $85 \cdot 72$
25. $709 \cdot 48$
26. $602 \cdot 34$
27. 273×11
28. 607×65
29. 301×12
30. 513×34
31. $3{,}001 \times 19$
32. $4{,}005 \times 72$
33. $5{,}072 \times 48$
34. $8{,}801 \times 25$
35. $5{,}003 \times 40$
36. $2{,}881 \times 70$

Find the product and check by estimation.

37. (372)(403)
38. (699)(101)
39. $8{,}500 \times 17$
40. $7{,}200 \times 27$
41. 406×305
42. 702×509
43. $46 \cdot 8 \cdot 9$
44. $13 \cdot 11 \cdot 5$

45. $81 \times 2 \times 13$

46. $3 \times 15 \times 88$

47. (10)(10)(400)

48. (20)(80)(30)

49. $57 \times 81 \times 5$

50. $73 \times 4 \times 33$

51. $\begin{array}{r} 8{,}972 \\ \times \ \ 365 \\ \hline \end{array}$

52. $\begin{array}{r} 7{,}552 \\ \times \ \ 841 \\ \hline \end{array}$

53. $\begin{array}{r} 18{,}650 \\ \times \ 2{,}949 \\ \hline \end{array}$

54. $\begin{array}{r} 21{,}320 \\ \times \ 7{,}159 \\ \hline \end{array}$

In each group of three products, one is wrong. Use estimation to identify which product is incorrect.

55. **a.** $802 \times 755 = 605{,}510$ **b.** $39 \times 4{,}722 = 184{,}158$ **c.** $77 \times 6{,}005 = 46{,}385$

56. **a.** $618 \times 555 = 342{,}990$ **b.** $86{,}331 \times 21 = 18{,}129{,}511$ **c.** $380 \times 772 = 293{,}360$

57. **a.** $9 \times 37{,}118 = 334{,}062$ **b.** $82 \times 961 = 7{,}882$ **c.** $13 \times 986 = 12{,}818$

58. **a.** $3{,}002 \times 9 = 2{,}718$ **b.** $58 \times 891 = 51{,}678$ **c.** $106 \times 68 = 7{,}208$

Mixed Practice

Multiply and check by estimation.

59. $48 \cdot 5 \cdot 12$

60. $89 \times 10{,}000$

61. $\begin{array}{r} 9{,}605 \\ \times \ \ \ 24 \\ \hline \end{array}$

62. (809)(201)

63. $357{,}000 \times 3$

64. $301 \cdot 34$

65. (50)(60)(100)

66. 495×21

Applications

B *Solve. Then check by estimation.*

67. Underwater explorers in the eastern Mediterranean Sea found the wreck of an Egyptian ship that had sunk 33 centuries earlier. How long ago in years did the ship sink? (*Hint:* 1 century = 100 years)

68. Each day, an athlete in training takes two capsules. If each capsule contains 1,600 international units (IU) of vitamin A, how much vitamin A does he take daily?

69. The walls of a human heart are made of muscles that contract about 100,000 times a day. (*Source: American Heart Association's Your Heart: An Owner's Manual*)

a. How many contractions are there in 30 days?

b. How many more contractions are there in 40 days than in 30 days?

70. It is estimated that the United States needs to produce about 125,000 new jobs a month to maintain the present unemployment rate. (*Source:* theatlantic.com)

a. How many new jobs altogether would have to be created in a year for the present unemployment rate to remain constant?

b. If 800,000 new jobs are created in a year, how short of the goal is this?

71. The 2010 Honda Civic Hybrid gets about 45 miles per gallon of gasoline. If the fuel tank holds about 12 gallons of gasoline, can a person drive from San Francisco to Los Angeles, a distance of 276 miles, without refilling the car's fuel tank? (*Source:* automobiles.honda.com)

72. A Canadian football field is 330 ft long and 195 ft wide. By contrast, a football field in the National Football League measures 360 ft by 160 ft. Which field has a larger area? (*Source:* wikipedia.org)

73. Find the area of the countertop shown in the diagram.

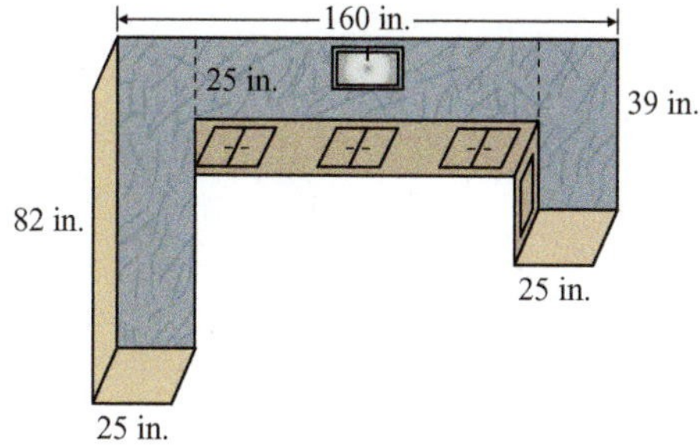

74. Calculate the area of the deck shown in the diagram.

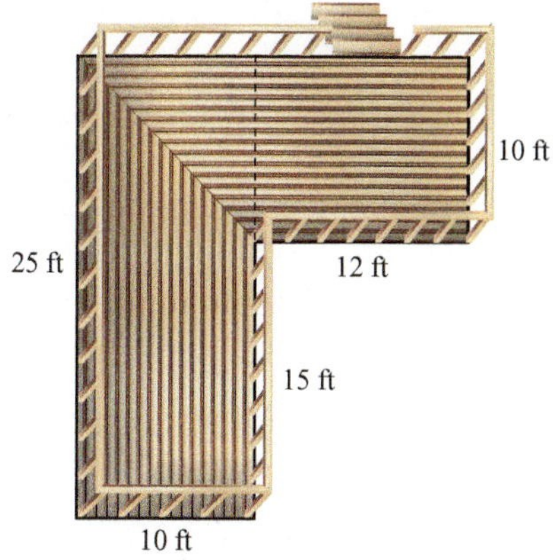

75. On the following map, 1 inch corresponds to 250 miles in the real world. How many miles actually separate towns A and B?

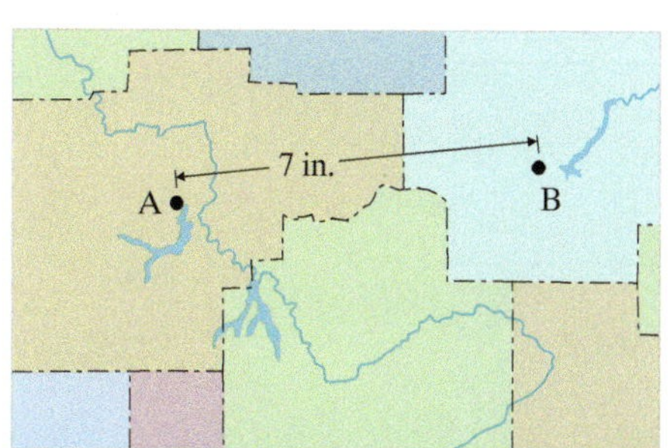

76. Angles are measured in either degrees (°) or radians (rad). A radian is about 57°. Express in degrees the measure of the angle shown.

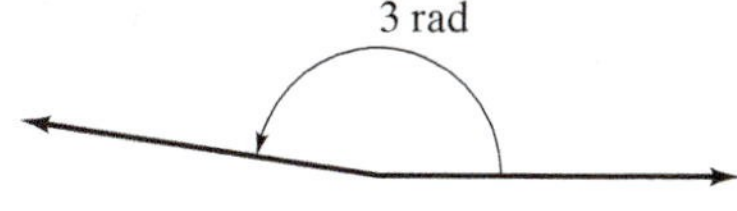

77. It costs $130 to join a local health club and $26 for each month of the membership. How much does a 1-year membership cost?

78. A customer bought an LCD television on the installment plan offered by an electronics store. If the total cost of the television is $1,689 and the customer pays $145 each month, how much does she have left to pay off after 8 months of payments?

79. During a 5-day week, a truck driver daily drove 42 miles an hour for 7 hours.

a. How far did she drive in one day?

b. How far did she drive during the week?

80. A young couple took out a mortgage on a condo. They paid $790 per month for 15 years.

a. How much did they pay annually toward the mortgage?

b. How much did it cost them to pay off the mortgage altogether?

81. The state of Colorado is approximately rectangular in shape, as shown. If the area of Kansas is about 82,000 sq mi, which state is larger? (*Source: The Columbia Gazeteer of the World*)

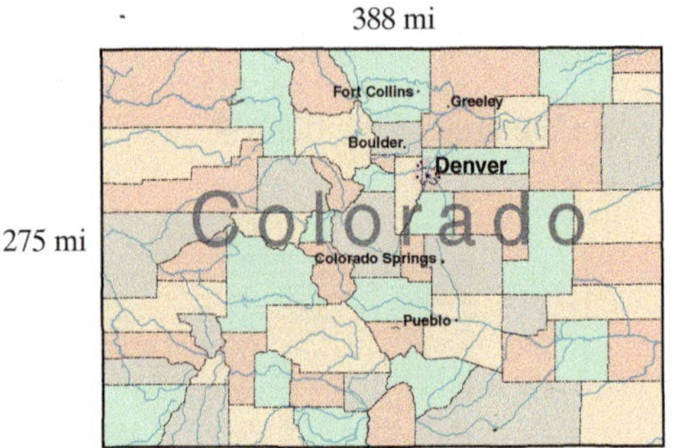

82. Tuition at a state college for full-time in-state residents is approximately $1,750 per semester. In a fall semester, there were 30,963 of these students. How much revenue did the full-time in-state resident students generate?

• Check your answers on page A-1.

MINDStretchers

Writing

1. Study the following diagram. Explain how it justifies the Distributive Property.

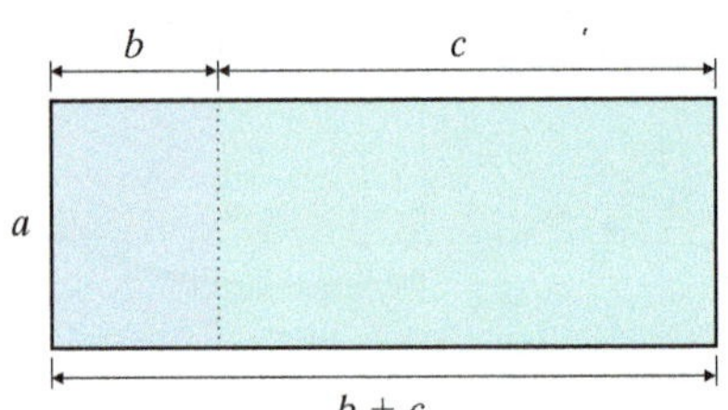

Mathematical Reasoning

2. Consider the six digits 1, 3, 5, 7, 8, and 9. Fill in the blanks with these digits, using each digit only once, so as to form the largest possible product.

Historical

3. Centuries ago in India and Persia, the **lattice method** of multiplication was popular. The following example, in which we multiply 57 by 43, illustrates this method. Explain how it works.

Cultural Note

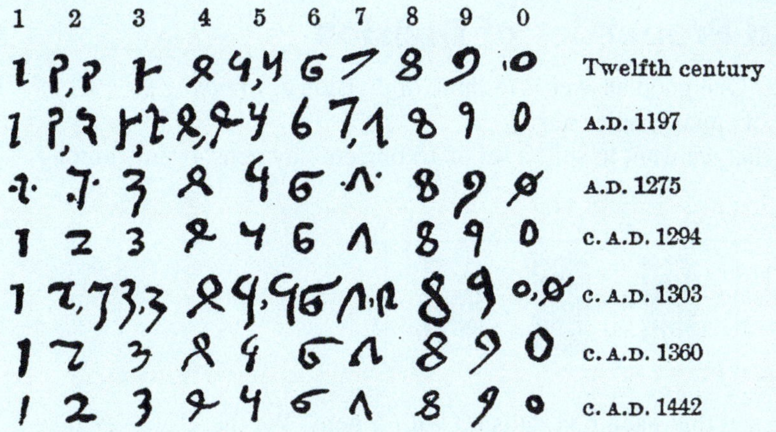

1 2 3 4 5 6 7 8 9 0

Twelfth century

A.D. 1197

A.D. 1275

c. A.D. 1294

c. A.D. 1303

c. A.D. 1360

c. A.D. 1442

The way the ten digits are written has evolved over time. Early Hindu symbols found in a cave in India date from more than two thousand years ago. About twelve hundred years ago, an Indian manuscript on arithmetic, which had been translated into Arabic, was carried by merchants to Europe where it was later translated into Latin.

This table shows European examples of digit notation from the twelfth to the fifteenth century, when the printing press led to today's standardized notation. Through international trade, these symbols became known throughout the world.

Source: David Eugene Smith and Jekuthiel Ginsburg, *Numbers and Numerals, A Story Book for Young and Old* (New York: Bureau of Publications, Teachers College, Columbia University, 1937)

1.4 Dividing Whole Numbers

OBJECTIVES

- **A** To divide whole numbers
- **B** To solve applied problems involving the division of whole numbers

The Meaning and Properties of Division

What does it mean to divide? One good answer is to think of division as *breaking up a set of objects* into a given number of equal smaller sets.

For instance, suppose that we want to split a set of 15 objects, say pens, evenly among 3 boxes.

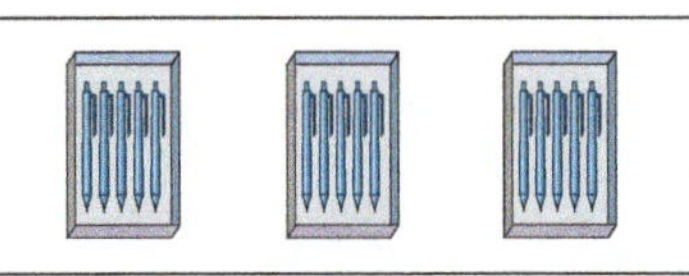

From the diagram we see that each box ends up with 5 pens. We therefore say that 15 divided by 3 is 5, which we can write as follows:

$$\begin{array}{rll} & 5 & \textbf{Quotient} \\ \textbf{Divisor} & 3\overline{)15} & \textbf{Dividend} \end{array}$$

In a division problem, the number that is being used to divide another number is called the *divisor*. The number being divided is the *dividend*. The result is the *quotient*.

We can also think of division as the *opposite* (*inverse*) of multiplication. Consider the following pair of problems that illustrate this point.

$$\begin{array}{ccc} \begin{array}{r} 5 \\ 3\overline{)15} \end{array} & \text{because} & 5 \times 3 = 15 \\ \textbf{Division} & & \textbf{Related multiplication} \end{array}$$

The following relationship connects multiplication and division.

$$\text{Quotient} \times \text{Divisor} = \text{Dividend}$$

Note that this relationship allows us to check our answer to a division problem by multiplying.

There are several common ways to write a division problem.

$$\begin{array}{r} 5 \\ 3\overline{)15}, \end{array} \quad \frac{15}{3} = 5, \quad \text{or} \quad 15 \div 3 = 5$$

Usually, we use the first of these to compute the answer. However, no matter which way we write this problem, 3 is the divisor, 15 is the dividend, and 5 is the quotient.

TIP When reading a division problem, we say that we are dividing either the divisor *into* the dividend or the dividend *by* the divisor. For instance, $3\overline{)15}$ is read either "3 divided into 15" or "15 divided by 3."

When calculating a quotient, we frequently use the following properties of division.

	Division	Related Multiplication
• Any whole number (except 0) divided by itself is 1.	$6\overline{)6}$ with quotient 1	$1 \times 6 = 6$
• Any whole number divided by 1 is the number itself.	$1\overline{)12}$ with quotient 12	$12 \times 1 = 12$
• Zero divided by any whole number (other than 0) is 0.	$8\overline{)0}$ with quotient 0	$0 \times 8 = 0$
• Division by 0 is not permitted.	$0\overline{)5}$ with quotient ?	$? \times 0 = 5$ **There is no number that when multiplied by 0 equals 5.**

Dividing Whole Numbers

Multiplication is the opposite of division. So in the simple division problem $3\overline{)15}$, we know that the answer is 5 because we have memorized that $5 \cdot 3$ is 15. But what should we do when the dividend is a larger number?

Consider the following problem: Divide 9 into 5,112 and check the answer.

- We start with the greatest place (thousands) in the dividend. We consider the dividend to be 5 thousands and think $9\overline{)5}$. Since $9 \cdot 1 = 9$ and 9 is larger than 5, there are no thousands in the quotient.

$$\begin{array}{r} 0 \leftarrow \text{Thousands} \\ 9\overline{)5{,}112} \end{array}$$

- So we go to the hundreds place in the dividend. We consider the dividend to be 51 hundreds and think $9\overline{)51}$. Since $9 \cdot \mathbf{5} = 45$, we position the **5** in the hundreds place of the quotient.

$$\begin{array}{rl} 5 & \leftarrow \text{Hundreds} \\ 9\overline{)5{,}112} & \\ -4{,}500 & \leftarrow 500 \cdot 9 = 4500 \\ \hline 612 & \leftarrow \text{Difference} \end{array}$$

- Next, we move to the tens place of the difference, 612. We consider the new dividend to be 61 tens and think $9\overline{)61}$. Since $9 \cdot \mathbf{6} = 54$, we position the **6** in the tens place of the quotient.

$$\begin{array}{rl} 56 & \leftarrow \text{Tens} \\ 9\overline{)5{,}112} & \\ -4{,}500 & \\ \hline 612 & \\ -540 & \leftarrow 60 \cdot 9 = 540 \\ \hline 72 & \leftarrow \text{Difference} \end{array}$$

- Finally, we go to the ones place of the difference, 72. We consider the new dividend to be 72 ones. So we think $9\overline{)72}$. Since $9 \cdot \mathbf{8} = 72$, we position the **8** in the ones place of the quotient.

  So 568 is our answer.

$$\begin{array}{rl} 568 & \leftarrow \text{Ones} \\ 9\overline{)5{,}112} & \\ -4{,}500 & \\ \hline 612 & \\ -540 & \\ \hline 72 & \\ -72 & \leftarrow 8 \cdot 9 = 72 \\ \hline 0 & \leftarrow \text{Difference} \end{array}$$

Instead of writing 0's as placeholders, we can use the following shortcut.

$$\begin{array}{r} 568 \\ 9\overline{)5{,}112} \\ \underline{-4\,5} \\ 61 \\ \underline{-54} \\ 72 \\ \underline{-72} \\ 0 \end{array}$$

← These arrows help us to keep track of which digit we have brought down.

Check

$$\begin{array}{r} 568 \\ \underline{\times\ \ 9} \\ 5{,}112 \end{array}$$

← The product equals the dividend, so our answer is correct.

Note that each time we subtract in a division problem, the difference is less than the divisor. Why must that be true?

EXAMPLE 1

Divide and check: $4{,}263 \div 7$

Solution

Think $7\overline{)42}$.

$$\begin{array}{r} 609 \\ 7\overline{)4{,}263} \\ \underline{-4\,2} \\ 06 \\ \underline{-0} \\ 63 \\ \underline{-63} \\ 0 \end{array}$$

- $-4\,2$ ← $6 \times 7 = 42$. Subtract.
- 06 ← Think $7\overline{)6}$. There are zero 7's in 6.
- -0 ← $0 \times 7 = 0$. Subtract.
- 63 ← Think $7\overline{)63}$.
- -63 ← $9 \times 7 = 63$. Subtract.

Check

$$\begin{array}{r} 609 \\ \underline{\times\ \ 7} \\ 4{,}263 \end{array}$$

The product agrees with our dividend. Note the 0 in the quotient. Can you explain why the 0 is needed?

PRACTICE 1

Compute $9\overline{)7{,}263}$ and then check your answer.

TIP In writing your answer to a division problem, position the first digit of the quotient over the *rightmost digit* of the number into which you are dividing (the 6 over the 2 in Example 1).

$$\begin{array}{r} \mathbf{6}09 \\ \downarrow\downarrow\downarrow \\ 7\overline{)4{,}\mathbf{2}63} \end{array}$$

EXAMPLE 2

Compute $\frac{2{,}709}{9}$. Then check your answer.

Solution

$$\begin{array}{r} 301 \\ 9\overline{)2{,}709} \\ \underline{-2\,7} \\ 00 \\ \underline{-0} \\ 09 \\ \underline{-9} \\ 0 \end{array}$$

Check

$$\begin{array}{r} 301 \\ \underline{\times\ \ 9} \\ 2{,}709 \end{array}$$

PRACTICE 2

Carry out the following division and check your answer.

$$8\overline{)56{,}016}$$

In Examples 1 and 2, note that the remainder is 0; that is, the divisor goes evenly into the dividend. However, in some division problems, that is not the case. Consider, for instance, the problem of dividing 16 pens *equally* among 3 boxes.

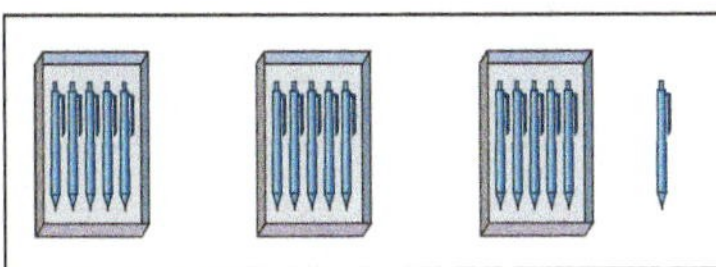

From the diagram, we see that each box contains 5 pens *but* that 1 pen—the *remainder*—is left over.

$$\begin{array}{rl} 5 & \leftarrow \text{Number of pens in each box} \\ \text{Number of boxes} \rightarrow 3\overline{)16} & \leftarrow \text{Total number of pens} \\ \underline{-15} & \leftarrow \text{Total number of pens in the boxes} \\ 1 & \leftarrow \text{Number of pens remaining} \end{array}$$

We write the answer to this problem as 5 R1 (read "5 Remainder 1"). Note that $(3 \times 5) + 1 = 16$. The following relationship is always true.

(Quotient × Divisor) + Remainder = Dividend

When a division problem results in a remainder as well as a quotient, we use this relationship for checking.

EXAMPLE 3

Find the quotient of 55,811 and 6. Then check.

Solution

$$\begin{array}{r} 9{,}301\text{ R5} \\ 6\overline{)55{,}811} \\ \underline{-54} \\ 1\,8 \\ \underline{-1\,8} \\ 01 \\ \underline{-0} \\ 11 \\ \underline{-6} \\ 5 \end{array}$$

Our answer is therefore 9,301 R5.

(Quotient × Divisor) + Remainder = Dividend

Check $(9{,}301 \times 6) + 5 =$

$55{,}806 + 5 = 55{,}811$

Since this matches the dividend, our answer checks.

PRACTICE 3

Compute $8\overline{)42{,}329}$ and check.

Now, let's consider division problems in which a divisor has more than one digit. Notice that such problems involve rounding.

EXAMPLE 4

Compute $\frac{2{,}574}{34}$ and check.

Solution In order to estimate the first digit of the quotient, we round 34 to 30 and 257 to 260.

$$\begin{array}{r} 8 \\ 34\overline{)2{,}574} \\ -2\,72 \end{array}$$

← Think 260 ÷ 30, or 26 ÷ 3. The quotient 8 goes over the 7 because we are dividing 34 into 257.

← 8 × 34 = 272. Try to subtract.

Because 272 is too large, we reduce our estimate in the quotient by 1 and try 7.

$$\begin{array}{r} 76 \\ 34\overline{)2{,}574} \\ -2\,38 \\ \hline 194 \\ -204 \\ \hline \end{array}$$

← 7 × 34 = 238. Subtract.

← Think 190 ÷ 30 or 19 ÷ 3.

← 6 × 34 = 204. Try to subtract.

Because 204 is too large, we reduce our estimate in the quotient by 1 and try 5.

$$\begin{array}{r} 75 \text{ R24} \\ 34\overline{)2{,}574} \\ -2\,38 \\ \hline 194 \\ -170 \\ \hline 24 \end{array}$$

So our answer is 75 R24.

Check (75 × 34) + 24 = 2,574

Since 2,574 is the dividend, our answer checks.

PRACTICE 4

Divide 23 into 1,818. Then check.

EXAMPLE 5

Divide $26\overline{)1{,}849}$ and then check.

Solution First, we round 26 to 30 and 184 to 180. Think 180 ÷ 30 = 6.

$$\begin{array}{r} 6 \\ 26\overline{)1{,}849} \\ -1\,56 \\ \hline 28 \end{array}$$

← This difference is larger than the divisor, so we increase the 6 in the quotient by 1.

$$\begin{array}{r} 71 \\ 26\overline{)1{,}849} \\ -182 \\ \hline 29 \\ -26 \\ \hline 3 \end{array}$$

Our answer is therefore 71 R3.

Check (71 × 26) + 3 = 1,849

PRACTICE 5

Compute and check: $15\overline{)1{,}420}$

TIP If the divisor has more than one digit, estimate each digit in the quotient by rounding and then dividing. If the product is too large or too small, adjust it up or down by 1 and then try again.

EXAMPLE 6

Find the quotient of 13,559 and 44. Then check.

Solution

```
      308 R7
 44)13,559
   -13 2
      35   ← This number is smaller than the divisor, so
     -0      the next digit in the quotient is 0.
     359
    -352
       7
```

Check $(308 \times 44) + 7 = 13{,}559$

PRACTICE 6

Divide 16,999 by 28. Then check your answer.

EXAMPLE 7

Divide and check: $6{,}000 \div 20$

Solution We set up the problem as before.

```
    300
20)6,000
  -60
    00
   -00        Check     300
    00                × 20
   -00                6,000
     0
```

Because the divisor and dividend both end in zero, a quicker way to do Example 7 is by dropping zeros.

```
2Ø)6,00Ø ← Drop one 0 from both
           the divisor and the
           dividend.

 300
2)600 ← Then divide.
```

PRACTICE 7

Compute 40)8,000 and then check.

TIP Dropping the same number of zeros at the right end of both the divisor and the dividend does not change the quotient.

Estimating Quotients

As for other operations, estimating is an important skill for division. Checking a quotient by estimation is faster than checking it by multiplication, although less exact. And in some division problems, we need only an approximate answer.

How do we estimate a quotient? A good way is to round the divisor to its greatest place. The new divisor then contains only one nonzero digit and so is relatively easy to divide by mentally. Then, we round the dividend to the place of our choice. Finally, we compute the estimated quotient by calculating its first digit and then attaching the appropriate number of zeros.

EXAMPLE 8

Calculate $\frac{7{,}004}{34}$ and then check by estimation.

Solution

$$\begin{array}{r} 206 \leftarrow \text{Exact quotient} \\ 34\overline{)7{,}004} \\ \underline{-6\,8} \\ 204 \\ \underline{-204} \end{array}$$

Check $34\overline{)7{,}004}$ Round 34 to 30 and round 7,004 to 7,000.

$\downarrow \quad \downarrow$

$30\overline{)7{,}000}$ Think 70 ÷ 30 or 7 ÷ 3.

$$\begin{array}{r} 200 \leftarrow \text{Estimated quotient} \\ 30\overline{)7{,}000} \end{array}$$

Note that, to the right of the 2 in the estimated quotient, we added a 0 over each of the digits in the dividend. Our answer (206) is close to our estimate (200), and so our answer is reasonable.

PRACTICE 8

Compute 100,568 ÷ 104 and use estimation to check.

EXAMPLE 9

Sound travels at about 340 meters per second, whereas light travels at 299,792,458 meters per second. Estimate how many times as fast as the speed of sound is the speed of light.

Solution To estimate a quotient, we first round the divisor and the dividend to their largest place value.

$340\overline{)299{,}792{,}458}$

$\downarrow \quad \downarrow$

$300\overline{)300{,}000{,}000}$ Round 340 to 300 and 299,792,458 to 300,000,000.

Then, we divide.

$$\begin{array}{r} 1{,}000{,}000 \\ 300\overline{)300{,}000{,}000} \end{array}$$

So the speed of light is about 1,000,000 times faster than the speed of sound.

PRACTICE 9

Based on population projections, China will have a population of 1,394,638,699 in the year 2025. In that same year, Brazil will have a population of 231,886,946. Estimate how many times the population of Brazil will the population of China be in 2025. (*Source:* sasweb.ssd.census.gov)

Dividing Whole Numbers on a Calculator

When using a calculator to divide, we must enter the numbers in the correct order to get the correct answer. We first enter the number into *which we are dividing (the dividend) and then the number* by *which we are dividing (the divisor).*

EXAMPLE 10

Use a calculator to divide $18\overline{)11{,}718}$.

Solution

Press	Display
11718 [÷] 18 [ENTER =]	11718 / 18 651.

A reasonable estimate is 10,000 ÷ 20, or 500, which is fairly close to 651.

PRACTICE 10

Find the following quotient with a calculator:

$$\frac{47{,}034}{78}$$

1.4 Exercises

FOR EXTRA HELP MyMathLab MathXL PRACTICE WATCH READ REVIEW

Mathematically Speaking

Fill in each blank with the most appropriate term or phrase from the given list.

subtraction	divided	divisor	multiplication
quotient	product	increased	

1. When dividing, the dividend is divided by the __________.

2. The result of dividing is called the __________.

3. The opposite operation of division is ______________.

4. Any whole number __________ by 1 is equal to the number itself.

A *Divide and check.*

5. $5\overline{)2{,}000}$
6. $5\overline{)6{,}000}$
7. $5\overline{)12{,}800}$
8. $8\overline{)12{,}504}$
9. $9\overline{)2{,}709}$
10. $2\overline{)5{,}780}$
11. $7\overline{)21{,}021}$
12. $5\overline{)27{,}450}$
13. $3\overline{)24{,}132}$
14. $2\overline{)30{,}534}$
15. $9\overline{)4{,}500}$
16. $3\overline{)4{,}512}$

Find the quotient and check.

17. $300 \div 10$
18. $400 \div 20$
19. $700 \div 50$
20. $6{,}000 \div 20$
21. $\frac{8{,}400}{200}$
22. $\frac{7{,}500}{300}$
23. $\frac{16{,}000}{40}$
24. $\frac{48{,}000}{20}$
25. $6{,}996 \div 44$
26. $9{,}660 \div 92$
27. $80{,}295 \div 15$
28. $31{,}031 \div 13$
29. $39{,}078 \div 39$
30. $49{,}497 \div 21$
31. $249{,}984 \div 36$
32. $499{,}992 \div 24$
33. $52\overline{)52{,}052}$
34. $24\overline{)48{,}072}$
35. $12\overline{)36{,}600}$
36. $36\overline{)25{,}560}$
37. $25\overline{)22{,}675}$
38. $15\overline{)30{,}480}$
39. $49\overline{)58{,}849}$
40. $19\overline{)38{,}570}$
41. $6{,}512 \div 10$
42. $8{,}922 \div 25$
43. $304 \div 27$
44. $206 \div 45$
45. $\frac{10{,}175}{87}$
46. $\frac{21{,}109}{25}$
47. $\frac{63{,}002}{90}$
48. $\frac{12{,}509}{61}$
49. $47\overline{)34{,}000}$
50. $66\overline{)99{,}980}$
51. $14\overline{)6{,}000}$
52. $32\overline{)3{,}007}$
53. $65\overline{)65{,}660}$
54. $39\overline{)30{,}009}$
55. $42\overline{)39{,}000}$
56. $97\overline{)13{,}502}$
57. $537\overline{)387{,}177}$
58. $265\overline{)197{,}160}$
59. $638\overline{)98{,}890}$
60. $152\overline{)34{,}048}$

In each group of three quotients, one is wrong. Use estimation to identify which quotient is incorrect.

61. **a.** $18{,}473 \div 91 = 203$ **b.** $43{,}364 \div 74 = 586$ **c.** $14{,}562 \div 18 = 8{,}009$

62. **a.** $43{,}710 \div 93 = 47$ **b.** $71{,}048 \div 107 = 664$ **c.** $11{,}501 \div 31 = 371$

63. **a.** $455{,}260 \div 65 = 704$ **b.** $11{,}457 \div 57 = 201$ **c.** $10{,}044 \div 93 = 108$

64. **a.** $178{,}267 \div 89 = 2{,}003$ **b.** $350{,}007 \div 21 = 1{,}667$ **c.** $37{,}185 \div 37 = 1{,}005$

Mixed Practice

Divide and check.

65. 38,095 ÷ 42

66. $\dfrac{63,147}{21}$

67. $6\overline{)12,000}$

68. 4,907 ÷ 7

69. $\dfrac{48,000}{20}$

70. $36\overline{)249,986}$

71. $\dfrac{3,330}{9}$

72. 4,090 ÷ 91

Applications

B *Solve and check.*

73. A part-time student is taking 9 credit-hours this semester at a local community college. If her tuition bill is $1,215, how much does each credit-hour cost?

74. A car used 15 gallons of gas on a 300-mile trip. How many miles per gallon (mpg) of gas did the car get?

75. The area of the Pacific Ocean is about 64 million square miles, and the area of the Atlantic Ocean is approximately 32 million square miles. The Pacific is how many times as large as the Atlantic? (*Source: The New Encyclopedia Britannica*)

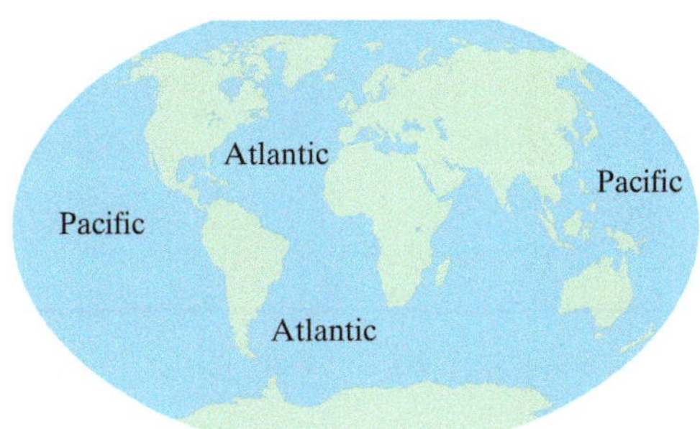

76. The diameter of Earth is about 8,000 miles, whereas the diameter of the Moon is about 2,000 miles. How many times the Moon's diameter is Earth's? (*Source: The New Encyclopedia Britannica*)

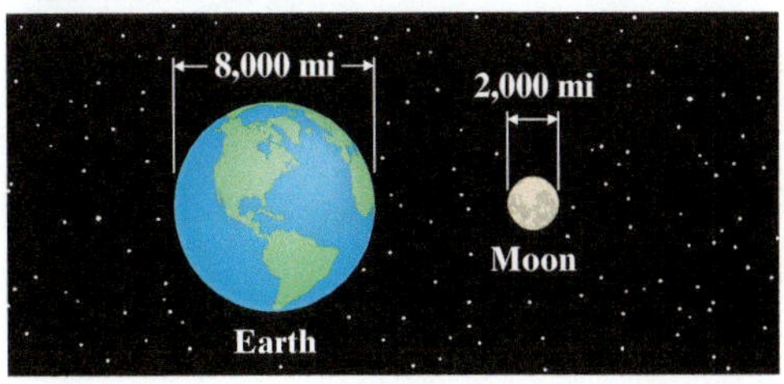

77. In the year 2030, Ohio is projected to have a population of about 12,300,000 people. If Ohio has a total land area of about 41,000 square miles, how many people per square mile will there be in 2030? (*Source:* Ohio Department of Development)

78. A certified medical assistant has an annual salary of $26,472. What is her gross monthly income?

79. A 150-pound person can burn about 360 calories in 1 hour doing yoga. How many calories are burned in 1 minute? (*Source:* American Cancer Society)

80. Ryan Howard signed a 5-year, $125,000,000 contract extension with the Philadelphia Phillies in 2010. On average, what is his pay per year from the contract extension? (*Source:* sportsillustrated.com)

81. A homeowner is remodeling a bathroom with dimensions 96 inches and 114 inches. For the floor, she has selected tiles that measure 6 inches by 6 inches.

a. How many tiles must she purchase?

b. The tiles come in boxes of 12. How many boxes of tiles must she purchase?

c. If each box of tiles costs $18, how much will she spend on the tiles for the floor?

82. The group admission rate for 15 or more people at Six Flags Great Adventure amusement park is $30 per person. A student group hosted a field trip to the park and charged $47 per ticket, covering both the cost of admission to the park and the bus transportation. (*Source:* Six Flags Great Adventure, 2010)

a. If the total amount the group collected for tickets was $1,739, how many students went on the field trip?

b. Calculate the total cost of admissions for the students on the field trip.

c. What was the cost of the bus transportation?

• Check your answers on page A-2.

MINDStretchers

Writing

1. Use the problem $10 \div 2 = 5$ to help explain why division can be thought of as repeated subtraction.

Mathematical Reasoning

2. Consider the following pair of problems.

   a. $2\overline{)7}$ **b.** $4\overline{)13}$

   Are the answers the same? Explain.

Groupwork

3. In the following division problem, A, B, and C each stand for a different digit. Working with a partner, identify all the digits. (*Hint:* There are two answers.)

$$\begin{array}{r} \text{ABA} \\ \text{AB}\overline{)\text{CACAB}} \\ \underline{-\text{CAB}} \\ \text{CA} \\ \underline{-\text{B}} \\ \text{CAB} \\ \underline{-\text{CAB}} \end{array}$$

1.5 Exponents, Order of Operations, and Averages

OBJECTIVES

- **A** To evaluate expressions involving exponents
- **B** To evaluate expressions using the rule for order of operations
- **C** To compute averages
- **D** To solve applied problems involving exponents, order of operations, or averages

Exponents

There are many mathematical situations in which we multiply a number by itself repeatedly. Writing such expressions in **exponential form** provides a shorthand method for representing this repeated multiplication of the same factor.

For instance, we can write $5 \cdot 5 \cdot 5 \cdot 5$ in exponential form as

$$5^4 \leftarrow \text{Exponent}$$

$$\uparrow \text{Base}$$

This expression is read "5 to the fourth *power*" or simply "5 to the fourth."

DEFINITION

An **exponent** (or **power**) is a number that indicates how many times another number (called the **base**) is used as a factor.

We read the power 2 or the power 3 in a special way. For instance, 5^2 is usually read "5 *squared*" rather than "5 to the second power." Similarly, we usually read 5^3 as "5 *cubed*" instead of "5 to the third power."

Let's look at a number written in exponential form—namely, 2^4. To evaluate this expression, we multiply 4 factors of 2.

$$\begin{aligned} 2^4 &= 2 \cdot 2 \cdot 2 \cdot 2 \\ &= 4 \cdot 2 \cdot 2 \\ &= 8 \cdot 2 \\ &= 16 \end{aligned}$$

In short, $2^4 = 16$. Do you see the difference between 2^4 and $2 \cdot 4$?

Sometimes we prefer to shorten expressions by writing them in exponential form. For instance, we can write $3 \cdot 3 \cdot 4 \cdot 4 \cdot 4$ in terms of powers of 3 and 4.

$$\underbrace{3 \cdot 3}_{\text{2 factors of 3}} \cdot \underbrace{4 \cdot 4 \cdot 4}_{\text{3 factors of 4}} = 3^2 \cdot 4^3$$

EXAMPLE 1

Rewrite

$$6 \cdot 6 \cdot 6 \cdot 10 \cdot 10 \cdot 10 \cdot 10$$

in exponential form.

Solution

$$\underbrace{6 \cdot 6 \cdot 6}_{\text{3 factors of 6}} \cdot \underbrace{10 \cdot 10 \cdot 10 \cdot 10}_{\text{4 factors of 10}} = 6^3 \cdot 10^4$$

PRACTICE 1

Write

$$5 \cdot 5 \cdot 5 \cdot 5 \cdot 5 \cdot 2 \cdot 2$$

in terms of powers.

EXAMPLE 2

Compute:

a. 1^5

b. 22^2

Solution

a. $1^5 = \underbrace{1 \cdot 1 \cdot 1 \cdot 1 \cdot 1}_{} = 1$

Note that 1 raised to any power is 1.

b. $22^2 = 22 \cdot 22 = 484$

After considering this example, can you explain the difference between squaring and doubling a number?

PRACTICE 2

Calculate:

a. 1^8

b. 11^3

EXAMPLE 3

Write $4^3 \cdot 5^3$ in standard form.

Solution

$$4^3 \cdot 5^3 = (4 \cdot 4 \cdot 4) \cdot (5 \cdot 5 \cdot 5)$$
$$= 64 \cdot 125$$
$$= 8{,}000$$

From this example, do you see the difference between cubing and tripling a number?

PRACTICE 3

Express $7^2 \cdot 2^4$ in standard form.

It is especially easy to compute *powers of 10.*

$$10^2 = 10 \cdot 10 = 1\underbrace{00}_{\text{2 zeros}}, \qquad 10^3 = 10 \cdot 10 \cdot 10 = 1{,}\underbrace{000}_{\text{3 zeros}}$$

$$10^4 = 10 \cdot 10 \cdot 10 \cdot 10 = 1\underbrace{0{,}000}_{\text{4 zeros}}$$

and so on.

Note the pattern.

EXAMPLE 4

The Milky Way, the galaxy to which the Sun and Earth belong, contains about 100 billion stars. Express this number in terms of a power of 10. (*Source: The New York Times Almanac 2010*)

Solution

$$\underbrace{100{,}000{,}000{,}000}_{\text{11 zeros}} = 10^{11}$$

So the Milky Way contains about 10^{11} stars.

PRACTICE 4

In 1804, the world population reached the milestone of one billion. Represent this number as a power of 10. (*Source:* U.S. Census Bureau)

Order of Operations

Some mathematical expressions involve more than one mathematical operation. For instance, consider $5 + 3 \cdot 2$. This expression seems to have two different values, depending on the order in which we perform the given operations.

Adding first	Multiplying first
$5 + 3 \cdot 2$	$5 + 3 \cdot 2$
$= 8 \cdot 2$	$= 5 + 6$
$= 16$	$= 11$

How are we to know which operation to carry out first? By consensus we agree to follow the rule called the **order of operations** so that everyone always gets the same value for an answer.

Order of Operations Rule

To evaluate mathematical expressions, carry out the operations *in the following order*.

1. First, perform the operations within any grouping symbols, such as parentheses ( ) or brackets [ ].
2. Then, raise any number to its power $\blacksquare^{\blacksquare}$.
3. Next, perform all multiplications and divisions as they appear from left to right.
4. Finally, do all additions and subtractions as they appear from left to right.

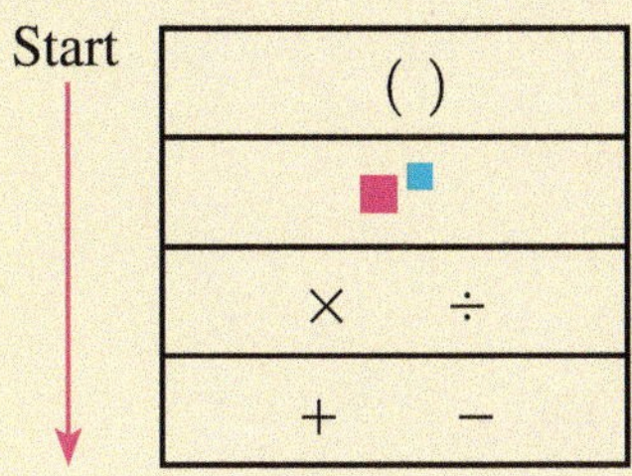

Applying this rule to the preceding example gives us the following result.

$$\begin{aligned} & 5 + 3 \cdot 2 && \text{Multiply first.} \\ = \; & 5 + 6 && \text{Then add.} \\ = \; & 11 \end{aligned}$$

So 11 is the correct answer.

Let's consider more examples that depend on the order of operations rule.

EXAMPLE 5

Simplify: $18 - 7 \cdot 2$

Solution Applying the rule, we multiply first, and then subtract.

$$18 - 7 \cdot 2 =$$
$$18 - 14 = 4$$

PRACTICE 5

Evaluate: $2 \cdot 8 + 4 \cdot 3$

EXAMPLE 6

Find the value of $3 + 2 \cdot (8 + 3^2)$.

Solution

$$3 + 2 \cdot (8 + 3^2) = 3 + 2 \cdot (8 + 9)$$ First, perform the operations in parentheses: square the 3.

$$= 3 + 2 \cdot 17$$ Then, add 8 and 9.

$$= 3 + 34$$ Next, multiply 2 by 17.

$$= 37$$ Finally, add 3 and 34.

PRACTICE 6

Simplify: $(4 + 1)^2 \times 6 - 4$

TIP When a division problem is written in the format $\frac{\square}{\square}$, parentheses are understood to be around both the dividend and the divisor. For instance,

$$\frac{10 - 2}{3 + 1} \text{ means } \frac{(10 - 2)}{(3 + 1)}.$$

EXAMPLE 7

Evaluate: $6 \cdot 2^3 - \frac{21 - 11}{2}$

Solution

$$6 \cdot 2^3 - \frac{21 - 11}{2} = 6 \cdot 2^3 - \frac{10}{2}$$ First, simplify the dividend by subtracting.

$$= 6 \cdot 8 - \frac{10}{2}$$ Then, cube.

$$= 48 - 5$$ Next, multiply and divide.

$$= 43$$ Finally, subtract.

PRACTICE 7

Simplify: $10 + \frac{24}{12 - 8} - 3 \times 4$

Some arithmetic expressions contain not only parentheses but also brackets. When simplifying expressions containing these grouping symbols, first perform the operations within the innermost grouping symbols and then continue to work outward.

EXAMPLE 8

Simplify: $5 \cdot [4(10 - 3^2) - 2]$

Solution

$$5 \cdot [4(10 - 3^2) - 2] = 5 \cdot [4(10 - 9) - 2]$$ Perform the operation in parentheses: square the 3.

$$= 5 \cdot [4 \cdot 1 - 2]$$ Subtract 9 from 10.

$$= 5 \cdot [4 - 2]$$ Multiply.

$$= 5 \cdot 2$$ Subtract.

$$= 10$$ Multiply.

PRACTICE 8

Evaluate: $[4 + 3(2^3 - 5)] \cdot 10$

EXAMPLE 9

Young's Rule is a rule of thumb for calculating the dose of medicine recommended for a child of a given age. According to this rule, the dose of acetaminophen in milligrams (mg) for a child who is eight years old can be calculated using the expression.

$$\frac{8 \times 500}{8 + 12}$$

What is the recommended dose?

Solution

$$\frac{8 \times 500}{8 + 12} = \frac{4{,}000}{20} \quad \text{First, simplify the dividend and the divisor.}$$

$$= 200 \quad \text{Then, divide.}$$

So the recommended dose is 200 milligrams.

PRACTICE 9

The minimum distance (in feet) that it takes a car to stop if it is traveling on a particular road surface at a speed of 30 miles per hour is given by the expression.

$$\frac{10 \times 30^2}{30 \times 5}$$

What is this minimum stopping distance?

Averages

We use an **average** to represent a set of numbers. Averages allow us to compare two or more sets. (For example, do the men or the women in your class spend more time studying?) Averages also allow us to compare an individual with a set. (For example, is the amount of time you spend studying above or below the class average?) The following definition shows how to compute an average.

DEFINITION

The **average** (or **mean**) of a set of numbers is the sum of those numbers divided by however many numbers are in the set.

EXAMPLE 10

What is the average of 100, 94, and 100?

Solution The average equals the sum of these three numbers divided by 3.

$$\frac{100 + 94 + 100}{3} = \frac{294}{3} = 98$$

PRACTICE 10

Find the average of \$30, \$0, and \$90.

EXAMPLE 11

The following map shows the five Great Lakes. The maximum depth of each of these lakes is given in the table. (*Source:* U.S. Environmental Protection Agency)

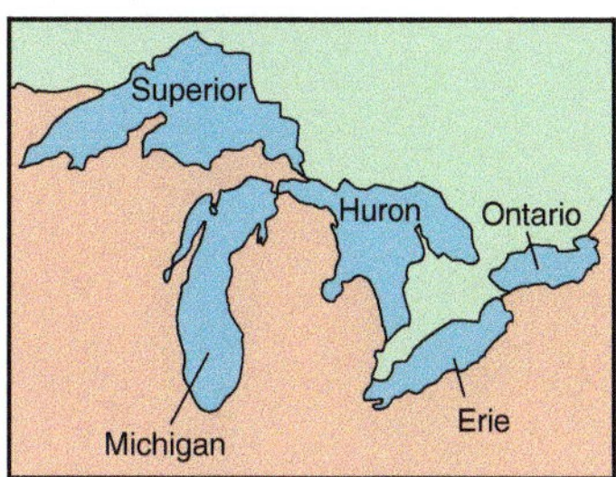

Lake	Maximum Depth (in meters)
Erie	64
Huron	229
Michigan	282
Ontario	244
Superior	406

PRACTICE 11

The table shown gives the number of at-home fatalities due to tornadoes in the United States each year from 2006 through 2009. (*Source:* spc.noaa.gov)

Year	Number of At-Home Fatalities
2006	58
2007	68
2008	99
2009	19

EXAMPLE 11 (continued)

a. What is the average maximum depth of the Great Lakes?

b. Which of the Great Lakes is deeper than the average?

Solution

a.
$$\frac{\text{The sum of the depths}}{\text{The number of lakes}} = \frac{64 + 229 + 282 + 244 + 406}{5} = \frac{1{,}225}{5} = 245$$

So the average maximum depth is 245 meters.

b. Lake Michigan and Lake Superior are deeper than the average.

a. What was the average annual number of at-home fatalities for these years?

b. In which years was the number of fatalities below the average?

Powers and Order of Operations on a Calculator

Let's use a calculator to carry out computations that involve either powers or the order of operations rule.

EXAMPLE 12

Calculate 23^3.

Solution

Press	Display
23 [^] 3 [ENTER]	23^3 12167.

PRACTICE 12

Use a calculator to compute 375^2.

EXAMPLE 13

Combine: $2 + 3 \times 4$

Solution

Press	Display
2 [+] 3 [×] 4 [ENTER]	2+3*4 14.

Note that some calculators do not follow the order of operations rule. When using this kind of calculator, enter the operations in the order specified by the order of operations rule to get the correct answer.

PRACTICE 13

On a calculator, compute $135 - 44 \div 11$.

1.5 Exercises

FOR EXTRA HELP MyMathLab      

Mathematically Speaking

Fill in each blank with the most appropriate term or phrase from the given list.

product	sum	adding	listing
subtracting	grouping	power	base

1. An exponent indicates how many times the ________ is used as a factor.

2. Parentheses and brackets are examples of __________ symbols.

3. An average of numbers on a list is found by ________ the numbers and then dividing by how many numbers there are on the list.

4. In evaluating an expression involving both a sum and a product, the ________ is evaluated first.

Complete each table by squaring the numbers given.

5.

n	0	2	4	6	8	10	12
n^2							

6.

n	1	3	5	7	9	11	13
n^2							

A *Complete each table by cubing the numbers given.*

7.

n	0	2	4	6	8
n^3					

8.

n	1	3	5	7	9
n^3					

Express each number as a power of 10.

9. $100 = 10^{\square}$

10. $1{,}000 = 10^{\square}$

11. $10{,}000 = 10^{\square}$

12. $100{,}000 = 10^{\square}$

13. $1{,}000{,}000 = 10^{\square}$

14. $10{,}000{,}000 = 10^{\square}$

Write each number in terms of powers.

15. $2 \cdot 2 \cdot 3 \cdot 3 = 2^{\square} \cdot 3^{\square}$

16. $2 \cdot 2 \cdot 5 \cdot 2 \cdot 5 = 2^{\square} \cdot 5^{\square}$

17. $5 \cdot 4 \cdot 4 \cdot 4 = 4^{\square} \cdot 5^{\square}$

18. $6 \cdot 7 \cdot 6 \cdot 7 \cdot 6 \cdot 7 = 6^{\square} \cdot 7^{\square}$

Write each number in standard form.

19. $6^2 \cdot 5^2$

20. $10^3 \cdot 9^2$

21. $2^5 \cdot 7^2$

22. $3^4 \cdot 4^3$

B *Evaluate.*

23. $8 + 5 \cdot 2$

24. $9 + 10 \cdot 2$

25. $8 - 12 \div 3$

26. $12 - 6 \div 2$

27. $18 \div 2 + 4$

28. $30 \div 3 + 6$

29. $6 \cdot 3 - 16 \div 4$

30. $7 \cdot 3 - 12 \div 4$

31. $10 + 5^2$

32. $9 + 2^3$

33. $(10 + 5)^2$

34. $(9 + 2)^3$

35. 10×5^2

36. 12×2^2

37. $(12 \div 2)^2$

38. $(10 \div 5)^2$

39. $15 \div (6 - 3)$

40. $24 \div (4 + 2)$

41. $2^6 - 6^2$

42. $3^5 - 5^3$

43. $8 + 5 - 3 - 2 \times 2$

44. $7 - 1 + 2 + 3 \cdot 2$

45. $(10 - 1)(10 + 1)$

46. $(8 - 1)(8 + 1)$

47. $10^2 - 1$

48. $8^2 - 1$

49. $\left(\frac{8 + 2}{7 - 2}\right)^2$

50. $\left(\frac{9 - 1}{3 + 5}\right)^3$

51. $\dfrac{5^3 - 2^3}{3}$

52. $\dfrac{3^2 + 5^2}{2}$

53. $4 + 12(10 - 3^2)$

54. $3 + 10(20 - 2^3)$

55. $10 \cdot 3^2 + \dfrac{10 - 4}{2}$

56. $\dfrac{3^3 + 1^3 + 2^3}{4}$

57. $(21 \div 7) + [(9 - 5) \cdot 2]^2$

58. $(30 \div 3) + [(7 - 4) \cdot 3]^2$

59. $[9 + 2(3^2 - 8)] + 7$

60. $15 + [3(8 - 2^2) - 6]$

61. $[2 \cdot (3 + 4)^2 - 3 \cdot (7 - 2)^2]^2$

62. $[3 \cdot (2 + 1)^2 - 2 \cdot (5 - 4)^2]^2$

63. $32 + 9 \cdot 215 \div 5$

64. $84 \cdot 27 + 32 \cdot 27^2 \div 2$

65. $48(48 - 31)(48 - 24)(48 - 41)$

66. $137^2 - 4(36)(22)$

In each exercise, the three squares stand for the numbers 4, 6, and 8 in some order. Fill in the squares to make true statements.

67. $\square \cdot 3 + \square \cdot 5 + \square \cdot 7 = 98$

68. $\square + 10 \times \square - \dfrac{\square}{2} = 42$

69. $(\square)(3 + \square) - 2 \cdot \square = 44$

70. $\square \cdot 3 + \square \cdot 5 + \square \cdot 7 = 82$

71. $\square + 10 \times \square - \square \div 2 = 45$

72. $\dfrac{48}{\square} - \dfrac{\square}{2} + (3 + \square)^2 = 127$

Insert parentheses, if needed, to make the expression on the left equal to the number on the right.

73. $5 + 2 \cdot 4^2 = 112$

74. $5 + 2 \cdot 4^2 = 69$

75. $5 + 2 \cdot 4^2 = 169$

76. $5 + 2 \cdot 4^2 = 37$

77. $8 - 4 \div 2^2 = 1$

78. $8 - 4 \div 2^2 = 7$

Find the area of each shaded region.

79.

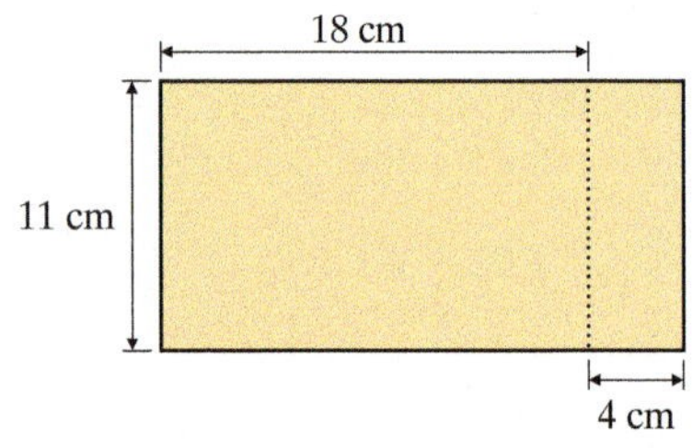

80.

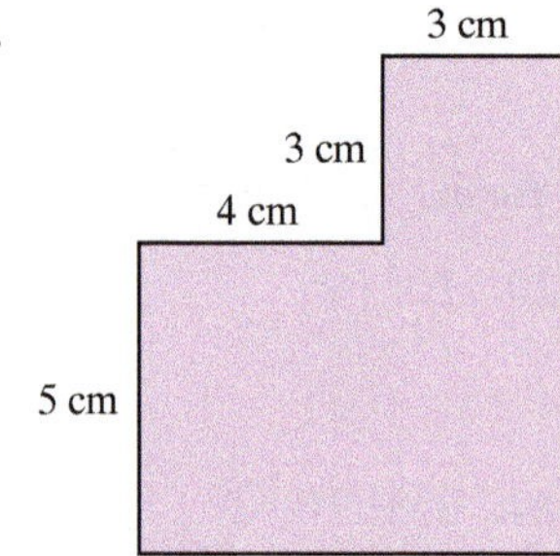

81.

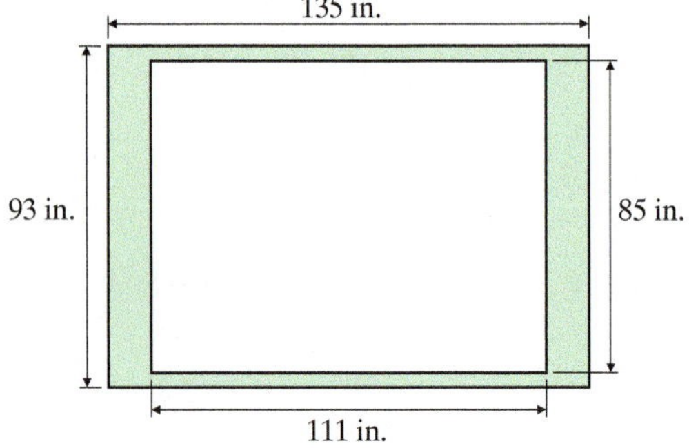

82.

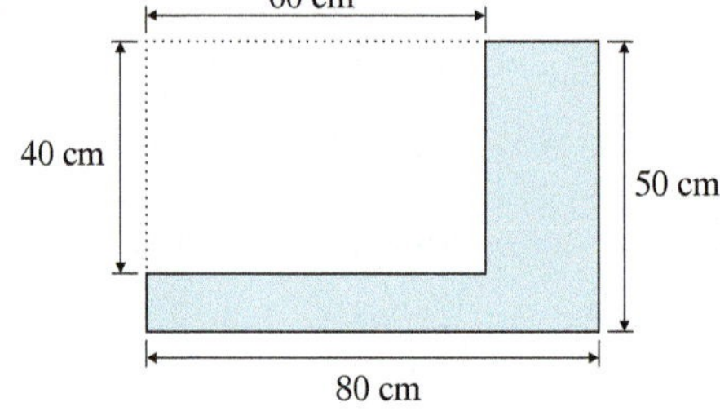

Complete each table.

83.

Input	Output
0	$21 + 3 \times \mathbf{0} =$
1	$21 + 3 \times \mathbf{1} =$
2	$21 + 3 \times \mathbf{2} =$

84.

Input	Output
0	$14 - 5 \times \mathbf{0} =$
1	$14 - 5 \times \mathbf{1} =$
2	$14 - 5 \times \mathbf{2} =$

C *Find the average of each set of numbers.*

85. 20 and 30

86. 10 and 50

87. 30, 60, and 30

88. 17, 17, and 26

89. 10, 0, 3, and 3

90. 5, 7, 7, and 17

91. 3,527 miles, 1,788 miles, and 1,921 miles

92. 3,432 miles, 1,822 miles, and 1,436 miles

93. Six 10's and four 5's

94. Sixteen 5's and four 0's

Mixed Practice

Solve.

95. Express 100,000,000 as a power of 10.

96. Find the area of the shaded region.

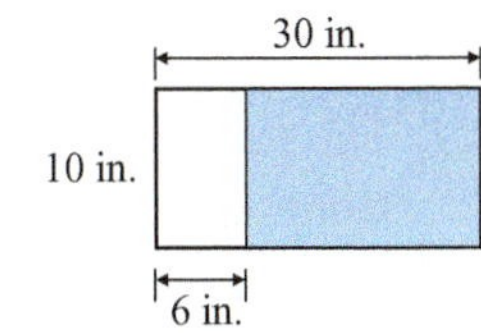

97. Square 17.

98. Rewrite in terms of powers of 2 and 7:
$2 \cdot 2 \cdot 2 \cdot 7 \cdot 7 = 2 \cdot 7$

99. Simplify: $50 - 2(10 - 3^2)$.

100. Cube 10.

101. Find the average of 10, 10, and 4.

102. Evaluate: 6×4^2

Applications

D *Solve and check.*

103. A 40-story office building has 25,000 square feet of space to rent. What is the average rental space on a floor?

104. The total area of the 50 states in the United States is about 3,700,000 square miles. If the state of Georgia's area is about 60,000 square miles, is its size above the average of all the states? Explain. (*Source: The New Encyclopedia Britannica*)

105. In a branch of mathematics called number theory, the numbers 3, 4, and 5 are called a *Pythagorean triple* because $3^2 + 4^2 = 5^2$ (that is, $9 + 16 = 25$). Show that 5, 12, and 13 are a Pythagorean triple.

106. If an object is dropped off a cliff, after 10 seconds it will have fallen $\frac{32 \cdot 10^2}{2}$ feet, ignoring air resistance. Express this distance in standard form, without exponents.

107. The solar wind streams off the Sun at speeds of about 1,000,000 miles per hour. Express this number as a power of 10. (*Source:* NASA)

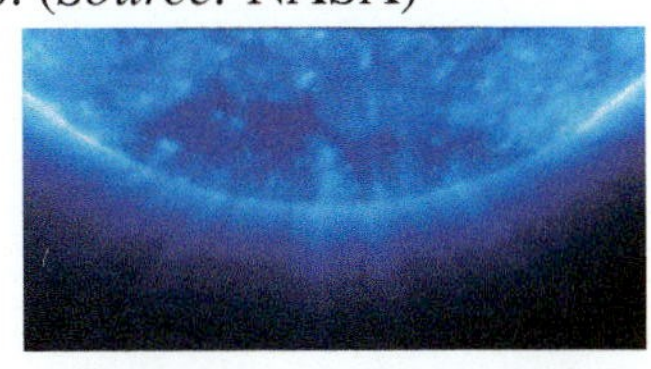

108. It has been estimated that there are 100,000,000,000,000 cells in the adult human body. Represent this number as a power of 10. (*Source:* ehd.org)

109. The following table shows a lab assistant's salary in various years.

Year	1	2	3
Salary	$19,400	$21,400	$23,700

a. Find the average salary for the three years.

b. How much greater was her average salary for the last two years than for all three years?

110. The following grade book shows a student's math test scores.

Test 1	Test 2	Test 3	Test 4
85	63	98	82

a. What is the average of his math scores?

b. If he were to get a 92 on the next math test, by how much would his average score increase?

111. In the last four home games, a college basketball team had scores of 68, 79, 57, and 72.

a. What was the average score for these games?

b. The average score for the last four away games was 64. On average, did the team score more at home or away? Explain.

112. A small theater company's production of *Romeo and Juliet* had 10 performances over two weekends. The attendance for each performance during the second weekend was 171, 297, 183, 347, and 232.

a. What was the average attendance for the performances during the second weekend?

b. If the average attendance at a performance during the first weekend was 272, was the average greater in the first or second weekend? Explain.

113. The following table gives the work stoppages for stoppages involving 1,000 or more workers in the United States in selected years.

Year	Total Number of Workers Involved	Total Number of Days Idle
2005	100,000	1,736,000
2006	70,000	2,688,000
2007	189,000	1,265,000
2008	72,000	1,954,000

(*Source:* U.S. Bureau of Labor Statistics)

a. Find the average number of workers per year involved in the work stoppages for the given four years, rounded to the nearest thousand.

b. Find the average annual number of days idle for the years 2006 through 2008. Was the number of days idle in 2008 above or below this average?

114. The chart below shows the countries that, in a recent year, had the largest military expenditures.

Country	Spending Level (in billions of U.S. dollars)	Per Capita Spending Level (in U.S. dollars)
United States	529	1,756
United Kingdom	59	990
France	53	875
China	50	37
Japan	44	341
Germany	37	447
Russia	35	244

(*Source:* infoplease.com)

a. Find the average military spending level for the five countries with the largest military spending levels.

b. What is the average per capita spending level for the seven countries shown?

115. Owners of a restaurant agreed to invest in redecorating the restaurant if the average number of customers per month for the coming 12 months is more than 500. The monthly tallies of customers for the restaurant during this period turned out to be: 372, 618, 502, 411, 638, 465, 572, 377, 521, 488, 458, and 602. Will the restaurant be redecorated?

116. The hospital chart shown is a record of a patient's temperature for two days.

Time	Temp. (°F)	Time	Temp. (°F)
6 A.M.	98	6 A.M.	101
10 A.M.	100	10 A.M.	102
2 P.M.	98	2 P.M.	101
6 P.M.	100	6 P.M.	102
10 P.M.	98	10 P.M.	100
2 A.M.	100	2 A.M.	100

What was her average temperature for this period of time?

• Check your answers on page A-2.

MINDStretchers

Writing

1. Evaluate the expressions in parts (a) and (b).

a. $7^2 + 4^2$ ___

b. $(7 + 4)^2$ ____

c. Are the answers to parts (a) and (b) the same? ___ If not, explain why not.

Mathematical Reasoning

2. The square of any whole number (called a **perfect square**) can be represented as a geometric square, as follows:

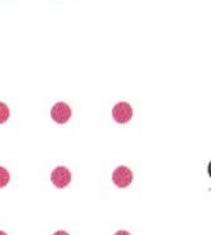

Try to represent the numbers 16, 25, 5, and 8 the same way.

16 25 5 8

Critical Thinking

3. Find the average of the whole numbers from 1 through 999.

1.6 More on Solving Word Problems

OBJECTIVE

A To solve applied problems involving the addition, subtraction, multiplication, or division of whole numbers using various problem-solving strategies

What Word Problems Are and Why They Are Important

In this section, we consider some general tips to help solve word problems.

Word problems can deal with any subject—from shopping to physics and geography to business. Each problem is a brief story that describes a particular situation and ends with a question. Our job, after reading and thinking about the problem, is to answer that question by using the given information.

Although there is no magic formula for solving word problems, you should keep the following problem-solving steps in mind.

To Solve Word Problems

- Read the problem carefully.
- Choose a strategy (such as drawing a picture, breaking up the question, substituting simpler numbers, or making a table).
- Decide which basic operation(s) are relevant and then translate the words into mathematical symbols.
- Perform the operations.
- Check the solution to see if the answer is reasonable. If it is not, start again by rereading the problem.

Reading the Problem

In a math problem, each word counts. So it is important to read the problem slowly and carefully, and not to scan it as if it were a magazine or newspaper article.

When reading a problem, we need to understand the problem's key points: *What information is given* and *what question is posed*. Once these points are clear, jot them down so as to help keep them in mind.

After taking notes, decide on a plan of action that will lead to the answer. For many problems, just thinking back to the meaning of the four basic operations will be helpful.

Operation	Meaning
$+$	Combining
$-$	Taking away
$\times$	Adding repeatedly
$\div$	Splitting up

Many word problems contain *clue words* that suggest performing particular operations. If we spot a clue word in a problem, we should consider whether the operation indicated in the table on the opposite page will lead us to a solution.

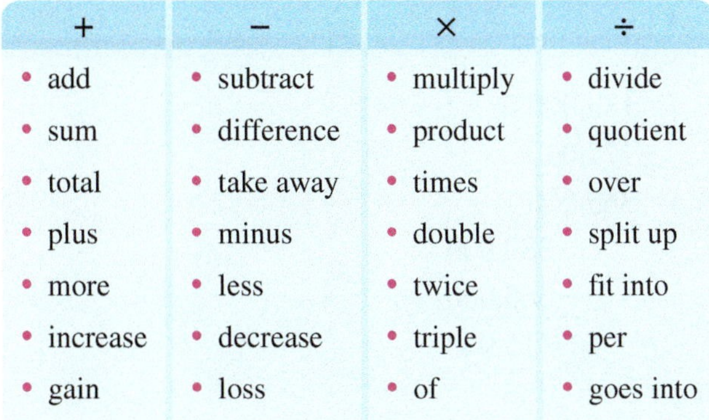

+	−	×	÷
• add	• subtract	• multiply	• divide
• sum	• difference	• product	• quotient
• total	• take away	• times	• over
• plus	• minus	• double	• split up
• more	• less	• twice	• fit into
• increase	• decrease	• triple	• per
• gain	• loss	• of	• goes into

However, be on guard—a clue word can be misleading. For instance, in the problem *What number increased by 2 is 6?*, we solve by subtracting, not adding.

Consider the following "translations" of these clues.

The patient's fever **increased by 5°**.	**+ 5**
The number of unemployed people **tripled**.	**× 3**
The length of the bedroom is **8 feet less** than the kitchen's.	**− 8**
The company's earnings were **split** among the **four** partners.	**÷ 4**

Choosing a Strategy

If no method of solution comes to mind after reading a problem, there are a number of problem-solving strategies that may help. Here we discuss four of these strategies: drawing a picture, breaking up the question, substituting simpler numbers, and making a table.

Drawing a Picture

Sketching even a rough representation of a problem—say, a diagram or a map—can provide insight into its solution, if the sketch accurately reflects the given information.

EXAMPLE 1

In an election, everyone voted for one of three candidates. The winner received 188,000 votes, and the second-place candidate got 177,000 votes. If 380,000 people voted in the election, how many people voted for the third candidate?

Solution To help us understand the given information, let's draw a diagram to represent the situation.

Candidate

1 2 3

188,000 177,000

380,000

We see from this diagram that to find the answer we need to do two things.

PRACTICE 1

A company slashed its workforce by laying off 1,150 employees during one month and laying off 2,235 employees during another month. Afterward, 7,285 employees remained. How many employees worked for the company before the layoffs began?

EXAMPLE 1 (continued)

- First, we need to add 188,000 to 177,000.

$$\begin{array}{r} 188{,}000 \\ +177{,}000 \\ \hline 365{,}000 \end{array}$$

- Then, we need to subtract this sum from 380,000.

$$\begin{array}{r} 380{,}000 \\ -365{,}000 \\ \hline 15{,}000 \end{array}$$

A good way to check our answer here is by adding.

$$188{,}000 + 177{,}000 + 15{,}000 = 380{,}000$$

Our answer checks, so 15,000 people voted for the third candidate.

Breaking Up the Question

Another effective problem-solving strategy is to break up the given question into a chain of simpler questions.

EXAMPLE 2

A student took a math test consisting of 20 questions, answering 3 questions wrong. How many more questions did she get right than wrong?

Solution Say that we do not know how to answer this question directly. Try to split it into several easier questions that lead to a solution.

- How many questions did the student get *right*? $20 - 3 = 17$
- How many questions did she get *wrong*? 3
- How many *more* questions did she get right than wrong? $17 - 3 = 14$

So the student had 14 more questions right than wrong.

This answer seems reasonable, because it must be less than 17, the number of questions that the student answered correctly.

PRACTICE 2

Teddy and Franklin Roosevelt were both U.S. presidents. Teddy was born in 1858 and died in 1919. Franklin was born in 1882 and died in 1945. How much longer did Franklin live than Teddy? (*Source:* Foner and Garraty, *The Reader's Companion to American History*)

Substituting Simpler Numbers

A word problem involving large numbers often seems difficult just because of the numbers. A good problem-solving strategy here is to consider first the identical problem but with simpler numbers. Solve the revised problem and then return to the original problem.

EXAMPLE 3

Raffle tickets cost $4 each. How many tickets must be sold for the raffle to break even if the prizes total $4,736?

Solution Suppose that we are not sure which operation to perform to solve this problem. Let's try substituting a simpler amount (say, $8) for the break-even amount of $4,736 and see if we can solve the resulting problem.

PRACTICE 3

A college has 47 sections of Math 110. If 33 students are enrolled in each section, how many students are taking Math 110?

The question would then become: How many \$4 tickets must be sold to make back \$8? Because it is a "fit-in" question, we must *divide* the \$8 by the \$4. Going back to the original problem, we see that we must divide \$4,736 by 4.

$$\$4{,}736 \div 4 = 1{,}184 \text{ tickets}$$

Is this answer reasonable? We can check either by estimating ($5{,}000 \div 4 = 1{,}250$, which is close to our answer) or by multiplying ($1{,}184 \times 4 = 4{,}736$, which also checks).

Making a Table

Finally, let's consider a strategy for solving word problems that involve many numbers. Organizing these numbers into a table often leads to a solution.

EXAMPLE 4

A borrower promises to pay back \$50 per month until a \$1,000 loan is settled. What is the remaining loan balance at the end of 5 months?

Solution We can solve by organizing the information in a table.

After Month	Remaining Balance
1	$1{,}000 - 50 = 950$
2	$950 - 50 = 900$
3	$900 - 50 = 850$
4	$850 - 50 = 800$
5	$800 - 50 = 750$

From the table, we see that the remaining balance after 5 months is \$750.

We can also solve this problem by breaking up the question into simpler questions.

- How much money did the borrower pay after 5 months? $5 \cdot 50 = \$250$
- How much money did the borrower still owe after 5 months? $1{,}000 - 250 = \$750$

Again, the remaining balance after 5 months is \$750.

PRACTICE 4

An athlete weighs 210 pounds and decides to go on a diet. If he loses 2 pounds a week while on the diet, how much will he weigh after 15 weeks?

1.6 Exercises

FOR EXTRA HELP MyMathLab   PRACTICE  WATCH  READ  REVIEW

Applications

A *Choose a strategy. Solve and check.*

1. In retailing, the difference between the gross sales and customer returns and allowances is called the net sales. If a store's gross sales were \$2,538 and customer returns and allowances amounted to \$388, what was the store's net sales?

2. The population of the United States in 1800 was 5,308,483. Ten years later, the population had grown to 7,239,881. During this period of time, did the country's population double? Justify your answer. (*Source: The Time Almanac 2000*)

3. A delivery van travels 27 miles west, 31 miles east, 45 miles west, and 14 miles east. How far is the van from its starting point?

4. Recycling one aluminum can saves enough energy to run a television for three hours. The average American watches 3,048 hours of television a year. For a year, how many aluminum cans would it take to power a television for the average American? (*Sources:* recycling-revolution.com and tvb.org)

5. A blue whale weighs about 300,000 pounds, and a great white shark weighs about 4,000 pounds. How many times the weight of a great white shark is the weight of a blue whale? (*Source:* wikipedia.com)

6. Two major naval disasters of the twentieth century involved the sinking of British ships—the *Titanic* and the *Lusitania.* The *Titanic,* which weighed about 93,000,000 pounds, was the most luxurious liner of its time; it struck an iceberg on its maiden voyage in 1912. The *Lusitania,* which weighed about 63,000,000 pounds, was sunk by a German submarine in 1915. How much heavier was the *Titanic* than the *Lusitania?* (*Source: The Oxford Companion to Ships and the Sea*)

7. A sales representative flew from Los Angeles to Miami (2,339 miles), then to New York (1,092 miles), and finally back to LA (2,451 miles). How many total miles did he fly?

8. A movie fan installed shelves for his collection of 400 DVDs. If 36 DVDs fit on each shelf, how many shelves did he need to house his entire collection?

9. Immigrants from all over the world came to the United States between 1931 and 1940 in the following numbers: 348,289 (Europe), 15,872 (Asia), 160,037 (Americas, outside the United States), 1,750 (Africa), and 2,231 (Australia/New Zealand). What was the total number of immigrants? (*Source:* George Thomas Kurian, *Datepedia of the United States*)

10. In 2008, there were 38,834 movie screens in the United States. A year later, there were 39,233 movie screens. What was the increase in the number of movie screens from 2008 to 2009? (*Source:* natoonline.org)

11. For each 4×6 print of a digital photo, a lab usually charges 10¢. During a promotion, the first twenty prints of each order are free. How much does the lab charge for fifty 4×6 prints during the promotion?

12. Because of a noisy neighbor, a young man decided to put acoustical tiles on the living room ceiling, which measures 21 feet by 18 feet. The tiles are square, with a side length of 1 foot. If the tiles cost $3 apiece, what is the total cost to cover the ceiling?

13. After a house was on the market for 4 months, the sellers reduced the asking price by $14,000. After 6 months, they reduced the asking price a second and final time. If the original asking price had been $229,000 and the final asking price was $198,000, by how much did the sellers reduce the price the final time?

14. A nurse sets the drip rate for an IV medication at 25 drops each minute. How many drops does a patient receive in 2 hours?

15. A car dealer offered to lease a car for $1,500 down and $189 per month. If a customer accepted a lease contract of 2 years, how much did the customer have to pay over the lease period?

16. A doctor instructed a patient to take 100 milligrams of a medication daily for 4 weeks. The local pharmacy dispensed 120 tablets, each containing 25 milligrams of the medication. After taking the tablets for 4 weeks, how many remained?

17. The part-time tuition rates per credit-hour at a community college were $95 for in-state residents and $257 for out-of-state residents. To take 9 credit-hours, how much more than an in-state resident does an out-of-state resident pay?

18. A scarf placed on sale is marked down by $16. At the register, the customer receives an additional discount of $6. If the final sale price of the scarf was $18, what was the original price?

19. An office manager needs to order 1,000 pens from an office supply catalog. If the catalog sells pens by the gross (that is, in sets of 144) and 7 gross were ordered, how many extra pens did the manager order?

20. At a sale, a shopper decides to buy three shirts costing $39 apiece and two pairs of shoes at $62 per pair. If he has $300 with him, is that enough money to pay for these items? Explain.

21. Dwight Eisenhower beat Adlai Stevenson in the 1952 and 1956 presidential elections. In 1952, Eisenhower received 442 electoral votes and Stevenson 89. In 1956, Eisenhower got 457 electoral votes and Stevenson 73. Which election was closer? By how many electoral votes? (*Source: World Almanac*)

22. A garden is rectangular in shape—26 feet in length and 14 feet in width. If fencing costs $13 a foot, how much will it cost to enclose the garden with this fencing?

23. A couple agrees to pay the seller of the house of their dreams $165,000. They put down $23,448 and promise to pay the balance in 144 equal installments. How much money will each installment be?

24. Earth revolves around the Sun in 365 days, but the planet Mercury does so in only 88 days. Compared to Earth, how many more complete revolutions will Mercury make in 1,000 days?

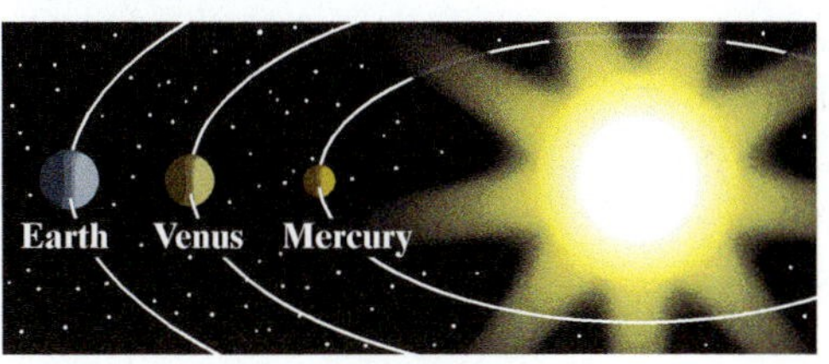

• Check your answers on page A-2.

Key Concepts and Skills

CONCEPT SKILL

Concept/Skill	Description	Example
[1.1] **Place value**	**Thousands**: Hundreds Tens Ones | **Ones**: Hundreds Tens Ones	846,120 ↑ 4 is in the ten thousands place.
[1.1] **To read a whole number**	Working from left to right, • read the number in each period, and then • name the period in place of the comma.	71,400 is read "seventy-one thousand, four hundred".
[1.1] **To write a whole number**	Working from left to right, • write the number named in each period, and • replace each period name with a comma.	"Five thousand, twelve" is written 5,012.
[1.1] **To round a whole number**	• Underline the place to which you are rounding. • The digit to the right of the underlined digit is called the *critical digit.* Look at the critical digit—if it is 5 or more, add 1 to the underlined digit; if it is less than 5, leave the underlined digit unchanged. • Replace all the digits to the right of the underlined digit with zeros.	$3\underline{8}6 \approx 390$ $4{,}\underline{8}17 \approx 4{,}800$
[1.2] **Addend, sum**	In an addition problem, the numbers being added are called *addends.* The result is called their *sum.*	6 + 4 = 10 ↑ ↑ ↑ Addend Addend Sum
[1.2] **The identity property of addition**	The sum of a number and zero is the original number.	$4 + 0 = 4$ $0 + 7 = 7$
[1.2] **The commutative property of addition**	Changing the order in which two numbers are added does not affect their sum.	$7 + 8 = 8 + 7$
[1.2] **The associative property of addition**	When adding three numbers, regrouping addends gives the same sum.	$(5 + 4) + 1 = 5 + (4 + 1)$
[1.2] **To add whole numbers**	• Write the addends vertically, lining up the place values. • Add the digits in the ones column, writing the right-most digit of the sum on the bottom. If the sum has two digits, carry the left digit to the top of the next column on the left. • Add the digits in the tens column as in the preceding step. • Repeat this process until you reach the last column on the left, writing the entire sum of that column on the bottom.	$\begin{array}{r} {}^{1}\ \ {}^{1}{}^{1}\ \ \\ 7{,}385 \\ 92{,}551 \\ +\ \ 2{,}007 \\ \hline 101{,}943 \end{array}$
[1.2] **Minuend, subtrahend, difference**	In a subtraction problem, the number that is being subtracted from is called the *minuend.* The number that is being subtracted is called the *subtrahend.* The answer is called the *difference.*	Difference ↓ 10 − 6 = 4 ↑ ↑ Minuend Subtrahend

Concept/Skill	Description	Example
[1.2] **To subtract whole numbers**	• On top, write the number *from which* we are subtracting. On the bottom, write the number that is being *taken away,* lining up the place values. Subtract in each column separately. • Start with the ones column. **a.** If the digit on top is *larger* than or *equal* to the digit on the bottom, subtract and write the difference below. **b.** If the digit on top is *smaller* than the digit on the bottom, borrow from the digit to the left on top. Then subtract and write the difference below the bottom digit. • Repeat this process until the last column on the left is finished, subtracting and writing its difference below.	8 ¹4 1 7,9 5 2 −1, 8 8 3 6, 0 6 9
[1.3] **Factor, product**	In a multiplication problem, the numbers being multiplied are called *factors*. The result is called their *product.*	Factor Product $4 \times 5 = 20$
[1.3] **The identity property of multiplication**	The product of any number and 1 is that number.	$1 \times 6 = 6$ $7 \times 1 = 7$
[1.3] **The multiplication property of 0**	The product of any number and 0 is 0.	$51 \times 0 = 0$
[1.3] **The commutative property of multiplication**	Changing the order in which two numbers are multiplied does not affect their product.	$3 \times 2 = 2 \times 3$
[1.3] **The associative property of multiplication**	When multiplying three numbers, regrouping the factors gives the same product.	$(4 \times 5) \times 6 = 4 \times (5 \times 6)$
[1.3] **The distributive property**	Multiplying a factor by the sum of two numbers gives the same result as multiplying the factor by each of the two numbers and then adding.	$2 \times (4 + 3)$ $= (2 \times 4) + (2 \times 3)$
[1.3] **To multiply whole numbers**	• Multiply the top factor by the ones digit in the bottom factor and write this product. • Multiply the top factor by the tens digit in the bottom factor and write this product leftward, beginning with the tens column. • Repeat this process until all the digits in the bottom factor are used. • Add the partial products, writing this sum.	693 × 71 693 48 51 49,203
[1.4] **Divisor, dividend, quotient**	In a division problem, the number that is being used to divide another number is called the *divisor.* The number into which it is being divided is called the *dividend.* The result is called the *quotient.*	Quotient 3 4)12 Divisor Dividend

continued

CONCEPT SKILL

Concept/Skill	Description	Example
[1.4] **To divide whole numbers**	• Divide 17 into 39, which gives 2. Multiply the 17 by 2 and subtract the result (34) from 39. Beside the difference (5), bring down the next digit (3) of the dividend. • Repeat this process, dividing the divisor (17) into 53. • At the end, there is a remainder of 2. Write it beside the quotient on top.	23 R2 17)393 34 53 51 2
[1.5] **Exponent (or power), base**	An *exponent* (or *power*) is a number that indicates how many times another number (called the *base*) is used as a factor.	Exponent $5^3 = 5 \times 5 \times 5$ Base
[1.5] **Order of operations rule**	To evaluate mathematical expressions, carry out the operations *in the following order.* **1.** First, perform the operations within any grouping symbols, such as parentheses ( ) or brackets [ ]. **2.** Then, raise any number to its power ■■. **3.** Next, perform all multiplications and divisions as they appear from left to right. **4.** Finally, do all additions and subtractions as they appear from left to right. ( ) ■■ × ÷ + −	$8 + 5 \cdot (3 + 1)^2 = 8 + 5 \cdot 4^2$ $= 8 + 5 \cdot 16$ $= 8 + 80$ $= 88$
[1.5] **Average (or mean)**	The *average* (or *mean*) of a set of numbers is the sum of those numbers divided by however many numbers are in the set.	The average of 3, 4, 10, and 3 is 5 because $\frac{3 + 4 + 10 + 3}{4} = \frac{20}{4} = 5$
[1.6] **To solve word problems**	• Read the problem carefully. • Choose a strategy (such as drawing a picture, breaking up the question, substituting simpler numbers, or making a table). • Decide which basic operation(s) are relevant and then translate the words into mathematical symbols. • Perform the operations. • Check the solution to see if the answer is reasonable. If it is not, start again by rereading the problem.	

CHAPTER 1 Review Exercises

Say Why

Fill in each blank.

1. The place values of 4 in 410 and of 6 in 7,699 ________ (are/are not) the same because ______________________ ______________________.

2. 5,605 rounded to the nearest thousand ________ (is/is not) 5,000 because ______________________.

3. The perimeter of a four-sided figure for which all sides have length 5 ________ (is/is not) 25 because ____________ ______________________.

4. The product of 8 and 7 ________ (is/is not) 15 because ______________________.

5. The area of a rectangle with length 10 and width 7 ________ (is/is not) 70 because ______________________ ______________________.

6. In the expression 9^2, 9 ________ (is/is not) the exponent because ______________________.

7. $2(3 + 5)$ ________ (is/is not) equal to $2 \cdot 3 + 2 \cdot 5$ because ______________________.

8. The quotient of 10 and 5 is ________ (is/is not) 2 because ______________________.

9. The average of 7, 11, and 0 ________ (is/is not) $\dfrac{7 + 11 + 0}{2}$ because ______________________ ______________________.

10. In evaluating the expression $9 + 3 \cdot 4$, we multiply ________ (before/after) adding because ______________________ ______________________.

[1.1] *In each whole number, identify the place that the digit 3 occupies.*

11. 23

12. 30,802

13. 385,000,000

14. 30,000,000,000

Write each number in words.

15. 497

16. 2,050

17. 3,000,007

18. 85,000,000,000

Write each number in standard form.

19. Two hundred fifty-one

20. Nine thousand, two

21. Fourteen million, twenty-five

22. Three billion, three thousand

Express each number in expanded form.

23. 2,500,000

24. 42,707

Round each number to the place indicated.

25. 571 to the nearest hundred

26. 938 to the nearest thousand

27. 384,056 to the nearest ten thousand

28. 68,332 to its largest place

[1.2] *Find the sum and check.*

29. $\begin{array}{r} 102 \\ 4{,}251 \\ +\ 5{,}133 \\ \hline \end{array}$

30. $\begin{array}{r} 53{,}569 \\ 10{,}000 \\ +\ 2{,}123 \\ \hline \end{array}$

31. $\begin{array}{r} 48{,}758 \\ 37{,}226 \\ +\ 87{,}559 \\ \hline \end{array}$

32. $\begin{array}{r} 95{,}000 \\ 25{,}895 \\ +\ 30{,}000 \\ \hline \end{array}$

Add and check.

33. 972,558 + 87,055 + 36,488 + 861,724

34. \$138,865 + \$729 + \$8,002 + \$75,471

Find the difference and check.

35. $\begin{array}{r} 876 \\ -\ 431 \\ \hline \end{array}$

36. $\begin{array}{r} 56{,}000 \\ -\ 45{,}984 \\ \hline \end{array}$

37. $\begin{array}{r} 98{,}118 \\ -\ 87{,}009 \\ \hline \end{array}$

38. $\begin{array}{r} 7{,}100 \\ -\ 1{,}590 \\ \hline \end{array}$

39. 60,000,000 − 48,957,777

40. \$5,000,000 − \$2,937,148

41. From 67,502 subtract 56,496.

42. Subtract 89,724 from 92,713.

[1.3] *Find the product and check.*

43. $\begin{array}{r} 72 \\ \times\ \ 6 \\ \hline \end{array}$

44. $\begin{array}{r} 400 \\ \times\ \ 3 \\ \hline \end{array}$

45. $\begin{array}{r} 2{,}923 \\ \times\ \ 51 \\ \hline \end{array}$

46. $\begin{array}{r} 6{,}000 \\ \times\ 2{,}000 \\ \hline \end{array}$

47. $\begin{array}{r} 14{,}921 \\ \times\ \ 32 \\ \hline \end{array}$

48. $\begin{array}{r} 8{,}152 \\ \times\ \ 125 \\ \hline \end{array}$

Multiply and check.

49. $2{,}751 \cdot 508$

50. (681)(498)(555)

[1.4] *Divide and check.*

51. $\dfrac{975}{25}$

52. $21\overline{)6{,}450}$

53. $13\overline{)491}$

54. $7{,}488 \div 11$

Find the quotient and check.

55. $8\overline{)205{,}000}$

56. $347\overline{)332{,}079}$

[1.5] *Compute.*

57. 7^3

58. 1^{10}

59. $2^3 \cdot 3^2$

60. $3 \cdot 10^5$

61. $20 - 3 \times 5$

62. $(9 + 4)^2$

63. $10 - \dfrac{6 + 4}{2}$

64. $3 + (5 - 1)^2$

65. $5 + [4^2 - 3(2 + 1)]$

66. $17 + [2(3^2 - 6) - 5]$

67. $98(50 - 1)(50 - 2)(50 - 3)$

68. $\dfrac{28^3 + 29^3 + 37^3 - 10}{(7 - 1)^2}$

Rewrite each expression, using exponents.

69. $7 \cdot 7 \cdot 5 \cdot 5 = 7^{■} \cdot 5^{■}$

70. $5 \cdot 2 \cdot 5 \cdot 2 \cdot 5 = 2^{■} \cdot 5^{■}$

Find the average.

71. 34 and 44

72. 20, 0, and 1

73. 5, 8, and 5

74. 4, 6, 3, and 7

Mixed Applications

Solve and check.

75. Beetles about the size of a pinhead destroyed 2,400,000 acres of forest. Express this number in words.

76. Scientists in Utah found a dinosaur egg one hundred fifty million years old. Write this number in standard form.

77. The following graph shows the consolidated assets of the six largest banks in the United States.

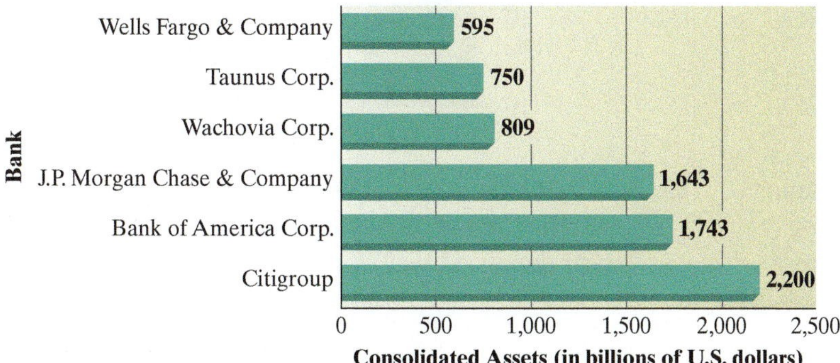

(*Source:* infoplease.com)

Find combined assets of Citigroup and Wachovia.

78. Halley's Comet is expected to visit Earth next in 2061. If the comet has a 76-year orbit, in what year did it last visit Earth? (*Source:* science.nasa.gov)

79. What is the land area of Texas to the nearest hundred thousand square miles? (*Source: Time Almanac 2010*)

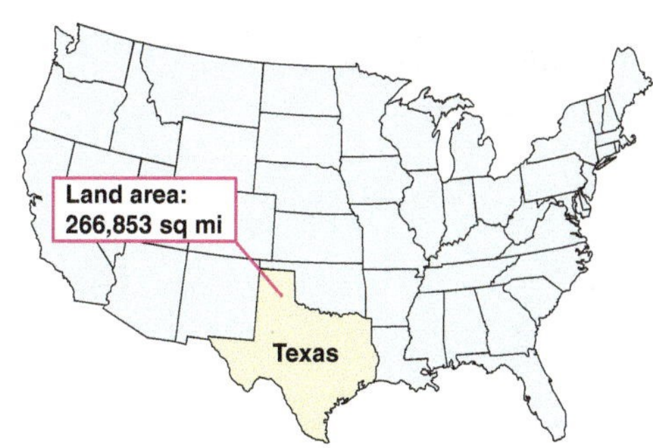

80. Apple Computer sold 31,855,000 iPods in the first two quarters of its 2010 fiscal year. How many iPods is this to the nearest million? (*Source:* wikipedia.org)

81. The Empire State Building is 1,250 feet high, and the Statue of Liberty is 152 feet in height. What is the minimum number of Statues of Liberty that would have to be stacked to be taller than the Empire State Building?

82. Both a singles tennis court and a football field are rectangular in shape. A tennis court measures 78 feet by 27 feet, whereas a football field measures 360 feet by 160 feet. About how many times the area of a tennis court is that of a football field?

83. The tallest building in the United States is Chicago's Willis Tower, which is 1,450 ft high. By contrast, the Dubai Tower, the tallest building in the world, is 1,267 ft higher. Find the height of the Dubai Tower.

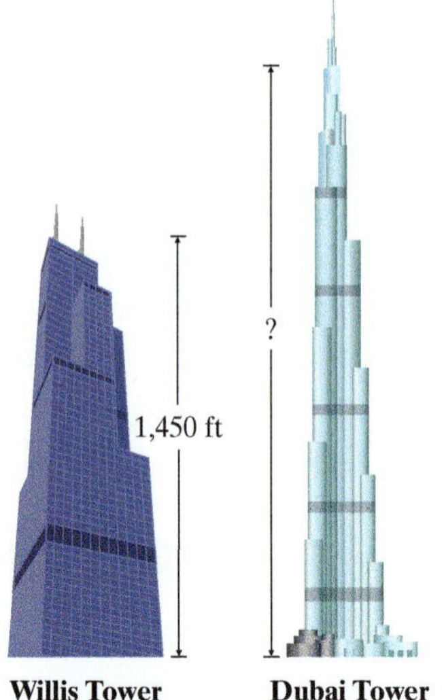

84. A landscaper needs 550 flower plants for a landscaping project. If a local garden center sells flats containing 24 plants, how many flats should the landscaper buy?

85. In a part-time job, a graduate student earned \$15,964 a year. How much money did she earn per week? (*Hint:* 1 year equals 52 weeks.)

86. Compute a company's net profit by completing the following business *skeletal profit and loss statement.*

Net sales	\$430,000
− Cost of merchandise sold	− 175,000
Gross margin	\$
− Operating expenses	− 135,000
Net profit	\$

87. In Giza, Egypt, the pyramid of Khufu has a base that measures 230 meters by 230 meters, whereas the pyramid of Khafre has a base that measures 215 meters by 215 meters. In area, how much larger than the base of the pyramid of Khafre is that of Khufu? (*Source:* pbs.org)

88. A millipede—a small insect with 68 body segments—has 4 legs per segment. How many legs does a millipede have?

89. Richard Nixon ran for the U.S. presidency three times. According to the table below, which was greater—the increase from 1960 to 1972 in the number of votes he got or the increase from 1968 to 1972? (*Source: The New York Times Almanac 2010*)

Year	Number of Votes for Nixon
1960	34,106,671
1968	31,785,148
1972	47,170,179

90. On a business trip, a sales representative flew from Chicago to Los Angeles to Boston. The chart below shows the air distances in miles between these cities.

Air Distance	Chicago	Los Angeles	Boston
Chicago	—	1,745	1,042
Los Angeles	1,745	—	2,596
Boston	1,042	2,596	—

If the sales rep earned a frequent flier point for each mile flown, how many points did he earn?

91. The Tour de France is a 20-stage bicycle race held in France annually. The chart shows the distances for the first 10 stages of the 2010 Tour de France. (*Source:* Le Tour de France)

Stage	Distance (in kilometers)
1	223
2	201
3	216
4	154
5	188
6	228
7	166
8	189
9	205
10	179

a. What was the total distance covered in the first 10 stages?

b. The entire race covered a distance of 3,632 kilometers. How many kilometers were covered in the last 10 stages?

92. The projected enrollment for public colleges in the United States from 2014 through 2018 is given in the table.

Year	Projected Enrollment
2014	14,758,000
2015	14,874,000
2016	14,981,000
2017	15,116,000
2018	15,241,000

(*Source:* census.gov)

a. What is the projected average enrollment in the years 2014 through 2018?

b. If a projected enrollment of 15,366,000 for the year 2019 were included, how would the average annual enrollment change?

93. Find the area of the figure.

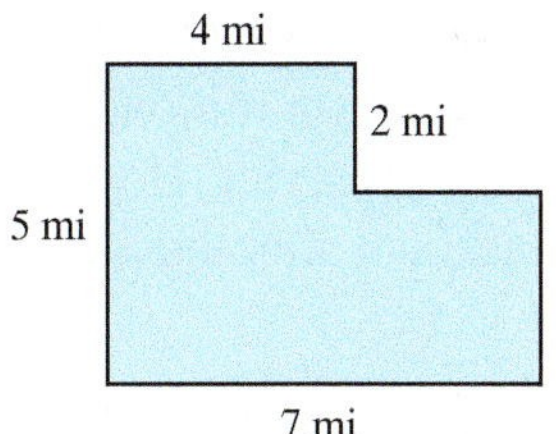

94. Find the perimeter of the figure.

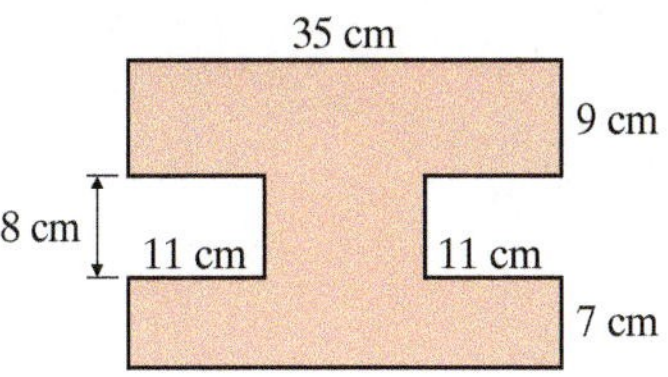

• Check your answers on page A-2.

CHAPTER 1 Posttest

FOR EXTRA HELP — Chapter Test Prep Videos. The Chapter Test Prep Videos with test solutions are available on DVD, in MyMathLab, and on YouTube (search "AkstDevMath" and click on "Channels").

To see if you have already mastered the topics in this chapter, take this test.

1. Write two hundred twenty-five thousand, sixty-seven in standard form.

2. Underline the digit that occupies the ten thousands place in 1,768,405.

3. Write 1,205,007 in words.

4. Round 196,593 to the nearest hundred thousand.

5. Find the sum of 398 and 1,496.

6. Subtract 398 from 1,005.

7. Subtract: $2{,}000 - 1{,}853$

8. Multiply: 328×907

9. Compute: $\dfrac{23{,}923}{47}$

10. Find the quotient: $59\overline{)36{,}717}$

11. Evaluate: 5^4

12. Write $5 \cdot 5 \cdot 4 \cdot 4 \cdot 4$ using exponents.

Simplify.

13. $4 \cdot 9 + 3 \cdot 4^2$

14. $29 - 3^3 \cdot (10 - 9)$

Solve and check.

15. The two largest continents in the world are Asia and Africa. To the nearest hundred thousand square miles, Asia's area is 17,200,000 and Africa's is 11,600,000. How much larger is Asia than Africa? (*Source: National Geographic Atlas of the World*)

16. In the year 2009, the state of Kansas had about 65,500 farms with an average size of 705 acres. How many acres of land in Kansas were devoted to farming? (*Source:* nossausda.gov)

17. A part-time student had \$1,679 in his checking account. He wrote a \$625 check for tuition, a \$546 check for rent, and a \$39 check for groceries. How much money remained in the account after these checks cleared?

18. A part-time taxi driver worked 4 days last week and made the following amounts of money: Monday \$95, Tuesday \$110, Wednesday \$132, and Friday \$155. How much money on the average did he make per work day?

19. A homeowner wishes to carpet the hallway shown below. If the cost of carpeting is about \$10 per square foot, approximately how much will the carpeting cost?

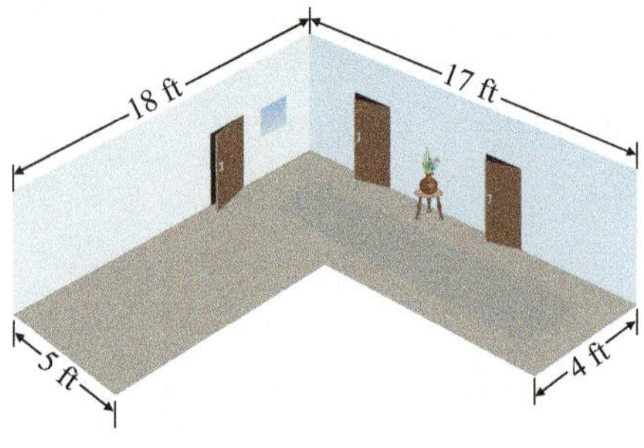

20. Many fast food chains display the caloric intake and the number of grams of fat associated with the chain's dishes.

Food	Calories	Grams of Fat
Original-recipe chicken whole wing	150	9
Original-recipe chicken breast	380	19
Original-recipe drumstick	140	8
Original-recipe thigh	360	25

How many more grams of fat are there in 2 original-recipe thighs than in 3 original-recipe drumsticks? (*Source: Washington Post*)

• Check your answers on page A-2.

Jimmy Stewart made the practice of filibustering famous in the Hollywood film *Mr. Smith Goes to Washington.*

CHAPTER 2

Fractions

Fractions and Filibustering

The rules of the United States Senate allow one or more members to *filibuster*, that is, to talk for as long as they wish and on any topic they choose. So a single member can in effect block the vote on a proposal.

The use of the filibuster to obstruct legislative action in the Senate has a long history going back to the early years of Congress. Unlimited debate was condoned on the grounds that any senator should have the right to address any issue for as long as necessary. However, in the course of time, the *cloture* procedure was developed to end a filibuster. In 1917, the Senate adopted the rule that invoking cloture requires a $\frac{2}{3}$ vote of the current 100 senators. This new rule was first put to the test in 1919, when a motion of cloture was passed ending a filibuster against the Treaty of Versailles. As a result, the state of war between Germany and the Allied Powers was over. In 1975, the Senate changed the rule, reducing the number of votes needed to stop a filibuster from $\frac{2}{3}$ to $\frac{3}{5}$ of the senators.

(*Sources:* senate.gov and wikipedia.org)

CHAPTER 2 PRETEST

To see if you have already mastered the topics in this chapter, take this test.

1. Find all the factors of 20.

2. Express 72 as the product of prime factors.

3. What fraction does the shaded part of the diagram represent?

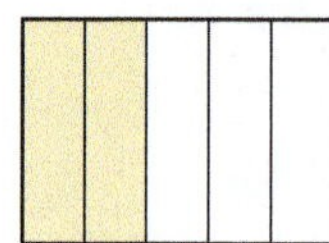

4. Write $20\frac{1}{3}$ as an improper fraction.

5. Express $\frac{31}{30}$ as a mixed number.

6. Write $\frac{9}{12}$ in simplest form.

7. What is the least common multiple of 10 and 4?

8. Which is greater, $\frac{1}{8}$ or $\frac{1}{9}$?

Add.

9. $\frac{1}{2} + \frac{7}{10}$

10. $7\frac{1}{3} + 5\frac{1}{2}$

Subtract.

11. $8\frac{1}{4} - 6$

12. $12\frac{1}{2} - 7\frac{7}{8}$

Multiply.

13. $2\frac{1}{3} \times 1\frac{1}{2}$

14. $\frac{5}{8} \times 96$

15. Divide: $3\frac{1}{3} \div 5$

16. Calculate: $2 + 1\frac{1}{3} \div \frac{4}{5}$

Solve. Write your answer in simplest form.

17. The St. Louis Cardinals, a National League baseball team, won 10 World Series. If 105 World Series have been played, what fraction of these series has this team won? (*Source:* wikipedia.org)

18. In a biology class, three-fourths of the students received a passing grade. If there are 24 students in the class, how many students received failing grades?

19. Find the perimeter of Central Park using the map below.

(*Source:* centralpark.nyc.org)

20. According to the nutrition information given, one serving of Total® Cereal contains 23 grams of carbohydrates. If one serving is $\frac{3}{4}$ cup, what amount of carbohydrates is contained in $2\frac{1}{4}$ cups of Total Cereal? (*Source:* General Mills)

• Check your answers on page A-2.

2.1 Factors and Prime Numbers

OBJECTIVES

- **A** To find the factors of a whole number
- **B** To identify prime and composite numbers
- **C** To find the prime factorization of a whole number
- **D** To find the least common multiple of two or more numbers
- **E** To solve applied problems using factoring or the least common multiple

What Factors Mean and Why They Are Important

Recall that in a multiplication problem, the whole numbers that we are multiplying are called **factors** of the product. For instance, 2 is said to be a factor of 8 because $2 \cdot 4 = 8$. Likewise, 4 is a factor of 8.

Another way of expressing the same idea is in terms of division: We say that 8 is **divisible** by 2, meaning that there is a remainder 0 when we divide 8 by 2.

$$\frac{8}{2} = 4 \text{ R}0$$

Note that: 1, 2, 4, and 8 are all factors of 8.

Although we factor whole numbers, a major application of factoring involves working with fractions, as we demonstrate in the next section.

Finding Factors

To identify the factors of a whole number, we divide the whole number by the numbers 1, 2, 3, 4, 5, 6, and so on, looking for remainders of 0.

EXAMPLE 1

Find all the factors of 6.

Solution Starting with 1, we divide each whole number into 6.

$\frac{6}{1} = 6 \text{ R}0$	$\frac{6}{2} = 3 \text{ R}0$	$\frac{6}{3} = 2 \text{ R}0$	$\frac{6}{4} = 1 \text{ R}2$	$\frac{6}{5} = 1 \text{ R}1$	$\frac{6}{6} = 1 \text{ R}0$
A factor	A factor	A factor	Not a factor	Not a factor	A factor

In finding the factors of 6, we do not need to divide 6 by the numbers 7 or greater. The reason is that no number larger than 6 could divide evenly into 6, that is, divide into 6 with no remainder.

So the factors of 6 are 1, 2, 3, and 6. Note that

- 1 is a factor of 6 and
- 6 is a factor of 6.

PRACTICE 1

What are the factors of 7?

TIP For any whole number, both *the number itself* and *1* are always factors. Therefore, all whole numbers (except 1) have at least two factors.

When checking to see if one number is a factor of another, it is generally faster to use the following **divisibility tests** than to divide.

The number is divisible by	if
2	the ones digit is 0, 2, 4, 6, or 8, that is, if the number is even.
3	the sum of the digits is divisible by 3.
4	the number named by the last two digits is divisible by 4.
5	the ones digit is either 0 or 5.
6	the number is even and the sum of the digits is divisible by 3.
9	the sum of the digits is divisible by 9.
10	the ones digit is 0.

EXAMPLE 2

What are the factors of 45?

Solution Let's see if 45 is divisible by 1, 2, 3, and so on, using the divisibility tests wherever they apply.

Is 45 divisible by	Answer
1?	Yes, because 1 is a factor of any number; $\frac{45}{1} = 45$, so 45 is also a factor.
2?	No, because the ones digit is not even.
3?	Yes, because the sum of the digits, $4 + 5 = 9$, is divisible by 3; $\frac{45}{3} = 15$, so 15 is also a factor.
4?	No, because 4 will not divide into 45 evenly.
5?	Yes, because the ones digit is 5; $\frac{45}{5} = 9$, so 9 is also a factor.
6?	No, because 45 is not even.
7?	No, because $45 \div 7$ has remainder 3.
8?	No, because $45 \div 8$ has remainder 5.
9?	We already know that 9 is a factor.
10?	No, because the ones digit is not 0.

The factors of 45 are, therefore, 1, 3, 5, 9, 15, and 45.

Note that we really didn't have to check to see if 9 was a factor—we learned that it was when we checked for divisibility by 5. Also, because the factors were beginning to repeat with 9, there was no need to check numbers greater than 9.

PRACTICE 2

Find all the factors of 75.

EXAMPLE 3

Identify all the factors of 60.

Solution Let's check to see if 60 is divisible by 1, 2, 3, 4, and so on.

Is 60 divisible by	Answer
1?	Yes, because 1 is a factor of all numbers; $\frac{60}{1} = 60$, so 60 is also a factor.
2?	Yes, because the ones digit is even; $\frac{60}{2} = 30$, so 30 is also a factor.
3?	Yes, because the sum of the digits, $6 + 0 = 6$, is divisible by 3; $\frac{60}{3} = 20$, so 20 is also a factor.
4?	Yes, because 4 will divide into 60 evenly; $\frac{60}{4} = 15$, so 15 is also a factor.
5?	Yes, because the ones digit is 0; $\frac{60}{5} = 12$, so 12 is also a factor.
6?	Yes, because the number is even, and the sum of the digits is divisible by 3; $\frac{60}{6} = 10$, so 10 is also a factor.
7?	No, because $60 \div 7$ has remainder 4.
8?	No, because $60 \div 8$ has remainder 4.
9?	No, because the sum of the digits, $6 + 0 = 6$, is not divisible by 9.
10?	We already know that 10 is a factor.

The factors of 60 are, therefore, 1, 2, 3, 4, 5, 6, 10, 12, 15, 20, 30, and 60.

PRACTICE 3

What are the factors of 90?

EXAMPLE 4

A presidential election takes place in the United States every year that is a multiple of 4. Was there a presidential election in 1866? Explain.

Solution The question is: Does 4 divide into 1866 evenly? Using the divisibility test for 4, we check whether 66 is a multiple of 4.

$$\frac{66}{4} = 16 \text{ R2}$$

Because $\frac{66}{4}$ has remainder 2, 4 is not a factor of 1866. So there was no presidential election in 1866.

PRACTICE 4

The doctor instructs a patient to take a pill every 3 hours. If the patient took a pill at 8:00 this morning, should she take one tomorrow at the same time? Explain.

Identifying Prime and Composite Numbers

Now, let's discuss the difference between prime numbers and composite numbers.

DEFINITIONS

A **prime number** is a whole number that has exactly two different factors: itself and 1.

A **composite number** is a whole number that has more than two factors.

Note that the numbers 0 and 1 are neither prime nor composite. But every whole number greater than 1 is either prime or composite, depending on its factors.

For instance, 5 is prime because its only factors are 1 and 5. But 8 is composite because it has more than two factors (it has four factors: 1, 2, 4, and 8).

Let's practice distinguishing between primes and composites.

EXAMPLE 5

Indicate whether each number is prime or composite.

a. 2 **b.** 78 **c.** 51 **d.** 19 **e.** 31

Solution

a. The only factors of 2 are 1 and 2. Therefore, 2 is prime.

b. Because 78 is even, it is divisible by 2. Having 2 as an "extra" factor—in addition to 1 and 78—means that 78 is composite. Do you see why all even numbers, except for 2, are composite?

c. Using the divisibility test for 3, we see that 51 is divisible by 3, because the sum of the digits 5 and 1, or 6, is divisible by 3. Because 51 has more than two factors, it is composite.

d. The only factors of 19 are itself and 1. Therefore, 19 is prime.

e. Because 31 has no factors other than itself and 1, it is prime.

PRACTICE 5

Decide whether each number is prime or composite.

a. 3 **b.** 57 **c.** 29

d. 34 **e.** 17

Finding the Prime Factorization of a Number

Every composite number can be written as the product of prime factors. This product is called its **prime factorization.** For instance, the prime factorization of 12 is $2 \cdot 2 \cdot 3$.

DEFINITION

The **prime factorization** of a whole number is the number written as the product of its prime factors.

Being able to find the prime factorization of a number is an important skill to have for working with fractions, as we show later in this chapter. A good way to find the prime factorization of a number is by making a **factor tree**, as illustrated in Example 6.

EXAMPLE 6

Write the prime factorization of 72.

Solution We start building a factor tree for 72 by dividing 72 by the smallest prime, 2. Because 72 is $2 \cdot 36$, we write both 2 and 36 underneath the 72. Then, we circle the 2 because it is prime.

Next, we divide 36 by 2, writing both 2 and 18, and circling 2 because it is prime.

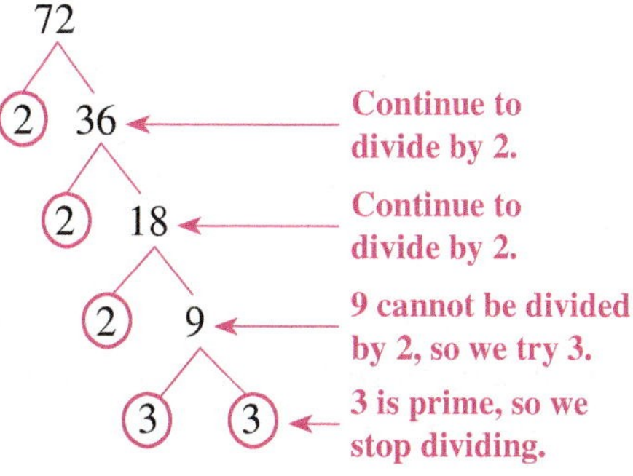

Below the 18, we write 2 and 9, again circling the 2. Because 9 is not divisible by 2, we divide it by the next smallest prime, 3. We stop this process because all the factors in the bottom row are prime. The prime factorization of 72 is the product of the circled factors.

$$72 = 2 \times 2 \times 2 \times 3 \times 3$$

We can also write this prime factorization as $2^3 \times 3^2$.

PRACTICE 6

Write the prime factorization of 56, using exponents.

EXAMPLE 7

Express 80 as the product of prime factors.

Solution The factor tree method for 80 is as shown.

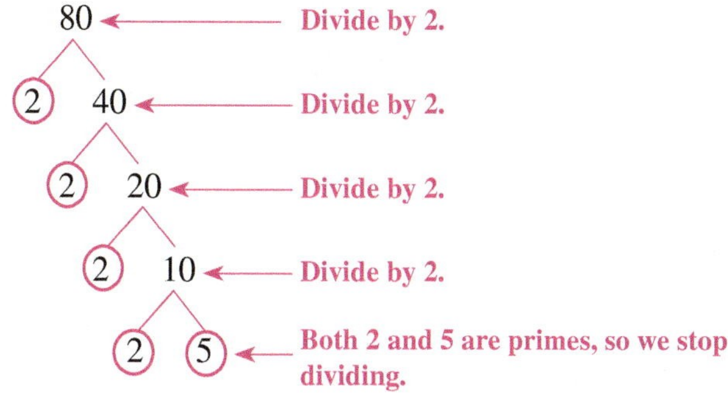

The prime factorization of 80 is $2 \times 2 \times 2 \times 2 \times 5$, or $2^4 \times 5$.

PRACTICE 7

What is the prime factorization of 75?

Finding the Least Common Multiple

The *multiples* of a number are the products of that number and the whole numbers. For instance, some multiples of 5 are the following:

0	5	10	15
0×5	1×5	2×5	3×5

A number that is a multiple of two or more numbers is called a *common multiple* of these numbers. To find the common multiples of 6 and 8, we first list the multiples of 6 and the multiples of 8 separately.

- The multiples of 6 are 0, 6, 12, 18, 24, 30, 36, 42, 48, 54, 60, ...
- The multiples of 8 are 0, 8, 16, 24, 32, 40, 48, 56, 64, ...

So the common multiples of 6 and 8 are 0, 24, 48, ... Of the nonzero common multiples, the *least* common multiple of 6 and 8 is 24.

DEFINITION

The **least common multiple (LCM)** of two or more whole numbers is the smallest nonzero whole number that is a multiple of each number.

A shortcut for finding the LCM—often faster than listing multiples—involves prime factorization.

To Compute the Least Common Multiple (LCM)

- Find the prime factorization of each number.
- Identify the prime factors that appear in each factorization.
- Multiply these prime factors, using each factor the greatest number of times that it occurs in any of the factorizations.

EXAMPLE 8

Find the LCM of 8 and 12.

Solution We first find the prime factorization of each number.

$$8 = 2 \times 2 \times 2 = 2^3 \qquad 12 = 2 \times 2 \times 3 = 2^2 \times 3$$

The factor 2 appears *three times* in the factorization of 8 and *twice* in the factorization of 12, so it must be included three times in forming the least common multiple. Also, the factor 3 appears once in the prime factorization of 12.

The highest power of 2

$$\text{LCM} = 2^3 \times 3 = 8 \times 3 = 24$$

As always, it is a good idea to check that our answer makes sense. We do so by verifying that 8 and 12 really are factors of 24.

PRACTICE 8

What is the LCM of 9 and 6?

EXAMPLE 9

Find the LCM of 5 and 9.

Solution First, we write each number as the product of primes.

$$5 = 5 \qquad 9 = 3 \times 3 = 3^2$$

To find the LCM, we multiply the highest power of each prime.

$$\text{LCM} = 3^2 \times 5 = 9 \times 5 = 45$$

So the LCM of 5 and 9 is 45. Note that 45 is also the product of 5 and 9.

PRACTICE 9

Find the LCM for 3 and 22.

TIP If two or more numbers have no common factor (other than 1), the LCM is their product. If one number is a multiple of another number, then their LCM is the larger of the two numbers.

Now, let's find the LCM of three numbers.

EXAMPLE 10

Find the LCM of 3, 5, and 6.

Solution First, we find the prime factorizations of these three numbers.

$$3 = 3 \qquad 5 = 5 \qquad 6 = 2 \times 3$$

The LCM is therefore the product $2 \times 3 \times 5$, which is 30. Note that 30 is a multiple of 3, 5, and 6, which supports our answer.

PRACTICE 10

Find the LCM of 2, 3, and 4.

EXAMPLE 11

A gym that is open every day of the week offers aerobics classes every third day and Tai Chi classes every fourth day. A student took both classes this morning. In how many days will the gym offer both classes on the same day?

Solution To answer this question, we ask: What is the LCM of 3 and 4? As usual, we begin by finding prime factorizations.

$$3 = 3 \qquad 4 = 2 \times 2 = 2^2$$

To find the LCM, we multiply 3 by 2^2.

$$\text{LCM} = 2^2 \times 3 = 12$$

So both classes will be offered again on the same day in 12 days.

PRACTICE 11

Suppose that a Senate seat and a House of Representatives seat were both filled this year. If the Senate seat is filled every 6 years and the House seat every 2 years, in how many years will both seats be up for election?

2.1 Exercises

FOR EXTRA HELP MyMathLab MathXL PRACTICE WATCH READ REVIEW

Mathematically Speaking

Fill in each blank with the most appropriate term or phrase from the given list.

division	least common multiple	composite	factor tree
divisibility	prime	remainders	common multiple
prime factorization	factors	multiples	

1. 1, 2, 3, 5, 6, 10, 15, and 30 are _______ of 30.

2. A(n) ___________ number is a whole number that has more than two factors.

3. A(n) _______ number has exactly two different factors: itself and 1.

4. The __________________ of two or more numbers is the smallest nonzero number that is a multiple of each number.

5. A number written as the product of its prime factors is called its _________________.

6. The ___________ test for 10 is to check if the ones digit is 0.

A *List all the factors of each number.*

7. 21 **8.** 10 **9.** 17 **10.** 9

11. 12 **12.** 15 **13.** 31 **14.** 47

15. 36 **16.** 35 **17.** 29 **18.** 73

19. 100 **20.** 98 **21.** 28 **22.** 48

B *Indicate whether each number is prime or composite. If it is composite, identify a factor other than the number itself and 1.*

23. 13 **24.** 7 **25.** 16 **26.** 24 **27.** 49

28. 75 **29.** 11 **30.** 31 **31.** 81 **32.** 45

C *Write the prime factorization of each number.*

33. 8 **34.** 10 **35.** 49 **36.** 14 **37.** 24

38. 18 **39.** 50 **40.** 40 **41.** 77 **42.** 63

43. 51 **44.** 57 **45.** 25 **46.** 49 **47.** 32

48. 64 **49.** 21 **50.** 22 **51.** 104 **52.** 105

53. 121 **54.** 169 **55.** 142 **56.** 62 **57.** 100

58. 200 **59.** 125 **60.** 90 **61.** 135 **62.** 400

D *Find the LCM in each case.*

63. 3 and 15 **64.** 9 and 12 **65.** 8 and 10 **66.** 4 and 6

67. 9 and 30 **68.** 20 and 21 **69.** 10 and 11 **70.** 15 and 60

71. 18 and 24 **72.** 30 and 150 **73.** 40 and 180 **74.** 100 and 90

75. 12, 5, and 50

76. 2, 8, and 10

77. 4, 7, and 12

78. 2, 3, and 5

79. 3, 5, and 7

80. 6, 8, and 12

81. 5, 15, and 20

82. 8, 24, and 56

Mixed Practice

Solve.

83. Write the prime factorization of 75.

84. Is 63 prime or composite? If it is composite, identify a factor other than the number itself and 1.

85. List all the factors of 72.

86. Find the LCM of 5, 10, and 12.

Applications

E *Solve.*

87. The federal government conducts a census every year that is a multiple of 10. Explain whether there will be a census in

a. 2015.

b. 2020.

88. Because of production considerations, the number of pages in a book that you are writing must be a multiple of 4. Can the book be

a. 196 pages long?

b. 198 pages long?

89. In 2006, the men's World Cup soccer tournament was held in Munich, Germany. If the tournament is held every 4 years, will there be a tournament in 2036? (*Source:* FIFA World Cup; Soccer Hall of Fame)

90. A car manufacturer recommends changing the oil every 3,000 miles. Would an oil change be recommended at 21,000 miles? Explain.

91. There are 15 players on a rugby team and 10 players on a men's lacrosse team. What is the smallest number of male students in a college that can be split evenly into either rugby or lacrosse teams? (*Source:* wikipedia.org)

92. The Fields Medal, the highest scientific award for mathematicians, is awarded every 4 years. The Dantzig Prize, an achievement award in the field of mathematical programming, is awarded every 3 years. If both were given in 2006, in what year will both be given again? (*Sources:* mathunion.org and siam.org.)

93. Two friends work in a hospital. One gets a day off every 5 days, and the other every 6 days. If they were both off today, in how many days will they again both be off?

94. A family must budget for life insurance premiums every 6 months, car insurance premiums every 3 months, and payments for a home security system every 4 months. If all these bills were due this month, in how many months will they again all fall due?

• Check your answers on page A-3.

MINDStretchers

Historical

1. The eighteenth-century mathematician Christian Goldbach made several famous conjectures (guesses) about prime numbers. One of these conjectures states: Every odd number greater than 7 can be expressed as the sum of three odd prime numbers. For instance, 11 can be expressed as 3 + 3 + 5. Write the following odd numbers as the sum of three odd primes.

 a. 57 = ■ + ■ + ■ **b.** 81 = ■ + ■ + ■

Mathematical Reasoning

2. What is the smallest whole number divisible by every whole number from 1 to 10?

Critical Thinking

3. Choose a three-digit number, say, 715. Find three prime numbers so that, when 715 is multiplied by the product of the three prime numbers, the product of all four numbers is 715,715.

2.2 Introduction to Fractions

OBJECTIVES

- **A** To read or write fractions or mixed numbers
- **B** To write improper fractions as mixed numbers or mixed numbers as improper fractions
- **C** To find equivalent fractions or to write fractions in simplest form
- **D** To compare fractions
- **E** To solve applied problems involving fractions

What Fractions Are and Why They Are Important

A fraction can mean *part of a whole*. Just as a whole number answers the question "How many?", a fraction answers the question "What part of?". Every day we use fractions in this sense. For example, we can speak of *two-thirds* of a class (meaning two of every three students) or *three-fourths* of a dollar (indicating that we have split a dollar into four equal parts and have taken three of these parts).

A fraction can also mean *the quotient of two whole numbers*. In this sense, the fraction $\frac{3}{4}$ tells us what we get when we divide the whole number 3 by the whole number 4.

DEFINITION

A **fraction** is any number that can be written in the form $\frac{a}{b}$, where a and b are whole numbers and b is nonzero.

From this definition, $\frac{1}{2}, \frac{3}{9}, \frac{6}{5}, \frac{8}{2}$, and $\frac{0}{1}$ are all fractions.

When written as $\frac{a}{b}$, a fraction has three components: $\frac{\text{Numerator}}{\text{Denominator}}$ ← Fraction line

- The **denominator** (on the bottom) stands for the number of parts into which the whole is divided.
- The **numerator** (on top) tells us how many parts of the whole the fraction contains.
- The **fraction line** separates the numerator from the denominator and stands for "out of" or "divided by."

Alternatively, a fraction can be represented as either a decimal or a percent. We discuss decimals and percents in Chapters 3 and 6.

Fraction Diagrams and Proper Fractions

Diagrams help us work with fractions. The fraction three-fourths, or $\frac{3}{4}$, is represented by the shaded part in each of the following diagrams:

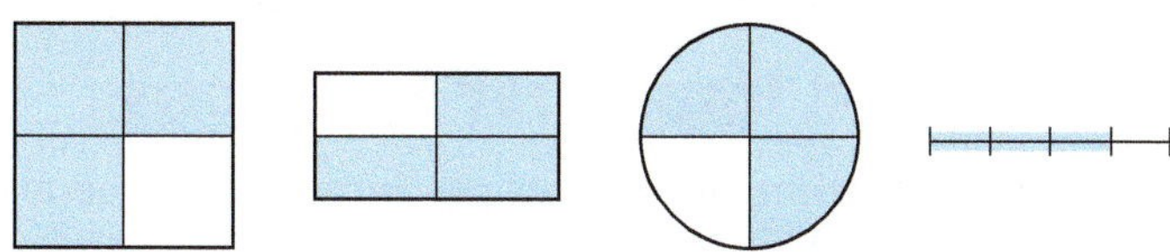

Note that in each diagram the whole has been divided into 4 *equal* parts, with 3 of the parts shaded.

The number $\frac{3}{4}$ is an example of a **proper fraction** because its numerator is smaller than its denominator. Let's consider some other examples of proper fractions.

EXAMPLE 1

In the diagram, what does the shaded portion represent?

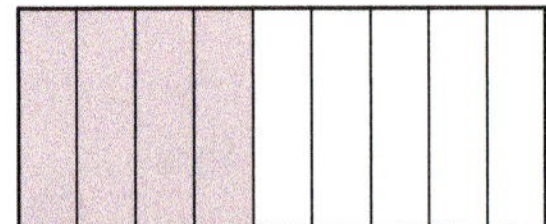

Solution In this diagram, the whole is divided into nine equal parts, so the denominator of the fraction shown is 9. Four of these parts are shaded, so the numerator is 4. The diagram represents the fraction $\frac{4}{9}$.

PRACTICE 1

The diagram illustrates what fraction?

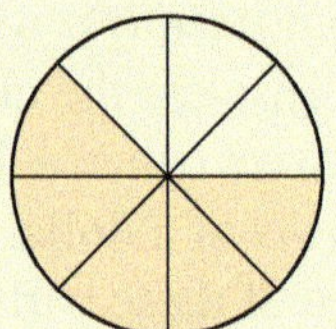

EXAMPLE 2

A college accepted 147 out of 341 applicants for admission into the nursing program. What fraction of the applicants were accepted into this program?

Solution Since there was a total of 341 applicants, the denominator of our fraction is 341. Because 147 of the applicants were accepted, 147 is the numerator. So the college accepted $\frac{147}{341}$ of the applicants into the nursing program.

PRACTICE 2

During a 30-minute television program, 7 minutes were devoted to commercials. What fraction of the time was for commercials?

EXAMPLE 3

The U.S. Senate approved a foreign-aid spending bill by a vote of 83 to 17. What fraction of the senators voted against the bill?

Solution First, we find the total number of senators. Because 83 senators voted for the bill and 17 voted against it, the total number of senators is $83 + 17$, or 100.

So $\frac{17}{100}$ of the senators voted against the bill.

PRACTICE 3

A college conducted an online survey of how many students send text messages. Of the 800 students who were surveyed, about 200 did *not* respond. Express the response rate as a fraction.

Mixed Numbers and Improper Fractions

On many jobs, if you work overtime, the rate of pay increases to one-and-a-half times the regular rate. A number such as $1\frac{1}{2}$, with a whole number part and a proper fraction part, is called a mixed number. A mixed number can also be expressed as an improper fraction, that is, a fraction whose numerator is greater than or equal to its denominator. The number $\frac{3}{2}$ is an example of an improper fraction.

Diagrams help us understand that mixed numbers and improper fractions are different forms of the same numbers, as Example 4 illustrates.

EXAMPLE 4

Draw diagrams to show that $2\frac{1}{3} = \frac{7}{3}$.

Solution First, represent the mixed number and the improper fraction in diagrams.

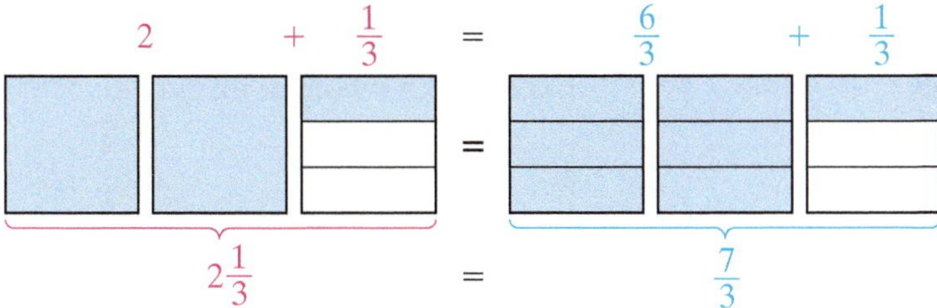

Both diagrams represent $2 + \frac{1}{3}$, so the numbers $2\frac{1}{3}$ and $\frac{7}{3}$ must be equal.

PRACTICE 4

By means of diagrams, explain why $1\frac{2}{3} = \frac{5}{3}$.

In Example 4, each unit (or square) corresponds to 1 whole, which is also three-thirds. That is why the total number of *thirds* in $2\frac{1}{3}$ is $(2 \times 3) + 1$, or 7. The number of *wholes* in $\frac{7}{3}$ is 2 wholes, with $\frac{1}{3}$ of a whole left over. We can generalize these observations into two rules.

To Change a Mixed Number to an Improper Fraction

- Multiply the denominator of the fraction by the whole-number part of the mixed number.
- Add the numerator of the fraction to this product.
- Write this sum over the denominator to form the improper fraction.

EXAMPLE 5

Write each of the following mixed numbers as an improper fraction.

a. $3\frac{2}{9}$ **b.** $12\frac{1}{4}$

Solution

a. $3\frac{2}{9} = \frac{(9 \times 3) + 2}{9}$ Multiply the denominator 9 by the whole number 3, adding the numerator 2. Place over the denominator.

$= \frac{27 + 2}{9} = \frac{29}{9}$ Simplify the numerator.

b. $12\frac{1}{4} = \frac{(4 \times 12) + 1}{4}$

$= \frac{48 + 1}{4} = \frac{49}{4}$

PRACTICE 5

Express each mixed number as an improper fraction.

a. $5\frac{1}{3}$ **b.** $20\frac{2}{5}$

To Change an Improper Fraction to a Mixed Number

- Divide the numerator by the denominator.
- If there is a remainder, write it over the denominator.

EXAMPLE 6

Write each improper fraction as a mixed or whole number.

a. $\frac{11}{2}$ **b.** $\frac{20}{20}$ **c.** $\frac{42}{5}$

Solution

a. $\frac{11}{2} = 2\overline{)11}$ with quotient 5 R1 — Divide the numerator by the denominator.

$\frac{11}{2} = 5\frac{1}{2}$ — Write the remainder over the denominator.

In other words, 5 R1 means that in $\frac{11}{2}$ there are 5 wholes with $\frac{1}{2}$ of a whole left over.

b. $\frac{20}{20} = 1$

c. $\frac{42}{5} = 8\frac{2}{5}$

PRACTICE 6

Express as a whole or mixed number.

a. $\frac{4}{2}$ **b.** $\frac{50}{9}$ **c.** $\frac{8}{3}$

Changing an improper fraction to a mixed number is important when we are dividing whole numbers: It allows us to express any remainder as a fraction. Previously, we would have said that the problems $2\overline{)7}$ and $4\overline{)13}$ both have the answer 3 R1. But by interpreting these problems as improper fractions, we see that their answers are different.

$$\frac{7}{2} = 3\frac{1}{2} \quad \text{but} \quad \frac{13}{4} = 3\frac{1}{4}$$

When a number is expressed as a mixed number, we know its size more readily than when it is expressed as an improper fraction. For instance, consider the mixed number $11\frac{7}{8}$. We immediately see that it is larger than 11 and smaller than 12 (that is, between 11 and 12). We could not reach this conclusion so easily if we were to examine only $\frac{95}{8}$, its improper form. However, there are situations—when we multiply or divide fractions—in which the use of improper fractions is preferable.

Equivalent Fractions

Some fractions that at first glance appear to be different from one another are really the same.

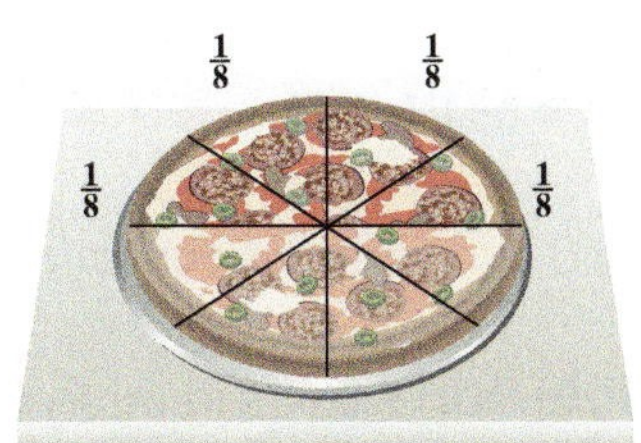

For instance, suppose that we cut a pizza into 8 equal slices, and then eat 4 of the slices. The shaded portion of the diagram at the right represents the amount eaten. Can you explain why in this diagram the fractions $\frac{4}{8}$ and $\frac{1}{2}$ describe the same part of the whole pizza? We say that these fractions are **equivalent**.

Any fraction has infinitely many equivalent fractions. To see why, let's consider the fraction $\frac{1}{3}$. We can draw different diagrams representing one-third of a whole.

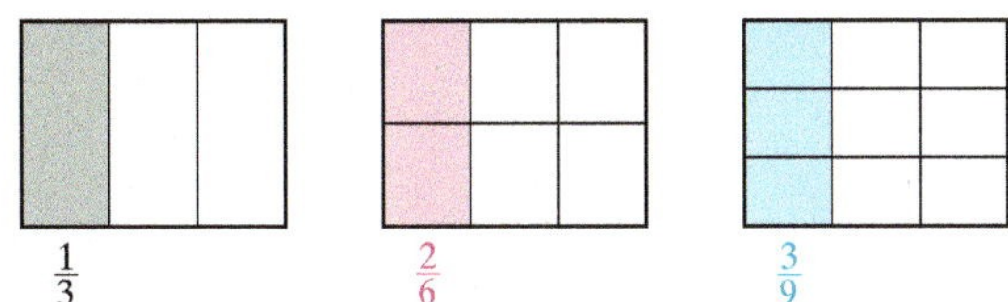

All the shaded portions of the diagrams are identical, so $\frac{1}{3} = \frac{2}{6} = \frac{3}{9}$.

A faster way to generate fractions equivalent to $\frac{1}{3}$ is to multiply both its numerator and denominator by the *same* whole number. Any whole number except 0 will do.

$$\frac{1}{3} = \frac{1 \cdot 2}{3 \cdot 2} = \frac{2}{6}$$
$$\frac{1}{3} = \frac{1 \cdot 3}{3 \cdot 3} = \frac{3}{9}$$
$$\frac{1}{3} = \frac{1 \cdot 4}{3 \cdot 4} = \frac{4}{12}$$
$$\frac{1}{3} = \frac{1 \cdot 5}{3 \cdot 5} = \frac{5}{15}$$

So $\frac{1}{3} = \frac{2}{6} = \frac{3}{9} = \frac{4}{12} = \frac{5}{15} = \ldots.$

Can you explain how you would generate fractions equivalent to $\frac{3}{5}$?

To Find an Equivalent Fraction

Multiply the numerator and denominator of $\frac{a}{b}$ by the same whole number n,

$$\frac{a}{b} = \frac{a \cdot n}{b \cdot n},$$

where both b and n are nonzero.

An important property of equivalent fractions is that their **cross products** are always equal.

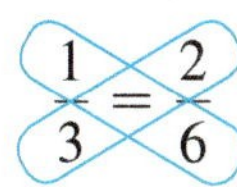

In this case, $1 \cdot 6 = 3 \cdot 2 = 6$

EXAMPLE 7

Find two fractions equivalent to $\frac{1}{7}$.

Solution Let's multiply the numerator and denominator by 2 and then by 6.

$$\frac{1}{7} = \frac{1 \cdot 2}{7 \cdot 2} = \frac{2}{14} \quad \text{and} \quad \frac{1}{7} = \frac{1 \cdot 6}{7 \cdot 6} = \frac{6}{42}$$

We use cross products to check.

$$\frac{1}{7} \stackrel{?}{=} \frac{2}{14}$$
$$1 \cdot 14 \stackrel{?}{=} 7 \cdot 2$$
$$14 \stackrel{\checkmark}{=} 14$$

So $\frac{1}{7}$ and $\frac{2}{14}$ are equivalent.

$$\frac{1}{7} \stackrel{?}{=} \frac{6}{42}$$
$$1 \cdot 42 \stackrel{?}{=} 7 \cdot 6$$
$$42 \stackrel{\checkmark}{=} 42$$

So $\frac{1}{7}$ and $\frac{6}{42}$ are equivalent.

PRACTICE 7

Identify three fractions equivalent to $\frac{2}{5}$.

EXAMPLE 8

Write $\frac{3}{7}$ as an equivalent fraction whose denominator is 35.

Solution $\frac{3}{7} = \frac{3 \cdot 5}{7 \cdot 5} = \frac{15}{35}$ Multiply the numerator and denominator by 5.

Therefore, $\frac{15}{35}$ is equivalent to $\frac{3}{7}$. To check, we find the cross products: Both $3 \cdot 35$ and $7 \cdot 15$ equal 105.

PRACTICE 8

Express $\frac{5}{8}$ as a fraction whose denominator is 72.

Writing a Fraction in Simplest Form

Previously in this section, we saw that $\frac{4}{8}$ and $\frac{1}{2}$ are equivalent fractions: $\frac{1}{2} = \frac{1 \cdot 4}{2 \cdot 4} = \frac{4}{8}$. We also can start with $\frac{4}{8}$ and find the equivalent fraction $\frac{1}{2}$ by dividing its numerator and denominator by the common factor 4:

$$\frac{4}{8} = \frac{\overset{1}{\cancel{4}}}{\underset{2}{\cancel{8}}} = \frac{1}{2}$$

Note that in the fraction $\frac{1}{2}$, the only common factor of the numerator and denominator is 1. A fraction is said to be **simplified** or **written in lowest terms** when the only common factor of its numerator and denominator is 1. The simplified form of a fraction is the equivalent fraction with the smallest numerator and denominator.

A good way to simplify a fraction is to first express both the numerator and denominator as the product of their prime factors. We can then divide out or cancel all common factors as the following example illustrates:

EXAMPLE 9

Simplify $\frac{9}{15}$.

Solution To express this fraction in simplest form, we write both the numerator and denominator as the product of primes.

$$\frac{9}{15} = \frac{\overset{1}{\cancel{3}} \cdot 3}{\underset{1}{\cancel{3}} \cdot 5} = \frac{3}{5}$$

Note that the only common factor of 3 and 5 is 1. To be sure that we have not made an error, let's check whether the cross products are equal: $9 \cdot 5 = 3 \cdot 15 = 45$.

PRACTICE 9

Write $\frac{14}{21}$ in lowest terms.

EXAMPLE 10

Write $\frac{42}{28}$ in lowest terms.

Solution

$$\frac{42}{28} = \frac{2 \cdot 3 \cdot 7}{2 \cdot 2 \cdot 7}$$ Express the numerator and denominator as the product of primes.

$$= \frac{\overset{1}{\cancel{2}} \cdot 3 \cdot \overset{1}{\cancel{7}}}{\underset{1}{\cancel{2}} \cdot 2 \cdot \underset{1}{\cancel{7}}}$$ Divide out the common factors, noting that 1 remains.

$$= \frac{3}{2}$$ Multiply the remaining factors.

PRACTICE 10

Simplify $\frac{42}{18}$.

EXAMPLE 11

A couple's annual income is $75,000. If they pay $9,000 for rent and $3,000 for food per year, rent and food account for what fraction of their income? Simplify the answer.

Solution First, we must find the part of the income that is paid for rent and food per year.

$$\underset{\text{Rent}}{\$9{,}000} + \underset{\text{Food}}{\$3{,}000} = \underset{\text{Part}}{\$12{,}000}$$

PRACTICE 11

An acre is a unit of area approximately equal to 4,800 square yards. A developer is selling parcels of land of 50 yards by 30 yards. What fraction of an acre is each parcel? Simplify the answer.

EXAMPLE 11 (continued)

The part is \$12,000 and the whole is \$75,000, so the fraction is $\frac{12{,}000}{75{,}000}$. We can simplify this fraction by cancelling common factors:

$$\frac{12{,}000}{75{,}000} = \frac{12,\cancel{000}}{75,\cancel{000}} = \frac{12}{75}$$

Note that canceling a 0 is the same as dividing by 10.

$$= \frac{3 \cdot 4}{3 \cdot 25} = \frac{\overset{1}{\cancel{3}} \cdot 4}{\underset{1}{\cancel{3}} \cdot 25} = \frac{4}{25}$$

Therefore, $\frac{4}{25}$ of the couple's income goes for rent and food.

Comparing Fractions

Some situations require us to *compare* fractions, that is, to rank them in order of size.

For instance, suppose that $\frac{5}{8}$ of one airline's flights arrive on time, in contrast to $\frac{3}{5}$ of another airline's flights. To decide which airline has a better record for on-time arrivals, we need to compare the fractions.

Or to take another example, suppose that the drinking water in your home, according to a lab report, has 2 parts per million (ppm) of lead. Is the water safe to drink if the federal limit on lead in drinking water is 15 parts per billion (ppb)? Again, we need to compare fractions.

One way to handle such problems is to draw diagrams corresponding to the fractions in question. The larger fraction corresponds to the larger shaded region.

For instance, the diagrams to the right show that $\frac{3}{4}$ is greater than $\frac{1}{4}$. The symbol $>$ stands for "greater than."

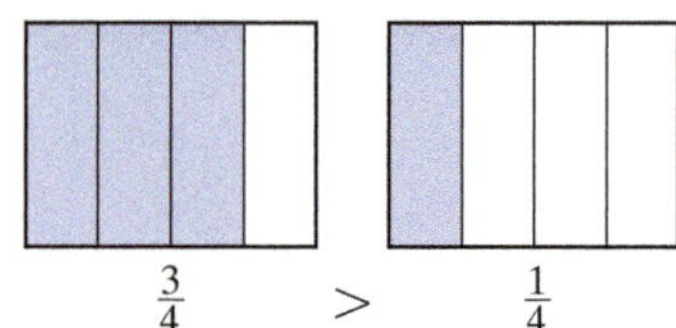

Both $\frac{3}{4}$ and $\frac{1}{4}$ have the same denominator, so we can rank them simply by comparing their numerators.

$$\frac{3}{4} > \frac{1}{4} \quad \text{because} \quad 3 > 1$$

For **like fractions**, the fraction with the larger numerator is the larger fraction.

DEFINITIONS

Like fractions are fractions with the same denominator.

Unlike fractions are fractions with different denominators.

To Compare Fractions

- If the fractions are like, compare their numerators.
- If the fractions are unlike, write them as equivalent fractions with the same denominator, and then compare their numerators.

EXAMPLE 12

Compare $\frac{7}{15}$ and $\frac{4}{9}$.

Solution These fractions are unlike because they have different denominators. Therefore, we need to express them as equivalent fractions having the same denominator. But what should that denominator be?

One common denominator that we can use is the *product of the denominators*: $15 \cdot 9 = 135$.

$$\frac{7}{15} = \frac{7 \cdot 9}{15 \cdot 9} = \frac{63}{135}$$ **$135 = 15 \cdot 9$, so the new numerator is $7 \cdot 9$ or 63.**

$$\frac{4}{9} = \frac{4 \cdot 15}{9 \cdot 15} = \frac{60}{135}$$ **$135 = 9 \cdot 15$, so the new numerator is $4 \cdot 15$ or 60.**

Next, we compare the numerators of the like fractions that we just found.

Because $63 > 60$, $\frac{63}{135} > \frac{60}{135}$. Therefore, $\frac{7}{15} > \frac{4}{9}$.

Another common denominator that we can use is the least common multiple of the denominators.

$$15 = 3 \times 5 \qquad 9 = 3 \times 3 = 3^2$$

The LCM is $3^2 \times 5 = 9 \times 5 = 45$. We then compute the equivalent fractions.

$$\frac{7}{15} = \frac{7 \cdot 3}{15 \cdot 3} = \frac{21}{45}$$ **Multiply the numerator and denominator by 3.**

$$\frac{4}{9} = \frac{4 \cdot 5}{9 \cdot 5} = \frac{20}{45}$$ **Multiply the numerator and denominator by 5.**

Because $\frac{21}{45} > \frac{20}{45}$, we know that $\frac{7}{15} > \frac{4}{9}$.

PRACTICE 12

Which is larger, $\frac{13}{24}$ or $\frac{11}{16}$?

Note that in Example 12 we computed the LCM of the two denominators. This type of computation is used frequently in working with fractions.

DEFINITION

For two or more fractions, their **least common denominator** (LCD) is the least common multiple of their denominators.

In Example 13, pay particular attention to how we use the LCD.

EXAMPLE 13

Order from smallest to largest: $\frac{3}{4}$, $\frac{7}{10}$, and $\frac{29}{40}$

Solution Because these fractions are unlike, we need to find equivalent fractions with a common denominator. Let's use their LCD as that denominator.

$$4 = 2 \times 2 = 2^2$$
$$10 = 2 \times 5$$
$$40 = 2 \times 2 \times 2 \times 5 = 2^3 \times 5$$

The LCD $= 2^3 \times 5 = 8 \times 5 = 40$. Check: 4 and 10 are both factors of 40.

We write each fraction with a denominator of 40.

$$\frac{3}{4} = \frac{3 \cdot 10}{4 \cdot 10} = \frac{30}{40} \qquad \frac{7}{10} = \frac{7 \cdot 4}{10 \cdot 4} = \frac{28}{40} \qquad \frac{29}{40} = \frac{29}{40}$$

Then, we order the fractions from smallest to largest. (The symbol $<$ stands for "less than.")

$$\frac{28}{40} < \frac{29}{40} < \frac{30}{40} \quad \text{or} \quad \frac{7}{10} < \frac{29}{40} < \frac{3}{4}$$

PRACTICE 13

Arrange $\frac{9}{10}$, $\frac{23}{30}$, and $\frac{8}{15}$ from smallest to largest.

EXAMPLE 14

About $\frac{39}{50}$ of Earth's atmosphere is made up of nitrogen, and $\frac{21}{100}$ is made up of oxygen. Does nitrogen or oxygen make up more of Earth's atmosphere? (*Source: The New York Times Almanac, 2010*)

Solution We need to compare $\frac{39}{50}$ with $\frac{21}{100}$. The LCD is 100.

$$\frac{39}{50} = \frac{78}{100}$$
$$\frac{21}{100} = \frac{21}{100}$$

Since $\frac{78}{100} > \frac{21}{100}$, $\frac{39}{50} > \frac{21}{100}$. Therefore, nitrogen makes up more of Earth's atmosphere than oxygen does.

PRACTICE 14

In a recent year, about $\frac{3}{20}$ of the commercial radio stations in the United States had an adult contemporary format and $\frac{1}{5}$ had a country format. In that year, were there more adult contemporary stations or country stations? (*Source:* musicbizacademy.com)

2.2 Exercises

FOR EXTRA HELP MyMathLab MathXL PRACTICE WATCH READ REVIEW

Mathematically Speaking

Fill in each blank with the most appropriate term or phrase from the given list.

improper fraction	proper fraction	like fractions
greatest common factor	simplify	mixed
convert	least common denominator	composite
equivalent		

1. A fraction whose numerator is smaller than its denominator is called a(n) ______________.

2. The improper fraction $\frac{5}{2}$ can be expressed as a(n) _______ number.

3. The fractions $\frac{6}{8}$ and $\frac{3}{4}$ are ____________.

4. Divide the numerator and denominator of a fraction by the same whole number in order to ___________ it.

5. Fractions with the same denominator are said to be _____________.

6. The _____________________ of two or more fractions is the least common multiple of their denominators.

A *Identify a fraction or mixed number represented by the shaded portion of each figure.*

7. 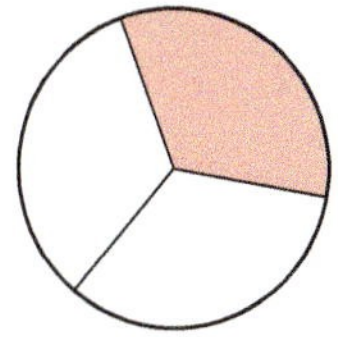

8. 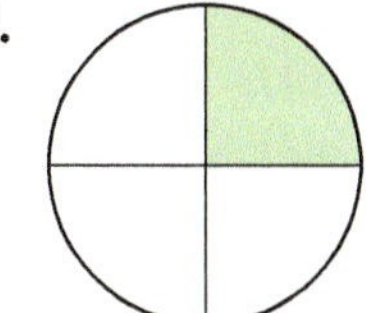

9. 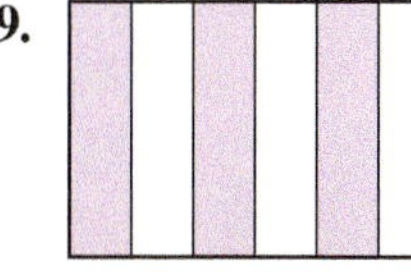

10.

11. 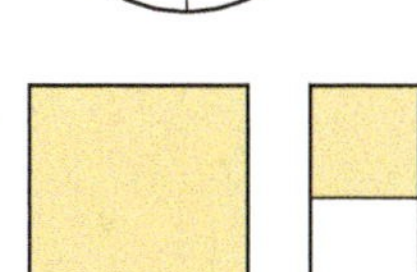

12. 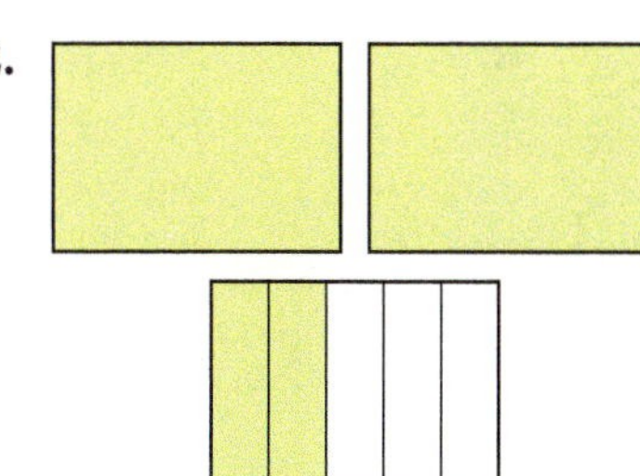

13. 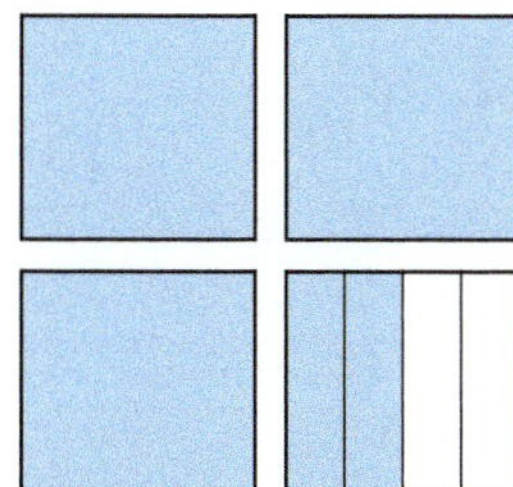

14. 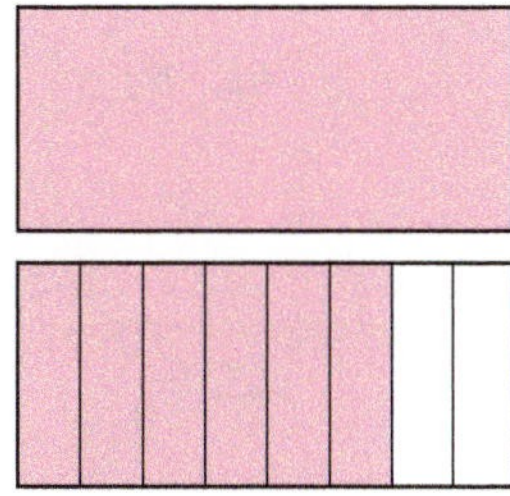

Draw a diagram to represent each fraction or mixed number.

15. $\frac{5}{7}$

16. $\frac{6}{11}$

17. $\frac{2}{9}$

18. $\frac{4}{10}$

19. $\frac{6}{6}$

20. $\frac{11}{11}$

21. $\frac{6}{5}$

22. $\frac{8}{3}$

23. $2\frac{1}{2}$

24. $4\frac{1}{5}$

25. $2\frac{1}{3}$

26. $3\frac{4}{9}$

B *Indicate whether each number is a proper fraction, an improper fraction, or a mixed number.*

27. $\frac{2}{5}$

28. $\frac{7}{12}$

29. $\frac{10}{9}$

30. $\frac{11}{10}$

31. $16\frac{2}{3}$

32. $12\frac{1}{2}$

33. $\frac{5}{5}$

34. $\frac{4}{4}$

35. $\frac{4}{9}$

36. $\frac{5}{6}$

37. $66\frac{2}{3}$

38. $10\frac{3}{4}$

Write each number as an improper fraction.

39. $2\frac{3}{5}$

40. $1\frac{1}{3}$

41. $6\frac{1}{9}$

42. $10\frac{2}{3}$

43. $11\frac{2}{5}$

44. $12\frac{3}{4}$

45. 5

46. 8

47. $7\frac{3}{8}$

48. $6\frac{5}{6}$

49. $9\frac{7}{9}$

50. $10\frac{1}{2}$

51. $13\frac{1}{2}$

52. $20\frac{1}{8}$

53. $19\frac{3}{5}$

54. $11\frac{5}{7}$

55. 14

56. 10

57. $4\frac{10}{11}$

58. $2\frac{7}{13}$

59. $8\frac{3}{14}$

60. $4\frac{1}{6}$

61. $8\frac{2}{25}$

62. $14\frac{1}{10}$

Express each fraction as a mixed or whole number.

63. $\frac{4}{3}$

64. $\frac{6}{5}$

65. $\frac{10}{9}$

66. $\frac{12}{5}$

67. $\frac{9}{3}$

68. $\frac{12}{12}$

69. $\frac{15}{15}$

70. $\frac{100}{100}$

71. $\frac{99}{5}$

72. $\frac{31}{2}$

73. $\frac{82}{9}$

74. $\frac{62}{3}$

75. $\frac{45}{45}$

76. $\frac{40}{3}$

77. $\frac{74}{9}$

78. $\frac{41}{8}$

79. $\frac{27}{2}$ **80.** $\frac{58}{11}$ **81.** $\frac{100}{9}$ **82.** $\frac{38}{3}$

83. $\frac{27}{1}$ **84.** $\frac{72}{9}$ **85.** $\frac{56}{7}$ **86.** $\frac{19}{1}$

C *Find two fractions equivalent to each fraction.*

87. $\frac{1}{8}$ **88.** $\frac{3}{10}$ **89.** $\frac{2}{11}$ **90.** $\frac{1}{10}$

91. $\frac{3}{4}$ **92.** $\frac{5}{6}$ **93.** $\frac{1}{9}$ **94.** $\frac{3}{5}$

Write an equivalent fraction with the given denominator.

95. $\frac{3}{4} = \frac{}{12}$ **96.** $\frac{2}{9} = \frac{}{18}$ **97.** $\frac{5}{8} = \frac{}{24}$ **98.** $\frac{7}{10} = \frac{}{20}$

99. $4 = \frac{}{10}$ **100.** $5 = \frac{}{15}$ **101.** $\frac{3}{5} = \frac{}{60}$ **102.** $\frac{4}{9} = \frac{}{63}$

103. $\frac{5}{8} = \frac{}{64}$ **104.** $\frac{3}{10} = \frac{}{40}$ **105.** $3 = \frac{}{18}$ **106.** $2 = \frac{}{21}$

107. $\frac{4}{9} = \frac{}{81}$ **108.** $\frac{7}{8} = \frac{}{24}$ **109.** $\frac{6}{7} = \frac{}{49}$ **110.** $\frac{5}{6} = \frac{}{48}$

111. $\frac{2}{17} = \frac{}{51}$ **112.** $\frac{1}{3} = \frac{}{90}$ **113.** $\frac{7}{12} = \frac{}{84}$ **114.** $\frac{1}{4} = \frac{}{100}$

115. $\frac{2}{3} = \frac{}{48}$ **116.** $\frac{7}{8} = \frac{}{56}$ **117.** $\frac{3}{10} = \frac{}{100}$ **118.** $\frac{5}{6} = \frac{}{144}$

Simplify, if possible.

119. $\frac{6}{9}$ **120.** $\frac{9}{12}$ **121.** $\frac{10}{10}$ **122.** $\frac{21}{21}$

123. $\frac{5}{15}$ **124.** $\frac{4}{24}$ **125.** $\frac{9}{20}$ **126.** $\frac{25}{49}$

127. $\frac{25}{100}$ **128.** $\frac{75}{100}$ **129.** $\frac{125}{1{,}000}$ **130.** $\frac{875}{1{,}000}$

131. $\frac{20}{16}$ **132.** $\frac{15}{9}$ **133.** $\frac{66}{32}$ **134.** $\frac{30}{18}$

135. $\frac{18}{32}$ **136.** $\frac{36}{45}$ **137.** $\frac{7}{24}$ **138.** $\frac{19}{51}$

139. $\frac{27}{9}$ **140.** $\frac{36}{144}$ **141.** $\frac{12}{84}$ **142.** $\frac{21}{36}$

143. $3\frac{38}{57}$ **144.** $11\frac{51}{102}$ **145.** $2\frac{100}{100}$ **146.** $1\frac{144}{144}$

D *Between each pair of numbers, insert the appropriate sign: <, =, or >.*

147. $\frac{7}{20}$ $\frac{11}{20}$

148. $\frac{5}{10}$ $\frac{3}{10}$

149. $\frac{1}{8}$ $\frac{1}{9}$

150. $\frac{5}{6}$ $\frac{7}{8}$

151. $\frac{2}{3}$ $\frac{6}{9}$

152. $\frac{9}{12}$ $\frac{3}{4}$

153. $2\frac{1}{3}$ $2\frac{9}{15}$

154. $2\frac{3}{7}$ $1\frac{1}{2}$

Arrange in increasing order.

155. $\frac{1}{2}, \frac{1}{3}, \frac{1}{4}$

156. $\frac{3}{2}, \frac{3}{3}, \frac{3}{4}$

157. $\frac{2}{3}, \frac{7}{12}, \frac{5}{6}$

158. $\frac{3}{4}, \frac{5}{6}, \frac{7}{8}$

159. $\frac{3}{5}, \frac{2}{3}, \frac{8}{9}$

160. $\frac{5}{8}, \frac{1}{2}, \frac{4}{11}$

Mixed Practice

Solve.

161. Choose the number whose value is between the other two: $\frac{7}{10}, \frac{8}{9}, \frac{5}{6}$.

162. Express $\frac{32}{6}$ as a mixed number.

163. Find two fractions equivalent to $\frac{2}{9}$.

164. Draw a diagram to represent $\frac{9}{10}$.

165. Write an equivalent fraction for $\frac{4}{5}$ with denominator 15.

166. Write $2\frac{3}{8}$ as an improper fraction.

Applications

E *Solve. Write your answer in simplest form.*

167. During the last 5 days, a student spent 11 hours studying mathematics at home. On the average, how much time is this per day?

168. A recipe for pasta with garlic and oil calls for 6 garlic cloves, peeled and chopped. If the recipe serves 4, how many garlic cloves on the average are in each serving?

169. As of 2009, the Nobel Prize was awarded to 40 women and 762 men. (*Source:* nobelprize.org)

a. What fraction of the Nobel prize winners were women?

b. What fraction were men?

170. It is projected that in 2015 there will be 182,000 physical therapists and 94,000 respiratory therapists employed as health care practitioners in the United States. (*Source: U.S. Census Statistical Abstract 2010*)

a. What fraction of these practitioners will be physical therapists?

b. What fraction will be respiratory therapists?

171. Of the 206 bones in the human skeleton, 106 are in the hands and feet. What fraction of these bones are *not* in the hands and feet? (*Source:* Henry Gray, *Anatomy of the Human Body*)

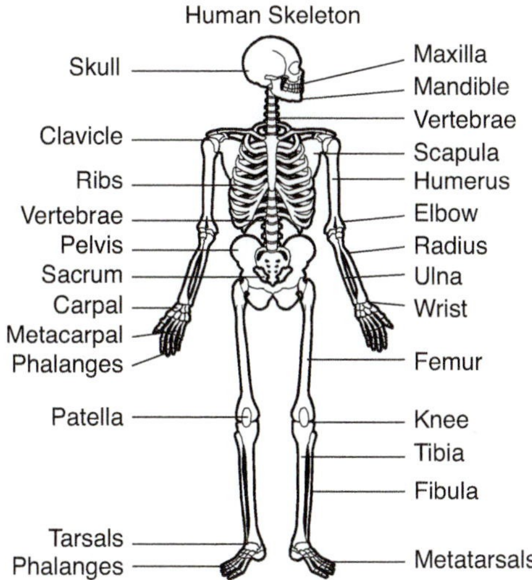

172. In a recent year of a National Basketball Association season, the Los Angeles Lakers won 65 of their 82 games led by the team captain, Kobe Bryant. Of the total games played, what fraction did the team *not* win? (*Sources:* Wikipedia.org and *Sports Illustrated Almanac 2010*)

173. A serving of frozen yogurt provides approximately $\frac{1}{10}$ of the recommended dietary allowance (RDA) of calcium, while a serving of plain yogurt provides $\frac{2}{5}$ of the RDA. Which of the two kinds of yogurt provides more calcium? Explain. (*Source:* dietary-supplements.info.nih.gov)

174. In a course on probability and statistics, a student learns that when rolling a pair of dice, the probability of getting a 5 is $\frac{1}{9}$, and the probability of getting a 6 is $\frac{5}{36}$. Does getting a 5 or a 6 have a greater probability? Explain.

175. In a recent year, of 64 million visits to other countries by U.S. tourists, approximately 19 million visits were to Mexico, 13 million were to Canada, and 4 million were to the United Kingdom. (*Source:* U.S. Department of Commerce)

a. The visits to the United Kingdom were what fraction of these 64 million visits?

b. What fraction of the visits were to either Mexico or Canada?

176. When fog rolled into the New York City area, visibility was reduced to one-sixteenth mile at JFK Airport, one-eighth mile at LaGuardia Airport, and one-half mile at Newark Airport.

a. Which of the three airports had the best visibility?

b. Which of the three airports had the worst visibility?

177. In a recent year, the weights of the first six draft picks for the Pittsburgh Steelers football team are given in the table below.

Player	Pouncey	Worilds	Sanders	Gibson	Scott	Butler
Weight (in pounds)	312	240	180	240	346	185

(*Source:* sports.yahoo.com)

What was their average weight?

178. The following chart gives the age of the first six American presidents at the time of their inauguration.

President	Washington	J. Adams	Jefferson	Madison	Monroe	J. Q. Adams
Age (in years)	57	61	57	57	58	57

What was their average age at inauguration? (*Source: Significant American Presidents of the United States*)

• Check your answers on page A-3.

MINDStretchers

Mathematical Reasoning

1. Identify the fraction that the shaded portion of the figure to the right represents.

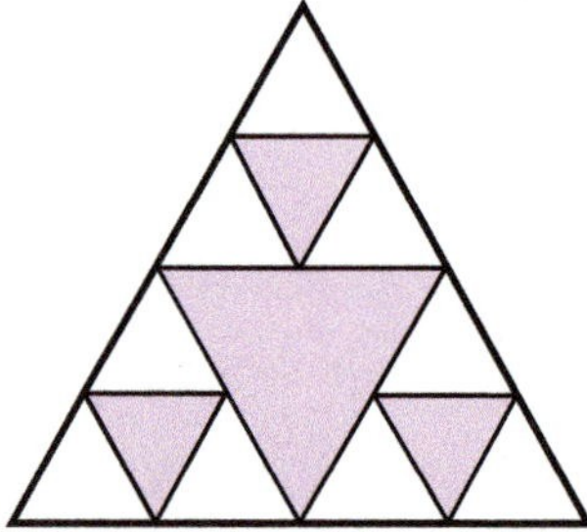

Groupwork

2. Working with a partner, determine how many fractions there are between the numbers 1 and 2.

Critical Thinking

3. Consider the three equivalent fractions shown. Note that the numerators and denominators are made up of the digits 1, 2, 3, 4, 5, 6, 7, 8, and 9—each appearing once.

$$\frac{3}{6} = \frac{7}{14} = \frac{29}{58}$$

a. Verify that these fractions are equivalent by making sure that their cross products are equal.

b. Write another trio of equivalent fractions that use the same nine digits only once.

$$\frac{2}{4} = \frac{}{} = \frac{}{}$$

2.3 Adding and Subtracting Fractions

OBJECTIVES

- **A** To add or subtract fractions or mixed numbers
- **B** To solve applied problems involving the addition or subtraction of fractions or mixed numbers

In Section 2.2 we examined what fractions mean, how they are written, and how they are compared. In the rest of this chapter, we discuss computations involving fractions, beginning with sums and differences.

Adding and Subtracting Like Fractions

Let's first discuss how to add and subtract like fractions. Suppose that an employee spends $\frac{1}{7}$ of his weekly salary for food and $\frac{2}{7}$ for rent. What part of his salary does he spend for food and rent combined? A diagram can help us understand what is involved. First, we shade one-seventh of the diagram, then another two-sevenths. We see in the diagram that the total shaded area is three-sevenths, $\frac{1}{7} + \frac{2}{7} = \frac{3}{7}$. Note that we added the original numerators to get the numerator of the answer but that *the denominator stayed the same*.

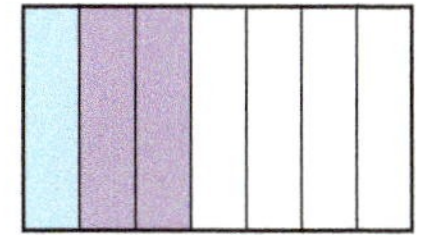

The diagram at the right illustrates the subtraction of like fractions, namely, $\frac{3}{7} - \frac{1}{7}$. If we shade three-sevenths of the diagram and then remove the shading in one-seventh, two-sevenths remain shaded. Therefore, $\frac{3}{7} - \frac{1}{7} = \frac{2}{7}$. Note that we could have gotten this answer simply by subtracting numerators without changing the denominator.

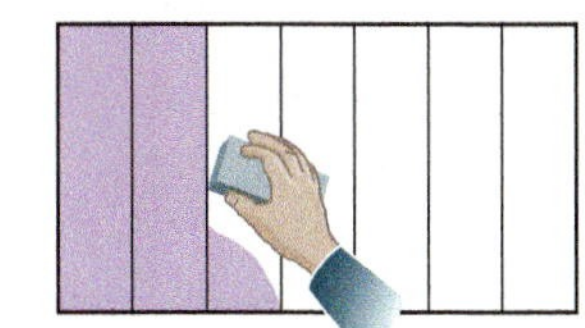

The following rule summarizes how to add or subtract fractions, *provided that they have the same denominator*.

To Add (or Subtract) Like Fractions

- Add (or subtract) the numerators.
- Use the given denominator.
- Write the answer in simplest form.

EXAMPLE 1

Add: $\frac{7}{12} + \frac{2}{12}$

Solution Applying the rule, we get $\frac{7}{12} + \frac{2}{12} = \frac{7+2}{12} = \frac{9}{12}$

$= \frac{3 \cdot 3}{4 \cdot 3} = \frac{3 \cdot \cancel{3}^{1}}{4 \cdot \cancel{3}_{1}} = \frac{3}{4}$.

Add the numerators.

Keep the same denominator.

Simplest form

PRACTICE 1

Find the sum of $\frac{7}{15}$ and $\frac{3}{15}$.

TIP Be careful *not* to add the denominators when adding fractions.

EXAMPLE 2

Find the sum of $\frac{12}{16}$, $\frac{3}{16}$, and $\frac{9}{16}$.

Solution

Answer as a mixed number

$$\frac{12}{16} + \frac{3}{16} + \frac{9}{16} = \frac{24}{16} = \frac{3}{2}, \text{ or } 1\frac{1}{2}$$

So the sum of $\frac{12}{16}$, $\frac{3}{16}$, and $\frac{9}{16}$ is $1\frac{1}{2}$.

PRACTICE 2

Add: $\frac{13}{40}$, $\frac{11}{40}$ and $\frac{23}{40}$

EXAMPLE 3

Find the difference between $\frac{11}{7}$ and $\frac{3}{7}$.

Solution

Subtract the numerators.

$$\frac{11}{7} - \frac{3}{7} = \frac{11 - 3}{7} = \frac{8}{7}, \text{ or } 1\frac{1}{7}$$

Keep the same denominator.

PRACTICE 3

Subtract: $\frac{19}{20} - \frac{11}{20}$

EXAMPLE 4

In the following diagram,

a. how far is it from the Administration Building to the Library via the Science Center?

b. which route from the Administration Building to the Library is shorter—via the Science Center or via the Student Center? By how much?

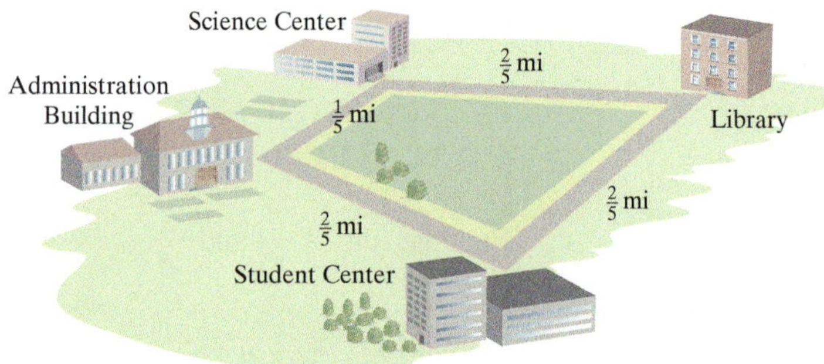

Solution

a. Examining the diagram, we see that

- the distance from the Administration Building to the Science Center is $\frac{1}{5}$ mile, and
- the distance from the Science Center to the Library is $\frac{2}{5}$ mile.

PRACTICE 4

A pediatrician prescribed $\frac{9}{20}$ gram of pain medication for a patient to take every 4 hours.

a. If the dosage were increased by $\frac{3}{20}$ gram, what would the new dosage be?

b. If the original dosage were decreased by $\frac{1}{20}$ gram, find the new dosage.

To find the distance from the Administration Building to the Library via the Science Center, we add.

$$\frac{1}{5} + \frac{2}{5} = \frac{3}{5}$$

So this distance is $\frac{3}{5}$ mile.

b. To find the distance from the Administration Building to the Library via the Student Center, we again add.

$$\frac{2}{5} + \frac{2}{5} = \frac{4}{5}$$

So this distance is $\frac{4}{5}$ mile. Since $\frac{3}{5} < \frac{4}{5}$, the route from the Administration Building to the Library via the Science Center is shorter than the route via the Student Center. Now we find the difference.

$$\frac{4}{5} - \frac{3}{5} = \frac{1}{5}$$

Therefore, the route via the Science Center is $\frac{1}{5}$ mile shorter than the route via the Student Center.

Adding and Subtracting Unlike Fractions

Adding (or subtracting) **unlike fractions** is more complicated than adding (or subtracting) like fractions. An extra step is required: changing the unlike fractions to equivalent like fractions. For instance, suppose that we want to add $\frac{1}{10}$ and $\frac{2}{15}$. Even though we can use any common denominator for these fractions, let's use their *least* common denominator to find equivalent fractions.

$$\begin{aligned} 10 &= 2 \cdot 5 \\ 15 &= 3 \cdot 5 \\ \text{LCD} &= 2 \cdot 3 \cdot 5 = 30 \end{aligned}$$

Let's rewrite the fractions vertically as equivalent fractions with the denominator 30.

$$\begin{aligned} \frac{1}{10} &= \frac{1 \cdot 3}{10 \cdot 3} = \frac{3}{30} \\ +\frac{2}{15} &= \frac{2 \cdot 2}{15 \cdot 2} = +\frac{4}{30} \end{aligned}$$

Now, we add the equivalent like fractions.

$$\begin{array}{r} \frac{3}{30} \\ +\frac{4}{30} \\ \hline \frac{7}{30} \end{array}$$

So $\frac{1}{10} + \frac{2}{15} = \frac{7}{30}$.

We can also add and subtract unlike fractions horizontally.

$$\frac{1}{10} + \frac{2}{15} = \frac{3}{30} + \frac{4}{30} = \frac{3+4}{30} = \frac{7}{30}$$

To Add (or Subtract) Unlike Fractions

- Write the fractions as equivalent fractions with the same denominator, usually the LCD.
- Add (or subtract) the numerators, keeping the same denominator.
- Write the answer in simplest form.

EXAMPLE 5

Add: $\frac{5}{12} + \frac{5}{16}$

Solution First, we find the LCD, which is 48. After finding equivalent fractions, we add the numerators, keeping the same denominator.

$$\begin{aligned} \frac{5}{12} &= \frac{5 \cdot 4}{12 \cdot 4} = \frac{20}{48} \\ +\frac{5}{16} &= \frac{5 \cdot 3}{16 \cdot 3} = +\frac{15}{48} \\ & \qquad\qquad\quad \frac{35}{48} \end{aligned}$$

← Already in lowest terms

PRACTICE 5

Add: $\frac{11}{12} + \frac{3}{4}$

EXAMPLE 6

Subtract $\frac{1}{12}$ from $\frac{1}{3}$.

Solution Because 3 is a factor of 12, the LCD is 12. Again, let's set up the problem vertically.

$$\begin{aligned} \frac{1}{3} &= \frac{4}{12} \\ -\frac{1}{12} &= -\frac{1}{12} \\ & \quad \frac{3}{12} = \frac{1}{4} \end{aligned}$$

Subtract the numerators, keeping the same denominator.

Write $\frac{3}{12}$ in lowest terms.

PRACTICE 6

Calculate: $\frac{4}{5} - \frac{1}{2}$

EXAMPLE 7

Combine: $\frac{1}{3} + \frac{1}{6} - \frac{3}{10}$

Solution First, we find the LCD of all three fractions. The LCD is 30.

$$\frac{1}{3} = \frac{10}{30}, \quad \frac{1}{6} = \frac{5}{30}, \quad \text{and} \quad \frac{3}{10} = \frac{9}{30}.$$

So $\frac{1}{3} + \frac{1}{6} - \frac{3}{10} = \frac{10}{30} + \frac{5}{30} - \frac{9}{30} = \frac{10 + 5 - 9}{30} = \frac{6}{30} = \frac{1}{5}$.

PRACTICE 7

Combine: $\frac{1}{3} - \frac{2}{9} + \frac{7}{8}$

EXAMPLE 8

A forest ranger leaves his lodge, hiking to three fire camps before returning to his lodge, as shown below. How far did he hike?

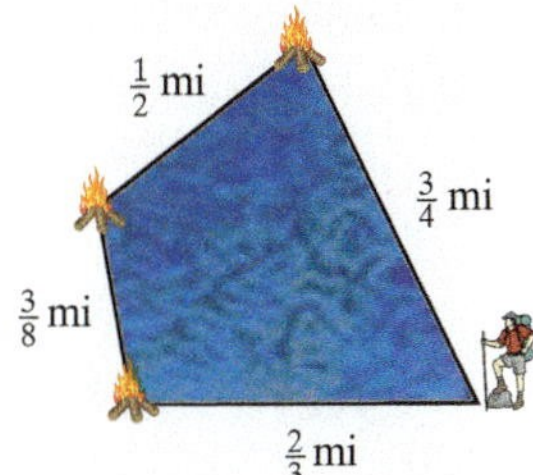

Solution The distance that the ranger hiked is the perimeter of the figure shown, that is, the sum of the lengths of its sides.

$$\text{Perimeter} = \frac{2}{3} + \frac{3}{8} + \frac{1}{2} + \frac{3}{4}$$

$$\begin{aligned} \frac{2}{3} &= \frac{16}{24} \leftarrow \text{LCD} \\ \frac{3}{8} &= \frac{9}{24} \\ \frac{1}{2} &= \frac{12}{24} \\ +\frac{3}{4} &= +\frac{18}{24} \\ \hline & \quad \frac{55}{24}, \text{ or } 2\frac{7}{24} \end{aligned}$$

So the ranger hiked $2\frac{7}{24}$ miles.

PRACTICE 8

Vet's Park Triangle is a neighborhood in Ann Arbor, Michigan, shown below. Find the perimeter. (*Source:* City-data.com)

Adding Mixed Numbers

Now, let's consider how to add **mixed numbers**, starting with those that have the same denominator.

Suppose, for instance, that we want to add $1\frac{1}{5}$ and $2\frac{1}{5}$. Let's draw a diagram to represent this sum.

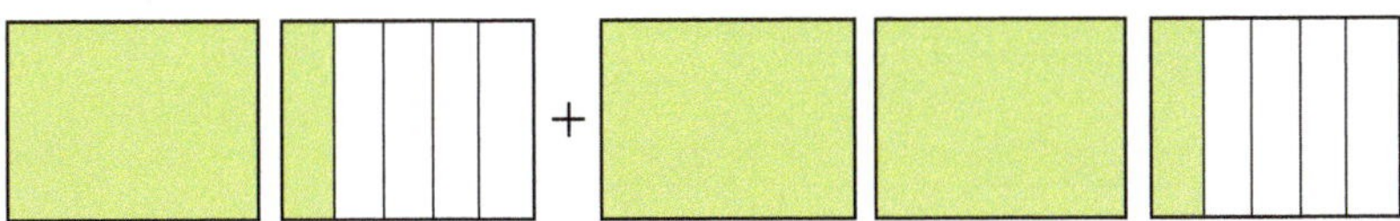

We can rearrange the elements of the diagram by combining the whole numbers and the fractions separately.

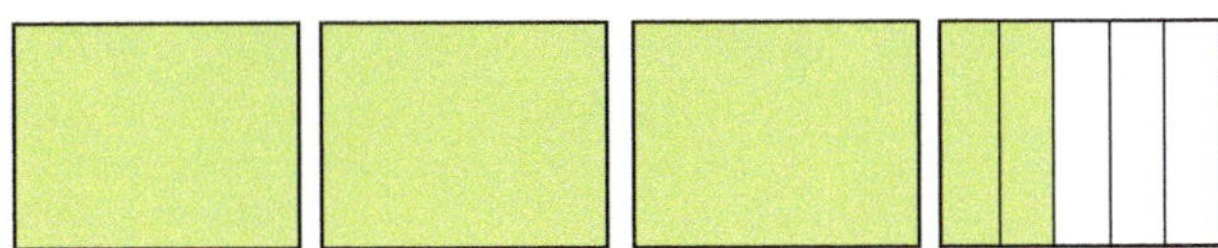

This diagram shows that the sum is $3\frac{2}{5}$.

Note that we can also write and solve this problem vertically.

$$\begin{array}{r} 1\frac{1}{5} \\ +2\frac{1}{5} \\ \hline 3\frac{2}{5} \end{array}$$

← Sum of the fractions

Sum of the whole numbers

EXAMPLE 9

Add: $8\frac{5}{9} + 10\frac{1}{9}$

Solution

$$\begin{array}{r} 8\frac{5}{9} \\ +10\frac{1}{9} \\ \hline 18\frac{6}{9} \end{array} = 18\frac{2}{3}$$

PRACTICE 9

Add: $25\frac{3}{10} + 9\frac{1}{10}$

EXAMPLE 10

Find the sum of $3\frac{3}{5}$, $2\frac{4}{5}$, and 6.

Solution Add the fractions, and then add the whole numbers.

$$\begin{array}{r} 3\frac{3}{5} \\ 2\frac{4}{5} \\ +\ 6 \\ \hline 11\frac{7}{5} \end{array} = 12\frac{2}{5}$$

Since $\frac{7}{5} = 1\frac{2}{5}$, we get $11\frac{7}{5} = 11 + 1\frac{2}{5} = 12\frac{2}{5}$.

So the sum is $12\frac{2}{5}$.

PRACTICE 10

Find the sum of $2\frac{5}{16}$, $1\frac{3}{16}$, and 4.

EXAMPLE 11

Two movies are shown back-to-back on TV without commercial interruption. The first runs $1\frac{3}{4}$ hours, and the second $2\frac{1}{4}$ hours. How long will it take to watch both movies?

Solution

We need to add $1\frac{3}{4}$ and $2\frac{1}{4}$.

$$\begin{array}{r} 1\frac{3}{4} \\ +2\frac{1}{4} \\ \hline 3\frac{4}{4} \end{array} = 3 + 1 = 4$$

Therefore, it will take 4 hours to watch the two movies.

PRACTICE 11

In a horse race, the winner beat the second-place horse by $1\frac{1}{2}$ lengths, and the second-place horse finished $2\frac{1}{2}$ lengths ahead of the third-place horse. By how many lengths did the third-place horse lose?

We have previously shown that when we add fractions with different denominators, we must first change the unlike fractions to equivalent like fractions. The same applies to adding mixed numbers that have different denominators.

To Add Mixed Numbers

- Write the fractions as equivalent fractions with the same denominator, usually the LCD.
- Add the fractions.
- Add the whole numbers.
- Write the answer in simplest form.

EXAMPLE 12

Find the sum of $3\frac{1}{5}$ and $7\frac{2}{3}$.

Solution The LCD is 15. Add the fractions, and then add the whole numbers.

$$\begin{array}{rcr} 3\frac{1}{5} & = & 3\frac{3}{15} \\ +7\frac{2}{3} & = & +7\frac{10}{15} \\ \hline & & 10\frac{13}{15} \end{array}$$

The sum of $3\frac{1}{5}$ and $7\frac{2}{3}$ is $10\frac{13}{15}$.

PRACTICE 12

Add $4\frac{1}{8}$ to $3\frac{1}{2}$.

EXAMPLE 13

Find the sum of $1\frac{2}{3}$, $8\frac{1}{4}$, and $3\frac{4}{5}$.

Solution Set up the problem vertically and use the LCD, which is 60. Add the fractions, and then add the whole numbers.

$$\begin{aligned} 1\frac{2}{3} &= 1\frac{40}{60} \\ 8\frac{1}{4} &= 8\frac{15}{60} \\ +3\frac{4}{5} &= +3\frac{48}{60} \end{aligned}$$

$$12\frac{103}{60} = 12 + 1\frac{43}{60} = 13\frac{43}{60}$$

PRACTICE 13

What is the sum of $5\frac{5}{8}$, $3\frac{1}{6}$, and $2\frac{5}{12}$?

Subtracting Mixed Numbers

Now, let's discuss how to subtract mixed numbers, beginning with those that have the same denominator.

For instance, suppose that we want to subtract $2\frac{1}{5}$ from $3\frac{2}{5}$. We draw a diagram to represent $3\frac{2}{5}$.

   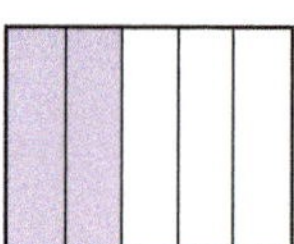

If we remove the shading from $2\frac{1}{5}$, then $1\frac{1}{5}$ remains shaded.

  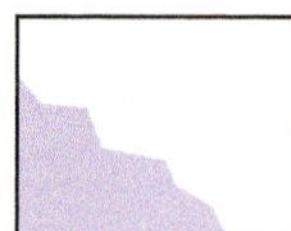 

So the difference is $1\frac{1}{5}$.

We can also write and solve this problem vertically.

$$\begin{array}{r} 3\frac{2}{5} \\ -2\frac{1}{5} \\ \hline 1\frac{1}{5} \end{array}$$

$\frac{1}{5}$ ← Difference of the fractions

1 ← Difference of the whole numbers

EXAMPLE 14

Subtract: $4\frac{5}{6} - 2\frac{1}{6}$

Solution We set up the problem vertically. Subtract the fractions, and then subtract the whole numbers.

$$\begin{array}{r} 4\frac{5}{6} \\ -2\frac{1}{6} \\ \hline 2\frac{4}{6} = 2\frac{2}{3} \end{array}$$

Therefore, the difference is $2\frac{2}{3}$.

PRACTICE 14

Subtract $5\frac{3}{10}$ from $9\frac{7}{10}$.

EXAMPLE 15

A construction job was scheduled to last $5\frac{3}{4}$ days, but was finished in $4\frac{1}{4}$ days. How many days ahead of schedule was the job?

Solution

This question asks us to subtract $4\frac{1}{4}$ from $5\frac{3}{4}$.

$$\begin{array}{r} 5\frac{3}{4} \\ -4\frac{1}{4} \\ \hline 1\frac{2}{4} = 1\frac{1}{2} \end{array}$$

So the job was $1\frac{1}{2}$ days ahead of schedule.

PRACTICE 15

A student is typing a report using a word processing program. To make room for the binding, the left margin is set at $2\frac{7}{10}$ inches and the right margin at $1\frac{3}{10}$ inches. The left margin is how much greater than the right margin?

When subtracting mixed numbers with different denominators, we begin by finding the least common denominator.

EXAMPLE 16

Subtract $2\frac{7}{100}$ from $5\frac{9}{10}$.

Solution As usual, we use the LCD (which is 100) to find equivalent fractions. Then we subtract the equivalent mixed numbers with the same denominator. Again, let's set up the problem vertically. Subtract the fractions, and then subtract the whole numbers.

$$\begin{array}{rcr} 5\frac{9}{10} & = & 5\frac{90}{100} \\ -2\frac{7}{100} & = & -2\frac{7}{100} \\ \hline & & 3\frac{83}{100} \end{array}$$

The answer is $3\frac{83}{100}$.

PRACTICE 16

Calculate: $8\frac{2}{3} - 4\frac{1}{12}$

EXAMPLE 17

Find the length of the pool shown below.

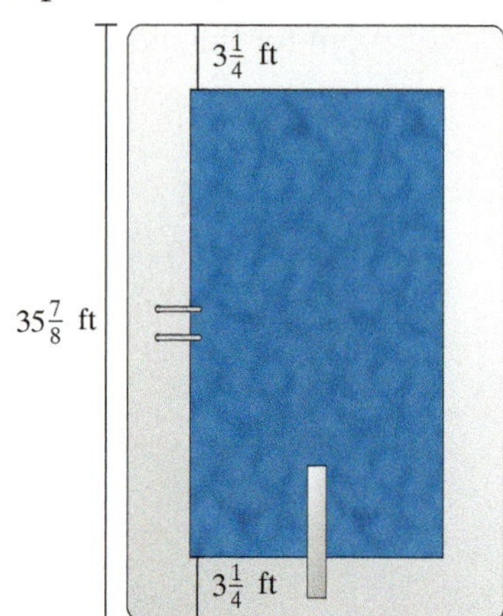

Solution The total length of the pool and the walkway is $35\frac{7}{8}$ feet. To find the length of the pool, we first add $3\frac{1}{4}$ feet and $3\frac{1}{4}$ feet. Then we subtract this sum from $35\frac{7}{8}$ feet.

$$\begin{array}{r} 3\frac{1}{4} \\ +3\frac{1}{4} \\ \hline 6\frac{2}{4} = 6\frac{1}{2} \end{array} \qquad \begin{array}{rcr} 35\frac{7}{8} & = & 35\frac{7}{8} \\ -6\frac{1}{2} & = & -6\frac{4}{8} \\ \hline & & 29\frac{3}{8} \end{array}$$

So the length of the pool is $29\frac{3}{8}$ feet. We can check this answer by adding $3\frac{1}{4}$, $29\frac{3}{8}$, and $3\frac{1}{4}$, getting $35\frac{7}{8}$.

PRACTICE 17

The outline of the state of Nevada is roughly in the shape of a *trapezoid*. The diagram shows the lengths of the borders that Nevada shares with each neighboring state. What is the total length of Nevada's border? (*Source:* econ.umn.edu)

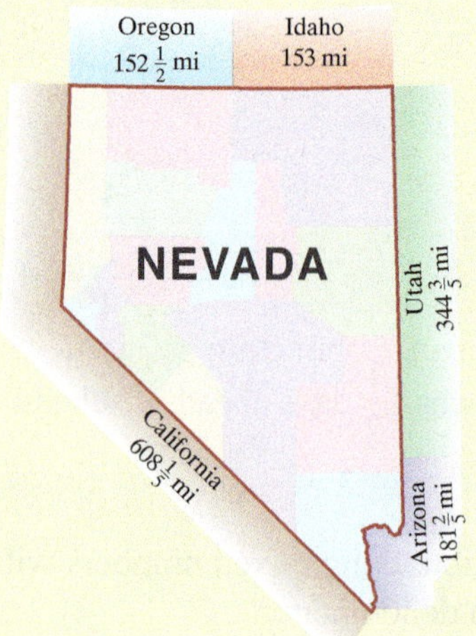

Recall from our discussion of subtracting whole numbers that, in problems in which a digit in the subtrahend is larger than the corresponding digit in the minuend, we need to regroup.

$$\begin{array}{r} \overset{2\,1}{\cancel{3}}\,2\,9 \\ -\;8\,7 \\ \hline 2\,4\,2 \end{array}$$

A similar situation can arise when we are subtracting mixed numbers. If the fraction on the bottom is larger than the fraction on top, we *regroup*, or *borrow from* the whole number on top.

EXAMPLE 18

Subtract: $6 - 1\frac{1}{3}$

Solution Let's rewrite the problem vertically.

$$\begin{array}{r} 6 \\ -1\frac{1}{3} \\ \hline \end{array}$$ There is no fraction on top from which to subtract $\frac{1}{3}$.

$$\begin{array}{r} 5\frac{3}{3} \\ -1\frac{1}{3} \\ \hline \end{array}$$ Regrouping, we express 6 as $5 + 1$, or $5 + \frac{3}{3}$, or $5\frac{3}{3}$.

$$\begin{array}{r} 5\frac{3}{3} \\ -1\frac{1}{3} \\ \hline 4\frac{2}{3} \end{array}$$ Now subtract.

So $6 - 1\frac{1}{3} = 4\frac{2}{3}$.

As in any subtraction problem, we can check our answer by addition.

$$4\frac{2}{3} + 1\frac{1}{3} = 5\frac{3}{3} = 6$$

PRACTICE 18

Subtract: $9 - 7\frac{5}{7}$

In Example 18, the answer is $4\frac{2}{3}$. Would we get the same answer if we compute $6\frac{1}{3} - 1$?

We have already discussed subtracting mixed numbers without regrouping as well as subtracting a mixed number from a whole number. Now, let's consider the general rule for subtracting mixed numbers.

To Subtract Mixed Numbers

- Write the fractions as equivalent fractions with the same denominator, usually the LCD.
- Regroup, or borrow from the whole number on top if the fraction on the bottom is larger than the fraction on top.
- Subtract the fractions.
- Subtract the whole numbers.
- Write the answer in simplest form.

EXAMPLE 19

Compute: $13\frac{2}{9} - 7\frac{8}{9}$

Solution First, we write the problem vertically.

$$\begin{array}{r} 13\frac{2}{9} \\ -7\frac{8}{9} \\ \hline \end{array}$$

Because $\frac{8}{9}$ is larger than $\frac{2}{9}$, we need to regroup as follows:

$$13\frac{2}{9} = 12 + 1 + \frac{2}{9} = 12 + \frac{9}{9} + \frac{2}{9} = 12\frac{11}{9}$$

$$\begin{array}{r} 12\frac{11}{9} \\ -7\frac{8}{9} \\ \hline \end{array}$$

Finally, we subtract, and then write the answer in simplest form.

$$\begin{array}{rcl} 13\frac{2}{9} & = & 12\frac{11}{9} \\ -7\frac{8}{9} & = & -7\frac{8}{9} \\ \hline & & 5\frac{3}{9}, \text{ or } 5\frac{1}{3} \end{array}$$

PRACTICE 19

Find the difference between $15\frac{1}{12}$ and $9\frac{11}{12}$.

EXAMPLE 20

Find the difference between $10\frac{1}{4}$ and $1\frac{5}{12}$.

Solution First, we write the equivalent fractions, using the LCD.

$$\begin{array}{rcr} 10\frac{1}{4} & = & 10\frac{3}{12} \\ -1\frac{5}{12} & = & -1\frac{5}{12} \\ \hline \end{array}$$

Then, we subtract.

$$\begin{array}{rcr} 10\frac{3}{12} & = & 9\frac{15}{12} \\ -1\frac{5}{12} & = & -1\frac{5}{12} \\ \hline & & 8\frac{10}{12} = 8\frac{5}{6} \end{array}$$

We regroup: $10\frac{3}{12} = 9 + \frac{12}{12} + \frac{3}{12} = 9\frac{15}{12}$

PRACTICE 20

Find the difference between $16\frac{1}{10}$ and $3\frac{3}{5}$.

EXAMPLE 21

In Oregon's Columbia River Gorge, a hiker walks along the Eagle Creek Trail, headed for Punchbowl Falls $2\frac{1}{10}$ miles away. After reaching Metlako Falls, does he have more or less than $\frac{1}{2}$ mile left to go? (*Source:* USDA Forest Service)

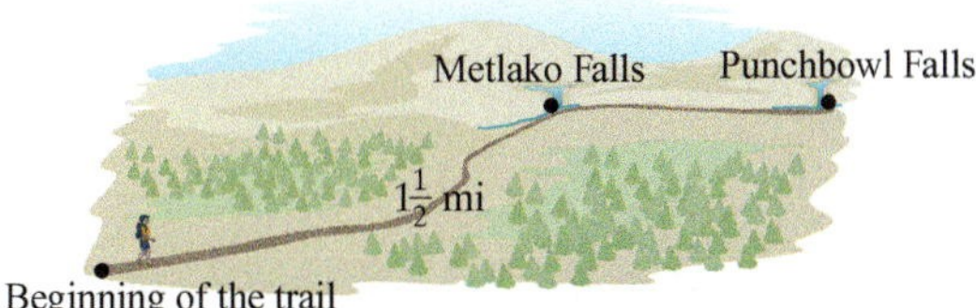

Eagle Creek Trail

Solution First, we must find the difference between the length of the trail from its begining to Punchbowl Falls, namely, $2\frac{1}{10}$ miles, and the distance already hiked, $1\frac{1}{2}$ miles.

$$\begin{array}{rcrcr} 2\frac{1}{10} & = & 2\frac{1}{10} & = & 1\frac{11}{10} \\ -1\frac{1}{2} & = & -1\frac{5}{10} & = & -1\frac{5}{10} \\ \hline & & & & \frac{6}{10} = \frac{3}{5} \end{array}$$

PRACTICE 21

A homeowner purchased the American single roll of wallpaper that is shown below and used $2\frac{7}{8}$ yards of the roll to paper a door panel. Is there enough paper left on the roll for a job that requires 2 yards of paper?

EXAMPLE 21 (continued)

So the distance left to hike is $\frac{3}{5}$ mile. Finally, we compare $\frac{3}{5}$ mile and $\frac{1}{2}$ mile.

$$\frac{3}{5} = \frac{6}{10} \qquad \frac{1}{2} = \frac{5}{10}$$

Because $6 > 5$, $\frac{6}{10} > \frac{5}{10}$. Therefore, $\frac{3}{5} > \frac{1}{2}$, and the hiker has more than $\frac{1}{2}$ mile left to go from Metlako Falls to Punchbowl Falls.

Another Method of Adding and Subtracting Mixed Numbers

Recall that any mixed number can be rewritten as an improper fraction. So when adding or subtracting mixed numbers, we can first express them as improper fractions. In a subtraction problem, this method has an advantage over the method previously discussed; namely, we never have to regroup. However, expressing mixed numbers as improper fractions may have the disadvantage of involving unnecessarily large numbers, as the following examples show.

EXAMPLE 22

Add: $14\frac{1}{6} + 8\frac{2}{3}$

Solution We begin by writing each mixed number as an improper fraction.

$14\frac{1}{6} + 8\frac{2}{3} = \frac{85}{6} + \frac{26}{3}$ Express $14\frac{1}{6}$ and $8\frac{2}{3}$ as improper fractions.

$= \frac{85}{6} + \frac{52}{6}$ Write the fractions to be added as equivalent fractions.

$= \frac{137}{6}$ Add the like fractions.

$= 22\frac{5}{6}$ Express the improper fraction as a mixed number.

PRACTICE 22

Find the sum: $7\frac{4}{5} + 2\frac{3}{4}$

Can you show that this method gives the same sum that we would have gotten if we had not expressed the mixed numbers as improper fractions? Explain.

EXAMPLE 23

Find the difference: $8\frac{5}{6} - 4\frac{9}{10}$

Solution $8\frac{5}{6} - 4\frac{9}{10} = \frac{53}{6} - \frac{49}{10}$ Write as improper fractions.

$= \frac{265}{30} - \frac{147}{30}$ Write as equivalent fractions.

$= \frac{118}{30}$ Subtract the like fractions.

$= 3\frac{28}{30}$ Express as a mixed number.

$= 3\frac{14}{15}$ Simplify.

Check that this answer is the same as we would have gotten without changing the mixed numbers to improper fractions.

PRACTICE 23

Subtract: $13\frac{1}{4} - 11\frac{7}{8}$

Estimating Sums and Differences of Mixed Numbers

When adding or subtracting mixed numbers, we can check by *estimating*, determining whether our estimate and our answer are close. Note that when we round mixed numbers, we round to the nearest whole number.

Checking a Sum by Estimating

$1\frac{1}{5} \rightarrow 1$ Because $\frac{1}{5} < \frac{1}{2}$, round *down* to the whole number 1.

$+2\frac{3}{5} \rightarrow +3$ Because $\frac{3}{5} > \frac{1}{2}$, round *up* to the whole number 3.

$3\frac{4}{5}$ 4 Our answer, $3\frac{4}{5}$, is close to 4, the sum of the rounded addends (1 and 3).

Checking a Difference by Estimating

$3\frac{2}{5} \rightarrow 3$ Because $\frac{2}{5} < \frac{1}{2}$, round *down* to 3.

$-1\frac{1}{5} \rightarrow -1$ Round *down* to 1.

$2\frac{1}{5}$ 2 Our answer, $2\frac{1}{5}$, is close to 2, the difference of the rounded numbers (3 and 1).

EXAMPLE 24

Combine and check: $5\frac{1}{3} - \left(2\frac{4}{5} + 1\frac{1}{10}\right)$

Solution Following the order of operations rule, we begin by adding the two mixed numbers in parentheses.

$$\begin{array}{rcr} 2\frac{4}{5} & = & 2\frac{8}{10} \\ +1\frac{1}{10} & = & +1\frac{1}{10} \\ \hline & & 3\frac{9}{10} \end{array}$$

Next, we subtract this sum from $5\frac{1}{3}$.

$$\begin{array}{rcrcr} 5\frac{1}{3} & = & 5\frac{10}{30} & = & 4\frac{40}{30} \\ -3\frac{9}{10} & = & -3\frac{27}{30} & = & -3\frac{27}{30} \\ \hline & & & & 1\frac{13}{30} \end{array}$$

So $5\frac{1}{3} - \left(2\frac{4}{5} + 1\frac{1}{10}\right) = 1\frac{13}{30}$.

Now, let's check this answer by estimating:

$$5\frac{1}{3} - \left(2\frac{4}{5} + 1\frac{1}{10}\right)$$
$$\downarrow \quad \downarrow \quad \downarrow$$
$$5 - (3 + 1) = 5 - 4 = 1$$

The estimate, 1, is close to our answer, $1\frac{13}{30}$.

PRACTICE 24

Calculate and check:

$8\frac{1}{4} - \left(3\frac{2}{5} - 1\frac{9}{10}\right)$

2.3 Exercises

FOR EXTRA HELP MyMathLab MathXL PRACTICE WATCH READ REVIEW

Mathematically Speaking

Fill in each blank with the most appropriate term or phrase from the given list.

denominators	regroup	equivalent
add	numerators	improper

1. To add like fractions, add the __________.

2. To subtract unlike fractions, rewrite them as __________ fractions with the same denominator.

3. When subtracting $2\frac{4}{5}$ from 7, __________ by writing 7 as $6\frac{5}{5}$.

4. Fractions with equal numerators and ______________ are equivalent to 1.

A *Add and simplify.*

5. $\frac{5}{8} + \frac{5}{8}$

6. $\frac{7}{10} + \frac{9}{10}$

7. $\frac{11}{12} + \frac{7}{12}$

8. $\frac{71}{100} + \frac{79}{100}$

9. $\frac{1}{5} + \frac{1}{5} + \frac{2}{5}$

10. $\frac{1}{7} + \frac{3}{7} + \frac{2}{7}$

11. $\frac{3}{20} + \frac{1}{20} + \frac{8}{20}$

12. $\frac{1}{10} + \frac{3}{10} + \frac{1}{10}$

13. $\frac{2}{3} + \frac{1}{2}$

14. $\frac{1}{4} + \frac{2}{5}$

15. $\frac{1}{2} + \frac{3}{8}$

16. $\frac{1}{6} + \frac{2}{3}$

17. $\frac{7}{10} + \frac{7}{100}$

18. $\frac{5}{6} + \frac{1}{12}$

19. $\frac{4}{5} + \frac{1}{8}$

20. $\frac{3}{4} + \frac{3}{7}$

21. $\frac{4}{9} + \frac{5}{6}$

22. $\frac{9}{10} + \frac{4}{5}$

23. $\frac{87}{100} + \frac{3}{10}$

24. $\frac{7}{20} + \frac{3}{4}$

25. $\frac{1}{3} + \frac{1}{4} + \frac{1}{6}$

26. $\frac{1}{5} + \frac{1}{6} + \frac{1}{3}$

27. $\frac{3}{8} + \frac{1}{10} + \frac{3}{16}$

28. $\frac{3}{10} + \frac{1}{3} + \frac{1}{9}$

29. $\frac{2}{9} + \frac{5}{8} + \frac{1}{4}$

30. $\frac{1}{2} + \frac{1}{3} + \frac{1}{4}$

31. $\frac{7}{8} + \frac{1}{5} + \frac{1}{4}$

32. $\frac{1}{10} + \frac{2}{5} + \frac{5}{6}$

Add and simplify. Then check by estimating.

33. $1 + 2\frac{1}{3}$

34. $4\frac{1}{5} + 2$

35. $8\frac{1}{10} + 7\frac{3}{10}$

36. $6\frac{1}{12} + 4\frac{1}{12}$

37. $7\frac{3}{10} + 6\frac{9}{10}$

38. $8\frac{2}{3} + 6\frac{2}{3}$

39. $5\frac{1}{6} + 9\frac{5}{6}$

40. $2\frac{3}{10} + 7\frac{9}{10}$

41. $5\frac{1}{4} + 5\frac{1}{6}$

42. $17\frac{3}{8} + 20\frac{1}{5}$

43. $3\frac{1}{3} + \frac{2}{5}$

44. $4\frac{7}{10} + \frac{7}{20}$

45. $8\frac{1}{5} + 5\frac{2}{3}$

46. $4\frac{1}{9} + 20\frac{7}{10}$

47. $\frac{2}{3} + 6\frac{1}{8}$

48. $\frac{1}{6} + 3\frac{2}{5}$

49. $9\frac{2}{3} + 10\frac{7}{12}$

50. $20\frac{3}{5} + 4\frac{1}{2}$

51. $6\frac{1}{10} + 3\frac{93}{100}$

52. $4\frac{8}{9} + 5\frac{1}{3}$

53. $4\frac{1}{2} + 6\frac{7}{8}$

54. $10\frac{5}{6} + 8\frac{1}{4}$

55. $30\frac{21}{100} + 5\frac{17}{20}$

56. $8\frac{3}{10} + 2\frac{321}{1,000}$

57. $80\frac{1}{3} + \frac{3}{4} + 10\frac{1}{2}$

58. $\frac{1}{3} + 25\frac{7}{24} + 100\frac{1}{2}$

59. $2\frac{1}{3} + 2 + 2\frac{1}{6}$

60. $4\frac{1}{8} + 4\frac{3}{16} + \frac{5}{4}$

61. $6\frac{7}{8} + 2\frac{3}{4} + 1\frac{1}{5}$

62. $1\frac{2}{3} + 5\frac{5}{6} + 3\frac{1}{4}$

63. $2\frac{1}{2} + 5\frac{1}{4} + 3\frac{5}{8}$

64. $4\frac{2}{3} + 2\frac{11}{36} + 1\frac{1}{2}$

Subtract and simplify.

65. $\frac{4}{5} - \frac{3}{5}$

66. $\frac{7}{9} - \frac{5}{9}$

67. $\frac{7}{10} - \frac{3}{10}$

68. $\frac{11}{12} - \frac{5}{12}$

69. $\frac{23}{100} - \frac{7}{100}$

70. $\frac{3}{2} - \frac{1}{2}$

71. $\frac{3}{4} - \frac{1}{4}$

72. $\frac{7}{9} - \frac{4}{9}$

73. $\frac{12}{5} - \frac{2}{5}$

74. $\frac{1}{8} - \frac{1}{8}$

75. $\frac{3}{4} - \frac{2}{3}$

76. $\frac{2}{5} - \frac{1}{6}$

77. $\frac{4}{9} - \frac{1}{6}$

78. $\frac{9}{10} - \frac{3}{100}$

79. $\frac{4}{5} - \frac{3}{4}$

80. $\frac{5}{6} - \frac{1}{8}$

81. $\frac{4}{7} - \frac{1}{2}$

82. $\frac{2}{5} - \frac{2}{9}$

83. $\frac{4}{9} - \frac{3}{8}$

84. $\frac{11}{12} - \frac{1}{3}$

85. $\frac{3}{4} - \frac{1}{2}$

86. $\frac{5}{6} - \frac{2}{3}$

Subtract and simplify. Then, check either by adding or by estimating.

87. $5\frac{3}{7} - 1\frac{1}{7}$

88. $6\frac{2}{3} - 1\frac{1}{3}$

89. $3\frac{7}{8} - 2\frac{1}{8}$

90. $10\frac{5}{6} - 2\frac{5}{6}$

91. $20\frac{1}{2} - \frac{1}{2}$

92. $7\frac{3}{4} - \frac{1}{4}$

93. $8\frac{1}{10} - 4$

94. $2\frac{1}{3} - 2$

95. $6 - 2\frac{2}{3}$

96. $4 - 1\frac{1}{5}$

97. $8 - 4\frac{7}{10}$

98. $2 - 1\frac{1}{2}$

99. $10 - 3\frac{2}{3}$

100. $5 - 4\frac{9}{10}$

101. $6 - \frac{1}{2}$

102. $9 - \frac{3}{4}$

103. $7\frac{1}{4} - 2\frac{3}{4}$

104. $5\frac{1}{10} - 2\frac{3}{10}$

105. $6\frac{1}{8} - 2\frac{7}{8}$

106. $3\frac{1}{5} - 1\frac{4}{5}$

107. $12\frac{2}{5} - \frac{3}{5}$

108. $3\frac{7}{10} - \frac{9}{10}$

109. $8\frac{1}{3} - 1\frac{2}{3}$

110. $2\frac{1}{5} - \frac{4}{5}$

111. $13\frac{1}{2} - 5\frac{2}{3}$

112. $7\frac{1}{10} - 2\frac{1}{7}$

113. $9\frac{3}{8} - 5\frac{5}{6}$

114. $2\frac{1}{10} - 1\frac{27}{100}$

115. $20\frac{2}{9} - 4\frac{5}{6}$

116. $9\frac{13}{100} - 6\frac{7}{10}$

117. $3\frac{4}{5} - \frac{5}{6}$

118. $1\frac{2}{8} - \frac{2}{6}$

119. $1\frac{3}{4} - 1\frac{1}{2}$

120. $2\frac{1}{2} - 1\frac{3}{4}$

121. $10\frac{1}{12} - 4\frac{2}{3}$

122. $7\frac{1}{4} - 1\frac{5}{16}$

123. $22\frac{7}{8} - 8\frac{9}{10}$

124. $9\frac{1}{10} - 3\frac{1}{2}$

125. $3\frac{1}{8} - 2\frac{3}{4}$

126. $3\frac{1}{4} - 2\frac{5}{16}$

Combine and simplify.

127. $\frac{5}{8} + \frac{9}{10} - \frac{1}{4}$

128. $\frac{2}{3} - \frac{1}{5} + \frac{1}{2}$

129. $12\frac{1}{6} + 5\frac{9}{10} - 1\frac{3}{10}$

130. $7\frac{1}{3} - 2\frac{4}{5} - 1\frac{1}{3}$

131. $15\frac{1}{2} - 3\frac{4}{5} - 6\frac{1}{2}$

132. $4\frac{1}{10} + 2\frac{9}{10} - 3\frac{3}{4}$

133. $20\frac{1}{10} - \left(\frac{1}{20} + 1\frac{1}{2}\right)$

134. $19\frac{1}{6} - \left(8\frac{9}{10} - \frac{1}{5}\right)$

Mixed Practice

Perform the indicated operations and simplify.

135. Subtract $1\frac{7}{8}$ from 6.

136. Add: $6\frac{1}{10} + 3\frac{7}{15}$

137. Calculate: $12\frac{2}{3} - \left(8\frac{5}{6} - 4\frac{1}{2}\right)$

138. Find the sum of $\frac{3}{8}, \frac{1}{2},$ and $\frac{1}{3}$.

139. Find the difference between $4\frac{3}{5}$ and $1\frac{2}{3}$.

140. Subtract: $\frac{9}{10} - \frac{1}{4}$

Applications

B *Solve. Write the answer in simplest form.*

141. A $\frac{7}{8}$-inch nail was hammered through a $\frac{3}{4}$-inch door. How far did it extend from the door?

142. A building occupies $\frac{1}{4}$ acre on a $\frac{7}{8}$-acre plot of land. What is the area of the land not occupied by the building?

143. The Kentucky Derby, Belmont Stakes, and the Preakness Stakes are three prestigious horse races that comprise the Triple Crown. (*Source:* http://infoplease.com)

a. Horses run $1\frac{3}{16}$ mile in the Preakness Stakes. If the Preakness Stakes is $\frac{5}{16}$ mile shorter than the Belmont Stakes, how far do horses run in the Belmont Stakes?

b. Horses run $1\frac{1}{4}$ miles in the Kentucky Derby. How much farther do horses run in the Belmont Stakes than in the Kentucky Derby?

144. In the year 2030, the total amount of electricity generated worldwide is projected to be approximately 32 trillion kilowatt-hours. Of this amount, $\frac{1}{32}$ is expected to be generated by liquid fuels, $\frac{1}{8}$ by nuclear power, and $\frac{7}{16}$ by coal. (*Source:* eia.doe.gov)

a. According to these projections, the combined amount of electricity generated by liquid fuels and nuclear power will be what fraction of the total world electricity?

b. As a fraction of the electricity generated worldwide, the amount of electricity generated by coal will be how much greater than the combined amount generated by liquid fuel and nuclear power?

145. The first game of a baseball doubleheader lasted $2\frac{1}{4}$ hours. The second game began after a $\frac{1}{4}$-hour break and lasted $2\frac{1}{2}$ hours. How long did the doubleheader take to play?

146. Three student candidates competed in a student government election. The winner got $\frac{5}{8}$ of the votes, and the second-place candidate got $\frac{1}{4}$ of the votes. If the rest of the votes went to the third candidate, what fraction of the votes did that student get?

147. In 1912, the *Titanic*, the largest passenger ship in the world, sank on its maiden voyage. The steamship had a length of $882\frac{3}{4}$ feet. Its width at the widest point was $92\frac{1}{2}$ feet. How much greater was its length than its width? (*Source:* titanic-titanic.com)

148. A shopper purchased two boxes of chocolates: one box containing a sugar-free assortment weighs $20\frac{5}{8}$ ounces, and the other box, containing milk toffee sticks, weighs $10\frac{1}{2}$ ounces. Find the total weight of the boxes. (*Source:* Russellstover.com)

149. In testing a new drug, doctors found that $\frac{1}{2}$ of the patients given the drug improved, $\frac{2}{5}$ showed no change in their condition, and the remainder got worse. What fraction got worse?

150. The size of a child's shoe is related to the length of his or her foot. The following table shows the relationship between shoe size and foot length for a variety of sizes.

Size	Foot Length (in inches)	Size	Foot Length (in inches)
4	$5\frac{3}{4}$	8	$6\frac{3}{4}$
5	6	9	7
6	$6\frac{1}{4}$	10	$7\frac{1}{4}$
7	$6\frac{1}{2}$	11	$7\frac{1}{2}$

(*Source:* kidbean.com/sizecharts.html)

Is the difference in foot length greater when comparing sizes 4 and 7 or when comparing sizes 7 and 10?

151. Suppose that four packages are placed on a scale, as shown. If the scale balances, how heavy is the small package on the right?

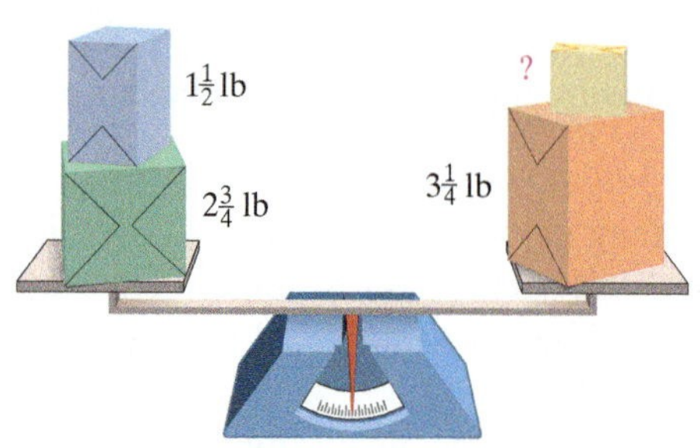

152. If the scale pictured balances, how heavy is the small package on the left?

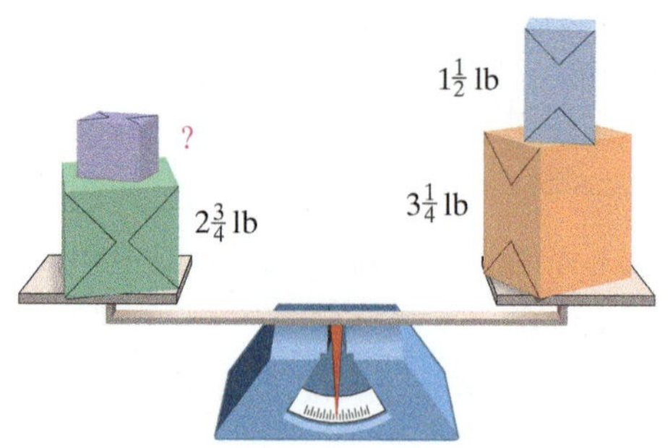

• Check your answers on page A-3.

MIND Stretchers

Groupwork

1. Working with a partner, complete the following magic square in which each row, column, and diagonal adds up to the same number.

$1\frac{1}{4}$		
	1	
$\frac{11}{12}$		$\frac{3}{4}$

Mathematical Reasoning

2. A fraction with 1 as the numerator is called a **unit fraction**. For example, $\frac{1}{7}$ is a unit fraction. Write $\frac{3}{7}$ as the sum of three unit fractions, using no unit fraction more than once.

$$\frac{3}{7} = \frac{1}{\square} + \frac{1}{\square} + \frac{1}{\square}$$

Writing

3. Consider the following two ways of subtracting $2\frac{4}{5}$ from $4\frac{1}{5}$.

Method 1

$$\begin{array}{rcr} 4\frac{1}{5} = 3 + \frac{5}{5} + \frac{1}{5} = & & 3\frac{6}{5} \\ -2\frac{4}{5} & = & -2\frac{4}{5} \\ \hline & & 1\frac{2}{5} \end{array}$$

Method 2

$$\begin{array}{rcr} 4\frac{1}{5} \rightarrow 4 + \frac{1}{5} + \frac{1}{5} = & & 4\frac{2}{5} \\ -2\frac{4}{5} \rightarrow 2 + \frac{4}{5} + \frac{1}{5} = & & -3 \\ \hline & & 1\frac{2}{5} \end{array}$$

a. Explain the difference between the two methods.

b. Explain which method you prefer.

c. Explain why you prefer that method.

2.4 Multiplying and Dividing Fractions

OBJECTIVES

A To multiply or divide fractions or mixed numbers

B To solve applied problems involving the multiplication or division of fractions or mixed numbers

This section begins with a discussion of multiplying fractions. We then move on to multiplying mixed numbers and conclude with dividing fractions and mixed numbers.

Multiplying Fractions

Many situations require us to multiply fractions. For instance, suppose that a mixture in a chemistry class calls for $\frac{4}{5}$ gram of sodium chloride. If we make only $\frac{2}{3}$ of that mixture, we need

$$\frac{2}{3} \text{ of } \frac{4}{5}$$
$$\downarrow$$
$$\frac{2}{3} \times \frac{4}{5}$$

that is, $\frac{2}{3} \times \frac{4}{5}$ gram of sodium chloride.

To illustrate how to find this product, we diagram these two fractions.

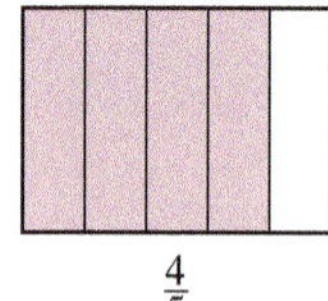

$\frac{4}{5}$

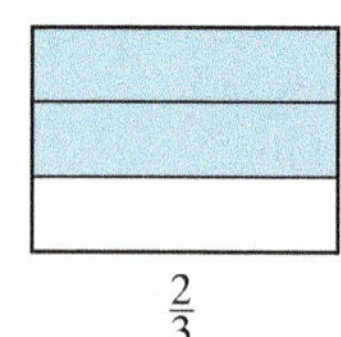

$\frac{2}{3}$

In the following diagram, we are taking $\frac{2}{3}$ of the $\frac{4}{5}$.

Note that we divided the whole into 15 parts and that our product, containing 8 of the 15 small squares, represents the double-shaded region. The answer is, therefore, $\frac{8}{15}$ of the original whole, which we can compute as follows:

$$\frac{2}{3} \times \frac{4}{5} = \frac{8}{15}$$

The numerator and denominator of the answer are the products of the original numerators and denominators.

To Multiply Fractions

- Multiply the numerators.
- Multiply the denominators.
- Write the answer in simplest form.

EXAMPLE 1

Multiply: $\frac{7}{8} \cdot \frac{9}{10}$

Solution

Multiply the numerators.

$$\frac{7}{8} \cdot \frac{9}{10} = \frac{7 \cdot 9}{8 \cdot 10} = \frac{63}{80}$$

Multiply the denominators.

PRACTICE 1

Find the product of $\frac{3}{4}$ and $\frac{5}{7}$.

EXAMPLE 2

Calculate: $\left(\frac{4}{5}\right)^2$

Solution

$$\left(\frac{4}{5}\right)^2 = \frac{4}{5} \cdot \frac{4}{5} = \frac{4 \cdot 4}{5 \cdot 5} = \frac{16}{25}$$

PRACTICE 2

Square $\frac{9}{10}$.

EXAMPLE 3

What is $\frac{3}{8}$ of 10?

Solution Finding $\frac{3}{8}$ of 10 means multiplying $\frac{3}{8}$ by 10.

$$\frac{3}{8} \times 10 = \frac{3}{8} \times \frac{10}{1} = \frac{3 \times 10}{8 \times 1} = \frac{30}{8} = \frac{15}{4}, \text{ or } 3\frac{3}{4}$$

PRACTICE 3

What is $\frac{2}{3}$ of 30?

In Example 3, we multiplied the two fractions first, and then simplified the answer. It is preferable, however, to reverse these steps: Simplify first, and then multiply. By first simplifying, sometimes referred to as *canceling*, we divide *any* numerator and *any* denominator by a common factor. Canceling before multiplying allows us to work with smaller numbers and still gives us the same answer.

EXAMPLE 4

Find the product of $\frac{4}{9}$ and $\frac{5}{8}$.

Solution

$$\frac{4}{9} \times \frac{5}{8} = \frac{\overset{1}{\cancel{4}}}{9} \times \frac{5}{\underset{2}{\cancel{8}}}$$ Divide the numerator 4 and the denominator 8 by 4.

$$= \frac{1 \times 5}{9 \times 2}$$ Multiply the resulting fractions.

$$= \frac{5}{18}$$

PRACTICE 4

Multiply: $\frac{7}{10} \cdot \frac{5}{11}$

EXAMPLE 5

Multiply: $\frac{9}{8} \times \frac{6}{5} \times \frac{7}{9}$

Solution We simplify and then multiply.

$$\frac{9}{8} \times \frac{6}{5} \times \frac{7}{9} = \frac{\overset{1}{\cancel{9}}}{\underset{4}{\cancel{8}}} \times \frac{\overset{3}{\cancel{6}}}{5} \times \frac{7}{\underset{1}{\cancel{9}}}$$

Divide the numerator 9 and the denominator 9 by 9. Divide the numerator 6 and the denominator 8 by 2.

$$= \frac{21}{20}, \text{ or } 1\frac{1}{20}$$

PRACTICE 5

Multiply: $\frac{7}{27} \cdot \frac{9}{4} \cdot \frac{8}{21}$

EXAMPLE 6

At a college, $\frac{3}{5}$ of the students take a math course. Of these students, $\frac{1}{6}$ take elementary algebra. What fraction of the students in the college take elementary algebra?

Solution We must find $\frac{1}{6}$ of $\frac{3}{5}$.

$$\frac{1}{6} \times \frac{3}{5} = \frac{1}{\underset{2}{\cancel{6}}} \times \frac{\overset{1}{\cancel{3}}}{5} = \frac{1 \times 1}{2 \times 5} = \frac{1}{10}$$

One-tenth of the students in the college take elementary algebra.

PRACTICE 6

A flight from New York to Los Angeles took 7 hours. With the help of the jet stream, the return trip took $\frac{3}{4}$ the time. How long did the trip from Los Angeles to New York take?

EXAMPLE 7

Of the 639 employees at a company, $\frac{4}{9}$ responded to a voluntary survey distributed by the human resources department. How many employees did not respond to the survey?

Solution Apply the strategy of breaking the problem into two parts.

- First, find $\frac{4}{9}$ of 639.
- Then, subtract the result from 639.

In short, we can solve this problem by computing $639 - \left(\frac{4}{9} \times 639\right)$.

$$639 - \left(\frac{4}{9} \times 639\right) = 639 - \left(\frac{4}{\underset{1}{\cancel{9}}} \times \frac{\overset{71}{\cancel{639}}}{1}\right) = 639 - 284 = 355$$

So 355 employees did not respond to the survey.

PRACTICE 7

The state sales tax on a car in Wisconsin is $\frac{1}{20}$ of the price of the car. What is the total amount a consumer would pay for a $19,780 car? (*Source:* revenue.wi.gov)

Multiplying Mixed Numbers

Some situations require us to multiply mixed numbers. For instance, suppose that your regular hourly wage is $\$7\frac{1}{2}$ and that you make time-and-a-half for working overtime. To find your overtime hourly wage, you need to multiply $1\frac{1}{2}$ by $7\frac{1}{2}$. The key here is to first rewrite each mixed number as an improper fraction.

$$1\frac{1}{2} \times 7\frac{1}{2} = \frac{3}{2} \times \frac{15}{2} = \frac{45}{4}, \text{ or } 11\frac{1}{4}$$

So you make $\$11\frac{1}{4}$ per hour overtime.

To Multiply Mixed Numbers

- Write the mixed numbers as improper fractions.
- Multiply the fractions.
- Write the answer in simplest form.

EXAMPLE 8

Multiply $2\frac{1}{5}$ by $1\frac{1}{4}$.

Solution $2\frac{1}{5} \times 1\frac{1}{4} = \frac{11}{5} \times \frac{5}{4}$ Write each mixed number as an improper fraction.

$= \frac{11 \times \overset{1}{\cancel{5}}}{\underset{1}{\cancel{5}} \times 4}$ Simplify and multiply.

$= \frac{11}{4}, \text{ or } 2\frac{3}{4}$

PRACTICE 8

Find the product of $3\frac{3}{4}$ and $2\frac{1}{10}$.

EXAMPLE 9

Multiply: $\left(4\frac{3}{8}\right)(4)\left(2\frac{2}{5}\right)$

Solution

$$\left(4\frac{3}{8}\right)(4)\left(2\frac{2}{5}\right) = \left(\frac{35}{8}\right)\left(\frac{4}{1}\right)\left(\frac{12}{5}\right)$$

$$= \left(\frac{\overset{7}{\cancel{35}}}{\underset{\underset{1}{\cancel{2}}}{\cancel{8}}}\right)\left(\frac{\overset{1}{\cancel{4}}}{1}\right)\left(\frac{\overset{6}{\cancel{12}}}{\underset{1}{\cancel{5}}}\right) = 42$$

Note in this problem that, although there are several ways to simplify, the answer always comes out the same.

PRACTICE 9

Multiply: $\left(1\frac{3}{4}\right)\left(5\frac{1}{3}\right)(3)$

EXAMPLE 10

A lawn surrounding a garden is to be installed, as depicted in the following drawing.

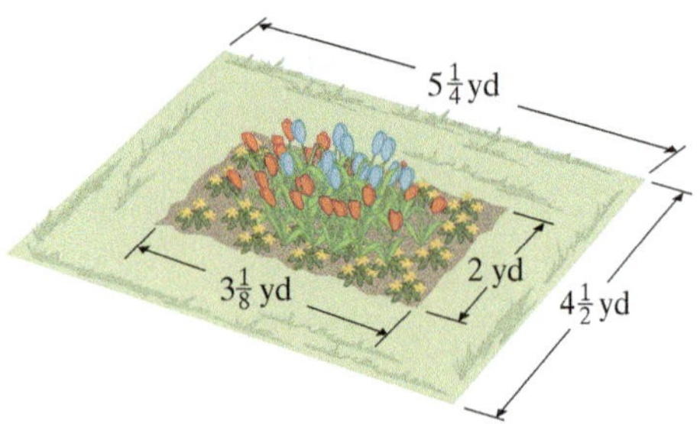

How many square yards of turf will we need to cover the lawn?

Solution Let's break this problem into three steps. First, we find the area of the rectangle with dimensions $5\frac{1}{4}$ yards and $4\frac{1}{2}$ yards. Then, we find the area of the small rectangle whose length and width are $3\frac{1}{8}$ yards and 2 yards, respectively.

$$5\frac{1}{4} \times 4\frac{1}{2} = \frac{21}{4} \times \frac{9}{2}$$

$$= \frac{189}{8}, \text{ or } 23\frac{5}{8}$$ The area of the large rectangle is $23\frac{5}{8}$ square yards.

$$3\frac{1}{8} \times 2 = \frac{25}{8} \times \frac{2}{1}$$

$$= \frac{25}{4}, \text{ or } 6\frac{1}{4}$$ The area of the small rectangle is $6\frac{1}{4}$ square yards.

Finally, we subtract the area of the small rectangle from the area of the large rectangle.

$$\begin{array}{rcr} 23\frac{5}{8} & = & 23\frac{5}{8} \\ -6\frac{1}{4} & = & -6\frac{2}{8} \\ \hline & & 17\frac{3}{8} \end{array}$$

The area of the lawn is, therefore, $17\frac{3}{8}$ square yards. So we will need $17\frac{3}{8}$ square yards of turf for the lawn.

PRACTICE 10

Two student club posters are shown below. How much greater is the area of the legal-size poster than the letter-size poster?

$8\frac{1}{2}$ in. × 11 in.
Letter-size paper

$8\frac{1}{2}$ in. × 14 in.
Legal-size paper

EXAMPLE 11

Simplify: $16\frac{1}{4} - 2 \cdot 4\frac{3}{5}$

Solution We use the order of operations rule, multiplying before subtracting.

$$\begin{aligned} 16\frac{1}{4} - 2 \cdot 4\frac{3}{5} &= 16\frac{1}{4} - \frac{2}{1} \cdot \frac{23}{5} \\ &= 16\frac{1}{4} - \frac{46}{5} \\ &= 16\frac{1}{4} - 9\frac{1}{5} \\ &= 16\frac{5}{20} - 9\frac{4}{20} \\ &= 7\frac{1}{20} \end{aligned}$$

PRACTICE 11

Calculate: $6 + \left(3\frac{1}{2}\right)^2$

Dividing Fractions

We now turn to quotients, beginning with dividing a fraction by a whole number. Suppose, for instance, that you want to share $\frac{1}{3}$ of a pizza with a friend, that is, to divide the $\frac{1}{3}$ into two equal parts. What part of the whole pizza will each of you receive?

This diagram shows $\frac{1}{3}$ of a pizza.

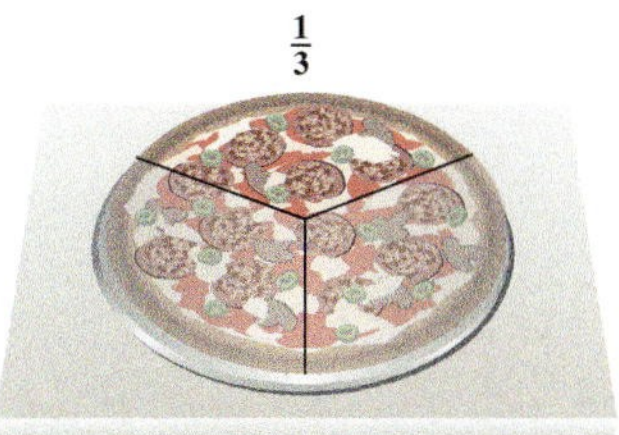

If we split each third into two equal parts, each part is $\frac{1}{6}$ of the pizza.

You and your friend will each get $\frac{1}{6}$ of the whole pizza, which you can compute as follows.

$$\frac{1}{3} \div 2 = \frac{1}{6}$$

Note that dividing a number by 2 is the same as taking $\frac{1}{2}$ of it. This equivalence suggests the procedure for dividing fractions shown next.

Divisor

$$\frac{1}{3} \div 2 = \frac{1}{3} \div \frac{2}{1} = \frac{1}{3} \times \frac{1}{2} = \frac{1 \times 1}{3 \times 2} = \frac{1}{6}$$

$\frac{2}{1}$ and $\frac{1}{2}$ are reciprocals.

This procedure involves *inverting*, or finding the *reciprocal* of the divisor. The reciprocal is found by switching the numerator and denominator.

To Divide Fractions

- Change the divisor to its reciprocal, and multiply the resulting fractions.
- Write the answer in simplest form.

EXAMPLE 12

Divide: $\frac{4}{5} \div \frac{3}{10}$

Solution $\frac{4}{5} \div \frac{3}{10} = \frac{4}{5} \times \frac{10}{3}$ Change the divisor to its reciprocal and multiply.

$= \frac{4}{\cancel{5}_1} \times \frac{\cancel{10}^2}{3}$ Divide the numerator 10 and the denominator 5 by 5.

$= \frac{4 \times 2}{1 \times 3}$ Multiply the fractions.

$= \frac{8}{3}$ Simplify.

$= 2\frac{2}{3}$

As in any division problem, we can check our answer by multiplying it by the divisor. $\frac{\cancel{8}^4}{\cancel{3}_1} \times \frac{\cancel{3}^1}{\cancel{10}_5} = \frac{4}{5}$

Because $\frac{4}{5}$ is the dividend, we have confirmed our answer.

PRACTICE 12

Divide: $\frac{3}{4} \div \frac{1}{8}$

TIP In a division problem, the fraction to the right of the division sign is the divisor. Always invert the divisor (the second fraction) and not the dividend (the first fraction).

EXAMPLE 13

What is $\frac{4}{7}$ divided by 20?

Solution $\frac{4}{7} \div 20 = \frac{4}{7} \times \frac{1}{20}$ Invert $\frac{20}{1}$ and multiply.

$= \frac{\overset{1}{\cancel{4}}}{7} \times \frac{1}{\underset{5}{\cancel{20}}}$ Divide the numerator 4 and the denominator 20 by 4.

$= \frac{1 \times 1}{7 \times 5} = \frac{1}{35}$

PRACTICE 13

Compute the following quotient:

$5 \div \frac{5}{8}$

EXAMPLE 14

To stop the developing process, photographers use a chemical called stop bath. Suppose that a photographer needs $\frac{1}{4}$ bottle of stop bath for each roll of film. If the photographer has $\frac{2}{3}$ bottle of stop bath left, can he develop three rolls of film?

Solution We want to find out how many $\frac{1}{4}$'s there are in $\frac{2}{3}$, that is, to compute $\frac{2}{3} \div \frac{1}{4}$.

$$\frac{2}{3} \div \frac{1}{4} = \frac{2}{3} \times \frac{4}{1} = \frac{8}{3} \quad \text{or} \quad 2\frac{2}{3}$$

↑ Find the reciprocal of the divisor, $\frac{1}{4}$, and then multiply.

So the photographer cannot develop three rolls of film.

PRACTICE 14

A cabin is built on ground that is sinking $\frac{3}{4}$ inch per year. How many years will it take the cabin to sink 2 inches?

Dividing Mixed Numbers

Dividing mixed numbers is similar to dividing fractions, except that there is an additional step.

To Divide Mixed Numbers

- Write the mixed numbers as improper fractions.
- Divide the fractions.
- Write the answer in simplest form.

EXAMPLE 15

Find: $9 \div 2\frac{7}{10}$

Solution $9 \div 2\frac{7}{10} = \frac{9}{1} \div \frac{27}{10}$ Write the whole number and the mixed number as improper fractions.

$= \frac{\overset{1}{\cancel{9}}}{1} \times \frac{10}{\underset{3}{\cancel{27}}}$ Invert and multiply.

$= \frac{10}{3}$, or $3\frac{1}{3}$

PRACTICE 15

Divide: $6 \div 3\frac{3}{4}$

EXAMPLE 16

What is $2\frac{1}{2} \div 4\frac{1}{2}$?

Solution $2\frac{1}{2} \div 4\frac{1}{2} = \frac{5}{2} \div \frac{9}{2} = \frac{5}{\underset{1}{\cancel{2}}} \times \frac{\overset{1}{\cancel{2}}}{9} = \frac{5}{9}$

Invert and multiply.

PRACTICE 16

Divide $2\frac{3}{8}$ by $5\frac{3}{7}$.

EXAMPLE 17

There are $6\frac{3}{4}$ yards of silk in a roll. If it takes $\frac{3}{4}$ yard to make one designer tie, how many ties can be made from the roll?

Solution The question is: How many $\frac{3}{4}$'s fit into $6\frac{3}{4}$? It tells us that we must divide.

$6\frac{3}{4} \div \frac{3}{4} = \frac{\overset{9}{\cancel{27}}}{\underset{1}{\cancel{4}}} \times \frac{\overset{1}{\cancel{4}}}{\underset{1}{\cancel{3}}} = 9$

So nine ties can be made from the roll of silk.

PRACTICE 17

According to a newspaper advertisement for a "diet shake," a man lost 33 pounds in $5\frac{1}{2}$ months. How much weight did he lose per month?

Estimating Products and Quotients of Mixed Numbers

As with adding or subtracting mixed numbers, it is important to check our answers when multiplying or dividing. We can check a product or a quotient of mixed numbers by estimating the answer and then confirming that our estimate and answer are reasonably close.

Checking a Product by Estimating

$$2\frac{1}{5} \times 7\frac{2}{3} = \frac{11}{5} \times \frac{23}{3} = \frac{253}{15}, \text{ or } 16\frac{13}{15}$$

$$\downarrow \quad \downarrow$$

$$2 \times 8 = 16$$

Our answer, $16\frac{13}{15}$, is close to 16, the product of the rounded factors.

Because $16\frac{13}{15}$ is near 16, $16\frac{13}{15}$ is a reasonable answer.

Checking a Quotient by Estimating

$$6\frac{1}{4} \div 2\frac{7}{10} = \frac{25}{4} \div \frac{27}{10} = \frac{\overset{}{25}}{\underset{2}{\cancel{4}}} \times \frac{\overset{5}{\cancel{10}}}{27} = \frac{125}{54}, \text{ or } 2\frac{17}{54}$$

$$\downarrow \quad \downarrow$$

$$6 \div 3 = 2$$

Our answer, $2\frac{17}{54}$, is close to 2, the quotient of the rounded dividend and divisor.

Because 2 is near $2\frac{17}{54}$, $2\frac{17}{54}$ is a reasonable answer.

EXAMPLE 18

Simplify and check: $3\frac{3}{4} \times 5\frac{1}{3} \div 2\frac{7}{9}$

Solution Following the order of operations rule, we work from left to right, multiplying the first two mixed numbers.

$$3\frac{3}{4} \times 5\frac{1}{3} = \frac{\overset{5}{\cancel{15}}}{\underset{1}{\cancel{4}}} \times \frac{\overset{4}{\cancel{16}}}{\underset{1}{\cancel{3}}} = 20$$

Then, we divide 20 by $2\frac{7}{9}$ to get the answer.

$$20 \div 2\frac{7}{9} = \frac{20}{1} \div \frac{25}{9} = \frac{\overset{4}{\cancel{20}}}{1} \times \frac{9}{\underset{5}{\cancel{25}}} = \frac{36}{5}, \text{ or } 7\frac{1}{5}$$

Now, let's check by estimating.

$$3\frac{3}{4} \times 5\frac{1}{3} \div 2\frac{7}{9}$$

$$\downarrow \quad \downarrow \quad \downarrow$$

$$4 \times 5 \div 3 = 20 \div 3 \approx 7$$

The answer, $7\frac{1}{5}$, and the estimate, 7, are reasonably close, confirming the answer.

PRACTICE 18

Compute and check:

$5\frac{3}{5} \div 2\frac{1}{10} \times 2\frac{1}{4}$

EXAMPLE 19

Calculate and check: $12 \div 1\frac{2}{3} + 5 \cdot 2\frac{9}{10}$

Solution According to the order of operations rule, we divide and multiply before adding.

$$\begin{aligned}
12 \div 1\frac{2}{3} + 5 \cdot 2\frac{9}{10} &= \frac{12}{1} \div \frac{5}{3} + \frac{5}{1} \cdot \frac{29}{10} \\
&= \frac{12}{1} \cdot \frac{3}{5} + \frac{5}{1} \cdot \frac{29}{10} \\
&= \frac{12}{1} \cdot \frac{3}{5} + \frac{\overset{1}{\cancel{5}}}{1} \cdot \frac{29}{\underset{2}{\cancel{10}}} \\
&= \frac{36}{5} + \frac{29}{2} \\
&= 7\frac{1}{5} + 14\frac{1}{2} \\
&= 7\frac{2}{10} + 14\frac{5}{10} \\
&= 21\frac{7}{10}
\end{aligned}$$

Now, we estimate the answer in order to check.

$$\begin{array}{ccccccc}
12 \div & 1\frac{2}{3} & + & 5 \cdot & 2\frac{9}{10} & & \\
\downarrow & \downarrow & & \downarrow & \downarrow & & \\
12 \div & 2 & + & 5 \cdot & 3 & \approx & 21
\end{array}$$

The estimate and the answer are close, confirming the answer.

PRACTICE 19

Compute and check:

$14\frac{1}{3} \div 2 - 6 \div 2\frac{1}{4}$

2.4 Exercises

FOR EXTRA HELP     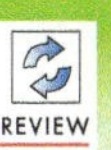

Mathematically Speaking

Fill in each blank with the most appropriate term or phrase from the given list.

reverse	proper fraction	multiply
divide	simplify	reciprocal
invert	factor	improper fraction

1. To find the product of the fractions $\frac{1}{7}$ and $\frac{5}{8}$, __________ 1 and 5, and 7 and 8.

2. To multiply mixed numbers, change each mixed number to its equivalent _______________.

3. The fraction $\frac{2}{3}$ is said to be the ________ of the fraction $\frac{3}{2}$.

4. To _______ fractions, change the divisor to its reciprocal, and multiply the resulting fractions.

5. To _______ a fraction is to find its reciprocal.

6. When multiplying fractions, we can divide any numerator and any denominator by a common _______.

A *Multiply and simplify.*

7. $\frac{1}{3} \times \frac{2}{5}$

8. $\frac{7}{8} \times \frac{1}{2}$

9. $\left(\frac{5}{8}\right)\left(\frac{2}{3}\right)$

10. $\left(\frac{3}{10}\right)\left(\frac{1}{4}\right)$

11. $\left(\frac{3}{4}\right)^2$

12. $\left(\frac{1}{8}\right)^2$

13. $\frac{4}{5} \times \frac{2}{5}$

14. $\frac{1}{2} \times \frac{3}{2}$

15. $\frac{7}{8} \times \frac{5}{4}$

16. $\frac{20}{3} \times \frac{2}{7}$

17. $\frac{5}{2} \cdot \frac{9}{8}$

18. $\frac{11}{10} \cdot \frac{9}{5}$

19. $\left(\frac{2}{5}\right)\left(\frac{5}{9}\right)$

20. $\left(\frac{4}{5}\right)\left(\frac{1}{4}\right)$

21. $\frac{7}{9} \times \frac{3}{4}$

22. $\frac{4}{5} \times \frac{1}{2}$

23. $\left(\frac{1}{8}\right)\left(\frac{6}{10}\right)$

24. $\left(\frac{4}{6}\right)\left(\frac{3}{8}\right)$

25. $\frac{10}{9} \times \frac{93}{100}$

26. $\frac{12}{5} \times \frac{15}{4}$

27. $\frac{2}{3} \times 20$

28. $\frac{5}{6} \times 5$

29. $\left(\frac{10}{3}\right)(4)$

30. $\left(\frac{5}{3}\right)(7)$

31. $\frac{2}{3} \times 24$

32. $\frac{3}{4} \times 12$

33. $\frac{2}{3} \cdot 6$

34. $100 \cdot \frac{2}{5}$

35. $18 \cdot \frac{2}{9}$

36. $20 \cdot \frac{4}{5}$

37. $\frac{7}{8} \times 10$

38. $\frac{5}{8} \times 12$

39. $\left(\frac{7}{8}\right)\left(1\frac{1}{2}\right)$

40. $\left(4\frac{1}{3}\right)\left(\frac{1}{5}\right)$

41. $\frac{1}{4} \cdot 8\frac{1}{2}$

42. $\frac{1}{3} \cdot 2\frac{1}{5}$

43. $\left(\frac{5}{6}\right)\left(1\frac{1}{9}\right)$

44. $\left(\frac{9}{10}\right)\left(2\frac{1}{7}\right)$

45. $\frac{1}{2} \times 5\frac{1}{3}$

46. $4\frac{1}{2} \times \frac{2}{3}$

47. $\frac{4}{5} \cdot 1\frac{1}{4}$

48. $\frac{3}{8} \cdot 5\frac{1}{3}$

49. $\left(\frac{3}{16}\right)\left(4\frac{2}{3}\right)$

50. $\left(\frac{7}{9}\right)\left(2\frac{1}{4}\right)$

51. $1\frac{1}{7} \times 1\frac{1}{5}$

52. $2\frac{1}{3} \times 1\frac{1}{2}$

53. $\left(2\frac{1}{10}\right)^2$

54. $\left(1\frac{1}{2}\right)^2$

55. $3\frac{9}{10} \cdot 2$

56. $5 \cdot 1\frac{1}{2}$

57. $100 \times 3\frac{3}{4}$

58. $1\frac{5}{6} \times 20$

59. $1\frac{1}{2} \times 5\frac{1}{3}$

60. $5\frac{1}{4} \times 1\frac{1}{9}$

61. $\left(2\frac{1}{2}\right)\left(1\frac{1}{5}\right)$

62. $\left(1\frac{3}{10}\right)\left(2\frac{4}{9}\right)$

63. $12\frac{1}{2} \cdot 3\frac{1}{3}$

64. $5\frac{1}{10} \cdot 1\frac{2}{3}$

65. $66\frac{2}{3} \times 1\frac{7}{10}$

66. $37\frac{1}{2} \times 1\frac{3}{5}$

67. $1\frac{5}{9} \times \frac{3}{8} \times 2$

68. $\frac{1}{8} \times 2\frac{1}{4} \times 6$

69. $\left(\frac{1}{2}\right)^2\left(2\frac{1}{3}\right)$

70. $\left(1\frac{1}{4}\right)^2\left(\frac{1}{5}\right)$

71. $\frac{4}{5} \times \frac{7}{8} \times 1\frac{1}{10}$

72. $8\frac{1}{3} \times \frac{3}{10} \times \frac{5}{6}$

73. $\left(1\frac{1}{2}\right)^3$

74. $\left(2\frac{1}{2}\right)^3$

Divide and simplify.

75. $\frac{3}{5} \div \frac{2}{3}$

76. $\frac{2}{3} \div \frac{3}{5}$

77. $\frac{4}{5} \div \frac{7}{8}$

78. $\frac{7}{8} \div \frac{4}{5}$

79. $\frac{1}{2} \div \frac{1}{7}$

80. $\frac{1}{7} \div \frac{1}{2}$

81. $\frac{5}{9} \div \frac{1}{8}$

82. $\frac{1}{8} \div \frac{5}{9}$

83. $\frac{4}{5} \div \frac{8}{15}$

84. $\frac{3}{10} \div \frac{6}{5}$

85. $\frac{7}{8} \div \frac{3}{8}$

86. $\frac{10}{3} \div \frac{5}{6}$

87. $\frac{9}{10} \div \frac{3}{4}$

88. $\frac{5}{6} \div \frac{1}{3}$

89. $\frac{1}{10} \div \frac{2}{5}$

90. $\frac{3}{4} \div \frac{6}{5}$

91. $\frac{2}{3} \div 7$

92. $\frac{7}{10} \div 10$

93. $\frac{2}{3} \div 6$

94. $\frac{1}{20} \div 2$

95. $8 \div \frac{1}{5}$

96. $8 \div \frac{2}{9}$

97. $7 \div \frac{3}{7}$

98. $10 \div \frac{2}{5}$

99. $4 \div \frac{3}{10}$

100. $10 \div \frac{2}{3}$

101. $1 \div \frac{1}{7}$

102. $3 \div \frac{1}{8}$

103. $2\frac{5}{6} \div \frac{3}{7}$

104. $5\frac{1}{9} \div \frac{2}{3}$

105. $1\frac{1}{3} \div \frac{4}{5}$

106. $7\frac{1}{10} \div \frac{1}{2}$

107. $8\frac{5}{6} \div \frac{9}{10}$

108. $6\frac{1}{2} \div \frac{1}{2}$

109. $20\frac{1}{10} \div \frac{1}{5}$

110. $15\frac{2}{3} \div \frac{5}{6}$

111. $\frac{1}{6} \div 2\frac{1}{7}$

112. $\frac{2}{7} \div 1\frac{1}{3}$

113. $\frac{1}{2} \div 2\frac{3}{5}$

114. $\frac{3}{4} \div 3\frac{1}{9}$

115. $4 \div 1\frac{1}{4}$

116. $7 \div 1\frac{9}{10}$

117. $2\frac{1}{10} \div 20$

118. $5\frac{6}{7} \div 14$

119. $2\frac{1}{2} \div 3\frac{1}{7}$

120. $3\frac{1}{7} \div 2\frac{1}{2}$

121. $8\frac{1}{10} \div 5\frac{3}{4}$

122. $1\frac{7}{10} \div 5\frac{1}{8}$

123. $2\frac{1}{3} \div 4\frac{1}{2}$

124. $8\frac{1}{6} \div 2\frac{1}{2}$

125. $6\frac{3}{8} \div 2\frac{5}{6}$

126. $1\frac{2}{3} \div 1\frac{2}{5}$

Simplify.

127. $\frac{1}{2} + \frac{2}{3} \times 1\frac{1}{3}$

128. $\frac{9}{10} + \frac{4}{5} \cdot 8$

129. $5 - \frac{1}{3} \times \frac{2}{5}$

130. $3 \div \frac{2}{5} - 2\frac{1}{3}$

131. $2\frac{3}{4} \times \frac{1}{8} + \frac{1}{5}$

132. $\frac{3}{8} \cdot \frac{1}{2} - \frac{1}{10}$

133. $4 - \frac{2}{9} \div \frac{3}{4}$

134. $6 \div 5 \times \frac{1}{4}$

135. $3\frac{1}{2} \times 6 \div 5$

136. $4 \cdot \frac{2}{3} - 1\frac{1}{8}$

137. $10 \times \frac{1}{8} \times 2\frac{1}{2}$

138. $\frac{1}{3} \div \frac{1}{6} \times \frac{2}{3}$

139. $8 \div 1\frac{1}{5} + 3 \cdot 1\frac{1}{2}$

140. $3\frac{1}{8} \div 5 + 4 \div 2\frac{1}{2}$

141. $\frac{5}{6} \cdot \frac{9}{10} - \frac{3}{5} \div \frac{6}{7}$

142. $\frac{6}{11} \div \frac{18}{55} - \frac{7}{26} \cdot \frac{13}{14}$

143. $\left(\frac{1}{4}\right)^2 \cdot \left(3 - 1\frac{2}{3}\right)^2$

144. $\left(1 - \frac{2}{5}\right)^2 \div \left(1\frac{1}{2}\right)^2$

145. $\left(1\frac{1}{2} \div \frac{1}{3}\right)^2 + \left(1 - \frac{1}{4}\right)^2$

146. $\left(3\frac{1}{2}\right)^2 + 2\left(1\frac{1}{2} - 1\frac{1}{3}\right)$

Mixed Practice

147. Divide $6\frac{1}{8}$ by $2\frac{3}{4}$.

148. Compute: $14 - 3 \div \left(\frac{4}{5}\right)^2$

149. Find the product of $\frac{3}{5}$ and $\frac{7}{8}$.

150. Find the quotient of $\frac{9}{10}$ and $\frac{2}{5}$.

151. Multiply $\frac{2}{3}$ by 12.

152. Calculate: $\left(4\frac{1}{2}\right)\left(6\frac{2}{3}\right)$

Applications

B *Solve. Write the answer in simplest form.*

153. In a local town, $\frac{5}{6}$ of the voting-age population is registered to vote. If $\frac{7}{10}$ of the registered voters voted in the election for mayor, what fraction of the voting-age population voted?

154. Last year, $\frac{1}{8}$ of the emergency room visits at a hospital were injury related. Of these, $\frac{2}{5}$ were due to motor vehicle accidents. What fraction of the emergency room visits were due to motor vehicle accidents?

155. A couple would like to add on an extension to their house. One of the construction companies that bid on the job would charge \$40,200 with $\frac{1}{30}$ down. How much money does the couple need to put down if they choose to use this company?

156. There is a rule of thumb to not spend more than $\frac{1}{4}$ of one's income on rent. If someone makes \$24,000 a year, what is the most he or she should spend per month on rent according to this rule?

157. A tile store charges $\$46\frac{1}{2}$ per square foot for granite countertops, including installation. Find the cost of buying and installing a granite countertop on the kitchen island shown.

158. Which of these rooms has the larger area?

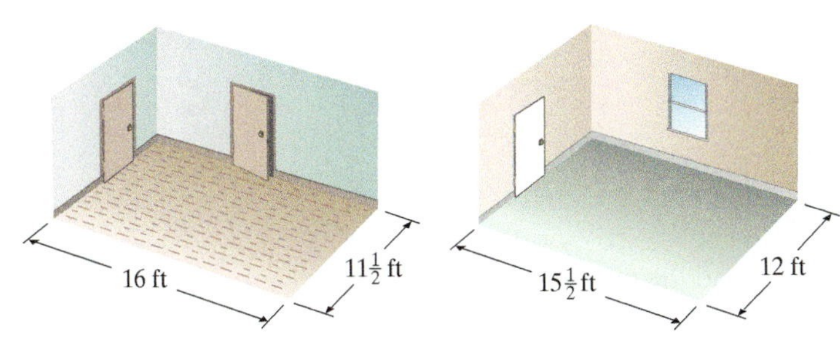

159. Students in an astronomy course learn that a first-magnitude star is $2\frac{1}{2}$ times as bright as a second-magnitude star, which in turn is $2\frac{1}{2}$ times as bright as a third-magnitude star. How many times as bright as a third-magnitude star is a first-magnitude star?

160. Some people believe that gasohol is superior to gasoline as an automotive fuel. Gasohol is a mixture of gasoline $\left(\frac{9}{10}\right)$ and ethyl alcohol $\left(\frac{1}{10}\right)$. How much more gasoline than ethyl alcohol is there in $10\frac{1}{2}$ gallons of gasohol?

161. Because of evaporation, a pond loses $\frac{1}{4}$ of its remaining water each month of summer. If it is full at the beginning of summer, what fraction of the original amount will the pond contain after three summer months?

162. A scientist is investigating the effects of cold on human skin. In one of the scientist's experiments, the temperature starts at 70°F and drops by $\frac{1}{10}$° every 2 minutes. What is the temperature after 6 minutes?

163. A trip to a nearby island takes $3\frac{1}{2}$ hours by boat and $\frac{1}{2}$ hour by airplane. How many times as fast as the boat is the plane?

164. Each dose of aspirin weighs $\frac{3}{4}$ grain. If a hospital pharmacist has 9 grains of aspirin on hand, how many doses can he provide?

165. A store sells two types of candles. The scented candle is 8 inches tall and burns $\frac{1}{2}$ inch per hour, whereas the unscented candle is 10 inches tall and burns $\frac{1}{3}$ inch per hour.

a. In an hour, which candle will burn more?

b. Which candle will last longer?

166. A college-wide fund-raising campaign collected \$3 million in $1\frac{1}{2}$ years for student scholarships.

a. What was the average amount collected per year?

b. By how much would this average increase if an additional \$1 million were collected?

• Check your answers on page A-3.

MINDStretchers

Writing

1. Every number except 0 has a reciprocal. Explain why 0 does not have a reciprocal.

Groupwork

2. In the following magic square, the *product* of every row, column, and diagonal is 1. Working with a partner, complete the square.

$\frac{2}{3}$		$1\frac{1}{2}$
$\frac{1}{2}$		

Patterns

3. Find the product: $1\frac{1}{2} \cdot 1\frac{1}{3} \cdot 1\frac{1}{4} \cdot \cdots \cdot 1\frac{1}{99} \cdot 1\frac{1}{100}$

Key Concepts and Skills

CONCEPT SKILL

Concept/Skill	Description	Example
[2.1] Prime number	A whole number that has exactly two different factors: itself and 1.	2, 3, 5
[2.1] Composite number	A whole number that has more than two factors.	4, 8, 9
[2.1] Prime factorization of a whole number	The number written as the product of its prime factors.	$30 = 2 \cdot 3 \cdot 5$
[2.1] Least common multiple (LCM) of two or more whole numbers	The smallest nonzero whole number that is a multiple of each number.	The LCM of 30 and 45 is 90.
[2.1] To compute the least common multiple (LCM)	• Find the prime factorization of each number. • Identify the prime factors that appear in each factorization. • Multiply these prime factors, using each factor the greatest number of times that it occurs in any of the factorizations.	$20 = 2 \cdot 2 \cdot 5$ $= 2^2 \cdot 5$ $30 = 2 \cdot 3 \cdot 5$ The LCM of 20 and 30 is $2^2 \cdot 3 \cdot 5$, or 60.
[2.2] Fraction	Any number that can be written in the form $\frac{a}{b}$, where a and b are whole numbers and b is nonzero.	$\frac{3}{11}, \frac{9}{5}$
[2.2] Proper fraction	A fraction whose numerator is smaller than its denominator.	$\frac{2}{7}, \frac{1}{2}$
[2.2] Mixed number	A number with a whole-number part and a proper fraction part.	$5\frac{1}{3}, 4\frac{5}{6}$
[2.2] Improper fraction	A fraction whose numerator is greater than or equal to its denominator.	$\frac{9}{4}, \frac{5}{5}$
[2.2] To change a mixed number to an improper fraction	• Multiply the denominator of the fraction by the whole-number part of the mixed number. • Add the numerator of the fraction to this product. • Write this sum over the denominator to form the improper fraction.	$4\frac{2}{3} = \frac{3 \times 4 + 2}{3}$ $= \frac{14}{3}$
[2.2] To change an improper fraction to a mixed number	• Divide the numerator by the denominator. • If there is a remainder, write it over the denominator.	$\frac{14}{3} = 4\frac{2}{3}$
[2.2] To find an equivalent fraction	Multiply the numerator and denominator of $\frac{a}{b}$ by the same whole number; that is, $\frac{a}{b} = \frac{a \cdot n}{b \cdot n}$, where both b and n are nonzero.	$\frac{3}{4} = \frac{3 \cdot 2}{4 \cdot 2} = \frac{6}{8}$

continued

CONCEPT SKILL

Concept/Skill	Description	Example
[2.2] **To simplify a fraction**	• First, express both the numerator and denominator as the product of their prime factors. • Then, divide out or cancel all common factors.	$\frac{30}{84} = \frac{\overset{1}{\cancel{2}} \cdot \overset{1}{\cancel{3}} \cdot 5}{\underset{1}{\cancel{2}} \cdot 2 \cdot \underset{1}{\cancel{3}} \cdot 7} = \frac{5}{14}$
[2.2] **Like fractions**	Fractions with the same denominator.	$\frac{2}{5}, \frac{3}{5}$
[2.2] **Unlike fractions**	Fractions with different denominators.	$\frac{3}{5}, \frac{3}{10}$
[2.2] **To compare fractions**	• If the fractions are like, compare their numerators. • If the fractions are unlike, write them as equivalent fractions with the same denominator and then compare their numerators.	$\frac{6}{8}, \frac{7}{8}$ $\quad 6 < 7$, so $\frac{6}{8} < \frac{7}{8}$ $\frac{2}{3}, \frac{12}{15}$ or $\frac{10}{15}, \frac{12}{15}$ $\quad 12 > 10$, so $\frac{12}{15} > \frac{2}{3}$
[2.2] **Least common denominator (LCD) of two or more fractions**	The least common multiple of their denominators.	The LCD of $\frac{11}{30}$ and $\frac{7}{45}$ is 90.
[2.3] **To add (or subtract) like fractions**	• Add (or subtract) the numerators. • Use the given denominator. • Write the answer in simplest form.	$\frac{1}{8} + \frac{1}{8} = \frac{2}{8} = \frac{1}{4}$ $\frac{3}{8} - \frac{1}{8} = \frac{2}{8} = \frac{1}{4}$
[2.3] **To add (or subtract) unlike fractions**	• Write the fractions as equivalent fractions with the same denominator, usually the LCD. • Add (or subtract) the numerators, keeping the same denominator. • Write the answer in simplest form.	$\frac{2}{3} + \frac{1}{2} = \frac{4}{6} + \frac{3}{6}$ $= \frac{7}{6}$, or $1\frac{1}{6}$ $\frac{5}{12} - \frac{1}{6} = \frac{5}{12} - \frac{2}{12}$ $= \frac{3}{12}$, or $\frac{1}{4}$
[2.3] **To add mixed numbers**	• Write the fractions as equivalent fractions with the same denominator, usually the LCD. • Add the fractions. • Add the whole numbers. • Write the answer in simplest form.	$4\frac{1}{2} = 4\frac{3}{6}$ $+6\frac{2}{3} = +6\frac{4}{6}$ $10\frac{7}{6} = 11\frac{1}{6}$
[2.3] **To subtract mixed numbers**	• Write the fractions as equivalent fractions with the same denominator, usually the LCD. • Regroup or borrow from the whole number on top if the fraction on the bottom is larger than the fraction on top. • Subtract the fractions. • Subtract the whole numbers. • Write the answer in simplest form.	$4\frac{1}{5} = 3\frac{6}{5}$ $-1\frac{2}{5} = -1\frac{2}{5}$ $2\frac{4}{5}$

Concept/Skill	Description	Example
[2.4] **To multiply fractions**	• Multiply the numerators. • Multiply the denominators. • Write the answer in simplest form.	$\frac{1}{2} \cdot \frac{3}{5} = \frac{3}{10}$
[2.4] **To multiply mixed numbers**	• Write the mixed numbers as improper fractions. • Multiply the fractions. • Write the answer in simplest form.	$2\frac{1}{2} \cdot 1\frac{2}{3} = \frac{5}{2} \cdot \frac{5}{3}$ $= \frac{25}{6}$, or $4\frac{1}{6}$
[2.4] **Reciprocal of $\frac{a}{b}$**	The fraction $\frac{b}{a}$ formed by switching the numerator and denominator.	The reciprocal of $\frac{4}{3}$ is $\frac{3}{4}$.
[2.4] **To divide fractions**	• Change the divisor to its reciprocal, and multiply the resulting fractions. • Write the answer in simplest form.	$\frac{2}{5} \div \frac{3}{7} = \frac{2}{5} \cdot \frac{7}{3}$ $= \frac{14}{15}$
[2.4] **To divide mixed numbers**	• Write the mixed numbers as improper fractions. • Divide the fractions. • Write the answer in simplest form.	$2\frac{1}{2} \div 1\frac{1}{3} =$ $\frac{5}{2} \div \frac{4}{3} =$ $\frac{5}{2} \cdot \frac{3}{4} = \frac{15}{8}$, or $1\frac{7}{8}$

Cultural Note

In societies throughout the world and across the centuries, people have written fractions in strikingly different ways. In ancient Greece, for example, the fraction $\frac{1}{4}$ was written Δ″ where Δ (read "delta") is the fourth letter of the Greek alphabet.

At one time, people wrote the numerator and denominator of fractions in Roman numerals, as shown at the left in a page from a sixteenth-century German book. In today's notation, the last fraction shown on the page is $\frac{200}{460}$.

I/IIII Dieſſe figur iſt vñ bedeüt ain fiertel von ainez ganzen/alſo mag man auch ain fünfftail/ayn ſechſtail/ain ſybentail oder zwai ſechſtail 2c. vnd alle ander brüch beſchreiben/Als I/V | I/VI | I/VII | II/VI 2c.

VI/VIII Diß ſein Sechs achtail/das ſein ſechstail der acht ain gantz machen.

IX/XI Diß Figur bezaigt ann newn ayilfftail das ſeyn IX tail/der XI ain gantz machen.

XX/XXXI Diß Figur bezaichet/zwentzigk ainundreyſigk tail/das ſein zwentzigk tail der ainsundreiſſigk ain gantz machen.

IIC/IIIIC.LX Diß ſein zwaihundert tail/der Fierhundert vnd ſechzigk ain gantz machen.

Source: David Eugene Smith and Jekuthiel Ginsburg, *Numbers and Numerals, a Story Book for Young and Old* (New York: Bureau of Publications, Teachers College, Columbia University, 1937).

CHAPTER 2 Review Exercises

Say Why *Fill in each blank.*

1. Twenty-seven $\underset{\text{is/is not}}{\rule{2cm}{0.4pt}}$ a composite number because ______________________________.

2. The prime factorization of 180 $\underset{\text{is/is not}}{\rule{2cm}{0.4pt}}$ $3^2 \times 4 \times 5$ because ______________________________.

3. The expression $\frac{3}{0}$ $\underset{\text{is/is not}}{\rule{2cm}{0.4pt}}$ a fraction because ______________________________.

4. The expression $\frac{12}{11}$ $\underset{\text{is/is not}}{\rule{2cm}{0.4pt}}$ an improper fraction because ______________________________.

5. The expression $\frac{16}{48}$ $\underset{\text{is/is not}}{\rule{2cm}{0.4pt}}$ equivalent to $\frac{1}{3}$ because ______________________________.

6. The expressions $\frac{12}{15}$ and $\frac{12}{16}$ $\underset{\text{are/are not}}{\rule{2cm}{0.4pt}}$ unlike fractions because ______________________________.

7. The least common denominator of $\frac{5}{8}$ and $\frac{7}{12}$ $\underset{\text{is/is not}}{\rule{2cm}{0.4pt}}$ 24 because ______________________________.

8. The reciprocal of $\frac{6}{8}$ $\underset{\text{is/is not}}{\rule{2cm}{0.4pt}}$ $\frac{3}{4}$ because ______________________________.

[2.1] *Find all the factors of each number.*

9. 150 **10.** 180 **11.** 57 **12.** 70

Indicate whether each number is prime or composite.

13. 23 **14.** 33 **15.** 87 **16.** 67

Write the prime factorization of each number, using exponents.

17. 36 **18.** 75 **19.** 99 **20.** 54

Find the LCM.

21. 6 and 14 **22.** 5 and 10 **23.** 18, 24, and 36 **24.** 10, 15, and 20

[2.2] *Identify the fraction or mixed number represented by the shaded portion of each figure.*

25. 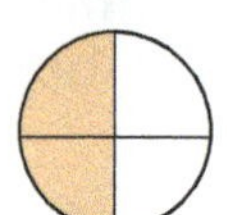

26. 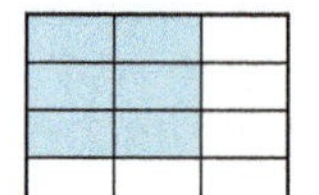

27. 

28.  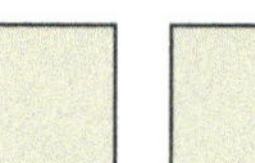 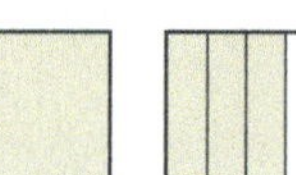

Indicate whether each number is a proper fraction, an improper fraction, or a mixed number.

29. $4\frac{1}{8}$ **30.** $\frac{5}{6}$ **31.** $\frac{3}{2}$ **32.** $\frac{7}{1}$

Write each mixed number as an improper fraction.

33. $7\frac{2}{3}$
34. $1\frac{4}{5}$
35. $9\frac{1}{10}$
36. $8\frac{3}{7}$

Write each fraction as a mixed number or a whole number.

37. $\frac{13}{2}$
38. $\frac{14}{3}$
39. $\frac{11}{4}$
40. $\frac{12}{12}$

Write an equivalent fraction with the given denominator.

41. $7 = \frac{}{12}$
42. $\frac{2}{7} = \frac{}{14}$
43. $\frac{1}{2} = \frac{}{10}$
44. $\frac{9}{10} = \frac{}{30}$

Simplify.

45. $\frac{14}{28}$
46. $\frac{15}{21}$
47. $\frac{30}{45}$
48. $\frac{54}{72}$
49. $5\frac{2}{4}$
50. $8\frac{10}{15}$
51. $6\frac{12}{42}$
52. $8\frac{45}{63}$

Insert the appropriate sign: <, =, or >.

53. $\frac{5}{8} \quad \frac{3}{8}$
54. $\frac{5}{6} \quad \frac{1}{6}$
55. $\frac{2}{3} \quad \frac{4}{5}$
56. $\frac{9}{10} \quad \frac{7}{8}$
57. $\frac{3}{4} \quad \frac{5}{8}$
58. $\frac{7}{10} \quad \frac{5}{9}$
59. $3\frac{1}{5} \quad 1\frac{9}{10}$
60. $5\frac{1}{8} \quad 5\frac{1}{9}$

Arrange in increasing order.

61. $\frac{2}{7}, \frac{3}{8}, \frac{1}{2}$
62. $\frac{1}{5}, \frac{1}{3}, \frac{2}{15}$
63. $\frac{4}{5}, \frac{9}{10}, \frac{3}{4}$
64. $\frac{7}{8}, \frac{7}{9}, \frac{13}{18}$

[2.3] *Add and simplify.*

65. $\frac{2}{5} + \frac{4}{5}$
66. $\frac{7}{20} + \frac{8}{20}$
67. $\frac{5}{8} + \frac{7}{8} + \frac{3}{8}$
68. $\frac{3}{10} + \frac{1}{10} + \frac{2}{10}$
69. $\frac{1}{3} + \frac{2}{5}$
70. $\frac{7}{8} + \frac{5}{6}$
71. $\frac{9}{10} + \frac{1}{2} + \frac{2}{5}$
72. $\frac{3}{8} + \frac{4}{5} + \frac{3}{4}$
73. $2 + 3\frac{7}{8}$
74. $6\frac{1}{4} + 3\frac{1}{4}$
75. $8\frac{7}{10} + 1\frac{9}{10}$
76. $5\frac{5}{6} + 2\frac{1}{6}$
77. $2\frac{1}{3} + 4\frac{1}{3} + 5\frac{2}{3}$
78. $1\frac{3}{10} + \frac{9}{10} + 2\frac{1}{10}$
79. $5\frac{2}{5} + \frac{3}{10}$
80. $9\frac{1}{6} + 8\frac{3}{8}$
81. $10\frac{2}{3} + 12\frac{3}{4}$
82. $20\frac{1}{2} + 25\frac{7}{8}$
83. $10\frac{3}{5} + 7\frac{9}{10} + 2\frac{1}{4}$
84. $20\frac{7}{8} + 30\frac{5}{6} + 4\frac{1}{3}$

Subtract and simplify.

85. $\frac{3}{8} - \frac{1}{8}$
86. $\frac{7}{9} - \frac{1}{9}$
87. $\frac{5}{3} - \frac{2}{3}$
88. $\frac{4}{6} - \frac{4}{6}$

89. $\frac{3}{10} - \frac{1}{20}$

90. $\frac{1}{2} - \frac{1}{8}$

91. $\frac{3}{5} - \frac{1}{4}$

92. $\frac{1}{3} - \frac{1}{10}$

93. $12\frac{1}{2} - 5$

94. $4\frac{3}{10} - 2$

95. $8\frac{7}{8} - 5\frac{1}{8}$

96. $20\frac{3}{4} - 2\frac{1}{4}$

97. $12 - 5\frac{1}{2}$

98. $4 - 2\frac{3}{10}$

99. $7 - 4\frac{1}{3}$

100. $1 - \frac{4}{5}$

101. $6\frac{1}{10} - 4\frac{3}{10}$

102. $2\frac{5}{8} - 1\frac{7}{8}$

103. $5\frac{1}{4} - 2\frac{3}{4}$

104. $7\frac{1}{6} - 3\frac{5}{6}$

105. $3\frac{1}{10} - 2\frac{4}{5}$

106. $7\frac{1}{2} - 4\frac{5}{8}$

107. $5\frac{1}{12} - 4\frac{1}{2}$

108. $6\frac{2}{9} - 2\frac{1}{3}$

109. $\frac{1}{3} + \frac{5}{6} - \frac{1}{2}$

110. $7\frac{9}{10} - 1\frac{1}{5} + 2\frac{3}{4}$

[2.4] *Multiply and simplify.*

111. $\frac{3}{4} \times \frac{1}{4}$

112. $\frac{1}{2} \times \frac{7}{8}$

113. $\left(\frac{5}{6}\right)\left(\frac{3}{4}\right)$

114. $\left(\frac{2}{3}\right)\left(\frac{1}{4}\right)$

115. $\frac{2}{3} \cdot 8$

116. $\frac{1}{10} \cdot 7$

117. $\left(\frac{1}{5}\right)^3$

118. $\left(\frac{2}{3}\right)^3$

119. $\frac{1}{2} \times \frac{2}{3} \times \frac{3}{4}$

120. $\frac{7}{8} \times \frac{2}{5} \times \frac{1}{6}$

121. $\frac{4}{5} \times 1\frac{1}{5}$

122. $\frac{2}{3} \times 2\frac{1}{3}$

123. $5\frac{1}{3} \cdot \frac{1}{2}$

124. $\frac{1}{10} \cdot 6\frac{2}{3}$

125. $1\frac{1}{3} \cdot 4\frac{1}{2}$

126. $3\frac{1}{4} \cdot 5\frac{2}{3}$

127. $6\frac{3}{4} \times 1\frac{1}{4}$

128. $8\frac{1}{2} \times 2\frac{1}{2}$

129. $\frac{7}{8} \times 1\frac{1}{5} \times \frac{3}{7}$

130. $1\frac{3}{8} \times \frac{10}{11} \times 1\frac{1}{4}$

131. $\left(3\frac{1}{3}\right)^3$

132. $\left(1\frac{1}{2}\right)^3$

133. $\frac{5}{8} + \frac{1}{2} \cdot 5$

134. $1\frac{9}{10} - \left(\frac{2}{3}\right)^2$

135. $4\left(\frac{2}{5}\right) + 3\left(\frac{1}{6}\right)$

136. $6\left(1\frac{1}{2} - \frac{3}{10}\right)$

Find the reciprocal.

137. $\frac{2}{3}$

138. $1\frac{1}{2}$

139. 8

140. $\frac{1}{4}$

Divide and simplify.

141. $\frac{7}{8} \div 5$

142. $\frac{5}{9} \div 9$

143. $\frac{2}{3} \div 5$

144. $\frac{1}{100} \div 2$

145. $\frac{1}{2} \div \frac{2}{3}$

146. $\frac{2}{3} \div \frac{1}{2}$

147. $6 \div \frac{1}{5}$

148. $7 \div \frac{4}{5}$

149. $\frac{7}{8} \div \frac{3}{4}$

150. $\frac{9}{10} \div \frac{1}{2}$

151. $\frac{3}{5} \div \frac{3}{10}$

152. $\frac{2}{3} \div \frac{1}{6}$

153. $3\frac{1}{2} \div 2$

154. $2 \div 3\frac{1}{2}$

155. $6\frac{1}{3} \div 4$

156. $4 \div 6\frac{1}{3}$

157. $8\frac{1}{4} \div 1\frac{1}{2}$

158. $3\frac{2}{5} \div 1\frac{1}{3}$

159. $4\frac{1}{2} \div 2\frac{1}{4}$

160. $7\frac{1}{5} \div 2\frac{2}{5}$

161. $\left(5 - \frac{2}{3}\right) \div \frac{4}{9}$

162. $6\frac{1}{2} \div \left(\frac{1}{2} + 4\frac{1}{2}\right)$

163. $7 \div 2\frac{1}{4} + 5 \div \left(1\frac{1}{2}\right)^2$

164. $\left(1\frac{2}{3}\right)^2 \times 2 + 9 \div 4\frac{1}{2}$

Mixed Applications

Solve. Write the answer in simplest form.

165. The Summer Olympic Games are held during each year divisible by 4. Were the Olympic Games held in 1990?

166. What is the smallest amount of money that you can pay in both all quarters and all dimes?

167. Eight of the 32 human teeth are incisors. What fraction of human teeth are incisors? (*Source:* Ilsa Goldsmith, *Human Anatomy for Children*)

168. The planets in the solar system (including the "dwarf planet" Pluto) consist of Earth, two planets closer to the Sun than Earth, and six planets farther from the Sun than Earth. What fraction of the planets in the solar system are closer than Earth to the Sun? (*Source:* Patrick Moore, *Astronomy for the Beginner*)

169. A Filmworks camera has a shutter speed of $\frac{1}{8{,}000}$ second and a Lensmax camera has a shutter speed of $\frac{1}{6{,}000}$ second. Which shutter is faster? (*Hint:* The faster shutter has the smaller shutter speed.)

170. In a recent year, among Americans who were 65 years of age or older 15 million were male and 21 million female. What fraction of this population was female? (*Source:* U. S. Census Bureau)

171. An insurance company reimbursed a patient \$275 on a dental bill of \$700. Did the patient get more or less than $\frac{1}{3}$ of the money paid back? Explain.

172. A union goes on strike if at least $\frac{2}{3}$ of the workers voting support the strike call. If 23 of the 32 voting workers support a strike, should a strike be declared? Explain.

173. A grand jury has 23 jurors. Sixteen jurors are needed for a quorum, and a vote of 12 jurors is needed to indict.

a. What fraction of the full jury is needed to indict?

b. Suppose that 16 jurors are present. What fraction of those present is needed to indict?

174. In a tennis match, Lisa Gregory went to the net 12 times, winning the point 7 times. By contrast, Monica Yates won the point 4 of the 6 times that she went to the net.

a. Which player went to the net more often?

b. Which player had a better rate of winning points at the net?

175. In a math course, $\frac{3}{5}$ of a student's grade is based on four in-class exams, and $\frac{3}{20}$ of the grade is based on homework. What fraction of a student's grade is based on in-class exams and homework?

176. A metal alloy is made by combining $\frac{1}{4}$ ounce of copper with $\frac{2}{3}$ ounce of tin. Find the alloy's total weight.

177. The weight of a diamond is measured in carats. What is the difference in weight between a $\frac{3}{4}$-carat and a $\frac{1}{2}$-carat diamond?

178. During a sale, the price of a sweater was marked $\frac{1}{4}$ off the original price of \$45. Using a coupon, a customer received an additional $\frac{1}{5}$ off the sale price. What fraction of the original price was the final sale price?

179. In a math class, $\frac{3}{8}$ of the students are chemistry majors and $\frac{2}{3}$ of those students are women. If there are 48 students in the math class, how many women are chemistry majors?

180. Three-eighths of the undergraduate students at a two-year college receive financial aid. If the college has 4,296 undergraduate students, how many undergraduate students do not receive financial aid?

181. Of the first 10 artists to receive the Grammy Lifetime Achievement Award, $\frac{4}{5}$ were men. How many of these awardees were women? (*Source: Top 10 of Everything 2010*)

182. A sea otter eats about $\frac{1}{5}$ of its body weight each day. How much will a 35-pound otter eat in a day? (*Source:* Karl W. Kenyon, *The Sea Otter in the Eastern Pacific Ocean*)

183. An investor bought \$1,000 worth of a technology stock. At the beginning of last year, it had increased in value by $\frac{2}{5}$. During the year, the value of the stock declined by $\frac{1}{4}$. What was the value of the stock at the end of last year?

184. In Roseville, 40 of every 1,000 people who want to work are unemployed, in contrast to 8 of every 100 people in Georgetown. How many times as great as the unemployment rate in Roseville is the unemployment rate in Georgetown?

185. A brother and sister want to buy as many goldfish as possible for their new fish tank. A rule of thumb is that the total length of fish (in inches) should be less than the capacity of the tank (in gallons). If they have a 10-gallon tank and goldfish average $\frac{1}{2}$ inch in length, how many fish should they buy?

186. A commuter is driving to the city of Denver 15 miles away. If he has already driven $3\frac{1}{4}$ miles, how far is he from Denver?

187. The wingspread of a Boeing 777-300 jet is $199\frac{11}{12}$ feet, whereas the wingspread of a Boeing 747-400 is $211\frac{5}{12}$ feet. How much longer is the wingspread of a Boeing 747-400 jet? (*Source:* boeing.com)

188. A Chicago family plans to take a vacation traveling by express bus either to Kansas City or to Indianapolis. The trip to Kansas City takes $11\frac{3}{4}$ hours, in contrast to a trip to Indianapolis that takes only $4\frac{1}{3}$ hours. How much longer is the first trip? (*Source:* greyhound.com)

189. An airplane is flying $1\frac{1}{2}$ times the speed of sound. If sound travels at about 1,000 feet per second, at what speed is the plane flying?

190. When standing upright, the pressure per square inch on a person's hip joint is about $2\frac{1}{2}$ times his or her body weight. If the person weighs 200 pounds, what is that pressure? (*Source:* pnas.org)

191. A cubic foot of water weighs approximately $62\frac{1}{2}$ pounds. If a basin contains $4\frac{1}{2}$ cubic feet of water, how much does the water weigh?

192. In 2009, the attendance at Fenway Park in Boston was about $1\frac{1}{4}$ times the attendance ten years earlier. If attendance in 1999 was approximately 2,500,000, what was the attendance in 2009? Round to the nearest hundred thousand. (*Source:* ballparks.com)

193. It took the space shuttle Endeavor $1\frac{1}{2}$ hours to orbit Earth. How many orbits did the Endeavor make in 12 hours? (*Source:* NASA)

194. An offshore wind farm measures $10\frac{1}{4}$ miles by $2\frac{1}{2}$ miles. Find the area of the wind farm.

195. The following chart is a record of the amount of time (in hours) that two employees spent working the past weekend. Complete the chart.

Employee	Saturday	Sunday	Total
L. Chavis	$7\frac{1}{2}$	$4\frac{1}{4}$	
R. Young	$5\frac{3}{4}$	$6\frac{1}{2}$	
Total			

196. Complete the following chart.

Worker	Hours per Day	Days Worked	Total Hours	Wage per Hour	Gross Pay
Maya	5	3		\$7	
Noel	$7\frac{1}{4}$	4		\$10	
Alisa	$4\frac{1}{2}$	$5\frac{1}{2}$		\$9	

197. According to a newspaper advertisement, a man on a diet lost 60 pounds in $5\frac{1}{2}$ months. On the average, how much weight did he lose per month?

198. According to the nutrition label on a box of cereal, one serving is $1\frac{1}{4}$ cups. If the box contains 18 servings of cereal, how many cups of cereal does it contain?

199. The American explorers Lewis and Clark traveled about 8,000 miles in $2\frac{1}{2}$ years, mapping an overland route to the Pacific Ocean. Approximately how many miles did they travel per year? (*Source:* nps.gov)

200. One of the largest yachts in the world is the German-built motor yacht Arctic P. If the yacht's length is $87\frac{2}{3}$ meters and its beam (its width at the widest point) is $14\frac{3}{4}$ meters, how many times its width is its length, to the nearest whole number? (*Source:* superyachts.com)

• Check your answers on page A-3.

CHAPTER 2 Posttest

FOR EXTRA HELP

The Chapter Test Prep Videos with test solutions are available on DVD, in MyMathLab, and on YouTube (search "AkstDevMath" and click on "Channels").

To see if you have mastered the topics in this chapter, take this test.

1. List all the factors of 63.
2. Write 54 as the product of prime factors.
3. What fraction of the diagram is shaded?

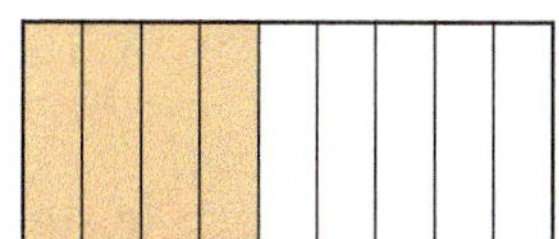

4. Write 12 as an improper fraction.
5. Express $\frac{41}{4}$ as a mixed number.
6. Write $\frac{875}{1{,}000}$ in simplest form
7. Which is smaller, $\frac{2}{3}$ or $\frac{5}{10}$?
8. What is the LCD for $\frac{3}{8}$ and $\frac{1}{12}$?

Add.

9. $\frac{2}{3} + \frac{1}{8} + \frac{3}{4}$
10. $6\frac{7}{8} + 1\frac{3}{10}$

Subtract.

11. $6 - 1\frac{5}{7}$
12. $10\frac{1}{6} - 4\frac{2}{5}$

Multiply.

13. $\left(\frac{1}{9}\right)^2$
14. $2\frac{2}{3} \times 4\frac{1}{2}$
15. Divide: $2\frac{1}{3} \div 3$
16. Calculate: $14\frac{1}{2} - 5 \cdot 1\frac{1}{3}$

Solve. Write your answer in simplest form.

17. Four of the 44 men who have served as vice-president of the United States were born in Indiana. What fraction of these men were *not* born in Indiana? (*Source: The New York Times Almanac 2010*)
18. In an Ironman triathlon, an athlete completed the 112-mile bike ride in $5\frac{5}{6}$ hours. What was the average number of miles she bicycled each hour?
19. The film *Super Troopers* lasts $1\frac{2}{3}$ hours. If the film is half over, how much remains to the running time of the film?
20. A college student has a part-time job, working $5\frac{1}{2}$ hours on Monday and $6\frac{1}{4}$ hours on Tuesday. If the student makes \$8 an hour, how much money did the student make for the two days of work?

• Check your answers on page A-4.

Cumulative Review Exercises

To help you review, solve the following:

1. Write in words: 5,000,315

2. Round 1,876,529 to the nearest hundred thousand.

3. Multiply: $5{,}814 \times 100$

4. Find the quotient: $89\overline{)80{,}812}$

5. Evaluate: $24 \div (2 + 4) - 3$

6. Evaluate: $\left(\frac{6 - 5}{2 + 3}\right)^2$

7. Write the prime factorization of 84 using exponents.

8. Find the least common multiple of 20 and 24.

9. Write $\frac{75}{100}$ in simplest form.

10. Which is larger, $\frac{1}{4}$ or $\frac{3}{8}$?

11. Add: $\frac{5}{8} + \frac{5}{6}$

12. Subtract: $8 - 1\frac{3}{5}$

13. Find the product: $1\frac{1}{2} \cdot 4\frac{2}{3}$

14. Divide: $4 \div 2\frac{3}{4}$

15. In a recent year, the two U.S. corporations with largest revenues were ExxonMobil and Walmart Stores. How much greater were the ExxonMobil revenues? (*Source: Fortune*)

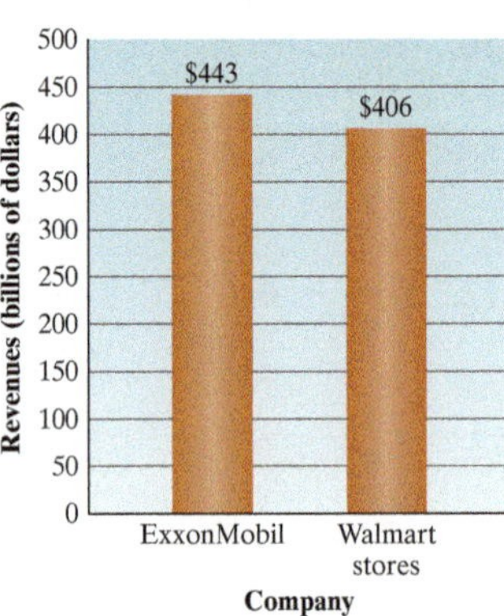

16. A jury decided on punishments in an oil spill case. The jury ordered the captain of the oil barge to pay $5,000 in punitive damages and the oil company to pay $5 billion. The amount that the company had to pay is how many times the amount that the captain had to pay?

17. The following table gives the number of named storms (tropical storms, hurricanes, and subtropical storms) that hit the United States in recent years:

Year	Number of Named Storms
2005	28
2006	10
2007	15
2008	18
2009	9

(*Source:* aoml.noaa.gov)

How far above the five-year average was the number of named storms in 2005?

18. A gallon of paint covers approximately 400 square feet. A room has two walls that are 8 feet high and 13 feet wide, and two walls that are 8 feet high and 15 feet wide. The doors and windows along those walls total 78 square feet. Will one gallon of paint cover the walls? Why or why not?

19. A homeowner wants to refinish his basement over three weekends. He completes $\frac{1}{4}$ of the job the first weekend and $\frac{5}{12}$ of the job the second weekend. What fraction of the job remains to be completed?

20. In a theater program, a student purchases a 7-inch-long piece of trim for costumes. From this purchase, how many $1\frac{3}{4}$-inch-long pieces of trim can be made?

• Check your answers on page A-4.

CHAPTER 3

Decimals

Decimals and Blood Tests

Blood tests reveal a great deal about a person's health—whether to reduce the cholesterol level to lower the risk of heart disease, or raise the red blood cell count to prevent anemia. And blood tests identify a variety of diseases, for example, AIDS and mononucleosis.

Blood analyses are typically carried out in clinical laboratories. Technicians in these labs operate giant machines that perform thousands of blood tests per hour.

What these blood tests, known as "chemistries," actually do is to analyze blood for a variety of substances, such as creatinine and calcium.

In any blood test, doctors look for abnormal levels of the substance being measured. For instance, the normal range on the creatinine test is typically from 0.7 to 1.5 milligrams (mg) per unit of blood. A high level may mean kidney disease; a low level, muscular dystrophy.

The normal range on the calcium test may be 9.0 to 10.5 mg per unit of blood. A result outside this range is a clue for any of several diseases.

(*Source:* Dixie Farley, "Top 10 Laboratory Tests: Blood Will Tell," *FDA Consumer*, Vol. 23)

CHAPTER 3 PRETEST

To see if you have already mastered the topics in this chapter, take this test.

1. In the number 27.081, what place does the 8 occupy?

2. Write in words: 4.012

3. Round 3.079 to the nearest tenth.

4. Which is largest: 0.00212, 0.0029, or 0.000888?

Perform the indicated operations.

5. $7.02 + 3.5 + 11$

6. $2.37 + 5.0038$

7. $13.79 - 2.1$

8. $9 - 2.7 + 3.51$

9. $8.3 \times 1{,}000$

10. 8.01×2.3

11. $(0.12)^2$

12. $5 + 3 \times 0.7$

13. $6.05 \div 1{,}000$

14. $\dfrac{9.81}{0.3}$

Express as a decimal.

15. $\dfrac{7}{8}$

16. $2\dfrac{5}{6}$, rounded to the nearest hundredth

Solve.

17. In a science course, a student learns that an acid is stronger if it has a lower pH value. Which is stronger, an acid with a pH value of 3.7 or an acid with a pH value of 2.95?

18. The following table shows the quarterly revenues for Microsoft Corporation in a recent fiscal year.

	First	Second	Third	Fourth
Revenue (in billions of dollars)	15.06	16.63	13.65	13.1

What was Microsoft's total revenue for the four quarters? (*Source:* microsoft.com)

19. A serving of iceberg lettuce contains 3.6 milligrams (mg) of vitamin C, whereas romaine lettuce contains 11.9 mg of vitamin C. Romaine lettuce is how many times as rich in vitamin C as iceberg lettuce? Round the answer to the nearest whole number. (*Source: The Concise Encyclopedia of Foods and Nutrition*)

20. A tourist visiting Orlando makes a long-distance telephone call that costs \$0.85 for the first 3 minutes and \$0.17 for each additional minute. What is the cost of a 20-minute call?

• Check your answers on page A-4.

3.1 Introduction to Decimals

OBJECTIVES

- **A** To read or write decimals
- **B** To find the fraction equivalent to a decimal
- **C** To compare decimals
- **D** To round decimals
- **E** To solve applied problems involving decimals

What Decimals Are and Why They Are Important

Decimal notation is in common use. When we say that the price of a book is $32.75, that the length of a table is 1.8 meters, or that the answer displayed on a calculator is 5.007, we are using decimals.

A number written as a **decimal** has

- a whole-number part, which *precedes* the decimal point, and
- a fractional part, which *follows* the decimal point.

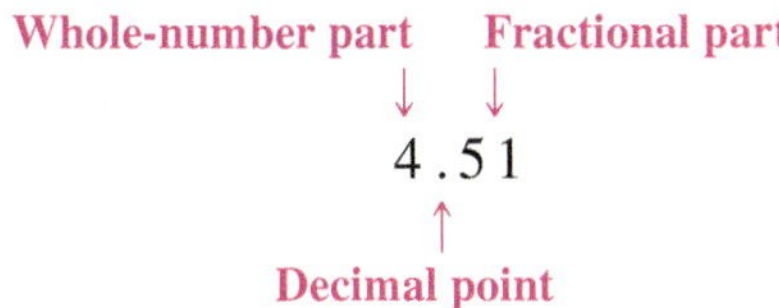

A decimal without a decimal point shown is understood to have one at the right end and is the same as a whole number. For instance, 3 and 3. are the same number.

The fractional part of any decimal has as its denominator a power of 10, such as 10, 100, or 1,000. The use of the word *decimal* reminds us of the importance of the number 10 in this notation, just as decade means 10 years or December meant the 10th month of the year (which it was for the early Romans).

In many problems, we can choose to work with either decimals or fractions. Therefore, we need to know how to work with both if we are to use the easier approach to solve a particular problem.

Decimal Places

Each digit in a decimal has a place value. The place value system for decimals is an extension of the place value system for whole numbers.

The places to the right of the decimal point are called **decimal places**. For instance, the number 64.149 is said to have three decimal places.

For a whole number, place values are powers of 10: 1, 10, 100, ... By contrast, each place value for the fractional part of a decimal is the reciprocal of a power of 10: $\frac{1}{10}, \frac{1}{100}, \frac{1}{1{,}000}, \ldots$.

The first decimal place after the decimal point is the tenths place. Working to the right, the next decimal places are the hundredths place, the thousandths place, and so on.

The following table shows the place values in the decimals 0.54 and 0.30716.

Ones	•	Tenths	Hundredths	Thousandths	Ten-thousandths	Hundred-thousandths
1	and	$\frac{1}{10}$	$\frac{1}{100}$	$\frac{1}{1{,}000}$	$\frac{1}{10{,}000}$	$\frac{1}{100{,}000}$
0	.	5	4			
0	.	3	0	7	1	6

The next table shows the place values for the decimals 7,204.5 and 513.285.

Thousands	Hundreds	Tens	Ones	•	Tenths	Hundredths	Thousandths
1,000	100	10	1	and	$\frac{1}{10}$	$\frac{1}{100}$	$\frac{1}{1,000}$
7	2	0	4	.	5		
	5	1	3	.	2	8	5

EXAMPLE 1

In each number, identify the place that the digit 3 occupies.

a. 0.134 **b.** 92.388 **c.** 0.600437

Solution

a. The hundredths place

b. The tenths place

c. The hundred-thousandths place

PRACTICE 1

What place does the digit 1 occupy in each number?

a. 566.184

b. 43.57219

c. 0.921

Changing Decimals to Fractions

Knowing the place value system is the key to understanding what decimals mean, how to read them, and how to write them.

- The decimal 0.9, or .9, is another way of writing $(0 \times 1) + \left(9 \times \frac{1}{10}\right)$, or $\frac{9}{10}$. This decimal is read the same as the equivalent fraction: "nine tenths."
- The decimal 0.21 represents 2 tenths + 1 hundredth. This expression simplifies to the following:

$$\left(2 \times \frac{1}{10}\right) + \left(1 \times \frac{1}{100}\right) = \frac{2}{10} + \frac{1}{100} = \frac{20}{100} + \frac{1}{100}, \text{ or } \frac{21}{100}$$

So 0.21 is read "twenty-one hundredths."

- The decimal 0.149 stands for $\frac{149}{1,000}$.

$$\left(1 \times \frac{1}{10}\right) + \left(4 \times \frac{1}{100}\right) + \left(9 \times \frac{1}{1,000}\right) = \frac{1}{10} + \frac{4}{100} + \frac{9}{1,000}$$
$$= \frac{100}{1,000} + \frac{40}{1,000} + \frac{9}{1,000}$$
$$= \frac{149}{1,000}$$

So 0.149 is read "one hundred forty-nine thousandths."

Let's summarize these examples.

Decimal	Equivalent Fraction	Read as
0.9	$\frac{9}{10}$	Nine tenths
0.21	$\frac{21}{100}$	Twenty-one hundredths
0.149	$\frac{149}{1{,}000}$	One hundred forty-nine thousandths

Note that in each of these decimals, the fractional part is the same as the numerator of the equivalent fraction: $0.149 = \frac{149}{1{,}000}$.

We can use the following rule to rewrite any decimal as a fraction or a mixed number.

To Change a Decimal to the Equivalent Fraction or Mixed Number

- Copy the nonzero whole-number part of the decimal and drop the decimal point.
- Place the fractional part of the decimal in the numerator of the equivalent fraction.
- Make the denominator of the equivalent fraction 1 followed by as many zeros as the decimal has decimal places.
- Simplify the resulting fraction, if possible.

EXAMPLE 2

Express 0.75 in fractional form and simplify.

Solution The whole-number part of the decimal is 0. We drop the decimal point. The fractional part (75) of the decimal becomes the numerator of the equivalent fraction. Since the decimal has two decimal places, we make the denominator of the equivalent fraction 1 followed by two zeros (100). So we can write 0.75 as $\frac{75}{100}$, which simplifies to $\frac{3}{4}$.

PRACTICE 2

Write 0.875 as a fraction in lowest terms.

EXAMPLE 3

Express 1.87 as a mixed number.

Solution This decimal is equivalent to a mixed number whose whole-number part is 1. The fractional part (87) of the decimal is the numerator of the equivalent fraction. The decimal has two decimal places, so the fraction's denominator has two zeros (that is, it is 100).

$$1.87 = 1\frac{87}{100}$$

Do you see that the answer can also be written as $\frac{187}{100}$?

PRACTICE 3

The decimal 2.03 is equivalent to what mixed number?

EXAMPLE 4

Find the equivalent fraction of each decimal.

a. 3.2 **b.** 3.200

Solution

a. 3.2 represents $3\frac{2}{10}$, or $3\frac{1}{5}$.

b. 3.200 equals $3\frac{200}{1{,}000}$, or $3\frac{1}{5}$.

PRACTICE 4

Express each decimal in fractional form.

a. 5.6 **b.** 5.6000

TIP Adding zeros in the rightmost decimal places does not change a decimal's value. However, generally decimals can be written without these extra zeros.

EXAMPLE 5

Write each decimal as a mixed number.

a. 1.309 **b.** 1.39

Solution

a. $1.309 = 1\frac{309}{1{,}000}$

b. $1.39 = 1\frac{39}{100}$

PRACTICE 5

What mixed number is equivalent to each decimal?

a. 7.003 **b.** 4.1

Knowing how to change a decimal to its equivalent fraction also helps us read the decimal.

EXAMPLE 6

Express each decimal in words.

a. 0.319 **b.** 2.71 **c.** 0.08

Solution

a. $0.319 = \frac{319}{1{,}000}$

We read the decimal as "three hundred nineteen thousandths."

b. $2.71 = 2\frac{71}{100}$

The decimal point is read as "and." We read the decimal as "two and seventy-one hundredths."

c. $0.08 = \frac{8}{100}$

We read the decimal as "eight hundredths." Note that we *do not simplify* the equivalent fraction when reading the decimal.

PRACTICE 6

Express each decimal in words.

a. 0.61

b. 4.923

c. 7.05

EXAMPLE 7

Write each number in decimal notation.

a. Seven tenths **b.** Five and thirty-two thousandths

Solution

a. Since 7 is in the tenths place, the decimal is written as 0.7.

b. The whole number preceding *and* is in the ones place. The last digit of 32 is in the thousandths place.

5.032

We replace *and* with the decimal point. We need a 0 to hold the tenths place.

The answer is 5.032.

PRACTICE 7

Write each number in decimal notation.

a. Forty-three thousandths

b. Ten and twenty-six hundredths

EXAMPLE 8

For hay fever, an allergy sufferer takes a decongestant pill that has a tablet strength of three hundredths of a gram. Write the equivalent decimal.

Solution "Three hundredths" is written 0.03, with the digit 3 in the hundredths place.

PRACTICE 8

The number pi (usually written π) is approximately three and fourteen hundredths. Write this approximation as a decimal.

Comparing Decimals

Suppose that we want to compare two decimals—say, 0.6 and 0.7. The key is to rethink the problem in terms of fractions.

$$0.6 = \frac{6}{10} \qquad 0.7 = \frac{7}{10}$$

Because $\frac{7}{10} > \frac{6}{10}$, $0.7 > 0.6$.

The following procedure provides another way to compare decimals that is faster than converting the decimals to fractions.

To Compare Decimals

- Rewrite the numbers vertically, lining up the decimal points.
- Working from left to right, compare the digits that have the same place value. At the first place value where the digits differ, the decimal which has the largest digit with this place value is the largest decimal.

EXAMPLE 9

Which is larger, 0.729 or 0.75?

Solution First, let's line up the decimal point.

↓
0.729
0.75
↑

We see that both decimals have a 0 in the ones place. We next compare the digits in the tenths place and see that, again, they are the same. Looking to the right in the hundredths place, we see that $5 > 2$. Therefore, $0.75 > 0.729$. Note that the decimal with more digits is not necessarily the larger decimal.

PRACTICE 9

Which is smaller, 0.83 or 0.8297?

EXAMPLE 10

Rank from smallest to largest: 2.17, 2.1, and 0.99

Solution First, we position the decimals so that the decimal points are aligned.

↓
2.17
2.1
0.99
↑

Working from left to right, we see that in the ones place, the first two decimals have a 2 and the third decimal has a 0, so the third decimal is the smallest of the three. To decide which of the first two decimals is smaller, we compare the digits in the tenths place. Since both of these decimals have a 1 in the tenths place, we proceed to the hundredths place. A 0 is understood to the right of the 1 in 2.1, so we compare 0 and 7.

↓
2.17
2.10
0.99
↑

Since $0 < 7$, we conclude that $2.10 < 2.17$. Therefore, the three decimals from smallest to largest are 0.99, 2.1, and 2.17.

PRACTICE 10

Rewrite in decreasing order: 3.5, 3.51, and 3.496

EXAMPLE 11

Plastic garbage bags come in three thicknesses (or gauges): 0.003 inch, 0.0025 inch, and 0.002 inch. The three gauges are called lightweight, regular weight, and heavyweight. Which is the lightweight gauge?

Solution To find the smallest of the decimals, we first line up the decimal points.

$$\begin{array}{l} 0.003 \\ 0.0025 \\ 0.002 \end{array}$$

Working from left to right, we see that the three decimals have the same digits until the thousandths place, where $3 > 2$. Therefore, 0.003 must be the heavyweight gauge. To compare 0.0025 and 0.002, we look at the ten-thousandths place. The 5 is greater than the 0 that is understood to be there (0.0020). So 0.0025 inch must be the regular-weight gauge, and 0.002 the lightweight gauge.

PRACTICE 11

The higher the energy efficiency rating (EER) of an air conditioner, the more efficiently it uses electricity. Which of the following air conditioners is least efficient? (*Source:* Consumer Guide)

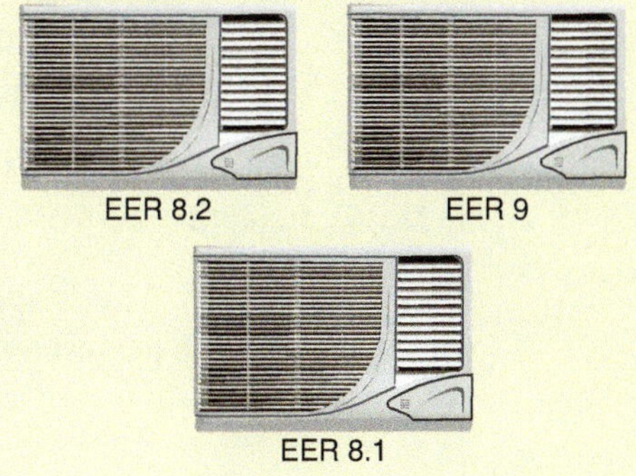

Rounding Decimals

As with whole numbers, we can round decimals to a given place value. For instance, suppose that we want to round the decimal 1.38 to the nearest tenth. The decimal 1.38 lies between 1.3 and 1.4, so one of these two numbers will be our answer—but which? To decide, let's take a look at a number line.

1.31 1.33 1.35 1.37 1.39
1.3 1.32 1.34 1.36 1.38 1.4

Do you see from this diagram that 1.38 is closer to 1.4 than to 1.3?

$$1.38 \approx 1.4$$

Tenths place

Rounding a decimal to the nearest tenth means that the last digit lies in the tenths place.

The following table shows the relationship between the place to which we are rounding and the number of decimal places in our answer.

Rounding to the Nearest	Means That the Rounded Decimal Has
tenth $\left(\frac{1}{10}\right)$	one decimal place.
hundredth $\left(\frac{1}{100}\right)$	two decimal places.
thousandth $\left(\frac{1}{1,000}\right)$	three decimal places.
ten-thousandth $\left(\frac{1}{10,000}\right)$	four decimal places.

Note that the number of decimal places is the same as the number of zeros in the corresponding denominator.

The following rule can be used to round decimals.

To Round a Decimal to a Given Decimal Place

- Underline the digit in the place to which the number is being rounded.
- The digit to the right of the underlined digit is called the *critical digit*. Look at the critical digit—if it is 5 or more, add 1 to the underlined digit; if it is less than 5, leave the underlined digit unchanged.
- Drop all decimal places to the right of the underlined digit.

Let's apply this rule to the problem that we just considered—namely, rounding 1.38 to the nearest tenth.

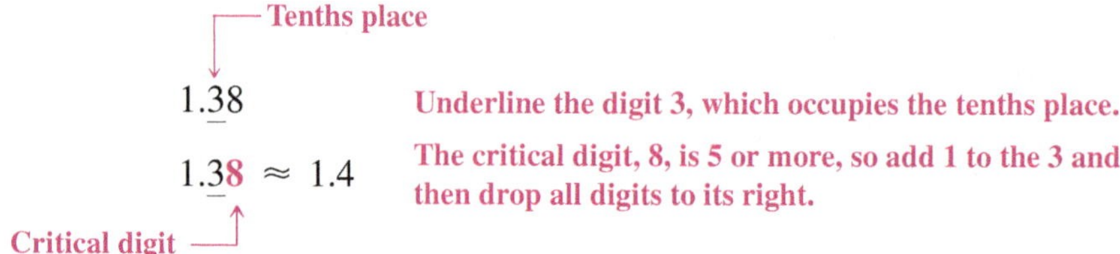

The following examples illustrate this method of rounding.

EXAMPLE 12

Round 94.735 to

a. the nearest tenth.

b. two decimal places.

c. the nearest thousandth.

d. the nearest ten.

e. the nearest whole number.

Solution

a. First, we underline the digit 7 in the tenths place: 94.7̲35. The critical digit, 3, is less than 5, so we do not add 1 to the underlined digit. Dropping all digits to the right of the 7, we get 94.7. Note that our answer has only one decimal place because we are rounding to the nearest tenth.

b. We need to round 94.735 to two decimal places (to the nearest hundredth).

94.73̲5 ≈ 94.74

The critical digit is 5 or more. Add 1 to the underlined digit and drop the decimal place to the right.

c. 94.735̲ ≈ 94.735 because the critical digit to the right of the 5 is understood to be 0.

PRACTICE 12

Round 748.0772 to

a. the nearest tenth.

b. the nearest hundredth.

c. three decimal places.

d. the nearest whole number.

e. the nearest hundred.

d. We are rounding 94.735 to the nearest ten (*not tenth*), which is a whole-number place.

94.735 ≈ 90

Because 4 < 5, keep 9 in the tens place, insert 0 in the ones place and drop all decimal places.

e. Rounding to the nearest whole number means rounding to the nearest 1.

94.735 ≈ 95

Because 7 > 5, change the 4 to 5 and drop all decimal places.

EXAMPLE 13

Round 3.982 to the nearest tenth.

Solution First, we underline the digit 9 in the tenths place and identify the critical digit: 3.982. The critical digit, 8, is more than 5, so we add 1 to the 9, get 10, and write down the 0. We add the 1 to 3, getting 4, and drop the 8 and 2.

3.982 ≈ 4.0

Drop

The answer is 4.0. Note that we do not drop the 0 in the tenths place of the answer to indicate that we have rounded to that place.

PRACTICE 13

Round 7.2962 to two decimal places.

EXAMPLE 14

The rate of exchange between currencies varies with time. On a particular day the euro, the currency used in many countries of Western Europe, could have been exchanged for 1.23502 dollars. What was this price to the nearest cent? (*Source:* xe.com)

Solution A cent is one-hundredth of a dollar. Therefore, we need to round 1.23502 to the nearest hundredth.

1.23502 ≈ 1.24

So the price to the nearest cent was $1.24.

PRACTICE 14

Mount Waialeale on the Hawaiian island of Kauai is one of the world's wettest places, with an average annual rainfall of 11.68 meters. What is the amount of rainfall to the nearest tenth of a meter? (*Source:* wikipedia.org)

3.1 Exercises

FOR EXTRA HELP MyMathLab

PRACTICE

WATCH

READ REVIEW

Mathematically Speaking

Fill in each blank with the most appropriate term or phrase from the given list.

less	greater	increasing	left
decreasing	ten	hundredths	multiple
thousandths	power	right	tenth

1. A decimal place is a place to the ________ of the decimal point.

2. The fractional part of a decimal has as its denominator a ________ of 10.

3. The decimal 0.17 is equivalent to the fraction seventeen __________.

4. The decimal 209.95 rounded to the nearest ________ is 210.0.

5. The decimal 0.371 is ________ than the decimal 0.3499.

6. The decimals 0.48, 0.4, and 0.371 are written in __________ order.

A *Underline the digit that occupies the given place.*

7. 2.78 Tenths place

8. 6.835 Tenths place

9. 9.01 Hundredths place

10. 0.772 Hundredths place

11. 2.00175 Ten-thousandths place

12. 4.00189 Ten-thousandths place

13. 823.001 Thousandths place

14. 829.006 Thousandths place

Identify the place occupied by the underlined digit.

15. $25.\underline{7}1$

16. $3.00\underline{2}$

17. $8.1\underline{8}3$

18. $\underline{4}9.771$

19. $1{,}077.04\underline{2}$

20. $2.8371\underline{0}7$

21. $\$25\underline{3}.72$

22. $\$7{,}571.3\underline{9}$

Write each decimal in words.

23. 0.53

24. 0.72

25. 0.305

26. 0.849

27. 0.6

28. 0.3

29. 5.72

30. 3.89

31. 24.002

32. 370.081

Write each number in decimal notation.

33. Eight tenths

34. Six tenths

35. One and forty-one thousandths

36. Eighteen and four thousandths

37. Sixty and one hundredth

38. Ninety-two and seven hundredths

39. Four and one hundred seven thousandths

40. Five and sixty-three thousandths

41. Three and two tenths meters

42. Ninety-eight and six tenths degrees

B *For each decimal, find the equivalent fraction or mixed number, written in lowest terms.*

43. 0.6

44. 0.8

45. 0.39

46. 0.27

47. 1.5

48. 9.8

49. 8.000

50. 6.700

51. 5.012

52. 20.304

C *Between each pair of numbers, insert the appropriate sign, <, =, or >, to make a true statement.*

53. 3.21 2.5

54. 8.66 4.952

55. 0.71 0.8

56. 1.2 1.38

57. 9.123 9.11

58. 0.72 0.7

59. 4 4.000

60. 7.60 7.6

61. 8.125 feet 8.2 feet

62. 2.45 pounds 2.5 pounds

Rearrange each group of numbers from smallest to largest.

63. 7.1, 7, 7.07

64. 0.002, 0.2, 0.02

65. 5.001, 4.9, 5.2

66. 3.85, 3.911, 2

67. 9.6 miles, 9.1 miles, 9.38 miles

68. 2.7 seconds, 2.15 seconds, 2 seconds

D *Round as indicated.*

69. 17.36 to the nearest tenth

70. 8.009 to two decimal places

71. 3.5905 to the nearest thousandth

72. 3.5902 to the nearest thousandth

73. 37.08 to one decimal place

74. 3.08 to one decimal place

75. 0.396 to the nearest hundredth

76. 0.978 to the nearest hundredth

77. 7.0571 to two decimal places

78. 3.038 to one decimal place

79. 8.7 miles to the nearest mile

80. $3.57 to the nearest dollar

Round to the indicated place.

81.

To the Nearest	8.0714	0.9916
Tenth		
Hundredth		
Ten		

82.

To the Nearest	0.8166	72.3591
Tenth		
Hundredth		
Ten		

Mixed Practice

Solve.

83. In the decimal 0.024, underline the digit in the tenths place.

84. What is the equivalent fraction of 3.8?

85. Round 870.062 to the nearest hundredth.

86. Write four and thirty-one thousandths in decimal notation.

87. Write in increasing order: 2.14 meters, 2.4 meters, and 2.04 meters.

88. Write 0.05 in words.

Applications

The following statements involve decimals. Write all decimals in words.

89. It takes the Earth 23.934 hours to rotate once about its axis. (*Source:* NASA)

90. Male Rufous hummingbirds weigh an average of 0.113 ounce. (*Source:* Lanny Chambers, *Facts about Hummingbirds*)

91. Over two years, the average score on a college admissions exam increased from 18.7 to 18.8.

92. A chemistry text gives 55.85 as the atomic mass of iron and 63.55 as the atomic mass of copper.

93. The following table shows the average amount of time spent on selected daily household activities each year by U.S. civilians.

Activity	Average Amount of Time (in number of hours per year)
Housework	211.7
Lawn and garden care	69.4
Food preparation and cleanup	189.8
Caring for household members	193.5
Household management	47.5

(*Source:* bls.gov)

94. The coefficient of friction is a measure of the amount of friction produced when one surface rubs against another. The following table gives these coefficients for various surfaces.

Materials	Coefficient of Friction
Wood on wood	0.3
Steel on steel	0.15
Steel on wood	0.5
A rubber tire on a dry concrete road	0.7
A rubber tire on a wet concrete road	0.5

(*Source: CRC Handbook of Chemistry and Physics*)

95. Bacteria are single-celled organisms that typically measure from 0.00001 inch to 0.00008 inch across.

96. In one month, the consumer confidence index rose from 71.9 points to 80.2 points.

The following statements involve numbers written in words. Write each number in decimal notation.

97. The area of a plot of land is one and two tenths acres.

98. The lead in many mechanical pencils is seven tenths millimeter thick.

99. At the first Indianapolis 500 auto race in 1911, the winning speed was seventy-four and fifty-nine hundredths miles per hour. (*Source:* Jack Fox, *The Indianapolis 500*)

100. In 1796, there was a U.S. coin in circulation, the half cent, worth five thousandths of a dollar.

101. At sea level, the air pressure on each square inch of surface area is fourteen and seven tenths pounds.

102. A doctor prescribed a dosage of one hundred twenty-five thousandths milligram of Prolixin.

103. According to the owner's manual, the voltage produced by a camcorder battery is nine and six tenths volts (V).

104. In preparing an injection, a nurse measured out one and eight tenths milliliters of sterile water.

105. The electrical usage in a tenant's apartment last month amounted to three hundred fifty-two and one tenth kilowatt hours (kWh).

106. In one day, the Dow Jones Industrial Average fell by three and sixty-three hundredths points.

Solve.

107. The following table shows the three medalists in the men's skating short program at the 2010 Winter Olympics in Vancouver, Canada.

Country	Skater	Score
United States	Evan Lysacek	90.3
Japan	Daisuke Takahashi	90.25
Russian Federation	Evgeni Plushenko	90.85

(*Source:* sports.yahoo.com/olympics/vancouver)

Which of these three top skaters earned the highest score for the short program?

108. The more powerful an earthquake is, the higher its magnitude is on the Richter scale. Great earthquakes, such as the 1906 San Francisco earthquake, have magnitudes of 8.0 or higher. Is an earthquake with magnitude 7.8 considered to be a great earthquake? (*Source: The New Encyclopedia Britannica*)

109. Last winter, a homeowner's average daily heating bill was for 8.75 units of electricity. This winter, it was for 8.5 units. During which winter was the average higher?

110. In order to qualify for the dean's list at a community college, a student's grade point average (GPA) must be 3.5 or above. Did a student with a GPA of 3.475 make the dean's list?

111. The following table shows estimates of the lead emissions in the United States for two given years.

Year	Amount of Lead (in millions of tons)
1985	0.022
2005	0.003

(*Source:* Environmental Protection Agency)

In which year was the amount of lead emissions less?

112. As part of her annual checkup, a patient had a blood test. The normal range for a particular substance is 1.1 to 2.3. If she scored 0.95, was her blood in the normal range?

113. The following table shows the amount of money that a jury awarded a husband and wife who were plaintiffs in a lawsuit.

Plaintiff	Award (in millions of dollars)
Husband	1.875
Wife	1.91

Whose award was less than the $1.9 million that each plaintiff had demanded?

114. The table below shows the amount of toxic emissions released into the air from three factories during the same time in a recent year. Which of the factories released the most toxic emissions?

	Electronics Factory	Food Factory	Chemical Factory
Toxic emissions (in millions of pounds)	1.5	1.4	1.48

Round to the indicated place.

115. A bank pays interest on all its accounts to the nearest cent. If the interest on an account is $57.0285, how much interest does the bank pay?

116. A city's sales tax rate, expressed as a decimal, is 0.0825. What is this rate to the nearest hundredth?

117. According to the organizers of a lottery, the probability of winning the lottery is 0.0008. Round this probability to three decimal places.

118. One day last week, a particular foreign currency was worth $0.7574. How much is this currency worth to the nearest tenth of a dollar?

119. In terms of land area, North America is 1.36 times as large as South America. Round this decimal to the nearest tenth. (*Source: National Geographic Atlas of the World*)

120. The length of the Panama Canal is 50.7 miles. Round this length to the nearest mile. (*Source: The New Encyclopedia Britannica*)

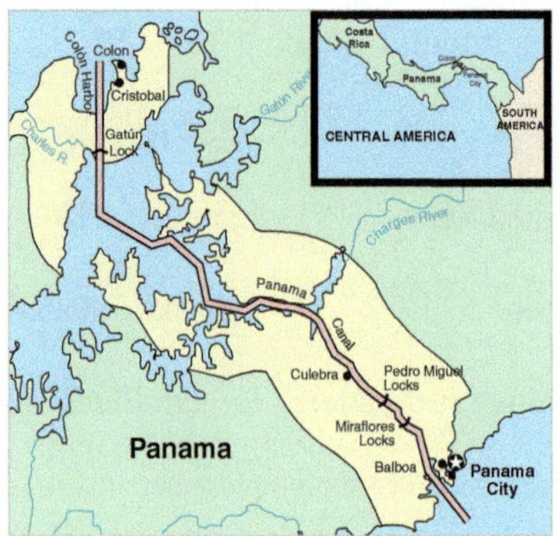

• Check your answers on page A-4.

MINDStretchers

Critical Thinking

1. For each question, either give the answer or explain why there is none.

a. Find the *smallest* decimal that when rounded to the nearest tenth is 7.5.

b. Find the *largest* decimal that when rounded to the nearest tenth is 7.5.

Writing

2. The next whole number after 7 is 8. What is the next decimal after 0.7? Explain.

Groupwork

3. Working with a partner, list fifteen numbers between 2.5 and 2.6.

Cultural Note

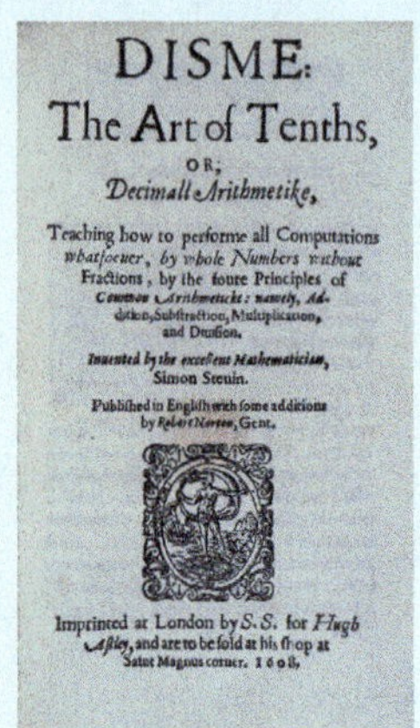

DISME:
The Art of Tenths,
OR,
Decimall Arithmetike,
Teaching how to performe all Computations *whatsoever*, *by whole Numbers without* Fractions, by the foure Principles of *Common Arithmeticke*: *namely*, Addition, Substraction, Multiplication, and Diuision.
Inuented by the excellent Mathematician, Simon Steuin.
Published in English with some additions by *Robert Norton*, Gent.
Imprinted at London by *S. S.* for *Hugh Astley*, and are to be sold at his shop at Saint Magnus corner. 1608.

In 1585, Simon Stevin, a Dutch engineer, published a book entitled *The Art of Tenths* (*La Disme in French*) in which he presented a thorough account of decimals. Stevin sought to teach everyone "with an ease unheard of, all computations necessary between men by integers without fractions."

Stevin did not invent decimals; their history dates back thousands of years to ancient China, medieval Arabia, and Renaissance Europe. However, Stevin's writings popularized decimals and also supported the notion of decimal coinage—as in American currency, where there are 10 dimes to the dollar.

Source: Morris Kline, *Mathematics, a Cultural Approach* (Reading, Massachusetts: Addison-Wesley Publishing Company, 1962), p. 614.

3.2 Adding and Subtracting Decimals

OBJECTIVES

- **A** To add or subtract decimals
- **B** To solve applied problems involving the addition or subtraction of decimals

In Section 3.1 we discussed the meaning of decimals and how to compare and round them. Now, we turn our attention to computing with decimals, starting with addition and subtraction.

Adding Decimals

Adding decimals is similar to adding whole numbers: We add the digits in each place value position, regrouping when necessary. Suppose that we want to find the sum of two decimals: 1.2 + 3.5. First, we rewrite the problem vertically, lining up the decimal points in the addends. Then, we add as usual, inserting the decimal point below the other decimal points.

$$\begin{array}{r} 1.2 \\ +3.5 \\ \hline 4.7 \end{array} \quad \text{This addition is equivalent to} \quad \begin{array}{r} 1\frac{2}{10} \\ +3\frac{5}{10} \\ \hline 4\frac{7}{10} \end{array}$$

Note that when we added the mixed numbers corresponding to the decimals, we got $4\frac{7}{10}$, which is equivalent to 4.7. This example suggests the following rule:

To Add Decimals

- Rewrite the numbers vertically, lining up the decimal points.
- Add.
- Insert a decimal point in the sum below the other decimal points.

EXAMPLE 1

Find the sum: 2.7 + 80.13 + 5.036

Solution

$$\begin{array}{r} 2.7 \\ 80.13 \\ +5.036 \\ \hline 87.866 \end{array}$$

Rewrite the addends with decimal points lined up vertically.

Add.

Insert the decimal point in the sum.

PRACTICE 1

Add: 5.12 + 4.967 + 0.3

EXAMPLE 2

Compute: 2.367 + 5 + 0.143

Solution Recall that 5 and 5. are equivalent.

$$\begin{array}{r} 2.367 \\ 5. \\ +0.143 \\ \hline 7.510 \end{array} = 7.51$$

Line up the decimal points and add.

Insert the decimal point in the sum.

We can drop the extra 0 at the right end.

PRACTICE 2

What is the sum of 7.31, 8, and 23.99?

EXAMPLE 3

A runner's time was 0.06 second longer than the world record of 21.71 seconds. What was the runner's time?

Solution We need to compute the sum of the two numbers. The runner's time was 21.77 seconds.

$$\begin{array}{r} 0.06 \\ +21.71 \\ \hline 21.77 \end{array}$$

PRACTICE 3

A child has the flu. This morning, his body temperature was 99.4°F. What was his temperature after it went up by 2.7°?

Subtracting Decimals

Now, let's discuss subtracting decimals. As with addition, subtracting decimals is similar to subtracting whole numbers. To compute the difference between 12.83 and 4.2, we rewrite the problem vertically, lining up the decimal points. Then, we subtract as usual, inserting a decimal point below the other decimal points.

$$\begin{array}{r} \downarrow \\ 12.83 \\ -4.2 \\ \hline 8.63 \\ \uparrow \end{array} \quad \text{is equivalent to} \quad \begin{array}{rcr} 12\frac{83}{100} & = & 12\frac{83}{100} \\ -4\frac{2}{10} & = & -4\frac{20}{100} \\ \hline & & 8\frac{63}{100}, \end{array} \quad \text{or} \quad 8.63$$

Again, note that when we subtracted the equivalent mixed numbers, we got the same difference.

As in any subtraction problem, we can check this answer by adding the subtrahend (4.2) to the difference (8.63), confirming that we get the original minuend (12.83). This example suggests the following rule:

$$\begin{array}{r} 8.63 \\ +4.2 \\ \hline 12.83 \end{array}$$

To Subtract Decimals

- Rewrite the numbers vertically, lining up the decimal points.
- Subtract, inserting extra zeros in the minuend if necessary for regrouping.
- Insert a decimal point in the difference, below the other decimal points.

EXAMPLE 4

Subtract and check: 5.038 − 2.11

Solution

$$\begin{array}{r} \downarrow \\ 5.038 \\ -2.11 \\ \hline 2.928 \\ \uparrow \end{array}$$

Rewrite the problem with decimal points lined up vertically.

Subtract. Regroup when necessary.

Insert the decimal point in the answer.

Check To verify that our difference is correct, we check by addition.

$$\begin{array}{r} 2.928 \\ +2.11 \\ \hline 5.038 \end{array}$$

PRACTICE 4

Find the difference and check: 71.3825 − 25.17

EXAMPLE 5

65 is how much larger than 2.04?

Solution Recall that 65 and 65. are equivalent.

Insert the zeros needed for regrouping.

$$\begin{array}{r} 65.00 \\ -2.04 \\ \hline 62.96 \end{array}$$

Line up the decimal points.
Subtract.
Insert the decimal point in the answer.

Check

$$\begin{array}{r} 62.96 \\ +\ 2.04 \\ \hline 65.00 \end{array} = 65$$

PRACTICE 5

How much greater is \$735 than \$249.57?

EXAMPLE 6

A Burger King Whopper Jr.® contains 1.02 grams of sodium, whereas a McDonald's Quarter Pounder ® contains 0.73 grams. How much more sodium does the Whopper Jr.® contain? (*Sources:* nutrition.mcdonalds.com and bk.com)

Solution We need to find the difference between 1.02 and 0.73.

$$\begin{array}{r} 1.02 \\ -0.73 \\ \hline 0.29 \end{array}$$

So a Burger King Whopper Jr.® contains 0.29 grams more sodium.

PRACTICE 6

The New York Marathon has timing mats located every 3.1 miles throughout its 26.2-mile course. A runner passes one of the mats 9.3 miles into the race. How much further does the runner have to go to finish the race? (*Source*: nycmarathon.org)

EXAMPLE 7

Suppose that a part-time employee's salary is \$350 a week, less deductions. The following table shows these deductions.

Deduction	Amount
Federal, state, and city taxes	\$100.80
Social Security	26.78
Union dues	8.88

What is the employee's take-home pay?

Solution Let's use the strategy of breaking the question into two simpler questions.

- *How much money is deducted per week?* The weekly deductions (\$100.80, \$26.78, and \$8.88) add up to \$136.46.
- *How much of the salary is left after subtracting the total deductions?* The difference between \$350 and \$136.46 is \$213.54, which is the employee's take-home pay.

PRACTICE 7

A sales rep, working in Ohio, wants to drive from Circleville to Columbus. How much shorter is it to drive directly to Columbus instead of going by way of Lancaster? (*Source:* mapquest.com)

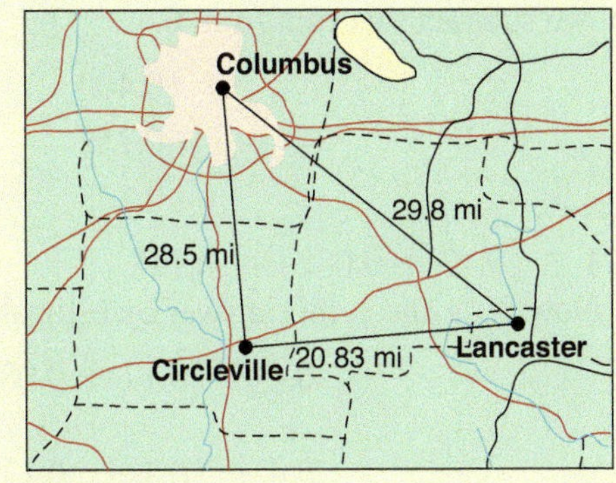

Estimating Sums and Differences

Being able to estimate in your head the sum or difference between two decimals is a useful skill, for either checking or approximating an exact answer. To estimate, simply round the numbers to be added or subtracted and then carry out the operation on the rounded numbers.

EXAMPLE 8

Compute the sum 0.17 + 0.4 + 0.083. Use estimation to check.

Solution First, we add. Then, to check, we round the addends—say, to the nearest tenth—and add the rounded numbers.

$$\begin{array}{rcr} 0.17 & \approx & 0.2 \\ 0.4 & \approx & 0.4 \\ +0.083 & \approx & +0.1 \\ \hline \text{Exact sum} \rightarrow 0.653 & & 0.7 \leftarrow \text{Estimated sum} \end{array}$$

Our exact sum is close to our estimated sum, and in fact, rounds to it.

PRACTICE 8

Add 0.093, 0.008, and 0.762. Then, check by estimating.

EXAMPLE 9

Subtract 0.713 − 0.082. Then check by estimating.

Solution First, we find the exact answer and then round the given numbers to get an estimate.

$$\begin{array}{rcr} 0.713 & \approx & 0.7 \\ -0.082 & \approx & -0.1 \\ \hline \text{Exact difference} \rightarrow 0.631 & & 0.6 \leftarrow \text{Estimated difference} \end{array}$$

Our exact answer, 0.631, is close to 0.6.

PRACTICE 9

Compute: 0.17 − 0.091. Use estimation to check.

EXAMPLE 10

Combine and check: 0.4 − (0.17 + 0.082)

Solution Following the order of operations rule, we begin by adding the two decimals in parentheses.

$$\begin{array}{r} 0.17 \\ +0.082 \\ \hline 0.252 \end{array}$$

Next, we subtract this sum from 0.4.

$$\begin{array}{r} 0.400 \\ -0.252 \\ \hline 0.148 \end{array}$$

So 0.4 − (0.17 + 0.082) = 0.148.

Now, let's check this answer by estimating:

$$\begin{array}{l} 0.4 - (0.17 + 0.082) \\ \downarrow \quad\quad \downarrow \quad\quad \downarrow \\ 0.4 - (0.2 + 0.1) = 0.4 - 0.3 = 0.1 \end{array}$$

The estimate, 0.1, is close to 0.148.

PRACTICE 10

Calculate and check:
0.813 − (0.29 − 0.0514)

In the following examples, we estimate a sum or difference to approximate the correct answer, not to check it.

EXAMPLE 11

A movie budgeted at \$7.25 million ended up costing \$1.655 million more. Estimate the final cost of the movie.

Solution Let's round each number to the nearest million dollars.

$$\begin{array}{lcr} 1.655 & \approx & 2 \text{ million} \\ 7.25 & \approx & +7 \text{ million} \\ \hline & & 9 \text{ million} \end{array}$$

Adding the rounded numbers, we see that the movie cost approximately \$9 million.

PRACTICE 11

From the deposit ticket shown below, estimate the total amount deposited.

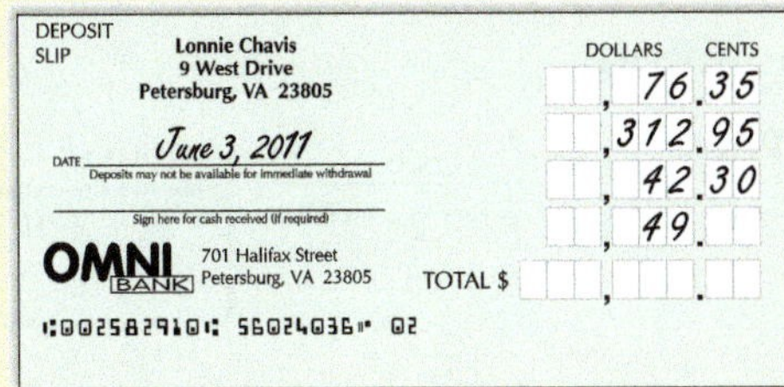

DEPOSIT SLIP

Lonnie Chavis
9 West Drive
Petersburg, VA 23805

DATE June 3, 2011
Deposits may not be available for immediate withdrawal

Sign here for cash received (if required)

OMNI BANK 701 Halifax Street Petersburg, VA 23805

DOLLARS	CENTS
76	35
312	95
42	30
49	
TOTAL $	

Estimate: ____________________

EXAMPLE 12

When the underwater tunnel connecting the United Kingdom and France was built, French and British construction workers dug from their respective countries. They met at the point shown on the map.

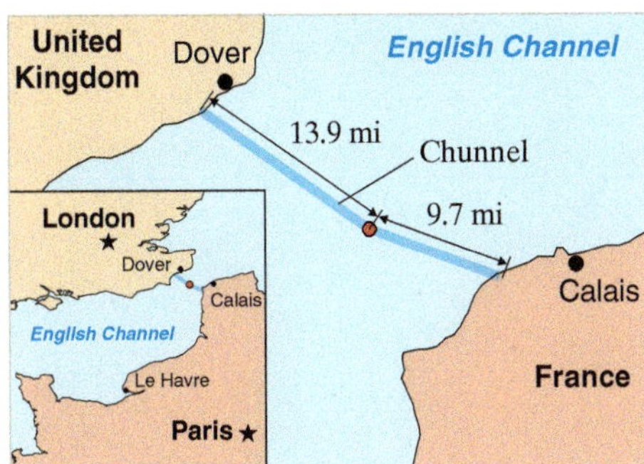

Estimate how much farther the British workers had dug than the French workers. (*Source: The New York Times*)

Solution We can round 13.9 to 14 and 9.7 to 10. The difference between 14 and 10 is 4, so the British workers dug about 4 miles farther than the French workers.

PRACTICE 12

An art collector bought a painting for \$2.3 million. A year later, she sold the painting for \$4.1 million. Estimate her profit on the sale.

Adding and Subtracting Decimals on a Calculator

When adding or subtracting decimals, press the ⊡ key to enter the decimal point. If a sum or difference ends with a 0 in the rightmost decimal place, does your calculator drop the 0? If a sum or difference has no whole-number part, does your calculator insert a 0?

EXAMPLE 13

Compute: $2.7 + 4.1 + 9.2$

Solution

Press	Display
2.7 [+] 4.1 [+] 9.2 [ENTER]	2.7 + 4.1 + 9.2 16.

We can check this sum by estimating: $3 + 4 + 9 = 16$, which is the same as the answer calculated.

PRACTICE 13

Find the sum: $3.82 + 9.17 + 66.24$

EXAMPLE 14

Find the difference: $83.71 - 83.70002$

Solution

Press	Display
83.71 [−] 83.70002 [ENTER]	83.71 - 83.70002 0.00998

We can check this difference by adding: $0.00998 + 83.70002 = 83.71$.

PRACTICE 14

Compute: $5.00003 - 5.00001$

3.2 Exercises

FOR EXTRA HELP MyMathLab  PRACTICE  WATCH  READ 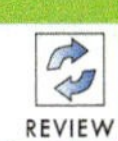 REVIEW

Mathematically Speaking

Fill in each blank with the most appropriate term or phrase from the given list.

sum	decimal points	difference
any number	rightmost digits	zeros

1. When adding decimals, rewrite the numbers vertically, lining up the ______________.

2. Inserting _______ at the right end of a decimal does not change its value.

3. To estimate the _______ of 0.31 and 0.108, add 0.3 and 0.1.

4. To estimate the ___________ between 0.31 and 0.108, subtract 0.1 from 0.3.

A *Find the sum. Check by estimating.*

5. 3.89 + 5.44
6. 2.17 + 4.29
7. 0.6 + 0.3
8. 0.2 + 0.6
9. 6.03 + 2.1
10. 1.4 + 3.96
11. 13.05 + 8.4
12. 21.07 + 5.1
13. 2.67 + 5
14. 8 + 4.99
15. \$74 + \$3.21
16. \$8.77 + \$62
17. 0.49023 + 0.5997
18. 1.002 + 0.20013
19. 8.01 + 6.7 + 9.45
20. 9.73 + 5.99 + 3.688
21. 34.7 + 5.84 + 3 + 0.882
22. 75.285 + 2 + 3.871 + 0.5
23. 7 millimeters + 3.5 millimeters + 9.82 millimeters
24. 10.35 inches + 32 inches + 54.9 inches
25. 4.7 kilograms + 2.98 kilograms + 9.002 kilograms
26. 0.85 second + 1.72 seconds + 3.009 seconds
27. 3.861 + 2.89 + 3.775 + 9.00813 + 3.77182
28. \$8.99 + \$3.99 + \$17.83 + \$15 + \$201.75

Find the difference. Check either by estimating or by adding.

29. 0.8 − 0.1
30. 0.9 − 0.3
31. 20.72 − 3.92
32. 12.98 − 5.73
33. 23.81 − 5.4
34. 17.49 − 10.2
35. 80.2 − 4.57
36. 97.1 − 3.23
37. 25.99 − 3.666
38. 32.99 − 7.555
39. 0.27 − 0.1
40. 0.29 − 0.2
41. 1.032 − 0.9178
42. 0.01 − 0.0001
43. 13.2 − 7
44. 9.6 − 4
45. 20 − 4.63
46. 8 − 2.55
47. 10 − 4.1
48. 13 − 7.2
49. 8 − 1.79
50. 9 − 4.63
51. 3.2 pounds − 1.35 pounds
52. 23.5 seconds − 2.8 seconds
53. 103.7°F − 98.8°F
54. 32.5 grams − 19.27 grams

Compute.

55. $35.2 - 2.86 + 9.07 - 1.658$

56. $10 - 2.38 + 4.92 - 6.02$

57. 30 milligrams − 0.5 milligram − 1.6 milligrams

58. \$20.93 + \$1.07 − \$19.58

59. $5.21 - (1.03 + 0.975)$

60. $6.953 - (4.09 + 0.008)$

61. $41.075 - 2.87104 - 17.005$

62. $0.00661 + 1.997 - 0.05321$

In each group of three computations, one answer is wrong. Use estimation to identify which answer is incorrect.

63. a. $\begin{array}{r} 0.059 \\ 0.00234 \\ +0.036 \\ \hline 0.09734 \end{array}$ **b.** $\begin{array}{r} 0.1903 \\ 0.074 \\ +0.2051 \\ \hline 0.4694 \end{array}$ **c.** $\begin{array}{r} 0.00441 \\ 0.06882 \\ +0.0103 \\ \hline 0.8353 \end{array}$

64. a. $\begin{array}{r} \$32.71 \\ 43.09 \\ +\ 8.27 \\ \hline \$74.07 \end{array}$ **b.** $\begin{array}{r} \$19.37 \\ 2. \\ +\ 7.22 \\ \hline \$28.59 \end{array}$ **c.** $\begin{array}{r} \$139.26 \\ 82.87 \\ +\ 3.01 \\ \hline \$225.14 \end{array}$

65. a. $\begin{array}{r} 0.35 \\ -0.1007 \\ \hline 0.2493 \end{array}$ **b.** $\begin{array}{r} 0.072 \\ -0.0056 \\ \hline 0.664 \end{array}$ **c.** $\begin{array}{r} 0.03 \\ -0.008 \\ \hline 0.022 \end{array}$

66. a. $\begin{array}{r} 8.551 \\ -2.9995 \\ \hline 5.5515 \end{array}$ **b.** $\begin{array}{r} 78.328 \\ -\ 5.5 \\ \hline 7.2828 \end{array}$ **c.** $\begin{array}{r} 65 \\ -\ 2.778 \\ \hline 62.222 \end{array}$

Mixed Practice

Solve.

67. Calculate: $4.78 + 13 - 10.009$

68. Find the difference between 90.1 and 12.58.

69. Add: 0.5 pound + 3 pounds + 4.25 pounds

70. Subtract: \$20 − \$6.95

71. Compute: $8 - 2.4 + 6.0013$

72. What is the sum of 1.265, 7, and 0.14?

Applications

B *Solve and check.*

73. A paperback book that normally sells for \$13 is now on sale for \$11.97. What is the discount in dollars and cents?

74. During a drought, the mayor of a city attempted to reduce daily water consumption to 3.1 million gallons. If daily water consumption fell to 1.948 million gallons above that goal, estimate the city's consumption.

75. A skeleton was found at an archaeological dig. Radiocarbon dating—a technique used for estimating age—indicated that the skeleton was 56 centuries old, plus or minus 0.8 centuries. According to this estimate, what is the greatest possible age of the skeleton?

76. A college launched a campaign to collect \$3 million to build a new technology complex. If \$1.316 million has been collected so far, how much more money, to the nearest million dollars, is needed?

77. As an investment, a couple bought an apartment house for \$2.3 million. Two years later, they sold the apartment house for \$4 million. What was their profit?

78. A woman sued her business partner and was awarded \$1.5 million. On appeal, however, the award was reduced to \$0.75 million. By how much was the award reduced?

79. In setting up a page in word processing, the margins of a page are usually expressed in decimal parts of an inch. How long is each typed line on the page?

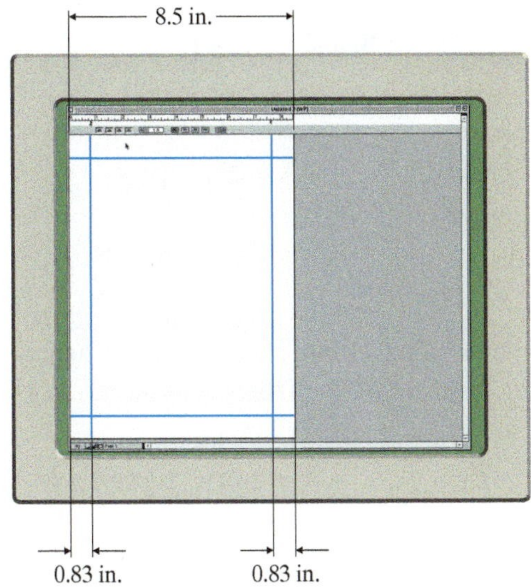

80. In the picture below, how much clearance will there be between the top of the roof cargo carrier and the top of the garage door?

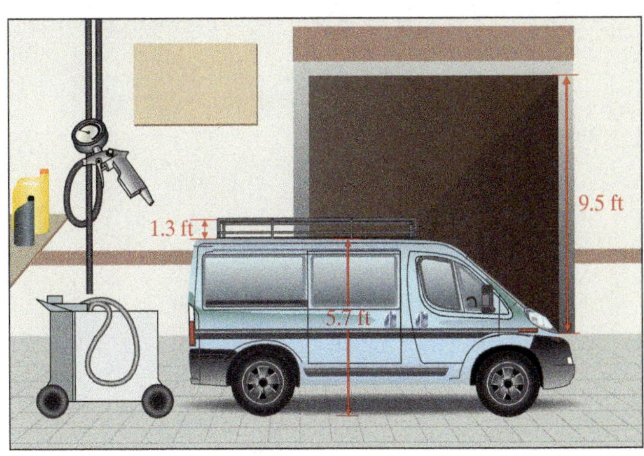

81. A radio disc jockey wants to choose among compact disc tracks that last 3.5, 2.8, 2.9, 2.6, and 1.6 minutes. Can he select tracks so as to get between 9.8 and 10 minutes of music? Explain.

82. A shopper plans to buy three items that cost \$4.99, \$7.99, and \$2.99 each. If she has \$15 with her, will she have enough money to pay for all three items? Explain.

83. When gymnasts compete, they receive scores in four separate events: vault (VT), uneven bar (UB), balance beam (BB), and floor exercises (FX). The total of these four event scores is called the all-around score (AA). The following chart shows the qualifying scores earned by three gymnasts at the Beijing, China 2008 Summer Olympics.

Gymnast	VT	UB	BB	FX	AA
Nastia Liukin (U.S.)	15.1	15.95	15.975	15.35	
Yang Yilin (China)	15.2	16.65	15.5	15	
Shawn Johnson (U.S.)	16	15.325	15.975	15.425	

(*Source*: wikipedia.org)

a. Calculate the all-around scores for these three competitors.

b. Which competitor had the highest all-around score?

84. The graph shown gives the number of households that watched particular TV programs in a given week, according to the Nielsen Top 20 ratings.

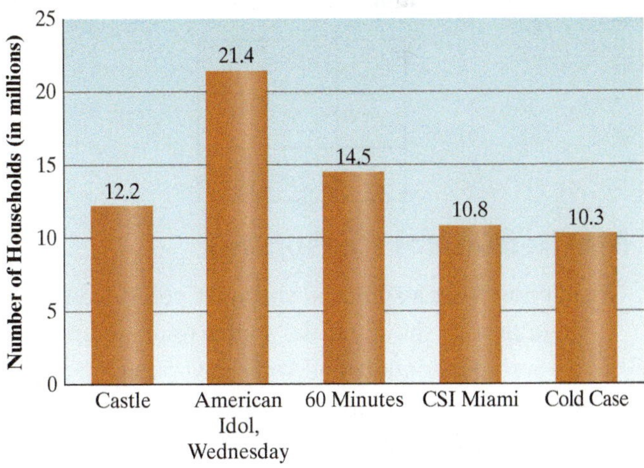

(*Source:* usatoday.com)

a. How many more households watched *American Idol-Wednesday* than *60 Minutes*?

b. Write the names of the five programs in increasing order of viewership.

85. To prevent anemia, a doctor advises his patient to take at least 18 milligrams of iron each day. The following table shows the amount of iron in the food that the patient ate yesterday. Did she get enough iron? If not, how much more does she need?

Food	Iron (in milligrams)
Chicken, breast	1.1
Canned tuna	0.8
Fortified oatmeal	10
Seedless raisins	1.5
Frozen spinach	1.9
Whole wheat bread	0.9

(*Source:* ods.od.nih.gov)

86. When filling a prescription, buying a generic drug rather than a brand-name drug can often save money. The following table shows the prices of various brand-name and generic drugs.

Drug	Brand-Named Price	Generic Price
Fexofenadine	\$30	\$8.80
Furosemide	\$35.35	\$17.94
Omeprazole	\$48.47	\$37.44
Metformin	\$186.89	\$78.39
Synthroid	\$61.56	\$38.02

How much money will a shopper save if he buys all five generic drugs rather than the brand-name drugs?

• Check your answers on page A-4.

MINDStretchers

Groupwork

1. Working with a partner, find the missing entries in the following magic square, in which 3.75 is the sum of every row, column, and diagonal.

0.75	1.25	
2.		

Mathematical Reasoning

2. Suppose that a spider is sitting at point *A* on the rectangular web shown. If the spider wants to crawl along the web horizontally and vertically to munch on the delicious fly caught at point *B*, how long is the shortest route that the spider can take?

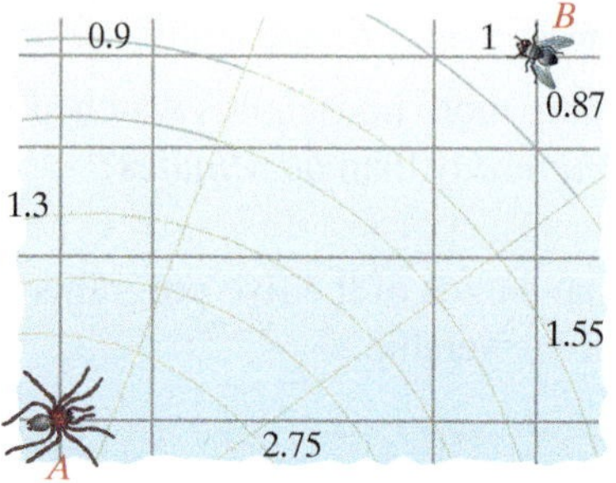

Writing

3. a. How many pairs of whole numbers are there whose sum is 7?

b. How many pairs of decimals are there whose sum is 0.7?

c. Explain why (a) and (b) have different answers.

3.3 Multiplying Decimals

OBJECTIVES

- **A** To multiply decimals
- **B** To solve applied problems involving the multiplication of decimals

In this section, we discuss how to multiply two or more decimals, finding both the exact product and an estimated product.

Multiplying Decimals

To find the product of two decimals—say, 1.02 and 0.3—we multiply the same way we multiply whole numbers. But with decimals we need to know where the decimal point goes in the product. To find out, let's change each decimal to its fractional equivalent.

$$\begin{array}{r} 1.02 \\ \times\ 0.3 \\ \hline 306 \end{array} \quad \text{is equivalent to} \rightarrow \quad 1\frac{2}{100} \times \frac{3}{10}$$

$$= \frac{102}{100} \times \frac{3}{10} = \frac{306}{1{,}000}, \quad \text{or} \quad 0.306$$

Where should we place the decimal point?

The product is in thousandths, so it has three decimal places.

Looking at the multiplication problem with decimals, note that *the product has as many decimal places as the total number of decimal places in the factors*. This example illustrates the following rule for multiplying decimals.

$$\begin{array}{r} 1.0\,2 \\ 0.3 \\ \hline 0.3\,0\,6 \end{array}$$

To Multiply Decimals

- Multiply the factors as if they were whole numbers.
- Find the total number of decimal places in the factors.
- Count that many places from the right end of the product, and insert a decimal point.

EXAMPLE 1

Multiply: 6.1 × 3.7

Solution First, multiply 61 by 37, ignoring the decimal points.

$$\begin{array}{r} 6.1 \\ \times\ 3.7 \\ \hline 4\,2\,7 \\ 1\,8\,3 \\ \hline 2\,2\,5\,7 \end{array}$$

Then, count the total number of decimal places in the factors.

$$\begin{array}{r} 6.1 \leftarrow \text{One decimal place} \\ \times\ 3.7 \leftarrow \text{One decimal place} \\ \hline 4\,2\,7 \\ 1\,8\,3 \\ \hline 2\,2.5\,7 \leftarrow \text{Two decimal places in the product} \end{array}$$

Insert the decimal point two places from the right end. So, 22.57 is the product.

PRACTICE 1

Find the product: 2.81 × 3.5

EXAMPLE 2

Find the product of 0.75 and 4.

Solution Let's multiply 0.75 by 4, ignoring the decimal point.

$$\begin{array}{r} 0.75 \\ \times \quad 4 \\ \hline 300 \end{array}$$

Count the total number of decimal places.

$$\begin{array}{rl} 0.75 & \leftarrow \text{Two decimal places} \\ \times \quad 4 & \leftarrow \text{Zero decimal places (4 is a whole number)} \\ \hline 3.00 & \leftarrow \text{Two decimal places in the product} \end{array}$$

So the product is 3.00, which simplifies to 3.

PRACTICE 2

Multiply: 0.28×5

EXAMPLE 3

Multiply 0.03 and 0.25, rounding the answer to the nearest thousandth.

Solution

$$\begin{array}{r} 0.25 \\ \times \ 0.03 \\ \hline 075 \\ 000 \\ \hline 0.0075 \end{array}$$

Rounding to the nearest thousandth, we get 0.008.

PRACTICE 3

What is the product of 0.44 and 0.03, rounded to the nearest hundredth?

EXAMPLE 4

Multiply: $(1.1)(3.5)(0.8)$

Solution To find the product of three factors, we can first multiply the two left factors and then multiply this product by the third factor:

$$\underbrace{(1.1)(3.5)}(0.8) =$$
$$(3.85)(0.8) = 3.08$$

So 3.08 is the final product.

PRACTICE 4

Evaluate: $(0.2)(0.3)(0.4)$

EXAMPLE 5

Simplify: $3 + (1.2)^2$

Solution Recall that, according to the order of operations rule, we first must find $(1.2)^2$ and then add 3.

$$3 + (1.2)^2 =$$
$$3 + 1.44 = 4.44$$

PRACTICE 5

Evaluate: $10 - (0.3)^2$

EXAMPLE 6

Multiply: 8.274×100

Solution

$$\begin{array}{r} 8.274 \leftarrow \text{Three decimal places} \\ \times\ 100 \leftarrow \text{Zero decimal places} \\ \hline 827.400 \leftarrow \text{Three decimal places in the product} \end{array}$$

So the product is 827.400, or 827.4 after we drop the extra zeros.

Note that the second factor (100) is a power of 10 ending in **two** zeros and that the product is identical to the first factor except that the decimal point is moved to the right **two** places.

PRACTICE 6

Compute: $0.325 \times 1{,}000$

Example 6 suggests the following shortcut:

TIP To multiply a decimal by a power of 10, move the decimal point to the right the same number of places as the power of 10 has zeros.

EXAMPLE 7

Find the product: $2.89 \times 1{,}000$

Solution We see that 1,000 is a power of 10 and has three zeros. So to multiply 2.89 by 1,000, we simply move the decimal point in 2.89 to the right three places.

Add a 0 to move three places. $2.89 \times 1{,}000 = 2890.$

So the product is 2,890, with the 0 serving as a placeholder.

PRACTICE 7

Multiply 32.7 by 10,000.

EXAMPLE 8

The popularity of a television show is measured in ratings, where each rating point represents 1,149,000 homes in which the show is watched. After examining the table at the right, answer each question. (*Source:* wikipedia.org)

Show	Rating
1	10.3
2	9.1

a. In how many homes was Show 1 watched?

b. In how many more homes was Show 1 watched than Show 2?

Solution

a. To find the number of homes in which Show 1 was watched, we multiply its rating, 10.3, by 1,149,000, which gives us 11,834,700.

b. To compare the popularity of Show 1 and Show 2, we compute the number of homes in which Show 2 was viewed: $9.1 \times 1{,}149{,}000 = 10{,}455{,}900$. The number of Show 1 homes exceeds the number of Show 2 homes by $11{,}834{,}700 - 10{,}455{,}900$, or 1,378,800 homes.

PRACTICE 8

A chemistry student learns that a molecule is made up of atoms. For instance, the water molecule H_2O consists of two atoms of hydrogen H and one atom of oxygen O. Each of these atoms has an atomic weight.

Atom	Atomic Weight
H	1.008
O	15.999

a. After examining the chart above, compute the weight of the water molecule.

b. Round this weight to the nearest whole number.

Estimating Products

Being able to estimate mentally the product of two decimals is useful for either checking or approximating an exact answer. To estimate, round each factor so that it has only one nonzero digit. Then, multiply the rounded factors.

EXAMPLE 9

Multiply 0.703 by 0.087 and check the answer by estimating.

Solution First, we multiply the factors to find the exact product. Then, we round each factor and multiply them.

$$\begin{array}{rcrl} & 0.703 & \approx & 0.7 \leftarrow \text{Rounded to have one nonzero digit} \\ & \times\ 0.087 & \approx & \times\,0.09 \leftarrow \text{Rounded to have one nonzero digit} \\ \text{Exact product} \rightarrow & 0.061161 & & 0.063 \leftarrow \text{Estimated product} \end{array}$$

We see that the exact product and the estimated product are fairly close.

PRACTICE 9

Find the product of 0.0037×0.092, estimating to check.

EXAMPLE 10

Calculate and check: $(4.061)(0.72) + (0.91)(0.258)$

Solution Following the order of operations rule, we begin by finding the two products.

$$(4.061)(0.72) = 2.92392 \qquad (0.91)(0.258) = 0.23478$$

Then, we add these two products.

$$2.92392 + 0.23478 = 3.1587$$

So $(4.061)(0.72) + (0.91)(0.258) = 3.1587$.

Now, let's check this answer by estimating.

$$\begin{array}{l} (4.061)(0.72) + (0.91)(0.258) \\ \quad\downarrow \qquad \downarrow \qquad\quad \downarrow \qquad \downarrow \\ \quad(4) \quad (0.7) + (0.9) \quad (0.3) = 2.8 + 0.27 = 3.07 \approx 3 \end{array}$$

The estimate 3 is close to 3.1587.

PRACTICE 10

Compute and check:
$(0.488)(9.1) - (3.5)(0.227)$

EXAMPLE 11

The sound waves of an elephant call can travel through both the ground and the air. Through the air, the waves may travel 6.63 miles. If they travel 1.5 times as far through the ground, what is the estimated ground distance? (*Source:* wikipedia.org)

Solution We know that the waves may travel 6.63 miles through the air and 1.5 times as far through the ground. To find the estimated ground distance, we compute this product.

$$\begin{array}{rcr} 6.63 & \approx & 7 \\ \times\,1.5 & \approx & \times\,2 \\ \hline & & 14 \end{array}$$

So the estimated ground distance of the sound waves of an elephant call is about 14 miles.

PRACTICE 11

Earth travels through space at a speed of 18.6 miles per second. Estimate how far Earth travels in 60 seconds. (*Source:* The Diagram Group, *Comparisons*)

Multiplying Decimals on a Calculator

Multiply decimals on a calculator by entering each decimal as you would enter a whole number, but insert a decimal point as needed. If there are too many decimal places in your answer to fit in the display, investigate how your calculator displays the answer.

EXAMPLE 12

Compute 8,278.55 × 0.875, rounding your answer to the nearest hundredth. Then, check the answer by estimating.

Solution

Now, 7,243.73125 rounded to the nearest hundredth is 7,243.73. Checking by estimating, we get 8,000 × 0.9, or 7,200, which is close to our exact answer.

PRACTICE 12

Find the product of 2,471.66 and 0.33, rounding to the nearest tenth. Check the answer.

EXAMPLE 13

Find $(1.9)^2$

Solution

Now, let's check by estimating. Since 1.9 rounded to the nearest whole number is 2, $(1.9)^2$ should be close to 2^2, or 4, which is close to our exact answer, 3.61.

PRACTICE 13

Calculate: $(2.1)^3$

3.3 Exercises

FOR EXTRA HELP MyMathLab MathXL PRACTICE WATCH READ REVIEW

Mathematically Speaking

Fill in each blank with the most appropriate term or phrase from the given list.

add	three	factors	five
first factor	four	multiplication	
square	two	division	

1. The operation understood in the expression (3.4) (8.9) is_____________.

2. When multiplying decimals, the number of decimal places in the product is equal to the total number of decimal places in the _______.

3. To multiply a decimal by 100, move the decimal point _______ places to the right.

4. The product of 0.27 and 8.18 has _______ decimal places.

5. To compute the expression $(8.5)^2 + 2.1$, first _______.

6. To multiply a decimal by 1,000, move the decimal point _______ places to the right.

A *Insert a decimal point in each product. Check by estimating.*

7. $2.356 \times 1.27 = 299212$

8. $97.26 \times 5.3 = 515478$

9. $3{,}144 \times 0.065 = 204360$

10. $837 \times 0.15 = 12555$

11. $71.2 \times 35 = 24920$

12. $0.002 \times 37 = 0074$

13. $0.0019 \times 0.051 = 969$

14. $0.0089 \times 0.0021 = 1869$

15. $2.87 \times 1{,}000 = 287000$

16. $492.31 \times 10 = 492310$

17. $\$4.25 \times 0.173 = \$73525$

18. 11.2 feet $\times 0.75 = 8400$ feet

Find the product. Check by estimating.

19. 0.6×0.9

20. 0.8×0.7

21. 0.5×0.8

22. 0.6×0.8

23. 0.1×0.2

24. 0.9×0.5

25. 0.04×0.07

26. 0.03×0.01

27. 2.55×0.3

28. 8.07×0.6

29. 0.96×2.1

30. 0.87×3.1

31. 38.01×0.2

32. 12.02×0.05

33. 125×0.004

34. 135×0.006

35. 3.8×1.54

36. 9.51×2.7

37. 13.74×11

38. 12.45×11

39. 12.459×0.3

40. 72.558×0.2

41. $(0.675)(2.66)$

42. $(4.003)(0.59)$

43. 83.127×100 **44.** 49.247×100 **45.** $0.0023 \times 10{,}000$ **46.** $0.0135 \times 10{,}000$

47. $(1.5)(0.6)(0.1)$ **48.** $(12)(3.5)(0.2)$ **49.** $(0.03)(1.4)(25)$ **50.** $(2.6)(0.5)(0.9)$

51. $(0.001)^3$ **52.** $(0.1)^4$ **53.** 17 feet $\times$ 2.5 **54.** 15 hours $\times$ 7.5

55. 3.5 miles $\times$ 0.4 **56.** 9.1 meters $\times$ 1,000

57. $\begin{array}{r} 43.87 \\ \times\ 0.975 \\ \hline \end{array}$ **58.** $\begin{array}{r} 18{,}275.33 \\ \times\ \quad 0.39 \\ \hline \end{array}$ **59.** $\begin{array}{r} 99{,}125 \\ \times\ \quad 2.75 \\ \hline \end{array}$ **60.** $\begin{array}{r} 3.512 \\ \times\ 1.47 \\ \hline \end{array}$

Simplify.

61. 0.7×10^2 **62.** 0.6×10^4 **63.** $30 - 2.5 \times 1.7$

64. $18 - 3.4 \times 1.6$ **65.** $1 + (0.3)^2$ **66.** $6 + (1.2)^2$

67. $0.8(1.3 + 2.9) - 0.5$ **68.** $4 - 2.1(3.5 - 1.8)$ **69.** $(5.2 - 3.9)(0.9 + 2.14)$

70. $(8 + 4.5)(8 - 4.5)$ **71.** $0.4(3 - 2.9)(2 + 1.5)$ **72.** $0.5(1 + 0.2)(1 - 0.2)$

Complete each table.

73.

Input	Output
1	$3.8 \times \mathbf{1} - 0.2 =$
2	$3.8 \times \mathbf{2} - 0.2 =$
3	$3.8 \times \mathbf{3} - 0.2 =$
4	$3.8 \times \mathbf{4} - 0.2 =$

74.

Input	Output
1	$7.5 \times \mathbf{1} + 0.4 =$
2	$7.5 \times \mathbf{2} + 0.4 =$
3	$7.5 \times \mathbf{3} + 0.4 =$
4	$7.5 \times \mathbf{4} + 0.4 =$

Each product is rounded to the nearest hundredth. In each group of three products, one is wrong. Use estimation to explain which product is incorrect.

75. **a.** $51.6 \times 0.813 \approx 419.51$ **b.** $2.93 \times 7.283 \approx 21.34$ **c.** $(5.004)^2 \approx 25.04$

76. **a.** $0.004 \times 3.18 \approx 0.01$ **b.** $2.99 \times 0.287 \approx 0.86$ **c.** $(1.985)^3 \approx 10.82$

77. **a.** $4.913 \times 2.18 \approx 10.71$ **b.** $0.023 \times 0.71 \approx 0.16$ **c.** $(8.92)(1.0027) \approx 8.94$

78. **a.** $\$138.28 \times 0.075 \approx \$10.37$ **b.** $0.19 \times \$487.21 \approx \$92.57$ **c.** $0.77 \times \$6{,}005.79 \approx \$462.45$

Mixed Practice

Solve.

79. Simplify: $9 - (0.5)^2$

80. Compute: $2.1 + 5 \times 0.6$

81. Multiply 0.75 and 0.09, rounding the answer to the nearest thousandth.

82. Multiply: (2.3)(4.5)(0.6)

83. Find the product of 0.56 and 8.

84. Find the product: $3.01 \times 1{,}000$

Applications

B *Solve. Check by estimating.*

85. Sound travels at approximately 1,000 feet per second (fps). If a jet is flying at Mach 2.9 (that is, 2.9 times the speed of sound), what is its speed?

86. If insurance premiums of \$323.50 are paid yearly for 10 years for a life insurance policy, how much did the policy holder pay altogether in premiums?

87. The planet in the solar system closest to the Sun is Mercury. The average distance between these two bodies is 57.9 million kilometers. Express this distance in standard form. (*Source:* Jeffrey Bennett et al., *The Cosmic Perspective*)

88. According to the first American census in 1790, the population of the United States was approximately 3.9 million. Write this number in standard form. (*Source: The Statistical History of the United States*)

89. A construction company builds custom-designed swimming pools, including the circular pool shown below. The area of the bottom of this pool is approximately 3.14×9^2 square feet. Find this area to the nearest tenth. (*Source:* Pritchett Construction Co.)

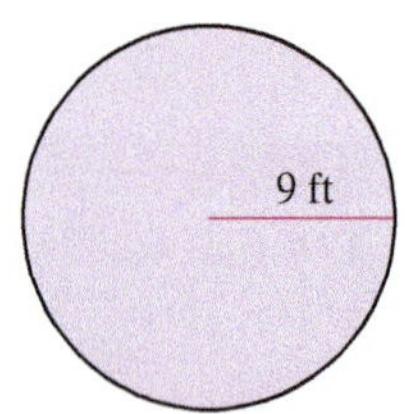

90. Find the area (in square meters) of the floor pictured.

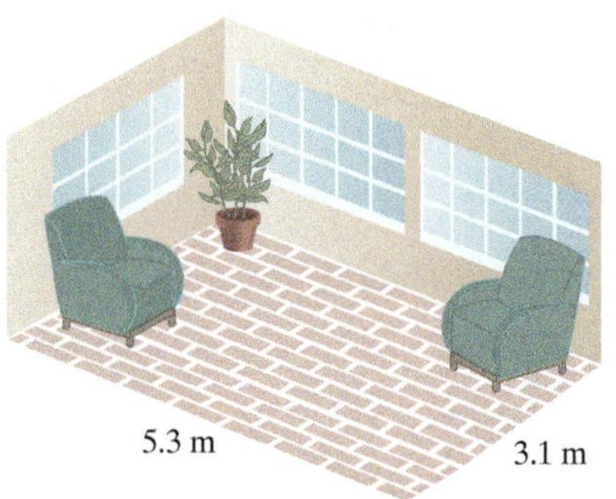

91. Over a 5-day period, a nurse administered 10 tablets to a patient. If each tablet contained 0.125 milligram of the drug Digoxin, how much Digoxin did the nurse administer?

92. Water weighs approximately 62.5 pounds per cubic foot (lb/ft^3). If a bathtub contains about 30 ft^3 of water, how much does the water in the bathtub weigh?

93. A tennis player weighing 180 pounds burns 10.9 calories per minute while playing singles tennis. How many calories would he burn in 2 hours? (*Source:* caloriesperhour.com)

94. A plumber is paid \$37.50 per hour for the first 40 hours worked. She gets time and a half, \$56.25, for any time over her 40-hour week. If she works 49 hours in a week, how much is her pay?

95. The sales receipt for a shopper's purchases is as follows:

Purchase	Quantity	Unit Price	Price
Belt	1	\$11.99	\$__.__
Shirt	3	\$16.95	\$__.__
Total Price			\$__.__

a. Complete the table.

b. If the shopper pays for these purchases with four \$20 bills, how much change should he get?

96. On an electric bill, *usage* is the difference between the meter's *current reading* and the *previous reading* in kilowatt hours (kWh). The *amount due* is the product of the usage and the *rate per kWh*. Find the two missing quantities in the table, rounding to the nearest hundredth.

Previous Reading	**750.07 kWh**
Current Reading	**1,115.14 kWh**
Usage	**kWh**
Rate per kWh	**\$0.10**
Amount Due	**\$**

97. Scientists have discovered a relationship between the length of a person's bones and the person's overall height. For instance, an adult male's height (in inches) can be predicted from the length of his femur bone by using the formula $(1.9 \times \text{femur}) + 32.0$. With this formula, find the height of the German giant Constantine, whose femur measured 29.9 inches. (*Source: Guinness World Records*)

98. In order to buy a \$125,000 house, a couple puts down \$25,000 and takes out a mortgage on the balance. To pay off the mortgage, they pay \$877.57 per month for the following 360 months. How much more will they end up paying for the house than the original price of \$125,000?

• Check your answers on page A-4.

MINDStretchers

Patterns

1. When $(0.001)^{100}$ is multiplied out, how many decimal places will it have?

Mathematical Reasoning

2. Give an example of two decimals

a. whose sum is greater than their product, and

b. whose product is greater than their sum.

Groupwork

3. In the product to the right, each letter stands for a different digit. Working with a partner, identify all the digits.

$$\begin{array}{r} A.B \\ \times\, B.A \\ \hline C\,D \end{array}$$

3.4 Dividing Decimals

OBJECTIVES

- **A** To find the decimal equivalent to a fraction
- **B** To divide decimals
- **C** To solve applied problems involving the division of decimals

In this section, we first consider changing a fraction to its decimal equivalent, which involves both division and decimals. We then move on to our main concern—the division of decimals.

Changing a Fraction to the Equivalent Decimal

Earlier in this chapter, we discussed how to change a decimal to its equivalent fraction. Now let's consider the opposite problem—how to change a fraction to its equivalent decimal.

When the denominator of a fraction is already a power of 10, the problem is simple. For example, the decimal equivalent of $\frac{43}{100}$ is 0.43.

But what about the more difficult problem in which the denominator is *not* a power of 10? A good strategy is to find an equivalent fraction that does have a power of 10 as its denominator. Consider, for instance, the fraction $\frac{3}{4}$. Since 4 is a factor of 100, which is a power of 10, we can easily find an equivalent fraction having a denominator of 100.

$$\frac{3}{4} = \frac{3 \cdot 25}{4 \cdot 25} = \frac{75}{100} = 0.75$$

So 0.75 is the decimal equivalent of $\frac{3}{4}$.

There is a faster way to show that $\frac{3}{4}$ is the same as 0.75, without having to find an equivalent fraction. Because $\frac{3}{4}$ can mean $3 \div 4$, we divide the numerator (3) by the denominator (4). Note that if we continue to divide to the hundredths place, there is no remainder.

```
       0.75     ← Insert the decimal point directly above
                  the decimal point in the dividend.
    4)3.00      ← The decimal point is after the 3.
      0           Insert enough 0's to continue
      —           dividing as far as necessary.
      3 0
      2 8
      ———
        20
        20
        ——
```

So this division also tells us that $\frac{3}{4}$ equals 0.75.

To Change a Fraction to the Equivalent Decimal

- Divide the denominator of the fraction into the numerator, inserting to its right both a decimal point and enough zeros to get an answer either without a remainder or rounded to a given decimal place.
- Place a decimal point in the quotient directly above the decimal point in the dividend.

EXAMPLE 1

Express $\frac{1}{2}$ as a decimal.

Solution To find the decimal equivalent, we divide the fraction's numerator by its denominator.

$$\begin{array}{r} 0.5 \\ 2\overline{)1.0} \\ \underline{0} \\ 1\,0 \\ \underline{1\,0} \end{array}$$

Add a decimal point and a 0 to the right of the 1.

So 0.5 is the decimal equivalent of $\frac{1}{2}$.

Check We verify that the fractional equivalent of 0.5 is $\frac{1}{2}$.

$$0.5 = \frac{5}{10} = \frac{1}{2}$$

The answer checks.

PRACTICE 1

Write the fraction $\frac{3}{8}$ as a decimal.

EXAMPLE 2

Convert $2\frac{3}{5}$ to a decimal.

Solution Let's first change this mixed number to an improper fraction: $2\frac{3}{5} = \frac{13}{5}$. We can then change this improper fraction to a decimal by dividing its numerator by its denominator.

$$\begin{array}{r} 2.6 \\ 5\overline{)13.0} \\ \underline{10} \\ 3\,0 \\ \underline{3\,0} \end{array}$$

So 2.6 is the decimal form of $2\frac{3}{5}$.

Check We convert this answer back from a decimal to its mixed number form.

$$2.6 = 2\frac{6}{10} = 2\frac{3}{5}$$ The answer checks.

PRACTICE 2

Write $7\frac{5}{8}$ as a decimal.

When converting some fractions to decimal notation, we keep getting a remainder as we divide. In this case, we round the answer to a given decimal place.

EXAMPLE 3

Convert $4\frac{8}{9}$ to a decimal, rounded to the nearest hundredth.

Solution First, we change this mixed number to an improper fraction. Then, we convert it to a decimal.

$$4\frac{8}{9} = \frac{44}{9} = 9\overline{)44.000} = 4.888$$

In order to round to the nearest hundredth, we must continue to divide to the thousandths place. So we insert three 0's.

```
    4.888
9)44.000
  36
   8 0
   7 2
     80
     72
      80
      72
```

Finally, we round to the nearest hundredth: $4.888 \approx 4.89$

PRACTICE 3

Express $83\frac{1}{3}$ as a decimal, rounded to the nearest tenth.

In Example 3, note that if instead of rounding we had continued to divide we would have gotten as our answer the **repeating decimal** 4.88888 . . . (also written $4.\overline{8}$). Can you think of any other fraction that is equivalent to a repeating decimal?

Let's now consider some word problems in which we need to convert fractions to decimals.

EXAMPLE 4

Reformulated gasoline (RFG) is a "cleaner" gasoline required to be used in nine major metropolitan areas of the United States with the worst ozone air pollution problem. In a particular week, the average retail regular gas price for RFG areas was $\$2\frac{7}{8}$. Express this amount in dollars and cents, to the nearest cent. (*Source:* epa.gov and eia.doe.gov)

Solution To solve, we must convert the mixed number $2\frac{7}{8}$ to a decimal.

$$2\frac{7}{8} = \frac{23}{8} = 8\overline{)23.000} = 2.875 \approx 2.88$$

```
   2.875
8)23.000
  16
   70
   64
    60
    56
     40
     40
```

So the average price for RFG was \$2.88, to the nearest cent.

PRACTICE 4

About $\frac{13}{125}$ of the world's land surface is covered with ice. Express this fraction as a decimal, rounded to the nearest tenth. (*Source:* enotes.com)

Dividing Decimals

Before we turn our attention to dividing one decimal by another, let's consider simpler problems in which we are dividing a decimal by a whole number. An example of such a problem is $0.6 \div 2$. We can write this expression as the fraction $\frac{0.6}{2}$, which can be rewritten as the quotient of two whole numbers by multiplying the numerator and denominator by 10 as follows:

$$\frac{0.6}{2} = \frac{0.6 \times 10}{2 \times 10} = \frac{6}{20}$$

We then convert the fraction $\frac{6}{20}$ to the equivalent decimal, as we have previously discussed.

$$\begin{array}{r} 0.3 \\ 20\overline{)6.0} \\ \underline{0} \\ 6\,0 \\ \underline{6\,0} \end{array}$$

So $\frac{0.6}{2} = 0.3$

Note that we get the same quotient if we divide the number in the original problem as follows:

$$\begin{array}{r} 0.3 \leftarrow \textbf{Quotient} \\ \textbf{Divisor} \rightarrow 2\overline{)0.6} \leftarrow \textbf{Dividend} \\ \underline{0} \\ 6 \\ \underline{6} \end{array}$$

It is important to write the decimal point in the quotient directly above the decimal point in the dividend.

Next, let's consider a division problem in which we are dividing one decimal by another: $0.006 \div 0.02$. Writing this expression as a fraction, we get $\frac{0.006}{0.02}$. Since we have already discussed how to divide a decimal by a whole number, the goal here is to find a fraction equivalent to $\frac{0.006}{0.02}$ where the denominator is a whole number. Multiplying the numerator and denominator by 100 will do just that.

$$0.006 \div 0.02 = \frac{0.006}{0.02} = \frac{0.006 \times 100}{0.02 \times 100} = \frac{0.6}{2}$$

We know from the previous problem that $\frac{0.6}{2} = 0.3$. Since $\frac{0.006}{0.02} = \frac{0.6}{2}$, we conclude $\frac{0.006}{0.02} = 0.3$.

A shortcut to multiplying by 100 in both the divisor and the dividend is to move the decimal point two places to the right.

$$\text{So } 0.02\overline{)0.006} \text{ (quotient } 0.3\text{) is equivalent to } 2\overline{)0.6} \text{ (quotient } 0.3\text{)}$$

As in any division problem, we can check our answer by confirming that the product of the quotient and the *original divisor* equals the *original dividend.*

Division Problem	**Check**
$0.6 \div 2 = 0.3$	$\begin{array}{r} 0.3 \\ \underline{\times 2} \\ 0.6 \end{array}$
$0.006 \div 0.02 = 0.3$	$\begin{array}{r} 0.3 \\ \underline{\times 0.02} \\ 0.006 \end{array}$

These examples suggest the following rule:

To Divide Decimals

- Move the decimal point in the divisor to the right end of the number.
- Move the decimal point in the dividend the same number of places to the right as in the divisor.
- Insert a decimal point in the quotient directly above the decimal point in the dividend.
- Divide the new dividend by the new divisor, inserting zeros at the right end of the dividend as necessary.

EXAMPLE 5

What is 0.035 divided by 0.25?

Solution Move the decimal point to the right end, making the divisor a whole number.

$$0.25\overline{)0.035}$$

Move the decimal point in the dividend the same number of places.

Now, we divide 3.5 by 25, which gives us 0.14.

$$\begin{array}{r} 0.14 \\ 25\overline{)3.50} \\ \underline{2\ 5} \\ 1\ 00 \\ \underline{1\ 00} \end{array}$$

Check We see that the product of the quotient and the original divisor is equal to the original dividend.

$$\begin{array}{r} 0.14 \\ \underline{\times 0.25} \\ 070 \\ \underline{028} \\ 0.0350 = 0.035 \end{array}$$

PRACTICE 5

Divide and check: 2.706 ÷ 0.15

EXAMPLE 6

Find the quotient: 6 ÷ 0.0012. Check the answer.

Solution

The decimal point is moved four places to the right.

$$0.0012\overline{)6.0000}$$

To move the decimal point four places to the right, we must add four 0's as placeholders.

$$\begin{array}{r} 5{,}000 \\ 12\overline{)60{,}000} \end{array}$$

Check

$$\begin{array}{r} 5{,}000 \\ \underline{\times\ 0.0012} \\ 6.0000 = 6 \end{array}$$

The answer checks.

PRACTICE 6

Divide $\frac{8.2}{0.004}$ and then check.

EXAMPLE 7

Divide and round to the nearest hundredth: $0.7\overline{)40.2}$
Then, check.

Solution $0.7\overline{)40.2}$

$$\begin{array}{r} 57.428 \approx 57.43 \text{ to the nearest hundredth} \\ 7\overline{)402.000} \\ \underline{35} \\ 52 \\ \underline{49} \\ 3\,0 \\ \underline{2\,8} \\ 20 \\ \underline{14} \\ 60 \\ \underline{56} \\ 4 \end{array}$$

Check

$$\begin{array}{r} 57.43 \\ \times\ \ 0.7 \\ \hline 40.201 \approx 40.2 \end{array}$$

Because we rounded our answer, the check gives us a product only approximately equal to the original dividend.

PRACTICE 7

Find the quotient of 8.07 and 0.11, rounded to the nearest tenth.

EXAMPLE 8

Compute and check: $8.319 \div 1{,}000$

Solution

$$\begin{array}{r} 0.008319 \\ 1{,}000\overline{)8.319000} \\ \underline{8\,000} \\ 3190 \\ \underline{3000} \\ 1900 \\ \underline{1000} \\ 9000 \\ \underline{9000} \end{array}$$

Check

$0.008319 \times 1{,}000 = 0008.319 = 8.319$

Note that the divisor (1,000) is a power of 10 ending in three zeros, and that the quotient is identical to the dividend except that the decimal point is moved to the left three places.

$$\frac{8.319}{1{,}000} = 0.008319$$

PRACTICE 8

Divide: $100\overline{)3.41}$

Example 8 suggests the following shortcut.

TIP To divide a decimal by a power of 10, move the decimal point to the left the same number of places as the power of 10 has zeros.

Can you explain the difference between the shortcuts for multiplying and for dividing by a power of 10?

EXAMPLE 9

Compute: $\dfrac{7.2}{100}$

Solution Since we are dividing by 100, a power of 10 with two zeros, we can find this quotient simply by moving the decimal point in 7.2 to the left two places.

$$\frac{7.2}{100} = .072, \quad \text{or} \quad 0.072$$

So the quotient is 0.072.

PRACTICE 9

Calculate: $0.86 \div 1{,}000$

Now, let's try using these skills in some applications.

EXAMPLE 10

The following table gives the area of each of the world's three largest oceans.

Ocean	Area (in millions of square kilometers)
Pacific Ocean	155.6
Atlantic Ocean	76.8
Indian Ocean	68.6

(*Source:* cia.gov)

The area of the Pacific Ocean is how many times as great as the area of the Atlantic Ocean, rounded to the nearest tenth?

Solution The area of the Pacific Ocean is 155.6 and that of the Atlantic Ocean is 76.8 (both in millions of square kilometers). To find how many times as great 155.6 is when compared to 76.8, we find the quotient of these numbers.

$$76.8\overline{)155.6} = 768\overline{)1556}$$

$$\begin{array}{r} 2.02 \\ 768\overline{)1556.00} \\ \underline{1536} \\ 20\ 0 \\ \underline{0} \\ 2000 \\ \underline{1536} \\ 464 \end{array}$$

Rounding to the nearest tenth, we find that the area of the Pacific Ocean is 2.0 times that of the Atlantic Ocean.

PRACTICE 10

The table gives the amount of selected foods consumed per capita in the United States in a recent year.

Food	Annual Per Capita Consumption (in pounds)
Red meat	112.0
Poultry	72.7
Fish and shellfish	16.5

The amount of red meat consumed was how many times as great as the amount of poultry, rounded to the nearest tenth? (*Source:* U.S. Department of Agriculture)

Estimating Quotients

As we have shown, one way to check the quotient of two decimals is by multiplying. Another way is by estimating.

To check a decimal quotient by estimating, we can round each decimal to one nonzero digit and then mentally divide the rounded numbers. But we must be careful to position the decimal point correctly in our estimate.

Mental estimation is also a useful skill for approximating a quotient.

EXAMPLE 11

Divide and check by estimating: $3.36 \div 0.021$

Solution $0.021\overline{)3.360}$

We compute the exact answer.

$$\begin{array}{r} 160 \\ 21\overline{)3,360} \\ \underline{2\,1} \\ 1\,26 \\ \underline{1\,26} \\ 00 \\ \underline{00} \end{array}$$

So 160 is our quotient.

Now, let's check by estimating. Because $3.36 \approx 3$ and $0.021 \approx 0.02$, $3.36 \div 0.021 \approx 3 \div 0.02$. We mentally divide to get the estimate.

$$\begin{array}{r} 150 \\ 0.02\overline{)3.00} \end{array}$$

Our estimate 150 is reasonably close to our exact answer, 160.

PRACTICE 11

Compute and check by estimating:
$8.229 \div 0.39$

EXAMPLE 12

Calculate and check: $(9.13) \div (0.2) + (4.6)^2$

Solution Following the order of operations rule, we begin by finding the square and then the quotient.

$$(4.6)^2 = 21.16$$
$$(9.13) \div (0.2) = 45.65$$

Then we add these two results.

$$21.16 + 45.65 = 66.81$$

So $(9.13) \div (0.2) + (4.6)^2 = 66.81$.

Now, let's check this answer by estimating.

$$\begin{array}{c} (9.13) \div (0.2) + (4.6)^2 \\ \downarrow \quad\quad \downarrow \quad\quad\quad \downarrow \\ \underbrace{9 \div 0.2}_{45} \quad + \quad 25 \quad \approx 45 + 25 \approx 70 \end{array}$$

The estimate 70 is close to our answer, 66.81.

PRACTICE 12

Compute and check:
$13.07 + (8.4 \div 0.5)^2$

EXAMPLE 13

The water in a filled aquarium weighs 638.25 pounds. If 1 cubic foot of water weighs 62.5 pounds, estimate how many cubic feet of water there are in the aquarium.

Solution We know that the water in the aquarium weighs 638.25 pounds. Since 1 cubic foot of water weighs 62.5 pounds, we can estimate the number of cubic feet of water in the aquarium by computing the quotient $638.25 \div 62.5$, which is approximately $600 \div 60$, or 10. So a reasonable estimate for the amount of water in the aquarium is 10 cubic feet.

PRACTICE 13

The following graph shows the number of farms, in a recent year, in five states.

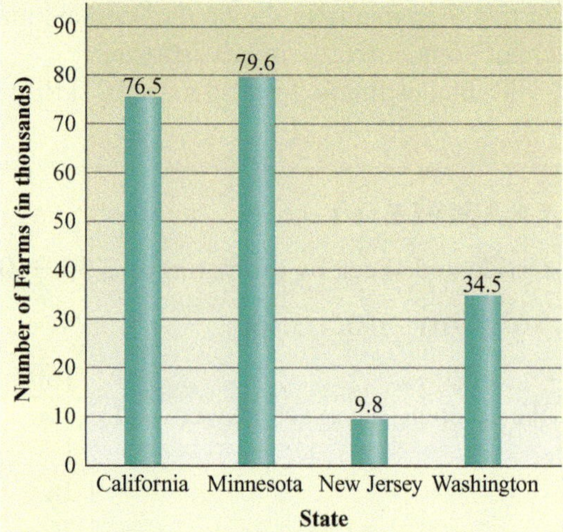

The number of farms in Minnesota is about how many times as great as the number in New Jersey? (*Source:* U.S. Department of Agriculture)

Dividing on a Calculator

When dividing decimals on a calculator, be careful to enter the dividend first and then the divisor. Note that when the dividend is larger than the divisor, the quotient is greater than 1 and when the dividend is smaller than the divisor, the quotient is less than 1.

EXAMPLE 14

Calculate $8.6 \div 1.6$ and round to the nearest tenth.

Solution

The answer, when rounded to the nearest tenth, is 5.4. As expected, the answer is greater than 1, because $8.6 > 1.6$.

PRACTICE 14

Compute the quotient $8.6\overline{)1.6}$ and round to the nearest tenth.

EXAMPLE 15

Divide $0.3\overline{)0.07}$, rounding to the nearest hundredth.

Solution

The answer, when rounded to the nearest hundredth, is 0.23. As expected, the answer is less than 1, because $0.07 < 0.3$.

PRACTICE 15

Find the quotient, rounding to the nearest hundredth: $0.3 \div 0.07$

3.4 Exercises

FOR EXTRA HELP MyMathLab MathXL PRACTICE WATCH READ REVIEW

Mathematically Speaking

Fill in each blank with the most appropriate term or phrase from the given list.

quotient	three	divisor	decimal
dividend	right	terminating	four
fraction	product	left	repeating

1. To change a fraction to the equivalent ________, divide the numerator of the fraction by its denominator.
2. An example of a(n) ________ decimal is 0.3333 . . . .
3. When dividing decimals, move the decimal point in the divisor to the ________ end.
4. To divide a decimal by 1,000, move the decimal point ________ places to the left.
5. To estimate the ________ of 0.813 and 0.187, divide 0.8 by 0.2.
6. When dividing a decimal by a whole number, the decimal point in the quotient is placed above the decimal point in the ________.

A *Change to the equivalent decimal. Then, check.*

7. $\frac{1}{2}$
8. $\frac{3}{5}$
9. $\frac{3}{8}$
10. $\frac{1}{8}$
11. $\frac{37}{10}$
12. $\frac{57}{10}$
13. $1\frac{5}{8}$
14. $2\frac{7}{8}$
15. $6\frac{1}{5}$
16. $8\frac{2}{5}$
17. $21\frac{3}{100}$
18. $60\frac{17}{100}$

Change to the equivalent decimal. Round to the nearest hundredth.

19. $\frac{2}{3}$
20. $\frac{5}{6}$
21. $\frac{7}{9}$
22. $\frac{1}{3}$
23. $3\frac{1}{9}$
24. $2\frac{4}{7}$
25. $5\frac{1}{16}$
26. $10\frac{11}{32}$

B *Divide and check.*

27. $4\overline{)17}$
28. $2\overline{)35}$
29. $5\overline{)21}$
30. $6\overline{)33}$
31. $8\overline{)11}$
32. $6\overline{)9}$
33. $18\overline{)153}$
34. $14\overline{)217}$

Divide. Express any remainder as a decimal rounded to the nearest thousandth.

35. $7\overline{)23}$
36. $9\overline{)41}$
37. $11\overline{)3}$
38. $13\overline{)2}$
39. $7\overline{)46}$
40. $6\overline{)82}$
41. $13\overline{)911}$
42. $12\overline{)208}$

Insert the decimal point in the appropriate place.

43. $\begin{array}{r} 5\,8\,8\,2 \\ 0.7\overline{)4\,1.1\,7\,4} \end{array}$
44. $\begin{array}{r} 5\,7 \\ 3\overline{)0.0\,1\,7\,1} \end{array}$
45. $\begin{array}{r} 6\,6\,3 \\ 0.5\,8\overline{)0.0\,3\,8\,4\,5\,4} \end{array}$
46. $\begin{array}{r} 6\,9 \\ 3.9\overline{)2\,6.9\,1} \end{array}$

Divide. Check, either by multiplying or by estimating.

47. $8\overline{)23.1}$
48. $8\overline{)24.6}$
49. $7\overline{)2.002}$
50. $6\overline{)4.002}$

51. $\frac{17.2}{4}$ 52. $\frac{18.5}{5}$ 53. $\frac{0.003}{2}$ 54. $\frac{0.009}{5}$

55. $8.65 \div 5$ 56. $7.74 \div 6$ 57. $11.5 \div 4$ 58. $16.5 \div 4$

59. $0.2\overline{)0.8}$ 60. $0.3\overline{)0.6}$ 61. $0.05\overline{)3.52}$ 62. $0.04\overline{)1.92}$

63. $\frac{47}{0.5}$ 64. $\frac{86}{0.2}$ 65. $\frac{5}{0.4}$ 66. $\frac{9}{0.6}$

67. $0.03 \div 0.1$ 68. $0.04 \div 0.2$ 69. $0.38 \div 1.9$ 70. $0.75 \div 2.5$

71. $95.2 \div 100$ 72. $81.6 \div 100$ 73. $0.082 \div 100$ 74. $0.093 \div 100$

Divide, rounding to the nearest hundredth. Check, either by multiplying or by estimating.

75. $0.8\overline{)307.1}$ 76. $0.6\overline{)401.8}$ 77. $0.9\overline{)0.0057}$ 78. $0.2\overline{)0.0063}$

79. $\frac{3.69}{0.4}$ 80. $\frac{3.98}{0.8}$ 81. $\frac{87}{0.009}$ 82. $\frac{23}{0.006}$

83. $41 \div 0.021$ 84. $91 \div 0.071$ 85. $35.77 \div 0.11$ 86. $29.11 \div 0.17$

87. $49.071 \div 0.728$ 88. $18.032 \div 0.796$ 89. $3 \div 0.0721$ 90. $4 \div 0.0826$

Perform the indicated operations.

91. $\frac{3.06}{4} + 2$ 92. $\frac{2.04}{3} + 1$ 93. $\frac{18.27 - 8.4}{0.3}$ 94. $\frac{26.77 - 10.1}{0.4}$

95. $\frac{13.05}{7.27 - 7.02}$ 96. $\frac{14.07}{8.41 - 8.01}$ 97. $\frac{8.1 \times 0.2}{0.4}$ 98. $\frac{4.7 \times 5.6}{0.8}$

99. $(82.9 - 3.6) \div (0.21 - 0.01)$ 100. $(3.21 - 0.207) \div (2.08 - 2.072)$

101. $8.73 \div 0.2 + (2.5)^2$ 102. $4.86 \div 0.2 + (3.1)^2$

Complete each table.

103.

Input	Output
1	**1** $\div 5 - 0.2 =$
2	**2** $\div 5 - 0.2 =$
3	**3** $\div 5 - 0.2 =$
4	**4** $\div 5 - 0.2 =$

104.

Input	Output
1	**1** $\div 4 + 0.4 =$
2	**2** $\div 4 + 0.4 =$
3	**3** $\div 4 + 0.4 =$
4	**4** $\div 4 + 0.4 =$

Each of the following quotients is rounded to the nearest hundredth. In each group of three quotients, one is wrong. Use estimation to identify which quotient is incorrect.

105. **a.** $5.7 \div 89 \approx 0.06$ **b.** $0.77 \div 0.0019 \approx 405.26$ **c.** $31.5 \div 0.61 \approx 516.39$

106. **a.** $\frac{9.83}{4.88} \approx 0.20$ **b.** $\frac{2.771}{0.452} \approx 6.13$ **c.** $\frac{389.224}{1.79} \approx 217.44$

107. **a.** $61.27 \div 0.057 \approx 1{,}074.91$ **b.** $0.614 \div 2.883 \approx 2.13$ **c.** $0.0035 \div 0.00481 \approx 0.73$

108. **a.** \$365 ÷ \$4.89 ≈ 7.46 **b.** \$17,358.27 ÷ \$365 ≈ 47.56 **c.** \$3,000 ÷ \$2.54 ≈ 1,181.10

Mixed Practice

Solve.

109. Express $\frac{4}{5}$ as a decimal.

110. Divide $1.6\overline{)8.5}$ and round to the nearest tenth.

111. Change $1\frac{1}{6}$ to a decimal, rounded to the nearest hundredth.

112. Simplify: $81.5 - \frac{32}{0.4}$

113. What is 0.063 divided by 0.14?

114. Find the quotient: $9 \div 0.0072$

Applications

C *Solve and check.*

115. A stalactite is an icicle-shaped mineral deposit that hangs from the roof of a cave. If it took a thousand years for a stalactite to grow to a length of 3.7 inches, how much did it grow per year?

116. In a strong storm, the damage to 100 houses was estimated at \$12.7 million. What was the average damage per house?

117. The women's softball team won 21 games and lost 14. The men's softball team won 22 games and lost 18.

a. The women's team won what fraction of the games that it played, expressed as a decimal?

b. The men's team won what fraction of its games, expressed as a decimal?

c. Which team has a better record? Explain.

118. Yesterday, 0.08 inches of rain fell. Today, $\frac{1}{4}$ inch of rain fell.

a. How much rain fell today, expressed as a decimal?

b. Which day did more rain fall? Explain.

119. The table shown gives the highest gasoline mileage for three small SUVs.

a. For each SUV, compute how many miles it gets per gallon, rounded to the nearest whole number.

b. Which SUV gives the highest mileage?

SUV	Distance Driven (in miles)	Gasoline Used (in gallons)	Gasoline Mileage (in miles per gallon)
Honda CR-V	40.5	1.9	
Ford Escape Hybrid	62.4	2.4	
GMC Terrain	42.6	2.4	

(*Source: Consumer Reports*)

120. The table shown gives the total number of games played and assists made by three basketball players over the same period of time.

a. Compute the average number of assists per game for each player, expressed as a decimal rounded to the nearest tenth.

b. Decide which player had the highest average.

Player	Number of Games	Number of Assists	Average
Steve Nash	236	2,507	
Chris Paul	203	2,266	
Deron Williams	226	2,385	

(*Source:* nba.com/statistics)

121. On the stock market, shares of Citigroup, Inc. common stock were traded at \$3.63 per share. How many shares could have been bought for \$7,260? (*Source:* finance.yahoo.com)

122. A light microscope can distinguish two points separated by 0.0005 millimeter, whereas an electron microscope can distinguish two points separated by 0.0000005 millimeter. The electron microscope is how many times as powerful as the light microscope?

123. Typically, the heaviest organ in the body is the skin, weighing about 9 pounds. By contrast, the heart weighs approximately 0.7 pound. About how many times the weight of the heart is that of the skin? (*Source: World of Scientific Discovery*)

124. At a community college, each student enrolled pays a \$19.50 student fee per semester. In a given semester, if the college collected \$39,000 in student fees, how many students were enrolled?

125. A shopper buys four organic chickens. The chickens weigh 3.2 pounds, 3.5 pounds, 2.9 pounds, and 3.6 pounds. How much less than the average weight of the four chickens was the weight of the lightest one?

126. A dieter joins a weight-loss club. Over a 5-month period, she loses 8 pounds, 7.8 pounds, 4 pounds, 1.5 pounds, and 0.8 pound. What was her average monthly weight loss, to the nearest tenth of a pound?

127. Babe Ruth got 2,873 hits in 8,399 times at bat, resulting in a batting average of $\frac{2{,}873}{8{,}399}$, or approximately .342. Another great player, Ty Cobb, got 4,189 hits out of 11,434 times at bat. What was his batting average, expressed as a decimal rounded to the nearest thousandth? (Note that batting averages don't have a zero to the left of the decimal point because they can never be greater than 1.) (*Source:* baseball-reference.com)

128. On January 23, 1960, the Trieste became the only manned deep-diving research vessel ever to reach the bottom of the Mariana Trench. This trench is the deepest part of any ocean on Earth, with a depth of 10,911 meters. At a descent rate of 3,290.4 meters per hour, approximately how long did it take for the Trieste to reach the bottom? Round the answer to the nearest tenth of an hour. (*Source:* absoluteastronomy.com)

• Check your answers on page A-5.

MINDStretchers

Patterns

1. In the *repeating decimal* 0.142847142847142847 . . . , identify the 994th digit to the right of the decimal point.

Groupwork

2. In the following magic square, 3.375 is the *product* of the numbers in every row, column, and diagonal. Working with a partner, fill in the missing numbers.

	0.25	3
		0.5

Writing

3. **a.** $0.5 \div 0.8 = ?$

   b. $0.8 \div 0.5 = ?$

   c. Find the product of your answers in parts (a) and (b). Explain how you could have predicted this product.

Key Concepts and Skills

CONCEPT SKILL

Concept/Skill	Description	Example
[3.1] **Decimal**	A number written with two parts: a whole number, which precedes the decimal point, and a fractional part, which follows the decimal point.	Whole-number part → 3.721 ← Fractional part; ↑ Decimal point
[3.1] **Decimal place**	A place to the right of the decimal point.	Decimal places: 8.**035** (Tenths, Hundredths, Thousandths)
[3.1] **To change a decimal to the equivalent fraction or mixed number**	• Copy the nonzero whole-number part of the decimal and drop the decimal point. • Place the fractional part of the decimal in the numerator of the equivalent fraction. • Make the denominator of the equivalent fraction 1 followed by as many zeros as the decimal has decimal places. • Simplify the resulting fraction, if possible.	The decimal 3.25 is equivalent to the mixed number $3\frac{25}{100}$ or $3\frac{1}{4}$.
[3.1] **To compare decimals**	• Rewrite the numbers vertically, lining up the decimal points. • Working from left to right, compare the digits that have the same place value. At the first place value where the digits differ, the decimal which has the largest digit with this place value is the largest decimal.	1.073 1.06999 In the ones place and the tenths place, the digits are the same. But in the hundredths place, $7 > 6$, so $1.073 > 1.06999$.
[3.1] **To round a decimal to a given decimal place**	• Underline the digit in the place to which the number is being rounded. • The digit to the right of the underlined digit is called the *critical digit*. Look at the critical digit—if it is 5 or more, add 1 to the underlined digit; if it is less than 5, leave the underlined digit unchanged. • Drop all decimal places to the right of the underlined digit.	$23.9\underline{3}81 \approx 23.94$ ↑ Critical digit
[3.2] **To add decimals**	• Rewrite the numbers vertically, lining up the decimal points. • Add. • Insert a decimal point in the sum below the other decimal points.	0.035 0.08 + 0.00813 0.12313
[3.2] **To subtract decimals**	• Rewrite the numbers vertically, lining up the decimal points. • Subtract, inserting extra zeros in the minuend if necessary for regrouping. • Insert a decimal point in the difference below the other decimal points.	0.90370 − 0.17052 0.73318

continued

CONCEPT SKILL

Concept/Skill	Description	Example
[3.3] **To multiply decimals**	• Multiply the factors as if they were whole numbers. • Find the total number of decimal places in the factors. • Count that many places from the right end of the product, and insert a decimal point.	21.07 ← Two decimal places × 0.18 ← Two decimal places 3.7926 ← Four decimal places
[3.4] **To change a fraction to the equivalent decimal**	• Divide the denominator of the fraction into the numerator, inserting to its right both a decimal point and enough zeros to get an answer either without a remainder or rounded to a given decimal place. • Place a decimal point in the quotient directly above the decimal point in the dividend.	$\frac{7}{8} = 8\overline{)7.000}$ with quotient 0.875
[3.4] **To divide decimals**	• Move the decimal point in the divisor to the right end of the number. • Move the decimal point in the dividend the same number of places to the right as in the divisor. • Insert a decimal point in the quotient directly above the decimal point in the dividend. • Divide the new dividend by the new divisor, inserting zeros at the right end of the dividend as necessary.	$3.5\overline{)71.05}$ = 20.3 $35\overline{)710.5}$ 70 10 0 10 5 10 5

CHAPTER 3 Review Exercises

Say Why *Fill in each blank.*

1. The expression $7\frac{3}{10}$ $\underset{\text{is/is not}}{\underline{\hspace{2cm}}}$ considered to be a decimal because ________________________________.

2. In the decimal 2.781, the 2 $\underset{\text{is/is not}}{\underline{\hspace{2cm}}}$ in a decimal place because ________________________________.

3. The number 48.726 rounded to the nearest hundredth $\underset{\text{is/is not}}{\underline{\hspace{2cm}}}$ 48.72 because ________________________________.

4. We $\underset{\text{can/cannot}}{\underline{\hspace{2cm}}}$ add, subtract, or compare decimals by rewriting them vertically and then lining up the decimal points because ________________________________.

5. The number 9.1313 $\underset{\text{is/is not}}{\underline{\hspace{2cm}}}$ a repeating decimal because ________________________________.

6. Without calculating, we know that $0.04 \div 0.23$ $\underset{\text{is/is not}}{\underline{\hspace{2cm}}}$ less than 1 because ________________________________.

[3.1] *Name the place that each underlined digit occupies.*

7. $8.3\underline{5}9$ **8.** $13.\underline{0}05$ **9.** $8{,}024.\underline{5}$ **10.** $0.000\underline{3}$

Express each number as a fraction, mixed number, or whole number.

11. 0.35 **12.** 8.2 **13.** 4.007 **14.** 10.000

Write each decimal in words.

15. 0.72 **16.** 5.6

17. 3.0009 **18.** 510.036

Write each decimal in decimal notation.

19. Seven thousandths **20.** Two and one tenth

21. Nine hundredths **22.** Seven and forty-one thousandths

Between each pair of numbers, insert the appropriate sign, $<$, $=$, or $>$, to make a true statement.

23. 0.037   0.04 **24.** 2.031   2.0301 **25.** 5.12   4.71932 **26.** 2   1.8

Rearrange each group of numbers from largest to smallest.

27. 0.72, 0.8, 1.002 **28.** 0.003, 0.00057, 0.004

Round as indicated.

29. 7.31 to the nearest tenth **30.** 0.0387 to the nearest thousandth

31. 4.3868 to two decimal places **32.** $899.09 to the nearest dollar

[3.2] *Perform the indicated operations. Check by estimating.*

33. $8.2 + 3.91$ **34.** $50 + 2.7 + 0.05$ **35.** $\$8 + \$3.25 + \$12.88$ **36.** $8.4\text{ m} + 3.6\text{ m}$

37. $30.7 - 1.92$

38. $93 - 5.248$

39. $2.5 - (0.72 - 0.054)$

40. $54.17 - (8 - 2.731)$

41. $5.398 + 8.72 + 92.035 + 0.7723 - 3.714 - 5.008$

42. $\$87{,}259.39 + \$2{,}098.35 + \$1{,}387.92 + \$203.14$

[3.3] *Find the product. Check by estimating.*

43. 7.28×0.4

44. $(288)(3.5)$

45. 0.005×0.002

46. $(3.7)^2$

47. $2.71 \cdot 1{,}000$

48. 0.0034×10

49. $8 - (1.5)^2$

50. $3(2.4) + 7(0.9)$

51. $18{,}772.35 \times 0.0836$

52. $(74.862)(5.901)$

[3.4] *Change to the equivalent decimal.*

53. $\frac{5}{8}$

54. $90\frac{1}{5}$

55. $4\frac{1}{16}$

56. $\frac{45}{1{,}000}$

Express each fraction as a decimal. Round to the nearest hundredth.

57. $\frac{1}{6}$

58. $\frac{2}{7}$

59. $8\frac{1}{3}$

60. $11\frac{2}{9}$

Divide and check.

61. $2\overline{)1.3}$

62. $\frac{4.8}{3}$

63. $0.7 \div 4$

64. $\frac{2.77}{10}$

Divide. Round to the nearest tenth.

65. $4.67 \div 0.9$

66. $\frac{2.35}{0.73}$

67. $\frac{7.11}{0.3}$

68. $0.06\overline{)981.5}$

69. $18.74 \div 9.7$

70. $220 \div 0.61$

71. $81.37\overline{)247.062}$

72. $247.062\overline{)81.37}$

Simplify.

73. $\frac{(1.3)^2 - 1.1}{0.5}$

74. $\frac{2.5 - (0.4)^2}{0.02}$

75. $\frac{13.75}{9.6 - 9.2}$

76. $(2.5)(3.5) \div 6.25$

Mixed Applications

Solve.

77. Recently, a champion swimmer swam 50 meters in 25.2 seconds and then swam 100 meters farther in 29.29 seconds. How long did she take to swim the 150 meters?

78. On a certain day, the closing price of one share of Home Depot was \$14.57. If the closing price for the day was \$0.43 higher than the opening price, what was the opening price? (*Source:* finance.yahoo.com)

79. The venom of a certain South American frog is so poisonous that 0.0000004 ounce of the venom can kill a person. How is this decimal read?

80. A carpenter is constructing steps leading to a terrace. The cross-section of the steps is shown below. Find the missing dimension.

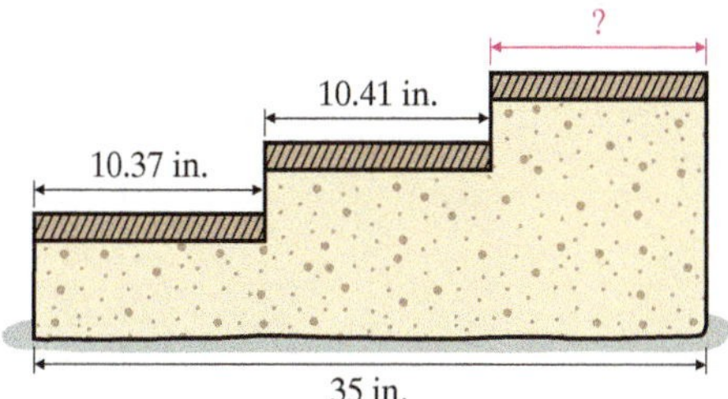

81. Astronomers use the term *astronomical unit* (or AU) for the average distance between Earth and the Sun. The distance 1 AU is about 93,000,000 miles. The average distance from the Sun to Mars is about 140,000,000 miles. Express in astronomical units, rounded to the nearest tenth, the average distance between Mars and the Sun. (*Source*: Jeffrey Bennett et al., *The Cosmic Perspective*)

82. In the United States Congress, there are 100 senators and 435 representatives. How many times as many representatives as senators are there?

83. A supermarket sells a 4-pound package of ground meat for $5.20 and a 5-pound package of ground meat for $6.20. What is the difference between the two prices per pound?

84. A homeowner is making plans to build a patio in her backyard and would like to save money by doing the work herself. The dimensions of the proposed patio are 9.67 yards by 5.33 yards. The labor costs would be about $5 per square yard. Estimate how much money the homeowner would save in labor by building the patio herself.

85. In a chemistry lab, a student weighs a compound three times, getting 7.15 grams, 7.18 grams, and 7.23 grams. What is the average of these weights, to the nearest hundredth of a gram?

86. A team of geologists scaled a mountain. At the base of the mountain, the temperature had been 11°C. The temperature fell 0.75 degrees for every 100 meters the team climbed. After they climbed 1,000 meters, what was the temperature?

87. The following form was adapted from the *U.S. Individual Income Tax Return*. Find the total income in line 22.

7	Wages, salaries, tips, etc.	7	28,774.71
8	Taxable interest income	8	
9	Dividend income	9	232.55
10	Taxable refunds, credits, or offsets of state and local income taxes	10	349.77
11	Alimony received	11	
12	Business income or (loss)	12	
13	Capital gain or (loss)	13	511.74
14	Other gains or (losses)	14	5,052.71
15	IRA distributions: taxable amount	15	
16	Pensions and annuities: taxable amount	16	
17	Rents, royalties, partnerships, estates, trusts, etc.	17	1,240.97
18	Farm income or (loss)	18	
19	Unemployment compensation	19	
20	Social Security benefits	20	
21	Other income	21	
22	Add the amounts shown in the far right column	22	

88. The following table shows the quarterly revenues, in billions of dollars, for Google and Yahoo! for a recent year.

Ending	Google	Yahoo!
June 30	5.523	1.573
September 30	5.945	1.575
December 31	6.674	1.732
March 31	6.775	1.597

(*Source:* finance.yahoo.com)

How much more were Google's earnings than Yahoo!'s for the year?

• Check your answers on page A-5.

CHAPTER 3 Posttest

To see if you have mastered the topics in this chapter, take this test.

1. In the number 0.79623, which digit occupies the thousandths place?

2. Write in words: 5.102

3. Round 320.1548 to the nearest hundredth.

4. Which is smallest, 0.04, 0.0009, or 0.00028?

5. Express 3.04 as a mixed number.

6. Write as a decimal: four thousandths

Perform the indicated operations.

7. $2.3 + 0.704 + 1.35$

8. $\$5.27 + \$9 - \$8.61$

9. 2.09×10

10. 5.2×1.1

11. $(0.1)^3$

12. $\dfrac{3.52}{2} + \dfrac{4.8}{3}$

13. $2.9 \div 1{,}000$

14. $\dfrac{9.81}{0.3}$

Express as a decimal.

15. $\dfrac{3}{8}$

16. $4\dfrac{1}{6}$, rounded to the nearest hundredth

Solve.

17. The element hydrogen is so light that 1 cubic foot of hydrogen weighs only 0.005611 pound. Round this weight to the nearest hundredth of a pound.

18. Historically, a mile was the distance that a Roman soldier covered when he took 2,000 steps. If a mile is 5,280 feet, how many feet, to the nearest tenth of a foot, was a Roman's step?

19. The Triple Crown consists of three horse races—the Kentucky Derby (1.25 miles), the Preakness Stakes (1.1875 miles), and the Belmont Stakes (1.5 miles). Which race is longest? (*Source: World Almanac 2010*)

20. A part of the real estate tax in Berkeley, California provides funds for the maintenance and servicing of traffic signals and other public lighting. This amount of tax (in dollars) on a house is 0.0108 times the area of the house (in square feet). What is this tax amount on a house that is 3,000 square feet in area? (*Source:* ci.berkeley.ca.us)

• Check your answers on page A-5.

Cumulative Review Exercises

To help you review, solve the following:

1. Round 591,622 to the nearest million.

2. Subtract: 6,063 − 3,987

3. Multiply: (409)(67)

4. Find the area of the shaded region.

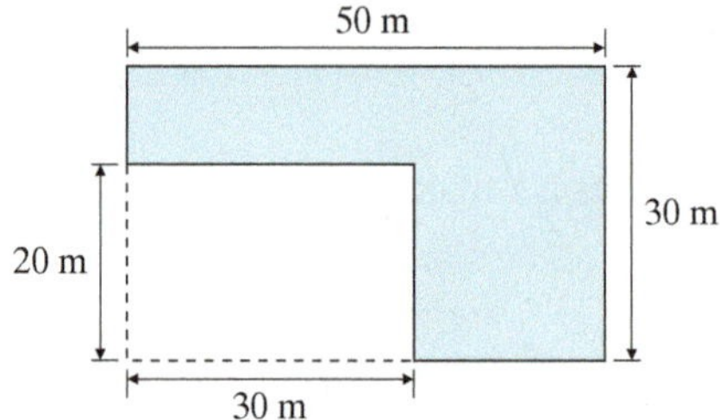

5. Simplify: $22 + 4(9 - 2^2)$

6. List all the factors of 60.

7. Which is larger, $1\frac{1}{2}$ or $1\frac{3}{8}$?

8. Subtract: $5 - 2\frac{1}{3}$

9. Estimate: $7\frac{9}{10} \times 4\frac{1}{13}$

10. Calculate: $\frac{2}{5} + \frac{1}{3} \cdot \frac{1}{2}$

11. Write the equivalent fraction or mixed number for 4.72 in lowest terms.

12. Round 38.363 to one decimal place.

13. Find the difference: 64.99 − 4.777

14. Divide: 29.89 ÷ 0.049

15. How much more than \$25 will a shopper need in order to buy items costing, including tax, \$8.39, \$7.34, and \$9.44?

16. A community garden occupies $\frac{1}{5}$ acre on a $\frac{1}{3}$-acre plot of land. What is the area of the land not occupied by the garden?

17. A dating service advertises that it has been introducing thousands of singles for 20 years. On the average, how many successful marriages were arranged per year?

18. A satellite orbiting Earth travels at 16,000 miles per hour. An orbit takes 1.6 hours. How far will the satellite travel once around, to the nearest thousand miles?

19. An electric company charges \$0.09693 per kilowatt hour. If a restaurant used 2,000 kilowatt hours of electricity in a certain week, what was its weekly cost?

20. In a recent year, there were 1.596 billion Internet users worldwide. In hundreds of millions, the number of users in Asia was 6.57 and in North America, 2.51. How many times the number of Internet users in North America was the number in Asia, rounded to the nearest tenth? (*Source:* internetworldstats.com)

• Check your answers on page A-5.

CHAPTER 4

Basic Algebra: Solving Simple Equations

Algebra and Pressure

Perhaps you have noticed that a woman's high heels sink into soft ground even when the much larger heels of a man's shoes do not. The pressure that a heel exerts on the ground depends not only on the weight of the person walking but also on the area of the heel. Smaller heels result in greater pressure.

Physicists have observed that when a man and a woman have equal weights, the product of the area of their heels and the pressure that they exert on the ground is the same. In this situation, suppose that the man's heels are 10 square inches in area, in contrast to only $\frac{1}{4}$ square inch for the woman's high heels and that the man's heels each exert 15 pounds per square inch on the ground. We can conclude that:

$$\frac{1}{4}p = 10 \cdot 15$$

Algebra allows us to solve this equation and to find p, the pressure exerted by each of the woman's high heels.

(*Source:* W. Thomas Griffith, *The Physics of Everyday Phenomena*, Dubuque, IA: Wm. C. Brown Publishers, 1992)

CHAPTER 4 PRETEST

To see if you have already mastered the topics in this chapter, take this test.

Write each algebraic expression in words.

1. $t - 4$

2. $\frac{y}{3}$

Translate each phrase to an algebraic expression.

3. 8 more than m

4. Twice n

Evaluate each algebraic expression.

5. $\frac{x}{4}$, for $x = 16$

6. $5 - y$, for $y = 3\frac{1}{2}$

Translate each sentence to an equation.

7. The sum of x and 3 equals 5.

8. The product of 4 and y is 12.

Solve and check.

9. $x + 4 = 10$

10. $t - 1 = 9$

11. $2n = 26$

12. $\frac{a}{4} = 3$

13. $8 = m + 1.9$

14. $15 = 0.5n$

15. $m - 3\frac{1}{2} = 10$

16. $\frac{n}{10} = 1.5$

Write an equation. Solve and check.

17. The planet Jupiter has 36 more moons than the planet Uranus. If Jupiter has 63 moons, how many does Uranus have? (*Source:* NASA)

18. Tickets for all movies shown before 5:00 P.M. at a local movie theater qualify for the bargain matinee price, which is \$2.75 less than the regular ticket price. If the bargain price is \$6.75, what is the regular price?

19. In Michigan, about two-fifths of the area is covered with water. This portion of the state represents about 39,900 square miles. What is the area of Michigan? (*Source: The New York Times Almanac, 2010*)

20. An 8-ounce cup of regular tea has about 40 milligrams of caffeine, which is 10 times the amount of caffeine in a cup of decaffeinated tea. How much caffeine is in a cup of decaffeinated tea?

• Check your answers on page A-6.

4.1 Introduction to Basic Algebra

OBJECTIVES

- **A** To translate phrases to algebraic expressions and vice versa
- **B** To evaluate an algebraic expression for a given value of the variable
- **C** To solve applied problems involving algebraic expressions

What Algebra Is and Why It Is Important

In this chapter, we discuss some of the basic ideas in algebra. These ideas will be important throughout the rest of this book.

In algebra, we use letters to represent unknown numbers. The expression $2 + 3$ is arithmetic, whereas the expression $x + y$ is algebraic, since x and y represent numbers whose values are not known. With *algebraic expressions*, such as $x + y$, we can make general statements about numbers and also find the value of unknown numbers.

We can think of algebra as a *language*: The idea of translating ordinary words to algebraic notation and vice versa is the key. Often, just writing a problem algebraically makes the problem much easier to solve. We present ample proof of this point repeatedly in the chapters that follow.

We begin our discussion of algebra by focusing on what algebraic expressions mean and how to translate and evaluate them.

Translating Phrases to Algebraic Expressions and Vice Versa

To apply mathematics to a real-world situation, we often need to be able to express that situation algebraically. Consider the following example of this kind of translation.

Suppose that you are enrolled in a college course that meets 50 minutes a day for 3 days a week. The course therefore meets for $50 \cdot 3$, or 150 minutes, a week.

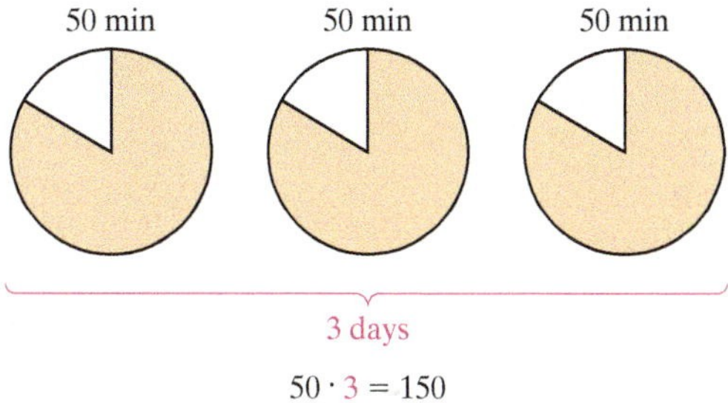

Now, suppose that in a semester the 50-minute class meets d days but that we do not know what number the letter d represents. How many minutes per semester does the class meet? Do you see that we can express the answer as $50d$, that is, 50 times d days?

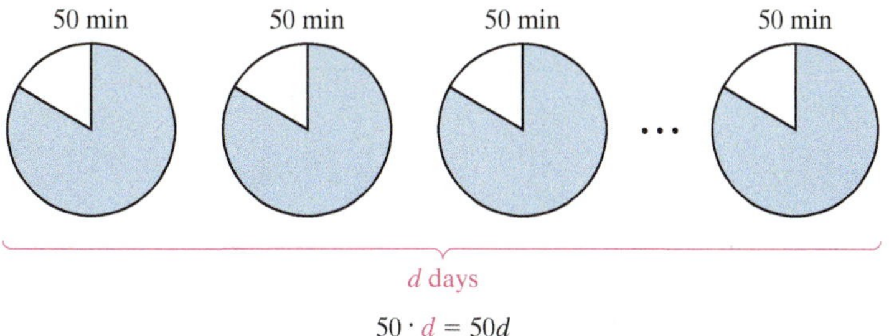

In algebra, a *variable* is a letter, or other symbol, used to represent an unknown number. In the algebraic expression $50d$, for instance, d is a variable and 50 is a *constant*. Note that in writing an algebraic expression, we usually omit any multiplication symbol: $50d$ means $50 \cdot d$.

DEFINITIONS

A **variable** is a letter that represents an unknown number.

A **constant** is a known number.

An **algebraic expression** is an expression that combines variables, constants, and arithmetic operations.

There are many translations of an algebraic expression to words, as the following table indicates.

$x + 4$ translates to	$n - 3$ translates to	$\frac{3}{4} \cdot y$ or $\frac{3}{4}y$ translates to	$z \div 5$ or $\frac{z}{5}$ translates to
• x plus 4 • x increased by 4 • the sum of x and 4 • 4 more than x • 4 added to x	• n minus 3 • n decreased by 3 • the difference between n and 3 • 3 less than n • 3 subtracted from n	• $\frac{3}{4}$ times y • the product of $\frac{3}{4}$ and y • $\frac{3}{4}$ of y	• z divided by 5 • the quotient of z and 5 • z over 5

EXAMPLE 1

Translate each algebraic expression in the table to words.

Solution

Algebraic Expression	Translation
a. $\frac{p}{3}$	p divided by 3
b. $x - 4$	4 less than x
c. $5f$	5 times f
d. $2 + y$	the sum of 2 and y
e. $\frac{2}{3}a$	$\frac{2}{3}$ of a

PRACTICE 1

Translate each algebraic expression to words.

Algebraic Expression	Translation
a. $\frac{1}{2}p$	
b. $5 - x$	
c. $y \div 4$	
d. $n + 3$	
e. $\frac{3}{5}b$	

EXAMPLE 2

Translate each word phrase in the table to an algebraic expression.

Solution

Word Phrase	Translation
a. 16 more than m	$m + 16$
b. the product of 5 and b	$5b$
c. the quotient of 6 and z	$6 \div z$
d. a decreased by 4	$a - 4$
e. $\frac{3}{8}$ of t	$\frac{3}{8}t$

PRACTICE 2

Express each word phrase as an algebraic expression.

Word Phrase	Translation
a. x plus 9	
b. 10 times y	
c. the difference between n and 7	
d. p divided by 5	
e. $\frac{2}{5}$ of v	

As we have seen, any letter or symbol can be used to represent a variable. For example, *five less than a number* can be translated to $n - 5$, where n represents the number.

Let's consider the following example.

EXAMPLE 3

Express each phrase as an algebraic expression.

Solution

Word Phrase	Translation
a. 2 less than a number	$n - 2$, where n represents the number
b. an amount divided by 10	$\frac{a}{10}$, where a represents the amount
c. $\frac{3}{8}$ of a price	$\frac{3}{8}p$, where p represents the price

PRACTICE 3

Translate each word phrase to an algebraic expression.

Word Phrase	Translation
a. a quantity increased by 12	
b. the quotient of 9 and an account balance	
c. a cost multiplied by $\frac{2}{7}$	

Now, let's look at word problems that involve translations.

EXAMPLE 4

Suppose that p partners share equally in the profits of a business. What is each partner's share if the profit was \$2,000?

Solution Each partner should get the quotient of 2,000 and p, which can be written algebraically as $\frac{2{,}000}{p}$ dollars.

PRACTICE 4

Next weekend, a student wants to study for his four classes. If he has h hours to study in all and he wants to devote the same amount of time to each class, how much time will he study per class?

EXAMPLE 5

At registration, n out of 100 classes are closed. How many classes are not closed?

Solution Since n classes are closed, the remainder of the 100 classes are not closed. So we can represent the number of classes that are not closed by the algebraic expression $100 - n$.

PRACTICE 5

Of s shrubs in front of a building, 3 survived the winter. How many shrubs died over the winter?

Evaluating Algebraic Expressions

In this section, we look at how to evaluate algebraic expressions. Let's begin with a simple example.

Suppose that the balance in a savings account is \$200. If d dollars is then deposited, the balance will be $(200 + d)$ dollars.

To evaluate the expression $200 + d$ for a particular value of d, we replace d with that number. If \$50 is deposited, we replace d by 50:

$$200 + d = 200 + 50 = 250$$

So the new balance will be \$250.

The following rule is helpful for evaluating expressions:

To Evaluate an Algebraic Expression

- Substitute the given value for each variable.
- Carry out the computation.

Now, let's consider some more examples.

EXAMPLE 6

Evaluate each algebraic expression.

Solution

Algebraic Expression	Value
a. $n + 8$, if $n = 15$	$15 + 8 = 23$
b. $9 - z$, if $z = 7.89$	$9 - 7.89 = 1.11$
c. $\frac{2}{3}r$, if $r = 18$	$\frac{2}{3} \cdot 18 = 12$
d. $y \div 4$, if $y = 3.6$	$3.6 \div 4 = 0.9$

PRACTICE 6

Find the value of each algebraic expression.

Algebraic Expression	Value
a. $\frac{s}{4}$, if $s = 100$	
b. $0.2y$, if $y = 1.9$	
c. $x - 4.2$, if $x = 9$	
d. $25 + z$, if $z = 1.6$	

The following examples illustrate how to write and evaluate expressions to solve word problems.

EXAMPLE 7

Power consumption for a period of time is measured in watt-hours, where a watt-hour means 1 watt of power for 1 hour. How many watt-hours of energy will a 60-watt bulb consume in h hours? In 3 hours?

Solution The expression that represents the number of watt-hours used in h hours is $60h$. So for $h = 3$, the number of watt-hours is $60 \cdot 3$, or 180. Therefore, 180 watt-hours of energy will be consumed in 3 hours.

PRACTICE 7

When deciding how much money to spend on a new car, a good rule of thumb to follow is to budget about one-fifth of your monthly net income for a car payment. How much should you set aside for the monthly car payment if your net income is n dollars per month? If your net income is $3,750? (*Source:* automotive.com)

EXAMPLE 8

Suppose that there are 180 days in the local school year. How many days was a student present at school if she was absent d days? 9 days?

Solution If d represents the number of days that the student was absent, the expression $180 - d$ represents the number of days that she was present. If she was absent 9 days, we substitute 9 for d in the expression:

$$180 - d = 180 - 9 = 171$$

So the student was present 171 days.

PRACTICE 8

At a coffee shop, a lunch bill was $18.45 plus the tip. What was the total amount of the lunch, including a tip of t dollars? A tip of $3?

4.1 Exercises

FOR EXTRA HELP MyMathLab  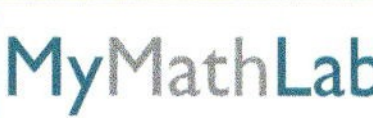    

PRACTICE WATCH READ REVIEW

Mathematically Speaking

Fill in each blank with the most appropriate term or phrase from the given list.

arithmetic	constant	evaluate
translate	variable	algebraic

1. A(n) __________ is a letter that represents an unknown number.

2. A(n) __________ is a known number.

3. A(n) __________ expression combines variables, constants, and arithmetic operations.

4. To __________ an algebraic expression, replace each variable with the given number, and carry out the computation.

A *Translate each algebraic expression to two different word phrases.*

5. $t + 9$

6. $8 + r$

7. $c - 12$

8. $x - 5$

9. $c \div 3$

10. $\frac{z}{7}$

11. $10s$

12. $11t$

13. $y - 10$

14. $w - 1$

15. $7a$

16. $4x$

17. $x \div 6$

18. $\frac{y}{5}$

19. $x - \frac{1}{2}$

20. $x - \frac{1}{3}$

21. $\frac{1}{4}w$

22. $\frac{4}{5}y$

23. $2 - x$

24. $8 - y$

25. $1 + x$

26. $n + 7$

27. $3p$

28. $2x$

29. $n - 1.1$

30. $x - 6.5$

31. $y \div 0.9$

32. $\frac{n}{2.4}$

Translate each word phrase to an algebraic expression.

33. x plus 10

34. d plus 12

35. 1 less than n

36. 9 less than b

37. the sum of y and 5

38. the sum of x and 11

39. t divided by 6

40. r divided by 2

41. the product of 10 and y

42. the product of 5 and p

43. the difference between w and 5

44. the difference between n and 5

45. n increased by $\frac{4}{5}$

46. x increased by $\frac{2}{3}$

47. the quotient of z and 3

48. The quotient of n and 10

49. $\frac{2}{7}$ of x

50. $\frac{2}{3}$ of y

51. 6 subtracted from k

52. 8 subtracted from z

53. 12 more than a number

54. 18 more than a number

55. the difference between a number and 5.1

56. the difference between a number and 8.2

B ***Evaluate each algebraic expression.***

57. $y + 7$, if $y = 19$

58. $3 + n$, if $n = 2.9$

59. $7 - x$, if $x = 4.5$

60. $19 - y$, if $y = 6.7$

61. $\frac{3}{4}p$, if $p = 20$

62. $\frac{4}{5}n$, if $n = 30$

63. $x \div 2$, if $x = 2\frac{1}{3}$

64. $\frac{n}{3}$, if $n = 7.5$

65. $p - 7.9$, if $p = 9$

66. $y - 20.1$, if $y = 30$

67. $x \div \frac{5}{6}$, if $x = \frac{1}{6}$

68. $\frac{1}{3}y$, if $y = \frac{1}{2}$

Complete each table.

69.

x	$x + 8$
1	
2	
3	
4	

70.

x	$x + 10$
1	
2	
3	
4	

71.

n	$n - 0.2$
1	
2	
3	
4	

72.

b	$b - 0.4$
1	
2	
3	
4	

73.

x	$\frac{3}{4}x$
4	
8	
12	
16	

74.

n	$\frac{2}{3}n$
3	
6	
9	
12	

75.

z	$\frac{z}{2}$
2	
4	
6	
8	

76.

y	$\frac{y}{5}$
5	
10	
15	
20	

Mixed Practice

Solve.

77. Translate the phrase "7 less than x" to an algebraic expression.

78. Evaluate the algebraic expression $0.5t$, if $t = 8$.

79. Translate the algebraic expression $\frac{n}{2}$ to two different phrases.

80. Evaluate the algebraic expression $\frac{1}{4}y$, if $y = \frac{2}{3}$.

81. Translate the phrase "the product of 3.5 and t" to an algebraic expression.

82. Evaluate the algebraic expression $x + 1$, if $x = 4$.

83. Translate the algebraic expression $x + 6$ to two different phrases.

84. Evaluate the algebraic expression $n - 20$, if $n = 30$.

Applications

C *Solve.*

85. A patient receives m milligrams of medication per dose. Her doctor orders her medication to be decreased by 25 milligrams. How much medication will she then receive per dose?

86. When a borrower takes out a mortgage, each monthly payment has two parts. One part goes toward the principal and the other toward the interest. If the principal payment is \$344.86 and the interest payment is i, write an algebraic expression for the total payment.

87. The top of the Flatiron Building in New York City, so called because it is shaped like a clothing iron, is in a form similar to the triangle pictured below. Write an expression for the sum of the measures of the three angles. (*Source:* flatironbid.org)

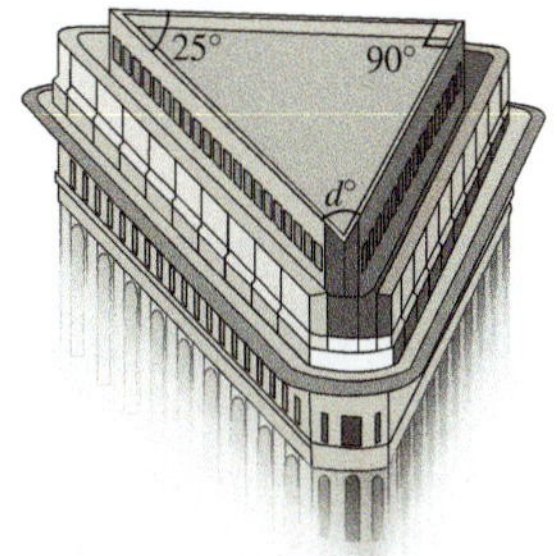

88. Professional land surveyors establish official land, air space, and water boundaries. Below is a sample of a typical lot survey. Write an expression for the sum of the lengths of the sides of the lot. (*Source:* lsrp.com)

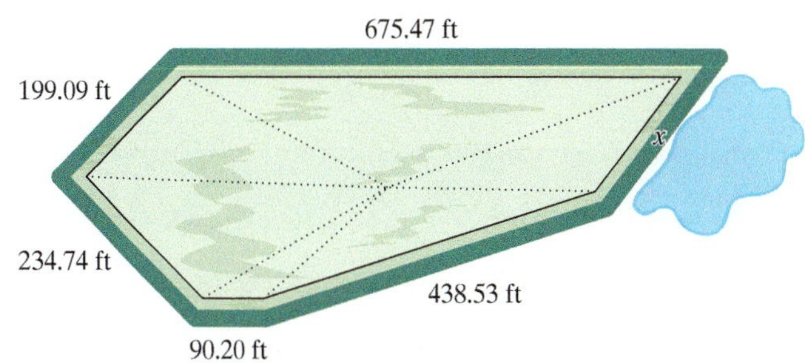

89. If a long-distance trucker drives at a speed of r miles per hour for t hours, she will travel a distance of $r \cdot t$ miles. How far will she travel at a speed of 55 miles per hour in 4 hours?

90. If a basketball player makes b baskets in a attempts, his field goal average is defined to be $\frac{b}{a}$. Find the field goal average of a player who made 12 baskets in 25 attempts.

91. A bank charges customers a fee of \$2.50 for each withdrawal made at its ATMs.
    a. Write an expression for the total fee charged to a customer for w of these withdrawals
    b. Find the total fee if the customer makes 9 withdrawals.

92. A computer network technician charges \$80 per hour for labor.
    a. Write an expression for the cost of h hours of work.
    b. Find the cost of a networking job that takes $2\frac{1}{2}$ hours.

• Check your answers on page A-6.

MINDStretchers

Mathematical Reasoning

1. Consider the expression $x + x$.

   a. Why does this expression mean the same as the expression $2x$?

   b. What does the expression $\underbrace{x + x + x + \cdots + x}_{n \text{ times}}$ mean in terms of multiplication?

Groupwork

2. Working with a partner, consider the areas of the following rectangles. For some values of x, the rectangle on the left has a larger area; for other values of x, the rectangle on the right is larger.

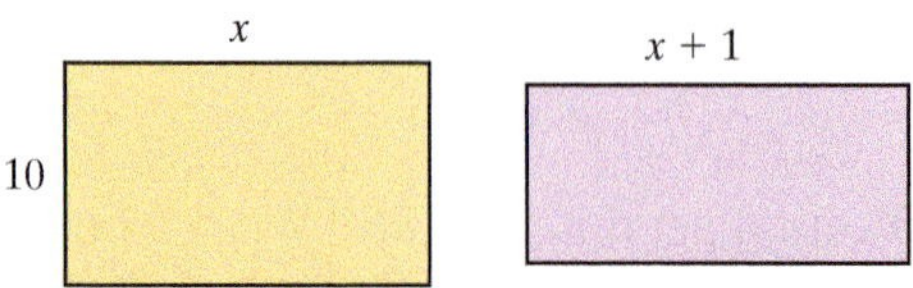

   a. Find a value of x for which the rectangle on the left has the larger area.

   b. Find a value of x for which the area of the rectangle on the right is larger.

Writing

3. Algebra is universal; that is, it is used in all countries of the world regardless of the language spoken. If you know how to speak a language other than English, translate each of the following algebraic expressions to that language.

   a. $7x$ b. $x - 2$ c. $3 + x$ d. $\dfrac{x}{3}$

Cultural Note

Solving an equation to identify an unknown number is similar to using a balance scale to determine an unknown weight. Egyptians 3,400 years ago used balance scales to weigh objects such as gold rings.

The balance scale is an ancient measuring device. These scales were used by Sumerians for weighing precious metals and gems at least 9,000 years ago.

Source: O. Dilke, *Mathematics and Measurement* (Berkeley: University of California Press/British Museum, 1987).

4.2 Solving Addition and Subtraction Equations

OBJECTIVES

- **A** To translate sentences involving addition or subtraction to equations
- **B** To solve addition or subtraction equations
- **C** To solve applied problems involving addition or subtraction equations

What Equations Are and Why They Are Important

An equation contains two expressions separated by an equal sign.

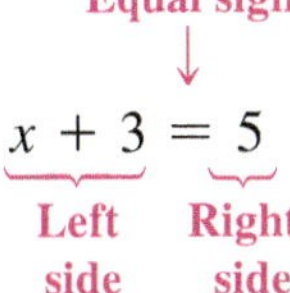

DEFINITION

An **equation** is a mathematical statement that two expressions are equal.

For example,

$$1 + 2 = 3$$
$$x - 5 = 6$$
$$2 + 7 + 3 = 12$$
$$3x = 9$$

are all equations.

Equations are used to solve a wide range of problems. A key step in solving a problem is to translate the sentences that describe the problem to an equation that models the problem. In this section, we focus on equations that involve either addition or subtraction. In the next section, we consider equations involving multiplication or division.

Translating Sentences to Equations

In translating sentences to equations, certain words and phrases mean the same as the equal sign:

- equals
- is
- is equal to
- is the same as
- yields
- results in

Let's look at some examples of translating sentences to equations that involve addition or subtraction and vice versa.

EXAMPLE 1

Translate each sentence in the table to an equation.

Solution

Sentence	Equation
a. The sum of y and 3 is equal to $7\frac{1}{2}$.	$y + 3 = 7\frac{1}{2}$
b. The difference between x and 9 is the same as 14.	$x - 9 = 14$
c. Increasing a number by 1.5 yields 3.	$n + 1.5 = 3$
d. 6 less than a number is 10.	$n - 6 = 10$

PRACTICE 1

Write an equation for each word phrase or sentence.

Sentence	Equation
a. n decreased by 5.1 is 9.	
b. y plus 2 is equal to 12.	
c. The difference between a number and 4 is the same as 11.	
d. 5 more than a number is $7\frac{3}{4}$.	

EXAMPLE 2

In a savings account, the previous balance P plus a deposit of \$7.50 equals the new balance of \$43.25. Write an equation that represents this situation.

Solution

The previous balance plus the deposit equals the new balance.

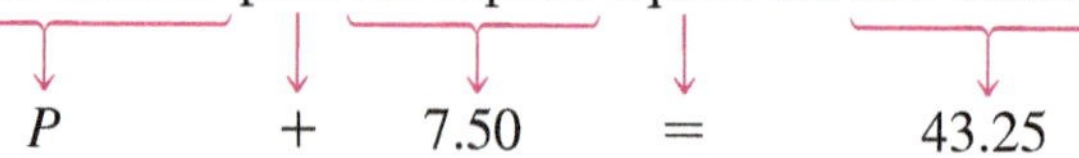

$$P \quad + \quad 7.50 \quad = \quad 43.25$$

So the equation is $P + 7.50 = 43.25$.

PRACTICE 2

The sale price of a jacket is \$49.95. This amount is \$6 less than the regular price p. Write an equation that represents this situation.

Equations Involving Addition and Subtraction

Suppose that you are told that five *more than some number* is equal to seven. You can find that number by solving the *addition equation* $x + 5 = 7$. To solve this equation means to find a number that, when substituted for the variable x, makes the equation a true statement. Such a number is called a *solution* of the equation.

To solve the equation $x + 5 = 7$, we can think of a balance scale like the one shown.

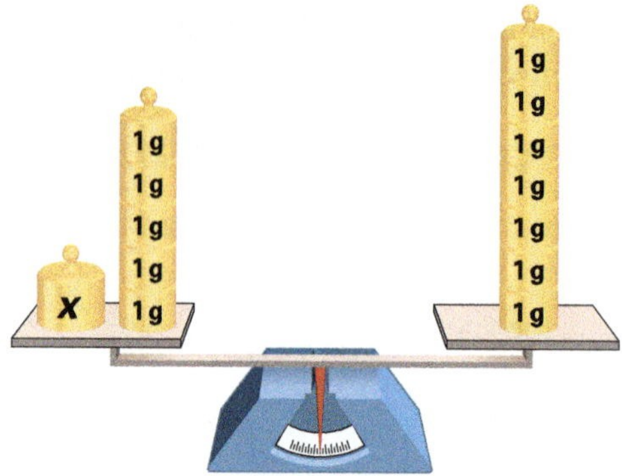

For the balance to remain level, whatever we do to one side, we must also do to the other side. In this case, if we subtract 5 grams from each side of the balance, we can conclude that the unknown weight, x, must be 2 grams, as shown below. So 2 is the solution of the equation $x + 5 = 7$.

Similarly in the *subtraction equation* $x - 5 = 7$, if we add 5 to each side of the equation, we find that x equals 12.

In solving these and other equations, the key is to **isolate the variable**, that is, to get the variable alone on one side of the equation.

These examples suggest the following rule:

To Solve Addition or Subtraction Equations

- For an addition equation, *subtract* the same number from each side of the equation in order to isolate the variable on one side.
- For a subtraction equation, *add* the same number to each side of the equation in order to isolate the variable on one side.
- In either case, check the solution by substituting the value of the unknown in the original equation to verify that the resulting equation is true.

Because addition and subtraction are **opposite operations**, one operation "undoes" the other. The following examples illustrate how to perform an opposite operation to each side of an equation when you are solving for the unknown.

EXAMPLE 3

Solve and check: $y + 9 = 17$

Solution

$$y + 9 = 17$$

$$y + 9 - 9 = 17 - 9 \quad \text{Subtract 9 from each side of the equation.}$$

$$y + 0 = 8$$

$$y = 8 \quad \text{Any number added to 0 is the number.}$$

The solution is 8.

Check

$$y + 9 = 17$$

$$8 + 9 \stackrel{?}{=} 17 \quad \text{Substitute 8 for } y \text{ in the original equation.}$$

$$17 \stackrel{\checkmark}{=} 17 \quad \text{The equation is true, so 8 is the solution to the equation.}$$

Note that, because 9 was added to y in the original equation, we solved by subtracting 9 from both sides of the equation in order to isolate the variable.

PRACTICE 3

Solve and check: $x + 5 = 14$

EXAMPLE 4

Solve and check: $n - 2.5 = 0.7$

Solution

$$n - 2.5 = 0.7$$

$$n - 2.5 + 2.5 = 0.7 + 2.5 \quad \text{Add 2.5 to each side of the equation.}$$

$$n - 0 = 3.2$$

$$n = 3.2$$

The solution is 3.2.

Check

$$n - 2.5 = 0.7$$

$$3.2 - 2.5 \stackrel{?}{=} 0.7 \quad \text{Substitute 3.2 for } n \text{ in the original equation.}$$

$$0.7 \stackrel{\checkmark}{=} 0.7$$

Can you explain why checking an answer is important?

PRACTICE 4

Solve and check: $t - 0.9 = 1.8$

EXAMPLE 5

Solve and check: $x + \frac{1}{3} = 3\frac{1}{2}$

Solution

$$x + \frac{1}{3} = 3\frac{1}{2}$$

$$x + \frac{1}{3} - \frac{1}{3} = 3\frac{1}{2} - \frac{1}{3}$$ Subtract $\frac{1}{3}$ from each side of the equation.

$$x = 3\frac{1}{6}$$

The solution is $3\frac{1}{6}$.

Check

$$x + \frac{1}{3} = 3\frac{1}{2}$$

$$3\frac{1}{6} + \frac{1}{3} \stackrel{?}{=} 3\frac{1}{2}$$ Substitute $3\frac{1}{6}$ for x in the original equation.

$$3\frac{1}{2} \stackrel{\checkmark}{=} 3\frac{1}{2}$$

PRACTICE 5

Solve and check: $m + \frac{1}{4} = 5\frac{1}{2}$

Equations are often useful **mathematical models** of real-world situations, as the following examples show. To derive these models, we need to be able to translate word sentences to algebraic equations, which we then solve.

EXAMPLE 6

Write each sentence as an equation. Then, solve and check.

Solution

Sentence	Equation	Check
a. 15 is equal to y increased by 9.	$15 = y + 9$ $15 - 9 = y + 9 - 9$ $6 = y$, or $y = 6$	$15 = y + 9$ $15 \stackrel{?}{=} 6 + 9$ $15 \stackrel{\checkmark}{=} 15$
b. 10 is equal to m decreased by 8.	$10 = m - 8$ $10 + 8 = m - 8 + 8$ $18 = m$, or $m = 18$	$10 = m - 8$ $10 \stackrel{?}{=} 18 - 8$ $10 \stackrel{\checkmark}{=} 10$

Note that we isolated the variable on the right side of the equation instead of the left side. The result is the same.

PRACTICE 6

Translate each sentence to an algebraic equation. Then, solve and check.

a. 11 is 4 less than m.

b. The sum of 12 and n equals 21.

EXAMPLE 7

Suppose that a chemistry experiment requires students to find the weight of the water in a flask. If the weight of the flask with water is 21.49 grams and the weight of the empty flask is 9.56 grams, write an equation to find the weight of the water. Then, solve and check.

Solution Recall that some problems can be solved by representing the given information in a diagram. Let's use that strategy here.

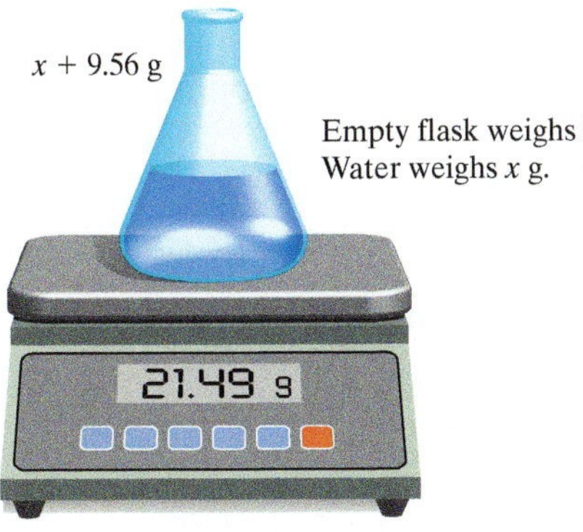

The diagram suggests the equation $21.49 = x + 9.56$, where x represents the weight of the water. Solving this equation, we get:

$$21.49 = x + 9.56$$
$$21.49 - 9.56 = x + 9.56 - 9.56$$
$$11.93 = x, \text{ or } x = 11.93$$

Check

$$21.49 = x + 9.56$$
$$21.49 \stackrel{?}{=} 11.93 + 9.56$$
$$21.49 \stackrel{\checkmark}{=} 21.49$$

So the weight of the water is 11.93 grams.

PRACTICE 7

An online discount book retailer charges a shipping fee of \$3.99. The total cost of a book, including the shipping fee, was \$27.18. Write an equation to determine the cost of the book without the shipping fee. Then, solve and check.

EXAMPLE 8

Harvard College (in Cambridge, Massachusetts) and the College of William and Mary (in Williamsburg, Virginia) are the two oldest institutions of higher learning in the United States. Harvard, founded in 1636, is 57 years older than William and Mary. Write an equation to determine when William and Mary was founded. Then, solve and check. (*Source: National Center for Education Statistics*)

Solution Let y represent the year in which William and Mary was founded. We know that 57 years earlier than the year y is 1636, the year in which Harvard was founded. This gives us the equation:

$$y - 57 = 1636$$

Now, we solve for the unknown.

$$y - 57 + 57 = 1636 + 57$$
$$y = 1693$$

Check

$$y - 57 = 1636$$
$$1693 - 57 \stackrel{?}{=} 1636$$
$$1636 \stackrel{\checkmark}{=} 1636$$

So William and Mary was founded in 1693.

PRACTICE 8

The two U.S. states with the largest area are Alaska and Texas. The area of Texas, 269,000 square miles, is approximately 394,000 square miles smaller than that of Alaska. Write an equation to determine Alaska's area. Then, solve and check. (*Source: The New York Times Almanac, 2010*)

4.2 Exercises

FOR EXTRA HELP MyMathLab MathXL PRACTICE WATCH READ REVIEW

Mathematically Speaking

Fill in each blank with the most appropriate term or phrase from the given list.

constant	subtract	equation
translates	simplifies	variable
add	sentence	

1. A(n) __________ is a mathematical statement that two expressions are equal.

2. A solution of an equation is a number that, when substituted for the __________, makes the equation a true statement.

3. In the equation $x + 2 = 5$, __________ from each side of the equation in order to isolate the variable.

4. The equation $x - 1 = 6$ __________ to the sentence "The difference between x and 1 is 6."

A *Translate each sentence to an equation.*

5. z minus 9 is 25.

6. x minus 7 is 29.

7. The sum of 7 and x is 25.

8. The sum of m and 19 is 34.

9. t decreased by 3.1 equals 4.

10. r decreased by 5.1 equals 6.4.

11. $\frac{3}{2}$ increased by a number yields $\frac{9}{2}$.

12. $\frac{8}{3}$ increased by a number yields $\frac{13}{3}$.

13. $3\frac{1}{2}$ less than a number is equal to 7.

14. $1\frac{1}{2}$ less than a number is equal to $7\frac{1}{4}$.

B *By answering yes or no, indicate whether the value of x shown is a solution of the given equation.*

15.

Equation	Value of x	Solution?
a. $x + 1 = 9$	8	
b. $x - 3 = 4$	5	
c. $x + 0.2 = 5$	4.8	
d. $x - \frac{1}{2} = 1$	$\frac{1}{2}$	

16.

Equation	Value of x	Solution?
a. $x - 39 = 5$	44	
b. $x - 2 = 6$	4	
c. $x + 2.8 = 4$	1.2	
d. $x - \frac{2}{3} = 1$	$1\frac{2}{3}$	

Identify the operation to perform on each side of the equation to isolate the variable.

17. $x + 4 = 6$

18. $x + 10 = 17$

19. $x - 6 = 9$

20. $x - 11 = 4$

21. $x - 7 = 24$

22. $10 = x - 3$

23. $x + 21 = 25$

24. $3 = x + 2$

Solve and check.

25. $a - 7 = 24$

26. $x - 9 = 13$

27. $y + 19 = 21$

28. $z + 23 = 31$

29. $x - 2 = 10$

30. $t - 4 = 19$

31. $n + 9 = 13$

32. $d + 12 = 12$

33. $5 + m = 7$

34. $17 + d = 20$

35. $39 = y - 51$

36. $44 = c - 3$

37. $z + 2.4 = 5.3$

38. $t + 2.3 = 6.7$

39. $n - 8 = 0.9$

40. $c - 0.7 = 6$

41. $y + 8.1 = 9$

42. $a + 0.7 = 2$

43. $x + \frac{1}{3} = 9$

44. $z + \frac{2}{5} = 11$

45. $m - 1\frac{1}{3} = 4$

46. $s - 4\frac{1}{2} = 8$

47. $x + 3\frac{1}{4} = 7$

48. $t + 1\frac{1}{2} = 5$

49. $c - 14\frac{1}{5} = 33$

50. $a - 9\frac{7}{10} = 27\frac{2}{3}$

51. $x - 3.4 = 9.6$

52. $m - 12.5 = 13.7$

53. $5 = y - 1\frac{1}{4}$

54. $3 = t - 1\frac{2}{3}$

55. $5\frac{3}{4} = a + 2\frac{1}{3}$

56. $4\frac{1}{3} = n + 3\frac{1}{2}$

57. $2.3 = x - 5.9$

58. $4.1 = d - 6.9$

59. $y - 7.01 = 12.9$

60. $x - 3.2 = 5.23$

61. $x + 3.443 = 8$

62. $x + 0.035 = 2.004$

63. $2.986 = y - 7.265$

64. $3.184 = y - 1.273$

Translate each sentence to an equation. Solve and check.

65. 3 more than *n* is 11.

66. 15 more than *x* is 33.

67. 6 less than *y* equals 7.

68. 4 less than *t* equals 1.

69. If 10 is added to *n*, the sum is 19.

70. 25 added to a number *m* gives a result of 53.

71. *x* increased by 3.6 is equal to 9.

72. *n* increased by 3.5 is equal to 7.

73. A number minus $4\frac{1}{3}$ is the same as $2\frac{2}{3}$.

74. A number minus $5\frac{1}{2}$ is the same as $2\frac{1}{2}$.

Choose the equation that best describes the situation.

75. After 6 months of dieting and exercising, an athlete lost $8\frac{1}{2}$ pounds. If she now weighs 135 pounds, what was her original weight?

a. $w + 8\frac{1}{2} = 135$ **b.** $w - 126\frac{1}{2} = 8\frac{1}{2}$

c. $w - 8\frac{1}{2} = 135$ **d.** $w + 135 = 143\frac{1}{2}$

76. A teenager has *d* dollars. After buying an Xbox 360 Elite for \$299.99, he has \$6.01 left. How many dollars did he have at first?

a. $d + 299.99 = 306$ **b.** $d - 299.99 = 6.01$

c. $d - 299.99 = 306$ **d.** $d + 6.01 = 299.99$

77. According to a 30-day sample, the two most downloaded English-language authors are the British novelist Charles Dickens and the American humorist Mark Twain. In the sample, there were 37,541 downloads of Dickens and 5,268 fewer of Twain. How many downloads of Twain were there? (*Source:* Project Gutenberg)

a. $x + 5{,}268 = 37{,}541$ **b.** $x - 5{,}268 = 37{,}541$

c. $x + 5{,}268 = 32{,}273$ **d.** $x - 37{,}541 = 5{,}268$

78. The CN Tower in Canada and Canton Tower in China are two of the tallest telecommunications towers in the world. Of these structures, the CN Tower is 555 meters tall, which is 55 meters shorter than the tower in China. How tall is the Canton Tower? (*Source:* gztvtower.info.com)

a. $x + 55 = 555$ **b.** $x + 555 = 610$

c. $x - 55 = 555$ **d.** $x - 555 = 55$

Mixed Practice

Solve and check.

79. $10 = a - 4.5$

80. $x + \frac{1}{2} = 6$

Solve.

81. The life expectancy in the United States of a female born in the year 2000 was 79.3 years. For a female born two decades later, it is projected to be 2.6 years greater. Choose the equation to find the life expectancy of a female born in the year 2020. (*Source:* U.S. Census Bureau)

a. $x + 79.3 = 2.6$ **b.** $x - 2.6 = 79.3$

c. $x + 2.6 = 79.3$ **d.** $x - 20 = 2.6$

82. The hygienist at a dentist's office cleaned a patient's teeth. The total bill came to $125, which was partially covered by dental insurance. If the patient paid $60 out of pocket toward the bill, choose the equation to find how much of the bill was covered by insurance.

a. $x + 60 = 125$ **b.** $x - 60 = 125$

c. $x + 125 = 60$ **d.** $x - 125 = 60$

83. Is 3 a solution to the equation $10 - x = 7$?

84. Is 6 a solution to the equation $x + 4.5 = 7.5$?

85. Translate the sentence "The sum of 4.2 and n is 8" to an equation.

86. Translate the sentence "x decreased by 4 is 10" to an equation.

87. Identify the operation to perform on each side of the equation $y - 1.9 = 5$ to isolate the variable.

88. Identify the operation to perform on each side of the equation $n + 2 = 10$ to isolate the variable.

Applications

C *Write an equation. Solve and check.*

89. A local community college increased the cost of a credit hour by $12 for this year. If the cost of a credit hour for this year is $106, what was the cost last year?

90. The first algebra textbook was written by the Arab mathematician Muhammad ibn Musa al-Khwarazmi. The title of that book, which gave rise to the word *algebra*, was *Aljabr wa'lmuqabalah*, meaning "the art of bringing together unknowns to match a known quantity." If the book appeared in the year 825, how many years ago was this? (*Source:* R.V. Young, *Notable Mathematicians*)

91. In the triangle shown, angles A and B are complementary, that is, the sum of their measures is 90°. Find x, the number of degrees in angle B.

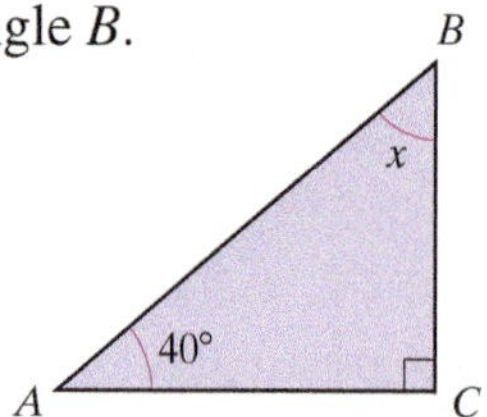

92. In the following diagram, angles ABD and CBD are supplementary, that is, the sum of their measures is 180°. Find y.

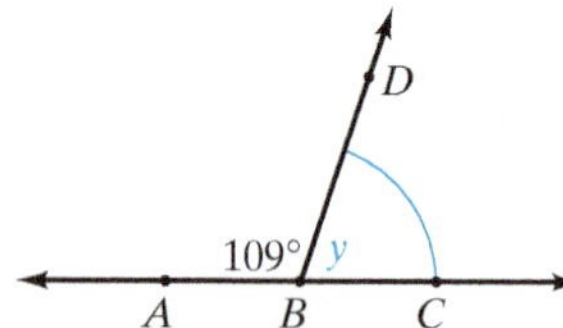

93. An article on Broadway shows reported that this week the box office receipts for a particular show were $621,000. If that amount was $13,000 less than last week's, how much money did the show take in last week?

94. Mount Kilimanjaro, the highest elevation on the continent of Africa, is 299 meters lower than Mount McKinley, the highest elevation on the continent of North America. If Mount Kilimanjaro is 5,895 meters high, how high is Mount McKinley? (*Source: The World Factbook, 2010*)

95. On a state freeway, the minimum speed limit is 45 miles per hour. This is 20 miles per hour lower than the maximum speed limit. What is the maximum speed limit?

96. The melting point of silver is 1,763 degrees Fahrenheit. This is 185 degrees less than the melting point of gold. What is the melting point of gold? (*Source: The New York Times Almanac, 2010*)

97. In a recent year, the U.S. charities that received the greatest private support were two organizations headquartered in Alexandria, Virginia—United Way Worldwide ($4,023,362,895) and the Salvation Army ($1,876,674,000). How much more money did United Way Worldwide receive? (*Source:* philanthropy.com)

98. During a recession, an automobile company laid off 18,578 employees, reducing its workforce to 46,894. How many employees did the company have before the recession?

• Check your answers on page A-6.

MINDStretchers

Groupwork

1. Working with a partner, compare the equations $x - 4 = 6$ and $x - a = b$.

a. Use what you know about the first equation to solve the second equation for x.

b. What are the similarities and the differences between the two equations?

Writing

2. Equations often serve as models for solving word problems.
Write two different word problems corresponding to each of the following equations.

a. $x + 4 = 9$

-
-

b. $x - 1 = 5$

-
-

Critical Thinking

3. In the magic square at the right, the sum of each row, column, and diagonal is the same. Find that sum and write and solve equations to get the values of f, g, h, r, and t.

f	6	11
g	10	h
r	14	t

4.3 Solving Multiplication and Division Equations

OBJECTIVES

- **A** To translate sentences involving multiplication or division to equations
- **B** To solve multiplication or division equations
- **C** To solve applied problems involving multiplication or division equations

Translating Sentences to Equations

In order to translate sentences involving multiplication or division to equations, we must recall the key words that indicate when to multiply and when to divide.

EXAMPLE 1

Translate each sentence in the table to an equation.

Solution

Sentence	Equation
a. The product of 3 and x is equal to 0.6.	$3x = 0.6$
b. The quotient of y and 4 is 15.	$\frac{y}{4} = 15$
c. Two-thirds of a number is 9.	$\frac{2}{3}n = 9$
d. One-half is equal to some number over 6.	$\frac{1}{2} = \frac{n}{6}$

PRACTICE 1

Write an equation for each sentence.

Sentence	Equation
a. Twice x is the same as 14.	
b. The quotient of a and 6 is 1.5.	
c. Some number divided by 0.3 is equal to 1.	
d. Ten is equal to one-half of some number.	

EXAMPLE 2

A house sold for \$125,000. This amount is twice its assessed value x. Write an equation to represent this situation.

Solution The selling price of the house is twice its assessed value x.

125,000 = $2x$

So the equation is $125{,}000 = 2x$.

PRACTICE 2

The area of a rectangle is equal to the product of its length (3 feet) and its width (w). The rectangle's area is 15 square feet. Represent this relationship in an equation.

Equations Involving Multiplication and Division

As with addition equations, we can also solve *multiplication equations* by thinking of a balance scale like the one shown below at the left.

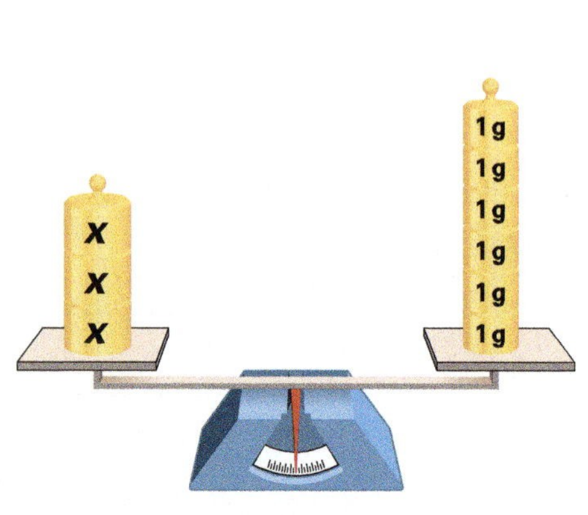

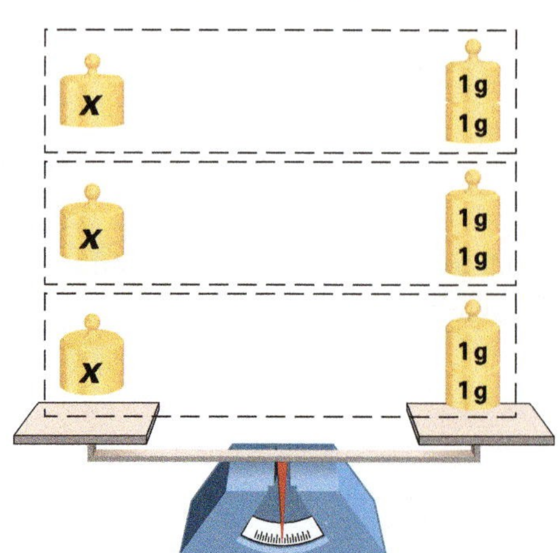

For example, consider the sentence "Three times some number x equals six," which translates to the multiplication equation $3x = 6$. We want to find the number that, when substituted for the variable x, makes this equation a true statement. To keep the balance level, whatever we do to one side we must do to the other side. In this case, dividing each side of the balance by 3 shows that in each group the unknown x must equal 2, as shown on the previous page.

Similarly, in the division equation $\frac{x}{4} = 3$, we can multiply each side of the equation by 4 and then conclude that x equals 12.

These examples suggest the following rule:

To Solve Multiplication or Division Equations

- For a multiplication equation, divide by the same number on each side of the equation in order to isolate the variable on one side.
- For a division equation, multiply by the same number on each side of the equation in order to isolate the variable on one side.
- In either case, check the solution by substituting the value of the unknown in the original equation to verify that the resulting equation is true.

Because multiplication and division are opposite operations, one "undoes" the other. The following examples show how to perform the opposite operation on each side of an equation to solve for the unknown.

EXAMPLE 3

Solve and check: $5x = 20$

Solution $5x = 20$

$\frac{5x}{5} = \frac{20}{5}$ Divide each side of the equation by 5: $\frac{5x}{5} = 1x$, or x.

$x = 4$

The solution is 4.

Check $5x = 20$

$5(4) \stackrel{?}{=} 20$ Substitute 4 for x in the original equation.

$20 \stackrel{\checkmark}{=} 20$ The equation is true, so 4 is the solution to the original equation.

PRACTICE 3

Solve and check: $6x = 30$

In Example 3, can you explain why $1x = x$?

EXAMPLE 4

Solve and check: $5 = \frac{y}{2}$

Solution $5 = \frac{y}{2}$

Multiply each side of the equation by 2:

$2 \cdot 5 = 2 \cdot \frac{y}{2}$ $2 \cdot \frac{y}{2} = \frac{2}{1} \cdot \frac{y}{2} = 1y$, or y.

$10 = y$, or $y = 10$

The solution is 10.

PRACTICE 4

Solve and check: $1 = \frac{a}{6}$

EXAMPLE 4 (continued)

Check $5 = \frac{y}{2}$

$5 \stackrel{?}{=} \frac{10}{2}$ Substitute 10 for y in the original equation.

$5 \stackrel{\checkmark}{=} 5$

EXAMPLE 5

Solve and check: $0.2n = 4$

Solution $0.2n = 4$

$\frac{0.2n}{0.2} = \frac{4}{0.2}$ Divide each side by 0.2: $0.2\overline{)4.0}$ gives $20.$ or 20.

$n = 20$

The solution is 20.

Check $0.2n = 4$

$0.2(20) \stackrel{?}{=} 4$ Substitute 20 for n in the original equation.

$4.0 \stackrel{?}{=} 4$

$4 \stackrel{\checkmark}{=} 4$

PRACTICE 5

Solve and check: $1.5x = 6$

EXAMPLE 6

Solve and check: $\frac{m}{0.5} = 1.3$

Solution $\frac{m}{0.5} = 1.3$

$(0.5)\frac{m}{0.5} = (0.5)(1.3)$ Multiply each side by 0.5.

$m = 0.65$

The solution is 0.65.

Check $\frac{m}{0.5} = 1.3$

$\frac{0.65}{0.5} \stackrel{?}{=} 1.3$ Substitute 0.65 for m in the original equation.

$1.3 \stackrel{\checkmark}{=} 1.3$

PRACTICE 6

Solve and check: $\frac{a}{2.4} = 1.2$

EXAMPLE 7

Solve and check: $\frac{2}{3}n = 6$

Solution

$$\frac{2}{3}n = 6$$

$$\frac{2}{3}n \div \frac{2}{3} = 6 \div \frac{2}{3} \quad \text{Divide each side by } \frac{2}{3}.$$

$$\left(\frac{2}{3}n\right)\left(\frac{3}{2}\right) = 6\left(\frac{3}{2}\right)$$

$$\left(\frac{2}{3}\right)\left(\frac{3}{2}\right)n = 6\left(\frac{3}{2}\right)$$

$$n = 9$$

The solution is 9.

Check

$$\frac{2}{3}n = 6$$

$$\frac{2}{3}(9) \stackrel{?}{=} 6 \quad \text{Substitute 9 for } n \text{ in the original equation.}$$

$$\frac{2}{3}\left(\frac{9}{1}\right) \stackrel{?}{=} 6$$

$$6 \stackrel{\checkmark}{=} 6$$

PRACTICE 7

Solve and check: $\frac{3}{4}x = 12$

As in the case of addition and subtraction equations, multiplication and division equations can be useful mathematical models of real-world situations. To derive these models, we translate word sentences to algebraic equations and solve.

EXAMPLE 8

Write each sentence as an algebraic equation. Then, solve and check.

Solution

Sentence	Equation	Check
a. Thirty-five is equal to the product of 5 and x.	$35 = 5x$ $\frac{35}{5} = \frac{5x}{5}$ $7 = x$, or $x = 7$	$35 = 5x$ $35 \stackrel{?}{=} 5(7)$ $35 \stackrel{\checkmark}{=} 35$
b. One equals p divided by 3.	$1 = \frac{p}{3}$ $3 \cdot 1 = 3 \cdot \frac{p}{3}$ $3 = p$, or $p = 3$	$1 = \frac{p}{3}$ $1 \stackrel{?}{=} \frac{3}{3}$ $1 \stackrel{\checkmark}{=} 1$

PRACTICE 8

Translate each sentence to an equation. Then, solve and check.

Sentence	Equation	Check
a. Twelve is equal to the quotient of z and 6.		
b. Sixteen equals twice x.		

EXAMPLE 9

A baseball player runs 360 feet when hitting a home run.

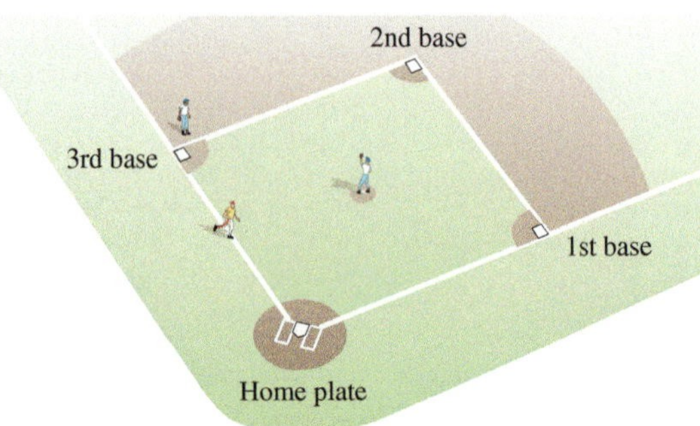

If the distances between successive bases on a baseball diamond are equal, how far is it from third base to home plate? Write an equation. Then, solve and check.

Solution Let x equal the distance between successive bases. Since these distances are equal, $4x$ represents the distance around the bases. But we know that the distance around the bases also equals 360 feet. So $4x = 360$. We solve this equation for x.

$$4x = 360$$

$$\frac{4x}{4} = \frac{360}{4} \quad \text{Divide each side by 4.}$$

$$x = 90$$

Check

$$4x = 360$$

$$4(90) \stackrel{?}{=} 360 \quad \text{Substitute 90 for } x \text{ in the original equation.}$$

$$360 \stackrel{\checkmark}{=} 360$$

So the distance from third base to home plate is 90 feet.

PRACTICE 9

The Pentagon is the headquarters of the U.S. Department of Defense.

The distance around the Pentagon is about 1.6 kilometers. If each side of the Pentagon is the same length, write an equation to find that length. Then, solve and check. (*Source:* Gene Gurney, *The Pentagon*)

EXAMPLE 10

At an annual meeting, town officials recommended allocating a sum of money for emergency training to be split evenly among three committees. What was this sum if each committee was to be allocated for emergency training $4,000?

Solution Let s equal the sum of money allocated. Each committee was allocated $4,000, which is equal to s divided by 3. So we write the following equation:

$$4{,}000 = \frac{s}{3}$$

PRACTICE 10

At a local club, rock concert tickets sold for $25.50 each. If 87 of them were sold, how much money did the club take in from these tickets? Write an equation. Then, solve and check.

We can solve this equation by multiplying both sides by 3.

$$(3)4{,}000 = (3)\frac{s}{3}$$

$$12{,}000 = s, \text{ or } s = 12{,}000$$

Check $4{,}000 = \frac{s}{3}$

$$4{,}000 \stackrel{?}{=} \frac{12{,}000}{3}$$

$$4{,}000 \stackrel{\checkmark}{=} 4{,}000$$

So the sum of money allocated for emergency training was \$12,000.

EXAMPLE 11

The life expectancy in Zambia is one of the lowest of any country in the world. Zambia's life expectancy is only about 39 years, which is approximately $\frac{1}{2}$ that in the United States. Write an equation to find the life expectancy in the United States. Then, solve and check. (*Source:* cia.gov)

Solution Let e equal the U.S. life expectancy. Zambia's life expectancy, 39 years, is equal to $\frac{1}{2}$ of e, so we write the following equation:

$$39 = \frac{1}{2}e$$

We can solve this equation by dividing both sides by $\frac{1}{2}$.

$$39 \div \frac{1}{2} = \frac{1}{2}e \div \frac{1}{2}$$

$$(39)\left(\frac{2}{1}\right) = \left(\frac{1}{2}e\right)\left(\frac{2}{1}\right)$$

$$(39)\left(\frac{2}{1}\right) = \left(\frac{1}{2}\right)\left(\frac{2}{1}\right)e$$

$$78 = e, \text{ or } e = 78$$

Check $39 = \frac{1}{2}e$

$$39 \stackrel{?}{=} \frac{1}{2}(78)$$

$$39 \stackrel{\checkmark}{=} 39$$

So the life expectancy in the United States is approximately 78 years.

PRACTICE 11

Last year, a junior partner in a small law firm received $\frac{3}{8}$ of the firm's profits. What were those profits if she got \$150,000? Write an equation. Then, solve and check.

4.3 Exercises

FOR EXTRA HELP MyMathLab Math XL PRACTICE WATCH READ REVIEW

Mathematically Speaking

Fill in each blank with the most appropriate term or phrase from the given list.

divide	expression	equation	evaluating
addition	division	checked	multiply
substituting	solved		

1. In the equation $2x = 6$, ________ each side of the equation by 2 in order to isolate the variable.

2. In the equation $\frac{x}{5} = 3$, ________ each side of the equation by 5 in order to isolate the variable.

3. Check whether a number is a solution of an equation by ________ the number for the variable in the equation.

4. An equation is ________ by finding its solution.

5. The equal sign separates the two sides of a(n) ________.

6. Multiplication and ________ are opposite operations.

A *Translate each sentence to an equation.*

7. $\frac{3}{4}$ of a number y is 12.

8. $\frac{2}{3}$ of a number x is 20.

9. A number x divided by 7 is equal to $\frac{7}{2}$.

10. A number z divided by 8 is equal to $\frac{8}{3}$.

11. $\frac{1}{3}$ of x is 2.

12. $\frac{1}{4}$ of x is 6.

13. The quotient of a number and 3 is equal to $\frac{1}{3}$.

14. The quotient of a number and 4 is equal to $\frac{1}{4}$.

15. The product of 9 and an amount is the same as 27.

16. The product of 8 and an amount is the same as 32.

B *By answering yes or no, indicate whether the value of x shown is a solution of the given equation.*

17.

Equation	Value of x	Solution?
a. $7x = 21$	3	
b. $3x = 12$	36	
c. $\frac{x}{4} = 8$	2	
d. $\frac{x}{0.2} = 4$	8	

18.

Equation	Value of x	Solution?
a. $\frac{x}{3} = 10$	30	
b. $2.5x = 5$	2	
c. $2x = \frac{1}{3}$	$\frac{1}{6}$	
d. $\frac{x}{0.4} = 3$	12	

Identify the operation to perform on each side of the equation to isolate the variable.

19. $3x = 15$

20. $6y = 18$

21. $\frac{x}{2} = 9$

22. $\frac{y}{6} = 1$

23. $\frac{3}{4}a = 21$

24. $\frac{2}{3}m = 14$

25. $1.5b = 15$

26. $2.6x = 52$

Solve and check.

27. $5x = 30$

28. $8y = 8$

29. $\frac{x}{2} = 9$

30. $\frac{n}{9} = 3$

31. $36 = 9n$

32. $125 = 5x$

33. $\frac{x}{7} = 13$

34. $\frac{w}{10} = 21$

35. $1.7y = 6.8$

36. $0.5a = 7.5$

37. $2.1b = 42$

38. $1.5x = 45$

39. $\frac{m}{15} = 10.5$

40. $\frac{p}{10} = 12.1$

41. $\frac{t}{0.4} = 1$

42. $\frac{n}{0.5} = 6$

43. $\frac{2}{3}x = 1$

44. $\frac{1}{8}n = 3$

45. $\frac{1}{4}x = 9$

46. $\frac{3}{7}t = 15$

47. $17t = 51$

48. $100x = 400$

49. $10y = 4$

50. $100n = 50$

51. $7 = \frac{n}{100}$

52. $40 = \frac{p}{10}$

53. $2.5 = \frac{x}{5}$

54. $4.6 = \frac{z}{2}$

55. $2 = 4x$

56. $3 = 5x$

57. $\frac{14}{3} = \frac{7}{9}m$

58. $\frac{4}{9} = \frac{2}{3}a$

Solve. Round to the nearest tenth. Check.

59. $3.14x = 21.3834$

60. $2.54x = 78.25$

61. $\frac{x}{1.414} = 3.5$

62. $\frac{x}{1.732} = 1.732$

Translate each sentence to an equation. Solve and check.

63. The product of 8 and n is 56.

64. The product of 12 and m is 3.

65. $\frac{3}{4}$ of a number y is equal to 18.

66. $\frac{1}{3}$ of a number x is 16.

67. A number x divided by 5 is 11.

68. A number y divided by 100 is 10.

69. Twice x is equal to 36.

70. 3 times m is 90.

71. $\frac{1}{2}$ of an amount is 4.

72. $\frac{5}{7}$ of a number is 10.

73. A number divided by 5 is equal to $1\frac{3}{5}$.

74. An amount divided by 14 is equal to $1\frac{1}{2}$.

75. The quotient of a number and 2.5 is 10.

76. A quantity divided by 15 equals 3.6.

Choose the equation that best describes each situation.

77. Suppose that a teenager spends \$20, which is $\frac{1}{4}$ of his total savings m. How much money did he have in the beginning?

a. $m - \frac{1}{4} = 20$ **b.** $4m = 20$

c. $m + \frac{1}{4} = 20$ **d.** $\frac{1}{4}m = 20$

78. Find the weight of a child if $\frac{1}{3}$ of her weight is 9 pounds.

a. $3x = 9$ **b.** $\frac{1}{3}x = 9$

c. $x + 3 = 9$ **d.** $x + \frac{1}{3} = 9$

79. A high school student plans to buy an MP3 player 8 weeks from now. If the MP3 player costs \$140, how much money must the student save each week in order to buy it?

a. $8c = 140$ **b.** $c + 8 = 140$

c. $\frac{c}{8} = 140$ **d.** $c - 8 = 140$

80. The student government at a college sold tickets to a play. From the ticket sales, they collected \$300, which was twice the cost of the play. How much did the play cost?

a. $\frac{n}{2} = 300$ **b.** $n - 2 = 300$

c. $2n = 300$ **d.** $n + 2 = 300$

Mixed Practice *Solve and check.*

81. $11 = 2x$

82. $\frac{x}{6} = 9$

Solve.

83. The cost of dinner at a restaurant was split evenly among 3 friends. If each friend paid \$25.75, choose the equation to find the amount on the check.

a. $x + 3 = 25.75$ **b.** $3x = 25.75$

c. $x - 3 = 25.75$ **d.** $\frac{x}{3} = 25.75$

84. According to an Internet speed test, the web connection on a college computer has a download speed of 15 megabits per second (Mbps) and an upload speed of 0.67 Mbps. Choose the equation to find how many times the download speed is the upload speed. (*Source:* compnetworking.about.com)

a. $\frac{x}{0.67} = 15$ **b.** $x + 0.67 = 15$

c. $15x = 0.67$ **d.** $x - 0.67 = 15$

85. Is 25 a solution of the equation $0.4x = 10$?

86. Is 2 a solution of the equation $\frac{n}{3} = 6$?

87. Translate the sentence "Twice x is 5" to an equation.

88. Translate the sentence "The quotient of y and 3 is 6" to an equation.

89. Identify the operation to perform on each side of the equation $\frac{n}{2} = 3$ to isolate the variable.

90. Identify the operation to perform on each side of the equation $4x = 7$ to isolate the variable.

Applications

C *Write an equation. Solve and check.*

91. For a square city block, the perimeter is 60 units. Find the length of one side of the square city block.

92. According to the nutrition label, one packet of regular instant oatmeal has five-eighths the calories of one packet of maple and brown sugar instant oatmeal. If the regular oatmeal has 100 calories, how many calories does the maple and brown sugar oatmeal have?

93. In an Ironman 70.3 triathlon, athletes must complete a 56-mile bike ride. This is one-half the distance of the bike ride in a regular Ironman triathlon. What distance must an athlete bike in a regular Ironman triathlon? (*Source:* World Triathlon Corporation)

94. The town of Ruidoso is located in southeastern New Mexico. During the summer, the population triples, growing to 25,000 as tourists arrive to enjoy the horse racing and the cool mountain air. What is the town's population in the off-season, rounded to the nearest thousand? (*Sources:* wikitravel.org and ruidoso.net)

95. An online media rental service charges members a monthly flat rate of \$8.99 to watch unlimited TV episodes and movies streamed to their televisions and computers. What does the service charge for 3 months?

96. One plan offered by a long-distance phone service provider charges \$0.07 per minute for long-distance phone calls. A customer using this plan was charged \$22.26 for long-distance calls this month. How many minutes of long-distance calls did she make this month?

97. A lab technician prepared an alcohol-and-water solution that contained 60 milliliters of alcohol. This was two-fifths of the total amount of the solution.

a. What was the total amount of solution the lab technician prepared?

b. How much water was in the solution?

98. A sales representative invested \$5,500 of his sales bonus in the stock market. This represents one-third of his total sales bonus.

a. What was his total sales bonus?

b. How much of his sales bonus was not invested in the stock market?

99. Unoccupied 25-story condo towers in Orange County, California sold for $\frac{2}{3}$ off their original value. If the towers sold for \$128 million, what was their original value? (*Source:* lansner.ocregister.com)

100. Attendance at the Brooklyn Museum in 2009 dropped by a quarter from the previous year to about 340,000. What was the attendance in 2008, to the nearest ten thousand? (*Source: New York Times*)

101. The population density of a country is the quotient of the country's population and its land area. The population density of the United States is approximately 86.8 persons per square mile. If the land area of the United States is 3,537,438 square miles, find the U.S. population to the nearest million. (*Source:* census.gov)

102. In a recent year, the top two U.S. airlines in terms of total passengers were Southwest Airlines and American Airlines. Southwest Airlines flew 101,300,000 passengers, which was about 1.18 times the number of passengers that American flew. To the nearest million, how many passengers did American fly? (*Source:* bts.gov)

• Check your answers on page A-6.

MINDStretchers

Writing

1. Write two different word problems that are applications of each equation.

a. $4x = 20$

•

•

b. $\frac{x}{2} = 5$

•

•

Groupwork

2. The equations $\frac{r}{7} = 2$ and $\frac{7}{r} = 2$ are similar in form. Working with a partner, answer the following questions.

a. How would you solve the first equation for r?

b. How can you use what you know about the first equation to solve the second equation for r?

c. What are the similarities and differences between the two equations?

Critical Thinking

3. In a magic square with four rows and four columns, the sum of the entries in each row, column, and diagonal is the same. If the entries are the consecutive whole numbers 1 through 16, what is the sum of the numbers in each diagonal?

Key Concepts and Skills

CONCEPT SKILL

Concept/Skill	Description	Example
[4.1] Variable	A letter that represents an unknown number.	x, y, t
[4.1] Constant	A known number.	$2, \frac{1}{3}, 5.6$
[4.1] Algebraic expression	An expression that combines variables, constants, and arithmetic operations.	$x + 3, \frac{1}{8}n$
[4.1] To evaluate an algebraic expression	• Substitute the given value for each variable. • Carry out the computation.	Evaluate $8 - x$ for $x = 3.5$: $8 - x = 8 - 3.5$, or 4.5
[4.2] Equation	A mathematical statement that two expressions are equal.	$2 + 4 = 6, x + 5 = 7$
[4.2] To solve addition or subtraction equations	• For an addition equation, subtract the same number from each side of the equation in order to isolate the variable on one side. • For a subtraction equation, add the same number to each side of the equation in order to isolate the variable on one side. • In either case, check the solution by substituting the value of the unknown in the original equation to verify that the resulting equation is true.	$y + 9 = 15$ $y + 9 - 9 = 15 - 9$ $y = 6$ Check $y + 9 = 15$ $6 + 9 \stackrel{?}{=} 15$ $15 \stackrel{\checkmark}{=} 15$ $w - 6\frac{1}{2} = 8$ $w - 6\frac{1}{2} + 6\frac{1}{2} = 8 + 6\frac{1}{2}$ $w = 14\frac{1}{2}$ Check $w - 6\frac{1}{2} = 8$ $14\frac{1}{2} - 6\frac{1}{2} \stackrel{?}{=} 8$ $8 \stackrel{\checkmark}{=} 8$
[4.3] To solve multiplication or division equations	• For a multiplication equation, divide by the same number on each side of the equation in order to isolate the variable on one side. • For a division equation, multiply by the same number on each side of the equation in order to isolate the variable on one side. • In either case, check the solution by substituting the value of the unknown in the original equation to verify that the resulting equation is true.	$1.3r = 26$ $\frac{1.3r}{1.3} = \frac{26}{1.3}$ $r = 20$ Check $1.3r = 26$ $1.3(20) \stackrel{?}{=} 26$ $26 \stackrel{\checkmark}{=} 26$ $\frac{x}{7} = 8$ $7 \cdot \frac{x}{7} = 7 \cdot 8$ $x = 56$ Check $\frac{x}{7} = 8$ $\frac{56}{7} \stackrel{?}{=} 8$ $8 \stackrel{\checkmark}{=} 8$

CHAPTER 4 Review Exercises

Say Why

Fill in each blank.

1. In the expression $7x + 5$, x ________ a variable
   is/is not
   because ____
   ____.

2. In the expression $6t - 5$, $6t$ ________ a constant
   is/is not
   because ____
   ____.

3. An algebraic expression __________ include a division
   can/cannot
   symbol because ____
   ____.

4. An equation __________ contain exactly one expression
   can/cannot
   because ____
   ____.

5. The number 28 ________ a solution of the equation
   is/is not
   $72 - x = 44$ because ____
   ____.

6. Rewriting $x - 3 = 10$ as $x = 10 + 3$ ________
   is/is not
   an example of isolating the variable because
   ____.

[4.1] *Translate each algebraic expression to words.*

7. $x + 1$
8. $y + 4$
9. $w - 1$
10. $s - 3$
11. $\frac{c}{7}$
12. $\frac{a}{10}$
13. $2x$
14. $6y$
15. $y \div 0.1$
16. $n \div 1.6$
17. $\frac{1}{3}x$
18. $\frac{1}{10}w$

Translate each word phrase to an algebraic expression.

19. Nine more than m
20. The sum of b and $\frac{1}{2}$
21. y decreased by 1.4
22. Three less than z
23. The quotient of 3 and x
24. n divided by 2.5
25. The product of an amount and 3
26. Twelve times some number

Evaluate each algebraic expression.

27. $b + 8$, for $b = 4$
28. $d + 12$, for $d = 7$
29. $a - 5$, for $a = 5$
30. $c - 9$, for $c = 15$
31. $1.5x$, for $x = 0.2$
32. $1.3t$, for $t = 5$
33. $\frac{1}{2}n$, for $n = 3$
34. $\frac{1}{6}a$, for $a = 2\frac{1}{2}$
35. $w - 9.6$, for $w = 10$
36. $v - 3\frac{1}{2}$, for $v = 8$
37. $\frac{m}{1.5}$, for $m = 2.4$
38. $\frac{x}{0.2}$, for $x = 1.8$

[4.2] *Solve and check.*

39. $x + 11 = 20$

40. $y + 15 = 24$

41. $n - 19 = 7$

42. $b - 12 = 8$

43. $a + 2.5 = 6$

44. $c + 1.6 = 9.1$

45. $x - 1.8 = 9.2$

46. $y - 1.4 = 0.6$

47. $w + 1\frac{1}{2} = 3$

48. $s + \frac{2}{3} = 1$

49. $c - 1\frac{1}{4} = 5\frac{1}{2}$

50. $p - 6 = 5\frac{2}{3}$

51. $7 = m + 2$

52. $10 = n + 10$

53. $39 = c - 39$

54. $72 = y - 18$

55. $38 + n = 49$

56. $37 + x = 62$

57. $4.0875 + x = 35.136$

58. $24.625 = m - 1.9975$

[4.2–4.3] *Translate each sentence to an equation.*

59. n decreased by 19 is 35.

60. 37 less than a equals 234.

61. 9 increased by a number is equal to $15\frac{1}{2}$.

62. 26 more than a number is $30\frac{1}{3}$.

63. Twice y is 16.

64. The product of t and 25 is 175.

65. 34 is equal to n divided by 19.

66. 17 is the quotient of z and 13.

67. $\frac{1}{3}$ of a number equals 27.

68. $\frac{2}{5}$ of a number equals 4.

By answering yes or no, indicate whether the value of x shown is a solution to the given equation.

69.

Equation	Value of x	Solution?
a. $0.3x = 6$	2	
b. $x - \frac{1}{2} = 1\frac{2}{3}$	$2\frac{1}{6}$	
c. $\frac{x}{0.5} = 7$	3.5	
d. $x + 0.1 = 3$	3.1	

70.

Equation	Value of x	Solution?
a. $0.2x = 6$	30	
b. $x + \frac{1}{2} = 1\frac{2}{3}$	$\frac{5}{6}$	
c. $\frac{x}{0.2} = 4.1$	8.2	
d. $x + 0.5 = 7.4$	6.9	

[4.3] *Solve and check.*

71. $2x = 10$

72. $8t = 16$

73. $\frac{a}{7} = 15$

74. $\frac{n}{6} = 9$

75. $9y = 81$

76. $10r = 100$

77. $\frac{w}{10} = 9$

78. $\frac{x}{100} = 1$

79. $1.5y = 30$

80. $1.2a = 144$

81. $\frac{1}{8}n = 4$

82. $\frac{1}{2}b = 16$

83. $\frac{m}{1.5} = 2.1$

84. $\frac{z}{0.3} = 1.9$

85. $100x = 40$

86. $10t = 5$

87. $0.3 = \frac{m}{4}$

88. $1.4 = \frac{b}{7}$

89. $0.866x = 10.825$

90. $\frac{x}{0.707} = 2.1$

Mixed Applications

Write an algebraic expression for each problem. Then, evaluate the expression for the given amount.

91. The temperature increases 2 degrees an hour. By how many degrees will the temperature increase in h hours? In 3 hours?

92. During the fall term, a math tutor works 20 hours per week. What is the tutor's hourly wage if she earns d dollars per week? $191 per week?

93. The local supermarket sells a certain fruit for 89¢ per pound. How much will p pounds cost? 3 pounds?

94. After having borrowed $3,000 from a bank, a customer must pay the amount borrowed plus a finance charge. How much will he pay the bank if the finance charge is d dollars? $225?

Write an equation. Then, solve and check.

95. After depositing $238 in a checking account, the balance will be $517. What was the balance before the deposit?

96. A bowler's final score is the sum of her handicap and scratch score (actual score). If a bowler has a final score of 225 and a handicap of 50, what was her scratch score?

97. Drinking bottled water is more popular in some countries than in others. In a recent year, the per capita consumption of bottled water in the United States was about 100 liters, or approximately 2.9 times as much as it was in the United Kingdom. Find the per capita consumption in the United Kingdom, to the nearest liter. (*Source:* britishbottledwater.org)

98. Hurricane Gilbert was one of the strongest storms to hit the Western Hemisphere in the twentieth century. A newspaper reported that the hurricane left 500,000 people, or about one-fourth of the population of Jamaica, homeless. Approximately what was the population of Jamaica? (*Source:* J. B. Elsner and A. B. Kara, *Hurricanes of the North Atlantic*)

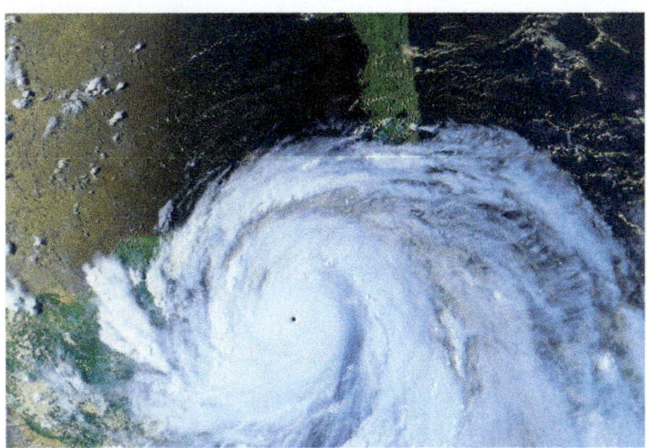

99. On the Moon, a person weighs about one-sixth of his or her weight on Earth. What is the weight on Earth of an astronaut who weighs 30 pounds on the Moon?

100. During an economic recession, a U.S. senator proposed a recovery plan that would cost $3 trillion. Opponents criticized the plan for being too expensive and said that it was 2.5 times the cost of an alternative plan. What was the cost of the alternative plan?

101. The normal body temperature is 98.6 degrees Fahrenheit. An ill patient had a temperature of 101°F. This temperature is how many degrees above normal?

102. This year, a community college received 8,957 applications for admission, which amounts to 256 fewer than were received last year. How many applications did the community college receive last year?

• Check your answers on page A-6.

CHAPTER 4 Posttest

FOR EXTRA HELP

CHAPTER Test Prep VIDEOS

The Chapter Test Prep Videos with test solutions are available on DVD, in MyMathLab, and on YouTube (search "AkstDevMath" and click on "Channels").

To see if you have mastered the topics in this chapter, take this test.

Write each algebraic expression in words.

1. $x + \frac{1}{2}$

2. $\frac{a}{3}$

Translate each word phrase to an algebraic expression.

3. 10 less than a number

4. The quotient of 8 and p

Evaluate each algebraic expression.

5. $a - 1.5$, for $a = 1.5$

6. $\frac{b}{9}$, for $b = 2\frac{1}{4}$

Translate each sentence to an equation.

7. The difference between x and 6 is $4\frac{1}{2}$.

8. The quotient of y and 8 is 3.2.

Solve and check.

9. $x + 10 = 10$

10. $y - 6 = 6$

11. $81 = 3n$

12. $82 = \frac{a}{9}$

13. $m - 1.8 = 6$

14. $1.5n = 75$

15. $10x = 5\frac{1}{2}$

16. $\frac{n}{100} = 7.6$

Write an equation. Then, solve and check.

17. A recipe for seafood stew requires $2\frac{1}{4}$ pounds of fish. After buying $1\frac{3}{4}$ pounds of bluefish, a chef decides to fill out the recipe with codfish. How many pounds of codfish should he buy?

18. In a recent year, the federal government removed the Rocky gray wolf from the list of endangered species, resulting in the first Rocky gray wolf hunting season in decades. During this season, about one-fourth of the population, or an estimated 500 wolves, were killed. Approximately how many Rocky gray wolves had lived in the wild before the hunt began? (*Source:* treehugger.com)

19. According to official estimates, the world population in 2045 is projected to be about one-and-a-half times what it was in 1999. If the projected population is 9 billion, what was the population in 1999? (*Source:* census.gov)

20. In chemistry, an *endothermic reaction* is one that absorbs heat. As a result of an endothermic reaction, the temperature of a solution dropped by 19.8 degrees Celsius to 7.6°C. What was the temperature of the solution before the reaction took place? (*Source:* Timberlake, *Chemistry: An Introduction to General, Organic, and Biological Chemistry*)

• Check your answers on page A-7.

Cumulative Review Exercises

To help you review, solve the following:

1. Round 314,159 to the nearest hundreds.

2. One of the three differences below is wrong. Use estimation to identify which difference is incorrect.

a. $\begin{array}{r} 675{,}029 \\ -126{,}384 \\ \hline 548{,}645 \end{array}$ **b.** $\begin{array}{r} 539{,}324 \\ -126{,}384 \\ \hline 412{,}940 \end{array}$ **c.** $\begin{array}{r} 954{,}736 \\ -365{,}976 \\ \hline 488{,}760 \end{array}$

3. Multiply: 804×29

4. Find the quotient and check: $35{,}020 \div 34$

5. Write the equivalent fraction with the given denominator: $\frac{7}{8} = \frac{\quad}{96}$

6. Write as a mixed number in simplified form: $\frac{56}{40}$

7. Subtract: $8\frac{1}{4} - 2\frac{7}{8}$

8. Write the decimal 5.239 in words.

9. Insert the appropriate sign, $<$, $=$, or $>$, to make a true statement. 6.356 6.36

10. Compute: $12 - (3.2 + 4.91)$

11. Find the quotient: $7.5 \div 1{,}000$

12. Decide whether 2 is a solution of the equation $w + 3 = 5$

13. Solve and check: $n - 3.8 = 4$

14. Solve and check: $\frac{x}{2} = 16$

15. In animating a cartoon, artists had to draw 24 images to appear during 1 second of screen time. How many images did they have to draw to produce a 5-minute cartoon?

16. Dental insurance reimbursed a patient \$200 on a bill of \$700. Did the patient get less or more than $\frac{1}{3}$ of his money back? Explain.

17. In a recent review of world energy, the annual oil production of the United States was approximately 300 million tons, or three-fifths of Saudi Arabia's production. Write an equation to find Saudi Arabia's annual oil production. Then solve and check. (*Source: Top 10 of Everything 2010*)

18. According to the Environmental Protection Agency, U.S. residents produce about 4.4 pounds of garbage per person per day. At this rate, how much garbage does a family of 3 produce in a week? (*Source:* epa.gov)

19. Farmers depend on bees to pollinate many crop plants, such as apples and cherries. In the American Midwest, the acreage of crops is large as compared with the number of bees, so farmers are especially concerned if the number of beehives decline. When the number of beehives in the state of Illinois dropped from 101,000 to 46,000, how big a drop was this? (*Source:* ag.uiuc.edu)

20. Due to World War I, the number of U.S. military personnel on active duty in 1918—roughly 2.9 million—was about 4.5 times what it had been the year before. (*Source: The New York Times Almanac 2010*)

a. Write an equation to find the number of personnel in 1917.

b. Solve, rounding the solution to the nearest hundred thousand.

• Check your answers on page A-7.

CHAPTER 5

Ratio and Proportion

Ratio, Proportion, and the Connecticut Compromise

One of the most contested issues at the 1787 Constitutional Convention in Philadelphia was how the U.S. Congress was to be structured. The convention delegates from large states, such as Massachusetts, supported the Virginia Plan in which the number of representatives from a state would be *proportional* to its population so that a state with double the population of another would have twice as many representatives. By contrast, the delegates from smaller states, such as Delaware, favored the New Jersey Plan, in which every state has the same number of representatives in Congress.

Finally, after months of heated debate, the convention delegates agreed to the Connecticut Compromise under which Congress would have two chambers whose approval would be required for legislation to pass. In one chamber, the Senate, big and small states alike would have the same number of representatives. In the other chamber, the House of Representatives, representation would be proportional so that for all states the *ratio* of the population to the number of House representatives would be roughly the same. To this day, the Constitution provides that seats in the House are apportioned among the states by population, as determined by the census conducted every ten years.

(*Source:* Douglas Brinkley, *American Heritage History of the United States*, Viking, 1998)

CHAPTER 5 PRETEST

To see if you have already mastered the topics in this chapter, take this test.

Write each ratio or rate in simplest form.

1. 6 to 8

2. 40 to 100

3. \$30 to \$18

4. 19 feet to 51 feet

5. 48 gallons of water in 15 minutes

6. 10 milligrams every 6 hours

Find the unit rate.

7. 12 dental assistants for every 6 dentists

8. 35 calculators for 35 students

Determine the unit price.

9. \$690 for 3 boxes of ceramic tiles

10. 12 bottles of lemon iced tea for \$6.00

Determine whether each proportion is true or false.

11. $\frac{2}{3} = \frac{16}{24}$

12. $\frac{32}{20} = \frac{8}{3}$

Solve and check.

13. $\frac{6}{8} = \frac{x}{12}$

14. $\frac{21}{x} = \frac{2}{3}$

15. $\frac{\frac{1}{2}}{4} = \frac{2}{x}$

16. $\frac{x}{6} = \frac{8}{0.3}$

Solve.

17. A contractor combines 80 pounds of sand with 100 pounds of gravel. In this mixture, what is the ratio of sand to gravel?

18. A machine at a potato chip factory can peel 12,000 pounds of potatoes in 60 minutes. At this rate, how many pounds of potatoes can it peel per minute?

19. The *aspect ratio* of an image is the ratio of the image's width to its height. In digital camera photos, a common aspect ratio is 4 to 3. With this ratio, how high is a photo that is 6 inches wide? (*Source:* wikipedia.org)

20. Suppose the scale on a Louisiana map is 3 inches to 31 miles. If two cities, Baton Rouge and New Orleans, are 7.4 inches apart on the map, what is the actual distance, to the nearest mile, between the two cities? (*Source:* ersys.com)

• Check your answers on page A-7.

5.1 Introduction to Ratios

OBJECTIVES

- **A** To write ratios of like quantities in simplest form
- **B** To write rates in simplest form
- **C** To solve applied problems involving ratios

What Ratios Are and Why They Are Important

We frequently need to compare quantities. Sports, medicine, and business are just a few areas where we use **ratios** to make comparisons. Consider the ratios in the following examples.

- The volleyball team won 4 games for every 3 they lost.
- A physician assistant prepared a 1-to-25 boric acid solution.
- The stock's price-to-earnings ratio was 13 to 1.

Can you think of other examples of ratios in your daily life?

The preceding examples illustrate the following definition of a ratio.

DEFINITION

A **ratio** is a comparison of two quantities expressed as a quotient.

There are several ways to write a ratio. For instance, we can write the ratio 3 to 10 as

$$3 \text{ to } 10 \qquad 3:10 \qquad \frac{3}{10}$$

No matter which notation we use for this ratio, it is read "3 to 10."

Simplifying Ratios

Because a ratio can be written as a fraction, we can say that, as with any fraction, a ratio is in simplest form (or written in lowest terms) when 1 is the only common factor of the numerator and denominator.

Let's consider some examples of writing ratios in simplest form.

EXAMPLE 1

Write the ratio 10 to 5 in simplest form.

Solution The ratio 10 to 5 expressed as a fraction is $\frac{10}{5}$.

$$\frac{10}{5} = \frac{10 \div 5}{5 \div 5} = \frac{2}{1}$$

So the ratio 10 to 5 is the same as the ratio 2 to 1. Note that the ratio 2 to 1 means that the first number is twice as large as the second number.

PRACTICE 1

Write the ratio 8:12 in simplest form.

Frequently, we deal with quantities that have units, such as months or feet. When both quantities in a ratio have the same unit, they are called **like quantities**. In a ratio of like quantities, the units drop out.

EXAMPLE 2

Express the ratio 5 months to 3 months in simplest form.

Solution The ratio 5 months to 3 months expressed as a fraction is $\frac{5 \text{ months}}{3 \text{ months}}$. Simplifying, we get $\frac{5}{3}$, which is already in lowest terms.

PRACTICE 2

Express in simplest form the ratio 9 feet to 5 feet.

TIP Note that with ratios we do not rewrite improper fractions as mixed numbers because our answer must be a comparison of *two* numbers. So in Example 2, we write the ratio as $\frac{5}{3}$ rather than as $1\frac{2}{3}$.

EXAMPLE 3

24-karat gold is pure gold. By contrast, 14-karat gold, commonly used to make jewelry, consists of 14 parts out of 24 parts pure gold; the rest is another metal such as copper or silver added for hardness. In 14-karat gold, what is the ratio of gold to the other metal? (*Source:* essortment.com)

Solution In 14-karat gold, 14 parts of 24 parts are pure gold. First, we calculate the number of parts that are not pure gold.

$$24 - 14 = 10$$

So 10 of 24 parts are the other metal. Next, let's write the ratio of pure gold to the other metal.

$$\frac{\text{Number of parts of pure gold}}{\text{Number of parts of the other metal}} = \frac{14}{10} = \frac{7}{5}$$

We conclude that the ratio of gold to the other metal is 7 to 5.

PRACTICE 3

The Waist to Hip Ratio (WHR) is a ratio, commonly expressed as a decimal, that has been shown to be a good predictor of possible cardiovascular problems in both men and women. If a male has a WHR greater than 1, then he is considered to be at high risk for these problems. Calculate the WHR of a male with a waist measurement of 40 inches and a hip measurement 2 inches less. Is he at high risk? (*Source: The Medical Journal of Australia*)

Now, let's compare **unlike quantities**, that is, quantities that have different units or are different kinds of measurement. Such a comparison is called a **rate**.

DEFINITION
A **rate** is a ratio of unlike quantities.

For instance, suppose that your rate of pay is \$52 for each 8 hours of work. Simplifying this rate, we get:

$$\frac{\$52}{8 \text{ hours}} = \frac{\$13}{2 \text{ hours}}$$

So you are paid \$13 for every 2 hours that you worked. Note that the units are expressed as part of the answer.

EXAMPLE 4

Simplify each rate.

a. 350 miles to 18 gallons of gas

b. 18 trees to produce 2,000 pounds of paper

Solution

a. 350 miles to 18 gallons $= \dfrac{350 \text{ miles}}{18 \text{ gallons}} = \dfrac{175 \text{ miles}}{9 \text{ gallons}}$

b. 18 trees to 2,000 pounds $= \dfrac{18 \text{ trees}}{2{,}000 \text{ pounds}} = \dfrac{9 \text{ trees}}{1{,}000 \text{ pounds}}$

PRACTICE 4

Express each rate in simplest form.

a. 150 milliliters of medication infused every 60 minutes

b. 18 pounds lost in 12 weeks

Examples 1, 2, 3, and 4 illustrate the following rule for simplifying a ratio or rate:

To Simplify a Ratio or Rate

- Write the ratio or rate as a fraction.
- Express the fraction in simplest form.
- If the quantities are like, drop the units. If the quantities are unlike, keep the units.

Frequently, we want to find a particular kind of rate called a *unit rate*. In the rate $\dfrac{\$13}{2 \text{ hours}}$, for instance, it would be useful to know what is earned for each hour (that is, the hourly wage). We need to rewrite $\dfrac{\$13}{2 \text{ hours}}$ so that the denominator is 1 hour.

$$\frac{\$13}{2 \text{ hours}} = \frac{\$13 \div 2}{2 \text{ hours} \div 2} = \frac{\$6.50}{1 \text{ hour}} = \$6.50 \text{ per hour, or } \$6.50/\text{hr}$$

Note that "per" means "divided by."

Here, we divided the numbers in both the numerator and denominator by the number in the denominator.

DEFINITION
A **unit rate** is a rate in which the number in the denominator is 1.

EXAMPLE 5

Write as a unit rate.

a. 275 miles in 5 hours

b. \$3,453 for 6 weeks

Solution First, we write each rate as a fraction. Then, we divide numbers in the numerator and denominator by the number in the denominator, getting 1 in the denominator.

PRACTICE 5

Express as a unit rate.

a. a fall of 192 feet in 4 seconds

b. 15 hits in 40 times at bat

EXAMPLE 5 (continued)

a. 275 miles in 5 hours $= \dfrac{275 \text{ miles}}{5 \text{ hours}} = \dfrac{275 \text{ miles} \div 5}{5 \text{ hours} \div 5} = \dfrac{55 \text{ miles}}{1 \text{ hour}}$, or 55 mph

b. \$3,453 for 6 weeks $= \dfrac{\$3{,}453}{6 \text{ weeks}} = \dfrac{\$3{,}453 \div 6}{6 \text{ weeks} \div 6} = \dfrac{\$575.50}{1 \text{ week}}$, or \$575.50 per week

EXAMPLE 6

To reduce expenses, a commuter buys a fuel-efficient car. If the car goes 60 miles on 2.5 gallons of gas, what is its fuel economy, that is, how many miles per gallon does it get?

Solution To find the car's fuel economy, we calculate the ratio of the distance that it travels (60 miles) to the amount of gas that it uses (2.5 gallons).

$$\text{Fuel economy} = \frac{60 \text{ miles}}{2.5 \text{ gallons}}$$

Next, we simplify by dividing both the numerator and denominator by the number in the denominator, 2.5:

$$\frac{60 \text{ miles}}{2.5 \text{ gallons}} = \frac{24 \text{ miles}}{1 \text{ gallon}}$$

So the car gets 24 miles per gallon.

PRACTICE 6

Because of heavy traffic, a bus took 30 minutes to cover a distance of 20 city blocks. How many minutes per city block did the bus move?

In order to get the better buy, we sometimes compare prices by computing the price of a single item. This **unit price** is a type of unit rate.

DEFINITION
A **unit price** is the price of one item, or one unit.

To find a unit price, we write the ratio of the total price of the units to the number of units and then, simplify.

$$\text{Unit price} = \frac{\text{Total price}}{\text{Number of units}}$$

Let's consider some examples of unit pricing.

EXAMPLE 7

Find the unit price.

a. \$300 for 12 months of membership

b. 6 credits for \$234

c. 10-ounce box of wheat flakes for \$2.70

PRACTICE 7

Determine the unit price.

a. 4 supersaver flights for \$696

b. \$22 for 8 hours of parking

c. \$19.80 for 20 song downloads

Solution

a. $\dfrac{\$300}{12 \text{ months}} = \$25/\text{month}$

b. $\dfrac{\$234}{6 \text{ credits}} = \$39/\text{credit}$

c. $\dfrac{\$2.70}{10 \text{ ounces}} = \$0.27/\text{ounce}$

In the next example, we apply the concept of unit price to determine which is the better deal.

EXAMPLE 8

For the following two boxes of bandages, which is the better buy?

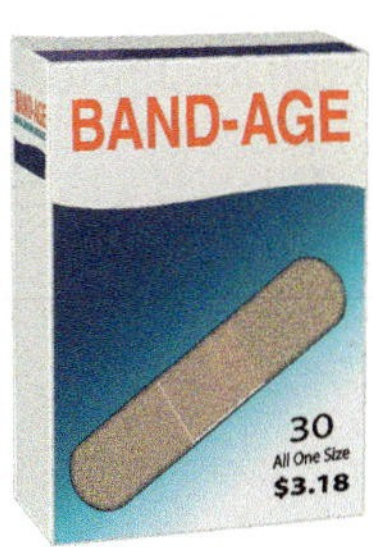

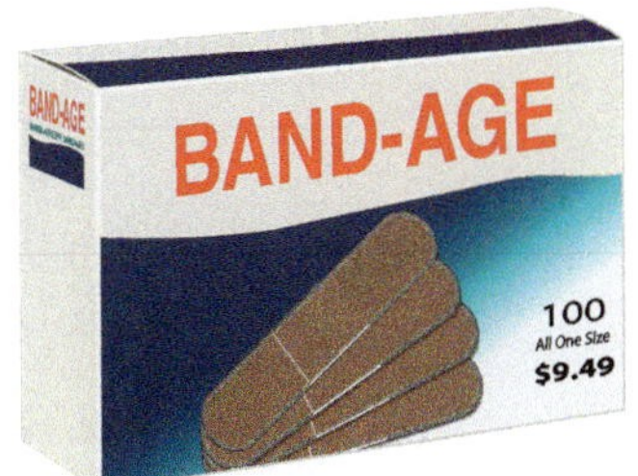

Solution First, we find the unit price for each box of bandages. Then, we round to the nearest cent and compare the prices.

$$\text{Unit price} = \frac{\text{Total price}}{\text{Number of units}} = \frac{\$3.18}{30} = \$0.106 \approx \$0.11 \text{ per bandage}$$

$$\text{Unit price} = \frac{\text{Total price}}{\text{Number of units}} = \frac{\$9.49}{100} = \$0.0949 \approx \$0.09 \text{ per bandage}$$

Since $\$0.09 < \$0.11$, the better buy is the box with 100 bandages.

PRACTICE 8

Which bottle of vitamin C has the lower unit price?

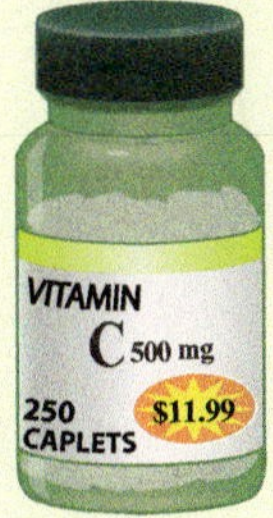

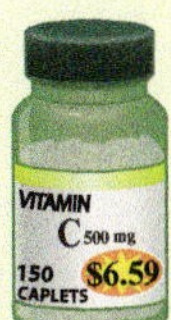

Cultural Note

The shape of a grand piano is dictated by the length of its strings. When a stretched string vibrates, it produces a particular pitch, say C. A second string of comparable tension will produce another pitch, which depends on the ratio of the string lengths. For instance, if the ratio of the second string to the first string is 18 to 16, then plucking the second string will produce the pitch B.

Around 500 B.C., the followers of the mathematician Pythagoras learned to adjust string lengths in various ratios so as to produce an entire scale. Thus the concept of ratio is central to the construction of pianos, violins, and many other musical instruments.

Sources: John R. Pierce, *The Science of Musical Sound* (New York: Scientific American Library, 1983)
David Bergamini, *Mathematics* (New York: Time-Life Books, 1971)

5.1 Exercises

FOR EXTRA HELP MyMathLab 

Mathematically Speaking

Fill in each blank with the most appropriate term or phrase from the given list.

weight of a unit	difference	number of units
like	fractional form	denominator
simplest form	unlike	
numerator	quotient	

1. A ratio is a comparison of two quantities expressed as a(n) ____________.

2. A rate is a ratio of ____________ quantities.

3. A ratio is said to be in ____________ when 1 is the only common factor of the numerator and denominator.

4. Quantities that have the same units are called ____________ quantities.

5. A unit rate is a rate in which the number in the ____________ is 1.

6. To find the unit price, divide the total price of the units by the ____________.

A *Write each ratio in simplest form.*

7. 6 to 9
8. 9 to 12
9. 10 to 15
10. 21 to 27
11. 55 to 35
12. 8 to 10
13. 2 to $1\frac{1}{3}$
14. 25 to $1\frac{1}{4}$
15. 2.5 to 10
16. 1.25 to 100
17. 60 minutes to 45 minutes
18. \$40 to \$25
19. 10 feet to 10 feet
20. 75 tons to 75 tons
21. 30¢ to 18¢
22. 66 years to 32 years
23. 7 miles per hour to 24 miles per hour
24. 21 gallons to 20 gallons
25. 1,000 acres to 50 acres
26. 2,000 miles to 25 miles
27. 8 grams to 7 grams
28. 19 ounces to 51 ounces
29. 24 seconds to 30 seconds
30. 28 milliliters to 42 milliliters

B *Write each rate in simplest form.*

31. 25 telephone calls in 10 days
32. 42 gallons in 4 minutes
33. 288 calories burned in 40 minutes
34. 190 e-mails in 25 days
35. 2 million hits on a website in 6 months
36. 50 million troy ounces of gold produced in 12 months
37. 68 baskets in 120 attempts
38. 18 boxes of cookies for \$45
39. 296 points in 16 games
40. 12 knockouts in 16 fights
41. 500 square feet of carpeting for \$1,645
42. 300 full-time students to 200 part-time students

43. 48 males for every 9 females
44. 3 case workers for every 80 clients
45. 40 Democrats for every 35 Republicans
46. $12,500 in 6 months
47. 2 pounds of zucchini for 16 servings
48. 57 hours of work in 9 days
49. 1,535 flights in 15 days
50. 25 pounds of plaster for 2,500 square feet of wall
51. 3 pounds of grass seeds for 600 square feet of lawn
52. 684 parts manufactured in 24 hours

Determine the unit rate.

53. 3,375 revolutions in 15 minutes
54. 3,000 houses to 1,500 acres of land
55. 120 gallons of heating oil for 15 days
56. 48 yards in 8 carries
57. 3 tanks of gas to cut 10 acres of lawn
58. 192 meters in 6 seconds
59. 8 yards of material for 5 dresses
60. 648 heartbeats in 9 minutes
61. 20 hours of homework in 10 days
62. $200 for 8 hours of work
63. A run of 5 kilometers in 20 minutes
64. 56 calories in 4 ounces of orange juice
65. 140 fat calories in 2 tablespoons of peanut butter
66. 60 children for every 5 adults

Find the unit price, rounding to the nearest cent if necessary.

67. 12 bars of soap for $5.40
68. 4 credit hours for $200
69. 6 rolls of film that cost $17.70
70. 2 notebooks that cost $6.90
71. 3 plants for $200
72. $240,000 for a 30-second prime time television commercial spot
73. 5 nights in a hotel for $495
74. 60 minutes of Internet access for $3

Complete each table, rounding if necessary. Determine which is the better buy.

75. Honey lemon cough drops

Number of Units	Total Price	Unit Price
30	$1.69	
100	$5.49	

76. Stretchable disposable diapers

Number of Units	Total Price	Unit Price
36	$8.69	
60	$14.99	

77. Staples® Bright White Inkjet paper

Number of Units (Sheets)	Total Price	Unit Price
500	$9.69	
2,500	$42.99	

78. Colgate® Total toothpaste

Number of Units (Ounces)	Total Price	Unit Price
6	$3.59	
7	$4.19	

Fill in the table. Which is the best buy?

79. Glad® trash bags, large, with drawstring

Number of Units	Total Price	Unit Price
14	$8.49	
25	$11.49	
28	$7.49	

80. CVS® AA alkaline batteries

Number of Units	Total Price	Unit Price
4	$2.57	
10	$3.89	
24	$6.89	

Mixed Practice

Solve.

81. To the nearest cent, find the unit price of an 18-ounce jar of creamy peanut butter that costs $2.89.

82. Complete the table. Then, find the best buy. Starbucks® Cappuccino

Number of units (Fluid ounces)	Total Price	Unit Price
12	$3.15	
16	$3.95	
24	$4.25	

83. Simplify the rate: 4 tutors for every 30 students.

84. Write as a unit rate: 50 lots to 0.2 square mile.

85. Write the ratio 20 to 4 in simplest form.

86. Express $\frac{30 \text{ centimeters}}{45 \text{ centimeters}}$ in simplest form.

Applications

C *Solve. Simplify if possible.*

87. The number line shown is marked off in equal units. Find the ratio of the length of the distance x to the distance y.

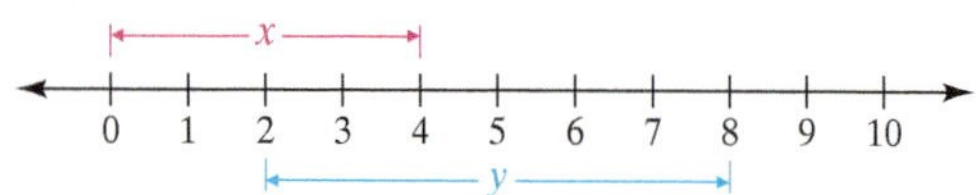

88. In the following rectangle, what is the ratio of the width to the length?

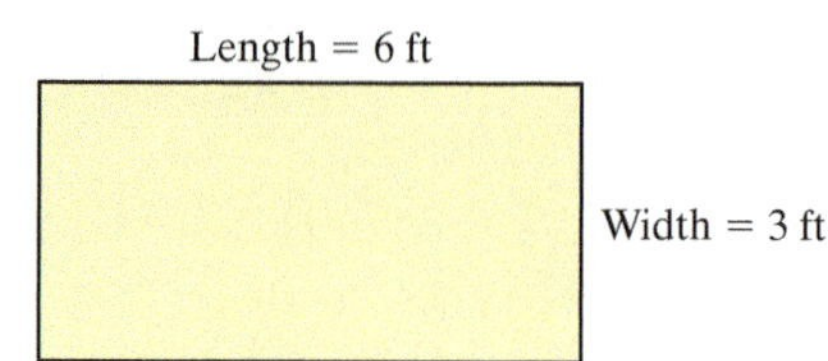

89. In 10 ounces of cashew nuts, there are 1,700 calories. How many calories are there per ounce?

90. For a building valued at $200,000 the property tax is $4,000. Find the ratio of the tax to the building's value.

91. The *cartridge yield* of a computer printer is the number of pages that it will print before the toner runs out. A cartridge that sells for $69.99 has a yield of 2,000 pages. Rounded to the nearest cent, what is the cost per page of printing? (*Source:* smartcomputing.com)

92. A bathtub contains 20 gallons of water. If the tub empties in 4 minutes, what is the rate of flow of the water per minute?

93. In a student government election, 1,000 students cast a vote for the incumbent, 900 voted for the opponent, and 100 cast a protest vote. What was the ratio of the incumbent's vote to the total number of votes?

94. At a college, 4,500 of the 7,500 students are female. What is the ratio of females to males at the college?

95. In finance, the *return on investment* (ROI) is the ratio of profit or loss on an investment relative to the amount of money invested. ROI is commonly calculated to compare the performance of one investment relative to another. Find this ratio for an investment of $9,000 with a profit of $1,500. (*Sources:* investopedia.com and ehow.com)

96. About 15,600 people can ride El Toro, a roller coaster at Six Flags Great Adventure in New Jersey, in 12 hours. Approximately how many people per hour can ride El Toro? (*Source:* wikipedia.org)

97. At the beginning of the 112th U.S. Congress, there were 193 Democrats in the House of Representatives and 51 Democrats in the Senate. For Republicans, there were 242 in the House and 47 in the Senate. Was the ratio of Democrats to Republicans higher in the House or in the Senate? (*Source:* wikipedia.org)

98. The table below shows the breakdown of the number of patients in two hospital units at a local city hospital. Is the ratio of nurses to patients in the intensive care unit higher or lower than the ratio of nurses to patients in the medical unit?

	Intensive Care Unit	Medical Unit
Patients	25	65
Nurses	8	11

99. The following table deals with five of the longest-reigning monarchs in history.

Monarch	Country	Reign	Length of Reign (in years)
King Louis XIV	France	1643–1715	72
King John II	Liechtenstein	1858–1929	71
Emperor Franz-Josef	Austria-Hungary	1848–1916	68
Queen Victoria	United Kingdom	1837–1901	64
Emperor Hirohito	Japan	1926–1989	63

(*Source: The Top 10 of Everything 2006*)

a. What is the ratio of Emperor Hirohito's length of reign to that of Emperor Franz-Josef?

b. What is the ratio of Queen Victoria's length of reign to that of King Louis XIV?

100. The following bar graph deals with popular singers and the number of their albums that "went platinum," that is, sold more than 1 million copies.

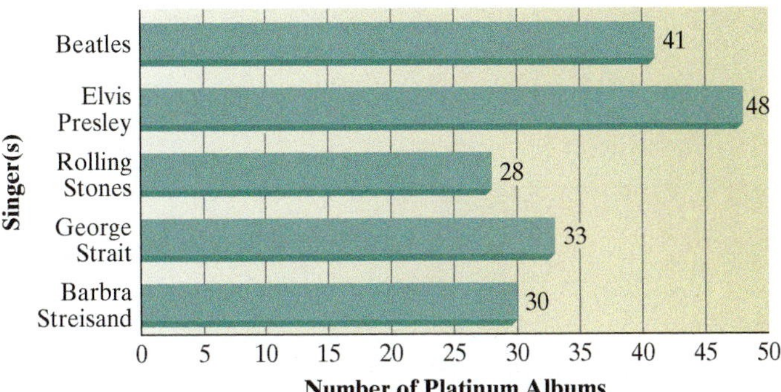

(*Source:* riaa.com)

a. Find the ratio of the number of platinum albums for Elvis Presley as compared to George Strait.

b. What is the ratio of the number of platinum albums for the Rolling Stones as compared to Barbra Streisand?

101. In the insurance industry, a loss ratio is the ratio of total losses paid out by an insurance company to total premiums collected for a given time period.

$$\text{Loss ratio} = \frac{\text{Losses paid}}{\text{Premiums collected}}$$

In 2 months, a certain insurance company paid losses of \$6,400,000 and collected premiums of \$12,472,000. What is the loss ratio, rounded to the nearest hundredth?

102. Analysts for a brokerage firm prepare research reports on companies with stocks traded in various stock markets. One statistic that an analyst uses is the price-to-earnings (P.E.) ratio.

$$\text{P.E. ratio} = \frac{\text{Market price per share}}{\text{Earnings per share}}$$

Find the P.E. ratio, rounded to the nearest hundredth, for a stock that had a per-share market price of \$70.75 and earnings of \$5.37 per share.

• Check your answers on page A-7.

MINDStretchers

History

1. For a **golden rectangle**, the ratio of its length to its width is approximately 1.618 to 1 (the **golden ratio**).

To the ancient Egyptians and Greeks, a golden rectangle was considered to be the ratio most pleasing to the eye. Show that index cards in either of the two standard sizes (3×5 and 5×8) are close approximations to the golden rectangle.

Investigation

2. The distance around a circle is called its **circumference** (C). The distance across the circle through its center is called its **diameter** (d).

a. Use a string and ruler to measure C and d for both circles shown.

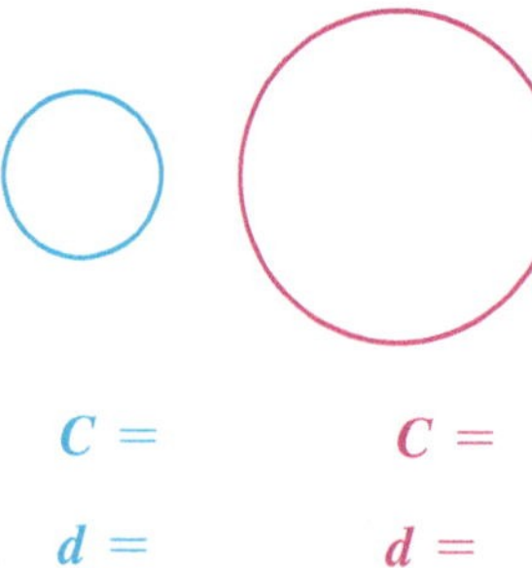

$C =$ $C =$

$d =$ $d =$

b. Compute the ratio of C to d for each circle. Are the ratios approximately equal?

$\frac{C}{d} =$ $\frac{C}{d} =$

Writing

3. Sometimes we use *differences* rather than *quotients* to compare two quantities. Give an example of each kind of comparison and any advantages and disadvantages of each approach.

5.2 Solving Proportions

OBJECTIVES

A To write or solve proportions

B To solve applied problems involving proportions

Writing Proportions

When two ratios—for instance, 1 to 2 and 4 to 8—are equal, they are said to be *in proportion*. We can write "1 is to 2 as 4 is to 8" as $\frac{1}{2} = \frac{4}{8}$. Such an equation is called a **proportion**.

Proportions are common in daily life and are used in many situations, such as finding the distance between two cities from a map with a given scale or the amount of a worker's pay for four weeks given the weekly pay.

DEFINITION

A **proportion** is a statement that two ratios are equal.

One way to see if a proportion is true is to determine whether the *cross products* of the ratios are equal. For example, we see that the proportion

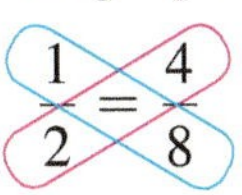

is true, because $2 \cdot 4 = 1 \cdot 8$, or $8 = 8$. However, the proportion $\frac{3}{5} = \frac{9}{10}$ is not true, since $5 \cdot 9 \neq 3 \cdot 10$.

EXAMPLE 1

Determine whether the proportion 4 is to 3 as 16 is to 12 is true.

Solution First, we write the ratios in fractional form: $\frac{4}{3} = \frac{16}{12}$

$$\frac{4}{3} = \frac{16}{12}$$

$3 \cdot 16 \stackrel{?}{=} 4 \cdot 12$ Set the cross products equal.

$48 \stackrel{\checkmark}{=} 48$

So the proportion 4 is to 3 as 16 is to 12 is true.

PRACTICE 1

Are the ratios 10 to 4 and 15 to 6 in proportion?

EXAMPLE 2

Is $\frac{15}{9} = \frac{8}{5}$ a true proportion?

Solution $\frac{15}{9} \stackrel{?}{=} \frac{8}{5}$

$9 \cdot 8 \stackrel{?}{=} 15 \cdot 5$ Set the cross products equal.

$72 \neq 75$

The cross products are not equal. So the proportion is not true.

PRACTICE 2

Determine whether $\frac{15}{6} = \frac{8}{3}$ is a true proportion.

EXAMPLE 3

A college claims that the student-to-faculty ratio is 13 to 1. If there are 96 faculty for 1,248 students, is the college's claim true?

Solution The college claims a student-to-faculty ratio of $\frac{13}{1}$, and the actual ratio of students to faculty is $\frac{1,248}{96}$. We want to know if these two ratios are equal.

$$\begin{array}{rl} \text{Students} \rightarrow 13 & 1,248 \leftarrow \text{Students} \\ \text{Faculty} \rightarrow \overline{\ 1\ } \stackrel{?}{=} & \overline{\ 96\ } \leftarrow \text{Faculty} \end{array}$$

$$1 \cdot 1,248 \stackrel{?}{=} 13 \cdot 96 \quad \text{Set the cross products equal.}$$

$$1,248 \stackrel{\checkmark}{=} 1,248$$

Since the cross products are equal, the college's claim is true.

PRACTICE 3

A company has a policy making the compensation of its CEO proportional to the dividends that are paid to shareholders. If the dividends increase from \$72 to \$80 and the CEO's compensation is increased from \$360,000 to \$420,000, was the company's policy followed?

Solving Proportions

Suppose that you make \$840 for working 4 weeks in a book shop. At this rate of pay, how much money will you make in 10 weeks? To solve this problem, we can write a proportion in which the rates compare the amount of pay to the time worked. We want to find the amount of pay corresponding to 10 weeks, which we call x.

$$\begin{array}{c} \text{Pay} \rightarrow \\ \text{Time} \rightarrow \end{array} \frac{840}{4} = \frac{x}{10} \begin{array}{c} \leftarrow \text{Pay} \\ \leftarrow \text{Time} \end{array}$$

After setting the cross products equal, we find the missing value.

$$\frac{840}{4} = \frac{x}{10}$$

$$4 \cdot x = 840 \cdot 10$$

$$4x = 8,400$$

$$\frac{4x}{4} = \frac{8,400}{4} \quad \text{Divide each side of the equation by 4.}$$

$$x = 2,100$$

So you will make \$2,100 in 10 weeks.

We can check our solution by substituting 2,100 for x in the original proportion.

$$\frac{840}{4} = \frac{x}{10}$$

$$\frac{840}{4} \stackrel{?}{=} \frac{2,100}{10}$$

$$4 \cdot 2,100 \stackrel{?}{=} 840 \cdot 10 \quad \text{Set the cross products equal.}$$

$$8,400 \stackrel{\checkmark}{=} 8,400$$

Our solution checks.

To Solve a Proportion

- Find the cross products, and set them equal.
- Solve the resulting equation.
- Check the solution by substituting the value of the unknown in the original equation to be sure that the resulting proportion is true.

EXAMPLE 4

Solve and check: $\dfrac{2}{3} = \dfrac{x}{15}$

Solution

$$\frac{2}{3} = \frac{x}{15}$$

$$3 \cdot x = 2 \cdot 15 \quad \text{Set the cross products equal.}$$

$$3x = 30$$

$$\frac{3x}{3} = \frac{30}{3} \quad \text{Divide each side by 3.}$$

$$x = 10$$

The solution is 10.

Check

$$\frac{2}{3} = \frac{x}{15}$$

$$\frac{2}{3} \stackrel{?}{=} \frac{10}{15} \quad \text{Substitute 10 for } x.$$

$$3 \cdot 10 \stackrel{?}{=} 2 \cdot 15 \quad \text{Set the cross products equal.}$$

$$30 \stackrel{\checkmark}{=} 30$$

PRACTICE 4

Solve and check: $\dfrac{x}{6} = \dfrac{12}{9}$

EXAMPLE 5

Solve and check: $\dfrac{\frac{1}{4}}{12} = \dfrac{x}{96}$

Solution

$$\frac{\frac{1}{4}}{12} = \frac{x}{96}$$

$$12 \cdot x = \frac{1}{4} \cdot 96 \quad \text{Set the cross products equal.}$$

$$12x = 24$$

$$\frac{12x}{12} = \frac{24}{12} \quad \text{Divide each side by 12.}$$

$$x = 2$$

The solution is 2.

Check

$$\frac{\frac{1}{4}}{12} = \frac{x}{96}$$

$$\frac{\frac{1}{4}}{12} \stackrel{?}{=} \frac{2}{96} \quad \text{Substitute 2 for } x.$$

$$12(2) \stackrel{?}{=} \frac{1}{4} \cdot (96) \quad \text{Set the cross products equal.}$$

$$24 \stackrel{\checkmark}{=} 24$$

PRACTICE 5

Solve and check: $\dfrac{\frac{1}{2}}{2} = \dfrac{3}{x}$

EXAMPLE 6

Forty pounds of sodium hydroxide are needed to neutralize 49 pounds of sulfuric acid. At this rate, how many pounds of sodium hydroxide are needed to neutralize 98 pounds of sulfuric acid? (*Source:* Peter Atkins and Loretta Jones, *Chemistry*)

Solution Let n represent the number of pounds of sodium hydroxide needed. We set up a proportion to compare the amount of sodium hydroxide to the amount of sulfuric acid.

$$\frac{40}{49} = \frac{n}{98} \quad \begin{array}{l}\text{Sodium hydroxide} \\ \text{Sulfuric acid}\end{array}$$

$$49n = 40 \cdot 98 \quad \text{Set the cross products equal.}$$

$$49n = 3{,}920$$

$$\frac{49n}{49} = \frac{3{,}920}{49} \quad \text{Divide each side by 49.}$$

$$n = 80$$

Check

$$\frac{40}{49} = \frac{n}{98}$$

$$\frac{40}{49} \stackrel{?}{=} \frac{80}{98} \quad \text{Substitute 80 for } n.$$

$$49 \cdot 80 \stackrel{?}{=} 40 \cdot 98 \quad \text{Set the cross products equal.}$$

$$3{,}920 \stackrel{\checkmark}{=} 3{,}920$$

So 80 pounds of sodium hydroxide are needed to neutralize 98 pounds of sulfuric acid.

PRACTICE 6

Saffron is a powder made from crocus flowers and is used in the manufacture of perfume. Some 8,000 crocus flowers are required to make 2 ounces of saffron. How many flowers are needed to make 16 ounces of saffron? (*Source: The World Book Encyclopedia*)

TIP A good way to set up a proportion is to write quantities of the same kind in the numerators and their corresponding quantities of the other kind in the denominators.

EXAMPLE 7

If St. Louis and Cincinnati on the map below are 1.6 inches apart, what is the actual distance between them?

$\frac{1}{2}$ in. = 100 mi

PRACTICE 7

The first aircraft to fly faster than the speed of sound was a research plane piloted by Major Charles E. Yeager of the U.S. Air Force on Oct. 14, 1947. Aircraft speed, especially supersonic, is commonly expressed as a Mach number—the ratio of the speed of the aircraft to the speed of sound (1,066 kilometers per hour). If Yeager's plane reached a top speed of 1,126 kilometers per hour exceeding the speed of sound, what was the Mach speed of the plane, rounded to the nearest hundredth? (*Source:* concorde-jet.com)

Solution We know that $\frac{1}{2}$ inch corresponds to 100 miles. Let's set up a proportion that compares inches to miles, letting m represent the unknown number of miles.

$$\frac{\frac{1}{2}\text{ inch}}{100\text{ miles}} = \frac{1.6\text{ inches}}{m\text{ miles}}$$

$$\frac{\frac{1}{2}}{100} = \frac{1.6}{m}$$

$$\frac{1}{2}m = (100)(1.6)$$ Set the cross products equal.

$$\frac{1}{2}m = 160$$

$$\frac{1}{2}m \div \frac{1}{2} = 160 \div \frac{1}{2}$$ Divide each side by $\frac{1}{2}$.

$$\frac{1}{2}m \times \frac{2}{1} = 160 \times \frac{2}{1}$$

$$m = 320$$

So the cities are 320 miles apart.

Check

$$\frac{\frac{1}{2}}{100} = \frac{1.6}{m}$$

$$\frac{\frac{1}{2}}{100} \stackrel{?}{=} \frac{1.6}{320}$$

$$100(1.6) \stackrel{?}{=} \frac{1}{2} \cdot (320)$$

$$160 \stackrel{\checkmark}{=} 160$$

EXAMPLE 8

In the following diagram, the heights and shadow lengths of the two objects shown are in proportion. Find the height of the tree h.

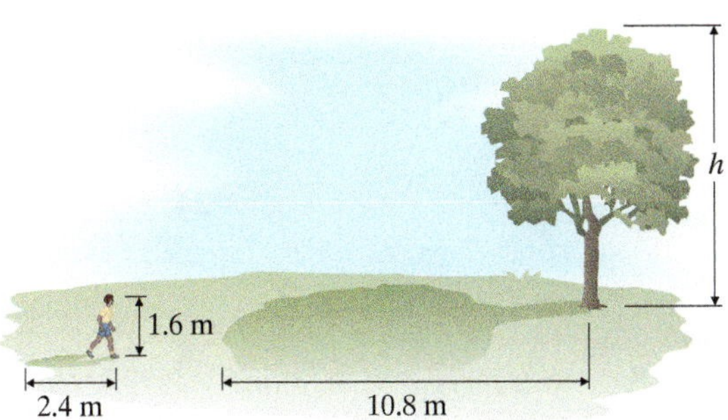

Solution The heights and shadow lengths are in proportion, so we write the following:

$$\frac{\text{Height} \rightarrow h\text{ meters}}{\text{Shadow} \rightarrow 10.8\text{ meters}} = \frac{1.6\text{ meters} \leftarrow \text{Height}}{2.4\text{ meters} \leftarrow \text{Shadow}}$$

$$\frac{h}{10.8} = \frac{1.6}{2.4}$$

$$2.4h = (10.8)(1.6)$$

$$\frac{2.4h}{2.4} = \frac{17.28}{2.4}$$

$$h = 7.2$$

So the height of the tree is 7.2 meters.

Check

$$\frac{h}{10.8} = \frac{1.6}{2.4}$$

$$\frac{7.2}{10.8} \stackrel{?}{=} \frac{1.6}{2.4}$$

$$(10.8)(1.6) \stackrel{?}{=} (7.2)(2.4)$$

$$17.28 \stackrel{\checkmark}{=} 17.28$$

PRACTICE 8

A community college in Hawaii has a student-to-faculty ratio of 14 to 1. How many faculty members are at the college if it has 4,200 students? (*Source:* hawaii.edu)

5.2 Exercises

FOR EXTRA HELP MyMathLab  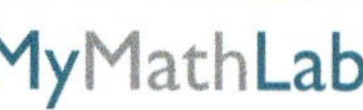    

PRACTICE WATCH READ REVIEW

Mathematically Speaking

Fill in each blank with the most appropriate term or phrase from the given list.

equation	check	like
products	solve	cross products
as	proportion	

1. A(n) __________ is a statement that two ratios are equal.

2. To determine if a proportion is true, check whether the __________ of the ratios are equal.

3. The proportion $\frac{4}{5} = \frac{8}{10}$ can be read "4 is to 5 __________ 8 is to 10."

4. To __________ the proportion $\frac{x}{2} = \frac{4}{6}$, find the value of x that makes the proportion true.

A *Indicate whether each statement is true or false.*

5. Thirty is to 9 as 40 is to 12.

6. Nine is to 12 as 12 is to 16.

7. Two is to 3 as 7 is to 16.

8. Three is to 8 as 10 is to 27.

9. One and one-tenth is to 0.3 as 44 is to 12.

10. One and one-half is to 2 as 0.6 is to 0.8.

11. $\frac{3}{6} = \frac{2}{5}$

12. $\frac{4}{7} = \frac{5}{8}$

13. $\frac{12}{28} = \frac{18}{42}$

14. $\frac{28}{24} = \frac{7}{6}$

15. $\frac{6}{1} = \frac{3}{\frac{1}{2}}$

16. $\frac{5}{30} = \frac{\frac{1}{3}}{2}$

Solve and check.

17. $\frac{4}{8} = \frac{10}{x}$

18. $\frac{3}{2} = \frac{42}{x}$

19. $\frac{x}{19} = \frac{10}{5}$

20. $\frac{x}{78} = \frac{1}{6}$

21. $\frac{5}{x} = \frac{15}{12}$

22. $\frac{15}{x} = \frac{6}{10}$

23. $\frac{4}{1} = \frac{52}{x}$

24. $\frac{17}{1} = \frac{51}{x}$

25. $\frac{7}{4} = \frac{14}{x}$

26. $\frac{18}{15} = \frac{6}{x}$

27. $\frac{x}{8} = \frac{3}{6}$

28. $\frac{x}{35} = \frac{5}{7}$

29. $\frac{70}{x} = \frac{21}{6}$

30. $\frac{4}{x} = \frac{92}{23}$

31. $\frac{x}{12} = \frac{25}{20}$

32. $\frac{x}{45} = \frac{20}{25}$

33. $\frac{28}{x} = \frac{36}{27}$

34. $\frac{24}{x} = \frac{27}{63}$

35. $\frac{x}{10} = \frac{4}{3}$

36. $\frac{x}{2} = \frac{6}{5}$

37. $\frac{4}{x} = \frac{\frac{2}{5}}{10}$

38. $\frac{\frac{3}{4}}{6} = \frac{3}{x}$

39. $\frac{x}{27} = \frac{1.6}{24}$

40. $\frac{x}{1.8} = \frac{28}{24}$

41. $\frac{10.5}{x} = \frac{5}{10}$

42. $\frac{9}{x} = \frac{7.2}{32}$

43. $\frac{7}{0.9} = \frac{x}{36}$

44. $\frac{56}{4.8} = \frac{x}{18}$

45. $\frac{600}{x} = \frac{3}{1\frac{1}{2}}$

46. $\frac{12}{x} = \frac{5}{2\frac{1}{3}}$

47. $\frac{15}{2} = \frac{x}{2\frac{2}{3}}$

48. $\frac{20}{3} = \frac{x}{1\frac{1}{2}}$

49. $\frac{\frac{1}{2}}{\frac{1}{5}} = \frac{x}{4}$

50. $\frac{2}{\frac{4}{5}} = \frac{\frac{2}{3}}{x}$

51. $\frac{3}{2.7} = \frac{6}{x}$

52. $\frac{1.5}{6} = \frac{x}{1.2}$

53. $\frac{\frac{1}{3}}{x} = \frac{2}{1.2}$

54. $\frac{2.5}{x} = \frac{\frac{1}{4}}{50}$

55. $\frac{x}{0.16} = \frac{0.15}{4.8}$

56. $\frac{1.5}{1.25} = \frac{x}{0.5}$

Mixed Practice

57. Solve and check: $\frac{\frac{3}{4}}{5} = \frac{x}{8}$

58. Solve and check: $\frac{1.6}{x} = \frac{2.4}{27}$

59. Solve and check: $\frac{3}{2} = \frac{2\frac{2}{5}}{x}$

60. Determine whether the proportion 8 is to 1 as 2 is to $\frac{1}{4}$ is true.

61. Is $\frac{4}{9} = \frac{3}{8}$ a true or false statement?

62. Solve and check: $\frac{x}{9} = \frac{5}{6}$

B Applications

Solve and check.

63. An average adult's heart beats 8 times every 6 seconds, whereas a newborn baby's heart beats 7 times every 3 seconds. Determine whether these rates are the same. (*Source: Mosby's Medical, Nursing, and Allied Health Dictionary*)

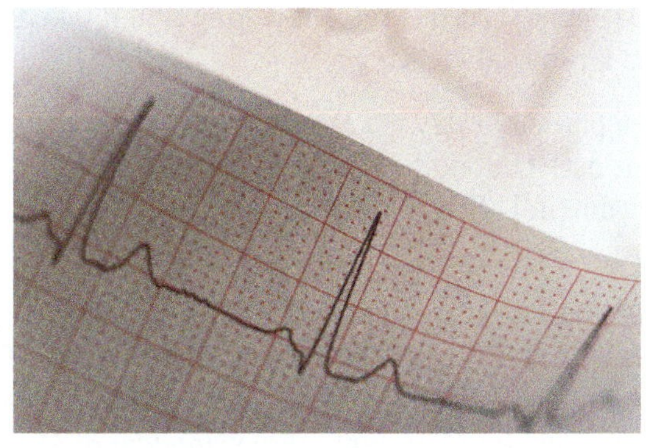

64. An intravenous fluid is infused at a rate of 2.5 milliliters per minute. How many milliliters are infused per hour?

65. A dripping faucet wastes about 15 gallons of water daily. About how much water is wasted in 3 hours? (*Hint:* 1 day = 24 hours)

66. A full-time college student paid tuition of \$1,296 for 12 credits, and a part-time student paid \$1,008 for 9 credits. Were the tuition rates the same?

67. The recommended daily allowance of protein for adults is 0.8 grams for every 2.2 pounds of body weight. If you weigh 150 pounds, how many grams of protein to the nearest tenth should you consume each day? (*Source: The Nutrition Desk Reference*)

68. A homeowner is preparing a solution of insecticide and water to spray her house plants. The directions on the insecticide bottle instruct her to mix 1 part of insecticide with 50 parts of water. How much water must she mix with 2 tablespoons of insecticide?

69. In carbon dioxide molecules, for every 2 oxygen atoms there is 1 carbon atom. How many oxygen atoms combine with 50 carbon atoms to form carbon dioxide molecules?

70. The scale on a map is $\frac{1}{4}$ centimeter to 50 kilometers. Find the actual distance between two towns represented by 10 centimeters on the map.

71. The following rectangular photo is to be enlarged so that the width of the enlargement is 25 inches. If the dimensions of the photo are to remain in proportion, what should the length of the enlargement be?

72. Architects use computers to render their designs. If the actual length of the kitchen shown on the computer-generated floor plan is 10 feet, what is the actual length of the dining area?

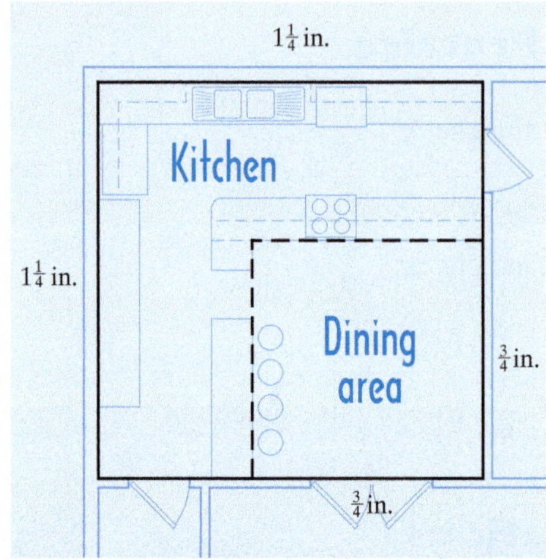

73. A popular scale for building model railroads is the N scale, in which model trains are $\frac{1}{160}$ the size of actual trains. Using this scale, what is the model size of a boxcar that is actually 40 feet long?

74. Thirty gallons of oil flow through a pipe in 4 hours. At this rate, how long will it take for 280 gallons to flow through this same pipe?

75. The ratio of your federal income tax to state income tax is 10 to 3. How much is your state income tax if your federal income tax is \$2,000?

76. A student's computer can download a 4-megabyte song in 3 seconds. At this rate, how long will it take to download a 6-megabyte song?

77. To determine the number of fish in a lake, researchers tagged 150 of them. In a later sample, they found that 6 of 480 fish were tagged. About how many fish were in the lake?

78. On a particular day, 89 Japanese yen were worth the same as 1 U.S. dollar. If a shirt cost 1,780 yen, what was its value in U.S. dollars?

79. A 5-speed bicycle has a chain linking the pedal sprocket and the gears on the rear wheel. The ratio of pedal turns to rear-wheel turns in first gear is 9 to 14. How many times in first gear does the rear wheel turn if the pedals turn 180 times?

80. The tallest land animal is the giraffe. How tall is a giraffe that casts a shadow 320 centimeters long, if a man nearby who is 180 centimeters tall casts a shadow 100 centimeters long? (*Source: Encyclopedia of Mammals*)

81. A tablet of medication consists of two substances in the ratio of 9 to 5. If the tablet contains 140 milligrams of medication, how much of each substance is in the tablet?

82. A certain metal is 5 parts tin and 2 parts lead. How many kilograms of each are there in 28 kilograms of the metal?

83. The nutrition label from a box of General Mills Total cereal indicates that a $\frac{3}{4}$-cup serving contains the following:

Nutrition Facts
Fat 0.5 g
Cholesterol 0 mg
Sodium 190 mg
Potassium 90 mg
Carbohydrates 23 g
Protein 2 g

a. How many grams of carbohydrates are there in 3 cups of cereal?

b. What is the amount of protein in $1\frac{1}{2}$ cups of cereal?

84. The following recipe is for Nilla Apple Crisp:

Nilla Apple Crisp — *Serves 12*

4 large Granny Smith apples (2 pounds), peeled, thinly sliced
1/2 cup of packed brown sugar
2 teaspoons of ground cinnamon
1/3 cup of old-fashioned or quick-cooking oats
1/4 cup of cold margarine
25 Reduced Fat Nilla Wafers, crushed (about 1 cup of crumbs)
1 1/2 cups of thawed Cool Whip Lite Whipped Topping

(*Source:* Kraftrecipes.com)

a. How many cups of thawed Cool Whip Lite Whipped Topping are needed for 18 servings?

b. What is the number of servings if $1\frac{1}{2}$ cups of packed brown sugar are used and the other ingredients are increased proportionately?

85. A senator reported that 640 metric tons of spent nuclear fuel had produced 660,000 gallons of nuclear waste. At this rate, how many gallons of nuclear waste, to the nearest thousand, would be produced by 810 metric tons of fuel?

86. A car uses 0.16 gallon of gas to travel through a tunnel 3.6 miles long. At this rate, how many gallons of gas, to the nearest whole number, are needed to travel 2,885 miles across country?

• Check your answers on page A-7.

MINDStretchers

Mathematical Reasoning

1. Pictorial comparisons (called *analogies*) are used on many standardized tests. Fill in the blank.

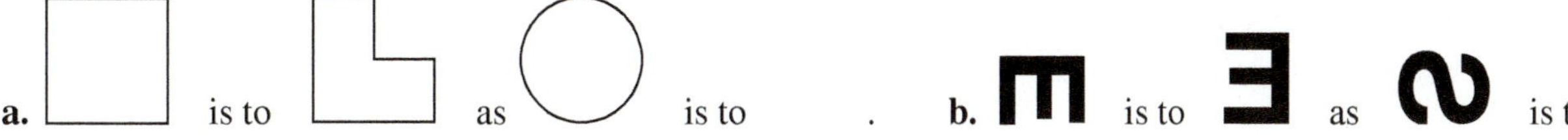

Groupwork

2. Work with a partner on the following.

a. Complete the following table.

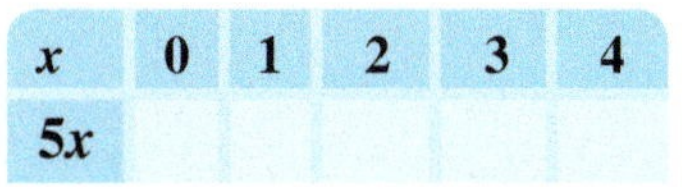

x	0	1	2	3	4
$5x$					

b. Write as many true proportions as you can, based on the values in the table.

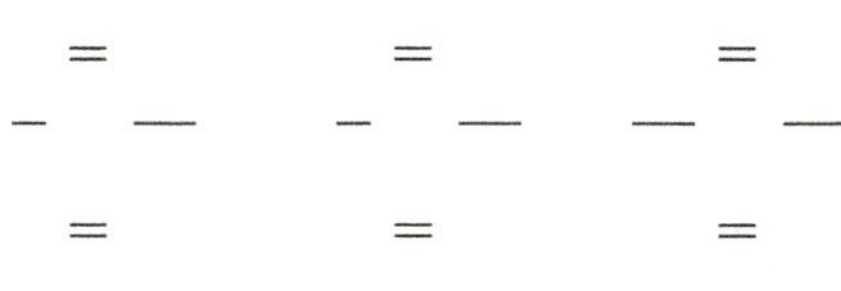

Technology

3. On the web, there are many currency calculators that convert a given amount of a first currency into the equivalent amount of a second currency. Locate one such calculator. Use your knowledge of proportions to confirm that the currency calculator is working correctly.

Key Concepts and Skills

CONCEPT SKILL

Concept/Skill	Description	Example
[5.1] **Ratio**	A comparison of two quantities expressed as a quotient.	3 to 4, $\frac{3}{4}$, or 3:4
[5.1] **Rate**	A ratio of unlike quantities.	$\frac{10 \text{ students}}{3 \text{ tutors}}$
[5.1] **To simplify a ratio**	• Write the ratio as a fraction. • Express the fraction in simplest form. • If the quantities are alike, drop the units. If the quantities are unlike, keep the units.	9:27 is the same as 1:3, because $\frac{9}{27} = \frac{1}{3}$ 21 hours to 56 hours $= \frac{21 \text{ hours}}{56 \text{ hours}} = \frac{21}{56} = \frac{3}{8}$ 175 miles per 7 gallons $= \frac{175 \text{ miles}}{7 \text{ gallons}} = \frac{25 \text{ miles}}{1 \text{ gallon}}$, or 25 mpg
[5.1] **Unit rate**	A rate in which the number in the denominator is 1.	$\frac{180 \text{ calories}}{1 \text{ ounce}}$, or 180 calories per ounce, or 180 cal/oz
[5.1] **Unit price**	The price of one item, or one unit.	\$0.69 per can, or \$0.69/can
[5.2] **Proportion**	A statement that two ratios are equal.	$\frac{5}{8} = \frac{15}{24}$
[5.2] **To solve a proportion**	• Find the cross products, and set them equal. • Solve the resulting equation. • Check the solution by substituting the value of the unknown in the original equation to verify that the resulting proportion is true.	$\frac{6}{9} = \frac{2}{x}$ $6x = 18$ $x = 3$ **Check** $\frac{6}{9} = \frac{2}{x}$ $\frac{6}{9} \stackrel{?}{=} \frac{2}{3}$ $6 \cdot 3 \stackrel{?}{=} 9 \cdot 2$ $18 \stackrel{\checkmark}{=} 18$

CHAPTER 5 Review Exercises

Say Why *Fill in each blank.*

1. The ratio $\frac{\$25}{\$225}$ ________ (is/is not) written in simplified form as $\frac{\$1}{\$9}$ because ____.

2. The ratio $\frac{\$51}{6.8 \text{ hours}}$ ________ (is/is not) a rate because ____.

3. The rate $\frac{1 \text{ mile}}{5{,}280 \text{ feet}}$ ________ (is/is not) a unit rate because ____.

4. \$10 per foot ________ (is/is not) an example of a unit price because ____.

5. $\frac{2}{3} - \frac{x}{5}$ ________ (is/is not) a proportion because ____.

6. In the proportion $\frac{x}{2} = \frac{3}{5}$, $2 \cdot 3$ and $5x$ ________ (are/are not) cross products because ____.

[5.1] *Write each ratio or rate in simplest form.*

7. 10 to 15
8. 28 to 56
9. 3 to 4
10. 50 to 16
11. 10,400 votes to 6,500 votes
12. 9 cups to 12 cups
13. 88 feet in 10 seconds
14. 45 applicants for 10 positions

Write each ratio as a unit rate.

15. 4 pounds of grass seed to plant in 1,600 square feet of lawn
16. 75 billion telephone calls in 150 days
17. 48 yards in 6 downs
18. 3,200 square feet covered by 8 gallons of paint
19. 21,000,000 vehicles produced in 2 years
20. 532,000 commuters traveled in 7 days

Find the unit price for each item.

21. \$475 for 4 nights
22. \$19.45 for 5 DVD movie rentals
23. \$80,000 for 64 computer stations
24. \$9,364 for 100 shares of stock

Fill in each table. Which is the better buy?

25. *The New Yorker*® magazine issues

Number of Units	Total Price	Unit Price
47	\$39.95	
94	\$69.95	

26. Custom laser checks

Number of Units	Total Price	Unit Price
100	\$13.95	
250	\$37.95	

Complete each table. Determine the best buy.

27. Stop Aging Now® green tea extract capsules

Number of Units	Total Price	Unit Price
30	\$16.95	
90	\$44.85	
180	\$77.70	

28. Johnson's Baby Oil®

Number of Units (fluid ounces)	Total Price	Unit Price
4	\$2.78	
14	\$3.92	
20	\$3.74	

[5.2] *Indicate whether each proportion is true or false.*

29. $\frac{15}{25} = \frac{3}{5}$

30. $\frac{3}{1} = \frac{1}{3}$

31. $\frac{50}{45} = \frac{10}{8}$

32. $\frac{15}{6} = \frac{5}{2}$

Solve and check.

33. $\frac{1}{2} = \frac{x}{12}$

34. $\frac{9}{12} = \frac{x}{4}$

35. $\frac{12}{x} = \frac{3}{8}$

36. $\frac{x}{72} = \frac{5}{12}$

37. $\frac{1.6}{7.2} = \frac{x}{9}$

38. $\frac{x}{12} = \frac{1.2}{1.8}$

39. $\frac{5}{\frac{1}{2}} = \frac{7}{x}$

40. $\frac{3}{5} = \frac{x}{\frac{2}{3}}$

41. $\frac{2\frac{1}{4}}{x} = \frac{1}{30}$

42. $\frac{3}{1\frac{3}{5}} = \frac{x}{24}$

43. $\frac{\frac{5}{6}}{x} = \frac{2}{1.8}$

44. $\frac{\frac{2}{3}}{4} = \frac{x}{0.9}$

45. $\frac{0.36}{4.2} = \frac{2.4}{x}$

46. $\frac{x}{0.21} = \frac{0.12}{0.18}$

Mixed Applications

Solve and check.

47. An airplane has 12 first-class seats and 180 seats in coach. What is the ratio of first-class seats to coach seats?

48. A computer store sells $23,000 worth of desktop computers and $45,000 worth of laptops in a given month. What is the ratio of desktop computer to laptop sales?

49. If a personal care attendant earns $540 for a 6-day workweek, how much does she earn per day?

50. A glacier in Alaska moves about 2 inches in 16 months. How far does the glacier move per month?

51. In a recent year, approximately 200,000,000 of the 300,000,000 people in the United States were Internet users. What is the ratio of Internet users to the total population? (*Source:* internetworldstats.com)

52. A city's public libraries spend about $9.50 in operating expenses for every book they circulate. If their operating expenses amount to $475,000, how many books circulate?

53. In a college's day-care center, the required staff-to-child ratio is 2 to 5. If there are 60 children and 12 staff in the day-care center, is the center in compliance with the requirement?

54. Despite the director's protests, the 1924 silent film *Greed* was edited down from about 42 reels of film to 10 reels. If the original version was about 9 hours long, about how long was the edited version? (*Source: The Film Encyclopedia*)

55. A sports car engine has an 8-to-1 compression ratio. Before compression, the fuel mixture in a cylinder takes up 440 cubic centimeters of space. How much space does the fuel mixture occupy when fully compressed?

56. On an architectural drawing of a planned community, a measurement of 25 feet is represented by 0.5 inches. If two houses are actually 62.5 feet apart, what is the distance between them on the drawing?

57. The *density of a substance* is the ratio of its mass to its volume. To the nearest hundredth, find the density of gasoline if a volume of 317.45 cubic centimeters has a mass of 216.21 grams.

58. The admission rate at a college is the ratio of the number of admitted students to the number of applicants. At Harvard College for the class of 2014, there were 30,489 applicants of whom 2,110 were admitted. Find Harvard's admission rate, expressed as a decimal rounded to the nearest hundredth. (*Source:* news.harvard.edu)

• Check your answers on page A-7.

CHAPTER 5 Posttest

FOR EXTRA HELP

Chapter Test Prep Videos

The Chapter Test Prep Videos with test solutions are available on DVD, in MyMathLab, and on YouTube (search "AkstDevMath" and click on "Channels").

To see if you have mastered the topics in this chapter, take this test.

Write each ratio or rate in simplest form.

1. 8 to 12

2. 15 to 42

3. 55 ounces to 31 ounces

4. 180 miles to 15 miles

5. 65 revolutions in 60 seconds

6. 3 centimeters for every 75 kilometers

Find the unit rate.

7. 340 miles in 5 hours

8. 200-meter dash in 25 seconds

Determine the unit price.

9. \$4,080 for 30 days

10. 25 greeting cards for \$20

Determine whether each proportion is true or false.

11. $\frac{8}{21} \stackrel{?}{=} \frac{16}{40}$

12. $\frac{7}{3} \stackrel{?}{=} \frac{63}{27}$

Solve and check.

13. $\frac{15}{x} = \frac{6}{10}$

14. $\frac{102}{17} = \frac{36}{x}$

15. $\frac{0.9}{36} = \frac{0.7}{x}$

16. $\frac{\frac{1}{3}}{4} = \frac{x}{12}$

Solve.

17. To advertise his business, an owner can purchase 3 million e-mail addresses for \$120 or 5 million e-mail addresses for \$175. Which is the better buy?

18. The Association of American Medical Colleges has called for increasing the number of students attending medical schools so as to reduce projected physician shortages. In a recent year, the entering class of medical schools was about 18,000 students as compared to 16,000 students five years earlier. What is the ratio of the later enrollment to the earlier enrollment? (*Source:* aamc.org)

19. A man $6\frac{1}{4}$ feet tall casts a 5-foot shadow. A nearby tree casts a 20-foot shadow. If the heights and shadow lengths of the man and tree are proportional, how tall is the tree?

20. A nurse takes his patient's pulse. What is the patient's pulse per minute if it beats 12 times in 15 seconds?

• Check your answers on page A-8.

Cumulative Review Exercises

To help you review, solve the following:

1. Add: $\$93{,}281 + \$8{,}429 + \$6{,}701$

2. Divide: $\dfrac{5{,}103}{27}$

3. Calculate: $7 \cdot 2^3 - \dfrac{21 - 13}{2}$

4. Write the prime factorization of 168.

5. Find the difference: $3\frac{1}{10} - 2\frac{7}{10}$

6. Divide: $\frac{1}{3} \div 3\frac{1}{4}$

7. Simplify: $\frac{2}{3} \times 1\frac{1}{2} - \frac{1}{4}$

8. Multiply: $8.2 \times 1{,}000$

9. Estimate: $12\frac{1}{7} \div 3\frac{9}{10}$

10. Solve and check: $x + 6.5 = 9$

11. Solve and check: $\frac{3}{10}n = 21$

12. Simplify the ratio: 2.5 to 10

13. Find the unit price: 3 yards for \$12

14. Solve and check: $\dfrac{\frac{1}{2}}{4} = \dfrac{x}{6}$

15. What is the area of the singles tennis court shaded in the diagram?

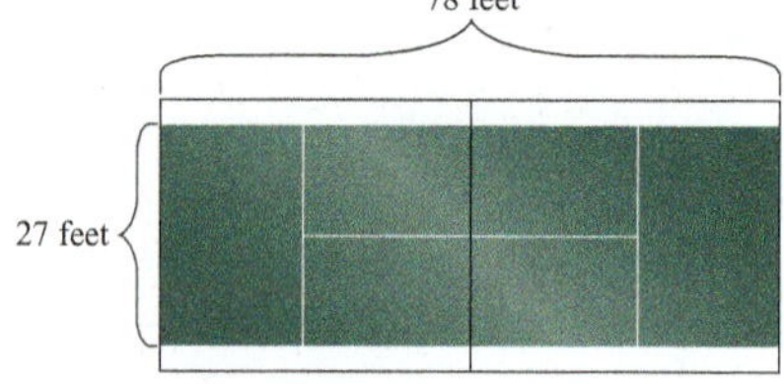

16. A college graduate looking for a teaching position takes a temporary substitute-teaching job. He works two days and makes \$178.35. At this rate of pay, how much would he make, to the nearest cent, for teaching five days? (*Source:* okaloosaschools.com)

17. The barometric pressure fell from 30.02 inches to 29.83 inches. By how many inches did it fall?

18. Write the algebraic expression for the number of miles a driver travels if she drives at a speed of r miles per hour for t hours. How far will she travel in 4 hours at a speed of 65 miles per hour?

19. A rule of thumb for growing lily bulbs is to plant them 3 times as deep as they are wide. How deep should a gardener plant a lily bulb that is 2.5 inches wide?

20. In 2010, air traffic between the United States and Western Europe was disrupted because of ash spewing from a volcano in Iceland. On one day alone, 1,000 of 29,000 scheduled flights were cancelled. What fraction of the flights were cancelled? (*Source:* cnn.com)

• Check your answers on page A-8.

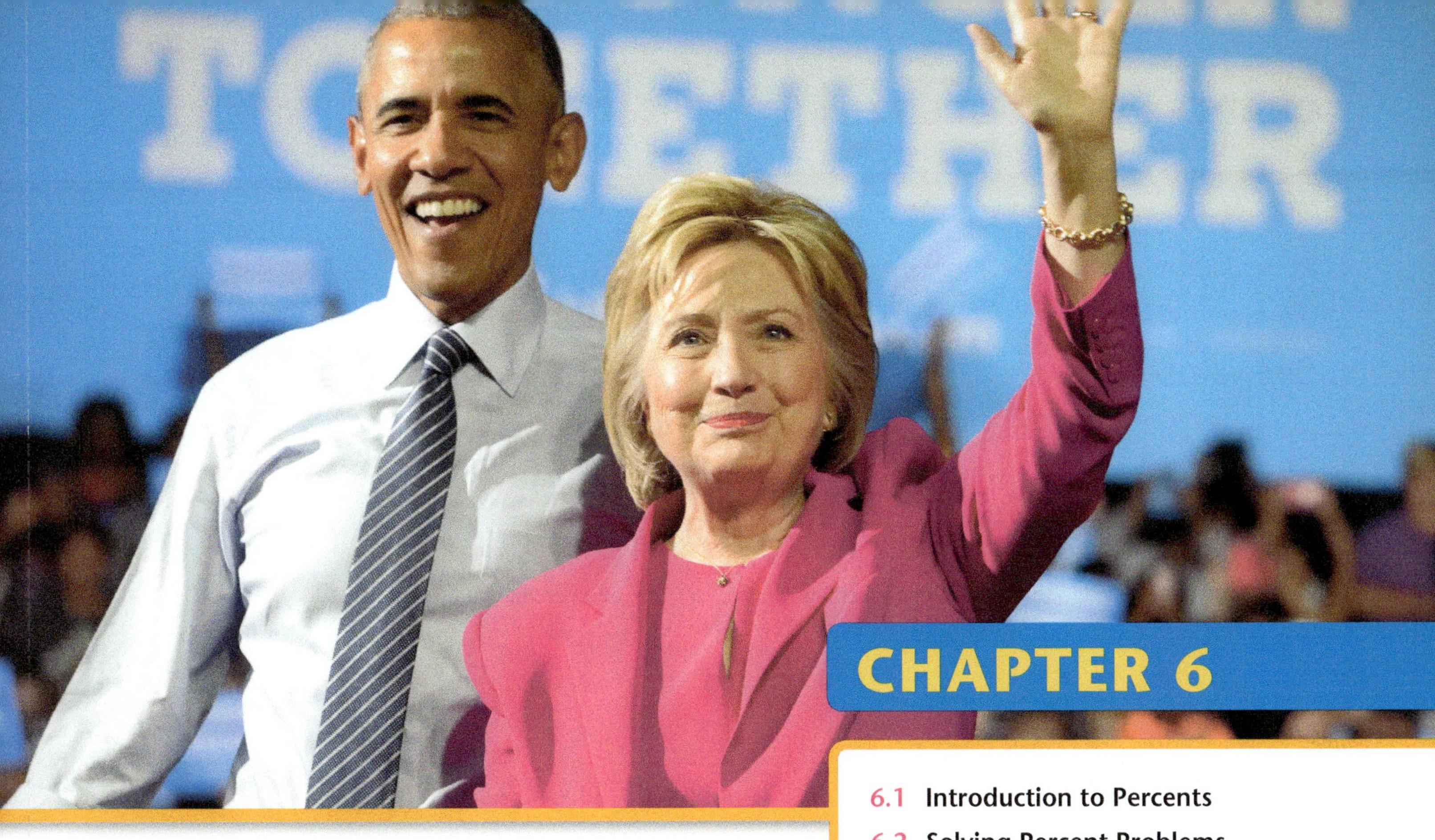

CHAPTER 6

Percents

Percents and Political Polls

In 2007, Barack Obama and Hillary Clinton competed in Iowa for the Democratic nomination for President of the United States. When asked which candidate they would vote for if the Democratic caucus were held that day, 30% of interviewees said Mr. Obama and 26% Mrs. Clinton. But every poll and survey has a *margin of error*. The Iowa poll's margin of error was plus or minus 5%, meaning that support for Mr. Obama was likely between 25% and 35%, and support for Mrs. Clinton between 21% and 31%. The results in Iowa were therefore too close to call—there was essentially a tie. Barack Obama went on to be nominated and to win the election.

(*Source*: *New York Times*, November 28, 2007)

CHAPTER 6 PRETEST

To see if you have already mastered the topics in this chapter, take this test.

Rewrite.

1. 5% as a fraction

2. $37\frac{1}{2}\%$ as a fraction

3. 250% as a decimal

4. 3% as a decimal

5. 0.007 as a percent

6. 8 as a percent

7. $\frac{2}{3}$ as a percent, rounded to the nearest whole percent

8. $1\frac{1}{10}$ as a percent

Solve.

9. What is 75% of 50 feet?

10. Find 110% of 50.

11. 80% of what number is 25.6?

12. 2% of what number is 5?

13. What percent of 10 is 4?

14. What percent of 4 is 10?

15. In a municipal savings account, a city employee earned 3% interest on $350. How much money did the employee earn in interest for 1 year?

16. The number of students enrolled at a community college rose from 2,475 last year to 2,673 this year. What was the percent increase in the college's enrollment?

17. In the depths of the Great Depression, 24% of the U.S. civilian labor force was unemployed. Write this percent as a simplified fraction. (*Source:* census.gov)

18. In a chemistry lab, a student dissolved 10 milliliters of acid in 30 milliliters of water. What percent of the solution was acid?

19. For parties of 8 or more, a restaurant automatically adds an 18% tip to the restaurant check. What tip would be added to a dinner check for a party of 10 if the total bill was $339.50?

20. A patient's health insurance covered 80% of the cost of her operation. She paid the remainder, which came to $2,000. Find the total cost of the operation.

• Check your answers on page A-8.

6.1 Introduction to Percents

OBJECTIVES

- **A** To find the fraction or the decimal equivalent of a given percent
- **B** To find the percent equivalent of a given fraction or decimal
- **C** To solve applied problems involving percent conversions

What Percents Are and Why They Are Important

Percent means divided by 100. So 50% (read "fifty percent") means 50 divided by 100 (or 50 out of 100).

A percent can also be thought of as a ratio or a fraction with denominator 100. For example, we can look at 50% either as the ratio of 50 parts to 100 parts or as the fraction $\frac{50}{100}$, or $\frac{1}{2}$. Since a fraction can be written as a decimal, we can also think of 50% as 0.50, or 0.5.

In the diagram at the right, 50 of the 100 squares are shaded. This shaded portion represents 50%.

We can use diagrams to represent other percents.

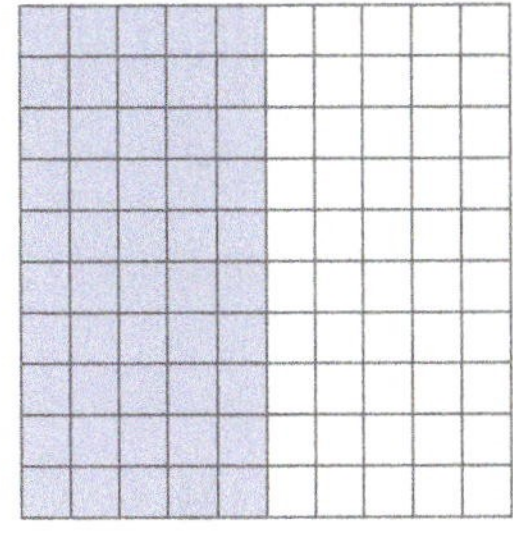

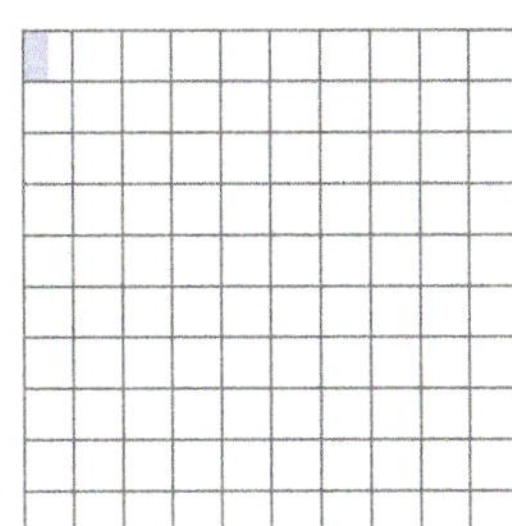

In the diagram to the left, $\frac{1}{2}$% is equivalent to the shaded portion,

$$\frac{\frac{1}{2}}{100}, \text{ or } \frac{1}{200}.$$

The entire diagram at the right is shaded, so 100% means $\frac{100}{100}$, or 1.

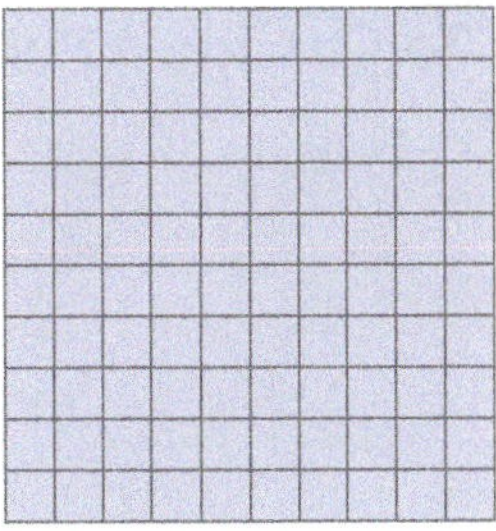

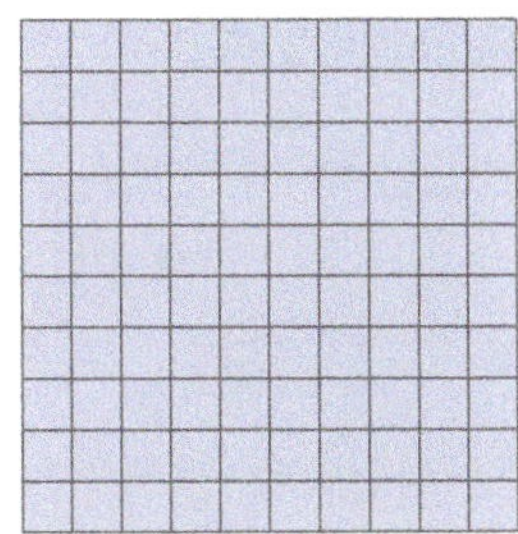

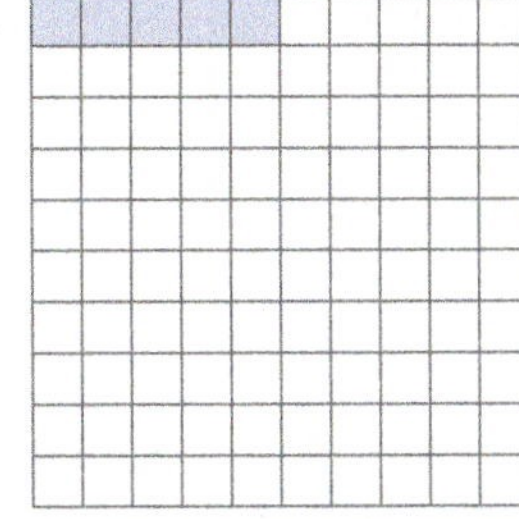

We can express 105% as $\frac{105}{100}$, or $1\frac{1}{20}$, as shown by the shaded portions of the diagrams.

Percents are commonly used, as the following statements taken from a single page of a newspaper illustrate.

- About 10% of the city's budget goes to sanitation.
- Blanket Sale—30% to 40% off!
- The number of victims of the epidemic increased by 125% in just 6 months.

A key reason for using percents so frequently is that they are easy to compare. For instance, we can tell right away that a discount of 30% is larger than a discount of 22%, simply by comparing the whole numbers 30 and 22.

To see how percents relate to fractions and decimals, let's consider finding equivalent fractions, decimals, and percents. In Chapter 3, we discussed two of the six types of conversions:

- changing a decimal to a fraction, and
- changing a fraction to a decimal.

Here, we consider the remaining four types of conversions:

- changing a percent to a fraction,
- changing a percent to a decimal,
- changing a decimal to a percent, and
- changing a fraction to a percent.

Note that each type of conversion changes the way the number is written—but not the number itself.

Changing a Percent to a Fraction

Suppose that we want to rewrite a percent—say, 30%—as a fraction. Because percent means divided by 100, we simply drop the % sign, place 30 over 100, and simplify.

$$30\% = \frac{30}{100} = \frac{3}{10}$$

Therefore, the fraction $\frac{3}{10}$ is just another way of writing the percent 30%. This result suggests the following rule:

To Change a Percent to the Equivalent Fraction

- Drop the % sign from the given percent and place the number over 100.
- Simplify the resulting fraction, if possible.

EXAMPLE 1

Write 7% as a fraction.

Solution To change this percent to a fraction, we drop the percent sign and write the 7 over 100. The fraction is already in lowest terms.

$$7\% = \frac{7}{100}$$

PRACTICE 1

Find the fractional equivalent of 21%.

EXAMPLE 2

Express 150% as a fraction.

Solution $150\% = \frac{150}{100} = \frac{3}{2}$, or $1\frac{1}{2}$

Note that the answer is larger than 1 because the original percent was more than 100%.

PRACTICE 2

What is the fractional equivalent of 225%?

EXAMPLE 3

Express $\frac{1}{10}\%$ as a fraction.

Solution To find the equivalent fraction, we drop the % sign and then put the number over 100.

$$\frac{\frac{1}{10}}{100} = \frac{1}{10} \div 100 = \frac{1}{10} \div \frac{100}{1} = \frac{1}{10} \times \frac{1}{100} = \frac{1}{1{,}000}$$

So $\frac{1}{10}\%$ expressed as a fraction is $\frac{1}{1{,}000}$.

PRACTICE 3

Change $\frac{2}{3}\%$ to a fraction.

EXAMPLE 4

Express $33\frac{1}{3}\%$ as a fraction.

Solution To find the equivalent fraction, we first drop the % sign and then put the number over 100.

$$\frac{33\frac{1}{3}}{100} = 33\tfrac{1}{3} \div 100 = 33\tfrac{1}{3} \div \frac{100}{1} = \frac{100}{3} \div \frac{100}{1} = \frac{\cancel{100}^{1}}{3} \times \frac{1}{\cancel{100}_{1}} = \frac{1}{3}$$

So $33\frac{1}{3}\%$ expressed as a fraction is $\frac{1}{3}$.

PRACTICE 4

Change $12\frac{1}{2}\%$ to a fraction.

EXAMPLE 5

The Ring of Fire contains 75% of the volcanoes on Earth. Express this percent as a fraction. (*Source:* nationalgeographic.com)

Ring of Fire

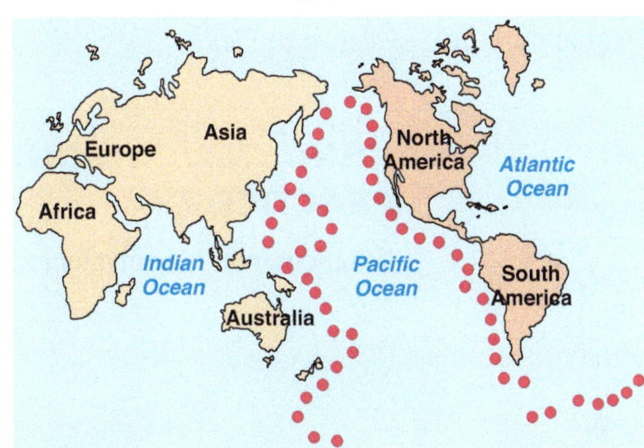

Solution $75\% = \frac{75}{100} = \frac{3}{4}$

So $\frac{3}{4}$ of the volcanoes on Earth are located in the Ring of Fire.

PRACTICE 5

Rwanda is the first country where more than half of the members of the legislature (56%) are women. Express this percent as a fraction. (*Source: Top 10 of Everything, 2010*)

Changing a Percent to a Decimal

Now, let's consider rewriting a percent as a decimal. For instance, take 75%. We begin by writing this percent as a fraction.

$$75\% = \frac{75}{100}$$

Converting this fraction to a decimal, we divide:

$$100\overline{)75.00}\quad 0.75$$

Note that we could have gotten this answer simply by moving the decimal point two places to the left and dropping the % sign.

$$75\% = 75.\% = .75, \text{ or } 0.75$$

This example suggests the following rule:

To Change a Percent to the Equivalent Decimal

- Drop the % sign from the given percent and divide the number by 100.

EXAMPLE 6

Change 42% to a decimal.

Solution Drop the % sign from 42% and divide by 100. Recall that to divide a decimal by 100, we can simply move the decimal point two places to the left.

$$42\% = 42.\% = .42, \text{ or } 0.42$$

So the decimal equivalent of 42% is 0.42.

PRACTICE 6

Express 31% as a decimal.

TIP A shortcut for changing a percent to its equivalent decimal is dropping the percent sign and moving the decimal point *two places* to the *left*.

EXAMPLE 7

Find the decimal equivalent of 1%.

Solution The unwritten decimal point lies to the right of the 1. Moving the decimal point two places to the left and dropping the % sign, we get:

$$1\% = 01.\% = .01, \text{ or } 0.01$$

Note that we inserted a 0 as a placeholder, because there was only a single digit to the left of the decimal point.

PRACTICE 7

What decimal is equivalent to 5%?

EXAMPLE 8

Convert 37.5% to a decimal.

Solution 37.5% = .375, or 0.375

In this problem, the given number is a percent even though it involves a decimal point.

PRACTICE 8

Rewrite 48.2% as a decimal.

EXAMPLE 9

Change $12\frac{1}{2}\%$ to a decimal.

Solution To find the decimal equivalent of $12\frac{1}{2}\%$, we begin by converting the fraction $\frac{1}{2}$ in the mixed number to its decimal equivalent 0.5. Then, we move the decimal point to the left two places, dropping the % sign.

$$\frac{1}{2} = 2\overline{)1.0} \quad 0.5, \text{ or } .5$$

$$12\frac{1}{2}\% = 12.5\% = .125, \text{ or } 0.125$$

PRACTICE 9

Express the following percent as a decimal: $62\frac{1}{4}\%$

EXAMPLE 10

In 1945 at the end of World War II, the public debt of the United States was 602% of what it had been in 1940. Write this percent as a decimal.

Solution 602% = 602.%, or 6.02

The 1945 U.S. debt was 6.02 times what it had been 5 years earlier.

PRACTICE 10

In 1991, the movie sequel Highlander II was released and grossed 163.7% of what Highlander I had made five years earlier. Express this percent as a decimal. (*Source: Top 10 of Everything, 2010*)

Changing a Decimal to a Percent

Suppose that we want to change 0.75 to a percent. Because 100% is the same as 1, we can multiply this number by 100% without changing its value.

$$0.75 \times 100\% = 75\%$$

We could have gotten this answer simply by moving the decimal point to the right two places and adding the % sign.

$$0.75 = 075.\% = 75\%$$

Note that we dropped the decimal point in the answer because it was to the right of the units digit.

To Change a Decimal to the Equivalent Percent

- Multiply the number by 100 and insert a % sign.

EXAMPLE 11

Write 0.425 as a percent.

Solution We multiply 0.425 by 100 and add a % sign.

$$0.425 = 0.425 \times 100\% = 42.5\%, \text{ or } 42\frac{1}{2}\%$$

PRACTICE 11

What percent is equivalent to the decimal 0.025?

TIP A shortcut for changing a decimal to its equivalent percent is inserting a % sign and moving the decimal point *two places* to the *right*.

EXAMPLE 12

Convert 0.03 to a percent.

Solution $0.03 = 003.\% = 3\%$

PRACTICE 12

Change 0.09 to a percent.

EXAMPLE 13

Express 0.1 as a percent.

Solution In the given number, only a single digit is to the right of the decimal point. So to move the decimal point two places to the right, we need to insert a 0 as a placeholder.

$$0.1 = 0.10 = 10.\% = 10\%$$

PRACTICE 13

What percent is equivalent to 0.7?

EXAMPLE 14

What percent is equivalent to 2?

Solution Recall that a whole number such as 2 has a decimal point understood to its right. We move the decimal point two places to the right.

$$2 = 2. = 2.00 = 200.\% = 200\%$$

So the answer is 200%, which makes sense: 200% is double 100%, just as 2 is double 1.

PRACTICE 14

Rewrite 3 as a percent.

EXAMPLE 15

Express 0.2483 as a percent, rounded to the nearest whole percent.

Solution First, we obtain the exact percent equivalent.

$$0.2483 = 24.83\%$$

To round this number to the nearest whole percent, we underline the digit 4. Then, we check the critical digit immediately to its right. This digit is 8, so we round up.

$$2\underline{4}.83\% \approx 25.\% = 25\%$$

PRACTICE 15

Convert 0.714 to a percent, rounded to the nearest whole percent.

EXAMPLE 16

Red blood cells make up about 0.4 of the total blood volume in the human body, whereas 55% of the total blood volume is plasma. Which makes up more of the blood volume—red blood cells or plasma? Explain. (*Source:* Mayo Clinic)

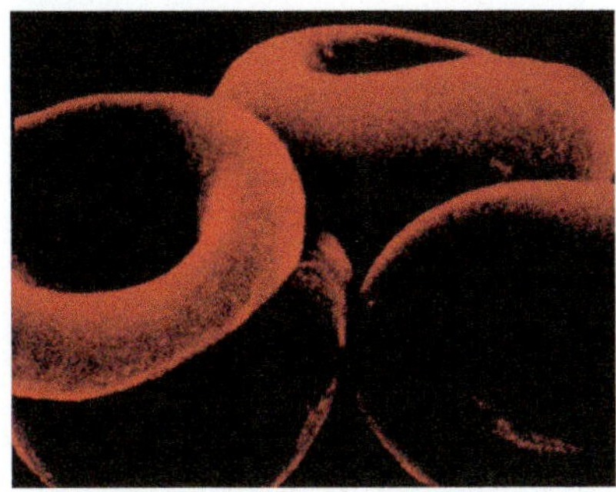

Solution We want to compare the decimal 0.4 and the percent 55%. One way is to change the decimal to a percent.

$$0.4 = 0.40 = 40.\% = 40\%$$

Since 40% is less than 55%, we conclude that plasma makes up more of the blood volume.

PRACTICE 16

Air is a mixture of many gases. For example, 0.78 of air is nitrogen, and 0.93% is argon. Is there more nitrogen or argon in air? Explain.

Changing a Fraction to a Percent

Now, let's change a fraction to a percent. Consider, for instance, the fraction $\frac{1}{5}$. To convert this fraction to a percent, multiply $\frac{1}{5}$ by 100%, which is equal to 1.

$$\frac{1}{5} = \frac{1}{5} \times 100\% = \frac{1}{\cancel{5}_1} \times \frac{\cancel{100}^{20}}{1}\% = 20\%$$

To Change a Fraction to the Equivalent Percent

- Multiply the fraction by 100%.

EXAMPLE 17

Rewrite $\frac{7}{20}$ as a percent.

Solution To change the given fraction to a percent, we multiply by 100%.

$$\frac{7}{20} = \frac{7}{20} \times 100\% = \frac{7}{\cancelto{1}{20}} \times \frac{\cancelto{5}{100}}{1}\% = 35\%$$

PRACTICE 17

Convert $\frac{4}{25}$ to a percent.

EXAMPLE 18

Which is larger: 130% or $1\frac{3}{8}$?

Solution To compare, let's express $1\frac{3}{8}$ as a percent.

$$1\frac{3}{8} = 1\frac{3}{8} \times 100\% = \frac{11}{8} \times \frac{100}{1}\%$$

$$= \frac{11}{\cancelto{2}{8}} \times \frac{\cancelto{25}{100}}{1}\% = \frac{275}{2}\% = 137\frac{1}{2}\%$$

Because $137\frac{1}{2}\%$ is larger than 130%, so is $1\frac{3}{8}$.

PRACTICE 18

True or false: $\frac{2}{3} > 60\%$. Justify your answer.

EXAMPLE 19

A student got 28 of 30 questions correct on a test. If all the questions were equal in value, what was the student's grade, rounded to the nearest whole percent?

Solution The student answered $\frac{28}{30}$ of the questions right. To find the student's grade, we change this fraction to a percent.

$$\frac{28}{30} = \frac{28}{30} \times 100\% = \frac{28}{\cancelto{3}{30}} \times \frac{\cancelto{10}{100}}{1}\%$$

$$= \frac{280}{3}\% = 93\frac{1}{3}\% = 93.3\ldots\% \approx 93\%$$

Note that the critical digit is 3, so we round down. The rounded grade was therefore 93%.

PRACTICE 19

An administrative assistant spends $490 out of her monthly salary of $1,834 on rent. What percent of her monthly salary is spent on rent, rounded to the nearest whole percent?

6.1 Exercises

FOR EXTRA HELP MyMathLab 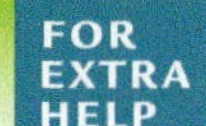 PRACTICE  WATCH  READ  REVIEW 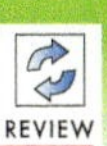

Mathematically Speaking

Fill in each blank with the most appropriate term or phrase from the given list.

right	fraction	percent	divide
decimal	left	whole number	multiply

1. A(n) __________ is a ratio or fraction with denominator 100.

2. To change a percent to the equivalent __________, drop the % sign from the given percent, and place the number over 100.

3. To change a percent to the equivalent decimal, move the decimal point two places to the __________ and drop the % sign.

4. To change a fraction to the equivalent percent, __________ the fraction by 100 and insert a % sign.

A *Change each percent to a fraction or mixed number. Simplify.*

5. 8% **6.** 3% **7.** 250% **8.** 110%

9. 33% **10.** 41% **11.** 18% **12.** 44%

13. 14% **14.** 45% **15.** 65% **16.** 92%

17. $\frac{3}{4}\%$ **18.** $\frac{1}{10}\%$ **19.** $\frac{3}{10}\%$ **20.** $\frac{1}{5}\%$

21. $7\frac{1}{2}\%$ **22.** $2\frac{1}{2}\%$ **23.** $14\frac{2}{7}\%$ **24.** $28\frac{4}{7}\%$

Convert each percent to a decimal.

25. 6% **26.** 9% **27.** 72% **28.** 25%

29. 0.1% **30.** 0.2% **31.** 102% **32.** 113%

33. 42.5% **34.** 10.5% **35.** 500% **36.** 400%

37. $106\frac{9}{10}\%$ **38.** $201\frac{1}{10}\%$ **39.** $3\frac{1}{2}\%$ **40.** $2\frac{4}{5}\%$

41. $\frac{9}{10}\%$ **42.** $\frac{7}{10}\%$ **43.** $\frac{3}{4}\%$ **44.** $\frac{1}{4}\%$

B *Express each decimal as a percent.*

45. 0.31 **46.** 0.37 **47.** 0.17 **48.** 0.18

49. 0.3 **50.** 0.4 **51.** 0.04 **52.** 0.05

53. 0.125 **54.** 0.875 **55.** 1.29 **56.** 1.07

57. 2.9 **58.** 3.5 **59.** 2.87 **60.** 3.62

61. 1.016 **62.** 1.003 **63.** 9 **64.** 7

Change each fraction to a percent.

65. $\frac{3}{10}$ **66.** $\frac{1}{2}$ **67.** $\frac{1}{10}$ **68.** $\frac{3}{20}$

69. $\frac{4}{25}$ **70.** $\frac{6}{25}$ **71.** $\frac{9}{10}$ **72.** $\frac{7}{10}$

73. $\frac{3}{50}$ **74.** $\frac{1}{50}$ **75.** $\frac{5}{9}$ **76.** $\frac{2}{9}$

77. $\frac{1}{9}$ **78.** $\frac{4}{7}$ **79.** 6 **80.** 8

81. $1\frac{1}{2}$ **82.** $2\frac{3}{5}$ **83.** $2\frac{1}{6}$ **84.** $1\frac{1}{3}$

Replace □ *with* < *or* >.

85. $2\frac{1}{4}$ □ 240% **86.** $3\frac{5}{6}$ □ 380% **87.** $\frac{1}{2}\%$ □ 50% **88.** $\frac{1}{40}$ □ $\frac{1}{4}\%$

Express as a percent, rounded to the nearest whole percent.

89. $\frac{4}{9}$ **90.** $\frac{3}{7}$ **91.** 2.2469 **92.** 1.1633

Complete each table.

93.

Fraction	Decimal	Percent
		$33\frac{1}{3}$
	0.666 …	
	0.25	
		75%
		20%
$\frac{2}{5}$		
	0.6	

94.

Fraction	Decimal	Percent
	0.8	
$\frac{1}{6}$		
$\frac{5}{6}$		
		12.5%
	0.375	
		$62\frac{1}{2}\%$
	0.875	

Mixed Practice

Solve.

95. Change 104% to a mixed number.

96. What percent is equivalent to $\frac{2}{5}$?

97. Express $3\frac{1}{6}$ as a percent.

98. Express $62\frac{1}{2}\%$ as a fraction.

99. Convert 27.5% to a decimal.

100. Find the decimal equivalent to $\frac{3}{8}\%$.

101. What percent is equivalent to 3.1?

102. Change 0.003 to a percent.

103. Which is smaller, $2\frac{5}{9}$ or 254%?

104. Express 1.2753 to the nearest whole percent.

Applications

C *Solve.*

105. It is estimated that 96% of all e-mail messages received are *spam* (unsolicited junk e-mail). Express this percent as a decimal. (*Source:* govtech.com)

106. According to a recent study, 65% of children have had an imaginary companion by age 7. Express this percent as a fraction. (*Source:* uwnews.org)

107. The following graph shows the percent of people in the United States who get their local news regularly from various sources.

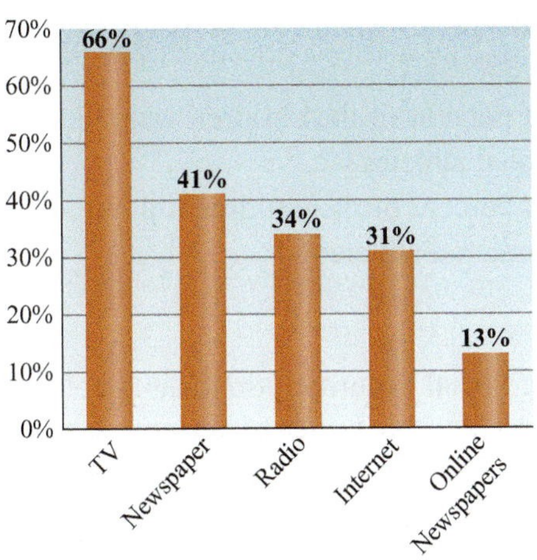

Regular Sources of Local News

(*Source:* people-press.org)

What fraction of people in the United States do *not* get their news regularly from TV?

108. The following graph shows the distribution of investments for a retiree. Express as a decimal the percent of investments that are in equities.

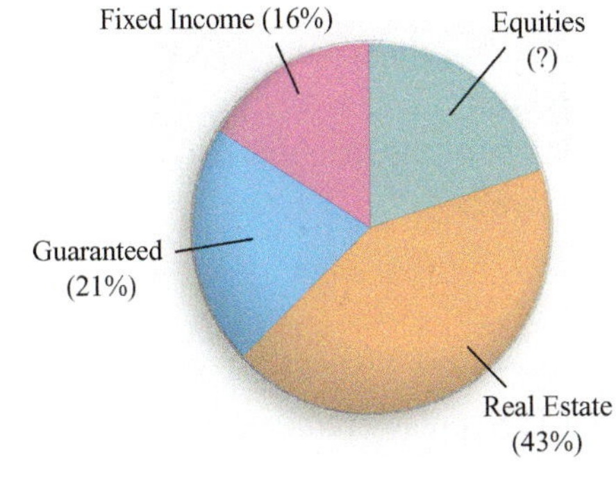

109. According to the nutrition label, one large egg contains 6 grams of protein. This is 10% of the daily value (DV) for protein. Express this percent as a fraction.

110. A bank offers a Visa credit card with a fixed annual percentage rate (APR) of 16.99%. Express the APR as a decimal.

111. Los Angeles has an area 10.05 times that of San Francisco. Express this decimal as a percent. (*Source:* U.S. Census Bureau)

112. In France, the main source of electricity is nuclear power. In a recent year, more than $\frac{3}{4}$ of the country's total electrical production was nuclear. Write this fraction as a percent. (*Source:* International Atomic Energy Agency)

113. When the recession ended, the factory's output grew by 135%. Write this percent as a simplified mixed number.

114. According to a survey, 78% of the arguments that couples have are about money. Express this percent as a decimal.

115. According to a recent U.S. Bureau of Labor Statistics report, 51.5% of all union members are government employees. Convert this percent to a decimal. (*Source:* rsc.tomprice.house.gov)

116. After an oil spill, 15% of the wildlife survived. Express this percent as a fraction.

117. The World Health Organization estimates that in developed countries, $\frac{1}{4}$ of women and 42% of men smoke. In these countries, is smoking more common among women or among men? Explain. (*Source:* americanheart.org)

118. The state sales tax rate in Indiana is 7%, whereas in Iowa it is $\frac{3}{50}$. Which state has a lower sales tax rate? Explain. (*Source:* taxadmin.org)

119. A quality control inspector found 2 defective machine parts out of 500 manufactured.

a. What percent of the machine parts manufactured were defective?

b. What percent of the machine parts manufactured were not defective?

120. In a survey of several hundred children, 3 out of every 25 children indicated that they wanted to become professional athletes when they grow up. (*Source: National Geographic Kids*)

a. What percent of the children wanted to become professional athletes?

b. What percent of the children did not want to become professional athletes?

121. In the 2008 U.S. presidential election, 131,257,328 people turned out to vote. At the time, there were 189,844,867 registered voters. What percent of the registered voters, to the nearest whole percent, voted? (*Sources:* fec.gov and eac.gov)

122. The first Social Security retirement benefits were paid in 1940 to Ida May Fuller of Vermont. She had paid in a total of \$24.85 and got back \$20,897 before her death in 1975. Express the ratio of what she got back to what she put in as a percent, rounded to the nearest whole percent. (*Source:* James Trager, *The People's Chronology*)

• Check your answers on page A-8.

MINDStretchers

Mathematical Reasoning

1. By mistake, you move the decimal point to the right instead of to the left when changing a percent to a decimal. Your answer is how many times as large as the correct answer?

Writing

2. A study of the salt content of seawater showed that the average salt content varies from 33‰ to 37‰, where the symbol ‰ (read "per mil") means "for every thousand." Explain why you think the scientist who wrote this study did not use the % symbol.

Critical Thinking

3. What percent of the region shown is shaded in?

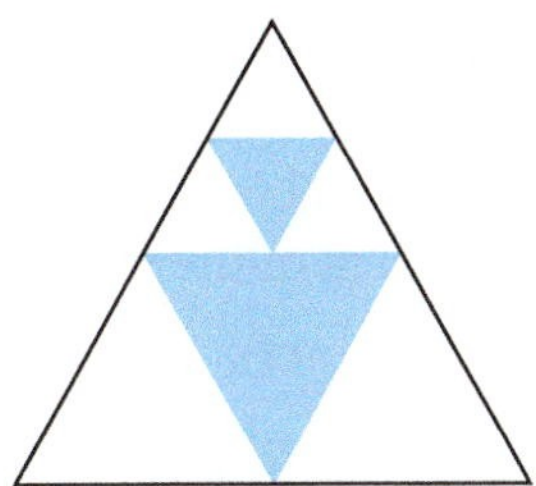

Cultural Note

Throughout history, the concepts of percent and taxation have been interrelated. At the peak of the Roman Empire, the Emperor Augustus instituted an inheritance tax of 5% to provide retirement funds for the military. Another emperor, Julius Caesar, imposed a 1% sales tax on the population. And in Roman Asia, tax collectors exacted a tithe of 10% on crops. If landowners could not pay, the collectors offered to lend them funds at interest rates that ranged from 12% up to 48%.

Roman taxation served as a model for modern countries when these countries developed their own systems of taxation many centuries later.

Sources: Frank J. Swetz, *Capitalism and Arithmetic: The New Math of the 15th Century,* Open Court, 1987; Carolyn Webber and Aaron Wildavsky, *A History of Taxation and Expenditure in the Western World,* Simon and Schuster, 1986.

6.2 Solving Percent Problems

OBJECTIVES

A To find the amount, the base, or the percent in a percent problem

B To solve applied problems involving percents

The Three Basic Types of Percent Problems

Frequently, we think of a percent not in isolation but rather in connection with another number. In other words, we take *a percent of a number*.

Consider, for example, the problem of taking 50% of 8. This problem is equivalent to finding $\frac{1}{2}$ of 8, which gives us 4.

Note that this percent problem, like all others, involves three numbers.

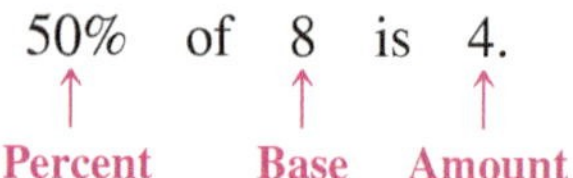

- The 50% is called the **percent** (or the **rate**). The percent always contains the % sign.
- The 8 is called the **base**. The base of a percent—the number that we are taking the percent of—always follows the word *of* in the statement of the problem.
- The remaining number 4 is called the **amount** (or the **part**).

Percent problems involve finding one of the three numbers. For example, if we omit the 4 in "50% of 8 is 4," we ask the question: What is 50% of 8? Omitting the 8, we ask: 50% of what number is 4? And omitting the 50%, we ask: What percent of 8 is 4?

There are several ways to solve these three basic percent questions. In this section, we discuss two ways—the translation method and the proportion method.

The Translation Method

In the translation method, a percent problem has the form

The percent of the base is the amount.

The percent problem gives only two of the three quantities. To find the missing quantity using the translation method, we translate to a simple equation that we then solve.

This method depends on translating the words in the given problem to the appropriate mathematical symbols.

Word(s)	Math Symbol
What, what number, what percent	x (or some other letter)
is	$=$
of	$\times$ or $\cdot$
percent, %	Percent value expressed as a decimal or fraction

Let's translate several percent problems to equations.

EXAMPLE 1

Translate each question to an equation, using the translation method.

a. What is 10% of 2? **b.** 20% of what number is 5?

c. What percent of 8 is 4?

Solution

a. In this problem, we are looking for the amount.

What is 10% of 2?
$$x = 0.1 \cdot 2$$

b. Here, we are looking for the base, that is, the number after the word *of*.

20% of what number is 5?
$$0.2 \cdot x = 5$$

c. This problem asks "what percent?" So we are looking for the percent.

What percent of 8 is 4?
$$x \cdot 8 = 4$$

PRACTICE 1

Use the translation method to set up an equation.

a. What is 70% of 80?

b. 50% of what number is 10?

c. What percent of 40 is 20?

Finding an Amount

Now, let's apply the translation method to solve the type of percent problem in which we are given both the percent and the base and are looking for the amount.

EXAMPLE 2

What is 25% of 8?

Solution First, we translate the question to an equation.

What is 25% of 8?
$$x = \frac{1}{4} \cdot 8$$

Then, we solve this equation:

$$x = \frac{1}{4} \cdot 8 = \frac{1}{\cancel{4}_1} \cdot \frac{\cancel{8}^2}{1} = \frac{2}{1}$$
$$= 2$$

So 2 is 25% of 8. Would we have gotten the same answer if we had translated 25% to 0.25?

PRACTICE 2

What is 20% of 40?

EXAMPLE 3

Find 200% of 30.

Solution We can reword the problem as a question.

What is 200% of 30?
$$x = 2 \cdot 30$$

Solving this equation, we get $x = 60$. So 200% of 30 is 60.

PRACTICE 3

150% of 8 is what number?

TIP When the percent is less than 100%, the amount is *less* than the base. When the percent is more than 100%, the amount is *more* than the base.

EXAMPLE 4

What is $66\frac{2}{3}\%$ of 15?

Solution First, let's change the percent to a fraction:

$$66\frac{2}{3}\% = \frac{66\frac{2}{3}}{100}$$

$$= 66\frac{2}{3} \div \frac{100}{1}$$

$$= \frac{\overset{2}{\cancel{200}}}{3} \times \frac{1}{\underset{1}{\cancel{100}}} = \frac{2}{3}$$

Translating the question to an equation, we get:

$$\begin{array}{ccccc} \text{What} & \text{is} & 66\frac{2}{3}\% & \text{of} & 15? \\ \downarrow & \downarrow & \downarrow & \downarrow & \downarrow \\ x & = & \frac{2}{3} & \cdot & 15 \end{array}$$

$$= \frac{2}{\underset{1}{\cancel{3}}} \cdot \overset{5}{\cancel{15}} = \frac{10}{1} = 10$$

So $66\frac{2}{3}\%$ of 15 is 10.

PRACTICE 4

Find $33\frac{1}{3}\%$ of 600.

EXAMPLE 5

A marketing account manager has 3.5% of her monthly salary put into a 401(k) plan. How much did she put into the 401(k) plan if her monthly salary is \$3,200?

Solution We are looking for the monthly amount placed into the 401(k) plan, which is 3.5% of \$3,200.

$$\begin{array}{ccccc} \text{What} & \text{is} & 3.5\% & \text{of} & \$3{,}200? \\ \downarrow & \downarrow & \downarrow & \downarrow & \downarrow \\ x & = & (0.035) & \cdot & (3{,}200) \\ & = & 112 & & \end{array}$$

So she has \$112 per month put into the 401(k) plan. Note that this amount has the same unit (dollars) as the base.

PRACTICE 5

Of the 600 workers at a factory, 8.5% belong to a union. How many workers are in the union?

Finding a Base

Now, let's consider some examples of using the translation method to find the base when we know the percent and the amount.

EXAMPLE 6

4% of what number is 8?

Solution We begin by writing the appropriate equation.

4%	of	what number	is	8?
↓	↓	↓	↓	↓
0.04	·	x	=	8

Next, we solve this equation.

$$0.04x = 8$$

$$\frac{0.04}{0.04}x = \frac{8}{0.04} \quad \text{Divide each side by 0.04.}$$

$$x = \frac{8}{0.04} = 200 \qquad 0.04\overline{)8.00} = 4\overline{)800.} \;\; 200.$$

So 4% of 200 is 8.

PRACTICE 6

6 is 12% of what number?

EXAMPLE 7

108 is 120% of what number?

Solution We consider the following question:

120%	of	what number	is	108?
↓	↓	↓	↓	↓
1.2	·	x	=	108

Solving, we get:

$$1.2x = 108$$

$$\frac{1.2}{1.2}x = \frac{108}{1.2}$$

$$x = 90$$

So 108 is 120% of 90.

PRACTICE 7

250% of what number is 18?

EXAMPLE 8

A college awarded financial aid to 3,843 students, which was 45% of the total number of students enrolled at the college. What was the student enrollment at the college?

Solution We must answer the following question:

45%	of	what number	is	3,843?
↓	↓	↓	↓	↓
0.45	·	x	=	3,843

Next, we solve the equation.

$$0.45x = 3{,}843$$

$$\frac{0.45x}{0.45} = \frac{3{,}843}{0.45}$$

$$x = 8{,}540$$

So 8,540 students were enrolled at the college.

PRACTICE 8

There was a glut of office space in a city, with 400,000 square feet, or 16% of the total office space, vacant. How much office space did the city have?

Finding a Percent

Finally, let's look at the third type of percent problem, in which we are given the base and the amount and are looking for the percent.

EXAMPLE 9

What percent of 80 is 60?

Solution We begin by writing the appropriate equation.

What percent of 80 is 60?

$$x \cdot 80 = 60$$

$$80x = 60 \quad \text{Write the equation in standard form.}$$

$$\frac{80}{80}x = \frac{60}{80} \quad \text{Divide each side by 80.}$$

$$x = \frac{60}{80} = \frac{3}{4} \quad \text{Simplify.}$$

Since we are looking for a percent, we change $\frac{3}{4}$ to a percent. So 75% of 80 is 60.

$$x = \frac{3}{4} = \frac{3}{4} \cdot \frac{100}{1}\% = 75\%$$

PRACTICE 9

What percent of 6 is 5?

EXAMPLE 10

What percent of 60 is 80?

Solution We begin by writing the appropriate equation, as shown to the right.

What percent of 60 is 80?

$$x \cdot 60 = 80$$

$$60x = 80$$

$$\frac{60}{60}x = \frac{80}{60}$$

$$x = \frac{80}{60} = \frac{4}{3}$$

Finally, we want to change $\frac{4}{3}$ to a percent.

$$x = \frac{4}{3} = \frac{4}{3} \cdot \frac{100}{1}\% = \frac{400}{3}\% = 133\frac{1}{3}\%$$

So $133\frac{1}{3}\%$ of 60 is 80.

PRACTICE 10

What percent of 8 is 9?

EXAMPLE 11

A young couple buys a house for \$125,000, making a down payment of \$25,000 and paying the difference over time with a mortgage. What percent of the cost of the house was the down payment?

Solution We write the question as shown to the right.

$$\begin{array}{ccccc} \text{What percent} & \text{of} & \$125{,}000 & \text{is} & \$25{,}000? \\ \downarrow & \downarrow & \downarrow & \downarrow & \downarrow \\ x & \cdot & 125{,}000 & = & 25{,}000 \end{array}$$

$$125{,}000x = 25{,}000$$

$$\frac{125{,}000}{125{,}000}x = \frac{25{,}000}{125{,}000}$$

$$x = \frac{25}{125} = \frac{1}{5}$$

Next, we change $\frac{1}{5}$ to a percent. $x = \frac{1}{5} = \frac{1}{5} \cdot \frac{100}{1}\% = 20\%$

So the down payment was 20% of the total cost of the house.

PRACTICE 11

Of the 400 acres on a farm, 120 were used to grow corn. What percent of the total acreage was used to grow corn?

The Proportion Method

So far, we have used the translation method to solve percent problems. Now, let's consider an alternative approach, the proportion method.

Using the proportion method, we view a percent relationship in the following way.

$$\frac{\textbf{Amount}}{\textbf{Base}} = \frac{\textbf{Percent}}{\textbf{100}}$$

If we are given two of the three quantities, we set up this proportion and then solve it to find the third quantity.

EXAMPLE 12

What is 60% of 35?

Solution The base (the number after the word *of*) is 35. The percent (the number followed by the % sign) is 60. The amount is unknown. We set up the proportion, substitute into it, and solve.

$$\frac{\text{Amount}}{\text{Base}} = \frac{\text{Percent}}{100}$$

$$\frac{x}{35} = \frac{60}{100}$$

$$100x = 60 \cdot 35 \quad \text{Set cross products equal.}$$

$$\frac{100}{100}x = \frac{2{,}100}{100} \quad \text{Divide each side by 100.}$$

$$x = 21$$

So 60% of 35 is 21.

PRACTICE 12

Find 108% of 250.

EXAMPLE 13

15% of what number is 21?

Solution Here, the number after the word *of* is missing, so we are looking for the base. The amount is 21, and the percent is 15. We set up the proportion, substitute into it, and solve.

$$\frac{\text{Amount}}{\text{Base}} = \frac{\text{Percent}}{100}$$

$$\frac{21}{x} = \frac{15}{100}$$

$$15x = 2{,}100 \quad \text{Set cross products equal.}$$

$$\frac{\cancel{15}}{\cancel{15}}x = \frac{2{,}100}{15} \quad \text{Divide each side by 15.}$$

$$x = 140$$

So 15% of 140 is 21.

PRACTICE 13

2% of what number is 21.6?

EXAMPLE 14

What percent of \$45 is \$30?

Solution We know that the base is 45, the amount is 30, and we are looking for the percent.

$$\frac{30}{45} = \frac{x}{100}$$

$$45x = 3{,}000$$

$$\frac{\cancel{45}}{\cancel{45}}x = \frac{3{,}000}{45}$$

$$x = 66\frac{2}{3}$$

So we conclude that $66\frac{2}{3}\%$ of \$45 is \$30.

PRACTICE 14

What percent of 63 is 21?

EXAMPLE 15

A car depreciated, that is, dropped in value, by 20% during its first year. By how much did the value of the car decline if it cost \$30,500 new?

Solution The question here is: What is 20% of \$30,500? So the percent is 20, the base is \$30,500 and we are looking for the amount. We set up the proportion and solve.

$$\frac{x}{30{,}500} = \frac{20}{100}$$

$$100x = 610{,}000$$

$$\frac{\cancel{100}}{\cancel{100}}x = \frac{610{,}000}{100}$$

$$x = 6{,}100$$

So the value of the car depreciated by \$6,100.

PRACTICE 15

A credit card company requires a minimum payment of 4% of the balance. What is the minimum payment if the credit card balance is \$2,450?

EXAMPLE 16

Each day, an adult takes tablets containing 24 milligrams of zinc. If this amount is 160% of the recommended daily allowance, how many milligrams are recommended? (*Source: Podiatry Today*)

Solution Here, we are looking for the base. The question is: 160% of what amount is 24 milligrams? We set up the proportion and solve.

$$\frac{24}{x} = \frac{160}{100}$$
$$160x = 2{,}400$$
$$\frac{\cancel{160}}{\cancel{160}}x = \frac{2{,}400}{160}$$
$$x = 15$$

Therefore, the recommended daily allowance of zinc is 15 milligrams. Note that this base is less than the amount (24 milligrams). Why must that be true?

PRACTICE 16

According to a newspaper article, a Nobel Prize winner had to pay the Internal Revenue Service $129,200—or 38% of his prize—in taxes. How much was his Nobel Prize worth?

EXAMPLE 17

A college accepted 1,620 of the 4,500 applicants for admission. What was the acceptance rate, expressed as a percent?

Solution The question is: What percent of 4,500 is 1,620?

$$\frac{1{,}620}{4{,}500} = \frac{x}{100}$$
$$4{,}500x = 162{,}000$$
$$\frac{\cancel{4{,}500}}{\cancel{4{,}500}}x = \frac{162{,}00\cancel{0}}{4{,}50\cancel{0}}$$
$$x = 36$$

So the college's acceptance rate was 36%.

PRACTICE 17

A bookkeeper's annual salary was raised from $38,000 to $39,900. What percent of her original annual salary is her new annual salary?

Percents on a Calculator

Many calculators have a percent key (%), sometimes used with the 2nd function (2nd). However, the percent key functions differently on different models. Check to see if the following approach works on your machine. If it does not, experiment to find an approach that does.

EXAMPLE 18

Use a calculator to find 50% of 8.

Solution

Press	Display
50 [2nd] [%] [×] 8 [ENTER]	50% * 8 4.

PRACTICE 18

What is 8.25% of $72.37, to the nearest cent?

6.2 Exercises

FOR EXTRA HELP MyMathLab  PRACTICE  WATCH   READ  REVIEW

Mathematically Speaking

Fill in each blank with the most appropriate term or phrase from the given list.

amount	of	base
is	what	percent

1. The ________ is the number that we are taking the percent of.
2. The ________ is the result of taking the percent of the base.
3. The ________ of the base is the amount.
4. In the translation method of solving a percent problem, ________ is replaced by a multiplication symbol.

A *Find the amount.*

5. What is 75% of 8?
6. What is 50% of 48?
7. Compute 100% of 23.
8. Compute 200% of 6.
9. Find 41% of 7.
10. Find 6% of 9.
11. What is 35% of \$400?
12. What is 40% of 10 miles?
13. What is 3.1% of 20?
14. What is 0.5% of 7?
15. Compute 1.8% of 2.5.
16. Compute 3.5% of 4.6.
17. Compute $\frac{1}{2}\%$ of 20.
18. Compute $\frac{1}{10}\%$ of 35.
19. What is $12\frac{1}{2}\%$ of 32?
20. What is $66\frac{2}{3}\%$ of 33?
21. What is $7\frac{1}{8}\%$ of \$257.13, rounded to the nearest cent?
22. What is 8.9% of 7,325 miles, rounded to the nearest mile?

Find the base.

23. 25% of what number is 8?
24. 30% of what number is 120?
25. \$12 is 10% of how much money?
26. \$195 is 1% of what salary?
27. 5 is 200% of what number?
28. 14 is 200% of what number?
29. 2% of what amount of money is \$5?
30. 20% of how many meters is 8 meters?
31. 15 is $33\frac{1}{3}\%$ of what number?
32. 85 is $8\frac{1}{2}\%$ of what number?
33. 3.5 is 200% of what number?
34. 8.1 is 150% of what number?
35. 0.5% of what number is 23?
36. 0.75% of what number is 24?
37. 0.12% of what number is 3.6?
38. 0.25% of what number is 100.4?
39. 6.5% of how much money is \$3,200, rounded to the nearest cent?
40. 4,718 is $2\frac{1}{8}\%$ of what number?

Find the percent.

41. 50 is what percent of 100?

42. 13 is what percent of 52?

43. What percent of 8 is 6?

44. What percent of 50 is 20?

45. What percent of 12 is 10?

46. What percent of 15 is 5?

47. 2 miles is what percent of 8 miles?

48. \$16 is what percent of \$20?

49. \$30 is what percent of \$20?

50. 10 is what percent of 8?

51. 9 feet is what percent of 8 feet?

52. 35¢ is what percent of 21¢?

53. 2.5 is what percent of 4?

54. 1.4 is what percent of 8?

55. \$1.80 is what percent of \$3.60?

56. What percent of 0.3 is 1.5?

57. What percent of 251,749 is 76,801, rounded to the nearest percent?

58. 8,422 is what percent of 11,630, to the nearest percent?

Mixed Practice

Solve.

59. Compute $37\frac{1}{2}\%$ of 160.

60. Calculate 0.01% of 55, rounded to the nearest hundredth.

61. What percent of 15 is 10?

62. What percent of 20 is 30?

63. 20% of what distance is 35 miles?

64. 2.5% of what is 32?

65. 3 feet is what percent of 60 feet?

66. Find 7.2% of \$300.

67. 4% of what amount of money is \$20?

68. \$24 is what percent of \$300?

69. What is 40% of 25?

70. $\frac{3}{4}\%$ of what number is 60?

Applications

B *Solve.*

71. During a tournament, a golfer made par on 12 of 18 holes. On what percent of the holes on the course did she make par?

72. In the first quarter of last year, a steel mill produced 300 tons of steel. If this was 20% of the year's output, find that output.

73. A student answered 90% of the questions on a math exam correctly. If she answered 36 questions correctly, how many questions were on the exam?

74. A property management company sold 80% of the condominium units in a new building with 90 units. How many units were sold?

75. Payroll deductions comprise 40% of the gross income of a student working part-time. If his deductions total \$240, what is his gross income?

76. In 1862, the U.S. Congress enacted the nation's first income tax, at the rate of 3%. How much income tax would you have paid if you made \$2,500? (*Source:* U.S. Bureau of the Census)

77. According to the report on a country's economic conditions, 1.5 million people, or 8% of the workforce, were unemployed. How large was the workforce?

78. A recipe for cattle feed calls for 1,200 pounds of corn, 400 pounds of oats, 200 pounds of protein, 100 pounds of beet pulp, 75 pounds of cottonseed hulls, and 25 pounds of molasses. What percent of this mixture is corn? (*Source:* cattlepages.com)

79. In basketball, a foul shot is called a *free throw*. The recipient of the most valuable player award on a college basketball team made 75% of 96 free throw attempts. How many free throws did he make? (*Source:* wikipedia.org)

80. According to a recent telephone survey, 424 thousand out of 673 thousand adults interviewed in the United States were either overweight or obese. What percent of interviewees, to the nearest whole percent, were either overweight or obese? (*Source:* webmd.com)

81. Flexible-fuel vehicles run on E85, an alternative fuel that is a blend of ethanol and gasoline containing 85% ethanol. How much ethanol is in 12 gallons of E85?

82. Of the 80 classrooms on a community college campus, 75% are equipped with whiteboards. How many classrooms have whiteboards?

83. A company's profits amounted to 10% of its sales. If the profits were \$3 million, compute the company's sales.

84. In a recent survey of U.S. colleges, the average tuition and fees at private four-year colleges was approximately \$20 thousand, in contrast to about \$2 thousand for public two-year colleges. The second figure is what percent of the first figure? (*Source: The Chronicle of Higher Education Almanac*)

85. A math lab coordinator is willing to spend up to 25% of her income on housing. What is the most she can spend if her annual income is \$36,000?

86. An office supply warehouse shipped 648 cases of copy paper. If this represents 72% of the total inventory, how many cases of paper did the warehouse have in its inventory?

87. A lab technician mixed 36 milliliters of alcohol with 84 milliliters of water to make a solution. What percent of the solution was alcohol?

88. A shopper lives in a town where the sales tax is 5%. Across the river, the tax is 4%. If it costs her \$6 to make the round trip across the river, should she cross the river to buy a \$250 television set?

89. The following graph shows the breakdown of the projected U.S population by gender in the year 2020. If the population is expected to be 340 million people, how many more women than men will there be in 2020? (*Source:* census.gov)

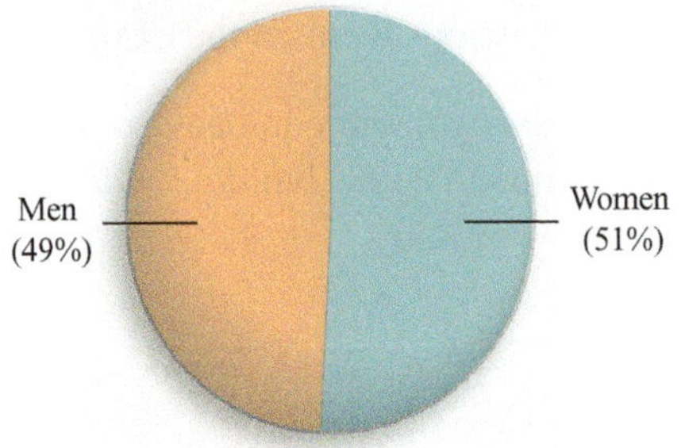

90. The graph gives the use of the Internet by U.S. adults, according to a recent national survey. All percents are rounded to the nearest whole percent.

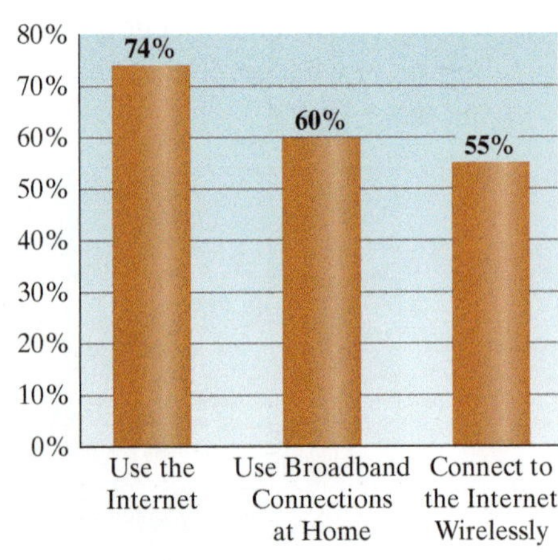

The survey involved tracking a sample of 2,258 adults. How many more adults in this sample use broadband connections at home than those who connect to the internet wirelessly? (*Source:* pewinternet.org)

91. In a company, 85% of the employees are female. If 765 males work for the company, what is the total number of employees?

92. A quarterback completed 15 passes or 20% of his attempted passes. How many of his attempted passes did he *not* complete?

93. The state of Michigan has an urban population of about 8 million people, with the remaining population of 2 million people living in rural areas. (*Source:* ers.usda.gov)

a. Approximately what is the total population of Michigan?

b. Approximately what percent of Michigan's population is urban?

94. A homeowner builds a family room addition on his 1,650-square-foot house, increasing the area of the house by 495 square feet.

a. Calculate the total area of the house with the addition.

b. What percent of the original area is the total area?

• Check your answers on page A-8.

MINDStretchers

Writing

1. Do you prefer solving percent problems using the translation method or the proportion method? In a few sentences, explain why.

Critical Thinking

2. At a college, 20% of the women commute, in contrast to 30% of the men. Yet more women than men commute. Explain how this result is possible.

Technology

3. On the web, go to the U.S. Bureau of the Census home page (census.gov). Write a percent problem of interest to you involving data from the site, and solve the problem.

6.3 More on Percents

OBJECTIVES

- **A** To solve percent increase or decrease problems
- **B** To solve percent problems involving taxes, commissions, or discounts
- **C** To solve simple or compound interest problems
- **D** To solve applied problems involving percents

Finding a Percent Increase or Decrease

Next, let's consider a type of "what percent" problem that deals with a *changing quantity*. If the quantity is increasing, we speak of a *percent increase*; if it is decreasing, we speak of a *percent decrease*.

Here is an example: Last year, a family paid \$2,000 in health insurance, and this year, their health insurance bill was \$2,500. By what percent did this expense increase?

Note that this problem states the value of a quantity at two points in time. We are asked to find the percent increase between these two values.

To solve, we first compute the difference between the values, that is, between the *new value* and the *original value*.

$$\underset{\text{New value}}{\underset{\downarrow}{2{,}500}} - \underset{\text{Original value}}{\underset{\downarrow}{2{,}000}} = \underset{\text{Change in value}}{\underset{\downarrow}{500}}$$

The question posed is expressed as follows:

$$\underset{x}{\underset{\downarrow}{\text{What percent}}} \quad \underset{\cdot}{\underset{\downarrow}{\text{of}}} \quad \underset{2{,}000}{\underset{\downarrow}{2{,}000}} \quad \underset{=}{\underset{\downarrow}{\text{is}}} \quad \underset{500}{\underset{\downarrow}{500?}}$$

It is important to note that the *base* here—as in all percent change problems—is the original value of the quantity.

Next, we solve the equation.

$$2{,}000x = 500$$

$$\frac{\cancel{2{,}000}}{\cancel{2{,}000}}x = \frac{500}{2{,}000}$$

$$x = \frac{1}{4} = 0.25, \text{ or } 25\%$$

So we conclude that the family's health insurance expense *increased* by 25%.

To Find a Percent Increase or Decrease

- Compute the difference between the new and the original values.
- Compute what percent this difference is of the original value.

EXAMPLE 1

The cost of a marriage license had been \$10. Later it rose to \$15. What percent increase was this?

PRACTICE 1

To accommodate a flood of tourists, businesses in town boosted the number of hotel beds from 25 to 100. What percent increase is this?

Solution The original cost of the license was \$10, and the new cost was \$15. The change in cost is, therefore, \$15 − \$10, or \$5. So the question is as follows:

What percent of \$10 is \$5?

$$x \cdot 10 = 5$$

$$10x = 5$$

$$\frac{\not{10}}{\not{10}}x = \frac{\not{5}^{1}}{\not{10}_{2}}$$

$$x = \frac{1}{2} = 0.5, \text{ or } 50\%$$

So the cost of the license increased by 50%.

EXAMPLE 2

Suppose that an animal species is considered to be endangered if its population drops by more than 60%. If a species' population fell from 40 to 18, should we consider the animal endangered?

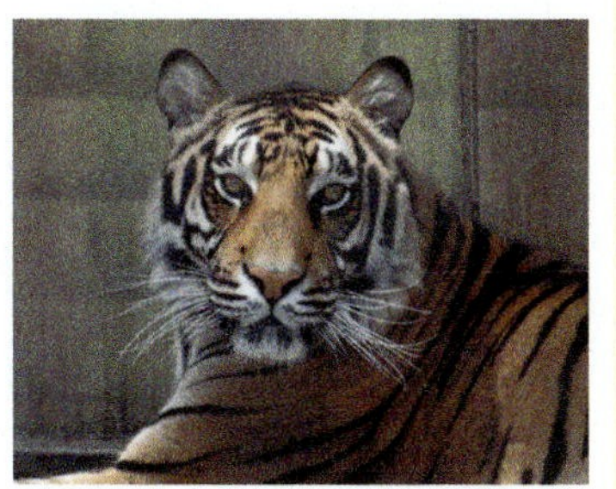

Solution The population dropped from 40 to 18, that is, by 22. The question is how the percent decrease compares with 60%. We compute.

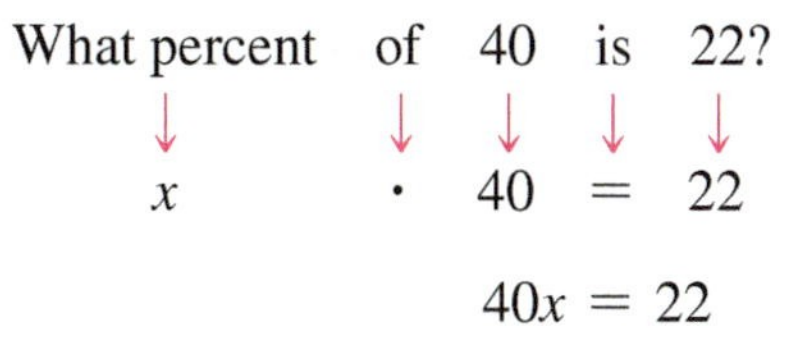

What percent of 40 is 22?

$$x \cdot 40 = 22$$

$$40x = 22$$

$$x = \frac{22}{40} = \frac{11}{20} = 0.55, \text{ or } 55\%$$

Since the population decreased by less than 60%, the species is not considered to be endangered.

PRACTICE 2

Major financial crashes took place on both Tuesday, October 29, 1929, and Monday, October 19, 1987. On the earlier date, the stock index dropped from 300 to 230. On the latter date, it dropped from 2,250 to 1,750. As a percent, did the stock index drop more in 1929 or in 1987? (*Source: The Wall Street Journal*)

Business Applications of Percent

The idea of percent is fundamental to business and finance. Percent applications are part of our lives whenever we buy or sell merchandise, pay taxes, and borrow or invest money.

Taxes

Governments levy taxes to pay for a variety of services, from supporting schools to paving roads. There are many kinds of taxes, including sales, income, property, and import taxes.

In general, the amount of a tax that we pay is a percent of a related value. For instance, sales tax is usually computed as a percent of the price of merchandise sold. Thus, in a town where the sales tax rate is 7%, we could compute the tax on any item sold by computing 7% of the price of that item.

Similarly, property tax is commonly computed by taking a given percent (the tax rate) of the property's assessed value. And an import tax is calculated by taking a specified percent of the market value of the imported item.

EXAMPLE 3

The sales tax on a \$950 digital camcorder is \$71.25. What is the sales tax rate, expressed as a percent?

Solution We must consider the following question:

71.25 is what percent of 950?

$$71.25 = x \cdot 950$$

$$950x = 71.25$$

$$\frac{\cancel{950}x}{\cancel{950}} = \frac{71.25}{950}$$

$$x = 0.075, \text{ or } 7.5\%$$

So the rate of the sales tax is 7.5%, or $7\frac{1}{2}\%$.

PRACTICE 3

When registering a new car, the owner paid a 2.5% import tax on the purchase price of \$18,500. How much import tax did he pay? (*Source:* justlanded.com)

Commission

To encourage salespeople to make more sales, many of them, instead of receiving a fixed salary, are paid on *commission*. Working on commission means that the amount of money that they earn is a specified percent—say, 10%—of the total sales for which they are responsible. Often salespeople make a flat fee in addition to a commission based on sales.

EXAMPLE 4

The owner of a condo in San Diego sold it for \$222,000. On this amount, she paid a real estate agent a commission of 6%.

a. Find the amount of the commission.

b. How much money did the owner make from the sale after paying the agent's fee?

Solution

a. The commission is 6% of \$222,000.

What is 6% of \$222,000?

$$x = 0.06 \cdot 222{,}000 = 13{,}320$$

So the commission amounted to \$13,320.

b. The seller made \$222,000 − \$13,320, or \$208,680.

PRACTICE 4

A sales associate at a furniture store is paid a base monthly salary of \$1,500. In addition, she earns a 9% commission on her monthly sales. If her total sales this month is \$12,500, calculate

a. her commission, and

b. her total monthly income.

Discount

In buying or selling merchandise, the term *discount* refers to a reduction on the merchandise's original price. The rate of discount is usually expressed as a percent of the original price.

EXAMPLE 5

A drugstore gives senior citizens a 10% discount. If some pills normally sell for \$16 a bottle, how much will a senior citizen pay?

Solution Note that, because senior citizens get a discount of 10%, they pay 100% − 10%, or 90%, of the normal price.

The question then becomes:

$$\begin{array}{ccccc} \text{What} & \text{is} & 90\% & \text{of} & \$16? \\ \downarrow & \downarrow & \downarrow & \downarrow & \downarrow \\ x & = & 0.9 & \cdot & 16 = 14.4 \end{array}$$

So a senior citizen will pay \$14.40 for a bottle of the pills.

Note that another way to solve this problem is first to compute the amount of the discount (10% of \$16) and then to subtract this discount from the original price. With this approach, do we get the same answer?

PRACTICE 5

Find the sale price. \$69.60

Have students write their responses in a journal.

Simple Interest

Anyone who has been late in paying a credit card bill or who has deposited money in a savings account knows about *interest*. When we lend or deposit money, we make interest. When we borrow money, we pay interest.

Interest depends on the amount of money borrowed (the *principal*), the annual rate of interest (usually expressed as a percent), and the length of time the money is borrowed (usually expressed in years). We can compute the amount of interest by multiplying the principal by the rate of interest and the number of years. This type of interest is called *simple interest* to distinguish it from *compound interest* (which we discuss later).

EXAMPLE 6

How much simple interest is earned in 1 year on a principal of \$900 at an annual interest rate of 6.5%?

Solution To compute the interest, we multiply the principal by the rate of interest and the number of years.

$$\begin{array}{ccccccc} & & \text{Principal} & & \text{Rate of Interest} & & \text{Number of Years} \\ & & \downarrow & & \downarrow & & \downarrow \\ \text{Interest} & = & 900 & \times & 0.065 & \times & 1 \\ & = & 58.5 & & & & \end{array}$$

So \$58.50 in interest is earned.

PRACTICE 6

What is the simple interest on an investment of \$20,000 for 1 year at an annual interest rate of 7.25%? \$1,450

EXAMPLE 7

A customer deposited \$825 in a savings account that each year pays 5% in simple interest, which is credited to his account. What is the account balance after 2 years?

Solution To solve this problem, let's break it into two questions:

- How much interest did the customer make after 2 years?
- What is the sum of the original deposit and that interest?

First, let's find the interest. To do this, we multiply the principal by the rate of interest and the number of years.

Principal ↓ Rate of Interest ↓ Number of Years ↓

$$\text{Interest} = (825)\ (0.05)\ (2)$$
$$= 82.50$$

The customer made \$82.50 in interest.

Now, let's find the account balance by adding the amount of the original deposit to the interest made.

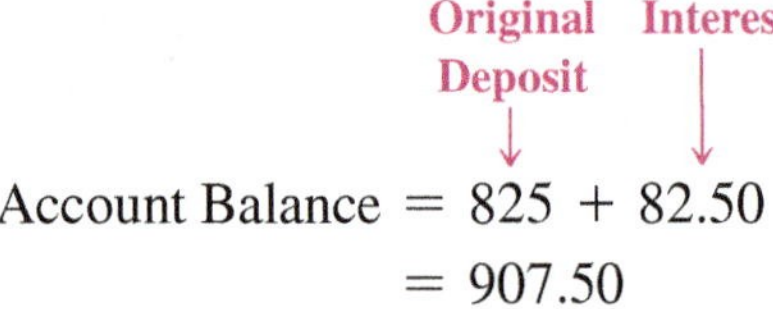

$$\text{Account Balance} = 825 + 82.50$$
$$= 907.50$$

So the account balance after 2 years is \$907.50

PRACTICE 7

A bank account pays 6% simple interest on \$1,600 for 2 years. Compute the account balance after 2 years.

Compound Interest

As we have seen, simple interest is based on the principal. Most banks, however, pay their customers *compound interest*, which is based on both the principal and the previous interest generated.

For instance, suppose that a bank customer has \$1,000 deposited in a savings account that pays 5% interest compounded annually. There were no withdrawals or other deposits. Let's compute the balance in the account at the end of the third year.

The following table shows the account balance after the customer has left the money in the account for 3 years. After 1 year, the account will contain \$1,050 (that is, 100% of the original \$1,000 added to 5% of \$1,000, giving us 105% of \$1,000).

Year	Balance at the End of the Year
0	\$1,000
1	$\$1{,}000 + 0.05 \times \$1{,}000 = \$1{,}050.00$
2	$\$1{,}050 + 0.05 \times \$1{,}050 = \$1{,}102.50$
3	$\$1{,}102.50 + 0.05 \times \$1{,}102.50 \approx \$1{,}157.63$

The balance in the account after the third year is \$1,157.63, rounded to the nearest cent.

Alternatively, for each year we can multiply the account balance by 1.05 to compute the balance at the end of the next year.

Year	Balance at the End of the Year
0	\$1,000
1	$1.05 \times \$1{,}000 = \$1{,}050.00$
2	$(1.05)^2 \times \$1{,}000 = \$1{,}102.50$
3	$(1.05)^3 \times \$1{,}000 \approx \$1{,}157.63$

Note that the balance at the end of the third year is $(1.05)^3 \times \$1{,}000$, or \$1,157.63, in agreement with our previous computation. What would the balance be at the end of the fourth year? What is the relationship between the number of years the money has been invested and the power of 1.05?

In computing the preceding answer, we needed to raise the number 1.05 to a power. Before scientific calculators became available, compound interest problems were commonly solved by use of a compound interest table that contained information such as the following:

Number of Years	4%	5%	6%	7%
1	1.04000	1.05000	1.06000	1.07000
2	1.08160	1.10250	1.12360	1.14490
3	1.12486	1.15763	1.19102	1.22504

When using such a table to calculate a balance, we simply multiply the principal by the number in the table corresponding to the rate of interest and the number of years for which the principal is invested. For instance, after 3 years a principal of \$1,000 compounded at 5% per year results in a balance of $1.15763 \times 1{,}000$, or \$1,157.63, as we previously noted.

Today, problems of this type are generally solved on a calculator.

EXAMPLE 8

A couple deposited \$7,000 in a bank account and did not make any withdrawals or deposits in the account for 3 years. The interest is compounded annually at a rate of 3.5%. What will be the amount in their account at the end of this period?

Solution Each year, the amount in the account is 100% + 3.5%, or 1.035 times the previous year's balance. So at the end of 3 years, the number of dollars in the account is calculated as follows:

Principal	First Year	Second Year	Third Year
↓	↓	↓	↓
7,000 ×	1.035 ×	1.035 ×	1.035

$$7{,}000 \times 1.035 \times 1.035 \times 1.035$$

It makes sense to use a calculator to carry out this computation. One way to key in this computation on a calculator is as follows.

Press	Display
7000 [×] 1.035 [^] 3 [ENTER]	7000 * 1.035 ^ 3 7761.025125

So at the end of 3 years, they have \$7,761.03 in the account, rounded to the nearest cent.

PRACTICE 8

Find the balance after 4 years on a principal amount of \$2,000 invested at a rate of 6% compounded annually.

6.3 Exercises

FOR EXTRA HELP MyMathLab MathXL PRACTICE WATCH READ REVIEW

Mathematically Speaking

Fill in each blank with the most appropriate term or phrase from the given list.

discount	on salary	on commission
interest	simple	compound
final	original	

1. When computing a percent increase or decrease, the ___________ value is used as the base of the percent.

2. Sellers who are paid a fixed percent of the sales for which they are responsible are said to work ___________.

3. A reduction on the price of merchandise is called a(n) ___________.

4. When interest is paid on both the principal and the previous interest generated, it is called ___________ interest.

A *Find the percent increase or decrease.*

5.

Original Value	New Value	Percent Increase or Decrease
\$10	\$12	
\$10	\$8	
\$6	\$18	
\$35	\$70	
\$14	\$21	
\$10	\$1	
\$8	\$6.50	
\$6	\$5.25	

6.

Original Value	New Value	Percent Increase or Decrease
\$5	\$6	
\$12	\$10	
\$4	\$9	
\$25	\$45	
\$10	\$36	
\$100	\$20	
4 ft	3 ft	
8 lb	4.5 lb	

B *Compute the sales tax. Round to the nearest cent.*

7.

Selling Price	Rate of Sales Tax	Sales Tax
\$30.00	5%	
\$24.88	3%	
\$51.00	$7\frac{1}{2}\%$	
\$196.23	4.5%	

8.

Selling Price	Rate of Sales Tax	Sales Tax
\$40.00	6%	
\$16.98	4%	
\$85.00	$5\frac{1}{2}\%$	
\$286.38	5%	

Compute the commission. Round to the nearest cent.

9.

Sales	Rate of Commission	Commission
$700	10%	
$450	2%	
$870	$4\frac{1}{2}\%$	
$922	7.5%	

10.

Sales	Rate of Commission	Commission
$400	1%	
$670	3%	
$610	$6\frac{1}{2}\%$	
$2,500	8.25%	

Compute the discount and sale price. Round to the nearest cent.

11.

Original Price	Rate of Discount	Discount	Sale Price
$700.00	25%		
$18.00	10%		
$43.50	20%		
$16.99	5%		

12.

Original Price	Rate of Discount	Discount	Sale Price
$200.00	30%		
$21.00	50%		
$88.88	10%		
$72.50	40%		

C *Calculate the simple interest and the final balance. Round to the nearest cent.*

13.

Principal	Interest Rate	Time (in years)	Interest	Final Balance
$300	4%	2		
$600	7%	2		
$500	8%	2		
$375	10%	4		
$1,000	3.5%	3		
$70,000	6.25%	30		

14.

Principal	Interest Rate	Time (in years)	Interest	Final Balance
$100	6%	5		
$800	4%	5		
$500	3%	10		
$800	6%	10		
$250	1.5%	2		
$300,000	4.25%	20		

Calculate the final balance after compounding the interest annually. Round to the nearest cent.

15.

Principal	Interest Rate	Time (in years)	Final Balance
$500	4%	2	
$6,200	3%	5	
$300	1%	8	
$20,000	4%	2	
$145	3.8%	3	
$810	2.9%	10	

16.

Principal	Interest Rate	Time (in years)	Final Balance
$300	6%	1	
$2,900	5%	4	
$800	3%	5	
$10,000	3%	4	
$250	1.1%	2	
$200	3.3%	5	

Mixed Practice

Complete each table. Round to the nearest whole percent.

17.

Original Value	New Value	Percent Decrease
$5	$4.50	

18.

Original Value	New Value	Percent Increase
$220	$300	

Complete each table. Round to the nearest cent.

19.

Original Price	Rate of Discount	Discount	Sale Price
$87.33	40%		

20.

Assessed Value of a House	Rate of County Property Tax	County Property Tax
$150,000	0.9475%	

21.

Selling Price	Rate of Sales Tax	Sales Tax
$200	7.25%	

22.

Sales	Rate of Commission	Commission
$537.14	10%	

23.

Principal	Interest Rate	Kind of Interest	Time (in years)	Interest	Final Balance
$3,000	5%	simple	5		

24.

Principal	Interest Rate	Kind of Interest	Time (in years)	Final Balance
$259.13	5.8%	compound annually	12	

Applications

D *Solve.*

25. An upscale department-store chain reported that total sales this year were $2.3 billion—up from $1.8 billion last year. Find the percent increase in sales, to the nearest whole percent.

26. Last year, a local team won 20 games. This year, it won 15 games. What was the percent decrease of games won?

27. In 9 years, the number of elderly nursing home residents rose from 200,000 to 1.3 million. By what percent did the number of residents increase?

28. Due to a decrease in demand, a manufacturing plant decreased its production from 2,400 to 1,800 units per day. What was the percent decrease in the number of units produced per day?

29. The first commercial telephone exchange was set up in New Haven, Connecticut, in 1878. Between 1880 and 1890, the number of telephones in the United States increased from 50 thousand to 200 thousand, in round numbers. What percent increase was this? (*Source:* census.gov)

30. A patient's medication was decreased from 250 milligrams to 200 milligrams per dose. What was the percent decrease in the dosage?

31. A customer paid 5% sales tax on a notebook computer that sold for $1,699. Calculate the amount of sales tax that she paid.

32. Last year, a town assessed the value of a residential property at $272,000. If the homeowner paid property tax of $3,264, what was the property tax rate?

33. A customer bought a cell phone for $150. The total selling price of the phone, including sales tax, was $159.75. What was the sales tax rate?

34. In a town, the sales tax rate is 7%. It costs $5 to travel to a nearby town and back where the sales tax is only 5.8%. Is it worthwhile to make this trip to purchase an item that sells in both towns for $800?

35. A pharmaceutical sales representative earns a 12% monthly commission on all her sales above $5,000. Find her commission if her sales this month totaled $27,500.

36. A sales assistant earns a flat salary of $150 plus a 10% commission on sales of $3,000. What were his total earnings?

37. On a restaurant table, a customer leaves $1.35 as a tip. Assuming that the customer left a 15% tip, how much was the bill before the tip?

38. A salesperson receives a 5% commission on the first $2,000 in sales and a 7% commission on sales above $2,000. How much commission does she earn on sales of $3,500?

39. A coupon for an online sporting goods company gives customers a discount on any order. The discounted price on a tennis racquet that normally sells for $180 is $153. What percent is the discount? (*Source:* sportsauthority.com)

40. A car rental company gives members of the organization AARP a 5% discount on standard rates. For a car that ordinarily rents for $43.84 per day, what is the daily rate, to the nearest cent, after the AARP discount? (*Source:* enterprise.com)

41. A store sells a television that lists for $399 at a 35% discount rate. What is the sale price?

42. An appliance store has a sale on all its appliances. A washing machine that originally sold for $800 is on sale for $680. What is the discount rate?

43. A bank customer borrowed $3,000 for 1 year at 5% simple interest to buy a computer. How much interest did the customer pay?

44. How much simple interest is earned on $600 at an 8% annual interest rate for 2 years?

45. A couple deposited $5,000 in a bank. How much interest will they have earned after 1 year if the interest rate is 5%?

46. A student borrowed $2,000 from a friend, agreeing to pay her 4% simple interest. If he promised to repay her the entire amount at the end of 3 years, how much money must he pay her?

47. At a home goods store, a customer bought a down comforter that originally cost $180.

a. What was the sale price of the comforter?
b. Calculate the total amount the customer paid after 6% sales tax was added to the purchase.

48. During a sale, a shoe store marked down the price of a pair of sneakers that originally cost $80 by 40%.

a. What was the sale price of the sneakers?
b. After two weeks, the store marked down the sale price by another 60%. What percent off the original price was the sale price after the second discount was applied?

49. An investor put $3,000 in an account that pays 4% interest, compounded annually. Find the amount in the account after 2 years.

50. A bank pays 4.5% interest, compounded annually, on a 2-year certificate of deposit (CD) that initially costs $500. What is the value of the CD at the end of the 2 years, rounded to the nearest cent?

51. A city had a population of 4,000. If the city's population increased by 10% per year, what was the population 4 years later?

52. An art dealer bought a painting for $10,000. If the value of the painting increased by 50% per year, what was its value 4 years later?

• Check your answers on page A-8.

MINDStretchers

Writing

1. Explain the difference between simple interest and compound interest.

Technology

2. Using a spreadsheet, construct a three-column table showing the original price, the 10% discount, and the selling price for items with an original price of any whole number of dollars between $1 and $100.

Mathematical Reasoning

3. If a quantity increases by a given percent and then decreases by the same percent, will the final value be the same as the original value? Explain.

Key Concepts and Skills

CONCEPT SKILL

Concept/Skill	Description	Example
[6.1] **Percent**	A ratio or fraction with denominator 100. It is written with the % sign, which means divided by 100.	$7\% = \frac{7}{100}$ ↑ Percent
[6.1] **To change a percent to the equivalent fraction**	• Drop the % sign from the given percent and place the number over 100. • Simplify the resulting fraction, if possible.	$25\% = \frac{25}{100} = \frac{1}{4}$
[6.1] **To change a percent to the equivalent decimal**	• Drop the % sign from the given percent and divide the number by 100.	$23.5\% = .235$, or 0.235
[6.1] **To change a decimal to the equivalent percent**	• Multiply the number by 100 and insert a % sign.	$0.125 = 12.5\%$
[6.1] **To change a fraction to the equivalent percent**	• Multiply the fraction by 100%.	$\frac{1}{5} = \frac{1}{5} \times 100\% = \frac{1}{\cancel{5}_1} \times \frac{\cancel{100}^{20}}{1}\%$ $= 20\%$
[6.2] **Base**	The number that we are taking the percent of. It always follows the word *of* in the statement of a percent problem.	50% of 8 is 4. ↑ (8) Base
[6.2] **Amount**	The result of taking the percent of the base.	50% of 8 is 4. ↑ (4) Amount
[6.2] **To solve a percent problem using the translation method**	• Translate as follows: What number, what percent ⟶ x is ⟶ = of ⟶ × or · % ⟶ decimal or fraction • Set up the equation. **The percent of the base is the amount.** • Solve.	What is 50% of 8? ↓ ↓ ↓ ↓ ↓ $x = 0.5 \cdot 8$ $x = 4$ 30% of what number is 6? ↓ ↓ ↓ ↓ ↓ $0.3 \cdot x = 6$ $\frac{\cancel{0.3}x}{\cancel{0.3}} = \frac{6}{0.3}$ $x = \frac{6}{0.3} = 20$ What percent of 8 is 2? ↓ ↓ ↓ ↓ ↓ $x \cdot 8 = 2$ $8x = 2$ $x = \frac{2}{8} = \frac{1}{4} = 25\%$

continued

CONCEPT SKILL

Concept/Skill	Description	Example
[6.2] **To solve a percent problem using the proportion method**	• Identify the amount, the base, and the percent, if known. • Set up and substitute into the proportion. $\frac{\text{Amount}}{\text{Base}} = \frac{\text{Percent}}{100}$ • Solve for the unknown quantity.	50% of 8 is what number? $\frac{x}{8} = \frac{50}{100}$ $100x = 400$ $x = 4$ 30% of what number is 6? $\frac{6}{x} = \frac{30}{100}$ $30x = 600$ $x = 20$ What percent of 8 is 2? $\frac{2}{8} = \frac{x}{100}$ $8x = 200$ $x = 25$ So the answer is 25%.
[6.3] **To find a percent increase or decrease**	• Compute the difference between the new and the original values. • Determine what percent this difference is of the original value.	Find the percent increase for a quantity that changes from 4 to 5. Difference: $5 - 4 = 1$ What percent of 4 is 1? ↓ ↓ ↓ ↓ ↓ $x \cdot 4 = 1$ $x = \frac{1}{4} = 0.25$, or 25%

CHAPTER 6 Review Exercises

Say Why

Fill in each blank.

1. $129\frac{1}{2}\%$ and $\frac{129\frac{1}{2}}{100}$ ________ equivalent because
are/are not
____
____.

2. $\frac{1}{2}$ and $\frac{1}{2}\%$ ________ equivalent because ________
are/are not
____
____.

3. In "50% of 8 is 4", 8 ________ the base because
is/is not
____
____.

4. In "50% of 8 is 4", 50 ________ the amount because
is/is not
____
____.

5. The base of a percent ________ always larger than the
is/is not
amount because ____
____.

6. If 5% of a number is 20, then that number ________
is/is not
greater than 20 because ____
____
____.

7. 100% of a number ________ equal to that number
is/is not
because ____
____.

8. If a quantity doubles in value between two points in time, then the percent increase ________ 200%
is/is not
because ____
____.

[6.1] *Complete the following tables.*

9.

Fraction	Decimal	Percent
$\frac{1}{4}$		
	0.7	
		$\frac{3}{4}\%$
$\frac{5}{8}$		
		41%
$1\frac{1}{100}$		
		260%
	3.3	
	0.12	
		$66\frac{2}{3}\%$
$\frac{1}{6}$		

10.

Fraction	Decimal	Percent
$\frac{3}{8}$		
	0.49	
		0.1%
		150%
	0.875	
		$83\frac{1}{3}\%$
$2\frac{3}{4}$		
	1.2	
	0.75	
		10%
$\frac{1}{3}$		

[6.2] *Solve.*

11. What is 40% of 30?

12. What percent of 5 is 6?

13. 2 feet is what percent of 4 feet?

14. 30% of what number is 6?

15. What percent of 8 is 3.5?

16. Find 55% of 10.

17. $12 is 200% of what amount of money?

18. 2 is what percent of 10?

19. What is 1.2% of 25?

20. Find 115% of 400.

21. 35% of $200 is what?

22. $\frac{1}{2}$% of what number is 5?

23. 15 is what percent of 0.75?

24. 4.5 is what percent of 18?

25. Calculate $33\frac{1}{3}$% of $600.

26. What percent of $9 is $4?

27. Find 60% of $20.

28. 2.5% of how much money is $40?

29. What percent of $7.99 is $1.35, to the nearest whole percent?

30. 3.5 is $8\frac{1}{4}$% of what number, to the nearest hundredth?

[6.3] *Complete the following tables.*

31.

Original Value	New Value	Percent Decrease
24	16	

32.

Original Value	New Value	Percent Decrease
360 mi	300 mi	

33.

Selling Price	Rate of Sales Tax	Sales Tax
$50	6%	

34.

Sales	Rate of Commission	Commission
$600	4%	

35.

Original Price	Rate of Discount	Discount	Sale Price
$200	15%		

36.

Principal	Interest Rate	Time (in years)	Simple Interest	Final Balance
$200	4%	2		

Mixed Applications

Solve.

37. A compact fluorescent light bulb (CFL) will last up to 8,000 hours. If another CFL lasts 25% longer, what is the life of this bulb? (*Sources:* smarthome.com and bulbs.com)

38. Jonas Salk developed the polio vaccine in 1954. The number of reported polio cases in the United States dropped from 29,000 to 15,000 between 1955 and 1956. What was the percent drop, to the nearest whole percent? (*Source:* census.gov)

39. For their fees, one real estate agent charges 11% of a year's rent and another charges the first month's rent. Which agent charges more?

40. A particular community bank makes available loans with simple interest. How much interest is due on a five-year car loan of $24,000 based on a simple interest rate of 6%?

41. According to a city survey, 49% of respondents approve of how the mayor is handling his job and 31% disapprove. What percent neither approved nor disapproved?

42. Plastics make up about 11% and paper makes up $\frac{9}{25}$ of the solid municipal waste in the United States. Which makes up more of the solid municipal waste? (*Source:* Energy Information Administration)

43. According to a study, 25% of employees do not take all of their vacation time due to the demands of their jobs. Express this percent as a fraction. (*Source:* Families and Work Institute)

44. In a recent year, 16 out of every 25 high school graduates in Ohio took the ACT college entrance exam. Express this fraction as a percent. (*Source: The World Almanac, 2010*)

45. It takes a worker 50 minutes to commute to work. If he has been traveling for 20 minutes, what percent of his trip has been completed?

46. Among left-handed people are a number of U.S. presidents, including Ronald Reagan, George H.W. Bush, Bill Clinton, and Barack Obama. About 3 out of every 20 people are lefties. Express as a percent. (*Sources:* indiana.edu and scientificamerican.com)

47. A couple financed a 30-year mortgage at a fixed interest rate of 6.29%. Express this rate as a decimal.

48. The length of a person's thigh bone is usually about 27% of his or her height. Estimate someone's height whose thigh bone is 20 inches long. (*Source: American Journal of Physical Anthropology*)

49. The following table deals with estimates of the recoverable coal reserves in two states that are leading coal producers.

State	Coal Reserves (in billions of short tons)
West Virginia	17
Illinois	38

The size of West Virginia's reserves is what percent of the size of Illinois' reserves rounded to the nearest percent? (*Source*: nma.org)

50. A clothing store places the following ad in a local newspaper:

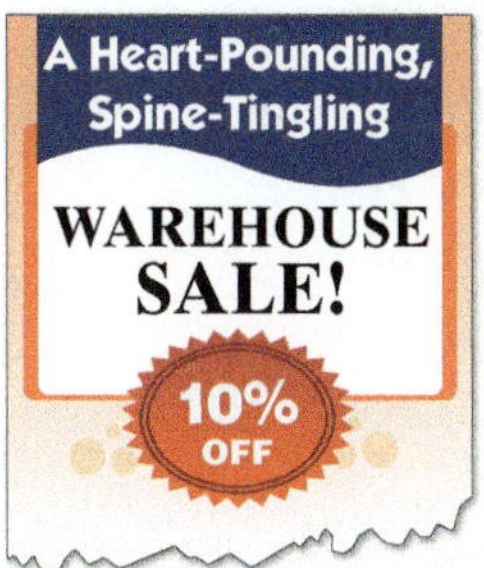

At the store, what is the sale price of a suit that regularly sells for $230?

51. The salary of an executive assistant had been $30,000 before she got a raise of $1,000. If the rate of inflation is 5%, has her salary kept pace with inflation?

52. In a scientific study that relates weight to health, people are considered overweight if their actual weight is at least 20% above their ideal weight. If you weigh 160 pounds and have an ideal weight of 130 pounds, are you considered overweight?

53. An airline oversold a flight to Los Angeles by nine seats, or 5% of the total number of seats available on the airplane. How many seats does the airplane have?

54. When an assistant became editor, the magazine's weekly circulation increased from 50,000 to 60,000. By what percent did the circulation increase?

55. The winner of a men's U.S. Open tennis match got $87\frac{1}{2}\%$ of his first serves in. If he had 72 first serves, how many went in?

56. According to a recent survey, there are approximately 78 million owned dogs and 94 million owned cats in the United States. The number of dogs is what percent of the number of cats, to the nearest percent? (*Source:* humanesociety.org)

57. At an auction, a bidder bought a table for $150. The auction house also charged a "buyer's premium"—an extra fee—of 10%. How much did the bidder pay in all?

58. According to the news report, 80 tons of food met only 20% of the food needs in the refugee camp. How much additional food was needed?

59. A traveler needs 14,000 more frequent-flier miles to earn a free trip to Hawaii, which is 20% of the total number needed. How many frequent-flier miles in all does this award require?

60. The sales tax rate on a flat panel TV bought in New Orleans is 9%. What was the selling price (not including the sales tax) if the sales tax amounted to $53.91? (*Source:* forbes.com)

61. A sales representative for wholesale products earned $49,000 per year plus 10% commission on sales totaling $25,000. What was his total income for the year?

62. At the end of the year, the receipts of a retail store amounted to $200,000. Of these receipts, 85% went for expenses; the rest was profit. How much profit did the store make?

63. If a bank customer deposits $7,000 in a bank account that pays a 5.5% rate of interest compounded annually, what will be the balance after 2 years?

64. Suppose that a country's economy expands by 2% per year. By what percent will it expand in 10 years, to the nearest whole percent?

65. Complete the following table which shows the net income for Texas Instruments Inc. in four consecutive quarters.

Quarter Ending	Income (in millions of dollars)	Percent of Total Income (rounded to the nearest tenth of a percent)
Jun 30, 2009	260	
Sep 30, 2009	538	
Dec 31, 2009	655	
Mar 31, 2010	658	
Total		

(*Source:* finance.yahoo.com)

66. The following graph shows the sources from which the federal government received income in a recent year:

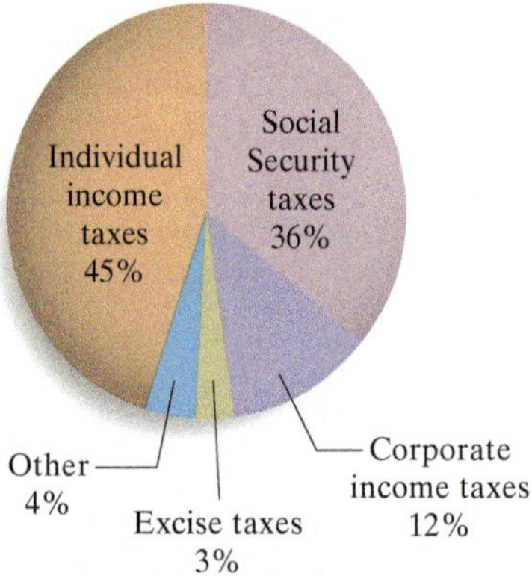

If the total amount of money taken in was $2,500 billion, compute how much money was received from each source, to the nearest billion dollars. (*Source:* taxpolicycenter.org)

• Check your answers on page A-9.

CHAPTER 6 Posttest

FOR EXTRA HELP

CHAPTER Test Prep VIDEOS

The Chapter Test Prep Videos with test solutions are available on DVD, in MyMathLab, and on YouTube* (search "AkstDevMath" and click on "Channels").

To see if you have mastered the topics in this chapter, take this test.

Rewrite.

1. 4% as a fraction

2. $27\frac{1}{2}\%$ as a fraction

3. 174% as a decimal

4. 8% as a decimal

5. 0.009 as a percent

6. 10 as a percent

7. $\frac{5}{6}$ as a percent, rounded to the nearest whole percent

8. $2\frac{1}{5}$ as a percent

Solve.

9. What is 25% of 30 miles?

10. Find 120% of 40.

11. 3% of what number is 9?

12. 8% of what number is 16?

13. What percent of 10 is 6?

14. What percent of 4 is 10?

15. To pay for tuition, a college student borrows \$2,000 from a relative for 2 years at 5% simple interest. Find the amount of simple interest that is due.

16. In a parking lot that has 150 spaces, 4% are for handicap parking. How many handicap spaces are in the lot?

17. A customer paid \$14.95 in sales tax on an iPhone that sells for \$299 (before tax). What was the sales tax rate?

18. Milk is approximately 50% cream. How much milk is needed to produce 2 pints of cream?

19. For some Broadway and off-Broadway performances, TKTS discount booths sell tickets at 30% off the full price plus a \$4 per ticket service charge. What is the total cost of three tickets that sell for \$98 at full price? (*Source*: tdf.org)

20. A college ended six straight years of tuition increases by raising its tuition from \$3,000 to \$3,100. Find the percent increase.

• Check your answers on page A-10.

Cumulative Review Exercises

To help you review, solve the following.

1. Express 10,000,000 as a power of 10.

2. Divide: $1{,}962 \div 18$

3. Find the sum of $3\frac{4}{5}$ and $1\frac{9}{10}$.

4. Find the difference: $32.25 - 4.68$

5. Multiply: 0.2×3.5

6. Express $\frac{5}{6}$ as a decimal, rounded to the nearest hundredth.

7. Divide, rounding to the nearest hundredth: $5.122 \div 0.7$

8. Translate the phrase "the difference between a number and 6.7" into an algebraic expression.

9. Solve and check: $w + 17\frac{2}{5} = 41$

10. Solve for x: $\frac{x}{3} = 2.5$

11. Write as a unit rate: $327.60 for 40 hours.

12. Solve and check: $\frac{1.2}{x} = \frac{1.8}{21}$

13. Change $18\frac{2}{11}\%$ to a fraction.

14. 20% of what amount is $200?

Solve.

15. The government withdrew $\frac{1}{4}$ million of its 2 million troops. What fraction of the total is this?

16. Three FM stations are highlighted on the radio dial shown. These stations have frequencies 99.5 (WBAI), 104.3 (WAXQ), and 105.9 (WQXR). Label the three stations on the dial.

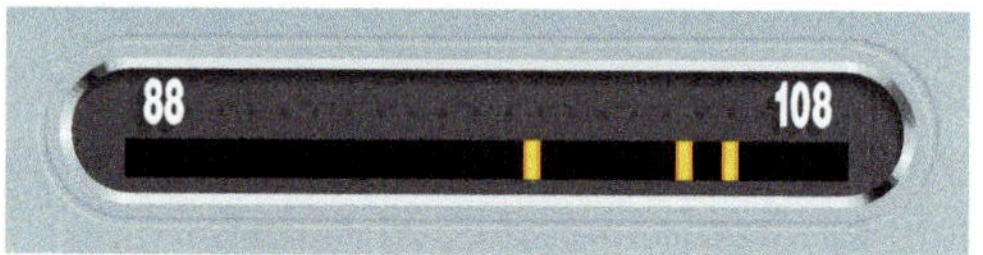

17. At the Westminster Dog Show, the Best in Show prize has been won by terriers three times as often as working group breeds (such as boxers and Great Danes). If terriers have won 45 out of the 103 times that the prize has been awarded, determine how many times working group dogs have won. (*Source*: wikipedia.org)

18. Twitter is a popular social networking service used for tweeting, that is, for sending brief messages. According to a recent survey, 11 of 100 adults who use the internet have tweeted. At this rate among 6,500 such adults, how many would be expected to have tweeted? (*Sources*: wikipedia.org and socialmediatoday.com)

19. In a recent year, about 28% of the 992 thousand U.S. doctors were female. How many female doctors were there, to the nearest thousand? (*Source:* ama.assn.org)

20. Between the years 2010 and 2050, the U.S. population is projected to double. The portion of this population 85 and over is projected to grow more rapidly, increasing from 4 million to 21 million. What percent increase is this? (*Source*: census.gov)

• Check your answers on page A-10.

CHAPTER 7

Signed Numbers

Signed Numbers and Chemistry

In chemistry, a valence is assigned to each element in a compound. Valences help us study the ways in which the elements combine to form the compound.

The valence is a positive or negative whole number that expresses the combining capacity of the element. For example in the compound $CaCl_2$ (calcium chloride), the element calcium (Ca) has a valence of $+2$, whereas the element chlorine (Cl) has a valence of -1.

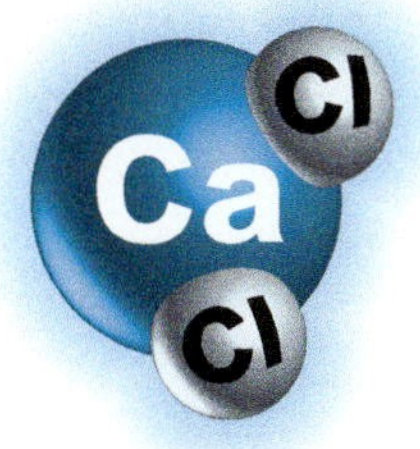

The valences in any chemical compound add up to 0. So if you know how to perform signed number computations, you can predict the chemical formula of any compound.

(*Source:* Karen C. Timberlake, *Basic Chemistry*, Prentice Hall, 2011)

CHAPTER 7 PRETEST

To see if you have already mastered the topics in this chapter, take this test.

1. Which is larger, -23 or $+7$?

2. A negative number has an absolute value of 4. What is the number?

Compute.

3. $-8 + (-9)$

4. $-20 + 20$

5. $34 - 41$

6. $-9 - (-9)$

7. -5×15

8. -8^2

9. $-\frac{3}{4} \times \frac{2}{3}$

10. $\left(-\frac{1}{2}\right)^2$

11. $-18 \div (-9)$

12. $-1.8 \div (-0.9)$

13. $-2 + 5 + (-3) + 8$

14. $10 + (-3) - (-1)$

15. $-9 - 3^2 \times (-5)$

16. $8 \div (-2) + 3 \cdot (-1)$

Solve.

17. The temperature at noon was 74°F. A cold front moved into the region, causing the temperature to drop an average of 4°F per hour over the next 4 hours. What was the temperature at 4:00 P.M.?

18. A fee of \$2.50 is deducted from a student's account each time she uses an ATM on campus. In 1 month, she uses the ATM 7 times. Express as a signed number the impact on her account as a result of the ATM fees.

19. The graph below gives the average December temperatures in degrees Celsius for various cities around the world.

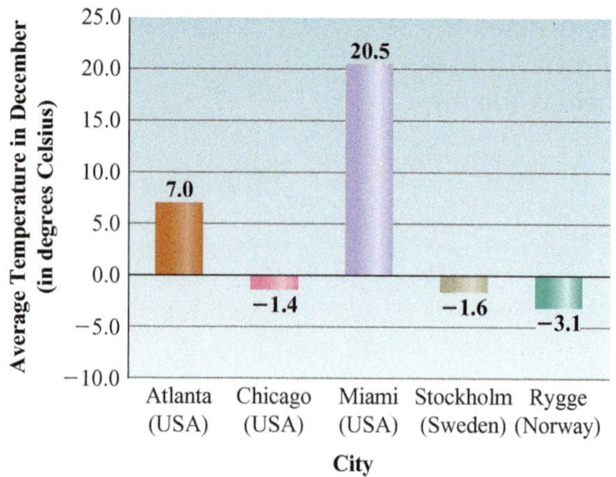

(*Source:* worldclimate.com)

How much greater is the temperature of the warmest city as compared to the coolest city?

20. The Greek mathematician Pythagoras died about 500 B.C. In the year A.D. 2000, how many centuries had passed since his death? (*Source:* wikipedia.org)

• Check your answers on page A-10.

7.1 Introduction to Signed Numbers

OBJECTIVES

- **A** Represent signed numbers on the number line
- **B** To find the opposite of a signed number
- **C** To find the absolute value of a signed number
- **D** To compare signed numbers
- **E** To solve applied problems involving signed numbers

What Signed Numbers Are and Why They Are Important

In this chapter, we discuss negative numbers and show how they relate to positive numbers—the numbers greater than 0. Negative and positive numbers together are referred to as *signed numbers*.

Here are a few applications of signed numbers.

- In football, a positive number represents yards gained; a negative number, yards lost.
- In terms of time, positive applies to a time after an event took place; negative, to a time before that event.
- In the study of electricity, positive represents one kind of electric charge; negative, the opposite kind of electric charge.

These applications can help you develop intuition in working with signed numbers and understand what they represent.

The Number Line

Our previous discussion of the number line on page 3 included only 0 and the positive numbers. However, the number line can be extended to represent the negative numbers also. If we label the numbers to the right of 0 as positive and extend the line leftward past 0, then we label the numbers to the left of 0 as negative.

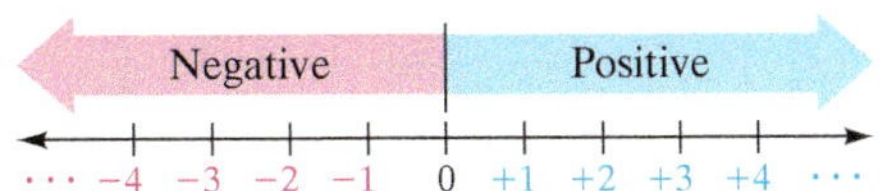

Note that we write "negative two" as −2 and "positive three" as +3, or just 3. However, we write no sign before 0 because 0 is neither negative nor positive.

DEFINITIONS

A **positive number** is a number greater than 0.

A **negative number** is a number less than 0.

A **signed number** is a number with a sign that is either positive or negative.

In drawing the number line, we usually label only the integers, that is, the whole numbers and the corresponding negatives.

DEFINITION

The **integers** are the numbers . . . , −4, −3, −2, −1, 0, +1, +2, +3, +4, . . . , continuing indefinitely in both directions.

Positive fractions and decimals and their corresponding negatives can also be represented on the number line. Let's look at how to locate the points on the number line that correspond to the numbers $\frac{3}{4}$, 3.8, and -2.4. The following number line shows these locations.

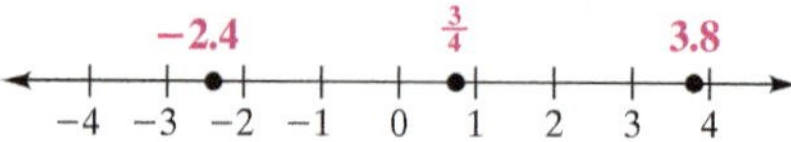

EXAMPLE 1

Locate $\frac{1}{2}$, -2.8, $-\frac{1}{8}$, and 1.2 on the number line.

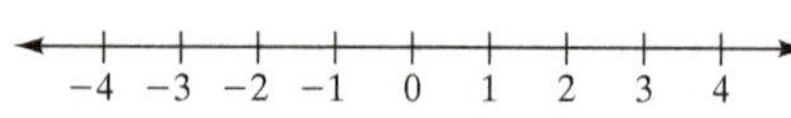

Solution

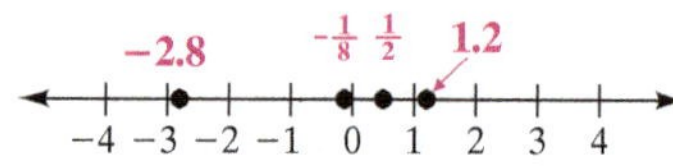

PRACTICE 1

Locate $1\frac{9}{10}$, -1, -3.1, and 0 on the number line.

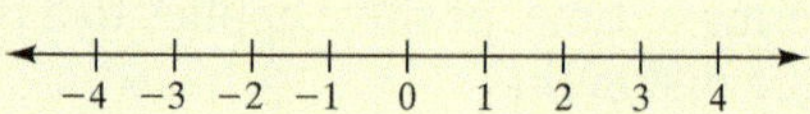

On the number line, we say that -1 and $+1$ (or 1) are the *opposites* of each other. Similarly -50 and $+50$ (or 50) are opposites. What is the opposite of 0?

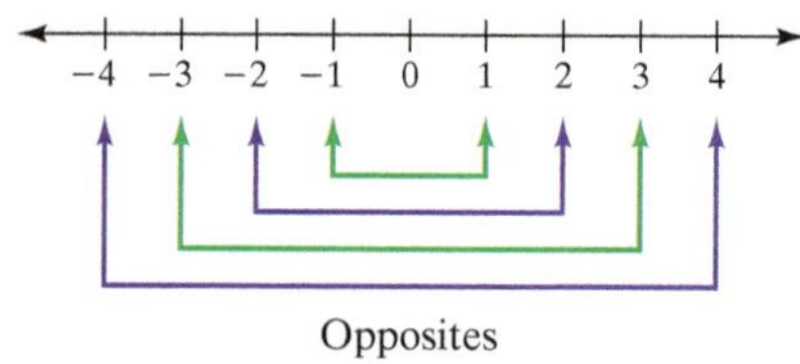

Opposites

DEFINITION

Two numbers that are the same distance from 0 on the number line but on opposite sides of 0 are called **opposites.**

From this definition, we see that opposite numbers have opposite signs.

EXAMPLE 2

Find the opposite of each number in the table.

Solution

Number	Opposite
a. 5	-5
b. $-\frac{1}{2}$	$\frac{1}{2}$
c. 1.5	-1.5
d. -100	100

PRACTICE 2

Find the opposite of each number.

Number	Opposite
a. 9	
b. $-4\frac{9}{10}$	
c. -2.9	
d. 31	

Because the number -2 is negative, it lies 2 units to the left of 0. The number 2, which is positive, lies in the opposite direction: 2 units to the right of 0.

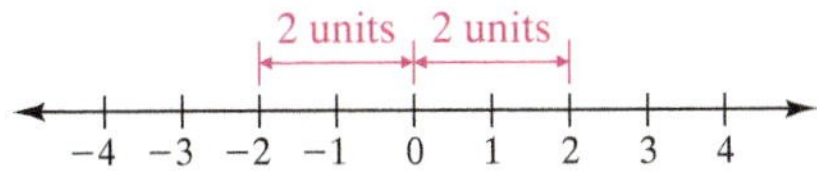

When you locate a number on the number line, the *distance* of that number from 0 is called its *absolute value*. Thus, the absolute value of $+2$ is 2, and the absolute value of -2 is 2.

DEFINITION

The **absolute value** of a number is its distance from 0 on the number line. The absolute value of a number is represented by the symbol $|\ |$.

For example, we write the absolute value of -2 as $|-2|$.

Several properties of absolute value follow from this definition.

- The absolute value of a positive number is the number itself.
- The absolute value of a negative number is its opposite.
- The absolute value of 0 is 0.
- The absolute value of a number is always positive or 0.

These properties help us find the absolute value of any number.

EXAMPLE 3

Compute.

a. $|-8|$ **b.** $|0|$ **c.** $\left|-\frac{1}{2}\right|$ **d.** $|5.3|$

Solution

a. Because -8 is negative, its absolute value is its opposite, or 8.

b. The absolute value of 0 is 0.

c. $\frac{1}{2}$ **d.** 5.3

PRACTICE 3

Compute.

a. $|9|$ **b.** $|1\frac{3}{4}|$ **c.** $|-4.1|$ **d.** $|-5|$

EXAMPLE 4

Determine the sign and the absolute value of the number.

a. 25 **b.** -1.9

Solution

a. Sign: +; absolute value: 25

b. Sign: −; absolute value: 1.9

PRACTICE 4

What are the sign and the absolute value of the number?

	Sign	Absolute value
a. -4		
b. $6\frac{1}{2}$		

Comparing Signed Numbers

The number line helps us compare two signed numbers, that is, to decide which number is larger and which is smaller. On the number line, a number to the right is the larger number.

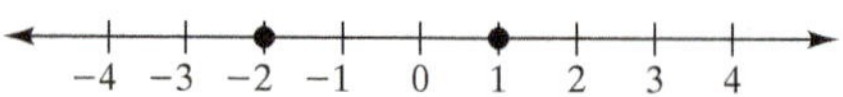

So $1 > -2$.

To Compare Signed Numbers

- Locate the points being compared on the number line; a number to the right is larger than a number to the left.

When comparing signed numbers, remember the following:

- Zero is greater than any negative number because all negative numbers lie to the left of 0.
- Zero is less than any positive number because all positive numbers lie to the right of 0.
- Any positive number is greater than any negative number because all positive numbers lie to the right of all negative numbers.

EXAMPLE 5

Which is larger?

a. $\frac{1}{2}$ or 0 **b.** -1 or -3 **c.** 1.4 or -3

Solution

a. Because $\frac{1}{2}$ (or $+\frac{1}{2}$) is to the right of 0 on the number line, $\frac{1}{2}$ is greater than 0.

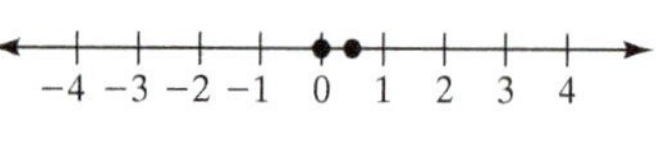

b. Because -1 is to the right of -3, $-1 > -3$, that is, -1 is larger.

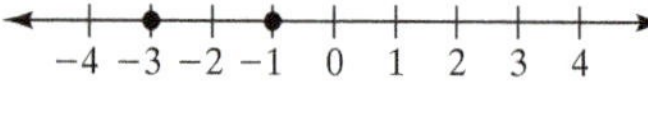

c. Because 1.4 is to the right of -3, $1.4 > -3$, that is, 1.4 is the larger of the two numbers.

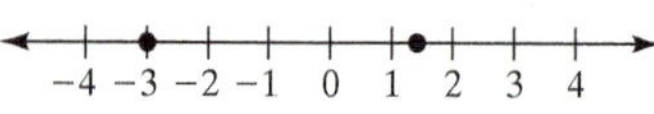

PRACTICE 5

Which is smaller?

a. 0 or 2

b. -5 or -2

c. 2.3 or -4

Now, let's try some practical applications of comparing signed numbers. The key is to be able to determine if a number is negative or positive. You should become familiar with the following words that indicate the sign of a number.

Negative	Positive
Loss	Gain
Below	Above
Decrease	Increase
Down	Up
Withdrawal	Deposit
Past	Future
Before	After

EXAMPLE 6

Express as a signed number: Badwater Basin in Death Valley is the lowest elevation in the Western Hemisphere at 282 feet below sea level. (*Source:* National Park Service)

Solution The number in question represents an elevation below sea level, so we write it as a negative number: −282 feet.

PRACTICE 6

Represent as a signed number: The New York Giants gained 2 yards on a play.

EXAMPLE 7

The following table shows the temperature below which various plants freeze and die.

Plant	Asters	Carnations	Mums
Hardy to	−20°F	−5°F	−30°F

In a very cold climate, which would be planted?
(*Source: The American Horticultural Society A–Z Encyclopedia of Garden Plants*)

Solution First, we compare the temperatures of the asters and the carnations. Because $-20° < -5°$, the asters are hardier than the carnations. Next, we compare the temperatures of the asters and the mums. Because $-20° > -30°$, the mums are hardier. So the mums would be the best of the three to plant.

PRACTICE 7

A student studying astronomy learns that *apparent magnitude* is how bright a star appears when viewed from Earth. The lower the apparent magnitude, the brighter the star appears. The table shows the apparent magnitude for various stars. (*Source: Encyclopedia Britannica Almanac*)

Star	Canopus	Sirius	Alpha Centauri
Apparent Magnitude	−0.72	−1.46	−0.01

Which of these stars is the brightest?

7.1 Exercises

FOR EXTRA HELP MyMathLab MathXL PRACTICE WATCH READ REVIEW

Mathematically Speaking

Fill in each blank with the most appropriate term or phrase from the given list.

larger	smaller	opposites	right
positive number	left	absolute value	integers
signed number	negative number	whole numbers	

1. A number greater than 0 is a(n) ________________.

2. A number less than 0 is a(n) ________________.

3. A number with a sign that is either positive or negative is called a(n) __________________.

4. The __________ are the numbers . . . , $-4, -3, -2, -1, 0, 1, 2, 3, 4, \ldots$, continuing indefinitely in both directions.

5. Two numbers that are the same distance from 0 on the number line but on opposite sides of 0 are called __________.

6. The ________________ of a number is its distance from 0 on the number line.

7. For two numbers on the number line, the number on the left is __________ than the number on the right.

8. For two numbers on the number line, the number on the __________ is larger.

A *Mark the corresponding point for each number on the number line.*

9. $-2, 2.5, 3$

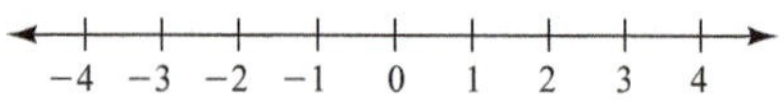

10. $4, -1, 1.1$

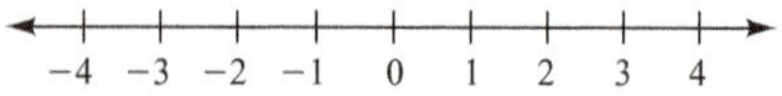

11. $0, -\frac{1}{2}, 1\frac{4}{5}$

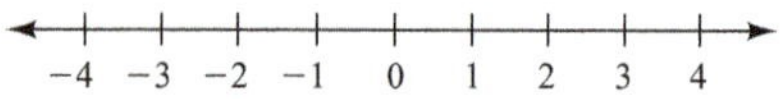

12. $-\frac{1}{3}, 2, -3\frac{9}{10}$

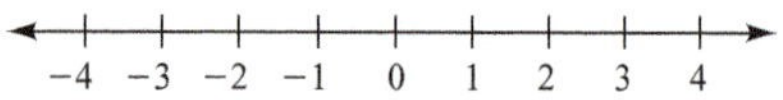

B *Find the opposite of each number.*

13. 8

14. 3

15. 10.2

16. 8.4

17. -25

18. -5

19. $-5\frac{1}{2}$

20. $-2\frac{1}{3}$

21. -4.1

22. -1.2

23. 0.5

24. 0.8

C *Evaluate.*

25. $|-6|$

26. $|-39|$

27. $\left|-\frac{4}{5}\right|$

28. $|-1\frac{2}{3}|$

29. $|2|$

30. $|8|$

31. $|-0.6|$

32. $|-5.8|$

Determine the sign and the absolute value of each number.

	Sign	Absolute Value		Sign	Absolute Value
33. 8			**34.** 11		
35. −4.3			**36.** −9.2		
37. −7			**38.** −30		
39. $\frac{1}{5}$			**40.** $\frac{3}{4}$		

Solve.

41. How many numbers have an absolute value of 5?

42. How many numbers have an absolute value of 0.5?

43. Is there a number whose absolute value is −1?

44. Are there three different numbers that have the same absolute value?

D *Circle the larger number in each pair.*

45. −4 and −7

46. −6 and −9

47. 12 and 0

48. 0 and 87

49. −3 and 2

50. −3 and 14

51. −4 and $-2\frac{1}{3}$

52. −6 and $-3\frac{1}{4}$

53. −29 and −2

54. −4 and −27

55. 9 and −22

56. 8 and −15

57. −8 and −2

58. −3 and −14

59. −7 and $-7\frac{1}{4}$

60. −8 and $-8\frac{1}{2}$

61. −8.3 and −8.5

62. −3.9 and −3.4

63. $-3\frac{1}{2}$ and $-3\frac{2}{3}$

64. $-7\frac{1}{2}$ and $-7\frac{1}{4}$

Indicate whether each inequality is true or false.

65. $-5 > -7$

66. $-8 > -9$

67. $-1 < 3.4$

68. $-4 < 3.6$

69. $0 > -2\frac{3}{4}$

70. $0 > -5\frac{3}{4}$

71. $2 > -2$

72. $8 > -8$

73. $-3.5 > -3.4$

74. $-1.6 < -1.7$

75. $-4\frac{1}{3} < 0$

76. $-8\frac{2}{3} < 0$

Arrange the numbers in each group from smallest to largest.

77. 3, −3, 0

78. 3.5, −3.1, −3, 0, 4

79. −9, 9, −4.5

80. $-2\frac{1}{2}$, −2, 3, −2.7

Express each quantity as a signed number.

81. A withdrawal of $150 from an account

82. 6 kilometers below sea level

83. A rise in temperature of 14.5°C

84. A gain of $3\frac{1}{4}$ pounds while on a diet

Mixed Practice

Solve.

85. Locate $2\frac{1}{2}$ and −3.9 on the number line.

−4 −3 −2 −1 0 1 2 3 4

86. Find the opposite of $-2\frac{1}{4}$.

87. Write as a signed number:
a. a withdrawal of \$10.98 from a bank account
b. a deposit of \$100 into the account

88. What are the sign and absolute value of the following numbers?
a. 4 **b.** $-\frac{2}{3}$

	Sign	**Absolute Value**
a.		
b.		

89. Evaluate: **a.** $|0.5|$ **b.** $|-11|$

90. Which number is larger, -4.95 or -4?

91. Complete using the symbol $<$ or $>$.
a. $-9 \square -6$ **b.** $0 \square -8\frac{2}{3}$

92. Rewrite -1.7, -2, and $-\frac{3}{4}$ from largest to smallest.

Applications

E *Solve.*

93. The Mariana Trench, the deepest point in the Pacific Ocean, is 11,033 meters below sea level, and the Puerto Rico Trench, the deepest point in the Atlantic Ocean, is 8,648 meters below sea level. Which trench is deeper? (*Source:* marianatrench.com)

94. A small toy company shows a loss of \$0.3 million for the second quarter of its business and a loss of \$0.9 million for the third quarter. In which quarter did the company show the greater loss?

95. Would a patient be receiving more medication if his dosage is decreased by 50 milligrams or if it is decreased by 25 milligrams?

96. Would a group of passengers be higher if they took the elevator down 2 floors or if they took it down 5 floors?

97. A bone density test is used to determine whether a person has osteoporosis (brittle bone disease). If the result of a bone density test, called the T-score, is below -2.5, then a person has osteoporosis. Does a patient whose T-score is -1.8 have osteoporosis? (*Source:* mayoclinic.com)

98. A bank customer has a checking account with overdraft privileges. The account is currently overdrawn by \$109.45. If the customer pays off \$100 of the overdraft, will his account still be overdrawn?

99. The following table shows the average surface temperature on several planets.

Planet	Temperature (in degrees Fahrenheit)
Mars	-81
Saturn	-218
Uranus	-323

Which planet is the warmest? (*Source:* nasa.gov)

100. The following graph gives the boiling point (in degrees Celsius) of three liquids.

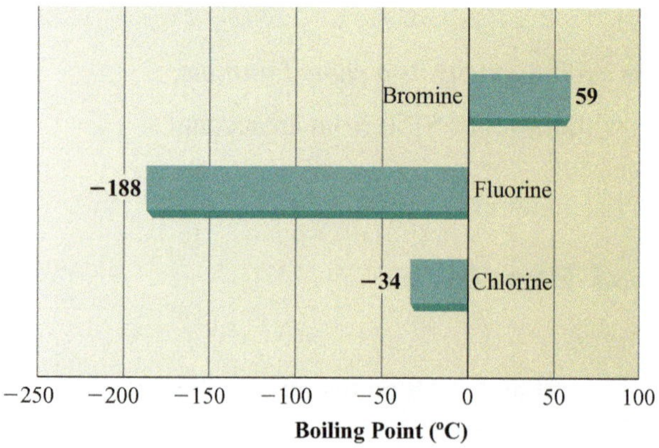

Which of these liquids has the lowest boiling point? (*Source: CRC Handbook of Chemistry and Physics*)

101. In ice hockey, the *plus/minus* statistic is used to rate individual players. If a player is on the ice when *his* team scores a goal, then he gets a *plus* point. If he is on the ice when the *other* team scores a goal, he gets a *minus* point. In theory, the higher a player's plus/minus rating, the better the player. The following table shows an individual hockey player's plus/minus rating for each of the three periods of a certain game. (*Source:* wiki.answers.com)

Period	Plus/Minus Rating
First	−3
Second	1
Third	−2

a. Locate the scores on the number line below. What does 0 on the number line represent?

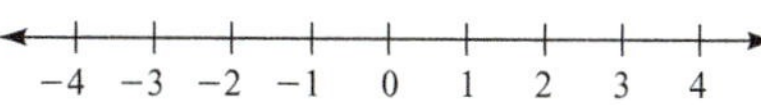

b. In which period were the most goals scored by the other team while the player was on the ice?

102. The following graph shows the estimated change in population of four Midwestern cities during the first decade of the 21st century.

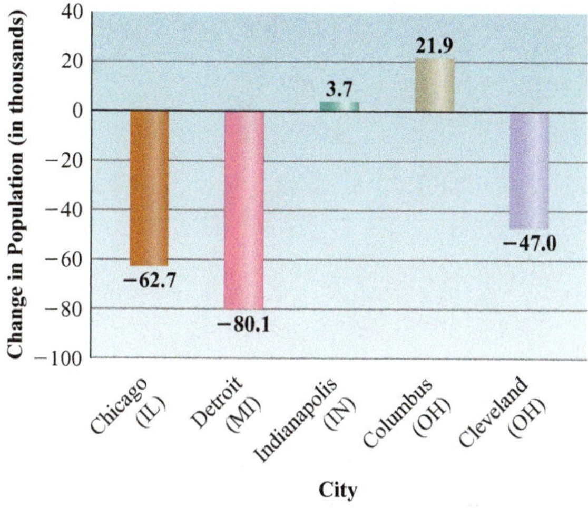

(*Source:* census.gov)

a. Which cities grew in population?

b. Which of the cities had the largest decline in population?

• Check your answers on page A-10.

MINDStretchers

Groupwork

1. a. List several numbers between −2 and −3.

b. How many numbers are there between −2 and −3?

Mathematical Reasoning

2. On the thermometer at the right, highlight all temperatures within 4 degrees of −1°.

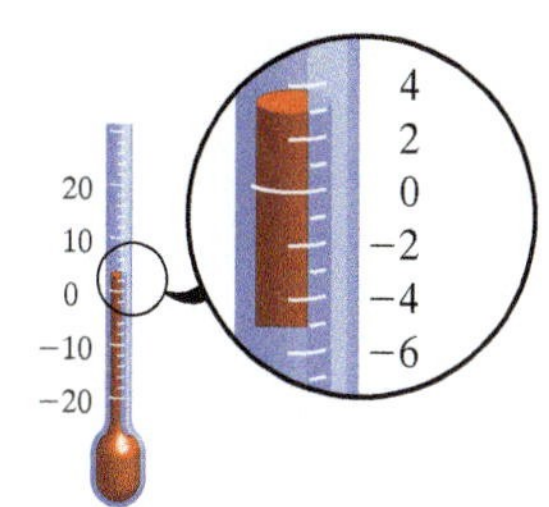

Technology

3. Using a computer spreadsheet application such as Microsoft Excel®, enter the numbers

$$-4, 9, 2, 0, -5, -1, 7, 2, 0, 9, 9, 7, -3, -4, 6, 1, 4, 3, 0$$

and sort these numbers in ascending order. Then, sort them in descending order. Is the same number in the middle with both sorts? Explain how you could have predicted this result.

Cultural Note

Until the work of sixteenth-century Italian physicists, no one was able to measure temperature. Liquid-in-glass thermometers were invented around 1650, when glassblowers in Florence, Italy, were able to create the intricate shapes that thermometers require. Thermometers from the seventeenth and eighteenth centuries provided a model for working with negative numbers that led to their wider acceptance in the mathematical and scientific communities. Numbers above and below 0 represented temperatures above and below the freezing point of water, just as they do on the Celsius scale today. Before the introduction of these thermometers, a number such as -1 was difficult to interpret for those who believed that the purpose of numbers is to count or to measure.

By contrast, the early Greek mathematicians had rejected negative numbers, calling them absurd. A thousand years later, in the seventh century A.D., the Indian mathematician Brahmagupta argued for accepting negative numbers and put down the first comprehensive rules for computing with them.

Sources: Lancelot Hogben, *Mathematics in the Making* (London: Galahad Books, 1960).

Henri Michel, *Scientific Instruments in Art and History* (New York: Viking Press, 1966).

Calvin C. Clawson, *The Mathematical Traveler* (New York and London: Plenum Press, 1994).

7.2 Adding Signed Numbers

OBJECTIVES

- **A** To add signed numbers
- **B** To solve applied problems involving the addition of signed numbers

Our previous work in addition, subtraction, multiplication, and division was restricted to positive numbers—whether those positive numbers happened to be whole numbers, fractions, or decimals. Now, we consider computations involving *any* signed numbers, starting with the operation of addition.

Suppose that we want to add two negative numbers, say -1 and -2. It is helpful to look at this problem in terms of money. If your hourly wage went down \$1 and down again \$2, altogether it went down \$3. In terms of signed numbers, this example can be expressed as:

$$-1 + (-2) = -3$$

We can also look at this problem on the number line. To add -1 and -2, we start at the point corresponding to the first number, -1. The second number, -2, is *negative*, so we move 2 units to the *left*. We end at -3, which is the answer we expected.

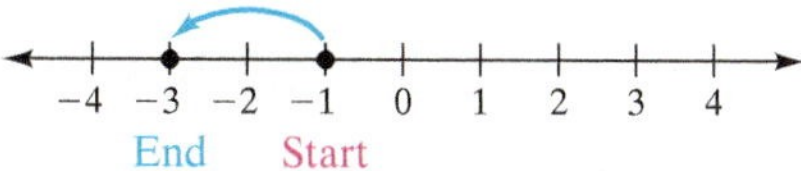

Now, let's consider adding the signed numbers -1 and $+3$. Thinking of this problem in terms of money may make it clearer. If your hourly wage went down \$1 and then up \$3, your total hourly increase is \$2. Using signed numbers, this example can be written as:

$$-1 + 3 = 2$$

We can picture this problem by starting at -1 on the number line. The second number, 3, is *positive*, so we move 3 units to the *right*. We end at 2, which is the answer.

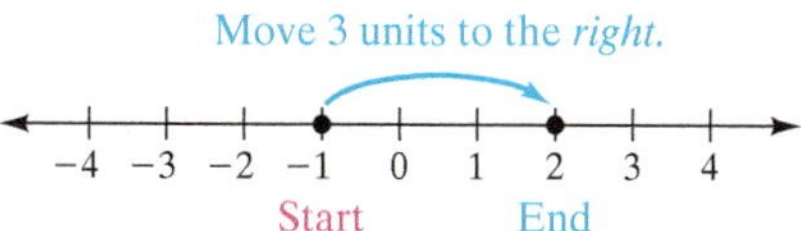

The following rule provides a shortcut for adding signed numbers:

To Add Two Signed Numbers

- If the numbers have the same sign, add the absolute values and keep the sign.
- If the numbers have different signs, subtract the smaller absolute value from the larger and take the sign of the number with the larger absolute value.

EXAMPLE 1

Add: -3 and -2

Solution The sum of the absolute values is 5.

$$|-3| + |-2| = 3 + 2 = 5$$

Both -3 and -2 are negative, so their sum is negative.

$$(-3) + (-2) = -5$$

Check

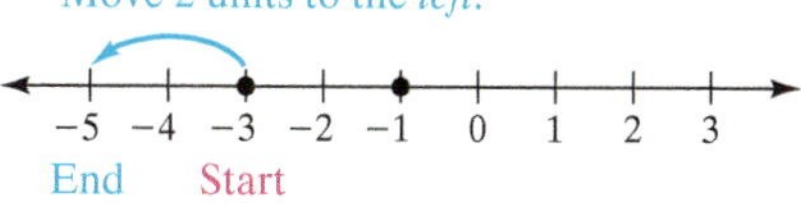

PRACTICE 1

Combine: -8 and -17

EXAMPLE 2

Find the sum: $(-3.9) + (-0.5)$

Solution $|-3.9| = 3.9$ and $|-0.5| = 0.5$

$$3.9 + 0.5 = 4.4$$

The sum of two negative numbers is negative, so

$$(-3.9) + (-0.5) = -4.4$$

Check Move 0.5 units to the *left*.

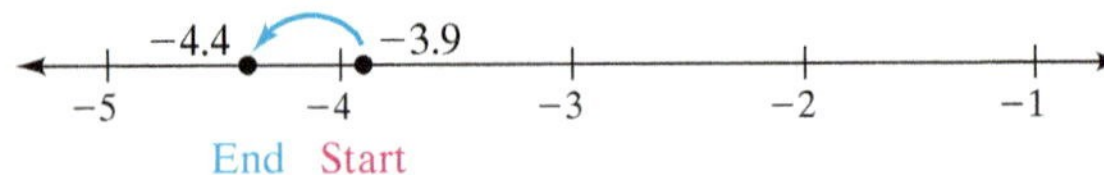

PRACTICE 2

Add: $-3 + (-1\frac{1}{2})$

EXAMPLE 3

Add: $(-1) + 2$

Solution Here, we are adding numbers with different signs. First, we find the absolute values.

$$|-1| = 1 \quad \text{and} \quad |2| = 2$$

Next, we subtract the smaller absolute value from the larger.

$$2 - 1 = 1$$

Because 2 has the larger absolute value and its sign is positive, the sum is also positive. Our answer is 1, or +1.

$$(-1) + 2 = 1$$

Check Move 2 units to the *right*.

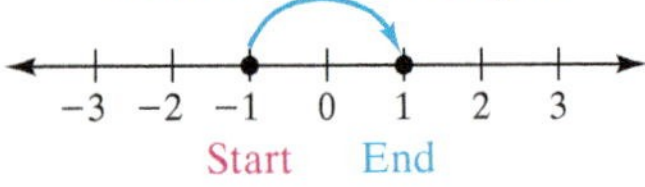

PRACTICE 3

Find the sum of -2 and 9.

Note in Example 3 that, when we added a negative number to 2, we got a smaller result—namely, 1.

EXAMPLE 4

Combine: $(-2) + (+2)$

Solution $|-2| = 2$ and $|+2| = 2$ **Find the absolute values.**

$2 - 2 = 0$ **Subtract the absolute values.**

Zero is neither positive nor negative, for it has no sign.

$$(-2) + (+2) = 0$$

Check Move 2 units to the *right*.

−3 −2 −1 0 1 2 3

Start End

Do you see why the sum of -44 and $+44$ is 0? How about $2.77 + (-2.77)$ or $2\frac{5}{8} + (-2\frac{5}{8})$?

PRACTICE 4

Find the sum: $-35 + 35$

EXAMPLE 5

Add $3\frac{4}{5}$ to $(-1\frac{1}{5})$.

Solution $|-1\frac{1}{5}| = 1\frac{1}{5}$ and $|3\frac{4}{5}| = 3\frac{4}{5}$ Find the absolute values.

$$\begin{array}{r} 3\frac{4}{5} \\ -1\frac{1}{5} \\ \hline 2\frac{3}{5} \end{array}$$ Subtract the absolute values.

Because $|3\frac{4}{5}|$ is greater than $|-1\frac{1}{5}|$, the answer is positive. So

$$3\frac{4}{5} + (-1\frac{1}{5}) = 2\frac{3}{5}$$

PRACTICE 5

Add: $-2.1 + 0.8$

Some addition problems involve the sum of three or more signed numbers. Rearranging the signed numbers to add the positives and negatives separately can make the addition easier. Note that this rearrangement does not affect the sum because addition is a commutative and associative operation.

EXAMPLE 6

Find the sum: $3 + (-1) + (-8) + 2 + (-11)$

Solution We are adding two positive and three negative numbers. Rearranging the numbers by sign, we get the following.

$$\underbrace{3 + 2}_{\text{Positives}} + \underbrace{(-1) + (-8) + (-11)}_{\text{Negatives}}$$

First, we add the positives. $3 + 2 = 5$

Then, we add the negatives. $(-1) + (-8) + (-11) = -20$

Finally, we combine the positive and the negative subtotals.

$$5 + (-20) = -15$$

So $3 + (-1) + (-8) + 2 + (-11) = -15$.

PRACTICE 6

$-3 + 1 + 8 + (-6) = ?$

EXAMPLE 7

The most famous of all comets is Halley's Comet, which passes by Earth every 76 years. For other comets, however, the length of time between visits is much longer. The Great Comet, for example, comes near Earth only once every 3,000 years. If this comet visited Earth about 1200 B.C., approximately when was its next visit? (*Source:* Mark R. Kidger, "Some Thoughts on Comet Hale-Bopp")

PRACTICE 7

Lake Baikal in Russia is the deepest lake in the world. The deepest point in the lake is 1,187 meters below sea level. If the surface is 1,643 meters above this point, what is the elevation of the surface? (*Source:* bww.irk.ru)

EXAMPLE 7 (continued)

Solution To help you understand this problem, we draw the number line. Any number line involving time is called a *time line*. On a time line, positive years are A.D. and negative years are B.C.

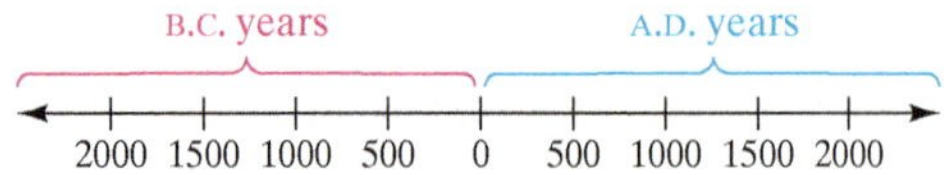

So 1200 B.C. is represented by -1200. We must add -1200 and $+3000$.

$|-1200| = 1200$ and $|3000| = 3000$ **Find the absolute values.**

$3000 - 1200 = 1800$ **Subtract the absolute values.**

Since the absolute value of 3000 is greater than the absolute value of -1200, the answer is positive.

$$(-1200) + 3000 = 1800$$

Therefore, the Great Comet came near Earth in about A.D. 1800.

Check

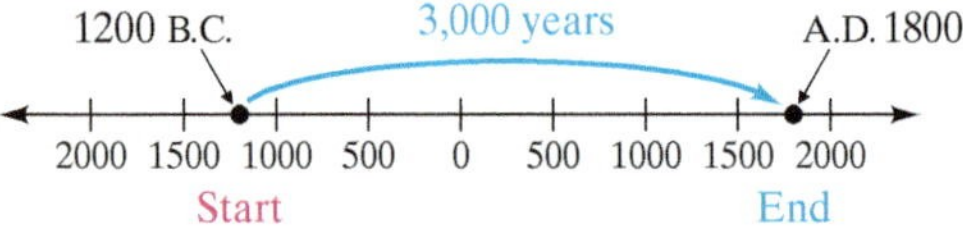

Signed Numbers on a Calculator

The numbers that we have entered so far on a calculator have been positive numbers. To enter a negative number, we need to hit a special key that indicates that the sign of the number is negative. Some calculators have a negative sign key, [(−)]*. Others have a change of sign key,* [(+/−)]*. Be careful not to confuse either of these keys with the subtraction key,* [−].

EXAMPLE 8

Calculate: $-1.3 + (-5.8)$

Solution

Press	Display
[(−)] 1.3 [+] [(] [(−)] 5.8 [)] [ENTER =]	−1.3 + (−5.8) −7.1

PRACTICE 8

Calculate: $-1.3 + (-5.891) + (4.713)$

7.2 Exercises

FOR EXTRA HELP MyMathLab MathXL PRACTICE WATCH READ REVIEW

Mathematically Speaking

Fill in each blank with the most appropriate term or phrase from the given list.

commutative	right	larger
absolute values	left	distributive
smaller	numbers	

1. To add (-6) and $(+6)$ on the number line, start at (-6) and move 6 units to the ________.

2. To find the sum of two signed numbers with the same sign, add the ____________ and keep the sign.

3. To find the sum of two signed numbers with different signs, subtract the smaller absolute value from the larger and take the sign of the number with the ________ absolute value.

4. Rearranging signed numbers to add the positives and negatives separately does not affect the sum, because the operation of addition is associative and ____________.

A *Find the sum of each pair of numbers. Use the number line as a visual check.*

5. $6 + (-5)$

6. $3 + (-9)$

7. $-2 + 5$

8. $-4 + 9$

9. $-9 + (-2)$

10. $-6 + (-4)$

11. $7 + (-7)$

12. $3 + (-2)$

Find the sum.

13. $67 + (-67)$

14. $2 + (-2)$

15. $-10 + 5$

16. $-12 + 7$

17. $-100 + 300$

18. $-20 + 60$

19. $8 + (-2)$

20. $5 + (-3)$

21. $-60 + (-90)$

22. $-50 + (-40)$

23. $-7 + 2$

24. $-9 + 4$

25. $-27 + 0$

26. $-13 + 0$

27. $-9 + 9$

28. $-2 + 2$

29. $5.2 + (-0.3)$

30. $-0.6 + 1$

31. $-0.2 + 0.3$

32. $-5.5 + 0$

33. $60 + (-0.5)$

34. $-0.7 + 0.7$

35. $-9.8 + 3.9$

36. $6.1 + (-5.9)$

37. $(-5.6) + (-8.9)$

38. $(-0.8) + (-0.5)$

39. $\left(-\frac{1}{2}\right) + \left(-5\frac{1}{2}\right)$

40. $-1\frac{1}{3} + \left(-2\frac{2}{3}\right)$

41. $-1\frac{1}{5} + \frac{3}{5}$

42. $2\frac{1}{6} + \left(-\frac{5}{6}\right)$

43. $-\frac{2}{5} + 2$

44. $-14 + \frac{1}{3}$

45. $1\frac{1}{2} + \left(-1\frac{3}{5}\right)$

46. $1\frac{3}{8} + \left(-2\frac{1}{4}\right)$

47. $(-24) + 20 + (-98)$

48. $35 + (-17) + (-18)$

49. $25 + (-19) + (-16)$

50. $20 + (-8) + (-12)$

51. $-27 + 50 + (-14)$

52. $-34 + (-9) + 15$

53. $12 + (-7) + (-12\frac{1}{2})$

54. $-8 + (-4) + (-8\frac{1}{4})$

55. $(-7) + 12 + 0 + (-7) + 9$

56. $(-3) + 8 + (-9) + 3 + (-4)$

57. $10 + (-9) + (-1) + 0 + 3$

58. $9 + (-3) + 8 + (-4) + 5$

59. $(-5) + (-2) + 6 + (-4) + 5$

60. $-6 + 18 + (-15) + 7 + (-3)$

61. $-0.3 + (-2.6) + (-4)$

62. $-5.25 + (-0.4) + 3$

63. $-12 + 7.58 + 12$

64. $-3.7 + (-1.88) + 5$

65. $8.756 + (-9.08) + (-4.59)$

66. $-5.405 + 6 + (-6.89)$

67. $-3.001 + (-0.59) + 8$

68. $-10 + 5.17 + (-10.002)$

Mixed Practice

Solve.

69. Find the sum of $-6 + 9$ on the number line.

−6 −5 −4 −3 −2 −1 0 1 2 3 4 5 6

70. Combine 16 and (-24).

71. Add: $9.6 + (-9.6)$

72. Combine: $-4\frac{2}{9} + \left(-2\frac{1}{9}\right)$

73. Add -8, 14, and -10.

74. Find the sum of -1.7, -3.95, and 10.

Applications

B ***Solve. Express each answer as a signed number.***

75. The lowest elevation in Africa is Lake Assal at 512 feet below sea level. The highest elevation, Mount Kilimanjaro, is 19,852 feet above Lake Assal. What is the elevation of Mount Kilimanjaro? (*Source: The New York Times Almanac 2010*)

76. A student owes $2,456 on her credit card. After making a payment of $350, what is the balance on her credit card?

77. During a recession, a manufacturer laid off 182 employees. A year later, another 56 employees were laid off. What was the change in the number of employees working for the manufacturer as a result of the two layoffs?

78. A computer retailer decreases the price of a laptop computer by $150 during a sale. As a special promotion, the price is decreased another $75 for customers who trade in their old laptop. What is the total price change for a customer who trades in his old laptop?

79. In a physics class, students study the properties of atomic particles, including protons and electrons. They learn that a proton has an electric charge of $+1$, whereas an electron has an electric charge of -1. What is the total charge of a collection of 3 protons and 4 electrons?

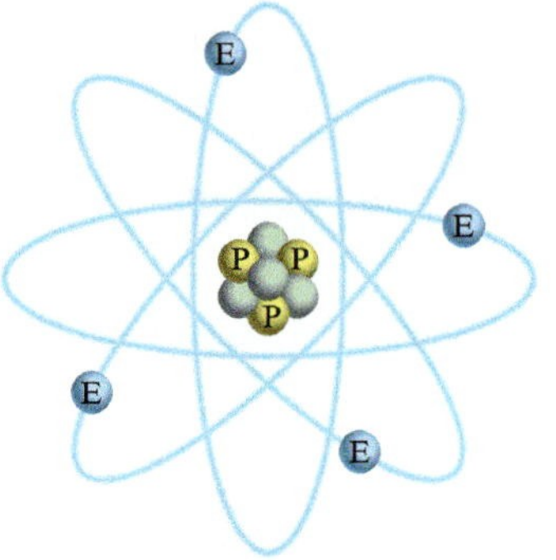

80. Cleopatra became queen of Egypt in 51 B.C. She left the throne 20 years later. In what year was that?

81. Ten years ago, a couple got married. Four years later, they got divorced. When was their divorce?

82. To conduct an experiment, a chemist cooled a substance to $-10°C$. During the experiment, a chemical reaction took place that raised the temperature of the substance by $15°$. What was the final temperature?

83. A football team gained 5 yards on its first down, lost 7 yards on second down, and lost 4 yards on third down. What was the overall change in position after third down?

84. In the last 4 months, a dieter lost 5 pounds, gained 2 pounds, lost 1 pound, and maintained his weight, respectively. What was his overall change in weight?

85. The table below shows for recent years the change from the previous year in the number of union members in the United States.

Year	Change in the Number of Union Members (in millions)
2006	-0.3
2007	$+0.3$
2008	$+0.4$
2009	-0.8
2010	-0.6

(*Source:* Barry T. Hirsch and David A. Macpherson; unionstats.com)

a. What was the net change in the number of union members for 2006–2010?

b. If the number of union members in 2006 was 15.7 million, what was the corresponding number in 2010?

86. The following graph shows a company's bottom line (gain or loss) for various quarters, in millions of dollars.

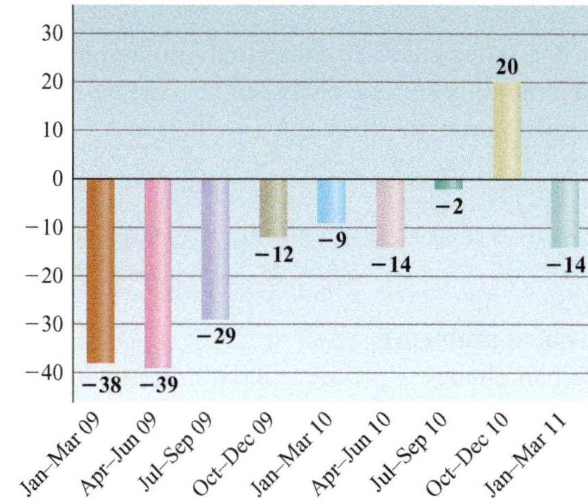

a. What was the company's bottom line for 2010?

b. Was the company's bottom line greater in the last quarter of 2009 or in the first quarter of 2011?

• Check your answers on page A-10.

MINDStretchers

Groupwork

1. Work with a partner on the following.

a. Fill in the following addition table.

+	+3	−2	−1
+4			
−3			
−1			

b. Why do the nine numbers that you entered sum to 0?

Writing

2. For signed numbers, does *adding* always mean *increasing*? Explain.

Patterns

3. Find the missing numbers in the following sequence: $-5, -7, -6, -8, -7, -9, -8,$ ______, ______, ______

7.3 Subtracting Signed Numbers

OBJECTIVES

A To subtract signed numbers

B To solve applied problems involving the subtraction of signed numbers

The subtraction of signed numbers is based on two topics previously discussed—adding signed numbers and finding the opposite of a signed number.

Let's first consider a subtraction problem involving money. Suppose that you have \$10 in your bank account and withdraw \$3. Then \$7 will be left in the account. In terms of signed numbers, this can be expressed as

$$10 - (+3) = 7$$

On the other hand, suppose that you had started with \$10 in the account and the bank imposed a monthly service charge of \$3. The balance in your account once more would be \$7. Using signed numbers, this can be written as

$$10 + (-3) = 7$$

The answers to the two problems are the same; so

$$10 - (+3) = 7 \text{ and } 10 + (-3) = 7$$

are equivalent problems.

We can change a problem in subtracting signed numbers to an equivalent problem in adding signed numbers by adding the *opposite* of the number that we want to subtract. The following rule provides a shortcut for subtracting signed numbers:

To Subtract Two Signed Numbers

- Change the operation of subtraction to addition. Then, add the first number and the opposite of the second number.
- Follow the rule for adding signed numbers.

To see if this rule works when the number we are subtracting is negative, consider $4 - (-1)$. Recall that every subtraction problem has a related addition problem. Therefore, $4 - (-1) = 5$ because $5 + (-1) = 4$. Note that we get the same result using the rule for subtracting signed numbers.

$$4 \underset{\text{Subtract}}{\underset{\uparrow}{-}} \underset{\text{Negative 1}}{\underset{\uparrow}{(-1)}} = 4 \underset{\text{Add}}{\underset{\uparrow}{+}} \underset{\text{Positive 1}}{\underset{\uparrow}{(+1)}} = 5$$

EXAMPLE 1

Find the difference: $-2 - (-4)$

Solution We change the operation of subtraction to addition. Then, we add the first number and the opposite of the second number.

$$\begin{aligned} &-2 - (-4) \\ &\quad\ \downarrow\ \ \downarrow \\ &= -2 + (+4) \end{aligned}$$

We already know how to add a negative and a positive number. So we get:

$$-2 + 4 = +2, \text{ or } 2$$

PRACTICE 1

Find the difference: $-4 - (-2)$

Note that in Example 1 and Practice 1, the numbers in the differences are the same except for their order. But the answers are quite different: $-2 - (-4) = 2$, whereas $-4 - (-2) = -2$. Why do you think this is so?

EXAMPLE 2

Compute: $3 - (-9)$

Solution Change negative 9 to its opposite, positive 9.

$$3 - (-9) = 3 + (+9) = +12, \text{ or } 12$$

Change subtraction to addition.

Note that when we subtracted -9 from 3, we got an answer larger than 3.

PRACTICE 2

Subtract: $9 - (-9)$

EXAMPLE 3

Subtract: $-2 - 8\frac{1}{3}$

Solution $-2 - 8\frac{1}{3} = -2 + \left(-8\frac{1}{3}\right) = -10\frac{1}{3}$

PRACTICE 3

Find the difference: $-9 - 12.1$

EXAMPLE 4

Calculate: $5 + (-6) - (-11)$

Solution This problem involves addition and subtraction. According to the order of operations rule, we work from left to right.

$$\begin{aligned} 5 + (-6) - (-11) &= -1 - (-11) && \text{Add 5 and } (-6). \\ &= -1 + 11 && \text{Subtract } -11. \\ &= 10 && \text{Add 11.} \end{aligned}$$

PRACTICE 4

$-2 - 3 + (-5) = ?$

EXAMPLE 5

Simplify: $4 - [2 - (-3)]$

Solution This problem involves an operation within brackets. According to the order of operations rule, first we work within the brackets.

$$\begin{aligned} 4 - [2 - (-3)] &= 4 - [2 + 3] && \text{Subtract } -3 \text{ from 2.} \\ &= 4 - 5 && \text{Add 2 and 3.} \\ &= -1 && \text{Subtract 5 from 4.} \end{aligned}$$

PRACTICE 5

Calculate: $7 - [3 + (-2)]$

EXAMPLE 6

Normally we think of oxygen as a gas. However, when cooled to $-183°C$ (its boiling point), oxygen becomes a liquid. If it is cooled further to $-218°C$ (its melting point), oxygen becomes a solid. How much higher is the boiling point of oxygen than its melting point? (*Source: Handbook of Chemistry & Physics*)

Solution We need to compute how much greater is -183 than -218.

$$\begin{aligned} (-183) - (-218) &= (-183) + (+218) \\ &= +35 \end{aligned}$$

The boiling point of oxygen is 35°C higher than its melting point.

PRACTICE 6

Julius Caesar, the Roman general and statesman, was born in about 100 B.C. How many years ago was that, to the nearest hundred years? (*Source: wikipedia.org*)

7.3 Exercises

FOR EXTRA HELP MyMathLab MathXL PRACTICE WATCH READ REVIEW

Mathematically Speaking

Fill in each blank with the most appropriate term or phrase from the given list.

absolute value	order of operations	addition
sum	multiplication	signed numbers
difference	opposite	

1. To subtract two signed numbers, change the operation of subtraction to addition, and change the number being subtracted to its __________. Then, follow the rule for adding signed numbers.

2. Every subtraction problem has a related __________ problem.

3. When a signed number problem involves addition and subtraction, work from left to right according to the ________________ rule.

4. When subtracting a negative number, the __________ is greater than the original number.

A *Find the difference.*

5. $5 - (-2)$
6. $7 - (-3)$
7. $4 - 8$
8. $5 - 9$
9. $-9 - 5$
10. $-44 - 2$
11. $42 - (-2)$
12. $36 - (-4)$
13. $50 - 75$
14. $44 - 83$
15. $-20 - (-1)$
16. $-18 - (-3)$
17. $3 - (-3)$
18. $4 - (-4)$
19. $0 - 38$
20. $0 - 56$
21. $-13 - 13$
22. $-15 - 15$
23. $13 - (-13)$
24. $14 - (-14)$
25. $8 - 23$
26. $7 - 34$
27. $800 - (-200)$
28. $300 - (-100)$
29. $7 - 8.52$
30. $9.1 - 10.84$
31. $9.2 - (-0.5)$
32. $8.6 - (-0.7)$
33. $-5.2 - (-5.2)$
34. $-0.5 - (-0.5)$
35. $8.6 - (-1.9)$
36. $7.4 - (-3.1)$
37. $-10 - (-9.5)$
38. $-6 - (-8.7)$
39. $4\frac{1}{2} - 9\frac{1}{2}$
40. $6\frac{1}{5} - 8\frac{1}{5}$
41. $10 - 2\frac{1}{4}$
42. $12 - 5\frac{2}{3}$
43. $-7 - \frac{1}{4}$
44. $-9 - \frac{1}{8}$
45. $5\frac{3}{4} - \left(-1\frac{1}{2}\right)$
46. $6\frac{1}{2} - \left(-1\frac{1}{3}\right)$

Combine.

47. $4 + (-6) - (-9)$
48. $10 + (-6) - (-8)$
49. $7 - 7 + (-5)$
50. $8 - 8 + (-9)$
51. $-12 + 3.6 - (-6.5)$
52. $4.6 - (-5) + (-3.6)$
53. $6 + \left(-4\frac{1}{5}\right) + \left(-2\frac{3}{10}\right)$
54. $-2\frac{1}{2} - (-3) + 5\frac{1}{4}$
55. $-8 + (-4) - 9 + 7 + (-1)$
56. $-5 - (-1) + 6 + (-3) - 4$
57. $6 - [5 - (-4)]$
58. $2 - [3 + (-5)]$
59. $7.043 - 9.002 - 1.883$
60. $-6.192 - 0.337 - (-23.94)$
61. $-8.722 + (-3.913) - 3.86$
62. $2.884 - 0.883 + (-6.125)$

Mixed Practice

Solve.

63. Subtract: $-16 - 9$

64. Find the difference: $8.1 - 10.46$

65. Subtract $-19\frac{3}{4}$ from $-19\frac{3}{4}$.

66. From 6 subtract -5.

67. Combine: $-4 + (-5) + 6 - (-4)$

68. Simplify: $3 - [2 + (-9)]$

Applications

B *Solve. Express each answer as a signed number.*

69. Two airplanes take off from the same airport. One flies west and the other east, as shown. How far apart are they?

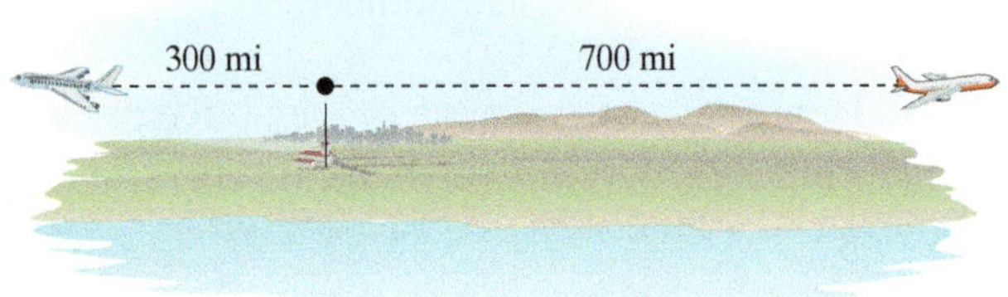

70. Two friends get on different elevators at the same floor. One goes up 2 floors, the other goes down 4 floors. How many floors are they apart?

71. The highest point on the continent of South America is Mt. Aconcagua at an elevation of 22,834 feet above sea level. The lowest point is the Valdes Peninsula at an elevation of 131 feet below sea level. How much higher is Mt. Aconcagua than the Valdes Peninsula? (*Source:* National Geographic Society)

72. Ethiopia was founded around 1,000 B.C., and the United States in A.D. 1789. About how much older is Ethiopia than the United States? (*Source: The Concise Columbia Encyclopedia*)

73. In business, net income is calculated by subtracting the costs from the revenue. What is a company's net income if its revenues were \$2.3 million and its costs were \$3.7 million? Express as a signed number.

74. Rapid City, South Dakota, holds the U.S. record for a 2-hour temperature change. On January 12, 1911, the temperature at 6 A.M. was 49°F. If it was 62° colder by 8 A.M., what was the temperature at 8 A.M.? (*Source:* crh.noaa.gov)

75. In St. Louis, Missouri, the water level of the Mississippi River is measured in feet above or below a zero level. On January 16, 1940, the river reached its lowest level, at 6.2 feet below the zero level.

  a. Express this measurement as a signed number.

  b. The highest level of the Mississippi River in St. Louis, recorded on August 1, 1993, was 49.6 feet above the zero level. What was the difference between the highest and lowest levels? (*Source:* ams.usda.gov)

76. In any year, the U.S. budget surplus or deficit is the difference between what the federal government takes in and what it spends. In 2001 there was a surplus of \$127.4 billion, whereas in 2002 there was a deficit of \$157.8 billion. (*Source:* gpoaccess.gov)

  a. Write each quantity as a signed number.

  b. Find the difference between the surplus and the deficit.

77. The bar graph shows the annual precipitation for Phoenix, Arizona for the years from 2005 through 2009. (*Source:* azwater.gov)

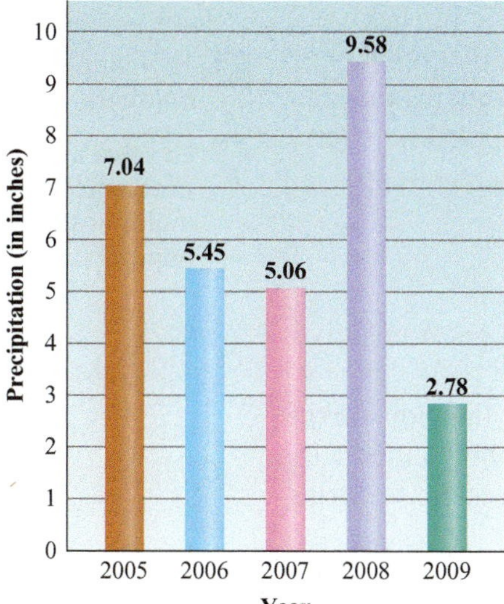

a. The precipitation in 2009 was 5.52 inches below the normal annual precipitation. What is the normal annual precipitation?

b. Find the change in annual precipitation from 2008 to 2009.

78. The bar graph shows the closing price of a share of Whole Foods Market, Inc. stock for a 5-day period in July of 2010. (*Source:* finance.yahoo.com)

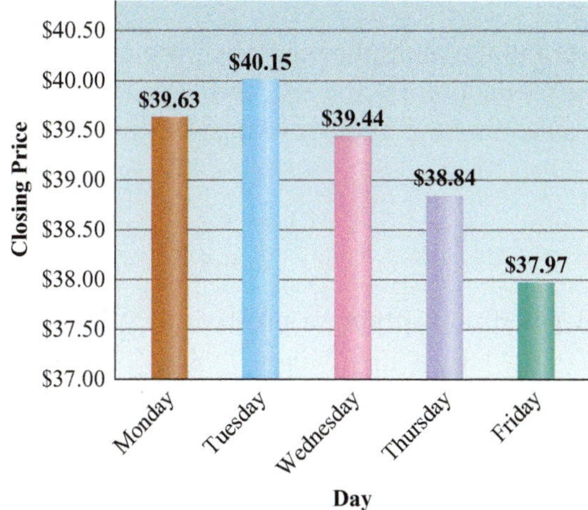

a. Calculate the change in the closing price per share from the previous day for Tuesday through Friday. Express each change as a signed number.

b. If the change in the closing price on Monday was $0.86, what was the closing price the previous day?

• Check your answers on page A-10.

MINDStretchers

Groupwork

1. Working with a partner, rearrange the numbers in the square on the left so that the sum of every row, column, and diagonal is -6.

-3	$+2$	-2
-5	-4	-6
0	-1	$+1$

		-2

Mathematical Reasoning

2. The following two columns of numbers add up to the same sum.

3	7
7	?
1	7
9	8
5	9

What number does the question mark represent?

Writing

3. Consider the following two problems:

$$8 - (-2) = 8 + 2 = 10$$

$$8 \div \frac{4}{7} = 8 \times \frac{7}{4} = 14$$

Explain in what way the two problems are similar.

7.4 Multiplying Signed Numbers

OBJECTIVES

A To multiply signed numbers

B To solve applied problems involving the multiplication of signed numbers

We now turn to the multiplication of signed numbers. Consider, for example, the problem of finding the product of +4 and −2, or $4(-2)$. We know that multiplying a number by 4 means the same as adding the number to itself 4 times. Using the rule for adding signed numbers, we get:

$$\begin{aligned} 4(-2) &= -2 + (-2) + (-2) + (-2) \\ &= -8 \end{aligned}$$

Note that when multiplying a positive number by a negative number, we get a negative product.

Let's take another look at this same problem in practical terms. Suppose that you are on a diet and you lose 2 pounds per month. Compared to your current weight, how much will you weigh 4 months from now? The answer is that you will weigh 8 pounds less. So we write

$$4(-2) = -8$$

Now, we examine a different question. Again assume that you lose 2 pounds per month by dieting. Four months ago, you were heavier than you are now. How much heavier were you? To answer this question, note that each month you lost 2 pounds but that you are going back in time 4 months. So you weighed 8 pounds more than you do now, which can be expressed as

$$-4(-2) = +8, \text{ or } 8$$

Note that when we multiply two negative numbers, we get a positive number.

We can use the following rule to multiply signed numbers:

To Multiply Two Signed Numbers

- Multiply the absolute values of the numbers.
- If the numbers have the same sign, their product is positive; if the numbers have different signs, their product is negative.

Another way to think of multiplying signed numbers is as follows:

Positive · Positive = Positive	Positive · Negative = Negative
Negative · Negative = Positive	Negative · Positive = Negative

EXAMPLE 1

Find the product of −2 and −1.

Solution First, we find the absolute values.

$$|-2| = 2 \text{ and } |-1| = 1$$

Next, we multiply the absolute values.

$$2 \cdot 1 = 2$$

Since the numbers have the same sign, the product is positive. The answer is +2, or 2. So we can write

$$(-2)(-1) = 2$$

PRACTICE 1

Compute: $-8(-4)$

EXAMPLE 2

Calculate: $(5)(-10)$

Solution $|5| = 5$ and $|-10| = 10$ Find the absolute value of each factor.

$5 \cdot 10 = 50$ Multiply the absolute values.

The factors have different signs, so the product is negative.

$$(5)(-10) = -50$$

PRACTICE 2

Multiply: $(-5)(2)$

The next example involves raising signed numbers to a power. Note that in an expression such as $(-7)^2$, the parentheses around the -7 indicate that this negative number is to be squared. By contrast, in the expression -7^2 without parentheses, the positive number 7 is to be squared and the result is preceded by a negative sign.

EXAMPLE 3

Evaluate: **a.** $(-7)^2$ **b.** -7^2

Solution

a. Recall that $(-7)^2$ means $(-7)(-7)$. Because the two factors have the same sign, their product is positive. So $(-7)^2 = 49$.

b. The expression -7^2 means $-(7 \cdot 7)$. So

$$-7^2 = -(7 \cdot 7) = -(49)$$
$$= -49$$

PRACTICE 3

Simplify:

a. $(-1)^2$ **b.** -1^2

Note that in Example 3a, we squared a negative number so that the answer was positive. On the other hand, in Example 3b, we squared a positive number. The square was preceded by a negative sign, giving us a negative answer.

EXAMPLE 4

Find the product of $2\frac{1}{5}$ and -5.

Solution $2\frac{1}{5} \cdot 5 = \frac{11}{5} \cdot \frac{5}{1} = 11$ Multiply the absolute values.

Since the factors $2\frac{1}{5}$ and -5 have different signs, their product is negative.

$$2\frac{1}{5} \cdot (-5) = -11$$

PRACTICE 4

Multiply: $\left(-1\frac{1}{3}\right)\left(-\frac{1}{5}\right)$

EXAMPLE 5

Multiply: $(-1.4)(-0.6)$

Solution $(1.4)(0.6) = 0.84$ Multiply the absolute values.

Since the factors have the same sign, we get:

$$(-1.4)(-0.6) = 0.84$$

PRACTICE 5

Find the product of -2.5 and 8.

EXAMPLE 6

Calculate: $8(-2)(-3)$

Solution We multiply from left to right.

$$8(-2)(-3) \quad \text{Positive} \cdot \text{Negative} = \text{Negative}$$
$$= -16 \cdot (-3) \quad \text{Negative} \cdot \text{Negative} = \text{Positive}$$
$$= 48$$

PRACTICE 6

Multiply: $-8(-2)(-3)$

Comparing Example 6 and Practice 6, we note that both problems have the same absolute values but different signs. The product in Example 6 is positive, because there are two negative factors. By contrast, the answer to Practice 6 is negative, because there are three negative factors. Can you explain why a product is positive if it has an even number of negative factors, whereas a product is negative if it has an odd number of negative factors?

EXAMPLE 7

Simplify: $[8 - 3(-2)]^2$

Solution Use the order of operations rule.

$$[8 - 3(-2)]^2 = [8 - (-6)]^2 \quad \text{Multiply 3 by } -2.$$
$$= [8 + 6]^2 \quad \text{Subtract } -6 \text{ from 8.}$$
$$= [14]^2 \quad \text{Add.}$$
$$= 196 \quad \text{Square.}$$

PRACTICE 7

Calculate: $[-3 + 2(-5)]^2$

EXAMPLE 8

An oil company is drilling for oil. Each day, the workers drill down 20 feet farther until they hit a pool of oil, as shown. Will they reach oil by the end of the fifth day?

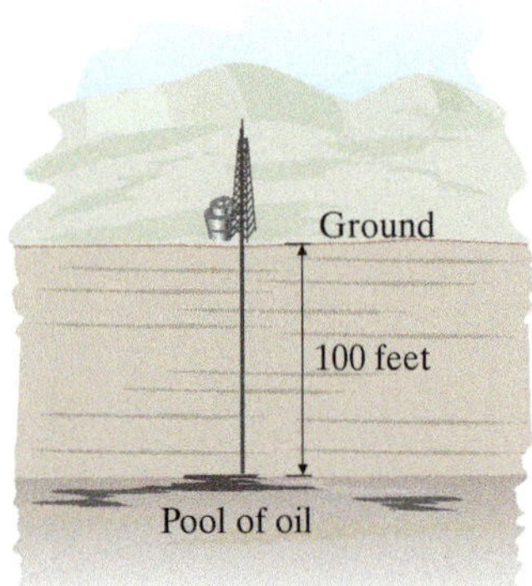

Solution Let's represent movement downward by a negative number. Since each of the 5 days they drill 20 feet farther down, we compute $5 \cdot (-20)$. Using the rule for multiplying signed numbers, we get -100. Therefore, the drill will reach 100 feet below ground level by the fifth day—the depth of the pool of oil.

PRACTICE 8

Alvin is a deep submergence vehicle operated by the Woods Hole Oceanographic Institution for marine research. On a research mission, it descends from the surface to the ocean floor 2,000 meters below at a rate of 30 meters per minute. Will *Alvin* reach the ocean floor in 60 minutes? (*Source:* American Geophysical Union)

7.4 Exercises

FOR EXTRA HELP MyMathLab MathXL PRACTICE WATCH READ REVIEW

Mathematically Speaking

Fill in each blank with the most appropriate term or phrase from the given list.

odd	positive	negative	even
product	sum	prime	

1. The product of two numbers with the same sign is ________.

2. The ________ of two numbers with different signs is negative.

3. The product of a(n) ________ number of negative factors is positive.

4. The product of a(n) ________ number of negative factors is negative.

A *Find the product.*

5. $(2)(-5)$
6. $-4 \cdot 9$
7. $-2 \cdot 5$
8. $-3 \cdot 7$

9. $-5 \cdot (-5)$
10. $-4 \cdot (-3)$
11. $-34(-9)$
12. $-35(-7)$

13. $2 \cdot (-8)$
14. $-1 \cdot 5$
15. $907 \cdot (-9)$
16. $-5 \cdot (812)$

17. $5(-8)$
18. $8 \cdot (-53)$
19. $-88 \cdot 2$
20. $20 \cdot (-30)$

21. $(-200)(-4)$
22. $-4 \cdot (-200)$
23. $-80 \cdot 90$
24. $-60 \cdot 40$

25. $(2.5)(-2)$
26. $(0.3)(-0.2)$
27. $(0.2)(-50)$
28. $3 \cdot (-0.3)$

29. $(-1.2)(-4.6)$
30. $(-0.7)(-1.8)$
31. $(5)(-1.6)$
32. $(-40)(2.7)$

33. $-\frac{1}{3} \cdot \frac{5}{9}$
34. $\left(-\frac{5}{6}\right) \cdot \left(\frac{2}{3}\right)$
35. $1\frac{1}{4}\left(-\frac{2}{3}\right)$
36. $-\frac{1}{5} \cdot 2\frac{1}{2}$

Evaluate.

37. -5^2
38. -6^2
39. $(-100)^2$
40. $(-300)^2$

41. $(-0.5)^2$
42. $(-0.4)^2$
43. $(-0.1)^3$
44. $(-0.2)^3$

45. $\left(-\frac{3}{4}\right)^2$
46. $\left(-\frac{1}{5}\right)^2$
47. $(-1)^3$
48. $(-4)^3$

49. $(-0.308)^2$
50. $(-7.96)^2$

Multiply.

51. $(9)(12)(-2)$
52. $(2)(-3)(-200)$
53. $(5)(-2)(-1)(3)(-2)$

54. $(-5)(-2)(-1)(3)(-2)$
55. $(-5)(-3)(0)$
56. $(-7)(0)(-10)$

57. $10 \cdot \left(-\frac{1}{2}\right) \cdot (-1)$
58. $\left(-\frac{1}{2}\right)(-4)\left(-\frac{1}{2}\right)$
59. $\frac{4}{5} \cdot \left(-\frac{8}{9}\right) \cdot \frac{1}{3}$

60. $-\frac{3}{4} \cdot \frac{1}{2} \cdot \left(-\frac{5}{7}\right)$
61. $(-2.64)(0.03)(-1.85)$
62. $(5.24)(-0.18)(-2.4)$

Simplify.

63. $(-3)^2 + (-4)$

64. $(-4)^2 + (-5)$

65. $-7 + 3(-3) - 10$

66. $-5 + 2(-8) - 2$

67. $-3(4) + (-6)(-2)$

68. $8 \cdot (-2) + (-3)(-1)$

69. $2(-8) + 3(-4)$

70. $3(-9) + 2(-7)$

71. $(-0.5)^2 + 1^2$

72. $(-0.3)^2 + 0.3^2$

73. $\frac{3}{5}(-10) - 6$

74. $\frac{1}{5}(-15) + 32$

75. $-5 \cdot (-3 + 1.2)$

76. $-6 \cdot (-4 + 2.7)$

77. $(6 - 8.4) \cdot (3 + 1.5)$

78. $(4.6 - 5) \cdot (-1 + 0.6)$

79. $[-2 + 3(-5)]^2$

80. $[3 - 4(-2)]^2$

81. $-2(8 - 12) + 24 + (-3)^2$

82. $5^2 + 4(-6 - 4) + (-9)(-8)$

Complete each table. Express each answer as a signed number.

83.

Input	Output
a. −2	$(-3)(\mathbf{-2}) - 1 =$
b. −1	$(-3)(\mathbf{-1}) - 1 =$
c. 0	$(-3)(\mathbf{0}) - 1 =$
d. +1	$(-3)(\mathbf{+1}) - 1 =$
e. +2	$(-3)(\mathbf{+2}) - 1 =$

84.

Input	Output
a. −2	$(-5)(\mathbf{-2}) + 1 =$
b. −1	$(-5)(\mathbf{-1}) + 1 =$
c. 0	$(-5)(\mathbf{0}) + 1 =$
d. +1	$(-5)(\mathbf{+1}) + 1 =$
e. +2	$(-5)(\mathbf{+2}) + 1 =$

Mixed Practice

Solve.

85. Multiply: $805(-6)$

86. Find the product of $-1\frac{1}{2}$ and $-1\frac{1}{3}$.

87. Calculate: $-(0.01)^2$

88. Compute: $\left(-\frac{2}{3}\right)\left(-\frac{4}{5}\right)\left(-\frac{9}{10}\right)$

89. Simplify: $(-4 + 5) - (-3)^2$

90. Evaluate: $\frac{2}{25}(-10)^2 - 6(4 - 7)$

Applications

B *Solve. Express each answer as a signed number.*

91. Tidal gauge measurements show that the sea level at Kodiak Island in Alaska is dropping at a rate of 12 millimeters per year. At this rate, how much will the sea level change in 6 years? (*Source:* National Oceanic and Atmospheric Administration)

92. A patient's dosage of medication is decreased 25 milligrams per day for 1 week. What is the change in her medication at the end of the week?

93. A piece of real estate property dropped \$1,475 in value each month. What was the change in value for 3 months?

94. Temperatures can be measured in both the Fahrenheit and Celsius scales. To find the Celsius equivalent of the temperature $-4°F$, we need to compute $\frac{5}{9} \cdot (-4 - 32)$. Simplify this expression.

95. The melting point of a substance is the temperature at which it changes from solid to liquid at standard atmospheric pressure. The melting point of mercury is $-40°C$. Find the melting point of krypton if it is 4 times as great as that of mercury. (*Source:* EnvironmentalChemistry.com)

96. In the 10 basketball games the Panthers played this season, they won 3 games by 2 points, won 2 games by 1 point, lost 4 games by 1 point, and tied in the final game. In these games, what was the difference between the number of points they scored and the number scored by the opposing teams?

97. Two seconds after release, the elevation of an object is $\frac{1}{2}(-32)(2)^2$ feet with respect to the point of release. What is this elevation?

98. During a drought, the water level in a reservoir fell 2 inches per week for 6 straight weeks. What was the change in the water level in the reservoir at the end of this period?

99. The balance in a student's bank account is \$1,000. Each month, \$150 is withdrawn.

a. What is the change in the account balance after 6 months?

b. What is the balance in the account after 6 months?

100. A man lost 1.8 pounds per week through a diet and exercise program.

a. What was his net change in weight after 15 weeks?

b. If the man weighed 183 pounds at the start of the program, how much did he weigh after 15 weeks?

• Check your answers on page A-11.

MINDStretchers

Groupwork

1. Ask a partner to think of two negative numbers. Then, you decide which is larger—the product of these numbers or their sum. Switch roles with your partner and repeat the exercise.

Critical Thinking

2. Fill in the following times table.

×	−1	3	−2
2			
−3			
−2			

Verify that the sum of the nine entries is 0. Why is this so?

Writing

3. A salesman says that he loses a little money on each item sold but makes it up in volume. Explain if this is possible.

7.5 Dividing Signed Numbers

OBJECTIVES

- **A** To divide signed numbers
- **B** To solve applied problems involving the division of signed numbers

Now, let's consider an example of division, the last of the four basic operations. Suppose that you and a friend together owe \$8 and you both agree to split the debt evenly. Then, each of you will owe \$4.

A debt is considered negative, so this problem requires us to calculate $-8 \div 2$. Recall that every division problem has a related multiplication problem. We see that $-8 \div 2 = -4$ because $-4 \cdot 2 = -8$. Note that when we divide a negative number by a positive number, we get a negative quotient.

Let's look at an example in which we divide one negative number by another negative number. Suppose that your friend owes you \$8 and agrees to repay the debt in installments of \$2 each. How many installments must your friend pay? The answer, of course, is 4.

This problem asks us to calculate $(-8) \div (-2)$. We know that $4 \cdot (-2) = -8$, so it follows that $(-8) \div (-2) = 4$. This example illustrates that dividing two negative numbers gives a positive quotient.

We can use the following rule for dividing signed numbers:

To Divide Two Signed Numbers

- Divide the absolute values of the numbers.
- If the numbers have the same sign, their quotient is positive; if the numbers have different signs, their quotient is negative.

Another way to think of dividing signed numbers is as follows:

Positive ÷ Positive = Positive   Positive ÷ Negative = Negative
Negative ÷ Negative = Positive   Negative ÷ Positive = Negative

EXAMPLE 1

Find the quotient: $-16 \div (-8)$

Solution First, find the absolute values.

$$|-16| = 16 \text{ and } |-8| = 8$$

Next, divide the absolute values.

$$16 \div 8 = 2$$

The numbers have the same sign, so the quotient is positive.

$$-16 \div (-8) = 2$$

PRACTICE 1

Divide: $-24 \div (-2)$

EXAMPLE 2

Simplify: $\dfrac{-8}{16}$

Solution $|-8| = 8$ and $|16| = 16$ Find the absolute values.

$\dfrac{8}{16} = \dfrac{1}{2}$ Express the quotient of the absolute values as a fraction.

Because the numbers have different signs, the answer is negative.

$$\frac{-8}{16} = -\frac{1}{2}$$

PRACTICE 2

Simplify: $\dfrac{9}{-15}$

TIP When a fraction has a negative sign in its numerator or denominator, we often rewrite the fraction as a negative number. For instance, we write both $\dfrac{-1}{2}$ and $\dfrac{1}{-2}$ as $-\dfrac{1}{2}$.

EXAMPLE 3

$7.4 \div (-2) = ?$

Solution $|7.4| = 7.4$ and $|-2| = 2$ Find the absolute values.

$7.4 \div 2 = 3.7$ Divide the absolute values.

The numbers have different signs, so their quotient is negative.

$$7.4 \div (-2) = -3.7$$

PRACTICE 3

Divide: $-1.5 \div 5$

EXAMPLE 4

Divide: $-8 \div \left(-1\frac{3}{5}\right)$

Solution We divide the absolute values of -8 and $-1\frac{3}{5}$.

$$8 \div 1\frac{3}{5} = 8 \div \frac{8}{5} = \cancel{8} \times \frac{5}{\cancel{8}} = 5$$

The quotient of two negative numbers is positive.

$$-8 \div \left(-1\frac{3}{5}\right) = 5$$

PRACTICE 4

Find the quotient: $-\dfrac{1}{2} \div 3$

We use the order of operations rule to simplify the following expressions in Example 5.

EXAMPLE 5

Simplify. **a.** $-10 + (-8) \div (-2)$ **b.** $\dfrac{-7 + (-3)^2}{2}$

Solution

a. $-10 + (-8) \div (-2) = -10 + 4$ Perform division before addition. Divide -8 by -2.

$= -6$ Add.

b. $\dfrac{-7 + (-3)^2}{2} = \dfrac{-7 + 9}{2}$ Simplify the numerator first.

$= \dfrac{2}{2}$ Add -7 and 9.

$= 1$ Divide.

PRACTICE 5

Simplify.

a. $6 - (-12) \div (-2)$

b. $\dfrac{5 - (-1)^2}{-4}$

EXAMPLE 6

The federal deficit in 1910 was about \$20 million. Five years later, it was \$60 million. How many times greater was the deficit of 1915 than that of 1910? (*Source:* infoplease.com)

Solution The problem asks us to compute $-60 \div (-20)$. The quotient of numbers with the same sign is positive, so the answer is 3. That is, the 1915 deficit was 3 times as great as the deficit of 1910.

PRACTICE 6

About $\frac{3}{4}$ of Earth's coral reefs are located in the Indian and Pacific Oceans. These reefs are declining at a rate of approximately 600 square miles per year. By how many square miles per month are the reefs declining, expressed as a signed number? (*Source:* news.bbc.co.uk)

EXAMPLE 7

The table shows the change in the closing price of a share of a software company's stock each day over a 5-day period.

Day	Change in Closing Price (in cents)
Monday	+32
Tuesday	−18
Wednesday	−21
Thursday	+16
Friday	−54

What was the average daily change in the closing price of a share of the stock?

Solution To compute the average change, we add the five changes and divide the sum by 5.

$$\frac{+32 + (-18) + (-21) + (+16) + (-54)}{5}$$

Recall from the order of operations rule that we must find the sum in the numerator before dividing by the denominator.

$$\frac{+48 + (-93)}{5} = \frac{-45}{5} = -9$$

So the average daily change over the 5-day period was down 9 cents per share.

PRACTICE 7

A young girl has a fever. The following chart shows how her temperature changed each day this week.

Monday	Up 2°
Tuesday	Up 1°
Wednesday	Down 1°
Thursday	Up 1°
Friday	Down 3°

What was the average daily change in her temperature?

7.5 Exercises

FOR EXTRA HELP MyMathLab  PRACTICE  WATCH  READ  REVIEW 

Mathematically Speaking

Fill in each blank with the most appropriate term or phrase from the given list.

addition	positive	negative
unequal	equal	multiplication

1. The quotient of two numbers with the same signs is __________.

2. The quotient of two numbers with different signs is __________.

3. The fractions $\frac{-2}{3}$, $\frac{2}{-3}$, and $-\frac{2}{3}$, are __________ in value.

4. Every division problem has a related ____________ problem.

A *Find the quotient. Simplify.*

5. $-20 \div (-4)$
6. $-7 \div (-1)$
7. $0 \div 5$
8. $0 \div 3$
9. $10 \div (-2)$
10. $-9 \div 3$
11. $16 \div (-8)$
12. $-12 \div 4$
13. $-250 \div (-10)$
14. $-300 \div (-3)$
15. $-200 \div 8$
16. $-20 \div 10$
17. $-35 \div (-5)$
18. $-8 \div (-4)$
19. $6 \div (-3)$
20. $-8 \div 2$
21. $-17 \div (-1)$
22. $-20 \div (-2)$
23. $-72 \div (-12)$
24. $-440 \div (-10)$
25. $-2.4 \div 8$
26. $-0.26 \div 2$
27. $-4 \div 0.2$
28. $9 \div (-0.6)$
29. $-4.8 \div (-0.3)$
30. $-2.6 \div (-0.2)$
31. $\left(-\frac{2}{3}\right) \div \frac{4}{5}$
32. $\left(-\frac{5}{6}\right) \div \left(-\frac{5}{6}\right)$
33. $7 \div \left(-\frac{1}{3}\right)$
34. $9 \div \left(-\frac{1}{3}\right)$
35. $-40 \div 2\frac{1}{2}$
36. $-30 \div 1\frac{1}{2}$
37. $(-15.1214) \div (-2.45)$
38. $-0.749 \div -0.214$
39. $-12.25 \div 3.5$
40. $50.8369 \div (-7.13)$

Simplify.

41. $\frac{-1}{5}$
42. $\frac{-1}{7}$
43. $\frac{-11}{-11}$
44. $\frac{-3}{-11}$
45. $\frac{4}{-10}$
46. $\frac{5}{-10}$
47. $\frac{-11}{-2}$
48. $\frac{-2}{-11}$
49. $\frac{-17}{-4}$
50. $\frac{-26}{-5}$
51. $\frac{-9}{-12}$
52. $\frac{-14}{-16}$
53. $-8 \div (-2)(-2)$
54. $-3(-4) \div (-2)$
55. $(3 - 7)^2 \div (-4)$
56. $(4 - 6)^2 \div (1 - 5)^2$
57. $\frac{2^2 - (-6)}{2}$
58. $\frac{3^2 - (-7)}{2}$
59. $\frac{2^2 + (-6)}{-2}$
60. $\frac{3^2 + (-7)}{-1}$
61. $\left(\frac{-8}{-2}\right)\left(\frac{8}{-2}\right)$
62. $\frac{-10}{2} \cdot \frac{-6}{5}$
63. $\frac{-9 - (-3)}{2}$
64. $\frac{-5 - (-7)}{2}$

65. $\dfrac{3(-0.2)^2}{-2}$

66. $\dfrac{(-16)(1.5)^2}{-1}$

67. $(-15) + (-3)^2 - 2 \cdot (-1)$

68. $-12 \cdot 2 + (-2)^2 - (-5) \cdot 3$

69. $24 \div (-8) + (-5) \cdot 6$

70. $-49 \div (-7)^2 - 4 \cdot (-3)$

71. $(-13 - 3) \div (-2 - 6)$

72. $10 + (-8) \div (-4)(-5)$

73. $[18 + (-4)] \div (-2)$

74. $[25 + (-10)] \div (-3)$

Insert parentheses, if needed, to make the expression on the left equal to the number on the right.

75. $9 \div 1 - 4 = -3$

76. $-2 + 8(-12) = -72$

77. $6 \div 3 - 1 - 4 = -1$

78. $-10 + 8 \div 2 - 5 \cdot 3 = -16$

79. $8 - 10 \cdot 2 - (-5) + 13 \div 4 = -6$

80. $12 \div (-5) + 1 + (-6)(-1) + 2 = -9$

Mixed Practice

Solve.

81. Divide: $-\dfrac{4}{5} \div \dfrac{2}{3}$

82. Divide -0.75 by -0.5.

83. Simplify: $\dfrac{19}{-6}$

84. Find the quotient: $-0.06 \div (-0.3)$

85. Evaluate: $(5 - 3)^2 \div (1 - 4)^3$

86. Simplify: $-4 - 9 \div 3(-5) + 2$

Applications

B *Solve. Express each answer as a signed number.*

87. The population of a certain city decreased by 60,989 in 10 years. Find the average annual change in population.

88. A new computer purchased for $1,800 will have a salvage value of $400. If its value decreases $280 per year, in how many years will it reach its salvage value?

89. In the decade between 1990 and 2000, the population of Washington, D.C., dropped from 607 thousand to 572 thousand, rounded to the nearest thousand. During this decade, what was the change in population per year? (*Source:* census.gov)

90. A football running back lost 4 yards on each of several plays. His total yardage lost was 24 yards. How many plays were involved?

91. The altitude of a plane decreased from 25,000 feet to 19,000 feet in 6 minutes. At what rate did the altitude of the plane change?

92. Over a 5-year period, the height of a cliff eroded by 3.5 feet. By how many feet did it change per year?

93. A meteorologist is expected to accurately predict the average high temperature for the next 5 days. This week the high temperatures were 3°, 0°, −8°, −11°, and 1°. If her prediction for these days was −3°, was it correct?

94. In a statistics course, a student needs to carry out the following computation.

$$\frac{(-0.5)^2 + (0.3)^2 + (0.2)^2}{3}$$

Find this number, rounded to the nearest hundredth.

95. The bar graph shows the daily high temperature in degrees Fahrenheit in Fairbanks, Alaska, for the first week of January in a recent year. (*Source:* climate.gi.alaska.edu)

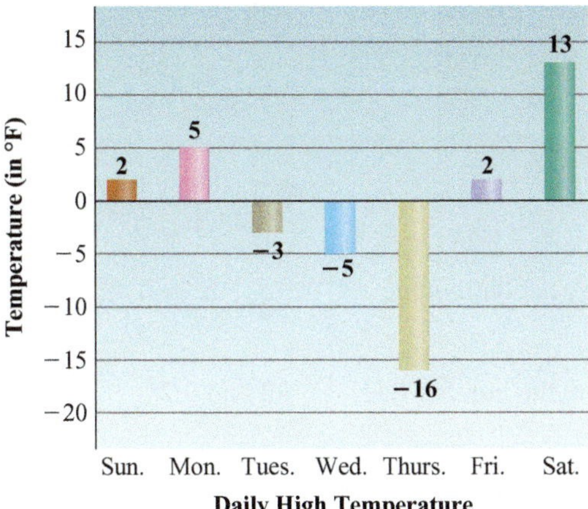

a. Which day of the first week was the coldest? The warmest?

b. To the nearest degree, what was the average daily high temperature that week?

96. The table below shows the stockholders' equity for Ford Motor Co. in each of four consecutive quarters.

Quarter	Stockholders' Equity (in billions)
1	−\$11
2	−\$9
3	−\$8
4	−\$6

(*Sources*: investopedia.com and finance.yahoo.com)

a. Find the average quarterly stockholders' equity for the first two quarters.

b. If the stockholders' equity had been \$2 billion more in the fourth quarter, what would the average for the four quarters have been?

• Check your answers on page A-11.

MINDStretchers

Patterns

1. Find the missing numbers in the following sequence.

$$+1296, +648, -216, -108, +36, +18, -6, \_\_\_\_, \_\_\_\_, \_\_\_\_$$

Groupwork

2. Do the following with a partner.
   - Take your partner's age in years.
   - Square it.
   - Subtract 9.
   - Divide the result by 3 less than your partner's age.
   - Subtract 53.
   - Add your partner's age.
   - Divide by 2.
   - Add 5^2.

   Verify that you wind up where you started—with your partner's age.

Writing

3. Explain the difference between the *opposite* of a number and the *reciprocal* of a number.

Key Concepts and Skills

CONCEPT SKILL

Concept/Skill	Description	Example
[7.1] **Positive number**	A number greater than 0.	$5, \frac{1}{3}, 2.7$
[7.1] **Negative number**	A number less than 0.	$-5, -\frac{1}{3}, -2.7$
[7.1] **Signed number**	A number with a sign that is either positive or negative.	$5, -5, \frac{1}{3}, -\frac{1}{3}, 2.7, -2.7$
[7.1] **Integers**	The numbers . . . , −4, −3, −2, −1, 0, 1, 2, 3, 4, . . . continuing indefinitely in both directions.	$+5, -5$
[7.1] **Opposites**	Two numbers that are the same distance from 0 on the number line but on opposite sides of 0.	$+2$ and -2
[7.1] **Absolute value**	The distance of a number from 0 on the number line, represented by the symbol | |.	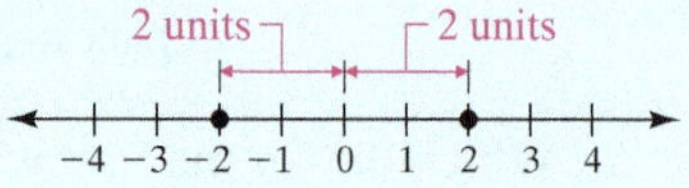 $\lvert -2 \rvert = 2, \quad \lvert +2 \rvert = 2$
[7.1] **To compare signed numbers**	• Locate the points being compared on the number line. A number to the right is larger than a number to the left.	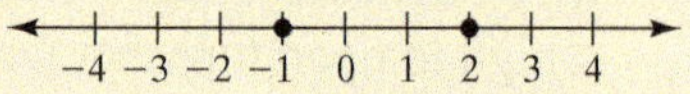 $2 > -1$
[7.2] **To add two signed numbers**	• If the numbers have the same sign, add the absolute values and keep the sign. • If the numbers have different signs, subtract the smaller absolute value from the larger and take the sign of the number with the larger absolute value.	$-0.5 + (-1.7) = -2.2$ because $\lvert -0.5 \rvert + \lvert -1.7 \rvert = 0.5 + 1.7 = 2.2$ $3\frac{1}{2} + (-9) = -5\frac{1}{2}$ because $\lvert -9 \rvert > \left\lvert +3\frac{1}{2} \right\rvert$ and $9 - 3\frac{1}{2} = 5\frac{1}{2}$
[7.3] **To subtract two signed numbers**	• Change the operation of subtraction to addition. Then add the first number and the opposite of the second number. • Follow the rule for adding signed numbers.	$-2 - (-5) = -2 + 5$ $= +3$, or 3
[7.4] **To multiply two signed numbers**	• Multiply the absolute values of the numbers. • If the numbers have the same sign, their product is positive; if the numbers have different signs, their product is negative.	$(-8)\left(-\frac{1}{2}\right) = +4$, or 4 $-0.2 \times 4 = -0.8$
[7.5] **To divide two signed numbers**	• Divide the absolute values of the numbers. • If the numbers have the same sign, their quotient is positive; if the numbers have different signs, their quotient is negative.	$\frac{-8}{-4} = +2$, or 2 $18 \div (-2) = -9$

CHAPTER 7 Review Exercises

Say Why

Fill in each blank.

1. The number $\frac{7}{9}$ $\underset{\text{is/is not}}{\underline{\hspace{2cm}}}$ a signed number because ____.

2. The number -17 $\underset{\text{is/is not}}{\underline{\hspace{2cm}}}$ less than 0 because ____.

3. The number -8.2 $\underset{\text{is/is not}}{\underline{\hspace{2cm}}}$ an integer because ____ ____.

4. The numbers -28.7 and 28.7 $\underset{\text{are/are not}}{\underline{\hspace{2cm}}}$ opposites because ____ ____.

5. The number -3 $\underset{\text{is/is not}}{\underline{\hspace{2cm}}}$ the absolute value of any number because ____ ____.

6. When we add -1 to a signed number, the sum $\underset{\text{is/is not}}{\underline{\hspace{2cm}}}$ larger than that signed number because ____ ____.

[7.1] *Mark the corresponding point for each number on the number line.*

7. -3

8. 1.5

−4 −3 −2 −1 0 1 2 3 4

Find the opposite signed number.

9. $+6$

10. -4

11. $-7\frac{1}{2}$

12. 10.1

Find the absolute value.

13. $|10|$

14. $|+2.5|$

15. $\left|-1\frac{1}{5}\right|$

16. $|-7|$

Circle the larger number.

17. -11 and -15

18. -15 and 10

19. 9 and $-5\frac{1}{3}$

20. -6.75 and -2

Arrange the numbers in each group from smallest to largest.

21. $-8, 8, -3.5$

22. $9, -6, -9.7$

23. $-2\frac{1}{2}, 0, -2.9$

24. $-4, -1\frac{1}{4}, 0$

Express each quantity as a signed number.

25. Ten feet above sea level

26. A loss of $350 on an investment

[7.2] *Find the sum.*

27. $-10 + (-10)$

28. $8 + (-10)$

29. $-5\frac{1}{2} + 12$

30. $-\frac{1}{4} + \left(-\frac{3}{4}\right)$

31. $0.9 + (-5)$

32. $-1.2 + (-0.8)$

33. $-8 + 5 + (-4)$

34. $12 + (-12) + \left(-\frac{1}{4}\right)$

[7.3] *Find the difference.*

35. $-10 - (-10)$

36. $14 - (-14)$

37. $5 - 15$

38. $-2 - 9$

39. $2.5 - (-0.5)$

40. $-\frac{1}{8} - 4$

[7.4] *Find the product.*

41. $-10(-10)$ **42.** $-15 \cdot 3$ **43.** $\frac{-2}{-3}\left(\frac{+10}{-11}\right)$

44. $3.5 \times (-2.1)$ **45.** $4(-3)(-6)$ **46.** $-2(-3)(-5)$

Evaluate.

47. $\left(\frac{1}{4}\right)^2$ **48.** $(-0.7)^2$ **49.** $(-6)^2$ **50.** -9^2

[7.5] *Find the quotient.*

51. $-35 \div (-7)$ **52.** $-80 \div 8$ **53.** $20 \div (-4)$

54. $-\frac{1}{8} \div (-4)$ **55.** $15 \div (-0.3)$ **56.** $\frac{-10}{-5}$

[7.2–7.5] *Simplify.*

57. $-8 - (-3) + 20$ **58.** $12 \cdot (-3)^2 - (-6)$ **59.** $(-7 + 3) \cdot (-5)^2$ **60.** $(20 - 30) \div (-10)$

61. $\frac{(-9.1)(-0.6)}{2}$ **62.** $\frac{-8 - 5.1}{5}$ **63.** $10^2 + \frac{-8 - 2}{2} + (-3)^2$ **64.** $\frac{10}{2} - (5 - 9)^2(-1)$

65. $6 - [5 + (-9)]$ **66.** $8 - [4 - (-6)]$ **67.** $[10 - 5(-1)]^2$ **68.** $[-2 + 3(-6)]^2$

Mixed Applications

Solve.

69. After answering a question incorrectly, a contestant on *Jeopardy* had \$1,000 deducted from his score of \$600. What was his new score?

70. The Chou dynasty ruled China between 1027 B.C. and 256 B.C. The philosopher Confucius was born in about 551 B.C. and died in about 479 B.C. Was the Chou dynasty in power throughout Confucius's lifetime? (*Source: Asian History on File*)

71. A customer has his monthly car payment automatically deducted from his checking account. If his monthly car payment is \$235, express the annual change in balance in his checking account as a signed number.

72. An administrative assistant had a balance of \$1,498.56 on her credit card. What is her new balance after charges totaling \$378.12, a payment of \$250, and a finance charge of \$23.15 are included?

73. A meteorologist reports that today's low temperature was $-5°F$ and that the normal low for the day is 23°F. How far below the normal low temperature is the low temperature today?

74. An instructor deducts 4 points for each incorrect answer on an exam. If a student received 92 out of a possible 120 points on the exam, how many questions did he answer incorrectly?

75. An investor bought 100 shares of a media company's stock for $3,500. The value of the stock was $3,380 after one month. Express the investor's change in value per share as a signed number.

76. Two of the most influential math books in history were *The Elements*, which Euclid wrote in 323 B.C., and *The Principia*, which Isaac Newton wrote in A.D 1687. How many years apart were these books written? (*Source: Notable Mathematicians from Ancient Times to the Present*)

77. Buddha was born in 563 B.C. and died in 483 B.C. Was he alive in 500 B.C.? (*Source: Compton's Encyclopedia*)

78. On a diver's first day of scuba diving, he dove to a depth of 30 feet below the surface of the sea. If on the next day he dove to a depth 3 times as great, how deep did he dive on that day?

79. Golf scores are commonly expressed as over or under *par*—the number of expected strokes on each hole. To *birdie* a hole is to take 1 less stroke than par, to *eagle* is to take 2 fewer strokes than par, and to *bogie* is to take 1 more stroke than par. With 3 birdies, 2 eagles, 1 par, and 2 bogies, how far over or under par is the golfer altogether?

80. Physicists have shown that, if an object is thrown upward at a speed of 100 feet per second, its elevation after 5 seconds will be

$$-16 \times 5^2 + 100 \times 5$$

feet relative to the point at which the object was thrown. How far above or below that point will the object be at that time?

81. The following bar graph shows the record low temperatures for selected states.

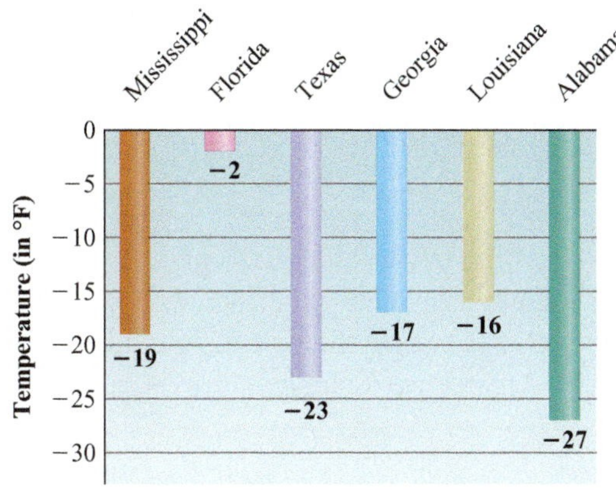

(*Source: National Climatic Data Center*)

a. Which state had the coldest record low temperature?

b. How much higher than the record low temperature in Louisiana was the record low temperature in Florida?

82. The following table shows the net income for American International Group, Inc. (AIG) for three consecutive years.

Year	Net Income (in billions of dollars)
2007	−6.2
2008	−99.3
2009	−10.9

a. What was the company's total net income for the three years?

b. What was the average annual net income for the company?

• Check your answers on page A-11.

CHAPTER 7 Posttest

FOR EXTRA HELP

CHAPTER Test Prep VIDEOS

The Chapter Test Prep Videos with test solutions are available on DVD, in MyMathLab, and on YouTube (search "AkstDevMath" and click on "Channels").

To see if you have mastered the topics in this chapter, take this test.

1. Which is smaller, -10 or -4?

2. A number has an absolute value of $\frac{1}{2}$ and is negative. What is the number?

Evaluate.

3. $-8 + 8$

4. $4.5 + (-5)$

5. $42 - 91$

6. $-12 - (-12)$

7. -23×9

8. -0.5×0.2

9. -12^2

10. $\left(-\frac{1}{4}\right)^2$

11. $-64 \div 16$

12. $\frac{1}{2} \div (-4)$

13. $-4 + 6 + (-7) + 9$

14. $15 - (-7) + (-1)$

15. $-8 - 4^2 \cdot (-3)$

16. $(2 - 8)^2 \div (-2)$

Solve.

17. A dieter on a weight-loss program lost 12 pounds in the first month. Over the next 5 months, he lost an additional 39 pounds. Express as a signed number the total change in weight in 6 months.

18. A copier that was purchased new for \$6,900 is worth \$4,700 four years later. If the copier changes in value by the same amount each year, what is the rate at which its value changes, expressed as a signed number?

19. On April 20, 2010, Deepwater Horizon, an oil rig owned by British Petroleum (BP), exploded off the coast of Louisiana, sending oil gushing into the Gulf of Mexico from the well nearly a mile below the water's surface. The graph below shows the percent change in stock prices for the top five oil companies in the world from the day of the explosion to two weeks after.

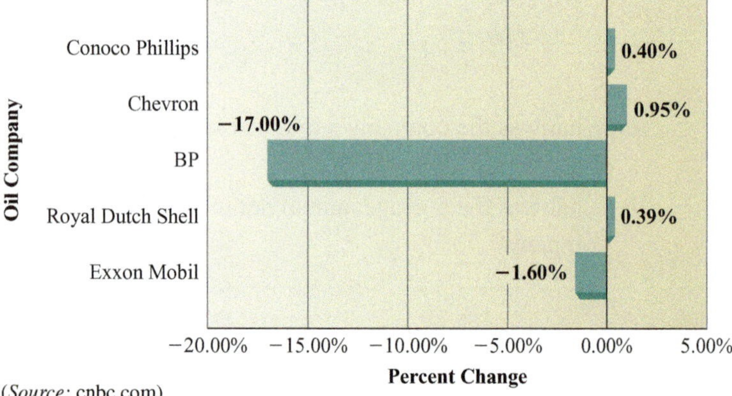

(*Source:* cnbc.com)

Rank these companies in order from biggest loss to biggest gain in their stock prices.

20. At 36,000 feet, the temperature outside an airplane had been $-56°C$. When the plane landed, the temperature outside rose to $20°C$. How much greater was the temperature outside the plane on the ground than in the air?

• Check your answers on page A-11.

Cumulative Review Exercises

To help you review, solve the following:

1. Round 2,891 to the nearest thousand.

2. $(5 - 9)^2 \div (-6 + 2)$

3. Multiply: $(4)\left(2\frac{1}{2}\right)$

4. Add: $8 + 2.1 + 3.9$

5. Find the product: $(3.7)(0.4)(2.1)$

6. Perform the indicated operations: $\frac{32.44 - 11.8}{6}$

7. Solve for x: $x + 7.5 = 9$

8. Translate the sentence "$\frac{5}{9}$ of x is 40" to an equation.

9. Solve for n: $\frac{1.4}{7} = \frac{13}{n}$

10. Change 1.125 to a percent.

11. What percent of 2.5 is 0.5?

12. Find the sum: $-7.84 + (-0.3) + 2.1$

13. Simplify: $(-0.5)^2 + (0.5)^2$

14. Divide: $\frac{5}{12} \div \left(-\frac{3}{4}\right)$

15. In a varsity baseball game, the starting pitcher pitched $6\frac{1}{3}$ innings before he was relieved. The relief pitcher lasted until the end of the 9th inning. How many innings was the relief pitcher in the game? (*Source: Kern Valley Sun*)

Solve.

16. A patient is to be given a total of 480 milligrams of medication per day. If the medication is to be administered every 4 hours, how much medication should be administered with each dose?

17. A patient has a total cholesterol level of 200 and an HDL (also called "good") cholesterol level of 50, both in terms of milligrams per deciliter. Find the ratio of total cholesterol to HDL cholesterol. (*Source*: americanheart.org)

18. When mortgage rates dropped, the number of housing starts rose from 4,000 to 5,000. What percent increase is this?

19. Three of the coldest temperature readings ever recorded on Earth were −89°C, −68°C, and −63°C. Of these three temperatures, which was the coldest? (*Source: Time Almanac, 2010*)

20. Nitrogen gas, when cooled, becomes liquid at −196°C. This liquid becomes solid at −210°C. At −200°C, is nitrogen a gas, a liquid or a solid? (*Source:* nasa.gov)

• Check your answers on page A-11.

CHAPTER 8

Basic Statistics

Statistics and the Law

Lawyers make frequent use of statistical evidence to win their cases.

Statistics on the distribution of blood and hair types in the general population are commonly used as evidence in physical assault and robbery trials. For those plaintiffs who claim that they are suffering from exposure to a toxic agent, their lawyers often present statistical evidence about the general incidence of their illness. And cases of race and sex discrimination typically focus on such statistics as the proportion of people who are admitted to a college or hired by a company or the average length of time that employees have spent in positions before being promoted.

This use of statistics in the U.S. legal system goes back to a landmark nineteenth-century trial, wherein the claim was made that the signature of Sylvia Howland on her will was forged. The turning point of the case was the testimony of an expert witness—a Harvard mathematician—who developed a system of statistically analyzing the degree of similarity among 42 signatures of the deceased. On the basis of these statistics, he testified that the signature on the will was unreasonably similar to another of Ms. Howland's from which it had probably been traced.

(*Source:* Jack B. Weinstein, "Litigation and Statistics," *Statistical Science*, 3, (3), 1988, pp. 286–297)

CHAPTER 8 PRETEST

To see if you have already mastered the topics in this chapter, take this test.

1. In the 2009 World Series, the Philadelphia Phillies, champions of the National League, played the New York Yankees, champions of the American League. The Yankees defeated the Phillies four games to two, as indicated in the following table of runs scored in the six games.

Team	Game 1	Game 2	Game 3	Game 4	Game 5	Game 6
Philadelphia Phillies	6	1	5	4	8	3
New York Yankees	1	3	8	7	6	7

(*Source:* mlb.com)

Was the range of runs that the Phillies scored in a Series game the same as that for the Yankees?

2. What is the mode of the number of days in a month? (*Reminder:* February has 28 or 29 days; April, June, September, and November have 30 days; and January, March, May, July, August, October, and December have 31 days.)

3. A local fire department tracks its emergency response times. Last month, the response times were: 12 minutes, 7 minutes, 20 minutes, 10 minutes, 6 minutes, 15 minutes, 8 minutes, 12 minutes, and 9 minutes. What was the mean response time?

4. The following table shows the amount of rainfall (in millimeters) each month for a recent year in London. (*Source:* worldweather.org)

Jan	Feb	Mar	Apr	May	June
53	36	48	47	51	50

July	Aug	Sept	Oct	Nov	Dec
48	54	53	57	57	57

What was the median amount of monthly rainfall?

5. Late in the spring term, your grades were: Spanish I (3 credits)—A; Music (2 credits)—A; Social Science (4 credits)—C; and Physical Education (1 credit)—B. The grades are assigned the following points: A = 4, B = 3, C = 2, D = 1, and F = 0. Calculate your GPA.

In each case, use the given table or graph to answer the question.

6. The following mortality table gives estimates for the life spans of individuals (in years), taking into account such factors as year of birth and gender. (*Source:* U.S. National Center for Health Statistics)

Year of Birth	1920	1930	1940	1950	1960	1970	1980	1990	2000	2010
Male	53.6	58.1	60.8	65.6	66.6	67.1	70.0	71.8	74.3	75.6
Female	54.6	61.6	65.2	71.1	73.1	74.7	77.5	78.8	79.7	81.4

For which years was the difference between male and female life spans greater than in 1950?

7. The following graph shows the percent of American cancer patients surviving 5 or more years, during various periods of time.

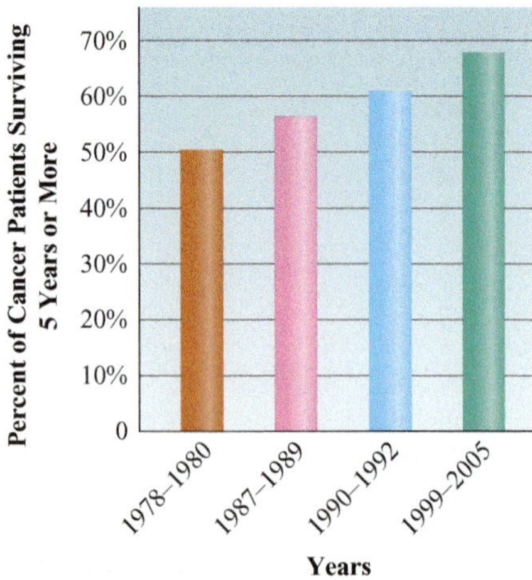

(*Source:* National Cancer Institute)

In which period(s) of time shown did more than three-fifths of the cancer patients survive 5 or more years?

8. The following pictograph shows the average circulation of some major newspapers across the United States in a recent year.

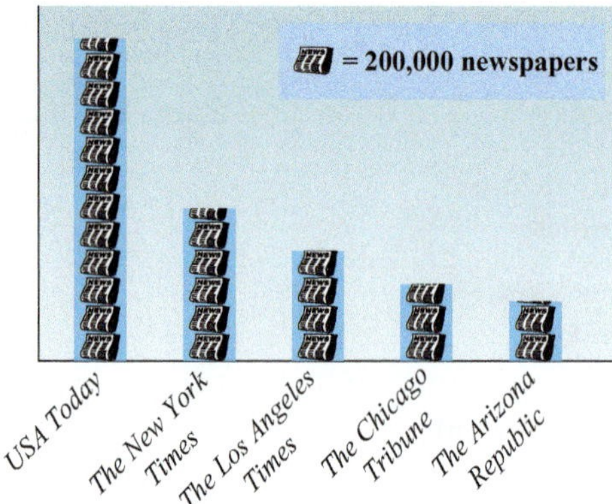

(*Source: Top Ten of Everything, 2010*)

What was the approximate daily circulation of *The New York Times*?

9. The first automated teller machine (ATM) in the United States was installed in 1971 at the Citizens & Southern National Bank in Atlanta. Overall, the number of ATMs has grown rapidly. The following graph shows the number in recent years.

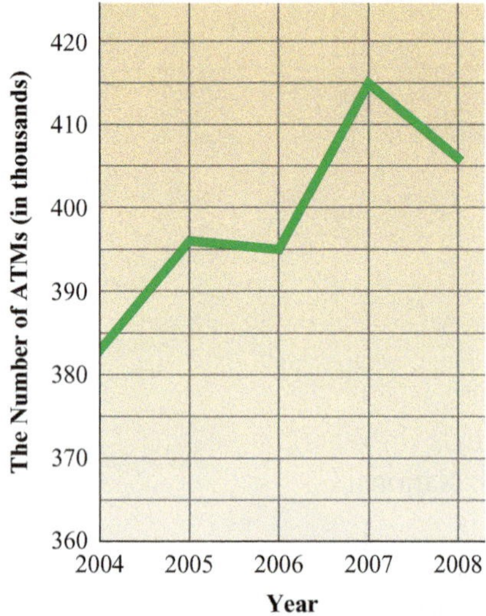

(*Source:* Insurance Information Institute)

In the year 2008, approximately how many ATMs were there in the United States?

10. The graph shows the breakdown of days of school missed in a recent 12-month period due to illness or injury for U.S. children 5–17 years of age.

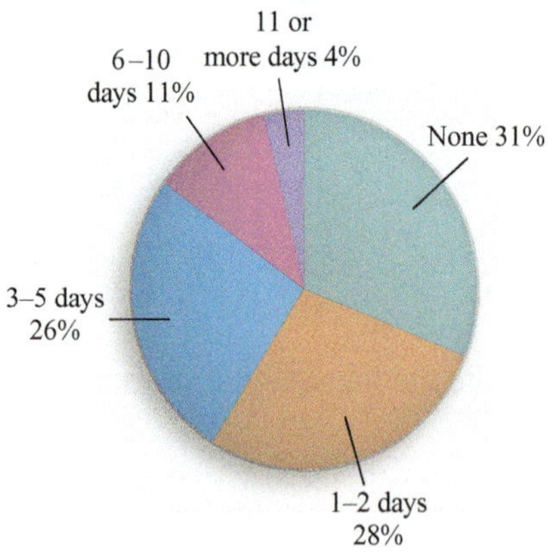

(*Source:* cdc.gov)

What percent of children missed 3 or more days of school?

• Check your answers on page A-11.

Cultural Note

MEMENTO MORI

LONDON'S Dreadful Viſitation:
Or, A COLLECTION of All the
Bills of Mortality
For this Preſent Year:
Beginning the 27th of *December* 1664. and
ending the 19th. of *December* following:
As alſo, *The GENERAL or whole years BILL:*
According to the Report made to the
KING's Moſt Excellent Majeſty,
By the Company of Pariſh-Clerks of London. &c

LONDON:
Printed and are to be ſold by *E. Cotes* living in *Alderſgate-ſtreet.*
Printer to the ſaid Company 1665.

A seventeenth-century English clothing salesman named John Graunt had the insight to apply a numerical approach to major social problems. In 1662, he published a book entitled *Natural and Political Observations upon the Bills of Mortality*, and thus founded the science of statistics.

Graunt was curious about the periodic outbreaks of the bubonic plague in London, and his book analyzed the number of deaths in London each week due to various causes. He was the first to discover that, at least in London, the number of male births exceeded the number of female births. He also found that there was a higher death rate in urban areas than in rural areas and that more men than women died violent deaths. Graunt summarized large amounts of information to make it understandable and made conjectures about large populations based on small samples. Graunt was also a pioneer in examining expected life span—a statistic that became vital to the insurance companies formed at the end of the seventeenth century.

Sources: Morris Kline, *Mathematics, a Cultural Approach* (Reading, Mass.: Addison-Wesley Publishing Company, 1962), p. 614.

F. N. David, *Games, Gods and Gambling* (New York: Hafner Publishing Company, 1962).

8.1 Introduction to Basic Statistics

OBJECTIVES

- **A** To find the mean, median, or mode(s) of a set of numbers
- **B** To find the range of a set of numbers
- **C** To solve applied problems involving basic statistics

What Basic Statistics Is and Why It Is Important

Statistics is the branch of mathematics that deals with ways of handling large quantities of information. The goal is to make this information easier to interpret.

With unorganized data, spotting trends and making comparisons is difficult. The study of statistics teaches you how to organize data in various ways in order to make the data more understandable.

One approach is to calculate special numbers, also called statistics, which describe the data. In this section, we consider four statistics: the mean, the median, the mode, and the range.

You have already seen that another way to organize data is to display the information in the form of a table or graph. We will discuss tables and graphs in greater detail in the next section of this chapter.

Many situations lend themselves to the application of statistical techniques. Wherever there are large quantities of information—from sports to business—statistics can help us to find meaning where, at first glance, there seems to be none, and to become more quantitatively literate.

Averages

We begin our introduction to statistics by revisiting the meaning of "average." Previously, we defined the average of a set of numbers to be the sum of the numbers divided by however many numbers are in the set. This statistic, which is more precisely called the *arithmetic mean*, or just the **mean**, is what most people think of as the average. However, it is not the only kind of average used to represent the numbers in a set.

A second average, the *median*, may describe a set of numbers better than the mean when there is an unusually large or unusually small number in the set to be averaged. The third average, the *mode*, has a special property—unlike the mean and the median, it is always in the set of numbers being averaged.

Mean

Let's look at an example of the mean.

EXAMPLE 1

The area of the United States is about 3,800,000 square miles.

a. Approximately what is the average area of each of the 50 states?

b. Is Michigan above or below average in area?

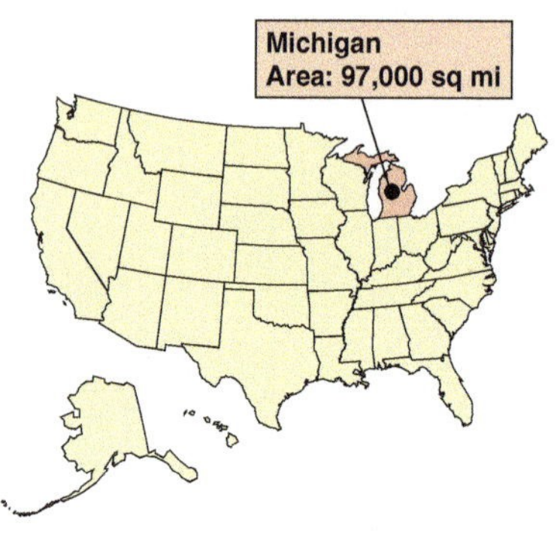

Solution

a. The mean area of a state is $\frac{3,800,000}{50}$, or 76,000 square miles.

b. Since 97,000 is greater than 76,000, Michigan is above average in area.

PRACTICE 1

Reggie Jackson hit five home runs in the 1977 World Series, which lasted six games. By contrast, Lou Gehrig hit four home runs in the 1928 World Series, a four-game series. On the average, which baseball player hit fewer home runs per game?

Note that a property of the mean is that it is changed substantially if even a single number in a set of numbers is replaced by one much larger or much smaller. For instance, if five people each make \$10 per hour, then their mean hourly wage is \$10. However, if the hourly wage of one of these individuals jumps to \$500, then the mean wage skyrockets to \$108, more than 10 times the previous mean.

Another kind of mean, called the *weighted average*, is used when some numbers in a set count more heavily than others. Weighted average comes into play, for instance, if you want to compute the average of your test scores in a class and the final exam counts twice as much as any of the other tests. Or, if you are computing your grade point average (GPA) and some courses carry more credits than others.

EXAMPLE 2

Last term, a student's grades were as follows.

Course	Credits	Grade	Grade Equivalent
Psychology	4	A	4
English	4	C	2
Art	3	B	3
Physical Education	1	B	3

If the student's GPA is 3.5 this term, did she have a higher or lower GPA last term?

Solution To calculate the GPA for last term, we first multiply the number of credits each course carries by the numerical grade equivalent received. We then add these products to find the total number of grade points. Finally, we divide this sum by the total number of credits.

Number of credits for the first course

Grade equivalent of the first course

$$\text{GPA} = \frac{4 \cdot 4 + 4 \cdot 2 + 3 \cdot 3 + 1 \cdot 3}{12}$$

Total number of credits

$$= \frac{16 + 8 + 9 + 3}{12} = \frac{36}{12} = 3$$

A GPA of 3 is less than a GPA of 3.5. So the student had a lower GPA last term.

PRACTICE 2

The following table shows the test scores that a classmate earned.

Exam	Score
1	95
2	80
3	80
Final	90

If the final exam is equivalent to two other exams, did the classmate earn an exam average above or below 85?

Median

As we have seen, a very large number can affect the mean of a set of numbers to such an extent that it is not representative of the set. Another kind of average, the *median*, is used when we wish to reduce the impact of an extreme number in the set, for instance, in computing average salary.

DEFINITION

In a set of numbers arranged in numerical order, the **median** of the numbers is the number in the middle. If there are two numbers in the middle, the median is the mean of the two middle numbers.

EXAMPLE 3

Find the median.

a. 6, 8, 2, 1, and 5 **b.** 6, 8, 2, and 5

Solution

a. We arrange the numbers from smallest to largest.

The middle number
↓
1 2 **5** 6 8

So the median is 5. If we arrange the numbers from largest to smallest, we get the same answer.

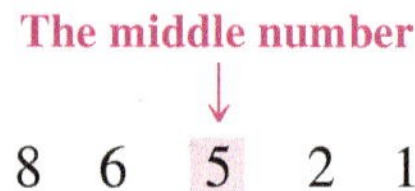

The median is still 5.

b. We order the numbers from smallest to largest.

2 5 6 8

Since four numbers are on the list and four is even, no single number is in the middle. In this case, the median is the mean of the two middle numbers.

Two middle numbers
↓
2 **5 6** 8
↓

$$\frac{5 + 6}{2} = 5.5$$

So 5.5 is the median.

PRACTICE 3

Compute the median.

a. 7, 2, 8, 5, 10, 7, 9, 10, 2, 5, 8, and 6

b. 0, 4, 1, 5, 7, 2, 5, 9, and 3

EXAMPLE 4

The first 11 justices on the U.S. Supreme Court served the following numbers of years: 5, 8, 1, 20, 5, 9, 0, 13, 4, 15, and 30. (*Source: Time Almanac 2010*)

a. What was the median number of years that they served on the court?

b. The next appointed justice served on the court for 3 years. With the addition of this new justice, by how much did the median number of years that the justices served change?

Solution

a. To compute the median of the years of service, let's first arrange these numbers in increasing order:

The middle number
↓
0 1 4 5 5 **8** 9 13 15 20 30

Because 8 is the middle number, the median number of years of service was 8.

PRACTICE 4

The weekend box office receipts (in millions of dollars) for five leading movies was:

25 16 9 12 26

a. What was the median box office receipt for these movies?

b. A sixth movie also took in \$9 million at the box office. How does the median box office receipt of the six movies compare with that of the five movies? (*Source: Variety*)

b. Let's arrange the 12 numbers in increasing order:

Two middle numbers
↓

0 1 3 4 5 **5 8** 9 13 15 20 30

The numbers 5 and 8 are in the middle. So the median is $\frac{5 + 8}{2}$, or 6.5. Since the median decreased from 8 to 6.5, it dropped by 1.5, that is, by 1.5 years.

Mode

The last type of average that we consider is the mode. Note that a set of numbers can have one mode, more than one mode, or even no mode.

DEFINITION

The **mode** of a set of numbers is the number (or numbers) occurring most frequently in the set.

EXAMPLE 5

Compute the mode(s).

a. 8, 6, 10, 8, 10, 8, 9, and 6

b. 2, 9, 3, 5, 7, 12, 3, 2, 18, 12, 2, and 3

c. 4, 10, 1, 5, 12, and 7

Solution

a. When we count how often each number occurs on this list, we see that 6 occurs twice, 8 occurs three times, 9 once, and 10 twice. Because there are more 8's than any other number, 8 is the mode.

b. Here, 2 occurs three times, 3 occurs three times, 5 once, 7 once, 9 once, 12 twice, and 18 once. So both 2 and 3, occurring most frequently, are modes.

c. No number occurs more than once. There is no mode.

PRACTICE 5

Find the mode(s).

a. 7, 2, 5, 1, 2, 5, and 2

b. 9, 1, 0, 4, 9, 4, 1, 5, 9, and 4

c. 8, 13, 9, and 2

EXAMPLE 6

Every U.S. state has a maximum posted speed limit for driving passenger vehicles. One state has a limit of 45 miles per hour, another one has a limit of 50 miles per hour, 26 have a limit of 55 miles per hour, 3 have a limit of 60 miles per hour, 16 have a limit of 65 miles per hour, and 3 have a limit of 70 miles per hour. What is the mode of these speed limits? (*Source:* iihs.org)

Solution The speed limit 55 miles per hour occurs more frequently than any other limit (26 times). So 55 miles per hour is the mode of the state speed limits.

PRACTICE 6

Students in a class discussed the number of hours in their college schedules. One student had a weekly schedule consisting of 3 hours of classes, 2 students had 6 hours, 15 had 12 hours, 1 had 13 hours, and 1 had 14 hours. Find the mode of the number of hours in these schedules.

Range

The last statistic that we consider is called the *range*. The range is not an average because it does not represent a typical number in the set. Instead, the range is a measure of the spread of the numbers in the set.

DEFINITION

The **range** of a set of numbers is the *difference* between the largest and the smallest number in the set.

EXAMPLE 7

Find the range of the numbers 3, 13, 2, 5, 9, and 2.

Solution The largest number in the set is 13, and the smallest is 2. So the range is 13 − 2, or 11.

PRACTICE 7

What is the range of 8, 10, 3, and 8?

EXAMPLE 8

Americans have seen a great deal of variability in gasoline prices at the pumps. The U.S. average retail prices (in dollars) for all grades of gasoline, including taxes, during the first six months of a recent year were as follows:

Jan	Feb	Mar	Apr	May	June
3.10	3.08	3.29	3.51	3.82	4.11

The average retail prices for gasoline during the last six months of the year were:

July	Aug	Sept	Oct	Nov	Dec
4.11	3.83	3.76	3.11	2.21	1.75

(*Source:* tonto.eia.doe.gov)

In which of the two six-month periods was the spread on gasoline prices greater?

Solution During the first six-month period, the highest average gas price was \$4.11 and the lowest was \$3.08. Therefore the range of gas prices was \$4.11 – \$3.08, or \$1.03. For the second period, the highest average gas price was \$4.11 and the lowest was \$1.75. So the range was \$4.11 – \$1.75, or \$2.36. Since the second period had a larger range, its spread was greater.

PRACTICE 8

In a couple of years, each of the 45 states in the U.S. with a state minimum hourly wage rate had one of the following rates: \$8.06, \$7.25, \$7.30, \$8.55, \$5.15, \$7.24, \$7.75, \$6.25, \$6.15, \$8.25, \$8.40, \$7.50, \$8.00, \$7.40. What is the range of these rates? (*Source:* dol.gov)

8.1 Exercises

FOR EXTRA HELP MyMathLab MathXL PRACTICE WATCH READ REVIEW

Mathematically Speaking

Fill in each blank with the most appropriate term or phrase from the given list.

range	weighted	mean
median	statistics	arithmetic
mode	algebra	

1. The branch of mathematics that deals with handling large quantities of information is called ____________.

2. The sum of the numbers in a set divided by however many numbers are in the set is called the ____________.

3. When some numbers in a set count more heavily than other numbers in the set, the average is said to be ____________.

4. The middle number in a set of numbers arranged in numerical order is called the ____________.

5. The ____________ is the number (or numbers) occurring most frequently in a set of numbers.

6. The ____________ is the difference between the largest number and the smallest number in a set of numbers.

A *Compute the indicated statistics. Round to the nearest tenth, where necessary.*

7.

Numbers	Mean	Median	Mode(s)
a. 8, 2, 9, 4, 8			
b. 3, 0, 0, 3, 10			
c. 4, 6, 9, 1, 1, 3			
d. 7.5, 9, 8.5, 5.5, 8.1			
e. $3\frac{1}{2}, 3\frac{3}{4}, 4, 3\frac{1}{2}, 3\frac{1}{4}$			
f. 4, −2, −1, 0, −1			

8.

Numbers	Mean	Median	Mode(s)
a. 5, 3, 5, 5, 3			
b. 7, 0, 7, 6, 0			
c. 5, 1, 0, 7, 7, 4			
d. 2.1, 3.6, 1.4, 2.5, 2.4			
e. $4\frac{1}{2}, 3\frac{3}{4}, 4, 4\frac{1}{2}, 4\frac{1}{4}$			
f. −1, −3, −3, −2, 4			

9. Calculate the mean, rounded to the nearest cent.
$9,125.88 $11,724.87 $12,705 $11,839.75
$13,500.79 $14,703.71

10. Find the mean, rounded to the nearest foot, of the following measurements.
3,725 ft 3,719 ft 3,740 ft 3,726 ft 3,729 ft
3,734 ft 3,725 ft

B *Complete each table.*

11.

Numbers	Range
a. 20, 11, 3, 4, 16	
b. 2.3, 5.7, 10.2, 6.1, 0.9	
c. $6\frac{5}{6}, 5\frac{1}{2}, \frac{1}{3}, 8, 5\frac{3}{4}$	
d. −2, 6, −4, −1, −4	

12.

Numbers	Range
a. 8, 0, 14, 3, 11	
b. 7.4, 2.9, 3.5, 8.6, 4.1	
c. $3\frac{1}{4}, 4\frac{2}{3}, 6\frac{1}{2}, \frac{9}{10}, 3\frac{7}{8}$	
d. −2, 4, −5, −2, 0	

Applications

C *Solve and check.*

13. Here are a student's grades last term: A in College Skills (2 credits), B in World History (4 credits), C in Music (2 credits), A in Spanish (3 credits), and B in Physical Education (1 credit). Did the student make the Dean's List, which requires a GPA of 3.5? Explain. (*Reminder:* A = 4, B = 3, C = 2, and D = 1.)

14. On a test, 9 students earned 80, 10 students earned 70, and 1 student earned 75. Was the grade of 75 below the class average (mean), exactly average, or above the class average? Explain.

15. A grandmother leaves a total of $1,000,000 to her 10 grandchildren. What is the mean amount left to each grandchild? Can you compute the median amount with the given information? Explain.

16. A woman and four men are riding in an elevator. Two men are taller than the woman, and two are shorter. Who has the median height of the people in the elevator?

17. In the U.S. House of Representatives, 435 members of Congress represent the 50 states. The table below shows the number of representatives of 8 states.

State	Number of Representatives
Maine	2
Indiana	9
Wisconsin	8
Hawaii	2
Colorado	7
North Carolina	13
Tennessee	9
Nebraska	3

(*Source:* 2010.census.gov)

Which of these 8 states has representation that is above the average for all 50 states?

18. The table shows the quarterly revenues for Ford Motor Company in a recent year.

Quarter	Revenue (in billions)
1	$31.7
2	$35.4
3	$31.6
4	$35.1

(*Source:* finance.yahoo.com)

What was the median quarterly revenue?

19. The table shows the salary of six teachers based on the number of years of service at a local school.

Years of Service	Salary
6	$44,424
10	$57,418
1	$37,925
4	$42,656
13	$58,358
18	$70,852

a. Find the median salary.

b. What is the range?

20. The table shows the prime interest rate on June 1st of the years 2000 through 2010.

2000	2001	2002	2003	2004	2005
9.50%	7.00%	4.75%	4.25%	4.00%	6.00%

2006	2007	2008	2009	2010
8.00%	8.25%	5.00%	3.25%	3.25%

(*Source:* moneycafe.com)

a. What is the mode of the interest rates for the given years?

b. Find the range.

21. The diameters for the eight planets of the solar system, rounded to the nearest 1,000 miles, are as follows:

Planet	Diameter (in thousands of miles)
Mercury	3
Venus	8
Earth	8
Mars	4
Jupiter	89
Saturn	75
Uranus	32
Neptune	31

(*Source: Encyclopedia Americana*)

Find each of the following distances, rounded to the nearest 1,000 miles:

a. mean diameter

b. median diameter

c. mode(s) of the diameters

d. range of the diameters

22. Consider the following utility bills for the past 10 months:

Month	Utility Bill
January	$90
February	$80
March	$90
April	$70
May	$100
June	$110
July	$140
August	$140
September	$100
October	$90

Find each of the following:

a. the mean bill

b. the median bill

c. the mode(s) of the bills

d. the range of the bills

23. In the year 1990, when the number of U.S. residents was about 249 million, the U.S. Postal Service delivered some 166 billion pieces of mail. By 2003, when the population had grown to 292 million, the Service delivered approximately 202 billion pieces of mail. On the average, how many more pieces of mail, to the nearest whole number, did a resident receive in 2003 than in 1990? (*Source:* census.gov)

24. Students earned the following grades on a college math test:

85	90	60	45	95	70	60	90	100	25	85	70	80
75	55	85	100	40	95	50	75	65	90	75	60	50

Using the mean as the average, how far above or below the class average, to the nearest whole number, was the test score of 75?

• Check your answers on page A-11.

MINDStretchers

Groupwork

1. Working with a partner, construct an example of a set of 10 numbers

a. whose mean, median, and mode are equal.

b. whose mean is less than its median.

c. that has two modes.

Mathematical Reasoning

2. Can the range of a set of numbers be equal to a negative number? Explain.

Investigation

3. In your college library or on the web, research the legal drinking age in 10 countries of your choosing. Then, determine the mean, median, mode, and range of these ages.

8.2 Tables and Graphs

OBJECTIVES

A To solve applied problems involving tables

B To solve applied problems involving graphs

What Tables and Graphs Are and Why They Are Important

We frequently present data in the form of tables or graphs. A **table** is a rectangular display of data. A **graph** is a picture or diagram of the data.

Organizing data in a table or graph makes it easier for a reader to make comparisons, to understand relationships, to spot trends, and to gain a sense of the data.

Graphs generally provide less accurate information than tables because we often have to read a graph by estimating. However, graphs are pictorial, so they make a more lasting impression than a table.

Tables and graphs are used in many different situations. Train schedules, insurance premium charts, and accountants' spreadsheets are common examples of tables. Graphs, such as a bar graph of a changing population, a line graph of fluctuating stock prices, and a circle graph of budget allocations, appear regularly in newspapers, magazines, and reports.

Tables

A table consists of rows and columns. Rows run horizontally, and columns run vertically. The nature of the entries in a row or column is described with labels called *headings*.

To read a table, first identify a particular row and column, and then locate the entry at their intersection. Consider the following table that shows the typical heartbeat rates for people of various ages:

Column

Heading →

Person's Age	Heartbeats per Minute
Newborn	135
2	110
6	95
10	87
20	71
40	72
60	74

Row → (Newborn, 135)

Entry → (60)

As the headings indicate, the entries in the first column are a person's age. The second column gives the number of times per minute that the heart of a person of that age typically beats.

As given in the table, the heart of a 20-year-old beats 71 times a minute. Does this table suggest that a child's heart beats more quickly than the heart of an adult? How many times per minute would you estimate that the heart of an 8-year-old beats?

Example 1 illustrates both reading and drawing conclusions from tables.

EXAMPLE 1

The following is the schedule of math classes this term:

Course	Section	Day/Time	Room	Professor
010	090	Tu W Th 9–10:50	516	Einstein
011	091	M W Th 9–9:50	518	Banneker
011	611	Tu Th 6–7:15	516	Kovalevski
051	111	M Tu W Th 11–11:50	523	Noether
051	711	M W 7–8:40	518	Hilbert
056	131	M W Th 1–2:50	516	Banneker
100	121	M Tu W Th 12–12:50	511	Hilbert
104	081	M W Th 8–8:50	511	Einstein
104	111	M W Th 11–11:50	511	Newton
150	091	M Tu W Th 9–9:50	511	Newton
150	511	M W 5:25–7:05	511	Kovalevski
206	131	M Tu W Th 1–1:50	523	Noether
301	141	M Tu W Th 2–2:50	523	Hilbert
302	511	Tu Th 5:25–7:05	520	Gauss

a. In what room does Math 301, Section 141 meet?

b. Is Professor Kovalevski teaching a section of Math 056 this term?

c. Today is Monday, and a student needs to speak to Professor Einstein. When and where is he teaching?

Solution

a. Math 301, Section 141 meets in room 523.

b. No, Professor Kovalevski is not teaching Math 056.

c. Professor Einstein is in room 511 from 8 to 8:50.

PRACTICE 1

A mail-order catalog contains the following chart for determining shipping and handling (S&H) charges:

Amount of Merchandise	Up to $5	$5.01–$15	$15.01–$25*
Charges	$2.95	$3.95	$4.95

*Add $0.10 for each additional $1 of merchandise over $25.

a. What are the S&H charges on $23.45 worth of merchandise?

b. How much must a customer pay in all for merchandise that, excluding S&H charges, sells for $3?

c. How much are the S&H charges on merchandise selling for $30?

Graphs

Now, let's discuss displaying data in the form of graphs. We deal with five kinds of graphs: pictographs, bar graphs, histograms, line graphs, and circle graphs.

Pictographs

A **pictograph** is a kind of graph in which images such as people, books, or coins are used to represent and to compare quantities. A *key* is given to explain what each image represents.

Pictographs are visually appealing. However, they make it difficult to distinguish between small differences—say, between a half and a third of an image.

EXAMPLE 2

The following graph shows the number of degrees awarded in the United States in a recent year:

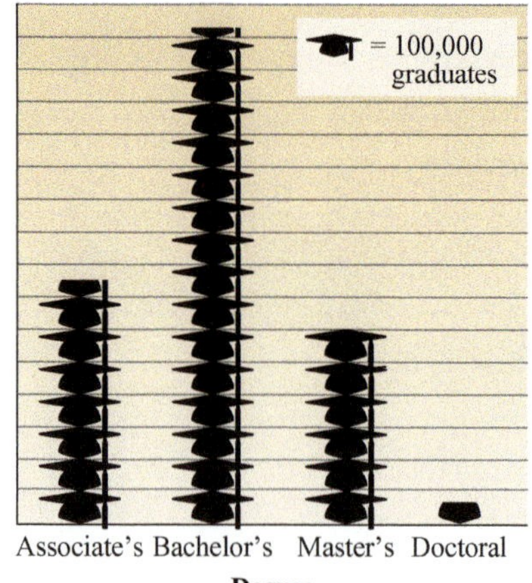

(*Source:* census.gov)

a. What does the symbol 🎓 mean in the key at the top of the graph?

b. About how many master's degrees were awarded?

c. About how many more bachelor's degrees than associate's degrees were awarded?

Solution

a. According to the key, the symbol 🎓 represents 100,000 graduates.

b. The number of master's degrees awarded was about 6(100,000), or 600,000.

c. About 15(100,000), or 1,500,000, bachelor's degrees were awarded, in contrast to about $7\frac{1}{2}(100{,}000)$, or 750,000, associate's degrees. So there were approximately 750,000 more bachelor's degrees awarded.

PRACTICE 2

The following pictograph shows the number of passengers who departed from or arrived at four busy U.S. airports in a recent year.

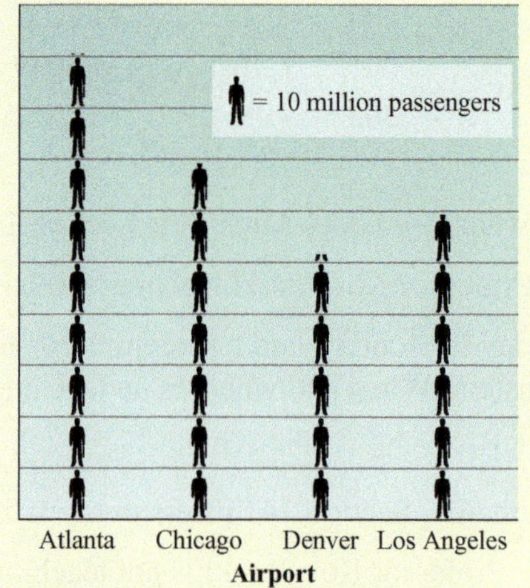

(*Source: The World Almanac 2010*)

a. What does the symbol 🚶 represent?

b. Which of the four airports was the busiest in terms of passengers?

c. Approximately how many passengers did the Los Angeles airport serve?

Bar Graphs

On a **bar graph**, quantities are represented by thin, parallel rectangles called bars. The length of each bar is proportional to the quantity that it represents.

On some graphs, the bars extend to the right. On other graphs, they extend upward or downward. Sometimes, bar lengths are labeled. Other times, they are read against an *axis*—a straight line parallel to the bars and similar to a number line.

Bar graphs are especially useful for making comparisons or contrasts among a few quantities, as the following example illustrates.

EXAMPLE 3

The graph shows the net income of Delta Air Lines Inc. in recent years.

a. What was the approximate net income of the company in fiscal year 2009?

b. About how much greater was the net income in fiscal year 2007 than in fiscal year 2006?

c. Describe the information shown by the graph.

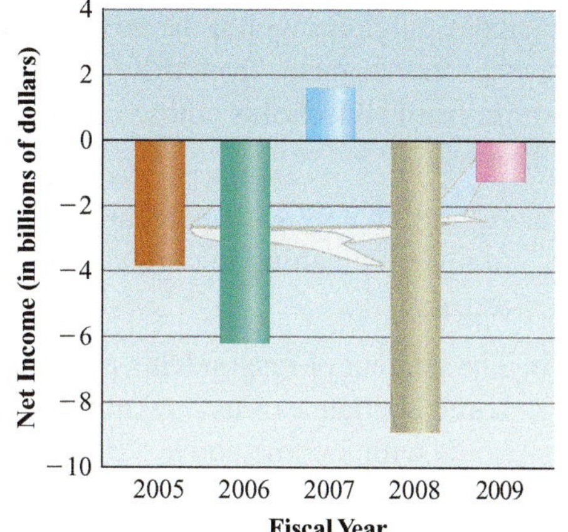

(*Source:* dailyfinance.com)

Solution

a. In 2009, the net income of the company was about −\$1.2 billion, that is, a loss of about \$1.2 billion.

b. In 2006, the net income was about −\$6.2 billion. The next year, it was about \$1.6 billion. So the net income for 2007 was approximately \$7.8 billion greater than in 2006.

c. The company operated at a profit in 2007. In other years, however, it operated at a loss, especially in 2008.

PRACTICE 3

The following graph shows the value of five leading farm commodities in California for a recent year.

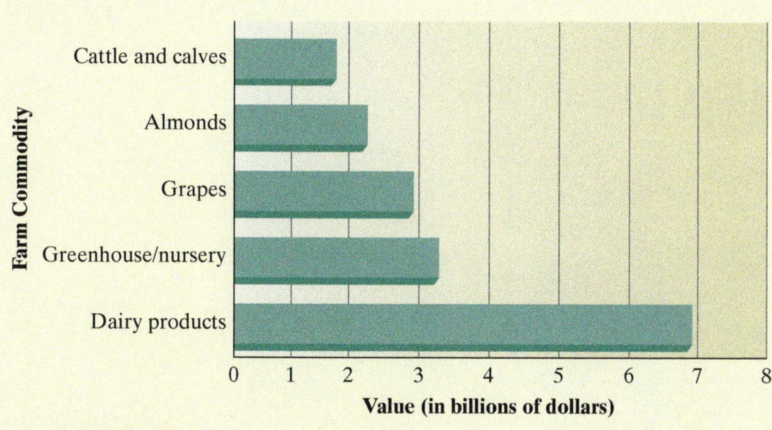

(*Source:* ers.usda.gov)

a. Which commodity had the greatest value?

b. What was the approximate value of grapes?

c. About how much greater was the value of greenhouse/nursery than almonds?

The next graph is an example of a *double-bar graph*. This kind of graph is used to compare two sets of data in various ways, as Example 4 illustrates.

EXAMPLE 4

The graph shows the percent of U.S. respondents according to a recent poll who were planning to make a major purchase within the next six months. The poll compares responses from the general population with those from households whose incomes exceeded $100 thousand.

a. Approximately what percent of the general population were planning to purchase home improvements/repairs?

b. The percent of respondents planning to purchase home appliances was how much greater from households with income above $100 thousand than from the general population?

Household income > $100K
General population
Percent of Respondents Planning to Make Purchase < 6 mo.
20%
18%
16%
14%
12%
10%
8%
6%
4%
2%
0%
Digital camera
Home appliance
Home improvements/ repairs
TV set
Car/Truck
Computer
Purchase

(*Source: Adweek*)

Solution

a. The percent of respondents from the general population who were planning to purchase home improvements/repairs was approximately 7%.

b. The percent of respondents planning to purchase home appliances was about 10% for households with income above $100 thousand and about 6% for the general population. So the difference was approximately 4%.

PRACTICE 4

The following graph shows the unemployment rate of U.S. youth (16- to 24-year-olds) versus adults (ages 25 and over) for five years.

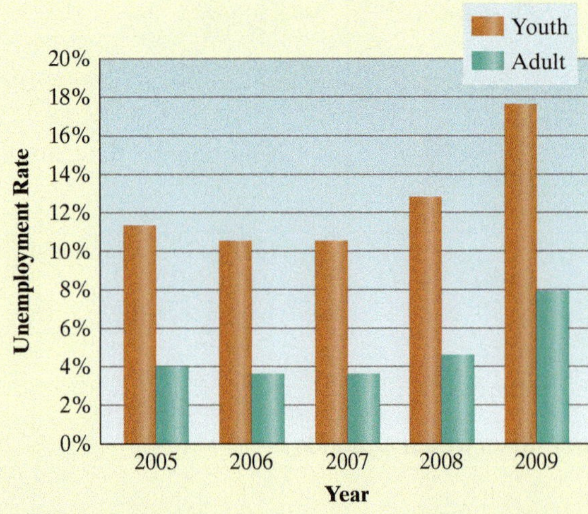

(*Source*: bls.gov)

a. For the year 2006, approximately what was the unemployment rate for youth?

b. For the year 2009, approximately what was the difference between the unemployment rate for youth and that for adults?

Histograms

Now, let's consider another kind of bar graph called a **histogram**. To understand what a histogram is, we consider an example.

Suppose that in a math class, 24 students take a final exam. The results are organized into a *frequency table*, as shown to the right. In the left column, note that the scores are grouped into *class intervals* (ranges of numbers) all of the same width and that these class intervals are written in increasing order. The right column shows the *class frequencies*, that is, the number of students who earned scores that fall into the class interval on the left. How would you have predicted the sum of the class frequencies in the right column?

Score (Class Interval)	Frequency (Class Frequency)
40–49	2
50–59	2
60–69	6
70–79	4
80–89	6
90–99	4

A histogram is a graph of a frequency table. In a histogram, adjacent bars touch. For each bar, the width represents a class interval, and the height stands for the corresponding class frequency. Consider the following histogram, which corresponds to the frequency table, as shown below.

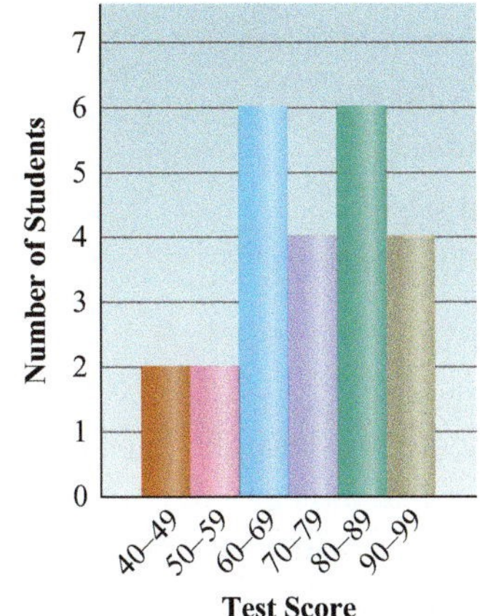

Note that according to the histogram, 4 students scored between 70 and 79, as given in the frequency table. Can you explain how the histogram shows that 10 students scored 80 or above?

EXAMPLE 5

The following graph shows the number of earthquakes worldwide in a recent year with magnitude 1.0 or greater.

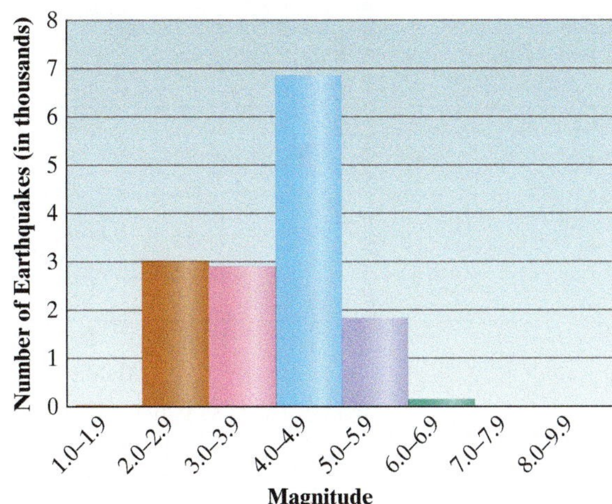

(*Source:* earthquake.usgs.gov)

PRACTICE 5

The following histogram summarizes the ages of the first 44 U.S. presidents at the time of their initial inauguration.

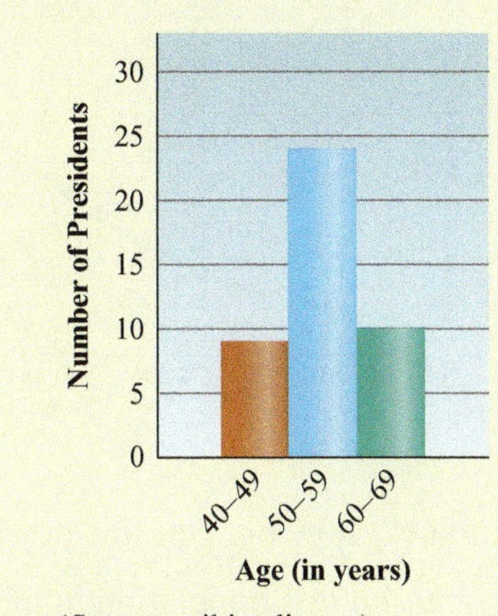

(*Source:* wikipedia.org)

EXAMPLE 5 (continued)

a. Approximately how many earthquakes were there with magnitude between 5.0 and 5.9?

b. To the nearest thousand, how many earthquakes were there with magnitude 4.9 or below?

c. What is the approximate ratio in simplified form of the number of earthquakes with magnitude 4.0–4.9 to those with magnitude 3.0–3.9?

Solution

a. The height of the bar for the class interval 5.0–5.9 is approximately 1,800. So there were about 1,800 earthquakes with magnitude between 5.0 and 5.9.

b. The number of earthquakes with magnitude 4.9 or below is the sum of the height of the bar for the class interval 4.0–4.9, as well as the height of all bars to the left. To the nearest thousand, this sum is 13,000.

c. The ratio of the number of earthquakes with magnitude 4.0–4.9 to those with magnitude 3.0–3.9 is about 7 to 3.

a. Approximately how many presidents were in their fifties when they were first inaugurated?

b. Were any presidents younger than 40 at their initial inauguration?

c. About how many presidents were 59 or younger when they were first inaugurated?

Line Graphs

On a **line graph**, points are connected by straight-line segments. The position of any point on a line graph is read against the vertical axis and the horizontal axis.

A line graph, also called a **broken-line graph**, is commonly used to highlight changes and trends over a period of time. Especially when we have data for many points in time, we are more likely to use a line graph than a bar graph.

EXAMPLE 6

The following graph shows the number of Americans 65 years of age and older from 1900 to 2010.

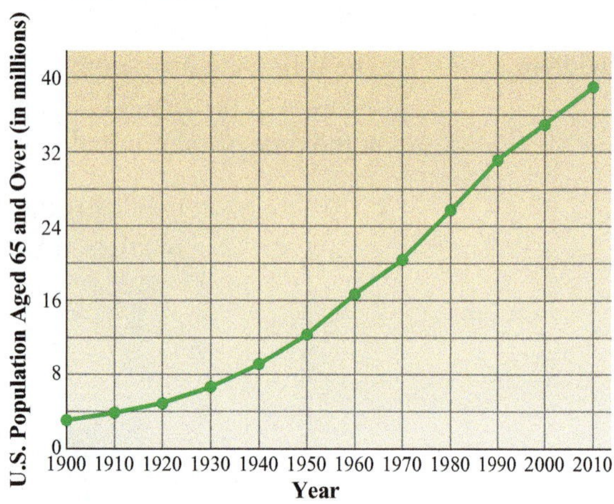

(*Source:* census.gov)

a. Approximately how big was this population in the year 2000?

b. In what year did this population number about 21 million?

c. In the year 2000, the U.S. population overall was approximately 4 times as large as it had been in the year 1900. Did the population shown in the graph grow more quickly?

PRACTICE 6

The following graph shows the mean temperatures in Chicago over a 30-year period for each month of the year.

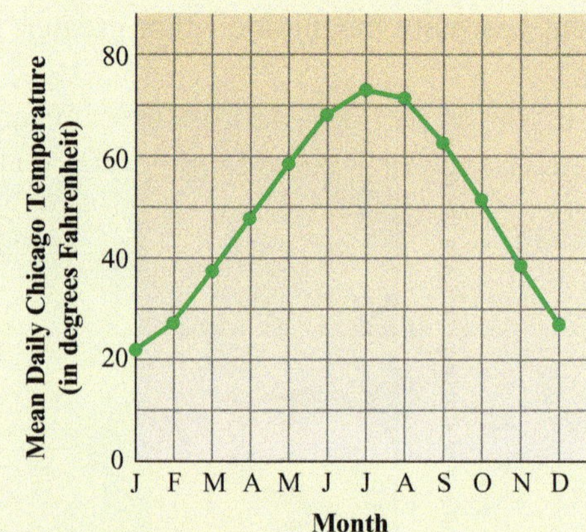

(*Source:* The U.S. National Climatic Data Center)

a. Which month in Chicago has the highest mean temperature?

Solution

a. In 2000, there were about 35 million Americans aged 65 and above.

b. There were approximately 21 million Americans aged 65 and above in the year 1970.

c. In 2000, the overall U.S. population was 4 times what it had been in 1900. But the population shown in the graph grew by a factor of about 10 and so grew more quickly.

b. Approximately what is the mean temperature in February?

c. What trend does the graph show?

Comparison line graphs show two or more changing quantities, as Example 7 illustrates.

EXAMPLE 7

The graph gives the number of male and female participants in high school athletic programs in the U.S. for selected years.

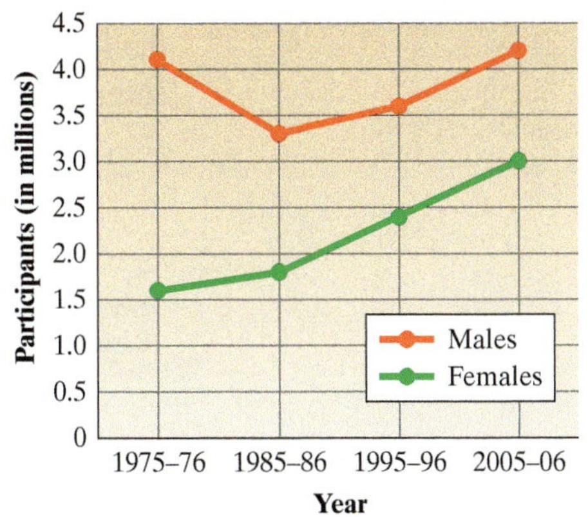

(*Source:* infoplease.com)

a. Estimate the number of females who participated in high school athletic programs in 2005–06.

b. Was the difference between the number of male and the number of female participants greater in 1985–86 or 1995–96?

c. Describe the trend that this graph shows.

Solution

a. In 2005–06, the number of female participants in high school athletic programs was approximately 3 million.

b. In 1985–86, the number of male participants was about 3.3 million and the number of female participants about 1.8 million. So the difference was about 1.5 million. By contrast in 1995–96, there were about 3.6 million males and 2.4 million females, with a difference of approximately 1.2 million. Since 1.5 million is larger than 1.2 million, the difference was greater in 1985–86.

c. Each year, there were more male participants than female participants. The number of female participants increased each year. The number of male participants decreased in 1985–86 and then increased in subsequent years.

PRACTICE 7

The following graph shows the mean precipitation for the cities of Seattle, Washington, and Orlando, Florida, in selected months.

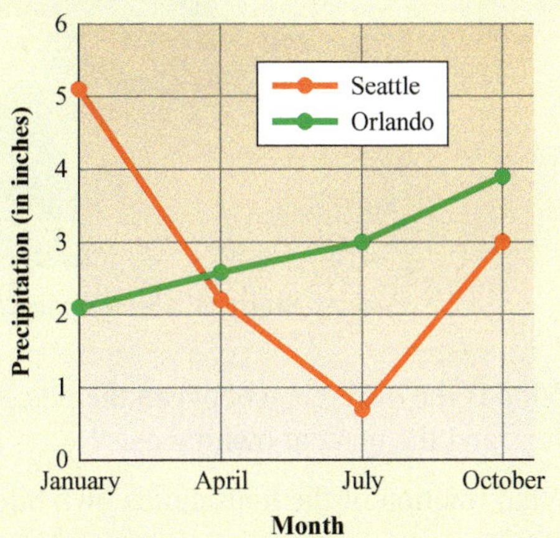

(*Source:* weatherbase.com)

a. On the average, does it rain more in Seattle or in Orlando during the month of October?

b. Approximately what is the average precipitation in Seattle for the month of January?

c. What trend does this graph show?

Circle Graphs

Circle graphs are commonly used to show how a whole amount—say, an entire budget or population—is broken into its parts. The graph resembles a pie (the whole amount) that has been cut into slices (the parts).

Each slice (or *sector*) is proportional in size to the part of the whole that it represents. Each slice is appropriately labeled with either its actual count or the percent of the whole that it represents.

The following example illustrates how to read and interpret the information given by a circle graph.

EXAMPLE 8

The following graph shows the percents of American households that own a single kind of pet:

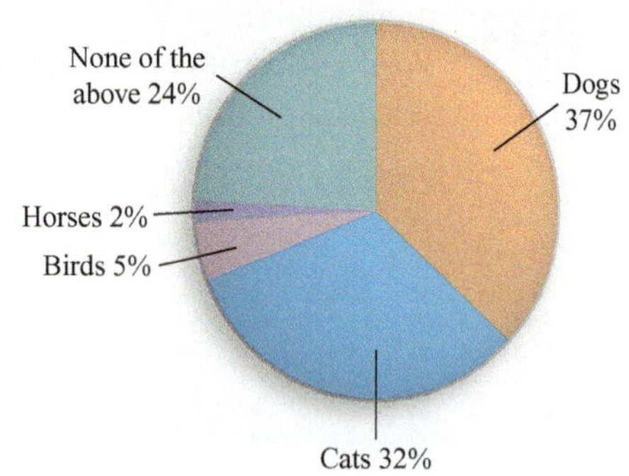

(*Source: Statistical Abstract of the United States, 2010*)

a. What is the difference between the percent of households owning dogs and the percent owning cats?

b. What fraction of the households own birds?

c. How many times as great is the percent of households owning cats as the percent owning horses?

Solution

a. Dog owners comprise 37% of the households, in contrast to 32% for cats. So the difference is 5%.

b. 5%, or 5 out of every 100, of the households own birds, which is equivalent to $\frac{5}{100}$, or $\frac{1}{20}$.

c. 32% of the households own cats, and 2% own horses. So the percent of households that owns cats is 16 times the percent that owns horses.

PRACTICE 8

A sample of U.S. e-mail users were asked how frequently they check their e-mail. The following graph summarizes their responses.

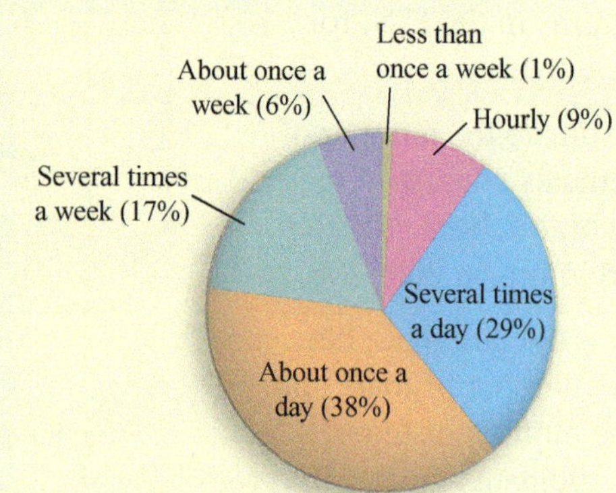

(*Source:* digitalcenter.org)

a. What fraction of the respondents check their e-mail about once a week?

b. What percent check their e-mail at least several times a day?

c. How many times as great is the percent of users who check their e-mail hourly than the percent who check their e-mail less than once a week?

8.2 Exercises

FOR EXTRA HELP MyMathLab  PRACTICE  WATCH  READ  REVIEW

Mathematically Speaking

Fill in each blank with the most appropriate term or phrase from the given list.

line graph	circle graph	heading
rows	columns	histogram
bar graph	graph	
pictograph	table	

1. A ________ is a rectangular display of data.

2. A ________ is a picture or diagram of data.

3. In a table, ________ run horizontally.

4. On a __________, images such as people, books or coins are used to represent quantities.

5. On a ________, quantities are represented by thin, parallel rectangles.

6. A __________ is a graph of a frequency table.

7. A __________ is commonly used to highlight changes and trends over a period of time.

8. A __________ resembles a pie (the whole) that has been cut into slices (the parts).

Applications

A *Read each table and solve.*

9. The following table shows how to determine a stockbroker's commission in a stock transaction. The commission depends on both the number of shares sold and the price per share.

	Number of Shares				
Price per Share	**100**	**200**	**300**	**400**	**500**
\$1–\$20	\$40	\$50	\$60	\$70	\$80
>\$20	\$40	\$60	\$80	\$90	\$100

a. What is the broker's commission on a sale of 300 shares of stock at \$15.75 a share?

b. What is the commission on a sale of 500 shares of stock at \$30 a share?

c. Will an investor pay her broker a lower commission if she sells 400 shares of stock in a single deal or 200 shares of stock in each of two deals?

10. The table shows the federal income tax schedule in a recent year for single filers.

If taxable income is over	But not over	The tax is
\$0	\$8,350	10% of the amount over \$0
\$8,350	\$33,950	\$835 plus 15% of the amount over \$8,350
\$33,950	\$82,250	\$4,675 plus 25% of the amount over \$33,950
\$82,250	\$171,550	\$16,750 plus 28% of the amount over \$82,250
\$171,550	\$372,950	\$41,754 plus 33% of the amount over \$171,550
\$372,950	No limit	\$108,216 plus 33% of the amount over \$372,950

(*Source:* irs.gov)

a. What is the tax for a person whose taxable income was \$33,950?

b. Compute the tax for a person whose taxable income is \$27,000.

11. An atlas contains a table of road distances (in miles) between various U.S. cities.

	Los Angeles	Chicago	Houston
Los Angeles	—	2,112	1,556
Chicago	2,112	—	1,092
Houston	1,556	1,092	—

(*Source: Rand McNally Road Atlas*)

a. According to this table, how far is Los Angeles from Houston?

b. Chicago is how much closer to Houston than to Los Angeles?

c. What is the meaning of the blanks in the table?

12. At a college, the math courses that students take depend on their scores on the Arithmetic and Algebra Placement Test. Test scores get translated into course placements as follows.

		Arithmetic Score		
		0–10	**11–15**	**16–28**
Algebra Score	**0–10**	Math 1 and then Math 3	Math 2 and then Math 3	Math 3
	11–15	Math 1 and then Math 3	Math 4	Math 3
	16–20	Math 1	Math 2	Math 5

a. If you score 12 in arithmetic and 8 in algebra, what should you take?

b. Suppose that you score 8 in arithmetic and 12 in algebra. What is your placement?

c. What must you score to take Math 5?

B *Read each graph and solve.*

13. The pictograph shows the projected employment for five major U.S. industries in the year 2018.

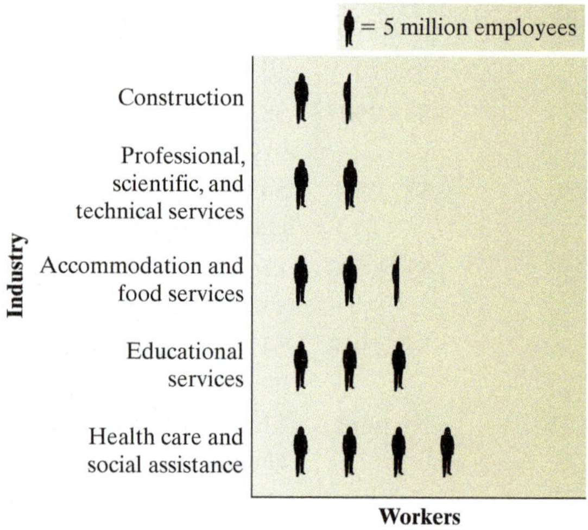

(*Source:* bls.gov)

a. Which of the industries will employ the most people in 2018?

b. Approximately how many workers will be employed in the accommodation and food services industry in 2018?

14. The following pictograph shows the percent of adults by age group who have accessed the internet wirelessly, according to a recent survey.

(*Source:* pewinternet.org)

a. Approximately what percent of 30–49 year olds have accessed the internet wirelessly?

b. For those adults who accessed the internet wirelessly, estimate the ratio of the percent of those 65 and over to the percent of 50–64 year olds.

15. In chemistry, the pH scale measures how acidic or basic a solution is. A solution with a pH of 7 is considered neutral. Solutions with a pH less than 7 are acids, and solutions with a pH greater than 7 are bases. The graph below shows the pH of various solutions.

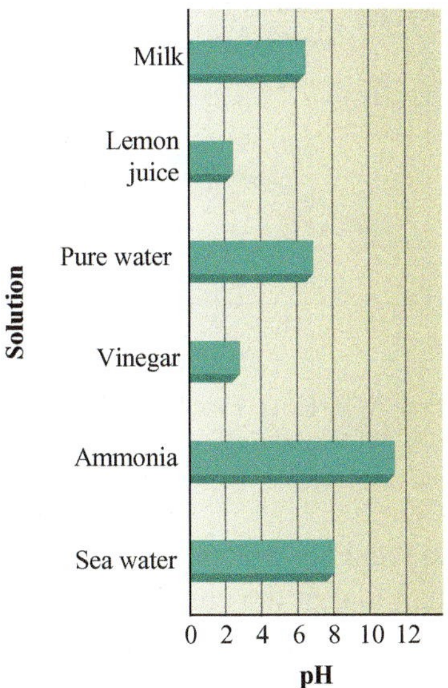

a. Which of the solutions are acids?

b. Approximate the pH of sea water.

c. Which solution is neutral?

16. The following graph shows the percent changes in the U.S. consumer price index (CPI) for energy types between November and December 2009 and between May and June 2010.

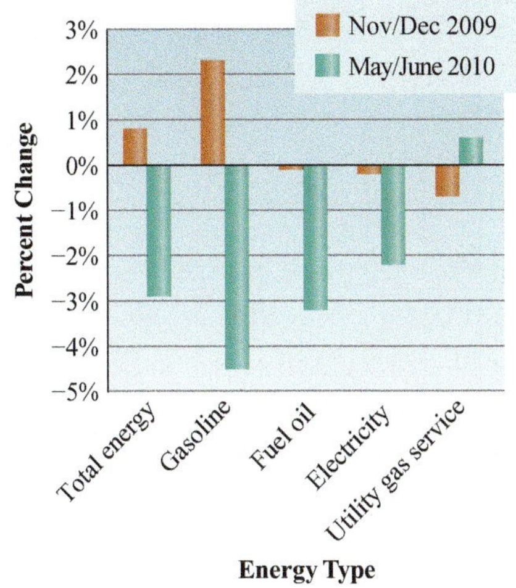

(*Source:* bls.gov)

a. Estimate the percent increase in the CPI for total energy between November and December 2009.

b. Between May and June 2010, the CPI for which type of energy increased?

c. According to the graph, which type of energy between which consecutive months had the greatest percent increase in CPI?

17. The following histogram shows the waiting time for applicants while at a passport office:

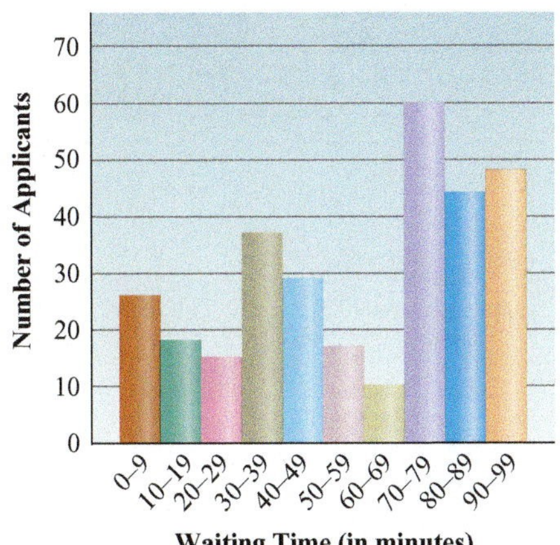

a. Did more applicants wait less than 10 minutes or more than 89 minutes?

b. Approximately how many applicants waited 80 minutes or more?

c. About how many applicants waited between 60 and 79 minutes?

18. The histogram shows the number of 2009 college-bound high school seniors taking the SAT by family income.

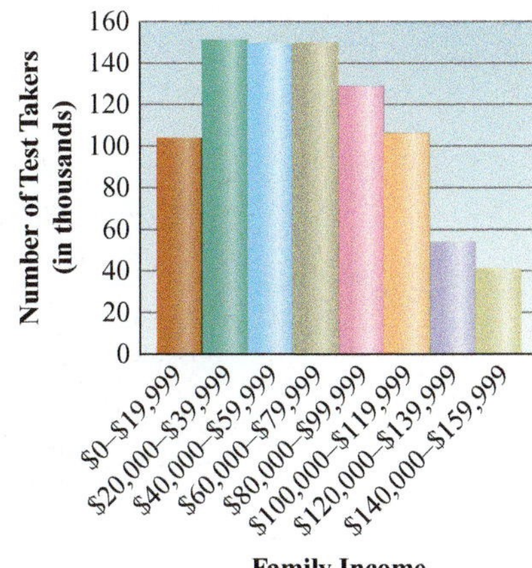

(*Source:* professionals.collegeboard.com)

a. About how many 2009 college-bound seniors with family incomes less than $60,000 took the SAT?

b. About how many 2009 college-bound seniors with family incomes between $80,000 and $119,999 took the SAT?

c. Describe the trend in this histogram.

19. The following graph shows the number of cell phone subscribers (in millions) for recent years.

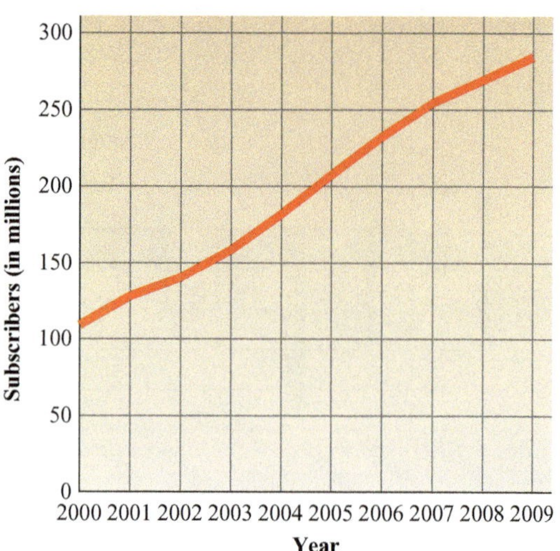

(*Source:* CTIA)

a. About how many subscribers were there in the year 2008?

b. In what year did the number of subscribers first exceed 200 million?

c. Describe the trend shown in the graph.

20. A human child and a chimp were raised together. Scientists graphed the number of words that the child and the chimp understood at different ages.

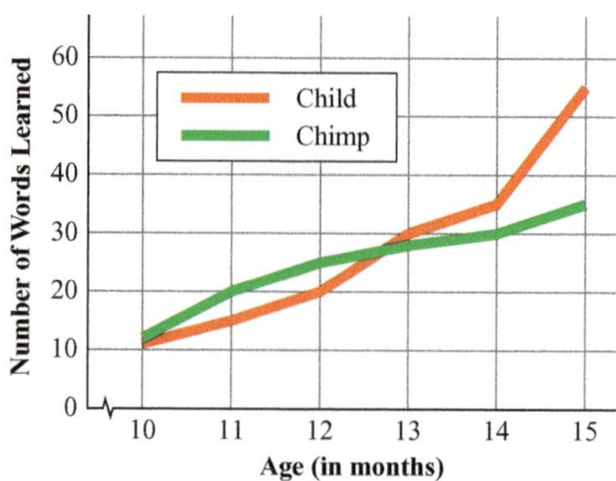

(*Sources:* A. H. Kritz, *Problem Solving in the Sciences;* W. N. Kellogg and L. A. Kellogg, *The Ape and the Child*)

a. At about what age was the child's vocabulary first better than that of the chimp?

b. At age 15 months, about how many more words did the child know than the chimp?

21. The graph shows the percent of Americans affiliated with various political parties, according to a recent poll.

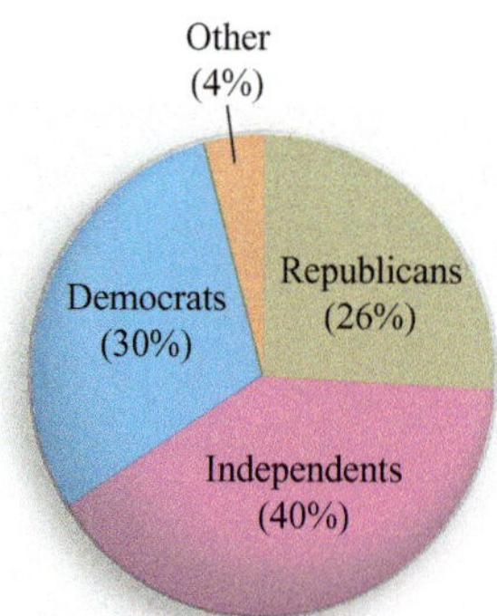

(*Source:* gallup.com)

a. According to this graph, are there more Americans who consider themselves Democrats or Republicans?

b. What fraction of Americans consider themselves Independent?

c. If 15,000 people were polled, how many of these considered themselves neither Republicans, Democrats, nor Independents?

22. The following circle graph shows the percent of social security recipients receiving various types of payment in a recent month:

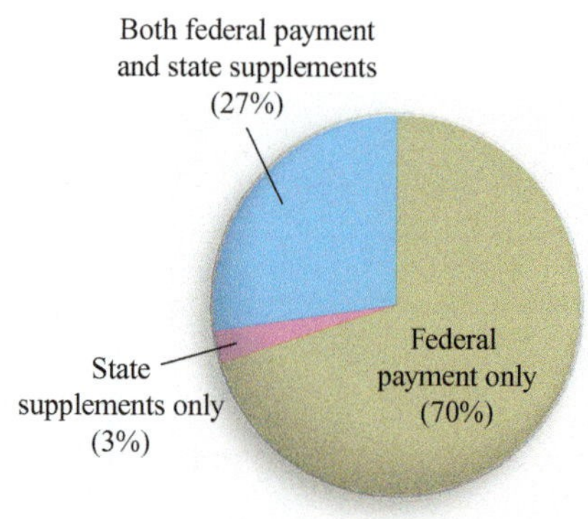

(*Source:* ssa.gov)

a. What percent of the recipients received both federal payment and state supplements?

b. What percent of the recipients received state supplements?

c. Is it true that the number of recipients who received federal payment only was more than ten times as great as the number of recipients who received state supplements only?

• Check your answers on page A-11.

MINDStretchers

Technology

1. On a computer, use a spreadsheet program to draw a circle graph that represents the following data, showing the percent of men and the percent of women at a party:

Gender	Number of Guests at a Party
Men	12
Women	8

Writing

2. A *stacked bar graph* not only allows comparisons between quantities but also shows how each quantity is divided into parts. For example, the following stacked bar graph deals, for each month of a year, with a company's total revenue, which is generated partly in store and partly online. In a few sentences, describe some of the main trends that this graph implies.

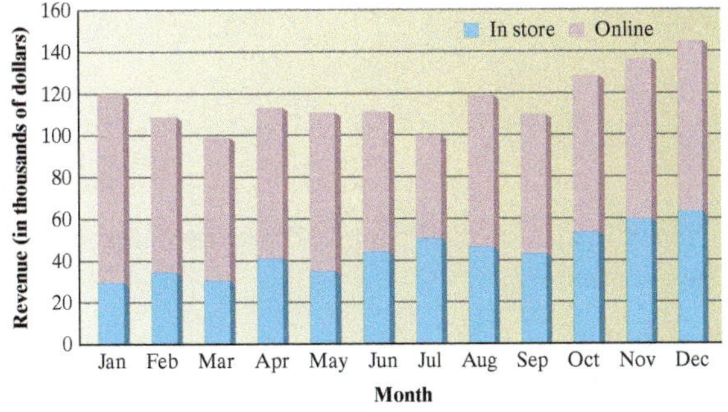

Mathematical Reasoning

3. Consider the following two bar graphs. They both represent the same data, namely the percent of voters in the presidential election of 2008 who voted either Democratic or Republican. (*Source:* wikipedia.org)

   a. Explain the difference between the impression that the top graph gives and the impression the bottom graph gives.

   b. How is this difference in impression achieved?

   c. Why would someone prefer to give one impression more than the other?

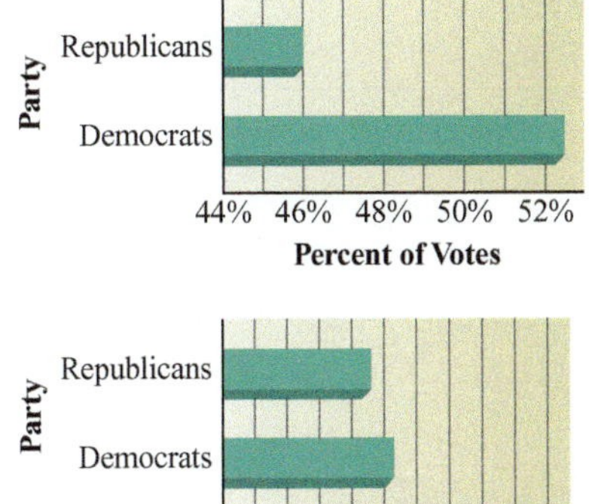

Key Concepts and Skills

CONCEPT SKILL

Concept/Skill	Description	Example
[8.1] **Mean**	Given a set of numbers, the sum of the numbers divided by however many numbers are in the set.	For 0, 0, 1, 3, and 5, the mean is: $\frac{0+0+1+3+5}{5} = \frac{9}{5} = 1.8$
[8.1] **Median**	Given a set of numbers arranged in numerical order, the number in the middle. If there are two numbers in the middle, the mean of the two middle numbers.	For 0, 0, 1, 3, and 5, the median is 1.
[8.1] **Mode**	Given a set of numbers, the number (or numbers) occurring most frequently in the set.	For 0, 0, 1, 3, and 5, the mode is 0.
[8.1] **Range**	Given a set of numbers, the difference between the largest and the smallest number in the set of numbers.	For 0, 0, 1, 3, and 5, the range is $5 - 0$, or 5.
[8.2] **Table**	A rectangular display of data.	Column ↓; Row → | Dems. | Reps. Men | 37 | 45 Women | 14 | 15
[8.2] **Pictograph**	A graph in which images such as people, books, or coins are used to represent the quantities.	= 10 robots; A B C D; Factory
[8.2] **Bar graph**	A graph in which quantities are represented by thin, parallel rectangles called bars. The length of each bar is proportional to the quantity that it represents.	Decade Ending in: 1990, 1970, 1950, 1930, 1910; Fatalities: 0, 2,000, 4,000, 6,000, 8,000
[8.2] **Histogram**	A graph of a frequency table.	Number of Students: 0–7; Test Score: 40–49, 50–59, 60–69, 70–79, 80–89, 90–99

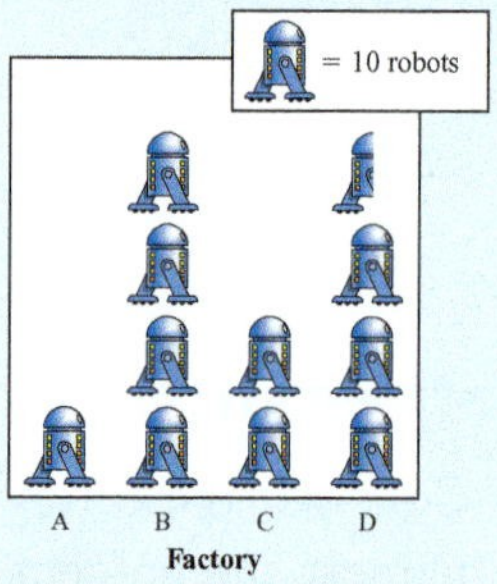

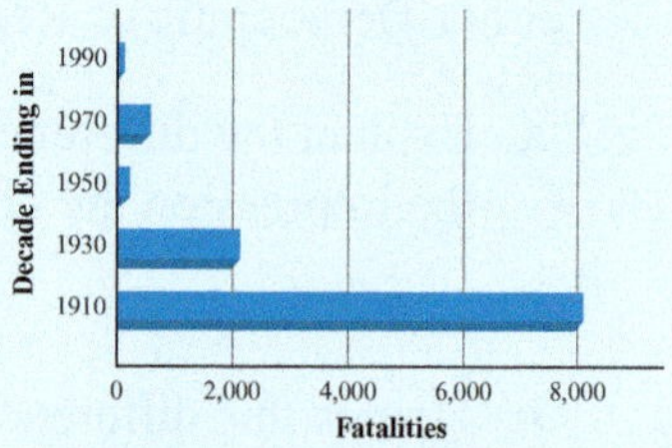

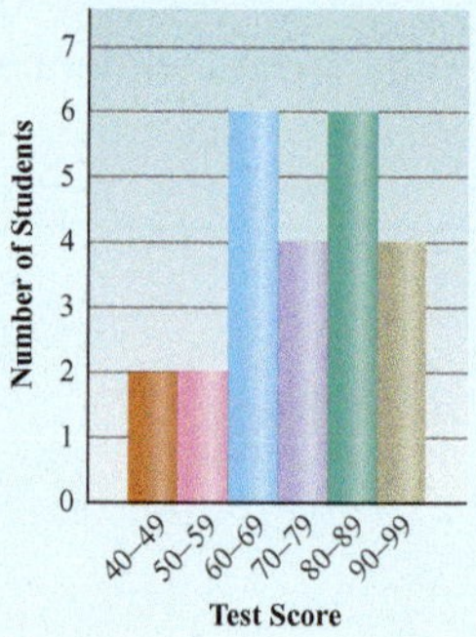

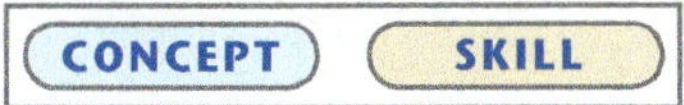

Concept/Skill	Description	Example
[8.2] **Line graph**	A graph in which points are connected by straight-line segments. The position of any point on a line graph is read against the vertical axis and the horizontal axis.	Number of Cases of Measles (in thousands) 0, 5, 10, 15, 20, 25, 30 1986 1987 1988 1989 1990 1991 1992 1993 Year
[8.2] **Circle graph**	A graph that resembles a pie (a whole amount) that has been cut into slices (the parts).	Other $197 Europe $364 Western Hemisphere $283 Japan $92 Canada $132

CHAPTER 8 Review Exercises

Say Why

Fill in each blank.

1. A set of numbers ________ have more than one mean
   can/cannot
   because ______________________________________
   ______________________________________.

2. A set of numbers ________ have more than one mode
   can/cannot
   because ______________________________________
   ______________________________________.

3. The range in the set 1, 4, 4, 4 ________ the same as
   is/is not
   the range in the set 401, 404, 404, 404 because
   ____
   ______________________________________.

4. A table __________ provide more exact information
   does/does not
   than a graph because ______________________
   ______________________________________.

5. A pictograph ________ typically used to represent
   is/is not
   small differences because ______________________
   ______________________________________.

6. A circle graph ________ used to make comparisons or
   is/is not
   contrasts among quantities because ___________
   ______________________________________.

[8.1] *Compute the desired statistic for each list of numbers.*

7.	List of Numbers	Mean	Median	Mode	Range
a.	6, 7, 4, 10, 4, 5, 6, 8, 7, 4, 5				
b.	1, 3, 4, 4, 2, 3, 1, 4, 5, 1				

Mixed Applications

8. The following table shows how long the first five American presidents and their wives lived:

President	Age	President's Wife	Age
George Washington	67	Martha Washington	70
John Adams	90	Abigail Adams	74
Thomas Jefferson	83	Martha Jefferson	34
James Madison	85	Dolley Madison	81
James Monroe	73	Eliza Monroe	62

(*Source: Presidents, First Ladies, and Vice Presidents*)

   a. Using the median as the average, did the husbands or the wives live longer?

   b. By how many years?

   c. What was the range of the ages of the presidents?

9. The median age of National Public Radio (NPR) listeners is 55 years. (*Source:* npr.org)

   a. Explain what this statement means.

   b. By how many years is the age of a 25-year-old listener above or below the median?

10. A soda machine is considered reliable if the range of the amounts of soda that it dispenses is less than 2 fluid ounces. In 10 tries, a particular machine dispensed the following amounts (in fluid ounces).

8.1 7.8 8.6 8.1 8.4 7.8 8 7.7 6.9 8.4

Is the machine reliable? Explain.

11. In order for a small local zoo to be profitable, it must average a minimum of at least 12,500 paying visitors per month. For the past 5 months, the number of paying visitors was as follows:

14,912 9,873 11,025 15,207 14,528

Using the mean as the average, determine whether the zoo was profitable.

12. In a recent year, the amount of cargo (in tons) that the five busiest U.S. ports handled was as follows:

	Cargo (in tons)		
Port	**Total**	**Domestic**	**Foreign**
Port of South Louisiana, LA	229,040,085	121,549,984	107,490,101
Houston, TX	216,064,325	70,721,886	145,342,439
New York, NY and NJ	157,202,043	65,780,088	91,421,955
Long Beach, CA	85,939,895	15,383,519	70,556,376
Beaumont, TX	81,383,531	24,339,637	57,043,894

(*Source:* U.S. Army Corps of Engineers)

a. To the nearest 10,000 tons, how much domestic cargo did Houston handle?

b. Which of these five ports handled the least foreign cargo?

c. To the nearest 10,000 tons, how much more domestic cargo did New York handle than Beaumont?

d. Which of these ports handled more foreign than domestic cargo?

13. The following pictograph shows the number of monthly domestic flights in the U.S. during recent years.

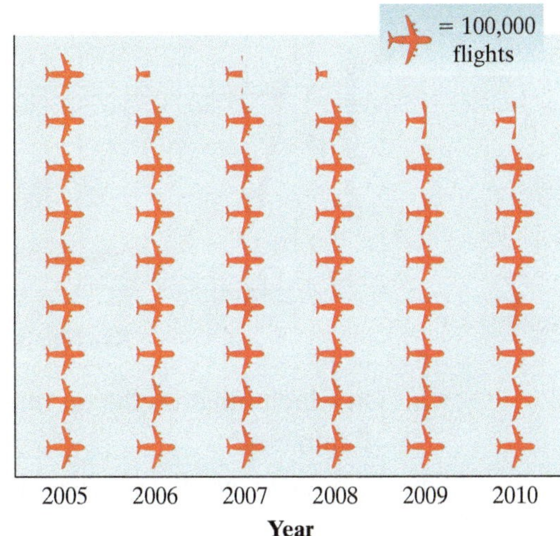

(*Source: Newsweek*)

a. About how many monthly domestic flights were there in 2007?

b. In which year shown were there the most monthly domestic flights?

c. About how many more monthly domestic flights were there in 2008 than in 2010?

14. The following graph shows the seven largest libraries in the United States and the number of volumes held by each.

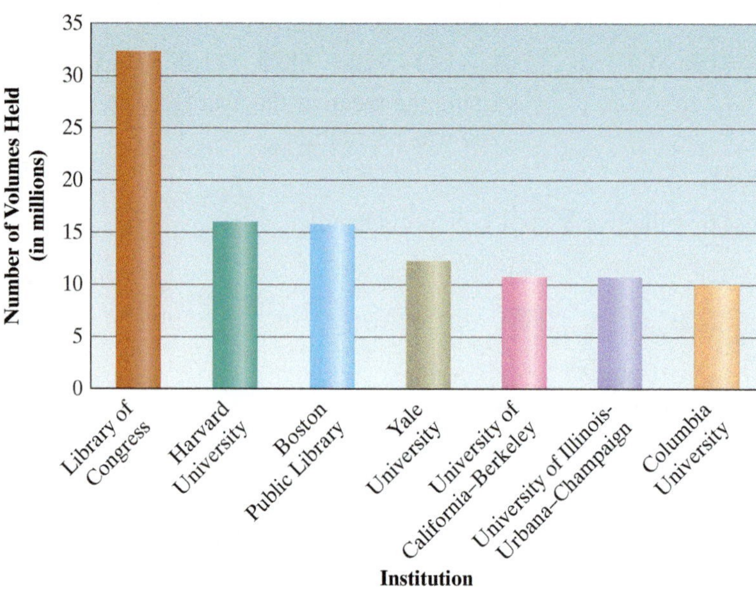

(*Source:* ala.org)

a. Approximately how many volumes are held in the library at Columbia University?
b. About how many more volumes does the Boston Public Library hold than the University of California–Berkeley?
c. Which libraries appear to hold half the number of volumes that the Library of Congress holds?

15. The following histogram displays the number of U.S. licensed drivers in a recent year, aged 20 through 84, according to their age.

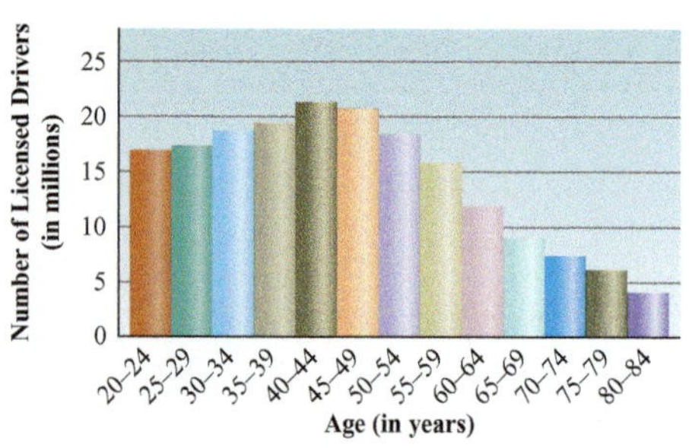

(*Source:* fhwa.dot.gov)

a. About how many licensed drivers were between 35 and 49 years of age?
b. If the total number of licensed drivers was about 200 million, approximately what percent of these were between the ages of 35 and 39?
c. Describe the trend in this graph.

16. Consider the following learning curve that shows how long a rat running through a maze takes on each run:

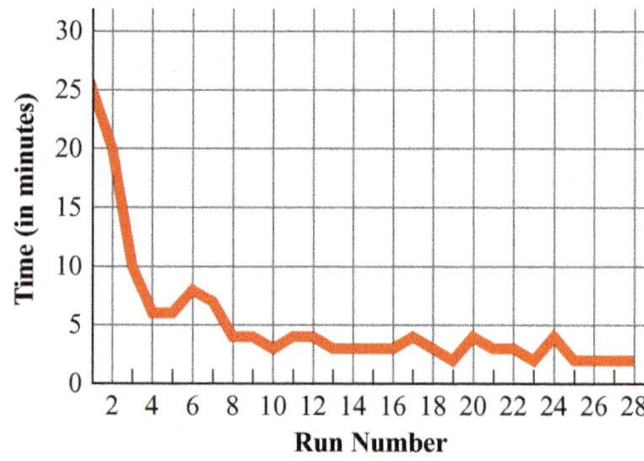

a. On which run does the rat run through the maze in 10 minutes?
b. How long does the rat take on the tenth run?
c. What general conclusion can you draw from this learning curve?

17. The graph shows the health care expenditures from out-of-pocket and insurance sources.

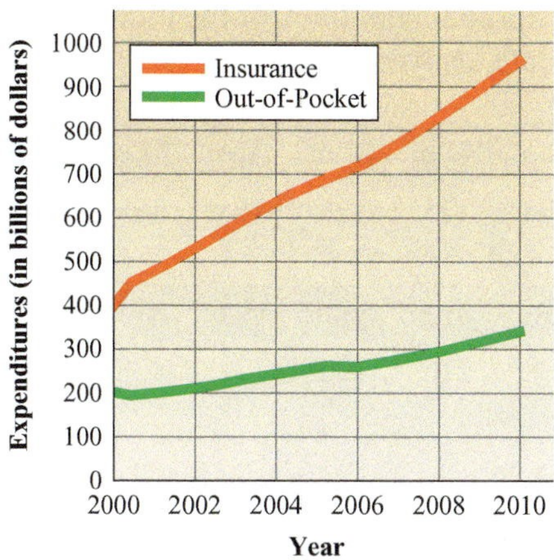

(*Source:* Centers for Medicare and Medicaid Services)

a. In what year were the insurance expenditures approximately $600 billion?

b. In the year 2000, what was the approximate ratio of out-of-pocket expenditures to insurance expenditures?

c. Express in words the trend the graph is illustrating.

18. The following graph shows the number of organ transplants performed in the United States for a recent year.

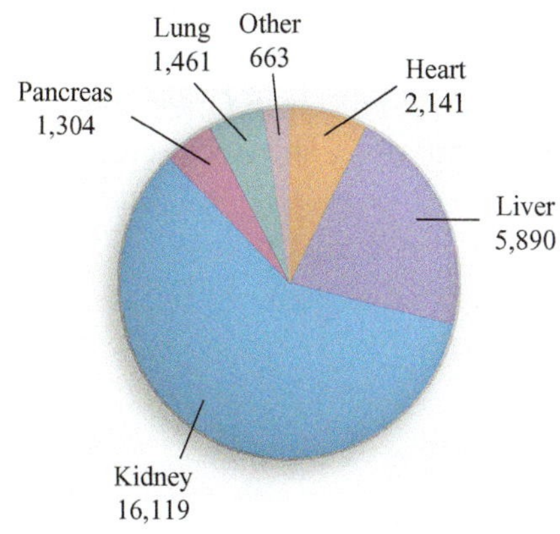

(*Source:* ustransplant.org)

a. How many transplants in all were performed?

b. How many more liver transplants than lung transplants were performed?

c. What percent of all transplants performed were kidney transplants, to the nearest whole percent?

19. In a college algebra course, each of two tests counts 20% of a student's course average, whereas the final exam counts 60% of the course average. Find the course average of a student who scored 80 and 90 on her tests and 70 on the final exam.

20. A survey was conducted on thousands of households throughout the United States comparing the views of internet users with non-users on whether communication technology has made the world a better place. The graph below summarizes their responses.

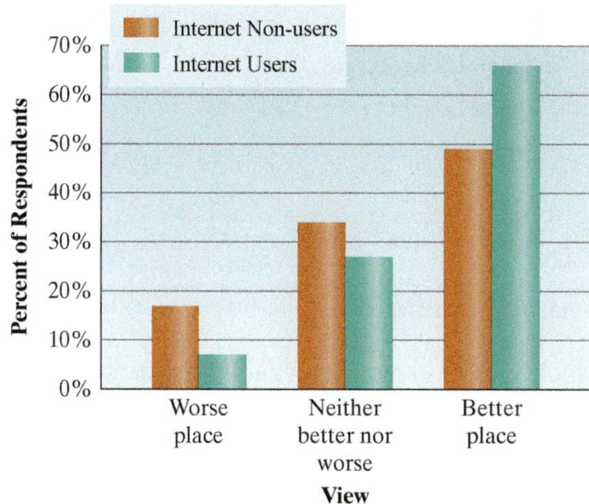

(*Source:* digitalcenter.org)

Did a greater percent of Internet users or non-users indicate that communication technology has made the world a better place?

• Check your answers on page A-12.

CHAPTER 8 Posttest

FOR EXTRA HELP

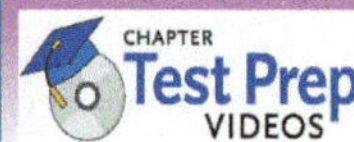

The Chapter Test Prep Videos with test solutions are available on DVD, in MyMathLab, and on YouTube (search "AkstDevMath" and click on "Channels").

To see whether you have mastered the topics in this chapter, take this test.

1. At the end of your first term, your grades were as follows: English Composition I (4 credits)—A, College Skills and Freshman Orientation (2 credits)—B, History (3 credits)—C, and Art (1 credit)—A. If A = 4, B = 3, C = 2, D = 1, and F = 0, calculate your GPA.

2. A local hospital kept track of the number of babies born each month last year.

Jan	Feb	Mar	Apr	May	June
106	115	138	165	189	202
July	**Aug**	**Sept**	**Oct**	**Nov**	**Dec**
208	216	190	172	138	105

What was the mean number of babies born at the hospital in a month last year?

3. A math instructor records the following exam scores for students in her Math 110 course:

86 78 96 82 74 56 72 76
88 60 48 76 100 98 64 80

What was the median exam score?

4. The weights (in pounds) of the 25 players on the active roster for the San Diego Padres during a recent season were as follows:

195 250 200 200 210 200 225 215 200 240
205 230 205 200 175 225 190 210 205 210
195 220 200 160 210

(*Source:* mlb.com)

What are the mode and range of the weights?

5. Shown in the following table is the fuel economy in miles per gallon (mpg) of the six most fuel-efficient vehicles for a recent year.

Vehicle	City (mpg)	Highway (mpg)
Toyota Prius	51	48
Honda Civic Hybrid	40	45
Honda Insight	40	43
Ford Fusion Hybrid	41	36
Smart for Two	33	41
Nissan Altima Hybrid	35	33

(*Source:* thedailygreen.com)

Which vehicles have better fuel efficiency for city driving than for highway driving?

6. A sample of adults were asked how they try to avoid catching the flu. The following graph summarizes their responses.

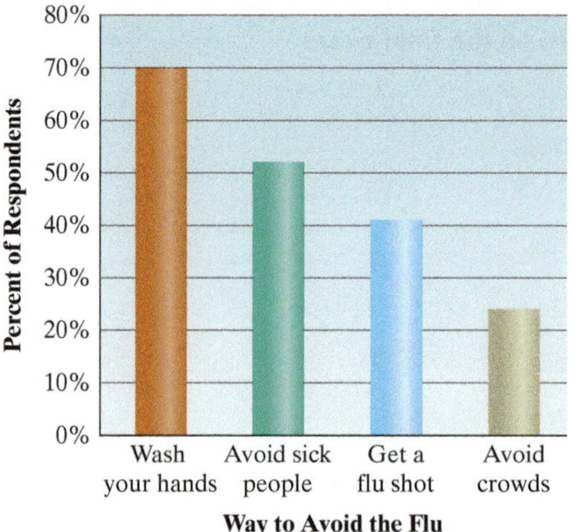

(*Source:* usatoday.com)

Approximately how much greater was the percent of respondents who avoid the flu by washing their hands than by getting a flu shot?

7. The following graph shows the number of U.S. adult victims of identity fraud for the years 2003 through 2009.

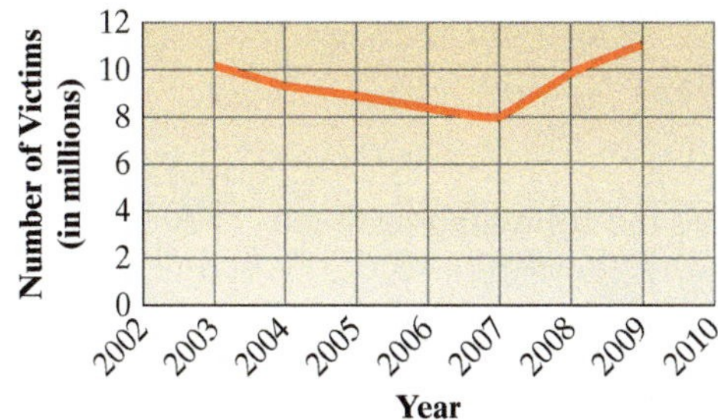

(*Source:* identitytheftassistance.org)

About how many more victims of identity fraud were there in 2009 than in 2007?

8. The graph below shows music sales for the years 2000–2009, including album sales, digital track sales, and overall music purchases.

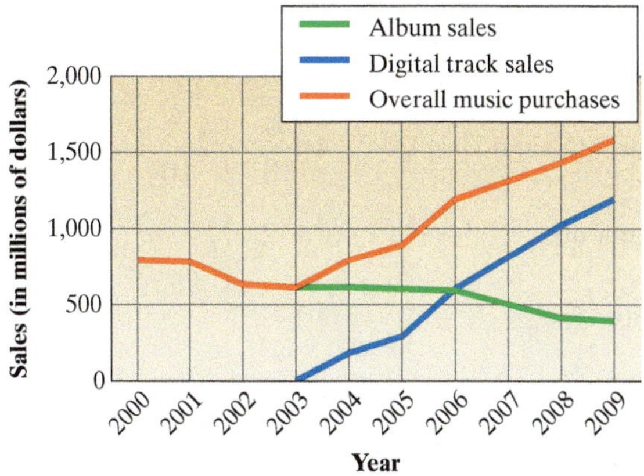

(*Source:* narm.com)

a. In what year were digital track sales first higher than album sales?

b. Describe the relationship between album, digital track, and overall music purchases. Use this relationship to describe overall music purchases for the year 2006.

9. The following circle graph shows the breakdown of the U.S. coastline (in miles) by region:

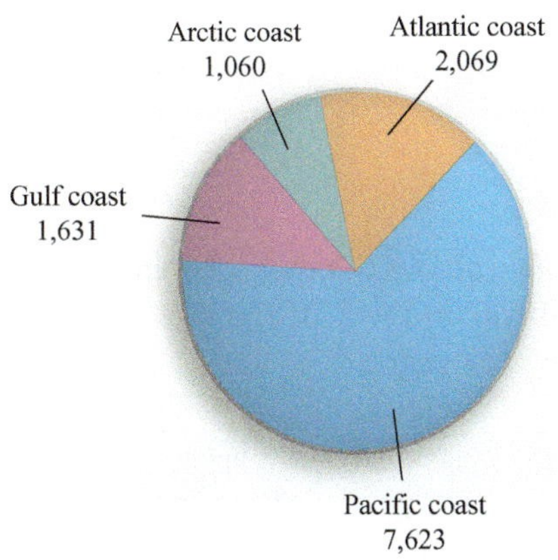

(*Source:* National Oceanic and Atmospheric Administration, U.S. Department of Commerce)

To the nearest 10%, what percent of the U.S. coastline is on the Pacific coast?

10. The following graph shows the number of Nobel prizes awarded in the sciences for research conducted in the United States and in the United Kingdom.

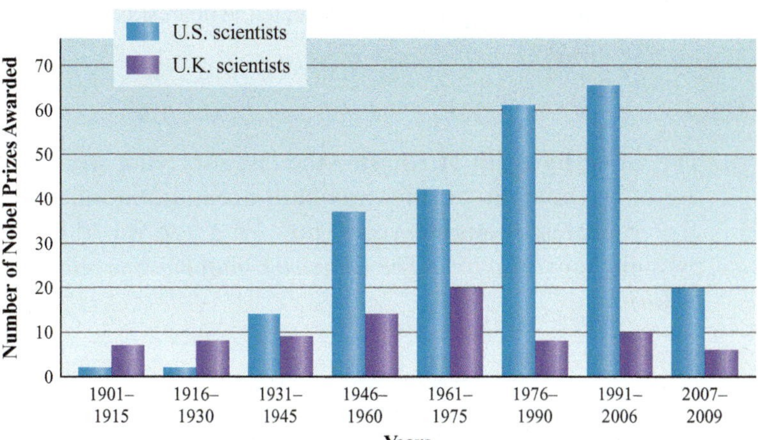

(*Source:* wikipedia.org)

In which period of time did scientists in the United States receive about twice as many prizes as scientists in the United Kingdom?

• Check your answers on page A-12.

Cumulative Review Exercises

To help you review, solve the following.

1. Simplify: $3 + 4(7 - 2)$

2. Simplify: $\frac{20}{25}$

3. Combine and simplify: $4\frac{17}{100} - 2\frac{3}{10}$

4. Estimate the following product: $\left(8\frac{1}{10}\right)\left(4\frac{9}{10}\right)$

5. Compute: $7.34 - (2.46 + 2.07)$

6. $(3.01)(1{,}000) = ?$

7. Solve for x: $x - 7\frac{1}{4} = 10$

8. Is 24 a solution to the equation $\frac{x}{0.4} = 6$?

9. Solve for n: $\frac{7}{10} = \frac{n}{30}$

10. Which is larger, 462% or $4\frac{5}{8}$?

11. What is $12\frac{1}{2}\%$ of 16?

12. What are the sign and the absolute value of $-\frac{7}{4}$?

13. Find the quotient and simplify: $-8 \div \left(\frac{4}{9}\right)$

14. Find the mean, median, mode(s), and range of 3.6, 3.9, 3.3, 3.8, and 3.9.

15. Recently in the United States, there were 14.9 million students enrolled in public high schools and 1.4 million students enrolled in private high schools. How many times the second enrollment is the first, rounded to the nearest tenth? (*Source:* National Center for Educational Statistics)

16. There were 480 unemployed people for 60 job openings. How many jobless people were there per opening?

17. Between July 2009 and July 2010, the number of full-time workers in the United States dropped from 114 million to 112 million. Find the percent decrease, to the nearest percent. (*Source:* bls.gov)

Solve.

18. The Great Pyramid of Khufu—the last surviving wonder of the ancient world—was built around 2680 B.C. To the nearest thousand years, how long ago was this pyramid built? (*Source: The Concise Columbia Encyclopedia*)

19. The following table shows the maximum number of home runs hit by a player in each of the two major baseball leagues during recent years.

Year	National League	American League
2005	48	51
2006	54	58
2007	54	50
2008	37	48
2009	39	47

(*Source:* baseball-almanac.com)

Was the median number of home runs higher in the National League or in the American League?

20. Many Americans vacation in Australia. The graph below shows the age and gender of visitors to Australia from the United States for a recent year.

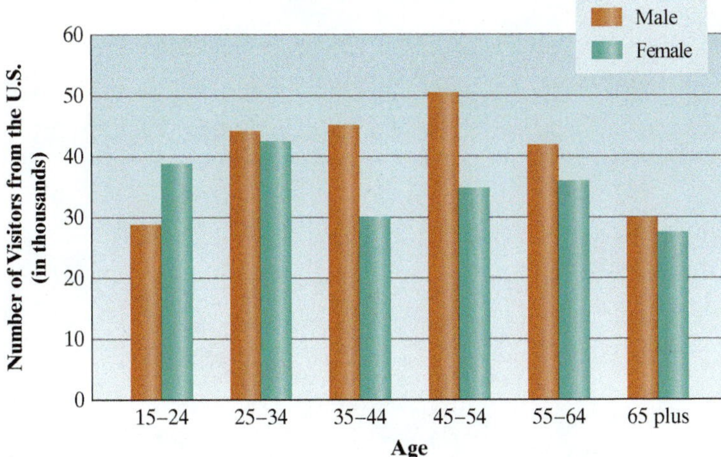

(*Source:* ret.gov.au)

In terms of gender and age, which was the largest group of visitors from the U.S. to Australia?

• Check your answers on page A-12.

CHAPTER 9

Measurement and Units

Units and Space Exploration

The *Mars Climate Orbiter* was a spacecraft designed to study Martian weather and launched by the National Aeronautics and Space Administration (NASA) in 1998. The craft's mission was to orbit the red planet at an altitude of between 140 and 150 kilometers (460,000–500,000 feet). However, because of a navigational error, the $125 million craft dropped down to 57 kilometers (190,000 feet) above the surface. At this low altitude, the orbiter overheated and was torn apart by atmospheric stresses and friction. The disaster occurred because NASA, which uses metric units, misinterpreted the data sent to them by engineers who had built the craft's thrusters. Instead of metric units, the engineers, who were working for a private contractor, had used U.S. customary units.

(*Sources:* spaceref.com and cnn.com)

CHAPTER 9 PRETEST

To see if you have already mastered the topics in this chapter, take this test.

Solve.

1. 6 in. = _____ ft

2. 6 ft = _____ in.

3. 3 pt = _____ fl oz

4. Compute: 5 hr 2 min − 1 hr 10 min

5. Which of the following units is a measure of length?

a. a gram **b.** a meter **c.** a liter **d.** a second

6. The width of a dime is about _____.

a. a meter **b.** a kilometer
c. a millimeter **d.** a centimeter

7. 3.5 kg = _____ g

8. 2,100 mm = _____ m

9. 8,000 mg = _____ g

10. 1.5 mm = ________ km

11. 2,000 mL = _____ kL

12. Round to the nearest tenth: 10 lb ≈ _____ kg

13. 78 in. ≈ _____ cm

14. Express 2 L in quarts.

15. In a recent year, Americans generated about 250 million tons of trash of which 83 million tons were either recycled or composted. How many pounds of trash were neither recycled nor composted? (*Source:* epa.gov)

16. A medicated shampoo to fight psoriasis comes in bottles with capacity 200 mL and 1 L. The larger bottle contains how many times the amount of shampoo that the smaller contains?

17. The weight of a standard gold bar held in the U.S. Bullion Depository in Fort Knox, Kentucky, is approximately 27 lb 8 oz. How many ounces does one gold bar weigh? (*Sources:* usmint.gov and goldinfo.net)

18. The possible placement that a weather sensor can be mounted is (1.5 ± 0.3) m—that is, 1.5 meters plus or minus 0.3 meters—above ground level. What is the maximum height in centimeters that the sensor can be mounted above ground? (*Source:* ofcm.gov)

19. The driving distance between Montreal and Toronto is about 539 km. Express this distance to the nearest hundred miles. (*Source:* trailcanada.com)

20. Hollywood has released major 3D films, such as *The Polar Express*, shot in 70 mm IMAX format. Express the format of this film in centimeters. (*Source:* wikipedia.org)

• Check your answers on page A-12.

9.1 U.S. Customary Units

OBJECTIVES

A To change a measurement from one U.S. customary unit to another

B To add or subtract measurements expressed in U.S. customary units

C To solve applied problems involving U.S. customary units

What Measurement and Units Are and Why They Are Important

When measuring, we express quantities in standardized units such as pounds and yards, which enable us to compare characteristics of physical objects. With standardized units, we can decide which of two packages is heavier or how many times longer one room is than another.

Throughout this chapter, we focus on four kinds of measure: length, weight, capacity, and time. The units that we use to express these measurements come from two systems of measurement: the U.S. customary system and the metric system.

U.S. Customary Units of Length, Weight, Capacity, and Time

We begin with U.S. customary units (also called U.S. units). These are the units that we use most often in everyday situations.

Sometimes, we need to change the unit in which a measurement is expressed. To convert units, we must know how different units are related.

In measuring *length*, the main U.S. units are miles (mi), yards (yd), feet (ft), and inches (in.). The key conversion relationships among these units are as shown in the table to the right.

Length
5,280 ft = 1 mi
3 ft = 1 yd
12 in. = 1 ft

The main U.S. units of *weight* are tons, pounds (lb), and ounces (oz). These relationships are shown in the table to the right.

Weight
2,000 lb = 1 ton
16 oz = 1 lb

Next, let's consider the main U.S. units of *capacity* (*liquid volume*): gallons (gal), quarts (qt), pints (pt), cups (c), and fluid ounces (fl oz). Here, the table to the right shows how these units are related.

Capacity
4 qt = 1 gal
2 pt = 1 qt
2 c = 1 pt
8 fl oz = 1 c

Finally, there are the main U.S. units of *time*: years (yr), months (mo), weeks (wk), days, hours (hr), minutes (min), and seconds (sec). Some of the key relationships among these units of time are as shown in the table to the right.

Time
365 days = 1 yr
12 mo = 1 yr
52 wk = 1 yr
7 days = 1 wk
24 hr = 1 day
60 min = 1 hr
60 sec = 1 min

Study these relationships so that you can use them to solve problems involving units.

Note that the abbreviations of the units are the same regardless of whether they are singular or plural. For example, the abbreviation for foot (ft) is the same as the abbreviation for feet (ft).

Changing Units

Suppose that you are ordering from a catalog the placemat pictured below. If you know that the length of the space to be covered is 2 ft, how long is it in inches? To change a length given in feet to inches, we need to know that there are 12 in. in 1 ft.

$$\begin{aligned} 2 \text{ ft} &= 2 \times (\mathbf{1\ ft}) \quad \textbf{Substitute 12 in. for 1 ft.} \\ &= 2 \times \mathbf{12\ in.} \\ &= 24 \text{ in.} \end{aligned}$$

So the placemat is 24 in. long.

Another way of solving this problem is to multiply the original measurement by the *unit factor* $\frac{12 \text{ in.}}{1 \text{ ft}}$. The unit factor method, commonly used in the physical and health sciences, is particularly helpful in solving complex conversion problems. Because the numerator 12 in. and the denominator 1 ft represent the same length, the unit factor $\frac{12 \text{ in.}}{1 \text{ ft}}$ is equivalent to 1.

$$\begin{aligned} 2 \text{ ft} &= 2 \cancel{\text{ft}} \times \frac{12 \text{ in.}}{1 \cancel{\text{ft}}} \\ &= \frac{2 \times 12 \text{ in.}}{1} = 24 \text{ in.} \end{aligned}$$

Note how we simplified the answer by canceling common units, as if the units were numbers.

Or suppose that we wanted to change 24 in. to feet. We can solve this problem by multiplying the original measurement by the unit factor $\frac{1 \text{ ft}}{12 \text{ in.}}$.

$$\begin{aligned} 24 \text{ in.} &= \overset{2}{\cancel{24}} \cancel{\text{in.}} \times \frac{1 \text{ ft}}{\underset{1}{\cancel{12}} \cancel{\text{in.}}} \\ &= 2 \text{ ft} \end{aligned}$$

Let's look back at the two problems that we just solved. Both involved inches and feet. In the first problem, we multiplied by the unit factor $\frac{12 \text{ in.}}{1 \text{ ft}}$; in the second problem, we multiplied by its reciprocal, $\frac{1 \text{ ft}}{12 \text{ in.}}$. (Note that both unit factors are equivalent to 1.)

TIP When converting from one unit to another unit, multiply the original measurement by the unit factor that has the *desired unit in its numerator* and the *original unit in its denominator.*

EXAMPLE 1

5 qt = _____ pt

Solution From the capacity chart at the beginning of this section, we know that 2 pt = 1 qt. We want to change from *quarts* to *pints*, so we use the unit factor $\frac{2\text{ pt}}{1\text{ qt}}$, where the desired unit is in the numerator and the original unit is in the denominator.

$$5\text{ qt} = 5\text{ qt} \times \frac{2\text{ pt}}{1\text{ qt}}$$
$$= 10\text{ pt}$$

PRACTICE 1

32 oz = _____ lb

EXAMPLE 2

Express 30 min in hours.

Solution $30\text{ min} = \overset{1}{30}\text{ min} \times \frac{1\text{ hr}}{\underset{2}{60}\text{ min}}$

$$= \frac{1}{2}\text{ hr}$$

PRACTICE 2

Change 5 ft to yards.

TIP When we change *a large unit to a small unit,* the numerical part of the answer is larger than the original number. But when we change *a small unit to a large unit,* the numerical part of the answer is smaller than the original number.

EXAMPLE 3

How many cups are equivalent to 3 gal?

Solution To solve this problem, we must use several steps.

Step 1. Change gallons to quarts.

Step 2. Change quarts to pints.

Step 3. Change pints to cups.

To combine these three steps, we multiply the original measurement by a chain of appropriate unit factors to get *cups* in the final answer.

$$3\text{ gal} = 3\text{ gal} \times \frac{4\text{ qt}}{1\text{ gal}} \times \frac{2\text{ pt}}{1\text{ qt}} \times \frac{2\text{ c}}{1\text{ pt}}$$
$$= \frac{3 \times 4 \times 2 \times 2}{1}\text{ c}$$
$$= 48\text{ c}$$

PRACTICE 3

How many seconds are there in a day?

EXAMPLE 4

The following sign is displayed on a bridge:

If a car weighs about 3,000 lb, can it cross the bridge safely?

Solution To compare the weight limit of the bridge and the weight of the car, let's express the two measurements in the same unit.

$$\begin{aligned} 2 \text{ tons} &= 2 \text{ tons} \times \frac{2{,}000 \text{ lb}}{1 \text{ ton}} \\ &= 4{,}000 \text{ lb} \end{aligned}$$

Because the bridge can support 4,000 lb—a lot more than the car weighs—it can safely cross the bridge.

PRACTICE 4

Rhubarb or "pie plant" is used in making pies, tarts, and sauces. If a cook makes 2 pints of rhubarb sauce, will there be at least 5 cups of sauce made? (*Source:* rhubarbinf.com)

In Example 4, we used the unit factor $\frac{2{,}000 \text{ lb}}{1 \text{ ton}}$ instead of $\frac{1 \text{ ton}}{2{,}000 \text{ lb}}$. Explain why.

Adding and Subtracting Mixed Units

Some measurements involve more than a single unit. For instance, we may express in *mixed units* the height of a child as 3 ft 6 in. or the weight of a package as 10 lb 7 oz.

When a measurement is expressed in mixed units, we write the larger unit with the largest possible whole number. For instance, the length of a movie is written as 3 hr 5 min and not as 2 hr 65 min, even though the two lengths of time are equal.

EXAMPLE 5

Convert 2 lb 15 oz to ounces.

Solution

$$\begin{aligned} 2 \text{ lb } 15 \text{ oz} &= 2 \text{ lb} \times \frac{16 \text{ oz}}{1 \text{ lb}} + 15 \text{ oz} \\ &= (32 + 15) \text{oz} \\ &= 47 \text{ oz} \end{aligned}$$

PRACTICE 5

Write 1 ft 7 in. in inches.

EXAMPLE 6

Change 271 min to hours and minutes.

Solution

$$\begin{aligned} 271 \text{ min} &= 4 \times 60 \text{ min} + 31 \text{ min} \\ &= \left(4 \times 60 \text{ min} \times \frac{1 \text{ hr}}{60 \text{ min}}\right) + 31 \text{ min} \\ &= 4 \text{ hr } 31 \text{ min} \end{aligned}$$

PRACTICE 6

What is 20 mo expressed in years and months?

There are many practical situations in which we need to add or subtract measurements written in mixed units—for instance, when we want to find out how much longer one car is than another or how much paint there will be if we combine the contents of several cans.

In addition or subtraction problems, only quantities having the same unit can be added or subtracted.

$$\begin{array}{r} 1\text{ ft }\ 4\text{ in.} \\ +1\text{ ft }\ 2\text{ in.} \\ \hline 2\text{ ft }\ 6\text{ in.} \end{array} \qquad \begin{array}{r} 2\text{ hr }\ 10\text{ min} \\ -1\text{ hr }\ \ 3\text{ min} \\ \hline 1\text{ hr }\ \ 7\text{ min} \end{array}$$

Often addition or subtraction problems involve changing units.

EXAMPLE 7

Find the sum:

$$\begin{array}{r} 7\text{ ft }\ 9\text{ in.} \\ +2\text{ ft }\ 5\text{ in.} \\ \hline \end{array}$$

Solution

$$\begin{array}{r} 7\text{ ft }\ 9\text{ in.} \\ +2\text{ ft }\ 5\text{ in.} \\ \hline 14\text{ in.} \end{array}$$

Start with the inches column. (The smaller unit is always in the column on the right.) Add the numbers in the inches column.

$$\begin{array}{r} \scriptstyle 1\text{ ft} \qquad\quad \\ 7\text{ ft }\ 9\text{ in.} \\ +2\text{ ft }\ 5\text{ in.} \\ \hline \not{14}\text{ in.} \\ \scriptstyle 2\text{ in.} \end{array}$$

Because 14 in. = 1 ft 2 in., replace the 14 in. by 2 in. and carry the 1 ft to the feet column.

$$\begin{array}{r} \scriptstyle 1\text{ ft} \qquad\quad \\ 7\text{ ft }\ 9\text{ in.} \\ +2\text{ ft }\ 5\text{ in.} \\ \hline 10\text{ ft }\ 2\text{ in.} \end{array}$$

Add the numbers in the feet column.

The sum is 10 ft 2 in.

PRACTICE 7

Add 3 lb 10 oz and 1 lb 14 oz.

EXAMPLE 8

A local theater was showing a double feature of two films—*Pirates of the Caribbean: At World's End* and *Pirates of the Caribbean: Dead Man's Chest*. The first film ran 2 hr 49 min and the second, 2 hr 30 min. How long was the double feature? (*Source:* wikipedia.org)

Solution

$$\begin{array}{r} \scriptstyle 1\text{ hr} \qquad\qquad \\ 2\text{ hr }\ 49\text{ min} \\ 2\text{ hr }\ 30\text{ min} \\ \hline 5\text{ hr }\ \not{79}\text{ min} \\ \scriptstyle 19\text{ min} \end{array}$$

Start with the minutes column. Because 79 min = 1 hr 19 min, replace the 79 min with 19 min and carry the 1 hr to the hours column. Add the numbers in the hours column.

So the double feature ran 5 hr 19 min.

PRACTICE 8

An online camping gear company ships two camping tents to a customer. If each tent weighs 6 lb 10 oz, what is their combined weight?

Now, let's look at some examples of subtraction.

EXAMPLE 9

Subtract 1 yd 2 ft from 3 yd 1 ft.

Solution

$\overset{2}{\cancel{3}}$ yd $\overset{4}{\cancel{1}}$ ft -1 yd 2 ft		Start with the feet column. Because 2 ft is larger than 1 ft, replace the 3 yd with 2 yd, and the borrowed yard with 3 ft, which when added to 1 ft gives 4 ft.
2 yd 4 ft -1 yd 2 ft 1 yd 2 ft		Subtract the numbers within each column.

Check Remember that we can check a subtraction by using addition.

1 yd (carried)
1 yd 2 ft
+1 yd 2 ft
3 yd $\cancel{4}$ ft
1 ft

Adding, we get 3 yd 1 ft, so our answer checks.

PRACTICE 9

How much greater is 6 gal than 5 gal 1 qt?

EXAMPLE 10

The *Mona Lisa* is one of the best known paintings in the world.

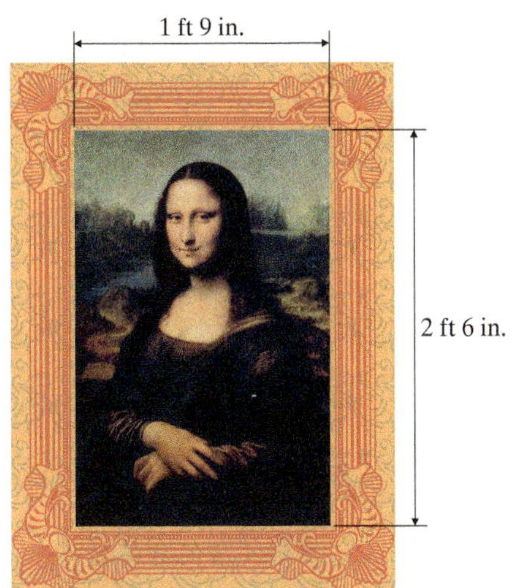

What is the difference between the height and width of the painting?

Solution

$\overset{1}{\cancel{2}}$ ft $\overset{18}{\cancel{6}}$ in.
-1 ft 9 in.
0 ft 9 in.

Because 9 in. is larger than 6 in., replace the 2 ft with 1 ft, and the borrowed foot with 12 in., which when added to 6 in. gives 18 in.

So the difference between the height and the width is 9 in.

PRACTICE 10

Two of the most famous horses in history were Citation and Secretariat. In 1948, Citation won the Kentucky Derby with a time of 2 min 5 sec. Twenty-five years later, Secretariat won in 1 min 59 sec. What was the difference between the two times? (*Source: Facts and Dates of American Sports*)

9.1 Exercises

FOR EXTRA HELP MyMathLab MathXL PRACTICE WATCH READ REVIEW

Mathematically Speaking

Fill in each blank with the most appropriate term or phrase from the given list.

smaller	pound	gallon	weight
larger	numerator	denominator	unit factor
sign	length	unit	

1. In the U.S. customary system, the yard is a unit of ______________.

2. In the U.S. customary system, the ______________ is a unit of capacity.

3. When we change a large unit to a small unit, the numerical part of the answer is ______________ than the original number.

4. When changing from miles to feet, we multiply the original measurement by $\frac{5{,}280 \text{ ft}}{1 \text{ mi}}$, which is called a ______________.

5. In addition or subtraction problems, only quantities having the same ______________ can be added or subtracted.

6. When converting from one unit to another unit, multiply the original measurement by the unit factor that has the desired unit in its ______________.

A *Change each quantity to the indicated unit.*

7. 48 in. = ____ ft

8. 6 pt = ____ qt

9. 9 ft = ____ yd

10. 48 mo = ____ yr

11. 60 ft = ____ in.

12. 8 qt = ____ pt

13. 7 min = ____ sec

14. 4 lb = ____ oz

15. 10 yd = ____ ft

16. 2 yr = ____ mo

17. 32 oz = ____ lb

18. 30 sec = ____ min

19. 2 mi = _____ yd

20. 5 hr = ____ min

21. 32 pt = ____ fl oz

22. $\frac{1}{4}$ day = ____ hr

23. $\frac{1}{2}$ gal = ____ qt

24. 32 fl oz = ____ pt

25. $2\frac{1}{2}$ qt = ____ pt

26. $1\frac{1}{2}$ qt = ____ gal

27. 7 pt = ____ qt

28. 36 hr = ____ days

29. $1\frac{1}{2}$ tons = _____ lb

30. 7,000 lb = ____ tons

31. 45 min = ____ hr

32. 7,920 yd = ____ mi

33. $\frac{1}{2}$ day = ____ hr

34. $\frac{1}{2}$ hr = ____ day

35. 5 min 10 sec = ____ sec

36. 1 lb 2 oz = ____ oz

37. 90 in. = ____ ft ____ in.

38. 50 hr = ____ days ____ hr

Complete each table.

	Length	Inches	Feet	Yards
39.	Giraffe (height)		16	
40.	Baseball (diameter)	3		
41.	Dog pen (width)			5
42.	Pond (depth)		12	

	Weight	Ounces	Pounds	Tons
43.	Ostrich egg		3	
44.	Baseball bat	29		
45.	Tongue of a blue whale			4
46.	Pony		200	

	Capacity	Fluid Ounces	Pints	Quarts
47.	Can of paint			1
48.	Case of cream		24	
49.	Bottle of mouthwash	16		
50.	Container of milk			2

	Time	Seconds	Minutes	Hours
51.	Rocket blast	50		
52.	Baseball game			3
53.	News report		22	
54.	Standing ovation		8	

B *Compute.*

55. 4 lb 7 oz
−2 lb 9 oz

56. 2 hr
−1 hr 2 min

57. 20 lb 5 oz
+ 9 lb 10 oz

58. 5 lb 10 oz
+1 lb 8 oz

59. 5 yr 7 mo + 3 yr 11 mo

60. 4 yr − 2 yr 3 mo

61. 5 gal 1 qt − 2 gal 2 qt

62. 1 pt 10 fl oz + 3 pt 8 fl oz

63. 2 qt 1 pt + 1 qt 1 pt

64. 2 ft − 5 in.

65. 6 min 2 sec
1 min 10 sec
+1 min 3 sec

66. 20 ft 5 in.
5 ft 7 in.
+ 9 ft 10 in.

Mixed Practice *Solve.*

67. Find the sum of 10 lb 12 oz and 3 lb 5 oz.

68. 2 gal = ________ pt

69. Express 4 min 7 sec in seconds.

70. Compute: 4 yr − 1 yr 3 mo

71. Convert 15 ft to yards.

72. How many pints are equivalent to 20 fl oz?

Applications

C *Solve.*

73. The longest field goal made by Tom Dempsey while playing football for the New Orleans Saints was 189 ft. The longest field goal made by Jason Elam while playing for the Denver Broncos was 63 yd. Determine if Elam's field goal was less than, greater than, or equal to that of Demsey. (*Source:* mmbolding.com)

74. The Tyrannosaurus rex (T. rex) was a dinosaur that lived in what is now western North America some 65 million years ago. Fossils show that the T. rex measured up to 42 ft in length. Express this length in yards. (*Source:* wikipedia.org)

75. In 1940, Cornelius Warmerdam used a bamboo pole to vault 15 ft 8 in., setting a record. In 1962, Dave Tork, using a fiberglass pole, vaulted 16 ft 2 in. How much higher was Tork's vault than Warmerdam's? (*Source: Facts and Dates of American Sports*)

76. One of the tallest women who ever lived was an American named Sandy Allen, who, at age 22, was 91 in. tall. What was her height in feet and inches? (*Source: Guinness Book of World Records*)

77. The record for a person holding his or her breath under water without special equipment is 823 sec. Express this time in minutes and seconds. (*Source: Atlantic Monthly*)

78. In an Olympic marathon, an athlete runs 26 mi plus an additional 385 yd. How many total yards does an athlete run in a marathon?

79. Born in 1934, the Dionne sisters were Canadians who became world famous as the first quintuplets to survive beyond infancy. At birth, the tiniest of these babies weighed 1 lb 15 oz, and the largest weighed 3 lb 4 oz. What was the difference between their weights? (*Source: Encyclopaedia Brittanica*)

80. Abraham Lincoln spoke of "four score and seven years ago." If a score is 20, how many months are there in four score and seven years?

81. The lease on a tenant's apartment runs for 3 yr. If he has lived in the apartment already for 1 yr 2 mo, does he have more or less than $1\frac{1}{2}$ yr left on the lease?

82. A United Airlines nonstop flight from New Orleans to Los Angeles is scheduled to last 4 hr 20 min. According to the schedule, how much time remains $1\frac{1}{2}$ hr into the flight? (*Source:* expedia.com)

83. In the spring of 2009, a first annual fishing competition was held in Southern California in which the anglers who caught the largest halibut (by weight) shared prize money. Two of the heaviest halibut caught in the contest weighed 31 lb and 17 lb 2 oz. What was their difference in weight? (*Source:* danawharf.com)

84. Evergreen trees come in different sizes. In one year, a fast-growing evergreen that had been 5 ft 10 in. tall grew by 4 ft 7 in. At the end of the year, what was the height of the tree?

85. As part of a kitchen remodel, a homeowner bought a new refrigerator.

a. The refrigerator is 36 in. wide and 30 in. deep. Express the width and depth in feet.

b. Using the answer from part (a), calculate the area of floor space the refrigerator will occupy.

86. A recipe calls for 8 fl oz of chicken broth.

a. How many pints of chicken broth are needed for the recipe?

b. If 1 qt of chicken broth is available, will there be enough to triple the recipe?

87. The highest mountain in the world is Mt. Everest. If its peak is 29,035 ft above sea level, find the height of the mountain to the nearest tenth of a mile.

88. Earth is made up of three main layers: the crust, the mantle, and the core. The core, which is composed of a liquid outer core and a solid inner core, is about 2,156 mi thick. How many feet thick is the core, rounded to the nearest million? (*Source:* U.S. Geological Survey)

• Check your answers on page A-12.

MINDStretchers

Groupwork

1. The U.S. customary system includes units that measure not only length, weight, capacity or time but also energy. For example, the British Thermal Unit (BTU) is commonly used to measure the amount of energy put out by a furnace or the cooling capacity of an air conditioner. Working with a partner, identify some other units in the U.S. customary system not discussed in this section. What do these units measure?

Mathematical Reasoning

2. In measuring, we often introduce errors. Suppose that each of two measurements could be as much as an inch off. If we then add the two measurements, how far from the truth could our sum be? Explain.

History

3. The foot is not the only body part used as a measure. For instance, the ancient Egyptians used the *mouthful* as a unit of measure of volume. In your college library or on the Web, investigate other examples.

Cultural Note

Joseph Louis Lagrange (1736–1813) was a French mathematician who was influential during the years following the French Revolution of 1789 in developing the metric system of measures based on decimals and powers of 10. Since then, the United States has been resistant to adopting the metric system, although in 1790, Thomas Jefferson argued that the country should adopt a decimal system of weights and measures.

Source: Gullberg, *Mathematics: From the Birth of Numbers* (New York: W. W. Norton, 1997), p. 52.

9.2 Metric Units and Metric/U.S. Customary Unit Conversions

OBJECTIVES

- **A** To identify units in the metric system
- **B** To change a measurement from one metric unit to another
- **C** To change a measurement from a metric unit to a U.S. customary unit, and vice versa
- **D** To solve applied problems involving metric units or U.S. customary units

Metric Units of Length, Weight, and Capacity

Now, we turn to metric units. Developed by French scientists over 200 years ago, the metric system (formally known as the International System of Units, or SI) has become standard in most countries of the world. Even in the United States, metric units predominate in many important fields, including scientific research, medicine, the film industry, food and drink packaging, sports, and the import–export industry.

As in Section 9.1, which dealt with the U.S. customary system, here we consider measurements of length, weight, and capacity. Time units are identical in both systems, so we do not discuss them in this section. Again, abbreviations of units in the singular and plural are the same.

We begin this discussion of the metric system by considering the basic metric units:

- the **meter (m)**, a unit of length, which gives the metric system its name;
- the **gram (g)**, a unit of weight, or more precisely of mass; and
- the **liter (L)**, a unit of capacity, that is, of liquid volume.

There are quite a few other metric units as well. The names for many of the other units are formed by combining a basic unit with one of the metric prefixes. The following table contains a list of metric prefixes, with those most commonly used in bold:

METRIC PREFIXES

Prefix	Symbol	Meaning
Kilo-	k	One thousand (1,000)
Hecto-	h	Hundred (100)
Deka-	da	Ten (10)
Deci-	d	One tenth $\left(\frac{1}{10}\right)$
Centi-	c	One hundred**th** $\left(\frac{1}{100}\right)$
Milli-	m	One thousand**th** $\left(\frac{1}{1,000}\right)$

Next, let's see how the three basic metric units combine with the metric prefixes to form new units. We begin with units of length.

Length

The table shows the four most commonly used metric units of length: kilometers (km), meters (m), centimeters (cm), and millimeters (mm). Memorize the table to the right, noting what each unit means as well as its symbol.

METRIC UNITS OF LENGTH

Unit	Symbol	Meaning
Kilometer	km	1,000 meters
Meter	m	1 meter
Centimeter	cm	$\frac{1}{100}$ meter
Millimeter	mm	$\frac{1}{1,000}$ meter

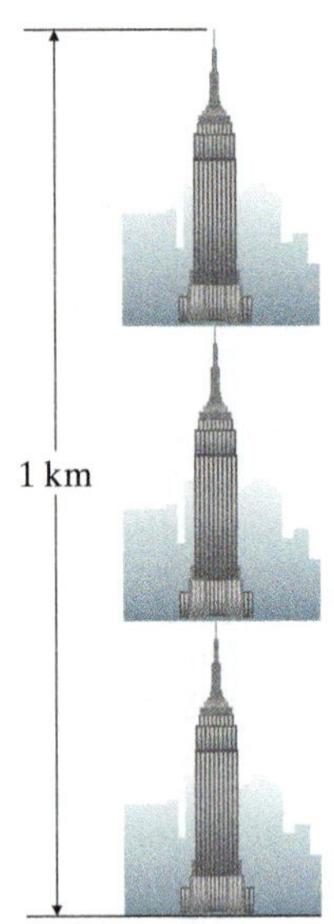

In the preceding table, the largest unit of length is the *kilometer*. A kilometer is a little more than half a mile, or about 3 times the height of the Empire State Building, as shown to the left. Great lengths, such as the distance between two cities, are expressed in kilometers.

The *meter*, the basic metric unit of length, is a little longer than a yard, or about the width of a twin bed. In the metric system, we use meters to measure medium-size lengths—say, the length of a room.

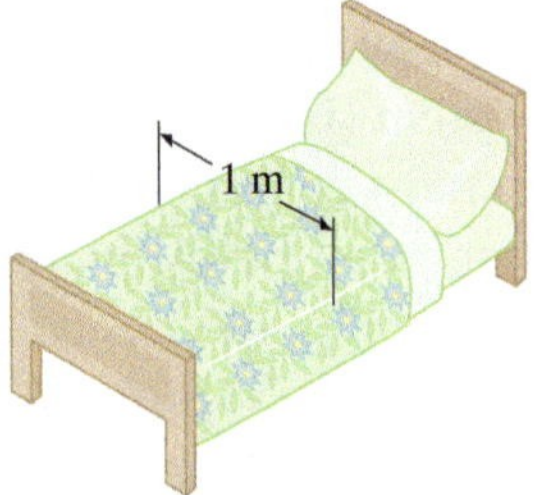

The next unit of length, the *centimeter*, is approximately the width of your little finger, or somewhat less than half an inch. In the metric system, the width of an envelope is expressed in centimeters.

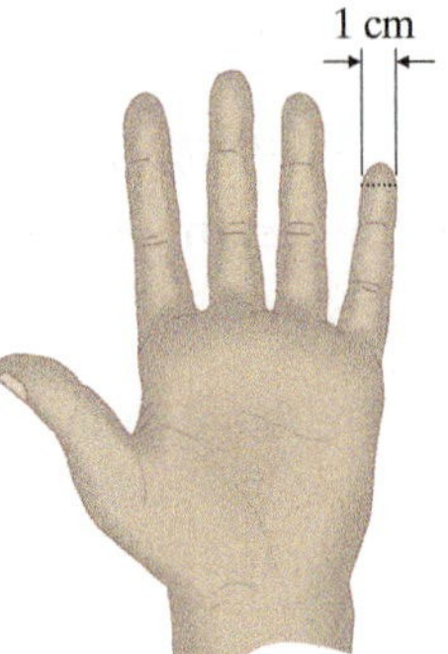

The smallest unit of length in the table is the *millimeter*. A millimeter is about the thickness of a dime. We use the millimeter to measure short lengths—say, the dimensions of an insect.

Weight

Now, we turn to the metric units of weight, shown in the following table: kilograms (kg), grams (g), and milligrams (mg). Memorize this table, which deals with the three most commonly used metric units of weight.

METRIC UNITS OF WEIGHT

Unit	Symbol	Meaning
Kilogram	kg	1,000 grams
Gram	g	1 gram
Milligram	mg	$\frac{1}{1{,}000}$ gram

The largest unit of weight in the table is the *kilogram*. A kilogram is approximately 2 lb, or about the weight of this textbook. Large weights—say, that of a car or of a person—are expressed in kilograms.

This textbook

The next unit, the *gram*, is smaller, only about $\frac{1}{30}$ oz, or about the weight of a raisin.

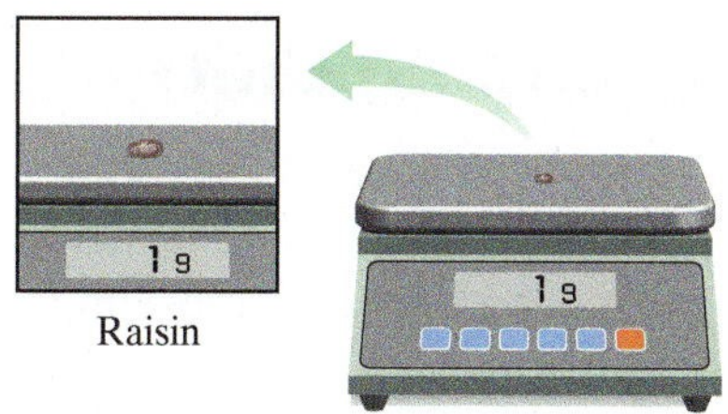

Raisin

The smallest unit, the *milligram*, is tiny—about the weight of a strand of hair. It is therefore used to measure small weights—say, that of a pill.

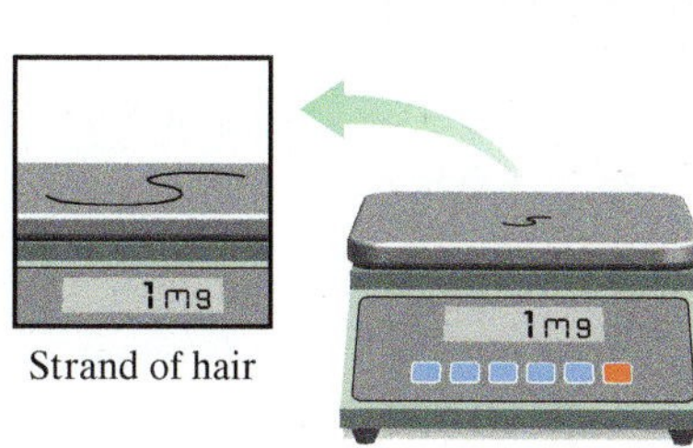

Strand of hair

Capacity (Liquid Volume)

Amounts of liquid are commonly measured in terms of liquid volume, or equivalently, the capacity of containers that hold the liquid. The following table deals with three metric units of capacity: kiloliters (kL), liters (L), and milliliters (mL). Memorize this table, which describes the three primary metric units of capacity.

METRIC UNITS OF CAPACITY

Unit	Symbol	Meaning
Kiloliter	kL	1,000 liters
Liter	L	1 liter
Milliliter	mL	$\frac{1}{1,000}$ liter

In this table, the largest unit of capacity is the *kiloliter*. The amount of water that a typical collapsible swimming pool holds is about 1 kiloliter. Kiloliters are used to measure large volumes of liquid, for instance, the capacity of an oil barge or the amount of soda that a factory produces annually.

The second unit of liquid volume, the *liter*, is slightly more than a quart, a typical size for a bottle of mouthwash. Liters are used in measuring larger quantities of liquid, such as the amount of water that a sink will hold or, in some countries, the amount of gasoline purchased at the pump.

The smallest unit in this table is the *milliliter*, which represents a very small amount of liquid—about as much as an eyedropper contains. Milliliters are used in measuring small volumes of liquid—say, the amount of perfume in a tiny bottle.

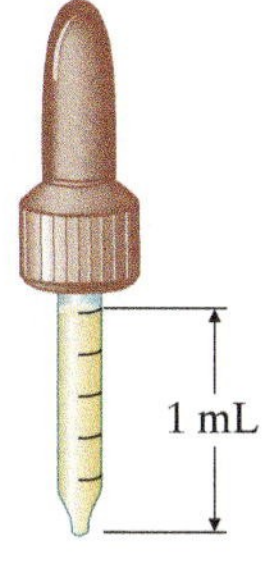

Changing Units

As we have already seen, sometimes we need to change the unit in which a measurement is expressed. One reason the metric system is widely used is that unit conversions in this system are much easier than those in the U.S. customary system. Such conversions simply involve multiplying or dividing by a power of 10, such as 100 or 1,000. The following table shows several metric conversion relationships.

Length	Weight	Capacity
1,000 m = 1 km	1,000 g = 1 kg	1,000 L = 1 kL
100 cm = 1 m	1,000 mg = 1 g	1,000 ml = 1 L
1,000 mm = 1 m		

From these relationships, we can set up unit factors to carry out unit conversions.

EXAMPLE 1

1.5 g = ________ mg

Solution To convert to *milligrams*, we use the unit factor $\frac{1{,}000 \text{ mg}}{1 \text{ g}}$ because 1,000 mg = 1 g.

$$\begin{aligned} 1.5 \text{ g} &= 1.5 \cancel{\text{g}} \times \frac{1{,}000 \text{ mg}}{1 \cancel{\text{g}}} \\ &= 1.5 \times 1{,}000 \text{ mg} \\ &= 1{,}500 \text{ mg} \end{aligned}$$

To multiply 1.5 by 1,000, move the decimal point in 1.5 three places to the right.

PRACTICE 1

3,100 mg = ____ g

As illustrated in Example 1, a part of a metric unit is usually expressed as a decimal, not as a fraction. That is, we write 1.5 g, not $1\frac{1}{2}$ g.

EXAMPLE 2

500 m = ________ km

Solution To convert meters to kilometers, we multiply the original measurement by the unit factor $\frac{1 \text{ km}}{1{,}000 \text{ m}}$.

$$\begin{aligned} 500 \text{ m} &= 500 \cancel{\text{m}} \times \frac{1 \text{ km}}{1000 \cancel{\text{m}}} \\ &= \frac{500}{1{,}000} \text{ km} \\ &= 0.5 \text{ km} \end{aligned}$$

To divide 500 by 1,000, move the decimal point in 500 three places to the left.

As a quick check, note that the number of the larger unit (0.5 km) is less than the number of the smaller unit (500 m) for an equivalent measurement.

PRACTICE 2

2,500 cm = ____ m

So far we have used unit factors to convert between units of measure in the metric system. An alternative method to unit factors is moving the decimal point to the right or to the left using the *metric conversion line.*

kilo- hecto- deka- *unit* deci- centi- milli-

For example, let's revisit Example 1, where we changed 1.5 g to milligrams. Here we consider the metric conversion line for weight.

kg hg dag g dg cg mg

Note that each unit is equivalent to 10 times the unit to its right. We locate the given unit grams (g) on the line. To change to milligrams (mg), we need to move along the line 3 units to the right.

kg hg dag g dg cg mg

Similarly, in 1.5 we move the decimal point 3 places to the right.

$$1.5 \text{ g} = 1\,500. \text{ mg} = 1{,}500 \text{ mg}$$

So, as before, we find that 1.5 g is equivalent to 1,500 mg.

In the next example, we revisit Example 2, where we used the unit factor method to convert 500 m to kilometers. This time, however, we use the metric conversion line.

EXAMPLE 3

Change 500 m to kilometers.

Solution First, we list the metric units of length from largest to smallest. Then, we locate meters (m) on the metric conversion line and move to kilometers (km).

km hm dam m dm cm mm

Similarly, we move the decimal point 3 places to the left in 500.

$$500. \text{ m} = 0.500 \text{ km} = 0.5 \text{ km}$$

As with the unit factor method, we conclude that 500 m = 0.5 km.

PRACTICE 3

Use the metric conversion line to change 2,500 cm to meters.

TIP When using the metric conversion line to change one metric unit to another, locate the original unit on the line and move to the desired unit. Then, in the numerical part of the original measurement, move the decimal point the same number of places in the same direction as you moved on the metric conversion line.

EXAMPLE 4

Express 3 kL in milliliters.

Solution First, we consider the unit factor method. Because the table of capacity on page 413 does not indicate how many milliliters are equivalent to a kiloliter, we need to solve the problem in steps.

Step 1. Change kiloliters to liters.

Step 2. Change liters to milliliters.

To combine these two steps, we multiply the original measurement by a chain of appropriate unit factors to get *milliliters* for the final answer.

$$\begin{aligned} 3 \text{ kL} &= 3 \cancel{\text{kL}} \times \frac{1{,}000 \cancel{\text{L}}}{1 \cancel{\text{kL}}} \times \frac{1{,}000 \text{ mL}}{1 \cancel{\text{L}}} \\ &= 3 \times 1{,}000 \times 1{,}000 \text{ mL} \\ &= 3{,}000{,}000 \text{ mL} \end{aligned}$$

Alternatively, let's use the metric conversion line. We list the units of capacity in decreasing order from left to right, and then move from kiloliters to milliliters on the metric conversion line.

kL hL daL L dL cL mL

6 units to the right

$$3 \text{ kL} = 3. \text{ kL} = 3\,000\,000. \text{ mL} = 3{,}000{,}000 \text{ mL}$$

Note that with either method, we get 3,000,000 mL.

PRACTICE 4

Change 5,000,000 mL to kiloliters.

EXAMPLE 5

For a healthy 20-year-old female, the U.S. recommended dietary allowances (RDAs) for calcium and iron are 1 g and 18 mg, respectively. Which RDA is higher? (*Source:* iron.edu)

Solution When comparing quantities expressed in different units, we convert them to the same unit, usually the smaller unit. Here, we change 1 g to milligrams.

$$1 \text{ g} = 1 \text{ g} \times \frac{1{,}000 \text{ mg}}{1 \text{ g}}$$
$$= 1{,}000 \text{ mg}$$

Using the metric conversion line gives us:

kg hg dag g dg cg mg

3 units to the right

1. g = 1 000. mg = 1,000 mg

So the RDA for calcium is 1,000 mg, which is higher than the 18-mg RDA for iron.

PRACTICE 5

The small intestine is 6 m long, and the large intestine is 150 cm long. Which is shorter? (*Source: Webster's New World Book of Facts*)

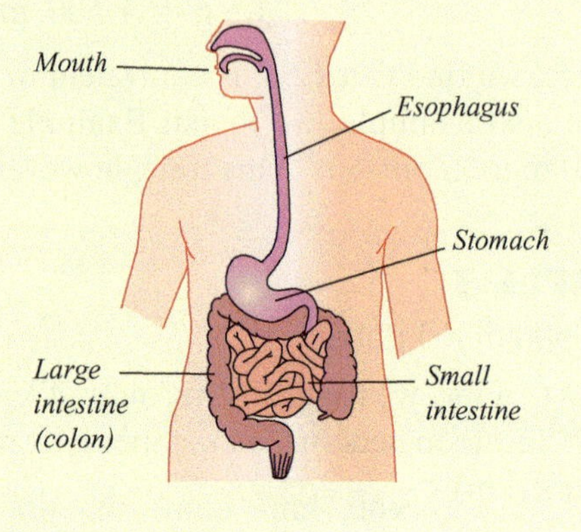

Changing Extreme Units

In addition to the metric prefixes already discussed, there are others that are in common use. With these prefixes, we can create very large or very small units. Often the need for such units has arisen because of advances in technology, science, or medicine. For example, the prefix "mega-" meaning 1,000,000 makes it convenient to express the large memories and high speeds that today's computers possess. At the other end of the spectrum, the prefix "micro-", meaning $\frac{1}{1{,}000{,}000}$, is helpful in measuring tiny doses of medicines or vaccines. The following table lists some of these metric prefixes. Commit these to memory.

OTHER METRIC PREFIXES

Prefix	Symbol	Meaning
Giga-	G	One billion (1,000,000,000)
Mega-	M	One million (1,000,000)
Micro-	mc	One million**th** $\left(\frac{1}{1{,}000{,}000}\right)$

We can combine these prefixes with the metric units. For instance, the microgram is a tiny unit of weight and the microsecond is a tiny unit of time.

These prefixes are also sometimes combined with units that, while in common use, are technically not part of the metric system. For example, the megabyte and the gigabyte are large units of computer memory based on the byte (B).

Consider the following table.

OTHER METRIC UNITS

Unit	Symbol	Meaning	Relationship
Gigabyte	GB	1,000,000,000 bytes	1,000,000,000 B = 1 GB
Megabyte	MB	1,000,000 bytes	1,000,000 B = 1 MB
Microsecond	mcsec	$\frac{1}{1{,}000{,}000}$ second	1,000,000 mcsec = 1 sec
Microgram	mcg	$\frac{1}{1{,}000{,}000}$ gram	1,000,000 mcg = 1 g

We can use the unit factor method to convert the units in this table to other units, as the following examples illustrate.

EXAMPLE 6

2,000 mcg = ________ g

Solution Because 1 g = 1,000,000 mcg, to convert micrograms to grams we use the unit factor $\frac{1\text{ g}}{1{,}000{,}000\text{ mcg}}$.

$$2{,}000\text{ mcg} = 2{,}000\text{ mcg} \times \frac{1\text{ g}}{1{,}000{,}000\text{ mcg}}$$
$$= \frac{2{,}000}{1{,}000{,}000}\text{ g}$$
$$= 0.002\text{ g}$$

Move the decimal point in 2,000 six places to the left.

PRACTICE 6

5.3 sec = ________ mcsec

EXAMPLE 7

Many computers come equipped with 4 GB of RAM (random access memory) to run today's demanding operating systems. Express this amount of RAM in terms of megabytes.

Solution

$$4\text{ GB} = 4\text{ GB} \times \frac{1{,}000{,}000{,}000\text{ B}}{1\text{ GB}} \times \frac{1\text{ MB}}{1{,}000{,}000\text{ B}}$$
$$= 4 \times 1{,}000\text{ MB}$$
$$= 4{,}000\text{ MB}$$

So the amount of RAM is 4,000 MB.

PRACTICE 7

Vitamin D is important for the proper absorption of calcium from food. The U.S. Institute of Medicine recommends that all individuals under the age of 50 take 5 mcg of Vitamin D daily. Express this amount in milligrams. (*Source:* mayoclinic.com)

Metric/U.S. Customary Unit Conversions

In some situations, we need to change a measurement expressed in a U.S. unit to a metric unit, or vice versa. For example, if Americans were driving in Ireland and saw a road sign giving the distance to the next town in kilometers, they might want to express that distance in miles.

Or suppose that shoppers wanted to buy mouthwash and wondered how many pint bottles are equal in capacity to a 750-mL bottle. Here, they might want to change pints to milliliters.

To convert, we must either have memorized or have access to metric/U.S. unit conversion relationships. The following table shows some of these key relationships:

METRIC/U.S. UNIT CONVERSION RELATIONSHIPS

Length	Weight	Capacity
1.6 km ≈ 1 mi	910 kg ≈ 1 ton	260 gal ≈ 1 kL
1,600 m ≈ 1 mi	2.2 lb ≈ 1 kg	3.8 L ≈ 1 gal
3,300 ft ≈ 1 km	450 g ≈ 1 lb	1.1 qt ≈ 1 L
3.3 ft ≈ 1 m	28 g ≈ 1 oz	2.1 pt ≈ 1 L
39 in. ≈ 1 m		470 mL ≈ 1 pt
30 cm ≈ 1 ft		
2.5 cm ≈ 1 in.		

EXAMPLE 8

Express 2 oz in grams.

Solution According to the conversion table, 1 oz ≈ 28 g.

To express in *grams*, we multiply 2 oz by the unit factor $\frac{28\text{ g}}{1\text{ oz}}$.

$$2\text{ oz} \approx 2\,\cancel{\text{oz}} \times \frac{28\text{ g}}{1\,\cancel{\text{oz}}} \approx 56\text{ g}$$

So 2 oz is about 56 g.

PRACTICE 8

Express 10 gal in terms of liters.

Note that our answer in Example 8 is only an approximation because the unit factor is not exact. Also note that the number of grams is more than the number of ounces because an ounce is larger than a gram.

EXAMPLE 9

Since 1866, the capacity of an oil barrel in the United States has been standardized at 42 gal. How many liters of oil does an oil barrel hold, rounded to the nearest liter? (*Source:* Daniel Yergin, *The Prize*)

Solution The conversion table indicates that 3.8 L ≈ 1 gal.

To convert to liters, we use the conversion factor $\frac{3.8\text{ L}}{1\text{ gal}}$.

$$42\text{ gal} \approx 42\,\cancel{\text{gal}} \times \frac{3.8\text{ L}}{1\,\cancel{\text{gal}}} \approx 159.6\text{ L}$$

So an oil barrel holds approximately 160 L of oil.

PRACTICE 9

An iron-nickel meteorite from Argentina was for sale on an auction website. If the meteorite weighs 1,760 g, find its weight in pounds to the nearest whole number. (*Source:* astromart.com)

9.2 Exercises

FOR EXTRA HELP MyMathLab MathXL PRACTICE WATCH READ REVIEW

Mathematically Speaking

Fill in each blank with the most appropriate term or phrase from the given list.

kilo-	centi-	milli-
weight	deci-	quart
liter	length	

1. In the metric system, the meter is a unit of _______.

2. In the metric system, the _______ is a unit of capacity.

3. The prefix _______ means one thousand.

4. The prefix _______ means one-thousandth.

5. The prefix _______ means one-hundredth.

6. The prefix _______ means one-tenth.

A *Choose the unit that would most likely be used to measure each quantity.*

7. The volume of liquid in a test tube
a. millimeter **b.** milligram **c.** milliliter

8. The weight of a television set
a. milligram **b.** gram **c.** kilogram

9. The width of a street
a. millimeter **b.** meter **c.** kilometer

10. The length of a river
a. kilometer **b.** kilogram **c.** millimeter

Choose the best estimate in each case.

11. The capacity of a large bottle of soda
a. 1 mL **b.** 1 L **c.** 1 kg

12. The width of film for slides
a. 35 mm **b.** 35 cm **c.** 35 m

13. The height of the Washington Monument
a. 170 cm **b.** 170 m **c.** 170 km

14. The length of a pencil
a. 20 mm **b.** 20 cm **c.** 20 m

15. The capacity of a bottle of hydrogen peroxide
a. 400 mL **b.** 400 L **c.** 400 g

16. The weight of an adult
a. 70 mg **b.** 70 g **c.** 70 kg

B *Change each quantity to the indicated unit.*

17. 1,000 mg = _____ g

18. 253 mm = _____ m

19. 750 g = _____ kg

20. 2 L = _____ mL

21. 0.08 kL = _____ L

22. 4.3 kg = _____ g

23. 3.5 m = _____ mm

24. 900 m = _____ km

25. 5 mL = _____ L

26. 250 mg = _____ g

27. 4,000 mm = _____ m

28. 5 L = _____ mL

29. 7,000 L = _____ kL

30. 2,500 mg = _____ g

31. 413 cm = _____ m

32. 2.8 m = _____ cm

33. 0.002 kg = _____ mg

34. 3,000 mm = _____ cm

35. 7,500 mL = _____ kL

36. 2.1 km = _______ cm

37. 0.03 g = _______ mcg

38. 5,000 mcsec = _____ sec

39. 4 MB = _______ B

40. 8,000,000 B = _____ GB

41. 7,000,000 mcg = _____ g

42. 0.09 sec = _____ mcsec

43. 1,280 MB = _____ GB

44. 0.04 mg = _____ mcg

Complete each table.

	Length	Millimeters	Centimeters	Meters
45.	Swan's wingspan		238	
46.	Mouse pad		5	
47.	Jumping spider	10		
48.	Power cord			3

	Weight	Milligrams	Grams	Kilograms
49.	Capsule	300		
50.	Human liver		1,560	
51.	Pastry		450	
52.	Kangaroo cub			1

	Capacity	Milliliters	Liters	Kiloliters
53.	Container of tile cleaner	709		
54.	Bottle of spring water		3	
55.	Gas station fuel tank			17
56.	Aquarium		110	

Compute.

57. 3 km + 250 m

58. 5 L − 600 mL

59. 98 kg + 25.6 g

60. 30 cm + 2 m

C *Change each quantity to the indicated unit. If needed, round the answer to the nearest tenth of the unit.*

61. 30 oz ≈ ___ g

62. 4 mi ≈ ____ km

63. 10 cm ≈ ___ in.

64. 900 g ≈ ___ lb

65. 48 in. ≈ ____ m

66. 6 qt ≈ ____ L

67. 5 pt ≈ ___ L

68. 6 ft ≈ ____ cm

Mixed Practice

Solve.

69. Combine 3 m and 50 cm.

70. Change 500 g to kilograms.

71. 2.5 m = ____ mm

72. Express 2,000 mL in liters.

73. The distance between a student's home and her college would most likely be measured in

a. kilometers **b.** kilograms **c.** millimeters

74. Express 60 cm in feet.

75. Change 3 L to quarts.

76. The best estimate for the capacity of a bottle of olive oil is

a. 500 L **b.** 500 kg **c.** 500 mL

Applications

D *Solve. If needed, round the answer to the nearest tenth of the unit.*

77. According to a medical journal, the average daily U.S. diet contains 6,000 mg of sodium. How many grams is this? (*Source: Journal of the American Medical Association*)

78. Some road signs in the United States give the speed limit in both miles per hour (mph) and kilometers per hour (kph). Confirm that the speed limits in the South Dakota sign to the right are approximately the same. (*Source:* worldofstock.com)

79. A nurse must administer a 4-mL dose of a drug daily to a patient. If there is 1 L of this drug on hand, will it last the patient 120 days?

80. Vitamin C commonly comes in pills with a strength of 500 mg. How many of these pills will an adult need to take if she wants a dosage of half a gram?

81. A student in a physics lab measured the length of a pendulum string as 7.5 cm. Express this length in inches.

82. In the Summer Olympics, a major track-and-field event is the 100-m dash. How long is this race in kilometers?

83. The diameter of the primary mirror of the Hubble Space Telescope is 2.4 m. Express this diameter in millimeters. (*Source:* hubblesite.org)

84. The Willis Tower, formerly the Sears Tower, in Chicago is 442 m tall. Express this height in kilometers. (*Source:* Emporis Buildings)

85. The side of a square tile is 75 mm long. If 100 of these tiles are placed on the floor side by side, what is their total length in centimeters?

86. In 1948, a medical researcher studied the possible use of large doses of ascorbic acid to cure tuberculosis (TB). He reported administering about 3,000,000 mg of ascorbic acid to patients with TB. Express this amount in kilograms. (*Source:* vitamincfoundation.org)

87. A dairy equipment company sells refrigerated tanks, called milk coolers. These tanks are available in the following sizes: 0.2 kL, 0.5 kL, 1 kL, 2 kL, 3 kL, 5 kL, and 10 kL. Find the capacity of the smallest of these tanks, expressed in milliliters. (*Source:* tolcontrols.com)

88. The trading system for the New York Stock Exchange takes 650 mcsec to process a trade. Express this amount of time in seconds. (*Source:* reuters.com)

89. Folic acid is a B-complex vitamin needed by the body to produce red blood cells. A health-conscious college student takes 800 mcg tablets of folic acid daily. How much folic acid, in milligrams, does she take in 7 days? (*Source:* ncbi.nlm.nih.gov)

90. Two hours before surgery, a patient was told to drink 300 mL of a clear fluid. Express this amount of fluid in quarts.

91. One of the heaviest babies ever born was an Italian boy who at birth weighed 360 oz. What was the baby's weight in kilograms? (*Source: Guinness Book of World Records*)

92. A prehistoric bird had a wingspan of 8 m. Express this wingspan in feet. (*Source: Guinness Book of World Records*)

93. The average weight of an adult human's brain is about 3 lb. Express this weight in grams. (*Source: The Top 10 of Everything, 2006*)

94. A passenger car is generally considered small if the distance between its front and back wheels is less than 95 in. What is this distance expressed in meters?

95. A chemistry professor mixes the contents of two beakers containing 2.5 L and 700 mL of a liquid. What is the combined amount in liters?

96. A can contains 355 mL of soda. Express in liters the amount of soda in a six-pack of these cans.

• Check your answers on page A-12.

MINDStretchers

History

1. Either in your college library or on the web, investigate the history of decimal coinage in the United States. In what respect are decimal coinage and the metric system similar? Explain.

Writing

2. Consider the following measurement expressed in the metric system.

$$\underbrace{37}_{\text{km}},\underbrace{568}_{\text{m}}.\underbrace{251}_{\text{mm}}\text{ meters}$$

Note how we can split this measurement in meters into other metric units. Would this work with U.S. units? Explain.

Groupwork

3. A liter is about 10% more than a quart. Work with a partner to answer the following questions.

a. Do you think that, as the United States goes metric, containers will increase in size? Explain.

b. Suppose containers were to increase in size. What do you think the economic consequence of this increase would be?

Key Concepts and Skills

	U.S. Customary Units	Relationships
[9.1] **Length**	Mile (mi), yard (yd), foot (ft), and inch (in.)	5,280 ft = 1 mi 3 ft = 1 yd 12 in. = 1 ft
[9.1] **Weight**	Ton, pound (lb), and ounce (oz)	2,000 lb = 1 ton 16 oz = 1 lb
[9.1] **Capacity (liquid volume)**	Gallon (gal), quart (qt), pint (pt), cup (c), and fluid ounce (fl oz)	4 qt = 1 gal 2 pt = 1 qt 2 c = 1 pt 8 fl oz = 1 c
[9.1] **Time**	Year (yr), month (mo), week (wk), day, hour (hr), minute (min), and second (sec)	365 days = 1 yr 12 mo = 1 yr 52 wk = 1 yr 7 days = 1 wk 24 hr = 1 day 60 min = 1 hr 60 sec = 1 min

METRIC PREFIXES

Prefix	Symbol	Meaning
Kilo-	k	One thousand (1,000)
Hecto-	h	Hundred (100)
Deka-	da	Ten (10)
Deci-	d	One tenth $\left(\frac{1}{10}\right)$
Centi-	c	One hundredth $\left(\frac{1}{100}\right)$
Milli-	m	One thousandth $\left(\frac{1}{1,000}\right)$

	Metric Units	Relationships
[9.2] **Length**	Kilometer (km), meter (m), centimeter (cm), and millimeter (mm)	1,000 m = 1 km 100 cm = 1 m 1,000 mm = 1 m
[9.2] **Weight**	Kilogram (kg), gram (g), and milligram (mg)	1,000 g = 1 kg 1,000 mg = 1 g
[9.2] **Capacity (liquid volume)**	Kiloliter (kL), liter (L), and milliliter (mL)	1,000 L = 1 kL 1,000 mL = 1 L
[9.2] **Time**	Same as U.S. customary units.	

OTHER METRIC PREFIXES

Prefix	Symbol	Meaning
Giga-	G	One billion (1,000,000,000)
Mega-	M	One million (1,000,000)
Micro-	mc	One million**th** $\left(\frac{1}{1{,}000{,}000}\right)$

OTHER METRIC UNITS

Unit	Symbol	Meaning	Relationship
Gigabyte	GB	1,000,000,000 bytes	1,000,000,000 B = 1 GB
Megabyte	MB	1,000,000 bytes	1,000,000 B = 1 MB
Microsecond	mcsec	$\frac{1}{1{,}000{,}000}$ second	1,000,000 mcsec = 1 sec
Microgram	mcg	$\frac{1}{1{,}000{,}000}$ gram	1,000,000 mcg = 1 g

Key Metric/U.S. Unit Conversion Relationships

[9.2] **Length**	1.6 km ≈ 1 mi 1,600 m ≈ 1 mi 3,300 ft ≈ 1 km 3.3 ft ≈ 1 m 39 in. ≈ 1 m 30 cm ≈ 1 ft 2.5 cm ≈ 1 in.
[9.2] **Weight**	910 kg ≈ 1 ton 2.2 lb ≈ 1 kg 450 g ≈ 1 lb 28 g ≈ 1 oz
[9.2] **Capacity (liquid volume)**	260 gal ≈ 1 kL 3.8 L ≈ 1 gal 1.1 qt ≈ 1 L 2.1 pt ≈ 1 L 470 mL ≈ 1 pt

CHAPTER 9 Review Exercises

Say Why

Fill in each blank.

1. To change 6 days to hours, we _______ (do/do not) multiply by $\frac{1 \text{ day}}{24 \text{ hr}}$ because ____.

2. A container _______ (can/cannot) weigh 1 quart because ____.

3. A centimeter _______ (is/is not) longer than a millimeter because ____.

4. 7.3 km _______ (is/is not) equivalent to 7,300 m because ____.

5. For a digital camera, the unit *megapixel* _______ (is/is not) equivalent to one million pixels because ____.

6. In chemistry, the unit *microliter* _______ (is/is not) equivalent to $\frac{1}{1{,}000{,}000{,}000}$ of a liter because ____.

[9.1] *Change each quantity to the indicated unit.*

7. 5 yd = ____ ft
8. 20 mo = ____ yr
9. 32 oz = ____ lb
10. 10 ft = ____ yd
11. $1\frac{1}{2}$ tons = ____ lb
12. $8\frac{1}{2}$ lb = ____ oz
13. 3 pt = ____ fl oz
14. 150 sec = ____ min
15. 7 hr 15 min = ____ min
16. 50 in. = ____ ft ____ in.
17. 10,560 ft = ____ mi
18. 2,000 oz = ____ lb

Compute the given sum or difference.

19. 4 hr 20 min + 3 hr 50 min
20. 20 ft − 1 ft 3 in.
21. 3 gal 2 qt − 1 gal 3 qt
22. 3 lb 6 oz + 2 lb 9 oz + 1 lb 3 oz

[9.2] *Choose the unit that you would most likely use to measure each quantity.*

23. The weight of a car
    a. milligrams **b.** grams **c.** kilograms
24. The width of a pencil's point
    a. millimeters **b.** centimeters **c.** meters
25. The capacity of an oil barrel
    a. milliliters **b.** liters **c.** meters
26. The distance a commuter drives
    a. millimeters **b.** centimeters **c.** kilometers

Choose the best estimate in each case.

27. The width of a piece of typing paper
    a. 16 mm **b.** 16 cm **c.** 16 km
28. The capacity of a bottle of mouthwash
    a. 100 mL **b.** 100 L **c.** 100 g
29. The weight of an aspirin pill
    a. 200 mg **b.** 200 g **c.** 200 kg
30. The length of an athlete's long jump
    a. 6.72 mm **b.** 6.72 cm **c.** 6.72 m

Change each quantity to the indicated unit.

31. 37 mg = ____ g

32. 4 kL = ____ L

33. 8 m = ____ cm

34. 2.1 km = ____ m

35. 600 mm = __ m

36. 5,100 g = __ kg

37. 5,000 mcsec = ____ sec

38. 4 GB = ____ MB

Change each quantity to the indicated unit, rounding to the nearest unit.

39. 4 oz ≈ ____ g

40. 5 cm ≈ ____ in.

41. 32 km ≈ ____ mi

42. 4 gal ≈ ____ L

Mixed Applications

Solve.

43. A DVD plays for 72 min. Express this playing time in hours.

44. In a recent year, a typical U.S. resident used about 1,430 gal of water a day for residential, agricultural, and industrial purposes. How many pints is this? (*Source:* usgs.gov)

45. *Frankenstein* (130 min) and *Dracula* (1 hr 15 min) are two classic horror films made in 1931. Which film is longer?

46. Some doctors recommend that athletes drink about 600 mL of fluid each hour. Express this amount in liters.

47. A teaspoon of common table salt contains about 2,000 mg of sodium. How many grams of sodium is this?

48. In a factory, a chemical process produced 3 mg of a special compound each hour. How many grams were produced in 24 hr?

49. A high blood level of cholesterol is a risk factor for heart disease. Having a low level of cholesterol, less than 200 mg per deciliter of blood, is considered desirable. Express this amount of cholesterol in grams. (*Source:* ext.colostate.edu)

50. It is estimated that 750 kg of pesticide is sprayed on a typical U.S. golf course each year. How many grams is this? (*Source: Journal of Pesticide Reform*)

51. A computer virus checker took 349 min to scan each file on a 320 GB hard drive. Express this length of time to the nearest hour.

52. A daily reference value (DRV) is a reference point that serves as a general guideline for a healthy diet. For a 2,000-calorie diet, the DRV for fiber is 25 g. Express this DRV in milligrams. (*Source:* U.S. Food and Drug Administration)

53. In Olympic gymnastics, the floor exercise is performed on a square mat measuring 10 m on a side. What is the length of one side to the nearest foot?

54. The weight of a precious stone is given in carats, where 1 carat is equal to 200 mg. The Hope Diamond weighs 45.52 carats. Express this weight in grams. (*Source:* Smithsonian Institution)

55. A website for bodybuilders recommends taking the dietary supplement melatonin to promote sleep without the hazards of prescription sleeping pills. The site notes that successful results can be achieved with dosages ranging from 100 mcg to 200 mg. What is the difference between these dosages in micrograms? (*Source:* bodybuilding.com)

56. A computer has a hard drive with capacity 50 GB. Express this capacity in megabytes.

57. In pairs figure skating, the free skate is 1 min 40 sec longer than the short program. If the short program is 2 min 50 sec, how long is the free skate? (*Source:* usfsa.org)

58. The following diagram shows the heights of an average U.S. woman and an average 10-year-old U.S. girl.

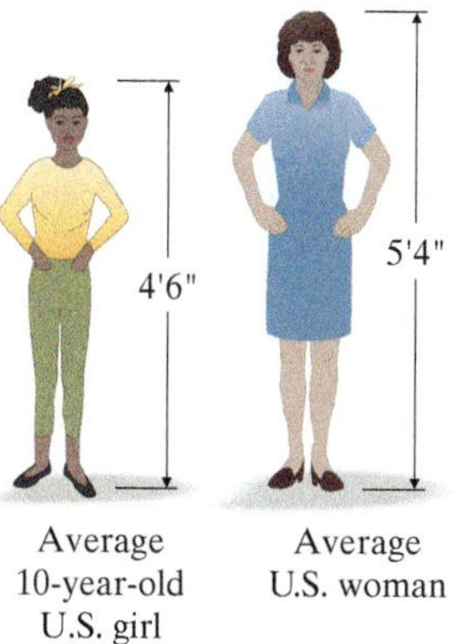

What is the difference in their heights? (*Source: Archives of Pediatrics and Adolescent Medicine*)

59. Thimerosal is a compound widely used since the 1930s as a preservative in a variety of drugs, including vaccines. Because of concerns about this use of thimerosal, a study was conducted in which individuals received up to 26,000 mcg of thimerosal. No toxic effects were found. What was the largest quantity of thimerosal in milligrams that these individuals received? (*Source:* fda.gov)

60. In 1956, IBM shipped the first hard drive, about the size of two refrigerators. This hard drive held 5,000,000 bytes of data. Express this quantity in megabytes. (*Source:* pcworld.com)

61. One of the shortest dinosaurs that ever lived was only 60 cm long when fully grown. What was the length of this dinosaur, rounded to the nearest inch? (*Source: Encarta Learning Zone Encyclopedia*)

62. About 43,000 pt of donated blood are used each day in the United States and Canada. How many liters is this, to the nearest 10,000 L? (*Source:* americasblood.org)

63. The table shows the average gestation period, in days, for various mammals. (*Source: The World Almanac and Book of Facts, 2010*)

Mammal	Gestation Period
Polar bear	240
Cow	284
Hippopotamus	238
Gorilla	258
Sea lion	350

How many more weeks is the gestation period for a sea lion than a hippopotamus?

64. The table shows the Saffir-Simpson Hurricane Scale, which is used to rate a hurricane based on its intensity. (*Source:* National Weather Service, NOAA)

Category	Wind Speed (mph)
1	74–95
2	96–110
3	111–130
4	131–155
5	156 and above

Express the range of wind speeds for a category 3 hurricane to the nearest kilometer per hour.

• Check your answers on page A-13.

CHAPTER 9 Posttest

FOR EXTRA HELP

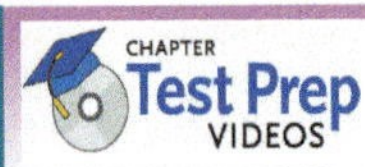

The Chapter Test Prep Videos with test solutions are available on DVD, in MyMathLab, and on YouTube (search "AkstDevMath" and click on "Channels").

To see if you have mastered the topics in this chapter, take this test.

1. 120 sec = ____ min

2. 7 yd = ____ ft

3. 3 gal = ____ pt

4. Add: 1 ft 9 in. and 1 ft 6 in.

5. Subtract 1 hr 25 min from 3 hr 10 min.

6. Which of the following units is a measure of capacity?
a. a gram **b.** a meter **c.** a liter **d.** an hour

7. The weight of a baby is measured in
a. milligrams **b.** grams **c.** kilograms

8. 400 cm = ____ m

9. 6 m = ____ mm

10. 4.6 sec = ______ mcsec

11. 2 mg = ____ mcg

12. 5 GB = ____ MB

13. 3 oz ≈ ____ g

14. 3 L ≈ ____ qt

15. A scientist dissecting a snake found in its stomach a centipede 140 mm long. Express the length of the centipede in centimeters. (*Source:* scienceblogs.com)

16. The cruising altitude of a passenger jet is 33,000 ft. Express this altitude in miles, rounded to the nearest whole number.

17. An adult humpback whale weighs about 60,000 lb and an adult finback whale weights about 40 tons. Which whale weighs more? (*Source:* Whale Center of New England)

18. On December 26, 2004, a powerful earthquake created huge waves that raced across the Indian Ocean. By the end of the day, more than 150,000 people in 11 countries were dead or missing, making this perhaps the most destructive tsunami in history and resulting in Earth spinning 3 mcsec faster. Express in seconds the change in spin time. (*Source:* nytimes.com)

19. A seven-year-old computer came with 960 MB of memory, in contrast to a more recent computer with 2 GB of memory. The new computer had how many times as much memory as the old, to the nearest tenth?

20. In order for an orbiting satellite to remain in the same spot over Earth, it must be approximately 35,786 km above the surface of Earth. Express this distance to the nearest thousand miles. (*Source:* Marshall Space Flight Center, NASA)

• Check your answers on page A-13.

Cumulative Review Exercises

To help you review, solve the following:

1. Compute: 9^3

2. Express as a mixed number: $\frac{11}{2}$

3. Combine and simplify: $\frac{5}{6} + \frac{1}{3} - \frac{11}{12}$

4. Divide: $\frac{2.8}{0.2}$

5. Solve and check: $\frac{20}{27} = \frac{5}{9}k$

6. Determine the unit rate: 120 mg of sodium in 5 crackers

7. \$36,000 is what percent of \$54,000?

8. Find the price of a book that normally sells for \$24 but is on sale at a 25% discount.

9. Find the difference: $-8 - (-3\frac{2}{5})$

10. Find the product: $\left(-\frac{2}{5}\right)\left(-1\frac{2}{3}\right)$

11. Find the range: 8, 6, 2, 9, 1, and 6

12. Solve: $3(x - 2) = -1$

13. Fill in the blank: 7 ft = _____ in.

14. Change 0.04 kg to grams.

15. On an Indiana farm, 160 bushels of wheat were raised on $5\frac{1}{3}$ acres. How many bushels per acre of wheat were raised?

16. The closing price of a share of technology stock on Monday was \$23.86. On Tuesday, the closing price was \$22.39. What was the change in the price of the stock?

17. In Finland, police levied traffic fines proportional to an offender's income. For speeding, a wealthy driver was fined about \$70,000, and a student, with a monthly income of approximately \$700, was fined about \$100. Estimate, to the nearest hundred thousand dollars, the monthly income of the wealthy driver. (*Source:* Steve Stecklow, "Helsinki on Wheels: Fast Finns Find Fines Fit Their Finances," *Wall Street Journal*)

18. The bar graph shows the number of American Kennel Club registrations for various breeds in a recent year.

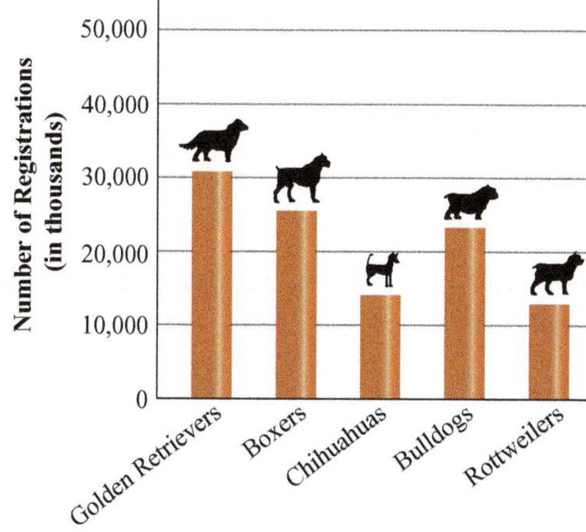

About how many more golden retrievers were registered than bulldogs? (*Source:* akc.org)

19. Fishermen commonly want to find the weight of a fish that they have caught, even when they are going to throw it back into the water. They can approximate this weight W (in pounds) by measuring the length of the fish L and the girth of the fish G (both in inches). Here girth means the distance around the body of the fish at its largest point.

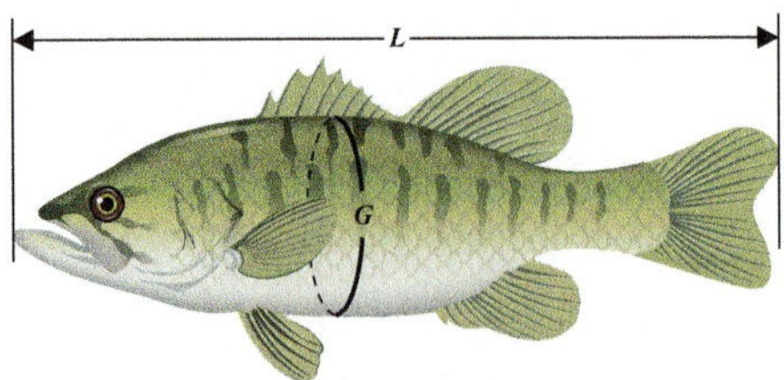

The formula $W = \dfrac{L \cdot G^2}{800}$ approximates the weight of a trout. Find, to the nearest pound, the weight of a trout 30 in. long with girth 15 in. (*Source:* dnr.wi.gov)

20. The distance from Memphis to Atlanta is given on the map shown below. Express this distance to the nearest hundred miles. (*Source: The World Almanac 2010*)

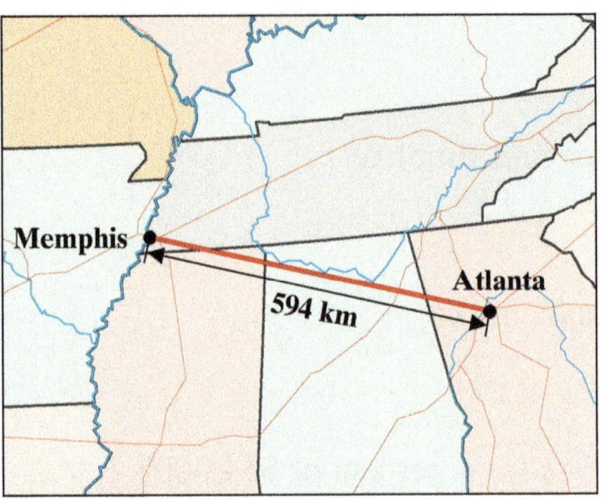

• Check your answers on page A-13.

Basic Geometry

Geometry and Architecture

Students of geometry study abstract figures in space, whereas architects design real structures in space. The two fields, geometry and architecture, are, therefore, closely related.

The simplest architectural structures have basic geometric shapes. An igloo in the far north and a dome that graces a state capitol are shaped like hemispheres. A tepee is in the shape of a cone, and the peak of a roof is a pyramid.

The rectangle plays an especially important role in architectural design. Bricks, windows, doors, rooms, buildings, lots, city blocks, and street grids are all based on the rectangle—one of the most adaptable shapes for human needs.

Of all the rectangles with a given area, the square has the smallest perimeter. As a result, warehouses are often built in the form of squares. On the other hand, houses, hotels, and hospitals—for which daylight and a long perimeter are more important—are seldom square in shape.

(*Source:* William Blackwell, *Geometry in Architecture,* John Wiley and Sons, 1984)

CHAPTER 10 PRETEST

To see if you have already mastered the topics in this chapter, take this test.

1. Sketch and label an example of an obtuse $\angle PQR$

2. Find: $\sqrt{121}$

3. Find the supplement of 100°.

4. Find the complement of 36°.

Find each perimeter or circumference. Use $\pi \approx 3.14$, when needed.

5.

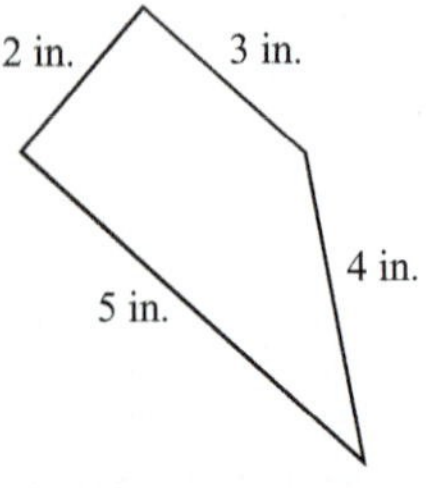

6.

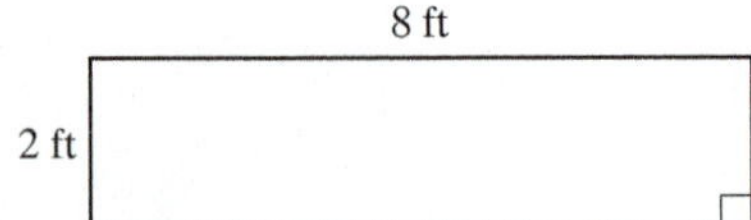

7. A circle with a diameter of 4 in.

8. A square with side 2.6 m.

Find each area. Use $\pi \approx 3.14$, when needed.

9.

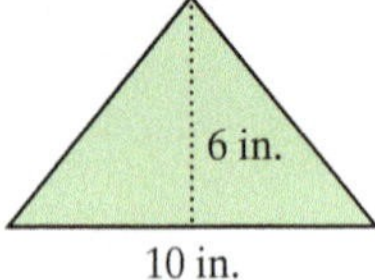

10.

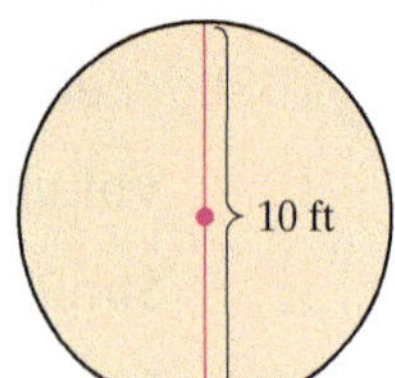

Find each volume. Use $\pi \approx 3.14$, when needed.

11.

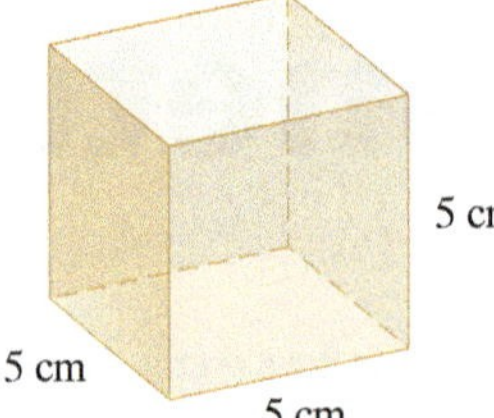

12.

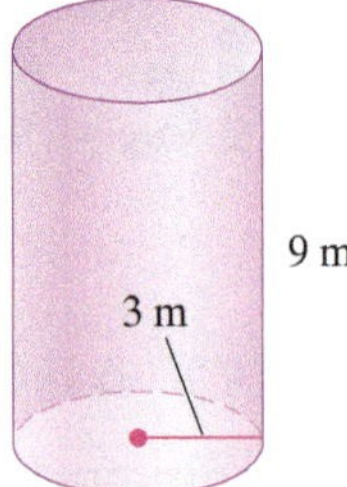

Find the value of each unknown.

13.

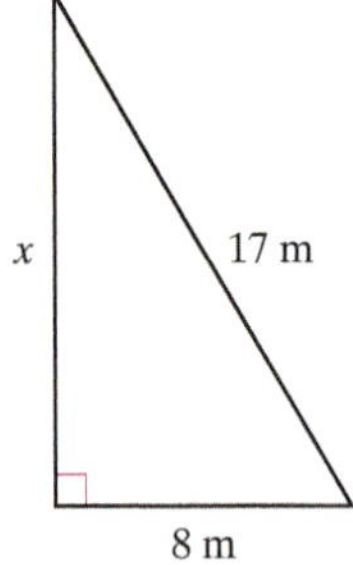

14.

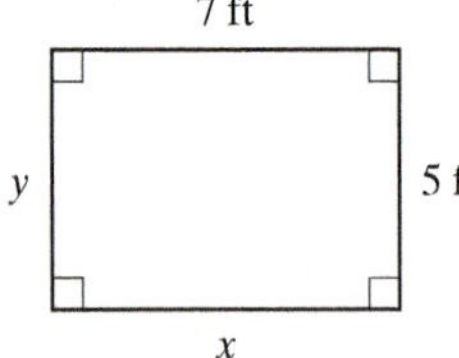

15.

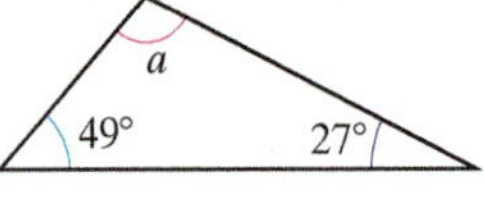

16. For the diagrams shown, ΔABC is similar to ΔDEF. Find y.

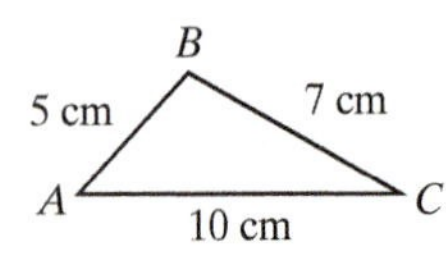

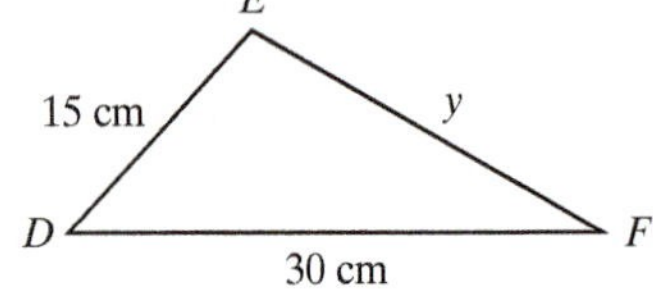

17. On the front of the brick shown, does each corner form an acute angle?

Solve.

18. Rescuers are searching for survivors of a shipwreck that took place within a mile of a rock. What is the area of the region that would be most appropriate to search for survivors? Round to the nearest square mile.

19. An entertainment center has just enough room to hold a 52-inch television, measured diagonally. Will an LCD HDTV with sides 44.4 in. and 30.2 in. fit in the entertainment center?

20. In constructing the foundation for a house, a contractor digs a hole 6 ft deep, 54 ft long, and 25 ft wide. How many cubic feet of earth are removed?

• Check your answers on page A-13.

Cultural Note

A *tessellation* is any repeating pattern of interlocking shapes. Some shapes, depending on their geometric properties, will tessellate, that is, go on indefinitely, covering the plane without overlapping and without gaps. These shapes include squares and equilateral triangles. Many other shapes will not tessellate. Tessellations are commonly found in the home—for instance, in the design of wall, ceiling, and floor coverings. More elaborate tessellations are found in mosaics that survive from ancient times. The tile mosaic shown to the left is found at the Alhambra, the summer residence of Moorish kings built in the fourteenth century in Granada, Spain.

(*Source:* The Metropolitan Museum of Art, *Islamic Art and Geometric Design: Activities for Learners*, New York: The Metropolitan Museum of Art, 2004)

10.1 Introduction to Basic Geometry

OBJECTIVES

- **A** To identify basic geometric concepts or figures
- **B** To find missing sides or angles
- **C** To solve applied problems involving basic geometric concepts or figures

What Geometry Is and Why It Is Important

The word *geometry*, which dates back thousands of years, means "measurement of the Earth." Today, we use the term to mean the branch of mathematics that deals with concepts such as point, line, angle, perimeter, area, and volume.

Ancient peoples, including the Egyptians, used the principles of geometry in their construction projects. They understood these principles because of observations they made in their daily lives and their studies of the physical forms in nature.

Geometry also has many practical applications in such diverse fields as art and design, multimedia, architecture, physics, and engineering. In city planning, geometric concepts, relationships, and notation are often used when designing the layout of a city. Note below how the use of geometric thinking helps to transform the street plan on the left to the geometric diagram on the right, making it easier to focus on the key features of the street plan.

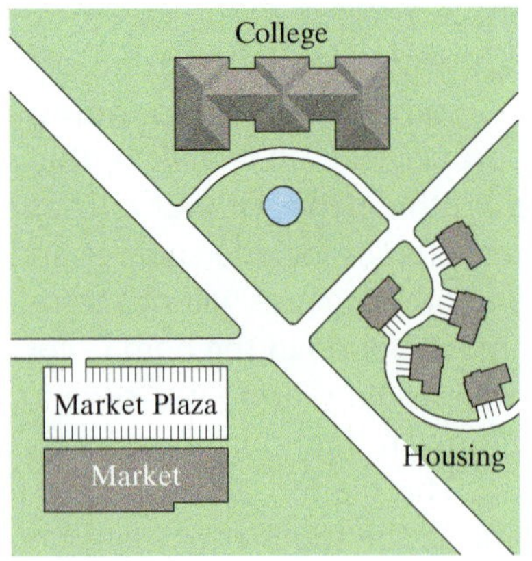

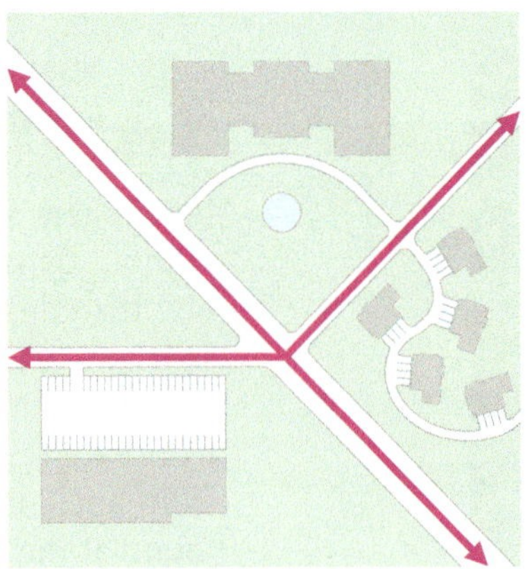

Basic Geometric Concepts

Let's first consider some of the basic concepts that underlie the study and application of geometry. The following table lists and explains some basic geometric terms illustrated in the preceding street plan. We use these terms throughout this chapter.

Definition	Example
A **point** is an exact location in space. A point has no dimension.	A • (read "point A")
A **line** is a collection of points along a straight path that extends endlessly in both directions. A line has only one dimension.	$\overleftrightarrow{CB}$ (read "line CB")
A **line segment** is a part of a line having two endpoints. Every line segment has a length.	$\overline{AB}$ (read "line segment AB") The length of $\overline{AB}$ is denoted AB.
A **ray** is a part of a line having only one endpoint.	$\overrightarrow{CD}$ (read "ray CD") (The endpoint is always the first letter.)

Definition	Example
An **angle** consists of two rays that have a common endpoint called the **vertex** of the angle.	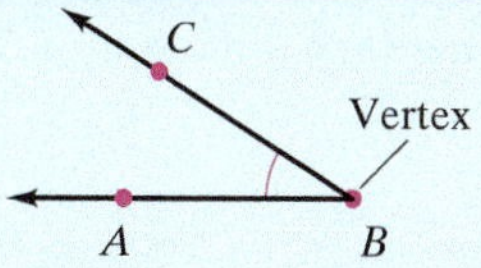 $\angle ABC$ (read "angle ABC") (The vertex is always the middle letter.) $\angle ABC$ can also be written as $\angle CBA$ or just $\angle B$.
A **plane** is a flat surface that extends endlessly in all directions.	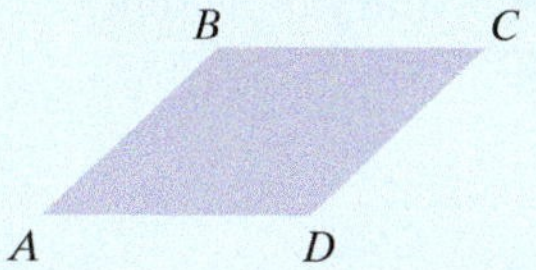 Plane $ABCD$

The unit in which angles are commonly measured is the degree (°). Angles are classified according to their measures. To indicate the measure of $\angle ABC$, we write $m\angle ABC$.

Definition	Example
A **straight angle** is an angle whose measure is 180°.	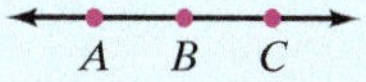 $\angle ABC$ is a straight angle, $m\angle ABC = 180°$.
A **right angle** is an angle whose measure is 90°.	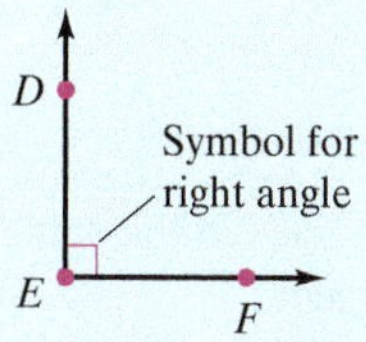 $\angle DEF$ is a right angle; $m\angle DEF = 90°$.
An **acute angle** is an angle whose measure is less than 90°.	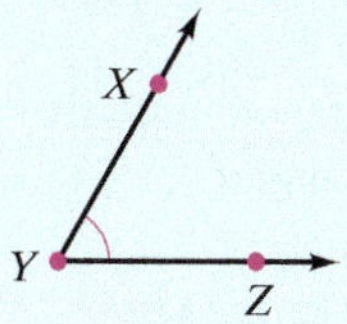 $\angle XYZ$ is an acute angle.
An **obtuse angle** is an angle whose measure is more than 90° and less 180°.	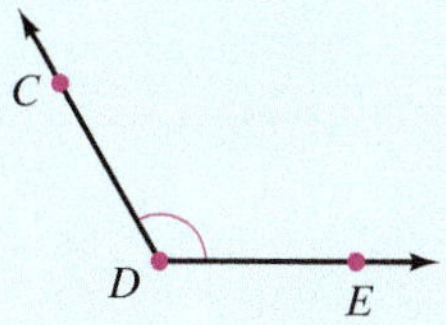 $\angle CDE$ is an obtuse angle.

continued

Definition	Example
Two angles are complementary if the sum of their measures is 90°.	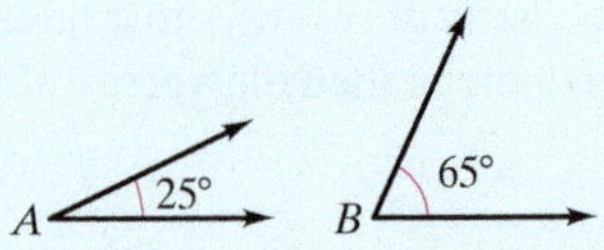 $m\angle A + m\angle B = 25° + 65° = 90°$ $\angle A$ and $\angle B$ are complementary angles.
Two angles are supplementary if the sum of their measures is 180°.	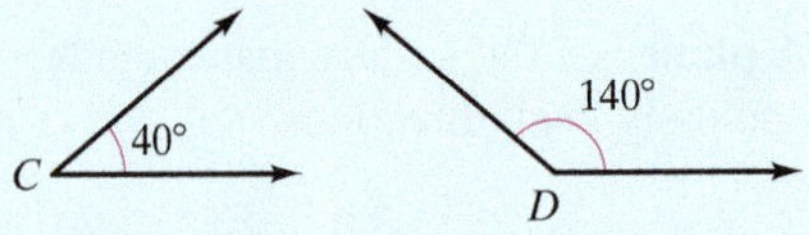 $m\angle C + m\angle D = 40° + 140° = 180°$ $\angle C$ and $\angle D$ are supplementary angles.

Lines in a plane are either intersecting or parallel.

Definition	Example
Intersecting lines are two lines that cross.	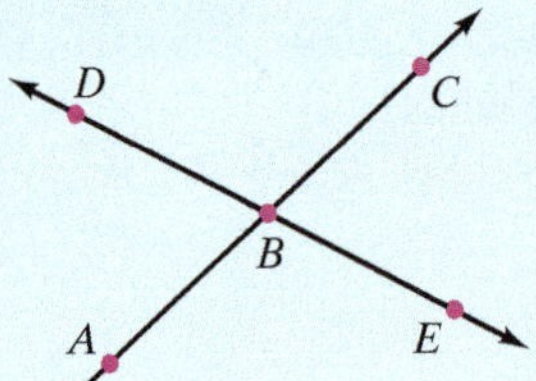 $\overleftrightarrow{AC}$ intersects $\overleftrightarrow{DE}$ at point B.
Parallel lines are two lines in the same plane that do not intersect.	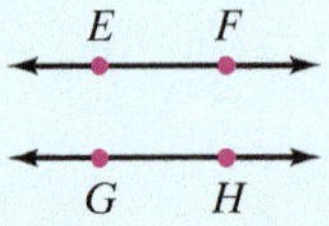 $\overleftrightarrow{EF} \parallel \overleftrightarrow{GH}$ is read "$\overleftrightarrow{EF}$ is parallel to $\overleftrightarrow{GH}$."
Perpendicular lines are two lines that intersect to form right angles.	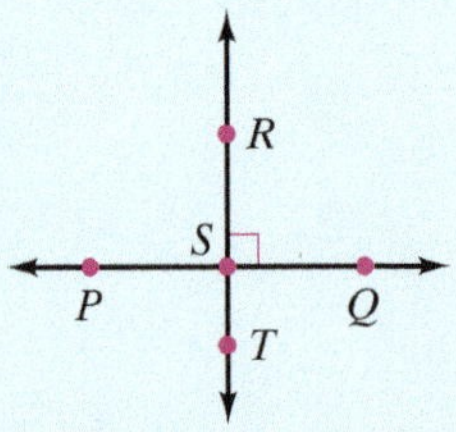 $\overleftrightarrow{RT} \perp \overleftrightarrow{PQ}$ is read "$\overleftrightarrow{RT}$ is perpendicular to $\overleftrightarrow{PQ}$." $\angle RSP$, $\angle RSQ$, $\angle PST$, and $\angle QST$ are all right angles.

When two lines intersect, two special pairs of angles are formed.

Definition	Example
Vertical angles are two opposite angles with equal measure formed by two intersecting lines.	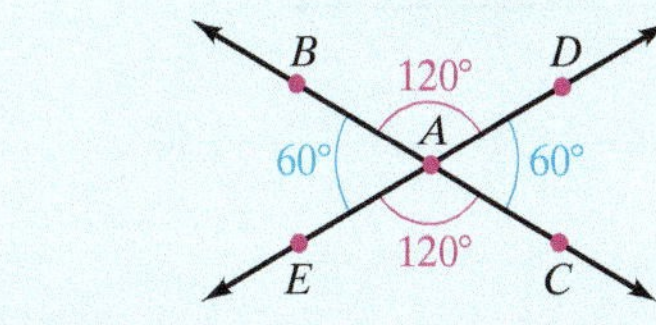 $\angle BAE$ and $\angle DAC$ are vertical angles. $\angle BAD$ and $\angle EAC$ are vertical angles.

Try drawing another pair of vertical angles. Do you think that their measures are equal? Can you describe the pair of angles formed by intersecting lines that are not vertical angles?

Now, let's consider more examples involving these basic geometric terms.

EXAMPLE 1

Sketch and label $\overline{EF}$.

Solution First, sketch a line segment.

Then, label the line segment. 

This line segment is written $\overline{EF}$ and is read "line segment EF."

PRACTICE 1

Draw $\angle ABC$.

EXAMPLE 2

$\angle A$ and $\angle B$ are complementary angles. Find the measure of $\angle A$ if $m\angle B = 69°$.

Solution Because $\angle A$ and $\angle B$ are complementary angles, $m\angle A + m\angle B = 90°$.

$$m\angle A + m\angle B = 90°$$
$$m\angle A + 69° = 90°$$
$$m\angle A + 69° - 69° = 90° - 69° \quad \text{Subtract 69° from each side.}$$
$$m\angle A = 21°$$

PRACTICE 2

What is the measure of the angle complementary to 37°?

EXAMPLE 3

Find the measure of the angle that is supplementary to 89°.

Solution To find the measure of the angle that is supplementary to 89°, we write the following equation:

$$89° + x = 180°$$

where x represents the measure of the supplementary angle.

$$89° + x = 180°$$
$$89° - 89° + x = 180° - 89°$$
$$x = 91°$$

So an angle with measure 91° is supplementary to one with measure 89°.

PRACTICE 3

What is the measure of the angle supplementary to 15°?

EXAMPLE 4

In the diagram to the right, $\angle ABC$ is a straight angle. Find y.

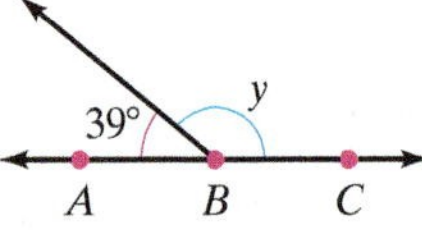

Solution Because $\angle ABC$ is a straight angle, $y + 39° = 180°$. We solve this equation for y.

$$y + 39° = 180°$$
$$y + 39° - 39° = 180° - 39°$$
$$y = 141°$$

PRACTICE 4

In the diagram shown, find x.

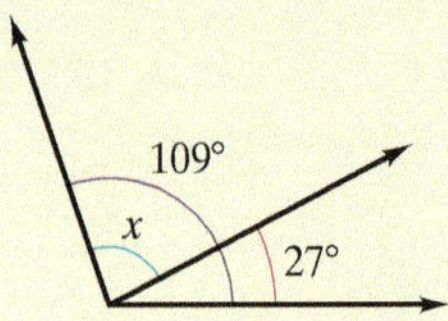

EXAMPLE 5

Find the values of x and y in the diagram to the right.

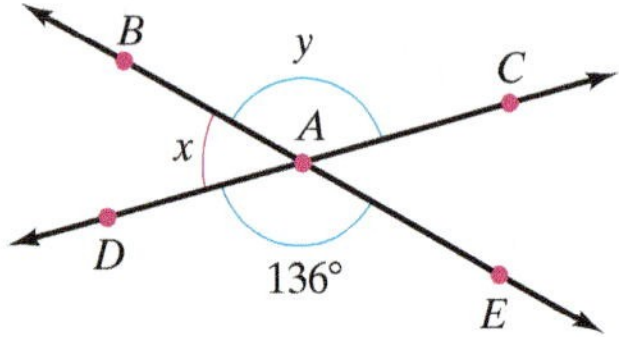

Solution Because $\angle BAC$ and $\angle DAE$ are vertical angles and $\angle DAE = 136°$, $\angle BAC = 136°$, or $y = 136°$. Because $\angle DAC$ is a straight angle, the sum of x and y is 180°.

$$x + y = 180°$$
$$x + 136° = 180°$$
$$x + 136° - 136° = 180° - 136°$$
$$x = 44°$$

So $x = 44°$ and $y = 136°$.

PRACTICE 5

In the following diagram, what are the values of a and b?

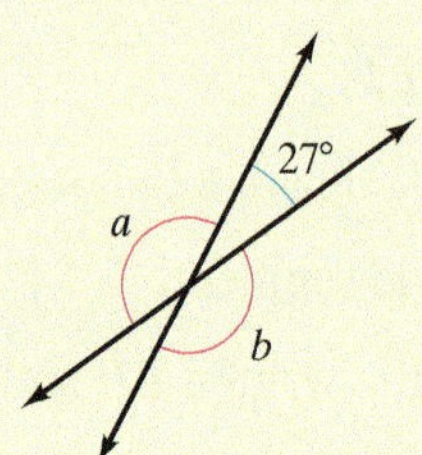

Basic Geometric Figures

Here, we use the concepts just discussed to define some basic geometric figures: triangles, trapezoids, parallelograms, rectangles, squares, and circles. Except for circles, these figures are *polygons*.

DEFINITION

A **polygon** is a closed plane figure made up of line segments.

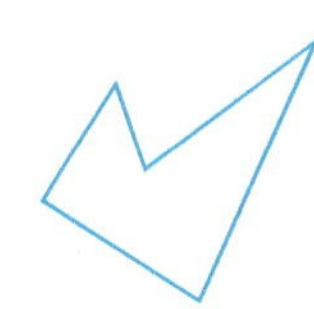

Closed: A polygon

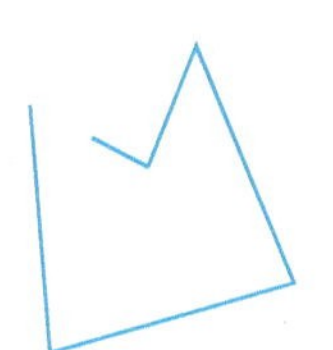

Not closed: Not a polygon

Polygons are classified according to the number of their sides. Here we examine two types of polygons—triangles and quadrilaterals.

Definition	Example
A **triangle** is a polygon with three sides.	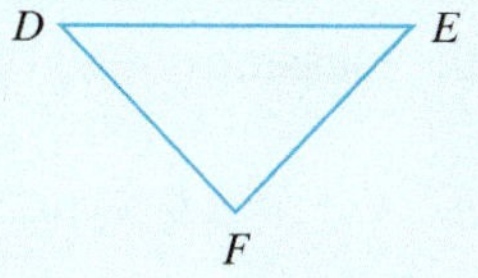 D E F — ΔDEF (read "triangle DEF") ΔDEF has three *vertices* (plural of *vertex*)—points D, E, and F. ΔDEF has sides $\overline{DE}$, $\overline{EF}$, and $\overline{DF}$.
A **quadrilateral** is a polygon with four sides.	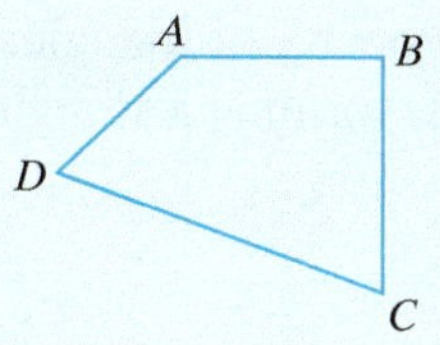 A B D C — Quadrilateral $ABCD$ has four vertices—points A, B, C, and D.

Triangles are classified by the lengths of their sides or the measures of their angles.

Definition	Example
An **equilateral triangle** is a triangle with three sides equal in length.	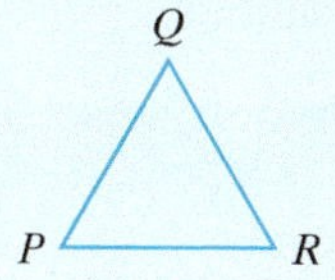 Q P R — $\overline{PQ}$, $\overline{QR}$, and $\overline{PR}$ have equal lengths.
An **isosceles triangle** is a triangle with two or more sides equal in length.	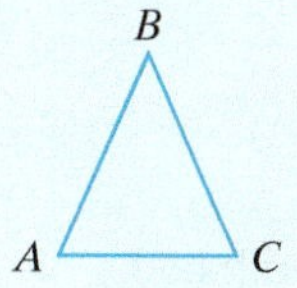 B A C — $\overline{AB}$ and $\overline{BC}$ have equal lengths.
A **scalene triangle** is a triangle with no sides equal in length.	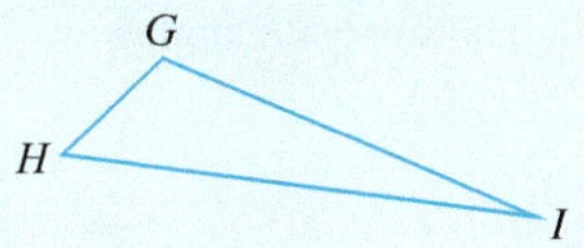 G H I — The three sides have unequal lengths.
An **acute triangle** is a triangle with three *acute* angles.	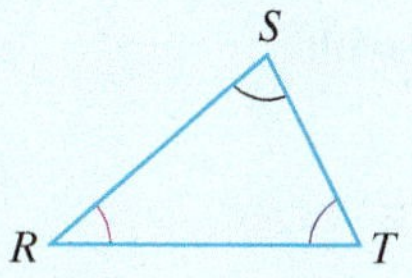 S R T — $\angle R$, $\angle S$, and $\angle T$ are acute angles.
A **right triangle** is a triangle with one right angle.	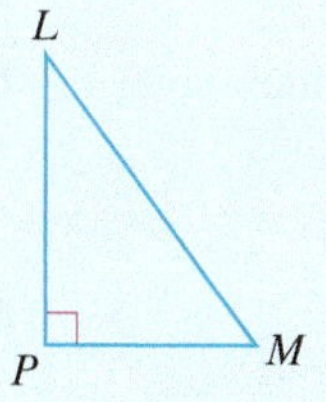 L P M — $\angle P$ is a right angle.
An **obtuse triangle** is a triangle with one obtuse angle.	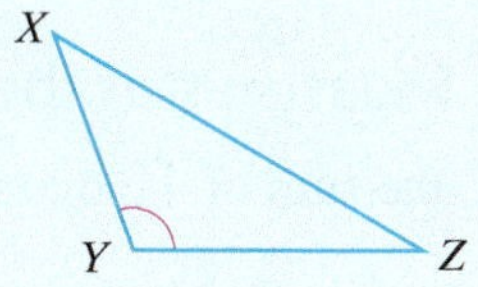 X Y Z — $\angle Y$ is an obtuse angle.

The Sum of the Measures of the Angles of a Triangle

In any triangle, the sum of the measures of all three angles is 180°, that is, for any ΔABC,

$$m\angle A + m\angle B + m\angle C = 180°$$

To demonstrate that this property of triangles is reasonable, we can put the three angles of any triangle next to each other forming a straight angle.

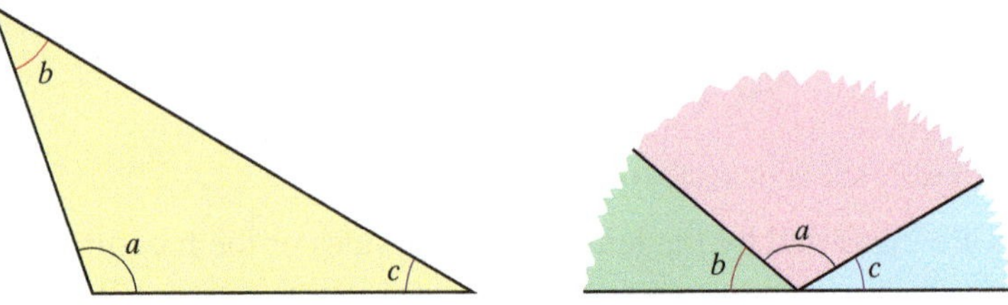

We have already seen that a polygon with four sides is called a *quadrilateral*. Let's consider special types of quadrilaterals.

Definition	Example
A **trapezoid** is a quadrilateral with only one pair of opposite sides parallel.	A B C D $\overline{AB} \parallel \overline{CD}$
A **parallelogram** is a quadrilateral with both pairs of opposite sides parallel. Opposite sides are equal in length, and opposite angles have equal measures.	L M P O $\overline{LM} \parallel \overline{PO}$ and $\overline{LP} \parallel \overline{MO}$ $\overline{LM}$ and $\overline{PO}$ have equal lengths, and $\overline{LP}$ and $\overline{MO}$ have equal lengths. $m\angle L = m\angle O$ and $m\angle P = m\angle M$
A **rectangle** is a parallelogram with four right angles.	R S T U $m\angle R = m\angle T = m\angle U = m\angle S = 90°$
A **square** is a rectangle with four sides equal in length.	D E F G $\overline{DE}$, $\overline{EG}$, $\overline{FG}$, and $\overline{DF}$ have equal lengths.

The Sum of the Measures of the Angles of a Quadrilateral

In any quadrilateral, the sum of the measures of the angles is 360°. That is, for any quadrilateral $ABCD$,

$$m\angle A + m\angle B + m\angle C + m\angle D = 360°$$

We can see why this property of quadrilaterals is true by cutting a quadrilateral into two triangles. In each triangle, the sum of the measures of the three angles is 180°.

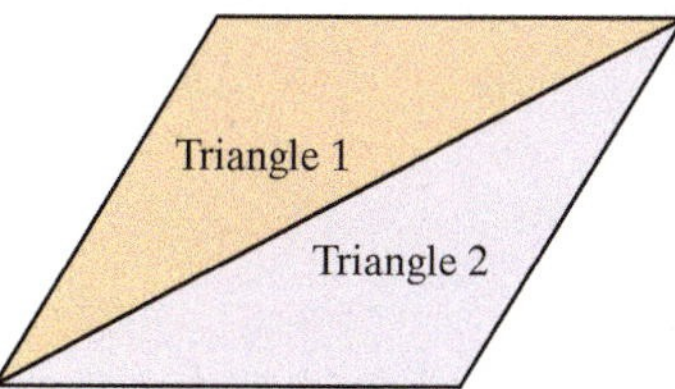

The last basic geometric figure we consider here is the circle.

Definition	Example
A **circle** is a closed plane figure made up of points that are all the same distance from a fixed point called the **center**.	O Circle with center O
A **diameter** is a line segment that passes through the center of a circle and has both endpoints on the circle.	A O B Diameter $\overline{AB}$
A **radius** is a line segment with one endpoint on the circle and the other at the center.	O B Radius $\overline{OB}$

Note that the diameter (d) of a circle is twice the radius (r), or $d = 2r$.

EXAMPLE 6

Sketch and label isosceles triangle ABC. Name the equal sides.

Solution

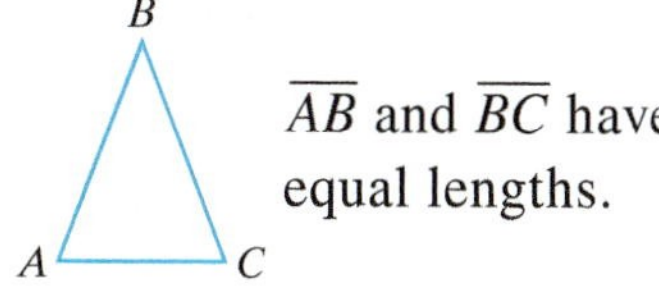

$\overline{AB}$ and $\overline{BC}$ have equal lengths.

PRACTICE 6

Draw and label quadrilateral $ABCD$ that has at least one right angle with opposite sides equal and parallel. Name both pairs of parallel sides.

EXAMPLE 7

In ΔDEF, $m\angle D = 45°$ and $m\angle E = 65°$. Find $m\angle F$.

Solution First, we draw a diagram.

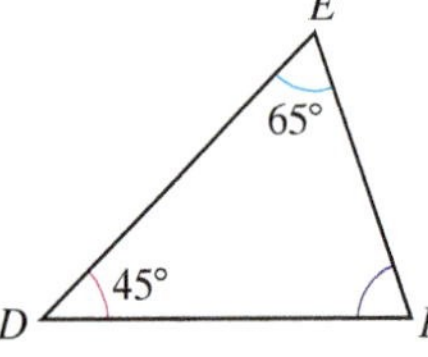

The sum of the measures of the angles is 180°, so we write the following:

$$
\begin{aligned}
m\angle D + m\angle E + m\angle F &= 180° \\
45° + 65° + m\angle F &= 180° \\
110° + m\angle F &= 180° \\
110° - 110° + m\angle F &= 180° - 110° \\
m\angle F &= 70°
\end{aligned}
$$

PRACTICE 7

In ΔRST, $\angle S$ is a right angle and $m\angle T = 30°$. What is $m\angle R$?

EXAMPLE 8

In the quadrilateral shown, what is $m\angle D$?

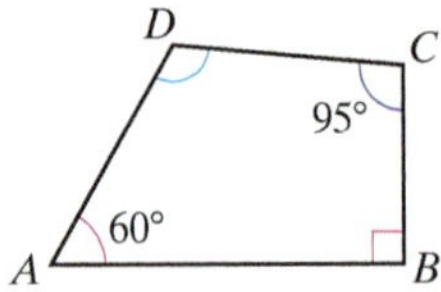

Solution The sum of the measures of the four angles is 360°. Note that $\angle B$ is a right angle, so $m\angle B = 90°$. We write the following:

$$
\begin{aligned}
m\angle A + m\angle B + m\angle C + m\angle D &= 360° \\
60° + 90° + 95° + m\angle D &= 360° \\
245° + m\angle D &= 360° \\
245° - 245° + m\angle D &= 360° - 245° \\
m\angle D &= 115°
\end{aligned}
$$

PRACTICE 8

In the trapezoid shown, what is $m\angle U$?

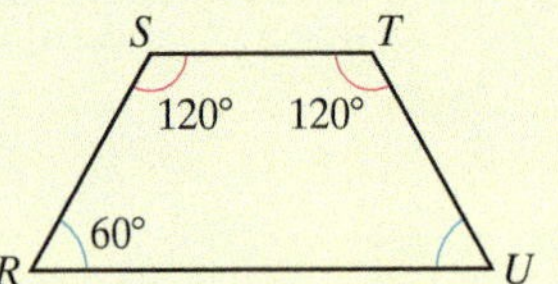

EXAMPLE 9

A dartboard has a diameter of 18 in. What is the radius of the dartboard?

Solution The diameter of the dartboard is 18 in. The distance from the center to a point on the edge of the board is the radius. To find the radius, we divide the diameter by 2.

$$18 \div 2 = 9$$

So the radius of the dartboard is 9 in.

PRACTICE 9

A manhole cover has a radius of 12 in. What is the diameter of the manhole cover?

10.1 Exercises

FOR EXTRA HELP     

Mathematically Speaking

Fill in each blank with the most appropriate term or phrase from the given list.

radius	supplementary	parallelogram	complementary	trapezoid
scalene	vertical	isosceles	line segment	parallel
acute	perpendicular	ray	obtuse	diameter

1. A(n) __________ is a part of a line having two endpoints.

2. The measure of a(n) __________ angle is more than $90°$ and less than $180°$.

3. Two angles are ____________ if the sum of their measures is $90°$.

4. Two angles are ____________ if the sum of their measures is $180°$.

5. Lines that intersect to form right angles are called ____________.

6. Opposite angles with equal measure formed by two intersecting lines are called __________.

7. Lines in the same plane that do not intersect are called __________.

8. A(n) __________ triangle has two or more sides equal in length.

9. A(n) __________ triangle has no sides equal in length.

10. A(n) __________ is a quadrilateral with only one pair of opposite sides parallel.

11. A(n) ___________ is a quadrilateral with both pairs of opposite sides parallel.

12. A(n) __________ is a line segment that passes through the center of a circle and has both endpoints on the circle.

A *Sketch and label each geometric object. Where appropriate, use symbols to express your answer.*

13. Point P

14. $\overleftrightarrow{AB}$

15. $\overline{BC}$

16. $\overrightarrow{AB}$

17. Parallel lines $\overleftrightarrow{MN}$ and $\overleftrightarrow{ST}$

18. Perpendicular lines $\overleftrightarrow{UV}$ and $\overleftrightarrow{WX}$

19. Equilateral ΔABC

20. Isosceles ΔPQR

21. Circle with center O

22. Trapezoid

23. Scalene ΔABC

24. Right ΔWXY

25. Acute $\angle FGH$

26. Vertical angles

Solve.

27. Find x.

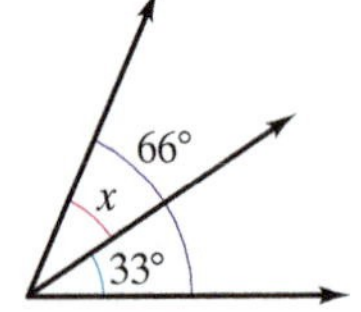

28. $\angle PQR$ is a straight angle. Find the measure of $\angle PQS$.

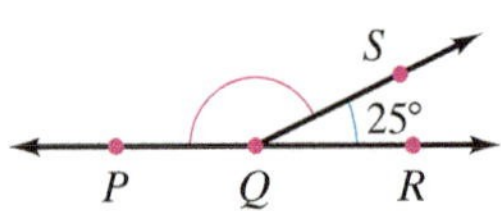

29. $\angle DEF$ is a right angle. Find the measure of $\angle DEG$.

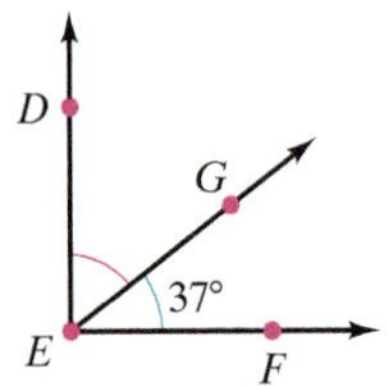

30. Solve for x and y.

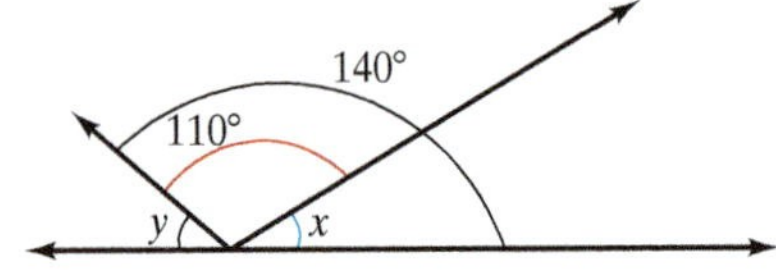

In the diagram shown, $\overleftrightarrow{AB} \perp \overleftrightarrow{CD}$ and $m\angle CPE = 35°$. Find the measure of each angle.

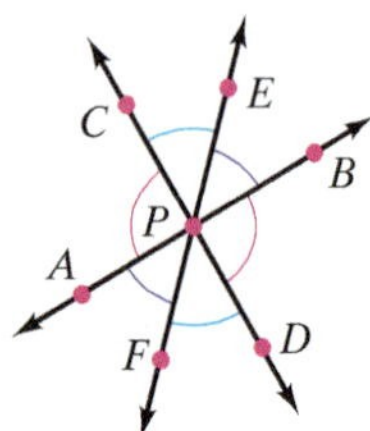

31. $\angle CPD$

32. $\angle APD$

33. $\angle BPD$

34. $\angle CPB$

35. $\angle APB$

36. $\angle BPE$

37. $\angle FPD$

38. $\angle APF$

In each figure, find the measure of the unknown angle(s).

39.

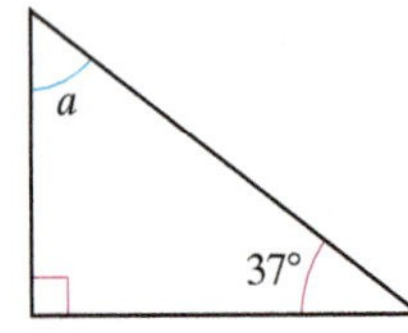

40.

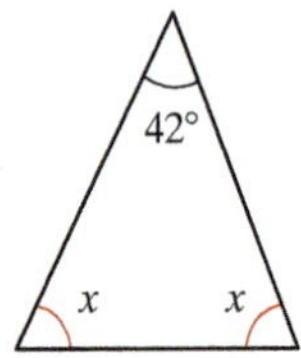

41.

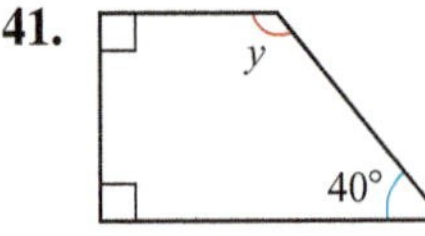

42.

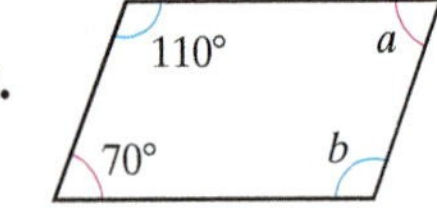

Solve.

43. Find the complement of 35°.

44. What is the measure of an angle that is complementary to itself?

45. Find the supplement of 105°.

46. What is the measure of an angle that is supplementary to 88°?

47. In ΔABC, $m\angle A = 35°$ and $m\angle B = 75°$. Find $m\angle C$.

48. In ΔDEF, $m\angle E = 90°$ and $m\angle F = 19°$. Find $m\angle D$.

49. In a parallelogram, the sum of three of the angles is 275°. What is the measure of the fourth angle?

50. In a triangle in which all angles are equal, what is the measure of each angle?

Mixed Practice

Solve.

51. Sketch and label the diameter of a circle

52. In the following diagram, find the measure of $\angle DBE$.

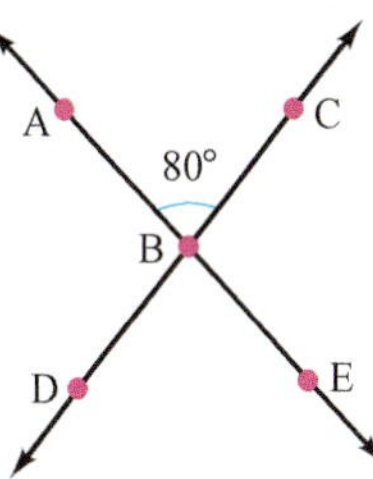

53. Find x.

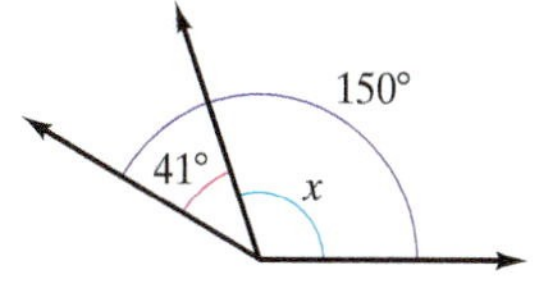

54. Find x.

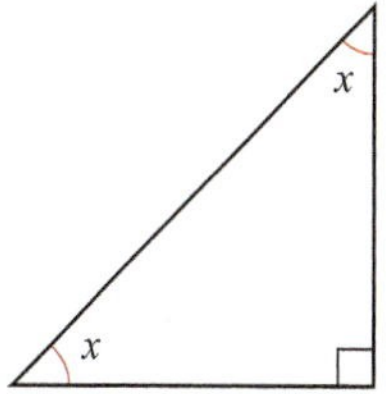

55. Sketch and label obtuse ΔRST.

56. What is the measure of an angle that is complementary to 20°?

Applications

C *Solve.*

57. An ancient circular medicine wheel made with rocks, as shown in the diagram, was built by Native Americans in Wyoming. What is the wheel's radius? (*Source:* Works Projects Administration, *Wyoming: A Guide to Its History, Highways, and People*)

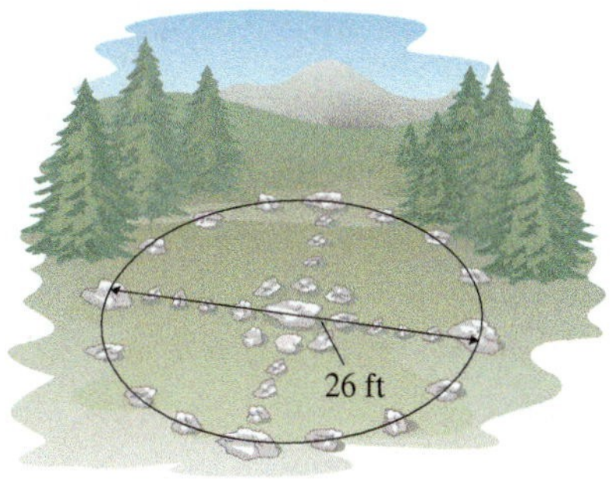

58. The Human Centrifuge, located outside of Philadelphia, is a sphere at the end of a 50-foot-long metal pole spinning in a circular room. The centrifuge was used to sling around early astronauts in order to understand the effect on their bodies of high pressure. What is the diameter of the circle formed by the spinning centrifuge? (*Sources:* nadcmuseum.org and hq.nasa.gov)

59. In a physics class, the instructor connects two blocks with a cord that passes over a small, frictionless pulley, as shown. What is the measure of the missing angle?

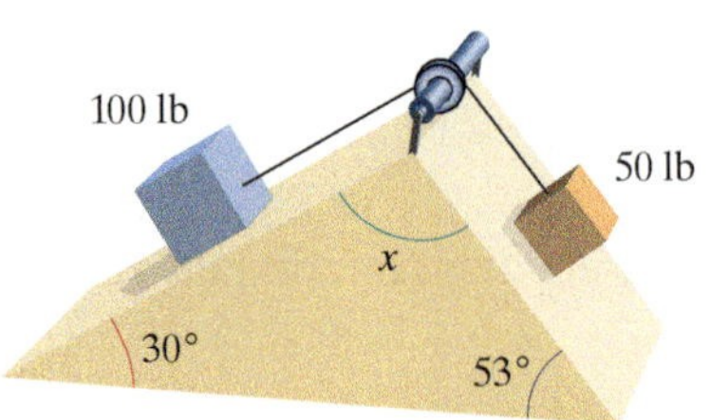

60. In the diagram below, when a beam of light strikes a plane mirror, $m\angle 1 = m\angle 2$. If $m\angle 1 = 40°$, what is $m\angle 3$.

61. The Bermuda triangle, mapped below, is a region of the Atlantic Ocean famous because many people, aircraft, and ships have disappeared within its bounds. Is this triangle acute, right, or obtuse?

62. The 1996 U.S. presidential election was a contest between the Democrat Bill Clinton, the Republican Bob Dole, and the Reform Party's Ross Perot. In the circle graph below, which candidate's popular vote is represented by an acute angle? (*Source:* wikipedia.org)

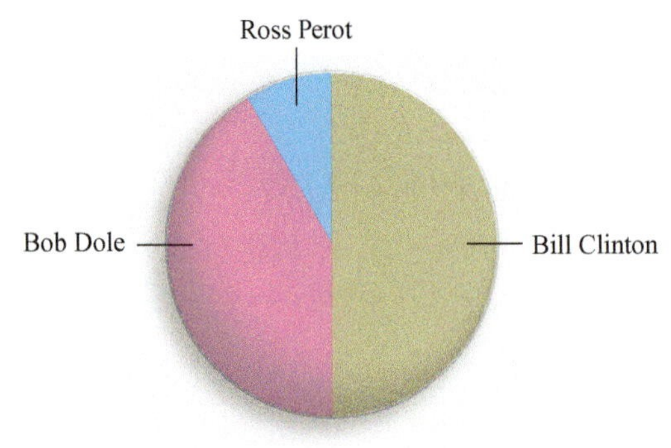

63. The State Dining Room in the White House is a rectangle approximately 49 ft by 36 ft. What would be the dimensions of a new rug for the room if a strip of 2 ft of wood floor needs to be seen around the outside of the rug? (*Source:* whitehousemuseum.org)

64. Hikers walk due west and then veer off on a path at an angle 23° to the north. To head due north what angle must they make with the path?

65. A ceramic tile in the shape of a parallelogram used for a bathroom remodeling project is shown in the diagram.

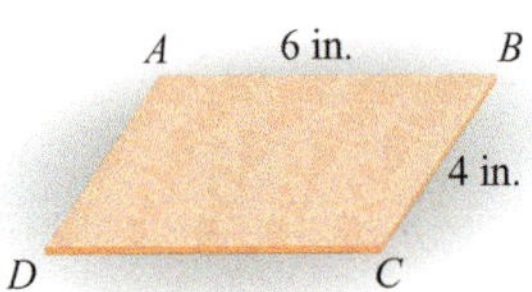

a. What is the length of the side parallel to $\overline{AB}$?

b. If $m\angle C = 120°$, then what is $m\angle D$?

66. A circular mirror shown in the diagram has a radius of 10 in.

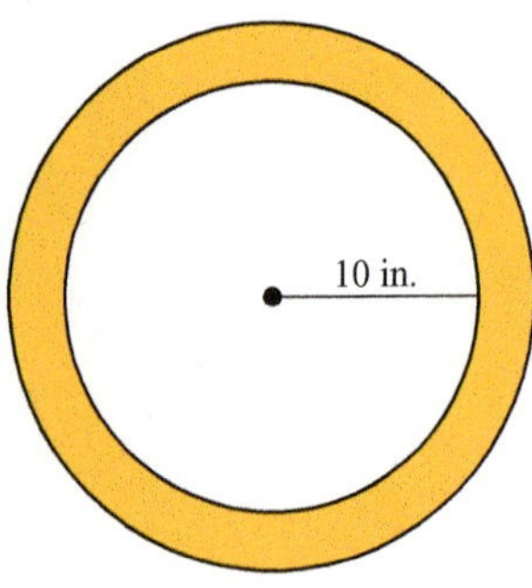

a. What is the diameter of the mirror, *excluding* the circular frame?

b. The diameter of the mirror, *including* the frame, is 28 in. How wide is the frame?

67. The top and bottom edges of the planter shown below are parallel.

a. What is the shape of each side of the planter?

b. For all sides, the measure of each of the lower angles is 100°. Find the measure of the upper angles, which are also equal to one another.

68. WORT-FM is a noncommercial radio station in south central Wisconsin. It broadcasts a 2000-watt signal in every direction from a transmitter in Madison, Wisconsin.

a. Describe the station's coverage area.

b. If the coverage area of the station has a 100-mile diameter, what is the radius of this area? (*Source:* wort-fm.org)

• Check your answers on page A-13.

MINDStretchers

Patterns

1. How many rectangles can you find in the diagram below?

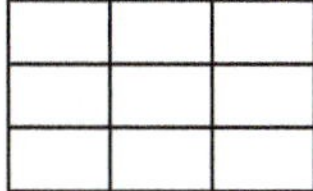

Writing

2. Can a triangle contain one right angle and one obtuse angle? Explain.

Groupwork

3. A *hexagon* is a polygon with 6 sides and 6 angles. Working with a partner, show how you can find the sum of the measures of the angles in the hexagon.

10.2 Perimeter and Circumference

OBJECTIVES

- **A** To find the perimeter of a polygon or the circumference of a circle
- **B** To find the perimeter or circumference of a composite geometric figure
- **C** To solve applied problems involving perimeter or circumference

The Perimeter of a Polygon

One of the most basic features of a plane geometric figure is its *perimeter*. The length of a fence around a plot of land, the length of a state's border, and the length of a picture frame are examples of perimeters.

DEFINITION

The **perimeter** of a polygon is the distance around it.

To find the perimeter of any polygon, we add the lengths of its sides. Note that perimeters are measured in linear units such as feet or meters.

Suppose that we want to build a fence around the garden shown below:

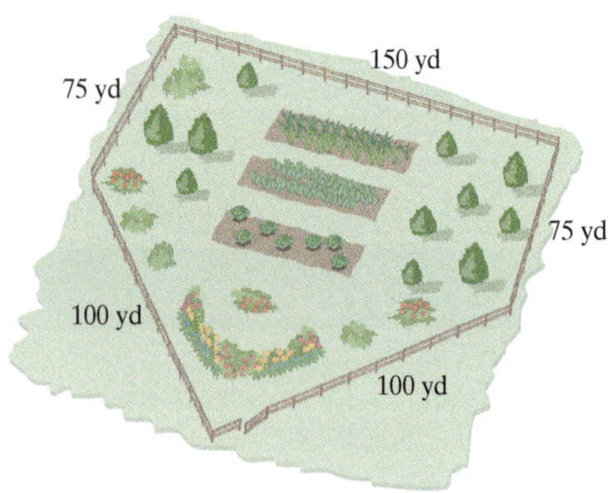

How much fencing do we need? Using the definition of perimeter, we obtain the distance around this garden.

$$75 + 150 + 75 + 100 + 100 = 500$$

So we need 500 yd of fencing.

For some polygons, we can also use a *formula* to find the perimeter. Let's consider the formulas for the perimeter of a triangle, a rectangle, and a square.

Figure	Formula	Example
Triangle	$P = a + b + c$ Perimeter equals the sum of the lengths of the three sides.	$a = 12$ cm, $b = 20$ cm, $c = 24$ cm $P = a + b + c$ $= 12 + 20 + 24$ $= 56$, or 56 cm

Figure	Formula	Example
Rectangle	$P = 2l + 2w$ Perimeter equals twice the length plus twice the width.	$l = 10$ m, $w = 5$ m $P = 2l + 2w$ $= 2 \cdot 10 + 2 \cdot 5$ $= 20 + 10$ $= 30$, or 30 m
Square	$P = 4s$ Perimeter equals 4 times the length of a side.	$s = 6$ ft $P = 4s$ $= 4 \cdot 6$ $= 24$, or 24 ft

EXAMPLE 1

Find the perimeter of the polygon shown.

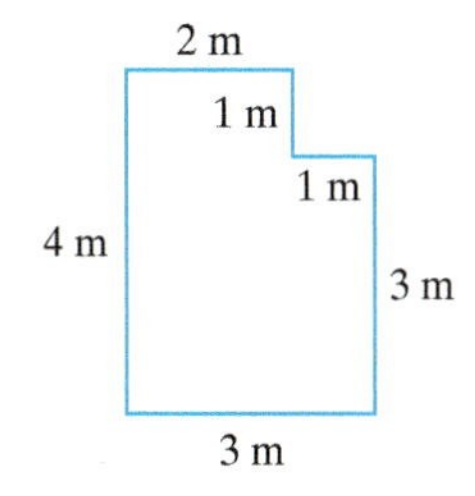

Solution To find the perimeter, we add the lengths of the sides.

$$2 + 1 + 1 + 3 + 3 + 4 = 14$$

So the perimeter is 14 m.

PRACTICE 1

What is the perimeter of this polygon?

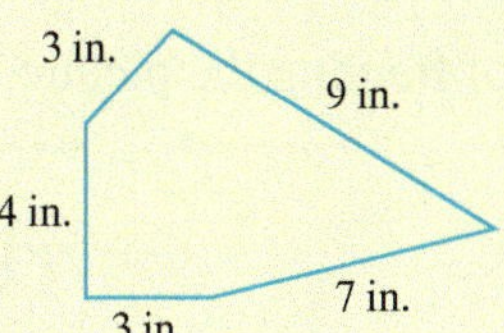

EXAMPLE 2

Find the perimeter of an equilateral triangle with side 1.4 m long.

Solution Recall that all three sides of an equilateral triangle are equal.

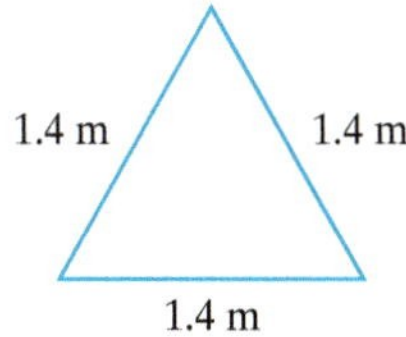

We use the formula for the perimeter of a triangle.

$$\begin{aligned} P &= a + b + c \\ &= 1.4 + 1.4 + 1.4 \\ &= 4.2 \end{aligned}$$

Therefore, the perimeter of the triangle is 4.2 m. Because all three sides are equal in length, we could have used the formula

$$P = 3s = 3(1.4) = 4.2, \text{ or } 4.2 \text{ m}$$

PRACTICE 2

Find the perimeter of a square with side $\frac{3}{4}$ mi long.

EXAMPLE 3

A rectangular picture is 35 in. long and 25 in. wide. To frame the picture costs \$2.50 per inch. What is the cost of framing the picture?

Solution Let's draw a diagram.

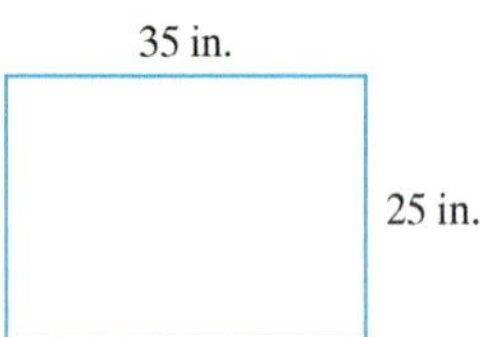

The picture is rectangular, so we use the formula $P = 2l + 2w$ to find its perimeter.

$$\begin{aligned} P &= 2l + 2w \\ &= 2(35) + 2(25) \\ &= 70 + 50 \\ &= 120 \end{aligned}$$

The distance around the picture is 120 in. To find the cost of framing the picture, we multiply this perimeter by the cost per inch.

$$\begin{aligned} \text{Cost} &= 120\ \text{in.} \times \frac{\$2.50}{\text{in.}} \\ &= \$300 \end{aligned}$$

So the cost of framing the picture is \$300.

PRACTICE 3

A square garden has sides 10 ft long. How much will it cost to install a fence around the garden if the fence costs \$1.75 per foot?

The Circumference of a Circle

We refer to the perimeter of a polygon but the *circumference* of a circle.

DEFINITION

The distance around a circle is called its **circumference.**

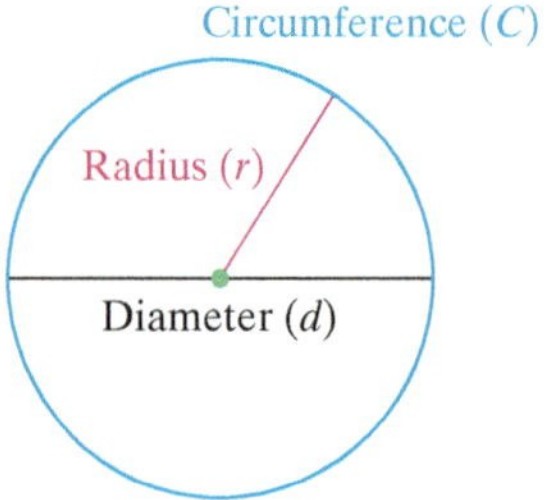

For every circle, the ratio of the circumference C to the diameter d is the same number, which is written as π (read "pi"). This relationship $\frac{C}{d} = \pi$ can also be written as $C = \pi d$ or $C = 2\pi r$. Do you see why πd and $2\pi r$ are equal? Explain.

The value of π is 3.1415926. . . . It is an *irrational number*, so that when π is expressed as a decimal, the digits go on indefinitely without any pattern being repeated. For convenience, we often use an approximate value of π, such as 3.14 or $\frac{22}{7}$, when calculating a circumference by hand.

Figure	Formula	Example
Circumference	$C = \pi d$, or $C = 2\pi r$ Circumference equals π times the diameter, or 2 times π times the radius r.	10 cm $C = \pi d$ $\approx 3.14(10)$ ≈ 31.4, or 31.4 cm

EXAMPLE 4

Find the circumference of the circle shown. Use $\pi \approx 3.14$.

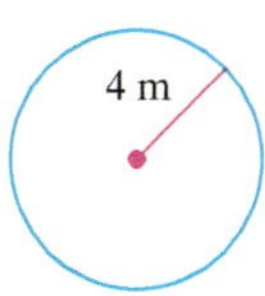

Solution The radius of the circle is 4 m. We use the formula for the circumference of a circle in terms of the radius.

$$\begin{aligned} C &= 2\pi r \\ &\approx 2(3.14)(4) \quad \text{Substitute 4 for } r. \\ &\approx 25.12 \end{aligned}$$

So the circumference is approximately 25.12 m.

PRACTICE 4

What is the circumference of the circle shown? Use $\pi \approx \frac{22}{7}$.

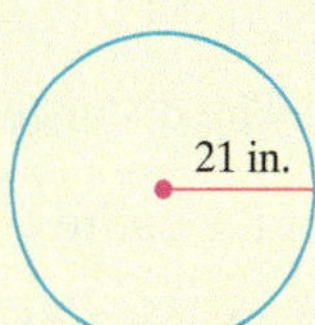

EXAMPLE 5

The diameter of a rolling wheel is 20 in. How far does it travel in one complete turn?

Solution First, let's draw a diagram.

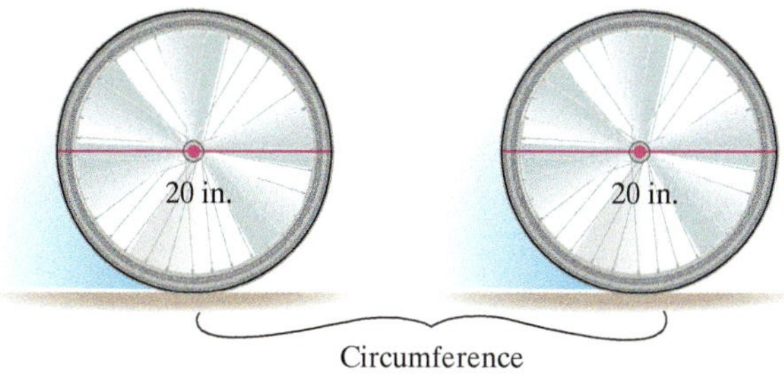

PRACTICE 5

A circular swimming pool has a radius of 18 ft. If a metal rail is to be placed around the edge of the pool, how many feet of railing are needed?

EXAMPLE 5 (continued)

The wheel makes one complete turn, so we know that it travels a distance equal to its circumference. We use the formula for the circumference in terms of the diameter.

$$\begin{aligned} C &= \pi d \\ &\approx 3.14(20) \quad \text{Substitute 20 for } d. \\ &\approx 62.8 \end{aligned}$$

So the wheel travels approximately 62.8 in. in one turn. Do you see how to solve this problem using the formula $C = 2\pi r$?

Composite Figures

Two or more basic geometric figures may be combined to form a **composite figure.**

EXAMPLE 6

Find the perimeter of the following figure, which consists of a semicircle and a rectangle.

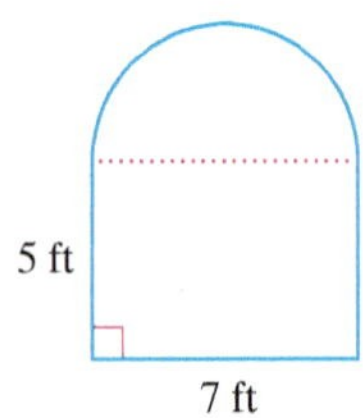

Solution The upper part of the figure is a semicircle with a diameter of 7 ft. The distance around the semicircle is $\frac{1}{2}$ the circumference of the entire circle. So let's compute this circumference and then divide by 2. Here, we use $\pi \approx \frac{22}{7}$.

$$\begin{aligned} C &= \pi d \\ &\approx \frac{22}{\cancel{7}}(\cancel{7}) \\ &\approx 22 \end{aligned}$$

The distance around the semicircle is approximately $\frac{22}{2}$, or 11 ft.

Now, we find the perimeter of three sides of the rectangle at the lower part of the figure.

$$\begin{aligned} P &= 5 + 7 + 5 \\ &= 17 \end{aligned}$$

The perimeter of the lower part is 17 ft.

$$\begin{aligned} \text{Total perimeter} &= \text{Circumference of top} + \text{Perimeter of bottom} \\ &\approx 11 + 17 \\ &\approx 28 \end{aligned}$$

So the perimeter of the composite figure is approximately 28 ft.

PRACTICE 6

What is the perimeter of the figure shown, which is composed of a square and a semicircle?

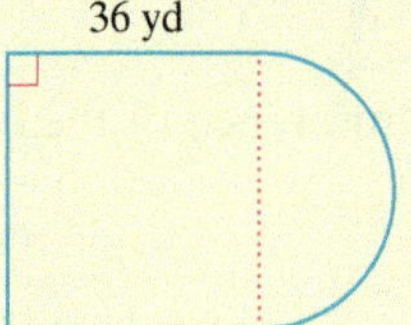

EXAMPLE 7

To prevent a cellar from being flooded, a plumber puts drainage pipes around the outside of a building complex. The outline of the building, consisting of two rectangles and a square, is shown.

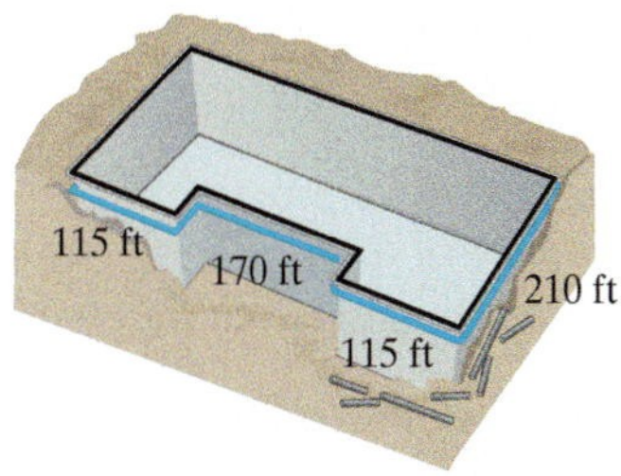

Ignoring the distance between the pipes and the walls, what is the length of the pipes required to go around the complex?

Solution We want to find the perimeter of the building. So we break the composite figure into three figures—two rectangles and one square.

We find the length of the indented part of each rectangular shape by subtracting 170 ft from 210 ft to get 40 ft.

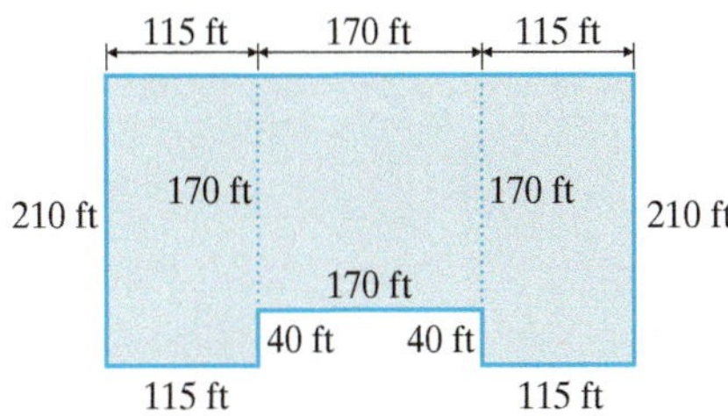

Now, we know all the lengths that make up the building complex, so we can find its perimeter.

$$P = \underbrace{(115 + 210 + 115 + 40)}_{\text{Left rectangle}} + \underbrace{(170 + 170)}_{\text{Center square}}$$
$$+ \underbrace{(115 + 210 + 115 + 40)}_{\text{Right rectangle}}$$
$$= 1{,}300$$

Note that the dotted lines are not part of the perimeter because they lie inside the figure.

So 1,300 ft of drainage pipe is needed to go around the building complex.

PRACTICE 7

A commuter drives from home to work and then returns home, as shown below. What is the total length of the trip?

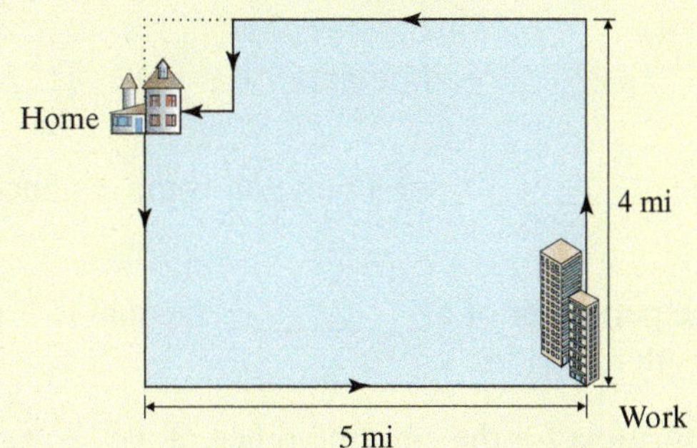

10.2 Exercises

FOR EXTRA HELP MyMathLab MathXL PRACTICE WATCH READ REVIEW

Mathematically Speaking

Fill in each blank with the most appropriate term or phrase from the given list.

simple	rectangle	composite
circle	circumference	square
perimeter	length	

1. The __________ of a polygon is the distance around it.

2. The perimeter of a __________ is equal to the sum of twice the length and twice the width.

3. The perimeter of a __________ is equal to 4 times the length of a side.

4. The distance around a circle is called its __________.

5. A formula for the circumference of a __________ is $C = 2\pi r$.

6. Two or more basic geometric figures are combined in a __________ figure.

A *Find the perimeter or circumference of each figure. Use* $\pi \approx 3.14$, *when needed.*

7. 2 in., 6 in., 3 in., 1 in., 5 in.

8. 2 cm, 3 cm, 3 cm, 4 cm, 4 cm, 3 cm, 3 cm, 2 cm

9. 2.5 m, 0.5 m

10. 6.5 ft, 6.5 ft

11. $3\frac{1}{2}$ yd, $3\frac{1}{2}$ yd, $3\frac{1}{2}$ yd

12. 5 m, 13 m, 12 m

13. 10 m

14.

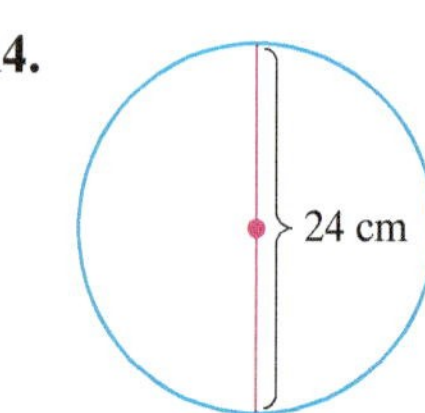

15. 7 ft

16.

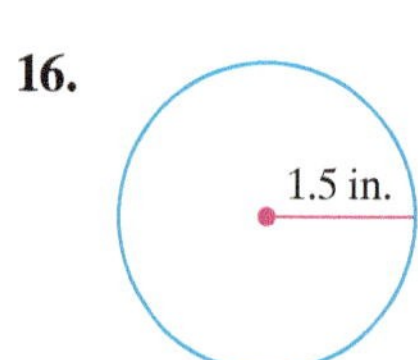

Find the perimeter or circumference. Use $\pi \approx 3.14$, *when needed.*

17. A square with side $5\frac{1}{4}$ yd long

18. A circle whose radius is 20 in. long

19. A rectangle of length $5\frac{3}{4}$ ft and width $3\frac{1}{4}$ ft

20. A triangle whose side lengths are 2 in., $1\frac{1}{2}$ in., and $\frac{7}{8}$ in.

21. An isosceles triangle whose equal sides are $7\frac{1}{2}$ cm long and whose third side is 4 cm long

22. A rectangle with length 8 m and width $4\frac{1}{2}$ m

23.

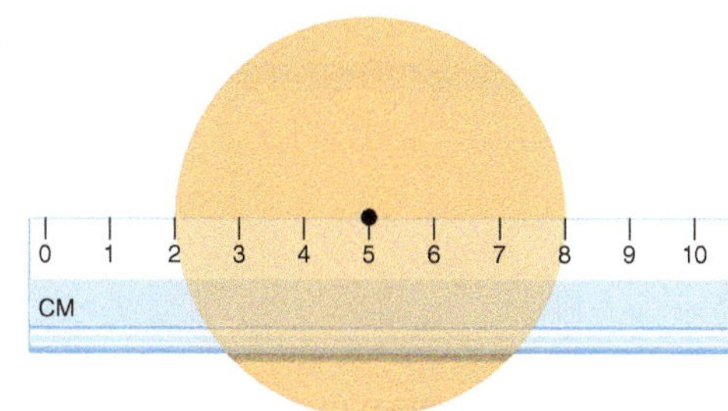

24.

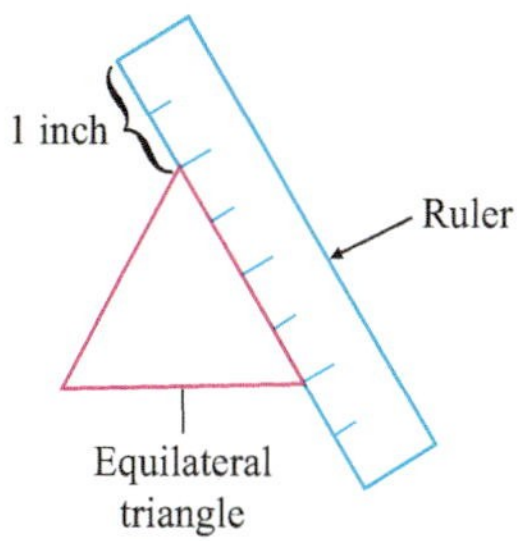

25. A circle whose diameter is 3.54 m long

26. A polygon whose side lengths are 22.75 ft, 25.73 ft, 15.94 ft, 18.23 ft, 21.65 ft, and 34.98 ft

B *Find the perimeter of each composite geometric figure. Use* $\pi \approx 3.14$*, when needed.*

27.

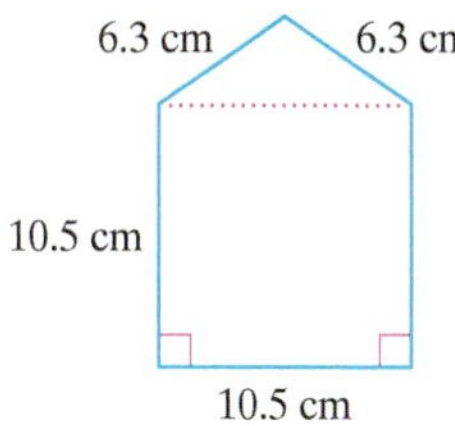

28.

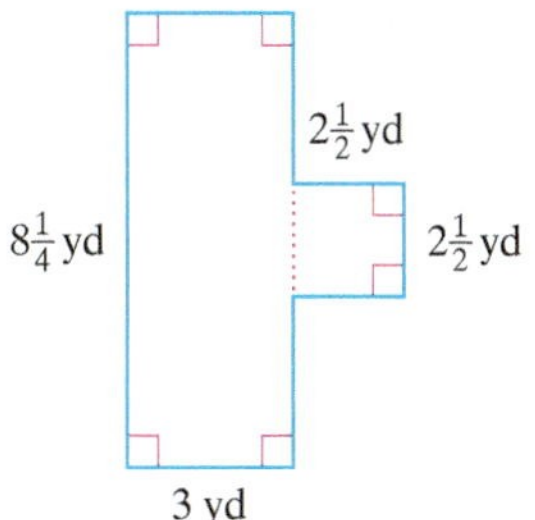

29.

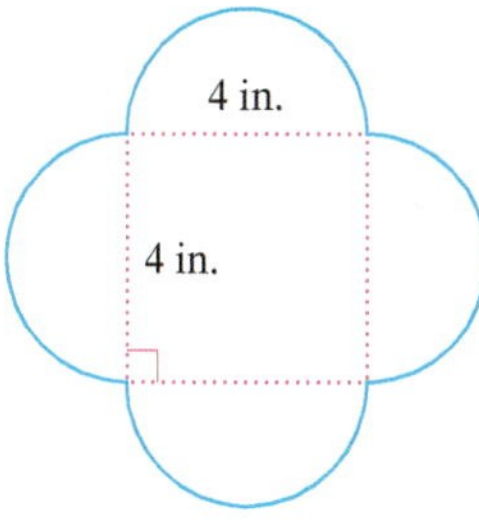

30.

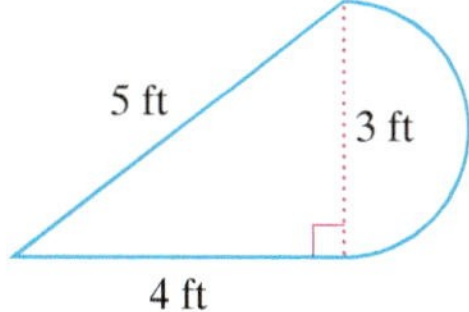

31.

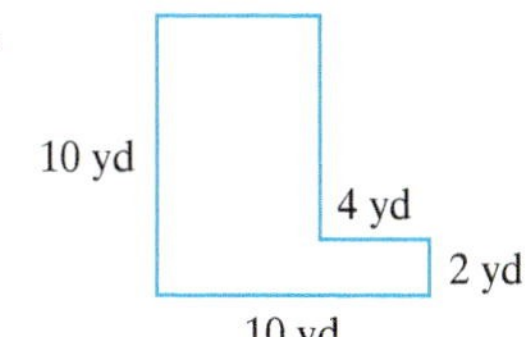

32.

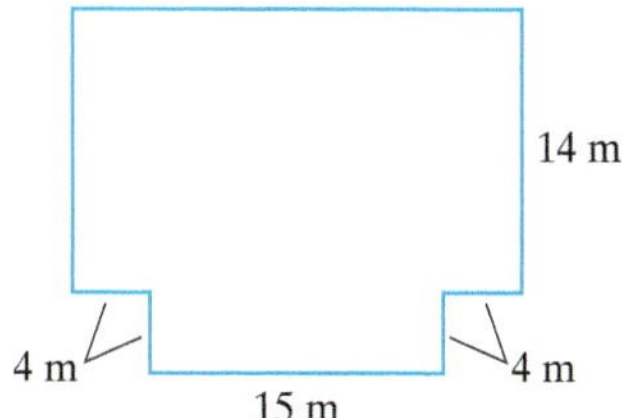

Mixed Practice

Find the perimeter or circumference of each figure. Use $\pi \approx 3.14$*, when needed.*

33.

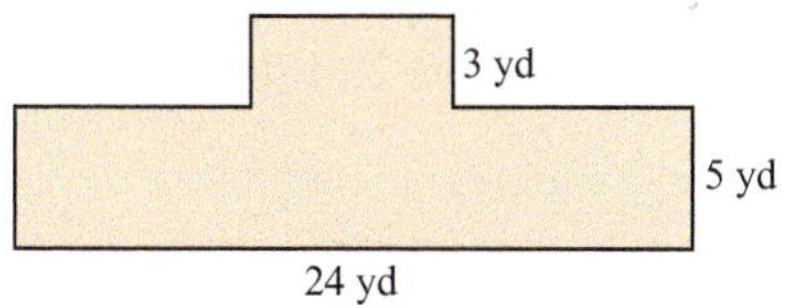

34.

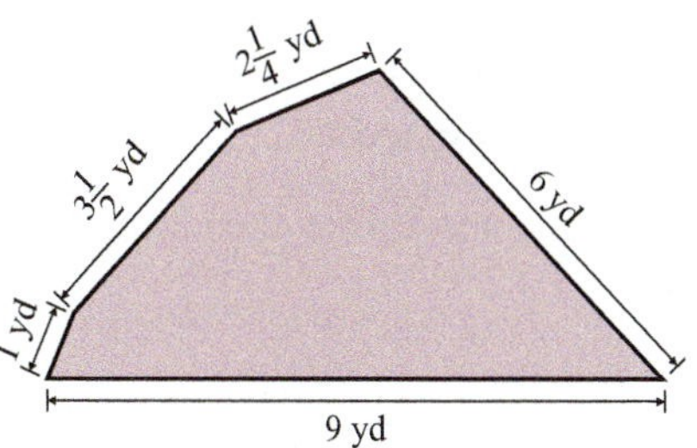

35. An equilateral triangle with side length 9 cm

36.

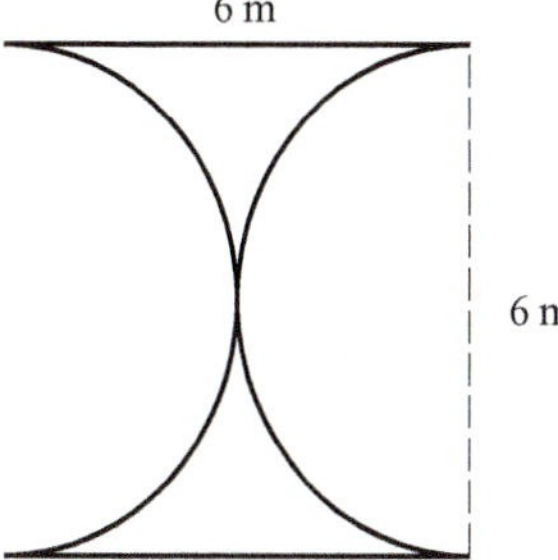

37. 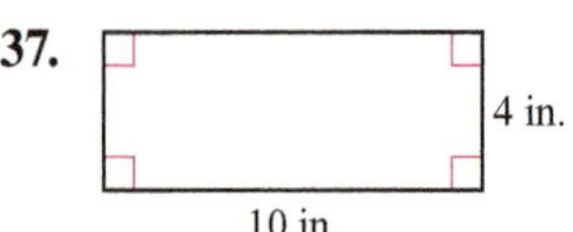

38. 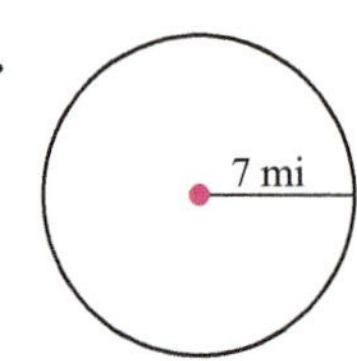

Applications

C *Solve.*

39. Find the perimeter of the doubles tennis court shown below.

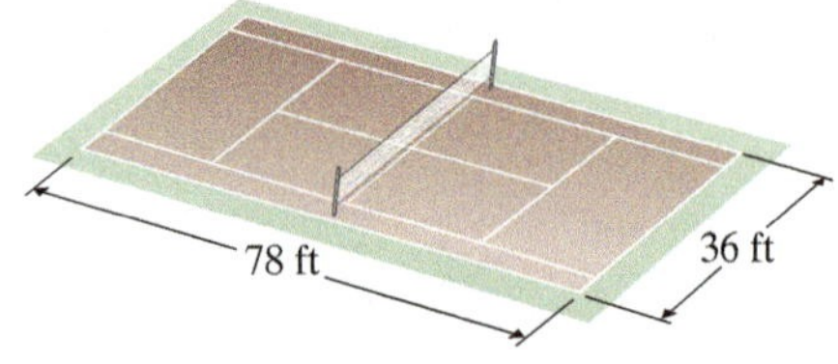

40. If a student drives from Atlanta to New York City to Chicago and back to Atlanta, what is the total mileage?

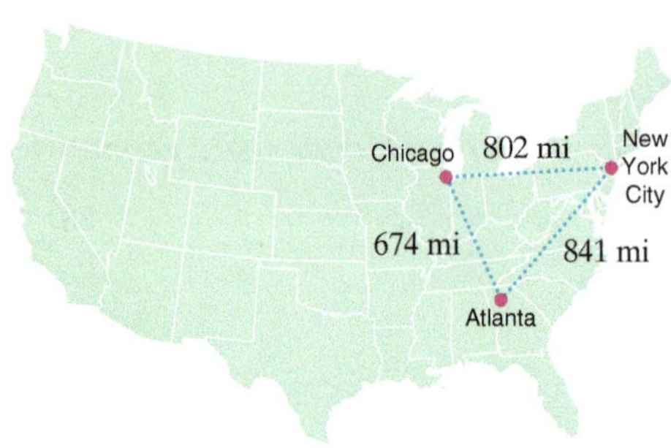

41. As the following diagram shows, bicycle wheels come in different diameters.

In one wheel rotation, how much farther does the 27-inch bicycle wheel go than the 25-inch bicycle wheel?

42. The Texas Star at Fair Park in Dallas is the largest ferris wheel in North America. The diameter of the wheel is 212 ft. How many feet does a rider travel in one revolution of the wheel? (*Source:* bigtex.com)

43. A field 50 m wide and 100 m long is to be enclosed with a fence. If fence posts are placed every 10 m, how many posts are needed?

44. If rug binding costs $1.95 per foot, what is the cost of binding a rectangular rug that is 21 ft long and 12 ft wide?

45. Find the length of line needed for the clothesline pulley.

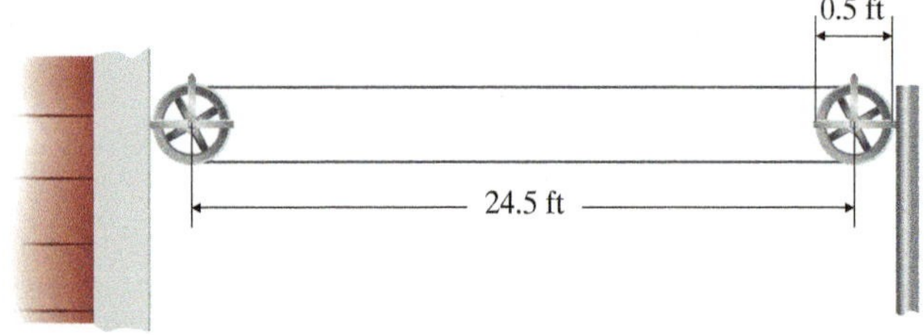

46. A carpenter plans to lay a wood molding in the room shown. If the room has three doors, each 3 ft wide, what is the total length of floor molding required?

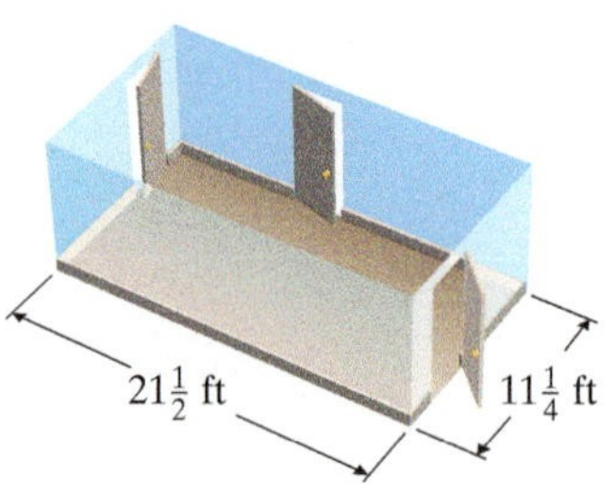

47. The radius of Earth is about 6,400 km. If a satellite is orbiting 400 km above Earth, find the distance that the satellite travels in one orbit.

Orbit
Earth
400 km
Satellite
6,400 km

48. A circular crater on the Moon has a circumference of about 214.66 mi. What is the radius of the crater?

• Check your answers on page A-13.

MINDStretchers

Investigation

1. Draw three triangles. Label the sides of each triangle a, b, and c. Measure each side, writing the measurements in the following table. Compare the sum of any two sides of a triangle with its third side.

	a	b	c	$a + b$	$a + c$	$b + c$
Triangle 1						
Triangle 2						
Triangle 3						

How does the length of the side of a triangle compare to the sum of the lengths of the other two sides?

Mathematical Reasoning

2. Consider the cart pictured. Which wheel do you think is likely to wear out more quickly? Justify your answer.

Groupwork

3. Explain how you can approximate the circumference of a circular room with a ruler. Compare your method with those of other members of the group.

10.3 Area

OBJECTIVES

- **A** To find the area of a polygon or a circle
- **B** To find the area of a composite figure
- **C** To solve applied problems involving area

The Area of a Polygon and a Circle

Area is a measure of the size of a plane geometric figure. The size of a piece of paper, the size of a volleyball court, and the size of a lawn are all examples of areas.

To find the area of the rectangle shown, we split it into little 1-inch by 1-inch squares, each representing 1 square inch.

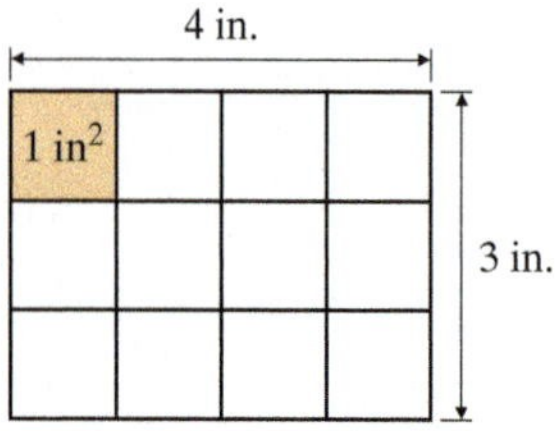

Then, we count the number of square inches within the rectangle, which is 12 square inches.

Each row of the rectangle contains 4 square inches, and the rectangle has 3 rows. So a shortcut to counting the total number of square inches is to multiply 3 by 4, getting 12 square inches in all. Note that areas are measured in square units, such as square inches (sq in. or in^2), square miles (sq mi or mi^2), or square meters (m^2).

DEFINITION
Area is the number of square units that a figure contains.

In this section, we focus on finding the area of common polygons and circles. First, we consider the areas of polygons.

Figure	Formula	Example
Rectangle	$A = lw$ Area equals the length times the width. 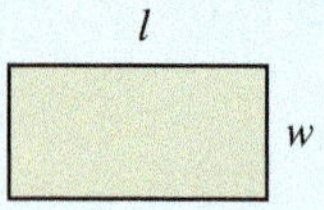	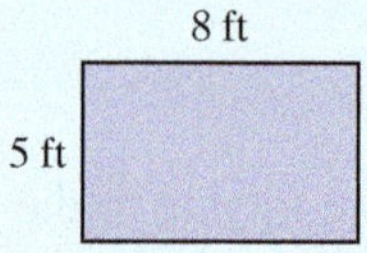 $A = lw$ $= 8 \cdot 5$ $= 40$, or 40 ft^2
Square	$A = s \cdot s = s^2$ Area equals the square of a side. 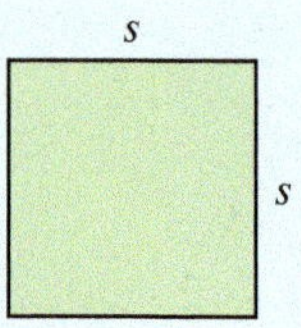	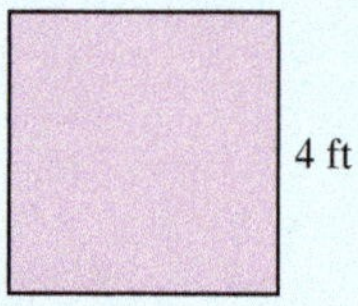 $A = s^2$ $= (4)^2$ $= 4 \cdot 4$ $= 16$, or 16 ft^2

continued

Figure	Formula	Example
Triangle	$A = \frac{1}{2}bh$ Area equals one-half the base times the height. 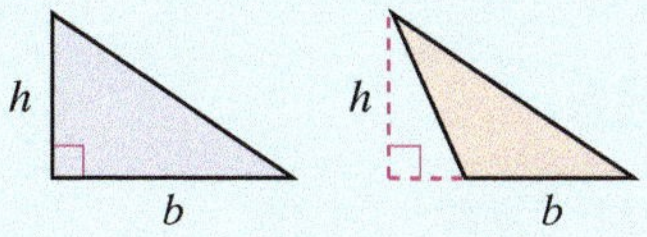	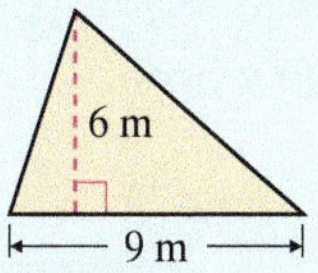 6 m, 9 m $A = \frac{1}{2}\boldsymbol{bh}$ $= \frac{1}{2} \cdot 9 \cdot 6$ $= \frac{1}{2} \cdot 54$ $= 27$, or 27 m^2
Parallelogram	$A = bh$ Area equals the base times the height. 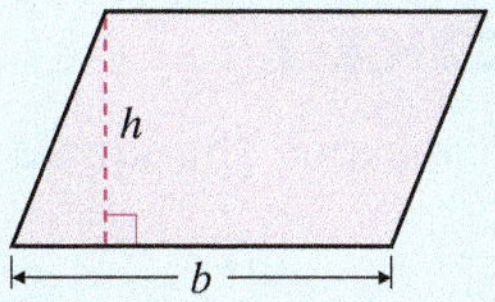	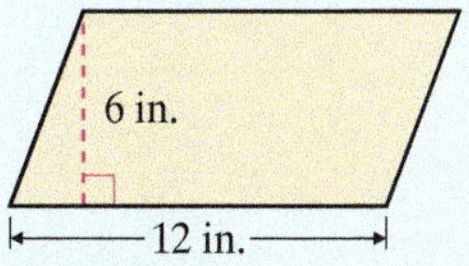 6 in., 12 in. $A = \boldsymbol{bh}$ $= 12 \cdot 6$ $= 72$, or 72 in^2
Trapezoid	$A = \frac{1}{2}h(b + B)$ Area equals one-half the height times the sum of the bases. 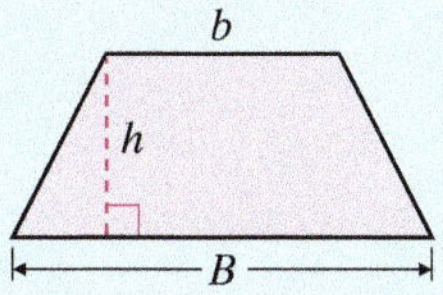	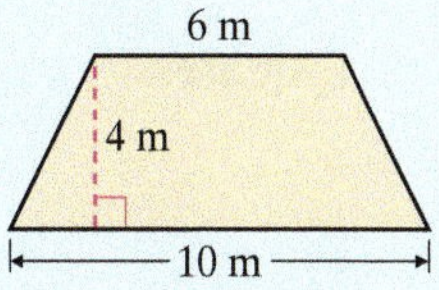 6 m, 4 m, 10 m $A = \frac{1}{2}\boldsymbol{h}(\boldsymbol{b} + \boldsymbol{B})$ $= \frac{1}{2} \cdot 4(6 + 10)$ $= \frac{1}{2} \cdot 4 \cdot 16$ $= 32$, or 32 m^2

Now, let's consider the area of a circle. As in the case of the circumference, the area of a circle is expressed in terms of π. Recall that π is approximately 3.14 or $\frac{22}{7}$.

Figure	Formula	Example
Circle	$A = \pi r^2$ Area equals π times the square of the radius. 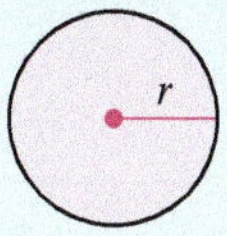	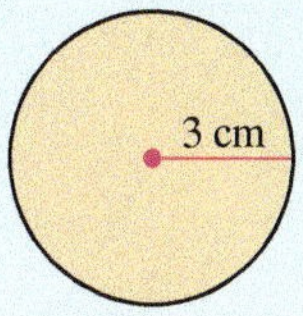 3 cm $A = \pi r^2$ $\approx 3.14\,(3)^2$ $\approx 3.14\,(9)$ ≈ 28.26, or 28.26 cm^2

EXAMPLE 1

Find the area of a rectangle whose length is 5 ft and whose width is 3 ft.

Solution First, we draw a diagram to visualize the problem.

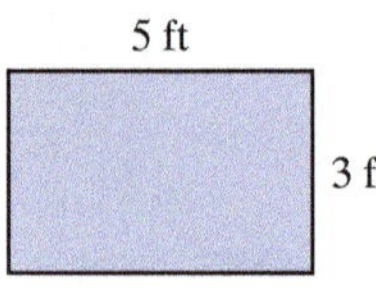

Then, we use the formula for the area of a rectangle.

$$\begin{aligned} A &= lw \\ &= (5)(3) \\ &= 15 \end{aligned}$$

So the area of the rectangle is 15 ft^2.

PRACTICE 1

A rectangle has length 6 cm and width 2 cm. Find its area.

EXAMPLE 2

Find the area of the square.

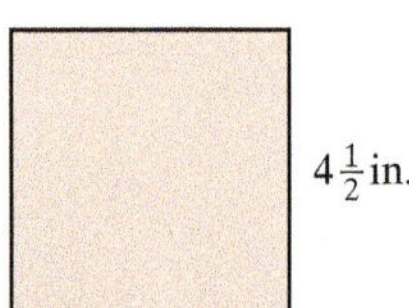

Solution We use the formula for the area of a square.

$$\begin{aligned} A &= s^2 \\ &= \left(4\frac{1}{2}\right)^2 \\ &= \left(4\frac{1}{2}\right)\left(4\frac{1}{2}\right) \\ &= \frac{9}{2} \cdot \frac{9}{2} \\ &= \frac{81}{4}, \text{ or } 20\frac{1}{4} \end{aligned}$$

So the area of the square is $20\frac{1}{4}$ in^2.

PRACTICE 2

What is the area of a square with side 3.6 cm?

EXAMPLE 3

Find the area of a triangle with base 8 cm and height 5.9 cm.

Solution First, we draw a diagram.

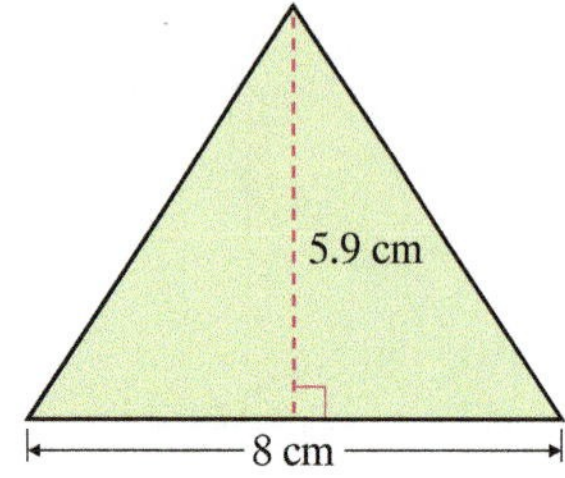

PRACTICE 3

A triangle has a height of 3 in. and a base of 5 in. What is its area?

Next, we use the formula for finding the area of a triangle.

$$A = \frac{1}{2}bh$$
$$= \frac{1}{\cancel{2}_1}(\cancel{8}^4)(5.9)$$
$$= 23.6$$

So the area of the triangle is 23.6 cm^2.

EXAMPLE 4

What is the area of a parallelogram with base $6\frac{1}{2}$ m and height 3 m?

Solution We draw a diagram and then use the formula for the area of a parallelogram.

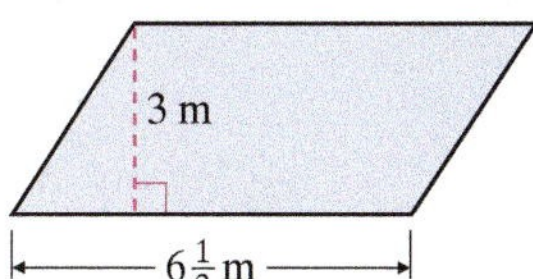

$$A = bh$$
$$= \left(6\frac{1}{2}\right)(3)$$
$$= \frac{13}{2} \times \frac{3}{1}$$
$$= 19\frac{1}{2}$$

So the area of the parallelogram is $19\frac{1}{2}$ m^2.

PRACTICE 4

Find the area of a parallelogram whose base is 5 ft and height is $2\frac{1}{2}$ ft.

EXAMPLE 5

What is the area of the trapezoid shown?

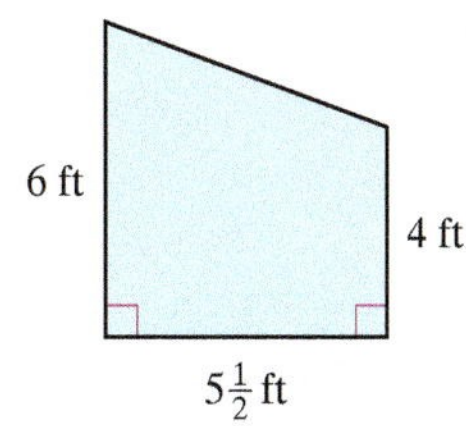

Solution This polygon is a trapezoid, so we use the following formula to find its area.

$$A = \frac{1}{2}h(b + B)$$
$$= \frac{1}{2} \cdot 5\frac{1}{2}(6 + 4)$$
$$= \frac{1}{2} \cdot \frac{11}{\cancel{2}_1} \cdot \cancel{10}^5$$
$$= \frac{55}{2}, \text{ or } 27\frac{1}{2}$$

So the area of the trapezoid is $27\frac{1}{2}$ ft^2.

PRACTICE 5

Find the area of the following trapezoid.

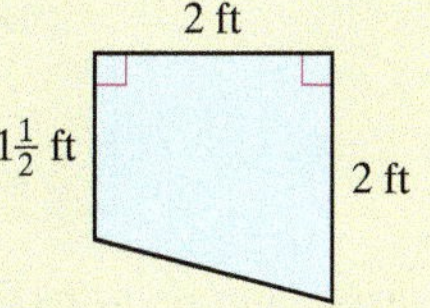

EXAMPLE 6

What is the area of a circle whose diameter is 8 m?

Solution First, we draw a diagram.

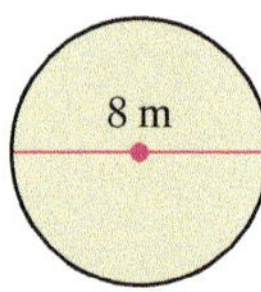

We know that the radius is one-half of 8 m, or 4 m, which we substitute in the formula for the area of a circle.

$$\begin{aligned} A &= \pi r^2 \\ &\approx 3.14(4)^2 \\ &\approx 3.14(16) \\ &\approx 50.24 \end{aligned}$$

So the area of the circle is approximately 50.24 m^2.

PRACTICE 6

Find the area of a circle whose radius is 5 yd.

EXAMPLE 7

An artist wants to buy an ad in a magazine that charges $1,000 per square inch for advertising space. Will the cost of the ad be greater than $6,000?

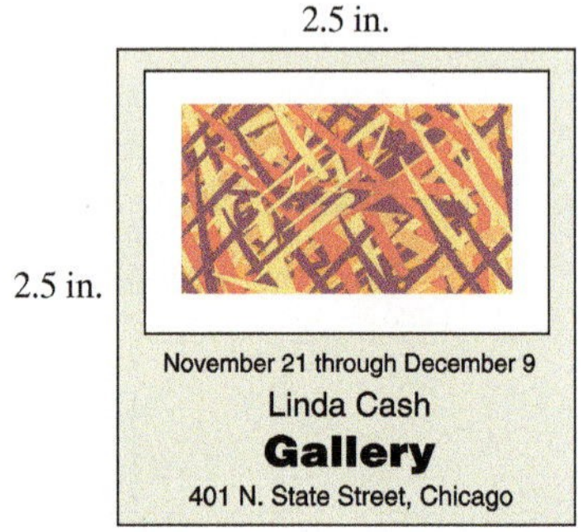

Solution First, we need to find the area of the ad, which is square.

$$\begin{aligned} A &= s^2 \\ &= (2.5)^2 \\ &= (2.5)(2.5) \\ &= 6.25\text{, or } 6.25 \text{ square inches} \end{aligned}$$

To find the cost of the ad, we multiply 6.25 by 1,000, getting 6,250. The cost of the ad is $6,250. So the cost is greater than $6,000.

PRACTICE 7

In a flooring store, a customer wants to buy tile for a 9-foot by 12-foot room using 1 ft^2 tiles that sell for $4.99 apiece. Will $500 be enough to pay for the tiles?

EXAMPLE 8

In a certain town, only students living outside a 2-mile radius of their school must pay a fee for bus transportation. To the nearest square mile, what is the area of the region in which students do not pay a fee for bus transportation?

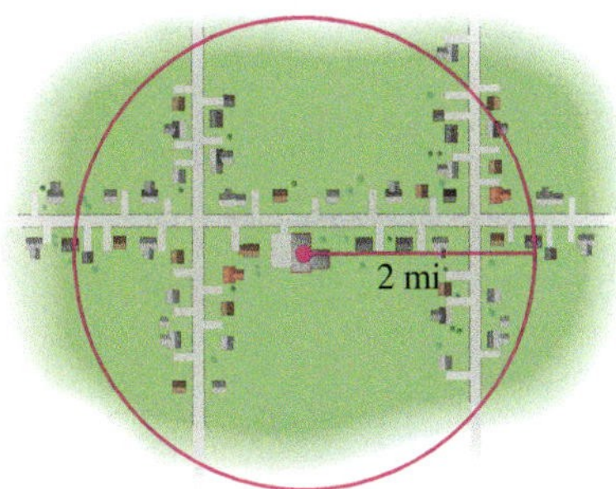

Solution We need to find the area of the region, which is a circle.

$$\begin{aligned} A &= \pi r^2 \\ &\approx \pi(2)^2 \\ &\approx 3.14\,(4) \\ &\approx 12.56 \end{aligned}$$

So the area of the region is approximately 13 mi^2.

PRACTICE 8

The following diagram shows the region in which the beam of light from a lighthouse can be seen in any direction in the fog. What is the area of this region?

Composite Figures

Recall that a composite figure comprises two or more simple figures. Let's consider finding areas of such figures.

EXAMPLE 9

Find the area of the shaded portion of the figure.

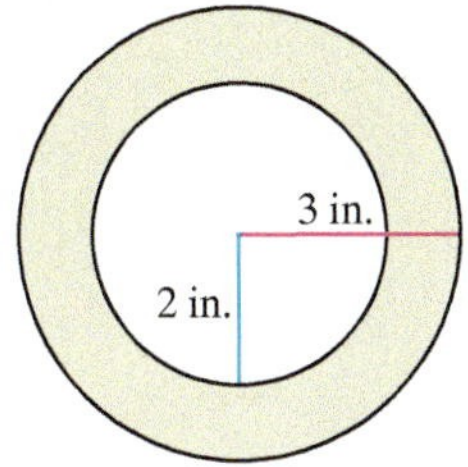

Solution To find the area of the shaded portion, we subtract the area of the small (inner) circle from the area of the large (outer) circle.

$$\begin{aligned} \text{Shaded Area} &= \text{Area of large circle} - \text{Area of small circle} \\ &\approx 3.14\,(3)^2 - 3.14\,(2)^2 \\ &\approx 3.14\,(9) - 3.14\,(4) \\ &\approx 28.26 - 12.56 \\ &\approx 15.70 \end{aligned}$$

So the area of the shaded figure is approximately 15.7 in^2.
How could the distributive property be used to solve this problem?

PRACTICE 9

Find the shaded area.

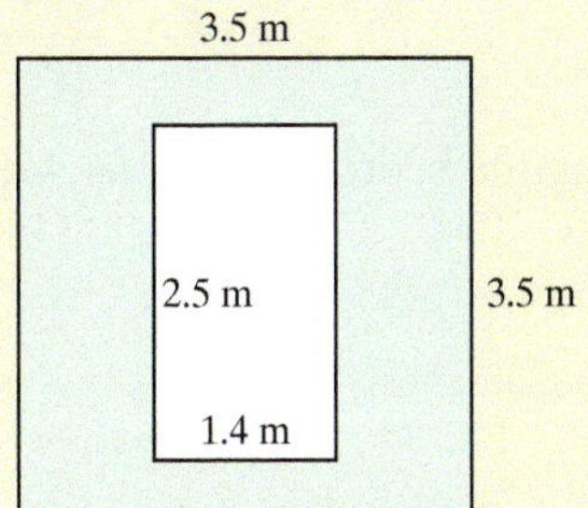

EXAMPLE 10

At \$19 per square foot, how much will it cost to carpet the bedroom pictured?

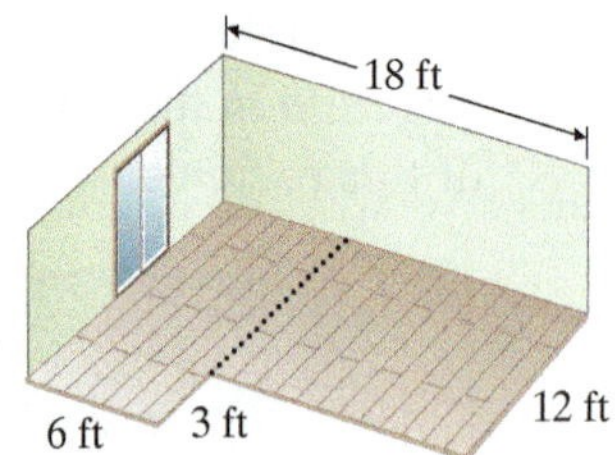

Solution First, we must find the area of the room. Note that the room consists of a 15-foot by 6-foot rectangle, and a square 12 ft on a side.

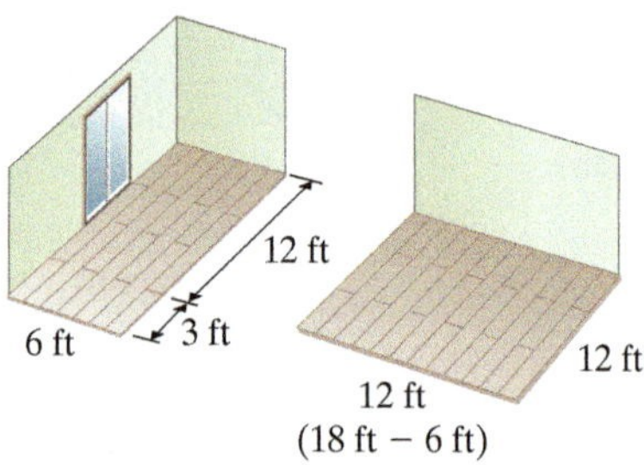

$$\begin{aligned} \text{Total area} &= \text{Area of rectangle} + \text{Area of square} \\ &= l \cdot w + s^2 \\ &= 15 \cdot 6 + (12)^2 \\ &= 90 + 144 \\ &= 234, \text{ or } 234 \text{ ft}^2 \end{aligned}$$

The total area of the room is 234 square feet.

Since the carpet costs \$19 per square foot, we calculate the total cost as follows:

$$234 \text{ ft}^2 \times \frac{\$19}{\text{ft}^2} = \$4{,}446$$

So carpeting the bedroom costs \$4,446.

PRACTICE 10

A coating of polyurethane is applied to the central circle on the gymnasium floor shown below. What is the area of the part of the floor that still needs coating?

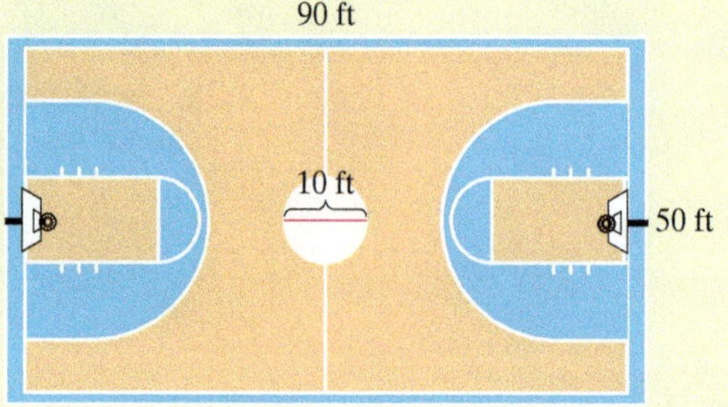

10.3 Exercises

FOR EXTRA HELP MyMathLab MathXL PRACTICE WATCH READ REVIEW

Mathematically Speaking

Fill in each blank with the most appropriate term or phrase from the given list.

trapezoid	volume	square meters
circle	meters	triangle
square	area	

1. The number of square units that a figure contains is called its __________.

2. Areas are measured in square units, such as __________.

3. The area of a(n) __________ is equal to one-half the product of the base and the height.

4. The area of a(n) __________ is equal to the square of a side.

5. A formula for the area of a(n) __________ is $A = \pi r^2$.

6. The formula $A = \frac{1}{2}h(b + B)$ is used to find the area of a(n) __________.

A ***Find the area of each figure. Use $\pi \approx 3.14$, when needed.***

7. 25 m, 5 m

8. 10 in., 10 in.

9. 5 ft, 12 ft

10. 9 cm, 4 cm

11. 10 yd, 29 yd

12. 25 in., 55 in.

13. 15 cm

14. 16 ft

15. 7 m, 4 m, 9 m

16.

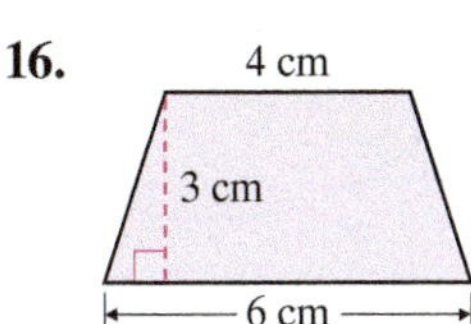

17. A parallelogram with base 4 m and height 3.9 m

18. A parallelogram with base 6.5 in. and height 4 in.

19. A circle with diameter 20 in.

20. A circle with radius 100 ft

21. A triangle with height 2.5 ft and base 5 ft

22. A triangle with base 8 in. and height $6\frac{1}{2}$ in.

23. A trapezoid with height 4.2 yd and bases 7 yd and 14 yd

24. A trapezoid with height 3.5 m and bases 4 m and 6.5 m

25. A rectangle with length 2.6 m and width 1.4 m

26. A rectangle with length $\frac{1}{2}$ ft and width $\frac{2}{3}$ ft

27. A square with side $\frac{1}{4}$ yd long

28. A square with side 15.5 cm long

B *Find the shaded area.*

29.

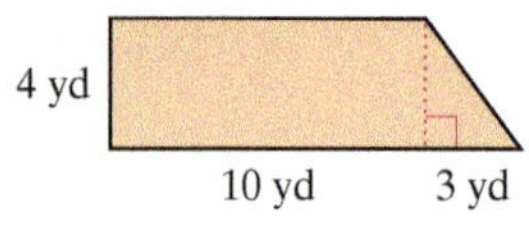

30.

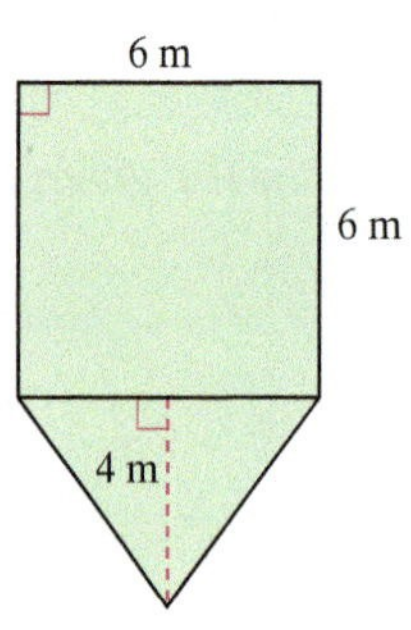

31.

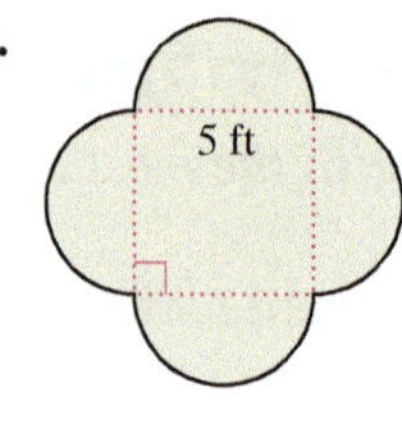

32.

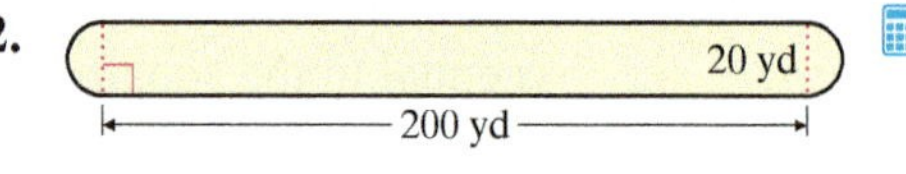

33.

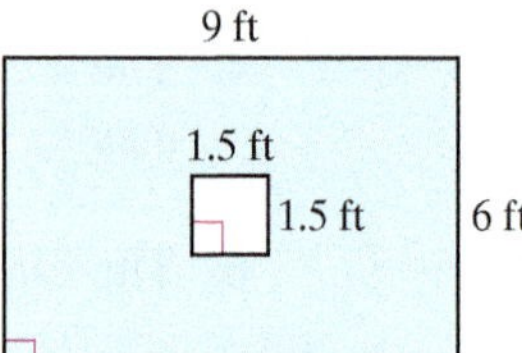

34.

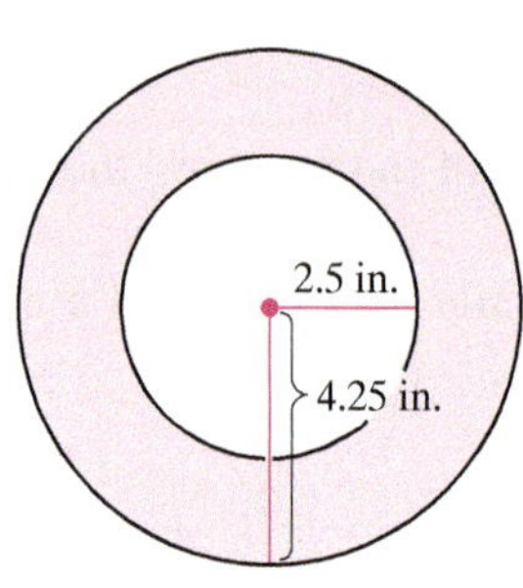

Mixed Practice

Find the area of each shaded region or described figure. Use $\pi \approx 3.14$*, when needed.*

35.

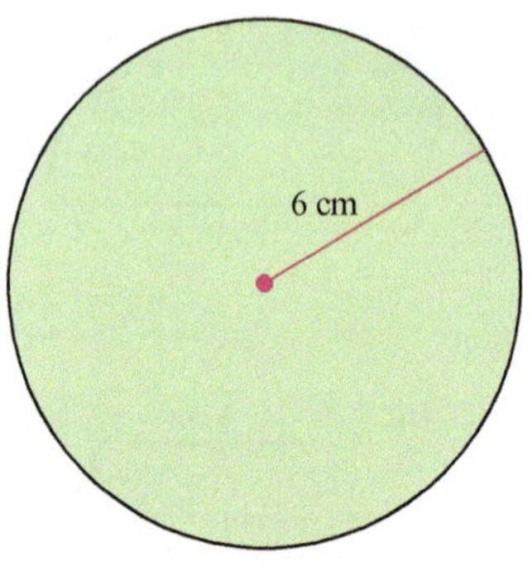

36.

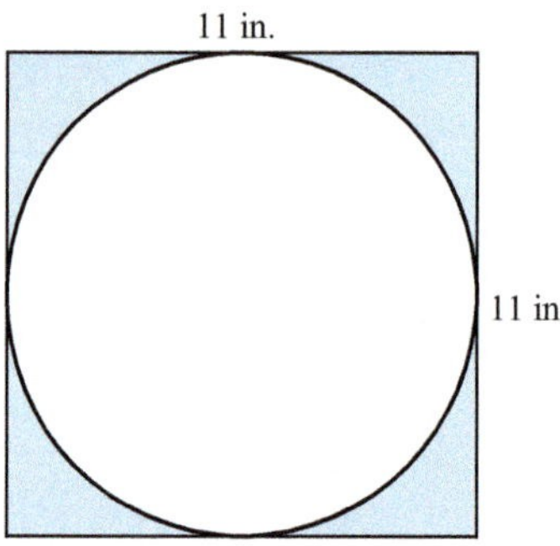

37.

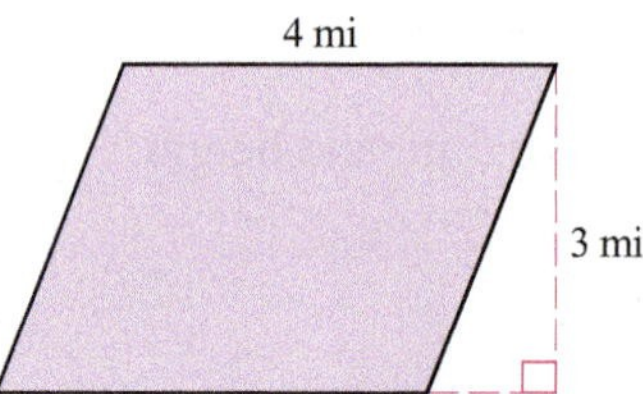

38.

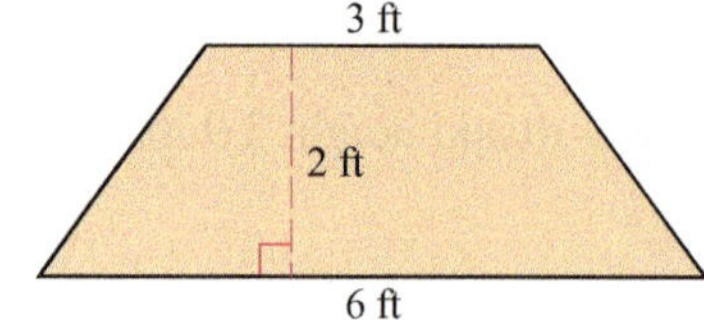

39. A right triangle with legs 6.5 m and 10 m

40. A parallelogram with base 8.5 yd and height 7 yd

Applications

C *Solve. Use* $\pi \approx 3.14$, *when needed.*

41. The boxing ring below is a square that measures 18 ft on a side inside the ropes. The area outside the ropes, called the apron, extends 2 ft beyond the ropes on each side. What is the area of the apron?

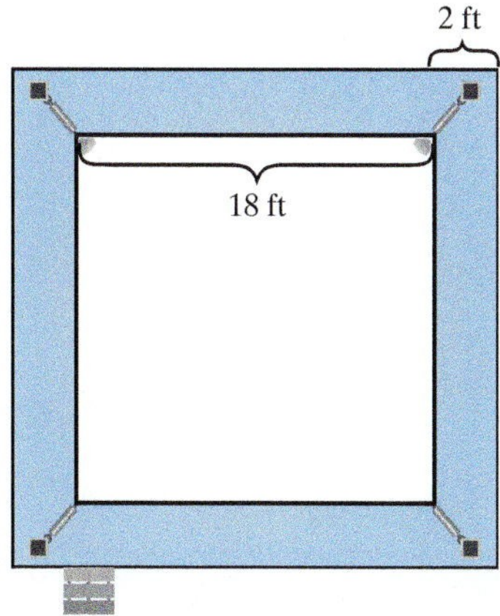

42. The base of the United Nations Secretariat building is a rectangle with length 88 m and width 22 m. The Empire State Building has a rectangular base measuring 129 m by 57 m. What is the difference in the area between the two bases? (*Sources:* docomomo-us.org and newyork-transportation.com)

43. A microscope allows a scientist to see a circular region that is 0.25 mm in diameter. What is the area of this region?

44. An air-traffic control tower can identify an airplane within 10 mi of the tower in any direction. What area does the tower cover?

45. Suppose that an L-shaped house is located on the rectangular lot shown. How much yard space is there?

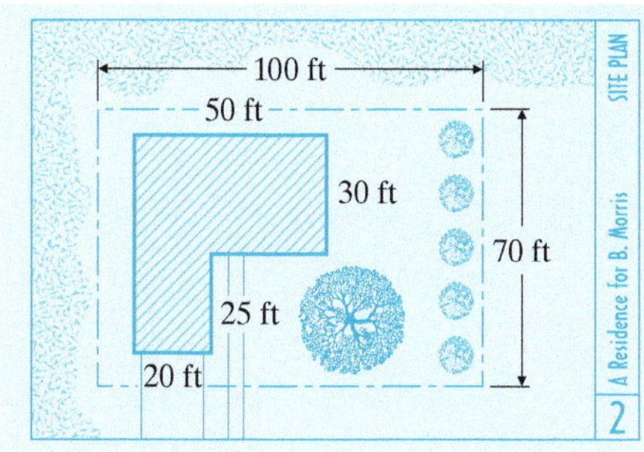

46. A walkway 2 yd wide, shown below, is built around the entire building below. Find the area of the walkway.

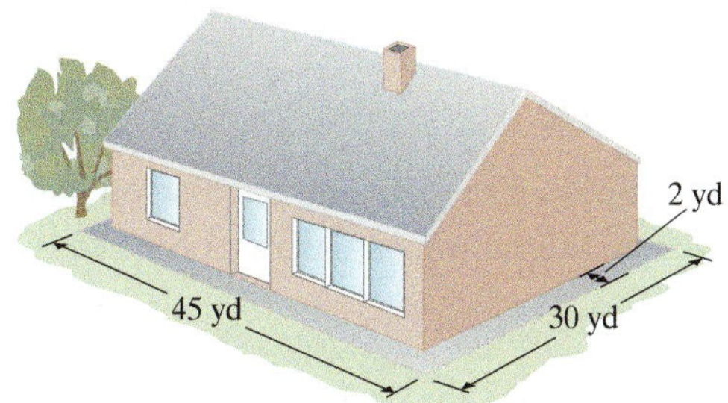

47. An online furniture company is advertising two sizes of a trapezoid-shaped computer workstation. For the smaller workstation, the two parallel sides are 24 in. and 48 in., and the depth is 24 in. The larger has parallel sides of length 30 in. and 60 in. and a depth of 30 in. What is the difference between the area of the two workstations? (*Source:* csnlibraryfurniture.com)

48. The John Hancock Tower, the tallest skyscraper in New England, has a floor plan shaped like a parallelogram. If the parallelogram's base is 293 ft and its height is 107 ft, find the area of the floor plan. (*Source:* pcfandp.com)

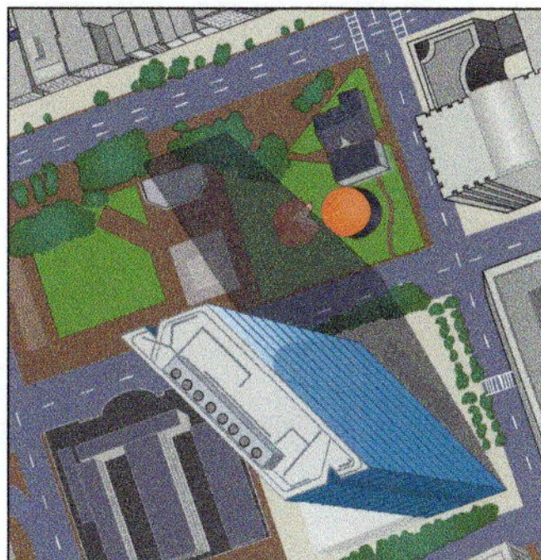

49. A circular rug comes in two sizes. In one, the diameter is 41 in., and in the other only 32 in. If the smaller rug sells for \$79.95 and the prices are proportional to the areas, find the selling price of the larger rug, rounded to the nearest cent.

50. Consider the two tables pictured. How much larger in area to the nearest square meter is the semicircular table than the rectangular table?

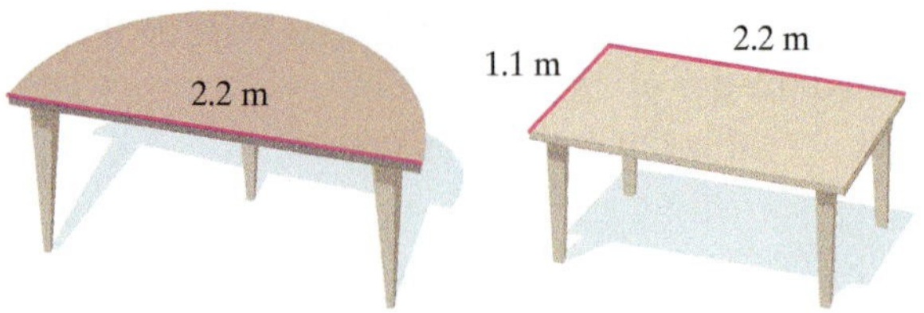

• Check your answers on page A-13.

MINDStretchers

Investigation

1. • Draw a square.
   • Measure its side lengths.
   • Find its area.
   • Double the side length of the square.
   • Find the area of the new square.
   • Start with another square and repeat this process several times.
   • How does doubling the side length of a square affect its area?

Groupwork

2. In the diagram to the right, each small square represents 1 square inch. Working with a partner, estimate the area of the oval.

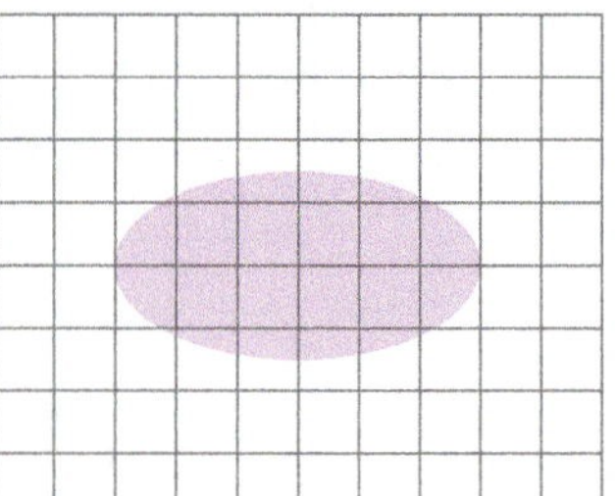

Mathematical Reasoning

3. In the diagram below, $\overline{AC}$ is parallel to $\overline{ED}$.

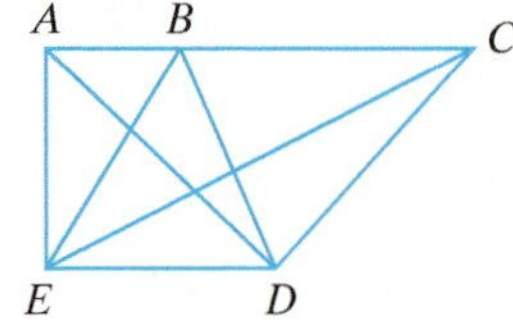

What relationship do you see between the areas of ΔEAD, ΔEBD, and ΔECD? Justify your answer.

10.4 Volume

OBJECTIVES

- **A** To find the volume of a geometric solid
- **B** To find the volume of a composite geometric solid
- **C** To solve applied problems involving volume

The Volume of a Geometric Solid

Volume is a measure of the amount of space inside a three-dimensional figure. The amount of water in an aquarium, the amount of juice in a can, or the amount of grain in a bin are all examples of volumes.

To find the volume of the box shown, we can split it into little 1-inch by 1-inch by 1-inch cubes, each representing 1 cubic inch. Then, we count the number of cubic inches within the box, which is 24 cubic inches.

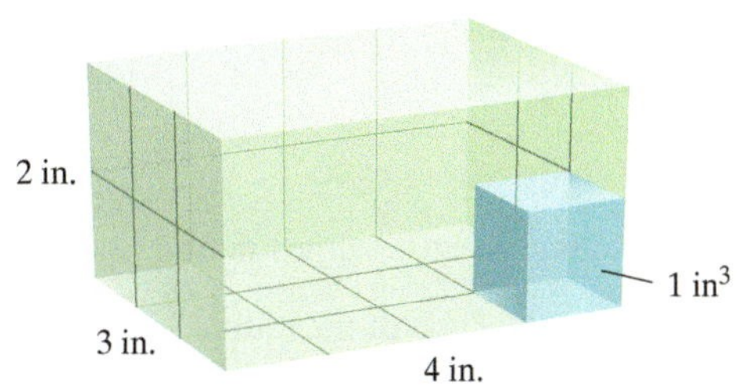

A shortcut to counting the total number of cubic inches is to multiply the length, the width, and the height: $4 \times 3 \times 2$, getting 24, or 24 cubic inches. Note that volumes are measured in cubic units, such as cubic inches (cu in. or in^3), cubic miles (cu mi or mi^3), or cubic meters (cu m or m^3).

DEFINITION

Volume is the number of cubic units required to fill a three-dimensional figure.

In this section, we consider basic three-dimensional objects and find their volume by using the following formulas:

Definition	Formula	Example
A **rectangular solid** is a solid in which all six faces are rectangles.	$V = lwh$ Volume equals length times width times height. (h, face, w, l)	(3 ft, 4 ft, 6 ft) $V = lwh$ $= 6 \cdot 4 \cdot 3$ $= 72$, or $72\ \text{ft}^3$
A **cube** is a solid in which all six faces are squares.	$V = e^3$ Volume equals the cube of the edge. (edge, e, e, e)	(3 cm, 3 cm, 3 cm) $V = e^3$ $= (3)^3$ $= 27$, or $27\ \text{cm}^3$

continued

Definition	Formula	Example
A **cylinder** is a solid in which the bases are circles and are perpendicular to the height.	$V = \pi r^2 h$ Volume equals π times the square of the radius times the height.	$V = \pi r^2 h$ $\approx 3.14(2)^2(5)$ $\approx 3.14(4)(5)$ ≈ 62.8, or 62.8 in^3
A **sphere** is a three-dimensional figure made up of all points a given distance from the center.	$V = \frac{4}{3}\pi r^3$ Volume equals $\frac{4}{3}$ times π times the cube of the radius.	$V = \frac{4}{3}\pi r^3$ $\approx \frac{4}{3}(3.14)(3)^3$ $\approx \frac{4}{3}(3.14)(27)$ $\approx \frac{339.12}{3}$ ≈ 113.04, or 113.04 yd^3

EXAMPLE 1

Find the volume of the rectangular solid shown.

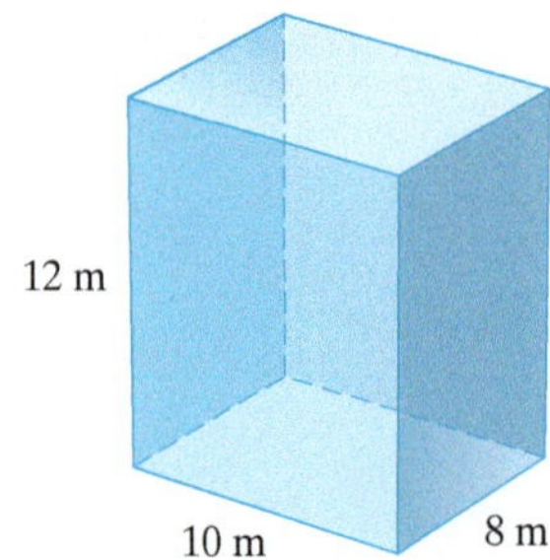

Solution To find the volume of a rectangular solid, use the formula $V = lwh$.

$$V = lwh$$
$$= 10 \cdot 8 \cdot 12$$
$$= 960$$

The volume of the rectangular solid is 960 m^3.

PRACTICE 1

What is the volume of this box?

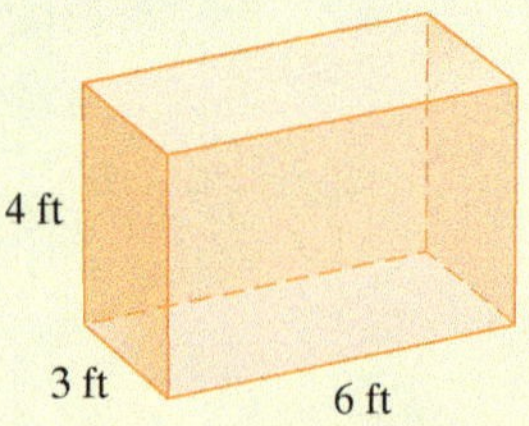

EXAMPLE 2

Find the volume of a cube with length 11 in. on a side.

Solution Use the formula for the volume of a cube, $V = e^3$.

$$\begin{aligned} V &= e^3 \\ &= (11)^3 \\ &= 11 \cdot 11 \cdot 11 \\ &= 1{,}331 \end{aligned}$$

So the volume is 1,331 in^3.

PRACTICE 2

What is the volume of a cube with an edge 15 cm long?

EXAMPLE 3

What is the volume of the cylinder shown?

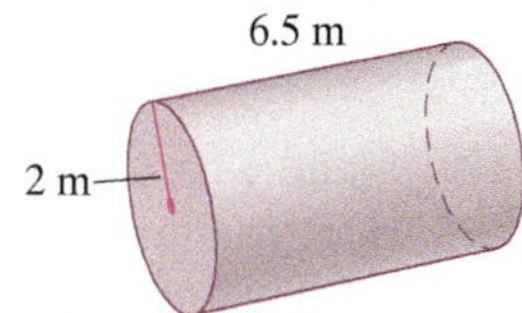

Solution In this cylinder, the radius of the base is 2 m and the height is 6.5 m. We substitute these quantities into the formula for the volume of a cylinder.

$$\begin{aligned} V &= \pi r^2 h \\ &\approx 3.14\,(2)^2(6.5) \\ &\approx 3.14\,(4)(6.5) \\ &\approx 81.64 \end{aligned}$$

So the volume is approximately 82 m^3.

PRACTICE 3

Find the volume enclosed by this pipe.

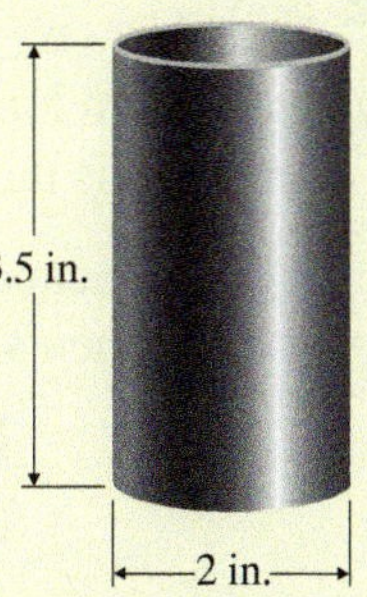

EXAMPLE 4

What is the volume of the sphere shown? Use $\pi \approx \dfrac{22}{7}$.

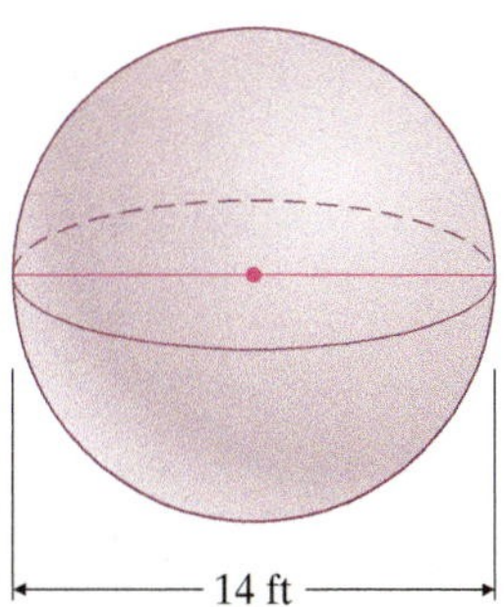

PRACTICE 4

What is the volume of a ball whose diameter is 6 in.?

EXAMPLE 4 (continued)

Solution We use the formula for the volume of a sphere. Since the diameter of the sphere is 14 ft, the radius is 7 ft.

$$V = \frac{4}{3}\pi r^3$$

$$V \approx \frac{4}{3} \cdot \frac{22}{7}(7)^3$$

$$\approx \frac{4}{3} \cdot \frac{22}{\not{7}} \cdot \frac{\not{7} \cdot 7 \cdot 7}{1}$$

$$\approx \frac{4{,}312}{3}, \text{ or } 1{,}437\frac{1}{3}$$

So the volume is about 1,437 ft^3.

EXAMPLE 5

A shipping van is 10.2 m long and 2.5 m wide. If it can be filled to a height of 1.8 m, what is the capacity of the van?

Solution To find the capacity or volume of the van, we use the formula $V = lwh$ and then substitute the given values.

$$V = lwh$$
$$= (10.2)(2.5)(1.8)$$
$$= 45.9$$

So the capacity of the van is 45.9 m^3.

PRACTICE 5

During an experiment, a meteorologist fills a spherical weather balloon with helium. If she fills the weather balloon until its diameter is 2 m, what is its volume?

Composite Geometric Solids

Now, let's find the volume of composite geometric solids made up of two or more basic solid figures.

EXAMPLE 6

Find the volume of the solid pictured, which is a cube with a cylinder deleted from its center.

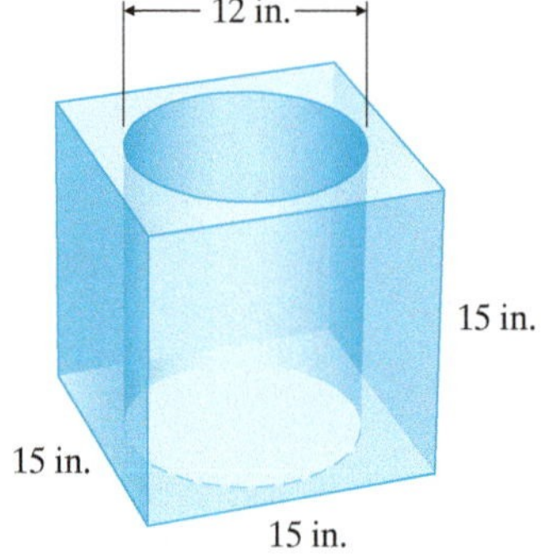

PRACTICE 6

A ball is packaged in a cube-shaped box touching all its sides, as pictured.

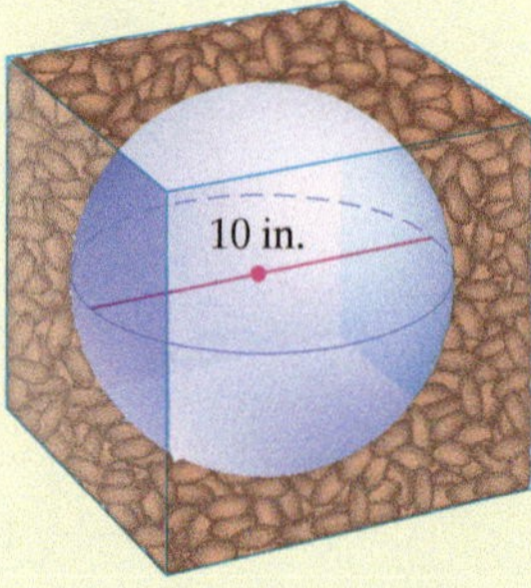

Find the amount of space occupied by the peanut packing.

Solution To find the volume of this solid, we subtract the volume of the cylinder from the volume of the cube.

$$\begin{aligned}\text{Volume of the solid} &= \text{Volume of cube} - \text{Volume of cylinder}\\ &= e^3 - \pi r^2 h\\ &\approx (15)^3 - (3.14)(6)^2(15)\\ &\approx 3{,}375 - (3.14)(36)(15)\\ &\approx 3{,}375 - 1{,}695.6\\ &\approx 1{,}679.4\end{aligned}$$

So the volume of the solid is approximately 1,679 in^3.

EXAMPLE 7

A pharmaceutical company produces a medicine capsule that is shaped like a cylinder with half a sphere at each end, shown below. What is the volume of the capsule?

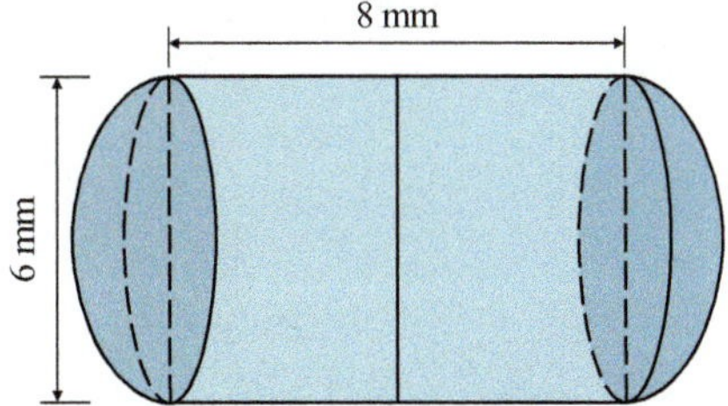

Solution To find the volume of the capsule, we add the volume of the cylinder to the volume of the sphere formed by the two ends.

Volume of capsule = Volume of cylinder + Volume of sphere

$$\begin{aligned}V &= \pi r^2 h + \frac{4}{3}\pi r^3\\ &\approx 3.14(3)^2(8) + \frac{4}{3}(3.14)(3)^3\\ &\approx 3.14(9)(8) + \frac{4}{3}(3.14)(27)\\ &\approx 226.08 + 113.04\\ &\approx 339.12\end{aligned}$$

So the volume of the capsule is approximately 339 mm^3.

PRACTICE 7

Each tier of a wedding cake is a 4-inch-high rectangular solid. What is the total volume of the cake?

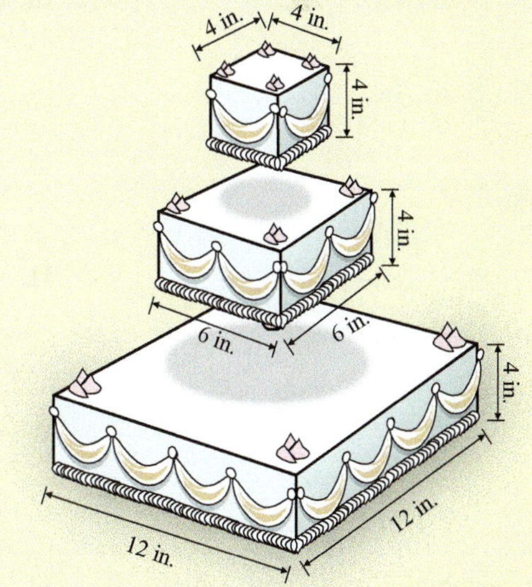

10.4 Exercises

FOR EXTRA HELP MyMathLab Math XL PRACTICE WATCH READ REVIEW

Mathematically Speaking

Fill in each blank with the most appropriate term or phrase from the given list.

circle	composite	volume
cube	rectangular solid	simple
cylinder	area	sphere

1. The number of cubic units required to fill a three-dimensional figure is called its ________.

2. Combining two or more basic solid figures results in a(n) __________ geometric solid.

3. The formula $V = lwh$ is used the find the volume of a(n) ______________.

4. A formula for the volume of a(n) ________ is $V = e^3$.

5. A(n) ________ is a solid in which the bases are circles and are perpendicular to the height.

6. A(n) __________ is a three-dimensional figure made up of all points a given distance from the center.

A *Find the volume of each solid. Use* $\pi \approx 3.14$*, when needed.*

7.
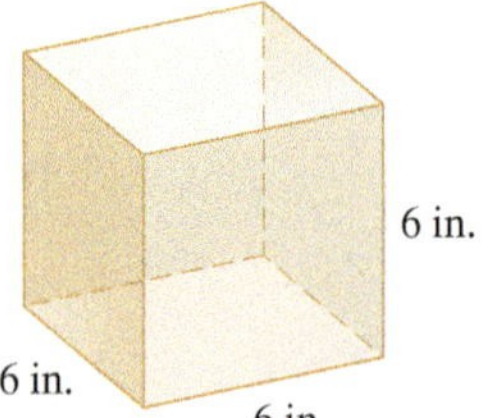

8.
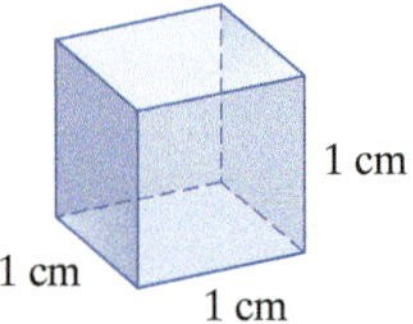

9.
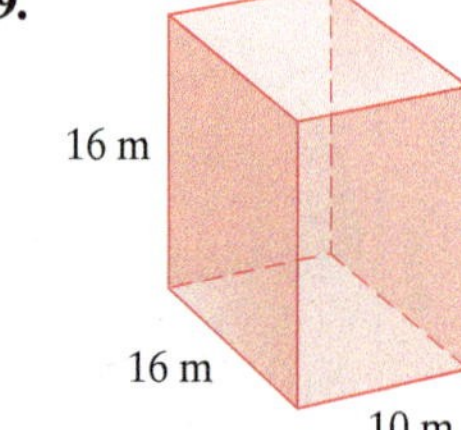

10.
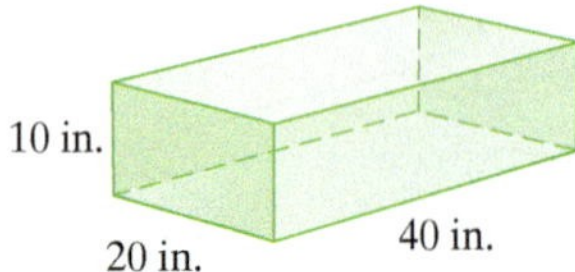

11.
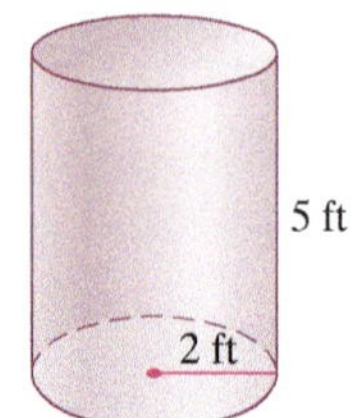

12.
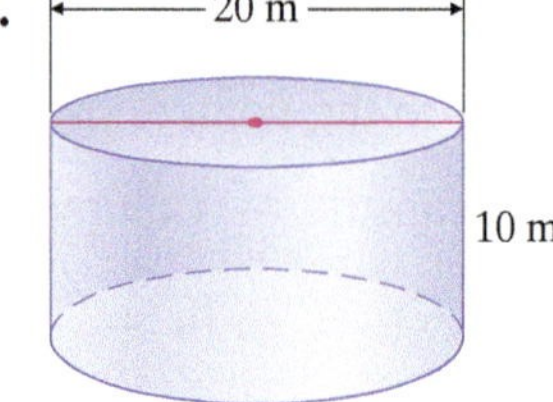

13. A rectangular solid with length 3.5 ft, width 5.5 ft, and height 6.5 ft

14. A cube with length 45 m on a side

15.
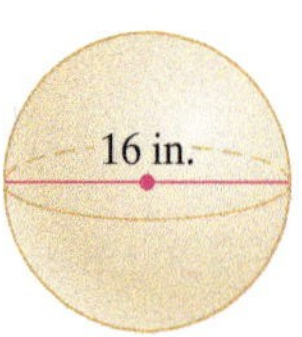

16.
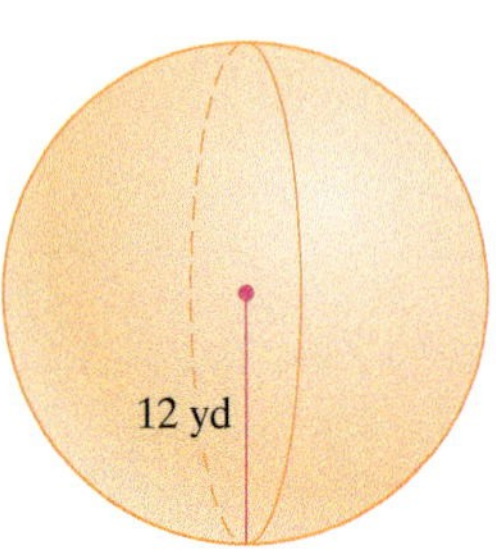

17.
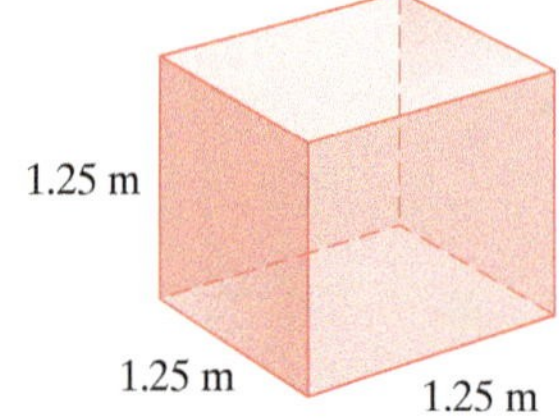

18.
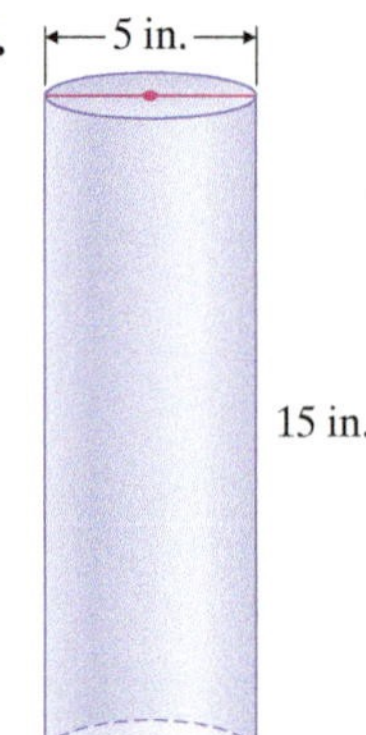

B *Find the volume of each composite geometric solid. Use $\pi \approx 3.14$, when needed.*

19.

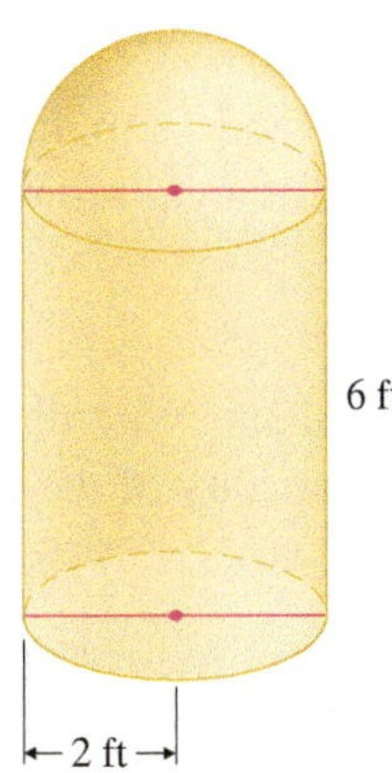

20.

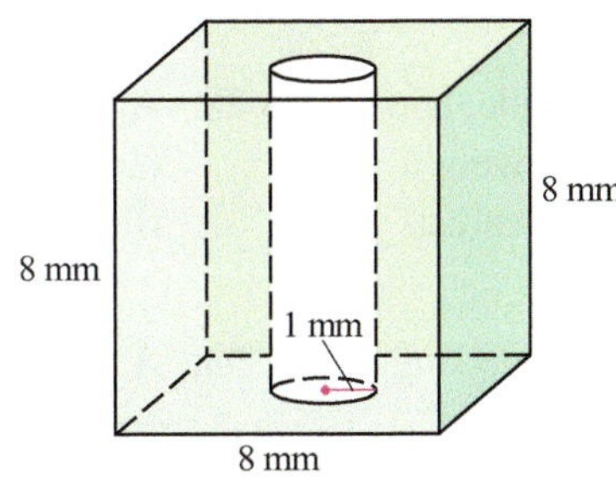

Mixed Practice

Find the volume of each solid. Use $\pi \approx 3.14$, when needed.

21.

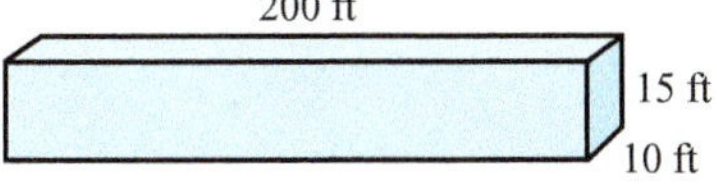

22.

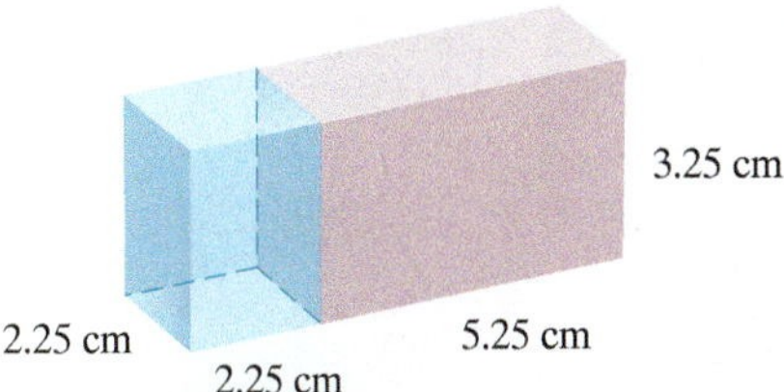

23.

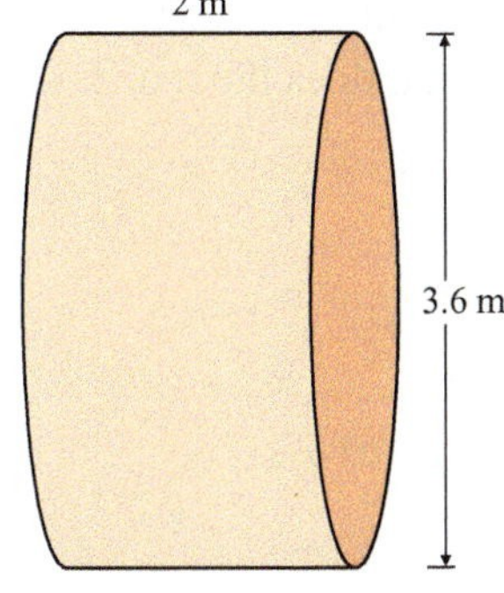

24.

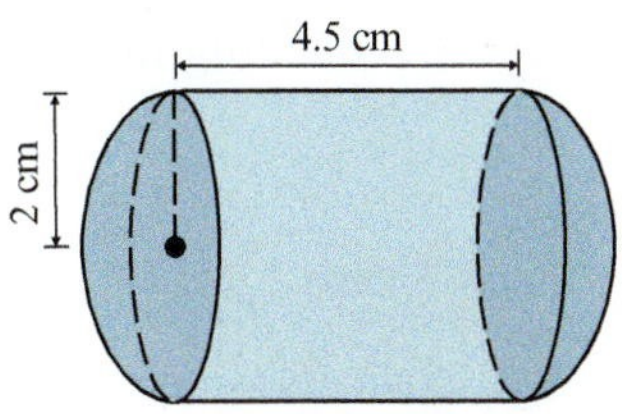

25.

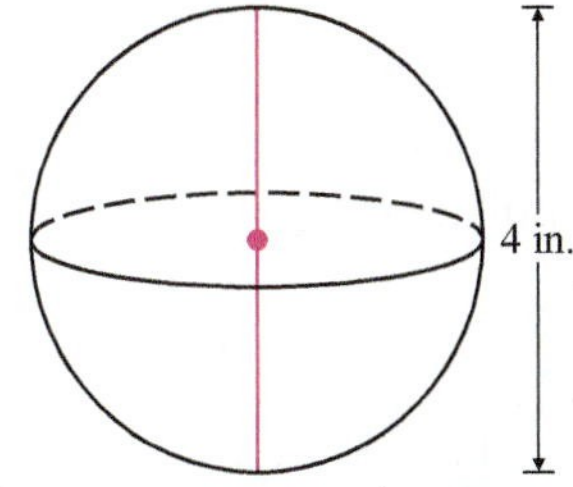

26. A cube with side length 9 in.

Applications

Solve. Use $\pi \approx 3.14$, when needed.

27. A rectangular gold bar is 20 cm long, 10 cm wide, and 6 cm high. If the bar weighs 13 kg, what is the weight of the bar per cubic centimeter? Round to the nearest hundredth.

28. A fish tank is 12 in. wide, 18 in. long, and 16 in. high. How much water, to the nearest tenth of a gallon, will fill the tank if 1gal equals 231 in^3?

29. The crew of a spaceship began a search for a missing shuttlecraft 1,000 mi in every direction. How many cubic miles did the crew search? Round to the nearest million.

30. A steel storage tank is cylindrical in shape, with diameter 12 ft and height 35 ft. Find the capacity of the tank. (*Source:* dktanks.com)

31. Bubbles occur when air is trapped inside a liquid such as soda or a solid such as glass. The surface of a bubble is spherical. Find the volume of a bubble with radius 1.5 in. Round to the nearest cubic inch. (*Source:* eHow.com)

32. A regulation basketball in the National Basketball Association has a radius of approximately 4.7 in. What is the volume of such a basketball? (*Source:* wikipedia.org)

33. A large box is placed in a delivery truck, as pictured. What is the volume of the space remaining in the truck?

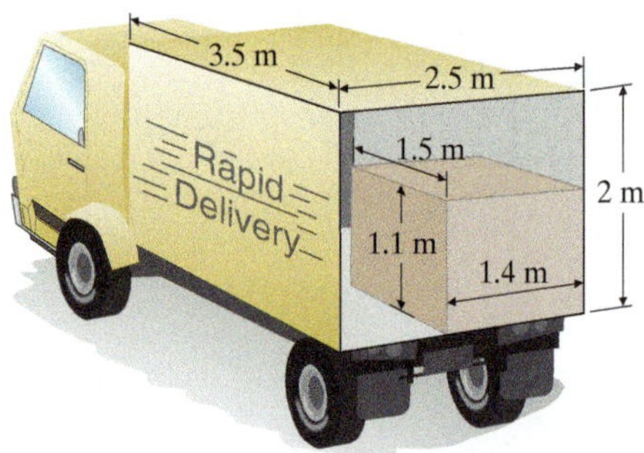

34. The metal machine part shown is cylindrical in shape, with a smaller cylinder drilled out of its center. How many cubic millimeters of metal does the machine part contain?

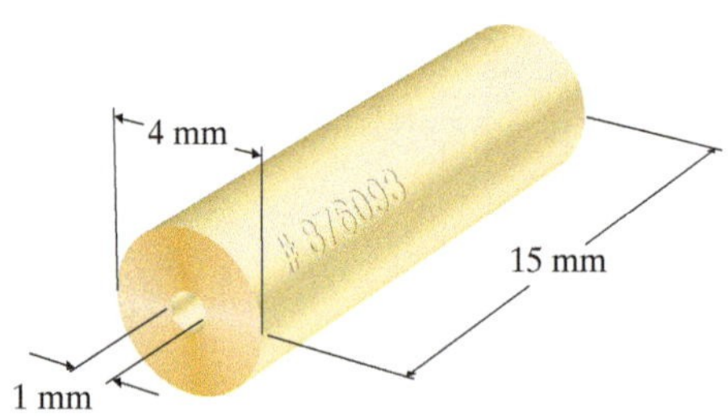

35. The planet Jupiter is the third brightest object in the night sky, after the Moon and Venus. It is the largest of the planets, with a radius of about 71,000 km in contrast to only 6,400 km for Earth. How many times the volume of Earth is that of Jupiter? Round to the nearest hundred. (*Source:* Bennett et al., *The Cosmic Perspective*)

36. The metal recycling containers are 15 in. deep and 15 in. wide. Their heights, from left to right, are 24 in., 26 in., and 22 in. Find the total volume of the containers. (*Source:* twinsupply.com)

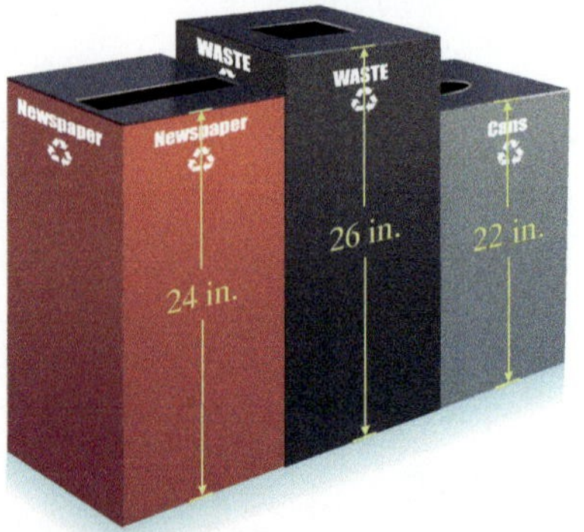

37. In a chemistry course, students learn that a substance's density is found by dividing its mass by its volume. If a 300-gram block of wood is 7 cm by 5 cm by 3 cm, find the density of the wood.

38. A cylindrical wheat silo has a height of 35 yd and a diameter of 5 yd. What is the maximum amount of wheat that can be stored in this silo?

• Check your answers on page A-13.

MINDStretchers

Groupwork

1. Working with a partner, decide which three-dimensional figure can be formed by folding each pattern.

a.

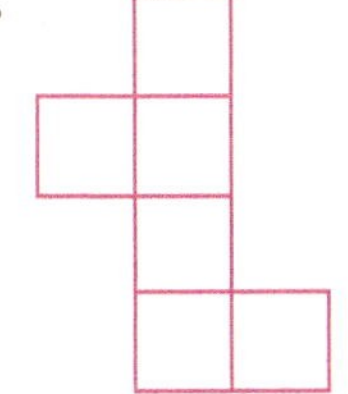

b.

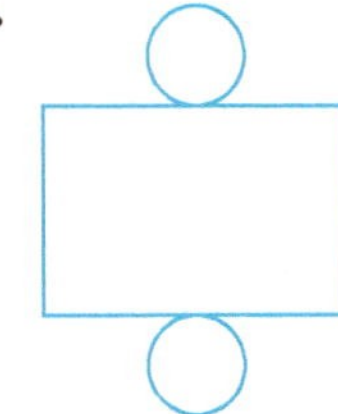

Mathematical Reasoning

2. Geometric solids have not only volume but also *surface area*. Find the total surface area of a cube with volume x^3.

Patterns

3. Consider the following table:

Figure	Measure	Number of Dimensions	Formula
Circle	Circumference	1	$C = 2\pi r$
Circle	Area	2	$A = \pi r^2$
Sphere	Volume	3	$V = \frac{4}{3}\pi r^3$

Describe the pattern that you observe.

10.5 Similar Triangles

OBJECTIVES

- **A** To find the missing sides of similar triangles
- **B** To solve applied problems involving similar triangles

Identifying Corresponding Sides of Similar Triangles

Some figures have the same shape but a different size. For example, when a photograph is enlarged, everything in the enlargement is the same shape as in the original—only larger. In this section, we focus on triangles that have this relationship, which are called *similar triangles.*

DEFINITION

Similar triangles are triangles that have the same shape but not necessarily the same size.

When two triangles are similar, for each angle of the first triangle there corresponds an angle of the second triangle with the same measure. The sides opposite these *corresponding angles* are called *corresponding sides*.

In similar triangles, the measures of corresponding angles are equal and corresponding sides are in proportion. For example, the following triangles ABC and DEF are similar:

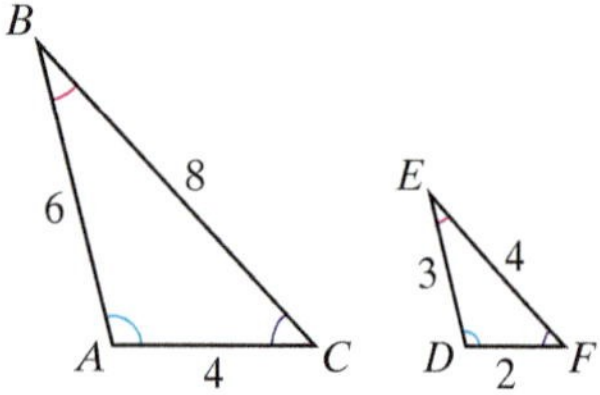

Since these triangles are similar, the measures of their corresponding angles are equal. So we write:

$$m\angle A = m\angle D$$
$$m\angle B = m\angle E$$
$$m\angle C = m\angle F$$

Also, the lengths of the corresponding sides are in proportion, that is:

$$\frac{AB}{DE} = \frac{BC}{EF} = \frac{AC}{DF}$$
$$\frac{6}{3} = \frac{8}{4} = \frac{4}{2} = \frac{2}{1}$$

The ratio of the corresponding sides is $\frac{2}{1}$.

When we write that two triangles are similar, we name them so that the order of corresponding angles in both triangles is the same. In this case,

$$\Delta ABC \sim \Delta DEF$$

where the symbol "$\sim$" means "is similar to."

EXAMPLE 1

$\Delta RST \sim \Delta XYZ$. Name the corresponding sides of these triangles.

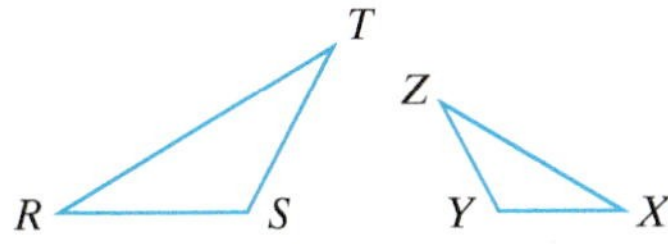

Solution Because $\Delta RST \sim \Delta XYZ$, $m\angle R = m\angle X$, $m\angle S = m\angle Y$, and $m\angle T = m\angle Z$. We know that the corresponding sides are opposite the angles with equal measure. So we write the following:

Because $m\angle R = m\angle X$, $\overline{ST}$ corresponds to $\overline{YZ}$. — $\overline{ST}$ is opposite $\angle R$, and $\overline{YZ}$ is opposite $\angle X$.

Because $m\angle S = m\angle Y$, $\overline{RT}$ corresponds to $\overline{XZ}$. — $\overline{RT}$ is opposite $\angle S$, and $\overline{XZ}$ is opposite $\angle Y$.

Because $m\angle T = m\angle Z$, $\overline{RS}$ corresponds to $\overline{XY}$. — $\overline{RS}$ is opposite $\angle T$, and $\overline{XY}$ is opposite $\angle Z$.

PRACTICE 1

$\Delta ABC \sim \Delta GHI$. List the corresponding sides of these triangles.

Finding the Missing Sides of Similar Triangles

Since corresponding sides in similar triangles are in proportion, we can use these proportions to find the length of a missing side.

To Find a Missing Side of a Similar Triangle

- Write the ratios of the lengths of the corresponding sides.
- Write a proportion using a ratio with known terms and a ratio with an unknown term.
- Solve the proportion for the unknown term.

EXAMPLE 2

In the following diagram, $\Delta TAP \sim \Delta RUN$. Find x.

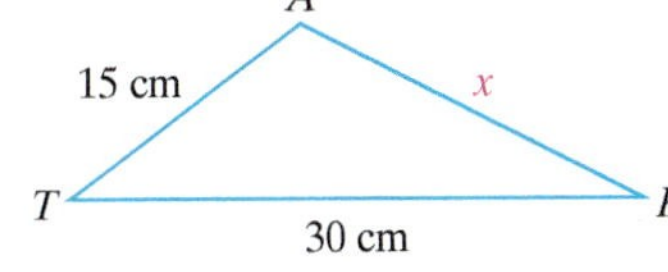

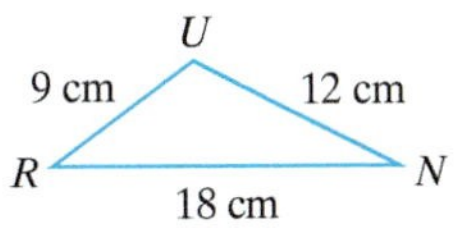

Solution Because $\Delta TAP \sim \Delta RUN$, we write the ratios of the lengths of the corresponding sides as follows:

$$\frac{TA}{RU} = \frac{AP}{UN} = \frac{TP}{RN}$$

$$\frac{15}{9} = \frac{x}{12} = \frac{30}{18}$$

PRACTICE 2

In the following diagram, $\Delta DOT \sim \Delta PAN$. Find y.

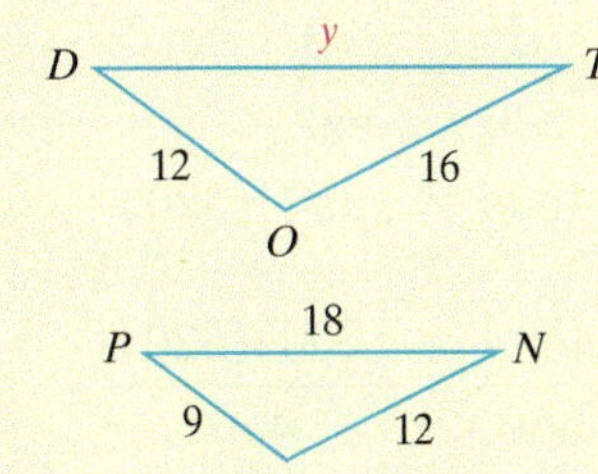

EXAMPLE 2 (continued)

To solve for x, we can consider either the proportion $\frac{15}{9} = \frac{x}{12}$ or the proportion $\frac{x}{12} = \frac{30}{18}$. Note that each proportion contains the unknown term x. If we choose the first proportion and solve for x, we get the following:

$$\frac{15}{9} = \frac{x}{12}$$
$$9x = 180$$
$$x = 20$$

So x is 20 cm.

Similar triangles are useful in finding lengths that cannot be measured directly, as Example 3 illustrates.

EXAMPLE 3

A surveyor took the measurements shown. If $\Delta ABC \sim \Delta EFC$, find d, the distance across the river.

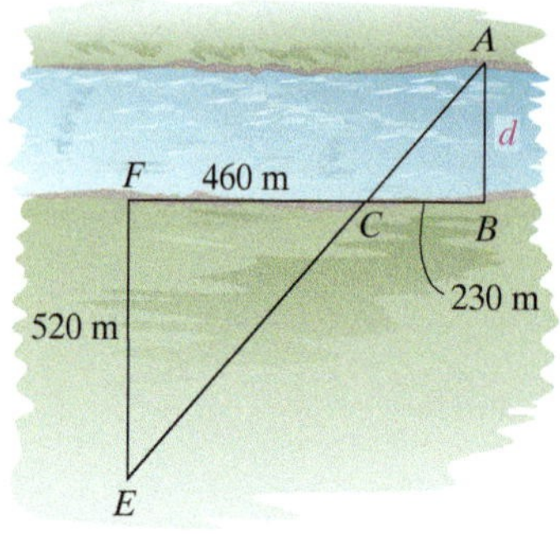

Solution Here, $\Delta ABC \sim \Delta EFC$, so we write the proportion $\frac{AB}{EF} = \frac{BC}{FC}$.

Then, we substitute the values given in the diagram.

$$\frac{d}{520} = \frac{230}{460}$$
$$460d = (230)(520) \quad \text{Cross multiply.}$$
$$\frac{\overset{1}{\cancel{460}}d}{\underset{1}{\cancel{460}}} = \frac{(\overset{1}{\cancel{230}})(520)}{\underset{2}{\cancel{460}}} \quad \text{Divide both sides by 460.}$$
$$d = 260$$

So the distance across the river is 260 m.

PRACTICE 3

The height of a man and his shadow form a triangle similar to that formed by a nearby tree and its shadow. What is the height of the tree?

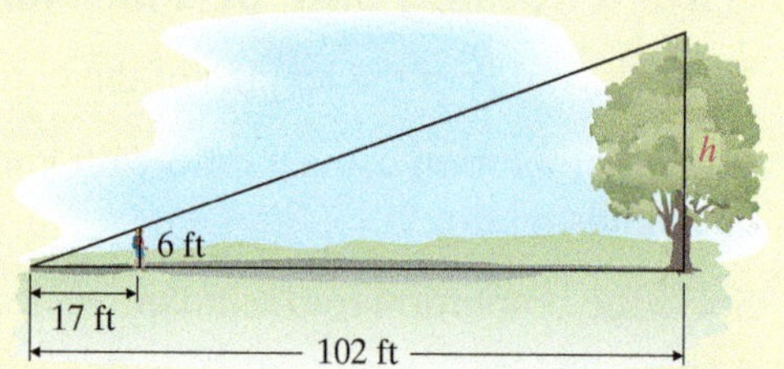

10.5 Exercises

FOR EXTRA HELP MyMathLab   PRACTICE  WATCH  READ 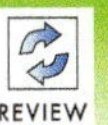 REVIEW

Mathematically Speaking

Fill in each blank with the most appropriate term or phrase from the given list.

equal	shape	similar
area	in proportion	corresponding

1. The symbol ~ is used to indicate that triangles are _____________.

2. Similar triangles have the same _____________ but not necessarily the same size.

3. In similar triangles, _____________ sides are opposite angles with equal measure.

4. Corresponding sides of similar triangles are _____________.

A *Find the value of each unknown.*

5. $\triangle DEF \sim \triangle ABC$

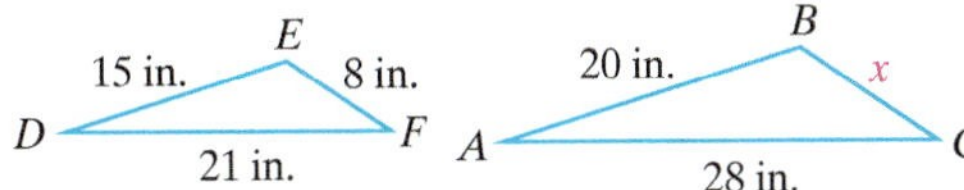

6. $\triangle LOM \sim \triangle RST$

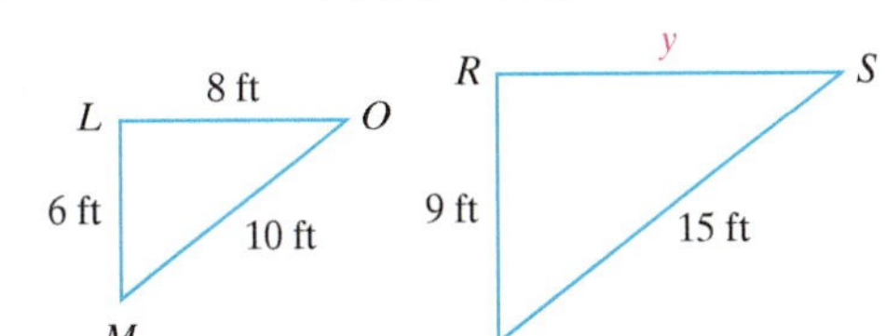

7. $\triangle DOT \sim \triangle PAN$

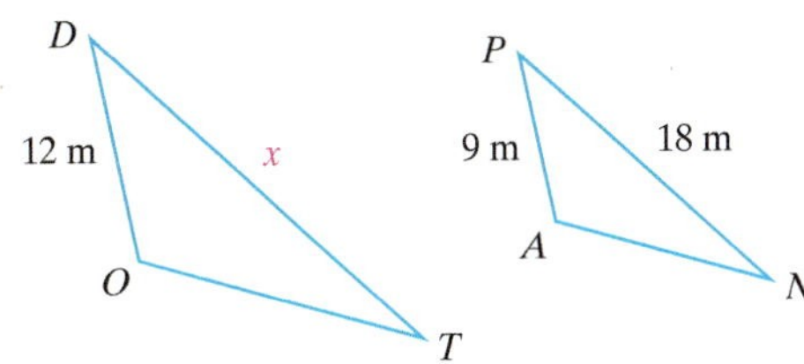

8. $\triangle ACT \sim \triangle MLK$

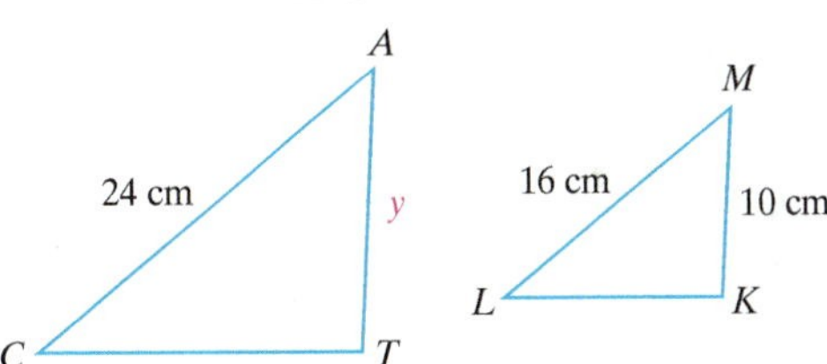

9. $\triangle DEF \sim \triangle ABC$

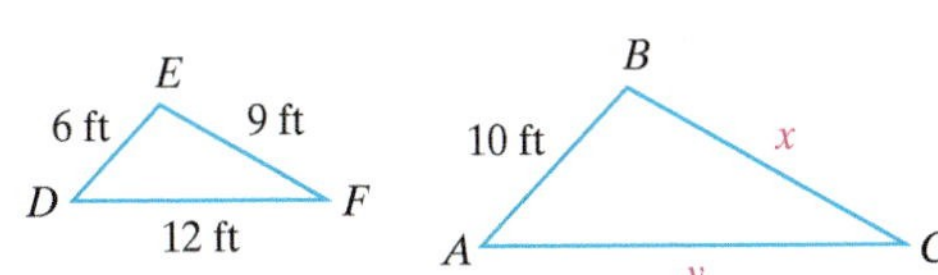

10. $\triangle DOT \sim \triangle PIN$

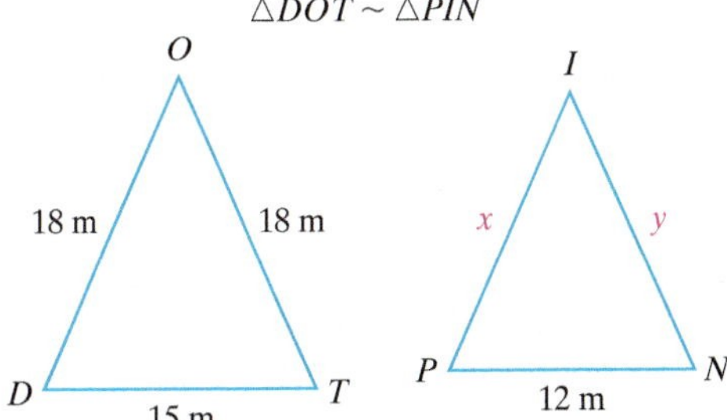

11. $\triangle TAP \sim \triangle RON$

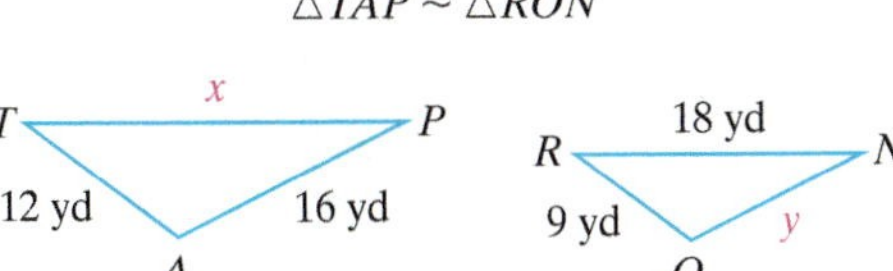

12. $\triangle FEG \sim \triangle CBD$

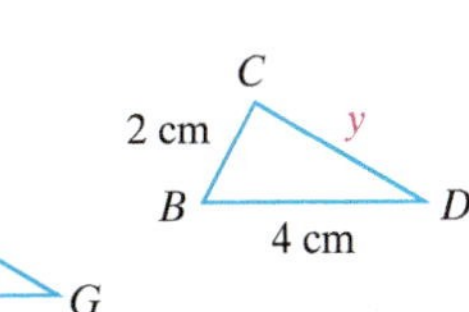

13.

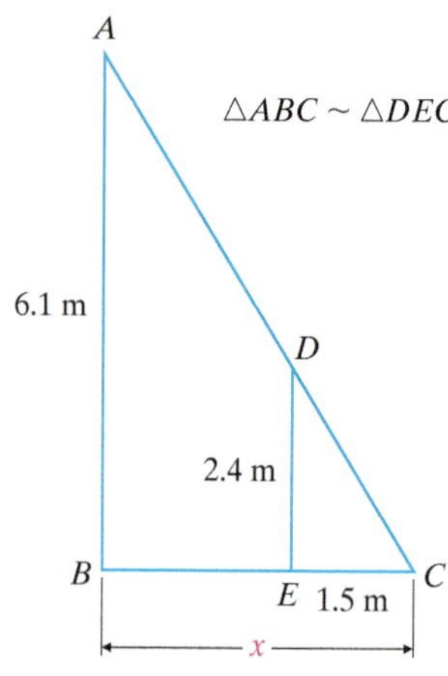

14.

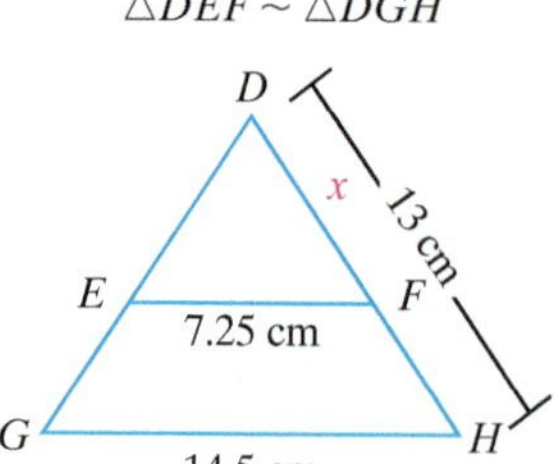

Mixed Practice

Find the value of each unknown.

15. $\Delta ABC \sim \Delta DEF$

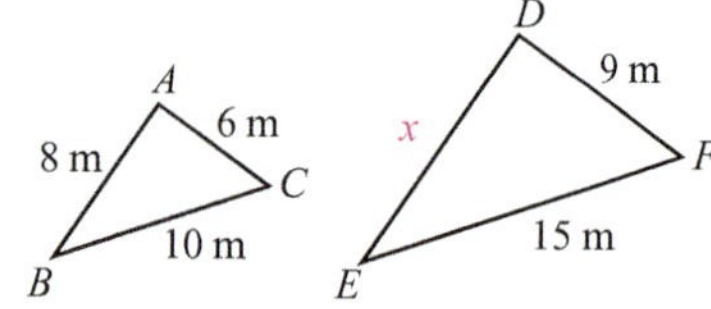

16. $\Delta PQR \sim \Delta STU$

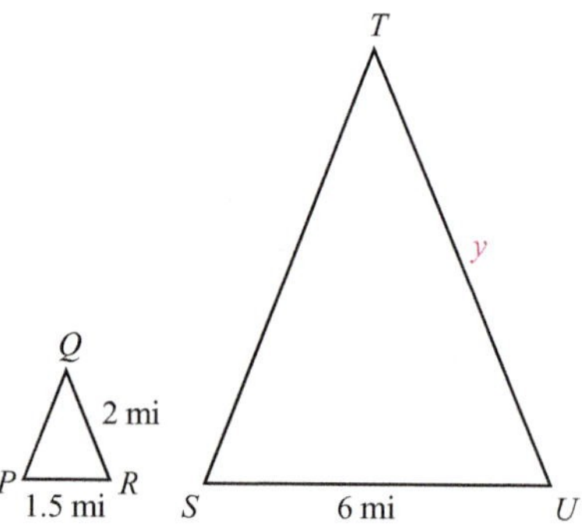

17. $\Delta ABC \sim \Delta ADE$

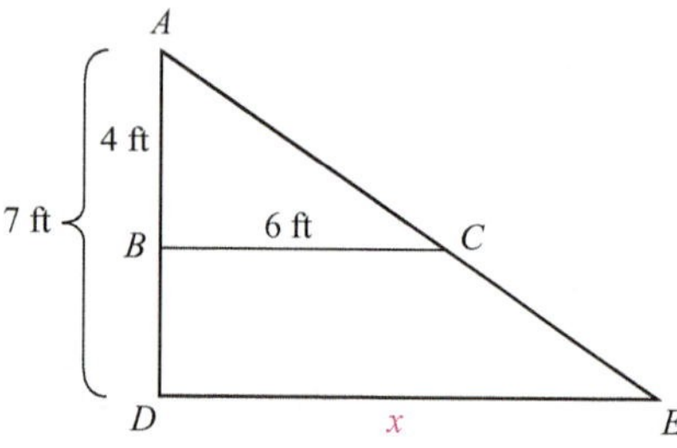

18. $\Delta PQR \sim \Delta TSR$

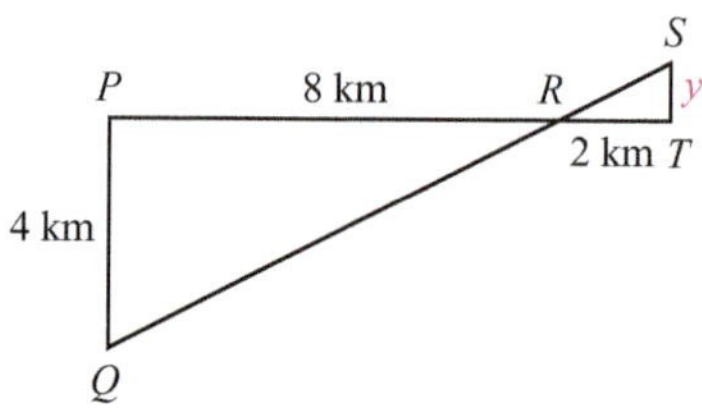

Applications

B *Solve. Assume the triangles are similar.*

19. The diagram below (not drawn to scale) shows the shadows cast by a column and a ruler. Find the height of the column.

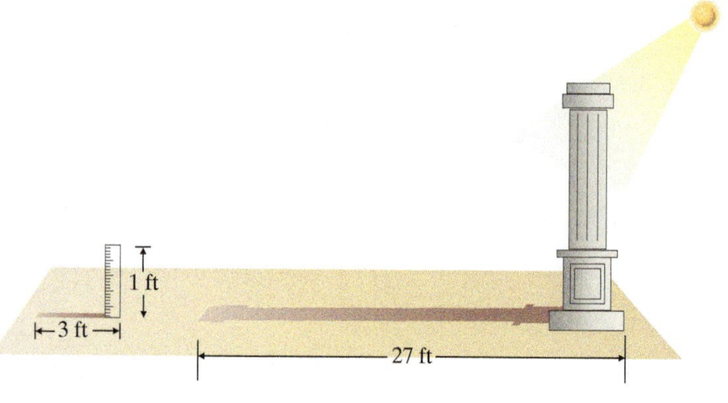

20. Light from a flashlight shines through a transparent dragon puppet onto a screen behind it, as shown below. Find the height of the puppet's image.

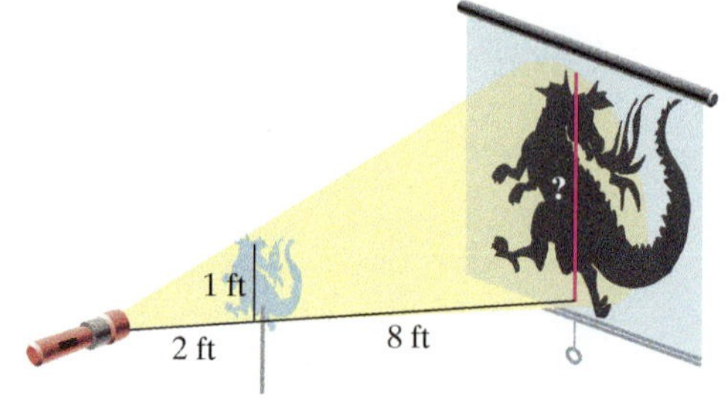

21. A Coast Guard observer sees a boat out on the ocean and wants to know how far it is from the shore. Use the diagram to find that distance.

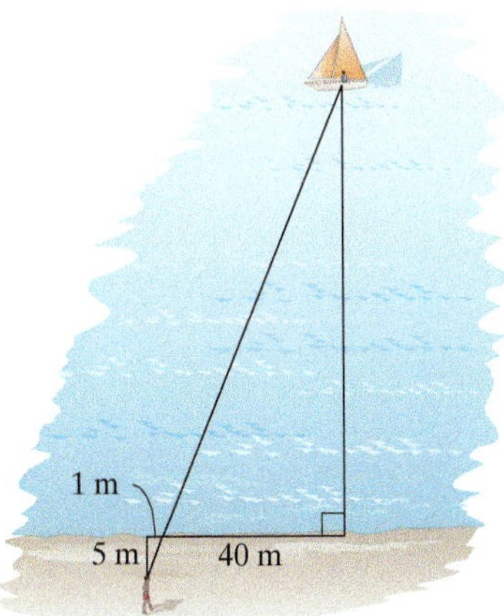

22. One way to measure the height of a building is to position a mirror on the ground so that the top of the building's reflection can be seen. Find the height of the building.

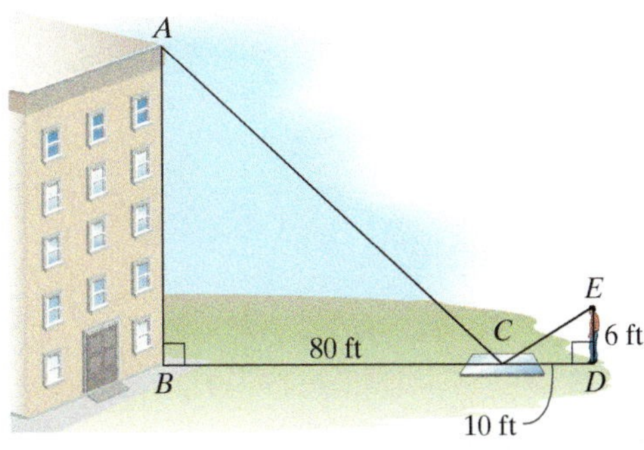

23. To make paintings appear to be three-dimensional, artists draw parallel lines such as railroad tracks getting closer to one another the further away they get. The "vanishing point" refers to the point in the distance where the lines appear to meet, forming the vertex of a triangle. Find the missing dimension in the image at right. (*Source:* math.vcu.edu)

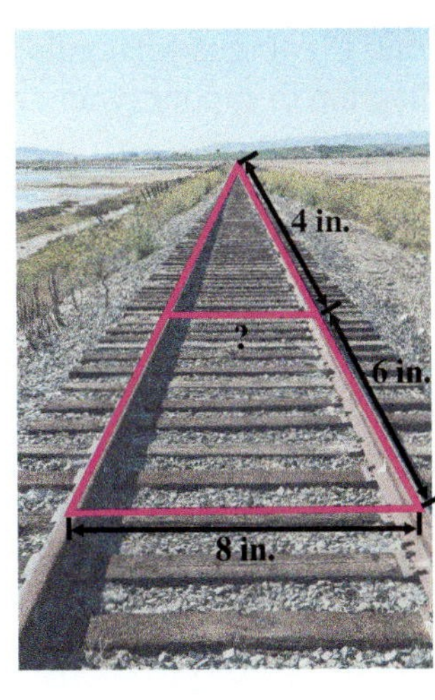

24. The diagram below shows the side view of an escalator, where ΔABC is similar to ΔCBD. Find the length of $\overline{CD}$.

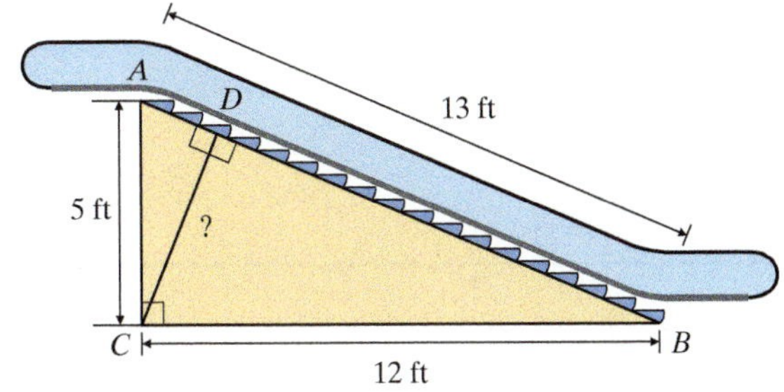

25. Two support wires are attached to a utility pole as shown in the diagram. If $\Delta ABC \sim \Delta ADE$, find AD.

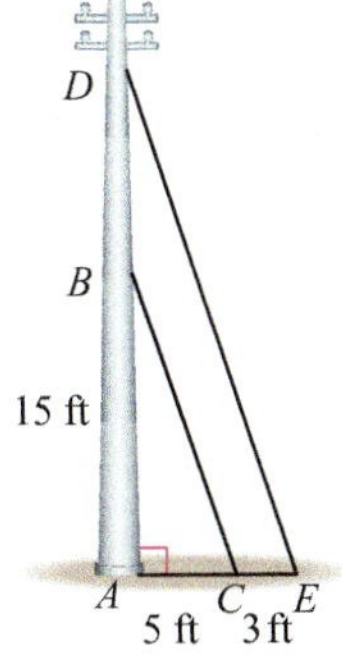

26. To measure $\overline{AB}$, the distance across a certain lake, the length of line segments $\overline{BC}$, $\overline{DC}$, and $\overline{ED}$ were "staked out," as shown in the diagram below. If $\Delta ABC \sim \Delta EDC$, how wide is the lake?

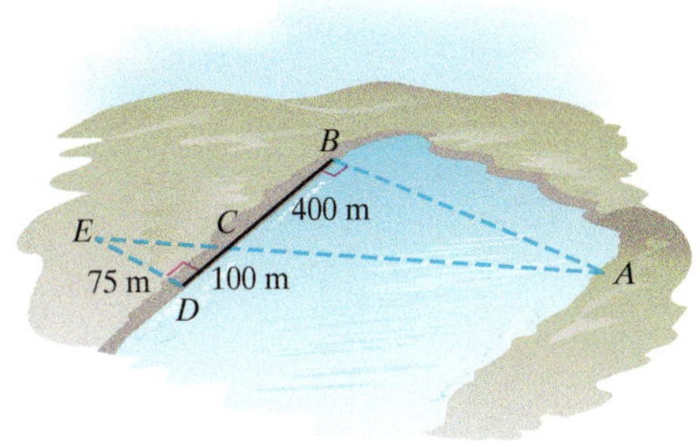

• Check your answers on page A-14.

MINDStretchers

Patterns

1. List 10 pairs of similar triangles in the following square.

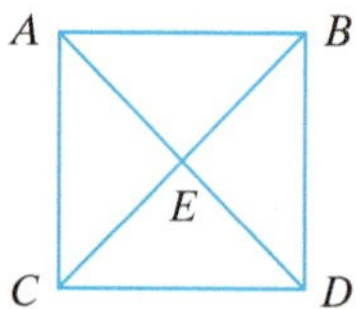

Writing

2. On a computer, you can shrink an image without changing its shape so that the new image is similar to the old one. Give some other everyday examples of similarity.

Groupwork

3. Working with a partner, decide whether this statement is true: If two quadrilaterals have corresponding angles equal, the two quadrilaterals must have the same shape. Draw a diagram to support your answer.

10.6 Square Roots and the Pythagorean Theorem

OBJECTIVES

- **A** To find the square root of a number
- **B** To find the unknown side of a right triangle, using the Pythagorean theorem
- **C** To solve applied problems involving a square root or the Pythagorean theorem

In Section 10.3, we found the area of a square by squaring the length of one of its sides. In this section, we look at the opposite of squaring a number, that is, finding its *square root.*

Finding the Square Root of a Number

Consider the following problem. Suppose that a square has an area of 25 ft^2. What is the length s of each side?

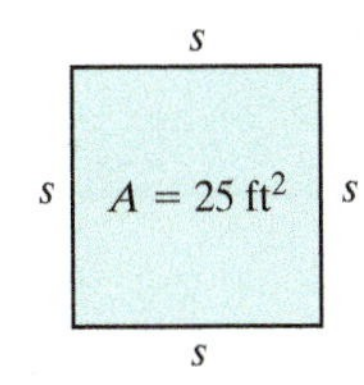

Recall that, for the area of a square, $A = s^2$. So for $25 = s^2$, we need to determine what whole number when multiplied by itself equals 25. Because $25 = 5 \cdot 5$, the whole number is 5. So a square with an area of 25 ft^2 has sides of length 5 feet.

Squaring the whole number 5 gives 25. So we say that 25 is the *square* of 5, or that 5 is the (principal) *square root* of 25 (written $\sqrt{25} = 5$).

Since 25 is the square of a whole number, it is called a *perfect square.* Perfect squares play a special role in the discussion of square roots.

DEFINITIONS

A **perfect square** is a number that is the square of a whole number.

The (principal) **square root** of a number n, written $\sqrt{n}$, is the positive number whose square is n.

Since $36 = 6^2$, $\sqrt{36} = 6$. And since $4 = 2^2$, $\sqrt{4} = 2$. Squaring and taking a square root are opposite operations, since one operation undoes the other.

EXAMPLE 1

Find the square root.

a. $\sqrt{64}$ **b.** $\sqrt{100}$

Solution In each case, we need to find the whole number that when squared gives us the number under the square root sign.

a. $\sqrt{64} = 8$ because $8 \cdot 8 = 64$.

b. $\sqrt{100} = 10$ because $10 \cdot 10 = 100$.

PRACTICE 1

What is the square root of each perfect square?

a. $\sqrt{49}$

b. $\sqrt{144}$

In Example 1, note that $100 > 25$ and $\sqrt{100} > \sqrt{25}$. In general, larger numbers have larger square roots.

Many numbers are not perfect squares. For instance, 28 is not a perfect square because there is no whole number that when multiplied by itself equals 28. If a number is not a perfect square, we can either estimate or use a calculator to find its approximate square root.

EXAMPLE 2

$\sqrt{28}$ lies between which two consecutive whole numbers?

Solution To begin, let's find the two consecutive perfect squares that 28 lies between. The number 28 is more than 25 and less than 36, which are consecutive perfect squares.

Because 28 lies between 5^2 and 6^2, $\sqrt{28}$ lies between 5 and 6.

PRACTICE 2

Between which two consecutive whole numbers does $\sqrt{47}$ lie?

EXAMPLE 3

Using a calculator, approximate each square root. Round to the nearest tenth.

a. $\sqrt{75}$ **b.** $\sqrt{21}$

Solution

a. Press | Display

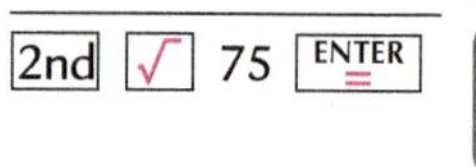

So $\sqrt{75} \approx 8.7$.

b. Press | Display

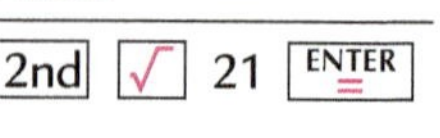

Therefore, $\sqrt{21} \approx 4.6$.

PRACTICE 3

Using a calculator, approximate each square root. Round to the nearest hundredth.

a. $\sqrt{56}$

b. $\sqrt{12}$

The Pythagorean Theorem

Recall that a right triangle is a triangle that has one 90° angle. In a right triangle, the side opposite the right angle is called the **hypotenuse**. The other two sides are called **legs**.

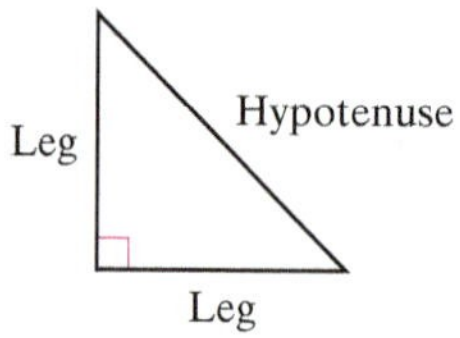

The lengths of the three sides of a right triangle are related in a special way. To understand this relationship, consider the areas of the squares on the legs and on the hypotenuse, as in the following example:

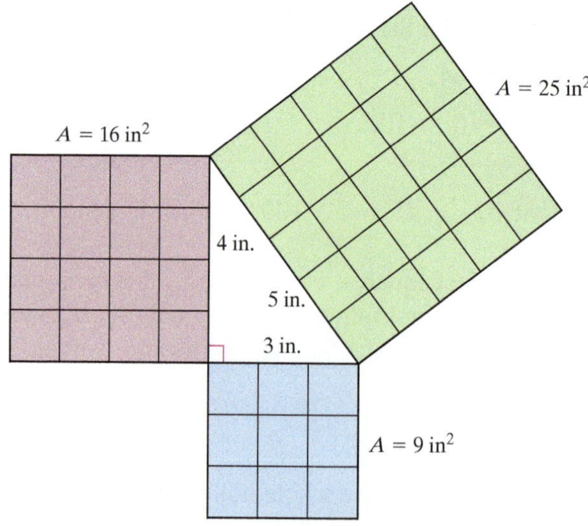

$$9 \text{ in}^2 + 16 \text{ in}^2 = 25 \text{ in}^2$$

Area of the square on one leg + Area of the square on the other leg = Area of the square on the hypotenuse

In general, if we let a and b represent the lengths of the legs and c represent the length of the hypotenuse, then $a^2 + b^2 = c^2$.

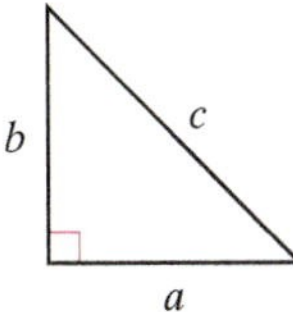

This relationship is called the *Pythagorean theorem.*

The Pythagorean Theorem

For every right triangle, the sum of the squares of the lengths of the two legs equals the square of the length of the hypotenuse, that is,

$$a^2 + b^2 = c^2$$

where a and b are the lengths of the legs, and c is the length of the hypotenuse.

We can use the Pythagorean theorem to find the third side of a right triangle if we know the other two sides.

EXAMPLE 4

Find the length of the hypotenuse.

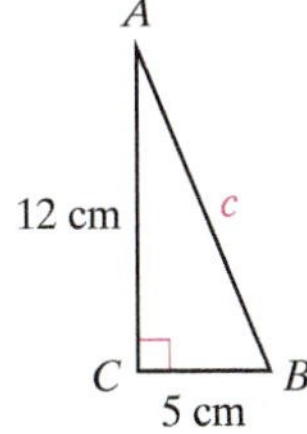

Solution To find the length of the hypotenuse, we use the Pythagorean theorem.

$$a^2 + b^2 = c^2$$
$$5^2 + 12^2 = c^2$$
$$25 + 144 = c^2$$
$$169 = c^2$$
$$\sqrt{169} = c$$ Taking a square root is the opposite of squaring.
$$13 = c, \text{ or } c = 13$$

So the hypotenuse is 13 cm long.

PRACTICE 4

Find the length of the unknown side in ΔABC.

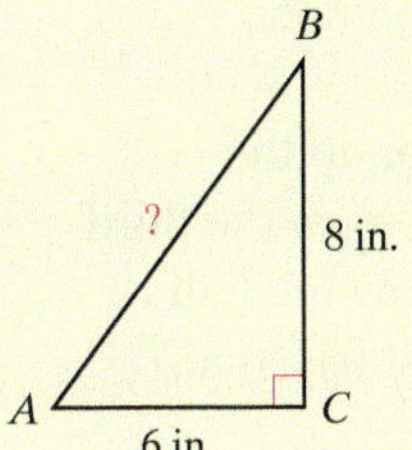

EXAMPLE 5

If b equals 1 cm and c equals 2 cm, what is a in a right triangle, where a and b are the lengths of the legs and c is the length of the hypotenuse? Round the answer to the nearest tenth of a centimeter.

PRACTICE 5

In a right triangle, one leg equals 2 ft and the hypotenuse equals 4 ft. Approximate the length of the missing leg. Round the answer to the nearest tenth of a foot.

EXAMPLE 5 (continued)

Solution First, we draw a diagram.

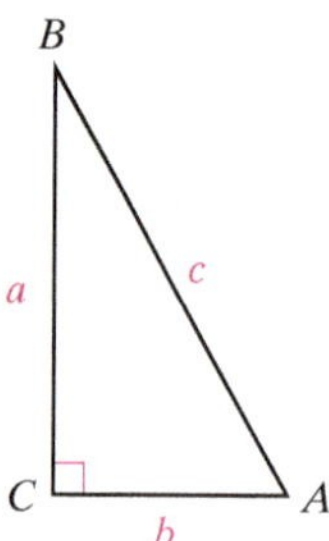

Then, we use the Pythagorean theorem and substitute the given values to obtain the following:

$$a^2 + b^2 = c^2$$
$$a^2 + 1^2 = 2^2$$
$$a^2 + 1 = 4$$
$$a^2 + 1 - 1 = 4 - 1$$
$$a^2 = 3$$
$$a = \sqrt{3}$$ Taking a square root is the opposite of squaring.

To express the answer as a decimal, we use a calculator. Rounding to the nearest tenth, we find that $\sqrt{3}$ is 1.7. So $a \approx 1.7$ cm.

EXAMPLE 6

A baseball diamond is a square with sides 90 ft long. How far, to the nearest foot, must the third baseman throw the ball to reach the first baseman?

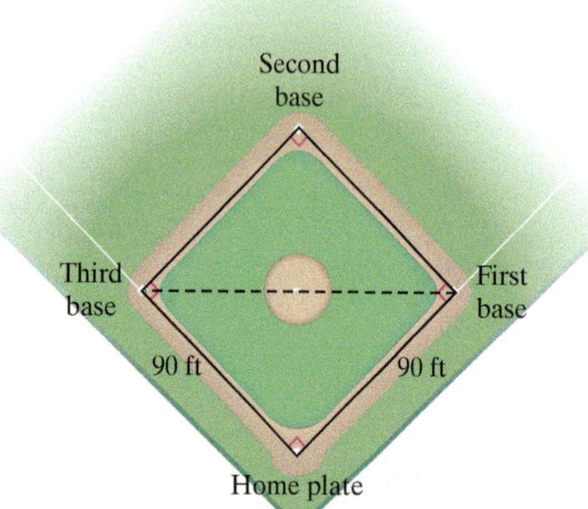

Solution The sides of the diamond together with the diagonal from third base to first base form a right triangle. To find the distance from third base to first base, we use the Pythagorean theorem.

$$a^2 + b^2 = c^2$$
$$90^2 + 90^2 = c^2$$
$$8{,}100 + 8{,}100 = c^2$$
$$16{,}200 = c^2$$
$$\sqrt{16{,}200} = c$$
$$c = \sqrt{16{,}200}$$

We use a calculator and round to find that $\sqrt{16{,}200}$ is approximately 127 ft. So the third baseman must throw the ball approximately 127 ft to reach the first baseman.

PRACTICE 6

Stair stringers, the structural supporting parts of staircases, are used by carpenters in building stairs. What is the length, to the nearest tenth of a foot, of the stair stringer shown in the diagram below? (*Source:* wikipedia.org)

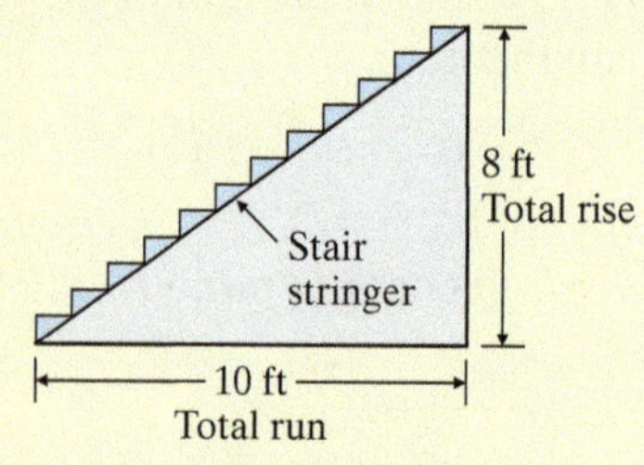

10.6 Exercises

FOR EXTRA HELP MyMathLab MathXL PRACTICE WATCH READ REVIEW

Mathematically Speaking

Fill in each blank with the most appropriate term or phrase from the given list.

squaring	leg	hypotenuse
multiple	Area of three squares	prime
consecutive		perfect square
doubling	Pythagorean theorem	square root

1. The number 5 is the __________ of 25.

2. Finding a square root is the opposite of __________ the number.

3. The square of a whole number is said to be a(n) ______________.

4. The whole numbers 5 and 6 are ______________.

5. In a right triangle, the longest side is called the __________.

6. If a and b are the lengths of the legs of a right triangle and c is the length of the hypotenuse, then the ______________ states that $a^2 + b^2 = c^2$.

A *Find each square root.*

7. $\sqrt{9}$
8. $\sqrt{4}$
9. $\sqrt{16}$
10. $\sqrt{36}$
11. $\sqrt{81}$
12. $\sqrt{64}$
13. $\sqrt{169}$
14. $\sqrt{121}$
15. $\sqrt{400}$
16. $\sqrt{225}$
17. $\sqrt{256}$
18. $\sqrt{900}$

Determine between which two consecutive whole numbers each square root lies.

19. $\sqrt{50}$
20. $\sqrt{7}$
21. $\sqrt{80}$
22. $\sqrt{31}$
23. $\sqrt{39}$
24. $\sqrt{2}$
25. $\sqrt{14}$
26. $\sqrt{105}$

Approximate each square root. Round to the nearest tenth, if needed.

27. $\sqrt{5}$
28. $\sqrt{11}$
29. $\sqrt{37}$
30. $\sqrt{74}$
31. $\sqrt{139}$
32. $\sqrt{165}$
33. $\sqrt{9{,}801}$
34. $\sqrt{8{,}649}$

B *Find each missing length. Round to the nearest tenth, if needed.*

35. 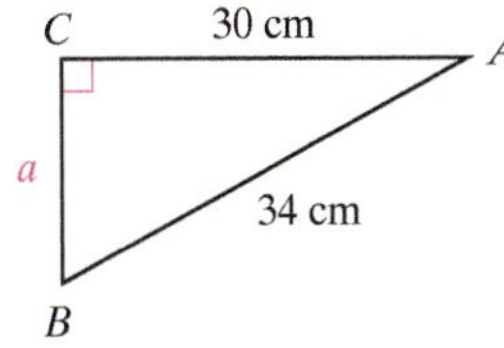

36. 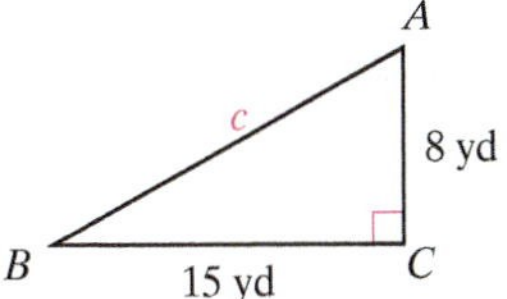

37. 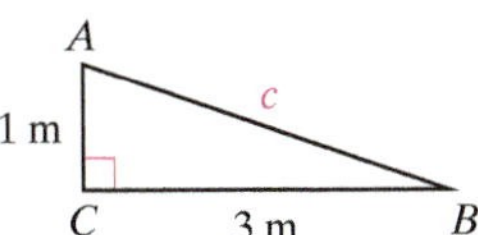

38. 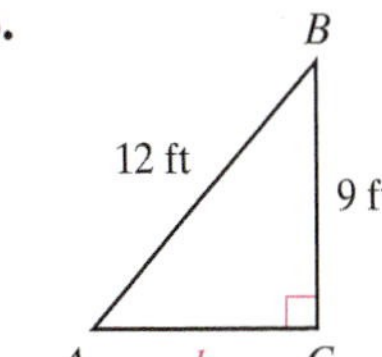

Given a right triangle with legs a and b, and hypotenuse c, find the missing side. Round to the nearest tenth, if needed.

	a	*b*	*c*
39.	24 m		25 m
40.	5 in.	12 in.	
41.	6 ft		10 ft
42.		4 cm	5 cm
43.	12 m	16 m	
44.		9 in.	15 in.
45.	7 cm	9 cm	
46.	2 yd	5 yd	
47.		18 ft	20 ft
48.	2 in.	2 in.	

Mixed Practice

Solve.

49. Find $\sqrt{196}$.

50. Determine between which two consecutive whole numbers $\sqrt{95}$ lies.

51. Find the missing length.

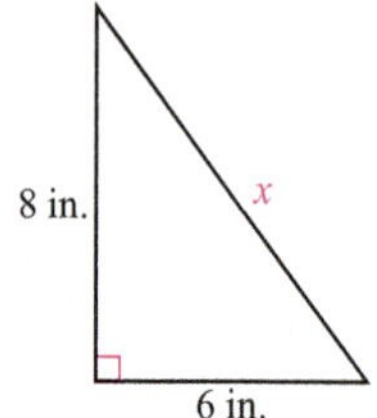

52. Find the missing length.

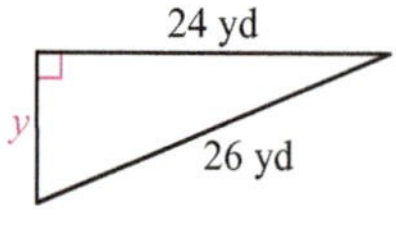

53. Find the missing length. Round to the nearest tenth.

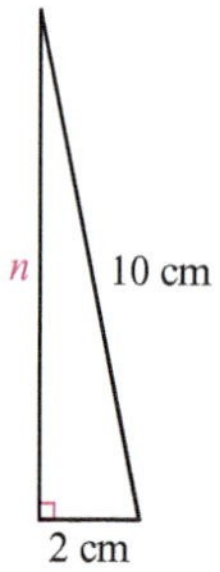

54. Find $\sqrt{41}$ to the nearest tenth.

Applications

C *Solve. Use a calculator, if needed.*

55. A contractor leans a ladder against the side of a building. How high up the building does the ladder reach?

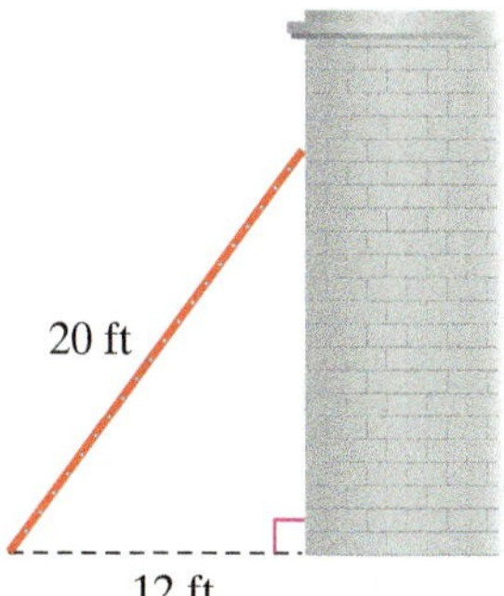

56. A scuba diver swims away from the boat and then dives, as shown. How far from the boat, to the nearest foot, will he be?

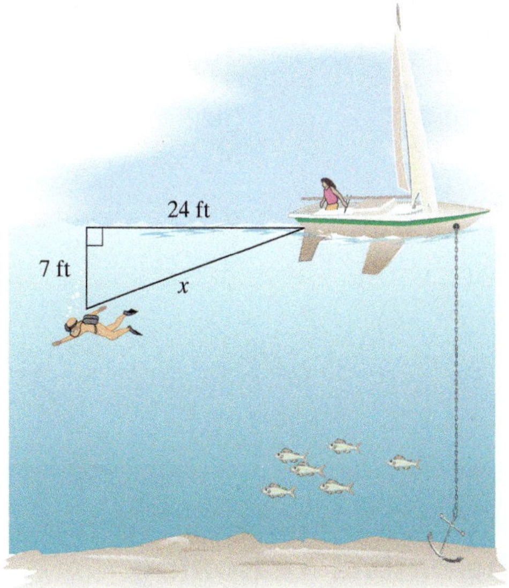

57. What is the length of the rectangular plot of land shown?

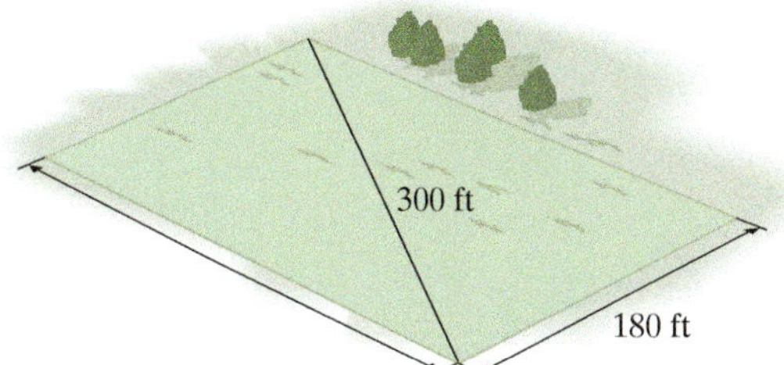

58. *ABCE* is a rectangular picnic area, with a picnic table at point *B* and the entrance at point *D*. The lengths *BC*, *CD*, and *BE* are as shown below:

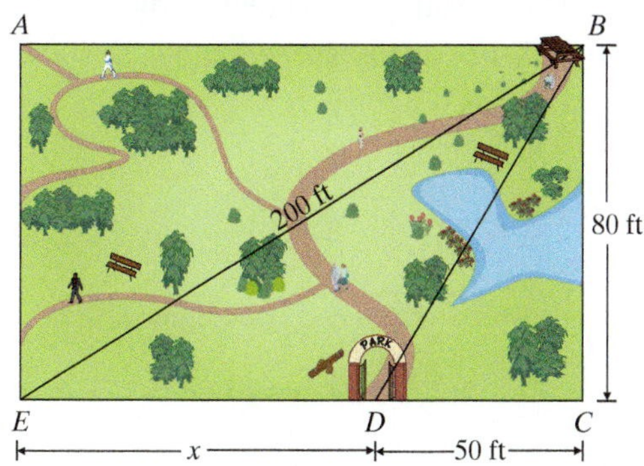

Find the distance between the entrance and point *E*.

59. A builder constructed a roof of wooden beams. According to the diagram, what is the length of the sloping beam?

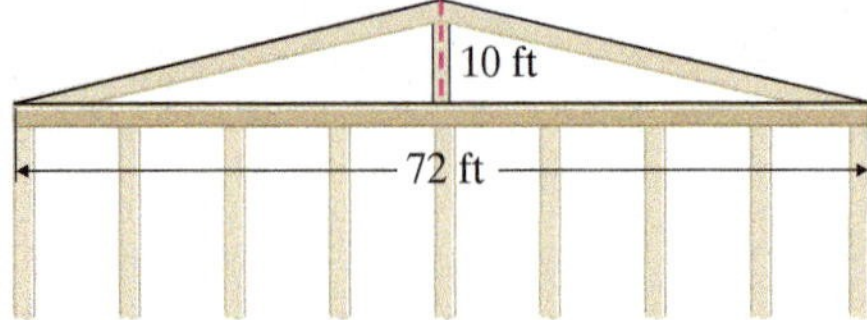

60. A college is constructing an access ramp to a door in one of its buildings, as shown. Find the length of the ramp.

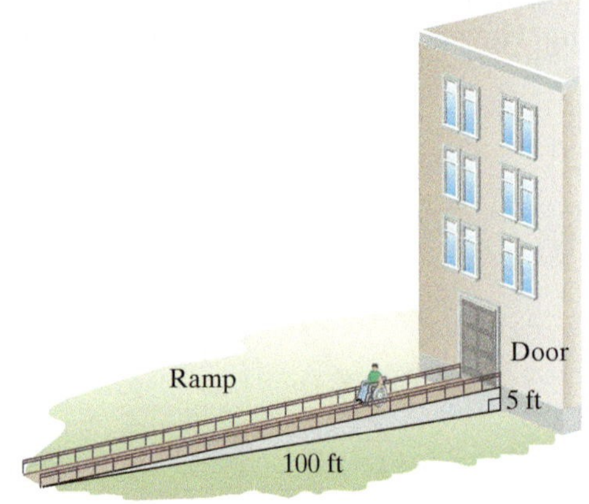

• Check your answers on page A-14.

MINDStretchers

Mathematical Reasoning

1. Give an example of a number that is smaller than its square root.

Writing

2. Thousands of years ago, the ancient Egyptians used a clever way of creating a right angle for their construction projects. For example, to create a right triangle with side lengths 3, 4, and 5, they would use a rope tying it in a circle with 12 equally spaced knots, as shown:

Explain why this procedure would create the right triangle. (*Source:* Peter Tompkins, *Secrets of the Great Pyramids*)

Investigation

3. Choose a whole number. Use a calculator to determine whether it is a perfect square.

Key Concepts and Skills

CONCEPT SKILL

Concept/Skill	Description	Example
[10.1] **Point**	An exact location in space, with no dimension.	•A
[10.1] **Line**	A collection of points along a straight path that extends endlessly in both directions.	A B $\overleftrightarrow{AB}$
[10.1] **Line segment**	A part of a line having two endpoints.	B C $\overline{BC}$
[10.1] **Ray**	A part of a line having only one endpoint.	A B $\overrightarrow{AB}$
[10.1] **Angle**	Two rays that have a common endpoint called the *vertex* of the angle.	A B C $\angle ABC$
[10.1] **Plane**	A flat surface that extends endlessly in all directions.	B C A D
[10.1] **Straight angle**	An angle whose measure is 180°.	180° A B C
[10.1] **Right angle**	An angle whose measure is 90°.	D E F
[10.1] **Acute angle**	An angle whose measure is less than 90°.	X 65° Y Z
[10.1] **Obtuse angle**	An angle whose measure is more than 90° and less than 180°.	C 120° D E
[10.1] **Complementary angles**	Two angles the sum of whose measures is 90°.	25° A 65° B
[10.1] **Supplementary angles**	Two angles the sum of whose measures is 180°.	40° C 140° D

continued

CONCEPT SKILL

Concept/Skill	Description	Example
[10.1] **Intersecting lines**	Two lines that cross.	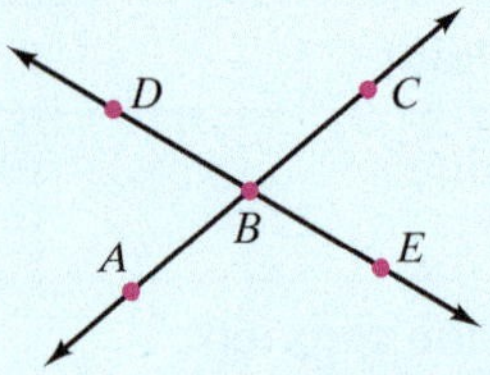 D C B A E
[10.1] **Parallel lines**	Two lines on the same plane that do not intersect.	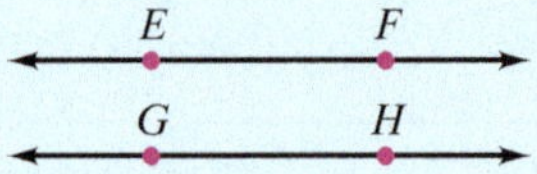 E F G H $\overleftrightarrow{EF} \parallel \overleftrightarrow{GH}$
[10.1] **Perpendicular lines**	Two lines that intersect to form right angles.	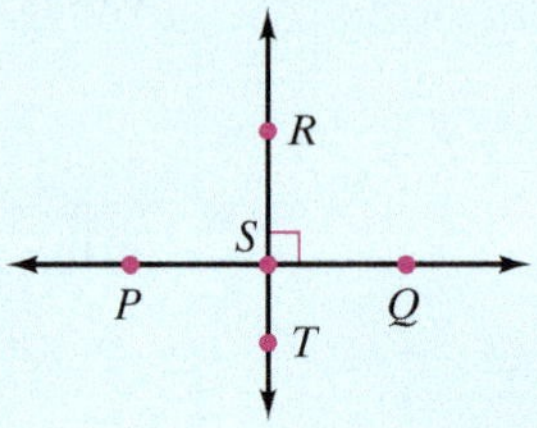 R S P Q T $\overleftrightarrow{RT} \perp \overleftrightarrow{PQ}$
[10.1] **Vertical angles**	Two opposite angles with equal measure formed by two intersecting lines.	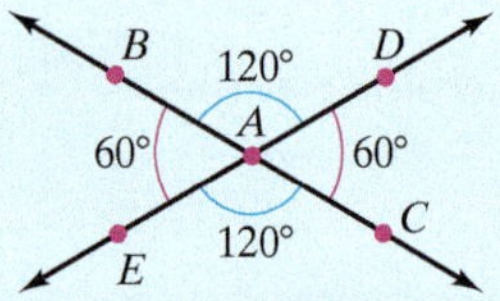 B 120° D 60° A 60° C E 120°
[10.1] **Polygon**	A closed plane figure made up of line segments.	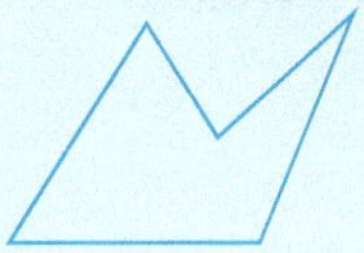
[10.1] **Triangle**	A polygon with three sides.	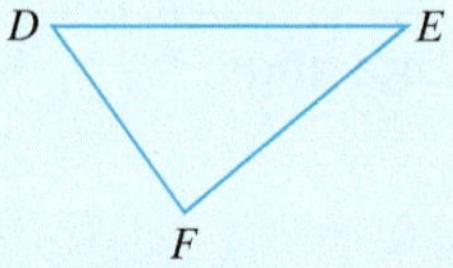 D E F
[10.1] **Quadrilateral**	A polygon with four sides.	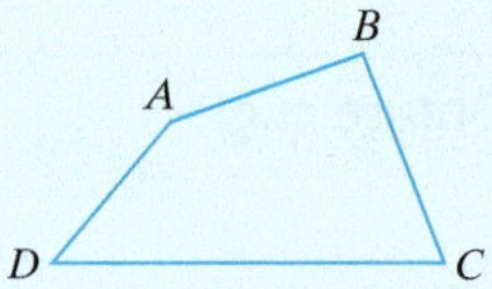 B A D C
[10.1] **Equilateral triangle**	A triangle with three sides equal in length.	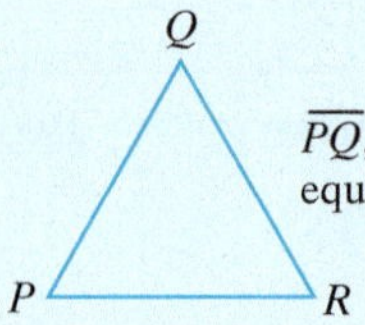 Q P R $\overline{PQ}$, $\overline{QR}$, and $\overline{PR}$ have equal lengths.
[10.1] **Isosceles triangle**	A triangle with two or more sides equal in length.	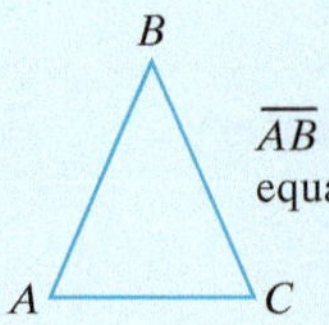 B A C $\overline{AB}$ and $\overline{BC}$ have equal lengths.

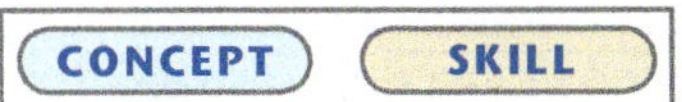

Concept/Skill	Description	Example
[10.1] **Scalene triangle**	A triangle with no sides equal in length.	$\overline{HG}$, $\overline{GI}$, and $\overline{HI}$ have unequal lengths.
[10.1] **Acute triangle**	A triangle with three acute angles.	Triangle RST
[10.1] **Right triangle**	A triangle with one right angle.	Triangle LPM
[10.1] **Obtuse triangle**	A triangle with one obtuse angle.	Triangle XYZ
[10.1] **The sum of the measures of the angles of a triangle**	In any triangle, the sum of the measures of all three angles is 180°.	$m\angle A + m\angle B + m\angle C = 180°$
[10.1] **Trapezoid**	A quadrilateral with only one pair of opposite sides parallel.	$\overline{AB} \parallel \overline{CD}$
[10.1] **Parallelogram**	A quadrilateral with both pairs of opposite sides parallel. Opposite sides are equal in length, and opposite angles have equal measures.	$\overline{LM} \parallel \overline{PO}$ $\overline{LP} \parallel \overline{MO}$ $\overline{LM}$ and $\overline{PO}$ have equal lengths, and $\overline{LP}$ and $\overline{MO}$ have equal lengths.
[10.1] **Rectangle**	A parallelogram with four right angles.	Rectangle $RSTU$

continued

CONCEPT SKILL

Concept/Skill	Description	Example
[10.1] **Square**	A rectangle with four sides equal in length.	D E F G $\overline{DE}, \overline{EG}, \overline{FG}$, and $\overline{DF}$ have equal lengths.
[10.1] **The sum of the measures of the angles of a quadrilateral**	In any quadrilateral, the sum of the measures of the angles is 360°.	B C A D $m\angle A + m\angle B + m\angle C + m\angle D = 360°$
[10.1] **Circle**	A closed plane figure made up of points that are all the same distance from a fixed point called the center.	O
[10.1] **Diameter**	A line segment that passes through the center of a circle and has both endpoints on the circle.	A O B Diameter $\overline{AB}$
[10.1] **Radius**	A line segment with one endpoint on the circle and the other at the center.	O B Radius $\overline{OB}$
[10.2] **Perimeter**	The distance around a polygon.	7 cm, 3 cm, 2 cm, 6 cm, 5 cm $P = 3 + 7 + 2 + 5 + 6 = 23$ $P = 23$ cm
[10.2] **Circumference**	The distance around a circle.	10 in. $C = 10\pi \approx 31.4$ $C \approx 31.4$ in.

Concept/Skill	Description	Example
[10.3] **Area**	The number of square units that a figure contains.	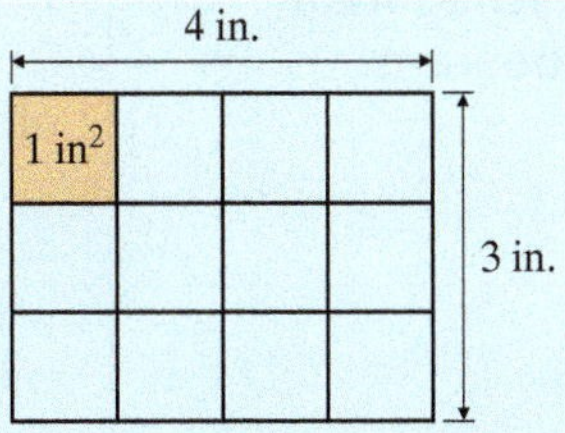 4 in. 1 in² 3 in. $A = 4 \cdot 3 = 12$ $A = 12 \text{ in}^2$
[10.4] **Volume**	The number of cubic units required to fill a three-dimensional figure.	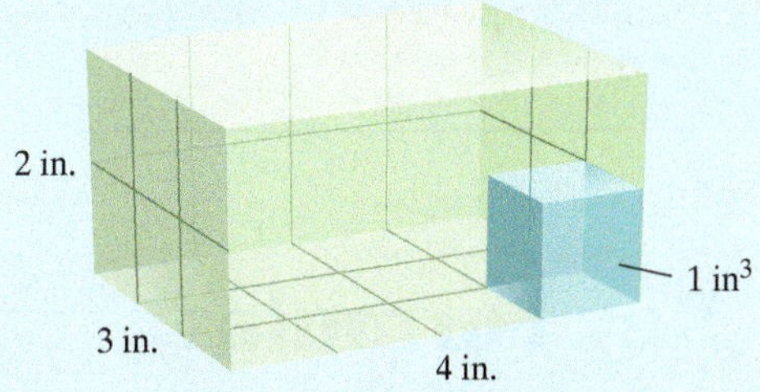 2 in. 3 in. 4 in. 1 in³ $V = 2 \cdot 3 \cdot 4 = 24$ $V = 24 \text{ in}^3$
[10.5] **Similar triangles**	Triangles that have the same shape but not necessarily the same size.	$\Delta ABC \sim \Delta DEF$ 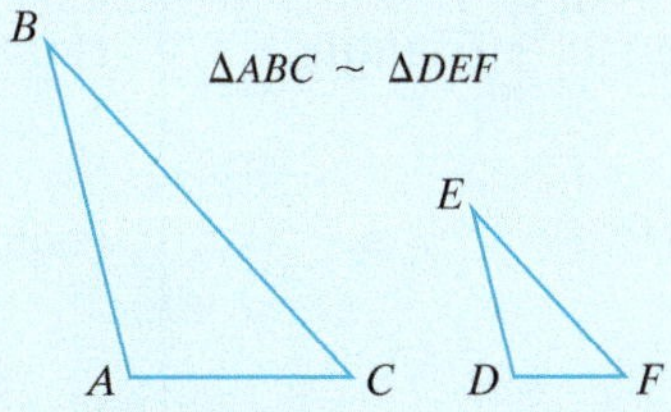
[10.5] **Corresponding sides**	In similar triangles, the sides opposite the equal angles.	In the similar triangles pictured, $\overline{AB}$ corresponds to $\overline{DE}$, $\overline{BC}$ corresponds to $\overline{EF}$, and $\overline{AC}$ corresponds to $\overline{DF}$.
[10.5] **To find a missing side of a similar triangle**	• Write the ratios of the lengths of the corresponding sides. • Write a proportion using a ratio with known terms and a ratio with an unknown term. • Solve the proportion for the unknown term.	$\Delta TRS \sim \Delta XYW$ Find a. 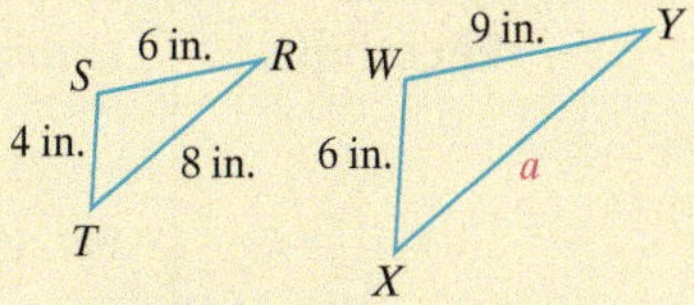 6 in. 4 in. 8 in. 9 in. 6 in. a $\frac{ST}{WX} = \frac{TR}{XY}$ $\frac{4}{6} = \frac{8}{a}$ $4a = 48$ $a = 12$, or 12 in.
[10.6] **Perfect square**	A number that is the square of a whole number.	49 and 144
[10.6] **(Principal) square root of *n***	The positive number, written $\sqrt{n}$, whose square is n.	$\sqrt{36}$ and $\sqrt{8}$

continued

Concept/Skill	Description	Example
[10.6] **Pythagorean theorem**	For every right triangle, the sum of the squares of the lengths of the two legs equals the square of the length of the hypotenuse, that is, $$a^2 + b^2 = c^2$$ where a and b are the lengths of the legs, and c is the length of the hypotenuse.	a; 25 yd; 24 yd Find a. $a^2 + b^2 = c^2$ $a^2 + (24)^2 = (25)^2$ $a^2 + 576 = 625$ $a^2 + 576 - 576 = 625 - 576$ $a^2 = 49$ $a = \sqrt{49}$ $= 7$, or 7 yd

Key Formulas

Figure	Formula	Example
[10.2]–[10.3] **Triangle**	*Perimeter* $$P = a + b + c$$ Perimeter equals the sum of the lengths of the three sides. *Area* $$A = \frac{1}{2}bh$$ Area equals one-half the base times the height.	6 m; 8 m; 4.8 m; 10 m $P = a + b + c$ $= 6 + 10 + 8$ $= 24$, or 24 m $A = \frac{1}{2}bh$ $= \frac{1}{2} \cdot 10 \cdot 4.8$ (10 and 2 cancelled to 5 and 1) $= 24$, or 24 m^2
[10.2]–[10.3] **Rectangle**	*Perimeter* $$P = 2l + 2w$$ Perimeter equals twice the length plus twice the width. *Area* $$A = lw$$ Area equals the length times the width.	3 in.; 7 in. $P = 2l + 2w$ $= 2(7) + 2(3)$ $= 14 + 6$ $= 20$, or 20 in. $A = lw$ $= 7 \cdot 3$ $= 21$, or 21 in^2

Figure	Formula	Example
[10.2]–[10.3] **Square**	*Perimeter* $P = 4s$ Perimeter equals four times the length of a side. *Area* $A = s^2$ Area equals the square of a side.	$\frac{1}{2}$ in. $P = 4s$ $= 4 \cdot \frac{1}{2}$ $= 2$, or 2 in. $A = s^2$ $= \left(\frac{1}{2}\right)^2$ $= \frac{1}{4}$, or $\frac{1}{4}$ in^2
[10.3] **Parallelogram**	*Area* $A = bh$ Area equals the base times the height.	3 ft 6 ft $A = bh$ $= 6 \cdot 3$ $= 18$, or 18 ft^2
[10.3] **Trapezoid**	*Area* $A = \frac{1}{2}h(b + B)$ Area equals one-half the height times the sum of the bases.	3 in. 4 in. 5 in. $A = \frac{1}{2}h(b + B)$ $= \frac{1}{2} \cdot 4\,(3 + 5)$ $= \frac{1}{2} \cdot 4 \cdot 8$ (4 and 2 cancelled to 2 and 1) $= 16$, or 16 in^2
[10.2]–[10.3] **Circle**	*Circumference* $C = \pi d$, or $C = 2\pi r$ Circumference equals π times the diameter, or 2 times π times the radius. *Area* $A = \pi r^2$ Area equals π times the square of the radius.	8 cm $C = \pi d$ $\approx 3.14(8)$ ≈ 25.12, or 25.12 cm $A = \pi r^2$ $\approx 3.14\,(4)^2$ $\approx 3.14\,(16)$ ≈ 50.24, or 50.24 cm^2 Note: $d = 8$ cm, so $r = 4$ cm.

continued

Figure	Formula	Example
[10.4] **Rectangular solid**	*Volume* $V = lwh$ Volume equals length times width times height.	5 cm, 7 cm, 15 cm $V = lwh$ $= 15 \cdot 7 \cdot 5$ $= 525$, or $525\ \text{cm}^3$
[10.4] **Cube**	*Volume* $V = e^3$ Volume equals the cube of the edge.	2 in. $V = e^3$ $= (2)^3$ $= 2 \cdot 2 \cdot 2$ $= 8$, or $8\ \text{in}^3$
[10.4] **Cylinder**	*Volume* $V = \pi r^2 h$ Volume equals π times the square of the radius times the height.	12 m, 4 m $V = \pi r^2 h$ $\approx 3.14\,(4)^2(12)$ $\approx 3.14\,(16)(12)$ ≈ 603, or $603\ \text{m}^3$
[10.4] **Sphere**	*Volume* $V = \frac{4}{3}\pi r^3$ Volume equals $\frac{4}{3}$ times π times the cube of the radius.	2 ft $V = \frac{4}{3}\pi r^3$ $\approx \frac{4}{3}(3.14)(2)^3$ $\approx \frac{4}{3}(3.14)(8)$ $\approx \frac{100.48}{3}$ ≈ 33, or $33\ \text{ft}^3$

CHAPTER 10 Review Exercises

Say Why *Fill in each blank.*

1. Consider the diagram below.

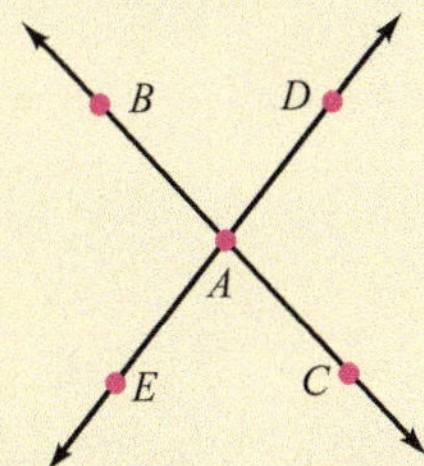

$m\angle BAE$ ________ equal to $m\angle DAC$ because
is/is not

____

____.

2. Consider the diagram below.

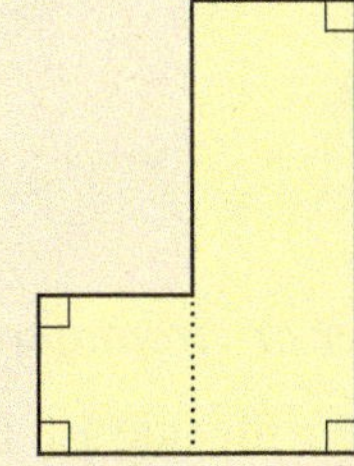

The figure ________ composite because
is/is not

____

____.

3. The area of a figure ________ be expressed in feet
can/cannot

because ____

____.

4. A face of a cube ________ have 6 edges because
does/does not

____

____.

5. Consider the diagram below.

The two triangles shown ________ similar because
are/are not

____

____.

6. Consider the diagram below.

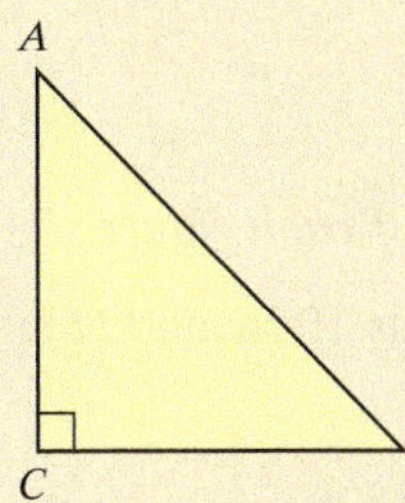

$\overline{AB}$ ________ the hypotenuse of ΔABC because
is/is not

____.

[10.1] *Sketch and label an example of each of the following.*

7. $\overline{AB}$

8. $\angle PQR$

9. Parallel lines $\overleftrightarrow{ST}$ and $\overleftrightarrow{UV}$

10. Obtuse ΔABC

Find each missing angle.

11.

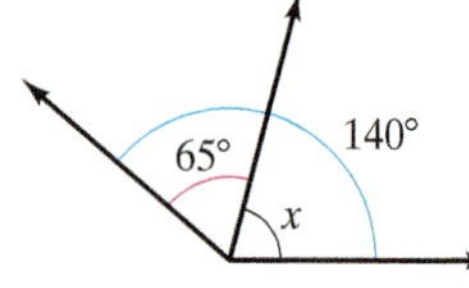

12.

13.

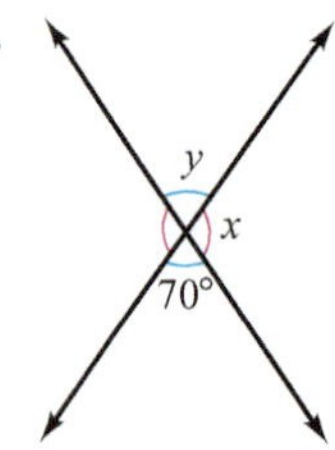

14.

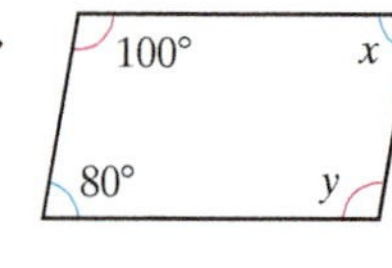

[10.2] *Find each perimeter or circumference. Use* $\pi \approx$ **3.14,** *when needed.*

15. An equilateral triangle with side 1.8 m

16. A polygon whose side lengths are 4.5 ft, 9 ft, 7.5 ft, 3 ft, and 6 ft

17.

18.

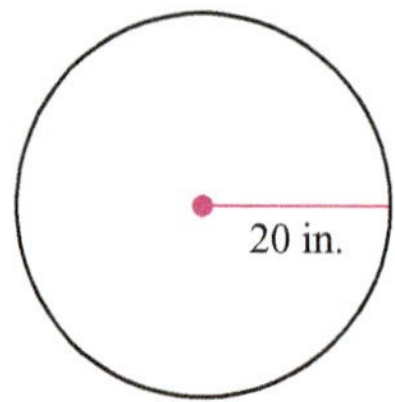

[10.3] *Find the area of each figure. Use* $\pi \approx$ **3.14,** *when needed.*

19. A square with side 15 yd

20. A circle with radius 14 ft

21.

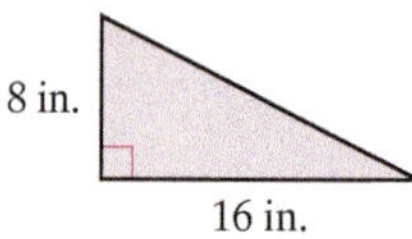

22.

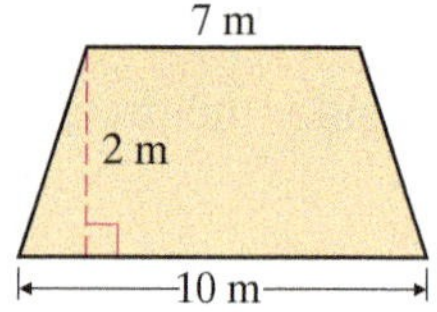

[10.4] *Find the volume of each figure. Use* $\pi \approx$ **3.14,** *when needed.*

23. A cylinder with radius 10 in. and height 4.2 in.

24. A rectangular solid with length 16 ft, width $4\frac{1}{2}$ ft, and height 3 ft

25. A cube with edge 1.25 m

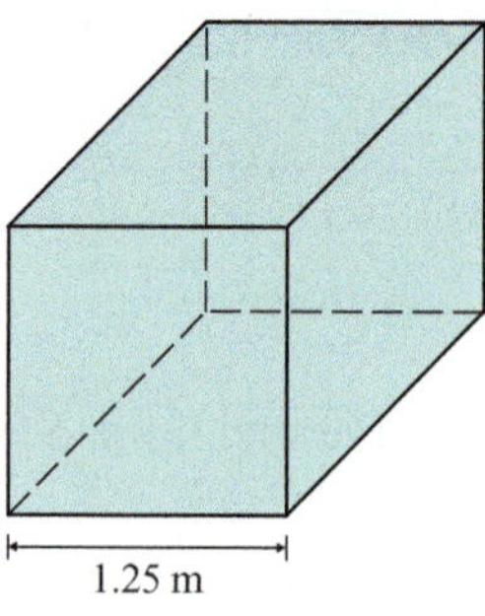

26. A sphere with diameter 2.5 cm

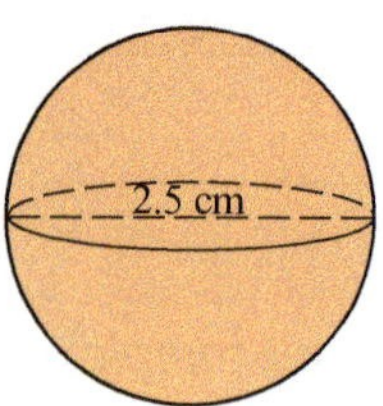

[10.2]–[10.4] *Solve.*

27. Find the perimeter of the figure shown, which is made up of a semicircle and a trapezoid. Use $\pi \approx 3.14$.

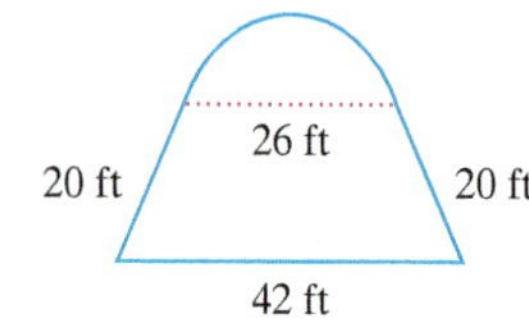

28. Find the area of the shaded portion of the figure. Use $\pi \approx 3.14$.

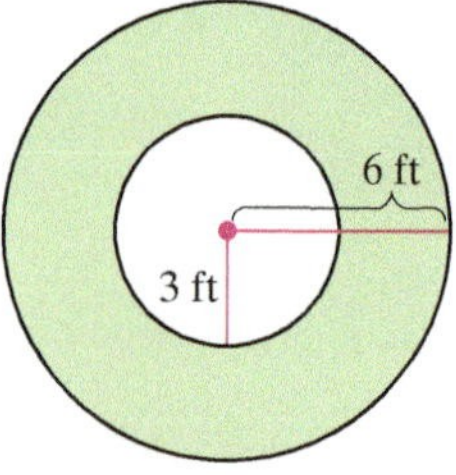

29. What is the area of the figure that consists of a square and two semicircles?

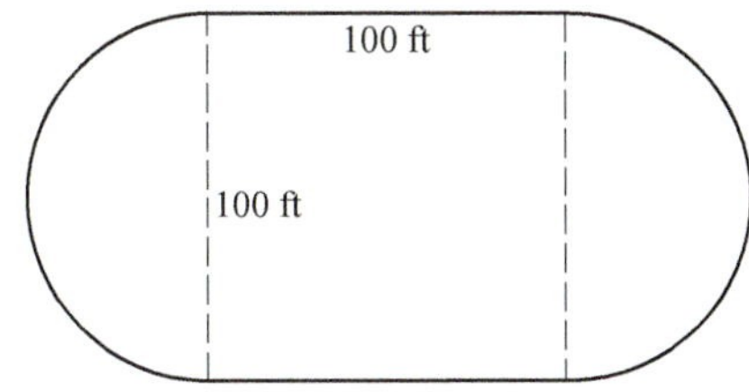

30. Find the volume of the shaded region between the sphere and the cube.

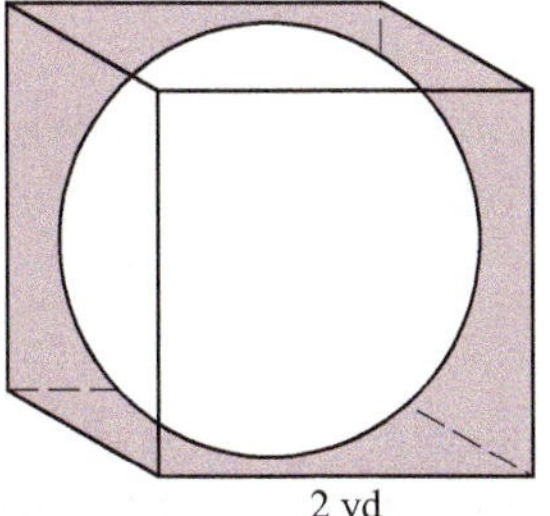

[10.5] ***Find the value of each unknown.***

31.

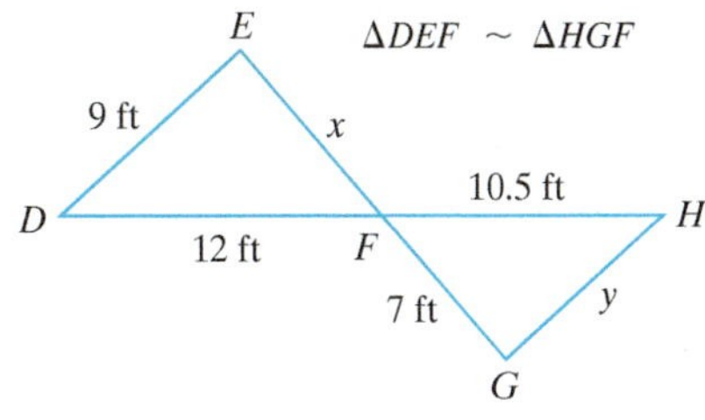

32.

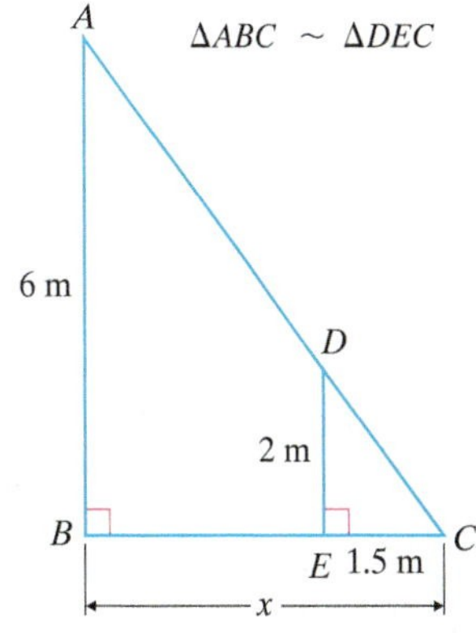

[10.6] ***Find the square root.***

33. $\sqrt{9}$ **34.** $\sqrt{64}$ **35.** $\sqrt{121}$ **36.** $\sqrt{900}$

Determine between which two consecutive whole numbers each square root lies.

37. $\sqrt{3}$ **38.** $\sqrt{84}$ **39.** $\sqrt{40}$ **40.** $\sqrt{10}$

Find the square root. Round to the nearest hundredth.

41. $\sqrt{8}$ **42.** $\sqrt{1,235}$ **43.** $\sqrt{195}$ **44.** $\sqrt{29}$

For a given triangle, the lengths of the legs are a and b, and the length of the hypotenuse is c. Find the length of the missing side. Round to the nearest tenth, if needed.

	a	*b*	*c*
45.	9 ft		15 ft
46.		24 in.	26 in.
47.	8 yd	5 yd	
48.	2 ft	2 ft	

Mixed Applications

Solve.

49. A roll of aluminum foil is 12 in. wide and 2,400 in. long. Find the area of the roll of aluminum foil.

50. Six weeks after an underwater oil well exploded, oil pouring into the Gulf of Mexico spread throughout a circular region with radius 200 mi. How big was the affected area? (*Source:* myfoxdc.com)

51. In a couple's apartment, an air conditioner can cool a room up to 3,000 ft^3 in volume. Based on the floor plan of their living room and a ceiling height of 10 ft, can the air conditioner cool the room?

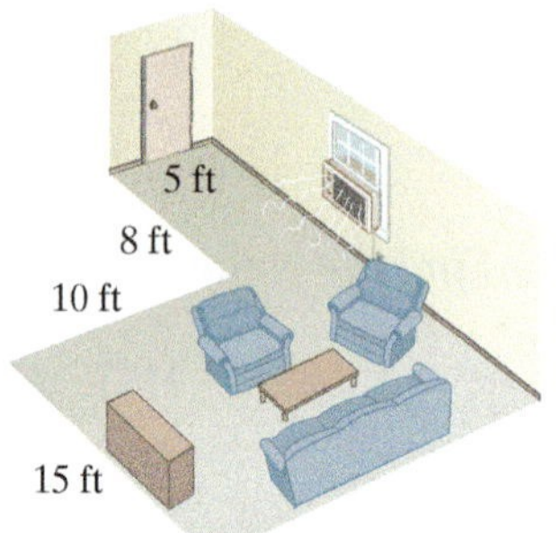

52. Of the two high-definition TV screens shown below, how much greater is the area of the larger screen?

53. A pilot flies 12 mi west from city A to city B. Then, he flies 5 mi south from city B to city C. What is the straight-line distance from city A to city C?

54. How high up on a wall will a 12-foot ladder reach if the bottom of the ladder is placed 6 ft from the wall? Round to the nearest foot.

55. On the pool table shown, a player hits the ball at point E. It ricochets off point C and winds up in the pocket at point A. If $\Delta ABC \sim \Delta EDC$, find CD.

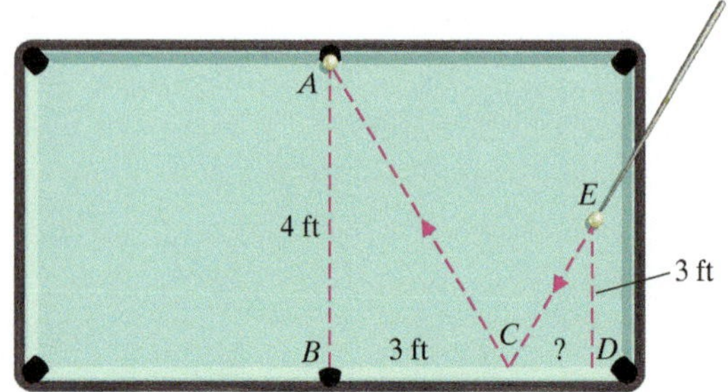

56. On the campus map below, the two triangles are similar. Find the distance between the Athletic Center and the Student Center.

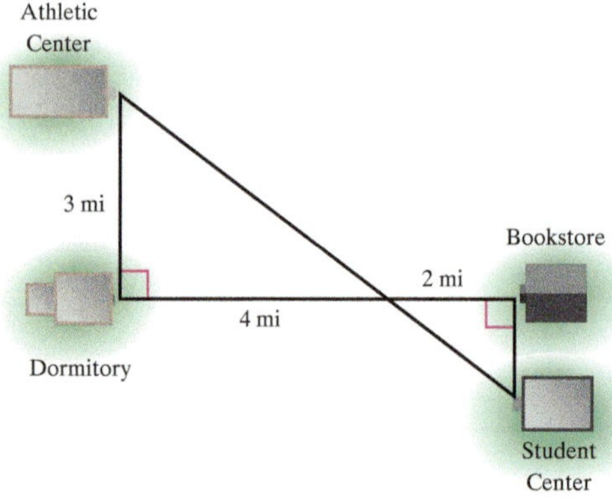

57. From the following drawing, find the total length of the building's walls.

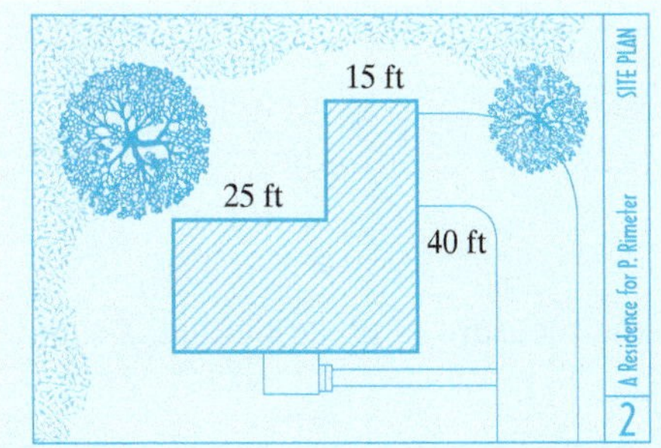

58. According to interior designers, the distances between the refrigerator, stove, and sink usually form a *work triangle*. To be efficient, the perimeter of a work triangle must be no more than 22 ft. Determine whether the model kitchen shown is efficient.

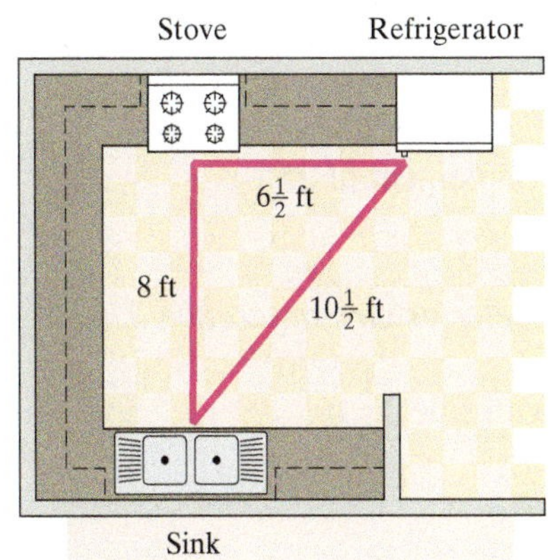

59. The coffee in this cylindrical can weighs 13 oz.

What is the weight of a cubic inch of coffee, to the nearest tenth of an ounce?

60. How much soil is needed to fill the flower box shown to 1 cm from the top? Round to the nearest cubic centimeter.

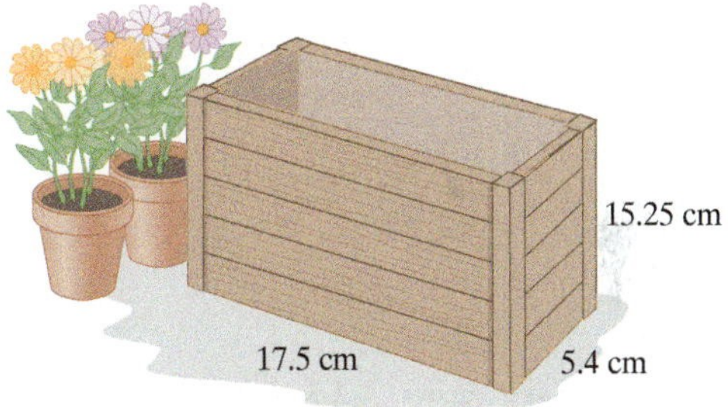

61. Find the area of the picture matting shown.

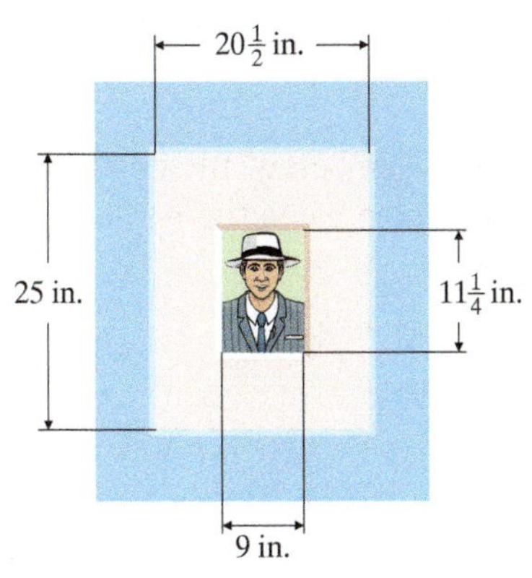

62. The game of racquetball is played with a small, hollow rubber ball, as shown.

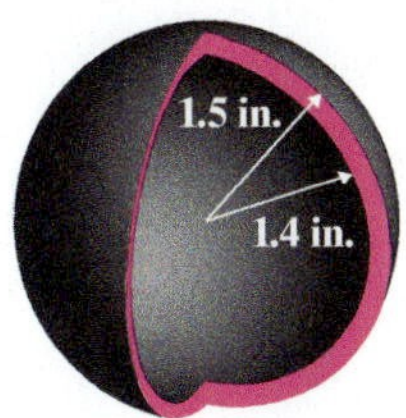

How much rubber, to the nearest tenth of a cubic inch, does a racquetball contain?

• Check your answers on page A-14.

CHAPTER 10 Posttest

FOR EXTRA HELP

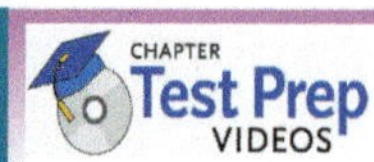

The Chapter Test Prep Videos with test solutions are available on DVD, in MyMathLab, and on YouTube (search "AkstDevMath" and click on "Channels").

To see if you have mastered the topics in this chapter, take this test.

1. Sketch and label an example of acute $\angle PQR$.

2. Find: $\sqrt{225}$

3. Find the complement of 25°.

4. What is the measure of an angle that is supplementary to 91°?

Find each perimeter or circumference. Use $\pi \approx 3.14$*, when needed.*

5. A square with side $3\frac{1}{2}$ ft

6. An equilateral triangle with side 1.5 m

7.

8 cm

8.

$5\frac{1}{2}$ ft

2 ft

Find each area. Use $\pi \approx 3.14$*, when needed.*

9.

6 ft

9 ft

10.

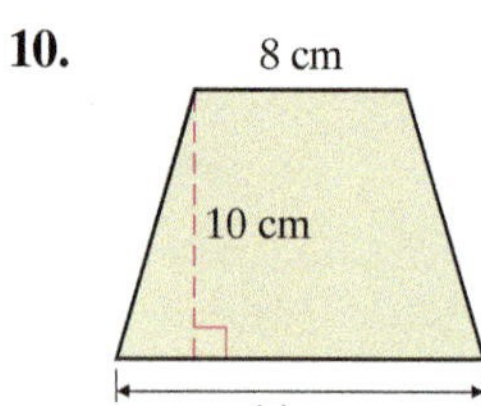

Find each volume. Use $\pi \approx 3.14$*, when needed.*

11.

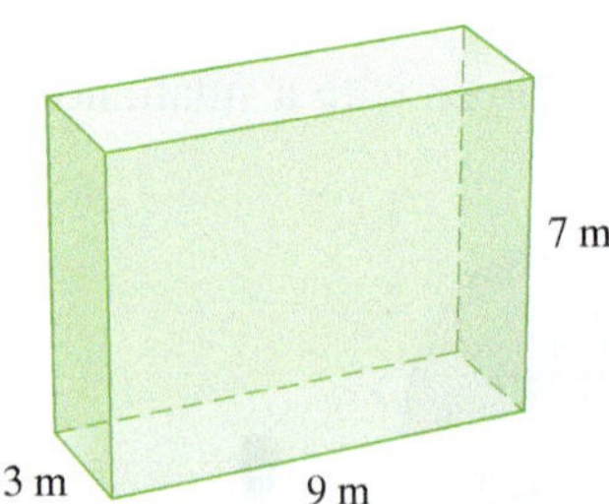

12.

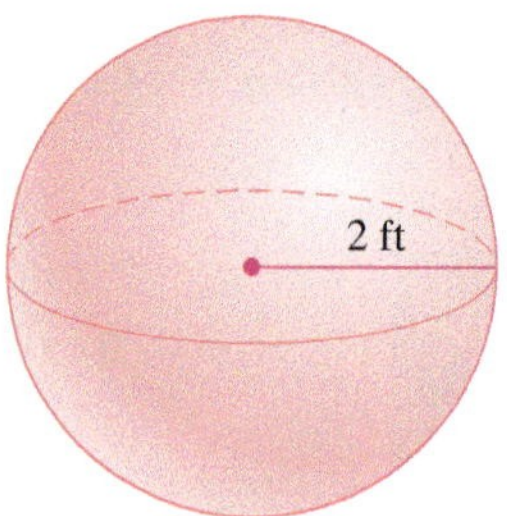

Find the value of each unknown.

13.

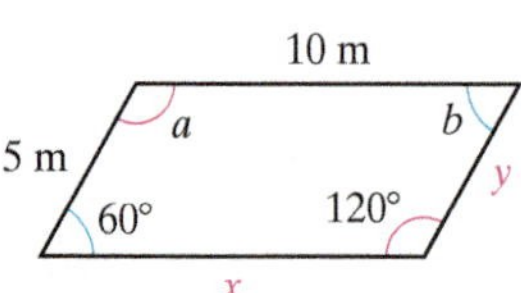

14.

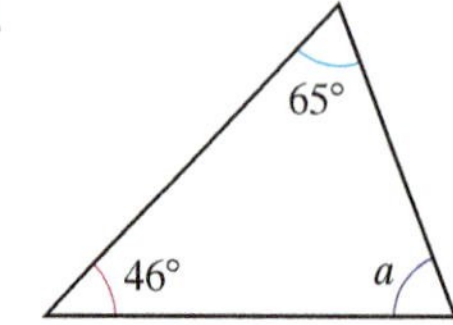

15.

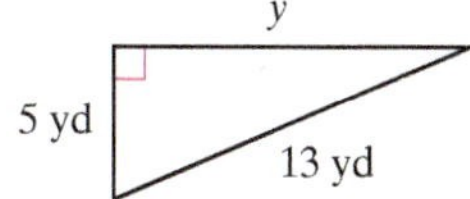

16. In the following diagram, $\Delta ABC \sim \Delta DEF$. Find x and y.

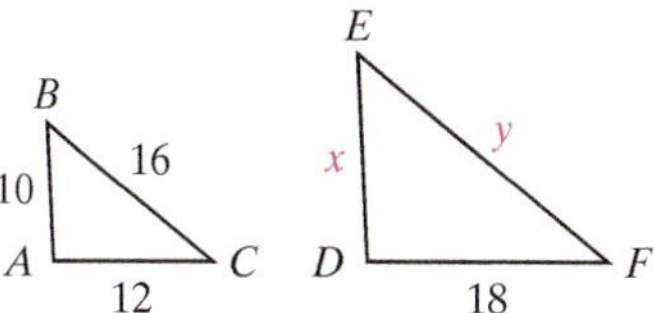

17. Dupont Circle, a major traffic circle and park located in northwest Washington, DC, is formed by the intersection of Massachusetts Avenue NW, Connecticut Avenue NW, New Hampshire Avenue NW, P Street NW, and 19th Street NW.

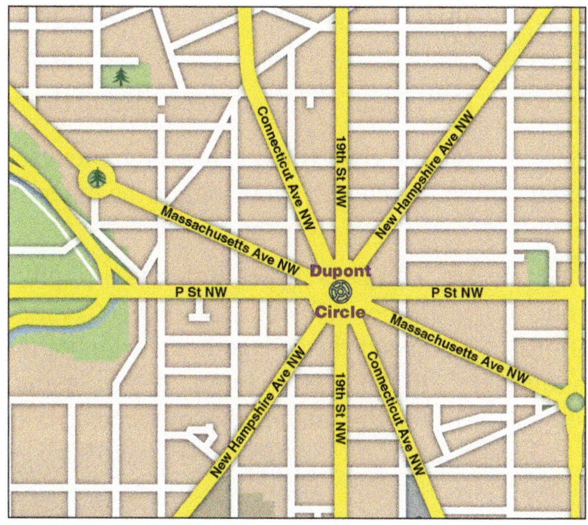

Is the angle formed by Massachusetts Avenue NW and New Hampshire Avenue NW below P Street NW acute? (*Source:* maps.google.com)

Solve.

18. A board foot is a special measure of volume used in the lumber industry. If a board foot contains 144 in^3 of wood, how many board feet rounded to the nearest tenth are there in the board shown?

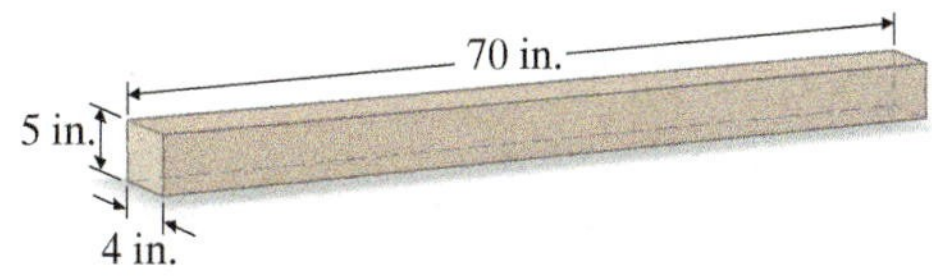

19. Suppose that the leash on a dog is 5 m long. To the nearest square meter, what is the area of the dog's "run"? Use $\pi \approx 3.14$.

20. An airplane flying due north is 200 mi from the airport. At the same time, another airplane flying due east is 150 mi from the airport. How far apart are the two airplanes at that time?

• Check your answers on page A-14.

Cumulative Review Exercises

To help you review, solve the following.

1. Divide and round to the nearest whole number: 23,802 ÷ 396

2. Simplify: $\frac{24}{30}$

3. Round to the nearest tenth: 3.061

4. Find the product: 38.759×0.4

5. Solve for y: $\frac{y}{12} = \frac{2}{3}$

6. 40% of what number is 20?

7. Calculate the final balance after compounding the interest annually if the principal is \$500, the interest rate is 8%, and the time is 2 yr.

8. Evaluate: $(-4)(-2) + 7$

9. Find the median of the following salaries: \$63,500, \$209,800, \$59,300, and \$57,100.

10. Solve the equation $32 - \frac{b}{8} = 29$.

11. Rewrite 64 fl oz as quarts.

12. 15 km ≈ ___ mi, rounded to the nearest tenth.

13. A common liquid insecticide for houseplants requires mixing 50 parts of water with every part of insecticide. What fraction of the mixture is insecticide?

14. If a right triangle has a hypotenuse of length 13 and a leg of length 5, how long is the other leg?

15. The Dermon Building is a 1925 office building located in downtown Memphis, Tennessee. Because of its architectural and historic significance, it is listed on the National Register of Historic Places. The building is rectangular in shape, 149 ft wide and 75 ft deep. Find the building's perimeter. (*Source:* dermonbuilding.com)

16. The Dow Jones index is a popular measure of how 30 large U.S. companies are trading. The following table shows the change in the value of this index from one year to the next, in terms of the closing value of the index on the first trading day of the year.

Years	Change in the Dow Jones Index
2007/2008	+569
2008/2009	−4,009
2009/2010	+1,529

What is the average of these changes? (*Sources:* wikipedia.org and pbs.org)

17. A typical hurricane has a calm, circular center, called the *eye*, which is approximately 30 mi across. What is the area that the eye covers?

18. One spouse had a yearly income \$12,000 more than the other. If the spouses' combined yearly income was \$60,000, determine their individual *monthly* incomes.

19. The graph shows the number of identity-theft complaints reported in the United States for the years from 2001 to 2008. About how many more complaints were there in 2005 than in 2001? (*Source:* Federal Trade Commission)

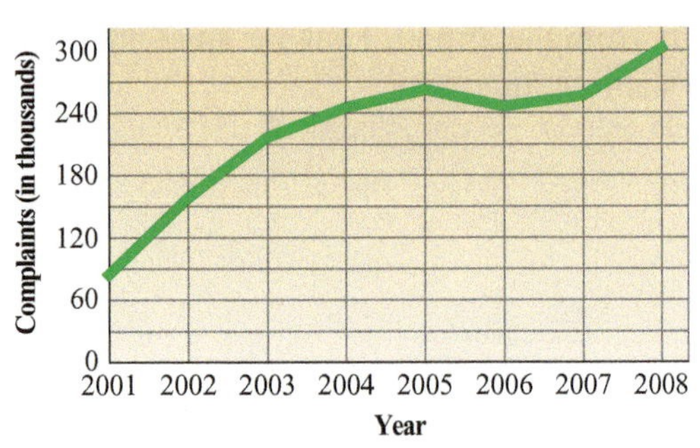

20. Find the width of the river pictured. The triangles are similar.

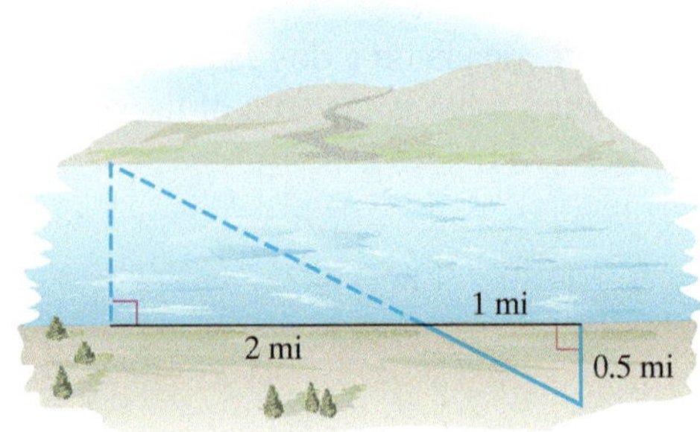

• Check your answers on page A-14.

Normal bone structure

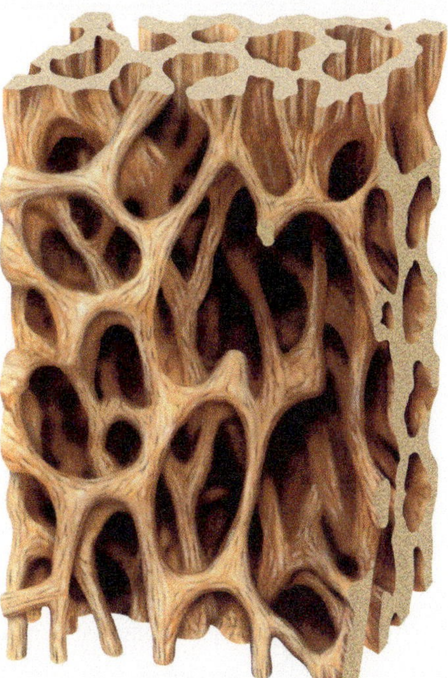
Osteoporotic bone

CHAPTER 11

Introduction to Real Numbers and Algebraic Expressions

Osteoporosis and Real Numbers

Approximately 10 million Americans have osteoporosis, a disease in which bones become weak and are likely to break. To diagnose osteoporosis, doctors commonly employ a bone mineral density (BMD) test. A person's BMD score is compared to a norm based on the optimal density of a healthy 30-year-old adult. Scores below the norm are indicated in negative numbers. For instance, a score of -2 indicates a low bone mass and a score of -2.5 or less is diagnosed as osteoporosis. Generally, a score of -1 is equivalent to a 10% loss of bone density.

The National Osteoporosis Foundation recommends treatment if you have a result that is:

- less than -1.5 with risk factors, or
- less than -2 with no risk factors.

The results of many medical tests are reported in terms of positive and negative numbers.

(*Source:* myhealth.gov)

CHAPTER 11 PRETEST

To see if you have already mastered the topics in this chapter, take this test.

1. Express as a positive or negative integer: a profit of \$2000

2. Is 0 a rational number?

3. Graph the number $-\frac{1}{2}$ on the following number line.

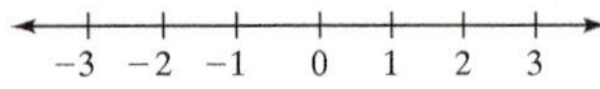

4. Use $>$ or $<$ to fill in the box to make a true statement: $-31 \square -1$

5. Add: $9 + (-4) + 2 + (-9)$

6. What is the opposite of -5?

7. Subtract: $-6 - 7$

8. Simplify: $3(-7) - 5$

9. What is the reciprocal of $\frac{1}{4}$?

10. Divide: $(-72) \div (-8)$

11. Express in exponential form: $-6 \cdot 6 \cdot 6 \cdot 6$

12. Write using exponents: $-2a \cdot b \cdot b \cdot b$

13. Evaluate: $2x - 4y + 8$, if $x = -2$ and $y = 3$

14. Apply the distributive property: $-3(a + b)$

15. Combine like terms: $3n - 7 + n$

16. Simplify: $-5(2 - x) + 9x$

17. Mount Kilimanjaro, Africa's highest point, is 5895 meters (m) above sea level. Lake Assal, its lowest point, is 156 m below sea level. What is the difference in the elevations of these two points? (*Source: National Geographic Family Reference Atlas of the World*)

18. In the 10 games played this season, a team won 3 games by 4 points, won 2 games by 1 point, lost 4 games by 3 points, and tied in the final game. In these games, did the team score more or fewer points than its opposing teams?

19. The formula for the perimeter of a rectangle is $P = 2l + 2w$. Calculate the perimeter of the following rectangle.

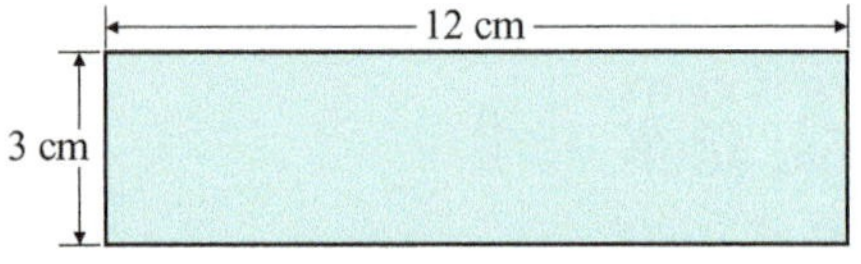

20. In golf, scores are given in terms of par; scores above par are positive, and scores below par are negative. The table shows Dustin Johnson's scores for each round of a recent PGA Championship tournament.

Round	First	Second	Third	Fourth
Score	−1	−4	−5	+1

(*Source:* espn.com)

In which round did he have the lowest score?

• Check your answers on page A-14.

11.1 Real Numbers

OBJECTIVES

- **A** To identify different kinds of real numbers—integers, rational numbers, and irrational numbers
- **B** To graph real numbers on the number line
- **C** To compare real numbers
- **D** To solve applied problems involving the comparison of real numbers

What Real Numbers Are and Why They Are Important

Real numbers are numbers that can be represented as points on the number line. All the numbers discussed in Chapter 7 were real numbers. They extend the numbers used in arithmetic and allow us to solve problems that we could not otherwise solve.

Let's begin our discussion by looking at different kinds of real numbers.

Integers

As we have seen in Chapter 7, the integers are the numbers . . . , $-4, -3, -2, -1, 0, 1, 2, 3, 4$ . . . , continuing indefinitely in both directions.

EXAMPLE 1

The Dow Jones Industrial Average on the stock market declined 4 points today. Express this situation as an integer.

Solution The number in question represents a decline (or loss), so we write a negative integer, namely -4.

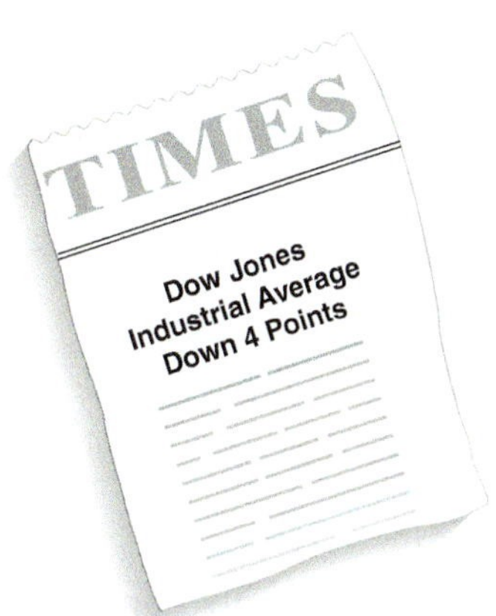

PRACTICE 1

Represent as an integer: A passion flower plant freezes and dies at a temperature of 18° below 0° Fahrenheit (°F).

Rational Numbers

Suppose that we want to represent the following situation: *The New York Giants lost one-half yard on a play.* To express a loss of one-half yard, we need a kind of number other than whole numbers or integers. We need a *rational number*—a number that can be written as the quotient of two integers, where the denominator is not equal to zero.

DEFINITION

Rational numbers are numbers that can be written in the form $\frac{a}{b}$, where a and b are integers and $b \neq 0$.

Here are some examples of rational numbers:

- $\frac{2}{3}$, since it is the ratio of integers.
- 5, since it can be written in the form $\frac{5}{1}$. In general, any integer is also a rational number.
- $7\frac{1}{4}$, since it can be written $\frac{29}{4}$.
- 0.03, since it can be written in the form $\frac{3}{100}$.
- $-\frac{1}{2}$, since it can be written $\frac{-1}{2}$.

A rational number has a decimal representation that either *terminates* or *repeats*. Here are some examples:

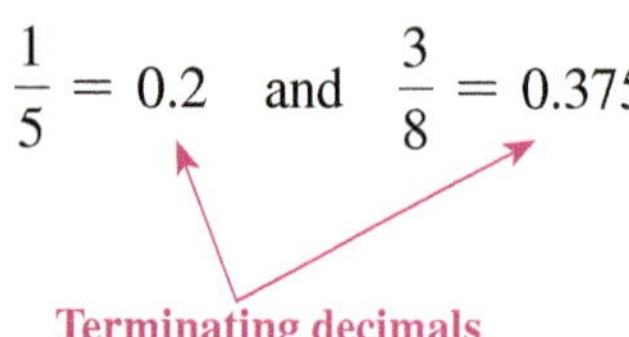
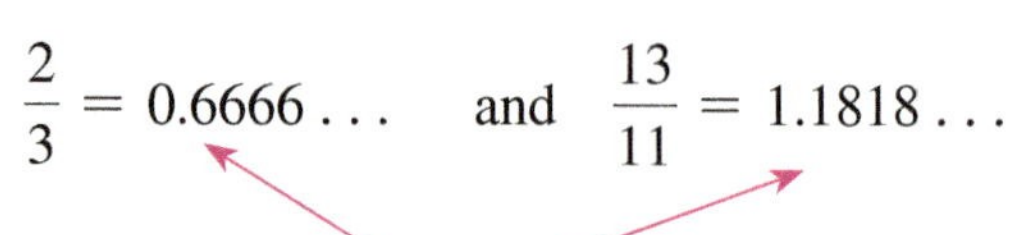

$\frac{1}{5} = 0.2$ and $\frac{3}{8} = 0.375$ $\qquad$ $\frac{2}{3} = 0.6666\ldots$ and $\frac{13}{11} = 1.1818\ldots$

Terminating decimals $\qquad$ Repeating decimals

Just as with integers, we can picture rational numbers as points on a number line. To *graph* a rational number, we locate the point on the number line and mark it as shown.

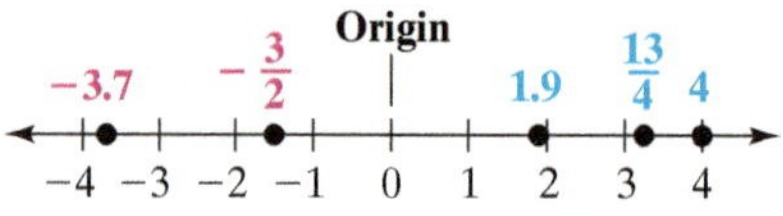

On the above number line, note that

- the point at 0 is called the *origin*,
- numbers to the right of 0 are positive, and numbers to the left of 0 are negative,
- the number 0 is neither positive nor negative.

Until Chapter 18 of this text, most of the numbers with which we work will be rational numbers.

EXAMPLE 2

Graph each number on the same number line.

a. $-\frac{7}{2}$ $\qquad$ **b.** 2.1

Solution

a. Because $-\frac{7}{2}$ can be expressed as $-3\frac{1}{2}$, the point $-\frac{7}{2}$ is graphed halfway between -3 and -4.

b. The number 2.1, or $2\frac{1}{10}$, is between 2 and 3 but is closer to 2.

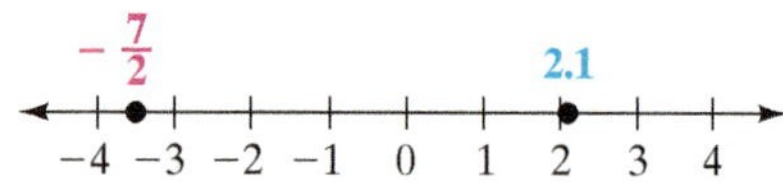

PRACTICE 2

Graph each number on the number line.

a. -1.7 $\qquad$ **b.** $\frac{5}{4}$

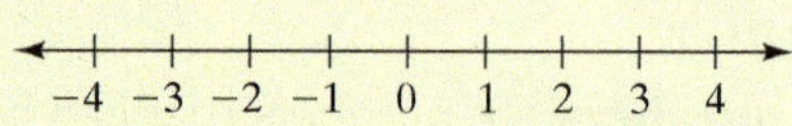

Irrational Numbers

Recall that any rational number can be written as the quotient of two integers. However, there are other real numbers that cannot be written in this form. Such numbers are called *irrational numbers*, and, as we have discussed in Section 10.2, their corresponding decimal representations continue indefinitely and have no repeating pattern. Examples of irrational numbers are

$\sqrt{2} = 1.4142\ldots$ The square root of 2

$-\sqrt{3} = -1.7320\ldots$ The negative square root of 3

$\pi = 3.1415\ldots$ Pi, the ratio of the circumference of a circle to its diameter

In many computations with irrational numbers, we use decimal approximations that are rounded to a certain number of decimal places, for instance:

$$\sqrt{2} \approx 1.41 \qquad -\sqrt{3} \approx -1.73 \qquad \pi \approx 3.14$$

Recall that the symbol $\approx$ means "is approximately equal to."

Irrational numbers, like rational numbers, can be graphed on the number line. The rational numbers and the irrational numbers together make up the *real numbers.*

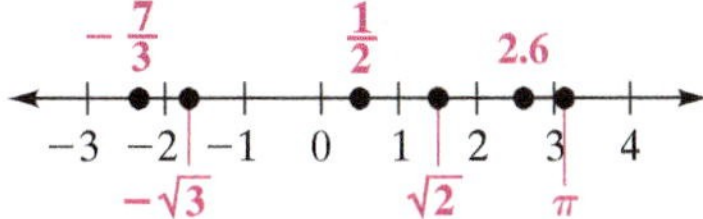

Comparing Real Numbers

Recall from Section 7.1 that the number line helps us to compare two real numbers, that is, to determine which number is larger.

Given two numbers on the number line, *the number to the right is larger than the number to the left*. Similarly, *the number to the left is smaller than the number to the right.*

−4 −1

−4 −3 −2 −1 0 1 2 3 4

The *equal sign* (=) and the *inequality symbols* ($\neq$, <, $\leq$, > and $\geq$) are used to compare numbers.

$=$	means *is equal to*	$\frac{5}{2} = 2\frac{1}{2}$ is read "$\frac{5}{2}$ is equal to $2\frac{1}{2}$."
$\neq$	means *is not equal to*	$3 \neq -3$ is read "3 is not equal to -3."
$<$	means *is less than*	$-1 < 0$ is read "-1 is less than 0."
$\leq$	means *is less than or equal to*	$4 \leq 7$ is read "4 is less than or equal to 7."
$>$	means *is greater than*	$-2 > -5$ is read "-2 is greater than -5."
$\geq$	means *is greater than or equal to*	$3 \geq 1$ is read "3 is greater than or equal to 1."

The statements $3 \neq -3$, $-1 < 0$, $4 \leq 7$, $-2 > -5$, and $3 \geq 1$ are *inequalities.*

EXAMPLE 3

Indicate whether each inequality is true or false. Explain.

a. $0 > -1$ **b.** $0 \geq \frac{1}{2}$ **c.** $-3.5 < 1.5$

d. $-10 \leq -10$ **e.** $-4 > -2$

Solution

a. $0 > -1$ True, because 0 is to the right of -1.

b. $0 \geq \frac{1}{2}$ False, because 0 is to the left of $\frac{1}{2}$.

c. $-3.5 < 1.5$ True, because -3.5 is to the left of 1.5.

d. $-10 \leq -10$ True, because -10 is equal to -10.

e. $-4 > -2$ False, because -4 is to the left of -2.

PRACTICE 3

Determine whether each inequality is true or false. Explain.

a. $-2 < -1$

b. $0 \leq -5$

c. $\frac{10}{4} \geq \frac{5}{2}$

d. $0.3 > 0$

e. $-2.4 > 1.6$

We noted in Example 3(a) that $0 > -1$. Is $-1 < 0$? Explain why.

EXAMPLE 4

Graph the numbers $-2, \frac{1}{2}, -1$, and $-\frac{1}{4}$. Then, list them in order from least to greatest.

Solution

Reading the graph, we see that the numbers in order from the least to the greatest are $-2, -1, -\frac{1}{4}$, and $\frac{1}{2}$.

PRACTICE 4

Graph the numbers $-2.4, 3, -\frac{1}{2}$, and -1.6. Then, write them in order from largest to smallest.

EXAMPLE 5

The average temperature in Fairbanks, Alaska, in the month of December is -9°F. The average temperature in Barrow, Alaska, in the same month is -11°F. Which place is warmer in December? Explain. (*Source:* climatetemp.info)

Solution We need to compare -9 with -11. Because $-9 > -11$, it is warmer in Fairbanks than in Barrow.

PRACTICE 5

The table below shows the elevation of three lakes.

Lake	Elevation (in feet)
The Caspian Sea (Asia–Europe)	-92
Lake Maracaibo (South America)	0
Lake Eyre (Australia)	-52

(*Source: Geological Survey, U.S. Department of the Interior*)

Which lake has the lowest elevation? Explain.

EXAMPLE 6

Historians use number lines, called *timelines*, to show dates of historical events. On the timeline shown below, the B.C. dates are considered to be negative whereas the A.D. dates are positive.

a. Locate on the following timeline the world history events shown in the table.

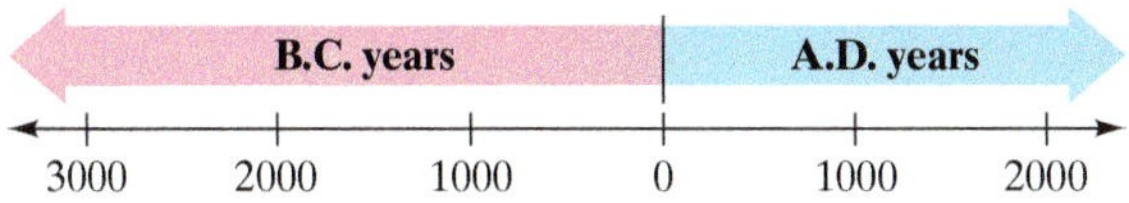

Event	Date
(A) Hieroglyphic writing developed in Egypt.	3200 B.C.
(B) Charlemagne (Charles the Great) was crowned emperor by Pope Leo III in Rome.	A.D. 800
(C) Hun invaders from Asia entered Europe.	A.D. 372
(D) Mayan civilization began to develop in Central America.	1500 B.C.
(E) Sweden seceded from the Scandinavian Union.	A.D. 1523
(F) In Greece, the Parthenon was built.	438 B.C.
(G) The city of Rome was founded, according to legend, by Romulus.	753 B.C.
(H) The evolution of England's unique political institutions began with the *Magna Carta.*	A.D. 1215

(Source: The World Almanac Book of Facts, 2010)

b. Order the events from the most recent to the earliest event.

Solution B.C. dates are considered to be negative numbers, and A.D. dates positive numbers. So we graph on a timeline similar to the way we graph on a number line. That is, we locate each year on the timeline and mark it as shown below:

a.

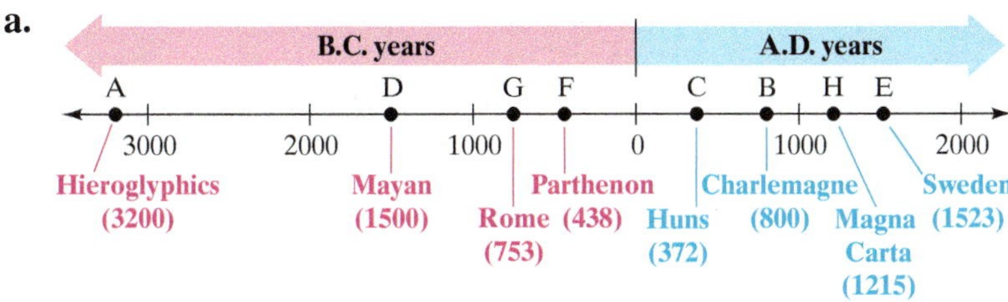

b. The events from the most recent to the earliest: E, H, B, C, F, G, D, and A.

PRACTICE 6

Consider the following table that shows the highlights in the development of algebra.

Event	Date
(A) Babylonian algebra found on cuneiform clay tablets dates back to the reign of King Hammurabi.	1700 B.C.
(B) Greek algebra (as formulated by the Pythagoreans) was geometric.	500 B.C.
(C) The Greek mathematician Diophantus introduced a style of writing equations.	A.D. 250
(D) Greek algebra (as formulated by Euclid) was geometric.	300 B.C.
(E) Bhaskara was one of the most prominent Hindu mathematicians in algebra.	A.D. 1100
(F) Mohammed ibn-Musa al-Khwarizmi wrote the book *Al-jabr* (translated as "algebra" in Latin).	A.D. 825
(G) Modern symbols and notation in algebra emerged.	A.D. 1500

(*Source:* NCTM, *Historical Topics for the Mathematics Classroom)*

a. Locate on the timeline below the events listed in the table.

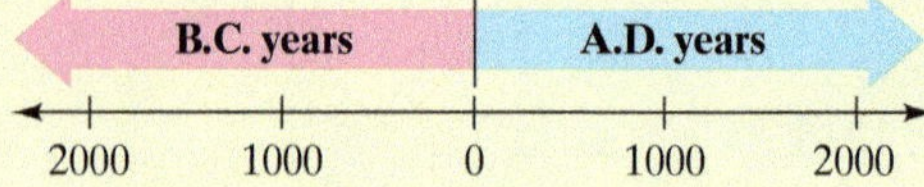

b. Then, order these highlights from the earliest to the most recent event.

11.1 Exercises

FOR EXTRA HELP MyMathLab MathXL PRACTICE WATCH READ REVIEW

Mathematically Speaking

Fill in each blank with the most appropriate term or phrase from the given list.

rational numbers	natural numbers	negative numbers
whole numbers	irrational numbers	real numbers
either terminate or	integers	neither terminate nor
center	origin	

1. The ________________ are 1, 2, 3, 4, . . .

2. The ________________ consist of 0 and the natural numbers.

3. Numbers to the left of 0 on the number line are ________________.

4. The ________________ are the numbers . . . , −4, −3, −2, −1, 0, +1, +2, +3, +4, . . . , continuing indefinitely in both directions.

5. The point at 0 on the number line is called the ________________.

6. Real numbers that cannot be written as the quotient of two integers are called ________________.

7. The corresponding decimal representations of irrational numbers ________________ repeat.

8. The rational and the irrational numbers together make up the ________________.

A ***Express each quantity as a positive or negative number.***

9. 5 km below sea level

10. A profit of \$1000

11. A temperature drop of 22.5°C

12. A gain of $6\frac{1}{2}$ lb

13. A withdrawal of \$160 from an account

14. A debt of \$1500.45

Classify each number by writing a check in the appropriate boxes.

	Whole Numbers	Integers	Rational Numbers	Real Numbers
15. −7				
16. −3				
17. $3\frac{1}{6}$				
18. $4\frac{1}{2}$				
19. 10				
20. 15				

B ***Graph each number on the number line.***

21. −3

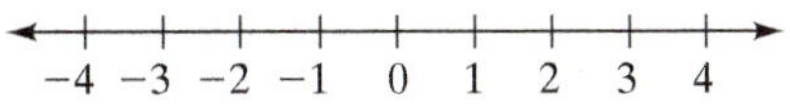

22. −1

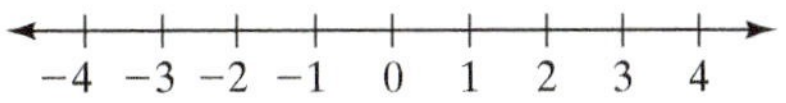

23. $-\frac{1}{2}$

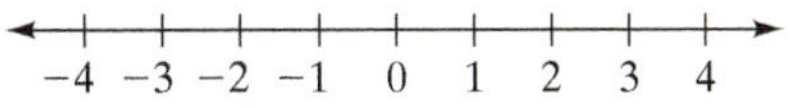

24. $-\frac{3}{8}$

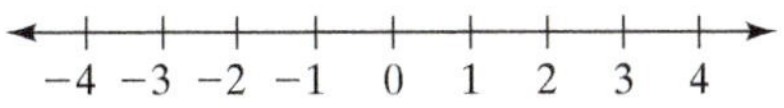

25. $3\frac{1}{4}$ (number line: −4 −3 −2 −1 0 1 2 3 4)

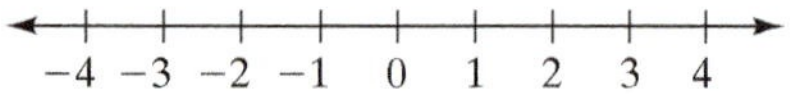

26. $2\frac{9}{10}$ (number line: −4 −3 −2 −1 0 1 2 3 4)

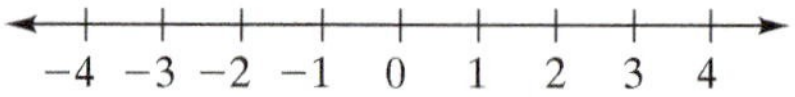

27. -2.9 (number line: −4 −3 −2 −1 0 1 2 3 4)

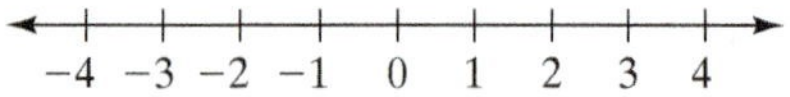

28. -3.2 (number line: −4 −3 −2 −1 0 1 2 3 4)

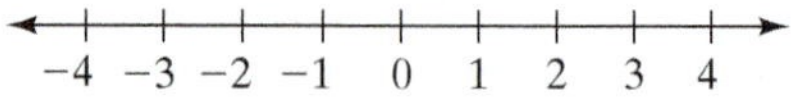

C *Indicate whether each inequality is true or false.*

29. $-7 < -5$

30. $-3 < -2$

31. $-1 > 2.5$

32. $-3 > 4.5$

33. $0 \geq -1\frac{1}{4}$

34. $-6 \leq -6$

Replace each ▭ with <, >, or = to make a true statement.

35. 0 ▭ -1

36. 4 ▭ -7

37 -1.5 ▭ -2

38. -1.6 ▭ -2

39. 2.5 ▭ $2\frac{1}{2}$

40. 3.25 ▭ $3\frac{1}{4}$

41. $|-4|$ ▭ $|4|$

42. $-|5|$ ▭ $|-5|$

43. 6.2 ▭ $|-7.1|$

44. 7.4 ▭ $|-8.6|$

Graph the numbers in each group on the number line. Then, write the numbers from largest to smallest.

45. $3\frac{1}{2}, -1.5, -\frac{1}{2}, 0;$ (number line: −4 −3 −2 −1 0 1 2 3 4)

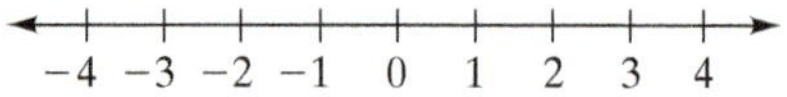

46. $2\frac{1}{2}, -4, 3, -2.5$ (number line: −4 −3 −2 −1 0 1 2 3 4)

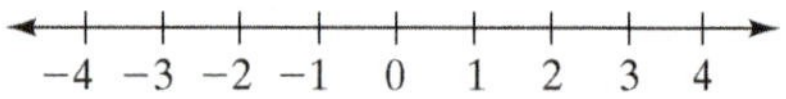

47. $-1, 2, -2, -3, 1$ (number line: −4 −3 −2 −1 0 1 2 3 4)

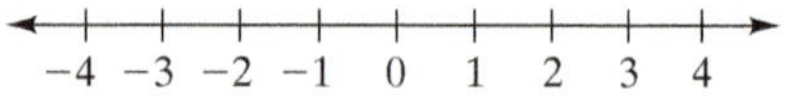

48. $-3, 3, -3.5, 3.5$ (number line: −4 −3 −2 −1 0 1 2 3 4)

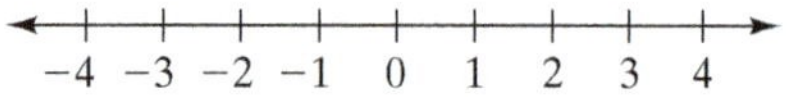

Mixed Practice

Solve.

49. Express the quantity as a positive or negative number: a loss of \$53.

50. Graph the number $-1\frac{5}{8}$ on the number line.

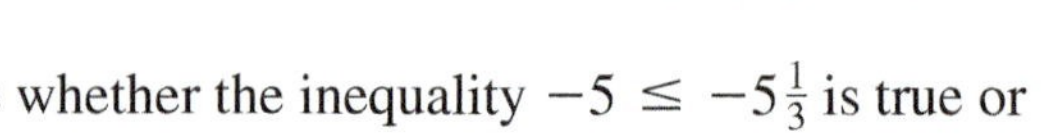

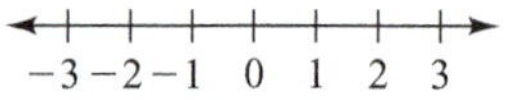

51. Classify the number by writing a check in the appropriate boxes.

	Whole Numbers	Integers	Rational Numbers	Real Numbers
2.6				

52. Indicate whether the inequality $-5 \leq -5\frac{1}{3}$ is true or false.

Replace the ▭ with <, >, or = to make a true statement.

53. -7.8 ▭ -8.2

54. Graph the numbers $\frac{1}{2}, -2\frac{1}{2}, 1\frac{1}{2}, -2$ on a number line. Then, write the numbers from largest to smallest.

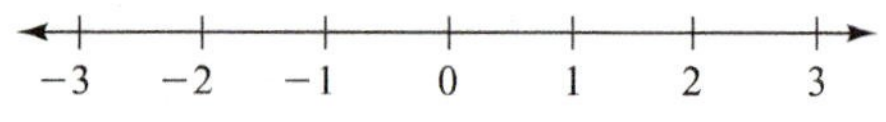

Applications

D *Solve.*

55. Today, a person in debt owes \$200. Last week, he owed \$2000. Was he better off financially last week, or is he better off today?

56. Three of the coldest temperature readings ever recorded on Earth were −90°F, −129°F, and −87°F. Of these three temperatures, which was the coldest? (*Source:* ncdc.noaa.gov)

57. Recall that astronomers use the term *apparent magnitude* to indicate the brightness of a star as seen from Earth. The following number line shows the apparent magnitude of various stars and other objects. For historical reasons, the brighter a star or object is, the farther to the left it is graphed on the number line.

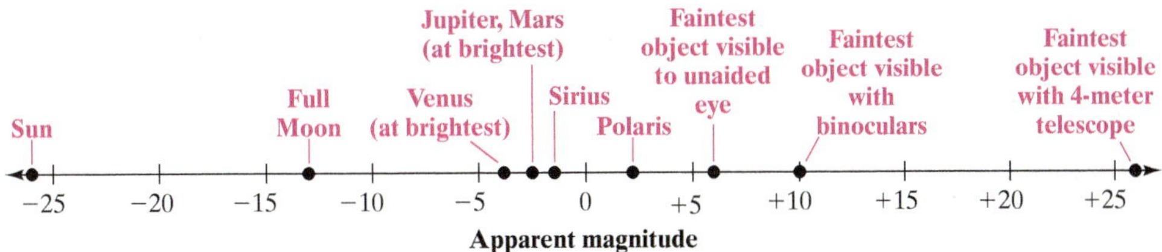

(*Sources:* absolute astronomy.com and topastronomy.com)

a. Which star is brighter as seen from Earth, Polaris or Sirius?

b. Use this number line to estimate the apparent magnitude of the full Moon.

c. The giant star Beta Sagittae lies hundreds of light-years from Earth, with an apparent magnitude of 4.4. Plot this star's apparent magnitude on the number line.

58. The following timeline shows the years of some major technological innovations.

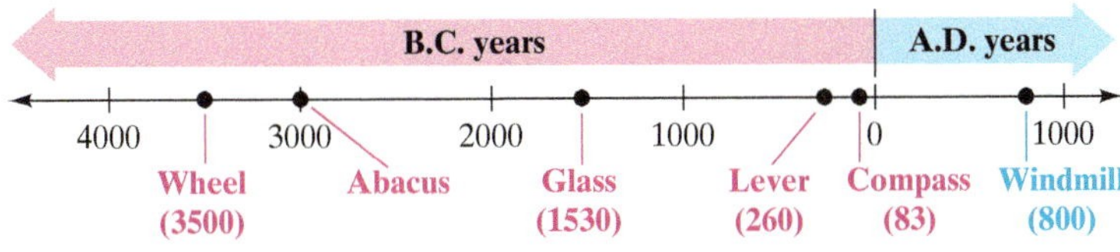

(*Source:* Bill Yenne, *100 Inventions That Shaped World History*)

a. Which was invented earlier, the lever or the compass?

b. According to the timeline, what was invented between 2000 B.C. and A.D. 1000?

c. The mechanical clock was invented around A.D. 950. Plot a point on the timeline to represent this invention.

59. Below are the birth years for a variety of famous people throughout history.

Leonardo da Vinci (Italian painter)	A.D. 1452	Aristotle (Greek philosopher/scientist)	384 B.C.
Keith Richards (English musician)	A.D. 1943	Wolfgang Amadeus Mozart (Austrian composer)	A.D. 1756
Charles Dickens (English writer)	A.D. 1812	Miles Davis (American jazz musician)	A.D. 1926
William Shakespeare (English playwright)	A.D. 1564	Marie Antoinette (French queen)	A.D. 1755
Tiger Woods (American golfer)	A.D. 1976	Oliver Cromwell (English statesman)	A.D. 1599
Attila (Hun king)	A.D. 406	Socrates (Greek philosopher)	469 B.C.

(*Source: Chambers Biographical Dictionary*)

a. Who was born later, Aristotle or Socrates?

b. Who was born later, Aristotle or Attila ?

c. Which of the individuals listed in the table was born the earliest?

d. Which of the individuals listed in the table was born most recently?

60. The following table gives the change from the previous month in the opening stock price per share for Delta Air Lines, Inc. (stock symbol: DAL) for each month of 2009.

Jan.	Feb.	Mar.	Apr.	May	Jun.
\$2.86	−\$4.22	−\$2.24	\$0.67	\$0.76	−\$0.19

Jul.	Aug.	Sept.	Oct.	Nov.	Dec.
−\$0.27	\$1.22	\$0.08	\$1.83	−\$1.81	\$1.17

(*Source:* finance.yahoo.com)

a. What was the greatest increase in the opening stock price?

b. What was the smallest increase in the opening stock price?

c. What was the greatest decrease in the opening stock price?

d. What was the smallest decrease in the opening stock price?

• Check your answers on page A-14.

MINDStretchers

Groupwork

1. Working with a partner, develop a diagram to show the relationship among the real numbers, the rational numbers, the irrational numbers, the integers, the noninteger rational numbers, the whole numbers, the negative integers, the natural numbers, and zero.

Mathematical Reasoning

2. Is there a largest number less than 5 that is
   a. an integer?
   b. a rational number?

Research

3. Using your college library or the Web, investigate the role the Pythagoreans played in discovering irrational numbers. Write a few sentences to summarize your findings.

Cultural Note

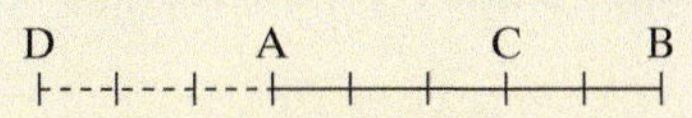

The seventeenth-century English mathematician and cryptographer John Wallis is generally credited with inventing the number line as shown above. Published in his *A Treatise of Algebra* in 1685, the number line gave meaning to negative numbers, the existence of which was controversial. Wallis was the leading English mathematician before Sir Isaac Newton. A professor of geometry at Oxford University, Wallis wrote on a variety of mathematical topics, and introduced the symbol ∞ for infinity. He was the first person to devise a system to teach deaf mutes, and was also one of the founders of the Royal Society, the oldest learned society for science in existence.

(*Sources:* Jan Gullberg, New York: W. W. Norton & Company, *Mathematics, From the Birth of Numbers,*1997; wikipedia.org; newworldencyclopedia.org)

11.2 Using Properties to Add and Subtract Real Numbers

OBJECTIVES

A To add or subtract real numbers

B To solve applied problems involving the addition or subtraction of real numbers

Recall our discussion in Sections 7.2 and 7.3 of adding and subtracting signed numbers. All the numbers in those sections were real numbers. Here, we consider some important properties that relate to addition.

Addition of Real Numbers

First, let's review addition. We can add two numbers on the number line, moving to the left if the second number is negative and to the right if it is positive.

EXAMPLE 1

Add -1 and 3 on the number line.

Solution

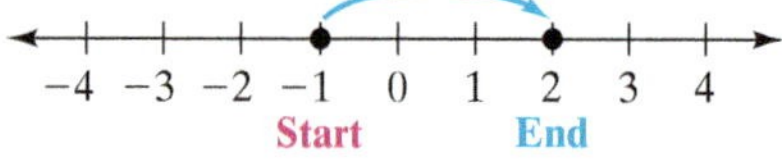

So $-1 + 3 = 2$.

PRACTICE 1

Add 2 and -3 on the number line.

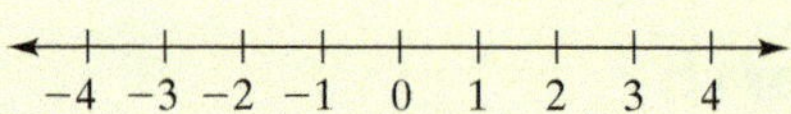

EXAMPLE 2

Add 2 and -2 on the number line.

Solution

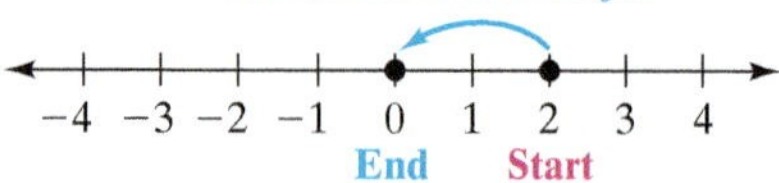

So $2 + (-2) = 0$. Note that 2 and (-2) are opposites, and their sum is 0.

PRACTICE 2

Add -5 and 5 on the number line.

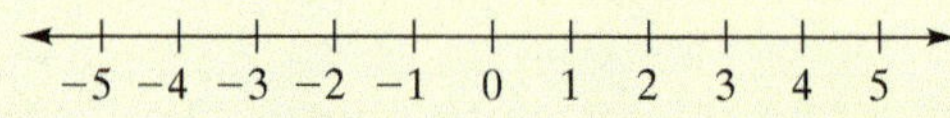

Example 2 suggests that when adding two numbers that are opposites, such as 2 and -2, the sum is 0. We call such numbers *additive inverses.* Every real number has an opposite, or additive inverse, as stated in the following property:

The Additive Inverse Property

For any real number a, there is exactly one real number $-a$ such that

$$a + (-a) = 0 \quad \text{and} \quad (-a) + a = 0.$$

In words, this property states that any number added to its opposite is zero.

EXAMPLE 3

Using a number line, add $-1\frac{1}{2}$ and 0.

Solution

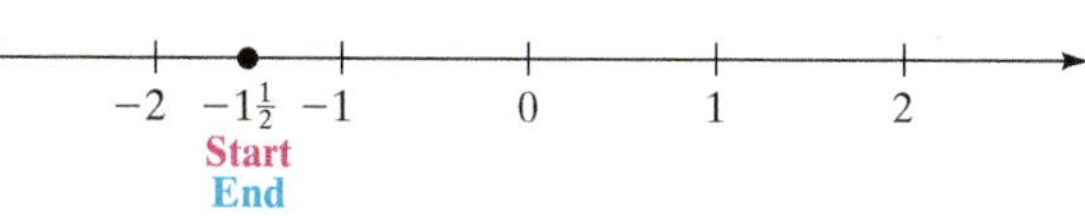

So $-1\frac{1}{2} + 0 = -1\frac{1}{2}$. Note that adding 0 to $-1\frac{1}{2}$ gives us $-1\frac{1}{2}$, the same number we started with.

PRACTICE 3

Add 0 and 0.5 on the following number line.

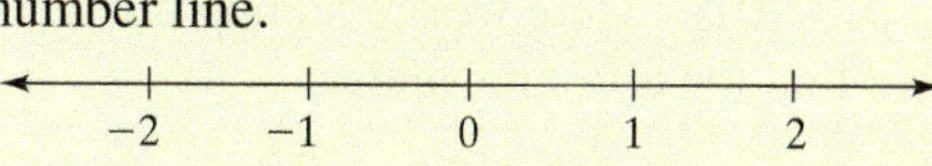

Example 3 illustrates another important property in adding real numbers—the *additive identity property*, also called the *identity property of addition*. This property, as well as other properties discussed in Chapter 1 with respect to whole numbers, applies to all real numbers.

The Additive Identity Property

For any real number a,

$$a + 0 = a \quad \text{and} \quad 0 + a = a.$$

In words, this property states that any number added to zero is the original number.

Instead of working on the number line, we have seen that real numbers can be added by using the following rule:

To Add Real Numbers

- If the numbers have the same sign, add their absolute values and keep their sign.
- If the numbers have different signs, find the difference between the larger absolute value and the smaller absolute value, and take the sign of the number with the larger absolute value.

EXAMPLE 4

Add −8 and −19.

Solution Because both numbers are negative, we find their absolute values and then add:

$$|-8| + |-19| = 8 + 19 = 27$$

The sum of two negative numbers is negative.

$$-8 + (-19) = -27$$

Negative numbers (−8 and −19); Negative sum (−27)

PRACTICE 4

Find the sum: $-13 + (-18)$

EXAMPLE 5

Find the sum of -6.7 and 5.2.

Solution Here we are adding numbers with *different* signs. First, we find the absolute values:

$$|-6.7| = 6.7 \qquad \text{and} \qquad |5.2| = 5.2$$

Then, we find the difference between the larger absolute value and the smaller one:

$$6.7 - 5.2 = 1.5$$

Because -6.7 has the larger absolute value and its sign is negative, the sum is also negative. So

$$-6.7 + (+5.2) = -6.7 + 5.2 = -1.5$$

Negative number (points to -6.7)

Positive number (points to $+5.2$)

The sum takes the sign of the number with the larger absolute value.

PRACTICE 5

Add: $10.1 + (-6.6)$

Can you use the number line to explain why the sum in Example 5 is negative?

EXAMPLE 6

Combine: $-\frac{2}{9} + \frac{2}{9}$

Solution

$$-\frac{2}{9} + \frac{2}{9} = 0$$

Note that $-\frac{2}{9}$ and $\frac{2}{9}$ are additive inverses. So their sum is 0.

PRACTICE 6

Combine: $-\frac{1}{3} + \frac{1}{3}$

EXAMPLE 7

Add: $\frac{3}{8} + \left(-\frac{5}{16}\right)$

Solution

First, we find the LCD of the fractions. The LCD is 16. So

$$\frac{3}{8} + \left(-\frac{5}{16}\right) = \frac{6}{16} + \frac{-5}{16} = \frac{1}{16}.$$

PRACTICE 7

Add: $-\frac{7}{18} + \frac{1}{9}$

The *commutative property of addition* allows us to add two numbers in any order, getting the same sum. For example, $2 + 6 = 8$ and $6 + 2 = 8$.

The Commutative Property of Addition

For any two real numbers a and b,

$$a + b = b + a.$$

In words, this property states that the sum of two numbers is the same regardless of order.

By contrast, the *associative property of addition* lets us regroup numbers that are added without affecting the sum. For example, $(2 + 6) + 3 = 8 + 3 = 11$ and $2 + (6 + 3) = 2 + 9 = 11$.

The Associative Property of Addition

For any three real numbers, a, b, and c,

$$(a + b) + c = a + (b + c).$$

In words, this property states that when adding three numbers, their sum is the same regardless of how they are grouped.

When adding three or more real numbers, it is usually easier to add the positives separately from the negatives. This rearrangement does not affect the sum because of the commutative and associative properties.

EXAMPLE 8

Find the sum: $5 + (-6) + (-9) + 3 + (-5)$

Solution Let's rearrange the numbers by sign.

$$\underbrace{5 + 3}_{\text{Positives}} + \underbrace{(-6) + (-9) + (-5)}_{\text{Negatives}}$$

$5 + 3 = 8$	**Add the positives.**
$(-6) + (-9) + (-5) = -20$	**Add the negatives.**
$8 + (-20) = -12$	**Find the sum of the positive and the negative sums.**

So $5 + (-6) + (-9) + 3 + (-5) = -12$.

PRACTICE 8

Find the sum:
$-8 + (-4) + 7 + (-8) + 3$

Can you think of another way to get the sum in Example 8?

EXAMPLE 9

In January 2005, the average price of a gallon of regular unleaded gasoline was \$1.82. In each of the next five Januaries, the average price increased by \$0.49, decreased by \$0.04, increased by \$0.77, decreased by \$1.26, and increased by \$0.94. What was the average price in January 2010? (*Source:* bls.gov)

Solution We can represent an increase in the average price of \$0.49 and a decrease in the average price of \$0.04 as $+0.49$ and -0.04, respectively. Similarly, an increase of \$0.77, a decrease of \$1.26, and an increase of \$0.94 can be represented by $+0.77$, -1.26, and $+0.94$.

To find the average price in January 2010, we start with the average price in January 2005. Then, we add each of the increases or decreases as follows:

$$+1.82 + (+0.49) + (-0.04) + (+0.77) + (-1.26) + (+0.94) = +2.72$$

So the average price of a gallon of regular unleaded gasoline in January 2010 was \$2.72.

PRACTICE 9

The price of a certain stock on Monday was \$37.50 per share. On Tuesday, the price of a share went up \$2; on Wednesday, it went down \$1; and on Thursday, it went down another \$2. What was the share price of the stock on Thursday?

Subtraction of Real Numbers

Now, let's review subtraction. As we have seen, subtracting two real numbers is based on adding the first number to the opposite of the second number, as the following rule indicates:

To Subtract Real Numbers

- Change the operation of subtraction to addition and change the number being subtracted to its opposite.
- Follow the rule for adding real numbers.

EXAMPLE 10

Find the difference: $2 - (-5)$

Solution

Change the operation from subtraction to addition.

$$2 - (-5) = 2 + (+5) = 7$$

Change the number being subtracted from -5 to $+5$.

PRACTICE 10

Find the difference: $4 - (-1)$

EXAMPLE 11

Evaluate: $-7 - (-1) + 5 + (-3)$

Solution

$$
\begin{aligned}
-7 - (-1) + 5 + (-3) \\
&= -7 + (+1) + 5 + (-3) && \text{Subtract } -1 \text{ from } -7. \\
&= -6 + 5 + (-3) && \text{Add } -7 \text{ and } +1. \\
&= -1 + (-3) && \text{Add } -6 \text{ and } 5. \\
&= -4 && \text{Add } -1 \text{ and } -3.
\end{aligned}
$$

PRACTICE 11

Simplify: $4 + (-6) - (-11) + 8$

EXAMPLE 12

Simplify: $10 - [0 - (-3)]$

Solution

$$
\begin{aligned}
10 - [0 - (-3)] &= 10 - [0 + 3] && \text{Subtract } -3 \text{ from } 0. \\
&= 10 - 3 && \text{Add 0 and 3.} \\
&= 7 && \text{Subtract 3 from 10.}
\end{aligned}
$$

PRACTICE 12

Simplify: $5 - [12 + (-7)]$

EXAMPLE 13

Egypt emerged as a nation in about 3100 B.C. and Ethiopia in about 3000 B.C. How much older is Egypt than Ethiopia? (*Source: The World Book Encyclopedia*)

Solution Recall that a B.C. year corresponds to a negative integer. So 3100 B.C. and 3000 B.C. are represented by -3100 and -3000, respectively. Because $-3000 > -3100$, we write -3000 first:

$$
\begin{aligned}
-3000 - (-3100) &= -3000 + (+3100) \\
&= 100
\end{aligned}
$$

So Egypt is about 100 years older than Ethiopia.

PRACTICE 13

Paper was invented in China in about 100 B.C., and wood block printing in about A.D. 770. How much older is the invention of paper than that of wood block printing? (*Source: The World Book Encyclopedia*)

11.2 Exercises

FOR EXTRA HELP MyMathLab MathXL PRACTICE WATCH READ REVIEW

Mathematically Speaking

Fill in each blank with the most appropriate term or phrase from the given list.

adding two numbers in any order	regrouping numbers that are added	associative property of addition
identities	subtracting	opposites
commutative property of addition	additive identity property	additive inverse property

1. The ______________________ states that for any real number a, there is exactly one real number $-a$ such that $a + (-a) = 0$ and $(-a) + a = 0$.

2. Additive inverses are also called ______________________.

3. The ______________________ states that for any real number a, $a + 0 = a$ and $0 + a = a$.

4. The commutative property of addition states that ______________________ results in the same sum.

5. When ______________________, change the second number to the opposite and then add.

6. The associative property of addition states that ______________________ gives us the same sum.

A ***Find the sum of each pair of numbers using the number line.***

7. $4 + (-3)$

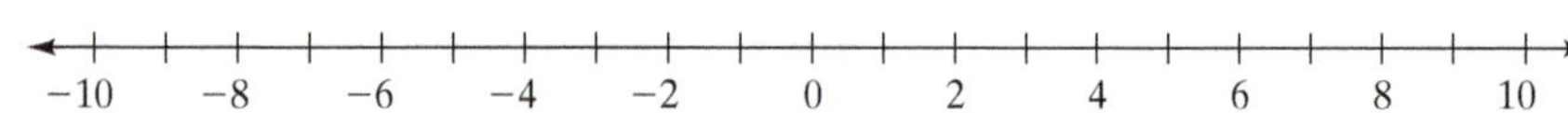

8. $6 + (-2)$

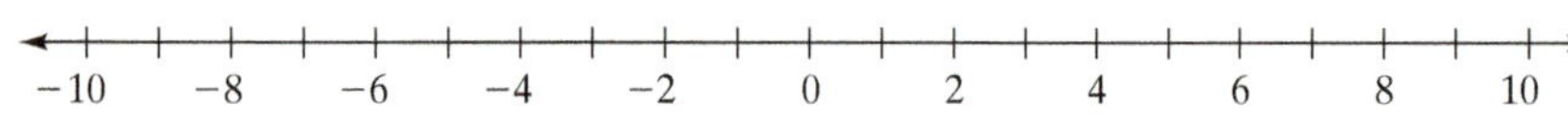

9. $8 + (-8)$

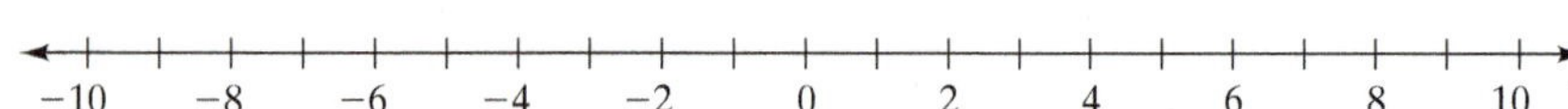

10. $3 + (-3)$

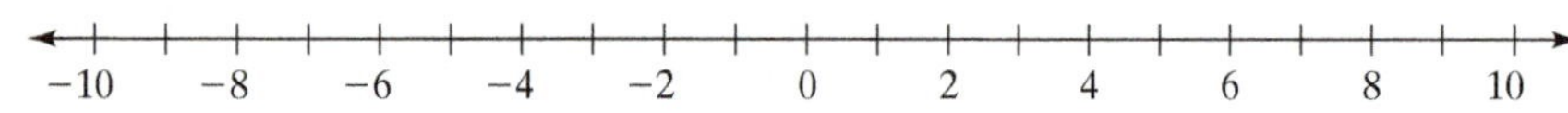

11. $-3 + (-5)$

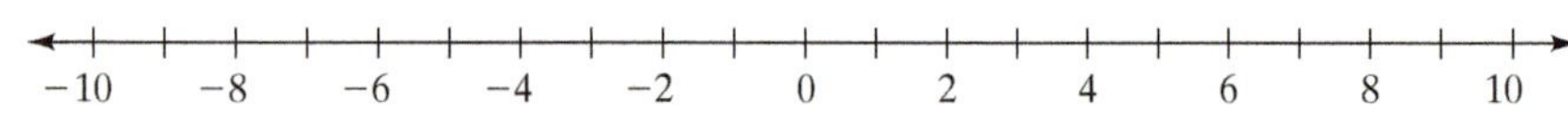

12. $-4 + (-2)$

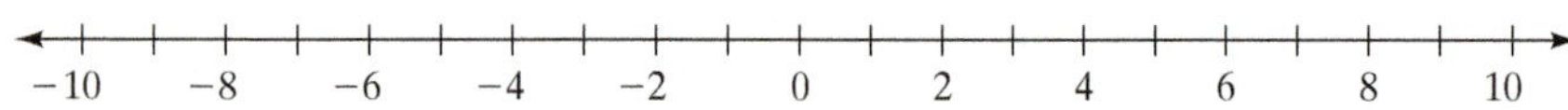

Name the property of addition illustrated.

13. $3 + (-3) = 0$

14. $(-100) + 100 = 0$

15. $5 + (-6) = (-6) + 5$

16. $-1.8 + 2.4 = 2.4 + (-1.8)$

17. $(-4) + 0 = -4$

18. $0 + 2\frac{1}{2} = 2\frac{1}{2}$

19. $(2 + 3) + 6 = 2 + (3 + 6)$

20. $(5 + 6) + (-1) = 5 + [6 + (-1)]$

21. $-a + 0 = -a$

22. $-a + (-b) = -b + (-a)$

Find the sum.

23. $24 + (-1)$ **24.** $10 + (-6)$ **25.** $-50 + (-30)$ **26.** $(-18) + (-18)$

27. $60 + (-90)$ **28.** $2 + (-10)$ **29.** $-18 + 18$ **30.** $-10 + 10$

31. $5.2 + (-0.9)$ **32.** $6.1 + (-5.9)$ **33.** $-10.5 + 0$ **34.** $0 + (-0.3)$

35. $-9.6 + 3.9$ **36.** $-7.2 + 2.8$ **37.** $(-9.8) + (-6.5)$ **38.** $-0.8 + (-0.9)$

39. $\frac{4}{15} + \left(-\frac{2}{3}\right)$ **40.** $-\frac{5}{6} + \frac{7}{12}$ **41.** $2\frac{1}{3} + (-1\frac{1}{2})$ **42.** $6\frac{1}{4} + (-1\frac{1}{3})$

Simplify.

43. $15 + (-9) + (-15) + 9$ **44.** $45 + (-27) + 0 + (-18)$ **45.** $-0.4 + (-2.6) + (-4)$

46. $(-6.25) + (-0.4) + 3$ **47.** $(-58) + 10.48 + 58$ **48.** $-3.7 + 3.7 + (-1.88)$

49. $107 + (-97) + (-45) + 23$ **50.** $-64 + 7 + (-10) + (-19)$

Find the difference.

51. $-24 - 7$ **52.** $-6 - 9$ **53.** $(-19) - 25$ **54.** $(-49) - 2$

55. $52 - (-19)$ **56.** $24 - (-31)$ **57.** $60 - 95$ **58.** $70 - 92$

59. $-34 - (-2)$ **60.** $-30 - (-1)$ **61.** $16 - (-16)$ **62.** $21 - (-21)$

63. $0 - 45$ **64.** $0 - 36$ **65.** $-31 - 31$ **66.** $-25 - 25$

67. $22 - (-22)$ **68.** $8 - (-19)$ **69.** $6 - 7.42$ **70.** $10.1 - 11.84$

71. $-7.3 - 0.5$ **72.** $-3 - 0.1$ **73.** $(-5.6) - (-5.6)$ **74.** $(-0.4) - (-0.4)$

75. $8.6 - (-1.7)$ **76.** $9.4 - (-2.5)$ **77.** $-\frac{1}{3} - \frac{5}{6}$ **78.** $-\frac{4}{5} - \frac{7}{8}$

79. $-18 - \left(-\frac{3}{4}\right)$ **80.** $-12 - \left(-\frac{1}{4}\right)$

Simplify.

81. $3 + (-6) - (-15)$ **82.** $12 + (-4) - (-9)$ **83.** $8 - 10 + (-5)$

84. $12 - 16 + (-5)$ **85.** $9 - 12 - 18$ **86.** $7 - 4 - 12$

87. $-9 + (-4) - 9 + 4$ **88.** $-6 + (-1) - 5 + 8$ **89.** $11 - (8 - 2)$

90. $4 - (6 - 1)$ **91.** $-12 + [9 - (-3)]$ **92.** $-4 - [2 + (-6)]$

93. $3 - [0 + (-15)]$ **94.** $-10 - [-8 + 0]$

Mixed Practice

Name the property of addition illustrated.

95. $(-2 + 5) + 8 = -2 + (5 + 8)$

96. $-5.9 + 5.9 = 0$

Solve.

97. Find the sum of 2 and (-9) using the number line.

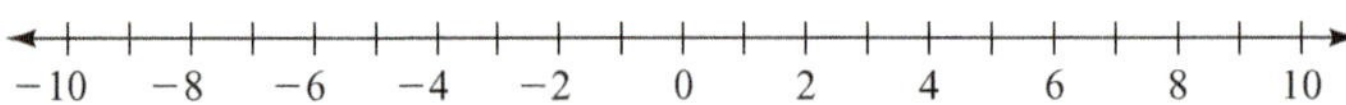

Find the sum.

98. $37 + 53 + (-38)$

99. $22 + (-15) + (-22)$

100. $-8.5 + 4.8$

Find the difference.

101. $17 - (-31)$

102. $-7.6 - 5.8$

103. $(-23) - 15$

104. $-28 - (-17)$

105. $\frac{5}{6} - \left(-\frac{3}{4}\right)$

Simplify.

106. $-5 - (-3) + 2 + (-9)$

107. $9 - (8 + 1)$

108. $18 - [(-13) - (-9)]$

Applications

B *Solve. Express your answer as a real number.*

109. The temperature on the top of a mountain was 2° below 0°. If it then got 7° warmer, what was the temperature?

110. A plane cruising at an altitude of 32,000 ft hit an air pocket and dropped 700 ft. What was its new altitude?

111. In the first quarter of Super Bowl XLII, the Pittsburgh Steelers outscored the Arizona Cardinals by three points. In the second quarter, the Cardinals were outscored by seven points. The Steelers scored three points more than the Cardinals in the third quarter. Finally in the fourth quarter, the Cardinals outscored the Steelers by nine points. Who won the game and by how many points? (*Source: Sports Illustrated Almanac 2010*)

112. Estimate the total annual profits for Dot.com Corporation according to the following chart. (Note that a red number written in parentheses is negative.)

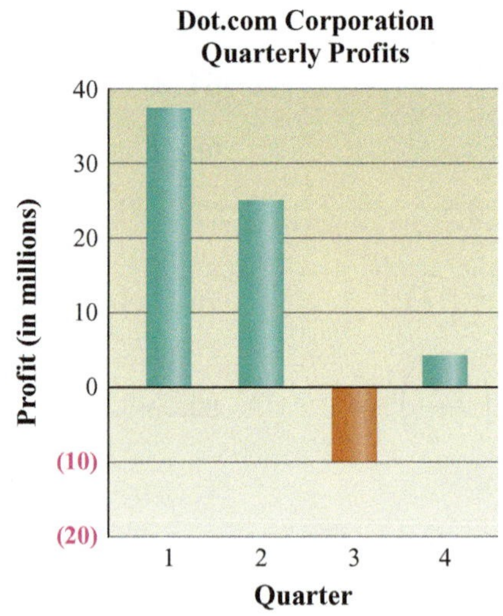

113. With $371.25 in his checking account, an artist writes checks for $71.33 and $51.66. He also deposits $35. After the two checks clear and his deposit is credited, will he still have enough money to cover a check of $250?

114. In 2009, the U.S. federal budget allotted $1.795 billion for Supporting Student Success. For the next year, the allotment decreased by $0.254 billion. In 2011, the allotment was $0.245 billion more than for the previous year. How much money was allotted for this program in 2011? (*Source:* whitehouse.gov)

115. The first Olympic Games occurred in 776 B.C. Approximately how many centuries were there between the first Olympic games and the Olympic games held in A.D. 2010?

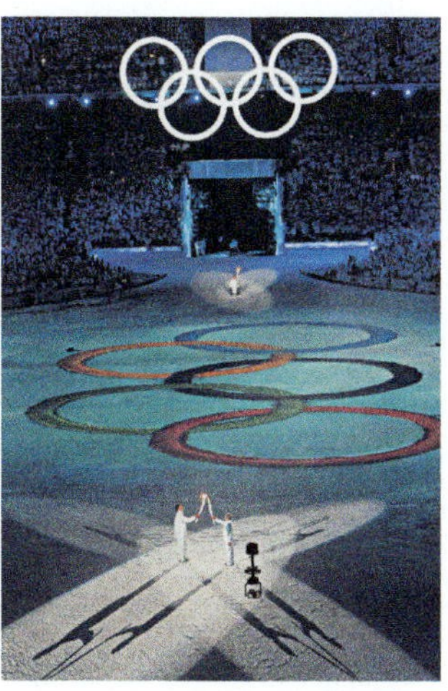

116. The following chart shows the record high and low temperatures (in degrees Fahrenheit) for a number of U.S. states.

State	Record High (°F)	Record Low (°F)
Alabama	114	−27
California	134	−45
Louisiana	114	−16
Minnesota	114	−64
New York	108	−57
Virginia	110	−30

(*Source:* wikipedia.org)

Which state had the greatest difference in record extreme temperatures?

117. A company reported a loss of $281,330 last year and a loss of $5291 this year. By how much money was the company's loss reduced?

118. The value of a computer company's stock rose by $0.50 per share and then dropped by $0.75. What was the overall change in value?

119. Matter is liquid when its temperature is between its melting point and its boiling point. The following table shows the melting and boiling points (in degrees Celsius) of various elements.

Element	Melting Point (°C)	Boiling Point (°C)
Krypton	−157	−153
Neon	−249	−246
Bromine	−7	59

(*Source: The New York Times Almanac*, 2010)

a. For each of these elements, find the difference between its boiling and melting points.

b. Which of the elements is liquid in the widest range of temperatures?

c. Which of the elements is liquid at 0°C?

120. Superconductors allow for very efficient passage of electric currents. For practical use, a superconductor must work above −196°C, which is the boiling point of nitrogen.

a. In 1986, the first high-temperature superconductor that was able to conduct electricity without resistance at a temperature of −238°C was discovered. How many degrees below the boiling point of nitrogen did this first high-temperature superconductor work?

b. In 1987, the researcher Paul Chu discovered a new class of materials that conduct electricity at −178°C. How many degrees above the boiling point of nitrogen did the new materials conduct electricity?

c. In 1990, other researchers created a miniature transistor that conducts electricity at a temperature that is 48°C above the boiling point of nitrogen. What is this temperature? (*Sources: World Book Encyclopedia* and uh.edu)

• Check your answers on page A-15.

MINDStretchers

Mathematical Reasoning

1. Consider the following addition table:

+	x	y	z
x	y	z	x
y	z	x	y
z	x	y	z

a. Use the table to find $x + y$.

b. What is the additive identity element in the addition table? Explain how you know.

c. What is the opposite of x in this table? Explain how you know.

Critical Thinking

2. Rearrange the numbers in the square on the left so that it becomes a magic square in which the sum of every row, column, and diagonal is -6.

-3	2	-2
-5	-4	-6
0	-1	1

Writing

3. Explain whether it is always true that the difference between two numbers is smaller than either of the numbers. Give an example to justify your answer.

11.3 Using Properties to Multiply and Divide Real Numbers

OBJECTIVES

A To multiply or divide real numbers

B To solve applied problems involving the multiplication or division of real numbers

Recall our discussion in Sections 7.4 and 7.5 of multiplying and dividing. Here, we consider some important properties that relate to these operations.

Multiplication of Real Numbers

First, let's review multiplication. We have seen that real numbers can be multiplied by using the following rule:

To Multiply Real Numbers

- Multiply their absolute values.
- If the numbers have the same sign, their product is positive; if they have different signs, their product is negative.

EXAMPLE 1

Multiply -2 by -9.

Solution

$|-2| = 2$ and $|-9| = 9$ — Find the absolute values.

$2 \cdot 9 = 18$ — Multiply the absolute values.

$-2(-9) = 18$ — The product of two negatives is positive.

PRACTICE 1

Find the product of -1 and -100.

EXAMPLE 2

Multiply: $3(-5)$

Solution

$|3| = 3$ and $|-5| = 5$ — Find the absolute values.

$3 \cdot 5 = 15$ — Multiply the absolute values.

$3(-5) = -15$ — The product of a positive and a negative is negative.

PRACTICE 2

Calculate: $-5 \cdot 3$

Comparing Example 2 and Practice 2, we see that $3(-5) = -5 \cdot 3$, which suggests a property of real numbers discussed in Chapter 1, the *commutative property of multiplication.*

The Commutative Property of Multiplication

For any two real numbers a and b,

$$a \cdot b = b \cdot a.$$

In words, this property states that the product of two numbers is the same regardless of order.

Another property involving the multiplication of real numbers is the *multiplicative identity property*, also called the *identity property of multiplication.*

The Multiplicative Identity Property

For any real number a,

$$a \cdot 1 = a \quad \text{and} \quad 1 \cdot a = a.$$

In words, this property states that the product of any number and one is the original number.

The next property also involves multiplication.

The Multiplication Property of 0

For any real number a,

$$a \cdot 0 = 0 \quad \text{and} \quad 0 \cdot a = 0.$$

In words, this property states that the product of any number and 0 is 0.

EXAMPLE 3

Find the product.

a. $-9\left(\frac{1}{3}\right)$ **b.** $-\frac{1}{5}(-25)$ **c.** $-4(0)$

d. $-1.5(-1.5)$ **e.** $0.4(-6)$ **f.** $-7(1)$

Solution

a. $-9\left(\frac{1}{3}\right) = \frac{\overset{3}{\cancel{-9}}}{1} \cdot \frac{1}{\underset{1}{\cancel{3}}} = -3$ Negative · Positive = Negative

b. $-\frac{1}{5}(-25) = -\frac{1}{\underset{1}{\cancel{5}}} \cdot \frac{\overset{5}{\cancel{-25}}}{1} = 5$ Negative · Negative = Positive

c. $-4(0) = 0$ Multiplication property of zero

d. $-1.5(-1.5) = 2.25$ Negative · Negative = Positive

e. $0.4(-6) = -2.4$ Positive · Negative = Negative

f. $-7(1) = -7$ Multiplicative identity property

PRACTICE 3

Multiply.

a. $\left(-\frac{2}{3}\right)(-12)$

b. $\left(-\frac{1}{3}\right)\left(\frac{5}{9}\right)$

c. $(-0.4)(-0.3)$

d. $2.5(-1.9)$

e. $0 \cdot (-2.8)$

f. $1 \cdot \frac{2}{3}$

The next property of real numbers—the *associative property of multiplication*—allows us to *regroup* the product of three numbers. For example, $(2 \cdot 3) \cdot 5 = 2 \cdot (3 \cdot 5)$.

The Associative Property of Multiplication

For any three real numbers, a, b, and c,

$$(a \cdot b)c = a(b \cdot c).$$

In words, this property states that when multiplying three real numbers, the product is the same regardless of how they are grouped.

EXAMPLE 4

Calculate: $-3(-2)(9)$

Solution

$$-3(-2)(9) = 6(9) \quad \text{Multiply } -3 \text{ by } -2.$$
$$= 54 \quad \text{Multiply 6 by 9.}$$

So $(-3)(-2)(9) = 54$.

PRACTICE 4

Multiply: $-8(4)(-2)$

EXAMPLE 5

Multiply: $5(-2)(-1)(3)(-2)$

Solution A good way to calculate this product is to rearrange the numbers by sign.

$$\underbrace{5(3)}_{\text{Positives}} \qquad \underbrace{-2(-1)(-2)}_{\text{Negatives}}$$

$5(3) = 15$ Find the product of the positives.

$-2(-1)(-2) = 2(-2) = -4$ Find the product of the negatives.

$15(-4) = -60$ Multiply the two products above.

So $5(-2)(-1)(3)(-2) = -60$.

PRACTICE 5

Find the product:

$(-6)(-1)(4)(2)(-5)$

In Example 4, the product was positive because there were *two* negative factors. By contrast, in Example 5 the answer was negative because there were *three* negative factors. Can you explain why a product is positive if it has an even number of negative factors, whereas a product is negative if it has an odd number of negative factors?

According to the order of operations rule given in Section 1.5, multiplication is performed before either addition or subtraction, working from left to right.

EXAMPLE 6

Simplify: $-2(24) - 5(-6)$

Solution We use the order of operations rule.

$$\begin{aligned} -2(24) - 5(-6) &= -48 - (-30) && \text{Multiply first.} \\ &= -48 + 30 && \text{Subtract } -30 \text{ from } -48. \\ &= -18 && \text{Add } -48 \text{ and } 30. \end{aligned}$$

So $-2(24) - 5(-6) = -18$.

PRACTICE 6

Calculate: $4(-25) - (-2)(36)$

Following the order of operations rule, we simplify mathematical expressions by first performing the operations within any grouping symbols such as parentheses ( ) or brackets [ ].

EXAMPLE 7

Calculate: $5 - 3(2 - 4)$

Solution

$$\begin{aligned} 5 - 3(2 - 4) &= 5 - 3(-2) && \text{Subtract within parentheses.} \\ &= 5 - (-6) && \text{Multiply.} \\ &= 11 && \text{Subtract.} \end{aligned}$$

So $5 - 3(2 - 4) = 11$.

PRACTICE 7

Simplify: $10 - 5(3 + 1)$

EXAMPLE 8

Simplify: $11 - 4(-3 + 1)^2$

Solution

$$\begin{aligned} 11 - 4(-3 + 1)^2 &= 11 - 4(-2)^2 && \text{Add within parentheses.} \\ &= 11 - 4(4) && \text{Square } -2. \\ &= 11 - 16 && \text{Multiply.} \\ &= -5 && \text{Subtract.} \end{aligned}$$

So $11 - 4(-3 + 1)^2 = -5$.

PRACTICE 8

Calculate: $9 - 5(6 - 7)^2$

EXAMPLE 9

Simplify: $2[(3 - 8)^2 - (-6)^2]$

Solution

$$\begin{aligned} 2[(3 - 8)^2 - (-6)^2] &= 2[(-5)^2 - (-6)^2] && \text{Subtract within parentheses.} \\ &= 2[25 - 36] && \text{Square } -5. \text{ Then square } -6. \\ &= 2[-11] && \text{Subtract within brackets.} \\ &= -22 && \text{Multiply.} \end{aligned}$$

PRACTICE 9

Simplify:

$-20 + 3[(1 - 4)^2 + (-2)]$

EXAMPLE 10

Temperatures can be measured in both the Fahrenheit and Celsius systems. The normal *melting point* of the element mercury is about $-37.9°$F. To find the Celsius equivalent of this temperature, we need to compute $\frac{5}{9}(-37.9 - 32)$. Simplify this expression.

(*Source: CRC Handbook of Chemistry and Physics*)

Solution

$$\frac{5}{9}(-37.9 - 32) = \frac{5}{9}(-69.9)$$
$$\approx -38.8$$

So $-37.9°$F is equivalent to $-38.8°$C.

PRACTICE 10

If a rock is thrown upward on the moon with an initial velocity of 10 ft/sec, the rock's velocity after 3 sec will be $[10 - (5.3)(3)]$ ft/sec. Simplify this expression and interpret the result. (Note: Objects moving upward have positive velocity and objects moving downward have negative velocity.) (*Source:* NASA)

Division of Real Numbers

Now, let's review division. As we have seen, real numbers can be divided using the following rule:

To Divide Real Numbers

- Divide their absolute values.
- If the numbers have the same sign, their quotient is positive; if the numbers have different signs, their quotient is negative.

EXAMPLE 11

Find the quotient.

a. $(-16) \div (-2)$ **b.** $\frac{-24}{6}$ **c.** $\frac{-2}{-8}$

d. $\frac{9.4}{-2}$ **e.** $\frac{-15}{-0.3}$

Solution In each problem, first we find the absolute values. Next, we divide them. Then, we attach the appropriate sign to this quotient.

a. $(-16) \div (-2)$

$|-16| = 16$ and $|-2| = 2$

$16 \div 2 = 8$

Since the numbers have the *same* signs, their quotient is positive. So $(-16) \div (-2) = 8$.

PRACTICE 11

Divide.

a. $40 \div (-5)$

b. $\frac{-42}{-6}$

c. $\frac{-5}{10}$

d. $\frac{-6.3}{9}$

e. $\frac{-24}{-0.4}$

EXAMPLE 11 (continued)

b. $\dfrac{-24}{6}$ $\quad |-24| = 24$ and $|6| = 6$

$$\frac{24}{6} = 4$$

Since the numbers have *different* signs, the quotient is negative.

So $\dfrac{-24}{6} = -4$.

c. $\dfrac{-2}{-8}$ $\quad |-2| = 2$ and $|-8| = 8$

$$\frac{2}{8} = \frac{1}{4}$$

Since the numbers have the *same* signs, the quotient is positive.

So $\dfrac{-2}{-8} = \dfrac{1}{4}$.

d. $\dfrac{9.4}{-2}$ $\quad |-9.4| = 9.4$ and $|-2| = 2$

$$\frac{9.4}{2} = 4.7$$

Since the numbers have *different* signs, the quotient is negative.

So $\dfrac{9.4}{-2} = -4.7$.

e. $\dfrac{-15}{-0.3}$ $\quad |-15| = 15$ and $|-0.3| = 0.3$

$$\frac{15}{0.3} = \frac{15.0}{0.3} = \frac{150}{3} = 50$$

Since the numbers have the *same* signs, the quotient is positive.

So $\dfrac{-15}{-0.3} = 50$.

Some division problems involve 0. For instance, $0 \div (-5) = 0$ because $(-5) \cdot 0 = 0$. On the other hand, $(-5) \div 0$ is *undefined* because there is no real number that when multiplied by 0 gives -5.

These two examples lead us to the following conclusion:

Division Involving Zero

For any nonzero real number a,

$$0 \div a = 0.$$

For any nonzero real number a,

$$a \div 0 \text{ is undefined.}$$

In words, these properties state that zero divided by any nonzero number is zero, whereas any number divided by zero is undefined.

Recall that in Section 2.4 we expressed the rule for dividing fractions in terms of the *reciprocal* of a number. In algebra, we also refer to the reciprocal of a number as its *multiplicative inverse*.

Multiplicative Inverse Property

For any nonzero real number a,

$$a \cdot \frac{1}{a} = 1 \quad \text{and} \quad \frac{1}{a} \cdot a = 1,$$

where a and $\frac{1}{a}$ are **multiplicative inverses** (or **reciprocals**) of each other.

In words, this property states that the product of a number and its multiplicative inverse is one. For example,

- $\frac{1}{3}$ and 3 are multiplicative inverses because $\frac{1}{3} \cdot 3 = 1$
- $-\frac{5}{6}$ and $-\frac{6}{5}$ are multiplicative inverses because $\left(-\frac{5}{6}\right)\left(-\frac{6}{5}\right) = 1$

EXAMPLE 12

Complete the following table:

Solution

Number	Reciprocal
a. 4	$\frac{1}{4}$ is the reciprocal of 4 because $4 \cdot \frac{1}{4} = 1$
b. $-\frac{3}{4}$	$-\frac{4}{3}$ is the reciprocal of $-\frac{3}{4}$ because $\left(-\frac{3}{4}\right) \cdot \left(-\frac{4}{3}\right) = 1$
c. -10	$-\frac{1}{10}$ is the reciprocal of -10 because $-10\left(-\frac{1}{10}\right) = 1$
d. $1\frac{1}{2}$	$\frac{2}{3}$ is the reciprocal of $1\frac{1}{2}$ because $1\frac{1}{2} \cdot \frac{2}{3} = \frac{3}{2} \cdot \frac{2}{3} = 1$

PRACTICE 12

Fill in the following table:

Number	Reciprocal
a. -5	
b. $\frac{1}{-8}$	
c. $1\frac{1}{3}$	
d. $-\frac{8}{5}$	

Now, let's consider division of real numbers by using reciprocals. Recall that we subtract by adding an opposite. Similarly, we can divide by multiplying by a reciprocal.

Division of Real Numbers

For any real numbers a and b, where b is nonzero,

$$a \div b = \frac{a}{b} = a \cdot \frac{1}{b}.$$

In words, this rule states that the quotient of two numbers is the product of the first number and the reciprocal of the second number.

EXAMPLE 13

Divide.

a. $-\frac{1}{3} \div \frac{5}{6}$ b. $-\frac{1}{2} \div 4$

Solution

a. $-\frac{1}{3} \div \frac{5}{6} = -\frac{1}{3} \cdot \frac{6}{5} = -\frac{1}{\cancel{3}_1} \cdot \frac{\cancel{6}^2}{5} = -\frac{2}{5}$

b. $-\frac{1}{2} \div 4 = -\frac{1}{2} \div \frac{4}{1} = -\frac{1}{2} \cdot \frac{1}{4} = -\frac{1}{8}$

PRACTICE 13

Divide.

a. $-\frac{8}{9} \div \frac{2}{3}$

b. $-10 \div \left(-\frac{2}{5}\right)$

We use the order of operations rule to simplify the following expressions.

EXAMPLE 14

Simplify.

a. $-8 \div (-2)(-2)$ b. $\frac{-5 + (-7)}{2}$

Solution

a. $-8 \div (-2)(-2) = 4(-2)$ Perform multiplications and divisions as they occur from left to right. Divide -8 by -2.

$= -8$ Multiply 4 by -2.

b. $\frac{-5 + (-7)}{2} = \frac{-12}{2}$ Parentheses are understood to be around the numerator. Add -5 and -7.

$= -6$ Divide -12 by 2.

PRACTICE 14

Simplify.

a. $(-3)(-4) \div (2)(-2)$

b. $\frac{-9 - (-3)}{2}$

EXAMPLE 15

During clinical practice, a student nurse took care of a patient with a fever. He recorded the patient's temperature at the same time every day for five days. The following table shows the change in the patient's temperature each day.

Day	Temperature Change
Monday	Up 2.5°
Tuesday	Down 2°
Wednesday	Down 1.5°
Thursday	Up 1°
Friday	Down 3°

What was the average daily change in the patient's temperature?

Solution To compute the average daily change, we add the five temperature changes and divide the sum by 5, the number of days the temperature was recorded.

$$\frac{2.5 + (-2) + (-1.5) + 1 + (-3)}{5}$$

Since parentheses are understood to be around both the numerator and the denominator, we find the sum in the numerator before dividing by the denominator.

$$\frac{2.5 + (-2) + (-1.5) + 1 + (-3)}{5} = \frac{-3}{5} = -0.6$$

So the average daily change in temperature during the five days was $-0.6°$.

PRACTICE 15

During the past four months, the number of cell phone minutes a customer used changed from the previous month as follows:

Month	Change (in minutes)
1	Down 300
2	Up 200
3	Down 500
4	Up 100

What was the average monthly change in minutes used?

11.3 Exercises

FOR EXTRA HELP MyMathLab MathXL PRACTICE WATCH READ REVIEW

Mathematically Speaking

Fill in each blank with the most appropriate term or phrase from the given list.

regroup the product of three numbers	multiply two numbers in either order	any number and 1
any number and 0		

1. The commutative property of multiplication allows us to _______________.

2. The multiplicative identity property tells us that the product of __________________ is the original number.

3. The associative property of multiplication allows us to ________________.

4. Any nonzero real number a divided by zero is ________________.

5. Any nonzero real number a has a multiplicative inverse, or __________________, which is written $\frac{1}{a}$.

6. To divide two real numbers, multiply the dividend by the reciprocal of the __________________.

A ***Name the property of multiplication illustrated.***

7. $2(5) = 5(2)$

8. $1.7(-3) = -3(1.7)$

9. $(-4 \cdot 6) \cdot 3 = -4 \cdot (6 \cdot 3)$

10. $(-8 \cdot 5) \cdot 4 = -8 \cdot (5 \cdot 4)$

11. $-9 \cdot 1 = -9$

12. $1 \cdot (-10) = -10$

13. $-8 \cdot 0 = 0$

14. $0 \cdot \frac{1}{2} = 0$

Find the product.

15. $6(-2)$

16. $7(-4)$

17. $-7(-3)$

18. $-5(-5)$

19. $-12\left(\frac{1}{4}\right)$

20. $-15\left(\frac{2}{3}\right)$

21. $\left(1\frac{1}{3}\right)\left(-\frac{4}{9}\right)$

22. $\left(2\frac{1}{5}\right)\left(-\frac{2}{7}\right)$

23. $-1.5(-0.6)$

24. $-1.7(-0.4)$

25. $1.2(-50)$

26. $1.5(-60)$

27. $3(-2)(-20)$

28. $-9(-12)(2)$

29. $-15(-3)(0)$

30. $-8.5(0)(2.6)$

31. $-6(1)(-2)(-3)(-4)$

32. $6(-1)(-2)(3)(-4)$

33. $\left(-\frac{1}{3}\right)\left(-\frac{1}{3}\right)\left(-\frac{1}{3}\right)$

34. $\left(-\frac{1}{2}\right)\left(-\frac{1}{2}\right)\left(-\frac{1}{2}\right)$

Simplify.

35 $-7 + 3(-2) - 10$

36. $-4 + 2(-5) - 3$

37. $-3 - 5(-6)$

38. $-10 - 2(-8)$

39. $\left(\frac{3}{5}\right)(-15) + 6$

40. $\left(\frac{3}{4}\right)(-16) + 20$

41. $-5 \cdot (-3 + 4)$

42. $(-10 + 7) \cdot (-3)$

43. $-6 - 3(5 - 9)$

44. $5 - 2(4 - 10)$

45. $7 - 3(-2 + 5)^2$

46. $5(4 - 6)^2 - 9$

47. $3[5(2^3 - 10) + 6]$

48. $-2[3 + 6(4 - 5)^3]$

Complete each table.

49.

	a.	b.	c.	d.	e.
Number	$-\frac{1}{2}$	5	$-\frac{3}{4}$	$3\frac{1}{5}$	-1
Reciprocal					

50.

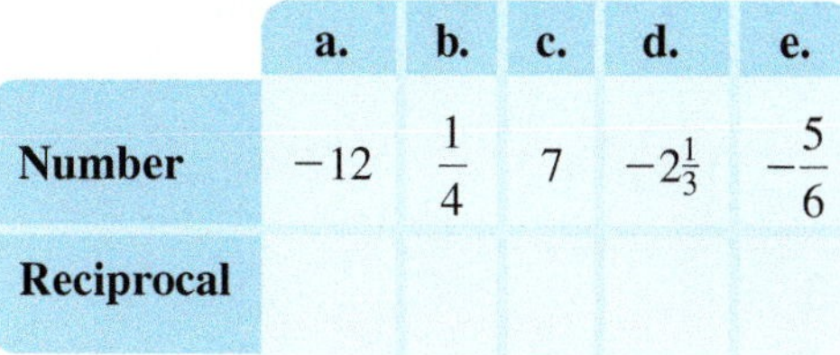

	a.	b.	c.	d.	e.
Number	-12	$\frac{1}{4}$	7	$-2\frac{1}{3}$	$-\frac{5}{6}$
Reciprocal					

Divide.

51. $-8 \div (-1)$

52. $-12 \div (-1)$

53. $-63 \div 7$

54. $-16 \div 4$

55. $\frac{0}{-9}$

56. $0 \div (-10)$

57. $-250 \div (-10)$

58. $-300 \div (-10)$

59. $-200 \div (-8)$

60. $-400 \div (-5)$

61. $\frac{-2}{16}$

62. $\frac{2}{-10}$

63. $\frac{4}{5} \div \left(-\frac{2}{3}\right)$

64. $\frac{7}{12} \div \left(-\frac{1}{6}\right)$

65. $8 \div \left(-\frac{1}{4}\right)$

66. $5 \div \left(-\frac{1}{6}\right)$

67. $2\frac{1}{2} \div (-20)$

68. $3\frac{1}{2} \div (-10)$

69. $(-3.5) \div 7$

70. $(-5.6) \div (8)$

71. $\frac{-3}{-0.3}$

72. $\frac{-1.8}{-0.6}$

Simplify.

73. $-16 \div (-2)(-2)$

74. $-36 \div (-3)(-2)$

75. $(3 - 7) \div (-4)$

76. $(5 - 8) \div (-3)$

77. $\frac{2 + (-6)}{-2}$

78. $\frac{10 + (-4)}{3}$

79. $(4 - 6) \div (1 - 5)$

80. $(-15 - 3) \div (-2 - 4)$

81. $-56 \div 7 - 4 \cdot (-3)$

82. $32 \div (-8) + (-5) \cdot 6$

83. $(-4)\left(\frac{1}{2}\right) - 2 \div \left(-\frac{1}{8}\right)$

84. $(-6) \div \left(\frac{2}{3}\right) + (-10)\left(\frac{2}{5}\right)$

MIXED PRACTICE

Find the product.

85. $-2.8(-1.3)$

86. $\frac{2}{5}\left(-\frac{3}{4}\right)$

87. $3(-5)(1)(-4)(-2)$

Solve.

88. Name the property of multiplication that $1.7(-6.3) = -6.3(1.7)$ illustrates.

Simplify.

89. $-5 - 6(-2) + (-3)$

90. $\left(\frac{4}{9}\right)(-18) - (-3)$

91. $-2[(3 - 8)^2 \div (-5)]$

Divide.

92. $\left(-4\frac{1}{2}\right) \div 3$

93. $\frac{5}{6} \div \left(-\frac{3}{8}\right)$

94. $(-0.72) \div (-6)$

95. $-65 \div (-13)$

Solve.

96. Complete the table.

	a.	b.	c.	d.	e.
Number	8	$\frac{2}{3}$	$-\frac{1}{4}$	-6	$-1\frac{1}{3}$
Reciprocal					

Simplify.

97. $-12 \div (5 - 7)$

98. $\frac{4 - (-6)}{-2}$

Applications

Solve.

99. During a drought, the water level in a reservoir dropped 3 in. each week for 5 straight weeks. Express the overall change in water level as a signed number.

100. On January 31, the high temperature in Chicago was 40°F. The high temperature then dropped 3°F per day for 3 days. If 32°F is freezing on the Fahrenheit scale, was it below freezing on February 3?

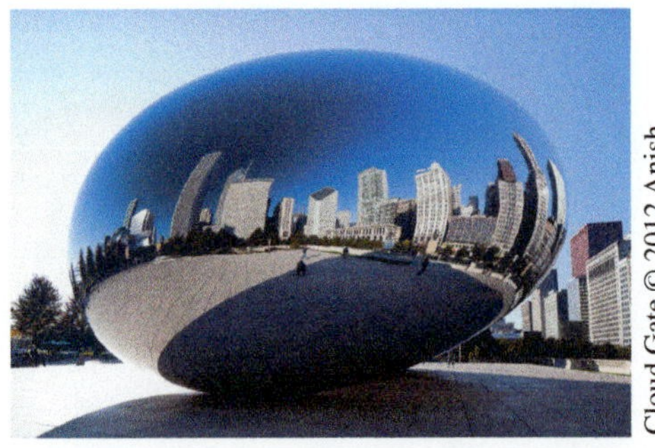

101. The following tables show the number of calories in servings of various foods and the number of calories burned by various activities:

Food	Number of Calories Per Serving
Apple	80
Banana	105
Pretzel, stick	30
Ginger ale, can	125
Donut	210

(*Source: The World Almanac and Book of Facts*)

Activity (1 hr)	Number of Calories Burned*
Swimming	−288
Bicycling	−612
Football	−498
Basketball	−450
Scrubbing floors	−440

*For a 150-lb person.
(*Source: Exercise & Weight Control*, President's Council on Physical Fitness and Sports)

Find the net number of calories in each situation.

a. A weight watcher eats 3 servings of pretzel sticks and then plays basketball for $\frac{1}{2}$ hour.

b. A dieter swims for 30 min and drinks 2 cans of ginger ale.

c. An athlete eats 3 servings of apples and 2 servings of donuts. Later, he goes bicycling for 2 hr and then swims for 1 hr.

102. To discourage guessing on a test, an instructor takes off for wrong answers, grading according to the following scheme:

Performance on a Test Item	Score
Correct	5
Incorrect	−2
Blank	0

What score would the instructor give to each of the following tests?

Number of Items Correct	Number of Items Incorrect	Number of Items Blank	Test Grade
a. 17	1	2	
b. 19	1	0	
c. 12	7	1	

103. The following table shows the change from the previous year in the number of customers at a restaurant for each of five years:

Year	Change in the Number of Customers
2007	An increase of 5700
2008	No change
2009	A decrease of 2600
2010	A decrease of 900
2011	A decrease of 1200

For the five years, what was the average change per year in the number of customers?

104. The following bar graph shows the quarterly net income for the company American International Group, Inc. (AIG) for four consecutive quarters. (Note that a red number written in parentheses is negative.)

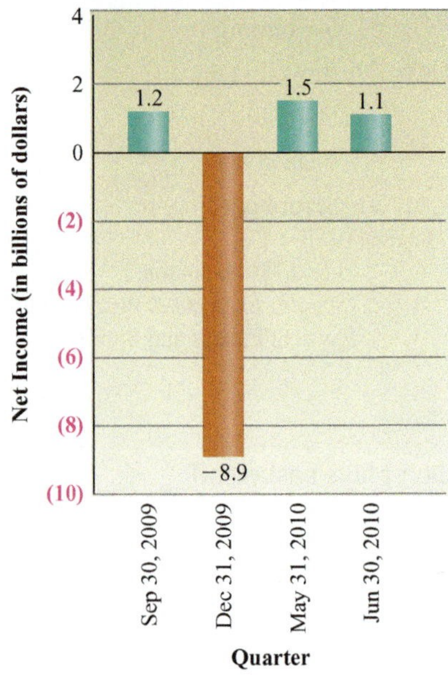

(*Source:* finance.yahoo.com)

To the nearest billion dollars, what was the average net income for a quarter?

105. A football running back lost a total of 24 yd in 6 plays. What was the average number of yards he lost on each play?

106. The federal deficit in 1940 was about \$3 billion. Five years later at the end of World War II, it was about \$48 billion. How many times the deficit of 1940 was that of 1945? (*Source: Budget of the United States Government, Fiscal Year 2000*, 1999)

• Check your answers on page A-15.

MIND*Stretchers*

Mathematical Reasoning

1. Explain how the number line can be used to find the product of two integers.

Historical

2. At a very early age, the eighteenth-century mathematician Carl Friedrich Gauss found the sum of the first 100 positive integers within a few minutes by using the following method. First, he wrote the sum both forward and backward.

$$1 + 2 + 3 + \cdots + 98 + 99 + 100$$
$$100 + 99 + 98 + \cdots + 3 + 2 + 1$$

Then, he added the 100 vertical pairs, getting a sum of 101 for each pair. He concluded that the product 100(101) was twice the correct answer, which turns out to be $\frac{100(101)}{2} = 50(101) = 5050$. Show how to find the sum of the first 1000 *negative* integers using Gauss's method. Explain your work.

Groupwork

3. Consider the following five integers: $-2, 6, -9, 18, -36$. Working with a partner, explain which two of the five integers you would choose

a. to find the smallest quotient.

b. to find the largest quotient.

11.4 Algebraic Expressions and Exponents

OBJECTIVES

- **A** To find the number of terms in an algebraic expression
- **B** To write algebraic expressions using exponents
- **C** To solve applied problems involving algebraic expressions

Algebraic Expressions

Algebraic expressions consist of one or more *terms*, separated by addition signs. If there are subtraction signs, we can rewrite the expression in an equivalent form using addition. For instance, we can think of the algebraic expression

$$2x + \frac{y}{3} - 4 \quad \text{as} \quad 2x + \frac{y}{3} + (-4).$$

This algebraic expression is made up of three terms.

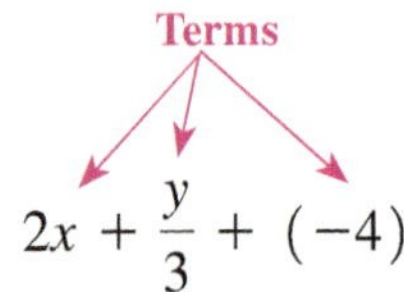

DEFINITION
A **term** is a number, a variable, or the product or quotient of numbers and variables.

EXAMPLE 1

Find the number of terms in each expression.

a. $3y + 1$ **b.** $\frac{a}{b}$

Solution

a. The expression $3y + 1$ has two terms.

b. The expression $\frac{a}{b}$ has one term.

PRACTICE 1

Determine how many terms are in each expression.

a. $2a + 3 - b$ **b.** $-4xy$

Exponents

Recall that we can use *exponential notation* as a shorthand method for representing repeated multiplication of the same factor.

DEFINITION
For any real number x and any positive integer a,

$$x^a = \underbrace{x \cdot x \cdots x \cdot x}_{a \text{ factors}},$$

where x is called the **base** and a is called the **exponent** (or **power**).

In exponential notation, the exponent indicates how many times the base is used as a factor.

The expression x^a is read "x to the ath power," or "x to the power a." However, the exponents 2 and 3 are usually read in a special way. For instance, we generally read 5^2 as "5 *squared*" rather than "5 to the second power." Similarly, we read 5^3 as "5 *cubed*" instead of "5 to the third power."

A number raised to the power 1 is that number. For example, $5^1 = 5$ and $x^1 = x$.

In Section 1.5, we evaluated 2^4. Now, let's consider the expression $(-2)^4$. To evaluate this expression, we multiply 4 factors of -2:

$$\underset{\text{Base}\ \uparrow}{(-2)}^{\overset{\text{Exponent}}{\downarrow}4} = \underbrace{(-2)(-2)(-2)(-2)}_{\text{4 factors of } -2}$$

$$\begin{aligned}(-2)^4 &= \underbrace{(-2)(-2)}(-2)(-2)\\ &= \underbrace{4(-2)}(-2)\\ &= \underbrace{(-8)(-2)}\\ &= 16\end{aligned}$$

In short, $(-2)^4 = 16$.

Next, let's consider the expression -2^4. To evaluate this expression, we multiply 4 factors of 2. Then, we take the opposite:

$$-\underset{\text{Base}\ \uparrow}{2}^{\overset{\text{Exponent}}{\downarrow}4} = -\underbrace{(2)(2)(2)(2)}_{\text{4 factors of 2}}$$

$$= -16$$

Note that in the expression $(-2)^4$ the base is -2, whereas in -2^4 the base is 2.

EXAMPLE 2

Evaluate.

a. $(-3)^4$ **b.** $-3^4(-2)^2$

Solution

a. $(-3)^4 = (-3)(-3)(-3)(-3) = 81$

b. $-3^4(-2)^2 = -(3)(3)(3)(3)(-2)(-2)$
$= -(81)(4)$
$= -324$

PRACTICE 2

Evaluate.

a. -6^2

b. $(-6)^2(-3)^2$

Sometimes we put an expression into exponential form. Such expressions may involve more than one base. For instance, the expression $(-4)(-4)(-4)(-3)(-3)$ can be rewritten in terms of powers of -4 and -3:

$$\underbrace{(-4)(-4)(-4)}_{\text{3 factors of } -4}\underbrace{(-3)(-3)}_{\text{2 factors of } -3} = (-4)^3(-3)^2$$

Consider the following examples.

EXAMPLE 3

Express in exponential form.

a. $(6)(6)(-10)(-10)(-10)(-10)$

b. $(4)(4)(-3)(-3)(4)$

Solution

a. $\underbrace{(6)(6)}_{\text{2 factors of 6}}\underbrace{(-10)(-10)(-10)(-10)}_{\text{4 factors of } -10} = 6^2(-10)^4$

b. $(4)(4)(-3)(-3)(4) = (4)(4)(4)(-3)(-3) = 4^3(-3)^2$

PRACTICE 3

Write using exponents.

a. $(2)(2)(2)(2)(-5)(-5)$

b. $(-6)(-6)(8)(8)(-6)(-6)$

EXAMPLE 4

Rewrite each expression using exponents.

a. $-2n \cdot n$ **b.** $-3x \cdot x \cdot y \cdot y \cdot y \cdot y$

Solution

a. $-2n \cdot n = -2n^2$

b. $-3x \cdot x \cdot y \cdot y \cdot y \cdot y = -3x^2y^4$

PRACTICE 4

Express in exponential form.

a. $-x \cdot x \cdot x \cdot x \cdot x$

b. $2m \cdot m \cdot m \cdot n \cdot n \cdot n \cdot n$

Can you explain the difference between the expressions $2n$ and n^2?

EXAMPLE 5

The population of a small town doubles every 5 yr. If the town's population started with n people, what is its population after 20 yr?

Solution We know that the town's population started with n people and doubles every 5 yr. To find the population after 20 yr, consider the following table.

Time Passed (in years)	Number of 5-yr Time Periods	Population of the Town
5	1	$2^1 \cdot n$, or $2n$
10	2	$2^2 \cdot n$, or $4n$
15	3	$2^3 \cdot n$, or $8n$
20	4	$2^4 \cdot n$, or $16n$

Each time period, the population is doubled, that is, multiplied by 2.

So the population after 20 yr is 2^4n or $16n$.

PRACTICE 5

A bacteriologist observes that the population of a bacteria growing in a petri dish triples in size every 2 hr. If x cells were present in the initial population, what was the population after 10 hr? Write the answer using exponents.

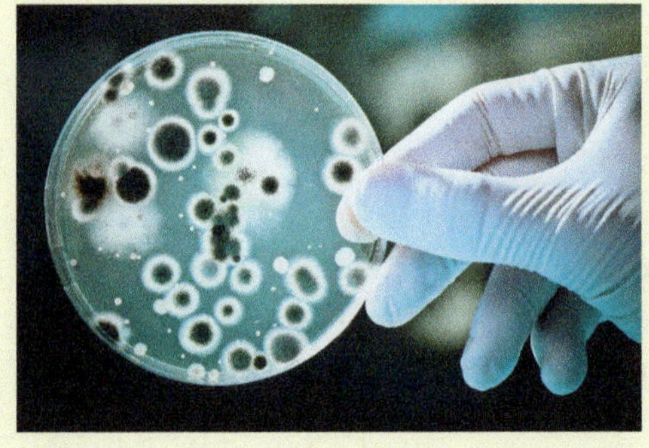

11.4 Exercises

FOR EXTRA HELP MyMathLab     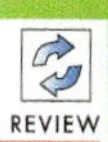

PRACTICE WATCH READ REVIEW

Mathematically Speaking

Fill in each blank with the most appropriate term or phrase from the given list.

base	coefficients	algebraic expression	x to the ath power
exponent	a to the xth power	terms	

1. In the expression $8x^2 + 11$, there are two ________________.

2. A(n) ________________ consists of one or more terms, separated by addition signs.

3. In the expression x^a, x is called the ________________.

4. The expression x^a can be read as ________________.

A *Determine the number of terms in each expression.*

5. $-5x$

6. $-7y$

7. $a + b$

8. $10 + 2y$

9. $xy + \frac{x}{y} - z$

10. $x - y + z$

Evaluate.

11. -3^2

12. $(-4)^2$

13. $(-3)^3(-4)^2$

14. $-4^2(-3)^2$

B *Write using exponents.*

15. $(-2)(-2)(-2)(4)(4)$

16. $(-5)(-5)(-5)(-5)(2)(2)$

17. $(6)(6)(-3)(-3)(-3)$

18. $(-2)(-2)(4)(4)(4)(4)$

19. $(2)(-1)(-1)(2)(2)(2)$

20. $(-10)(3)(3)(3)(3)(-10)$

21. $3(n)(n)(n)$

22. $2(x)(x)(x)(x)$

23. $-4a \cdot a \cdot a \cdot b \cdot b$

24. $5r \cdot s \cdot s \cdot s \cdot s$

25. $-y \cdot y \cdot y$

26. $(-y)(-y)(-y)$

27. $10a \cdot a \cdot a \cdot b \cdot b \cdot c$

28. $-5x \cdot y \cdot y \cdot z$

29. $-x \cdot x \cdot y \cdot y \cdot y$

30. $(-x)(-x)(-x)(-y)(-y)$

Mixed Practice

Write using exponents.

31. $(-4)(-4)(5)(5)(5)$

32. $(-1)(-1)(-1)(-1)(7)(7)$

Write using exponents.

33. $-3p \cdot p \cdot p \cdot q$

34. $a \cdot a(-b)(-b)$

Determine the number of terms in each expression.

35. $a - \frac{b}{2}$

36. $x + \frac{x}{y} - z$

Evaluate.

37. $-3^2(-2)^2$

38. $-1^3(-4)^3$

Applications

C *Solve.*

39. An initial investment of 5000 dollars doubles every 10 yr. Write in exponential form the value of the investment after 30 yr.

40. A colony of bacteria *E. coli* doubles in size every 20 min when grown in a rich medium. If the colony started with m bacteria, how many bacteria were in the colony 2 hr later? Express the answer using exponents. (*Source:* eb.com)

41. The area of a square can be found by squaring the length of a side. Write an expression in exponential form to represent the area of the square shown.

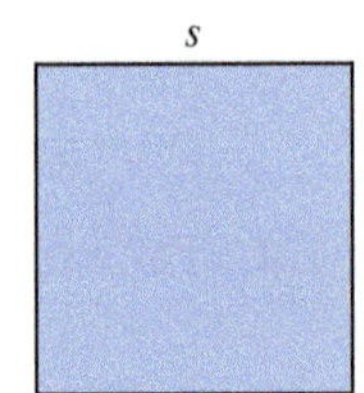

42. The volume of a cube can be found by cubing the length of an edge. Write an expression in exponential form to represent the volume of this cube.

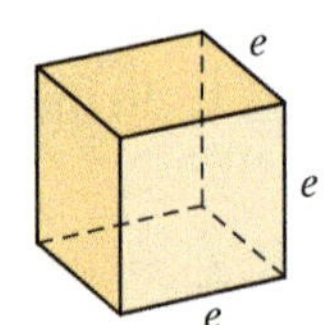

• Check your answers on page A-15.

MINDStretchers

Writing

1. Explain the difference between an *arithmetic expression* and an *algebraic expression.*

Mathematical Reasoning

2. Can there be two different numbers a and b for which $a^b = b^a$? Justify your answer.

Groupwork

3. The *algebra tiles* shown below represent the expressions $3x - 2$ and $x^2 + 1$, respectively.

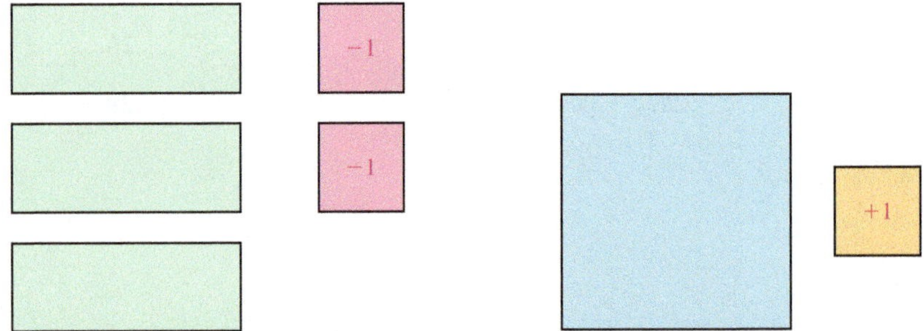

Working with a partner, represent each of the following expressions by algebra tiles.

a. $x - 4$ **b.** $2x + 3$ **c.** $x^2 + x + 1$

11.5 Evaluating Algebraic Expressions and Formulas

OBJECTIVES

- **A** To evaluate algebraic expressions
- **B** To evaluate formulas
- **C** To solve applied problems involving algebraic expressions or formulas

To review our discussion of evaluating algebraic expressions from Section 4.1, lets look at an example.

EXAMPLE 1

Evaluate each algebraic expression.

Solution

Algebraic Expression	Value
a. $9 - z$, if $z = -2$	$9 - z = 9 - (-2) = 9 + 2 = 11$
b. $-2cd$, if $c = -1$ and $d = 2$	$-2cd = -2(-1)(2) = 4$

PRACTICE 1

Find the value of each algebraic expression.

Algebraic Expression	Value
a. $25 + m$, if $m = -10$	
b. $-3xy$, if $x = -2$ and $y = 5$	

In evaluating some algebraic expressions, we need to use the order of operations rule.

EXAMPLE 2

Find the value of each expression for $x = -3$, $y = 2$, $w = -5$, and $z = 4$.

a. $2x + 5y$ **b.** $3(w + z)$

c. $2w^2 - 3y^3$ **d.** $x^2 - 5x + 1$

Solution

a. $2x + 5y = 2(-3) + 5(2)$ Replace x with -3 and y with 2. Then, multiply.
$= -6 + 10$ Add.
$= 4$

b. $3(w + z) = 3(-5 + 4) = 3(-1) = -3$

c. $2w^2 - 3y^3 = 2(-5)^2 - 3(2)^3 = 2(25) - 3(8) = 50 - 24 = 26$

d. $x^2 - 5x + 1 = (-3)^2 - 5(-3) + 1$
$= 9 + 15 + 1$
$= 25$

PRACTICE 2

Evaluate each expression for $a = 2$, $b = -3$, $c = -4$, and $d = 5$.

a. $5a - 2c$

b. $2(d - b)$

c. $2a^3 + 4b^2$

d. $c^2 + 4c - 2$

EXAMPLE 3

Evaluate each expression if $a = 3$, $b = 4$, and $c = -2$.

a. $\frac{c}{5 - a}$ **b.** $\frac{b + c}{b - c}$ **c.** $(-b)^2$ **d.** $-b^2$

Solution

a. $\frac{c}{5 - a} = \frac{-2}{5 - 3} = \frac{-2}{2} = -1$

b. $\frac{b + c}{b - c} = \frac{4 + (-2)}{4 - (-2)} = \frac{2}{6} = \frac{1}{3}$

c. $(-b)^2 = (-4)^2 = (-4)(-4) = 16$

d. $-b^2 = -4^2 = -(4)(4) = -16$

PRACTICE 3

Find the value of each expression when $x = -5$, $y = -3$, and $z = 1$.

a. $\frac{x - 2z}{y}$ **b.** $\frac{x - z}{x + y}$

c. $(-y)^4$ **d.** $-y^4$

Consider the expressions in Examples 3(c) and 3(d). Explain why they are not equal.

EXAMPLE 4

Physicists have shown that if an object is shot straight upward at a speed of 100 ft/sec, its location after t sec will be

$$-16t^2 + 100t$$

feet above the point of release. How far above or below the point of release will the object be after 5 sec?

Solution To determine the position of the object after 5 sec, we must substitute 5 for t in the expression $-16t^2 + 100t$.

$$\begin{aligned} -16t^2 + 100t &= -16(5)^2 + 100(5) \\ &= -16(25) + 100(5) \\ &= -400 + 500 \\ &= 100 \end{aligned}$$

So after 5 sec, the object will be 100 ft above the point of release. Note that a position above the point of release is positive, whereas a position below the point of release is negative.

PRACTICE 4

The U.S. per capita consumption of bottled water (in gallons) can be approximated by the expression $2.0x + 15.7$, where x represents the number of years since 2003. Estimate the per capita consumption in 2015. (*Sources: Statistical Abstract of the United States, 2010* and census.gov)

Formulas

A **formula** is an equation that indicates how variables are related to one another. Some formulas express geometric relationships; others express physical laws. Just as with algebraic expressions, the letters and mathematical symbols in a formula represent numbers and words.

EXAMPLE 5

To predict the temperature T at a particular altitude a, meteorologists subtract $\frac{1}{200}$ of the altitude from the temperature g on the ground. Here, T and g are in degrees Fahrenheit and a is in feet. Translate this rule to a formula.

Solution Stating the rule briefly in words, the temperature at a particular altitude equals the difference between the temperature on the ground and $\frac{1}{200}$ times the altitude. Now, we translate this rule to mathematical symbols.

$$T = g - \frac{1}{200}a,$$

which is the desired formula.

PRACTICE 5

To convert a temperature C expressed in Celsius degrees to the temperature F expressed in Fahrenheit degrees, we multiply the Celsius temperature by $\frac{9}{5}$ and then add 32. Write a formula that expresses this relationship.

The method of evaluating formulas is similar to that of evaluating algebraic expressions. We substitute all the given numbers for the variables and then carry out the computations using the order of operations rule.

EXAMPLE 6

The formula for finding simple interest is $I = Prt$, where I is the interest in dollars, P is the principal (the amount invested) in dollars, r is the annual rate of interest, and t is the time in years that the principal has been on deposit. Find the amount of interest on a principal of \$3000 that has been on deposit for 2 years at a 6% annual rate of interest.

Solution We know that $P = 3000$, $r = 6\%$, and $t = 2$. Converting 6% to its decimal form, we get 0.06. Substituting into the formula gives us:

$$\begin{aligned} I &= Prt \\ &= 3000(0.06)(2) \\ &= 360 \end{aligned}$$

So the interest earned is \$360.

PRACTICE 6

Given the distance formula $d = rt$, find the value of d if the rate r is 50 mph and the time t is 1.6 hr.

EXAMPLE 7

The markup M on an item is its selling price S minus its cost C.

a. Express this relationship as a formula.

b. If a digital camera cost a retailer \$399.95 and was then sold for \$559, how much was the markup on the camera?

Solution

a. We write the formula $M = S - C$.

b. To find the markup, we substitute for S and C.

$$\begin{aligned} M &= S - C \\ &= 559 - 399.95 \\ &= 159.05 \end{aligned}$$

So the markup was \$159.05.

PRACTICE 7

Kelvin and Celsius temperature scales are commonly used in science. To convert a temperature expressed in Celsius degrees C to degrees Kelvin, K, add 273 to the Celsius temperature.

a. Write this relationship as a formula.

b. Suppose that in a chemistry experiment, C equals -6. What is the value of K?

11.5 Exercises

MyMathLab    

A *Evaluate each algebraic expression, if $a = 4$, $b = 3$, and $c = -2$.*

1. $-5 + b$
2. $-6 + a$
3. $-2ac$
4. $-4cb$
5. $-2a^2$
6. $-4b^2$
7. $2a - 15$
8. $3a - 18$
9. $a + 2c$
10. $4b + 3a$
11. $2(a - c)$
12. $4(a - b)$
13. $-a + b^2$
14. $-b + c^2$
15. $3a^2 - c^3$
16. $c^3 - 2a^2$
17. $\dfrac{a + b}{b - a}$
18. $\dfrac{a - c}{c + a}$
19. $\dfrac{3}{5}(a + b + c)^2$
20. $\dfrac{5}{9}(a - b - c)^2$

Evaluate each algebraic expression if $w = -0.5$, $x = 2$, $y = -3$, and $z = 1.5$.

21. $2w^2 - 3x + y - 4z$
22. $2x - 3w - y^2 + z$
23. $w - 7z - \dfrac{1}{4}(x - 6y)$
24. $5x - \dfrac{2}{5}(8w + 2y) - 4z$
25. $\dfrac{-10xy}{(w - z)^2}$
26. $\dfrac{-(2z - y)^2}{9wx}$

Complete each table.

27.

x	0	1	2	−1	−2
$2x + 5$					

28.

x	0	1	2	−1	−2
$3x + 1$					

29.

y	0	1	2	3	4
$y - 0.5$					

30.

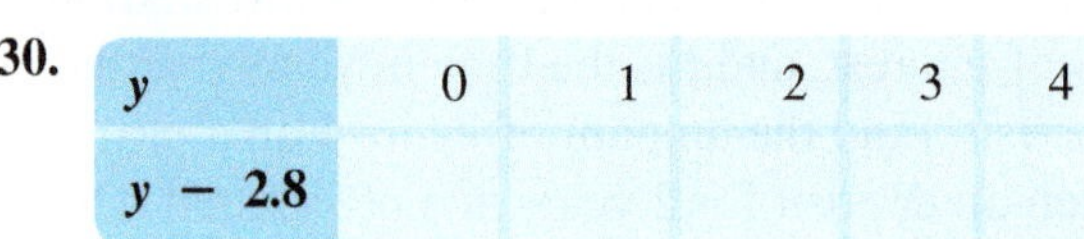

y	0	1	2	3	4
$y - 2.8$					

31.

x	0	2	4	−2	−4
$-\frac{1}{2}x$					

32.

x	0	5	10	−5	−10
$-\frac{3}{5}x$					

33.

n	2	4	6	−2	−4
$\frac{n}{2}$					

34.

n	5	10	−5	−10	−15
$\frac{n}{5}$					

35.

g	0	1	2	−1	−2
$-g^2$					

36.

g	0	1	2	−1	−2
g^2					

37.

a	0	1	2	−1	−2
$a^2 + 2a - 2$					

38.

a	0	1	2	−1	−2
$-a^2 - 2a + 2$					

B *Evaluate each formula for the quantity requested.*

Formula	Given	Find
39. $C = \frac{5}{9}(F - 32)$	$F = -4°$	C
40. $C = \frac{5}{9}(F - 32)$	$F = -8°$	C
41. $A = P(1 + rt)$	$P = \$2000$, $r = 5\%$, and $t = 2$ yr	A
42. $A = P(1 + rt)$	$P = \$3000$, $r = 6\%$, and $t = 3$ yr	A
43. $P = 2l + 2w$	$l = 2\frac{1}{2}$ ft and $w = 1\frac{1}{4}$ ft	P
44. $P = 2l + 2w$	$l = 3\frac{1}{2}$ ft and $w = 2\frac{3}{4}$ ft	P
45. $A = \frac{a + b + c + d}{4}$	$a = -8$, $b = -6$, $c = 4$ and $d = -2$	A
46. $A = \frac{a + b + c + d}{4}$	$a = -10$, $b = -8$, $c = 6$, and $d = -4$	A
47. $C = \pi d$	$\pi \approx 3.14$ and $d = 100$ m	C
48. $C = \pi d$	$\pi \approx 3.14$ and $d = 120$ m	C
49. $C = A \cdot \frac{W}{150 \text{ lb}}$	$A = 100$ mg and $W = 30$ lb	C
50. $C = A \cdot \frac{W}{150 \text{ lb}}$	$A = 120$ mg and $W = 40$ lb	C
51. $A = 6e^2$	$e = 1.5$ cm	A
52. $A = 6e^2$	$e = 2.5$ cm	A
53. $V = \pi r^2 h$	$\pi = 3.14$, $r = 10$ in., and $h = 10$ in.	V
54. $V = \pi r^2 h$	$\pi = 3.14$, $r = 2$ in., and $h = 25$ in.	V

Mixed Practice

Solve.

55. Evaluate the algebraic expression if $w = 3$, $x = -1.5$, $y = -2$, and $z = 0.5$.

$$y^2 - 2z + \frac{1}{3}(2x - w)$$

Complete each table.

56.

x	0	8	12	-8	-12
$-\frac{3}{4}x$					

57.

x	0	1	2	-1	-2
$-2x + 4$					

Evaluate each algebraic expression if $a = 3, b = -4,$ and $c = 2$.

58. $-5(c + b)$

59. $b^2 - 3a^2$

60. $\frac{c + a}{b - a}$

Evaluate each formula for the quantity requested.

Formula	Given	Find
61. $A = \frac{1}{2}bh$	$b = 7$ in. and $h = 3$ in.	A
62. $F = \frac{9}{5}C + 32$	$C = -10°$	F

Applications

C *Write each relationship as a formula.*

63. The average A of three numbers a, b, and c is the sum of the numbers divided by 3.

64. The perimeter P of a rectangle is twice the sum of the length l and the width w.

65. For every right triangle, the sum of the squares of the two legs a and b equals the square of the hypotenuse c.

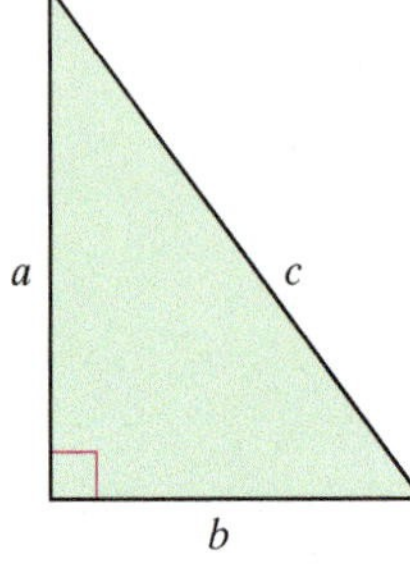

66. The *aspect ratio* of a hang glider is a measure of how well it can glide and soar. The aspect ratio R is the square of the glider's wingspan s divided by the wing area A.

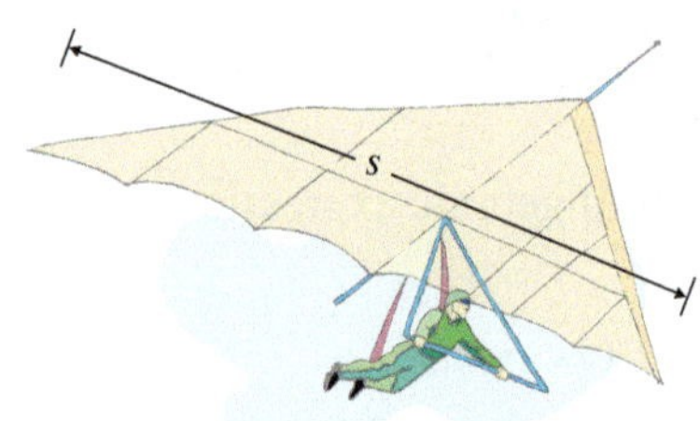

67. The equivalent energy E of a mass equals the product of the mass m and the square of the speed of light c.

68. The weight of an object depends on the gravitational pull of the planet or moon it is on. For instance, the weight E of an astronaut on Earth is 6 times the astronaut's weight m on the moon.

69. The length l of a certain spring in centimeters is 25 more than 0.4 times the weight w (in grams) of the object hanging from it.

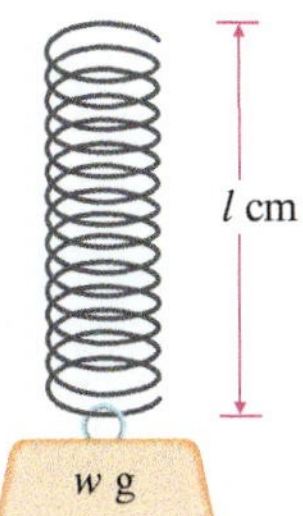

70. In electronics, when two resistors R_1 and R_2 are connected in parallel, the total resistance R between the points X and Y can be founded by dividing the product of the resistances by the sum of the resistances.

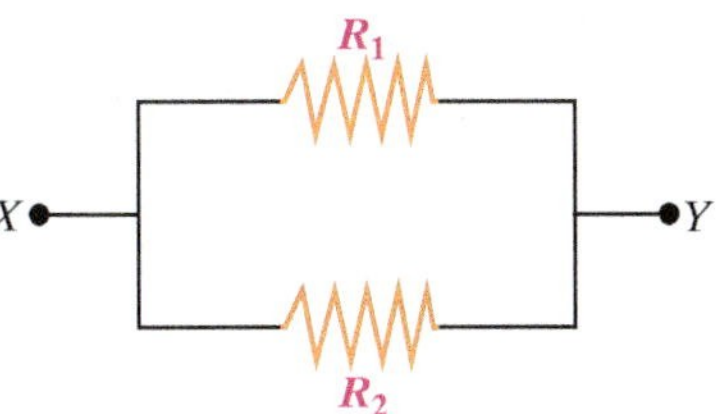

Solve.

71. The distance that a free-falling object drops, ignoring friction, is given by the formula

$$S = \frac{1}{2}gt^2,$$

where S is the distance (in feet), g is the acceleration due to gravity, and t is time (in seconds). Find the distance that an object falls if $g = 32 \text{ ft/sec}^2$ and $t = 2$ sec.

72. The volume of a right circular cylinder is given by the formula

$$V = \pi r^2 h,$$

where r is the radius of the base of the cylinder and h is the height of the cylinder. For the cylinder shown, find the volume in cubic centimeters. Use 3.14 for π.

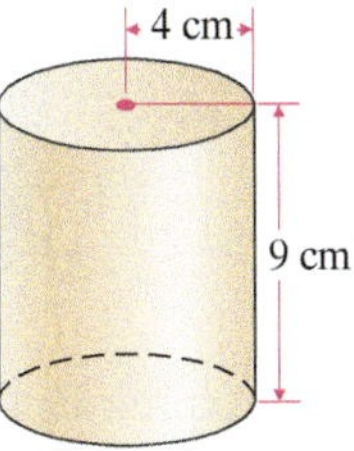

73. The percent markup m of an item based on cost is equal to 100 times the difference of the selling price s of the item and its cost c, divided by the cost.

a. Express this relationship as a formula.

b. If the selling price of a bottle of vitamin E is $8.75 and the cost is $6.25, what is the percent markup on the vitamin E?

74. To calculate the speed f of an object (in feet per second), we multiply its speed m (in miles per hour) by $\frac{22}{15}$.

a. Write this relationship as a formula.

b. If an object is moving at 60 mph, what is its speed in feet per second?

75. The total calories C in a piece of food is the sum of its protein, carbohydrate, and fat calories. A particular piece of food has p grams of protein, f grams of fat, and c grams of carbohydrates. A gram of fat contains 9 calories, and there are 4 calories per gram of carbohydrate or protein. (*Source:* Duyff, *American Dietetic Association Complete Food and Nutrition Guide*)

a. Express C in terms of p, f, and c.

b. How many calories are there in 5 g of fat, 0 g of carbohydrates, and 2 g of protein?

76. A simple approximation of the amount A of water in a human body is 60% of the total weight w.

a. Write this relationship as a formula.

b. If a person weighs 145 lb, approximately how much of the person's weight is water?

• Check your answers on page A-15.

MINDStretchers

Mathematical Reasoning

1. Complete the first three rows of the following table. Make up your own values of a and b in the 4th and 5th rows and complete the table.

a	b	$a + b$	$\lvert a\rvert$	$\lvert b\rvert$	$\lvert a\rvert + \lvert b\rvert$	$\lvert a + b\rvert$	T or F, $\lvert a\rvert + \lvert b\rvert = \lvert a + b\rvert$	T or F, $\lvert a\rvert + \lvert b\rvert \geq \lvert a + b\rvert$
−3	4							
1	−2							
−5	−4							

What general relationship do these examples suggest? Explain why this relationship is true.

Patterns

2. Consider the following table.

Odd number n	1	3	5	7	...	1999
Counting number C	1	2	3	4	...	1000

Write a formula that expresses C in terms of n.

Groupwork

3. Some expressions that appear to be different are, in fact, equivalent. Confirm that the expressions $\dfrac{x^2 - 9}{x + 3}$ and $x - 3$ are equal for various values of x.

11.6 Simplifying Algebraic Expressions

OBJECTIVES

- **A** To identify coefficients, like terms, and unlike terms
- **B** To use the distributive property
- **C** To combine like terms
- **D** To simplify algebraic expressions involving parentheses or brackets
- **E** To solve applied problems involving algebraic expressions

In an algebraic expression, the numerical factor of each variable term is said to be its **numerical coefficient** or, simply, its **coefficient**. For instance, the term $3y$ has coefficient 3 and the term $-4y$ has coefficient -4. Note that the coefficient of y is 1 because $y = 1y$. Likewise, the coefficient of $-y$ is -1 because $-y = -1y$.

The terms $3y$, $-4y$, and y are said to be *like* terms because they all contain the variable y raised to the power 1. By contrast, the terms $3x$ and $3x^2$ are *unlike* terms because the variable x is raised to different powers. Similarly, the terms $3x^2$ and $5y^2$ are unlike because they involve different variables.

DEFINITION

Like terms are terms that have the same variables with the same exponents. Terms that are not like are called **unlike terms.**

EXAMPLE 1

For each algebraic expression, identify the terms and name the coefficient of the terms. Then, indicate whether the terms are like or unlike.

a. $3a + 5b$ **b.** $n - 6n$

c. $4x^2 + 2x$ **d.** $2a^2b - 3a^2b$

Solution

a. $3a + 5b$ has terms $3a$ and $5b$; unlike

(coefficient; different variables)

b. $n - 6n$ has terms n and $-6n$; like

(coefficient; same variable)

c. $4x^2 + 2x$ has terms $4x^2$ and $2x$; unlike

(coefficient; same variable but different exponents)

d. $2a^2b - 3a^2b$ has terms $2a^2b$ and $-3a^2b$; like

(coefficient; same variables with same exponents)

PRACTICE 1

Identify and name the coefficients of the terms of each algebraic expression. Then, state whether the terms are like or unlike.

a. $m - 3m$

b. $5x + 7$

c. $2x^2y - 3xy^2$

d. $m + 2m - 4m$

Combining Like Terms

An important property used to simplify algebraic expressions is the *distributive property*.

The Distributive Property

For any real numbers a, b, and c,

$$a \cdot (b + c) = a \cdot b + a \cdot c.$$

In words, this property states that a number times the sum of two quantities is equal to the number times one quantity plus the number times the other quantity.

Another way to express the distributive property is $(b + c)a = ba + ca$. Can you explain why?

The distributive property also holds for $a(b - c)$. Since $b - c = b + (-c)$, it follows that $a(b - c) = ab - ac$ and $(b - c)a = ba - ca$.

EXAMPLE 2

Use the distributive property.

a. $2(x + y)$ **b.** $(3 + 7) \cdot n$

c. $(-5)(a - 2b)$ **d.** $0.6(x - 5)$

Solution

a. $2(x + y) = 2 \cdot x + 2 \cdot y = 2x + 2y$

b. $(3 + 7) \cdot n = 3 \cdot n + 7 \cdot n = 3n + 7n$

c. $(-5)(a - 2b) = (-5) \cdot a + (-5) \cdot (-2b) = -5a + 10b$

d. $0.6(x - 5) = 0.6(x) - 0.6(5) = 0.6x - 3$

PRACTICE 2

Rewrite using the distributive property.

a. $(-10)(4r + s)$

b. $(5 + 1) \cdot w$

c. $3(g - 3h)$

d. $1.5(y + 2)$

We can use the distributive property to simplify algebraic expressions involving like terms. Adding or subtracting like terms using the distributive property is called *combining like terms*. Unlike terms, such as $4x^2$ and $2x$, cannot be combined.

EXAMPLE 3

Simplify.

a. $2x + 6x$ **b.** $a - 8a$

c. $6y - y + 5$ **d.** $3b + 2b - 5b$

Solution

a. $2x + 6x = (2 + 6)x$ Use the distributive property.
$= 8x$ Add 2 and 6.

b. $a - 8a = (1 - 8)a$ Recall that the coefficient of a is 1.
$= -7a$ Use the distributive property.

c. $6y - y + 5 = (6 - 1)y + 5$ Recall that the coefficient of $-y$ is -1. Use the distributive property.
$= 5y + 5$

d. $3b + 2b - 5b = (3 + 2 - 5)b$ Use the distributive property.
$= 0 \cdot b$ Recall that $0 \cdot b = 0$.
$= 0$

PRACTICE 3

Simplify.

a. $5x + x$

b. $-5y - y$

c. $a - 3a + b$

d. $-9t + 3t + 6t$

EXAMPLE 4

Combine like terms, if possible.

a. $2x^2 + x$ **b.** $8m^2n^2 + 2m^2n^2$ **c.** $-6a^2b + 5a^2b$

Solution

a. $2x^2 + x$ This algebraic expression cannot be simplified because the terms are unlike.

Unlike terms

b. $8m^2n^2 + 2m^2n^2 = (8 + 2)m^2n^2 = 10m^2n^2$

Like terms

c. $-6a^2b + 5a^2b = (-6 + 5)a^2b = -1a^2b = -a^2b$

Like terms

PRACTICE 4

Combine like terms, if possible.

a. $y^2 - 3y^2$

b. $7a^2b + 3ab^2$

c. $4xy^2 - xy^2$

Simplifying Algebraic Expressions Involving Parentheses

Some algebraic expressions involve parentheses. We can use the distributive property to remove the parentheses in order to simplify these expressions.

EXAMPLE 5

Simplify: $\frac{1}{2}(x + 6) - 4$

Solution

$$\frac{1}{2}(x + 6) - 4 = \frac{1}{2}x + 3 - 4$$ Use the distributive property.

$$= \frac{1}{2}x - 1$$ Combine like terms.

PRACTICE 5

Simplify: $3(y - 4) + 2$

Let's consider simplifying algebraic expressions such as $-(4y - 6)$, in which a negative sign precedes an expression in parentheses. Just as $-y = -1$, so does $-(4y - 6) = -1(4y - 6)$.

EXAMPLE 6

Simplify: $-(4y - 6)$

Solution

$$-(4y - 6) = -1(4y - 6)$$

$$= -1 \cdot 4y + (-1)(-6)$$ Use the distributive property.

$$= -4y + 6$$

Because the terms in the expression $-4y + 6$ are unlike, it is not possible to simplify the expression further.

PRACTICE 6

Simplify: $-(2a - 3b)$

EXAMPLE 7

Simplify: $2x - 3 - (x + 4)$

Solution

$$\begin{aligned} 2x - 3 - (x + 4) &= 2x - 3 - 1(x + 4) \\ &= 2x - 3 - x - 4 && \text{Use the distributive property.} \\ &= 2x - x - 3 - 4 \\ &= x - 7 && \text{Combine like terms.} \end{aligned}$$

PRACTICE 7

Simplify: $5y - 6 - (y - 5)$

TIP

- When removing parentheses preceded by a *minus sign*, all the terms in parentheses change to the opposite sign.
- When removing parentheses preceded by a *plus sign*, all the terms in parentheses keep the same sign.

EXAMPLE 8

Simplify: $(14a - 9) - 2(3a + 4)$

Solution

$$\begin{aligned} (14a - 9) - 2(3a + 4) &= 14a - 9 - 6a - 8 \\ &= 8a - 17 \end{aligned}$$

PRACTICE 8

Simplify: $(y + 3) - 3(y + 7)$

Some algebraic expressions contain not only parentheses but also brackets. When simplifying expressions containing both types of grouping symbols, first remove the innermost grouping symbols by using the distributive property and then continue to work outward.

EXAMPLE 9

Simplify: $5 + 2[-4(x - 3) + 8x]$

Solution

$$\begin{aligned} 5 + 2[-4(x - 3) + 8x] &= 5 + 2[-4x + 12 + 8x] && \text{Use the distributive property inside the brackets.} \\ &= 5 + 2[4x + 12] && \text{Combine like terms inside the brackets.} \\ &= 5 + 8x + 24 && \text{Use the distributive property.} \\ &= 8x + 29 && \text{Combine like terms.} \end{aligned}$$

PRACTICE 9

Simplify: $10 - [4y + 3(2y - 1)]$

Now, we consider applied problems that involve simplifying expressions. To solve these problems, first we translate word phrases to algebraic expressions, and then simplify.

EXAMPLE 10

When a hospital is filled to capacity, it has p patients in private rooms and 20 more patients in semiprivate rooms than in private rooms. The daily rate for a patient in a private room is \$300, and in a semiprivate room, the daily rate is \$200. Write an algebraic expression to represent the total amount of money that the hospital takes in per day when all of the rooms are full. Then, simplify the expression.

Solution We know that p represents the number of patients in private rooms. Since the hospital has 20 more patients in semiprivate rooms than in private rooms, $p + 20$ represents the number of patients in semiprivate rooms. A patient in a private room pays \$300 per day, and a patient in a semiprivate room pays \$200 per day.

Organizing the information into a table can help us to clarify the relationship between the key quantities in the problem.

Type of Room	Cost per Patient	Number of Patients in Rooms	Total Amount of Money
Private	\$300	p	$300p$
Semiprivate	\$200	$p + 20$	$200(p + 20)$

So when the hospital is full, the total amount of money that the hospital takes in per day is represented by the algebraic expression $300p + 200(p + 20)$.

Simplifying $300p + 200(p + 20)$, we get:

$$\begin{aligned} 300p + 200(p + 20) &= 300p + 200p + 4000 \\ &= 500p + 4000 \end{aligned}$$

So the total amount of money that the hospital receives per day is $(500p + 4000)$ dollars.

PRACTICE 10

For a concert, a local performing arts center sold c tickets for children and 40 fewer tickets for adults. Write an algebraic expression to represent the total income received by the center if the cost of a ticket is \$12 for adults and \$5 for children. Then, simplify the expression.

11.6 Exercises

FOR EXTRA HELP MyMathLab MathXL PRACTICE WATCH READ REVIEW

Mathematically Speaking

Fill in each blank with the most appropriate term or phrase from the given list.

coefficient	unlike	negative
outermost	number	like
positive	innermost	
associative property of addition	distributive property	

1. The ________________ of the term $-x$ is -1.

2. Terms that have the same variables with the same exponents are called ________________ terms.

3. The ________________ states that for any real numbers a, b, and c, $a \cdot (b + c) = a \cdot b + a \cdot c$.

4. We cannot combine ________________ terms.

5. When removing parentheses preceded by a(n) ________________ sign, change all the terms in parentheses to the opposite sign.

6. To simplify expressions containing brackets and parentheses, begin by removing the ________________ grouping symbols.

A ***Name the coefficient in each term.***

7. $7x$ **8.** $100a$ **9.** $-5y$ **10.** $-18t$

11. ab **12.** m **13.** $-x^2$ **14.** $-xy^3$

15. $-0.1n$ **16.** $2.5s$ **17.** $\frac{2}{3}a^2b$ **18.** $-\frac{1}{4}mn$

19. $2x - 5y$ **20.** $-x + 10y$

Identify the terms, and indicate whether they are like or unlike.

21. $2a - a$ **22.** $10r + r$

23. $5p + 3$ **24.** $30 - 2A$

25. $4x^2 - 6x^2$ **26.** $x^2 + 7x^3$

27. $-20n - 3n$ **28.** $-15n - 8n$

B ***Use the distributive property.***

29. $-7(x - y)$ **30.** $-3(x - y)$ **31.** $(1 - 10) \cdot a$

32. $(3 - 12) \cdot x$ **33.** $-0.5(r + 3)$ **34.** $-0.4(p + 5)$

C ***Simplify, if possible, by combining like terms.***

35. $3x + 7x$ **36.** $8p + 3p$ **37.** $-10n - n$

38. $-y - 5y$ **39.** $20a - 10a + 4a$ **40.** $2r - 5r + r$

41. $3y - y + 2$ **42.** $7y - 2y + 1$ **43.** $8b^3 + b^3 - 9b^3$

44. $6x^2 + 2x^2 - 14x^2$ **45.** $-b^2 + ab^2$ **46.** $-3x^2 - 5x$

47. $3r^2t^2 + r^2t^2$

48. $m^2n^2 + 2m^2n^2$

49. $3x^2y - 5xy^2$

50. $10ab^2 + a^2b$

D *Simplify.*

51. $2(x + 3) - 4$

52. $3(a - 5) - 1$

53. $(7x + 1) + 2(2x - 1)$

54. $(3x - 4) + 2(2x - 18)$

55. $-(3y - 10)$

56. $-(2x + 5)$

57. $5x - 3 - (x + 6)$

58. $5x - 9 - (x + 4)$

59. $-4(n - 9) + 3(n + 1)$

60. $5(2b - 1) + 4(c + b)$

61. $x - 4 - 2(x - 1) + 3(2x + 1)$

62. $a + 1 - 3(a + 1) + 8(4a + 5)$

63. $7 + 3[x - 2(x - 1)]$

64. $2 - [n - 4(n + 5)]$

65. $10 - 3[4(a + 2) - 3a]$

66. $5 + 2[-(z + 7) + 4z]$

Mixed Practice

Simplify, if possible, by combining like terms.

67. $4pq^2 - 6q^2$

68. $12m + 2 - m$

Solve.

69. Identify the terms in the expression $3a + 3a^2$ and indicate whether they are like or unlike.

70. Use the distributive property: $-3(x - 2y)$

71. Name the coefficient in the term xy.

Simplify.

72. $2n - 7 - (3n + 2)$

73. $y - 3 - 4(y - 2) + 2(3y + 1)$

74. $(5a - 6) + 3(2a + 1)$

Applications

E ***Write an algebraic expression. Then, simplify.***

75. What is the sum of the angles in the triangle shown?

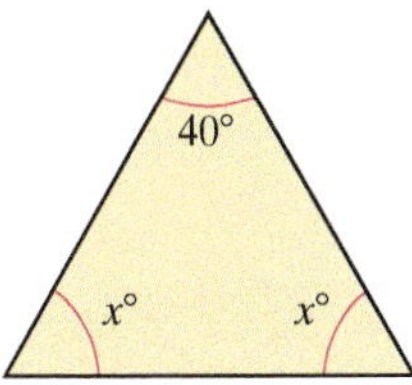

76. According to a will, an estate is to be divided among 2 children and 2 grandchildren. Each grandchild is to receive one-half the amount c that each child receives from the estate. What is the total value of the estate?

77. A baseball fan buys a ticket for a game for d dollars. Two of his friends decide to go to the game at the last minute and purchase tickets for \$4 more per ticket. What is the total cost of the 3 tickets?

78. The shape of Colorado is approximately a rectangle. Its length is 100 mi more than its width, w. What is the approximate perimeter of Colorado?

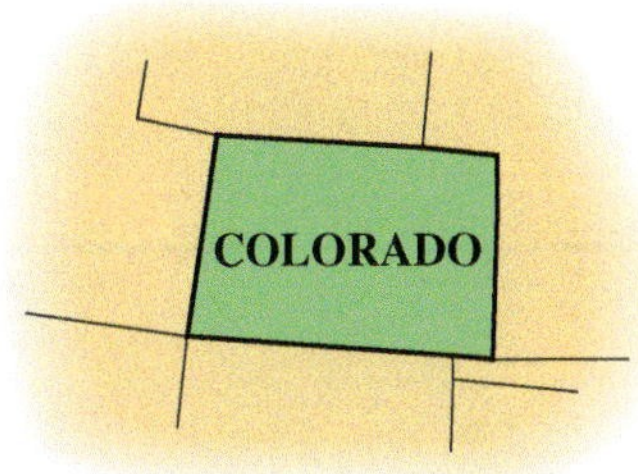

79. If n is the first of 3 consecutive integers, what is the sum of the 3 consecutive integers? (*Hint*: Consecutive integers are integers that differ by 1 unit.)

80. On Tuesday, a company's stock fell by 6% from its value the day before. If v represents the value of the stock on Monday, what was its value on Tuesday?

81. There are 54 questions on the Mathematics section of the Scholastic Aptitude Test (SAT). The raw score for this portion of the test is found as follows: 1 point is awarded to a correct answer, 0 points are given if the question is left unanswered, and $\frac{1}{4}$ of a point is deducted for an incorrect answer. What is the Mathematics raw score of a student who answer c questions correctly and leaves b questions blank? (*Source:* sat.collegeboard.com)

82. A condo building contains 100 apartments. Of these, y are 4-room apartments, and the others are 3-room apartments. How many rooms are there in the condo building?

• Check your answers on page A-15.

MIND*Stretchers*

Groupwork

1. Parentheses have different meanings in different situations. Working with a partner, explain the meaning of parentheses in each context.

   a. $(-2)(-3)$

   b. $5(x + 2)$

   c.

Income	$713,014
Expenditures	$961,882
Profit	($248,868)

   d. I am studying algebra (my favorite subject!).

Research

2. By searching the web or checking in your college's math learning center, investigate how *algebra tiles* can be used to combine like terms. Summarize your findings.

Writing

3. Explain the meaning of *combining like terms* in the following examples:

   a. 5 ft 3 in. +7 ft 6 in.

   b. $16\frac{2}{3} - 5\frac{1}{3}$

   c. $10x - 1 + 4 + 2x$

Key Concepts and Skills

CONCEPT SKILL

Concept/Skill	Description	Example
[11.1] **Real numbers**	Numbers that can be represented as points on the number line.	$\frac{2}{3}, -\frac{1}{2}, 5, 7.4$, and $\sqrt{2}$
[11.1] **Integers**	The numbers ..., $-4, -3, -2, -1, 0, +1, +2, +3, +4, \ldots$ continuing indefinitely in both directions.	$-20, -7, 0, +5$, and $+17$
[11.1] **Rational numbers**	Numbers that can be written in the form $\frac{a}{b}$, where a and b are integers and $b \neq 0$.	$-1.6, \frac{2}{3}, -\frac{1}{2}, 5$, and 0.04
[11.2] **The additive inverse property**	For any real number a, there is exactly one real number $-a$, such that $a + (-a) = 0$ and $-a + a = 0$, where a and $-a$ are **additive inverses (opposites)** of each other.	$7 + (-7) = 0$ and $-7 + 7 = 0$
[11.2] **The additive identity property**	For any real number a, $a + 0 = a$ and $0 + a = a$.	$7 + 0 = 7$ and $0 + 7 = 7$
[11.2] **To add real numbers**	• If the numbers have the same sign, add the absolute values and keep the sign. • If the numbers have different signs, find the difference between the larger absolute value and the smaller absolute value, and take the sign of the number with the larger absolute value.	$-5 + (-2) = -7$ $+5 + (+2) = +7$, or 7 $-5 + (+2) = -3$ $+5 + (-2) = +3$, or 3
[11.2] **The commutative property of addition**	For any two real numbers a and b, $a + b = b + a$.	$6 + 2 = 2 + 6$
[11.2] **The associative property of addition**	For any three real numbers, a, b, and c, $(a + b) + c = a + (b + c)$.	$(8 + 4) + 1 = 8 + (4 + 1)$
[11.2] **To subtract real numbers**	• Change the operation of subtraction to addition and change the number being subtracted to its opposite. • Follow the rule for adding real numbers.	$2 - (-5)$ $= 2 + (+5) = 7$
[11.3] **To multiply real numbers**	• Multiply their absolute values. • If the numbers have the same sign, their product is positive; if they have different signs, their product is negative.	$(-3)(-8) = 24$ $3(-8) = -24$
[11.3] **The commutative property of multiplication**	For any two real numbers a and b, $a \cdot b = b \cdot a$.	$(-2)(-7) = (-7)(-2)$
[11.3] **The multiplicative identity property**	For any real number a, $a \cdot 1 = a$ and $1 \cdot a = a$.	$3 \cdot 1 = 3$ and $1 \cdot 3 = 3$
[11.3] **The multiplication property of zero**	For any real number a, $a \cdot 0 = 0$ and $0 \cdot a = 0$.	$2 \cdot 0 = 0$ and $0 \cdot 2 = 0$
[11.3] **The associative property of multiplication**	For any three real numbers, a, b, and c, $(a \cdot b)c = a(b \cdot c)$.	$(-1 \cdot 2)3 = -1(2 \cdot 3)$

continued

CONCEPT SKILL

Concept/Skill	Description	Example
[11.3] **The order of operations rule**	To evaluate mathematical expressions, carry out the operations *in the following order*: **1.** First, perform the operations within any grouping symbols, such as parentheses ( ) or brackets [ ]. **2.** Then, raise any number to its power. **3.** Next, perform all multiplications and divisions as they appear from left to right. **4.** Finally, do all additions and subtractions as they appear from left to right.	$2 + 3 \cdot 5 = 17$ $-3(8 - 4)^2 = -48$
[11.3] **To divide real numbers**	• Divide their absolute values. • If the numbers have the same sign, their quotient is positive; if they have different signs, their quotient is negative.	$-16 \div (-2) = 8$ $-24 \div 3 = -8$
[11.3] **Division involving zero**	For any nonzero real number a, $$0 \div a = 0.$$ For any nonzero real number a, $$a \div 0 \text{ is undefined.}$$	$0 \div 5 = 0$ and $5 \div 0$ is undefined.
[11.3] **Division of real numbers**	For any real numbers a and b where b is nonzero, $$a \div b = \frac{a}{b} = a \cdot \frac{1}{b}.$$	$-\frac{3}{5} \div 6 = -\frac{3}{5} \div \frac{6}{1}$ $= -\frac{\overset{1}{\cancel{3}}}{5} \cdot \frac{1}{\underset{2}{\cancel{6}}} = -\frac{1}{10}$
[11.3] **The multiplicative inverse property**	For any nonzero real number a, $$a \cdot \frac{1}{a} = 1 \text{ and } \frac{1}{a} \cdot a = 1,$$ where a and $\frac{1}{a}$ are **multiplicative inverses (reciprocals)** of each other.	$-\frac{3}{4}$ and $-\frac{4}{3}$ are multiplicative inverses because $-\frac{3}{4} \cdot \left(-\frac{4}{3}\right) = 1.$
[11.4] **Algebraic expression**	An expression in which constants and variables are combined using standard arithmetic operations.	$\frac{2}{5}n + 9$
[11.4] **Term**	A number, a variable, or the product or quotient of numbers and variables.	$-11, 3x, \frac{n}{4}$
[11.4] **Exponential notation**	For any real number x and any positive integer a, $$x^a = \underbrace{x \cdot x \cdots x \cdot x}_{a \text{ factors}},$$ where x is called the **base** and a is called the **exponent** (or **power**).	Exponent ↓ $(-2)^3 = (-2)(-2)(-2)$ ↑ $= -8$ Base
[11.6] **Coefficient**	The numerical factor of a variable term.	Coefficient ↙ ↓ ↘ $3x$, y or $1y$, $-2x^2$
[11.6] **Like terms**	Terms that have the same variable and the same exponent.	n and $6n$ $5x^3$ and $-2x^3$
[11.6] **Unlike terms**	Terms that are not like.	x and $3y$ a^2 and a

Concept/Skill	Description	Example
[11.6] **The distributive property**	For any real numbers a, b, and c, $$a \cdot (b + c) = a \cdot b + a \cdot c$$ and $$(b + c) \cdot a = b \cdot a + c \cdot a.$$	$2(x + y) = 2x + 2y$ and $(3 + 7) \cdot n = 3 \cdot n + 7 \cdot n$
[11.6] **To combine like terms**	• Use the distributive property. • Add or subtract.	$2x + 6x = (2 + 6)x$ $= 8x$

CHAPTER 11 Review Exercises

Say Why

Fill in each blank.

1. The decimal 0.5 __________ (is/is not) a rational number because ____ ____.

2. The additive inverse of 17 __________ (is/is not) -17 because ____.

3. The expression $-\frac{a}{3}$ and $\frac{3}{a}$ __________ (are/are not) multiplicative inverses because ________________________________.

4. The expression $-(6a)^2$ __________ (is/is not) equivalent to $(-6a)(-6a)$ because __________________________ ____.

5. The number 125 __________ (is/is not) a power of 5 because ____.

6. The expression $15p^2q^3$ and $-9p^3q^2$ __________ (are/are not) like terms because ________________________________ ____.

[11.1] ***Express each quantity as a signed number.***

7. 3 mi above sea level

8. A withdrawal of \$160 from an account

Graph each number on the number line.

9. -1

−4 −3 −2 −1 0 1 2 3 4

10. $2\frac{9}{10}$

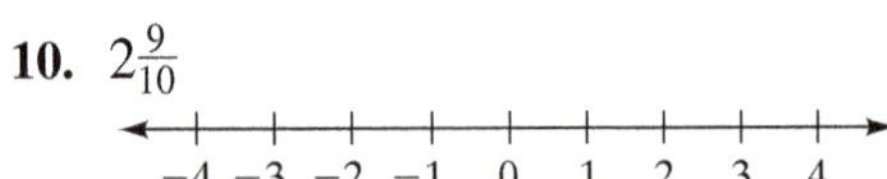

11. 0.5

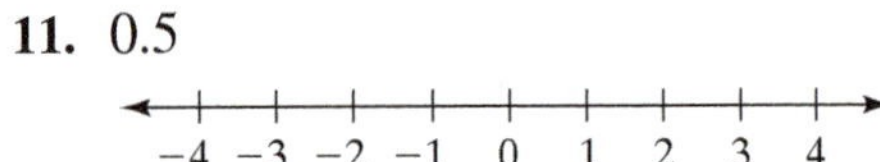

12. -3.75

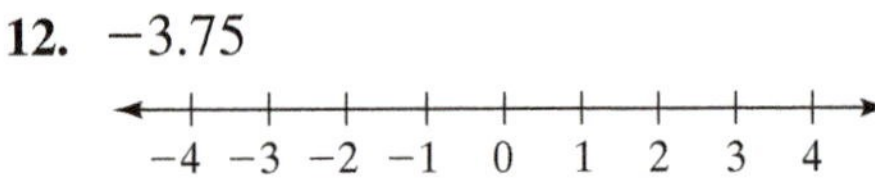

Indicate whether each inequality is true or false.

13. $-7 < -5$

14. $-1 > 3$

[11.2] ***Find the sum of each pair of numbers using the number line.***

15. $-4 + (-1)$

−5 −4 −3 −2 −1 0 1 2 3 4 5

16. $3 + (-7)$

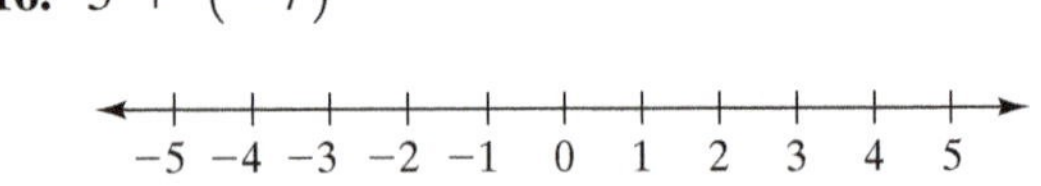

Name the property of addition illustrated by each statement.

17. $5 + (-6) = (-6) + 5$

18. $0 + 25 = 25$

19. $(1 + 3) + 6 = 1 + (3 + 6)$

20. $-10 + 10 = 0$

Find the sum.

21. $9 + (-9)$

22. $4 + (-2)$

23. $-3 + 5$

24. $0 + (-15)$

25. $-3 + (-2)$

26. $-3 + 7 + (-89)$

27. $-0.5 + (-3.6) + (-4)$

28. $-2 + 5.3 + 12$

Find the difference.

29. $12 - 3$

30. $36 - 47$

31. $-52 - 3$

32. $2 - 5$

33. $-19 - 8$

34. $24 - (-3)$

35. $8 - (-8)$

36. $0 - 5$

37. $6 - 7.42$

38. $-9 - \left(-\frac{3}{8}\right)$

Combine.

39. $2 + (-4) - (-7)$

40. $-3 - (-1) + 12$

[11.3] ***Name the property of multiplication illustrated by each statement.***

41. $-3(5) = 5(-3)$

42. $(-8 \cdot 6) \cdot 2 = -8 \cdot (6 \cdot 2)$

43. $-9 \cdot 1 = -9$

44. $-7 \cdot 0 = 0$

Find the product.

45. $2(-5)$

46. $-3 \cdot 7$

47. $-60 \cdot 90$

48. $-8(-300)$

49. $(-2.7)(-10)$

50. $\left(\frac{3}{4}\right)\left(-\frac{1}{3}\right)$

51. $5(-4)(-300)$

52. $(-1)(-12)(3)$

Simplify.

53. $-8 + 3(-2) - 9$

54. $3 - 2(-3) - (-5)$

55. $-9 - 5(-7)$

56. $20 - 3(-6)$

57. $-4(-2 + 5)$

58. $(-12 + 6)(-1)$

Find the reciprocal.

59. $-\frac{2}{3}$

60. 8

Find the quotient. Simplify.

61. $-30 \div (-10)$

62. $6 \div (-1)$

63. $-\frac{11}{5}$

64. $\frac{4}{5} \div \left(-\frac{2}{3}\right)$

Simplify.

65. $-16 \div 2(-4)$

66. $(9 - 23) \div (-13 + 6)$

67. $\frac{3 + (-1)}{-2}$

68. $\frac{5(7 - 3)}{-8 - 2}$

69. $(-3) + 8 - 2 \cdot (-4)$

70. $10 \div (-2) + (-3) \cdot 5$

[11.4] ***Determine the number of terms in each expression.***

71. $-x + y - 3z$

72. $\frac{m}{n} + 4$

73. $-9t$

74. $3a - 1 + 7b + c$

Write each using exponents.

75. $-3(-3)(-3)(-3)$

76. $-5(-5)(-5)(3)(3)$

77. $4(x)(x)(x)$

78 $-5a \cdot a \cdot b \cdot b \cdot b \cdot c$

[11.5] ***Evaluate each algebraic expression if $a = 2$, $b = 5$, and $c = -1$.***

79. $30 + c$

80. $-\frac{4}{9}b$

81. $-5a^2$

82. $10(b - c)$

83. $\frac{1 - a}{c}$

84. $4a^2 - 4ab + b^2$

[11.6]

85. Apply the distributive property: $-5(x - y)$

86. Combine like terms: $4x + 10x - 2y$

87. Simplify: $3x^2 - x^2 - 4x^2$

88. Combine: $2r^2t^2 - r^2t^2$

89. Simplify: $2(a - 5) + 1$

90. Simplify: $-(3x + 2)$

91. Combine like terms: $-3x - 5 - (x + 10)$

92. Simplify: $(2a - 4) + 2(a - 5) - 3(a + 1)$

Mixed Applications

Solve.

93. Express as a signed number: a gain of \$700

94. The price of a share of stock fell by \$0.50 each week for 4 wk in a row. What was the overall change in the price?

95. In *exothermic* chemical reactions, the surrounding temperature rises; in *endothermic* chemical reactions, the surrounding temperature drops. If a chemical reaction causes the surrounding temperature to change from −7°C to −4°C, was it exothermic or endothermic?

96. For each 100 people in the United States in a recent year, there were approximately 1.383 births, 0.838 deaths, and a net migration of +0.425 people. How many more people per 100 were there at the end of the year than at the beginning? Express as a percent. (*Source:* cia.gov)

97. The following graph shows five countries and their lowest point on land, measured in meters. The countries with negative altitudes have land below sea level.

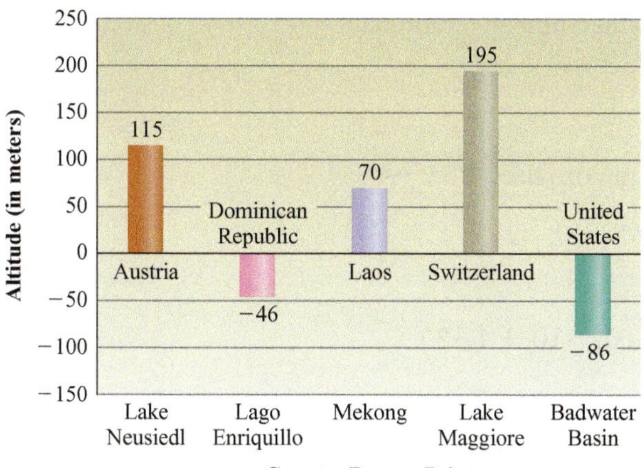

(*Source:* wikipedia.org)

What is the approximate difference in altitude between the country with the highest altitude and the one with the lowest altitude?

98. The bar graph shows the closing price of a share of General Electric Co. stock for a five-day period in November 2010.

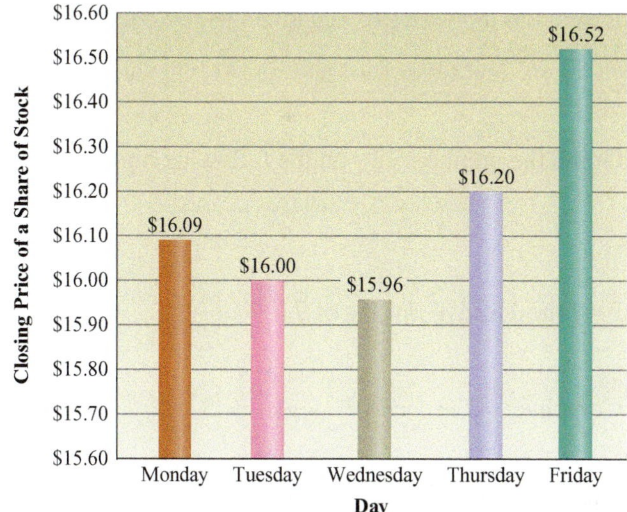

(*Source:* finance.yahoo.com)

Calculate the change in the closing price per share from the previous day for Tuesday through Friday, expressing each change as a signed number.

99. The length of a rectangle is double its width w. Find the rectangle's perimeter.

100. One year, a company's loss was $60,000. The next year, it was only $20,000. How many times the second loss is the first loss?

101. A colony of bacteria triples in size every 6 hr. At one point, there are 10 bacteria in the colony. Write in exponential form the number of bacteria in the colony 18 hr later.

102. Firefighters use the formula $S = 0.5N + 26$ to compute the maximum horizontal range S (in feet) of water from a particular hose, where N is the hose's nozzle pressure (in pounds). Calculate S if $N = 90$ lb. (*Source:* firedistrict7.com)

103. On June 1, 2009, the interest rate on a one-year Treasury bill (T-bill) was 0.48%. A year later, the rate dropped to 0.35%. How much less interest (in dollars) would be earned on a one-year T-bill starting at $\$P$ on June 1, 2010 than a one-year T-bill starting at $\$P$ on June 1, 2009? (*Source:* treas.gov)

104. Find the annual interest earned on an investment of $600 if x dollars is invested at an interest rate of 5% per year and the remainder is invested at a rate of 7% per year.

105. A homeowner needs to have sufficient current to operate the electrical appliances in the home. Electricians use the formula $I = \frac{P}{E}$ to compute the current I (in amperes) needed in terms of the power P (in watts) and the energy E (in volts). Find I if $P = 2300$ watts and $E = 115$ volts.

106. A checking account has a balance of $410. If $900 is deposited, a check for $720 is written, and two withdrawals of $300 each are made through an ATM, by how much money is the account overdrawn?

• Check your answers on page A-16.

CHAPTER 11 Posttest

FOR EXTRA HELP  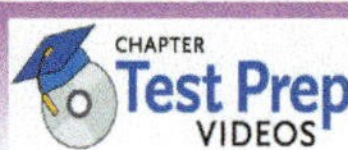

The Chapter Test Prep Videos with test solutions are available on DVD, in MyMathLab, and on YouTube (search "AkstDevMath" and click on "Channels").

To see if you have mastered the topics in this chapter, take this test.

1. Express as a positive or negative integer: a loss of 10,000 jobs

2. Is the number $\frac{2}{5}$ rational?

3. Graph the number $-2\frac{1}{2}$ on the following number line.

−3 −2 −1 0 1 2 3

4. True or false? $1 > -4$

5. Find the additive inverse of 7.

6. Add: $10 + (-3)$

7. Combine: $2 + (-3) + (-1) + 5$

8. Simplify: $4 + (-1)(-6)$

9. What is the reciprocal of 12?

10. Divide: $-15 \div 0.3$

11. Write in exponential form: $-5(-5)(-5)$

12. Write using exponents. $-3a \cdot b \cdot b \cdot c \cdot c \cdot c$

13. Evaluate $-b^3$ if $b = -2$

14. Evaluate $3a + b - c$ if $a = -1$, $b = 0$, and $c = 2$.

15. Simplify: $4y + 3 - 7y + 10y + 1$

16. Combine: $8t + 1 - 2(3t - 1)$

17. On a double-or-nothing wager, a gambler bets \$5 and wins. Express a signed number the amount of money he won.

18. Last year, a company suffered a loss of \$20,000. This year, it showed a profit of \$50,000. How big an improvement was this?

19. The balance in a checking account was d dollars. Some time later, the balance was 5% higher. What was the new balance?

20. In football, a team has 4 plays, called "downs," to move the football 10 yd or more toward their goal, or else they lose the ball to their opponents. A team had a series of downs in which they gained no ground, lost 10 yd, lost 8 yd, and gained 37 yd. Did the team succeed in keeping the ball?

• Check your answers on page A-16.

Cumulative Review Exercises

To help you review, solve the following.

1. Find the product: $63 \times 2 \times 12$

2. Insert parentheses, if needed, to make $8 + 4 \cdot 2^2$ equal to 48.

3. Subtract: $5\frac{1}{3} - 2\frac{1}{2}$

4. Simplify: $6 - \frac{2}{3} \div \frac{8}{9}$

5. Compute: $(0.1)^3$

6. Translate, and then solve: A number increased by 15.7 is equal to 84.6.

7. Solve for t: $\frac{3}{8}t = 24$

8. What percent of 30 is 25?

9. Combine: $-6 + (-3) - 7 + 2 + (-5)$

10. Calculate: $(-1)^2 - 3$

11. Compute the mean, median, mode(s), and range of $2\frac{1}{2}, 3\frac{1}{4}$, 2, 3, and $2\frac{1}{4}$.

12. 8 min = _______ sec

13. Determine between which two consecutive whole numbers $\sqrt{50}$ lies.

14. Pirates are terrorizing a growing number of shipping lanes. According to the International Chamber of Commerce, ships reported 445 attacks in 2010, up 10% from 2009. Rounding to the nearest whole number, calculate the number of ship attacks reported in 2009. (*Source:* icc-ccs.org)

15. A doctor changed a patient's dosage of thyroxine from 0.075 milligrams to 0.1 milligrams. Explain whether this change represented an increase or a decrease.

16. A sailor serves on a large military ship on which there are 75 officers and 1,275 enlisted personnel. What is the ratio of officers to enlisted personnel among the ship's crew?

17. On a given stock market trading day, the "advance/decline ratio" is a measure of market performance. This ratio compares the number of stocks whose value increases with the number whose value decreases. If on a particular day the advance/decline ratio was 3 to 2 and 1,800 stocks increased in value, how many stocks decreased in value? (*Source:* stock-market-investors.com)

18. The following table shows the net income for Zale Corporation in four consecutive quarters.

Quarter	1	2	3	4
Net Income (in millions)	−$89.8	−$59.7	$6.6	−$12.1

(*Source:* finance.yahoo.com)

What was the average quarterly net income?

19. The following graph shows the growth of active users on the social network Facebook in recent years.

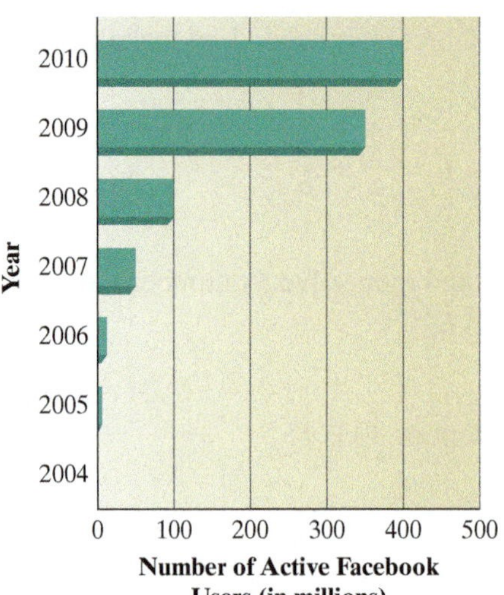

(*Source:* facebook.com)

Approximately how many more active Facebook users were there in 2010 than in 2008?

20. "Barefooting" is a kind of water skiing in which the skier is barefoot, having no water skis. Invented in Cypress Gardens, Florida, in 1947, the sport requires the motor boat pulling the skier to travel faster than in regular water skiing, at a speed that depends on the skier's weight. The required boat speed s (in miles per hour) can be approximated using the formula

$$s = \frac{w}{10} + 20,$$

where weight w is in pounds. For a 200-pound barefooted skier, how fast must the motor boat go? (*Sources:* usawaterski.org and wikipedia.org)

• Check your answers on page A-16.

CHAPTER 12

Solving Linear Equations and Inequalities

Cell Phones and Algebra

Today's generation takes wireless communication for granted. The cellular telephone, one of the more important wireless devices, was first introduced to the public in the mid-1980s. Since then, cell phones have shrunk both in size and in price.

Using a cell phone generally entails choosing a calling plan with a monthly service charge. As a rule, this charge consists of two parts: a specified flat monthly fee and a per-minute usage charge varying with the number of minutes of airtime used.

The choice among calling plans can be confusing, especially with the lure of promotional incentives. One nationally advertised calling plan requires a monthly charge of \$44.99 for 450 min of free local calls and 45 cents per minute for additional local calls. A competing plan is for \$64.99 a month for 900 min of free local calls plus 40 cents a minute for additional local calls. Under what circumstances is one deal better than the other? The key to answering this question is solving inequalities such as:

$$44.99 + 0.45(x - 450) < 64.99 + 0.4(x - 900)$$

(*Source:* Tom Farley, "The Cell-Phone Revolution," *American Heritage of Invention and Technology*, American Heritage, 2007)

CHAPTER 12 PRETEST

To see if you have already mastered the topics in this chapter, take this test.

1. Is 4 a solution of the equation $7 - 2x = 3x - 11$?

2. What must be done to the following equation in order to isolate the variable? $n + 2 = -6$

Solve and check.

3. $6 + y = -5$

4. $n - \frac{1}{2} = -5$

5. $\frac{y}{-5} = 1$

6. $-n = 8$

7. $\frac{2}{3}x - 3 = -9$

8. $4x - 8 = -10$

9. $9x + 13 = 7x + 19$

10. $8 - [2 + 3(x - 5)] = 3$

11. $-2(3n - 1) = -7n$

12. $14x - (8x - 13) = 12x + 3$

13. Solve $v - 5u = w$ for v in terms of u and w.

14. Draw the graph of $x \leq 2$.

−3 −2 −1 0 1 2 3

15. Solve and graph: $x + 3 > 3$

−3 −2 −1 0 1 2 3

16. An office photocopier makes 30 copies per minute. How long will it take to copy a 360-page document?

17. A florist charges a flat fee of \$100 plus \$70 for each centerpiece for a wedding. If the total bill for the centerpieces was \$1500, how many centerpieces did the florist make?

18. For 3 hr, a family on vacation drove on a highway. When traffic increased, they drove 10 mph more slowly for an additional 2 hr. If the total length of the trip was 175 mi, how fast were they driving at first?

19. The formula for finding the amount of kinetic energy used is $E = \frac{1}{2}mv^2$. Solve this equation for m in terms of E and v.

20. A gym offers two membership options: Option A is \$55 per month for unlimited use of the gym and Option B is \$10 per month plus \$3 for each hour a member uses the gym. For how many hours of use per month will Option A be a better deal?

• Check your answers on page A-16.

12.1 Solving Linear Equations: The Addition Property

OBJECTIVES

- **A** To determine whether a given number is a solution of a given equation
- **B** To solve linear equations using the addition property
- **C** To solve applied problems using the addition property

Solving Equations

Recall from Section 4.2 that an *equation* is a mathematical statement that two expressions are equal to one another. In this chapter, we continue our discussion of equations, but consider:

- equations involving both negative as well as positive numbers
- equations that are more complicated, and
- an approach to solving equations based on several key algebraic ideas

An equation may be either true or false. The equation $x + 4 = 9$ is true if 5 is substituted for the variable x.

$$x + 4 = 9$$
$$5 + 4 = 9 \quad \text{True}$$

However, this equation is false if 6 is substituted for the variable x.

$$x + 4 = 9$$
$$6 + 4 = 9 \quad \text{False}$$

How many values of x will make $x + 4 = 9$ a true equation? How many values of x will make this equation false?

The number 5 is called a *solution* of the equation $x + 4 = 9$ because we get a true statement when we substitute 5 for x.

EXAMPLE 1

Is 2 a solution of the equation $3x + 1 = 11 - 2x$?

Solution

$$3x + 1 = 11 - 2x$$
$$3(2) + 1 \stackrel{?}{=} 11 - 2(2) \quad \text{Substitute 2 for } x.$$
$$6 + 1 \stackrel{?}{=} 11 - 4 \quad \text{Evaluate each side of the equation.}$$
$$7 = 7 \quad \text{True}$$

Since $7 = 7$ is a true statement, 2 is a solution of $3x + 1 = 11 - 2x$.

PRACTICE 1

Determine whether 4 is a solution of the equation $5x - 4 = 2x + 5$.

EXAMPLE 2

Determine whether -1 is a solution of the equation $2x - 4 = 6(x + 2)$.

Solution

$$2x - 4 = 6(x + 2)$$
$$2(-1) - 4 \stackrel{?}{=} 6(-1 + 2) \quad \text{Substitute } -1 \text{ for } x.$$
$$-2 - 4 \stackrel{?}{=} 6(1) \quad \text{Evaluate each side of the equation.}$$
$$-6 = 6 \quad \text{False}$$

Since $-6 = 6$ is a false statement, -1 is *not* a solution of the equation.

PRACTICE 2

Is -8 a solution of the equation $5(x + 3) = 3x - 1$?

In this chapter, we will work mainly with equations in one variable that have one solution. These equations are called *linear equations* or *first-degree equations* because the exponent of the variable is 1.

DEFINITION

A **linear equation in one variable** is an equation that can be written in the form

$$ax + b = c$$

where a, b, and c are real numbers and $a \neq 0$.

Some linear equations have a special relationship. Consider the equations $x = 2$ and $x + 1 = 3$:

$x = 2$	The solution is 2.	$x + 1 = 3$	By inspection, we see the solution is 2.
		$2 + 1 \stackrel{?}{=} 3$	
		$3 = 3$	True

The equations $x = 2$ and $x + 1 = 3$ have the same solution and so are *equivalent*.

DEFINITION

Equivalent equations are equations that have the same solution.

Recall that to solve an equation means to find the number or constant that, when substituted for the variable, makes the equation a true statement. This solution is found by changing the equation to an equivalent equation of the following form:

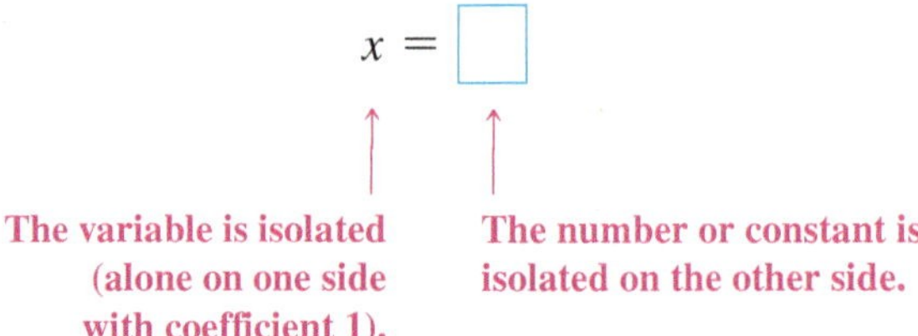

Using the Addition Property to Solve Linear Equations

One of the properties that we use in solving equations involves adding. Consider the equation $\frac{6}{3} = 2$. Adding 4 to each side of the equation gives us $\frac{6}{3} + 4 = 2 + 4$. Using mental arithmetic, we see that $\frac{6}{3} + 4 = 2 + 4$ is also a true statement. In general, adding the same number to each side of a true equation results in another true equation. So adding -3 to both sides of the equation $x + 3 = 7$ gives us the equivalent equation $x + 3 + (-3) = 7 + (-3)$, or $x = 4$. This example suggests the following property:

The Addition Property of Equality

For any real numbers a, b, and c, $a = b$ and $a + c = b + c$ are equivalent.

In words, this property states that when adding any real number to each side of an equation, the result is an equivalent equation. Now, let's apply this property to solving equations.

EXAMPLE 3

Solve and check: $x - 5 = -11$

Solution To solve this equation, we isolate the variable by adding 5, which is the additive inverse of -5, to each side of the equation.

$$x - 5 = -11$$

$$x - 5 + 5 = -11 + 5$$ Add 5 to each side of the equation, getting an equivalent equation.

$$x + 0 = -6$$

$$x = -6$$ Recall that $x + 0 = x$.

Check

$$x - 5 = -11$$

$$-6 - 5 \stackrel{?}{=} -11$$ Substitute -6 for x in the original equation.

$$-11 = -11$$ True

So the solution is -6.

PRACTICE 3

Solve and check: $y - 12 = -7$

Can you explain why checking a solution is important?

EXAMPLE 4

Solve and check: $-9 = a + 6$

Solution

$$-9 = a + 6$$

$$-9 + (-6) = a + 6 + (-6)$$ Add -6 to each side of the equation.

$$-15 = a + 0$$

$$-15 = a$$

or $$a = -15$$

Check

$$-9 = a + 6$$

$$-9 \stackrel{?}{=} (-15) + 6$$ Substitute -15 for a.

$$-9 = -9$$ True

So the solution is -15.

PRACTICE 4

Solve and check: $-2 = n + 15$

Are the solutions to the equations $-9 = a + 6$ and $a + 6 = -9$ the same? Explain.

Because subtracting a number is the same as adding its opposite, the addition property allows us to subtract the same value from each side of an equation. For instance, suppose $a = b$. Subtracting c from each side gives us $a - c = b - c$. So an alternative approach to solving Example 4 is to subtract the same number, namely 6, from each side of the equation, as follows:

$$-9 = a + 6$$

$$-9 - 6 = a + 6 - 6$$

$$-15 = a, \text{ or } a = -15$$

Note that this approach gives us the same solution, namely -15, that we got using the approach in Example 4.

EXAMPLE 5

Solve and check: $r + \frac{1}{5} = -\frac{3}{5}$

Solution

$$r + \frac{1}{5} = -\frac{3}{5}$$

$$r + \frac{1}{5} - \frac{1}{5} = -\frac{3}{5} - \frac{1}{5} \quad \text{Subtract } \frac{1}{5} \text{ from each side of the equation.}$$

$$r + 0 = -\frac{4}{5}$$

$$r = -\frac{4}{5}$$

Check

$$r + \frac{1}{5} = -\frac{3}{5}$$

$$-\frac{4}{5} + \frac{1}{5} \stackrel{?}{=} -\frac{3}{5} \quad \text{Substitute } -\frac{4}{5} \text{ for } r.$$

$$-\frac{3}{5} = -\frac{3}{5} \quad \text{True}$$

PRACTICE 5

Solve and check: $-\frac{3}{8} = s + \frac{1}{8}$

EXAMPLE 6

Solve and check: $m - (-26.1) = 32$

Solution

$$m - (-26.1) = 32$$

$$m + 26.1 = 32$$

$$m + 26.1 - 26.1 = 32 - 26.1 \quad \text{Subtract 26.1 from each side of the equation.}$$

$$m = 5.9$$

Check

$$m - (-26.1) = 32$$

$$5.9 - (-26.1) \stackrel{?}{=} 32 \quad \text{Substitute 5.9 for } m.$$

$$5.9 + 26.1 \stackrel{?}{=} 32$$

$$32 = 32 \quad \text{True}$$

So the solution is 5.9.

PRACTICE 6

Solve and check: $5 = 4.9 - (-x)$

Before we consider an application of solving a linear equation, recall our discussion from Chapter 4 of how to translate word phrases to algebraic expressions. Now, let's look at some examples of translating word sentences to equations, namely those involving addition or subtraction.

Sentence: A number increased by 1.1 equals 8.6.

↓ ↓ ↓ ↓ ↓

Equation: $n \quad + \quad 1.1 \quad = \quad 8.6$

Sentence:	A number	minus	one-half	equals	five.
	↓	↓	↓	↓	↓
Equation:	y	$-$	$\frac{1}{2}$	$=$	5

Note that in both examples we used a variable to represent the unknown number.

In solving applied problems, first we translate the given word sentences to equations, and then we solve the equations.

EXAMPLE 7

The mean distance between the planet Venus and the Sun is 31.2 million mi more than the mean distance between the planet Mercury and the Sun. (*Source: The New York Times Almanac 2010*)

a. Using the following diagram, write an equation to find Mercury's mean distance from the Sun.

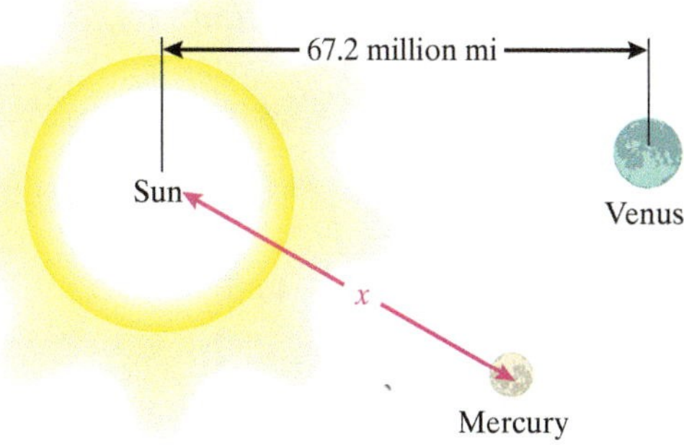

b. Solve this equation.

Solution

a. Let x represent Mercury's mean (or average) distance from the Sun. From the diagram, we see that the mean distance between Venus and the Sun is 67.2 million mi. This distance is 31.2 million mi more than Mercury's mean distance to the Sun. So we translate this sentence to an equation:

Sentence:	67.2	is	31.2	plus	x
	↓	↓	↓	↓	↓
Equation:	67.2	$=$	31.2	$+$	x

b. Solve the equation:

$$67.2 = x + 31.2$$

$$67.2 - 31.2 = x + 31.2 - 31.2$$ **Subtract 31.2 from each side of the equation.**

$$36.0 = x$$

Check

$$67.2 = x + 31.2$$

$$67.2 \stackrel{?}{=} 36.0 + 31.2$$ **Substitute 36.0 for x.**

$$67.2 = 67.2$$ **True**

So we conclude that Mercury's mean distance from the Sun is 36.0 million mi.

PRACTICE 7

As a result of a chemical reaction, the temperature of a substance rose. The original temperature had been 23.7°C, and the final temperature was 36.0°C. Find the change in temperature.

12.1 Exercises

FOR EXTRA HELP MyMathLab MathXL PRACTICE WATCH READ REVIEW

Mathematically Speaking

Fill in each blank with the most appropriate term or phrase from the given list.

equal equations	linear equations in one variable	equivalent equations
isolate	addition property of equality	solve
additive identity property		

1. Equations that can be written in the form $ax + b = c$, where a, b, and c are real numbers and $a \neq 0$, are called ____________.

2. Equations that have the same solution are called ____________.

3. To ____________ an equation means to find the value that, when substituted for the variable, makes the equation true.

4. The ____________ states that for any real numbers a, b, and c, $a = b$ and $a + c = b + c$ are equivalent.

A ***Indicate whether it is true or false that the value of x shown is a solution to the given equation.***

5.

Value of x	Equation	True or False
a. -8	$3x + 13 = -11$	
b. 7	$28 - x = 7 - 4x$	
c. 9	$2(x - 3) = 12$	
d. $\frac{2}{3}$	$12x - 2 = 6x + 2$	

6.

Value of x	Equation	True or False
a. -2	$3x - 2 = 10$	
b. 4	$2x + 3 = 5x - 9$	
c. 1	$3(5 - x) = 18$	
d. $-\frac{1}{2}$	$6 - 5x = 5x + 11$	

B ***Indicate what must be done to each side of the equation in order to isolate the variable.***

7. $x + 4 = -6$

8. $x + 3 = -3$

9. $z - (-3.5) = 5$

10. $x - (-15) = 10$

11. $-2 = -1 + x$

12. $-6 = -5 + x$

13. $9 = x - 2\frac{1}{5}$

14. $7 = x - 3\frac{1}{4}$

Solve and check.

15. $y + 9 = -14$

16. $x + 2 = -10$

17. $t - 4 = -4$

18. $m - 6 = -1$

19. $9 + a = -3$

20. $10 + x = -4$

21. $z - 4 = -10$

22. $n - 4 = -1$

23. $-30 = x + 12$

24. $-25 = y + 21$

25. $-6 = t - 12$

26. $-19 = r - 19$

27. $-4 = -10 + r$

28. $-7 = -2 + m$

29. $-15 + n = -2$

30. $-14 + c = -6$

31. $x + \frac{2}{3} = -\frac{1}{3}$

32. $z + \frac{1}{8} = -\frac{3}{8}$

33. $8 + y = 4\frac{1}{2}$

34. $9 + z = 6\frac{1}{4}$

35. $m + 2.4 = 5.3$

36. $n + 3.2 = 8.4$

37. $-2.3 + t = -5.9$

38. $-3.4 + r = -9.5$

39. $a - (-35) = 30$

40. $x - (-25) = 24$

41. $m - \left(-\frac{1}{4}\right) = -\frac{1}{4}$

42. $a - \left(-\frac{1}{5}\right) = -\frac{1}{5}$

Translate each sentence to an equation. Then, solve and check.

43. Two more than a number is -12.

44. 3.2 more than a number is -20.

45. If $4\frac{1}{3}$ is subtracted from a number, the result is $21\frac{1}{2}$.

46. A number minus $2\frac{1}{7}$ is the same as $1\frac{1}{2}$.

47. If -3 is added to a number, the result is -1.

48. If -5.2 is added to a number, the result is 12.

49. -17 subtracted from a number equals 11.

50. -5 subtracted from a number equals -8.

Choose the equation that best describes the situation.

51. After dieting and exercising, a featherweight boxer lost 6 lb. If the boxer now weighs 127 lb, what was his original weight?

a. $x - 127 = -6$ **b.** $x + 6 = 127$
c. $x + 127 = 6$ **d.** $x - 6 = 127$

52. After paying a bill of \$5.25, a customer has \$2.75 left. How much money did she have prior to paying the bill?

a. $d - 5.25 = 2.75$ **b.** $d + 2.75 = 5.25$
c. $d + 5.25 = 2.75$ **d.** $d - 2.75 = -5.25$

53. A digital picture that takes up 4.7 megabytes (Mb) of memory is saved on a computer. If 250 Mb of memory are free after the picture is saved, how many megabytes of free memory did the computer have before the picture was saved?

a. $x - 4.7 = 250$ **b.** $x - 250 = -4.7$
c. $x + 4.7 = 250$ **d.** $x + 250 = 4.7$

54. At a college, tuition costs students \$3000 a semester. If a student received \$2250 in financial aid, how much more money does the student need to pay the balance of the semester's tuition?

a. $m - 2250 = 3000$ **b.** $m + 2250 = 3000$
c. $m + 3000 = 2250$ **d.** $m - 3000 = 2250$

Mixed Practice

Indicate what must be done to each side of the equation in order to isolate the variable.

55. $\frac{2}{3} + b = 3$

56. $-5 = a - 3$

Solve.

57. Indicate whether the value of x shown is a solution to the given equation by answering true or false.

Value of x	Equation	True or False
a. -4	$6x + 15 = -11$	
b. 9	$4x - 21 = 2x - 3$	
c. 7	$-2(x - 5) = 4$	
d. $\frac{3}{4}$	$5 - 8x = -4x + 2$	

Translate each sentence to an equation. Then, solve and check.

58. The difference between 0 and a number is 32.

59. If 2.5 is subtracted from a number, the result is -3.8.

Solve and check.

60. $x - 12 = -7$

61. $-19 = m + 4$

62. $-3.8 + t = -6.6$

63. $7 + n = 2\frac{2}{3}$

64. $x - (-2.1) = 1$

Applications

C *Write an equation. Solve and check.*

65. If the speed of a car is increased by 10 mph, the speed will be 44 mph. What is the speed of the car now?

66. The difference between the boiling point of gold and the boiling point of aluminum is 337°C. If aluminum boils at 2519°C, the lower temperature of the two metals, what is the boiling point of gold? (*Source: The New York Times Almanac 2010*)

67. A student uses about 370 calories after 1 hr of hiking. This is 190 calories more than is used after 1 hr of stretching. How many calories does the student use in an hour of stretching? (*Source:* U.S. Department of Agriculture)

68. A patient's temperature dropped by 2.4°F to 98.6°F. What had the patient's temperature been?

69. After descending 170 m, a traffic helicopter is 215 m above the ground, as illustrated in the following diagram. What was the original height of the helicopter?

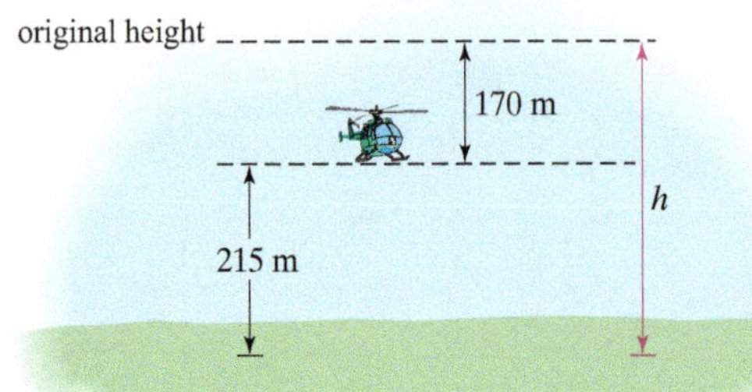

70. In the following diagram, $\angle ABC$ and $\angle CBD$ are complementary angles, that is, the sum of their measures is 90°. Find the number of degrees in $\angle CBD$.

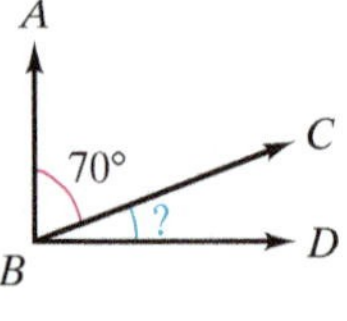

71. In the triangle shown, the sum of the measures of the three angles is $180°$. Find y.

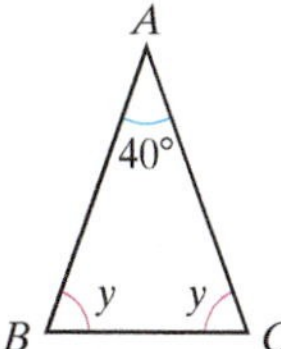

72. At the close of business today, the price per share of a stock was \$21.63. This was down \$0.31 from the closing price per share yesterday. What was the closing price per share of the stock yesterday?

• Check your answers on page A-16.

MINDStretchers

Mathematical Reasoning

1. Suppose that $x - a = b$.

   a. What happens to x if a decreases and b remains the same?

   b. What happens to x if a remains the same and b decreases?

Writing

2. Are the equations $a = b$ and $a - c = b - c$ equivalent? Explain.

Research

3. In your college library or on the Web, investigate the origin of the word *algebra*. Write a few sentences to show what you have learned.

Cultural Note

The picture on the left reflects the mathematical tradition of representing an unknown by the letter *x*. The painting of an unnamed woman, entitled *Madame X* (1884), is by the American artist John Singer Sargent. About a dozen films made throughout the twentieth century dealing with enigmatic women bear the same name.

The practice of using an *x* and other letters from the end of the alphabet to represent mathematical unknowns goes back to the seventeenth-century French mathematician René Descartes, who made major contributions to the development of algebra.

Source: Florian Cajori, *A History of Mathematical Notations* (Chicago: The Open Court Publishing Company, 1929).

12.2 Solving Linear Equations: The Multiplication Property

OBJECTIVES

A To solve linear equations using the multiplication property

B To solve applied problems using the multiplication property

In the previous section, we used the addition property of equality to solve equations involving addition or subtraction. Here, we consider a second property to solve equations involving multiplication or division.

The Multiplication Property of Equality

For any real numbers a, b, and c, $c \neq 0$, $a = b$ and $a \cdot c = b \cdot c$ are equivalent.

In words, this property states that when multiplying each side of an equation by any nonzero real number, the result is an equivalent equation. Let's apply the multiplication property to solving equations.

EXAMPLE 1

Solve and check: $\frac{x}{4} = 11$

Solution In this equation, note that $\frac{x}{4}$ is the same as $\frac{1}{4} \cdot x$. To solve this equation, we isolate the variable by multiplying each side of the equation by 4, the reciprocal of $\frac{1}{4}$.

$$\frac{x}{4} = 11$$

$$4 \cdot \frac{x}{4} = 4 \cdot 11$$ Mutiply each side of the equation by 4, getting an equivalent equation.

$$1x = 44$$ $4 \cdot \frac{x}{4} = 4 \cdot \frac{1}{4}x = 1x$

$$x = 44$$ $1x = x$

Check $\frac{x}{4} = 11$

$$\frac{44}{4} \stackrel{?}{=} 11$$ Substitute 44 for x.

$$11 = 11$$ True

So the solution is 44.

PRACTICE 1

Solve and check: $\frac{y}{3} = 21$

EXAMPLE 2

Solve and check: $9x = -72$

Solution

$$9x = -72$$

$$\left(\frac{1}{9}\right)9x = \left(\frac{1}{9}\right)(-72) \quad \text{Multiply each side of the equation by } \frac{1}{9}.$$

$$1x = -8 \quad \frac{1}{9} \cdot 9 = 1$$

$$x = -8$$

Check

$$9x = -72$$

$$9(-8) \stackrel{?}{=} -72 \quad \text{Substitute } -8 \text{ for } x.$$

$$-72 = -72 \quad \text{True}$$

So the solution is -8.

PRACTICE 2

Solve and check: $-7y = -63$

Because dividing by a number is the same as multiplying by its reciprocal, the multiplication property allows us to divide each side of an equation by a non-zero number. For instance, suppose $a = b$. Multiplying each side by $\frac{1}{c}$ gives us $a \cdot \frac{1}{c} = b \cdot \frac{1}{c}$, for $c \neq 0$. This equation is equivalent to $\frac{a}{c} = \frac{b}{c}$. So an alternative approach to solving Example 2 is to divide each side of the equation by the same number, namely 9, as shown:

$$9x = -72$$

$$\frac{9x}{9} = \frac{-72}{9}$$

$$x = -8$$

Note that this approach gives us the same solution, -8, that we got using the approach in Example 2.

EXAMPLE 3

Solve and check: $-y = -15$

Solution

$$-y = -15$$

$$-1y = -15 \quad \text{The coefficient of } -y \text{ is } -1.$$

$$\frac{-1y}{-1} = \frac{-15}{-1} \quad \text{Divide each side of the equation by } -1.$$

$$y = 15$$

Check

$$-y = -15$$

$$-1(15) \stackrel{?}{=} -15 \quad \text{Substitute 15 for } y.$$

$$-15 = -15 \quad \text{True}$$

So the solution is 15.

PRACTICE 3

Solve and check: $-x = 10$

EXAMPLE 4

Solve and check: $46 = -4.6n$

Solution

$$46 = -4.6n$$

$$\frac{46}{-4.6} = \frac{-4.6n}{-4.6} \quad \text{Divide each side of the equation by } -4.6.$$

$$-10 = n \quad \text{or}$$

$$n = -10$$

Check

$$46 = -4.6n$$

$$46 \stackrel{?}{=} -4.6(-10) \quad \text{Substitute } -10 \text{ for } n.$$

$$46 = 46 \quad \text{True}$$

So the solution is -10.

PRACTICE 4

Solve and check: $-11.7 = -0.9z$

EXAMPLE 5

Solve and check: $\frac{2w}{3} = 8$

Solution

$$\frac{2w}{3} = 8$$

$$\frac{2}{3}w = 8$$

$$\frac{3}{2} \cdot \frac{2}{3}w = \frac{3}{2} \cdot 8 \quad \text{Multiply each side of the equation by } \frac{3}{2}.$$

$$w = 12 \quad \frac{3}{2} \cdot \frac{2}{3} = 1 \text{ and } 1w = w.$$

Check

$$\frac{2}{3}w = 8$$

$$\frac{2}{3}(12) \stackrel{?}{=} 8 \quad \text{Substitute 12 for } w.$$

$$2 \cdot 4 \stackrel{?}{=} 8 \quad \text{Simplify.}$$

$$8 = 8 \quad \text{True}$$

So the solution is 12.

PRACTICE 5

Solve and check: $\frac{6y}{7} = -12$

Can you show another way to solve Example 5? Explain.

Let's now consider applied problems involving the multiplication property. To solve these problems, we translate word sentences to equations involving multiplication or division as follows:

Sentence: Three times a number x equals -12.

$3 \cdot x = -12$

Equation: $3x = -12$

Sentence: A number d divided by -2 equals 8.

Equation: $d \div -2 = 8$

$$\frac{d}{-2} = 8$$

EXAMPLE 6

A student applies for a job that pays an hourly overtime wage of \$15.90. The overtime wage is 1.5 times the regular hourly wage. What is the regular hourly wage for the job?

Solution Letting w represent the regular hourly wage, we write the word sentence, and then translate it to an equation.

Sentence:

The overtime wage \$15.90 is 1.5 times the regular hourly wage.

Equation: $15.90 = 1.5 \cdot w$

$15.90 = 1.5w$

Next, we solve the equation for w.

$$15.9 = 1.5w$$

$$\frac{15.9}{1.5} = \frac{1.5}{1.5}w$$

$$10.6 = w, \text{ or } w = 10.6$$

Check $15.90 = 1.5w$

$15.90 \stackrel{?}{=} 1.5(10.6)$ Substitute 10.6 for w.

$15.90 = 15.90$ True

So the regular wage is \$10.60.

PRACTICE 6

A mechanic billed a customer \$189.50 for labor to repair his car. If one-fourth of the bill was for labor, how much was the total bill?

Let's now turn to a particular kind of applied problem—a problem involving *motion*. To solve such problems, we need to use the equation $d = rt$, where d is distance, r is the average rate or speed, and t is time.

EXAMPLE 7

One of the fastest pitchers in the history of Japanese baseball was Yoshinori Sato, whose pitches were clocked at about 147 ft/sec. If the distance from the pitcher's mound to home plate is 60.5 ft, approximate the time it took Sato's pitches to reach home plate, to the nearest hundredth of a second. (*Source:* japantoday.com)

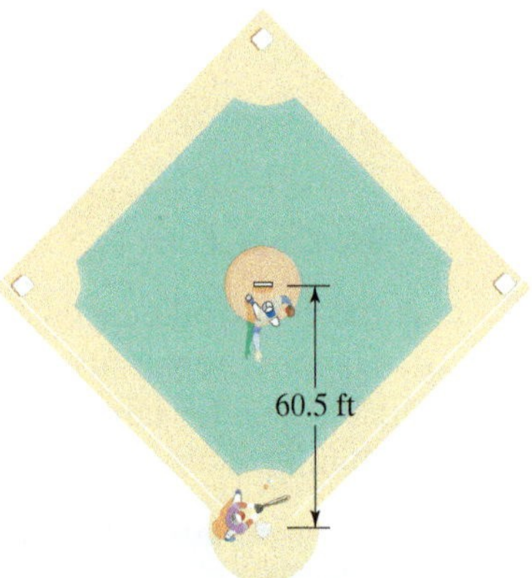

Solution The distance between the pitcher's mound and home plate is 60.5 ft and the rate of the ball thrown is 147 ft/sec. Using the equation $d = rt$, we can find the time.

$$d = rt$$

$$60.5 = 147t \quad \text{Substitute 60.5 for } d \text{ and 147 for } r.$$

$$\frac{60.5}{147} = \frac{147t}{147}$$

$$0.41 \approx t, \text{ or}$$

$$t \approx 0.41$$

Check Since the solution of the original equation is a rounded value, the check will not result in an exact equality. To verify the solution, check that the expressions on each side of the equation are approximately equal to one another.

$$60.5 = 147t$$

$$60.5 \stackrel{?}{\approx} 147 \cdot 0.41 \quad \text{Substitute 0.41 for } t.$$

$$60.5 \approx 60.3 \quad \text{True}$$

So Sato's pitches took approximately 0.41 sec to reach home plate.

PRACTICE 7

A driver makes a trip from Washington, D.C., the U.S. capital, to Philadelphia, Pennsylvania, the home of Independence Hall. If she averages 60 mph, how long, to the nearest tenth of an hour, will it take her to get to Philadelphia? (*Source:* mapquest.com)

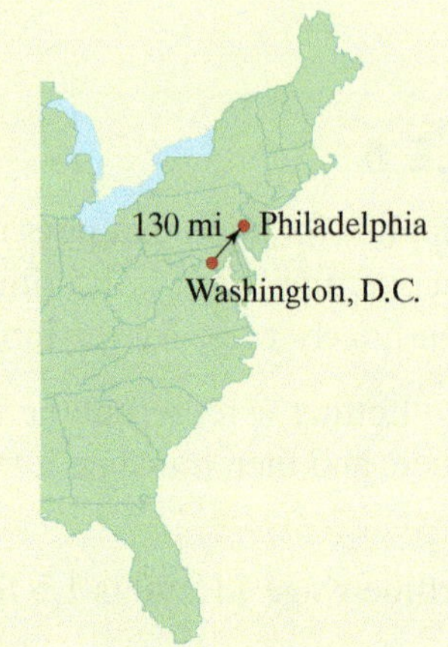

Some percent problems involve *interest*. When you lend, deposit, or invest money, you *earn* interest. When you borrow, you *pay* interest.

The amount of interest earned (or paid) I depends on the *principal* P (the amount of money invested or borrowed), the annual rate of interest r (usually expressed as a percent), and the length of time t the money is invested or borrowed (generally expressed in years). We compute the amount of interest by multiplying the principal by the rate of interest and the number of years: $I = Prt$. This kind of interest is called *simple interest*.

EXAMPLE 8

At a local bank, a customer earned $94.50 in interest on a principal of $900 in an account with an annual interest rate of 5.25%. How many years did it take the customer to earn the interest?

Solution To compute the number of years, we use the formula $I = Prt$.

$$I = P \cdot r \cdot t$$

$$94.5 = (900)(0.0525)\,t$$

$$94.5 = 47.25t$$

$$\frac{94.5}{47.25} = \frac{47.25t}{47.25}$$

$$2 = t, \quad \text{or} \quad t = 2$$

So it took the customer 2 yr to earn the interest.

PRACTICE 8

In 1 yr, a bank paid $130 in simple interest on an initial balance of $2000. What annual rate of interest was paid?

12.2 Exercises

FOR EXTRA HELP MyMathLab MathXL PRACTICE WATCH READ REVIEW

A ***Indicate what must be done to each side of the equation in order to isolate the variable.***

1. $\frac{x}{3} = -4$ **2.** $\frac{a}{-6} = 1$ **3.** $-5x = 20$ **4.** $-4x = 30$ **5.** $-2.2n = 4$

6. $1.5x = -6$ **7.** $\frac{3}{4}x = 12$ **8.** $\frac{2}{3}x = -6$ **9.** $-\frac{5y}{2} = 15$ **10.** $-\frac{8n}{5} = 4$

Solve and check.

11. $6x = -30$ **12.** $-8y = 8$ **13.** $\frac{n}{2} = 9$ **14.** $\frac{w}{10} = -21$

15. $\frac{a}{4} = 1.2$ **16.** $\frac{n}{7} = -1.3$ **17.** $-5x = 2.5$ **18.** $-2y = 0.08$

19. $42 = -6c$ **20.** $50 = -2x$ **21.** $11 = -\frac{r}{2}$ **22.** $4 = \frac{-m}{3}$

23. $\frac{5}{6}x = 10$ **24.** $\frac{3}{4}d = -3$ **25.** $-\frac{2}{5}y = 1$ **26.** $\frac{2}{3}r = -8$

27. $\frac{3n}{4} = 6$ **28.** $\frac{5a}{6} = 5$ **29.** $\frac{4c}{3} = -4$ **30.** $-\frac{2z}{7} = 8$

31. $-\frac{x}{2.4} = -1.2$ **32.** $-\frac{n}{0.5} = -1.3$ **33.** $-2.5a = 5$ **34.** $-2.25 = -1.5t$

35. $\frac{2}{3}y = \frac{4}{9}$ **36.** $\frac{5}{6}c = \frac{2}{3}$

Translate each sentence to an equation. Then, solve and check.

37. The product of -4 and a number is 56.

38. The product of -8 and a number is 72.

39. A number divided by 0.2 is 1.1.

40. A number divided by 0.6 is 1.8.

41. The quotient of a number and -3.5 is 30.

42. The quotient of a number and -2.5 is 40.

43. $\frac{1}{6}$ of a number is $2\frac{4}{5}$.

44. $\frac{5}{8}$ of a number is 20.

45. A shopper used half his money to buy a backpack. If the backpack cost \$20, how much money did he have prior to this purchase?

a. $20 = 2m$ **b.** $20m = \frac{1}{2}$

c. $20 = \frac{m}{2}$ **d.** $\frac{m}{20} = \frac{1}{2}$

46. Four college students divided the cost of lunch evenly among themselves. If the check was for \$20.25, how much did each student pay?

a. $x + 4 = 20.25$ **b.** $4x = 20.25$

c. $\frac{x}{4} = 20.25$ **d.** $x - 4 = 20.25$

47. A student plans to buy a DVD player 6 wk from now. If the DVD player costs \$150, how much money must she save per week in order to buy it?

a. $6p = 150$ **b.** $150p = 6$

c. $\frac{p}{6} = 150$ **d.** $\frac{p}{150} = 6$

48. The student government at a college sold tickets to a play. From ticket sales, it collected \$800, which was twice the cost of the play. How much did the play cost?

a. $\frac{c}{800} = 2$ **b.** $800c = 2$

c. $\frac{c}{2} = 800$ **d.** $2c = 800$

Mixed Practice

Solve and check.

49. $5x = -20$

50. $\frac{2n}{7} = 4$

51. $8 = -\frac{a}{3}$

52. $-\frac{y}{3.8} = -0.3$

Translate each sentence to an equation. Then, solve and check.

53. The quotient of a number and 5 is equal to 2.

54. $\frac{3}{8}$ of a number is -12.

Indicate what must be done to each side of the equation in order to isolate the variable.

55. $-5.2m = 4$

56. $\frac{b}{7} = 3$

Applications

B ***Write an equation that best describes the situation. Then, solve and check.***

57. According to a geologist, sediment at the bottom of a local lake accumulated at the rate of 0.02 cm/yr. How long did it take to create a layer of sediment 10.5 cm thick?

58. Because of evaporation, the water level in an aquarium drops at a rate of $\frac{1}{10}$ in./hr. In how many hours will the level drop 2 in.?

59. The bus trip from Miami to San Francisco takes 70 hr. What is the average speed of the bus rounded to the nearest mile per hour? (*Source:* Greyhound)

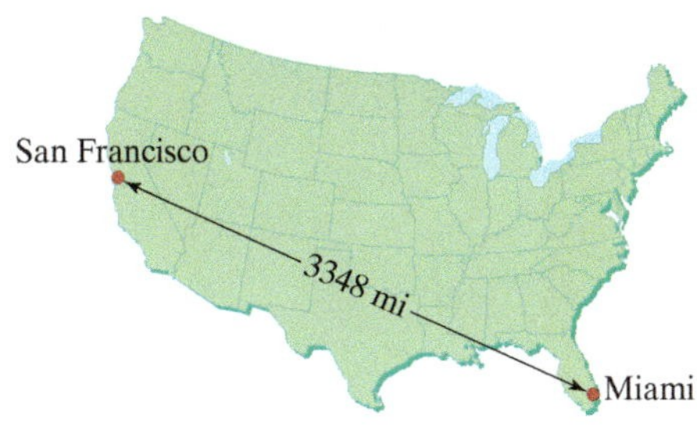

60. The coat in the ad shown is selling at $\frac{1}{4}$ the regular price. What was the regular price?

61. A customer has only \$20 to spend at a local print shop that charges \$0.05 per copy (which includes sales tax). At this rate, how many copies can she afford to make?

62. Consider the two parcels of land shown—one a rectangle and the other a square. For which value of x do the two parcels have the same area?

x mi; $\frac{1}{2}$ mi; 6 mi; 6 mi

63. A city offers to pay a disposal company \$40 per ton to bury 20,000 tons of toxic waste. If this deal represents $\frac{2}{3}$ of the disposal company's projected income, what is that income?

64. The diameter of a tree trunk increases as the tree ages and adds rings. Suppose a trunk's diameter increases by 0.2 in./yr. How many years will it take the tree to increase in diameter from 4 in. to 12 in.?

65. The maximum depth of the Caspian Sea is approximately 1000 m. If this depth is about $\frac{1}{5}$ the maximum depth of the Mediterranean Sea, what is the maximum depth of the Mediterranean Sea? (*Source:* wikipedia.org)

66. A top-secret plane flew 3000 mi in $1\frac{1}{2}$ hr. What was the average speed of this plane?

67. A student takes a job in the college's student center so that she can buy books that cost \$187.50. How many hours must she work to make this amount if she earns \$7.50/hr?

68. A double-trailer truck is driven at an average speed of 54 mph from Atlanta to Cincinnati, a driving distance of 457 mi. To the nearest tenth of an hour, how long did the trip take? (*Source:* mapquest.com)

69. Last year, a young couple paid a total of \$10,020 in rent for their apartment. How much money did they pay per month in rent?

70. An equilateral triangle is a triangle that has three sides equal in length. If the perimeter of an equilateral triangle is $10\frac{1}{2}$ ft, how long is each side?

71. After 1 yr, a student paid \$100 in simple interest on a loan to buy a computer and a printer. If she had borrowed \$2000, what was the annual rate of interest?

72. How long would it take a bank account to generate \$24 in simple interest on \$600 at 8% annual interest?

73. How much money should an investor put into a savings account in order to earn \$250 in simple interest after 1 yr, if the annual interest rate is 5%?

74. An aunt made a loan to her nephew, who agreed to pay her 4% simple interest. If he owed her \$2240 in all at the end of 3 yr, how much money had she lent him?

• Check your answers on page A-16.

MINDStretchers

Mathematical Reasoning

1. In the course of solving a linear equation, you reach the step $5x = 3x$.
   If you then divide both sides by x, you get $5 = 3$, which is impossible. Did you make an error? Explain.

Critical Thinking

2. If you multiply each side of the equation $0.24r = -12.48$ by 100, the result is an equivalent equation. Explain why it is helpful to carry out this multiplication in solving the equation.

Writing

3. Consider the equation $ax = b$, where a and b are both negative numbers. Could x also be a negative number? Explain.

12.3 Solving Linear Equations by Combining Properties

OBJECTIVES

- **A** To solve linear equations using both the addition and multiplication properties
- **B** To solve linear equations involving parentheses and combining like terms
- **C** To solve applied problems using the addition and multiplication properties, or involving parentheses and combining like terms

In the previous sections, we solved simple equations involving either the addition property or the multiplication property. We now turn our attention to solving equations that involve both properties.

Solving Equations Using Both the Addition and Multiplication Properties

We need to use both the addition property and the multiplication property to solve equations such as:

$$3x + 4 = 7 \quad \text{and} \quad \frac{r}{2} - 5 = 9$$

To solve these equations, we first use the addition property to get the variable term alone on one side. Then, we use the multiplication property to isolate the variable.

EXAMPLE 1

Solve and check: $3x + 4 = 7$

Solution

$$3x + 4 = 7$$

$$3x + 4 - 4 = 7 - 4 \quad \text{Subtract 4 from each side of the equation.}$$

$$3x = 3$$

$$\frac{3x}{3} = \frac{3}{3} \quad \text{Divide each side of the equation by 3.}$$

$$x = 1$$

Check

$$3x + 4 = 7$$

$$3(1) + 4 \stackrel{?}{=} 7 \quad \text{Substitute 1 for } x.$$

$$3 + 4 \stackrel{?}{=} 7$$

$$7 = 7 \quad \text{True}$$

So the solution is 1.

PRACTICE 1

Solve and check: $2y + 1 = 9$

In solving Example 1, would we get the same solution if we divided before subtracting? Explain.

EXAMPLE 2

Solve and check: $\frac{r}{2} - 5 = 9$

Solution

$$\frac{r}{2} - 5 = 9$$

$$\frac{r}{2} - 5 + 5 = 9 + 5 \quad \text{Add 5 to each side of the equation.}$$

$$\frac{r}{2} = 14$$

$$2 \cdot \frac{r}{2} = 2 \cdot 14 \quad \text{Multiply each side of the equation by 2.}$$

$$r = 28$$

Check

$$\frac{r}{2} - 5 = 9$$

$$\frac{28}{2} - 5 \stackrel{?}{=} 9 \quad \text{Substitute 28 for } r.$$

$$14 - 5 \stackrel{?}{=} 9$$

$$9 = 9 \quad \text{True}$$

So the solution is 28.

PRACTICE 2

Solve and check: $\frac{c}{5} - 1 = 8$

EXAMPLE 3

Solve: $-4s + 7 = 3$

Solution

$$-4s + 7 = 3$$

$$-4s + 7 - 7 = 3 - 7 \quad \text{Subtract 7 from each side of the equation.}$$

$$-4s = -4$$

$$\frac{-4s}{-4} = \frac{-4}{-4} \quad \text{Divide each side by } -4.$$

$$s = 1$$

So the solution is 1.

PRACTICE 3

Solve: $-6b - 5 = 13$

Solving Equations by Combining Like Terms

Now, let's consider an equation that has like terms on the same side of the equation. In order to solve this kind of equation, we combine all like terms before using the addition and multiplication properties.

EXAMPLE 4

Solve and check: $2x - 5x = 12$

Solution

$$2x - 5x = 12$$

$$-3x = 12 \quad \text{Combine like terms.}$$

$$\frac{-3x}{-3} = \frac{12}{-3} \quad \text{Divide each side of the equation by } -3.$$

$$x = -4$$

Check

$$2x - 5x = 12$$

$$2(-4) - 5(-4) \stackrel{?}{=} 12 \quad \text{Substitute } -4 \text{ for } x.$$

$$-8 + 20 \stackrel{?}{=} 12$$

$$12 = 12 \quad \text{True}$$

So the solution is -4.

PRACTICE 4

Solve and check: $8n + 10n = 24$

EXAMPLE 5

Solve and check: $7x - 3x - 6 = 6$

Solution

$$7x - 3x - 6 = 6$$

$$4x - 6 = 6 \quad \text{Combine like terms.}$$

$$4x - 6 + 6 = 6 + 6 \quad \text{Add 6 to each side of the equation.}$$

$$4x = 12$$

$$\frac{4x}{4} = \frac{12}{4} \quad \text{Divide each side of the equation by 4.}$$

$$x = 3$$

Check

$$7x - 3x - 6 = 6$$

$$7(3) - 3(3) - 6 \stackrel{?}{=} 6 \quad \text{Substitute 3 for } x.$$

$$21 - 9 - 6 \stackrel{?}{=} 6$$

$$6 = 6 \quad \text{True}$$

So the solution is 3.

PRACTICE 5

Solve and check: $5 - t - t = -1$

Suppose an equation has like terms that are on opposite sides of the equation. To solve, we use the addition property to get the like terms together on the same side so that they can be combined.

EXAMPLE 6

Solve: $13z + 5 = -z + 12$

Solution

$$13z + 5 = -z + 12$$

$$z + 13z + 5 = z + (-z) + 12 \quad \text{Add } z \text{ to each side of the equation.}$$

$$14z + 5 = 12 \quad \text{Combine like terms.}$$

$$14z + 5 - 5 = 12 - 5 \quad \text{Subtract 5 from each side of the equation.}$$

$$14z = 7$$

$$\frac{14z}{14} = \frac{7}{14} \quad \text{Divide each side of the equation by 14.}$$

$$z = \frac{1}{2}$$

So the solution is $\frac{1}{2}$.

PRACTICE 6

Solve: $3f - 12 = -f - 15$

Some equations have fractional terms. The key to solving a fractional equation is to *clear the equation of the fractional terms.* We do this by first determining their *least common denominator* (LCD), and then by multiplying both sides of the equation by the LCD.

EXAMPLE 7

Solve and check: $\frac{1}{2}x + \frac{1}{5}x = \frac{7}{10}$

Solution

The fractional terms in this equation are $\frac{1}{2}x$, $\frac{1}{5}x$ and $\frac{7}{10}$. The denominators of these terms are 2, 5, and 10, so the LCD is 10. To clear the equation of the fractional terms, we multiply each side of the equation by 10.

$$\frac{1}{2}x + \frac{1}{5}x = \frac{7}{10}$$

$$10 \cdot \left(\frac{1}{2}x + \frac{1}{5}x\right) = 10 \cdot \frac{7}{10} \quad \text{Multiply each side of the equation by the LCD.}$$

$$10 \cdot \frac{1}{2}x + 10 \cdot \frac{1}{5}x = 10 \cdot \frac{7}{10} \quad \text{Use the distributive property.}$$

$$5x + 2x = 7 \quad \text{Simplify.}$$

$$7x = 7 \quad \text{Combine like terms.}$$

$$x = 1 \quad \text{Divide each side of the equation by 7.}$$

PRACTICE 7

Solve and check: $\frac{1}{3}w - \frac{1}{6}w = \frac{2}{3}$

EXAMPLE 7 (continued)

Check $\frac{1}{2}x + \frac{1}{5}x = \frac{7}{10}$

$\frac{1}{2}(1) + \frac{1}{5}(1) \stackrel{?}{=} \frac{7}{10}$ Substitute 1 for x.

$\frac{1}{2} + \frac{1}{5} \stackrel{?}{=} \frac{7}{10}$

$\frac{5}{10} + \frac{2}{10} \stackrel{?}{=} \frac{7}{10}$

$\frac{7}{10} = \frac{7}{10}$ True

So the solution is 1.

Solving Equations Containing Parentheses

Some equations contain parentheses. To solve these equations, we first remove the parentheses using the distributive property. Then, we proceed as in previous examples.

EXAMPLE 8

Solve and check: $2c = -3(c - 5)$

Solution $2c = -3(c - 5)$

$2c = -3c + 15$ Use the distributive property.

$3c + 2c = 3c - 3c + 15$ Add $3c$ to each side of the equation.

$5c = 15$ Combine like terms.

$\frac{5c}{5} = \frac{15}{5}$ Divide each side of the equation by 5.

$c = 3$

Check $2c = -3(c - 5)$

$2(3) \stackrel{?}{=} -3(3 - 5)$ Substitute 3 for c.

$6 \stackrel{?}{=} -3(-2)$

$6 = 6$ True

So the solution is 3.

PRACTICE 8

Solve and check: $-5(z + 6) = z$

As Example 8 suggests, we use the following procedure to solve a linear equation:

To Solve Linear Equations

- Use the distributive property to clear the equation of parentheses, if necessary.
- Combine like terms where appropriate.
- Use the addition property to isolate the variable term.
- Use the multiplication property to isolate the variable.
- Check by substituting the solution in the original equation.

Let's apply this procedure in the next example. Recall from Section 11.6 that when removing parentheses preceded by a minus sign, all the terms in the parentheses change to the opposite sign.

EXAMPLE 9

Solve: $2(x + 5) - (x - 2) = 4x + 6$

Solution

$2(x + 5) - (x - 2) = 4x + 6$

$2x + 10 - x + 2 = 4x + 6$ — Use the distributive property.

$x + 12 = 4x + 6$ — Combine like terms.

$x - 4x + 12 = 4x - 4x + 6$ — Subtract $4x$ from each side of the equation.

$-3x + 12 = 6$ — Combine like terms.

$-3x + 12 - 12 = 6 - 12$ — Subtract 12 from each side of the equation.

$-3x = -6$

$\frac{-3x}{-3} = \frac{-6}{-3}$ — Divide each side of the equation by -3.

$x = 2$

So the solution is 2.

PRACTICE 9

Solve: $2(t - 3) - 3(t - 2) = t + 8$

EXAMPLE 10

Solve: $13 - [4 + 2(x - 1)] = 3(x + 2)$

Solution

$13 - [4 + 2(x - 1)] = 3(x + 2)$

$13 - [4 + 2x - 2] = 3(x + 2)$ — Working from the innermost pair of grouping symbols, use the distributive property.

$13 - [2 + 2x] = 3(x + 2)$

$13 - 2 - 2x = 3x + 6$

$11 - 2x = 3x + 6$

$11 - 2x - 3x = 3x - 3x + 6$

$11 - 5x = 6$

$11 - 11 - 5x = 6 - 11$

$-5x = -5$

$\frac{-5x}{-5} = \frac{-5}{-5}$

$x = 1$

So the solution is 1.

PRACTICE 10

Solve: $4[5y - (y - 1)] = 7(y - 2)$

Now, let's consider applications that lead to equations like those that we have discussed in this section.

EXAMPLE 11

An insurance company settles a claim by multiplying the claim by a certain factor and then subtracting the deductible. A payment of \$3500 is made on a claim filed for \$5000. If the company has a \$500 deductible, what factor was used to settle the claim?

Solution First, let x represent the factor used to settle a claim. Then, write the equation:

The claim times the factor less the deductible is the payment.

$$5000 \quad \cdot \quad x \quad - \quad 500 \quad = \quad 3500$$

Next, we solve for x.

$$5000x - 500 = 3500$$

$$5000x - 500 + 500 = 3500 + 500 \quad \text{Add 500 to each side of the equation.}$$

$$5000x = 4000$$

$$\frac{5000x}{5000} = \frac{4000}{5000} \quad \text{Divide each side of the equation by 5000.}$$

$$x = \frac{4}{5}, \text{ or } 0.8$$

So the factor used to compute the \$5000 claim is $\frac{4}{5}$, or 0.8.

PRACTICE 11

A car is purchased for \$12,000. The car's value depreciates \$1100 in value per year for each of the first 6 yr of ownership. At what point will the car have a value of \$6500?

Recall our discussion of motion problems in the previous section. Many motion problems lead to equations of the type that we have discussed in this section, in particular, to equations containing parentheses. Some of these problems involve one or more objects traveling the same distance, at different rates, and for different lengths of time. As in other motion problems, we use the formula $d = rt$.

EXAMPLE 12

A bus leaves St. Petersburg traveling at 45 mph. An hour later, a second bus leaves the same city traveling at 55 mph in the same direction. In how many hours will the second bus overtake the first bus?

Solution Let t represent the number of hours traveled by the second bus. Since the first bus left an hour earlier than the second, $t + 1$ represents the number of hours traveled by the first bus.

Recall from Section 12.2 that the distance that an object travels is the product of its average rate of travel (speed) and the time that it has traveled. Putting these quantities in a table clarifies their relationship.

	Rate	· Time	= Distance
First bus	45	$t + 1$	$45(t + 1)$
Second bus	55	t	$55t$

PRACTICE 12

Two friends plan to meet in Boston. A half an hour after one friend took a local train, the other friend takes an Amtrak express train on a parallel track. If both friends leave from the same station, how long will it take the express train to catch up with the local train if their speeds are 60 mph and 50 mph, respectively?

Since the two buses travel the same distance to the point where they meet, we can write the following equation and then solve:

$$45(t + 1) = 55t$$
$$45t + 45 = 55t$$
$$45t - 45t + 45 = 55t - 45t$$
$$45 = 10t$$
$$\frac{45}{10} = \frac{10}{10}t$$
$$4.5 = t, \quad \text{or}$$
$$t = 4.5$$

So the second bus will overtake the first bus in 4.5 hr, or $4\frac{1}{2}$ hr.

EXAMPLE 13

A shopper walks from home to the market at a rate of 3 mph. After shopping, he returns home following the same route walking at 2 mph. If the walk back from the market takes 10 min more than the walk to the market, how far away is the market?

Solution Let t represent the time of the walk to the market. Since the rates are given in miles per hour, we express the time in hours.

The return trip takes 10 min more time. We change 10 min to $\frac{10}{60}$ hr, or $\frac{1}{6}$ hr, getting $t + \frac{1}{6}$ for the time of the return trip.

	Rate	· Time	= Distance
Going	3	t	$3t$
Returning	2	$t + \frac{1}{6}$	$2\left(t + \frac{1}{6}\right)$

Since the walk to and from the market followed the same route, the distances each way are equal.

$$3t = 2\left(t + \frac{1}{6}\right)$$
$$3t = 2t + \frac{1}{3}$$
$$3t - 2t = 2t - 2t + \frac{1}{3}$$
$$t = \frac{1}{3}$$

It takes $\frac{1}{3}$ hr to walk to the market. So the market is $3 \cdot \frac{1}{3}$, or 1 mi away from home.

PRACTICE 13

On a round trip over the same roads, a car averaged 25 mph going and 30 mph returning. If the entire trip took 5 hr 30 min, what is the distance each way?

In some motion problems, an object travels at different speeds for different parts of the trip. The total distance traveled is the sum of the partial distances.

EXAMPLE 14

A car is driven for 3 hr in a rainstorm. After the weather clears, the car is driven 10 mph faster for 2 more hours, completing a 250-mile trip. How fast was the car driven during the storm?

Solution Let x represent the speed of the car during the storm. The speed of the car after the storm passes can be represented by $x + 10$. Drawing a diagram helps us to visualize the problem and to see that the total distance driven is the sum of the two partial distances.

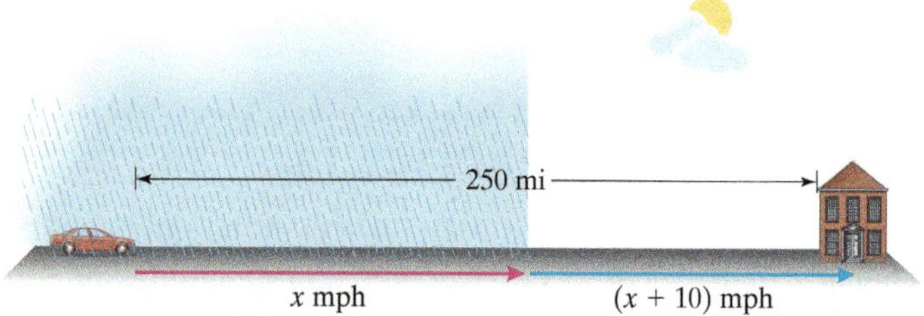

Next, let's complete a table in order to organize the relevant information. Recall that distance is the product of rate and time.

	Rate	· Time	= Distance
Storm	x	3	$3x$
Clear	$x + 10$	2	$2(x + 10)$

We are told that the sum of the two partial distances is 250 mi, giving us an equation to solve.

$$\begin{aligned} 3x + 2(x + 10) &= 250 \\ 3x + 2x + 20 &= 250 \\ 5x + 20 &= 250 \\ 5x &= 230 \\ x &= \frac{230}{5} = 46 \end{aligned}$$

So the car was driven at a speed of 46 mph during the storm.

PRACTICE 14

A cyclist pedals uphill for 2 hr. She then continues downhill 10 mph faster for another hour. If the entire trip was 40 mi in length, what was her downhill speed?

Some applications of percent involve *multiple investments,* as the following example illustrates.

EXAMPLE 15

A book editor invested part of her \$5000 bonus in a fund that paid 4% simple interest and the rest in a CD that paid 6% simple interest. Find the amount invested at each rate if the overall interest earned in 1 yr was \$260.

Solution Let's set up a table to organize the given information. Note that the interest earned for each investment is the product of each principal, the rate of interest, and the time.

	Principal ·	Rate of Interest ·	Time =	Interest Earned
Fund	x	0.04	1	$0.04x$
CD	$5000 - x$	0.06	1	$0.06(5000-x)$
Total Interest				260

The interest earned on the total investment is the sum of the interest earned on the fund and the interest earned on the CD.

$$0.04x + 0.06(5000 - x) = 260$$
$$0.04x + 300 - 0.06x = 260$$
$$-0.02x = -40$$
$$\frac{-0.02x}{-0.02} = \frac{-40}{-0.02}$$
$$x = 2000$$

So \$2000 was invested in the fund and \$5000 − \$2000, or \$3000, was invested in the CD.

PRACTICE 15

The amount of money a broker invested in bonds was double what she invested in a mutual fund. After 1 yr, the investment in bonds gained 10% and the investment in the mutual fund lost 10%. If the net gain was \$700, how much was each of her investments?

Another type of problem involving percent is a *mixture* or *solution problem.* In this kind of problem, the amount of a particular ingredient in the solution or mixture is often expressed as a percent of the total. Chemists and pharmacists commonly need to solve mixture and solution problems.

EXAMPLE 16

A chemist added water to 8 L of a 50% alcohol solution. How much water must be added so that the alcohol concentration of the new solution is 40%?

Solution We need to add a quantity of water, with 0% alcohol, to dilute 8 L of 50% alcohol solution so that the new solution has a 40% alcohol concentration. Let x represent the desired amount of water to be added. It is helpful to organize the given information into a table, as follows:

Action	Percent of Alcohol in the Solution (expressed as a decimal)	Amount of Solution (in liters)	Amount of Alcohol (in liters)
Start with	0.5	8	4
Add	0	x	0
Finish with	0.4	$x + 8$	$0.4(x + 8)$

In each row, the amount of alcohol is the product of the percent of alcohol in the solution and the amount of the solution. Also in the new solution, the amount of alcohol is equal to the sum of the amount of alcohol in the initial solution and the amount of alcohol in the water added.

$$0.4(x + 8) = 4$$
$$0.4x + 3.2 = 4$$
$$0.4x = 0.8$$
$$\frac{0.4}{0.4}x = \frac{0.8}{0.4}$$
$$x = 2$$

So 2 L of water must be added.

PRACTICE 16

How many grams of salt must be added to 30 g of a solution that is 20% salt to make a solution that is 25% salt?

How would you explain each entry in the table in Example 16?

12.3 Exercises

FOR EXTRA HELP MyMathLab MathXL PRACTICE WATCH READ REVIEW

A *Solve and check.*

1. $3x - 1 = 8$

2. $7r - 8 = 13$

3. $9t + 17 = -1$

4. $2y + 1 = 9$

5. $20 - 5m = 45$

6. $25 - 3c = 34$

7. $\frac{n}{2} - 1 = 5$

8. $\frac{s}{3} - 2 = -4$

9. $\frac{x}{5} + 15 = 0$

10. $\frac{y}{3} + 3 = 42$

11. $3 - t = 1$

12. $2 - x = 2$

13. $-8 - b = 11$

14. $5 - y = 8$

15. $\frac{2}{3}x - 9 = 17$

16. $\frac{4}{5}d - 3 = 13$

17. $\frac{4}{5}r + 20 = -20$

18. $\frac{3y}{8} + 14 = -10$

19. $3y + y = -8$

20. $4a + 3a = -21$

21. $7z - 2z = -30$

22. $4x - x = 18$

23. $28 - a + 4a = 7$

24. $5 - 8x - 2x = -25$

25. $1 = 1 - 6t - 4t$

26. $-1 = 5 - z - z$

27. $3y + 2 = -y - 2$

28. $3n + 6 = -n - 6$

29. $5r - 4 = 2r + 6$

30. $7 - m = 5 + 3m$

B *Solve and check.*

31. $4(x + 7) = 7 + x$

32. $3t - 2 = 4(t - 2)$

33. $5(y - 1) = 2y + 1$

34. $5a - 4 = 7(a + 2)$

35. $3a - 2(a - 9) = 4 + 2a$

36. $5 - 2(3x - 4) = 3 - x$

37. $5(2 - t) - (1 - 3t) = 6$

38. $\frac{3}{5}(15y + 10) - 3(4y + 3) = 0$

39. $2y - 3(y + 1) = -(5y + 3) + y$

40. $9n + 5(n + 3) = -(n + 13) - 2$

41. $2[3z - 5(2z - 3)] = 3z - 4$

42. $5[2 - (2n - 4)] = 2(5 - 3n)$

43. $-8m - [2(11 - 2m) + 4] = 9m$

44. $7 - [4 + 2(a - 3)] = 11(a + 2)$

Choose the equation that best describes the situation.

45. A car leaves Seattle traveling at a rate of 45 mph. One hour later, a second car leaves from the same place, along the same road, at 54 mph. If the first car travels for t hr, in how many hours will the second car overtake the first car?

a. $54(t - 1) = 45t$ **b.** $45(t + 1) = 54t$
c. $54(t + 1) = 45t$ **d.** $45(t - 1) = 54t$

46. A company budgets \$600,000 for an advertising campaign. It must pay \$4000 for each television commercial and \$1000 per radio commercial. If the company plans to air 50 fewer radio commercials than television commercials, find the number of television commercials t that will be in the advertising campaign.

a. $4000t + 1000(t + 50) = 600{,}000$ **b.** $4000t + 1000(t - 50) = 600{,}000$
c. $4000t + 1000t - 50 = 600{,}000$ **d.** $1000t + 4000(t - 50) = 600{,}000$

47. A taxi fare is \$3.00 for the first mile and \$1.25 for each additional mile. If a passenger's total cost was \$5.50, how far did she travel in the taxi?

a. $3.00x + 1.25 = 5.50$ **b.** $3.00 + 1.25x = 5.50$
c. $3.00 + 1.25(x + 1) = 5.50$ **d.** $3.00 + 1.25(x - 1) = 5.50$

48. A family's budget allows $\frac{1}{3}$ of the family's monthly income for housing and $\frac{1}{4}$ of its monthly income for food. If a total of \$1050 a month is budgeted for housing and food, what is the family's monthly income?

a. $\frac{1}{3}x = 1050 + \frac{1}{4}x$ **b.** $\frac{1}{3}x - \frac{1}{4}x = 1050$
c. $\frac{1}{3}x + \frac{1}{4}x = 1050$ **d.** $\frac{1}{4}x - \frac{1}{3}x = 1050$

Mixed Practice

Solve and check.

49. $5 - 5x + x = 21$

50. $16 - 3t = 31$

51. $4z + 3(5 - z) = -(z - 3) + 8$

52. $\frac{r}{7} + 3 = 11$

53. $0.8x - 1.5(x - 1) = 5$

54. $7y - 6 = 3(y - 6)$

Applications

C *Write an equation and solve.*

55. A part-time student at a college pays a student fee of \$45 plus \$135 per credit. How many credits is a part-time student carrying who pays \$1260 in all?

56. A health club charges members \$10/mo plus \$5/hr to use the facilities. If a member was charged \$55 this month, how many hours did she use the facilities?

57. In a local election, a newspaper reported that one candidate received twice as many votes as the other. Altogether, they received a total of 3690 votes. How many votes did each candidate receive?

58. Calcium carbonate (chalk) consists of 10 parts of calcium for each 3 parts of carbon and 12 parts of oxygen by weight. Find the amount of calcium in 75 lb of chalk.

59. A parking garage charges \$3 for the first hour and \$2 for each additional hour or fraction thereof. If \$9 was paid for parking, how many hours was the car parked in the garage?

60. A machinist earns \$11.50 an hour for the first 35 hr and \$15.30 for each hour over 35 per week. How many hours did he work this week if he earned \$555.50?

61. The office manager of an election campaign office needs to print 5000 postcards. It costs 2 cents to print a large postcard and 1 cent to print a small postcard. If \$85 is allocated for printing postcards, how many of each type of postcard can be printed?

62. The owner of a small factory has 8 employees. Some of the employees make \$10/hr, whereas the others make \$15/hr. If the total payroll is \$105/hr, how many employees make the higher rate of pay?

63. Twenty minutes after a father left for work on the bus, he noticed that he had left his briefcase at home. His son left home, driving at 36 mph, to catch the bus that was traveling at 24 mph. How long did it take the son to catch the bus?

64. In the rush hour, a commuter drives to work at 30 mph. Returning home off-peak, she takes $\frac{1}{4}$ hr less time driving at 40 mph. What is the distance between the commuter's work and her home?

65. The two snails shown below crawl toward each other at rates that differ by 2 cm/min. If it takes the snails 27 min to meet, how fast is each snail crawling?

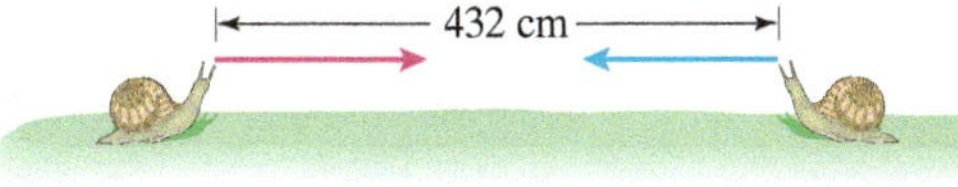

66. A signal is sent from a station on the ground to a satellite. The signal bounces off the satellite and then is received at a second ground station. If the signal traveled 2400 mi in all, at what speed was it traveling?

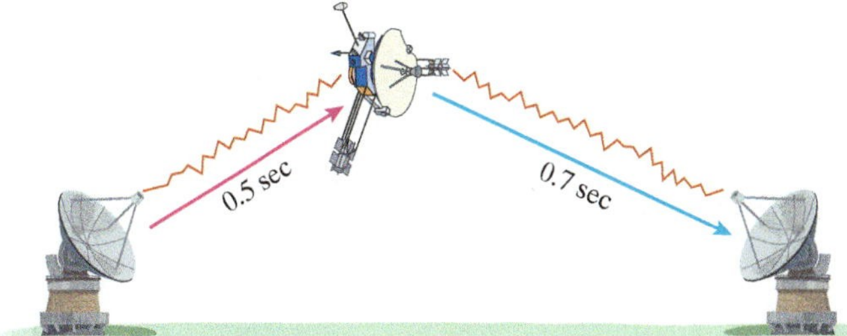

67. Two trucks leave a depot at the same time, traveling in opposite directions. One truck goes 4 mph faster than the other. After 2 hr, the trucks are 212 mi apart. What is the speed of the slower truck?

68. If a student drives from college to home at 40 mph, then he is 15 min late. However, if he makes the same trip at 50 mph, he is 12 min early. What is the distance between his college and his home?

69. Part of \$34,000 is invested at an interest rate of 8% and the rest at an interest rate of 10%. If the total interest earned in 1 yr was \$3000, how much money was invested at each rate?

70. The amount of money invested at 5% simple interest was half the amount invested at 7% simple interest. If the total yearly interest earned was \$380, how much money was invested at each rate?

71. An amount of \$20,000 was invested in a fund with a return of 8%. How much money was invested in a fund with a 5% return if the total return on both investments was \$2100?

72. A beneficiary split her \$100,000 inheritance between two investments. One of the investments gained 12% and the other lost 8%. If she broke even on the two investments, how much money did she invest in each?

73. A basic lemon vinaigrette salad dressing can be made by mixing 1 cup of olive oil with $\frac{1}{3}$ cup of lemon juice along with pinches of salt and white pepper. How much more olive oil should be added to make a dressing that is 20% lemon juice? (The salt and pepper contribution is negligible and can be left out of the computation.) (*Source: The Sauce Bible*)

74. A brand of antifreeze states that a radiator containing a solution that is 50% antifreeze and 50% water provides protection for a temperature as low as −34°F, whereas a solution that is 70% antifreeze provides protection down to −84°F. A car's radiator has 4 qt of a 50% solution. If the capacity of the radiator is 6 qt and the rest of the radiator is filled with pure antifreeze, what percent of the resulting solution is antifreeze? (*Source:* Prestone II Antifreeze)

75. How many ounces of a 40% acetic acid solution should a photographer add to 30 oz of a 4% acetic acid solution to obtain a 10% solution?

76. A pharmacist has 10 L of a 5% drug solution. How many liters of 2% solution should be added to produce a solution that is 3%?

• Check your answers on page A-16.

MINDStretchers

Mathematical Reasoning

1. Give an example of an equation that involves combining like terms and that has:
   - **a.** no solution.
   - **b.** an infinite number of solutions.

Patterns

2. Tables can be useful in solving equations.
   - **a.** Complete the following table. After examining your results, identify the solution to the following equation: $3(x - 2) = 2(x + 1)$.

x	0	2	4	6	8	10	12
$3(x - 2)$							
$2(x + 1)$							

   - **b.** Try a similar approach to solving the equation $7x = 5x + 11$. What conclusion can you draw about the solution?

x	0	2	4	6	8	10	12
$7x$							
$5x + 11$							

Groupwork

3. Working with a partner, choose a month on a calendar.
   - **a.** Ask your partner to select four days of the month that form a 2×2 square, but only to tell you the sum of the four days. Determine the four days.
   - **b.** Reverse roles with your partner and repeat part (a).
   - **c.** Compare how you and your partner responded to part (a).

12.4 Solving Literal Equations and Formulas

OBJECTIVES

- **A** To solve literal equations
- **B** To solve or evaluate formulas
- **C** To solve applied problems involving formulas

In many situations, equations describe the relationship between two or more variables. Such equations are called *literal* equations and can be used to describe situations such as how the amount you pay depends on what you buy, how the distance you walk determines how long the walk takes, and how the dosage of a medicine you need to take relates to your weight.

DEFINITION

A **literal equation** is an equation involving two or more variables.

Consider, for instance, the following literal equation:

$$x = y + z$$

Since there is more than one variable, we can solve for one of the variables in terms of the others. For instance, in this equation we can solve for y in terms of x and z.

EXAMPLE 1

Solve $x = y + z$ for y in terms of x and z.

Solution

$$x = y + z$$

$$x - z = y + z - z \quad \text{Subtract } z \text{ from each side of the equation.}$$

$$x - z = y, \text{ or } y = x - z$$

PRACTICE 1

Solve $q = 1 - p$ for p in terms of q.

We see from Example 1 that the solution to a literal equation is not a number, but an algebraic expression.

EXAMPLE 2

Solve $2a + b = c$ for a in terms of b and c.

Solution

$$2a + b = c$$

$$2a + b - b = c - b \quad \text{Subtract } b \text{ from each side of the equation.}$$

$$2a = c - b \quad \text{Simplify.}$$

$$\frac{2a}{2} = \frac{c - b}{2} \quad \text{Divide each side of the equation by 2.}$$

$$a = \frac{c - b}{2}$$

So $a = \dfrac{c - b}{2}$.

PRACTICE 2

Solve $3r - s = t$ for r in terms of s and t.

Note that we can solve literal equations by using the addition and multiplication properties that we have already used to solve other equations. Can you explain how to check your solution to a literal equation?

EXAMPLE 3

Solve $\frac{2K}{3m} = n$ for K.

Solution

$$\frac{2K}{3m} = n$$

$$3m \cdot \frac{2K}{3m} = 3m \cdot n \quad \text{Multiply each side of the equation by } 3m.$$

$$2K = 3mn$$

$$\frac{2K}{2} = \frac{3mn}{2} \quad \text{Divide each side of the equation by 2.}$$

$$K = \frac{3mn}{2}$$

So $K = \frac{3mn}{2}$.

PRACTICE 3

Solve $\frac{4x}{5a} = c$ for x.

EXAMPLE 4

Consider the equation $Ax + By = C$. Solve for y in terms of A, B, C, and x.

Solution

$$Ax + By = C$$

$$Ax - Ax + By = C - Ax \quad \text{Subtract } Ax \text{ from each side of the equation.}$$

$$By = C - Ax \quad \text{Simplify.}$$

$$\frac{By}{B} = \frac{C - Ax}{B} \quad \text{Divide each side of the equation by } B.$$

$$y = \frac{C - Ax}{B}$$

So $y = \frac{C - Ax}{B}$.

PRACTICE 4

Consider the equation $y = mx + b$. Solve for x in terms of y, m, and b.

Recall that in Section 11.5 we discussed formulas—literal equations in a real-world context. Here we focus on solving a formula for one variable in terms of the other variables. When we need to use a formula repeatedly to find the value of a particular variable, the computation can be simplified by first solving for that variable in the formula.

EXAMPLE 5

$P = 2(l + w)$ is the formula for the perimeter P of a rectangle in terms of its length l and width w.

a. Solve for l in terms of P and w.

b. Using the formula found in part (a), find the length of a rectangle with a perimeter of 20 cm and a width of 6 cm.

Solution

a.

$$P = 2(l + w)$$

$$P = 2l + 2w \quad \text{Use the distributive property.}$$

$$P - 2w = 2l + 2w - 2w \quad \text{Subtract } 2w \text{ from each side of the equation.}$$

$$P - 2w = 2l$$

$$\frac{P - 2w}{2} = \frac{2l}{2} \quad \text{Divide each side of the equation by 2.}$$

$$\frac{P - 2w}{2} = l, \text{ or}$$

$$l = \frac{P - 2w}{2}$$

So $l = \frac{P - 2w}{2}$.

b. $P = 20$ cm and $w = 6$ cm. We substitute in the formula $l = \frac{P - 2w}{2}$ to find the value of l.

$$l = \frac{P - 2w}{2} = \frac{20 - 2 \cdot 6}{2} = \frac{20 - 12}{2} = \frac{8}{2} = 4$$

So the length of the rectangle is 4 cm.

PRACTICE 5

$A = P(1 + rt)$ is the formula for computing the amount in an account earning simple interest. In the formula, A stands for the amount, P for the original principal, r for the annual rate of interest, and t for time.

a. Find a formula for r in terms of A, P, and t.

b. Using the equation found in part (a), evaluate r if $A = \$2100$, $P = \$2000$, and $t = 2$ years.

EXAMPLE 6

$V = lwh$ is the formula for finding the volume of a rectangular solid, where l represents the length of the solid, w the width, and h the height.

a. Solve the formula for h.

b. Using the equation found in part (a), find the value of h for $V = 48$ cu ft, $l = 8$ ft, and $w = 2$ ft.

Solution

a. Solving $V = lwh$ for h, we get:

$$V = lwh$$

$$\frac{V}{lw} = \frac{lwh}{lw}$$

$$\frac{V}{lw} = h$$

So $h = \frac{V}{lw}$.

PRACTICE 6

$A = \frac{1}{2}bh$ is the formula for finding the area of a triangle, where b is the base and h is the height.

a. Express h in terms of A and b.

b. Using the formula found in part (a), find the value of h for $A = 63$ in^2 and $b = 9$ in.

b. To find the value of h, we substitute 48 for V, 8 for l, and 2 for w in the formula $h = \frac{V}{lw}$.

$$h = \frac{48}{8(2)} = 3$$

The height is 3 ft.

EXAMPLE 7

To convert a temperature expressed in Fahrenheit degrees F to Celsius degrees C, a meteorologist multiplies $\frac{5}{9}$ by the difference between the Fahrenheit temperature and 32.

a. Write a formula for this relationship.

b. Solve the formula for F.

c. What Fahrenheit temperature corresponds to a Celsius temperature of 30°?

Solution

a. Stating the rule in words, we get:

Celsius temperature, C, is equal to $\frac{5}{9}$ times the quantity Fahrenheit temperature, F, minus 32.

Then, we translate this relationship to a formula.

$$C = \frac{5}{9}(F - 32)$$

b. Now, we solve the formula found in part (a) for F.

$$C = \frac{5}{9}(F - 32)$$

$$\frac{9}{5}C = \frac{9}{5} \cdot \frac{5}{9}(F - 32)$$

$$\frac{9}{5}C = F - 32$$

$$\frac{9}{5}C + 32 = F$$

So $F = \frac{9}{5}C + 32$.

c. To find the Fahrenheit temperature that corresponds to a Celsius temperature of 30°, we substitute 30 for C in the formula.

$$F = \frac{9}{5}C + 32 = \frac{9}{5}(30) + 32 = 54 + 32 = 86$$

The corresponding Fahrenheit temperature is, therefore, 86°.

PRACTICE 7

To find the area A of a trapezoid, multiply $\frac{1}{2}$ its height h by the sum of the trapezoid's upper base b and lower base B.

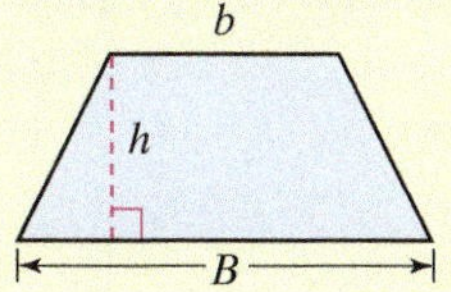

a. Translate this relationship to a formula.

b. Solve the formula for b.

c. What is the upper base of a trapezoid whose area is 32 cm^2, height is 4 cm, and lower base is 11 cm?

12.4 Exercises

FOR EXTRA HELP MyMathLab MathXL PRACTICE WATCH READ REVIEW

Mathematically Speaking

Fill in each blank with the most appropriate term or phrase from the given list.

number	expression	variable
linear equation	literal equation	formula
constant	algebraic expression	

1. A(n) ________________ is an equation involving two or more variables.

2. Literal equations can be solved for one ______________ in terms of the others.

3. The solution to a literal equation is a(n) ________________.

4. A(n) ________________ is a literal equation in a real-world context.

A ***Solve each equation for the indicated variable.***

5. $y + 10 = x$ for y

6. $b + 13 = a$ for b

7. $d - c = 4$ for d

8. $x - z = -5$ for x

9. $-3y = da$ for d

10. $ax = 5b$ for x

11. $\frac{1}{2}n = 2p$ for n

12. $\frac{3}{2}m = -4l$ for m

13. $a = \frac{1}{2}xyz$ for z

14. $w = \frac{2}{3}rst$ for r

15. $3x + y = 7$ for x

16. $x + 2y = 5$ for y

17. $3x + 4y = 12$ for y

18. $5a + 2b = 10$ for a

19. $y - 4t = 0$ for y

20. $6 = p - 4z$ for p

21. $-5b + p = r$ for b

22. $-7a + 3b = c$ for a

23. $h = 2(m - 2l)$ for l

24. $3(a - 2b) = c$ for b

B ***Solve each formula for the indicated variable.***

25. Uniform motion: $d = rt$ for r

26. Electrical power: $P = iV$ for i

27. Perimeter of a triangle: $P = a + b + c$ for b

28. Perimeter of a rectangle: $P = 2l + 2w$ for w

29. Circumference of a circle: $C = \pi d$ for d

30. Aspect ratio of a hang glider: $R = \frac{s^2}{a}$ for a

31. Power: $P = I^2R$ for R

32. Centripetal force: $F = \frac{mv^2}{r}$ for m

33. Average of three numbers: $A = \frac{a + b + c}{3}$ for a

34. Distance of a free-falling object: $S = \frac{1}{2}gt^2$ for g

35. Arithmetic progression: $S = a + (n - 1)d$ for a

36. Simple interest: $A = P(1 + rt)$ for t

Solve each formula for the indicated variable. Then, find the value of this variable using the given information.

37. Simple interest

a. $I = Prt$ for r

b. $I = \$12$, $P = \$200$, and $t = 2$ yr

38. Perimeter of a square

a. $P = 4s$ for s

b. $P = 60$ ft

39. Area of a parallelogram

a. $A = bh$ for b

b. $A = 30\text{ m}^2$ and $h = 6$ m

40. Average of two numbers

a. $A = \dfrac{a + b}{2}$ for b

b. $A = 11.5$ and $a = 10$

Mixed Practice

Solve each formula for the indicated variable.

41. Volume of a cylinder: $V = \pi r^2 h$ for h

42. Area of a rectangle: $A = lw$ for w

Solve each equation for the indicated variable.

43. $m = \dfrac{2}{5}abc$ for b

44. $-10x = yz$ for x

45. $4w + 9z = 3$ for z

46. $a + b = -3$ for a

Solve for the variable shown in color. Then, find the value of this variable for the given values of the other variables.

47. Profit: $P = R - C$ when $P = \$500$ and $C = \$2000$

48. Per capita income: $C = \dfrac{T}{P}$ when $C = \$40{,}000$ and $P = 300{,}000{,}000$

Applications

C ***Solve.***

49. In physics, Charles's Law describes the relationship between the volume of a gas and its temperature. The law states that the volume V divided by the temperature T is equal to a constant K.

a. Write an equation to express this relationship

b. Solve the relationship for V.

50. In geometry, the volume V of a cylinder is the product of the area B of the circular base and the height h.

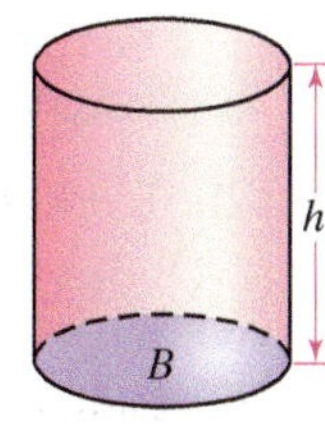

a. Write a formula for this relationship.

b. Solve the formula for h.

51. In nursing, Clark's Rule for medication expresses a relationship between the recommended dosages for a child and for an adult. The rule states that a child's dosage C equals the product of the weight W of the child in pounds divided by 150 and the adult's dosage A.

a. Write a formula for this relationship.

b. Solve the formula for A.

52. The Scholastic Aptitude Test (SAT) is a well-known test taken by high school seniors applying for college admission. Scores on such tests are often converted to standardized scores, or *z-scores*. In 2010, the formula for determining a z-score on the mathematics portion of the SAT was $z = \dfrac{x - 516}{116}$, where x is an individual score on the test. The value 516 is the national average score on the test whereas 116 is a measure of how spread out the scores were. (*Source:* professionals.collegeboard.com)

a. Solve for x in terms of z.

b. If your z-score is 2.2, what was your score on the test, rounded to the nearest whole number?

53. During a storm, the number of miles away m a bolt of lightning strikes can be estimated by first counting the number of seconds s between the bolt of lightning and the associated clap of thunder and then dividing by 5.

a. Translate this relationship to a formula.

b. Solve for s in terms of m.

c. If lightning strikes 2.5 mi away, how many seconds will elapse before the thunder is heard?

54. To estimate a man's shoe size S, triple his foot length l (expressed in inches) and subtract 22.

a. Express this relationship as a formula.

b. Estimate the shoe size of a man with a 12-in. foot.

c. Estimate the foot length of a man with shoe size 12.

55. The circumference C of a circle can be found by doubling the product of the constant π and the circle's radius r.

a. Express this relationship as a formula.

b. Solve this formula for r in terms of C and π.

c. Find the value of r rounded to the nearest tenth if C is 5 ft and π is approximately 3.14.

56. According to Newton's second law of motion, the force F (in newtons) applied to an object equals the product of the object's mass m (in kilograms) and its acceleration a (in m/sec^2).

a. Translate this relationship to a formula.

b. Solve for a in terms of m and F.

c. Find a (in m/sec^2) if m is 3 kg and F is 10 newtons.

• Check your answers on page A-17.

MINDStretchers

Research

1. By examining books in your college library or websites, find several examples of literal equations that relate two or more variables. Write the equation and explain what the variables represent.

Groupwork

2. Working with a partner, give an example of a situation that the following formula might describe.

$$y = mx + b$$

Explain what each variable represents in your example.

Patterns

3. Consider the following table.

x	0	1	2	3	4	⋯	10
y	1	3	5	7	9	⋯	21

Write an equation for y in terms of x.

12.5 Solving Linear Inequalities

OBJECTIVES

- **A** To determine if a number is a solution of an inequality
- **B** To graph the solutions of linear inequalities on the number line
- **C** To solve linear inequalities using the addition and multiplication properties of inequalities
- **D** To solve applied problems involving linear inequalities

In Section 11.1, we used the symbols $<$, $\le$, $>$, $\ge$, $=$, and $\neq$ to compare two real numbers. For example, with the real numbers -5 and 4, we can write the following statements:

$$-5 < 4 \qquad 4 > -5 \qquad -5 \neq 4$$

DEFINITION

An **inequality** is any mathematical statement containing $<$, $\le$, $>$, $\ge$, or $\neq$.

Solutions of Inequalities

Now, consider an inequality that involves a variable, say $x < 2$. Let's look at the values of x that make this inequality true.

	Values for x	$x < 2$	True or False?
Values for x that are less than 2	1	$1 < 2$	True
	$\frac{1}{2}$	$\frac{1}{2} < 2$	True
	0	$0 < 2$	True
	-1	$-1 < 2$	True
Values for x that are not less than 2	2	$2 < 2$	False
	3	$3 < 2$	False
	$3\frac{1}{2}$	$3\frac{1}{2} < 2$	False
	4	$4 < 2$	False

Note that there are many values of x that make $x < 2$ true. Can you name them all? Explain.

DEFINITION

A **solution of an inequality** is any value of the variable that makes the inequality true. To **solve an inequality** is to find all of its solutions.

EXAMPLE 1

Determine whether -3 is a solution of the inequality $2x + 5 \ge -3$.

Solution To determine if -3 is a solution of the inequality, we substitute -3 for x and simplify.

$$2x + 5 \ge -3$$

$$2(-3) + 5 \overset{?}{\ge} -3 \qquad \text{Substitute } -3 \text{ for } x.$$

$$-6 + 5 \overset{?}{\ge} -3 \qquad \text{Multiply.}$$

$$-1 \ge -3 \qquad \text{True}$$

Because $-1 \ge -3$ is a true statement, -3 is a solution of the inequality.

PRACTICE 1

Is 4 a solution of the inequality $\frac{1}{2}x - 2 < -1$?

For any inequality, we can draw on the number line a picture of its solutions—the *graph* of the inequality. Graphing the solutions of an inequality can be clearer than describing the solutions in symbols or words.

EXAMPLE 2

Draw the graph of $x < 2$.

Solution The graph of $x < 2$ includes all points on the number line to the left of 2. The open circle on the graph shows that 2 is *not* a solution.

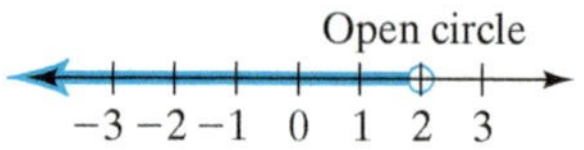

PRACTICE 2

Draw the graph of $x > 1$.

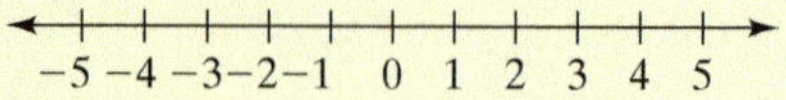

Note that solving an inequality generally results in a range of numbers rather than in a single number.

EXAMPLE 3

Draw the graph of $x \geq -\frac{1}{2}$.

Solution The graph of $x \geq -\frac{1}{2}$ includes all points on the number line to the right of $-\frac{1}{2}$ and also $-\frac{1}{2}$. The *closed circle* shows that $-\frac{1}{2}$ is a solution of $x \geq -\frac{1}{2}$.

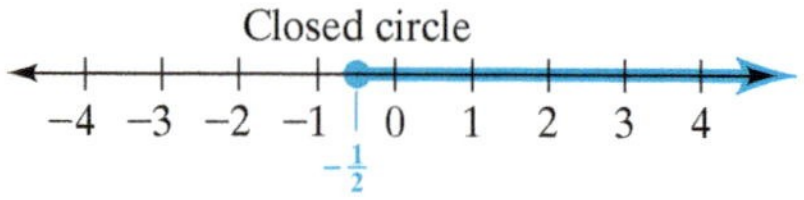

PRACTICE 3

Draw the graph of $x \leq -1\frac{1}{2}$.

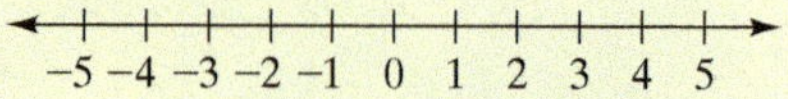

EXAMPLE 4

Draw the graph of $-1 \leq x < 2$.

Solution This inequality is read either "-1 is less than or equal to x *and* x is less than 2" or "x is greater than or equal to -1 *and* x is less than 2." The solutions of this inequality are all values of x that satisfy both $-1 \leq x$ and $x < 2$, and its graph is the overlap of the graphs of $-1 \leq x$ and $x < 2$.

Graph of $-1 \leq x$:

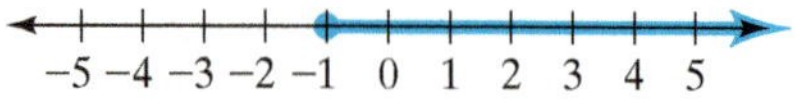

Graph of $x < 2$:

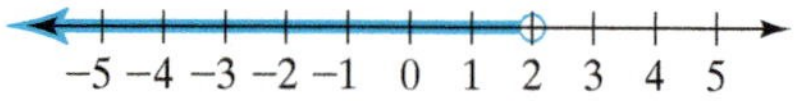

PRACTICE 4

Draw the graph of $-3 < x < 4$.

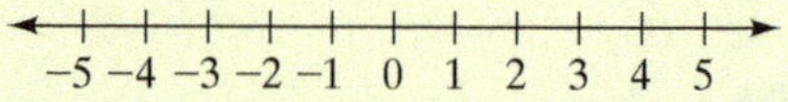

Graph of $-1 \leq x < 2$:

−5 −4 −3 −2 −1 0 1 2 3 4 5

Overlap of the graphs of $-1 \leq x$ and $x < 2$

Note that the graph includes −1 and all points *between* −1 and 2 on the number line. A closed circle at −1 means that −1 is a solution of the inequality, whereas an open circle at 2 means that 2 is not a solution of the inequality.

Solving Inequalities Using the Addition Property

Now, let's consider what happens when we perform the same operation on each side of an inequality. In the following inequality, we add 4 to each side of the inequality $5 < 12$.

$$5 < 12 \quad \text{True}$$

$$5 + 4 \stackrel{?}{<} 12 + 4 \quad \text{Add 4 to each side of the inequality.}$$

$$9 < 16 \quad \text{True}$$

Note that $5 < 12$ and $5 + 4 < 12 + 4$ are both true. In the same way, adding −3 to both sides of $x + 3 < 7$ gives us the *equivalent* inequality $x + 3 + (-3) < 7 + (-3)$, or $x < 4$. Alternatively, we could subtract 3 from both sides of the original inequality, getting the same inequality as before. Inequalities are said to be equivalent if they have the same solutions. For instance, 2 is a solution of $x + 3 < 7$ just as it is a solution of $x < 4$.

This example suggests the addition property of inequalities.

The Addition Property of Inequalities

For any real numbers a, b, and c:

- $a < b$ and $a + c < b + c$ are equivalent.
- $a > b$ and $a + c > b + c$ are equivalent.

Similar statements hold for $\leq$ and $\geq$.

TIP When the same number is added to or subtracted from each side of an inequality, the *direction* of the inequality is unchanged.

The way we solve inequalities is similar to the way we solve equations.

Equation	Inequality
$x + 3 = 5$	$x + 3 < 5$
$x + 3 - 3 = 5 - 3$	$x + 3 - 3 < 5 - 3$
$x = 2$ Solution	$x < 2$ Solution
$x + 3 = 5$ is equivalent to $x = 2$.	$x + 3 < 5$ is equivalent to $x < 2$.

We solve inequalities by expressing them as equivalent inequalities in which the variable term is isolated on one side.

EXAMPLE 5

Solve and graph: $y + 5 > 9$

Solution

$$y + 5 > 9$$
$$y + 5 - 5 > 9 - 5 \quad \text{Subtract 5 from each side of the inequality.}$$
$$y > 4$$

The graph of $y > 4$ is:

−1 0 1 2 3 4 5 6 7

Note that an open circle is drawn at 4 to show that 4 is not a solution.

Because an inequality has many solutions, we cannot check all of the solutions as we did with an equation. However, we can do a partial check of the solutions of an inequality by substituting points on the graph in the original inequality. For instance, to check that all values for y greater than 4 are the solutions of $y + 5 > 9$, we replace y in the original inequality with some points on the graph and some points not on the graph.

Values for y	$y + 5 > 9$	True or False
6	$6 + 5 > 9$	True
5	$5 + 5 > 9$	True
4	$4 + 5 > 9$	False
3	$3 + 5 > 9$	False

The table confirms that the solution of $y + 5 > 9$ is $y > 4$. That is, any number greater than but not equal to 4 is a solution.

PRACTICE 5

Solve and graph: $n + 5 > 4$

EXAMPLE 6

Solve and graph: $z - 2 \leq -3\frac{1}{2}$

Solution

$$z - 2 \leq -3\tfrac{1}{2}$$
$$z - 2 + 2 \leq -3\tfrac{1}{2} + 2 \quad \text{Add 2 to each side of the inequality.}$$
$$z \leq -1\tfrac{1}{2}$$

So all numbers less than or equal to $-1\frac{1}{2}$ are solutions. The graph of $z \leq -1\frac{1}{2}$ is:

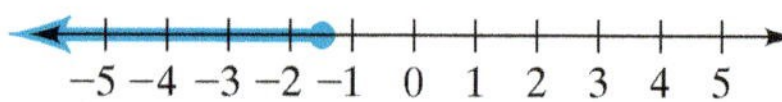

PRACTICE 6

Solve and graph: $x - 4 \leq 1\frac{1}{2}$

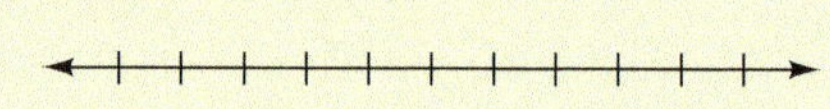

EXAMPLE 7

Solve and graph: $6y - 3 < 5y + 4$

Solution $6y - 3 < 5y + 4$

$6y - 5y - 3 < 5y - 5y + 4$ Subtract $5y$ from each side of the inequality.

$y - 3 < 4$

$y - 3 + 3 < 4 + 3$ Add 3 to each side of the inequality.

$y < 7$

So all numbers less than or equal to 7 are solutions of $6y - 3 < 5y + 4$. The graph of $y < 7$ is:

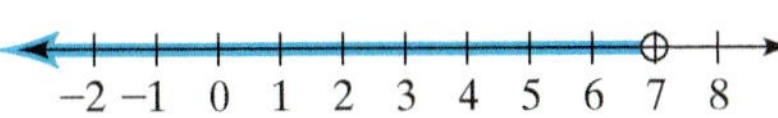

PRACTICE 7

Solve and graph: $4x + 5 > 3x - 2$

Solving Inequalities Using the Multiplication Property

Consider the inequality $12 < 15$. Let's look at what happens when we multiply this inequality by a *positive* number:

$12 < 15$

$12 \cdot 3 \overset{?}{<} 15 \cdot 3$ Multiply each side of the inequality by 3.

$36 < 45$ True

Note that the direction of the last inequality is the same as that of the first inequality. Now, we consider multiplying each side of the original inequality by a *negative* number:

$12 < 15$

$12(-3) \overset{?}{<} 15(-3)$ Multiply each side of the inequality by -3.

$-36 \overset{?}{<} -45$ False, unless the direction of the inequality sign is reversed

$-36 > -45$ True

Observe that the direction of the last inequality is the reverse of that of the first inequality.

These examples suggest the *multiplication property of inequalities.*

The Multiplication Property of Inequalities

For any real numbers a, b, and c:

- If c is positive, then $a < b$ and $ac < bc$ are equivalent.
- If c is negative, then $a < b$ and $ac > bc$ are equivalent.

Similar statements hold for $>$, $\leq$, and $\geq$.

We can demonstrate a similar property for division.

$12 < 15$

$\frac{12}{3} \overset{?}{<} \frac{15}{3}$ Divide each side of the inequality by 3.

$4 < 5$ True

$12 < 15$

$\frac{12}{-3} \overset{?}{<} \frac{15}{-3}$ Divide each side of the inequality by -3.

$-4 \overset{?}{<} -5$ False, so we need to reverse the direction of the inequality sign.

$-4 > -5$ True

Note that when we multiply or divide each side of an inequality by a positive number, the direction of the inequality remains the same. But when we multiply or divide each side of an inequality by a negative number, the direction of the inequality is *reversed.*

EXAMPLE 8

Solve and graph: $\frac{x}{2} < 3$

Solution $\frac{x}{2} < 3$

$2 \cdot \frac{x}{2} < 2 \cdot 3$ Multiply each side of the inequality by 2.

$x < 6$

So any number less than 6 is a solution. The graph of $x < 6$ is:

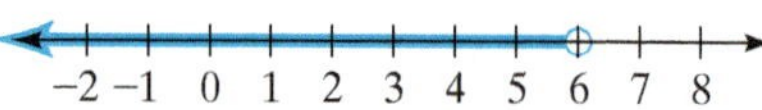

PRACTICE 8

Solve and graph: $\frac{x}{3} \leq 1$

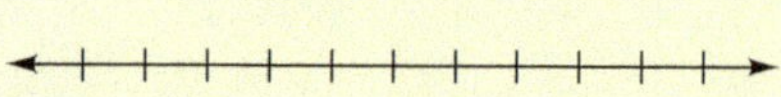

EXAMPLE 9

Solve and graph: $-4z \leq 12$

Solution $-4z \leq 12$

$\frac{-4z}{-4} \geq \frac{12}{-4}$ Divide each side of the inequality by -4 and reverse the direction of the inequality.

$z \geq -3$

The solution of $-4z \leq 12$ is $z \geq -3$, so all numbers greater than or equal to -3 are solutions. The graph of $z \geq -3$ is:

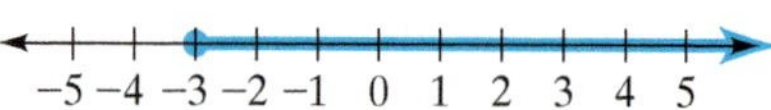

PRACTICE 9

Solve and graph: $-3x > 15$

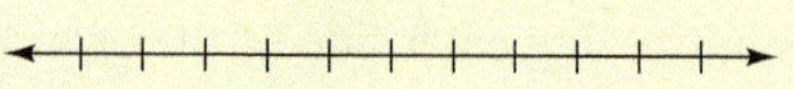

EXAMPLE 10

Solve and graph: $-21 < 6y - 9y$

Solution $-21 < 6y - 9y$

$-21 < -3y$ Combine like terms.

$\frac{-21}{-3} > \frac{-3y}{-3}$ Divide each side of the inequality by -3 and reverse the direction of the inequality.

$7 > y$, or $y < 7$

So all numbers less than 7 are solutions. The graph is:

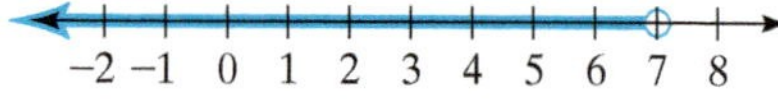

PRACTICE 10

Solve and graph: $10 > 5x - 7x$

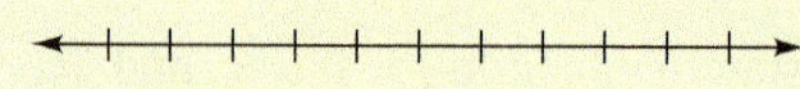

As in solving equations, we may need to use more than one property of inequalities to solve some inequalities.

EXAMPLE 11

Solve: $5y - 4 - 6y \leq -8$

Solution $5y - 4 - 6y \leq -8$

$-y - 4 \leq -8$ Combine like terms.

$-y - 4 + 4 \leq -8 + 4$ Add 4 to each side of the inequality.

$-y \leq -4$ Simplify.

$\frac{-y}{-1} \geq \frac{-4}{-1}$ Divide each side of the inequality by -1, and reverse the direction of the inequality.

$y \geq 4$

So all numbers greater than or equal to 4 are solutions.

PRACTICE 11

Solve: $-6 \geq 3z + 4 - z$

EXAMPLE 12

Solve: $3n - 2(n + 3) < 14$

Solution $3n - 2(n + 3) < 14$

$3n - 2n - 6 < 14$ Use the distributive property.

$n - 6 < 14$ Combine like terms.

$n - 6 + 6 < 14 + 6$ Add 6 to each side of the inequality.

$n < 20$

So all numbers less than 20 are solutions.

PRACTICE 12

Solve: $7x - (9x + 1) > -5$

Some common word phrases used in applied problems involving inequalities and their translations are shown in the following table:

Word Phrase	Translation
x is less than a	$x < a$
x is less than or equal to a	$x \leq a$
x is greater than a	$x > a$
x is greater than or equal to a	$x \geq a$
x is at most a	$x \leq a$
x is no more than a	$x \leq a$
x is at least a	$x \geq a$
x is no less than a	$x \geq a$

EXAMPLE 13

In geometry, the triangle inequality states that the sum of the lengths of any two sides of a triangle is greater than the length of the third side. In the isosceles triangle shown, write and solve an inequality to find the possible side lengths a. Graph the inequality.

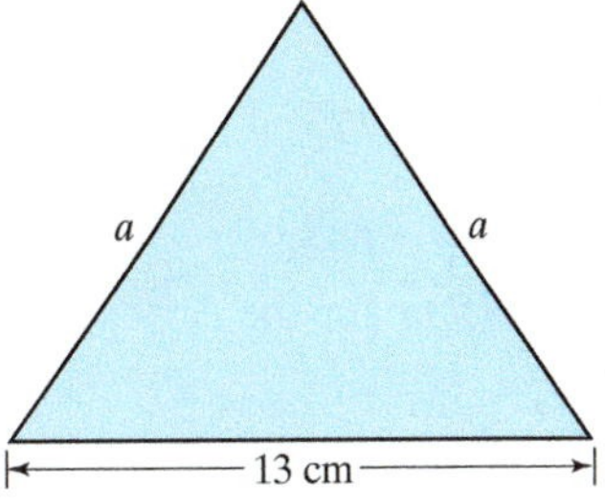

PRACTICE 13

In the triangle shown, for which values of x will the perimeter be greater than or equal to 14 in.?

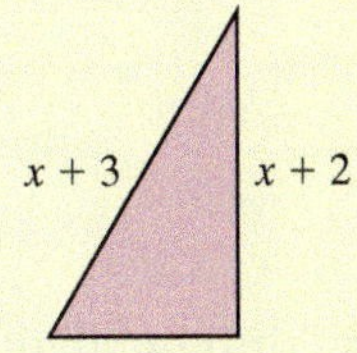

EXAMPLE 13 (continued)

Solution The sum of the lengths of the two equal sides is greater than the length of the third side.

$$a + a > 13$$
$$2a > 13$$

Solving the inequality, we get:

$$\frac{2a}{2} > \frac{13}{2} \quad \text{Divide each side by 2.}$$
$$a > 6.5$$

The graph is:

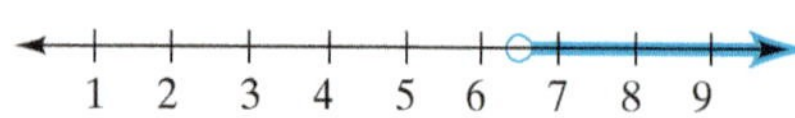

We conclude that the length of each side a is any number greater than 6.5 cm.

EXAMPLE 14

A factory's quality-control department randomly selects a sample of 5 lightbulbs to test. In order to meet quality-control standards, the lightbulbs in the sample must last an average of at least 950 hr. Four of the selected lightbulbs lasted 925 hr, 1000 hr, 950 hr, and 900 hr. How many hours must the fifth lightbulb last if the sample is to meet quality-control standards?

Solution The average number of hours that the 5 lightbulbs must last is the sum of the hours each lightbulb lasts divided by the number of lightbulbs in the sample, which is 5.

$$\text{Average number of hours} = \frac{925 + 1000 + 950 + 900 + x}{5}$$

where x represents the number of hours the fifth bulb lasts. In order for the average to be *at least* 950, it must be greater than or equal to 950. So we write and solve the inequality.

$$\frac{925 + 1000 + 950 + 900 + x}{5} \geq 950$$
$$5\left(\frac{925 + 1000 + 950 + 900 + x}{5}\right) \geq 950 \cdot 5$$
$$925 + 1000 + 950 + 900 + x \geq 4750$$
$$3775 + x \geq 4750$$
$$3775 - 3775 + x \geq 4750 - 3775$$
$$x \geq 975$$

So the fifth lightbulb in the sample must last at least 975 hr for the sample to meet quality control standards.

PRACTICE 14

A student has two part-time jobs. On the first job, she works 15 hr a week at \$8.50 an hour. The second job pays only \$7.50 an hour, but she can work as many hours as she wants. To make at least \$300, how many hours should she work on the second job?

12.5 Exercises

FOR EXTRA HELP MyMathLab MathXL PRACTICE WATCH READ REVIEW

Mathematically Speaking

Fill in each blank with the most appropriate term or phrase from the given list.

equation	reversed	unchanged
graph	inequality	positive
negative	solution	open
closed		

1. A(n) __________ is any mathematical statement containing $<$, $\leq$, $>$, $\geq$, or $\neq$.

2. A(n) _______ of an inequality is any value of the variable that makes the inequality true.

3. In the graph of the inequality $x > -7$, the circle is _______.

4. In the graph of the inequality $x \geq -7$, the circle is _______.

5. When the same number is added to or subtracted from each side of an inequality, the direction of the inequality is ____________.

6. According to the multiplication property of inequalities, if c is __________, then $a \geq b$ and $ac \geq bc$ are equivalent.

7. According to the multiplication property of inequalities, if c is __________, then $a \geq b$ and $ac \leq bc$ are equivalent.

8. When each side of an inequality is divided by a negative number, the direction of the inequality is _______.

A ***Indicate whether the value of x shown is a solution of the given inequality by answering true or false.***

9.

Value of x	Inequality	True or False
a. 1	$8 - 3x > 5$	
b. 4	$4x - 7 \leq 2x + 1$	
c. -7	$6(x + 6) < -9$	
d. $-\frac{3}{4}$	$8x + 10 \geq 12x + 15$	

10.

Value of x	Inequality	True or False
a. 3	$2x + 1 < -10$	
b. -5	$1 - 3x > 5x + 12$	
c. -2	$-(4x + 8) \geq 0$	
d. $\frac{1}{3}$	$9x - 7 \leq 6x - 11$	

B ***Graph on the number line.***

11. $x > 3$

−2 −1 0 1 2 3 4 5 6 7 8

12. $x > -1$

−5 −4 −3 −2 −1 0 1 2 3 4 5

13. $x < -5$

−8 −7 −6 −5 −4 −3 −2 −1 0 1 2

14. $x < 1$

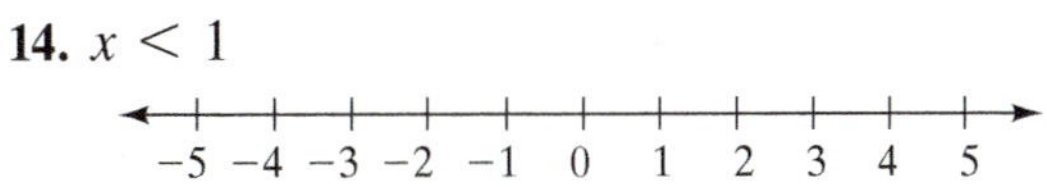

15. $x \geq 2$

−5 −4 −3 −2 −1 0 1 2 3 4 5

16. $x \geq 1$

−5 −4 −3 −2 −1 0 1 2 3 4 5

17. $x < -2.5$

−5 −4 −3 −2 −1 0 1 2 3 4 5

18. $x < -1.5$

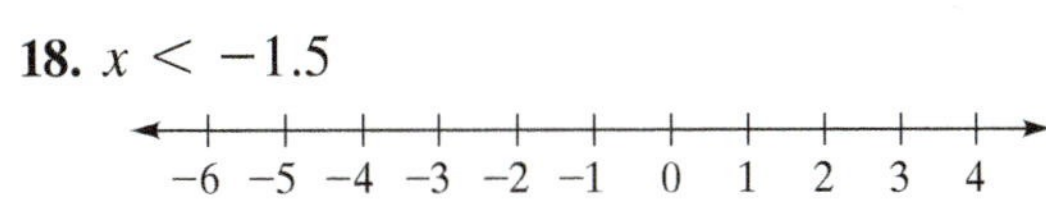

19. $x > -2\frac{1}{2}$

−5 −4 −3 −2 −1 0 1 2 3 4 5

20. $x \geq 4\frac{1}{2}$

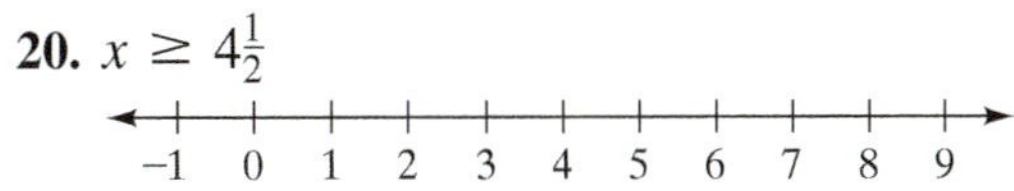

21. $x \geq -\frac{1}{3}$

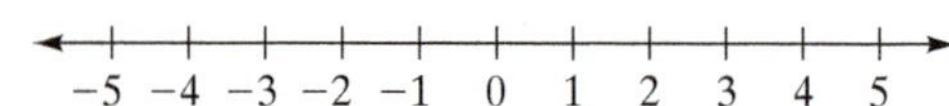

22. $x < \frac{3}{4}$

−5 −4 −3 −2 −1 0 1 2 3 4 5

23. $-3 < x < 1$

−5 −4 −3 −2 −1 0 1 2 3 4 5

24. $0 < x \leq 4$

−5 −4 −3 −2 −1 0 1 2 3 4 5

25. $-\frac{1}{2} \leq x < 2$

−5 −4 −3 −2 −1 0 1 2 3 4 5

26. $-4 \leq x \leq 1\frac{1}{2}$

−5 −4 −3 −2 −1 0 1 2 3 4 5

C ***Solve and graph.***

27. $v + 2 < -5$

28. $s + 1 \leq 1$

29. $y - 5 > -5$

30. $x - 5 > -1$

31. $y + 2 \leq 5.5$

32. $t + 3 \leq 6.5$

33. $v - 17 \leq -15$

34. $n - 25 > -30$

35. $-2 \geq x - 4$

36. $4 < a - 3$

37. $\frac{1}{3}a < -1$

38. $\frac{1}{2}x \leq 3$

39. $-5y > 10$

40. $-7t \leq -21$

41. $2x \geq 0$

42. $-3.5m < 0$

43. $-\frac{3}{4}a \geq 3$

44. $-\frac{x}{3} \geq 2$

−7 −6 −5 −4 −3 −2 −1 0 1 2 3

45. $6 \leq -\frac{2}{3}n$

46. $4 \leq -\frac{y}{2}$

Solve.

47. $\frac{n}{3} + 2 > 3$

48. $\frac{1}{2}y + 4 \geq -1$

49. $3x - 12 \leq 6$

50. $2v - 9 > -7$

51. $-21 - 3y > 0$

52. $-36 - 4y > 0$

53. $5n - 11 \geq 2n + 28$

54. $24 - 9s < -13s + 8$

55. $-4m + 8 \leq -3m + 1$

56. $-6y + 13 > y + 6$

57. $-7x + 4x + 23 < 2$

58. $5t - 7 + 2t \leq -14$

59. $-3(z + 5) > -15$

60. $2(8 + w) < 22$

61. $0.5(2x + 1) \geq 3x$

62. $-0.2(10d - 5) > 9$

63. $2(x - 2) - 3x \geq -1$

64. $2(4z - 3) < 3(z + 4)$

65. $7y - (9y + 1) < 5$

66. $-(6m - 2) \leq 0$

67. $0.4(5x + 1) \geq 3x$

68. $0.2(y - 3) > 8$

69. $5x + 1 < 3x - 2(4x - 3)$

70. $-2(0.5 - 4t) > -3(4 - 3.5t)$

71. $3 + 5n \leq 6(n - 1) + n$

72. $3(4 - 2m) \geq 2(3m - 6)$

73. $-\frac{4}{3}x - 16 > x + \frac{1}{3}x$

74. $\frac{2}{3}z - 4 < z + \frac{1}{3} + \frac{1}{3}z$

75. $0.2y > 1500 + 2.6y$

76. $x + 1.6x \leq 52$

Choose the inequality that best describes the situation.

77. To vote in the United States, a citizen must be at least 18 years old.

a. $a < 18$ **b.** $a > 18$
c. $a \leq 18$ **d.** $a \geq 18$

78. The number of people seated in a theater is at most 650.

a. $n < 650$ **b.** $n \leq 650$
c. $n > 650$ **d.** $n \geq 650$

79. It is generally accepted that a person has a fever if the person's temperature is above 98.6°F. To convert a Celsius temperature to its Fahrenheit equivalent, we use the formula $F = \frac{9}{5}C + 32$. For which Celsius temperatures C does a person have a fever?

a. $\frac{9}{5}C + 32 \geq 98.6$ **b.** $\frac{9}{5}C + 32 \leq 98.6$
c. $\frac{9}{5}C + 32 < 98.6$ **d.** $\frac{9}{5}C + 32 > 98.6$

80. A teenager has a $20 gift certificate at the iTunes store. He decides to purchase three episodes of his favorite television show at $2.99 each, and is also browsing the music store where songs can be downloaded at $1.29 apiece. What is the maximum number of songs he can purchase without exceeding the amount of the gift certificate?

a. $8.97 + 1.29n > 20$ **b.** $8.97 + 1.29n < 20$
c. $8.97 + 1.29n \geq 20$ **d.** $8.97 + 1.29n \leq 20$

Mixed Practice

Graph on the number line.

81. $-1 < x \leq 3\frac{1}{2}$

−2 −1 0 1 2 3 4 5 6

82. $x > -0.5$

−4 −3 −2 −1 0 1 2 3 4

Solve.

83. Indicate whether the value of x shown is a solution of the given inequality by answering true or false.

Value of x	Inequality	True or False
a. 2	$4 - 2x < -1$	
b. 6	$3x - 5 \geq 21 - 2x$	
c. −8	$-(2x + 4) \leq 12$	
d. $-\frac{1}{2}$	$8x - 5 > 4x - 8$	

Solve and graph.

84. $-3 \leq x - 7$

−2 −1 0 1 2 3 4 5 6

85. $-\frac{2}{3}m \geq 4$

−8 −6 −4 −2 0 2 4 6 8

86. $-8t > -24$

−4 −3 −2 −1 0 1 2 3 4

Solve.

87. $3(a + 4) \geq 2(3a - 2)$

88. $-5m + 6 \geq -4m + 4$

89. $8x + 7 - 3x < 32$

90. $0.3m - 1200 < 2.8m$

D Applications

Write an inequality and solve.

91. On the last three chemistry exams, a student scored 81, 85, and 91. What score must the student earn on the next exam to have an average above 85?

92. A rectangular deck that is 14 ft long is to be built onto the back of a house. What would be the area of the deck if it is at least 12 ft wide?

93. A novelty store, open Monday through Saturday, must sell an average of \$200 worth of merchandise per day to break even on expenses. The following table shows sales for one week:

Day	Monday	Tuesday	Wednesday	Thursday	Friday
Amount of Sales	\$250	\$250	\$150	\$130	\$180

How much must the store make in sales on Saturday to at least break even for the week?

94. An American tourist about to travel abroad wants to buy euros from one of two currency exchange booths. The first booth charges \$1.38 per euro plus a fixed surcharge of \$3. The second charges \$1.50 per euro with no surcharge. For what kind of transaction does the second booth offer a better deal?

95. A telemarketer claims that every call made to a customer costs at least \$2. If a call costs \$0.50 plus \$0.10 for each minute, how long does each call last?

96. One side of a triangular garden is 2 ft longer than another side. The third side is 4 ft long. What are the maximum lengths of the other two sides if the perimeter of the garden can be no longer than 12 ft?

97. A real estate agent gets $2\frac{1}{2}\%$ on every house she sells plus a \$1,000 bonus. Her supervisor offers an alternative deal of 3% on the sale of every house and no bonus. Should she accept the deal? Under what circumstances?

98. A parking garage offers two payment options: a \$20 flat fee for the whole day, or \$5 plus \$2 per hour for each hour or part thereof that a customer parks. Which is the better option for the customer? Explain why.

99. A person weighing 200 lb volunteers for a clinical trial of a new diet pill. If he loses 2.5 lb per month using the diet pill combined with regular exercise, when will he weigh less than 180 lb?

100. A diver scored 9.2, 9.6, 9.7, 9.4, and 9.3 in the first 5 dives in a diving competition, where all scores are given in tenths. To win first place, she must beat a total score of 56.6. What is the lowest score she can get on her last dive to win the competition?

101. A Gallup poll found that more than 3 and less than 9 percent of the U.S. public believed that the Apollo moon landing of July 1969 was faked. Express in terms of inequalities how many millions of Americans p among 200 million members of the public were of this opinion. (*Source:* gallup.com)

102. At least 10,000 hours of study and practice are needed for someone to become an "expert" in a particular field, according to one estimate. If a person has studied a subject for 10 hr/wk, 36 wk per school year, through 12 yr of schooling, how many more hours are needed for him to be an expert? (*Source:* Malcolm Gladwell, *Outliers*)

• Check your answers on page A-17.

MINDStretchers

Mathematical Reasoning

1. Explain for which values of x the following inequality holds: $x + 5 > x + 2$.

Groupwork

2. Working with a partner, explore whether each of the following statements is always, sometimes, or never true. Give examples.

a. If $a < b$ and $c < d$, then $ac < bd$.

b. $a - b \leq a + b$

Writing

3. Clearly identifying the variables, give examples of inequalities that

a. you wish were true.

b. you wish were false.

Key Concepts and Skills

CONCEPT SKILL

Concept/Skill	Description	Example
[12.1] **Equation**	A mathematical statement that two expressions are equal.	$y - 6 = 9$
[12.1] **Solution of an equation**	A value of the variable that makes the equation a true statement.	2 is a solution of the equation $3x + 1 = 11 - 2x$.
[12.1] **Linear equation in one variable**	An equation that can be written in the form $ax + b = c$, where a, b, and c are real numbers and $a \neq 0$.	$4x + 1 = 3$
[12.1] **Equivalent equations**	Equations that have the same solution.	$x = 2$ and $x + 1 = 3$
[12.1] **The addition property of equality**	For any real numbers a, b, and c, $a = b$ and $a + c = b + c$ are equivalent.	If $x - 3 = 5$, then $(x - 3) + 3 = 5 + 3$
[12.2] **The multiplication property of equality**	For any real numbers a, b, and c, $c \neq 0$, $a = b$ and $a \cdot c = b \cdot c$ are equivalent.	If $\frac{n}{2} = 8$, then $\frac{n}{2} \cdot 2 = 8 \cdot 2$
[12.3] **To solve linear equations**	• Use the distributive property to clear the equation of parentheses, if necessary. • Combine like terms where appropriate. • Use the addition property to isolate the variable term. • Use the multiplication property to isolate the variable. • Check by substituting the solution in the original equation.	$2(x + 5) = 6$ $2x + 10 = 6$ $2x + 10 - 10 = 6 - 10$ $2x = -4$ $x = -2$ **Check** $2(x + 5) = 6$ $2(-2 + 5) \stackrel{?}{=} 6$ $6 = 6$ True
[12.4] **Literal equation**	An equation involving two or more variables.	$2t + b = c$
[12.5] **Inequality**	Any mathematical statement containing $<, \leq, >, \geq$, or $\neq$.	$x \geq -4$
[12.5] **Solution of an inequality**	Any value of the variable that makes the inequality true.	0 is a solution of $x < 7$.
[12.5] **The addition property of inequalities**	For any real numbers a, b, and c: • $a < b$ and $a + c < b + c$ are equivalent. • $a > b$ and $a + c > b + c$ are equivalent. Similar statements hold for $\leq$ and $\geq$.	If $x < 1$, then $x + 5 < 1 + 5$.

[12.5] **The multiplication property of inequalities**	For any real numbers a, b, and c:
	• If c is positive, then $a < b$ and $ac < bc$ are equivalent.
	• If c is negative, then $a < b$ and $ac > bc$ are equivalent.
	Similar statements hold for $>$, $\leq$, and $\geq$.
		If $\frac{x}{2} > 4$, then
		$2 \cdot \frac{x}{2} > 2 \cdot 4$.
		If $-2x < 4$, then
		$\frac{-2x}{-2} > \frac{4}{-2}$.

CHAPTER 12 Review Exercises

Say Why

Fill in each blank.

1. The number -7 ________ (is/is not) a solution to the equation $5x + 1 = 36$ because ______________________________.

2. The equation $2 = -\frac{3}{4}x$ ________ (is/is not) linear because ______________________________.

3. When 5 is subtracted from each side of the equation $2x + 5 = 10$, the result ________ (is/is not) an equivalent equation because ______________________________.

4. The equations $5m + 2 = \frac{1}{2}$ and $10m + 4 = 1$ ________ (are/are not) equivalent because ______________________________.

5. The number 8 ________ (is/is not) a solution of $x - 6.5 > 1.5$ because ______________________________.

6. If $-15x < 3$, then $\frac{-15x}{-15}$ ________ (is/is not) less than $\frac{3}{-15}$ because ______________________________.

[12.1]

7. Is 2 a solution of the equation $5x + 3 = 7 - 4x$?

8. Determine whether 0 is a solution of the equation $4x - 15 = 5(x - 3)$.

Solve and check.

9. $x - 3 = -12$
10. $t + 10 = 8$
11. $-9 = a + 5$
12. $4 = n - 7$
13. $y - (-3.1) = 11$
14. $r + 4.8 = 20$

[12.2] *Solve and check.*

15. $\frac{x}{3} = -2$
16. $\frac{z}{2} = -5$
17. $2x = -20$
18. $-5d = 15$
19. $-y = -4$
20. $-x = 3$
21. $20.5 = 0.5n$
22. $30 = -0.2r$
23. $\frac{2t}{3} = -6$
24. $\frac{5y}{6} = -10$

[12.3] *Solve and check.*

25. $2x + 1 = 7$
26. $-t - 4 = 5$
27. $\frac{a}{2} - 3 = -10$
28. $\frac{r}{3} - 6 = 12$
29. $-y + 7 = -2$
30. $-2t + 3 = 1$
31. $4x - 2x - 5 = 7$
32. $3y - y + 12 = 6$
33. $z + 1 = -2z + 10$
34. $n - 3 = -n + 7$
35. $c = -2(c + 1)$
36. $p = -(p - 5)$

37. $2(x + 1) - (x - 8) = -x$

38. $-(x + 2) - (x - 4) = -5x$

39. $3[2n - 4(n + 1)] = 6n - 12$

40. $-4(2x - 6) = 7[x - (3x - 1)]$

41. $10 - [3 + (2x - 1)] = 3x$

42. $x - [5 + (3x - 4)] = -x$

[12.4]

43. Solve $a - 5b = 2c$ for a in terms of b and c.

44. Solve $\frac{2a}{b} = n$ for a in terms of b and n.

45. An isosceles triangle is a triangle with two sides equal in length. $P = 2a + b$ is a formula for the perimeter P of an isosceles triangle in terms of its equal sides a and the third side b. Solve for a in terms of P and b.

46. $V = \frac{1}{3}Bh$ is a formula for the volume V of a cone in terms of the area of its base B and its height h. Solve for h in terms of B and V.

[12.5] ***Graph each inequality.***

47. $x < 2$

−4 −3 −2 −1 0 1 2 3 4

48. $x \geq -4.5$

−6 −5 −4 −3 −2 −1 0 1 2

49. $-2 < x \leq 2$

−4 −3 −2 −1 0 1 2 3 4

50. $-0.5 < x < 5$

−2 −1 0 1 2 3 4 5 6

Solve. Then, graph.

51. $-n \leq 2$

−3 −2 −1 0 1 2 3 4

52. $y + 1 > 6$

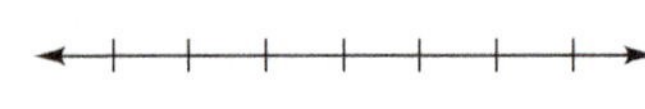

53. $-\frac{1}{2}t + 3 \leq 3$

54. $8y - 2 \leq 6y + 2$

55. $\frac{1}{2}(8 - 12x) \leq x - 10$

56. $0.5n - 0.3 < 0.2(2n + 1)$

Mixed Applications

Solve.

57. An air conditioner's energy efficiency ratio (EER) is the quotient of its British thermal unit (Btu) rating and its wattage. What is the Btu rating of a 2000-watt air conditioner if its EER is 8?

58. For a wedding, the reception costs will be $5000 plus $50 per guest. The bride and groom have budgeted $12,000 for the reception. How many guests can the bride and groom invite to the wedding?

59. A polygon is a closed geometric figure with straight sides. In any polygon with n sides, the sum of the measures of its angles is $180(n - 2)$ degrees. If the measures of the angles of a polygon add up to 540°, how many sides does the polygon have?

60. A newspaper reported that a candidate received 15,360 more votes than her opponent, and that 39,210 votes were cast in the election. How many votes were cast for each candidate?

61. To print b books, it costs a publisher $(100{,}000 + 25b)$ dollars. The publisher receives $50b$ dollars in revenue for selling this number of books. How many books must be sold for the revenue to equal the cost?

62. The road connecting two factories is 380 mi long. A truck leaves one of the factories traveling toward the other factory at 45 mph, while at the same time a second truck leaves the other factory heading at 50 mph toward the first. How long after the departure will the trucks meet?

63. A plane flying between two cities at 400 mph arrives half an hour behind schedule. If the plane had flown at a speed of 500 mph, it would have been on time. Find the distance between the cities.

64. Two friends leave a party at 10 P.M. driving in opposite directions. One drives at a speed of 40 mph, whereas the other drives at a speed of 32 mph. At what time are the two friends 18 mi apart?

65. How many pints of a 1% solution of disinfectant must be combined with 4 pt of a 10% solution to make a 5% solution?

66. How much pure alcohol must be mixed with 6 L of a 60% alcohol solution to make a 70% alcohol solution?

67. To convert a weight expressed in kilograms k to the equivalent number of pounds p, multiply the number of kilograms by 2.2.

a. Write a formula for this relationship.

b. Solve this formula for k.

68. To send a telegram, it costs \$2 for the first 10 words in the telegram and y cents for each additional word.

a. Write an equation to find the cost C of a telegram 26 words long.

b. Solve this equation for y.

69. In planning for the future, a young couple invested \$2000, partly in an insured fund and partly in an uninsured fund. The profit on the insured fund was 2%, whereas on the uninsured fund it was 4%. If the total profit earned was \$72, how much money was invested in the insured fund?

70. The financial manager of a company had \$100,000 in cash to invest. He chose to deposit some of these funds in corporate bonds paying 8% simple interest and the remainder in a money market fund that earns 4% annual simple interest. If the interest after 1 yr generated by the corporate investment was \$4880 greater than that from the money market fund, how much money was invested in the corporate bonds?

71. In 2 yr, a certificate of deposit in the amount of \$2000 paid \$120 altogether in simple interest. What was the annual rate of interest, expressed as a percent?

72. How long would it take a savings account paying 2% annually in simple interest to generate \$500 in interest on a deposit of \$5000?

• Check your answers on page A-17.

CHAPTER 12 Posttest

FOR EXTRA HELP

CHAPTER Test Prep VIDEOS

The Chapter Test Prep Videos with test solutions are available on DVD, in MyMathLab, and on YouTube (search "AkstDevMath" and click on "Channels").

To see if you have mastered the topics in this chapter, take this test.

1. Determine whether -2 is a solution of the equation $3x - 4 = 6(x + 2)$.

2. What must be done to the equation $x - 1 = -10$ in order to isolate the variable?

Solve and check.

3. $s - 4 = -7$

4. $\frac{1}{4} = y + \frac{3}{4}$

5. $\frac{n}{2} = -3$

6. $-y = -11$

7. $\frac{3y}{4} = 6$

8. $2x + 5 = 11$

9. $10x + 1 = -x + 23$

10. $\frac{1}{4}x + \frac{1}{5}x = \frac{9}{5}$

11. $16a = -4(a - 5)$

12. $2(x + 5) - (x + 4) = 7x + 1$

13. Solve $5n + p = t$ for p in terms of n and t.

14. Draw the graph of $-1 \leq x < 3$.

−3 −2 −1 0 1 2 3

15. Solve $-2z \leq 6$. Then, graph.

−3 −2 −1 0 1 2 3

16. In a recent year, the per capita debt of Wisconsin was about \$4 thousand. If the total debt of the state was approximately \$24 billion, estimate the population of the state. [*Source: The World Almanac and Book of Facts, 2011*]

17. A taxi charges \$4.00 for the first mile plus \$1.25 for each additional mile. On a fare of \$16.50, how long was the ride?

18. A woman's shoe size S is given by the formula $S = 3L - 21$, where L (in inches) is her foot length. Solve for L in terms of S.

19. Two friends live 33 mi apart. They cycled from their homes, riding toward one another and meeting $1\frac{1}{2}$ hr later. If one friend cycled 2 mph faster than the other, what were their two rates?

20. A cell phone service offers two calling plans. Plan A costs \$39.99 per month plus \$0.79 per minute (or part thereof) for calls outside the network. Plan B costs \$54.99 per month plus \$0.59 per minute (or part thereof) for calls outside the network. Under what conditions will the monthly cost of Plan A exceed the monthly cost of Plan B?

• Check your answers on page A-18.

Cumulative Review Exercises

To help you review, solve the following:

1. Which is larger, $\frac{1}{2}$ or $\frac{5}{8}$?

2. Express 4.007 in words.

3. Find the product: $(-0.5)(24)(-0.2)$

4. True or false: $1 > -2$

5. Simplify: $2 + (-1) + (-4) + 4$

6. True or false: 3 is to 5 as 8 is to 15.

7. Solve and check: $\dfrac{\frac{1}{4}}{6} = \dfrac{x}{12}$

8. Find the mean, median, mode(s), and range of 5.6, 5.9, 5.3, 5.8, and 5.9.

9. A mother gives birth to quintuplets—four boys and one girl. Two boys are longer than the girl, and two are shorter. Which of the babies has the median length?

10. Fill in the blank: 6000 MB = __ GB

11. Subtract 2 hr 35 min from 4 hr.

12. Find the perimeter of a square with side $4\frac{1}{2}$ ft.

13. Find the missing length.

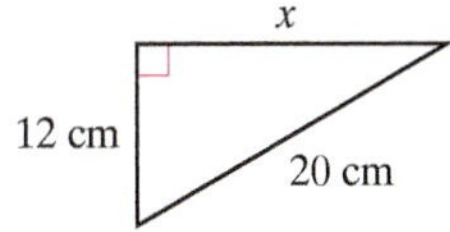

14. 24 is 75% of what number?

15. In a recent year, males experienced 18,400,000, or 40%, of the common operations and procedures performed in the United States. To the nearest million, how many were performed in all? (*Source: The New York Times Almanac 2010*)

16. An initial investment of \$1000 triples in value every 15 yr. The value of this investment after 60 yr is how many times the initial investment? Write in exponential form.

17. For the first 2 hr of a trip, a driver averaged 55 mph. For the remainder of the trip, he drove at an average of 60 mph. If the entire trip covered 140 mi, how many hours did the trip last?

18. A computer technician charges \$50 for the first hour of work and \$25 for each additional hour of work.

a. Write a formula that expresses the amount of money A that the technician charges in terms of the time t that he works.

b. Solve this equation for t.

c. If the technician sent you a bill for \$125, how many hours did he claim to work?

19. During what is called the "Group Stage" of soccer's World Cup tournament, games are scored as follows: a win is worth 3 points, a draw 1 point, and a loss 0 points. Each team plays 3 games in the Group Stage. (For example in 2010, the U.S. had 1 win, 2 draws, and 0 losses, giving them a total of 5 points.) If a team wins w games, gets a draw in d games, and loses l games, what is its score? (*Source:* fifa.com)

20. The fuel bill for a condo building was less than 8% of the building's operating expenses. If the fuel bill was \$74,000, find the building's operating expenses.

• Check your answers on page A-18.

CHAPTER 13

Graphing Linear Equations and Inequalities

Get Rich with Graphing?

If financial analysts could predict trends on Wall Street, they would know when to buy and when to sell stock in order to maximize profits. To arrive at educated predictions of future trends, analysts use the statistical tool of linear regression.

When using linear regression, an analyst plots points that show a particular stock's recent selling prices. Then, the analyst sketches the straight line that is closest to passing through these points. This **regression line** provides an estimate of how much the stock is likely to increase or decrease in price, and at what pace. If the actual selling prices differ considerably from the predicted prices for a lengthy period of time, the analyst suspects a new trend and recomputes the regression line.

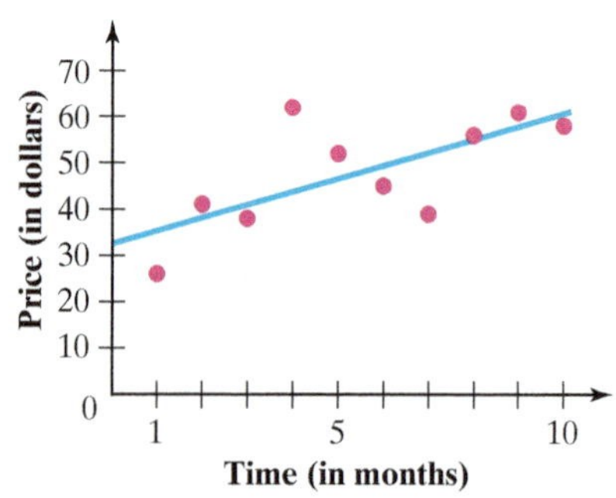

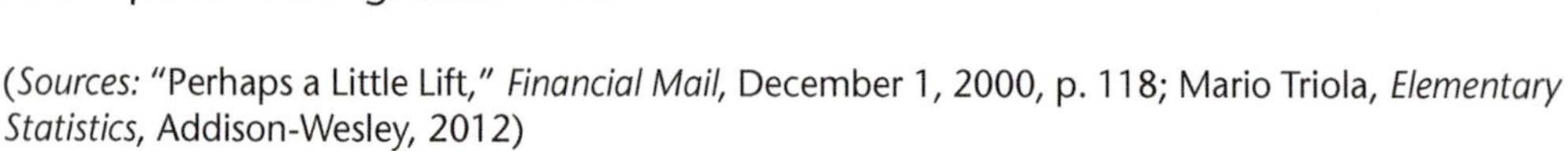
(*Sources:* "Perhaps a Little Lift," *Financial Mail*, December 1, 2000, p. 118; Mario Triola, *Elementary Statistics*, Addison-Wesley, 2012)

CHAPTER 13 PRETEST

To see if you have already mastered the topics in this chapter, take this test.

1. On the coordinate plane below, plot the points $A(1, 4)$ and $B(-3, -6)$.

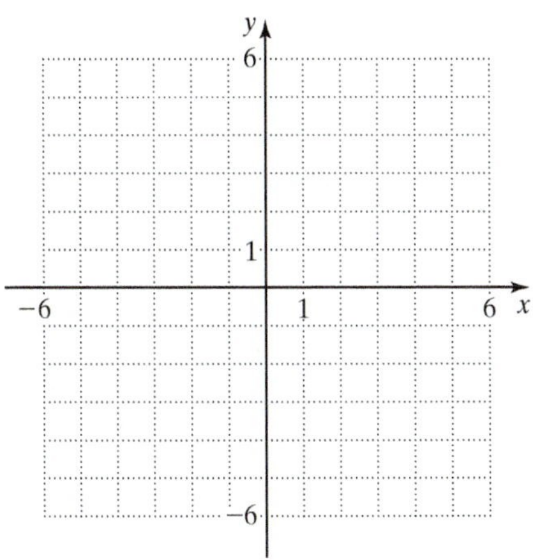

2. In which quadrant is the point $(3, -5)$ located?

3. Given two points $C(7, 5)$ and $D(1, 2)$, compute the slope of the line that passes through the points.

4. For the points $A(2, 1)$, $B(0, 5)$, $C(1, 7)$, and $D(4, 1)$, indicate whether $\overleftrightarrow{AB}$ is parallel to $\overleftrightarrow{CD}$. Explain.

5. For the points $P(-3, 3)$, $Q(1, -1)$, $R(-2, -2)$, and $S(4, 4)$, indicate whether $\overleftrightarrow{PQ}$ is perpendicular to $\overleftrightarrow{RS}$. Explain.

6. In the graph shown, find the x-intercept and the y-intercept.

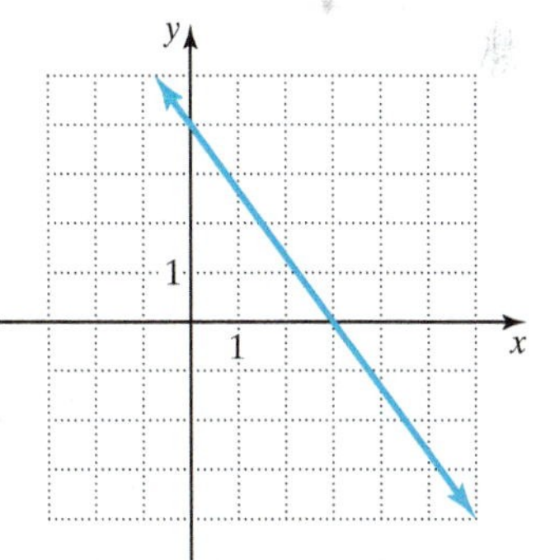

7. For the equation $y = 2x - 5$, complete the following table:

x	4	7		
y			0	−1

Graph each equation.

8. $y = -3$

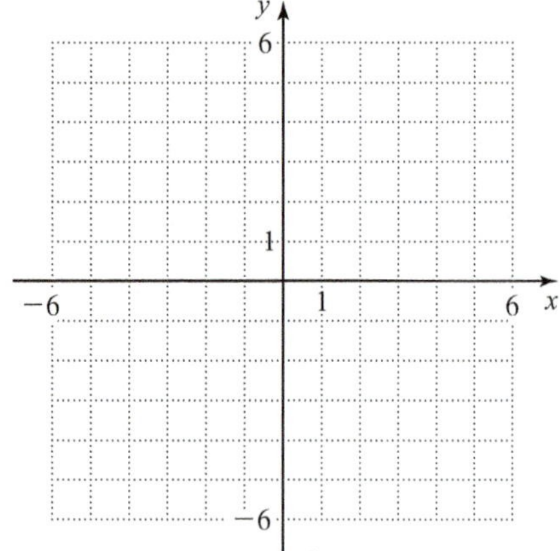

9. $x = 2$

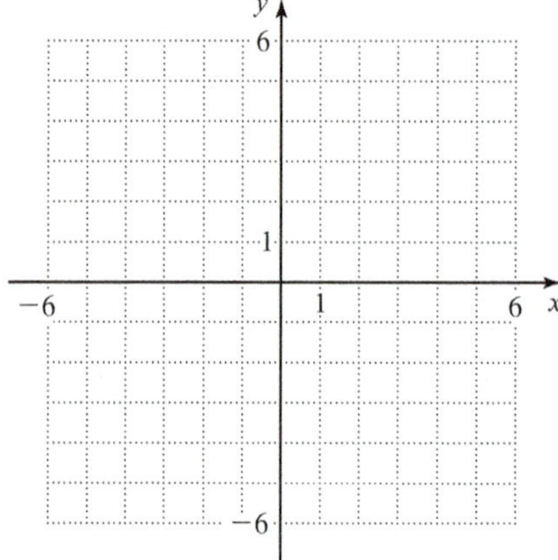

10. $y = -3x + 2$

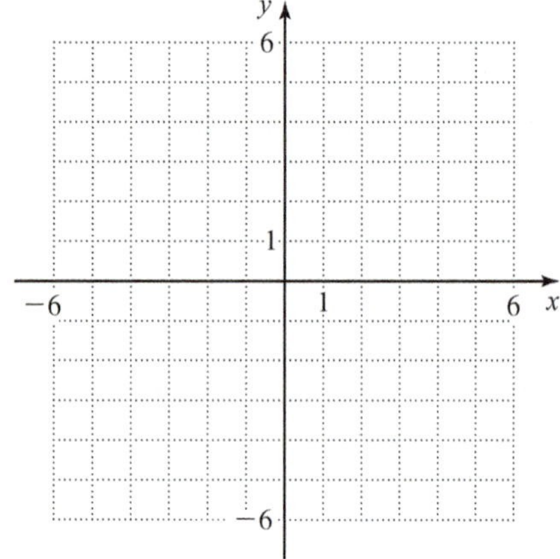

11. $2x - 3y = 6$

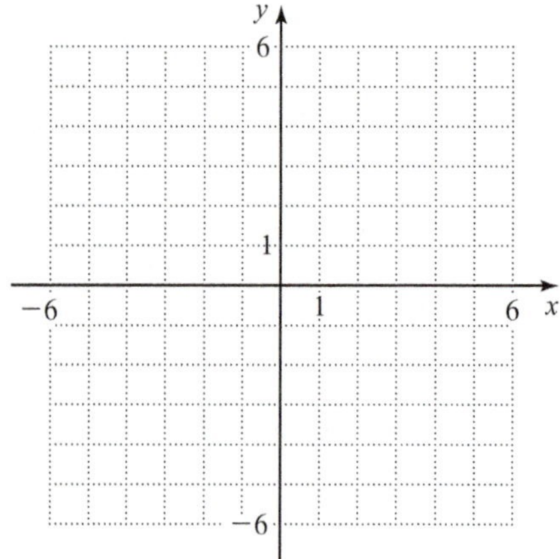

12. What are the slope and the y-intercept of the graph of $y = 2x - 5$?

13. Write the equation $5x - y = 8$ in slope-intercept form.

14. Find an equation of the line with slope 2 that passes through the point (0, 8).

15. Find an equation of the line that passes through the points (4, 1) and $(2, -1)$.

16. Graph: $y > 3x + 1$

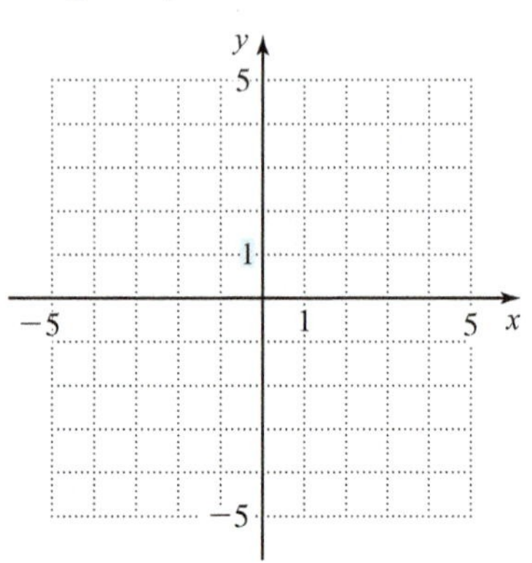

17. In the graph shown, the y-axis stands for the number of congressional representatives from a state and the x-axis stands for the state's population according to a recent U.S. census.

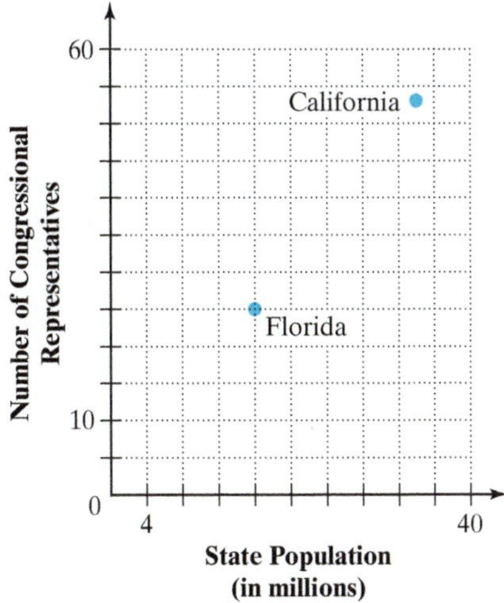

Draw the line that passes through the points. Is this line's slope positive, negative, or zero? How does the state population relate to the number of representatives from that state?

18. Farmers want to develop varieties of wheat that grow at a faster rate. The graph shown displays the growth pattern of two new varieties of wheat. Which variety grows more quickly?

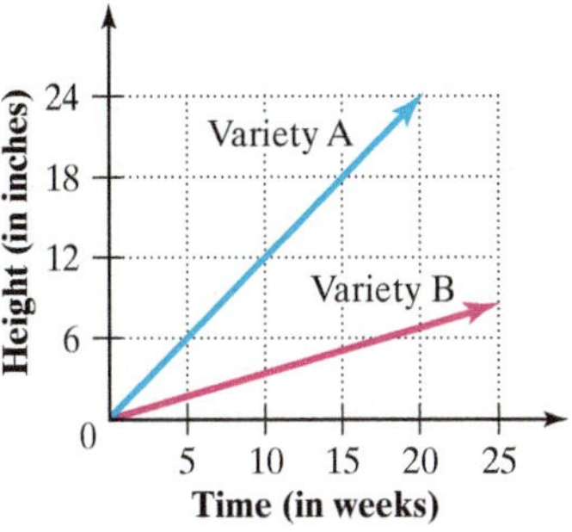

19. A local video store charges a daily rental fee of \$2.50 per movie.
   a. Express the daily cost c of renting x movies.
   b. Choose an appropriate scale for the axes and graph this relationship.

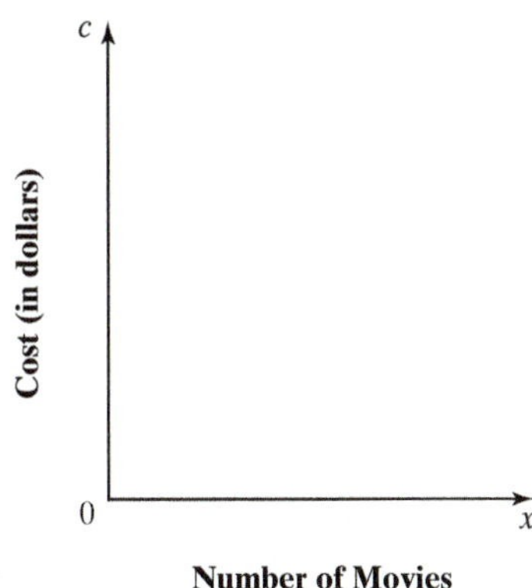

   c. What is the slope of the line in part (b)? Explain its significance in this context.

20. A textbook sales representative is driving to a college 200 mi away at a speed of 50 mph.
   a. Express the distance d (in miles) the sales representative travels in terms of the time t (in hours) he has been driving.
   b. Choose an appropriate scale for the axes and graph this relationship.

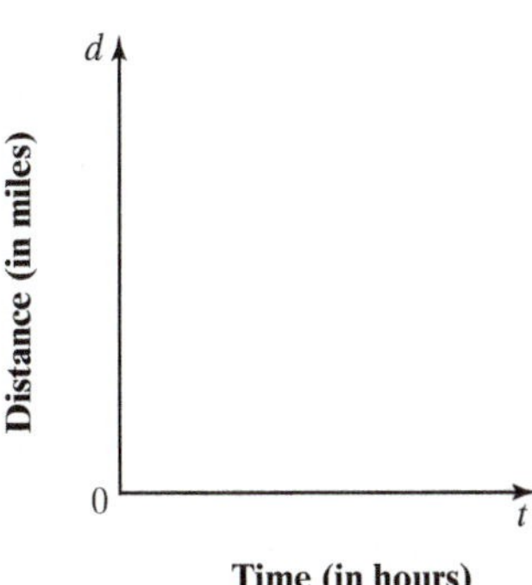

   c. What is the slope of the graph? Explain its significance in terms of the trip.

• Check your answers on page A-18.

13.1 Introduction to Graphing

OBJECTIVES

- **A** To identify and plot points on a coordinate plane
- **B** To identify the quadrants of a coordinate plane
- **C** To solve applied problems involving graphs

What Graphing Is and Why It Is Important

Many mathematical relationships can be expressed as equations, inequalities, or their graphs. Although lacking the precision of an equation, a graph can clarify at a glance patterns and trends in a relationship, helping us to understand that relationship better.

In the past, the graphing approach to problem solving was usually more time-consuming than the traditional algebraic approach. Today, the use of graphing calculators and computer software packages has made graphing easier. But to utilize these graphing tools, you must first understand the concepts and skills involved in graphing, which are discussed in this chapter.

In this chapter, the relationships that we graph are relatively simple. In Chapters 14 and 19, we will apply graphing techniques to more complex relationships.

Plotting Points

If you were to enter a theater or sports arena with a ticket for row 5, seat 3, you would know exactly where to sit that is, at $(5, 3)$. Such a system of **coordinates** in which we associate a pair of numbers in a given order with a corresponding location is commonplace.

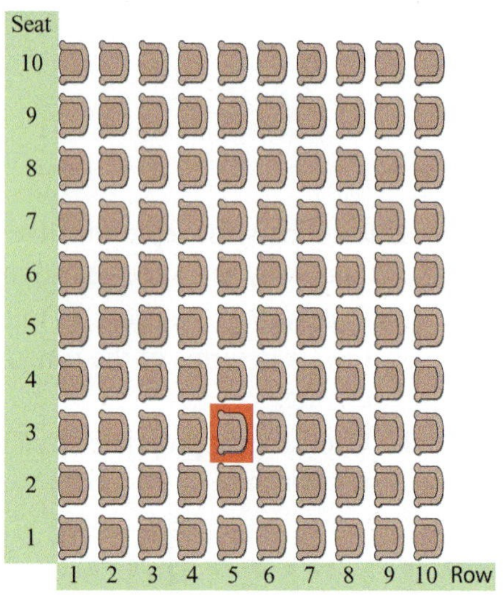

The flat surface on which we draw graphs is called a **coordinate plane**. To create a coordinate plane, we first sketch two perpendicular number lines—one horizontal, the other vertical—that intersect at their zeros. Each number line is called an **axis**. The point where the axes intersect is called the **origin**. It is common practice to refer to the horizontal number line as the ***x*-axis** and the vertical number line as the ***y*-axis**.

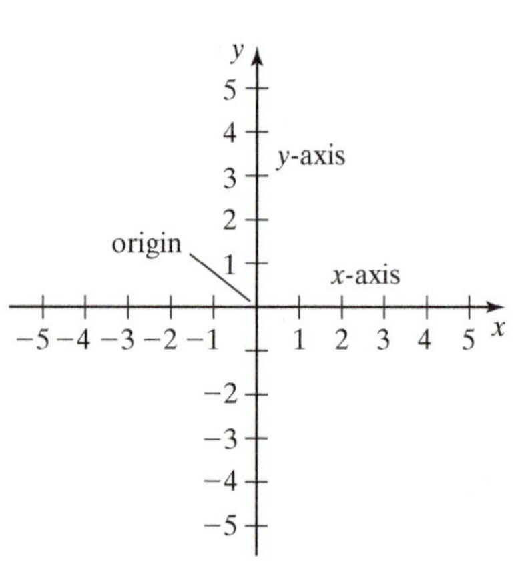

Each point in a coordinate plane is represented by a pair of numbers called an **ordered pair.** For example, the origin is the point (0, 0). The first number in an ordered pair represents a horizontal distance and is called the ***x*-coordinate.** The second number represents a vertical distance and is called the ***y*-coordinate.**

To **plot** a point in the coordinate plane, we find its location represented by its ordered pair. For example, to plot the point (3, 1), we start at the origin and go 3 units *to the right,* then go *up* 1 unit. For this point, we say that $x = 3$ and $y = 1$.

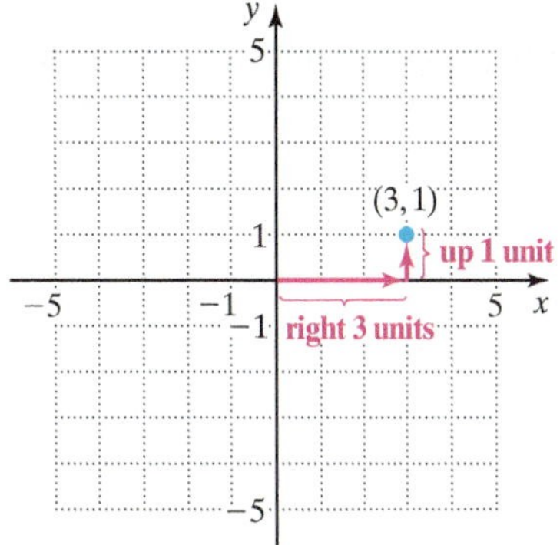

Notice that the two numbers in an ordered pair are written in parentheses, separated by a comma. Do the ordered pairs (3, 1) and (1, 3) correspond to different points? Why?

When an ordered pair has a negative x-coordinate, the corresponding point is to the left of the y-axis, as shown in the following coordinate plane. Similarly, when an ordered pair's y-coordinate is negative, the point is below the x-axis.

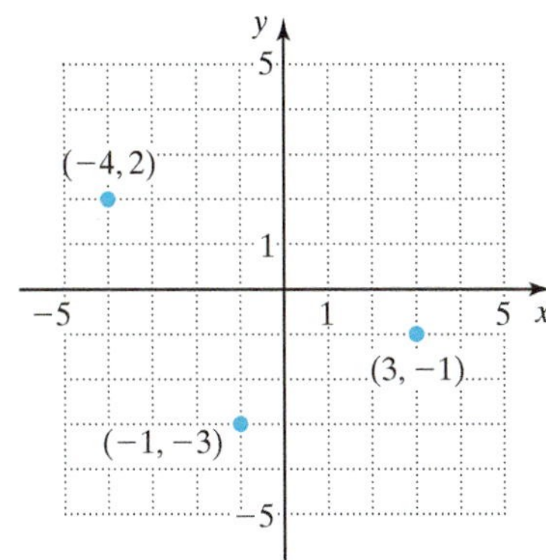

Any ordered pair whose y-coordinate is 0 corresponds to a point that is on the x-axis. For instance, the points $(-4, 0)$, $(0, 0)$, and $(3, 0)$ are on the x-axis, as shown in the graph below on the left. Similarly, any ordered pair whose x-coordinate is 0 corresponds to a point that is on the y-axis. For instance, the points $(0, 2)$, $(0, -1)$, and $(0, -4)$ are on the y-axis, as shown in the middle graph below.

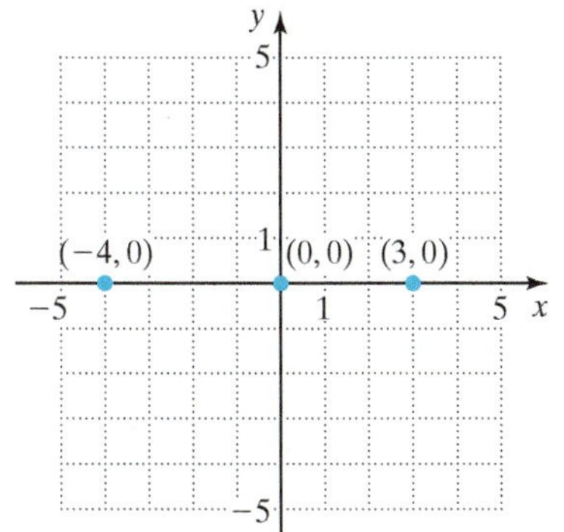

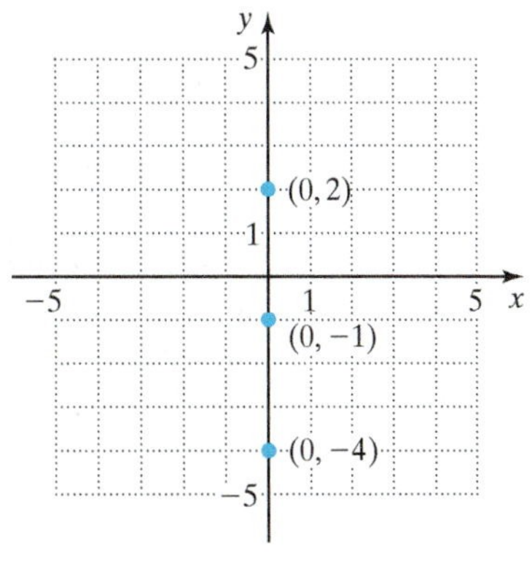

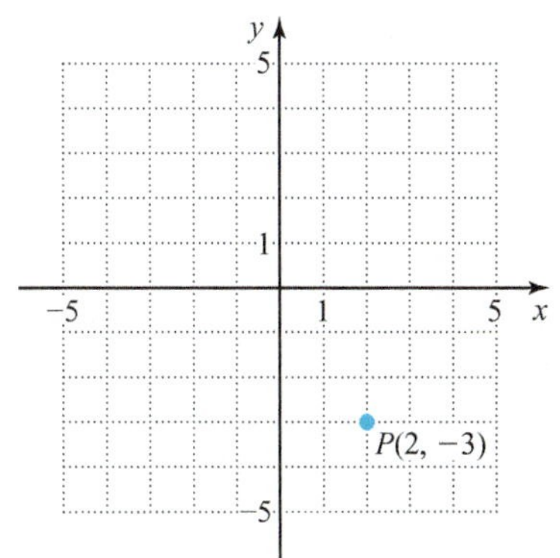

Sometimes we name points with letters. We can refer to a point as P or A or any other letter that we choose. Generally, a capital letter is used. If we want to emphasize that point P has coordinates $(2, -3)$, we can write it as $P(2, -3)$, as shown in the graph above on the right.

If there is a point whose coordinates we do not know, we can refer to it as (x, y) or $P(x, y)$, where x and y are the unknown coordinates.

Now, let's look at some examples of plotting points.

EXAMPLE 1

Plot the following points on a coordinate plane.

a. $(5, 2)$ **b.** $(3, -4)$ **c.** $(-1, 1)$
d. $(0, 3)$ **e.** $(-4, -2)$ **f.** $(0, 0)$

Solution The points are plotted on the coordinate plane as shown.

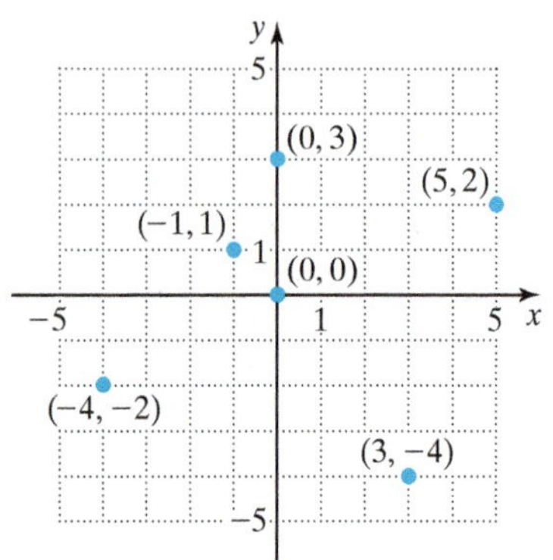

PRACTICE 1

On a coordinate plane, plot the points corresponding to each ordered pair.

a. $(0, -5)$ **b.** $(4, 4)$
c. $(-2, 2)$ **d.** $(5, 0)$
e. $(-3, -2)$ **f.** $(2, -4)$

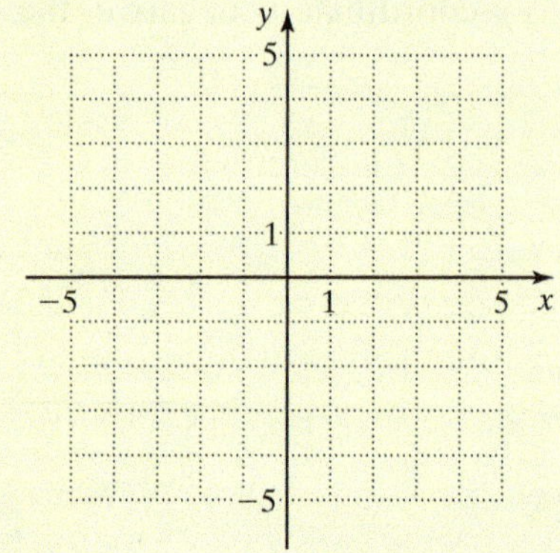

The x- and y-axes are boundaries that separate a coordinate plane into four regions called *quadrants*. These quadrants are named in counterclockwise order starting with Quadrant I, as shown below:

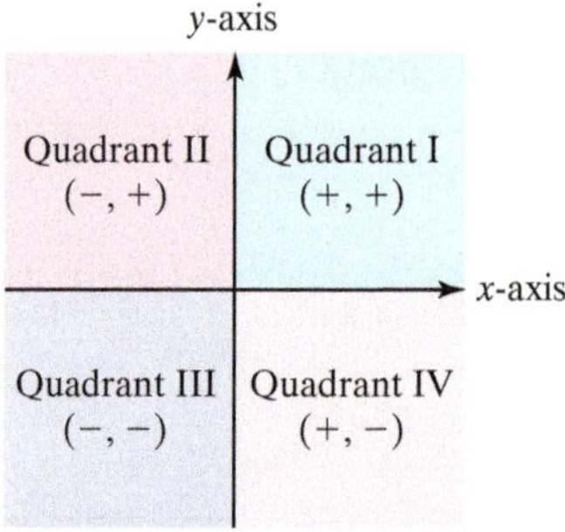

The quadrant in which a point is located tells us something about its coordinates. For instance, any point in Quadrant I is to the right of the y-axis and above the x-axis, so both its coordinates must be positive. Points in this quadrant are of particular interest in applied problems in which all quantities are positive.

Any point in Quadrant II lies to the left of the y-axis and above the x-axis, so its x-coordinate must be negative and its y-coordinate positive. Points in Quadrant III are to the left of the y-axis and below the x-axis, so both coordinates are negative. And finally, any point in Quadrant IV is to the right of the y-axis and below the x-axis, so its x-coordinate must be positive and its y-coordinate negative. Points that lie on an axis are not in any quadrant.

EXAMPLE 2

Determine the quadrant in which each point is located.

a. $(-5, 5)$

b. $(7, 20)$

c. $(1.3, -4)$

d. $(-4, -5)$

Solution

a. $(-5, 5)$ is in Quadrant II.

b. $(7, 20)$ is in Quadrant I.

c. $(1.3, -4)$ is in Quadrant IV.

d. $(-4, -5)$ is in Quadrant III.

PRACTICE 2

In which quadrant is each point located?

a. $\left(-\frac{1}{2}, 3\right)$

b. $(6, -7)$

c. $(-1, -4)$

d. $(2, 9)$

Often the points that we are to plot affect how we draw the axes on a coordinate plane. For instance, in the following coordinate planes we choose for each axis an appropriate **scale**—the length between adjacent tick marks—to conveniently plot all points in question.

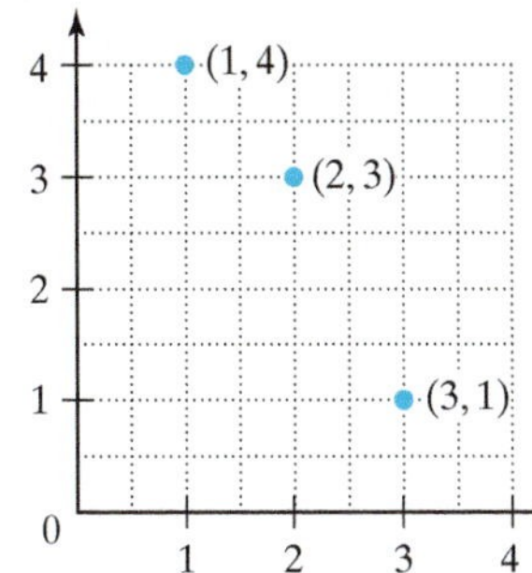

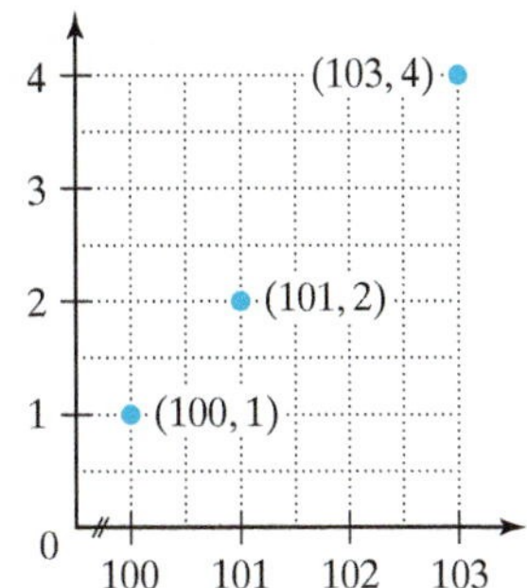

TIP Depending on the location of the points to be plotted, we can choose to show only part of a coordinate plane.

EXAMPLE 3

A young entrepreneur started a dot-com company that made a profit of \$10,000 in its first year of business. In the second year, the company's profit grew to \$15,000. In the third year, however, the company lost \$5000. Plot points on a coordinate plane to display this information.

Solution We let x represent the year of business and y represent the company's profit in dollars that year. The three points to be plotted are:

$(1, 10{,}000)$, $(2, 15{,}000)$, and $(3, -5000)$.

PRACTICE 3

The following table shows the average monthly temperatures for the first four months of the year (where month 1 represents January) for Chicago, Illinois. (*Source:* U.S. National Climatic Data Center)

Month *m*	1	2	3	4
Temperature *t* (°F)	22	27	37	49

EXAMPLE 3 (continued)

Notice that in the third year, the company's loss is represented by a negative profit. Because the y-coordinates are large, we use 5000 as the scale on the y-axis. Then, we plot the points on the following coordinate plane.

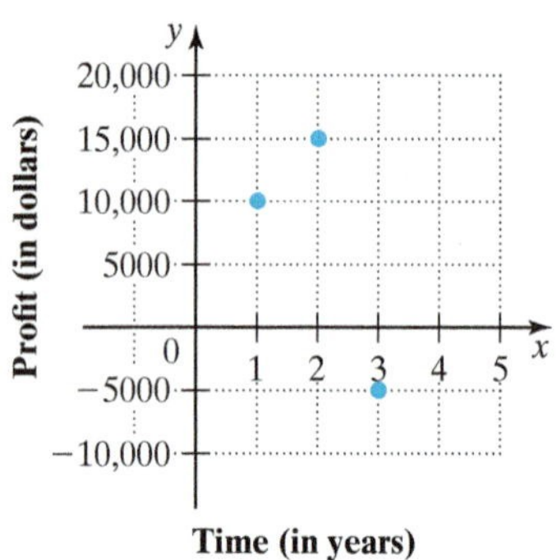

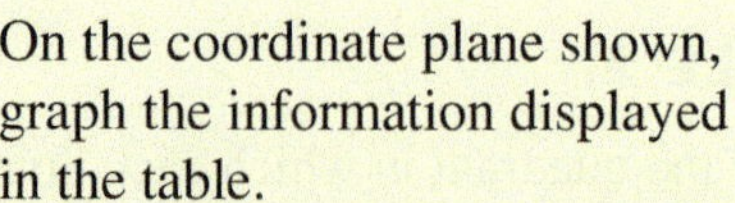
On the coordinate plane shown, graph the information displayed in the table.

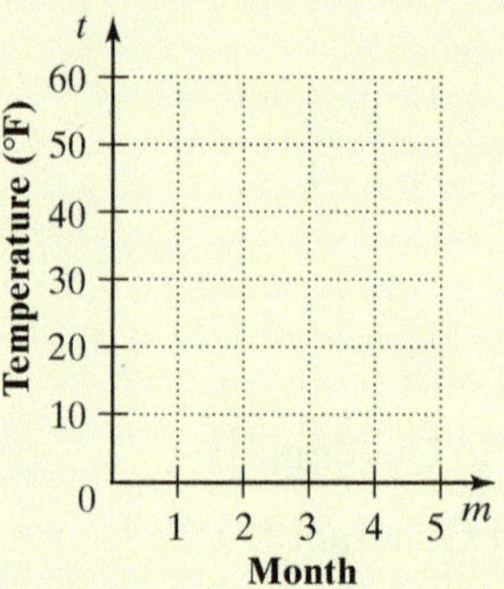

We end this discussion of plotting points with a final comment about variables. On a coordinate plane, the first coordinate of each point is a value of one quantity (or variable), and the second coordinate is a value of another quantity. For instance, in Example 3 we considered the profit that a company makes at various times. One variable represents time and the other variable the profit. Notice that the profit made by the company depends on the time rather than the other way around. So we refer to time as the *independent* variable and the profit as the *dependent* variable. It is customary when plotting points to assign the independent variable to the horizontal axis and the dependent variable to the vertical axis. However, as shown in Practice 3, letters other than x and y can be used to represent quantities.

Interpreting Graphs

Points plotted on a coordinate plane are merely dots on a piece of paper. However, their significance comes to life when we understand the information that they convey.

Describing the trend on a coordinate plane tells the story of that trend. When key points are missing, as is frequently the case, the story is incomplete. In such cases, we may want to make a prediction, that is, to extend the observed pattern so as to estimate the missing data. Such predictions, while not certain, at least allow us to make decisions based on the best available evidence. We may also want to speculate about the conditions that underlie an observed pattern of plotted points.

The trend among plotted points on a coordinate plane shows a relationship between the two variables. To highlight the relationship, it is common practice either to draw a line that passes through the plotted points or to connect adjacent points with short line segments.

Consider the following examples that involve interpreting trends on a coordinate plane:

EXAMPLE 4

The graph shows the cost C of parking a car at a lot for time t hr. Describe the trend that you observe in terms of both the coordinate plane and the cost of parking.

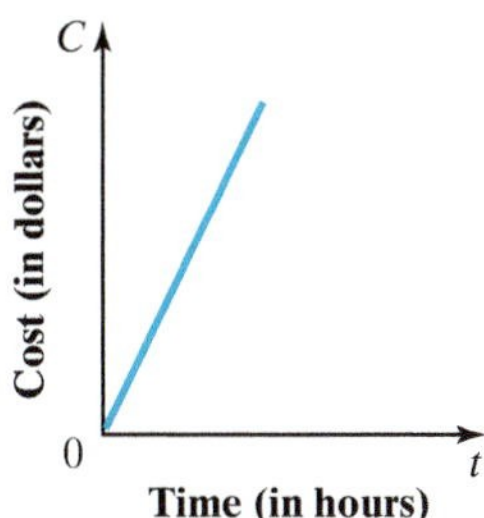

Solution On the graph, the larger C-values correspond to the larger t-values. In terms of parking, we see that the longer a car is parked in the lot, the more it costs to park the car.

PRACTICE 4

The value V of a new car after t years is displayed on the following graph. Describe the line graph in terms of the changing value of the car.

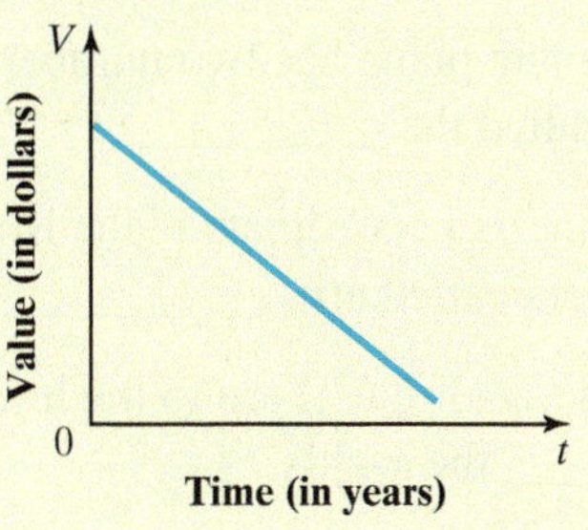

EXAMPLE 5

The following graph shows the cost that a shipping company charges to send a package, depending on the package's weight.

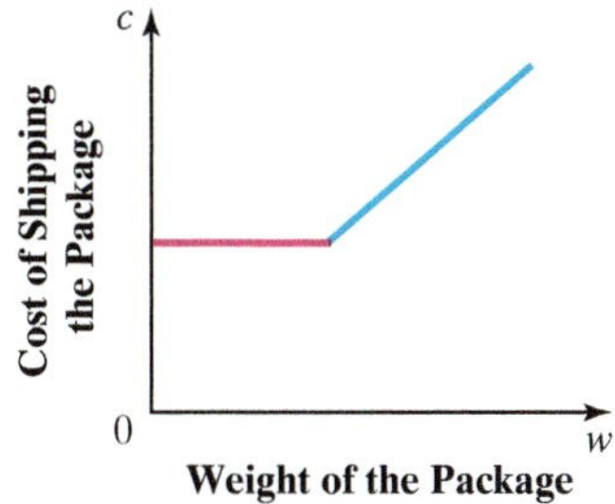

Write a brief story describing the displayed relationship. What business practice does the horizontal line segment reflect?

Solution From the horizontal line segment, we see that the cost of shipping is constant (that is, a flat rate) for lighter packages up to a certain weight. The horizontal line segment indicates that the company established a minimum cost for sending lightweight packages. Since the slanted line segment goes upward to the right, the cost of shipping increases with the weight of heavier packages.

PRACTICE 5

The following graph shows the number of times per minute that a runner's heart beats. Describe in a sentence or two the pattern you observe. Use this pattern to write a scenario as to what the runner might be doing.

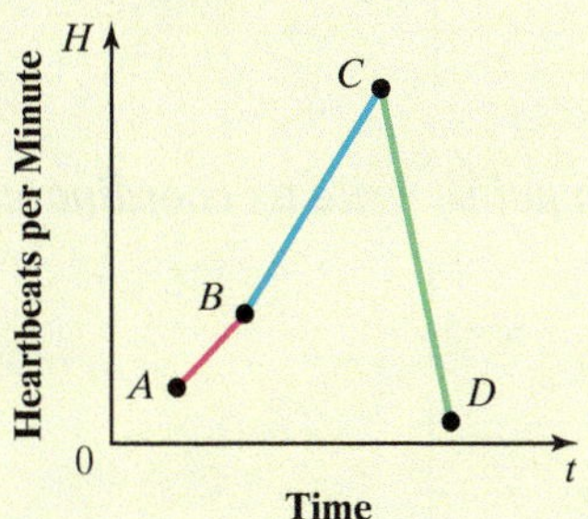

Mathematically Speaking

Fill in each blank with the most appropriate term or phrase from the given list.

above	horizontal	below
dependent	x-axis	independent
origin	coordinate	center
ordered pair	vertical	y-axis

1. A coordinate plane has two number lines that intersect at a point called the __________.

2. The horizontal number line on a coordinate plane is usually referred to as the __________.

3. Each point in a coordinate plane is represented by a pair of numbers called a(n) ______________.

4. The y-coordinate of a point on a coordinate plane represents a(n) __________ distance.

5. Points in Quadrant III are to the left of the y-axis and __________ the x-axis.

6. The _______________ variable is usually assigned to the horizontal axis.

A *On the coordinate plane below, plot the points with the given coordinates.*

7. $A(0, 5)$ $B(-1, -5)$ $C(1, 4)$ $D(3, -3)$ $E(-4, 2)$ $F(5, 0)$

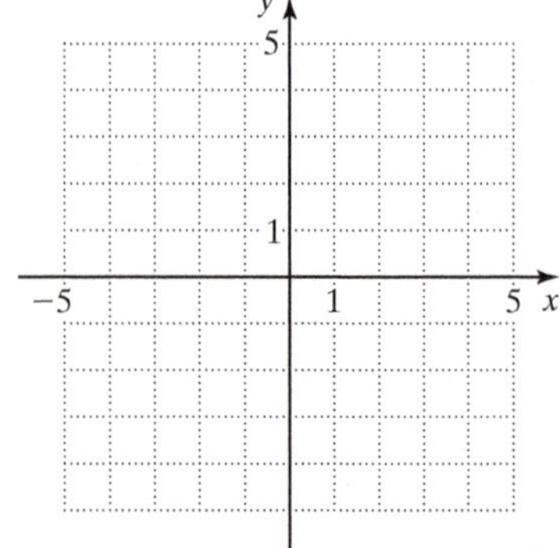

8. $A(-2, 4)$ $B(0, 0)$ $C(2, -1)$ $D(-3, 0)$ $E(3, 4)$ $F(-4, -2)$

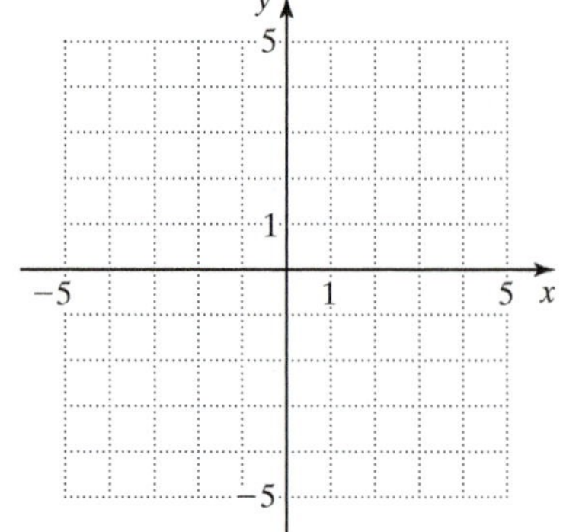

Next to each point, write its coordinates.

9.

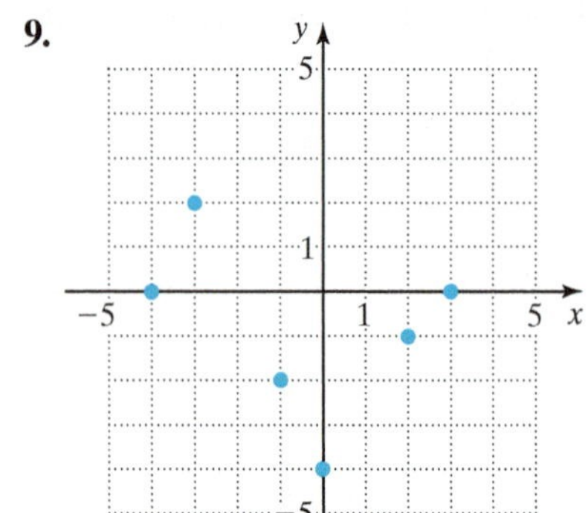

10.

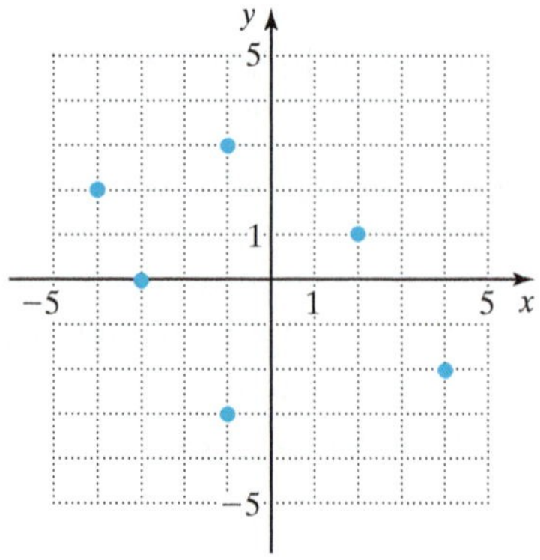

B *Identify the quadrant in which each point is located.*

11. $(-2, -3)$

12. $(-13, -24)$

13. $(-9, 5)$

14. $(-5.1, 4)$

15. $\left(3, -\frac{1}{2}\right)$

16. $\left(3\frac{1}{2}, -8\right)$

17. $(65, 11)$

18. $(8, 6.2)$

Mixed Practice

Solve.

19. Plot the points with the given coordinates on the coordinate plane.
$A(3, 2)$ $B(4, 0)$ $C(-4, -1)$ $D(-2, 3)$ $E(0, -3)$

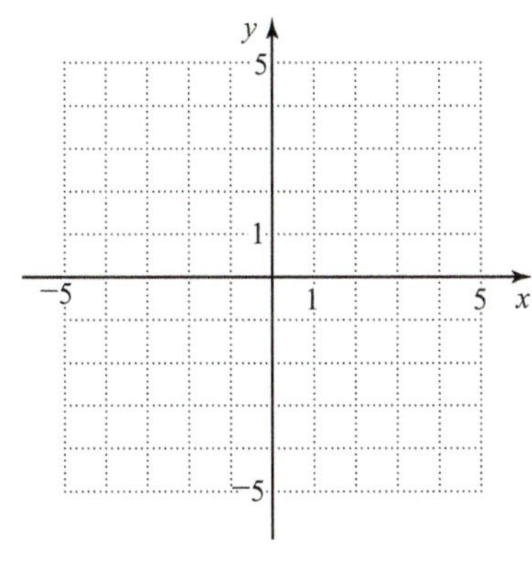

20. Write the coordinates next to each point.

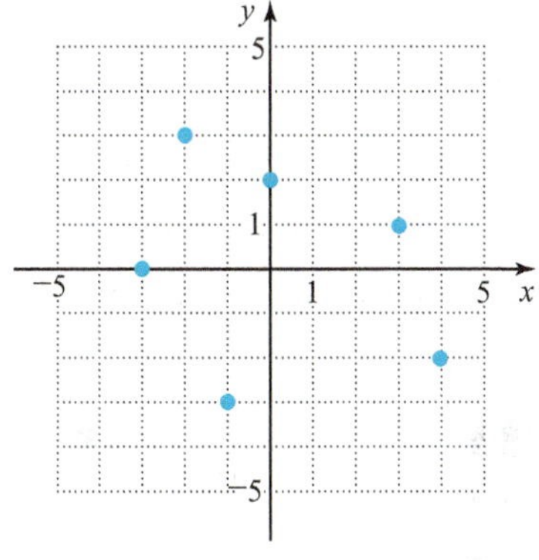

Identify the quadrant in which each point is located.

21. $(-3, 7)$

22. $\left(6, -\frac{3}{4}\right)$

23. $(27, 39)$

24. $(-4.2, -3.8)$

Applications

C *Solve.*

25. College students coded *A*, *B*, *C*, and *D* took placement tests in mathematics and in English. The following coordinate plane displays their scores.

a. Estimate the coordinates of the plotted points.

b. Which students scored higher in English than in mathematics?

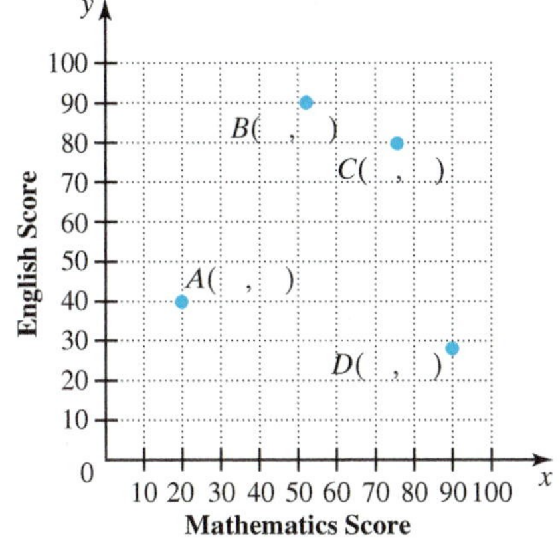

26. Suppose that a financier owns shares of stock in three companies—Dearborn, Inc. (*D*), Ellsworth Products (*E*), and Fairfield Publications (*F*). On the following coordinate plane, the *x*-value of a point represents the change in value of a share of the indicated stock from the previous day. The *y*-value stands for the number of shares of that stock owned by the financier.

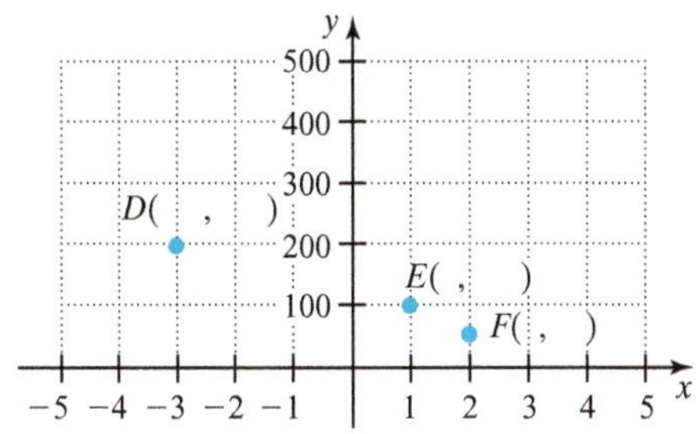

a. Name the coordinates of the plotted points.

b. For each point, explain the significance of the product of the point's coordinates.

27. The following table gives the percent of the U.S. adult population that smoked in various years. (*Source:* cdc.gov)

Year	1980	1990	2000	2010
Percent of the Population That Smoked	33%	26%	23%	20%

Plot this information on the coordinate plane below:

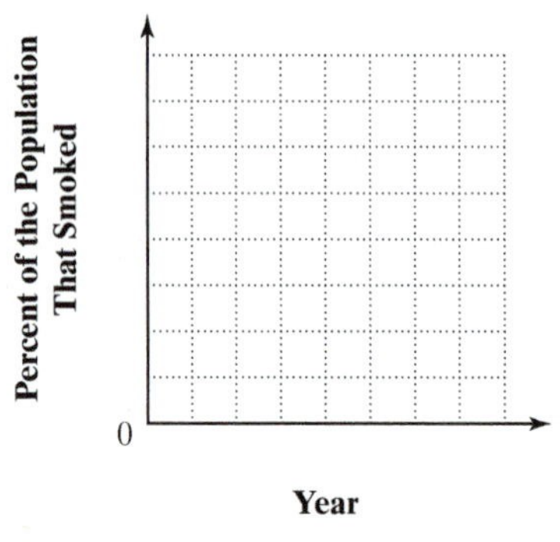

28. The number of electoral votes cast for the winning candidate in recent presidential elections is displayed in the following table:

Year	1996	2000	2004	2008
Electoral Votes	379	271	286	365

(*Source: The New York Times*)

Plot this information in the coordinate plane shown:

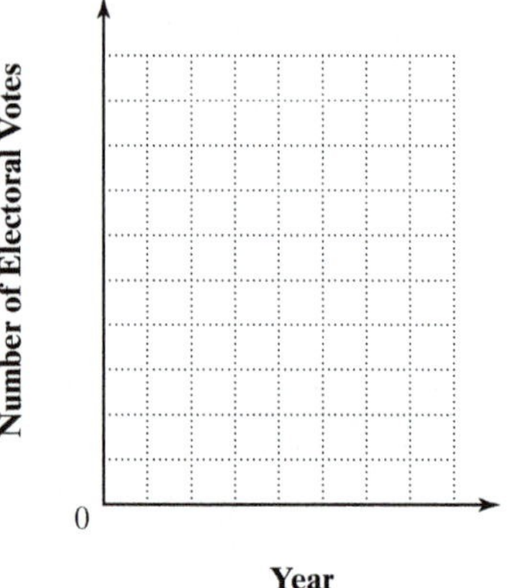

29. A chemist conducts an experiment to measure the melting and boiling points of four substances, as indicated in the following table:

Substance	Symbol	Melting Point (°C)	Boiling Point (°C)
Chlorine	Cl	−101	−35
Oxygen	O	−218	−183
Bromine	Br	−7	59
Phosphorous	P	44	280

On the coordinate plane shown below, an x-value represents a substance's melting point and a y-value stands for its boiling point, both in degrees Celsius.

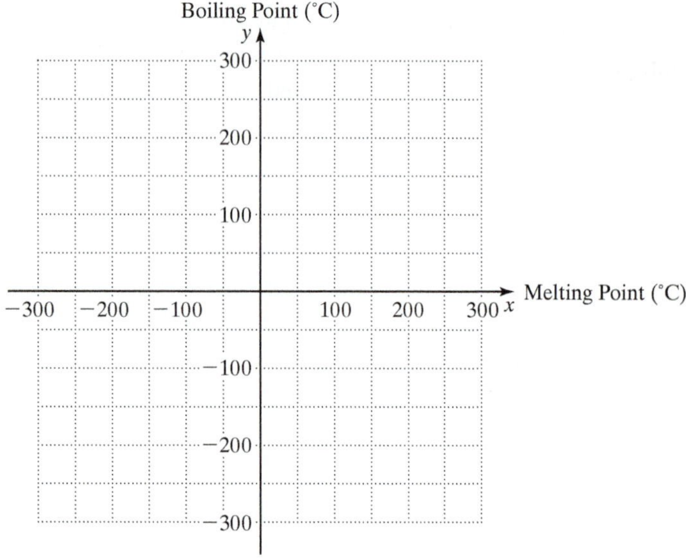

a. Plot points for the four substances. Label each point with the appropriate substance symbol.

b. For each point, which of its coordinates is larger—the x-value or the y-value? In a sentence, explain this pattern.

30. Meteorologists use the windchill index to determine the windchill temperature (how cold it feels outside) relative to the actual temperature when the wind speed is considered. The following table shows the actual temperatures in degrees Fahrenheit and the related windchill temperatures when the wind speed is 5 mph:

Actual Temperature T	-10	-5	0	5	10	15
Windchill Temperature W	-22	-16	-11	-5	1	7

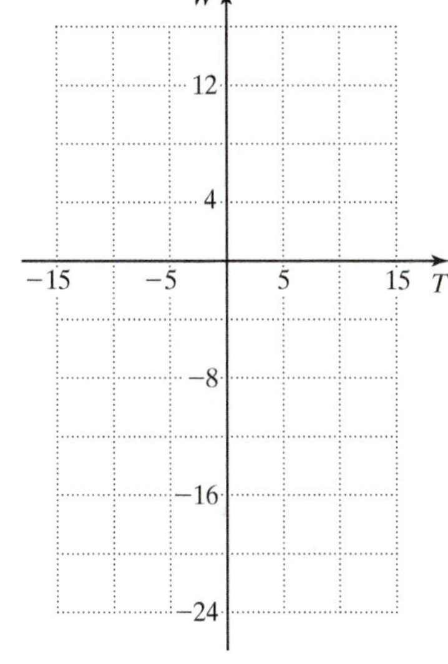

a. Plot points (T, W) on the given coordinate plane.

b. For the plotted points, describe the pattern that you observe.

31. On the following coordinate plane, a y-coordinate stands for the number of senators from a state. The corresponding x-coordinate represents that state's population according to a recent U.S. census. Describe the pattern that you observe. (*Source:* census.gov)

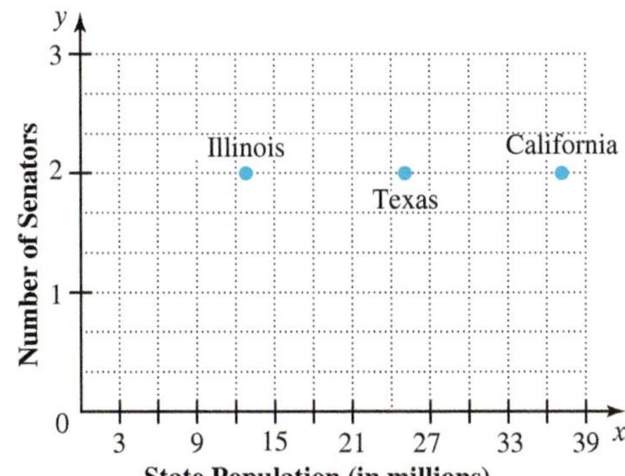

32. Last year's daily closing values (in dollars) of a share of a technology stock are plotted on the graph below. What story is this graph telling?

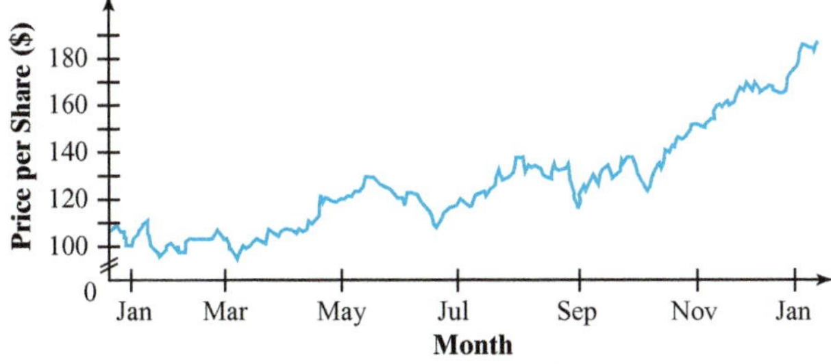

33. A child walks away from a wall, stands still, and then approaches the wall. In a couple of sentences, explain which of the graphs below could describe this motion.

a.

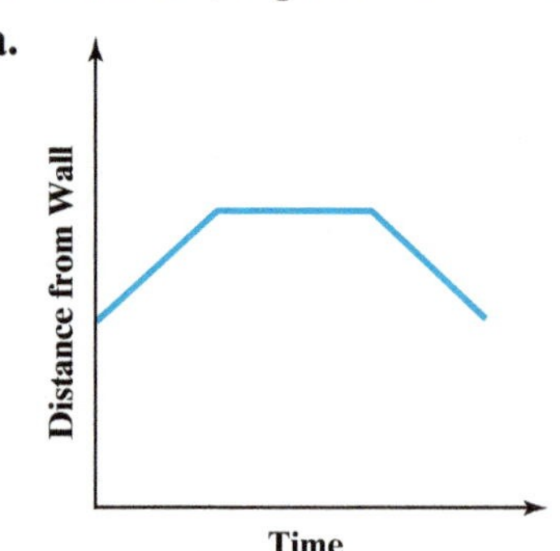

b.

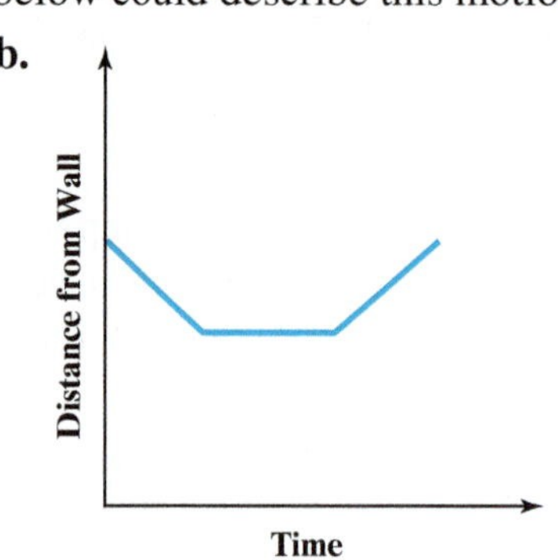

c.

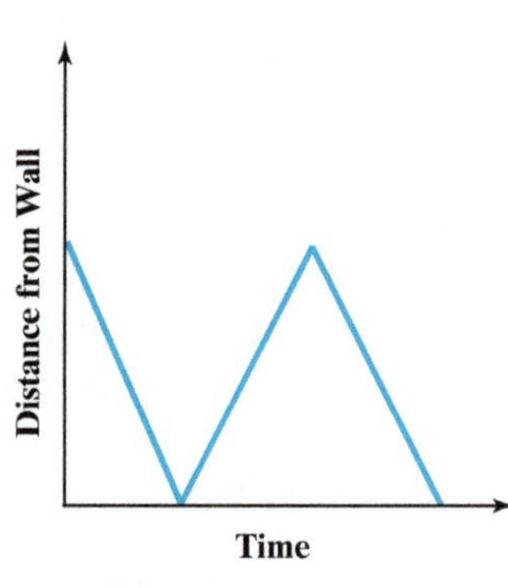

34. The following graph shows the temperature of a patient on a particular day. Describe the overall pattern you observe in the patient's temperature over the time period.

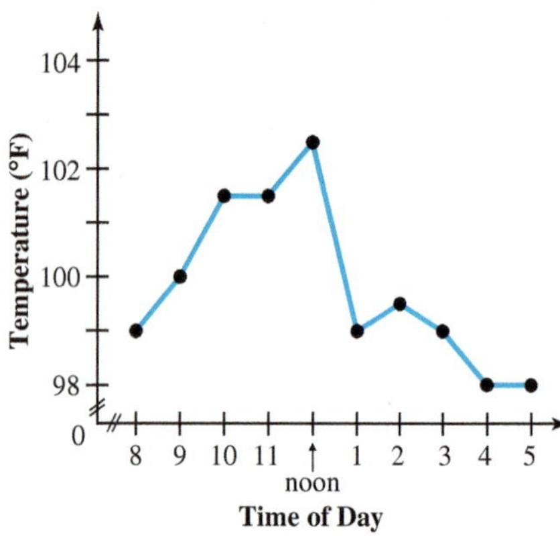

• Check your answers on page A-19.

MINDStretchers

Writing

1. Many situations involve using a coordinate system to identify positions. Two such situations are given below:
   - a chessboard
   - a map

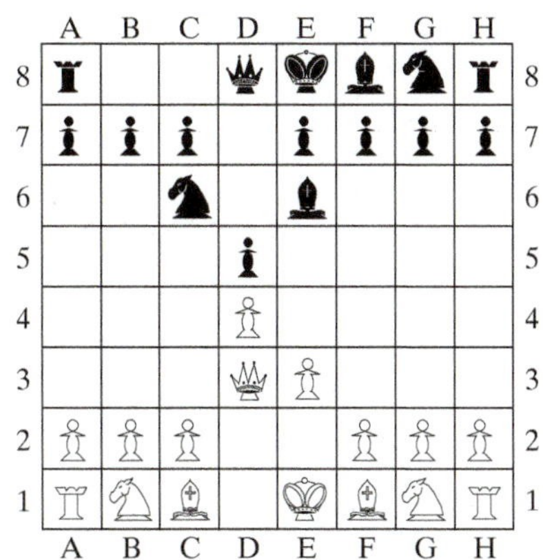

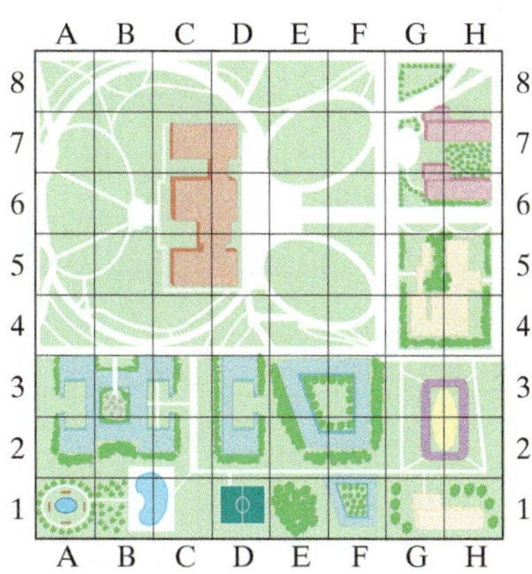

   a. Explain to what extent a chessboard and an atlas map are coordinate systems.

   b. Identify some other examples of coordinate systems in everyday life.

Critical Thinking

2. The map to the right shows a square section of a city. You want to walk along the horizontal and vertical streets from point $(-2, -2)$ to point (2, 2). One possible route is $(-2, -2)\rightarrow(-1, -2)\rightarrow(0, -2)\rightarrow(0, -1)\rightarrow(0, 0)\rightarrow(0, 1)\rightarrow(0, 2)\rightarrow(1, 2)\rightarrow(2, 2)$ as pictured below.

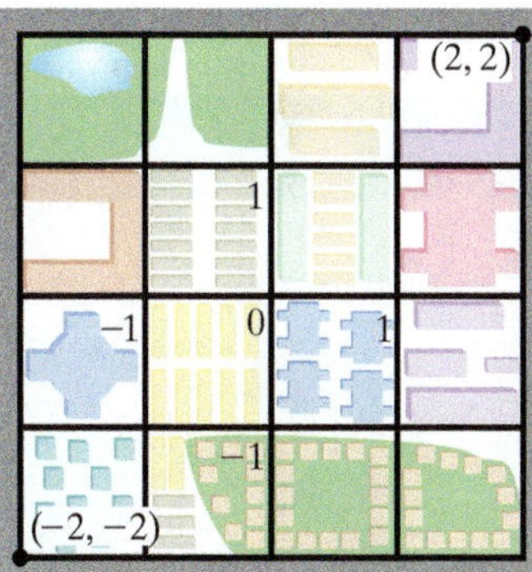

(continued)

This route is 8 blocks long. List four other 8-block routes from $(-2, -2)$ to $(2, 2)$.

$(-2, -2) \to (\_\_, \_\_) \to (\_\_, \_\_) \to (\_\_, \_\_) \to (\_\_, \_\_)$
$\to (\_\_, \_\_) \to (\_\_, \_\_) \to (\_\_, \_\_) \to (2, 2)$

$(-2, -2) \to (\_\_, \_\_) \to (\_\_, \_\_) \to (\_\_, \_\_) \to (\_\_, \_\_)$
$\to (\_\_, \_\_) \to (\_\_, \_\_) \to (\_\_, \_\_) \to (2, 2)$

$(-2, -2) \to (\_\_, \_\_) \to (\_\_, \_\_) \to (\_\_, \_\_) \to (\_\_, \_\_)$
$\to (\_\_, \_\_) \to (\_\_, \_\_) \to (\_\_, \_\_) \to (2, 2)$

$(-2, -2) \to (\_\_, \_\_) \to (\_\_, \_\_) \to (\_\_, \_\_) \to (\_\_, \_\_)$
$\to (\_\_, \_\_) \to (\_\_, \_\_) \to (\_\_, \_\_) \to (2, 2)$

Groupwork

3. Two points are plotted on a coordinate plane. Discuss with a partner what is special about a third point whose x-coordinate is the average of the first two x-coordinates, and whose y-coordinate is the average of the first two y-coordinates.

Cultural Note

It was the seventeenth-century French mathematician and philosopher René Descartes (pronounced day-KART) who developed the concepts that underlie graphing. The story goes that one morning Descartes, who liked to stay in bed and meditate, began to eye a fly crawling on his bedroom ceiling. In a flash of insight, he realized that it was possible to express mathematically the fly's position in terms of its distance to the two adjacent walls.

13.2 Slope

In the previous section, we discussed points on a coordinate plane. Now, let's look at (straight) lines that pass through points. A key characteristic of a line is its slope. In this section, we focus on the slope of a line and its relationship to the corresponding equation.

OBJECTIVES

A To find the slope of a line that passes through two given points

B To determine whether the slope of a given line is positive, negative, zero, or undefined

C To graph a line that passes through a given point and has a given slope

D To determine whether two given lines are parallel or perpendicular

E To solve applied problems involving slope

Slope

On an airplane, would you rather glide downward gradually or drop like a stone? Would you rather ski down a run that drops precipitously or ski across a gently inclined snowfield? These questions relate to *slope*, the extent to which a line is slanted. In other words, slope is a measure of a line's steepness.

Slope, also called **rate of change**, is an important concept in the study of graphing. Examining the slope of a line can tell us if the quantity being graphed increases or decreases, as well as how fast the quantity is changing. For example, in one application, the slope of a line can represent the speed of a moving object. In another application, the slope can stand for the rate at which a share of stock is changing in value.

To understand exactly what slope means, let's suppose that a straight line on a coordinate plane passes through two arbitrary points. We can call the coordinates of the first point (x_1, y_1), read "x sub 1" and "y sub 1," and the coordinates of the second point (x_2, y_2). These coordinates are written with *subscripts* in order to distinguish them from one another. We can plot these two points on a coordinate plane, and then graph the line passing through them.

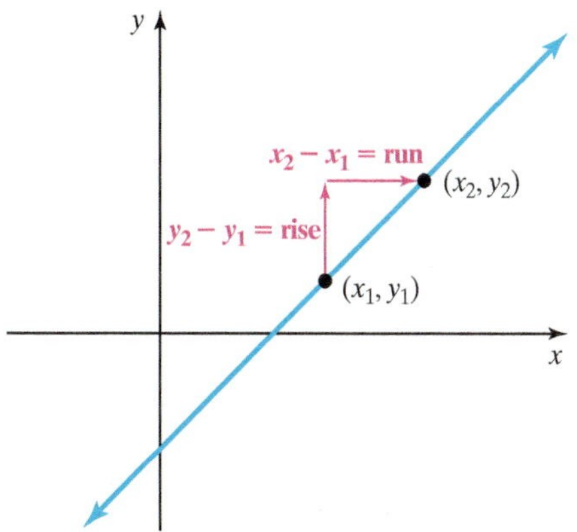

We usually represent the slope of a line by the letter m and define slope to be the ratio of the change in the y-values to the change in the x-values. Using the coordinates of the points (x_1, y_1) and (x_2, y_2) shown in the previous graph gives us the following formula:

$$m = \frac{\text{change in } y\text{-values}}{\text{change in } x\text{-values}} = \frac{y_2 - y_1}{x_2 - x_1}, \quad \text{where } x_1 \neq x_2$$

In this formula, the numerator of the fraction is the vertical change called the *rise* and the denominator is the horizontal change called the *run*. So another way of writing the formula for slope is $m = \frac{\text{rise}}{\text{run}}$.

DEFINITION

The **slope** m of a line passing through the points (x_1, y_1) and (x_2, y_2) is defined to be

$$m = \frac{y_2 - y_1}{x_2 - x_1}, \quad \text{where } x_1 \neq x_2.$$

Can you explain why in the definition of slope, x_1 and x_2 must not be equal?

Note that when using the formula for slope, it does not matter which point is chosen for (x_1, y_1) and which point for (x_2, y_2) as long as the order of subtraction of the coordinates is the same in both the numerator and denominator.

EXAMPLE 1

Find the slope of the line that passes through the points $(2, 1)$ and $(4, 2)$. Plot the points, and then sketch the line.

Solution Let $(2, 1)$ stand for (x_1, y_1) and $(4, 2)$ for (x_2, y_2).

$$\begin{array}{cc} (2, 1) & (4, 2) \\ \uparrow\uparrow & \uparrow\uparrow \\ x_1\, y_1 & x_2\, y_2 \end{array}$$

Substituting into the formula for slope, we get:

$$m = \frac{y_2 - y_1}{x_2 - x_1} = \frac{2 - 1}{4 - 2} = \frac{1}{2}$$

Now, let's plot $(2, 1)$ and $(4, 2)$, and then sketch the line passing through them.

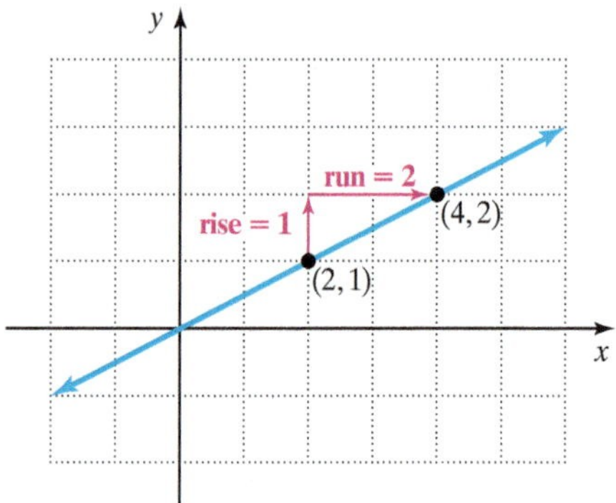

We can also find the slope of a line using its graph. From $(2, 1)$ and $(4, 2)$, we see that the change in y-values (the rise) is $2 - 1$, or 1. The change in x-values (the run) is $4 - 2$, or 2. So $m = \frac{\text{rise}}{\text{run}} = \frac{1}{2}$.

Therefore, we get the same answer whether we use the formula

$$m = \frac{y_2 - y_1}{x_2 - x_1} \quad \text{or} \quad m = \frac{\text{rise}}{\text{run}}.$$

PRACTICE 1

Find the slope of a line that contains the points, $(1, 2)$ and $(4, 3)$. Plot the points, and then sketch the line.

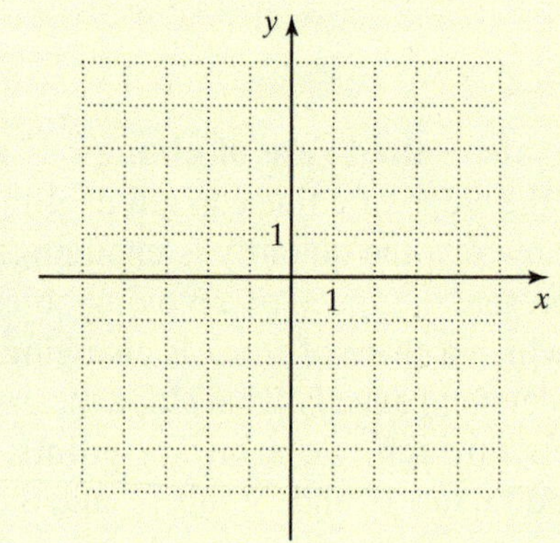

A line rising to the right as shown in Example 1 has a **positive** slope. We say that such a line is **increasing** because as the x-values get larger, the corresponding y-values also get larger.

EXAMPLE 2

Sketch the line passing through the points $(-3, 1)$ and $(2, -2)$. Find the slope.

Solution First, we plot the points $(-3, 1)$ and $(2, -2)$. Then, we draw a line passing through them.

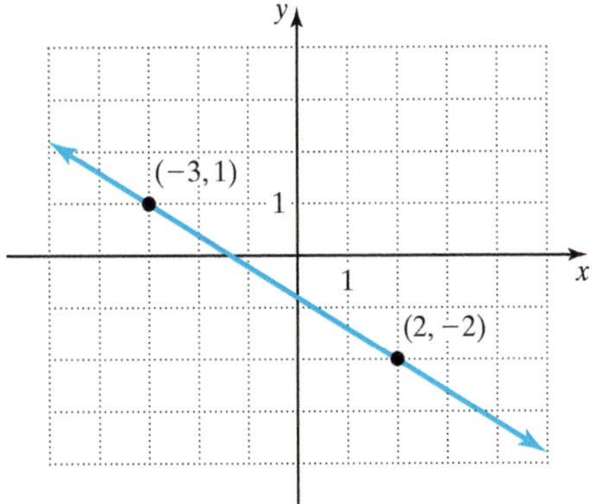

Next, find the slope.

$$\begin{array}{cccc} (-3, & 1) & (2, & -2) \\ \uparrow & \uparrow & \uparrow & \uparrow \\ x_1 & y_1 & x_2 & y_2 \end{array}$$

$$m = \frac{y_2 - y_1}{x_2 - x_1} = \frac{-2 - 1}{2 - (-3)} = \frac{-3}{5} = -\frac{3}{5}$$

PRACTICE 2

Sketch the line that contains the points $(-2, 1)$ and $(3, -5)$. Find the slope.

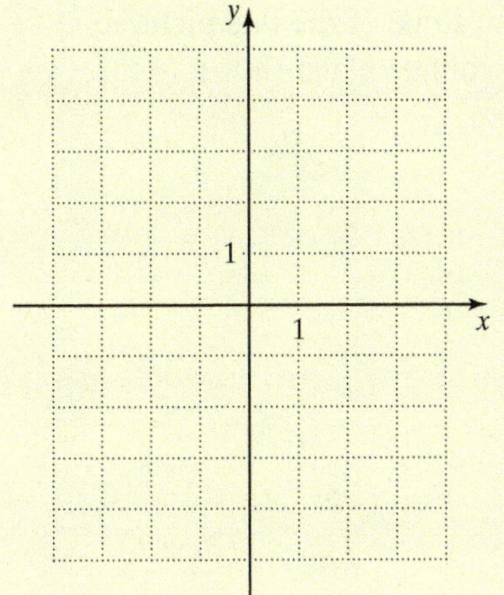

A line falling to the right as shown in Example 2 has a **negative** slope. We say that such a line is **decreasing** because as the x-values get larger, the corresponding y-values get smaller.

EXAMPLE 3

Find the slope of a line that passes through the points $(7, 5)$ and $(-1, 5)$. Plot the points, and then sketch the line.

Solution First, we find the slope of the line.

$$\begin{array}{cccc} (7, & 5) & (-1, & 5) \\ \uparrow & \uparrow & \uparrow & \uparrow \\ x_1 & y_1 & x_2 & y_2 \end{array}$$

$$m = \frac{y_2 - y_1}{x_2 - x_1} = \frac{5 - 5}{-1 - 7} = \frac{0}{-8} = 0$$

So the slope of this line is 0.

Next, we plot the points, and then sketch the line passing through them.

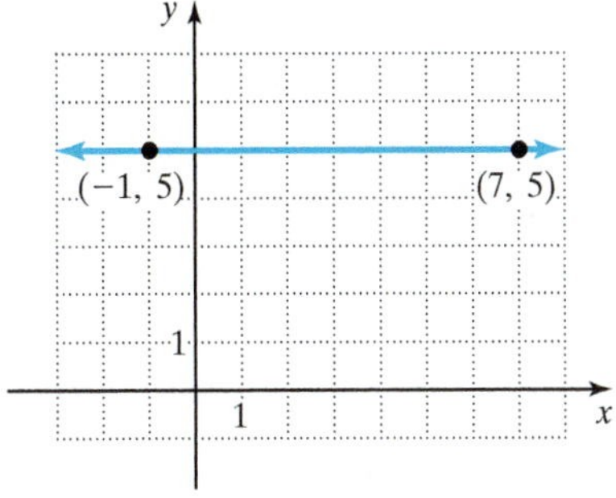

PRACTICE 3

On the following coordinate plane, plot the points $(2, -1)$ and $(6, -1)$. Sketch the line, and then compute its slope.

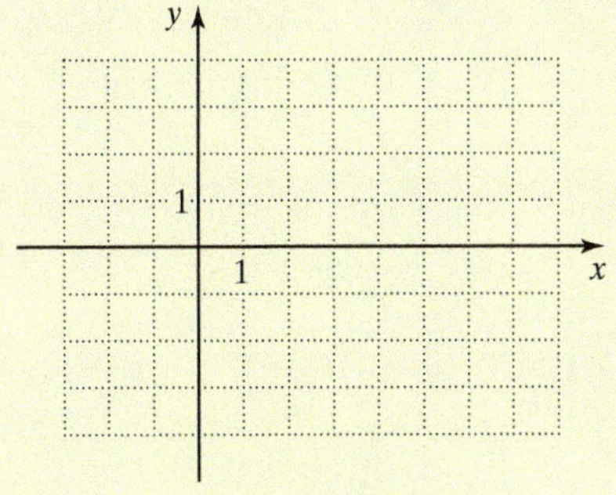

When the slope of a line is 0, its graph is a **horizontal line** as shown in Example 3. All points on a horizontal line have the same y-coordinate, that is, the y-values are constant for all x-values.

EXAMPLE 4

What is the slope of the line pictured on the coordinate plane shown to the right?

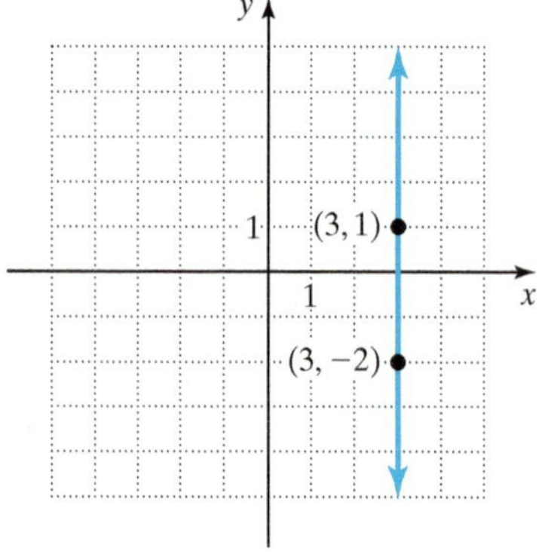

Solution

$$\begin{matrix} (3, & -2) & (3, & 1) \\ \uparrow & \uparrow & \uparrow & \uparrow \\ x_1 & y_1 & x_2 & y_2 \end{matrix}$$

$$m = \frac{y_2 - y_1}{x_2 - x_1} = \frac{1 - (-2)}{3 - 3} = \frac{3}{0}$$

Since division by 0 is undefined, the slope of this line is undefined.

PRACTICE 4

Find the slope of the line that passes through the points $(-2, 7)$ and $(-2, 0)$.

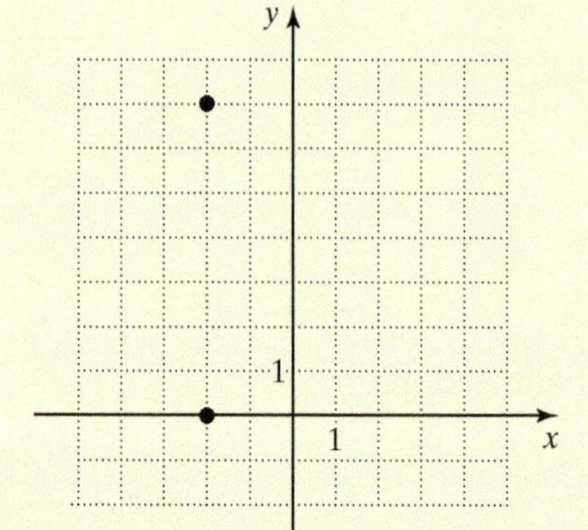

When the slope of a line is undefined, its graph is a **vertical line** as shown in Example 4. All points on a vertical line have the same x-coordinate, that is, the x-values are constant for all y-values.

As we have seen in Examples 1 through 4, the sign of the slope of a line tells us a lot about the line. As we continue graphing lines, it will be helpful to keep in mind the following graphs:

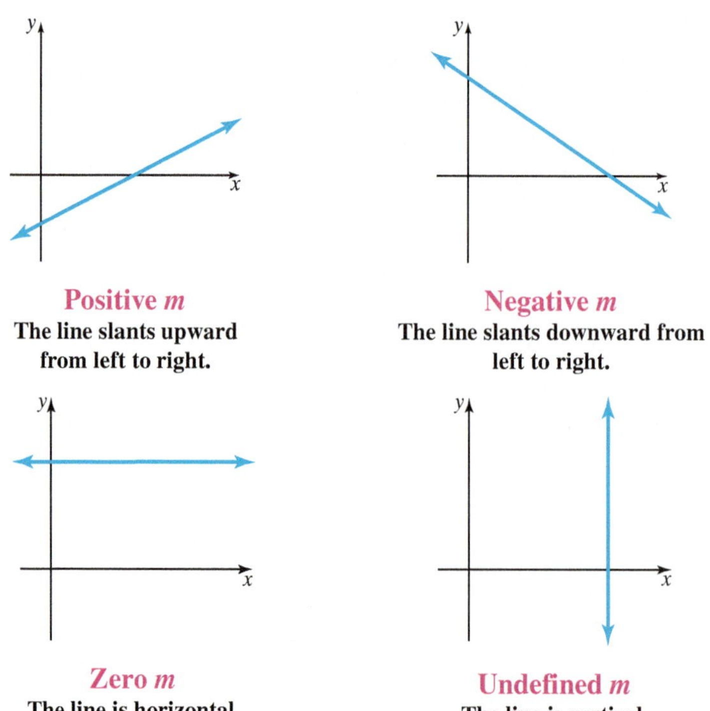

In the next example, we graph two lines on a coordinate plane.

EXAMPLE 5

Calculate the slopes for the lines shown.

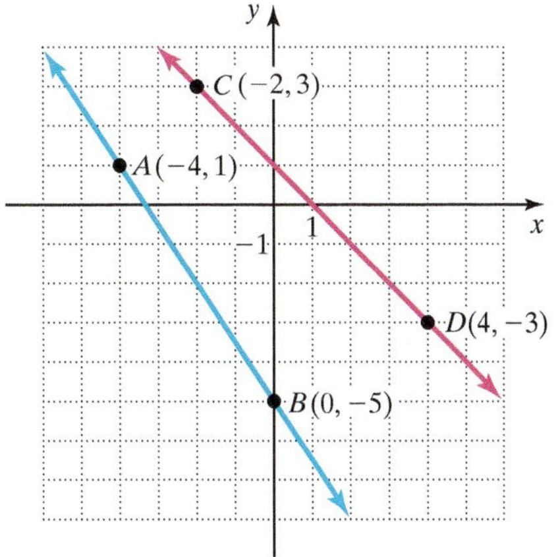

Solution Line AB, written $\overleftrightarrow{AB}$, passes through $A(-4, 1)$ and $B(0, -5)$. Its slope is:

$$m = \frac{y_2 - y_1}{x_2 - x_1} = \frac{1 - (-5)}{(-4) - 0} = \frac{1 + 5}{-4} = \frac{6}{-4} = -\frac{3}{2}$$

For $\overleftrightarrow{CD}$ passing through $C(-2, 3)$ and $D(4, -3)$, the slope is:

$$m = \frac{y_2 - y_1}{x_2 - x_1} = \frac{3 - (-3)}{(-2) - 4} = \frac{3 + 3}{(-2) - 4} = \frac{6}{-6} = -1$$

Note that both lines have negative slopes and slant downward from left to right.

PRACTICE 5

Compute the slopes for the lines shown.

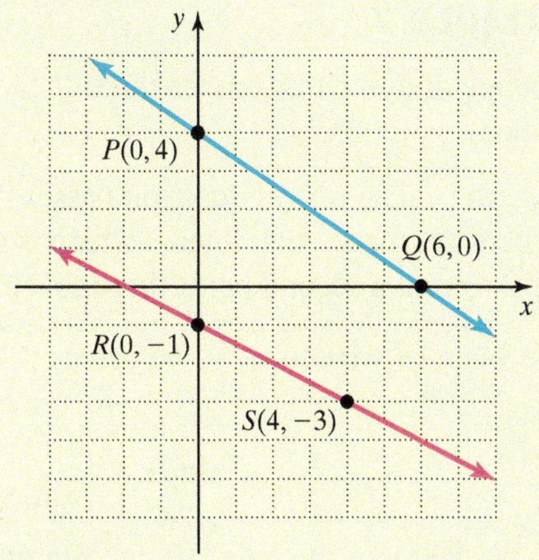

As shown in the next example, how a line slants often helps us to interpret the information given in the graph.

EXAMPLE 6

The graph to the right shows the amount of money that your dental insurance reimburses you, depending on the amount of your dental bill. Is the slope of the graphed line positive or negative? Explain how you know. What does this mean in terms of insurance reimbursement?

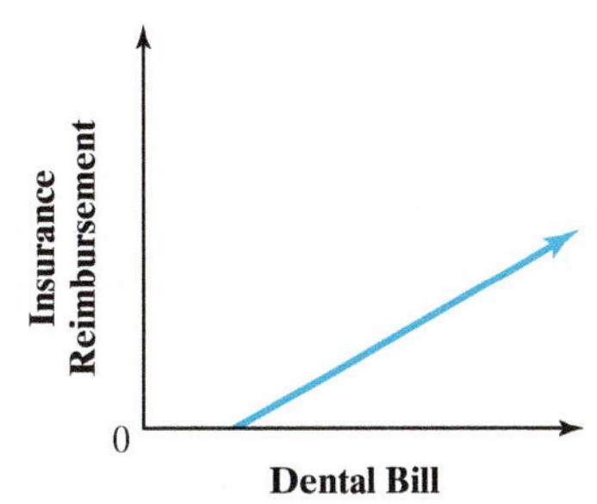

Solution Since the graphed line slants upward from left to right, its slope is positive. According to this graph, larger x-values correspond to larger y-values. So your dental insurance reimburses you more for larger dental bills.

PRACTICE 6

A doctor is trying to help eliminate an epidemic. Explain, in terms of slope, which scenario would be the most desirable.

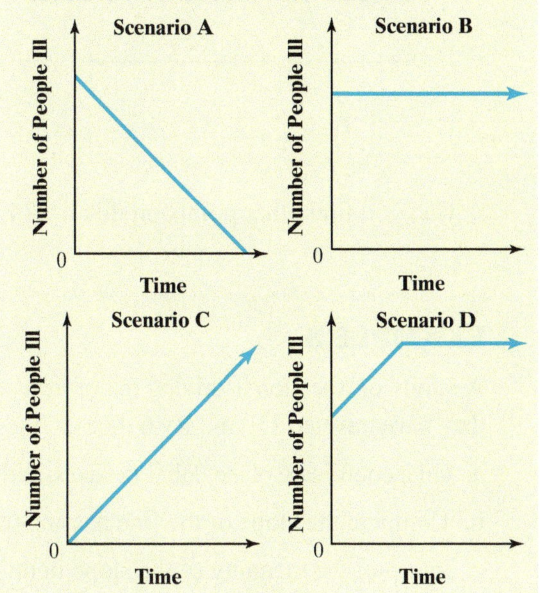

We have already graphed a line by plotting two points and drawing the line passing through them. Now, let's look at graphing a line when given the slope of the line and a point on the line.

EXAMPLE 7

The slope of a line that passes through the point $(2, 5)$ is 3. Graph the line.

Solution The line in question passes through the point $(2, 5)$. But there are many such lines—which is the right one? We use the slope 3 to find a second point through which the line also passes.

Since 3 can be written as $\frac{3}{1}$, we have:

$$\text{slope} = \frac{\text{rise}}{\text{run}} = \frac{3}{1}$$

We first plot the point $(2, 5)$. Starting at $(2, 5)$, we move 3 units up (for a rise of 3) and then 1 unit to the right (for a run of 1). Arriving at the point $(3, 8)$, we sketch the line passing through the points $(2, 5)$ and $(3, 8)$, as shown in the graph on the left below.

Since $\frac{3}{1} = \frac{-3}{-1}$, we could have started at $(2, 5)$ and moved down 3 units (for a rise of -3) and then 1 unit to the left (for a run of -1). In this case, we would arrive at $(1, 2)$, which is another point on the same line, as shown in the graph on the right below.

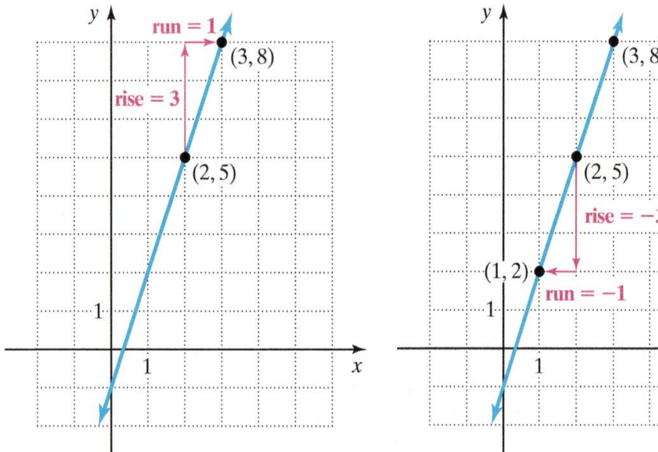

Can you find other points on this line? Explain.

PRACTICE 7

Graph the line with slope 4 that passes through the point $(1, -2)$.

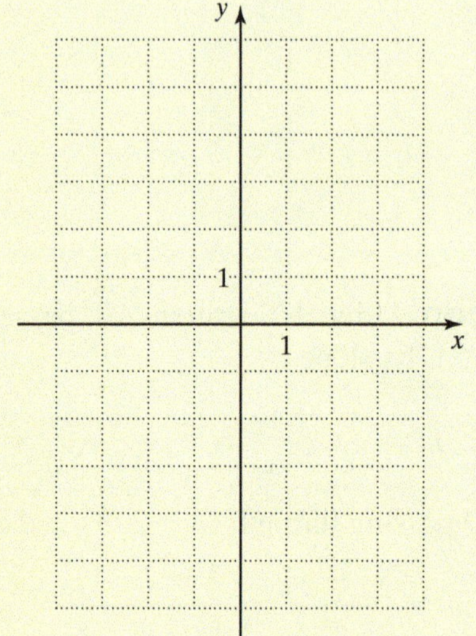

EXAMPLE 8

A family on vacation is driving out of town at a constant speed. At 2 o'clock, they have traveled 110 mi. By 6 o'clock, they have traveled 330 mi.

a. On a coordinate plane, label the axes, and then plot the appropriate points.

b. Compute the slope of the line passing through the points.

c. Interpret the meaning of the slope in this situation.

PRACTICE 8

In 1985, the streetcar running along Boston's Arborway Corridor had a daily ridership of 28,000. The streetcar was then replaced by a bus. Twenty-five years later, daily ridership had dropped to 14,000. (*Source:* arborway.org)

Solution

a. Label the axes on the coordinate plane. Then, plot the points $(2, 110)$ and $(6, 330)$.

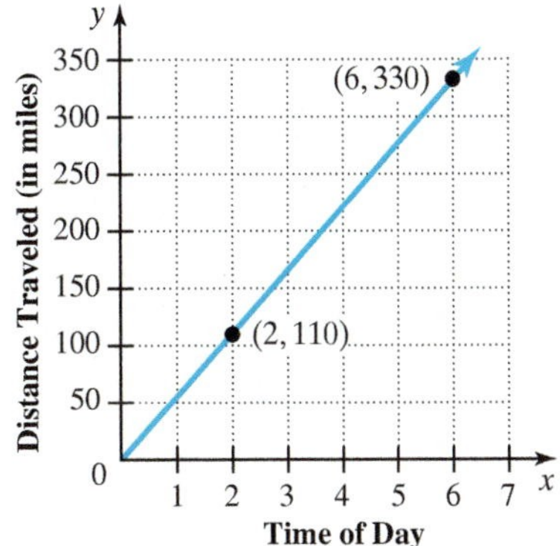

b. The slope of the line through the two points is:

$$\begin{aligned} m &= \frac{y_2 - y_1}{x_2 - x_1} \\ &= \frac{330 - 110}{6 - 2} \\ &= \frac{220}{4} \\ &= 55 \end{aligned}$$

c. Here the slope is the change in distance divided by the change in time. In other words, the slope is the average speed the family traveled, which is 55 mph.

a. On a coordinate plane, label the axes, and then plot the appropriate points for daily ridership r in y years after 1985.

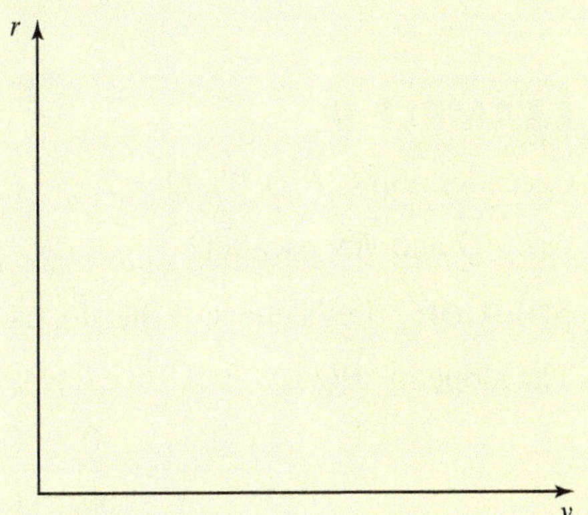

b. Compute the slope of the line that passes through the points.

c. Interpret the meaning of the slope in this situation.

Parallel and Perpendicular Lines

By examining the slopes of straight lines on a coordinate plane, we can solve problems that require us to determine if:

- two given lines are parallel or
- two given lines are perpendicular.

Let's consider parallel lines first.

Since the slope of a line measures its slant, lines in a coordinate plane with equal slopes are parallel, that is, do not intersect. So on the coordinate plane shown below, if we knew the coordinates of points P, Q, R, and S, we could verify that $\overleftrightarrow{PQ}$ and $\overleftrightarrow{RS}$ are parallel by computing their slopes and then checking that these slopes are equal.

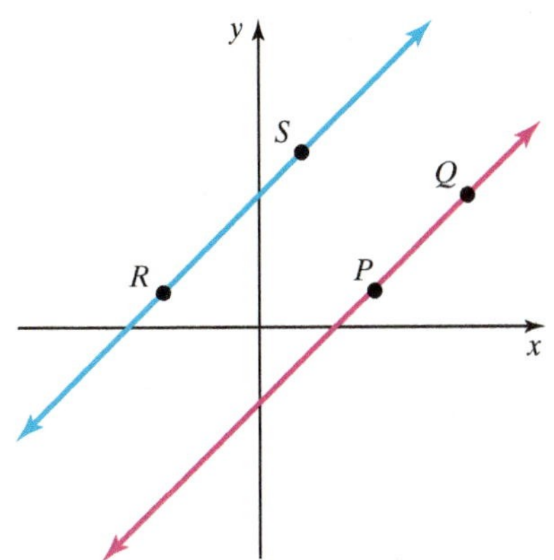

Parallel Lines

Two nonvertical lines are **parallel** if and only if their slopes are equal. That is, if the slopes are m_1 and m_2, then $m_1 = m_2$.

EXAMPLE 9

Consider points $P(0, 0)$, $Q(-2, -5)$, $R(0, 5)$, and $S(-2, 0)$. Are $\overleftrightarrow{PQ}$ and $\overleftrightarrow{RS}$ parallel?

Solution Let's check if the slopes of $\overleftrightarrow{PQ}$ and $\overleftrightarrow{RS}$ are equal.

The slope of $\overleftrightarrow{PQ}$ is:

$$m = \frac{y_2 - y_1}{x_2 - x_1} = \frac{0 - (-5)}{0 - (-2)} = \frac{0 + 5}{0 + 2} = \frac{5}{2}$$

The slope of $\overleftrightarrow{RS}$ is:

$$m = \frac{y_2 - y_1}{x_2 - x_1} = \frac{5 - 0}{0 - (-2)} = \frac{5}{0 + 2} = \frac{5}{2}$$

Since the slopes of $\overleftrightarrow{PQ}$ and $\overleftrightarrow{RS}$ are equal, $\overleftrightarrow{PQ}$ and $\overleftrightarrow{RS}$ are parallel.

PRACTICE 9

Decide whether $\overleftrightarrow{EF}$ and $\overleftrightarrow{GH}$ are parallel, given points $E(0, 4)$, $F(4, -1)$, $G(0, 8)$, and $H(8, -2)$.

EXAMPLE 10

A pediatric nurse kept track of the weights of two children who are twins. Use the graph to answer the following questions:

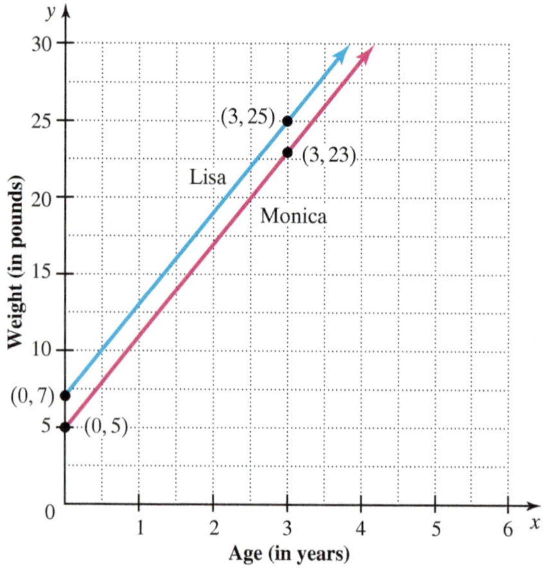

a. Which twin was heavier at birth?

b. Are the two lines parallel?

c. Which twin grew at a faster rate?

d. Assuming that the rates of growth of the twins remain constant, could this graph be used to project the weight of the twins at age 8? Explain.

PRACTICE 10

The graph shown gives the income of a computer lab technician and a multimedia designer as related to the number of years that they have been employed.

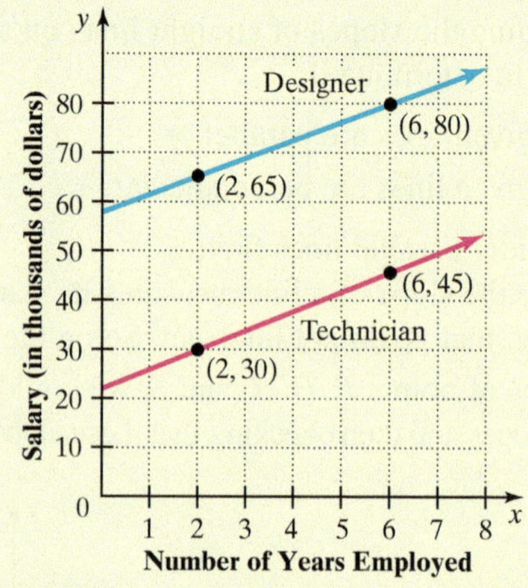

Use this graph to answer the following questions.

Solution

a. At birth, Lisa weighed 7 lb and Monica weighed 5 lb. So Lisa was heavier.

b. To determine if the two lines are parallel, we begin by computing the slope of the line passing through the points $(0, 7)$ and $(3, 25)$.

$$m = \frac{y_2 - y_1}{x_2 - x_1} = \frac{7 - 25}{0 - 3} = \frac{-18}{-3} = 6$$

Now, we compute the slope for the line passing through the points $(0, 5)$ and $(3, 23)$.

$$m = \frac{y_2 - y_1}{x_2 - y_1} = \frac{5 - 23}{0 - 3} = \frac{-18}{-3} = 6$$

Since the slopes of the two lines are equal, the graphed lines are parallel.

c. Since the two lines are parallel, the twins grew at the same rate.

d. If we extend the x- and y-axes, we could project the weight of each child at age 8 by reading the corresponding y-coordinate for x equal to 8.

a. Are the two lines parallel?

b. Does your answer to part (a) agree with your observation of the graph? Explain.

c. Which employee's salary increased at a faster rate?

d. From the graph, estimate the starting salary of the multimedia designer.

Now, let's consider the problem of determining whether two given lines are perpendicular, that is, they intersect at a right, or 90°, angle.

On a coordinate plane, two lines are perpendicular to one another when the product of their slopes is -1. For instance, the slope of $\overleftrightarrow{PQ}$ in the following graph is:

$$m = \frac{7 - 2}{3 - 1} = \frac{5}{2}$$

The slope of $\overleftrightarrow{QR}$ is:

$$m = \frac{2 - 0}{1 - 6} = \frac{2}{-5} = -\frac{2}{5}$$

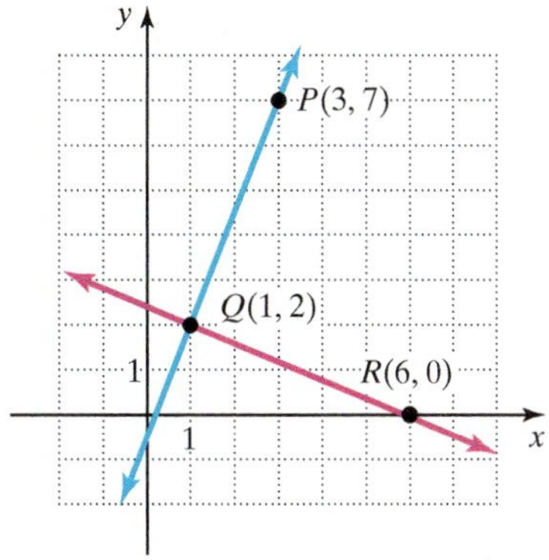

The product of these slopes is:

$$m = \left(\frac{5}{2}\right)\left(-\frac{2}{5}\right) = -1$$

Since the slopes are *negative reciprocals* of each other, the two lines must be perpendicular to one another.

Perpendicular Lines

Two nonvertical lines are **perpendicular** if and only if the product of their slopes is -1. That is, if the slopes are m_1 and m_2, then $m_1 \cdot m_2 = -1$.

EXAMPLE 11

Determine if $\overleftrightarrow{AB}$ and $\overleftrightarrow{BC}$ shown in the graph are perpendicular to one another.

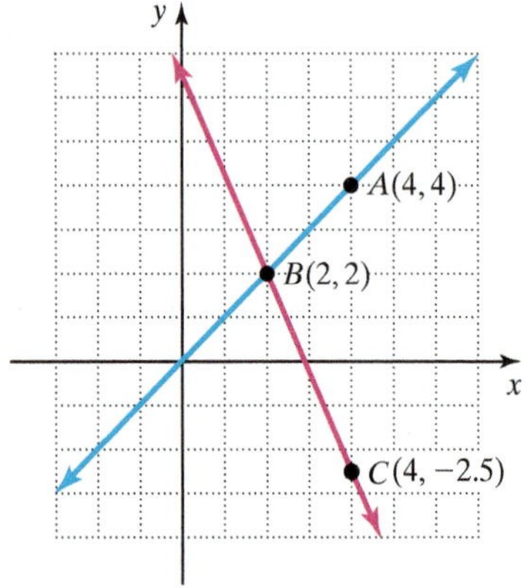

Solution First, let's compute the slopes of the two lines in question.

The slope of $\overleftrightarrow{AB}$ is:

$$m = \frac{y_2 - y_1}{x_2 - x_1} = \frac{4 - 2}{4 - 2} = \frac{2}{2} = 1$$

The slope of $\overleftrightarrow{BC}$ is:

$$m = \frac{y_2 - y_1}{x_2 - x_1} = \frac{2 - (-2.5)}{2 - 4} = \frac{2 + 2.5}{2 - 4} = \frac{4.5}{-2} = -2.25$$

To check if $\overleftrightarrow{AB}$ and $\overleftrightarrow{BC}$ are perpendicular, we find the product of their slopes:

$$(1)(-2.25) = -2.25$$

The lines are not perpendicular to one another since this product is not equal to -1.

PRACTICE 11

Consider points $A(1, 3)$, $B(2, 5)$, and $C(-1, 2)$. Decide if $\overleftrightarrow{AB}$ is perpendicular to $\overleftrightarrow{AC}$.

EXAMPLE 12

On the coordinate plane shown, x-values represent streets and y-values represent avenues. A road is to be constructed running straight from 4th Street and 3rd Avenue to 9th Street and 9th Avenue. A second road will run from 2nd Street and 10th Avenue to 8th Street and 5th Avenue. Will the roads meet at right angles?

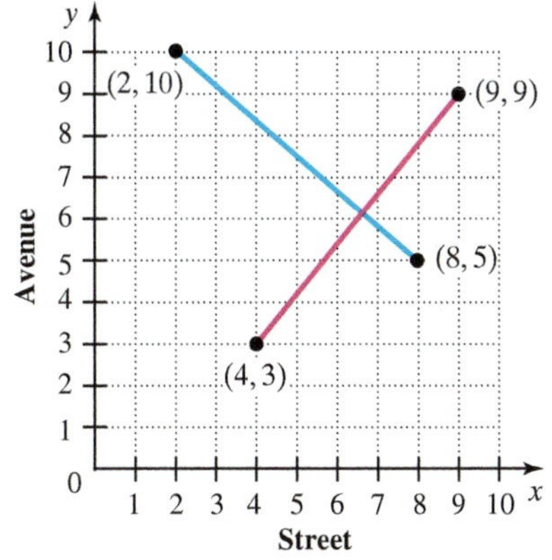

Solution Let's find the slopes of the two roads. The slope of the road from 9th Street and 9th Avenue to 4th Street and 3rd Avenue is:

$$m = \frac{y_2 - y_1}{x_2 - x_1} = \frac{9 - 3}{9 - 4} = \frac{6}{5}$$

The slope of the road from 2nd Street and 10th Avenue to 8th Street and 5th Avenue is:

$$m = \frac{y_2 - y_1}{x_2 - x_1} = \frac{10 - 5}{2 - 8} = \frac{5}{-6} = -\frac{5}{6}$$

The product of these slopes is:

$$\left(\frac{6}{5}\right)\left(-\frac{5}{6}\right) = -1$$

Therefore, the roads will be perpendicular to one another.

PRACTICE 12

Consider the square 6 units on each side shown in the diagram below. By examining their slopes, determine whether the diagonals of the square are perpendicular to one another.

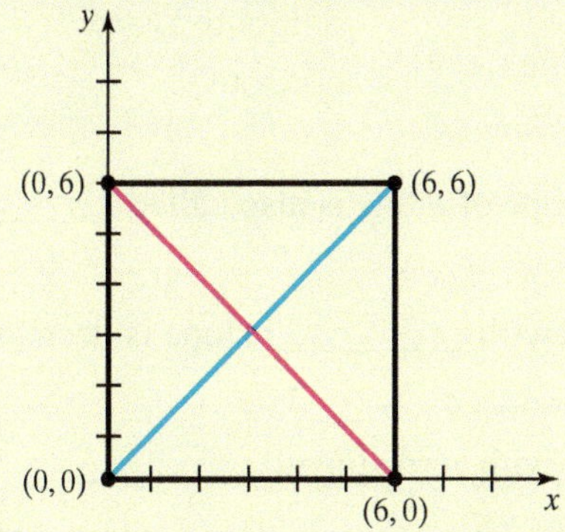

13.2 Exercises

FOR EXTRA HELP MyMathLab MathXL PRACTICE WATCH READ REVIEW

Mathematically Speaking

Fill in each blank with the most appropriate term or phrase from the given list.

y-coordinate	parallel	negative
positive	vertical	x-coordinate
perpendicular	rate of change	run
horizontal	rise	

1. The slope of a line is also called its ______________.

2. In the slope formula, the vertical change is called the __________.

3. A line with __________ slope is decreasing.

4. If all points on a line have the same ____________, then the line is vertical.

5. A line with zero slope is __________.

6. A line with undefined slope is __________.

7. Two nonvertical lines are __________ if and only if their slopes are equal.

8. Two nonvertical lines are ____________ if and only if the product of their slopes is -1.

A *Compute the slope m of the line that passes through the given points. Plot these points on the coordinate plane, and sketch the line that passes through them.*

9. $(2, 3)$ and $(-2, 0)$, $m =$

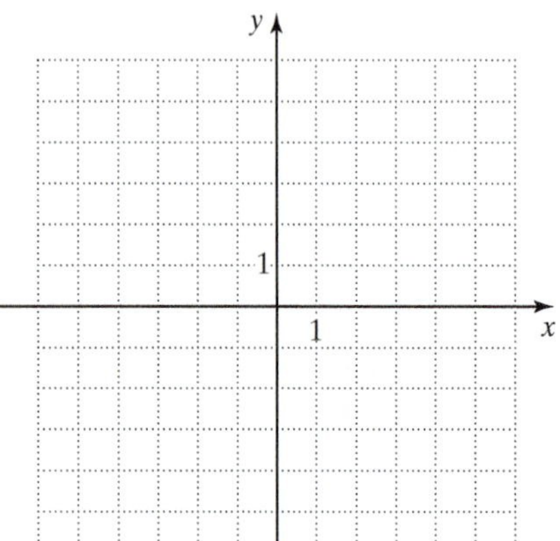

10. $(-1, 4)$ and $(0, 5)$, $m =$

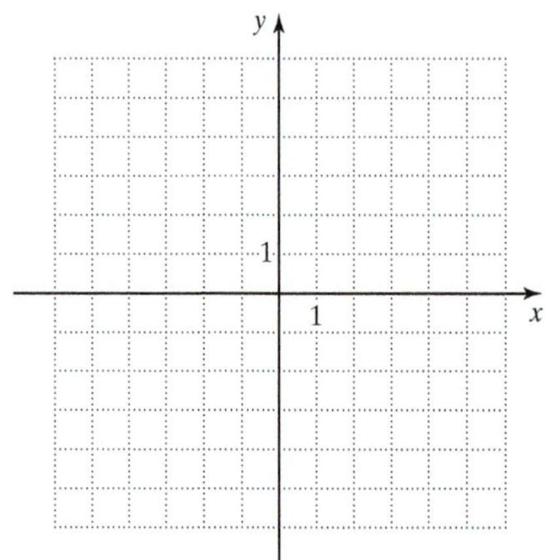

11. $(6, -4)$ and $(6, 1)$, $m =$

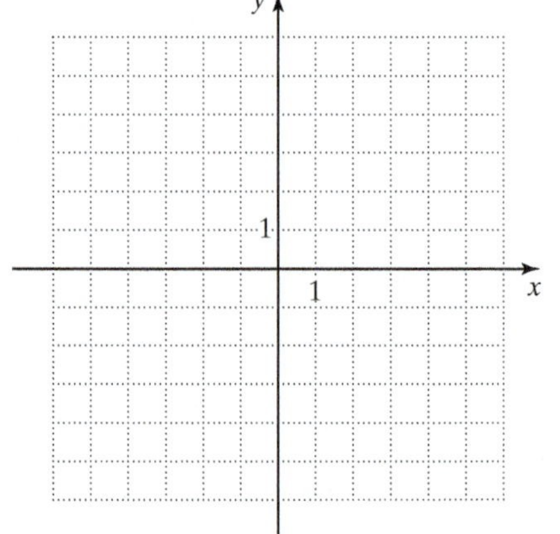

12. $(1, 1)$ and $(1, -3)$, $m =$

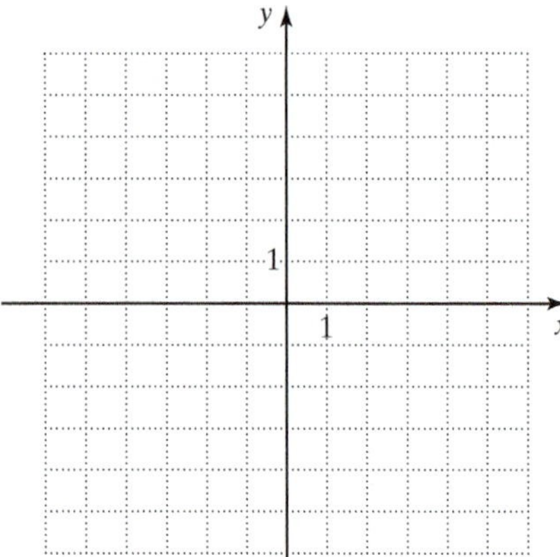

13. $(-2, 1)$ and $(3, -1)$, $m =$

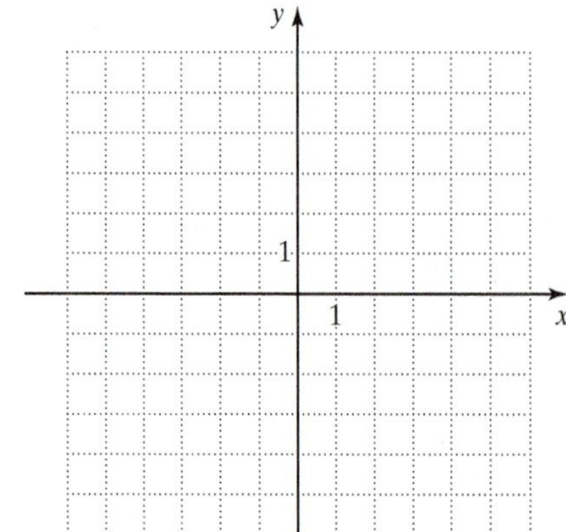

14. $(0, 0)$ and $(-2, 5)$, $m =$

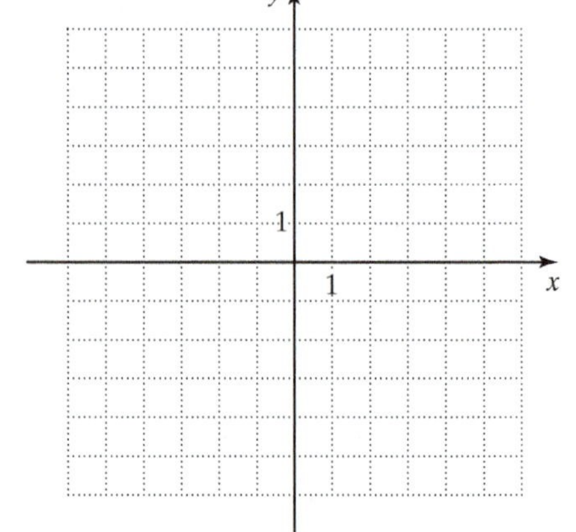

15. $(-1, -4)$ and $(3, -4)$, $m =$

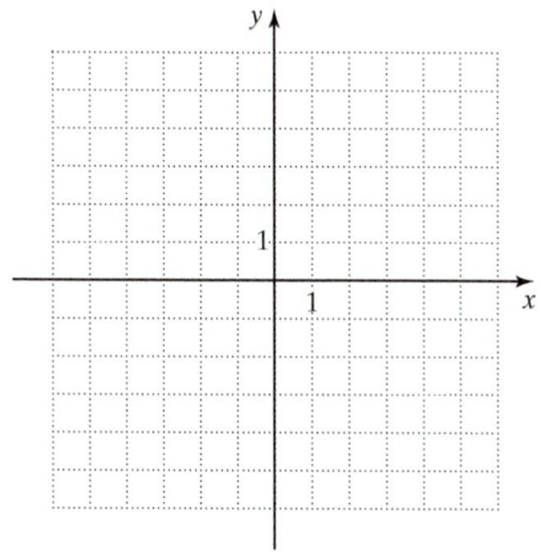

16. $(3, 0)$ and $(5, 0)$, $m =$

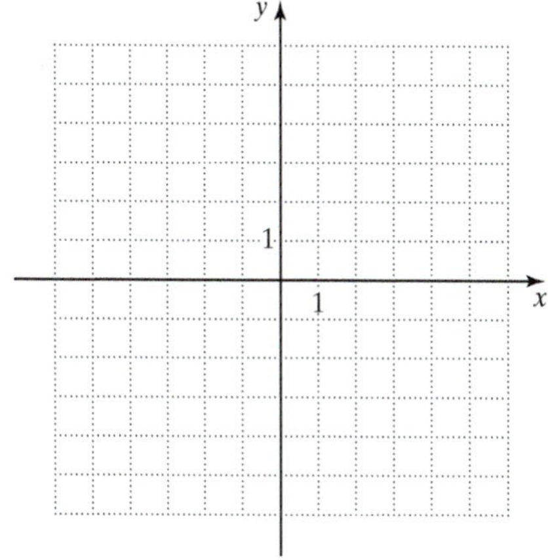

17. $(0.5, 0)$ and $(0, 3.5)$, $m =$

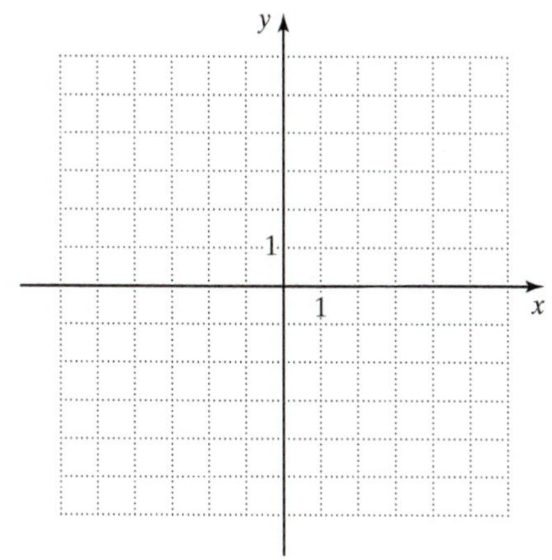

18. $(4, 4.5)$ and $(1, 2.5)$, $m =$

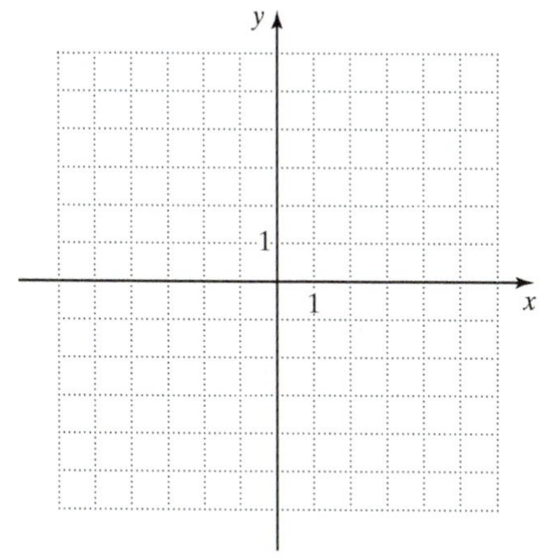

On each graph, calculate the slopes for the lines shown.

19.

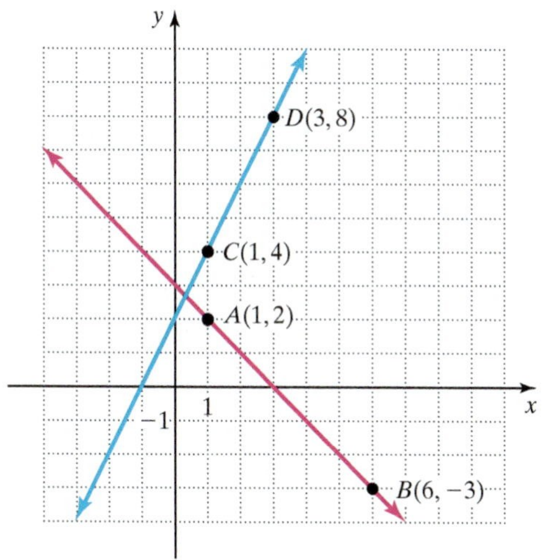

20.

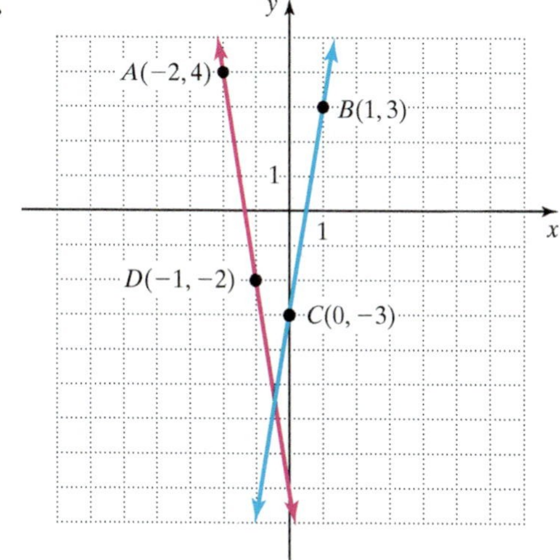

B *Indicate whether the slope of each graph is positive, negative, zero, or undefined. Then, state whether the line is horizontal, vertical, or neither.*

21.

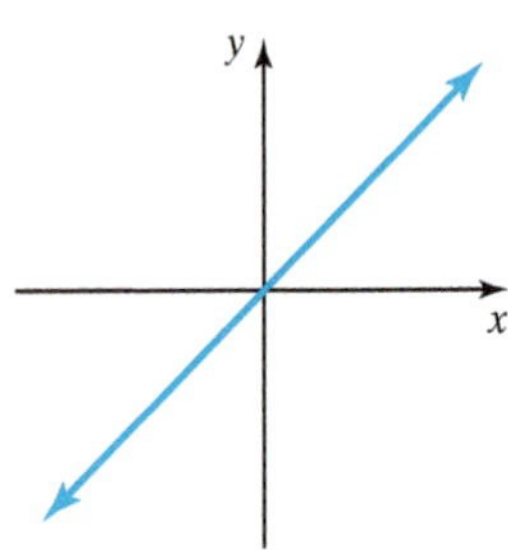

22.

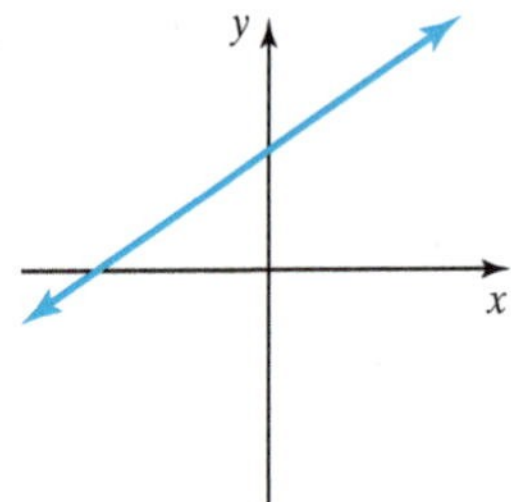

23.

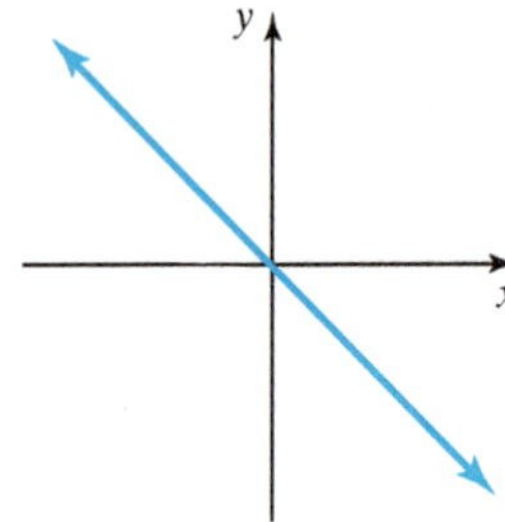

24.

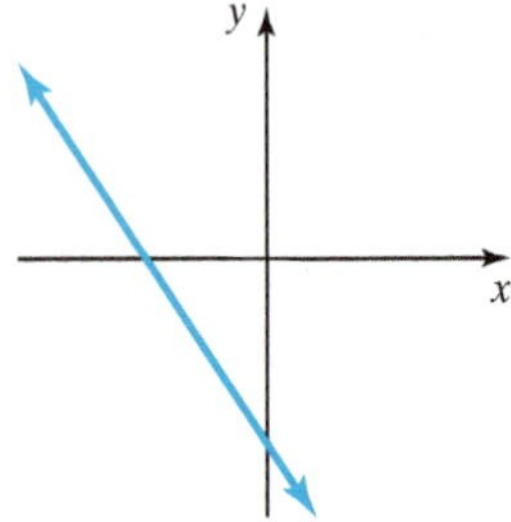

25.

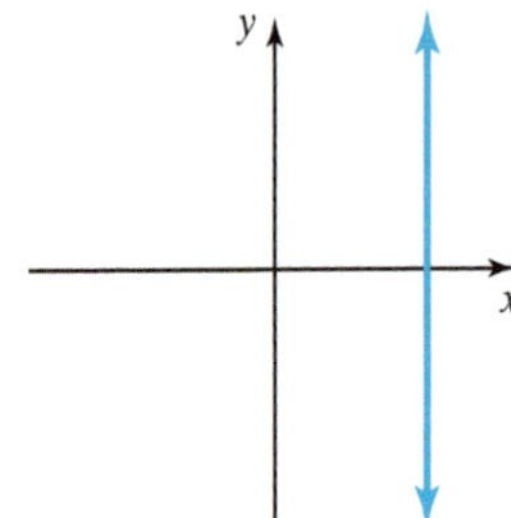

26.

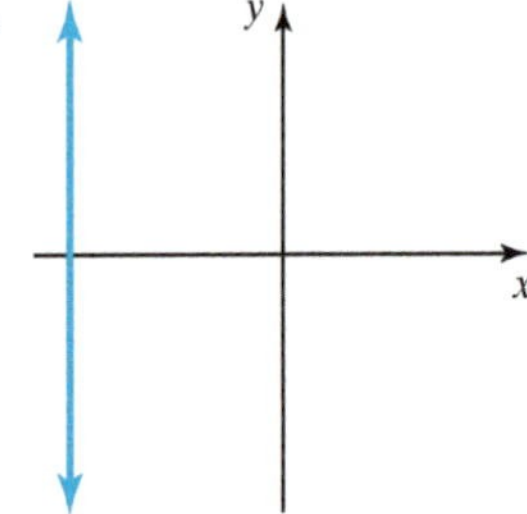

27.

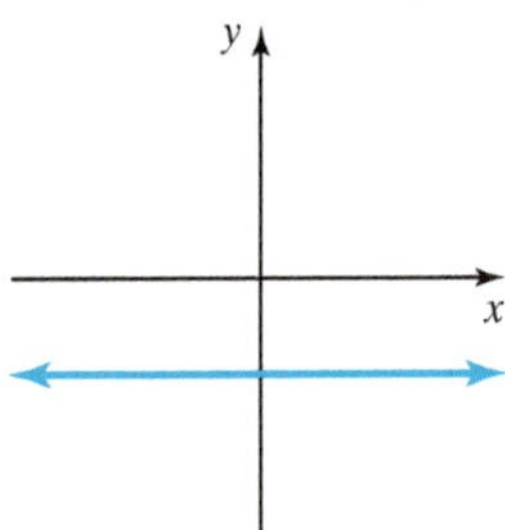

28.

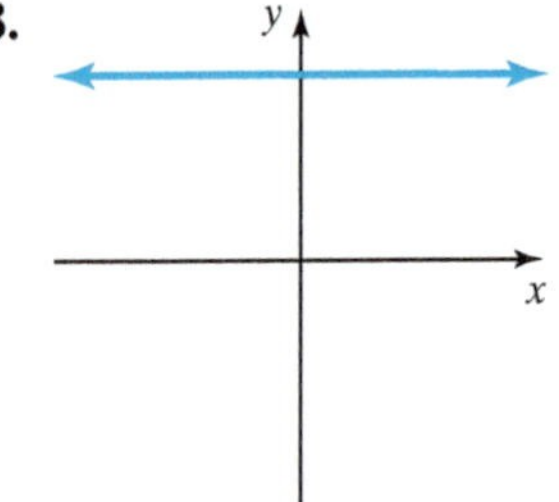

C *Graph the line on the coordinate plane using the given information.*

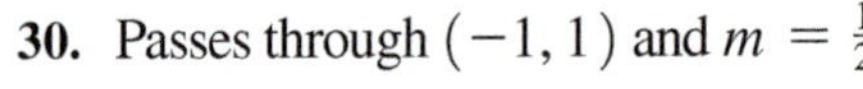

29. Passes through $(2, 5)$ and $m = 4$

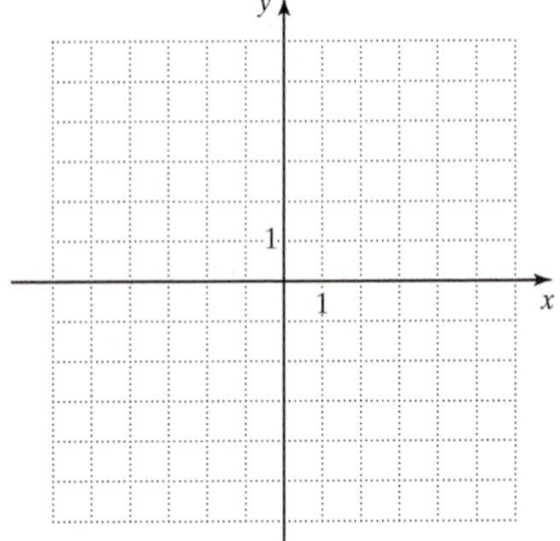

30. Passes through $(-1, 1)$ and $m = \frac{1}{2}$

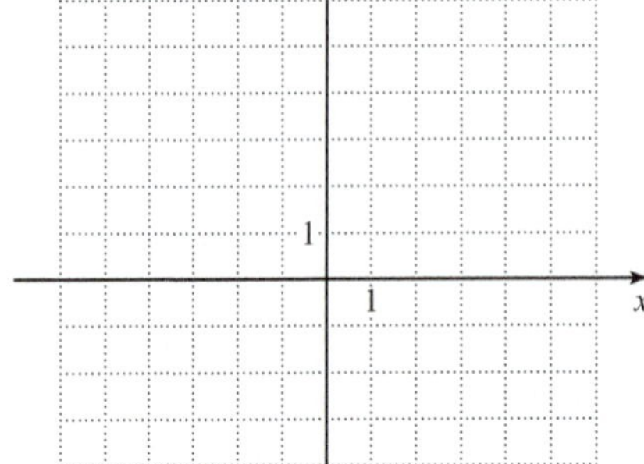

31. Passes through $(2, 5)$ and $m = -\frac{4}{3}$

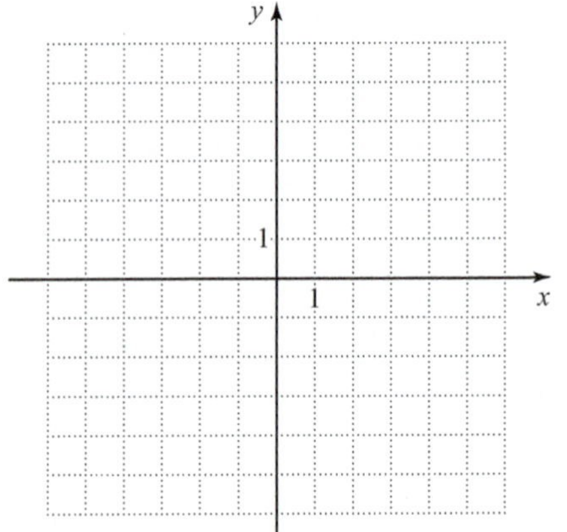

32. Passes through $(1, 5)$ and $m = -3$

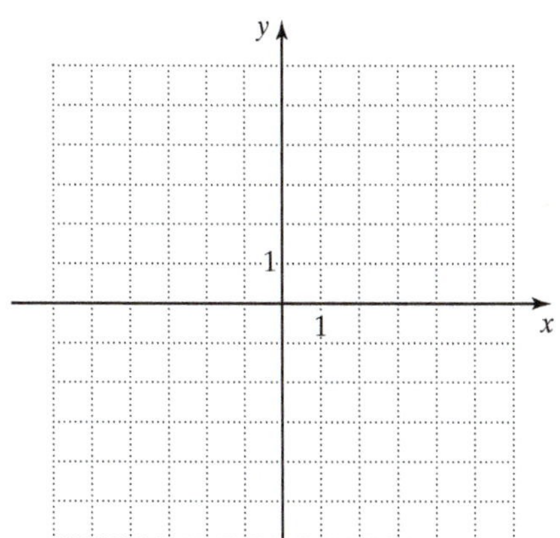

33. Passes through $(0, -6)$ and $m = 0$

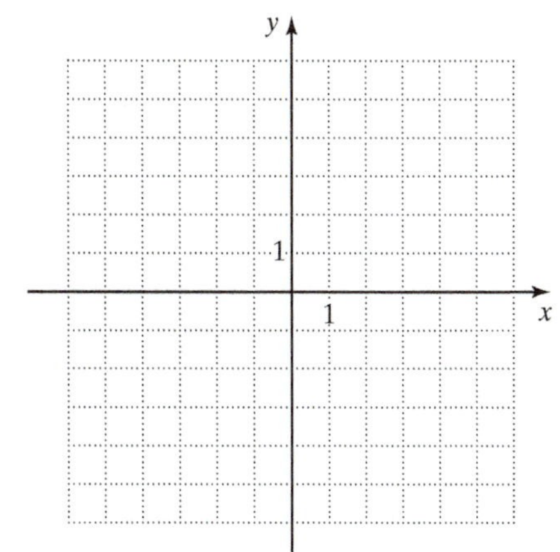

34. Passes through $(0, -2)$ and $m = 0$

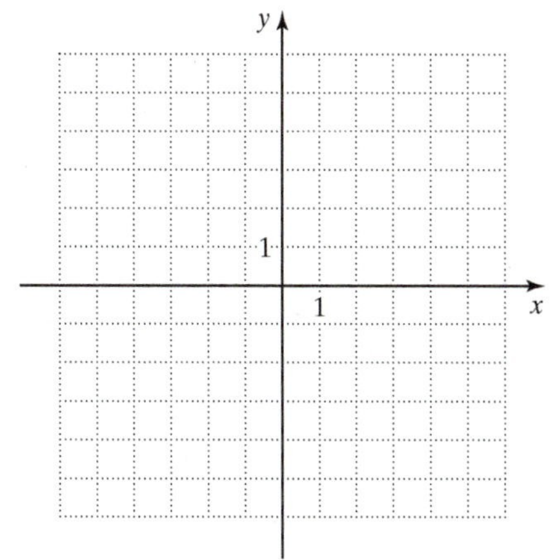

35. Passes through $(-4, 0)$ and the slope is undefined

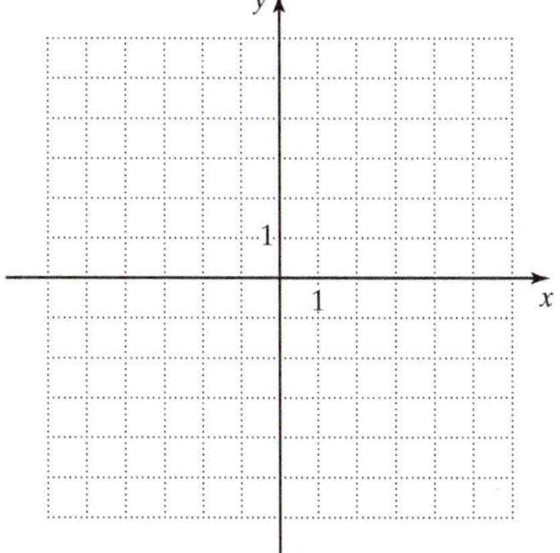

36. Passes through $(-6, 2)$ and the slope is undefined

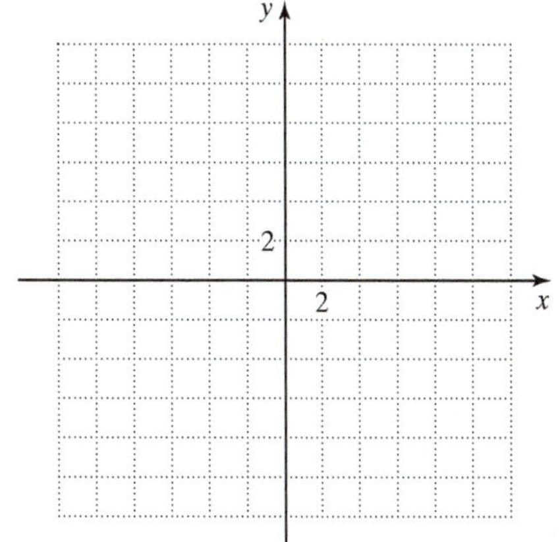

D *Determine whether $\overleftrightarrow{PQ}$ and $\overleftrightarrow{RS}$ are parallel or perpendicular.*

37.

	P	*Q*	*R*	*S*
a.	$(0, -1)$	$(1, 3)$	$(5, 0)$	$(7, 8)$
b.	$(9, 1)$	$(7, 4)$	$(0, 0)$	$(6, 4)$

38.

	P	*Q*	*R*	*S*
a.	$(3, 3)$	$(7, 7)$	$(-5, 5)$	$(2, -2)$
b.	$(8, 0)$	$(0, 4)$	$(0, -4)$	$(-12, 2)$

Mixed Practice

Indicate whether the slope of each graph is positive, negative, zero, or undefined. Then, state whether the line is horizontal, vertical, or neither.

39.

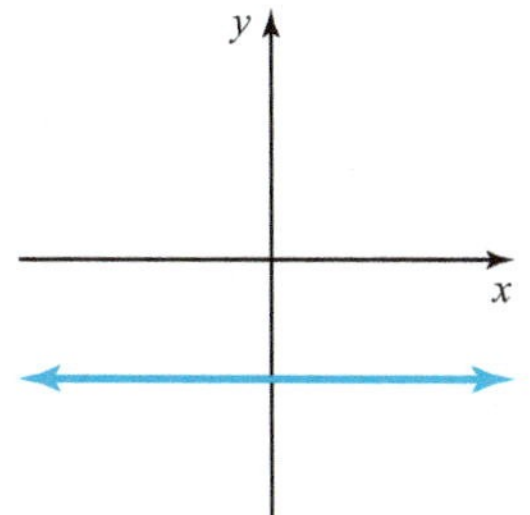

40.

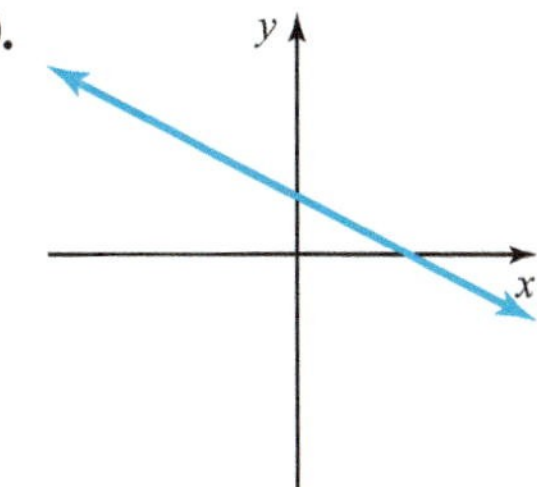

41. Graph the line on the coordinate plane if the line passes through $(-2, 3)$ and $m = 4$.

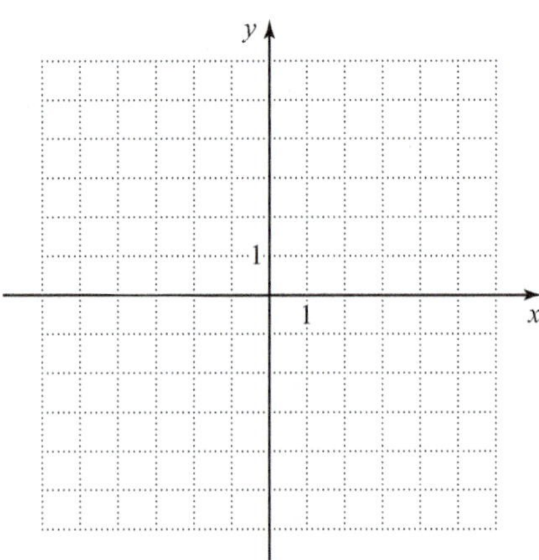

42. Calculate the slopes for the lines shown in the graph.

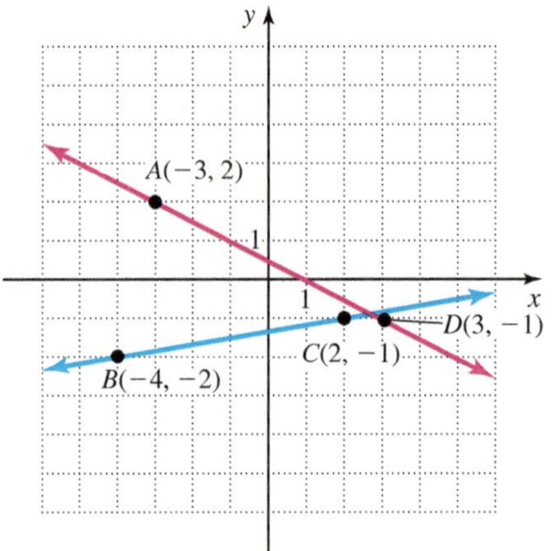

43. Determine whether $\overleftrightarrow{AB}$ and $\overleftrightarrow{CD}$ are parallel or perpendicular.

	A	B	C	D
a.	$(-3, 2)$	$(5, -2)$	$(1, -4)$	$(5, 4)$
b.	$(-1, 5)$	$(5, -3)$	$(2, 2)$	$(5, -2)$

Compute the slope m of the line that passes through the given points. Plot these points on the coordinate plane, and sketch the line that passes through them.

44. $(-4, -5)$ and $(2, -1)$, $m =$

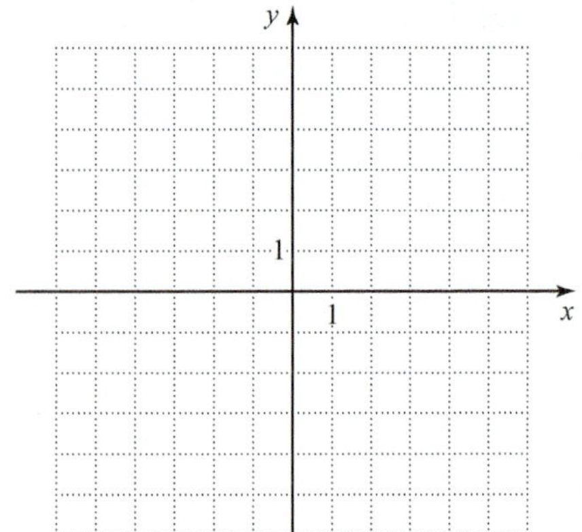

45. $(-3, -2)$ and $(-3, 4)$, $m =$

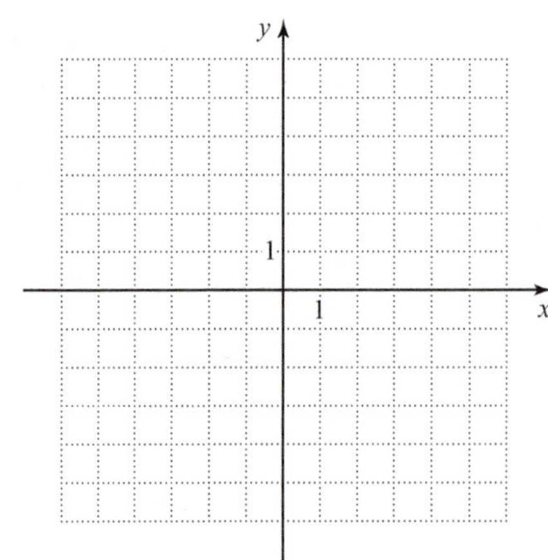

46. $(-4, 4.5)$ and $(-2, -1.5)$, $m =$

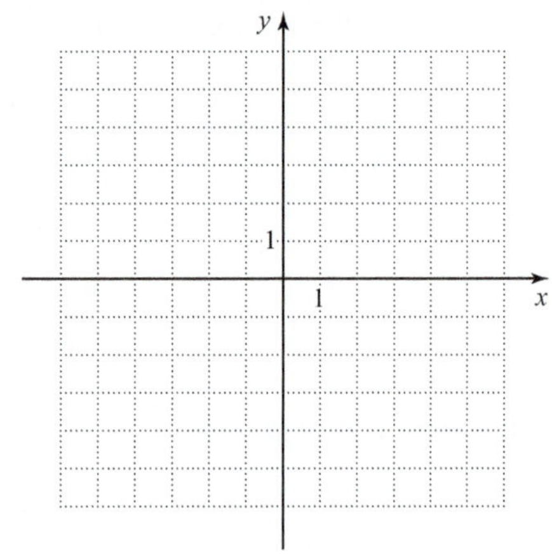

Applications

E *Solve.*

47. A chemist conducts an experiment on the gas contained in a sealed tube. The experiment is to heat the gas and then to measure the resulting pressure in the tube. In the lab manual, points are plotted and the line is sketched to show the gas pressure for various temperatures.

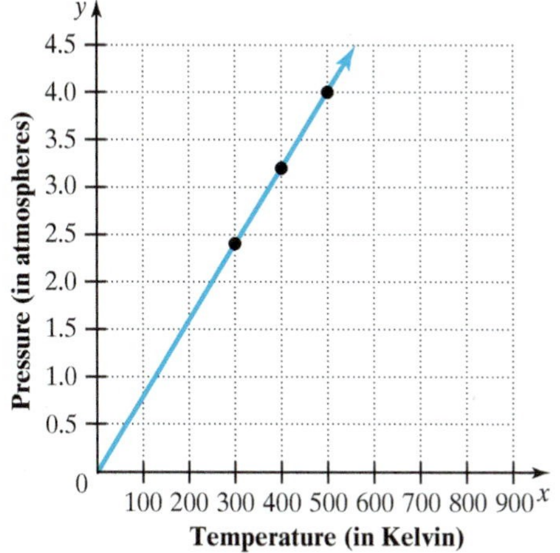

a. Is the slope of this line positive, negative, zero, or undefined?

b. In a sentence, explain the significance of the answer to part (a) in terms of temperature and pressure.

48. To reduce their taxes, many businesses use *the straight-line method of depreciation* to estimate the change in the value over time of equipment that they own. The graph shows the value of equipment owned from the time of purchase to 7 yr later.

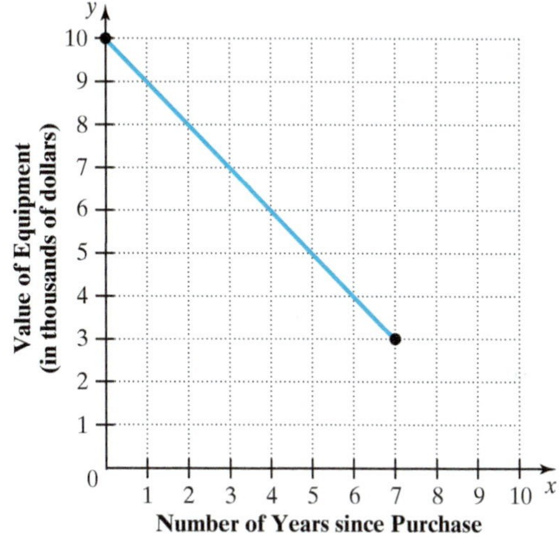

a. Is the slope of this line positive, negative, zero, or undefined?

b. In a sentence or two, explain the significance of the answer to part (a) in terms of the value of the equipment over time.

49. Two motorcyclists leave at the same time, racing down a road. Consider the lines in the graph that show the distance traveled by each motorcycle at various times.

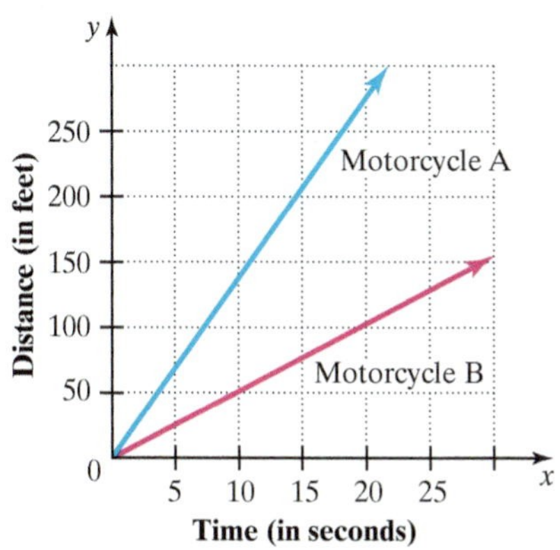

a. Which motorcycle first travels 100 ft?

b. Which motorcycle is traveling more slowly?

c. Explain what the slopes mean in this situation.

50. Many day-care centers charge parents additional fees for arriving late to pick up their children. The following graph shows the late fee for two day-care centers:

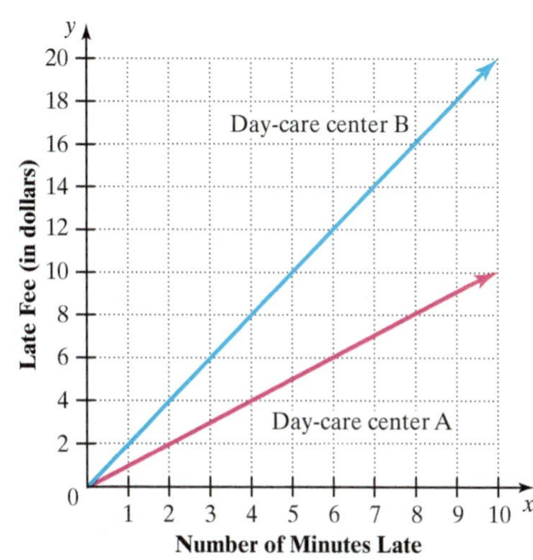

a. Which day-care center charges a higher late fee?

b. Explain what the slopes represent in this situation.

51. The following graph records the amount of garbage deposited in landfills A and B after they are opened.

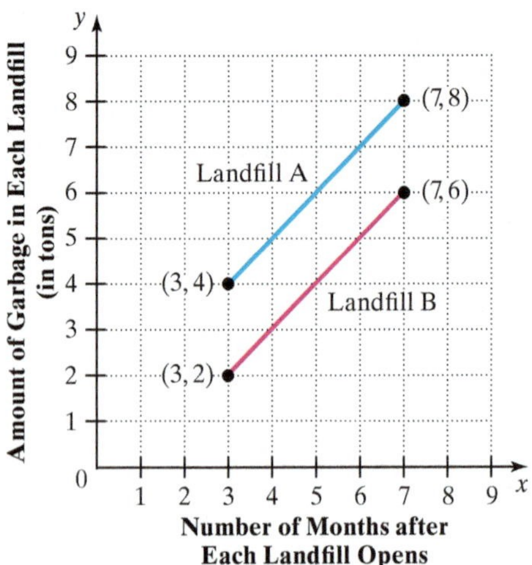

Are the garbage deposits at the two landfills growing at the same rate? Explain in a sentence or two how you know.

52. The weights of a brother and sister from age 3 yr to 7 yr are recorded in the graph. Did their weights increase at the same rate? Explain.

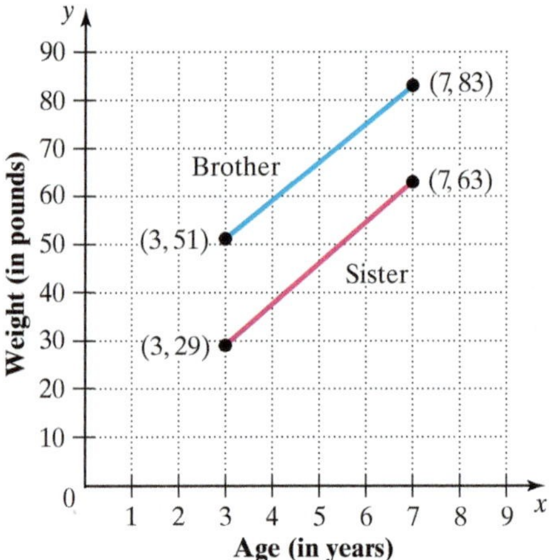

53. After a stock split, the per-share value of the stock increased, as shown in the following table:

Number of Days after the Split	Per-share Stock Value (in dollars)	Point
2	27	$P(2, 27)$
4	51	$Q(4, 51)$
8	79	$R(8, 79)$

a. Choose appropriate scales and label each axis. Plot the points, and then sketch $\overleftrightarrow{PQ}$ and $\overleftrightarrow{QR}$.

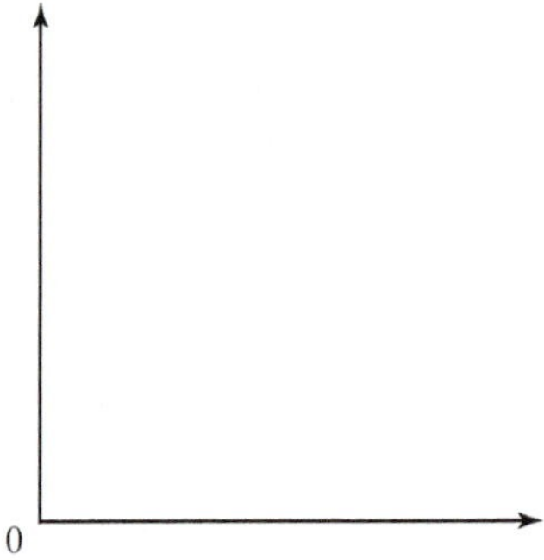

b. Determine whether the rate of increase changed over time. Explain.

54. The position of a dropped object for various times after the object is released is given in the table below:

Time After Release (in seconds)	Position (in feet)	Point
1	−16	$A\,(1, -16)$
2	−64	$B\,(2, -64)$
3	−144	$C\,(3, -144)$

a. Choose appropriate scales and label each axis. Plot the points, and then sketch $\overleftrightarrow{AB}$ and $\overleftrightarrow{BC}$.

b. Compute the slopes of $\overleftrightarrow{AB}$ and $\overleftrightarrow{BC}$.

c. Was the rate of fall for the dropped object constant throughout the experiment? Explain.

55. A hiker is at point A as shown in the graph and wants to take the shortest route through a field to reach a nearby road represented by $\overleftrightarrow{BC}$. The shortest route will be to walk perpendicular to the road. Is $\overleftrightarrow{AD}$ the shortest route? Explain.

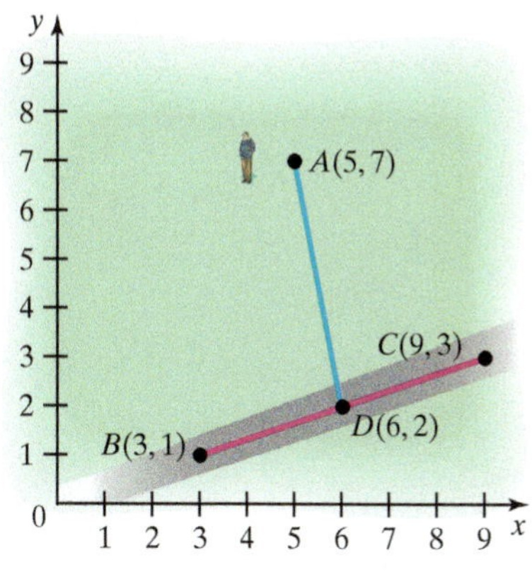

56. The coordinates of the vertices of triangle ABC are shown. Is triangle ABC a right triangle? Explain.

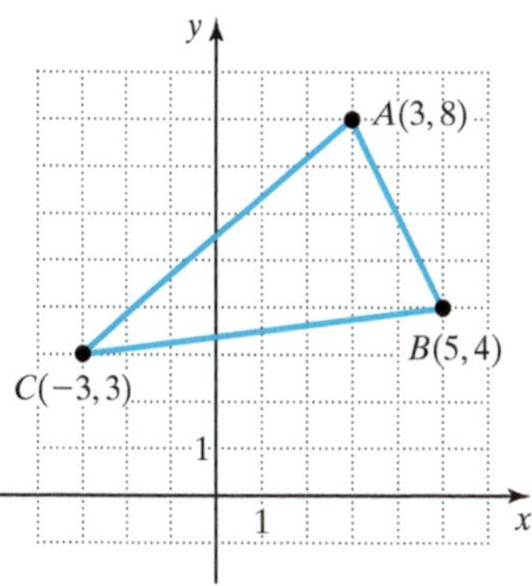

57. On a highway, a driver sets a car's cruise control for a constant speed of 55 mph.

a. Of the following, which graph shows the distance the car travels? Using the slope of the line, explain.

Graph I

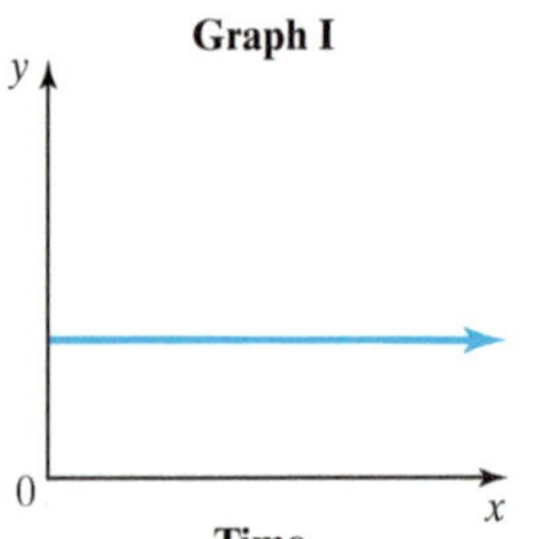

Graph II

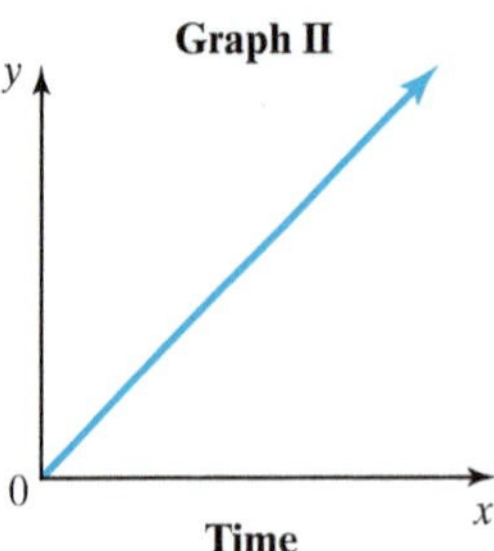

b. Which of the above graphs shows the speed of the car? Using the slope of the line, explain.

58. City leaders consider imposing an income tax on residents whose income is above a certain amount, as pictured below.

Plan A

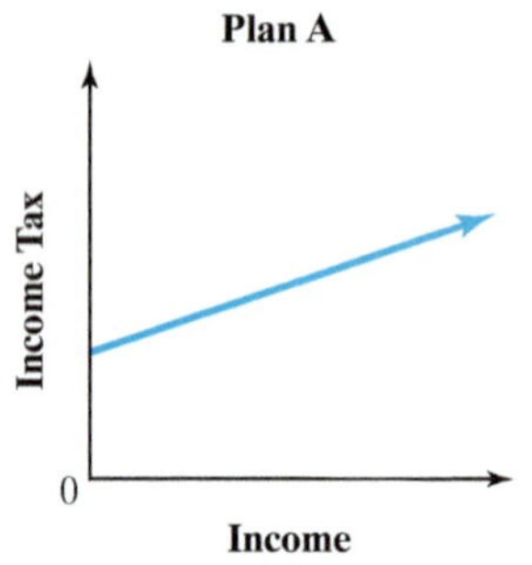

Plan B

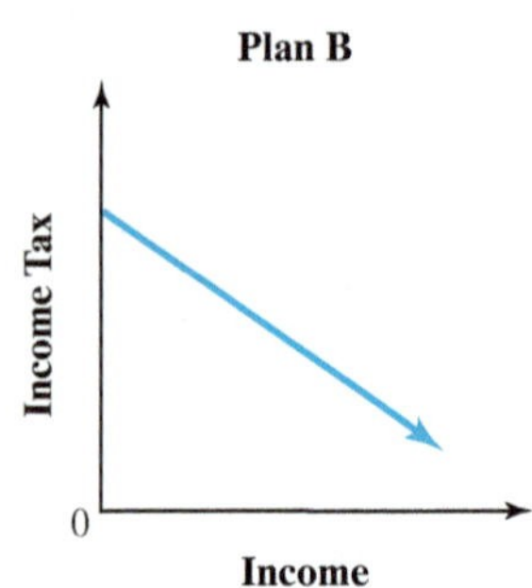

Plan C

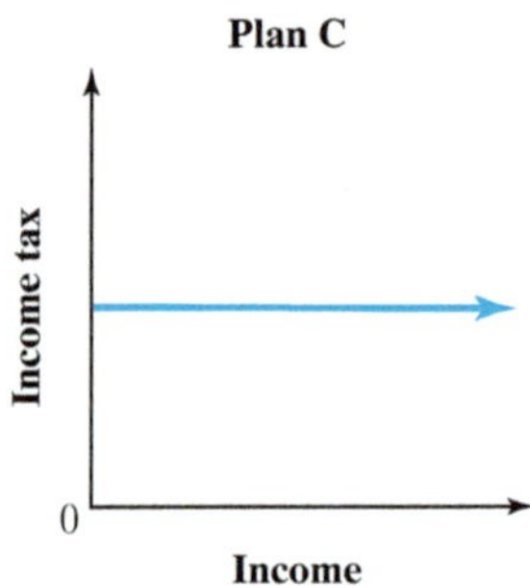

a. Using the slopes of the lines, describe each plan.

b. Which plan do you think is the fairest? Explain.

• Check your answers on page A-20.

MINDStretchers

Groupwork

1. A geoboard is a square flat surface with pegs forming a grid pattern. You can stretch rubber bands around the pegs to explore geometric questions such as, "What is the slope of $\overline{AB}$?" Pictured is a geoboard with rubber bands forming a series of steps.

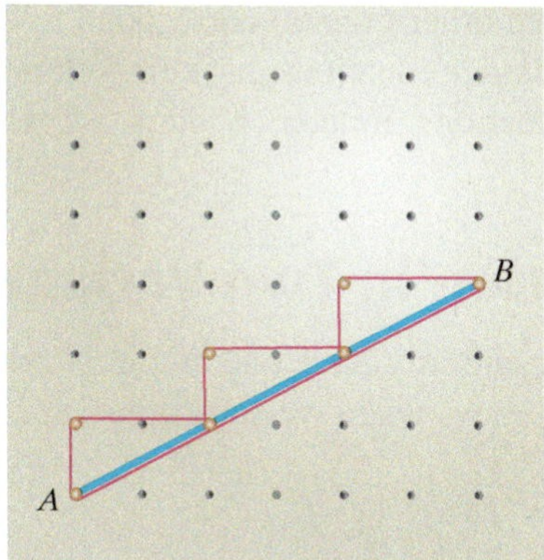

Using a geoboard or graph paper, determine whether the slope of $\overleftrightarrow{AB}$ increases or decreases if:

a. the rise of each step increases by one peg.

b. the run of each step increases by one peg.

Writing

2. Describe a real-world situation that each of the following graphs might illustrate. Explain the significance of slope in the situation that you have described.

a. 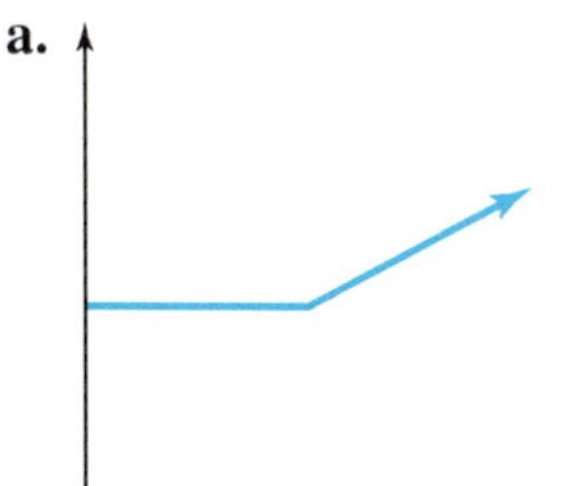

b. 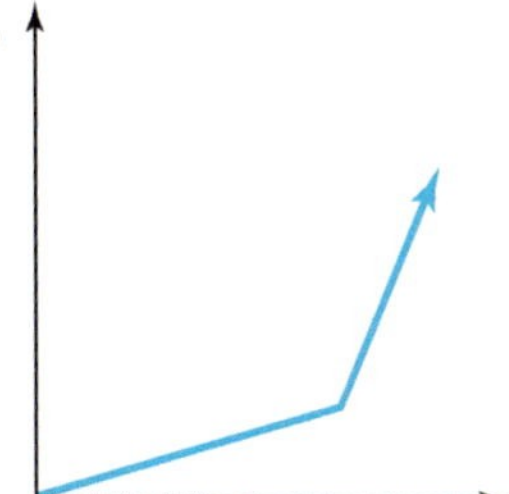

c. 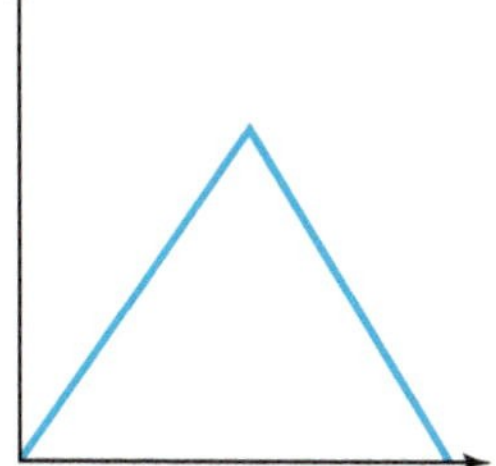

Mathematical Reasoning

3. If the slope of a line is positive, explain why any line perpendicular to it must have a negative slope.

13.3 Linear Equations and Their Graphs

OBJECTIVES

- **A** To identify the coordinates of points that satisfy a given linear equation in two variables
- **B** To graph given linear equations in two variables
- **C** To solve applied problems involving graphs of linear equations

In the first section of this chapter, we developed the idea of coordinates—an ordered pair of numbers associated with a point on the plane. In the second section, we shifted our attention from points to lines and their slopes. In this section, we focus on Descartes' most startling idea—on a coordinate plane, lines (and indeed other graphs) correspond to equations. Line graphs and their associated equations, called *linear* equations, have important applications in the real world, allowing us to model many situations. In later chapters, we will discuss other types of graphs.

Solutions of a Linear Equation in Two Variables

Recall that linear equations in one variable generally have one and only one solution. For instance, the solution of $2x + 5 = 11$ is 3 because when we substitute 3 for x, the equation is true.

$$2x + 5 = 11$$
$$2 \cdot 3 + 5 \stackrel{?}{=} 11$$
$$11 = 11 \quad \text{True}$$

We now consider linear equations in *two* variables, for instance $y = 2x + 5$. We can also express $y = 2x + 5$ as $-2x + y = 5$.

$$y = 2x + 5$$
$$-2x + y = 5 \quad \text{Subtract } 2x \text{ from each side of the equation.}$$

DEFINITION

A **linear equation in two variables,** x and y, is an equation that can be written in the *general form* $Ax + By = C$, where A, B, and C are real numbers and A and B are not both 0.

Note that in the general form of a linear equation, the two variable terms are on one side of the equation and the constant term is on the other. The equation $-2x + y = 5$, for instance, is in general form, with $A = -2$, $B = 1$, and $C = 5$.

Here are some other examples of linear equations in general form:

Equation	A	B	C
$5x + 3y = 10$	5	3	10
$x - 2y = 6 \rightarrow 1x + (-2)y = 6$	1	−2	6
$x = -4 \rightarrow 1x + 0y = -4$	1	0	−4
$y = -4 \rightarrow 0x + 1y = -4$	0	1	−4

Now, let's look at what we mean by a solution of a linear equation in two variables.

DEFINITION

A **solution** of an equation in two variables is an ordered pair of numbers that when substituted for the variables makes the equation true.

Applying this definition to the equation $-2x + y = 5$, we observe that $x = 3$ and $y = 11$ is a solution of the equation because substituting 3 for x and 11 for y makes the equation true.

$$\begin{aligned} -2x + y &= 5 \\ -2 \cdot 3 + 11 &\stackrel{?}{=} 5 \\ 5 &= 5 \quad \text{True} \end{aligned}$$

There are many other solutions of this equation as well. For instance, $x = 0$ and $y = 5$ is another solution.

$$\begin{aligned} -2x + y &= 5 \\ -2 \cdot 0 + 5 &\stackrel{?}{=} 5 \\ 5 &= 5 \quad \text{True} \end{aligned}$$

Unlike a linear equation in one variable that has at most one solution, linear equations in two variables have an infinite number of solutions. Can you explain why this is the case?

Now, how do we *find* solutions of a linear equation in two variables? Typically, we start with the value of one of the variables, and then compute the corresponding value of the other, as the following example illustrates.

EXAMPLE 1

For the equation $4x + y = -5$, find five solutions by completing the following table:

x	-2	3	0		
y				0	1

Solution In the first column of this table, we substitute -2 for x in the given equation, and then solve for y.

$$\begin{aligned} 4x + y &= -5 \\ 4(-2) + y &= -5 \\ -8 + y &= -5 \\ y &= 3 \end{aligned}$$

To find the corresponding y in the second column, we substitute 3 for x:

$$\begin{aligned} 4x + y &= -5 \\ 4(3) + y &= -5 \\ 12 + y &= -5 \\ y &= -17 \end{aligned}$$

And in the third column, we substitute 0 for x:

$$\begin{aligned} 4x + y &= -5 \\ 4(0) + y &= -5 \\ 0 + y &= -5 \\ y &= -5 \end{aligned}$$

PRACTICE 1

For the equation $-2x + y = 1$, find the missing values in the following table:

x	0	5	-3		
y				0	-3

EXAMPLE 1 (continued)

The next two columns are different from the earlier ones: they give us y-values and require us to solve for x. First, we see that in the fourth column y is 0.

$$\begin{aligned} 4x + y &= -5 \\ 4x + 0 &= -5 \quad \text{Substitute 0 for } y. \\ 4x &= -5 \\ x &= -\frac{5}{4} \end{aligned}$$

In the final column of the table, $y = 1$.

$$\begin{aligned} 4x + y &= -5 \\ 4x + 1 &= -5 \quad \text{Substitute 1 for } y. \\ 4x &= -6 \\ x &= -\frac{6}{4} = -\frac{3}{2} \end{aligned}$$

We have found the missing values, so the table reads:

x	-2	3	0	$-\frac{5}{4}$	$-\frac{3}{2}$
y	3	-17	-5	0	1

The Graph of a Linear Equation in Two Variables

The graph of an equation, more precisely the graph of the *solutions* of that equation, is a kind of picture of the equation.

DEFINITION

The **graph** of a linear equation in two variables consists of all points whose coordinates make the equation true.

Given a linear equation, how do we find its graph? A general strategy is to first isolate one of the variables, unless it is already done. Then, we identify several solutions of the equation, keeping track of the x- and y-values in a table. We plot the points and then sketch the line passing through them. That line is the graph of the given equation, as the following example illustrates.

Let's graph the equation $y = 3x + 1$. The variable y is already isolated, so we find y-values by substituting arbitrary values of x. For instance, let x equal 0. To find y, we substitute 0 for x.

$$y = 3x + 1 = 3 \cdot 0 + 1 = 0 + 1 = 1$$

So $x = 0$ and $y = 1$ is a solution of this equation. We say that the ordered pair $(0, 1)$ is a solution of $y = 3x + 1$.

Let's choose three other values of x, say -1, 1, and 2. Substituting -1 for x in the equation, we get: $y = 3x + 1 = 3(-1) + 1 = -2$

Substituting 1 for x, we get: $y = 3x + 1 = 3 \cdot 1 + 1 = 4$

Substituting 2 for x gives us: $y = 3x + 1 = 3 \cdot 2 + 1 = 7$

Next, we enter these results in a table.

x	-1	0	1	2
y	-2	1	4	7

Then, we plot on a coordinate plane the four points $A(-1, -2)$, $B(0, 1)$, $C(1, 4)$, and $D(2, 7)$. If we have not made a mistake, the points will all lie on the same line. We know that the line segments $\overline{AB}$, $\overline{BC}$, and $\overline{CD}$ are part of the same line because they all have equal slopes. The graph of the equation $y = 3x + 1$ is the line passing through these points. So any point on this line satisfies the equation $y = 3x + 1$.

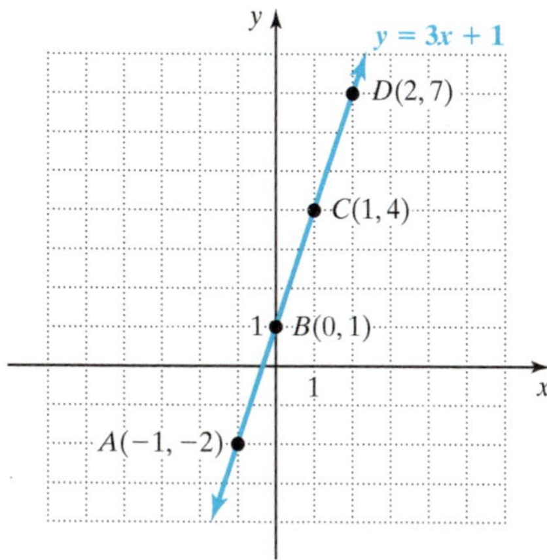

Do you think that if we had chosen three other x-values we would have gotten the same graph? Check to see that this is the case.

This example suggests the following procedure for graphing a linear equation:

To Graph a Linear Equation in Two Variables

- Isolate one of the variables—usually y—if it is not already done.
- Choose three x-values, entering them in a table.
- Complete the table by calculating the corresponding y-values.
- Plot the three points—two to draw the line and the third to serve as a *checkpoint*.
- Check that the points seem to lie on the same line.
- Draw the line passing through the points.

A couple of observations about the preceding example are worth making:

- The slope of the line graphed is 3. We can see this by taking any pair of points on the line, for example $(0, 1)$ and $(1, 4)$, and computing the slope of the line between them.

$$m = \frac{y_2 - y_1}{x_2 - x_1} = \frac{1 - 4}{0 - 1} = \frac{-3}{-1} = 3$$

- This slope is identical to the coefficient of x in the equation $y = 3x + 1$.

We also see that the point where the graph crosses the y-axis is $(0, 1)$. This point is called the *y-intercept*. Note that the constant term in the equation $y = 3x + 1$ is also 1.

These relationships are more than a coincidence, and we will say more about them in Section 13.4.

EXAMPLE 2

Graph the equation $y = -\frac{3}{2}x$ by choosing three points whose coordinates satisfy the equation.

Solution We begin by choosing x-values. In this case, we choose multiples of 2 for the x-values. Then, we find the corresponding y-values.

x	$y = -\frac{3}{2}x$	(x, y)
0	$y = -\frac{3}{2}(0) = 0$	$(0, 0)$
2	$y = -\frac{3}{2}(2) = -3$	$(2, -3)$
4	$y = -\frac{3}{2}(4) = -6$	$(4, -6)$

We plot the points, and then draw a line passing through them to get the desired line.

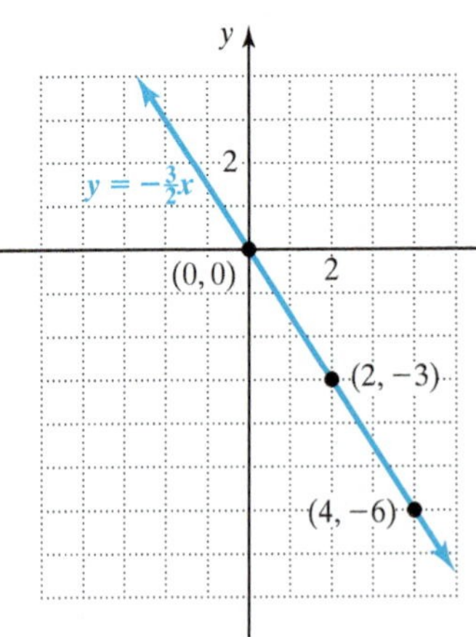

PRACTICE 2

Graph the equation $y = -\frac{3}{5}x$.

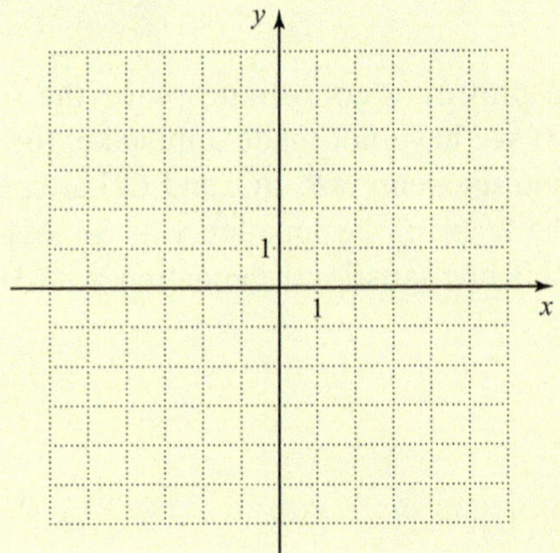

Note that in Example 2, we chose multiples of 2 for the x-values. Can you explain why?

EXAMPLE 3

Consider the equation $2x + y = 3$.

a. Graph the equation.

b. Find the slope of the line.

Solution

a. We begin by solving the equation for y.

$$2x + y = 3$$
$$y = -2x + 3 \quad \text{Subtract } 2x \text{ from each side.}$$

Next, we choose three values for x, for instance, -1, 0, and 2. Then, we enter them into a table and find their corresponding y-values as follows:

x	$y = -2x + 3$	(x, y)
-1	$y = -2(-1) + 3 = 2 + 3 = 5$	$(-1, 5)$
0	$y = -2(0) + 3 = 0 + 3 = 3$	$(0, 3)$
2	$y = -2(2) + 3 = -4 + 3 = -1$	$(2, -1)$

PRACTICE 3

Consider the equation $-2x + y = -5$.

a. Graph the equation.

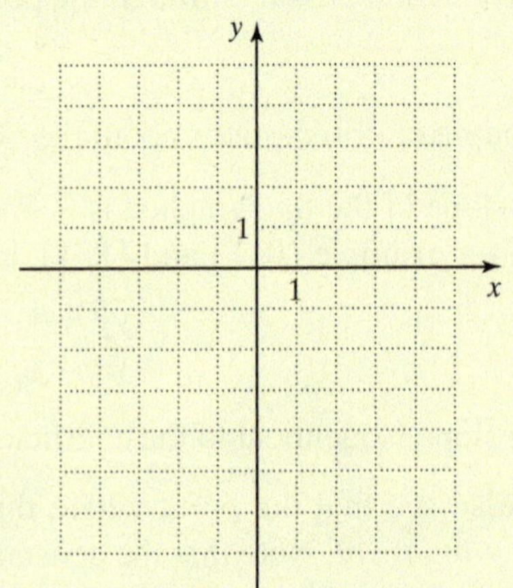

b. Find the slope of the line.

Now, we plot the points on the coordinate plane. Since the points seem to lie on the same line, we draw a line passing through the points.

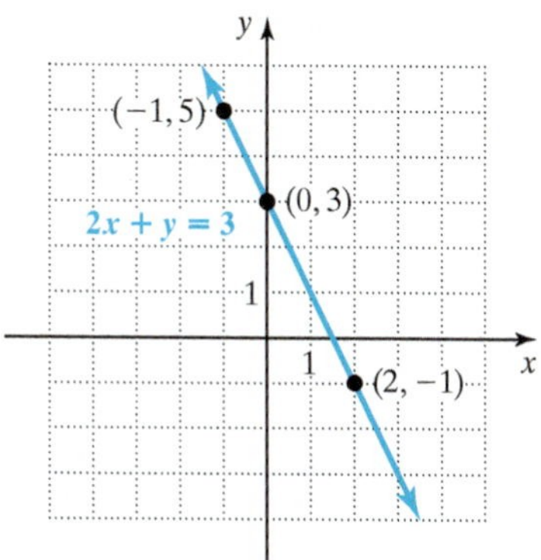

b. To find the slope of the line, we can consider the points $(0, 3)$ and $(2, -1)$.

$$m = \frac{3 - (-1)}{0 - 2} = \frac{4}{-2} = -2$$

Note that the slope is the coefficient of x in the equation $y = -2x + 3$.

We have graphed equations in general form by first isolating y and then computing y-values for arbitrary x-values. Now, we graph equations in general form with a different approach using x- and y-intercepts. Note that intercepts stand out on a graph and are easy to plot. Since an x-intercept lies on the x-axis, its y-value must be 0. Similarly, since a y-intercept lies on the y-axis, the x-value of a y-intercept must be 0.

DEFINITION

The ***x*-intercept** of a line is the point where the graph crosses the *x*-axis. The ***y*-intercept** is the point where the graph crosses the *y*-axis.

The following graph shows a line passing through two points, $(0, 4)$ and $(3, 0)$, which are both intercepts.

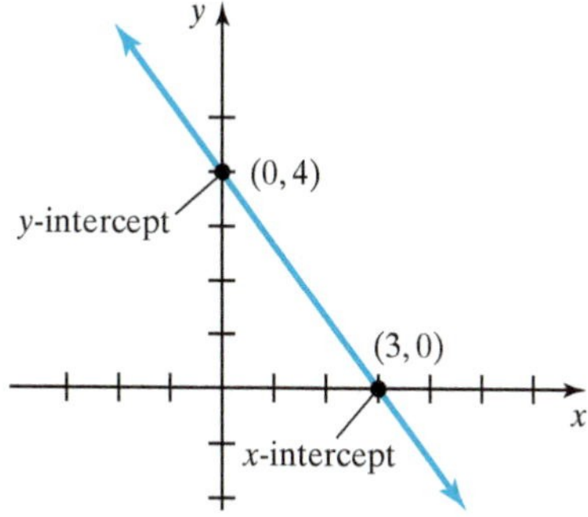

Using the x- and y-intercepts to graph equations can save work, especially when the coefficients of the two variables are factors of the constant term, as shown in the next example.

EXAMPLE 4

Consider the equation $2x + 3y = 6$. Find the x- and y-intercepts. Then, graph.

Solution Since the y-intercept has x-value 0, we let $x = 0$ and then solve for y.

For $x = 0$:

$$2x + 3y = 6$$
$$2 \cdot 0 + 3y = 6 \quad \text{Substitute 0 for } x.$$
$$3y = 6$$
$$\frac{3y}{3} = \frac{6}{3}$$
$$y = 2$$

So the y-intercept is $(0, 2)$.

Similarly, the x-intercept has y-value 0. So we let $y = 0$ and then solve for x.

For $y = 0$:

$$2x + 3y = 6$$
$$2x + 3 \cdot 0 = 6 \quad \text{Substitute 0 for } y.$$
$$2x = 6$$
$$\frac{2x}{2} = \frac{6}{2}$$
$$x = 3$$

So the x-intercept is $(3, 0)$.

Before graphing, we choose a third point to be used as a checkpoint.

For $2x + 3y = 6$, let $x = 6$:

$$2x + 3y = 6$$
$$2 \cdot 6 + 3y = 6 \quad \text{Substitute 6 for } x.$$
$$12 + 3y = 6$$
$$3y = -6$$
$$y = -2$$

So the checkpoint is $(6, -2)$.

Plotting the points $(0, 2)$, $(3, 0)$, and $(6, -2)$ on a coordinate plane, we confirm that they seem to lie on the same line. Finally, we draw a line through the points, getting the desired graph.

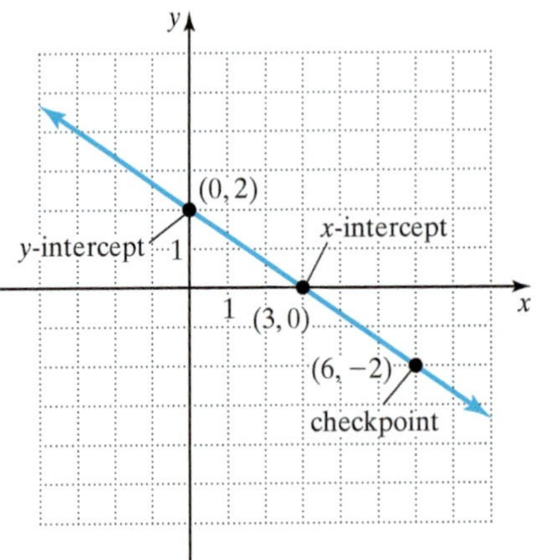

PRACTICE 4

Consider the equation $x - 2y = 4$. Find the x- and y-intercepts. Then, graph.

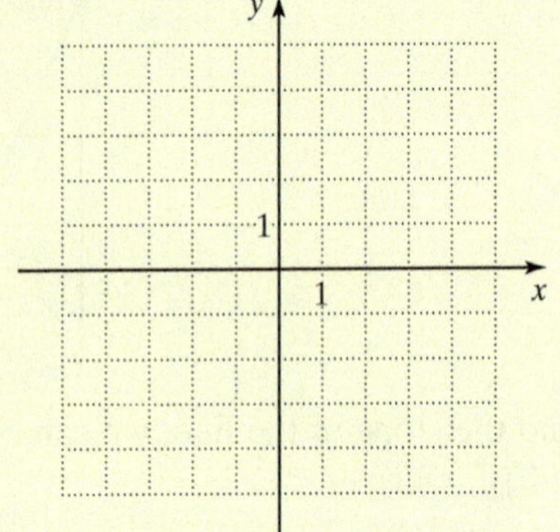

Example 4 suggests the following procedure:

To Graph a Linear Equation in Two Variables Using the *x*- and *y*-intercepts

- Let $x = 0$, and find the y-intercept.
- Let $y = 0$, and find the x-intercept.
- Find a checkpoint.
- Plot the three points.
- Check that the points seem to lie on the same line.
- Draw the line passing through the points.

We know that the general form of a linear equation in two variables is $Ax + By = C$. Sometimes in a linear equation one of the two variables is missing, that is, the coefficient of one of the two variables is zero. Consider the following equations:

$$y = 9 \rightarrow 0x + y = 9$$
$$x = -5.8 \rightarrow x + 0y = -5.8$$

Let's look at the graphs of these equations.

EXAMPLE 5

Graph:

a. $y = 9$ **b.** $x = -5.8$

Solution

a. For the line $y = 9$, the coefficient of the x-term is 0. The x-value can be any real number and the y-value is always 9. So the graph is as follows:

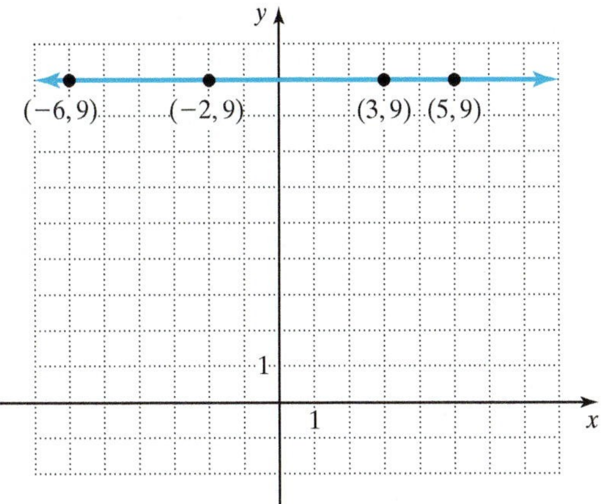

The graph of this equation is a horizontal line. Recall that the slope of a horizontal line is 0.

PRACTICE 5

On the given coordinate plane, graph:

a. $y = -1$ **b.** $x = 2.5$

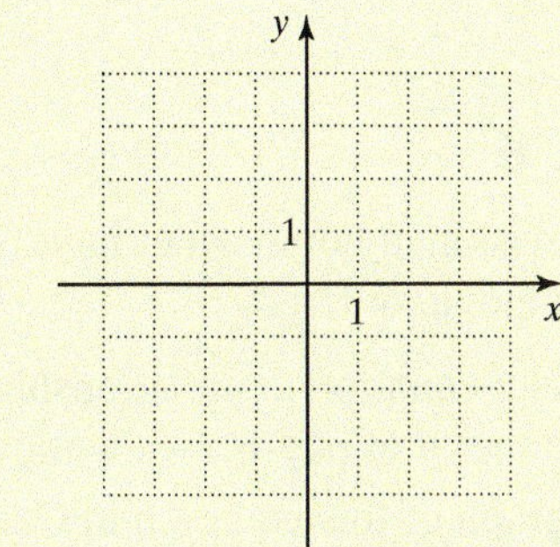

EXAMPLE 5 (continued)

b. For the line $x = -5.8$, the coefficient of the y-term is 0. The y-value can be any real number and the x-value is always -5.8. So the graph is as follows:

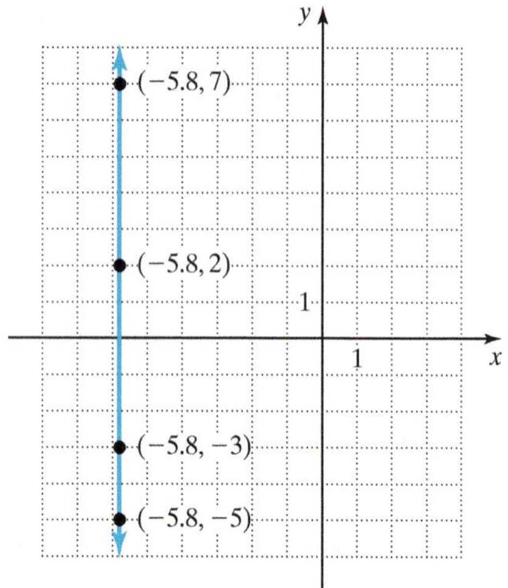

The graph of this equation is a vertical line. The slope of this line is undefined.

In general, the graph of the equation $x = a$ is a vertical line passing through the point $(a, 0)$, and the graph of the equation $y = b$ is a horizontal line passing through the point $(0, b)$. What are the equations of the x- and y-axes?

Now, let's use our knowledge of graphing linear equations to solve applied problems.

EXAMPLE 6

For cable television, a homeowner pays \$30 per month plus \$5 for each pay-per-view movie ordered.

a. Express as an equation the relationship between the monthly bill B and the number n of pay-per-view movies.

b. Draw the graph of this equation in Quadrant I of a coordinate plane.

c. Compute the slope of this graph. In terms of the cable TV bill, explain the significance of the slope.

d. In terms of the cable TV bill, explain the significance of the B-intercept of the graph.

e. From the graph in part (b), estimate what the cable bill would be if the homeowner had ordered 15 pay-per-view movies that month.

Solution

a. The monthly bill (in dollars) amounts to the sum of 30 and 5 times the number of pay-per-view movies which the homeowner ordered. So

$$B = 5n + 30.$$

PRACTICE 6

A stockbroker charges as her commission on stock transactions \$40 plus 3% of the value of the sale.

a. Write as an equation the commission C in terms of the sales s.

b. To draw the graph of this equation in Quadrant I, we enter several nonnegative n-values, say 0, 10, and 20, in a table and then compute the corresponding B-values.

n	$B = 5n + 30$	(n, B)
0	$B = 5(0) + 30 = 30$	$(0, 30)$
10	$B = 5(10) + 30 = 80$	$(10, 80)$
20	$B = 5(20) + 30 = 130$	$(20, 130)$

Since the monthly bill B depends on the number n of pay-per-view movies ordered each month, we label the horizontal axis with the independent variable n and the vertical axis with the dependent variable B. Now, we choose an appropriate scale for each axis and plot $(0, 30)$, $(10, 80)$, and $(20, 130)$. Then, we draw the line passing through these points.

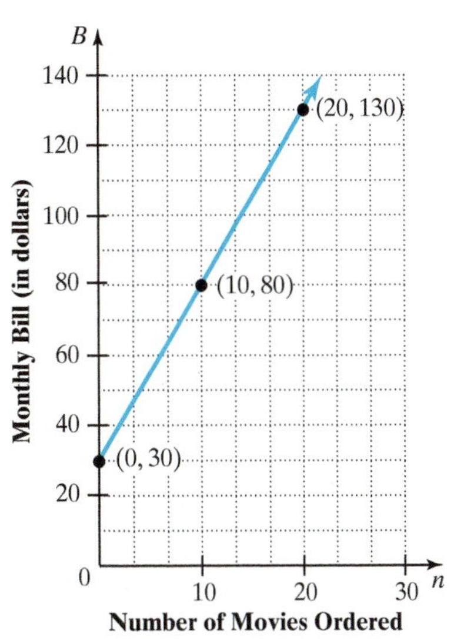

We restricted the graph to Quadrant I because the number of movies ordered n and the corresponding bill B are always nonnegative.

c. Substituting n and B for x and y, respectively, in the slope formula, we get the following slope:

$$m = \frac{B_2 - B_1}{n_2 - n_1} = \frac{80 - (30)}{10 - (0)} = \frac{50}{10} = 5.$$

Note that we could have predicted this answer since we know that the bill increases by \$5 for every additional pay-per-view movie ordered.

d. Since the B-intercept is the point $(0, 30)$, \$30 would be the amount of the bill if the homeowner had watched no pay-per-view movies at all during the month.

e. From the graph, it appears that if $n = 15$, then $B = 105$, that is, the cable bill would be \$105.

b. Draw a graph showing this relationship on sales up to \$1000. Be sure to choose an appropriate scale for each axis.

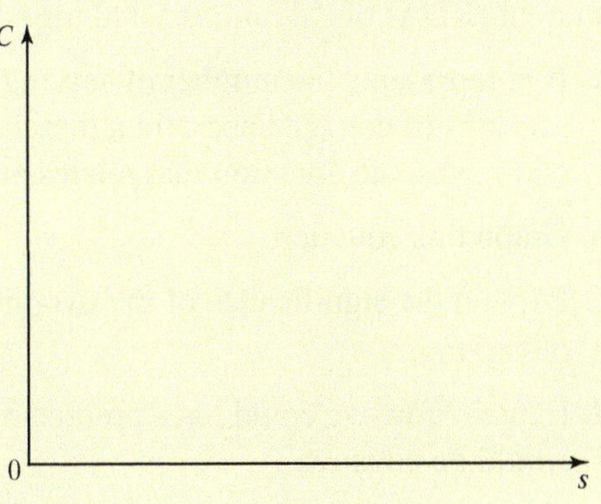

c. Compute the slope of this graph. In terms of the stock broker's commission, explain the significance of the slope.

d. If the value of a sale is \$500, estimate from the graph in part (b) the broker's commission.

EXAMPLE 7

A dietician uses milk and cottage cheese as sources of calcium in her diet. One serving of milk contains 300 mg of calcium, and one serving of cottage cheese contains 100 mg of calcium. The recommended daily amount (RDA) of calcium is 1000 mg.

a. If m represents the number of servings of milk and c the number of servings of cottage cheese in a diet that contains the RDA of calcium, write an equation that relates m and c.

b. Graph this equation.

c. Explain the significance of the two intercepts in terms of the number of servings.

d. Explain how we could have predicted that the slope of this graph would be negative.

Solution

a. The amount of calcium in the milk is $300m$, and the amount of calcium in the cottage cheese is $100c$. Since the RDA of calcium is 1000 mg, the following equation holds:

$$300m + 100c = 1000, \text{ or } 3m + c = 10$$

b. To graph, we identify the m- and c-intercepts, as well as a third point, say with $m = 1$.

m	0	$3\frac{1}{3}$	1
c	10	0	7

Next, we choose an appropriate scale for each axis and plot the three points, checking that the points all lie on the same line. Then, we draw the line passing through them.

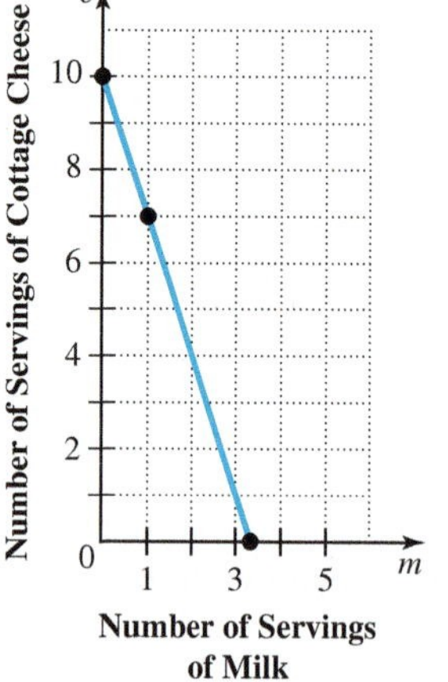

c. The m-intercept represents the number of servings she would need to meet the daily minimum requirement if she uses only milk as her source of calcium. The c-intercept represents the number of servings she would need to meet the RDA if she uses only cottage cheese as her source of calcium.

d. Even without drawing this line, we know that its slope has to be negative for the following reason: The RDA of calcium is a fixed amount (1000 mg). So larger values of m must correspond to smaller values of c. The line will therefore have to be decreasing, falling to the right and with a negative slope.

PRACTICE 7

An athlete has just signed a contract with a total value of $10 million. According to the terms of the contract, she earns $2 million in some years and $1 million in other years.

a. Let x stand for the number of years in which she earns $2 million and y the number of $1 million years. Write an equation that relates x and y.

b. Choose an appropriate scale for each axis on the coordinate plane. Graph the equation found in part (a).

c. Describe in terms of the contract the significance of the slope of the graph.

d. Describe in terms of the contract the significance of the x- and y-intercepts of the graph.

13.3 Exercises

FOR EXTRA HELP MyMathLab MathXL PRACTICE WATCH READ REVIEW

Mathematically Speaking

Fill in each blank with the most appropriate term or phrase from the given list.

three x-values	vertical	x-intercept
graph	solution	horizontal
y-intercept	three points	

1. A(n) ____________ of an equation in two variables is an ordered pair of numbers that when substituted for the variables makes the equation true.

2. The ____________ of a linear equation in two variables consists of all points whose coordinates satisfy the equation.

3. One way of graphing a linear equation in two variables is to plot ____________.

4. The ____________ of a line is the point where the graph crosses the x-axis.

5. One way to graph a linear equation using intercepts is to first let $x = 0$ and then find the ____________.

6. The graph of the equation $y = c$ is a(n) ____________ line passing through the point $(0, c)$.

A *Complete each table so that the ordered pairs are solutions of the given equation.*

7. $y = 3x - 8$

x	4	7	
x			0

8. $y = 2x - 5$

x	0		
y		15	17

9. $y = -5x$

x	3.5	6		
y			$\frac{1}{2}$	-8

10. $y = -10x$

x	$\frac{1}{5}$	2.9		
y			-6	-1

11. $3x + 4y = 12$

x	0	-4		
y			-3	0

12. $4x + y = 8$

x	5	0		
y			0	16

13. $y = \frac{1}{3}x - 1$

x	3	6	-3	
y				-1

14. $y = -\frac{3}{2}x + 2$

x	$\frac{4}{3}$	6	-2	
y				2

B *Graph each equation by finding three points whose coordinates satisfy the equation.*

15. $y = x$

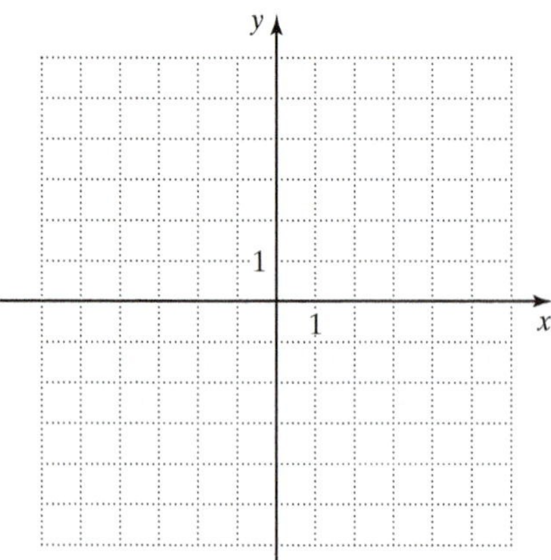

16. $y = 3x$

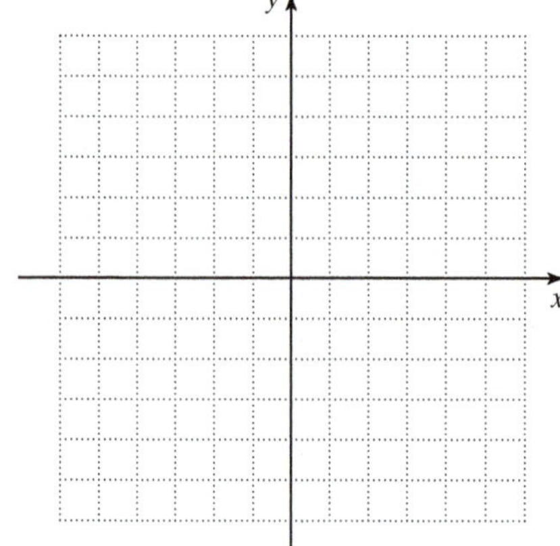

17. $y = \frac{1}{2}x$

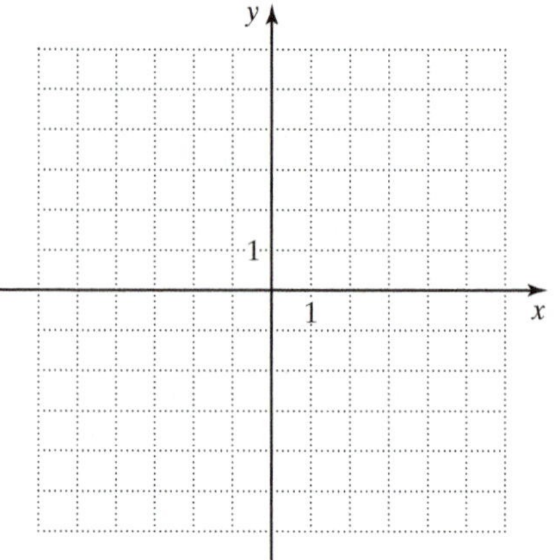

18. $y = \frac{1}{4}x$

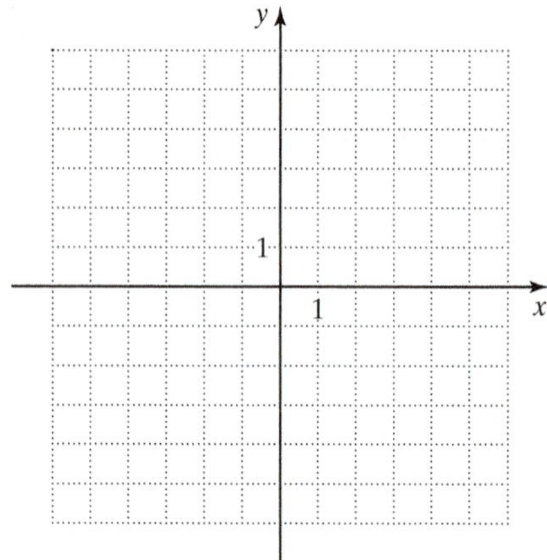

19. $y = -\frac{5}{4}x$

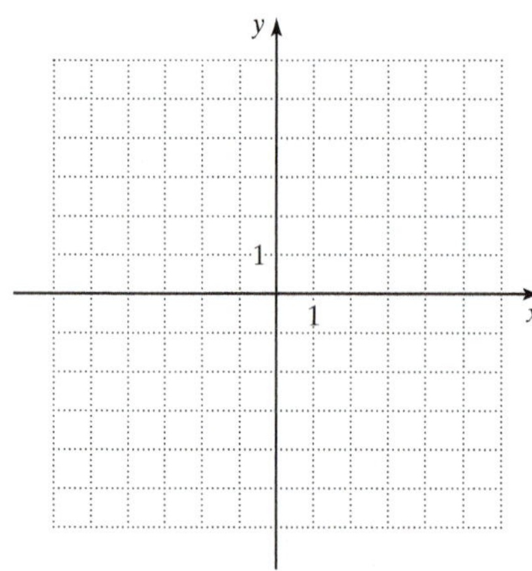

20. $y = -\frac{3}{2}x$

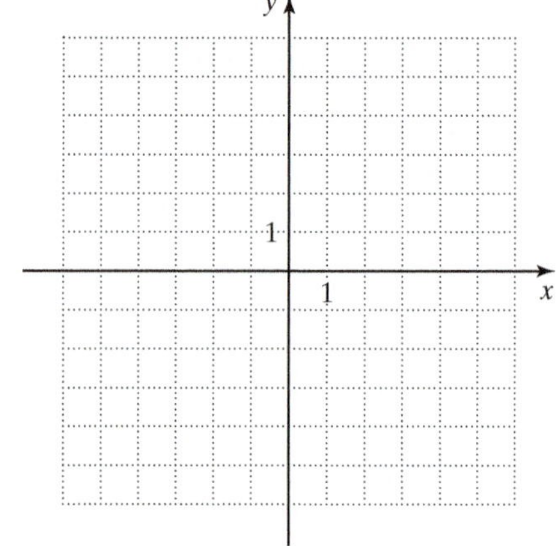

21. $y = 2x + 1$

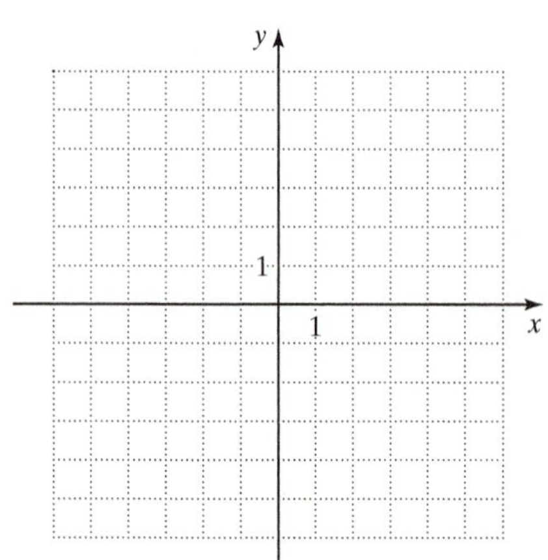

22. $y = 3x + 1$

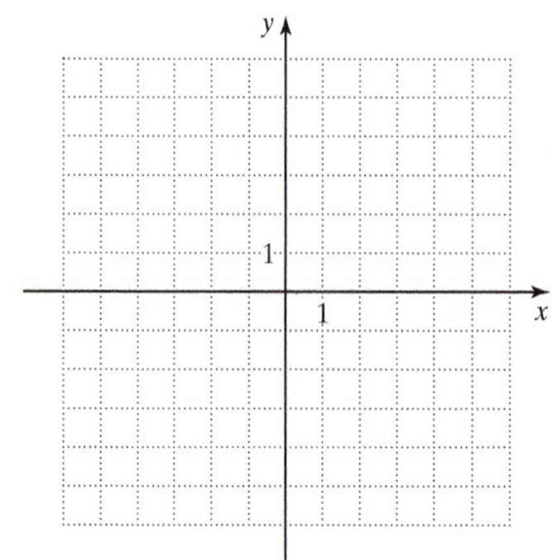

23. $y = -\frac{1}{3}x + 1$

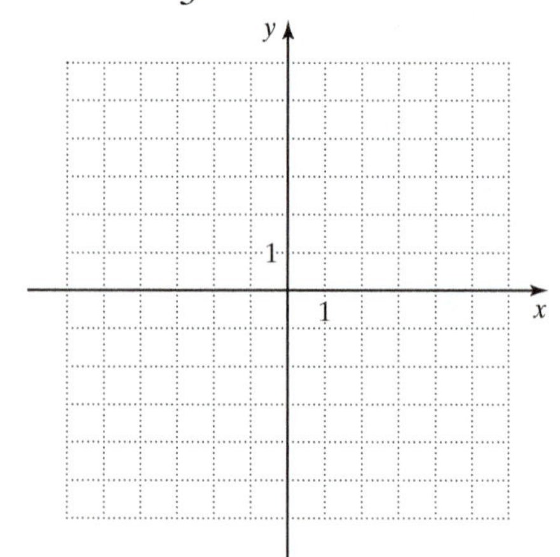

24. $y = -\frac{3}{4}x + 2$

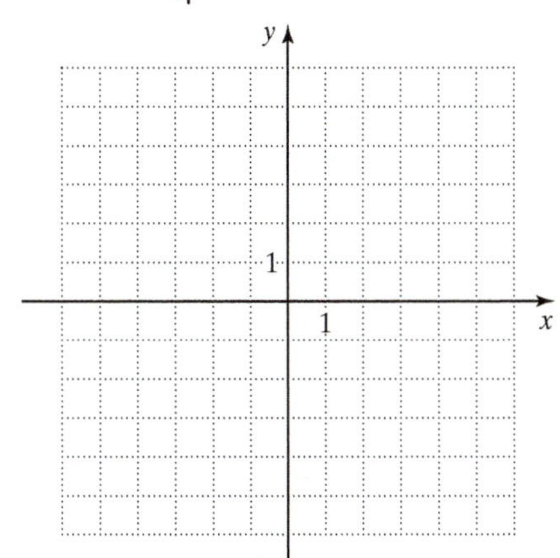

25. $y - 2x = -3$

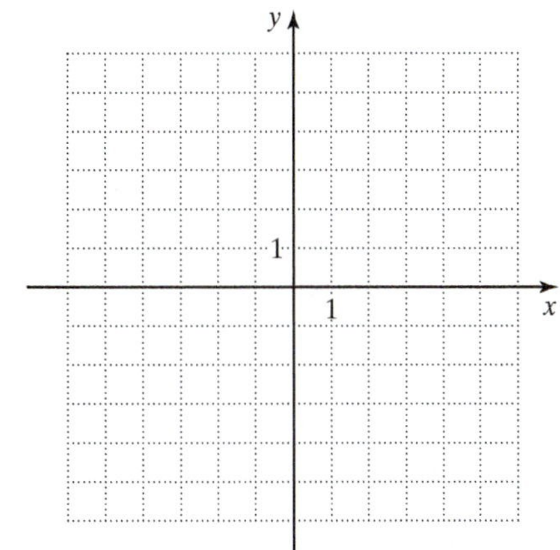

26. $y - 3x = 2$

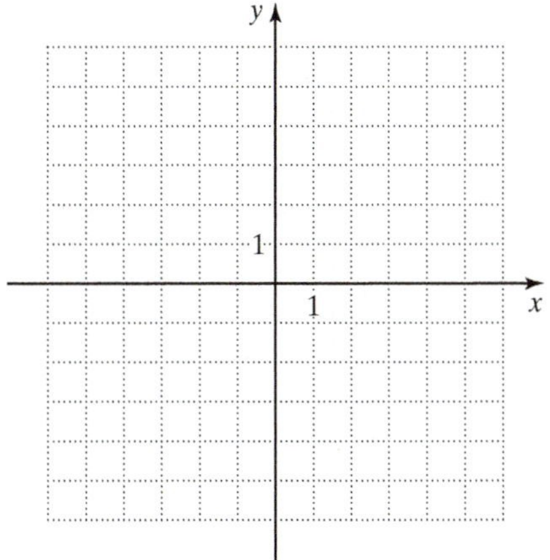

27. $x + y = 6$

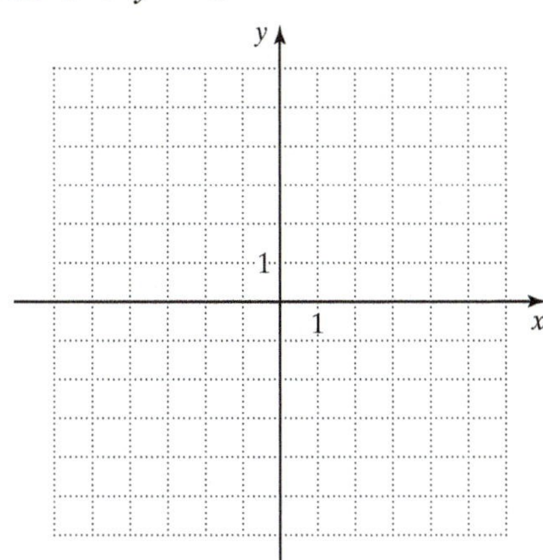

28. $3x + y = 4$

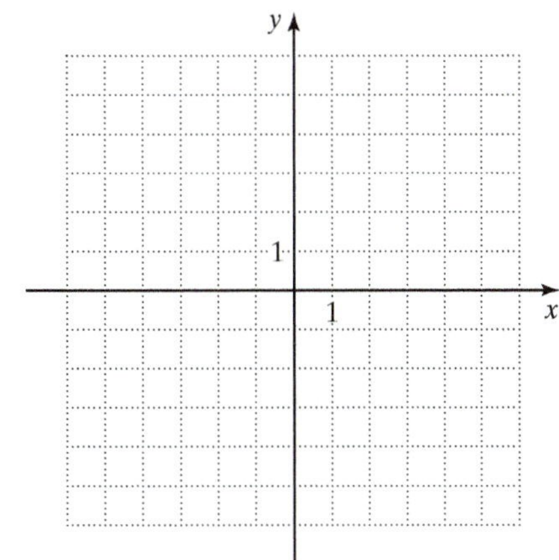

29. $x - 2y = 4$

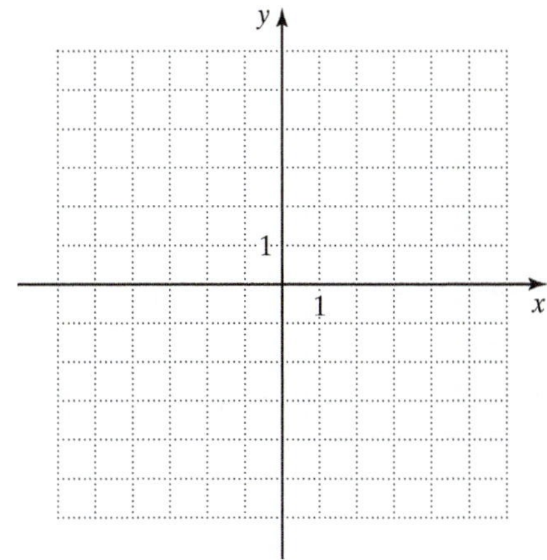

30. $x - 3y = 15$

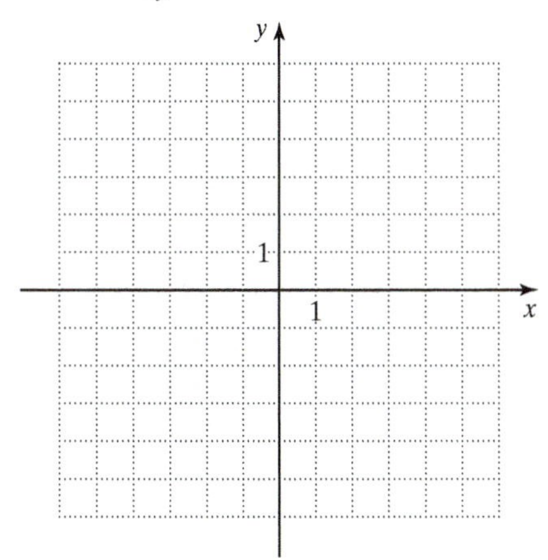

For each equation, find the x- and y-intercepts. Then, use the intercepts to graph the equation.

31. $5x + 3y = 15$

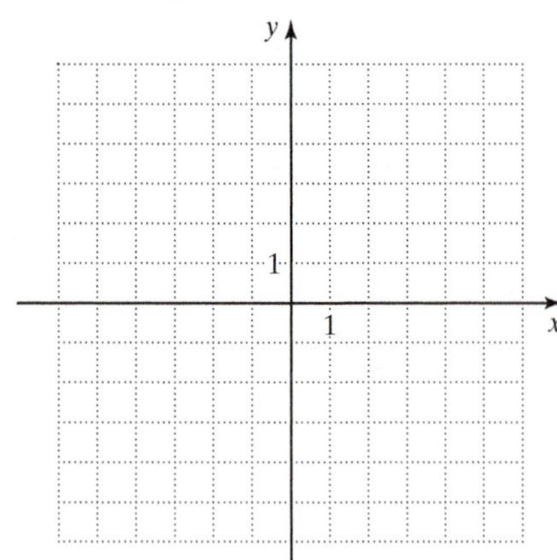

32. $4x + 5y = 20$

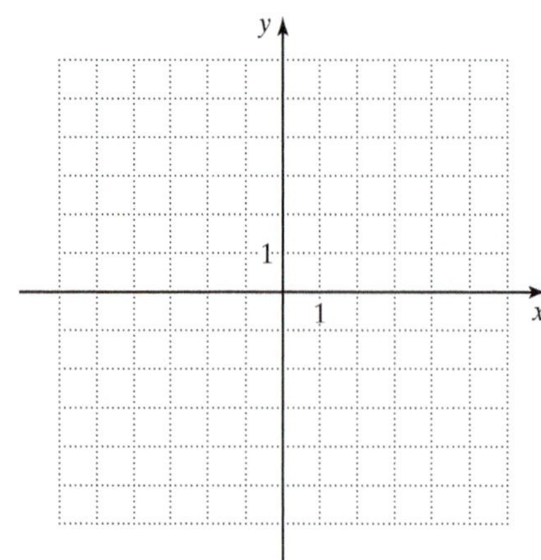

33. $3x - 6y = 18$

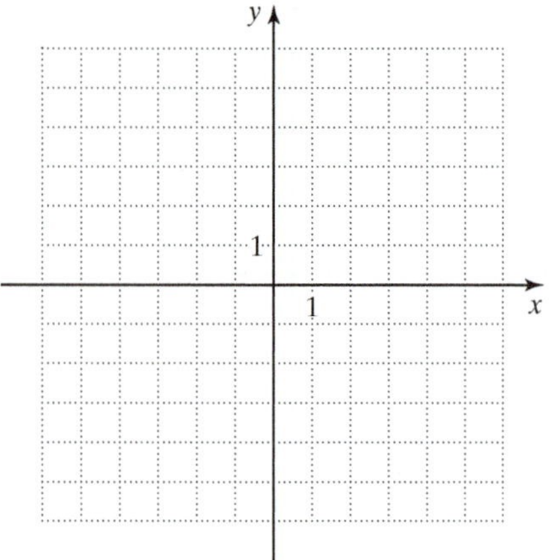

34. $7x - 2y = -7$

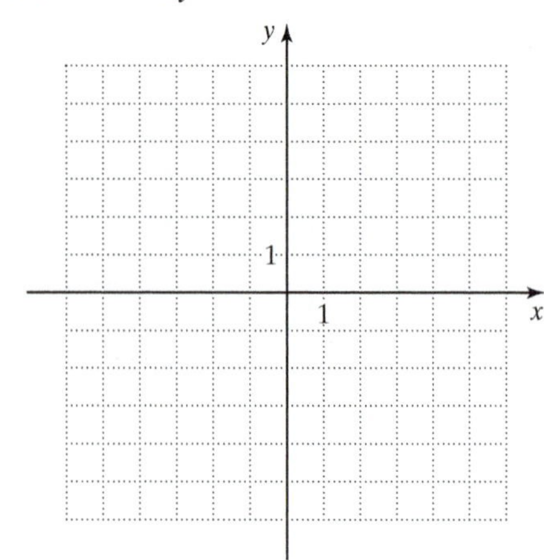

35. $3y - 2x = -6$

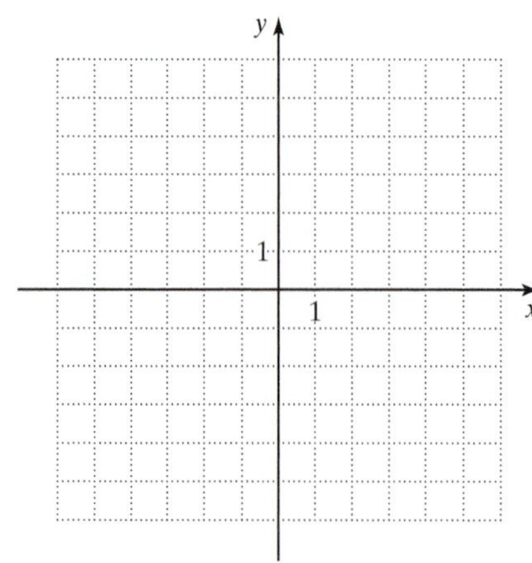

36. $4y - 5x = 10$

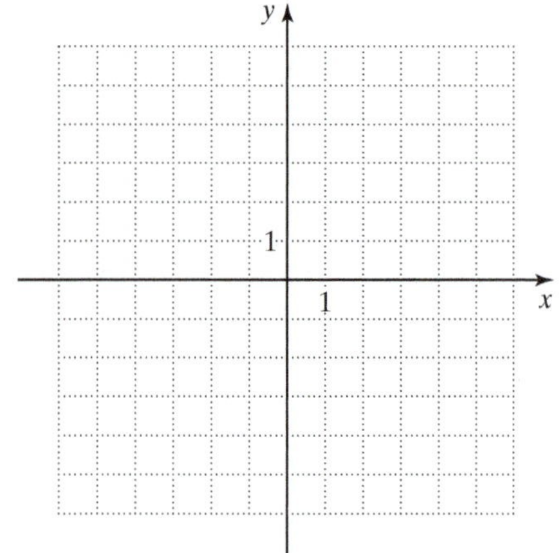

37. $9y + 6x = -9$

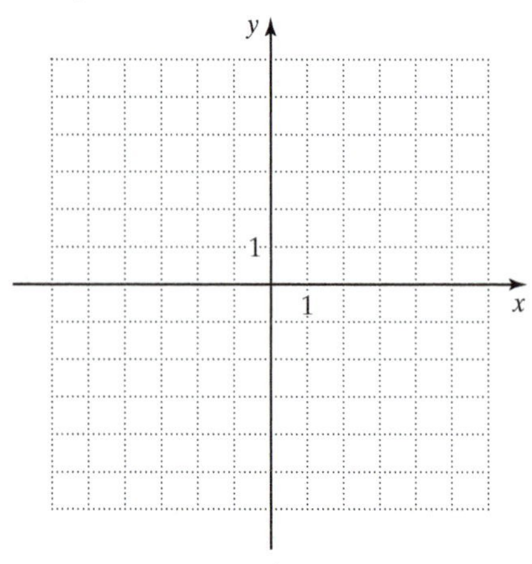

38. $4y + 8x = -4$

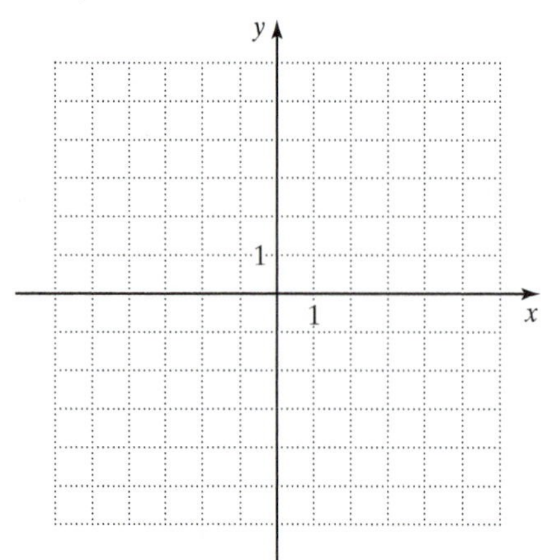

39. $y = \frac{1}{2}x + 2$

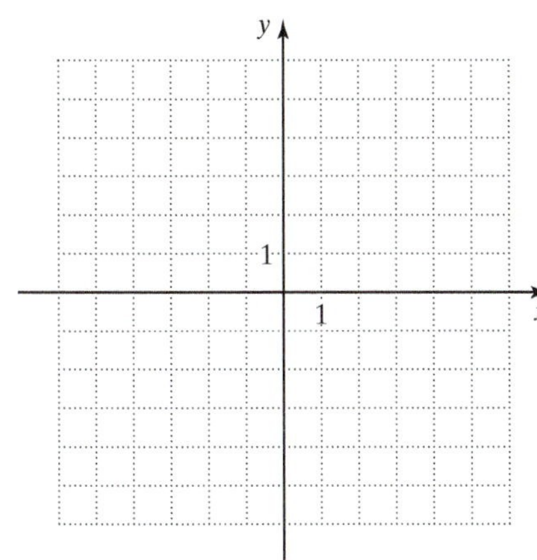

40. $y = \frac{5}{4}x - 5$

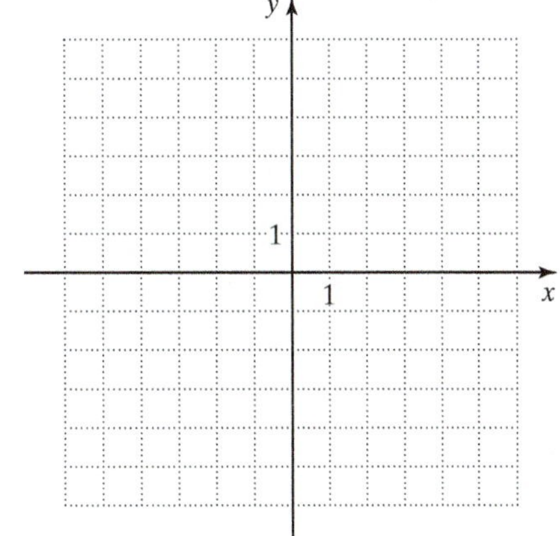

Graph.

41. $y = -2$

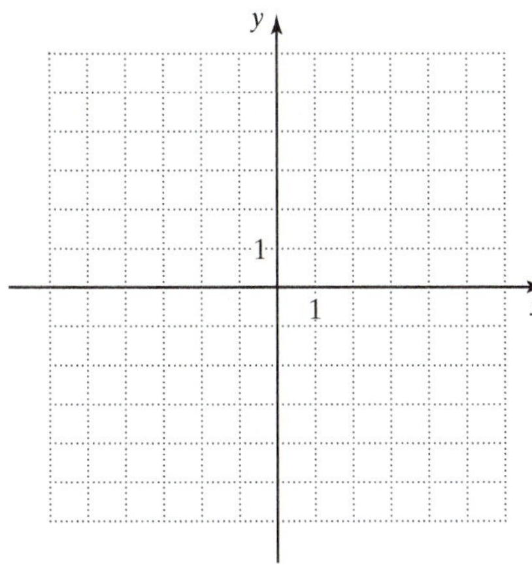

42. $y = 0$

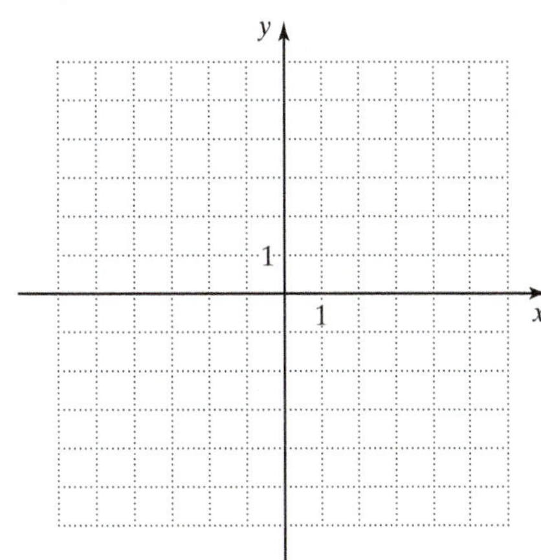

43. $x = 3$

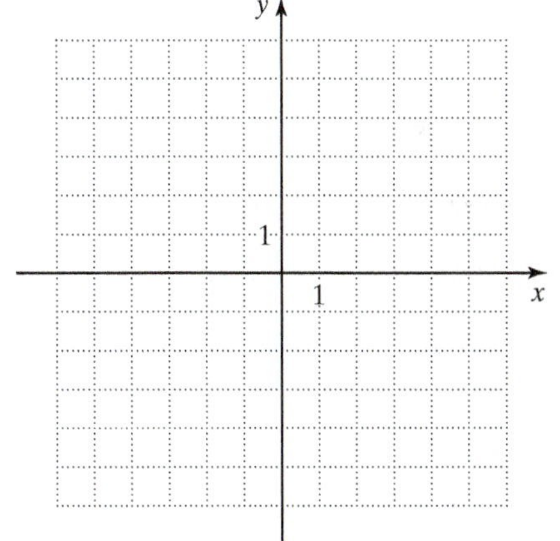

44. $x = 0$

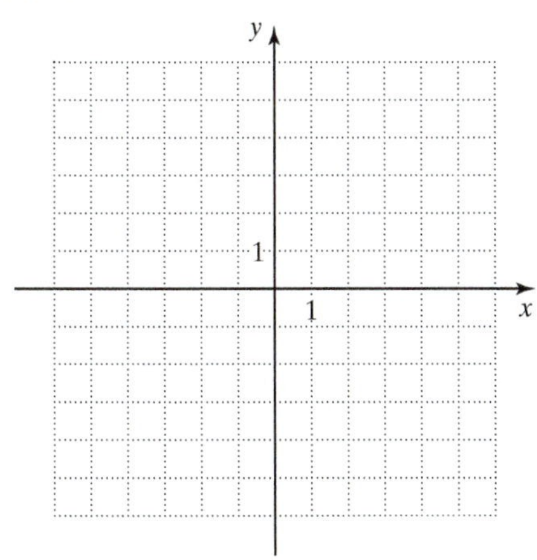

45. $x = -5.5$

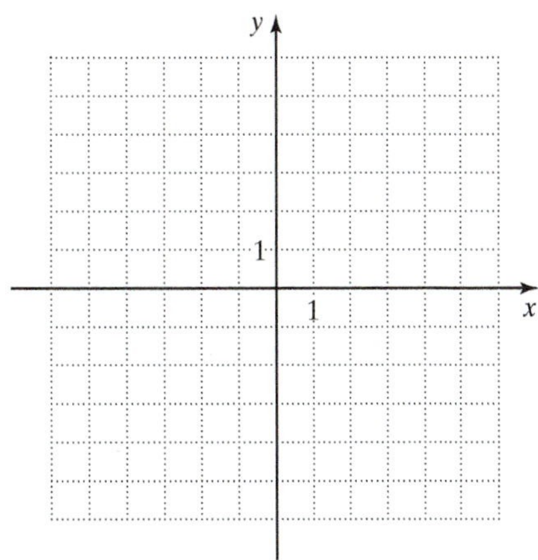

46. $y = -0.5$

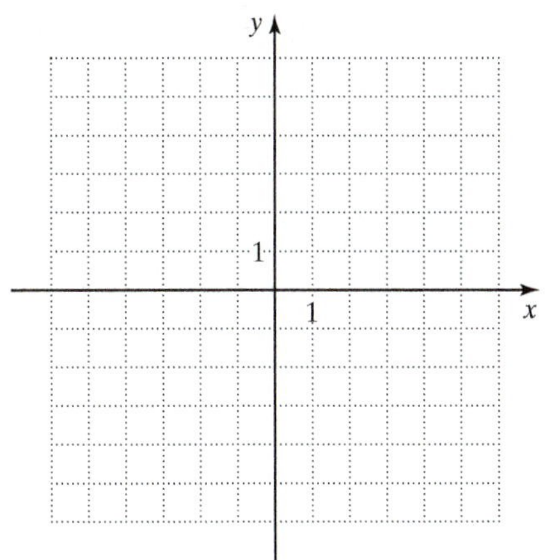

47. $y = \frac{1}{2}x + 3$

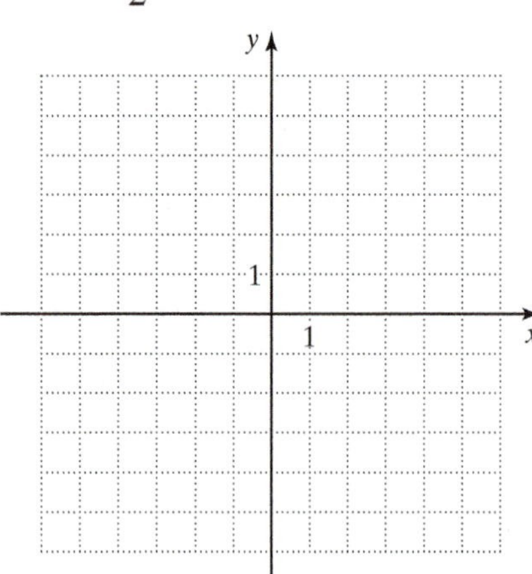

48. $y = \frac{1}{2}x + 6$

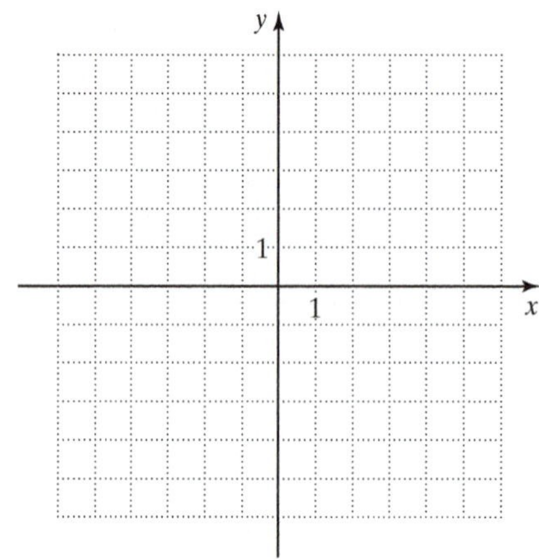

49. $3x + 5y = -15$

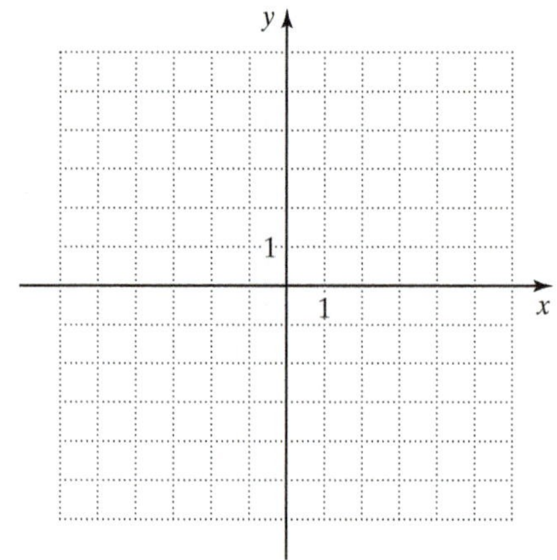

50. $3y - 5x = 15$

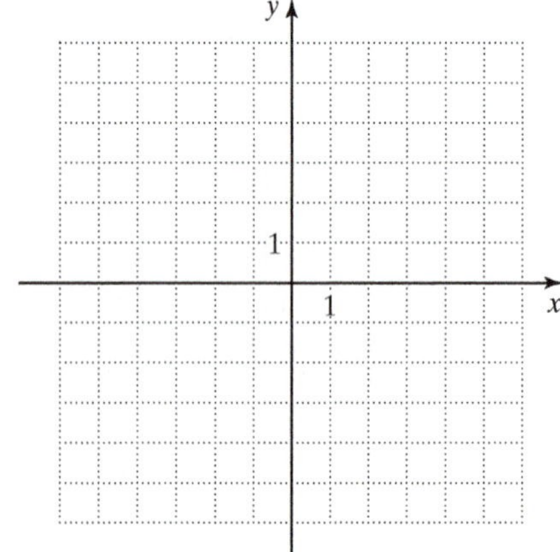

51. $y = -\frac{3}{5}x + \frac{2}{5}$

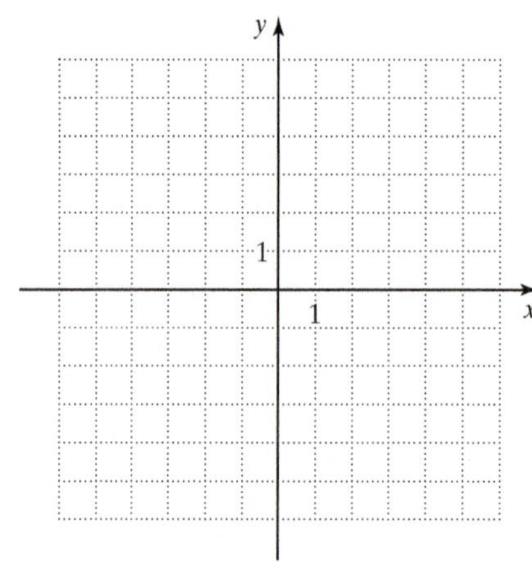

52. $y = -\frac{1}{4}x + \frac{3}{4}$

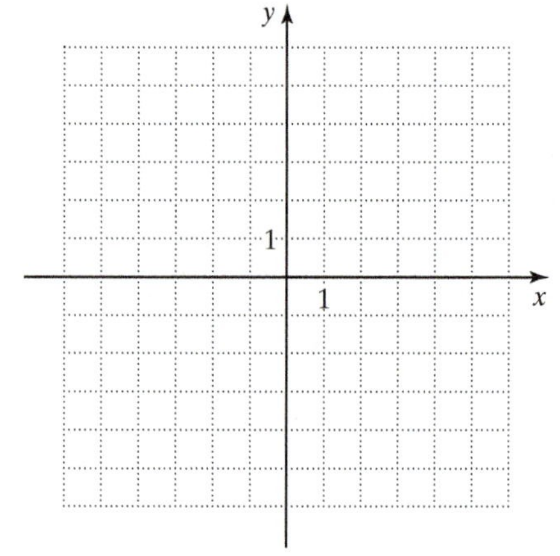

Mixed Practice

Complete each table so that the ordered pairs are solutions of the given equation.

53. $y = -2x + 6$

x	-3	$\frac{5}{2}$		
y			-10	4

54. $3x - 4y = 6$

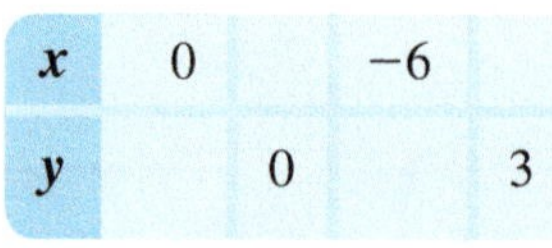

x	0		-6	
y		0		3

For each equation, find the x- and y-intercepts, and another point whose coordinates satisfy the equation. Then, use these points to graph the equation.

55. $y = -\frac{1}{2}x + 2$

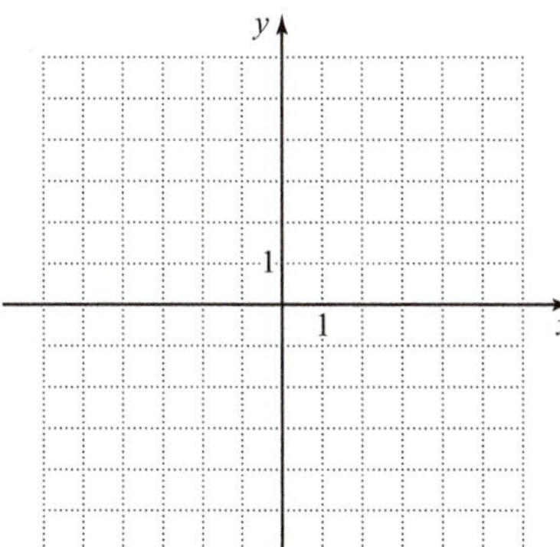

56. $2x - y = 4$

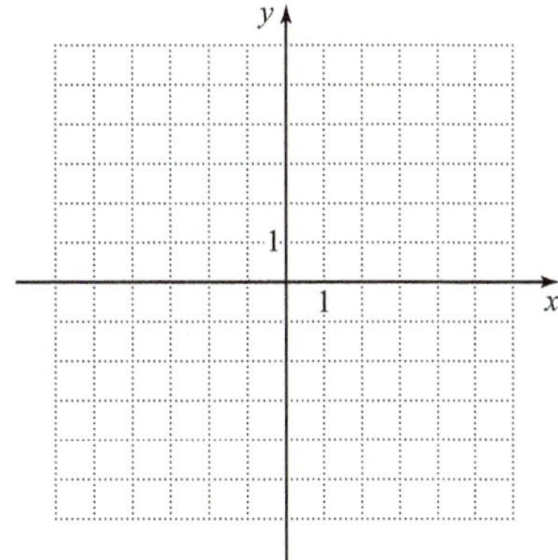

57. $6x - 2y = -12$

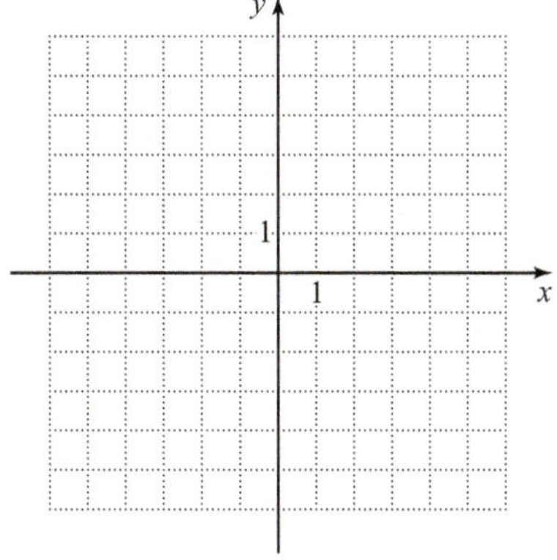

58. $y = \frac{3}{2}x - 3$

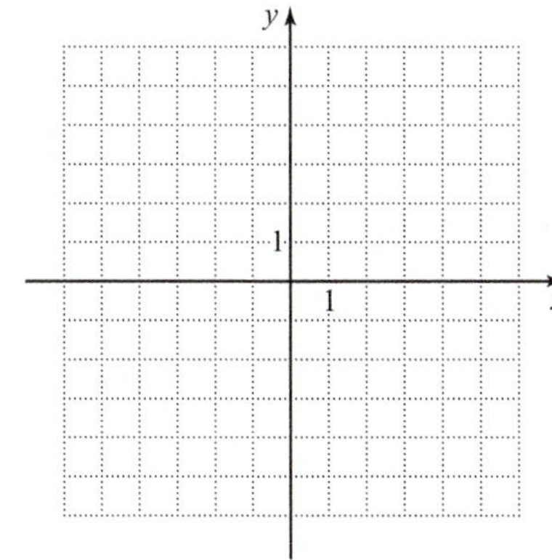

59. $y = -4x$

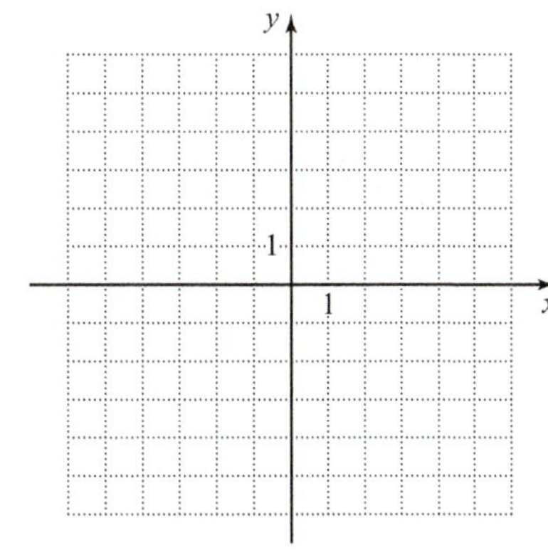

60. $y = \frac{3}{4}x$

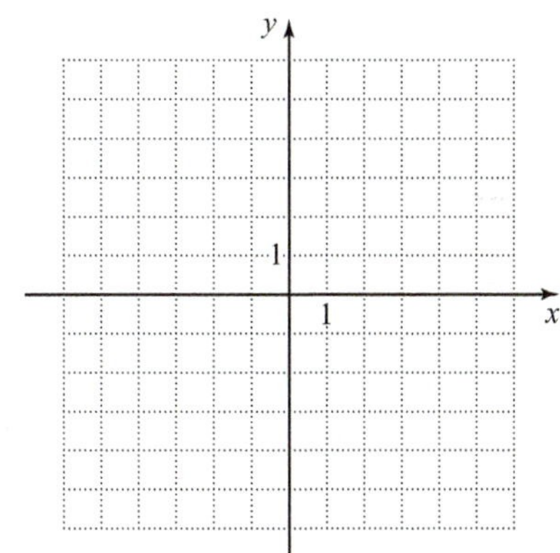

61. $3y + 6x = -3$

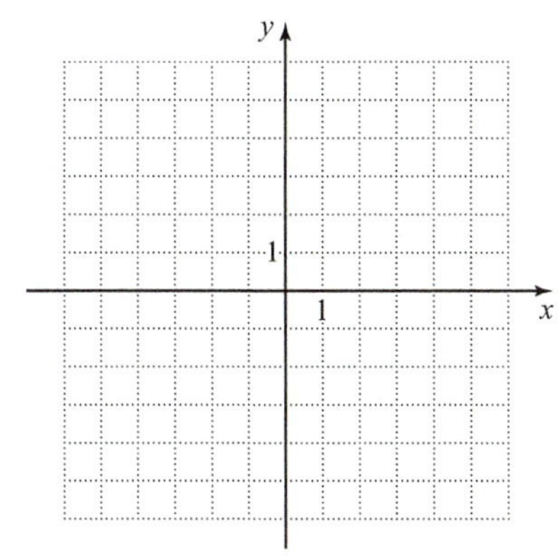

Graph.

62. $x = -3\frac{1}{2}$

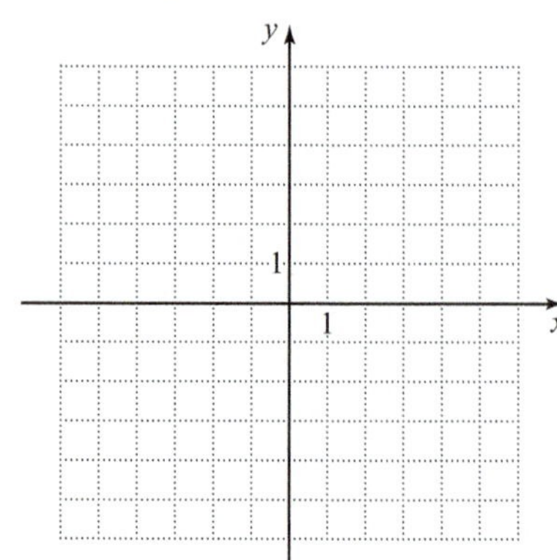

63. $y = -0.5x + 3$

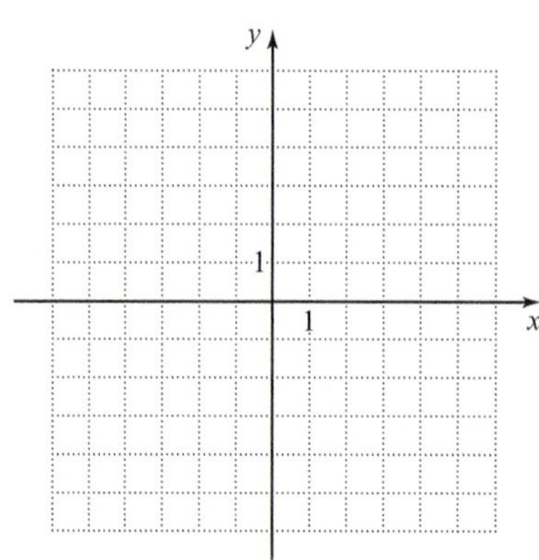

64. $2y - 4x = 8$

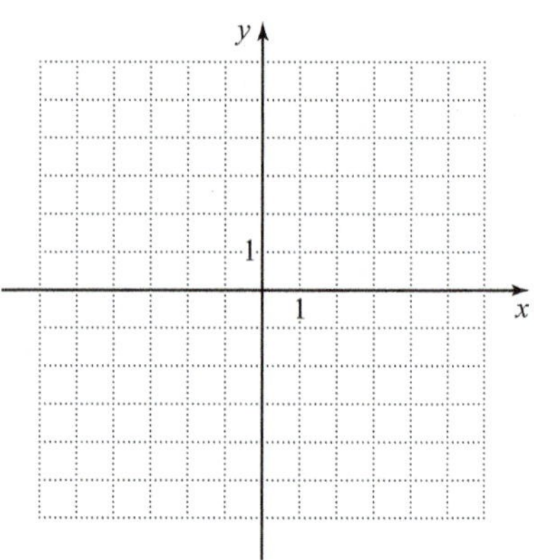

Applications

C *Solve.*

65. In a physics lab, students study the mathematics of motion. They learn that if an object is tossed straight upward with an initial velocity of 10 ft/sec, then after t sec the object will be traveling at a velocity of v ft/sec, where

$$v = 10 - 32t.$$

a. Complete the table at the right.

t	0		1	1.5	2
v		−6			

Explain what a positive value of v means. What does a negative value of v mean?

b. Choose an appropriate scale for each axis, and then graph this equation.

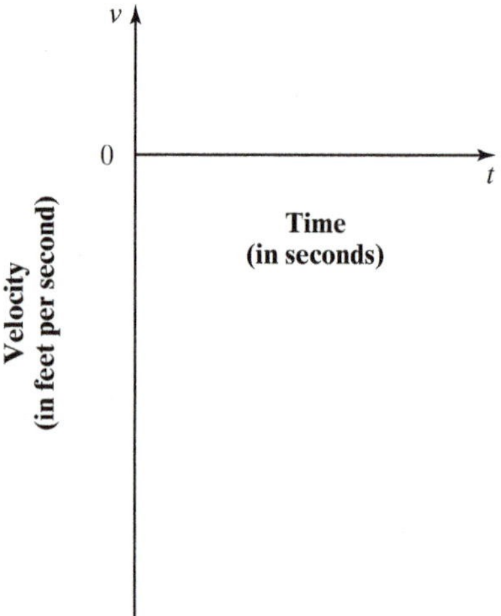

c. In terms of the object's motion, explain the significance of the v-intercept.

d. In terms of the object's motion, explain the significance of the t-intercept.

66. Each day, a local grocer varies the price p of an item in dollars, and then keeps track of the number s of items sold. According to his records, the following equation describes the relationship between s and p:

$$s = -2p + 12$$

a. Complete the table shown at the right.

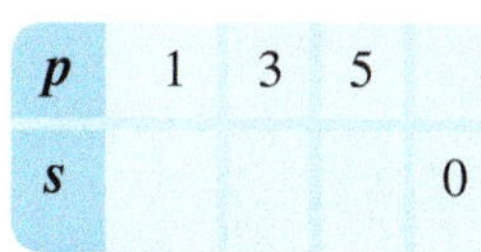

p	1	3	5	
s				0

b. Choose appropriate scales for the axes, and then graph the equation.

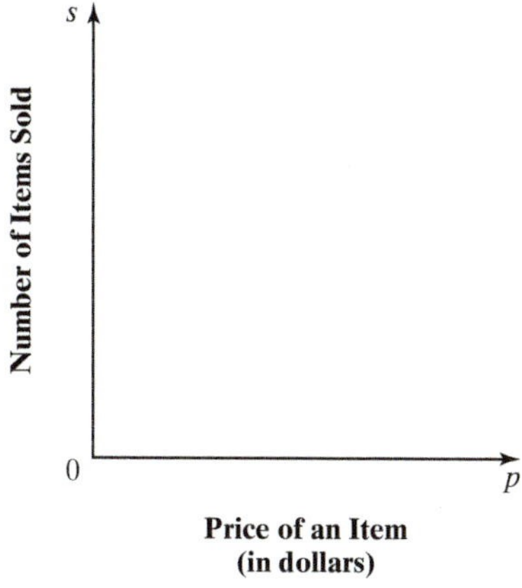

c. Explain why it makes sense to consider the graph only in Quadrant I.

d. From the graph, estimate the price required to sell 4 items.

67. A young couple buys furniture for $2000, agreeing to pay $500 down and $100 at the end of each month until the entire debt is paid off.

a. Express the amount P paid off in terms of the number m of monthly payments.

b. Complete the table shown at the right.

m	1	2	3
P			

c. Choose an appropriate scale for the axes, and then graph this equation.

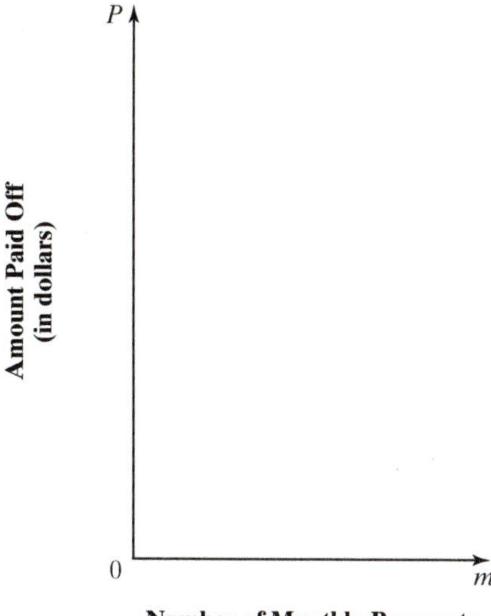

68. Students studying forensic science know that when the femur bone of an adult female is unearthed, a good estimate of her height h is 29 more than double the length l of the femur bone, where all measurements are in inches. (*Source:* nsbri.org)

a. Express this relationship as a formula.

b. Complete the table shown at the right.

l	20	25	30
h			

c. Choose an appropriate scale for the axes, and then graph this relationship.

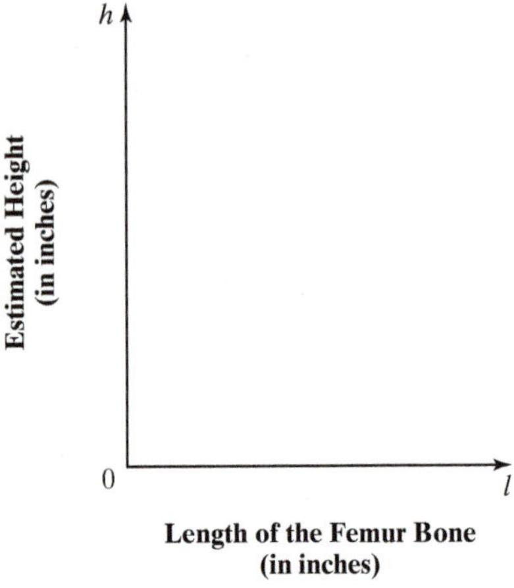

d. Use the graph to estimate the height of a woman whose femur bone was 14 in. in length.

69. The coins in a cash register, with a total value of \$2, consist of n nickels and d dimes.

a. Represent this relationship as an equation.

b. Graph the equation found in part (a).

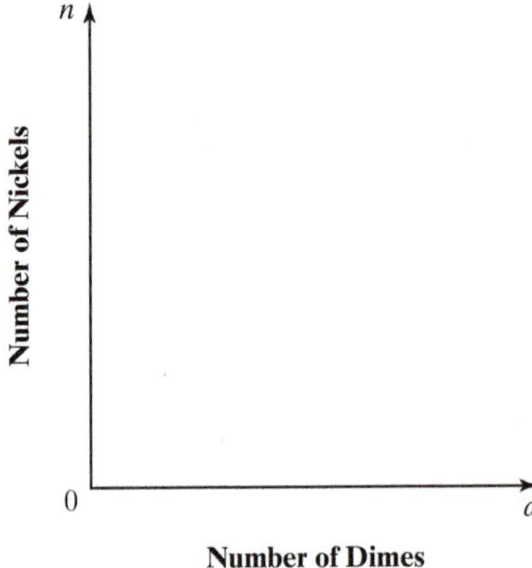

c. Explain in a sentence or two why not every point on this graph in Quadrant I is a reasonable solution to the problem.

70. On the first leg of a trip, a truck driver drove for x hr at a constant speed of 50 mph. On the second leg of the trip, he drove for y hr consistently at 40 mph. In all, he drove 1000 mi.

a. Translate this information into an equation.

b. Choose appropriate scales for the axes, and then graph the equation found in part (a).

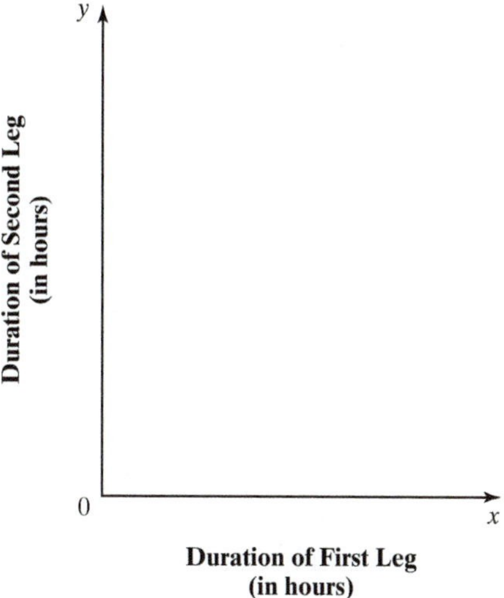

c. What are the x- and y-intercepts of this graph? Explain their significance in terms of the trip.

d. Find the slope of the line. Explain whether you would have expected the slope to be positive or negative, and why.

71. At a computer rental company, the fee F for renting a laptop is \$40 plus \$5 for each of the d days that the laptop is rented.

a. Express this relationship as an equation.

b. Choose appropriate scales for the axes, and then graph the equation expressed in part (a).

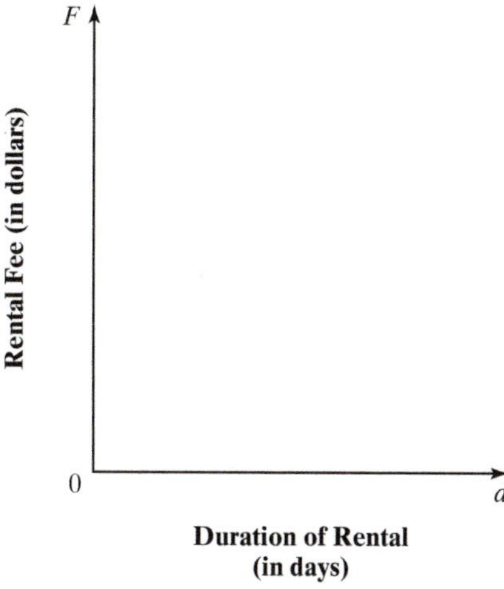

c. How much more does it cost to rent a laptop for 12 days than for 6 days?

72. At a local community center, the annual cost c to use the swimming pool includes an annual membership fee of \$75 plus \$5 per hour for h hr of pool time.

a. Write an equation for the annual cost of swimming at the community center in terms of the number of hours of pool time.

b. Choose appropriate scales for the axes, and then graph the equation for up to and including 150 hr.

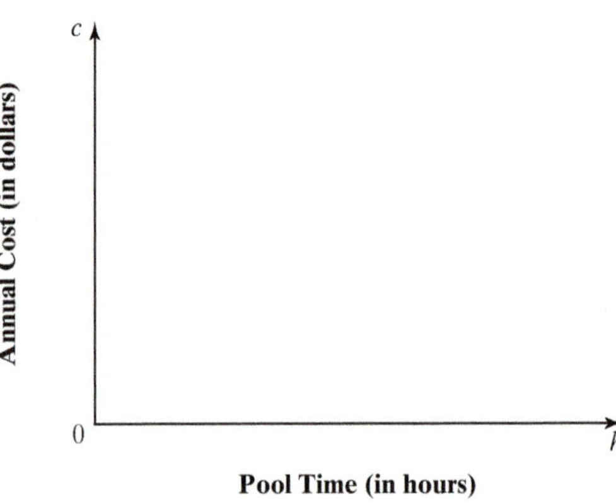

c. Use the graph to estimate the annual cost of using the pool for 25 hr.

d. Suppose the annual cost for swimming was \$500. Estimate the number of hours of pool time.

• Check your answers on page A-21.

MINDStretchers

Groupwork

1. Not all graphs are linear. For example, the graph of the equation $y = x^2 - 4$ is nonlinear, as the graph at the right illustrates:

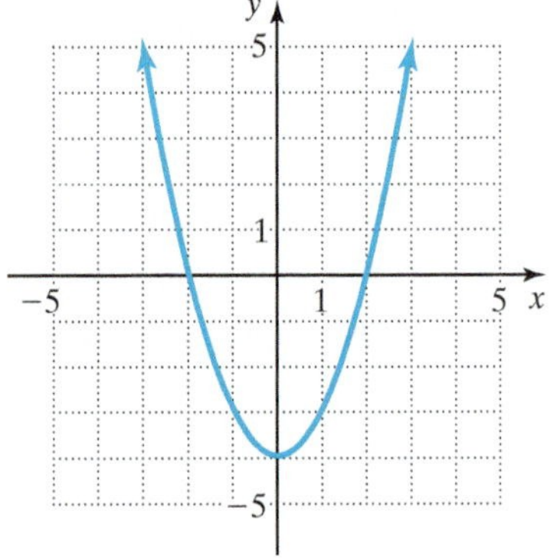

a. Identify the x- and y-intercepts for the graph shown.

b. Show that the x- and y-intercepts found in part (a) satisfy the equation $y = x^2 - 4$.

Writing

2. Give some advantages and disadvantages of graphing a linear equation by finding three arbitrary points versus using the intercepts.

Mathematical Reasoning

3. Recall that for a linear equation to be in general form, it must be written as $Ax + By = C$, where A, B, and C are real numbers and A and B are not both 0. What would the graph of this equation look like if A and B were both 0?

13.4 More on Linear Equations and Their Graphs

OBJECTIVES

- **A** To identify the slope and y-intercept of an equation written in slope-intercept form
- **B** To write given linear equations in slope-intercept form
- **C** To graph a line using its slope and y-intercept
- **D** To find the equation of a line, given two points on the line or its slope and one point
- **E** To solve applied problems involving graphs of linear equations

In the previous section of this chapter, we considered linear equations written in the general form and how to graph them. Now, we will discuss two other forms of linear equations: the *slope-intercept form* and the *point-slope form*. Using these two forms, we will continue to show how line graphs and their corresponding equations have real-world applications.

Slope-Intercept Form

Recall from Section 13.3 that one approach to graphing an equation written in general form is to isolate y.

$$-5x + y = 2 \quad \text{An equation in general form}$$
$$y = 5x + 2 \quad \text{Solve for } y.$$

The linear equation $y = 5x + 2$ is said to be in slope-intercept form. The graph of this equation has slope 5, which is equal to the coefficient of x, and y-intercept $(0, 2)$, where 2 is the constant term of the equation.

DEFINITION

A linear equation is in **slope-intercept form** if it is written as

$$y = mx + b,$$

where m and b are constants. In this form, m is the slope and $(0, b)$ is the y-intercept of the graph of the equation.

This form is used for identifying the slope and y-intercept of the graph of a linear equation without drawing the graph of the equation. The coefficient of x is the slope m, and the constant term b is the y-coordinate of the y-intercept $(0, b)$.

The following table gives additional examples of equations written in slope-intercept form:

Equation	Slope m	y-intercept $(0, b)$
$y = \frac{1}{2}x + 1$	$\frac{1}{2}$	$(0, 1)$
$y = 2x - 1$	2	$(0, -1)$
$y = -7x \rightarrow y = -7x + 0$	-7	$(0, 0)$
$y = 5 \rightarrow y = 0x + 5$	0	$(0, 5)$

EXAMPLE 1

Find the slope and y-intercept of the equation $y = 3x - 5$.

Solution The equation $y = 3x - 5$ or $y = 3x + (-5)$ is already in slope-intercept form, $y = mx + b$. The slope m is 3, and the y-intercept is $(0, -5)$ since the equation has constant term -5.

PRACTICE 1

Find the slope and y-intercept of the equation $y = -2x + 3$.

EXAMPLE 2

For the graph of $y = 3x$, find the slope and y-intercept.

Solution We can rewrite $y = 3x$ as $y = 3x + 0$. Now, the equation $y = 3x + 0$ is in slope-intercept form, with slope $m = 3$ and $b = 0$. Since $b = 0$, the y-intercept is $(0, 0)$. That is, the graph passes through the origin.

PRACTICE 2

Find the slope and y-intercept of the graph of $y = -x$.

EXAMPLE 3

Express $3x + 5y = 6$ in slope-intercept form.

Solution Since the slope-intercept form of an equation is $y = mx + b$, we need to solve the given equation for y.

$$\begin{aligned} 3x + 5y &= 6 \\ 5y &= -3x + 6 \\ \frac{5y}{5} &= \frac{-3}{5}x + \frac{6}{5} \\ y &= -\frac{3}{5}x + \frac{6}{5} \end{aligned}$$

So $y = -\frac{3}{5}x + \frac{6}{5}$ is the equation written in slope-intercept form, where m is $-\frac{3}{5}$ and b is $\frac{6}{5}$.

PRACTICE 3

Express $3x - 2y = 4$ in slope-intercept form.

EXAMPLE 4

Write $y - 1 = 5(x - 1)$ in slope-intercept form.

Solution To get the equation in the form $y = mx + b$, we must solve for y.

$$\begin{aligned} y - 1 &= 5(x - 1) \\ y - 1 &= 5x - 5 \\ y &= 5x - 5 + 1 \\ y &= 5x - 4 \end{aligned}$$

So $y = 5x - 4$ is the equation written in slope-intercept form, where m is 5 and b is -4.

PRACTICE 4

Change the equation $y - 2 = 4(x + 1)$ to slope-intercept form.

We can use the slope-intercept form to write the equation of a line when given its slope and y-intercept.

EXAMPLE 5

Write the equation of the line with slope $-\frac{4}{5}$ and y-intercept $(0, -3)$.

Solution We are given that the slope is $-\frac{4}{5}$ and the y-intercept is $(0, -3)$. So $m = -\frac{4}{5}$ and $b = -3$. We substitute these values in the slope-intercept form:

$$y = mx + b$$

$$y = -\frac{4}{5}x + (-3), \quad \text{or } y = -\frac{4}{5}x - 3$$

PRACTICE 5

A line on a coordinate plane has slope 1 and intersects the y-axis 2 units above the origin. Write its equation in slope-intercept form.

Recall our discussion in Section 13.2 of parallel and perpendicular lines. The next two examples deal with the equations of lines that are parallel or perpendicular.

EXAMPLE 6

Find an equation of the line that is parallel to the graph of $y = 4x + 1$ and has y-intercept $(0, 3)$.

Solution The line $y = 4x + 1$ is in slope-intercept form. The slope of this line is 4. Since parallel lines have the same slope, the line we want will also have slope $m = 4$. Since its y-intercept is $(0, 3)$, $b = 3$. Therefore, the desired equation is $y = 4x + 3$.

PRACTICE 6

What is the equation of the line parallel to the graph of $y = -2x + 3$ with y-intercept $(0, -1)$?

EXAMPLE 7

What is the equation of the line that is perpendicular to the graph of $y = 3x - 1$ and has y-intercept $(0, 1)$?

Solution The line $y = 3x - 1$ is written in slope-intercept form. So its slope must be 3. We know that the slopes of two perpendicular lines are negative reciprocals of each other. Therefore, the slope of the line we want has slope $m = -\frac{1}{3}$. Since its y-intercept is $(0, 1)$, $b = 1$.

So the desired equation is $y = -\frac{1}{3}x + 1$.

PRACTICE 7

Find the equation of the line that is perpendicular to the graph of $y = 2x + 5$ and has y-intercept $(0, -2)$.

EXAMPLE 8

At her college, a student pays $75 per credit-hour plus a flat student fee of $100. Find an equation in slope-intercept form that relates the amount A that she pays to the number h of credit-hours in her program.

Solution The amount A in dollars that the student pays is the sum of 75 times the number h of credit-hours in her program and 100. So we have:

$$A = 75h + 100$$

This equation is written in slope-intercept form.

PRACTICE 8

A bathtub, which has a capacity of 45 gal, is filled to the top with water. The tub starts to drain at a rate of 3 gal/min. Write an equation in slope-intercept form that expresses the amount of water w left in the tub (in gallons) in terms of the time t that the tub has been draining (in minutes).

We have already graphed equations of the form $y = mx + b$ by finding three points whose coordinates satisfy the equation. Now, let's focus on graphing such equations using the slope and the y-intercept.

Consider the equation $y = 3x - 1$, which has slope 3 and y-intercept $(0, -1)$. Since 3 is $\frac{3}{1}$, we know from the definition of slope that the *rise* is 3 and the *run* is 1. Starting at $(0, -1)$, we move up 3 units and then 1 unit to the right to find a second point $(1, 2)$ on the line. Then, we draw the line through the two points $(0, -1)$ and $(1, 2)$.

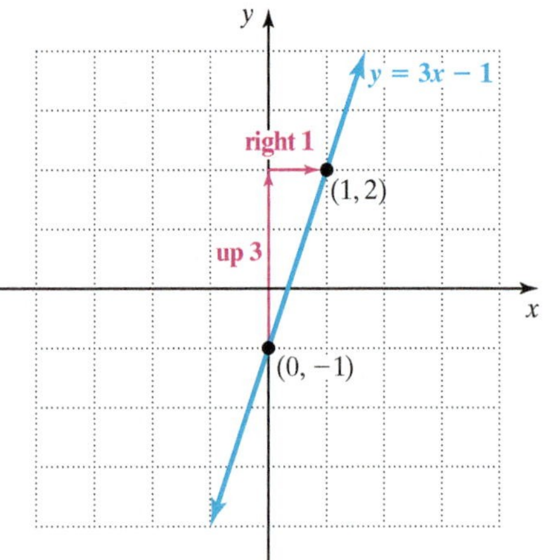

This example leads us to the following rule:

To Graph a Linear Equation in Two Variables Using the Slope and the *y*-intercept

- First, locate the y-intercept.
- Then, use the slope to find a second point on the line.
- Finally, draw the line through the two points.

EXAMPLE 9

Graph $y = -\frac{3}{2}x + 2$ using the slope and y-intercept.

Solution Since $y = -\frac{3}{2}x + 2$ is in slope-intercept form, the slope is $-\frac{3}{2}$ and the y-intercept is $(0, 2)$. First, we locate the y-intercept $(0, 2)$. Since the slope $-\frac{3}{2}$ equals $\frac{-3}{2}$, from the point $(0, 2)$ we move *down* 3 units and then 2 units to the *right* to find the second point $(2, -1)$. Then, we draw the line through the points $(0, 2)$ and $(2, -1)$ as shown in the graph.

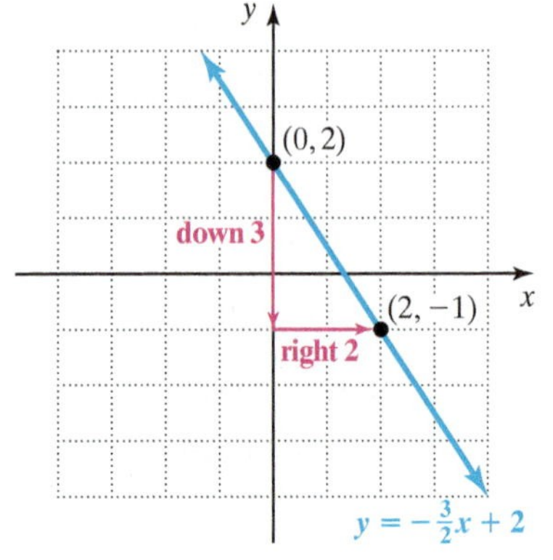

PRACTICE 9

Use the slope and y-intercept to graph $y = -\frac{1}{3}x - 4$.

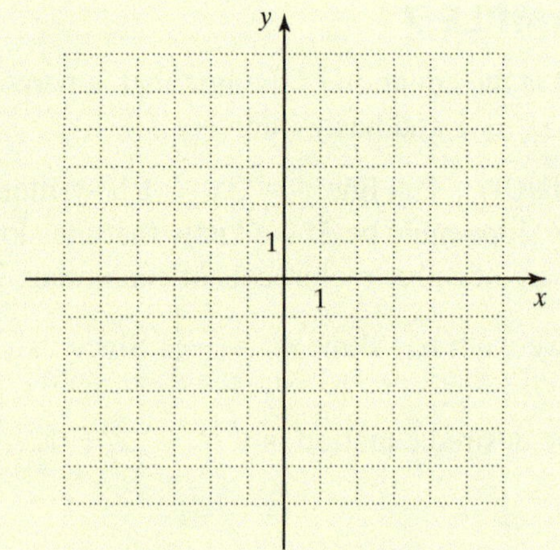

Note that since $\frac{-3}{2} = \frac{3}{-2}$, from the point $(0, 2)$ we could have moved *up* 3 units, and then 2 units to the *left* to find the second point $(-2, 5)$. Then, we could have drawn the line through $(0, 2)$ and $(-2, 5)$ to obtain the same graph of the equation, as shown on the coordinate plane to the right.

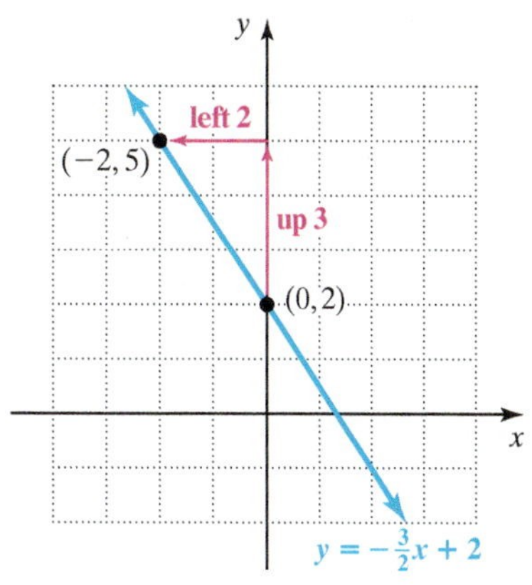

Recall that in Section 13.3 we graphed $2x + 3y = 6$ using the intercepts. We can also graph this equation using the slope and y-intercept. Which method do you prefer? Explain why.

Point-Slope Form

The last form of a linear equation that we will discuss is called the point-slope form.

DEFINITION

The **point-slope form** of a linear equation is written as

$$y - y_1 = m(x - x_1),$$

where x_1, y_1, and m are constants. In this form, m is the slope and (x_1, y_1) is a point that lies on the graph of the equation.

This form is useful for finding the equation of a line in two particular situations:

- when we know the slope of the line and a point on it, or
- when we know two points on the line.

EXAMPLE 10

A line with slope -2 passes through the point $(-1, 5)$. Find the equation of this line written in point-slope form.

Solution Since we know a point on the line and the slope of the line, we can substitute directly into the point-slope form, where $x_1 = -1$, $y_1 = 5$, and $m = -2$.

$$y - y_1 = m(x - x_1)$$
$$y - 5 = -2[x - (-1)]$$
$$y - 5 = -2(x + 1) \quad \text{Point-slope form}$$

We can leave this equation in point-slope form, or we can simplify the equation and write it in either general form

$$y - 5 = -2x - 2$$
$$2x + y = 3 \quad \text{General form}$$

or in slope-intercept form

$$y = -2x + 3 \quad \text{Slope-intercept form}$$

PRACTICE 10

A line passing through the point $(7, 0)$ has slope 2. Find its equation in point-slope form.

EXAMPLE 11

What is the equation of the line passing through the points $(3, 5)$ and $(2, 1)$?

Solution Since we know the coordinates of two points on the line, we can find its slope.

$$m = \frac{y_2 - y_1}{x_2 - x_1} = \frac{5 - 1}{3 - 2} = \frac{4}{1} = 4$$

The line with slope $m = 4$ passing through the point $(3, 5)$ is:

$$y - y_1 = m(x - x_1)$$
$$y - 5 = 4(x - 3)$$

PRACTICE 11

Find the equation in point-slope form of the line passing through $(7, 7)$ and the origin.

The equation found in Example 11 is $y = 4x - 7$ in slope-intercept form. If we substitute the point $(2, 1)$ rather than the point $(3, 5)$ into the point-slope form, will the resulting equation be the same?

EXAMPLE 12

An accountant's computer decreases in value by \$400 a year. The computer was worth \$1600 one year after he bought it. Write an equation that gives the value V of the computer in terms of the number of years t since the purchase.

Solution Each year that passes, t increases by 1 and V decreases by 400. Therefore, the graph of the equation we are seeking has slope -400. Because the computer is worth \$1600 one year after purchase, the graph passes through the point $(1, 1600)$. Since we know a point on the line as well as its slope, we can find the point-slope form of the equation.

$$V - V_1 = m(t - t_1)$$
$$V - 1600 = -400(t - 1)$$

If we like, we can simplify the equation and write it in slope-intercept form:

$$V - 1600 = -400t + 400$$
$$V = -400t + 2000$$

PRACTICE 12

The total weight of a box used for shipping baseballs increases by 5 oz for each baseball that is packed in the box. A box with 4 balls weighs 27 oz. Write an equation that expresses the total weight w of the box in terms of the number of baseballs b packed in the box.

Can you explain why the slope of the line in Example 12 is negative?

Using a Calculator or Computer to Graph Linear Equations

Calculators with graphing capabilities and computers with graphing software allow us to graph equations at the push of a key, even those with complicated coefficients. Although they vary somewhat in terms of features and commands, these machines all graph the equation of your choice on a coordinate plane.

To graph an equation, begin by making certain that the equation is in slope-intercept form. On many graphers, pressing the [Y =] *key results in a screen being displayed on which you enter the equation. For instance, if you wanted to graph* $2x - y = 1$, *you would first solve for y, resulting in* $y = 2x - 1$, *and then enter* $2x - 1$ *to the right of \Y1 = on the screen. Pressing the* [GRAPH] *key displays a coordinate plane in which the graph of* $y = 2x - 1$ *is sketched, as we see on the screen to the right.*

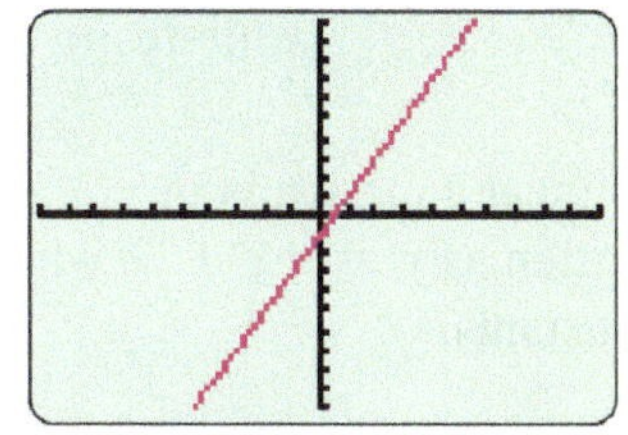

Many graphers have a **TRACE** *feature that highlights a point on the graph and displays its coordinates. As you hold down an arrow key, you can see how the coordinates change as the highlighted point moves along the graph.*

TIP The viewing window allows you to set the range and scales for the axes. Before you graph an equation, be sure to set the viewing window in which you would like to display the graph.

EXAMPLE 13

Graph $y - x = 2$, and then use the **TRACE** feature to identify the y-intercept.

Solution First, solve for y: $y = x + 2$. Next, press [Y =] and enter $x + 2$ to the right of \Y1 =. Then, set the viewing window in which you want to display the graph. Finally, press [GRAPH] to display the graph of the equation. If the graph of $y = x + 2$ does not appear, check your grapher's instruction manual.

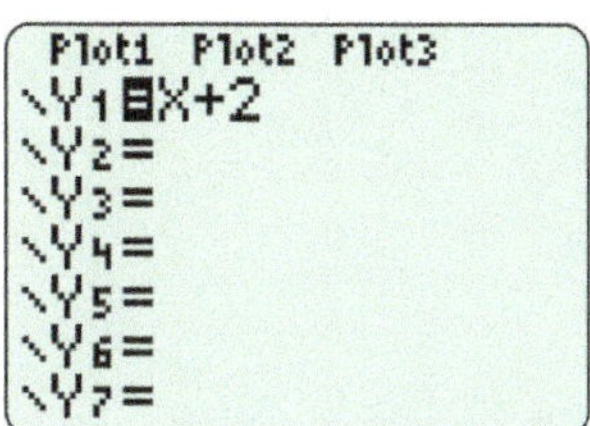

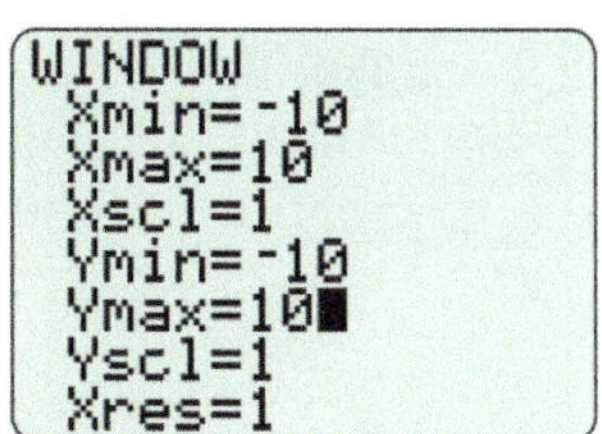

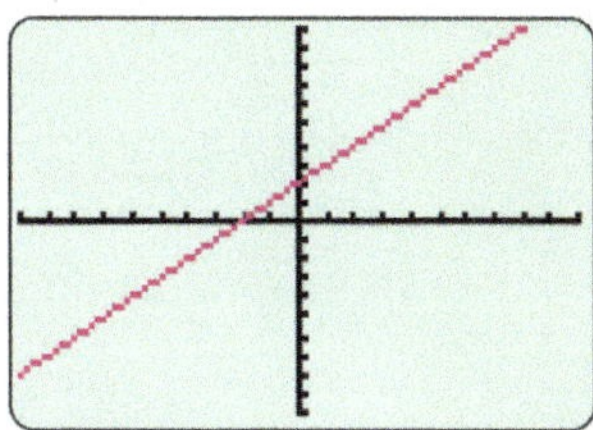

With the **TRACE** feature, run the cursor along the graph until it appears to be on the y-axis. The displayed coordinates of this y-intercept are approximately $x = 0$ and $y = 2$.

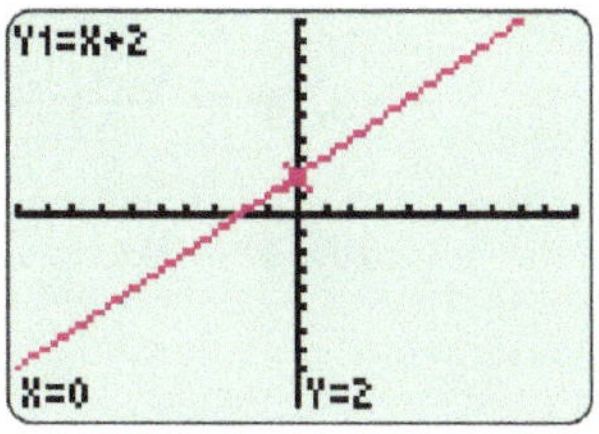

PRACTICE 13

Graph $2y + x = 3$, and then find the y-intercept with the **TRACE** feature.

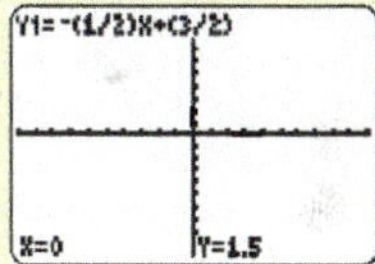

13.4 Exercises

FOR EXTRA HELP MyMathLab MathXL PRACTICE WATCH READ REVIEW

Mathematically Speaking

Fill in each blank with the most appropriate term or phrase from the given list.

standard	point-slope	slope
x-intercept	y-intercept	slope-intercept

1. A linear equation is in ___________ form if it is written as $y = mx + b$, where m and b are constants.

2. For an equation of a line written in slope-intercept form, m is the ___________ of the line.

3. For an equation of a line written in slope-intercept form, $(0, b)$ is the ___________ of the line.

4. The ___________ form of a linear equation is written as $y - y_1 = m(x - x_1)$, where x_1, y_1, and m are constants.

A *Complete each table.*

5.

Equation	Slope m	y-intercept $(0, b)$	Which Graph Type Best Describes the Line? ╱ ╲ — |	x-intercept
$y = 3x - 5$				
$y = -2x$				
$y = 0.7x + 3.5$				
$y = \frac{3}{4}x - \frac{1}{2}$				
$6x + 3y = 12$				
$y = -5$				
$x = -2$				

6.

Equation	Slope m	y-intercept $(0, b)$	Which Graph Type Best Describes the Line? ╱ ╲ — |	x-intercept
$y = -3x + 5$				
$y = 2x$				
$y = 1.5x + 6$				
$y = \frac{2}{3}x + \frac{1}{2}$				
$4x + 6y = 24$				
$y = 0.3$				
$x = 2$				

Find the slope and y-intercept of each equation.

7. $y = -x + 2$

8. $y = -\frac{1}{2}x + 3$

9. $y = 3x - 4$

10. $y = 4x - 2$

B *Write the following equations in slope-intercept form.*

11. $x - y = 10$

12. $3x - y = 15$

13. $x + 10y = 10$

14. $x + y = 7$

15. $6x + 4y = 1$

16. $3x + 5y = 15$

17. $2x - 5y = 10$

18. $4x - 8y = 12$

19. $y + 1 = 3(x + 5)$

20. $y - 1 = 3(x - 5)$

C *Match the equation to its graph.*

21. $4x - 2y = 6$

22. $-2x + 4y = 8$

23. $2y - x = 8$

24. $6x + 3y = -9$

a.

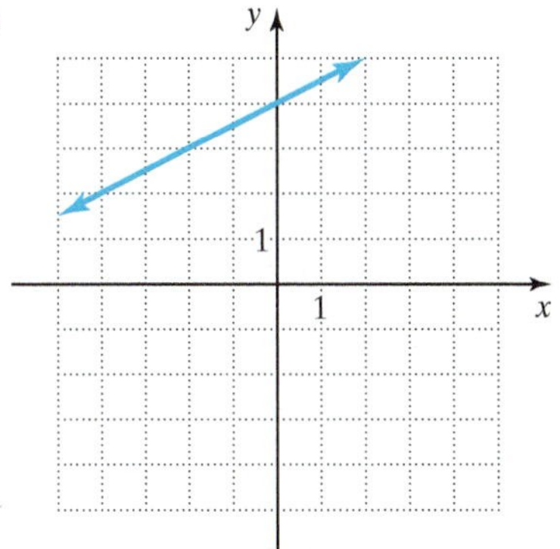

b.

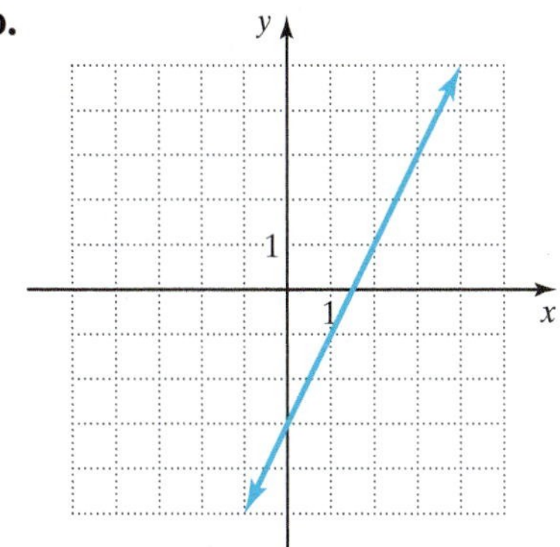

c.

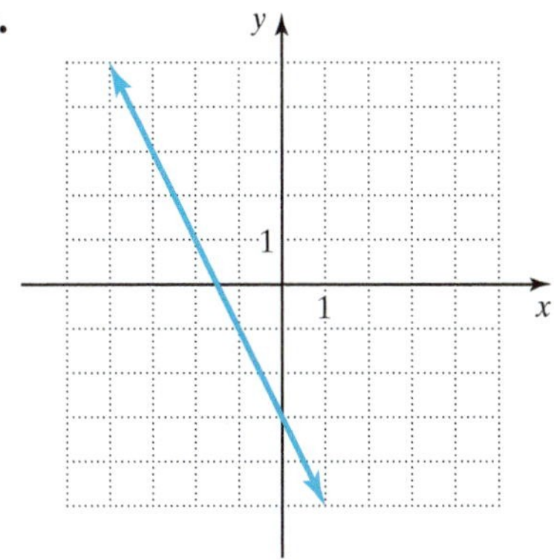

d.

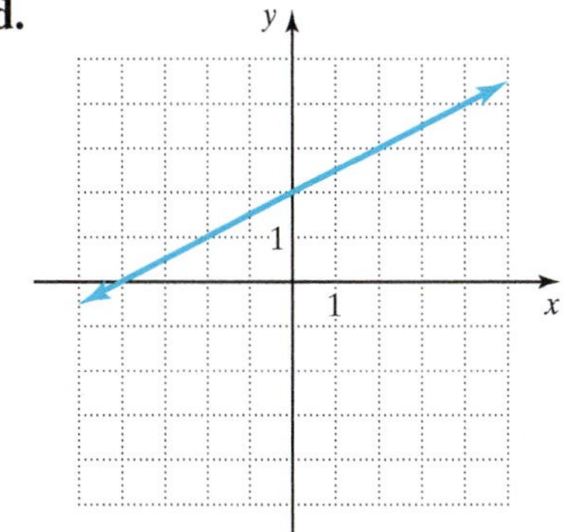

Graph the following equations using the slope and y-intercept.

25. $y = 2x + 1$

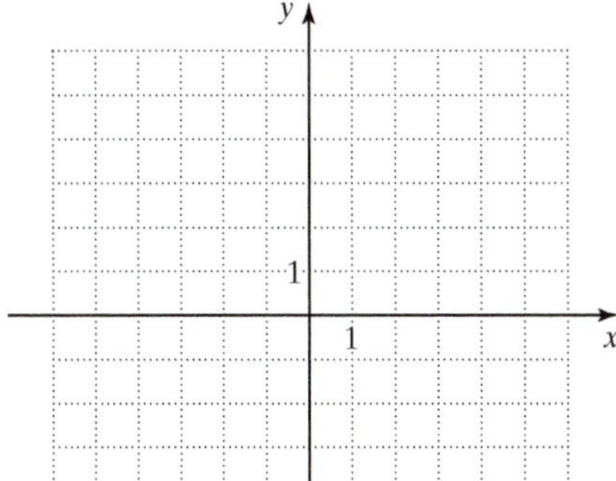

26. $y = 3x + 1$

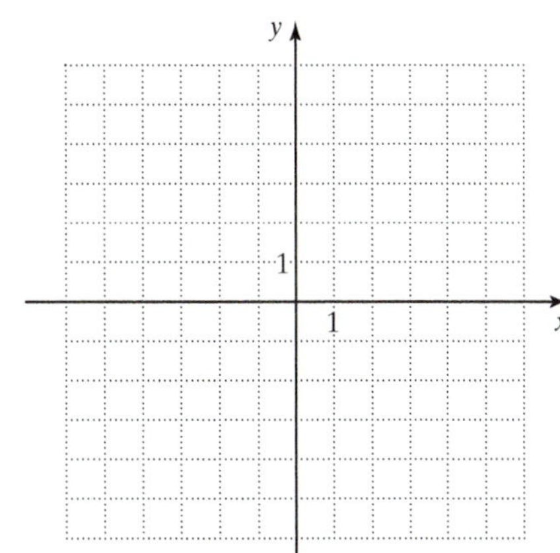

27. $y = -\frac{2}{3}x + 6$

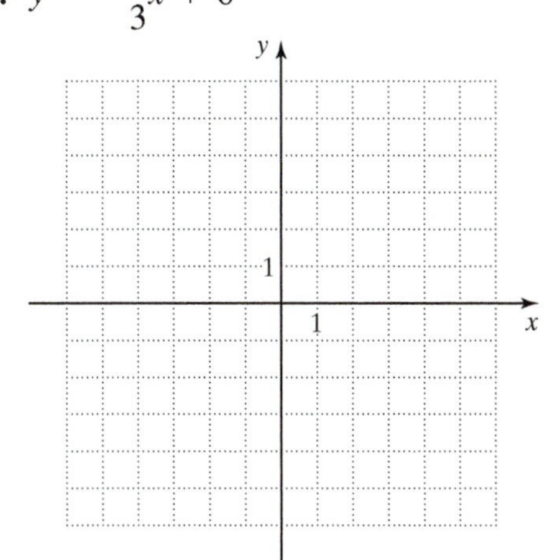

28. $y = -\frac{3}{2}x - 6$

29. 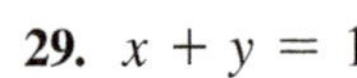
$x + y = 1$

30. 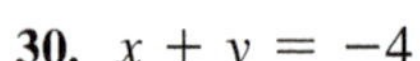
$x + y = -4$

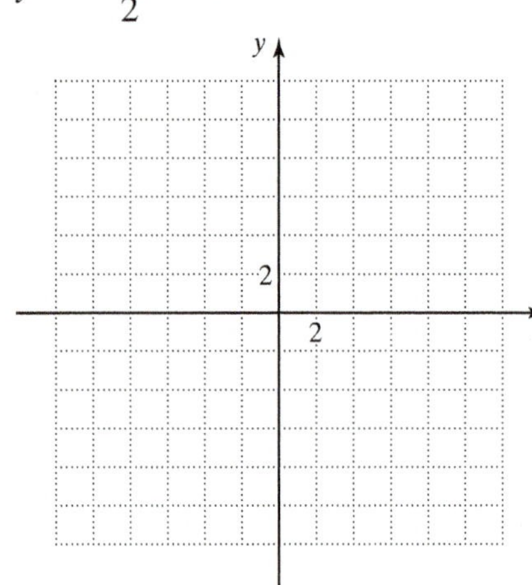

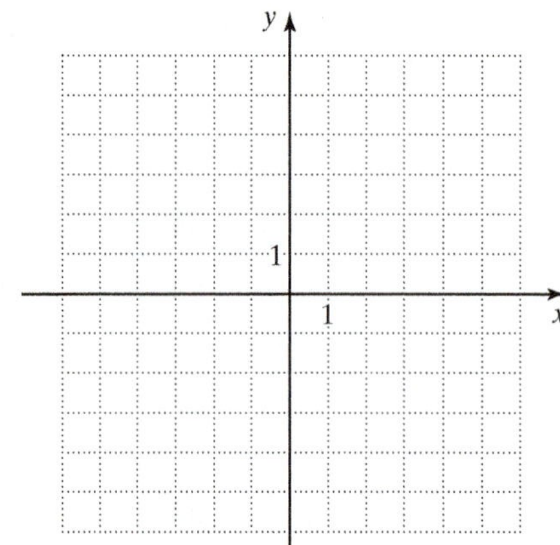

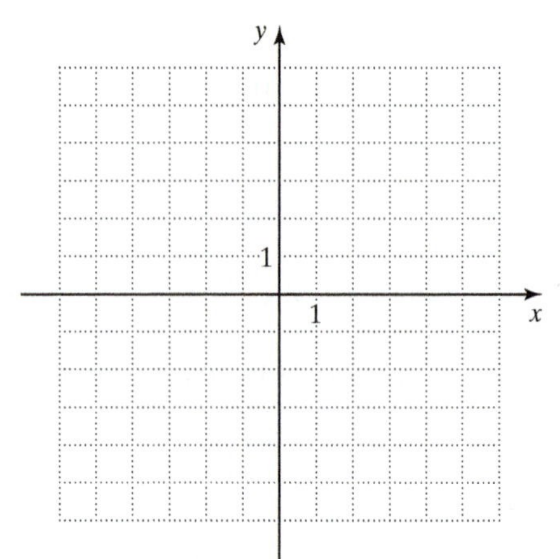

31. $y = -\frac{3}{4}x$

32. 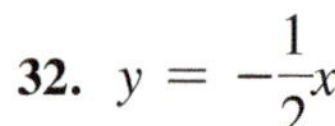
$y = -\frac{1}{2}x$

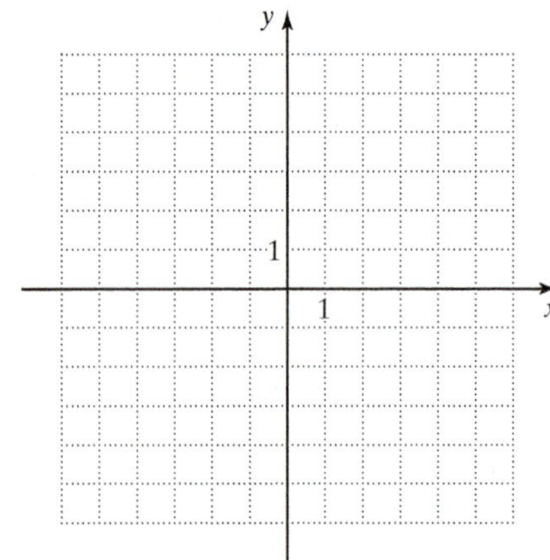

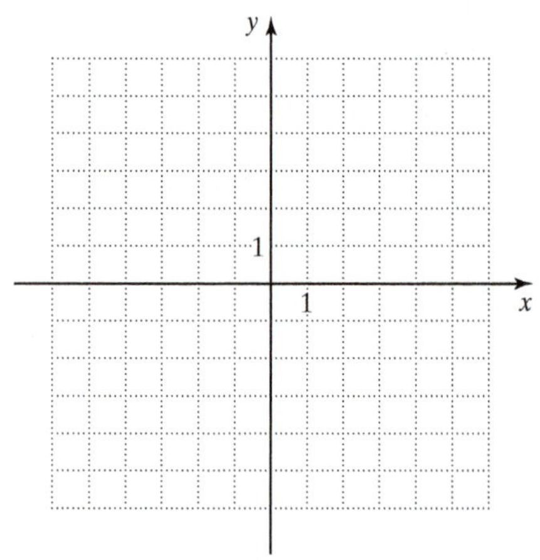

33. $x + 2y = 4$

34. $2x + 3y = 12$

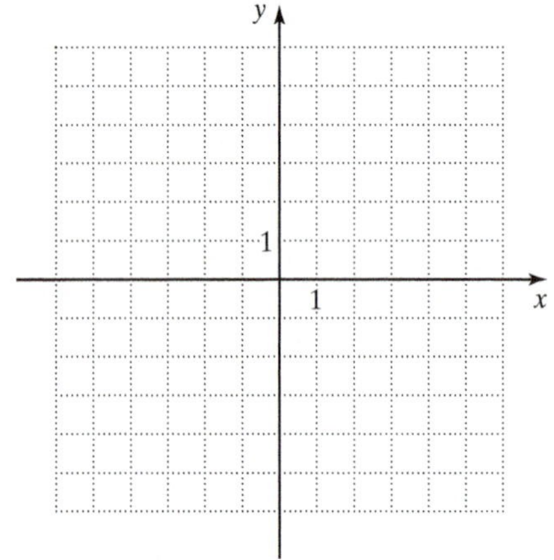

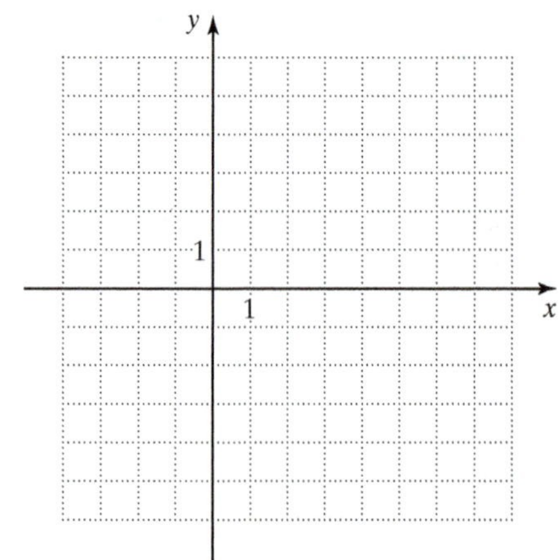

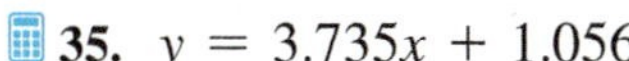

35. $y = 3.735x + 1.056$

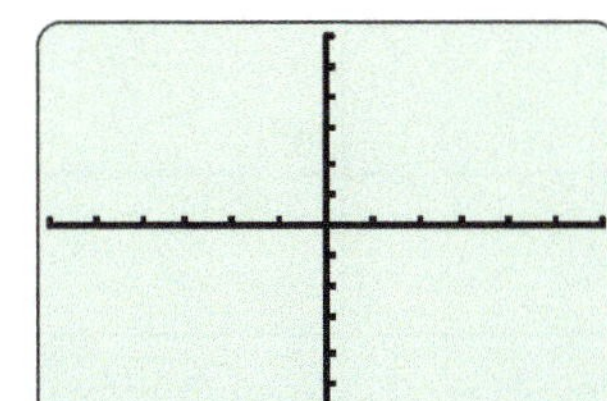

36. $y = -0.875x + 2.035$

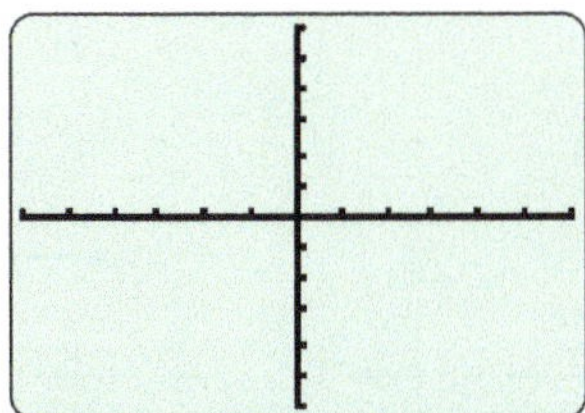

D *Solve.*

37. Find the equation of the line with slope 3 that passes through the point $(0, 7)$.

38. What is the equation of the line that has slope -1 with y-intercept $(0, -2)$?

39. Find the equation of the line that is parallel to the graph of $y = 5x - 1$ and has y-intercept $(0, -20)$.

40. What is the equation of the line that is parallel to the graph $y = \frac{1}{3}x - 1$ and has y-intercept $(0, 4)$?

41. What is the equation of the line that is perpendicular to the graph of $y = 2x$ and that passes through $(-2, 5)$?

42. Find the equation of the line that is perpendicular to the graph $y = -x$ and that passes through the point $(1, -3)$.

43. What is the equation of the line passing through the points $(2, 1)$ and $(1, 2)$?

44. The points $(5, 1)$ and $(2, -3)$ lie on a line. Find its equation.

45. Find the equation of the line passing through points $(-1, -5)$ and $(-7, -6)$.

46. What is the equation of the line passing through the origin and the point $(3, 5)$?

47. Write the equation of the vertical line that passes through the point $(-3, 5)$.

48. What is the equation of the vertical line passing through the point $(1, -8)$?

49. What is the equation of the horizontal line passing through the point $(2, -6)$?

50. What is the equation of the horizontal line passing through the point $(-4, 7)$?

Find the equation of each graph.

51.

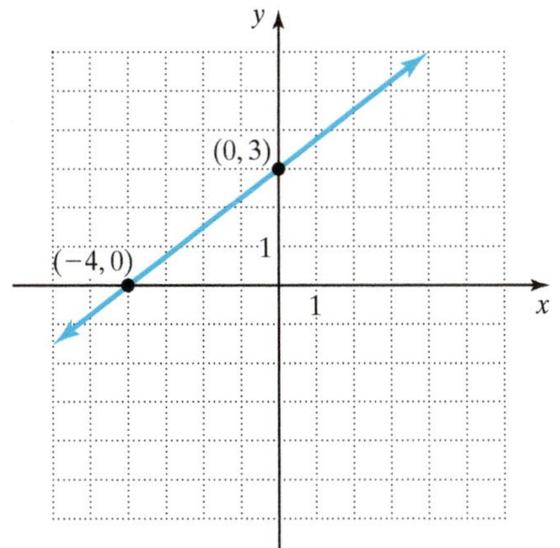

52.

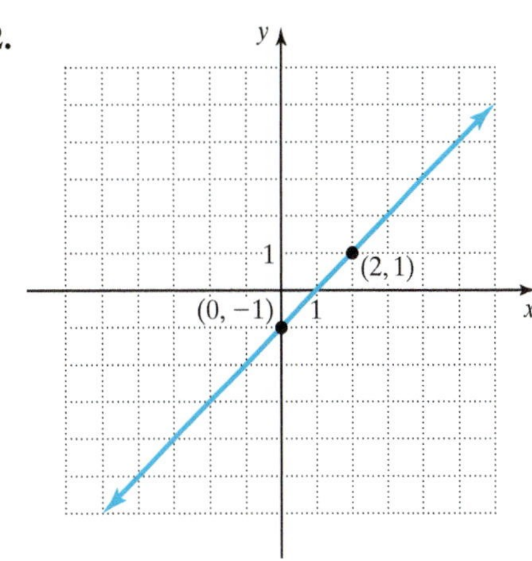

53.

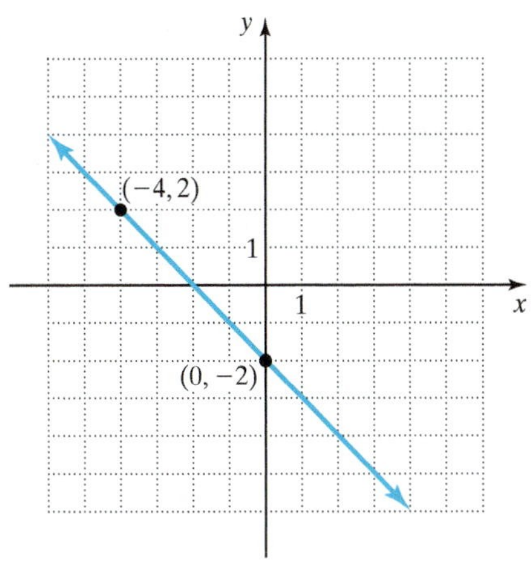

54.

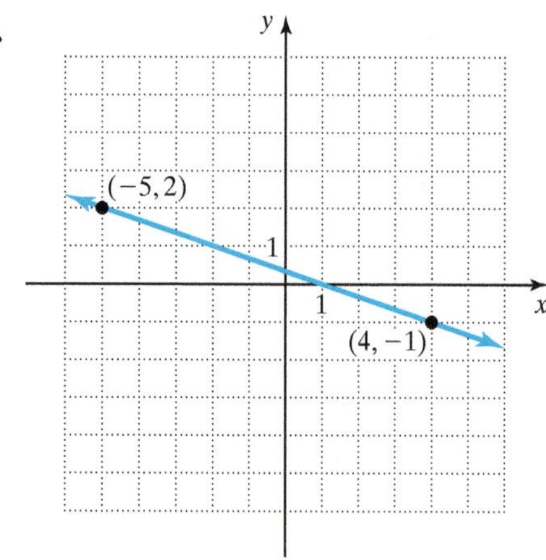

55.

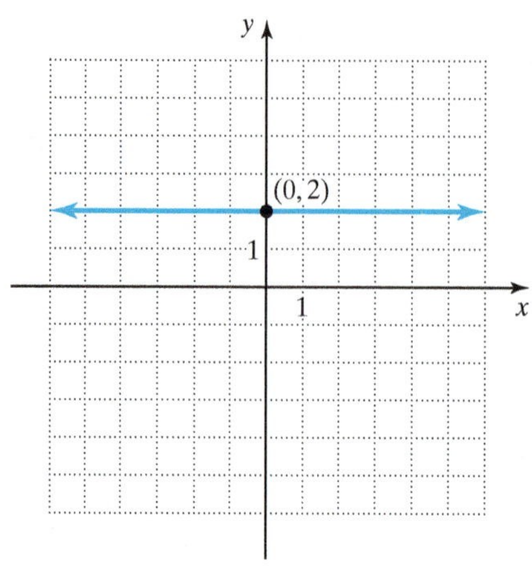

56.

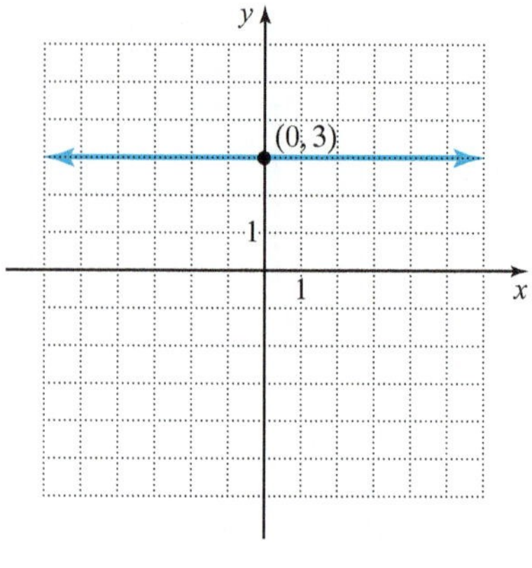

57.

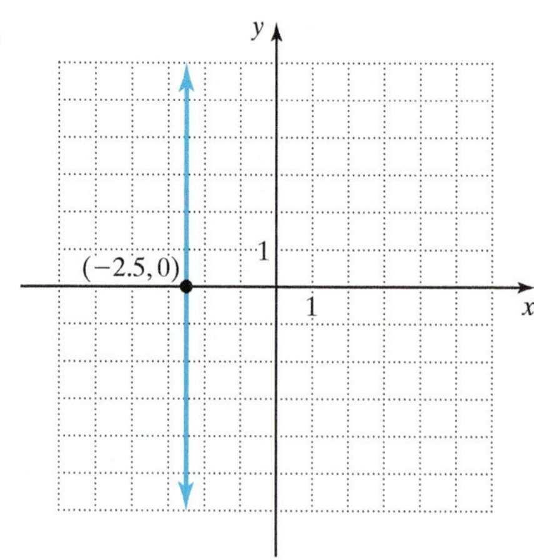

58.

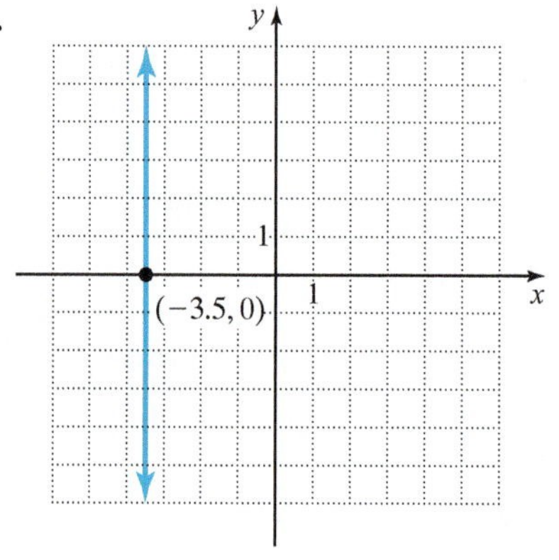

Mixed Practice

59. Complete the table.

Equation	Slope m	y-intercept $(0, b)$	Which Graph Type Best Describes the Line? ╱ ╲ — |	x-intercept
$y = -7x + 2$				
$y = 4x$				
$y = 2.5x + 10$				
$y = \frac{2}{3}x - \frac{1}{4}$				
$5x + 4y = 20$				
$x = 9$				
$y = -3.2$				

60. Find the slope and y-intercept of $y = -\frac{2}{5}x + 3$.

Write the following equations in slope-intercept form.

61. $4x - y = 5$

62. $3x - 6y = 8$

63. Which equation describes the graph?

a. $-4x + y = 5$

b. $4x + y = -5$

c. $-5x + y = -4$

d. $5x + y = -4$

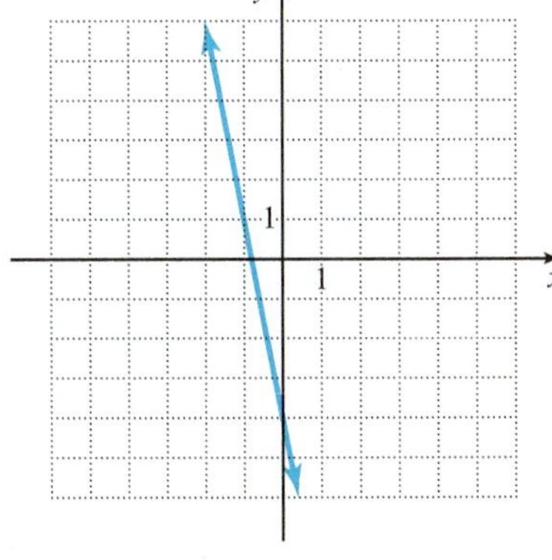

64. What is the equation of the line that is parallel to the graph $y = \frac{1}{2}x + 2$ and has x-intercept $(3, 0)$?

65. What is the equation of the line that is perpendicular to the graph of $y = -2x + 1$ that passes through the point $(4, 1)$?

66. What is the equation of the line that passes through the points $(2, -2)$ and $(-2, 1)$?

Graph the following equations using the slope and y-intercept.

67. $y = \frac{3}{5}x - 2$

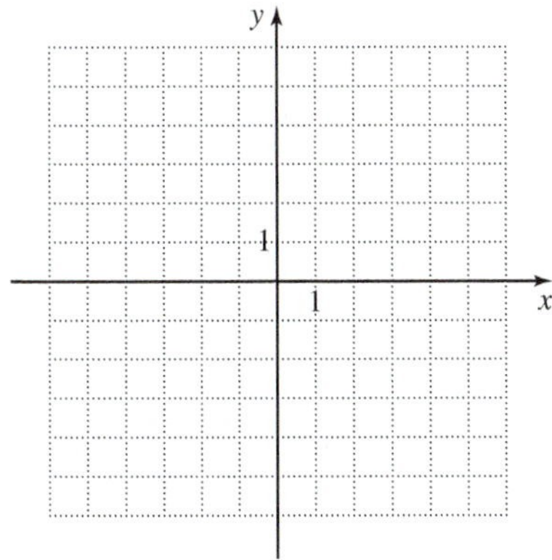

68. $3x + 2y = 4$

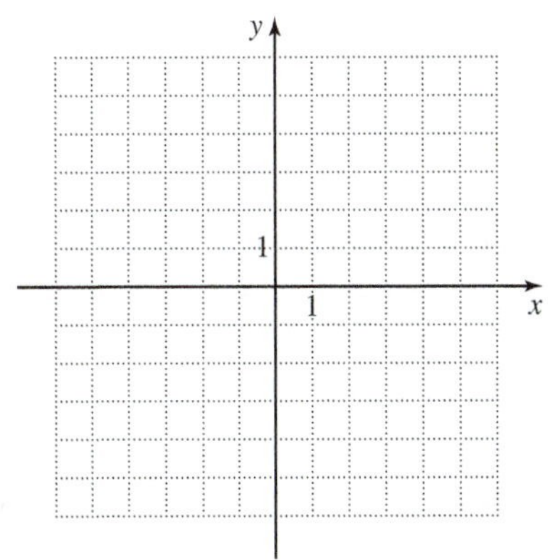

Find the equation of each graph.

69.

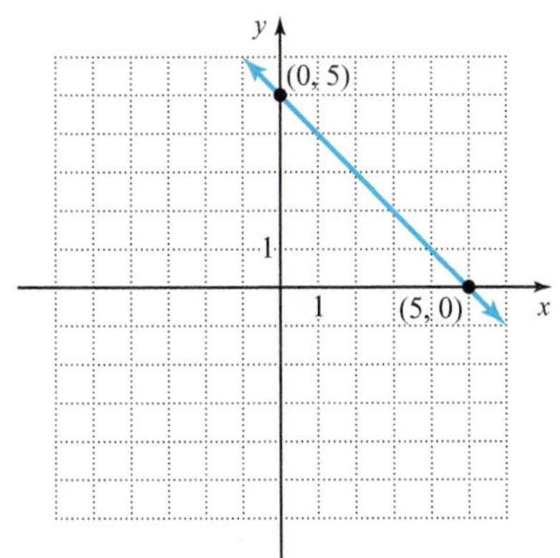

70.

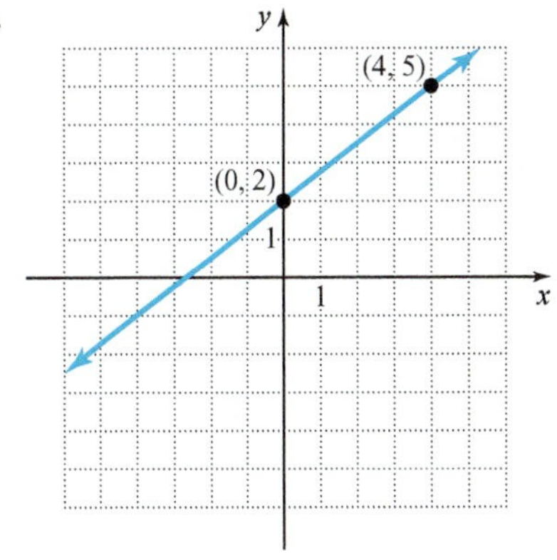

Applications

E *Solve.*

71. The following graph describes the relationship between Fahrenheit temperature F and Celsius temperature C.

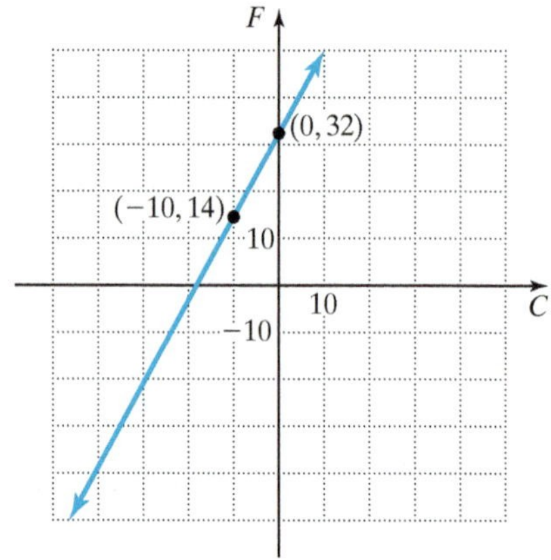

a. Find the slope of the line.

b. Find the equation of the line in slope-intercept form.

c. Water boils at 212°F. Use part (b) to find the Celsius temperature at which water boils.

72. The owner of a shop buys a piece of machinery for \$1500. The value V of the machinery declines by \$150 per year.

a. Write an equation for V after t years in slope-intercept form.

b. Graph the equation found in part (a).

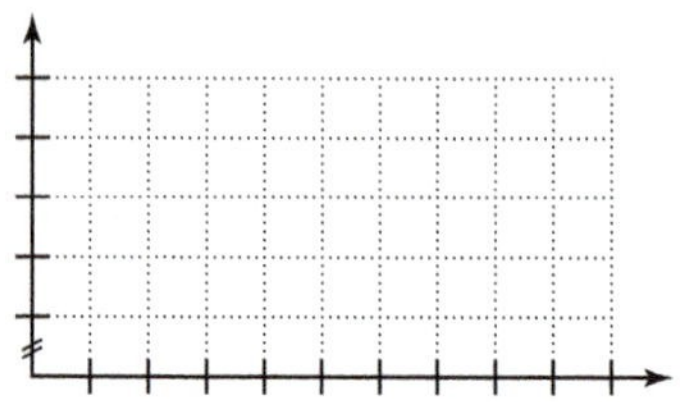

c. Explain the significance of the two intercepts in this context.

73. Each month, a utility company charges its residential customers a flat fee for electricity plus 6 cents per kilowatt-hour (kWh) consumed. Last month, a customer used 500 kWh of electricity, and his bill amounted to \$45.

a. Express as an equation in point-slope form the relationship between the customer's monthly bill y in cents and the number x of kilowatt-hours of electricity consumed.

b. Express the equation found in part (a) in slope-intercept form.

c. What does the y-intercept represent in this situation?

74. A condo unit has been appreciating in value at \$5000/yr. Three years after it was purchased, it was worth \$65,000.

a. Find an equation that expresses the value y of the condo in terms of the number x of years since it was purchased.

b. Graph the equation found in part (a).

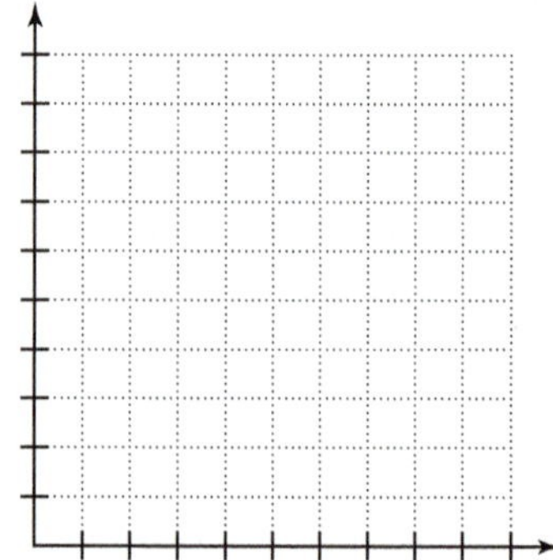

c. What is the significance of the y-intercept in this situation?

75. A salesperson earns a salary of \$1500 per month plus a commission of 3% of the total monthly sales.

a. Write a linear equation giving the salesperson's total monthly income I in terms of sales S.

b. Graph the equation found in part (a).

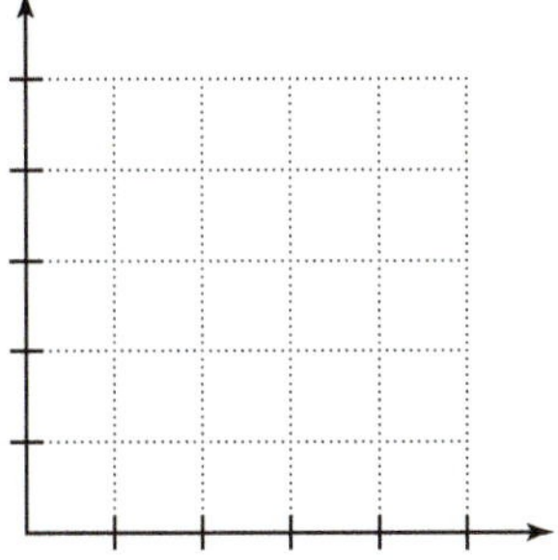

c. Find the salesperson's income on monthly sales of \$6200.

76. When the brakes on a train are applied, the speed of the train decreases by the same amount every second. Two seconds after applying the brakes, the train's speed is 88 mph. After 4 sec, its speed is 60 mph.

a. Write an equation expressing the speed s of the train in terms of the time t seconds after applying the brakes.

b. Graph the equation found in part (a).

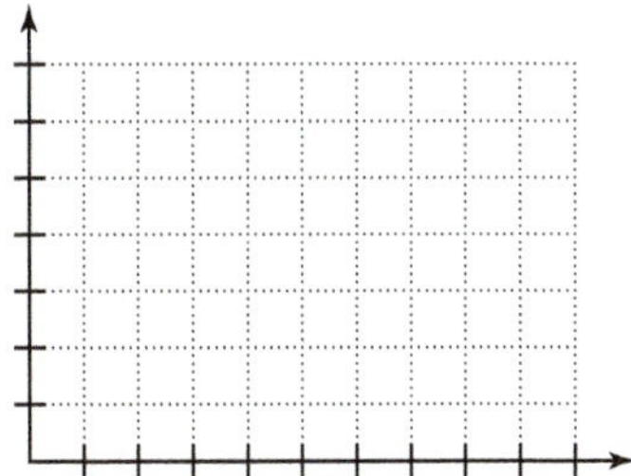

c. What was the speed of the train when the brakes were first applied?

77. Pressure under water increases with greater depth. The pressure P on an object and the depth d below sea level are related by a linear equation. The pressure at sea level is 1 atmosphere (atm), whereas 33 ft below sea level the pressure is 2 atm. Find the equation expressing P in terms of d.

78. The length of a heated object and the temperature of the object are related by a linear equation. A rod at 0° Celsius is 10 m long, and at 25° Celsius it is 10.1 m long. Write an equation for length in terms of temperature.

79. When a force is applied to a spring, its length changes. The length L and the force F are related by a linear equation. The spring shown here was initially 10 in. long when no force was applied. What is the equation that expresses length in terms of force?

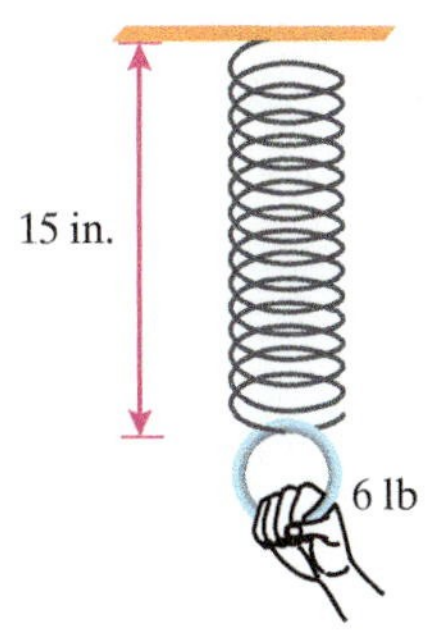

80. A company purchased a computer workstation for \$8000. After 3 yr, the estimated value of the workstation was \$4400. If the value V in dollars and the age a of the workstation are related by a linear equation, find an equation that expresses V in terms of a.

• Check your answers on page A-23.

MINDStretchers

Technology

1. Consider $2x - 7 = 0$, which is a linear equation in x.
   a. Solve the equation.
   b. On a graphing calculator or computer with graphing software, graph $y = 2x - 7$. Then, find the x-intercept of the line. Explain in a sentence or two how you can use this approach to solve the equation $2x - 7 = 0$.

Critical Thinking

2. Consider the equation $y = mx + b$. Explain under what circumstances its graph lies completely in Quadrants I and II.

Mathematical Reasoning

3. What kind of line corresponds to an equation that can be written in *general form* but in neither slope-intercept form nor point-slope form?

13.5 Linear Inequalities and Their Graphs

OBJECTIVES

A To identify the coordinates of points that satisfy a given linear inequality in two variables

B To graph linear inequalities in two variables

C To solve applied problems involving graphs of linear inequalities

In Section 12.5, we showed how to graph inequalities in one variable on a number line. In such inequalities, the solutions are real numbers. For instance, the graph of $x \le 2$ is shown below.

Now, we consider graphing inequalities in two variables on a coordinate plane. In this case, the solutions are ordered pairs of real numbers. For instance, if we want to graph the solutions to $x \le 2$ in the coordinate plane, we would shade the region to the left of the vertical line $x = 2$ because every ordered pair in this region has an x-coordinate that is less than 2. Every point on the line is also a solution to $x \le 2$ because each pair on this line has an x-coordinate equal to 2.

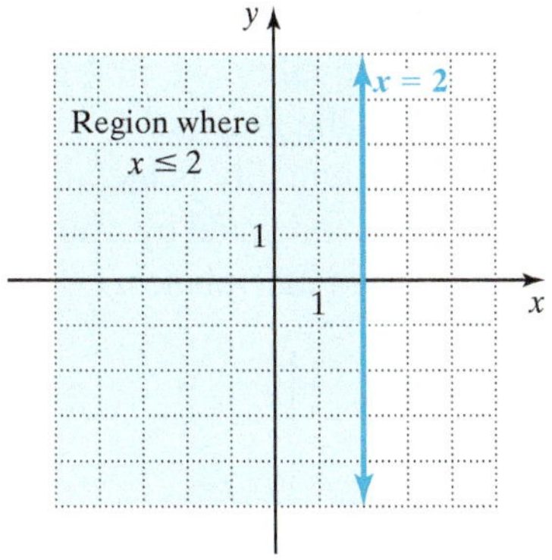

In this section, we focus our attention on graphing linear inequalities in *two* variables, such as $2x + 3y > 1$.

DEFINITION

A **linear inequality in two variables** is an inequality that can be written in the form $Ax + By < C$, where A, B, and C are real numbers and A and B are not both 0. The inequality symbol can be $<$, $>$, $\le$, or $\ge$.

As we will see, a linear inequality has as its graph a half-plane, which is a region of the coordinate plane bounded on one side by a straight line.

Consider some situations that give rise to such inequalities:

The number of men m and women w invited to a party is at most 20. → $m + w \le 20$

Your income i exceeds your expenses e by more than \$1000. → $i > e + 1000$

Let's look at what we mean by a *solution* of such inequalities.

DEFINITION

A **solution of an inequality in two variables** is an ordered pair of numbers that when substituted for the variables makes the inequality a true statement.

EXAMPLE 1

Is the ordered pair $(2, 5)$ a solution to the inequality $y \geq x + 1$?

Solution When we substitute 2 for x and 5 for y in the given inequality, it becomes:

$$y \geq x + 1$$
$$5 \overset{?}{\geq} 2 + 1$$
$$5 \geq 3 \quad \text{True}$$

Since $5 \geq 3$ is true, the ordered pair $(2, 5)$ is a solution to $y \geq x + 1$. Is the ordered pair $(6, 5)$ a solution to this inequality?

PRACTICE 1

Is $(1, 3)$ a solution to the inequality $y < x - 1$?

By the *graph* of an inequality in two variables, we mean the set of all points on the plane whose coordinates satisfy the inequality. To explore what such a graph looks like, let's consider $y \geq 2x$. To find the graph of this inequality, we first graph the corresponding equation $y = 2x$. Since the inequality symbol is $\geq$, the graph of $y \geq 2x$ includes points for which y is either greater than or equal to $2x$. Because equality is included, points on the line $y = 2x$ are part of the graph of the inequality, which is indicated by drawing the *boundary line* $y = 2x$ as a solid line. This line cuts the plane into two half-planes.

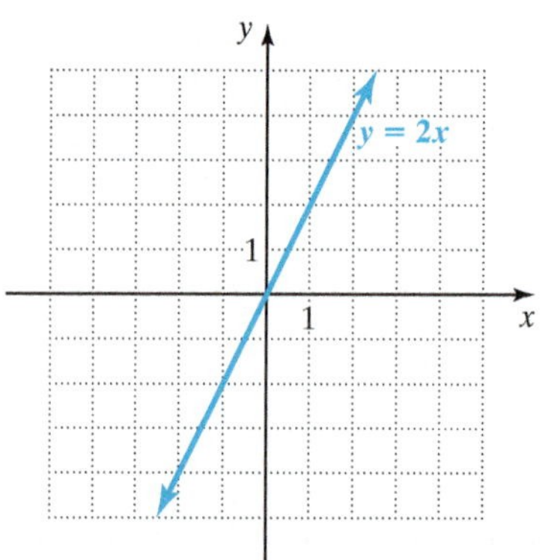

Next, we take an arbitrary point on either side of the boundary line—a *test point*. If the coordinates of the test point satisfy the inequality, then the desired graph contains the half-plane in which the test point lies. Otherwise, the desired graph contains the other half-plane.

Suppose that we take $(4, 0)$ as our test point.

$$y \geq 2x$$
$$0 \overset{?}{\geq} 2(4)$$
$$0 \geq 8 \quad \text{False}$$

Since the inequality does not hold for the point $(4, 0)$, the half-plane that we want is the region above the graph of $y = 2x$, so we shade this region. Therefore, the graph of $y \geq 2x$ is the boundary line and the shaded region.

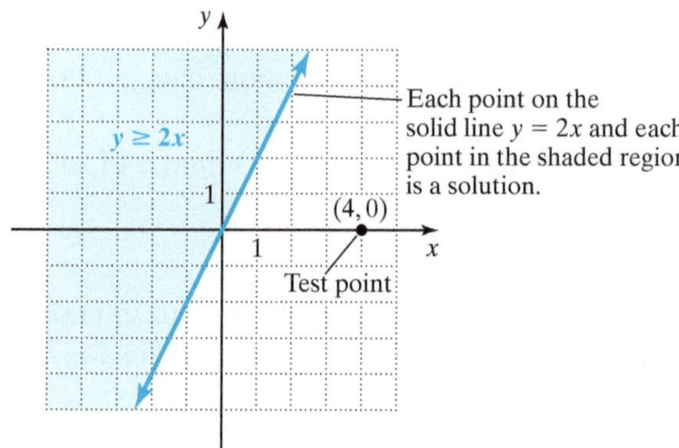

If the inequality had been $y > 2x$ instead of $y \geq 2x$, the boundary line would not have been part of the graph. We would have indicated the exclusion of the boundary with a broken line, as in the following diagram:

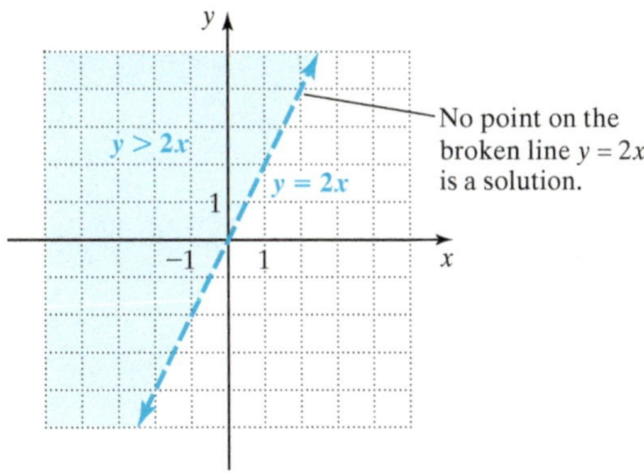

To Graph a Linear Inequality in Two Variables

- Graph the corresponding linear equation. For an inequality with the symbol $\leq$ or $\geq$, draw a solid line; for an inequality with the symbol $<$ or $>$, draw a broken line. This line is the boundary between two half-planes.
- Choose a test point in either half-plane and substitute the coordinates of this point in the inequality. If the resulting inequality is true, then the graph of the inequality is the half-plane containing the test point. If it is not true, then the other half-plane is the graph. A solid line is part of the graph, and a broken line is not.

EXAMPLE 2

Find the graph of $y - 2x < 4$.

Solution First, we graph the equation $y - 2x = 4$. Solving for y gives $y = 2x + 4$. We draw a broken line since the original inequality symbol is $<$. Then, we choose in either half-plane a test point, say $(0, 0)$.

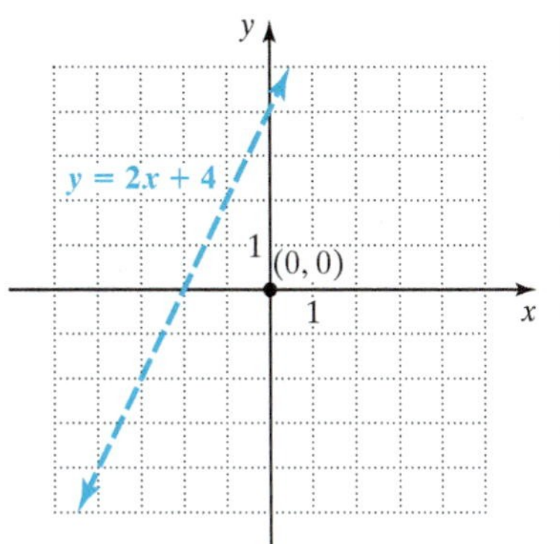

We substitute $x = 0$ and $y = 0$ into the inequality:

$$y - 2x < 4$$
$$0 - 2 \cdot 0 \stackrel{?}{<} 4$$
$$0 < 4 \quad \text{True}$$

Since the inequality is true, the graph of the inequality is the half-plane containing the test point. So the half-plane below the line is

PRACTICE 2

Graph the inequality $y + 3x \geq 6$.

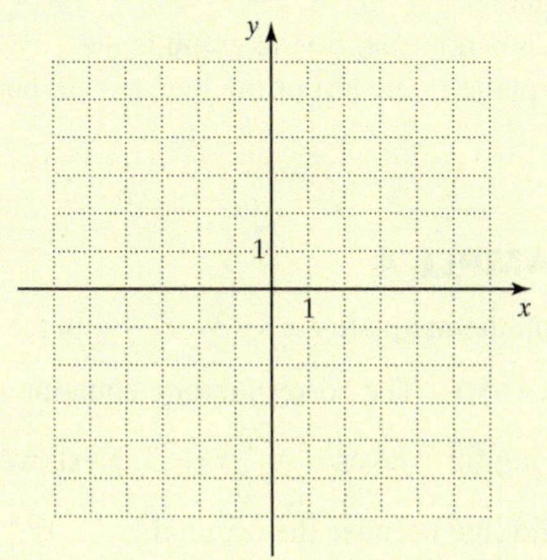

EXAMPLE 2 (continued)

our graph. Note that the graph does not include the boundary line since the original inequality symbol is $<$.

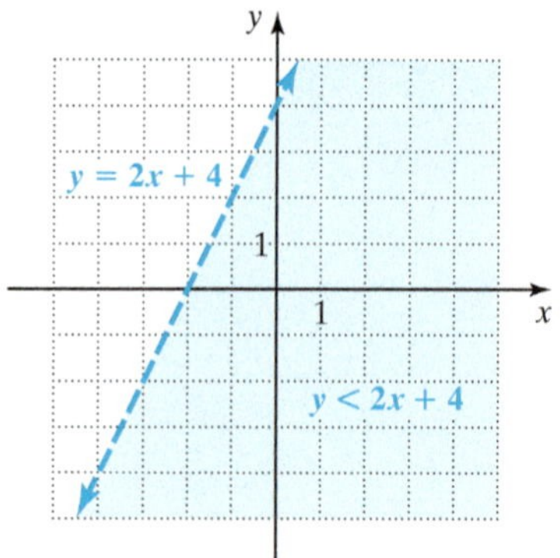

We can test any point on the graph. Why is $(0, 0)$ a good choice?

EXAMPLE 3

Graph $x < 0$.

Solution The boundary line $x = 0$ is the y-axis. It is drawn as a broken line, since the original inequality symbol is $<$. We need to select a test point on either side of the boundary line. Let's take the point $(1, 0)$ as the test point, which is in the half-plane to the right of the line $x = 0$. Substituting into the inequality $x < 0$, we get $1 < 0$, which is not true. So the graph is the half-plane to the left of the line $x = 0$, but not including this line.

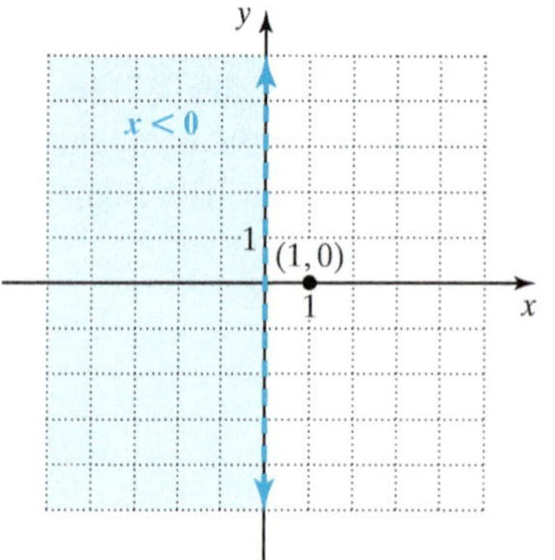

PRACTICE 3

Find the graph of $y \geq -5$.

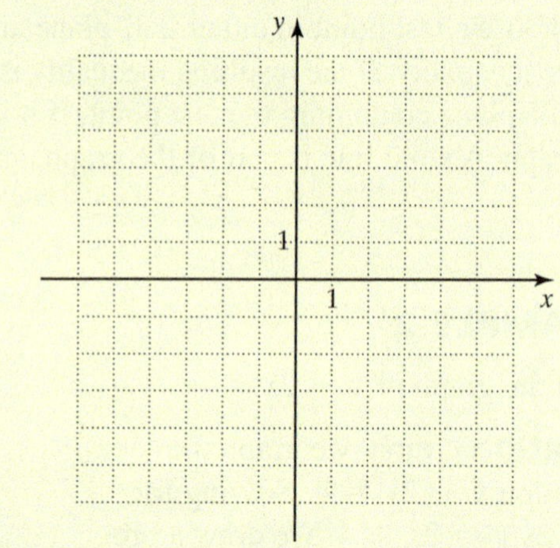

EXAMPLE 4

Graph the inequality $x - 3y \leq -9$ in the first quadrant.

Solution The corresponding equation is $x - 3y = -9$. Solving for y gives $y = \frac{1}{3}x + 3$. Next, we graph this line, drawing a solid line because the original inequality symbol is $\leq$.

Taking $(0, 0)$ as the test point, we check whether $0 - 3(0) \leq -9$. Since this inequality does not hold, the test point is not part of the graph. So the graph in Quadrant I is the region in the quadrant above the line $y = \frac{1}{3}x + 3$, and including it.

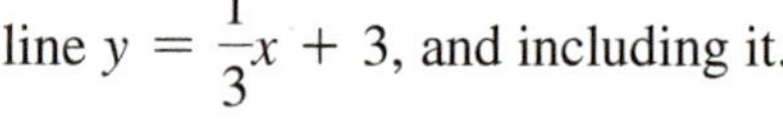

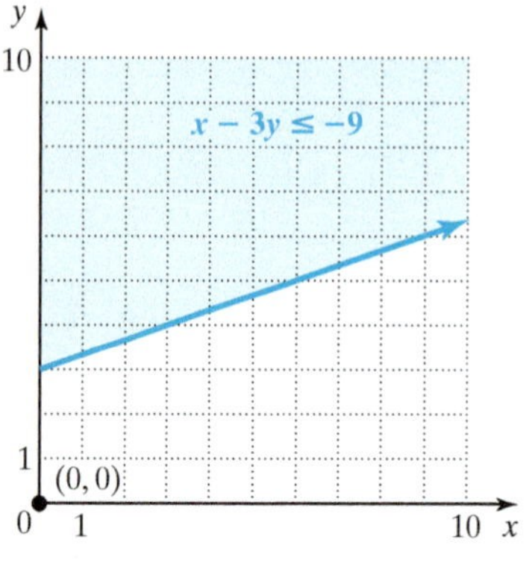

PRACTICE 4

Find the graph of $x - 2y > -6$ in Quadrant I.

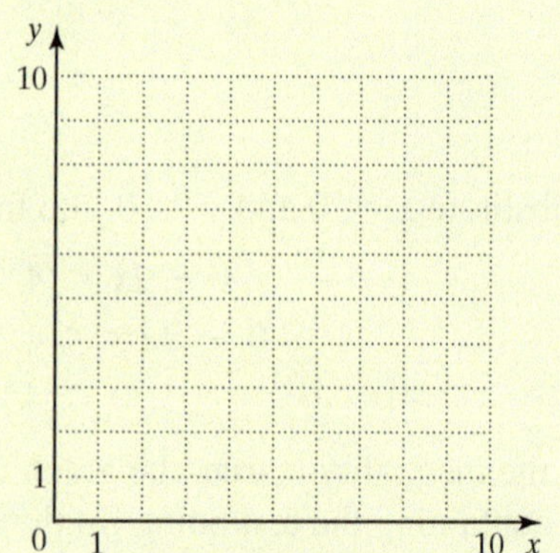

EXAMPLE 5

An online perfumery makes \$50 on each bottle of perfume and \$20 on each bottle of cologne. To earn a profit, the total amount of money made daily must exceed the perfumery's daily overhead of \$1000.

a. Express as an inequality: The amount of money made on selling p bottles of perfume and c bottles of cologne is greater than the overhead.

b. Graph this inequality.

c. Use this graph to decide whether selling 10 bottles of perfume and 12 bottles of cologne results in a profit.

Solution

a. The factory makes \$50 on each bottle of perfume sold. Therefore, the factory makes $50p$ dollars by selling p bottles of perfume. Similarly, the factory makes $20c$ dollars by selling c bottles of cologne. Since the total amount of money made must be greater than the overhead of \$1000, the inequality is $50p + 20c > 1000$. If we divide both sides of this inequality by the common factor 10, we get $5p + 2c > 100$.

b. To graph the inequality $5p + 2c > 100$, we first graph the corresponding equation, $5p + 2c = 100$. We restrict our attention to the portion of the graph in Quadrant I, since the variables can only assume nonnegative values. The boundary line is not included since the symbol in the linear inequality is $>$. Using $(0, 0)$ as a test point, we see that the inequality $5 \cdot 0 + 2 \cdot 0 > 100$ is false. So the graph of $5p + 2c > 100$ is the region in Quadrant I above the boundary line, but not including the line.

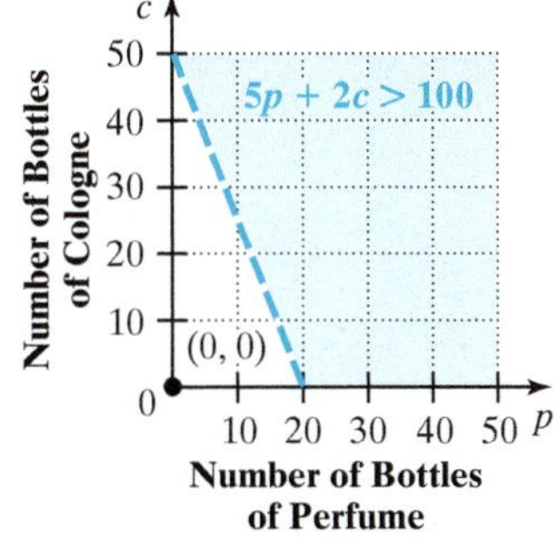

c. To decide whether there was a profit when selling 10 bottles of perfume and 12 bottles of cologne, we plot the point $(10, 12)$. Since the point lies *outside* the graph of our inequality, there was no profit.

PRACTICE 5

Each day, a refinery can produce both diesel fuel and gasoline, with a total maximum output of 3000 gal.

a. Express this relationship as an inequality, where d represents the amount of diesel fuel produced and g the amount of gasoline produced, both in gallons.

b. Graph this inequality.

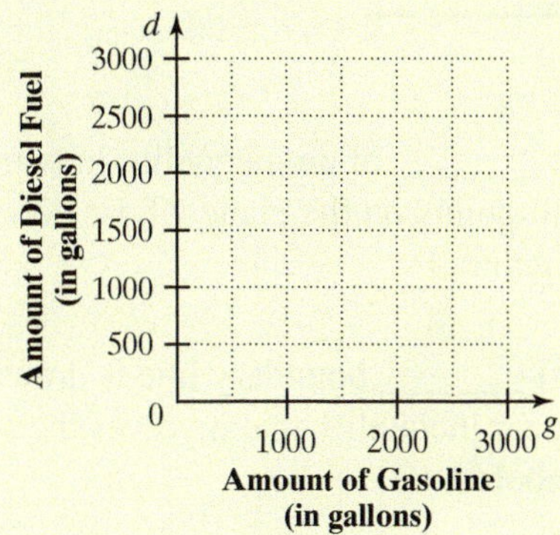

c. Explain the significance of the intercepts of this graph.

13.5 Exercises

FOR EXTRA HELP MyMathLab MathXL PRACTICE WATCH READ REVIEW

Mathematically Speaking

Fill in each blank with the most appropriate term or phrase from the given list.

broken	solution	solid
is	graph	is not
half-plane	ray	

1. The graph of a linear inequality in two variables is a(n) ____________.

2. A(n) _______ of an inequality in two variables is an ordered pair of numbers that when substituted for the variables makes the inequality a true statement.

3. The _______ of an inequality in two variables is the set of all points on the plane whose coordinates satisfy the inequality.

4. A(n) _______ boundary line is drawn when graphing a linear inequality that involves the symbol $\leq$ or the symbol $\geq$.

5. A(n) _______ boundary line is drawn when graphing a linear inequality that involves the symbol $<$ or the symbol $>$.

6. If the boundary line is broken, it _______ part of the graph.

A *Decide if the given ordered pair is a solution to the inequality.*

7. $y < 3x \quad (0, 0)$

8. $y > -5x \quad (-1, 4)$

9. $y \geq 2x - 1 \quad (-\frac{1}{2}, -2)$

10. $y \leq -\frac{2}{3}x + 5 \quad (6, 1)$

11. $2x - 3y > 10 \quad (10, 8)$

12. $5x + 3y \geq 12 \quad (0, -2)$

B *Each solid or broken line is the graph of $y = x$. Shade in the graph of the given inequality.*

13. $y > x$

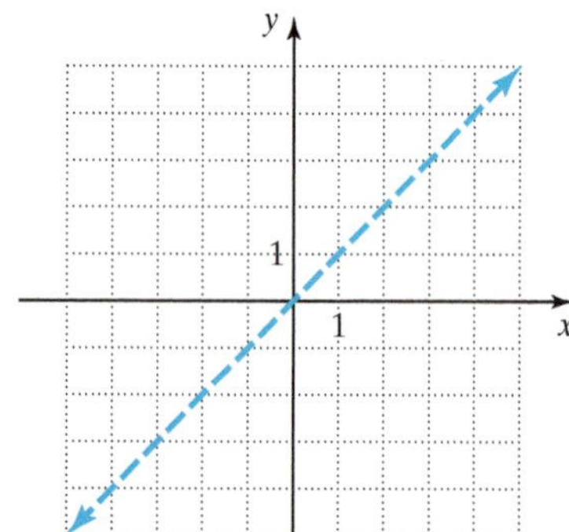

14. $y \geq x$

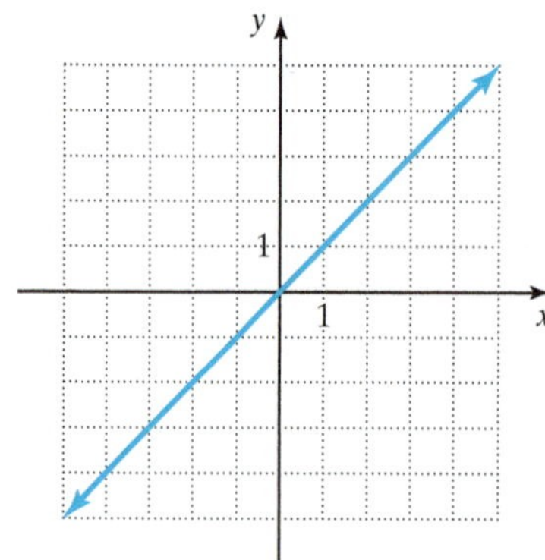

15. $x \leq y$

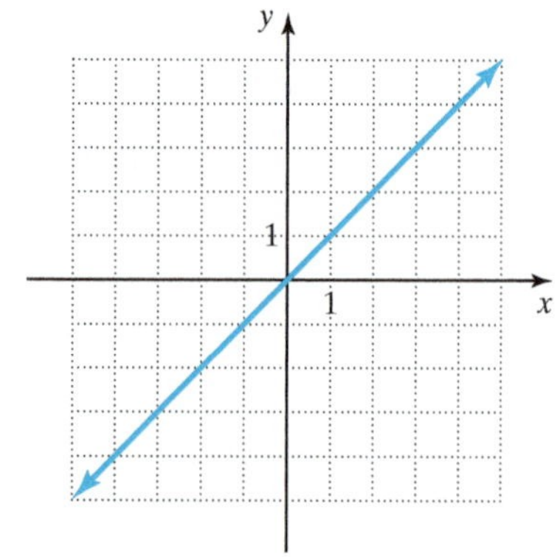

16. $x < y$

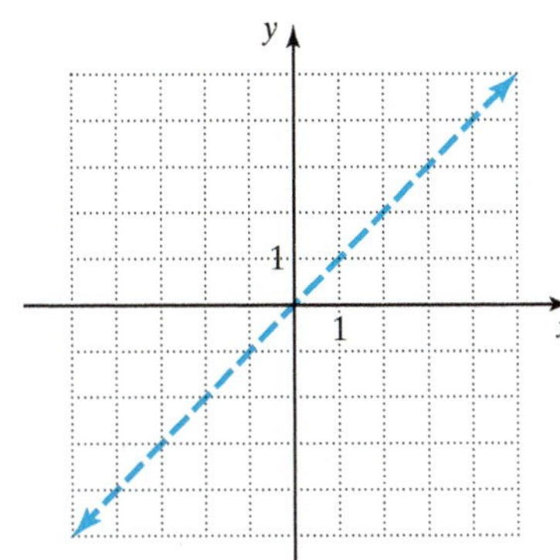

17. $y < x$

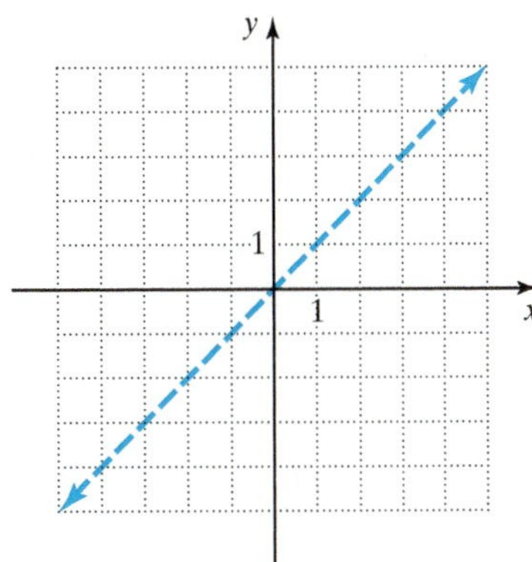

18. $y \leq x$

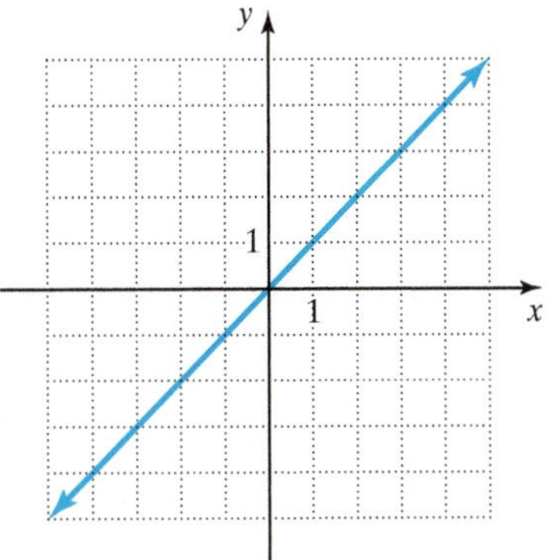

Match each inequality to its graph.

19. $y < \frac{1}{4}x - 1$

20. $y > -2x + 3$

21. $x - 4y \leq 4$

22. $2x + y \geq 3$

a.

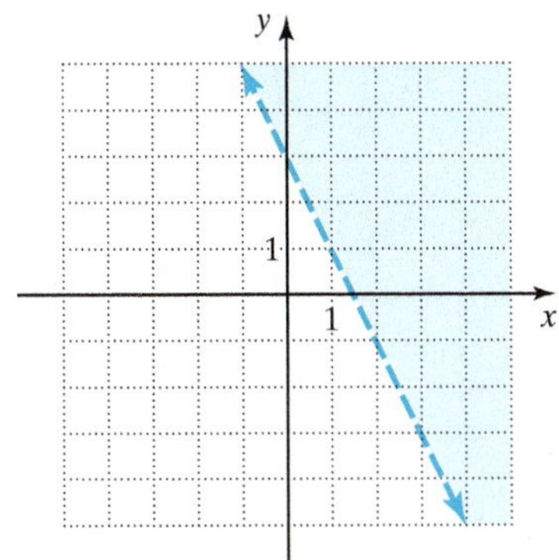

b.

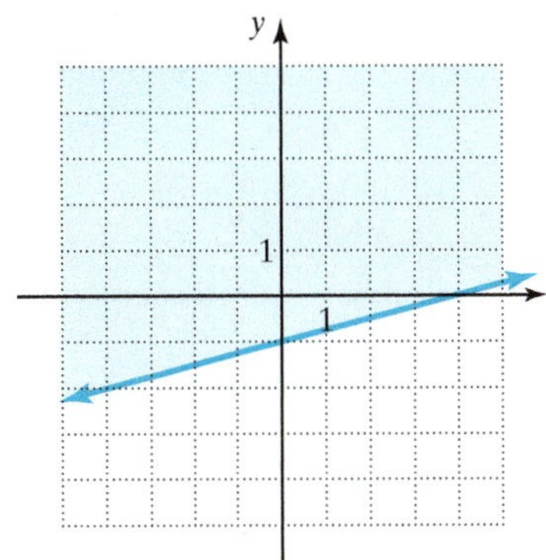

c.

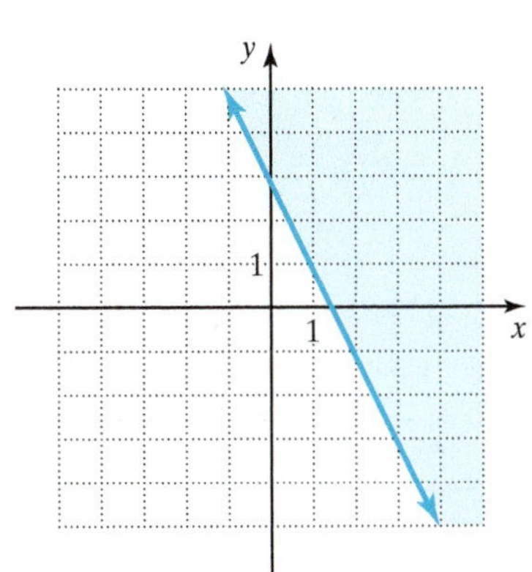

d.

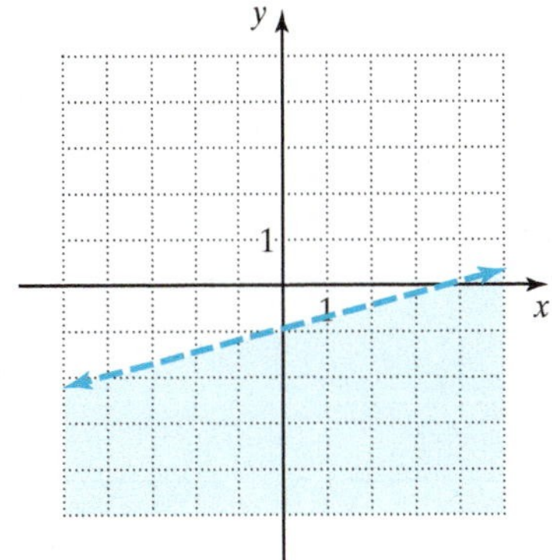

Graph the linear inequality.

23. $x > -5$

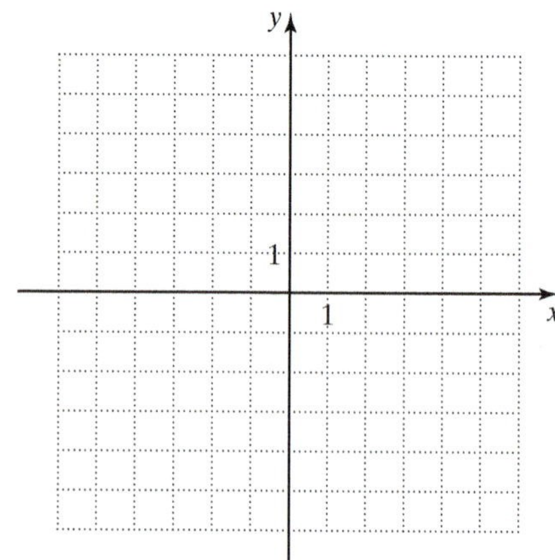

24. $x > 3$

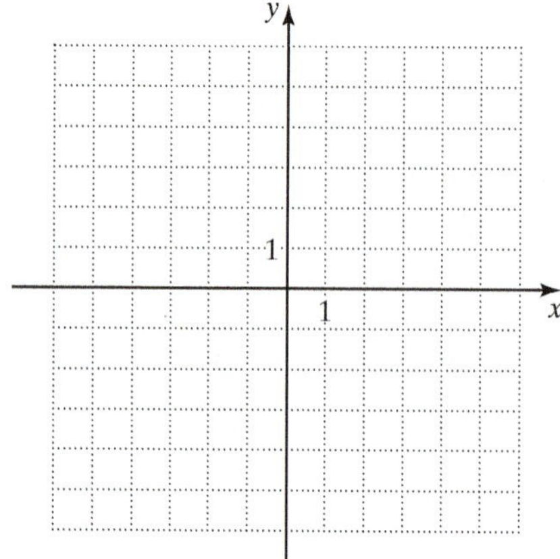

25. $y < 0$

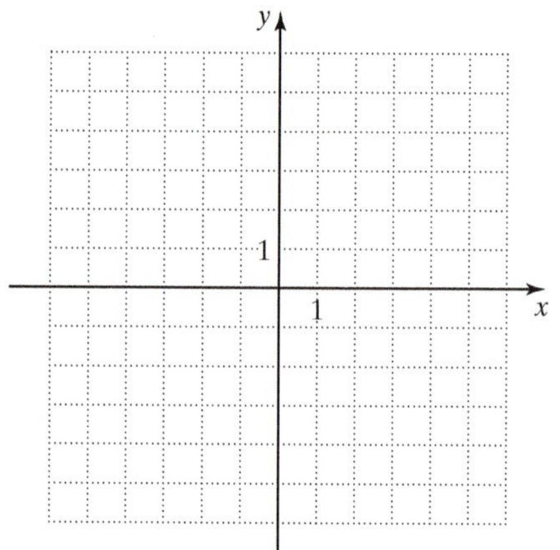

26. $y < 4$

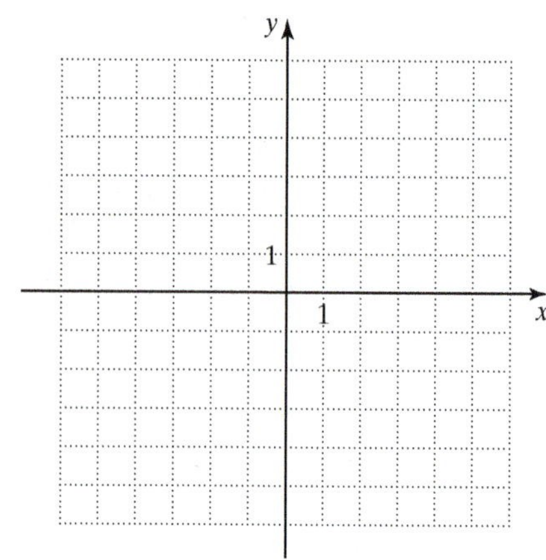

27. $y \leq 3x$

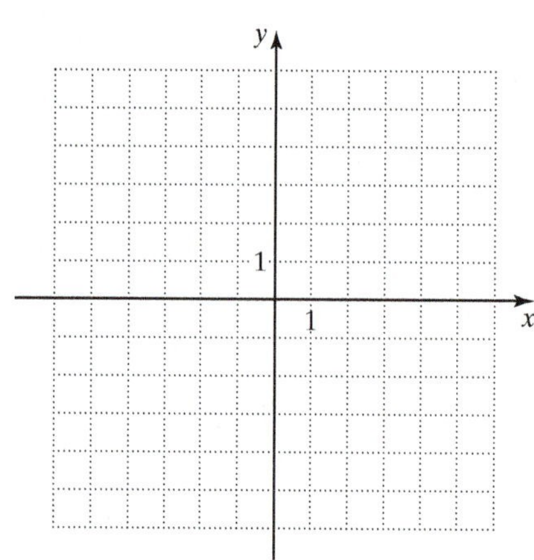

28. $y \leq -x$

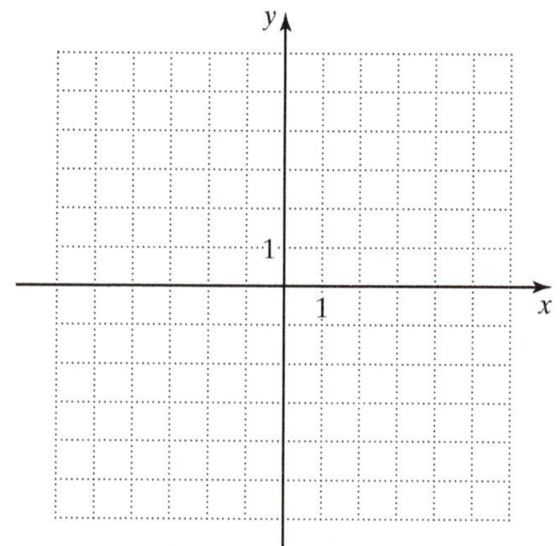

29. $y \geq -2x$

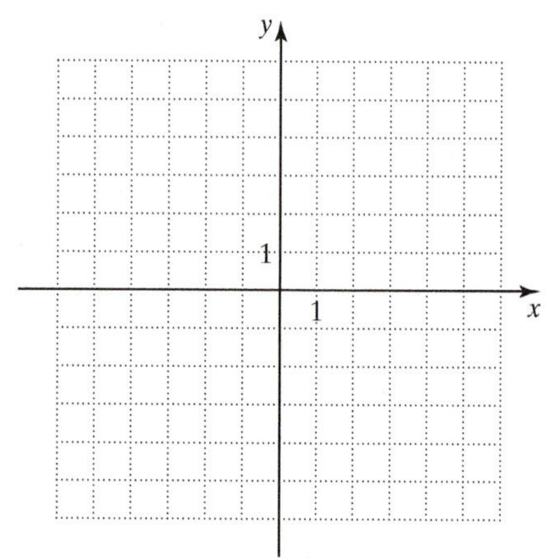

30. $y \geq 4x$

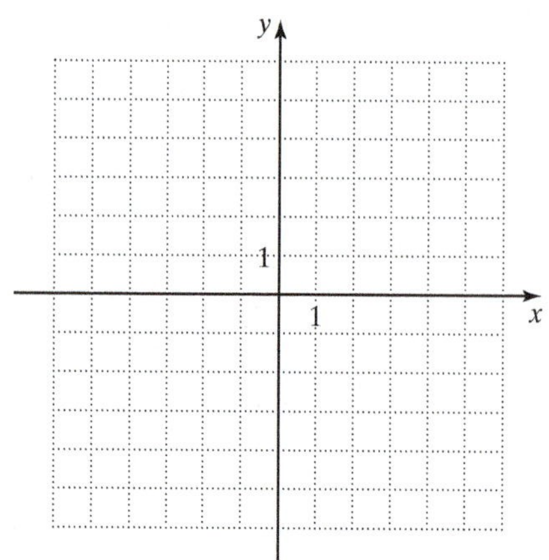

31. $y \leq \frac{1}{2}x$

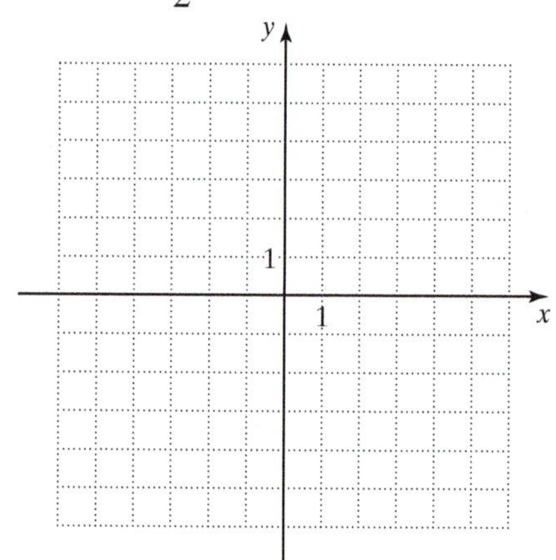

32. $y > -\frac{2}{3}x$

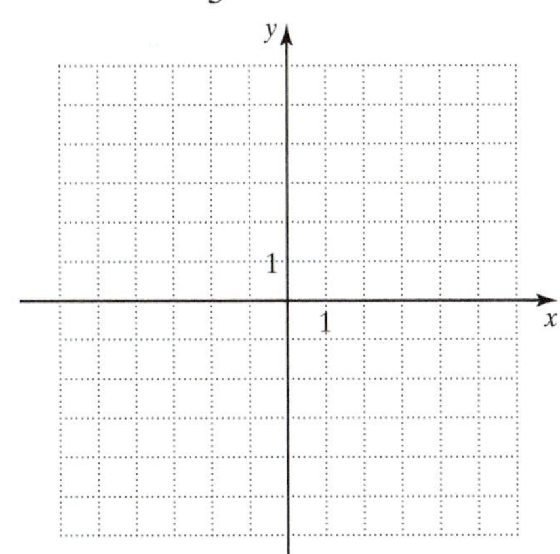

33. $y > 3x + 5$

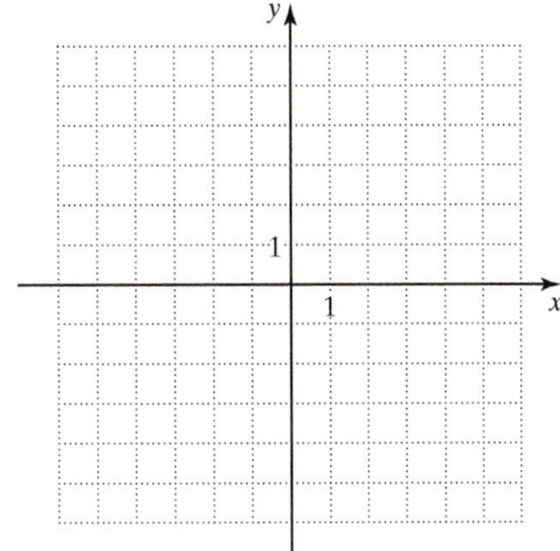

34. $y \geq -x - 1$

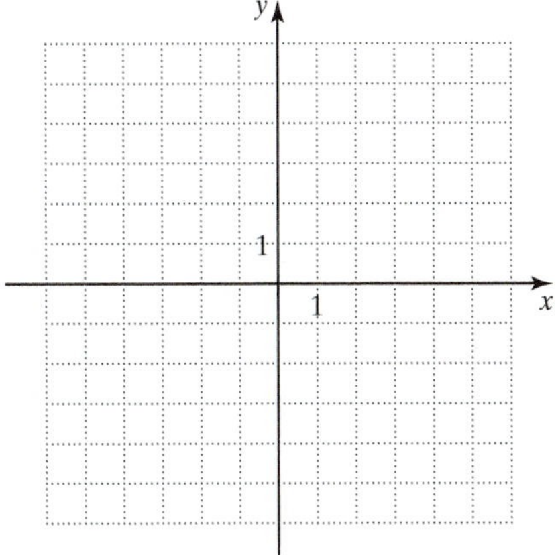

35. $5y - x > 10$

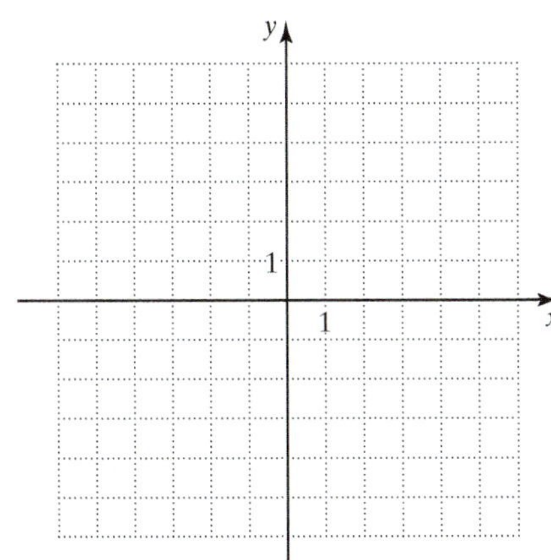

36. $4y + x < -12$

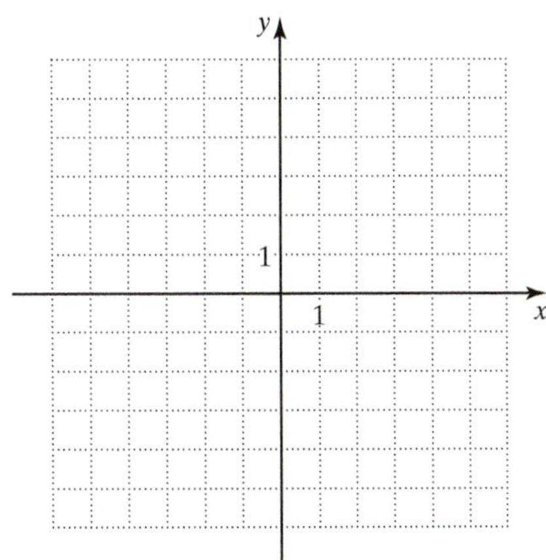

37. $2x - 3y \geq 3$

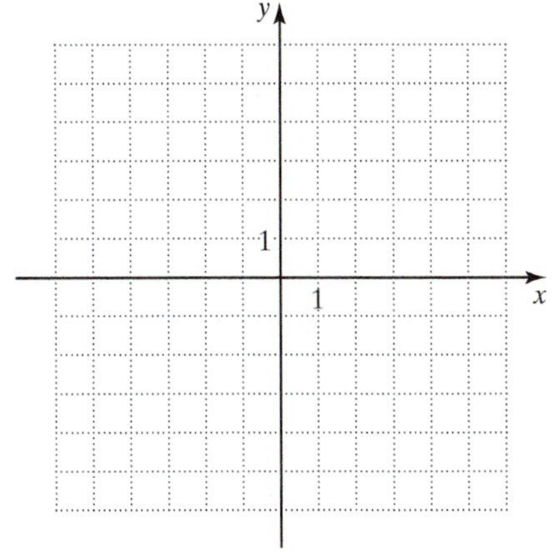

38. $3x - 2y \leq 4$

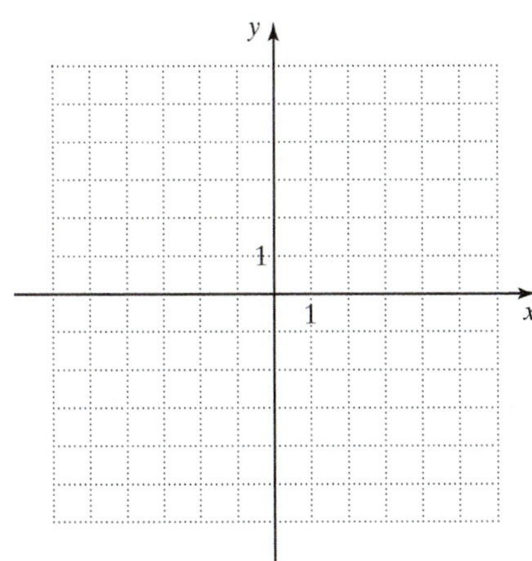

39. $4x - 5y \leq 10$

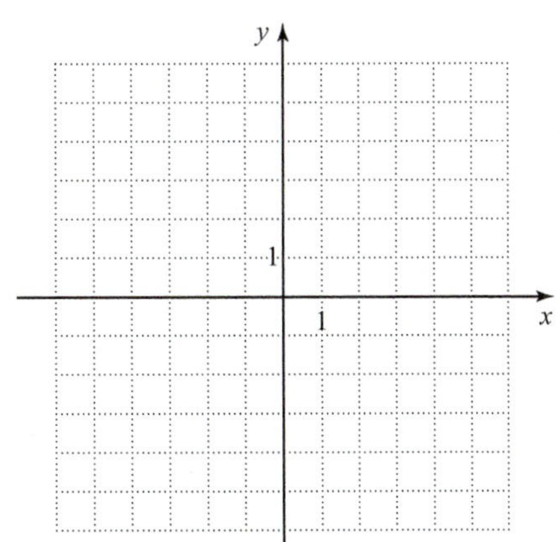

40. $-3x + 2y < 1$

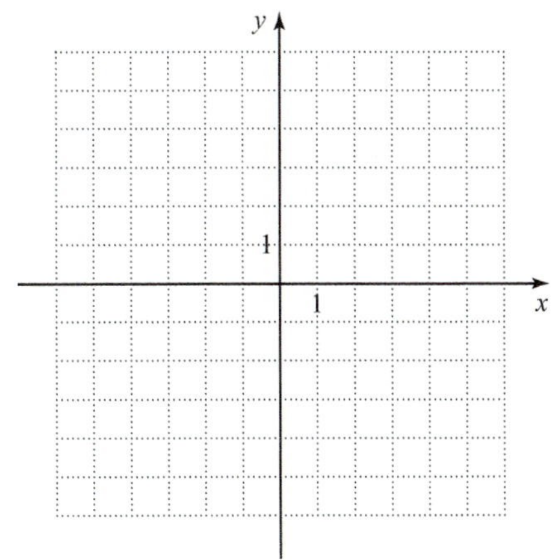

Mixed Practice

Decide if the given ordered pair is a solution to the inequality.

41. $y > -\frac{1}{2}x + 2 \quad (4, 0)$

42. $x - 2y \leq -2 \quad (8, 6)$

43. Which equation describes the graph?

a. $y + 4x \geq 2$

b. $y - 4x \geq 2$

c. $y > 4x + 2$

d. $y < 4x + 2$

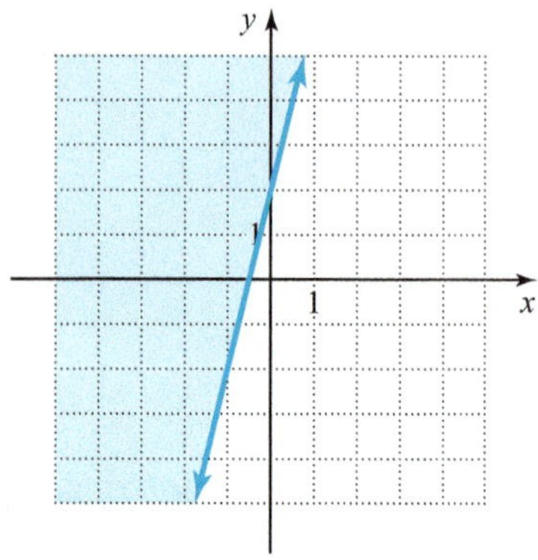

44. Shade in the graph of $x > y$.

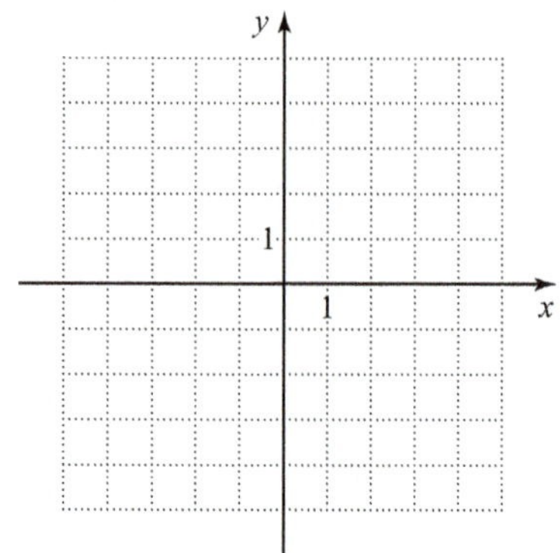

Graph the linear inequality.

45. $y < -3$

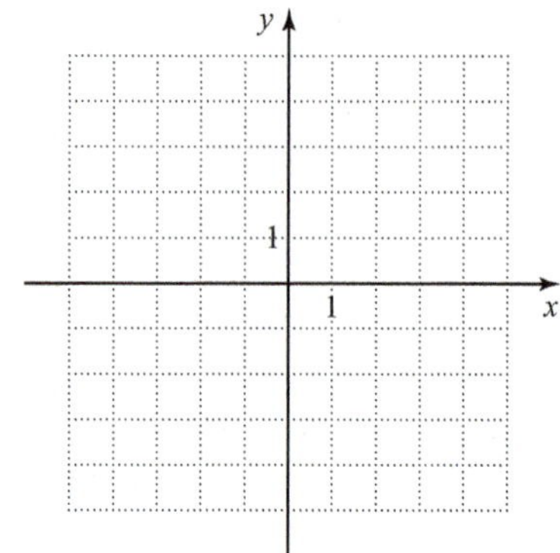

46. $y \leq \frac{2}{3}x$

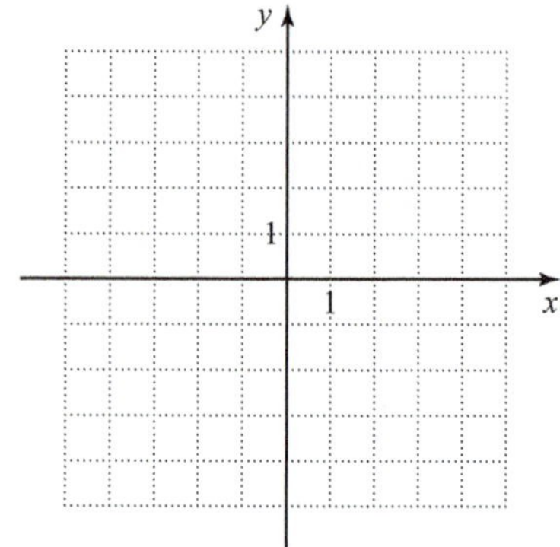

47. $y > -2x - 2$

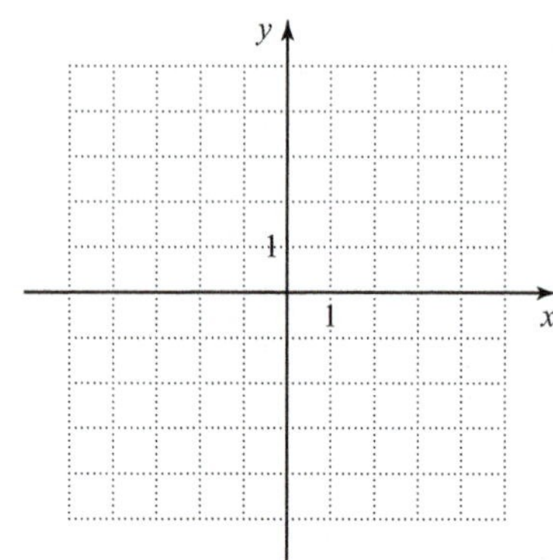

48. $2x - 3y < 6$

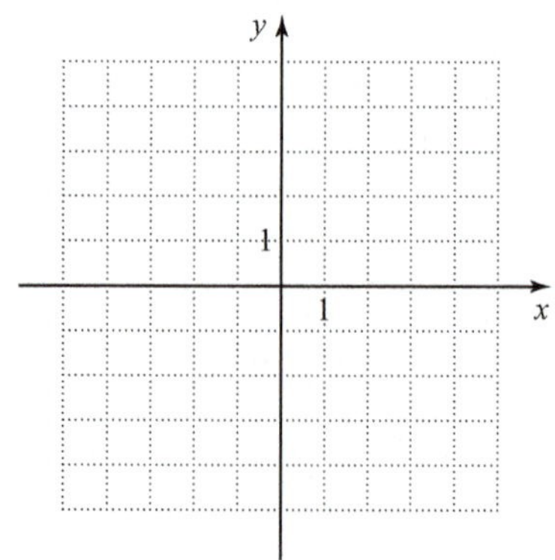

Applications

C *Solve.*

49. According to a guideline, a family's housing expenses h should be less than $\frac{1}{4}$ of the family's combined income i.

a. Express this guideline as an inequality.

b. Graph this inequality.

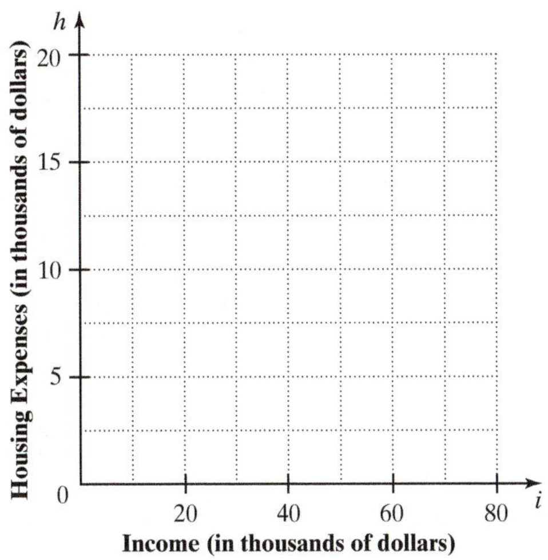

c. Choose a point on the graph. For this point, explain why the guideline holds.

50. To purchase an apartment in a particular building, a buyer is allowed to take out a mortgage m that is at most 75% of the apartment's selling price s.

a. Express this relationship as an inequality.

b. Graph this inequality.

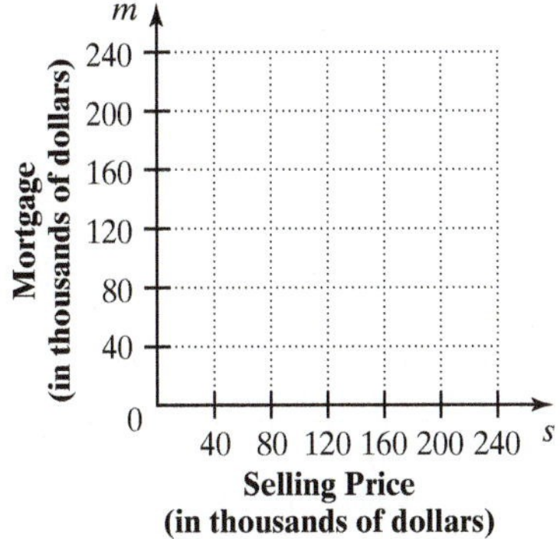

c. Plot the point $(100, 90)$. For this point, explain in terms of the mortgage policy of the building whether the buyer will be able to buy the apartment.

51. A printing company ships x copies of a college's student handbook to the uptown campus and y copies to the downtown campus. The company must ship a total of at least 200 copies to these two locations.

a. Express this relationship as an inequality.

b. Graph this inequality.

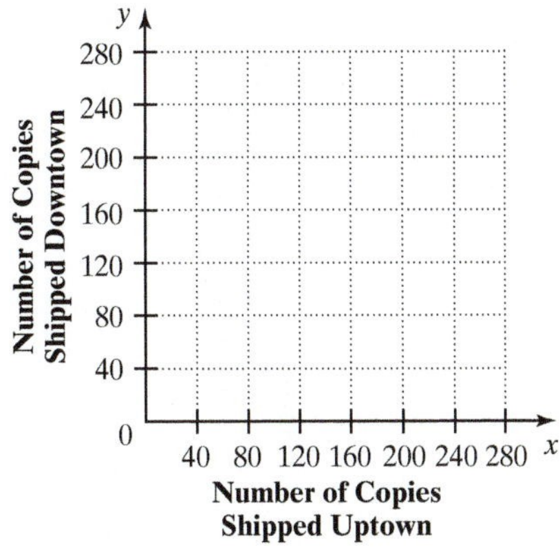

c. Give the coordinates of a point that satisfies the inequality. Check that the coordinates satisfy the company's shipping requirement.

52. In September 2010, an investor had a maximum of $1500 to purchase stocks. She wanted to buy x shares of General Electric (GE) and y shares of Ford Motor Company (F). General Electric was selling at approximately $16 per share and Ford at approximately $12 per share. (*Source:* NYSE.com)

a. Write as an inequality the possible amount of each stock that she could have bought.

b. Solve this problem graphically.

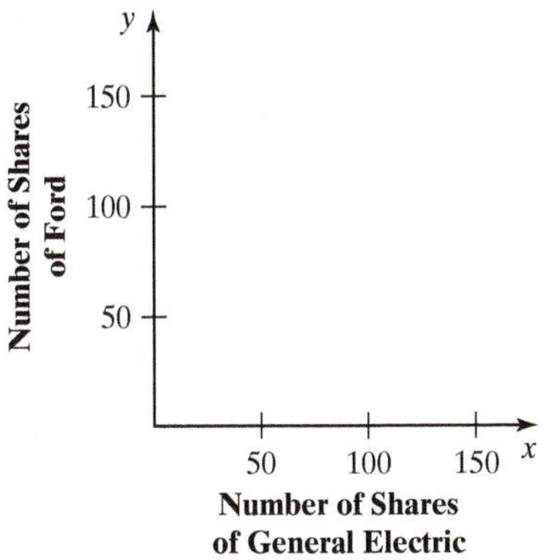

53. A local gourmet coffee shop sells small and large gift baskets. A small gift basket sells for \$30 and a large gift basket sells for \$75. The coffee shop would like a revenue of at least \$1500 per month on the sale of gift baskets.

a. Write an inequality, where x is the number of small gift baskets sold in a month and y is the number of large gift baskets sold, to represent this situation.

b. Graph this inequality.

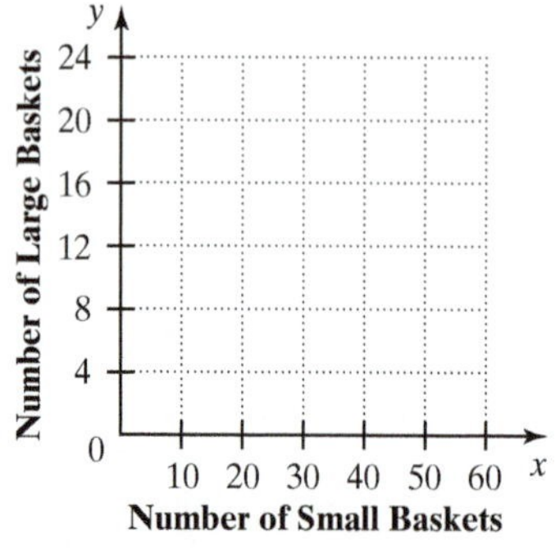

c. Use the graph to determine if selling 20 small gift baskets and 20 large gift baskets will generate the desired revenue.

54. In moving into a new apartment, a young couple needs to borrow money for both furniture and a car. The loan for furniture has a 10% annual interest rate, whereas the car loan has a 5% annual interest rate. The couple can afford at most \$2000 in interest payments for the year.

a. Express the given information as an inequality, representing the car loan amount by c and the furniture loan amount by f.

b. Graph this inequality.

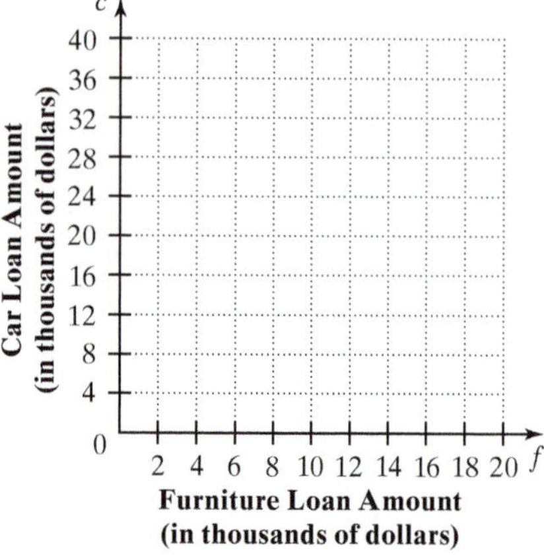

c. What is the maximum car loan amount that the couple can afford?

d. If the couple borrows \$20,000 for the car loan, what is the most that they can afford to borrow for furniture?

55. A plane is carrying bottled water and cases of medicine to victims of a flood. Each bottle of water weighs 10 lb, and each case of medicine weighs 15 lb. The plane can carry a maximum of 50,000 lb of cargo.

a. Express this weight limitation of the cargo as an inequality in terms of the number of bottles of water w and the number of cases of medicine m in the plane.

b. Graph this inequality.

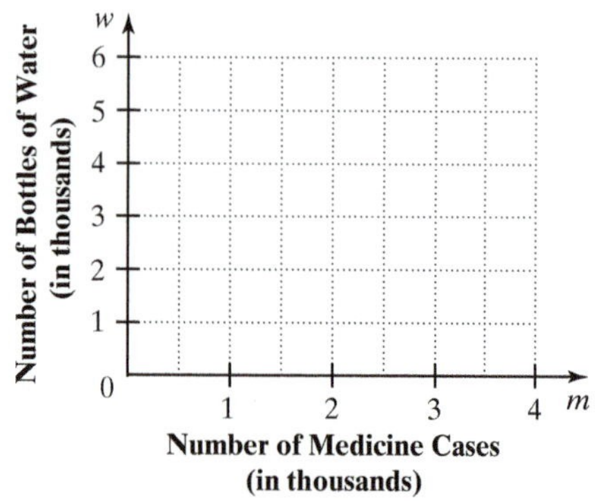

c. Identify several quantities of water and medicine that the plane can carry.

56. An elevator has a maximum capacity of 1600 lb. Suppose that the average weight of an adult is 160 lb and the average weight of a child is 40 lb.

a. Write an inequality that relates the number of adults a and the number of children c who can ride an elevator without overloading it.

b. Graph this inequality.

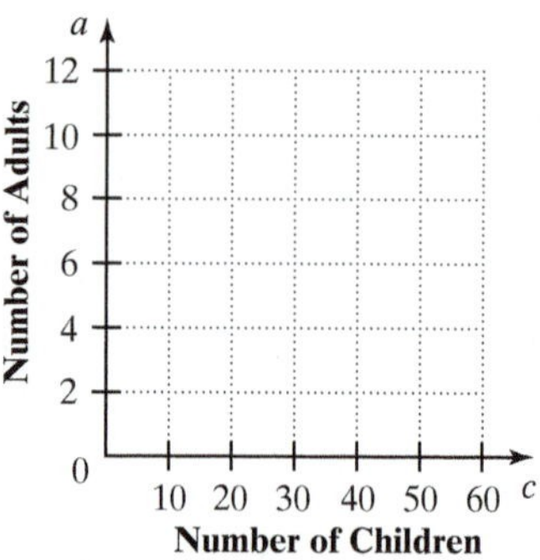

c. What is an example of a group of people who will overload the elevator?

57. A student has two part-time jobs. One pays \$8/hr, and the other \$10/hr. Between the two jobs, the student needs to earn at least \$200/wk.

a. Write an inequality that shows the number of hours that the student can work at each job.

b. Graph this inequality.

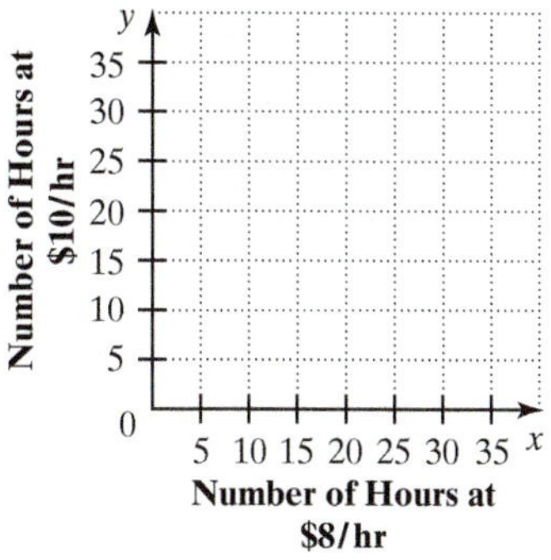

c. Give some examples of the number of hours that the student can work at each job.

58. Scientists who study weather have developed linear models that relate a region's weather conditions to the kind of vegetation that grows in the region. One such model predicts desert conditions if $3t - 35p > 140$, where t represents the average annual temperature (in degrees Fahrenheit) and p the annual precipitation (in inches).

a. Graph this relationship on the coordinate plane.

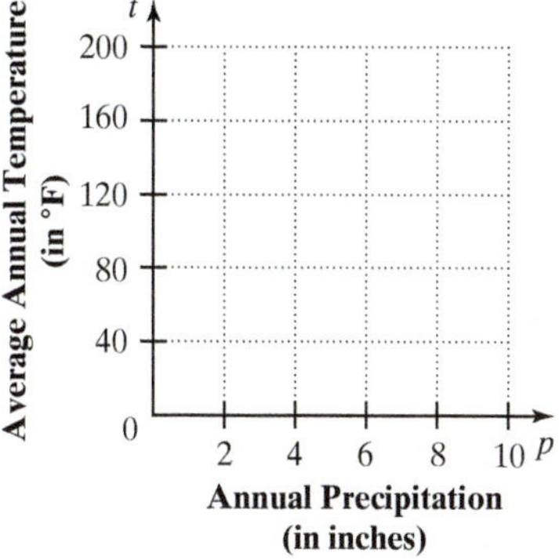

b. Give some examples of weather conditions that this model predicts will lead to desert conditions.

59. While on a diet, a model wants to snack on fresh apples and bananas. An apple contains 60 calories and a banana contains 100 calories. If she wants to consume fewer than 300 calories, find the number of apples a and the number of bananas b that she can eat. Solve this problem graphically.

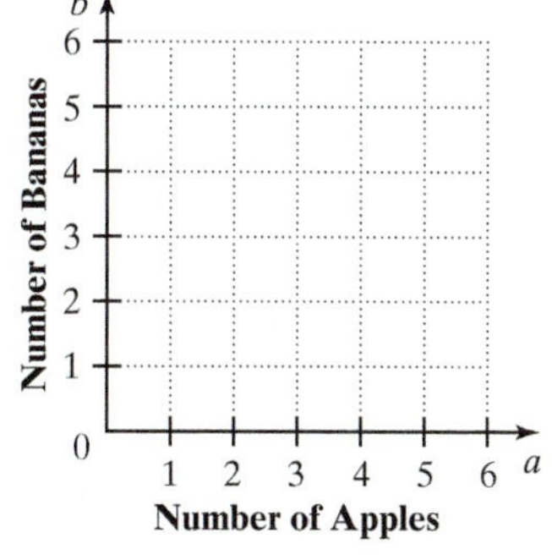

60. A student on spring break wants to drive no more than 300 mi in one day. The trip is along two highways. On the first highway, the student drives at an average speed of 60 mph for x hr and on the second highway at a speed of 50 mph for y hr. What are some possible times that the student can drive? Solve this problem graphically.

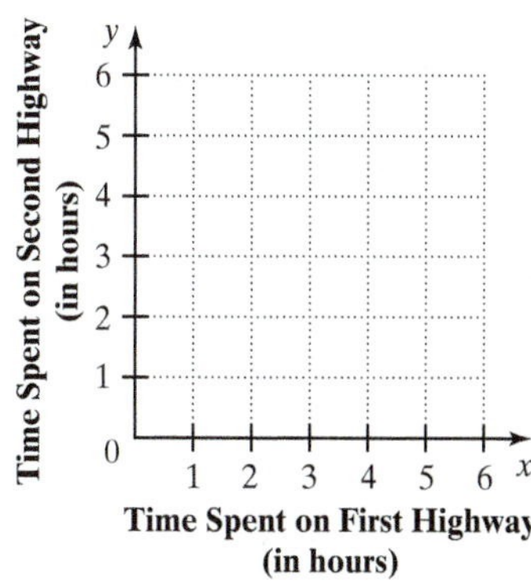

• Check your answers on page A-24.

MINDStretchers

Mathematical Reasoning

1. Consider the three graphs: $y < b$, $y = b$, and $y > b$, where b is a positive number. If you were to show these graphs on the same coordinate plane, what would you get? Would you get the same answer if b were negative?

Groupwork

2. In playing a carnival game, you roll a pair of dice—a red die and a blue die—each with six faces numbered 1, 2, 3, 4, 5, and 6. The grid below shows all the possible outcomes when you roll the two dice.

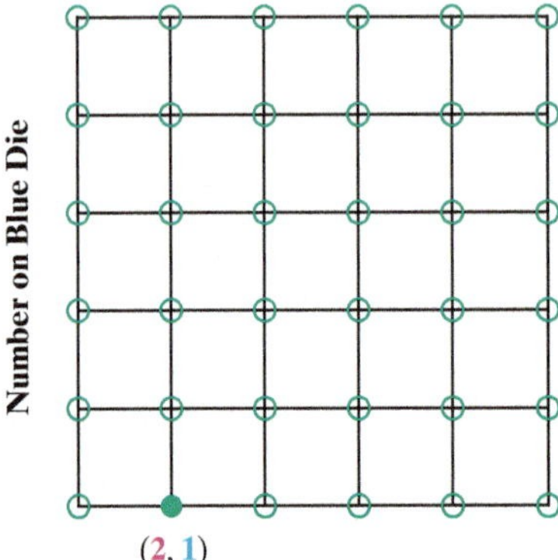

For each point on the grid, the first coordinate represents the roll on the red die and the second coordinate represents the roll on the blue die. For instance, the point $(2, 1)$ corresponds to rolling a 2 on the red and a 1 on the blue.

a. How many points in all are there on the grid?

b. If the number on the blue die is greater than the number on the red die, you will win a prize. Fill in the points on the grid that correspond to winning a prize. How many points did you fill in?

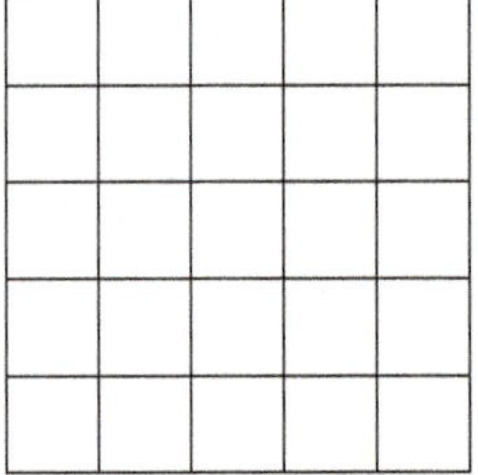

c. What fraction of the total number of points on the grid did you fill in? What does this fraction represent?

Writing

3. Compare solving a linear *equation* in two variables by graphing with solving a linear *inequality* in two variables by graphing. What are the similarities? What are the differences?

Key Concepts and Skills

CONCEPT SKILL

Concept/Skill	Description	Example
[13.1] **Coordinate plane**	A flat surface on which we draw graphs. The **coordinate plane** is formed by two perpendicular lines called **axes** which intersect at the **origin**.	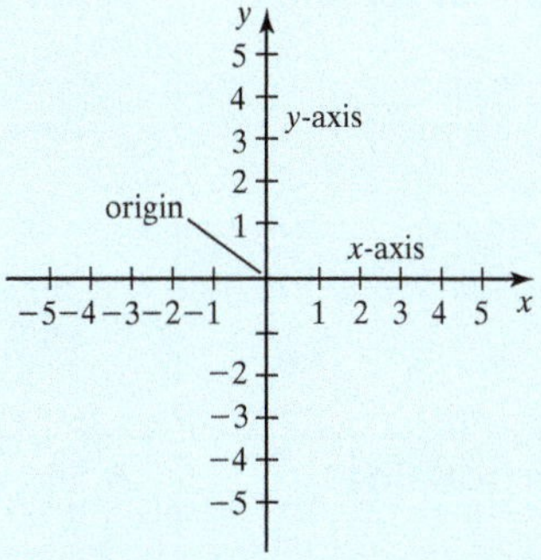 y-axis; origin; x-axis; −5 −4 −3 −2 −1, 1 2 3 4 5
[13.1] **Coordinates**	An **ordered pair** of numbers that represents a point in the coordinate plane.	(−4, 2); (3, −1); (−1, −3)
[13.1] **Quadrant**	One of four regions of a coordinate plane separated by axes.	Quadrant II (−,+); Quadrant I (+,+); Quadrant III (−,−); Quadrant IV (+,−)
[13.1] **Scale**	The length between adjacent tick marks on an axis in a coordinate plane.	Scale; −200 −100, 100 200
[13.2] **Slope**	The **slope** m of a line passing through the points (x_1, y_1) and (x_2, y_2): $m = \frac{y_2 - y_1}{x_2 - x_1}$, where $x_1 \neq x_2$	For $(1, 5)$ and $(-2, 6)$, $m = \frac{6-5}{-2-1} = \frac{1}{-3} = -\frac{1}{3}$

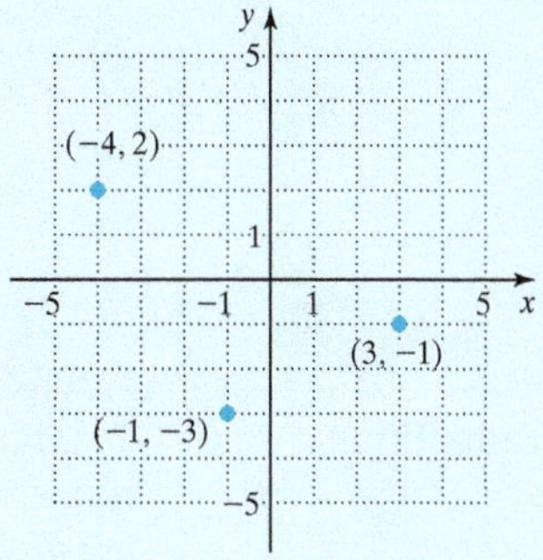

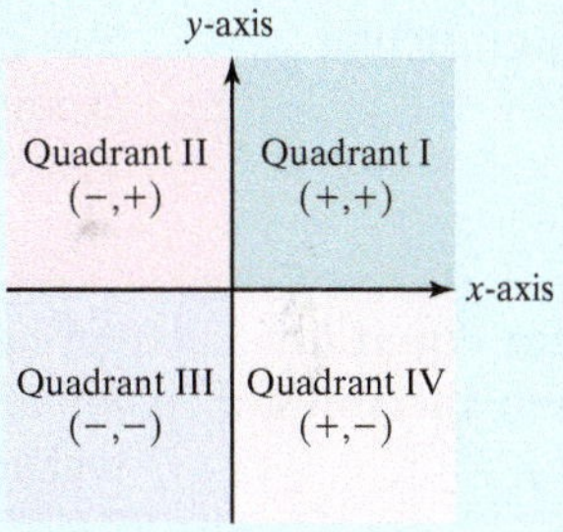

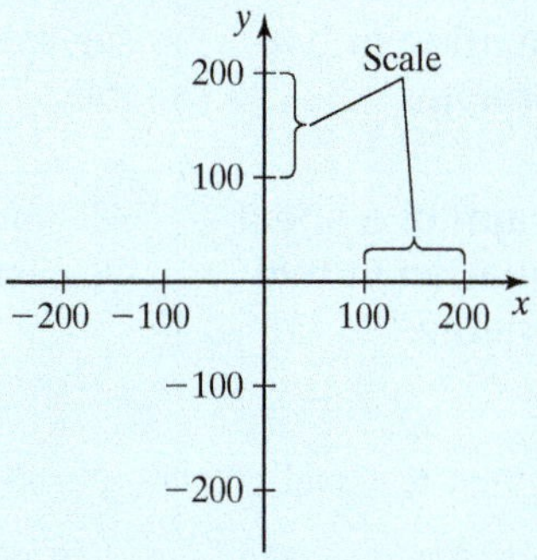

continued

Concept/Skill	Description	Example
[13.2] **Horizontal line**	A line whose slope is 0.	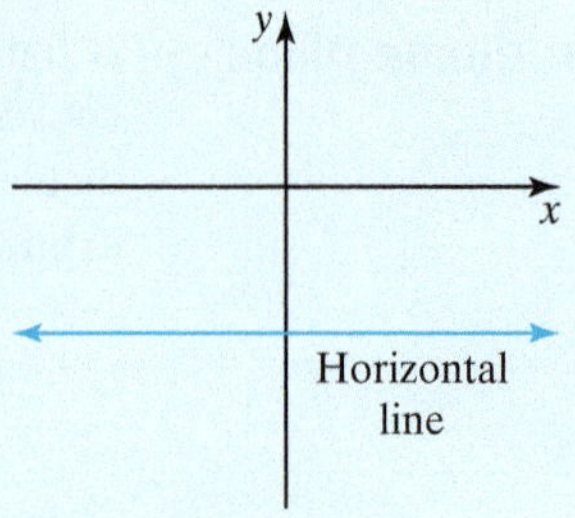
[13.2] **Vertical line**	A line whose slope is undefined.	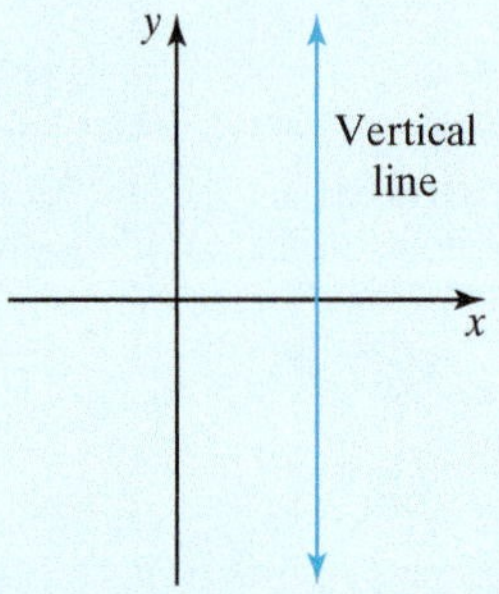
[13.2] **Parallel lines**	Two nonvertical lines are **parallel** if and only if their slopes are equal. That is, if the slopes are m_1 and m_2, then $m_1 = m_2$.	The line passing through $(0, 1)$ and $(2, 5)$ and the line passing through $(3, 6)$ and $(1, 2)$ are parallel since both lines have slope 2.
[13.2] **Perpendicular lines**	Two nonvertical lines are **perpendicular** if and only if the product of their slopes is -1. That is, if the slopes are m_1 and m_2, then $m_1 \cdot m_2 = -1$.	The line passing through $(0, 3)$ and $(1, 4)$ and the line passing through $(2, 8)$ and $(3, 7)$ are perpendicular since the product of their slopes, 1 and -1, is -1.
[13.3] **Linear equation in two variables, *x* and *y***	An equation that can be written in the *general form* $Ax + By = C$, where A, B, and C are real numbers and A and B are not both 0.	$3x + 5y = 7$
[13.3] **Solution of an equation in two variables**	An ordered pair of numbers that when substituted for the variables makes the equation true.	$(1, 5)$ is a solution of the equation $y = x + 4$: $5 \stackrel{?}{=} 1 + 4$; $5 = 5$ True
[13.3] **Graph of a linear equation in two variables**	All points whose coordinates satisfy the equation.	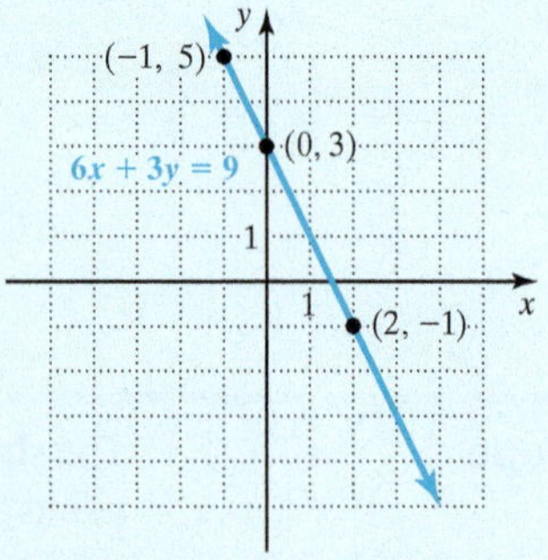

Concept/Skill	Description	Example
[13.3] **To graph a linear equation in two variables**	• Isolate one of the variables—usually y—if it is not already done. • Choose three x-values, entering them in a table. • Complete the table by calculating the corresponding y-values. • Plot the three points—two to draw the line and the third to serve as a *checkpoint*. • Check that the points seem to lie on the same line. • Draw the line passing through the points.	To graph $y - 3x = 1$: $y = 3x + 1$ (see table and graph below)

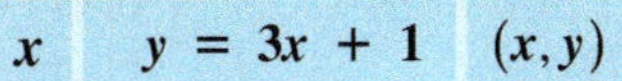

x	$y = 3x + 1$	(x, y)
0	$y = 3(0) + 1$	$(0, 1)$
1	$y = 3(1) + 1$	$(1, 4)$
2	$y = 3(2) + 1$	$(2, 7)$

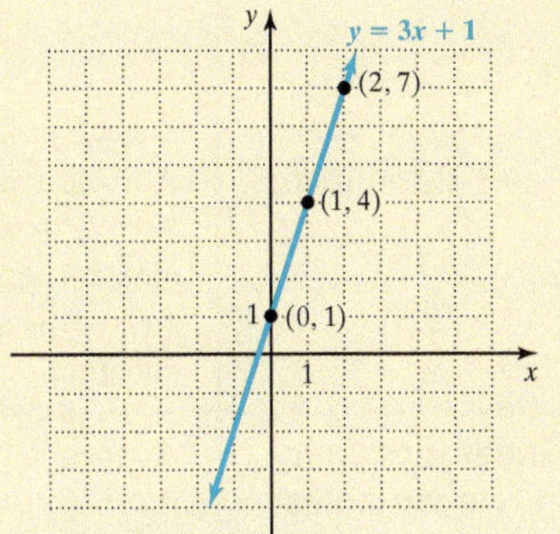

Concept/Skill	Description	Example
[13.3] **Intercepts of a line**	The x-intercept: the point where the graph crosses the x-axis. The y-intercept: the point where the graph crosses the y-axis.	(see graph below)

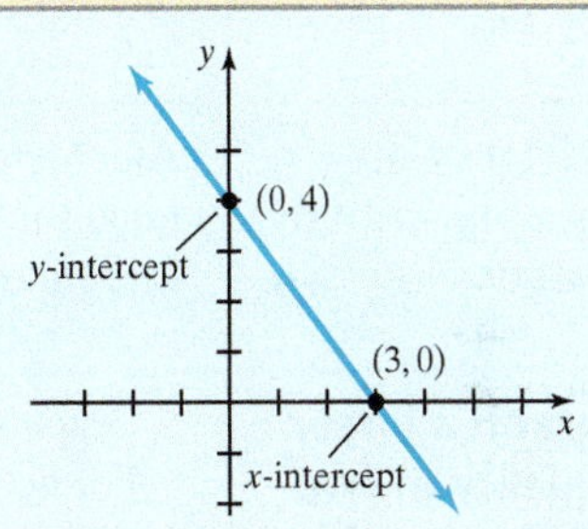

Concept/Skill	Description	Example
[13.3] **To graph a linear equation in two variables using the x- and y-intercepts**	• Let $x = 0$, and find the y-intercept. • Let $y = 0$, and find the x-intercept. • Find a checkpoint. • Plot the three points. • Check that the points seem to lie on the same line. • Draw the line passing through the points.	To graph $2x + 3y = 6$: (see table and graph below)

	$2x + 3y = 6$	(x, y)
$x = 0$	$2 \cdot 0 + 3y = 6; y = 2$	$(0, 2)$
$y = 0$	$2x + 3 \cdot 0 = 6; x = 3$	$(3, 0)$
$x = 6$	$2 \cdot 6 + 3y = 6; y = -2$	$(6, -2)$

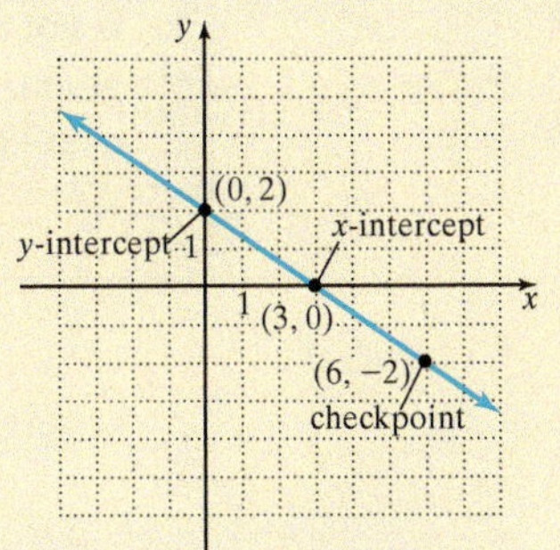

Concept/Skill	Description	Example
[13.4] **Slope-intercept form**	A linear equation written as $y = mx + b$, where m and b are constants. In this form, m is the slope and $(0, b)$ is the y-intercept of the graph of the equation.	The line with slope 5 and y-intercept $(0, 2)$: $y = mx + b$ $y = 5x + 2$

continued

CONCEPT SKILL

Concept/Skill	Description	Example
[13.4] **To graph a linear equation in two variables using the slope and y-intercept**	• First, locate the y-intercept. • Then, use the slope to find a second point on the line. • Finally, draw the line through the two points.	To graph $y = 3x - 1$: The y-intercept is $(0, -1)$ and the slope is 3. 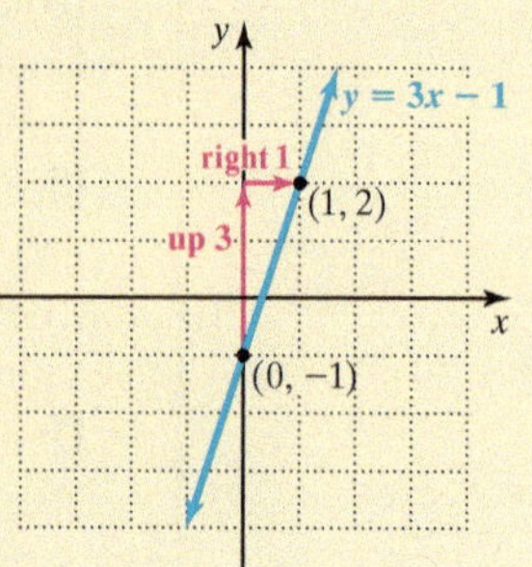
[13.4] **Point-slope form**	A linear equation written as $y - y_1 = m(x - x_1)$, where x_1, y_1, and m are constants. In this form, m is the slope and (x_1, y_1) is a point that lies on the graph of the equation.	The line with slope 5 passing through the point (2, 1): $y - y_1 = m(x - x_1)$ $y - 1 = 5\,(x - 2)$
[13.5] **Linear inequality in two variables**	An inequality that can be written in the form $Ax + By < C$, where A, B, and C are real numbers and A and B are not both 0. The inequality symbol can be $<, >, \geq$, or $\leq$.	$5x + 3y < 1$
[13.5] **Solution of an inequality in two variables**	An ordered pair of numbers that when substituted for the variables makes the inequality a true statement.	$(3, 1)$ is a solution to the inequality $x < 5y$: $3 < 5(1)$ $3 < 5$ True
[13.5] **To graph a linear inequality in two variables**	• Graph the corresponding linear equation. For an inequality with the symbol $\leq$ or $\geq$, draw a solid line; for an inequality with the symbol $<$ or $>$, draw a broken line. This line is the boundary between two half-planes. • Choose a test point in either half-plane and substitute the coordinates of this point in the inequality. If the resulting inequality is true, then the graph of the inequality is the half-plane containing the test point. If it is not true, then the other half-plane is the graph. A solid line is part of the graph, and a broken line is not.	To graph $y > x + 1$, first graph the line $y = x + 1$. The inequality does not hold for the test point $(0, 0)$. The graph of $y > x + 1$ is the half-plane above and excluding the line $y = x + 1$. 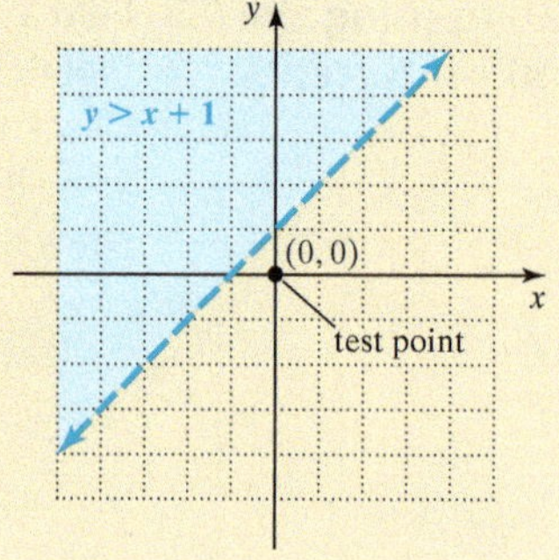

CHAPTER 13 Review Exercises

Say Why

Fill in each blank.

1. The points $(2, -5)$ and $(3, 5)$ ________ (are/are not) in the same quadrant because ________________________________.

2. The point $(3, -1)$ ________ (does/does not) lie on the graph of the equation $2x - 5y = 1$ because ________________________________.

3. The graph of the equation $y = -2x + 7$ ________ (is/is not) increasing because ________________________________.

4. The graph of the equation $y = 5$ ________ (is/is not) horizontal because ________________________________.

5. The x-intercept of $y = 2x + 6$ ________ (is/is not) -3 because ________________________________.

6. The slope of a vertical line ________ (is/is not) undefined because ________________________________.

7. A line with slope 5 ________ (is/is not) perpendicular to a line with slope $\frac{1}{5}$ because ________________________________.

8. $(0, 0)$ ________ (is/is not) a solution to the linear inequality $2x - 5y \geq 0$ because ________________________________.

[13.1]

9. Plot the points with the given coordinates.
$A(0, 5)$ $B(-1, -6)$ $C(3, -4)$ $D(-2, 2)$

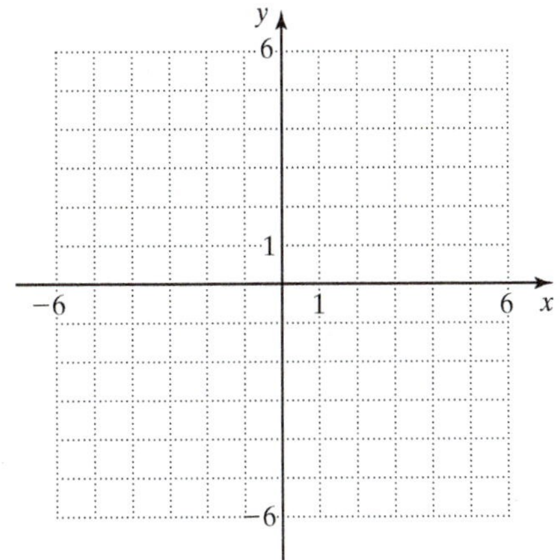

10. Fill in the coordinates of each point.

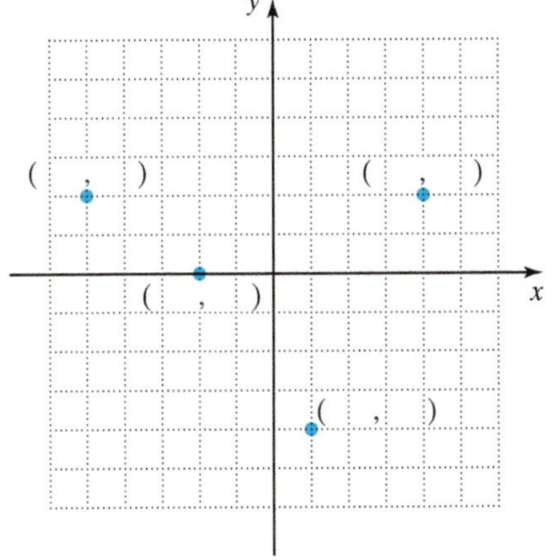

Identify the quadrant in which each point is located:

11. $(5, -1)$

12. $(-7, -2)$

[13.2] *Compute the slope m of the line that passes through the given points. Plot these points on the coordinate plane, and draw the line.*

13. $(2, 0)$ and $(3, 5)$, $m =$

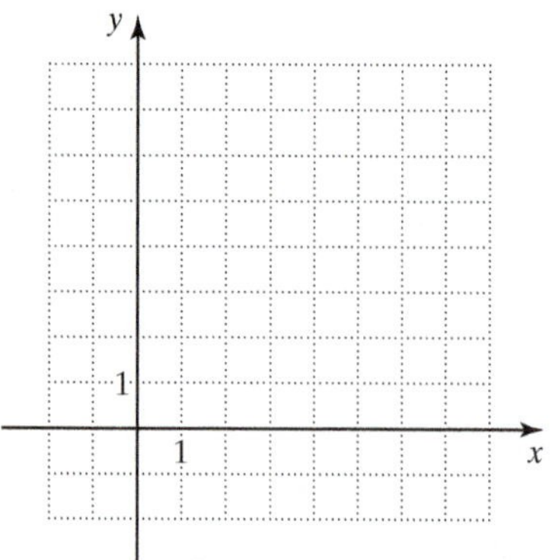

14. $(5, 7)$ and $(2, 7)$, $m =$

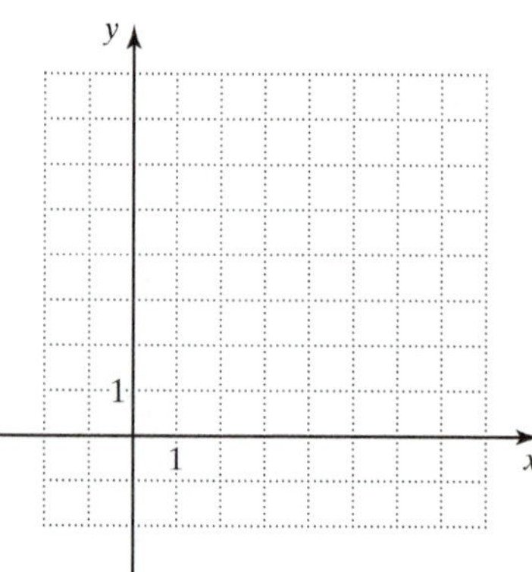

Draw the line on the coordinate plane based on the given information.

15. Passes through $(3, -1)$ and $m = 4$

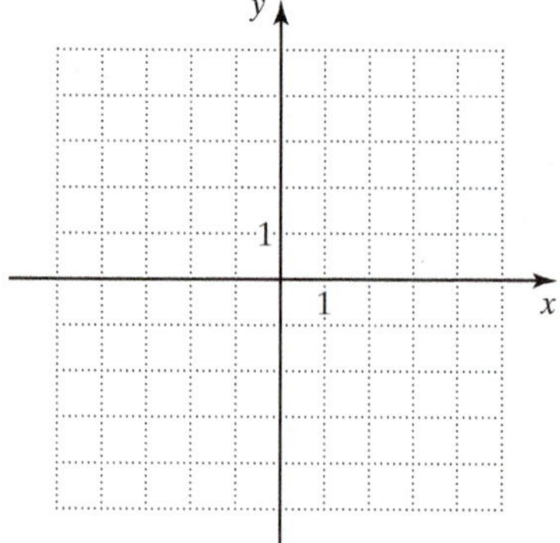

16. Passes through $(0, 0)$ and $m = -\frac{1}{2}$

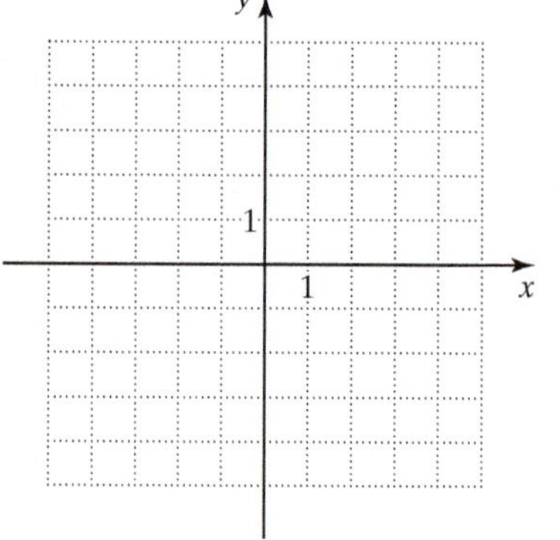

Determine whether the slope of each graph is positive, negative, zero, or undefined.

17.

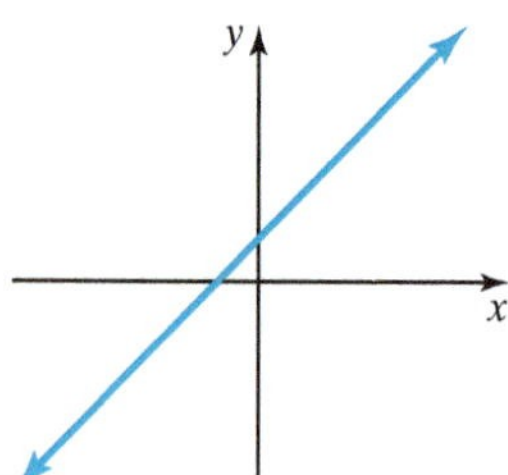

18.

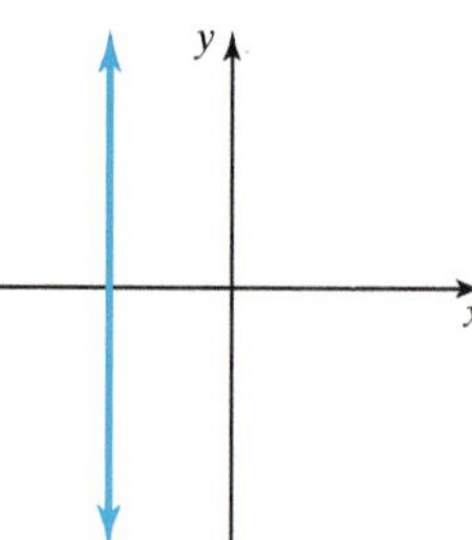

19.

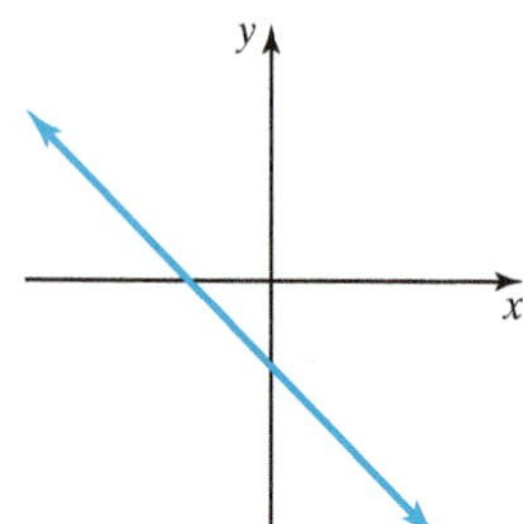

20.

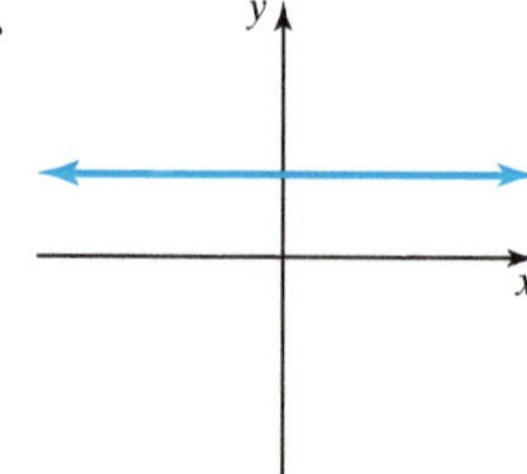

Determine whether $\overleftrightarrow{AB}$ and $\overleftrightarrow{CD}$ are parallel or perpendicular.

21. $A(5, 0)$ $B(3, 0)$ $C(-3, -2)$ $D(1, -2)$

22. $A(4, 8)$ $B(5, 9)$ $C(2, -3)$ $D(0, -1)$

For the following graphed line, find:

23. the x-intercept.

24. the y-intercept.

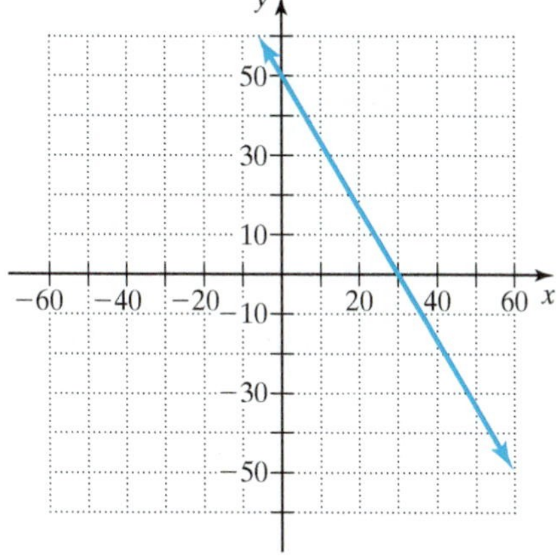

[13.3] *Complete each table of values for the given equation.*

25. $y = 2x - 5$

x	0	1		
y			0	1

26. $y = -x + 3$

x	2	5		
y			7	−5

Graph.

27. $4x - 3y = -12$

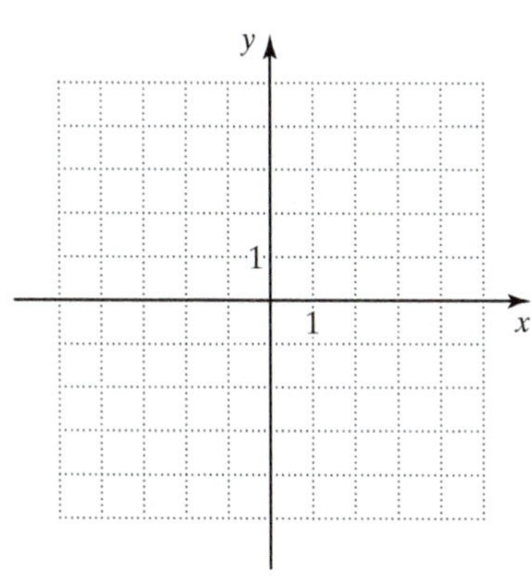

28. $x + 2y = -6$

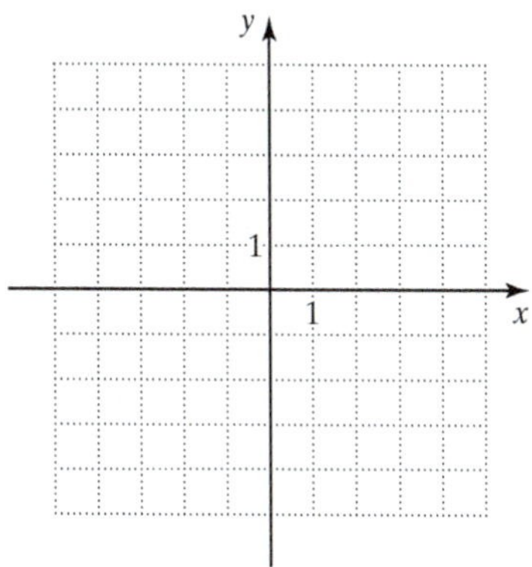

29. $y = \frac{1}{2}x$

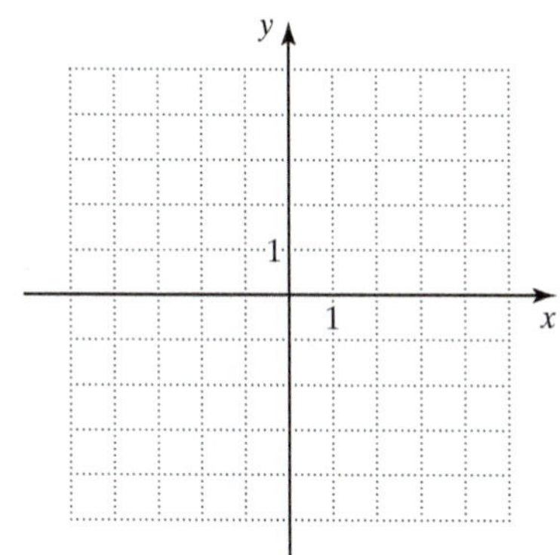

30. $y = -x + 2$

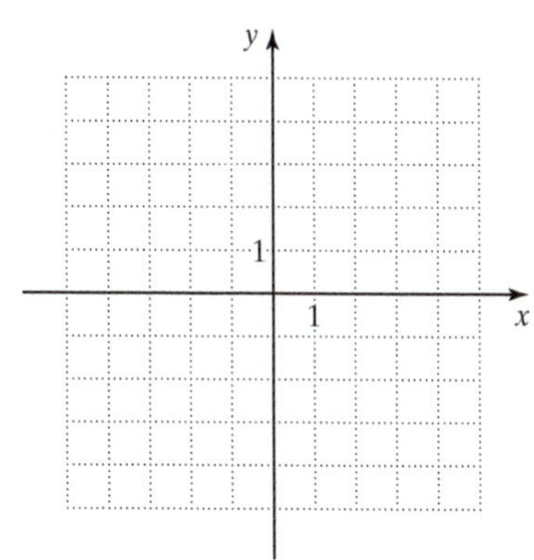

[13.4] *Write each equation in slope-intercept form.*

31. $x - y = 10$

32. $x + 2y = -1$

Complete the following table:

	Equation	Slope m	y-intercept $(0, b)$	Indicate Which Graph Type Best Describes the Line. ╱ ╲ — |	x-intercept
33.	$y = 4x - 16$				
34.	$y = -\frac{1}{3}x$				

35. Find the slope of a line perpendicular to the line $x - 2y = 4$.

36. Find the slope of a line parallel to the line $3x - y = 1$.

37. Find the equation of the line with slope -1 that passes through the point $(3, 5)$.

38. Write the equation of the horizontal line that passes through the point $(3, 0)$.

39. The points $(2, 0)$ and $(1, 5)$ lie on a line. Find its equation.

40. What is an equation of the line passing through the points $(3, 1)$ and $(-2, 0)$?

Find the equation of each graph.

41.

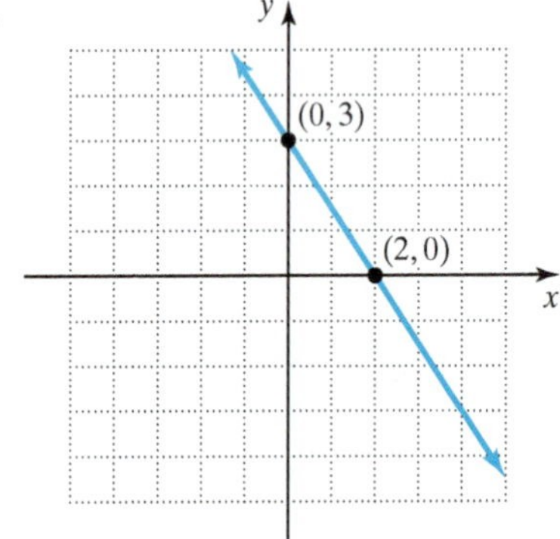

42.

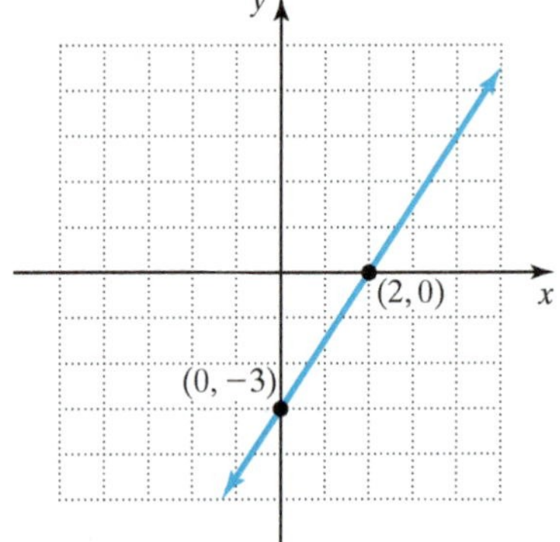

[13.5] *Decide if the ordered pair is a solution to the inequality.*

43. $(-2, 7)$, $x + y < 1$

44. $(1, -4)$, $2x - 3y \geq 14$

Each line is the graph of $y = -x$. Shade in the graph of the given inequality.

45. $y > -x$

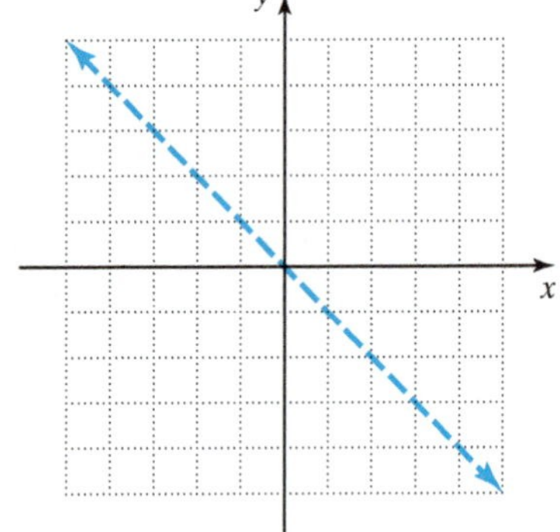

46. $y < -x$

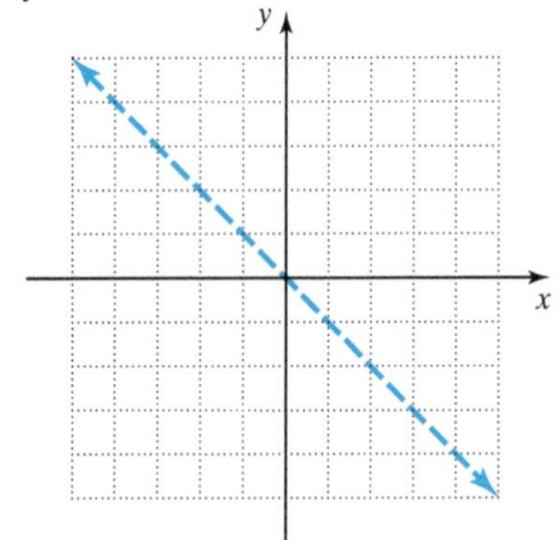

Graph the linear inequality.

47. $y \leq 2x$

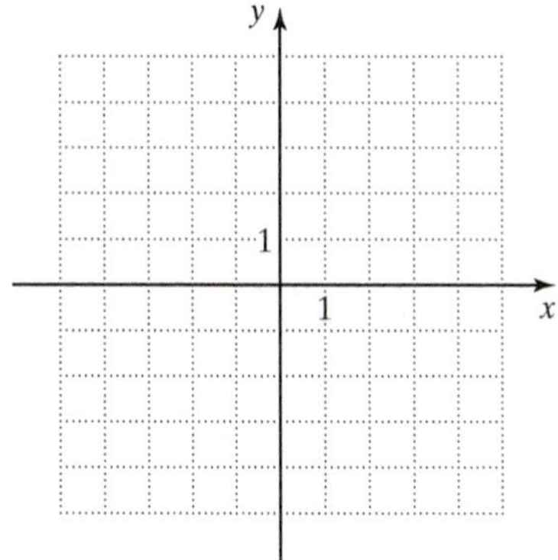

48. $y - x > -1$

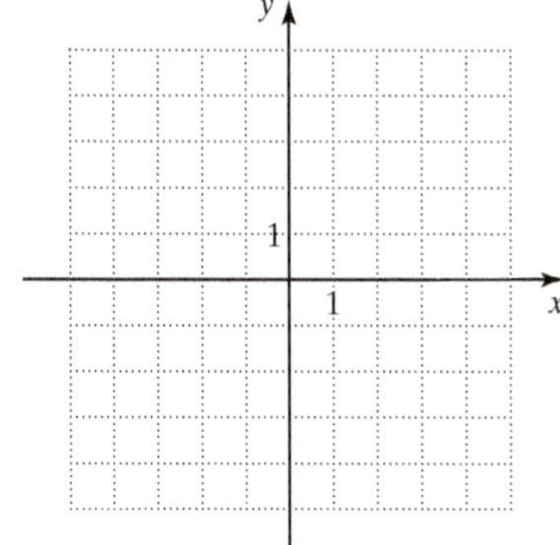

Mixed Applications

Solve.

49. The following table shows the amount A (in dollars) that a bed and breakfast charges when renting a room for s days.

Length of Stay s	Rental Amount A
2	180
5	450

a. Graph the points given in the table. Draw a line passing through the points.

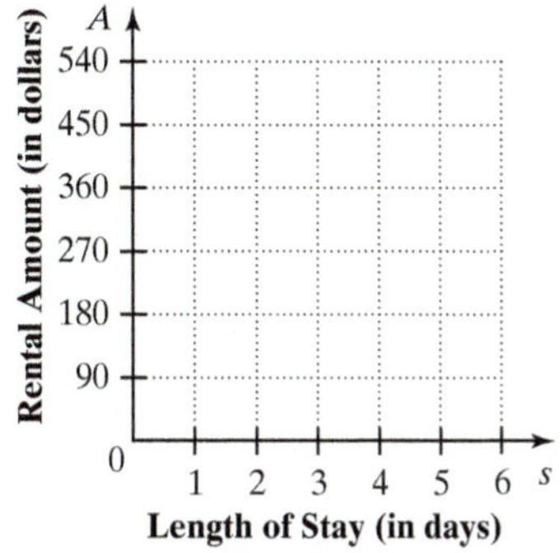

b. What is the A-intercept of this line? Explain its significance in terms of renting a room.

50. The following table shows the cost C (in cents) of duplicating q flyers at a print shop.

Quantity q	Cost C
1	4
10	40

a. Plot the points given in the table and draw the line passing through them.

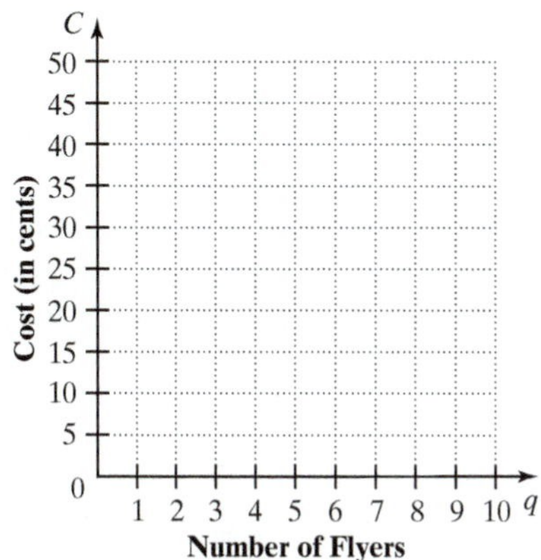

b. Calculate the slope of this line. Explain its significance in terms of the price structure at the print shop.

51. A man drives toward a town, stops, and then again drives toward the town. Which of the following graphs could describe this motion? Explain your answer.

a.

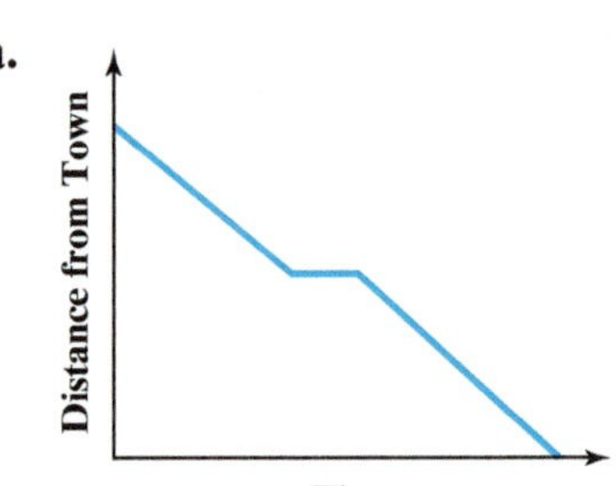

b.

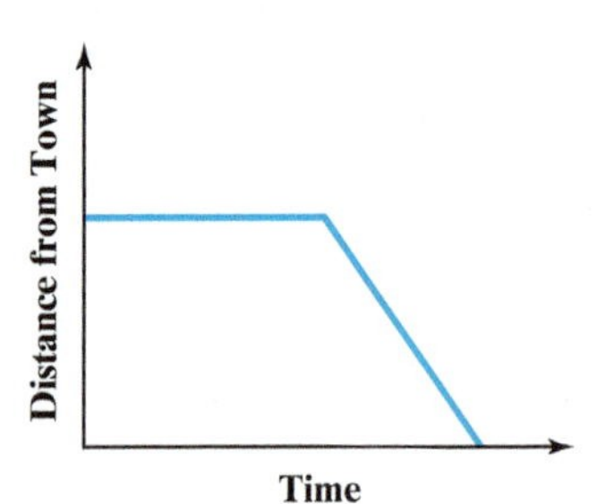

c.

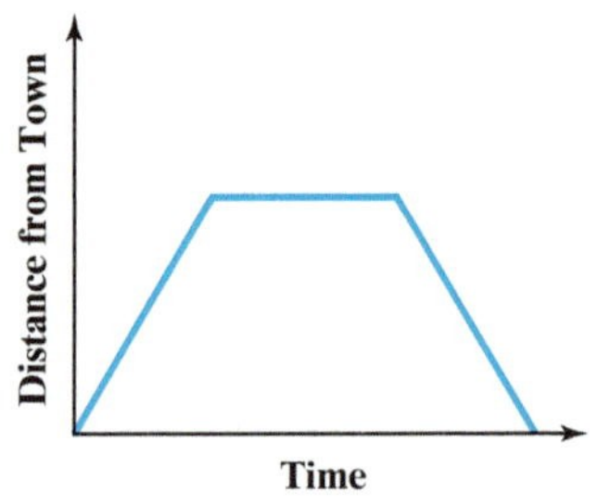

52. The graph below shows the altitude of an airplane during a flight. Write a brief story describing the altitude of the plane relative to the duration of the flight.

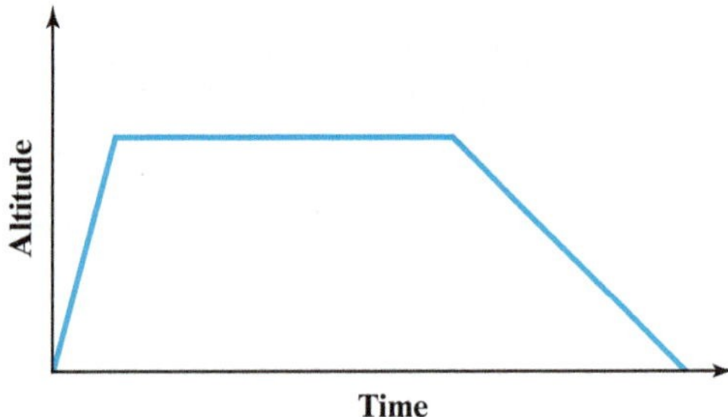

53. A novelist negotiated the following deal with her publisher: a \$20,000 bonus plus 9% of book sales.

a. Express her income i in terms of book sales s.

b. Draw a graph of this equation for sales up to \$500,000.

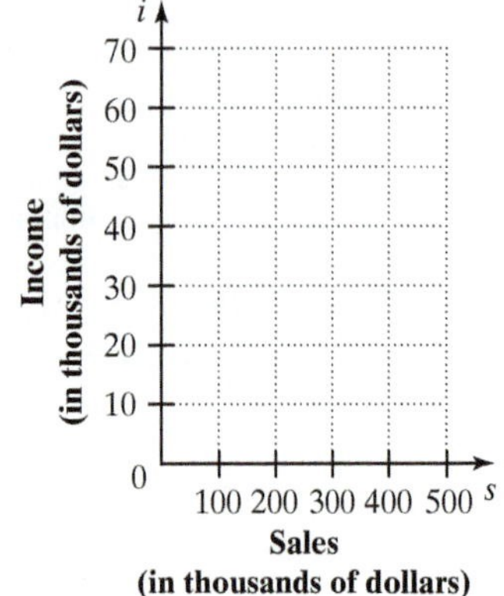

54. A bank account with an initial balance of \$100 earns simple interest at an annual rate of 4%. The amount A in the account after t years is given by:

$$A = 100 + 4t$$

a. Graph this equation.

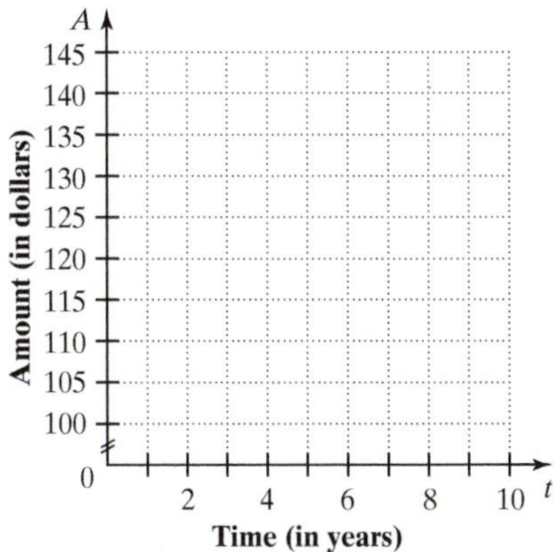

b. What is the A-intercept of this graph? Explain its significance in terms of the bank account.

55. The width of a jewel case is $\frac{1}{4}$ in. for a single compact disc and $\frac{1}{2}$ in. for double compact discs. If there are s single jewel cases and d double jewel cases on a shelf 30 in. long, then

$$\frac{1}{4}s + \frac{1}{2}d < 30.$$

a. Graph this inequality.

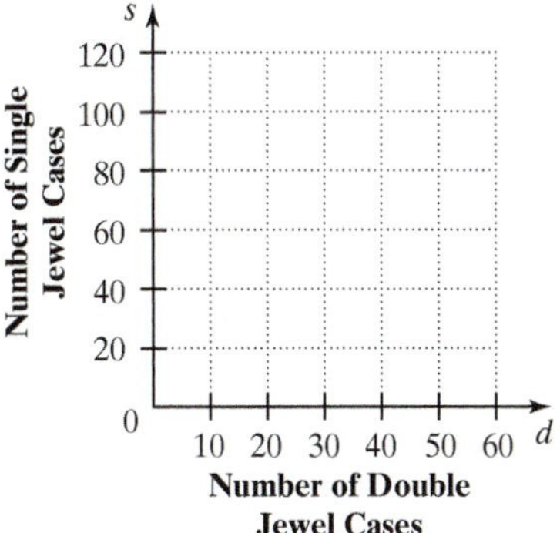

b. From this graph, identify one possible combination of single and double jewel cases that will fit on the shelf.

56. To be able to catch up and pass a friend driving away at a speed of 50 mph, it is necessary to cover a distance of d mi in t hr, where

$$d > 50t.$$

a. Graph this inequality.

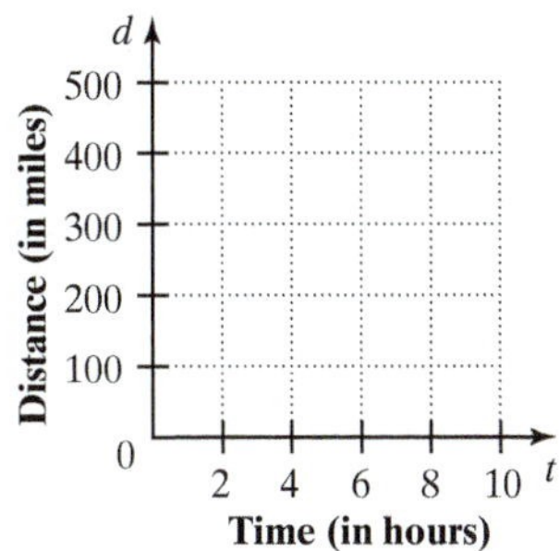

b. From this graph, choose a point in the shaded region. Explain what its coordinates mean in terms of catching up and passing the friend.

• Check your answers on page A-26.

CHAPTER 13 Posttest

To see whether you have mastered the topics in this chapter, take this test.

1. On the coordinate plane shown, plot the points $A(-2, 0)$ and $B(5, 3)$.

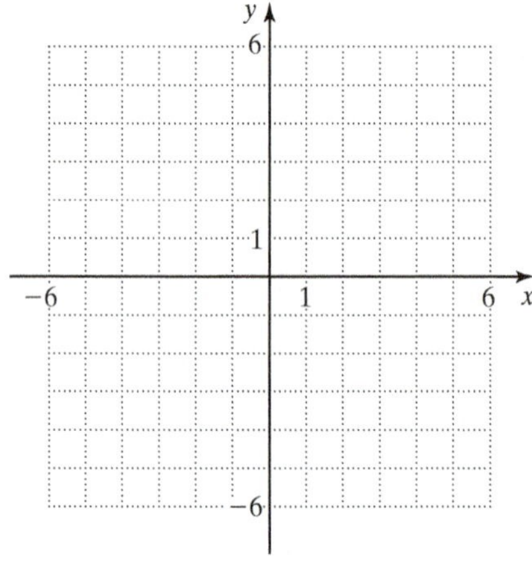

2. In which quadrant is the point $(-5, 3)$ located?

3. Given two points $C(8, 1)$ and $D(3, -4)$, compute the slope of the line that passes through the points.

4. Are the graphs of $y = 3x + 1$ and $y = 3x - 2$ parallel? Explain how you know.

5. For the points $A(0, 1)$, $B(2, 8)$, $C(0, 6)$, and $D(7, 4)$, indicate whether $\overleftrightarrow{AB}$ is perpendicular to $\overleftrightarrow{CD}$. Explain.

6. In the following graph, find the x-intercept and the y-intercept.

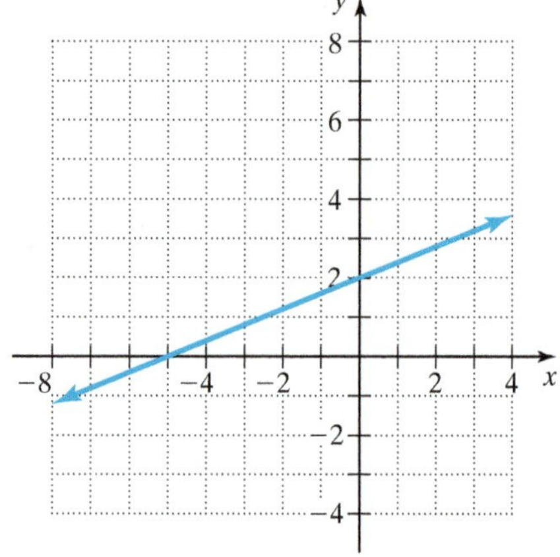

7. The graph on the following coordinate plane shows how the rental cost at a local car rental establishment relates to the number of miles that the car has been driven.

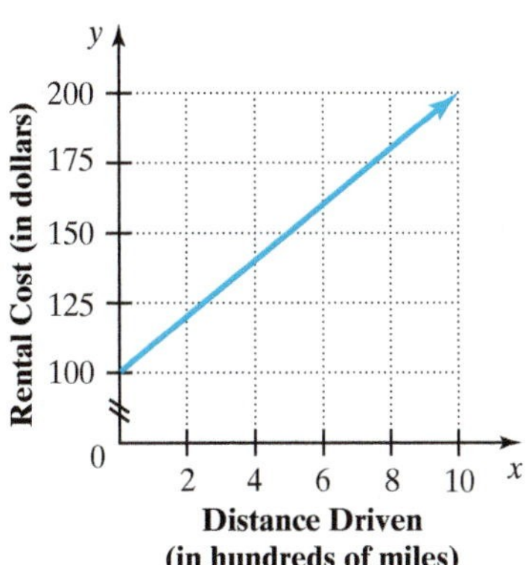

Is the slope of the graphed line positive, negative, zero, or undefined? Describe in a sentence or two the relationship between the rental cost and the number of miles driven.

8. Do the points $(0, 0)$, $(-2, -4)$, and $(1, 2)$ lie on the same line? Explain.

9. For the equation $y = -3x + 1$, complete the following table:

x	-3	5		
y			0	-2

Graph the equation.

10. $y = 2$

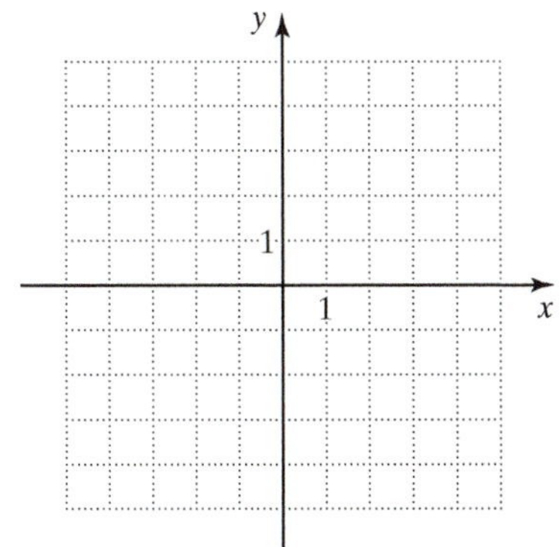

11. $x = -4$

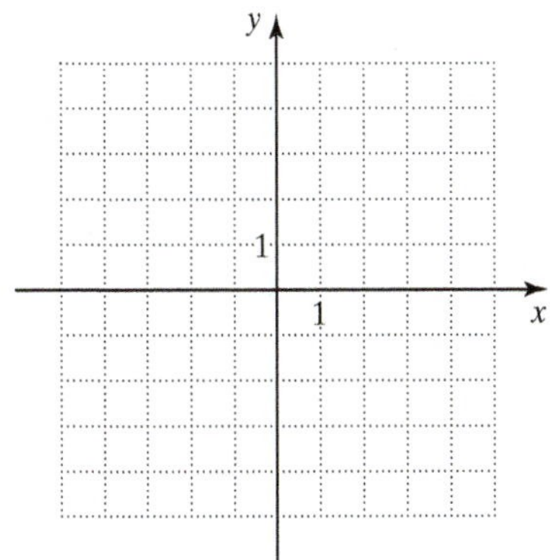

12. $y = -x - 3$

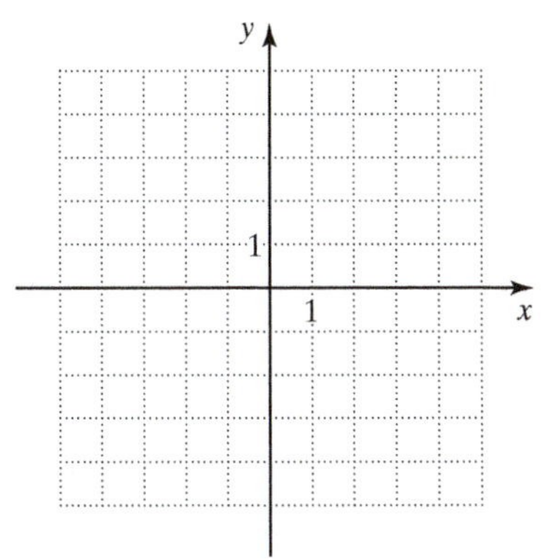

13. $3x - 2y = 6$

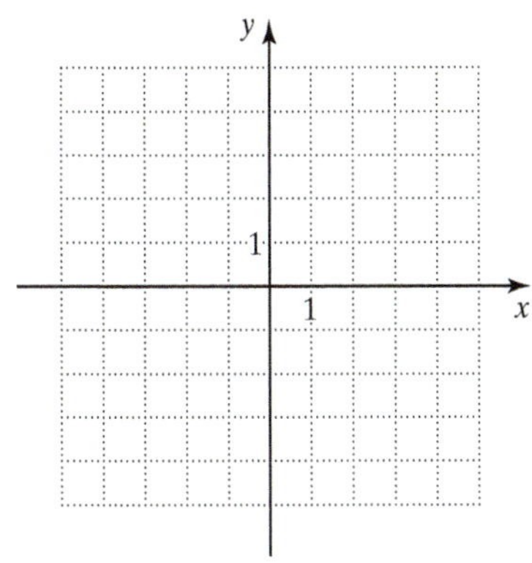

14. What are the slope and the y-intercept of the graph of $y = 3x + 1$?

15. Write the equation $2x - y = 5$ in slope-intercept form.

16. Find the equation of the line with slope -1 that passes through the point $(0, -3)$.

17. The points $(3, 5)$ and $(-4, 2)$ lie on a line. Find its equation.

18. Graph $y \le -\frac{1}{2}x + 1$.

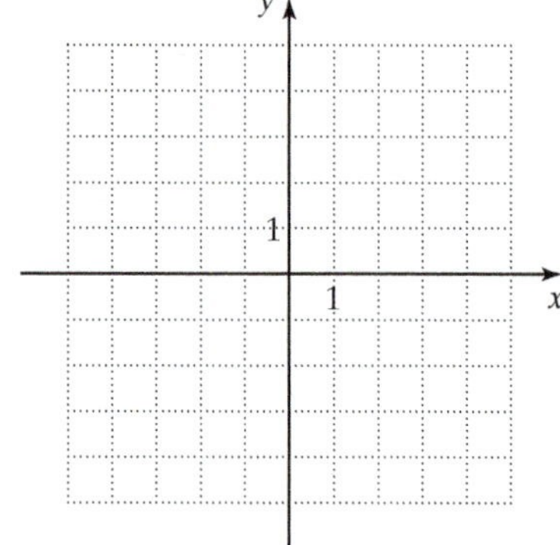

19. An entrepreneur is establishing a small business to manufacture leather bags. Her initial investment is \$1000, and her unit cost to manufacture each bag is \$30. Write an equation that gives the total cost C of manufacturing b bags. Plot the total cost of manufacturing 100, 200, and 300 bags.

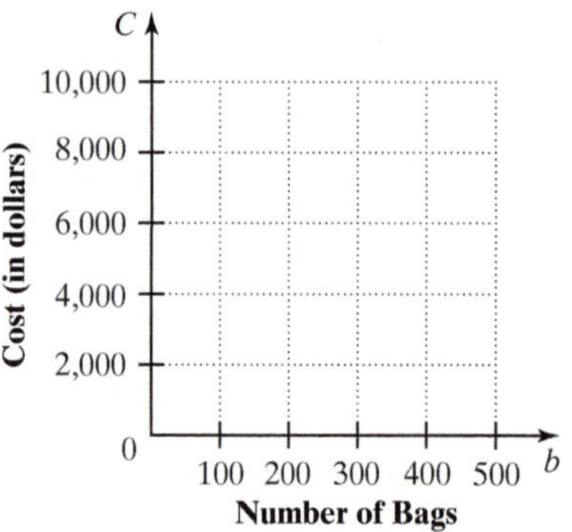

20. In 2010, the Library of Congress held over 124,000 telephone books and microfilmed city directories from across the U.S. In addition, the library acquires over 8,000 more of these holdings each year. Find the inequality relating the number of years x passed since 2010 to the number of these holdings y in the Library of Congress' collection, and graph this inequality. (*Source:* loc.gov)

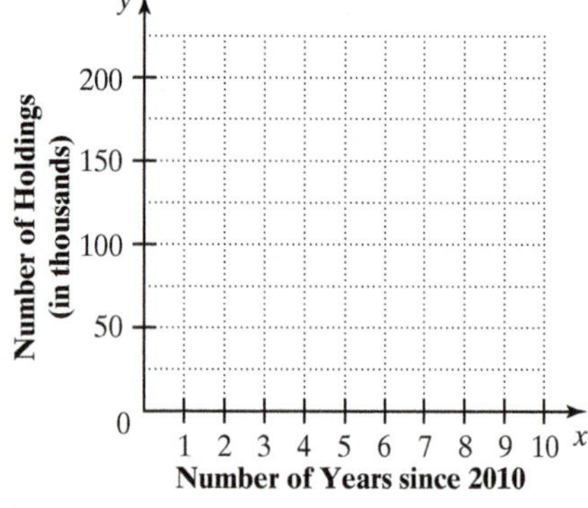

• Check your answers on page A-27.

Cumulative Review Exercises

To help you review, solve the following:

1. Calculate: $15\frac{1}{2} - 3 \cdot 2\frac{1}{4}$

2. Simplify: $3 + \frac{2}{3} \div \frac{2}{5}$

3. Replace ▢ with $>, <$, or $=$ to make a true statement: -2.3 ▢ $-2\frac{1}{3}$

4. Find the difference: $3.9 - 8.2$

5. Solve and check: $\frac{4}{5} = \frac{x}{3}$

6. Calculate: $(7 - 9) \div (-2 - 6)$

7. Find the median: 10, 1, 3, 7, and 4

8. Calculate the range: -3, 4, 8, 0, and 10

9. Express 3 min 10 sec in seconds.

10. Fill in the blank: 511 cm = 5.11 m

11. Find the missing angle:

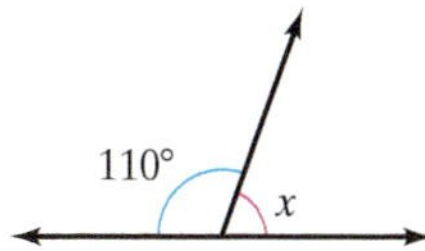

12. Find the volume of a cube with edge 5 m.

13. Find the value of $x^2 - 4x + 1$ if $x = -3$.

14. Solve for x: $2x - 1 = 5x + 11$

15. An air conditioner can reduce the temperature in a room by 8°F every 5 min. The temperature in the room was 62°F after the air conditioner had been running for 10 min. Write a linear equation that expresses the temperature in the room in terms of the time that the air conditioner has been running.

16. In a recent year, the combined cost of Medicare and Medicaid accounted for 13% of \$4 trillion, the total of all federal budget outlays. Find to the nearest hundred billion dollars the combined cost of Medicare and Medicaid. (*Source*: cnn.com)

17. The speed of sound S (in meters per second) in air at temperature T can be approximated by the following formula:

$$S = 0.6T + 331.$$

Solve for T in terms of S. (*Source:* Peter J. Nolan, *Fundamentals of College Physics*)

18. The following graph shows the average weight for American males, age 20 and over. Describe the overall pattern in this population's weight over a lifespan. (*Source:* cdc.gov)

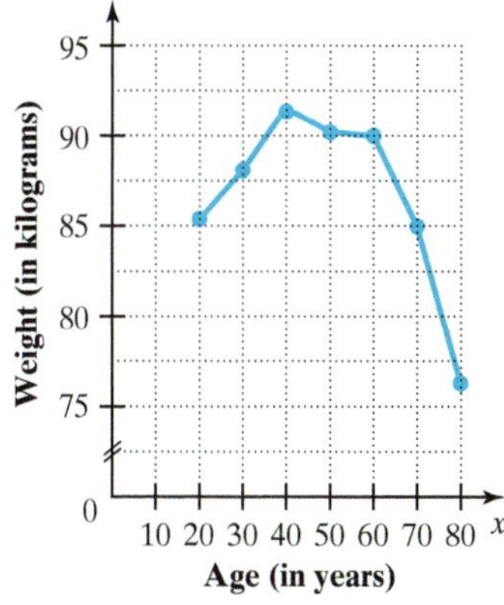

19. St. Augustine grass is a type of grass found in the southern United States. Its growth rate varies with the season. After this grass is mowed at the beginning of the season, its average height in the summer and the fall is graphed over time as shown below. In which season does the grass grow more quickly? (*Source:* aggie-horticulture.tamu.edu)

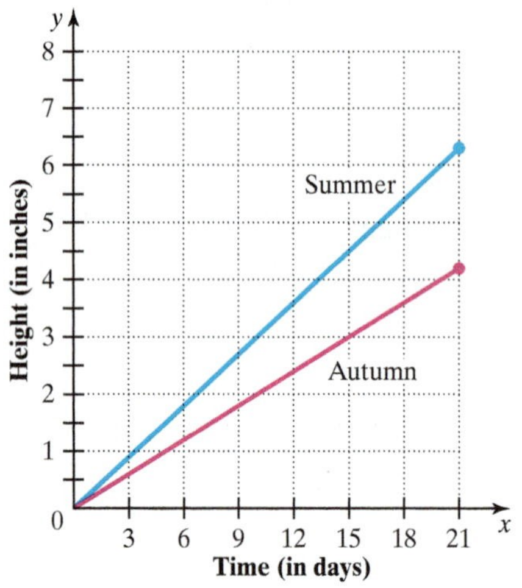

20. A drama club washes cars as a fund-raising activity. The club charges \$6 to wash each car.

a. Write an equation that expresses the club's income y in terms of the number x of cars they wash.

b. Graph the equation.

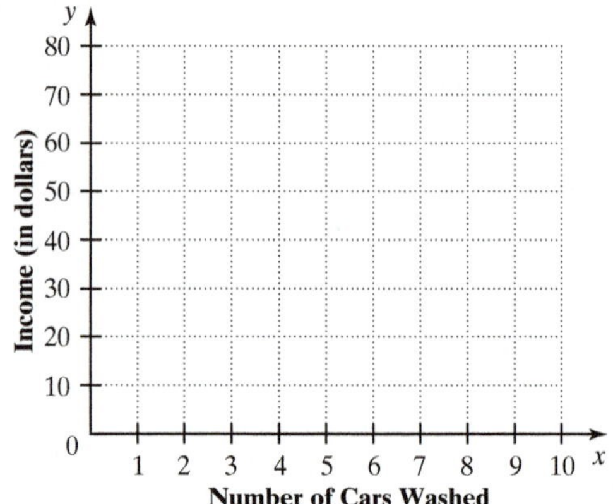

c. What are the x- and y-intercepts of the graph?

• Check your answers on page A-27.

CHAPTER 14

Solving Systems of Linear Equations

Economics and Linear Curves

In a market, sellers can set the price at which their goods are offered for sale. How does this price affect the number of goods that buyers are willing to purchase? How does it affect the number of goods that producers are willing to supply the sellers?

Generally, as the *price* of an item increases, the *quantity* of items sold declines. This trend is captured in a **demand curve**—commonly approximated by a straight line with a negative slope. The coordinates of each point on this line correspond to the price at which retailers sell items and the quantity of items they sell.

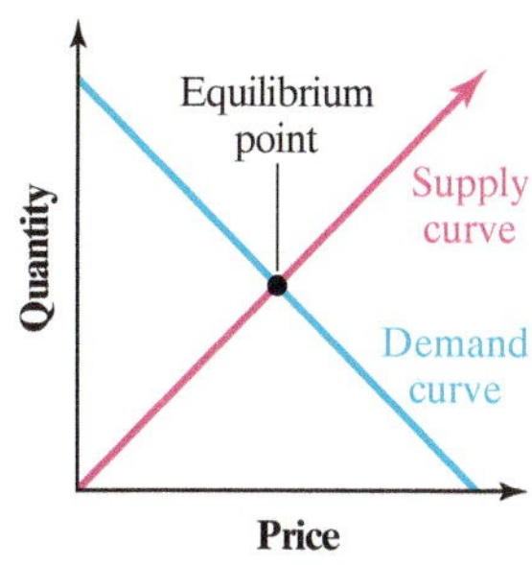

By contrast, the **supply curve** has a positive slope, meaning that as selling prices increase, wholesalers are inclined to make more goods available to retailers. The coordinates of a point on this line represent the price at which retailers sell items and the quantity of items that wholesalers supply to the retailers.

Graphing both the supply curve and the demand curve on the same coordinate plane shows the price at which the market is **at equilibrium**. At this point of equilibrium, the quantity supplied is equal to the quantity demanded.

(*Source:* Michael Parkin, *Economics*, Pearson Addison-Wesley, 2010)

CHAPTER 14 PRETEST

To see if you have already mastered the topics in this chapter, take this test.

1. Determine which ordered pair is a solution of the following system:

$$\begin{aligned} x + 2y &= 5 \\ 5x - y &= -8 \end{aligned}$$

   a. $(5, 0)$
   b. $(-1, 3)$
   c. $(1, -3)$

2. For the system graphed, indicate the number of solutions.

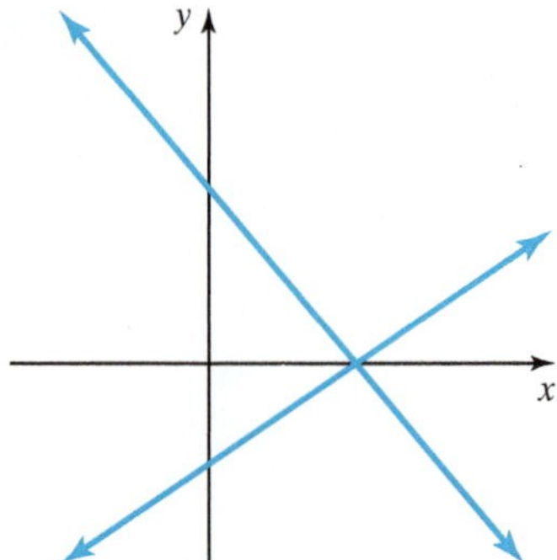

Solve each system by graphing.

3. $x + y = -2$
   $y = x + 4$

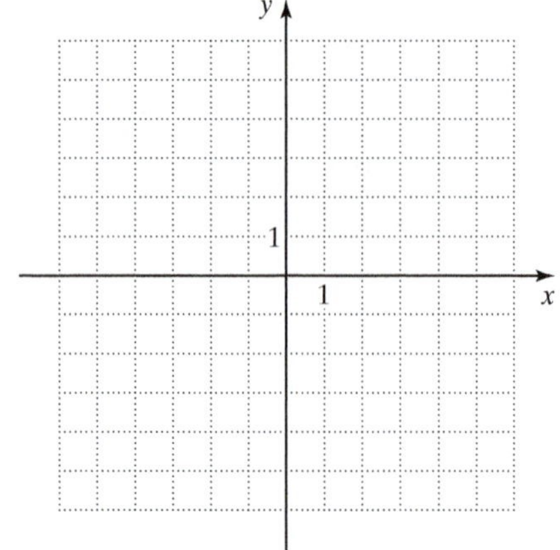

4. $x - 2y = 1$
   $y = 2$

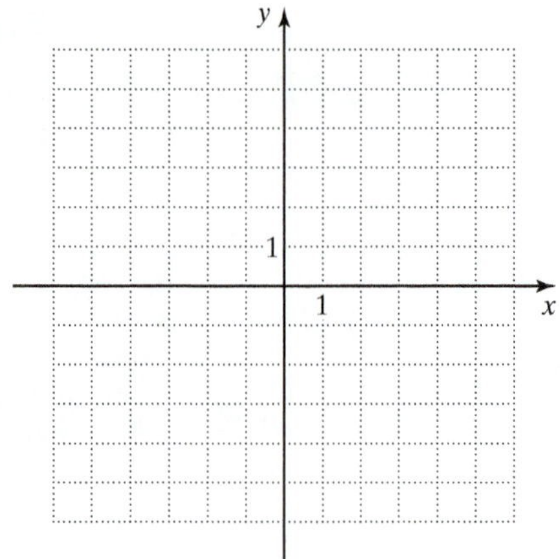

5. $x - 2y = -4$
   $4y = 2x + 8$

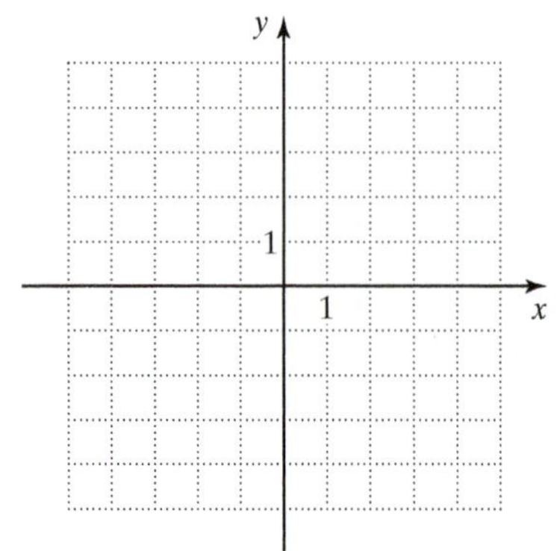

6. $3y = 6 - x$
   $x + 3y = 3$

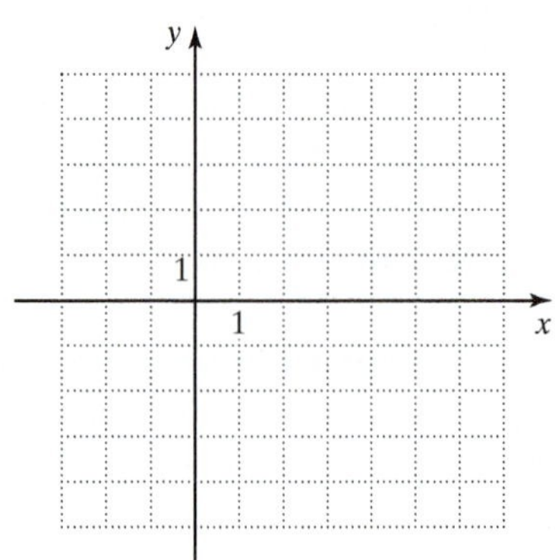

Solve each system by substitution.

7. $x - 2y = 7$
$y = -11 - x$

8. $7x - 4y = 10$
$x - 2y = 0$

9. $a + 3b = -2$
$a = 2b - 7$

Solve each system by elimination.

10. $2x + 5y = -13$
$-2x + 6y = -20$

11. $6x - 8y = 36$
$1.5x - 2y = 9$

12. $3x - 7y = -19$
$2x + 3y = -5$

Solve each system.

13. $4x + y = 0$
$5y = 12 - 8x$

14. $-3n + 5m = 10$
$-4m = -2(n + 1)$

15. $x + 9 = 2y$
$8y - 13 = 4x$

16. $6x + 10y - 12 = 0$
$3x + 2.5y - 6 = 0$

Solve.

17. At a college commencement, four times as many bachelor's degrees as associate's degrees were awarded to graduating students. If 3095 students graduated, how many bachelor's degrees and associate's degrees were awarded?

18. For a student club fund-raiser, the number of \$2 raffle tickets printed was three times the number of \$5 tickets. If all of the tickets are sold, receipts from the \$5 tickets will be \$50 less than those from the \$2 tickets. How many \$5 tickets were printed?

19. A lottery winner invested \$200,000 of her winnings in two funds earning 5% and 6% simple interest, respectively. If after one year she earned \$11,200 in interest, how much did she invest in each fund?

20. On a boating trip, it took 2 hr to travel 13 mi with the current. It took the same amount of time to travel 11 mi against the current on the return trip. Find the speed of the boat and the speed of the current.

• Check your answers on page A-28.

14.1 Introduction to Systems of Linear Equations: Solving by Graphing

OBJECTIVES

- **A** To decide whether an ordered pair is a solution of a system of linear equations in two variables
- **B** To determine the number of solutions of a system of linear equations
- **C** To solve systems of linear equations by graphing
- **D** To solve applied problems involving systems of linear equations

What Systems of Linear Equations Are and Why They Are Important

Recall from previous chapters that some situations are described by a linear equation in one variable, say, $2x + 1 = 10$, whereas others are described by a single linear equation in *two* variables, for instance, $y = 3x - 5$.

Now, let's consider situations in which the relationship between two variables is described by a *pair* of linear equations, for instance:

$$x + y = 7$$
$$x - y = 3$$

Groups of equations, called *systems*, serve as a model for a wide variety of applications in fields such as business and science. A system can represent the conditions that must be satisfied in a particular situation. For instance, the system might describe how the cost of products relate to one another or the motion of an airplane in various wind conditions.

In this chapter, we deal with three approaches to solving systems of linear equations, namely by *graphing*, *substitution*, and *elimination*.

Introduction to Systems of Linear Equations

We begin by focusing on the meaning of a system of equations.

DEFINITION

A **system of equations** is a group of two or more equations solved simultaneously.

Systems of equations are sometimes written with large braces:

$$\begin{Bmatrix} x + y = 7 \\ x - y = 3 \end{Bmatrix} \quad \text{or} \quad \begin{cases} x + y = 7 \\ x - y = 3 \end{cases}$$

Braces are used to emphasize that any solution of a system must satisfy *all* the equations in the system. For instance, $x = 5$ and $y = 2$ is a solution of the system above, because when we substitute 5 for x and 2 for y into the equations, *both* equations are true:

$$x + y = 7 \quad \rightarrow \quad 5 + 2 \stackrel{?}{=} 7 \quad \text{True}$$
$$x - y = 3 \quad \rightarrow \quad 5 - 2 \stackrel{?}{=} 3 \quad \text{True}$$

A solution of a system of two equations is commonly represented as an ordered pair of numbers. For instance, the solution of the system just mentioned can be written as $(5, 2)$. Can you explain why $(2, 5)$ is not a solution of this system?

DEFINITION

A **solution** of a system of two linear equations in two variables is an ordered pair of numbers that makes both equations in the system true.

EXAMPLE 1

Consider the following system:

$$x - 2y = 6$$
$$2x + 5y = 3$$

a. Is $(4, -1)$ a solution of the system?

b. Is $(2, -2)$ a solution of the system?

Solution

a. To decide if the ordered pair $(4, -1)$ is a solution of this system, we substitute the x-coordinate 4 for x and the y-coordinate -1 for y in the equations and check if both equations are true.

$$x - 2y = 6 \rightarrow 4 - 2(-1) \stackrel{?}{=} 6 \xrightarrow{\text{Simplifies to}} 6 = 6 \quad \text{True}$$
$$2x + 5y = 3 \rightarrow 2(4) + 5(-1) \stackrel{?}{=} 3 \xrightarrow{\text{Simplifies to}} 3 = 3 \quad \text{True}$$

So the ordered pair $(4, -1)$ satisfies both equations and so is a solution of the system.

b. To see if $(2, -2)$ is a solution, we substitute 2 for x and -2 for y in the equations and check if they are both true.

$$x - 2y = 6 \rightarrow 2 - 2(-2) \stackrel{?}{=} 6 \xrightarrow{\text{Simplifies to}} 6 = 6 \quad \text{True}$$
$$2x + 5y = 3 \rightarrow 2(2) + 5(-2) \stackrel{?}{=} 3 \xrightarrow{\text{Simplifies to}} -6 = 3 \quad \text{False}$$

So the ordered pair $(2, -2)$ is not a solution of the system because it does not satisfy both equations.

PRACTICE 1

Consider the following system:

$$3x + 2y = 5$$
$$4x - 2y = -5$$

Determine whether the following ordered pairs are solutions of the system.

a. $(0, 2.5)$

b. $(1, -1)$

Number of Solutions of a System

In solving a system of linear equations, a question that immediately comes to mind is how many solutions the system has. Let's consider this question graphically. Since each equation is linear, both graphs are lines. Now, suppose that we graph the two equations on the same coordinate plane. Any point at which the two graphs of the system intersect is a solution of the system, because that point must satisfy both equations.

For instance, let's reconsider the system discussed on the previous page

$$x + y = 7$$
$$x - y = 3$$

and solve it by graphing both equations.

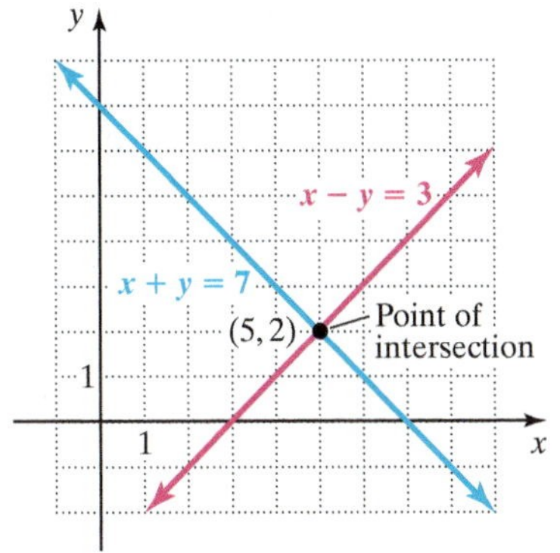

Note that the lines intersect at $(5, 2)$—precisely the ordered pair that we have previously determined to be a solution of this system. Since the two lines meet at a single point, the system has *exactly one* solution.

Not all systems have one solution. For instance, consider the following system in which there are *no* solutions:

$$\begin{aligned} 3x - y &= 2 \\ 3x - y &= 4 \end{aligned}$$

Graphing this system, we get:

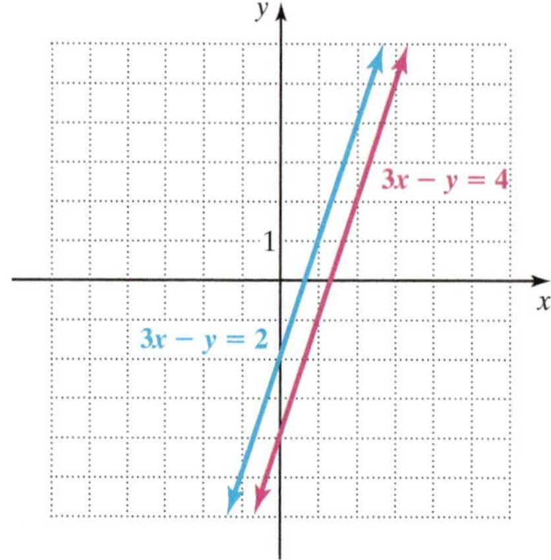

We observe that the lines are parallel and, therefore, do not intersect. So in this case, the system has no solutions.

Finally, let's examine a system that has more than one solution.

$$\begin{aligned} 2x - 2y &= 6 \\ y &= x - 3 \end{aligned}$$

The graph of this system is as follows:

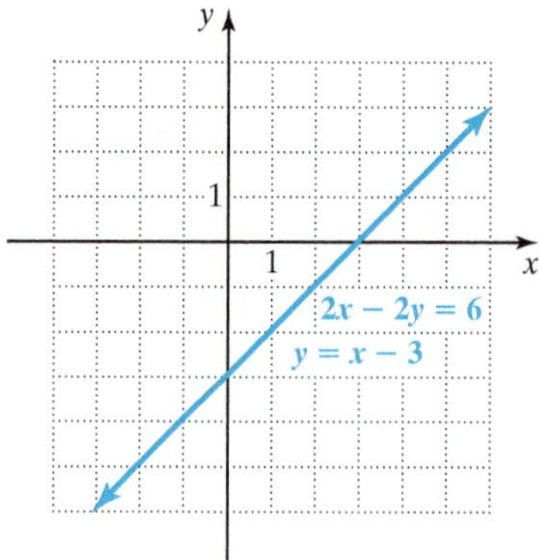

Note that only one line is shown. The reason is that both equations in the system have the same graph. Any point on this line, for instance $(3, 0)$, is a solution of both equations. A system such as this one has *infinitely many* solutions, namely every point on the line.

Every system of linear equations has one solution, no solution, or infinitely many solutions. The following table summarizes the main features of the three types of systems:

Number of Solutions	Description of the System's Graph	Possible Graph
One solution	The lines intersect at exactly one point.	
No solution	The lines are parallel.	
Infinitely many solutions	The lines coincide, that is, they are the same line.	

EXAMPLE 2

For each system graphed, determine the number of solutions.

a.

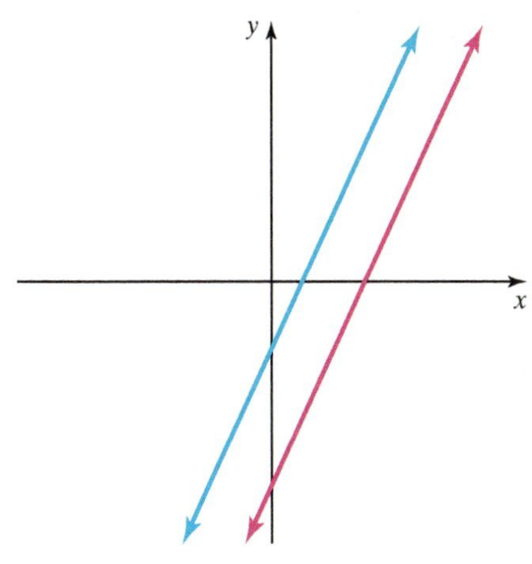

PRACTICE 2

Determine the number of solutions of each of the following systems:

a.

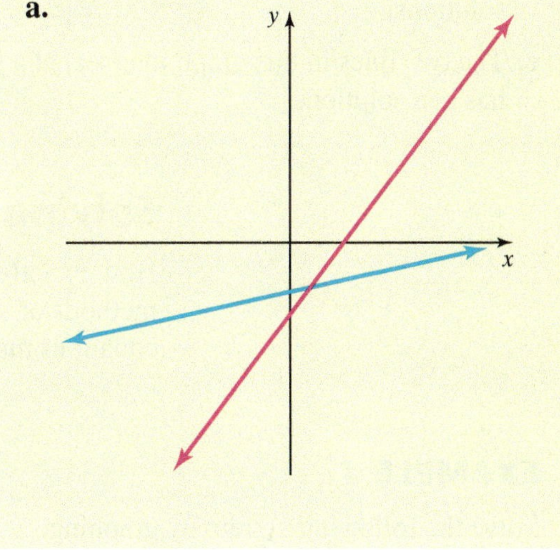

EXAMPLE 2 (continued)

b.

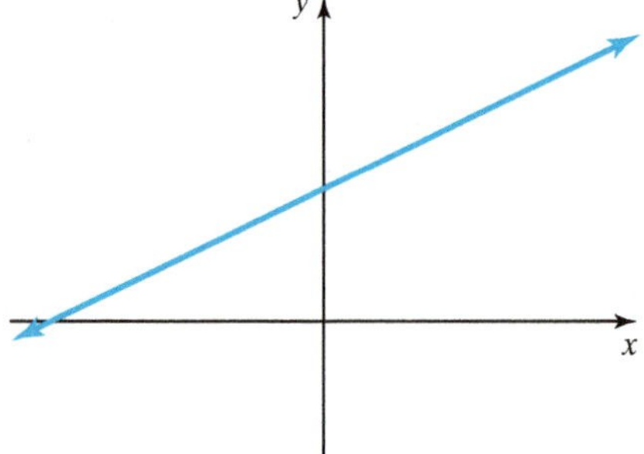

c.

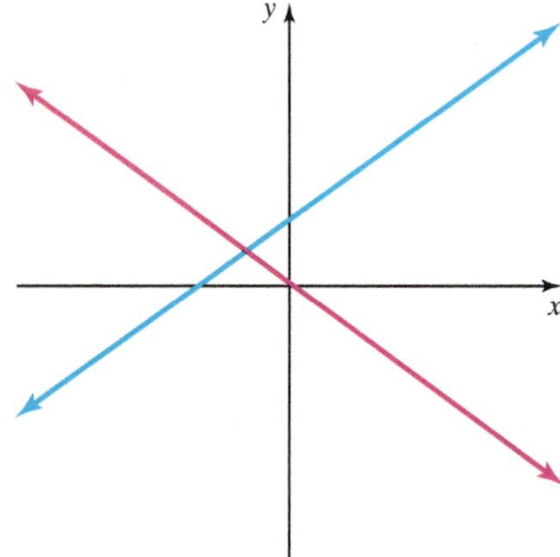

Solution

a. The graph of the system consists of two lines that appear to be parallel. Since the lines do not intersect, the system has no solution.

b. This graph is a single line. Therefore, the system has infinitely many solutions.

c. The two lines in this graph intersect at a single point. So the system has one solution.

b.

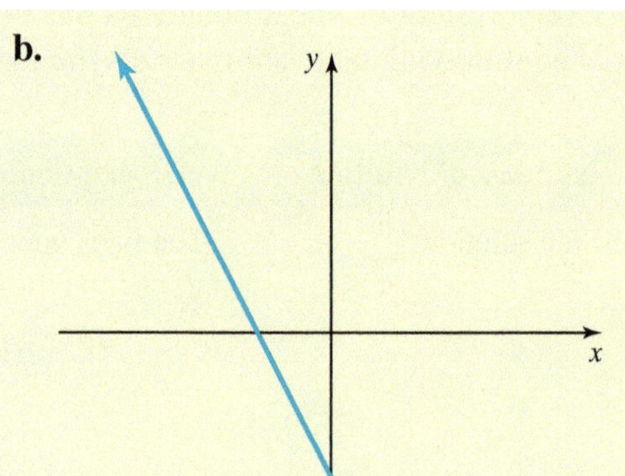

c.

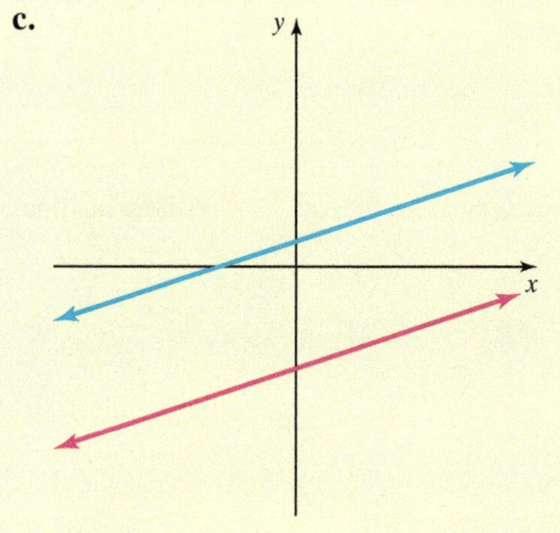

Solving Systems by Graphing

How exactly are systems of linear equations solved? In this chapter, we consider several methods of solving systems. Let's first discuss the **graphing method** in which we graph the equations that make up the system. Any point of intersection is a solution of the system.

EXAMPLE 3

Solve the following system by graphing:

$$\begin{aligned} x + y &= 6 \\ x - y &= -4 \end{aligned}$$

Solution Let's graph each linear equation by using the x- and y-intercept method and then sketching the line that passes through these points.

PRACTICE 3

On the given coordinate plane, solve the following system by graphing:

$$\begin{aligned} x + y &= 2 \\ x - y &= 4 \end{aligned}$$

$x + y = 6$ | $x - y = -4$

x	y
0	6
6	0
3	3

x	y
0	4
−4	0
2	6

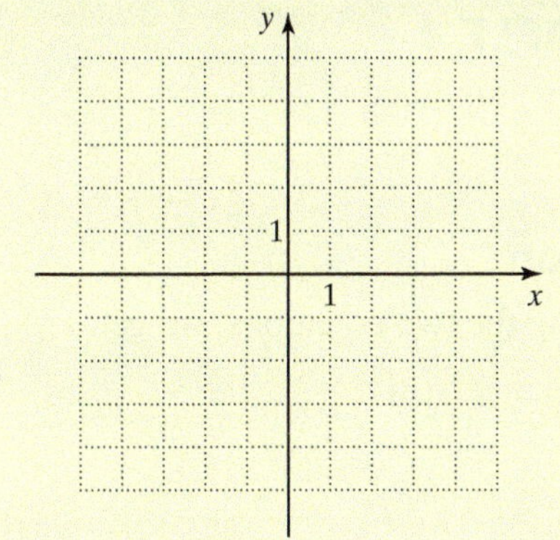

On the same coordinate plane, we plot the points, and then graph both equations.

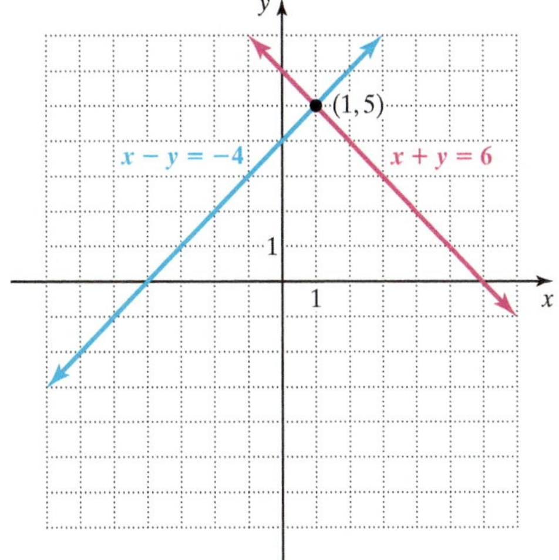

The corresponding lines appear to intersect at the point $(1, 5)$, giving $x = 1$ and $y = 5$.

Check Since solving a system of equations by graphing may result in approximate solutions, we confirm that $(1, 5)$ is the solution by substituting these values into the original equations:

$$x + y = 6 \xrightarrow{\text{Substitute 1 for } x \text{ and 5 for } y.} 1 + 5 \stackrel{?}{=} 6$$
$$6 = 6 \quad \text{True}$$

$$x - y = -4 \xrightarrow{\text{Substitute 1 for } x \text{ and 5 for } y.} 1 - 5 \stackrel{?}{=} -4$$
$$-4 = -4 \quad \text{True}$$

So $(1, 5)$ is the solution of the system.

To Solve a System of Linear Equations by Graphing

- Graph both equations on the same coordinate plane.
- There are three possibilities:
  a. If the lines intersect at exactly one point, then the solution is the ordered pair of coordinates for the point of intersection. Check that these coordinates satisfy both equations.
  b. If the lines are parallel, then there is no solution of the system.
  c. If the lines coincide, then there are infinitely many solutions, namely all the ordered pairs of coordinates that represent points on the line.

EXAMPLE 4

Solve by graphing.

$$y = 2x + 5$$
$$2x - y = 2$$

Solution

Graphing the two equations, we get:

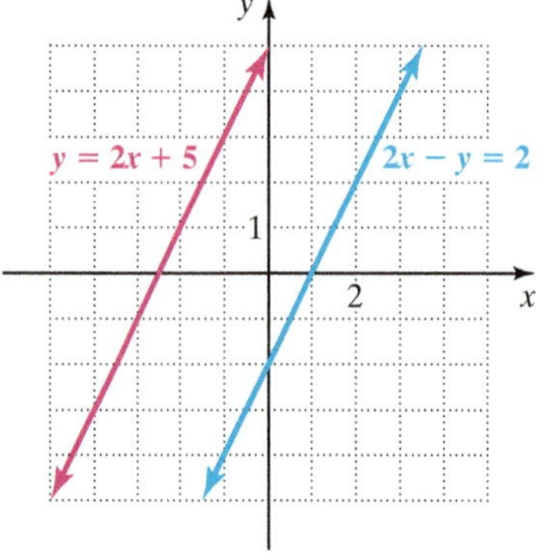

The lines appear to be parallel, which suggests that the system has no solution. To confirm that the lines are parallel, we can check that their slopes are equal. The graph of the first equation, $y = 2x + 5$, has slope 2. To find the slope of the second equation, we write $2x - y = 2$ in slope-intercept form, getting $y = 2x - 2$. The graph of this equation also has slope 2. Therefore, the lines are parallel and the system has no solution.

PRACTICE 4

Solve for x and y by graphing.

$$y = x - 6$$
$$x - y = 4$$

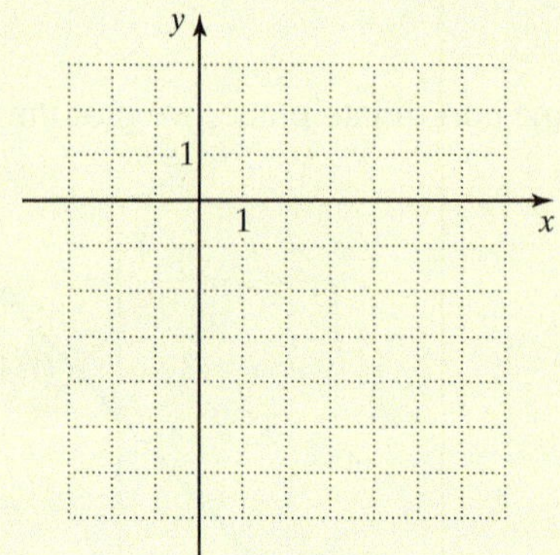

EXAMPLE 5

Solve by graphing.

$$2y = -8x + 2$$
$$-4x - y = -1$$

Solution

When we graph the two equations in this system, we get the same graph.

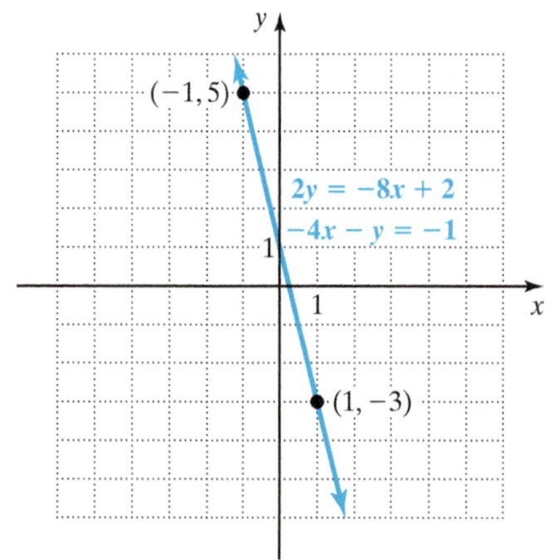

Note that when the two equations in the system are changed to slope-intercept form, the equations are identical.

$$2y = -8x + 2 \xrightarrow{\text{Isolate } y.} y = -4x + 1$$

$$-4x - y = -1 \xrightarrow{\text{Isolate } y.} y = -4x + 1$$

We conclude that the system has infinitely many solutions. All points on the line, some of which are indicated, are solutions. Can you identify another point on the graph and confirm that it is a solution to the system?

PRACTICE 5

Solve by graphing.

$$6x = 15 - 3y$$
$$y = 5 - 2x$$

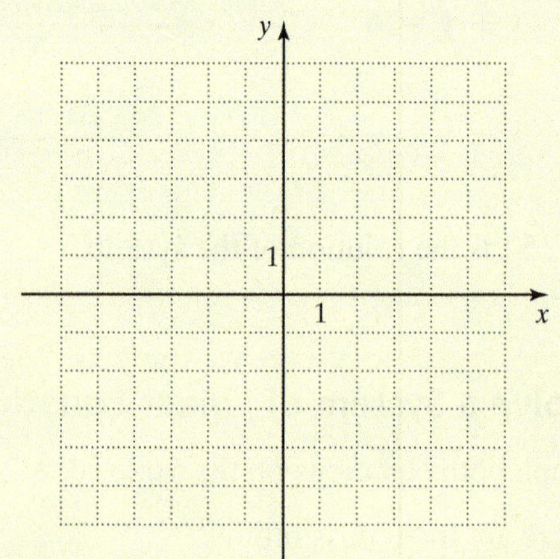

EXAMPLE 6

The U.S. House of Representatives has 435 members. On a certain bill, all the representatives voted, and 15 more representatives voted for the bill than against the bill. (There were no abstentions.)

a. If s represents the number of representatives who *supported* the bill, and n the number of representatives who did *not* support the bill, express the given information as a system of equations.

b. On a coordinate plane, graph the system found in part (a).

c. Find the coordinates of the point of intersection.

d. In this problem, what is the significance of the coordinates of the point of intersection?

Solution

a. The given information can be expressed algebraically as:

$$s + n = 435$$
$$s = n + 15$$

b. First, let's graph s along the vertical axis and n along the horizontal axis. Since the number of representatives voting for or against a bill is between 0 and 435, we then choose an appropriate scale and label the two axes accordingly. Next, we graph the two equations.

c. The lines intersect at the point $(210, 225)$, that is, $n = 210$ and $s = 225$.

d. We conclude that 210 representatives voted against the bill and 225 voted for the bill.

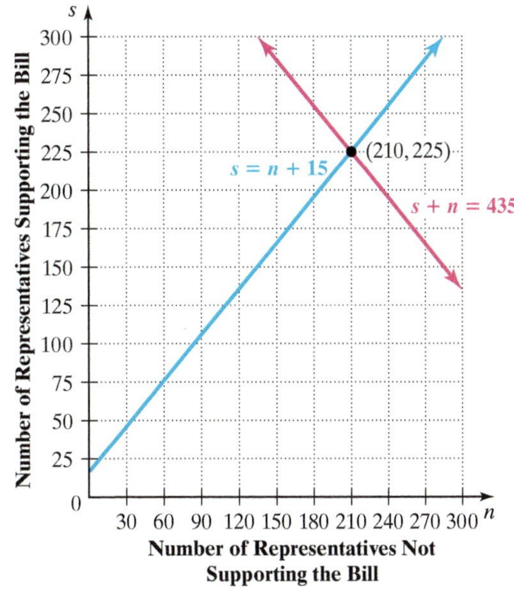

PRACTICE 6

A liberal arts student transferring to a four-year institution took a test of verbal skills and a test of mathematical skills. Her total score was 1150, and the verbal score v was 100 less than the math score m.

a. Express the given information as a system of equations.

b. On a coordinate plane, graph the system found in part (a).

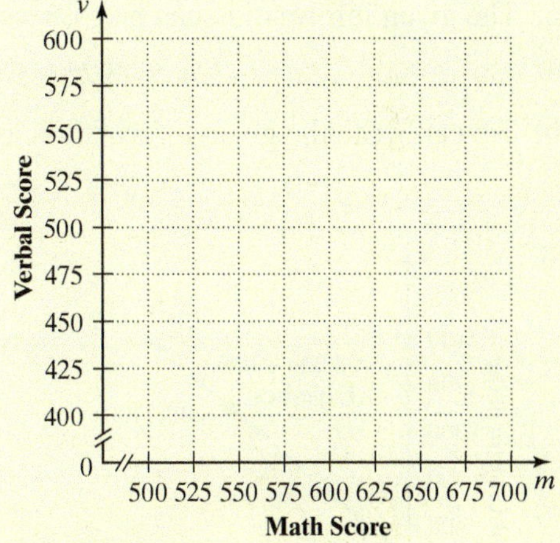

c. Identify the coordinates of the point of intersection.

d. In this situation, what is the significance of the coordinates of the point of intersection?

When running a business, it is important to determine both the income that the business makes and the expenses that it takes to run the business. The business' income and its expenses depend on the number of items produced and sold. These quantities can be graphed on a coordinate plane, as shown in the next example. The point at which the income for a business equals its expenses is called the *break-even point*.

EXAMPLE 7

For a start-up business, an entrepreneur determined that to produce computer-generated, silk-screen T-shirts it will cost \$3.25 a shirt plus \$450 in fixed overhead. Each shirt produced is sold at \$5.50.

a. If x represents the number of T-shirts sold and y the amount it costs to produce the T-shirts, write an equation that expresses y in terms of x.

b. If x represents the number of T-shirts produced and y the amount of income from selling the T-shirts, write an equation that relates x and y.

c. On a coordinate plane, graph the lines found in parts (a) and (b).

d. Find the break-even point for producing the T-shirts. Explain its significance in terms of the x- and y-coordinates.

Solution

a. The given information can be expressed as:

$$y = 3.25x + 450$$

b. We can write the given information as:

$$y = 5.50x$$

c.

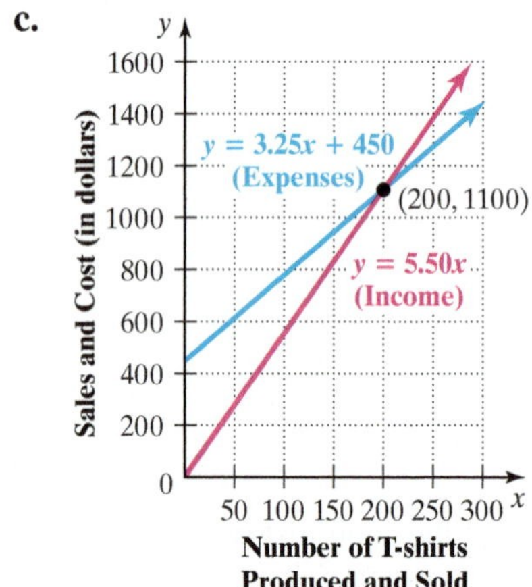

d. Since the break-even point is the point where the income equals the expenses, we must find the point of intersection of the lines $y = 3.25x + 450$ and $y = 5.50x$. The intersection of the lines is the point $(200, 1100)$, which is the break-even point. So when 200 shirts are produced and sold, the income and the expenses will be the same, \$1100. After 200 shirts are produced and sold, the business will start making a profit.

PRACTICE 7

To print a newsletter costs \$450 fixed overhead plus \$1.50 a copy. The newsletters sell for \$3 each.

a. If x represents the number of copies printed and y the amount of money it costs to print the newsletter, write an equation that expresses y in terms of x.

b. If x represents the number of copies printed and y the amount of income from newsletter sales, write an equation that relates x and y.

c. On the coordinate plane below, graph the lines found in parts (a) and (b).

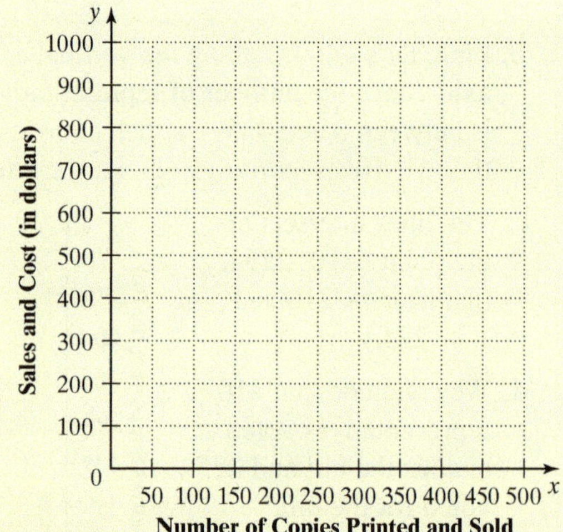

d. Find the break-even point for printing the newsletter.

In this section, we have examined the graphing method of solving systems of linear equations. A major advantage of this method over alternative methods is that it helps us to visualize the problem and its solution. However, a disadvantage of this approach is that our reading of a graph may be inaccurate, particularly when a coordinate of the point of intersection is not an integer or is very large. So a solution found by the graphing method may not be exact.

Solving Systems of Linear Equations on a Grapher

Both graphing calculators and computers with graphing software can help facilitate the process of solving systems of linear equations. As with the paper-and-pencil approach, a grapher displays the graphs of the equations that make up a system on the same coordinate plane. We can then use one of the special features of the grapher to read the coordinates of the point at which the graphed lines intersect, that is, the solution of the system.

The most common features of a grapher that help us to read the coordinates of the point of intersection are TRACE, ZOOM, and INTERSECT.

- *With the TRACE feature, the cursor runs along either of the graphed lines until it is positioned on or near the point of intersection; the coordinates of that point are then displayed.*
- *The ZOOM feature lets us position the cursor as close as we want to the point of intersection.*
- *The INTERSECT feature automatically calculates the point of intersection.*

Note that each of the features may give only an approximation for the point of intersection. However, the most accurate approximation of the point of intersection is given by the INTERSECT feature.

EXAMPLE 8

Use either a graphing calculator or graphing software to solve:

$$4x - y = 11$$
$$x = y + 6$$

Solution Begin by solving each equation for y.

$$4x - y = 11 \xrightarrow{\text{Isolate } y.} y = 4x - 11$$

$$x = y + 6 \xrightarrow{\text{Isolate } y.} y = x - 6$$

Then, press the $\boxed{Y=}$ key, and enter $4x - 11$ to the right of **Y1** = and $x - 6$ to the right of **Y2** =. Set the viewing window. Then, press the $\boxed{\text{GRAPH}}$ key to display the coordinate plane on which the two equations are graphed. The **TRACE** feature can be used to move a cursor along one of the lines toward the intersection of the graphs by holding down an arrow key. Note that as the cursor is moved, the changing coordinates of its position will be displayed on the screen. Once the cursor reaches the point of intersection, we can read the coordinates on the screen.

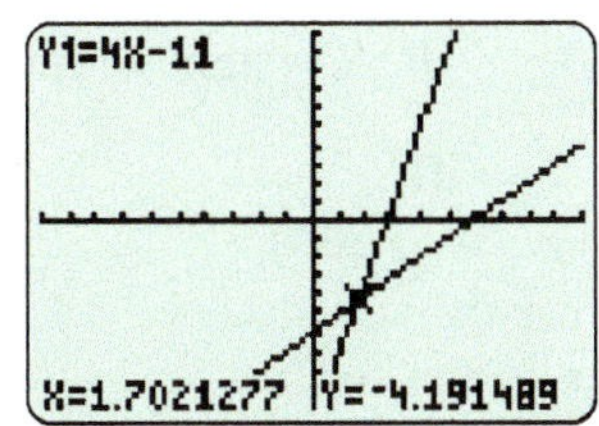

Using the TRACE feature

To get a better approximation of the solution, we can either activate the **ZOOM** feature to zoom in on the intersection point or activate the **INTERSECT** feature.

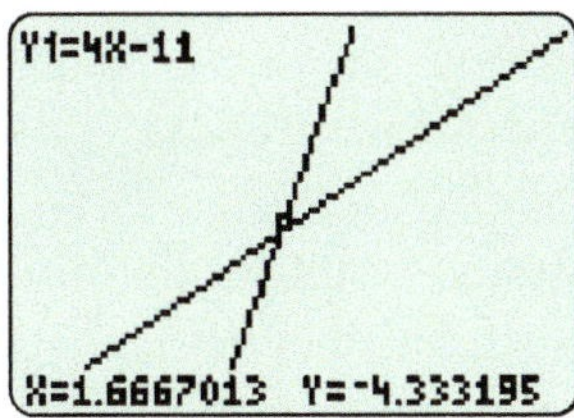

Using the ZOOM feature

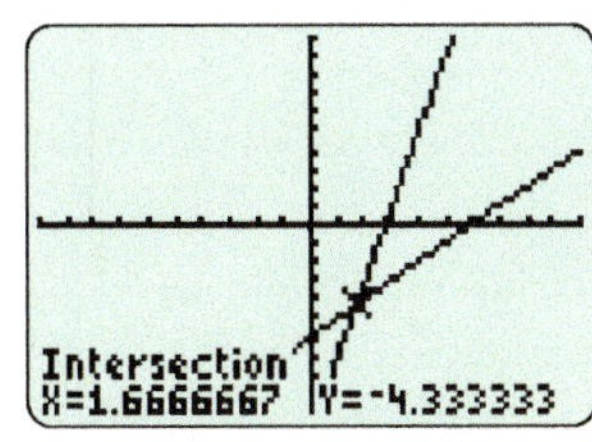

Using the INTERSECT feature

So the approximate solution is $(1.667, -4.333)$.

PRACTICE 8

Use a grapher to solve the following system of equations:

$$8x - y = 1$$
$$y = x + 5$$

14.1 Exercises

FOR EXTRA HELP MyMathLab MathXL PRACTICE WATCH READ REVIEW

Mathematically Speaking

Fill in each blank with the most appropriate term or phrase from the given list.

are parallel	solution	coincide
graph	set of equations	system of equations

1. A(n) ________________ is a group of two or more equations solved simultaneously.

2. A(n) ________________ of a system of two linear equations in two variables is an ordered pair of numbers that makes both equations in the system true.

3. If a system of linear equations has no solution, its graph consists of lines that ________________.

4. If a system has infinitely many solutions, its graph consists of lines that ________________.

A *Indicate whether each ordered pair is or is not a solution of the given system.*

5. $x + y = 3$
$2x - y = 6$

a. $(0, 3)$ ________
b. $(3, 3)$ ________
c. $(3, 0)$ ________

6. $x - 6y = 3$
$x - y = -7$

a. $(-2, -9)$ ________
b. $(-9, -2)$ ________
c. $(9, -2)$ ________

7. $4x + 5y = 0$
$7x - y = 0$

a. $(1, 7)$ _____
b. $(-5, 4)$ _____
c. $(0, 0)$ _____

8. $2x - 2y = 30$
$8x + 2y = -10$

a. $(1, -9)$ ________
b. $(16, 1)$ ________
c. $(2, -11)$ ________

B *Match each system with the appropriate graph.*

9. a. A system with solution $(1, 3)$

b. A system with solution $(-1, 3)$

c. A system with infinitely many solutions

d. A system with no solution

I

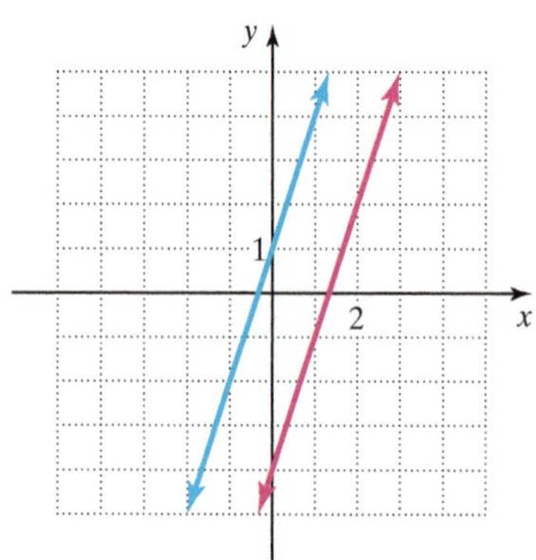

II

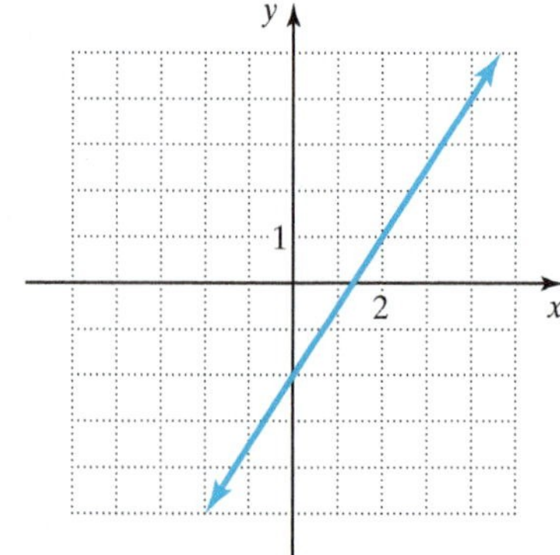

III

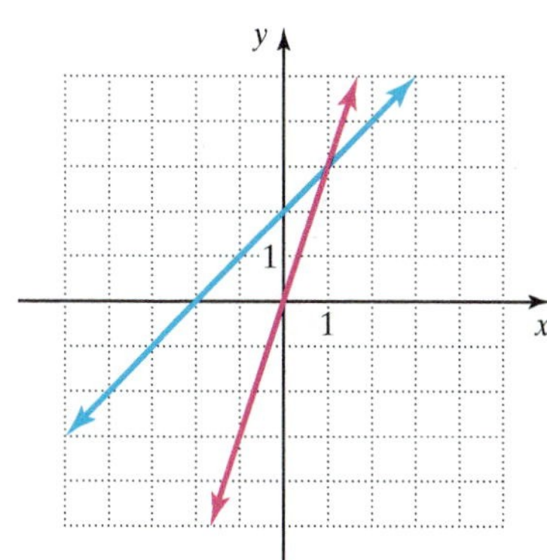

IV

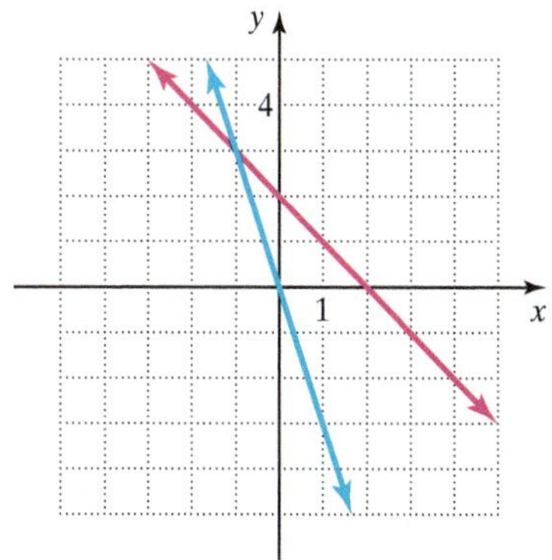

10. **a.** A system with solution $\left(-\frac{1}{4}, -1\frac{1}{4}\right)$

b. A system with solution $(-1, 3)$

c. A system with infinitely many solutions

d. A system with no solution

I

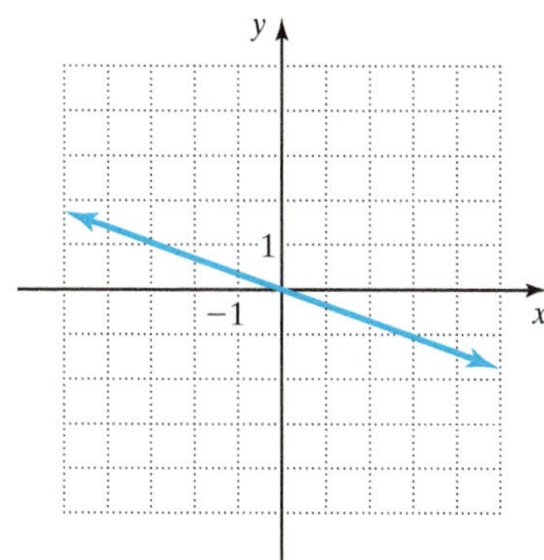

II

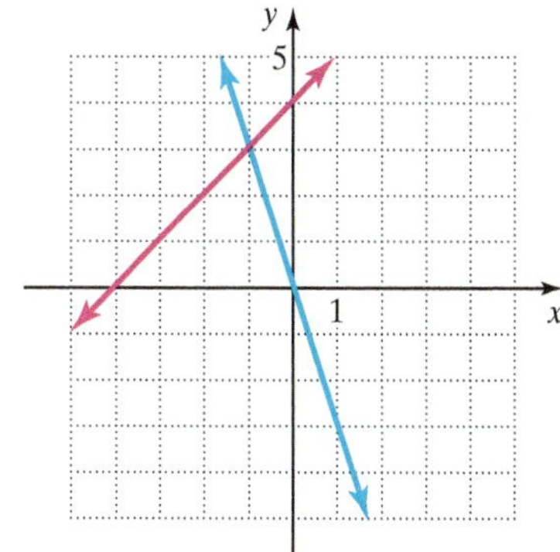

III

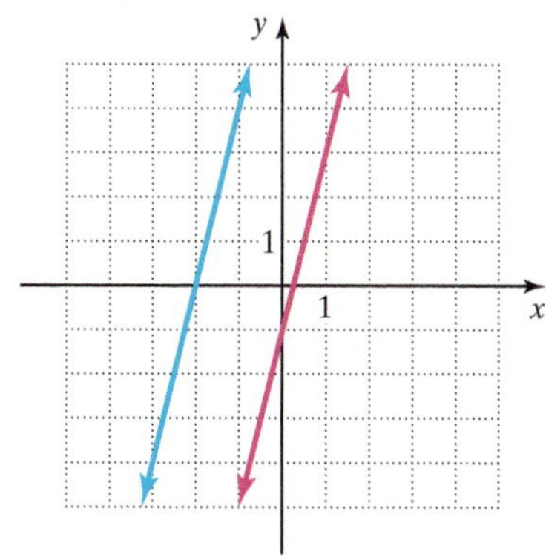

IV

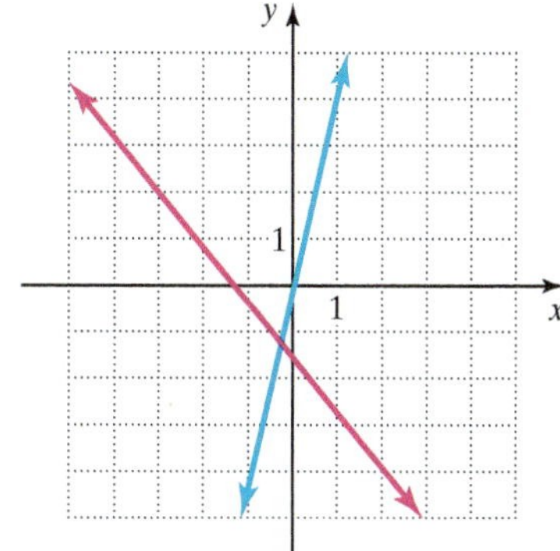

C *Solve by graphing.*

11. $x - y = 2$
$x + y = 4$

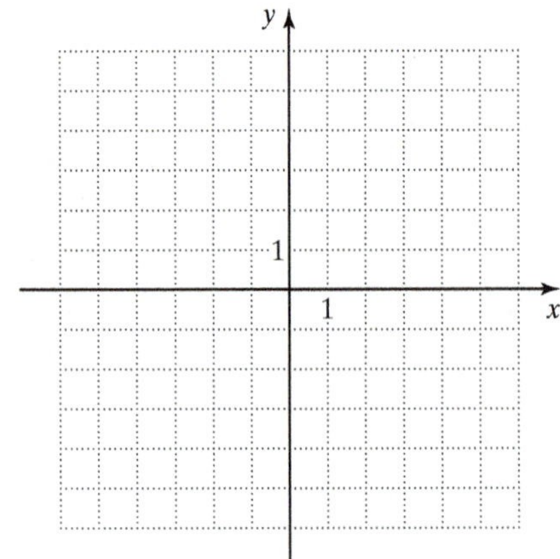

12. $x + 2y = 3$
$x + y = 2$

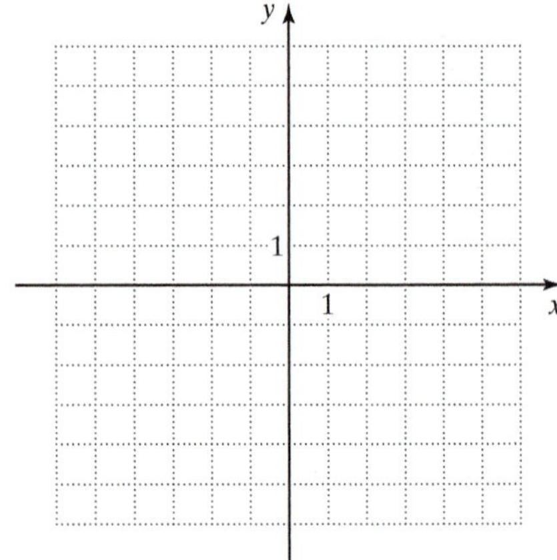

13. $y = x + 4$
$x + y = 4$

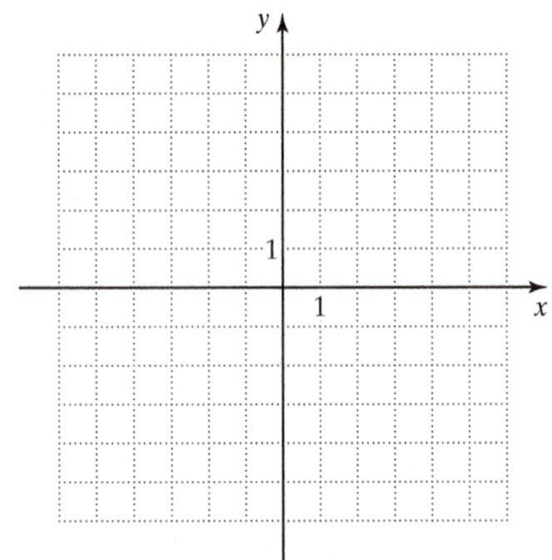

14. $-5 = 2x + y$
$y = -x$

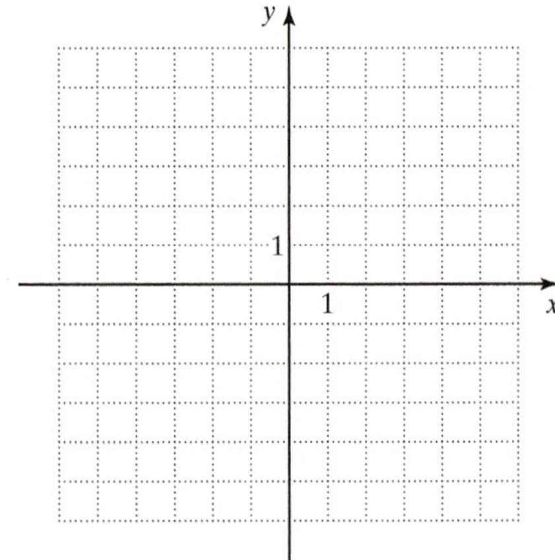

15. $y = -x + 6$
$y = -3x + 8$

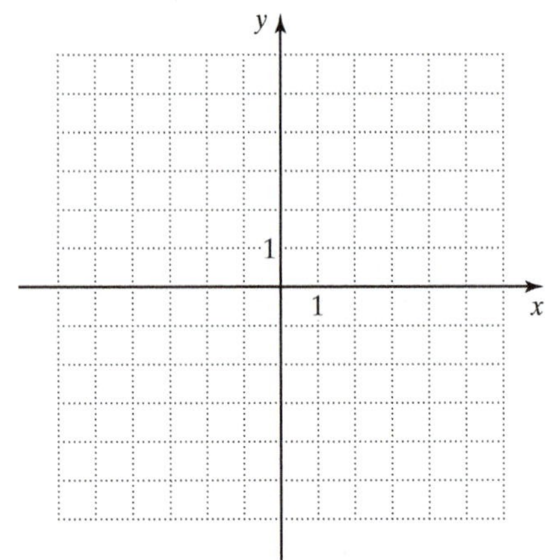

16. $y = x + 1$
$y = -x - 3$

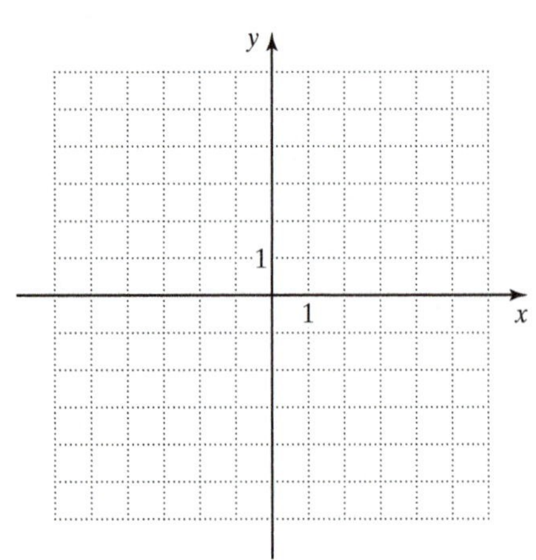

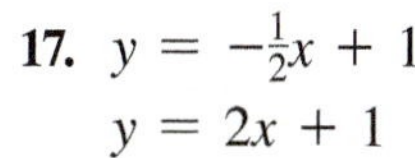

17. $y = -\frac{1}{2}x + 1$
$y = 2x + 1$

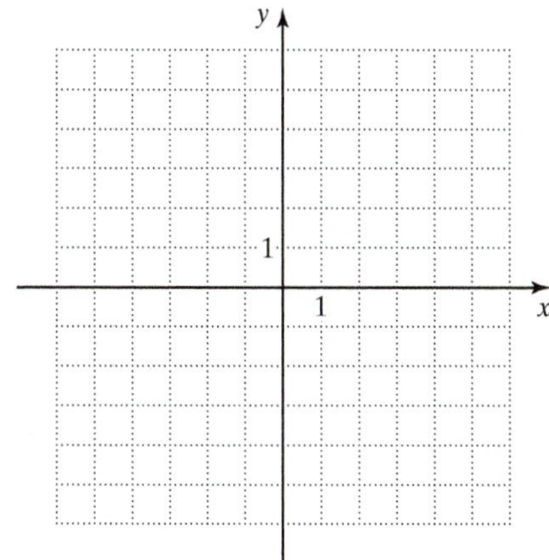

18. $y = 2x - 6$
$y = 3 - \frac{1}{4}x$

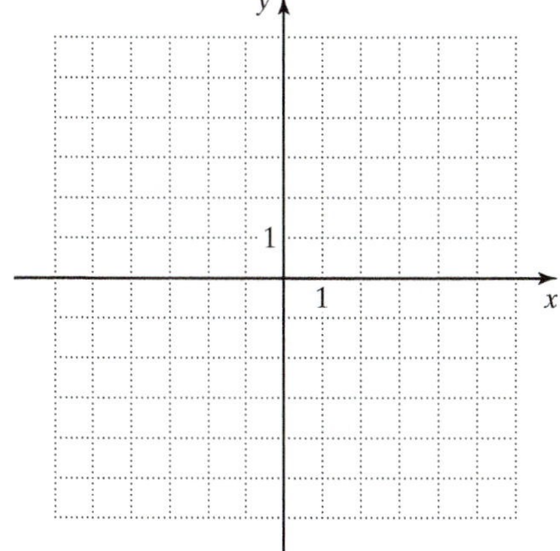

19. $y = 5 + 3x$
$x + y = -3$

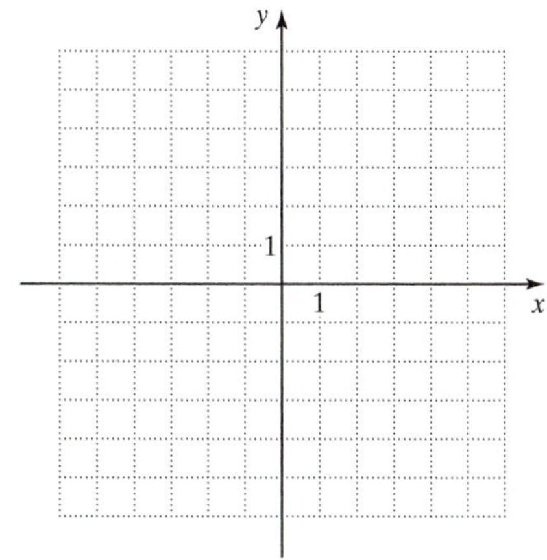

20. $2x + y = -3$
$y = -(x + 4)$

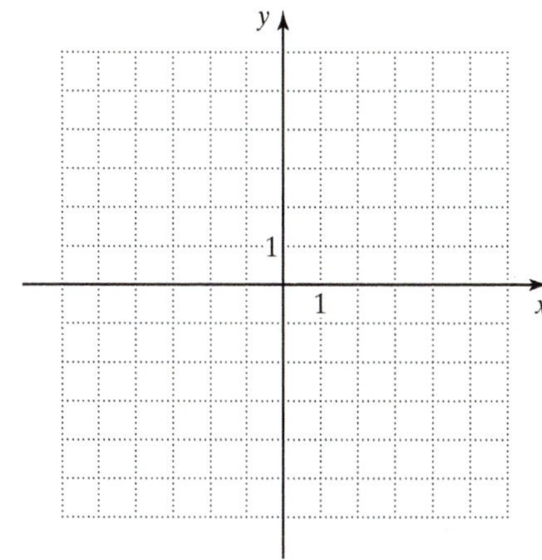

21. $2y = 6x + 2$
$3y - 9x = 3$

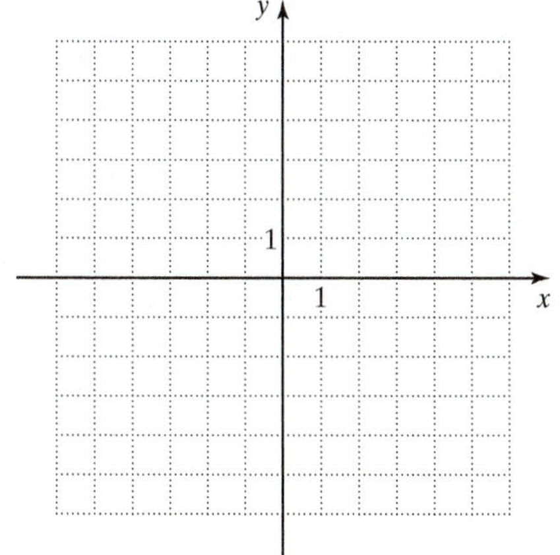

22. $4x = 8y + 4$
$5x - 10y = 5$

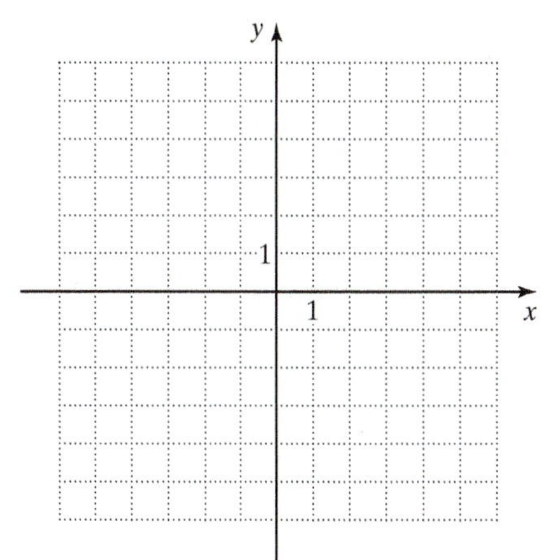

23. $2x + y = -4$
$y = -2x + 3$

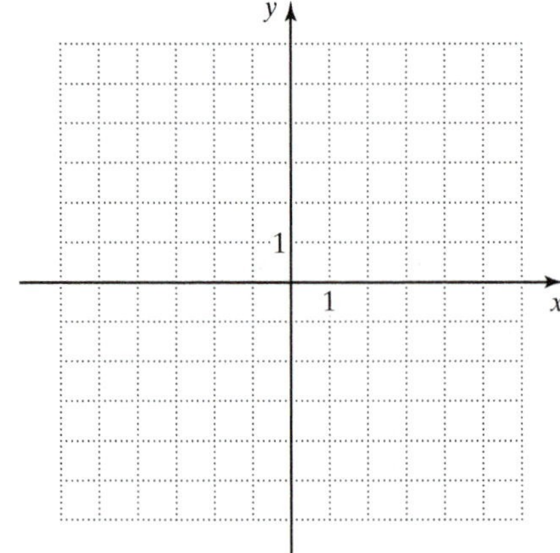

24. $3x - y = 1$
$6x + 4 = 2y$

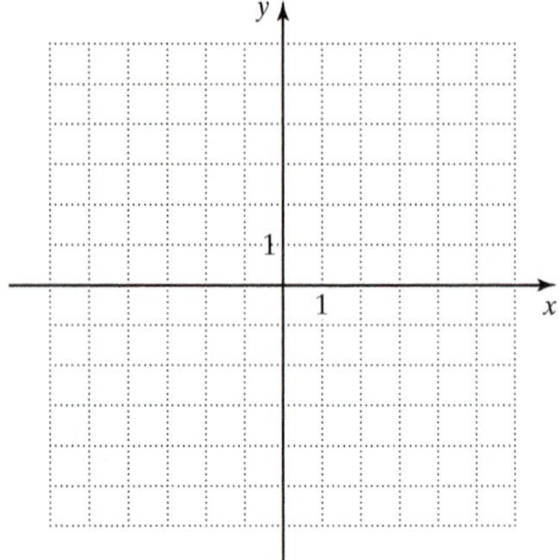

25. $x = 5 + y$
$-2x + 2y = -10$

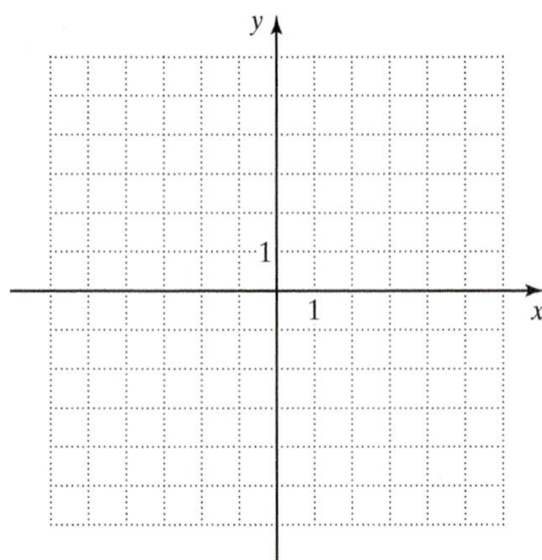

26. $x + y = 6$
$3y - 18 = -3x$

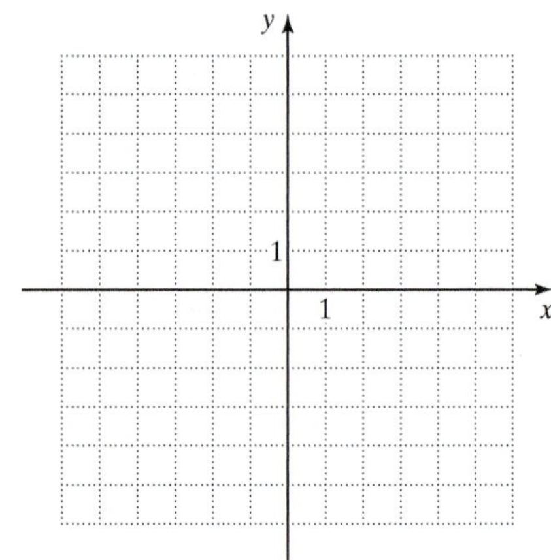

27. $x - y = -1$
$x - y = 4$

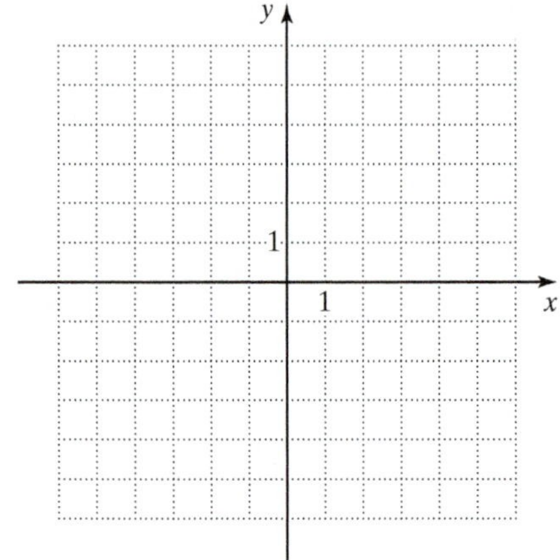

28. $x + 5y = 6$
$x + 5y = 0$

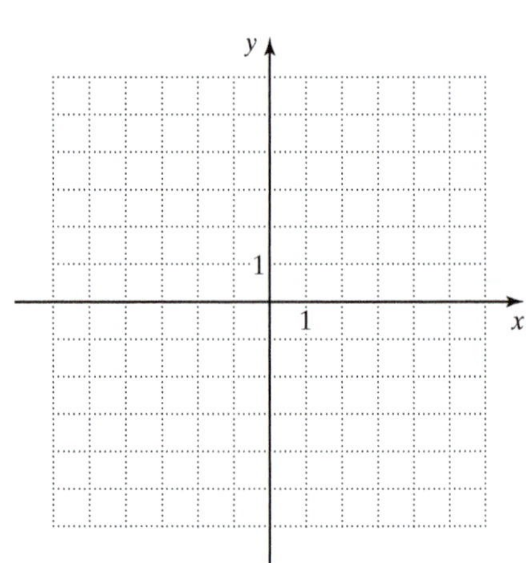

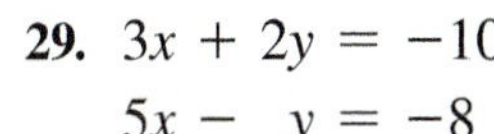

29. $\begin{aligned} 3x + 2y &= -10 \\ 5x - y &= -8 \end{aligned}$

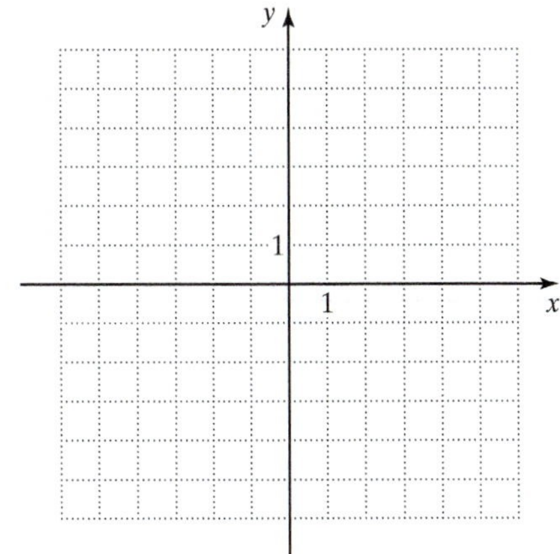

30. $\begin{aligned} -x - 2y &= 8 \\ -6x + 4y &= 0 \end{aligned}$

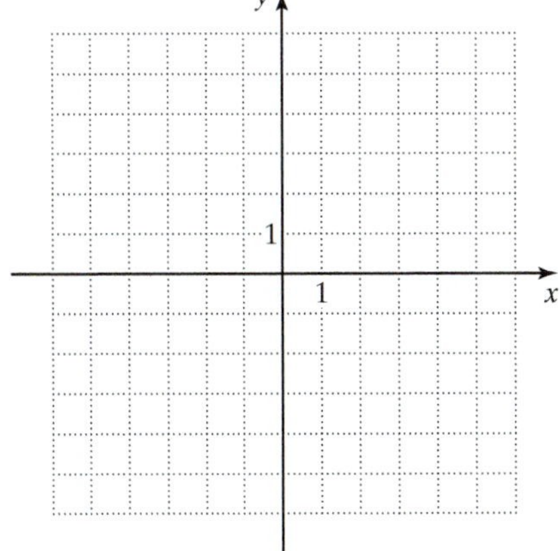

31. $\begin{aligned} -3x + y - 14 &= 0 \\ 3x - y - 11 &= 0 \end{aligned}$

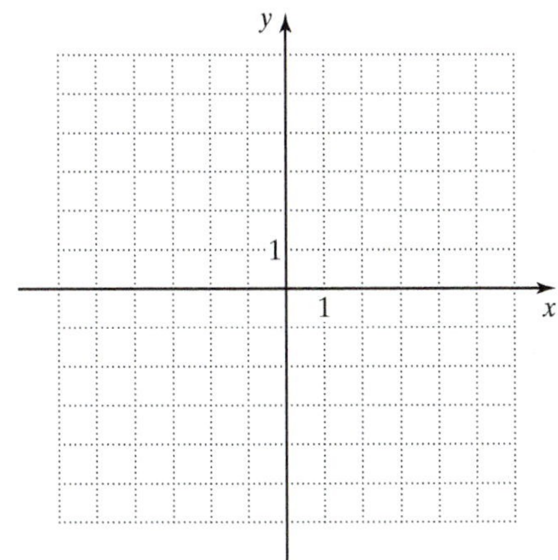

32. $\begin{aligned} 7x - y - 2 &= 0 \\ 14x - 2y - 12 &= 0 \end{aligned}$

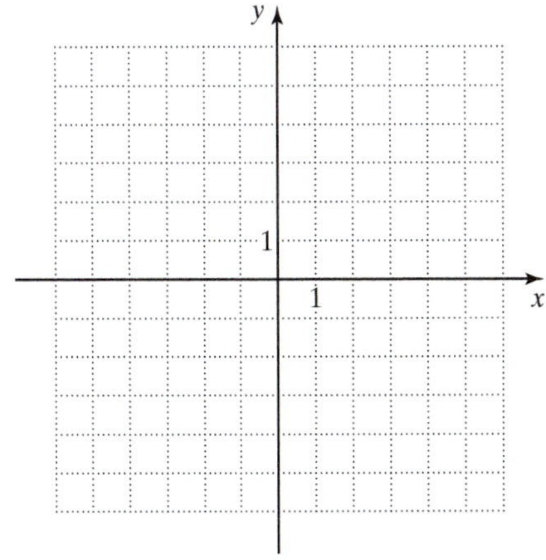

33. $\begin{aligned} 10x + 2y &= -6 \\ y &= 2 \end{aligned}$

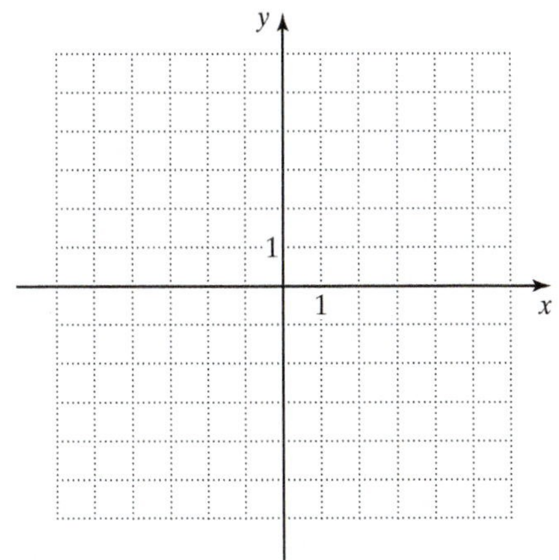

34. $\begin{aligned} x + 5y &= -15 \\ x &= 5 \end{aligned}$

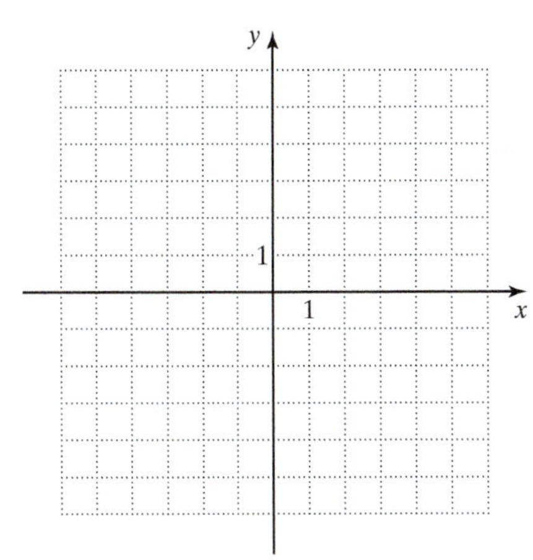

35. $x = 0$
$x - 2y = 4$

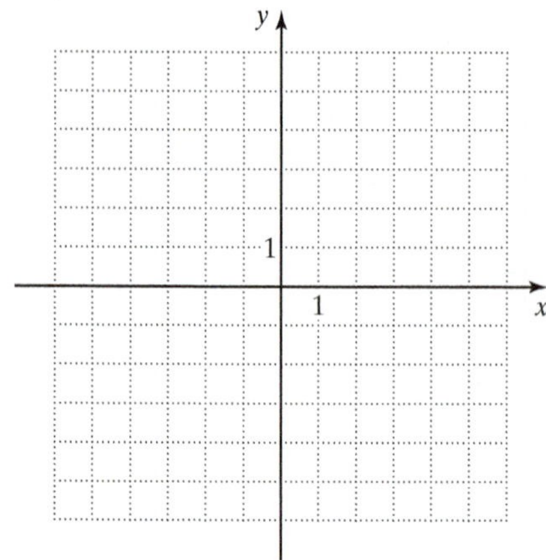

36. $y = -3$
$y = 2x + 3$

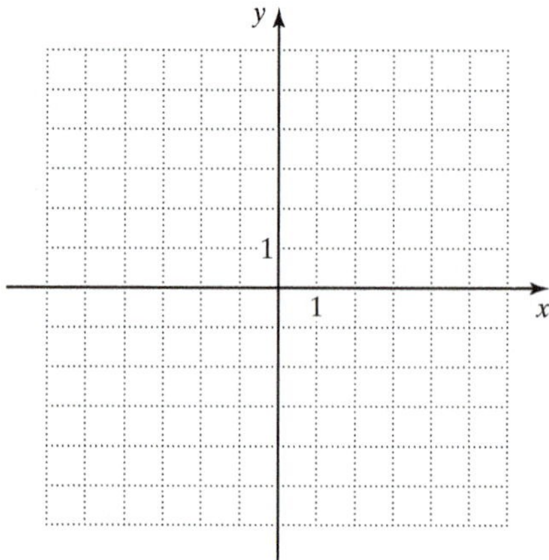

37. $4x - 4y = -8$
$y = \frac{2}{3}x$

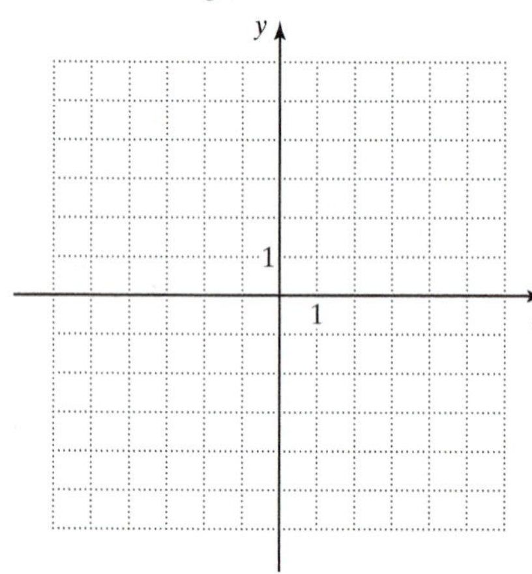

38. $3x + 2y = 6$
$y = -\frac{3}{4}x$

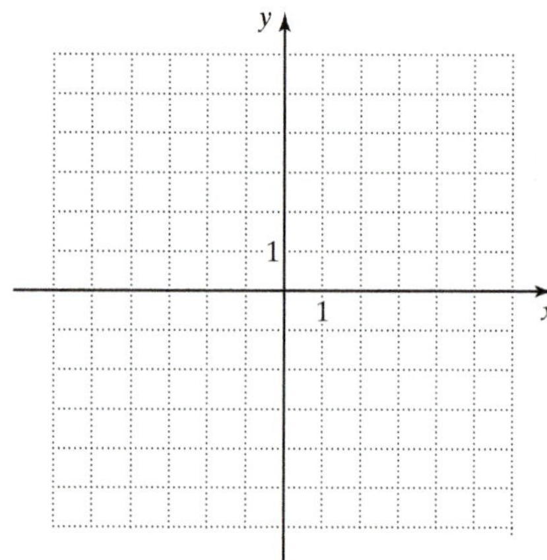

Mixed Practice

Solve by graphing.

39. $2y - 4x = -4$
$y = -2 + 2x$

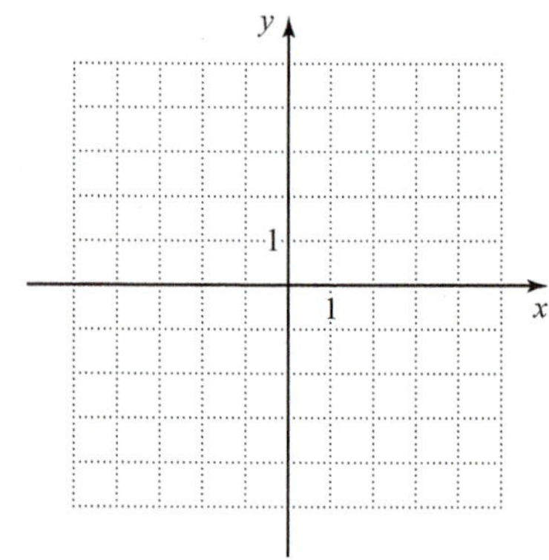

40. $4x + 2 = 2y$
$2x - y = 3$

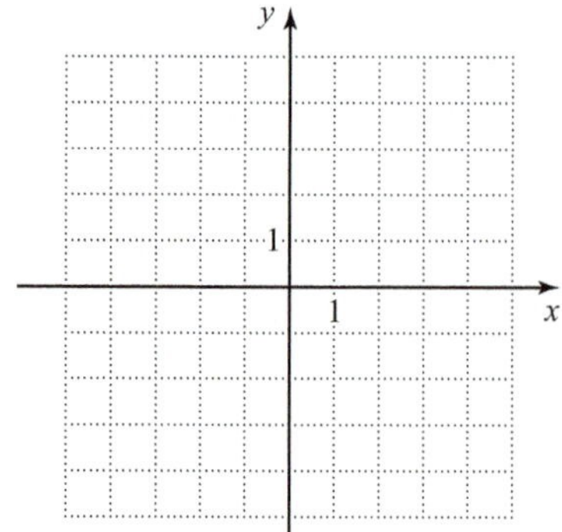

41. $y = -3 + \frac{1}{3}x$
$y = -2x + 4$

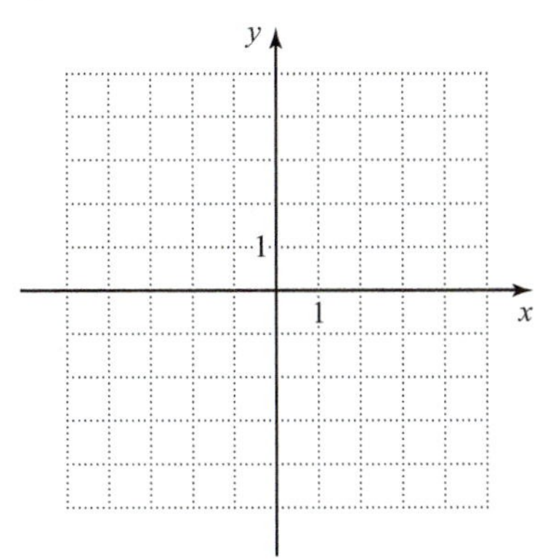

42. $-4 = 2x - y$
$x = 2y + 1$

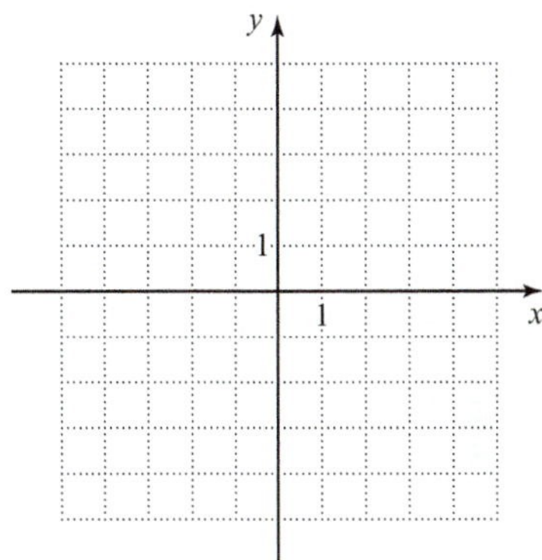

43. $3x - 2y = 2$
$4x + y = 10$

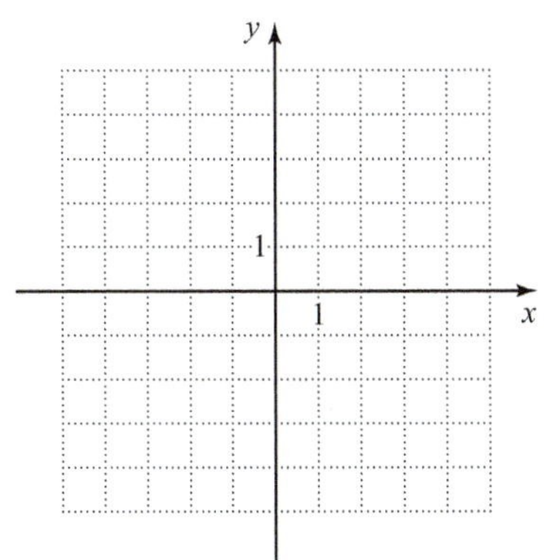

44. $x - 3y = -6$
$x = -3$

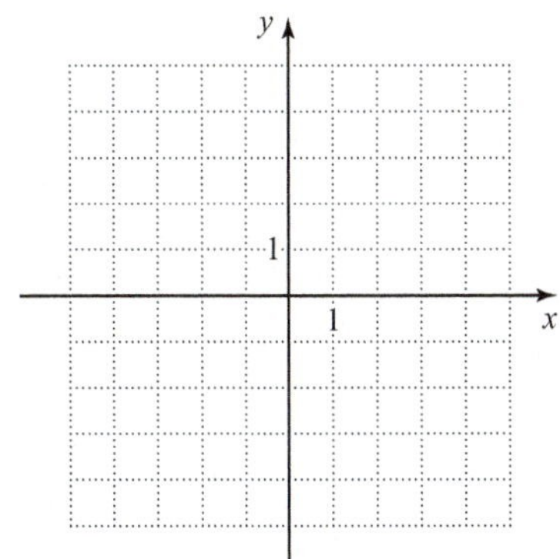

Solve.

45. Which system matches the given graph?

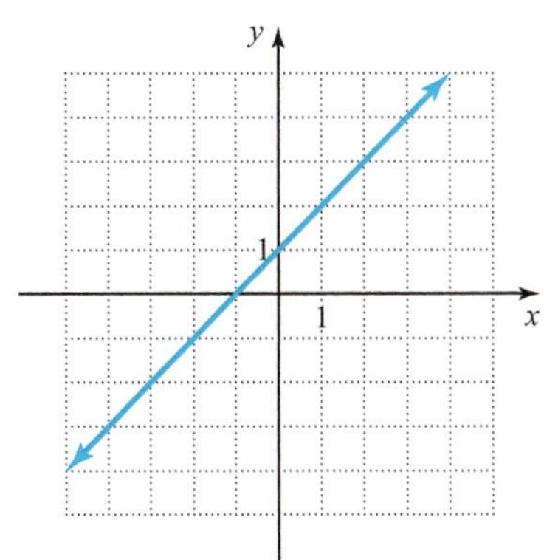

a. A system with no solution
b. A system with infinitely many solutions
c. A system with exactly one solution

46. Indicate whether each ordered pair is or is not a solution to the system.

$$x - 2y = -2$$
$$x - y = 4$$

a. $(-2, -6)$
b. $(12, 8)$
c. $(10, 6)$

Applications

D *Solve.*

47. A young married couple had a combined annual income of $57,000.

a. If the wife made $3000 more than the husband, write these relationships as a system of equations. Let x represent the husband's income and y the wife's income.

b. Graph the equations.

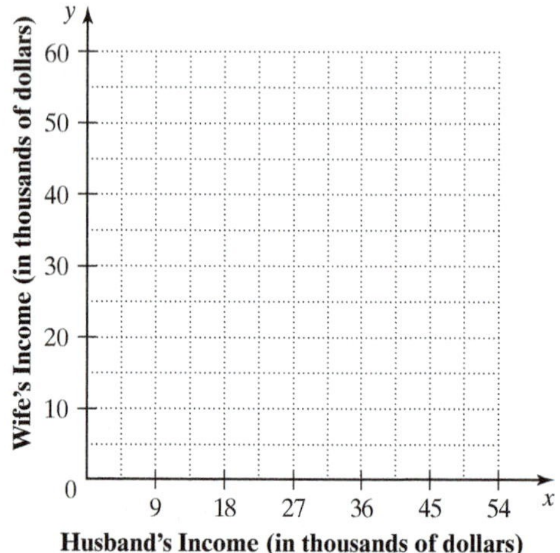

c. Find the two incomes.

48. An airplane flying with a tailwind flew at a speed of 450 mph, relative to the ground. When flying against the tailwind, it flew at a speed of 350 mph.

a. Express these relationships as equations, where x represents the speed of the airplane in calm air and y the speed of the wind.

b. Graph these equations.

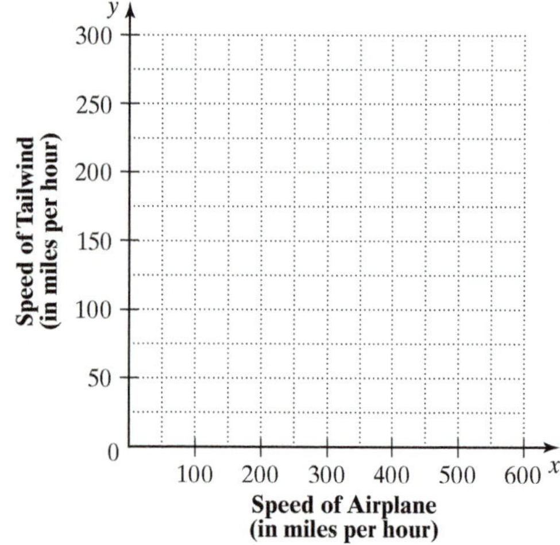

c. Find the speed of the airplane in calm air and the speed of the wind.

49. Mike the plumber charges \$75 for a house call and then \$40/hr for labor. Sally the plumber charges \$100 for a house call and then \$30/hr for labor.

a. Write a cost equation for each plumber, where y is the total cost of plumbing repairs and x is the number of hours of labor.

b. Graph the two equations.

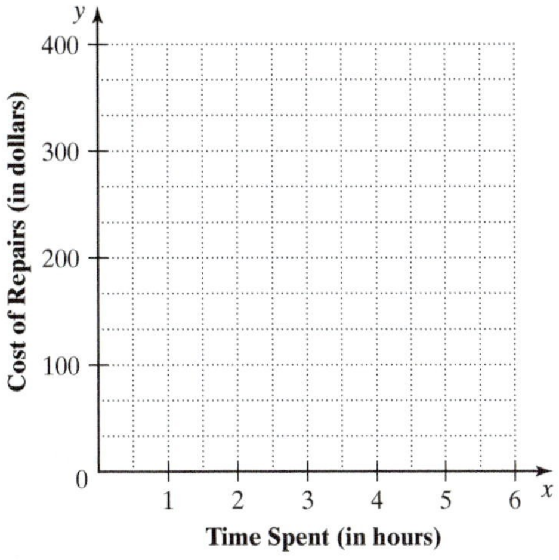

c. Determine the number of hours of plumbing repairs that would be required for the two plumbers to charge the same amount.

d. Determine from the graph which plumber charges less if the estimated amount of time to carry out the plumbing repairs is 5 hr.

50. To connect to the Web, customers must choose between two Internet service providers (ISPs). Flat ISP charges its customers a monthly flat fee of \$10 regardless of how many hours they connect to the Web. A competing company, Variable ISP, charges \$2.50 per month plus \$0.25 for each hour of connection time.

a. Express each company's price p in terms of hours connected h.

b. Draw a graph that shows how each company's price relates to connection time.

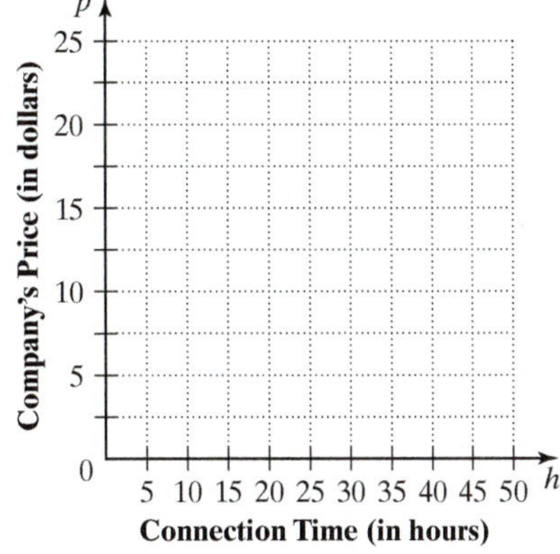

c. For which connection time do the two companies charge the same?

51. A small company duplicates DVDs. The cost of duplicating is \$30 fixed overhead plus \$0.25 per DVD duplicated. The company generates revenues of \$1.50 per DVD. Use a graph to determine the break-even point for duplicating DVDs.

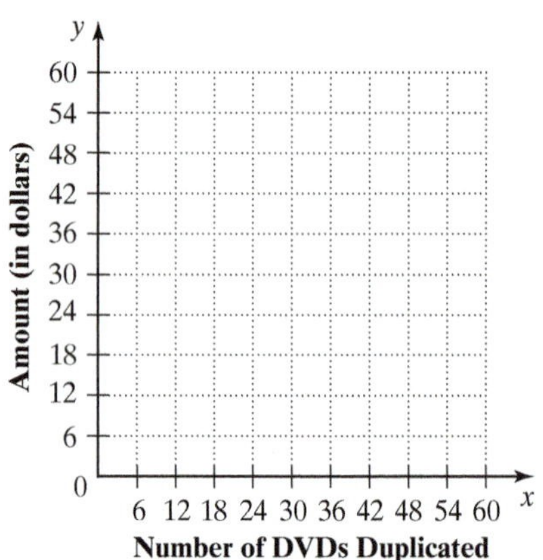

52. A clothing company sells jackets for \$140 per jacket. The company's fixed costs are \$9000 and the variable costs are \$50 per jacket. Use a graph to determine the break-even point for production.

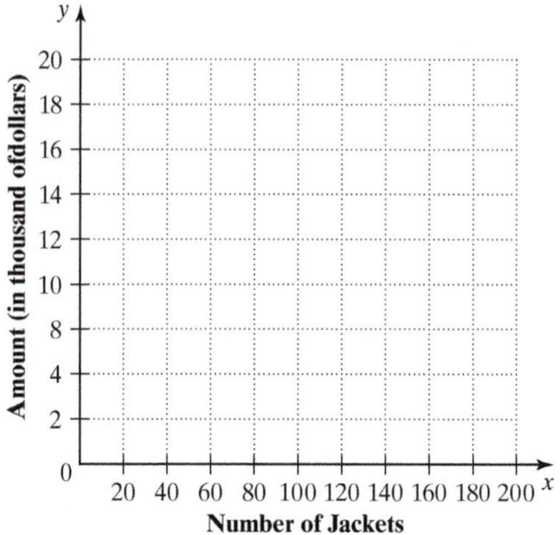

53. A movie fan rented 6 films at a local video store for one day. The daily rental charge was \$2 on some films and \$4 on others. If the total rental charge was \$22, use a graph to determine how many \$4 films were rented.

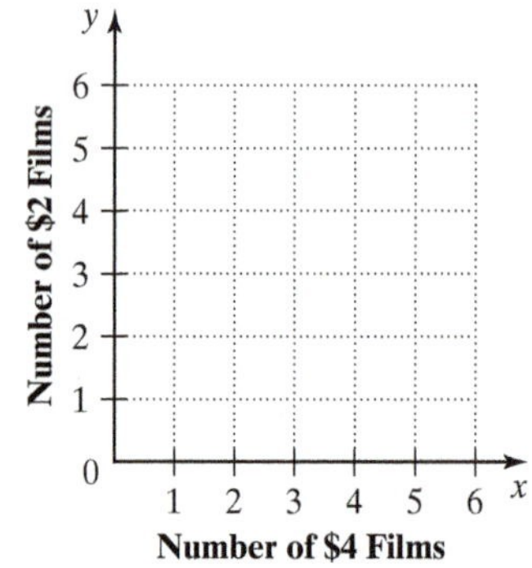

54. An appliance store sells washer-dryer combos for \$1500. If the washer costs \$200 more than the dryer, use a graph to find the cost of each appliance.

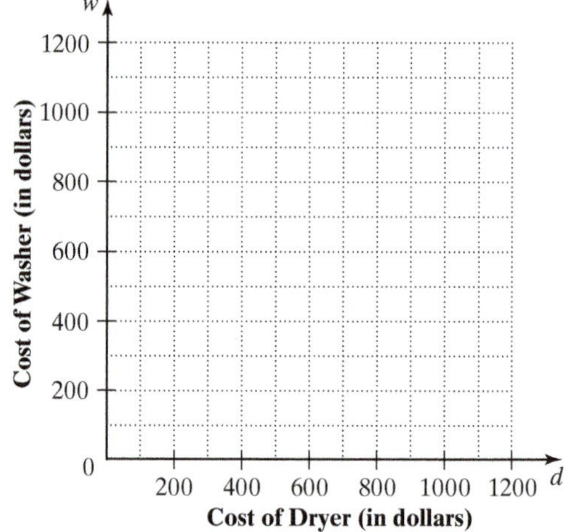

55. Silver's Gym charges a \$300 initiation fee plus \$30 per month. DeLuxe Fitness Center has an initial charge of \$400 but only charges \$25 per month. Use a graph to determine for what number of months both health clubs will charge the same amount.

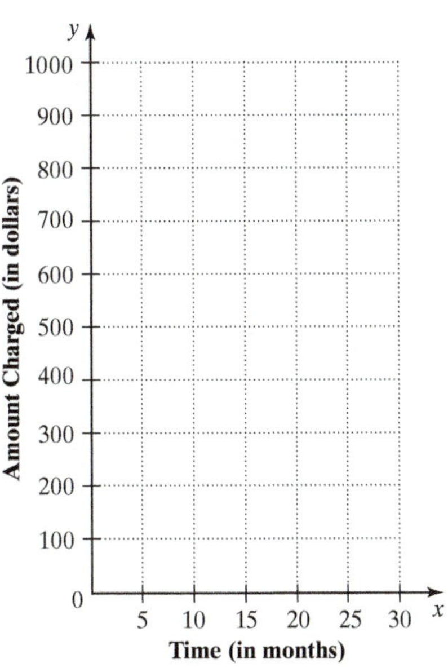

56. A plant nursery is selling a 7-foot specimen of a tree that grows about 1.5 ft/yr and a 6-foot specimen of a tree that grows 2 ft/yr. Use a graph to determine in how many years the two trees will be the same height.

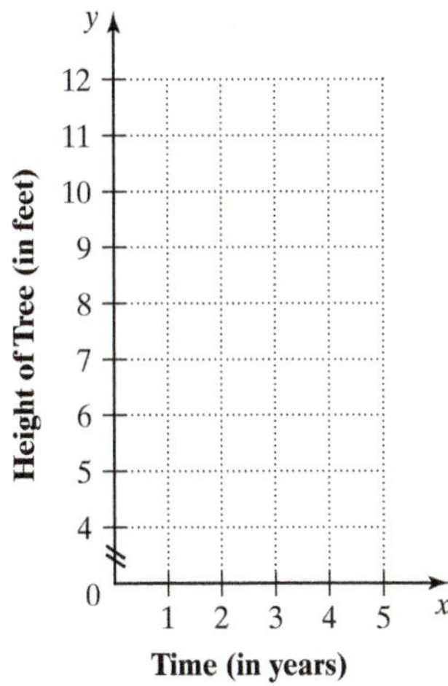

• Check your answers on page A-28.

MINDStretchers

Critical Thinking

1. Write a system of linear equations that has $(2, 5)$ as its only solution.

Writing

2. Is it possible for a system of two linear equations to have exactly two solutions? If not, explain why.

Mathematical Reasoning

3. Not every system of equations is linear. For example the system

$$y = 2x$$
$$y = x^2$$

has the graph shown to the right. How many solutions does this system have? Explain how you know.

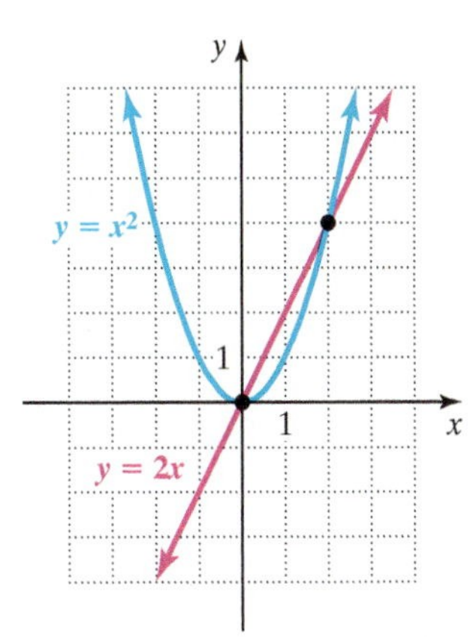

14.2 Solving Systems of Linear Equations by Substitution

OBJECTIVES

A To solve systems of linear equations by substitution

B To solve applied problems involving systems of linear equations

This section deals with solving a system of equations by the **substitution method**. As in the previous section, we restrict the discussion to systems of two linear equations in two variables. In applying the substitution method, we solve for one variable in terms of the other in one linear equation, getting a much simpler problem.

As compared with the graphing method, the substitution method has the advantage of being faster and also of giving us an exact solution. This consideration is especially important if the coordinates of the solution are not integral, or in an applied problem where precision counts. The substitution method particularly lends itself to solving systems in which a variable is isolated in one of the original equations.

To see how solving by substitution works, let's look at some examples.

EXAMPLE 1

Solve by substitution: **(1)** $x + y = 10$

(2) $y = 2x + 4$

Solution Notice that Equation (2) is solved for y in terms of x. So we can substitute the expression $2x + 4$ from Equation (2) for y in Equation (1).

(1) $x + y = 10$

$x + (2x + 4) = 10$ Substitute $2x + 4$ for y.

We now have an equation that contains only one variable, namely x, and so is easy to solve.

$$x + (2x + 4) = 10$$
$$3x + 4 = 10$$
$$3x = 6$$
$$x = 2$$

Now that we have found the value of x, let's substitute it in either of the original equations in order to determine the corresponding value of y. Substituting 2 for x in Equation (2), we get:

(2) $y = 2x + 4$

$= 2(2) + 4$

$= 8$

So the solution to the system is $(2, 8)$, that is, $x = 2$ and $y = 8$.

Check We can check the solution by substituting 2 for x and 8 for y in the original equations.

(1) $x + y = 10$

$2 + 8 \stackrel{?}{=} 10$

$10 = 10$ True

(2) $y = 2x + 4$

$8 \stackrel{?}{=} 2 \cdot 2 + 4$

$8 = 8$ True

The check confirms the solution $(2, 8)$.

PRACTICE 1

Use the substitution method to solve:

(1) $x - y = 7$

(2) $x = -y + 1$

In Example 1, would we have gotten the same solution if we had solved for y in the *first* equation and then replaced y in the second equation?

The preceding example suggests the following procedure for solving systems of linear equations in two variables.

To Solve a System of Linear Equations by Substitution

- In one of the equations, solve for either variable in terms of the other variable.
- In the other equation, substitute the expression equal to the variable found in the previous step. Then, solve the resulting equation for the remaining variable.
- Substitute the value found in the previous step in either of the original equations and solve for the other variable.
- Check by substituting the values in both equations of the original system.

EXAMPLE 2

Use the substitution method to solve the following system:

$$\begin{aligned} (1)\quad d &= 3q - 8 \\ (2)\quad 3d - 4q &= 10 \end{aligned}$$

Solution Since d is isolated in Equation (1), we can substitute the expression $3q - 8$ for d in Equation (2).

$$\begin{aligned} (2)\quad 3d - 4q &= 10 \\ 3(3q - 8) - 4q &= 10 \quad \text{Substitute } 3q - 8 \text{ for } d. \\ 9q - 24 - 4q &= 10 \\ 5q - 24 &= 10 \\ 5q &= 34 \\ q &= \frac{34}{5}, \text{ or } 6.8 \end{aligned}$$

Now, let's solve for d by substituting 6.8 for q in Equation (1).

$$\begin{aligned} (1)\quad d &= 3q - 8 \\ &= 3(6.8) - 8 \\ &= 20.4 - 8 \\ &= 12.4 \end{aligned}$$

The solution is $q = 6.8$ and $d = 12.4$. Note that neither value is an integer. So if we solve this system by the graphing method, we have to estimate the coordinates of a point.

Check We can confirm the solution by substituting these values in both of the original equations.

$$\begin{aligned} (1)\quad d &= 3q - 8 \\ 12.4 &\stackrel{?}{=} 3(6.8) - 8 \\ 12.4 &\stackrel{?}{=} 20.4 - 8 \\ 12.4 &= 12.4 \quad \text{True} \end{aligned}$$

$$\begin{aligned} (2)\quad 3d - 4q &= 10 \\ 3(12.4) - 4(6.8) &\stackrel{?}{=} 10 \\ 37.2 - 27.2 &\stackrel{?}{=} 10 \\ 10 &= 10 \quad \text{True} \end{aligned}$$

So the solution is confirmed.

PRACTICE 2

Solve for m and n by substitution:

$$\begin{aligned} (1)\quad m &= -5n + 1 \\ (2)\quad 2m + 3n &= 7 \end{aligned}$$

In Example 2, would we have gotten the same solution to the system if the variables had been called x and y instead of q and d? Explain.

EXAMPLE 3

Solve the system by substitution.

$$\begin{aligned} (1)\quad 5x - 3y &= 5 \\ (2)\quad 2x - y &= 1 \end{aligned}$$

Solution First, we need to solve for x or y in either of the equations. Let's solve for y in Equation (2) where the coefficient of y is -1.

$$\begin{aligned} (2)\quad 2x - y &= 1 \\ -y &= -2x + 1 \\ y &= 2x - 1 \quad \text{Divide each side by } -1. \end{aligned}$$

Next, we substitute the expression $2x - 1$ for y in Equation (1) and solve for x.

$$\begin{aligned} (1)\quad 5x - 3y &= 5 \\ 5x - 3(2x - 1) &= 5 \quad \text{Substitute } 2x - 1 \text{ for } y. \\ 5x - 6x + 3 &= 5 \\ -x + 3 &= 5 \\ -x &= 2 \\ x &= -2 \end{aligned}$$

Finally, we solve for y by substituting -2 for x in Equation (2).

$$\begin{aligned} (2)\quad 2x - y &= 1 \\ 2(-2) - y &= 1 \\ -4 - y &= 1 \\ -y &= 5 \\ y &= -5 \end{aligned}$$

So the solution is $(-2, -5)$. Check by substituting in the original system.

PRACTICE 3

Solve by substitution.

$$\begin{aligned} (1)\quad 2x - 7y &= 7 \\ (2)\quad 6x - y &= 1 \end{aligned}$$

EXAMPLE 4

Solve by substitution:

$$\begin{aligned} (1)\quad x - 4y &= 15 \\ (2)\quad -2x + 8y &= 5 \end{aligned}$$

Solution We begin by solving for either x or y in one of the equations. Let's solve for x in Equation (1) since the coefficient of x in this equation is 1.

$$\begin{aligned} (1)\quad x - 4y &= 15 \\ x &= 4y + 15 \end{aligned}$$

Next, we substitute the expression $4y + 15$ for x in equation (2).

$$\begin{aligned} (2)\quad -2x + 8y &= 5 \\ -2(4y + 15) + 8y &= 5 \quad \text{Substitute } 4y + 15 \text{ for } x. \\ -8y - 30 + 8y &= 5 \\ -30 &= 5 \quad \text{False} \end{aligned}$$

Getting a false statement means that no value of y makes this last equation true. So the original system has no solution.

PRACTICE 4

Use the substitution method to solve the following system:

$$\begin{aligned} (1)\quad 3x + y &= 10 \\ (2)\quad -6x - 2y &= 1 \end{aligned}$$

What do you think the graph of the system in Example 4 looks like?

EXAMPLE 5

Solve by substitution:

$$\begin{aligned} &(1) \quad 6x + 2y = 4 \\ &(2) \quad -y = 3x - 2 \end{aligned}$$

Solution Since the coefficient of y in Equation (2) is -1, let's solve this equation for y.

$$\begin{aligned} (2) \quad -y &= 3x - 2 \\ y &= -3x + 2 \quad \text{Divide each side by } -1. \end{aligned}$$

Now, we substitute $-3x + 2$ for y in Equation (1).

$$\begin{aligned} (1) \quad 6x + 2y &= 4 \\ 6x + 2(-3x + 2) &= 4 \quad \text{Substitute } -3x + 2 \text{ for } y. \\ 6x - 6x + 4 &= 4 \\ 4 &= 4 \quad \text{True} \end{aligned}$$

The original system has infinitely many solutions, since every x-value makes the last equation true. For each x-value, the corresponding y-value can be found by substituting it in either of the original equations and solving for y. Some solutions are $(0, 2)$ and $(1, -1)$.

PRACTICE 5

Solve for x and y:

$$\begin{aligned} &(1) \quad y = -2x + 4 \\ &(2) \quad 10x + 5y = 20 \end{aligned}$$

In Example 5, what do you think the graph of the system looks like? Can you identify a particular solution to the system?

TIP When solving a system of linear equations in two variables by substitution:

- if we get a false statement, then the system has no solution.
- if we get a true statement, then the system has infinitely many solutions.

Now, let's use the substitution method to solve some applications.

EXAMPLE 6

A car rental agency has two plans:

- In the Ambassador Plan, renting a car for one day costs \$35 plus \$0.25/mi driven.
- In the Diplomat Plan, a one-day car rental costs \$50 plus \$0.10/mi driven.

a. For each plan, write a linear equation that relates a day's price p for renting a car to the number of miles driven n. Express the given information as a system of equations.

b. Use the substitution method to solve the system of linear equations.

c. In the context of this problem, what is the significance of the solution?

PRACTICE 6

To watch movies on premium channels, a couple decides to choose between two television cable deals:

- the TV Deal that costs \$20 installation and \$35/mo, and
- the Movie Deal that costs \$30 installation and \$25/mo.

a. Write an equation for each deal, expressing the cost of a deal c in terms of the number of months n for which the couple signs up.

EXAMPLE 6 (continued)

Solution

a. The Ambassador Plan can be expressed as $p = 0.25n + 35$; the Diplomat Plan becomes $p = 0.10n + 50$. The system representing both plans is therefore:

$$\textbf{(1)} \quad p = 0.25n + 35$$
$$\textbf{(2)} \quad p = 0.10n + 50$$

b. To solve the system, we can set the two expressions for p equal to each other.

$$0.25n + 35 = 0.10n + 50$$

Solving for n gives us:

$$\begin{aligned} 0.25n + 35 &= 0.10n + 50 \\ 0.15n &= 15 \\ n &= \frac{15}{0.15} \\ n &= 100 \end{aligned}$$

To solve for p, we can substitute 100 for n in Equation (1).

$$\begin{aligned} \textbf{(1)} \quad p &= 0.25n + 35 \\ &= 0.25(\mathbf{100}) + 35 \\ &= 25 + 35 \\ &= 60 \end{aligned}$$

So the solution to the system is $n = 100$ and $p = 60$.

c. With the appropriate units, the solution is $n = 100$ mi and $p = \$60$. This means that the cost of a one-day rental on the two plans is the same amount of money, namely \$60, only when the car is driven 100 mi. For other distances driven, the plans charge different amounts.

b. Solve the system of linear equations by substitution.

c. In the context of this problem, what is the significance of the solution?

Recall that we discussed mixture problems involving a single equation in Section 12.3. The following example shows how we can apply our knowledge of solving systems of linear equations to these problems.

EXAMPLE 7

How much 30% alcohol solution and 50% alcohol solution must be mixed to get 10 gal of 42% solution?

Solution We solve this problem as we did earlier mixture problems, namely by organizing the given information in a table. Let's represent the amount of 30% solution by x and the amount of 50% solution by y.

Action	Percent of Alcohol	Amount of Solution (in gallons)	Amount of Alcohol (in gallons)
Start with	30%	x	$0.3x$
Add	50%	y	$0.5y$
Finish with	42%	10	0.42(10), or 4.2

PRACTICE 7

A chemist wishes to combine an alloy that is 20% copper with one that is 50% copper to obtain 15 oz of an alloy that is 25% copper. Find the quantities of the alloys required.

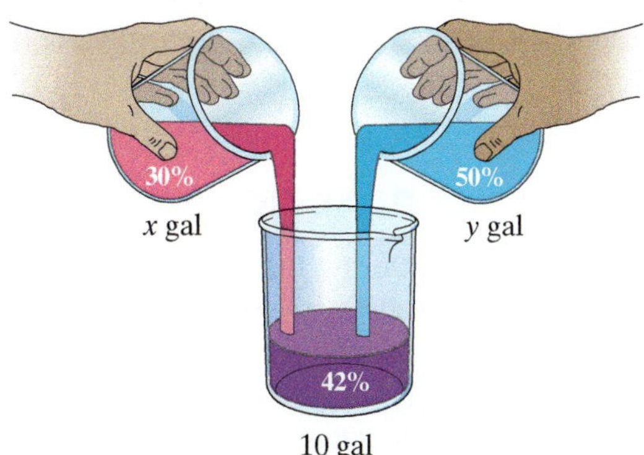

The amount of alcohol in the 30% solution is 30% of x, or $0.3x$. The amount of alcohol in the 50% solution is 50% of y, or $0.5y$. The amount of alcohol in the 42% solution is 42% of 10, or 4.2. Since the total amount of the combined solutions is 10 gal and the total amount of alcohol is 4.2 gal, we get the following system:

$$\begin{aligned} &\textbf{(1)} \quad & x + y &= 10 \\ &\textbf{(2)} \quad & 0.3x + 0.5y &= 4.2 \end{aligned}$$

In applying the substitution method, we begin by solving for y in Equation (1).

$$\begin{aligned} &\textbf{(1)} \quad & x + y &= 10 \\ && y &= -x + 10 \end{aligned}$$

We then substitute $-x + 10$ for y in Equation (2).

$$\begin{aligned} &\textbf{(2)} \quad & 0.3x + 0.5y &= 4.2 \\ && 0.3x + 0.5(\mathbf{-x + 10}) &= 4.2 \\ && 0.3x - 0.5x + 5 &= 4.2 \\ && -0.2x &= -0.8 \\ && x &= \frac{-0.8}{-0.2} \\ && x &= 4 \end{aligned}$$

After replacing x by 4 in Equation (1), we solve for y.

$$\begin{aligned} &\textbf{(1)} \quad & x + y &= 10 \\ && \mathbf{4} + y &= 10 \\ && y &= 6 \end{aligned}$$

So the solution to the system is $(4, 6)$. In other words, 4 gal of 30% solution and 6 gal of 50% solution are needed to produce 10 gal of the 42% solution.

A system of linear equations can also serve as a model for investment problems, as the following example illustrates:

EXAMPLE 8

A stockbroker had \$10,000 to invest for her client. The broker invested part of this amount at a low-risk, low-yield 5% rate of return per year and the rest at a high-risk, high-yield 7% rate. If the client earned a return of \$550 in one year, how much money did the broker invest at each rate?

Solution The following table reflects the given information. Here, x stands for the amount of the investment at a 5% return, and y the investment at a 7% return.

Rate of Return	Amount of Investment (\$)	Amount of Return (\$)
5%	x	$0.05x$
7%	y	$0.07y$
TOTAL	10,000	550

The amount of return on each investment is the product of the rate of return and the amount of the investment. We add the amount of the individual investments to find the total investment, and the amount of returns on each investment to find the total amount of return. Since the total investment is \$10,000 and the total return is \$550, we get the following system:

$$\begin{aligned} &\textbf{(1)} & x + y &= 10{,}000 \\ &\textbf{(2)} & 0.05x + 0.07y &= 550 \end{aligned}$$

Now, let's solve for y in Equation (1).

$$\begin{aligned} &\textbf{(1)} & x + y &= 10{,}000 \\ & & y &= 10{,}000 - x \end{aligned}$$

We then substitute $10{,}000 - x$ for y in Equation (2).

$$\begin{aligned} &\textbf{(2)} & 0.05x + 0.07y &= 550 \\ & & 0.05x + 0.07(10{,}000 - x) &= 550 \\ & & 0.05x + 700 - 0.07x &= 550 \\ & & -0.02x + 700 &= 550 \\ & & -0.02x &= -150 \\ & & x &= \frac{-150}{-0.02} \\ & & x &= 7500 \end{aligned}$$

After substituting 7500 for x in Equation (1), we solve for y.

$$\begin{aligned} &\textbf{(1)} & x + y &= 10{,}000 \\ & & 7500 + y &= 10{,}000 \\ & & y &= 2500 \end{aligned}$$

Therefore, the solution to the system is (7500, 2500). In other words, \$7500 was invested at 5% and \$2500 at 7%.

PRACTICE 8

During the recession of the early 1990s, the stock prices of fast-food companies suffered. Between January 5, 1990 and January 4, 1991, McDonald's stock fell approximately 15% and Wendy's fell about 77%. If a financier invested \$120,000 in these two companies in January 1990, and ended up with stock worth only \$77,200 in January 1991, how much did the financier invest in each company? (*Source:* google.com/finance)

14.2 Exercises

FOR EXTRA HELP MyMathLab  PRACTICE  WATCH  READ  REVIEW

A *Solve by substitution and check.*

1. $x + y = 10$
$y = 2x + 1$

2. $x - y = 7$
$x = 5y + 3$

3. $y = -3x - 15$
$y = -x - 7$

4. $y = -2x - 21$
$y = 5x$

5. $-x - y = 8$
$x = -3y$

6. $x - y = 15$
$y = -4x$

7. $4x + 2y = 10$
$x = 2$

8. $2y + x = 10$
$y = -5$

9. $-x + 20y = 0$
$x - y = 0$

10. $5x + 3y = 0$
$x + y = 0$

11. $6x + 4y = 2$
$2x + y = 0$

12. $x - 3y = 0$
$2x - 3y = 6$

13. $3x + 5y = -12$
$x + 2y = -6$

14. $3x + 5y = -1$
$3x + y = -5$

15. $m = 20 - 2n$
$2m + 4n = -22$

16. $x + 2y = 4$
$3x + 6y = 3$

17. $7x - 3y = 26$
$3x - y = 11$

18. $x + 3y = 1$
$-3x - 5y = -2$

19. $8x + 2y = -1$
$y = -4x + 1$

20. $4x + 2y = 4$
$y = 5 - 2x$

21. $6x - 2y = 2$
$y = 3x - 1$

22. $2x - 6y = -12$
$x = 3y - 6$

23. $2x + 6y = 12$
$x + 3y = 6$

24. $3u + 6v = 60$
$-2u = 4v - 40$

25. $p + 2q = 13$
$q + 7 = 4p$

26. $a - b = -1$
$6b = 5a$

27. $s - 3t + 5 = 0$
$-4s + t - 9 = 0$

28. $2l - 3w + 6 = 0$
$l - w - 10 = 0$

Mixed Practice

Solve by substitution and check.

29. $y = 2x + 12$
$y = -3x - 3$

30. $2y - 8x = -2$
$y = 4x - 1$

31. $6x + 2y = 2$
$y = -3x + 2$

32. $-x - y = 12$
$x = -5y$

33. $5x + 3y = -6$
$-3x + y = 5$

34. $2y - 3x = 6$
$y - 3x = 0$

Applications

B *Solve.*

35. Two taxi companies compete in the same neighborhood. One of these companies charges \$3 for the taxi drop plus \$1.25 for each mile driven, while the other charges \$2 for the taxi drop, plus \$1.50 for each mile driven.

a. Express these relationships as an algebraic system, where m represents miles driven and c represents cost.

b. Solve the system and interpret the results.

36. Two electricians make house calls. One charges \$75 for a visit plus \$50/hr of work. The other charges \$95 per visit plus \$40/hr of work.

a. Letting c represent cost and h represent hours worked, write a linear equation for each electrician that expresses the charge for a house call in terms of the length of a visit.

b. For how many hours of work do the two electricians charge the same?

37. On a particular airline route, a full-price coach ticket costs \$310 and a discounted coach ticket costs \$210. On one of these flights, there were 172 passengers in coach, which resulted in a total ticket income of \$44,120. How many full-price tickets were sold?

38. In 2010, the combined cost of two military aircraft—a Charger and a Fighter—was \$260 million. A year later, the cost of a Charger increased by 10% and the cost of a Fighter decreased by 10%. If the combined cost in 2011 was \$246 million, find the cost of each type of aircraft in 2010.

39. A laboratory technician needs to make a 10-liter batch of antiseptic that is 60% alcohol. How can she combine a batch of antiseptic that is 30% alcohol with another that is 70% to get the desired concentration?

40. A bottle of fruit juice contains 20% water. How much water must be added to this bottle to produce 8 L of fruit juice that is 50% water?

41. A corporation merged two departments into one. In one department, 5% of the employees were women, whereas in the other department, 80% were women. When the departments were merged, 50% of the 150 employees were women. How many women were in each department before the merger?

42. A hospital needs 30 L of a 10% solution of disinfectant. How many liters of a 20% solution and a 4% solution should be mixed to obtain this 10% solution?

43. A student took out two loans totaling \$5000. She borrowed the maximum amount she could at 6% and the remainder at 7% interest per year. At the end of the first year, she owed \$310 in interest. How much was loaned at each rate?

44. A man invested three times as much money in a bond fund that earned 8% in a year as he did in a mutual fund that returned 4% in the year. How much money did he invest in each fund if the total earnings for the year were \$112?

45. A \$40,000 investment was split so that part was invested at a 7% annual rate of interest and the rest at 9%. If the total annual earnings were \$3140, how much money was invested at each rate?

46. A financial adviser counseled a client to invest \$15,000, split between two stocks. At the end of one year, the investment in one stock increased in value by 4%, and the investment in the second stock increased in value by 8%. If the total increase in value of the investment was \$1120, how much money was invested in each stock?

• Check your answers on page A-30.

MIND*Stretchers*

Groupwork

1. Cramer's Rule is a formula that can be used to solve a system of linear equations for x and y.
Consider the following system:

$$ax + by = c$$
$$dx + ey = f$$

The formula states that $x = \dfrac{ce - bf}{ae - bd}$ and $y = \dfrac{af - cd}{ae - bd}$. Note that this mechanical approach allows machines to solve systems of equations.

a. Working with a partner, make up your own values for a, b, c, d, e, and f, and substitute these values in the system.

$$\_\_\_\_\_x + \_\_\_\_\_y = \_\_\_\_\_$$
$$\_\_\_\_\_x + \_\_\_\_\_y = \_\_\_\_\_$$

b. Use Cramer's Rule to calculate x and y.

$$x = \frac{ce - bf}{ae - bd} = \_\_\_\_\_$$
$$y = \frac{af - cd}{ae - bd} = \_\_\_\_\_$$

c. By substitution, check whether (x, y) is in fact a solution to the system.

Mathematical Reasoning

2. On the coordinate plane, consider the quadrilateral $ABCD$ shown below.
At what point do the diagonals $\overline{AC}$ and $\overline{BD}$ intersect? Explain how to find the answer exactly.

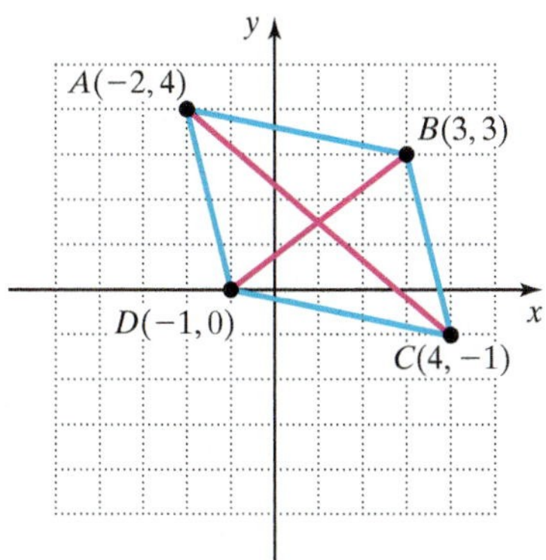

Critical Thinking

3. For what value of k will the system shown have infinitely many solutions? Explain why?

$$kx - 2y = 10$$
$$4x - y = 5$$

14.3 Solving Systems of Linear Equations by Elimination

OBJECTIVES

- **A** To solve systems of linear equations by elimination
- **B** To solve applied problems involving systems of linear equations

The final method of solving a system of linear equations that we consider is called the **elimination** (or **addition**) **method**. As in the substitution method discussed in the previous section, the elimination method involves changing a given system of two equations in two variables to one equation in one variable. A system in which the coefficients of a variable are opposites particularly lends itself to the elimination method.

Recall that in solving linear equations in one variable, we frequently used the addition property of equality: the equations $a = b$ and $a + c = b + c$ are equivalent. That is, if we add the same number to both sides of an equation, we get an equivalent equation.

The elimination method for solving *systems* is based on a closely related property of equality: If $a = b$ and $c = d$, then $a + c = b + d$. This property allows us to "add equations."

Let's look at examples of applying this property in the elimination method.

EXAMPLE 1

Solve the following system by the elimination method.

$$\begin{aligned} &(1)\quad x + y = 4 \\ &(2)\quad x - y = 2 \end{aligned}$$

Solution First, we decide which variable to eliminate. Since the coefficients of the y-terms in the two equations are opposites (namely $+1$ and -1), the y-terms are eliminated if we add the equations.

$$\begin{array}{rr} (1) & x + y = 4 \\ (2) & \underline{x - y = 2} \\ & 2x + 0y = 6 \\ & 2x = 6 \\ & x = 3 \end{array}$$

Next, to find y, we substitute 3 for x in either of the original equations. Substituting in Equation (1) we get:

$$\begin{array}{rr} (1) & x + y = 4 \\ & 3 + y = 4 \\ & y = 4 - 3 \\ & y = 1 \end{array}$$

So $x = 3$ and $y = 1$. That is, the solution is $(3, 1)$.

Check We check the solution by substituting these values for x and y in both of the original equations.

$$\begin{array}{rlrl} (1) & x + y = 4 & (2) & x - y = 2 \\ & 3 + 1 \stackrel{?}{=} 4 & & 3 - 1 \stackrel{?}{=} 2 \\ & 4 = 4 \quad \text{True} & & 2 = 2 \quad \text{True} \end{array}$$

So our solution $(3, 1)$ is confirmed.

PRACTICE 1

Solve for x and y.

$$\begin{aligned} &(1)\quad x + y = 6 \\ &(2)\quad x - y = -10 \end{aligned}$$

EXAMPLE 2

Use the elimination method to solve for x and y.

$$\begin{aligned} \textbf{(1)}\quad 3x + 2y &= 14 \\ \textbf{(2)}\quad 5x + 2y &= -8 \end{aligned}$$

Solution In this system, adding the two given equations does not eliminate either variable. However, note that the two y-terms have the same coefficient, namely 2. So if we multiply Equation (1) by -1 and then add the equations, the y-terms will cancel out.

$$\begin{array}{lrcl} \textbf{(1)} & 3x + 2y = 14 & \xrightarrow{\text{Multiply by } -1.} & -3x - 2y = -14 \\ \textbf{(2)} & 5x + 2y = -8 & & \underline{5x + 2y = -8} \quad \text{Add the equations.} \\ & & & 2x \qquad = -22 \\ & & & x = -11 \end{array}$$

Next, we substitute -11 for x in either of the original equations. Let's choose Equation (1), and then solve for y.

$$\begin{aligned} \textbf{(1)}\qquad 3x + 2y &= 14 \\ 3(-11) + 2y &= 14 \\ -33 + 2y &= 14 \\ 2y &= 14 + 33 \\ 2y &= 47 \\ y &= \frac{47}{2} = 23.5 \end{aligned}$$

So the solution is $(-11, 23.5)$.

PRACTICE 2

Solve the following system using the elimination method:

$$\begin{aligned} \textbf{(1)}\quad 4x + 3y &= -7 \\ \textbf{(2)}\quad 5x + 3y &= -5 \end{aligned}$$

EXAMPLE 3

Solve:

$$\begin{aligned} \textbf{(1)}\quad 3x - y &= -2 \\ \textbf{(2)}\quad x + 5y &= 10 \end{aligned}$$

Solution Note that the coefficient of x in Equation (2) is $+1$. By multiplying this equation by -3 and then adding the two equations, we eliminate the x-terms.

$$\begin{array}{lrcl} \textbf{(1)} & 3x - y = -2 & & 3x - y = -2 \\ \textbf{(2)} & x + 5y = 10 & \xrightarrow{\text{Multiply by } -3.} & \underline{-3x - 15y = -30} \\ & & & -16y = -32 \quad \text{Add the equations.} \\ & & & y = 2 \end{array}$$

Next, we substitute 2 for y in Equation (2) and then solve for x.

$$\begin{aligned} \textbf{(2)}\qquad x + 5y &= 10 \\ x + 5(2) &= 10 \\ x + 10 &= 10 \\ x &= 0 \end{aligned}$$

So the solution is $(0, 2)$.

PRACTICE 3

Solve for x and y:

$$\begin{aligned} \textbf{(1)}\quad x - 3y &= -18 \\ \textbf{(2)}\quad 5x + 2y &= 12 \end{aligned}$$

How could we have solved the system in Example 3 another way?

EXAMPLE 4

Use the elimination method to solve the following system of linear equations:

$$\begin{aligned} &\textbf{(1)} \quad 4x + 3y = -19 \\ &\textbf{(2)} \quad 3x - 2y = -10 \end{aligned}$$

Solution This system is more complicated to solve than the previous examples because there is no single integer that we can multiply either equation by that will eliminate a variable when we add the equations. Instead, we must multiply *both* equations by integers that lead to the elimination of a variable. There are a number of possible strategies to accomplish this. We can, for instance, multiply Equation (1) by 2 and Equation (2) by 3 to eliminate the y-terms when the equations are added.

$$\begin{aligned} &\textbf{(1)} \quad 4x + 3y = -19 \xrightarrow{\text{Multiply by 2.}} \quad 8x + 6y = -38 \\ &\textbf{(2)} \quad 3x - 2y = -10 \xrightarrow{\text{Multiply by 3.}} \quad 9x - 6y = -30 \\ &\hphantom{\textbf{(2)} \quad 3x - 2y = -10 \xrightarrow{\text{Multiply by 3.}}} \quad 17x = -68 \quad \text{Add the equations.} \\ &\hphantom{\textbf{(2)} \quad 3x - 2y = -10 \xrightarrow{\text{Multiply by 3.}}} \quad\quad\quad x = -4 \end{aligned}$$

Now, let's substitute -4 for x in Equation (1), and then solve for y.

$$\begin{aligned} \textbf{(1)} \quad 4x + 3y &= -19 \\ 4(-4) + 3y &= -19 \\ -16 + 3y &= -19 \\ 3y &= 16 + (-19) \\ 3y &= -3 \\ y &= -1 \end{aligned}$$

So the solution is $(-4, -1)$.

PRACTICE 4

Solve by elimination:

$$\begin{aligned} &\textbf{(1)} \quad 5x - 7y = 24 \\ &\textbf{(2)} \quad 3x - 5y = 16 \end{aligned}$$

How could we have solved the system in Example 4 by eliminating x instead of y?

To Solve a System of Linear Equations by Elimination

- Write both equations in the general form $Ax + By = C$.
- Choose the variable that you want to eliminate.
- If necessary, multiply one or both equations by appropriate numbers so that the coefficients of the variable to be eliminated are opposites.
- Add the equations. Then, solve the resulting equation for the remaining variable.
- Substitute the value found in the previous step in either of the original equations and solve for the other variable.
- Check by substituting the values in both equations of the original system.

EXAMPLE 5

Solve by elimination:

$$\begin{aligned} &(1) & 5x &= 3y \\ &(2) & -3x + 2y &= 9 \end{aligned}$$

Solution Equation (1) is not in the form $Ax + By = C$, so let's begin by rewriting it in general form.

$$\begin{aligned} &(1) & 5x &= 3y \xrightarrow{\text{Write in general form.}} & 5x - 3y &= 0 \\ &(2) & -3x + 2y &= 9 & -3x + 2y &= 9 \end{aligned}$$

Now, suppose we choose to eliminate the x-terms. To do this, we can multiply Equation (1) by 3, Equation (2) by 5, and then add the equations.

$$\begin{aligned} &(1) & 5x - 3y &= 0 \xrightarrow{\text{Multiply by 3.}} & 15x - 9y &= 0 \\ &(2) & -3x + 2y &= 9 \xrightarrow{\text{Multiply by 5.}} & -15x + 10y &= 45 \quad \text{Add the equations.} \\ & & & & y &= 45 \end{aligned}$$

To solve for x, let's substitute 45 for y in Equation (1).

$$\begin{aligned} (1) \quad 5x &= 3y \\ 5x &= 3(45) \\ 5x &= 135 \\ x &= 27 \end{aligned}$$

So the solution is $(27, 45)$.

PRACTICE 5

Use the elimination method to solve the following system:

$$\begin{aligned} &(1) & -2x + 5y &= 20 \\ &(2) & 3x &= 7y - 26 \end{aligned}$$

EXAMPLE 6

Solve by elimination:

$$\begin{aligned} &(1) & 4x - 6y + 12 &= 0 \\ &(2) & 2x - 3y &= -4 \end{aligned}$$

Solution We begin by writing Equation (1) in general form.

$$\begin{aligned} &(1) & 4x - 6y + 12 &= 0 \xrightarrow{\text{Write in general form.}} & 4x - 6y &= -12 \\ &(2) & 2x - 3y &= -4 & 2x - 3y &= -4 \end{aligned}$$

Now, let's eliminate the y-terms. To do this, we can multiply Equation (2) by -2, and then add the equations.

$$\begin{aligned} &(1) & 4x - 6y &= -12 & 4x - 6y &= -12 \\ &(2) & 2x - 3y &= -4 \xrightarrow{\text{Multiply by } -2.} & -4x + 6y &= 8 \\ & & & & 0 &= -4 \quad \text{False} \end{aligned}$$

Since adding the equations yields a false statement, the original system has no solution.

PRACTICE 6

Solve:

$$\begin{aligned} 3x &= 4 + y \\ 9x - 3y &= 12 \end{aligned}$$

Let's use our knowledge of the elimination method to solve some applied problems, beginning with a motion problem on the following page.

EXAMPLE 7

It takes a plane 3 hr to fly between two airports, traveling with a tailwind at a ground speed of 500 mph. The plane then takes 4 hr to make the return trip against the same wind. What is the speed of the plane in still air? What is the speed of the wind?

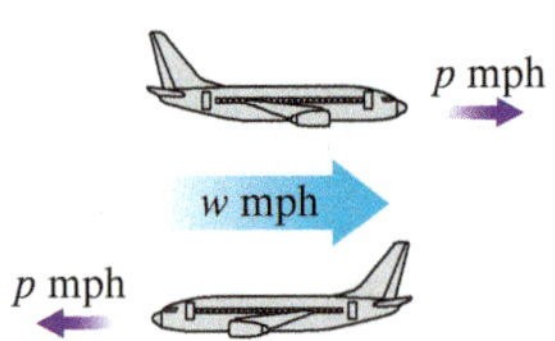

Solution Let p represent the speed of the plane in still air and w represent the speed of the wind. On the initial flight, the wind is with the plane so its speed relative to the ground is $p + w$. When returning, the wind is against the plane so its ground speed is $p - w$. We can organize the given information in the following table:

	Ground speed ·	Time =	Distance
Going	$p + w$	3	$3\,(p + w)$
Returning	$p - w$	4	$4\,(p - w)$

Note that since the distance the plane travels is the product of its ground speed and the time it travels, we can compute each entry in the distance column of the table by multiplying the corresponding entries in the ground speed and the time columns.

Since we are told that the speed going is 500 mph, we have:

$$p + w = 500$$

But the distance going and the distance returning are equal, so

$$3(p + w) = 4(p - w)$$

Now, we have a system of two equations, which we must solve.

$$\begin{aligned} \textbf{(1)}\quad p + w &= 500 \\ \textbf{(2)}\quad 3(p + w) &= 4(p - w) \end{aligned}$$

We can write Equation (2) in general form, by simplifying.

$$\begin{aligned} 3(p + w) &= 4(p - w) \\ 3p + 3w &= 4p - 4w \\ 3p - 4p + 3w + 4w &= 0 \\ -p + 7w &= 0 \end{aligned}$$

The system then becomes:

$$\begin{aligned} \textbf{(1)}\quad p + w &= 500 \\ \textbf{(2)}\quad -p + 7w &= 0 \end{aligned}$$

Adding the equations eliminates the p-terms.

$$\begin{aligned} 8w &= 500 \\ w &= 62.5 \end{aligned}$$

Finally, let's substitute 62.5 for w in Equation (1) and solve for p.

$$\begin{aligned} \textbf{(1)}\quad p + 62.5 &= 500 \\ p &= 437.5 \end{aligned}$$

So the speed of the plane in still air is 437.5 mph, and the wind speed is 62.5 mph.

PRACTICE 7

A whale swimming with the current traveled 80 mi in 2 hr. Swimming against the current, the whale traveled only 40 mi in the same amount of time. Find the whale's speed in calm water and the speed of the current.

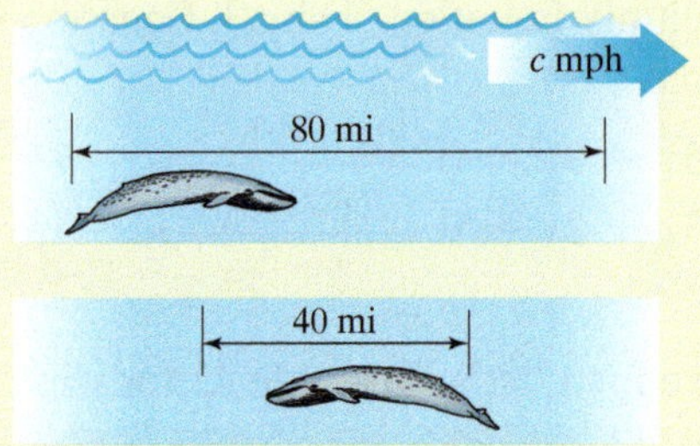

EXAMPLE 8

A student had two part-time jobs in a restaurant. One week she earned a total of \$306, working 12 hr as a cashier and 10 hr as a cook. The next week, she worked 14 hr as a cashier and 22 hr cooking, earning \$512. What is her hourly wage as a cashier? As a cook?

Solution Let x represent the student's hourly wage as a cashier and y represent the student's hourly wage as a cook. The first week, the student earned \$306, and the second week, \$512. So we must solve the following system:

$$\begin{aligned} &\textbf{(1)} \quad 12x + 10y = 306 \\ &\textbf{(2)} \quad 14x + 22y = 512 \end{aligned}$$

We can divide each equation by 2 to simplify.

$$\begin{aligned} &\textbf{(1)} \quad 6x + 5y = 153 \\ &\textbf{(2)} \quad 7x + 11y = 256 \end{aligned}$$

Let's eliminate the x-terms by multiplying Equation (1) by 7 and Equation (2) by -6. Then, we add the equations.

$$\begin{aligned} &\textbf{(1)} \quad 6x + 5y = 153 \xrightarrow{\text{Multiply by 7.}} 42x + 35y = 1071 \\ &\textbf{(2)} \quad 7x + 11y = 256 \xrightarrow{\text{Multiply by } -6.} -42x - 66y = -1536 \end{aligned}$$

Adding the equations eliminates the x-terms.

$$\begin{aligned} -31y &= -465 \\ y &= \frac{-465}{-31} \\ y &= 15 \end{aligned}$$

Finally, we substitute 15 for y in the original Equation (1) and solve for x.

$$\begin{aligned} \textbf{(1)} \quad 12x + 10y &= 306 \\ 12x + 10(15) &= 306 \\ 12x + 150 &= 306 \\ 12x &= 156 \\ x &= \frac{156}{12} \\ x &= 13 \end{aligned}$$

So the student earned \$13/hr as a cashier and \$15/hr as a cook.

PRACTICE 8

Admission prices at a high school football game were \$10 for adults and \$6 for students. The total value of the 175 tickets sold was \$1450. How many adults and how many students attended the game?

In this chapter, we have discussed three methods of solving a system of linear equations—the graphing method, the substitution method, and the elimination (or addition) method. To help in deciding which method to apply in a given problem, listed in the table below are some advantages and disadvantages of each method.

Method	Advantages	Disadvantages
Graphing Method	• Provides a picture that makes relationships understandable.	• Approximates solutions, particularly when they are not integers or are large. • Can be time consuming if not using a grapher.
Substitution Method	• Gives exact solutions. • Is easy to use when a variable in one of the original equations is isolated.	• No picture. • Can result in complicated equations with parentheses or fractions.
Elimination Method	• Gives exact solutions. • Is easy to use when the two coefficients of a variable are opposites.	• No picture.

14.3 Exercises

FOR EXTRA HELP MyMathLab Math XL PRACTICE    

A *Solve.*

1. $x + y = 3$
$x - y = 7$

2. $x - y = 10$
$x + y = -8$

3. $x + y = -4$
$-x + 3y = -6$

4. $5x - y = 8$
$2x + y = -1$

5. $10p - q = -14$
$-4p + q = -4$

6. $a + b = -4$
$-a + 2b = -8$

7. $3x + y = -3$
$4x + y = -4$

8. $x + 4y = -3$
$x - 7y = 19$

9. $3x + 5y = 10$
$3x + 5y = -5$

10. $8x + 2y = 3$
$4x + y = -9$

11. $9x + 6y = -15$
$-3x - 2y = 5$

12. $4x + y = -3$
$8x + 2y = -6$

13. $5x + 2y = -9$
$-5x + 2y = 11$

14. $7x + 4y = -6$
$-x + 4y = 10$

15. $2s + d = -2$
$5s + 3d = -6$

16. $-5x + 8y = -7$
$-6x + 9y = -9$

17. $3x - 5y = 1$
$7x - 8y = 17$

18. $3x + 2y = 9$
$-2x + 3y = -19$

19. $5x + 2y = -1$
$4x - 5y = -14$

20. $10x - 3y = 9$
$3x - 2y = -5$

21. $7p + 3q = 15$
$-5p - 7q = 16$

22. $8a + 2b = 18$
$4a - 3b = -15$

23. $6x + 5y = -8.5$
$8x + 10y = -3$

24. $6x - 6y = -3.6$
$-4x + 8y = -16$

25. $3.5x + 5y = -3$
$2x = -2y$

26. $3x - 3y = 0$
$1.5y = -6x + 30$

27. $2x - 4 = -y$
$x + 2y = 0$

28. $y = -3x + 7$
$4x + 2y = 11$

29. $8x + 10y = 1$
$-4x - 5y + 6 = 0$

30. $x - y = 6$
$3x = 3y + 10$

Mixed Practice

Solve

31. $4a - b = -10$
$-3a + b = 7$

32. $9x - 6y = 3$
$3x - 2y = 6$

33. $3x - 5y = 4$
$-6x + 10y = -8$

34. $7p + 4q = -12$
$4p + q = -3$

35. $5x + 3y = -3$
$-7x - 5y = 4$

36. $4x + 3y = -5.5$
$5x + 6y = -3.5$

Applications

B *Solve.*

37. A quarterback throws a pass that travels 40 yd with the wind in 2.5 sec. If he had thrown the same pass against the wind, the football would have traveled 20 yd in 2 sec. Find the speed of a pass that the quarterback would throw if there were no wind.

38. A crew team rows in a river with a current. When the team rows with the current, the boat travels 14 mi in 2 hr. Against the current, the team rows 6 mi in the same amount of time. At what speed does the team row in still water?

39. To enter a zoo, adult visitors must pay \$5, whereas children and seniors pay only half price. On one day, the zoo collected a total of \$765. If the zoo had 223 visitors that day, how many half-price admissions and how many full-price admissions did the zoo collect?

40. The molecular weight of a compound such as water or hydrogen peroxide is found by adding the atomic weights of all the atoms in the compound. A water molecule consists of two hydrogen atoms and one oxygen atom, with a molecular weight of approximately 18. A hydrogen peroxide molecule consists of two hydrogen atoms and two oxygen atoms, with a molecular weight of about 34. What is the atomic weight of hydrogen? Of oxygen? (*Source:* John T. Moore, *Chemistry Made Simple*)

41. The British Parliament consists of two houses—the House of Lords (consisting of peers who are appointed) and the House of Commons (consisting of Members of Parliament, or MPs, who are elected). In November 2010, there were a total of 1388 peers and MPs, with 88 more peers than MPs. How many MPs and how many peers were there? (*Source:* www.parliament.uk)

42. Compact discs are stored in single jewel cases, which are 0.375 in. thick, and in multiple-CD jewel cases, which have a thickness of 0.875 in. If 86 of these jewel cases exactly fit on the storage shelf shown, how many of the jewel cases are single?

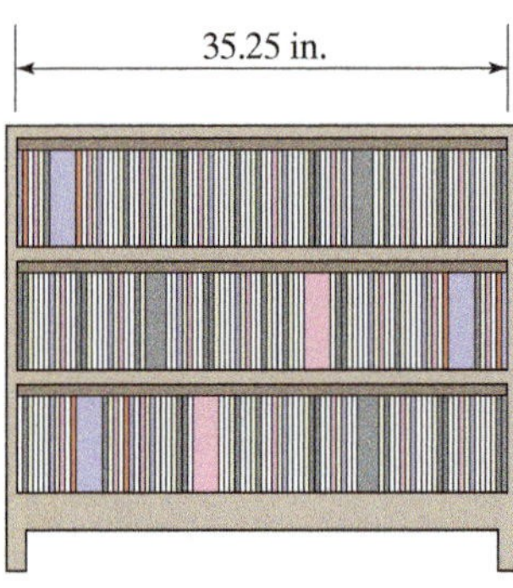

43. A particular computer takes 43 nanoseconds to carry out 5 sums and 7 products, and 42 nanoseconds to perform 2 sums and 9 products. How long does the computer take to carry out one sum? To carry out one product?

44. A wholesale novelty shop sells some embroidered scarves for \$12 each and others for \$15 each. A customer pays \$234 for seventeen scarves. How many scarves at each price did she buy?

45. One issue of a journal has 3 full-page ads and 5 half-page ads, generating \$6075 in advertising revenue. The next issue has 4 full-page ads and 4 half-page ads, resulting in advertising revenue of \$6380. Determine the advertising rates in this journal for full-page and half-page ads.

46. According to a law of physics, the lever shown below will balance when the products of each weight and the length of its force arm are equal.

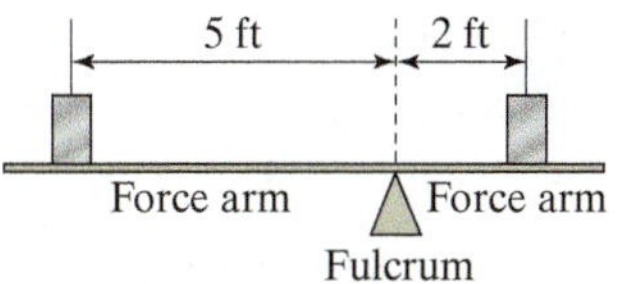

The weights shown above balance. If 10 lb are added to the left weight, which is then moved 1 ft closer to the fulcrum, the lever will again balance. Find the two original weights.

• Check your answers on page A-30.

MINDStretchers

Groupwork

1. Your friend performs the following magic trick: She asks you to think of two numbers but not to tell her what they are. Instead, you tell her the sum and the difference of the two numbers. She promptly tells you what the two original numbers were. Explain how your friend does the trick. Working with partners, try to perform this trick.

Writing

2. Consider the following system of equations:

$$\begin{aligned} &\textbf{(1)} \quad 5x - 8y = 4 \\ &\textbf{(2)} \quad 12x + 24y = 11 \end{aligned}$$

Explain how the concept of LCM relates to solving this system by elimination.

Mathematical Reasoning

3. The elimination method can be extended to three linear equations in three variables. Solve the following system:

$$\begin{aligned} 4x - y - 3z &= 30 \\ 3x - 2y - 6z &= -5 \\ x - z &= 5 \end{aligned}$$

Cultural Note

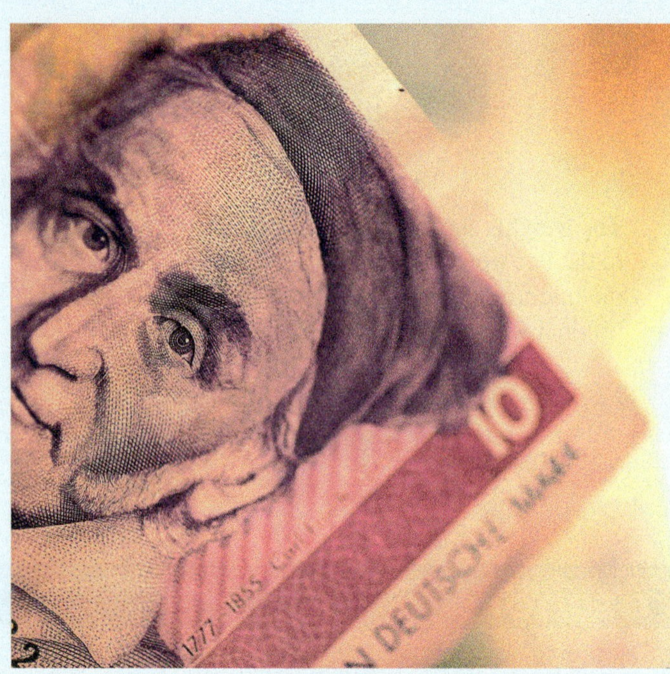

Carl Friedrich Gauss (1777–1855) is generally considered to be one of the greatest mathematicians in history. Among his many mathematical contributions was the *Gaussian elimination method* of solving systems of linear equations, discussed in this chapter. Gauss (rhymes with "house") is credited with being the first to prove the major mathematical result known as the Fundamental Theorem of Algebra. He was also an important scientist. In astronomy, he laid the theoretical foundation for predicting a planet's orbit. To honor this German's groundbreaking work in physics, his name is given to the unit (*gauss*) used today to express the strength of a magnetic field. Also, Gauss' portrait appears on the German ten-mark bill, as shown to the left.

(*Sources:* Roger Cooke, *The History of Mathematics*, John Wiley, 1997; D. E. Smith, *History of Mathematics*, Dover Publications, 1923)

Key Concepts and Skills

CONCEPT SKILL

Concept/Skill	Description	Example
[14.1] **System of equations**	Two or more equations considered simultaneously, that is, together.	$x + y = 7$ $x - y = 1$
[14.1] **Solution of a system of linear equations in two variables**	An ordered pair of numbers that makes both equations in the system true.	Is $(4, 3)$ a solution of the following system? $x + y = 7 \rightarrow 4 + 3 \stackrel{?}{=} 7$ True $x - y = 1 \rightarrow 4 - 3 \stackrel{?}{=} 1$ True Yes, $(4, 3)$ is a solution of the system.
[14.1] **To solve a system of linear equations by graphing**	• Graph both equations on the same coordinate plane. • There are three possibilities: **a.** If the lines intersect at exactly one point, then the solution is the ordered pair of coordinates for the point of intersection. Check that these coordinates satisfy both equations. **b.** If the lines are parallel, then there is no solution of the system. **c.** If the lines coincide, then there are infinitely many solutions, namely all the ordered pairs of coordinates that represent points on the line.	$x + y = 7$ $x - y = 3$ 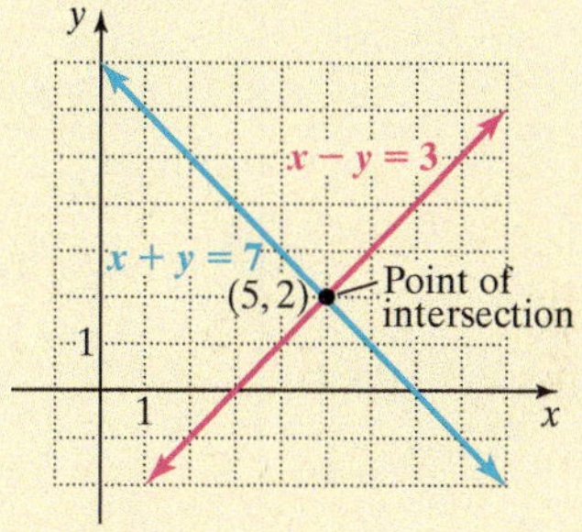 $3x - y = 2$ $3x - y = 4$ 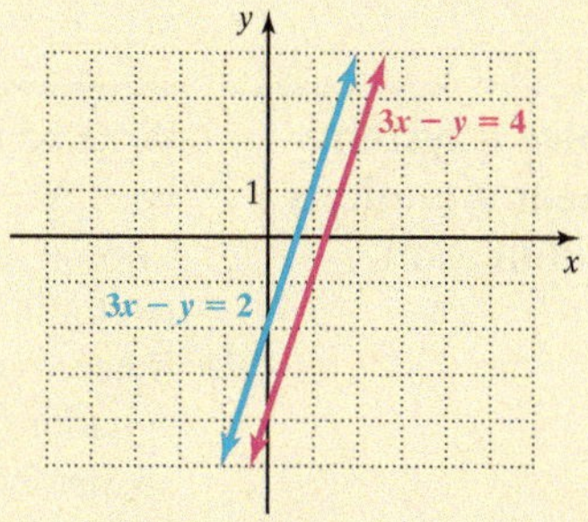 $x - y = 3$ $2x - 2y = 6$ 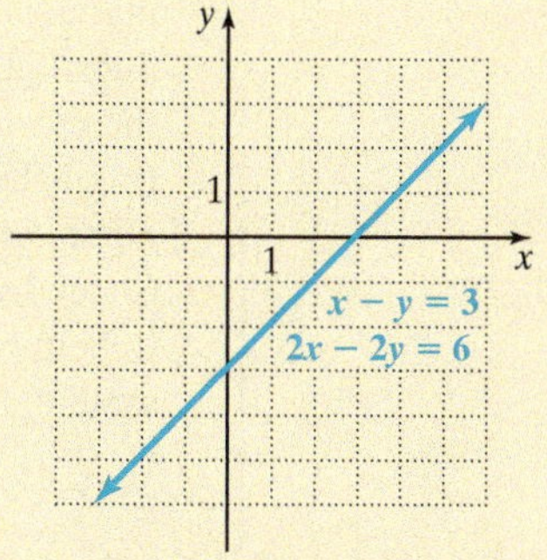
[14.2] **To solve a system of linear equations by substitution**	• In one of the equations, solve for either variable in terms of the other variable.	**(1)** $2x + 3y = 2$ **(2)** $x + 4y = 6$ **(2)** $x + 4y = 6$ $x = -4y + 6$

continued

Concept/Skill	Description	Example
	• In the other equation, substitute the expression equal to the variable found in the previous step. Then, solve the resulting equation for the remaining variable. • Substitute the value found in the previous step in either of the original equations and solve for the other variable. • Check by substituting the values in both equations of the original system.	(1) $2x + 3y = 2$ $2(-4y + 6) + 3y = 2$ $-8y + 12 + 3y = 2$ $-5y + 12 = 2$ $-5y = -10$ $y = 2$ (2) $x + 4y = 6$ $x + 4(2) = 6$ $x + 8 = 6$ $x = -2$ **Check** (1) $2x + 3y = 2$ $2(-2) + 3(2) \stackrel{?}{=} 2$ $-4 + 6 \stackrel{?}{=} 2$ $2 = 2$ True (2) $x + 4y = 6$ $-2 + 4(2) \stackrel{?}{=} 6$ $-2 + 8 \stackrel{?}{=} 6$ $6 = 6$ True So the solution is $(-2, 2)$.
[14.3] **To solve a system of linear equations by elimination**	• Write both equations in the general form $Ax + By = C$. • Choose the variable that you want to eliminate. • If necessary, multiply one or both equations by appropriate numbers so that the coefficients of the variable to be eliminated are opposites. • Add the equations. Then, solve the resulting equation for the remaining variable. • Substitute the value found in the previous step in either of the original equations and solve for the other variable. • Check by substituting the values in both equations of the original system.	(1) $3x - 2y = -5$ (2) $2x - 4y = 2$ Multiply Equation (1) by 2. $6x - 4y = -10$ Multiply Equation (2) by -1. $-2x + 4y = -2$ $4x = -12$ $x = -3$ (2) $2x - 4y = 2$ $2(-3) - 4y = 2$ $-6 - 4y = 2$ $-4y = 8$ $y = -2$ **Check** (1) $3x - 2y = -5$ $3(-3) - 2(-2) \stackrel{?}{=} -5$ $-9 + 4 \stackrel{?}{=} -5$ $-5 = -5$ True (2) $2x - 4y = 2$ $2(-3) - 4(-2) \stackrel{?}{=} 2$ $-6 + 8 \stackrel{?}{=} 2$ $2 = 2$ True So the solution is $(-3, -2)$.

CHAPTER 14 Review Exercises

Say Why

Fill in each blank.

1. The ordered pair $(-1, 2)$ ________ (is/is not) a solution of the system of equations $\begin{aligned} 3x + 4y &= 5 \\ -x + 2y &= 3 \end{aligned}$ because ____.

2. If the graph of a system of equations consists of parallel lines, then the system __________ (has/does not have) any solutions because ____.

3. If the graph of a system of equations consists of exactly one line, then the system __________ (has/does not have) infinitely many solutions because ____.

4. In solving the system of equations $\begin{aligned} 2x - y &= 7 \\ -2x + 3y &= 10 \end{aligned}$ adding the equations __________ (does/does not) eliminate the x-terms because ____.

[14.1]

5. Consider the following system:

$$\begin{aligned} x + 2y &= -4 \\ 3x - y &= 3 \end{aligned}$$

Is $(2, -3)$ a solution of the system?

6. For each of the systems graphed, determine the number of solutions.

a.

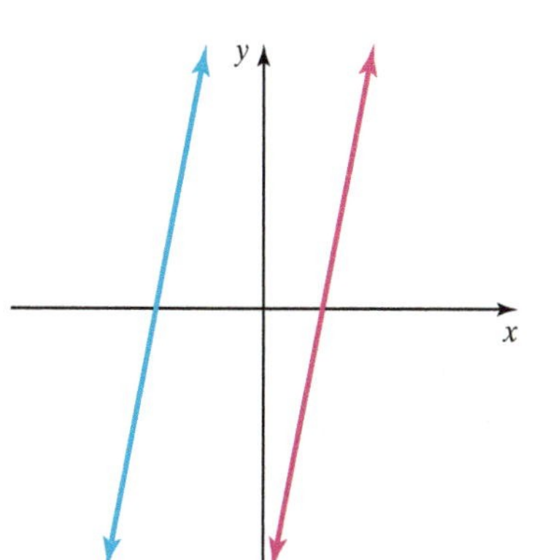

b.

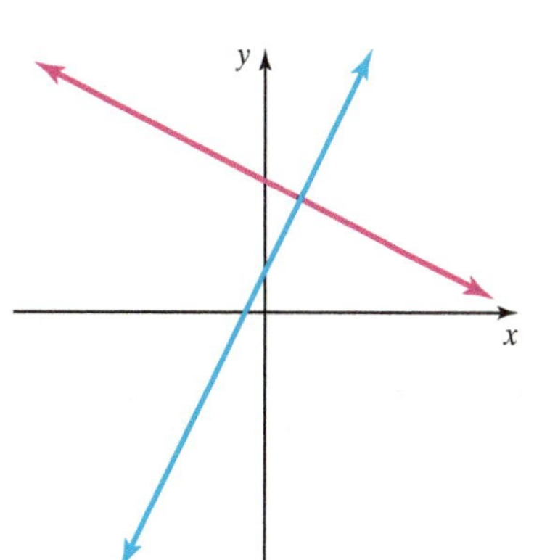

c.

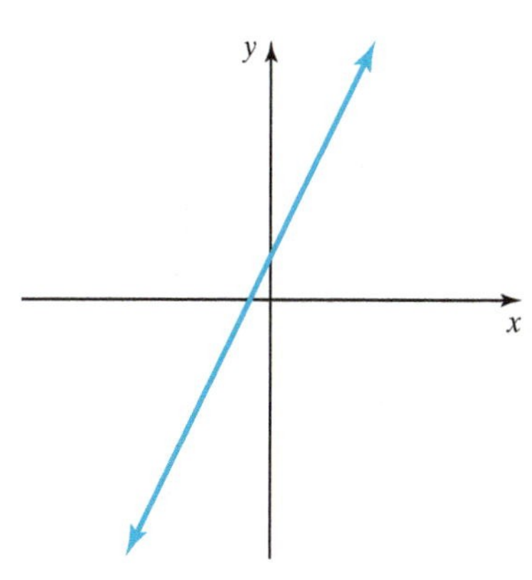

7. Match each system with the appropriate graph:
   a. A system with solution $(2, 1)$
   b. A system with solution $(1, 2)$
   c. A system with infinitely many solutions
   d. A system with no solution

I

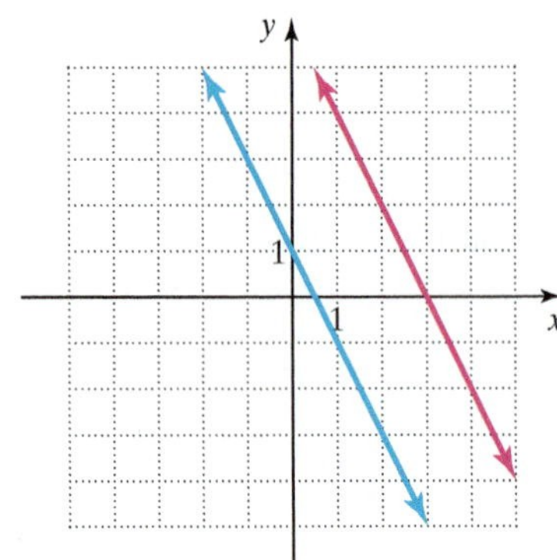

II

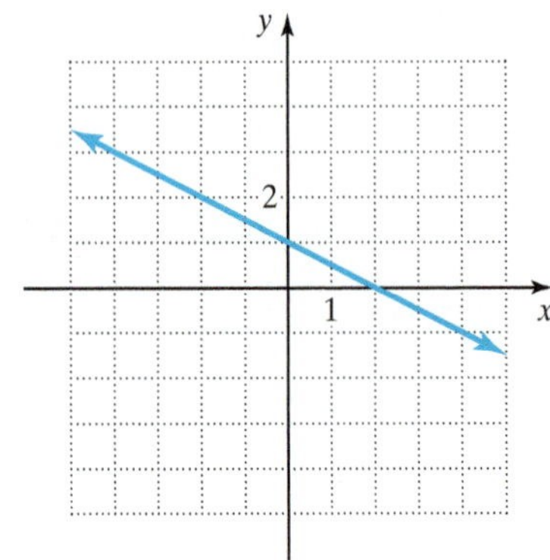

III

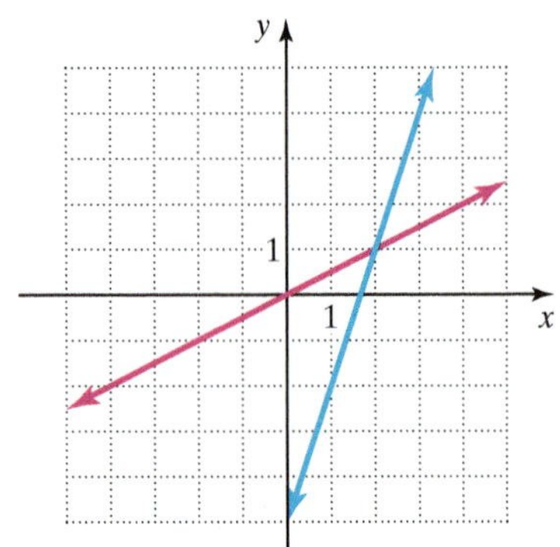

IV

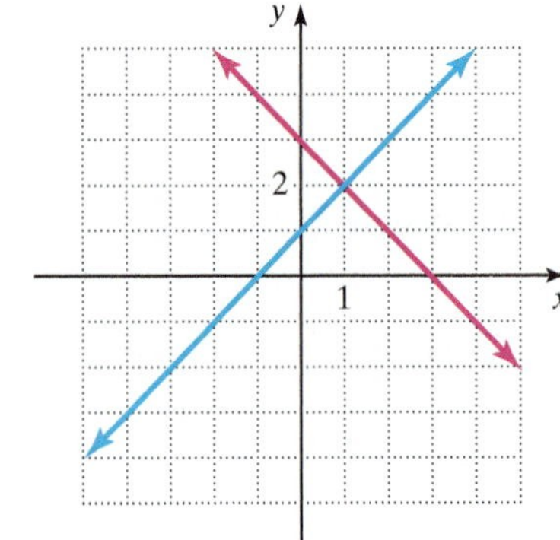

Solve each system by graphing.

8. $x + y = 6$
   $x - y = -4$

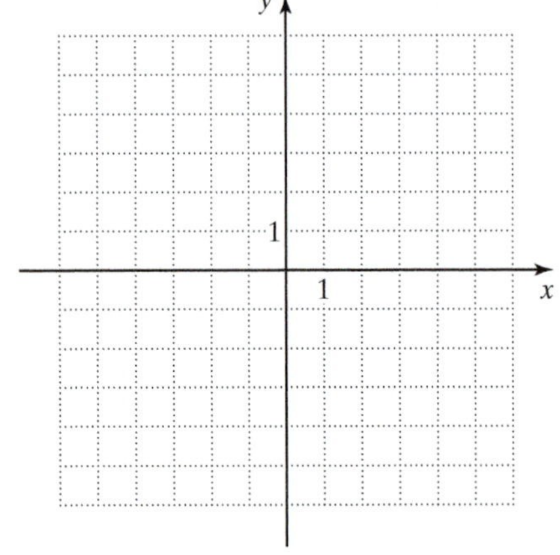

9. $y = 2x$
   $6x - 3y = 3$

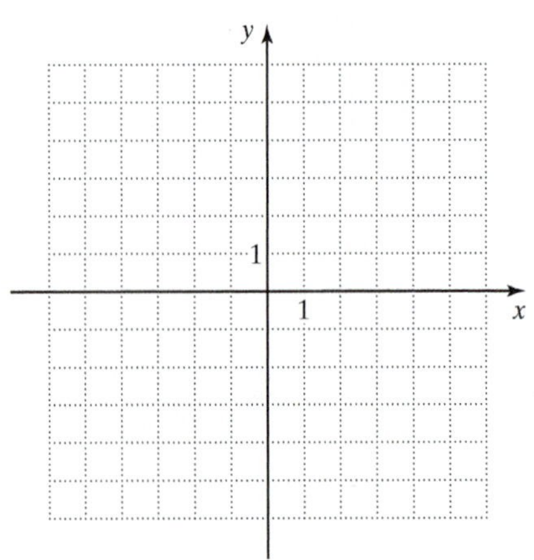

10. $\quad 2y = -8x + 2$
$-4x - y = -1$

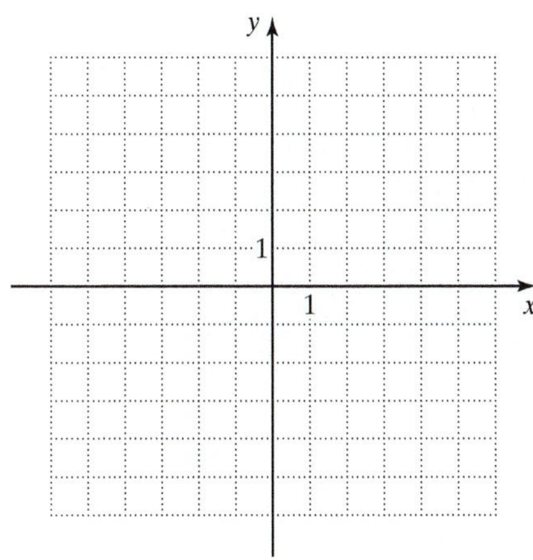

11. $x - 3y = -15$
$y - x = 5$

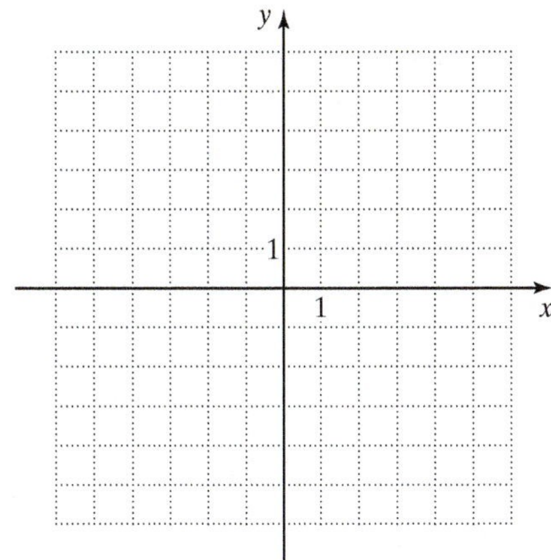

[14.2] *Solve each system by substitution.*

12. $x + y = 3$
$y = 2x + 6$

13. $a = 3b - 4$
$a + 4b = 10$

14. $x - 3y = 1$
$-2x + 6y = 7$

15. $10x + 2y = 14$
$-y = 5x - 7$

[14.3] *Solve each system by elimination.*

16. $x + y = 1$
$x - y = 7$

17. $2x + 3y = 8$
$4x + 6y = 16$

18. $4x = 9 - 5y$
$2x + 3y = 3$

19. $3x + 2y = -4$
$4x - 3y = 23$

Mixed Applications

Solve.

20. A student starts a typing service. He buys a computer and a printer for \$1750, and then charges \$5.50 per page for typing. Expenses for ink, paper, and electricity amount to \$0.50 per page.
    a. Write the given information as a system of equations.

    b. How many pages must the student type to break even?

21. A job applicant must choose between two sales positions. One position pays \$10/hr, and the other \$8/hr plus a base pay of \$50/wk.

a. Express the weekly salaries of the two positions as a system of equations.

b. How many hours would the applicant have to work per week in order to earn the same amount of money at each position?

22. The movie screen shown has a perimeter of 332 ft. If the length is 26 ft more than the width, find the area of the screen.

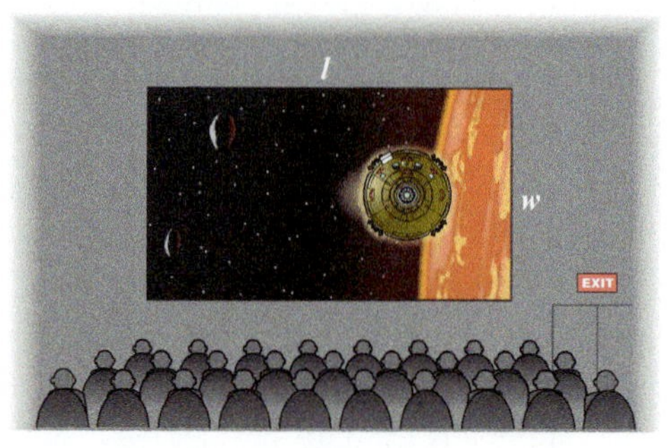

23. A doubles tennis court is 42 ft longer than it is wide. If the court's perimeter is 228 ft, find its dimensions.

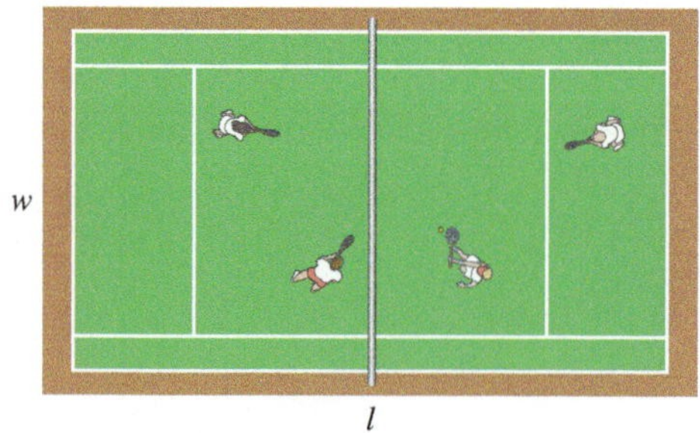

24. The coin box of a vending machine contains only nickels and dimes. There are 350 coins worth \$25. How many nickels and how many dimes did the box contain?

25. Two trains start 500 mi apart and speed toward each other. The difference between the average speeds of the trains is 5 mph. If they pass one another after 4 hr, find the rate of each train.

26. During a season, a college basketball team scored 2437 points on a combination of three-point and two-point baskets. If the team made 1085 baskets, how many two-point baskets and how many three-point baskets did the team make?

27. A pharmacist has 10% and 30% alcohol solutions in stock. To prepare 200 mL of a 25% solution, how much of each should the pharmacist mix?

28. A farmer is preparing an insecticide by mixing a 50% solution with water. How much of this solution and how much water are needed to fill a 2000-liter tank with a 35% solution?

29. Last year, a financial adviser recommended that her client invest part of his \$50,000 in secure municipal bonds that paid 6% and the rest in corporate stocks that paid 8%. How much money did the client put into each type of investment if the total annual return was \$3200?

30. An investor split \$10,000 between a high-risk mutual fund and a low-risk mutual fund. Last year, the high-risk fund paid 12% and the low-risk fund paid 2%, for a total of \$900. How much money was invested in each fund?

31. Two airplanes leave an airport at the same time, one flying 100 mph faster than the other. The planes travel in opposite directions, and after 2 hr they are 1800 mi apart. Determine the speed of the slower plane.

32. Flying with the wind, a bird flew 13 mi in half an hour. On the return trip against the wind, it was able to travel only 8 mi in the same amount of time. Find the speed of the bird in still air and the speed of the wind.

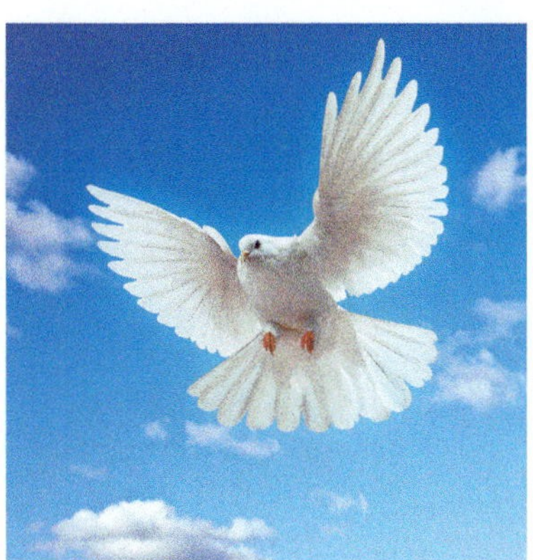

33. The U.S. Senate has 100 members. After debating the merits of a treaty, all the senators voted, and 14 more voted for the treaty than against. None of the senators abstained.

a. How many senators voted *for* the treaty?

b. How many senators voted *against* the treaty?

34. In a chemistry lab, a piece of copper starting at 2°C is heated at the rate of 3°/min. At the same time, a piece of iron starting at 86°C is being cooled at the rate of 4°/min.

a. After how much time will the two metals be at the same temperature?

b. After how much time will the iron be 14° colder than the copper?

• Check your answers on page A-30.

CHAPTER 14 Posttest

To see if you have mastered the topics in this chapter, take this test.

1. Indicate which ordered pair is a solution of the following system:

$$\begin{aligned} x + y &= -1 \\ 3x - y &= 1 \end{aligned}$$

a. $(0, -1)$

b. $(-1, 0)$

c. $(2, -3)$

2. How many solutions does the graphed system appear to have?

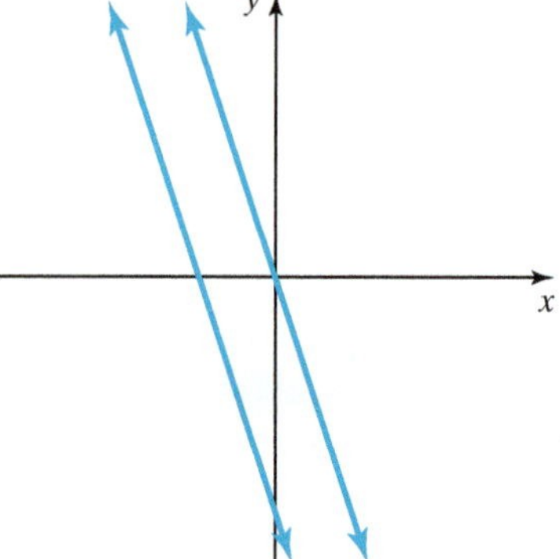

Solve each system by graphing.

3. $x - y = 3$
$x + y = 3$

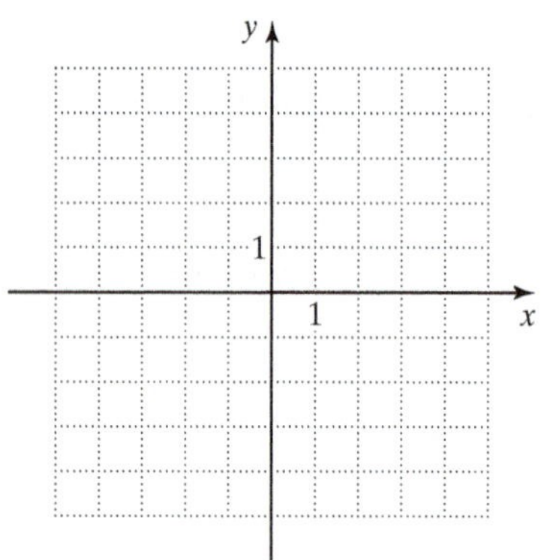

4. $5 = 2x - y$
$y = 2x$

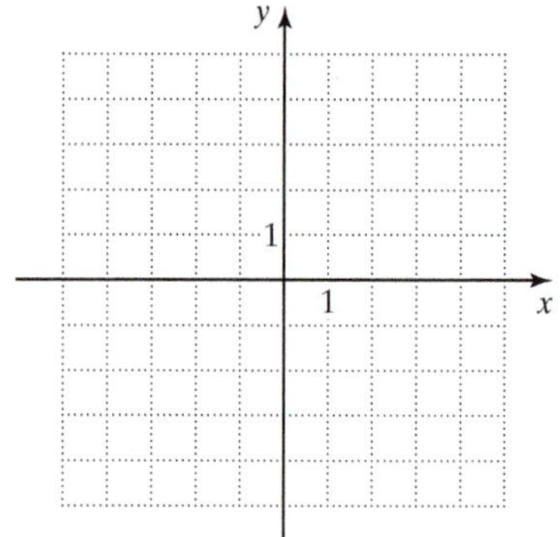

5. $2x + 3y = 4$
$3x - y = -5$

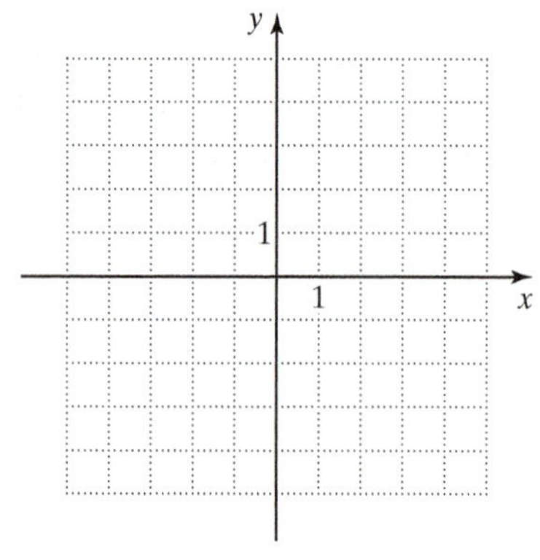

6. $6y = 4 - 2x$
$x + 3y = 2$

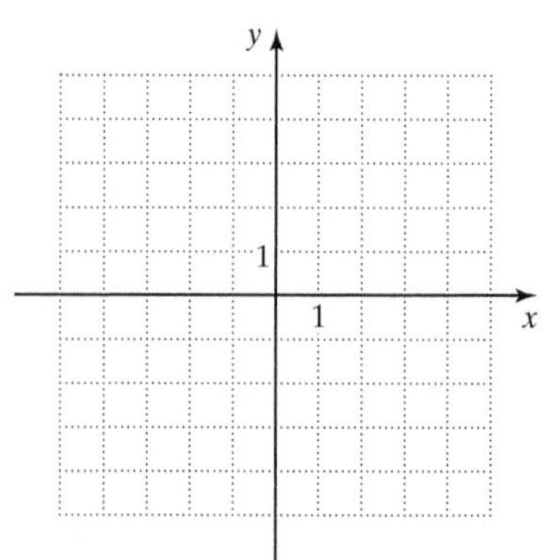

Solve each system by substitution.

7. $x = 3y - 7$
$y = x + 5$

8. $3x - 5y = -12$
$x + 2y = 7$

9. $u - 3v = -12$
$5u + v = 8$

Solve each system by elimination.

10. $4x + y = 3$
$7x - y = 19$

11. $x - y = 5$
$2x - 2y = 5$

12. $-5p + 2q = 1$
$4p + 3q = 1.5$

Solve each system.

13. $7x = 2y$
$-4x + y = 1$

14. $4l = -(m + 3)$
$8l + 2m = -6$

15. $5x + 2y = -1$
$x - 1 = y$

16. $5x + 3y - 9 = 0$
$2x - 7y - 20 = 0$

Solve.

17. In a local election, the ratio of votes for the winning candidate to the losing candidate was 2 to 1. If 6306 votes were cast in the election, how many votes did the winning candidate get?

18. Nutritional information for 3-oz servings of turkey and of salmon is given in the following table:

	Turkey, light meat	Salmon
Amount of Fat (in grams)	3	3
Number of Calories	135	99

How many servings of turkey and of salmon would it take to get 9 g of fat and 333 cal?

19. A 20% iodine solution is mixed with a 60% iodine solution to produce 4 gal of a 50% iodine solution. How many gallons of each solution are needed?

20. A small airplane traveled 170 mph with a tailwind and 130 mph with a headwind. Find the speed of the wind and the speed of the airplane in still air.

• Check your answers on page A-31.

Cumulative Review Exercises

To help you review, solve the following:

1. Arrange in decreasing order: $\frac{2}{3}, \frac{5}{6}, \frac{7}{12}$

2. Compute: $8 - 2.4 + 5.003$

3. Solve and check: $10n = 3\frac{1}{2}$

4. Is 5 a solution of the equation $3p + 1 = 9 - p$?

5. Is $\frac{1.8}{27} = \frac{1.6}{24}$ a true or false statement?

6. True or false: $-6 < -5$

7. Compute: 3 yr − 1 yr 4 mo

8. Calculate: $-4 \div 2 + 3(-1)(8)$

9. Simplify: $2(4x - 1) - 3(y + x)$

10. Solve: $5(x + 1) - (x - 2) = x - 2$

11. Solve and graph: $3x - 7 < 4(x - 2)$

12. Find the slope and y-intercept of the graph whose equation is $3x + 6y = 12$.

13. Graph the linear inequality: $4y - x \leq 8$

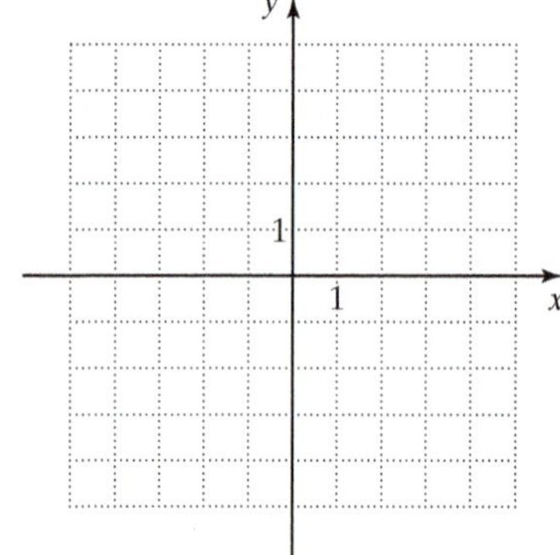

14. Solve by elimination:

$$3x + 2y = -10$$
$$-3x + 2y = 14$$

15. In 2009, the amount of Social Security tax withheld from a paycheck was 6.2%. The amount of Medicare tax withheld was 1.45%. For a 2009 gross income of g, how much was withheld for both Social Security and Medicare? (*Source:* irs.gov)

16. The maximum speed of a supersonic airplane S (in miles per hour) is commonly represented by its Mach number M, where

$$M = \frac{S}{740}.$$

What is the maximum speed of an airplane flying at Mach 2.1?

17. In renting a car for a day, you must choose between two local car agencies. One agency charges \$35/day plus \$0.20/mi, and the other charges \$50/day plus \$0.15/mi. Under what circumstances do the two agencies charge the same amount for a one-day rental?

18. An average adult male has about 5 L of blood, consisting of plasma and cells. About 90% of the plasma is water, which accounts for half of the total amount of blood. How much blood, to the nearest liter, is plasma, and how much is cells? (*Source:* americasblood.org)

19. The following table displays the approximate number of medical doctors in the United States for various years.

Year	1980	1990	2000	2007
Number of Doctors (in hundreds of thousands)	5	6	8	9

(*Source:* American Medical Association, *Physician Characteristics and Distribution in the U.S., 2010*)

Plot this information in the coordinate plane below.

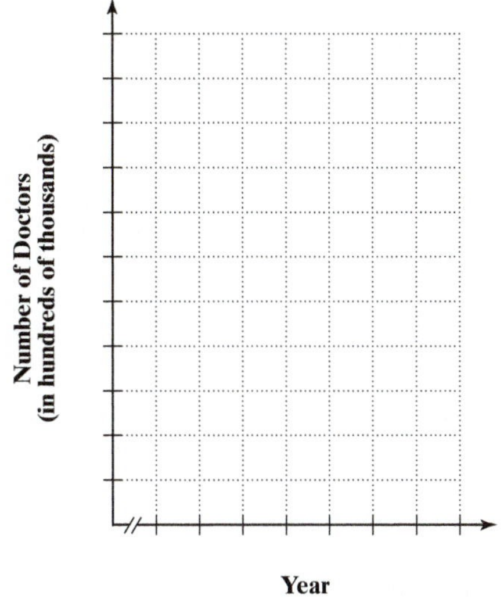

20. The state of Colorado is roughly shaped like a rectangle, with borders that are latitude and longitude lines. The north-south border is about 100 mi shorter than the east-west border, and the perimeter is approximately 1320 mi. Estimate the area of the state. (*Source:* census.gov)

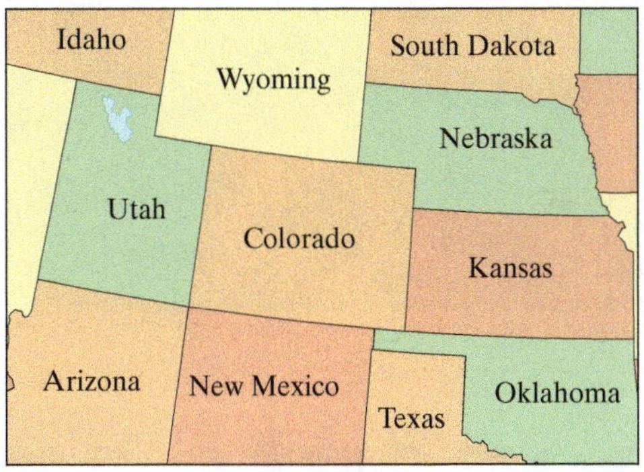

• Check your answers on page A-31.

CHAPTER 15

Exponents and Polynomials

Layoffs and Polynomials

Layoffs are common whenever the economy contracts. For instance, in 1933, the worst year of the Great Depression, so many American workers were laid off that one-fourth of the U.S. labor force was unemployed.

Suppose that 100 employees work in a factory. If the fraction f of these employees are laid off, the factory will still employ $100(1 - f)$, or $100 - 100f$, workers. If the factory again lays off the same fraction of employees, the remaining number of workers can be represented by $100(1 - f)(1 - f)$, an expression that can also be written as the *polynomial* $100 - 200f + 100f^2$.

(*Source:* Michael Parkin, *Economics*, Addison-Wesley, 1999)

CHAPTER 15 PRETEST

To see if you have already mastered the topics in this chapter, take this test.

Simplify.

1. $x^5 \cdot x^4$

2. $y^7 \div y^3$

3. $-3a^0$

4. $(4x^4y^3)^2$

5. $\left(\frac{a}{b^5}\right)^3$

6. $(5x^{-1}y^4)^{-2}$

7. For the polynomial $6x^4 + 5x^3 + x^2 - 7x + 8$, identify:

a. the terms ________________

b. the coefficients ________________

c. the degree ________________

d. the constant term ________________

8. Find the sum: $(2n^2 + 7n - 10) + (n^2 - 6n + 12)$

9. Subtract: $(8x^2 - 9) - (7x^2 - x - 5)$

10. Combine: $(6a^2b + ab - a^2) + (2a^2 + 3a^2b - 5b^2) - (7a^2 - 2ab - b^2)$

Multiply.

11. $3x^2(x^2 - 4x + 9)$

12. $(n + 3)(2n^2 + n - 6)$

13. $(4x + 9)(x - 3)$

14. $(3y - 7)(3y + 7)$

15. $(5 - 2n)^2$

Divide.

16. $\dfrac{9t^4 - 18t^3 - 45t^2}{9t^2}$

17. $(4x^2 - 3x - 10) \div (x - 2)$

Solve.

18. In chemistry, a mole (mol) of any material contains 6×10^{23} molecules. Express in scientific notation the number of molecules that 200 mol of hydrogen will contain. (*Source:* Karen Timberlake, *Chemistry: An Introduction to General Organic and Biological Chemistry*)

19. A student deposits \$100 in an account at interest rate r (in decimal form) compounded annually. The amount in the account after 2 yr is given by:

$$A = 100(1 + r)^2$$

Write A as a polynomial in r without parentheses.

20. The polynomial $0.08x + 6.07$ approximates the world population (in billions) for a particular year, where x represents the number of years after 2000. According to this model, find the world population in 2030 to the nearest billion. (*Source:* census.gov)

• Check your answers on page A-31.

15.1 Laws of Exponents

OBJECTIVES

- **A** To simplify expressions with exponents, including the exponent zero or one
- **B** To simplify expressions by using the product or quotient rule of exponents
- **C** To simplify expressions with negative exponents
- **D** To solve applied problems involving exponents

What Exponents Are and Why They Are Important

In Chapter 11, we considered algebraic expressions. In this chapter, we discuss a particular kind of expression called a *polynomial* and operations involving polynomials. Central to our discussion of these operations is a thorough understanding of the laws of exponents.

Exponents play an important role in arithmetic and algebra. Powers of 10 are key to the decimal place-value system that underlies the reading, writing, and computation of real numbers. There are many other applications of exponents in the sciences and in business, such as compound interest, population growth, and scientific notation as used in astronomy and physics.

Exponents

Recall the definition of an exponent from Section 11.4: An exponent (or power) is a number that indicates how many times another number (called the *base*) is used as a factor. For instance,

$$5^3 = \underbrace{5 \cdot 5 \cdot 5}_{\text{3 factors of 5}}$$

(Exponent: 3; Base: 5)

In general, if a is a positive integer, the expression x^a means

$$x^a = \underbrace{x \cdot x \cdot x \cdot \cdots \cdot x}_{a \text{ factors}},$$

where the exponent a indicates that there are a factors of x.

It follows from the definition of an exponent that raising a base to the power 1 has 1 factor of the base.

Exponent 1

For any real number x, $x^1 = x$.

In words, any number raised to the power 1 is the number itself. For instance, $7^1 = 7$, $(-5)^1 = -5$, and $x^1 = x$.

EXAMPLE 1

Multiply.

a. 2^5 **b.** $\left(-\frac{1}{3}\right)^4$ **c.** $(-x)^1$ **d.** $(-x)^2$

Solution

a. $2^5 = 2 \cdot 2 \cdot 2 \cdot 2 \cdot 2 = 32$

b. $\left(-\frac{1}{3}\right)^4 = \left(-\frac{1}{3}\right)\left(-\frac{1}{3}\right)\left(-\frac{1}{3}\right)\left(-\frac{1}{3}\right) = \frac{1}{81}$

c. $(-x)^1 = -x$

d. $(-x)^2 = (-x)(-x) = x^2$

PRACTICE 1

Multiply.

a. 10^4

b. $\left(-\frac{1}{2}\right)^5$

c. $(-y)^6$

d. $(-y)^1$

The expression 10^0 involves raising a number to the 0 power. To understand the value of this expression, consider the following pattern:

$$\begin{aligned} 10^5 &= 10 \cdot 10 \cdot 10 \cdot 10 \cdot 10 = 100{,}000 \\ 10^4 &= 10 \cdot 10 \cdot 10 \cdot 10 = 10{,}000 \\ 10^3 &= 10 \cdot 10 \cdot 10 = 1000 \\ 10^2 &= 10 \cdot 10 = 100 \\ 10^1 &= 10 \\ 10^0 &= ? \end{aligned}$$

Note that to go from 10^5 to 10^4 we divide the first number by 10. The pattern continues, and going from 10^2 to 10^1, we divide 10^2 by 10. So to go from 10^1 to 10^0, it seems reasonable for us to divide 10^1 by 10, which is 1. Thus, we take 10^0 to be equal to 1. Would we have gotten the same result had we considered powers of 2 instead of powers of 10 in order to determine the value of 2^0?

Exponent 0

For any nonzero real number x, $x^0 = 1$.

In words, any nonzero number raised to the power 0 is equal to 1. For example, $8^0 = 1$ and $(-100)^0 = 1$. Note that the expression 0^0 is undefined. Throughout the remainder of this text, we will assume that any variable raised to the 0 power represents a nonzero number.

EXAMPLE 2

Simplify.

a. 25^0 **b.** $(-3.5)^0$ **c.** a^0

d. $-a^0$ **e.** $4x^0$

Solution

a. $25^0 = 1$ **b.** $(-3.5)^0 = 1$ **c.** $a^0 = 1$

d. $-a^0 = -1 \cdot a^0 = -1 \cdot 1 = -1$

e. $4x^0 = 4 \cdot 1 = 4$

PRACTICE 2

Simplify.

a. 8^0

b. $\left(-\frac{2}{3}\right)^0$

c. y^0

d. $-y^0$

e. $(4x)^0$

Laws of Exponents

The laws of exponents are rules that apply to exponents. These rules are not arbitrary but follow logically from the definition of exponent.

Let's first discuss the *product rule*. This rule applies when we multiply two powers with the same base. Consider the expression $x^4 \cdot x^6$. Using the definition of an exponent we get:

$$x^4 \cdot x^6 = \underbrace{\overbrace{(x \cdot x \cdot x \cdot x)}^{\text{4 factors}}\overbrace{(x \cdot x \cdot x \cdot x \cdot x \cdot x)}^{\text{6 factors}}}_{\text{10 factors}} = x^{10}$$

Since 4 factors of x and 6 additional factors of x make 10 factors of x, it follows that

$$x^4 \cdot x^6 = x^{4+6} = x^{10}.$$

This result can be generalized as follows:

The Product Rule of Exponents

For any nonzero real number x and for any nonnegative integers a and b,

$$x^a \cdot x^b = x^{a+b}.$$

In words, to multiply powers of the same base, add the exponents and keep the base the same.

EXAMPLE 3

Simplify using the product rule, if possible.

a. $2^3 \cdot 2^5$ **b.** $(-5)^2 \cdot (-5)^8$ **c.** $x^8 \cdot x^{10}$

d. $m^4 \cdot m$ **e.** $x^2 \cdot y^3$

Solution When we multiply powers of the same base, the product rule tells us to add the exponents but *not to change the base.*

a. $2^3 \cdot 2^5 = 2^{3+5} = 2^8$

b. $(-5)^2 \cdot (-5)^8 = (-5)^{2+8} = (-5)^{10}$

c. $x^8 \cdot x^{10} = x^{8+10} = x^{18}$

d. $m^4 \cdot m = m^4 \cdot m^1 = m^{4+1} = m^5$

e. In the expression $x^2 \cdot y^3$, we cannot apply the product rule because the bases are not the same.

PRACTICE 3

Simplify using the product rule, if possible.

a. $10^8 \cdot 10^4$

b. $(-4)^3 \cdot (-4)^3$

c. $n^3 \cdot n^7$

d. $y^5 \cdot y^0$

e. $a \cdot b^4$

Now, we discuss another law of exponents—the *quotient rule*. This rule applies when we divide two powers of the same base. Consider the expression $\frac{x^6}{x^2}$. Using the definition of exponent, we get:

$$\frac{x^6}{x^2} = \frac{\overbrace{x \cdot x \cdot x \cdot x \cdot \overset{1}{\cancel{x}} \cdot \overset{1}{\cancel{x}}}^{\text{6 factors}}}{\underbrace{\underset{1}{\cancel{x}} \cdot \underset{1}{\cancel{x}}}_{\text{2 factors}}} = \underbrace{x \cdot x \cdot x \cdot x}_{\text{4 factors}} = x^4$$

So

$$\frac{x^6}{x^2} = x^{6-2} = x^4,$$

which suggests the following rule:

The Quotient Rule of Exponents

For any nonzero real number x and for any positive integers a and b,

$$\frac{x^a}{x^b} = x^{a-b}.$$

In words, to divide powers of the same base, subtract the exponent in the denominator from the exponent in the numerator, and keep the base the same.

EXAMPLE 4

Simplify using the quotient rule, if possible.

a. $\frac{10^5}{10^2}$ **b.** $(-2)^5 \div (-2)^5$ **c.** $\frac{p^{12}}{p^7}$

d. $\frac{y^{13}}{y}$ **e.** $\frac{x^4}{y^2}$

Solution When we divide powers of the same base, the quotient rule tells us to subtract the exponents but *not to change the base.*

a. $\frac{10^5}{10^2} = 10^{5-2} = 10^3$

b. $(-2)^5 \div (-2)^5 = \frac{(-2)^5}{(-2)^5} = (-2)^{5-5} = (-2)^0 = 1$

c. $\frac{p^{12}}{p^7} = p^{12-7} = p^5$

d. $\frac{y^{13}}{y} = \frac{y^{13}}{y^1} = y^{13-1} = y^{12}$

e. In the expression $\frac{x^4}{y^2}$, we cannot apply the quotient rule because the bases are not the same.

PRACTICE 4

Simplify using the quotient rule, if possible.

a. $\frac{7^7}{7^2}$

b. $(-9)^6 \div (-9)^5$

c. $\frac{s^{10}}{s^{10}}$

d. $\frac{r^8}{r}$

e. $\frac{a^5}{b^3}$

Note in Example 4(b) that the quotient rule confirms the fact that any nonzero real number raised to the 0 power is 1. That is,

$$\frac{(-2)^5}{(-2)^5} = \frac{-32}{-32} = 1 \quad \text{and} \quad \frac{(-2)^5}{(-2)^5} = (-2)^{5-5} = (-2)^0 = 1$$

EXAMPLE 5

Simplify.

a. $x^3 \cdot x \cdot x^5$ **b.** $(a^2b)(ab^4)$ **c.** $\frac{t^3 \cdot t^5}{t^2}$

Solution

a. $x^3 \cdot x \cdot x^5 = x^9$ — Use the product rule.

b. $(a^2b)(ab^4) = a^2 \cdot a^1 \cdot b^1 \cdot b^4$ — Rearrange the factors.

$= a^3b^5$ — Use the product rule.

c. $\frac{t^3 \cdot t^5}{t^2} = \frac{t^8}{t^2} = t^6$ — Use the product rule in the numerator. Then, use the quotient rule.

PRACTICE 5

Simplify.

a. $y^2 \cdot y^3 \cdot y^4$

b. $(x^3y^3)(x^2y^3)$

c. $\frac{a^7}{a \cdot a^4}$

Negative Exponents

Until now, we have considered only exponents that were either positive integers or 0. What meaning should we give to *negative exponents*?

The quotient rule is the key to answering this question. Consider, for instance, the quotient $\frac{6^4}{6^7}$. On the one hand, we can simplify this fraction by using the definition of exponent and canceling the common factors, getting:

$$\frac{6^4}{6^7} = \frac{\overset{1}{\cancel{6}} \cdot \overset{1}{\cancel{6}} \cdot \overset{1}{\cancel{6}} \cdot \overset{1}{\cancel{6}}}{\underset{1}{\cancel{6}} \cdot \underset{1}{\cancel{6}} \cdot \underset{1}{\cancel{6}} \cdot \underset{1}{\cancel{6}} \cdot 6 \cdot 6 \cdot 6} = \frac{1}{6^3}$$

On the other hand, we can simplify by using the quotient rule, getting:

$$\frac{6^4}{6^7} = 6^{4-7} = 6^{-3}$$

Since

$$\frac{6^4}{6^7} = 6^{-3} \quad \text{and} \quad \frac{6^4}{6^7} = \frac{1}{6^3},$$

we conclude that $6^{-3} = \frac{1}{6^3}$, which suggests the following rule:

Negative Exponent

For any nonzero real number x and for any integer a,

$$x^{-a} = \frac{1}{x^a}.$$

In words, to evaluate x^{-a}, take the reciprocal of x^{-a} and change the sign of the exponent.

In general, an expression with exponents is considered *simplified* when it is written with only positive exponents.

EXAMPLE 6

Simplify.

a. 5^{-2} **b.** p^{-8} **c.** $-(8x)^{-1}$ **d.** $(-4)^{-2}$

Solution

a. $5^{-2} = \frac{1}{5^2} = \frac{1}{25}$ **b.** $p^{-8} = \frac{1}{p^8}$

c. $-(8x)^{-1} = -1(8x)^{-1} = \frac{-1}{(8x)^1} = -\frac{1}{8x}$

d. $(-4)^{-2} = \frac{1}{(-4)^2} = \frac{1}{16}$

PRACTICE 6

Simplify.

a. 9^{-2}

b. n^{-5}

c. $-(3y)^{-1}$

d. $(5)^{-3}$

TIP A negative exponent indicates a reciprocal. For example, $5^{-2} = \frac{1}{25}$.

The product rule of exponents and the quotient rule of exponents, which were defined for positive-integer exponents, also hold for negative-integer exponents.

EXAMPLE 7

Simplify by writing each expression using only positive exponents.

a. $2^{-1}q$ **b.** $5x^{-2}$ **c.** $\frac{y^6}{y^{10}}$

d. $4^{-1} \cdot x^{-5} \cdot x^2$ **e.** $\frac{1}{x^{-2}}$

Solution

a. $2^{-1}q = \frac{1}{2} \cdot q = \frac{q}{2}$

b. $5x^{-2} = 5 \cdot \frac{1}{x^2} = \frac{5}{x^2}$

c. $\frac{y^6}{y^{10}} = y^{6-10} = y^{-4} = \frac{1}{y^4}$

d. $4^{-1} \cdot x^{-5} \cdot x^2 = 4^{-1} \cdot x^{-5+2} = \frac{1}{4}x^{-3} = \frac{1}{4} \cdot \frac{1}{x^3} = \frac{1}{4x^3}$

e. $\frac{1}{x^{-2}} = 1 \div x^{-2} = 1 \div \frac{1}{x^2} = 1 \cdot \frac{x^2}{1} = x^2$

PRACTICE 7

Write as expressions using only positive exponents.

a. $8^{-1}s$

b. $3x^{-1}$

c. $\frac{r^3}{r^9}$

d. $3^2 \cdot g^{-1} \cdot g^{-4}$

e. $\frac{1}{x^{-3}}$

The definition of a negative exponent and Example 7(e), in which we saw that $\frac{1}{x^{-2}} = x^2$, suggest the following:

Reciprocal of x^{-a}

For any nonzero real number x and any integer a,

$$\frac{1}{x^{-a}} = x^a.$$

In words, the reciprocal of x^{-a} is x^a.

EXAMPLE 8

Write as an expression using positive exponents.

a. $\frac{5}{x^{-4}}$ **b.** $\frac{1}{3y^{-1}}$ **c.** $\frac{a^2}{b^{-3}}$

Solution

a. $\frac{5}{x^{-4}} = 5 \cdot \frac{1}{x^{-4}} = 5 \cdot x^4 = 5x^4$

b. $\frac{1}{3y^{-1}} = \frac{1}{3} \cdot \frac{1}{y^{-1}} = \frac{1}{3} \cdot y^1 = \frac{y}{3}$

c. $\frac{a^2}{b^{-3}} = a^2 \cdot \frac{1}{b^{-3}} = a^2 \cdot b^3 = a^2b^3$

PRACTICE 8

Write as an expression using positive exponents.

a. $\frac{1}{a^{-3}}$

b. $\frac{2}{5x^{-2}}$

c. $\frac{r^3}{2s^{-1}}$

EXAMPLE 9

Physicists study different kinds of electromagnetic waves, including radio waves, X-rays, and gamma rays. The diagram below, called the *electromagnetic spectrum*, shows the relationship among the wavelengths of these waves.

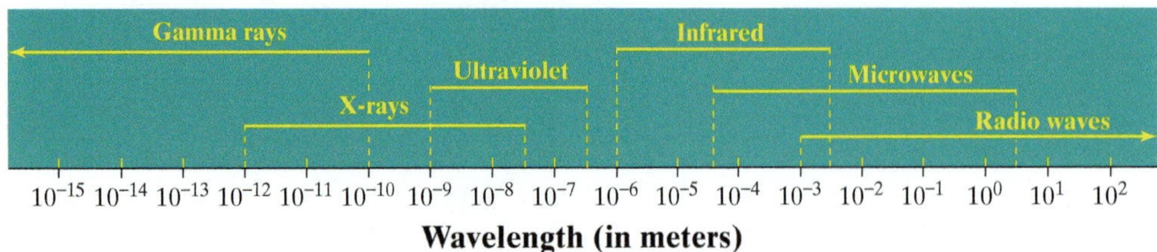

Consider a particular X-ray whose wavelength is 10^{-10} m and a gamma ray whose wavelength is 10^{-15} m. (*Source:* Arthur Beiser, *The Mainstream of Physics*)

a. Which of these rays has a greater length?

b. What is the ratio of the longer to the shorter wavelength?

Solution

a. The wavelengths are 10^{-10} m for the X-ray and 10^{-15} m for the gamma ray. Converting these expressions with negative exponents to equivalent expressions with positive exponents, the wavelengths can be written as $\frac{1}{10^{10}}$ m and $\frac{1}{10^{15}}$ m, respectively. Since $\frac{1}{10^{10}}$ has the smaller denominator, it is the larger number. Therefore, the X-ray has the greater wavelength.

b. We use the quotient rule to compute the ratio of the greater wavelength, 10^{-10} m, to the shorter wavelength, 10^{-15} m:

$$\frac{10^{-10}}{10^{-15}} = \frac{10^{15}}{10^{10}} = 10^{15-10} = 10^5$$

So the ratio is $\frac{10^5}{1}$, or 10^5 to 1.

PRACTICE 9

A computer's memory is often measured in *bits*, *bytes*, and *megabytes*. A bit (short for "binary digit") is the smallest unit of data in the memory of the computer. A byte is equal to 2^3 bits, whereas a megabyte (MB) is equal to 2^{20} bytes.

a. How many bits are in a megabyte? Write the answer as a power of 2.

b. In most computers, the hard drive's capacity is expressed in *gigabytes*. A gigabyte (GB) is 2^{10} megabytes. How many bytes are there in a gigabyte? Express the result as a power of 2.

15.1 Exercises

FOR EXTRA HELP MyMathLab MathXL PRACTICE WATCH READ REVIEW

Mathematically Speaking

Fill in each blank with the most appropriate term or phrase from the given list.

the opposite of	added	power of 1
multiplied	0 power	subtracted
1 divided by	divided	

1. The product rule of exponents states that when powers of the same base are multiplied, the exponents are ____________ and the base is left the same.

2. The quotient rule of exponents states that when powers of the same base are divided, the exponents are ____________ and the base is left the same.

3. Any nonzero real number raised to the ____________ is 1.

4. For any nonzero real number x and for any integer a, x^{-a} can be written as ____________ x^a.

A *Multiply.*

5. 5^3
6. 1^8
7. $\left(\frac{3}{4}\right)^2$
8. $\left(\frac{7}{8}\right)^2$
9. $(0.4)^2$
10. $(0.2)^2$
11. $-(0.5)^2$
12. $-(0.3)^2$
13. $(-2)^3$
14. $(-4)^3$
15. $(-3)^4$
16. $(-6)^4$
17. $\left(-\frac{1}{2}\right)^3$
18. $\left(-\frac{4}{5}\right)^3$
19. $(-x)^4$
20. $(-y)^3$

Simplify.

21. $(pq)^1$
22. $(xy)^1$
23. $(-3)^0$
24. $(-4)^0$
25. $-a^0$
26. $-b^0$

B *Simplify using the product rule, if possible.*

27. $10^9 \cdot 10^2$
28. $5^2 \cdot 5^3$
29. $4^3 \cdot 4^5$
30. $6^2 \cdot 6^8$
31. $a^4 \cdot a^2$
32. $x \cdot x^8$
33. $x^3 \cdot y^5$
34. $a^4 \cdot b^5$
35. $n^6 \cdot n$
36. $y \cdot y^0$
37. x^2y
38. a^4b^3

Simplify using the quotient rule, if possible.

39. $\frac{8^5}{8^3}$
40. $\frac{3^8}{3^2}$
41. $\frac{5^4}{5}$
42. $\frac{2^9}{2^3}$
43. $\frac{y^6}{y^5}$
44. $\frac{x^{12}}{x^{12}}$
45. $\frac{a^{10}}{a^4}$
46. $\frac{x^7}{x^5}$
47. $\frac{y^8}{x^4}$
48. $\frac{a^5}{b}$
49. $\frac{x^6}{x^6}$
50. $r^4 \div r^0$

Simplify.

51. $y^2 \cdot y^3 \cdot y$
52. $t \cdot t \cdot t^2$
53. $(p^2q^3)(p^5q^2)$
54. $(xy^2)(x^2y^6)$
55. $(yx^2)(xz^2)(yz)$
56. $a(a^4b^2)(bc^3)$
57. $\frac{a^2 \cdot a^3}{a^4}$
58. $\frac{t^4}{t^3 \cdot t}$
59. $\frac{x^2 \cdot x^4}{x^3 \cdot x}$
60. $\frac{y^5 \cdot y^5}{y^2 \cdot y^3}$

C *Write as an expression using only positive exponents.*

61. 5^{-1}
62. 7^{-1}
63. x^{-1}
64. a^{-1}
65. $(-3a)^{-1}$
66. $-(5y)^{-1}$
67. 2^{-4}
68. 7^{-3}
69. -3^{-4}
70. -5^{-2}
71. $8n^{-3}$
72. $2y^{-3}$
73. $(-x)^{-2}$
74. $(-a)^{-4}$
75. $-3^{-2}x$
76. $-4^{-1}y$
77. $x^{-2}y^3$
78. xy^{-3}
79. qr^{-1}
80. rs^{-1}
81. $4x^{-1}y^2$
82. $-5a^2b^{-4}$
83. $p^{-2} \cdot p^{-3}$
84. $t^{-3} \cdot t^{-3}$
85. $p^{-1} \cdot p^4$
86. $s^4 \cdot s^{-2}$
87. $\frac{a^3}{a^4}$
88. $\frac{n}{n^5}$
89. $\frac{2}{n^{-4}}$
90. $\frac{3}{n^{-1}}$
91. $\frac{p^4}{q^{-1}}$
92. $\frac{a}{b^{-6}}$
93. $\frac{t^{-2}}{t^3}$
94. $\frac{x^{-2}}{x^5}$
95. $\frac{x^5}{x^{-2}}$
96. $\frac{n^7}{n^{-1}}$
97. $\frac{a^{-4}}{a^{-5}}$
98. $\frac{y^{-1}}{y^{-6}}$
99. $\frac{a^{-3}}{b^{-3}}$
100. $\frac{x^{-4}}{y^{-2}}$

Mixed Practice

Simplify.

101. $(s^2t^4)(st^2)$
102. $\frac{a^6 \cdot a}{a^2 \cdot a^3}$

Solve.

103. Multiply: $\left(-\frac{2}{3}\right)^3$
104. Simplify $\frac{a^5}{a^4}$ using the quotient rule, if possible.
105. Simplify x^2y^3 using the product rule, if possible.
106. Simplify: -5^0

Write as an expression using only positive exponents.

107. $(-y)^{-6}$
108. $\frac{3}{4k^{-2}}$
109. $q^{-2} \cdot q^{-3}$
110. $a^{-3}b^2$
111. $\frac{x^4}{y^{-3}}$
112. $\frac{m^{-5}}{n^{-3}}$

Applications

D *Solve.*

113. The first day of an epidemic, 35 people got sick. Each day thereafter, the number of people who got ill doubled.

a. How many people were ill on the sixth day of the epidemic? On the tenth day?

b. How many times as great was the number of people ill on the tenth day as compared to the number ill on the sixth day?

114. The value of a new car t years after it is purchased is given by the expression $28{,}000(1.25)^{-t}$.

a. What is the value of the car one year after it was purchased?

b. Evaluate the expression for $t = 0$. Explain the significance of this value.

115. The concentration of a pollutant in a pond is 60 parts per million (ppm). The pollution level drops by 5% each month, so the amount of pollutant each month is 95% of the amount in the previous month, as shown in the following table:

Month	Pollution Level (ppm)
1	60
2	60×0.95
3	$60 \times (0.95)^2$
4	$60 \times (0.95)^3$
5	$60 \times (0.95)^4$
6	$60 \times (0.95)^5$
7	$60 \times (0.95)^6$

What will the pollution level be in the twelfth month?

116. The population of the United States in 1820 was approximately 10^7. One hundred years later, it was approximately 10^8. By what factor did the U.S. population grow during this century? (*Source:* census.gov)

117. A small, cube-shaped box is packed within a larger, cube-shaped box and is surrounded by Styrofoam peanuts. Since the side length of the larger box is $\frac{5}{2}$ that of the smaller box, we can represent the two side lengths as $5x$ and $2x$.

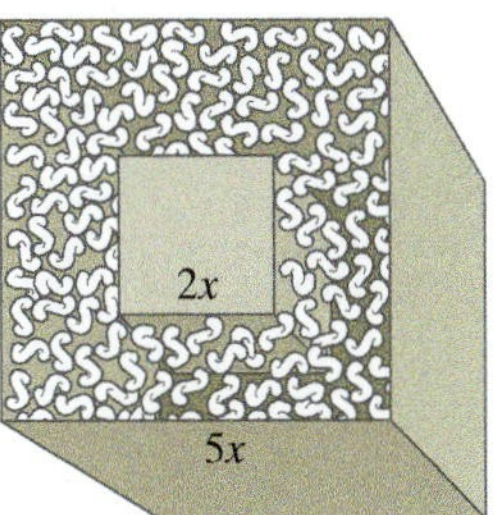

a. Write an expression for the volume of each box.

b. How many times the volume of the small box is the volume of the large box?

118. The *multiplication principle* states that the number of possible ways to do multiple things is the product of the number of ways to do each individual thing. For instance, if one die has 6 possible results, then two dice have 6×6 or 36 possible results. While playing a board game, a player rolls 3 red dice and 2 white dice. Each of the 6 faces of a die has a 1, 2, 3, 4, 5, or 6.

a. How many different possible rolls are there for the 5 dice?

b. How many different possible rolls are there where all the red dice come up 6, and both of the white dice come up as something other than 6?

• Check your answers on page A-32.

MINDStretchers

Writing

1. Identify the errors that were made resulting in the following false statements:

a. $4^{-3} = -\frac{1}{64}$

b. $x^{-2} \cdot x^{-3} = x^{6}$

c. $4^{3} \cdot 4^{-5} = 16^{-2}$

d. $5^{4} \div 5 = 1^{4}$

Groupwork

2. Which is larger: x^2 or x^{-2}? Explain your answer, and give some examples.

Mathematical Reasoning

3. Explain why a thousand million is the same as a billion.

15.2 More Laws of Exponents and Scientific Notation

OBJECTIVES

- **A** To simplify expressions by using the power rule
- **B** To simplify expressions by raising a product to a power
- **C** To simplify expressions by raising a quotient to a power
- **D** To write numbers in scientific notation or standard notation
- **E** To solve applied problems involving laws of exponents or scientific notation

In this section, we consider several additional laws of exponents, as well as an important application of exponents known as scientific notation.

Additional Laws of Exponents

In the previous section, we discussed the product rule for the product of powers and the quotient rule for the quotient of powers. We now consider a third rule known as the *power rule*. The power rule deals with expressions in which a power is raised to a power.

Let's consider, for instance, the expression $(x^2)^3$. Using the definition of an exponent gives us:

$$(x^2)^3 = \underbrace{x^2 \cdot x^2 \cdot x^2}_{\text{3 factors of } x^2} = x^{2+2+2} = x^6$$

So $(x^2)^3 = x^6$. We can generalize this result as follows:

The Power Rule of Exponents

For any nonzero real number x and any integers a and b,

$$(x^a)^b = x^{ab}.$$

In words, to raise a power to a power, *multiply* the exponents and *keep the base the same.*

EXAMPLE 1

Simplify using the power rule of exponents.

a. $(5^2)^2$ **b.** $(2^{-3})^2$ **c.** $-(p^4)^5$ **d.** $(q^2)^{-1}$

Solution We apply the power rule and then simplify.

a. $(5^2)^2 = 5^{2\cdot 2} = 5^4 = 625$

b. $(2^{-3})^2 = 2^{-3\cdot 2}$

$= 2^{-6}$

$= \dfrac{1}{2^6}$ Take the reciprocal of 2^{-6}, and change the exponent from -6 to 6.

$= \dfrac{1}{64}$

c. $-(p^4)^5 = -(p^{4\cdot 5}) = -p^{20}$

d. $(q^2)^{-1} = q^{2\cdot(-1)} = q^{-2} = \dfrac{1}{q^2}$

PRACTICE 1

Simplify using the power rule of exponents.

a. $(2^3)^2$

b. $(7^3)^{-1}$

c. $(q^2)^4$

d. $-(p^3)^{-5}$

TIP Be sure to distinguish between the *product* rule and the *power* rule.

Product rule: $x^a \cdot x^b = x^{a+b}$ — Add the exponents.

Power rule: $(x^a)^b = x^{ab}$ — Multiply the exponents.

Another law of exponents has to do with *raising a product to a power*. For instance, consider the expression $(5x)^3$.

Rearrange the factors.

$$(5x)^3 = (5x)(5x)(5x) = (5 \cdot 5 \cdot 5)(x \cdot x \cdot x) = 5^3 \cdot x^3 = 125x^3$$

So we see that $(5x)^3$ is the same as 5^3 times x^3. We can generalize this result as follows:

Raising a Product to a Power

For any nonzero real numbers x and y and any integer a,

$$(xy)^a = x^a \cdot y^a.$$

In words, to raise a product to a power, raise each factor to that power.

EXAMPLE 2

Simplify using the rule for raising a product to a power.

a. $(2y)^4$ **b.** $(-3a)^2$ **c.** $-(3a)^2$

Solution We apply the rule for raising a product to a power, and then simplify.

a. $(2y)^4 = 2^4 \cdot y^4 = 16y^4$

b. $(-3a)^2 = (-3)^2(a)^2 = 9a^2$

c. $-(3a)^2 = -3^2 \cdot a^2 = -9a^2$

PRACTICE 2

Simplify.

a. $(7a)^2$

b. $(-4x)^3$

c. $-(4x)^3$

EXAMPLE 3

Simplify.

a. $(2x^4)^5$ **b.** $(p^3q^5)^3$ **c.** $-7(m^5n^{10})^2$ **d.** $(5a^{-2}c^4)^{-2}$

Solution

a. $(2x^4)^5 = 2^5(x^4)^5$ Use the rule for raising a product to a power.

$= 32x^{20}$ Use the power rule.

b. $(p^3q^5)^3 = (p^3)^3(q^5)^3 = p^9q^{15}$

c. $-7(m^5n^{10})^2 = -7(m^5)^2(n^{10})^2 = -7m^{10}n^{20}$

d. $(5a^{-2}c^4)^{-2} = 5^{-2}(a^{-2})^{-2}(c^4)^{-2}$ Use the rule for raising a product to a power.

$= 5^{-2}(a^4)(c^{-8})$ Use the power rule.

$= \frac{1}{25} \cdot a^4 \cdot \frac{1}{c^8}$ Use the definition of a negative exponent.

$= \frac{a^4}{25c^8}$

PRACTICE 3

Simplify.

a. $(-6a^9)^2$

b. $(q^8r^{10})^2$

c. $-2(ab^7)^3$

d. $(7a^{-1}c^{-5})^2$

The next law of exponents that we discuss is *raising a quotient to a power*. For instance, consider the expression $\left(\frac{a}{b}\right)^4$, where a divided by b is raised to the fourth power. By the definition of exponent, we get:

$$\left(\frac{a}{b}\right)^4 = \frac{a}{b}\cdot\frac{a}{b}\cdot\frac{a}{b}\cdot\frac{a}{b} = \frac{a\cdot a\cdot a\cdot a}{b\cdot b\cdot b\cdot b} = \frac{a^4}{b^4}$$

So $\left(\frac{a}{b}\right)^4 = \frac{a^4}{b^4}$. We generalize this result as follows:

Raising a Quotient to a Power

For any nonzero real numbers x and y and any integer a,

$$\left(\frac{x}{y}\right)^a = \frac{x^a}{y^a}.$$

In words, to raise a quotient to a power, raise both the numerator and the denominator to that power.

EXAMPLE 4

Simplify by using the rule for raising a quotient to a power.

a. $\left(\frac{x}{5}\right)^3$ **b.** $\left(\frac{-a}{b}\right)^4$ **c.** $\left(\frac{5}{x}\right)^{-3}$ **d.** $\left(\frac{-3r^2}{st^4}\right)^3$ **e.** $\left(\frac{9u}{v^{-1}}\right)^2$

Solution Here, we use the rule for raising a quotient to a power, and then simplify.

a. $\left(\frac{x}{5}\right)^3 = \frac{x^3}{5^3} = \frac{x^3}{125}$ **b.** $\left(\frac{-a}{b}\right)^4 = \frac{(-a)^4}{b^4} = \frac{a^4}{b^4}$

c. $\left(\frac{5}{x}\right)^{-3} = \frac{5^{-3}}{x^{-3}}$

$= 5^{-3}\cdot\frac{1}{x^{-3}}$

$= \frac{1}{5^3}\cdot x^3$

$= \frac{x^3}{125}$

d. $\left(\frac{-3r^2}{st^4}\right)^3 = \frac{(-3r^2)^3}{(st^4)^3} = \frac{(-3)^3(r^2)^3}{s^3(t^4)^3} = \frac{-27r^6}{s^3t^{12}}$

e. $\left(\frac{9u}{v^{-1}}\right)^2 = \frac{(9u)^2}{(v^{-1})^2}$

$= \frac{9^2u^2}{v^{-2}}$

$= 81u^2\cdot\frac{1}{v^{-2}}$

$= 81u^2v^2$

PRACTICE 4

Simplify.

a. $\left(\frac{y}{3}\right)^2$ **b.** $\left(\frac{-u}{v}\right)^{10}$

c. $\left(\frac{3}{y}\right)^{-2}$ **d.** $\left(\frac{-10a^5}{3b^2c}\right)^2$

e. $\left(\frac{5x}{y^{-2}}\right)^3$

Note that the simplified form of the expression in Example 4(a) is the same as the simplified form of the expression in Example 4(c). Since $\left(\frac{5}{x}\right)^{-3} = \frac{x^3}{125}$ and $\left(\frac{x}{5}\right)^3 = \frac{x^3}{125}$, we conclude that $\left(\frac{5}{x}\right)^{-3} = \left(\frac{x}{5}\right)^3$. This conclusion leads us to another law of exponents—*raising a quotient to a negative power.*

Raising a Quotient to a Negative Power

For any nonzero real numbers x and y and any integer a,

$$\left(\frac{x}{y}\right)^{-a} = \left(\frac{y}{x}\right)^a.$$

In words, to raise a quotient to a negative power, take the reciprocal of the quotient and change the sign of the exponent.

EXAMPLE 5

Simplify.

a. $\left(\frac{2}{x}\right)^{-3}$ b. $\left(\frac{3r}{10s}\right)^{-1}$ c. $\left(\frac{x^4}{y^2}\right)^{-2}$

Solution Use the rule for raising a quotient to a negative power.

a. $\left(\frac{2}{x}\right)^{-3} = \left(\frac{x}{2}\right)^3 = \frac{x^3}{2^3} = \frac{x^3}{8}$

b. $\left(\frac{3r}{10s}\right)^{-1} = \left(\frac{10s}{3r}\right)^1 = \frac{10s}{3r}$

c. $\left(\frac{x^4}{y^2}\right)^{-2} = \left(\frac{y^2}{x^4}\right)^2 = \frac{(y^2)^2}{(x^4)^2} = \frac{y^4}{x^8}$

PRACTICE 5

Simplify.

a. $\left(\frac{5}{a}\right)^{-2}$

b. $\left(\frac{4u}{v}\right)^{-1}$

c. $\left(\frac{a^5}{b^3}\right)^{-2}$

Scientific Notation

Scientific notation is an important application of exponents—whether they are positive, negative, or zero. Scientists use this notation to abbreviate very large or very small numbers. Note that scientific notation is based on powers of 10.

Example	Standard Notation	Scientific Notation
The speed of light	983,000,000 ft/sec	9.83×10^8 ft/sec
The length of a virus	0.000000000001 m	1×10^{-12} m

Scientific notation has several advantages over standard notation. For example, when a number contains a long string of 0's, writing it in scientific notation can take fewer digits. Also, numbers written in scientific notation can be relatively easy to multiply or divide.

DEFINITION

A number is in **scientific notation** if it is written in the form

$$a \times 10^n$$

where n is an integer and a is greater than or equal to 1 but less than 10 ($1 \le a < 10$).

Note that any value of a that satisfies the inequality $1 \leq a < 10$ must have *one nonzero digit* to the left of the decimal point. For instance, 7.3×10^5 is written in scientific notation. Do you see why the numbers 0.83×10^2, 5×3^7, and 13.8×10^{-4} are *not* written in scientific notation?

TIP When written in scientific notation, large numbers have positive powers of 10, whereas small numbers have negative powers of 10. For instance, 3×10^{23} is large, whereas 3×10^{-23} is small.

Let's now consider how to change a number from scientific notation to standard notation.

EXAMPLE 6

Change the number 2.41×10^5 from scientific notation to standard notation.

Solution To express this number in standard notation, we need to multiply 2.41 by 10^5. Since $10^5 = 100{,}000$, multiplying 2.41 by 100,000 gives:

$$2.41 \times 10^5 = 2.41 \times 100{,}000 = 241{,}000.00 = 241{,}000$$

The number 241,000 is written in standard notation.

Note that the power of 10 here is *positive* and that the decimal point is moved five places *to the right*. So a shortcut for expressing 2.41×10^5 in standard notation is to move the decimal point in 2.41 five places to the right.

$$2.41 \times 10^5 = 2\,4\,1\,0\,0\,0. = 241{,}000$$

PRACTICE 6

Express 2.539×10^2 in standard notation.

EXAMPLE 7

Convert 3×10^{-5} to standard notation.

Solution Using the definition of a negative exponent, we get:

$$3 \times 10^{-5} = 3 \times \frac{1}{10^5}, \text{ or } \frac{3}{10^5}$$

Since $10^5 = 100{,}000$, dividing 3 by 100,000 gives us:

$$\frac{3}{10^5} = \frac{3}{100{,}000} = 0.00003$$

Here we note that the power of 10 is *negative* and that the decimal point, which is understood to be at the right end of a whole number, is moved five places *to the left*. So a shortcut for expressing 3×10^{-5} in standard notation is to move the decimal point in 3. five places to the left.

$$3 \times 10^{-5} = 3. \times 10^{-5} = .0\,0\,0\,0\,3 = .00003, \text{ or } 0.00003$$

PRACTICE 7

Change 4.3×10^{-9} to standard notation.

TIP When converting a number from scientific notation to standard notation, move the decimal point to the *right* if the power of 10 is *positive* and to the *left* if the power of 10 is *negative.*

Now, let's consider the reverse situation, namely changing a number in standard notation to scientific notation.

EXAMPLE 8

Express 37,000,000,000 in scientific notation.

Solution For a number to be written in scientific notation, it must be of the form

$$a \times 10^n$$

where n is an integer and $1 \leq a < 10$. We know that 37,000,000,000 and 37,000,000,000. are the same. We move the decimal point *to the left* so that there is one nonzero digit to the left of the decimal point. The power of 10 by which we multiply is the same as the number of places moved.

$$37{,}000{,}000{,}000 = 3.7000000000 \times 10^{10}$$

Move 10 places to the *left*.

$$= 3.7 \times 10^{10}$$

Since 3.7 and 3.7000000000 are equivalent, we can drop the trailing zeros. So 37,000,000,000 expressed in scientific notation is 3.7×10^{10}.

PRACTICE 8

Write 8,000,000,000,000 in scientific notation.

EXAMPLE 9

Convert 0.00000000000000002 to scientific notation.

Solution We must write the number 0.00000000000000002 in the form

$$a \times 10^n$$

where n is an integer and $1 \leq a < 10$. We move the decimal point *to the right* so that there is one nonzero digit to the left of the decimal point. The power of 10 by which we multiply is the number of places moved, preceded by a *negative* sign.

$$0.00000000000000002 = 000000000000000002. \times 10^{-17}$$

Move 17 places to the *right*.

$$= 2. \times 10^{-17} = 2 \times 10^{-17}$$

PRACTICE 9

Express 0.000000000071 in scientific notation.

Next, let's consider calculations involving numbers written in scientific notation. We focus on the operations of multiplication and division.

EXAMPLE 10

Calculate, writing the result in scientific notation.

a. $(4 \times 10^{-1})(2.1 \times 10^6)$

b. $(1.2 \times 10^5) \div (2 \times 10^{-4})$

PRACTICE 10

Calculate, writing the result in scientific notation.

a. $(7 \times 10^{-2})(3.52 \times 10^3)$

b. $(2.4 \times 10^3) \div (6 \times 10^{-9})$

EXAMPLE 10 (continued)

Solution

a. $(4 \times 10^{-1})(2.1 \times 10^{6})$

$= (4 \times 2.1)(10^{-1} \times 10^{6})$ Regroup the factors.

$= 8.4 \times 10^{-1+6}$ Use the product rule.

$= 8.4 \times 10^{5}$

b. $(1.2 \times 10^{5}) \div (2 \times 10^{-4})$

$= \dfrac{1.2 \times 10^{5}}{2 \times 10^{-4}}$

$= \dfrac{1.2}{2} \times \dfrac{10^{5}}{10^{-4}}$ Rewrite the quotient as a product of quotients.

$= 0.6 \times 10^{5-(-4)}$

$= 0.6 \times 10^{9}$ Use the quotient rule.

Note that 0.6×10^9 is not in scientific notation because 0.6 is not between 1 and 10, that is, it does not have one nonzero digit to the left of the decimal point. To write 0.6×10^9 in scientific notation, we convert 0.6 to scientific notation and then simplify the product.

$0.6 \times 10^{9} = (6 \times 10^{-1}) \times 10^{9}$ Convert 0.6 to scientific notation.

$= 6 \times (10^{-1} \times 10^{9})$

$= 6 \times 10^{8}$ Use the product rule.

So the answer is 6×10^8.

EXAMPLE 11

There are about 5×10^6 red blood cells per cubic millimeter of blood. Each of these red blood cells contains about 2×10^8 hemoglobin molecules. Calculate the approximate number of hemoglobin molecules per cubic millimeter of blood, writing the result in scientific notation. (*Source:* Sylvia Mader, *Inquiry into Life*)

Solution We need to find the product of the number of red blood cells per cubic millimeter of blood and the number of hemoglobin molecules per red blood cell:

$$(5 \times 10^{6})(2 \times 10^{8}) = \underbrace{(5 \times 2)}\underbrace{(10^{6} \times 10^{8})}$$
$$= 10 \times 10^{6+8}$$
$$= 10 \times 10^{14}$$

To write this number in scientific notation, we convert 10 to scientific notation and then simplify the product.

$$10 \times 10^{14} = (1.0 \times 10^{1}) \times 10^{14}$$
$$= 1.0 \times (10^{1} \times 10^{14})$$
$$= 1.0 \times 10^{15}$$

So there are approximately 1×10^{15} hemoglobin molecules per cubic millimeter of blood.

PRACTICE 11

The number of hairs on the average human head is estimated to be about 1.5×10^5. If there are approximately 6×10^9 people in the world, estimate the number of human head hairs in the world. (*Source: Time Almanac 2011*)

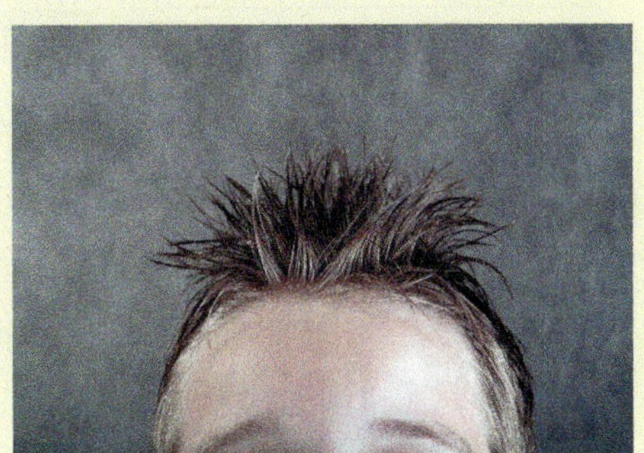

EXAMPLE 12

A certain DVD holds 9.4×10^9 bytes of information. How many files, each containing 9.4×10^6 bytes, can the DVD hold?

Solution To determine the number of files that will fit on the DVD, we divide:

$$\begin{aligned}(9.4 \times 10^9) \div (9.4 \times 10^6) &= \frac{9.4 \times 10^9}{9.4 \times 10^6}\\ &= \frac{9.4}{9.4} \times \frac{10^9}{10^6}\\ &= 1 \times 10^{9-6}\\ &= 1 \times 10^3\end{aligned}$$

So the DVD can hold 1×10^3, or 1000 files.

PRACTICE 12

At the very best, a light microscope can distinguish points 2×10^{-7} m apart, whereas an electronic microscope can distinguish points that are 2×10^{-10} m apart. The second number is how many times the first number? (*Source:* Sylvia Mader, *Inquiry into Life*)

Calculators and Scientific Notation

Calculators vary as to how numbers are displayed or entered in scientific notation.

Display

In order to avoid an overflow error, many calculator models change to scientific notation an answer that is either too small or too large to fit into the calculator's display. Calculators generally use the base 10 without displaying it. Some calculators display scientific notation with either an E or e, and others show a space. For example, 3.1E–4 or 3.1 –4 can represent 3.1×10^{-4}. What other differences do you see between written scientific notation and displayed scientific notation?

EXAMPLE 13

Multiply 1,000,000,000 by 2,000,000,000.

Solution

Press	Display
1000000000 [×] 2000000000 [ENTER]	1000000000*2000000000 2 E 18

Does your calculator display the product in scientific notation, that is, as 2E18, 2e18, or 2.   18?

PRACTICE 13

Square 0.000000005. How is the answer displayed?

Enter

Some calculators give the wrong answer to a computation if very large or very small numbers are entered *in standard form rather than in scientific notation. To enter a number in scientific notation, many calculators have a key labeled* EE, EXP, *or* EEX. *For a negative exponent, a key labeled* +/− *or* (−) *must be pressed either before or after the exponent key, depending on the calculator.*

EXAMPLE 14

Enter the number 5,000,000,000,000 in scientific notation.

Solution

PRACTICE 14

In your calculator, enter in scientific notation the number 0.00000000073.

EXAMPLE 15

Multiply 3.5×10^4 by 2.1×10^7 on a calculator.

Solution

So the answer is 7.35×10^{11}. If your calculator has enough places in the display, it may give the answer to this problem in standard form: 735,000,000,000.

PRACTICE 15

Use a calculator to divide 9.2×10^{12} by 2×10^4.

The following table summarizes the laws of exponents considered in Sections 15.1 and 15.2. For any nonzero numbers x and y, and any integers a and b:

Exponent 1	$x^1 = x$ (x can be any real number.)
Exponent 0	$x^0 = 1$
Product rule of exponents	$x^a \cdot x^b = x^{a+b}$
Quotient rule of exponents	$\frac{x^a}{x^b} = x^{a-b}$
Negative exponents	$x^{-a} = \frac{1}{x^a}$
Reciprocal of x^{-a}	$\frac{1}{x^{-a}} = x^a$
Power rule of exponents	$(x^a)^b = x^{ab}$
Raising a product to a power	$(xy)^a = x^a \cdot y^a$
Raising a quotient to a power	$\left(\frac{x}{y}\right)^a = \frac{x^a}{y^a}$
Raising a quotient to a negative power	$\left(\frac{x}{y}\right)^{-a} = \left(\frac{y}{x}\right)^a$

15.2 Exercises

FOR EXTRA HELP MyMathLab MathXL PRACTICE WATCH READ REVIEW

Mathematically Speaking

Fill in each blank with the most appropriate term or phrase from the given list.

add the factors	raise both the numerator and the denominator	left
right		factors
terms	raise each factor to that power	multiply the exponents
power form	add the powers	raise the reciprocal of the quotient
scientific notation		

1. The expression $(x^3)^2$ contains two ____________ of x^3.

2. The power rule of exponents states that to raise a power to a power, ____________ and leave the base the same.

3. To raise a product to a power, ____________.

4. To raise a quotient to a power, ____________ ____________ to that power.

5. To raise a quotient to a negative power, ____________ ____________ to the opposite of the given power.

6. A number is written in ____________ if it is in the form $a \times 10^n$, where n is an integer and a is greater than or equal to 1 but less than 10.

7. To convert a number from scientific to standard notation, move the decimal point to the ____________ if the power of 10 is negative.

8. To convert a number from standard to scientific notation, move the decimal point to the ____________ if the number is less than 1.

A *Simplify.*

9. $(2^2)^4$
10. $(3^3)^2$
11. $(5^2)^2$
12. $(2^3)^3$
13. $(10^5)^2$
14. $(0^5)^3$
15. $(4^{-2})^2$
16. $(2^{-3})^4$
17. $(x^4)^6$
18. $(p^2)^{10}$
19. $(y^4)^2$
20. $(n^3)^3$
21. $(x^{-2})^3$
22. $(y^{-5})^6$
23. $(n^{-2})^{-2}$
24. $(a^{-5})^{-4}$

B *Express in simplest form.*

25. $(4x)^3$
26. $(2y)^5$
27. $(-8y)^2$
28. $(-7a)^2$
29. $-(4n^5)^3$
30. $-(5x^3)^3$
31. $4(-2y^2)^4$
32. $2(-3t)^3$
33. $(3a)^{-2}$
34. $-(5t)^{-3}$
35. $(pq)^{-7}$
36. $(mn)^{-6}$
37. $(r^2t)^6$
38. $(a^3b^5)^4$
39. $(-2p^5q)^2$
40. $(-3a^2b^3)^4$
41. $-2(m^4n^8)^3$
42. $4(x^2y^3)^2$
43. $(-4m^5n^{-10})^3$
44. $(3a^{-3}c^8)^2$
45. $(a^3b^2)^{-4}$
46. $(p^3q^4)^{-2}$
47. $(4x^{-2}y^3)^2$
48. $(2x^{-2}y^3)^2$

C *Simplify.*

49. $\left(\frac{5}{b}\right)^3$
50. $\left(\frac{x}{4}\right)^2$
51. $\left(\frac{c}{b}\right)^2$
52. $\left(\frac{t}{s}\right)^5$
53. $-\left(\frac{a}{b}\right)^7$
54. $-\left(\frac{x}{y}\right)^3$
55. $\left(\frac{a^2}{3}\right)^3$
56. $\left(\frac{y^6}{4}\right)^3$

57. $\left(-\frac{p^3}{q^2}\right)^5$

58. $\left(-\frac{x^2}{y^3}\right)^4$

59. $\left(\frac{a}{4}\right)^{-1}$

60. $\left(\frac{b}{3}\right)^{-1}$

61. $\left(\frac{2x^5}{y^2}\right)^3$

62. $\left(\frac{n^2}{3w^5}\right)^2$

63. $\left(\frac{pq}{p^2q^2}\right)^5$

64. $\left(\frac{s^2t^3}{st^2}\right)^2$

65. $\left(\frac{3x}{y^{-3}}\right)^4$

66. $\left(\frac{p^{-1}}{5q^5}\right)^2$

67. $\left(\frac{-u^2v^3}{4vu^4}\right)^2$

68. $\left(\frac{2xy^3}{xy^2}\right)^4$

69. $\left(-\frac{x^{-2}y}{2z^{-4}}\right)^4$

70. $-\left(\frac{4a^{-4}}{bc^{-2}}\right)^2$

71. $\left(\frac{r^5}{t^6}\right)^{-2}$

72. $\left(\frac{y^3}{x^3}\right)^{-2}$

73. $\left(\frac{2a^4}{b^2}\right)^{-3}$

74. $\left(\frac{q^5}{5p^4}\right)^{-2}$

D *Express in standard notation.*

75. 3.17×10^8

76. 9.1×10^5

77. 1×10^{-6}

78. 8.33×10^{-4}

79. 6.2×10^6

80. 7.55×10^{10}

81. 4.025×10^{-5}

82. 2.1×10^{-3}

Express in scientific notation.

83. 420,000,000

84. 100,000,000

85. 0.0000035

86. 0.00017

87. 217,000,000,000

88. 154,800,000,000

89. 0.00000000731

90. 0.00000005672

Complete each table.

91.

Standard Notation	Scientific Notation (written)	Scientific Notation (displayed on a calculator)
975,000,000		
	4.87×10^8	
		1.652E−10
0.000000067		
	1×10^{-13}	
		3.281E9

92.

Standard Notation	Scientific Notation (written)	Scientific Notation (displayed on a calculator)
975,000,000,000		
	5×10^8	
		4.988E−7
0.0000048		
	9.34×10^{-9}	
		9.772E6

Calculate, writing the result in scientific notation.

93. $(3 \times 10^2)(3 \times 10^5)$

94. $(5 \times 10^4)(7.1 \times 10^3)$

95. $(2.5 \times 10^{-2})(8.3 \times 10^{-3})$

96. $(9.1 \times 10^{-13})(6.3 \times 10^{-10})$

97. $(2.1 \times 10^4)(8 \times 10^{-4})$

98. $(8.6 \times 10^9)(4.4 \times 10^{-12})$

99. $(2.5 \times 10^8) \div (2 \times 10^{-2})$

100. $(3.0 \times 10^4) \div (1 \times 10^3)$

101. $(6 \times 10^5) \div (2 \times 10^3)$

102. $(4.8 \times 10^{-3}) \div (8 \times 10^2)$

103. $(9.6 \times 10^{20}) \div (3.2 \times 10^{12})$

104. $(8.4 \times 10^6) \div (4.2 \times 10^7)$

Mixed Practice

Simplify.

105. Express 3.067×10^{-4} in standard notation.

106. Express 895,600,000 in scientific notation.

107. Complete the table.

Standard Notation	Scientific Notation (written)	Scientific Notation (displayed on a calculator)
428,000,000,000		
	3.24×10^6	
		5.224E−6
0.000000057		
	6.82×10^{-7}	
		4.836E7

Simplify, using only positive exponents.

108. $(a^{-3})^5$

109. $\left(\dfrac{a^2}{3b^5}\right)^{-2}$

110. $(2r^7s^{-3})^3$

111. $-(4y)^{-3}$

112. $\left(-\dfrac{m^2}{n^3}\right)^5$

113. $-\left(\dfrac{2x^{-2}}{y^{-3}z}\right)^4$

114. $2(-3x^4)^3$

Calculate, writing the result in scientific notation.

115. $(6.3 \times 10^{-4}) \div (9 \times 10^3)$

116. $(4.1 \times 10^{-3})(2.7 \times 10^{-2})$

Applications

E *Solve.*

117. Consider the two boxes shown. How many times the volume of the smaller box is the volume of the larger box?

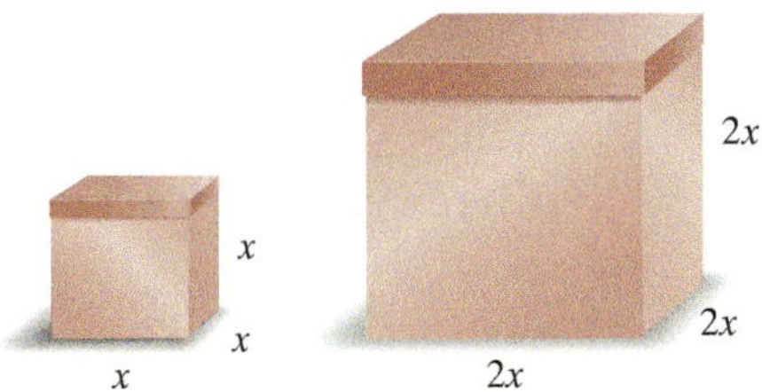

118. The infectious part of a virus is typically between 2.5×10^{-8} m and 2×10^{-7} m in size. Express these quantities in standard notation. (*Source:* Sylvia Mader, *Inquiry into Life*)

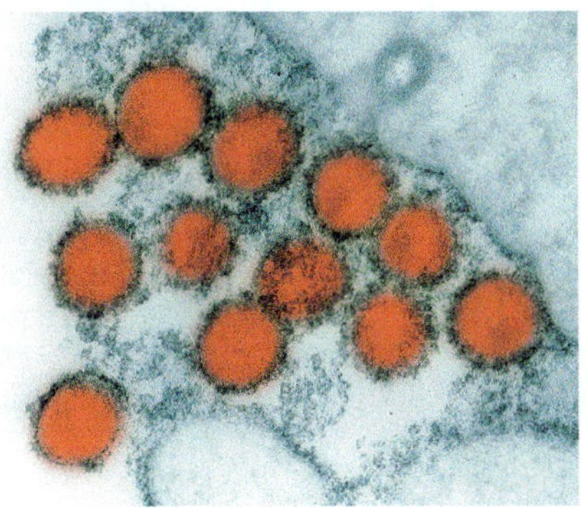

119. A DVD holds between 4×10^{9} and 1.7×10^{10} bytes of data. Express these quantities in standard notation.

120. After a flood, the radius of a circular pond doubles. How does the area of the pond change?

121. The wavelength of red light is 0.0000007 m. Write this length in scientific notation.

122. In a recent year, the U.S. federal budget had a deficit of $1,400,000,000,000. Express this amount in scientific notation. (*Source: The 2011 Statistical Abstract of the United States*)

123. The cell is considered the basic unit of life. Each day, the body destroys and replaces more than 200 billion cells. Write this quantity in scientific notation.

124. The diameter of an atom is about 1.1×10^{-10} m. What is this quantity in standard notation? (*Source:* Peter J. Nolan, *Fundamentals of College Physics*)

125. The mass of a proton is about 1.7×10^{-24} g. Rewrite this quantity in standard notation. (*Source:* Karen Timberlake, *Chemistry*)

126. To measure vast distances, astronomers use a unit called a *parsec*, which is equal to about 3.086×10^{18} cm. Express this quantity in standard form. (*Source:* Derek McNally, *Positional Astronomy*)

127. The world population is projected to be 7.6×10^{9} in 2020. What is this population expressed in standard notation? (*Source:* census.gov)

128. The diameter of a water molecule is about 2.8 angstroms, where one angstrom is 0.00000001 cm. What is the radius of a water molecule in centimeters, expressed in scientific notation? (*Source:* answers.com)

129. For each pound of body weight, a human body contains about 3.2×10^{4} microliters (μL) of blood. In turn, a microliter of blood contains about 5×10^{6} red blood cells. A person weighing 100 lb has approximately how many red blood cells?

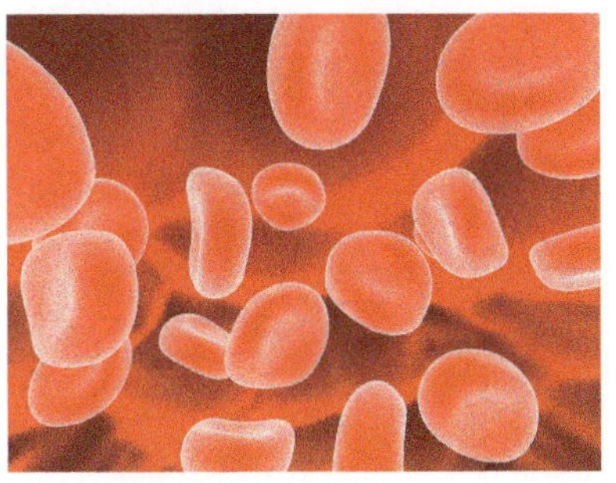

130. On the television series *Star Trek: The Next Generation*, the android Data could carry out 60 trillion operations per second. Express this rate in scientific notation.

131. Light travels through a vacuum at a speed of 186,000 mi/sec.

a. Express this speed in scientific notation.

b. How long will it take for light to travel to Earth from the star Vega, which is 1.58×10^{14} mi from Earth? (*Source: The New York Times Almanac 2011*)

132. There are 26,890,000,000,000,000,000 molecules of a gas in a cubic meter.

a. Rewrite this quantity in scientific notation.

b. What volume is required for 3.4×10^{20} molecules of the gas?

• Check your answers on page A-32.

MINDStretchers

Investigation

1. On a scientific calculator, enter the number 2. Double that number. Then, keep doubling the result. After how many doublings does your calculator display the number in scientific notation? Explain how you could have predicted that result.

Critical Thinking

2. What is the mathematical relationship between $(a^m)^n$ and $(a^m)^{-n}$? Justify your answer.

Research

3. In your college library or on the Web, determine the annual national debt for the United States for 5 consecutive years. Would you use scientific notation or standard notation to express these amounts? Explain why.

15.3 Basic Concepts of Polynomials

OBJECTIVES

- **A** To identify and classify polynomials
- **B** To simplify polynomials
- **C** To evaluate polynomials
- **D** To solve applied problems involving polynomials

What Polynomials Are and Why They Are Important

In Chapter 11, we discussed algebraic expressions in general. We now consider a particular kind of algebraic expression called a *polynomial*.

Just as whole numbers are fundamental to arithmetic, polynomials play a similarly key role in algebra.

Whole number: $3 \cdot 10^2 + 7 \cdot 10^1 + 8 \cdot 10^0 = 378$

Polynomial: $3 \cdot x^2 + 7 \cdot x^1 + 8 \cdot x^0 = 3x^2 + 7x + 8$

In fact, a good deal of algebra is devoted to studying properties of polynomials and operations on polynomials.

Many phenomena in the sciences and business can be described by polynomial expressions. Even when the description is only approximate, the polynomial approximation is often a good one.

Monomials

We begin by considering algebraic expressions called *monomials*.

DEFINITION

A **monomial** is an expression that is the product of a real number and variables raised to nonnegative integer powers.

Some examples of monomials are:

$$5x \quad -7t^4 \quad \frac{4}{5}x^2 \quad -2p^2q^5$$

Recall that in the expression $5x$, 5 is called the *coefficient*.

Note that a constant such as -12 can be thought of as $-12x^0$. So any constant is also considered a monomial.

The expression $5x + 3$ is not a monomial since it is the sum of $5x$ and 3, which are called *terms*. A monomial consists only of factors. Can you explain whether $\frac{2}{x}$ is a monomial?

Monomials serve as building blocks (or terms) for the larger set of polynomials. Polynomials are formed by adding and subtracting monomial terms.

DEFINITION

A **polynomial** is an algebraic expression with one or more monomials added or subtracted.

Here are some examples of polynomials:	Here are some examples of algebraic expressions that are *not* polynomials:
$3x^2 - 5x + 7$	$2x^{-1}$
$4t^2 - 3$	$\frac{t^2}{3} + \frac{4}{t}$
$-8x^2$	$\frac{n+1}{n}$
$17x^4 + 5x^3 - 8x^2 + x - 1$	$\frac{5x^2}{2x+9}$
$20pq - p^2 - 7q^2 + 6$	$2\sqrt{x} + 1$

EXAMPLE 1

Consider the polynomial $3x^5 + 2x^3 - 8$.

a. Identify the terms of the polynomial.

b. For each term, identify its coefficient.

Solution

a. The terms are $3x^5$, $2x^3$, and -8.

b. The coefficients of the terms are 3, 2, and -8, respectively.

PRACTICE 1

For the polynomial $-10x^2 + 4x + 20$, find (a) the terms and (b) their coefficients.

Classification of Polynomials

There are several ways to classify polynomials. One way is according to the number of variables in the polynomial, and another is according to the number of terms in the polynomial. Finally, a third kind of classification is according to the degree of the polynomial. We consider each of these classifications in turn.

Number of Variables

A polynomial such as $3x^2 - 6x + 9$ is said to be *in one variable,* namely in x. The polynomial $t^5 + 11t^4 - 7t^3 - t^2 + 10t - 50$ is also in one variable, namely in t. On the other hand, the polynomial $3x^4y - 5x^2y^3 + 9$ is in *two* variables, x and y. Throughout this text, we focus on polynomials in one variable.

Number of Terms

As we have seen, a polynomial with just one term is called a *monomial.* A polynomial with *two* terms is called a **binomial**. A **trinomial** is a polynomial with *three* terms. Polynomials with four or more terms are simply called *polynomials.*

$10x$ ⟵ Monomial

$3x - 2$ ⟵ Binomial

$x^2 + 9x + 6$ ⟵ Trinomial

$-x^3 + 8x^2 + x - 19$ ⟵ Polynomial

EXAMPLE 2

Classify each polynomial according to the number of terms.

Polynomial	Monomial	Binomial	Trinomial
$3x - 5$			
$7x^2 - 3x + 10$			
$10a^2$			

Solution

Polynomial	Monomial	Binomial	Trinomial
$3x - 5$		✓	
$7x^2 - 3x + 10$			✓
$10a^2$	✓		

PRACTICE 2

Classify each polynomial according to the number of terms.

Polynomial	Monomial	Binomial	Trinomial
$5x^2 + 2x + 9$			
$-4x^2$			
$12p - 1$			

Degree and Order of Terms

Let's consider a monomial in one variable. The **degree** of the monomial is the power of the variable in the monomial.

$3x^4 \longleftarrow$ Of the fourth degree or of degree 4

$-7y^2 \longleftarrow$ Of the second degree or of degree 2

Recall that a constant, such as 3, can be thought of as $3x^0$ and so is considered to be of degree 0.

8 is a monomial of degree 0.

-12 is a monomial of degree 0.

Polynomials are also classified by their degree. The degree of a polynomial is the highest degree of any of its terms. For instance, the degree of $8x^3 + 9x^2 - 7x - 1$ is 3 since 3 is the highest degree of any term.

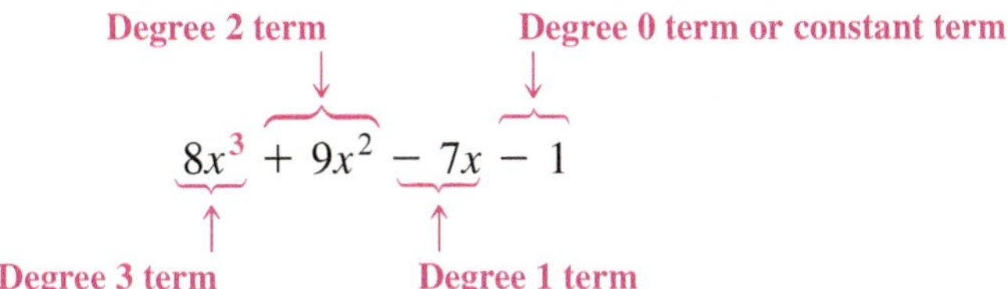

EXAMPLE 3

Identify the degree of each polynomial.

a. $-20x^3$ **b.** $5 + x$ **c.** $-7x^2 + 3x + 10$ **d.** -36

Solution

a. $-20x^3$ is of degree 3.

b. $5 + x = 5 + x^1$, which is of degree 1.

c. $-7x^2 + 3x + 10$ is of degree 2.

d. $-36 = -36x^0$, which is of degree 0.

PRACTICE 3

Indicate the degree of each polynomial.

a. $7n$

b. $x - x^2$

c. $2 + 4x^2 - 10x^3$

d. -8

The **leading term** of a polynomial is the term in the polynomial with the highest degree, and the coefficient of that term is called the **leading coefficient**. The term of degree 0 is called the **constant term**. So in the polynomial $3t^2 - 8t + 1$, the leading term is $3t^2$, the leading coefficient is 3, and the constant term is 1.

EXAMPLE 4

Complete the table.

Polynomial	Constant Term	Leading Term	Leading Coefficient
$2x^{10}$			
$4x + 25$			
$3 - 10x + 8x^2$			
$7x^3 - x - 8$			

PRACTICE 4

Complete the table.

Polynomial	Constant Term	Leading Term	Leading Coefficient
$-3x^7 + 9$			
x^5			
$x^4 - 7x - 1$			
$3x + 5x^3 + 20$			

Solution

Polynomial	Constant Term	Leading Term	Leading Coefficient
$2x^{10}$	0	$2x^{10}$	2
$4x + 25$	25	$4x$	4
$3 - 10x + 8x^2$	3	$8x^2$	8
$7x^3 - x - 8$	-8	$7x^3$	7

The terms of a polynomial are usually arranged in *descending order of degree*. That is, we write the leading term on the left, then the term of the next highest degree, and so forth.

$x^3 + 5x^2 - 3x - 10$ **Descending order of degree—the exponents get smaller from left to right.**

Occasionally, however, the terms of a polynomial are written in *ascending order of degree*.

$-10 - 3x + 5x^2 + x^3$ **Ascending order of degree—the exponents get larger from left to right.**

EXAMPLE 5

Rearrange each polynomial in descending order.

a. $3x^4 + 9x^2 - 7x^3 + x + 10$

b. $-7x + 6x^3 + 10$

Solution

a. We rewrite the polynomial so that the term with the highest exponent of x is on the left, the next highest exponent comes second, and so on.

$$3x^4 + 9x^2 - 7x^3 + x + 10 = 3x^4 - 7x^3 + 9x^2 + x^1 + 10x^0$$
$$= 3x^4 - 7x^3 + 9x^2 + x + 10$$

b. We rewrite the polynomial so that the exponents get smaller from left to right.

Degree 3 Degree 1

$$-7x + 6x^3 + 10 = 6x^3 + \boxed{} - 7x + 10$$

Degree 2 Degree 0

Note that a term with coefficient 0 is usually not written. The unwritten term is said to be a *missing term*. For instance, in the polynomial

$$6x^3 - 7x + 10$$

$0x^2$ is the missing term. So we can write this polynomial as $6x^3 + 0x^2 - 7x + 10$.

PRACTICE 5

Write each polynomial in descending order.

a. $-8x + 9x^5 - 7x^4 + 9x^2 - 6$

b. $x^3 + 7x^5 - 3x^2 + 8$

The concept of missing terms is important in the division of polynomials, as we will see later in this chapter.

Simplifying and Evaluating Polynomials

Recall that in Section 11.6, we discussed how to simplify algebraic expressions by combining like terms.

EXAMPLE 6

Simplify, and then put in descending order.

$$8 + 3x^3 + 9x + 1 - 8x + 7x^2 - 3x^2$$

Solution

$$\begin{aligned} &8 + 3x^3 + 9x + 1 - 8x + 7x^2 - 3x^2 \\ &= 8 + 3x^3 + 9x + 1 - 8x + 7x^2 - 3x^2 \\ &= 9 + 3x^3 + x + 4x^2 \quad \text{Combine like terms.} \\ &= 3x^3 + 4x^2 + x + 9 \end{aligned}$$

PRACTICE 6

Combine like terms, and then write in descending order.

$$2x^2 + 3x - x^2 + 5x^3 + 3x - 5x^3 + 20$$

Recall, also, our discussion of evaluating algebraic expressions in Section 11.5. Polynomials, like other algebraic expressions, are evaluated by replacing each variable with the given number and then carrying out the computation.

EXAMPLE 7

Find the value of $2x^2 - 8x - 5$ when:

a. $x = 3$ **b.** $x = -3$

Solution

a. $$\begin{aligned} 2x^2 - 8x - 5 &= 2(3)^2 - 8(3) - 5 \quad \text{Substitute 3 for } x. \\ &= 2(9) - 8(3) - 5 \\ &= 18 - 24 - 5 = -11 \end{aligned}$$

b. $$\begin{aligned} 2x^2 - 8x - 5 &= 2(-3)^2 - 8(-3) - 5 \quad \text{Substitute } -3 \text{ for } x. \\ &= 2(9) - 8(-3) - 5 \\ &= 18 + 24 - 5 \\ &= 37 \end{aligned}$$

PRACTICE 7

Find the value of $x^2 - 5x + 5$ when:

a. $x = 2$

b. $x = -2$

EXAMPLE 8

If \$1000 is deposited in a savings account that pays compound interest at a rate r compounded annually, then after 2 yr the balance in the account will be represented by the polynomial $(1000r^2 + 2000r + 1000)$ dollars. Find the balance if $r = 0.05$.

Solution We need to replace r by 0.05 in the polynomial.

$$\begin{aligned} 1000r^2 + 2000r + 1000 &= 1000(0.05)^2 + 2000(0.05) + 1000 \\ &= 1000(0.0025) + 2000(0.05) + 1000 \\ &= 2.5 + 100 + 1000 \\ &= 1102.5 \end{aligned}$$

So the account balance is \$1102.50.

PRACTICE 8

If an object is dropped from a height of 500 ft above the ground, its height (in feet above the ground) after t sec is given by the expression $500 - 16t^2$. How high above the ground is the object after 3 sec?

EXAMPLE 9

The polynomial $-322x^2 + 3027x + 15{,}530$ models the total number of adoptions for a particular year to the United States from other countries, where x represents the number of years after 1999. According to this model, how many of these intercountry adoptions were there, to the nearest thousand, in 2009? (*Source:* adoption.state.gov)

Solution Since the year 2009 is 10 years after 1999, we substitute 10 for x in the polynomial.

$$\begin{aligned} -322x^2 + 3027x + 15{,}530 &= -322(10)^2 + 3027(10) + 15{,}530 \\ &= -32{,}200 + 30{,}270 + 15{,}530 \\ &= 13{,}600 \approx 14{,}000 \end{aligned}$$

So there were, to the nearest thousand, 14,000 adoptions in 2009 to the United States from other countries.

PRACTICE 9

The polynomial $0.02x^3 - 0.75x^2 + 6.89x + 45.17$ can be used to model the temperature (in degrees Fahrenheit) at a particular time in Anaheim, California on November 26, 2010, where x represents the number of hours past 8:00 A.M. According to this model, find the temperature in Anaheim at 11:00 A.M. to the nearest degree. (*Source:* forecast.weather.gov)

15.3 Exercises

FOR EXTRA HELP MyMathLab MathXL PRACTICE WATCH READ REVIEW

Mathematically Speaking

Fill in each blank with the most appropriate term or phrase from the given list.

ascending	constant term	descending
power	coefficient	leading term
degree	leading coefficient	
polynomial	monomial	

1. A(n) ________________ is an expression that is the product of a real number and variables raised to non-negative integer powers.
2. The ________________ of the expression $-\frac{3}{4}x^3$ is $-\frac{3}{4}$.
3. The degree of a monomial in one variable is the ________________ of the monomial's variable.
4. The ________________ of a polynomial is the highest degree of any of its terms.
5. The ________________ of a polynomial is the term in the polynomial with the highest degree.
6. The coefficient of the leading term is called the ________________.
7. The term of degree 0 in a polynomial is called the ________________.
8. Polynomial terms are usually written in ________________ order of degree.

A *Indicate whether each of the following is a polynomial.*

9. $7x^2$
10. $2x^2$
11. $x - 7\sqrt{x} + 1$
12. $\dfrac{2}{x + 3}$
13. $10p + q$
14. $4a - 3a^2$
15. $2x^{-1}y - x^2$
16. $3y^{-1} + y$

For each polynomial, find (a) the terms and (b) their coefficients.

17. $5x^4 - 2x^3 + 1$
18. $-6y^3 + 4y - 3$

Classify each polynomial according to the number of terms.

19.

Polynomial	Monomial	Binomial	Trinomial
$5x - 1$			
$-5a^2$			
$-6a + 3$			
$x^3 + 4x^2 + 2$			

20.

Polynomial	Monomial	Binomial	Trinomial
$5x + x^2$			
$3x$			
$12p - 1$			
$2x^5 - x^3 + x$			

Rearrange each polynomial in descending order, and then identify its degree.

21. $3x^2 - 2x + 8 - 4x^3$

22. $5x^3 + 7x + 1 - 7x^2$

23. $2 - 3y$

24. $7 - 5x$

25. $4p^2 - p^4 + 3p^3 + 10 - p$

26. $25 - y + 2y^2 + y^3 - 3y^4$

27. $-y^3 - 2y + 2 - 4y^5$

28. $5x^3 + 3x^5 + 8x + 3$

29. $5a^2 - a$

30. $-8x^2 + 6x - 2$

31. $3p^3$

32. $5x^2$

Complete each table.

33.

Polynomial	Constant Term	Leading Term	Leading Coefficient
$-x^7 + 2$			
$2x - 30$			
$-5x + 1 + x^2$			
$7x^3 - 2x - 3$			

34.

Polynomial	Constant Term	Leading Term	Leading Coefficient
$5x^3 + 8$			
$-x + 10$			
$2x^2 - 3x + 4$			
$-5x + x^4 - 9$			

B *Simplify.*

35. $9x^3 - 7x^2 + 1 + x^3 + 10x + 5$

36. $2y^3 - 7y^2 + 1 + 2y^2 + 3y + 8$

37. $r^3 + 2r^2 + 15 + r^2 - 8r - 1$

38. $n^4 - n^3 - 7n^2 + n^2 + 10n + 3$

Combine like terms. Then, write the polynomial in descending order of powers.

39. $4x^2 - 2x - x^2 - 10 - 3x + 4$

40. $x^3 + 5x - 7x^2 - 1 - x^3 + x$

41. $6n^3 + 20n - n^2 + 2 - 4n^3 + 15n^2 + 8$

42. $8y^2 + y^3 - 8y + 20 + 3y + 9y^2$

Identify the missing terms of each polynomial.

43. $x^3 - 7x - 2$

44. $n^2 + 7n$

45. $6x^3 + 8x^2 + 1$

46. $x^4 - 3x$

C *Find the value of each polynomial for the given values of the variable.*

47. $7x - 3$, for $x = 2$ and $x = -2$

48. $5a + 11$, for $a = 0$ and $a = 2$

49. $n^2 - 3n + 9$, for $n = 7$ and $n = -7$

50. $3y^2 + 2y + 1$, for $y = 2$ and $y = -1$

51. $2.1x^2 + 3.9x - 7.3$, for $x = 2.37$ and $x = -2.37$

52. $0.1x^3 + 4.1x - 9.1$, for $x = 3.14$ and $x = -3.14$

Mixed Practice

Rearrange each polynomial in descending order, and then identify its degree.

53. $-a^3 + 2a - 4 + 5a^4$

54. $8x + 6x^3$

55. Complete the following table:

Polynomial	Constant Term	Leading Term	Leading Coefficient
$4x - 20$			
$-7x^6 + 9$			
$3x + 2 + x^2$			
$6x - x^5 + 11$			

56. Classify each polynomial according to the number of terms.

Expression	Monomial	Binomial	Trinomial
$2p - 3p^2$			
$-8x^3$			
$a^2 - 4a + 5$			

57. Is $x^2 + 4x^{-2} - 2$ a polynomial?

Find the value of each polynomial for the given values of the variable.

58. $5y^2 - 8y - 9$, for $y = 3$ and $y = -2$

59. $5.3x^2 - 2.7x - 6.8$, for $x = 1.25$ and $x = -3.87$

60. Simplify: $4n^3 - 8n^4 + 3 + 6n^2 + 5n^4 - 4n^2$

61. Identify the missing terms of the polynomial $9x^4 + x^2 - 3x$.

62. Combine like terms; then, write the polynomial in descending order of powers: $3m^2 - m^3 + 9m - 11m^2 - 15 + 7m$

Applications

D *Solve.*

63. The polynomial $1 + x^2 + x^{15} + x^{16}$ is used by computer scientists to detect errors in computer data. Classify this polynomial in terms of its variables and its degree.

64. The owner of a factory estimates that her profit (in dollars) is

$$0.003x^3 - 1.4x^2 + 300x - 1000$$

where x is the number of items that the factory produces. Describe this polynomial in terms of its variables and its degree.

65. The polynomial $x + \dfrac{x^2}{20}$ is the *stopping distance* of a car in feet after the brakes are applied, where the variable x is the speed of the car in miles per hour before braking. Find the stopping distance for a car that had been traveling at 40 mph.

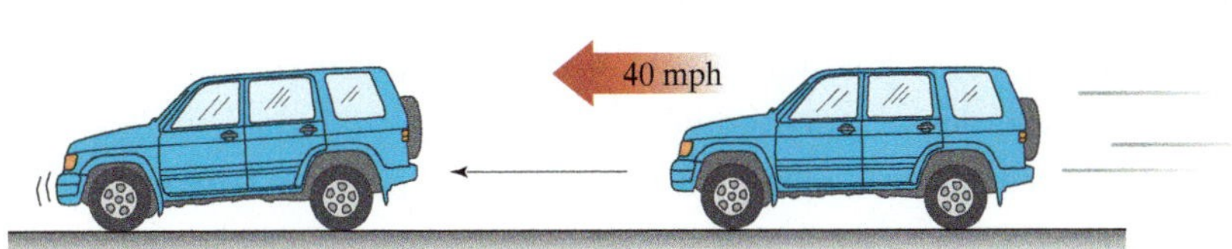

66. There are n teams that compete in a sports league, where each team plays every other team once. The following polynomial gives the total number of games that must be played:

$$0.5n^2 - 0.5n$$

If the league has 20 teams, how many games are played?

67. The average monthly cell telephone bill (in dollars) in the U.S. for a particular year is approximated by the polynomial $-0.18x^2 + 1.91x + 45.45$, where x represents the number of years after 2000. Find the average monthly cell telephone bill in 2003 to the nearest dollar. (*Source:* ctia.org)

68. The payroll (in millions of dollars) of the Philadelphia Flyers ice hockey team for a particular year is modeled by the polynomial $0.58x^2 + 4.96x + 41.48$, where x represents the number of years after 2007. According to this model, find the team payroll in 2015 to the nearest million dollars. (*Source:* usatoday.com)

69. The number of U.S. radio stations with a Spanish format for a particular year is approximated by the polynomial $-0.82x^3 + 6.09x^2 + 22.07x + 631.59$, where x represents the number of years after 2003. According to this model, find to the nearest hundred the number of these stations in 2005. (*Source:* mstreet.net)

70. The average basic cable television subscribers (in thousands) in the U.S. for a particular year can be approximated by the polynomial $-88.167x^3 + 184.500x^2 - 114.333x + 65{,}337$, where x represents the number of years after 2005. According to this model, how many basic cable television subscribers to the nearest million were there in 2006? (*Source:* census.gov)

• Check your answers on page A-32.

MINDStretchers

Research

1. The following prefixes are used with polynomials. Use a dictionary to fill in the table.

Prefix	Meaning of This Prefix	Three Words Beginning with This Prefix
Mono-		
Bi-		
Tri-		
Poly-		

Groupwork

2. There are some polynomials whose value is a prime number for many values of the variable. For instance, consider the second-degree polynomial $n^2 + n + 41$.
   a. Check that for n equal to a whole number between 0 and 39, the value of this polynomial is a prime number.

   b. Is the value of this polynomial a prime number for $n = 40$? Explain your answer.

Patterns

3. The degree of a monomial in more than one variable is the sum of the powers of the variables in that term. Recall that the degree of a polynomial is the highest degree of any of its terms. The following tables show polynomials that represent the area or the volume of various common geometric figures.

Area

Geometric Figure	Polynomial	Degree of the Polynomial
Square	s^2	
Triangle	$0.5bh$	
Trapezoid	$0.5hb + 0.5hB$	
Circle	$3.14r^2$	
Rectangle	lw	

Volume

Geometric Figure	Polynomial	Degree of the Polynomial
Cube	e^3	
Rectangular solid	lwh	
Sphere	$\frac{4}{3}\pi r^3$	
Cylinder	$\pi r^2 h$	

a. Complete the tables by finding the degree of each polynomial.

b. Describe the pattern you observe in the table for area. Explain your observation.

c. Describe the pattern you observe in the table for volume. Explain your observation.

Cultural Note

Muhammad ibn Musa al-Khwarizmi, a ninth-century mathematician, wrote *al-Kitab al-mukhtasar fi hisab al-jabr wa'l-muqabala* (*The Compendious Book on Calculation by Completion and Balancing*)—one of the earliest treatises on algebra and the source of the word *algebra*. This work dealt with solving equations as well as with practical applications of algebra. Al-Khwarizmi, from whose name the word *algorithm* derives, also wrote influential works on astronomy and on the Hindu numeration system.

(*Sources:* Jan Gullberg, *Mathematics From the Birth of Numbers*, W. W. Norton, 1997 and Morris Kline, *Mathematics, A Cultural Approach*, Addison-Wesley, 1962).

15.4 Adding and Subtracting Polynomials

OBJECTIVES

- **A** To add polynomials
- **B** To subtract polynomials
- **C** To solve applied problems involving the addition or subtraction of polynomials

In this section, we consider the addition and subtraction of polynomials and their applications to real-world situations. As our discussion proceeds, note the similarity between adding and subtracting whole numbers in arithmetic and these operations on polynomials in algebra.

Adding Polynomials

The key to adding polynomials is identifying like terms and then adding them.

To Add Polynomials

- Add the like terms.

As with whole numbers, we can add polynomials using either a horizontal or vertical format. To add polynomials horizontally, we simply remove the parentheses and combine like terms.

EXAMPLE 1

Find the sum: $(8x^2 + 3x + 4) + (-12x^2 + 7)$

Solution Recall that when a plus sign precedes terms in parentheses, we remove the parentheses and keep the sign of each term.

First polynomial: $(8x^2 + 3x + 4)$; Second polynomial: $(-12x^2 + 7)$

$$(8x^2 + 3x + 4) + (-12x^2 + 7)$$
$$= 8x^2 + 3x + 4 - 12x^2 + 7 \quad \text{Remove parentheses.}$$
$$= -4x^2 + 3x + 11 \quad \text{Combine like terms.}$$

PRACTICE 1

Add: $6x - 3$ and $9x^2 - 3x - 40$

EXAMPLE 2

Combine: $(3st^2 - 4st + t^2) + (8s^2t - 3t^2) + (10t^2 - 5st + 7s^2)$

Solution

$$(3st^2 - 4st + t^2) + (8s^2t - 3t^2) + (10t^2 - 5st + 7s^2)$$
$$= 3st^2 - 4st + t^2 + 8s^2t - 3t^2 + 10t^2 - 5st + 7s^2$$
$$= 3st^2 + 8s^2t - 9st + 7s^2 + 8t^2$$

PRACTICE 2

Combine: $(9p^2 + 4pq + 2q^2) + (-p^2 - 5q^2) + (2p^2 - 3pq - 7q^2)$

Now, let's look at how to add polynomials in a vertical format. Recall that in adding whole numbers, for instance 329 and 50, we position the addends so that digits with the same place value are in the same column.

Hundreds, Tens, Ones

$$\begin{array}{r} 329 \\ +\ 50 \\ \hline \end{array}$$

Similarly, when we add polynomials vertically, we position the polynomials so that like terms are in the same column. Suppose, for example, we want to add $7x^5 + x^3 - 3x^2 + 8$

and $2x^5 - 7x^4 + 9x - 9$. In general, a polynomial is considered to be in simplest form when each term is simplified and like terms are combined. Usually the terms are then rearranged in descending order. So, first we make sure that both polynomials are in descending order. Then, we write the polynomials as follows:

$$\begin{array}{r} 7x^5 + 0x^4 + x^3 - 3x^2 + 0x + 8 \\ \underline{2x^5 - 7x^4 + 0x^3 + 0x^2 + 9x - 9} \end{array}$$

Note that we could have just left a space for each missing term of the polynomials. Do you see that each column contains like terms? We then add the terms in each column, as shown.

$$\begin{array}{rrrrrr} 7x^5 & & +\,x^3 & -\,3x^2 & & +\,8 \\ 2x^5 & -\,7x^4 & & & +\,9x & -\,9 \\ \hline 9x^5 & -\,7x^4 & +\,x^3 & -\,3x^2 & +\,9x & -\,1 \end{array}$$

So the sum is $9x^5 - 7x^4 + x^3 - 3x^2 + 9x - 1$.

Let's consider some more examples.

EXAMPLE 3

Add vertically: $3x^2 - 5x - 6$ and $10x + 20$

Solution First, we check that both polynomials are in descending powers of x. Next, we rewrite the polynomials vertically, with like terms positioned in the same column:

Second-degree term; First-degree terms; Zero-degree (constant) terms

$$\begin{array}{rrr} 3x^2 & -\,5x & -\,6 \\ & 10x & +\,20 \\ \hline \end{array}$$

We add within columns.

$$\begin{array}{rrr} 3x^2 & -\,5x & -\,6 \\ & 10x & +\,20 \\ \hline 3x^2 & +\,5x & +\,14 \end{array}$$

So the sum is $3x^2 + 5x + 14$, which is in simplest form.

PRACTICE 3

Find the sum of $8n^2 + 2n - 1$ and $3n^2 - 2$ using a vertical format.

EXAMPLE 4

Find the sum of $7x^3 - 10x^2y + 8xy^2 + 13y^3$, $14xy^2 - 1$ and $3x^2y - 5xy^2 - y^3 + 2$ using a vertical format.

Solution

$$\begin{array}{rrrrr} 7x^3 & -\,10x^2y & +\,8xy^2 & +\,13y^3 & \\ & & 14xy^2 & & -\,1 \\ & 3x^2y & -\,5xy^2 & -\,y^3 & +\,2 \\ \hline 7x^3 & -\,7x^2y & +\,17xy^2 & +\,12y^3 & +\,1 \end{array}$$

PRACTICE 4

Add vertically:

$7p^3 - 8p^2q - 3pq^2 + 20$,
$10p^2q + pq^2 - q^3 + 5$, and $p^3 - q^3$

Is the sum in Example 4 in simplified form? Explain.

Subtracting Polynomials

Recall that when a minus sign precedes terms in parentheses, we remove the parentheses and change the sign of each term.

EXAMPLE 5

Remove the parentheses, and simplify.

a. $2x - (3x + 4y)$

b. $(5n + 2m) - (n + m) + (3n - 4m)$

Solution

a. In the expression $2x - (3x + 4y)$, note that $(3x + 4y)$ is preceded by a minus sign. So we remove the parentheses, and change the sign of each term in parentheses. Then, we combine like terms.

$$2x - (3x + 4y) = 2x - 3x - 4y$$
$$= -x - 4y$$

b. $(5n + 2m) \underbrace{- (n + m)}_{} \underbrace{+ (3n - 4m)}_{} = 5n + 2m - n - m + 3n - 4m = 7n - 3m$

For a polynomial preceded by a minus sign, *change* signs of terms.

For a polynomial preceded by a plus sign, *keep* signs of terms.

PRACTICE 5

Remove the parentheses, and simplify.

a. $-(4r - 3s) + 7r$

b. $(2p + 5q) + (p - 6q) - (3p + 2q)$

To subtract real numbers, we change the number being subtracted to its opposite, and then add. Subtraction of polynomials works very much in the same way.

To Subtract Polynomials

- Change the sign of each term of the polynomial being subtracted.
- Add the like terms.

For instance, suppose that we want to subtract the polynomial $2x^4 - 5x^3 + 4x^2 + x + 1$ from $3x^4 + x^3 - 4x^2 + 8x - 9$.

The polynomial from which we are subtracting — The polynomial being subtracted

$$(3x^4 + x^3 - 4x^2 + 8x - 9) - (2x^4 - 5x^3 + 4x^2 + x + 1)$$

$$= (3x^4 + x^3 - 4x^2 + 8x - 9) + (-2x^4 + 5x^3 - 4x^2 - x - 1)$$ Change the sign of each term of the polynomial being subtracted, and then add.

$$= 3x^4 + x^3 - 4x^2 + 8x - 9 - 2x^4 + 5x^3 - 4x^2 - x - 1$$ Remove parentheses.

$$= x^4 + 6x^3 - 8x^2 + 7x - 10$$ Combine like terms.

EXAMPLE 6

Subtract: $(5x^2 - 3x + 7) - (-2x^2 + 8x + 9)$

Solution

$$\begin{aligned}(5x^2 - 3x + 7) - (-2x^2 + 8x + 9) &= (5x^2 - 3x + 7) + (2x^2 - 8x - 9)\\ &= 5x^2 - 3x + 7 + 2x^2 - 8x - 9\\ &= 7x^2 - 11x - 2\end{aligned}$$

PRACTICE 6

Find the difference:
$(2x - 1) - (3x^2 + 15x - 1)$

Thus far, we have subtracted polynomials using a horizontal format. However, we can also subtract polynomials vertically, a skill that comes up when dividing polynomials. As in the case of the vertical addition of polynomials, the key in vertical subtraction is to position the polynomials so that like terms are in the same column. Then, we change the sign of each term of the polynomial being subtracted, and add.

EXAMPLE 7

Subtract using a vertical format: $(7x^2 - 3x + 7) - (x^2 + 8x - 9)$

Solution

$$\begin{array}{r} 7x^2 - 3x + 7 \\ -(x^2 + 8x - 9) \\ \hline \end{array}$$

Position like terms in the same columns.

$$\begin{array}{r} 7x^2 - 3x + 7 \\ -x^2 - 8x + 9 \\ \hline \end{array}$$

Change the sign of each term of the polynomial being subtracted.

$$\begin{array}{r} 7x^2 - 3x + 7 \\ -x^2 - 8x + 9 \\ \hline 6x^2 - 11x + 16 \end{array}$$

Add.

PRACTICE 7

Subtract vertically:
$(20x - 13) - (5x^2 - 12x + 13)$

EXAMPLE 8

Subtract $10x^2 + 8xy + y^2$ from $4y^2 - 9x^2$, using a vertical format.

Solution

$$\begin{array}{r} -9x^2 \qquad\quad + 4y^2 \\ -(10x^2 + 8xy + y^2) \\ \hline \end{array}$$

Position like terms in the same column.

$$\begin{array}{r} -9x^2 \qquad\quad + 4y^2 \\ -10x^2 - 8xy - y^2 \\ \hline \end{array}$$

Change the sign of each term in the polynomial being subtracted.

$$\begin{array}{r} -9x^2 \qquad\quad + 4y^2 \\ -10x^2 - 8xy - y^2 \\ \hline -19x^2 - 8xy + 3y^2 \end{array}$$

Add.

PRACTICE 8

Find the difference using a vertical format:
$(2p^2 - 7pq + 5q^2) - (3p^2 + 4pq - 12q^2)$

EXAMPLE 9

The polynomial $0.96x + 33.05$ approximates the number of residential natural gas consumers (in thousands) for a particular year in the state of Vermont x years after 2005. The corresponding polynomial for the state of Alaska is $0.35x + 10.85$. Find the polynomial that approximates the total number of these consumers for the two states. (*Source:* eia.gov)

Solution To determine the total number of residential natural gas consumers from either Vermont or Alaska, we add the number from Vermont to the number from Alaska. These numbers are approximated by the given polynomials.

$$\begin{array}{r} 0.96x + 33.05 \\ \underline{0.35x + 10.85} \\ 1.31x + 43.9 \end{array}$$

So $1.31x + 43.9$ represents the total number (in thousands) of these consumers for Vermont or Alaska.

PRACTICE 9

The polynomial $0.01x + 4.03$ approximates the yearly stadium attendance (in millions) at Yankee Stadium for a particular year x years after 2004. The corresponding polynomial for Fenway Park is $0.05x + 2.81$. Find the polynomial that approximates how much greater the attendance (in millions) is at Yankee Stadium than at Fenway Park. (*Source:* baseball-almanac.com)

15.4 Exercises

FOR EXTRA HELP MyMathLab MathXL PRACTICE  WATCH  READ REVIEW

A *Add horizontally.*

1. $3x^2 + 6x - 5$ and $-x^2 + 2x + 7$
2. $10x^2 + 3x + 9$ and $-x^2 - 5x + 1$
3. $2n^3 + n$ and $3n^3 + 8n$
4. $9y + 2y^2$ and $-y - 3y^2$
5. $10p + 3 + p^2$ and $p^2 - 7p - 4$
6. $x^2 + 3x - 8$ and $10 - 3x + 4x^2$
7. $8x^2 + 7xy - y^2$ and $3x^2 - 10xy + 3y^2$
8. $20p^2 + 15q^4 + 30pq$ and $-4pq + 10q^4 - p^2$
9. $2p^3 - p^2q - 5pq^2 + 1, 3p^2q + 2pq^2 - 4q^3 + 4,$ and $p^3 + q^3$
10. $2x^3 - 4x^2y + xy^2 + y^3, 3xy^2 - 6,$ and $2x^2y - xy^2 - 2y^3 + 3$

Add vertically.

11. $10x^2 - 3x - 8$ and $20x + 3$
12. $t^2 + 4t + 5$ and $-t + 10$
13. $5x^3 + 7x - 1$ and $x^2 + 2x + 3$
14. $2r^3 + r + 2$ and $-r^3 - 8r^2 + 5r - 6$
15. $5ab^2 - 3a^2 + a^3$ and $2ab^2 + 9a^2 - 4a^3$
16. $p^2q^3 - p^2 - q^3$ and $5p^2q^3 + p^2 + q^3$

B *Subtract horizontally.*

17. $2x^2 + 3x - 7$ from $x^2 + x + 4$
18. $2x^3 + 7x^2 + 3x$ from $8x^3 - 10x^2 + x$
19. $3x^3 + x^2 + 5x - 8$ from $x^3 + 10x^2 - 8x + 3$
20. $5t^2 - 7t - 1$ from $8t^2 - 3t + 2$
21. $5x + 9$ from $x^2 + 3x$
22. $3x - 7$ from $x^2 - x + 4$
23. $4y^2 - 6xy - 3$ from $1 - 6xy + 5x^2 - y^2$
24. $p^4 - 7p^2q^2 + q^4$ from $8p^4 - 3q^4$

Subtract vertically.

25. $7p^2 - 10p - 1$ from $2p^2 - 3p + 5$
26. $x^2 - 5x + 2$ from $3x^2 + 10x - 2$
27. $8t^3 - 5$ from $9t^3 - 12t^2 + 3$
28. $10x + 7$ from $x^2 + 2x - 6$
29. $r^3 - 3r^2s - 5$ from $4r^3 - 20r^2s - 7$
30. $-5x^3 - 2y^2$ from $13xy^3 + 7x^3 - 10y^2$

Remove the parentheses and simplify, if possible.

31. $7x - (8x + r)$
32. $8x - (9x + y)$
33. $2p - (3q + r)$
34. $t - (4r - s)$
35. $(4y - 1) + (3y^2 - y + 5)$
36. $(2x + 6) + (x^2 - 4x + 3)$
37. $(m^3 - 6m + 7) - (-9 + 6m)$
38. $(p^2 + 3p - 5) - (p^3 + 6 - p^2)$
39. $(2x^3 - 7x + 8) - (5x^2 + 3x - 1)$
40. $(n^3 - 4n + 2) - (n^2 - 8n + 1)$

41. $(8x^2 + 3x) + (x - 2) + (x^2 + 9)$

42. $(5y^2 + y) + (9y - 1) + (y^2 + 2)$

43. $(3x - 7) + (2x + 9) - (7x - 10)$

44. $(5n + 1) - (3n + 1) + (2n + 5)$

45. $(7x^2y^2 - 10xy + 4) - (2xy + 8) + (x^2y^2 - 10)$

46. $(2m^2n^2 - mn + 7) - (mn - 3) + (m^2n^2 + 7)$

Mixed Practice

Subtract horizontally.

47. $3m^2 - 4m + 2$ from $5m^2 - 9m - 7$

48. $x^2 - 5x$ from $3x - 4$

Remove the parentheses and simplify, if possible.

49. $(5x^3 - x - 8) - (-3x^2 - 3x + 4)$

50. $3a - (4b - c)$

51. $(3m - 1) - (4m + 5) + (6m - 2)$

52. $(6t - 3) + (2t^2 - 3t + 4)$

Add horizontally.

53. $3x^2 + 5x - 4$ and $-5x^2 + x + 7$

54. $6a^2 - 8b^3 - 7ab$ and $-5ab + 3b^3 - 11a^2$

Solve.

55. Subtract $p^2 - 4p + 3$ from $6p^2 + 7p - 4$ vertically.

56. Add $t^3 - 3t^2 + t + 9$ and $-8t^3 + t^2 - 2$ vertically.

Applications

C *Solve.*

57. The surface area of the cylindrical can shown is approximated by the polynomial expression $3.14r^2 + 3.14r^2 + 6.28rh$.

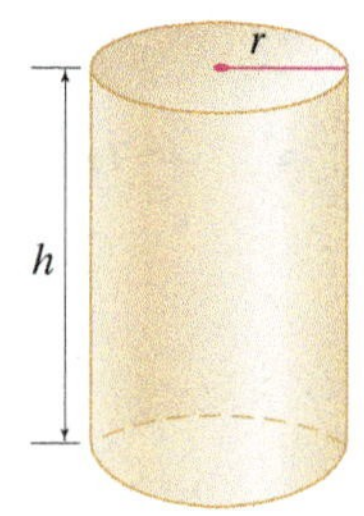

a. Simplify this polynomial expression.

b. Find a polynomial expression for the surface area of a can where the radius r and height h are equal.

58. The room shown below is in the shape of a cube.

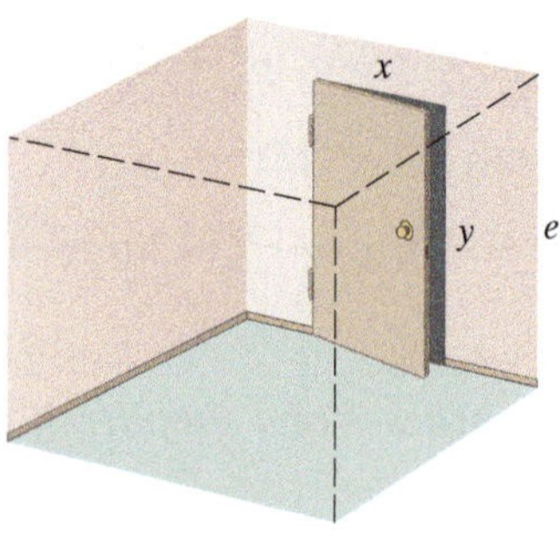

a. Write a simplified expression for the surface area of the four walls.

b. Find the surface area of the four walls when e is 9 ft, x is 3 ft, and y is 7 ft.

59. The total U.S. imports of petroleum (in thousands of barrels per day) in a given year are approximated by the polynomial $-37x^3 + 23x^2 - 13x + 13{,}718$, where x represents the number of years after 2005. The corresponding total of exports is modeled by the polynomial $-9x^3 + 84x^2 + 28x + 1175$. Write a polynomial that represents how many more thousands of barrels per day were imported than exported in a given year. (*Source:* eia.doe.gov)

60. The polynomial $327x^3 - 1651x^2 + 3x + 178{,}279$ models the number (in thousands) of male commissioned officers in the U.S. Department of Defense in a given year, where x represents the number of years after 2004. The corresponding polynomial for female commissioned officers is $149x^3 - 745x^2 + 310x + 34{,}022$. Find a polynomial that represents how many more thousands of male commissioned officers than female commissioned officers there were for a given year. (*Source:* prhome.defense.gov)

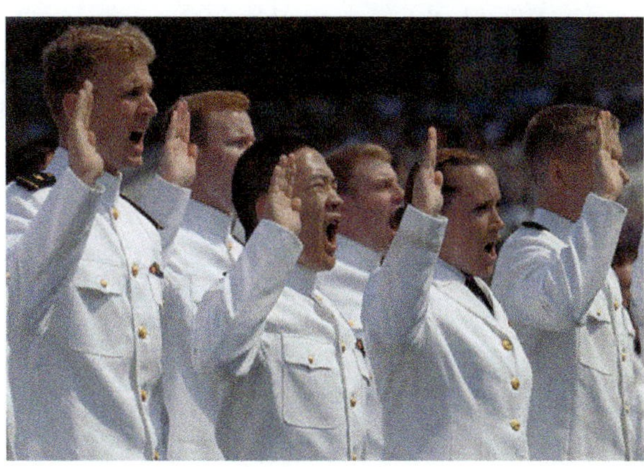

61. The polynomial $-2.4x^3 + 7.8x^2 - 6.0x + 68.3$ models the amount (in millions of dollars) spent on the Texas Rangers baseball team for a particular year, where x represents the number of years after 2007. The corresponding polynomial for the San Francisco Giants is $-1.6x^3 + 14.6x^2 - 26.6x + 90.2$. (*Source:* usatoday.com)

a. Find the polynomial that represents how much more the payroll is for the Giants than the Rangers.

b. Using the polynomial found in part (a), find how much more to the nearest million dollars the Giants payroll was than the Rangers in 2010.

62. The amount spent (in billions of dollars) on home health care in the U.S. for a particular year is modeled by the polynomial $0.1x^2 - 5.1x + 42.7$, where x represents the number of years after 2004. The corresponding polynomial for nursing home care is $0.3x^2 + 4.7x + 115.3$. Find the polynomial that represents the total expenditures for home health care and nursing home care during a given year. (*Source:* cms.gov)

• Check your answers on page A-33.

MINDStretchers

Patterns

1. A Fibonacci sequence is a list of numbers with the following property: After the first two numbers, every other number on the list is the sum of the two previous numbers. The following, for example, is a Fibonacci sequence:

$$\overset{\overset{7\,+\,11}{\downarrow}}{}$$
$$7, 11, 18, 29, 47, 76, 123, \ldots$$
$$\underset{\underset{11\,+\,18}{\uparrow}}{}$$

a. What is the next number in this sequence?

b. If the first two numbers in a Fibonacci sequence are a and b, find the third and fourth numbers. Check that the tenth number in the sequence is given by the polynomial $21a + 34b$.

Mathematical Reasoning

2. Is it possible to add two polynomials, each of degree 4, and have the sum be a polynomial of degree 2? If so, give an example. If not, explain why not.

Critical Thinking

3. Is the subtraction of polynomials a commutative operation? Give an example to support your answer.

15.5 Multiplying Polynomials

OBJECTIVES

- **A** To multiply monomials
- **B** To multiply a monomial by a polynomial
- **C** To multiply binomials
- **D** To multiply polynomials
- **E** To solve applied problems involving the multiplication of polynomials

In this section, we discuss how to multiply polynomials. We start by finding the product of two monomials.

Multiplying Monomials

When multiplying monomials such as $3x^2$ and $2x^5$, the product rule of exponents helps us to find the product.

$$\begin{aligned}(3x^2)(2x^5) &= (3 \cdot 2) \cdot (x^2 \cdot x^5) && \text{The commutative and associative properties of multiplication}\\ &= (3 \cdot 2) \cdot (x^{2+5}) && \text{The product rule of exponents}\\ &= 6x^7\end{aligned}$$

To Multiply Monomials

- Multiply the coefficients.
- Multiply the variables, using the product rule of exponents.

EXAMPLE 1

Multiply: $-2x \cdot 8x$

Solution
$$\begin{aligned}-2x \cdot 8x &= (-2 \cdot 8) \cdot (x \cdot x)\\ &= (-2 \cdot 8)(x^{1+1})\\ &= -16x^2\end{aligned}$$

PRACTICE 1

Find the product of $(-10x^2)$ and $(-4x^3)$.

EXAMPLE 2

Multiply: $(-2x^3y)(-4x^2y^4)(10xy)$

Solution

$$\begin{aligned}(-2x^3y)(-4x^2y^4)(10xy) &= (-2 \cdot -4 \cdot 10) \cdot (x^3 \cdot x^2 \cdot x) \cdot (y \cdot y^4 \cdot y)\\ &= 80(x^{3+2+1})(y^{1+4+1})\\ &= 80x^6y^6\end{aligned}$$

Note that the variables in a product are generally written in alphabetical order.

PRACTICE 2

Find the product: $(7ab^2)(10a^2b^3)(-5a)$

EXAMPLE 3

Simplify: $(-3p^2r)^3$

Solution

$$\begin{aligned}(-3p^2r)^3 &= (-3)^3(p^2)^3(r)^3 && \text{Use the rule for raising a product to a power.}\\ &= -27p^6r^3\end{aligned}$$

PRACTICE 3

Find the square of $-5xy^2$.

Multiplying a Monomial by a Polynomial

Now, let's use our knowledge of multiplying monomials to find the product of a monomial and a polynomial.

Consider, for instance, the product $(7x)(9x^2 + 5)$. We use the distributive property to find this product.

$$(7x)(9x^2 + 5) = (7x)(9x^2) + (7x)(5) = 63x^3 + 35x$$

Let's look at some more examples.

EXAMPLE 4

Multiply: $(-8x + 9)(-3x^2)$

Solution

$$\begin{aligned}(-8x + 9)(-3x^2) &= (-3x^2)(-8x + 9)\\ &= (-3x^2)(-8x) + (-3x^2)(9)\\ &= 24x^3 - 27x^2\end{aligned}$$

PRACTICE 4

Find the product: $(10s^2 - 3)(7s)$

EXAMPLE 5

Multiply: $3p^2q(5p^3 - 2pq + q^3)$

Solution

$$\begin{aligned}3p^2q(5p^3 - 2pq + q^3) &= 3p^2q(5p^3) + 3p^2q(-2pq) + 3p^2q(q^3)\\ &= 15p^5q - 6p^3q^2 + 3p^2q^4\end{aligned}$$

PRACTICE 5

Simplify:
$-2m^3n^2(-6m^3n^5 + 2mn^2 + n)$

EXAMPLE 6

Simplify: $8x^2(3x + 1) + x^2(5x^2 - 6x + 5)$

Solution

$$\begin{aligned}&8x^2(3x + 1) + x^2(5x^2 - 6x + 5)\\ &= 8x^2(3x) + 8x^2(1) + x^2(5x^2) + x^2(-6x) + x^2(5)\\ &= 24x^3 + 8x^2 + 5x^4 - 6x^3 + 5x^2\\ &= 5x^4 + 18x^3 + 13x^2 \quad \text{Combine like terms.}\end{aligned}$$

PRACTICE 6

Simplify:
$7s^3(-2s^2 + 5s + 4) - s^2(s^2 + 6s - 1)$

Multiplying Two Binomials

Now, we extend the discussion to the multiplication of binomials. As in the case of multiplying monomials and binomials, we use the distributive property.

Consider, for example, the product $(x + 4)(7x + 2)$. To apply the distributive property, we can think of the first factor $(x + 4)$ as a single number multiplied by the binomial $(7x + 2)$.

$$(a) \cdot (b + c) = (a) \cdot (b) + (a) \cdot (c)$$

$$\begin{aligned}(x + 4)(7x + 2) &= (x + 4)(7x) + (x + 4)(2) && \text{Use the distributive property.}\\ &= 7x(x + 4) + 2(x + 4) && \text{Use the commutative property.}\\ &= 7x \cdot x + 7x \cdot 4 + 2 \cdot x + 2 \cdot 4 && \text{Use the distributive property.}\\ &= 7x^2 + 28x + 2x + 8\\ &= 7x^2 + 30x + 8 && \text{Combine like terms.}\end{aligned}$$

Would we get the same answer if we multiplied $(7x + 2)$ by $(x + 4)$?

Let's consider some other examples.

EXAMPLE 7

Find the product: $(3x + 1)(5x - 2)$

Solution

$$\begin{aligned}(3x + 1)(5x - 2) &= (3x + 1)(5x) + (3x + 1)(-2)\\ &= 5x(3x + 1) + (-2)(3x + 1)\\ &= 5x \cdot 3x + 5x \cdot 1 + (-2) \cdot 3x + (-2) \cdot 1\\ &= 15x^2 + 5x - 6x - 2\\ &= 15x^2 - x - 2\end{aligned}$$

PRACTICE 7

Multiply: $(a - 1)(2a + 3)$

Another way to multiply two binomials is called the **FOIL method,** which is derived from the distributive property. With this method, we can memorize a formula that makes multiplying binomials quick and easy.

FOIL stands for **F**irst, **O**uter, **I**nner, and **L**ast. Let's see how this method works, applying it to the product $(x + 4)(7x + 2)$ that we discussed above.

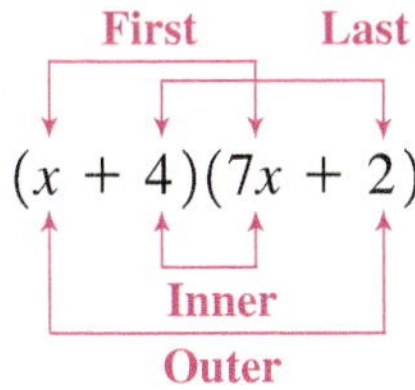

The two **first** terms in the binomials are x and $7x$, and their product is $7x^2$.

First

$(x + 4)(7x + 2)$ $\quad x \cdot 7x = 7x^2$

The two **outer** terms are x and 2, and their product is $2x$.

$(x + 4)(7x + 2)$ $\quad x \cdot 2 = 2x$

Outer

The two **inner** terms are 4 and $7x$, and their product is $28x$.

$(x + 4)(7x + 2)$ $\quad 4 \cdot 7x = 28x$

Inner

Next, the two **last** terms are 4 and 2, and their product is 8.

Last

$(x + 4)(7x + 2)$ $\quad 4 \cdot 2 = 8$

Finally, we add the four products, combining any like terms.

$$7x^2 + 2x + 28x + 8 = 7x^2 + 30x + 8$$

Note that, as expected, this answer is the same as the one found using the distributive property.

To Multiply Two Binomials Using the FOIL Method

Consider $(a + b)(c + d)$.

- Multiply the two *first* terms in the binomials.

$$(a + b)(c + d)$$ **Product is ac.** (F)

- Multiply the two *outer* terms.

$$(a + b)(c + d)$$ **Product is ad.** (O)

- Multiply the two *inner* terms.

$$(a + b)(c + d)$$ **Product is bc.** (I)

- Multiply the two *last* terms.

$$(a + b)(c + d)$$ **Product is bd.** (L)

- Find the sum of these four products.

$$(a + b)(c + d) = ac + ad + bc + bd$$

With some practice, the FOIL method can be done mentally.

EXAMPLE 8

Multiply: $(8x - 3)(2x - 1)$

Solution Using the FOIL method, we get:

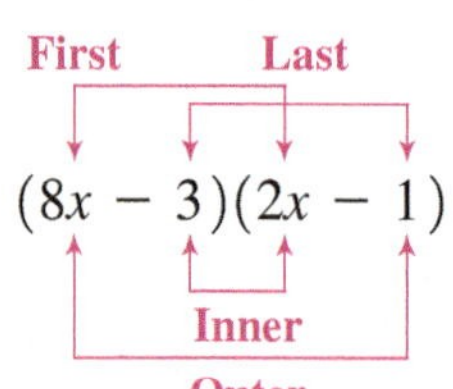

F: $(8x)(2x) = 16x^2$
O: $(8x)(-1) = -8x$
I: $(-3)(2x) = -6x$
L: $(-3)(-1) = 3$

So $(8x - 3)(2x - 1) = 16x^2 - 8x - 6x + 3 = 16x^2 - 14x + 3$. Note that the middle term, $-14x$, is the sum of the outer and inner products, which are like terms.

PRACTICE 8

Find the product of $8x + 3$ and $2x - 1$.

EXAMPLE 9

Multiply: $(3a + b)(2a - b)$

Solution

$$\begin{aligned}(3a + b)(2a - b) &= \overset{F}{(3a)(2a)} + \overset{O}{(3a)(-b)} + \overset{I}{(b)(2a)} + \overset{L}{(b)(-b)} \\ &= 6a^2 - 3ab + 2ab - b^2 \\ &= 6a^2 - ab - b^2\end{aligned}$$

PRACTICE 9

Find the product: $(7m - n)(2m + n)$

Multiplying Polynomials

Finally, let's consider multiplying polynomials in general. Here, we extend the previous discussion of multiplying binomials to multiplying polynomials that can have any number of terms. Let's consider the following example in which we multiply two polynomials written in a horizontal format.

EXAMPLE 10

Multiply: $(x^2 + 3x - 1)(x + 2)$

Solution

$(x^2 + 3x - 1)(x + 2)$

$= (x^2 + 3x - 1)(x) + (x^2 + 3x - 1)(2)$ Use the distributive property.

$= x^3 + 3x^2 - x + 2x^2 + 6x - 2$ Use the distributive property.

$= x^3 + 5x^2 + 5x - 2$ Combine like terms.

PRACTICE 10

Multiply $(n^2 + n - 2)(n - 4)$ using a horizontal format.

Instead of multiplying horizontally, we can multiply two polynomials in a vertical format similar to that used for multiplying whole numbers.

$$\begin{array}{r l}
x^2 + 3x - 1 & \\
x + 2 & \\
\hline
2x^2 + 6x - 2 & \text{Multiply } x^2 + 3x - 1 \text{ by } 2. \\
x^3 + 3x^2 - x \quad\;\; & \text{Multiply } x^2 + 3x - 1 \text{ by } x. \\
\hline
x^3 + 5x^2 + 5x - 2 & \text{Add like terms.}
\end{array}$$

Note that we positioned the terms of the two "partial products" in the shaded area so that each column contains like terms. Finally, to find the product of the original polynomials, we add down each column in the shaded area.

EXAMPLE 11

Find the product of $3y^2 + y + 5$ and $4y - 1$.

Solution We begin by rewriting the problem in a vertical format.

$$\begin{array}{r}
3y^2 + \;\; y + 5 \\
4y - 1 \\
\hline
-3y^2 - \;\; y - 5 \\
12y^3 + 4y^2 + 20y \quad\;\; \\
\hline
12y^3 + \;\; y^2 + 19y - 5
\end{array}$$

So we conclude:

$$(3y^2 + y + 5)(4y - 1) = 12y^3 + y^2 + 19y - 5$$

PRACTICE 11

Multiply: $(8n^2 - n + 3)(n + 2)$

EXAMPLE 12

Multiply: $(4x^3 - 2x + 1)(x + 5)$

Solution Let's multiply vertically.

Use $+0x^2$ for the missing x^2 term.

$$\begin{array}{r} 4x^3 + 0x^2 - 2x + 1 \\ x + 5 \\ \hline 20x^3 + 0x^2 - 10x + 5 \\ 4x^4 + 0x^3 - 2x^2 + x \quad \\ \hline 4x^4 + 20x^3 - 2x^2 - 9x + 5 \end{array}$$

Note that we wrote $0x^2$ for the missing second-degree term in the top polynomial. (Alternatively, we could have left a blank space there.) Similarly, we write a term with a 0 coefficient for each missing term in the partial products.

PRACTICE 12

Find the product of $8x^3 + 9x - 1$ and $3x + 7$.

EXAMPLE 13

Find the product of $a^2 + 2ab + b^2$ and $a + b$.

Solution

$$\begin{array}{r} a^2 + 2ab + b^2 \\ a + b \\ \hline a^2b + 2ab^2 + b^3 \\ a^3 + 2a^2b + ab^2 \quad \quad \\ \hline a^3 + 3a^2b + 3ab^2 + b^3 \end{array}$$

PRACTICE 13

Multiply $p^2 - 2pq + q^2$ by $p - q$.

EXAMPLE 14

A box factory makes an open-top box from a piece of cardboard by cutting out squares that are x ft by x ft from each corner, as shown below. The area of the base of the box is given by $(4 - 2x)(3 - 2x)$ ft^2. Rewrite the area of the base without parentheses.

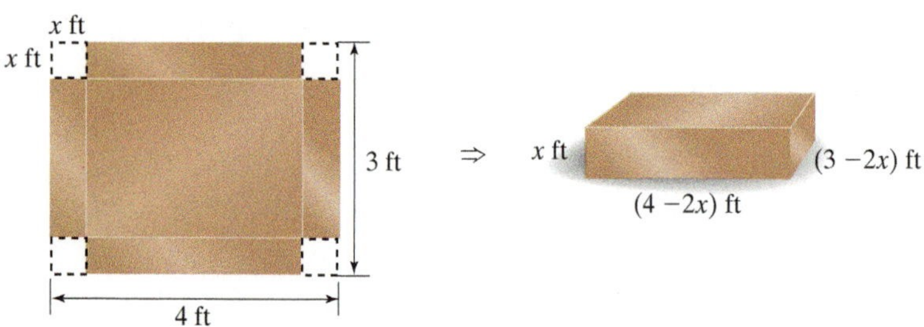

Solution To remove parentheses, we multiply the binomials using the FOIL method.

$$\begin{aligned} (4 - 2x)(3 - 2x) &= \overset{\text{F}}{(4)(3)} + \overset{\text{O}}{(4)(-2x)} + \overset{\text{I}}{(-2x)(3)} + \overset{\text{L}}{(-2x)(-2x)} \\ &= 12 - 8x - 6x + 4x^2 \\ &= 12 - 14x + 4x^2, \text{ or } 4x^2 - 14x + 12 \end{aligned}$$

So the area of the base is $(4x^2 - 14x + 12)$ ft^2.

PRACTICE 14

At the end of 2 yr, the amount of money in a savings account is given by

$$(P + Pr) + (P + Pr)r$$

where P is the initial balance and r is the annual rate of interest compounded annually. Rewrite this expression by removing parentheses and simplifying.

15.5 Exercises

FOR EXTRA HELP MyMathLab    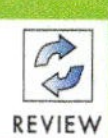

A *Multiply.*

1. $(6x)(-4x)$
2. $(-3x)(2x)$
3. $(9t^2)(-t^3)$
4. $(-6y^2)(7y^5)$
5. $(-5x^2)(-4x^4)$
6. $(-2h^2)(-3h^3)$
7. $(10x^3)(-7x^5)$
8. $(20a)(-5a^{99})$
9. $(4pq^2)(4p^2qr^2)$
10. $(8st^2)(6s^4t^7)$
11. $(-8x)^2$
12. $(-10p)^2$
13. $\left(\frac{1}{2}t^4\right)^3$
14. $\left(\frac{1}{3}n^3\right)^3$
15. $(7a)(10a^2)(-5a)$
16. $(-8n^3)(2n^3)(-n)$
17. $(2ab^2)(-3abc)(4a^2)$
18. $(4mn^3)(-m)(7mn^2)$

B *Find the product.*

19. $(7x - 5)x$
20. $(3y - 7)y$
21. $(9t + t^2)(5t)$
22. $(5x - x^2)(4x)$
23. $6a^3(4a^2 - 7a)$
24. $2y(8y - y^3)$
25. $4x^2(3x - 2)$
26. $-5p^7(9p - 3)$
27. $x^3(x^2 - 2x + 4)$
28. $t^2(t^2 + 8t + 1)$
29. $5x(3x^2 + 5x + 6)$
30. $4x(3x^2 - x + 2)$
31. $(5x^2 - 3x - 7)(-9x)$
32. $(10x^2 + x - 1)(2x)$
33. $6x^2(x^3 + 4x^2 - x - 1)$
34. $(x^3 + 6x^2 - 9x + 10)(-3x^4)$
35. $4p(7q - p^2)$
36. $(v + 3w^2)(-7v)$
37. $-pq^2(3p^3 - 9q)$
38. $(m^2 + n^2)\,3mn^5$
39. $2a^2b^3(3a^4b^2 + 10ab^5)$
40. $-4x^2y^2(7x^2y^3 - 4x^3y^4)$

Simplify.

41. $10x + 2x(-3x + 8)$
42. $-7x + 3x(5x - 9)$
43. $-x + 8x(x^2 - 2x + 1)$
44. $2x(8x^2 + 7x - 2) - 6x^2$
45. $9x(x^2 + 3x - 5) + 8x(-4x^2 + x)$
46. $x(x^2 + 11x) - 10x(2x^2 + 3)$
47. $-4xy(2x^2 + 4xy) + x^2y(7x^2 - 2y)$
48. $2s^3t(5s - t^2) - 7s^2t^2(9s^2 - 10t)$
49. $5a^2b^2(3ab^4 - a^3b^2) + 4a^2b^2(9ab^4 - 10a^3b^2)$
50. $(3p - 8pq)(5p^2) - (4p^3 + 1)(7q)$

C *Multiply.*

51. $(y + 2)(y + 3)$
52. $(x + 1)(x + 4)$
53. $(x - 3)(x - 5)$
54. $(n - 4)(n - 2)$
55. $(a - 2)(a + 2)$
56. $(x + 3)(x - 3)$
57. $(w + 3)(2w - 7)$
58. $(8x + 5)(x + 4)$
59. $(3 - 2y)(5y - 1)$
60. $(4u - 1)(3 - 2u)$
61. $(10p - 4)(2p - 1)$
62. $(7x + 1)(7x - 3)$
63. $(u + v)(u - v)$
64. $(x + y)(x - y)$
65. $(2p - q)(q - p)$
66. $(x + 4y)(x - y)$
67. $(3a - b)(a - 2b)$
68. $(5x + 4y)(x - y)$
69. $(p - 8)(4q + 3)$
70. $(x + 7)(6y - 1)$

D *Find the product.*

71. $(x - 3)(x^2 - 3x + 1)$
72. $(a + 2)(a^2 - 4a + 4)$
73. $(2x - 1)(x^2 + 3x - 5)$
74. $(8n + 3)(2n^2 - 9n - 1)$
75. $(a - b)(a^2 + ab + b^2)$
76. $(x^2 + xy + y^2)(y - x)$
77. $(3x)(x + 5)(x - 7)$
78. $(y^2)(8 - 3y)(8 + y)$
79. $(3n)(2n - 1)(2n + 1)$
80. $(-a)(a + 2b)(a - 3b)$

Mixed Practice

Simplify.

81. $m^2n(8m^2 - 6n) - 2mn(3m^2 - 7mn)$
82. $8t^2 + 5t(3t^2 - 2t + 1)$

Multiply.

83. $(-8s^3)(-5s^4)$
84. $3p(8p - p^2)$
85. $(9y^2 + 7y - 8)(-4y)$
86. $(u - 5)(u - 7)$
87. $(3x^4)(-8x^5)(x^2)$
88. $(2w - 3)(4 - w)$
89. $-a^2b(4a - 2b^3)$
90. $(3h^2k^7)(-9hk^5l)$
91. $(x - 1)(x^2 - 3x + 2)$
92. $(a + 6)(8a - 3)$

Applications

E *Solve.*

93. Backgammon is one of the world's oldest board games. The length and width of the distinctive board (shown below) differ by 30 mm. Find the area of the board in terms of x without using parentheses.

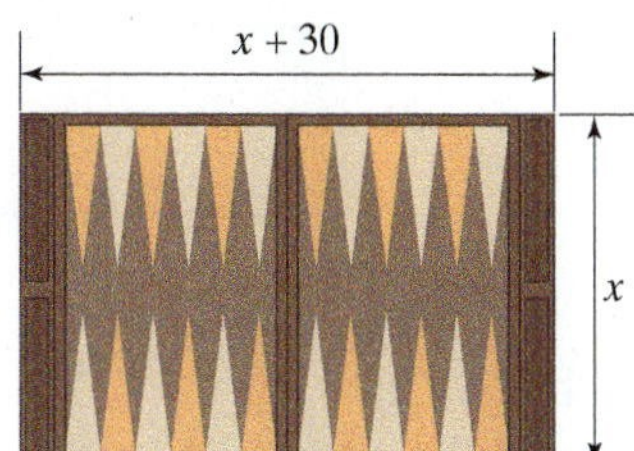

94. A raindrop with diameter d has a cross-sectional area A that can be modeled using the formula $6A - 25 = 2d(5d - 12)$. Solve this formula for A. (*Source:* Brian Lim, "Derivation of the Shape of a Raindrop," cs.cmu.edu)

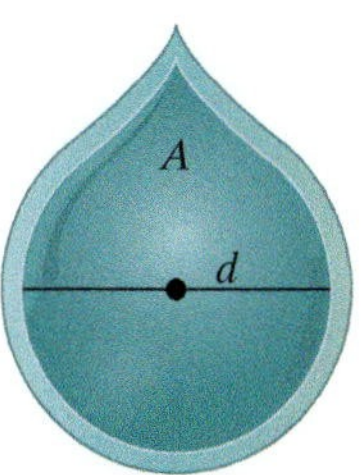

95. A factory has been selling 1000 color laser printers per year for \$1500 each. The company's market research indicates that for each \$100 that the price is raised, sales will fall by 30 units. The expression $(1500 + 100x)(1000 - 30x)$ gives the estimated revenue that the company will take in if it adjusts the price of a printer, where x represents the number of \$100 increases in the price. Rewrite this expression, multiplying out the factors.

96. A company's total revenue R is given by the equation

$$R = px$$

where p is the price of each item and x is the number of items sold. Write a polynomial for the revenue if $x = -\frac{1}{4}p + 100$.

97. Investment brokers use the formula $A = P(1 + r)^t$ for the amount of money A in a client's account that earns compound interest. In this formula, P is the principal (the original amount of money that the client invested), r is the rate of return per time period (in decimal form), and t is the number of time periods that the money has been invested.

a. If the client invested \$5000 for 3 periods, write this formula as a polynomial in r without parentheses.

b. If a client invested \$5000, how much greater is the amount of money in the account after 3 periods as compared with 2 periods? Write your answer without parentheses.

c. If the client's rate of return on the investment is 10%, how much money is represented by the expression in part (b)?

98. The expression for the volume of a sphere is $\frac{4}{3}\pi r^3$, where r is the radius of the sphere. Assume that the shape of the Earth is approximately a sphere of radius r mi.

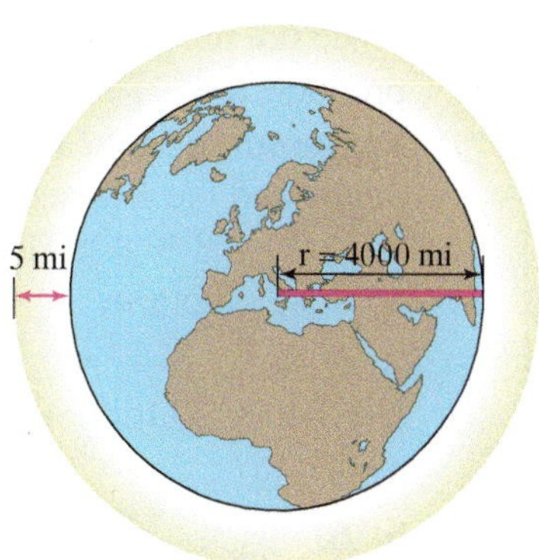

a. Find and simplify the expression for the volume of the sphere formed by everywhere rising 5 mi above the surface of the Earth.

b. Find and simplify the expression for the volume of the sphere formed by everywhere descending 5 mi below the surface of the Earth.

c. According to some scientists, all the life discovered so far in the universe is found in the layer around the Earth that extends 5 mi above the Earth's surface to 5 mi below. What is the volume of this layer?

• Check your answers on page A-33.

MINDStretchers

Mathematical Reasoning

1. We can draw rectangles to visualize the product of two binomials. Consider, for example, the product $(x + 9)(x + 2)$. Using the diagram at the right, explain how the product can be expressed as the sum of four areas.

	x	9
x	x^2	$9x$
2	$2x$	18

Compare your answer with the result of using the FOIL method to multiply the two binomials.

Patterns

2. Simplify the following polynomials:

$(x + y)^0 =$

$(x + y)^1 =$

$(x + y)^2 =$

$(x + y)^3 =$

In the following table, enter the coefficients from these polynomials. What pattern (known as Pascal's Triangle) do you observe in this table?

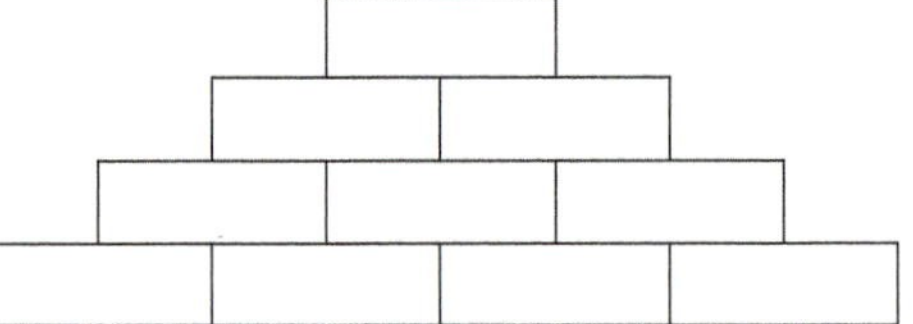

Historical

3. *Lattice multiplication* is a procedure that originated in India in the twelfth century. In this procedure, whole numbers are multiplied as if they were binomials. Each digit of each factor is multiplied separately. The products are recorded in little cells within a lattice, and then added along the diagonals. For instance, to find $29 \cdot 47$, the four products are placed in the small, diagonally split squares. The product of 2 and 4, shown in red, is 8.

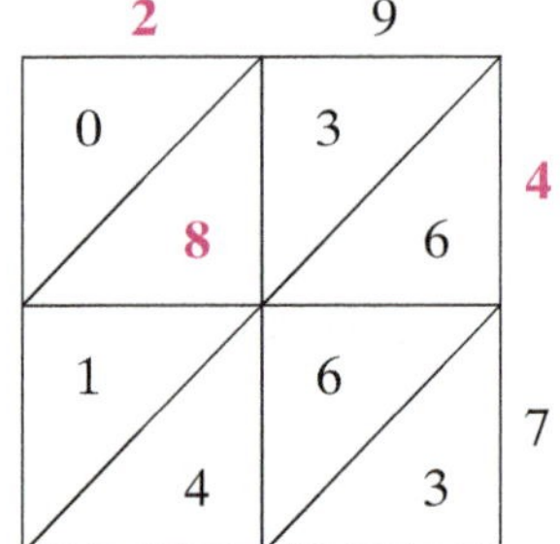

In the diagram to the right, the square in the upper left shows ${}^0/_8$, which represents 8. Since the product of 9 and 4 is 36, the square in the upper right contains ${}^3/_6$. The lower products are ${}^1/_4$ or 14 and ${}^6/_3$ or 63. These products are all added along the diagonal. For instance, in the diagonal shaded green, the sum $6 + 6 + 4$ is 16; the 6 is written below and the 1 is regrouped into the diagonal above and added into that diagonal: $1 + (3 + 8 + 1)$. The product, 1363, appears down the left side of the lattice and across the bottom.

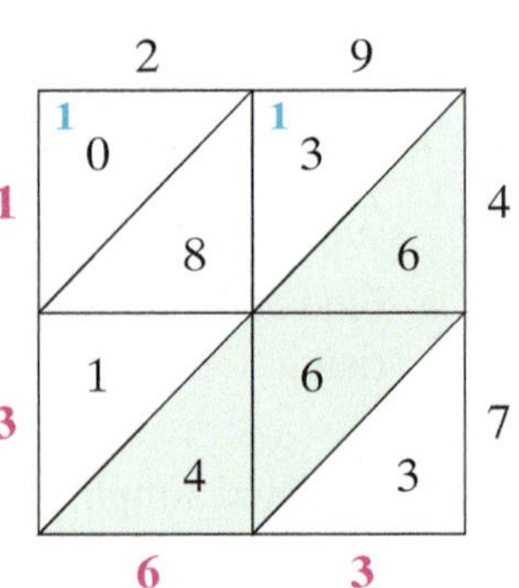

Use lattice multiplication to find the following products:

a. $53 \cdot 89$

b. $61 \cdot 94$

15.6 Special Products

OBJECTIVES

A To use the formulas for squaring a binomial

B To use the formula for multiplying the sum and the difference of two terms

C To solve applied problems involving formulas for special products

Recall that in the previous section, we discussed two ways of multiplying binomials—first by directly applying the distributive property, and second by using the FOIL method. In this section, we focus on yet a third approach—using formulas. These formulas provide a shortcut for finding special products, that is, special cases of multiplying binomials.

The Square of a Binomial

To square a binomial means to multiply that binomial by itself. Consider, for instance, $(x + 5)^2$, in which we square the *sum* of two terms. We can rewrite this expression as a binomial multiplied by itself and then use the FOIL method to find the product.

$$\begin{aligned}(x + 5)^2 &= (x + 5)(x + 5)\\ &= x^2 + 5x + 5x + 25\\ &= x^2 + 10x + 25\end{aligned}$$

Note that the expression $x^2 + 10x + 25$ is equal to $(x)^2 + 2(x)(5) + (5)^2$. So the square of the sum of x and 5 equals the square of x plus twice the product of x and 5 plus the square of 5. This observation suggests the following formula for squaring the sum of two terms.

The Square of a Sum

$$(a + b)^2 = a^2 + 2ab + b^2$$

In words, the square of the *sum* of two terms is equal to the square of the first term plus twice the product of the two terms plus the square of the second term.

Can you explain why this formula follows from the FOIL method of multiplying binomials?

EXAMPLE 1

Simplify: $(x + 7)^2$

Solution Since we are squaring the sum of x and 7, we can use the formula for the square of a sum.

First term ↓ $(x$; Second term ↓ $+ 7)^2$; The square of the first term ↓ x^2; Twice the product of the two terms $2(x)(7)$; The square of the second term ↓ $(7)^2$

$$\begin{aligned}(x + 7)^2 &= x^2 + 2(x)(7) + (7)^2\\ &= x^2 + 14x + 49\end{aligned}$$

PRACTICE 1

Simplify: $(p + 10)^2$

Note in Example 1 that $(x + 7)^2 = x^2 + 14x + 49$, whereas $x^2 + 7^2 = x^2 + 49$. So we see that $(x + 7)^2 \neq x^2 + (7)^2$.

TIP It is important to distinguish between the *square of the sum* of two terms and the *sum of the squares* of the terms: $(x + y)^2 \neq x^2 + y^2$

EXAMPLE 2

Simplify: $(p + q)^2$

Solution In $(p + q)^2$, the first term is p and the second term is q.

First term ↓, Second term ↓, The square of the first term ↓, The square of the second term ↓

$$(p + q)^2 = p^2 + \underbrace{2(p)(q)}_{\text{Twice the product of the two terms}} + q^2$$

$$= p^2 + 2pq + q^2$$

PRACTICE 2

Simplify: $(s + t)^2$

EXAMPLE 3

Simplify: $(2x + 3y)^2$

Solution

$$\begin{aligned}(2x + 3y)^2 &= (2x)^2 + 2(2x)(3y) + (3y)^2 \\ &= 4x^2 + 12xy + 9y^2\end{aligned}$$

PRACTICE 3

Simplify: $(4p + 5q)^2$

Now, let's examine a second and related formula, which involves the square of the *difference* of two terms rather than their sum. For instance, consider $(x - 5)^2$. Again, we rewrite the square of the binomial and then apply the FOIL method:

$$\begin{aligned}(x - 5)^2 &= (x - 5)(x - 5) \\ &= x^2 - 5x - 5x + 25 \\ &= x^2 - 10x + 25\end{aligned}$$

Note that the expression $x^2 - 10x + 25$ is equal to $x^2 - 2(x)(5) + 25$. So the square of the difference of x and 5 equals the square of x minus twice the product of x and 5 plus the square of 5. This example leads to the following formula for squaring the difference of two terms:

The Square of a Difference

$$(a - b)^2 = a^2 - 2ab + b^2$$

In words, the square of the difference of two terms is equal to the square of the first term minus twice the product of the two terms plus the square of the second term.

If we compare the formula for the square of a sum with the formula for the square of a difference, we see that the signs of the middle term of the resulting trinomial differ. That is, for $(a + b)^2$, the middle term is positive, whereas for $(a - b)^2$, the middle term is negative.

EXAMPLE 4

Simplify: $(8a - 1)^2$

Solution Here, we are squaring the difference of two terms, so we use the formula for the square of a difference.

First term ↓ $(8a$ − Second term ↓ $1)^2$ = The square of the first term ↓ $(8a)^2 - 2(8a)(1)$ + The square of the second term ↓ 1^2

Twice the product of the two terms

$$(8a - 1)^2 = (8a)^2 - 2(8a)(1) + 1^2$$
$$= 64a^2 - 16a + 1$$

PRACTICE 4

Simplify: $(5x - 2)^2$

EXAMPLE 5

Simplify: $(p - q)^2$

Solution

$$(p - q)^2 = p^2 - 2(p)(q) + q^2$$
$$= p^2 - 2pq + q^2$$

PRACTICE 5

Simplify: $(u - v)^2$

EXAMPLE 6

Simplify: $(3a - 4b)^2$

Solution

$$(3a - 4b)^2 = (3a)^2 - 2(3a)(4b) + (4b)^2$$
$$= 9a^2 - 24ab + 16b^2$$

PRACTICE 6

Simplify: $(2x - 9y)^2$

The Product of the Sum and Difference of Two Terms

The third special binomial formula relates to multiplying *the sum of two terms by the difference of the same two terms*. Explain why neither of the two previous formulas applies in this situation.

For example, consider the product $(x + 5)(x - 5)$. Using the FOIL method we get:

$$(x + 5)(x - 5) = x \cdot x + x \cdot (-5) + 5 \cdot x - 5 \cdot 5$$
$$= x^2 - 5x + 5x - 25$$

The middle terms cancel each other out.

$$= x^2 - 25$$

The Product of the Sum and Difference of Two Terms

$$(a + b)(a - b) = a^2 - b^2$$

In words, the product of the sum and difference of the *same* two terms is equal to the square of the first term minus the square of the second term.

EXAMPLE 7

Multiply: $(x + 11)(x - 11)$

Solution

First term	Second term		The square of the first term		The square of the second term
↓	↓		↓		↓

$$(x + 11)(x - 11) = x^2 - (11)^2$$
$$= x^2 - 121$$

PRACTICE 7

Find the product of $(t + 10)$ and $(t - 10)$.

EXAMPLE 8

Multiply.

a. $(p - q)(p + q)$ **b.** $(3m + 2n)(3m - 2n)$

Solution

a. Since $(p - q)(p + q) = (p + q)(p - q)$, the formula for finding the product of the sum and difference of two terms applies.

$$(p - q)(p + q) = p^2 - q^2$$

b. $(3m + 2n)(3m - 2n) = (3m)^2 - (2n)^2 = 9m^2 - 4n^2$

PRACTICE 8

Find the product.

a. $(r - s)(r + s)$

b. $(8s - 3t)(8s + 3t)$

EXAMPLE 9

Find the product: $(3a^2 - 5)(3a^2 + 5)$

Solution $(3a^2 - 5)(3a^2 + 5) = (3a^2)^2 - (5)^2 = 9a^4 - 25$

PRACTICE 9

Multiply: $(10 - 7k^2)(10 + 7k^2)$

EXAMPLE 10

The nineteenth-century French physician Jean Louis Poiseuille investigated the flow of blood in the smaller blood vessels of the body. He discovered that the speed of blood varies from point to point within a blood vessel. In a blood vessel of radius r at a point b units from the center of the blood vessel, the blood flow speed is given by the expression $k(r + b)(r - b)$, where k is a constant. Write this expression without parentheses.

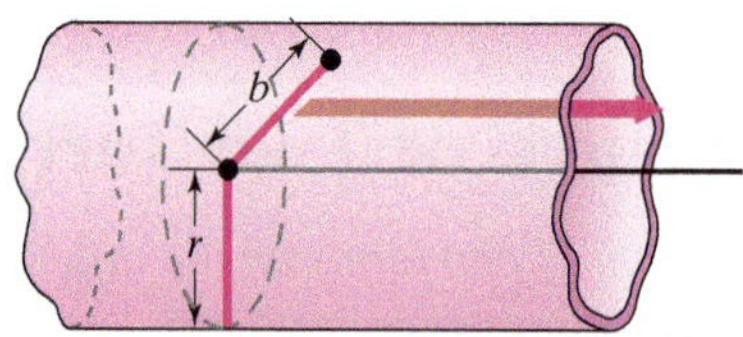

Solution We need to multiply out $k(r + b)(r - b)$.

$$k(r + b)(r - b) = k(r^2 - b^2)$$
$$= kr^2 - kb^2$$

So the expression is $kr^2 - kb^2$.

PRACTICE 10

The area of the square wooden frame shown can be represented by the polynomial $(S + s)(S - s)$, where s is the side length of the smaller square and S is the side length of the larger square. Write this expression without parentheses.

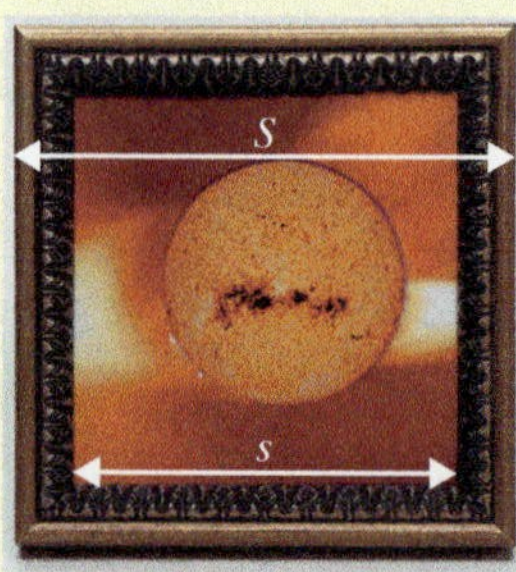

Memorize the formulas for the square of a sum, the square of a difference, and the product of the sum and difference of two terms. With sufficient practice, you will be able to find these special products mentally.

15.6 Exercises

FOR EXTRA HELP MyMathLab MathXL PRACTICE WATCH READ REVIEW

Mathematically Speaking

Fill in each blank with the most appropriate term or phrase from the given list.

minus	plus	positive
negative	times	divided by

1. The square of the sum of two terms is equal to the square of the first term _______ twice the product of the two terms plus the square of the second term.

2. In the formula for the square of the sum of two terms, the middle term of the trinomial is _______.

3. In the formula for the square of a difference of two terms, the middle term of the trinomial is _______.

4. The product of the sum and difference of the same two terms is equal to the square of the first term _______ the square of the second term.

A *Simplify.*

5. $(y + 2)^2$

6. $(a + 3)^2$

7. $(x + 4)^2$

8. $(n + 8)^2$

9. $(x - 11)^2$

10. $(b - 10)^2$

11. $(6 - n)^2$

12. $(9 - y)^2$

13. $(x + y)^2$

14. $(s + t)^2$

15. $(3x + 1)^2$

16. $(5x + 3)^2$

17. $(4n - 5)^2$

18. $(2x - 1)^2$

19. $(9x + 2)^2$

20. $(11m + 3)^2$

21. $\left(a + \frac{1}{2}\right)^2$

22. $\left(b + \frac{1}{5}\right)^2$

23. $(8b + c)^2$

24. $(3s + t)^2$

25. $(5x - 2y)^2$

26. $(3m - 4n)^2$

27. $(-x + 3y)^2$

28. $(-p + 2q)^2$

29. $(4x^3 + y^4)^2$

30. $(5a^2 - c^2)^2$

B *Multiply.*

31. $(a + 1)(a - 1)$

32. $(r + 8)(r - 8)$

33. $(4x - 3)(4x + 3)$

34. $(7y - 2)(7y + 2)$

35. $(10 + 3y)(3y - 10)$

36. $(-1 + 9x)(9x + 1)$

37. $\left(m - \frac{1}{2}\right)\left(m + \frac{1}{2}\right)$

38. $\left(m - \frac{1}{4}\right)\left(m + \frac{1}{4}\right)$

39. $(n + 0.3)(n - 0.3)$

40. $(y + 0.5)(y - 0.5)$

41. $(4a + b)(4a - b)$

42. $(p - 3q)(p + 3q)$

43. $(3x - 2y)(3x + 2y)$

44. $(10t + 3s)(10t - 3s)$

45. $(1 - 5n)(5n + 1)$

46. $(2s + 3)(3 - 2s)$

47. $x(x + 5)(x - 5)$

48. $y(y - 7)(y + 7)$

49. $5n^2(n + 7)^2$

50. $-8y^3(2y - 1)^2$

51. $(n^2 - m^4)(n^2 + m^4)$

52. $(x^3 + y^5)(x^3 - y^5)$

53. $(a - b)(a + b)(a^2 + b^2)$

54. $(x + y)(x - y)(x^2 + y^2)$

Mixed Practice

Simplify each expression.

55. $(6n + 4)^2$

56. $(2h - 7k)(2h + 7k)$

57. $(4p - 9)(4p + 9)$

58. $(w + 7)^2$

59. $(8 - a)^2$

60. $(3a + 8)(8 - 3a)$

61. $-2x^2(4x - 3)^2$

62. $(3x - 5y)^2$

Applications

C *Solve.*

63. The mayor of a city plans for a square-shaped park, as shown on the grid to the right. The area of the park can be modeled by the expression $(x - 5)^2 + (y - 1)^2$. Rewrite this expression by removing parentheses and simplifying.

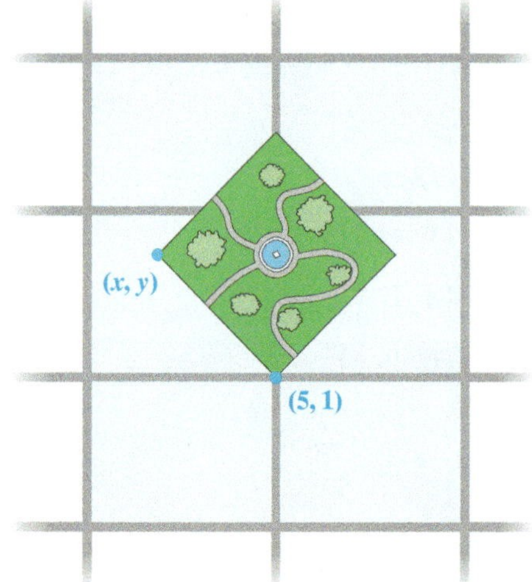

64. A Rubik's Cube is a three-dimensional mechanical puzzle considered to be the world's best-selling toy, with more than 300 million sold worldwide. On each of the 6 faces of the cube is an n by n grid of colored squares. (For instance, a cube with 3 by 3 grids is shown at the right.) Each face of the cube turns independently. The puzzle is solved when on each face all the colored squares are the same color, although different faces will have different colors. (*Source:* wikipedia.org)

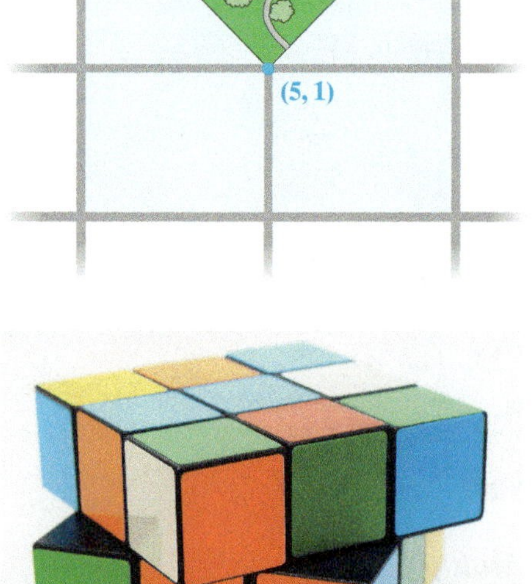

a. How many more colored squares does a Rubik's Cube with an $(n + 1)$ by $(n + 1)$ grid have than one with an n by n grid?

b. How many more colored squares in all does a Rubik's Cube with a 4×4 grid have than one with a 3×3 grid?

65. An investment of A dollars increases in value by $P\%$ for each of two years. The value of the investment at the end of the two years can be represented by

$$A\left(1 + \frac{P}{100}\right)\left(1 + \frac{P}{100}\right)$$

Multiply out this expression.

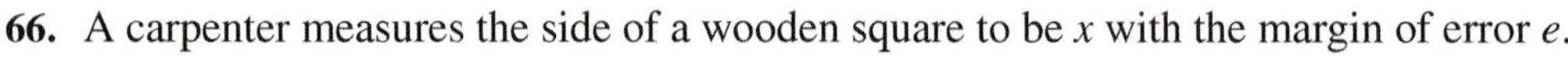

66. A carpenter measures the side of a wooden square to be x with the margin of error e.

a. What are the longest possible true dimensions of the wooden square? The shortest possible true dimensions?

b. What is the difference in area between the wooden square with the longest possible dimensions and the wooden square with the shortest possible dimensions in terms of e and x?

67. To measure the spread of data, statisticians compute the *sample variance* of the data. For a sample of size 3, they use the following formula:

$$\text{Sample variance} = \frac{(a-m)^2 + (b-m)^2 + (c-m)^2}{2}$$

Rewrite this formula without parentheses, combining like terms.

68. As the temperature of a lightbulb's filament changes from T_1 to T_2, the energy that the filament radiates changes by the quantity

$$a(T_1 - T_2)(T_1 + T_2)(T_1^2 + T_2^2)$$

Multiply to find this change.

69. Consider an $(x + y)$ by $(x - y)$ rectangle.
   a. What is the perimeter of this rectangle?
   b. Does a square with side length x have the same perimeter as the rectangle in part (a)?
   c. How much larger is the area of the square in part (b) than the area of the rectangle in part (a)?

70. When a radioactive substance decays, the remaining amount A of the substance after t min can be modeled by the formula $A = m(1 - r)^t$, where m is the initial mass of the substance, t is the elapsed time in minutes, and the rate of decay is r per minute. (*Source:* Atomic Mass Data Center)
   a. Write this formula without parentheses for the amount of 1000 micrograms of a radioactive substance that remains after 2 min.
   b. A substance decays at the rate of 3%. After 2 min, how much of 1000 micrograms of this substance is left, rounded to the nearest microgram? (*Hint:* Before substituting into the formula, change the rate from a percent to a decimal.)

• Check your answers on page A-33.

MINDStretchers

Patterns

1. Mentally compute each product. (*Hint:* Think of these computations as "special products.")

   a. $9999 \times 10{,}001$ **b.** $30\frac{1}{10} \times 29\frac{9}{10}$

Mathematical Reasoning

2. Suppose that you square two consecutive whole numbers and subtract the smaller square from the larger. Is it possible that the difference is an even number? Explain your answer.

Historical

3. By the year 2000 B.C., the astronomers of Mesopotamia knew the relationship $(a - b)(a + b) = a^2 - b^2$. They could demonstrate this relationship by a geometric model. Find the area of the remaining region if the yellow square is removed from the figure shown. Show that this model verifies the relationship $(a - b)(a + b) = a^2 - b^2$. (*Hint:* Find the area in two ways.)

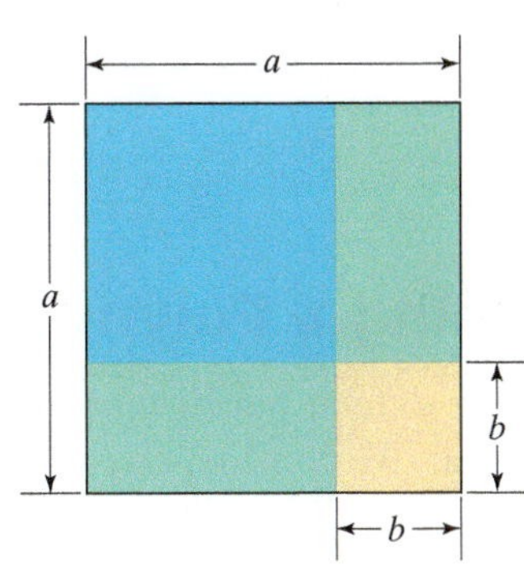

15.7 Dividing Polynomials

OBJECTIVES

- **A** To divide monomials
- **B** To divide a polynomial by a monomial
- **C** To divide a polynomial by a binomial
- **D** To solve applied problems involving the division of polynomials

The final section of this chapter deals with dividing polynomials. We progress from dividing a monomial by another monomial to dividing a polynomial by a binomial.

Dividing Monomials

The ability to divide monomials—the simplest of polynomial divisions—depends on our knowledge of both fractions and exponents, as the following example illustrates:

EXAMPLE 1

Simplify: $8x^6 \div 4x^2$

Solution We begin by rewriting the problem in fractional form.

$$8x^6 \div 4x^2 = \frac{8x^6}{4x^2}$$

$$\frac{8x^6}{4x^2} = \frac{8}{4} \cdot \frac{x^6}{x^2}$$

$$= 2x^{6-2}$$ Divide the coefficients and use the quotient rule of exponents.

$$= 2x^4$$

PRACTICE 1

Find the quotient: $-12n^6 \div 3n$

EXAMPLE 2

Divide: $\frac{-15a^2b^3c}{10ab^2}$

Solution $$\frac{-15a^2b^3c}{10ab^2} = \frac{-15}{10} \cdot \frac{a^2b^3c}{ab^2}$$

$$= \frac{-3}{2} \cdot a^{2-1}b^{3-2}c$$

$$= \frac{-3}{2}abc$$

$$= -\frac{3abc}{2}$$

PRACTICE 2

Find the quotient:

$(20p^3q^2r^4) \div (-5p^2q^2r)$

Dividing a Polynomial by a Monomial

Now, we extend our discussion of division to finding the quotient of a polynomial and a monomial. We consider an example, written in fractional form.

EXAMPLE 3

Simplify the quotient: $(10x^2 + 15x) \div (5x)$

Solution Let's write the given expression in fractional form.

$$(10x^2 + 15x) \div (5x) = \frac{\overbrace{10x^2 + 15x}^{\text{Polynomial}}}{\underbrace{5x}_{\text{Monomial}}}$$

Since we are dividing a polynomial by a monomial, we can rewrite the given fraction as the sum of two fractions with the same denominator. Then, we simplify each fraction.

$$\begin{aligned}\frac{10x^2 + 15x}{5x} &= \frac{10x^2}{5x} + \frac{15x}{5x}\\ &= 2x + 3\end{aligned}$$

PRACTICE 3

Divide: $(21x^3 - 14x^2) \div 7x$

Example 3 suggests that when dividing a polynomial by a monomial, we can divide each term in the polynomial by the monomial, and then add. Let's apply this shortcut in the following examples.

EXAMPLE 4

Divide: $\dfrac{9x^4 - 6x^3 + 12x^2}{-3x^2}$

Solution Divide each term in the numerator by the denominator. Then, add the quotients:

$$\begin{aligned}\frac{9x^4 - 6x^3 + 12x^2}{-3x^2} &= \frac{9x^4}{-3x^2} - \frac{6x^3}{-3x^2} + \frac{12x^2}{-3x^2}\\ &= -3x^2 + 2x - 4\end{aligned}$$

PRACTICE 4

Simplify: $\dfrac{14x^8 + 10x^5 - 8x^3}{-2x^3}$

EXAMPLE 5

Divide: $\dfrac{10p^4q^5 + 12p^2q^4 - pq^3}{2pq^2}$

Solution

$$\begin{aligned}\frac{10p^4q^5 + 12p^2q^4 - pq^3}{2pq^2} &= \frac{10p^4q^5}{2pq^2} + \frac{12p^2q^4}{2pq^2} - \frac{pq^3}{2pq^2}\\ &= 5p^3q^3 + 6pq^2 - \frac{q}{2}\end{aligned}$$

PRACTICE 5

Find the quotient:

$$\frac{-5a^7b^6 + a^2b^4 - 15ab^3}{5ab^3}$$

Dividing a Polynomial by a Binomial

Now, let's take a look at the general case—how to divide one polynomial by another polynomial. We will restrict our attention to dividing a polynomial by a *binomial*, but these remarks apply to dividing by polynomials with three or more terms as well.

The procedure for dividing a polynomial by a binomial is similar to that of dividing whole numbers, which is commonly called *long division*.

$$\begin{array}{r l}
12 & \longleftarrow \text{Quotient} \\
\text{Divisor} \longrightarrow 13\overline{)\ 158} & \longleftarrow \text{Dividend} \\
-\underline{13} & \\
28 & \\
-\underline{26} & \\
2 & \longleftarrow \text{Remainder}
\end{array}$$

We stop dividing because the remainder 2 is smaller than the divisor 13. So $158 \div 13 = 12$ with remainder 2.

Now, let's consider the following problem in dividing polynomials:

$$(x^2 - 5x - 24) \div (x + 3)$$

We set up the problem just as if we were dividing whole numbers, being careful to correctly distinguish between the dividend and the divisor. In this case, $x^2 - 5x - 24$ is the dividend and $x + 3$ is the divisor.

$$\begin{array}{r}
x \\
x + 3\overline{)x^2 - 5x - 24} \\
\underline{x^2 + 3x}
\end{array}$$

Divide the first term in the dividend by the first term in the divisor: Think $x^2 \div x = x$. Place x in the quotient above the x term in the dividend. Then, *multiply* x in the quotient by the divisor: $x(x + 3) = x^2 + 3x$.

$$\begin{array}{r}
x \\
x + 3\overline{)x^2 - 5x - 24} \\
\underline{x^2 + 3x} \\
-8x - 24
\end{array}$$

Subtract $(x^2 + 3x)$ from $(x^2 - 5x)$ by changing the signs of each term in $(x^2 + 3x)$ and then adding to get $-8x$. Bring down the next term, -24. We have to divide again because the degree of $-8x - 24$ is equal to the degree of $x + 3$. Both have degree 1.

$$\begin{array}{r}
x - 8 \\
x + 3\overline{)x^2 - 5x - 24} \\
\underline{x^2 + 3x} \\
-8x - 24 \\
\underline{-8x - 24} \\
0
\end{array}$$

The remainder is 0.

Divide x into $-8x$: Think $-8x \div x = -8$. Place -8 in the quotient above the constant term in the dividend. Then, *multiply* -8 in the quotient by the divisor: $-8(x + 3) = -8x - 24$. Finally, *subtract* $(-8x - 24)$ from $(-8x - 24)$ by changing the signs of each term in $(-8x - 24)$ and then adding to get 0.

Note that the degree of the remainder is less than the degree of the divisor because the remainder 0 is a constant, which has degree 0, and the divisor $x + 3$ has degree 1.

So $(x^2 - 5x - 24) \div (x + 3) = x - 8$. Can you explain how to check this quotient?

This example suggests the following general method for dividing a polynomial by a polynomial.

To Divide a Polynomial by a Polynomial

- Arrange each term of the dividend and divisor in descending order.
- Divide the first term of the dividend by the first term of the divisor. The result is the first term of the quotient.
- Multiply the first term of the quotient by the divisor and place the product under the dividend.
- Subtract the product, found in the previous step, from the dividend.
- Bring down the next term to form a new dividend.
- Repeat the process until the degree of the remainder is less than the degree of the divisor.

EXAMPLE 6

$$2x - 1\overline{)6x^2 + 9x - 6}$$

Solution The dividend and divisor are already in descending order.

$$\begin{array}{r} 3x \\ 2x - 1\overline{)6x^2 + 9x - 6} \\ \underline{6x^2 - 3x} \\ 12x - 6 \end{array}$$

Divide $6x^2$ by $2x$, getting $3x$.
Multiply $3x$ by $(2x - 1)$, getting $(6x^2 - 3x)$.
Subtract $(6x^2 - 3x)$ from $(6x^2 + 9x)$ and bring down -6.

$$\begin{array}{r} 3x + 6 \\ 2x - 1\overline{)6x^2 + 9x - 6} \\ \underline{6x^2 - 3x} \\ 12x - 6 \\ \underline{12x - 6} \\ 0 \end{array}$$

Divide $12x$ by $2x$, getting 6.
Multiply 6 by $(2x - 1)$, getting $(12x - 6)$.
Subtract $(12x - 6)$ from $(12x - 6)$.

Since the degree of 0 is less than the degree of $(2x - 1)$, the process stops.
So $(6x^2 + 9x - 6) \div (2x - 1) = 3x + 6$.

PRACTICE 6

Find the quotient:
$(10x^2 + 17x + 3) \div (5x + 1)$

Now, we focus on problems in dividing polynomials that have *remainders*.

Dividing whole numbers:

$$\begin{array}{r} 21 \\ 45\overline{)956} \\ \underline{90} \\ 56 \\ \underline{45} \\ 11 \end{array}$$

11 ⟵ Remainder

Check

$45 \cdot 21 + 11 \stackrel{?}{=} 956$
$945 + 11 = 956$ True

So $956 \div 45 = 21\frac{11}{45}$.

Dividing polynomials:

$$\begin{array}{r} x + 3 \\ x + 5\overline{)x^2 + 8x + 16} \\ \underline{x^2 + 5x} \\ 3x + 16 \\ \underline{3x + 15} \\ 1 \end{array}$$

1 ⟵ Remainder

Check

$(x + 5)(x + 3) + 1 \stackrel{?}{=} x^2 + 8x + 16$
$x^2 + 8x + 15 + 1 = x^2 + 8x + 16$ True

So $(x^2 + 8x + 16) \div (x + 5) = x + 3 + \frac{1}{x + 5}$.

Note that we can check a problem involving division of polynomials in the same way that we check division of whole numbers:

$$\text{Divisor} \cdot \text{quotient} + \text{remainder} = \text{Dividend}$$

EXAMPLE 7

Find the quotient: $(x^3 + 3x^2 - 8x + 2) \div (x + 5)$

Solution

$$\begin{array}{r} x^2 - 2x + 2 \\ x + 5 \overline{)x^3 + 3x^2 - 8x + 2} \\ \underline{x^3 + 5x^2 } \\ -2x^2 - 8x \\ \underline{-2x^2 - 10x } \\ 2x + 2 \\ \underline{2x + 10} \\ -8 \end{array}$$

Since the degree of the remainder is less than the degree of the divisor, we stop.

Check

$$(x + 5)(x^2 - 2x + 2) + (-8) \stackrel{?}{=} x^3 + 3x^2 - 8x + 2$$
$$x^3 + 3x^2 - 8x + 2 = x^3 + 3x^2 - 8x + 2 \quad \text{True}$$

So we write the answer as:

$$x^2 - 2x + 2 + \frac{-8}{x + 5}$$

PRACTICE 7

Divide $(3x^3 + 7x^2 + 11x + 5)$ by $(3x + 1)$.

Some problems in dividing polynomials involve terms that are not in *descending order*.

EXAMPLE 8

Divide $(6 + 8x^2 - 14x)$ by $(2x - 3)$.

Solution Before dividing, we place the terms in both the divisor and the dividend in descending order. Here we need to rearrange the terms in the dividend:

$$\begin{array}{r} 4x - 1 \\ 2x - 3 \overline{)8x^2 - 14x + 6} \\ \underline{8x^2 - 12x } \\ -2x + 6 \\ \underline{-2x + 3} \\ 3 \end{array} \quad \text{The remainder is 3.}$$

So $(8x^2 - 14x + 6) \div (2x - 3) = 4x - 1 + \dfrac{3}{2x - 3}$.

How would you check this solution?

PRACTICE 8

Divide: $(-21s + 10 + 9s^2) \div (3s - 2)$

In dividing polynomials, we may have *missing terms* in the dividend, as shown in the following example.

EXAMPLE 9

$x + 3\overline{)2x^3 + 7x^2 - 9}$

Solution Since there is no x-term in the dividend, we can insert $0x$ as a placeholder for the missing term.

$$\begin{array}{r} 2x^2 + x - 3 \\ x + 3\overline{)2x^3 + 7x^2 + 0x - 9} \\ \underline{2x^3 + 6x^2} \qquad\qquad \\ x^2 + 0x \qquad \\ \underline{x^2 + 3x} \qquad \\ -3x - 9 \\ \underline{-3x - 9} \\ 0 \end{array}$$

So $\dfrac{2x^3 + 7x^2 - 9}{x + 3} = 2x^2 + x - 3.$

PRACTICE 9

Divide: $\dfrac{4n^3 - 19n^2 - 4}{4n - 3}$

EXAMPLE 10

A homeowner wishes to increase the length of a flower bed by twice as much as the increase in the width. The area of the new flower bed is given by $(2x^2 + 20x + 48)$ ft^2.

a. Use long division to find the width of the new flower bed in terms of x.

b. If the homeowner increases the width by 1 ft, then what are the dimensions of the new flower bed?

Solution

a. To find the width of the new rectangular flower bed, we can use the formula $A = lw$, where the area A and the length l are given by $(2x^2 + 20x + 48)$ ft^2 and $(2x + 8)$ ft respectively. Since $A = lw,\ w = \dfrac{A}{l}$.

Using $w = \dfrac{A}{l}$, we conclude that $w = \dfrac{2x^2 + 20x + 48}{2x + 8}$. Dividing the numerator by the denominator, we find that $w = x + 6$. So the width is $(x + 6)$ ft.

b. So if the owner increases the width of the flower bed by 1 ft, the dimensions of the new flower bed are:

$$(1 + 6) \text{ ft by } (2 \cdot 1 + 8) \text{ ft, or } 7 \text{ ft by } 10 \text{ ft.}$$

PRACTICE 10

If \$10 is invested at an interest rate of r per year and compounded annually, the future value S in dollars at the end of the nth year is given by:

$$S = 10(1 + r)^n$$

a. What is the future value of the investment after 1 yr? After 2 yr?

b. Write the answers to part (a) without parentheses and in descending order.

c. Using your answer in part (b), determine how many times as great the future value of the investment is after 2 yr as compared to the future value after 1 yr.

15.7 Exercises

FOR EXTRA HELP MyMathLab     

Mathematically Speaking

Fill in each blank with the most appropriate term or phrase from the given list.

powers	polynomial by the monomial	quotient
dividend	coefficients	divisor
remainder	monomial by the polynomial	

1. To divide a monomial by a monomial, divide the ____________________ and use the quotient rule of exponents.

2. To divide a polynomial by a monomial, divide each term in the ____________________ and then add.

3. In dividing one polynomial by another, repeat the process until the degree of the ____________________ is less than the degree of the divisor.

4. To check if the quotient of two polynomials is correct, see if the ____________________ is the sum of the remainder and the product of the divisor and the quotient.

A *Simplify.*

5. $\dfrac{10x^4}{5x^2}$

6. $\dfrac{6x^3}{2x^2}$

7. $\dfrac{16a^8}{-4a}$

8. $\dfrac{-35y^2}{7y}$

9. $\dfrac{-8x^5}{-6x^4}$

10. $\dfrac{-9p^5}{-12p^2}$

11. $\dfrac{12p^2q^3}{3p^2q}$

12. $\dfrac{9a^4b}{3a^2b}$

13. $\dfrac{-24u^6v^4}{-8u^4v^2}$

14. $\dfrac{-4x^5y^4}{-2x^2y}$

15. $\dfrac{-15a^2b^5}{7ab^3}$

16. $\dfrac{21x^3y^5}{-10xy^2}$

17. $\dfrac{-6u^5v^3w^3}{4u^2vw^3}$

18. $\dfrac{-10x^3yz^2}{8x^2yz}$

B *Divide.*

19. $\dfrac{6n^2 + 10n}{2n}$

20. $\dfrac{12m^4 + 15m^3}{3m}$

21. $\dfrac{20b^4 - 10b}{10b}$

22. $\dfrac{2x^2 - 8x}{2x}$

23. $\dfrac{18a^2 + 12a}{-3a}$

24. $\dfrac{16x^2 + 10x}{-2x}$

25. $\dfrac{9x^5 - 6x^7}{3x^5}$

26. $\dfrac{6a^3 - 4a^2}{2a^2}$

27. $\dfrac{12a^4 - 18a^3 + 30a^2}{6a^2}$

28. $\dfrac{8x^5 + 4x^4 - 16x^3}{4x^2}$

29. $\dfrac{n^5 - 10n^4 - 5n^3}{-5n^3}$

30. $\dfrac{9y^6 - 3y^5 - 2y^4}{-3y^4}$

31. $\dfrac{20a^2b + 4ab^3}{8ab}$

32. $\dfrac{14xy^2 - 21x^3y^4}{7xy^2}$

33. $\dfrac{12x^2y^3 - 9xy - 3xy^2}{-3xy}$

34. $\dfrac{10ab^8 - 4ab^6 + 6ab^4}{-2ab^4}$

35. $\dfrac{8p^2q^3 - 4p^3q^3 + 6p^4q}{4p^2q}$

36. $\dfrac{6x^2y^3 - 18x^3y^4 + 9x^4y^5}{6x^2y^3}$

C *Find the quotient.*

37. $(x^2 - 4x - 21) \div (x + 3)$

38. $(x^2 + 6x - 40) \div (x + 10)$

39. $(56x^2 - 23x + 2) \div (8x - 1)$

40. $(30x^2 + 13x - 3) \div (6x - 1)$

41. $(6x^2 + 13x - 5) \div (2x + 5)$

42. $(10x^2 - x - 2) \div (5x + 2)$

43. $(-2x + 5x^2 - 3) \div (x - 1)$

44. $(19x + 2x^2 + 35) \div (x + 7)$

45. $(4 + 20x + 21x^2) \div (2 + 3x)$

46. $(-3 + x + 2x^2) \div (3 + 2x)$

Divide.

47. $\dfrac{x^2 + 2x + 5}{x + 2}$

48. $\dfrac{x^2 + 2x + 7}{x - 2}$

49. $\dfrac{-3 - 5x + 2x^2}{x - 3}$

50. $\dfrac{-2x + 5x^2 - 3}{x - 1}$

51. $\dfrac{8x^2 - 6x - 11}{4x + 3}$

52. $\dfrac{3x^2 - x - 8}{3x - 1}$

53. $\dfrac{-x + x^3 - 5x^2 + 5}{x + 1}$

54. $\dfrac{-5 + 11x - 7x^2 + x^3}{x - 5}$

55. $\dfrac{6x^3 - 11x^2 - 5x + 19}{3x - 4}$

56. $\dfrac{2x^3 + x^2 - 4x - 8}{2x + 1}$

57. $\dfrac{5x^2 - 2}{x - 4}$

58. $\dfrac{10x^2 - 2x}{x + 3}$

59. $\dfrac{4x^3 - x + 3}{2x - 3}$

60. $\dfrac{3x^3 + x^2 - 4}{x + 1}$

61. $\dfrac{x^3 + 27}{x + 3}$

62. $\dfrac{x^3 - 1}{x - 1}$

Mixed Practice

Simplify.

63. $\dfrac{56r^2}{-8r}$

64. $\dfrac{42x^5y^2}{7xy^2}$

65. $\dfrac{-18a^2b^3c}{27ab^2c}$

Divide.

66. $\dfrac{15a^8b - 21a^4b - 24a^3b}{3a^3b}$

67. $\dfrac{18n^6 - 48n^4 - 2n^3}{-6n^3}$

68. $(36x^2 + x - 2) \div (4x + 1)$

69. $(5 + 13x + 6x^2) \div (5 + 3x)$

70. $\dfrac{10t^5 - 6t^3}{2t}$

71. $\dfrac{4x^3 + x^2 - 4x + 2}{4x + 1}$

72. $\dfrac{4x^2 - 5x}{x - 3}$

73. $\dfrac{32m^3 - 72m}{-8m}$

74. $\dfrac{2x^2 - 13x - 24}{x - 8}$

Applications

D *Solve.*

75. The number of typed words that fit on a page depends on the font used as well as the font's point size. For the popular font Times New Roman, the average number of words that fit on a (single-spaced) page can be modeled by the polynomial $9x^2 - 340x + 3600$, where x is the point size of the font. (*Source:* writersservices.com)

a. Use this formula to approximate the number of words on a page for 10-point Times New Roman.

b. What is the average number of words on a page *per point* of font size?

c. What is the average number of words on a page per point for 10-point Times New Roman?

76. A prepaid phone card company charges a \$0.49 connection fee for each call plus \$0.01 for each minute.

a. Write an expression for the cost of a call in terms of x, the number of minutes the call lasts.

b. Write an expression for the cost per minute of a call.

c. Use division to rewrite the expression in part (b).

77. The formula $d = rt$ can be used to find the time t when given the distance d and the rate r.

a. Solve the equation $d = rt$ for t.

b. Use the answer from part (a) to find an expression for the time it takes to travel a distance of $(t^3 - 6t^2 + 7t + 14)$ mi at a rate of $(t + 1)$ mph.

78. A *geometric series* is a sum of terms where each term is formed by multiplying the previous term by a constant. For example, the series $5 + 5r + 5r^2$ has three terms where the first term is 5 and each of the other terms is r times the previous term. Use long division to show that the sum of the first three terms can be calculated from the formula $\dfrac{-5r^3 + 5}{-r + 1}$.

79. The polynomial $3x^3 + 348x^2 + 5250x + 45{,}000$ models the total expenditures on home health care (in millions of dollars) for a particular year in the United States, where x represents the number of years after 2004. The polynomial $3x + 300$ approximates the U.S. population (in millions) during the same years. Write a polynomial that approximates the total expenditures on home heath care per person. (*Source:* cms.gov)

80. The polynomial $-9x^3 - 750x^2 + 16{,}800x + 180{,}000$ models the total expenditures on prescription drugs (in millions of dollars) in the United States for a particular year, where x represents the number of years after 2004. The polynomial $3x + 300$ approximates the U.S. population (in millions) during the same years. Write a polynomial that approximates the total expenditures on prescription drugs per person. (*Source:* cms.gov)

• Check your answers on page A-33.

MIND*Stretchers*

Mathematical Reasoning

1. When you divide a polynomial by a trinomial, explain how you know if the trinomial is a factor of the polynomial.

Patterns

2. Consider the following table:

Divisor	Dividend	Quotient
$x + 1$	$x^2 - x + 1$	
$x + 1$	$x^3 - x^2 + x - 1$	
$x + 1$	$x^4 - x^3 + x^2 - x + 1$	

   a. Complete the table by dividing each dividend by the divisor.
   b. Predict the result of dividing $x^5 - x^4 + x^3 - x^2 + x - 1$ by $x + 1$. Verify your prediction.

Critical Thinking

3. Divide $(2y^2 + 5y + 3)$ by $(y + 1)$. For which values of y is the quotient larger in value than the divisor?

Key Concepts and Skills

CONCEPT SKILL

Concept/Skill	Description	Example
[15.1] **Exponent (or Power)**	A number that indicates how many times another number (called the *base*) is used as a factor.	$4^3 = \underbrace{4 \cdot 4 \cdot 4}_{\text{3 factors}}$ (4: Base; 3: Exponent)
[15.1] **Exponent 1**	For any real number x, $x^1 = x$.	$2^1 = 2$ $(ab)^1 = ab$
[15.1] **Exponent 0**	For any nonzero real number x, $x^0 = 1$.	$5^0 = 1$ $(ab)^0 = 1$
[15.1] **The product rule of exponents**	For any nonzero real number x and for any integers a and b, $x^a \cdot x^b = x^{a+b}$	$3^2 \cdot 3^3 = 3^{2+3} = 3^5 = 243$ $x \cdot x = x^{1+1} = x^2$
[15.1] **The quotient rule of exponents**	For any nonzero real number x and for any integers a and b, $\frac{x^a}{x^b} = x^{a-b}$	$2^5 \div 2^3 = \frac{2^5}{2^3} = 2^{5-3} = 2^2 = 4$ $\frac{x^6}{x} = x^{6-1} = x^5$
[15.1] **Negative exponents**	For any nonzero real number x and for any integer a, $x^{-a} = \frac{1}{x^a}$	$6^{-2} = \frac{1}{6^2} = \frac{1}{36}$ $(2x)^{-2} = \frac{1}{(2x)^2} = \frac{1}{4x^2}$
[15.1] **Reciprocal of x^{-a}**	For any nonzero real number x and for any integer a, $\frac{1}{x^{-a}} = x^a$	$\frac{1}{5^{-3}} = 5^3 = 125$ $\frac{x^3}{y^{-2}} = x^3y^2$
[15.2] **The power rule of exponents**	For any nonzero real number x and for any integers a and b, $(x^a)^b = x^{ab}$	$(2^3)^2 = 2^{3 \cdot 2} = 2^6 = 64$ $(p^7)^3 = p^{7 \cdot 3} = p^{21}$
[15.2] **Raising a product to a power**	For any nonzero real numbers x and y and any integer a, $(xy)^a = x^a \cdot y^a$	$(4y)^3 = 4^3y^3 = 64y^3$ $(c^4d^3)^5 = c^{4 \cdot 5}d^{3 \cdot 5}$ $= c^{20}d^{15}$
[15.2] **Raising a quotient to a power**	For any nonzero real numbers x and y and any integer a, $\left(\frac{x}{y}\right)^a = \frac{x^a}{y^a}$	$\left(\frac{2}{3}\right)^3 = \frac{2^3}{3^3} = \frac{8}{27}$ $\left(\frac{-5}{b}\right)^4 = \frac{(-5)^4}{b^4} = \frac{625}{b^4}$

Concept/Skill	Description	Example
[15.2] Raising a quotient to a negative power	For any nonzero real numbers x and y and any integer a, $\left(\frac{x}{y}\right)^{-a} = \left(\frac{y}{x}\right)^{a}$	$\left(\frac{2}{3}\right)^{-1} = \left(\frac{3}{2}\right)^{1} = \frac{3}{2}$ $\left(\frac{p}{q}\right)^{-3} = \left(\frac{q}{p}\right)^{3} = \frac{q^3}{p^3}$
[15.2] Scientific notation	A number is in scientific notation if it is written in the form $a \times 10^n$ where n is an integer and a is greater than or equal to 1 but less than 10 $(1 \leq a < 10)$.	5.3×10^9 and 2.41×10^{-5} are in scientific notation.
[15.3] Monomial	An expression that is the product of a real number and variables raised to nonnegative integer powers.	$3x^3$ $-4a^2b$
[15.3] Polynomial	An algebraic expression with one or more monomials added or subtracted.	$-5x$ ⟵ Monomial $2x + 1$ ⟵ Binomial $x^2 - 9x + 2$ ⟵ Trinomial $-x^3 + 7x^2 - x + 19$ (all: Polynomial)
[15.3] Degree of a monomial	The power of the variable in a monomial.	$3x^5$ is of degree 5.
[15.3] Degree of a polynomial	The highest degree of any of its terms.	$4x^2 - x + 3$ is of degree 2.
[15.3] Leading term of a polynomial	The term in a polynomial with the highest degree. Its coefficient is called the **leading coefficient** of the polynomial. The term of degree 0 is called the **constant term**.	In $-8x^2 - 7x + 3$, $-8x^2$ is the leading term, -8 is the leading coefficient, and 3 is the constant term.
[15.4] To add polynomials	• Add the like terms.	Find the sum horizontally: $(5x^2 + 6x - 9) + (-10x^2 + 7)$ $= 5x^2 + 6x - 9 - 10x^2 + 7$ $= -5x^2 + 6x - 2$ Add vertically: $(5x^2 + 6x - 9) + (-10x^2 + 7)$ $\begin{array}{r} 5x^2 + 6x - 9 \\ \underline{-10x^2 \qquad + 7} \\ -5x^2 + 6x - 2 \end{array}$

continued

CONCEPT SKILL

Concept/Skill	Description	Example
[15.4] **To subtract polynomials**	• Change the sign of each term of the polynomial being subtracted. • Add the like terms.	Subtract horizontally: $(2x^2 - 6x + 1) - (-x^2 + 4x + 5)$ $= 2x^2 - 6x + 1 + x^2 - 4x - 5$ $= 3x^2 - 10x - 4$ Subtract vertically: $(2x^2 - 6x + 1) - (-x^2 + 4x + 5)$ $\begin{array}{r} 2x^2 - 6x + 1 \\ \underline{-(-x^2 + 4x + 5)} \end{array} \quad \begin{array}{r} 2x^2 - 6x + 1 \\ \underline{x^2 - 4x - 5} \\ 3x^2 - 10x - 4 \end{array}$
[15.5] **To multiply monomials**	• Multiply the coefficients. • Multiply the variables, using the product rule of exponents.	Multiply: $(-3x^2y)(5x^3y^2)$ $= (-3 \cdot 5)(x^2 \cdot x^3)(y \cdot y^2) = -15x^5y^3$
[15.5] **To multiply two binomials using the FOIL method** 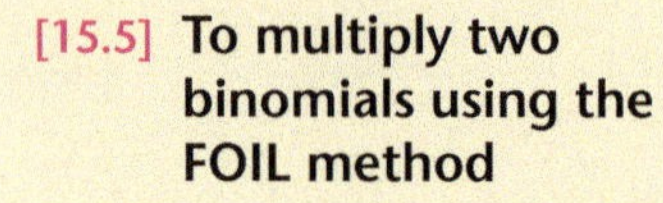	Consider $(a + b)(c + d)$. • Multiply the two first terms in the binomials. 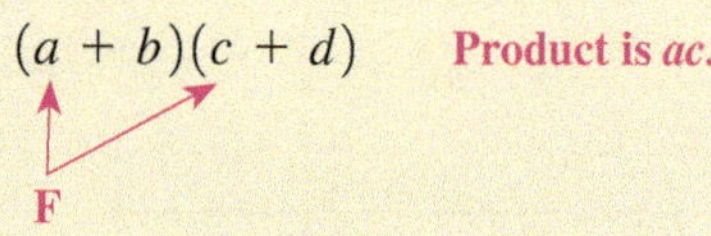 Product is ac. 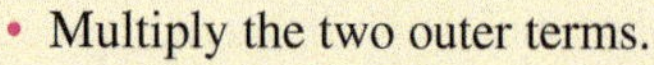 • Multiply the two outer terms. 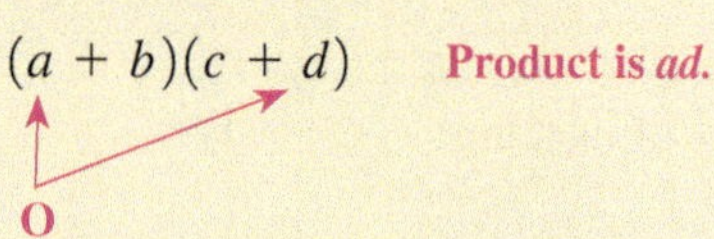 Product is ad. • Multiply the two inner terms. $(a + b)(c + d)$ Product is bc. I • Multiply the two last terms. 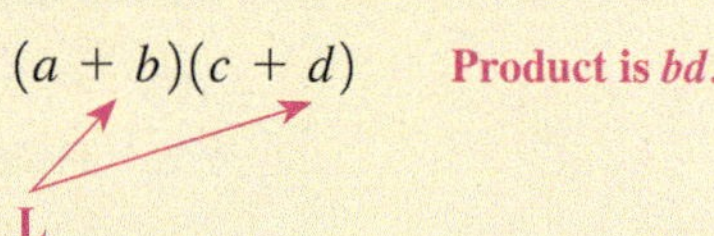 Product is bd. The product of the two binomials is the sum of these four products. $(a + b)(c + d) = ac + ad + bc + bd$	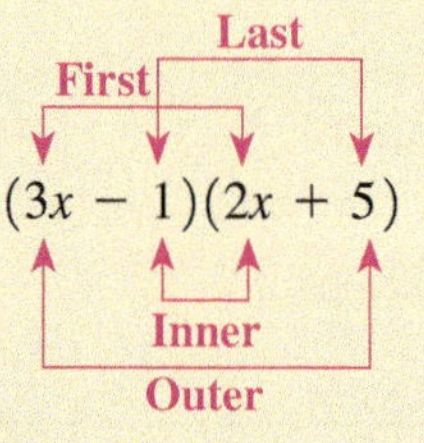 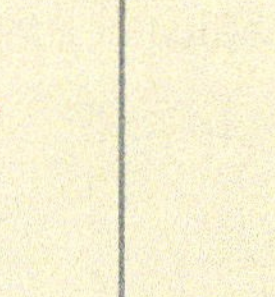 F $(3x)(2x) = 6x^2$ O $(3x)(5) = 15x$ I $(-1)(2x) = -2x$ L $(-1)(5) = -5$ The product $(3x - 1)(2x + 5)$ is the sum of these four products: $6x^2 + 15x - 2x - 5$ $= 6x^2 + 13x - 5$
[15.6] **The square of a sum**	$(a + b)^2 = a^2 + 2ab + b^2$	$(x + 3)^2 = x^2 + 2(x)(3) + (3)^2$ $= x^2 + 6x + 9$ $(7y + 5)^2 = (7y)^2 + 2(7y)(5) + (5)^2$ $= 49y^2 + 70y + 25$
[15.6] **The square of a difference**	$(a - b)^2 = a^2 - 2ab + b^2$	$(y - 6)^2 = y^2 - 2(y)(6) + (6)^2$ $= y^2 - 12y + 36$ $(4x - 1)^2 = (4x)^2 - 2(4x)(1) + (1)^2$ $= 16x^2 - 8x + 1$

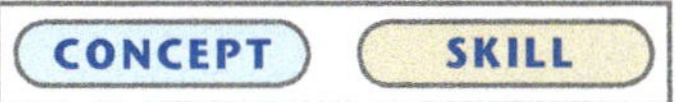

Concept/Skill	Description	Example
[15.6] **The product of the sum and difference of two terms**	$(a + b)(a - b) = a^2 - b^2$	$(2t + s)(2t - s) = (2t)^2 - (s)^2$ $= 4t^2 - s^2$
[15.7] **To divide a polynomial by a polynomial**	• Arrange each term of the dividend and divisor in descending order. • Divide the first term of the dividend by the first term of the divisor. The result is the first term of the quotient. • Multiply the first term of the quotient by the divisor and place the product under the dividend. • Subtract the product, found in the previous step, from the dividend. • Bring down the next term to form a new dividend. • Repeat the process until the degree of the remainder is less than the degree of the divisor.	$(3y^2 - 4y - 7) \div (y - 2)$ Quotient: $3y + 2$ Divisor: $y - 2$; Dividend: $3y^2 - 4y - 7$ $3y^2 - 6y$ $2y - 7$ $2y - 4$ Remainder $\longrightarrow -3$

CHAPTER 15 Review Exercises

Say Why

Fill in each blank.

1. $3^2 \cdot 3^4$ ________ (is/is not) equal to 3^8 because ____.

2. $(-5)^7 \div (-5)^4$ ________ (is/is not) equal to $(-5)^3$ because ____.

3. The expression $\frac{1}{a^3}$ ________ (can/cannot) be simplified as a^{-3} because ____.

4. The number 0.24×10^3 ________ (is/is not) written in scientific notation because ______________________________.

5. $-\frac{2}{3}x^{-3}$ ________ (is/is not) a monomial because ____.

6. In the polynomial $x^2 + 5$, the constant term ________ (does/does not) have degree 0 because ______________________________.

[15.1] *Simplify.*

7. $(-x)^3$
8. -31^0
9. $n^4 \cdot n^7$
10. $x^6 \cdot x$
11. $\frac{n^8}{n^5}$
12. $p^{10} \div p^7$
13. $y^4 \cdot y^2 \cdot y$
14. $(a^2b)(ab^2)$
15. x^0y
16. $\frac{n^4 \cdot n^7}{n^9}$

Write as an expression using only positive exponents.

17. $(5x)^{-1}$
18. $-3n^{-2}$
19. $8^{-2}v^4$
20. $\frac{1}{y^{-4}}$
21. $x^{-8} \cdot x^7$
22. $5^{-1} \cdot y^6 \cdot y^{-3}$
23. $\frac{a^5}{a^{-5}}$
24. $\frac{t^{-2}}{t^4}$
25. $\frac{x^{-2}}{y}$
26. $\frac{x^2}{y^{-1}}$

[15.2] *Simplify.*

27. $(10^2)^4$
28. $-(x^3)^3$
29. $(2x^3)^2$
30. $(-4m^5n)^3$
31. $3(x^{-2})^6$
32. $(a^3b^{-4})^{-2}$
33. $\left(\frac{x}{3}\right)^4$
34. $\left(\frac{-a}{b^3}\right)^2$
35. $\left(\frac{x}{y}\right)^{-6}$
36. $\left(\frac{x^2}{y^{-1}}\right)^5$
37. $\left(\frac{4a^3}{b^4c}\right)^2$
38. $\left(\frac{-u^{-5}v^2}{7w}\right)^2$

Express in standard notation.

39. 3.7×10^{10}
40. 1.63×10^9
41. 5.022×10^{-5}
42. 6×10^{-11}

Express in scientific notation.

43. 1,200,000,000,000

44. 427,000,000

45. 0.00000000000004

46. 0.00000056

Perform the indicated operation. Then, write the result in scientific notation.

47. $(1.4 \times 10^6)(4.2 \times 10^3)$

48. $(3 \times 10^{-2})(2.1 \times 10^5)$

49. $(1.8 \times 10^4) \div (3 \times 10^{-3})$

50. $(9.6 \times 10^{-4}) \div (1.6 \times 10^6)$

[15.3] *Indicate whether the expression is a polynomial.*

51. $3x^4 - 5x^3 + \dfrac{x^2}{4} - 8$

52. $-2x^2 - \dfrac{7}{x} + 1$

Classify each polynomial according to the number of terms.

53. $2x^5 + 7x^2 - 5$

54. $16 - 4t^2$

Write the polynomial in descending order. Then, identify the degree, leading term, and leading coefficient of the polynomial.

55. $8y - 3y^3 + y^2 - 1$

56. $n^4 - 6n^2 - 7n^3 + n$

Simplify. Then, write the polynomial in descending order of powers.

57. $10x - 8x^2 - 8x + 9x^2 - x^3 + 13$

58. $4n^3 - 7n + 9 - 3n^2 - n^3 + 7n^2 - 5 + n$

Evaluate the polynomial for the given values of the variable.

59. $2n^2 - 7n + 3$ for $n = -1$ and $n = 3$

60. $x^3 - 8$ for $x = 2$ and $x = -2$

[15.4] *Perform the indicated operations.*

61. $(4x^2 - x + 4) + (-3x^2 + 9)$

62. $(5y^4 - 2y^3 + 7y - 11) + (6 - 8y - y^2 - 5y^4)$

63. $(a^2 + 5ab + 6b^2) + (3a^2 - 9b^2) + (-7ab - 3a^2)$

64. $(5s^3t - 2st + t^2) + (s^2t - 5t^2) + (t^2 - 4st + 9s^2)$

65. $(x^2 - 5x + 2) - (-x^2 + 3x + 10)$

66. $(10n^3 + n^2 - 4n + 1) - (11n^3 - 2n^2 - 5n + 1)$

67.
$$\begin{array}{r} 5y^4 - 4y^3 \qquad\qquad + y - 6 \\ \underline{-(\qquad\quad y^3 - 2y^2 + 7y - 3)} \end{array}$$

68.
$$\begin{array}{r} -9x^3 + 8x^2 - 11x - 12 \\ \underline{+(11x^3 \qquad\quad - \quad x + 15)} \end{array}$$

Simplify.

69. $14t^2 - (10t^2 - 4t)$

70. $-(5x - 6y) + (3x - 7y)$

71. $(3y^2 - 1) - (y^2 + 3y + 2) + (-2y + 5)$

72. $(1 - 4x - 6x^2) - (7x - 8) - (-11x - x^2)$

[15.5] *Multiply.*

73. $-3x^4 \cdot 2x$

74. $(3ab)(8a^2b^3)(-6b)$

75. $2xy^2(4x - 5y)$

76. $(x^2 - 3x + 1)(-5x^2)$

77. $(n + 3)(n + 7)$

78. $(3x - 9)(x + 6)$

79. $(2x - 1)(4x - 1)$

80. $(3a - b)(3a + 2b)$

81. $(2x^3 - 5x + 2)(x + 3)$

82. $(y - 2)(y^2 - 7y + 1)$

Simplify.

83. $-y + 2y(-3y + 7)$

84. $4x^2(2x - 6) - 3x(3x^2 - 10x + 2)$

[15.6] ***Simplify.***

85. $(a - 1)^2$

86. $(s + 4)^2$

87. $(2x + 5)^2$

88. $(3 - 4t)^2$

89. $(5a - 2b)^2$

90. $(u^2 + v^2)^2$

Multiply.

91. $(m + 4)(m - 4)$

92. $(6 - n)(6 + n)$

93. $(7n - 1)(7n + 1)$

94. $(2x + y)(2x - y)$

95. $(4a - 3b)(4a + 3b)$

96. $x(x + 10)(x - 10)$

97. $-3t^2(4t - 5)^2$

98. $(p^2 - q^2)(p + q)(p - q)$

[15.7] ***Divide.***

99. $12x^4 \div 4x^2$

100. $\dfrac{-20a^3b^5c}{10ab^2}$

101. $(18x^3 - 6x) \div (3x)$

102. $\dfrac{10x^5 + 6x^4 - 4x^3 - 2x^2}{2x^2}$

103. $(3x^2 + 8x - 35) \div (x + 5)$

104. $\dfrac{13 - 5x^2 + 2x^3}{2x - 1}$

Mixed Applications

Solve.

105. The half-life of the element thorium-232 is 13,900,000,000 yr. Express this length of time in scientific notation. (*Source:* Peter J. Nolan, *Fundamentals of College Physics*)

106. Physicists use both the joule (J) and the electron volt (eV) as units of work, where 1 J is equal to 6.24×10^{18} eV. Rewrite this quantity in standard notation. (*Source:* Peter J. Nolan, *Fundamentals of College Physics*)

107. A grain of bee pollen is about 0.00003 m in diameter. Express this length in scientific notation.

108. The area of the United States is approximately 3.7×10^6 mi^2. Express this area in standard notation. (*Source: The New York Times Almanac*, 2011)

109. At a party there are n people present. If everyone shakes hands with everyone else, then the polynomial

$$\frac{n^2}{2} - \frac{n}{2}$$

gives the total number of handshakes. If 9 people are at the party, how many handshakes will there be?

110. An object falling from an altitude of 500 m will be $(-4.9t^2 + 500)$ m above the ground after t sec. What is the altitude of the object 2 sec into the fall?

111. The polynomial $0.02x^3 - 0.71x^2 + 5.95x + 65.08$ can be used to model the temperature (in degrees Fahrenheit) at a particular time in Orlando, Florida on November 26, 2010, where x represents the number of hours past 8:00 A.M. According to this model, find the temperature in Orlando at 11:00 A.M. to the nearest degree. (*Source:* forecast.weather.gov)

112. The polynomial $3.4x^3 - 20.1x^2 + 27.7x + 1683$ approximates the number of two-year colleges in the United States for a particular year, where x represents the number of years after 2005. Use this polynomial to approximate, to the nearest hundred, the number of two-year colleges in 2007. (*Source:* nces.ed.gov)

113. The area of a triangle can be represented by the expression $\frac{1}{2}bh$, where b is the length of the base and h, the height. The surface area of a pyramid is the sum of the area of its square base and the areas of the four (identical) triangular faces. In the pyramid located by the entrance to Paris' Louvre Museum, the triangular faces have a height 8 m less than the side length of its square base. (*Source:* Pei Cobb Fried & Partners)

a. Write a polynomial for the surface area of the Louvre Pyramid in terms of b, the side length of its square base.

b. The side length of the pyramid's base is 35 m. Find the surface area of the pyramid.

114. The polynomial $39x^2 + 4740x + 84{,}000$ approximates the total expenditures (in millions of dollars) on dental services in the U.S. for a particular year, where x represents the number of years after 2004. The polynomial $3x + 300$ approximates the U.S. population (in millions) during the same years. (*Source:* cms.gov)

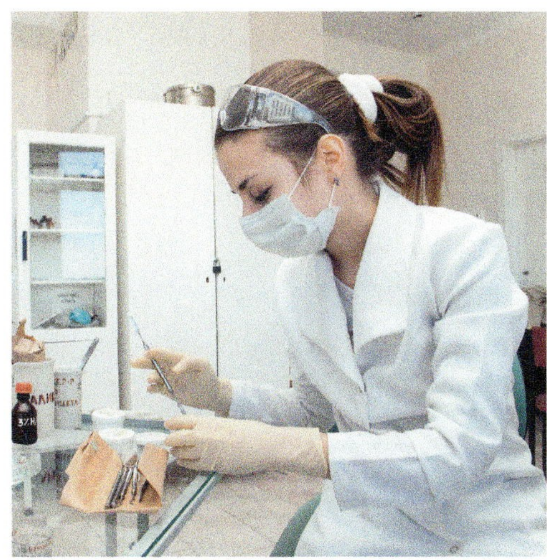

a. Write a polynomial that approximates the total annual expenditures on dental services per person.

b. Using this model, find the amount of expenditures on dental services per person in 2008 to the nearest hundred dollars.

115. The volume of a box is the product of its length, width, and height.

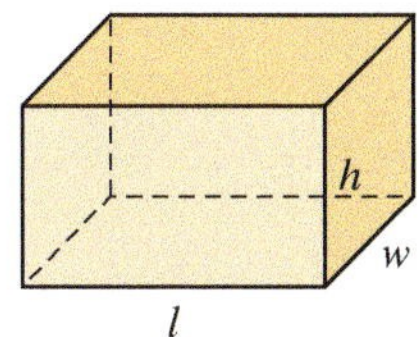

a. Write an expression for the volume (in cubic inches) of a box with length l in., width w in., and height h in.

b. Write an expression for the volume (in cubic inches) of a box with length l ft, width w ft, and height h ft. There are 12 in. in a foot.

116. The cylindrical storage vat pictured below has height r and a base with area πr^2.

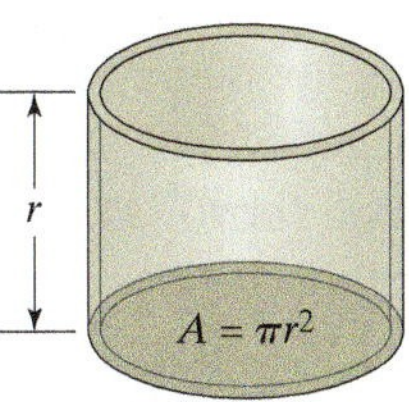

The volume V of the vat is the product of its height and the area of the base. Write a formula for V in terms of r.

• Check your answers on page A-34.

CHAPTER 15 Posttest

To see if you have mastered the topics in this chapter, take this test.

Simplify.

1. $x^6 \cdot x$

2. $n^{10} \div n^4$

3. $7a^{-1}b^0$

4. $(-3x^2y)^3$

5. $\left(\dfrac{x^2}{y^3}\right)^4$

6. $\left(\dfrac{3x^2}{y}\right)^{-3}$

7. For the polynomial $-x^3 + 2x^2 + 9x - 1$, identify:

a. the terms ____________________

b. the coefficients ____________________

c. the degree ____________________

d. the constant term ____________________

8. Find the sum: $(y^2 - 1) + (y^2 - y + 6)$

9. Subtract: $(x^2 - 7x - 4) - (2x^2 - 8x + 5)$

10. Combine: $(4x^2y^2 - 6xy - y^2) - (3x^2 + x^2y^2 - 2y^2) - (x^2 - 6xy + y^2)$

Multiply.

11. $(2mn^2)(5m^2n - 10mn + mn^2)$

12. $(y^3 - 2y^2 + 4)(y - 1)$

13. $(3x - 1)(2x + 7)$

14. $(7 - 2n)(7 + 2n)$

15. $(2m - 3)^2$

Divide.

16. $\dfrac{12s^3 + 15s^2 - 27s}{-3s}$

17. $(3t^3 - 5t^2 - t + 6) \div (3t - 2)$

Solve.

18. Medical X-rays, with a wavelength of about 10^{-10} m, can penetrate the flesh (but not the bones) of your body. Ultraviolet rays, which cause sunburn by penetrating only the top layer of skin, have a wavelength about 1000 times as long as X-rays. Find the length of ultraviolet rays. Write the answer in scientific notation. (*Source:* Peter J. Nolan, *Fundamentals of College Physics*)

19. A real estate broker sells two houses. The first house sells for \$140,000 and is expected to increase in value by \$1500/yr. The second house is purchased for \$90,000 and will likely appreciate by \$800/yr. Write an expression for the value of each house after x yr.

20. A young couple is saving up to purchase a car. They deposit \$1000 in an account that has an annual interest rate of r (in decimal form). At the end of 2 yr, the value of the account will be $1000(1 + r)^2$ dollars. Find the account balance at that time if the interest rate is 3%.

• Check your answers on page A-34.

Cumulative Review Exercises

To help you review, solve the following.

1. Round 208,711 to the nearest ten thousand.

2. Write the prime factorization of 60.

3. Express 3.2 as a mixed number.

4. Solve and check: $\dfrac{2.5}{2} = \dfrac{3.6}{x}$

5. Find the product: $-3.2(-2.3)$

6. Evaluate $3a^2 - 5ab + b^2$ for $a = -3$ and $b = 2$.

7. Simplify: $-4(y - 5) + 2(3y - 1)$

8. Solve $y = mx + b$ for m.

9. Solve: $-7x + 2(3x - 2) \geq 1$

10. Find the slope and y-intercept of the line $2x - 3y = 6$.

11. Graph the inequality: $y < 2x + 1$

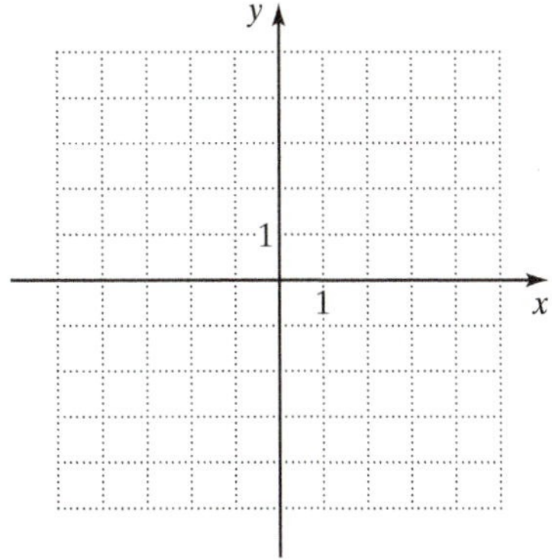

12. Solve by graphing: $\begin{aligned} y &= 1 + 3x \\ x + y &= -3 \end{aligned}$

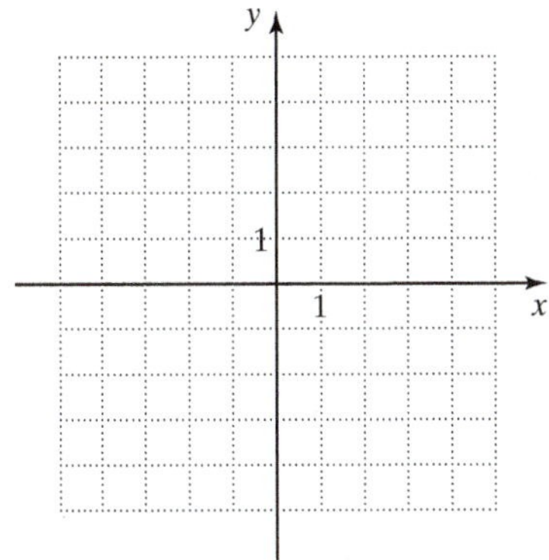

13. Solve by elimination:

$$\begin{aligned} 3x - 2y &= 10 \\ 2x + 3y &= -2 \end{aligned}$$

14. Find the difference: $(3m^2 - 8m + 7) - (2m^2 + 8m - 9)$

15. The Winter Olympics occur every 4 yr.

a. If x represents the first year that they took place, write expressions for the next three Winter Olympic years.

b. If the Winter Olympics were held in 1972, were they also held in 1980? Explain.

16. An executive goes out to dinner and leaves a 20% tip for the service.

a. The bill for the meal without tip is represented by b. Write an expression for the amount of the tip in terms of b.

b. The total cost c of the meal is the original bill plus the tip. Write an equation describing this situation.

17. On a travel website, the cost C of a vacation package is \$300 for airfare plus \$125/day for d days at the hotel.

a. Represent this relationship as an equation.

b. Choose appropriate scales for the axes, and then graph the equation found in part (a).

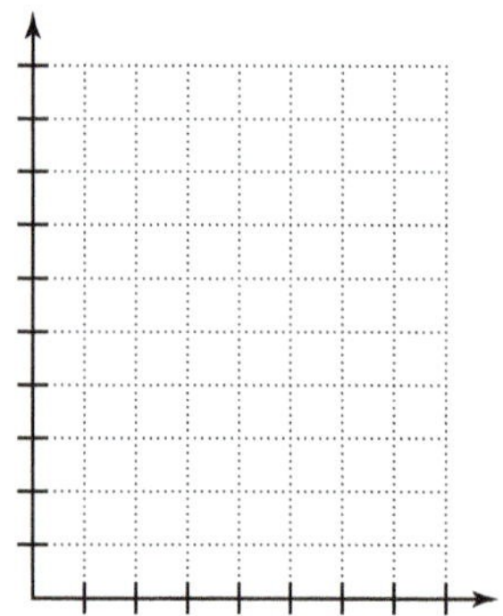

c. Explain the significance of the C-intercept in the context.

18. The cost of a small one-bedroom apartment is \$1500 security deposit and \$1000 rent per month. By contrast, a large studio apartment costs \$1800 security deposit and \$900 rent per month.

a. For each apartment, write a linear equation that expresses the cost of an apartment c (in dollars) in terms of the number of months m the apartment is rented.

b. Use the substitution method to solve the system of linear equations.

c. In the context of this problem, what is the significance of the solution?

19. According to Albert Einstein's famous equation $E = mc^2$, all objects, even resting ones, contain energy E. If the mass m of a raisin is 10^{-3} kg, and the speed c of light is about 3×10^8 m/sec, find the amount of energy the raisin contains. Write the answer in scientific notation.

20. The expression for the surface area of a cylinder is $2\pi rh + 2\pi r^2$, where r is the radius of the base and h is the height of the cylinder. If the radius of a cylindrical gift box is $(2x + 3)$ in. and its height is twice the radius, what is the surface area of the box? Write the answer as a polynomial in descending order of powers of x.

• Check your answers on page A-34.

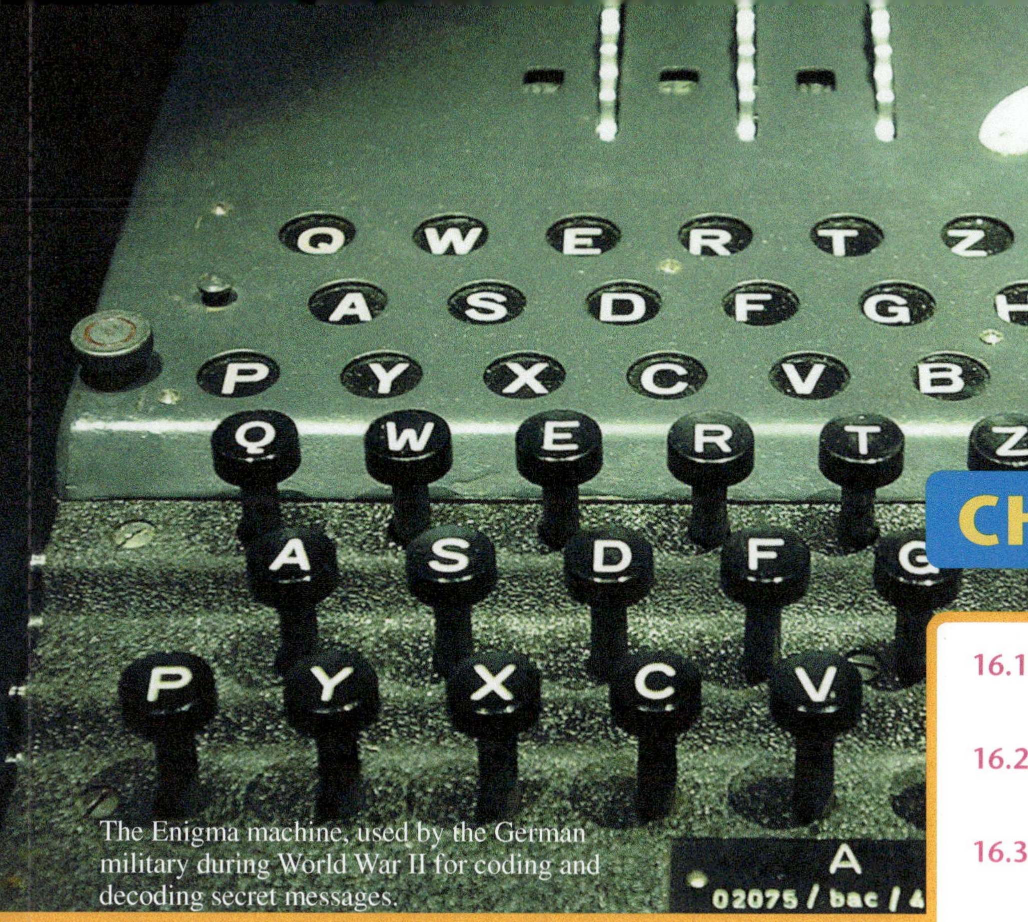

The Enigma machine, used by the German military during World War II for coding and decoding secret messages.

CHAPTER 16

Factoring Polynomials

Factoring and Cryptography

Cryptography, the science of coding and decoding messages, has been important throughout history. Sending secret messages in concealed form is particularly useful during wartime. For instance, the United States entered World War I in part because British intelligence intercepted and deciphered a message sent to the German minister in Mexico; the message called for a German–Mexican alliance against the United States. During World War II, the ability of the Allies to decode Nazi secrets allowed their commanders to eavesdrop on German plans and may have shortened the war by several years.

In peacetime, cryptographic techniques, used in electronic banking and in Web-based credit card purchases, play an increasingly important role in our lives.

Of the numerous cryptographic techniques for coding messages today, one important technique is *prime number encryption.* To crack a message coded in this way depends on finding the prime factors of a given, very large, whole number. Whereas multiplying two prime numbers is easy, reversing the process is difficult and drawn out. For instance, it has been estimated that finding the prime factorization of a five-digit whole number takes some 14 billion mathematical steps.

(*Sources:* Rudolf Kippenhahn, *Code Breaking: A History and Exploration,* The Overlook Press, 1999; F. H. Hinsley and Alan Stripp, Editors, *Codebreakers: The Inside Story of Bletchley Park,* Oxford University Press, 1994)

CHAPTER 16 PRETEST

To see if you have already mastered the topics in this chapter, take this test.

1. Find the greatest common factor of $18ab$ and $36a^4$.

Factor.

2. $4pq + 16p$

3. $10x^2y - 5x^3y^3 + 5xy^2$

4. $3x^2 + 6x + 2x + 4$

5. $n^2 - 11n + 24$

6. $4a + a^2 - 21$

7. $9y - 12y^2 + 3y^3$

8. $5a^2 + 6ab - 8b^2$

9. $-12n^2 + 38n + 14$

10. $4x^2 - 28x + 49$

11. $25n^2 - 9$

12. $x^2y - 4y^3$

13. $y^6 - 9y^3 + 20$

Solve.

14. $n(n - 6) = 0$

15. $3x^2 + x = 2$

16. $(y + 4)(y - 2) = 7$

17. The lateral surface area of a rectangular solid is given by the formula $A = 2lw + 2lh + 2wh$. Solve this formula for h in terms of A, w, and l.

18. A baseball player hits a pop-up fly ball with an initial velocity of 63 ft/sec from a height of 4 ft above the ground. The height of the ball (in feet) t sec after it is hit is given by the expression $-16t^2 + 63t + 4$. Write this expression in factored form.

19. A homeowner wants to fence off part of her yard to build a square play area for her children. Write an expression, in factored form, for the area of the yard *not* covered by the play area.

S ft

15 ft

15 ft

S ft

20. Find the dimensions of the LCD television screen shown if the length is 8 in. longer than the height.

- Check your answers on page A-34.

16.1 Common Factoring and Factoring by Grouping

OBJECTIVES

- **A** To find the greatest common factor (GCF) of two or more integers or terms
- **B** To factor out the greatest common factor from a polynomial
- **C** To factor by grouping
- **D** To solve applied problems involving factoring

What Factoring Is and Why It Is Important

In the previous chapter, we discussed how to *multiply* two factors in order to find their polynomial product.

Multiplying

$$\underbrace{(3x + 4)}_{\text{Factor}}\underbrace{(2x + 1)}_{\text{Factor}} = \underbrace{6x^2 + 11x + 4}_{\text{Product}}$$

In this chapter, we reverse the process, beginning with a polynomial and expressing it as a product of factors. Rewriting a polynomial as a product is called *factoring* the polynomial.

Factoring

$$\underbrace{6x^2 + 11x + 4}_{\text{Polynomial}} = \underbrace{(3x + 4)}_{\text{Factor}}\underbrace{(2x + 1)}_{\text{Factor}}$$

Just as factoring integers plays a key role in arithmetic, factoring polynomials is an important skill in algebra. Factoring polynomials helps us to simplify certain algebraic expressions and also to solve various types of equations.

Finding the Greatest Common Factor of Two or More Integers or Terms

We have already shown that every composite number can be written as the product of prime factors, called its *prime factorization*. For instance, the prime factorization of 15 is $3 \cdot 5$, and the prime factorization of 35 is $5 \cdot 7$. Since 5 appears in both factorizations, 5 is said to be a *common factor* of 15 and 35.

DEFINITION

A **common factor** of two or more integers is an integer that is a factor of each integer.

We can use the concept of common factor to find a *greatest common factor*.

DEFINITION

The **greatest common factor (GCF)** of two or more integers is the greatest integer that is a factor of each integer.

Let's consider an example of finding the greatest common factor of three numbers.

EXAMPLE 1

Find the GCF of 45, 63, and 81.

Solution First, we find the prime factorization of each number.

$$45 = 3 \cdot 3 \cdot 5 = 3^2 \cdot 5$$
$$63 = 3 \cdot 3 \cdot 7 = 3^2 \cdot 7$$
$$81 = 3 \cdot 3 \cdot 3 \cdot 3 = 3^4$$

Then, we look for the greatest common factor. We see that $3 \cdot 3 = 3^2$ is a factor of each number. Since no power of 3 higher than 3^2 and no prime number other than 3 is a factor of *all* three numbers, the GCF of 45, 63, and 81 is 3^2, or 9.

PRACTICE 1

What is the GCF of 24, 72, and 96?

We can extend the concept of greatest common factor to monomials. For instance, to find the greatest common factor of $-21x^2$ and $35xy^2$, we begin by writing the monomials in factored form.

$$-21x^2 = -1 \cdot 3 \cdot 7 \cdot x \cdot x$$
$$35xy^2 = 5 \cdot 7 \cdot x \cdot y \cdot y$$

From the factorizations we see that each monomial has a factor of 7 and a factor of x in common. Note that 7 is the greatest factor of the coefficients and x^1 is the highest power of x that is a factor of each monomial. So the greatest common factor of $-21x^2$ and $35xy^2$ is the product of 7 and x, or $7x$.

EXAMPLE 2

Find the GCF of x^4, x^3, and x^2.

Solution We write each monomial in factored form.

$$x^4 = x \cdot x \cdot x \cdot x$$
$$x^3 = x \cdot x \cdot x$$
$$x^2 = x \cdot x$$

From the factored forms, we see that the monomials have at most two factors of x in common. So the GCF of x^4, x^3, and x^2 is $x \cdot x$, or x^2.

PRACTICE 2

What is the GCF of a^3, a^2, and a?

Note that we could have written the factored forms of the monomials in Example 2 as:

$$x^4 = x^2 \cdot x^2$$
$$x^3 = x^2 \cdot x^1$$
$$x^2 = x^2$$

From these factored forms we see that x^2 is the highest power of the variable factor common to all three monomials. So the greatest common factor of the monomials is the lowest power of x in any of the monomials.

DEFINITION

The **greatest common factor (GCF) of two or more monomials** is the product of the greatest common factor of the coefficients and for each variable, the variable to the lowest power to which it is raised in any of the monomials.

Now, let's consider the greatest common factor of expressions involving more than one variable.

EXAMPLE 3

Identify the GCF of $3a^3b$ and $-15a^2b^4$.

Solution First, we write each monomial in factored form.

$$3a^3b = 3 \cdot a^2 \cdot a \cdot b$$
$$-15a^2b^4 = -1 \cdot 3 \cdot 5 \cdot a^2 \cdot b \cdot b^3$$

The greatest common factor of the coefficients is 3 and the common variable factors to the lowest powers are a^2 and b. So the GCF of $3a^3b$ and $-15a^2b^4$ is $3a^2b$.

PRACTICE 3

Find the GCF of $-18x^3y^4$ and $12xy^2$.

Factoring Out the Greatest Common Factor from a Polynomial

Recall from the previous chapter that we used the distributive property to multiply a monomial by a polynomial. For example, consider the product of $2x$ and $(x + 7)$.

$$2x(x + 7) = 2x \cdot x + 2x \cdot 7 = 2x^2 + 14x$$

When the terms of a polynomial have common factors, we can factor the polynomial by using the distributive property. So we can write $2x^2 + 14x$ in factored form by dividing out $2x$, the GCF of the terms $2x^2$ and $14x$.

$$2x^2 + 14x = 2x \cdot x + 2x \cdot 7 = \underset{\uparrow \text{ GCF}}{2x}(x + 7)$$

So the factored form of $2x^2 + 14x$ is $2x(x + 7)$.

EXAMPLE 4

Factor: $25x^3 + 10x^2$

Solution The GCF of $25x^3$ and $10x^2$ is $5x^2$.

$$25x^3 + 10x^2 = 5x^2(5x) + 5x^2(2) \quad \text{Factor out the GCF } 5x^2 \text{ from each term.}$$
$$= 5x^2(5x + 2) \quad \text{Use the distributive property.}$$

So the factorization of $25x^3 + 10x^2$ is $5x^2(5x + 2)$.

PRACTICE 4

Factor: $10y^2 + 8y^5$

Now, let's consider some examples of factoring polynomials in more than one variable.

EXAMPLE 5

Factor: $12c^2 - 2cd$

Solution The GCF of $12c^2$ and $-2cd$ is $2c$.

$$12c^2 - 2cd = 2c(6c) - 2c(d) \quad \text{Factor out the GCF } 2c \text{ from each term.}$$
$$= 2c(6c - d) \quad \text{Use the distributive property.}$$

PRACTICE 5

Factor: $21a^2b - 14a$

EXAMPLE 6

Factor: $3x^2y^4 + 9xy^2$

Solution

$$3x^2y^4 + 9xy^2 = 3xy^2(xy^2) + 3xy^2(3)$$
$$= 3xy^2(xy^2 + 3)$$

PRACTICE 6

Factor: $8a^2b^2 - 6ab^3$

EXAMPLE 7

Express in factored form: $20y^2 - 5y + 15$

Solution

$$20y^2 - 5y + 15 = 5(4y^2) - 5(y) + 5(3)$$
$$= 5(4y^2 - y + 3)$$

PRACTICE 7

Factor: $24a^2 - 48a + 12$

When solving some literal equations, it may be necessary to factor out common monomial factors.

EXAMPLE 8

Solve $ax + b = cx + d$ for x in terms of a, b, c, and d.

Solution To solve for x, we bring all the terms involving x to the left side of the equation and the other terms to the right side of the equation.

$$ax + b = cx + d$$
$$ax + b - b = cx + d - b \quad \text{Subtract } b \text{ from each side of the equation.}$$
$$ax = cx + d - b$$
$$ax - cx = cx - cx + d - b \quad \text{Subtract } cx \text{ from each side of the equation.}$$
$$ax - cx = d - b$$
$$x(a - c) = d - b \quad \text{Factor out } x \text{ on the left side of the equation.}$$
$$\frac{x(a - c)}{a - c} = \frac{d - b}{a - c} \quad \text{Divide each side of the equation by } (a - c)\text{, the coefficient of } x.$$
$$x = \frac{d - b}{a - c}$$

PRACTICE 8

Solve $ab = s^2 - ac$ for a in terms of b, c, and s.

Factoring by Grouping

Recall that the factorization of $2x^2 + 14x$ is $2x(x + 7)$. The factor $(x + 7)$ is called a *binomial* factor. The distributive property can be used to divide out not only a common monomial factor but also a common binomial factor, if there is one. For instance, let's consider the expression $x(x + 5) + 2(x + 5)$.

$$\underbrace{x(x + 5)}_{\text{First term}} + \underbrace{2(x + 5)}_{\text{Second term}}$$

In this polynomial, the binomial factor $(x + 5)$ is common to both terms. We can factor out $(x + 5)$, getting:

$$x(x + 5) + 2(x + 5) = (x + 5)(x + 2)$$

So the factored form of $x(x + 5) + 2(x + 5)$ is $(x + 5)(x + 2)$.

EXAMPLE 9

Factor: $x(x + 4) - 5(x + 4)$

Solution Factoring out $(x + 4)$, we get:

$$x(x + 4) - 5(x + 4) = (x + 4)(x - 5)$$

PRACTICE 9

Factor: $4(y - 3) + y(y - 3)$

In some algebraic expressions, such as $x(a - 7) + 3(7 - a)$, the binomial factors are opposites. In order to factor out a *common* binomial factor, we must rewrite one of the binomials by factoring out -1, as shown in the next example.

EXAMPLE 10

Factor: $x(a - 7) + 3(7 - a)$

Solution We factor out -1 from the binomial $(7 - a)$, getting $-1(a - 7)$.

$$x(a - 7) + 3(7 - a) = x(a - 7) + 3[-1(a - 7)]$$ Factor out -1 from $(7 - a)$.

$$= x(a - 7) - 3(a - 7)$$ Simplify.

$$= (a - 7)(x - 3)$$ Factor out $(a - 7)$.

PRACTICE 10

Factor: $3y(x - 1) + 2(1 - x)$

EXAMPLE 11

Factor: $5n(4n - 1) - (4n - 1)$

Solution

$$5n(4n - 1) - (4n - 1) = 5n(4n - 1) - 1(4n - 1)$$
$$= (4n - 1)(5n - 1)$$

PRACTICE 11

Factor: $(4 - 3x) + 2x(4 - 3x)$

When trying to factor a polynomial that has four terms, it may be possible to group pairs of terms in such a way that a common binomial factor can be found. This method is called **factoring by grouping**.

EXAMPLE 12

Factor by grouping: $xy - 4x + 3y - 12$

Solution

$$xy - 4x + 3y - 12 = (xy - 4x) + (3y - 12)$$ Group the first two terms and the last two terms.

$$= x(y - 4) + 3(y - 4)$$ Factor out the GCF from each group.

$$= (y - 4)(x + 3)$$ Write in factored form.

PRACTICE 12

Factor: $4b - 20 + ab - 5a$

EXAMPLE 13

Express in factored form: $6h - 6k - h^2 + hk$

Solution

$$6h - 6k - h^2 + hk = (6h - 6k) + (-h^2 + hk)$$ Group the first two terms and the last two terms.

$$= (6h - 6k) - (h^2 - hk)$$ Factor out -1 in the second group.

$$= 6(h - k) - h(h - k)$$ Factor out the GCF from each group.

$$= (h - k)(6 - h)$$ Write in factored form.

PRACTICE 13

Factor: $5y - 5z - y^2 + yz$

EXAMPLE 14

Each week, a sales associate receives a salary of d dollars as well as 5% commission on the value of the sales that she makes. Last week, sales amounted to x dollars, and this week, sales rose to y dollars. How much greater was the sales associate's total income this week than last week? Express this amount in factored form.

Solution Last week, the sales associate made d dollars in salary and $0.05x$ in commission. This week, the associate made d dollars in salary and $0.05y$ in commission. So the difference in total income is:

$$(d + 0.05y) - (d + 0.05x) = d + 0.05y - d - 0.05x$$
$$= 0.05y - 0.05x$$
$$= 0.05(y - x)$$

So the sales associate made $0.05(y - x)$ dollars more this week than last week.

PRACTICE 14

The distance an object under constant acceleration travels in time t is given by the expression $v_0t + \frac{1}{2}at^2$, where v_0 is the object's initial velocity and a is its acceleration. Factor this expression.

16.1 Exercises

FOR EXTRA HELP MyMathLab MathXL PRACTICE WATCH READ REVIEW

Mathematically Speaking

Fill in each blank with the most appropriate term or phrase from the given list.

sum	greatest common factor (GCF)	factoring
greatest factor	common factor	product
multiplying		

1. Rewriting a polynomial as a product is called ______________ the polynomial.
2. A(n) ______________ of two or more integers is an integer that is a factor of each integer.
3. The ______________ of two or more integers is the greatest integer that is a factor of each integer.
4. The greatest common factor of two or more monomials is the ______________ of the greatest common factor of the coefficients and, for each variable, the variable to the lowest power to which it is raised in any of the monomials.

A *Find the greatest common factor of each group of terms.*

5. 27, 54, and 81
6. 28, 35, 63
7. x^4, x^6, x^3
8. y^2, y, y^5
9. $16b, 8b^3, 12b^2$
10. $3a, 7a^2, 5a^4$
11. $-12x^5y^7, 4y^3$
12. $9m^3, 6m^2n$
13. $18a^5b^4, -6a^4b^3, 9a^2b^2$
14. $24mn, 32mn^2, 16m^2n$
15. $x(3x - 1)$ and $8(3x - 1)$
16. $6(5n + 2)$ and $n(5n + 2)$
17. $4x(x + 7)$ and $9x(x + 7)$
18. $y(y - 4)$ and $6y(y - 4)$

B *Factor out the greatest common factor.*

19. $3x + 6$
20. $10y + 15$
21. $24x^2 + 8$
22. $30y^2 - 6$
23. $27m - 9n$
24. $16r - 8t$
25. $7x^2 - 2x$
26. $3b^2 - 18b$
27. $5b^2 - 6b^3$
28. $4z^5 - 12z^2$
29. $10x^3 - 15x$
30. $12a^2 - 18a$
31. $a^2b^2 + ab$
32. $xy + x^2y^2$
33. $6xy^2 + 7x^2y$
34. $3p^3q^2 + 5p^2q^3$
35. $27pq^2 - 18p^2q$
36. $45c^2d - 15cd^2$
37. $2x^3y + 12x^3y^4$
38. $7a^2b^3 + 9a^4b^3$
39. $3c^3 + 6c^2 + 12$
40. $5y^2 - 20y + 10$
41. $9b^4 - 3b^3 + b^2$
42. $8y^5 - y^4 - 4y^2$
43. $2m^4 + 10m^3 - 6m^2$
44. $3x^3 - 9x^2 - 27x$
45. $5b^5 - 3b^3 + 2b^2$
46. $9c^4 + c^3 + 6c^2$
47. $15x^4 - 10x^3 - 25x$
48. $12m^6 + 9m^5 + 15m^3$
49. $4a^2b + 8a^2b^2 - 12ab$
50. $5m^2n - 15mn^2 + 10mn$
51. $9c^2d^2 + 12c^3d + 3cd^3$
52. $18x^2y^4 - 24xy^3 + 30x^3y$

C *Factor by grouping.*

53. $x(x-1)+3(x-1)$
54. $2(n+4)+n(n+4)$
55. $5a(a-1)-3(a-1)$
56. $4x(x+3)-7(x+3)$
57. $r(s+7)-2(7+s)$
58. $a(6+b)-7(b+6)$
59. $a(x-y)-b(x-y)$
60. $y(a-z)-x(a-z)$
61. $3x(y+2)-(y+2)$
62. $(n-1)-2m(n-1)$
63. $b(b-1)+5(1-b)$
64. $x(x-3)+2(3-x)$
65. $y(y-1)-5(1-y)$
66. $n(n-9)-4(9-n)$
67. $(t-3)-t(3-t)$
68. $w(w-4)+(4-w)$
69. $9a(b-7)+2(7-b)$
70. $2y(x-2)+3(2-x)$
71. $rs+3s+rt+3t$
72. $mn+2m+np+2p$
73. $xy+6y-4x-24$
74. $ab-5b-2a+10$
75. $15xy-9yz+20xz-12z^2$
76. $6ab+12ac-5bc-10c^2$
77. $2xz+8x+5yz+20y$
78. $3ab+9a+4bc+12c$

Solve for the indicated variable.

79. $TM = PC + PL$ for P
80. $S = a + Nd - d$ for d
81. $S = 2lw + 2lh + 2wh$ for l
82. $S = a + ar^n$ for a

Mixed Practice

Factor out the greatest common factor.

83. $16p^3 + 24p$
84. $4u^4 - 28u^2 + 36u$
85. $48rs^2 - 60r^2s$
86. $7m^3 - 4m^2$
87. $42j^2 - 6$

Solve.

88. $A = \frac{1}{2}hb_1 + \frac{1}{2}hb_2$ for h

Factor by grouping.

89. $st - 3t - 7s + 21$
90. $7b(b+2)-5(b+2)$
91. $3x(y-4)+5(4-y)$
92. $2bc + 8ab - 3ac - 12a^2$

Find the greatest common factor of each group of terms.

93. $16a^5b^3, -12a^2b^3, 20a^3b$
94. $14x^4, 21xy^3$

Applications

D *Solve.*

95. When an object with mass m increases in velocity from v_1 to v_2, its momentum increases by $mv_2 - mv_1$. Factor this expression.

96. One item sells for p dollars and another for q dollars. In addition, an 8% sales tax is charged on all items sold. An expression for the total selling price is $1.08p + 1.08q$. Write this expression in factored form.

97. In a meeting of diplomats, all diplomats must shake hands with one another. If there are n diplomats, the expression $0.5n^2 - 0.5n$ represents the total number of handshakes at the meeting. Factor this expression.

98. For an investment earning simple interest, the future value of the investment is represented by the expression $P + Prt$, where P is the present value of the investment, r is the annual interest rate, and t is the time in years. Factor this expression.

99. In a polygon with n sides, the number of diagonals is given by the expression $\frac{1}{2}n^2 - \frac{3}{2}n$. Write this expression in factored form.

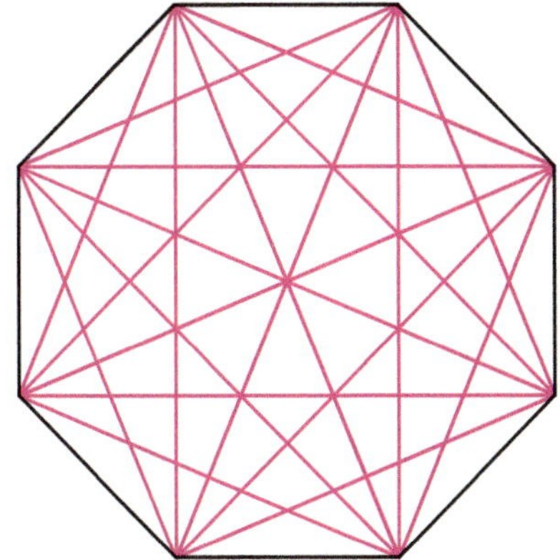

100. In a polygon with n sides, the interior angles (measured in degrees) add up to $180n - 360$. Find an equivalent expression by factoring out the GCF.

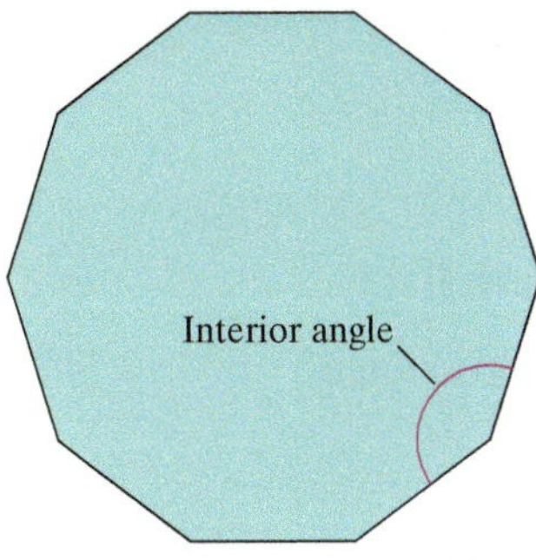

101. Consider the formula $P = nC + nT + D$, where P is the total price of a purchase, n is the number of items purchased, C is the cost per item, T is the tax on each item, and D is the total delivery charge. Solve for n in terms of P, C, T, and D.

102. The *harmonic mean* H of two numbers x and y is a kind of average. Solve for H in the formula $Hx + Hy = 2xy$.

103. For more than 30 yr, *Wheel of Fortune* has been a popular show on syndicated television. In one game, a player earned \$700 for each of n copies of the letter N and \$900 for each of s copies of the letter S, spending \$250 for each of e copies of the vowel E. Find in factored form the amount that the player won. (*Source: Wheel of Fortune*)

104. Many companies pay their hourly employees "time-and-a-half" during overtime hours, that is, during hours worked more than 40 per week. The regular hourly rate for employees is h dollars. During a particular week, x employees work 40 hours and y employees work 50 hours. What is the total cost of the company payroll for that week, expressed in factored form?

• Check your answers on page A-34.

MINDStretchers

Mathematical Reasoning

1. A four-digit whole number can be represented by the expression

$$1000d + 100c + 10b + a$$

where a is the digit in the units place, b is the digit in the tens place, c is the digit in the hundreds place, and d is the digit in the thousands place.

a. Consider several four-digit whole numbers (for instance, 8351). Then, for each of these numbers, reverse the order of the digits, forming a second number $1000a + 100b + 10c + d$ (here, 1538). Subtract the smaller number from the larger $(8351 - 1538 = 6813)$. Then, check whether this difference is divisible by 9.

b. Show that when you reverse the order of the digits of *any* four-digit whole number and subtract the two four-digit numbers, their difference must be divisible by 9.

Critical Thinking

2. Factor the expression $a^{n+2}b^n - a^nb^{n+1}$.

Groupwork

3. Working with a partner, for each polynomial list three numbers or monomials that when placed in the ☐ will make the polynomial factorable.

a. $2xy - 7x + \square - 14$

b. $xy^2 + \square + 3y^2 - 48$

16.2 Factoring Trinomials Whose Leading Coefficient Is 1

OBJECTIVES

A To factor trinomials of the form $ax^2 + bx + c$, where $a = 1$

B To solve applied problems involving factoring

In this section, we move on to another kind of factoring—factoring trinomials of the form $ax^2 + bx + c$, where $a = 1$. Recall that the coefficient a of the leading term ax^2 is called the *leading* coefficient. In other words, we are examining trinomials of the form $x^2 + bx + c$, where the leading coefficient is 1. First, we factor trinomials in which the constant term is positive, such as:

$$x^2 + 5x + \underset{\uparrow}{6} \qquad \text{and} \qquad x^2 - 8x + \underset{\uparrow}{16}$$

Constant term

Next, we factor trinomials in which the constant term is negative, for example:

$$x^2 + 3x - \underset{\uparrow}{10} \qquad \text{and} \qquad x^2 - 5x - \underset{\uparrow}{24}$$

Constant term

In Section 16.5, we will see how factoring trinomials helps us to solve related equations and applied problems.

Factoring $x^2 + bx + c$, for c Positive

Recall from Section 15.5 the FOIL method of multiplying two binomials.

$$\begin{aligned}(x + 2)(x + 3) &= \overset{F}{x^2} + \overset{O}{3x} + \overset{I}{2x} + \overset{L}{6} \\ &= x^2 + 5x + 6\end{aligned}$$

Note that when the leading coefficient of both binomials is 1, their product also has leading coefficient 1. This suggests that factoring a trinomial of the form $x^2 + bx + c$ gives the product of two binomials of the form $(x + ?)(x + ?)$, where each question mark represents an integer. To find the binomials, we apply the FOIL method in reverse.

For instance, let's factor $x^2 + 5x + 6$. Since the leading term of the trinomial is x^2, we apply the FOIL method to two binomial factors, placing x as the first term in each of the factors.

$$x^2 + 5x + 6 = (x + ?)(x + ?) = \overset{\substack{F\\\downarrow}}{x^2} + \overset{\substack{O\ I\\\downarrow}}{5x} + \overset{\substack{L\\\downarrow}}{6}$$

We see that the product of the constant terms of the binomial factors must be 6, the constant term of the trinomial. The sum of the outer and inner products must be $5x$. So we need to find two integers whose product is 6 and whose sum is 5.

To find these integers, consider all possible factors of 6, that is, of $+6$.

Factors of 6	Possible Binomial Factors	Sum of Outer and Inner Products
1, 6	$(x + 1)(x + 6)$	$6x + x = 7x$
−1, −6	$(x - 1)(x - 6)$	$-6x - x = -7x$
2, 3	$(x + 2)(x + 3)$	$3x + 2x = 5x$ ⟵ The correct middle term
−2, −3	$(x - 2)(x - 3)$	$-3x - 2x = -5x$

So $x^2 + 5x + 6 = (x + 2)(x + 3)$.

We check that the factors are correct by multiplying.

$$(x + 2)(x + 3) = x^2 + \underbrace{3x + 2x}_{\text{Sum of the outer and inner products}} + 6 = x^2 + \underbrace{5x}_{\text{Middle term}} + 6$$

This way of factoring a trinomial, listing all the possibilities, is sometimes called the *trial-and-error method.* Factoring a trinomial by the trial-and-error method involves recognizing patterns, looking for clues, and, once the factorization is found, multiplying to check. Note that by the commutative property of multiplication, we can also write the product $(x + 2)(x + 3)$ as $(x + 3)(x + 2)$.

EXAMPLE 1

Factor: $y^2 + 7y + 10$

Solution Applying the FOIL method in reverse, we know that the first term of each factor is y. So the factors are of the form $(y + ?)(y + ?)$. Using the trial-and-error method, we need to find two integers whose product is 10 and whose sum is 7. The following table shows how to test pairs of factors of 10, the constant term of the trinomial.

Factors of 10	Sum of Factors
1, 10	11
−1, −10	−11
2, 5	7 ⟵ Factors 2 and 5 have a sum of 7.
−2, −5	−7

So $y^2 + 7y + 10 = (y + 2)(y + 5)$, or $(y + 5)(y + 2)$.

Same sign

PRACTICE 1

Factor: $x^2 + 5x + 4$

Can you explain why in the preceding list of possible factors we need only have tested the positive factors of the constant term?

EXAMPLE 2

Factor: $x^2 - 10x + 16$

Solution The constant term of the trinomial is positive, and the coefficient of the x-term is negative. So the constant terms of the binomial factors must both be negative.

$$x^2 - 10x + 16 = (x - ?)(x - ?)$$

We need to find two negative integers whose product is 16 and whose sum is -10. In this case, we need to consider only negative factors of 16.

Factors of 16	Sum of Factors
$-1, -16$	-17
$-2, -8$	-10 ⟵ Factors -2 and -8 have a sum of -10.
$-4, -4$	-8

So $x^2 - 10x + 16 = (x - 2)(x - 8)$, or $(x - 8)(x - 2)$.

PRACTICE 2

Factor: $y^2 - 9y + 20$

TIP When the constant term c of the trinomial $x^2 + bx + c$ is positive,

- the constant terms of the binomial factors are both positive when b, the coefficient of the x-term in the trinomial, is positive, and
- the constant terms are both negative when b is negative.

Not every trinomial can be expressed as the product of binomial factors with coefficients that are integers. Polynomials that are not factorable are called **prime polynomials**. For instance, the trinomial $x^2 + 5x + 1$ is a prime polynomial. Can you explain why this trinomial is prime?

EXAMPLE 3

Factor: $x^2 + 4x + 6$

Solution Both the constant term and the coefficient of the x-term are positive. So if the trinomial is factorable, the constant terms of its binomial factors must both be positive as well.

$$x^2 + 4x + 6 = (x + ?)(x + ?)$$

We need to find two positive integers whose product is 6 and whose sum is 4.

Factors of 6	Sum of Factors
1, 6	7
2, 3	5

Since neither sum of the factors yields the correct coefficient of the x-term, we can conclude that the polynomial is not factorable. Therefore, this is a *prime polynomial.*

PRACTICE 3

Factor: $x^2 + 3x + 5$

When the terms of a trinomial are not in descending order, we usually rewrite the trinomial before factoring, as shown in the next example.

EXAMPLE 4

Factor: $12 - 8x + x^2$

Solution We begin by writing the terms in descending order.

$$12 - 8x + x^2 = x^2 - 8x + 12$$
$$= (x - ?)(x - ?)$$

Since the middle term of the trinomial is negative and the constant term is positive, we are looking for two negative integers. So we need to consider only negative factors of 12.

Factors of 12	Sum of Factors
−1, −12	−13
−2, −6	−8 ⟵ Factors −2 and −6 have a sum of −8.
−3, −4	−7

So $x^2 - 8x + 12 = (x - 2)(x - 6)$.

PRACTICE 4

Factor: $32 - 12y + y^2$

Note that in Example 4, we could also have factored $12 - 8x + x^2$ as $(2 - x)(6 - x)$, without rearranging the terms of the trinomial. Can you explain why the two solutions $(x - 2)(x - 6)$ and $(2 - x)(6 - x)$ are equal?

Now, let's consider a trinomial of the form $x^2 + bxy + cy^2$. Note that this trinomial contains more than one variable. When factoring these trinomials, we use the trial-and-error method where the binomial factors are of the form $(x + ?y)(x + ?y)$ and the question marks represent factors of c, the coefficient of the y^2 term, whose sum is b, the coefficient of the xy-term.

EXAMPLE 5

Factor: $x^2 + 3xy + 2y^2$

Solution $x^2 + 3xy + 2y^2 = (x + ?y)(x + ?y)$

Since the middle term and the last term are both positive, we look for only positive factors of 2, the coefficient of y^2, whose sum is 3, the coefficient of the xy-term.

Factors of 2	Sum of Factors
1, 2	3

So $x^2 + 3xy + 2y^2 = (x + 1y)(x + 2y) = (x + y)(x + 2y)$, or $(x + 2y)(x + y)$.

PRACTICE 5

Factor: $p^2 - 4pq + 3q^2$

Factoring $x^2 + bx + c$, for c Negative

Now, let's consider how to factor trinomials in which the coefficient of the first term is 1 and the sign of the constant term is negative.

EXAMPLE 6

Factor: $x^2 + 2x - 3$

Solution The goal is to find factors of -3 whose sum is 2.

Factors of -3	Sum of Factors	
$-1, 3$	2	⟵ Factors -1 and 3 have a sum of 2.
$1, -3$	-2	

So $x^2 + 2x - 3 = (x - 1)(x + 3)$.

Check We check by multiplying.

$$(x - 1)(x + 3) = x^2 + \underbrace{3x - x}_{\text{Sum of the outer and inner products}} - 3 = x^2 + \underbrace{2x}_{\text{Middle term}} - 3$$

PRACTICE 6

Factor: $x^2 + x - 6$

TIP When the constant term c of the trinomial $x^2 + bx + c$ is negative, the constant terms of the binomial factors have opposite signs.

EXAMPLE 7

Factor: $y^2 - 3y - 10$

Solution We must find two integers whose product is -10 and whose sum is -3. Since the constant term is negative, one of its factors must be positive and the other negative.

Factors of -10	Sum of Factors	
$1, -10$	-9	
$-1, 10$	9	
$2, -5$	-3	⟵ Factors 2 and -5 have a sum of -3.
$-2, 5$	3	

So $y^2 - 3y - 10 = (y + 2)(y - 5)$.

Note that since the sum of the two factors of -10 is negative, the negative factor must have a larger absolute value than the positive factor.

PRACTICE 7

Factor: $x^2 - 21x - 46$

EXAMPLE 8

Factor: $x^2 - 12 + x$

Solution

$x^2 - 12 + x = x^2 + x - 12$ — Write the terms in descending order.

$= (x + ?)(x - ?)$ — Since the constant term is negative, one of its factors is positive and the other factor is negative.

Now, we must find two factors of -12 whose sum is 1. So the positive factor must have a larger absolute value than the negative factor. Thus, we consider only factors of -12 for which the positive factor has the larger absolute value.

Factors of -12	Sum of Factors
$-1, 12$	11
$-2, 6$	4
$-3, 4$	1 ⟵ Factors -3 and 4 have a sum of 1.

So $x^2 + x - 12 = (x - 3)(x + 4)$.

PRACTICE 8

Factor: $y^2 - 24 + 2y$

Next, let's consider factoring a trinomial in two variables.

EXAMPLE 9

Factor: $x^2 - 5xy - 14y^2$

Solution $x^2 - 5xy - 14y^2 = (x + ?y)(x - ?y)$

We must find two factors of -14 whose sum is -5. Since the product of these factors is a negative number, one of the factors must be positive and the other negative. Since the sum of the factors is negative, the negative factor must have a larger absolute value than the positive factor. Thus, we consider only factors of -14 for which the negative factor has the larger absolute value.

Factors of -14	Sum of Factors
$1, -14$	-13
$2, -7$	-5 ⟵ Factors 2 and -7 have a sum of -5.

So $x^2 - 5xy - 14y^2 = (x + 2y)(x - 7y)$.

PRACTICE 9

Factor: $a^2 - 5ab - 24b^2$

Some trinomials have a common factor. When factoring such a trinomial, we factor out the GCF before trying to factor the trinomial into the product of two binomials, as shown in the next example.

EXAMPLE 10

Factor: $3x^2 + 6x - 24$

Solution Each term of the trinomial has a common factor of 3, so we begin by factoring out this factor.

$$3x^2 + 6x - 24 = 3(x^2 + 2x - 8)$$
$$= 3(x + ?)(x - ?)$$

Since the product of the two missing integers is negative, we need to consider only factors of -8, one positive and the other negative, where the positive factor has a larger absolute value than the negative factor.

Factors of -8	Sum of Factors
$-1, 8$	7
$-2, 4$	2 ⟵ Factors -2 and 4 have a sum of 2.

So $3x^2 + 6x - 24 = 3(x^2 + 2x - 8)$
$= 3(x - 2)(x + 4)$.

Note that after factoring out the GCF 3, neither of the remaining factors of the polynomial has a common factor.

PRACTICE 10

Factor: $y^3 - 9y^2 - 10y$

EXAMPLE 11

Express in factored form: $3a^4 - 21a^3 - 24a^2$

Solution

$3a^4 - 21a^3 - 24a^2 = 3a^2(a^2 - 7a - 8)$ Factor out the GCF $3a^2$ from each term.

$= 3a^2(a + 1)(a - 8)$ Factor $a^2 - 7a - 8$.

So $3a^4 - 21a^3 - 24a^2 = 3a^2(a + 1)(a - 8)$.

PRACTICE 11

Factor: $8x^3 + 8x^2 - 16x$

As in the previous examples, after factoring out the GCF the remaining trinomial can sometimes still be factored. A polynomial is **factored completely** when it is expressed as the product of a monomial and one or more prime polynomials. Throughout the remainder of this text, *factor* means to factor completely.

EXAMPLE 12

Factor: $-x^2 + 9x - 14$

Solution

$-x^2 + 9x - 14 = -1(x^2 - 9x + 14)$ Factor out -1 so that the leading coefficient is 1.

$= -1(x - 7)(x - 2)$
$= -(x - 7)(x - 2)$

PRACTICE 12

Write in factored form:
$-x^2 - 10x + 11$

EXAMPLE 13

Show that for any whole number n, the number represented by $n^2 + 7n + 12$ can be expressed as the product of two consecutive whole numbers.

Solution Since we want to represent the expression $n^2 + 7n + 12$ as a product, we factor it.

$$n^2 + 7n + 12 = (n + 3)(n + 4)$$

Note that the two factors, $n + 3$ and $n + 4$, are whole numbers that differ by 1.

$$(n + 4) - (n + 3) = n + 4 - n - 3 = 1$$

So $n^2 + 7n + 12$ can be expressed as the product of two consecutive whole numbers.

PRACTICE 13

A ball is tossed upward with a velocity of 32 ft/sec from a roof 48 ft above the ground. The expression $-16t^2 + 32t + 48$ approximates the height of the ball above the ground in feet after t sec. Factor this expression.

16.2 Exercises

FOR EXTRA HELP MyMathLab MathXL PRACTICE WATCH READ REVIEW

Mathematically Speaking

Fill in each blank with the most appropriate term or phrase from the given list.

are both negative	not factorable	opposite signs
a factor greater than 0	are both positive	ascending
factorable	descending	the same sign
	a common factor	

1. Polynomials that are ________________ are called prime polynomials.

2. Polynomials to be factored are generally written in ________________ order.

3. If $c < 0$ in the trinomial $x^2 + bx + c$, then the constant terms of the binomial factors have ________________.

4. If each term in a trinomial has ________________, factor out the GCF before trying to factor the trinomial into the product of two binomials.

A *Match each trinomial with its binomial factors.*

5. $x^2 - 16x + 28$ **a.** $(x - 1)(x + 28)$

6. $x^2 + 12x - 28$ **b.** $(x - 7)(x + 4)$

7. $x^2 + 29x + 28$ **c.** $(x - 4)(x - 7)$

8. $x^2 + 27x - 28$ **d.** $(x - 2)(x + 14)$

9. $x^2 - 3x - 28$ **e.** $(x + 28)(x + 1)$

10. $x^2 - 11x + 28$ **f.** $(x - 14)(x - 2)$

Find the missing factor.

11. $x^2 - 3x - 10 = (x + 2)(\quad)$

12. $x^2 - 11x + 18 = (x - 9)(\quad)$

13. $x^2 + 5x + 4 = (x + 1)(\quad)$

14. $x^2 - x - 12 = (x - 4)(\quad)$

15. $x^2 + 5x - 6 = (x + 6)(\quad)$

16. $x^2 + 4x - 21 = (x - 3)(\quad)$

Factor, if possible.

17. $x^2 + 6x + 8$

18. $x^2 + 9x + 8$

19. $x^2 + 5x - 6$

20. $x^2 + 2x - 3$

21. $x^2 + x + 2$

22. $x^2 + 13x + 7$

23. $x^2 + 2x - 8$

24. $x^2 + 4x - 45$

25. $x^2 - 4x + 3$

26. $x^2 - 6x + 5$

27. $y^2 - 12y + 32$

28. $m^2 - 5m + 4$

29. $t^2 - 4t - 5$

30. $s^2 - 4s - 12$

31. $x^2 + 2x - 1$

32. $x^2 + x + 7$

33. $y^2 - 9y + 20$

34. $a^2 - 7a + 10$

35. $y^2 - 13y - 48$

36. $x^2 - 3x - 18$

37. $b^2 + 11b + 28$

38. $p^2 + 13p + 22$

39. $x^2 - 14x + 49$

40. $x^2 - 12x + 36$

41. $-y^2 + 5y + 50$
42. $-x^2 + 5x + 84$
43. $x^2 + 64 - 16x$
44. $a^2 + 20 - 12a$
45. $16 - 10x + x^2$
46. $20 + b^2 - 21b$
47. $81 - 30w + w^2$
48. $y^2 + 72 - 17y$
49. $p^2 - 8pq + 7q^2$
50. $s^2 - 10st + 25t^2$
51. $p^2 - 4pq - 5q^2$
52. $x^2 - 4xy - 12y^2$
53. $m^2 - 12mn + 35n^2$
54. $r^2 - rs - 30s^2$
55. $x^2 + 9xy + 8y^2$
56. $a^2 + 12ab + 27b^2$
57. $5x^2 - 5x - 30$
58. $4y^2 + 12y - 40$
59. $2x^2 + 10x - 28$
60. $8r^2 - 56r - 64$
61. $12 - 18t + 6t^2$
62. $8 - 10x + 2x^2$
63. $3x^2 + 24 + 18x$
64. $5z^2 - 15 - 10z$
65. $y^3 + 3y^2 - 10y$
66. $x^3 - 6x^2 - 7x$
67. $a^3 + 8a^2 + 15a$
68. $q^3 - q^2 - 42q$
69. $t^4 - 14t^3 + 24t^2$
70. $x^4 + 6x^3 - 27x^2$
71. $4a^3 - 12a^2 + 8a$
72. $3y^3 - 18y^2 + 24y$
73. $2x^3 + 30x + 16x^2$
74. $5b^3 + 10b + 15b^2$
75. $4x^3 + 48x - 28x^2$
76. $3r^3 - 42r + 15r^2$
77. $-56s + 6s^2 + 2s^3$
78. $-20y - 16y^2 + 4y^3$
79. $2c^4 + 4c^3 - 70c^2$
80. $4t^4 + 24t^3 - 64t^2$
81. $ax^3 - 18ax^2 + 32ax$
82. $b^2y^3 - 5b^2y^2 - 36b^2y$

Mixed Practice

Factor, if possible.

83. $x^2 + 6x + 3$
84. $x^2 - xy - 12y^2$
85. $5x^4 - 15x^3 - 50x^2$
86. $3z^3 + 24z^2 + 36z$
87. $-w^2 + 6w + 40$
88. $b^2 - 6b + 5$
89. $6m^2 - 6m - 36$
90. $s^4 - 17s^3 + 72s^2$
91. $t^2 + 60 - 17t$
92. $n^2 - 11n + 18$

Solve.

93. Choose the correct binomial factors of the trinomial $x^2 - 5x - 24$.
    a. $(x + 8)$ and $(x - 3)$
    b. $(x + 3)$ and $(x - 8)$
    c. $(x - 6)$ and $(x + 4)$
    d. $(x - 4)$ and $(x + 6)$

94. Find the missing factor.
    $x^2 + 5x - 36 = (x - 4)(\quad)$

Applications

B *Solve.*

95. Scientists who study genetics use the equation $p^2 + 2pq + q^2 = 1$, where p represents a certain dominant gene and q represents a recessive gene. Rewrite the equation so that the left side is factored.

96. Show that for any whole number n, the number represented by $n^2 + 11n + 30$ can be expressed as the product of two consecutive whole numbers.

97. A child throws a stone downward with an initial velocity of 48 ft/sec from a height of 160 ft. One step in figuring out how long it takes for the stone to reach the ground is to factor the expression $16t^2 + 48t - 160$. What is its factorization?

98. A statistician found that the cost (in dollars) for a company to produce x units of a certain product can be approximated by $C = x^2 - 14x + 45$. Factor the expression on the right side of this equation.

99. According to specifications, a box manufacturer makes a closed box with a length and width that are 3 in. longer than the height. Let x equal the height of the box.

a. Find the surface area of the box.

b. Factor this polynomial.

100. For any whole number n, show that the number represented by $4n^2 + 20n + 24$ can be expressed as the product of two consecutive even numbers.

101. The Meteor Crater in Arizona is approximately circular, as shown in the photograph below.

The depth (in feet) of the crater below any point on a diameter can be modeled by the expression $\frac{57}{400{,}000}(x^2 - 2000x - 3{,}000{,}000)$, where x is the distance to the point on the diameter 1000 ft in from the edge. Write this expression in factored form. (*Source:* barringercrater.com)

102. The first field goal longer than 60 yd in National Football League history occurred on November 8, 1970. In one possible path the ball could have traveled, the height of the ball (in feet) is modeled by the polynomial $\left(-\frac{1}{18}\right)(x^2 + 57x - 180)$, where x is the ground distance (in yards) of the ball to the goal post. Factor this expression. (*Source:* nfl.com)

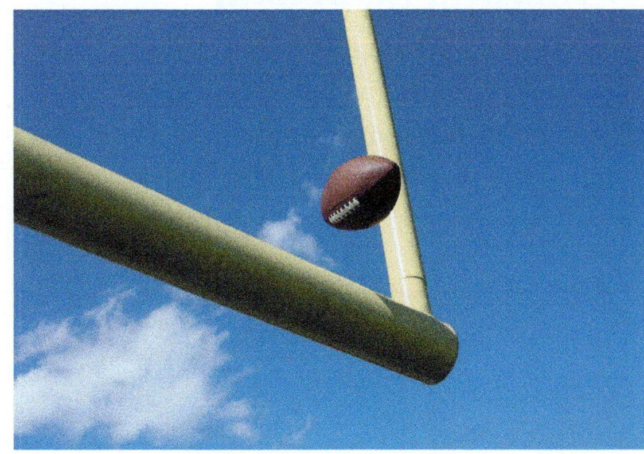

• Check your answers on page A-35.

MINDStretchers

Groupwork

1. Work with a partner. Next to each trinomial, list at least two integers that when inserted in the box will make the trinomial factorable.

   a. $x^2 + \square x + 60$

   b. $x^2 - x + \square$

Investigation

2. Algebra tiles give a physical representation of algebraic concepts. Below, the green tile (x by x) represents a square which models x^2, the square-term of the polynomial $x^2 + 5x + 6$. Each blue tile (x by 1) represents $1x$, or x. So the five blue tiles model $5x$, the linear term of the polynomial. Each orange tile (1 by 1) represents 1. Therefore, the six orange squares model the constant term 6. These tiles can be placed and moved according to the rules of algebra. Copy or trace the algebra tiles below onto a piece of paper. Then, cut out each tile separately. Finally, position all the tiles like a jigsaw puzzle so as to form a rectangle. (*Hint:* Factor the polynomial $x^2 + 5x + 6$.)

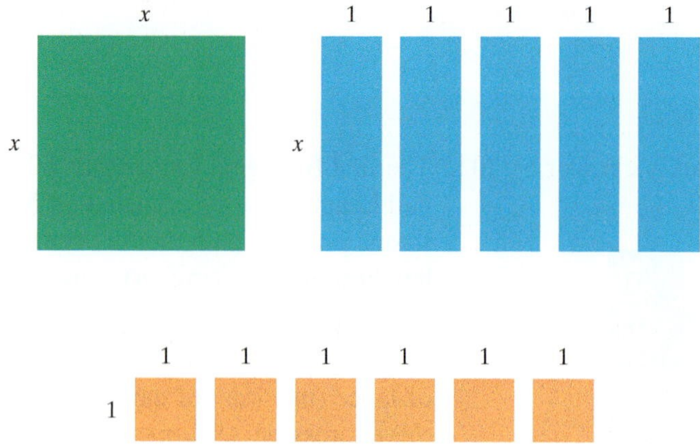

Mathematical Reasoning

3. In arithmetic, we simplify a fraction such as $\frac{20}{24}$ by writing the numerator and denominator in factored form and then dividing out common factors. For example, $\frac{20}{24} = \frac{\not{2} \cdot \not{2} \cdot 5}{\not{2} \cdot \not{2} \cdot 2 \cdot 3} = \frac{5}{6}$. Assuming $x \neq 1$ and $x \neq 3$, simplify $\frac{x^2 - 3x + 2}{x^2 - 4x + 3}$.

16.3 Factoring Trinomials Whose Leading Coefficient Is Not 1

OBJECTIVES

A To factor trinomials of the form $ax^2 + bx + c$, where $a \neq 1$

B To solve applied problems involving factoring

In the previous section, we discussed factoring trinomials of the form $ax^2 + bx + c$, where $a = 1$. Here, we consider polynomials whose leading coefficient is not 1, that is, trinomials such as:

$$2x^2 + 5x + 3 \quad \text{and} \quad 5x^2 - 13x - 6$$

The coefficient of the leading term is not 1.

The method of factoring that we use is, again, trial and error. However, we also discuss an alternative procedure that is based on factoring by grouping, a method covered in Section 16.1.

Factoring $ax^2 + bx + c$, for $a \neq 1$

Consider the product of two binomials:

$$\begin{aligned} & \quad\ \ \text{F} \qquad\quad \text{O} \qquad\quad \text{I} \qquad\quad \text{L} \\ (2x + 3)(x + 1) &= (2x)(x) + (2x)(1) + (3)(x) + (3)(1) \\ &= 2x^2 \quad + \quad 2x \quad + \quad 3x \quad + \quad 3 \\ &= 2x^2 + 5x + 3 \end{aligned}$$

Now, to reverse the process, we start with the product $2x^2 + 5x + 3$. To factor this polynomial, first check for common factors. Since there are no common factors other than 1 and -1, we use the FOIL method in reverse. Here, however, the leading coefficient is not 1, so we consider the factors of both 2 and 3. We then list and test all combinations of these factors to see if any will give us the desired middle term $5x$. In other words, we are looking for four integers so that:

$$2x^2 + 5x + 3 = (?x + ?)(?x + ?)$$

Factors of 2	Factors of 3	Possible Binomial Factors	Sum of Outer and Inner Products
2, 1	3, 1	$(2x + 3)(x + 1)$	$2x + 3x = 5x$ ← Correct middle term
		$(2x + 1)(x + 3)$	$6x + x = 7x$
2, 1	$-3, -1$	$(2x - 3)(x - 1)$	$-2x - 3x = -5x$
		$(2x - 1)(x - 3)$	$-6x - x = -7x$

Using this trial-and-error method, we find that $(2x + 3)(x + 1)$ is the correct factorization.

Note that this trial-and-error method for factoring a trinomial such as $2x^2 + 5x + 3$ is similar to the method for factoring trinomials with leading coefficient 1: We list and test all the possible factors of both the leading term and the constant term of the trinomial. We are looking for a combination of factors where the sum of the outer and inner products is the middle term of the trinomial. Practice and experience will shorten the process.

EXAMPLE 1

Factor: $3x^2 + 11x + 10$

Solution The terms of $3x^2 + 11x + 10$ have no common factors. So we proceed to use the trial-and-error method to factor this trinomial. Note that both the middle term $11x$ and the constant term 10 are positive. So we need to consider only combinations of the positive factors of 3 and positive factors of 10 that will give us a middle term with coefficient 11.

$$3x^2 + 11x + 10 = (?x + ?)(?x + ?)$$

Factors of 3	Factors of 10	Possible Binomial Factors	Middle Term
3, 1	2, 5	$(3x + 2)(x + 5)$	$15x + 2x = 17x$
		$(3x + 5)(x + 2)$	$6x + 5x = 11x$ ← Correct middle term
	10, 1	$(3x + 10)(x + 1)$	$3x + 10x = 13x$
		$(3x + 1)(x + 10)$	$30x + x = 31x$

So $3x^2 + 11x + 10 = (3x + 5)(x + 2)$.

PRACTICE 1

Factor: $5x^2 + 14x + 8$

EXAMPLE 2

Express in factored form: $15 - 17x + 4x^2$

Solution First, we rewrite the terms of the trinomial in descending order: $4x^2 - 17x + 15$. These terms have no common factor. To find the factorization of $4x^2 - 17x + 15$, we consider combinations of factors of 4 and factors of 15 that will result in a middle term with the coefficient -17.

Factors of 4	Factors of 15	Possible Binomial Factors	Middle Term
2, 2	$-3, -5$	$(2x - 3)(2x - 5)$	$-10x - 6x = -16x$
	$-15, -1$	$(2x - 15)(2x - 1)$	$-2x - 30x = -32x$
4, 1	$-3, -5$	$(4x - 3)(x - 5)$	$-20x - 3x = -23x$
		$(4x - 5)(x - 3)$	$-12x - 5x = -17x$ ← Correct middle term
	$-15, -1$	$(4x - 15)(x - 1)$	$-4x - 15x = -19x$
		$(4x - 1)(x - 15)$	$-60x - x = -61x$

So $4x^2 - 17x + 15 = (4x - 5)(x - 3)$.

PRACTICE 2

Factor: $21 - 25x + 6x^2$

Note that in Examples 1 and 2 we could have stopped the process of testing possible factorizations after finding the correct middle term.

EXAMPLE 3

Factor: $2y^2 + 19y - 10$

Solution The terms of $2y^2 + 19y - 10$ have no common factors. So we factor the trinomial by considering combinations of the factors of 2 and factors of -10 that will give us the middle term with coefficient 19.

Factors of 2	Factors of -10	Possible Binomial Factors	Middle Term
2, 1	2, -5	$(2y + 2)(y - 5)$	$-10y + 2y = -8y$
		$(2y - 5)(y + 2)$	$4y - 5y = -y$
	-2, 5	$(2y - 2)(y + 5)$	$10y - 2y = 8y$
		$(2y + 5)(y - 2)$	$-4y + 5y = y$
	10, -1	$(2y + 10)(y - 1)$	$-2y + 10y = 8y$
		$(2y - 1)(y + 10)$	$20y - y = 19y$ ← Correct middle term
	-10, 1	$(2y - 10)(y + 1)$	$2y - 10y = -8y$
		$(2y + 1)(y - 10)$	$-20y + y = -19y$

So $2y^2 + 19y - 10 = (2y - 1)(y + 10)$.

PRACTICE 3

Factor: $7y^2 + 47y - 14$

Can you explain why $(2y + 2)$, $(2y - 2)$, $(2y + 10)$, and $(2y - 10)$ can be immediately eliminated as possible factors of $2y^2 + 19y - 10$ in Example 3?

Consider the following possible binomial factors in Example 3.

The signs of the constant terms are reversed.

$$(2y - 1)(y + 10) = 2y^2 + 19y - 10$$
$$(2y + 1)(y - 10) = 2y^2 - 19y - 10$$

The sign of the middle term changes.

Comparing these possible factors suggest the following:

TIP Reversing the signs of the constant terms in binomial factors has the effect of switching the sign of the middle term in their product.

EXAMPLE 4

Factor: $5x^2 - 13x - 6$

Solution Since the terms of $5x^2 - 13x - 6$ have no common factors, let's look at the combinations of factors of 5 and factors of -6 that will give us the middle term with coefficient -13.

Factors of 5	Factors of -6	Possible Binomial Factors	Middle Term
5, 1	2, −3	$(5x + 2)(x - 3)$	$-15x + 2x = -13x$ ⟵ Correct middle term
		$(5x - 3)(x + 2)$	$10x - 3x = 7x$
	−2, 3	$(5x - 2)(x + 3)$	$15x - 2x = 13x$
		$(5x + 3)(x - 2)$	$-10x + 3x = -7x$
	−6, 1	$(5x - 6)(x + 1)$	$5x - 6x = -x$
		$(5x + 1)(x - 6)$	$-30x + x = -29x$
	6, −1	$(5x + 6)(x - 1)$	$-5x + 6x = x$
		$(5x - 1)(x + 6)$	$30x - x = 29x$

So $5x^2 - 13x - 6 = (5x + 2)(x - 3)$. Note that it was unnecessary to examine any factors after the first trial, since we found the correct combination for the middle term $-13x$.

PRACTICE 4

Factor: $2x^2 - x - 10$

EXAMPLE 5

Factor: $12y^3 + 2y^2 - 2y$

Solution Since $2y$ is the GCF of the trinomial, first we factor it out getting:

$$12y^3 + 2y^2 - 2y = 2y(6y^2 + y - 1)$$

Next, we factor $6y^2 + y - 1$, looking for a combination of factors of 6 and factors of -1 that will give us the middle term with coefficient 1.

Factors of 6	Factors of -1	Possible Binomial Factors	Middle Term
6, 1	1, −1	$(6y + 1)(y - 1)$	$-6y + y = -5y$
		$(6y - 1)(y + 1)$	$6y - y = 5y$
3, 2	1, −1	$(3y + 1)(2y - 1)$	$-3y + 2y = -y$
		$(3y - 1)(2y + 1)$	$3y - 2y = y$ ⟵ Correct middle term

So $12y^3 + 2y^2 - 2y = 2y(6y^2 + y - 1) = 2y(3y - 1)(2y + 1)$.

PRACTICE 5

Factor: $18x^3 - 21x^2 - 9x$

Now, we consider factoring a trinomial of the form $ax^2 + bxy + cy^2$. This type of trinomial contains more than one variable, so we need to look for a factorization of the form $(?x + ?y)(?x + ?y)$.

EXAMPLE 6

Factor: $12x^2 + 28xy + 8y^2$

Solution Since 4 is the GCF of the trinomial, let's first factor it out.

$$12x^2 + 28xy + 8y^2 = 4(3x^2 + 7xy + 2y^2)$$

Next, we factor $3x^2 + 7xy + 2y^2$. We look for the combination of factors of 3 and factors of 2 that will result in the middle term $7xy$.

$$4(3x^2 + 7xy + 2y^2) = 4(?x + ?y)(?x + ?y)$$

Factors of 3	Factors of 2	Possible Binomial Factors	Middle Term
3, 1	2, 1	$(3x + 2y)(x + y)$	$3xy + 2xy = 5xy$
		$(3x + y)(x + 2y)$	$6xy + xy = 7xy$ ← Correct middle term

So $12x^2 + 28xy + 8y^2 = 4(3x^2 + 7xy + 2y^2) = 4(3x + y)(x + 2y)$.

PRACTICE 6

Factor: $36c^2 - 12cd - 15d^2$

Now, let's consider an alternative procedure for factoring a trinomial $ax^2 + bx + c$ based on *grouping*. This method, which the next example illustrates, is sometimes called the *ac method*.

EXAMPLE 7

Factor: $2x^2 + 5x - 3$

Solution First, we check that the terms of $2x^2 + 5x - 3$ have no common factors. Next, instead of listing the factors of 2 and the factors of -3 as in the trial-and-error method, we begin by finding their product:

$$ac = (2)(-3) = -6$$

We then look for two factors of the number ac (that is, -6) that add up to b (that is, 5). The numbers 6 and -1 satisfy these conditions, since $(6)(-1) = -6$ and $(6) + (-1) = 5$. We use these numbers to split up the middle term in the original trinomial, and then rewrite the trinomial.

$$2x^2 + 5x - 3 = 2x^2 + 6x + (-1)x - 3$$ Split up the middle term: $5x = 6x + (-1)x$.

$$= [2x^2 + 6x] + [(-1)x - 3]$$ Group the first two terms and the last two terms.

$$= 2x(x + 3) + (-1)(x + 3)$$ Factor out the GCF from each group.

$$= (x + 3)(2x - 1)$$ Write in factored form.

So $2x^2 + 5x - 3 = (x + 3)(2x - 1)$. As usual, we can check that this factorization is correct by multiplication.

PRACTICE 7

Factor: $2x^2 - 7x - 4$

Now, let's solve some applied problems involving the factoring of trinomials.

EXAMPLE 8

A ball is thrown upward at 40 ft/sec from the top of a building 24 ft above the ground. The height of the ball above the ground (in feet) t sec after the ball is thrown is given by the expression $-16t^2 + 40t + 24$. Write this expression in factored form.

Solution We factor the expression $-16t^2 + 40t + 24$:

$$\begin{aligned} -16t^2 + 40t + 24 &= -8(2t^2 - 5t - 3) \\ &= -8(2t + 1)(t - 3) \end{aligned}$$

So the factorization of $-16t^2 + 40t + 24$ is $-8(2t + 1)(t - 3)$, which can also be written

$$8(-2t - 1)(t - 3),$$

or

$$8(2t + 1)(-t + 3).$$

PRACTICE 8

A bin is made from a 7-foot by 5-foot sheet of metal by cutting out squares of equal size from each corner and then turning up the sides. The volume of the resulting bin can be represented by the expression $4x^3 - 24x^2 + 35x$. Rewrite this expression in factored form.

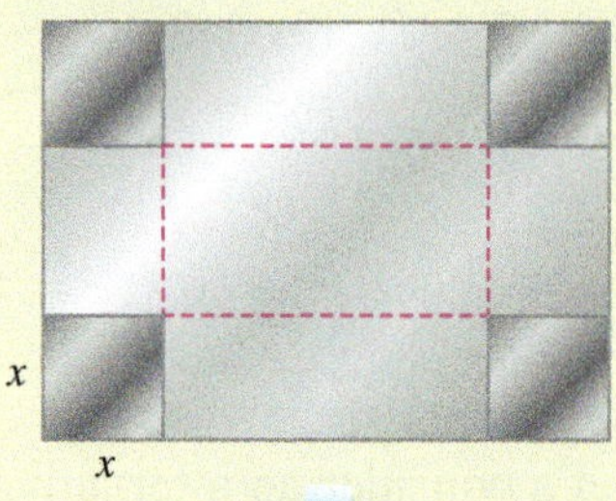

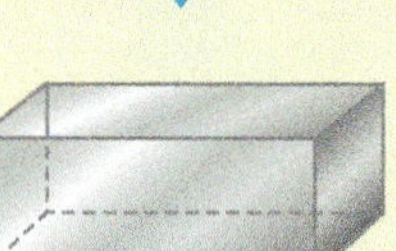

16.3 Exercises

FOR EXTRA HELP MyMathLab     

A *Match each trinomial with its binomial factors.*

1. $2x^2 + 3x - 9$ **a.** $(2x - 9)(x + 1)$

2. $2x^2 - 19x + 9$ **b.** $(2x - 9)(x - 1)$

3. $2x^2 - 11x + 9$ **c.** $(x - 9)(2x + 1)$

4. $2x^2 - 3x - 9$ **d.** $(x - 9)(2x - 1)$

5. $2x^2 - 17x - 9$ **e.** $(2x - 3)(x + 3)$

6. $2x^2 - 7x - 9$ **f.** $(x - 3)(2x + 3)$

Find the missing factor.

7. $3x^2 + 16x + 5 = (x + 5)(\quad)$

8. $2x^2 + 11x + 12 = (2x + 3)(\quad)$

9. $5x^2 - 13x - 6 = (5x + 2)(\quad)$

10. $2x^2 - x - 6 = (2x + 3)(\quad)$

11. $3x^2 - 11x + 6 = (3x - 2)(\quad)$

12. $6x^2 - 7x + 2 = (2x - 1)(\quad)$

Factor, if possible.

13. $3x^2 + 8x + 5$

14. $2x^2 + 15x + 7$

15. $2y^2 - 11y + 5$

16. $3y^2 - 10y + 7$

17. $3x^2 + 14x + 8$

18. $2x^2 + 11x + 9$

19. $5x^2 + 9x - 6$

20. $4y^2 - 16y - 7$

21. $6y^2 - y - 5$

22. $5x^2 + 17x - 12$

23. $2y^2 - 11y + 14$

24. $7y^2 - 19y + 10$

25. $9a^2 - 18a - 16$

26. $10m^2 - m - 21$

27. $4x^2 - 13x + 3$

28. $4n^2 - 9n + 2$

29. $6 + 17y + 12y^2$

30. $4 + 16n + 15n^2$

31. $-17m + 21 + 2m^2$

32. $-16x + 5 + 3x^2$

33. $-6a^2 - 7a + 3$

34. $-5b^2 - 14b + 3$

35. $8y^2 + 5y - 22$

36. $6x^2 + 5x - 25$

37. $7y^2 + 36y - 5$

38. $2y^2 + 27y + 14$

39. $8a^2 + 65a + 8$

40. $8n^2 + 33n + 4$

41. $6x^2 + 25x - 9$

42. $10x^2 + 21x - 10$

43. $8y^2 - 26y + 15$

44. $4m^2 - 16m - 9$

45. $14y^2 - 38y + 20$

46. $9y^2 - 24y + 15$

47. $28a^2 + 24a - 4$

48. $6x^2 + 40x - 14$

49. $-6b^2 + 40b + 14$

50. $-25m^2 + 65m + 30$

51. $12y^3 + 50y^2 + 28y$

52. $6x^3 + 45x^2 + 21x$

53. $14a^4 - 38a^3 + 20a^2$

54. $10n^4 - 35n^3 + 15n^2$

55. $2x^3y + 13x^2y + 15xy$

56. $3xy^3 + 10xy^2 + 3xy$

57. $6ab^3 - 44ab^2 + 14ab$

58. $12a^3b - 34a^2b + 24ab$

59. $20c^2 - 9cd + d^2$

60. $12a^2 - 25ab + 12b^2$

61. $2x^2 - 5xy - 3y^2$
62. $6s^2 - st - 12t^2$
63. $8a^2 - 6ab + b^2$
64. $3m^2 - 8mn + 5n^2$
65. $18x^2 + 3xy - 6y^2$
66. $4s^2 + 10st - 24t^2$
67. $16c^2 - 44cd + 30d^2$
68. $16a^2 - 48ab + 36b^2$
69. $27u^2 + 18uv + 3v^2$
70. $4a^2 + 26ab - 48b^2$
71. $42x^3 + 45x^2y - 27xy^2$
72. $60p^3 + 28p^2q - 16pq^2$
73. $-30x^4y + 35x^3y^2 + 15x^2y^3$
74. $-24x^3y^2 - 6x^2y^3 + 18xy^4$
75. $5ax^2 - 28axy - 12ay^2$
76. $3cx^2 + 7cxy - 20cy^2$

Mixed Practice

Factor, if possible.

77. $-5m + 2 + 3m^2$
78. $3y^2 + 2y + 5$
79. $8x^2 - 2x - 3$
80. $8x^2 + 36x - 20$
81. $14x^3 + 44x^2 + 6x$
82. $30a^3b - 55a^2b + 15ab$
83. $8m^2 - 18mn + 9n^2$
84. $24a^3 + 18a^2b - 15ab^2$
85. $7r^2 - 9r + 2$
86. $10a^2 + 11a - 6$

Solve.

87. Factor: $3x^2 - 22x + 7$.
   a. $(3x + 7)(x + 1)$
   b. $(x + 7)(3x + 1)$
   c. $(3x - 7)(x - 1)$
   d. $(x - 7)(3x - 1)$

88. Find the missing factor:
$5x^2 - 2x - 3 = (5x + 3)(\quad)$

Applications

B *Solve.*

89. An object is thrown upward so that its height (in meters above the ground) at time t sec is represented by the expression $-5t^2 - 21t + 20$. Factor this expression.

90. A box with width w has a volume that can be expressed as $3w^3 - 2w^2 - w$. Rewrite this expression in factored form.

91. A homeowner decides to increase the area of his 8-foot by 10-foot deck by increasing both the length and the width, as shown in the diagram below:

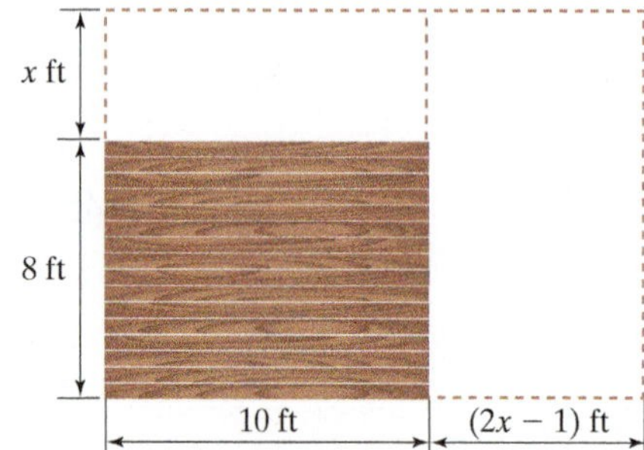

   a. Find the area of the expanded deck by adding the areas of the three rectangles.
   b. Express the area of the expanded deck in factored form.

92. The diagram below shows a circular pad with radius r on the square top of a dining room table.

   a. Find the area of the region of the table top not covered by the circular pad.
   b. Express the area in part (a) in factored form.

93. Show that the expression $4n^2 - 12n + 5$ can be written as the product of two integers that differ by 4, no matter what integer n represents.

94. The squares of the first n whole numbers add up to $\frac{2n^3 + 3n^2 + n}{6}$. Write this expression so that the numerator is in factored form.

• Check your answers on page A-35.

MINDStretchers

Groupwork

1. Work with a partner. Next to each polynomial, list at least two integers that, when inserted in the box, will make the polynomial factorable.

a. $2x^2 + \square x + 5$ **b.** $\square x^2 - 4x + 1$ **c.** $3x^2 - x + \square$

Investigation

2. Copy or trace the following algebra tiles onto a piece of paper. Then, cut out each tile separately. Finally, position all the tiles like a jigsaw puzzle so as to form a rectangle. (*Hint:* Factor the polynomial $2x^2 + 7x + 3$.)

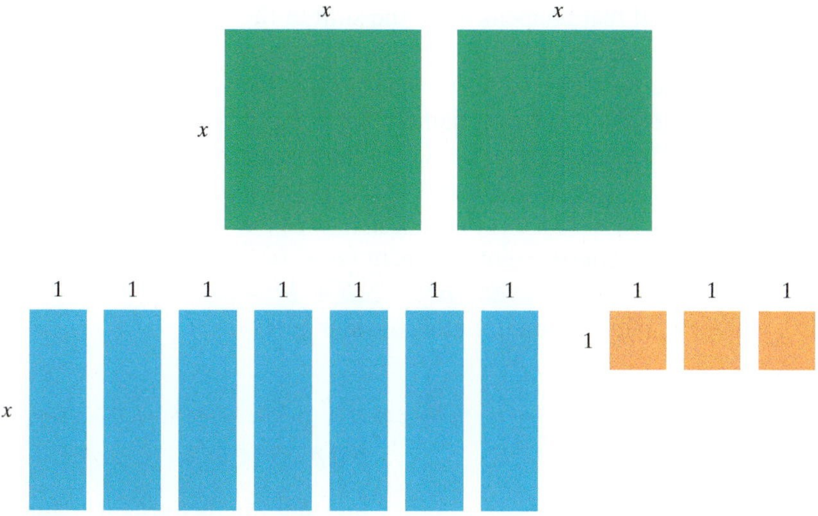

Writing

3. Do you prefer to factor trinomials using the trial-and-error method or the *ac* method? Explain why.

16.4 Factoring Perfect Square Trinomials and the Difference of Squares

OBJECTIVES

- **A** To determine if a polynomial is a perfect square trinomial or a difference of squares
- **B** To factor perfect square trinomials
- **C** To factor the difference of squares
- **D** To solve applied problems involving factoring

Recall that in Section 15.6, we considered formulas for multiplying binomials in certain special cases: the square of a sum, the square of a difference, and the product of the sum and the difference of the same two terms. In this section, we show that these formulas also allow us to factor special polynomials, called *perfect square trinomials* and the *difference of squares*. Recognizing these special polynomials makes it easier to factor them.

Factoring Perfect Square Trinomials

We have seen that squaring the sum or the difference of two terms gives us:

$$(a + b)^2 = a^2 + 2ab + b^2 \quad \text{and} \quad (a - b)^2 = a^2 - 2ab + b^2$$

Each of these products is called a **perfect square trinomial**. Such trinomials may be factored by reversing the multiplication process.

Factoring a Perfect Square Trinomial

$$a^2 + 2ab + b^2 = (a + b)^2$$
$$a^2 - 2ab + b^2 = (a - b)^2$$

The first formula shows us how to factor a trinomial that happens to be the sum of the squares of two terms *plus* twice their product. In this formula, the terms are a and b, the sum of their squares is $a^2 + b^2$, and $+2ab$ is twice their product. In words, the formula states that the factorization of such a trinomial is the square of the *sum* of the two terms, namely $(a + b)^2$.

The second formula applies when we want to factor a trinomial that is the sum of the squares of two terms *minus* twice their product. In this formula, the terms are again a and b, the sum of their squares is $a^2 + b^2$, and $-2ab$ is minus twice their product. In words, this formula states that the factorization of such a trinomial is the square of the *difference* of the two terms, namely $(a - b)^2$.

Keep in mind that the formulas $a^2 + 2ab + b^2 = (a + b)^2$ and $a^2 - 2ab + b^2 = (a - b)^2$ only apply when a polynomial is a perfect square trinomial. Let's consider an example of recognizing these trinomials.

EXAMPLE 1

Determine whether each polynomial is a perfect square trinomial.

a. $x^2 - 8x + 16$

b. $-x^2 + 2x + 1$

c. $x^2 + 4x + 1$

d. $n^2 - 6n - 9$

e. $9x^2 + 30xy + 25y^2$

PRACTICE 1

Indicate whether each trinomial is a perfect square.

a. $x^2 + 6x + 9$

b. $-4t^2 - 4t + 1$

c. $y^2 - 14y + 49$

d. $x^2 - 2x - 1$

e. $4p^2 - 4pq + q^2$

Solution

a. In the polynomial $x^2 - 8x + 16$, x^2 and 16 are perfect squares and correspond to a^2 and b^2, respectively, in the formula $a^2 - 2ab + b^2$. So a corresponds to x, and b corresponds to 4. Since the middle term, $-8x$, or $-2 \cdot x \cdot 4$, corresponds to $-2ab$, the polynomial is a perfect square trinomial.

$$x^2 - 8x + 16 = \underset{\underset{a^2}{\uparrow}}{x^2} - \underset{\underset{2ab}{\uparrow}}{\underline{2 \cdot x \cdot 4}} + \underset{\underset{b^2}{\uparrow}}{(4)^2}$$

b. In the polynomial $-x^2 + 2x + 1$, $-x^2$ is not a perfect square since its coefficient is negative. Therefore, $-x^2 + 2x + 1$ is not a perfect square trinomial.

c. For the polynomial $x^2 + 4x + 1$, x^2 and 1 are both perfect squares with a corresponding to x, and b to 1 in the formula $a^2 + 2ab + b^2$. However, the middle term $+4x$ is not twice the product of x and 1. So $x^2 + 4x + 1$ is not a perfect square trinomial.

d. The constant term of a perfect square trinomial, b^2, must be positive. So $n^2 - 6n - 9$ is not a perfect square, since its constant term -9 is negative.

e. In the polynomial $9x^2 + 30xy + 25y^2$, we know that $9x^2$ is equal to $(3x)^2$ and $25y^2$ equals $(5y)^2$. So a corresponds to $3x$ and b corresponds to $5y$ in the formula $a^2 + 2ab + b^2$. The middle term $+30xy$ can be expressed as $+2 \cdot 3x \cdot 5y$. Therefore, the trinomial is a perfect square.

$$9x^2 + 30xy + 25y^2 = \underset{\underset{a^2}{\uparrow}}{(3x)^2} + \underset{\underset{2ab}{\uparrow}}{\underline{2(3x)(5y)}} + \underset{\underset{b^2}{\uparrow}}{(5y)^2}$$

Now, let's practice *factoring* perfect square trinomials.

EXAMPLE 2

Factor: $x^2 + 12x + 36$

Solution For the trinomial $x^2 + 12x + 36$, the first term x^2 and the last term 36, or 6^2, are perfect squares. The middle term, $12x$ or $2 \cdot x \cdot 6$, is twice the product of x and 6. It follows that $x^2 + 12x + 36$ is a perfect square trinomial. Since its middle term is positive, we apply the formula for the square of a sum.

$$x^2 + 12x + 36 = \underset{\underset{a^2}{\uparrow}}{x^2} + \underset{\underset{2ab}{\uparrow}}{\underline{2 \cdot x \cdot 6}} + \underset{\underset{b^2}{\uparrow}}{(6)^2} = \underset{(a+b)^2}{(x + 6)^2}$$

Check We can confirm our answer by multiplying out $(x + 6)^2$.

$$(x + 6)^2 = (x + 6)(x + 6) = x^2 + 2 \cdot 6x + 36 = x^2 + 12x + 36$$

PRACTICE 2

Factor: $n^2 + 20n + 100$

EXAMPLE 3

Write as the square of a binomial: $x^2 + 9 - 6x$

Solution Let's begin by rewriting the trinomial in descending order: $x^2 - 6x + 9$.

$$x^2 - 6x + 9 = x^2 - \underbrace{2 \cdot x \cdot 3} + (3)^2 = (x - 3)^2$$

$$a^2 - 2ab + b^2 = (a - b)^2$$

PRACTICE 3

Factor: $t^2 + 4 - 4t$

EXAMPLE 4

Express $9x^2 - 6xy + y^2$ as the square of a binomial.

Solution Here, the trinomial contains more than one variable.

$$9x^2 - 6xy + y^2 = (3x)^2 - \underbrace{2 \cdot 3x \cdot y} + y^2 = (3x - y)^2$$

$$a^2 - 2ab + b^2 = (a - b)^2$$

PRACTICE 4

Write $25c^2 - 40cd + 16d^2$ as the square of a binomial.

EXAMPLE 5

Factor: $y^{10} + 16y^5 + 64$

Solution Since $y^{10} = (y^5)^2$ and $64 = (8)^2$, we know that y^{10} and 64 are perfect squares. Also, $16y^5 = 2 \cdot 8 \cdot y^5$, that is, $16y^5$ is twice the product of y^5 and 8. So we conclude that $y^{10} + 16y^5 + 64$ is a perfect square trinomial.

$$y^{10} + 16y^5 + 64 = (y^5)^2 + \underbrace{2 \cdot y^5 \cdot 8} + (8)^2 = (y^5 + 8)^2$$

$$a^2 + 2ab + b^2 = (a + b)^2$$

PRACTICE 5

Express $x^4 + 8x^2 + 16$ as the square of a binomial.

In factoring a perfect square trinomial, how do we know whether the binomial squared is the sum or the difference of two terms?

Factoring the Difference of Squares

Recall that when finding the product of the sum and the difference of the same terms, we get:

$$(a + b)(a - b) = a^2 - b^2$$

The product is a binomial that is called a **difference of squares**. We can factor such a binomial by reversing the multiplication process.

Factoring the Difference of Squares

$$a^2 - b^2 = (a + b)(a - b)$$

This formula is a shortcut for factoring a binomial equal to the square of one term *minus* the square of another term. The terms are a and b, and so their squares are a^2 and b^2. In words, the formula states that the factorization of a binomial that is the difference of the squares of two terms is the sum of the two terms times the difference of the same two terms, that is, $(a + b)(a - b)$.

EXAMPLE 6

Indicate whether each binomial is a difference of squares.

a. $x^2 - 81$ **b.** $x^2 + y^2$

c. $x^2 - y^3$ **d.** $4p^6 - q^2$

Solution

a. In the binomial $x^2 - 81$, both x^2 and 81 are perfect squares and correspond to a^2 and b^2, respectively, in our formula.

$$x^2 - 81 = x^2 - 9^2$$
$$\uparrow \quad \uparrow$$
$$a^2 - b^2$$

Here, a corresponds to x, and b corresponds to 9. So $x^2 - 81$ is a difference of squares.

b. In the expression $x^2 + y^2$, both x^2 and y^2 are perfect squares, where x corresponds to a and y to b. However, the binomial is the sum and not the difference of squares.

c. For $x^2 - y^3$, x^2 is a perfect square, but y^3 is not. So $x^2 - y^3$ is not a difference of squares.

d. The binomial $4p^6 - q^2$ can be rewritten as $(2p^3)^2 - q^2$, and so is a difference of squares.

PRACTICE 6

Determine whether each binomial is a difference of squares.

a. $x^2 - 64$

b. $x^2 + 49$

c. $x^3 - 16$

d. $r^4 - 9s^6$

Note, as the preceding example suggests, that even powers of a variable are perfect squares, whereas odd powers of a variable are not. Can you explain why?

EXAMPLE 7

Factor: $x^2 - 100$

Solution Since x^2 and 100 are perfect squares, $x^2 - 100$ is a difference of squares.

$$x^2 - 100 = x^2 - 10^2 = (x + 10)(x - 10)$$
$$\uparrow \quad \uparrow \quad \uparrow \quad \uparrow$$
$$a^2 - b^2 = (a + b)\ (a - b)$$

Check We can verify that $(x + 10)(x - 10)$ is the factorization of $x^2 - 100$ by multiplying.

$$(x + 10)(x - 10) = x^2 - 10x + 10x - 100 = x^2 - 100$$

PRACTICE 7

Factor: $y^2 - 121$

EXAMPLE 8

Write $16x^2 - 49y^2$ in factored form.

Solution Since $16x^2 = (4x)^2$ and $49y^2 = (7y)^2$, we know that $16x^2$ and $49y^2$ are perfect squares. So $16x^2 - 49y^2$ is a difference of squares.

$$16x^2 - 49y^2 = \underset{\substack{\uparrow \\ a^2}}{(4x)^2} \underset{\substack{\\ -}}{-} \underset{\substack{\uparrow \\ b^2}}{(7y)^2} \underset{\substack{\\ =}}{=} \underset{\substack{\uparrow \\ (a+b)}}{\underbrace{(4x + 7y)}}\underset{\substack{\uparrow \\ (a-b)}}{\underbrace{(4x - 7y)}}$$

PRACTICE 8

Express $9x^2 - 25y^2$ in factored form.

EXAMPLE 9

Factor: $4x^4 - 9y^6$

Solution Because $4x^4 = (2x^2)^2$ and $9y^6 = (3y^3)^2$, we see that $4x^4$ and $9y^6$ are both perfect squares. So $4x^4 - 9y^6$ is a difference of squares.

$$4x^4 - 9y^6 = \underset{\substack{\uparrow \\ a^2}}{(2x^2)^2} \underset{\substack{\\ -}}{-} \underset{\substack{\uparrow \\ b^2}}{(3y^3)^2} \underset{\substack{\\ =}}{=} \underset{\substack{\uparrow \\ (a+b)}}{\underbrace{(2x^2 + 3y^3)}}\underset{\substack{\uparrow \\ (a-b)}}{\underbrace{(2x^2 - 3y^3)}}$$

PRACTICE 9

Factor: $64x^8 - 81y^2$

EXAMPLE 10

Find an expression for the area of the cross section of the pipe pictured. Write this expression in factored form.

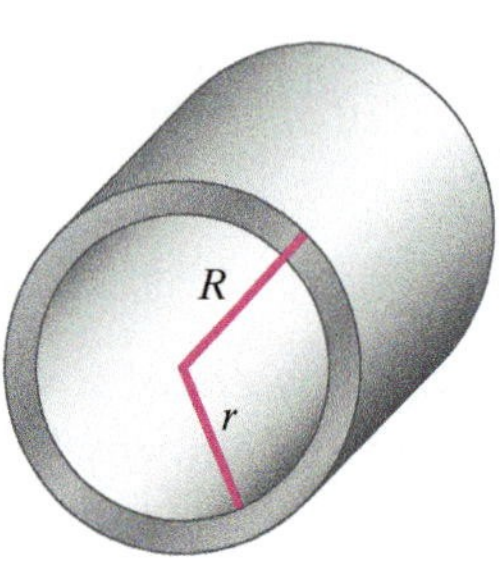

Solution The cross section is a ring-shaped region between two circles with the same center. The radius of the inner circle is r and the radius of the outer circle is R. Using the formula for the area of a circle helps us to find an expression for the area of the cross section:

Area of the cross section

$$\begin{aligned} &= \text{Area of the large circle} - \text{Area of the small circle} \\ &= \pi R^2 - \pi r^2 \\ &= \pi(R^2 - r^2) \\ &= \pi(R + r)(R - r) \end{aligned}$$

So in factored form, an expression for the area of the cross section of the pipe is $\pi(R + r)(R - r)$.

PRACTICE 10

A stone is dropped from a bridge 256 ft above a river. The height of the stone above the river (in feet) t sec after it is dropped is given by the expression $256 - 16t^2$. Factor this expression.

16.4 Exercises

FOR EXTRA HELP MyMathLab MathXL PRACTICE WATCH READ REVIEW

Mathematically Speaking

Fill in each blank with the most appropriate term or phrase from the given list.

odd	the square	even
perfect square trinomial	half the product	twice the product
square of the sum of two terms	difference of squares	

1. A(n) ____________________ of the form $a^2 + 2ab + b^2$ can be factored as $(a + b)^2$.

2. The trinomial $a^2 + 5ab + 25b^2$ is not a perfect square, because the middle term is not ____________________ of a and $5b$.

3. The binomial $a^2 - b^2$, a(n) ____________________, can be factored as $(a + b)(a - b)$.

4. Powers of a variable that are ____________________ are not perfect squares.

A *Determine whether each polynomial is a perfect square trinomial, a difference of squares, or neither.*

5. $x^2 + 2x + 1$
6. $y^2 + 8y + 16$
7. $-t^2 - 4t + 1$
8. $-y^2 + 2y + 9$
9. $x^2 - 6x + 9$
10. $n^2 - 2n + 1$
11. $x^2 - 25$
12. $y^2 - 49$
13. $81x^2 - 36y^2$
14. $4x^2 - 100y^2$
15. $25x^2 - 20x + 4$
16. $16x^2 - 24x + 9$
17. $y^2 + 1$
18. $y^2 + 3$
19. $25x^2 + 10xy + y^2$
20. $49x^2 + 14xy + y^2$
21. $x^3 - 1$
22. $x^4 + 9$

B *Factor, if possible.*

23. $x^2 - 12x + 36$
24. $y^2 - 14y + 49$
25. $y^2 + 20y + 100$
26. $x^2 + 2x + 1$
27. $a^2 - 4a + 4$
28. $b^2 - 22b + 121$
29. $x^2 - 6x - 9$
30. $y^2 - 10y - 25$
31. $4a^2 - 36a + 81$
32. $25b^2 - 20b + 4$
33. $49x^2 + 28x + 4$
34. $9y^2 + 24y + 16$
35. $36 - 60x + 25x^2$
36. $49 - 42y + 9y^2$
37. $m^2 + 26mn + 169n^2$
38. $4a^2 + 36ab + 81b^2$
39. $225a^2 - 30ab + b^2$
40. $25s^2 - 40st + 16t^2$
41. $y^4 + 2y^2 + 1$
42. $x^4 + 4x^2 + 4$
43. $6x^2 + 12x + 6$
44. $12y^2 + 24y + 12$
45. $27m^3 - 36m^2 + 12m$
46. $48y^3 - 24y^2 + 3y$
47. $4s^2t^3 + 80s^2t^2 + 400s^2t$
48. $2x^3y^2 - 52x^2y^2 + 338xy^2$

C *Factor, if possible.*

49. $m^2 - 64$

50. $n^2 - 1$

51. $y^2 - 81$

52. $x^2 - 16$

53. $144 - x^2$

54. $225 - t^2$

55. $100m^2 - 81$

56. $16n^2 - 25$

57. $36x^2 + 121$

58. $64n^2 + 169$

59. $1 - 9x^2$

60. $81 - 4y^2$

61. $x^2 - 4y^2$

62. $49c^2 - d^2$

63. $100x^2 - 9y^2$

64. $36a^2 - 121b^2$

65. $3k^3 - 147k$

66. $5m^3 - 125m$

67. $4y^4 - 36y^2$

68. $3t^5 - 300t^3$

69. $27x^2y - 3x^2y^3$

70. $50xy - 18x^3y$

71. $2a^2b^2 - 98$

72. $9x^4y^2 - 81$

73. $256 - r^4$

74. $625 - t^4$

75. $5x^4 - 80y^4$

76. $64s^4 - 4t^4$

77. $x^2(c - d) - 4(c - d)$

78. $y^2(a - b) - (a - b)$

79. $16(x - y) - a^2(x - y)$

80. $9(y - c) - x^2(y - c)$

Mixed Practice

Factor, if possible.

81. $9c^2 + 48cd + 64d^2$

82. $169m^2 - 49n^2$

83. $a^2 - 225b^4$

84. $8t^3 - 56t^2 + 98t$

85. $54a^4b^2 - 36a^2b + 6$

86. $12xy - 27xy^3$

87. $81 - w^4$

88. $196 - r^2$

89. $36u^2 + 60u + 25$

90. $121s^2 + 36$

Determine whether each polynomial is a perfect square trinominal, a difference of squares, or neither.

91. $j^2 - 169$

92. $81a^2 - 90ab + 25b^2$

Applications

D *Solve.*

93. If the radius of a balloon decreases from radius r_1 to radius r_2, then the drop in the balloon's surface area is given by the expression $4\pi r_1^2 - 4\pi r_2^2$. Write this expression in factored form.

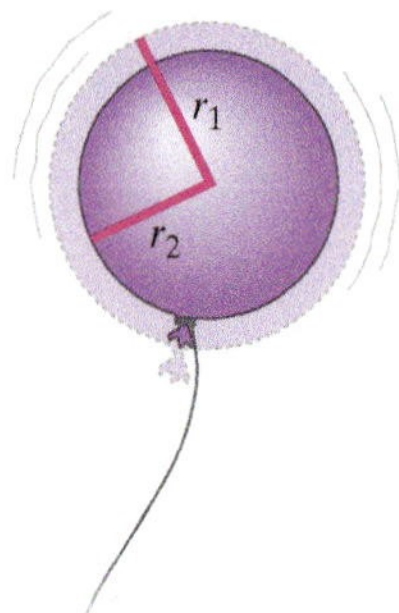

94. The height (in feet) of a stone dropped from a cliff 100 ft above a river, as shown in the illustration, is given by the expression $100 - 16t^2$, where t is time (in seconds). Factor this expression.

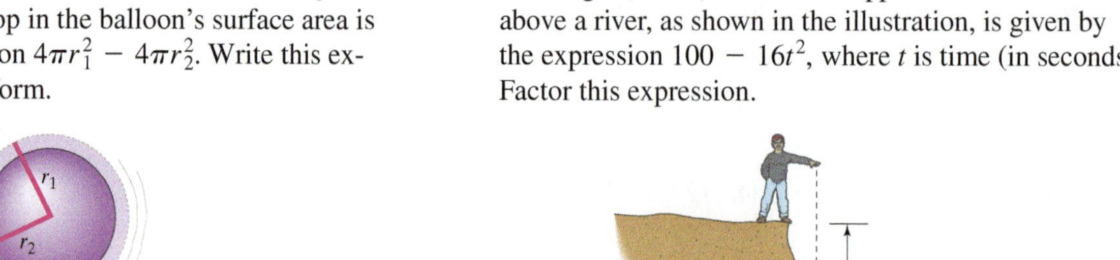

95. An open box is made from a 2-foot by 2-foot piece of cardboard by cutting out equal squares from each of the four corners and turning up the sides. The volume of the resulting box can be modeled by the polynomial $4x - 8x^2 + 4x^3$. Factor the expression.

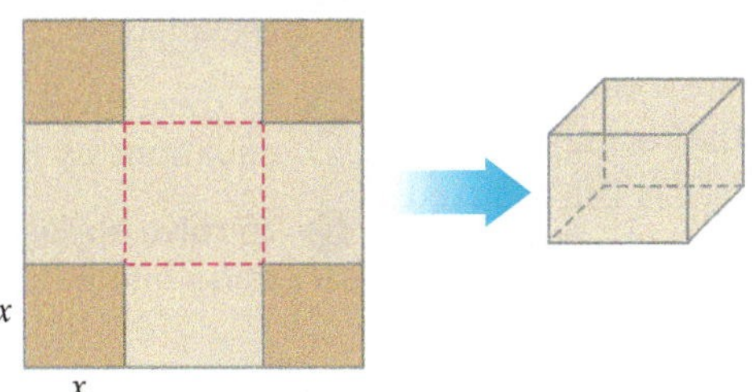

96. Find an expression, in factored form, for the area of the wooden border of the square picture frame shown.

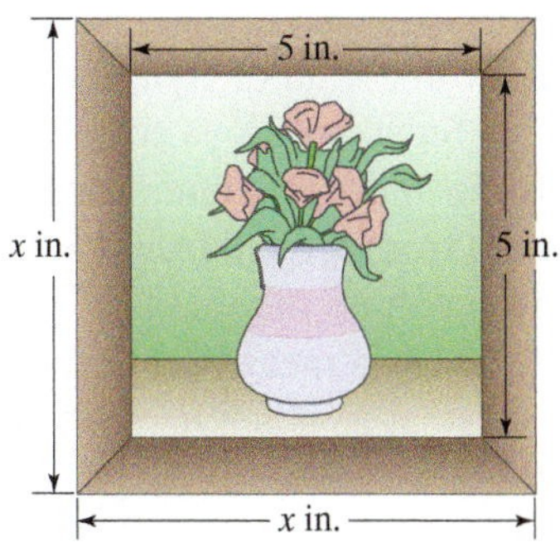

97. When the velocity of a rocket increases from v_1 to v_2, the force caused by air resistance increases by $kv_2^2 - kv_1^2$ for a constant k. Write this polynomial in factored form.

98. A \$16,000 investment grew by an average annual rate of return of r. After two years, the value of the investment in dollars was $16{,}000 + 32{,}000r + 16{,}000r^2$. What is the factorization of this expression?

• Check your answers on page A-35.

MINDStretchers

Patterns

1. Find two factors of 3599 both of which are greater than 1. (*Hint:* $3599 = 3600 - 1$.)

Groupwork

2. Try this trick with a partner:

   a. Take your partner's age in years.
   b. Square it.
   c. Subtract 9.
   d. Divide the result by 3 less than your partner's age.
   e. Subtract 53.
   f. Add your partner's age.
   g. Divide by 2.
   h. Add 5^2.
   i. Check that you wind up where you started—with your partner's age.

   In the table, record the results for three different ages. Then, in the fourth row, repeat the steps with a variable x representing your partner's age. Explain why this trick works.

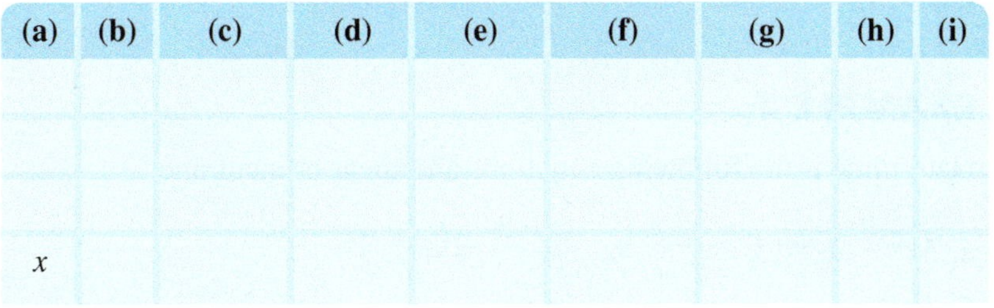

(a)	(b)	(c)	(d)	(e)	(f)	(g)	(h)	(i)
x								

Writing

3. In a few sentences, explain how the diagram at the right shows that $(a + b)^2 = a^2 + 2ab + b^2$.

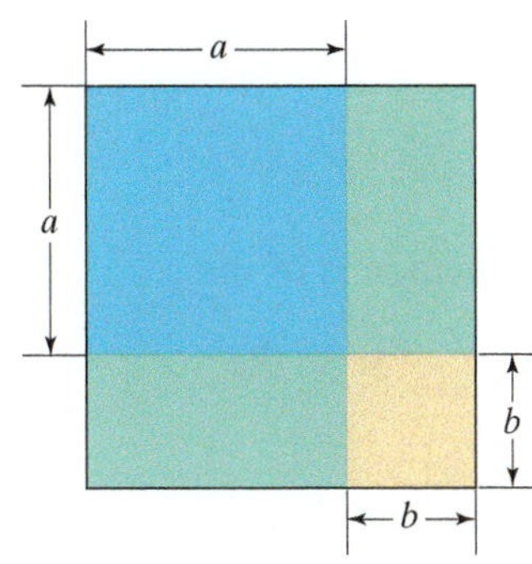

16.5 Solving Quadratic Equations by Factoring

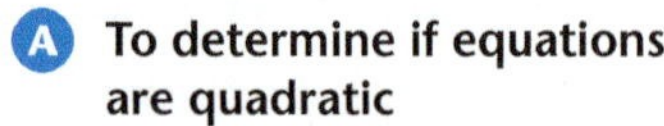

OBJECTIVES

- **A** To determine if equations are quadratic
- **B** To solve quadratic equations by factoring
- **C** To solve applied problems using quadratic equations

This section deals with a kind of equation that we have not previously considered, namely, a *quadratic equation*. Such equations come up in physics, finance, and other fields as well.

Consider, for instance, a situation involving the movement of a rocket. If the rocket is shot straight upward from ground level with an initial velocity of 80 ft/sec, the rocket's height above the ground h can be modeled by the expression $80t - 16t^2$, where t is the elapsed time (in seconds) and h is measured in feet. To find the time at which the rocket falls back and hits the ground (that is, when $h = 0$), we need to be able to solve the quadratic equation $80t - 16t^2 = 0$.

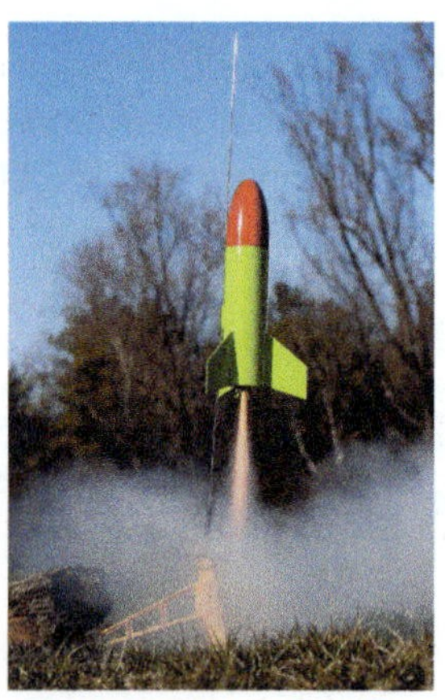

DEFINITION

A **second-degree** or **quadratic equation** is an equation that can be written in the form $ax^2 + bx + c = 0$, where a, b, and c are real numbers and $a \neq 0$.

Some examples of quadratic equations are:

$$x^2 - x + 6 = 0 \qquad 3x^2 - 12 = 0 \qquad (x - 1)^2 = 0$$

Can you explain why these polynomials are of the second degree?

EXAMPLE 1

Determine if the following equations are linear or quadratic.

a. $5x^2 - x = 3$ **b.** $\dfrac{x - 3}{2} = 4$

Solution

a. $5x^2 - x = 3$ is equivalent to $5x^2 - x - 3 = 0$. Since this equation is of the form $ax^2 + bx + c = 0$, where $a = 5$, $b = -1$, and $c = -3$, the equation is quadratic.

b. $\dfrac{x - 3}{2} = 4$ is equivalent to $x - 3 = 8$. This equation is of the form $ax + b = c$, where $a = 1$, $b = -3$, and $c = 8$. So the equation is linear.

PRACTICE 1

Determine whether these equations are quadratic or linear.

a. $3(x + 1) = 6$

b. $2x^2 = x^2 + 7x$

As in the case of a linear equation, a value is said to be a *solution* of a quadratic equation if substituting the value for the variable makes the equation a true statement.

Using Factoring to Solve Quadratic Equations

In this section, we consider those quadratic equations that can be solved by factoring. In Chapter 19, we will consider additional approaches to solving quadratic equations.

The key to solving quadratic equations by factoring is to apply the *zero-product property*.

The Zero-Product Property

If $ab = 0$, then $a = 0$ or $b = 0$, or both a and $b = 0$.

In words, this property states that if the product of two factors is zero, then either one or both of the factors must be zero.

Consider these examples of the zero-product property.

- If $2x = 0$, then x must be 0 (since $2 \neq 0$).
- If $x(3x - 1) = 0$, then either $x = 0$ or $3x - 1 = 0$.
- If $(x - 3)(x + 2) = 0$, then either $x - 3 = 0$ or $x + 2 = 0$.

Let's see how to use the zero-product property to solve a quadratic equation that is already in factored form.

EXAMPLE 2

Solve: $(2x - 1)(x + 6) = 0$

Solution Since $(2x - 1)(x + 6) = 0$, the zero-product property tells us that at least one of the factors must equal zero, that is either $2x - 1 = 0$ or $x + 6 = 0$.

$$(2x - 1)(x + 6) = 0$$

$$2x - 1 = 0 \quad \text{or} \quad x + 6 = 0 \qquad \text{Set each factor equal to 0.}$$

$$2x = 1 \qquad\qquad x = -6 \qquad \text{Solve each equation for } x.$$

$$x = \frac{1}{2}$$

Check We replace x by the values $\frac{1}{2}$ and -6 in the original equation.

Substitute $\frac{1}{2}$ for x.

$$(2x - 1)(x + 6) = 0$$

$$\left[2\cdot\left(\frac{1}{2}\right) - 1\right]\left(\frac{1}{2} + 6\right) \stackrel{?}{=} 0$$

$$(1 - 1)(6\tfrac{1}{2}) \stackrel{?}{=} 0$$

$$0 \cdot 6\tfrac{1}{2} \stackrel{?}{=} 0$$

$$0 = 0 \quad \text{True}$$

Substitute -6 for x.

$$(2x - 1)(x + 6) = 0$$

$$[(2)(-6) - 1](-6 + 6) \stackrel{?}{=} 0$$

$$(-13)(0) \stackrel{?}{=} 0$$

$$0 = 0 \quad \text{True}$$

So the solutions of the equation $(2x - 1)(x + 6) = 0$ are $\frac{1}{2}$ and -6.

PRACTICE 2

Solve: $(3x - 1)(x + 5) = 0$

In Example 2, note that we found the solutions of a quadratic equation by solving two linear equations: $2x - 1 = 0$ and $x + 6 = 0$. Can you explain how we know that these equations are linear?

When the quadratic expression in a second-degree equation is not given in factored form, we need to factor it before solving.

EXAMPLE 3

Solve: $y^2 - 5y = 0$

Solution

$$y^2 - 5y = 0$$

$$y(y - 5) = 0 \quad \text{Factor the left side of the equation.}$$

Next, we set each factor equal to 0. Then, we solve for y.

$$y = 0 \quad \text{or} \quad y - 5 = 0$$

$$y = 5$$

Check We verify our solutions in the original equation.

Substitute 0 for y.	Substitute 5 for y.
$y^2 - 5y = 0$	$y^2 - 5y = 0$
$(0)^2 - 5(0) \stackrel{?}{=} 0$	$(5)^2 - 5(5) \stackrel{?}{=} 0$
$0 - 0 \stackrel{?}{=} 0$	$25 - 25 \stackrel{?}{=} 0$
$0 = 0$ True	$0 = 0$ True

So the solutions are 0 and 5.

PRACTICE 3

Solve: $y^2 + 6y = 0$

In order to apply the zero-product property, the product of the factors of a quadratic expression must equal zero. This implies that a quadratic equation must be written in *standard form,* $ax^2 + bx + c = 0$, before it can be solved.

EXAMPLE 4

Solve: $2x^2 + x = 1$

Solution

$$2x^2 + x = 1$$

$$2x^2 + x - 1 = 0 \quad \text{Write in standard form by adding } -1 \text{ to each side.}$$

$$(2x - 1)(x + 1) = 0 \quad \text{Factor the left side of the equation.}$$

$$2x - 1 = 0 \quad \text{or} \quad x + 1 = 0 \quad \text{Set each factor to equal to 0.}$$

$$2x = 1 \qquad x = -1 \quad \text{Solve for } x.$$

$$x = \frac{1}{2}$$

PRACTICE 4

Solve: $4y^2 - 11y = 3$

Check We verify our solutions in the original equation.

Substitute $\frac{1}{2}$ for x.	Substitute -1 for x.
$2x^2 + x = 1$	$2x^2 + x = 1$
$2\left(\frac{1}{2}\right)^2 + \frac{1}{2} \stackrel{?}{=} 1$	$2(-1)^2 + (-1) \stackrel{?}{=} 1$
$2\left(\frac{1}{4}\right) + \frac{1}{2} \stackrel{?}{=} 1$	$2(1) + (-1) \stackrel{?}{=} 1$
$1 = 1$ True	$1 = 1$ True

So the solutions are $\frac{1}{2}$ and -1.

EXAMPLE 5

Solve: $2x(x - 3) = 8$

Solution We begin by writing the equation in standard form.

$$2x(x - 3) = 8$$

$$2x^2 - 6x = 8 \quad \text{Multiply.}$$

$$2x^2 - 6x - 8 = 0 \quad \text{Write in standard form by adding } -8 \text{ to each side.}$$

$$2(x^2 - 3x - 4) = 0 \quad \text{Factor out the GCF.}$$

$$2(x + 1)(x - 4) = 0 \quad \text{Write in factored form.}$$

Next, we set factors containing variables equal to 0. Then, we solve for x.

$$x + 1 = 0 \quad \text{or} \quad x - 4 = 0$$

$$x = -1 \qquad\qquad x = 4$$

Check

Substitute -1 for x.	Substitute 4 for x.
$2x(x - 3) = 8$	$2x(x - 3) = 8$
$2(-1)(-1 - 3) \stackrel{?}{=} 8$	$2(4)(4 - 3) \stackrel{?}{=} 8$
$(-2)(-4) \stackrel{?}{=} 8$	$8(1) \stackrel{?}{=} 8$
$8 = 8$ True	$8 = 8$ True

So the solutions are -1 and 4.

PRACTICE 5

Solve: $3t(t + 4) = 15$

These examples lead us to the following strategy (or rule) for solving a quadratic equation by factoring:

To Solve a Quadratic Equation by Factoring

- If necessary, rewrite the equation in standard form with 0 on one side.
- Factor the other side.
- Use the zero-product property to get two simple linear equations.
- Solve the linear equations.
- Check by substituting the solutions in the original quadratic equation.

EXAMPLE 6

A homeowner has a square garden that she wants to make longer. If she extends one side by 5 ft and the adjacent side by 2 ft, the resulting garden would be rectangular with an area of 130 ft^2. How much fencing will she need to enclose the enlarged garden?

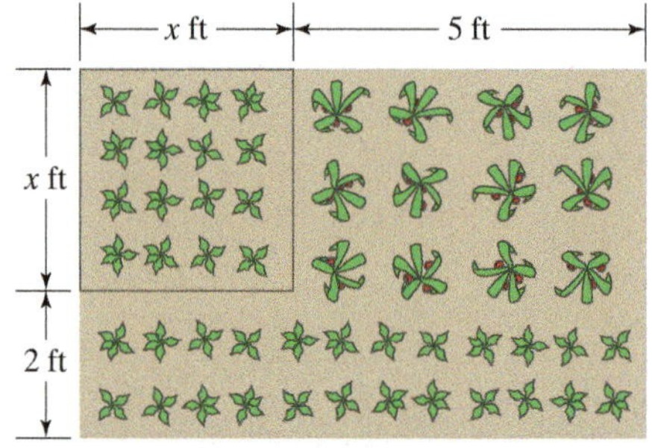

Solution Let's represent the length of each side of the square by x. The resulting rectangular garden will have dimensions $(x + 5)$ and $(x + 2)$. The area of a rectangle can be computed by multiplying its length and its width, and we are told that this area is 130.

$$(x + 5)(x + 2) = 130$$
$$x^2 + 7x + 10 = 130$$
$$x^2 + 7x - 120 = 0$$
$$(x + 15)(x - 8) = 0$$
$$x + 15 = 0 \quad \text{or} \quad x - 8 = 0$$
$$x = -15 \qquad\qquad x = 8$$

Check

Substitute -15 for x.

$$(x + 5)(x + 2) = 130$$
$$(-15 + 5)(-15 + 2) \stackrel{?}{=} 130$$
$$(-10)(-13) \stackrel{?}{=} 130$$
$$130 = 130 \quad \text{True}$$

Substitute 8 for x.

$$(x + 5)(x + 2) = 130$$
$$(8 + 5)(8 + 2) \stackrel{?}{=} 130$$
$$(13)(10) \stackrel{?}{=} 130$$
$$130 = 130 \quad \text{True}$$

So the two solutions of the equation are -15 and 8. Since x represents a length, we can reject the negative solution -15. We conclude that the length of each side of the square is 8 ft. To compute the perimeter of the rectangle, we can substitute into the following formula:

$$\begin{aligned} P &= 2l + 2w \\ &= 2(8 + 5) + 2(8 + 2) \\ &= 2(13) + 2(10) \\ &= 26 + 20 \\ &= 46 \end{aligned}$$

Therefore, 46 ft of fencing is needed to enclose the enlarged garden.

PRACTICE 6

A framemaker is planning to frame a rectangular painting that has an area of 80 in^2.

If she has 3 ft of framing to put around the picture, what should the dimensions of the frame be? (Ignore the frame's thickness.)

We can also apply what we know about solving quadratic equations to problems involving the Pythagorean theorem. This theorem relates to the lengths of the three sides of a right triangle.

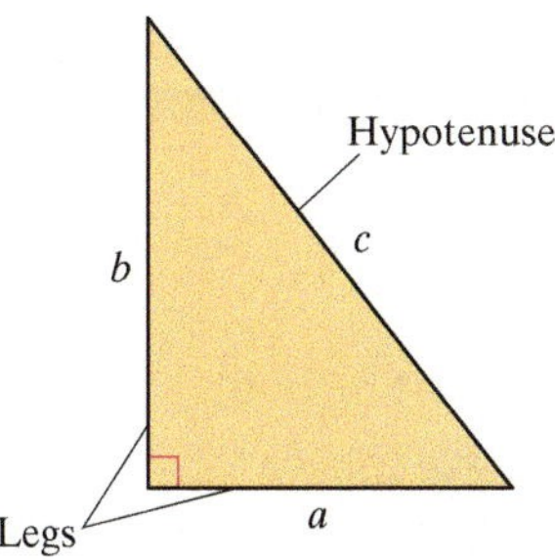

A right triangle has one 90° angle. The side opposite the 90° angle, the longest side, is called the *hypotenuse*. The other sides are called *legs*.

The Pythagorean theorem states that for every right triangle, the sum of the squares of the lengths of the legs equals the square of the length of the hypotenuse: $a^2 + b^2 = c^2$.

EXAMPLE 7

How far from the base of a building should a painter place a 17-foot ladder so that it reaches 15 ft up the building?

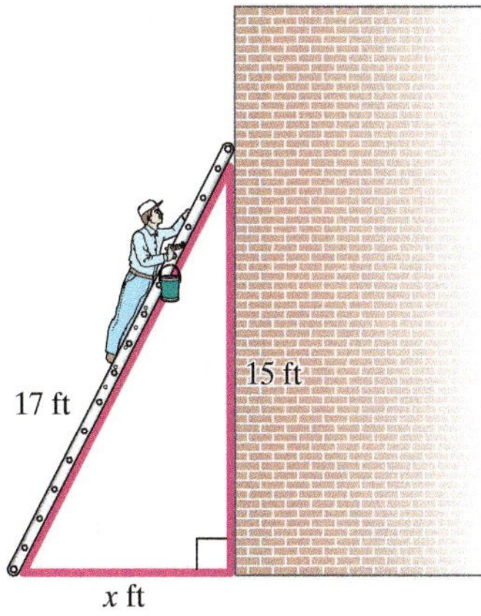

Solution Let x represent the distance from the base of the building to the bottom of the ladder. Note that the ladder, the ground, and the building form a right triangle. Using the Pythagorean theorem, we get:

$$a^2 + b^2 = c^2$$
$$x^2 + 15^2 = 17^2$$
$$x^2 + 225 = 289$$
$$x^2 - 64 = 0$$
$$(x + 8)(x - 8) = 0$$
$$x + 8 = 0 \quad \text{or} \quad x - 8 = 0$$
$$x = -8 \qquad\qquad x = 8$$

Since x represents a distance, we consider only the positive value of x, namely 8.

Check Substitute 8 for x: $x^2 + 15^2 = 17^2$

$$8^2 + 15^2 \stackrel{?}{=} 17^2$$
$$64 + 225 \stackrel{?}{=} 289$$
$$289 = 289 \quad \text{True}$$

So the painter should place the ladder 8 ft from the base of the building.

PRACTICE 7

Two scooters, traveling at constant rates, leave an intersection at the same time. One scooter travels north while the other travels east. When the scooter traveling east has gone 5 mi, the distance between the two scooters is 13 mi. How far has the scooter going north traveled?

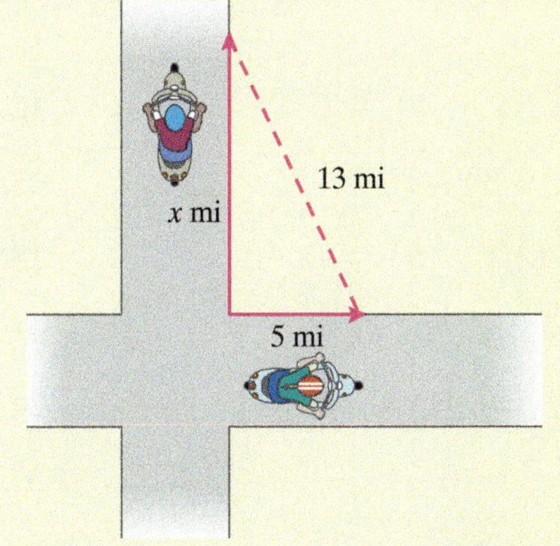

16.5 Exercises

FOR EXTRA HELP MyMathLab MathXL PRACTICE WATCH READ REVIEW

Mathematically Speaking

Fill in each blank with the most appropriate term or phrase from the given list.

sum of the squares of the legs	binomial equation	either one or both
in standard form	square of the sum of the legs	a perfect square
both	quadratic equation	

1. A second-degree or ______________ is an equation that can be written in the form $ax^2 + bx + c = 0$, where a, b, and c are real numbers and $a \neq 0$.

2. The zero-product property states that if the product of two factors is zero, then ______________ of the factors must be zero.

3. In using the zero-product property to solve a quadratic equation, the equation must be ______________.

4. The Pythagorean theorem states that for every right triangle, the ______________ equals the square of the hypotenuse.

A *Indicate whether each equation is linear or quadratic.*

5. $x^2 - 3x + 2 = 0$

6. $x^2 = 6x$

7. $(x + 3)(x - 4) = x^2$

8. $\dfrac{x + 1}{4} = 2$

9. $2x^2 + 12x = -10$

10. $(x + 4)(x - 1) = 14$

B *Solve.*

11. $(x + 3)(x - 4) = 0$

12. $(x - 2)(x - 1) = 0$

13. $4(x - 1) = 0$

14. $-7(2t + 3) = 0$

15. $y(3y + 5) = 0$

16. $2y(5y - 4) = 0$

17. $(2t + 1)(t - 5) = 0$

18. $(t - 3)(3t - 1) = 0$

19. $(2x + 3)(2x - 3) = 0$

20. $(5 - 4x)(5 + 4x) = 0$

21. $t(2 - 3t) = 0$

22. $t(1 + 2t) = 0$

23. $y^2 - 2y = 0$

24. $y^2 + 3y = 0$

25. $5x - 25x^2 = 0$

26. $3t + 6t^2 = 0$

27. $x^2 + 5x + 6 = 0$

28. $x^2 + x - 2 = 0$

29. $x^2 + x - 56 = 0$

30. $y^2 - y - 90 = 0$

31. $2x^2 - 5x - 3 = 0$

32. $2y^2 + 5y - 12 = 0$

33. $6x^2 - x - 2 = 0$

34. $4t^2 - 8t + 3 = 0$

35. $0 = 36x^2 - 12x + 1$

36. $0 = 25y^2 + 10y + 1$

37. $r^2 - 121 = 0$

38. $t^2 - 49 = 0$

39. $0 = (2x - 3)^2$

40. $0 = (4x + 1)^2$

41. $16x^2 - 16x + 4 = 0$

42. $9t^2 + 18t + 9 = 0$

43. $9m^2 + 15m - 6 = 0$

44. $6n^2 - 32n + 10 = 0$

45. $r^2 - r = 6$

46. $r^2 + 3r = 10$

47. $y^2 - 7y = -12$

48. $y^2 - y = 12$

49. $n^2 + 2n = 8$

50. $t^2 - 3t = 18$

51. $3y^2 + 4y = -1$

52. $3k^2 + 17k = -10$

53. $4x^2 + 6x = -2$

54. $12x^2 + 33x = 9$

55. $2n^2 = -10n$

56. $3m^2 = 6m$

57. $4x^2 = 1$

58. $9y^2 = 16$

59. $8y^2 = 2$

60. $12x^2 = 48$

61. $3r^2 + 6r = 2r^2 - 9$

62. $5n^2 + 36 = 12n + 4n^2$

63. $x(x - 1) = 12$

64. $r(r + 3) = 10$

65. $4t(t - 1) = 24$

66. $2n(n + 7) = -24$

67. $(y + 3)(y - 2) = 14$

68. $(m - 6)(m + 1) = -10$

69. $(3n - 2)(n + 5) = -14$

70. $(t - 5)(t + 2) = 18$

71. $(n + 2)(n + 4) = 12n$

72. $(3x + 5)(x - 1) = 16x$

73. $3x(2x - 5) = x^2 - 10$

74. $n^2 + 8 = 3n(n - 2)$

Mixed Practice

Solve.

75. $4r^2 + 11r - 3 = 0$

76. $0 = 9u^2 + 6u + 1$

77. $10x^2 + 25x - 15 = 0$

78. $n^2 + 3n = 28$

79. $5a^2 + 19a = 4$

80. $9b^2 = 27b$

81. $3s^2 + 64 = 16s + 2s^2$

82. $(x - 6)(x + 2) = 33$

83. $6t(3t - 5) = 0$

84. $(7 - 3m)(7 + 3m) = 0$

85. $4j + 8j^2 = 0$

Solve.

86. Is the equation $(x - 3)(x + 1) = 5$ linear or quadratic?

Applications

C *Solve.*

87. In a certain league, the teams play each other twice in a season. It can be shown that if there are n teams in the league, the teams must play $n^2 - n$ games. If the league plays 210 games in a season, how many teams were in the league?

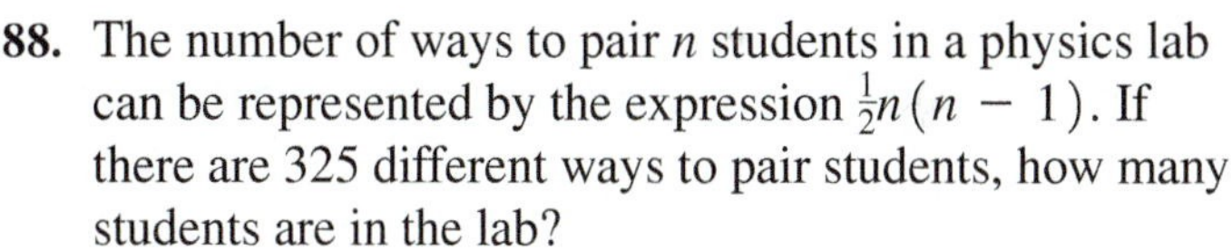

88. The number of ways to pair n students in a physics lab can be represented by the expression $\frac{1}{2}n(n - 1)$. If there are 325 different ways to pair students, how many students are in the lab?

89. Two cars leave an intersection, one traveling west and the other south. After some time, the faster car is 2 mi farther away from the intersection than the slower car. At that time, the two cars are 10 mi apart. How far did each car travel?

90. The sail on a sailboat is a right triangle in which the hypotenuse is called the leech. A 12-foot tall mainsail has a leech length of 13 ft. If sailcloth costs $\$10/\text{ft}^2$, what is the cost of a new mainsail?

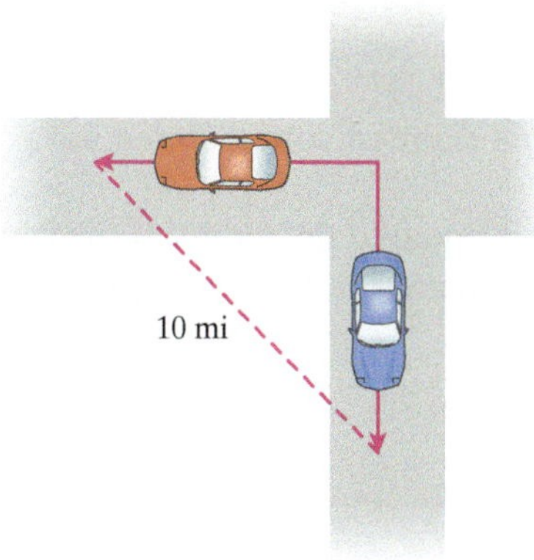

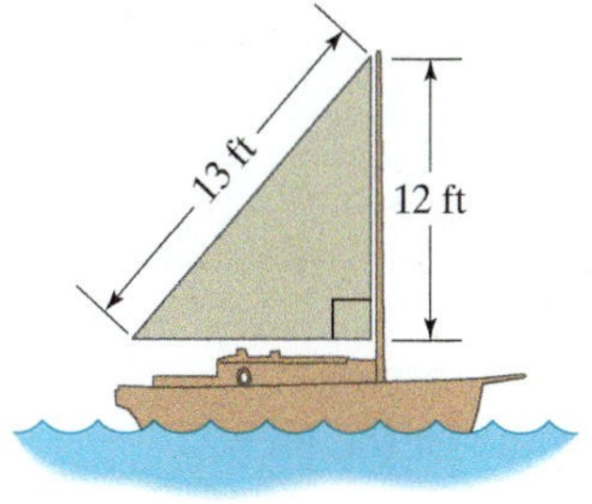

91. The base of 1 World Trade Center in New York City is a cube. The area of each side of the base is 40,000 ft^2. What is the height of the base of the tower? (*Source:* worldconstructionnetwork.com)

92. A diver jumps from a diving board 24 ft above a pool. After t sec, the diver's height h above the water (in feet) is given by the expression $-16t^2 + 8t + 24$. In how many seconds will the diver hit the water?

93. A businessman invested \$8000 in a high-risk growth fund that after two years was worth \$12,500. His broker used the equation $8000(1 + r)^2 = 12{,}500$ to find the average annual rate of return. What is this rate?

94. An artist wants to frame an 8-inch by 10-inch painting with a uniform border around the painting. She only has enough materials to cover 40 $in.^2$ of border. How wide can the border be based on her amount of materials?

95. The St. Louis Gateway Arch is the tallest national monument in the United States. At x yards (on the ground) from the center, the height of the arch (in yards) can be approximated by the expression $\frac{2}{105}[(105)^2 - x^2]$. (*Source:* gatewayarch.com)

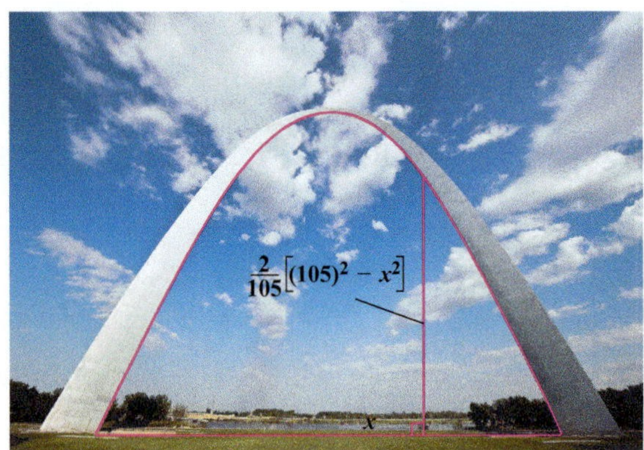

a. Factor this expression.
b. Find the width of the arch.

96. A standard DVD has a radius of 6 cm. The amount of data (in gigabytes) stored on the DVD from the center to r cm out can be modeled by the expression $0.04677\pi(r^2 - 4)$ except on the center ring on which the DVD spins, where no data can be stored. (*Source:* iso.org)

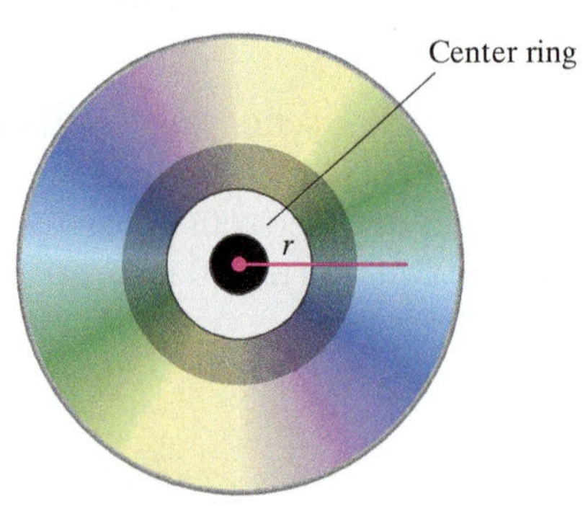

a. Write this expression in factored form.

b. When r is equal to the radius of the center ring, the amount of data stored is 0. What is the radius of the center ring?

• Check your answers on page A-35.

MINDStretchers

Mathematical Reasoning

1. Give an example of an equation:
   a. whose solutions are 2 and 3.
   b. whose only solution is 5.
   c. whose solutions have opposite signs.
   d. whose solutions are 2, 3, and 4.

Technology

2. Consider the quadratic equation $x^2 - 3x - 10 = 0$.
   a. Solve this equation.
   b. On a calculator or a computer, graph $y = x^2 - 3x - 10$. Use the graph to find the x-intercepts.
   c. Explain how you can use these intercepts to solve the original equation $x^2 - 3x - 10 = 0$.

Writing

3. Explain whether the zero-product property is true for more than two factors.

Key Concepts and Skills

CONCEPT SKILL

Concept/Skill	Description	Example
[16.1] **Common factor of two or more integers**	An integer that is a factor of each integer.	5 is a common factor of 15 and 50.
[16.1] **Greatest common factor (GCF) of two or more integers**	The greatest integer that is a factor of each integer.	The GCF of 45, 63, and 81 is 9.
[16.1] **Greatest common factor (GCF) of two or more monomials**	The product of the greatest common factor of the coefficients and, for each variable, the variable to the lowest power to which it is raised in any of the monomials.	The GCF of $6x^4$, $8x^3$, and $12x^2$ is $2x^2$.
[16.1] **Factoring by grouping**	Group pairs of terms and factor out a GCF in each group, if necessary. Then, factor out the common binomial factor.	$xy + x - 4y - 4$ $= (xy + x) + (-4y - 4)$ $= x(y + 1) - 4(y + 1)$ $= (y + 1)(x - 4)$
[16.2] **Factoring a trinomial of the form $ax^2 + bx + c$, where $a = 1$**	• List and test the factors of c to find two integers for $(x + ?)(x + ?)$ whose product is c and whose sum is b.	$x^2 - 8x + 12 = (x - 2)(x - 6)$ because Factors of 12 $(-2)\cdot(-6) = 12$ and $(-2) + (-6) = -8$ Sum of factors
[16.2] **Prime polynomial**	A polynomial that is not factorable.	$x^2 + 5x + 1$
[16.3] **Factoring a trinomial $ax^2 + bx + c$, where $a \neq 1$ (Trial-and-Error Method)**	• List and test the factors of a and of c to find four integers for $(?x + ?)(?x + ?)$ so that the product of the leading coefficients of the binomial factors is a, the product of the constant terms of the binomial factors is c, and the coefficients of the inner and outer products add up to b.	$2x^2 + 15x + 7 = (2x + 1)(x + 7)$ because $2 \cdot 1 = 2$, $1 \cdot 7 = 7$, and $2 \cdot 7 + 1 \cdot 1 = 15$.
[16.3] **Factoring a trinomial $ax^2 + bx + c$, where $a \neq 1$ (ac Method)**	• Form the product ac. • Find two factors of ac that add up to b. • Use these factors to split up the middle term in the original trinomial. • Group the first two terms and the last two terms. • From each group, factor out the common factor. • Factor out the common binomial factor.	For $2x^2 + 15x + 7$, $ac = 2 \cdot 7 = 14$ $1 \cdot 14 = 14$ and $1 + 14 = 15$ $2x^2 + 15x + 7$ $= 2x^2 + x + 14x + 7$ $= (2x^2 + x) + (14x + 7)$ $= x(2x + 1) + 7(2x + 1)$ $= (2x + 1)(x + 7)$
[16.4] **Factoring a perfect square trinomial**	$a^2 + 2ab + b^2 = (a + b)^2$ $a^2 - 2ab + b^2 = (a - b)^2$	$x^2 + 12x + 36 = (x + 6)^2$ $x^2 - 6x + 9 = (x - 3)^2$
[16.4] **Factoring the difference of squares**	$a^2 - b^2 = (a + b)(a - b)$	$x^2 - 100 = (x + 10)(x - 10)$

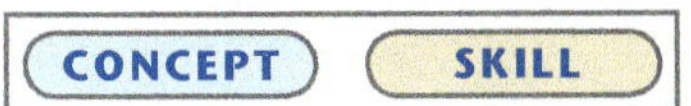

Concept/Skill	Description	Example
[16.5] **Second-degree or quadratic equation**	An equation that can be written in the form $ax^2 + bx + c = 0$, where a, b, and c are real numbers and $a \neq 0$.	$x^2 - x + 6 = 0$
[16.5] **The zero-product property**	If $ab = 0$, then $a = 0$ or $b = 0$, or both a and $b = 0$.	If $x(3x - 1) = 0$, then either $x = 0$ or $3x - 1 = 0$.
[16.5] **To solve a quadratic equation by factoring**	• If necessary, rewrite the equation in standard form with 0 on one side. • Factor the other side. • Use the zero-product property to get two simple linear equations. • Solve the linear equations. • Check by substituting the solutions in the original quadratic equation.	$x(x - 3) = 4$ $x^2 - 3x = 4$ $x^2 - 3x - 4 = 0$ $(x - 4)(x + 1) = 0$ $x - 4 = 0$ or $x + 1 = 0$ $x = 4$ $\quad$ $x = -1$ **Check** Substitute 4 for x. $4(4 - 3) \stackrel{?}{=} 4$ $4(1) \stackrel{?}{=} 4$ $4 = 4$ True Substitute -1 for x. $(-1)(-1 - 3) \stackrel{?}{=} 4$ $(-1)(-4) \stackrel{?}{=} 4$ $4 = 4$ True

CHAPTER 16 Review Exercises

Say Why

Fill in each blank.

1. The trinomial $x^2 + 2x + 2$ ________ (is/is not) a prime polynomial because ____.

2. The polynomial $4x^2 - 2x - 6$ ________ (is/is not) factored completely as $2(2x^2 - x - 3)$ because ____.

3. The trinomial $x^2 - 4xy + 4y^2$ ________ (is/is not) a perfect square trinomial because ____.

4. The binomial $9x^2 - 4y^2$ ________ (can/cannot) be factored as $(3x + 2y)(3x - 2y)$ because ____.

[16.1] *Find the greatest common factor.*

5. 48, 36, and 60

6. $9m^3n$, $24m^4$, and $15m^2n^2$

Factor.

7. $3x - 6y$

8. $16p^3q^2 + 18p^2q - 4pq^2$

9. $(n - 1) + n(n - 1)$

10. $xb - 5b - 2x + 10$

Solve for the indicated variable.

11. $d = rt_1 + rt_2$ for r

12. $ax + y = bx + c$ for x

[16.2] *Factor, if possible.*

13. $x^2 + x + 1$

14. $m^2 - m + 3$

15. $y^2 + 42 + 13y$

16. $m^2 - 7mn + 10n^2$

17. $24 - 8x - 2x^2$

18. $-15xy^2 + 3x^3 - 12x^2y$

[16.3] *Factor, if possible.*

19. $3x^2 + 5x - 2$

20. $5n^2 + 13n + 6$

21. $3n^2 - n - 1$

22. $6x^2 - x - 12$

23. $2a^2 + 3ab - 35b^2$

24. $16a - 4a^2 - 15$

25. $9y^3 - 21y + 60y^2$

26. $2p^2q - 3pq^2 - 2q^3$

[16.4] *Factor, if possible.*

27. $b^2 - 6b + 9$

28. $64 - x^2$

29. $25y^2 - 20y + 4$

30. $9a^2 + 24ab + 16b^2$

31. $81p^2 - 100q^2$

32. $4x^8 - 28x^4 + 49$

33. $48x^4 - 3y^4$

34. $x^2(x - 1) - 9(x - 1)$

[16.5] *Solve.*

35. $(x + 2)(x - 1) = 0$

36. $t(t - 4) = 0$

37. $3x^2 + 18x = 0$

38. $4x^2 + 4x + 1 = 0$

39. $y^2 - 10y = -16$

40. $3k^2 - k = 2$

41. $4n(2n + 3) = 20$

42. $(y - 1)(y + 2) = 10$

Mixed Applications

Solve.

43. The length of an object varies with its temperature. The expression $aLt_2 - aLt_1$ represents the change in the length of an object heated to temperature t_2, where L is its length at temperature t_1 and a is the *coefficient of linear expansion*, a constant that depends on the material that the object is made of. Write this expression in factored form.

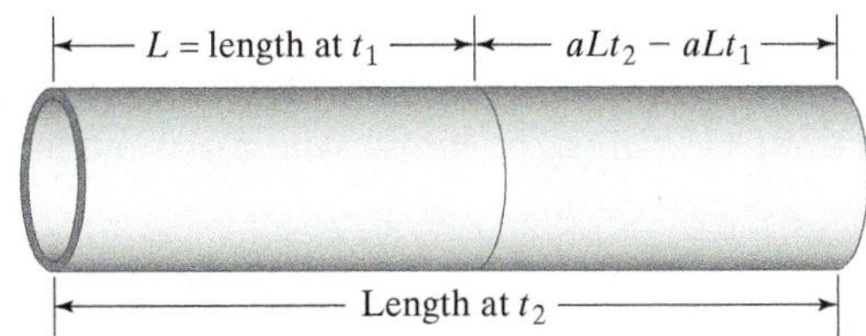

44. If a ball is thrown straight upward at v ft/sec, its height above the point of release is given by the expression $vt - 16t^2$, where t is the number of seconds after release. Write an expression in factored form for the distance between the object's location at t_1 and at t_2.

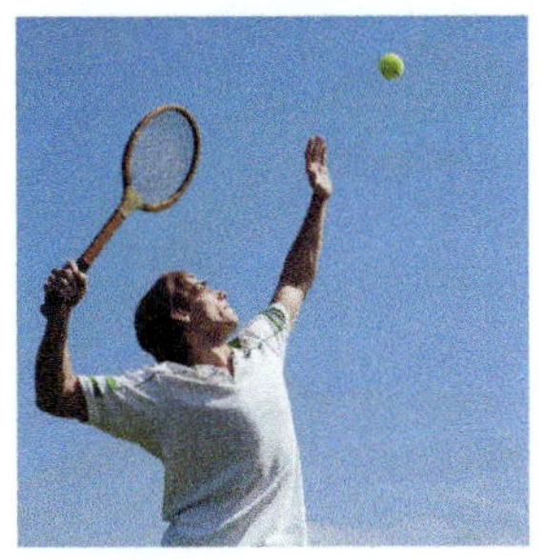

45. Find the distance between the two intersections shown on the city grid pictured if each block is 500 ft long.

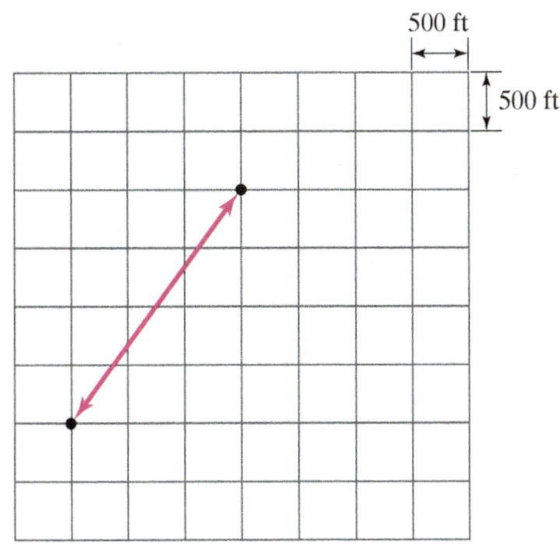

46. A kite maker designs the diamond-shaped kite shown. The diagonals of the kite cross at right angles. The vertical diagonal is 52 in. long. What is the length of the horizontal diagonal of the kite shown?

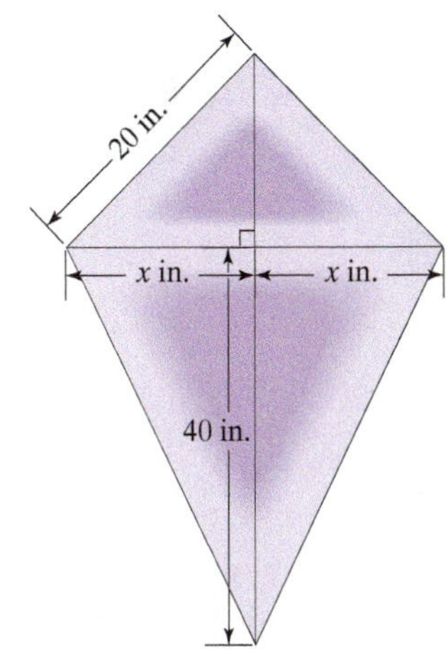

47. After t sec, the height of a rocket launched straight upward from ground level with an initial velocity of 76 ft/sec can be modeled by the polynomial $76t - 16t^2$. After how many seconds will the rocket reach a height of 18 ft above the launch (that is, equal to $+18$)?

48. The formula for the area of the trapezoid shown is $A = \frac{1}{2}hb + \frac{1}{2}hB$. Solve this formula for h in terms of A, b, and B.

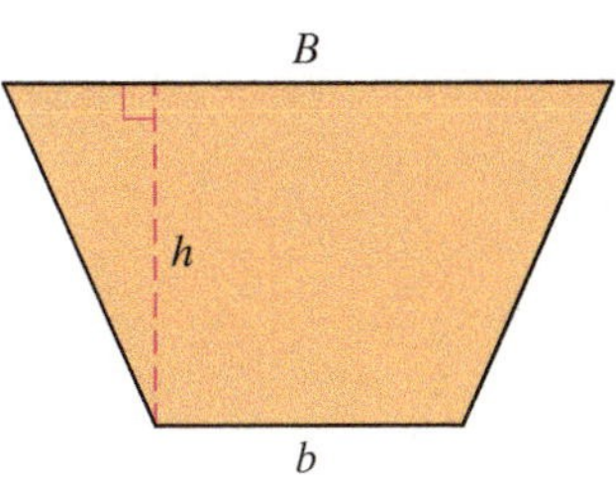

• Check your answers on page A-36.

CHAPTER 16 Posttest

To see whether you have mastered the topics in this chapter, take this test.

1. Find the greatest common factor of $12x^3$ and $15x^2$.

Factor.

2. $2xy - 14y$

3. $6pq^2 + 8p^3 - 16p^2q$

4. $ax - bx + by - ay$

5. $n^2 - 13n - 48$

6. $-8 + x^2 - 2x$

7. $15x^2 - 5x^3 + 20x$

8. $4x^2 + 13xy - 12y^2$

9. $-12x^2 + 36x - 27$

10. $9x^2 + 30xy + 25y^2$

11. $121 - 4x^2$

12. $p^2q^2 - 1$

13. $y^4 - 8y^2 + 16$

Solve.

14. $(n + 8)(n - 1) = 0$

15. $6x^2 + 10x = 4$

16. $(2n + 1)(n - 1) = 5$

17. The energy it takes to lift an object of mass m from level y_1 to level y_2 is $mgy_2 - mgy_1$, where g is a constant. Factor this expression.

18. In the right triangle shown, find the length of the missing side.

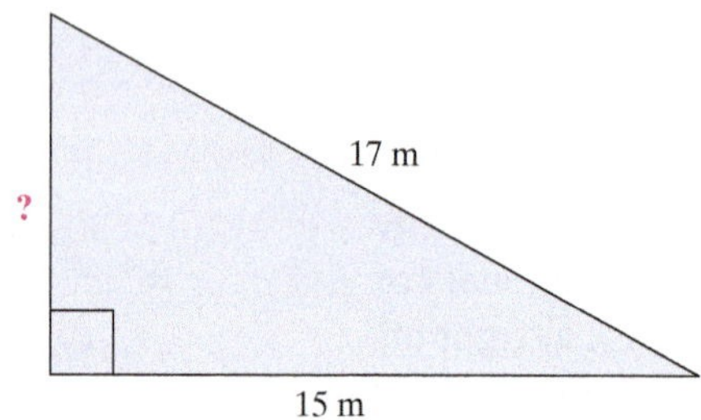

19. A rectangular garden measures 25 ft by 30 ft. The gardener wishes to surround the garden with a border of mulch x ft wide, as shown. Write an expression in factored form for the area of the mulch border in terms of x.

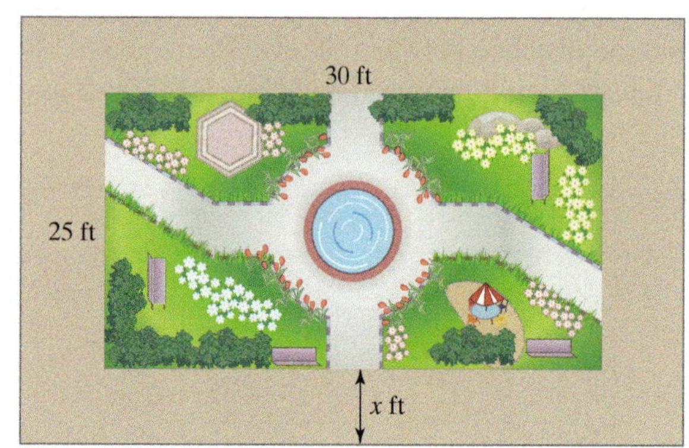

20. In a physics experiment, a weight is dropped from a platform 9 ft above the ground. The time t (in seconds) it takes the weight to reach the ground may be found by solving the equation $9 - 16t^2 = 0$. Solve for t.

• Check your answers on page A-36.

Cumulative Review Exercises

To help you review, solve the following:

1. Simplify: $2[11 + (5 + 4 \cdot 6^2)]$

2. Subtract: $4\frac{1}{2} - 2\frac{5}{6}$

3. Calculate $(0.4)^3$.

4. Find the unit price: 50 min of Web access for $4

5. 4 oz is what percent of 2 lb?

6. Solve: $z + 4 = 5z - 2(z + 6)$

7. Graph $2x + 4y = 12$.

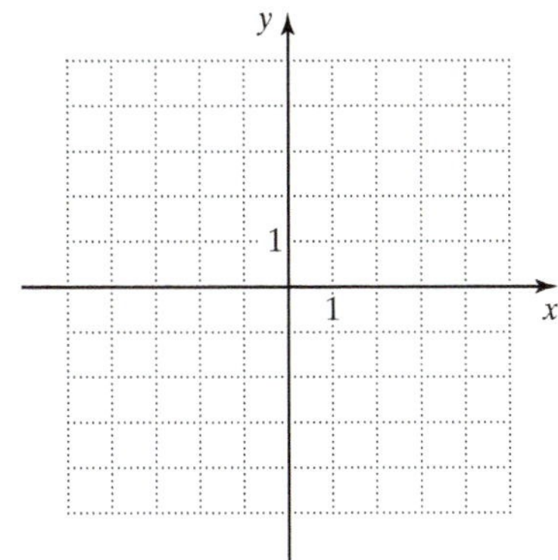

8. Graph the inequality $4x - 3y < 12$ on a coordinate plane.

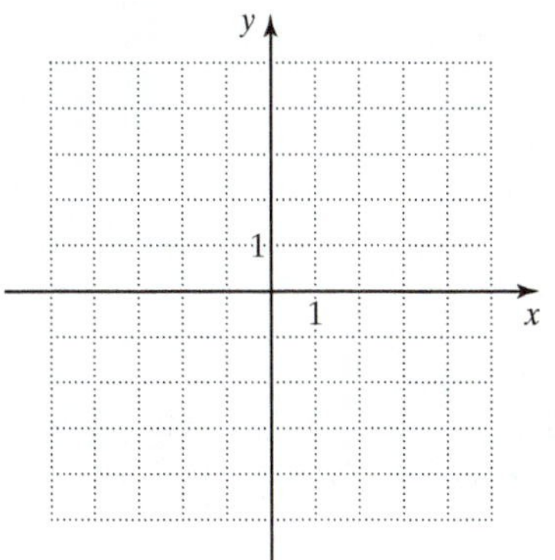

9. Solve by substitution and check:

$$4x + 2y = 1$$
$$y = 3 - 2x$$

10. Solve the system: **(1)** $x + y = 6$
(2) $y = 2x + 9$

11. Simplify: $\dfrac{n^2 \cdot n^3 \cdot n}{n^5}$

12. Find the product: $(3x + 1)(7x - 2)$

13. Factor, if possible: $16x^2 - 24x + 9$

14. Solve: $8x^2 + 22x = 6$

15. A share of Proctor & Gamble stock (ticker PG) opened at \$61.91. Five days later, it closed at \$61.16. What was the average daily change in the price of a share? (*Source: New York Times*)

16. A part-time college student pays a student fee of f dollars plus c dollars per credit. What is the charge for a student with an 8-credit schedule?

17. The owner of a shirt factory has fixed expenses of \$500 per day. It costs the factory \$15 to produce each shirt. If the shirts are sold wholesale at \$25 apiece, how many shirts must be sold per day to break even?

18. The following graph shows the average number of hours that American Internet users spent online per week in various years. (*Source: The World Almanac and Book of Facts, 2011*)

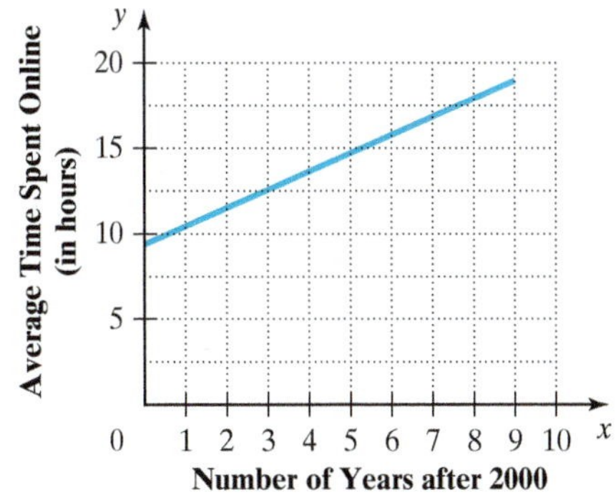

Is the slope of this line positive, negative, or zero?

19. The distance from the Earth to the Sun is approximately 150,000,000 km. Express this distance in scientific notation. (*Source:* Bennett et al., *The Cosmic Perspective*)

20. An expression for the total surface area of the tin can shown below is $2\pi r^2 + 2\pi rh$. Rewrite this expression in factored form.

• Check your answers on page A-36.

CHAPTER 17

Rational Expressions and Equations

Rational Expressions and Grapes

Science plays a role in viticulture, the art of growing grapes. For instance, scientists have used DNA fingerprinting—the same technique used in paternity suits and criminal trials—to trace the parentage of numerous grape varieties.

Viticulture also makes use of algebra. Since much of the flavor and color of grapes is in their skin, wine growers want to raise grapes with an increased surface-to-volume ratio. This ratio is larger for smaller grapes, which have proportionately more skin than larger grapes.

We can see why this is true by assuming that a grape is approximately a sphere of radius r. The ratio of a sphere's surface area to its volume is represented by the rational expression $\frac{4\pi r^2}{\frac{4}{3}\pi r^3}$, which simplifies to $\frac{3}{r}$. Substituting smaller values for the radius r in this expression gives larger values of the expression. So the smaller the grape is, the larger the surface-to-volume ratio.

(*Source:* David L. Wheeler, "Scholars Marry the Science and the Art of Winemaking," *Chronicle of Higher Education*, May 30, 1997)

CHAPTER 17 PRETEST

To see if you have already mastered the topics in this chapter, take this test.

1. Identify the values for which the rational expression $\frac{5}{x+6}$ is undefined.

2. Show that the rational expressions $\frac{n^2 - 2n}{3n}$ and $\frac{n-2}{3}$ are equivalent if $n \neq 0$.

Simplify.

3. $\frac{24x^2y^3}{6xy^5}$

4. $\frac{4a^2 - 8a}{a - 2}$

5. $\frac{w^2 - 6w}{36 - w^2}$

6. $\frac{\frac{5n}{8}}{\frac{n^3}{16}}$

7. Write $\frac{2}{3a}$ and $\frac{1}{9a - 18}$ in terms of their LCD.

Perform the indicated operation.

8. $\frac{12y}{y+1} - \frac{7y+2}{y+1}$

9. $\frac{1}{6x^2} + \frac{5}{4x}$

10. $\frac{3}{c-3} - \frac{1}{c+3}$

11. $\frac{1}{x^2 - 2x + 1} - \frac{2}{1 - x^2}$

12. $\frac{15a^3b}{20n^4} \cdot \frac{16n^2}{9ab}$

13. $\frac{y-4}{5y^2 + 10y} \cdot \frac{y^2 - 2y - 8}{y^2 - 16}$

14. $\frac{x^2 - x - 2}{x^2 + 5x + 4} \div \frac{x^2 - 7x + 10}{x - 5}$

Solve and check.

15. $\frac{3x - 8}{x^2 - 4} + \frac{2}{x - 2} = \frac{7}{x + 2}$

16. $\frac{x}{x+4} - 1 = \frac{2}{x-1}$

Solve.

17. A college student makes \$100 working at a part-time job. If the student's hourly rate of pay were doubled, then the student could work 5 hr less and make the same total amount of money. What is the student's hourly rate of pay?

18. An ice cream company has fixed costs of \$1500 and variable costs of \$2 for each gallon of ice cream it produces. The cost per gallon of producing x gal of ice cream is given by the expression $2 + \frac{1500}{x}$. Write this cost as a single rational expression.

19. On the first part of a 360-mile trip, a family drives 195 mi at an average speed of r mph. They drive the remainder of the trip at an average speed that is 10 mph less than their speed during the first part of the trip. If the entire trip took 6 hr, what was the average speed during each part of the trip?

20. It takes a photocopier t min to make 30 copies. If at the same rate it takes 5 min longer to make 90 copies, how long does it take the photocopier to make 30 copies?

• Check your answers on page A-36.

17.1 Rational Expressions

OBJECTIVES

- **A** To identify values for which a rational expression is undefined
- **B** To determine whether rational expressions are equivalent
- **C** To simplify rational expressions
- **D** To solve applied problems involving rational expressions

What Rational Expressions Are and Why They Are Important

In this chapter, we move beyond our previous discussion of polynomials to consider a type of algebraic expression called a *rational expression*. Rational expressions, sometimes called *algebraic fractions*, are useful in many disciplines, including the sciences, the social sciences, medicine, and business.

For instance, the work of some anthropologists who study the history of the human species involves rational expressions. In studying a fossil record, anthropologists examine ancient skulls and compute $\frac{W}{L}$, the ratio of each skull's width W to its length L. This expression is an example of a rational expression.

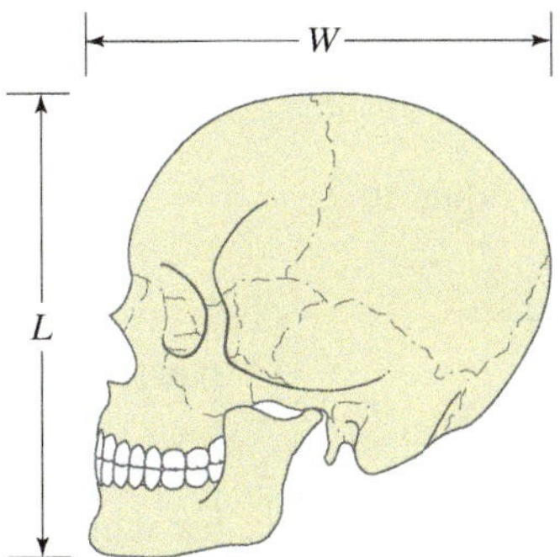

In this chapter, we discuss how to simplify rational expressions, as well as how to carry out operations on rational expressions. We also focus on how to solve equations involving rational expressions.

Introduction to Rational Expressions

Rational expressions in algebra are similar to fractions in arithmetic. In arithmetic, a fraction such as $\frac{3}{4}$ is the quotient of two integers, whereas in algebra, a rational expression such as $\frac{W}{L}$ is the quotient of two polynomials.

DEFINITION

A *rational expression* $\frac{P}{Q}$ is an algebraic expression that can be written as the quotient of two polynomials, P and Q, where $Q \neq 0$.

Other examples of rational expressions are:

$$\frac{5}{x-2} \qquad \frac{n^2+2n-1}{n+1} \qquad \frac{-a^2}{7bc}$$

A rational expression can be written in terms of division. For example,

$$\frac{5}{x-2} \text{ can be written as } 5 \div (x-2), \text{ and}$$

$$\frac{n^2+2n-1}{n+1} \text{ as } (n^2+2n-1) \div (n+1).$$

Since we can write a rational expression as division, we must be sure that its denominator does not equal 0. *When a variable is replaced with a value that makes the denominator 0, the rational expression is undefined.*

For instance, consider the rational expression $\frac{5}{x-2}$. For what values of x is this expression undefined? Setting the denominator $x - 2$ equal to 0 gives us:

$$x - 2 = 0$$
$$x = 2$$

When x is replaced with 2 in the rational expression, we get:

$$\frac{5}{x-2} = \frac{5}{2-2} = \frac{5}{0} \leftarrow \text{Undefined}$$

So $\frac{5}{x-2}$ is undefined when x is equal to 2.

EXAMPLE 1

Identify all numbers for which the following rational expressions are undefined.

a. $\frac{3}{x+5}$ **b.** $\frac{8x}{x^2-3x+2}$

Solution First, we find the values for which the rational expressions are undefined by setting each denominator equal to 0. Then, we solve for x.

a. $\frac{3}{x+5}$

$$x + 5 = 0$$
$$x = -5$$

So $\frac{3}{x+5}$ is undefined when x is equal to -5.

b. $\frac{8x}{x^2-3x+2}$

$$x^2 - 3x + 2 = 0$$
$$(x-2)(x-1) = 0 \quad \text{Factor.}$$
$$x - 2 = 0 \quad \text{or} \quad x - 1 = 0 \quad \text{Set each factor equal to 0.}$$
$$x = 2 \qquad\qquad x = 1 \quad \text{Solve for } x.$$

So $\frac{8x}{x^2-3x+2}$ is undefined when x is equal to 1 or 2.

PRACTICE 1

Indicate the values of the variable for which each rational expression is undefined.

a. $\frac{n+1}{n-4}$

b. $\frac{6}{n^2-9}$

Note that throughout the remainder of this book, *we will assume that no variable in a rational expression represents a value that makes the denominator 0.*

Equivalent Expressions

Recall from Chapter 2 that equivalent fractions, such as $\frac{1}{2}$ and $\frac{3}{6}$, are fractions that have the same value even though they are written differently. Similarly, **equivalent rational expressions** are rational expressions that have the same value, for any value of the variable that makes neither denominator 0.

To Find an Equivalent Rational Expression

Multiply the numerator and denominator of $\frac{P}{Q}$ by the same polynomial R.

$$\frac{P}{Q} = \frac{PR}{QR},$$

where $Q \neq 0$ and $R \neq 0$.

In words, this rule states that to find an equivalent rational expression, multiply the numerator and denominator of a rational expression by the same nonzero polynomial, as shown in Example 2 below.

EXAMPLE 2

Indicate whether each pair of rational expressions is equivalent.

a. $\frac{2}{3}$ and $\frac{2x}{3x}$ **b.** $\frac{d-5}{3}$ and $\frac{4d^2-20d}{12d}$

Solution

a. Multiplying both the numerator and denominator of $\frac{2}{3}$ by x, we see that $\frac{2}{3} = \frac{2x}{3x}$. So the expressions $\frac{2}{3}$ and $\frac{2x}{3x}$ are equivalent.

b. Multiplying both the numerator and denominator of $\frac{d-5}{3}$ by $4d$, we get:

$$\frac{(d-5)\cdot 4d}{3\cdot 4d} = \frac{4d^2-20d}{12d}$$

So $\frac{d-5}{3}$ and $\frac{4d^2-20d}{12d}$ are equivalent.

PRACTICE 2

Determine whether each pair of rational expressions is equivalent.

a. $\frac{b}{3}$ and $\frac{3b^2}{9b}$

b. $\frac{5}{a+7}$ and $\frac{5a}{a^2+7a}$

Note that we can also use the rule to find an equivalent rational expression by dividing the numerator and denominator of the rational expression by the same nonzero polynomial.

$$\frac{x^2 - 4}{x^2 + 5x + 6} = \frac{(x - 2)(x + 2)}{(x + 3)(x + 2)}$$ Factor the numerator and denominator.

$$= \frac{(x - 2)\overset{1}{\cancel{(x + 2)}}}{(x + 3)\underset{1}{\cancel{(x + 2)}}}$$ Divide out the common factor $(x + 2)$.

$$= \frac{x - 2}{x + 3}$$

So $\frac{x^2 - 4}{x^2 + 5x + 6}$ is equivalent to $\frac{x - 2}{x + 3}$.

TIP When simplifying a rational expression, do not divide out common terms in a sum or difference in the numerator and denominator. For instance, do not divide out the x's in $\frac{x - 2}{x + 3}$.

A rational expression is said to be *in simplest form* (or *reduced to lowest terms*) when its numerator and denominator have no common factor other than 1 or -1. Throughout the remainder of this book, we generally *simplify any answer that is a rational expression*. Can you explain the similarities and differences in simplifying a fraction in arithmetic and simplifying a rational expression in algebra?

To Simplify a Rational Expression

- Factor the numerator and denominator.
- Divide out any common factors in the numerator and denominator.

EXAMPLE 3

Write in simplest form.

a. $\frac{10x^2}{-5xy}$ b. $\frac{-4a^3b^2}{-2ab}$

Solution

a. $\frac{10x^2}{-5xy} = \frac{5x(2x)}{5x(-y)}$ Factor the numerator and denominator.

$$= \frac{\overset{1}{\cancel{5x}}(2x)}{\underset{1}{\cancel{5x}}(-y)}$$ Divide out the common factor $5x$.

$$= -\frac{2x}{y}$$ Simplify.

b. $\frac{-4a^3b^2}{-2ab} = \frac{-2ab(2a^2b)}{-2ab}$ Factor the numerator and denominator.

$$= \frac{\overset{1}{\cancel{-2ab}}(2a^2b)}{\underset{1}{\cancel{-2ab}}} = 2a^2b$$ Divide out the common factor $-2ab$, and simplify.

PRACTICE 3

Express in lowest terms.

a. $\frac{-12n^3}{-6mn}$

b. $\frac{-3x^4y}{2x^2y}$

EXAMPLE 4

Write in simplest form, if possible.

a. $\frac{3x - 6}{9x + 12}$ **b.** $\frac{n + 3}{2n + 6}$ **c.** $\frac{t + 3}{t - 1}$

Solution

a. $\frac{3x - 6}{9x + 12} = \frac{3(x - 2)}{3(3x + 4)}$ Factor the numerator and the denominator.

$= \frac{\overset{1}{\cancel{3}}(x - 2)}{\underset{1}{\cancel{3}}(3x + 4)}$ Divide out the common factor 3.

$= \frac{x - 2}{3x + 4}$

b. $\frac{n + 3}{2n + 6} = \frac{n + 3}{2(n + 3)}$ Factor the denominator.

$= \frac{\overset{1}{\cancel{n + 3}}}{2\underset{1}{\cancel{(n + 3)}}}$ Divide out the common factor $n + 3$.

$= \frac{1}{2}$

c. $\frac{t + 3}{t - 1}$ The numerator $t + 3$ and the denominator $t - 1$ have no common factor (other than 1). This expression cannot be simplified.

PRACTICE 4

Write in lowest terms.

a. $\frac{2y - 8}{4y + 6}$

b. $\frac{v - 4}{v + 3}$

c. $\frac{x + 2}{3x + 6}$

EXAMPLE 5

Simplify. **a.** $\frac{ab - ac}{ax + 2ay}$ **b.** $\frac{2t^2 - 2}{t^2 + t - 2}$ **c.** $\frac{3x^2 + 2x - 1}{3x^2 - 4x + 1}$

Solution

a. $\frac{ab - ac}{ax + 2ay} = \frac{a(b - c)}{a(x + 2y)}$

$= \frac{\overset{1}{\cancel{a}}(b - c)}{\underset{1}{\cancel{a}}(x + 2y)}$

$= \frac{b - c}{x + 2y}$

b. $\frac{2t^2 - 2}{t^2 + t - 2} = \frac{2(t^2 - 1)}{(t - 1)(t + 2)}$

$= \frac{2(t - 1)(t + 1)}{(t - 1)(t + 2)}$

$= \frac{2\overset{1}{\cancel{(t - 1)}}(t + 1)}{\underset{1}{\cancel{(t - 1)}}(t + 2)}$

$= \frac{2(t + 1)}{t + 2}$

c. $\frac{3x^2 + 2x - 1}{3x^2 - 4x + 1} = \frac{(3x - 1)(x + 1)}{(3x - 1)(x - 1)}$

$= \frac{\overset{1}{\cancel{(3x - 1)}}(x + 1)}{\underset{1}{\cancel{(3x - 1)}}(x - 1)} = \frac{x + 1}{x - 1}$

PRACTICE 5

Simplify.

a. $\frac{wt - wx}{wz - 3wg}$

b. $\frac{4n^2 - 4}{n^2 + 3n + 2}$

c. $\frac{2y^2 - y - 1}{2y^2 + 7y + 3}$

The following examples involve factors such as $a - b$ and $b - a$ that are opposites of each other; that is, they differ only in sign. Recall that since $(-1)(a - b) = b - a$,

it follows that $\frac{a - b}{b - a} = -1$ because $\frac{a - b}{b - a} = \frac{\overset{1}{\cancel{(a - b)}}}{-1\underset{1}{\cancel{(a - b)}}} = \frac{1}{-1} = -1$.

EXAMPLE 6

Write in lowest terms.

a. $\frac{s - 5}{5 - s}$ **b.** $\frac{2p - 10}{-p + 5}$

c. $\frac{3x - 6}{4 - x^2}$ **d.** $\frac{1 - x^2}{x^2 - 3x + 2}$

Solution

a. $\frac{s - 5}{5 - s} = \frac{s - 5}{-1(s - 5)}$ Write $5 - s$ as $-1(s - 5)$.

$= \frac{\overset{1}{\cancel{s - 5}}}{-1\underset{1}{\cancel{(s - 5)}}}$ Divide out the common factor $s - 5$.

$= \frac{1}{-1}$ Simplify.

$= -1$

b. $\frac{2p - 10}{-p + 5} = \frac{2(p - 5)}{-1(p - 5)} = \frac{2\overset{1}{\cancel{(p - 5)}}}{-1\underset{1}{\cancel{(p - 5)}}} = \frac{2}{-1} = -2$

c. $\frac{3x - 6}{4 - x^2} = \frac{3(x - 2)}{(2 - x)(2 + x)}$ Factor the numerator and denominator.

$= \frac{3(x - 2)}{-1(x - 2)(2 + x)}$ Write $(2 - x)$ as $-1(x - 2)$.

$= \frac{3\overset{1}{\cancel{(x - 2)}}}{-\underset{1}{\cancel{(x - 2)}}(2 + x)}$ Divide out the common factor $(x - 2)$.

$= -\frac{3}{x + 2}$ Simplify.

d. $\frac{1 - x^2}{x^2 - 3x + 2} = \frac{(1 + x)(1 - x)}{(x - 2)(x - 1)} = \frac{-(x + 1)(x - 1)}{(x - 2)(x - 1)}$

$= \frac{-(x + 1)\overset{1}{\cancel{(x - 1)}}}{(x - 2)\underset{1}{\cancel{(x - 1)}}} = -\frac{x + 1}{x - 2}$

PRACTICE 6

Simplify.

a. $\frac{-y - 1}{1 + y}$

b. $\frac{3x - 12}{-x + 4}$

c. $\frac{3n - 15}{25 - n^2}$

d. $\frac{4 - 9s^2}{3s^2 + s - 2}$

EXAMPLE 7

It costs a television manufacturer $\dfrac{95x + 10{,}000}{x}$ dollars per television to produce x televisions.

a. For which value of x is this rational expression undefined?

b. Explain in a sentence or two why you think that it makes sense for the cost per television set to be undefined for the value of x found in part (a).

Solution

a. A rational expression is undefined when its denominator is equal to 0. For the expression $\dfrac{95x + 10{,}000}{x}$, the denominator is 0 when $x = 0$.

b. When $x = 0$, the manufacturer is producing *no* television sets. So it makes no sense to speak of the cost per television set.

PRACTICE 7

For a circle of radius r, the ratio of its area to its circumference is given by the expression $\dfrac{\pi r^2}{2\pi r}$.

a. Simplify this expression.

b. For which value of r is the original rational expression undefined?

17.1 Exercises

FOR EXTRA HELP MyMathLab MathXL PRACTICE WATCH READ REVIEW

Mathematically Speaking

Fill in each blank with the most appropriate term or phrase from the given list.

exponential expression	undefined	rational expression
in simplest form	equivalent	equal to 0

1. A(n) ______________, $\frac{P}{Q}$, is an algebraic expression that can be written as the quotient of two polynomials, P and Q, where $Q \neq 0$.

2. A rational expression is ______________ if a variable is replaced with a value that makes its denominator 0.

3. Rational expressions are ______________ if they have the same value no matter what value replaces the variable.

4. A rational expression is said to be ______________ (or reduced to lowest terms) when its numerator and denominator have no common factor other than 1 or -1.

A *Identify the values for which the given rational expression is undefined.*

5. $\frac{7}{x}$

6. $\frac{-2}{c}$

7. $\frac{8}{y-2}$

8. $\frac{y-6}{y-1}$

9. $\frac{4}{x+2}$

10. $\frac{x-3}{x+5}$

11. $\frac{n+11}{2n-1}$

12. $\frac{x-4}{3x-2}$

13. $\frac{x^2+1}{x^2-1}$

14. $\frac{n+2}{n^2-16}$

15. $\frac{x^2+x+1}{x^2-x-20}$

16. $\frac{p^2+7}{p^2-4p-21}$

B *Indicate whether each pair of rational expressions is equivalent.*

17. $\frac{p}{q}$ and $\frac{pr}{qr}$

18. $\frac{8}{3y}$ and $\frac{16}{6y}$

19. $\frac{3t+5}{t+1}$ and $\frac{3t^2+5t}{t^2+t}$

20. $\frac{2x-1}{x-4}$ and $\frac{14x-7}{7x-28}$

21. $\frac{x-1}{x+3}$ and $-\frac{1}{3}$

22. $\frac{n+4}{2n+4}$ and $\frac{1}{2}$

23. $\frac{x-2}{2-x}$ and $\frac{2-x}{x-2}$

24. $\frac{y-x}{y+x}$ and $\frac{y+x}{y-x}$

25. $\frac{x^2+4}{x+2}$ and $x+2$

26. $\frac{(y+5)^2}{y+5}$ and $y+5$

C *Simplify.*

27. $\frac{10a^4}{12a}$

28. $\frac{4b}{6b^3}$

29. $\frac{3x^2}{12x^5}$

30. $\frac{-2y^6}{2y^4}$

31. $\frac{9s^3t^2}{6s^5t}$

32. $\frac{10r^3s^4}{5r^3s^4}$

33. $\frac{24a^4b^5}{3ab^2}$

34. $\frac{14yz^2}{21y^2z^2}$

35. $\frac{-2p^2q^3}{-10pq^4}$

36. $\frac{-3t^4s^5}{t^2s}$

37. $\frac{5x(x+8)}{4x(x+8)}$

38. $\frac{7n(n-1)}{2n(n-1)}$

39. $\frac{8x^2(5-2x)}{3x(2x-5)}$

40. $\frac{(a-b)ab^2}{3a^3(b-a)}$

41. $\frac{5x-10}{5}$

42. $\frac{3y+12}{3}$

43. $\frac{2x^2+2x}{4x^2+6x}$

44. $\frac{7y^3-5y^2}{3y^2+y}$

45. $\frac{a^2-4a}{ab-4b}$

46. $\frac{a^2+2b}{4b+2a^2}$

47. $\frac{6x-4y}{9x-6y}$

48. $\frac{10m-2n}{n-5m}$

49. $\frac{t^2-1}{t+1}$

50. $\frac{b^2-1}{b-1}$

51. $\frac{p^2-q^2}{q^2-p^2}$

52. $\frac{y^2-4x^2}{4x^2-y^2}$

53. $\frac{n-1}{2-2n}$

54. $\frac{x-3}{3-x}$

55. $\frac{(b-4)^2}{b^2-16}$

56. $\frac{m^2-1}{(m+1)^2}$

57. $\frac{2x^2+5x-3}{10x-5}$

58. $\frac{6p+12}{p^2-p-6}$

59. $\frac{a^3+9a^2+14a}{a^2-10a-24}$

60. $\frac{x^2+3x-4}{x^2+2x-3}$

61. $\frac{t^2-4t-5}{t^2-3t-10}$

62. $\frac{y^2+8y+15}{y^2-2y-15}$

63. $\frac{9-16d^2}{16d^2-24d+9}$

64. $\frac{6n^2+7n-10}{25-36n^2}$

65. $\frac{8s-2s^2}{2s^2-11s+12}$

66. $\frac{3x^2+2x-1}{12x^2-4x}$

67. $\frac{6x^2+5x+1}{6x^2-x-1}$

68. $\frac{2n^3+2n^2-4n}{n^3+2n^2-3n}$

69. $\frac{6y^2-7y+2}{6y^3+5y^2-6y}$

70. $\frac{3x^2+xy}{3x^2+7xy+2y^2}$

71. $\frac{2ab^2+4a^2b}{2b^2+5ab+2a^2}$

72. $\frac{p^2-4pq-12q^2}{2p^2-15pq+18q^2}$

73. $\frac{m^2+3mn-28n^2}{2m^2+4mn-48n^2}$

74. $\frac{3a^2+5ab-2b^2}{3a^2+8ab-3b^2}$

Mixed Practice

Simplify.

75. $\frac{12x^2y^6}{18x^3y^4}$

76. $\frac{-5a^7b^4}{a^3b^3}$

77. $\frac{3x^3-9x^2+6x}{x^3+x^2-6x}$

78. $\frac{x^2-3x-10}{x^2+4x+4}$

79. $\frac{w^2-2y^2}{2y^2-w^2}$

80. $\frac{h+3k^2}{12k^2+4h}$

81. $\frac{2x^2+x-3}{3x^2-3x}$

82. $\frac{r^2s(r-s)}{2rs^3(s-r)}$

Identify the values for which the given rational expression is undefined.

83. $\frac{p-5}{2p-3}$

84. $\frac{m+3}{m^2-25}$

Indicate whether each pair of rational expressions is equivalent.

85. $\frac{6y+7}{y+1}$ and $\frac{6y^2+7y}{y^2+y}$

86. $\frac{2u-9}{5u-9}$ and $\frac{2}{5}$

Applications

 *Solve.*

87. The surface area of a person's skin is important in both clothing design and medicine. A rough estimate of the skin area for someone with height h in. and waist circumference C in. is Ch in.^{2} Use this estimate to approximate the ratio of skin area to height. (*Source:* Weinstein and Adam, *Guesstimation: Solving the World's Problems on the Back of a Cocktail Napkin*)

88. The ratio of a company's stock price P to its earnings E, both per share, is called the company's *price-to-earnings* (or *P/E*) *ratio*. When is this ratio undefined? (*Source:* investopedia.com)

89. When a mathematics department had more faculty and staff members, each had an office with dimensions x ft by x ft. Now that the department has gotten smaller, its offices are being enlarged. Each faculty and each staff office will be made 2 ft wider, whereas each faculty office will also be made 5 ft longer.

a. Write an expression for the area (in square feet) of an enlarged faculty office.

b. Write an expression for the area (in square feet) of an enlarged staff office.

c. Write an expression for the ratio of the area of an enlarged faculty office to the area of an enlarged staff office. Simplify.

d. Find the value of the expression in part (c) if the length of each office had been 8 ft.

90. The baking time for bread depends on the ratio of its volume V to its surface area S. For each of the following types of bread, express the ratio $\frac{V}{S}$ in simplest form.

a. A rectangular loaf of bread

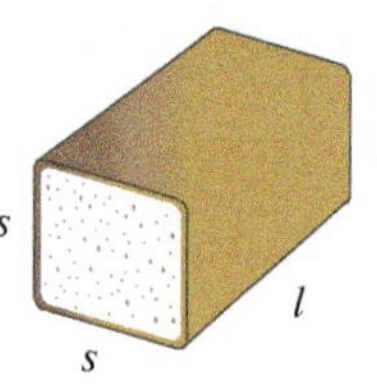

$V = s^2 l$

$S = 2s^2 + 4sl$

b. A cylindrical bread stick

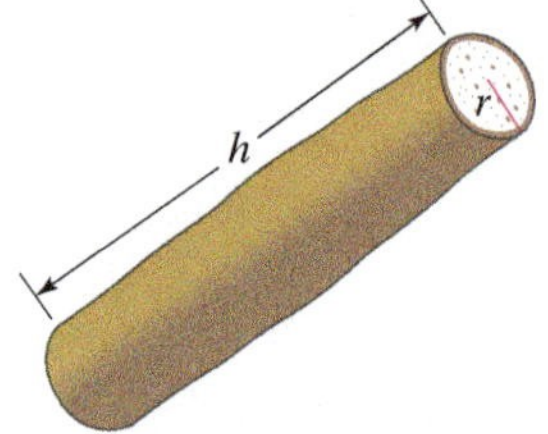

$V = \pi r^2 h$

$S = 2\pi r^2 + 2\pi rh$

91. An archer shoots an arrow at the target shown.

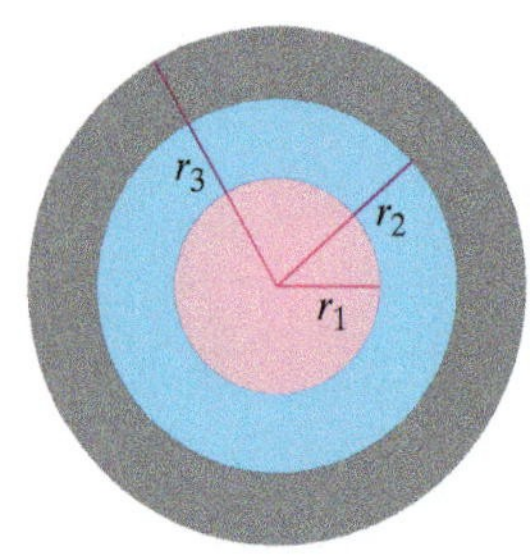

a. Find and simplify the ratio of the area of the inner circle to the area of the adjacent ring.

b. Find and simplify the ratio of the area of the smaller ring to the area of the larger ring.

92. A basketball with radius r is packed in a cubic box with side $2r$. The volume of the ball is $\frac{4}{3}\pi r^3$ and the volume of the box is $(2r)^3$.

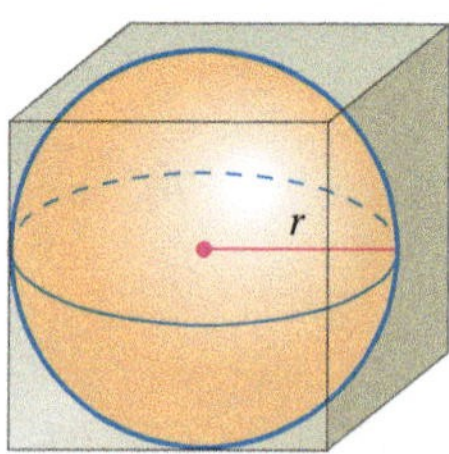

a. The smaller volume is what fraction of the larger volume? Simplify your answer.

b. Does the answer to part (a) depend on the value of r?

• Check your answers on page A-36.

MINDStretchers

Technology

1. Using a graphing calculator or computer software, display the graphs of $y = \dfrac{x^2 - 1}{x - 1}$ and $y = x + 1$ on the same screen. Compare the two graphs. What conclusion can you draw from this comparison?

Patterns

2. Some rational expressions, when simplified, are equivalent to polynomials. Check the simplifications of the following rational expressions:

   a. $\dfrac{x^2 - 1}{x - 1} = 1 + x$ **b.** $\dfrac{x^3 - 1}{x - 1} = 1 + x + x^2$ **c.** $\dfrac{x^4 - 1}{x - 1} = 1 + x + x^2 + x^3$

   Identify this pattern and extend it.

Writing

3. Sometimes, arithmetic fractions are written with a slanted fraction line, as in 3/4, rather than a horizontal fraction line, as in $\dfrac{3}{4}$. In a sentence or two, explain the disadvantage of writing a rational expression such as $\dfrac{x + 2}{x - 3}$ with a slanted fraction line.

17.2 Multiplying and Dividing Rational Expressions

OBJECTIVES

- **A** To multiply rational expressions
- **B** To divide rational expressions
- **C** To solve applied problems involving the multiplication or division of rational expressions

Multiplying Rational Expressions

Operations on rational expressions in algebra are similar to those on fractions in arithmetic. For instance, to multiply arithmetic fractions, we multiply their numerators to get the numerator of the product and multiply their denominators to get the denominator of the product.

$$\frac{2}{3}\cdot\frac{5}{7} = \frac{2\cdot 5}{3\cdot 7} = \frac{10}{21}$$

In algebra, rational expressions are multiplied in the same way.

$$\frac{2}{a}\cdot\frac{b}{y} = \frac{2b}{ay}$$

Recall from arithmetic that it is often easier to first divide out any common factors and then multiply the numerators and denominators.

$$\frac{2}{9}\cdot\frac{3}{10} = \frac{\overset{1}{\cancel{2}}}{\underset{3}{\cancel{9}}}\cdot\frac{\overset{1}{\cancel{3}}}{\underset{5}{\cancel{10}}} = \frac{1}{15}$$

Similarly, in algebra:

$$\frac{5a}{b}\cdot\frac{b}{2a^2} = \frac{5\overset{1}{\cancel{a}}}{\underset{1}{\cancel{b}}}\cdot\frac{\overset{1}{\cancel{b}}}{2\underset{a}{\cancel{a^2}}} = \frac{5}{2a}$$

EXAMPLE 1

Multiply.

a. $\frac{3}{x}\cdot\frac{5}{x}$ **b.** $\frac{4x^2}{7y^3}\cdot\frac{3y}{8x}$

Solution

a. There are no common factors in the numerators and denominators.

$\frac{3}{x}\cdot\frac{5}{x} = \frac{3\cdot 5}{x\cdot x}$ Multiply the numerators and multiply the denominators.

$= \frac{15}{x^2}$ Simplify.

b. $\frac{4x^2}{7y^3}\cdot\frac{3y}{8x} = \frac{\overset{1\ x}{\cancel{4}\cancel{x^2}}}{7\underset{y^2}{\cancel{y^3}}}\cdot\frac{3\overset{1}{\cancel{y}}}{\underset{2\ 1}{\cancel{8}\cancel{x}}}$ Divide out all common factors in the numerators and denominators.

$= \frac{x\cdot 3}{7y^2\cdot 2}$ Multiply the numerators and multiply the denominators.

$= \frac{3x}{14y^2}$ Simplify.

PRACTICE 1

Multiply.

a. $\frac{n}{7}\cdot\frac{2}{m}$

b. $\frac{p^2}{6q^2}\cdot\frac{2q}{5p}$

The following rule describes how we find the product of rational expressions, such as:

$$\frac{2n}{n^2 - 5n} \quad \text{and} \quad \frac{3n - 15}{4n}$$

To Multiply Rational Expressions

- Factor the numerators and denominators.
- Divide the numerators and denominators by all common factors.
- Multiply the remaining factors in the numerators and the remaining factors in the denominators.

EXAMPLE 2

Multiply.

a. $\dfrac{2n}{n^2 - 5n} \cdot \dfrac{3n - 15}{4n}$ b. $\dfrac{10}{9 - x^2} \cdot \dfrac{6 + 2x}{5}$

Solution

a. $\dfrac{2n}{n^2 - 5n} \cdot \dfrac{3n - 15}{4n} = \dfrac{2n}{n(n - 5)} \cdot \dfrac{3(n - 5)}{4n}$ Factor the numerator and the denominator.

$= \dfrac{\overset{1\,1}{\cancel{2}\cancel{n}}}{\underset{1}{\cancel{n}}\underset{1}{\cancel{(n - 5)}}} \cdot \dfrac{3\overset{1}{\cancel{(n - 5)}}}{\underset{2}{\cancel{4}}n}$ Divide out common factors.

$= \dfrac{3}{2n}$ Multiply the factors in the numerators and in the denominators.

b. $\dfrac{10}{9 - x^2} \cdot \dfrac{6 + 2x}{5} = \dfrac{10}{(3 - x)(3 + x)} \cdot \dfrac{2(3 + x)}{5}$

$= \dfrac{\overset{2}{\cancel{10}}}{(3 - x)\underset{1}{\cancel{(3 + x)}}} \cdot \dfrac{2\overset{1}{\cancel{(3 + x)}}}{\underset{1}{\cancel{5}}} = \dfrac{4}{3 - x}$

PRACTICE 2

Multiply.

a. $\dfrac{3t}{t^2 + 5t} \cdot \dfrac{3t + 15}{6t}$

b. $\dfrac{8}{36g^2 - 1} \cdot \dfrac{1 + 6g}{4}$

EXAMPLE 3

Find the product.

a. $\dfrac{x^2 - x - 20}{5x + 5} \cdot \dfrac{15x^2 - 15}{2x + 8}$

b. $\dfrac{16 + 6a - a^2}{a^2 - 10a - 24} \cdot \dfrac{a^2 - 6a - 27}{a^2 - 17a + 72}$

PRACTICE 3

Find the product.

a. $\dfrac{y^2 + 4y - 21}{3y - 9} \cdot \dfrac{6y^2 - 24}{y + 2}$

b. $\dfrac{x^2 - x - 30}{x^2 + 10x + 9} \cdot \dfrac{18 - 7x - x^2}{x^2 - 8x + 12}$

EXAMPLE 3 (continued)

Solution

a. $\dfrac{x^2 - x - 20}{5x + 5} \cdot \dfrac{15x^2 - 15}{2x + 8}$

$$= \frac{(x-5)(x+4)}{5(x+1)} \cdot \frac{15(x^2-1)}{2(x+4)}$$

$$= \frac{(x-5)(x+4)}{5(x+1)} \cdot \frac{15(x-1)(x+1)}{2(x+4)}$$

$$= \frac{(x-5)\overset{1}{\cancel{(x+4)}}}{\underset{1}{\cancel{5}}\underset{1}{\cancel{(x+1)}}} \cdot \frac{\overset{3}{\cancel{15}}(x-1)\overset{1}{\cancel{(x+1)}}}{2\underset{1}{\cancel{(x+4)}}}$$

$$= \frac{3(x-5)(x-1)}{2}$$

b. $\dfrac{16 + 6a - a^2}{a^2 - 10a - 24} \cdot \dfrac{a^2 - 6a - 27}{a^2 - 17a + 72}$

$$= \frac{(8-a)(2+a)}{(a-12)(a+2)} \cdot \frac{(a-9)(a+3)}{(a-8)(a-9)}$$

$$= \frac{(8-a)\overset{1}{\cancel{(2+a)}}}{(a-12)\underset{1}{\cancel{(a+2)}}} \cdot \frac{\overset{1}{\cancel{(a-9)}}(a+3)}{(a-8)\underset{1}{\cancel{(a-9)}}}$$

$$= \frac{-1(a-8)}{a-12} \cdot \frac{a+3}{a-8}$$

$$= \frac{-\overset{1}{\cancel{(a-8)}}}{a-12} \cdot \frac{a+3}{\underset{1}{\cancel{a-8}}}$$

$$= \frac{-(a+3)}{a-12}$$

$$= -\frac{a+3}{a-12}$$

Dividing Rational Expressions

In dividing arithmetic fractions, we take the reciprocal of the divisor and change the operation to multiplication. Recall that the reciprocal of a fraction is formed by interchanging its numerator and denominator.

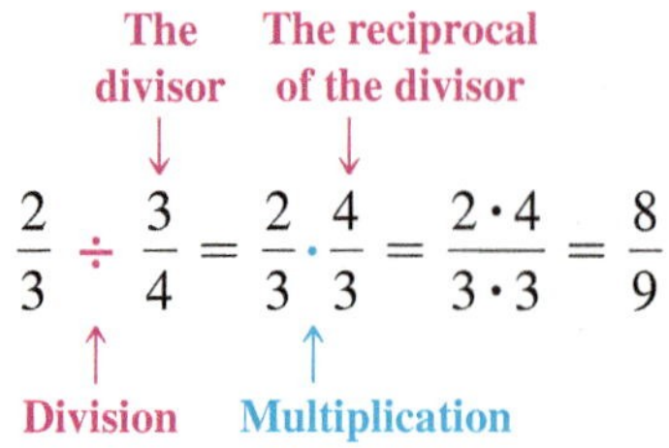

Rational expressions are divided in the same way.

The divisor — The reciprocal of the divisor

$$\frac{p}{q} \div \frac{r}{s} = \frac{p}{q} \cdot \frac{s}{r} = \frac{p \cdot s}{q \cdot r} = \frac{ps}{qr}$$

Division — Multiplication

In general, to divide rational expressions, we apply the following rule:

To Divide Rational Expressions

- Take the reciprocal of the divisor and change the operation to multiplication.
- Follow the rule for multiplying rational expressions.

EXAMPLE 4

Divide. **a.** $\frac{3}{x} \div \frac{y}{2}$ **b.** $\frac{6a}{b} \div \frac{3a^2}{b^3}$

Solution

a. $\frac{3}{x} \div \frac{y}{2} = \frac{3}{x} \cdot \frac{2}{y}$ Take the reciprocal of the divisor and change the operation to multiplication.

$= \frac{6}{xy}$ Multiply.

b. $\frac{6a}{b} \div \frac{3a^2}{b^3} = \frac{6a}{b} \cdot \frac{b^3}{3a^2}$

$= \frac{\overset{2\ 1}{\cancel{6}\cancel{a}}}{\underset{1}{\cancel{b}}} \cdot \frac{\cancel{b^3}^{\,b^2}}{\underset{1\ a}{\cancel{3}\cancel{a^2}}}$

$= \frac{2b^2}{a}$

PRACTICE 4

Find the quotient.

a. $\frac{a}{4} \div \frac{b}{6}$

b. $\frac{8q^3}{p^2} \div \frac{2q^4}{p^4}$

EXAMPLE 5

Find the quotient.

a. $\frac{a+1}{a-2} \div \frac{a+2}{a-1}$ **b.** $\frac{6x+3y}{4x-12y} \div \frac{10x+5y}{2x-6y}$

c. $\frac{x^2+3x+2}{3x+12} \div \frac{x^2-1}{7x}$

Solution

a. $\frac{a+1}{a-2} \div \frac{a+2}{a-1} = \frac{a+1}{a-2} \cdot \frac{a-1}{a+2}$ Take the reciprocal and change the operation to multiplication.

$= \frac{(a+1)(a-1)}{(a-2)(a+2)}$ Multiply.

PRACTICE 5

Find the quotient.

a. $\frac{x+3}{x-10} \div \frac{x+1}{x+3}$

b. $\frac{p+4q}{3p-6q} \div \frac{2p+8q}{2p-4q}$

c. $\frac{y^2+4y+3}{5y+10} \div \frac{y^2-1}{y}$

EXAMPLE 5 (continued)

b. $\dfrac{6x+3y}{4x-12y} \div \dfrac{10x+5y}{2x-6y}$

$= \dfrac{6x+3y}{4x-12y} \cdot \dfrac{2x-6y}{10x+5y}$ Take the reciprocal and change the operation to multiplication.

$= \dfrac{3(2x+y)}{4(x-3y)} \cdot \dfrac{2(x-3y)}{5(2x+y)}$ Factor the numerators and denominators.

$= \dfrac{3\overset{1}{\cancel{(2x+y)}}}{\underset{2}{\cancel{4}}\underset{1}{\cancel{(x-3y)}}} \cdot \dfrac{\overset{1}{\cancel{2}}\overset{1}{\cancel{(x-3y)}}}{5\underset{1}{\cancel{(2x+y)}}}$ Divide out the common factors.

$= \dfrac{3}{10}$ Multiply.

c. $\dfrac{x^2+3x+2}{3x+12} \div \dfrac{x^2-1}{7x} = \dfrac{x^2+3x+2}{3x+12} \cdot \dfrac{7x}{x^2-1}$

$= \dfrac{(x+1)(x+2)}{3(x+4)} \cdot \dfrac{7x}{(x+1)(x-1)}$

$= \dfrac{\overset{1}{\cancel{(x+1)}}(x+2)}{3(x+4)} \cdot \dfrac{7x}{\underset{1}{\cancel{(x+1)}}(x-1)}$

$= \dfrac{7x(x+2)}{3(x+4)(x-1)}$

EXAMPLE 6

According to the physicist Isaac Newton, the force of gravitation between two objects with mass m and M is given by the expression $G \cdot \dfrac{m}{d} \cdot \dfrac{M}{d}$, where d is the distance between them and G is the gravitational constant. Simplify this expression.

Solution

$$G \cdot \frac{m}{d} \cdot \frac{M}{d} = \frac{G}{1} \cdot \frac{m}{d} \cdot \frac{M}{d} = \frac{GmM}{d^2}$$

PRACTICE 6

When a body moves in a circular path, a force called the *centripetal force* is directed toward the center of the circle. This force has magnitude equal to the product of the object's mass m and its acceleration a. If $m = \dfrac{W}{g}$ and $a = \dfrac{v^2}{r}$, find the magnitude of the centripetal force in terms of W, g, v, and r, which represent the weight of the object, a constant due to gravity, the velocity of the object, and the radius of the circle, respectively.

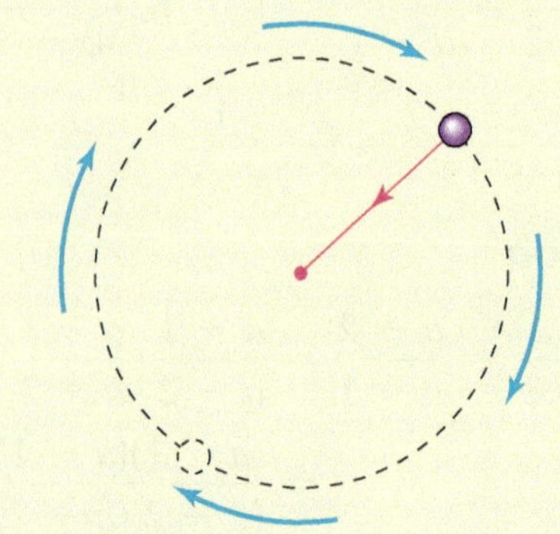

17.2 Exercises

FOR EXTRA HELP MyMathLab     

PRACTICE WATCH READ REVIEW

A *Multiply. Express the product in lowest terms.*

1. $\frac{1}{t^2} \cdot \frac{t}{4}$

2. $-\frac{y}{8} \cdot \frac{10}{y^3}$

3. $\frac{2}{a} \cdot \frac{3}{b}$

4. $\frac{5x}{2} \cdot \frac{2y}{3}$

5. $\frac{2x^4}{3x^5} \cdot \frac{5}{x^8}$

6. $\frac{4c}{d^3} \cdot \frac{3d}{8c^4}$

7. $-\frac{7x^2y}{3} \cdot \frac{6}{x^3y}$

8. $-\frac{5p^2}{10pq^2} \cdot \left(-\frac{6pq}{5p^3q}\right)$

9. $\frac{x}{x-2} \cdot \frac{5x-10}{x^4}$

10. $\frac{t}{t+3} \cdot \frac{4t+12}{t}$

11. $\frac{8n-3}{n^2} \cdot n$

12. $\frac{s+1}{s} \cdot s^2$

13. $\frac{8x-6}{5x+20} \cdot \frac{2x+8}{4x-3}$

14. $\frac{10y-1}{3y-6} \cdot \frac{5y-10}{10y-1}$

15. $\frac{5n-1}{6n+4} \cdot \frac{3n+2}{1-5n}$

16. $\frac{-4x-2}{10-x} \cdot \frac{6x+1}{2x+1}$

17. $\frac{x^2-4y^2}{x+y} \cdot \frac{3x+3y}{4x-8y}$

18. $\frac{a^2-4b^2}{6a-6b} \cdot \frac{10a-10b}{3a+6b}$

19. $\frac{p^4-1}{p^4-16} \cdot \frac{p^2+4}{p^2+1}$

20. $\frac{x^4-81}{x^2-x-12} \cdot \frac{x^2-16}{x^2+9}$

21. $\frac{n^2-2n-24}{n^2+6n+8} \cdot \frac{n^2+5n+6}{n^2-5n-6}$

22. $\frac{t^2+4t-21}{t^2+2t-15} \cdot \frac{t^2+t-20}{t^2+3t-28}$

23. $\frac{2y^2-y-6}{2y^2+y-3} \cdot \frac{2y^2-3y+1}{2y^2-9y+10}$

24. $\frac{2m+1}{2m^2+7m+3} \cdot \frac{m^2+2m-3}{2m+4m^2}$

25. $\frac{2}{x^3} \cdot \frac{4x}{5} \cdot \frac{10}{x^2}$

26. $\frac{a-3}{a^2} \cdot \frac{2a}{5} \cdot \frac{10a}{3}$

27. $\frac{x^2-7x+10}{2x-2} \cdot \frac{6x}{x^2-2x-15} \cdot \frac{x^2+2x-3}{x-2}$

28. $\frac{p^2-q^2}{p^2-pq} \cdot \frac{q}{2p^2-pq-q^2} \cdot \frac{6p+3q}{p}$

B *Divide. Express the quotient in lowest terms.*

29. $\frac{7}{a} \div \frac{14}{a}$

30. $\frac{-5}{n} \div \frac{n}{2}$

31. $\frac{p^3}{10} \div \frac{p^3}{20}$

32. $\frac{-s}{v^2} \div \frac{s^2}{v}$

33. $\frac{12}{x^3} \div \frac{6}{5x^2}$

34. $\frac{1}{a^2} \div \frac{2}{3a}$

35. $-\frac{3}{t} \div t$

36. $\frac{5}{s} \div s$

37. $\frac{9xy^2}{2x^3} \div \frac{3x^2y}{4y}$

38. $\frac{10a^3}{7ab^2} \div \frac{-6a^2b^2}{14ab}$

39. $\frac{c+3}{c-5} \div \frac{c+9}{c-7}$

40. $\frac{t+4}{t-6} \div \frac{2t-8}{2t-4}$

41. $\frac{6a-12}{8a+32} \div \frac{9a-18}{5a+20}$

42. $\frac{3x+6}{6x+18} \div \frac{2x+4}{x+3}$

43. $\frac{x+1}{10} \div \frac{1-x^2}{5}$

44. $\frac{4x+8}{8} \div \frac{4-x^2}{4}$

45. $\frac{p^2-1}{1-p} \div \frac{p+1}{p}$

46. $\frac{y+6}{3y} \div \frac{y^2-36}{6-y}$

47. $\frac{x^2y+3xy^2}{x^2-9y^2} \div \frac{5x^2y}{x^2-2xy-3y^2}$

48. $\frac{2x+1}{2x^2+7x+3} \div \frac{2x+4x^2}{x^2+2x-3}$

49. $\frac{2t^2 - 3t - 2}{2t + 1} \div (4 - t^2)$

50. $(y - x) \div \frac{x^2 - y^2}{x^2 + xy}$

51. $\frac{x^2 - 11x + 28}{x^2 - x - 42} \div \frac{x^2 - 2x - 8}{x^2 + 7x + 10}$

52. $\frac{a^2 - a - 56}{a^2 + 8a + 7} \div \frac{a^2 - 13a + 40}{a^2 - 4a - 5}$

53. $\frac{3p^2 - 3p - 18}{p^2 + 2p - 15} \div \frac{2p^2 + 6p - 20}{2p^2 - 12p + 16}$

54. $\frac{3y^2 + 13y + 4}{16 - y^2} \div \frac{3y^2 - 5y - 2}{3y - 12}$

Mixed Practice

Divide. Express the quotient in lowest terms.

55. $\frac{r^2}{s} \div \left(-\frac{r^4}{s}\right)$

56. $\frac{x^2 + 5x - 24}{x^2 + 9x + 8} \div \frac{x^2 - 10x + 21}{x^2 - 6x - 7}$

57. $\frac{3x + 9}{12} \div \frac{9 - x^2}{8}$

58. $\frac{h^2 + 2h - 8}{h + 4} \div (4 - h^2)$

59. $-\frac{6q^2}{15p^2q} \div \frac{12p^2q^2}{5p^2q}$

Multiply. Express the product in lowest terms.

60. $\frac{9a}{2b^2} \cdot \frac{b}{3a^3}$

61. $\frac{3y}{4} \cdot \frac{y + 3}{2y^2} \cdot \frac{6y}{3}$

62. $\frac{2c + 2d}{6c - 3d} \cdot \frac{4c^2 - d^2}{c + d}$

63. $\frac{x^2 + x - 2}{x + 1} \cdot \frac{3x + 3x^2}{3x^2 + 7x + 2}$

64. $\frac{u^2}{u + 5} \cdot \frac{5u + 25}{u}$

Applications

C *Solve.*

65. An investment of p dollars is growing at the annual simple interest rate r. The number of years that it will take the investment to be worth A dollars is given by the expression:

$$\left(\frac{A - p}{p}\right) \div r$$

Simplify this expression.

66. A store is having a sale, with each item selling at a discount of $x\%$.

a. With this discount, what percent of the normal price is a customer paying on each item?

b. What is the sale price of 10 items that normally sell for z dollars each?

67. A company has annual expenses totaling B dollars, of which $p\%$ goes toward rents. Of the rental expenses, $q\%$ goes toward the head office. Write as a rational expression the annual cost of the head office rental.

68. Physicists studying momentum may use the expression

$$\frac{W}{g} \cdot \frac{1}{t} \cdot (v_2 - v_1),$$

where W is the weight of an object, g is a constant, t is time, v_2 is the final velocity of the object, and v_1 is the initial velocity of the object. Write as a single rational expression.

69. Electricians use the following formula when studying the resistance in a heating element:

$$P = \frac{V^2}{R + r} \cdot \frac{r}{R + r}$$

Multiply out the right-hand side of this formula.

70. Suppose that the chance that one event occurs is $\frac{a}{b}$ and the chance that an independent (or unrelated) event occurs is $\frac{c}{d}$.

a. The chance that both events will occur is the product of their individual chances. Write this product as a rational expression.

b. The chance that neither event will occur can be represented by $\left(1 - \frac{a}{b}\right)\left(1 - \frac{c}{d}\right)$. Write this product as a single rational expression.

71. The following diagram shows a cylinder and its inscribed sphere, where the radius of the sphere is r and the height of the cylinder is h:

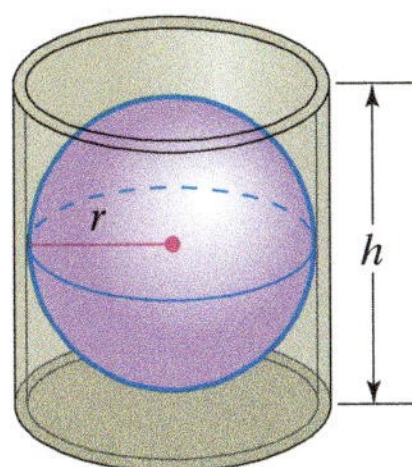

Volume of a sphere $= \frac{4}{3}\pi r^3$

Volume of a cylinder $= \pi r^2 h$

Surface area of a sphere $= 4\pi r^2$

Surface area of a cylinder $= 2\pi rh + 2\pi r^2$

Use these formulas to answer the following questions:

a. Find the ratio of the volume of the sphere to the volume of the cylinder.

b. Find the ratio of the surface area of the sphere to the surface area of the cylinder.

c. Divide the expression in part (a) by the expression in part (b).

72. In 1934, Harold Urey won the Nobel Prize in chemistry for discovering deuterium (also called *heavy hydrogen*). This discovery involved reducing the mass of an electron in an atom of deuterium.

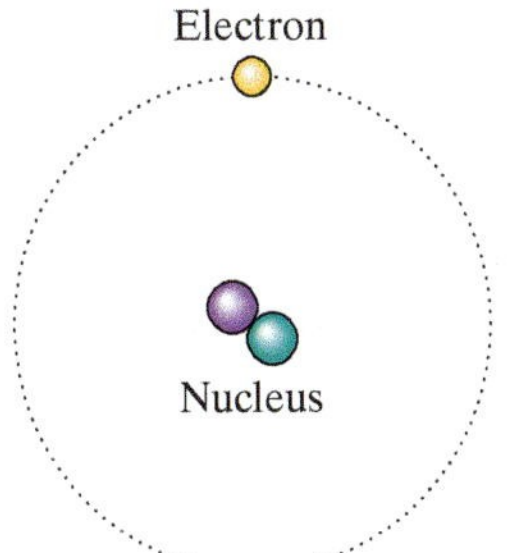

If m stands for the mass of an electron, then the reduced mass r can be represented by

$$r = \frac{mM}{M + m},$$

where M is the mass of the atom's nucleus. Find an expression for the quotient r divided by m in terms of m and M, and simplify. (*Source:* wikipedia.org)

• Check your answers on page A-36.

MINDStretchers

Critical Thinking

1. Even though the operation of division is not commutative, can you find four different polynomials P, Q, R, and S for which $\frac{P}{Q} \div \frac{R}{S} = \frac{R}{S} \div \frac{P}{Q}$? Explain.

Mathematical Reasoning

2. True or false, the expressions $\frac{\left(\frac{a}{b}\right)}{c}$ and $\frac{a}{\left(\frac{b}{c}\right)}$ are equivalent. Explain.

Groupwork

3. Working with a partner, find the following:
   a. Two different pairs of rational expressions whose quotient is $\frac{x + 4}{x + 6}$.
   b. Two different pairs of rational expressions whose product is $\frac{x^2 - x - 6}{2x^2 + 5x - 3}$.

17.3 Adding and Subtracting Rational Expressions

OBJECTIVES

- **A** To add or subtract rational expressions with the same denominator
- **B** To find the least common denominator (LCD) of two or more rational expressions
- **C** To add or subtract rational expressions with different denominators
- **D** To solve applied problems involving the addition or subtraction of rational expressions

In the last section, we discussed the multiplication and division of rational expressions. In this section, we consider how to find their sums and differences.

Adding and Subtracting Rational Expressions with the Same Denominator

Recall from arithmetic that to add like fractions, we add the numerators and keep the same denominator.

$$\frac{3}{5} + \frac{1}{5} = \frac{3 + 1}{5} = \frac{4}{5}$$

To subtract like fractions, we subtract the numerators and keep the same denominator.

$$\frac{3}{5} - \frac{1}{5} = \frac{3 - 1}{5} = \frac{2}{5}$$

In algebra, the process of adding and subtracting rational expressions with the same denominator is similar:

$$\frac{3}{x} + \frac{1}{x} = \frac{3 + 1}{x} = \frac{4}{x} \quad \text{and} \quad \frac{3}{x} - \frac{1}{x} = \frac{3 - 1}{x} = \frac{2}{x}$$

To Add (or Subtract) Rational Expressions with the Same Denominator

- Add (or subtract) the numerators and keep the same denominator.
- Simplify, if possible.

EXAMPLE 1

Add.

a. $\frac{3}{x - 1} + \frac{8}{x - 1}$ **b.** $\frac{3x}{4y} + \frac{x}{4y}$

c. $\frac{3x}{x + 1} + \frac{3}{x + 1}$ **d.** $\frac{2t^2 + 5t - 7}{3t - 1} + \frac{7t^2 - 5t + 6}{3t - 1}$

Solution

a. $\frac{3}{x - 1} + \frac{8}{x - 1} = \frac{3 + 8}{x - 1}$ Add the numerators and keep the same denominator.

$= \frac{11}{x - 1}$ Simplify.

b. $\frac{3x}{4y} + \frac{x}{4y} = \frac{4x}{4y}$ Add the numerators and keep the same denominator.

$= \frac{\overset{1}{\cancel{4}}x}{\underset{1}{\cancel{4}}y}$ Divide out the common factor.

$= \frac{x}{y}$ Simplify.

PRACTICE 1

Add.

a. $\frac{9}{y + 2} + \frac{1}{y + 2}$

b. $\frac{10r}{3s} + \frac{5r}{3s}$

c. $\frac{6t - 7}{t - 1} + \frac{1}{t - 1}$

d. $\frac{n^2 + 10n + 1}{n + 5} + \frac{2n^2 + 4n - 6}{n + 5}$

EXAMPLE 1 (continued)

c. $\frac{3x}{x+1} + \frac{3}{x+1} = \frac{3x+3}{x+1}$ Add the numerators and keep the same denominator.

$= \frac{3\cancel{(x+1)}}{\cancel{x+1}}$ Factor the numerator and divide out the common factor.

$= 3$ Simplify.

d. $\frac{2t^2 + 5t - 7}{3t - 1} + \frac{7t^2 - 5t + 6}{3t - 1} = \frac{(2t^2 + 5t - 7) + (7t^2 - 5t + 6)}{3t - 1}$

$= \frac{9t^2 - 1}{3t - 1}$ Combine like terms in the numerator.

$= \frac{\cancel{(3t-1)}(3t+1)}{\cancel{3t-1}}$

$= 3t + 1$

EXAMPLE 2

Subtract.

a. $\frac{6}{z} - \frac{4}{z}$ **b.** $\frac{5y}{8} - \frac{3y}{8}$ **c.** $\frac{x}{3y} - \frac{2x}{3y}$

Solution

a. $\frac{6}{z} - \frac{4}{z} = \frac{6-4}{z}$ Subtract the numerators and keep the same denominator.

$= \frac{2}{z}$ Simplify.

b. $\frac{5y}{8} - \frac{3y}{8} = \frac{5y - 3y}{8} = \frac{\overset{1}{\cancel{2}}y}{\underset{4}{\cancel{8}}} = \frac{y}{4}$

c. $\frac{x}{3y} - \frac{2x}{3y} = \frac{x - 2x}{3y} = \frac{-x}{3y} = -\frac{x}{3y}$

PRACTICE 2

Find the difference.

a. $\frac{12}{v} - \frac{7}{v}$

b. $\frac{7t}{10} - \frac{t}{10}$

c. $\frac{7p}{3q} - \frac{8p}{3q}$

EXAMPLE 3

Find the difference.

a. $\frac{x + 5y}{2y} - \frac{x - 11y}{2y}$ **b.** $\frac{3ax + bx}{a + 2b} - \frac{2ax - bx}{a + 2b}$

c. $\frac{6x + 12}{x^2 - x - 6} - \frac{x + 2}{x^2 - x - 6}$

PRACTICE 3

Subtract.

a. $\frac{7a - 4b}{3a} - \frac{a - 4b}{3a}$

b. $\frac{9xy - 5xz}{4y - z} - \frac{xy - 3xz}{4y - z}$

Solution

a. $\dfrac{x + 5y}{2y} - \dfrac{x - 11y}{2y}$

$= \dfrac{(x + 5y) - (x - 11y)}{2y}$ Subtract the numerators and keep the same denominator.

$= \dfrac{x + 5y - x + 11y}{2y}$ Remove the parentheses.

$= \dfrac{16y}{2y}$ Combine like terms.

$= \dfrac{\overset{8}{\cancel{16}}\overset{1}{\cancel{y}}}{\underset{1}{\cancel{2}}\underset{1}{\cancel{y}}}$ Divide out the common factors.

$= 8$ Simplify.

b. $\dfrac{3ax + bx}{a + 2b} - \dfrac{2ax - bx}{a + 2b} = \dfrac{(3ax + bx) - (2ax - bx)}{a + 2b}$

$= \dfrac{3ax + bx - 2ax + bx}{a + 2b}$

$= \dfrac{ax + 2bx}{a + 2b}$

$= \dfrac{x(a + 2b)}{a + 2b}$

$= \dfrac{x\overset{1}{\cancel{(a + 2b)}}}{\underset{1}{\cancel{a + 2b}}} = x$

c. $\dfrac{6x + 12}{x^2 - x - 6} - \dfrac{x + 2}{x^2 - x - 6} = \dfrac{(6x + 12) - (x + 2)}{x^2 - x - 6}$

$= \dfrac{6x + 12 - x - 2}{x^2 - x - 6}$

$= \dfrac{5x + 10}{x^2 - x - 6}$

$= \dfrac{5(x + 2)}{(x - 3)(x + 2)}$

$= \dfrac{5\overset{1}{\cancel{(x + 2)}}}{(x - 3)\underset{1}{\cancel{(x + 2)}}} = \dfrac{5}{x - 3}$

c. $\dfrac{2x + 13}{x^2 - 7x + 10} - \dfrac{5x + 7}{x^2 - 7x + 10}$

The Least Common Denominator of Rational Expressions

Recall that combining unlike arithmetic fractions involves finding their least common denominator (LCD). We write the fractions in terms of their LCD and then add these like fractions.

$$\frac{1}{2} + \frac{4}{5} = \frac{1}{2}\cdot\frac{5}{5} + \frac{4}{5}\cdot\frac{2}{2} = \frac{5}{10} + \frac{8}{10} = \frac{13}{10}$$

10 is the LCD of $\frac{1}{2}$ and $\frac{4}{5}$.

To find the LCD of rational expressions, we begin by factoring their denominators completely. For instance, consider the rational expressions $\frac{2}{9a^2}$ and $\frac{1}{12a^4}$, which we can write as:

$$\frac{2}{3 \cdot 3 \cdot a \cdot a} \quad \text{and} \quad \frac{1}{2 \cdot 2 \cdot 3 \cdot a \cdot a \cdot a \cdot a}$$

Denominators in factored form.

The LCD is the product of the different factors in the denominators, where the power of each factor is the greatest number of times that it occurs in any single denominator.

The highest power of this factor in any denominator

$$\text{LCD} = 2 \cdot 2 \cdot 3 \cdot 3 \cdot a \cdot a \cdot a \cdot a = 2^2 3^2 a^4$$

To Find the LCD of Rational Expressions

- Factor each denominator completely.
- Multiply these factors, using for the power of each factor the greatest number of times that it occurs in any of the denominators. The product of all these factors is the LCD.

EXAMPLE 4

Find the LCD of each group of rational expressions.

a. $\frac{1}{6}$ and $\frac{2}{n}$ **b.** $\frac{1}{10x}$ and $\frac{7}{12x^2}$

c. $\frac{3}{8p}, \frac{7}{10p^2q}$, and $\frac{1}{4pqr}$

Solution

a. We begin by factoring the denominators of $\frac{1}{6}$ and $\frac{2}{n}$.

Factor 6: $2 \cdot 3$

Factor n: n

No factor is repeated more than once in any denominator. So the LCD is the product of the factors.

$$\text{LCD} = 2 \cdot 3 \cdot n = 6n$$

b. $\frac{1}{10x}$ and $\frac{7}{12x^2}$

Factor $10x$: $2 \cdot 5 \cdot x$

Factor $12x^2$: $2^2 \cdot 3 \cdot x^2$

The factors 2 and x each appear at most twice in any denominator. The factors 3 and 5 each appear at most once in any denominator.

$$\text{LCD} = 2^2 \cdot 3 \cdot 5 \cdot x^2 = 60x^2$$

PRACTICE 4

For each group of rational expressions, determine the least common denominator.

a. $\frac{8}{y}$ and $\frac{x}{20}$

b. $\frac{1}{6t}$ and $\frac{4}{3t^2}$

c. $\frac{1}{5x}, \frac{7}{15xy^3}$, and $\frac{x+y}{2x^2}$

c. $\frac{3}{8p}, \frac{7}{10p^2q}$, and $\frac{1}{4pqr}$

$$\begin{aligned}&\text{Factor } 8p: \quad 2^3 \cdot p\\&\text{Factor } 10p^2q: \quad 2 \cdot 5 \cdot p^2 \cdot q\\&\text{Factor } 4pqr: \quad 2^2 \cdot p \cdot q \cdot r\\&\text{LCD} = 2^3 \cdot 5 \cdot p^2 \cdot q \cdot r = 40p^2qr\end{aligned}$$

In Example 4, we found the LCD of rational expressions whose denominators were monomials. Now, let's consider the case where the denominators may be trinomials or have binomial factors.

EXAMPLE 5

Find the LCD of each group of rational expressions.

a. $\frac{4x + 3}{3x^2 - 3x}$ and $\frac{x - 6}{x^2 - 2x + 1}$

b. $\frac{x + 4}{x + 7}, \frac{3}{x - 4}$, and $\frac{8x - 1}{x + 9}$

c. $\frac{2y}{x^2 - 2xy + y^2}, \frac{x}{x^2 - y^2}$, and $\frac{x - y}{x^2 + 2xy + y^2}$

Solution

a. First, we factor the denominators of $\frac{4x + 3}{3x^2 - 3x}$ and $\frac{x - 6}{x^2 - 2x + 1}$.

$$\begin{aligned}&\text{Factor } 3x^2 - 3x: \quad 3x(x - 1)\\&\text{Factor } x^2 - 2x + 1: \quad (x - 1)(x - 1) = (x - 1)^2\end{aligned}$$

The factor $3x$ appears at most once and the factor $(x - 1)$ appears at most twice in either denominator.

$$\text{LCD} = 3x(x - 1)^2$$

b. $\frac{x + 4}{x + 7}, \frac{3}{x - 4}$, and $\frac{8x - 1}{x + 9}$

$$\begin{aligned}&\text{Factor } x + 7: \quad x + 7\\&\text{Factor } x - 4: \quad x - 4\\&\text{Factor } x + 9: \quad x + 9\end{aligned}$$

No factor appears more than once in any denominator.

$$\text{LCD} = (x + 7)(x - 4)(x + 9)$$

c. $\frac{2y}{x^2 - 2xy + y^2}, \frac{x}{x^2 - y^2}$, and $\frac{x - y}{x^2 + 2xy + y^2}$

$$\begin{aligned}&\text{Factor } x^2 - 2xy + y^2: \quad (x - y)(x - y) = (x - y)^2\\&\text{Factor } x^2 - y^2: \quad (x - y)(x + y)\\&\text{Factor } x^2 + 2xy + y^2: \quad (x + y)(x + y) = (x + y)^2\\&\text{LCD} = (x - y)^2(x + y)^2\end{aligned}$$

PRACTICE 5

Determine the LCD for each group of rational expressions.

a. $\frac{9n + 1}{2n^2 + 2n}$ and $\frac{n - 5}{n^2 + 2n + 1}$

b. $\frac{7p + 1}{p + 2}, \frac{5p}{p - 1}$, and $\frac{3p + 1}{p + 5}$

c. $\frac{s - t}{s^2 + 4st + 4t^2}, \frac{3t}{s^2 - 4t^2}$, and $\frac{6s}{s^2 - 4st + 4t^2}$

Adding and Subtracting Rational Expressions with Different Denominators

In order to add or subtract rational expressions with different denominators, we will need to change each expression to an equivalent rational expression whose denominator is the LCD.

EXAMPLE 6

Write the following rational expressions in terms of their LCD:

a. $\frac{3}{5x^2}$ and $\frac{x+1}{6xy}$ **b.** $\frac{6x+1}{x^2-4}$ and $\frac{x}{x^2+5x+6}$

Solution

a. The LCD of $\frac{3}{5x^2}$ and $\frac{x+1}{6xy}$ is $30x^2y$. Since $5x^2 \cdot 6y = 30x^2y$, we multiply the numerator and denominator of $\frac{3}{5x^2}$ by $6y$ so that the denominator becomes the LCD.

$$\frac{3}{5x^2} = \frac{3 \cdot 6y}{5x^2 \cdot 6y} = \frac{18y}{30x^2y}$$

Note that multiplying the numerator and the denominator by $6y$ is the same as multiplying the expression by 1. So $\frac{3}{5x^2}$ and $\frac{18y}{30x^2y}$ are equivalent expressions.

Since $6xy \cdot 5x = 30x^2y$, we multiply the numerator and denominator of $\frac{x+1}{6xy}$ by $5x$ so that the denominator becomes the LCD.

$$\frac{x+1}{6xy} = \frac{(x+1) \cdot 5x}{6xy \cdot 5x} = \frac{5x(x+1)}{30x^2y}$$

b. To find the LCD, we begin by factoring the denominators of the expressions.

Factor $x^2 - 4$: $(x+2)(x-2)$
Factor $x^2 + 5x + 6$: $(x+2)(x+3)$

The LCD of $\frac{6x+1}{x^2-4}$ and $\frac{x}{x^2+5x+6}$ is $(x+2)(x-2)(x+3)$.

Since $(x^2-4)(x+3) = (x+2)(x-2)(x+3)$, we multiply the numerator and denominator of $\frac{6x+1}{x^2-4}$ by $(x+3)$ so that the denominator becomes the LCD.

$$\frac{6x+1}{x^2-4} = \frac{(6x+1)(x+3)}{(x+2)(x-2)(x+3)}$$

Since $(x^2+5x+6)(x-2) = (x+2)(x+3)(x-2)$, we multiply the numerator and denominator of $\frac{x}{x^2+5x+6}$ by $(x-2)$ so that the denominator becomes the LCD.

$$\frac{x}{x^2+5x+6} = \frac{x(x-2)}{(x+2)(x+3)(x-2)}$$

PRACTICE 6

Express each pair of rational expressions in terms of their LCD.

a. $\frac{2}{7p^3}$ and $\frac{p+3}{2p}$

b. $\frac{3y-2}{y^2-9}$ and $\frac{y}{y^2-6y+9}$

Now, let's look at how to add and subtract rational expressions with different denominators using the concept of equivalent rational expressions.

To Add (or Subtract) Rational Expressions with Different Denominators

- Find the LCD of the rational expressions.
- Write each rational expression with a common denominator, usually the LCD.
- Add (or subtract) the numerators.
- Simplify, if possible.

EXAMPLE 7

Perform the indicated operation.

a. $\frac{1}{2n} - \frac{3}{5n}$ **b.** $\frac{2}{3x^2} + \frac{5}{9x}$

Solution

a. The LCD of the rational expressions is $10n$.

$$\frac{1}{2n} - \frac{3}{5n} = \frac{1 \cdot 5}{2n \cdot 5} - \frac{3 \cdot 2}{5n \cdot 2}$$ Write as equivalent rational expressions with the LCD as the denominator.

$$= \frac{5}{10n} - \frac{6}{10n}$$ Simplify.

$$= \frac{-1}{10n}$$ Subtract the numerators.

$$= -\frac{1}{10n}$$ Simplify.

b. The LCD of the rational expressions is $9x^2$.

$$\frac{2}{3x^2} + \frac{5}{9x} = \frac{2 \cdot 3}{3x^2 \cdot 3} + \frac{5 \cdot x}{9x \cdot x}$$ Write as equivalent rational expressions with the LCD as the denominator.

$$= \frac{6}{9x^2} + \frac{5x}{9x^2}$$ Simplify.

$$= \frac{5x + 6}{9x^2}$$ Add the numerators.

PRACTICE 7

Combine.

a. $\frac{3}{4p} + \frac{1}{6p}$

b. $\frac{1}{5y} - \frac{2}{15y^2}$

EXAMPLE 8

Combine.

a. $\frac{y + 5}{y - 2} - \frac{y - 3}{y}$

b. $\frac{7a + 5}{a - 1} - \frac{a}{1 - a}$

PRACTICE 8

Combine.

a. $\frac{x + 2}{x} - \frac{x - 4}{x + 3}$

b. $\frac{3x - 4}{x - 1} + \frac{x + 1}{1 - x}$

EXAMPLE 8 (continued)

Solution

a. The LCD of the rational expression is $y(y - 2)$.

$$\frac{y+5}{y-2} - \frac{y-3}{y}$$

$$= \frac{(y+5)\cdot y}{(y-2)\cdot y} - \frac{(y-3)\cdot(y-2)}{y\cdot(y-2)}$$ Write in terms of the LCD.

$$= \frac{y^2+5y}{y(y-2)} - \frac{y^2-5y+6}{y(y-2)}$$ Multiply.

$$= \frac{(y^2+5y)-(y^2-5y+6)}{y(y-2)}$$ Subtract the numerators.

$$= \frac{y^2+5y-y^2+5y-6}{y(y-2)}$$ Remove the parentheses in the numerator.

$$= \frac{10y-6}{y(y-2)}$$ Simplify.

$$= \frac{2(5y-3)}{y(y-2)}$$ Factor the numerator.

b. The LCD of the rational expressions can be either $a - 1$ or $1 - a$, since $1 - a = -(a - 1)$. To get the LCD $(a - 1)$, we factor out -1 in the denominator of the expression $\frac{a}{1-a}$.

$$\frac{7a+5}{a-1} - \frac{a}{1-a}$$

$$= \frac{7a+5}{a-1} - \frac{a}{-(a-1)}$$ Write $1 - a$ as $-(a - 1)$.

$$= \frac{7a+5}{a-1} - \left(-\frac{a}{a-1}\right)$$ Write $\frac{a}{-(a-1)}$ as $-\frac{a}{(a-1)}$.

$$= \frac{7a+5}{a-1} + \frac{a}{a-1}$$ Simplify.

$$= \frac{7a+5+a}{a-1}$$ Add the numerators.

$$= \frac{8a+5}{a-1}$$ Simplify.

EXAMPLE 9

Combine.

a. $\frac{9}{2x+4} + \frac{x}{x^2-4}$

b. $\frac{2}{x-3} - \frac{x-1}{9-x^2}$

PRACTICE 9

Combine.

a. $\frac{1}{4x-16} + \frac{2x}{x^2-16}$

b. $\frac{3x-7}{x^2-1} + \frac{2}{1-x}$

Solution

a. $\dfrac{9}{2x+4} + \dfrac{x}{x^2-4}$

$= \dfrac{9}{2(x+2)} + \dfrac{x}{(x+2)(x-2)}$ Factor the denominators.

$= \dfrac{9 \cdot (x-2)}{2(x+2) \cdot (x-2)} + \dfrac{x \cdot 2}{(x+2)(x-2) \cdot 2}$ Write in terms of the LCD $2(x+2)(x-2)$.

$= \dfrac{9x-18}{2(x+2)(x-2)} + \dfrac{2x}{2(x+2)(x-2)}$ Multiply.

$= \dfrac{11x-18}{2(x+2)(x-2)}$ Add the numerators and combine like terms.

b. $\dfrac{2}{x-3} - \dfrac{x-1}{9-x^2}$

$= \dfrac{2}{x-3} - \dfrac{x-1}{(3-x)(3+x)}$

$= \dfrac{2}{x-3} - \dfrac{x-1}{-(x-3)(x+3)}$ Write $3 - x$ as $-(x-3)$.

$= \dfrac{2}{x-3} - \left[-\dfrac{x-1}{(x-3)(x+3)}\right]$

$= \dfrac{2}{x-3} + \dfrac{x-1}{(x-3)(x+3)}$

$= \dfrac{2 \cdot (x+3)}{(x-3) \cdot (x+3)} + \dfrac{x-1}{(x-3)(x+3)}$ Write in terms of the LCD $(x-3)(x+3)$.

$= \dfrac{2x+6}{(x-3)(x+3)} + \dfrac{x-1}{(x-3)(x+3)}$ Add the numerators and combine like terms.

$= \dfrac{(2x+6)+(x-1)}{(x-3)(x+3)}$

$= \dfrac{3x+5}{(x-3)(x+3)}$

EXAMPLE 10

Perform the indicated operations.

a. $\dfrac{7n}{n^2+4n+3} - \dfrac{3n-2}{n^2+2n+1}$ **b.** $\dfrac{y}{y-1} - \dfrac{y}{3} - \dfrac{4}{y+2}$

Solution

a. $\dfrac{7n}{n^2+4n+3} - \dfrac{3n-2}{n^2+2n+1}$

$= \dfrac{7n}{(n+3)(n+1)} - \dfrac{3n-2}{(n+1)^2}$ Factor the denominators.

$= \dfrac{7n \cdot (n+1)}{(n+3)(n+1) \cdot (n+1)}$

$- \dfrac{(3n-2) \cdot (n+3)}{(n+1)^2 \cdot (n+3)}$ Write in terms of the LCD $(n+3)(n+1)^2$.

PRACTICE 10

Perform the indicated operation.

a. $\dfrac{y}{y^2+5y+6} - \dfrac{4y+1}{y^2+3y+2}$

b. $\dfrac{2x}{5} - \dfrac{1}{x+1} - \dfrac{x-1}{4x}$

EXAMPLE 10 (continued)

$$= \frac{7n^2 + 7n}{(n + 3)(n + 1)^2} - \frac{3n^2 + 7n - 6}{(n + 3)(n + 1)^2}$$ Multiply.

$$= \frac{(7n^2 + 7n) - (3n^2 + 7n - 6)}{(n + 3)(n + 1)^2}$$ Subtract the numerators.

$$= \frac{7n^2 + 7n - 3n^2 - 7n + 6}{(n + 3)(n + 1)^2}$$ Remove the parentheses in the numerator.

$$= \frac{4n^2 + 6}{(n + 3)(n + 1)^2}$$ Combine like terms.

$$= \frac{2(2n^2 + 3)}{(n + 3)(n + 1)^2}$$ Factor the numerator.

b. The LCD is $3(y - 1)(y + 2)$.

$$\frac{y}{y - 1} - \frac{y}{3} - \frac{4}{y + 2}$$

$$= \frac{y \cdot 3(y + 2)}{(y - 1) \cdot 3(y + 2)} - \frac{y \cdot (y - 1)(y + 2)}{3 \cdot (y - 1)(y + 2)} - \frac{4 \cdot 3(y - 1)}{(y + 2) \cdot 3(y - 1)}$$

$$= \frac{3y^2 + 6y}{3(y - 1)(y + 2)} - \frac{y^3 + y^2 - 2y}{3(y - 1)(y + 2)} - \frac{12y - 12}{3(y - 1)(y + 2)}$$

$$= \frac{(3y^2 + 6y) - (y^3 + y^2 - 2y) - (12y - 12)}{3(y - 1)(y + 2)}$$

$$= \frac{3y^2 + 6y - y^3 - y^2 + 2y - 12y + 12}{3(y - 1)(y + 2)}$$

$$= \frac{-y^3 + 2y^2 - 4y + 12}{3(y - 1)(y + 2)}$$

EXAMPLE 11

A car gets c mpg in the city and d mpg more on the highway. Determine the amount of gas the car uses on a drive of x mi in the city and y mi on the highway, written as a single rational expression.

Solution In the city, the car is driven for x mi at c mpg. So the amount of gas consumed is $\frac{x}{c}$ gal. On the highway, the car is driven y mi at $(c + d)$ mpg. Therefore, $\frac{y}{c + d}$ gal are consumed. Altogether, the amount of gas used is:

$$\frac{x}{c} + \frac{y}{c + d} = \frac{x}{c} \cdot \frac{c + d}{c + d} + \frac{y}{c + d} \cdot \frac{c}{c}$$ The LCD is $c(c + d)$.

$$= \frac{x(c + d) + yc}{c(c + d)}$$

So the car uses $\frac{x(c + d) + yc}{c(c + d)}$ gal of gas.

PRACTICE 11

To find the percent change for the cost of an item, retailers can use the expression

$$100\left(\frac{C_1}{C_0} - 1\right),$$

where C_1 is the new cost and C_0 is the old cost. Write this percent as a single rational expression.

17.3 Exercises

FOR EXTRA HELP MyMathLab MathXL PRACTICE  WATCH  READ  REVIEW

A *Perform the indicated operation. Simplify, if possible.*

1. $\frac{5a}{12} + \frac{11a}{12}$
2. $\frac{y}{5} + \frac{4y}{5}$
3. $\frac{5t}{3} - \frac{2t}{3}$
4. $\frac{2x}{15} - \frac{8x}{15}$
5. $\frac{10}{x} + \frac{1}{x}$
6. $\frac{4}{a} + \frac{3}{a}$
7. $\frac{6}{7y} - \frac{1}{7y}$
8. $\frac{-7}{8x} + \frac{3}{8x}$
9. $\frac{5x}{2y} + \frac{x}{2y}$
10. $\frac{4a}{3b} + \frac{8a}{3b}$
11. $\frac{2p}{5q} - \frac{3p}{5q}$
12. $\frac{x}{6y} - \frac{11x}{6y}$
13. $\frac{2}{x+1} + \frac{7}{x+1}$
14. $\frac{8}{p+q} + \frac{1}{q+p}$
15. $\frac{5}{x+2} - \frac{9}{2+x}$
16. $\frac{10}{r+s} - \frac{14}{s+r}$
17. $\frac{a}{a+3} + \frac{1}{a+3}$
18. $\frac{p}{p-1} + \frac{3}{p-1}$
19. $\frac{3x}{x-8} + \frac{2x+1}{x-8}$
20. $\frac{4p}{p-5} + \frac{p-2}{p-5}$
21. $\frac{7x+1}{5x+2} - \frac{3x}{5x+2}$
22. $\frac{2x}{2x-1} - \frac{x+1}{2x-1}$
23. $\frac{9x+17}{2x+5} - \frac{3x+2}{2x+5}$
24. $\frac{5a+1}{2a+3} - \frac{a-5}{2a+3}$
25. $\frac{-7+5n}{3n-1} + \frac{7n+3}{3n-1}$
26. $\frac{-5+8x}{5x-1} + \frac{4-3x}{5x-1}$
27. $\frac{x^2-1}{x^2-4x-2} - \frac{x^2-x+3}{x^2-4x-2}$
28. $\frac{9+3x-x^2}{x^2+x+1} + \frac{x^2-5}{x^2+x+1}$
29. $\frac{x}{x^2-3x+2} + \frac{2}{x^2-3x+2} + \frac{x^2-4x}{x^2-3x+2}$
30. $\frac{a}{a^2-5a+4} + \frac{2}{a^2-5a+4} + \frac{a^2-4a}{a^2-5a+4}$
31. $\frac{2x-1}{3x^2-x+2} + \frac{8}{3x^2-x+2} - \frac{3x}{3x^2-x+2}$
32. $\frac{4y-3}{y^2+7y+1} - \frac{y}{y^2+7y+1} + \frac{6-y}{y^2+7y+1}$

B *The following expressions represent denominators of rational expressions. Find their LCD.*

33. $5(x+2)$ and $3(x+2)$
34. $9(4c-1)$ and $6(4c-1)$
35. $(p-3)(p+8)$ and $(p-3)(p-8)$
36. $(b+c)(5b)$ and $(b+c)(2b)$
37. $t, t+3$, and $t-3$
38. $s, 4s$, and $s+5$
39. $t^2+7t+10$ and t^2-25
40. n^2-1 and n^2+6n-7
41. $3s^2-11s+6$ and $3s^2+4s-4$
42. $2x^2+x-15$ and $-2x^2+9x-10$

Write each pair of rational expressions in terms of their LCD.

43. $\frac{1}{3x}$ and $\frac{5}{4x^2}$
44. $\frac{2}{5y^3}$ and $\frac{3}{y^2}$
45. $\frac{5}{2a^2}$ and $\frac{a-3}{7ab}$

46. $\frac{x+2}{4xy}$ and $\frac{7}{6x^2}$

47. $\frac{8}{n(n+1)}$ and $\frac{5}{(n+1)^2}$

48. $\frac{2}{c(c-3)^2}$ and $\frac{4}{(c-3)}$

49. $\frac{3n}{4n+4}$ and $\frac{2n}{n^2-1}$

50. $\frac{4y}{y^2-1}$ and $\frac{y}{2y+2}$

51. $\frac{2n}{n^2+6n+5}$ and $\frac{3n}{n^2+2n-15}$

52. $\frac{7t}{t^2-2t+1}$ and $\frac{4t}{t^2-5t+4}$

C *Perform the indicated operations. Simplify, if possible.*

53. $\frac{5}{3x}+\frac{1}{2x}$

54. $\frac{3}{4a}+\frac{2}{5a}$

55. $\frac{2}{3x^2}-\frac{5}{6x}$

56. $\frac{9}{7c^2}-\frac{3}{5c^3}$

57. $\frac{-2}{3x^2y}+\frac{4}{3xy^2}$

58. $\frac{6}{p^2q}+\frac{8}{pq^2}$

59. $\frac{1}{x+1}+\frac{1}{x-1}$

60. $\frac{2}{a+b}+\frac{2}{a-b}$

61. $\frac{p+6}{3}-\frac{2p+1}{7}$

62. $\frac{b+9}{5}-\frac{5-7b}{2}$

63. $x-\frac{10-4x}{2}$

64. $\frac{-2t+1}{8}-3t$

65. $\frac{3a+1}{6a}-\frac{a^2-2}{2a^2}$

66. $\frac{y^2-2}{2y^2}-\frac{2y-7}{4y}$

67. $\frac{a^2}{a-1}-\frac{1}{1-a}$

68. $\frac{6}{x-4}-\frac{x}{4-x}$

69. $\frac{4}{c-4}+\frac{c}{4-c}$

70. $\frac{a}{a-b}+\frac{b}{b-a}$

71. $\frac{x-5}{x+1}-\frac{x+2}{x}$

72. $\frac{p+4}{p}-\frac{p+3}{p-1}$

73. $\frac{4x-5}{x-4}+\frac{1-3x}{4-x}$

74. $\frac{5n}{1-n}+\frac{3n-2}{n-1}$

75. $\frac{5x}{x^2+x-2}+\frac{6}{x+2}$

76. $\frac{p}{p^2-3p+2}+\frac{4}{p-1}$

77. $\frac{4}{3n-9}-\frac{n}{n^2+2n-15}$

78. $\frac{-2x}{x^2+7x+12}+\frac{5}{4x+16}$

79. $\frac{2}{t+5}-\frac{t+6}{25-t^2}$

80. $\frac{4}{x-2}-\frac{x-1}{4-x^2}$

81. $\frac{4x}{x^2+2x+1}-\frac{2x+5}{x^2+4x+3}$

82. $\frac{y-1}{y^2-4y+4}+\frac{3y}{y^2-y-2}$

83. $\frac{2t-1}{2t^2+t-3}+\frac{2}{t-1}$

84. $\frac{4}{y-2}+\frac{3y-1}{y^2-6y+8}$

85. $\frac{4x}{x-1}+\frac{2}{3x}+\frac{x}{x^2-1}$

86. $\frac{c}{c+2}+\frac{3}{4c}+\frac{2c}{c^2-4}$

87. $\frac{5y}{3y-1}-\frac{3}{y-4}+\frac{y+1}{3y^2-13y+4}$

88. $\frac{6x-1}{2x+1}+\frac{x}{x-1}-\frac{4x}{2x^2-x-1}$

89. $\dfrac{a - 1}{(a + 3)^2} - \dfrac{2a - 3}{a + 3} - \dfrac{a}{4a + 12}$

90. $\dfrac{y}{8y - 16} + \dfrac{3y + 4}{y - 2} - \dfrac{y - 3}{(y - 2)^2}$

Mixed Practice

Write each pair of rational expressions in terms of their LCD.

91. $\dfrac{p + 3}{6p^2}$ and $\dfrac{5}{8pq}$

92. $\dfrac{3}{y(y - 2)}$ and $\dfrac{8}{(y - 2)^2}$

Perform the indicated operations. Simplify, if possible.

93. $\dfrac{9}{xy^2} + \dfrac{6}{x^2y}$

94. $\dfrac{7}{y - 3} - \dfrac{y}{3 - y}$

95. $\dfrac{c - 3}{c} - \dfrac{c - 2}{c + 1}$

96. $\dfrac{-2a}{a^2 + 7a + 10} + \dfrac{4}{3a + 15}$

97. $\dfrac{b + 3}{b^2 - 2b - 3} - \dfrac{4}{b^2 - 6b + 9}$

98. $\dfrac{2x}{x^2 - 1} + \dfrac{x}{x + 1} + \dfrac{2}{3x}$

99. $\dfrac{6}{m + n} - \dfrac{11}{n + m}$

100. $\dfrac{5x - 2}{3x - 1} - \dfrac{2x}{3x - 1}$

101. $\dfrac{r}{r^2 - r - 6} + \dfrac{3}{r^2 - r - 6} + \dfrac{r^2 - 5r}{r^2 - r - 6}$

102. If $j^2 - 4$ and $j^2 + 2j - 8$ represent denominators of rational expressions, find their LCD.

Applications

D *Solve.*

103. The position of an object thrown upward is given by the expression $vt + \dfrac{gt^2}{2}$. Here v is the initial velocity, t is the time, and g is a constant related to gravity. Write as a single rational expression.

104. Under certain conditions, the chances of an event happening can be represented by the rational expression $\dfrac{f}{t}$, whereas $\dfrac{t - f}{t}$ represents the chances of the event *not* happening. In this expression, f is the number of favorable outcomes and t is the total number of outcomes. What is the sum of these two expressions?

105. A bank pays an interest rate r compounded annually on all account balances. If a customer wanted the balance in an account to be \$1000 at the end of one year, she would need to have a current balance of $\dfrac{1000}{1 + r}$ dollars. However, if she were willing to wait two years for the balance to reach \$1000, then her current balance would need to be only $\dfrac{1000}{(1 + r)^2}$ dollars. Express the difference between the quantities $\dfrac{1000}{1 + r}$ dollars and $\dfrac{1000}{(1 + r)^2}$ dollars as a single rational expression.

106. An expression to find the length of base b of a trapezoid is $\dfrac{2A}{h} - a$. Another expression is $\dfrac{2A - ah}{h}$. Explain whether these two expressions are equivalent.

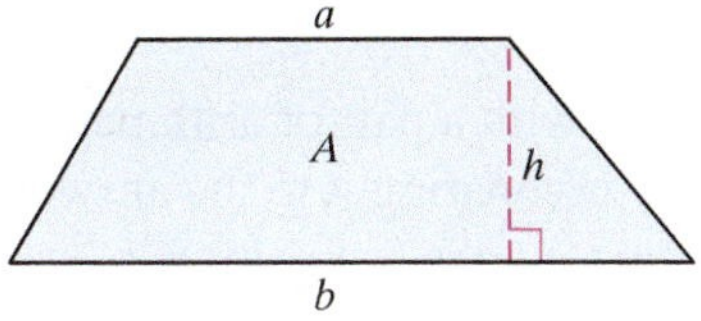

107. A trucker drove 20 mi at a speed r mph, and then returned at double the speed. How long did the whole trip take? Write the answer as a single rational expression.

108. A train traveled m mi at a speed of s mph. A bus following the same route traveled 5 mph slower. How much longer did the bus take than the train to make this trip? Write the answer as a single rational expression.

109. To maintain a checking account, a bank charges a customer \$3 per month and \$0.10 per check. For x checks, the cost per check is $\left(\frac{3}{x} + 0.1\right)$ dollars. Write this cost as a single rational expression.

110. In baseball, an expression for *runners per inning* is $\frac{H}{I} + \frac{W}{I}$, where H represents the number of hits, W the number of walks, and I the number of innings pitched. Explain whether the expression $\frac{H + W}{2I}$ is equivalent to $\frac{H}{I} + \frac{W}{I}$.

• Check your answers on page A-36.

MINDStretchers

Groupwork

1. When we break a given rational expression into *partial fractions*, we are expressing it as the sum of two other rational expressions. For instance, we could write the rational expression $\frac{4x + 1}{x^2 + x}$ as $\frac{3}{x + 1} + \frac{1}{x}$.

a. Confirm that $\frac{4x + 1}{x^2 + x} = \frac{3}{x + 1} + \frac{1}{x}$.

b. Find the missing numerators so that $\frac{5x + 1}{x^2 - 1} = \frac{?}{x - 1} + \frac{?}{x + 1}$.

Critical Thinking

2. Find the following product.

$$\left(1 + \frac{1}{n}\right)\left(1 + \frac{1}{n + 1}\right)\left(1 + \frac{1}{n + 2}\right)\left(1 + \frac{1}{n + 3}\right)\dots\left(1 + \frac{1}{n + 99}\right)\left(1 + \frac{1}{n + 100}\right)$$

Mathematical Reasoning

3. A formula sometimes used to add arithmetic fractions is:

$$\frac{p}{q} + \frac{r}{s} = \frac{ps + qr}{qs}$$

a. Working with a pair of arithmetic fractions of your choice, confirm that this formula gives the correct sum by comparing it to the answer derived using an alternative method.

b. Consider the left side of this formula as a sum of rational expressions. Explain why the right side of the formula is correct.

Cultural Note

The ancient Egyptians expressed most fractions as the sum of *unit fractions* (fractions whose numerators are 1). For instance, they would write the fraction $\frac{2}{5}$ as $\frac{1}{3} + \frac{1}{15}$. More generally, the rational expression $\frac{x + y}{xy}$ can be written as the sum of the unit fractions $\frac{1}{x}$ and $\frac{1}{y}$.

(*Source:* Jan Gullberg, *From the Birth of Numbers*, W. W. Norton, 1997)

17.4 Complex Rational Expressions

OBJECTIVES

A To simplify complex rational expressions

B To solve applied problems involving complex rational expressions

In arithmetic, a *complex fraction* is a fraction that in turn contains one or more fractions in its numerator, denominator, or both:

$$\frac{\frac{1}{3}}{\frac{4}{5}} \qquad \begin{array}{l}\leftarrow \text{Fraction in the numerator}\\ \leftarrow \text{Main fraction line}\\ \leftarrow \text{Fraction in the denominator}\end{array}$$

Such fractions can be put in standard form as the quotient of two integers by recalling that the main fraction line represents division:

$$\frac{\frac{1}{3}}{\frac{4}{5}} = \frac{1}{3} \div \frac{4}{5} = \frac{1}{3} \cdot \frac{5}{4} = \frac{5}{12}$$

A comparable expression in algebra is called *a complex rational expression* (or a *complex algebraic fraction*). Such an expression contains a rational expression in its numerator, denominator, or both. Some examples are shown below:

$$\frac{x^2 + 5x - 2}{\frac{3}{4}} \qquad \frac{7}{\frac{a + b}{2}} \qquad \frac{\frac{3n}{2} + \frac{1}{n}}{\frac{n - 4}{n^2}}$$

As in the case of an arithmetic complex fraction, a complex rational expression is usually simplified, that is, written as the quotient of two polynomials with no common factors.

We consider two methods of simplifying a complex rational expression: the *division method* and the *LCD method.*

The Division Method

In the division method, we begin by writing the numerator and denominator as single rational expressions, simplifying if necessary. Then, we write the complex rational expression as the numerator divided by the denominator.

For instance, suppose we want to simplify the complex rational expression below:

$$\frac{\frac{x}{2} + \frac{1}{2}}{\frac{x^2}{3} + \frac{x}{3}}$$

We first write the numerator and denominator as single rational expressions.

$$\frac{\frac{x}{2} + \frac{1}{2}}{\frac{x^2}{3} + \frac{x}{3}} = \frac{\frac{x + 1}{2}}{\frac{x^2 + x}{3}}$$

Add the rational expressions in the numerator.

Add the rational expressions in the denominator.

$$= \frac{x + 1}{2} \div \frac{x^2 + x}{3}$$

Write the expression as the numerator divided by the denominator.

$$= \frac{x+1}{2} \cdot \frac{3}{x^2+x}$$ Take the reciprocal of the divisor and change the operation to multiplication.

$$= \frac{\overset{1}{\cancel{x+1}}}{2} \cdot \frac{3}{x\underset{1}{\cancel{(x+1)}}}$$ Factor and divide out the common factor.

$$= \frac{3}{2x}$$ Simplify.

So we can conclude that the original complex rational expression $\dfrac{\frac{x}{2}+\frac{1}{2}}{\frac{x^2}{3}+\frac{x}{3}}$ is equivalent to the rational expression $\dfrac{3}{2x}$.

To Simplify a Complex Rational Expression: Division Method

- Write the numerator and denominator as single rational expressions in simplified form.
- Write the expression as the numerator divided by the denominator.
- Divide.
- Simplify, if possible.

Let's use the division method in the following examples.

EXAMPLE 1

Simplify.

a. $\dfrac{\frac{2}{x}}{\frac{6}{x^3}}$ **b.** $\dfrac{y}{\frac{y}{2}+\frac{y^2}{3}}$

Solution

a. Neither the numerator nor the denominator of the complex rational expression can be simplified. So we write the expression as the numerator divided by the denominator.

$$\frac{\frac{2}{x}}{\frac{6}{x^3}} = \frac{2}{x} \div \frac{6}{x^3}$$

$$= \frac{2}{x} \cdot \frac{x^3}{6}$$ Take the reciprocal of the divisor and change the operation to multiplication.

$$= \frac{\overset{1}{\cancel{2}}}{\underset{1}{\cancel{x}}} \cdot \frac{\overset{x^2}{\cancel{x^3}}}{\underset{3}{\cancel{6}}}$$ Divide out the common factors.

$$= \frac{x^2}{3}$$ Simplify.

PRACTICE 1

Simplify.

a. $\dfrac{\frac{3}{x^4}}{\frac{5}{x}}$

b. $\dfrac{2x}{\frac{x^2}{4}+\frac{x}{2}}$

EXAMPLE 1 (continued)

b. Let's express the denominator as a single rational expression.

$$\frac{y}{\frac{y}{2}+\frac{y^2}{3}} = \frac{y}{\frac{y\cdot 3}{2\cdot 3}+\frac{y^2\cdot 2}{3\cdot 2}}$$ Add the rational expressions in the denominator. The LCD is 6.

$$= \frac{y}{\frac{3y+2y^2}{6}}$$

$$= y \div \frac{3y+2y^2}{6}$$ Write the expression as the numerator divided by the denominator.

$$= y\cdot\frac{6}{3y+2y^2}$$ Take the reciprocal of the divisor and change the operation to multiplication.

$$= \overset{1}{\cancel{y}}\cdot\frac{6}{\underset{1}{\cancel{y}}(3+2y)}$$ Factor and divide out the common factor.

$$= \frac{6}{3+2y}$$ Simplify.

EXAMPLE 2

Simplify: $\dfrac{1+\frac{1}{x}}{1-\frac{1}{x}}$

Solution We begin by simplifying the numerator and denominator so that each is a single rational expression.

$$\frac{1+\frac{1}{x}}{1-\frac{1}{x}} = \frac{\frac{x+1}{x}}{\frac{x-1}{x}}$$

$$= \frac{x+1}{x} \div \frac{x-1}{x}$$

$$= \frac{x+1}{x}\cdot\frac{x}{x-1}$$

$$= \frac{x+1}{\underset{1}{\cancel{x}}}\cdot\frac{\overset{1}{\cancel{x}}}{x-1}$$

$$= \frac{x+1}{x-1}$$

PRACTICE 2

Simplify: $\dfrac{2-\frac{1}{n}}{2+\frac{1}{n}}$

The LCD Method

Now, we consider the LCD method of simplifying a complex rational expression. In this method, we multiply the numerator and denominator of the complex rational expression by the LCD of all rational expressions that appear within it.

Let's simplify the complex rational expression $\dfrac{\dfrac{x}{2}+\dfrac{1}{2}}{\dfrac{x^2}{3}+\dfrac{x}{3}}$ given on the first page of this section using the LCD method. First, we must find the LCD of all the rational expressions in the numerator and denominator. The denominators within the complex rational expression are 2 and 3, so the LCD is 6.

$$\frac{\dfrac{x}{2}+\dfrac{1}{2}}{\dfrac{x^2}{3}+\dfrac{x}{3}} = \frac{\left(\dfrac{x}{2}+\dfrac{1}{2}\right)\cdot 6}{\left(\dfrac{x^2}{3}+\dfrac{x}{3}\right)\cdot 6}$$ Multiply the numerator and denominator by the LCD 6.

$$= \frac{\dfrac{x}{2}\cdot 6+\dfrac{1}{2}\cdot 6}{\dfrac{x^2}{3}\cdot 6+\dfrac{x}{3}\cdot 6}$$ Use the distributive property in the numerator and in the denominator.

$$= \frac{3x+3}{2x^2+2x}$$ Simplify.

$$= \frac{3\overset{1}{\cancel{(x+1)}}}{2x\underset{1}{\cancel{(x+1)}}}$$ Factor and divide out the common factor.

$$= \frac{3}{2x}$$ Simplify.

Note that the division method and the LCD method result in the same answer.

To Simplify a Complex Rational Expression: LCD Method

- Find the LCD of all the rational expressions *within* both the numerator and denominator.
- Multiply the numerator and denominator of the complex rational expression by this LCD.
- Simplify, if possible.

EXAMPLE 3

Simplify. **a.** $\dfrac{\dfrac{3}{y^2}}{\dfrac{1}{y}}$ **b.** $\dfrac{n^2}{\dfrac{1}{n}+\dfrac{2}{n^2}}$

Solution

a. $$\frac{\dfrac{3}{y^2}}{\dfrac{1}{y}} = \frac{\dfrac{3}{y^2}\cdot y^2}{\dfrac{1}{y}\cdot y^2}$$ Multiply by the LCD y^2.

$$= \frac{3}{y}$$ Simplify.

PRACTICE 3

Simplify.

a. $\dfrac{\dfrac{4}{x}}{\dfrac{2}{x^3}}$

b. $\dfrac{\dfrac{1}{y}+\dfrac{3}{y^2}}{2y}$

EXAMPLE 3 (continued)

b. $$\frac{n^2}{\frac{1}{n}+\frac{2}{n^2}} = \frac{n^2 \cdot n^2}{\left(\frac{1}{n}+\frac{2}{n^2}\right)\cdot n^2}$$ Multiply by the LCD n^2.

$$= \frac{n^2 \cdot n^2}{\frac{1}{n}\cdot n^2 + \frac{2}{n^2}\cdot n^2}$$ Use the distributive property.

$$= \frac{n^4}{n+2}$$ Simplify.

EXAMPLE 4

Simplify.

a. $$\frac{3+\frac{1}{x^2}}{3-\frac{1}{x}}$$ **b.** $$\frac{\frac{1}{x^2}-\frac{1}{y^2}}{\frac{3}{x}-\frac{3}{y}}$$

Solution

a. Multiply the numerator and denominator by the LCD x^2.

$$\frac{3+\frac{1}{x^2}}{3-\frac{1}{x}} = \frac{\left(3+\frac{1}{x^2}\right)\cdot x^2}{\left(3-\frac{1}{x}\right)\cdot x^2}$$

$$= \frac{3\cdot x^2 + \frac{1}{x^2}\cdot x^2}{3\cdot x^2 - \frac{1}{x}\cdot x^2}$$

$$= \frac{3x^2+1}{3x^2-x}$$

b. Multiply the numerator and denominator by the LCD x^2y^2.

$$\frac{\frac{1}{x^2}-\frac{1}{y^2}}{\frac{3}{x}-\frac{3}{y}} = \frac{\left(\frac{1}{x^2}-\frac{1}{y^2}\right)\cdot x^2y^2}{\left(\frac{3}{x}-\frac{3}{y}\right)\cdot x^2y^2}$$

$$= \frac{\frac{1}{x^2}(x^2y^2) - \frac{1}{y^2}(x^2y^2)}{\frac{3}{x}(x^2y^2) - \frac{3}{y}(x^2y^2)}$$

$$= \frac{y^2-x^2}{3xy^2-3x^2y} = \frac{\overset{1}{\cancel{(y-x)}}(y+x)}{3xy\underset{1}{\cancel{(y-x)}}}$$

$$= \frac{y+x}{3xy}$$

PRACTICE 4

Express in simplest terms.

a. $$\frac{4+\frac{1}{y}}{4-\frac{1}{y^2}}$$

b. $$\frac{\frac{1}{2a^2}-\frac{1}{2b^2}}{\frac{5}{a}+\frac{5}{b}}$$

EXAMPLE 5

A resistor is an electrical device such as a lightbulb that offers resistance to the flow of electricity. When two lightbulbs with resistance R_1 and R_2 are connected in a certain kind of circuit, the combined resistance is given by the following expression:

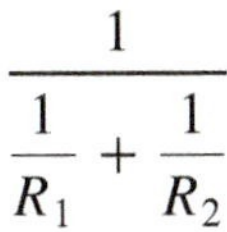

$$\frac{1}{\frac{1}{R_1}+\frac{1}{R_2}}$$

Simplify this expression for the combined resistance.

Solution

$$\frac{1}{\frac{1}{R_1}+\frac{1}{R_2}} = \frac{1 \cdot R_1R_2}{\left(\frac{1}{R_1}+\frac{1}{R_2}\right)\cdot R_1R_2}$$

$$= \frac{1 \cdot R_1R_2}{\frac{1}{\cancel{R_1}}\cdot \cancel{R_1}R_2 + \frac{1}{\cancel{R_2}}\cdot R_1\cancel{R_2}} = \frac{R_1R_2}{R_2 + R_1}$$

PRACTICE 5

The *harmonic mean* is a kind of average. The harmonic mean of the three numbers a, b, and c is given by the expression $\frac{3}{\frac{1}{a}+\frac{1}{b}+\frac{1}{c}}$.

Simplify this complex rational expression.

17.4 Exercises

FOR EXTRA HELP MyMathLab MathXL PRACTICE WATCH READ REVIEW

A *Simplify.*

1. $\dfrac{\frac{x}{5}}{\frac{x^2}{10}}$

2. $\dfrac{\frac{3}{s^2}}{\frac{s^3}{2}}$

3. $\dfrac{\frac{a+1}{2}}{\frac{a-1}{2}}$

4. $\dfrac{\frac{n-1}{n}}{\frac{n+1}{2n}}$

5. $\dfrac{3+\frac{1}{x}}{3-\frac{1}{x^2}}$

6. $\dfrac{1+\frac{1}{y^2}}{8-\frac{1}{y}}$

7. $\dfrac{\frac{1}{3d}-\frac{1}{d^2}}{d-\frac{9}{d}}$

8. $\dfrac{a-\frac{4}{a}}{\frac{1}{a^2}+\frac{1}{2a}}$

9. $\dfrac{1-\frac{4y^2}{x^2}}{3+\frac{6y}{x}}$

10. $\dfrac{\frac{p}{q}+2}{\frac{p^2}{q^2}-4}$

11. $\dfrac{\frac{2}{y}-\frac{1}{5}}{\frac{5}{y}-1}$

12. $\dfrac{\frac{2}{t}-\frac{3}{t^2}}{10+\frac{2}{t}}$

13. $\dfrac{1+\frac{4}{x}+\frac{4}{x^2}}{1+\frac{5}{x}+\frac{6}{x^2}}$

14. $\dfrac{1-\frac{1}{a^2}}{4-\frac{5}{a}+\frac{1}{a^2}}$

15. $\dfrac{3+\frac{1}{y+1}}{5-\frac{1}{y+1}}$

16. $\dfrac{2+\frac{1}{b-1}}{2-\frac{1}{b-1}}$

17. $\dfrac{\frac{x}{4}-\frac{x}{8}}{\frac{2}{y^2}+\frac{2}{y}}$

18. $\dfrac{\frac{n}{2}+\frac{n}{3}}{\frac{1}{m}-\frac{1}{m^2}}$

19. $\dfrac{\frac{x}{x+1}-\frac{2}{x}}{\frac{x}{3}}$

20. $\dfrac{\frac{y}{5}}{\frac{y}{y+2}+\frac{3}{y}}$

Mixed Practice

Simplify.

21. $\dfrac{\frac{m+2}{3m}}{\frac{m-1}{m}}$

22. $\dfrac{3-\frac{1}{d-2}}{3+\frac{1}{d-2}}$

23. $\dfrac{\frac{4}{u}-\frac{2}{u+1}}{\frac{u}{2}}$

24. $\dfrac{x-\frac{9}{x}}{\frac{1}{3x}-\frac{1}{x^2}}$

25. $\dfrac{\frac{3}{y^2}+\frac{4}{y}}{6+\frac{3}{y}}$

26. $\dfrac{3-\frac{1}{b}-\frac{2}{b^2}}{1-\frac{1}{b^2}}$

Applications

B *Solve.*

27. An expression from the study of electricity is

$$\frac{V}{\frac{1}{2R} + \frac{1}{2R + 2}},$$

where V represents voltage and R represents resistance. Simplify this expression.

28. At the beginning of the year, C dollars are invested in one business and D dollars in another. By the end of the year, the rate of return for the first investment was r and for the second, s. The value of the portfolio at the end of the year was therefore $C(1 + r) + D(1 + s)$. To show that the overall rate of return for the two investments was $\frac{Cr + Ds}{C + D}$, check that the expression $\frac{C(1 + r) + D(1 + s)}{1 + \frac{Cr + Ds}{C + D}}$ when simplified is the total initial investment.

29. The *earned run average* (ERA) is a statistic used in baseball to represent the average number of earned runs that a pitcher allows. A pitcher's ERA can be calculated using the expression

$$\frac{E}{\frac{I}{9}},$$

where E stands for the number of earned runs a pitcher gave up after pitching I innings. Simplify this complex rational expression.

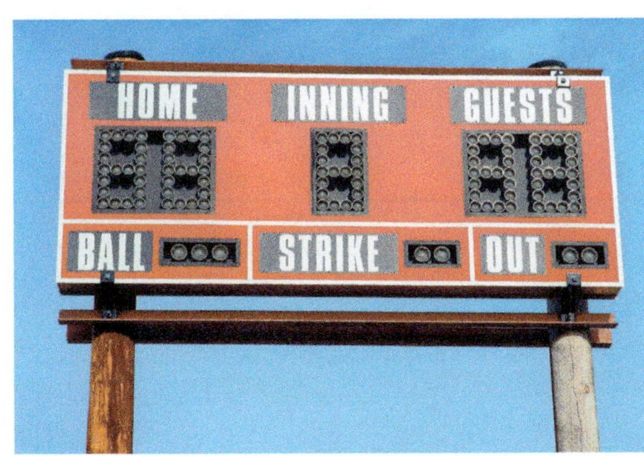

30. The expression

$$\frac{\frac{m}{c}}{1 - \frac{p^2}{c^2}}$$

is important in the design of airplanes. Write this expression in simplified form.

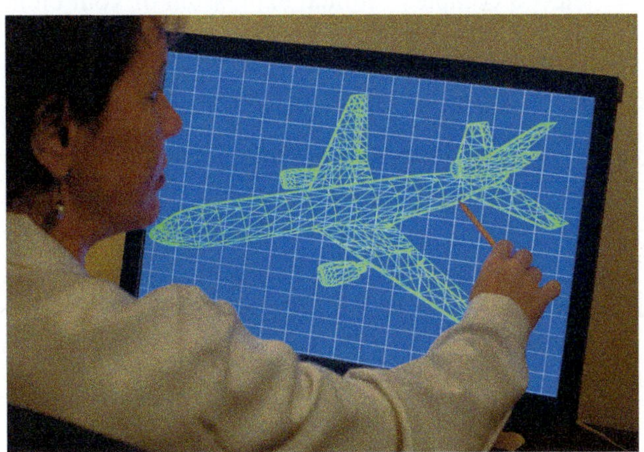

31. On a trip, an airplane flew at an average speed of a mph, returning on the same route at an average speed of b mph. The plane's average speed for the round trip is given by the following complex rational expression:

$$\frac{2}{\frac{1}{a} + \frac{1}{b}}$$

Simplify.

32. An object travels at speed s for a distance d and later travels at speed S for a distance D. The following expression represents the average speed for the entire trip:

$$\frac{d + D}{\frac{d}{s} + \frac{D}{S}}$$

Simplify this expression.

33. The weight of an object decreases as its distance from the Earth's surface increases. Suppose an object weighs w kg at sea level. Then at a height of h km above sea level, it will weigh (in kilograms)

$$\frac{w}{\left(1 + \frac{h}{6400}\right)^2}.$$

Show that $\dfrac{w}{\left(1 + \frac{h}{6400}\right)^2}$ and $\dfrac{6400^2 w}{(6400 + h)^2}$ are equivalent expressions.

34. The sound from a car horn as it approaches an observer seems to change its frequency. This is an example of the *Doppler effect*, which is studied in physics. The observer will hear a sound with frequency

$$\frac{S}{\frac{S - v}{f}},$$

where f is the actual frequency of the sound, S is the speed of sound, and v is the speed of the approaching object. Simplify the expression.

• Check your answers on page A-37.

MINDStretchers

Research

1. The harmonic mean is related to musical harmony. Either in your college library or on the Web, investigate the relationship between music and the harmonic mean. Summarize your findings in a few sentences.

Writing

2. Select a complex rational expression of your choice and simplify it using (a) the division method and (b) the LCD method. Which method do you prefer? Explain why.

Groupwork

3. Find a complex rational expression with numerator $\frac{4}{y}$ that simplifies to $\frac{3}{x + 5}$.

17.5 Solving Rational Equations

OBJECTIVES

A To solve equations involving rational expressions

B To solve applied problems involving rational equations

What Rational Equations Are and Why They Are Important

So far in this chapter, we have discussed rational *expressions*. Now, let's consider rational (or fractional) *equations*, that is, equations that contain one or more rational expressions. Here are some examples:

$$\frac{n}{2} - 3 = \frac{n}{5} \qquad \frac{1}{t} + \frac{1}{3} = \frac{1}{2} \qquad \frac{3}{x+1} - \frac{1}{8} = \frac{5}{x^2 - 1}$$

Many situations involving rates of work, motion, and proportions can be modeled by rational equations.

Solving Rational Equations

In solving rational equations, it is important not to confuse a rational expression with a rational equation.

$$\frac{4x}{3} + \frac{x}{3} \qquad \frac{4x}{3} + \frac{x}{3} = 5$$

Rational *expression* Rational *equation*

The key to solving a rational equation is to *clear the equation of rational expressions.* We do this by first determining their *least common denominator*, and then by multiplying both sides of the equation by the LCD.

EXAMPLE 1

Solve and check: $\frac{x}{2} + \frac{x}{6} = \frac{2}{3}$

Solution The rational expressions in this equation are $\frac{x}{2}$, $\frac{x}{6}$, and $\frac{2}{3}$. The denominators of these expressions are 2, 6, and 3, so the LCD is 6. To clear the equation of rational expressions, we multiply each side of the equation by 6.

$$\frac{x}{2} + \frac{x}{6} = \frac{2}{3}$$

$$6 \cdot \left(\frac{x}{2} + \frac{x}{6}\right) = 6 \cdot \frac{2}{3}$$ Multiply each side of the equation by the LCD.

$$6 \cdot \frac{x}{2} + 6 \cdot \frac{x}{6} = 6 \cdot \frac{2}{3}$$ Use the distributive property.

$$3x + x = 4$$ Simplify.

$$4x = 4$$ Combine like terms.

$$x = 1$$ Divide each side by 4.

PRACTICE 1

Solve and check: $\frac{y}{2} - \frac{y}{3} = \frac{1}{12}$

EXAMPLE 1 (continued)

Check

$$\frac{x}{2}+\frac{x}{6}=\frac{2}{3}$$

$$\frac{1}{2}+\frac{1}{6}\stackrel{?}{=}\frac{2}{3} \quad \text{Substitute 1 for } x.$$

$$\frac{3}{6}+\frac{1}{6}\stackrel{?}{=}\frac{2}{3}$$

$$\frac{4}{6}\stackrel{?}{=}\frac{2}{3}$$

$$\frac{2}{3}=\frac{2}{3} \quad \text{True}$$

So the solution to the given equation is 1.

Some rational equations, such as that in the next example, are in the form of a proportion. We can solve such equations either by setting the cross products equal or by multiplying both sides by the LCD. Here, we will use the LCD method.

EXAMPLE 2

Solve and check: $\frac{4}{5}=\frac{2}{x+8}$

Solution The denominators are 5 and $x + 8$, so the LCD is $5 \cdot (x + 8)$. We multiply each side of the equation by the LCD.

$$\frac{4}{5}=\frac{2}{x+8}$$

$$\overset{1}{\cancel{5}}\cdot(x+8)\frac{4}{\underset{1}{\cancel{5}}}=5\cdot\overset{1}{\cancel{(x+8)}}\frac{2}{\underset{1}{\cancel{x+8}}}$$

$$4\cdot(x+8)=5\cdot 2$$

$$4x+32=10$$

$$4x=-22$$

$$x=-\frac{22}{4}=-\frac{11}{2}=-5.5$$

Check

$$\frac{4}{5}\stackrel{?}{=}\frac{2}{-5.5+8} \quad \text{Substitute } -5.5 \text{ for } x.$$

$$\frac{4}{5}\stackrel{?}{=}\frac{2}{2.5}$$

$$\frac{4}{5}=\frac{4}{5} \quad \text{True.}$$

So the solution to the given equation is -5.5.

PRACTICE 2

Solve and check: $\frac{5}{x-1}=\frac{2}{3}$

The following rule describes the general procedure for solving equations of this type:

To Solve a Rational Equation

- Find the LCD of all rational expressions in the equation.
- Multiply each side of the equation by the LCD.
- Solve the equation.
- Check the solution(s) in the original equation.

EXAMPLE 3

A 125-pound adult gets a dosage of 2 mL of a particular drug. At this rate, how much additional drug does a 175-pound adult require?

Solution We begin by writing a proportion in which we use x to represent the amount of additional drug required. We then solve this equation either by multiplying by the LCD or by setting the cross products equal.

$$\frac{2}{125} = \frac{2 + x}{175} \quad \begin{matrix} \leftarrow \text{Amount of drug in milliliters} \\ \leftarrow \text{Weight in pounds} \end{matrix}$$

$$2 \cdot 175 = 125(2 + x) \quad \text{Cross multiply.}$$

$$350 = 250 + 125x \quad \text{Use the distributive property.}$$

$$125x = 100 \quad \text{Combine like terms.}$$

$$x = \frac{100}{125} \quad \text{Divide each side of the equation by 125.}$$

$$x = 0.8$$

Check

$$\frac{2}{125} = \frac{2 + x}{175}$$

$$\frac{2}{125} \stackrel{?}{=} \frac{2 + 0.8}{175} \quad \text{Substitute 0.8 for } x.$$

$$\frac{2}{125} \stackrel{?}{=} \frac{2.8}{175} \quad \text{Simplify.}$$

$$0.016 = 0.016 \quad \text{True}$$

So an additional 0.8 mL of the drug is required for a 175-pound adult.

PRACTICE 3

A homeowner pays $900 a month on a $100,000 mortgage. If instead she had a $75,000 mortgage, how much less money would she be paying per month at the same interest rate?

EXAMPLE 4

Solve and check: $\frac{4}{n} - \frac{n+1}{3} = 1$

Solution The LCD of the rational expressions in the equation is $3n$.

$$\frac{4}{n} - \frac{n+1}{3} = 1$$

$$3n \cdot \left(\frac{4}{n} - \frac{n+1}{3}\right) = 3n \cdot 1 \quad \text{Multiply each side of the equation by the LCD.}$$

$$3n \cdot \frac{4}{n} - 3n \cdot \frac{n+1}{3} = 3n \cdot 1 \quad \text{Distribute } 3n \text{ on the left side of the equation.}$$

$$\overset{1}{3\cancel{n}} \cdot \frac{4}{\underset{1}{\cancel{n}}} - \overset{1}{\cancel{3}n} \cdot \frac{n+1}{\underset{1}{\cancel{3}}} = 3n \cdot 1$$

$$12 - n(n+1) = 3n \quad \text{Simplify.}$$

$$12 - n^2 - n = 3n \quad \text{Use the distributive property.}$$

$$n^2 + n + 3n - 12 = 0$$

$$n^2 + 4n - 12 = 0 \quad \text{Write in standard form.}$$

$$(n+6)(n-2) = 0 \quad \text{Factor the left side of the equation.}$$

$$n + 6 = 0 \quad \text{or} \quad n - 2 = 0 \quad \text{Set each factor equal to 0.}$$

$$n = -6 \qquad\qquad n = 2$$

Check

Substitute -6 for n.

$$\frac{4}{n} - \frac{n+1}{3} = 1$$

$$\frac{4}{-6} - \frac{-6+1}{3} \stackrel{?}{=} 1$$

$$-\frac{2}{3} - \left(-\frac{5}{3}\right) \stackrel{?}{=} 1$$

$$-\frac{2}{3} + \frac{5}{3} \stackrel{?}{=} 1$$

$$\frac{3}{3} \stackrel{?}{=} 1$$

$$1 = 1 \quad \text{True}$$

Substitute 2 for n.

$$\frac{4}{n} - \frac{n+1}{3} = 1$$

$$\frac{4}{2} - \frac{2+1}{3} \stackrel{?}{=} 1$$

$$2 - 1 \stackrel{?}{=} 1$$

$$1 = 1 \quad \text{True}$$

Our check confirms that the solutions are -6 and 2.

PRACTICE 4

Solve and check: $\frac{x-2}{5} - 1 = -\frac{2}{x}$

EXAMPLE 5

Solve and check: $\dfrac{2}{x+3} + \dfrac{1}{x-3} = -\dfrac{6}{x^2-9}$

Solution First, we find the LCD of the rational expressions in this equation, which is $x^2 - 9$, or $(x+3)(x-3)$. Then, we multiply each side of the equation by the LCD.

$$\frac{2}{x+3} + \frac{1}{x-3} = -\frac{6}{x^2-9}$$

$$\frac{2}{x+3} + \frac{1}{x-3} = -\frac{6}{(x+3)(x-3)}$$

$$(x+3)(x-3)\cdot\frac{2}{x+3} + (x+3)(x-3)\cdot\frac{1}{x-3} = (x+3)(x-3)\cdot\frac{-6}{(x+3)(x-3)}$$

$$\overset{1}{\cancel{(x+3)}}(x-3)\cdot\frac{2}{\underset{1}{\cancel{x+3}}} + (x+3)\overset{1}{\cancel{(x-3)}}\cdot\frac{1}{\underset{1}{\cancel{x-3}}} = \overset{1}{\cancel{(x+3)(x-3)}}\cdot\frac{-6}{\underset{1}{\cancel{(x+3)(x-3)}}}$$

$$2(x-3) + (x+3) = -6$$
$$2x - 6 + x + 3 = -6$$
$$3x - 3 = -6$$
$$3x = -3$$
$$x = -1$$

Check

$$\frac{2}{x+3} + \frac{1}{x-3} = -\frac{6}{x^2-9}$$

$$\frac{2}{-1+3} + \frac{1}{-1-3} \stackrel{?}{=} -\frac{6}{(-1)^2-9} \quad \text{Substitute } -1 \text{ for } x.$$

$$\frac{2}{2} + \frac{1}{-4} \stackrel{?}{=} -\frac{6}{-8} \quad \text{Simplify.}$$

$$\frac{2}{2} + \left(-\frac{1}{4}\right) \stackrel{?}{=} -\left(-\frac{6}{8}\right)$$

$$1 - \frac{1}{4} \stackrel{?}{=} \frac{3}{4}$$

$$\frac{3}{4} = \frac{3}{4} \quad \text{True}$$

So the solution is -1.

PRACTICE 5

Solve and check: $\dfrac{4}{y+2} + \dfrac{2}{y-1} = \dfrac{12}{y^2+y-2}$

When multiplying each side of a rational equation by a variable expression, the resulting equation may have a solution that does not satisfy the original equation. If such a result makes a denominator in the original equation 0, then the rational expression is undefined. These **extraneous solutions** are *not* solutions of the original equation. So in solving rational equations, it is particularly important to check all possible solutions.

EXAMPLE 6

Solve and check: $\frac{x^2}{x-2} = \frac{4}{x-2}$

Solution The LCD of the rational expressions is $x - 2$.

$$\frac{x^2}{x-2} = \frac{4}{x-2}$$

$$(x-2)\cdot\frac{x^2}{x-2} = (x-2)\cdot\frac{4}{x-2}$$ Multiply each side of the equation by the LCD.

$$\overset{1}{\cancel{(x-2)}}\cdot\frac{x^2}{\underset{1}{\cancel{x-2}}} = \overset{1}{\cancel{(x-2)}}\cdot\frac{4}{\underset{1}{\cancel{x-2}}}$$

$$x^2 = 4$$

$$x^2 - 4 = 0$$

$$(x+2)(x-2) = 0$$ Factor the left side of the equation.

$x + 2 = 0$ or $x - 2 = 0$ Set each factor equal to 0.

$x = -2$ $\qquad x = 2$ Solve for x.

Check

Substitute -2 for x.

$$\frac{x^2}{x-2} = \frac{4}{x-2}$$

$$\frac{(-2)^2}{-2-2} \stackrel{?}{=} \frac{4}{-2-2}$$

$$-\frac{4}{4} = -\frac{4}{4}$$ True

Substitute 2 for x.

$$\frac{x^2}{x-2} = \frac{4}{x-2}$$

$$\frac{2^2}{2-2} \stackrel{?}{=} \frac{4}{2-2}$$

$$\frac{4}{0} = \frac{4}{0}$$ Undefined

Since we get undefined fractions when we substitute 2 for x in the original equation, 2 is *not* a solution. So the only solution is -2. Without solving, how could we have known that 2 is *not* a solution of the original equation? Explain.

PRACTICE 6

Solve and check: $x = \frac{9}{x+3} + \frac{3x}{x+3}$

Rational equations play an important role in **work problems**. In these problems, we typically want to compute how long it will take to complete a task.

The key to solving a work problem is to determine the *rate of work*, that is, the fraction of the task that is completed in one unit of time. For instance, if it takes a secretary 5 hr to type a report, then the secretary's rate of work—the fraction of the report typed in 1 hr—would be $\frac{1}{5}$. If it takes a painter 6 hr to paint a room, then $\frac{1}{6}$ of the room would be painted in an hour so that $\frac{1}{6}$ is the painter's rate of work.

Consider the following example.

EXAMPLE 7

Two company employees, one senior and the other junior, are responsible for carrying out a project. If the two employees had worked alone, the junior employee would have completed the project in 6 hr and the senior employee would have completed it in 4 hr. How long would it take the two employees to carry out the project working together?

Solution Since this is a work problem, we can use the following equation to determine how long it will take the two employees to carry out the project working together:

Rate of work · Time worked = Part of the task completed

Using this equation, we can set up a table. Let t represent the time it takes the two employees to complete the project working together. Note here that the task to be completed is the project.

	Rate of Work	· Time Worked	= Part of the Task Completed
Senior employee	$\frac{1}{4}$	t	$\frac{1}{4}\cdot t$
Junior employee	$\frac{1}{6}$	t	$\frac{1}{6}\cdot t$

Since the sum of the parts of the task completed must equal one complete task, we have:

$$\frac{1}{4}\cdot t+\frac{1}{6}\cdot t=1$$

$$\frac{t}{4}+\frac{t}{6}=1$$

To solve this equation, we multiply each side by the LCD 12.

$$12\cdot\frac{t}{4}+12\cdot\frac{t}{6}=12\cdot 1$$

$$3t+2t=12$$

$$5t=12$$

$$t=2\tfrac{2}{5}=2.4$$

So working together, it would take the two employees 2.4 hr (or 2 hr 24 min) to complete the project. We can confirm by checking, which is left as an exercise.

PRACTICE 7

A water tank has two pumps. Working alone, the less powerful pump can fill the tank in 10 hr, whereas the more powerful pump can fill it in 6 hr. How long will it take both pumps working together to fill the tank?

Another application of rational equations is **motion problems**. Recall that in these problems, an object moves at a constant rate r for time t and travels a distance d. These three quantities are related by the following formula:

$$d = rt$$

If we solve this equation for t, we get $t = \frac{d}{r}$, which is the formula that we apply in Example 8. This example involves the time a round trip took, which is the sum of the time going and the time returning.

EXAMPLE 8

A family on vacation drove 50 mi to a hotel, and then returned home following the same route. Because of lighter traffic, the family drove at twice the speed going to the hotel as compared to returning home. If the round trip took 3 hr, at what speed did the family return home?

Solution We are looking for the speed of the family's car returning home. Let's represent this unknown quantity by r. Since the family traveled twice as fast going to the hotel as returning home, their speed going to the hotel must have been $2r$. We use the formula $\frac{d}{r} = t$ to find the time traveled in each direction, and set up a table.

	Distance ÷	Rate =	Time
Going to the hotel	50	$2r$	$\frac{50}{2r}$
Returning home	50	r	$\frac{50}{r}$

Since it is given that the round trip took 3 hr in all, we write:

$$\frac{50}{2r} + \frac{50}{r} = 3$$

To solve this rational equation, we multiply each side of the equation by the LCD $2r$.

$$2r \cdot \frac{50}{2r} + 2r \cdot \frac{50}{r} = 2r \cdot 3$$

$$\overset{1}{\cancel{2r}} \cdot \frac{50}{\underset{1}{\cancel{2r}}} + 2\overset{1}{\cancel{r}} \cdot \frac{50}{\underset{1}{\cancel{r}}} = 2r \cdot 3$$

$$50 + 100 = 6r$$

$$6r = 150$$

$$r = 25$$

So the family returned home at 25 mph. We can confirm this solution by checking.

PRACTICE 8

A business executive traveled 1800 mi by jet, continuing the trip an additional 300 mi on a propeller plane. The speed of the jet was 3 times that of the propeller plane. If the entire trip took 6 hr, what was the speed of the propeller plane?

EXAMPLE 9

A boat travels 60 mph in still water. Find the speed of the river's current if the boat traveled 70 mi upriver in the same time that it took to travel 80 mi downriver.

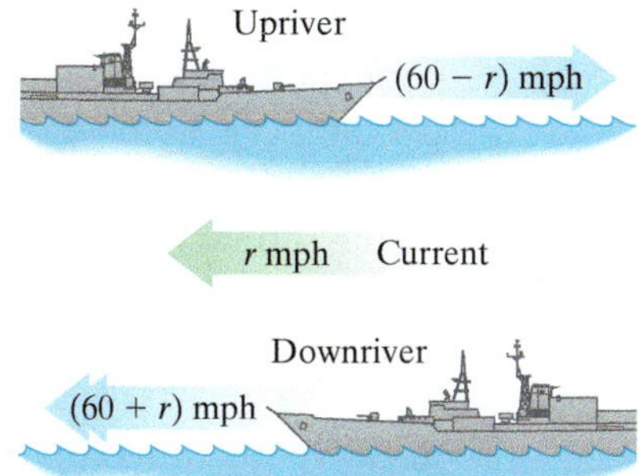

Solution Let r represent the speed of the river's current. When traveling downriver, the speed of the current is added to that of the boat. When traveling upriver, the speed of the current is subtracted from that of the boat. Applying the formula $t = \frac{d}{r}$, we can set up the following table:

	Distance (d)	÷ Rate (r)	= Time (t)
Upriver	70	$60 - r$	$\frac{70}{60 - r}$
Downriver	80	$60 + r$	$\frac{80}{60 + r}$

We are told that these two times are equal, so we write the following proportion and solve for r:

$$\frac{80}{60 + r} = \frac{70}{60 - r}$$

$$80(60 - r) = 70(60 + r) \quad \text{Cross multiply.}$$

$$4800 - 80r = 4200 + 70r$$

$$600 = 150r$$

$$r = 4$$

The speed of the river's current was 4 mph. How would you check that this answer is correct?

PRACTICE 9

The speed of the jet stream was 300 mph. When flying in the direction of the jet stream, a plane flies 1000 mi in the same time that it takes to fly 250 mi against the jet stream. Find the speed of the plane in still air.

Some rational equations are formulas or literal equations that relate two or more variables. Consider the next example.

EXAMPLE 10

The formula $\frac{1}{f} = \frac{1}{p} + \frac{1}{q}$ gives the focal length f of a lens, where p is the distance between the lens and an object and q is the distance between the image and the lens.

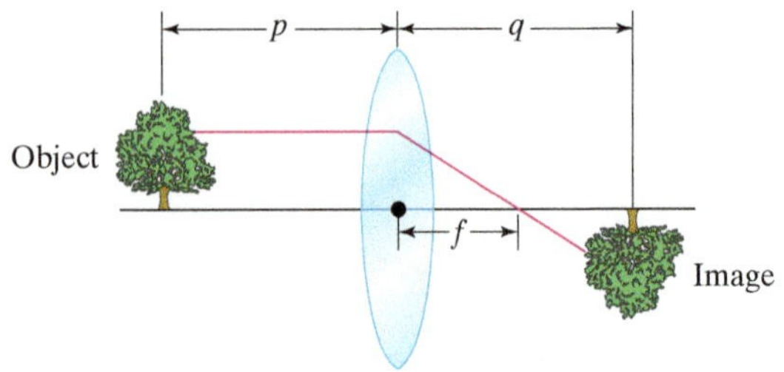

a. Solve this equation for f.

b. Use the formula found in part (a) to find the value of f when $p = 40$ cm and $q = 10$ cm.

Solution

a. Our task is to solve the equation for f.

$$\frac{1}{f} = \frac{1}{p} + \frac{1}{q}$$

$$fpq \cdot \frac{1}{f} = fpq \cdot \left(\frac{1}{p} + \frac{1}{q}\right)$$

$$fpq \cdot \frac{1}{f} = fpq \cdot \frac{1}{p} + fpq \cdot \frac{1}{q}$$

$$\overset{1}{f}pq \cdot \frac{1}{\underset{1}{\cancel{f}}} = f\overset{1}{\cancel{p}}q \cdot \frac{1}{\underset{1}{\cancel{p}}} + fp\overset{1}{\cancel{q}} \cdot \frac{1}{\underset{1}{\cancel{q}}}$$

$$pq = fq + fp$$

Since we are solving for f, we factor it out on the right side of the equation.

$$pq = f(q + p)$$

$$f = \frac{pq}{p + q}$$

b. Substituting 40 for p and 10 for q in the formula $f = \frac{pq}{p + q}$ we get:

$$f = \frac{40 \cdot 10}{40 + 10}$$

$$= \frac{400}{50}$$

$$= 8$$

So the focal length of the lens is 8 cm.

PRACTICE 10

Economists are interested in how the free market works. Suppose that a commodity sells at the price x dollars per unit. The number of units sold, called the *demand D*, can be modeled by the following equation:

$$D = \frac{500}{x} + 50$$

a. Solve for x in terms of D.

b. Find the value of x when the demand is 450 units.

17.5 Exercises

FOR EXTRA HELP MyMathLab  PRACTICE  WATCH  READ  REVIEW

A *Solve and check.*

1. $\frac{y}{2} + \frac{7}{10} = -\frac{4}{5}$

2. $\frac{x}{6} - \frac{1}{10} = \frac{2}{5}$

3. $\frac{1}{t} - \frac{7}{3} = -\frac{1}{3}$

4. $\frac{5}{p} + \frac{1}{3} = \frac{6}{p}$

5. $\frac{n}{8} - \frac{n}{2} = \frac{3}{4}$

6. $\frac{r}{2} + \frac{r}{3} = -\frac{5}{6}$

7. $x + \frac{1}{x} = 2$

8. $y + \frac{2}{y} = 3$

9. $\frac{t+1}{2t-1} - \frac{5}{7} = 0$

10. $\frac{b-3}{3b-2} - \frac{4}{5} = 0$

11. $\frac{t-2}{3} = 4$

12. $\frac{4}{x+1} = 3$

13. $\frac{2}{3} = \frac{4}{t-5}$

14. $\frac{1}{m+4} = \frac{2}{9}$

15. $\frac{8+x}{12} = \frac{22}{36}$

16. $\frac{75}{20} = \frac{30}{p-1}$

17. $\frac{y+3}{14} = \frac{y}{7}$

18. $\frac{n-6}{3} = \frac{n}{5}$

19. $\frac{x-1}{8} = \frac{x+1}{12}$

20. $\frac{3}{t-2} = \frac{9}{t+2}$

21. $\frac{x}{8} = \frac{2}{x}$

22. $\frac{4}{n} = \frac{n}{16}$

23. $\frac{2}{y} = \frac{y-4}{16}$

24. $\frac{5}{x+2} = \frac{x}{7}$

25. $\frac{a}{a+3} = \frac{4}{5a}$

26. $\frac{1}{3n} = \frac{n}{n+2}$

27. $\frac{y+1}{y+6} = \frac{y}{y+6}$

28. $\frac{2x+3}{x-3} = \frac{3x}{x-3}$

29. $\frac{2n}{n+1} = \frac{2}{n+1} + 1$

30. $\frac{4}{s-3} - \frac{3s}{s-3} = 2$

31. $\frac{5x}{x+1} = \frac{x^2}{x+1} + 2$

32. $\frac{x^2}{x-2} = \frac{2x}{x-2} + 3$

33. $\frac{s-5}{3} - \frac{8}{s} = -1$

34. $\frac{2}{c-2} + \frac{4}{c} = 2$

35. $\frac{m^2}{m+3} = \frac{9}{m+3}$

36. $\frac{n}{n+2} + \frac{2}{n-2} = \frac{-n+6}{n^2-4}$

37. $\frac{x}{x-3} - \frac{6}{x} = 1$

38. $\frac{2}{t} + \frac{t}{t+1} = 1$

39. $1 + \frac{4}{x^2} = \frac{4}{x}$

40. $\frac{1}{2x} + \frac{3}{x^2} = 1$

41. $\frac{2}{p+1} - \frac{1}{p-1} = \frac{2p}{p^2-1}$

42. $\frac{2}{y+2} - \frac{5}{2-y} = \frac{3y}{y^2-4}$

43. $\frac{3}{x} - \frac{1}{x+4} = \frac{5}{x^2+4x}$

44. $\frac{2}{y-2} + \frac{3}{y} = \frac{4}{-2y+y^2}$

45. $1 - \frac{6x}{(x-4)^2} = \frac{2x}{x-4}$

46. $\frac{4x}{x+1} - 2 = -\frac{3x}{(x+1)^2}$

47. $\frac{n+1}{n^2+2n-3} = \frac{n}{n+3} - \frac{1}{n-1}$

48. $\frac{7}{x-5} - \frac{5x+6}{x^2-3x-10} = \frac{x}{x+2}$

Mixed Practice

Solve and check.

49. $\frac{2}{b+3} = \frac{5b}{b^2-9} - \frac{3}{3-b}$

50. $\frac{3}{r} + \frac{1}{4} = \frac{12}{r}$

51. $\frac{6m}{m+1} - 3 = \frac{8m}{(m+1)^2}$

52. $\frac{a-3}{3(a-2)} - \frac{3}{8} = 0$

53. $2 - \frac{2x}{x+3} = \frac{6}{x+1}$

54. $\frac{8}{s} + \frac{s}{s+4} = 1$

55. $\frac{4}{w-3} = \frac{7}{w+3}$

56. $\frac{a}{a+3} = \frac{3}{2a}$

Applications

B *Solve.*

57. A clerical worker takes 4 times as long to finish a job as it does an executive secretary. Working together, it takes them 3 hr to finish the job. How long would it take the clerical worker, working alone, to finish the job?

58. One pipe can fill a tank in 1 min. A second pipe takes 2 min to fill the same tank. Working together, how long will it take both pipes to fill the tank?

59. A car traveled twice as fast on a dry road as it did making the return trip on a slippery road. The trip was 60 mi each way, and the round trip took 3 hr. What was the speed of the car on the dry road?

60. A cyclist rode uphill and then turned around and made the same trip downhill. The downhill speed was 3 times the uphill speed. If the trip each way was 18 mi and the entire trip lasted 2 hr, at what speed was the cyclist going uphill?

61. Steam exerts pressure on a pipe. In the formula

$$p = \frac{P}{LD},$$

p represents the pressure per square inch in the pipe, P stands for the total pressure in the pipe, L is the pipe's length, and D is the pipe's diameter. Solve this formula for D.

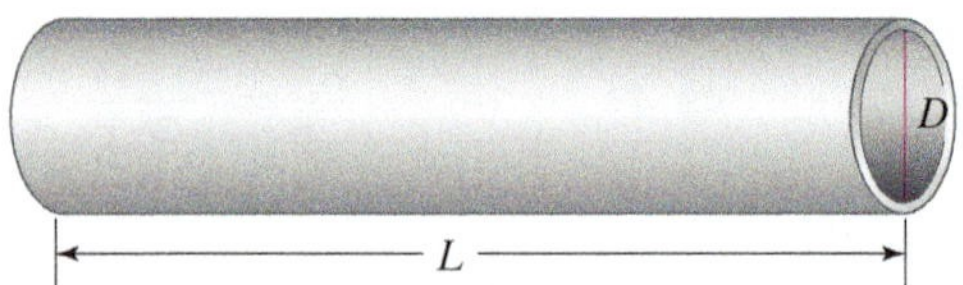

62. To approximate the appropriate dosage of medicine for a child, doctors sometimes use Young's rule:

$$C = \frac{aA}{a+12},$$

where C represents the child's dosage, a stands for the child's age, and A is the adult dosage. Solve this formula for a.

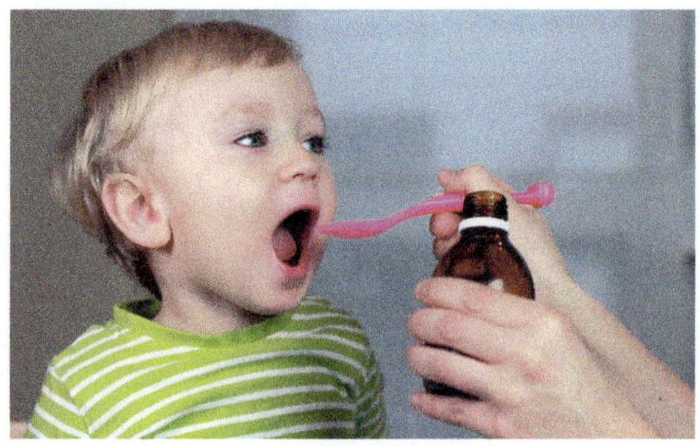

63. A local shop sells 5 bags of soil for $10. At this rate, how many fewer bags sell for $4?

64. It takes a laser printer 1 min 15 sec to print a 25-page report. At this rate, how much longer would it take the printer to print a 40-page report?

65. A chauffeur took 2 hr longer to drive 275 mi than he took to drive 165 mi. If his speed was the same on both trips, at what speed was he driving?

66. A train travels 225 mi in the same time that a bus travels 200 mi. If the speed of the train is 5 mph greater than that of the bus, find the speed of the bus.

67. A boat traveled 20 mi up the river, fighting a 6 mph current. The return trip, with the current, took half the time. How fast does the boat travel in still water?

68. A pilot flies 600 mi with a tailwind of 20 mph. Against the wind, he flies only 500 mi in the same amount of time. What is the speed of the plane in still air?

69. The city of Miami has two main interstate highways, I-75 and I-95. Suppose that an evacuation was planned for a portion of the city due to an impending hurricane. If all traffic moves smoothly, assume that I-75 alone could handle the evacuation traffic in x hr and I-95 in $(x + 3)$ hr, where x depends on the size of the evacuation.

a. How long would it take to carry out the evacuation using both highways?

b. How long would the evacuation take if $x = 21$ hr?

70. While you were in a car, the first 10 megabytes (MB) of a 170 MB download to your cell phone was on a 3G network. The remainder of the download, which altogether took m min, was on a 4G network that downloads 8 times as fast.

a. How fast (in megabytes per minute) was the 4G network?

b. How fast was this network if $m = 100$ min?

• Check your answers on page A-37.

MIND*Stretchers*

Groupwork

1. Working with a partner, show that there are no positive real numbers a and b for which the equation $\frac{1}{a} + \frac{1}{b} = \frac{1}{a + b}$ is true.

Mathematical Reasoning

2. Use your knowledge of systems of equations to solve the following system:

$$\textbf{(1)}\quad \frac{1}{x} + \frac{3}{y} = 1$$

$$\textbf{(2)}\quad -\frac{4}{x} + \frac{3}{y} = -3$$

Historical

3. A riddle from 1500 years ago deals with the age of the Greek mathematician Diophantus when he died. According to this riddle, "*. . . his boyhood lasted* $\frac{1}{6}$ *of his life; he married after* $\frac{1}{7}$ *more; his beard grew after* $\frac{1}{12}$ *more, and his son was born 5 years later; the son lived to half his father's age, and the father died 4 years after the son.*" Solve this riddle by determining Diophantus' age at his death. (*Source:* www.history.mcs.st-and.ac.uk)

Key Concepts and Skills

CONCEPT SKILL

Concept/Skill	Description	Example
[17.1] Rational expression	An algebraic expression $\frac{P}{Q}$ that can be written as the quotient of two polynomials, P and Q, where $Q \neq 0$.	$\frac{2x}{x+5}$, where $x + 5 \neq 0$
[17.1] Equivalent rational expressions	Rational expressions that have the same value, for any value of the variable that makes neither denominator 0.	$\frac{2}{3}$ and $\frac{2x}{3x}$, where $x \neq 0$
[17.1] To find an equivalent rational expression	• Multiply the numerator and denominator of $\frac{P}{Q}$ by the same polynomial R. $\frac{P}{Q} = \frac{PR}{QR}$, where $Q \neq 0$ and $R \neq 0$.	$\frac{3}{5y+1} = \frac{3 \cdot y}{(5y+1) \cdot y} = \frac{3y}{5y^2+y}$
[17.1] To simplify a rational expression	• Factor the numerator and denominator. • Divide out any common factors in the numerator and denominator.	$\frac{3n^2-3}{n^2+n-2}$ $= \frac{3\overset{1}{\cancel{(n-1)}}(n+1)}{\underset{1}{\cancel{(n-1)}}(n+2)} = \frac{3(n+1)}{n+2}$
[17.2] To multiply rational expressions	• Factor the numerators and denominators. • Divide the numerators and denominators by all common factors. • Multiply the remaining factors in the numerators and the remaining factors in the denominators.	$\frac{6x}{x^2-4x} \cdot \frac{x^2-16}{3x}$ $= \frac{6x}{x(x-4)} \cdot \frac{(x+4)(x-4)}{3x}$ $= \frac{\overset{2\ 1}{\cancel{6}\cancel{x}}}{\underset{1}{\cancel{x}}\underset{1}{\cancel{(x-4)}}} \cdot \frac{(x+4)\overset{1}{\cancel{(x-4)}}}{\underset{1}{\cancel{3}}x}$ $= \frac{2(x+4)}{x}$
[17.2] To divide rational expressions	• Take the reciprocal of the divisor, and change the operation to multiplication. • Follow the rule for multiplying rational expressions.	$\frac{x^2-x-6}{3x-9} \div \frac{4x+8}{5x}$ $= \frac{x^2-x-6}{3x-9} \cdot \frac{5x}{4x+8}$ $= \frac{\overset{1}{\cancel{(x+2)}}\overset{1}{\cancel{(x-3)}}}{3\underset{1}{\cancel{(x-3)}}} \cdot \frac{5x}{4\underset{1}{\cancel{(x+2)}}}$ $= \frac{5x}{12}$

continued

Concept/Skill	Description	Example
[17.3] **To add (or subtract) rational expressions with the same denominator**	• Add (or subtract) the numerators and keep the same denominator. • Simplify, if possible.	$\frac{3n-5}{n+1}+\frac{n+6}{n+1}$ $=\frac{(3n-5)+(n+6)}{n+1}$ $=\frac{3n-5+n+6}{n+1}=\frac{4n+1}{n+1}$ $\frac{2n^2+4n+6}{n-5}-\frac{n^2+10n+1}{n-5}$ $=\frac{(2n^2+4n+6)-(n^2+10n+1)}{n-5}$ $=\frac{2n^2+4n+6-n^2-10n-1}{n-5}$ $=\frac{n^2-6n+5}{n-5}$ $=\frac{(n-1)\overset{1}{\cancel{(n-5)}}}{\underset{1}{\cancel{(n-5)}}}=n-1$
[17.3] **To find the LCD of rational expressions**	• Factor each denominator completely. • Multiply these factors, using for the power of each factor the greatest number of times that it occurs in any of the denominators. The product of all these factors is the LCD.	$\frac{x}{x^2-4x+3}$ and $\frac{x+5}{2x-2}$ Factor x^2-4x+3: $(x-1)(x-3)$ Factor $2x-2$: $2(x-1)$ LCD $=2(x-1)(x-3)$
[17.3] **To add (or subtract) rational expressions with different denominators**	• Find the LCD of the rational expressions. • Write each rational expression with a common denominator, usually the LCD. • Add (or subtract) the numerators. • Simplify, if possible.	$\frac{n}{n^2+3n+2}+\frac{3}{2n+2}$ $=\frac{n}{(n+1)(n+2)}+\frac{3}{2(n+1)}$ $=\frac{n\cdot 2}{(n+1)(n+2)\cdot 2}$ $+\frac{3\cdot(n+2)}{2(n+1)\cdot(n+2)}$ $=\frac{2n}{2(n+1)(n+2)}+\frac{3n+6}{2(n+1)(n+2)}$ $=\frac{5n+6}{2(n+1)(n+2)}$ $\frac{5y}{y^2-9}-\frac{2}{y+3}$ $=\frac{5y}{(y+3)(y-3)}-\frac{2}{y+3}$ $=\frac{5y}{(y+3)(y-3)}-\frac{2\cdot(y-3)}{(y+3)\cdot(y-3)}$ $=\frac{5y}{(y+3)(y-3)}-\frac{2y-6}{(y+3)(y-3)}$ $=\frac{5y-(2y-6)}{(y+3)(y-3)}$ $=\frac{3y+6}{(y+3)(y-3)}=\frac{3(y+2)}{(y+3)(y-3)}$

Concept/Skill	Description	Example
[17.4] **Complex rational expression (complex algebraic fraction)**	A rational expression that contains rational expressions in its numerator, denominator, or both.	$\dfrac{x+1}{\frac{2}{3}}, \dfrac{\frac{a-b}{4}}{5}, \dfrac{\frac{6n}{5}-\frac{1}{n}}{\frac{n+3}{n^2}}$
[17.4] **To simplify a complex rational expression: division method**	• Write the numerator and denominator as single rational expressions in simplified form. • Write the expression as the numerator divided by the denominator. • Divide. • Simplify, if possible.	$\dfrac{\frac{x}{4}}{\frac{x^2}{3}+\frac{x}{2}} = \dfrac{\frac{x}{4}}{\frac{x^2}{3}\cdot\frac{2}{2}+\frac{x}{2}\cdot\frac{3}{3}}$ $= \dfrac{\frac{x}{4}}{\frac{2x^2+3x}{6}}$ $= \dfrac{x}{4} \div \dfrac{2x^2+3x}{6}$ $= \dfrac{x}{4}\cdot\dfrac{6}{2x^2+3x}$ $= \dfrac{\overset{1}{\cancel{x}}}{\underset{2}{\cancel{4}}}\cdot\dfrac{\overset{3}{\cancel{6}}}{\underset{1}{\cancel{x}}(2x+3)}$ $= \dfrac{3}{2(2x+3)}$
[17.4] **To simplify a complex rational expression: LCD method**	• Find the LCD of all the rational expressions *within* both the numerator and denominator. • Multiply the numerator and denominator of the complex rational expression by this LCD. • Simplify, if possible.	$\dfrac{1-\frac{1}{y}}{1-\frac{1}{y^2}} = \dfrac{\left(1-\frac{1}{y}\right)\cdot y^2}{\left(1-\frac{1}{y^2}\right)\cdot y^2}$ $= \dfrac{1\cdot y^2 - \frac{1}{\underset{1}{\cancel{y}}}\cdot \overset{y}{\cancel{y^2}}}{1\cdot y^2 - \frac{1}{\underset{1}{\cancel{y^2}}}\cdot \overset{1}{\cancel{y^2}}}$ $= \dfrac{y^2-y}{y^2-1}$ $= \dfrac{y\overset{1}{\cancel{(y-1)}}}{(y+1)\underset{1}{\cancel{(y-1)}}}$ $= \dfrac{y}{y+1}$

continued

Concept/Skill	Description	Example
[17.5] **To solve a rational equation**	• Find the LCD of all rational expressions in the equation. • Multiply each side of the equation by the LCD. • Solve the equation. • Check your solution(s) in the original equation.	$\frac{1}{x+3} + \frac{1}{x-3} = \frac{2}{x^2-9}$ $\frac{1}{x+3} + \frac{1}{x-3} = \frac{2}{(x+3)(x-3)}$ $(x+3)(x-3) \cdot \frac{1}{x+3}$ $+ (x+3)(x-3)\frac{1}{x-3}$ $= (x+3)(x-3) \cdot \frac{2}{(x+3)(x-3)}$ $x - 3 + x + 3 = 2$ $2x = 2$ $x = 1$ **Check** Substitute 1 for x. $\frac{1}{x+3} + \frac{1}{x-3} = \frac{2}{x^2-9}$ $\frac{1}{1+3} + \frac{1}{1-3} \stackrel{?}{=} \frac{2}{1^2-9}$ $\frac{1}{4} - \frac{1}{2} \stackrel{?}{=} -\frac{2}{8}$ $-\frac{1}{4} = -\frac{1}{4}$ True

CHAPTER 17 Review Exercises

Say Why

Fill in each blank.

1. The rational expression $\frac{2x-3}{x+4}$ ________ (is/is not) undefined when $x = \frac{3}{2}$ because ________________________________.

2. The expression $\frac{a^2(3-a)}{a(5+a)}$ ________ (is/is not) in simplest form because ________________________________.

3. The LCD of $\frac{5}{6x^3y}$ and $\frac{1}{2xy^2z}$ ________ (is/is not) $6x^3y^2z$ because ________________________________.

4. The expression $\frac{2}{\frac{x}{y}}$ ________ (is/is not) a complex rational expression because ________________________________.

5. The solution $a = 3$ ________ (is/is not) an extraneous solution to the equation $1 + \frac{9}{a^2} = \frac{6}{a}$ because ________________________________.

6. The rational expressions $\frac{3r}{s}$ and $\frac{9r^2s}{3rs^2}$ ________ (are/are not) equivalent because ________________________________.

[17.1]

7. Identify the values for which the given rational expression is undefined.

   a. $\frac{4}{x+1}$ b. $\frac{6x+12}{x^2-x-6}$

8. Determine whether each pair of rational expressions is equivalent.

   a. $\frac{2x}{y} \stackrel{?}{=} \frac{10x^2y}{5xy^2}$ b. $\frac{x^2-9}{x^2+6x+9} \stackrel{?}{=} \frac{x-3}{x+3}$

Simplify.

9. $\frac{12m}{20m^2}$

10. $\frac{15n-18}{9n+6}$

11. $\frac{x^2+2x-8}{4-x^2}$

12. $\frac{2x^2-3x-20}{3x^2-13x+4}$

[17.2] *Perform the indicated operation.*

13. $\frac{10mn}{3p^2} \cdot \frac{9np}{5m^2}$

14. $\frac{y-5}{4y+6} \cdot \frac{6y+9}{3y-15}$

15. $\frac{x+6}{x^2+x-30} \cdot \frac{x^2-10x+25}{2x+5}$

16. $\frac{2a^2-2a-4}{4-a^2} \cdot \frac{2a^2+a-6}{4a^2-2a-6}$

17. $\frac{x^2y}{2x} \div xy^2$

18. $\frac{5m+10}{2m-20} \div \frac{7m+14}{14m-20}$

19. $\frac{5y^2}{x^2-36} \div \frac{25xy-25y}{x^2-7x+6}$

20. $\frac{2x^2+x-1}{x^2+8x+7} \div \frac{6x^2+x-2}{x^2+14x+49}$

[17.3] *Write each pair of rational expressions in terms of their LCD.*

21. $\frac{1}{5x}$ and $\frac{3}{20x^2}$

22. $\frac{4}{n-1}$ and $\frac{n}{n+4}$

23. $\frac{1}{3x+9}$ and $\frac{x}{x^2+4x+3}$

24. $\frac{2}{3x^2-5x-2}$ and $\frac{1}{4-x^2}$

Perform the indicated operation.

25. $\frac{3t+1}{2t}+\frac{t-1}{2t}$

26. $\frac{5y}{y+7}-\frac{y-28}{y+7}$

27. $\frac{5y+4}{4y^2-2y}-\frac{2}{2y-1}$

28. $\frac{n}{3n+15}+\frac{n-2}{n^2+5n}$

29. $\frac{4}{x-3}-\frac{4x+1}{9-x^2}$

30. $\frac{y+3}{4-y^2}+\frac{1}{2-y}$

31. $\frac{2}{m+1}+\frac{6m-2}{m^2-2m-3}$

32. $\frac{3x-2}{x^2-x-12}-\frac{x+3}{x-4}$

33. $\frac{2x}{x^2+4x+4}-\frac{x-1}{x^2-2x-8}$

34. $\frac{n+4}{2n^2-3n+1}+\frac{n+1}{2n^2+5n-3}$

[17.4] *Simplify.*

35. $\dfrac{\frac{x}{2}}{\frac{3x^2}{7}}$

36. $\dfrac{1-\frac{9}{y}}{1-\frac{81}{y^2}}$

37. $\dfrac{\frac{1}{x}+\frac{1}{y}}{\frac{1}{2x}+\frac{1}{2y}}$

38. $\dfrac{4-\frac{3}{x}-\frac{1}{x^2}}{2-\frac{5}{x}+\frac{3}{x^2}}$

[17.5] *Solve and check.*

39. $\frac{2x}{x-4}=5-\frac{1}{x-4}$

40. $\frac{y+1}{y}+\frac{1}{2y}=4$

41. $\frac{5}{2x}+\frac{3}{x+1}=\frac{7}{x}$

42. $\frac{y-2}{y-4}=\frac{1}{y+2}+\frac{y+3}{y^2-2y-8}$

43. $\frac{x}{x+2}-\frac{2}{2-x}=\frac{x+6}{x^2-4}$

44. $\frac{3}{n^2-5n+4}-\frac{1}{n^2-4n+3}=\frac{n-3}{n^2-7n+12}$

45. $\frac{28}{x+3}=\frac{7}{9}$

46. $\frac{5}{3+y}=\frac{3}{7y+1}$

47. $\frac{3}{x-7}=\frac{x}{6}$

48. $\frac{11}{x-2}=\frac{x+7}{2}$

Mixed Applications

Solve.

49. A company found that the cost per booklet for printing x booklets can be represented by $\left(0.72 + \frac{200}{x}\right)$ dollars. Write this cost as a single rational expression.

50. Four friends decide to split the cost of renting a car equally. They discover that if they let one more friend share in the rental, the cost for each of the original four friends will be reduced by \$10. What is the total cost of the car rental?

51. A hiker walks a distance d mi at a speed of r mph, and then returns on the same path at a speed of s mph. The hiker's average speed for the entire trip can be represented by:

$$\frac{2d}{\frac{d}{r} + \frac{d}{s}}$$

Simplify this expression.

52. With the water running at full force, it takes 10 min to fill a bathtub. It then takes 15 min for the bathtub to drain. If by mistake the water is running at full force while the tub is draining, how long will it take the tub to fill?

53. A family on vacation drove 400 mi at two different speeds, 50 mph and 60 mph. The total driving time was 7 hr. How many miles did the family drive at 50 mph?

54. One student takes x hr to design a Web page. Another student takes an hour longer. What part of the job will be finished in an hour if the two students work together?

55. If t is the amount of time (in minutes) a runner needs to run a mile, then his or her average speed (in miles per hour) can be represented by $\frac{60}{t}$. For two runners, the ratio of their average speeds is $\frac{\frac{60}{t_1}}{\frac{60}{t_2}}$. Simplify this ratio.

56. Find a rational expression equal to the sum of the reciprocals of three consecutive integers, starting with n.

• Check your answers on page A-37.

CHAPTER 17 Posttest

FOR EXTRA HELP

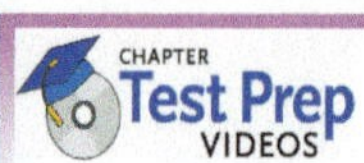

To see if you have mastered the topics in this chapter, take this test.

1. Identify the values for which the rational expression $\dfrac{3x}{x-8}$ is undefined.

2. Show that $\dfrac{y-3}{y}$ is equivalent to $-\dfrac{3y-y^2}{y^2}$.

Simplify.

3. $\dfrac{15a^3b}{12ab^2}$

4. $\dfrac{x^2-4x}{xy-4y}$

5. $\dfrac{3b^2-27}{b^2-4b-21}$

6. $\dfrac{\dfrac{3}{x^2}-\dfrac{1}{x}}{\dfrac{9}{x^2}-1}$

7. Write $\dfrac{4n-1}{n^2+6n-16}$, $\dfrac{2}{n+8}$, and $\dfrac{n}{4n-8}$ in terms of their LCD.

Perform the indicated operation.

8. $\dfrac{7x-10}{x+6}-\dfrac{5x-22}{x+6}$

9. $\dfrac{3}{2y-8}+\dfrac{2}{4y^2-16y}$

10. $\dfrac{5}{d-3}-\dfrac{d-4}{d^2-d-6}$

11. $\dfrac{5}{2x^2-3x-2}-\dfrac{x}{4-x^2}$

12. $\dfrac{n+1}{3n-18}\cdot\dfrac{n-6}{6n^3-6n}$

13. $\dfrac{a^2-25}{a^2-2a-24}\div\dfrac{a^2+a-30}{a^2-36}$

14. $\dfrac{x^2+6x+8}{x^2+x-2}\div\dfrac{x+4}{2x^2+12x+16}$

Solve and check.

15. $\dfrac{1}{y-5}+\dfrac{y+4}{25-y^2}=\dfrac{1}{y+5}$

16. $\dfrac{2y}{y-4}-2=\dfrac{4}{y+5}$

Solve.

17. Two rectangles each have area 28 ft^2. One rectangle has a width 3 ft greater and a length 3 ft shorter than the other rectangle. Find the shorter width.

18. Large files, which account for up to half of Internet traffic, can be downloaded to a computer from different sources and at different speeds at the same time. Suppose that there are three sources that contribute to the downloading of a large file. Working alone, one source would complete the downloading of the file in $2m$ min, the second in $3m$ min, and the third in $5m$ min. Working from the three sources together, how long would it take to download the file? (*Sources:* bittorrent.com and wikipedia.org)

19. A company owns two mail processing machines. The newer machine works twice as fast as the older one. Together, the two machines process 1000 pieces of mail in 20 min. How long does it take each machine, working alone, to process 1000 pieces of mail?

20. To get into shape, a cyclist bicycled for 20 mi slowly and then for 30 mi at 5 mph faster. Altogether she bicycled for 4 hr. What was her slow speed?

• Check your answers on page A-37.

Cumulative Review Exercises

To help you review, solve the following:

1. Which is larger, 0.6214 or 0.63?

2. Replace ▢ with $>$, $<$, or $=$ to make a true statement: $-|-4.6|$ ▢ -4.7

3. The following graph shows the percent changes in population of five American cities between the censuses in 2000 and 2010.

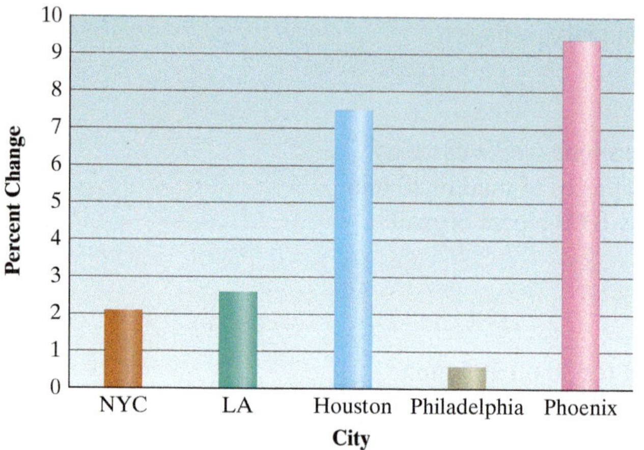

Estimate the difference between the percent change for Phoenix and that for Los Angeles. (*Source*: census.gov)

4. Find the area of the triangle shown below.

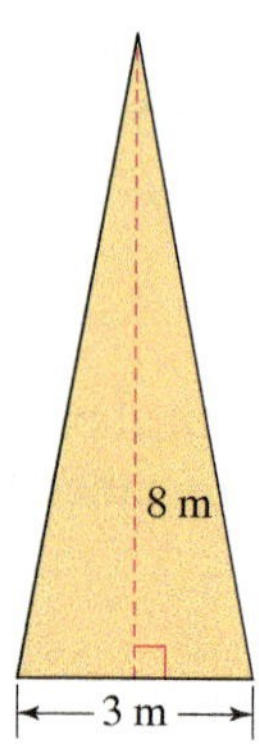

5. Subtract 2 ft 11 in. from 7 ft 2 in.

6. Simplify: $y - [2y - 3(y - 1)]$

7. Graph the inequality $y \geq 2$ on the coordinate plane.

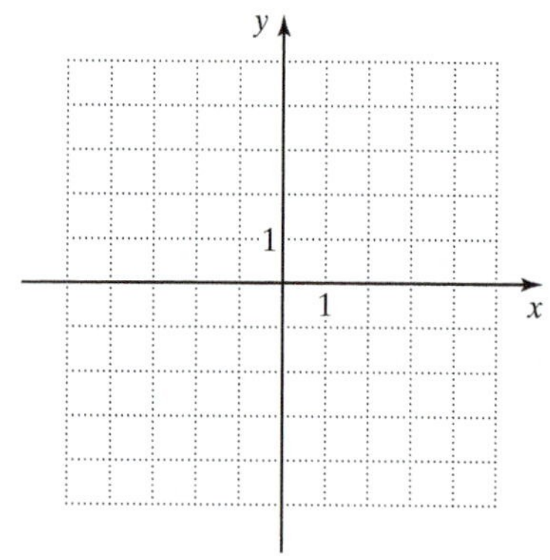

8. Solve by substitution, and then check.

$$x + y = 8$$
$$y = 2x - 1$$

9. True or false, the ordered pair $(2, -5)$ is a solution of the following system:

$$x + y = -3$$
$$3x - y = -11$$

10. Simplify: $(3k - 6l)^2$

11. Solve: $(n + 5)(n - 2) = 0$

12. Factor: $100y^2 - 81$

13. Simplify: $\dfrac{2 - \dfrac{6y}{x}}{1 - \dfrac{9y^2}{x^2}}$

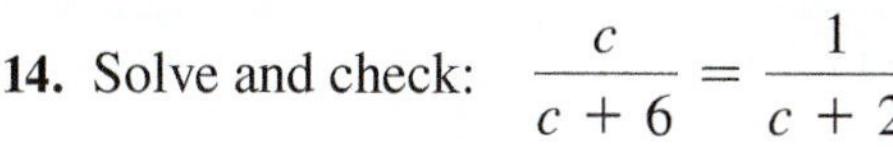

14. Solve and check: $\dfrac{c}{c + 6} = \dfrac{1}{c + 2}$

15. In round numbers, the population of the United States doubled between the years 1900 and 1950, and again between 1950 and 2000. If p represents the population in 1900, express in exponential form the population in 2000. (*Source:* census.gov)

16. A broker invested twice as much money in Stock A as in Stock B. Stock A had a return of 6% and Stock B a return of 4%. The total return was $1200. How much was invested in each stock?

17. The fastest train in the world is not France's TGV or the Japanese Bullet Train but, rather, the Pudong International Airport Express in Shanghai, China. This train travels at a maximum speed of 268 mph. At this rate, how far, to the nearest mile, would it travel in 2 min? (*Source:* beijing-visitor.com)

18. The U.S. annual per capita consumption of fruit in recent years can be modeled by the equation $y = 0.6x + 100$, where x is the number of years after 1970 and y is the number of pounds of fruit consumed per person. (*Source: The World Almanac and Book of Facts 2011*)

a. On a coordinate plane, label the axes, and then plot the graph of this equation.

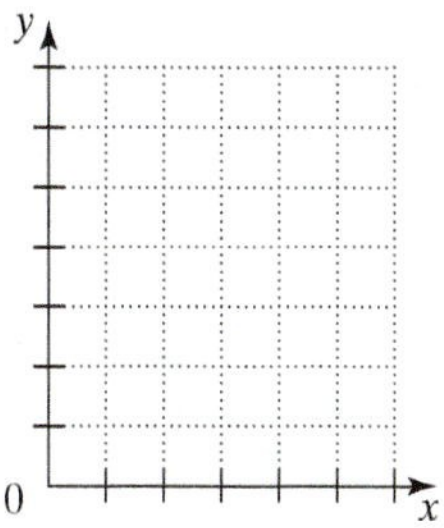

b. Identify the slope of this line and explain its significance according to this model.

c. Identify the y-intercept and explain its significance.

19. Two buses following the same route leave from the same station at different times. The Reston bus leaves 2 hr later than the Arlington bus. The Arlington bus is traveling at 40 mph and the Reston bus is traveling at 60 mph. How long will it take the Reston bus to overtake the Arlington bus?

20. A country's balance of payments can be found by subtracting its imports from its exports. If x is the number of years after 2000, the total exports of the United States for a particular year can be approximated by the polynomial $-5.6x^3 + 82.7x^2 - 227.6x + 1111.1$, and the total U.S. imports can be approximated by $-10.3x^3 + 132.8x^2 - 295.8x + 1494.8$ (both in billions of dollars). Find a polynomial to model the U.S. balance of payments. (*Source:* bea.gov)

• Check your answers on page A-37.

CHAPTER 18

Radical Expressions and Equations

Square Roots and Lighting

A theatrical lighting designer focuses the audience's attention on what is lit: actors, costumes, props, and scenery. In planning a production for the stage, the designer must take into account how much illumination a lighting fixture will deliver from a particular distance. An actor further from the fixture seems more dimly lit than the one who is closer.

In general, the distance d associated with an illumination of I on a person or object is given by the equation

$$d = \sqrt{\frac{k}{I}},$$

where k is a constant that depends on the fixture, d is in feet, and I is in foot-candles. It follows from this formula that if a lighting designer wants to increase the illumination on someone by a factor of four, the distance to the light source must be divided by the square root of 4, that is, by 2.

(*Source:* J. Michael Gillette, *Theatrical Design and Production*, McGraw-Hill, 1999)

CHAPTER 18 PRETEST

To see if you have already mastered the topics in this chapter, take this test.

Assume all variables and radicands represent nonnegative real numbers.

Simplify.

1. $\sqrt{81}$

2. $-\sqrt{27}$

3. $\sqrt{45a^2}$

4. $\sqrt{\dfrac{x}{64}}$

5. $6\sqrt{2} + \sqrt{2} - 3\sqrt{2}$

6. $\sqrt{12} + 2\sqrt{75}$

7. $\sqrt{9x^3} - 4x\sqrt{x} + x\sqrt{36x}$

Multiply or divide. Simplify, if possible.

8. $\sqrt{6} \cdot \sqrt{3}$

9. $\sqrt{2xy} \cdot \sqrt{10xy^3}$

10. $\dfrac{\sqrt{30}}{\sqrt{5}}$

11. $\sqrt{n}(\sqrt{n} + 2)$

12. $(\sqrt{3} - 1)(\sqrt{3} + 4)$

Simplify.

13. $\sqrt{\dfrac{5x}{6}}$

14. $\dfrac{\sqrt{40x^3}}{\sqrt{2x}}$

15. Rationalize the denominator: $\dfrac{8 + \sqrt{7}}{\sqrt{2}}$

Solve and check.

16. $\sqrt{x} - 1 = 5$

17. $y = \sqrt{4y - 3}$

Solve.

18. The velocity of a car (in meters per second) that starts from rest with a constant acceleration of a (in meters per second^2) can be found using the expression $\sqrt{2as}$, where s is the distance traveled (in meters). If the acceleration of the car is 2 m/sec^2, find its velocity after it has traveled 100 m.

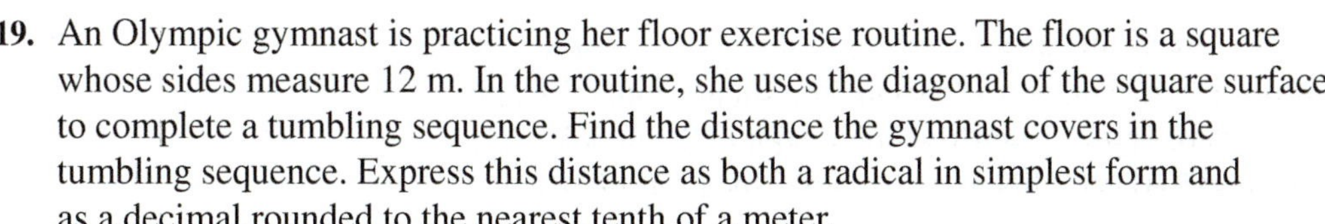

19. An Olympic gymnast is practicing her floor exercise routine. The floor is a square whose sides measure 12 m. In the routine, she uses the diagonal of the square surface to complete a tumbling sequence. Find the distance the gymnast covers in the tumbling sequence. Express this distance as both a radical in simplest form and as a decimal rounded to the nearest tenth of a meter.

20. Traffic accident investigators use the lengths of skid marks to determine the minimum speed a car was traveling before it skids to a stop. The minimum speed S (in miles per hour) can be approximated by the formula

$$S = \sqrt{30fL},$$

where f is the drag factor (or coefficient of friction) for the road surface and L is the length (in feet) of the skid marks. Solve this formula for L in terms of S and f.

• Check your answers on page A-38.

18.1 Introduction to Radical Expressions

OBJECTIVES

- **A** To evaluate radical expressions
- **B** To simplify radical expressions
- **C** To solve applied problems involving radical expressions

What Radicals Are and Why They Are Important

So far, we have considered two types of algebraic expressions, namely polynomials and rational expressions. In this chapter, we extend the discussion to a third type, called *radical expressions*. **Radical expressions**, such as $4\sqrt{y}$ or $5 - \sqrt{x}$, are algebraic expressions that involve *square roots*. Note that in this text we use the terms *radical* and *square root* interchangeably.

When we find a square root of a number, we ask ourselves: What number multiplied by itself gives the original number? This question arises naturally in many situations, notably in geometry. For instance, recall the formula for the area A of a square with side s.

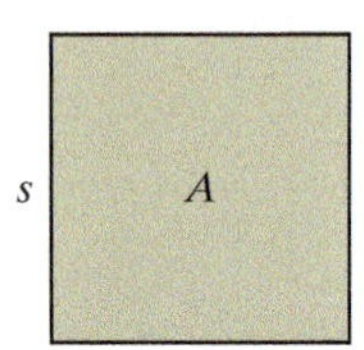

The formula $A = s^2$ states that to find the area of a square, we square the length of a side. If we know the value of A and wish to determine the value of s, we find the square root of A, that is, $s = \sqrt{A}$.

In this chapter, first we discuss evaluating and simplifying radical expressions, as well as carrying out the usual operations on them—adding, subtracting, multiplying, and dividing. Then, we consider how to solve equations involving radical expressions.

Introduction to Radicals

Let's begin with the definition of square root.

DEFINITION

A **square root** of a nonnegative real number a is a number that when squared is a.

Actually, every positive number has *two* square roots, one positive and the other negative. For example, the square roots of 4 are $+2$ and -2 because $(+2)^2 = 4$ and $(-2)^2 = 4$. The symbol $\sqrt{\ }$, called the **radical sign**, stands for the positive or *principal square root,* but is commonly referred to as "**the** square root." For instance, $\sqrt{9}$ is read "the square root of 9", or "radical 9", and represents $+3$, or 3. By contrast, the negative square root of 9, namely -3, is represented by $-\sqrt{9}$. Throughout the remainder of this text, when we speak of the square root of a number we mean its principal square root.

The number under the radical sign is called the **radicand**.

We read $\sqrt{n}$ as "the square root of n."

This chapter focuses on radicals with a nonnegative radicand, such as $\sqrt{0}$ and $\sqrt{3}$. We will not discuss radicals with a negative radicand, say $\sqrt{-4}$. Such a radical is not a real number, since no real number when squared results in a negative number.

Evaluating Radicals

Let's continue our discussion of radicals by finding the value of radicals with radicands that are whole numbers. A whole number is said to be a **perfect square** if it is the square of another whole number. For instance, 9 is a perfect square, since $9 = 3^2$. On the other hand, 8 is not a perfect square, since it is not the square of any whole number. Perfect squares play a special role in finding the value of square roots.

The square root of a perfect square is always a whole number. For instance, $\sqrt{9} = 3$, since $3^2 = 9$.

EXAMPLE 1

Find the value of the following radical expressions:

a. $\sqrt{16}$ **b.** $-3\sqrt{25}$

Solution

a. $\sqrt{16} = 4$ since $4^2 = 16$. **b.** $-3\sqrt{25} = -3 \cdot 5 = -15$

PRACTICE 1

Evaluate the following radical expressions:

a. $\sqrt{4}$

b. $-5\sqrt{49}$

TIP In evaluating radicals, remember that squaring, doubling, and halving a number are *not* the same as finding its square root.

The square root of a whole number that is not a perfect square is an *irrational number.* Recall from Section 11.1 that irrational numbers have decimal representations which neither terminate nor repeat. For instance, we have seen that $\sqrt{2}$ and $\sqrt{3}$ are irrational numbers.

$\sqrt{2} = 1.4142\ldots$, which is approximately 1.414.
$\sqrt{3} = 1.7320\ldots$, which is approximately 1.732.

A calculator can be used to approximate the value of these square roots by decimals rounded to a specific place value.

EXAMPLE 2

Evaluate $\sqrt{7}$, rounded to the nearest thousandth.

Solution Since the radicand 7 is not a perfect square, its square root is an irrational number. Using a calculator, we see that $\sqrt{7} = 2.6457\ldots$. Rounding to the nearest thousandth gives us 2.646.

PRACTICE 2

Find the value of $\sqrt{10}$, rounded to the nearest thousandth.

Simplifying Radicals

The operations of squaring and taking a square root undo each other. They are opposite operations in the same way that adding a number is the opposite of subtracting that number. The following two properties of radicals stem from this relationship.

Squaring a Square Root

For any nonnegative real number a,

$$(\sqrt{a})^2 = a.$$

In words, this property states that when we take the square root of a nonnegative number and then square the result, we get the original number. For instance, $(\sqrt{9})^2 = 9$ and $(\sqrt{8})^2 = 8$.

Taking the Square Root of a Square

For any nonnegative real number a,

$$\sqrt{a^2} = a.$$

In words, this property states that when we square a nonnegative number and then take the square root, the result is the original number. For instance, $\sqrt{3^2} = 3$.

We will use both of these properties repeatedly throughout this chapter. Do you see the difference between the two properties? Explain.

EXAMPLE 3

Simplify each radical expression.

a. $(\sqrt{25})^2$ **b.** $(\sqrt{2})^2$ **c.** $\sqrt{4^2}$ **d.** $\sqrt{9^2}$

Solution

a. $(\sqrt{25})^2 = 25$ Use the property of squaring a square root.

b. $(\sqrt{2})^2 = 2$

c. $\sqrt{4^2} = 4$ Use the property of taking the square root of a square.

d. $\sqrt{9^2} = 9$

PRACTICE 3

Simplify.

a. $(\sqrt{6})^2$

b. $(\sqrt{5})^2$

c. $\sqrt{7^2}$

d. $\sqrt{1^2}$

Some radicands contain variables. Consider, for instance, the expression $\sqrt{x}$. For negative values of x, the radicand is negative and so $\sqrt{x}$ is not a real number. *For purposes of simplicity, we will assume throughout the remainder of this text that all radicands and all variables are nonnegative.*

Let's look at some radicals that are perfect squares involving variables. For instance, the radicands for $\sqrt{x^8}$, $\sqrt{49y^6}$, and $\sqrt{25a^2b^4}$ are all perfect squares.

$$\sqrt{x^8} = \sqrt{(x^4)^2} = x^4$$
$$\sqrt{49y^6} = \sqrt{(7y^3)^2} = 7y^3$$
$$\sqrt{25a^2b^4} = \sqrt{(5ab^2)^2} = 5ab^2$$

Note that the exponent of a variable in any perfect square is always an even number, such as 8 or 10. When finding the square root of an *even* power, we can simply divide the power by 2. For instance,

$$\sqrt{x^8} = x^4 \leftarrow \frac{8}{2} = 4$$

EXAMPLE 4

Simplify each radical expression.

a. $\sqrt{n^6}$ **b.** $-\sqrt{16x^8}$ **c.** $\sqrt{100a^4b^2}$

Solution

a. The radicand n^6 is a perfect square, so we can use the property of taking the square root of a square:

$$\sqrt{n^6} = \sqrt{(n^3)^2} = n^3$$

b. $-\sqrt{16x^8} = -\sqrt{(4x^4)^2} = -4x^4$

c. $\sqrt{100a^4b^2} = \sqrt{(10a^2b)^2} = 10a^2b$

PRACTICE 4

Simplify.

a. $\sqrt{x^4}$

b. $\sqrt{64t^{10}}$

c. $-\sqrt{121x^2y^2}$

Just as we simplify fractions to express them in lowest terms, so we put radical expressions *in simplified form.* In this way, we can recognize that expressions that appear to be different are really the same. Also, simplified radical expressions are often easier to work with.

But what exactly does it mean to simplify a radical? It is easier to say when a radical expression is *not* simplified than to say when it *is* simplified. *A radical is not considered to be in simplified form if its radicand is divisible by a perfect square.* For instance, the radical $\sqrt{8}$ is not in simplified form because the perfect square 4 is a factor of the radicand 8. To simplify such a radical, we apply the *product rule of radicals.*

The Product Rule of Radicals

If a and b are any nonnegative real numbers, then

$$\sqrt{ab} = \sqrt{a} \cdot \sqrt{b}.$$

In words, this rule states that the square root of a product is the product of the square roots.

To check that the product rule works in a particular example, we consider the radical $\sqrt{4 \cdot 9}$.

$$\sqrt{4 \cdot 9} \stackrel{?}{=} \sqrt{4} \cdot \sqrt{9}$$
$$\sqrt{36} \stackrel{?}{=} 2 \cdot 3$$
$$6 = 2 \cdot 3 \quad \text{True}$$

So $\sqrt{4 \cdot 9} = \sqrt{4} \cdot \sqrt{9}$, as the product rule implies.

Now, let's use this rule to simplify $\sqrt{8}$.

$$\sqrt{8} = \sqrt{4 \cdot 2} \quad \text{Factor out the perfect square 4 from the radicand.}$$
$$= \sqrt{4} \cdot \sqrt{2} \quad \text{Use the product rule of radicals.}$$
$$= 2\sqrt{2} \quad \sqrt{4} = 2.$$

So $2\sqrt{2}$ is the simplified form of $\sqrt{8}$.

Some radicals contain radicands that have more than one possible perfect square factor. To simplify these radicals, we generally factor out the *largest* perfect square factor of the radicand, as shown in Example 5(a).

EXAMPLE 5

Simplify.

a. $\sqrt{48}$ **b.** $-4\sqrt{27}$ **c.** $\dfrac{\sqrt{24}}{6}$

Solution

a.
$$\sqrt{48} = \sqrt{16 \cdot 3} \quad \text{Factor out the perfect square 16.}$$
$$= \sqrt{16} \cdot \sqrt{3} \quad \text{Use the product rule of radicals.}$$
$$= 4\sqrt{3} \quad \text{Take the square root of the perfect square.}$$

b.
$$-4\sqrt{27} = -4\sqrt{9 \cdot 3} \quad \text{Factor out the perfect square 9.}$$
$$= -4\sqrt{9} \cdot \sqrt{3} \quad \text{Use the product rule of radicals.}$$
$$= -4 \cdot 3\sqrt{3} \quad \text{Take the square root of the perfect square.}$$
$$= -12\sqrt{3} \quad \text{Simplify.}$$

PRACTICE 5

Express in simplified form.

a. $\sqrt{72}$

b. $2\sqrt{40}$

c. $\dfrac{\sqrt{75}}{-15}$

EXAMPLE 5 (continued)

c. $\frac{\sqrt{24}}{6} = \frac{\sqrt{4 \cdot 6}}{6}$ Factor out the perfect square 4.

$= \frac{\sqrt{4} \cdot \sqrt{6}}{6}$ Use the product rule of radicals.

$= \frac{2\sqrt{6}}{6}$ Take the square root of the perfect square.

$= \frac{\overset{1}{\cancel{2}}\sqrt{6}}{\underset{3}{\cancel{6}}}$ Divide out the common factor.

$= \frac{\sqrt{6}}{3}$ Simplify.

Note that in Example 5(c), we cannot divide out the 6 in the numerator with the 6 in the denominator, since one is under a radical sign and the other is not.

TIP Be careful when dividing out common factors in radical expressions.

$$\frac{\sqrt{3}}{3} \neq \frac{\sqrt{\cancel{3}}}{\cancel{3}} \quad \text{but} \quad \frac{3\sqrt{3}}{3} = \frac{\cancel{3}\sqrt{3}}{\cancel{3}}$$

Now, let's apply the product rule to radical expressions involving variables. The key here is to factor out the largest possible even power of each variable. Note that these problems test our knowledge of exponents as well as of radicals.

EXAMPLE 6

Simplify.

a. $\sqrt{y^5}$ b. $-\sqrt{20a^6}$ c. $\sqrt{60x^5y}$

Solution

a. $\sqrt{y^5} = \sqrt{y^4 \cdot y}$ Factor out the perfect square y^4, using the product rule of exponents.

$= \sqrt{y^4} \cdot \sqrt{y}$ Use the product rule of radicals.

$= y^2\sqrt{y}$ Take the square root of the perfect square.

b. $-\sqrt{20a^6} = -\sqrt{4 \cdot 5 \cdot a^6}$ Factor out the perfect squares 4 and a^6.

$= -\sqrt{4} \cdot \sqrt{a^6} \cdot \sqrt{5}$ Use the product rule of radicals.

$= -2a^3\sqrt{5}$ Take the square root of the perfect squares.

c. $\sqrt{60x^5y} = \sqrt{4 \cdot 15 \cdot x^4 \cdot x \cdot y}$

$= \sqrt{4} \cdot \sqrt{x^4} \cdot \sqrt{15xy}$

$= 2x^2\sqrt{15xy}$

PRACTICE 6

Express in simplified form.

a. $\sqrt{x^3}$

b. $\sqrt{18n^4}$

c. $-\sqrt{50ab^2}$

Some radical expressions involve quotients. If the radicand is a perfect square, we can take the square root directly as shown in the next example.

EXAMPLE 7

Find the square root. **a.** $\sqrt{\frac{1}{4}}$ **b.** $\sqrt{\frac{x^2}{49}}$ **c.** $\sqrt{\frac{a^2}{b^8}}$

Solution

a. $\sqrt{\frac{1}{4}} = \frac{1}{2}$ since $\left(\frac{1}{2}\right)^2 = \frac{1}{4}$. **b.** $\sqrt{\frac{x^2}{49}} = \frac{x}{7}$ since $\left(\frac{x}{7}\right)^2 = \frac{x^2}{49}$.

c. $\sqrt{\frac{a^2}{b^8}} = \frac{a}{b^4}$ since $\left(\frac{a}{b^4}\right)^2 = \frac{a^2}{b^8}$.

PRACTICE 7

Find the square root.

a. $\sqrt{\frac{1}{16}}$ **b.** $\sqrt{\frac{y^2}{4}}$ **c.** $\sqrt{\frac{x^4}{y^6}}$

Some radical expressions contain quotients that are not perfect squares. To simplify these radicals, we use the following rule:

The Quotient Rule of Radicals

If a is a nonnegative real number and b is a positive real number, then

$$\sqrt{\frac{a}{b}} = \frac{\sqrt{a}}{\sqrt{b}}.$$

In words, this rule states that the square root of a quotient is the quotient of the square roots.

To check that the quotient rule works in a particular example, let's consider the radical $\sqrt{\frac{4}{9}}$.

$$\sqrt{\frac{4}{9}} \stackrel{?}{=} \frac{\sqrt{4}}{\sqrt{9}}$$

$$\sqrt{\left(\frac{2}{3}\right)^2} \stackrel{?}{=} \frac{2}{3}$$

$$\frac{2}{3} = \frac{2}{3} \quad \text{True}$$

So $\sqrt{\frac{4}{9}} = \frac{\sqrt{4}}{\sqrt{9}}$, which is in agreement with the quotient rule.

EXAMPLE 8

Simplify.

a. $\sqrt{\frac{7}{4}}$ **b.** $\sqrt{\frac{3x}{25}}$ **c.** $\sqrt{\frac{2a^5b^{10}}{9}}$

Solution

a. $\sqrt{\frac{7}{4}} = \frac{\sqrt{7}}{\sqrt{4}}$ Use the quotient rule of radicals.

$= \frac{\sqrt{7}}{2}$ Take the square root of the perfect square.

b. $\sqrt{\frac{3x}{25}} = \frac{\sqrt{3x}}{\sqrt{25}}$ Use the quotient rule of radicals.

$= \frac{\sqrt{3x}}{5}$ Take the square root of the perfect square.

PRACTICE 8

Simplify.

a. $\sqrt{\frac{3}{16}}$

b. $\sqrt{\frac{2y}{49}}$

c. $\sqrt{\frac{5x^7y^2}{4}}$

EXAMPLE 8 (continued)

c. $\sqrt{\dfrac{2a^5b^{10}}{9}} = \dfrac{\sqrt{2a^5b^{10}}}{\sqrt{9}}$ Use the quotient rule of radicals.

$= \dfrac{\sqrt{a^4b^{10}} \cdot \sqrt{2a}}{\sqrt{9}}$ Use the product rule of radicals.

$= \dfrac{a^2b^5\sqrt{2a}}{3}$ Take the square root of each perfect square.

Now, let's see if we can use our knowledge of radicals in solving some applied problems. The following problem deals with the Pythagorean theorem, a topic that we discussed in Section 16.5. Recall that this theorem describes the relationship between the lengths of the three sides of a right triangle, which is expressed as $c^2 = a^2 + b^2$. We can rewrite the theorem as

$$c = \sqrt{a^2 + b^2},$$

where c is the length of the hypotenuse, and a and b are the lengths of the other two sides.

TIP In evaluating a radical such as $\sqrt{a^2 + b^2}$ with arithmetic operations in the radicand, parentheses are understood around the radicand. So in evaluating $\sqrt{a^2 + b^2}$, we compute $a^2 + b^2$, take the square root, and then simplify, if possible.

EXAMPLE 9

A flat panel television is rectangular in shape, with dimensions 25 in. by 15 in. Find the length of the panel's diagonal, both as a radical and as a decimal rounded to the nearest inch.

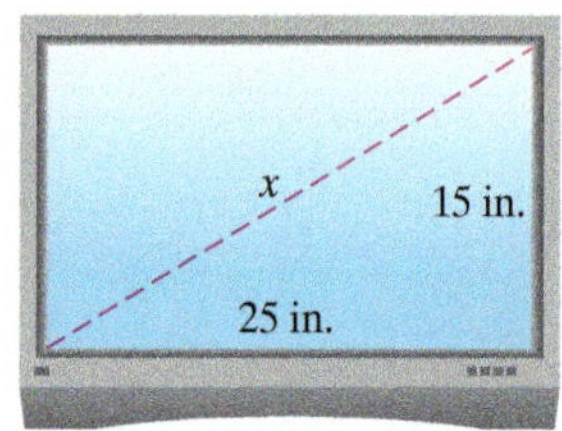

Solution Let x represent the length of the diagonal of the panel. The lengths of the other two sides of the triangle shown are 15 in. and 25 in. Using the Pythagorean theorem, we get:

$$\begin{aligned} x &= \sqrt{a^2 + b^2} \\ &= \sqrt{15^2 + 25^2} \\ &= \sqrt{225 + 625} \\ &= \sqrt{850} \\ &= \sqrt{25 \cdot 34} \\ &= \sqrt{25} \cdot \sqrt{34} \\ &= 5\sqrt{34} \end{aligned}$$

So the length of the diagonal of the panel is $5\sqrt{34}$ in., or approximately 29 in.

PRACTICE 9

Two Jeeps try to pull a car out of the snow, tugging at a 90° angle to each other.

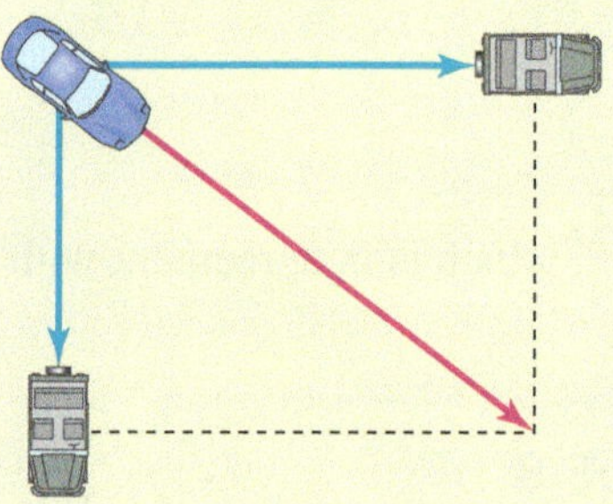

If one Jeep exerts a force of 600 lb and the other a force of 800 lb, the magnitude of the resulting force on the car (in pounds) can be found by computing $\sqrt{600^2 + 800^2}$. Calculate this force.

18.1 Exercises

FOR EXTRA HELP MyMathLab MathXL PRACTICE WATCH READ REVIEW

Mathematically Speaking

Fill in each blank with the most appropriate term or phrase from the given list.

odd	even	square root
product	divide by 2	square
quotient	positive square root	take the square root
double	irrational	

1. A(n) ________________ of a nonnegative real number a is a number that when squared is a.

2. The principal square root of 16, written $\sqrt{16}$, means the ________________ of 16.

3. The square root of a whole number that is not a perfect square is a(n) ________________ number.

4. If we take the square root of a nonnegative number and then ________________ the result, we get the original number.

5. When we square a nonnegative number and then ________________, the result is the original number.

6. The exponent of the variable in any perfect square is always a(n) ________________ number.

7. The square root of a product is the ________________ of the square roots.

8. The square root of a quotient is the ________________ of the square roots.

A *Find the value of each radical. When a radicand is not a perfect square, use a calculator to evaluate, rounding to the nearest thousandth.*

9. $\sqrt{36}$ **10.** $\sqrt{25}$ **11.** $\sqrt{1}$ **12.** $\sqrt{9}$

13. $-\sqrt{100}$ **14.** $-\sqrt{64}$ **15.** $3\sqrt{49}$ **16.** $2\sqrt{81}$

17. $\sqrt{5}$ **18.** $\sqrt{6}$ **19.** $3\sqrt{2}$ **20.** $4\sqrt{3}$

B *Simplify.*

21. $(\sqrt{16})^2$ **22.** $(\sqrt{5})^2$ **23.** $(\sqrt{11})^2$ **24.** $(\sqrt{10})^2$

25. $(\sqrt{5x})^2$ **26.** $(\sqrt{8y})^2$ **27.** $\sqrt{2^2}$ **28.** $\sqrt{3^2}$

29. $\sqrt{9^2}$ **30.** $\sqrt{14^2}$ **31.** $\sqrt{n^8}$ **32.** $\sqrt{s^2}$

33. $\sqrt{49y^2}$ **34.** $\sqrt{16t^4}$ **35.** $\sqrt{9x^4}$ **36.** $\sqrt{100y^6}$

37. $\sqrt{25x^2y^{10}}$ **38.** $\sqrt{49p^{12}q^4}$

Simplify each expression, factoring out any perfect square.

39. $\sqrt{32}$ **40.** $\sqrt{75}$ **41.** $-\sqrt{108}$ **42.** $-\sqrt{98}$

43. $6\sqrt{27}$ **44.** $3\sqrt{20}$ **45.** $\dfrac{\sqrt{48}}{12}$ **46.** $\dfrac{\sqrt{50}}{10}$

47. $\sqrt{11x^2}$ **48.** $\sqrt{7n^4}$ **49.** $\sqrt{n^5}$ **50.** $\sqrt{t^7}$

51. $\sqrt{20x^3}$ **52.** $\sqrt{32y^9}$ **53.** $-\sqrt{12p^2q}$

54. $-\sqrt{24a^3b^4}$ **55.** $9\sqrt{10x^3y^4}$ **56.** $2\sqrt{25a^7b^6}$

Simplify.

57. $\sqrt{\frac{4}{25}}$

58. $\sqrt{\frac{1}{36}}$

59. $-\sqrt{\frac{1}{4}}$

60. $-\sqrt{\frac{49}{9}}$

61. $\sqrt{\frac{81}{n^6}}$

62. $\sqrt{\frac{121}{x^8}}$

63. $\sqrt{\frac{x^4}{y^2}}$

64. $\sqrt{\frac{a^2}{b^6}}$

65. $-\sqrt{\frac{3}{4}}$

66. $-\sqrt{\frac{2}{9}}$

67. $\sqrt{\frac{5n}{16}}$

68. $\sqrt{\frac{3t}{25}}$

69. $\sqrt{\frac{3x^2y^6}{4}}$

70. $\sqrt{\frac{5a^4b^2}{9}}$

71. $-\sqrt{\frac{27x^6y}{16}}$

72. $-\sqrt{\frac{32ab^8}{25}}$

Mixed Practice

Simplify.

73. $\sqrt{\frac{36}{y^6}}$

74. $\sqrt{\frac{6m^4n^2}{49}}$

Find the value of each radical. When a radicand is not a perfect square, use a calculator to evaluate, rounding to the nearest thousandth.

75. $-\sqrt{144}$

76. $2\sqrt{8}$

Simplify each expression, factoring out any perfect square.

77. $5\sqrt{12}$

78. $\sqrt{11t^5}$

79. $-\sqrt{40a^4b^3}$

Simplify.

80. $(\sqrt{27x})^2$

81. $\sqrt{r^6}$

82. $\sqrt{64m^8}$

Applications

C *Solve.*

83. The *geometric mean m* of two numbers *a* and *b* is the square root of their product. The geometric mean can be thought of as a kind of average.
   a. Express this relationship as a formula.
   b. Find the geometric mean of 2 and 8.

84. To approximate the maximum speed *s* (in knots) of a sailboat, sailors multiply 1.3 by the square root of the length of the boat's waterline *w* (in feet).

   a. Express this relationship as a formula.
   b. Use this formula to approximate the maximum speed, to the nearest knot, of a sailboat with a 16-foot waterline.

85. When an object is dropped from a height h (in meters), it takes approximately $\sqrt{\frac{h}{5}}$ sec to reach the ground.

a. Find the time it takes for an object dropped from a height of 20 m to reach the ground.

b. If the same object is dropped from double the height, will it take twice as long to reach the ground? Explain.

86. The expression $2\pi\sqrt{\frac{L}{32}}$ approximates the time (in seconds) that a pendulum of length L (in feet) takes to make one complete swing back and forth.

a. How long does it take the pendulum shown to make one complete swing? Use 3.14 for π, and round to the nearest second.

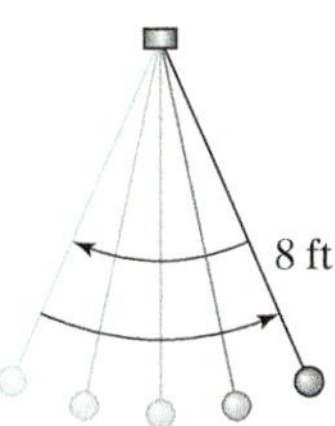

b. How long would it take a pendulum 4 times the length of that in part (a) to make a complete swing? Is this answer 4 times the answer to part (a)?

87. Under certain conditions, the expression $2\sqrt{5L}$ can be used to approximate the speed of a car (in miles per hour) that has left a skid mark of length L (in feet). If an investigating officer arrives at the scene of an accident where these conditions apply and finds a skid mark 180 ft long, at what speed was the car traveling at the time of the accident?

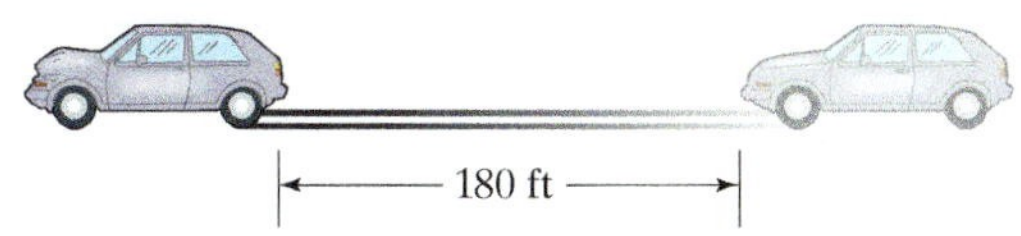

88. Firefighters can approximate the speed S at which water leaves a typical nozzle by using the formula

$$S = 12\sqrt{P},$$

where the pressure P is measured in pounds per square inch (psi) and the speed is in feet per second. Find the speed of the water if the pressure is 50 psi. Express your answer both as a simplified radical and as a decimal rounded to the nearest whole number. (*Source:* Jim Cottrell, "Fire Stream Physics," *Firefighter's News,* 1995)

89. A surveyor wishes to find the distance between the towns B and C across from one another on a lake, as shown in the illustration. Find this distance expressed both as a radical in simplified form and as a decimal rounded to the nearest mile.

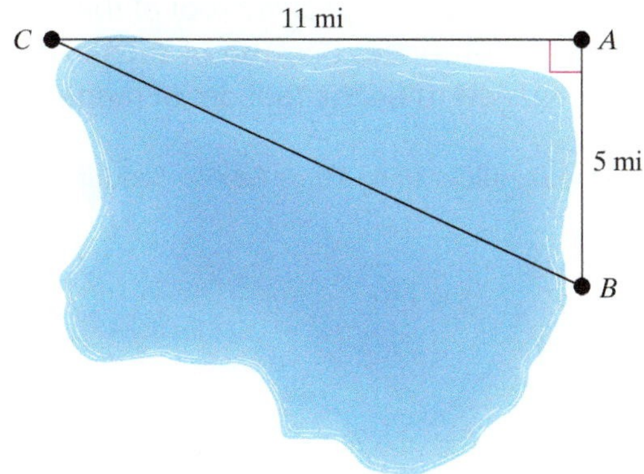

90. A plane is about to land at an airport. The flight path of the plane's descent is shown in the following diagram. What is the distance from the plane to the airport? Express this distance both as a radical and as a decimal rounded to the nearest multiple of 100 ft.

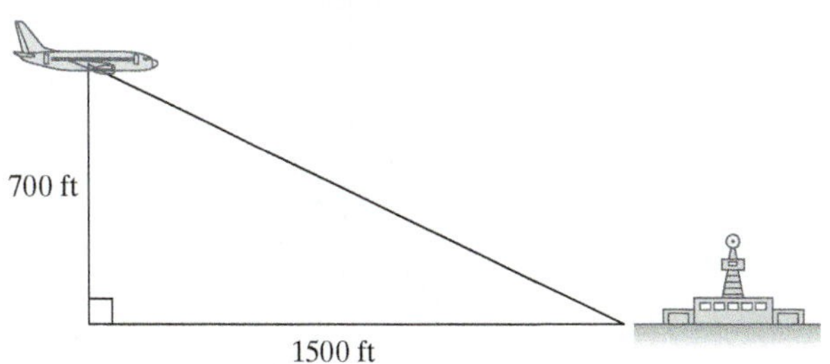

• Check your answers on page A-38.

MINDStretchers

Research

1. Using your college library or the Web, investigate whether there is a connection between the radical symbol and the symbol for a drugstore prescription.

Describe the results of your research.

Writing

2. Consider the following statement: *Just as with adding and subtracting, the operations of* ***squaring*** and ***taking a square root*** *are opposites*. In a few sentences, support or refute this statement.

Investigation

3. Consider the statement: For any positive number, the larger the number, the larger its square root.
   a. Use a calculator either to support this assertion with many examples or to give a counterexample. Sum up your findings.
   b. If you supported this assertion, do your examples prove the assertion true? If you gave a counterexample, does the counterexample prove the assertion false? Explain.

Cultural Note

The square root is related to one of the most revolutionary experiments in the history of science. Working some 500 years ago, Galileo Galilei discovered that contrary to ancient authority, the speed of falling objects, which he had dropped from the Leaning Tower of Pisa, was proportional to the square root of the height from which they were dropped. Because of such experiments, Galileo is considered to be the founder of modern physics.

(*Source:* Jeanne Bendick, *Along Came Galileo*, Beautiful Feat Books, 1999)

18.2 Adding and Subtracting Radical Expressions

OBJECTIVES

A To add or subtract radical expressions

B To solve applied problems involving the addition or subtraction of radical expressions

Sums and differences of many radicals, in contrast to other kinds of numbers such as fractions and decimals, cannot be simplified. For instance, there is no way to combine $\sqrt{2}$ and $\sqrt{3}$, so we cannot simplify the expression $\sqrt{2} + \sqrt{3}$. Similarly, we cannot simplify the expression $\sqrt{5} - 1$. However, we can approximate the value of these expressions in decimal form using a calculator.

Some other radicals *can* be combined. In this section, we discuss such radicals.

Adding and Subtracting Like Radicals

When adding or subtracting *like* radicals, we can simplify the result.

DEFINITIONS

Like radicals are radical expressions that have the same radicand. **Unlike radicals** are radical expressions with different radicands.

For instance, $4\sqrt{2}$ and $3\sqrt{2}$ are like radicals. By contrast, the radicals $7\sqrt{2}$ and $\sqrt{3}$ are unlike.

We use the distributive property to add or subtract like radicals, just as we do for adding or subtracting like terms.

Adding Like Terms

$$4x + 3x = (4 + 3)x = 7x$$

Like terms

Adding Like Radicals

$$4\sqrt{2} + 3\sqrt{2} = (4 + 3)\sqrt{2} = 7\sqrt{2}$$

Like radicals

Subtracting Like Terms

$$4x - 3x = (4 - 3)x = x$$

Like terms

Subtracting Like Radicals

$$4\sqrt{2} - 3\sqrt{2} = (4 - 3)\sqrt{2} = \sqrt{2}$$

Like radicals

EXAMPLE 1

Add or subtract.

a. $5\sqrt{3} + 2\sqrt{3}$

b. $7\sqrt{x} - 2\sqrt{x} - 4\sqrt{x}$

c. $4\sqrt{2y + 1} - \sqrt{2y + 1}$

d. $5\sqrt{3} + 3\sqrt{5}$

Solution

a. $5\sqrt{3} + 2\sqrt{3} = (5 + 2)\sqrt{3}$ Combine the coefficients using the distributive property.

$= 7\sqrt{3}$ Simplify.

b. $7\sqrt{x} - 2\sqrt{x} - 4\sqrt{x} = (7 - 2 - 4)\sqrt{x}$ Use the distributive property.

$= 1\sqrt{x}$ Simplify.

$= \sqrt{x}$

c. $4\sqrt{2y + 1} - \sqrt{2y + 1} = (4 - 1)\sqrt{2y + 1}$

$= 3\sqrt{2y + 1}$

d. $5\sqrt{3} + 3\sqrt{5}$ cannot be simplified because the radicals are not like.

PRACTICE 1

Add or subtract.

a. $8\sqrt{5} - 2\sqrt{5}$

b. $3\sqrt{n} + \sqrt{n} + 7\sqrt{n}$

c. $10\sqrt{t^2 - 3} + \sqrt{t^2 - 3}$

d. $4\sqrt{6} - 2\sqrt{2}$

To Add or Subtract *Like* Radicals

- Use the distributive property.
- Then, simplify.

Adding and Subtracting Unlike Radicals

Some *unlike* radicals become *like* when they are simplified. When this happens, they can be combined.

EXAMPLE 2

Combine. Simplify, if possible.

a. $\sqrt{12} + \sqrt{27}$

b. $7\sqrt{50} - \sqrt{72} + \sqrt{32}$

c. $-\sqrt{4x} + 3\sqrt{x}$

d. $a\sqrt{a^2b} - 5b\sqrt{b}$

Solution

a.
$$\begin{aligned} \sqrt{12} + \sqrt{27} &= \sqrt{4\cdot 3} + \sqrt{9\cdot 3} && \text{Factor out perfect squares.}\\ &= \sqrt{4}\cdot\sqrt{3} + \sqrt{9}\cdot\sqrt{3} && \text{Use the product rule of radicals.}\\ &= 2\sqrt{3} + 3\sqrt{3} && \text{Take the square root of the perfect squares.}\\ &= (2+3)\sqrt{3} && \text{Use the distributive property.}\\ &= 5\sqrt{3} && \text{Simplify.} \end{aligned}$$

b. $7\sqrt{50} - \sqrt{72} + \sqrt{32}$
$$\begin{aligned} &= 7\sqrt{25\cdot 2} - \sqrt{36\cdot 2} + \sqrt{16\cdot 2} && \text{Factor out perfect squares.}\\ &= 7\cdot\sqrt{25}\cdot\sqrt{2} - \sqrt{36}\cdot\sqrt{2} + \sqrt{16}\cdot\sqrt{2} && \text{Use the product rule of radicals.}\\ &= 7\cdot 5\sqrt{2} - 6\sqrt{2} + 4\sqrt{2} && \text{Take the square roots of the perfect squares.}\\ &= 35\sqrt{2} - 6\sqrt{2} + 4\sqrt{2} && \text{Simplify.}\\ &= (35 - 6 + 4)\sqrt{2} && \text{Use the distributive property.}\\ &= 33\sqrt{2} && \text{Simplify.} \end{aligned}$$

c. $-\sqrt{4x} + 3\sqrt{x} = -2\sqrt{x} + 3\sqrt{x} = (-2+3)\sqrt{x} = \sqrt{x}$

d.
$$\begin{aligned} a\sqrt{a^2b} - 5b\sqrt{b} &= a\cdot a\sqrt{b} - 5b\sqrt{b}\\ &= a^2\sqrt{b} - 5b\sqrt{b}\\ &= (a^2 - 5b)\sqrt{b} \end{aligned}$$

PRACTICE 2

Combine. Simplify, if possible.

a. $\sqrt{50} + \sqrt{98}$

b. $\sqrt{12} + 2\sqrt{75} - 6\sqrt{27}$

c. $-3\sqrt{16t} + \sqrt{9t}$

d. $\sqrt{25ab^4} + 7b^2\sqrt{a}$

To Add or Subtract *Unlike* Radicals

- Simplify the unlike radicals, if possible.
- If the result contains any like radicals, add or subtract them.

EXAMPLE 3

It takes $\sqrt{20}$ sec for an object to fall from the top of a 320-foot building, and $\sqrt{5}$ sec for the object to fall from the top of an 80-foot building. How much longer will it take the object to fall from the taller building? Express this quantity both in radical form and as a decimal rounded to the nearest second.

Solution We need to find the difference between $\sqrt{20}$ sec and $\sqrt{5}$ sec.

$$\begin{aligned}\sqrt{20} - \sqrt{5} &= \sqrt{4 \cdot 5} - \sqrt{5}\\ &= \sqrt{4} \cdot \sqrt{5} - \sqrt{5}\\ &= 2\sqrt{5} - \sqrt{5}\\ &= \sqrt{5}\end{aligned}$$

So it takes the object $\sqrt{5}$ sec, or approximately 2 sec, longer to fall from the taller building.

PRACTICE 3

A young couple wants to build a square cabin with an area of 200 m^2 on the corner of a square plot of land with area 1800 m^2. Find the length of the front yard shown in the illustration. Express this length as a decimal rounded to the nearest meter.

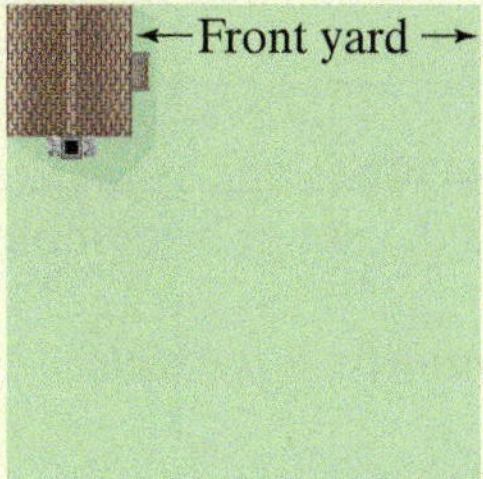

18.2 Exercises

FOR EXTRA HELP MyMathLab MathXL PRACTICE WATCH READ REVIEW

Mathematically Speaking

Fill in each blank with the most appropriate term or phrase from the given list.

distributive property	common	unlike
cannot	associative property of addition	can
like		

1. Radical expressions that have the same radicand are called ________________ radicals.
2. The ________________ is used to add or subtract like radicals.
3. Radicals that are ________________, even after being simplified, cannot be combined.
4. If we rewrite $2\sqrt{12} + 3\sqrt{75}$ as $4\sqrt{3} + 15\sqrt{3}$, we ________________ combine the terms.

A *Combine and simplify, if possible.*

5. $5\sqrt{7} + 3\sqrt{7}$
6. $7\sqrt{11} + \sqrt{11}$
7. $3\sqrt{2} - 8\sqrt{2}$
8. $\sqrt{3} - 10\sqrt{3}$
9. $6\sqrt{3} - 3\sqrt{6}$
10. $2\sqrt{7} + 7\sqrt{2}$
11. $-5\sqrt{11} - 10\sqrt{11} + 2\sqrt{11}$
12. $-4\sqrt{5} + 4\sqrt{5} + 2\sqrt{5}$
13. $7t\sqrt{3} + 2t\sqrt{3}$
14. $3y\sqrt{2} + 5y\sqrt{2}$
15. $13\sqrt{x} + 10\sqrt{x}$
16. $2\sqrt{3k} + 9\sqrt{3k}$
17. $2\sqrt{x} - 3\sqrt{y}$
18. $\sqrt{m} - 2\sqrt{n}$
19. $6\sqrt{x+1} - \sqrt{x+1}$
20. $\sqrt{2m-1} - 7\sqrt{2m-1}$
21. $\sqrt{8} - \sqrt{32}$
22. $\sqrt{12} - \sqrt{27}$
23. $\sqrt{50} + \sqrt{72}$
24. $\sqrt{48} + \sqrt{12}$
25. $-\sqrt{12} + 5\sqrt{3}$
26. $3\sqrt{2} - \sqrt{18}$
27. $6\sqrt{75} - 2\sqrt{12}$
28. $5\sqrt{72} - 4\sqrt{50}$
29. $5\sqrt{8} - 3\sqrt{12} + \sqrt{2}$
30. $2\sqrt{3} + \sqrt{20} + \sqrt{5}$
31. $2\sqrt{16y} + 3\sqrt{4y}$
32. $2\sqrt{9x} + 8\sqrt{4x}$
33. $\sqrt{9x} - \sqrt{16x^3}$
34. $2\sqrt{25a^3} - \sqrt{a}$
35. $\sqrt{25p} + \sqrt{64p} + \sqrt{p}$
36. $\sqrt{49a} - 2\sqrt{a} + 9\sqrt{a}$
37. $2\sqrt{16x} - \sqrt{x} - 3\sqrt{4x}$
38. $\sqrt{81y} + 6\sqrt{y} - 8\sqrt{25y}$
39. $-5x\sqrt{2x^3y^4} + x\sqrt{2x^5y^2}$
40. $-2\sqrt{9ab^5} - 3a\sqrt{4a^3b}$

Mixed Practice

Combine and simplify, if possible.

41. $\sqrt{27} - \sqrt{3} + \sqrt{6}$
42. $5\sqrt{7} + 7\sqrt{5}$
43. $6\sqrt{7p} - 2\sqrt{7p}$
44. $\sqrt{b} - 3\sqrt{36b^3}$
45. $\sqrt{75} + \sqrt{48}$
46. $8\sqrt{2} - \sqrt{72}$

Applications

B *Solve. Express the answer as a radical in simplified form.*

47. For the isosceles triangle shown, find:

a. the lengths of the missing sides.

b. the perimeter.

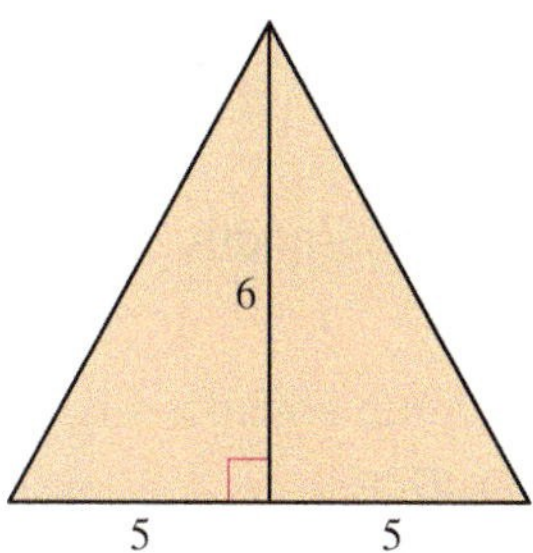

48. The size of a television set is commonly given by the length of the screen's diagonal. For the two sets pictured, find:

a. the lengths of their diagonals.

b. the difference of the diagonal lengths.

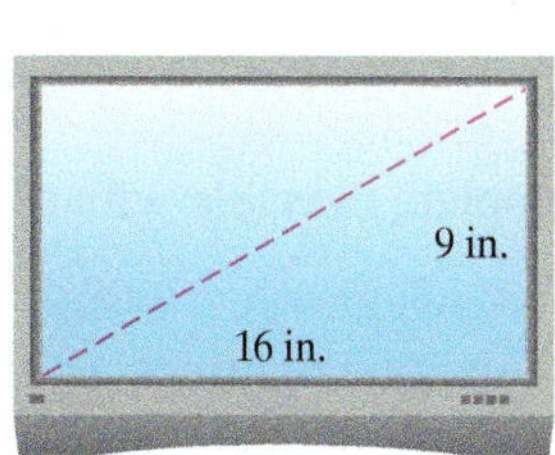

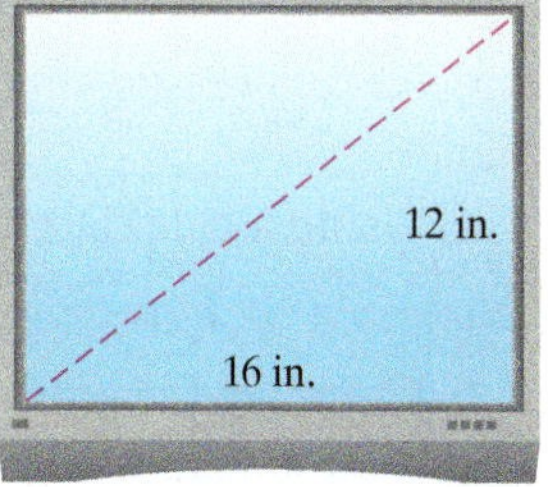

49. Two square tiles are pictured below:

Area = 90 in²

Area = 40 in²

a. Determine the side lengths of the two tiles.

b. How much longer is a side of the larger tile than the smaller tile?

50. The accompanying map shows the route that a college recruiter takes in calling on colleges each week. She starts at home (A), visits colleges in towns B, C, and D in that order, and then returns home.

a. How long is the road from A to B?

b. What is the length of the recruiter's weekly trip?

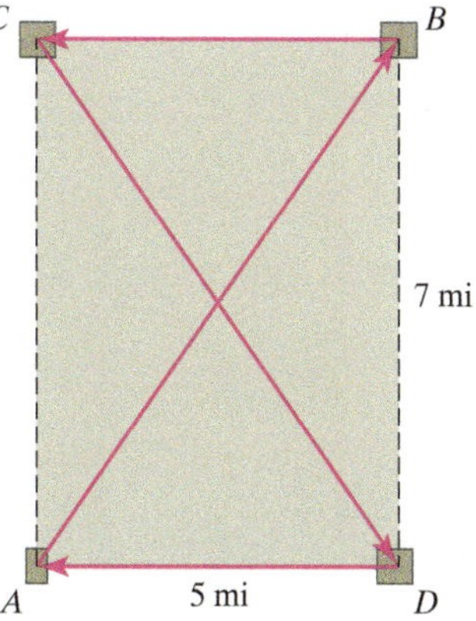

51. A manufacturer supplies machine parts to a retailer. The number n of machine parts and the total price P (in dollars) are related by the following equation:

$$P = 9\sqrt{n}$$

How much more will the manufacturer charge the retailer for 4000 machine parts than for 1000 machine parts?

52. It takes $\frac{\sqrt{50 - h}}{4}$ sec for an object to fall from a height of 50 ft to a height of h ft above the ground. From a height of 50 ft, how much longer does it take to drop to a height of 10 ft above the ground than to drop to a height of 40 ft above the ground?

• Check your answers on page A-38.

MINDStretchers

Groupwork

1. Working with a partner, choose several arbitrary nonnegative values for the variables a and b in the two left columns of the following table:

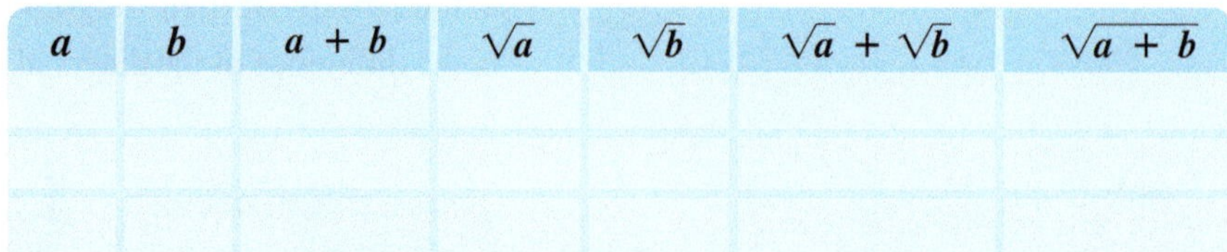

a	b	$a + b$	$\sqrt{a}$	$\sqrt{b}$	$\sqrt{a} + \sqrt{b}$	$\sqrt{a + b}$

a. Using a calculator, complete the table with each entry rounded to the nearest thousandth.

b. Compare the entries in the two rightmost columns. Are the entries in one column consistently larger than the entries in the other column? State a conjecture based on this observation.

Patterns

2. Consider the isosceles right triangle ABC shown below:

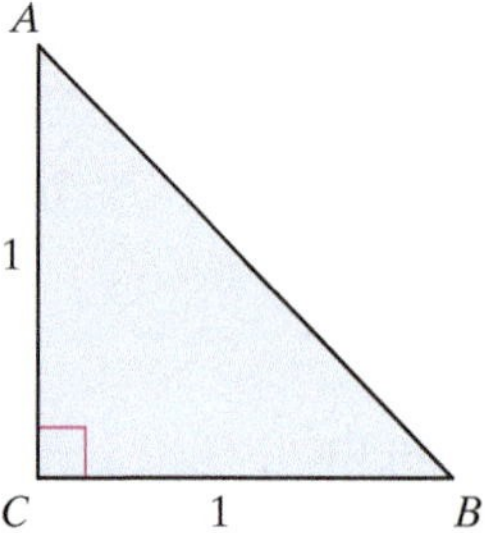

On side $\overline{AB}$, build right triangle ABD with $AD = 1$.

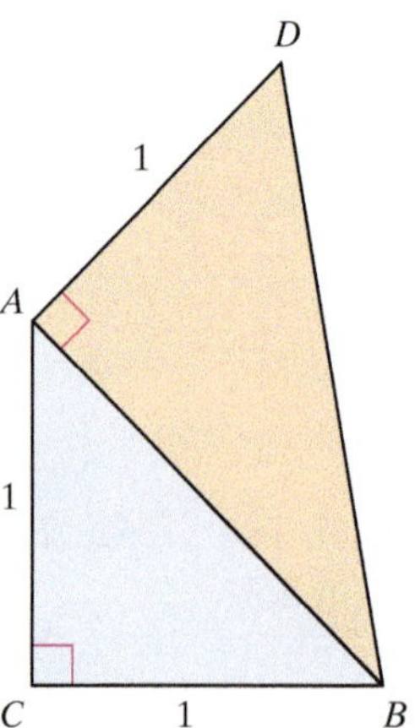

a. Find the lengths of $\overline{AB}$ and $\overline{DB}$.

b. How much longer is $\overline{DB}$ than $\overline{AB}$?

c. Now build another right triangle on $\overline{DB}$ with a leg of length 1. What is the length of $\overline{EB}$?

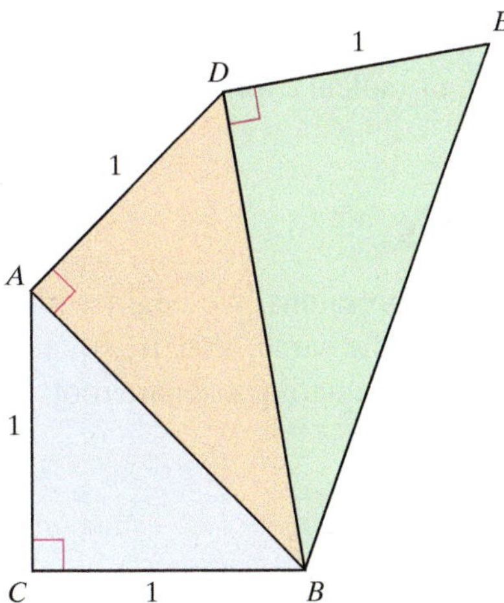

d. Describe the pattern that you observe in this *spiral* of triangles.

Writing

3. Recall that the radicals $\sqrt{2}$ and $5\sqrt{2}$ are called *like*, whereas the radicals $7\sqrt{2}$ and $\sqrt{3}$ are called *unlike*. List two other kinds of numbers where a distinction is made between like and unlike. Why are these distinctions made?

18.3 Multiplying and Dividing Radical Expressions

OBJECTIVES

- **A** To multiply radical expressions
- **B** To divide radical expressions
- **C** To rationalize the denominator in a radical expression
- **D** To solve applied problems involving the multiplication or division of radical expressions

Section 18.2 covered the addition and subtraction of radical expressions. In this section, we discuss how to find their products and quotients.

Multiplying Radical Expressions

In considering how to find the product of radical expressions, we begin with the simplest case—multiplying two radical expressions that are the same, that is, squaring a radical expression. Recall the property used in Section 18.1 for squaring a square root: $(\sqrt{a})^2 = a$.

EXAMPLE 1

Simplify. **a.** $\sqrt{3} \cdot \sqrt{3}$ **b.** $(\sqrt{5x})^2$ **c.** $(2\sqrt{n+1})^2$

Solution Here, we use the property of squaring a square root.

a. $\sqrt{3} \cdot \sqrt{3} = (\sqrt{3})^2 = 3$ **b.** $(\sqrt{5x})^2 = 5x$

c. $(2\sqrt{n+1})^2 = 2^2 \cdot (\sqrt{n+1})^2$ Use the rule of raising a product to a power.

$= 4(n+1)$

$= 4n + 4$

PRACTICE 1

Simplify:

a. $\sqrt{5} \cdot \sqrt{5}$

b. $(\sqrt{2y^3})^2$

c. $(3\sqrt{t+1})^2$

Now, let's look at multiplying two radical expressions that are different from one another. Here we apply the product rule of radicals, discussed in Section 18.1. Note that we can write this rule as

$$\sqrt{a} \cdot \sqrt{b} = \sqrt{ab},$$

where a and b are nonnegative numbers. In words, this rule states that the product of square roots is the square root of the product.

EXAMPLE 2

Multiply. Simplify, if possible.

a. $\sqrt{3} \cdot \sqrt{5}$ **b.** $(-5\sqrt{10})(6\sqrt{2})$ **c.** $\sqrt{12n^3} \cdot \sqrt{3n}$

Solution Here, we use the product rule of radicals before simplifying.

a. $\sqrt{3} \cdot \sqrt{5} = \sqrt{3 \cdot 5} = \sqrt{15}$

b. $(-5\sqrt{10})(6\sqrt{2}) = -5 \cdot 6 \cdot \sqrt{10 \cdot 2}$

$= -30\sqrt{20}$

$= -30\sqrt{4 \cdot 5}$

$= -30 \cdot 2\sqrt{5}$

$= -60\sqrt{5}$

c. $\sqrt{12n^3} \cdot \sqrt{3n} = \sqrt{12n^3 \cdot 3n}$

$= \sqrt{36n^4}$

$= 6n^2$

PRACTICE 2

Find the product. Simplify, if possible.

a. $\sqrt{7} \cdot \sqrt{10}$

b. $(9\sqrt{6})(-4\sqrt{3})$

c. $\sqrt{8y} \cdot \sqrt{2y^5}$

Next, we consider the multiplication of radical expressions that may contain more than one term. Just as with the multiplication of polynomials, the key here is to use the distributive property.

EXAMPLE 3

Find the product. Simplify, if possible.

a. $\sqrt{8}(2\sqrt{3} + \sqrt{2})$ **b.** $\sqrt{x}(5\sqrt{y} - 2)$

Solution

a. $\sqrt{8}(2\sqrt{3} + \sqrt{2})$

$= \sqrt{8}(2\sqrt{3}) + \sqrt{8}(\sqrt{2})$ Use the distributive property.

$= 2\sqrt{24} + \sqrt{16}$ Use the product rule of radicals.

$= 2\sqrt{4 \cdot 6} + 4$ Factor out a perfect square from the radicand 24.

$= 2 \cdot 2\sqrt{6} + 4$ Use the property of taking the square root of a square.

$= 4\sqrt{6} + 4$ Simplify.

b. $\sqrt{x}(5\sqrt{y} - 2)$

$= \sqrt{x}(5\sqrt{y}) - \sqrt{x}(2)$ Use the distributive property.

$= 5\sqrt{xy} - 2\sqrt{x}$ Use the product rule of radicals.

PRACTICE 3

Multiply. Simplify, if possible.

a. $\sqrt{6}(3\sqrt{3} - \sqrt{8})$

b. $\sqrt{a}(\sqrt{b} + 3)$

EXAMPLE 4

Find the product and simplify.

a. $(4\sqrt{2} - 1)(7\sqrt{2} + 3)$ **b.** $(2\sqrt{a} + 4)(\sqrt{a} - 1)$

Solution

a. $(4\sqrt{2} - 1)(7\sqrt{2} + 3)$

F L

$= (4\sqrt{2} - 1)(7\sqrt{2} + 3)$

I O

$= (4\sqrt{2})(7\sqrt{2}) + (4\sqrt{2})3 + (-1)(7\sqrt{2}) + (-1)(3)$ Use the FOIL method.

$= 4 \cdot 7 \cdot 2 + 4 \cdot 3\sqrt{2} - 7\sqrt{2} - 3$ Use the property of squaring a square root.

$= 56 + 12\sqrt{2} - 7\sqrt{2} - 3$ Simplify.

$= 53 + 5\sqrt{2}$ Combine like terms.

b. $(2\sqrt{a} + 4)(\sqrt{a} - 1)$

$= (2\sqrt{a})(\sqrt{a}) + (2\sqrt{a})(-1) + 4\sqrt{a} + 4(-1)$ Use the FOIL method.

$= 2a - 2\sqrt{a} + 4\sqrt{a} - 4$

$= 2a + 2\sqrt{a} - 4$

PRACTICE 4

Find the product and simplify.

a. $(2\sqrt{3} + 4)(\sqrt{3} - 1)$

b. $(\sqrt{x} + 2)(3\sqrt{x} - 2)$

Recall from Section 15.6 the formula for multiplying the sum and difference of the same two terms.

$$(a + b)(a - b) = a^2 - b^2$$

In the following example, the binomial factors are radical expressions.

EXAMPLE 5

Find the product and simplify.

a. $(\sqrt{10} + 5)(\sqrt{10} - 5)$ **b.** $(\sqrt{x} - \sqrt{y})(\sqrt{x} + \sqrt{y})$

Solution We apply the formula for the product of the sum and difference of two terms.

a. $(\sqrt{10} + 5)(\sqrt{10} - 5) = (\sqrt{10})^2 - (5)^2$
$= 10 - 25$
$= -15$

b. $(\sqrt{x} - \sqrt{y})(\sqrt{x} + \sqrt{y}) = (\sqrt{x})^2 - (\sqrt{y})^2 = x - y$

PRACTICE 5

Multiply and simplify:

a. $(\sqrt{7} - 3)(\sqrt{7} + 3)$

b. $(\sqrt{p} + \sqrt{q})(\sqrt{p} - \sqrt{q})$

Note that in Example 5, the products contain no radical sign. How would you explain why the radical signs drop out?

Recall from Section 15.6 that when squaring a binomial, the following formulas apply:

$(a + b)^2 = a^2 + 2ab + b^2$ The square of a sum

$(a - b)^2 = a^2 - 2ab + b^2$ The square of a difference

In the next example, the binomials being squared contain radicals.

EXAMPLE 6

Simplify.

a. $(\sqrt{3} + 5x)^2$ **b.** $(\sqrt{p} - 2)^2$

Solution

a. $(\sqrt{3} + 5x)^2 = (\sqrt{3})^2 + 2(\sqrt{3})(5x) + (5x)^2$
$= (\sqrt{3})^2 + 2(5)(\sqrt{3})(x) + (5)^2x^2$
$= 3 + 10x\sqrt{3} + 25x^2$
$= 25x^2 + 10x\sqrt{3} + 3$

Use the formula for squaring a binomial sum.

b. $(\sqrt{p} - 2)^2 = (\sqrt{p})^2 - 2(\sqrt{p})(2) + 2^2$
$= p - 2(\sqrt{p})(2) + 4$
$= p - 4\sqrt{p} + 4$

Use the formula for squaring a binomial difference.

PRACTICE 6

Simplify.

a. $(\sqrt{2} + b)^2$

b. $(\sqrt{x} - 6)^2$

EXAMPLE 7

In an *equilateral* triangle, all three sides are equal in length. The height h of this kind of triangle is given by the expression $\frac{s\sqrt{3}}{2}$, where s is the length of each side of the triangle. What is the height of the triangle shown?

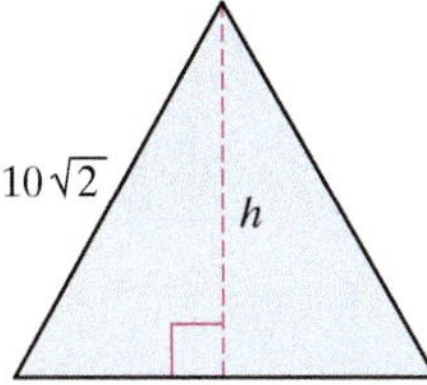

Solution In the triangle shown, $s = 10\sqrt{2}$. So we get:

$$h = \frac{s\sqrt{3}}{2}$$
$$= \frac{10\sqrt{2} \cdot \sqrt{3}}{2}$$
$$= \frac{10\sqrt{6}}{2} = 5\sqrt{6}$$

PRACTICE 7

The number of 24-hour days that it takes a planet in our solar system to revolve once around the Sun is approximated by the expression $0.2(\sqrt{R})^3$, where R is the average distance of the planet from the Sun (in millions of kilometers). For the planet Mercury, the average distance is about 60 million km. How many 24-hour days does it take Mercury to revolve around the Sun, to the nearest whole day?

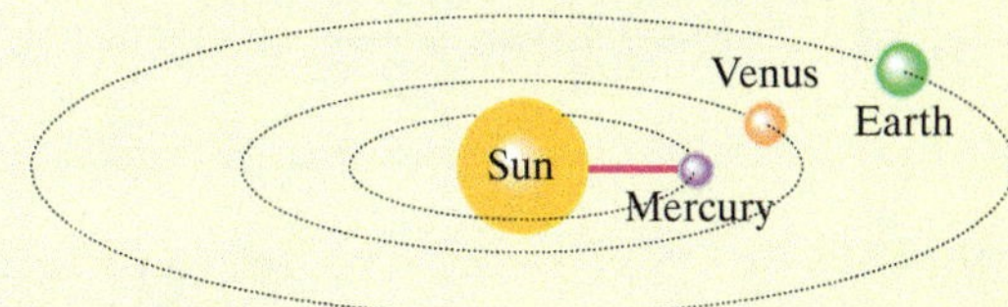

Dividing Radical Expressions

Now, let's consider dividing radical expressions. Here, we use the quotient rule of radicals discussed in Section 18.1. Note that we can write this rule as

$$\frac{\sqrt{a}}{\sqrt{b}} = \sqrt{\frac{a}{b}},$$

where a is nonnegative and b is positive. This rule allows us, in dividing two square roots, to bring the radicands under a single radical sign. We can then divide the radicands. This is particularly useful if their quotient happens to be a perfect square.

EXAMPLE 8

Find the quotient and simplify.

a. $\frac{\sqrt{26}}{\sqrt{2}}$ **b.** $\frac{\sqrt{16y^7}}{\sqrt{y}}$ **c.** $\frac{5\sqrt{10t}}{\sqrt{90t^3}}$

Solution Here, we use the quotient rule of radicals before simplifying.

a. $\frac{\sqrt{26}}{\sqrt{2}} = \sqrt{\frac{26}{2}} = \sqrt{13}$ **b.** $\frac{\sqrt{16y^7}}{\sqrt{y}} = \sqrt{\frac{16y^7}{y}} = \sqrt{16y^6} = 4y^3$

c. $\frac{5\sqrt{10t}}{\sqrt{90t^3}} = 5\sqrt{\frac{10t}{90t^3}}$ Use the quotient rule of radicals.

$= 5\sqrt{\frac{1}{9t^2}}$ Simplify the radicand.

$= 5 \cdot \frac{1}{3t}$ Use the property of taking the square root of a square.

$= \frac{5}{3t}$

PRACTICE 8

Divide and simplify.

a. $\frac{\sqrt{21}}{\sqrt{3}}$

b. $\frac{\sqrt{4x^5}}{\sqrt{x}}$

c. $\frac{\sqrt{2y}}{10\sqrt{8y^5}}$

When the radicand of a radical expression is a fraction, the expression is not considered to be simplified. To simplify, we can apply the quotient rule of radicals.

Radicands with perfect squares in the denominator lend themselves to this approach.

EXAMPLE 9

Simplify. **a.** $\sqrt{\dfrac{x^3}{y^8}}$ **b.** $\sqrt{\dfrac{9p^4}{q^{10}}}$ **c.** $\sqrt{\dfrac{7x^2}{4}}$

Solution Using the quotient rule of radicals, we get:

a. $\sqrt{\dfrac{x^3}{y^8}} = \dfrac{\sqrt{x^2 \cdot x}}{\sqrt{y^8}} = \dfrac{x\sqrt{x}}{y^4}$ **b.** $\sqrt{\dfrac{9p^4}{q^{10}}} = \dfrac{\sqrt{9p^4}}{\sqrt{q^{10}}} = \dfrac{3p^2}{q^5}$

c. $\sqrt{\dfrac{7x^2}{4}} = \dfrac{\sqrt{7x^2}}{\sqrt{4}} = \dfrac{x\sqrt{7}}{2}$

PRACTICE 9

Simplify.

a. $\sqrt{\dfrac{m^5}{n^4}}$ **b.** $\sqrt{\dfrac{y^6}{25x^2}}$

c. $\sqrt{\dfrac{5a^2}{9}}$

Rationalizing the Denominator

Some radical expressions are written as fractions with radicals in their denominators:

$$\frac{5}{\sqrt{2}} \qquad \frac{3x}{\sqrt{x-1}} \qquad \frac{v}{\sqrt{1-m^2}}$$

This type of expression, which can be difficult to evaluate without a calculator, is not considered to be simplified. However, we can always *rationalize the denominator* of such an expression, that is, rewrite the expression in an equivalent form that contains no radical in its denominator. To do this, we multiply the numerator and denominator by a square root that will make the radicand in the denominator a perfect square.

EXAMPLE 10

Rationalize the denominator.

a. $\dfrac{3}{\sqrt{5}}$ **b.** $\dfrac{\sqrt{6}}{\sqrt{x}}$ **c.** $\dfrac{\sqrt{25y^6}}{\sqrt{8}}$

Solution To rationalize the denominator, we multiply the numerator and denominator by a square root that makes the radicand in the denominator a perfect square.

a. $\dfrac{3}{\sqrt{5}} = \dfrac{3}{\sqrt{5}} \cdot \dfrac{\sqrt{5}}{\sqrt{5}}$ Multiply the numerator and denominator by $\sqrt{5}$.

$= \dfrac{3\sqrt{5}}{\sqrt{5^2}}$

$= \dfrac{3\sqrt{5}}{5}$

b. $\dfrac{\sqrt{6}}{\sqrt{x}} = \dfrac{\sqrt{6}}{\sqrt{x}} \cdot \dfrac{\sqrt{x}}{\sqrt{x}} = \dfrac{\sqrt{6x}}{x}$

PRACTICE 10

Rationalize the denominator.

a. $\dfrac{1}{\sqrt{2}}$

b. $\dfrac{\sqrt{5}}{\sqrt{s}}$

c. $\dfrac{\sqrt{49r^4}}{\sqrt{12}}$

c. $\dfrac{\sqrt{25y^6}}{\sqrt{8}} = \dfrac{5y^3}{\sqrt{8}}$ Take the square root in the numerator.

$$= \frac{5y^3}{\sqrt{8}} \cdot \frac{\sqrt{8}}{\sqrt{8}}$$

$$= \frac{5y^3\sqrt{8}}{8}$$

$$= \frac{5y^3(2\sqrt{2})}{8}$$

$$= \frac{5y^3\sqrt{2}}{4}$$

When the radicand of a radical expression is a fraction, we can simplify the radical by first applying the quotient rule and then rationalizing the denominator.

EXAMPLE 11

Simplify. **a.** $\sqrt{\dfrac{3}{5}}$ **b.** $\sqrt{\dfrac{x}{12}}$

Solution

a. $\sqrt{\dfrac{3}{5}} = \dfrac{\sqrt{3}}{\sqrt{5}} = \dfrac{\sqrt{3}}{\sqrt{5}} \cdot \dfrac{\sqrt{5}}{\sqrt{5}} = \dfrac{\sqrt{15}}{5}$

b. $\sqrt{\dfrac{x}{12}} = \dfrac{\sqrt{x}}{\sqrt{12}}$

$$= \frac{\sqrt{x}}{\sqrt{12}} \cdot \frac{\sqrt{3}}{\sqrt{3}}$$

$$= \frac{\sqrt{3x}}{\sqrt{36}}$$

$$= \frac{\sqrt{3x}}{6}$$

PRACTICE 11

Simplify.

a. $\sqrt{\dfrac{1}{6}}$

b. $\sqrt{\dfrac{n}{20}}$

To review, a radical expression is considered simplified if the following conditions are all met:

Condition	Simplified	Not Simplified
• The radicand has no factor that is a perfect square.	$5\sqrt{2}$	$\sqrt{50} = \sqrt{25 \cdot 2}$
• There are no fractions under the radical sign.	$\dfrac{\sqrt{3}}{2}$	$\sqrt{\dfrac{3}{4}}$
• There are no radicals in the denominator.	$\dfrac{\sqrt{6}}{3}$	$\dfrac{2}{\sqrt{6}}$

Some radical expressions are in the form of fractions, with more than one term in the numerator. We may need to rationalize the denominator of such an expression.

EXAMPLE 12

Rationalize the denominator.

a. $\dfrac{\sqrt{2} + 3}{\sqrt{5}}$ **b.** $\dfrac{\sqrt{x} - 1}{\sqrt{y}}$

Solution

a. $$\frac{\sqrt{2} + 3}{\sqrt{5}} = \frac{(\sqrt{2} + 3)}{\sqrt{5}} \cdot \frac{\sqrt{5}}{\sqrt{5}}$$
$$= \frac{(\sqrt{2} + 3)\sqrt{5}}{5}$$
$$= \frac{(\sqrt{2})(\sqrt{5}) + 3\sqrt{5}}{5}$$
$$= \frac{\sqrt{10} + 3\sqrt{5}}{5}$$

b. $$\frac{\sqrt{x} - 1}{\sqrt{y}} = \frac{\sqrt{x} - 1}{\sqrt{y}} \cdot \frac{\sqrt{y}}{\sqrt{y}}$$
$$= \frac{(\sqrt{x} - 1)\sqrt{y}}{y}$$
$$= \frac{(\sqrt{x})(\sqrt{y}) - \sqrt{y}}{y}$$
$$= \frac{\sqrt{xy} - \sqrt{y}}{y}$$

PRACTICE 12

Rationalize the denominator.

a. $\dfrac{\sqrt{5} - 1}{\sqrt{3}}$

b. $\dfrac{\sqrt{c} + 2}{\sqrt{b}}$

Thus far, we have rationalized denominators containing a single radical term. But suppose that there are two radical terms in a denominator. In this case, the key is to identify the *conjugate* of the denominator. The expressions $a + b$ and $a - b$ are called conjugates of one another. Note that conjugates come in pairs, since they are the sum and difference of the same two terms.

Recall that in Example 5(a) we multiplied two radical expressions, $\sqrt{10} + 5$ and $\sqrt{10} - 5$, which are conjugates of one another. We used the formula $(a + b)(a - b) = a^2 - b^2$, and then observed that the product -15 contains no radical sign. The elimination of radical signs suggests a procedure for rationalizing a denominator with two terms, namely, *multiplying both the numerator and the denominator by the conjugate of the denominator.*

EXAMPLE 13

Rationalize the denominator.

a. $\dfrac{4}{1 + \sqrt{3}}$

b. $\dfrac{x}{\sqrt{y} - \sqrt{2}}$

PRACTICE 13

Rationalize the denominator.

a. $\dfrac{8}{3 - \sqrt{2}}$

b. $\dfrac{a}{\sqrt{b} + \sqrt{5}}$

Solution

a. $\dfrac{4}{1+\sqrt{3}} = \dfrac{4}{1+\sqrt{3}} \cdot \dfrac{1-\sqrt{3}}{1-\sqrt{3}}$ Multiply the numerator and the denominator by the conjugate of the denominator.

$= \dfrac{4(1-\sqrt{3})}{(1+\sqrt{3})(1-\sqrt{3})}$

$= \dfrac{4-4\sqrt{3}}{1^2-(\sqrt{3})^2}$ Use the formula for findng the product of the sum and difference of two terms.

$= \dfrac{4-4\sqrt{3}}{1-3}$

$= \dfrac{2(2-2\sqrt{3})}{-2}$

$= \dfrac{\overset{1}{\cancel{2}}(2-2\sqrt{3})}{-\underset{1}{\cancel{2}}}$

$= -2 + 2\sqrt{3}$

b. $\dfrac{x}{\sqrt{y}-\sqrt{2}} = \dfrac{x}{\sqrt{y}-\sqrt{2}} \cdot \dfrac{\sqrt{y}+\sqrt{2}}{\sqrt{y}+\sqrt{2}}$ Multiply the numerator and the denominator by the conjugate of the denominator.

$= \dfrac{x(\sqrt{y}+\sqrt{2})}{(\sqrt{y}-\sqrt{2})(\sqrt{y}+\sqrt{2})}$

$= \dfrac{x\sqrt{y}+x\sqrt{2}}{y-2}$ Use the formula for finding the product of the sum and difference of two terms.

In Examples 13(a) and (b), note that when we multiplied the numerator and denominator by the conjugate of the denominator, the radical sign was eliminated in the denominator, as expected.

EXAMPLE 14

The velocity v (in kilometers per second) of a meteor streaking toward the Earth can be modeled by the expression $\dfrac{450}{\sqrt{d}}$, where d is its distance from the center of the Earth (in kilometers). Find the velocity of a meteor when it is 20,000 km from the Earth's center, written as a radical in simplified form. Using a calculator, also express this velocity as a decimal rounded to the nearest kilometer per second.

Solution We substitute 20,000 for d in $v = \dfrac{450}{\sqrt{d}}$.

$$v = \frac{450}{\sqrt{20{,}000}} = \frac{450}{\sqrt{10{,}000 \cdot 2}} = \frac{450}{\sqrt{10{,}000} \cdot \sqrt{2}}$$

$$= \frac{\overset{9}{\cancel{450}}}{\underset{2}{\cancel{100}}\sqrt{2}} = \frac{9}{2\sqrt{2}} = \frac{9}{2\sqrt{2}} \cdot \frac{\sqrt{2}}{\sqrt{2}} = \frac{9\sqrt{2}}{2 \cdot 2} = \frac{9\sqrt{2}}{4}$$

So the velocity of the meteor is $\dfrac{9\sqrt{2}}{4}$ km/sec, or approximately 3 km/sec.

PRACTICE 14

The formula $P = \dfrac{590}{\sqrt{t}}$ approximates the pulse rate (in beats per minute) for an adult who is t in. tall. Find the pulse rate of an adult 72 in. tall, written as a radical expression in simplified form. Using a calculator, also express this pulse rate rounded to the nearest whole number of beats per minute.

18.3 Exercises

FOR EXTRA HELP MyMathLab MathXL PRACTICE WATCH READ REVIEW

Mathematically Speaking

Fill in each blank with the most appropriate term or phrase from the given list.

simplify	associative property of multiplication	bring the radicands under a single radical sign
take the square roots before dividing	are different from one another	rationalize
distributive property		
contain more than one term		

1. To multiply two radical expressions that ______________, we use the product rule: $\sqrt{a} \cdot \sqrt{b} = \sqrt{ab}$.

2. The key to multiplying radical expressions that contain more than one term is to use the ______________.

3. The quotient rule of radicals, $\frac{\sqrt{a}}{\sqrt{b}} = \sqrt{\frac{a}{b}}$, where $a \geq 0$ and $b > 0$, allows us to ______________.

4. To ______________ the denominator of an expression means to rewrite the expression in an equivalent form that contains no radical in its denominator.

A *Multiply. Simplify, if possible.*

5. $\sqrt{21} \cdot \sqrt{21}$ **6.** $\sqrt{17} \cdot \sqrt{17}$ **7.** $\sqrt{15} \cdot \sqrt{15}$ **8.** $\sqrt{13} \cdot \sqrt{13}$

9. $(\sqrt{3n})^2$ **10.** $(\sqrt{4x})^2$ **11.** $(\sqrt{5y})^2$ **12.** $(\sqrt{7a})^2$

13. $(4\sqrt{x} - 1)^2$ **14.** $(3\sqrt{y} + 1)^2$ **15.** $(2\sqrt{t} + 5)^2$ **16.** $(5\sqrt{n} - 3)^2$

17. $\sqrt{18} \cdot \sqrt{3}$ **18.** $\sqrt{6} \cdot \sqrt{15}$ **19.** $(-2\sqrt{5})(7\sqrt{10})$ **20.** $(3\sqrt{24})(-5\sqrt{2})$

21. $\sqrt{8x^3} \cdot \sqrt{2x}$ **22.** $\sqrt{3x} \cdot \sqrt{12x}$ **23.** $\sqrt{3r} \cdot \sqrt{5r}$ **24.** $\sqrt{2n} \cdot \sqrt{7n}$

25. $\sqrt{2x} \cdot \sqrt{5} \cdot \sqrt{10y}$ **26.** $\sqrt{6y} \cdot \sqrt{7xy} \cdot \sqrt{2x}$ **27.** $\sqrt{3}(\sqrt{3} - 1)$ **28.** $\sqrt{2}(\sqrt{2} + 10)$

29. $\sqrt{x}(\sqrt{x} - 7)$ **30.** $\sqrt{y}(8 + \sqrt{y})$ **31.** $\sqrt{a}(4\sqrt{b} + 1)$ **32.** $\sqrt{p}(3\sqrt{q} - 5)$

33. $(\sqrt{5} + 3)(\sqrt{5} + 2)$ **34.** $(4 - \sqrt{7})(3 + \sqrt{7})$

35. $(\sqrt{7} - \sqrt{9})(\sqrt{7} + \sqrt{9})$ **36.** $(\sqrt{6} + \sqrt{4})(\sqrt{6} - \sqrt{4})$

37. $(8\sqrt{3} + 1)(5\sqrt{3} - 2)$ **38.** $(4\sqrt{2} + 1)(5\sqrt{2} - 3)$

39. $(\sqrt{n} + 5)(3\sqrt{n} - 1)$ **40.** $(2 - 5\sqrt{t})(4 + \sqrt{t})$

41. $(6 - \sqrt{3})(6 + \sqrt{3})$ **42.** $(\sqrt{2} - 1)(\sqrt{2} + 1)$

43. $(5 + 2\sqrt{3})(5 - 2\sqrt{3})$ **44.** $(4\sqrt{10} - 3)(4\sqrt{10} + 3)$

45. $(\sqrt{x} + 2)(\sqrt{x} - 2)$ **46.** $(5 - \sqrt{y})(5 + \sqrt{y})$

47. $(\sqrt{a} + \sqrt{b})(\sqrt{a} - \sqrt{b})$ **48.** $(\sqrt{m} - \sqrt{n})(\sqrt{m} + \sqrt{n})$

49. $(\sqrt{3x} - \sqrt{y})(\sqrt{3x} + \sqrt{y})$ **50.** $(\sqrt{q} + \sqrt{5p})(\sqrt{q} - \sqrt{5p})$

51. $(\sqrt{2} - x)^2$

52. $(4y + \sqrt{3})^2$

53. $(\sqrt{x} - 1)^2$

54. $(2 + \sqrt{n})^2$

B *Find the quotient and simplify.*

55. $\frac{\sqrt{15}}{\sqrt{3}}$

56. $\frac{\sqrt{10}}{\sqrt{2}}$

57. $\frac{\sqrt{5}}{\sqrt{125}}$

58. $\frac{\sqrt{6}}{\sqrt{24}}$

59. $\frac{\sqrt{4a^3}}{\sqrt{a}}$

60. $\frac{\sqrt{9d^7}}{\sqrt{d^3}}$

61. $\frac{4\sqrt{5y}}{\sqrt{45y^5}}$

62. $\frac{\sqrt{3x^4}}{5\sqrt{48x^8}}$

63. $\sqrt{\frac{a^4}{b^6}}$

64. $\sqrt{\frac{m^8}{n^2}}$

65. $\sqrt{\frac{16x^{12}}{y^8}}$

66. $\sqrt{\frac{49p^6}{q^{10}}}$

67. $\sqrt{\frac{5x^{10}}{36}}$

68. $\sqrt{\frac{7y^2}{100}}$

Simplify.

69. $\frac{2}{\sqrt{3}}$

70. $\frac{1}{\sqrt{2}}$

71. $\frac{\sqrt{5}}{\sqrt{y}}$

72. $\frac{\sqrt{7}}{\sqrt{a}}$

73. $\sqrt{\frac{2}{11}}$

74. $\sqrt{\frac{3}{7}}$

75. $\sqrt{\frac{x^2}{5}}$

76. $\sqrt{\frac{n^4}{3}}$

77. $\sqrt{\frac{t}{50}}$

78. $\sqrt{\frac{y}{32}}$

79. $\sqrt{\frac{a}{2}}$

80. $\sqrt{\frac{6y}{5}}$

C *Rationalize the denominator.*

81. $\frac{\sqrt{5} + 2}{\sqrt{3}}$

82. $\frac{6 - \sqrt{2}}{\sqrt{10}}$

83. $\frac{\sqrt{n} - 1}{\sqrt{m}}$

84. $\frac{4 - \sqrt{a}}{\sqrt{b}}$

85. $\frac{15}{4 + \sqrt{6}}$

86. $\frac{8}{\sqrt{3} + 1}$

87. $\frac{11}{4 - \sqrt{5}}$

88. $\frac{10}{\sqrt{7} - 2}$

89. $\frac{4}{\sqrt{5} - \sqrt{3}}$

90. $\frac{5}{\sqrt{6} + \sqrt{2}}$

91. $\frac{a}{\sqrt{b} - \sqrt{3}}$

92. $\frac{x}{\sqrt{5} + \sqrt{y}}$

Mixed Practice

Rationalize the denominator.

93. $\frac{\sqrt{c} - 6}{\sqrt{d}}$

94. $\frac{32}{5 - \sqrt{17}}$

Find the quotient and simplify, if possible.

95. $\frac{\sqrt{81a^5}}{\sqrt{a}}$

96. $\frac{\sqrt{5b^3}}{4\sqrt{45b^7}}$

97. $\sqrt{\frac{64s^8}{t^{12}}}$

Simplify.

98. $\frac{\sqrt{15}}{\sqrt{y}}$

99. $\sqrt{\frac{c}{72}}$

Multiply, and then simplify.

100. $\sqrt{20k} \cdot \sqrt{5k^3}$

101. $\sqrt{r}(1 + 3\sqrt{r})$

102. $(3\sqrt{11} - 4)(3\sqrt{11} + 4)$

103. $(\sqrt{7m} - \sqrt{n})(\sqrt{7m} + \sqrt{n})$

104. $(8x + \sqrt{6})^2$

Applications

D *Solve.*

105. Variance (V) and standard deviation (s) are two measures that statisticians use for studying data sets. These two measures are related by the formula $s = \sqrt{V}$. If one data set has double the variance of another, how many times the standard deviation of the second data set is that of the first?

106. The expression $\sqrt{2gh}$ can be used to determine the velocity of a free-falling object (in feet per second), where $g = 32\ \text{ft/sec}^2$, and the object has fallen h ft. If a ball has fallen 60 ft, is the ball's velocity double that of a ball that has fallen 30 ft?

107. The distance d between two points (x_1, y_1) and (x_2, y_2) on the coordinate plane is given by the formula

$$d = \sqrt{(x_2 - x_1)^2 + (y_2 - y_1)^2}.$$

Check that point $(6, 10)$ is twice as far from the origin as the point $(3, 5)$.

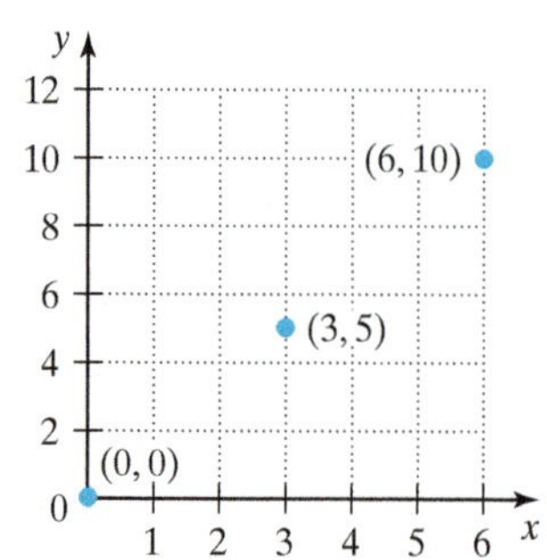

108. What is the area of the cross section of the pyramid pictured?

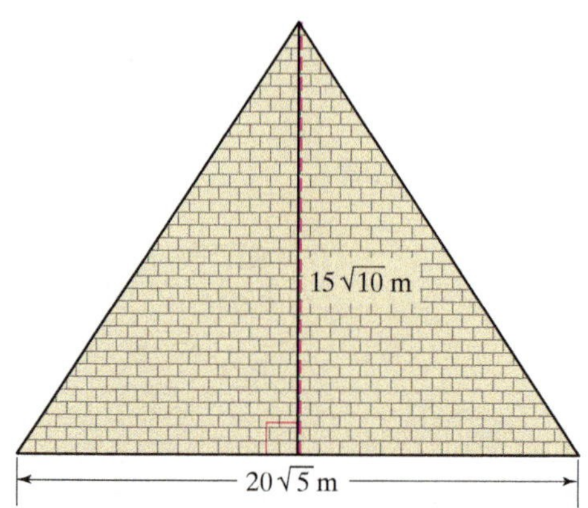

109. A hailstone will take t sec to drop d ft, where

$$t = \sqrt{\frac{d}{16}}.$$

How much time will it take a hailstone to drop 500 ft?

110. Chemists study the motion of gas molecules, called *diffusion*. The rates of diffusion (r_1 and r_2) of two gases are related by the formula

$$\frac{r_1}{r_2} = \frac{\sqrt{m_1}}{\sqrt{m_2}},$$

where m_1 and m_2 are the masses of the gas molecules. Find the ratio $\frac{r_1}{r_2}$, if $m_1 = 44$ units and $m_2 = 4$ units. Express the answer as a simplified radical.

111. An expression for finding the radius r of a cylindrical subwoofer is $\sqrt{\frac{V}{\pi h}}$, where V is the volume of the subwoofer, and h is its height. Simplify this expression, rationalizing the denominator.

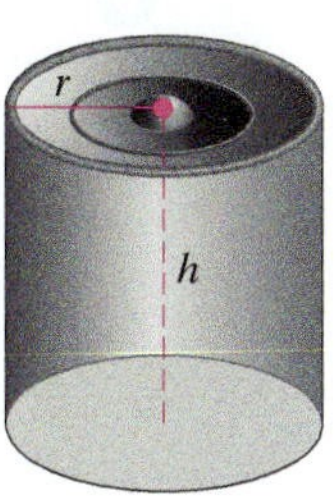

112. Suppose that an investment of P dollars grows at a fixed annual rate of return r. If at the end of two years the investment is worth A dollars, then the rate of return can be found from the formula

$$r = \frac{\sqrt{A} - \sqrt{P}}{\sqrt{P}}.$$

Rewrite this expression, rationalizing the denominator.

• Check your answers on page A-38.

MINDStretchers

Writing

1. Explain why each of the following radical expressions is *not* simplified.

a. $\sqrt{x^2 - 2ax + a^2}$ **b.** $\sqrt{\frac{x}{y}}$ **c.** $\frac{p}{\sqrt{q}}$

Mathematical Reasoning

2. Many quadratic equations have solutions that involve radical expressions. Determine whether $x = 1 - \sqrt{2}$ is a solution of the equation $x^2 - 2x = 1$.

Groupwork

3. Heron's formula states that the area of any triangle is equal to $\sqrt{s(s-a)(s-b)(s-c)}$, where the side lengths of the triangle are a, b, and c, and s is *half* the sum of the three side lengths, that is, $s = \frac{a+b+c}{2}$.

a. Working with a partner, apply Heron's formula to three triangles of your choice, filling in the four left columns in the following table.

Side Length a	Side Length b	Side Length c	Area of the Triangle	Area of the Triangle with Doubled Sides

b. For each of these triangles, form a new triangle whose sides are double the length of the sides of the original triangle. Use Heron's formula to compute the area of these new triangles and enter the results in the right column of the previous table.

c. In your three examples, does doubling all the side lengths of a triangle double its area?

18.4 Solving Radical Equations

OBJECTIVES

- **A** To solve radical equations
- **B** To solve applied problems involving radical equations

Finally, let's turn our attention to solving *radical equations*.

DEFINITION

A **radical equation** is an equation with a variable in one or more radicands.

Some examples of radical equations are:

$$\sqrt{3x} = 18 \qquad \sqrt{2n} = \sqrt{3n + 1} \qquad y - 2 = \sqrt{y} + 1$$

Can you explain why the equation $x + 3 = \sqrt{5}$ is not a radical equation?

The key to solving a radical equation is to find an equivalent equation with no square root. To do this, we can use the *squaring property of equality*.

The Squaring Property of Equality

For any real numbers a and b, if $a = b$, then $a^2 = b^2$.

In words, this property states that if two numbers are equal, then their squares are equal.

In solving a radical equation, we begin by *isolating the radical*. Applying the squaring property of equality, we then square each side of the equation. Squaring eliminates the radical and makes the equation easier to solve. However, the resulting equation may have solutions that are not solutions to the original equation. Recall that such "solutions" are called extraneous solutions. So when solving a radical equation, it is particularly important to check all possible solutions in the original equation.

EXAMPLE 1

Solve and check: $\sqrt{x} - 2 = 5$

Solution

$$\sqrt{x} - 2 = 5$$

$$\sqrt{x} - 2 + 2 = 5 + 2 \qquad \text{Add 2 to each side of the equation.}$$

$$\sqrt{x} = 7 \qquad \text{Simplify.}$$

$$(\sqrt{x})^2 = (7)^2 \qquad \text{Use the squaring property of equality.}$$

$$x = 49$$

Check

$$\sqrt{x} - 2 = 5$$

$$\sqrt{49} - 2 \stackrel{?}{=} 5 \qquad \text{Substitute 49 for } x \text{ in the original equation.}$$

$$7 - 2 \stackrel{?}{=} 5$$

$$5 = 5 \qquad \text{True}$$

So 49 is the solution.

PRACTICE 1

Solve and check: $\sqrt{y} + 3 = 7$

EXAMPLE 2

Solve and check: $\sqrt{x + 1} + 3 = 0$

Solution

$$\begin{aligned} \sqrt{x + 1} + 3 &= 0 \\ \sqrt{x + 1} &= -3 \\ (\sqrt{x + 1})^2 &= (-3)^2 \\ x + 1 &= 9 \\ x &= 8 \end{aligned}$$

Check

$$\begin{aligned} \sqrt{x + 1} + 3 &= 0 \\ \sqrt{8 + 1} + 3 &\stackrel{?}{=} 0 && \text{Substitute 8 for } x \text{ in the original equation.} \\ \sqrt{9} + 3 &\stackrel{?}{=} 0 \\ 3 + 3 &\stackrel{?}{=} 0 \\ 6 &\neq 0 && \text{False} \end{aligned}$$

Our check fails, so 8 is an extraneous solution, that is, it is *not* a solution to the original equation. Since 8 is the only possible solution, the equation has no solution.

PRACTICE 2

Solve and check: $\sqrt{2t - 5} + 7 = 0$

Some radical equations involve more than one square root, as in the following example:

EXAMPLE 3

Solve and check: $\sqrt{2n + 1} = \sqrt{5n - 2}$

Solution In this equation, the radicals are already isolated, so we begin by using the squaring property of equality to square each side of the equation.

$$\begin{aligned} \sqrt{2n + 1} &= \sqrt{5n - 2} \\ (\sqrt{2n + 1})^2 &= (\sqrt{5n - 2})^2 \\ 2n + 1 &= 5n - 2 \\ 1 &= 3n - 2 \\ 3 &= 3n \\ n &= 1 \end{aligned}$$

Check

$$\begin{aligned} \sqrt{2n + 1} &= \sqrt{5n - 2} \\ \sqrt{2 \cdot 1 + 1} &\stackrel{?}{=} \sqrt{5 \cdot 1 - 2} && \text{Substitute 1 for } n \text{ in the original equation.} \\ \sqrt{3} &= \sqrt{3} && \text{True} \end{aligned}$$

So 1 is the solution.

PRACTICE 3

Solve and check:
$\sqrt{4x + 7} = \sqrt{6x - 11}$

These examples suggest the following rule:

To Solve a Radical Equation

- Isolate a term with a radical.
- Square each side of the equation.
- Where possible, combine like terms.
- Solve the resulting equation.
- Check the possible solution(s) in the original equation.

Some radical equations are equivalent to quadratic equations with more than one solution.

EXAMPLE 4

Solve and check: $1 + \sqrt{1 - x} = x$

Solution

$$1 + \sqrt{1 - x} = x$$

$$\sqrt{1 - x} = x - 1 \qquad \text{Isolate the radical.}$$

$$(\sqrt{1 - x})^2 = (x - 1)^2 \qquad \text{Square each side of the equation.}$$

$$1 - x = x^2 - 2x + 1$$

$$0 = x^2 - x$$

$$x(x - 1) = 0 \qquad \text{Factor.}$$

$$x = 0 \quad \text{or} \quad x - 1 = 0 \qquad \text{Set each factor equal to 0.}$$

$$x = 1$$

Check

Substitute 0 for x.

$$1 + \sqrt{1 - x} = x$$

$$1 + \sqrt{1 - 0} \stackrel{?}{=} 0$$

$$1 + 1 \stackrel{?}{=} 0$$

$$2 \neq 0 \qquad \text{False}$$

Substitute 1 for x.

$$1 + \sqrt{1 - x} = x$$

$$1 + \sqrt{1 - 1} \stackrel{?}{=} 1$$

$$1 + 0 \stackrel{?}{=} 1$$

$$1 = 1 \qquad \text{True}$$

We see that $x = 1$ is a solution, whereas $x = 0$ is not. So the only solution to the original equation is 1.

PRACTICE 4

Solve and check: $y - \sqrt{y - 2} = 2$

EXAMPLE 5

Solve and check: $2\sqrt{x + 6} = \sqrt{x^2 + 19}$

Solution

$$2\sqrt{x + 6} = \sqrt{x^2 + 19}$$

$$(2\sqrt{x + 6})^2 = (\sqrt{x^2 + 19})^2 \qquad \text{Square each side of the equation.}$$

$$4(x + 6) = x^2 + 19$$

$$4x + 24 = x^2 + 19$$

$$0 = x^2 - 4x - 5$$

$$(x + 1)(x - 5) = 0$$

$$x + 1 = 0 \quad \text{or} \quad x - 5 = 0$$

$$x = -1 \qquad\qquad x = 5$$

Check

Substitute -1 for x.

$$2\sqrt{x + 6} = \sqrt{x^2 + 19}$$

$$2\sqrt{(-1) + 6} \stackrel{?}{=} \sqrt{(-1)^2 + 19}$$

$$2\sqrt{5} \stackrel{?}{=} \sqrt{20}$$

$$2\sqrt{5} = 2\sqrt{5} \qquad \text{True}$$

Substitute 5 for x.

$$2\sqrt{x + 6} = \sqrt{x^2 + 19}$$

$$2\sqrt{5 + 6} \stackrel{?}{=} \sqrt{(5^2) + 19}$$

$$2\sqrt{11} \stackrel{?}{=} \sqrt{44}$$

$$2\sqrt{11} = 2\sqrt{11} \qquad \text{True}$$

So both -1 and 5 are solutions.

PRACTICE 5

Solve and check:
$\sqrt{n^2 + 11} = 3\sqrt{n - 1}$

EXAMPLE 6

The approximate time t (in seconds) that it takes an object to fall freely a distance d (in feet) is given by the formula

$$t = \sqrt{\frac{d}{16}}.$$

If a skydiver had 3 sec of free fall, from what height did he leap?

Solution

$$t = \sqrt{\frac{d}{16}}$$

$$3 = \sqrt{\frac{d}{16}}$$

$$(3)^2 = \left(\sqrt{\frac{d}{16}}\right)^2$$

$$9 = \frac{d}{16}$$

$$144 = d, \quad \text{or} \quad d = 144$$

Check

$$t = \sqrt{\frac{d}{16}}$$

$$3 \stackrel{?}{=} \sqrt{\frac{144}{16}}$$

$$3 \stackrel{?}{=} \sqrt{9}$$

$$3 = 3 \quad \text{True}$$

So the skydiver must have leaped from a height of 144 ft.

PRACTICE 6

The sharper a road turns, the lower the speed a car can safely travel on the road without skidding. A formula used to find the maximum safe speed on a road is

$$s = \sqrt{2.5r},$$

where r is the radius of the road's curve (in feet) and s is the maximum safe speed of the car (in miles per hour). Find the radius r that will permit a maximum safe speed of 50 mph.

Recall from Section 12.4 that formulas state a relationship between two or more variables in a real-world context. In some cases, these formulas contain radicals.

EXAMPLE 7

To escape a planet's gravity (in meters per second^2), a spacecraft must achieve an initial velocity v (in meters per second) given by the formula

$$v = \sqrt{2gR},$$

where g is the planet's force of gravity and R is the planet's radius (in meters). Solve this formula for R.

Solution

$$v = \sqrt{2gR}$$

$$v^2 = 2gR$$

$$\frac{v^2}{2g} = R$$

$$R = \frac{v^2}{2g}$$

PRACTICE 7

The distance d (in miles) that a passenger can see from an airplane at altitude h (in feet) on a clear day can be approximated using the formula

$$d = \sqrt{\frac{3h}{2}}.$$

Solve this formula for h.

18.4 Exercises

FOR EXTRA HELP MyMathLab      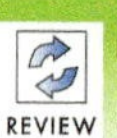

PRACTICE WATCH READ REVIEW

Mathematically Speaking

Fill in each blank with the most appropriate term or phrase from the given list.

rational equation	extraneous solution	their squares
radical solution	their square roots	square
isolate	radical equation	

1. A(n) ______________ is an equation in which a variable appears in one or more radicands.

2. The squaring property of equality states that if two numbers are equal, then ______________ are equal.

3. In solving a radical equation, the first step is to ______________ a term with a radical.

4. A possible solution of a radical equation that does not make the equation true is called a(n) ______________.

A *Solve and check.*

5. $\sqrt{x} = 3$

6. $\sqrt{n} = 7$

7. $\sqrt{2x} = 8$

8. $\sqrt{3x} = 9$

9. $\sqrt{x} + 6 = 0$

10. $\sqrt{4s} + 1 = 0$

11. $\sqrt{a} - 4 = 4$

12. $\sqrt{y} - 10 = -1$

13. $\sqrt{x + 3} = 3$

14. $\sqrt{n - 2} = 7$

15. $\sqrt{3t + 7} = 4$

16. $\sqrt{2t + 1} = 1$

17. $\sqrt{9t - 14} = \sqrt{2t}$

18. $\sqrt{3y + 2} = \sqrt{5y}$

19. $\sqrt{4x + 1} = \sqrt{2x + 7}$

20. $\sqrt{7x - 5} = \sqrt{4x + 13}$

21. $-2\sqrt{n - 2} = \sqrt{3n + 4}$

22. $\sqrt{5y + 2} = -3\sqrt{y - 2}$

23. $\sqrt{y - 1} + 4 = 6$

24. $1 + \sqrt{n + 4} = 11$

25. $3 - \sqrt{3x + 1} = 2$

26. $12 - \sqrt{2y - 1} = 7$

27. $\sqrt{x + 6} - x = 4$

28. $\sqrt{n + 8} - n = 2$

29. $7 + \sqrt{2x + 9} = x + 4$

30. $\sqrt{4y^2 + 5} = y + 4$

31. $7\sqrt{v} + v = -10$

32. $x + 6\sqrt{x} = -8$

33. $n - 3\sqrt{n + 2} = -4$

34. $y - 2\sqrt{y + 5} = -2$

35. $5\sqrt{y - 6} = \sqrt{y^2 - 14}$

36. $\sqrt{x^2 - 9} = 4\sqrt{x - 3}$

37. $\sqrt{2n^2 - 7} + 3 = n$

38. $\sqrt{4 - 3x} + 10 = x + 8$

39. $\sqrt{4x + 13} - 2x = -1$

40. $\sqrt{3x + 7} + 5 = 3x$

Mixed Practice

Solve and check.

41. $a - 4\sqrt{a + 5} = -8$

42. $\sqrt{2x^2 - 2x + 4} = x + 1$

43. $\sqrt{t} - 11 = -3$

44. $13 - \sqrt{3m - 5} = 9$

45. $\sqrt{7y + 8} = 6$

46. $\sqrt{9a - 13} = \sqrt{5a + 7}$

47. $\sqrt{9 - 4x} = x - 3$

48. $\sqrt{x^2 + 2} = 3\sqrt{x - 2}$

Applications

B *Solve.*

49. An electrical appliance with resistance R (in ohms) draws current I (in amps) and has power P (in watts). These quantities are related by the following formula:

$$I = \sqrt{\frac{P}{R}}$$

If an appliance has a resistance of 25 ohms and draws 10 amps, find its power.

50. For a sphere, the formula $r = \sqrt{\frac{S}{4\pi}}$ relates the radius r and the surface area S of the sphere. Find the surface area of a sphere with radius 8 in. Write your answer in terms of π.

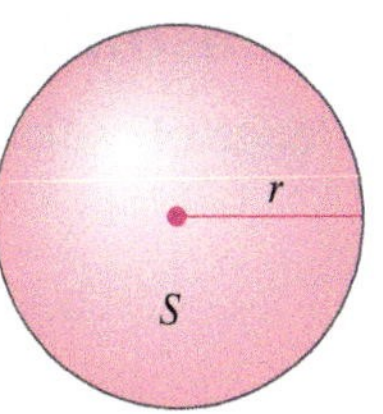

51. The period of a spring is the time it takes for the spring to stretch from one position, down, and then back again to its original position. The formula $T = 2\pi\sqrt{\frac{m}{k}}$, known as Hooke's law, expresses the period T (in seconds) in terms of the mass m (in grams) bobbing on the spring, where k is a constant. If $k = 8$, what mass will produce a period of 2 sec?

52. The formula for the diagonal of a rectangle d in terms of its length l and width w is:

$$d = \sqrt{l^2 + w^2}$$

Find the length of a rectangle if its width is 12 in. and its diagonal is 20 in.

53. The age of a tree t (in years) can be approximated using the formula

$$t = \frac{\sqrt{\frac{A}{\pi}} - b}{g},$$

where g is the width of each growth ring (in inches), b is the bark width (in inches), and A is the area of the tree's cross section (in square inches). (*Source:* National Climatic Data Center.)

a. Solve this formula for A.

b. Using this model, find the cross-section area of a 16-year-old tree with bark width $\frac{2}{3}$ in. and growth ring width $\frac{1}{3}$ in.

54. An investment pays an annual rate of return r on a principal of P dollars. The value V of the investment (in dollars) after 2 yr is related to the rate of return and the principal as follows:

$$r = \frac{\sqrt{V}}{\sqrt{P}} - 1$$

a. Solve this formula for V.

b. If the principal was \$100 and the rate of return is 0.05, what is the value of the investment?

• Check your answers on page A-38.

MINDStretchers

Groupwork

1. In general, $\sqrt{a} + \sqrt{b} \neq \sqrt{a + b}$. Working with a partner, determine if there are any positive numbers a and b for which $\sqrt{a} + \sqrt{b} = \sqrt{a + b}$. Justify your answer.

Mathematical Reasoning

2. Using your knowledge of radical equations, show how you would solve the following equation:

$$\sqrt{x - 16} + \sqrt{x + 11} = 9$$

Technology

3. On a coordinate plane, graph the radical equation $y = \sqrt{x}$

   a. by using a table.

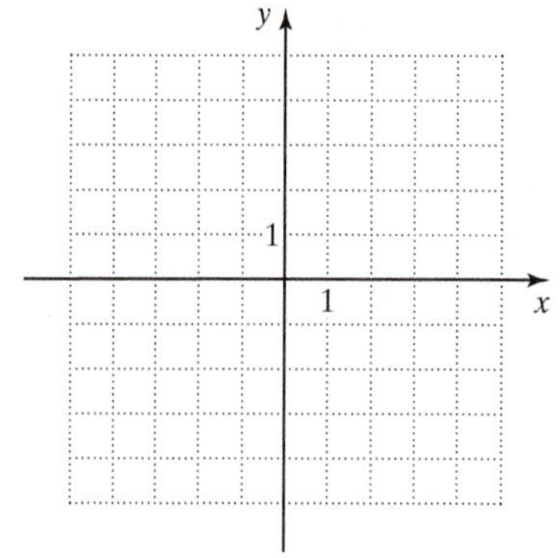

   b. by using a graphing calculator or graphing software.

Key Concepts and Skills

CONCEPT SKILL

Concept/Skill	Description	Example
[18.1] **Square root (of a nonnegative real number a)**	A number that when squared is a.	2 and -2 are square roots of 4 because $(2)^2 = 4$ and $(-2)^2 = 4$.
[18.1] **Radicand**	The number under a radical sign.	The radicand of $\sqrt{5}$ is 5.
[18.1] **Perfect square**	A whole number that is the square of another whole number.	9 is a perfect square because $9 = 3^2$.
[18.1] **Squaring a square root**	For any nonnegative real number a, $(\sqrt{a})^2 = a$.	$(\sqrt{8})^2 = 8$
[18.1] **Taking the square root of a square**	For any nonnegative real number a, $\sqrt{a^2} = a$.	$\sqrt{3^2} = 3$
[18.1] **The product rule of radicals**	If a and b are any nonnegative real numbers, then $\sqrt{ab} = \sqrt{a} \cdot \sqrt{b}$.	$\sqrt{4 \cdot 9} = \sqrt{4} \cdot \sqrt{9}$
[18.1] **The quotient rule of radicals**	If a is a nonnegative real number and b is a positive real number, then $\sqrt{\frac{a}{b}} = \frac{\sqrt{a}}{\sqrt{b}}$.	$\sqrt{\frac{4}{9}} = \frac{\sqrt{4}}{\sqrt{9}}$
[18.2] **Like radicals**	Radical expressions that have the same radicand.	$\sqrt{2}$ and $5\sqrt{2}$ are *like* radicals.
[18.2] **Unlike radicals**	Radical expressions with different radicands.	$7\sqrt{2}$ and $\sqrt{3}$ are *unlike* radicals.
[18.2] **To add or subtract *like* radicals**	• Use the distributive property. • Then, simplify.	$4\sqrt{2} + 3\sqrt{2} = (4 + 3)\sqrt{2} = 7\sqrt{2}$
[18.2] **To add or subtract *unlike* radicals**	• Simplify the unlike radicals, if possible. • If like radicals, then add or subtract.	$\sqrt{12} + \sqrt{27} = \sqrt{4 \cdot 3} + \sqrt{9 \cdot 3}$ $= 2\sqrt{3} + 3\sqrt{3}$ $= 5\sqrt{3}$
[18.3] **To multiply radicals**	• Apply the product rule of radicals. • Simplify, if possible.	$(\sqrt{10})(2\sqrt{2}) = 2 \cdot \sqrt{10 \cdot 2}$ $= 2\sqrt{20}$ $= 2 \cdot 2\sqrt{5}$ $= 4\sqrt{5}$
[18.3] **To divide radicals**	• Apply the quotient rule of radicals. • Simplify, if possible.	$\frac{\sqrt{20x^4}}{\sqrt{5x^2}} = \sqrt{\frac{20x^4}{5x^2}}$ $= \sqrt{4x^2}$ $= 2x$

continued

CONCEPT SKILL

Concept/Skill	Description	Example
[18.3] **To rationalize a denominator**	• If the denominator is a radical term, then multiply the numerator and denominator by the denominator. • If the denominator is a binomial with a radical term, then multiply the numerator and denominator by the conjugate of the denominator.	$\frac{\sqrt{x}}{\sqrt{8}} = \frac{\sqrt{x}}{\sqrt{8}} \cdot \frac{\sqrt{8}}{\sqrt{8}}$ $= \frac{\sqrt{8x}}{\sqrt{8^2}}$ $= \frac{\sqrt{4 \cdot 2x}}{8}$ $= \frac{2\sqrt{2x}}{8} = \frac{\sqrt{2x}}{4}$ $\frac{4}{1-\sqrt{3}} = \frac{4}{1-\sqrt{3}} \cdot \frac{1+\sqrt{3}}{1+\sqrt{3}}$ $= \frac{4(1+\sqrt{3})}{(1-\sqrt{3})(1+\sqrt{3})}$ $= \frac{4(1+\sqrt{3})}{1^2-(\sqrt{3})^2}$ $= \frac{4(1+\sqrt{3})}{1-3}$ $= \frac{4(1+\sqrt{3})}{-2}$ $= -2(1+\sqrt{3})$ $= -2-2\sqrt{3}$
[18.4] **Radical equation**	An equation with a variable in one or more radicands.	$\sqrt{3x} = 18$
[18.4] **The squaring property of equality**	For any real numbers a and b, if $a = b$, then $a^2 = b^2$.	If $\sqrt{x} = 7$, then $(\sqrt{x})^2 = (7)^2$.
[18.4] **To solve a radical equation**	• Isolate a term with a radical. • Square each side of the equation. • Where possible, combine like terms. • Solve the resulting equation. • Check the possible solution(s) in the original equation.	$\sqrt{x} + 1 = 5$ $\sqrt{x} = 4$ $(\sqrt{x})^2 = 4^2$ $x = 16$ **Check** $\sqrt{x} + 1 = 5$ $\sqrt{16} + 1 \stackrel{?}{=} 5$ $4 + 1 \stackrel{?}{=} 5$ $5 = 5$ True

CHAPTER 18 Review Exercises

Say Why

Fill in each blank.

1. The number $\sqrt{6}$ ________ (is/is not) an irrational number because ____.

2. The square of $\sqrt{27}$ ________ (is/is not) 27 because ____.

3. The radical $\sqrt{72}$ ________ (is/is not) in simplified form because ____.

4. The expression $\frac{\sqrt{5}}{\sqrt{3}}$ ________ (is/is not) simplified because ____.

5. The equation $x + \sqrt{27} = 4$ ________ (is/is not) a radical equation because ____.

6. The squaring property of equality ________ (is/is not) used to solve a radical equation because ____.

Assume that all variables and radicands represent nonnegative real numbers.

[18.1] *Simplify.*

7. $-\sqrt{49}$
8. $\sqrt{6^2}$
9. $(\sqrt{7x})^2$
10. $\sqrt{28}$
11. $-3\sqrt{18}$
12. $\sqrt{32x^3}$
13. $\sqrt{\frac{9}{25}}$
14. $-\sqrt{\frac{3t}{16}}$
15. $\sqrt{\frac{144}{x^{100}}}$
16. $2\sqrt{25a^5b^3}$

[18.2] *Combine.*

17. $2\sqrt{5} + \sqrt{5}$
18. $\sqrt{n} + 7\sqrt{n}$
19. $4x\sqrt{3} - 3x\sqrt{3}$
20. $\sqrt{27} - 2\sqrt{75}$
21. $x\sqrt{4x} + \sqrt{9x^3}$
22. $\sqrt{50a} - 3\sqrt{8a} + 8\sqrt{2a}$

[18.3] *Multiply. Simplify, if possible.*

23. $\sqrt{5} \cdot \sqrt{3}$
24. $\sqrt{8n} \cdot \sqrt{2n}$
25. $\sqrt{5a^2b^3} \cdot \sqrt{10ab^3}$
26. $\sqrt{x}(\sqrt{x} - 4)$
27. $(\sqrt{7} - 1)(\sqrt{7} + 1)$
28. $(\sqrt{y} + 1)(2\sqrt{y} - 3)$
29. $(\sqrt{y} + 5)^2$

Divide and simplify.

30. $\frac{\sqrt{54}}{\sqrt{6}}$
31. $\frac{\sqrt{3}}{\sqrt{12}}$
32. $\frac{\sqrt{48x}}{\sqrt{3x}}$
33. $\frac{\sqrt{24a^6}}{\sqrt{2a^3}}$

Rationalize the denominator.

34. $\dfrac{2}{\sqrt{11}}$

35. $\dfrac{\sqrt{3x^2}}{\sqrt{6x}}$

36. $\dfrac{4\sqrt{8} + \sqrt{2}}{\sqrt{2}}$

37. $\dfrac{10}{\sqrt{7} - 1}$

[18.4] *Solve and check.*

38. $\sqrt{n} - 3 = 5$

39. $\sqrt{2x + 1} = 4$

40. $\sqrt{4n - 5} = \sqrt{n + 10}$

41. $x - \sqrt{3x + 1} = 3$

Mixed Applications

Solve.

42. The length of one side of a square city block is $50\sqrt{2}$ ft. Find the area of the block.

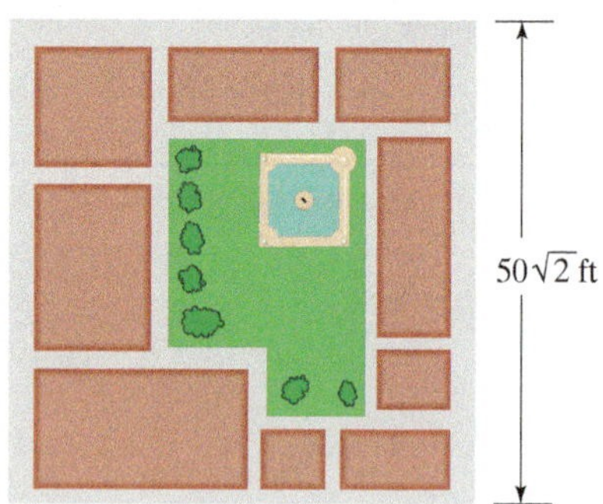

43. The formula for the distance that an astronaut can see on any planet is $d = \sqrt{\dfrac{rh}{2640}}$, where h is his height (in feet) and r is the radius of the planet (in miles). If the astronaut is 6 ft tall and the radius of Mars is about 2100 mi, how far can he see on Mars? Round the answer to the nearest mile.

44. The length of the diagonal of the metal box shown in the figure below is modeled by the expression $\sqrt{l^2 + w^2 + h^2}$, where l, w, and h represent the length, width, and height of the box, respectively. If the box is 5 in. long, 4 in. wide, and 3 in. high, will a screwdriver 8 in. long fit diagonally in the box?

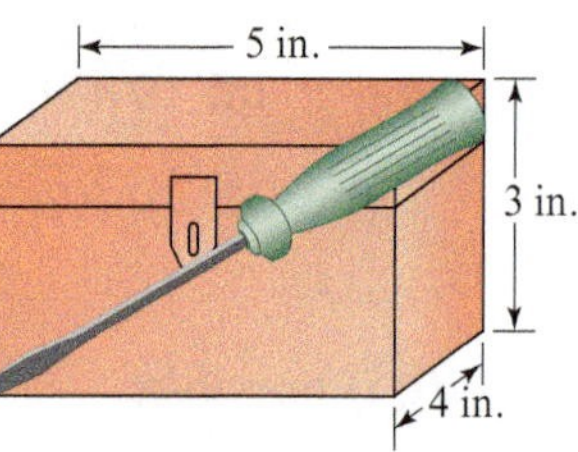

45. Find, in radical form, the perimeter of the kite shown below.

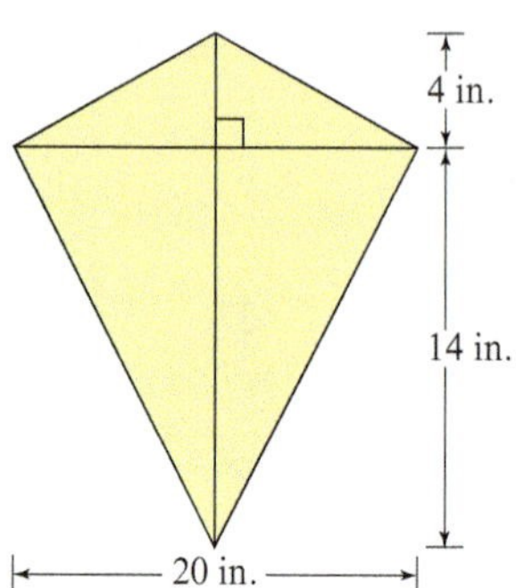

46. A formula for the radius of a sphere r is

$$r = \sqrt{\frac{S}{4\pi}},$$

where S is the surface area of the sphere. Write this formula, rationalizing the denominator.

47. Scientists who study lakes are concerned with their shape. These scientists have grouped lakes according to the concept of *shoreline development*, a measure of how closely a lake resembles a circle. Shoreline development D is defined by the formula

$$D = \frac{L}{2\sqrt{\pi A}},$$

where L is the length of the lake's shoreline and A is the area of the lake. (*Source:* David G. Frey, Ed., *Limnology in North America*)

a. Check that if a lake is perfectly circular, the shoreline development is equal to 1.

b. Rationalize the denominator in the formula.

c. Goslute Lake, which existed in the American southwest some 50 million years ago, had a shoreline development equal to 1.58. Solve for its area in terms of the length of its shoreline.

48. For the box shown with volume V, the length of a side x of the square base is $\sqrt{\frac{V}{15}}$. Simplify this expression.

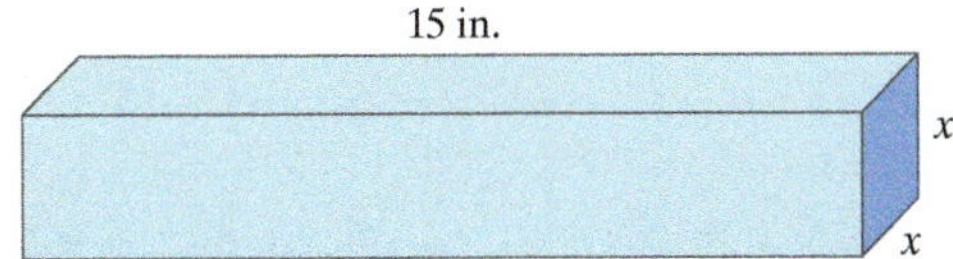

• Check your answers on page A-38.

CHAPTER 18 Posttest

FOR EXTRA HELP

CHAPTER Test Prep VIDEOS

The Chapter Test Prep Videos with test solutions are available on DVD, in MyMathLab, and on YouTube (search "AkstDevMath" and click on "Channels").

To see if you have mastered the topics in this chapter, take this test.

Assume all variables and radicands represent nonnegative real numbers.

Simplify.

1. $2\sqrt{36}$

2. $-3\sqrt{45}$

3. $\sqrt{32x^3y}$

4. $\sqrt{\frac{5n}{16}}$

5. $\sqrt{x} - 3\sqrt{x} + 5\sqrt{x}$

6. $\sqrt{8} - 4\sqrt{50} + \sqrt{18}$

7. $t\sqrt{4t} + 2\sqrt{16t^3}$

Multiply or divide. Simplify, if possible.

8. $(\sqrt{12})(\sqrt{2})$

9. $(\sqrt{5x^3y})(\sqrt{5x^2y})$

10. $\frac{\sqrt{75}}{\sqrt{3}}$

11. $\sqrt{y}(5\sqrt{y} - 1)$

12. $(\sqrt{3} + 4)^2$

Simplify.

13. $\sqrt{\frac{2p}{5}}$

14. $\frac{\sqrt{48x^4}}{\sqrt{2x}}$

15. Rationalize the denominator: $\frac{\sqrt{3} - 2\sqrt{6}}{\sqrt{3}}$

Solve and check.

16. $\sqrt{5x + 1} - 2 = 4$

17. $\sqrt{x + 11} = \sqrt{7x - 1}$

18. One of several expressions for the *windchill temperature* (WCT) is

$$91 + 0.08(3.7\sqrt{V} + 6 - 0.3V)(T - 91),$$

where T is the temperature (in degrees Fahrenheit) and V is the wind speed (in miles per hour). Find the WCT, rounded to the nearest whole number, if the temperature is 42°F and the wind speed is 16 mph.

19. Find the length of the ladder shown, expressed as a simplified radical.

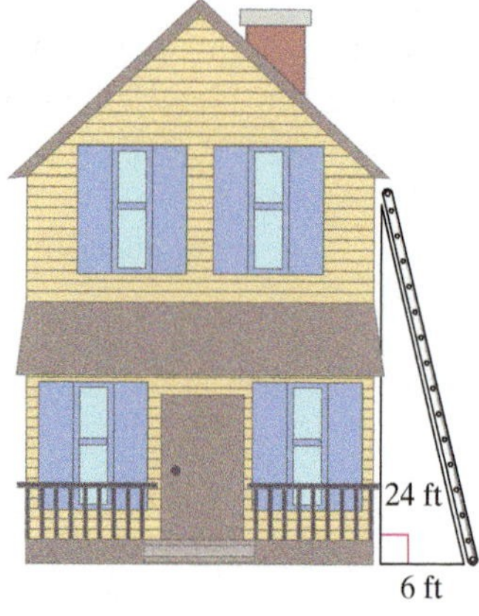

20. On a clear day, the distance d (in miles) that a lookout on a ship can see to the horizon from height h (in feet) can be approximated by the formula $d = 1.2\sqrt{h}$. How high must a lookout climb to see a ship 6 mi away?

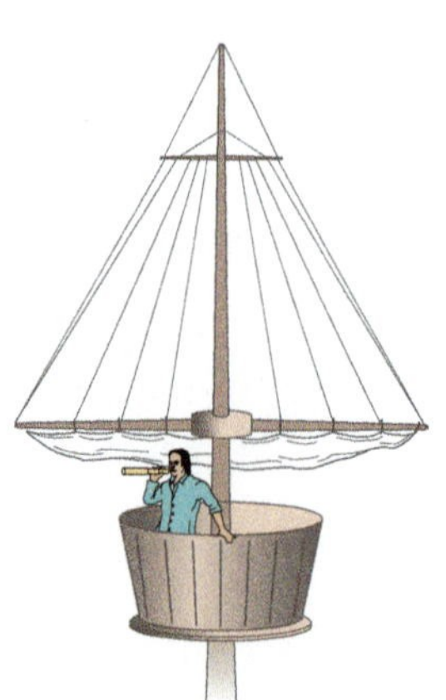

• Check your answers on page A-38.

Cumulative Review Exercises

To help you review, solve the following:

1. Rank from smallest to largest: 0.875, 0.83, 0.8625, and 0.083.

2. Solve and check: $\frac{2}{9} = \frac{c}{18}$

3. What percent is equivalent to 0.025?

4. Simplify: $\frac{2}{3} + \left(-\frac{4}{5}\right)$

5. Evaluate the expression $2a^2 - 3b + c - 4d$ if $a = -4.5$, $b = -2$, $c = 3$, and $d = 1.5$.

6. Solve and check: $c = -2(c - 1)$

7. Compute the slope m of the line that passes through the points $(4, 0)$ and $(3, 5)$. Plot these points on the coordinate plane and draw the line.

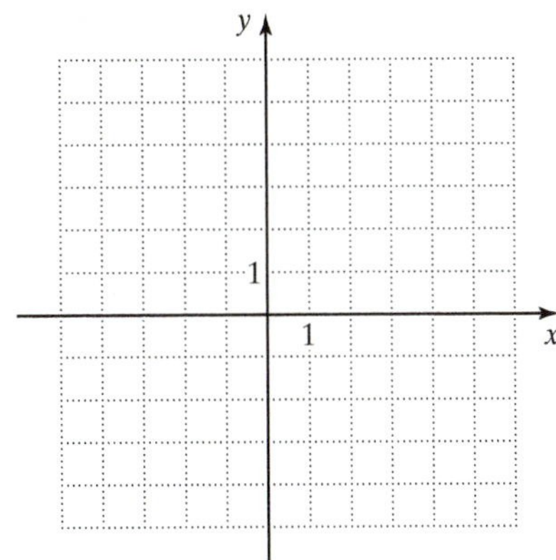

8. For the system of equations graphed, determine the number of solutions.

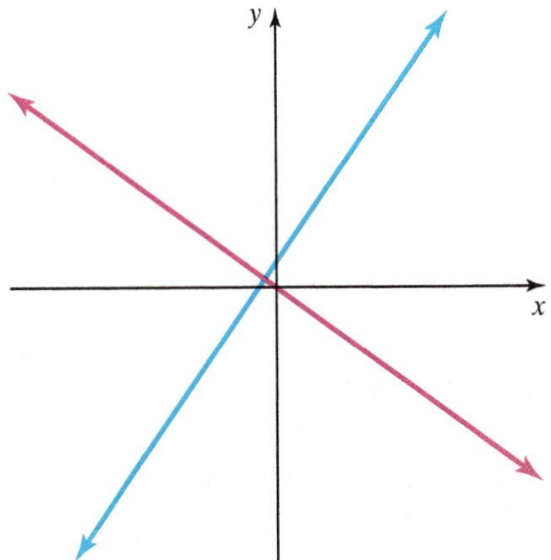

9. Find the product: $(4x^2 - 3)(4x - 1)$

10. Factor: $3x^3 + 24x^2 + 48x$

11. Add: $\frac{9}{c - 2} + \frac{c - 3}{c^2 - c - 2}$

12. Solve and check: $\frac{2}{m} = \frac{m - 3}{9}$

13. Solve and check: $\sqrt{n + 6} + 7 = 9$

14. Simplify $\frac{\sqrt{48}}{8}$.

15. The speed of sound in air s is modeled by the formula
$$s = 0.6t + 331,$$
where the speed is in meters per second and the temperature t is in degrees Celsius. At what temperature is this speed equal to 343 m/sec? (*Source:* Peter J. Nolan, *Fundamentals of College Physics*)

16. The average distance from the Sun is approximately 1.5×10^8 km to the Earth and about 6×10^9 km to the dwarf planet Pluto. The distance to Pluto is how many times the distance to the Earth, expressed in scientific notation? (*Source:* Jeffrey Bennett et al., *The Cosmic Perspective*)

17. A country's balance of payments can be found by subtracting its imports from its exports. If x is the number of years since 2000, the total exports of the United States can be approximated by the polynomial $-6x^3 + 83x^2 - 228x + 1111$, and the total U.S. imports can be approximated by $-10x^3 + 133x^2 - 296x + 1495$ (both polynomials in billions of dollars). Use this model to approximate the U.S. balance of payments. (*Source:* bea.gov)

18. The first week, a student with a part-time job made \$150 in x hr. The second week, he made \$175 working 1 more hour than the previous week. What was the ratio of the student's average pay per hour in the second week as compared to the first week?

19. At a studio that creates 3-D animation movies, a computer graphics server would take 72 hr working by itself to process the most intensive action sequences in a single animated movie frame. By contrast, a new desktop computer, although much more expensive, would take only 36 hr. (*Source:* tomshardware.com)

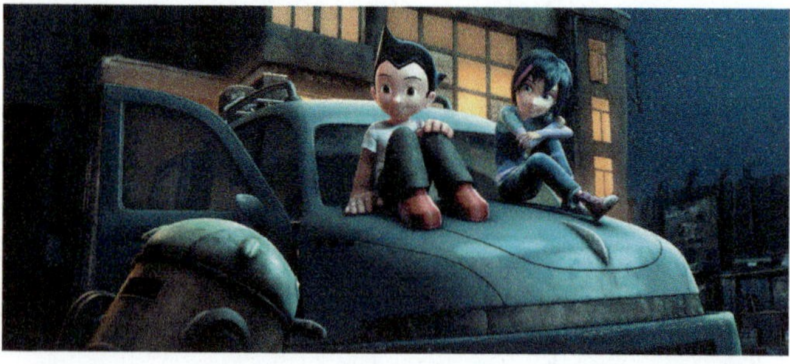

a. How many hours would it take for a computer graphics server and a new desktop computer working together to process one of these sequences?

b. One night, the studio used 5000 servers and 2000 desktop computers to process the animation. How many hours did it take to process a sequence?

20. Two hikers start a trip by walking due west from a campsite, as shown below. They then turn due north and walk to a waterfall. What is the distance d from the camp to the waterfall expressed in radical form? Also express this distance as a decimal, rounded to the nearest hundred meters.

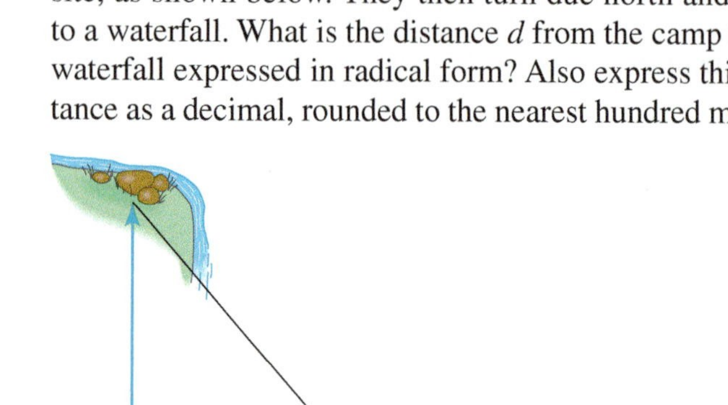

• Check your answers on page A-39.

CHAPTER 19

Quadratic Equations

Using Quadratic Equations to Test New Drugs

In testing new medicines or food additives, scientists try to determine their effect on the body by collecting and analyzing data. A method for analyzing data is to picture the data by plotting points on a coordinate plane.

For instance, the graph to the upper right illustrates the effect of various kinds of insulin on diabetic patients. It shows that some kinds of insulin have a longer-lasting effect than others.

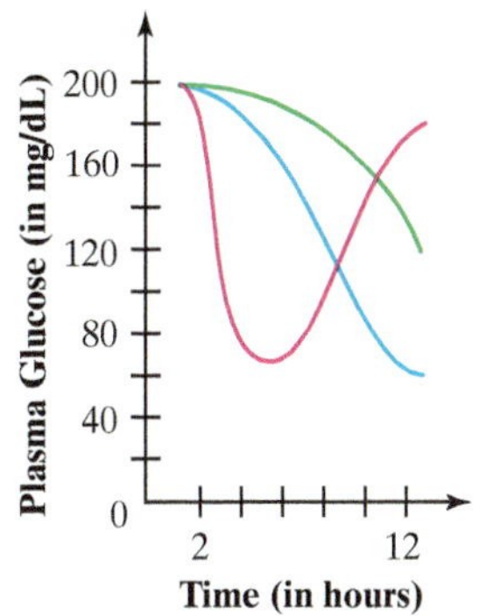

The *dose-effect curves* shown to the lower right deal with raising poultry. The curves show how the weight of turkeys changes when they are given a food additive called methionine. This substance, in the appropriate dosage, is thought to enhance turkey size and health.

Notice the shape of these graphs. Such curves can sometimes be approximated by *parabolas*, that is, by curves that correspond to quadratic equations. Scientists can then use these quadratic equations to summarize the observed data and to make predictions.

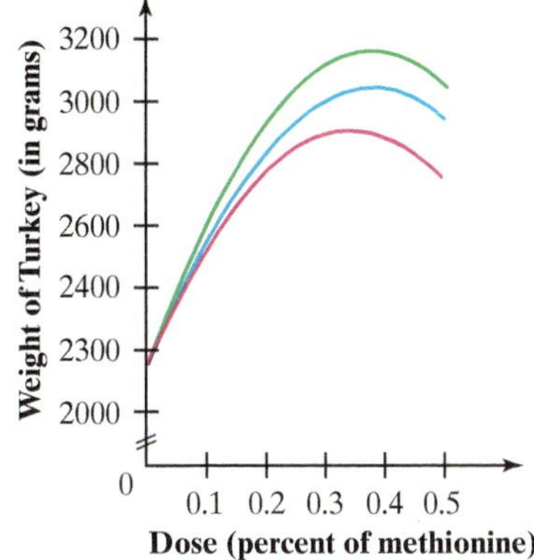

(*Sources:* Bertram G. Katzung, *Basic and Clinical Pharmacology,* 1995; Donald A. Berry, *Statistical Methodology in the Pharmaceutical Sciences,* 1990; David H. Tedeschi and Ralph E. Tedeschi, *Importance of Fundamental Principles in Drug Evaluation,* 1968)

CHAPTER 19 PRETEST

To see if you have already mastered the topics in this chapter, take this test.

1. Solve $3x^2 = 54$ using the square root property of equality.

2. Solve $A = 4\pi r^2$ for r. Assume that all variables are positive.

3. Fill in the blank to make a perfect square trinomial: $x^2 + 12x + [\ ]$.

4. Solve $3x^2 - 12x + 6 = 0$ by completing the square.

5. Solve $-2x^2 + 2x + 5 = 0$ by using the quadratic formula.

Solve.

6. $(x - 6)^2 = 25$

7. $n^2 + 7n + 12 = 0$

8. $y^2 - 6y = -4$

9. $4x^2 + 9 = 21$

10. $5n^2 - 10n - 4 = 0$

11. $(x + 2)(x + 2) = 18$

12. Sketch a parabola that has vertex $(0, 0)$ and passes through the points $(1, -1)$, $(-1, -1)$, $(2, -4)$, and $(-2, -4)$.

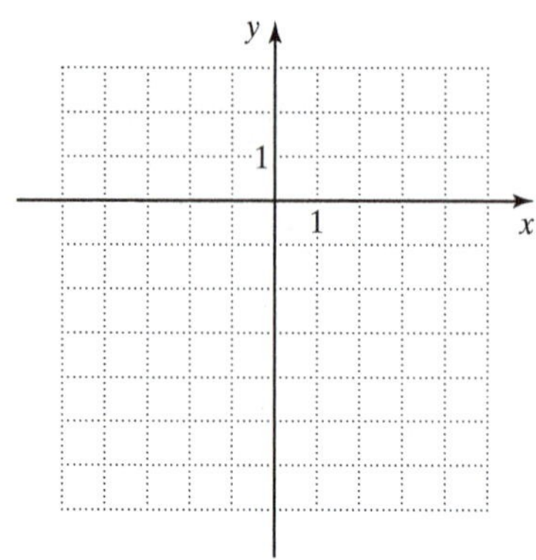

13. Find the vertex and axis of symmetry of the graph of $y = x^2 - 6x + 7$.

Sketch the graph of each equation.

14. $y = -x^2$

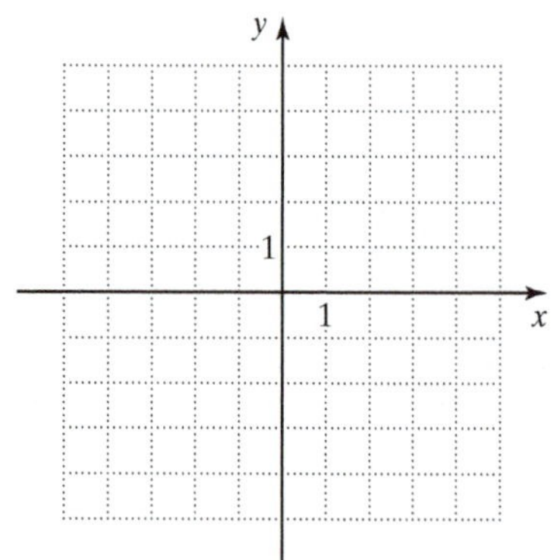

15. $y = x^2 - 2x + 1$

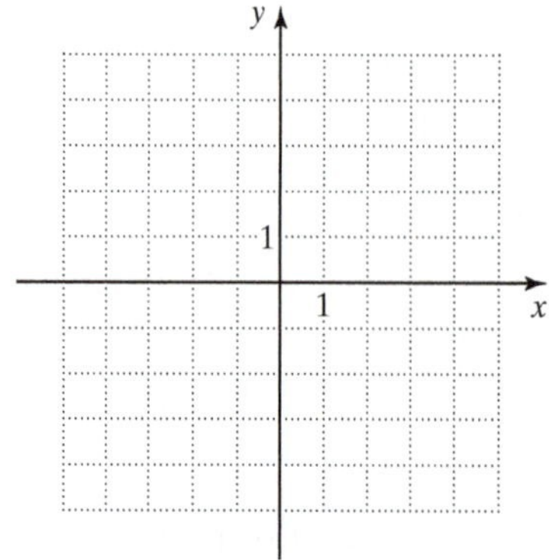

16. $y = 4x - 2x^2$

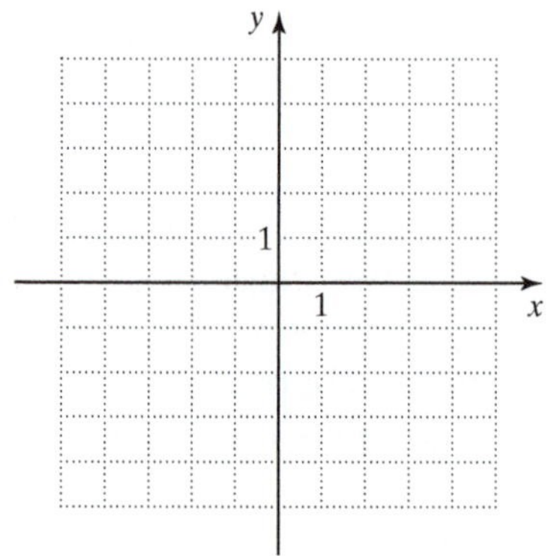

Solve.

17. The final velocity (in meters per second) of an object starting from rest that is accelerated at a constant rate is given by $v^2 = 2as$, where s is the distance traveled (in meters) during the acceleration a. Solve this equation for v.

18. A box manufacturer makes open boxes from 30 in. by 36 in. pieces of cardboard by cutting squares of equal size from the corners and turning up the sides. According to the manufacturing specifications, each box is to have a base with an area of 720 in^2. What size squares should be cut from the corners of the piece of cardboard in order to meet the box specifications?

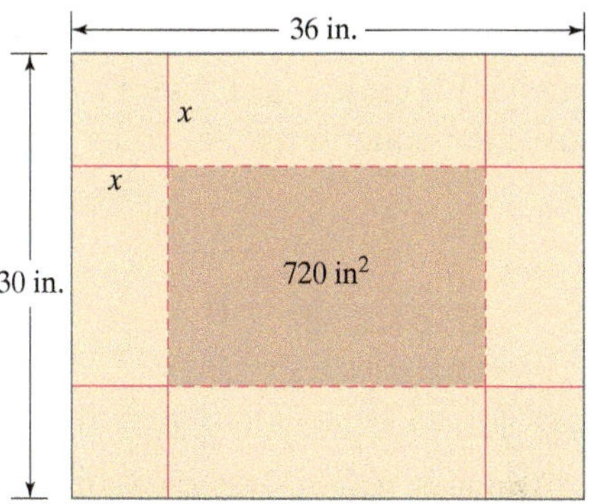

19. A Cessna airplane traveled 216 mi against a 15-mph headwind and 420 mi with a 25-mph tailwind. If the total flying time was 4 hr, what is the speed of the airplane in still air?

20. A rhythmic gymnast throws a ball straight up from a height of 5 ft with an initial velocity of 48 ft/sec. The equation $h = -16t^2 + 48t + 5$ models the ball's height h (in feet) after t sec. What is the maximum height the ball reaches?

• Check your answers on page A-39.

19.1 Solving Quadratic Equations by Using the Square Root Property

OBJECTIVES

- **A** To solve quadratic equations using the square root property
- **B** To solve literal equations or formulas using the square root property
- **C** To solve applied problems involving quadratic equations

Recall from Section 16.5 that a *quadratic equation* is an equation that can be written in the standard form $ax^2 + bx + c = 0$, where a, b, and c are real numbers and $a \neq 0$. In that section, we saw how quadratic equations can be used to solve problems related to physics, business, sports, and other fields. We solved such equations by factoring.

Factoring is only one method that can be used to solve quadratic equations. In this chapter, we consider three other methods: the square root property, completing the square, and the quadratic formula. We close the chapter by discussing how to graph quadratic equations.

Let's solve the equation $x^2 = 36$ by factoring, which is the method we discussed in Section 16.5.

$$x^2 = 36$$

$$x^2 - 36 = 0 \quad \text{Write the equation in standard form.}$$

$$(x + 6)(x - 6) = 0 \quad \text{Factor.}$$

$$x + 6 = 0 \quad \text{or} \quad x - 6 = 0 \quad \text{Set each factor equal to 0.}$$

$$x = -6 \qquad\qquad x = 6$$

The solutions are -6 and 6, which we can write as ± 6 (read "plus or minus 6").

The solutions ± 6 can be expressed as $\pm\sqrt{36}$. This suggests that we can also solve the equation $x^2 = 36$ simply by taking the square root of both sides.

The Square Root Property of Equality

If n is a nonnegative number and $x^2 = n$, then $x = \pm\sqrt{n}$, that is, $x = \sqrt{n}$ or $x = -\sqrt{n}$.

This property allows us to take the square root of each side of an equation, resulting in an equivalent equation, provided that each radicand is nonnegative.

Let's use this property to solve quadratic equations of the form $x^2 = n$ and $ax^2 = n$. Consider these examples.

EXAMPLE 1

Solve by taking square roots.

a. $x^2 = 49$ **b.** $3x^2 = 24$ **c.** $9 + 16t^2 = 0$

Solution

a. $x^2 = 49$

$x = \pm\sqrt{49}$ Use the square root property of equality.

$x = \pm 7$ Simplify.

Check

Substitute 7 for x.

$$x^2 = 49$$

$$(7)^2 \stackrel{?}{=} 49$$

$$49 = 49 \quad \text{True}$$

Substitute -7 for x.

$$x^2 = 49$$

$$(-7)^2 \stackrel{?}{=} 49$$

$$49 = 49 \quad \text{True}$$

So the solutions are 7 and -7.

PRACTICE 1

Use the square root property of equality to solve.

a. $x^2 = 100$

b. $2x^2 = 36$

c. $25 - 81t^2 = 0$

b. In this example, we must first divide each side of the equation by the coefficient of the x-squared term before using the square root property of equality.

$$3x^2 = 24$$
$$x^2 = 8 \quad \text{Divide each side by 3.}$$
$$x = \pm\sqrt{8} \quad \text{Use the square root property of equality.}$$
$$x = \pm 2\sqrt{2} \quad \text{Simplify.}$$

So the solutions are $2\sqrt{2}$ and $-2\sqrt{2}$.

c. $9 + 16t^2 = 0$

$$16t^2 = -9 \quad \text{Subtract 9 from each side of the equation.}$$
$$t^2 = -\frac{9}{16} \quad \text{Divide each side by 16.}$$

Recall from Section 18.1 that a radical with a negative radicand is not a real number. So there are no real-number solutions of the equation $9 + 16t^2 = 0$.

Now, let's look at how we use the square root property of equality to solve quadratic equations of the form $(ax + b)^2 = d$ and $a(x + c)^2 = d$, that is, equations containing the square of a binomial.

EXAMPLE 2

Solve using the square root property of equality.

a. $(y - 1)^2 = 25$ **b.** $4(x + 4)^2 = 8$ **c.** $(3x + 1)^2 = 18$

Solution

a. $(y - 1)^2 = 25$

$$y - 1 = \pm\sqrt{25} \quad \text{Use the square root property of equality.}$$
$$y - 1 = \pm 5 \quad \text{Simplify.}$$
$$y - 1 = 5 \quad \text{or} \quad y - 1 = -5$$
$$y = 6 \qquad\qquad y = -4$$

So the solutions are 6 and -4.

b. $4(x + 4)^2 = 8$

$$(x + 4)^2 = 2 \quad \text{Divide each side by 4.}$$
$$x + 4 = \pm\sqrt{2} \quad \text{Use the square root property of equality.}$$
$$x + 4 = \sqrt{2} \quad \text{or} \quad x + 4 = -\sqrt{2}$$
$$x = -4 + \sqrt{2} \qquad\qquad x = -4 - \sqrt{2}$$

So the solutions are $-4 + \sqrt{2}$ and $-4 - \sqrt{2}$.

PRACTICE 2

Solve using the square root property.

a. $(n + 2)^2 = 9$

b. $3(x - 4)^2 = 15$

c. $(4m - 1)^2 = 54$

EXAMPLE 2 (continued)

c. $(3x + 1)^2 = 18$

$$3x + 1 = \pm\sqrt{18} \quad \text{Use the square root property of equality.}$$

$$3x + 1 = \pm 3\sqrt{2}$$

$$3x + 1 = 3\sqrt{2} \quad \text{or} \quad 3x + 1 = -3\sqrt{2}$$

$$3x = -1 + 3\sqrt{2} \qquad 3x = -1 - 3\sqrt{2}$$

$$x = \frac{-1 + 3\sqrt{2}}{3} \qquad x = \frac{-1 - 3\sqrt{2}}{3}$$

So the solutions are $\frac{-1 + 3\sqrt{2}}{3}$ and $\frac{-1 - 3\sqrt{2}}{3}$.

In Example 2(c), note that since 3 is not a common factor of both terms in the numerator, we cannot cancel the 3 in the numerator with the 3 in the denominator.

EXAMPLE 3

Solve: $x - 5 = \frac{3}{x - 5}$

Solution

$$x - 5 = \frac{3}{x - 5}$$

$$(x - 5)^2 = 3 \quad \text{Multiply each side by } (x - 5).$$

$$x - 5 = \pm\sqrt{3} \quad \text{Use the square root property of equality.}$$

$$x - 5 = \sqrt{3} \quad \text{or} \quad x - 5 = -\sqrt{3}$$

$$x = 5 + \sqrt{3} \qquad x = 5 - \sqrt{3}$$

So the solutions are $5 + \sqrt{3}$ and $5 - \sqrt{3}$.

PRACTICE 3

Solve: $x + 4 = \frac{2}{x + 4}$

Recall that we solved linear literal equations in Section 12.4. Now, let's apply the square root property to solving quadratic literal equations. We assume that all variables and radicands are positive.

EXAMPLE 4

Solve for the indicated variable.

a. $V = \pi r^2 h$ for r (Volume formula for a cylinder)

b. $c^2 = a^2 + b^2$ for b (Pythagorean theorem)

Solution

a. Solving for r, we get:

$$V = \pi r^2 h$$

$$\frac{V}{\pi h} = r^2 \quad \text{Divide each side by } \pi h.$$

$$\pm\sqrt{\frac{V}{\pi h}} = r \quad \text{Use the square root property of equality.}$$

$$r = \pm\sqrt{\frac{V}{\pi h}} = \frac{\pm\sqrt{\pi h V}}{\pi h}$$

PRACTICE 4

Solve for the indicated variable.

a. $E = mc^2$ for c (c = speed of light)

b. $c^2 = a^2 + b^2$ for a

Since the length of the radius is positive, we accept only the positive square root: $r = \sqrt{\dfrac{V}{\pi h}}$

b. Solving for b, we get:

$$c^2 = a^2 + b^2$$

$$c^2 - a^2 = b^2 \qquad \text{Add } -a^2 \text{ to each side.}$$

$$\pm\sqrt{c^2 - a^2} = b \qquad \text{Use the square root property.}$$

So $b = \pm\sqrt{c^2 - a^2}$. Since b represents a side of a right triangle and its length is positive, we accept only the positive square root, namely

$$b = \sqrt{c^2 - a^2}.$$

Can the expression $\pm\sqrt{c^2 - a^2}$ in Example 4(b) be simplified? Explain.

EXAMPLE 5

A circular table top and a square table top have the same area. (Recall that the area of a square with side s is given by the formula $A = s^2$, whereas $A = \pi r^2$ represents the area of a circle with radius r.)

a. Express s in terms of r.

b. Approximating π by 3.14, find, to the nearest inch, the side of the square table if the radius of the circular table is 20 in.

Solution

a. Since the square and circular areas are equal, $s^2 = \pi r^2$. To express s in terms of r, we need to solve this equation for s.

$$s^2 = \pi r^2$$

$$s = \pm\sqrt{\pi}r$$

Since the side is positive in length, the solution is $s = \sqrt{\pi}r$, or $s = r\sqrt{\pi}$.

b. Next, we substitute 20 for r and simplify.

$$s = \sqrt{\pi}(20)$$

$$\approx \sqrt{3.14}(20)$$

$$\approx 35.44$$

So the side of the square table is approximately 35 in. long.

PRACTICE 5

A nuclear power plant on the Japanese coastline suffered a meltdown because of an earthquake and a tsunami. As a result, the government ordered a mandatory evacuation in a semicircular region centered around the power plant. (*Source:* latimes.com)

a. The area A of a semicircle with radius r is given by $A = \dfrac{1}{2}\pi r^2$. Solve this equation for r.

b. If the area of the region around the power plant was approximately 225 mi^2, find, to the nearest mile, the radius of the evacuated region.

EXAMPLE 6

If P dollars are invested at interest rate r (in decimal form) compounded annually, then at the end of 2 yr the amount will have grown to $A = P(1 + r)^2$.

a. Solve this equation for rate r.

b. What is the interest rate needed for an investment of \$1000 to grow to \$1210 in 2 yr?

Solution

a. First, we solve the equation $A = P(1 + r)^2$ for r.

$$A = P(1 + r)^2$$

$$P(1 + r)^2 = A$$

$$(1 + r)^2 = \frac{A}{P}$$

$$1 + r = \pm\sqrt{\frac{A}{P}}$$

$$r = -1 \pm \sqrt{\frac{A}{P}}$$

Since the interest rate is positive, we conclude that $r = -1 + \sqrt{\frac{A}{P}}$.

b. Substituting \$1000 for P and \$1210 for A, we get:

$$r = -1 + \sqrt{\frac{A}{P}}$$

$$= -1 + \sqrt{\frac{1210}{1000}}$$

$$= -1 + \sqrt{\frac{121}{100}}$$

$$= -1 + \frac{11}{10}$$

$$= 0.1$$

So the interest rate must be 10%.

PRACTICE 6

Loggers use *board feet* as a unit to measure the volume of timber. The formula

$$B = (d - 4)^2$$

gives an estimate of the number of board feet B in a 16-foot log, where d is the diameter of the log in inches and $d > 4$ in.

a. Solve the formula for diameter d.

b. What is the diameter of a log containing 16 board feet?

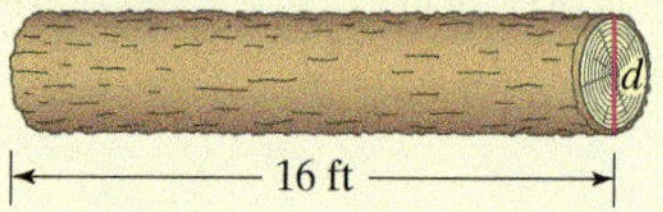

Would the value of r in Example 6 have changed if in the original equation we had substituted first and then solved for r? Explain.

19.1 Exercises

FOR EXTRA HELP MyMathLab MathXL PRACTICE WATCH READ REVIEW

Mathematically Speaking

Fill in each blank with the most appropriate term or phrase from the given list.

divide both sides by a	the coefficient of x^2	positive or negative square root of n
square of a binomial	square root of a binomial	take the square root of both sides
principal square root of n	binomial equation	
quadratic equation		

1. A(n) ________________________ is an equation that can be written in the standard form $ax^2 + bx + c = 0$, where a, b, and c are real numbers and $a \neq 0$.

2. The square root property of equality states that if p is a nonnegative number and $x^2 = n$, then x is equal to the ________________________.

3. Before using the square root property of equality to solve quadratic equations of the form $ax^2 = n$, we must ________________________.

4. We can use the square root property of equality to solve quadratic equations of the form $(ax + b)^2 = c$ or $a(x + c)^2 = d$, that is, equations containing the ________________________.

A *Solve by using the square root property.*

5. $y^2 = 9$

6. $x^2 = 16$

7. $p^2 = 2$

8. $x^2 = 5$

9. $n^2 = \frac{1}{4}$

10. $t^2 = \frac{16}{25}$

11. $5t^2 = 20$

12. $2n^2 = 72$

13. $4x^2 = 28$

14. $7n^2 = 21$

15. $3s^2 = 36$

16. $6x^2 = 48$

17. $\frac{1}{4}p^2 = 20$

18. $\frac{1}{3}s^2 = 7$

19. $5 + t^2 = 11$

20. $x^2 - 9 = 3$

21. $x^2 - 8 = 9$

22. $p^2 - 10 = 1$

23. $4n^2 - 9 = 0$

24. $9x^2 - 25 = 0$

25. $4x^2 + 18 = 8$

26. $9n^2 - 5 = -7$

27. $15 - 16x^2 = 3$

28. $6 - 18y^2 = 2$

29. $(n + 2)^2 = 9$

30. $(c - 5)^2 = 16$

31. $(x - 7)^2 = 49$

32. $(y + 9)^2 = 81$

33. $(x + 6)^2 = 5$

34. $(x - 2)^2 = 2$

35. $(5 - s)^2 = \frac{9}{16}$

36. $(1 - n)^2 = \frac{1}{4}$

37. $(y - 4)^2 = 9$

38. $(x + 2)^2 = 25$

39. $2(p - 5)^2 = 6$

40. $3(t + 4)^2 = 21$

41. $5(x + 1)^2 = 40$

42. $4(n - 3)^2 = 48$

43. $(3x + 1)^2 = 4$

44. $(2t - 3)^2 = 9$

45. $(4y + 5)^2 = 3$

46. $(8a - 1)^2 = 2$

47. $(2x - 7)^2 = 20$

48. $(3x + 4)^2 = 27$

49. $\left(\frac{1}{2}x - 5\right)^2 = 10$

50. $\left(\frac{1}{3}t + 1\right)^2 = 8$

51. $(x + 1)^2 + 49 = 0$

52. $(m - 5)^2 + 100 = 0$

53. $a + 8 = \frac{5}{a + 8}$

54. $x - 1 = \frac{2}{x - 1}$

55. $x - 3 = \frac{24}{x - 3}$

56. $n + 6 = \frac{27}{n + 6}$

57. $(y - 1)(y + 1) = 4$

58. $(n - 3)(n + 3) = -1$

B *Solve for the indicated variable.*

59. $ax^2 - b = 0$ for x

60. $h = 16t^2$ for t

61. $K = \frac{4\pi^2}{v^2 r}$ for v

62. $W = i^2Rt$ for i

63. $\frac{x^2}{16} - \frac{y^2}{25} = 1$ for y

64. $E = \frac{1}{2}mv^2$ for v

Mixed Practice

Solve by using the square root property.

65. $(5y - 6)^2 = 44$

66. $p^2 - 2 = 30$

67. $5 - 75m^2 = -4$

68. $(y - 4)^2 = 11$

69. $5(x + 2)^2 = 60$

70. $6t^2 = 42$

71. $a - 4 = \frac{63}{a - 4}$

72. Solve for x: $\frac{x^2}{4} + \frac{y^2}{9} = 1$

Applications

C *Solve.*

73. Two campus security officers communicate with one another using walkie-talkies. After meeting, one of the officers walks north at 45 ft/min and the other walks east at 60 ft/min. To the nearest minute, for how long can they hear one another if the walkie-talkies have a maximum range of 300 ft?

74. The length of a rectangular floor is twice the width. The area of the floor is 32 ft^2. What are the dimensions of the room?

75. A candy company wants to reduce the cost of its product. The candy is cylindrical in shape, with height h and with radius r. Management wants to reduce the size of the candies by 5%, while keeping the same shape. (Reminder: The formula for the volume of a cylinder is $V = \pi r^2 h$.)

a. If the company reduces only the height of the candy, what must the new height H be?

b. If the company reduces only the radius of the candy, what must the new radius R be?

76. A formula for the volume of the cone shown below is $V = \frac{1}{3}\pi r^2 h$.

a. Solve the formula for r.

b. Find the radius of the cone, rounded to the nearest tenth of an inch. Use $\pi \approx 3.14$.

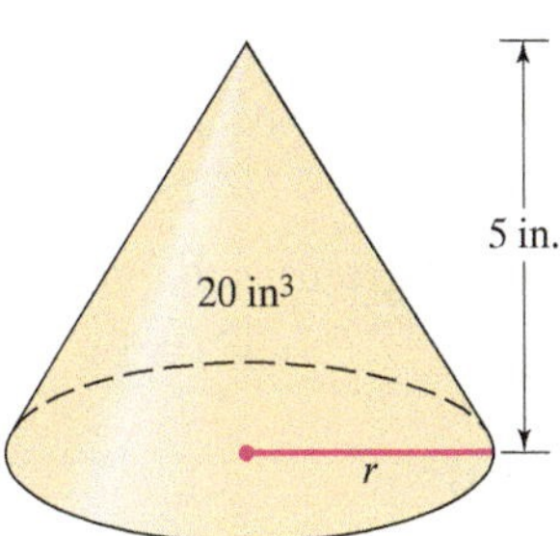

77. The graph of the equation $x^2 + y^2 = r^2$, where $r > 0$, is a circle whose center is the origin on the coordinate plane, as shown in the figure below. Solve this equation for y.

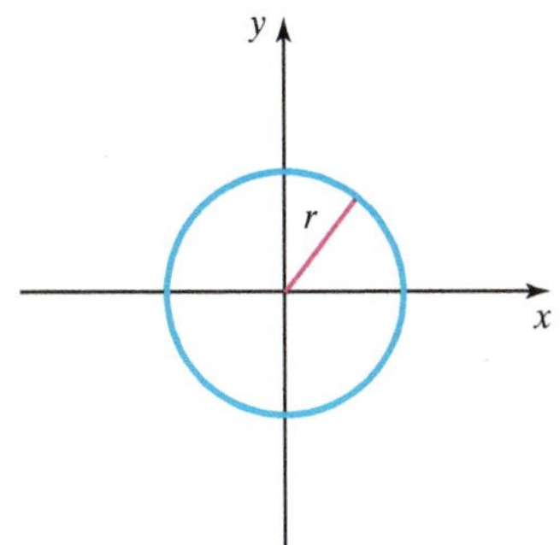

78. If a businessman deposits \$900 in an account with interest rate r compounded annually, the amount of money A in the account after 2 yr is given by the following equation:

$$A = 900(1 + r)^2$$

Solve this equation for r.

79. An object d ft from a point source of light receives I foot-candles of light, where

$$I = \frac{4050}{d^2}.$$

Solve this formula for the distance.

80. The owner of a pizzeria wants 20% of the surface of a pizza to be the crust around the edge (see the diagram below). If the radius of the pizza (including the crust) is r, find the appropriate width c of the pizza crust.

• Check your answers on page A-39.

MINDStretchers

Mathematical Reasoning

1. Consider the inequality $x^2 - 9 \leq 0$.

a. Identify three positive and three negative numbers that satisfy this inequality.

b. Solve the inequality.

Groupwork

2. Working in a small group, solve the following equations:

a. $x^4 = 16$ **b.** $t^3 = 9t$ **c.** $(n^2 - 1)^2 = 16$ **d.** $(y^2 + 5)^2 = 9$

Critical Thinking

3. Consider the equation $x^2 = a$. Solve for x. How many real solutions does this equation have? Explain.

Cultural Note

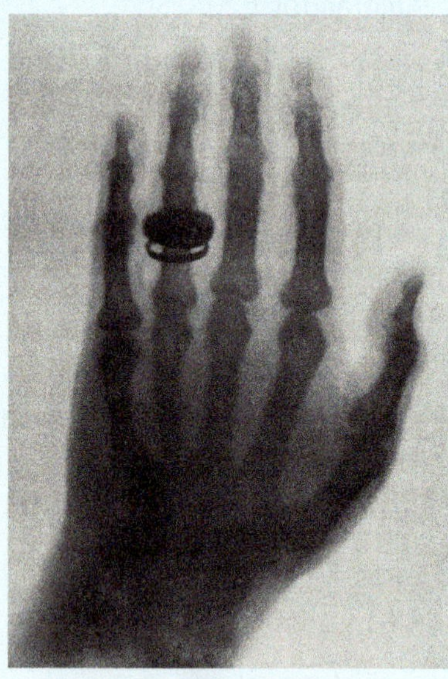

The photograph to the left is one of the first taken by the German physicist Wilhelm Roentgen in 1895 after he discovered strange rays that he called *X-rays* since they were a mysterious phenomenon. Roentgen's discovery allowed doctors to see inside the body for the first time without surgery and earned him the first Nobel Prize awarded for physics.

The practice of using an *x* and other letters from the end of the alphabet to represent mathematical unknowns goes back to the seventeenth-century French mathematician René Descartes, who made major contributions to the development of algebra.

(*Source:* Florian Cajori, *A History of Mathematical Notations*, The Open Court Publishing Company, 1929)

19.2 Solving Quadratic Equations by Completing the Square

OBJECTIVES

A To solve quadratic equations by completing the square

B To solve applied problems involving quadratic equations

In this section, we solve quadratic equations by **completing the square**. This method combines two previous approaches for solving quadratic equations, namely, taking square roots and factoring perfect square trinomials.

For instance, consider the equation $x^2 + 10x = 2$. This equation is different from those of the form solved in the previous section because the expression $x^2 + 10x$ is not a perfect square.

To use the square root property, we need to write the form of the left-hand side of the equation as a perfect square trinomial. Recall that in Section 15.6 we discussed perfect squares such as the following:

Perfect square ↓		The coefficient of the quadratic term is 1. ↓		The coefficient of the linear term ↓		The constant term ↓
$(x + 1)^2$	$=$	x^2	$+$	$2x$	$+$	1
$(x + 2)^2$	$=$	x^2	$+$	$4x$	$+$	4
$(x + 3)^2$	$=$	x^2	$+$	$6x$	$+$	9
$(x + 4)^2$	$=$	x^2	$+$	$8x$	$+$	16
⋮				⋮		

Note that in each perfect square trinomial, the coefficient of the quadratic term is 1, and the constant term is the square of one-half the coefficient of the linear term. For example, in $x^2 + 8x + 16$, we see that:

$$16 = \left(\frac{1}{2} \cdot 8\right)^2$$

More generally, when we square the binomial $x + c$ we get:

$$(x + c)^2 = x^2 + 2cx + c^2$$

where

$$c^2 = \left(\frac{1}{2} \cdot 2c\right)^2$$

↑ The constant term (under c^2); ↑ The coefficent of the linear term (under $2c$)

Observing this pattern leads us to a method of finding the value of the constant term that makes the general quadratic trinomial $x^2 + bx + c$ a perfect square.

To Complete the Square for the Expression $x^2 + bx$

- Find one-half of the coefficient of the linear term x: $\frac{1}{2}b$
- Square this value: $\left(\frac{1}{2}b\right)^2$
- Add the resulting value to the expression.

EXAMPLE 1

Fill in the blank to make each trinomial a perfect square.

a. $y^2 + 6y + \square$ **b.** $x^2 - 3x + \square$

Solution

a. $y^2 + 6y + \square$

The coefficient of the linear term y is 6.

$$\frac{1}{2}(6) = 3 \quad \text{Find one-half the coefficient of } y.$$

$$3^2 = 9 \quad \text{Square 3.}$$

The constant term to be added to $y^2 + 6y$ is 9. So we get the perfect square trinomial $y^2 + 6y + 9$, which is equal to $(y + 3)^2$.

b. $x^2 - 3x + \square$

The coefficient of the linear term x is -3.

$$\frac{1}{2}(-3) = -\frac{3}{2}$$

$$\left(-\frac{3}{2}\right)^2 = \frac{9}{4}$$

The constant term to be added to $x^2 - 3x$ is $\frac{9}{4}$. So we get:

$x^2 - 3x + \frac{9}{4}$, which is equal to $\left(x - \frac{3}{2}\right)^2$.

PRACTICE 1

Fill in the blank to make each trinomial a perfect square.

a. $x^2 - 12x + \square$

b. $n^2 + 5n + \square$

Now, let's apply the method of completing the square to solve a quadratic equation of the form $ax^2 + bx + c = 0$, where $a = 1$. Consider, for instance, the equation $x^2 + 8x = 5$.

We begin by completing the square on the left side of the equation $x^2 + 8x = 5$.

$$x^2 + 8x = 5 \quad \text{Take one-half the coefficient of the linear term } x: \frac{1}{2}(8) = 4. \text{ Then, square it: } 4^2 = 16.$$

We add 16 to the left side of the equation to complete the square, but to maintain equality we must also add 16 to the right side of the equation.

$$x^2 + 8x + 16 = 5 + 16$$
$$x^2 + 8x + 16 = 21$$

Next, we write the left side of the equation $x^2 + 8x + 16 = 21$ as the square of a binomial and then, solve the equation.

$$(x + 4)^2 = 21$$
$$x + 4 = \pm\sqrt{21} \quad \text{Use the square root property of equality.}$$
$$x + 4 = \sqrt{21} \quad \text{or} \quad x + 4 = -\sqrt{21}$$
$$x = -4 + \sqrt{21} \qquad x = -4 - \sqrt{21}$$

So the solutions are $-4 + \sqrt{21}$ and $-4 - \sqrt{21}$.

Let's look at some additional examples of solving quadratic equations by completing the square.

EXAMPLE 2

Solve by completing the square.

a. $y^2 - 6y = 7$ **b.** $x^2 - 5x + 3 = 0$

Solution

a.

$$y^2 - 6y = 7$$

$$y^2 - 6y + 9 = 7 + 9$$ Take one-half the coefficient of the linear term: $\frac{1}{2}(-6) = -3$. Then, square it: $(-3)^2 = 9$. Add 9 to each side of the equation.

$$y^2 - 6y + 9 = 16$$

$$(y - 3)^2 = 16$$ Write $y^2 - 6y + 9$ as the square of a binomial.

$$y - 3 = \pm 4$$ Use the square root property of equality.

$$y - 3 = 4 \quad \text{or} \quad y - 3 = -4$$

$$y = 7 \qquad\qquad y = -1$$

So the solutions of $y^2 - 6y = 7$ are 7 and -1. How would you check these solutions?

b. Before completing the square, we move all variable terms to one side of the equation and the constant term to the other side.

$$x^2 - 5x + 3 = 0$$

$$x^2 - 5x = -3$$ Add -3 to each side.

$$x^2 - 5x + \frac{25}{4} = -3 + \frac{25}{4}$$ Take one-half the coefficient of the linear term: $\frac{1}{2}(-5) = -\frac{5}{2}$. Then, square it: $\left(-\frac{5}{2}\right)^2 = \frac{25}{4}$. Add $\frac{25}{4}$ to each side.

$$x^2 - 5x + \frac{25}{4} = \frac{13}{4}$$ Simplify.

$$\left(x - \frac{5}{2}\right)^2 = \frac{13}{4}$$ Write $x^2 - 5x + \frac{25}{4}$ as the square of a binomial.

$$x - \frac{5}{2} = \pm\sqrt{\frac{13}{4}}$$ Use the square root property.

$$x - \frac{5}{2} = \sqrt{\frac{13}{4}} \quad \text{or} \quad x - \frac{5}{2} = -\sqrt{\frac{13}{4}}$$

$$x = \frac{5}{2} + \frac{\sqrt{13}}{2} \qquad x = \frac{5}{2} - \frac{\sqrt{13}}{2}$$

$$x = \frac{5 + \sqrt{13}}{2} \qquad x = \frac{5 - \sqrt{13}}{2}$$

So the solutions are $\frac{5 + \sqrt{13}}{2}$ and $\frac{5 - \sqrt{13}}{2}$.

PRACTICE 2

Solve by completing the square.

a. $y^2 + 4y = 21$

b. $n^2 + 7n + 5 = 0$

Next, we consider quadratic equations of the form $ax^2 + bx + c = 0$, where $a \neq 1$. Before completing the square, we must divide each side of the equation by a, the coefficient of the second-degree (or quadratic) term.

EXAMPLE 3

Solve by completing the square.

a. $2y^2 - 4y = 14$ **b.** $5m^2 - 10m + 4 = 0$

Solution

a.

$$2y^2 - 4y = 14$$

$$y^2 - 2y = 7$$ Divide each side of the equation by 2, the coefficient of the second-degree term.

$$y^2 - 2y + 1 = 7 + 1$$ Complete the square: add $\left(\frac{-2}{2}\right)^2 = 1$ to each side.

$$(y-1)^2 = 8$$

$$y - 1 = \pm\sqrt{8}$$

$$y - 1 = \sqrt{8} \quad \text{or} \quad y - 1 = -\sqrt{8}$$

$$y = 1 + 2\sqrt{2} \qquad y = 1 - 2\sqrt{2}$$

So the solutions are $1 + 2\sqrt{2}$ and $1 - 2\sqrt{2}$.

b. $5m^2 - 10m + 4 = 0$

$$m^2 - 2m + \frac{4}{5} = 0$$ Divide each side by 5, the coefficient of the second-degree term.

$$m^2 - 2m = -\frac{4}{5}$$ Subtract $\frac{4}{5}$ from each side.

$$m^2 - 2m + 1 = -\frac{4}{5} + 1$$ Complete the square: add $\left(\frac{-2}{2}\right)^2 = 1$ to each side.

$$(m-1)^2 = \frac{1}{5}$$

$$m - 1 = \pm\sqrt{\frac{1}{5}}$$

$$m - 1 = +\sqrt{\frac{1}{5}} \quad \text{or} \quad m - 1 = -\sqrt{\frac{1}{5}}$$

$$m = 1 + \sqrt{\frac{1}{5}} \qquad m = 1 - \sqrt{\frac{1}{5}}$$

$$m = 1 + \frac{\sqrt{5}}{5} \qquad m = 1 - \frac{\sqrt{5}}{5}$$

So the solutions are $1 + \frac{\sqrt{5}}{5}$ and $1 - \frac{\sqrt{5}}{5}$.

PRACTICE 3

Solve by completing the square.

a. $2y^2 - 16y = 8$

b. $4x^2 - 4x - 1 = 0$

The preceding examples suggest the following general method for solving a quadratic equation by completing the square.

To Solve the Quadratic Equation $ax^2 + bx + c = 0$ by Completing the Square

- If $a \neq 1$, then divide the equation by a, the coefficient of the second-degree term. If $a = 1$, then proceed to the next step.
- Move all terms with variables to one side of the equals sign and all constants to the other side.
- Take one-half the coefficient of the linear term. Then, square it and add this square to each side of the equation.
- Factor the side of the equation containing the variable terms, writing it as the square of a binomial.
- Use the square root property of equality.
- Solve the resulting equations.

EXAMPLE 4

Solve $4n^2 - 12n + 5 = 0$ by completing the square.

Solution

$$4n^2 - 12n + 5 = 0$$

$$n^2 - 3n + \frac{5}{4} = 0 \qquad \text{Divide each side by 4.}$$

$$n^2 - 3n = -\frac{5}{4}$$

$$n^2 - 3n + \frac{9}{4} = -\frac{5}{4} + \frac{9}{4} \qquad \text{Complete the square.}$$

$$\left(n - \frac{3}{2}\right)^2 = 1$$

$$n - \frac{3}{2} = \pm 1$$

$$n = \frac{3}{2} + 1 \quad \text{or} \quad n = \frac{3}{2} - 1$$

$$n = \frac{5}{2} \qquad\qquad n = \frac{1}{2}$$

So the solutions are $\frac{5}{2}$ and $\frac{1}{2}$.

PRACTICE 4

Solve $9x^2 - 9x - 4 = 0$ by completing the square.

EXAMPLE 5

The rectangular room pictured has an area of 280 ft^2 and a perimeter of 68 ft.

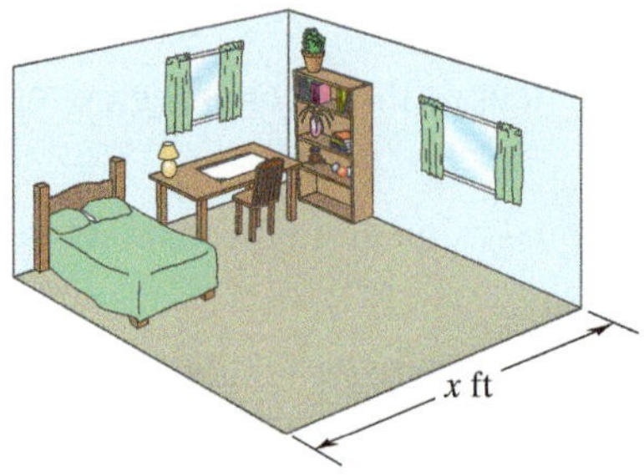

a. Express all four sides of the room in terms of x.

b. Find the dimensions of the room.

Solution

a. First, we must find the lengths of the four sides in terms of x. The area of the rectangle, which is given to be 280 ft^2, is the product of the length and the width. Since one side of the rectangle is x, the adjacent side is $\frac{280}{x}$ ft. So the dimensions of the rectangle are as pictured:

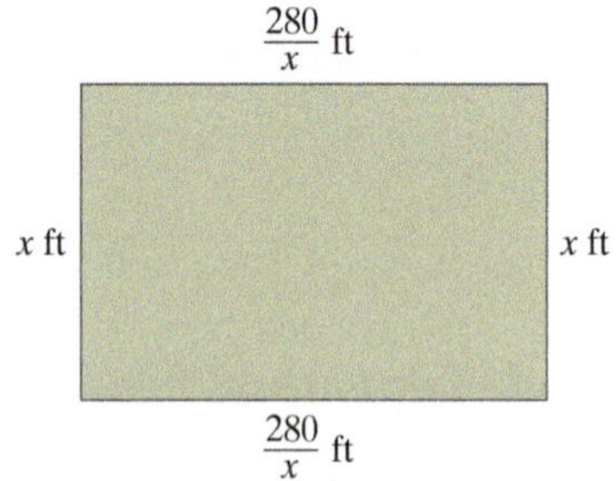

b. To find the dimensions of the room, we recall that the perimeter of the room is 68 ft. The perimeter is the sum of the side lengths:

$$2x + 2\left(\frac{280}{x}\right) = 68$$
$$x + \frac{280}{x} = 34$$
$$x^2 + 280 = 34x$$
$$x^2 - 34x = -280$$
$$x^2 - 34x + 289 = -280 + 289$$
$$(x - 17)^2 = 9$$
$$x - 17 = \pm 3$$
$$x = 17 \pm 3$$
$$x = 20 \quad \text{or} \quad x = 14$$

Note that if one side is 20, then the adjacent side, represented by $\frac{280}{x}$, is 14. And if one side is 14, then the adjacent side is 20. So the room is 20 ft by 14 ft.

PRACTICE 5

In ping-pong, the length of the top of the ping-pong table is 1 ft less than twice the width. The area of the ping-pong table is 36 ft^2.

a. Write an expression for the length of the ping-pong table in terms of the width w.

b. Find the length and width of the top of the table.

19.2 Exercises

FOR EXTRA HELP MyMathLab MathXL PRACTICE WATCH READ REVIEW

A *Fill in the blank to make each trinomial a perfect square.*

1. $x^2 + 6x +$ ▢

2. $t^2 + 4t +$ ▢

3. $x^2 - 2x +$ ▢

4. $n^2 - 10n +$ ▢

5. $x^2 + 5x +$ ▢

6. $x^2 - 3x +$ ▢

7. $y^2 - y +$ ▢

8. $n^2 + 9n +$ ▢

Solve by completing the square.

9. $x^2 + 4x = 0$

10. $p^2 + 6p = 0$

11. $b^2 - 10b = -1$

12. $x^2 - 20x = -50$

13. $y^2 + 14y - 15 = 0$

14. $t^2 + 10t + 21 = 0$

15. $x^2 - 6x - 4 = 0$

16. $x^2 - 4x - 2 = 0$

17. $n^2 + 8n + 20 = 0$

18. $x^2 - 2x + 2 = 0$

19. $x^2 - x - 3 = 0$

20. $y^2 + 7y + 8 = 0$

21. $2y^2 + 8y = 24$

22. $3x^2 + 12x = -12$

23. $6x^2 + 12x - 5 = 0$

24. $5a^2 + 10a - 3 = 0$

25. $4n^2 - 20n + 7 = 0$

26. $9y^2 - 18y - 7 = 0$

27. $4n^2 - 3n - 4 = 0$

28. $2x^2 - x - 2 = 0$

29. $2x^2 - 4x + 8 = 0$

30. $5x^2 - 20x = -30$

31. $(x - 3)(x + 1) = 1$

32. $(d + 2)(d + 3) = 8$

Mixed Practice

Solve by completing the square.

33. $2x^2 + x - 4 = 0$

34. $b^2 + 12b + 27 = 0$

35. $n^2 - 6n + 10 = 0$

36. $2x^2 - 4x - 38 = 0$

Fill in the blank to make each trinomial a perfect square.

37. $t^2 + 7t +$ ▢

38. $a^2 - 8a +$ ▢

Applications

B *Solve by completing the square.*

39. For the duration of an experiment, a biologist discovers that the number of insects N in a tank after d days is approximated by the model $N = d^2 + 12d + 6$. After how many days are there 51 insects?

40. An electronic company's revenue R (in dollars) is the product of the number of machines n that it sells and the price of a machine $(100 - n)$. This relationship can be expressed as:

$$R = n(100 - n)$$

How many machines must be sold for the revenue to be equal to \$1600?

41. If fireworks on the ground are shot straight upward at a speed of 32 ft/sec, then the height of the fireworks h (in feet) after t sec can be represented by the expression $-16t^2 + 32t$.

a. At what time will the fireworks be 16 ft above the ground?

b. At what time (to the nearest tenth of a second) will the fireworks be 8 ft above the ground?

42. Consider the rectangle shown below with the measurements given in feet.

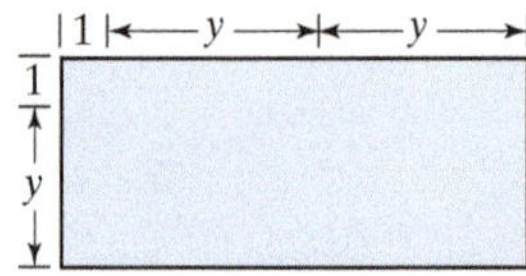

a. Write an expression for the width and length of the rectangle.

b. Write an expression for the area of the rectangle.

c. If the area of the rectangle is 25 ft^2, find the value of y to the nearest tenth of a foot.

43. Two friends leave a party, one driving north and the other east. They are 10 mi apart when one of them is 2 mi farther from the party than the other. At that time, how far were they each from the party?

44. A commuter walks to and from work, which is 2 mi from her apartment. Going home, she walks 1 mph slower than going to work. If it takes her 1 hr 10 min for the round trip, at what speed does she walk to work?

45. An office has two copying machines. Working alone, one machine takes 3 min longer to do a particular job than the other machine. When the machines are working together, it takes 4 min to do the job. To the nearest minute, how long would it take the faster machine to do the job alone?

46. Two painters are working on a house. If the first painter were doing the job by himself, it would take him 2 hr longer to paint the house than if the second painter were doing the job by himself. Working together, it takes the painters 5 hr to complete the job. To the nearest hour, how long would it take each painter to do the job working alone?

• Check your answers on page A-39.

MINDStretchers

Critical Thinking

1. A circular graph with radius r and centered around the point (h, k) can be represented by the equation $(x - h)^2 + (y - k)^2 = r^2$. Find the center and the radius of the circle represented by $x^2 - 4x + y^2 + 6y = -12$.

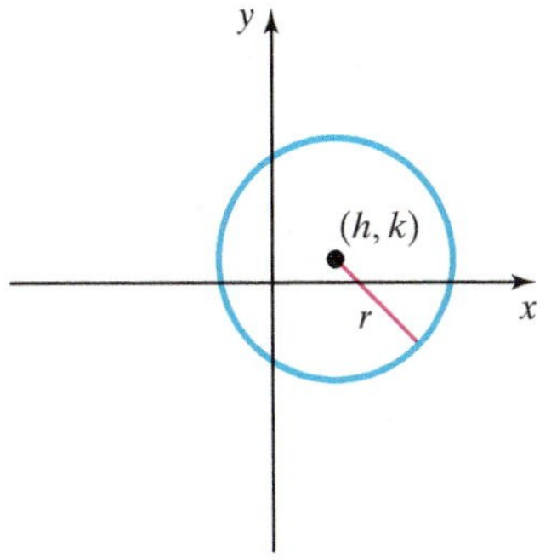

Mathematical Reasoning

2. We can represent the polynomial $x^2 + 10x$ geometrically by the area enclosed in the diagram below. In a few sentences, explain how we can use this diagram to "complete the square."

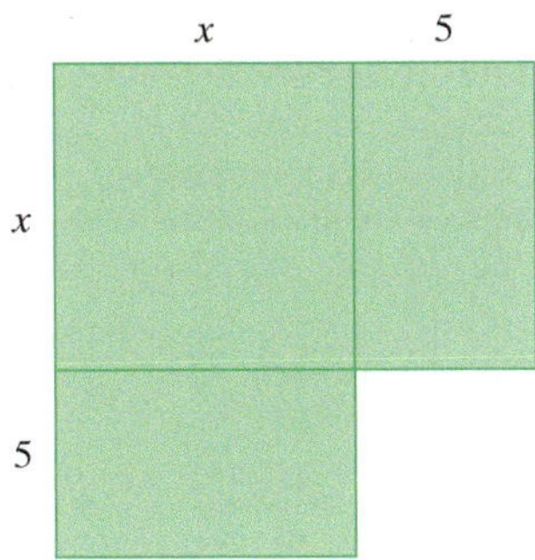

Groupwork

3. Working with a partner, consider the equation $x^2 + bx + c = 0$. The claim is made that this equation will have no real solution if $c > \frac{b^2}{4}$.

a. Choose some values of b and of c to check this claim. Record your findings.

b. Prove or disprove the claim by completing the square.

19.3 Solving Quadratic Equations by Using the Quadratic Formula

OBJECTIVES

- **A** To solve quadratic equations using the quadratic formula
- **B** To solve applied problems involving quadratic equations

Deriving the Quadratic Formula

In Section 19.2, we used the method of completing the square each time we solved a quadratic equation. So the question arises: Why not use this method and solve the general equation $ax^2 + bx + c = 0$, where $a \neq 0$, once and for all? By solving the general equation for x, we can derive the *quadratic formula*:

$$ax^2 + bx + c = 0$$

$$x^2 + \frac{b}{a}x + \frac{c}{a} = 0 \qquad \text{Divide each side by } a.$$

$$x^2 + \frac{b}{a}x = -\frac{c}{a} \qquad \text{Add } -\frac{c}{a} \text{ to each side.}$$

$$x^2 + \frac{b}{a}x + \frac{b^2}{4a^2} = -\frac{c}{a} + \frac{b^2}{4a^2} \qquad \text{Complete the square: } \left(\frac{1}{2}\cdot\frac{b}{a}\right)^2 = \frac{b^2}{4a^2}\text{; and add } \frac{b^2}{4a^2} \text{ to each side.}$$

$$\left(x + \frac{b}{2a}\right)^2 = \frac{b^2 - 4ac}{4a^2} \qquad \text{Write the left side as the square of a binomial, and write the right side as a single fraction.}$$

$$x + \frac{b}{2a} = \pm\sqrt{\frac{b^2 - 4ac}{4a^2}} \qquad \text{Assuming that the radicand is nonnegative, use the square root property of equality.}$$

$$x = -\frac{b}{2a} \pm \frac{\sqrt{b^2 - 4ac}}{2a} \qquad \text{Solve for } x \text{ by adding } -\frac{b}{2a} \text{ to each side.}$$

$$x = \frac{-b \pm \sqrt{b^2 - 4ac}}{2a} \qquad \text{Combine fractions.}$$

So the solutions of $ax^2 + bx + c = 0$ are $\frac{-b + \sqrt{b^2 - 4ac}}{2a}$ and $\frac{-b - \sqrt{b^2 - 4ac}}{2a}$.

This result gives us the *quadratic formula,* which can be used to solve any quadratic equation written in standard form.

The Quadratic Formula

If $ax^2 + bx + c = 0$, where a, b, and c are real numbers and $a \neq 0$, then

$$x = \frac{-b \pm \sqrt{b^2 - 4ac}}{2a}.$$

Solving Quadratic Equations Using the Quadratic Formula

The quadratic formula allows us to solve any quadratic equation of the form $ax^2 + bx + c = 0$, where $a \neq 0$. For instance, consider the equation $x^2 + 3x = 10$. To solve, we first write the equation in standard form: $ax^2 + bx + c = 0$.

$$x^2 + 3x = 10$$
$$x^2 + 3x - 10 = 0$$

In this equation $a = 1, b = 3$, and $c = -10$. So we substitute these values for a, b, and c in the formula, and then simplify.

$$x = \frac{-b \pm \sqrt{b^2 - 4ac}}{2a}$$

$$= \frac{-(3) \pm \sqrt{(3)^2 - 4(1)(-10)}}{2(1)}$$ Substitute 1 for a, 3 for b, and -10 for c.

$$= \frac{-3 \pm \sqrt{9 + 40}}{2}$$ Simplify.

$$= \frac{-3 \pm \sqrt{49}}{2}$$ Simplify.

$$= \frac{-3 \pm 7}{2}$$ Take the square root.

$$x = \frac{-3 + 7}{2} \quad \text{or} \quad x = \frac{-3 - 7}{2}$$

$$= \frac{4}{2} = 2 \qquad = \frac{-10}{2} = -5$$

So the solutions are 2 and -5.

Note that the equation $x^2 + 3x = 10$ can also be solved by factoring.

$$x^2 + 3x - 10 = 0$$

$$(x + 5)(x - 2) = 0$$

$$x + 5 = 0 \quad \text{or} \quad x - 2 = 0$$

$$x = -5 \qquad x = 2$$

So solving by factoring yields the same solutions as those we found by using the quadratic formula.

Can all equations that are solved by the quadratic formula also be solved by factoring? Explain.

To Solve a Quadratic Equation $ax^2 + bx + c = 0$ Using the Quadratic Formula

- Write the equation in standard form, if necessary.
- Identify the coefficients a, b, and c.
- Substitute values for a, b, and c in the formula $x = \dfrac{-b \pm \sqrt{b^2 - 4ac}}{2a}$.
- Simplify.

In showing how to apply the quadratic formula, first we consider quadratic equations of the form $ax^2 + bx + c = 0$, where $a = 1$.

EXAMPLE 1

Solve by using the quadratic formula.

a. $m^2 - 9m + 18 = 0$

b. $x^2 + 3x = -3$

PRACTICE 1

Solve by using the quadratic formula.

a. $n^2 - n = 0$

b. $m^2 + m = -1$

EXAMPLE 1 (continued)

Solution

a. $m^2 - 9m + 18 = 0$

This equation is already in standard form, so $a = 1$, $b = -9$, and $c = 18$. We substitute these values in the quadratic formula.

$$\begin{aligned} m &= \frac{-b \pm \sqrt{b^2 - 4ac}}{2a} \\ &= \frac{-(-9) \pm \sqrt{(-9)^2 - 4(1)(18)}}{2(1)} \\ &= \frac{9 \pm \sqrt{81 - 72}}{2} \\ &= \frac{9 \pm \sqrt{9}}{2} \\ &= \frac{9 \pm 3}{2} \end{aligned}$$

$$m = \frac{9 + 3}{2} = \frac{12}{2} = 6 \quad \text{or} \quad m = \frac{9 - 3}{2} = \frac{6}{2} = 3$$

So the solutions are 6 and 3.

b. $x^2 + 3x = -3$

First, let's write $x^2 + 3x = -3$ in standard form: $x^2 + 3x + 3 = 0$. We see that $a = 1$, $b = 3$, and $c = 3$. We substitute these values in the formula.

$$\begin{aligned} x &= \frac{-b \pm \sqrt{b^2 - 4ac}}{2a} \\ &= \frac{-(3) \pm \sqrt{(3)^2 - 4(1)(3)}}{2(1)} \\ &= \frac{-3 \pm \sqrt{9 - 12}}{2} \\ &= \frac{-3 \pm \sqrt{-3}}{2} \end{aligned}$$

In Section 18.1, we saw that a radical with a negative radicand is not a real number. Here, note that the radicand is negative. So there are no real solutions of the equation $x^2 + 3x = -3$.

TIP Before using the quadratic formula, make sure that the equation you are solving is in standard form.

Next, let's consider quadratic equations of the form $ax^2 + bx + c = 0$, where $a \neq 1$.

EXAMPLE 2

Use the quadratic formula to solve.

a. $4t^2 = 12t - 1$ **b.** $2y^2 + 3y = 1$

Solution

a.

$$4t^2 = 12t - 1$$

$$4t^2 - 12t + 1 = 0$$ Write the equation in standard form.

$$t = \frac{-b \pm \sqrt{b^2 - 4ac}}{2a}$$

$$= \frac{-(-12) \pm \sqrt{(-12)^2 - 4(4)(1)}}{2(4)}$$ Substitute 4 for a, −12 for b, and 1 for c.

$$= \frac{12 \pm \sqrt{144 - 16}}{8}$$

$$= \frac{12 \pm \sqrt{128}}{8}$$

$$x = \frac{12 + \sqrt{128}}{8} \quad \text{or} \quad x = \frac{12 - \sqrt{128}}{8}$$

$$= \frac{12 + 8\sqrt{2}}{8} \qquad = \frac{12 - 8\sqrt{2}}{8}$$

$$= \frac{4(3 + 2\sqrt{2})}{8} \qquad = \frac{4(3 - 2\sqrt{2})}{8}$$

$$= \frac{3 + 2\sqrt{2}}{2} \qquad = \frac{3 - 2\sqrt{2}}{2}$$

So the solutions are $\frac{3 + 2\sqrt{2}}{2}$ and $\frac{3 - 2\sqrt{2}}{2}$.

b.

$$2y^2 + 3y = 1$$

$$2y^2 + 3y - 1 = 0$$ Write the equation in standard form.

$$y = \frac{-b \pm \sqrt{b^2 - 4ac}}{2a}$$

$$= \frac{-3 \pm \sqrt{(3)^2 - 4(2)(-1)}}{2(2)}$$ Substitute 2 for a, 3 for b, and −1 for c.

$$= \frac{-3 \pm \sqrt{9 + 8}}{4}$$

$$= \frac{-3 \pm \sqrt{17}}{4}$$

So the solutions are $\frac{-3 + \sqrt{17}}{4}$ and $\frac{-3 - \sqrt{17}}{4}$.

PRACTICE 2

Use the quadratic formula to solve.

a. $5s^2 - 8s = 3$

b. $3m^2 = 7 - 3m$

Sometimes in solving a quadratic equation, we need to approximate the solutions using a calculator. This is often true when solving applied problems.

EXAMPLE 3

Solve $2t^2 + 4t = 7$. Round the solutions to the nearest tenth.

Solution $2t^2 + 4t = 7$

$2t^2 + 4t - 7 = 0$ Write the equation in standard form.

$t = \frac{-b \pm \sqrt{b^2 - 4ac}}{2a}$ Use the quadratic formula.

$= \frac{-4 \pm \sqrt{(4)^2 - 4(2)(-7)}}{2(2)}$ Substitute 2 for a, 4 for b, and -7 for c.

$= \frac{-4 \pm \sqrt{16 + 56}}{4}$

$= \frac{-4 \pm \sqrt{72}}{4}$

$= \frac{-4 \pm 6\sqrt{2}}{4}$

$t = \frac{-4 + 6\sqrt{2}}{4}$ or $t = \frac{-4 - 6\sqrt{2}}{4}$

$= \frac{-2 + 3\sqrt{2}}{2}$ $= \frac{-2 - 3\sqrt{2}}{2}$

≈ 1.121 ≈ -3.121

So the solutions are approximately 1.1 and -3.1.

PRACTICE 3

Solve $3x^2 = 6x + 4$. Round the solutions to the nearest tenth.

EXAMPLE 4

An experienced clerk can complete a data-entry project in 2 hr less than a new clerk can working alone. If the clerks work together, it would take 3 hr to complete the data-entry project. How long would it take each clerk to complete the data-entry project working alone? Round your answer to the nearest tenth of an hour.

Solution If it takes the experienced clerk n hr to complete the project working alone, then she works at the rate $\frac{1}{n}$ of the project per hour. If it takes the new clerk $(n + 2)$ hours working alone, then he works at the rate $\frac{1}{n + 2}$ of the project per hour.

Let's make a table using the fact that the product of rate and time worked is the amount of work completed.

	Rate of Work	Time Worked	Part of the Project Completed
Experienced Clerk	$\frac{1}{n}$	3	$\frac{3}{n}$
New Clerk	$\frac{1}{n + 2}$	3	$\frac{3}{n + 2}$

PRACTICE 4

Working together, two pumps can empty a tank in 3 hr. Working alone, one pump takes 5 hr more than the other to carry out this task. Using a calculator, determine how long, rounded to the nearest tenth of an hour, it takes each pump working alone to empty the tank.

Because the total amount of work completed in 3 hr is 1 project, we can write the equation $\frac{3}{n} + \frac{3}{n+2} = 1$. Now, we solve for n.

$$\frac{3}{n} + \frac{3}{n+2} = 1$$

$$n(n+2)\left(\frac{3}{n} + \frac{3}{n+2}\right) = n(n+2) \cdot 1$$

$$\cancel{n}(n+2)\frac{3}{\cancel{n}} + n\cancel{(n+2)}\frac{3}{\cancel{n+2}} = n(n+2)$$

$$3(n+2) + 3n = n(n+2)$$

$$3n + 6 + 3n = n^2 + 2n$$

$$6n + 6 = n^2 + 2n$$

$$0 = n^2 - 4n - 6, \text{ or } n^2 - 4n - 6 = 0$$

Using the quadratic formula to solve for n, we get:

$$n = \frac{-(-4) \pm \sqrt{(-4)^2 - 4(1)(-6)}}{2(1)}$$

$$= \frac{4 \pm \sqrt{16 + 24}}{2}$$

$$= \frac{4 \pm \sqrt{40}}{2}$$

$$= \frac{4 \pm 2\sqrt{10}}{2}$$

$$= \frac{2(2 \pm \sqrt{10})}{2}$$

$$= 2 \pm \sqrt{10}$$

$$n = 2 + \sqrt{10} \approx 5.162 \quad \text{or} \quad n = 2 - \sqrt{10} \approx -1.162$$

Because the amount of time it takes the experienced clerk to complete the project must be positive, we take n to be approximately 5.2. So working alone, it takes the experienced clerk about 5.2 hr to complete the data entry project. Because the new clerk's time working alone is 2 hr more than the experienced clerk's time, the new clerk's time is about 7.2 hr, rounded to the nearest tenth of an hour.

In this chapter, we have used four methods of solving quadratic equations in the form $ax^2 + bx + c = 0$. These methods are factoring, the square root property, the quadratic formula, and completing the square. The following table lists when to use each method:

Method	When to Use Method
Factoring	• Use when the polynomial can be factored.
The Square Root Property	• Use when $b = 0$.
The Quadratic Formula	• Use when the first two methods do not apply.
Completing the Square	• Use only when specified. This method is easier to use when $a = 1$ and b is even.

19.3 Exercises

FOR EXTRA HELP MyMathLab MathXL PRACTICE WATCH READ REVIEW

A *Write each quadratic equation in standard form. Then, identify the values of a, b, and c.*

Quadratic Equation	Standard Form	$a =$	$b =$	$c =$
1. $2x^2 + 9x - 1 = 0$				
2. $x^2 - 3x + 4 = 0$				
3. $-x^2 + 3x = 8$				
4. $x^2 + 4 = -x$				
5. $x^2 + 2x = 3x - 8$				
6. $5y + 7 = 2y^2 + 1$				
7. $\frac{1}{3}y^2 - \frac{1}{2}y = -\frac{1}{4}$				
8. $-1.04 + 0.001x + 0.002x^2 = 0$				
9. $(2x + 1)(3x - 5) = 10$				
10. $(4x - 9)(x) = 2$				

Solve by using the quadratic formula.

11. $t^2 + 2t - 3 = 0$

12. $x^2 + 8x + 7 = 0$

13. $n^2 + 3n + 6 = 0$

14. $x^2 - 5x + 7 = 0$

15. $7 + 6x - x^2 = 0$

16. $1 - x - x^2 = 0$

17. $y^2 - 4y = 1$

18. $x^2 + 6x = -9$

19. $3p^2 - 4p + 1 = 0$

20. $5p^2 + p - 3 = 0$

21. $4x^2 - 6x = -1$

22. $6y^2 + 8y = 2$

23. $2n^2 = 7n + 30$

24. $4t^2 = -4t + 9$

25. $x^2 + 3x + 2 = 0$

26. $x^2 - 6x + 5 = 0$

27. $3 + x^2 + 4x = 0$

28. $2y^2 + 15 = 11y$

29. $3p^2 = 2p$

30. $x^2 = -3x$

31. $4n^2 + 3n - 1 = 0$

32. $s^2 - 7s + 2 = 0$

33. $t^2 + 5 = 7t$

34. $y^2 = 4y + 2$

35. $(n + 7)(n - 8) = 5$

36. $(w - 3)(w - 2) = 12$

37. $5(x - 1) = x(x + 2)$

38. $x(x + 1) = -2(x + 5)$

39. $\frac{x^2}{3} - x = -\frac{1}{2}$

40. $\frac{3}{n - 2} - \frac{1}{n - 1} = 2$

Mixed Practice

Solve by using the quadratic formula.

41. $4m + 5 + m^2 = 0$

42. $4k^2 + 6k = 2$

43. $-6 + s^2 = s$

44. $(n + 3)(n - 5) = -10$

Write each quadratic equation in standard form. Then, identify the values of a, b, and c.

Quadratic Equation	Standard Form	$a =$	$b =$	$c =$
45. $\frac{2}{5}d^2 - \frac{4}{5}d = \frac{1}{5}$				
46. $(3x + 2)(2x - 1) = 4$				

Applications

B *Solve by using the quadratic formula. Use a calculator, if necessary.*

47. The sum of the consecutive whole numbers from 1 through n is given by the formula $S = \frac{n(n + 1)}{2}$. If this sum is 91, what is n?

48. The formula $d = \frac{n(n - 3)}{2}$ is used to find the number of diagonals in a polygon with n sides. How many sides does a polygon with 20 diagonals have?

49. A lab coordinator and a lab technician work in a mathematics laboratory. Working alone, the lab technician can set up the lab in 20 min less time than the lab coordinator. Together, they can set up the lab in 40 min. How long, to the nearest minute, does it take each of them to set up the lab working alone?

50. A college copy center has two photocopying machines. The older copier takes 2 hr longer than the newer copier to do a job. Together, the two copiers would take 5 hr to do the job. If the older machine breaks down, how long to the nearest hour will it take the newer machine to do the job?

51. A car travels 20 mph faster than a bicycle. The car travels 40 mi in 4 hr less than it takes the bicycle. Find the speed of the car, rounded to the nearest mile per hour.

52. Researchers investigating a new medicine found that the percent p (in decimal form) of the medicine in the blood after t sec can be modeled by $p = 0.08t - 0.01t^2$. After how many seconds, to the nearest tenth of a second, does the blood contain 10% of the medicine?

53. On a recent date, the Fahrenheit temperature y in Anaheim, California, between 11:00 A.M. and 6 P.M. could be modeled by

$$y = -0.4x^2 + 2.7x + 59.4,$$

where x represents the number of hours after 11:00 A.M. At approximately what times was the temperature 61° F?

54. The equation $y = 0.1x^2 + 2.4x + 291$ models the United States population y (in millions) between 2003 and 2010, where x represents the number of years since 2003. According to this model, in which year was the population approximately 305 million? (*Source:* cms.gov)

• Check your answers on page A-39.

MINDStretchers

Mathematical Reasoning

1. The radicand in the quadratic formula, $b^2 - 4ac$, is called the *discriminant*. Consider the following equations:

$$x^2 - 8x + 16 = 0 \qquad x^2 + 2x - 8 = 0 \qquad x^2 - 3x + 5 = 0$$

a. Compute the discriminant for each equation.

b. How many real solutions does an equation with a positive discriminant have?
An equation with a zero discriminant?
An equation with a negative discriminant?

Research

2. The equation $x^3 - x^2 + 4x = 3$ is said to be a *cubic equation*, because the highest power of the variable is 3. Investigate on the Web or in your college library whether mathematicians have discovered a *cubic formula* for solving cubic equations, just as the quadratic formula solves quadratic equations. Summarize your findings in a short paragraph.

Patterns

3. Find the solutions of each quadratic equation. Then, fill in the table.

Equation	Sum of Solutions	Product of Solutions
$x^2 + 5x + 6 = 0$		
$x^2 - 4x - 12 = 0$		
$2x^2 + 3x + 1 = 0$		
$3x^2 - x - 2 = 0$		

Consider the quadratic equation $ax^2 + bx + c = 0$. Based on the results in the table, identify a relationship between the coefficients of the terms of a quadratic equation and the sum and product of its solutions.

19.4 Graphing Quadratic Equations in Two Variables

OBJECTIVES

- **A** To graph quadratic equations in two variables
- **B** To solve applied problems involving quadratic equations

Recall from Chapter 13 our discussion of linear equations in two variables. There we graphed equations of the form $y = mx + b$. In this section, we graph quadratic equations in two variables. These equations are of the form $y = ax^2 + bx + c$, where $a \neq 0$. Examples of quadratic equations in two variables are:

$$y = x^2 \qquad y = 5 - 2x^2 \qquad y = x^2 - 4x + 3$$

Equations of this type, when graphed, have varied real-world applications. For example, they may provide insight into how best to achieve major business goals, such as maximizing profit or minimizing cost.

Graphing Quadratic Equations in Two Variables

Just as with the graphing of linear equations, we can graph quadratic equations by choosing values of one variable and then computing the corresponding values of the other variable to make a table of values.

For instance, let's consider the graph of the equation $y = x^2$. First, we select values for x, and then find the corresponding values for y.

x	$y = x^2$	(x, y)
-2	$y = (-2)^2 = 4$	$(-2, 4)$
-1	$y = (-1)^2 = 1$	$(-1, 1)$
0	$y = (0)^2 = 0$	$(0, 0)$
1	$y = (1)^2 = 1$	$(1, 1)$
2	$y = (2)^2 = 4$	$(2, 4)$

Next, we plot these points on a coordinate plane, and then draw a smooth curve through them. The result is the graph of $y = x^2$.

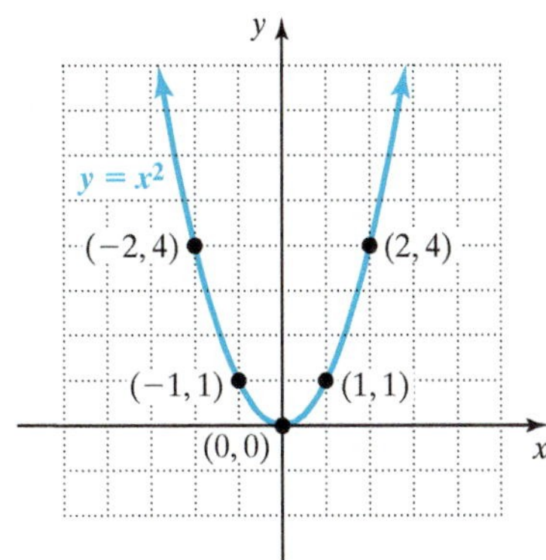

The graph of an equation of the form $y = ax^2 + bx + c$ is called a **parabola**. A parabola has the following characteristics:

- The graph is U-shaped, opening either upward or downward.
- The highest or lowest point of a parabola is called the **vertex** (or *turning point*) of the parabola.
- The graph has an axis of symmetry, which is a vertical line that passes through the vertex. This **axis of symmetry** divides the parabola into two parts that are mirror images of one another.
- The curve goes on indefinitely.

Consider the following graphs:

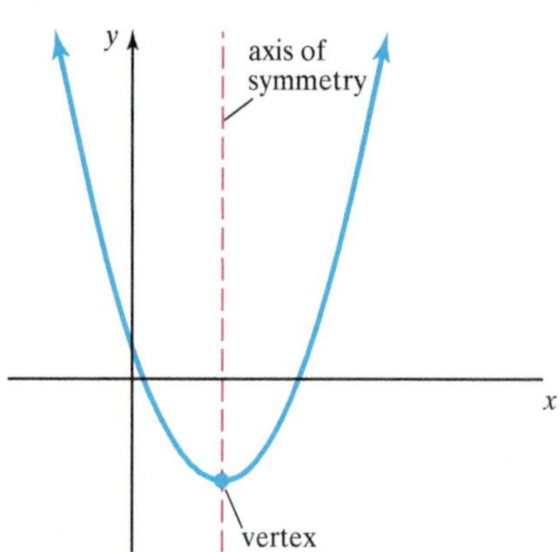

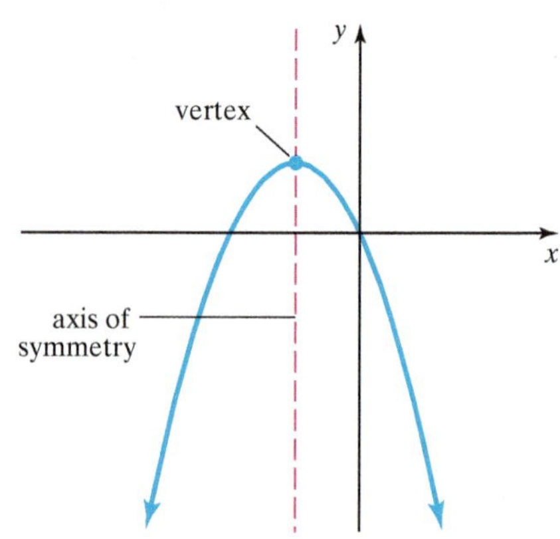

The parabola on the left opens upward, with the minimum y-value at the vertex. By contrast, the parabola on the right opens downward, with the maximum y-value at the vertex.

EXAMPLE 1

Graph: $y = 2x^2$

Solution First, we select values for x, and then find the corresponding values for y.

x	$y = 2x^2$	(x, y)
-2	$y = 2(-2)^2 = 2 \cdot 4 = 8$	$(-2, 8)$
-1	$y = 2(-1)^2 = 2 \cdot 1 = 2$	$(-1, 2)$
0	$y = 2(0)^2 = 2 \cdot 0 = 0$	$(0, 0)$
1	$y = 2(1)^2 = 2 \cdot 1 = 2$	$(1, 2)$
2	$y = 2(2)^2 = 2 \cdot 4 = 8$	$(2, 8)$

Next, we plot the points on a coordinate plane, and then draw a smooth curve through them. The result is the graph of the parabola $y = 2x^2$.

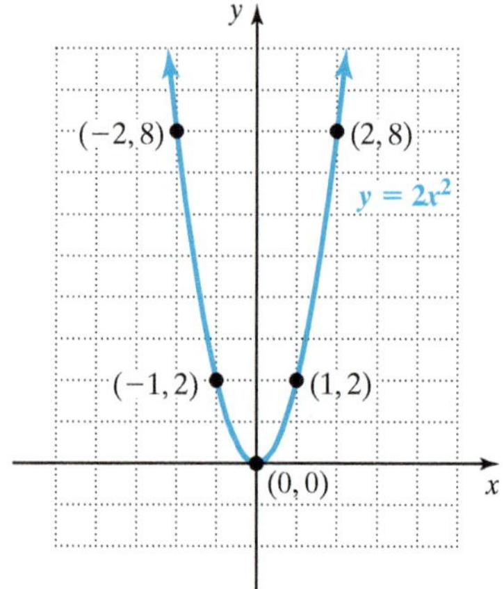

PRACTICE 1

Graph: $y = -2x^2$

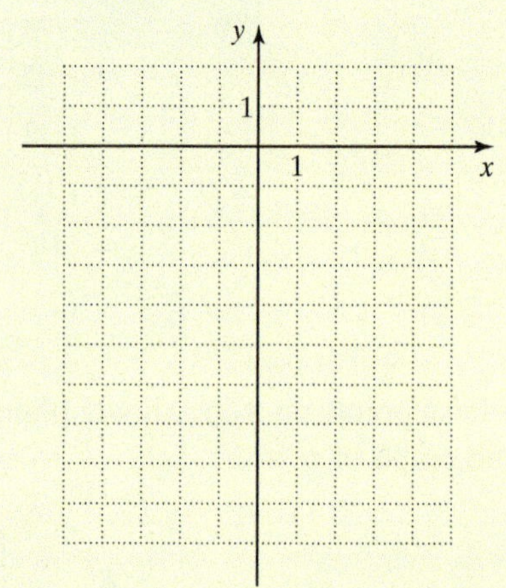

What are the differences and similarities between the graphs of $y = 2x^2$ and $y = -2x^2$?

Note that in Example 1, the vertex is the point $(0, 0)$ and the axis of symmetry is the y-axis. This is always true for the graph of any quadratic equation of the form $y = ax^2$.

Now, let's look at the graphs of equations that are of the more general form $y = ax^2 + bx + c$. Although we can always make a table of x- and y-values to graph a quadratic equation, we consider two other methods of graphing quadratic equations. One method is *graphing the vertex and several points on either side of the vertex.* The other method is *graphing the vertex and the x- and y-intercepts.*

Recall that the vertex is the point on the parabola where the curve turns and that it lies on the axis of symmetry.

The Equation of the Axis of Symmetry of a Parabola

For a parabola given by the quadratic equation $y = ax^2 + bx + c$, the equation of the axis of symmetry is $x = -\frac{b}{2a}$.

Since the axis of symmetry is a vertical line, its equation is of the form $x =$ a constant. From our discussion of vertical lines, we know that every point on the axis of symmetry, including the vertex, has x-coordinate $-\frac{b}{2a}$. We can use this fact to find the coordinates of the vertex as follows:

To Find the Vertex of a Parabola

For a parabola given by the quadratic equation $y = ax^2 + bx + c$,

- The x-coordinate of the vertex is $-\frac{b}{2a}$.
- The y-coordinate of the vertex is found by substituting $-\frac{b}{2a}$ for x into the equation and then computing y.

Now, let's use the vertex and several points on either side of the vertex to graph a quadratic equation.

EXAMPLE 2

Graph: $y = 5 - 2x^2$

Solution Writing the equation in standard form, $y = -2x^2 + 0x + 5$, we see that $a = -2$ and $b = 0$. The x-coordinate of the vertex is:

$$-\frac{b}{2a} = \frac{0}{2(-2)} = 0$$

Substituting 0 for x into the equation $y = 5 - 2x^2$, we find the y-coordinate of the vertex:

$$y = 5 - 2x^2 = 5 - 2(0)^2 = 5$$

So the vertex is $(0, 5)$.

PRACTICE 2

Graph: $y = -3x^2 + 1$

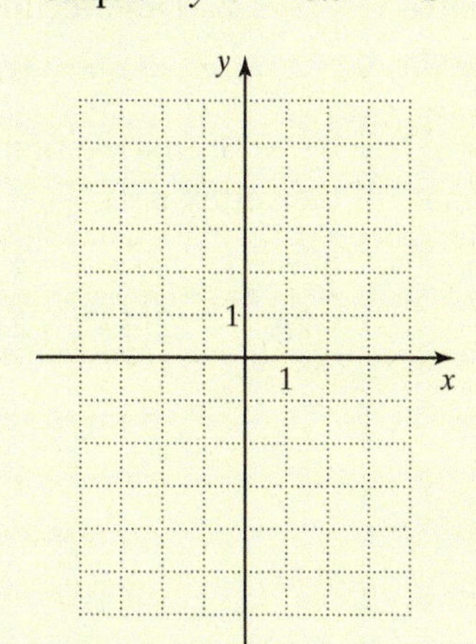

EXAMPLE 2 (continued)

Next, we find points on either side of the vertex.

x	$y = 5 - 2x^2$	(x, y)
-2	$y = 5 - 2x^2 = 5 - 2(-2)^2 = -3$	$(-2, -3)$
-1	$y = 5 - 2x^2 = 5 - 2(-1)^2 = 3$	$(-1, 3)$
0	$y = 5 - 2x^2 = 5 - 2(0)^2 = 5$	$(0, 5)$ ← vertex
1	$y = 5 - 2x^2 = 5 - 2(1)^2 = 3$	$(1, 3)$
2	$y = 5 - 2x^2 = 5 - 2(2)^2 = -3$	$(2, -3)$

Now, we plot the points and draw a smooth curve through them. The result is the graph of the parabola $y = 5 - 2x^2$. Note that the y-axis, with equation $x = 0$, is the axis of symmetry for the parabola.

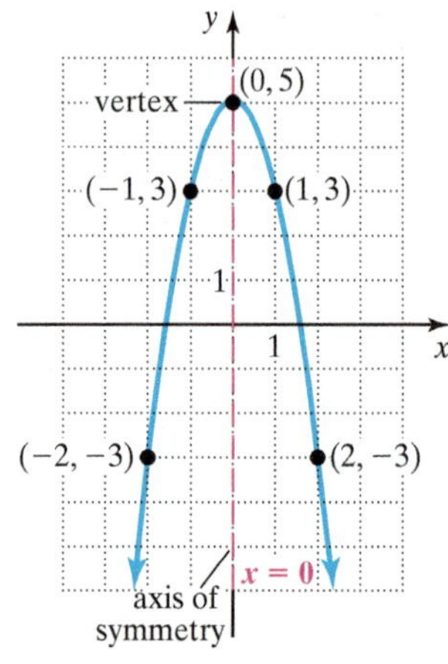

In Example 1, we saw that for $y = 2x^2$, where a is positive, the parabola opens upward. On the other hand, in Example 2, the graph of $y = 5 - 2x^2$, where a is negative, opens downward. Knowing the sign of a is useful in graphing $y = ax^2 + bx + c$.

TIP When a is positive, the parabola opens upward. When a is negative, the parabola opens downward.

Finally, we consider the graphing of quadratic equations of the form $y = ax^2 + bx + c$ using both the vertex and the x- and y-intercepts.

EXAMPLE 3

Consider $y = x^2 - 4x + 3$.

a. Sketch its graph. **b.** Describe the graph.

Solution

a. To graph $y = x^2 - 4x + 3$, first we find the vertex. The x-coordinate of the vertex is:

$$-\frac{b}{2a} = -\frac{-4}{2(1)} = \frac{4}{2} = 2$$

PRACTICE 3

Consider $y = -x^2 + 2x + 3$.

a. Sketch its graph.

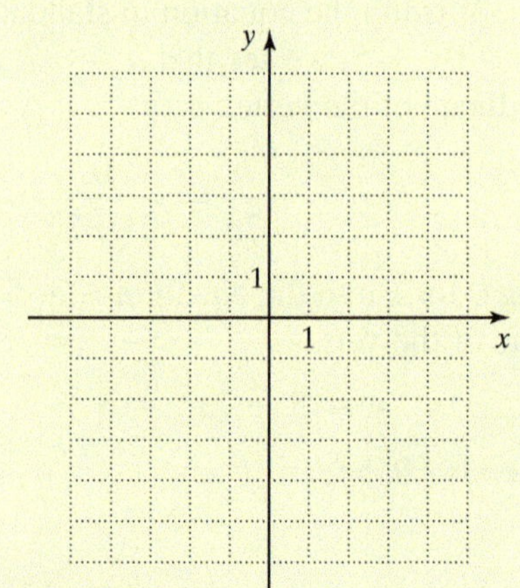

Substituting 2 for x, we get:

$$y = x^2 - 4x + 3 = (2)^2 - 4(2) + 3 = 4 - 8 + 3 = -1$$

So the vertex of the parabola is $(2, -1)$, and the axis of symmetry is $x = 2$.

Next, we find the x- and y-intercepts. Recall that to find the x-intercept of a graph, we let $y = 0$ and solve for x.

$$y = x^2 - 4x + 3$$
$$0 = x^2 - 4x + 3$$
$$0 = (x - 3)(x - 1)$$
$$x - 3 = 0 \quad \text{or} \quad x - 1 = 0$$
$$x = 3 \qquad\qquad x = 1$$

So the x-intercepts are $(3, 0)$ and $(1, 0)$.

Similarly, to find the y-intercept, we let $x = 0$ and solve for y.

$$y = x^2 - 4x + 3$$
$$= (0)^2 - 4(0) + 3 = 3$$

The y-intercept is therefore $(0, 3)$.

Finally, we plot the vertex, the x-intercepts, and the y-intercept on a coordinate plane. Then, we draw a smooth curve through the points. The result is the graph of the equation $y = x^2 - 4x + 3$.

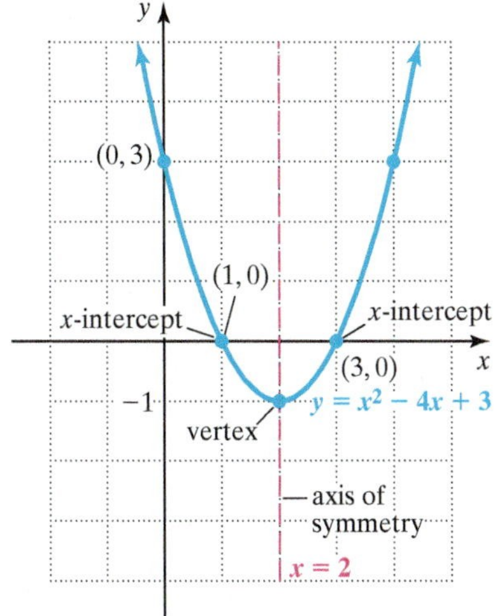

b. The parabola $y = x^2 - 4x + 3$ opens upward since $a = 1$ is positive. The curve turns at the vertex $(2, -1)$, the lowest point of the graph. The equation of the axis of symmetry is $x = 2$.

b. Describe the graph.

Can you explain the relationship between the y-intercept of a parabola and the constant term of the corresponding quadratic equation in the form $y = ax^2 + bx + c$?

EXAMPLE 4

A Roman candle is shot straight upward from the ground with an initial velocity of 96 ft/sec. After t sec, its distance s (in feet above the ground) is given by the equation $s = 96t - 16t^2$. Graph the equation. Find the maximum height reached by the Roman candle.

Solution Let's graph the equation $s = 96t - 16t^2$ by using the vertex and t- and s-intercepts. First, we find the vertex. The t-coordinate of the vertex is:

$$-\frac{b}{2a} = -\frac{96}{2(-16)} = 3$$

Substituting 3 for t, we get the value of s as follows:

$$s = 96t - 16t^2 = 96(3) - 16(3)^2 = 288 - 144 = 144$$

So the vertex has coordinates $(3, 144)$. Note that the line of symmetry is $t = 3$.

Next, we find the t- and s-intercepts of the graph. To find the t-intercept, we let $s = 0$ and solve for t.

$$s = 96t - 16t^2$$
$$0 = 96t - 16t^2$$
$$0 = 16t(6 - t)$$
$$16t = 0 \quad \text{or} \quad 6 - t = 0$$
$$t = 0 \qquad\qquad t = 6$$

So the t-intercepts are $(0, 0)$ and $(6, 0)$. Similarly, to find the s-intercept, we let $t = 0$ and solve for s.

$$s = 96t - 16t^2$$
$$s = 96(0) - 16(0)^2 = 0$$

The s-intercept is therefore $(0, 0)$.

Finally, we plot the vertex and the t- and s-intercepts on a coordinate plane to draw a smooth curve through the points.

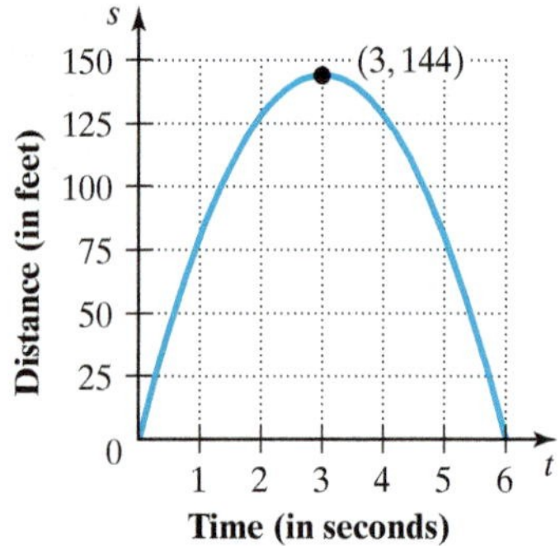

Since the maximum s-value is 144, the maximum height reached by the Roman candle is 144 ft. Notice that the distance above the ground increases between 0 and 3 sec and then decreases between 3 and 6 sec after the Roman candle is shot.

PRACTICE 4

The total profit or loss P (in dollars) made by a street vendor is given by the equation $P = x^2 - 10x$, where x is the number of items sold. Graph this equation. Find the number of items sold that results in the most money lost.

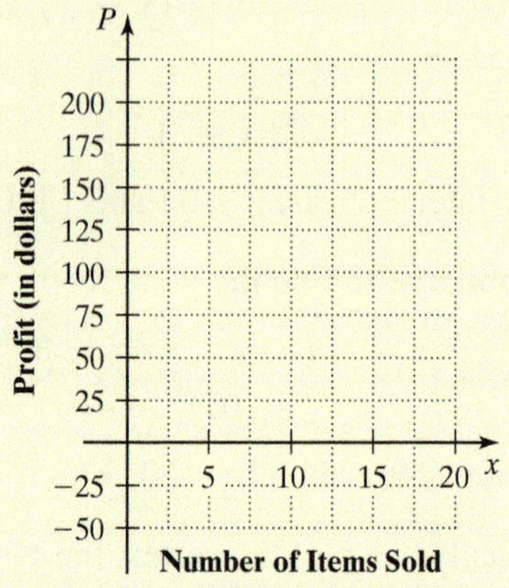

Note that in Example 4, the graph of $s = 96t - 16t^2$ does not go below the t-axis. Why is this the case? Explain.

19.4 Exercises

FOR EXTRA HELP MyMathLab MathXL PRACTICE WATCH READ REVIEW

Mathematically Speaking

Fill in each blank with the most appropriate term or phrase from the given list.

positive	vertex	negative
axis of symmetry	equation of the axis of symmetry	graph of an equation
equation		

1. The ________________ of the form $y = ax^2 + bx + c$ is called a parabola.

2. The highest or lowest point of a parabola is said to be its ________________.

3. The vertical line passing through a parabola's vertex is called its ________________.

4. For the graph of $y = ax^2 + bx + c$, the ________________________ is $x = -\frac{b}{2a}$.

5. The parabola corresponding to the equation $y = ax^2 + bx + c$ opens upward when a is ________________.

6. The parabola corresponding to the equation $y = ax^2 + bx + c$ opens downward when a is ________________.

A *Graph a parabola with the following characteristics.*

7. Passes through the points $(-2, 5)$, $(-1, 2)$, $(0, 1)$, $(1, 2)$, and $(2, 5)$.

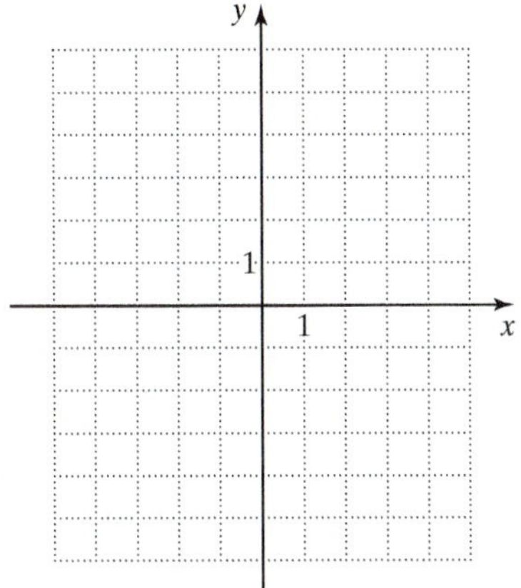

8. Passes through the points $(-2, -3)$, $(-1, 0)$, $(0, 1)$, $(1, 0)$, and $(2, -3)$.

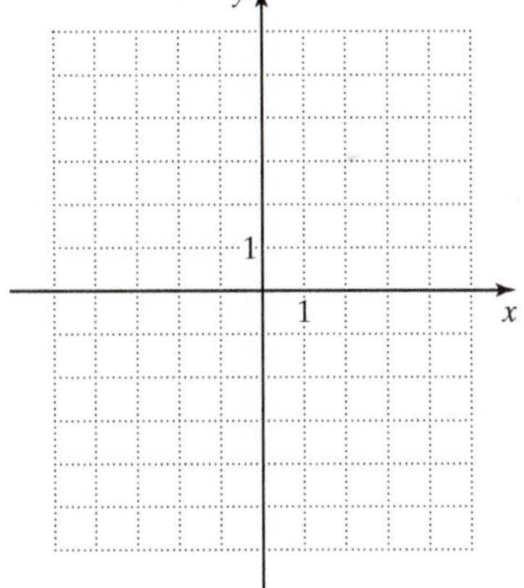

9. Has vertex $(0, 0)$ and also passes through the points $(-2, 4)$, $(-1, 1)$, $(1, 1)$, and $(2, 4)$.

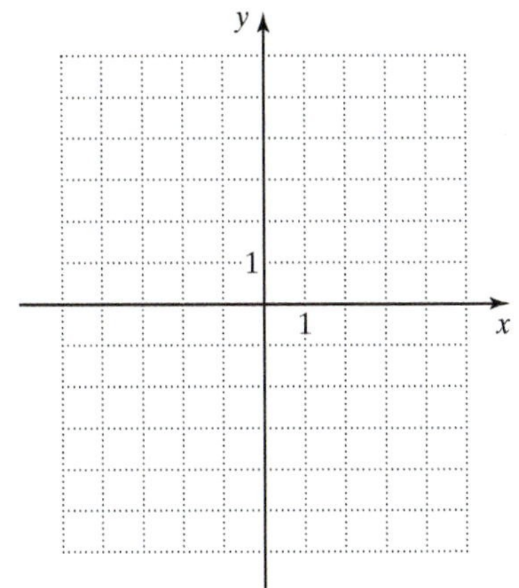

10. Has vertex $(0, 3)$ and also passes through the points $(2, -1)$, $(-2, -1)$, $(3, -6)$, and $(-3, -6)$.

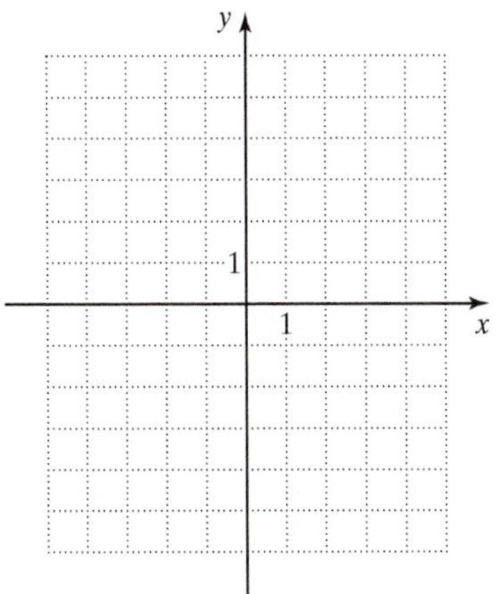

11. Has vertex $(2, -1)$, y-intercept $(0, 3)$, and x-intercepts $(1, 0)$ and $(3, 0)$.

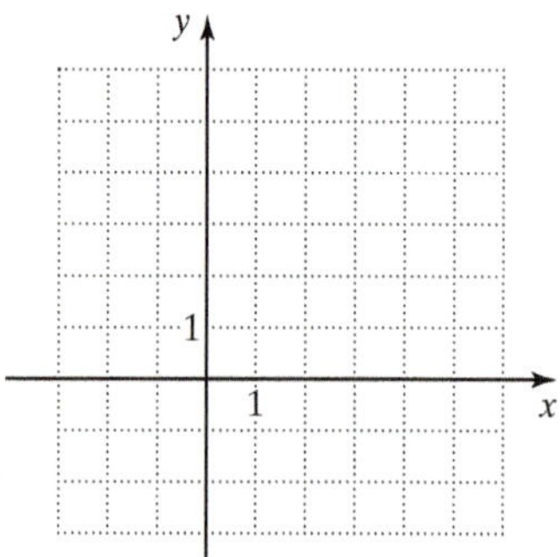

12. Has vertex $(-3, -1)$, y-intercept $(0, 8)$, and passes through the point $(-6, 8)$.

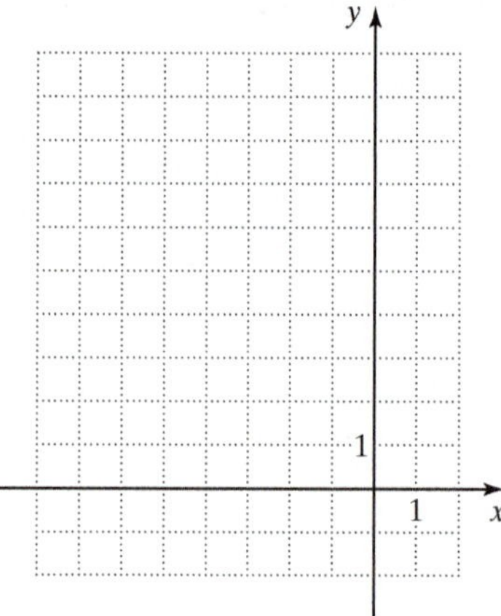

Find the axis of symmetry and the vertex of the graph of each equation.

13. $y = \frac{1}{2}x^2$

14. $y = \frac{1}{4}x^2$

15. $y = 5 - 4x + x^2$

16. $y = 2x^2 - 8x - 1$

Sketch the graph of each equation.

17. $y = 3x^2$

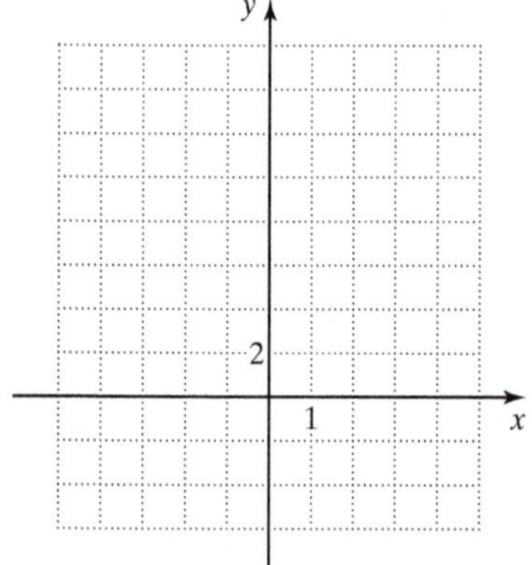

18. $y = 4x^2$

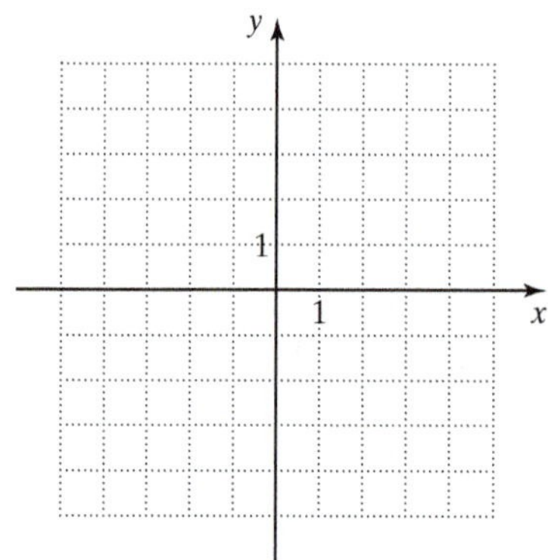

19. $y = -3x^2$

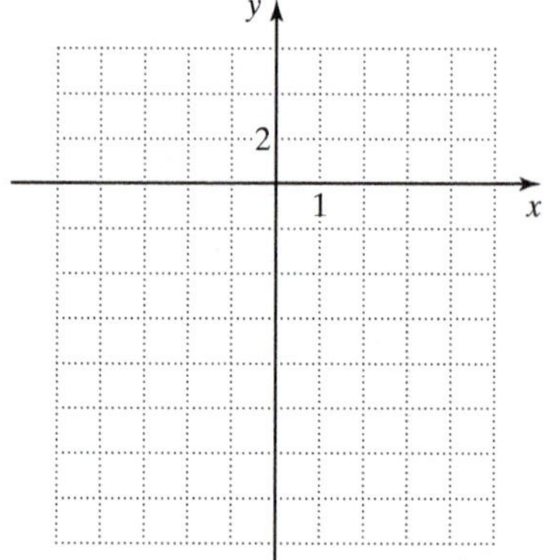

20. $y = -5x^2$

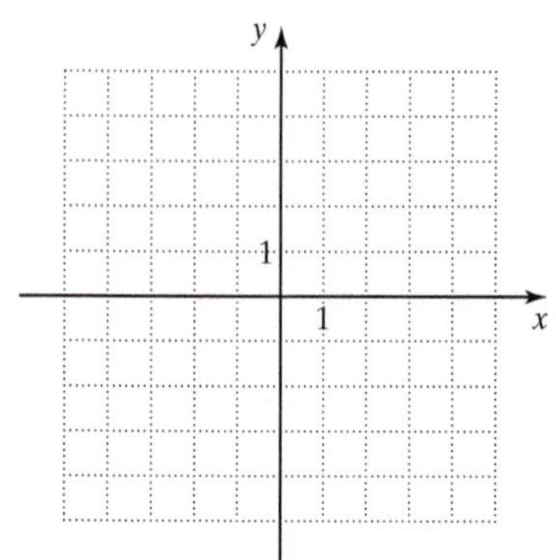

21. $y = 4 - x^2$

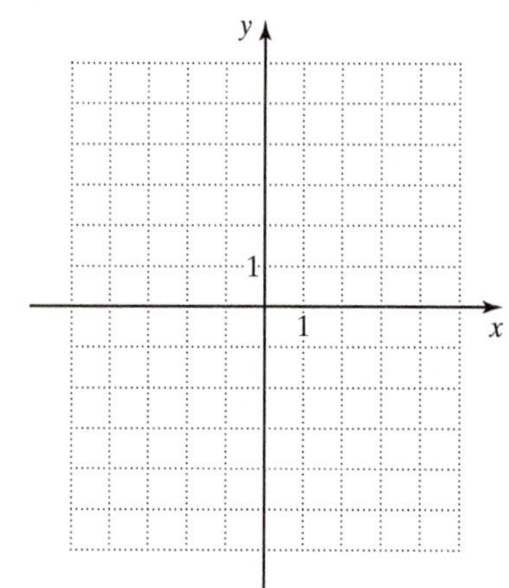

22. $y = 3 - x^2$

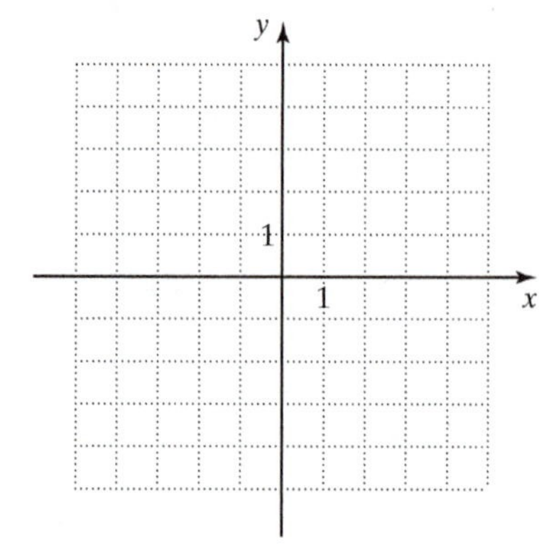

23. $y = x^2 - 9$

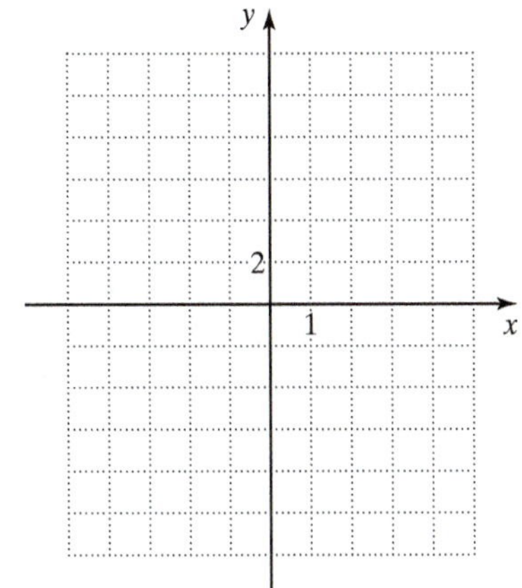

24. $y = x^2 - 4$

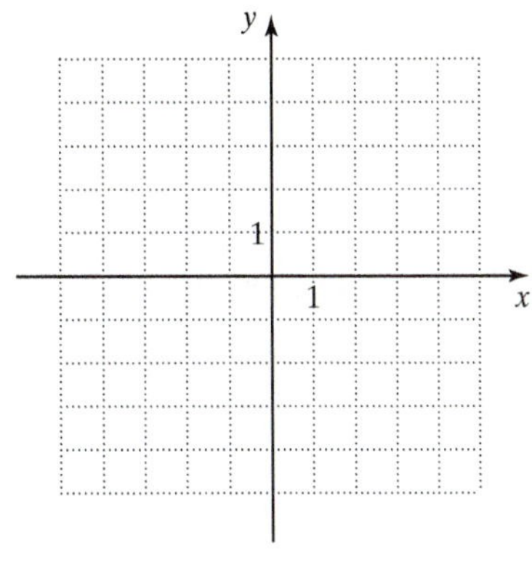

25. $y = x^2 - 2x$

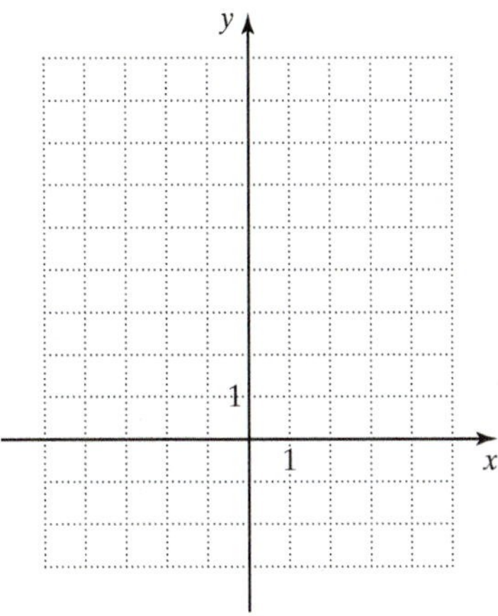

26. $y = x^2 - 4x$

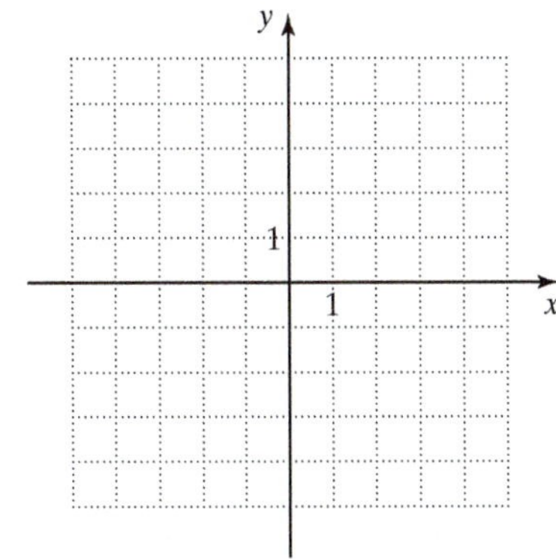

27. $y = -x^2 + 4x$

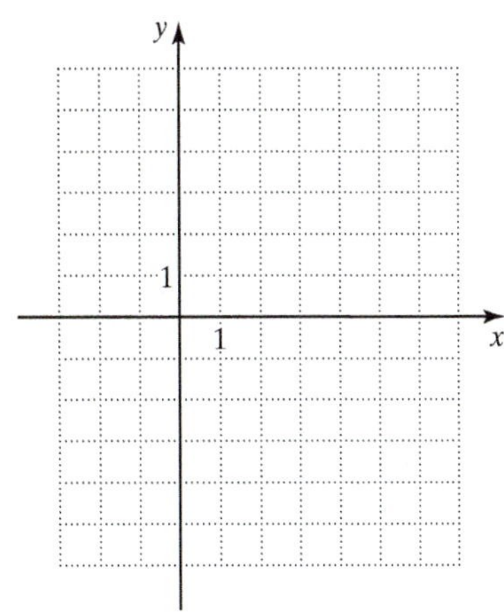

28. $y = -x^2 + 2x$

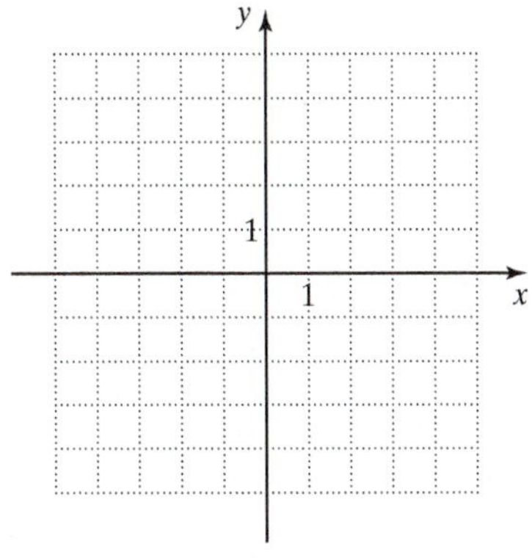

29. $y = -x^2 + 3x + 2$

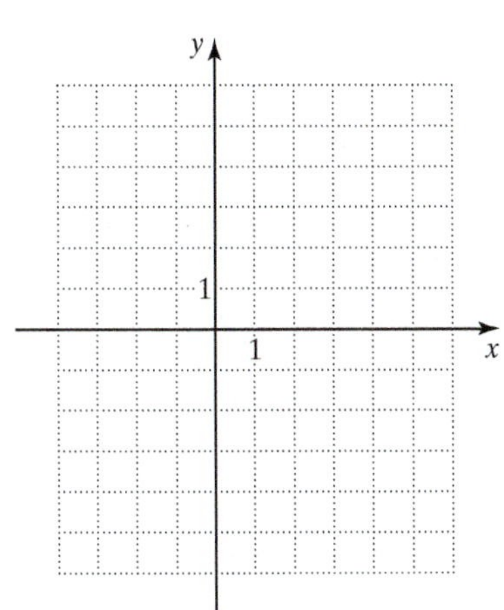

30. $y = -x^2 + 5x - 4$

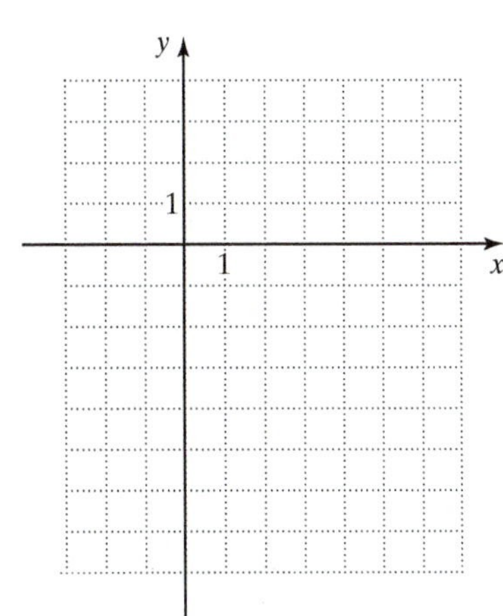

31. $y = x^2 - x - 6$

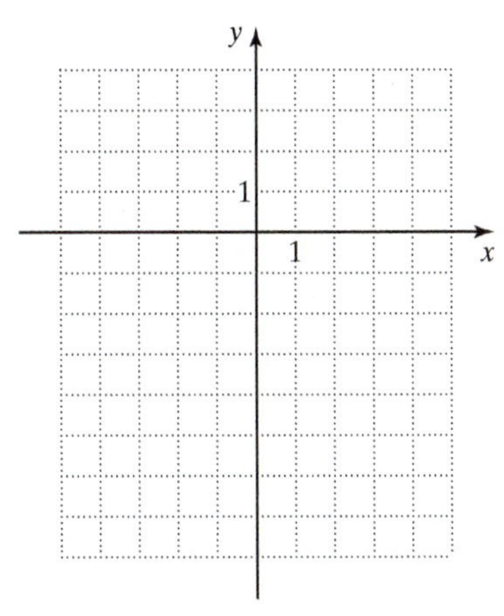

32. $y = x^2 - 4x + 4$

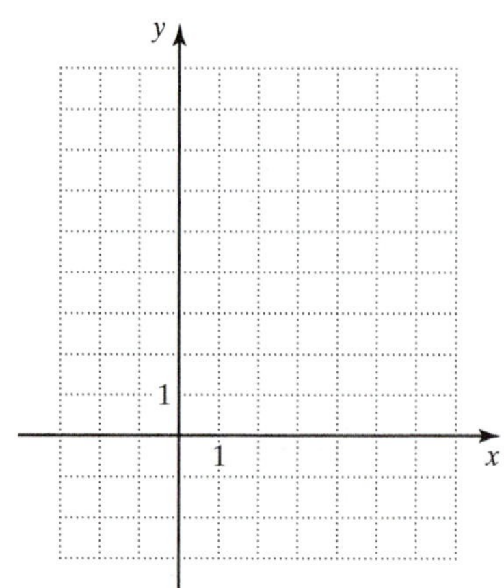

33. $y = 3x^2 - 4x + 1$

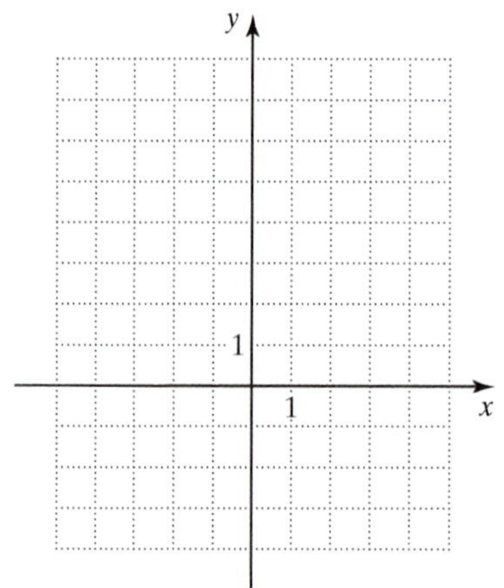

34. $y = 2x^2 + 3x - 2$

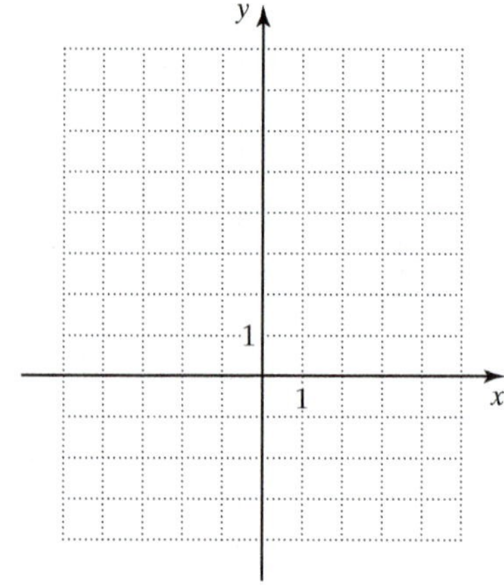

35. $y = (x + 2)^2$

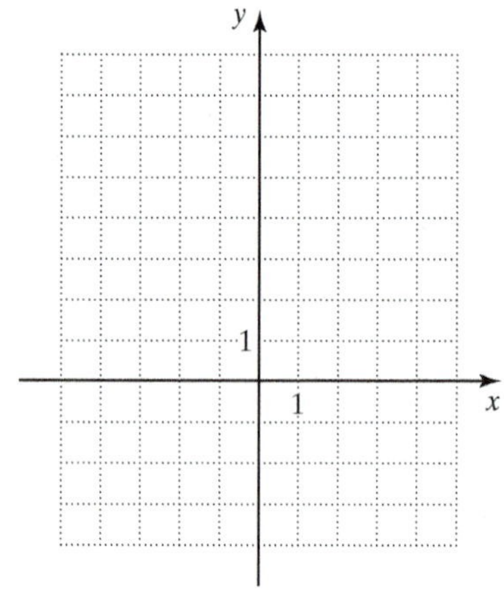

36. $y = (x + 3)^2$

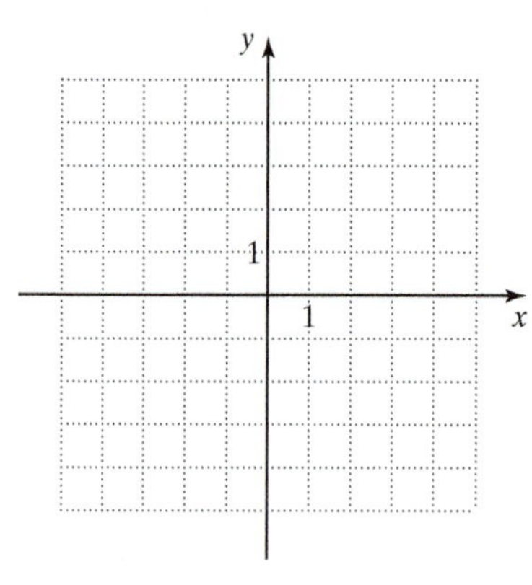

37. $y = (x - 2)^2$

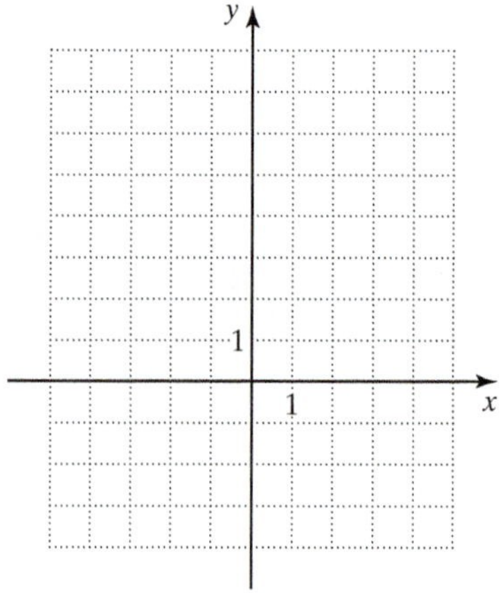

38. $y = (x - 3)^2$

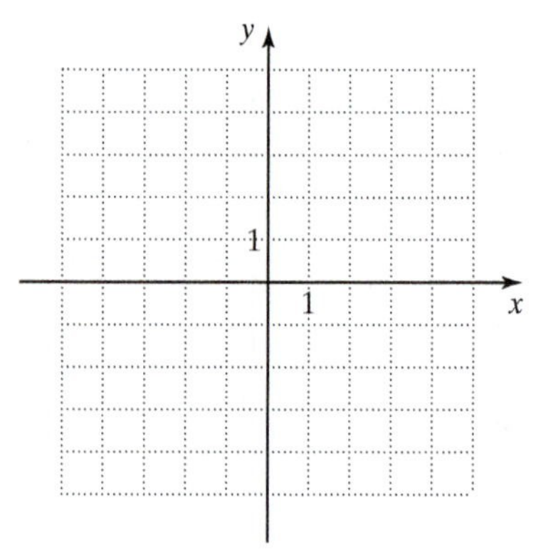

39. $y = 2 - 3x^2$

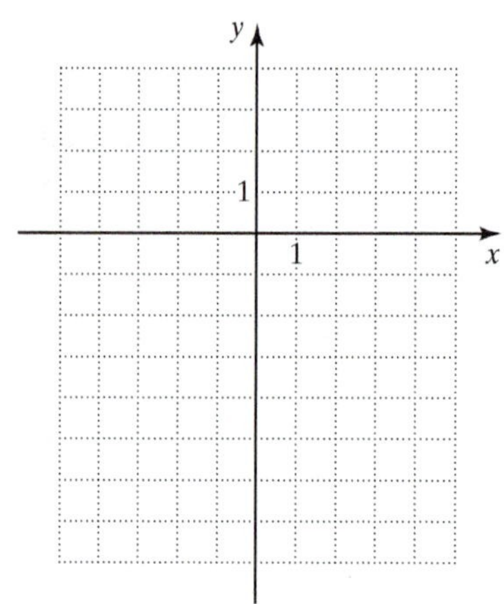

40. $y = 3x - 2x^2$

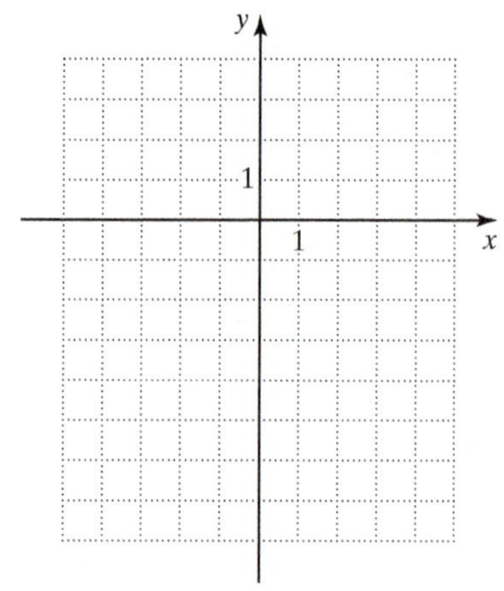

41. $y = -\frac{1}{2}x^2 - 5$

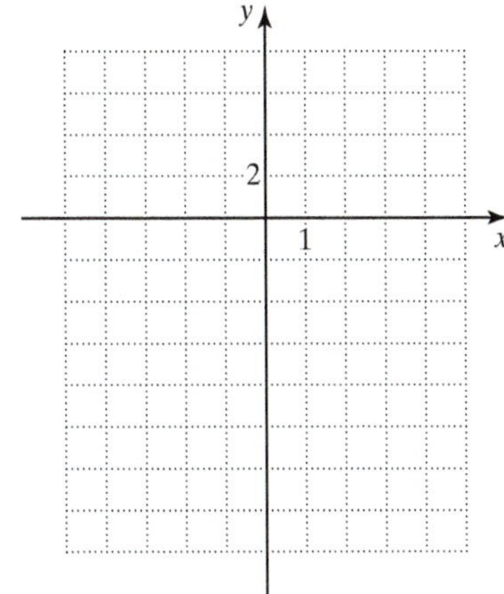

42. $y = -\frac{1}{4}x^2 - 4$

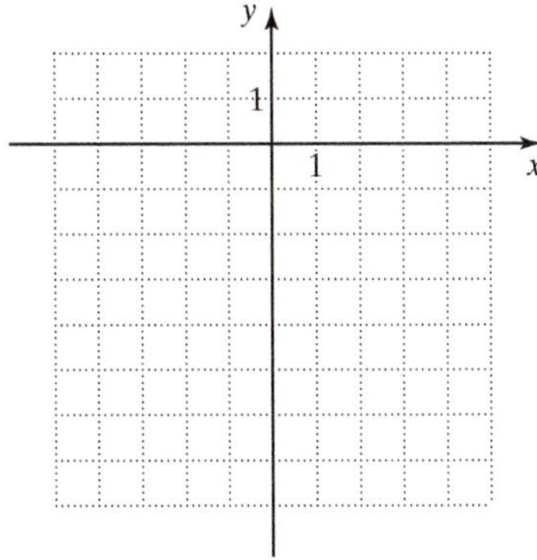

43. $y = \frac{3}{4}x^2 + 2$

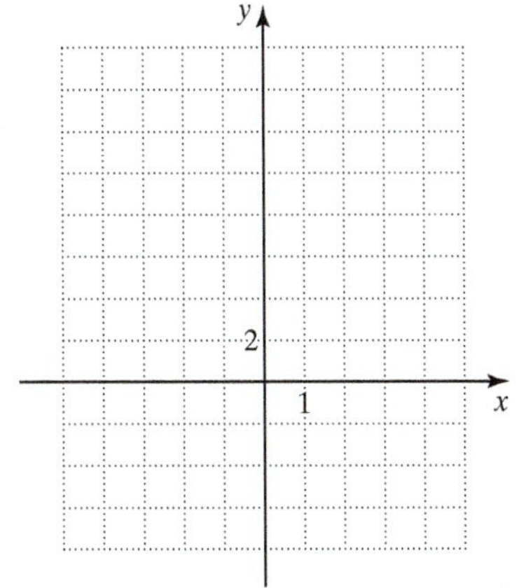

44. $y = \frac{1}{2}x^2 + 1$

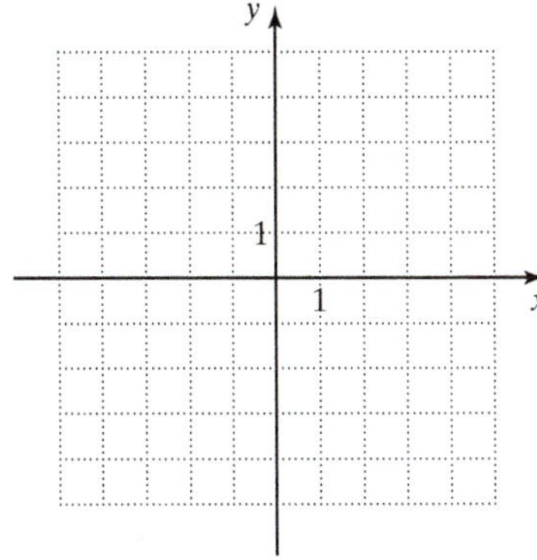

Mixed Practice

Solve.

45. Graph a parabola that has vertex $(0, -5)$ and also passes through the points $(-2, -1), (-1, -4), (1, -4), (2, -1)$.

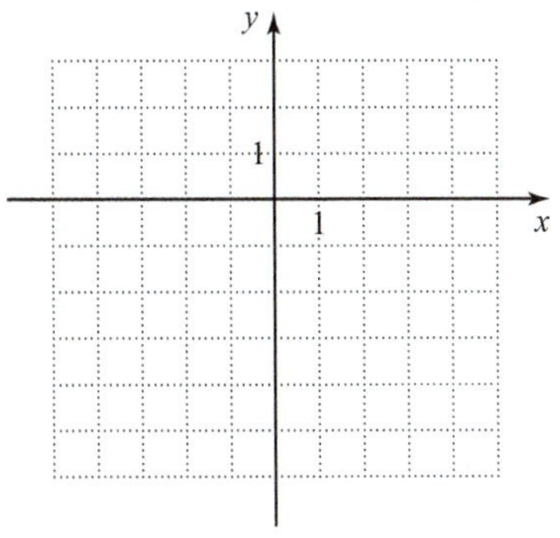

46. Find the axis of symmetry and the vertex of the graph of the equation $y = 4 - 3x^2 + 6x$.

Sketch the graph of each equation.

47. $y = 2x^2 - 3x + 1$

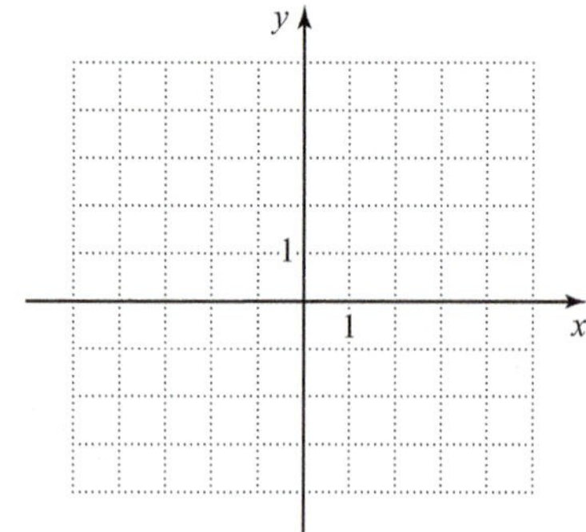

48. $y = -x^2 + 2x$

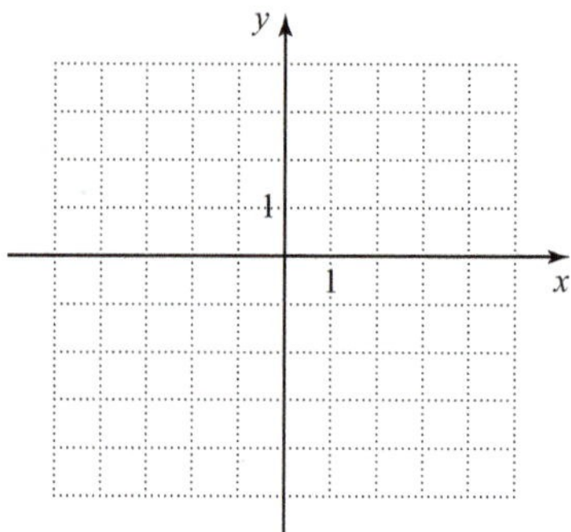

49. $y = -x^2 - 3x + 2$

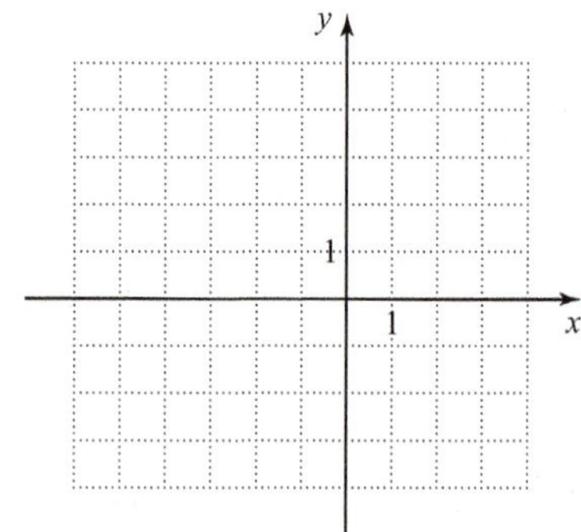
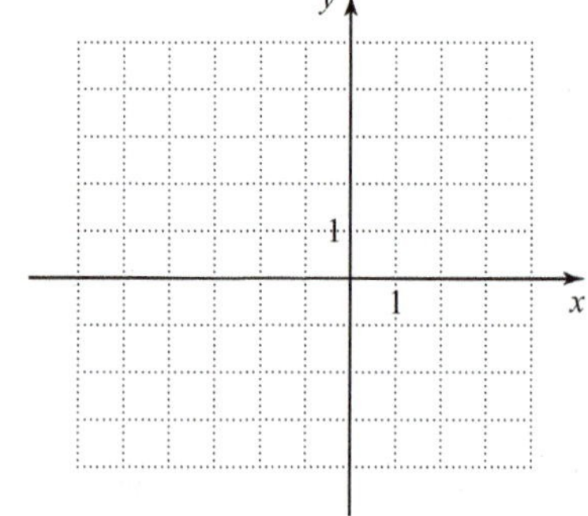

50. $y = \frac{1}{4}x^2 - 2$

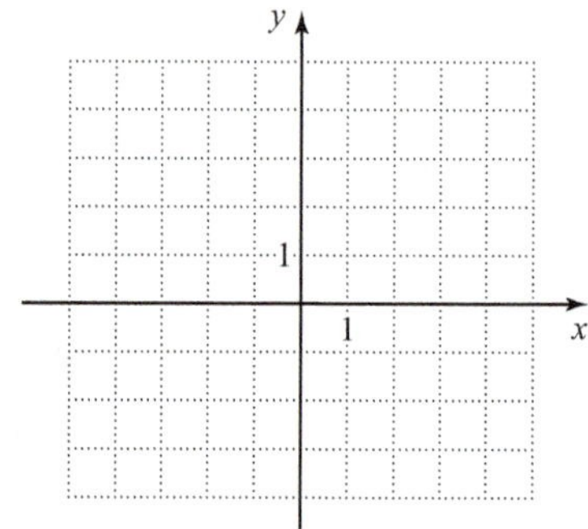

Applications

B *Solve.*

51. A company's annual payroll (in millions of dollars) in various years is modeled by $0.4x^2 - 2.4x + 23.9$, where x represents the number of years after 2005.

a. Graph the following equation:
$y = 0.4x^2 - 2.4x + 23.9$

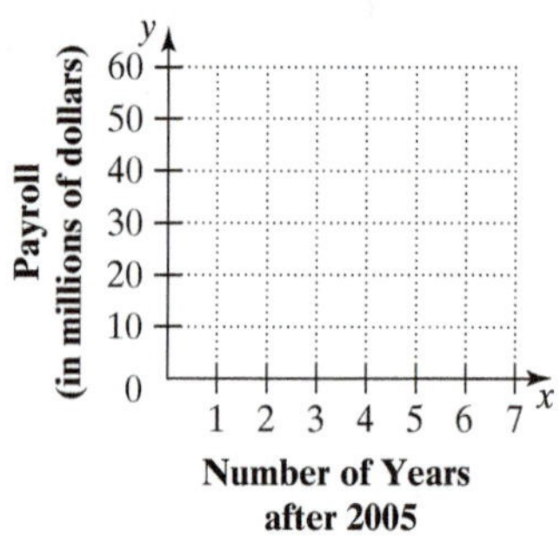

b. According to this model, in what year was the payroll minimized?

c. In a sentence, describe the trend in the team's payroll.

52. The manager of a coat factory found that the unit cost y (in dollars) of producing x coats can be approximated using the expression $0.002x^2 - 2.7x + 1000$.

a. Graph the equation $y = 0.002x^2 - 2.7x + 1000$.

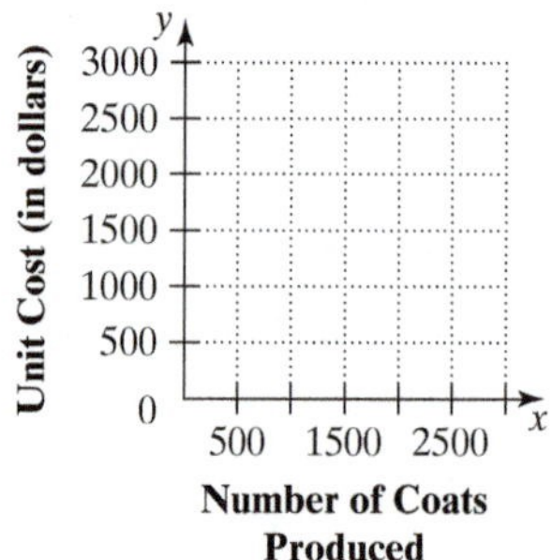

b. According to this model, the unit cost of producing coats will be lowest if how many coats are produced?

c. Describe how the unit cost varies with the number of coats produced.

53. The daily profit y of a company varies with the number of products x it sells per day. The daily profit can be approximated by the expression $-0.1x^2 + 10x$.

a. Graph the equation $y = -0.1x^2 + 10x$.

b. How many products must the company sell per day to maximize the profit?

c. Describe how the profit changes as sales increase.

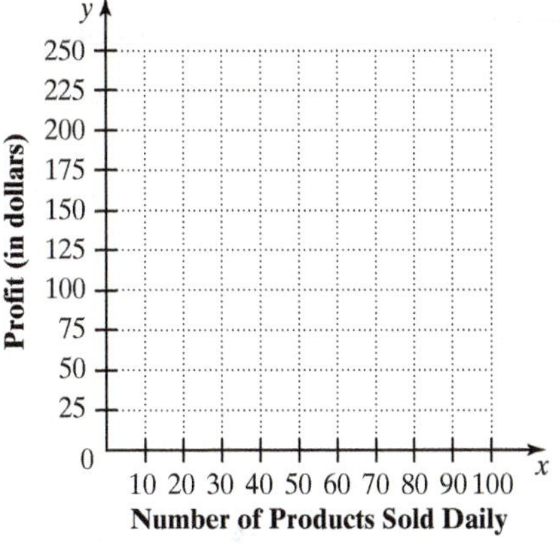

54. A child throws a ball straight upward from a height of 6 ft with an initial velocity of 80 ft/sec. The ball's height h ft above the ground after t sec is modeled by the following equation:

$$h = 6 + 80t - 16t^2$$

a. Graph this equation.

b. Will the ball reach a height of 4 ft above the ground?

c. Describe how the height of the ball changes over time.

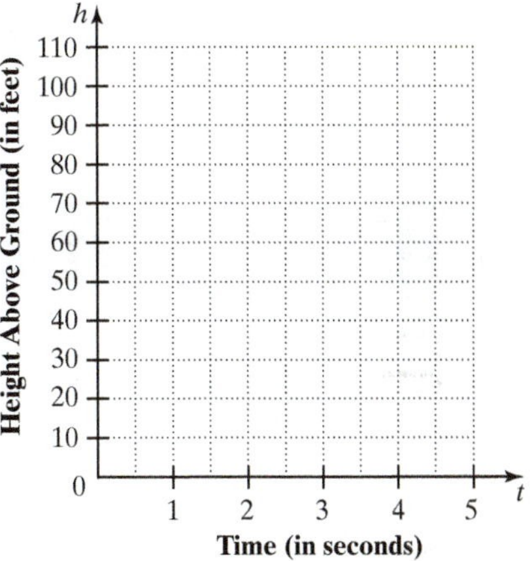

55. The manager of a store determines that the store can sell x items at a price of $100 - \frac{1}{4}x$ dollars per item.

a. Write a formula for the total revenue R that these sales generate.

b. Using the formula in part (a), complete a table of the revenue generated by the sale of 0, 100, 200, 300, and 400 items.

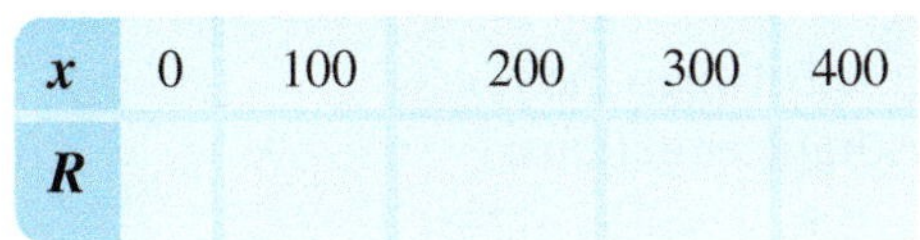

x	0	100	200	300	400
R					

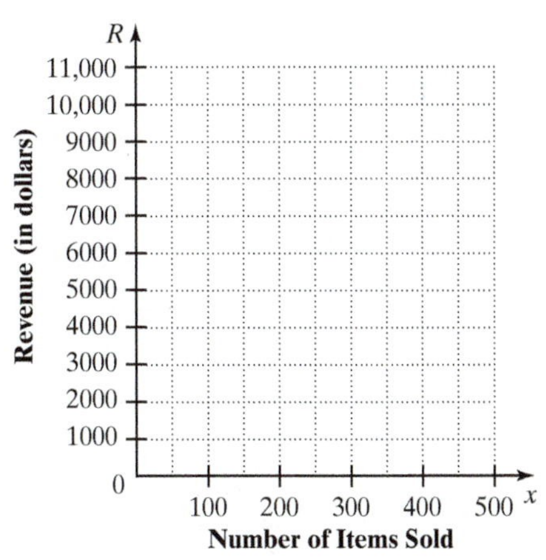

c. Graph the formula.

d. How many items should the store sell to maximize the revenue?

e. What is that maximum revenue?

56. The formula $h = 147t - 4.9t^2$ can be used to predict the height in meters of a small rocket launched straight upward from the ground with an initial velocity of 147 m/sec. The number of seconds after launch is represented by t.

a. Using the given formula, complete the table.

t	0	5	10	15	20	25	30
h							

b. From the preceding table, plot the ordered pairs (t, h). Draw a smooth curve passing through the points.

c. What are the t- and h-intercepts of the graph?

d. Approximate the maximum height reached by the rocket.

e. After how many seconds does the rocket hit the ground?

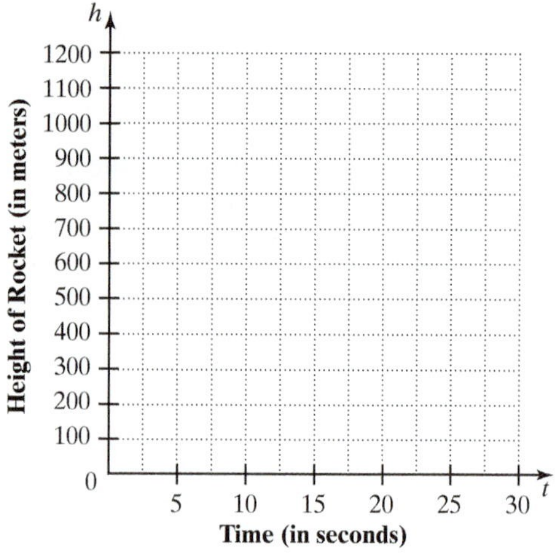

• Check your answers on page A-40.

MINDStretchers

Patterns

1. Consider the graph of $y = x^2$.

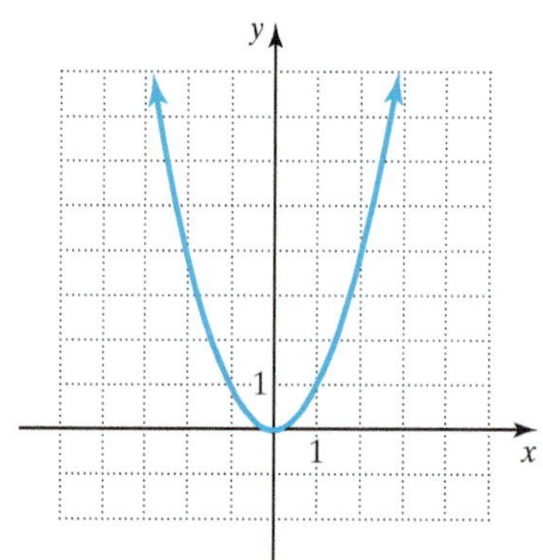

In a few sentences, describe any relationship that you observe between the graph of $y = x^2$ and the graph of each of the following quadratic equations.

a. $y = (x + 1)^2$

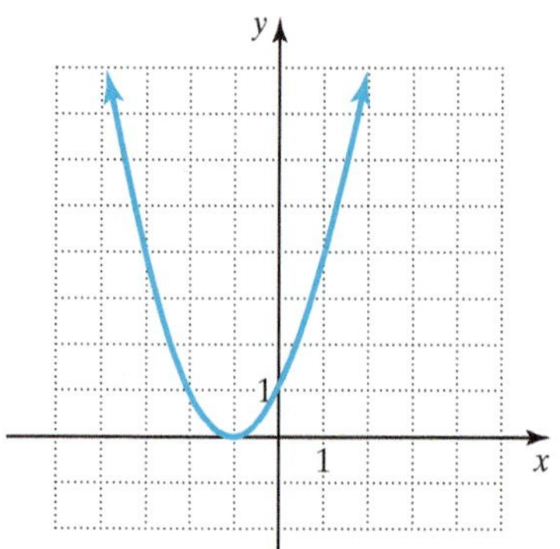

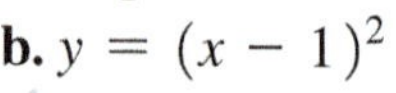

b. $y = (x - 1)^2$

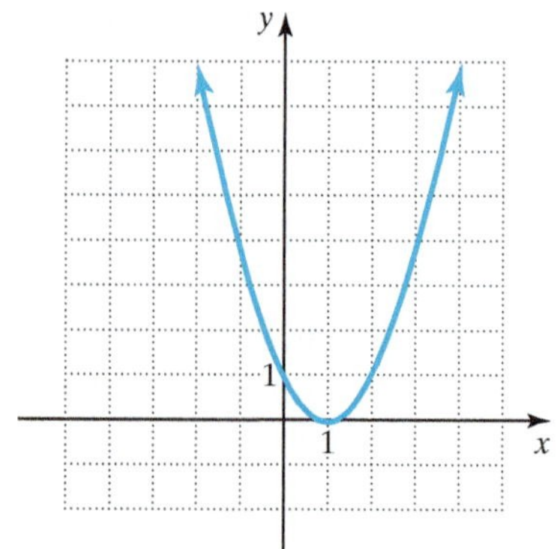

c. $y = x^2 + 1$

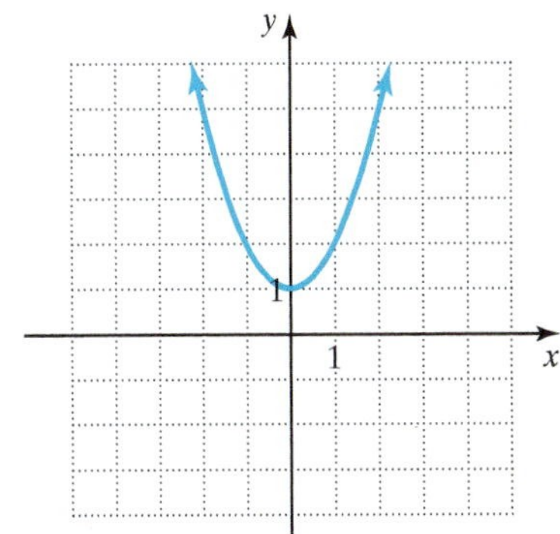

d. $y = x^2 - 1$

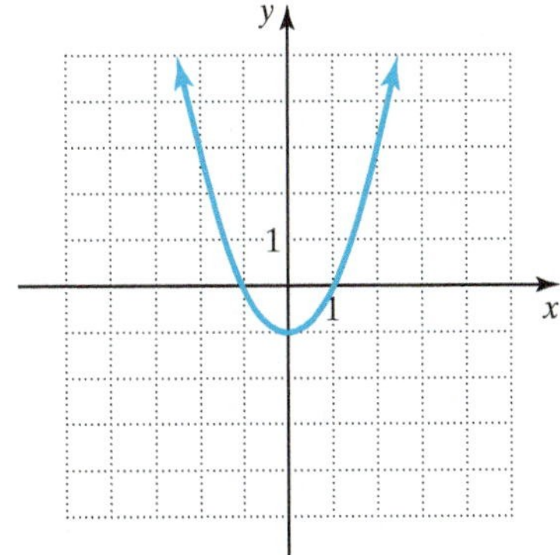

Research

2. Using the Web or your college library, find five situations in which parabolas are important in the real world. Write a sentence about each example.

Mathematical Reasoning

3. For the three graphs shown to the right, identify the graph(s) that fit each description below. Each graph is of the form $y = ax^2 + bx + c$.

a. $a > 0$

b. $a < 0$

c. The graph with the *greatest* value of $|a|$

d. The graph with the *least* value of $|a|$

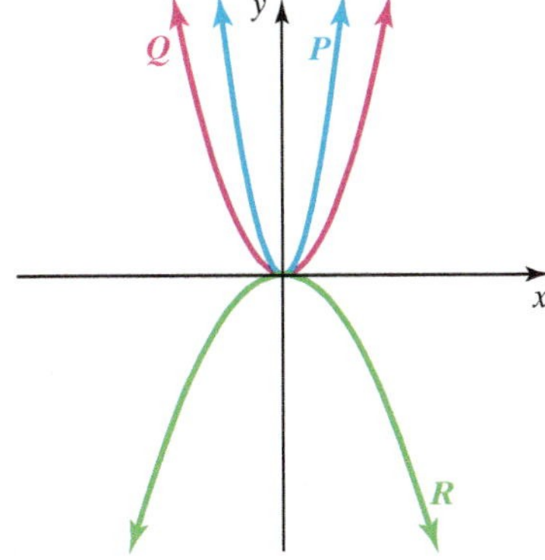

Key Concepts and Skills

CONCEPT SKILL

Concept/Skill	Description	Example
[19.1] **The square root property of equality**	If n is a nonnegative number and $x^2 = n$, then $x = \pm\sqrt{n}$, that is, $x = \sqrt{n}$ or $x = -\sqrt{n}$.	$x^2 = 15$ $x = \sqrt{15}$ or $x = -\sqrt{15}$
[19.2] **To complete the square for the expression $x^2 + bx$**	• Find one-half of the coefficient of the linear term x: $\frac{1}{2}b$ • Square this value: $\left(\frac{1}{2}b\right)^2$ • Add the resulting value to the expression.	$x^2 + 8x + c$: $\frac{1}{2}(8) = 4$ $4^2 = 16$ $x^2 + 8x + 16 = (x + 4)^2$
[19.2] **To solve the quadratic equation $ax^2 + bx + c = 0$ by completing the square**	• If $a \neq 1$, then divide the equation by a, the coefficient of the second-degree term. If $a = 1$, then proceed to the next step. • Move all terms with variables to one side of the equals sign and all constants to the other side. • Take one-half of the coefficient of the linear term. Then, square it and add this square to both sides of the equation. • Factor the side containing the variables, writing it as the square of a binomial. • Use the square root property of equality. • Solve the resulting equations.	$3x^2 + 18x + 12 = 0$ $x^2 + 6x + 4 = 0$ $x^2 + 6x = -4$ $\frac{1}{2}(6) = 3$ $3^2 = 9$ $x^2 + 6x + 9 = -4 + 9$ $(x + 3)^2 = 5$ $x + 3 = \pm\sqrt{5}$ $x = -3 \pm \sqrt{5}$ Solutions: $x = -3 + \sqrt{5}$ and $x = -3 - \sqrt{5}$
[19.3] **The quadratic formula**	If $ax^2 + bx + c = 0$, where a, b, and c are real numbers and $a \neq 0$, then $x = \dfrac{-b \pm \sqrt{b^2 - 4ac}}{2a}$.	$2x^2 + 4x - 7 = 0$, where $a = 2$, $b = 4$, and $c = -7$

CONCEPT SKILL

Concept/Skill	Description	Example
[19.3] **To solve a quadratic equation $ax^2 + bx + c = 0$ using the quadratic formula**	• Write the equation in standard form, if necessary. • Identify the coefficients a, b, and c. • Substitute values for a, b, and c in the formula $x = \frac{-b \pm \sqrt{b^2 - 4ac}}{2a}$. • Simplify.	$3x^2 = 6x + 4$ $3x^2 - 6x - 4 = 0$ $a = 3$, $b = -6$, and $c = -4$ $x = \frac{-b \pm \sqrt{b^2 - 4ac}}{2a}$ $= \frac{-(-6) \pm \sqrt{(-6)^2 - 4(3)(-4)}}{2(3)}$ $= \frac{6 \pm \sqrt{36 + 48}}{6}$ $= \frac{6 \pm \sqrt{84}}{6}$ $= \frac{6 \pm 2\sqrt{21}}{6}$ $x = \frac{3 \pm \sqrt{21}}{3}$ Solutions: $x = \frac{3 + \sqrt{21}}{3}$ and $x = \frac{3 - \sqrt{21}}{3}$
[19.4] **Parabola**	The graph of an equation of the form $y = ax^2 + bx + c$.	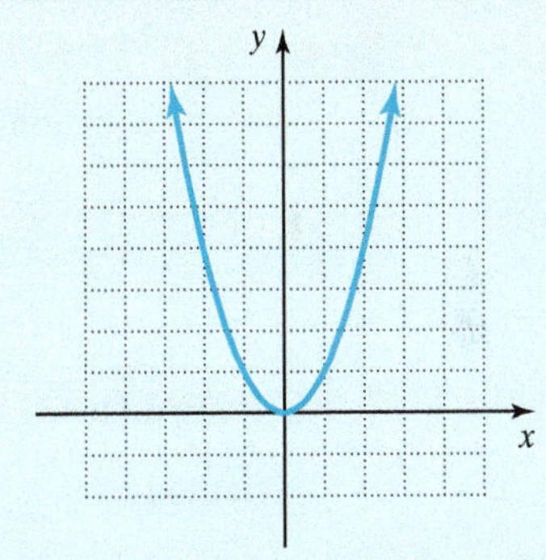
[19.4] **Vertex of a parabola**	The highest (maximum y-value) or lowest (minimum y-value) point of a parabola.	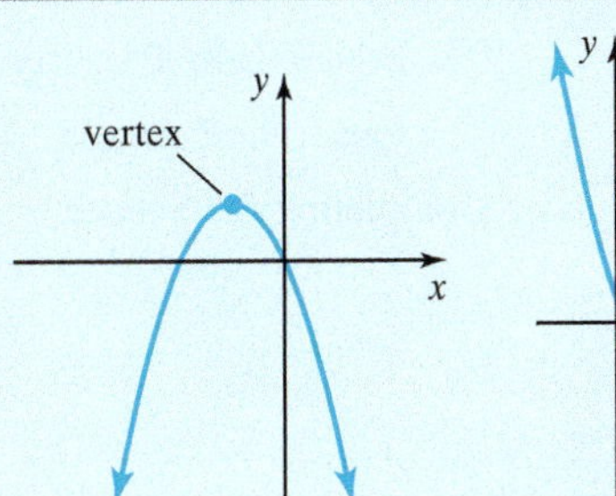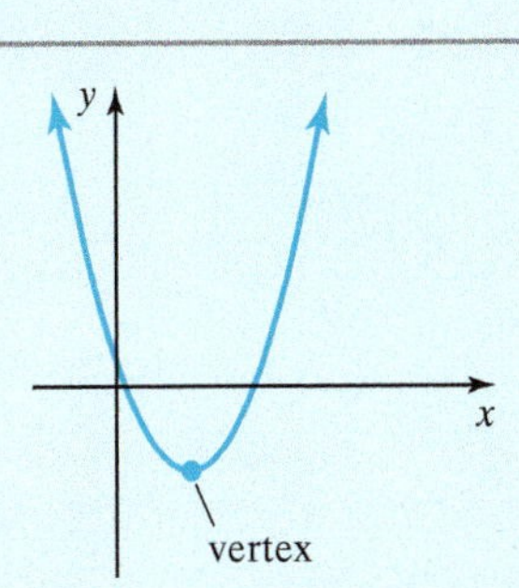

continued

CONCEPT SKILL

Concept/Skill	Description	Example
[19.4] **The axis of symmetry**	The axis of symmetry is a vertical line that passes through the vertex of a parabola. For a parabola given by the quadratic equation $y = ax^2 + bx + c$, the equation of the axis of symmetry is $x = -\frac{b}{2a}$.	For $y = x^2 - 4x + 3$, the equation of the axis of symmetry, where $a = 1$ and $b = -4$, is $x = -\frac{b}{2a} = -\frac{(-4)}{2(1)} = 2.$ 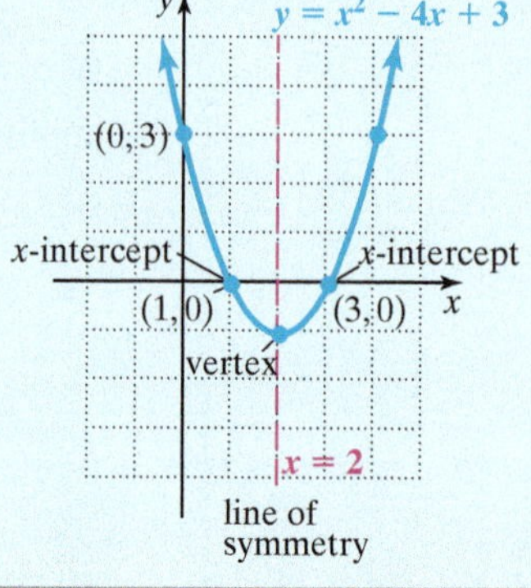
[19.4] **To find the vertex of a parabola**	For a parabola given by the quadratic equation $y = ax^2 + bx + c$: • The x-coordinate of the vertex is $-\frac{b}{2a}$. • The y-coordinate of the vertex is found by substituting $-\frac{b}{2a}$ for x into the equation and computing y.	For $y = x^2 + 2x - 3$, $a = 1$, $b = 2$, and $c = -3$. $x = -\frac{b}{2a} = -\frac{2}{2(1)} = -1$ $y = x^2 + 2x - 3$ $y = (-1)^2 + 2(-1) - 3 = -4$ Vertex: $(-1, -4)$

Solving a Quadratic Equation ($ax^2 + bx + c = 0$)

Method	When to Use Method
Factoring	• Use when the polynomial can be factored.
The Square Root Property	• Use when $b = 0$.
The Quadratic Formula	• Use when the first two methods do not apply.
Completing the Square	• Use only when specified. This method is easier to use when $a = 1$ and b is even.

CHAPTER 19 Review Exercises

Say Why

Fill in each blank.

1. If $a^2 = 14$, then a ________ (is/is not) equal to $\sqrt{14}$ or $-\sqrt{14}$ because ____.

2. The expression $x^2 + 2ax + a^2$ ________ (is/is not) a perfect square because ____.

3. The graph of $y = 2x^2 - 3x + 1$ ________ (is/is not) a parabola because ____.

4. The axis of symmetry of the graph of $y = -4x^2 + 5x - 3$ ________ (is/is not) a vertical line because ____.

5. The graph of the equation $y = -4x^2 + 5x - 3$ opens ____________ (upward/downward) because ____.

6. The parabola $y = 3x^2$ has a ____________ (minimum/maximum) value because ____.

[19.1] *Solve by using the square root property.*

7. $y^2 = 24$
8. $4x^2 = 12$
9. $x^2 + 5 = 0$
10. $(2n - 5)^2 = 18$

Solve for the indicated variable.

11. $A = \pi r^2$ for r (Area of a circle)
12. $d = \frac{1}{2}gt^2$ for t (t = time)

[19.2] *Fill in the blank to complete the square.*

13. $x^2 - 10x + \square$
14. $x^2 + 7x + \square$

Solve by completing the square.

15. $n^2 - 6n = 27$
16. $y^2 + 3y = 4$
17. $2x^2 + 8x = 4$
18. $4y^2 = 4y + 1$

[19.3] *Solve by using the quadratic formula.*

19. $y^2 - 2y - 1 = 0$
20. $-x^2 = 8x + 1$
21. $2n^2 + n = 5$
22. $(x + 3)(x - 2) = -10$
23. $y^2 = -5y$
24. $\frac{2}{x} + \frac{1}{x + 3} = 1$

[19.4] *Graph a parabola with the following characteristics.*

25. Passes through the points $(0, -3)$, $(1, -4)$, $(-1, 0)$, $(2, -3)$, and $(3, 0)$.

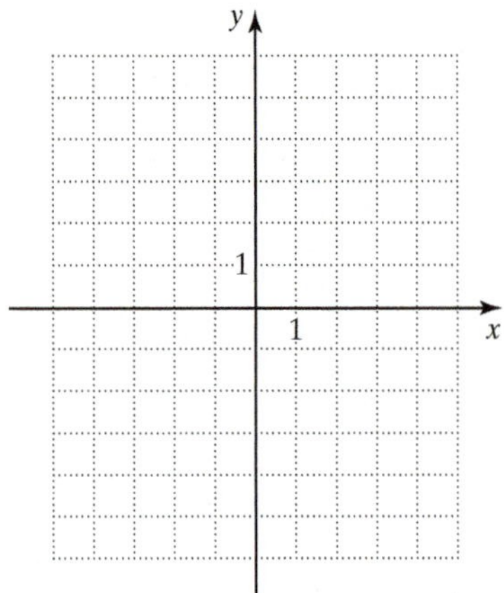

26. Has vertex $(-1, -4)$, y-intercept $(0, -3)$, and x-intercepts $(-3, 0)$, and $(1, 0)$.

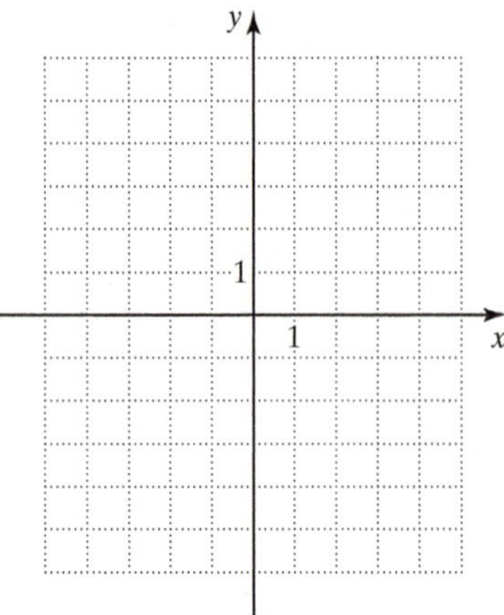

For each equation, find the axis of symmetry and the vertex of the corresponding parabola.

27. $y = 4x + x^2$

28. $y = -x^2 - 2x + 3$

Sketch the graph of each equation.

29. $y = \frac{1}{2}x^2$

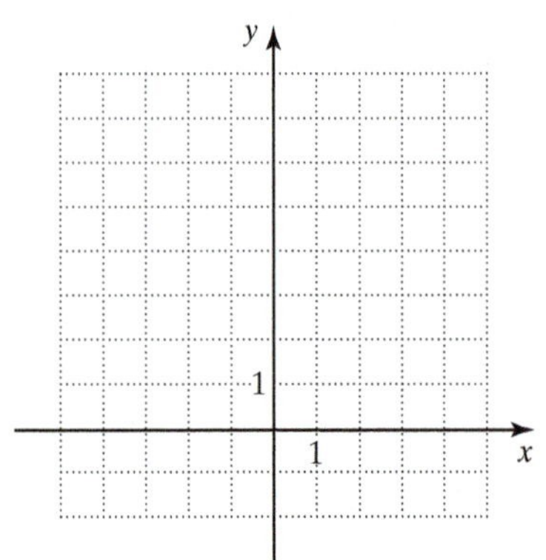

30. $y = x^2 - 5x - 6$

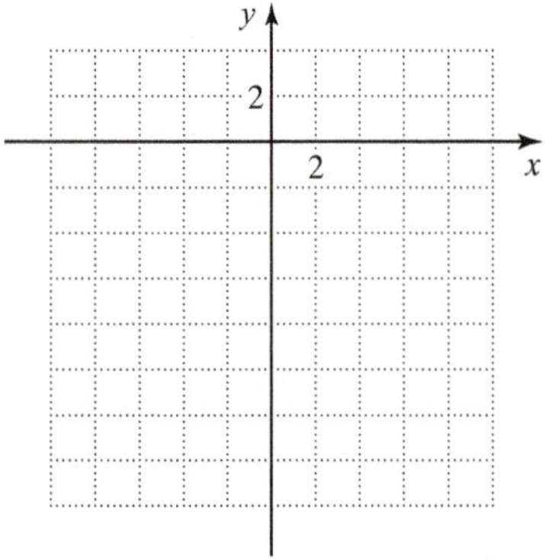

31. $y = -x^2 - 2x$

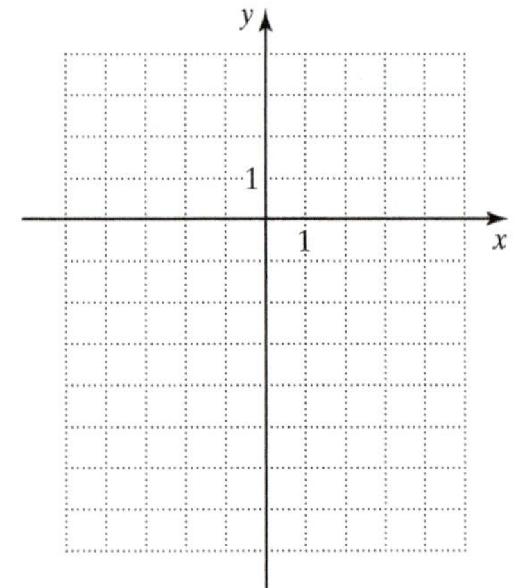

32. $y = 8 - 2x - 3x^2$

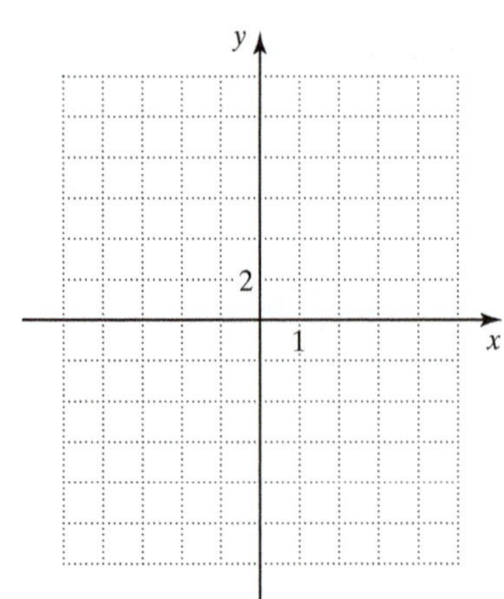

Mixed Applications

Solve. Use a calculator, if necessary.

33. A map company wants to make a globe with a surface area of 615 in^2. Use the formula $A = 4\pi r^2$, where A is the surface area, r is the radius of the sphere, and π is approximately 3.14. Find the radius to the nearest tenth of an inch.

34. A rug company uses the equation $P = 90x - x^2$ to determine its profit P for selling x rugs. If the company had a profit of \$2000, how many rugs did it sell?

35. The length of a rectangular banner is 5 ft more than twice its width. To the nearest tenth of a foot, find the dimensions if the banner has an area of 20 ft^2.

36. The formula $2D = n(n - 3)$ gives the number of diagonals D in a polygon of n sides. An architect designs a building with 14 diagonals. How many sides does the building have?

37. A travel agency offers a vacation package to the Bahamas at a discount rate. It determined that the amount of profit P per person can be modeled by the equation $P = 40n - n^2$, where n is the number of persons.

a. Graph $P = 40n - n^2$.

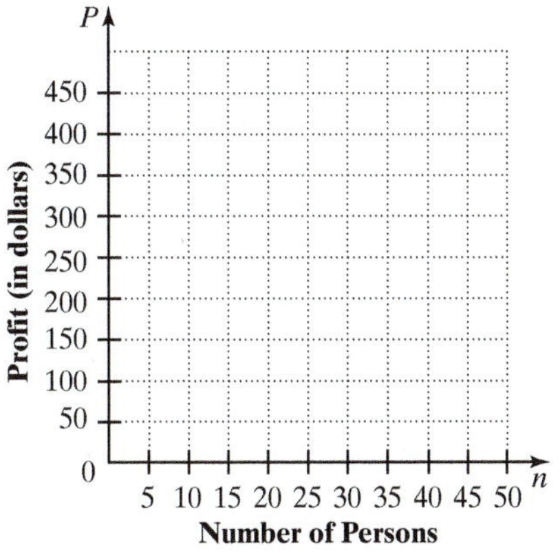

b. What is the agency's profit per person if 16 people go to the Bahamas?

c. If the profit was \$351 per person, how many people went to the Bahamas?

d. What is the maximum profit per person that the agency can make?

38. A ball is thrown straight up into the air from a height of 6 ft with an initial velocity of 40 ft/sec. The height h (in feet) after t sec is modeled by the equation $h = -16t^2 + 40t + 6$.

a. Graph the equation.

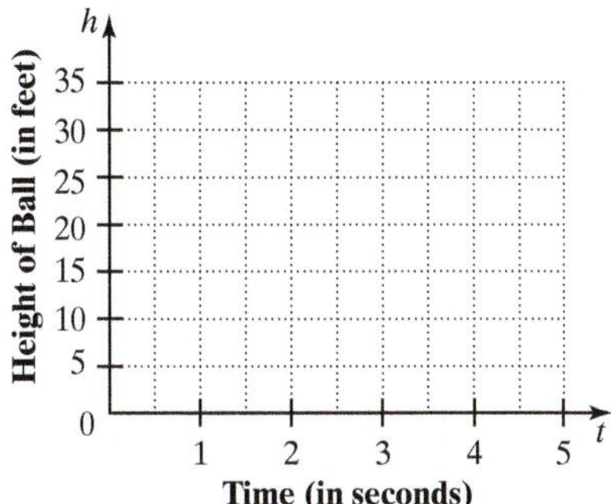

b. In how many seconds does the ball reach its maximum height?

c. What is the maximum height of the ball?

39. An inexperienced Web designer takes 8 days longer to develop a particular website than an experienced designer. Working together, the two designers can develop the site in 3 days. How long would it take each, working alone, to develop the site?

40. Because of traffic, a commuter drives home at an average speed that is 10 mph slower than her speed driving to work. The distance between her home and work is 15 mi. If it takes her 1 hr, round trip, what is her average speed driving to work to the nearest mile per hour?

• Check your answers on page A-41

CHAPTER 19 Posttest

To see if you have mastered the topics in this chapter, take this test.

1. Solve $4(x + 1)^2 = 32$ using the square root property of equality.

2. Solve $V = \frac{1}{3}\pi r^2 h$ for r. Assume that all variables are positive.

3. Fill in the blank to make the trinomial a perfect square: $x^2 - x +$ ___.

4. Solve $2y^2 - 6y - 2 = 0$ by completing the square.

5. Solve $2y^2 + 3y - 4 = 0$ using the quadratic formula.

Solve.

6. $6x^2 = 72$

7. $x^2 + 6x = 40$

8. $25(y - 4)^2 = 49$

9. $4x^2 - 8x - 3 = 0$

10. $(3x - 4)(x + 4) = -12$

11. $\frac{3}{n - 1} + \frac{5}{n + 1} = 1$

12. Sketch a parabola that has vertex $(-1, 1)$ and passes through the points $(-3, -3)$, $(-2, 0)$, $(0, 0)$, and $(1, -3)$.

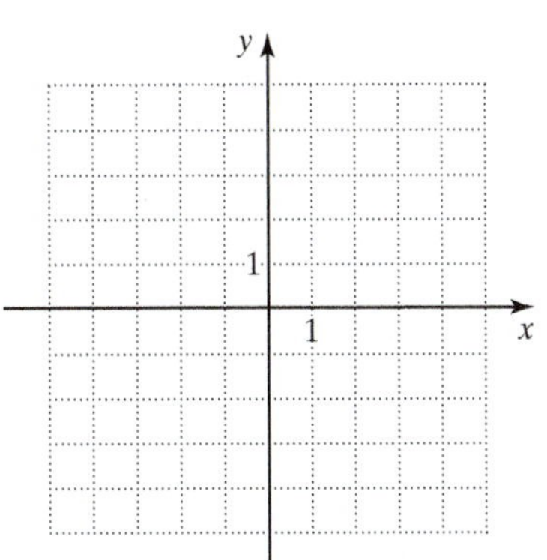
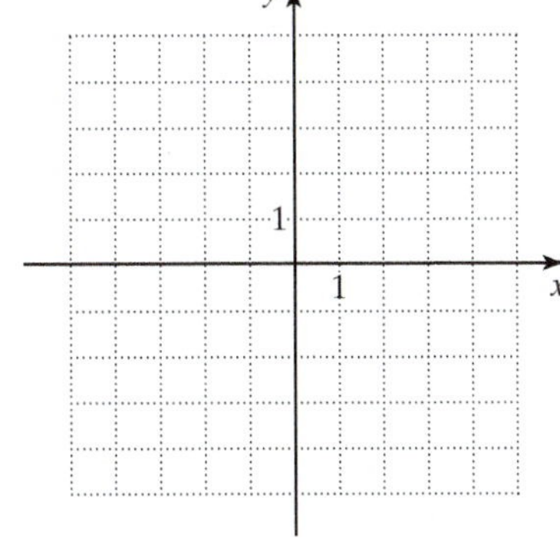

13. Find the vertex and the axis of symmetry of the graph of $y = x^2 + 3x + 4$.

Sketch the graph of each equation.

14. $y = x^2 - 2x - 3$

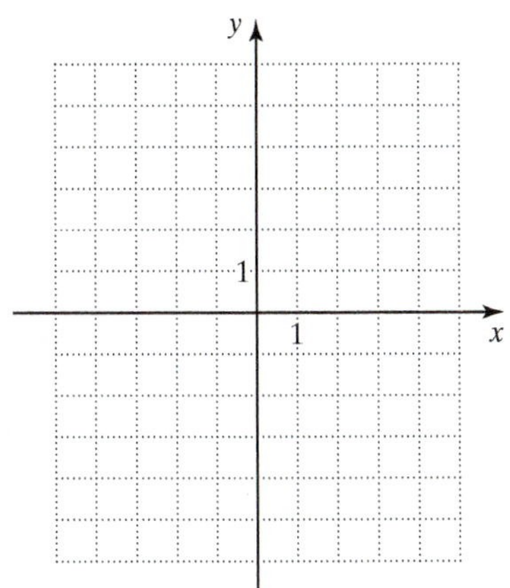

15. $y = 2x^2 + x - 15$

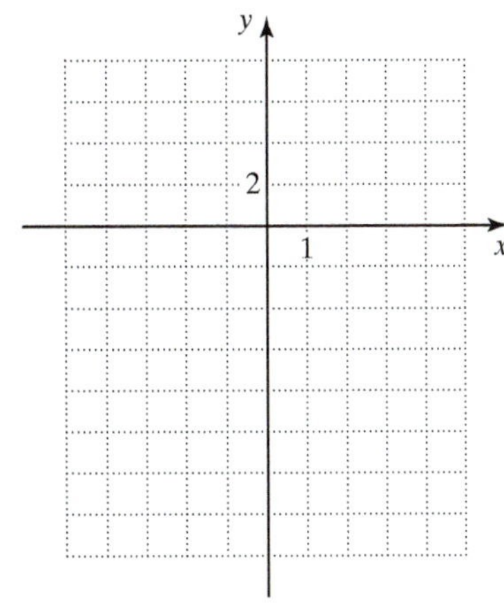

16. $y = (2 - x)^2$

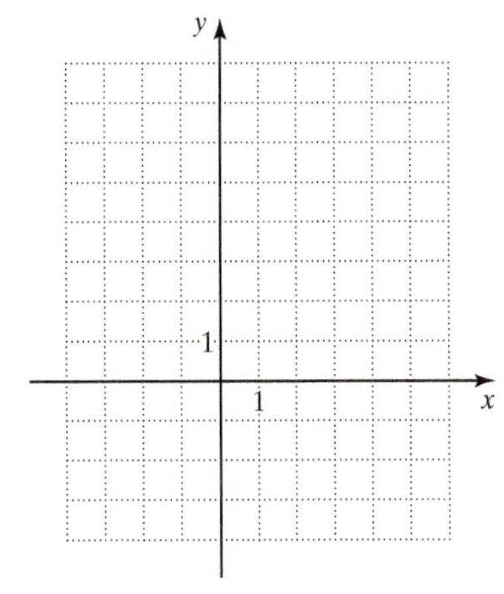

Solve. Use a calculator, if necessary.

17. The period of a pendulum is the time it takes for it to swing back and forth. The formula $l = 0.81t^2$ relates the length l (in feet) of a pendulum to the time t (in seconds) that the pendulum takes to swing back and forth. Solve this formula for t.

18. German federal highways, known as *autobahns*, have no mandatory speed limit. The equation $d = 0.0056s^2 + 0.14s$ models the distance d (in meters) that a car traveling at a speed of s km/hr needs to stop. Determine, to the nearest whole number, the maximum speed of a car that can stop within 75 m on an autobahn. (*Source:* wikipedia.org)

19. The recent U.S. average per capita personal income y (in dollars) can be roughly modeled by the quadratic expression $10x^2 - 860x + 14{,}560$, where x is the number of years since 1900. Using this model, in what year in the future would we expect the per capita income to reach \$75,000? (*Source: The New York Times Almanac, 2011*)

20. The equation $h = 80t - 16t^2$ models the height h (in feet) an object propelled straight up from the ground with an initial velocity of 80 ft/sec reaches in t sec. What is the maximum height the object reaches?

• Check your answers on page A-42.

Cumulative Review Exercises

To help you review, solve the following:

1. Evaluate: $(3x + y)^2 + z$ if $x = -2, y = 5$, and $z = -9$

2. Simplify: $7 + 9[11 - 4(3x + 6 - x)]$

3. Solve the equation $ax - by = c$ for y. Then, find the value of y when $a = 2, b = 5, c = 6$, and $x = 1$.

4. Graph: $2x - 3y < 6$

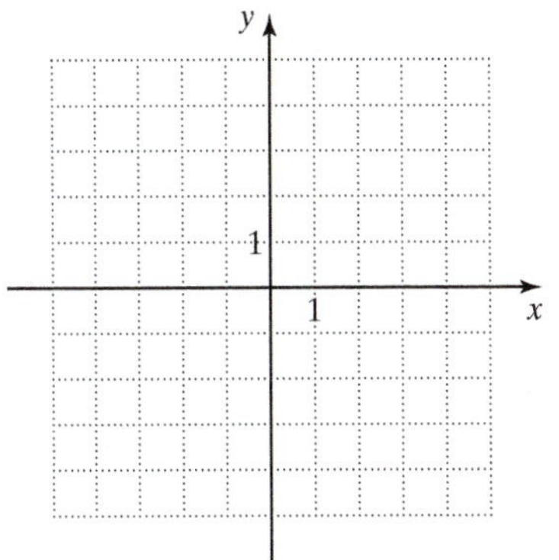

5. Solve: $\begin{aligned} -4x + 3y &= 13 \\ 3x + 4y &= 9 \end{aligned}$

6. Multiply: $(3n - 2)(5 + 6n)$

7. Divide: $\dfrac{4x^3}{y^2 - 4} \div \dfrac{6xy - 18x}{y^2 - y - 6}$

8. Factor: $9c^2d^2 + 6c^3d + 3cd^3$

9. Solve: $3x^2 - 19x - 14 = 0$

10. Simplify: $\dfrac{\dfrac{1}{4x} - \dfrac{1}{x^2}}{x - \dfrac{16}{x}}$

11. Solve: $\dfrac{c}{c + 6} = \dfrac{1}{c + 2}$

12. Simplify: $\dfrac{\sqrt{x^3y^4}}{\sqrt{xy}}$

13. Solve: $x^2 - 8x + 4 = 0$

14. Solve: $3x^2 + 5x = 2$

15. In 1862, the U.S. Congress enacted the nation's first income tax, at the rate of 3%. At this rate, what was the income of a taxpayer whose tax amounted to \$22.50? (*Source: The Seattle Times*)

16. A multimedia company started with 2 employees. In 6 months, the company had 7 employees. The number of employees increased at a steady rate.
    a. Write a linear equation that models the relationship between the number of employees n and number of months t since the company started.
    b. Graph the equation in Quadrant I.
    c. Explain the relationship between the equation in part (a) and the slope of the graph in part (b).

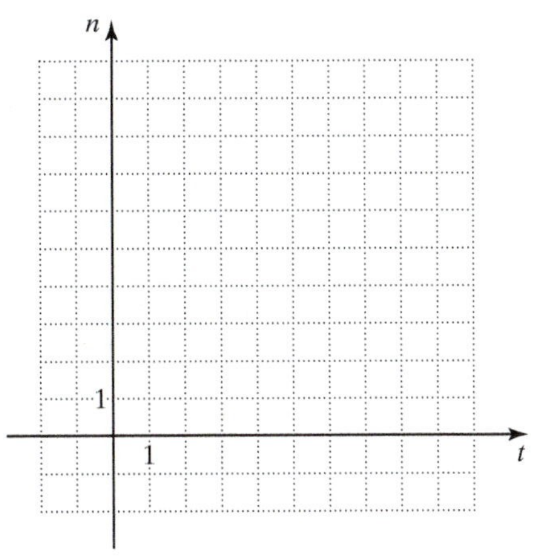

17. A certain bacteria population triples in size every day. Suppose a sample starts with 400 bacteria. The expression $400 \cdot 3^x$ models the number of bacteria in the sample after x days. Evaluate the expression for $x = 0, 1,$ and 2.

18. An average-sized audio file is 0.004 gigabytes (GB), and an average-sized TV episode video file is 0.350 GB. A certain portable multimedia player has 80 GB of storage. Express as an inequality the possible number of average-sized audio a and video files b that can be stored on this device. (*Source:* apple.com)

19. The speed of sound in air is approximately 300 m/sec. The sound from a firecracker exploding in an empty field expands outward in a growing circle. (*Source:* Everest, *Master Handbook of Acoustics*)

a. How far from the firecracker can its sound be heard in t sec?

b. In how many seconds, to the nearest second, will the sound of the firecracker be heard over an area of a square kilometer?

20. A sandbag is dropped from a hot-air balloon that is 128 ft above the ground. The height of the sandbag above the ground is given by the equation $h = -16t^2 + 128$, where h is the height (in feet) and t is the time (in seconds).

a. Graph the equation.

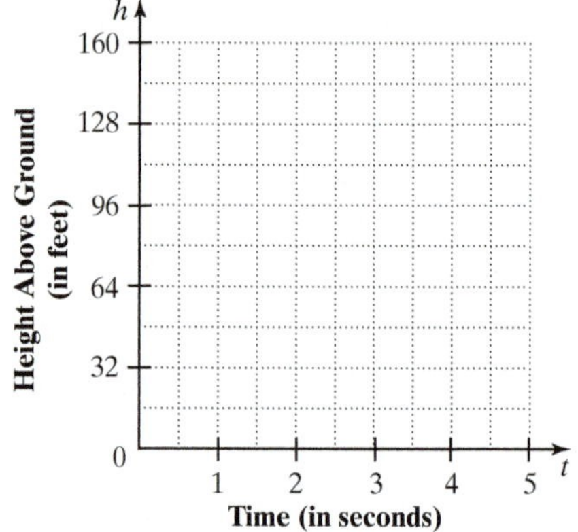

b. How far did the sandbag fall from second 0 to second 1? Explain.

• Check your answers on page A-42.

Appendix A

Table of Symbols

$\ldots$	and so forth
,	a separator for periods in a whole number
$+$	add
$-$	subtract
$\cdot, \times, (a)(b), 2y$	multiply
$\frac{a}{b}, \div, x+1\overline{)x^2-1}$	divide
$=$	is equal to
$\approx$	is approximately equal to
$\neq$	is not equal to
$<$	is less than
$\leq$	is less than or equal to
$>$	is greater than
$\geq$	is greater than or equal to
x^n	x raised to the power n
$(\ )$	parentheses (a grouping symbol)
$[\ ]$	brackets (a grouping symbol)
$\frac{a}{b}$	fraction
.	decimal point
π	pi (a number approximately equal to $\frac{22}{7}$ or 3.14)
%	percent
$-a$	the opposite, or additive inverse, of a
$\frac{1}{a}$	the reciprocal, or multiplicative inverse, of a
$\lvert a \rvert$	the absolute value of a
$\bullet A$	point A
$\overleftrightarrow{CB}$	line CB
$\overline{AB}$	line segment AB
$\overrightarrow{CD}$	ray CD
$\angle ABC$	angle ABC
$m\angle ABC$	the measure of angle ABC
$^\circ$	degree (for angles)
$\parallel$	parallel
$\perp$	perpendicular

Δ	triangle
$\sim$	is similar to
$\sqrt{a}$	the principal square root of *a*
$\sqrt[3]{a}$	the cube root of *a*
$\sqrt[n]{a}$	the *n*th root of *a*
(a, b)	the interval consisting of all real numbers between *a* and *b*, excluding *a* and *b*
$[a, b]$	the interval consisting of all real numbers between *a* and *b*, including *a* and *b*
$[a, b)$	the interval consisting of all real numbers between *a* and *b*, including *a* and excluding *b*
$(a, b]$	the interval consisting of all real numbers between *a* and *b*, excluding *a* and including *b*
∞	infinity
$\{a, b, c\}$	a set of elements
$\cup$	union
$\cap$	intersection
(x, y)	the ordered pair whose first coordinate is *x* and whose second coordinate is *y*
$f(x)$	the function *f* of *x*

Appendix B

Review of Basic Mathematics

To help you review basic mathematics (covered in Chapters 1–10), try the following selected exercises, folding the page to hide the answer column.

Topic	Exercise	Answer	Section
1. Writing whole numbers	Use digits and commas to write the amount seventy-three million, five dollars.	$73,000,005	[1.1]
2. Calculating with whole numbers	*Evaluate.* **a.** $(35 \div 7) + [(10 - 6)\cdot 3]^2$ **b.** $[13 + 2(3^2 - 8)] \div 5$	**a.** 149 **b.** 3	[1.2–1.5]
3. Using problem-solving strategies	*Draw a diagram to represent the following situation:* A college on trimesters admitted 4,285 students in the 1st trimester and 2,950 students in the 2nd trimester. For the three trimesters, 10,285 students were admitted to the college. How many students did the college admit in the 3rd trimester?	Students in Trimesters 1st | 2nd | 3rd 4,285 | 2,950 | ? 10,285 3,050 students	[1.6]
4. Comparing fractions	*Arrange in increasing order:* $\frac{5}{8}, \frac{1}{2}, \frac{4}{11}$	$\frac{4}{11}, \frac{1}{2}, \frac{5}{8}$	[2.1–2.2]
5. Adding and subtracting fractions	*Combine:* $10\frac{1}{2} - \left(4\frac{3}{8} + 3\frac{1}{4}\right)$	$2\frac{7}{8}$	[2.3]
6. Multiplying and dividing fractions	*Simplify:* $\left(3\frac{1}{8} \div 8\right)\cdot\left(4 \div 2\frac{1}{2}\right)$	$\frac{5}{8}$	[2.4]
7. Reading and writing decimals	Underline the digit that occupies the hundredths place in 509.772. Then, write the decimal in words.	$509.7\underline{7}2$ Five hundred nine and seven hundred seventy-two thousandths	[3.1]
8. Adding and subtracting decimals	*Calculate:* $0.5 - (0.19 + 0.086)$	0.224	[3.2]
9. Multiplying and dividing decimals	*Perform the indicated operations:* $\dfrac{4.7 \times 5.6}{0.4}$	65.8	[3.3–3.4]

continued

<table>
<tr><th>Topic</th><th>Exercise</th><th>Answer</th><th>Section</th></tr>
<tr><td>10. Translating and evaluating algebraic expressions</td><td>Translate each algebraic expression in the table to words. Then, find the value of the expression for the given value.
<table><tr><th>Algebraic Expression</th><th>Translation</th><th>Value</th></tr><tr><td>a. $r + 6$, if $r = 17$</td><td></td><td></td></tr><tr><td>b. $4 - x$, if $x = 1.2$</td><td></td><td></td></tr><tr><td>c. $n \div 10$, if $n = 5$</td><td></td><td></td></tr><tr><td>d. $\frac{1}{2}y$, if $y = \frac{2}{3}$</td><td></td><td></td></tr></table></td><td><table><tr><th>Algebraic Expression</th><th>Translation</th><th>Value</th></tr><tr><td>a. $r + 6$, if $r = 17$</td><td>6 more than r</td><td>23</td></tr><tr><td>b. $4 - x$, if $x = 1.2$</td><td>x less than 4</td><td>2.8</td></tr><tr><td>c. $n \div 10$, if $n = 5$</td><td>n divided by 10</td><td>$\frac{1}{2}$, or 0.5</td></tr><tr><td>d. $\frac{1}{2}y$, if $y = \frac{2}{3}$</td><td>$\frac{1}{2}$ of y</td><td>$\frac{1}{3}$</td></tr></table></td><td>[4.1]</td></tr>
<tr><td>11. Translating sentences to equations that involve solving by adding or subtracting</td><td>Translate each of the following sentences to an equation. Then, solve.
a. The sum of a number and $3\frac{1}{2}$ equals $4\frac{1}{3}$.
b. The difference between x and 1.6 is the same as 10.</td><td>a. $n + 3\frac{1}{2} = 4\frac{1}{3}$
$n = \frac{5}{6}$
b. $x - 1.6 = 10$
$x = 11.6$</td><td>[4.2]</td></tr>
<tr><td>12. Translating sentences to equations that involve solving by multiplying or dividing</td><td>Translate each of the following sentences to an equation. Then, solve.
a. Three-fourths of a number y is equal to 15.
b. The quotient of a number and 3.5 is 100.</td><td>a. $\frac{3}{4}y = 15$
$y = 20$
b. $\frac{n}{3.5} = 100$
$n = 350$</td><td>[4.3]</td></tr>
<tr><td>13. Writing ratios of like quantities and unlike quantities (rates)</td><td>Write each ratio or rate in simplest form.
a. 48 hours to 24 hours
b. 4 kilograms to 5 kilograms
c. $2,194 for 4 weeks</td><td>a. $\frac{2}{1}$
b. $\frac{4}{5}$
c. $\frac{\$548.50}{1 \text{ week}}$, or $548.50 per week</td><td>[5.1]</td></tr>
<tr><td>14. Writing and solving proportions</td><td>a. Determine whether the proportion 9 is to 1 as 3 is to $\frac{1}{3}$ is true.
b. Solve and check:
$\frac{2.5}{x} = \frac{\frac{1}{4}}{50}$</td><td>a. Yes
b. 500</td><td>[5.2]</td></tr>
</table>

<table>
<tr><th>Topic</th><th>Exercise</th><th>Answer</th><th>Section</th></tr>
<tr><td>15. Changing a percent to the equivalent fraction and vice versa</td><td>a. Convert $\frac{1}{2}$% to a fraction.
b. Express $\frac{1}{2}$ as a percent.</td><td>a. $\frac{1}{200}$
b. 50%</td><td>[6.1]</td></tr>
<tr><td>16. Changing a percent to the equivalent decimal and vice versa</td><td>a. Rewrite 106% as a decimal.
b. Find the percent equivalent of 0.375.</td><td>a. 1.06
b. 37.5%</td><td>[6.1]</td></tr>
<tr><td>17. Solving percent problems using the translation method or the proportion method</td><td>a. Find the amount:
What is $66\frac{2}{3}$% of 21?
b. Find the base:
$15 is 20% of how much money?
c. Find the percent:
What percent of 0.5 is 1.5?</td><td>a. 14
b. $75
c. 300%</td><td>[6.2]</td></tr>
<tr><td>18. Solving percent increase or decrease problems</td><td>Find the percent increase or decrease.
<table>
<tr><th>Original Value</th><th>New Value</th><th>Percent Increase or Decrease</th></tr>
<tr><td>a. $20</td><td>$5</td><td>?</td></tr>
<tr><td>b. $7</td><td>$21</td><td>?</td></tr>
<tr><td>c. $12</td><td>$18</td><td>?</td></tr>
<tr><td>d. 6 oz</td><td>2.5 oz</td><td>?</td></tr>
</table></td><td>
<table>
<tr><th>Original Value</th><th>New Value</th><th>Percent Increase or Decrease</th></tr>
<tr><td>a. $20</td><td>$5</td><td>75% decrease</td></tr>
<tr><td>b. $7</td><td>$21</td><td>200% increase</td></tr>
<tr><td>c. $12</td><td>$18</td><td>50% increase</td></tr>
<tr><td>d. 6 oz</td><td>2.5 oz</td><td>$58\frac{1}{3}$% decrease</td></tr>
</table></td><td>[6.3]</td></tr>
<tr><td>19. Representing signed numbers on a number line</td><td>Locate $-\frac{1}{2}$, 1.5, −3.1, and $1\frac{3}{4}$ on the number line. Write these numbers in order from smallest to largest.
−4 −3 −2 −1 0 1 2 3 4</td><td>−3.1 $-\frac{1}{2}$ 1.5 $1\frac{3}{4}$
−4 −3 −2 −1 0 1 2 3 4
$-3.1, -\frac{1}{2}$, 1.5, and $1\frac{3}{4}$</td><td>[7.1]</td></tr>
<tr><td>20. Adding and subtracting signed numbers</td><td>Combine.
a. $14 + (-8) + \left(-14\frac{1}{2}\right)$
b. $11 - [3 - (-9)]$</td><td>a. $-8\frac{1}{2}$
b. −1</td><td>[7.2–7.3]</td></tr>
</table>

continued

<table>
<tr><th>Topic</th><th>Exercise</th><th>Answer</th><th>Section</th></tr>
<tr><td>21. Multiplying signed numbers</td><td>a. Multiply: $(2)(-3)(-200)\left(-\frac{1}{4}\right)$
b. Evaluate: $(-0.3)^2$</td><td>a. −300
b. 0.09</td><td>[7.4]</td></tr>
<tr><td>22. Dividing signed numbers</td><td>Simplify:
a. $-5(-4) \div \left(-\frac{1}{2}\right)$
b. $(-2.25) \div (0.5)$</td><td>a. −40
b. −4.5</td><td>[7.5]</td></tr>
<tr><td>23. Finding the mean, median, mode(s), or range of a set of numbers</td><td>Compute the indicated statistics. Round to the nearest tenth, where necessary.
<table>
<tr><th>Numbers</th><th>Mean</th><th>Median</th><th>Mode(s)</th><th>Range</th></tr>
<tr><td>a. 6, 3, 6, 6, 3</td><td>?</td><td>?</td><td>?</td><td>?</td></tr>
<tr><td>b. 7, 3, 5, 3, 7</td><td>?</td><td>?</td><td>?</td><td>?</td></tr>
<tr><td>c. 2.3, 6.7, 7.1, 5.9</td><td>?</td><td>?</td><td>?</td><td>?</td></tr>
<tr><td>d. $2\frac{1}{2}, 2\frac{3}{4}, 3, 2\frac{1}{2}, 2\frac{1}{4}$</td><td>?</td><td>?</td><td>?</td><td>?</td></tr>
<tr><td>e. 5, −1, −2, 0, −2</td><td>?</td><td>?</td><td>?</td><td>?</td></tr>
</table></td><td>
<table>
<tr><th>Numbers</th><th>Mean</th><th>Median</th><th>Mode(s)</th><th>Range</th></tr>
<tr><td>a. 6, 3, 6, 6, 3</td><td>4.8</td><td>6</td><td>6</td><td>3</td></tr>
<tr><td>b. 7, 3, 5, 3, 7</td><td>5</td><td>5</td><td>3 and 7</td><td>4</td></tr>
<tr><td>c. 2.3, 6.7, 7.1, 5.9</td><td>5.5</td><td>6.3</td><td>None</td><td>4.8</td></tr>
<tr><td>d. $2\frac{1}{2}, 2\frac{3}{4}, 3, 2\frac{1}{2}, 2\frac{1}{4}$</td><td>$2\frac{3}{5}$</td><td>$2\frac{1}{2}$</td><td>$2\frac{1}{2}$</td><td>$\frac{3}{4}$</td></tr>
<tr><td>e. 5, −1, −2, 0, −2</td><td>0</td><td>−1</td><td>−2</td><td>7</td></tr>
</table></td><td>[8.1]</td></tr>
<tr><td>24. Changing measurements from one U.S. customary unit to another</td><td>Change each quantity to the indicated unit.
a. 12 ft = ____ yd
b. $3\frac{1}{2}$ qt = ____ pt
c. 18 oz = ____ lb ____ oz</td><td>a. 4
b. 7
c. 1, 2</td><td>[9.1]</td></tr>
<tr><td>25. Adding or subtracting measurements expressed in U.S. customary units</td><td>Compute.
a. 1 pt 10 fl oz
+4 pt 8 fl oz
b. 6 yr − 1 yr 5 mo</td><td>a. 6 pt 2 fl oz
b. 4 yr 7 mo</td><td>[9.1]</td></tr>
<tr><td>26. Changing a measurement from one metric unit to another</td><td>Change each quantity to the indicated unit.
a. 1,500 mg = ____ g
b. 600 m = ____ km
c. 7 kL = ____ L
d. 3.5 m = ____ cm
e. 6 MB = ____ B
f. 0.05 mg = ____ mcg</td><td>a. 1.5
b. 0.6
c. 7,000
d. 350
e. 6,000,000
f. 50</td><td>[9.2]</td></tr>
</table>

Topic	Exercise	Answer	Section
27. Finding the perimeter, area, or volume of a geometric figure	**a.** The composite figure below consists of a right triangle and a semi-circle. Find the perimeter and the area of the figure, rounded to the nearest hundredth. 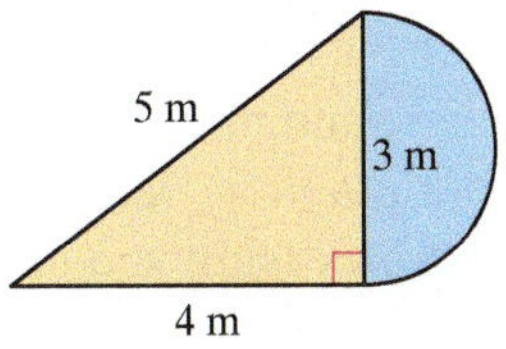 **b.** In the following diagram one cylinder has been cut out of another cylinder. Find the volume of the remaining solid, rounded to the nearest tenth. 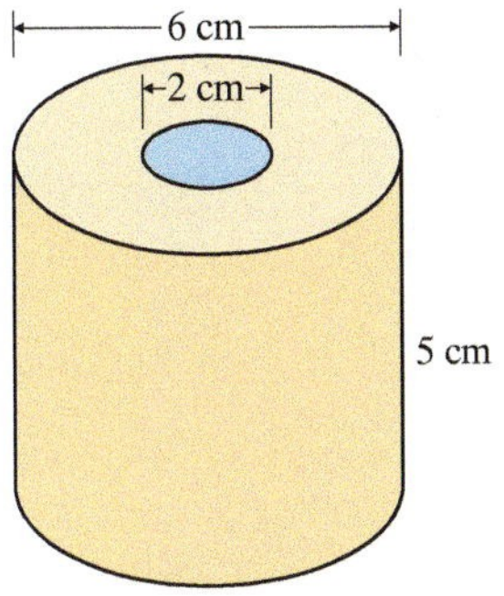 	**a.** $P \approx 13.71$ m $A \approx 9.53\ \text{m}^2$ **b.** $V \approx 125.6\ \text{cm}^3$	[10.2–10.4]
28. Finding the missing side of a similar triangle	Find the value of the unknown. 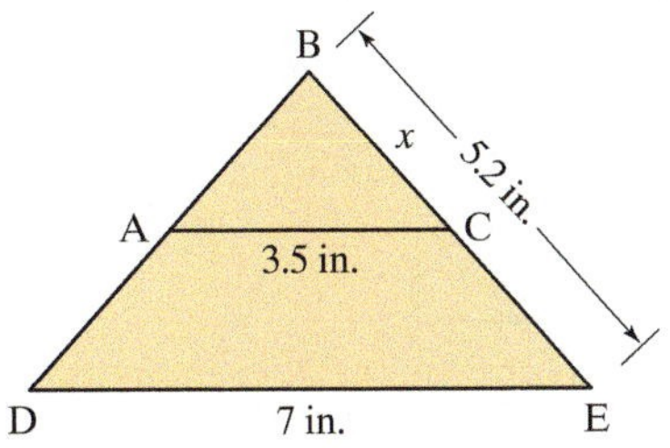 $\triangle DBE \sim \triangle ABC$	$x = 2.6$ in.	[10.5]
29. Finding the unknown side of a right triangle	Find the value of x. 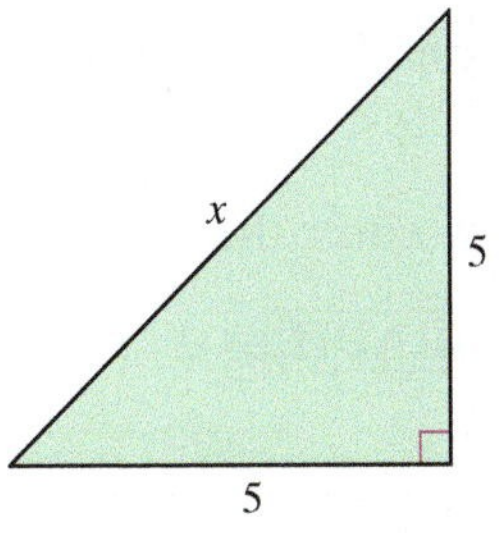 	$5\sqrt{2}$	[10.6]

Appendix C

Introduction to Graphing Calculators

This appendix covers the basic graphing features of a graphing calculator (or graphing software) used in this text. Note that the keystrokes and screens presented here and in the calculator inserts found throughout the text may be different from those on your graphing calculator. Refer to your user's manual for specific information and instructions on accessing and using the features of your particular model.

Graphing Equations

To graph an equation, it must be entered into the graphing calculator's *equation editor* in "$y =$ " form. On many graphing calculators, the equation editor can be displayed by pressing the **Y =** key. Note that you may enter more than one equation in the equation editor.

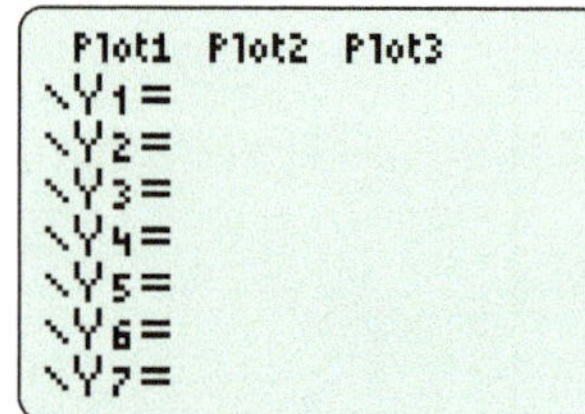

Equation Editor

For example, to graph the equation $4x + 2y = 6$, you must first solve the equation for y, getting $y = -2x + 3$. Then, enter the expression $-2x + 3$ to the right of **Y1** = in the equation editor. Finally, press the **GRAPH** key to display the graph on a coordinate plane.

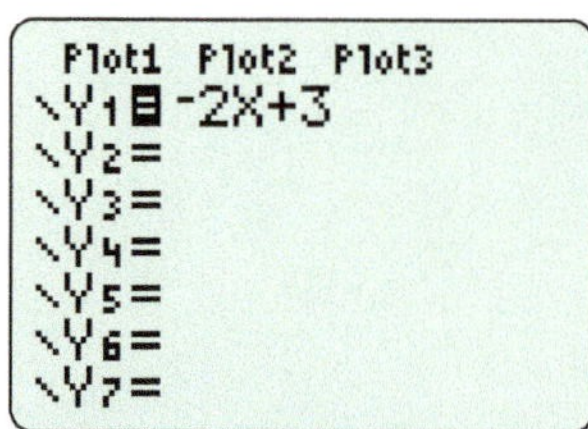

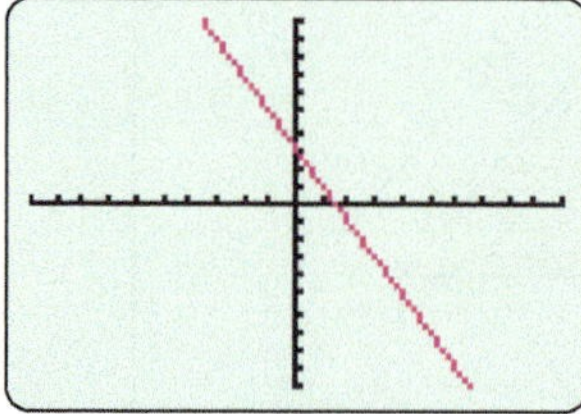

Note that most graphing calculators have two separate keys to distinguish between subtraction and negation. The **−** key is used for subtraction and the **(−)** key is used for negation.

You must be careful when entering equations containing fractions. To ensure that the graphing calculator interprets the input correctly, you may need to enclose all or part of the fraction in parentheses. For instance, to graph the equation $y = \frac{1}{2}x$, enter the expression $(1/2)x$ to the right of **Y1** = in the equation editor. Then, press **GRAPH** to display the graph of the equation.

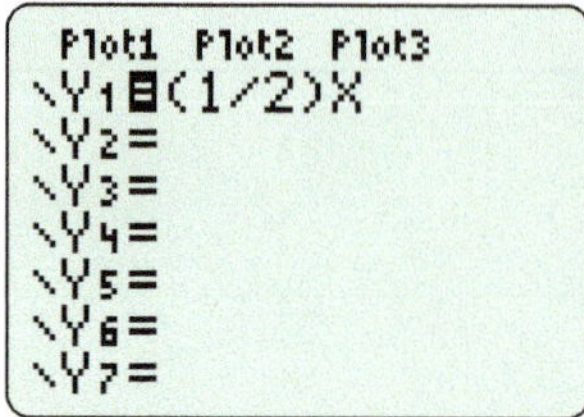

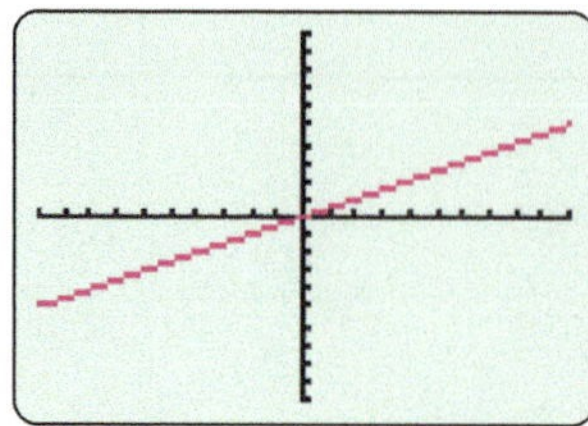

Viewing Windows

The screen on which a graph is displayed is called the *viewing window* and it represents a portion of a coordinate plane. The viewing window is defined by the following values:

Xmin: the minimum x value displayed on the x-axis

Xmax: the maximum x value displayed on the x-axis

Xscl: the distance between adjacent tick marks on the x-axis

Ymin: the minimum y value displayed on the y-axis

Ymax: the maximum y value displayed on the y-axis

Yscl: the distance between adjacent tick marks on the y-axis

You can set the viewing window by entering the minimum and maximum values and the scales for the axes in the *window editor*. The window editor can be displayed by pressing the **WINDOW** key. The screen on the left shows the window editor and displays the settings for the corresponding *standard viewing window* shown on the right.

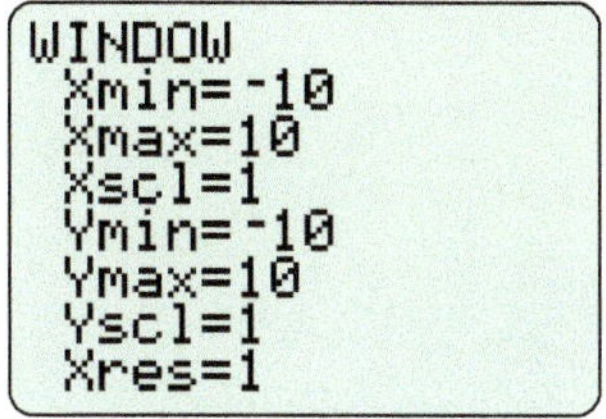

Window Editor

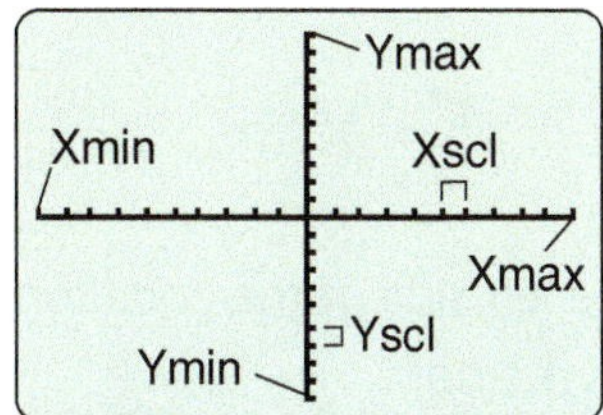

Viewing Window

The graph of an equation can be easily misinterpreted if an inappropriate viewing window is selected. For instance, compare the graph of the quadratic equation $y = x^2 - 15x + 14$, as shown in the following three viewing windows:

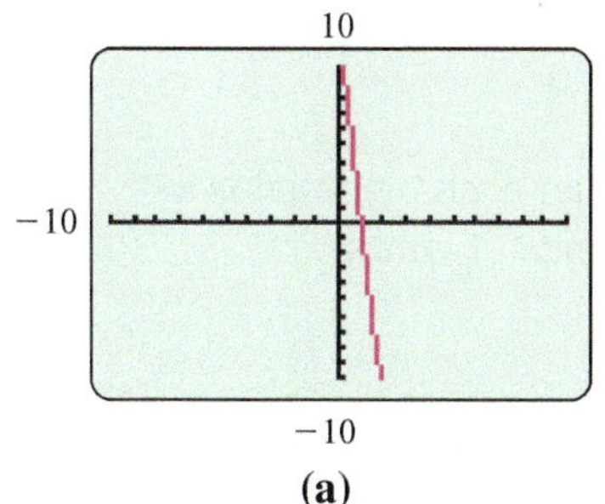

(a)

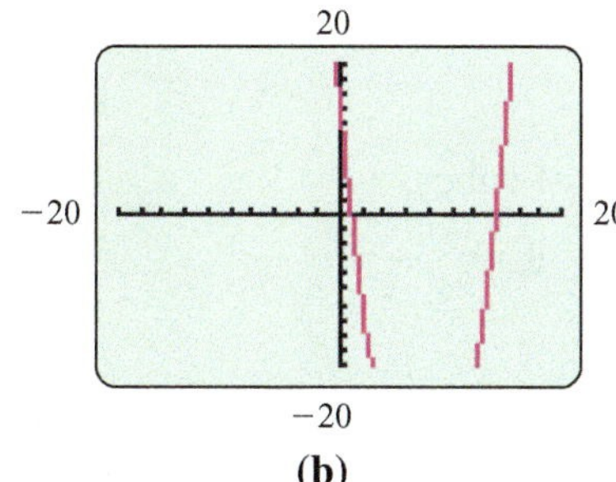

(b)

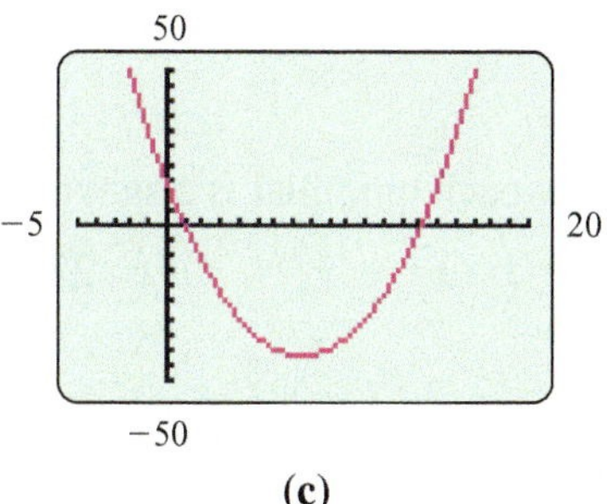

(c)

Although each of these viewing windows displays the graph of the equation, the viewing window in (c) is best because it shows all the key features of the graph, that is, the x- and y-intercepts and the vertex. Selecting an appropriate viewing window may require some practice, but familiarity with key features of the graphs of the equations discussed in the text will facilitate this process.

Other graphing calculator features, such as **TRACE**, **ZOOM**, and **INTERSECT**, are discussed in calculator inserts throughout the text.

Appendix D

Factoring the Sum of Cubes and the Difference of Cubes

OBJECTIVES

- **A** To factor a sum of cubes
- **B** To factor a difference of cubes
- **C** To solve applied problems involving the sum or difference of cubes

As we discussed in Section 16.4, there are formulas that make it easier to factor polynomials having a special form. Two formulas not covered in that section deal with the *sum of cubes* and the *difference of cubes*. These formulas provide a shortcut for factoring a binomial that is in the form of the cube of one term plus or minus the cube of another term. To state the formulas, we let the terms be a and b, with corresponding cubes a^3 and b^3.

Factoring the Sum of Cubes

$$a^3 + b^3 = (a + b)(a^2 - ab + b^2)$$

Factoring the Difference of Cubes

$$a^3 - b^3 = (a - b)(a^2 + ab + b^2)$$

TIP In these two formulas, the sign of b^3 on the left side of the equation is

- the *same* as that of b in the binomial factor on the right.
- the *opposite* of that of the middle term of the trinomial on the right.

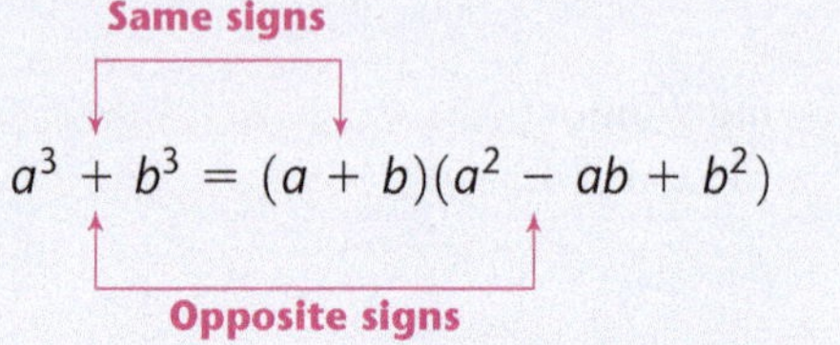

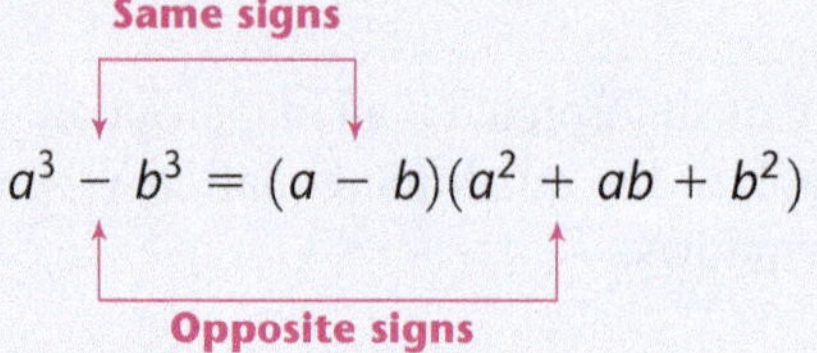

EXAMPLE 1

Indicate whether each binomial is a sum or difference of cubes.

a. $x^3 - 8$ **b.** $x^3 + y^3$ **c.** $27p^3 - q^6$ **d.** $x^2 + x^3$

Solution

a. In the expression $x^3 - 8$, both x^3 and 8 (or 2^3) are perfect cubes, where x corresponds to a and 2 to b.

$$x^3 - 8 = x^3 - 2^3$$

(↑ a^3 − ↑ b^3)

So $x^3 - 8$ is a *difference of cubes*.

b. In $x^3 + y^3$, both x^3 and y^3 are *perfect cubes* and correspond to a^3 and b^3, respectively in our formula. So $x^3 + y^3$ is a *sum of cubes*.

c. For $27p^3 - q^6$, both $27p^3$ and q^6 are perfect cubes because $27p^3 = (3p)^3$ and $q^6 = (q^2)^3$. So $27p^3 - q^6$ is a *difference of cubes*.

d. In $x^2 + x^3$, the first term x^2 is not a perfect cube. So the expression is *neither a sum of cubes nor a difference of cubes*.

PRACTICE 1

Indicate whether each binomial is a sum or difference of cubes.

a. $n^3 - 1$

b. $p^3 + q^4$

c. $x^3 + y^9$

d. $8x^6 - x^3$

EXAMPLE 2

Factor:

a. $x^3 - y^3$

b. $y^3 + 8$

Solution

a. The expression $x^3 - y^3$ is a difference of cubes, so we can apply that formula.

$$x^3 - y^3 = (x)^3 - (y)^3 = (x - y)(x^2 + xy + y^2)$$

$$a^3 \quad - \quad b^3 \qquad (a - b)(a^2 + ab + b^2)$$

So the factorization of $x^3 - y^3$ is $(x - y)(x^2 + xy + y^2)$. How can you check that this factorization is correct?

b. The expression $y^3 + 8$ is a sum of cubes.

$$y^3 + 8 = (y)^3 + (2)^3 = (y + 2)[y^2 - (y)(2) + (2)^2]$$

$$a^3 \quad + \quad b^3 \qquad (a + b)(a^2 - ab + b^2)$$

We conclude that the factorization of $y^3 + 8$ is $(y + 2)(y^2 - 2y + 4)$.

PRACTICE 2

Factor:

a. $r^3 + s^3$

b. $y^3 - 64$

EXAMPLE 3

Factor:

a. $x^3 + 27$

b. $2x^3 + 2$

c. $64n^3 - n^6$

Solution

a. Since $27 = 3^3$, $x^3 + 27$ is a sum of cubes.

$$x^3 + 27 = x^3 + 3^3 \quad \text{Use the sum-of-cubes formula.}$$
$$= (x + 3)(x^2 - x \cdot 3 + 3^2)$$
$$= (x + 3)(x^2 - 3x + 9)$$

So the factorization of $x^3 + 27$ is $(x + 3)(x^2 - 3x + 9)$.

b. We note that the two terms $2x^3$ and 2 of the binomial $2x^3 + 2$ have 2 as the greatest common factor.

$$2x^3 + 2 = 2(x^3 + 1) \quad \text{Factor out the GCF.}$$
$$= 2(x + 1)(x^2 - x + 1) \quad \text{Use the sum-of-cubes formula.}$$

So the factorization of $2x^3 + 2$ is $2(x + 1)(x^2 - x + 1)$.

c. We see that the two terms of the binomial $64n^3 - n^6$ have n^3 as the greatest common factor.

$$64n^3 - n^6 = n^3(64 - n^3) \quad \text{Factor out the GCF.}$$
$$= n^3(4 - n)(16 + 4n + n^2) \quad \text{Use the difference-of-cubes formula.}$$

So the factorization of $64n^3 - n^6$ is $n^3(4 - n)(16 + 4n + n^2)$.

PRACTICE 3

Factor:

a. $125 - x^3$

b. $5n^3 - 40$

c. $2n + 54n^4$

D Exercises

FOR EXTRA HELP MyMathLab  

Mathematically Speaking

Fill in each blank with the most appropriate term or phrase from the given list.

sum of cubes	prime	perfect square
difference of cubes	factorable	perfect cube

1. Both $x^3 + 1$ and $x^3 - 1$ are __________.

2. The expression $8x^9$ is a __________.

3. The product $(a - b)(a^2 + ab + b^2)$ can be used to factor the __________________.

4. The product $(a + b)(a^2 - ab + b^2)$ can be used to factor the ___________.

A *Factor, if possible.*

5. $p^3 + q^3$

6. $a^3 + b^3$

7. $n^3 + 8$

8. $x^3 + 27$

9. $x^3 + y^6$

10. $t^6 + 1$

11. $27n^3 + 1$

12. $1 + 64y^3$

13. $p^2 + q^2$

14. $8 + n^4$

15. $128x^3 + 2x^6$

16. $a^3 + 64a^6$

17. $2p^3 + 54$

18. $x^4 + xy^6$

B *Factor, if possible.*

19. $a^3 - b^3$

20. $s^3 - t^3$

21. $y^3 - 1$

22. $t^3 - 8$

23. $27 - n^6$

24. $a^3 - b^6$

25. $64 - 27y^6$

26. $27t^6 - 8s^3$

27. $8 - x^5$

28. $n^5 - 3$

29. $5x^3 - 5$

30. $2n^3 - 16$

31. $2y^6 - 2$

32. $t^2 - t^8$

Mixed Practice

Factor, if possible.

33. $p^3 + 1$

34. $a^3 - b^3$

35. $27x^6 - 8$

36. $2y^4 + 128y^7$

Applications

C *Solve.*

37. A furniture store sells two sizes of cube-shaped stackable storage cabinets. The dimensions of the interior storage space (in cubic inches) of the smaller cube is x-inches by x-inches by x-inches. The interior storage space of the larger cube is y-inches by y-inches by y-inches. Write an expression, in factored form, that shows the combined storage space of a large and small cube.

38. The total capacity of two barrels is given by the expression $\pi r^2 h + \pi s^2 h$, where h is the height of both vats and r and s are their radii. Write this expression in factored form.

39. On each side, a cube-shaped box is x-inches long. If each side length were increased by 1 in., then the volume would increase by $[(x + 1)^3 - x^3]$ in^3.

a. Simplify this expression by multiplying out.

b. Simplify the same expression, but by factoring.

40. The volume of a basketball is $\left(\frac{4}{3}\pi r_1^3 - \frac{4}{3}\pi r_2^3\right)$ in^3 greater than that of a soccer ball, where r_1 is the radius of a basketball and r_2 is the radius of a soccer ball.

a. Write this expression in factored form.

b. If r_1 and r_2 are approximately 5 in. and 4 in. respectively, show that both the original and the factored expressions have the same value, in terms of π. (*Source*: wikipedia.com)

• Check your answers on page A-42.

MIND *Stretchers*

Writing

1. Explain in words how to factor the whole number 1027.

Groupwork

2. Working with a partner, complete the following table:

Polynomial	$x^2 + y^2$	$x^3 + y^3$	$x^4 + y^4$	$x^5 + y^5$	$x^6 + y^6$
Prime?					

Mathematical Reasoning

3. Find the common factor for $x^3 - y^3$ and $(x - y)^3$.

Answers

CHAPTER 1

Chapter 1 Pretest, p. 2

1. Two hundred five thousand, seven **2.** 1,235,000 **3.** Hundred thousands **4.** 8,100 **5.** 8,226 **6.** 4,714 **7.** 185 **8.** 29,124 **9.** 260 **10.** 308 R6 **11.** 2^3 **12.** 36 **13.** 5 **14.** 43 **15.** 75 years old **16.** \$55 **17.** 68 **18.** 324 sec **19.** \$36 **20.** Room C, which measures 126 sq ft

Section 1.1 Practices, pp. 4–8

1, *p. 4:* **a.** Thousands **b.** Hundred thousands **c.** Ten millions **2, *p. 4:*** Eight billion, three hundred seventy-six thousand, fifty-two **3, *p. 4:*** \$7,372,050 Seven million, three hundred seventy-two thousand, fifty dollars **4, *p. 5:*** \$95,000,003 **5, *p. 5:*** \$375,000 **6, *p. 6:*** **a.** 2 ten thousands + 7 thousands + 0 hundreds + 1 ten + 3 ones = 20,000 + 7,000 + 0 + 10 + 3 or 20,000 + 7,000 + 10 + 3 **b.** 1 million + 2 hundred thousands + 7 ten thousands + 9 tens + 3 ones = 1,000,000 + 200,000 + 70,000 + 90 + 3 **7, *p. 7:*** **a.** 52,000 **b.** 50,000 **8, *p. 8:*** 420,000,000 **9, *p. 8:*** **a.** One million, six hundred ninety-nine thousand, two hundred **b.** 1,960,000

Exercises 1.1, pp. 9–13

1. whole numbers **3.** odd **5.** standard form **7.** placeholder **9.** expanded form **11.** 4,867 **13.** 316 **15.** 28,461,013 **17.** Hundred thousands **19.** Hundreds **21.** Billions **23.** Four hundred eighty-seven thousand, five hundred **25.** Two million, three hundred fifty thousand **27.** Nine hundred seventy-five million, one hundred thirty-five thousand **29.** Two billion, three hundred fifty-two **31.** One billion **33.** 10,120 **35.** 150,856 **37.** 6,000,055 **39.** 50,600,195 **41.** 400,072 **43.** 3 ones = 3 **45.** 8 hundreds + 5 tens + 8 ones = 800 + 50 + 8 **47.** 2 millions + 5 hundred thousands + 4 ones = 2,000,000 + 500,000 + 4 **49.** 670 **51.** 7,100 **53.** 30,000 **55.** 700,000 **57.** 30,000

59.

To the nearest	135,842	2,816,533
Hundred	135,800	2,816,500
Thousand	136,000	2,817,000
Ten thousand	140,000	2,820,000
Hundred thousand	100,000	2,800,000

61. 1 ten thousand + 2 thousands + 5 tens + 1 one = 10,000 + 2,000 + 50 + 1 **63.** 40,059 **65.** 1,056,100; one million, fifty-six thousand, one hundred **67.** Nine hundred thousand **69.** Forty-eight thousand, three hundred eighty-one **71.** Three hundred million **73.** 100,000,000,000 **75.** 3,288 **77.** 3,233,300,000,000 **79.** 150 ft **81.** 20,000 mi **83.** 1,900 **85.** **a.** Three million, six hundred thousand, nine hundred thirty sq mi **b.** 301,000 sq mi

Section 1.2 Practices, pp. 15–23

1, *p. 15:* 385 **2, *p. 16:*** 10,436 **3, *p. 17:*** 16 mi **4, *p. 18:*** 651 **5, *p. 19:*** 4,747 **6, *p. 20:*** 750 plant species **7, *p. 20:*** **a.** 286,000 **b.** 193,000 **c.** Less **8, *p. 22:*** 9,477 **9, *p. 22:*** 2,791 **10, *p. 23:*** 20,000 ft

Calculator Practices, pp. 23–24

11, *p. 23:* 49,532 **12, *p. 24:*** 31,899 **13, *p. 24:*** 2,499 ft

Exercises 1.2, pp. 25–31

1. right **3.** sum **5.** associative property of addition **7.** subtrahend **9.** 177,778 **11.** 14,710 **13.** 14,002 **15.** 56,188 **17.** 6,978 **19.** 4,820 **21.** 413 **23.** 14,865 **25.** 15,509 m **27.** 82 hr **29.** \$104,831 **31.** \$12,724 **33.** 31,200 tons **35.** 13,296,657 **37.** 1,662,757

39.

+	400	200	1,200	300	Total
300	700	500	1,500	600	3,300
800	1,200	1,000	2,000	1,100	5,300
Total	1,900	1,500	3,500	1,700	8,600

41.

+	389	172	1,155	324	Total
255	644	427	1,410	579	3,060
799	1,188	971	1,954	1,123	5,236
Total	1,832	1,398	3,364	1,702	8,296

43. a; possible estimate: 12,800 **45.** a; possible estimate: \$900,000 **47.** 217 **49.** 90 **51.** 362 **53.** 68,241 **55.** 2,285 **57.** 52,999 **59.** 2,943 **61.** 203,465 **63.** 368 **65.** 4,996 **67.** 982 **69.** 1,995 mi **71.** \$669 **73.** \$3,609 **75.** 273 books **77.** 209 m **79.** 2,001,000 **81.** 813,429 **83.** c; possible estimate: 40,000,000 **85.** a; possible estimate: \$200,000 **87.** 7,065 **89.** 1,676 **91.** 5,186 **93.** 281,000,000 **95.** 3,400,000 sq mi **97.** **a.** Austria, 16; Canada, 26; Germany 30; Norway, 23; United States, 37 **b.** United States **99.** About 43 years old **101.** No, the elevator is not overloaded. The total weight of passengers is 963 lb. **103.** 180°F **105.** 151 mi **107.** 19,403,000 **109.** 1,454 **111.** **a.** Less (2,804) **b.** 2,932 seats **113.** **a.** 280,000 species **b.** 1,060,000 species **c.** 260,000 species **115.** \$28,576

Section 1.3 Practices, pp. 34–37

1, *p. 34:* 608 **2, *p. 34:*** 4,230 **3, *p. 35:*** 480,000 **4, *p. 35:*** 205,296 **5, *p. 36:*** 107 sq ft **6, *p. 36:*** 112,840 **7, *p. 37:*** No; possible estimate = 20,000

Calculator Practices, p. 38

8, *p. 38:* 1,026,015 **9, *p. 38:*** 345,546

Exercises 1.3, pp. 39–43

1. product **3.** identity property of multiplication **5.** addition **7.** 400 **9.** 142,000 **11.** 170,000 **13.** 7,000,000 **15.** 12,700 **17.** 418 **19.** 3,248,000 **21.** 65,268 **23.** 817 **25.** 34,032 **27.** 3,003 **29.** 3,612 **31.** 57,019 **33.** 243,456 **35.** 200,120 **37.** 149,916 **39.** 144,500 **41.** 123,830 **43.** 3,312 **45.** 2,106 **47.** 40,000 **49.** 23,085 **51.** 3,274,780 **53.** 54,998,850 **55.** c; possible estimate: 480,000 **57.** b; possible estimate: 80,000 **59.** 2,880 **61.** 230,520 **63.** 1,071,000 **65.** 300,000 **67.** 3,300 yr **69.** **a.** 3,000,000 **b.** 1,000,000 **71.** Yes **73.** 5,775 sq in. **75.** 1,750 mi **77.** \$442 **79.** **a.** 294 mi **b.** 1,470 mi **81.** Colorado; area $\approx$ 106,700 sq mi

Section 1.4 Practices, pp. 46–50

1, *p. 46:* 807 **2, *p. 46:*** 7,002 **3, *p. 47:*** 5,291 R1 **4, *p. 48:*** 79 R1 **5, *p. 48:*** 94 R10 **6, *p. 49:*** 607 R3 **7, *p. 49:*** 200 **8, *p. 50:*** 967 **9, *p. 50:*** 5 times

Calculator Practice, p. 51

10, ***p. 51:*** 603

Exercises 1.4, pp. 52–54

1. divisor **3.** multiplication **5.** 400 **7.** 2,560 **9.** 301 **11.** 3,003 **13.** 8,044 **15.** 500 **17.** 30 **19.** 14 **21.** 42 **23.** 400 **25.** 159 **27.** 5,353 **29.** 1,002 **31.** 6,944 **33.** 1,001 **35.** 3,050 **37.** 907 **39.** 1,201 **41.** 651 R2 **43.** 11 R7 **45.** 116 R83 **47.** 700 R2 **49.** 723 R19 **51.** 428 R8 **53.** 1,010 R10 **55.** 928 R24 **57.** 721 **59.** 155 **61.** c; possible estimate: 800 **63.** a; possible estimate: 7,000 **65.** 907 R1 **67.** 2,000 **69.** 2,400 **71.** 370 **73.** $135 **75.** 2 times **77.** 300 people per square mile **79.** 6 calories **81. a.** 304 tiles **b.** 26 boxes **c.** $468

Section 1.5 Practices, pp. 55–59

1, ***p. 55:*** $5^5 \cdot 2^2$ **2,** ***p. 56:*** **a.** 1 **b.** 1,331 **3,** ***p. 56:*** 784 **4,** ***p. 56:*** 10^9 **5,** ***p. 57:*** 28 **6,** ***p. 58:*** 146 **7,** ***p. 58:*** 4 **8,** ***p. 58:*** 130 **9,** ***p. 59:*** 60 ft **10,** ***p. 59:*** $40 **11,** ***p. 59:*** **a.** 61 fatalities **b.** 2006 and 2009

Calculator Practices, p. 60

12, ***p. 60:*** 140,625; **13,** ***p. 60:*** 131

Exercises 1.5, pp. 61–65

1. base **3.** adding

5.

n	0	2	4	6	8	10	12
n^2	0	4	16	36	64	100	144

7.

n	0	2	4	6	8
n^3	0	8	64	216	512

9. 10^2 **11.** 10^4 **13.** 10^6 **15.** $2^2 \cdot 3^2$ **17.** $4^3 \cdot 5^1$ **19.** 900 **21.** 1,568 **23.** 18 **25.** 4 **27.** 13 **29.** 14 **31.** 35 **33.** 225 **35.** 250 **37.** 36 **39.** 5 **41.** 28 **43.** 6 **45.** 99 **47.** 99 **49.** 4 **51.** 39 **53.** 16 **55.** 93 **57.** 67 **59.** 18 **61.** 529 **63.** 419 **65.** 137,088 **67.** $\boxed{4} \cdot 3 + \boxed{6} \cdot 5 + \boxed{6} \cdot 7 = 98$ **69.** $(\boxed{8})(3 + \boxed{4}) - 2 \cdot \boxed{6} = 44$ **71.** $\boxed{8} + 10 \times \boxed{4} - \boxed{6} \div 2 = 45$ **73.** $(5 + 2) \cdot 4^2 = 112$ **75.** $(5 + 2 \cdot 4)^2 = 169$ **77.** $(8 - 4) \div 2^2 = 1$ **79.** 242 sq cm **81.** 3,120 sq in.

83.

Input	Output
0	$21 + 3 \times \mathbf{0} = 21$
1	$21 + 3 \times \mathbf{1} = 24$
2	$21 + 3 \times \mathbf{2} = 27$

85. 25 **87.** 40 **89.** 4 **91.** 2,412 mi **93.** 8 **95.** 10^8 **97.** 289 **99.** 48 **101.** 8 **103.** 625 sq ft **105.** $5^2 + 12^2 = 13^2$; $25 + 144 = 169$ **107.** 10^6 **109. a.** $21,500 **b.** $1,050 **111. a.** 69 **b.** At home; the average score for home games was higher than the average score for away games. **113. a.** 108,000 workers **b.** 1,969,000 workers; below average **115.** Yes, because the average number of customers is 502.

Section 1.6 Practices, pp. 67–69

1, ***p. 67:*** 10,670 employees **2,** ***p. 68:*** 2 yr **3,** ***p. 68:*** 1,551 students **4,** ***p. 69:*** 180 lb

Exercises 1.6, pp. 70–71

1. $2,150 **3.** 27 mi **5.** 75 times **7.** 5,882 mi **9.** 528,179 immigrants **11.** 300¢, or $3 **13.** $17,000 **15.** $6,036 **17.** $1,458 **19.** 8 extra pens **21.** 1952 was closer by 31 votes. **23.** $983

Chapter 1 Review Exercises, pp. 75–79

1. are; possible answer: both digits are in the hundreds place **2.** is not; possible answer: the critical digit 6 is greater than 5 **3.** is not; possible answer: the perimeter is the sum of the lengths of the figure's sides **4.** is not; possible answer: 8 times 7 is 56 **5.** is; possible answer: the area of a rectangle is the product of its length and its width **6.** is not; possible answer: 9 is the base which is raised to the power 2 **7.** is; possible answer: of the distributive property **8.** is; possible answer: 10 divided by 5 is 2 **9.** is not; possible answer: the sum of the numbers should be divided by 3 and not by 2 since there are three numbers **10.** before; possible answer: of the order of operations rule **11.** Ones **12.** Ten thousands **13.** Hundred millions **14.** Ten billions **15.** Four hundred ninety-seven **16.** Two thousand, fifty **17.** Three million, seven **18.** Eighty-five billion **19.** 251 **20.** 9,002 **21.** 14,000,025 **22.** 3,000,003,000 **23.** 2 millions + 5 hundred thousands = 2,000,000 + 500,000 **24.** 4 ten thousands + 2 thousands + 7 hundreds + 7 ones = 40,000 + 2,000 + 700 + 7 **25.** 600 **26.** 1,000 **27.** 380,000 **28.** 70,000 **29.** 9,486 **30.** 65,692 **31.** 173,543 **32.** 150,895 **33.** 1,957,825 **34.** $223,067 **35.** 445 **36.** 10,016 **37.** 11,109 **38.** 5,510 **39.** 11,042,223 **40.** $2,062,852 **41.** 11,006 **42.** 2,989 **43.** 432 **44.** 1,200 **45.** 149,073 **46.** 12,000,000 **47.** 477,472 **48.** 1,019,000 **49.** 1,397,508 **50.** 188,221,590 **51.** 39 **52.** 307 R3 **53.** 37 R10 **54.** 680 R8 **55.** 25,625 **56.** 957 **57.** 343 **58.** 1 **59.** 72 **60.** 300,000 **61.** 5 **62.** 169 **63.** 5 **64.** 19 **65.** 12 **66.** 18 **67.** 10,833,312 **68.** 2,694 **69.** $7^2 \cdot 5^2$ **70.** $2^2 \cdot 5^3$ **71.** 39 **72.** 7 **73.** 6 **74.** 5 **75.** Two million, four hundred thousand **76.** 150,000,000 **77.** $3,009,000,000,000 **78.** 1985 **79.** 300,000 sq mi **80.** 32,000,000 iPods **81.** 9 **82.** 27 times **83.** 2,717 **84.** 23 flats **85.** $307 per week

86.

Net sales	$430,000
− Cost of merchandise sold	− 175,000
Gross margin	$255,000
− Operating expenses	− 135,000
Net profit	$120,000

87. 6,675 sq m **88.** 272 legs **89.** 1968 to 1972 (15,385,031 votes) **90.** 4,341 points **91. a.** 1,949 km **b.** 1,683 km **92. a.** 14,994,000 **b.** The average would increase by 62,000. **93.** 29 sq mi **94.** 162 cm

Chapter 1 Posttest, p. 80

1. 225,067 **2.** 1,7$\underline{6}$8,405 **3.** One million, two hundred five thousand, seven **4.** 200,000 **5.** 1,894 **6.** 607 **7.** 147 **8.** 297,496 **9.** 509 **10.** 622 R19 **11.** 625 **12.** $4^3 \cdot 5^2$ **13.** 84 **14.** 2 **15.** 5,600,000 sq mi **16.** 46,177,500 acres **17.** $469 **18.** $123 **19.** $1,380 **20.** 26 g of fat

CHAPTER 2

Chapter 2 Pretest, p. 82

1. 1, 2, 4, 5, 10, 20 **2.** $2 \times 2 \times 2 \times 3 \times 3$, or $2^3 \times 3^2$ **3.** $\frac{2}{5}$ **4.** $\frac{61}{3}$ **5.** $1\frac{1}{30}$ **6.** $\frac{3}{4}$ **7.** 20 **8.** $\frac{1}{8}$ **9.** $1\frac{1}{5}$ **10.** $12\frac{5}{6}$ **11.** $2\frac{1}{4}$ **12.** $4\frac{5}{8}$ **13.** $3\frac{1}{2}$ **14.** 60 **15.** $\frac{2}{3}$ **16.** $3\frac{2}{3}$ **17.** $\frac{2}{21}$ **18.** 6 students **19.** 6 mi **20.** 69 g

Section 2.1 Practices, pp. 83–89

1, ***p. 83:*** 1, 7 **2,** ***p. 84:*** 1, 3, 5, 15, 25, 75 **3,** ***p. 85:*** 1, 2, 3, 5, 6, 9, 10, 15, 18, 30, 45, 90 **4,** ***p. 85:*** Yes; 24 is a multiple of 3. **5,** ***p. 86:*** **a.** Prime **b.** Composite **c.** Prime **d.** Composite **e.** Prime **6,** ***p. 87:*** $2^3 \times 7$ **7,** ***p. 87:*** 3×5^2 **8,** ***p. 88:*** 18 **9,** ***p. 89:*** 66 **10,** ***p. 89:*** 12 **11,** ***p. 89:*** 6 yr

Exercises 2.1, pp. 90–91

1. factors **3.** prime **5.** prime factorization **7.** 1, 3, 7, 21 **9.** 1, 17 **11.** 1, 2, 3, 4, 6, 12 **13.** 1, 31 **15.** 1, 2, 3, 4, 6, 9, 12, 18, 36 **17.** 1, 29 **19.** 1, 2, 4, 5, 10, 20, 25, 50, 100 **21.** 1, 2, 4, 7, 14, 28 **23.** Prime **25.** Composite (2, 4, 8) **27.** Composite (7) **29.** Prime **31.** Composite (3, 9, 27) **33.** 2^3 **35.** 7^2 **37.** $2^3 \times 3$ **39.** 2×5^2 **41.** 7×11 **43.** 3×17 **45.** 5^2 **47.** 2^5 **49.** 3×7 **51.** $2^3 \times 13$ **53.** 11^2 **55.** 2×71 **57.** $2^2 \times 5^2$ **59.** 5^3 **61.** $3^3 \times 5$ **63.** 15 **65.** 40 **67.** 90 **69.** 110 **71.** 72 **73.** 360 **75.** 300 **77.** 84 **79.** 105 **81.** 60 **83.** 3×5^2 **85.** 1, 2, 3, 4, 6, 8, 9, 12, 18, 24, 36, and 72 **87. a.** No, because 2015 is not a multiple of 10 **b.** Yes, because 2020 is a multiple of 10 **89.** No **91.** 30 students **93.** 30 days

Section 2.2 Practices, pp. 93–102

1, *p. 94:* $\frac{5}{8}$ **2, *p. 94:*** $\frac{7}{30}$ **3, *p. 94:*** $\frac{3}{4}$
4, *p. 95:*

5, *p. 95:* **a.** $\frac{16}{3}$ **b.** $\frac{102}{5}$ **6, *p. 96:*** **a.** 2 **b.** $5\frac{5}{9}$ **c.** $2\frac{2}{3}$ **7, *p. 98:*** Possible answer: $\frac{4}{10}, \frac{6}{15}, \frac{8}{20}$ **8, *p. 98:*** $\frac{45}{72}$ **9, *p. 99:*** $\frac{2}{3}$ **10, *p. 99:*** $\frac{7}{3}$ **11, *p. 99:*** $\frac{5}{16}$ **12, *p. 101:*** $\frac{11}{16}$ **13, *p. 102:*** $\frac{8}{15}, \frac{23}{30}, \frac{9}{10}$ **14, *p. 102:*** Country stations

Exercises 2.2, pp. 103–108

1. proper fraction **3.** equivalent **5.** like fractions **7.** $\frac{1}{3}$ **9.** $\frac{3}{6}$ **11.** $1\frac{1}{4}$
13. $3\frac{2}{4}$ **15.** **17.** **19.**

21. **23.**

25. **27.** Proper **29.** Improper

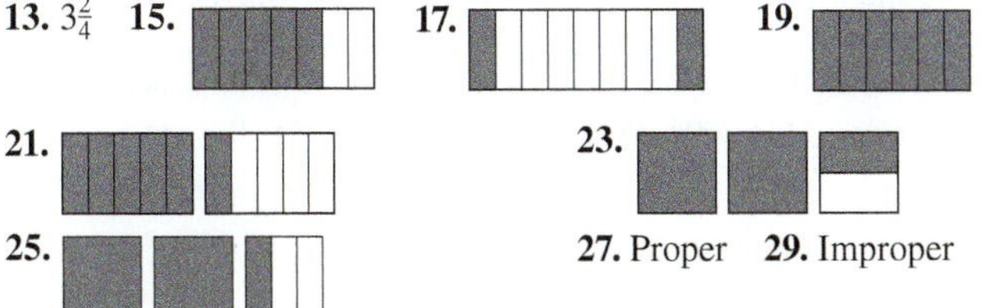

31. Mixed **33.** Improper **35.** Proper **37.** Mixed **39.** $\frac{13}{5}$ **41.** $\frac{55}{9}$ **43.** $\frac{57}{5}$ **45.** $\frac{5}{1}$ **47.** $\frac{59}{8}$ **49.** $\frac{88}{9}$ **51.** $\frac{27}{2}$ **53.** $\frac{98}{5}$ **55.** $\frac{14}{1}$ **57.** $\frac{54}{11}$ **59.** $\frac{115}{14}$ **61.** $\frac{202}{25}$ **63.** $1\frac{1}{3}$ **65.** $1\frac{1}{9}$ **67.** 3 **69.** 1 **71.** $19\frac{4}{5}$ **73.** $9\frac{1}{9}$ **75.** 1 **77.** $8\frac{2}{9}$ **79.** $13\frac{1}{2}$ **81.** $11\frac{1}{9}$ **83.** 27 **85.** 8 **87.** Possible answers: $\frac{2}{16}, \frac{3}{24}$ **89.** Possible answers: $\frac{4}{22}, \frac{6}{33}$ **91.** Possible answers: $\frac{6}{8}, \frac{9}{12}$ **93.** Possible answers: $\frac{2}{18}, \frac{3}{27}$ **95.** 9 **97.** 15 **99.** 40 **101.** 36 **103.** 40 **105.** 54 **107.** 36 **109.** 42 **111.** 6 **113.** 49 **115.** 32 **117.** 30 **119.** $\frac{2}{3}$ **121.** 1 **123.** $\frac{1}{3}$ **125.** $\frac{9}{20}$ **127.** $\frac{1}{4}$ **129.** $\frac{1}{8}$ **131.** $\frac{5}{4}$, or $1\frac{1}{4}$ **133.** $\frac{33}{16}$, or $2\frac{1}{16}$ **135.** $\frac{9}{16}$ **137.** $\frac{7}{24}$ **139.** 3 **141.** $\frac{1}{7}$ **143.** $3\frac{2}{3}$ **145.** 3 **147.** $<$ **149.** $>$ **151.** $=$ **153.** $<$ **155.** $\frac{1}{4}, \frac{1}{3}, \frac{1}{2}$ **157.** $\frac{7}{12}, \frac{2}{3}, \frac{5}{6}$ **159.** $\frac{3}{5}, \frac{2}{3}, \frac{8}{9}$ **161.** $\frac{5}{6}$ **163.** Possible answers: $\frac{4}{18}, \frac{6}{27}$ **165.** $\frac{12}{15}$ **167.** $2\frac{1}{5}$ hr per day **169. a.** $\frac{20}{401}$ **b.** $\frac{381}{401}$ **171.** $\frac{50}{103}$ **173.** The plain yogurt, because $\frac{2}{5}$ is greater than $\frac{1}{10}$. **175. a.** $\frac{1}{16}$ **b.** $\frac{1}{2}$ **177.** $250\frac{1}{2}$ lb

Section 2.3 Practices, pp. 109–124

1, *p. 109:* $\frac{2}{3}$ **2, *p. 110:*** $1\frac{7}{40}$ **3, *p. 110:*** $\frac{2}{5}$ **4, *p. 110:*** **a.** $\frac{3}{5}$ g **b.** $\frac{2}{5}$ g **5, *p. 112:*** $1\frac{2}{3}$ **6, *p. 112:*** $\frac{3}{10}$ **7, *p. 112:*** $\frac{71}{72}$ **8, *p. 113:*** $2\frac{1}{30}$ mi **9, *p. 114:*** $34\frac{2}{5}$ **10, *p. 114:*** $7\frac{1}{2}$ **11, *p. 115:*** 4 lengths **12, *p. 115:*** $7\frac{5}{8}$ **13, *p. 116:*** $11\frac{5}{24}$ **14, *p. 117:*** $4\frac{2}{5}$ **15, *p. 117:*** $1\frac{2}{5}$ in. **16, *p. 118:*** $4\frac{7}{12}$ **17, *p. 118:*** $1{,}439\frac{7}{10}$ mi **18, *p. 119:*** $1\frac{2}{7}$ **19, *p. 120:*** $5\frac{1}{6}$ **20, *p. 121:*** $12\frac{1}{2}$ **21, *p. 121:*** No, there will be only $1\frac{5}{8}$ yd left. **22, *p. 122:*** $10\frac{11}{20}$ **23, *p. 123:*** $1\frac{3}{8}$ **24, *p. 124:*** $6\frac{3}{4}$

Exercises 2.3, pp. 125–128

1. numerators **3.** regroup **5.** $1\frac{1}{4}$ **7.** $1\frac{1}{2}$ **9.** $\frac{4}{5}$ **11.** $\frac{3}{5}$ **13.** $1\frac{1}{6}$ **15.** $\frac{7}{8}$ **17.** $\frac{77}{100}$ **19.** $\frac{37}{40}$ **21.** $1\frac{5}{18}$ **23.** $1\frac{17}{100}$ **25.** $\frac{3}{4}$ **27.** $\frac{53}{80}$ **29.** $1\frac{7}{72}$ **31.** $1\frac{13}{40}$ **33.** $3\frac{1}{3}$ **35.** $15\frac{2}{5}$ **37.** $14\frac{1}{5}$ **39.** 15 **41.** $10\frac{5}{12}$ **43.** $3\frac{11}{15}$ **45.** $13\frac{13}{15}$ **47.** $6\frac{19}{24}$ **49.** $20\frac{1}{4}$ **51.** $10\frac{3}{100}$ **53.** $11\frac{3}{8}$ **55.** $36\frac{3}{50}$ **57.** $91\frac{7}{12}$ **59.** $6\frac{1}{2}$ **61.** $10\frac{33}{40}$ **63.** $11\frac{3}{8}$ **65.** $\frac{1}{5}$ **67.** $\frac{2}{5}$ **69.** $\frac{4}{25}$ **71.** $\frac{1}{2}$ **73.** 2 **75.** $\frac{1}{12}$ **77.** $\frac{5}{18}$ **79.** $\frac{1}{20}$ **81.** $\frac{1}{14}$ **83.** $\frac{5}{72}$ **85.** $\frac{1}{4}$ **87.** $4\frac{2}{7}$ **89.** $1\frac{3}{4}$ **91.** 20 **93.** $4\frac{1}{10}$ **95.** $3\frac{1}{3}$ **97.** $3\frac{3}{10}$ **99.** $6\frac{1}{3}$ **101.** $5\frac{1}{2}$ **103.** $4\frac{1}{2}$ **105.** $3\frac{1}{4}$ **107.** $11\frac{4}{5}$ **109.** $6\frac{2}{3}$ **111.** $7\frac{5}{6}$ **113.** $3\frac{13}{24}$ **115.** $15\frac{7}{18}$ **117.** $2\frac{29}{30}$ **119.** $\frac{1}{4}$ **121.** $5\frac{5}{12}$ **123.** $13\frac{39}{40}$ **125.** $\frac{3}{8}$ **127.** $1\frac{11}{40}$ **129.** $16\frac{23}{30}$ **131.** $5\frac{1}{5}$ **133.** $18\frac{11}{20}$ **135.** $4\frac{1}{8}$ **137.** $8\frac{1}{3}$ **139.** $2\frac{14}{15}$ **141.** $\frac{1}{8}$ in. **143. a.** $1\frac{1}{2}$ mi **b.** $\frac{1}{4}$ mi **145.** 5 hr **147.** $790\frac{1}{4}$ ft **149.** $\frac{1}{10}$ **151.** 1 lb

Section 2.4 Practices, pp. 131–140

1, *p. 131:* $\frac{15}{28}$ **2, *p. 131:*** $\frac{81}{100}$ **3, *p. 131:*** 20 **4, *p. 131:*** $\frac{7}{22}$ **5, *p. 132:*** $\frac{2}{9}$ **6, *p. 132:*** $5\frac{1}{4}$ hr **7, *p. 132:*** \$20,769 **8, *p. 133:*** $7\frac{7}{8}$ **9, *p. 133:*** 28 **10, *p. 134:*** $25\frac{1}{2}$ sq in. **11, *p. 135:*** $18\frac{1}{4}$ **12, *p. 136:*** 6 **13, *p. 137:*** 8 **14, *p. 137:*** $2\frac{2}{3}$ yr **15, *p. 138:*** $1\frac{3}{5}$ **16, *p. 138:*** $\frac{7}{16}$ **17, *p. 138:*** 6 lb **18, *p. 139:*** 6 **19, *p. 140:*** $4\frac{1}{2}$

Exercises 2.4, pp. 141–144

1. multiply **3.** reciprocal **5.** invert **7.** $\frac{2}{15}$ **9.** $\frac{5}{12}$ **11.** $\frac{9}{16}$ **13.** $\frac{8}{25}$ **15.** $\frac{35}{32} = 1\frac{3}{32}$ **17.** $\frac{45}{16} = 2\frac{13}{16}$ **19.** $\frac{2}{9}$ **21.** $\frac{7}{12}$ **23.** $\frac{3}{40}$ **25.** $\frac{31}{30} = 1\frac{1}{30}$ **27.** $\frac{40}{3} = 13\frac{1}{3}$ **29.** $\frac{40}{3} = 13\frac{1}{3}$ **31.** 16 **33.** 4 **35.** 4 **37.** $\frac{35}{4} = 8\frac{3}{4}$ **39.** $1\frac{5}{16}$ **41.** $2\frac{1}{8}$ **43.** $\frac{25}{27}$ **45.** $2\frac{2}{3}$ **47.** 1 **49.** $\frac{7}{8}$ **51.** $1\frac{13}{35}$ **53.** $4\frac{41}{100}$ **55.** $7\frac{4}{5}$ **57.** 375 **59.** 8 **61.** 3 **63.** $41\frac{2}{3}$ **65.** $113\frac{1}{3}$ **67.** $1\frac{1}{6}$ **69.** $\frac{7}{12}$ **71.** $\frac{77}{100}$ **73.** $3\frac{3}{8}$ **75.** $\frac{9}{10}$ **77.** $\frac{32}{35}$ **79.** $3\frac{1}{2}$ **81.** $4\frac{4}{9}$ **83.** $1\frac{1}{2}$ **85.** $2\frac{1}{3}$ **87.** $1\frac{1}{5}$ **89.** $\frac{1}{4}$ **91.** $\frac{2}{21}$ **93.** $\frac{1}{9}$ **95.** 40 **97.** $16\frac{1}{3}$ **99.** $13\frac{1}{3}$ **101.** 7 **103.** $6\frac{11}{18}$ **105.** $1\frac{2}{3}$ **107.** $9\frac{22}{27}$ **109.** $100\frac{1}{2}$ **111.** $\frac{7}{90}$ **113.** $\frac{5}{26}$ **115.** $3\frac{1}{5}$ **117.** $\frac{21}{200}$ **119.** $\frac{35}{44}$ **121.** $1\frac{47}{115}$ **123.** $\frac{14}{27}$ **125.** $2\frac{1}{4}$ **127.** $1\frac{7}{18}$ **129.** $4\frac{13}{15}$ **131.** $\frac{87}{160}$ **133.** $3\frac{19}{27}$ **135.** $4\frac{1}{5}$ **137.** $3\frac{1}{8}$ **139.** $11\frac{1}{6}$ **141.** $\frac{1}{20}$ **143.** $\frac{1}{9}$ **145.** $20\frac{13}{16}$ **147.** $2\frac{5}{22}$ **149.** $\frac{21}{40}$ **151.** 8 **153.** $\frac{7}{12}$ **155.** \$1,340 **157.** \$1,116 **159.** $6\frac{1}{4}$ **161.** $\frac{27}{64}$ **163.** 7 times **165. a.** The scented candle **b.** The unscented candle

Chapter 2 Review Exercises, pp. 148–154

1. is; possible answer: it has more than two factors: 1, 3, 9, and 27 **2.** is not; possible answer: 4 is not a prime number **3.** is not; possible answer: the denominator of a fraction must be nonzero **4.** is; possible answer: the numerator, 12, is greater than the denominator, 11 **5.** is; possible answer: if you divide both the numerator and the denominator by the same number, 16, you get $\frac{1}{3}$ **6.** are; possible answer: they have different denominators **7.** is; possible answer: 24 is the least common multiple of 8 and 12 **8.** is not; possible answer: the reciprocal, $\frac{8}{6}$, is formed by switching the numerator and the denominator **9.** 1, 2, 3, 5, 6, 10, 15, 25, 30, 50, 75, 150 **10.** 1, 2, 3, 4, 5, 6, 9, 10, 12, 15, 18, 20, 30, 36, 45, 60, 90, 180 **11.** 1, 3, 19, 57 **12.** 1, 2, 5, 7, 10, 14, 35, 70 **13.** Prime **14.** Composite **15.** Composite **16.** Prime **17.** $2^2 \times 3^2$ **18.** 3×5^2 **19.** $3^2 \times 11$ **20.** 2×3^3 **21.** 42 **22.** 10 **23.** 72 **24.** 60 **25.** $\frac{2}{4}$ **26.** $\frac{6}{12}$ **27.** $1\frac{1}{6}$ **28.** $2\frac{3}{5}$ **29.** Mixed **30.** Proper **31.** Improper **32.** Improper **33.** $\frac{23}{3}$ **34.** $\frac{9}{5}$ **35.** $\frac{91}{10}$ **36.** $\frac{59}{7}$ **37.** $6\frac{1}{2}$ **38.** $4\frac{2}{3}$ **39.** $2\frac{3}{4}$ **40.** 1 **41.** 84 **42.** 4 **43.** 5 **44.** 27 **45.** $\frac{1}{2}$ **46.** $\frac{5}{7}$ **47.** $\frac{2}{3}$ **48.** $\frac{3}{4}$ **49.** $5\frac{1}{2}$ **50.** $8\frac{2}{3}$ **51.** $6\frac{2}{7}$ **52.** $8\frac{5}{7}$ **53.** $>$ **54.** $>$ **55.** $<$ **56.** $>$ **57.** $>$ **58.** $>$ **59.** $>$ **60.** $>$ **61.** $\frac{2}{7}, \frac{3}{8}, \frac{1}{2}$ **62.** $\frac{2}{15}, \frac{1}{5}, \frac{1}{3}$ **63.** $\frac{3}{4}, \frac{4}{5}, \frac{9}{10}$ **64.** $\frac{13}{18}, \frac{7}{9}, \frac{7}{8}$ **65.** $\frac{6}{5} = 1\frac{1}{5}$ **66.** $\frac{3}{4}$ **67.** $\frac{15}{8} = 1\frac{7}{8}$ **68.** $\frac{3}{5}$ **69.** $\frac{11}{15}$ **70.** $1\frac{17}{24}$ **71.** $1\frac{4}{5}$ **72.** $1\frac{37}{40}$ **73.** $5\frac{7}{8}$ **74.** $9\frac{1}{2}$ **75.** $10\frac{3}{5}$ **76.** 8 **77.** $12\frac{1}{3}$ **78.** $4\frac{3}{10}$ **79.** $5\frac{7}{10}$ **80.** $17\frac{13}{24}$ **81.** $23\frac{5}{12}$ **82.** $46\frac{3}{8}$ **83.** $20\frac{3}{4}$ **84.** $56\frac{1}{24}$ **85.** $\frac{1}{4}$ **86.** $\frac{2}{3}$ **87.** 1 **88.** 0 **89.** $\frac{1}{4}$ **90.** $\frac{3}{8}$ **91.** $\frac{7}{20}$ **92.** $\frac{7}{30}$ **93.** $7\frac{1}{2}$ **94.** $2\frac{3}{10}$ **95.** $3\frac{3}{4}$ **96.** $18\frac{1}{2}$ **97.** $6\frac{1}{2}$ **98.** $1\frac{7}{10}$ **99.** $2\frac{2}{3}$ **100.** $\frac{1}{5}$ **101.** $1\frac{4}{5}$ **102.** $\frac{3}{4}$ **103.** $2\frac{1}{2}$ **104.** $3\frac{1}{3}$ **105.** $\frac{3}{10}$ **106.** $2\frac{7}{8}$ **107.** $\frac{7}{12}$ **108.** $3\frac{8}{9}$ **109.** $\frac{2}{3}$ **110.** $9\frac{9}{20}$ **111.** $\frac{3}{16}$ **112.** $\frac{7}{16}$ **113.** $\frac{5}{8}$ **114.** $\frac{1}{6}$ **115.** $5\frac{1}{3}$

116. $\frac{7}{10}$ **117.** $\frac{1}{125}$ **118.** $\frac{8}{27}$ **119.** $\frac{1}{4}$ **120.** $\frac{7}{120}$ **121.** $\frac{24}{25}$ **122.** $1\frac{5}{9}$ **123.** $2\frac{2}{3}$ **124.** $\frac{2}{3}$ **125.** 6 **126.** $18\frac{5}{12}$ **127.** $8\frac{7}{16}$ **128.** $21\frac{1}{4}$ **129.** $\frac{9}{20}$ **130.** $1\frac{9}{16}$ **131.** $37\frac{1}{27}$ **132.** $3\frac{3}{8}$ **133.** $3\frac{1}{8}$ **134.** $1\frac{41}{90}$ **135.** $2\frac{1}{10}$ **136.** $7\frac{1}{5}$ **137.** $\frac{3}{2}$ **138.** $\frac{2}{3}$ **139.** $\frac{1}{8}$ **140.** 4 **141.** $\frac{7}{40}$ **142.** $\frac{5}{81}$ **143.** $\frac{2}{15}$ **144.** $\frac{1}{200}$ **145.** $\frac{3}{4}$ **146.** $1\frac{1}{3}$ **147.** 30 **148.** $8\frac{3}{4}$ **149.** $1\frac{1}{6}$ **150.** $1\frac{4}{5}$ **151.** 2 **152.** 4 **153.** $1\frac{3}{4}$ **154.** $\frac{4}{7}$ **155.** $1\frac{7}{12}$ **156.** $\frac{12}{19}$ **157.** $5\frac{1}{2}$ **158.** $2\frac{11}{20}$ **159.** 2 **160.** 3 **161.** $9\frac{3}{4}$ **162.** $1\frac{3}{10}$ **163.** $5\frac{1}{3}$ **164.** $7\frac{5}{9}$ **165.** No **166.** 50¢ **167.** $\frac{1}{4}$ **168.** $\frac{2}{9}$ **169.** The Filmworks camera **170.** $\frac{7}{12}$ **171.** The patient got back more than $\frac{1}{3}$, because $\frac{275}{700} = \frac{11}{28} = \frac{33}{84}$, which is greater than $\frac{1}{3} = \frac{28}{84}$. **172.** Yes it should, because $\frac{23}{32}$ is greater than $\frac{2}{3}$. $\frac{23}{32} = \frac{69}{96}$, whereas $\frac{2}{3} = \frac{64}{96}$ **173. a.** $\frac{12}{23}$ **b.** $\frac{3}{4}$ **174. a.** Lisa Gregory **b.** Monica Yates **175.** $\frac{3}{4}$ **176.** $\frac{11}{12}$ oz **177.** $\frac{1}{4}$ carat **178.** $\frac{3}{5}$ **179.** 12 women **180.** 2,685 undergraduate students **181.** 2 awardees **182.** 7 lb **183.** $1,050 **184.** 2 times **185.** 19 fish **186.** $11\frac{3}{4}$ mi **187.** $11\frac{1}{2}$ ft **188.** $7\frac{5}{12}$ hr **189.** 1,500 fps **190.** 500 lb/sq in. **191.** $281\frac{1}{4}$ lb **192.** 3,100,000 **193.** 8 orbits **194.** $25\frac{5}{8}$ sq mi

195.

Employee	Saturday	Sunday	Total
L. Chavis	$7\frac{1}{2}$	$4\frac{1}{4}$	$11\frac{3}{4}$
R. Young	$5\frac{3}{4}$	$6\frac{1}{2}$	$12\frac{1}{4}$
Total	$13\frac{1}{4}$	$10\frac{3}{4}$	24

196.

Worker	Hours per Day	Days Worked	Total Hours	Wage per Hour	Gross Pay
Maya	5	3	15	$7	$105
Noel	$7\frac{1}{4}$	4	29	$10	$290
Alisa	$4\frac{1}{2}$	$5\frac{1}{2}$	$24\frac{3}{4}$	$9	$222\frac{3}{4}$

197. $10\frac{10}{11}$ lb **198.** $22\frac{1}{2}$ cups **199.** 3,200 mi **200.** 6 times

Chapter 2 Posttest, p. 155

1. 1, 3, 7, 9, 21, 63 **2.** 2×3^3 **3.** $\frac{4}{9}$ **4.** $\frac{12}{1}$ **5.** $10\frac{1}{4}$ **6.** $\frac{7}{8}$ **7.** $\frac{5}{10}$ **8.** 24 **9.** $1\frac{13}{24}$ **10.** $8\frac{7}{40}$ **11.** $4\frac{2}{7}$ **12.** $5\frac{23}{30}$ **13.** $\frac{1}{81}$ **14.** 12 **15.** $\frac{7}{9}$ **16.** $7\frac{5}{6}$ **17.** $\frac{10}{11}$ **18.** $19\frac{1}{5}$ mi **19.** $\frac{5}{6}$ hr **20.** $94

Chapter 2 Cumulative Review Exercises, p. 156

1. Five million, three hundred fifteen **2.** 1,900,000 **3.** 581,400 **4.** 908 **5.** 1 **6.** $\frac{1}{25}$ **7.** $2^2 \times 3 \times 7$ **8.** 120 **9.** $\frac{3}{4}$ **10.** $\frac{3}{8}$ **11.** $1\frac{11}{24}$ **12.** $6\frac{2}{5}$ **13.** 7 **14.** $1\frac{5}{11}$ **15.** $37 billion **16.** 1 million times **17.** 12 above **18.** Yes. The room has 370 square feet of wall area. **19.** $\frac{1}{3}$ **20.** 4 pieces

CHAPTER 3

Chapter 3 Pretest, p. 158

1. Hundredths **2.** Four and twelve thousandths **3.** 3.1 **4.** 0.0029 **5.** 21.52 **6.** 7.3738 **7.** 11.69 **8.** 9.81 **9.** 8,300 **10.** 18.423 **11.** 0.0144 **12.** 7.1 **13.** 0.00605 **14.** 32.7 **15.** 0.875 **16.** 2.83 **17.** One with a pH value of 2.95 **18.** $58.44 billion **19.** 3 times **20.** $3.74

Section 3.1 Practices: pp. 160–167

1, *p. 160:* **a.** The tenths place **b.** The ten-thousandths place **c.** The thousandths place **2, *p. 161:*** $\frac{7}{8}$ **3, *p. 161:*** $2\frac{3}{100}$ **4, *p. 162:*** **a.** $5\frac{3}{5}$ **b.** $5\frac{3}{5}$ **5, *p. 162:*** **a.** $7\frac{3}{1,000}$ **b.** $4\frac{1}{10}$ **6, *p. 162:*** **a.** Sixty-one hundredths **b.** Four and nine hundred twenty-three thousandths **c.** Seven and five hundredths **7, *p. 163:*** **a.** 0.043 **b.** 10.26 **8, *p. 163:*** 3.14 **9, *p. 164:*** 0.8297 **10, *p. 164:*** 3.51, 3.5, 3.496 **11, *p. 165:*** The one with the rating of 8.1, because 9 > 8.2 > 8.1 **12, *p. 166–167:*** **a.** 748.1 **b.** 748.08 **c.** 748.077 **d.** 748 **e.** 700 **13, *p. 167:*** 7.30 **14, *p. 167:*** 11.7 m

Exercises 3.1, pp. 168–172

1. right **3.** hundredths **5.** greater **7.** $2.\underline{7}8$ **9.** $9.0\underline{1}$ **11.** $2.001\underline{7}5$ **13.** $823.00\underline{1}$ **15.** Tenths **17.** Hundredths **19.** Thousandths **21.** Ones **23.** Fifty-three hundredths **25.** Three hundred five thousandths **27.** Six tenths **29.** Five and seventy-two hundredths **31.** Twenty-four and two thousandths **33.** 0.8 **35.** 1.041 **37.** 60.01 **39.** 4.107 **41.** 3.2 m **43.** $\frac{3}{5}$ **45.** $\frac{39}{100}$ **47.** $1\frac{1}{2}$ **49.** 8 **51.** $5\frac{3}{250}$ **53.** > **55.** < **57.** > **59.** = **61.** < **63.** 7, 7.07, 7.1 **65.** 4.9, 5.001, 5.2 **67.** 9.1 mi, 9.38 mi, 9.6 mi **69.** 17.4 **71.** 3.591 **73.** 37.1 **75.** 0.40 **77.** 7.06 **79.** 9 mi

81.

To the Nearest	8.0714	0.9916
Tenth	8.1	1.0
Hundredth	8.07	0.99
Ten	10	0

83. $0.\underline{0}24$ **85.** 870.06 **87.** 2.04 m, 2.14 m, 2.4 m **89.** Twenty-three and nine hundred thirty-four thousandths **91.** Eighteen and seven tenths; eighteen and eight tenths **93.** Two hundred eleven and seven tenths, sixty-nine and four tenths, one hundred eighty-nine and eight tenths, one hundred ninety-three and five tenths, forty-seven and five tenths **95.** One hundred-thousandth; eight hundred-thousandths **97.** 1.2 acres **99.** 74.59 mph **101.** 14.7 lb **103.** 9.6 V **105.** 352.1 kWh **107.** Evgeni Plushenko **109.** Last winter **111.** 2005 **113.** Husband's **115.** $57.03 **117.** 0.001 **119.** 1.4

Section 3.2 Practices: pp. 173–177

1, *p. 173:* 10.387 **2, *p. 173:*** 39.3 **3, *p. 174:*** 102.1°F **4, *p. 174:*** 46.2125 **5, *p. 175:*** $485.43 **6, *p. 175:*** 16.9 mi **7, *p. 175:*** 22.13 mi **8, *p. 176:*** 0.863 **9, *p. 176:*** 0.079 **10, *p. 176:*** 0.5744 **11, *p. 177:*** Possible estimate: $480 **12, *p. 177:*** Possible estimate: $2 million

Calculator Practices, p. 178

13, *p. 178:* 79.23; **14, *p. 178:*** 0.00002

Exercises 3.2, pp. 179–182

1. decimal points **3.** sum **5.** 9.33 **7.** 0.9 **9.** 8.13 **11.** 21.45 **13.** 7.67 **15.** $77.21 **17.** 1.08993 **19.** 24.16 **21.** 44.422 **23.** 20.32 mm **25.** 16.682 kg **27.** 23.30595 **29.** 0.7 **31.** 16.8 **33.** 18.41 **35.** 75.63 **37.** 22.324 **39.** 0.17 **41.** 0.1142 **43.** 6.2 **45.** 15.37 **47.** 5.9 **49.** 6.21 **51.** 1.85 lb **53.** 4.9°F **55.** 39.752 **57.** 27.9 mg **59.** 3.205 **61.** 21.19896 **63.** c; possible estimate: 0.084 **65.** b; possible estimate: 0.06 **67.** 7.771 **69.** 7.75 lb **71.** 11.6013 **73.** $1.03 **75.** 56.8 centuries **77.** $1.7 million **79.** 6.84 in. **81.** Yes; 2.8 + 2.9 + 2.6 + 1.6 = 9.9

83. a.

Gymnast	VT	UB	BB	FX	AA
Nastia Liukin (U.S.)	15.1	15.95	15.975	15.35	62.375
Yang Yilin (China)	15.2	16.65	15.5	15	62.35
Shawn Johnson (U.S)	16	15.325	15.975	15.425	62.725

b. Shawn Johnson **85.** A total of 16.2 mg of iron; no, she needs 1.8 mg more

Section 3.3 Practices: pp. 183–186

1, *p. 183:* 9.835 **2, *p. 184:*** 1.4 **3, *p. 184:*** 0.01 **4, *p. 184:*** 0.024 **5, *p. 184:*** 9.91 **6, *p. 185:*** 325 **7, *p. 185:*** 327,000 **8, *p. 185:*** **a.** 18.015 **b.** 18 **9, *p. 186:*** 0.0003404; possible estimate: $0.004 \times 0.09 = 0.00036$ **10, *p. 186:*** 3.6463 **11, *p. 186:*** Possible answer: 1,200 mi

Calculator Practices, p. 187

12, ***p. 187:*** 815.6 **13,** ***p. 187:*** 9.261

Exercises 3.3, pp. 188–191

1. multiplication **3.** two **5.** square **7.** 2.99212 **9.** 204.360 **11.** 2,492.0 **13.** 0.0000969 **15.** 2,870.00 **17.** $0.73525 **19.** 0.54 **21.** 0.4 **23.** 0.02 **25.** 0.0028 **27.** 0.765 **29.** 2.016 **31.** 7.602 **33.** 0.5 **35.** 5.852 **37.** 151.14 **39.** 3.7377 **41.** 1.7955 **43.** 8,312.7 **45.** 23 **47.** 0.09 **49.** 1.05 **51.** 0.000000001 **53.** 42.5 ft **55.** 1.4 mi **57.** 42.77325 **59.** 272,593.75 **61.** 70 **63.** 25.75 **65.** 1.09 **67.** 2.86 **69.** 3.952 **71.** 0.14

73.

Input	Output
1	$3.8 \times \mathbf{1} - 0.2 = 3.6$
2	$3.8 \times \mathbf{2} - 0.2 = 7.4$
3	$3.8 \times \mathbf{3} - 0.2 = 11.2$
4	$3.8 \times \mathbf{4} - 0.2 = 15$

75. a; possible estimate: 50 **77.** b; possible estimate: 0.014 **79.** 8.75 **81.** 0.068 **83.** 4.48 **85.** 2,900 fps **87.** 57,900,000 km **89.** 254.3 sq ft **91.** 1.25 mg **93.** 1,308 calories

95. a.

Purchase	Quantity	Unit Price	Price
Belt	1	$11.99	$11.99
Shirt	3	$16.95	$50.85
Total Price			$62.84

b. $17.16 **97.** 88.81 in.

Section 3.4 Practices: pp. 193–199

1, ***p. 193:*** 0.375 **2,** ***p. 193:*** 7.625 **3,** ***p. 194:*** 83.3 **4.** ***p. 194:*** 0.1 **5,** ***p. 196:*** 18.04 **6,** ***p. 196:*** 2,050 **7,** ***p. 197:*** 73.4 **8,** ***p. 197:*** 0.0341 **9,** ***p. 198:*** 0.00086 **10,** ***p. 198:*** 1.5 **11,** ***p. 199:*** 21.1; possible estimate: 20 **12,** ***p. 199:*** 295.31 **13,** ***p. 200:*** 8 times as great

Calculator Practices, p. 200

14, ***p. 200:*** 0.2 **15,** ***p. 200:*** 4.29

Exercises 3.4, pp. 201–204

1. decimal **3.** right **5.** quotient **7.** 0.5 **9.** 0.375 **11.** 3.7 **13.** 1.625 **15.** 6.2 **17.** 21.03 **19.** 0.67 **21.** 0.78 **23.** 3.11 **25.** 5.06 **27.** 4.25 **29.** 4.2 **31.** 1.375 **33.** 8.5 **35.** 3.286 **37.** 0.273 **39.** 6.571 **41.** 70.077 **43.** 58.82 **45.** 0.0663 **47.** 2.8875 **49.** 0.286 **51.** 4.3 **53.** 0.0015 **55.** 1.73 **57.** 2.875 **59.** 4 **61.** 70.4 **63.** 94 **65.** 12.5 **67.** 0.3 **69.** 0.2 **71.** 0.952 **73.** 0.00082 **75.** 383.88 **77.** 0.01 **79.** 9.23 **81.** 9,666.67 **83.** 1,952.38 **85.** 325.18 **87.** 67.41 **89.** 41.61 **91.** 2.765 **93.** 32.9 **95.** 52.2 **97.** 4.05 **99.** 396.5 **101.** 49.9

103.

Input	Output
1	$\mathbf{1} \div 5 - 0.2 = 0$
2	$\mathbf{2} \div 5 - 0.2 = 0.2$
3	$\mathbf{3} \div 5 - 0.2 = 0.4$
4	$\mathbf{4} \div 5 - 0.2 = 0.6$

105. c; possible estimate: 50 **107.** b; possible estimate: 0.2 **109.** 0.8 **111.** 1.17 **113.** 0.45 **115.** 0.0037 in. per yr **117. a.** 0.6 **b.** 0.55 **c.** The women's team has a better record. The team won $\frac{3}{5}$, or 0.6, of the games played, and the men's team won $\frac{11}{20}$, or 0.55, of the games played.

119. a.

SUV	Distance Driven (in miles)	Gasoline Used (in gallons)	Gasoline Mileage (miles per gallon)
Honda CR-V	40.5	1.9	21
Ford Escape Hybrid	62.4	2.4	26
GMC Terrain	42.6	2.4	18

b. Ford Escape Hybrid **121.** 2,000 shares **123.** 13 times **125.** 0.4 lb **127.** .366

Chapter 3 Review Exercises, pp. 207–210

1. is not; possible answer: it does not have a decimal point (or it is written in fractional form). **2.** is not; possible answer: decimal places are to the right of the decimal point, and this digit is to the left. **3.** is not; possible answer: the number to the right of the critical digit 2 is 6, so we round up, getting 48.73 **4.** can; possible answer: then each column will contain digits with the same place value. **5.** is not; possible answer: no digits repeat indefinitely. **6.** is; possible answer: the dividend is smaller than the divisor. **7.** Hundredths **8.** Tenths **9.** Tenths **10.** Ten-thousandths **11.** $\frac{7}{20}$ **12.** $8\frac{1}{5}$ **13.** $4\frac{7}{1{,}000}$ **14.** 10 **15.** Seventy-two hundredths **16.** Five and six tenths **17.** Three and nine ten-thousandths **18.** Five hundred ten and thirty-six thousandths **19.** 0.007 **20.** 2.1 **21.** 0.09 **22.** 7.041 **23.** $<$ **24.** $>$ **25.** $>$ **26.** $>$ **27.** 1.002, 0.8, 0.72 **28.** 0.004, 0.003, 0.00057 **29.** 7.3 **30.** 0.039 **31.** 4.39 **32.** $899 **33.** 12.11 **34.** 52.75 **35.** $24.13 **36.** 12 m **37.** 28.78 **38.** 87.752 **39.** 1.834 **40.** 48.901 **41.** 98.2033 **42.** $90,948.80 **43.** 2.912 **44.** 1,008 **45.** 0.00001 **46.** 13.69 **47.** 2,710 **48.** 0.034 **49.** 5.75 **50.** 13.5 **51.** 1,569.36846 **52.** 441.760662 **53.** 0.625 **54.** 90.2 **55.** 4.0625 **56.** 0.045 **57.** 0.17 **58.** 0.29 **59.** 8.33 **60.** 11.22 **61.** 0.65 **62.** 1.6 **63.** 0.175 **64.** 0.277 **65.** 5.2 **66.** 3.2 **67.** 23.7 **68.** 16,358.3 **69.** 1.9 **70.** 360.7 **71.** 3.0 **72.** 0.3 **73.** 1.18 **74.** 117 **75.** 34.375 **76.** 1.4 **77.** 54.49 sec **78.** $14.14 **79.** Four ten-millionths **80.** 14.22 in. **81.** 1.5 AU. **82.** 4.35 times **83.** $0.06 **84.** $250 **85.** 7.19 g **86.** 3.5°C **87.** 36,162.45

88.

Period Ending	Google	Yahoo!
June 30	5.523	1.573
September 30	5.945	1.575
December 31	6.674	1.732
March 31	6.775	1.597

$18.440 billion

Chapter 3 Posttest, p. 211

1. 6 **2.** Five and one hundred two thousandths **3.** 320.15 **4.** 0.00028 **5.** $3\frac{1}{25}$ **6.** 0.004 **7.** 4.354 **8.** $5.66 **9.** 20.9 **10.** 5.72 **11.** 0.001 **12.** 3.36 **13.** 0.0029 **14.** 32.7 **15.** 0.375 **16.** 4.17 **17.** 0.01 lb **18.** 2.6 ft **19.** Belmont Stakes **20.** $32.40

Chapter 3 Cumulative Review Exercises, p. 212

1. 1,000,000 **2.** 2,076 **3.** 27,403 **4.** 900 sq m **5.** 42 **6.** 1, 2, 3, 4, 5, 6, 10, 12, 15, 20, 30, 60 **7.** $1\frac{1}{2}$ **8.** $2\frac{2}{3}$ **9.** 32 **10.** $\frac{17}{30}$ **11.** $4\frac{18}{25}$ **12.** 38.4 **13.** 60.213 **14.** 610 **15.** $0.17 **16.** $\frac{2}{15}$ acre **17.** 325 **18.** 26,000 mi **19.** $193.86 **20.** 2.6

CHAPTER 4

Chapter 4 Pretest, p. 214

1. Possible answer: four less than t **2.** Possible answer: quotient of y and three **3.** $m + 8$ **4.** $2n$ **5.** 4 **6.** $1\frac{1}{2}$ **7.** $x + 3 = 5$ **8.** $4y = 12$ **9.** $x = 6$ **10.** $t = 10$ **11.** $n = 13$ **12.** $a = 12$ **13.** $m = 6.1$ **14.** $n = 30$ **15.** $m = 13\frac{1}{2}$, or 13.5 **16.** $n = 15$ **17.** $63 = x + 36$; 27 moons **18.** $6.75 = x - 2.75$; \$9.50 **19.** $\frac{2}{5}x = 39{,}900$; 99,750 sq mi **20.** $40 = 10x$; 4 mg

Section 4.1 Practices, pp. 216–218

1, *p. 216:* Answers may vary **a.** One-half of p **b.** x less than 5 **c.** y divided by 4 **d.** 3 more than n **e.** $\frac{3}{5}$ of b **2, *p. 216:*** **a.** $x + 9$ **b.** $10y$ **c.** $n - 7$ **d.** $p \div 5$ **e.** $\frac{2}{5}v$ **3, *p. 217:*** **a.** $q + 12$, where q represents the quantity **b.** $\frac{9}{a}$, where a represents the account balance **c.** $\frac{2}{7}c$, where c represents the cost **4, *p. 217:*** $\frac{h}{4}$ hr **5, *p. 217:*** $s - 3$ **6, *p. 218:*** **a.** 25 **b.** 0.38 **c.** 4.8 **d.** 26.6 **7, *p. 218:*** $\frac{1}{5}n$ dollars; \$750 **8, *p. 218:*** The total amount was $(18.45 + t)$ dollars; \$21.45 for $t = \$3$

Exercises 4.1, pp. 219–221

1. variable **3.** algebraic **5.** 9 more than t; t plus 9 **7.** c minus 12; 12 subtracted from c **9.** c divided by 3; the quotient of c and 3 **11.** 10 times s; the product of 10 and s **13.** y minus 10; 10 less than y **15.** 7 times a; the product of 7 and a **17.** x divided by 6; the quotient of x and 6 **19.** x minus $\frac{1}{2}$; $\frac{1}{2}$ less than x **21.** $\frac{1}{4}$ times w; $\frac{1}{4}$ of w **23.** 2 minus x; the difference between 2 and x **25.** 1 increased by x; x added to 1 **27.** 3 times p; the product of 3 and p **29.** n decreased by 1.1; n minus 1.1 **31.** y divided by 0.9; the quotient of y and 0.9 **33.** $x + 10$ **35.** $n - 1$ **37.** $y + 5$ **39.** $t \div 6$ **41.** $10y$ **43.** $w - 5$ **45.** $n + \frac{4}{5}$ **47.** $z \div 3$ **49.** $\frac{2}{7}x$ **51.** $k - 6$ **53.** $n + 12$ **55.** $n - 5.1$ **57.** 26 **59.** 2.5 **61.** 15 **63.** $1\frac{1}{6}$ **65.** 1.1 **67.** $\frac{1}{5}$

69.

x	$x + 8$
1	9
2	10
3	11
4	12

71.

n	$n - 0.2$
1	0.8
2	1.8
3	2.8
4	3.8

73.

x	$\frac{3}{4}x$
4	3
8	6
12	9
16	12

75.

z	$\frac{z}{2}$
2	1
4	2
6	3
8	4

77. $x - 7$ **79.** Possible answers: n over 2; n divided by 2 **81.** $3.5t$ **83.** Possible answers: 6 more than x; the sum of x and 6 **85.** $(m - 25)$ mg **87.** $25° + 90° + d°$, or $115° + d°$ **89.** 220 mi **91.** **a.** $2.5w$ dollars **b.** \$22.50

Section 4.2 Practices, pp. 223–227

1, *p. 223:* **a.** $n - 5.1 = 9$ **b.** $y + 2 = 12$ **c.** $n - 4 = 11$ **d.** $n + 5 = 7\frac{3}{4}$ **2, *p. 224:*** $p - 6 = 49.95$, where p is the regular price. **3, *p. 225:*** $x = 9$; **4, *p. 225:*** $t = 2.7$; **5, *p. 226:*** $m = 5\frac{1}{4}$; **6, *p. 226:*** **a.** $11 = m - 4$; $m = 15$ **b.** $12 + n = 21$; $n = 9$ **7, *p. 227:*** $x + 3.99 = 27.18$; \$23.19 **8, *p. 227:*** $269{,}000 = x - 394{,}000$; 663,000 sq mi

Exercises 4.2, pp. 228–231

1. equation **3.** subtract **5.** $z - 9 = 25$ **7.** $7 + x = 25$ **9.** $t - 3.1 = 4$ **11.** $\frac{3}{2} + y = \frac{9}{2}$ **13.** $n - 3\frac{1}{2} = 7$ **15.** **a.** Yes **b.** No **c.** Yes **d.** No **17.** Subtract 4. **19.** Add 6. **21.** Add 7. **23.** Subtract 21. **25.** $a = 31$ **27.** $y = 2$ **29.** $x = 12$ **31.** $n = 4$ **33.** $m = 2$ **35.** $y = 90$ **37.** $z = 2.9$ **39.** $n = 8.9$ **41.** $y = 0.9$ **43.** $x = 8\frac{2}{3}$ **45.** $m = 5\frac{1}{3}$ **47.** $x = 3\frac{3}{4}$ **49.** $c = 47\frac{1}{5}$ **51.** $x = 13$ **53.** $y = 6\frac{1}{4}$ **55.** $a = 3\frac{5}{12}$ **57.** $x = 8.2$ **59.** $y = 19.91$ **61.** $x = 4.557$ **63.** $y = 10.251$ **65.** $n + 3 = 11$; $n = 8$ **67.** $y - 6 = 7$; $y = 13$ **69.** $n + 10 = 19$; $n = 9$ **71.** $x + 3.6 = 9$; $x = 5.4$ **73.** $n - 4\frac{1}{3} = 2\frac{2}{3}$; $n = 7$ **75.** Equation c **77.** Equation a **79.** $a = 14.5$ **81.** Equation b **83.** Yes **85.** $4.2 + n = 8$ **87.** Add 1.9. **89.** $x + 12 = 106$; \$94 **91.** $40° + x = 90°$; 50° **93.** $621{,}000 = x - 13{,}000$; \$634,000 **95.** $45 = x - 20$; 65 mph **97.** $m + 1{,}876{,}674{,}000 = 4{,}023{,}362{,}895$; \$2,146,688,895

Section 4.3 Practices, pp. 232–237

1, *p. 232:* **a.** $2x = 14$ **b.** $\frac{a}{6} = 1.5$ **c.** $\frac{n}{0.3} = 1$ **d.** $10 = \frac{1}{2}n$ **2, *p. 232:*** $15 = 3w$ **3, *p. 233:*** $x = 5$ **4, *p. 233:*** $a = 6$ **5, *p. 234:*** $x = 4$ **6, *p. 234:*** $a = 2.88$ **7, *p. 235:*** $x = 16$ **8, *p. 235:*** **a.** $12 = \frac{z}{6}$, $z = 72$; $12 \stackrel{?}{=} \frac{72}{6}$, $12 \stackrel{\checkmark}{=} 12$ **b.** $16 = 2x$, $8 = x$, *or* $x = 8$; $16 \stackrel{?}{=} 2(8)$, $16 \stackrel{\checkmark}{=} 16$ **9, *p. 236:*** $1.6 = 5x$; 0.32 km **10, *p. 236:*** $\frac{x}{25.5} = 87$; \$2,218.50 **11, *p. 237:*** $\frac{3}{8}p = 150{,}000$; \$400.000

Exercises 4.3, pp. 238–241

1. divide **3.** substituting **5.** equation **7.** $\frac{3}{4}y = 12$ **9.** $\frac{x}{7} = \frac{7}{2}$ **11.** $\frac{1}{3}x = 2$ **13.** $\frac{n}{3} = \frac{1}{3}$ **15.** $9a = 27$ **17.** **a.** Yes **b.** No **c.** No **d.** No **19.** Divide by 3. **21.** Multiply by 2. **23.** Divide by $\frac{3}{4}$, or multiply by $\frac{4}{3}$. **25.** Divide by 1.5. **27.** $x = 6$ **29.** $x = 18$ **31.** $n = 4$ **33.** $x = 91$ **35.** $y = 4$ **37.** $b = 20$ **39.** $m = 157.5$ **41.** $t = 0.4$ **43.** $x = \frac{3}{2}$, or $1\frac{1}{2}$ **45.** $x = 36$ **47.** $t = 3$ **49.** $y = \frac{2}{5}$ **51.** $n = 700$ **53.** $x = 12.5$ **55.** $x = \frac{1}{2}$ **57.** $m = 6$ **59.** $x \approx 6.8$ **61.** $x \approx 4.9$ **63.** $8n = 56$; $n = 7$ **65.** $\frac{3}{4}y = 18$; $y = 24$ **67.** $\frac{x}{5} = 11$; $x = 55$ **69.** $2x = 36$; $x = 18$ **71.** $\frac{1}{2}a = 4$; $a = 8$ **73.** $\frac{n}{5} = 1\frac{3}{5}$; $n = 8$ **75.** $\frac{n}{2.5} = 10$; $n = 25$ **77.** Equation d **79.** Equation a **81.** $x = 5.5$ **83.** Equation d **85.** Yes **87.** $2x = 5$ **89.** Multiply by 2. **91.** $4s = 60$; 15 units **93.** $56 = \frac{1}{2}x$; 112 mi **95.** $\frac{c}{3} = 8.99$; \$26.97 **97.** **a.** $\frac{2}{5}x = 60$; 150 ml **b.** 90 ml **99.** $\frac{1}{3}x = 128$; \$384 million **101.** $\frac{p}{3{,}537{,}438} = 86.8$; 307,000,000 people

Chapter 4 Review Exercises, pp. 243–245

1. is; possible answer: x represents an unknown number **2.** is not; possible answer: constants are known numbers such as 6 or -5 **3.** can; possible answer: algebraic expressions combine constants, variables, and arithmetic operations **4.** cannot; possible answer: an equation is a mathematical statement that two expressions are equal **5.** is; possible answer: substituting 28 for x makes the equation $72 - 28 = 44$ a true statement **6.** is; possible answer: x is alone on one side of the equation **7.** x plus 1 **8.** Four more than y **9.** w minus 1 **10.** Three less than s **11.** c divided by 7 **12.** The quotient of a and 10 **13.** Two times x **14.** The product of 6 and y **15.** y divided by 0.1 **16.** The quotient of n and 1.6 **17.** One-third of x **18.** One-tenth of w **19.** $m + 9$ **20.** $b + \frac{1}{2}$ **21.** $y - 1.4$ **22.** $z - 3$ **23.** $\frac{3}{x}$ **24.** $n \div 2.5$ **25.** $3n$ **26.** $12n$ **27.** 12 **28.** 19 **29.** 0 **30.** 6 **31.** 0.3 **32.** 6.5 **33.** $1\frac{1}{2}$ **34.** $\frac{5}{12}$ **35.** 0.4 **36.** $4\frac{1}{2}$ **37.** 1.6 **38.** 9 **39.** $x = 9$ **40.** $y = 9$ **41.** $n = 26$ **42.** $b = 20$ **43.** $a = 3.5$ **44.** $c = 7.5$ **45.** $x = 11$ **46.** $y = 2$ **47.** $w = 1\frac{1}{2}$ **48.** $s = \frac{1}{3}$ **49.** $c = 6\frac{3}{4}$ **50.** $p = 11\frac{2}{3}$ **51.** $m = 5$ **52.** $n = 0$ **53.** $c = 78$ **54.** $y = 90$ **55.** $n = 11$ **56.** $x = 25$ **57.** $x = 31.0485$ **58.** $m = 26.6225$ **59.** $n - 19 = 35$ **60.** $a - 37 = 234$ **61.** $9 + n = 15\frac{1}{2}$ **62.** $n + 26 = 30\frac{1}{3}$ **63.** $2y = 16$ **64.** $25t = 175$ **65.** $34 = \frac{n}{19}$ **66.** $17 = \frac{z}{13}$ **67.** $\frac{1}{3}n = 27$ **68.** $\frac{2}{5}n = 4$ **69.** **a.** No **b.** Yes **c.** Yes **d.** No **70.** **a.** Yes **b.** No **c.** No **d.** Yes **71.** $x = 5$ **72.** $t = 2$ **73.** $a = 105$ **74.** $n = 54$ **75.** $y = 9$ **76.** $r = 10$ **77.** $w = 90$ **78.** $x = 100$ **79.** $y = 20$ **80.** $a = 120$ **81.** $n = 32$ **82.** $b = 32$ **83.** $m = 3.15$ **84.** $z = 0.57$ **85.** $x = \frac{2}{5}$, or 0.4 **86.** $t = \frac{1}{2}$, or 0.5 **87.** $m = 1.2$ **88.** $b = 9.8$ **89.** $x = 12.5$ **90.** $x = 1.4847$ **91.** $2h$ degrees; 6 degrees **92.** $\frac{d}{20}$ dollars per hr; \$9.55 per hr **93.** $\$0.89p$; \$2.67 **94.** $(3{,}000 + d)$ dollars; \$3,225 **95.** $x + 238 = 517$; \$279 **96.** $225 = x + 50$; 175 **97.** $2.9x = 100$; 34 L

98. $\frac{1}{4}x = 500{,}000$; 2,000,000 people **99.** $\frac{x}{6} = 30$; 180 lb
100. $2.5x = 3{,}000{,}000{,}000{,}000$; \$1,200,000,000,000 or \$1.2 trillion
101. $98.6 + x = 101$; 2.4°F **102.** $x - 256 = 8{,}957$; 9,213 applications

Chapter 4 Posttest, p. 246

1. Possible answer: x plus $\frac{1}{2}$ **2.** Possible answer: the quotient of a and 3 **3.** $n - 10$ **4.** $\frac{8}{p}$ **5.** 0 **6.** $\frac{1}{4}$ **7.** $x - 6 = 4\frac{1}{2}$ **8.** $\frac{y}{8} = 3.2$ **9.** $x = 0$ **10.** $y = 12$ **11.** $n = 27$ **12.** $a = 738$ **13.** $m = 7.8$ **14.** $n = 50$ **15.** $x = \frac{11}{20}$ **16.** $n = 760$ **17.** $1\frac{3}{4} + x = 2\frac{1}{4}$; $\frac{1}{2}$ lb **18.** $\frac{1}{4}x = 500$; 2,000 wolves **19.** $1.5x = 9$; 6 billion **20.** $x - 19.8 = 7.6$; 27.4°C

Chapter 4 Cumulative Review Exercises, pp. 247–248

1. 314,200 **2.** c **3.** 23,316 **4.** 1,030 **5.** $\frac{84}{96}$ **6.** $1\frac{2}{5}$ **7.** $5\frac{3}{8}$ **8.** Five and two hundred thirty-nine thousandths **9.** $<$ **10.** 3.89 **11.** 0.0075 **12.** Yes **13.** $n = 7.8$ **14.** $x = 32$ **15.** 7,200 images **16.** He got back $\frac{2}{7}$ of his money, which is less than $\frac{1}{3}$. **17.** $\frac{3}{5}x = 300{,}000{,}000$; 500 million tons **18.** 92.4 lb **19.** 55,000 beehives **20.** **a.** $4.5x = 2{,}900{,}000$ **b.** 600,000 personnel

CHAPTER 5

Chapter 5 Pretest, p. 250

1. $\frac{3}{4}$ **2.** $\frac{2}{5}$ **3.** $\frac{5}{3}$ **4.** $\frac{19}{51}$ **5.** $\frac{16 \text{ gal}}{5 \text{ min}}$ **6.** $\frac{5 \text{ mg}}{3 \text{ hr}}$ **7.** $\frac{2 \text{ dental assistants}}{1 \text{ dentist}}$ **8.** $\frac{1 \text{ calculator}}{1 \text{ student}}$ **9.** $\frac{\$230}{\text{box}}$ **10.** $\frac{\$0.50}{\text{bottle}}$ **11.** True **12.** False **13.** $x = 9$ **14.** $x = 31\frac{1}{2}$ **15.** $x = 16$ **16.** $x = 160$ **17.** $\frac{4}{5}$ **18.** 200 lb/min **19.** 4.5 in. **20.** 76 mi

Section 5.1 Practices, pp. 251–255

1, ***p. 251:*** $\frac{2}{3}$ **2,** ***p. 252:*** $\frac{9}{5}$ **3,** ***p. 252:*** $\frac{20}{19} \approx 1.05 > 1$; yes **4,** ***p. 253:*** **a.** $\frac{5 \text{ mL}}{2 \text{ min}}$ **b.** $\frac{3 \text{ lb}}{2 \text{ wk}}$ **5,** ***p. 253:*** **a.** 48 ft/sec **b.** 0.375 hit per time at bat **6,** ***p. 254:*** 1.5 min/city block **7,** ***p. 254:*** **a.** \$174/flight **b.** \$2.75/hr **c.** \$0.99/download **8,** ***p. 255:*** The 150-caplet bottle

Exercises 5.1, pp. 256–260

1. quotient **3.** simplest form **5.** denominator **7.** $\frac{2}{3}$ **9.** $\frac{2}{3}$ **11.** $\frac{11}{7}$ **13.** $\frac{3}{2}$ **15.** $\frac{1}{4}$ **17.** $\frac{4}{3}$ **19.** $\frac{1}{1}$ **21.** $\frac{5}{3}$ **23.** $\frac{7}{24}$ **25.** $\frac{20}{1}$ **27.** $\frac{8}{7}$ **29.** $\frac{4}{5}$ **31.** $\frac{5 \text{ calls}}{2 \text{ days}}$ **33.** $\frac{36 \text{ cal}}{5 \text{ min}}$ **35.** $\frac{1 \text{ million hits}}{3 \text{ mo}}$ **37.** $\frac{17 \text{ baskets}}{30 \text{ attempts}}$ **39.** $\frac{37 \text{ points}}{2 \text{ games}}$ **41.** $\frac{100 \text{ sq ft}}{\$329}$ **43.** $\frac{16 \text{ males}}{3 \text{ females}}$ **45.** $\frac{8 \text{ Democrats}}{7 \text{ Republicans}}$ **47.** $\frac{1 \text{ lb}}{8 \text{ servings}}$ **49.** $\frac{307 \text{ flights}}{3 \text{ days}}$ **51.** $\frac{1 \text{ lb}}{200 \text{ sq ft}}$ **53.** 225 revolutions/min **55.** 8 gal/day **57.** 0.3 tank/acre **59.** 1.6 yd/dress **61.** 2 hr/day **63.** 0.25 km/min **65.** 70 fat calories/tbsp **67.** \$0.45/bar **69.** \$2.95/roll **71.** \$66.67/plant **73.** \$99/night

75.

Number of Units	Total Price	Unit Price
30	\$1.69	\$0.06
100	\$5.49	\$0.05

100 cough drops

77.

Number of Units (Sheets)	Total Price	Unit Price
500	\$9.69	\$0.019
2,500	\$42.99	\$0.017

2,500 sheets of paper

79.

Number of Units	Total Price	Unit Price
14	\$8.49	\$0.61
25	\$11.49	\$0.46
28	\$7.49	\$0.27

28 trash bags

81. \$0.16/oz **83.** 2 tutors/15 students **85.** $\frac{5}{1}$ **87.** $\frac{2}{3}$ **89.** 170 cal/oz **91.** \$0.03 per page **93.** $\frac{1}{2}$ **95.** $\frac{1}{6}$ **97.** In the Senate **99.** **a.** $\frac{63}{68}$ **b.** $\frac{8}{9}$ **101.** 0.51 to 1

Section 5.2 Practices, pp. 261–265

1, ***p. 261:*** Yes **2,** ***p. 261:*** Not a true proportion **3,** ***p. 262:*** No **4,** ***p. 263:*** $x = 8$ **5,** ***p. 263:*** $x = 12$ **6,** ***p. 264:*** 64,000 flowers **7,** ***p. 264:*** Mach 1.06 **8,** ***p. 265:*** 300 faculty members

Exercises 5.2, pp. 266–269

1. proportion **3.** as **5.** True **7.** False **9.** True **11.** False **13.** True **15.** True **17.** $x = 20$ **19.** $x = 38$ **21.** $x = 4$ **23.** $x = 13$ **25.** $x = 8$ **27.** $x = 4$ **29.** $x = 20$ **31.** $x = 15$ **33.** $x = 21$ **35.** $x = 13\frac{1}{3}$ **37.** $x = 100$ **39.** $x = 1.8$ **41.** $x = 21$ **43.** $x = 280$ **45.** $x = 300$ **47.** $x = 20$ **49.** $x = 10$ **51.** $x = 5.4$ **53.** $x = \frac{1}{5}$ **55.** $x = 0.005$ **57.** $x = \frac{6}{5}$ **59.** $x = 1\frac{3}{5}$ **61.** False **63.** Not the same **65.** $1\frac{7}{8}$ gal **67.** 54.5 g **69.** 100 oxygen atoms **71.** $41\frac{2}{3}$ in. **73.** 0.25 ft **75.** \$600 **77.** 12,000 fish **79.** 280 times **81.** 90 mg and 50 mg **83.** **a.** 92 g **b.** 4 g **85.** 835,000 gal

Chapter 5 Review Exercises, pp. 271–272

1. is not; possible answer: when simplifying a ratio, drop the units if the ratio has like quantities **2.** is; possible answer: it is a comparison of two unlike quantities **3.** is not; possible answer: the denominator is not 1 **4.** is; possible answer: it is the price of one foot **5.** is not; possible answer: a proportion must have an equal sign **6.** are: possible answer: cross products are found by multiplying diagonally **7.** $\frac{2}{3}$ **8.** $\frac{1}{2}$ **9.** $\frac{3}{4}$ **10.** $\frac{25}{8}$ **11.** $\frac{8}{5}$ **12.** $\frac{3}{4}$ **13.** $\frac{44 \text{ ft}}{5 \text{ sec}}$ **14.** $\frac{9 \text{ applicants}}{2 \text{ positions}}$ **15.** 0.0025 lb/sq ft **16.** 500,000,000 calls/day **17.** 8 yd/down **18.** 400 sq ft/gal **19.** 10,500,000 vehicles/yr **20.** 76,000 commuters/day **21.** \$118.75/night **22.** \$3.89/rental **23.** \$1,250/station **24.** \$93.64/share

25.

Number of Units	Total Price	Unit Price
47	\$39.95	\$0.85
94	\$69.95	\$0.74

94 issues

26.

Number of Units	Total Price	Unit Price
100	\$13.95	\$0.14
250	\$37.95	\$0.15

100 checks

27.

Number of Units	Total Price	Unit Price
30	\$16.95	\$0.57
90	\$44.85	\$0.50
180	\$77.70	\$0.43

180 capsules

28.

Number of Units (fluid ounces)	Total Price	Unit Price
4	\$2.78	\$0.70
14	\$3.92	\$0.28
20	\$3.74	\$0.19

20 fl oz

29. True **30.** False **31.** False **32.** True **33.** $x = 6$ **34.** $x = 3$ **35.** $x = 32$ **36.** $x = 30$ **37.** $x = 2$ **38.** $x = 8$ **39.** $x = \frac{7}{10}$ **40.** $x = \frac{2}{5}$ **41.** $x = 67\frac{1}{2}$ **42.** $x = 45$ **43.** $x = \frac{3}{4}$, or 0.75 **44.** $x = \frac{3}{20}$, or 0.15 **45.** $x = 28$ **46.** $x = 0.14$ **47.** $\frac{1}{15}$ **48.** $\frac{23}{45}$ **49.** \$90/day **50.** 0.125 in./mo **51.** $\frac{2}{3}$ **52.** 50,000 books **53.** No **54.** $2\frac{1}{7}$ hr **55.** 55 cc **56.** 1.25 in. **57.** 0.68 g/cc **58.** 0.07 admitted students per applicant

Chapter 5 Posttest, p. 273

1. $\frac{2}{3}$ **2.** $\frac{5}{14}$ **3.** $\frac{55}{31}$ **4.** $\frac{12}{1}$ **5.** $\frac{13 \text{ revolutions}}{12 \text{ sec}}$ **6.** $\frac{1 \text{ cm}}{25 \text{ km}}$ **7.** 68 mph **8.** 8 m/sec **9.** $136/day **10.** $0.80/greeting card **11.** False **12.** True **13.** $x = 25$ **14.** $x = 6$ **15.** $x = 28$ **16.** $x = 1$ **17.** 5 million e-mail addresses **18.** $\frac{9}{8}$ **19.** 25 ft **20.** 48 beats/min

Chapter 5 Cumulative Review Exercises, p. 274

1. $108,411 **2.** 189 **3.** 52 **4.** $2^3 \cdot 3 \cdot 7$ **5.** $\frac{2}{5}$ **6.** $\frac{4}{39}$ **7.** $\frac{3}{4}$ **8.** 8,200 **9.** Possible answer: 3 **10.** $x = 2.5$ **11.** $n = 70$ **12.** $\frac{1}{4}$ **13.** $4 per yd **14.** $x = \frac{3}{4}$ **15.** 2,106 sq ft **16.** $445.88 **17.** 0.19 in. **18.** $r \cdot t$ miles; 260 mi **19.** 7.5 in. **20.** $\frac{1}{29}$

CHAPTER 6

Chapter 6 Pretest, p. 276

1. $\frac{1}{20}$ **2.** $\frac{3}{8}$ **3.** 2.5 **4.** 0.03 **5.** 0.7% **6.** 800% **7.** 67% **8.** 110% **9.** $37\frac{1}{2}$ ft **10.** 55 **11.** 32 **12.** 250 **13.** 40% **14.** 250% **15.** $10.50 **16.** 8% **17.** $\frac{6}{25}$ **18.** 25% **19.** $61.11 **20.** $10,000

Section 6.1 Practices, pp. 278–284

1, ***p. 278:*** $\frac{21}{100}$ **2,** ***p. 278:*** $\frac{9}{4}$, or $2\frac{1}{4}$ **3,** ***p. 279:*** $\frac{2}{300}$ **4,** ***p. 279:*** $\frac{1}{8}$ **5,** ***p. 279:*** $\frac{14}{25}$ **6,** ***p. 280:*** 0.31 **7,** ***p. 280:*** 0.05 **8,** ***p. 281:*** 0.482 **9,** ***p. 281:*** 0.6225 **10,** ***p. 281:*** 1.637 **11,** ***p. 282:*** 2.5% **12,** ***p. 282:*** 9% **13,** ***p. 282:*** 70% **14,** ***p. 282:*** 300% **15,** ***p. 283:*** 71% **16,** ***p. 283:*** Nitrogen; 78% > 0.93%, or 0.78 > 0.0093. **17,** ***p. 284:*** 16% **18,** ***p. 284:*** True. $\frac{2}{3} \approx 67\% > 60\%$ **19,** ***p. 284:*** 27%

Exercises 6.1, pp. 285–288

1. percent **3.** left **5.** $\frac{2}{25}$ **7.** $2\frac{1}{2}$ **9.** $\frac{33}{100}$ **11.** $\frac{9}{50}$ **13.** $\frac{7}{50}$ **15.** $\frac{13}{20}$ **17.** $\frac{3}{400}$ **19.** $\frac{3}{1,000}$ **21.** $\frac{3}{40}$ **23.** $\frac{1}{7}$ **25.** 0.06 **27.** 0.72 **29.** 0.001 **31.** 1.02 **33.** 0.425 **35.** 5 **37.** 1.069 **39.** 0.035 **41.** 0.009 **43.** 0.0075 **45.** 31% **47.** 17% **49.** 30% **51.** 4% **53.** 12.5% **55.** 129% **57.** 290% **59.** 287% **61.** 101.6% **63.** 900% **65.** 30% **67.** 10% **69.** 16% **71.** 90% **73.** 6% **75.** $55\frac{5}{9}\%$ **77.** $11\frac{1}{9}\%$ **79.** 600% **81.** 150% **83.** $216\frac{2}{3}\%$ **85.** < **87.** < **89.** 44% **91.** 225%

93.

Fraction	Decimal	Percent
$\frac{1}{3}$	0.333 . . .	$33\frac{1}{3}\%$
$\frac{2}{3}$	0.666 . . .	$66\frac{2}{3}\%$
$\frac{1}{4}$	0.25	25%
$\frac{3}{4}$	0.75	75%
$\frac{1}{5}$	0.2	20%
$\frac{2}{5}$	0.4	40%
$\frac{3}{5}$	0.6	60%

95. $1\frac{1}{25}$ **97.** $316\frac{2}{3}\%$ **99.** 0.275 **101.** 310% **103.** 254% **105.** 0.96 **107.** $\frac{17}{50}$ **109.** $\frac{1}{10}$ **111.** 1,005% **113.** $1\frac{7}{20}$ **115.** 0.515 **117.** Among men; $\frac{1}{4} = 25\% < 42\%$ **119. a.** 0.4% **b.** 99.6% **121.** 69%

Section 6.2 Practices, pp. 291–297

1, ***p. 291:*** **a.** $x = 0.7 \cdot 80$ **b.** $0.5 \cdot x = 10$ **c.** $x \cdot 40 = 20$ **2,** ***p. 291:*** 8 **3,** ***p. 291:*** 12 **4,** ***p. 292:*** 200 **5,** ***p. 292:*** 51 workers **6,** ***p. 293:*** 50 **7,** ***p. 293:*** 7.2 **8,** ***p. 293:*** 2,500,000 sq ft **9,** ***p. 294:*** $83\frac{1}{3}\%$ **10,** ***p. 294:*** $112\frac{1}{2}\%$ **11,** ***p. 295:*** 30% **12,** ***p. 295:*** 270 **13,** ***p. 296:*** 1,080 **14,** ***p. 296:*** $33\frac{1}{3}\%$ **15,** ***p. 296:*** $98 **16,** ***p. 297:*** $340,000 **17,** ***p. 297:*** 105% **18,** ***p. 297:*** $5.97

Exercises 6.2, pp. 298–301

1. base **3.** percent **5.** 6 **7.** 23 **9.** 2.87 **11.** $140 **13.** 0.62 **15.** 0.045 **17.** 0.1 **19.** 4 **21.** $18.32 **23.** 32 **25.** $120 **27.** 2.5 **29.** $250 **31.** 45 **33.** 1.75 **35.** 4,600 **37.** 3,000 **39.** $49,230.77 **41.** 50% **43.** 75% **45.** $83\frac{1}{3}\%$ **47.** 25% **49.** 150% **51.** $112\frac{1}{2}\%$ **53.** 62.5% **55.** 50% **57.** 31% **59.** 60 **61.** $66\frac{2}{3}\%$ **63.** 175 mi **65.** 5% **67.** $500 **69.** 10 **71.** $66\frac{2}{3}\%$ **73.** 40 questions **75.** $600 **77.** 18,750,000 people **79.** 72 free throws **81.** 10.2 gal **83.** $30,000,000 **85.** $9,000 **87.** 30% **89.** 6.8 million **91.** 5,100 employees **93. a.** 10 million **b.** 80%

Section 6.3 Practices, pp. 302–307

1, ***p. 302:*** 300% **2,** ***p. 303:*** 1929 **3,** ***p. 304:*** $462.50 **4,** ***p. 304:*** **a.** $1,125 **b.** $2,625 **5,** ***p. 305:*** $69.60 **6,** ***p. 305:*** $1,450 **7,** ***p. 306:*** $1,792 **8,** ***p. 307:*** $2,524.95

Exercises 6.3, pp. 308–312

1. original **3.** discount

5.

Original Value	New Value	Percent Increase or Decrease
$10	$12	20% increase
$10	$8	20% decrease
$6	$18	200% increase
$35	$70	100% increase
$14	$21	50% increase
$10	$1	90% decrease
$8	$6.50	$18\frac{3}{4}\%$ decrease
$6	$5.25	$12\frac{1}{2}\%$ decrease

7.

Selling Price	Rate of Sales Tax	Sales Tax
$30.00	5%	$1.50
$24.88	3%	$0.75
$51.00	$7\frac{1}{2}\%$	$3.83
$196.23	4.5%	$8.83

9.

Sales	Rate of Commission	Commission
$700	10%	$70.00
$450	2%	$9.00
$870	$4\frac{1}{2}\%$	$39.15
$922	7.5%	$69.15

11.

Original Price	Rate of Discount	Discount	Sale Price
$700.00	25%	$175.00	$525.00
$18.00	10%	$1.80	$16.20
$43.50	20%	$8.70	$34.80
$16.99	5%	$0.85	$16.14

ANSWERS

13.

Principal	Interest Rate	Time (in years)	Interest	Final Balance
$300	4%	2	$24.00	$324.00
$600	7%	2	$84.00	$684.00
$500	8%	2	$80.00	$580.00
$375	10%	4	$150.00	$525.00
$1,000	3.5%	3	$105.00	$1,105.00
$70,000	6.25%	30	$131,250.00	$201,250.00

15.

Principal	Interest Rate	Time (in years)	Final Balance
$500	4%	2	$540.80
$6,200	3%	5	$7,187.50
$300	1%	8	$324.86
$20,000	4%	2	$21,632.00
$145	3.8%	3	$162.17
$810	2.9%	10	$1,078.05

17.

Original Value	New Value	Percent Decrease
$5	$4.50	10%

19.

Original Price	Rate of Discount	Discount	Sale Price
$87.33	40%	$34.93	$52.40

21.

Selling Price	Rate of Sales Tax	Sales Tax
$200	7.25%	$14.50

23.

Principal	Interest Rate	Kind of Interest	Time (in years)	Interest	Final Balance
$3,000	5%	simple	5	$750.00	$3,750.00

25. 28% **27.** 550% **29.** 300% **31.** $84.95 **33.** 6.5% **35.** $2,700 **37.** $9 **39.** 15% **41.** $259.35 **43.** $150 **45.** $250 **47.** **a.** $144 **b.** $152.64 **49.** $3,244.80 **51.** 5,856

Chapter 6 Review Exercises, pp. 315–318

1. are; possible answer: a percent can be written as a fraction with denominator 100 **2.** are not; possible answer: $\frac{1}{2}$% is one-hundredth of $\frac{1}{2}$ **3.** is; possible answer: 8 is the number that we are taking the percent of **4.** is not; possible answer: the amount (4) is the percent (50%) of the base (8) **5.** is not; possible answer: when the percent is greater than 100%, the base is smaller than the amount **6.** is; possible answer: if the number were smaller than or equal to 20, then 5% of it would be still smaller **7.** is; possible answer: taking 100% of a number is the same as multiplying it by 1 **8.** is not; possible answer: the difference between the new and the original values is 100% of the original value

9.

Fraction	Decimal	Percent
$\frac{1}{4}$	0.25	25%
$\frac{7}{10}$	0.7	70%
$\frac{3}{400}$	0.0075	$\frac{3}{4}$%
$\frac{5}{8}$	0.625	62.5%
$\frac{41}{100}$	0.41	41%
$1\frac{1}{100}$	1.01	101%
$2\frac{3}{5}$	2.6	260%
$3\frac{3}{10}$	3.3	330%
$\frac{3}{25}$	0.12	12%
$\frac{2}{3}$	0.66 . . .	$66\frac{2}{3}$%
$\frac{1}{6}$	0.166 . . .	$16\frac{2}{3}$%

10.

Fraction	Decimal	Percent
$\frac{3}{8}$	0.375	37.5%
$\frac{49}{100}$	0.49	49%
$\frac{1}{1,000}$	0.001	0.1 %
$1\frac{1}{2}$	1.5	150%
$\frac{7}{8}$	0.875	87.5%
$\frac{5}{6}$	0.833 . . .	$83\frac{1}{3}$%
$2\frac{3}{4}$	2.75	275%
$1\frac{1}{5}$	1.2	120%
$\frac{3}{4}$	0.75	75%
$\frac{1}{10}$	0.1	10%
$\frac{1}{3}$	0.33 . . .	$33\frac{1}{3}$%

11. 12 **12.** 120% **13.** 50% **14.** 20 **15.** 43.75% **16.** 5.5 **17.** $6 **18.** 20% **19.** 0.3 **20.** 460 **21.** $70 **22.** 1,000 **23.** 2,000% **24.** 25% **25.** $200 **26.** $44\frac{4}{9}$% **27.** $12 **28.** $1,600 **29.** 17% **30.** 42.42

31.

Original Value	New Value	Percent Decrease
24	16	$33\frac{1}{3}$%

32.

Original Value	New Value	Percent Decrease
360 mi	300 mi	$16\frac{2}{3}$%

33.

Selling Price	Rate of Sales Tax	Sales Tax
$50	6%	$3.00

34.

Sales	Rate of Commission	Commission
$600	4%	$24

35.

Original Price	Rate of Discount	Discount	Sale Price
$200	15%	$30	$170

36.

Principal	Interest Rate	Time (in years)	Simple Interest	Final Balance
$200	4%	2	$16	$216

37. 10,000 hr **38.** 48% **39.** The agent that charges 11% **40.** $7,200 **41.** 20% **42.** Paper **43.** $\frac{1}{4}$ **44.** 64% **45.** 40% **46.** 15% **47.** 0.0629 **48.** Possible estimate: 80 in. **49.** 45% **50.** $207 **51.** No **52.** Yes **53.** 180 seats **54.** 20% **55.** 63 first serves **56.** 83% **57.** $165 **58.** 320 tons **59.** 70,000 mi **60.** $599 **61.** $51,500 **62.** $30,000 **63.** $7,791.18 **64.** 22%

65.

Quarter Ending	Income (in million of dollars)	Percent of Total Income (rounded to the nearest tenth of a percent)
June 30, 2009	260	12.3%
Sept 30, 2009	538	25.5%
Dec 31, 2009	655	31.0%
Mar 31, 2010	658	31.2%
Total	2,111	100.0%

66. Individual income taxes: $1,125 billion; Social Security taxes: $900 billion; corporate income taxes: $300 billion; excise taxes: $75 billion; other: $100 billion

Chapter 6 Posttest, p. 319

1. $\frac{1}{25}$ **2.** $\frac{11}{40}$ **3.** 1.74 **4.** 0.08 **5.** 0.9% **6.** 1,000% **7.** 83% **8.** 220% **9.** 7.5 mi **10.** 48 **11.** 300 **12.** 200 **13.** 60% **14.** 250% **15.** $200 **16.** 6 spaces **17.** 5% **18.** 4 pt **19.** $217.80 **20.** $3\frac{1}{3}$%

Chapter 6 Cumulative Review Exercises, p. 320

1. 10^7 **2.** 109 **3.** $5\frac{7}{10}$ **4.** 27.57 **5.** 0.7 **6.** 0.83 **7.** 7.32 **8.** $x - 6.7$ **9.** $23\frac{3}{5}$ **10.** 7.5 **11.** $8.19/hr **12.** 14 **13.** $\frac{2}{11}$ **14.** $1,000 **15.** $\frac{1}{8}$

16.

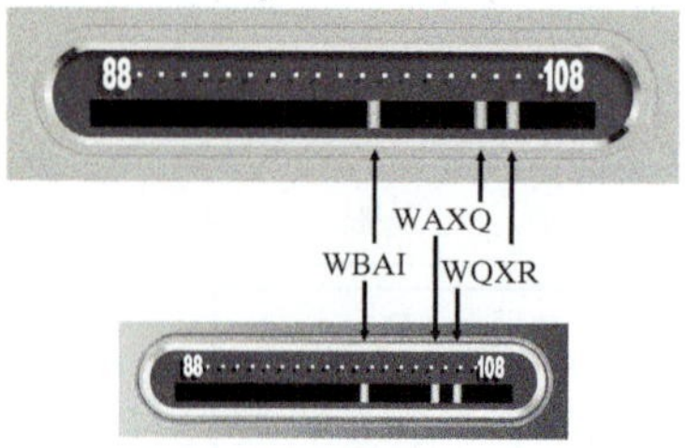

17. 15 times **18.** 715 adults **19.** 278 thousand **20.** 425%

CHAPTER 7

Chapter 7 Pretest, p. 322

1. +7 **2.** −4 **3.** −17 **4.** 0 **5.** −7 **6.** 0 **7.** −75 **8.** −64 **9.** $-\frac{1}{2}$ **10.** $\frac{1}{4}$ **11.** 2 **12.** 2 **13.** 8 **14.** 8 **15.** 36 **16.** −7 **17.** 58°F **18.** −$17.50 **19.** 23.6 degrees **20.** 25 centuries

Section 7.1 Practices, pp. 324–327

1, ***p. 324:***

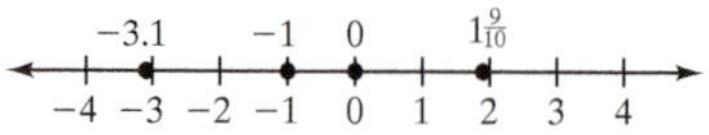

2, ***p. 324:*** **a.** −9 **b.** $4\frac{9}{10}$ **c.** 2.9 **d.** −31 **3,** ***p. 325:*** **a.** 9 **b.** $1\frac{3}{4}$ **c.** 4.1 **d.** 5 **4,** ***p. 325:*** **a.** Sign: −; absolute value: 4 **b.** Sign: +; absolute value: $6\frac{1}{2}$ **5,** ***p. 326:*** **a.** 0 **b.** −5 **c.** −4 **6,** ***p. 327:*** +2 yd **7,** ***p. 327:*** Sirius

Exercises 7.1, pp. 328–331

1. positive number **3.** signed number **5.** opposites **7.** smaller **9.**

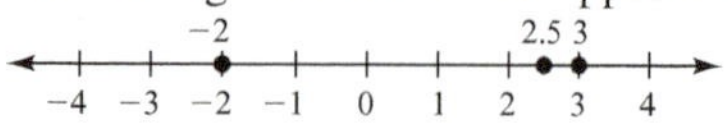

11.

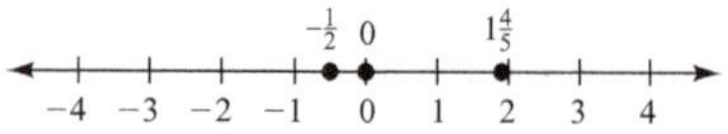

13. −8 **15.** −10.2 **17.** 25 **19.** $5\frac{1}{2}$ **21.** 4.1 **23.** −0.5 **25.** 6 **27.** $\frac{4}{5}$ **29.** 2 **31.** 0.6 **33.** Sign: +; Absolute value: 8 **35.** Sign: −; Absolute value: 4.3 **37.** Sign: −; Absolute value: 7 **39.** Sign: +; Absolute value: $\frac{1}{5}$ **41.** Two; −5 and 5 **43.** No **45.** −4 **47.** 12 **49.** 2 **51.** $-2\frac{1}{3}$ **53.** −2 **55.** 9 **57.** −2 **59.** −7 **61.** −8.3 **63.** $-3\frac{1}{2}$ **65.** T **67.** T **69.** T **71.** T **73.** F **75.** T **77.** −3, 0, 3 **79.** −9, −4.5, 9 **81.** −$150 **83.** +14.5°C **85.**

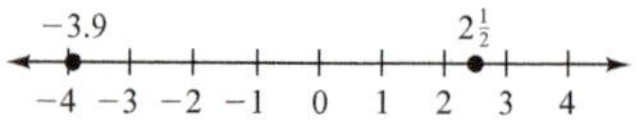

87. **a.** −$10.98 **b.** $100 **89.** **a.** 0.5 **b.** 11 **91.** **a.** < **b.** > **93.** Mariana Trench **95.** Decreased by 25 mg **97.** No **99.** Mars **101.** **a.**

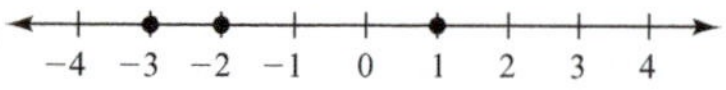

While the player is in the game, the number of goals scored by his team is equal to the number scored by the other team **b.** In the first period.

Section 7.2 Practices, pp. 333–336

1, ***p. 333:*** −25 **2,** ***p. 334:*** $-4\frac{1}{2}$ **3,** ***p. 334:*** 7 **4,** ***p. 334:*** 0 **5,** ***p. 335:*** −1.3 **6,** ***p. 335:*** 0 **7,** ***p. 335:*** 456 m **8,** ***p. 336:*** −2.478

Exercises 7.2, pp. 337–339

1. right **3.** larger **5.** 1 **7.** 3 **9.** −11 **11.** 0 **13.** 0 **15.** −5 **17.** 200 **19.** 6 **21.** −150 **23.** −5 **25.** −27 **27.** 0 **29.** 4.9 **31.** 0.1 **33.** 59.5 **35.** −5.9 **37.** −14.5 **39.** −6 **41.** $-\frac{3}{5}$ **43.** $1\frac{3}{5}$ **45.** $-\frac{1}{10}$ **47.** −102 **49.** −10 **51.** 9 **53.** $-7\frac{1}{2}$ **55.** 7 **57.** 3 **59.** 0 **61.** −6.9 **63.** 7.58 **65.** −4.914 **67.** 4.409 **69.** 3

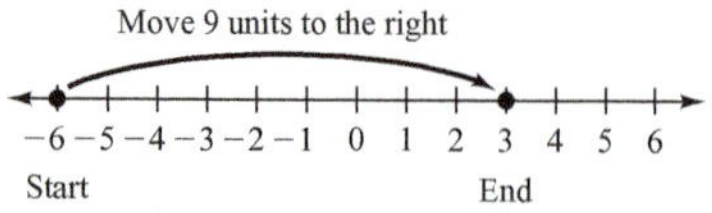

71. 0 **73.** −4 **75.** 19,340 ft **77.** −238 employees **79.** −1 **81.** −6 yr (6 yr ago) **83.** −6 yd **85.** **a.** −1.0 million **b.** 14.7 million

Section 7.3 Practices, pp. 340–342

1, ***p. 340:*** −2 **2,** ***p. 341:*** 18 **3,** ***p. 341:*** −21.1 **4,** ***p. 341:*** −10 **5,** ***p. 342:*** 6 **6,** ***p. 342:*** 2,100 yr

Exercises 7.3, pp. 343–345

1. opposite **3.** order of operations **5.** 7 **7.** −4 **9.** −14 **11.** 44 **13.** −25 **15.** −19 **17.** 6 **19.** −38 **21.** −26 **23.** 26 **25.** −15 **27.** 1,000 **29.** −1.52 **31.** 9.7 **33.** 0 **35.** 10.5 **37.** −0.5 **39.** −5 **41.** $7\frac{3}{4}$ **43.** $-7\frac{1}{4}$ **45.** $7\frac{1}{4}$ **47.** 7 **49.** −5 **51.** −1.9 **53.** $-\frac{1}{2}$ **55.** −15 **57.** −3 **59.** −3.842 **61.** −16.495 **63.** −25 **65.** 0 **67.** 1 **69.** 1,000 mi **71.** 22,965 ft **73.** −$1.4 million **75.** **a.** −6.2 ft **b.** 55.8 ft **77.** **a.** 8.30 in. **b.** −6.80 in.

Section 7.4 Practices, pp. 346–348

1, ***p. 346:*** 32 **2,** ***p. 347:*** −10 **3,** ***p. 347:*** **a.** 1 **b.** −1 **4,** ***p. 347:*** $\frac{4}{15}$ **5,** ***p. 347:*** −20 **6,** ***p. 348:*** −48 **7,** ***p. 348:*** 169 **8,** ***p. 348:*** No

Exercises 7.4, pp. 349–351

1. positive **3.** even **5.** −10 **7.** −10 **9.** 25 **11.** 306 **13.** −16 **15.** −8,163 **17.** −40 **19.** −176 **21.** 800 **23.** −7,200 **25.** −5 **27.** −10 **29.** 5.52 **31.** −8 **33.** $-\frac{5}{27}$ **35.** $-\frac{5}{6}$ **37.** −25 **39.** 10,000 **41.** 0.25 **43.** −0.001 **45.** $\frac{9}{16}$ **47.** −1 **49.** 0.094864 **51.** −216 **53.** −60 **55.** 0 **57.** 5 **59.** $-\frac{32}{135}$ **61.** 0.14652 **63.** 5 **65.** −26 **67.** 0 **69.** −28 **71.** 1.25 **73.** −12 **75.** 9 **77.** −10.8 **79.** 289 **81.** 41 **83.** **a.** 5 **b.** 2 **c.** −1 **d.** −4 **e.** −7 **85.** −4,830 **87.** −0.0001 **89.** −8 **91.** −72 mm **93.** −$4,425 **95.** −160° C **97.** −64 ft **99.** **a.** −$900 **b.** $100

Section 7.5 Practices, pp. 352–354

1, ***p. 352:*** 12 **2,** ***p. 353:*** $-\frac{3}{5}$ **3,** ***p. 353:*** −0.3 **4,** ***p. 353:*** $-\frac{1}{6}$ **5,** ***p. 354:*** **a.** 0 **b.** −1 **6.** ***p. 354:*** −50 sq mi/mo **7,** ***p. 354:*** 0 degrees

Exercises 7.5, pp. 355–357

1. positive **3.** equal **5.** 5 **7.** 0 **9.** −5 **11.** −2 **13.** 25 **15.** −25 **17.** 7 **19.** −2 **21.** 17 **23.** 6 **25.** −0.3 **27.** −20 **29.** 16 **31.** $-\frac{5}{6}$ **33.** −21 **35.** −16 **37.** 6.172 **39.** −3.5 **41.** $-\frac{1}{5}$ **43.** 1 **45.** $-\frac{2}{5}$ **47.** $5\frac{1}{2}$ **49.** $4\frac{1}{4}$ **51.** $\frac{3}{4}$ **53.** −8 **55.** −4 **57.** 5 **59.** 1 **61.** −16 **63.** −3 **65.** −0.06 **67.** −4 **69.** −33 **71.** 2 **73.** −7 **75.** $9 \div (1 - 4) = -3$ **77.** $6 \div (3 - 1) - 4 = -1$ **79.** $(8 - 10)\cdot 2 - (-5 + 13) \div 4 = -6$ **81.** $-\frac{6}{5}$ **83.** $-3\frac{1}{6}$ **85.** $-\frac{4}{27}$ **87.** −6,098.9 **89.** −3.5 thousand **91.** −1,000 ft/min **93.** Yes. **95.** **a.** Thursday; Saturday **b.** 0°F

Chapter 7 Review Exercises, pp. 359–361

1. is; possible answer: a positive sign is understood **2.** is; possible answer: it lies to the left of 0 on the number line **3.** is not; possible answer: the integers are the numbers . . . −4, −3, −2, −1, 0, 1, 2, 3, 4. . . continuing indefinitely in both directions **4.** are; possible answer: they are the same distance from 0 on the number line but on opposite sides of it **5.** is not; possible answer: all absolute values are either positive or 0 **6.** is not; possible answer: adding −1 on the number line means moving one unit to the left **7., 8.**

−3 1.5

−4 −3 −2 −1 0 1 2 3 4

9. −6 **10.** 4 **11.** $7\frac{1}{2}$ **12.** −10.1 **13.** 10 **14.** 2.5 **15.** $1\frac{1}{5}$ **16.** 7 **17.** −11 **18.** 10 **19.** 9 **20.** −2 **21.** −8, −3.5, 8 **22.** −9.7, −6, 9 **23.** −2.9, $-2\frac{1}{2}$, 0 **24.** −4, $-1\frac{1}{4}$, 0 **25.** +10 ft **26.** −$350 **27.** −20 **28.** −2 **29.** $6\frac{1}{2}$ **30.** −1 **31.** −4.1 **32.** −2 **33.** −7 **34.** $-\frac{1}{4}$ **35.** 0 **36.** 28 **37.** −10 **38.** −11 **39.** 3 **40.** $-4\frac{1}{8}$ **41.** 100 **42.** −45 **43.** $-\frac{20}{33}$ **44.** −7.35 **45.** 72 **46.** −30 **47.** $\frac{1}{16}$ **48.** 0.49 **49.** 36 **50.** −81 **51.** 5 **52.** −10 **53.** −5 **54.** $\frac{1}{32}$ **55.** −50 **56.** 2 **57.** 15 **58.** 114 **59.** −100 **60.** 1 **61.** 2.73 **62.** −2.62 **63.** 104 **64.** 21 **65.** 10 **66.** −2 **67.** 225 **68.** 400 **69.** −$400 **70.** Yes **71.** −$2,820 **72.** $1,649.83 **73.** 28°F **74.** 7 questions **75.** −$1.20 **76.** 2,010 years apart **77.** Yes **78.** 90 feet below the surface **79.** 5 under par **80.** 100 ft above the point of release (+100 ft) **81.** **a.** Alabama **b.** 14°F **82.** **a.** −$116.4 million **b.** −$38.8 billion

Chapter 7 Posttest, p. 362

1. −10 **2.** $-\frac{1}{2}$ **3.** 0 **4.** −0.5 **5.** −49 **6.** 0 **7.** −207 **8.** −0.1 **9.** −144 **10.** $\frac{1}{16}$ **11.** −4 **12.** $-\frac{1}{8}$ **13.** 4 **14.** 21 **15.** 40 **16.** −18 **17.** −51 lb **18.** −$550/yr **19.** BP, Exxon Mobil, Royal Dutch Shell, Conoco Phillips, Chevron **20.** 76°C

Chapter 7 Cumulative Review Exercises, p. 363

1. 3,000 **2.** −4 **3.** 10 **4.** 14 **5.** 3.108 **6.** 3.44 **7.** $x = 1.5$ **8.** $\frac{5}{9}x = 40$ **9.** $n = 65$ **10.** 112.5% **11.** 20% **12.** −6.04 **13.** 0.5 **14.** $-\frac{5}{9}$ **15.** $2\frac{2}{3}$ innings **16.** 80 mg **17.** 4:1 **18.** 25% **19.** −89°C **20.** A liquid

CHAPTER 8

Chapter 8 Pretest, pp. 365–366

1. Yes **2.** 31 **3.** 11 min **4.** 52 millimeters **5.** 3.1 **6.** 1960, 1970, 1980, 1990, and 2010 **7.** 1990–1992 and 1999–2005 **8.** 1,100,000 newspapers **9.** 406 thousand **10.** 41%

Section 8.1 Practices, pp. 368–372

1, ***p. 368:*** Reggie Jackson **2,** ***p. 369:*** Above; the exam average is 87. **3,** ***p. 370:*** **a.** 7 **b.** 4 **4,** ***p. 370:*** **a.** $16 million **b.** The median for the six movies was $2 million less. **5,** ***p. 371:*** **a.** 2 **b.** 4 and 9 **c.** No mode **6,** ***p. 371:*** 12 hr **7,** ***p. 372:*** 7 **8,** ***p. 372:*** $3.40

Exercises 8.1, pp. 373–375

1. statistics **3.** weighted **5.** mode

7.

Numbers	Mean	Median	Mode(s)
a. 8, 2, 9, 4, 8	6.2	8	8
b. 3, 0, 0, 3, 10	3.2	3	0 and 3
c. 4, 6, 9, 1, 1, 3	4	3.5	1
d. 7.5, 9, 8.5, 5.5, 8.1	7.7	8.1	None
e. $3\frac{1}{2}, 3\frac{3}{4}, 4, 3\frac{1}{2}, 3\frac{1}{4}$	$3\frac{3}{5}$	$3\frac{1}{2}$	$3\frac{1}{2}$
f. 4, −2, −1, 0, −1	0	−1	−1

9. $12,266.67

11.

Numbers	Range
a. 20, 11, 3, 4, 16	17
b. 2.3, 5.7, 10.2, 6.1, 0.9	9.3
c. $6\frac{5}{6}, 5\frac{1}{2}, \frac{1}{3}, 8, 5\frac{3}{4}$	$7\frac{2}{3}$
d. −2, 6, −4, −1, −4	10

13. No; the student's GPA was 3.25, which is less than 3.5. **15.** The mean amount is $100,000. We can't compute the median amount because we don't know the actual amount given to each grandchild. **17.** Indiana, North Carolina, and Tennessee **19.** **a.** $50,921 **b.** $32,927 **21.** **a.** 31,000 mi **b.** 20,000 mi **c.** 8,000 mi **d.** 86,000 mi **23.** 25 pieces of mail

Section 8.2 Practices, pp. 377–384

1, ***p. 377:*** **a.** $4.95 **b.** $5.95 **c.** $5.45 **2,** ***p. 378:*** **a.** 10 million passengers **b.** Atlanta **c.** 60 million passengers **3,** ***p. 379:*** **a.** Dairy products **b.** $2.9 billion **c.** $1 billion **4,** ***p. 380:*** **a.** 11% **b.** 10% **5,** ***p. 381:*** **a.** 24 presidents **b.** No **c.** 33 presidents **6,** ***p. 382:*** **a.** July **b.** 27°F **c.** In Chicago, mean temperatures increase from January through July and then decrease from July through December. **7,** ***p. 383:*** **a.** In Orlando **b.** 5 in. **c.** Possible answer: In Orlando, the average precipitation increases steadily from January through October. In Seattle, it peaks in January, dips in April and in July, and increases in October. **8,** ***p. 384:*** **a.** $\frac{3}{50}$ **b.** 38% **c.** 9 times as great

Exercises 8.2, pp. 385–388

1. table **3.** rows **5.** bar graph **7.** line graph **9.** **a.** $60 **b.** $100 **c.** She will pay a lower commission if she sells 400 shares in a single deal; 400 shares in a single deal will cost her $70 or $90, whereas two deals of 200 shares will cost her $100, $110, or $120. **11.** **a.** 1,556 mi **b.** 1,020 mi **c.** The blanks mean that there is no distance between a city and itself. **13.** **a.** Health care and social assistance **b.** 12.5 million workers **15.** **a.** Milk, lemon juice, and vinegar **b.** 8 **c.** Pure water **17.** **a.** More than 89 minutes **b.** 92 applicants **c.** 70 applicants **19.** **a.** 270 million subscribers **b.** 2005 **c.** Every year, the number of cell phone subscribers increased. **21.** **a.** Democrats **b.** $\frac{2}{5}$ **c.** 600

Chapter 8 Review Exercises, pp. 392–395

1. cannot; possible answer: the mean is the sum of the numbers divided by the number of numbers **2.** can; possible answer: there may be two or more numbers that occur with the highest frequency **3.** is; possible answer: the difference between the largest and smallest numbers in each set is the same **4.** does; possible answer: reading a graph generally involves estimating **5.** is not; possible answer: they are difficult to distinguish on a pictograph **6.** is not; possible answer: it is used to represent parts of a whole

7.

List of Numbers	Mean	Median	Mode	Range
6, 7, 4, 10, 4, 5, 6, 8, 7, 4, 5	6	6	4	6
1, 3, 4, 4, 2, 3, 1, 4, 5, 1	2.8	3	1 and 4	4

8. a. The husbands (83 yr) lived longer than the wives (70 yr). **b.** 13 yr **c.** 23 yr **9. a.** Half of the people who listen to NPR are younger than 55, and half are older. **b.** 30 years below the median **10.** This machine is reliable because the range is 1.7 fl oz. **11.** Since the average was 13,109, the zoo was profitable. **12. a.** 70,720,000 tons **b.** Beaumont, TX **c.** 41,440,000 tons more **d.** All except the Port of South Louisiana **13. a.** 830,000 monthly domestic flights **b.** In 2005 **c.** 80,000 more monthly domestic flights **14. a.** 10 million volumes **b.** 5 million volumes more **c.** Harvard University and the Boston Public Library **15. a.** 61 million licensed drivers **b.** 10% **c.** Possible answer: It increases for younger age groups, peaks for 40–44, and decreases for older age groups. **16. a.** Run number 3 **b.** Approximately 3 min **c.** With practice, the rat ran through the maze more quickly. **17. a.** 2003 **b.** $\frac{1}{2}$ **c.** Both out-of-pocket expenditures and insurance expenditures increased each year from 2000 through 2010 (with the exception of the period 2005–2006). **18. a.** 27,578 transplants **b.** 4,429 more liver transplants **c.** 58% **19.** 76 **20.** Internet users

Chapter 8 Posttest, pp. 396–397

1. 3.2 **2.** 162 **3.** 77 **4.** Mode: 200 lb; range: 90 lb **5.** Toyota Prius, Ford Fusion Hybrid, Mercury Milan Hybrid, and Nissan Altima Hybrid **6.** 30% greater **7.** 3 million more victims **8. a.** 2007 **b.** Overall music purchases are the sum of digital track and album sales. \$600 million + \$600 million = \$1,200 million, or \$1.2 billion, in sales for overall music purchases in the year 2006 **9.** 60% **10.** 1961–1975; U.S. scientists received approximately 40 and U.K. scientists received approximately 20 Nobel prizes.

Chapter 8 Cumulative Review Exercises, pp. 398–399

1. 23 **2.** $\frac{4}{5}$ **3.** $1\frac{87}{100}$ **4.** 40 **5.** 2.81 **6.** 3,010 **7.** $x = 17\frac{1}{4}$ **8.** No **9.** $n = 21$ **10.** $4\frac{5}{8}$ **11.** 2 **12.** Negative; $\frac{7}{4}$ **13.** −18 **14.** Mean: 3.7; median: 3.8; mode: 3.9; range: 0.6 **15.** 10.6 times **16.** 8 jobless people per opening **17.** 2% **18.** 5,000 yr **19.** Higher in the American League; the medians were 48 and 50. **20.** Males, age 45–54

CHAPTER 9

Chapter 9 Pretest, p. 401

1. $\frac{1}{2}$ **2.** 72 **3.** 48 **4.** 3 hr 52 min **5.** b **6.** d **7.** 3,500 **8.** 2.1 **9.** 8 **10.** 0.0000015 **11.** 0.002 **12.** 4.5 **13.** 195 **14.** About 2.2 qt **15.** 334 billion lb of trash **16.** 5 times **17.** 440 oz **18.** 180 cm **19.** 300 mi **20.** 7 cm

Section 9.1 Practices, pp. 404–407

1, ***p. 404:*** 2 **2,** ***p. 404:*** $1\frac{2}{3}$ yd **3,** ***p. 404:*** 86,400 sec **4,** ***p. 405:*** No **5,** ***p. 405:*** 19 in. **6,** ***p. 405:*** 1 yr 8 mo **7,** ***p. 406:*** 5 lb 8 oz **8,** ***p. 406:*** 13 lb 4 oz **9,** ***p. 407:*** 3 qt **10,** ***p. 407:*** 6 sec

Exercises 9.1, pp. 408–410

1. length **3.** larger **5.** unit **7.** 4 **9.** 3 **11.** 720 **13.** 420 **15.** 30 **17.** 2 **19.** 3,520 **21.** 512 **23.** 2 **25.** 5 **27.** $3\frac{1}{2}$ **29.** 3,000 **31.** $\frac{3}{4}$ **33.** 12 **35.** 310 **37.** 7 ft 6 in.

39.	192	$5\frac{1}{3}$
41.	180	15
43.	48	$\frac{3}{2000}$
45.	128,000	8,000
47.	32	2
49.	1	$\frac{1}{2}$
51.	$\frac{5}{6}$	$\frac{1}{72}$
53.	1,320	$\frac{11}{30}$

55. 1 lb 14 oz **57.** 29 lb 15 oz **59.** 9 yr 6 mo **61.** 2 gal 3 qt **63.** 4 qt **65.** 8 min 15 sec **67.** 14 lb 1 oz **69.** 247 sec **71.** 5 yd **73.** Equal to **75.** 6 in. **77.** 13 min 43 sec **79.** 1 lb 5 oz **81.** More; he has 1 yr 10 mo left on the lease. **83.** 13 lb 14 oz **85. a.** Width: 3 ft, depth: $2\frac{1}{2}$ ft **b.** $7\frac{1}{2}$ sq ft **87.** 5.5 mi

Section 9.2 Practices, pp. 414–418

1, ***p. 414:*** 3.1 **2,** ***p. 414:*** 25 **3,** ***p. 415:*** 25 m **4,** ***p. 415:*** 5 kL **5,** ***p. 416:*** Large intestine **6,** ***p. 417:*** 5,300,000 **7,** ***p. 417:*** 0.005 mg **8,** ***p. 418:*** 38 L **9,** ***p. 418:*** 4 lb

Exercises 9.2, pp. 419–422

1. length **3.** kilo- **5.** centi- **7.** c **9.** b **11.** b **13.** b **15.** a **17.** 1 **19.** 0.75 **21.** 80 **23.** 3,500 **25.** 0.005 **27.** 4 **29.** 7 **31.** 4.13 **33.** 2,000 **35.** 0.0075 **37.** 30,000 **39.** 4,000,000 **41.** 7 **43.** 1.28

45.	2,380	2.38
47.	1	0.01
49.	0.3	0.0003
51.	450,000	0.45
53.	0.709	0.000709
55.	17,000,000	17,000

57. 3,250 m, or 3.25 km **59.** 98,025.6 g, or 98.0256 kg **61.** 840 **63.** 4 **65.** 1.2 **67.** 2.4 **69.** 3.5 m or 350 cm **71.** 2,500 **73.** a **75.** Approximately 3.3 qt **77.** 6g **79.** Yes **81.** 3 in. **83.** 2,400 mm **85.** 750 cm **87.** 200,000 mL **89.** 5.6 mg **91.** 10.1 kg **93.** 1,350 g **95.** 3.2 L

Chapter 9 Review Exercises, pp. 425–427

1. do not; Possible answer: the unit factor must have the desired unit in the numerator and the original unit in the denominator: $\frac{24\text{ hr}}{1\text{ day}}$ **2.** cannot; Possible answer: a quart is a unit of capacity and not of weight **3.** is; Possible answer: the prefix "centi-" means $\frac{1}{100}$, the prefix "milli-" means $\frac{1}{1{,}000}$, and the first fraction is larger than the second **4.** is; Possible answer: on the metric conversion line, we move three units to the right to go from kilometers to meters **5.** is; Possible answer: the prefix "mega-" means one million **6.** is not; Possible answer: the prefix "micro-" means one millionth and not one billionth **7.** 15 **8.** $1\frac{2}{3}$ **9.** 2 **10.** $3\frac{1}{3}$ **11.** 3,000 **12.** 136 **13.** 48 **14.** $2\frac{1}{2}$ **15.** 435 **16.** 4 ft 2 in. **17.** 2

18. 125 **19.** 8 hr 10 min **20.** 18 ft 9 in. **21.** 1 gal 3 qt **22.** 7 lb 2 oz **23.** c **24.** a **25.** b **26.** c **27.** b **28.** a **29.** a **30.** c **31.** 0.037 **32.** 4,000 **33.** 800 **34.** 2,100 **35.** 0.6 **36.** 5.1 **37.** 0.005 **38.** 4,000 **39.** 112 **40.** 2 **41.** 20 **42.** 15 **43.** 1.2 hr **44.** 11,440 pt **45.** *Frankenstein* **46.** 0.6 L **47.** 2 g **48.** 0.072 g **49.** 0.2 g **50.** 750,000 g **51.** 6 hr **52.** 25,000 mg **53.** 33 ft **54.** 9.104 g **55.** 199,900 mcg **56.** 50,000 MB **57.** 4 min 30 sec **58.** 10 in. **59.** 26 mg **60.** 5 MB **61.** 24 in. **62.** 20,000 L **63.** 16 wk **64.** 178–208 km/hr

Chapter 9 Posttest, p. 428

1. 2 **2.** 21 **3.** 24 **4.** 3 ft 3 in. **5.** 1 hr 45 min **6.** c **7.** c **8.** 4 **9.** 6,000 **10.** 4,600,000 **11.** 2,000 **12.** 5,000 **13.** 84 **14.** 3.3 **15.** 14 cm **16.** 6 mi **17.** Finback whale **18.** 0.000003 sec **19.** 2.1 times as much memory **20.** 22,000 mi

Chapter 9 Cumulative Review Exercises, pp. 429–430

1. 729 **2.** $5\frac{1}{2}$ **3.** $\frac{1}{4}$ **4.** 14 **5.** $k = \frac{4}{3}$ **6.** 24 mg of sodium/cracker **7.** $66\frac{2}{3}\%$ **8.** \$18 **9.** $-4\frac{3}{5}$ **10.** $\frac{2}{3}$ **11.** 8 **12.** $\frac{5}{3}$ **13.** 84 **14.** 40 g **15.** 30 bushels per acre **16.** −\$1.47 **17.** \$500,000 **18.** 8,000 more golden retrievers **19.** 8 lb **20.** 400 mi

CHAPTER 10

Chapter 10 Pretest, pp. 432–433

1.

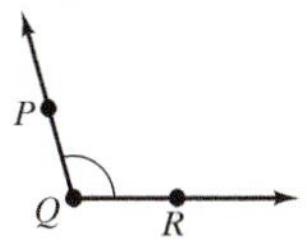

2. 11 **3.** 80° **4.** 54° **5.** 14 in. **6.** 20 ft **7.** Approximately 12.56 in. **8.** 10.4 m **9.** 30 in^2 **10.** Approximately 78.5 ft^2 **11.** 125 cm^3 **12.** Approximately 254.34 m^3 **13.** 15 m **14.** $x = 7$ ft; $y = 5$ ft **15.** $a = 104°$ **16.** 21 cm **17.** No, a right angle. **18.** 3 mi^2 **19.** No, because its diagonal is more than 53 in. **20.** 8,100 ft^3

Section 10.1 Practices, pp. 437–442

1, *p. 437:*

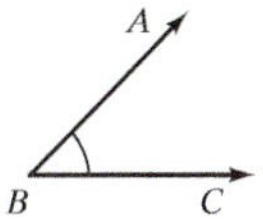

2, *p. 437:* 53° **3, *p. 437:*** 165°

4, *p. 438:* $x = 82°$ **5, *p. 438:*** $a = b = 153°$

6, *p. 441:*

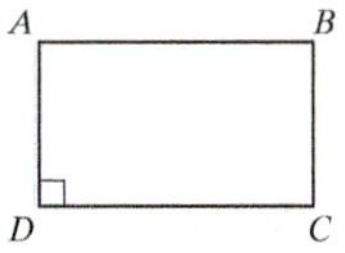

$\overline{AB} \| \overline{DC}$; $\overline{AD} \| \overline{BC}$

7, *p. 442:* 60° **8, *p. 442:*** $m\angle U = 60°$ **9, *p. 442:*** 24 in.

Exercises 10.1, pp. 443–447

1. line segment **3.** complementary **5.** perpendicular **7.** parallel **9.** scalene **11.** parallelogram **13.**

15.

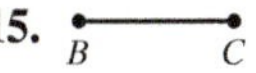

17.

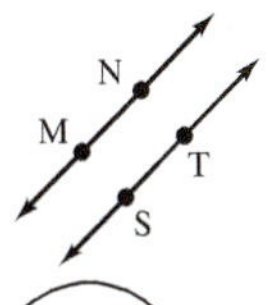

19.

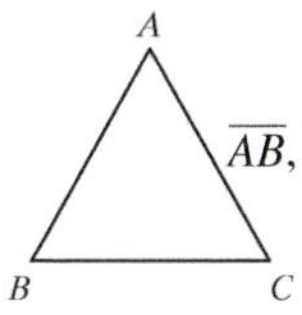

$\overline{AB}$, $\overline{AC}$, and $\overline{BC}$ are the same length.

21.

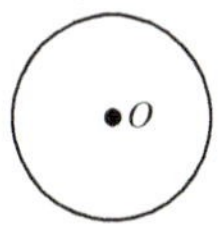

23.

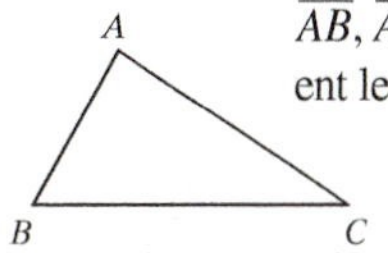

$\overline{AB}$, $\overline{AC}$, and $\overline{BC}$ have different lengths.

25.

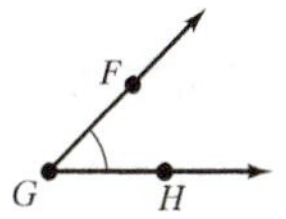

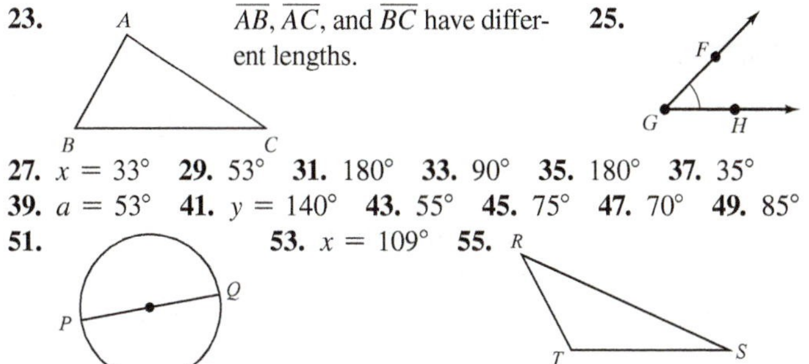

27. $x = 33°$ **29.** 53° **31.** 180° **33.** 90° **35.** 180° **37.** 35° **39.** $a = 53°$ **41.** $y = 140°$ **43.** 55° **45.** 75° **47.** 70° **49.** 85° **51.**

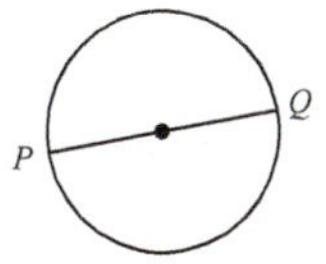

53. $x = 109°$ **55.**

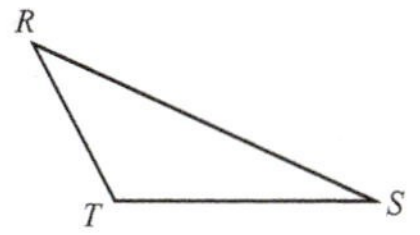

57. 13 ft **59.** 97° **61.** Acute **63.** 45 ft by 32 ft **65. a.** 6 in. **b.** 60° **67. a.** A trapezoid **b.** 80°

Section 10.2 Practices, pp. 449–453

1, *p. 449:* 26 in. **2, *p. 449:*** 3 mi **3, *p. 450:*** \$70 **4, *p. 451:*** Approximately 132 in. **5, *p. 451:*** Approximately 113 ft **6, *p. 452:*** Approximately 164.5 yd **7, *p. 453:*** 18 mi

Exercises 10.2, pp. 454–457

1. perimeter **3.** square **5.** circle **7.** 17 in **9.** 6 m **11.** $10\frac{1}{2}$ yd **13.** Approximately 62.8 m **15.** Approximately 21.98 ft **17.** 21 yd **19.** 18 ft **21.** 19 cm **23.** Approximately 18.84 cm **25.** Approximately 11.12 m **27.** 44.1 cm **29.** Approximately 25.12 in. **31.** 40 yd **33.** 64 yd **35.** 27 cm **37.** 28 in. **39.** 228 ft **41.** Approximately 6 in. **43.** 30 posts **45.** Approximately 50.57 ft **47.** Approximately 42,700 km

Section 10.3 Practices, pp. 460–464

1, *p. 460:* 12 cm^2 **2, *p. 460:*** 12.96 cm^2 **3, *p. 460:*** 7.5 in^2 **4, *p. 461:*** $12\frac{1}{2}$ ft^2 **5, *p. 461:*** $3\frac{1}{2}$ ft^2 **6, *p. 462:*** Approximately 78.5 yd^2 **7, *p. 462:*** No. A customer needs 108 tiles at a total cost of \$538.92. **8, *p. 463:*** Approximately 79 mi^2 **9, *p. 463:*** 8.75 m^2 **10, *p. 464:*** 4,421.5 ft^2

Exercises 10.3, pp. 465–468

1. area **3.** triangle **5.** circle **7.** 125 m^2 **9.** 30 ft^2 **11.** 290 yd^2 **13.** Approximately 706.5 cm^2 **15.** 32 m^2 **17.** 15.6 m^2 **19.** Approximately 314 in^2 **21.** 6.25 ft^2 **23.** 44.1 yd^2 **25.** 3.64 m^2 **27.** $\frac{1}{16}$ yd^2 **29.** 46 yd^2 **31.** Approximately 64.25 ft^2 **33.** 51.75 ft^2 **35.** Approximately 113.04 cm^2 **37.** 12 mi^2 **39.** 32.5 m^2 **41.** 160 ft^2 **43.** Approximately 0.05 mm^2 **45.** 5,000 ft^2 **47.** 486 in^2 **49.** \$131.25

Section 10.4 Practices, pp. 470–473

1, *p. 470:* 72 ft^3 **2, *p. 471:*** 3,375 cm^3 **3, *p. 471:*** Approximately 11 in^3 **4, *p. 471:*** Approximately 113 in^3 **5, *p. 472:*** Approximately 4.19 m^3 **6, *p. 472:*** The ball's volume is about 523 in^3, whereas the box's volume is 1,000 in^3. Therefore, the peanut packing occupies approximately 477 in^3. **7. *p. 473:*** 784 in^3

Exercises 10.4, pp. 474–477

1. volume **3.** rectangular solid **5.** cylinder **7.** 216 in^3 **9.** 2,560 m^3 **11.** Approximately 63 ft^3 **13.** Approximately 125.1 ft^3 **15.** Approximately 2,144 in^3 **17.** Approximately 1.95 m^3 **19.** Approximately 92.1 ft^3 **21.** 30,000 ft^3 **23.** Approximately 20.3 m^3 **25.** Approximately 33.5 in^3 **27.** 0.01 kg/cm^3 **29.** 4,187,000,000 mi^3 **31.** Approximately 14 in^3 **33.** 15.19 m^3 **35.** 1,400 times **37.** Approximately 2.9 g/cm^3

Section 10.5 Practices, pp. 479–480

1, *p. 479:* $\overline{AB}$ corresponds to $\overline{GH}$; $\overline{AC}$ corresponds to $\overline{GI}$; $\overline{BC}$ corresponds to $\overline{HI}$. **2, *p. 479:*** $y = 24$ **3, *p. 480:*** $h = 36$ ft

Exercises 10.5, pp. 481–483

1. similar **3.** corresponding **5.** $x = 10\frac{2}{3}$ in. **7.** $x = 24$ m **9.** $x = 15$ ft, $y = 20$ ft **11.** $x = 24$ yd, $y = 12$ yd **13.** $x = 3.8125$ m **15.** $x = 12$ m **17.** $x = 10.5$ ft **19.** 9 ft **21.** 200 m **23.** 3.2 in. **25.** 24 ft

Section 10.6 Practices, pp. 485–488

1, ***p. 485:*** **a.** 7 **b.** 12 **2,** ***p. 485:*** Between 6 and 7 **3,** ***p. 486:*** **a.** 7.48 **b.** 3.46 **4,** ***p. 487:*** 10 in. **5,** ***p. 487:*** 3.5 ft **6,** ***p. 488:*** Approximately 12.8 ft

Exercises 10.6, pp. 489–492

1. square root **3.** perfect square **5.** hypotenuse **7.** 3 **9.** 4 **11.** 9 **13.** 13 **15.** 20 **17.** 16 **19.** 7 and 8 **21.** 8 and 9 **23.** 6 and 7 **25.** 3 and 4 **27.** 2.2 **29.** 6.1 **31.** 11.8 **33.** 99 **35.** $a = 16$ cm **37.** $c \approx 3.2$ m **39. b.** 7 m **41. b.** 8 ft **43. c.** 20 m **45. c.** 11.4 cm **47. a.** 8.7 ft **49.** 14 **51.** $x = 10$ in. **53.** $n \approx 9.8$ cm **55.** 16 ft **57.** 240 ft **59.** Approximately 37.4 ft

Chapter 10 Review Exercises, pp. 501–505

1. is; possible answer: they are vertical angles **2.** is; possible answer: it is formed by combining a square and a rectangle **3.** cannot; possible answer: a foot is a unit of length and not of area **4.** does not; possible answer: it is a square and therefore has 4 sides **5.** are not; possible answer: they have different shapes **6.** is; possible answer: it is opposite right angle C **7.**

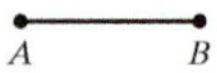

8.

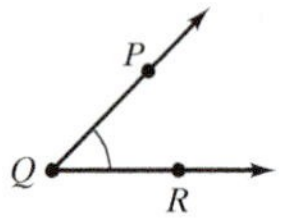

9.

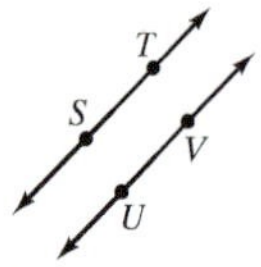

10.

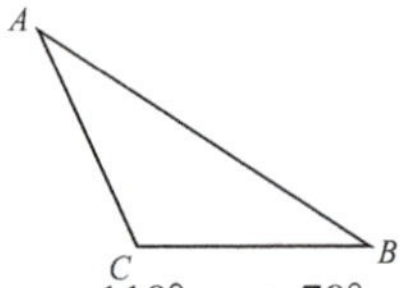

11. $x = 75°$ **12.** $x = 131°$

13. $x = 110°, y = 70°$ **14.** $x = 80°$; $y = 100°$ **15.** 5.4 m **16.** 30 ft **17.** 19 cm **18.** Approximately 125.6 in. **19.** 225 yd^2 **20.** Approximately 615 ft^2 **21.** 64 in^2 **22.** 17 m^2 **23.** Approximately 1,318.8 in^3 **24.** 216 ft^3 **25.** Approximately 1.95 m^3 **26.** Approximately 8.18 cm^3 **27.** Approximately 122.82 ft **28.** Approximately 84.78 ft^2 **29.** Approximately 17,850 ft^2 **30.** Approximately 3.81 yd^3 **31.** $x = 8$ ft, $y = 7.875$ ft **32.** $x = 4.5$ m **33.** 3 **34.** 8 **35.** 11 **36.** 30 **37.** 1 and 2 **38.** 9 and 10 **39.** 6 and 7 **40.** 3 and 4 **41.** 2.83 **42.** 35.14 **43.** 13.96 **44.** 5.39 **45. b.** 12 ft **46. a.** 10 in. **47. c.** 9.4 yd **48. c.** 2.8 ft **49.** 28,800 in^2 **50.** 125,600 mi^2 **51.** Yes, it can; the volume of the room is 2,650 ft^3. **52.** 220 in^2 **53.** 13 mi **54.** 10 ft **55.** $CD = 2\frac{1}{4}$ ft **56.** 7.5 mi **57.** 160 ft **58.** It is not efficient because the perimeter of the work triangle is 25 ft. **59.** 0.2 oz **60.** 1,347 cm^3 of soil **61.** 411.25 in^2 **62.** 2.6 in^3

Chapter 10 Posttest, pp. 506–507

1.

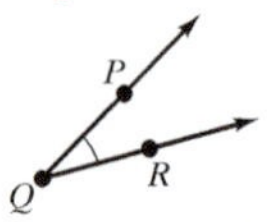

2. 15 **3.** 65° **4.** 89° **5.** 14 ft **6.** 4.5 m **7.** Approximately 25 cm **8.** 15 ft **9.** 54 ft^2 **10.** 110 cm^2 **11.** 189 m^3 **12.** Approximately 33 ft^3 **13.** $a = 120°$; $b = 60°$; $y = 5$ m; $x = 10$ m **14.** $a = 69°$ **15.** $y = 12$ yd **16.** $x = 15$; $y = 24$ **17.** No, it is obtuse. **18.** 9.7 board ft **19.** 79 m^2 **20.** 250 mi

Chapter 10 Cumulative Review Exercises, p. 508

1. 60 **2.** $\frac{4}{5}$ **3.** 3.1 **4.** 15.5036 **5.** $y = 8$ **6.** 50 **7.** \$583.20 **8.** 15 **9.** \$61,400 **10.** $b = 24$ **11.** 2 qt **12.** 9.4 **13.** $\frac{1}{51}$ **14.** 12 **15.** 448 ft **16.** −637, that is, 637 points down **17.** Approximately 707 mi^2 **18.** \$2,000 and \$3,000 **19.** Approximately 180,000 more complaints **20.** 1 mi **23.** 625.3 m^2 **24.** 0.2 **25.** 15 times

CHAPTER 11

Chapter 11 Pretest, p. 510

1. +\$2000 **2.** yes **3.**

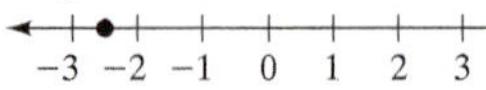

4. < **5.** −2 **6.** 5 **7.** −13 **8.** −26 **9.** $\frac{4}{1}$, or 4 **10.** 9 **11.** -6^4 **12.** $-2ab^3$ **13.** −8 **14.** $-3a - 3b$ **15.** $4n - 7$ **16.** $14x - 10$ **17.** 6051 m **18.** The team scored 2 more points than its opponents (+2). **19.** 30 cm **20.** The third round

Section 11.1 Practices, pp. 511–516

1, ***p. 511:*** −18°F **2,** ***p. 512:***

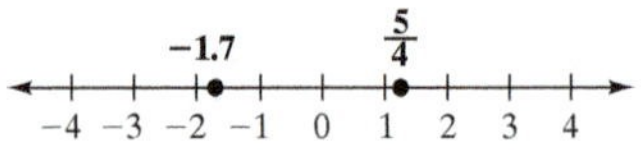

3, ***p. 514:*** **a.** True **b.** False **c.** True **d.** True **e.** False

4, ***p. 514:***

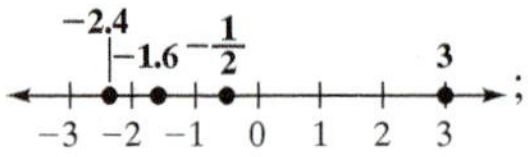

; 3, $-\frac{1}{2}$, −1.6, and −2.4 **5,** ***p. 514:*** The Caspian Sea; $-92 < -52 < 0$

6, ***p. 516:***

a.

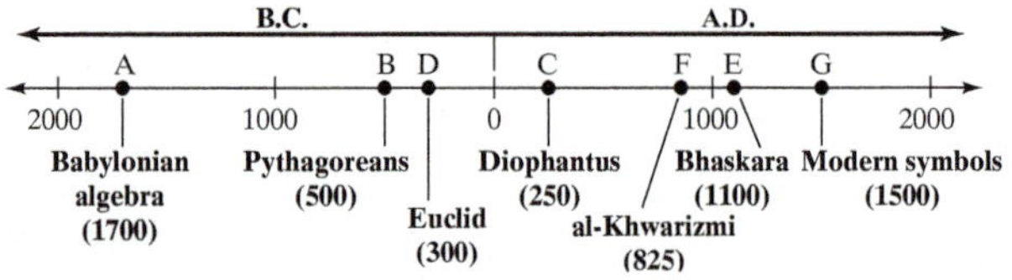

b. A, B, D, C, F, E, and G

Exercises 11.1, pp. 517–520

1. natural numbers **3.** negative numbers **5.** origin **7.** neither terminate nor **9.** −5 km **11.** −22.5°C **13.** −\$160

		Whole Numbers	Integers	Rational Numbers	Real Numbers
15.	−7		✓	✓	✓
17.	$3\frac{1}{6}$			✓	✓
19.	10	✓	✓	✓	✓

21.

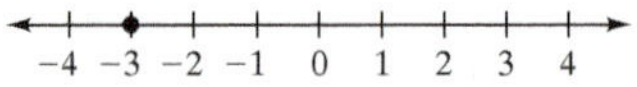

23.

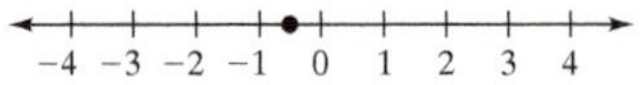

25.

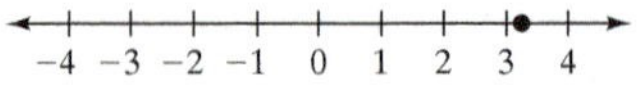

27.

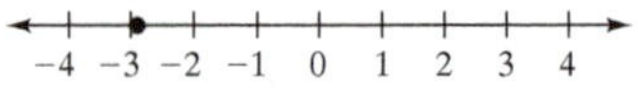

29. True **31.** False **33.** True **35.** > **37.** > **39.** = **41.** = **43.** < **45.**

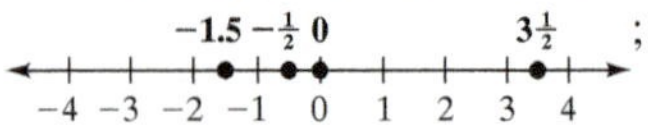

; $3\frac{1}{2}$, 0, $-\frac{1}{2}$, −1.5

47.

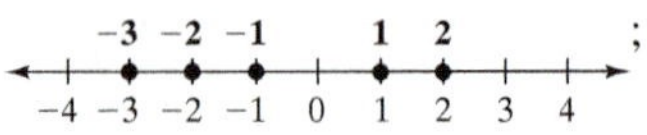

; 2, 1, −1, −2, −3 **49.** −\$53 **51.** Rational numbers and real numbers **53.** > **55.** Today **57. a.** Sirius **b.** −13

c.

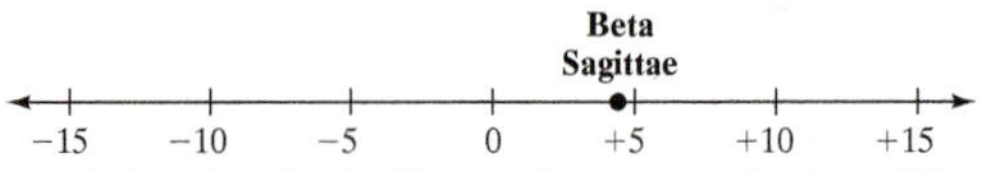

59. a. Aristotle **b.** Attila **c.** Socrates **d.** Tiger Woods

Section 11.2 Practices, pp. 522–527

1, *p. 522:* -1

−4 −3 −2 −1 0 1 2 3 4
End Start

2, *p. 522:* 0;

−5 −4 −3 −2 −1 0 1 2 3 4 5
Start End

3, *p. 523:* 0.5;

−2 −1 0 0.5 1 2
Start End

4, *p. 523:* -31 **5, *p. 524:*** 3.5 **6, *p. 524:*** 0 **7, *p. 524:*** $-\frac{5}{18}$ **8, *p. 525:*** -10 **9, *p. 526:*** \$36.50 **10, *p. 526:*** 5 **11, *p. 527:*** 17 **12, *p. 527:*** -3 **13, *p. 527:*** About 870 years older

Exercises 11.2, pp. 528–532

1. additive inverse property **3.** additive identity property **5.** subtracting **7.** 1 **9.** 0 **11.** -8 **13.** Additive inverse property **15.** Commutative property of addition **17.** Additive identity property **19.** Associative property of addition **21.** Additive identity property **23.** 23 **25.** -80 **27.** -30 **29.** 0 **31.** 4.3 **33.** -10.5 **35.** -5.7 **37.** -16.3 **39.** $-\frac{2}{5}$ **41.** $\frac{5}{6}$ **43.** 0 **45.** -7 **47.** 10.48 **49.** -12 **51.** -31 **53.** -44 **55.** 71 **57.** -35 **59.** -32 **61.** 32 **63.** -45 **65.** -62 **67.** 44 **69.** -1.42 **71.** -7.8 **73.** 0 **75.** 10.3 **77.** $-1\frac{1}{6}$ **79.** $-17\frac{1}{4}$ **81.** 12 **83.** -7 **85.** -21 **87.** -18 **89.** 5 **91.** 0 **93.** 18 **95.** Associative property of addition **97.** -7 **99.** -15 **101.** 48 **103.** -38 **105.** $1\frac{7}{12}$ **107.** 0 **109.** 5° above 0° $(+5)$ **111.** The Pittsburgh Steelers by four points **113.** Yes; he will have \$283.26 in the account. **115.** 28 centuries **117.** \$276,039 **119. a.** Krypton: 4°; neon: 3°; bromine: 66° **b.** bromine **c.** bromine

Section 11.3 Practices, pp. 533–541

1, *p. 533:* 100 **2, *p. 533:*** -15 **3, *p. 534:*** **a.** 8 **b.** $-\frac{5}{27}$ **c.** 0.12 **d.** -4.75 **e.** 0 **f.** $\frac{2}{3}$ **4, *p. 535:*** 64 **5, *p. 535:*** -240 **6, *p. 536:*** -28 **7, *p. 536:*** -10 **8, *p. 536:*** 4 **9, *p. 536:*** 1 **10, *p. 537:*** About -5.9; the rock is moving downward at a velocity of 5.9 ft/sec. **11, *p. 537:*** **a.** -8 **b.** 7 **c.** $-\frac{1}{2}$ **d.** -0.7 **e.** 60 **12, *p. 539:*** **a.** $-\frac{1}{5}$ **b.** -8 **c.** $\frac{3}{4}$ **d.** $-\frac{5}{8}$ **13, *p. 540:*** **a.** $-\frac{4}{3}$, or $-1\frac{1}{3}$ **b.** 25 **14, *p. 540:*** **a.** -12 **b.** -3 **15, *p. 541:*** -125 minutes (Down 125 minutes)

Exercises 11.3, pp. 542–546

1. multiply two numbers in either order **3.** regroup the product of three numbers **5.** reciprocal **7.** Commutative property of multiplication **9.** Associative property of multiplication **11.** Multiplicative identity property **13.** Multiplication property of zero **15.** -12 **17.** 21 **19.** -3 **21.** $-\frac{16}{27}$ **23.** 0.9 **25.** -60 **27.** 120 **29.** 0 **31.** 144 **33.** $-\frac{1}{27}$ **35.** -23 **37.** 27 **39.** -3 **41.** -5 **43.** 6 **45.** -20 **47.** -12 **49. a.** -2 **b.** $\frac{1}{5}$ **c.** $-\frac{4}{3}$ **d.** $\frac{5}{16}$ **e.** -1 **51.** 8 **53.** -9 **55.** 0 **57.** 25 **59.** 25 **61.** $-\frac{1}{8}$, or -0.125 **63.** $-\frac{6}{5}$ **65.** -32 **67.** $-\frac{1}{8}$ **69.** -0.5 **71.** 10 **73.** -16 **75.** 1 **77.** 2 **79.** $\frac{1}{2}$ **81.** 4 **83.** 14 **85.** 3.64 **87.** -120 **89.** 4 **91.** 10 **93.** $-\frac{20}{8}$ **95.** 5 **97.** 6 **99.** -15 in. (dropped 15 in.) **101. a.** -135 calories **b.** $+106$ calories **c.** -852 calories **103.** $+200$ customers (an increase of 200 customers per year) **105.** An average loss of 4 yd $(-4$ yd)

Section 11.4 Practices, pp. 548–550

1, *p. 548:* **a.** 3 **b.** 1 **2, *p. 549:*** **a.** -36 **b.** 324 **3, *p. 550:*** **a.** $2^4(-5)^2$ **b.** $(-6)^4(8)^2$ **4, *p. 550:*** **a.** $-x^5$ **b.** $2m^3n^4$ **5, *p. 550:*** The population after 10 hr was $243x$, or 3^5x.

Exercises 11.4, pp. 551–552

1. terms **3.** base **5.** 1 **7.** 2 **9.** 3 **11.** -9 **13.** -432 **15.** $(-2)^3(4)^2$ **17.** $6^2(-3)^3$ **19.** $(2)^4(-1)^2$ **21.** $3n^3$ **23.** $-4a^3b^2$ **25.** $-y^3$ **27.** $10a^3b^2c$ **29.** $-x^2y^3$ **31.** $(-4)^2(5)^3$ **33.** $-3p^3q$ **35.** 2 **37.** -36 **39.** $(2^3 \cdot 5000)$ dollars **41.** s^2

Section 11.5 Practices, pp. 553–555

1, *p. 553:* **a.** 15 **b.** 30 **2, *p. 553:*** **a.** 18 **b.** 16 **c.** 52 **d.** -2 **3, *p. 554:*** **a.** $\frac{7}{3}$ **b.** $\frac{3}{4}$ **c.** 81 **d.** -81 **4, *p. 554:*** 39.7 gal **5, *p. 555:*** $F = \frac{9}{5}C + 32$ **6, *p. 555:*** The distance d is 80 mi. **7, *p. 555:*** **a.** $K = C + 273$ **b.** K is 267.

Exercises 11.5, pp. 556–559

1. -2 **3.** 16 **5.** -32 **7.** -7 **9.** 0 **11.** 12 **13.** 5 **15.** 56 **17.** -7 **19.** 15 **21.** -14.5 **23.** -16 **25.** 15

27.

x	0	1	2	−1	−2
$2x + 5$	5	7	9	3	1

29.

y	0	1	2	3	4
$y - 0.5$	−0.5	0.5	1.5	2.5	3.5

31.

x	0	2	4	−2	−4
$-\frac{1}{2}x$	0	−1	−2	1	2

33.

n	2	4	6	−2	−4
$\frac{n}{2}$	1	2	3	−1	−2

35.

g	0	1	2	−1	−2
$-g^2$	0	−1	−4	−1	−4

37.

a	0	1	2	−1	−2
$a^2 + 2a - 2$	−2	1	6	−3	−2

39. $-20°$ **41.** \$2200 **43.** $7\frac{1}{2}$ ft **45.** -3 **47.** 314 m **49.** 20 mg **51.** 13.5 cm^2 **53.** 3140 in^3 **55.** 1

57.

x	0	1	2	−1	−2
$-2x + 4$	4	2	0	6	8

59. -11 **61.** 10.5 in^2 **63.** $A = \frac{a + b + c}{3}$ **65.** $a^2 + b^2 = c^2$ **67.** $E = mc^2$ **69.** $l = 0.4w + 25$ **71.** The object falls 64 ft. **73. a.** $m = \frac{100(s - c)}{c}$ **b.** 40% **75. a.** $C = 9f + 4(c + p)$ **b.** 53 calories

Section 11.6 Practices, pp. 561–565

1, *p. 561:* **a.** coefficients: 1 and -3; terms: m and $-3m$; like **b.** coefficient: 5; terms: $5x$ and 7; unlike **c.** coefficients: 2 and -3; terms: $2x^2y$ and $-3xy^2$; unlike **d.** coefficients: 1, 2, and -4; terms: m, $2m$, and $-4m$; like **2, *p. 562:*** **a.** $-40r - 10s$ **b.** $5w + w$ **c.** $3g - 9h$ **d.** $1.5y + 3$ **3, *p. 562:*** **a.** $6x$ **b.** $-6y$ **c.** $-2a + b$ **d.** 0 **4, *p. 563:*** **a.** $-2y^2$ **b.** Cannot be simplified **c.** $3xy^2$ **5, *p. 563:*** $3y - 10$ **6, *p. 563:*** $-2a + 3b$ **7, *p. 564:*** $4y - 1$ **8, *p. 564:*** $-2y - 18$ **9, *p. 564:*** $-10y + 13$ **10, *p. 565:*** $5c + 12(c - 40)$; $(17c - 480)$ dollars

Exercises 11.6, pp. 566–568

1. coefficient **3.** distributive property **5.** negative **7.** 7 **9.** -5 **11.** 1 **13.** -1 **15.** -0.1 **17.** $\frac{2}{3}$ **19.** 2; -5 **21.** Terms: $2a$ and $-a$; like

23. Terms: $5p$ and 3; unlike **25.** Terms: $4x^2$ and $-6x^2$; like **27.** Terms: $-20n$ and $-3n$; like **29.** $-7x + 7y$ **31.** $a - 10a$ **33.** $-0.5r - 1.5$ **35.** $10x$ **37.** $-11n$ **39.** $14a$ **41.** $2y + 2$ **43.** 0 **45.** Cannot be simplified **47.** $4r^2t^2$ **49.** Cannot be simplified **51.** $2x + 2$ **53.** $11x - 1$ **55.** $-3y + 10$ **57.** $4x - 9$ **59.** $-n + 39$ **61.** $5x + 1$ **63.** $-3x + 13$ **65.** $-3a - 14$ **67.** Cannot be simplified **69.** Terms; $3a$ and $3a^2$; unlike **71.** 1 **73.** $3y + 7$ **75.** $x + x + 40$; $(2x + 40)°$ **77.** $d + 2(d + 4)$; $(3d + 8)$ dollars **79.** $n + (n + 1) + (n + 2)$; $3n + 3$ **81.** $1c + 0b - 0.25(54 - b - c)$; $1.25c + 0.25b - 13.5$

Chapter 11 Review Exercises, pp. 572–575

1. is; possible answer: it can be expressed as the quotient of two integers, $\frac{1}{2}$ **2.** is; possible answer: their sum is 0 **3.** are not; possible answer: their product is not 1 **4.** is not; possible answer: the expression to be squared is $6a$ **5.** is; possible answer: it can be expressed as 5 to the third power **6.** are not; possible answer: they have different exponents of the variable p and different exponents of the variable q **7.** +3 mi **8.** −\$160

9. 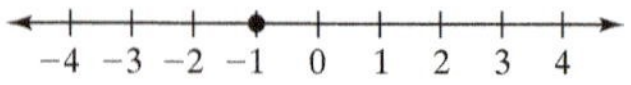

10. 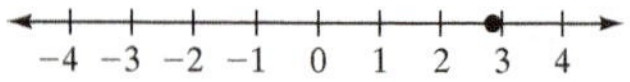

11. 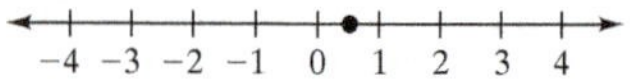

12. 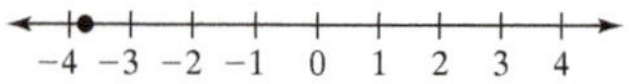 **13.** True **14.** False **15.** −5

16. −4 **17.** Commutative property of addition **18.** Additive identity property **19.** Associative property of addition **20.** Additive inverse property **21.** 0 **22.** 2 **23.** 2 **24.** −15 **25.** −5 **26.** −85 **27.** −8.1 **28.** 15.3 **29.** 9 **30.** −11 **31.** −55 **32.** −3 **33.** −27 **34.** 27 **35.** 16 **36.** −5 **37.** −1.42 **38.** $-8\frac{5}{8}$ **39.** 5 **40.** 10 **41.** Commutative property of multiplication **42.** Associative property of multiplication **43.** Multiplicative identity property **44.** Multiplication property of zero **45.** −10 **46.** −21 **47.** −5400 **48.** 2400 **49.** 27 **50.** $-\frac{1}{4}$ **51.** 6000 **52.** 36 **53.** −23 **54.** 14 **55.** 26 **56.** 38 **57.** −12 **58.** 6 **59.** $-\frac{3}{2}$ **60.** $\frac{1}{8}$ **61.** 3 **62.** −6 **63.** $-2\frac{1}{5}$ **64.** $-\frac{6}{5} = -1\frac{1}{5}$ **65.** 32 **66.** 2 **67.** −1 **68.** −2 **69.** 13 **70.** −20 **71.** 3 **72.** 2 **73.** 1 **74.** 4 **75.** $(-3)^4$ **76.** $(-5)^3 3^2$ **77.** $4x^3$ **78.** $-5a^2b^3c$ **79.** 29 **80.** $-\frac{20}{9}$ **81.** −20 **82.** 60 **83.** 1 **84.** 1 **85.** $-5x + 5y$ **86.** $14x - 2y$ **87.** $-2x^2$ **88.** r^2t^2 **89.** $2a - 9$ **90.** $-3x - 2$ **91.** $-4x - 15$ **92.** $a - 17$ **93.** +\$700 **94.** −\$2.00 **95.** Exothermic (+3°C) **96.** 0.97 per 100; 0.97% **97.** 281 m **98.** Tue: −\$0.09, Wed: −\$0.04, Thu: +\$0.24 Fri: +\$0.32 **99.** $6w$ **100.** The first loss is 3 times the second loss. **101.** $3^3 \cdot 10$ bacteria **102.** 71 ft **103.** $0.0013P$ **104.** $[0.05x + 0.07(600 - x)]$ dollars **105.** 20 amperes **106.** The account is overdrawn by \$10(−10).

Chapter 11 Posttest, p. 576

1. −10,000 **2.** Yes **3.** 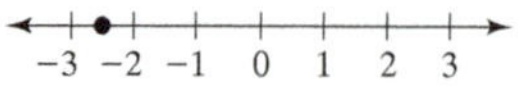 **4.** True **5.** −7 **6.** 7 **7.** 3 **8.** 10 **9.** $\frac{1}{12}$ **10.** −50 **11.** $(-5)^3$ **12.** $-3ab^2c^3$ **13.** 8 **14.** −5 **15.** $7y + 4$ **16.** $2t + 3$ **17.** +\$10(won \$10)d **18.** An improvement of \$70,000 **19.** $1.05d$ dollars **20.** Yes; the team kept the ball.

Chapter 11 Cumulative Review Exercises, pp. 577–578

1. 1,512 **2.** $(8 + 4) \cdot 2^2$ **3.** $2\frac{5}{6}$ **4.** $5\frac{1}{4}$ **5.** 0.001 **6.** $x + 15.7 = 84.6$; $x = 68.9$ **7.** $t = 64$ **8.** $83\frac{1}{3}\%$ **9.** −19 **10.** −2 **11.** Mean: $2\frac{3}{5}$; median: $2\frac{1}{2}$; no mode; range: $1\frac{1}{4}$ **12.** 480 **13.** 7 and 8 **14.** 405 **15.** This change represented an increase because 0.1 is larger than 0.075 (by 0.025). **16.** $\frac{1}{17}$ **17.** 1,200 stocks **18.** −\$38.75 million **19.** 300 million more users **20.** 40 mph

CHAPTER 12

Chapter 12 Pretest, p. 580

1. No **2.** Subtract 2 (or add −2). **3.** $y = -11$ **4.** $n = -4\frac{1}{2}$ **5.** $y = -5$ **6.** $n = -8$ **7.** $x = -9$ **8.** $x = -\frac{1}{2}$ **9.** $x = 3$ **10.** $x = 6$ **11.** $n = -2$ **12.** $x = \frac{5}{3}$ **13.** $v = w + 5u$ **14.** 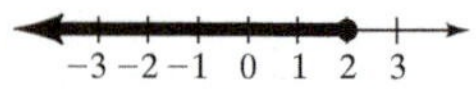 **15.** $x > 0$; 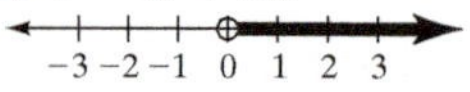 **16.** 12 min **17.** 20 centerpieces **18.** 39 mph **19.** $m = \frac{2E}{v^2}$ **20.** Option A is a better deal if the member uses the gym more than 15 hours per month $(x > 15)$.

Section 12.1 Practices: pp. 581–585

1, *p. 581:* No, 4 is not a solution. **2, *p. 581:*** Yes, −8 is a solution. **3, *p. 583:*** $y = 5$ **4, *p. 583:*** $n = -17$ **5, *p. 584:*** $s = -\frac{1}{2}$ **6, *p. 584:*** $x = 0.1$ **7, *p. 585:*** An increase of 12.3°C

Exercises 12.1, pp. 586–589

1. linear equations in one variable **3.** solve **5. a.** True **b.** False **c.** True **d.** True **7.** Subtract 4 (or add −4). **9.** Subtract 3.5 (or add −3.5). **11.** Add 1. **13.** Add $2\frac{1}{5}$. **15.** $y = -23$ **17.** $t = 0$ **19.** $a = -12$ **21.** $z = -6$ **23.** $x = -42$ **25.** $t = 6$ **27.** $r = 6$ **29.** $n = 13$ **31.** $x = -1$ **33.** $y = -3\frac{1}{2}$ **35.** $m = 2.9$ **37.** $t = -3.6$ **39.** $a = -5$ **41.** $m = -\frac{1}{2}$ **43.** $x + 2 = -12$; $x = -14$ **45.** $n - 4\frac{1}{3} = 21$; $n = 25\frac{5}{6}$ **47.** $x + (-3) = -1$; $x = 2$ **49.** $n - (-17) = 11$; $n = -6$ **51.** d **53.** a **55.** Subtract $\frac{2}{3}$ (or add $-\frac{2}{3}$). **57. a.** False **b.** True **c.** False **d.** True **59.** Possible answer: $x - 2.5 = -3.8$; $x = -1.3$ **61.** $m = -23$ **63.** $n = -4\frac{1}{3}$ **65.** $x + 10 = 44$; $x = 34$ mph **67.** $x + 190 = 370$; $x = 180$ calories **69.** $h - 170 = 215$; $h = 385$ m **71.** 70°

Section 12.2 Practices: pp. 590–595

1, *p. 590:* $y = 63$ **2, *p. 591:*** $y = 9$ **3, *p. 591:*** $x = -10$ **4, *p. 592:*** $z = 13$ **5, *p. 592:*** $y = -14$ **6, *p. 593:*** The total bill was \$758. **7, *p. 594:*** It will take about 2.2 hr. **8, *p. 595:*** 6.5% $(r = 0.065)$

Exercises 12.2, pp. 596–598

1. Multiply by 3. **3.** Divide by −5. **5.** Divide by −2.2. **7.** Multiply by $\frac{4}{3}$. **9.** Multiply by $-\frac{2}{5}$. **11.** $x = -5$ **13.** $n = 18$ **15.** $a = 4.8$ **17.** $x = -0.5$ **19.** $c = -7$ **21.** $r = -22$ **23.** $x = 12$ **25.** $y = -\frac{5}{2}$ **27.** $n = 8$ **29.** $c = -3$ **31.** $x = 2.88$ **33.** $a = -2$ **35.** $y = \frac{2}{3}$ **37.** $-4x = 56$; $x = -14$ **39.** $\frac{n}{0.2} = 1.1$; $n = 0.22$ **41.** $\frac{x}{-3.5} = 30$; $x = -105$ **43.** $\frac{1}{6}x = 2\frac{4}{5}$; $x = \frac{84}{5}$ **45.** c **47.** a **49.** $x = -4$ **51.** $a = -24$ **53.** $\frac{x}{5} = 2$; $x = 10$ **55.** Divide by −5.2. **57.** $0.02x = 10.5$; $x = 525$ yr **59.** $70r = 3348$; $r \approx 48$ mph **61.** $0.05c = 20$; $c = 400$ copies **63.** $\frac{2}{3}x = 800{,}000$; $x = \$1{,}200{,}000$ **65.** $\frac{1}{5}d = 1000$; $d = 5000$ m **67.** $7.50t = 187.50$; $t = 25$ hr **69.** $12x = 10{,}020$; $x = \$835$/mo **71.** 5% **73.** \$5000

Section 12.3 Practices: pp. 600–610

1, *p. 600:* $y = 4$ **2, *p. 601:*** $c = 45$ **3, *p. 601:*** $b = -3$ **4, *p. 602:*** $n = \frac{4}{3}$ **5, *p. 602:*** $t = 3$ **6, *p. 603:*** $f = -\frac{3}{4}$ **7, *p. 603:*** $w = 4$ **8, *p. 604:*** $z = -5$ **9, *p. 605:*** $t = -4$ **10, *p. 605:*** $y = -2$ **11, *p. 606:*** The car will have a value of \$6500 in 5 yr. **12, *p. 606:*** The express train will catch up with the local train in 2.5 hr, or $2\frac{1}{2}$ hr. **13, *p. 607:*** 75 mi **14, *p. 608:*** 20 mph **15, *p. 609:*** she invested \$7000 in a mutual fund and \$14,000 in bonds. **16, *p. 610:*** 2 g

Exercises 12.3, pp. 611–614

1. $x = 3$ **3.** $t = -2$ **5.** $m = -5$ **7.** $n = 12$ **9.** $x = -75$ **11.** $t = 2$ **13.** $b = -19$ **15.** $x = 39$ **17.** $r = -50$ **19.** $y = -2$ **21.** $z = -6$ **23.** $a = -7$ **25.** $t = 0$ **27.** $y = -1$ **29.** $r = \frac{10}{3}$ **31.** $x = -7$ **33.** $y = 2$ **35.** $a = 14$ **37.** $t = \frac{3}{2}$ **39.** $y = 0$ **41.** $z = 2$ **43.** $m = -2$ **45.** a **47.** d **49.** $x = -4$ **51.** $z = -2$ **53.** $x = -5$ **55.** $45 + 135x = 1260$; $x = 9$. The student is carrying 9 credits. **57.** $x + 2x = 3690$; $x = 1230$. One candidate received 1230 votes; the other candidate received 2460 votes. **59.** $3 + 2(t - 1) = 9$; $t = 4$. The car was parked in the garage for 4 hr. **61.** $0.02x + 0.01(5000 - x) = 85$; $x = 3500$. 3500 large postcards and 1500 small postcards can be printed. **63.** $24(t + \frac{1}{3}) = 36t$; $t = \frac{2}{3}$. It took $\frac{2}{3}$ hr, or 40 min, to catch the bus. **65.** $27r + 27(r + 2) = 432$; $r = 7$. One snail is crawling at a rate of 7 cm/min, the other is crawling at a rate of 9 cm/min. **67.** $2r + 2(r + 4) = 212$; $r = 51$. The speed of the slower truck is 51 mph. **69.** \$20,000 was invested at 8% and \$14,000 was invested at 10%. **71.** \$10,000 was invested at 5%. **73.** $\frac{1}{3}$ cup additional olive oil **75.** 6 oz

Section 12.4 Practices: pp. 616–619

1, *p. 616:* $p = 1 - q$ **2, *p. 616:*** $r = \frac{t + s}{3}$ **3, *p. 617:*** $x = \frac{5ac}{4}$ **4, *p. 617:*** $x = \frac{y - b}{m}$ **5, *p. 618:*** **a.** $r = \frac{A - P}{Pt}$ **b.** $r = 0.025$, or 2.5% **6, *p. 618:*** **a.** $h = \frac{2A}{b}$ **b.** $h = 14$ in. **7, *p. 619:*** **a.** $A = \frac{1}{2}h(b + B)$ **b.** $b = \frac{2A - hB}{h}$ **c.** $b = 5$ cm

Exercises 12.4, pp. 620–622

1. literal equation **3.** algebraic expression **5.** $y = x - 10$ **7.** $d = c + 4$ **9.** $d = \frac{-3y}{a}$ **11.** $n = 4p$ **13.** $z = \frac{2a}{xy}$ **15.** $x = \frac{7 - y}{3}$ **17.** $y = \frac{12 - 3x}{4}$ **19.** $y = 4t$ **21.** $b = \frac{p - r}{5}$ **23.** $l = \frac{2m - h}{4}$ **25.** $r = \frac{d}{t}$ **27.** $b = P - a - c$ **29.** $d = \frac{C}{\pi}$ **31.** $R = \frac{P}{I^2}$ **33.** $a = 3A - b - c$ **35.** $a = S - dn + d$ **37.** **a.** $r = \frac{I}{Pt}$ **b.** $r = 0.03$ or 3% per year **39.** **a.** $b = \frac{A}{h}$ **b.** $b = 5$ m **41.** $h = \frac{V}{\pi r^2}$ **43.** $b = \frac{5m}{2ac}$ **45.** $z = \frac{3 - 4w}{9}$ **47.** $R = P + C$; $R = \$2500$ **49.** **a.** $K = \frac{V}{T}$ **b.** $V = KT$ **51.** **a.** $C = \frac{W}{150} \cdot A$ **b.** $A = \frac{150C}{W}$ **53.** **a.** $m = \frac{S}{5}$ **b.** $s = 5m$ **c.** The thunder will be heard in 12.5 sec ($s = 12.5$). **55.** **a.** $C = 2\pi r$ **b.** $r = \frac{C}{2\pi}$ **c.** $r \approx 0.8$ ft

Section 12.5 Practices, pp. 623–630

1, *p. 623:* No, 4 is not a solution.

2, *p. 624:* −5 −4 −3 −2 −1 0 1 2 3 4 5

3, *p. 624:* −5 −4 −3 −2 −1 0 1 2 3 4 5

4, *p. 624:* −5 −4 −3 −2 −1 0 1 2 3 4 5

5, *p. 626:* $n > -1$; −5 −4 −3 −2 −1 0 1 2 3 4 5

6, *p. 626:* $x \le 5\frac{1}{2}$; −2 −1 0 1 2 3 4 5 6 7 8

7, *p. 627:* $x > -7$; −8 −7 −6 −5 −4 −3 −2 −1 0 1 2

8, *p. 628:* $x \le 3$; −5 −4 −3 −2 −1 0 1 2 3 4 5

9, *p. 628:* $x < -5$; −11 −10 −9 −8 −7 −6 −5 −4 −3 −2 −1

10, *p. 628:* $x > -5$; −8 −7 −6 −5 −4 −3 −2 −1 0 1 2

11, *p. 629:* $z \le -5$ **12, *p. 629:*** $x < 2$ **13, *p. 629:*** $(x + 3) + (x + 2) + x \ge 14$; $x \ge 3$; The perimeter will be greater than or equal to 14 in. for any value of x greater than or equal to 3. **14, *p. 630:*** $15(8.5) + 7.5t \ge 300$; $t \ge 23$. She should work at least 23 hr on the second job.

Exercises 12.5, pp. 631–635

1. inequality **3.** open **5.** unchanged **7.** negative **9.** **a.** False **b.** True **c.** False **d.** False

11. 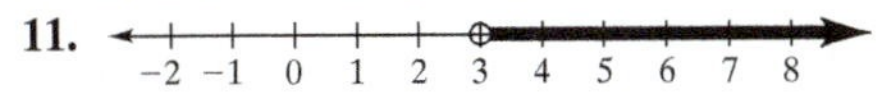

13. −8 −7 −6 −5 −4 −3 −2 −1 0 1 2

15. −5 −4 −3 −2 −1 0 1 2 3 4 5

17. 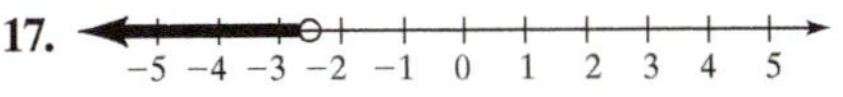

19. −5 −4 −3 −2 −1 0 1 2 3 4 5

21. −5 −4 −3 −2 −1 0 1 2 3 4 5

23. −5 −4 −3 −2 −1 0 1 2 3 4 5

25. −5 −4 −3 −2 −1 0 1 2 3 4 5

27. $v < -7$; −15 −14 −13 −12 −11 −10 −9 −8 −7 −6 −5

29. $y > 0$; −5 −4 −3 −2 −1 0 1 2 3 4 5

31. $y \le 3.5$; −5 −4 −3 −2 −1 0 1 2 3 4 5

33. $v \le 2$; −5 −4 −3 −2 −1 0 1 2 3 4 5

35. $2 \ge x$, or $x \le 2$; −5 −4 −3 −2 −1 0 1 2 3 4 5

37. $a < -3$; −10 −9 −8 −7 −6 −5 −4 −3 −2 −1 0

39. $y < -2$; −10 −9 −8 −7 −6 −5 −4 −3 −2 −1 0

41. $x \ge 0$; −5 −4 −3 −2 −1 0 1 2 3 4 5

43. $a \le -4$; −12 −11 −10 −9 −8 −7 −6 −5 −4 −3 −2

45. $-9 \ge n$, or $n \le -9$; −18 −17 −16 −15 −14 −13 −12 −11 −10 −9 −8

47. $n > 3$ **49.** $x \le 6$ **51.** $y < -7$ **53.** $n \ge 13$ **55.** $m \ge 7$ **57.** $x > 7$ **59.** $z < 0$ **61.** $x \le 0.25$ **63.** $x \le -3$ **65.** $y > -3$ **67.** $x \le 0.4$ **69.** $x < \frac{1}{2}$ **71.** $n \ge 4.5$ **73.** $x < -6$ **75.** $y < -625$ **77.** d **79.** d **81.** −2 −1 0 1 2 3 4 5

83. **a.** False **b.** True **c.** True **d.** True

85. $m \le -6$; −8 −6 −4 −2 0 2 4 6 8 **87.** $a \le \frac{16}{3}$

89. $x < 5$ **91.** $\frac{81 + 85 + 91 + x}{4} > 85$; $x > 83$. The student must score above 83. **93.** $\frac{250 + 250 + 150 + 130 + 180 + x}{6} \ge 200$; $x \ge 240$. The store must make at least \$240 in sales. **95.** $0.50 + 0.10x \ge 2$; $x \ge 15$. Each call lasts at least 15 min. **97.** $0.03h > 1000 + 0.025h$; $h > 200{,}000$. She should accept the deal on a house that she sells for more than \$200,000. **99.** $200 - 2.5x < 180$; $x > 8$. He will weigh less than 180 lb after 8 mo. **101.** $6 \le p \le 18$

Chapter 12 Review Exercises, pp. 637–639

1. is not; possible answer: when we substitute -7 for x we do not get a true statement **2.** is; possible answer: it can be written as $-\frac{3}{4}x + 0 = 2$, which is in the form $ax + b = c$ **3.** is; possible answer: the addition property of equality allows us to add -5 to both sides of the equation **4.** are; possible answer: multiplying each side of the first equation by 2 gives us the second equation **5.** is not; possible answer: $8 - 6.5$ is not greater than 1.5 **6.** is not; possible answer: if we divide each side of an inequality by a negative number, we reverse the inequality's direction **7.** No. **8.** 0 is a solution. **9.** $x = -9$ **10.** $t = -2$ **11.** $a = -14$ **12.** $n = 11$ **13.** $y = 7.9$ **14.** $r = 15.2$ **15.** $x = -6$ **16.** $z = -10$ **17.** $x = -10$ **18.** $d = -3$ **19.** $y = 4$ **20.** $x = -3$ **21.** $n = 41$ **22.** $r = -150$ **23.** $t = -9$ **24.** $y = -12$ **25.** $x = 3$ **26.** $t = -9$ **27.** $a = -14$ **28.** $r = 54$ **29.** $y = 9$ **30.** $t = 1$ **31.** $x = 6$ **32.** $y = -3$ **33.** $z = 3$ **34.** $n = 5$ **35.** $c = -\frac{2}{3}$ **36.** $p = \frac{5}{2}$ **37.** $x = -5$ **38.** $x = -\frac{2}{3}$ **39.** $n = 0$ **40.** $x = -\frac{17}{6}$ **41.** $x = \frac{8}{5}$ **42.** $x = -1$ **43.** $a = 2c + 5b$ **44.** $a = \frac{bn}{2}$ **45.** $a = \frac{P - b}{2}$ **46.** $h = \frac{3V}{B}$ **47.** −4 −3 −2 −1 0 1 2 3 4

48. 

49. −4 −3 −2 −1 0 1 2 3 4

50. −2 −1 0 1 2 3 4 5 6

51. $n \geq -2$; −3 −2 −1 0 1 2 3 4

52. $y > 5$; 4 5 6 7 8 9 10

53. $t \geq 0$; −1 0 1 2 3 4 5

54. $y \leq 2$; −3 −2 −1 0 1 2 3

55. $x \geq 2$; 0 1 2 3 4 5 6

56. $n < 5$; 1 2 3 4 5 6 7 **57.** 16,000 Btu

58. 140 guests **59.** 5 sides **60.** One candidate received 11,925 votes; the other received 27,285 votes. **61.** 4000 books **62.** The trucks will meet 4 hr after departure. **63.** 1000 mi **64.** 10:15 P.M. **65.** 5 pt **66.** 2 L **67. a.** $p = 2.2k$ **b.** $k = \frac{p}{2.2}$ **68. a.** $C = 2 + 16y$ **b.** $y = \frac{C - 2}{16}$ **69.** \$400 **70.** \$74,000 **71.** 3% **72.** 5 yr

Chapter 12 Posttest, p. 640

1. −2 is not a solution. **2.** Add 1 (or subtract −1) **3.** $s = -3$ **4.** $y = -\frac{1}{2}$ **5.** $n = -6$ **6.** $y = 11$ **7.** $y = 8$ **8.** $x = 3$ **9.** $x = 2$ **10.** $x = 4$ **11.** $a = 1$ **12.** $x = \frac{5}{6}$ **13.** $p = t - 5n$ **14.** −3 −2 −1 0 1 2 3

15. $z \geq -3$; −3 −2 −1 0 1 2 3 **16.** Approximately 6 million **17.** 11 mi **18.** $L = \frac{S + 21}{3}$ **19.** 10 mph and 12 mph **20.** The monthly cost of Plan A exceeds the monthly cost of Plan B if more than 75 min of calls are made outside the network.

Chapter 12 Cumulative Review Exercises, p. 641

1. $\frac{5}{8}$ **2.** Four and seven thousandths **3.** 2.4 **4.** True **5.** 1 **6.** False **7.** $x = \frac{1}{2}$ **8.** Mean: 5.7; median: 5.8; mode: 5.9; range: 0.6 **9.** The girl does. **10.** 6 **11.** 1 hr 25 min **12.** 18 ft **13.** 16 cm **14.** 32 **15.** 46,000,000 operations and procedures **16.** 3^4 **17.** 2.5 hr **18. a.** $A = 50 + 25(t - 1)$ **b.** $t = \frac{A - 25}{25}$ **c.** 4 hr **19.** $3w + d$ **20.** Greater than \$925,000

CHAPTER 13

Chapter 13 Pretest, pp. 643–645

1.

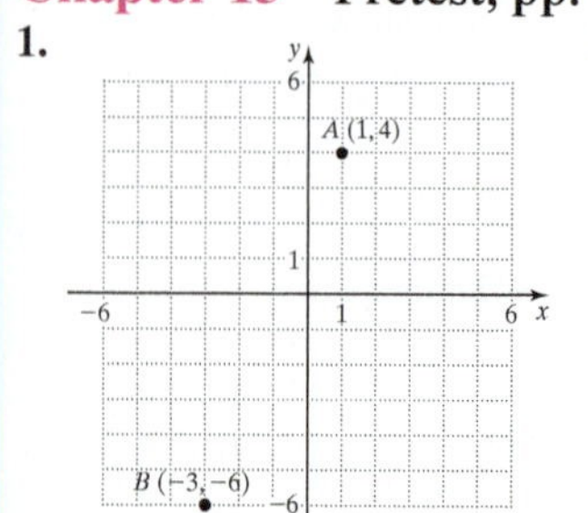

2. IV **3.** $m = \frac{1}{2}$ **4.** $\overleftrightarrow{AB}$ is parallel to $\overleftrightarrow{CD}$, since they have the same slope: −2. **5.** The slope of $\overleftrightarrow{PQ}$ is −1 and the slope of $\overleftrightarrow{RS}$ is 1. $\overleftrightarrow{PQ}$ is perpendicular to $\overleftrightarrow{RS}$, since the product of their slopes is −1. **6.** x-intercept: (3, 0); y-intercept: (0, 4) **7.**

x	4	7	$\frac{5}{2}$	2
y	3	9	0	−1

8.

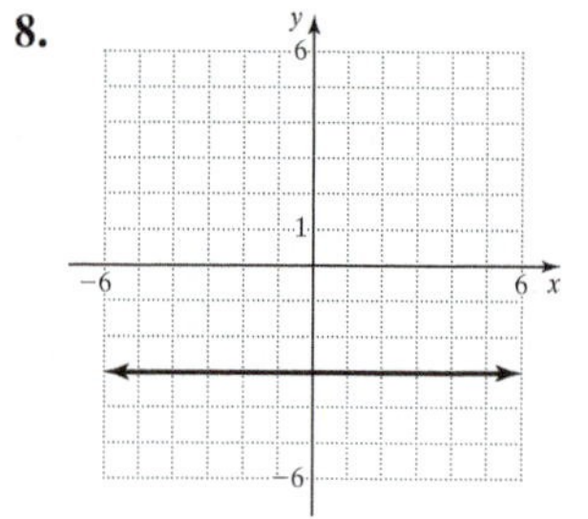

9.

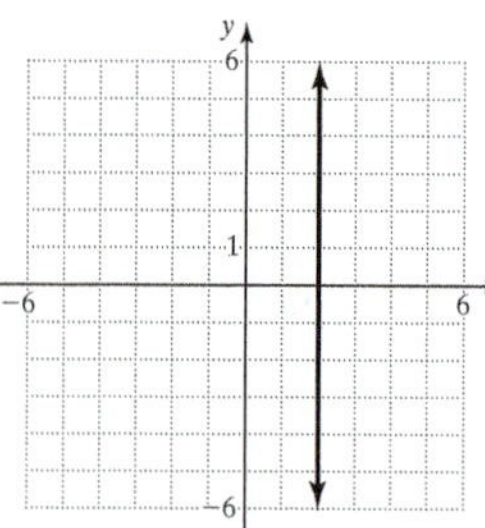

10.

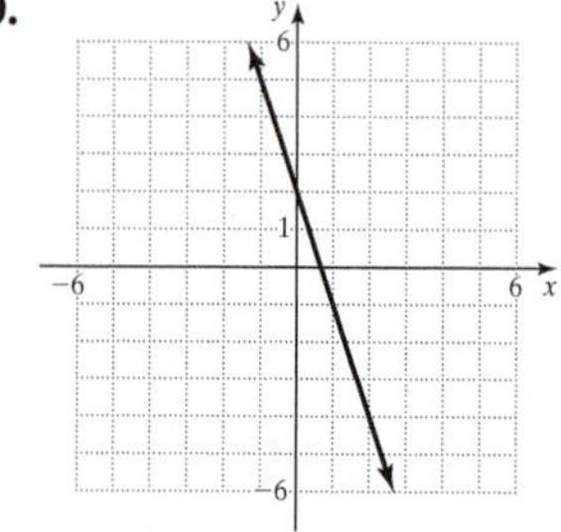

11.

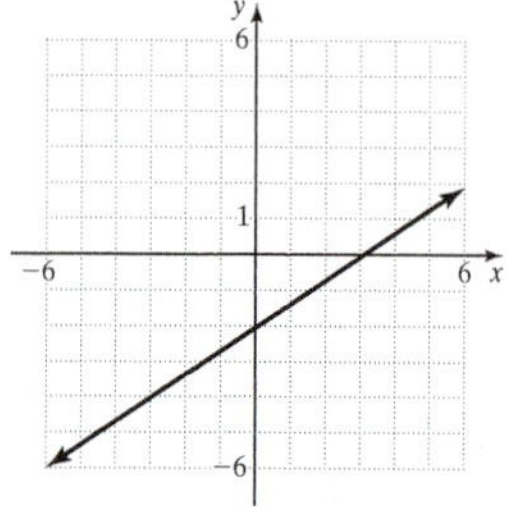

12. Slope is 2; y-intercept is (0, −5) **13.** $y = 5x - 8$ **14.** Slope-intercept form: $y = 2x + 8$; point-slope form: $y - 8 = 2(x - 0)$ **15.** Slope-intercept form: $y = x - 3$; Point-slope form: $y - 1 = (x - 4)$

16.

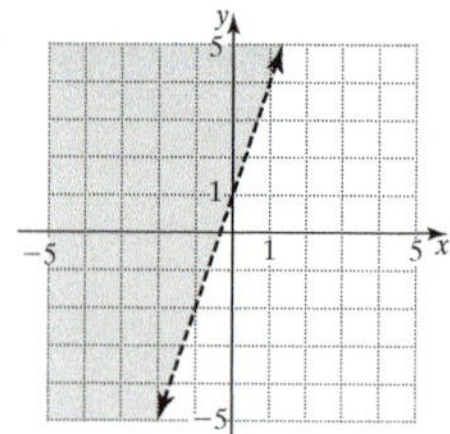

17. The slope of the line is positive. As the population of a state increases, the number of representatives in Congress from that state increases. **18.** Variety A grows faster.

19. a. $c = 2.5x$ **b.**

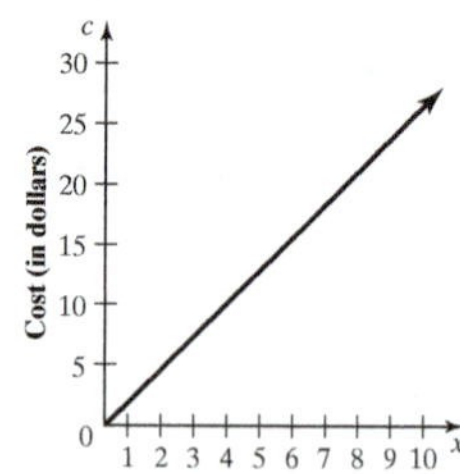

c. The slope of the line is 2.5. It represents the cost of renting a movie.

20. a. $d = 50t$ **b.**

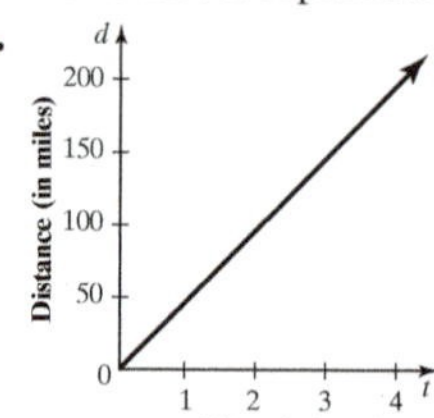

c. The slope of the graph is 50. It represents the speed the sales representative is driving.

Section 13.1 Practices, pp. 648–651

1, *p. 648:* **2, *p. 649:*** **a.** II **b.** IV **c.** III **d.** I

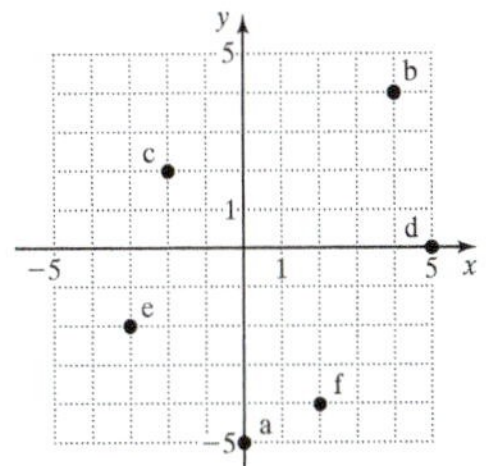

3, *p. 649:*

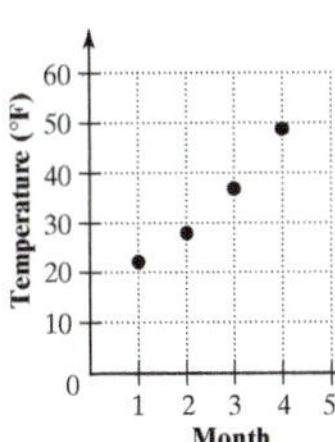

4, *p. 651:* The value of the car decreases as the number of years increases. **5, *p. 651:*** From *A* to *B* and *B* to *C*, the line segments slant up to the right, indicating that the runner's heartbeats per minute increase over this period of time. From *C* to *D*, the line segment slants downward, to the right, indicating that the runner's heartbeats per minute decrease. Possible scenario: The runner starts out warming up by jogging slowly for a certain length of time (*A* to *B*), then the runner jogs more quickly for some time (*B* to *C*), and finally, the runner jogs more slowly (*C* to *D*), resting at *D*.

Exercises 13.1, pp. 652–656

1. origin **3.** ordered pair **5.** below

7.

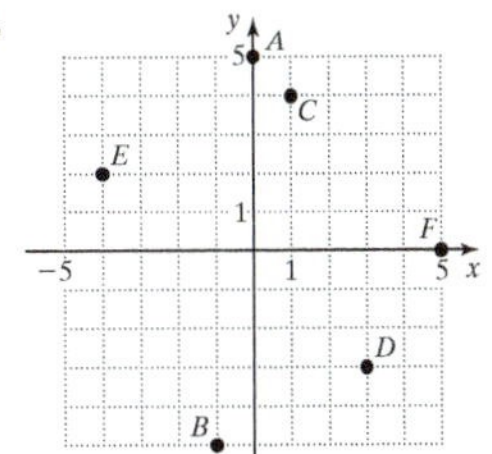

9.

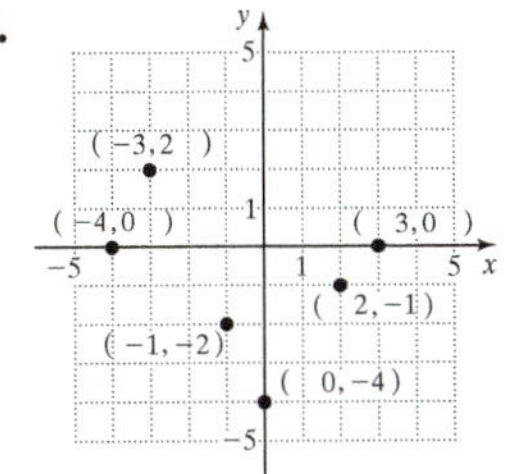

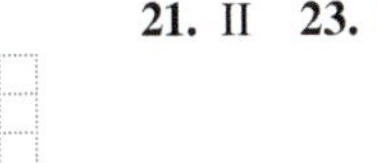

11. III **13.** II **15.** IV **17.** I

19.

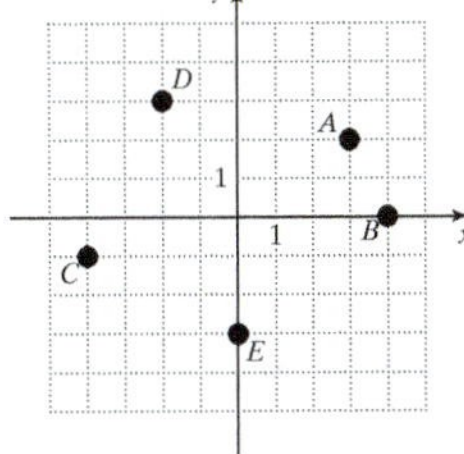

21. II **23.** I

25. a. $A(20, 40)$, $B(52, 90)$, $C(76, 80)$, and $D(90, 28)$ **b.** Students *A*, *B*, and *C* scored higher in English than in mathematics.

27.

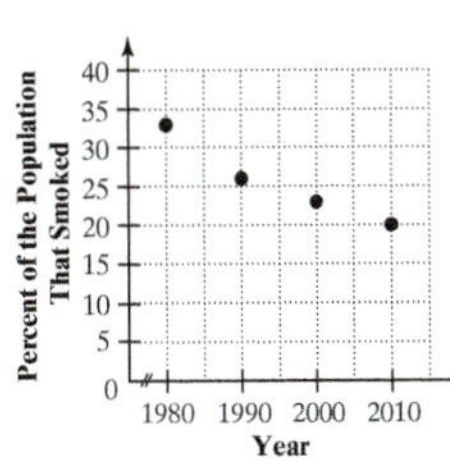

29. a.

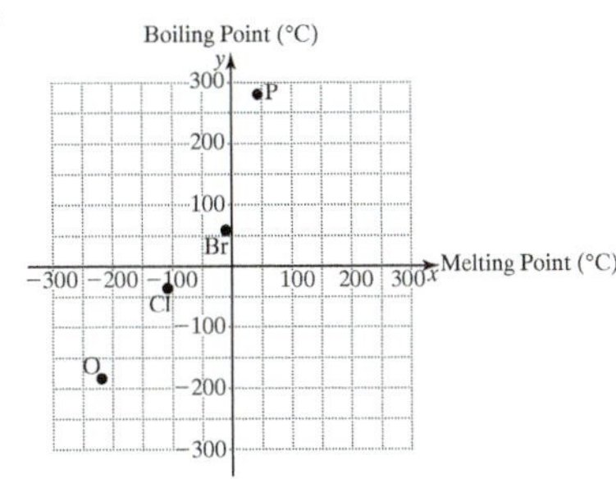

b. The *y*-coordinate is larger. The pattern shows that for each substance its boiling point is higher than its melting point. **31.** The number of senators from a state (2) is the same regardless of the size of the state's population. **33.** The graph in (a) could describe this motion. As the child moves away from the wall, the distance from the wall increases (line segment slants upward to the right). When the child stands still, the distance from the wall does not change (horizontal line segment). Finally, as the child moves toward the wall, the child's distance from the wall decreases (line segment slants downward to the right).

Section 13.2 Practices, pp. 660–669

1, *p. 660:* $m = \frac{1}{3}$ **2, *p. 661:*** $m = -\frac{6}{5}$ **3, *p. 661:***

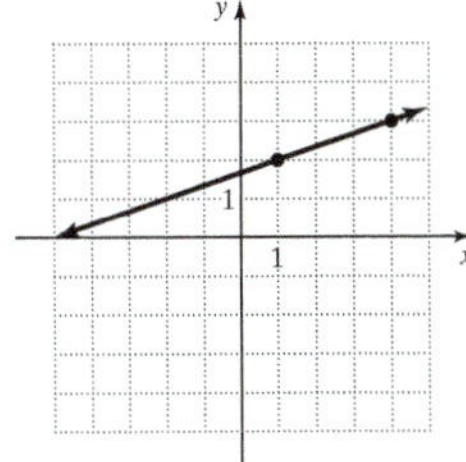

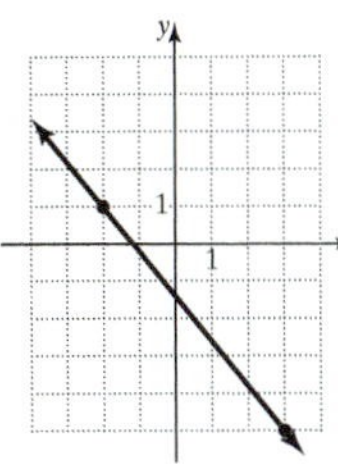

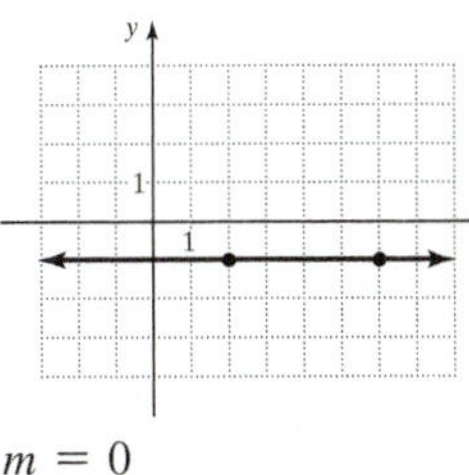

$m = 0$

4, *p. 662:* The slope is undefined.

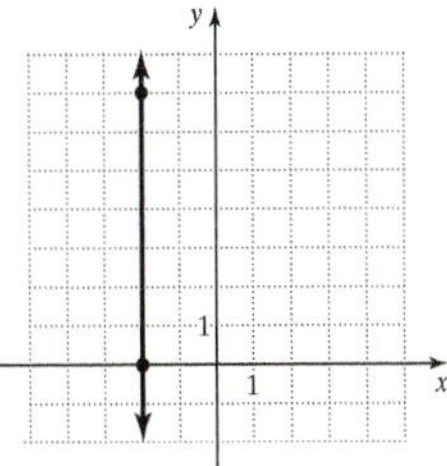

5, *p. 663:* Slope of $\overleftrightarrow{PQ}$: $-\frac{2}{3}$; slope of $\overleftrightarrow{RS}$: $-\frac{1}{2}$ **6, *p. 663:*** Scenario A is most desirable. The slope of the line is negative, which indicates a decrease in the number of people ill over time.

7, *p. 664:*

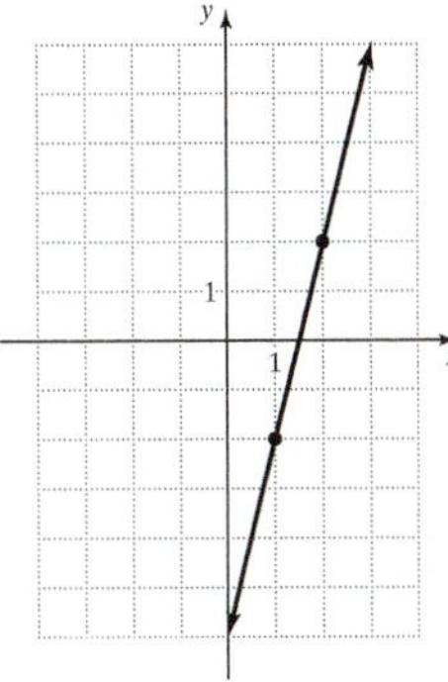

8, *p. 664:* **a.**

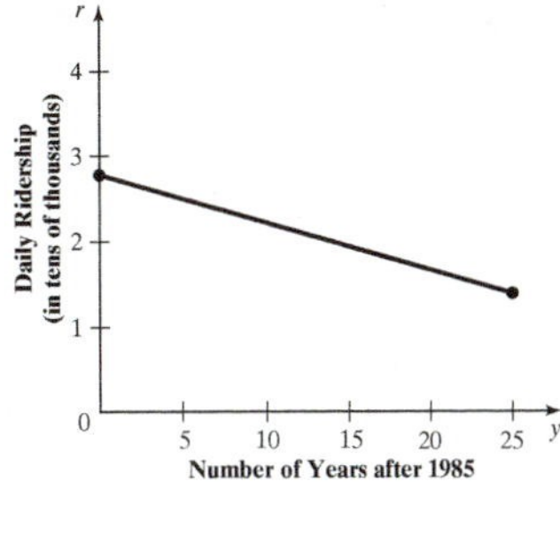

b. $m = -560$

c. The slope indicates that the daily ridership decreased by 560 per year. **9, *p. 666:*** The slope of $\overleftrightarrow{EF}$ is $-\frac{5}{4}$ and the slope of $\overleftrightarrow{GH}$ is $-\frac{5}{4}$. Since their slopes are equal, the lines are parallel. **10, *p. 666:*** **a.** Yes, the lines are parallel since their slopes are both $\frac{15}{4}$. **b.** Yes; the lines on the graph appear to be parallel. **c.** The salaries increased at the same rate. **d.** The starting salary of the multimedia designer was about $57,000. **11, *p. 668:*** The slope of $\overleftrightarrow{AB}$ is 2 and the slope of $\overleftrightarrow{AC}$ is $\frac{1}{2}$. Since the product of their slopes is not equal to -1, the lines are not perpendicular. **12, *p. 669:*** The slope of the diagonal from $(0, 0)$ to $(6, 6)$ is 1. The slope of the diagonal from $(0, 6)$ to $(6, 0)$ is -1. Since the product of the slopes is -1, the diagonals of the square are perpendicular.

Exercises 13.2, pp. 670–678

1. rate of change **3.** negative **5.** horizontal **7.** parallel
9. $\frac{3}{4}$

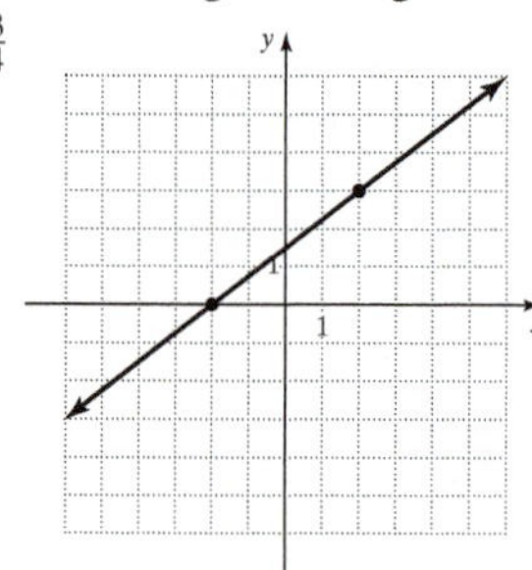

11. undefined

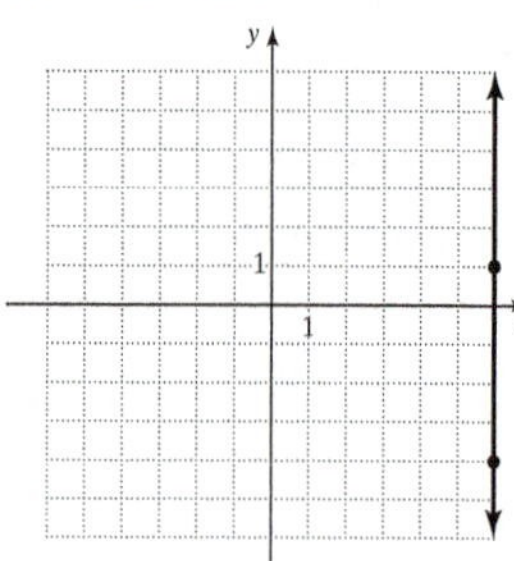

13. $-\frac{2}{5}$

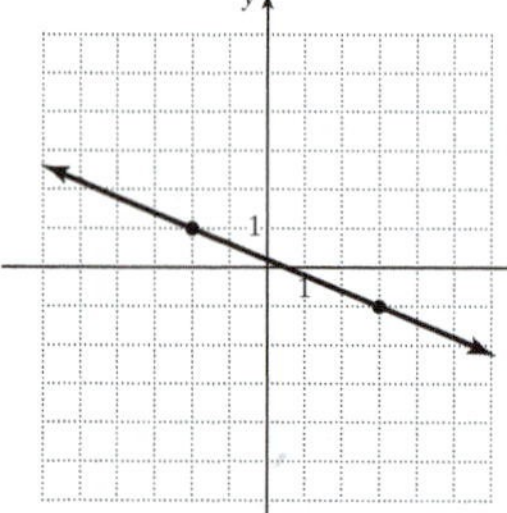

15. 0

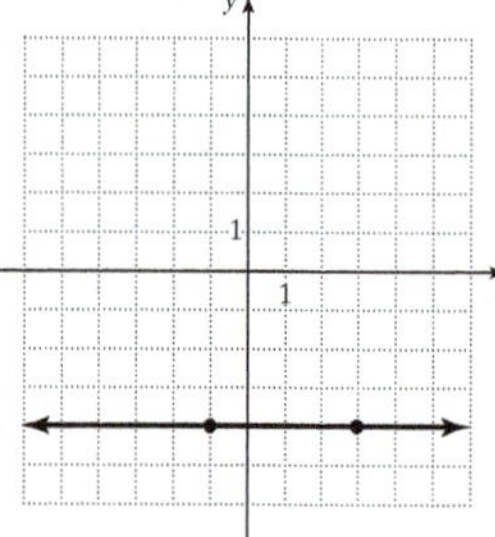

17. -7

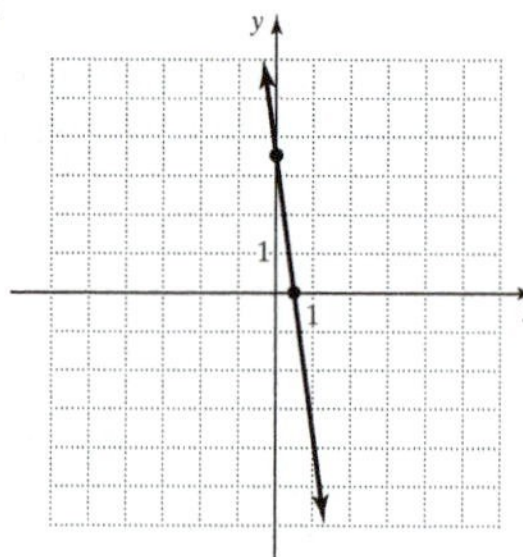

19. The slope of $\overleftrightarrow{AB}$ is -1. The slope of $\overleftrightarrow{CD}$ is 2. **21.** Positive slope; neither **23.** Negative slope; neither **25.** Undefined slope; vertical **27.** Zero slope; horizontal **29.**

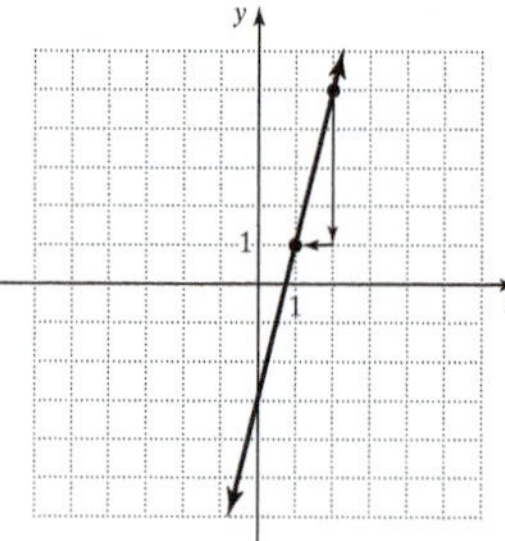

31.

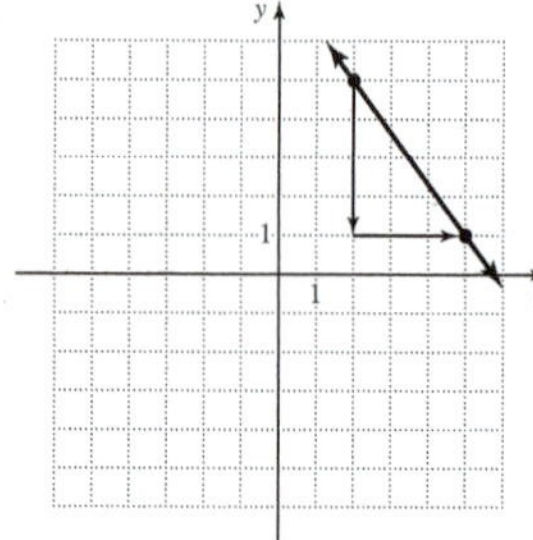

33.

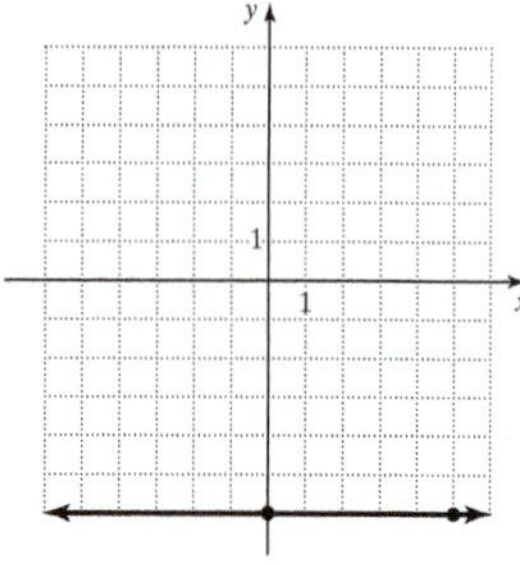

35.

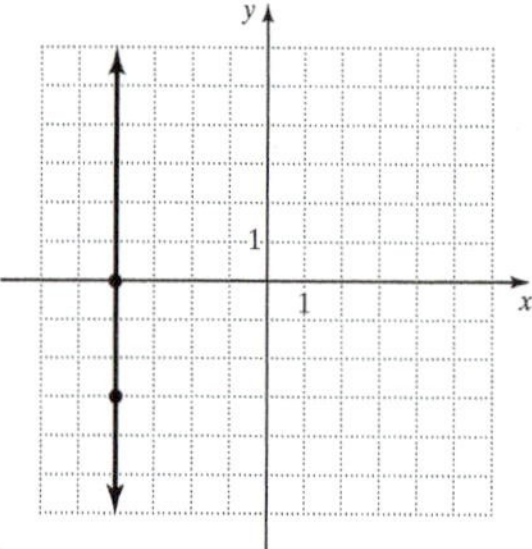

37. a. $\overleftrightarrow{PQ}: m = 4$; $\overleftrightarrow{RS}: m = 4$; the lines are parallel. **b.** $\overleftrightarrow{PQ}: m = -\frac{3}{2}$; $\overleftrightarrow{RS}: m = \frac{2}{3}$; the lines are perpendicular. **39.** Zero slope; horizontal
41.

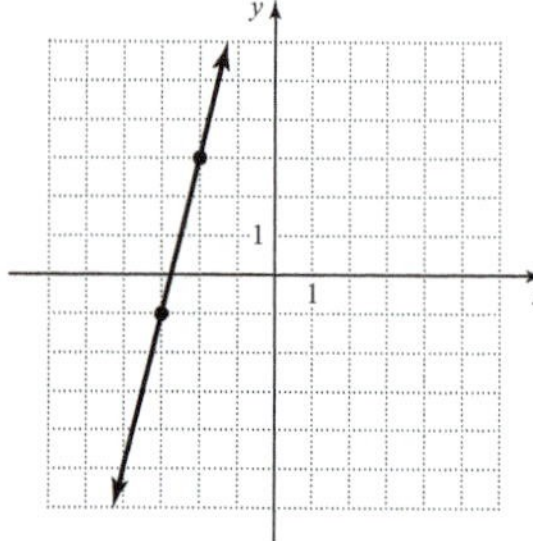

43. a. $\overleftrightarrow{AB}: m = -\frac{1}{2}$; $\overleftrightarrow{CD}: m = 2$; the lines are perpendicular.
b. $\overleftrightarrow{AB}: m = -\frac{4}{3}$; $\overleftrightarrow{CD}: m = -\frac{4}{3}$; the lines are parallel.
45. Undef.

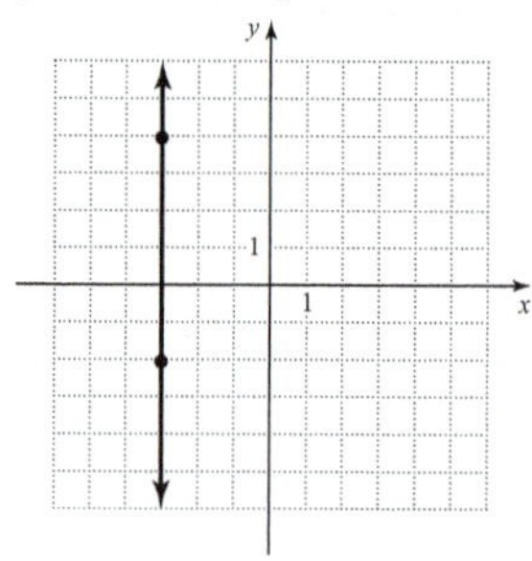

47. a. The slope is positive. **b.** A positive slope indicates that as the temperature of the gas increases, the pressure in the tube increases.
49. a. Motorcycle A **b.** Motorcycle B **c.** The slope is the change in distance over time, or the average speed of the motorcycles. **51.** The slope of each line is 1. Since the slopes of the lines are equal, the garbage deposits at the landfills are growing at the same rate. **53. a.**

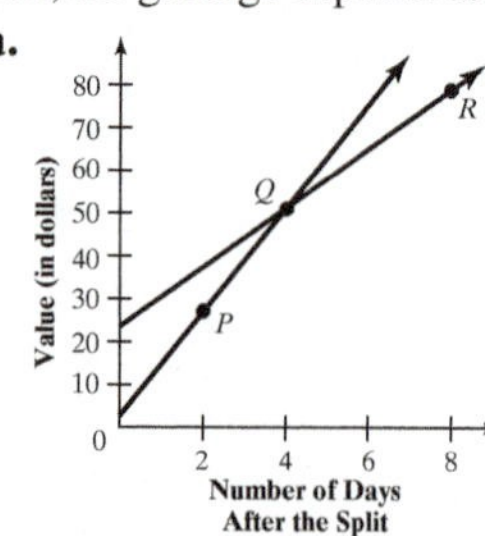

b. The slopes are 12 and 7. Since the slopes are not equal, the rate of increase did change over time. **55.** The product of the slopes of the two lines is $-5 \cdot \frac{1}{3}$, which is not equal to -1, so $\overleftrightarrow{AD}$ is not the shortest route.
57. a. Graph II; as the car travels, its distance increases with time. This implies a positive slope. **b.** Graph I; the car is set for a constant speed. The speed of the car does not change over time. This implies a 0 slope.

Section 13.3 Practices, pp. 681–690

1, ***p. 681:***

x	0	5	-3	$-\frac{1}{2}$	-2
y	1	11	-5	0	-3

2, ***p. 684:***

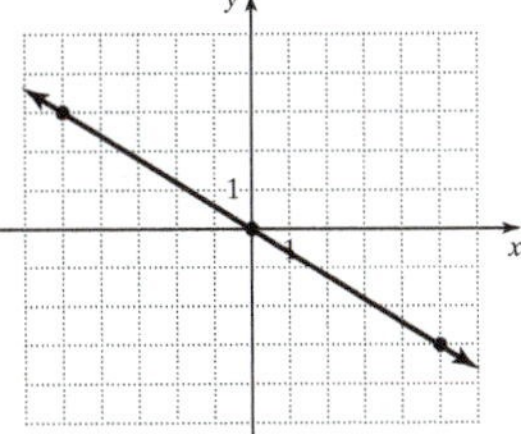

3, ***p. 684:*** **a.**

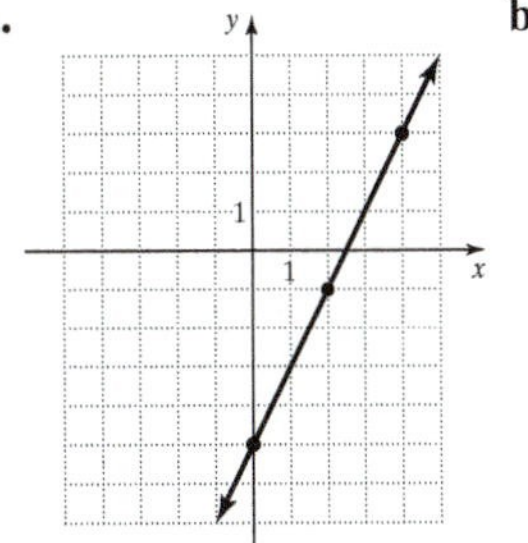

b. $m = 2$

4, ***p. 686:*** x-intercept: $(4, 0)$; y-intercept: $(0, -2)$;

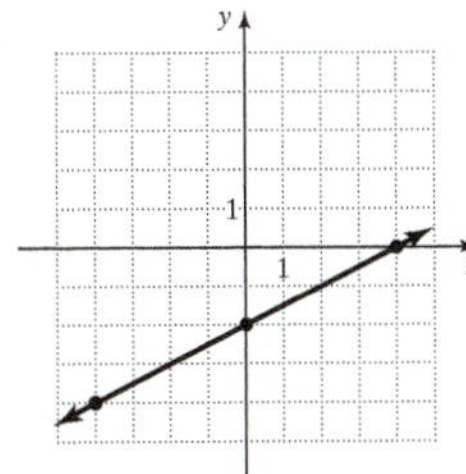

5, ***p. 687:***

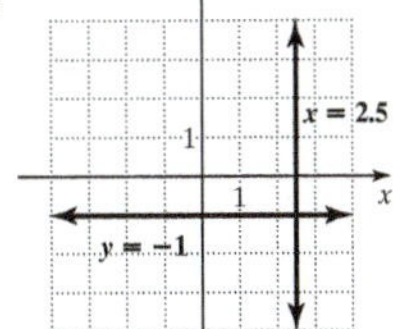

6, ***p. 688:*** **a.** $C = 0.03s + 40$

b.

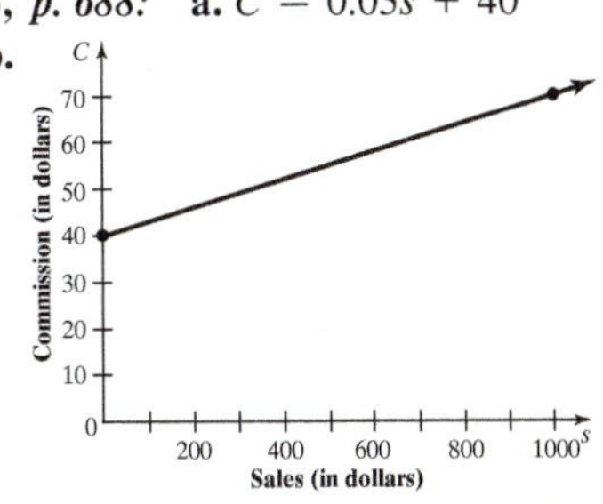

c. $m = 0.03$; for every sale, the commission increases by 0.03 times the increase in the value of the sale.

d. \$55 **7,** ***p. 690:*** **a.** $2x + y = 10$ **b.**

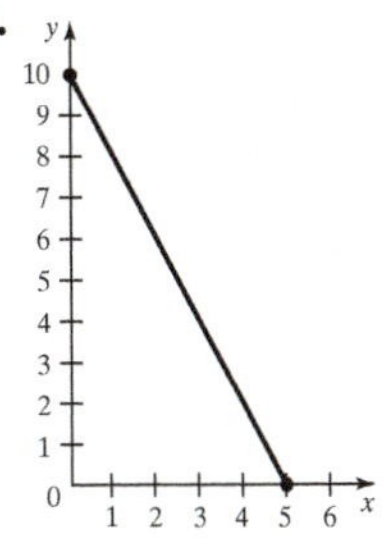

c. For each year the athlete is paid \$2 million, the number of years that she could be paid \$1 million decreases by 2 years. **d.** The x-intercept is the number of years of the contract if she was paid \$2 million in each year of the contract. The y-intercept is the number of years of the contract if she was paid \$1 million in each year of the contract.

Exercises 13.3, pp. 691–702

1. solution **3.** three points **5.** y-intercept

7.

x	4	7	$\frac{8}{3}$
y	4	13	0

9.

x	3.5	6	$-\frac{1}{10}$	$\frac{8}{5}$
y	-17.5	-30	$\frac{1}{2}$	-8

11.

x	0	-4	8	4
y	3	6	-3	0

13.

x	3	6	-3	0
y	0	1	-2	-1

15.

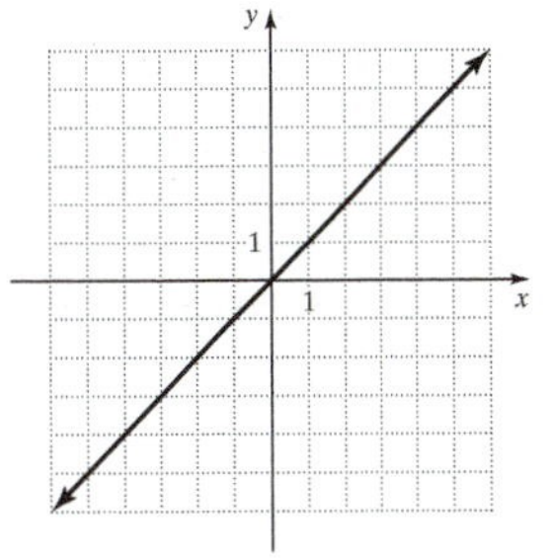

17.

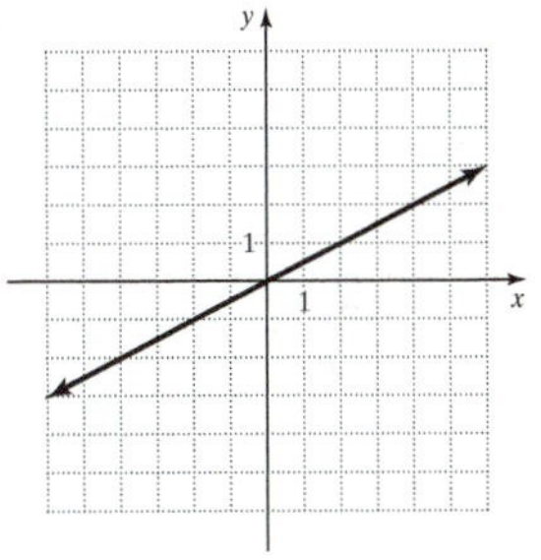

19.

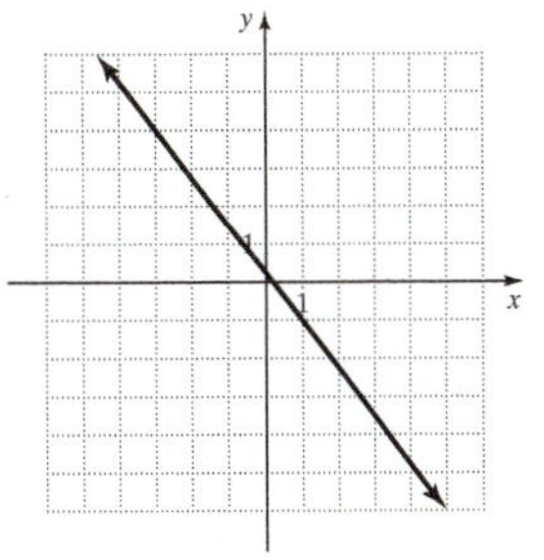

21.

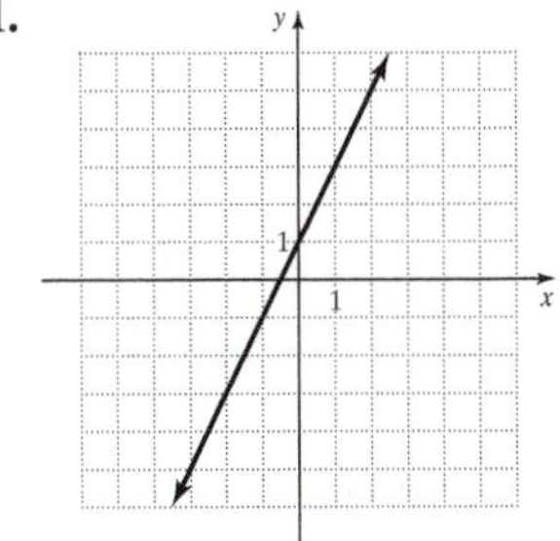

23.

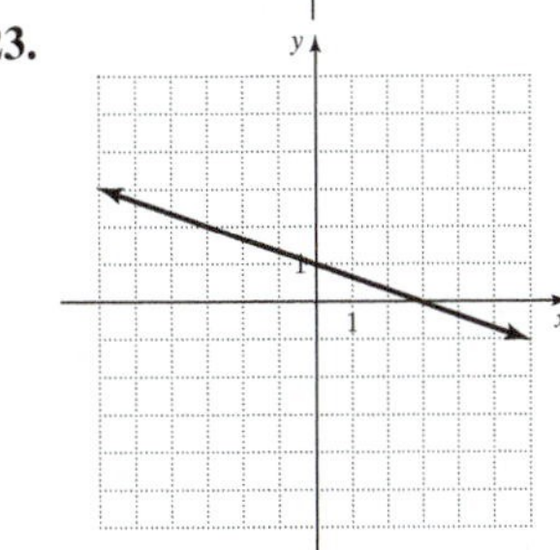

25.

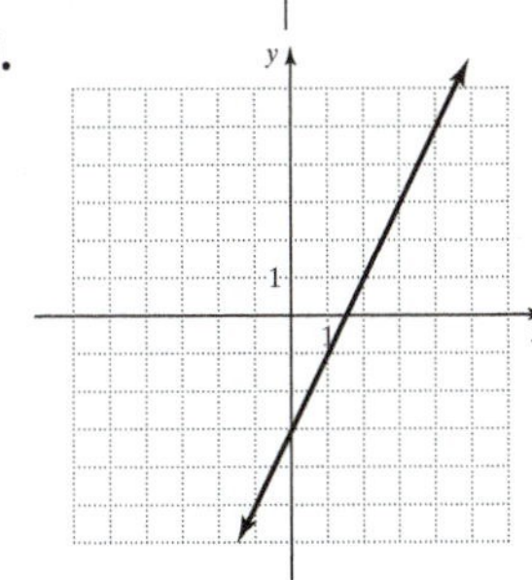

27.

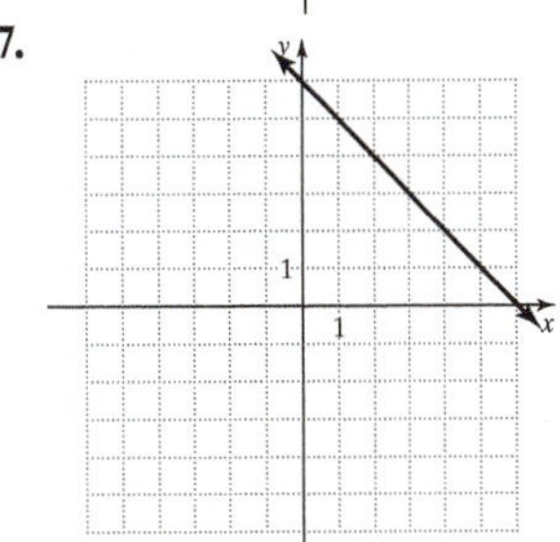

29.

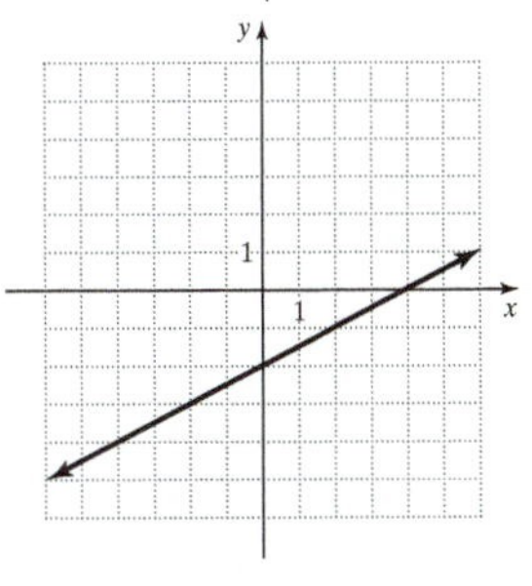

31.

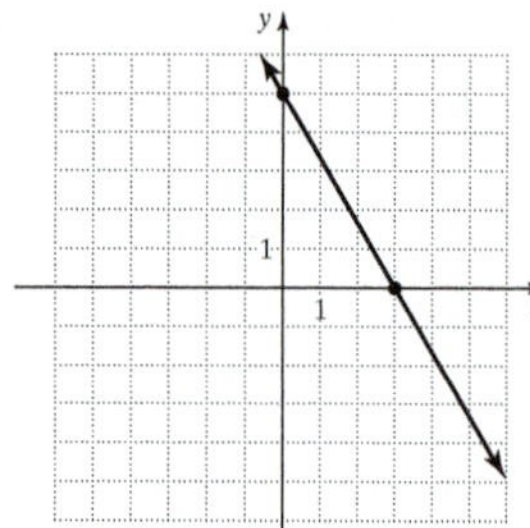

x-intercept: $(3, 0)$
y-intercept: $(0, 5)$

33.

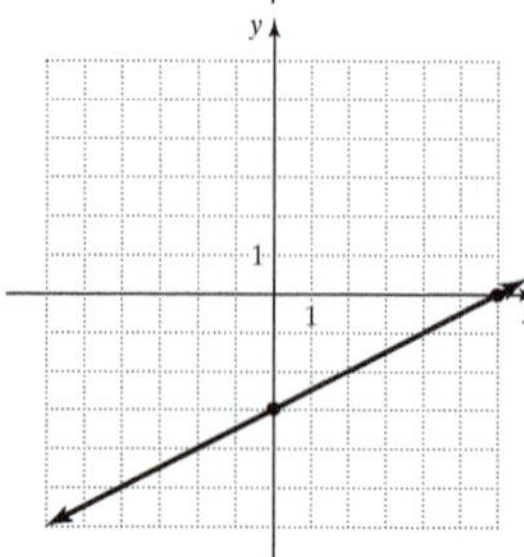

x-intercept: $(6, 0)$
y-intercept: $(0, -3)$

35.

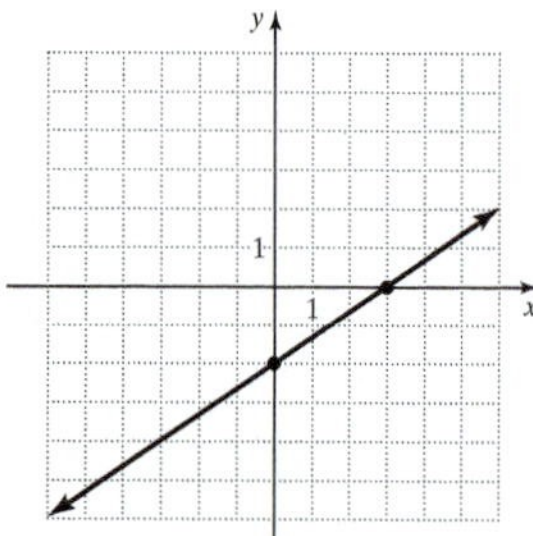

x-intercept: $(3, 0)$
y-intercept: $(0, -2)$

37.

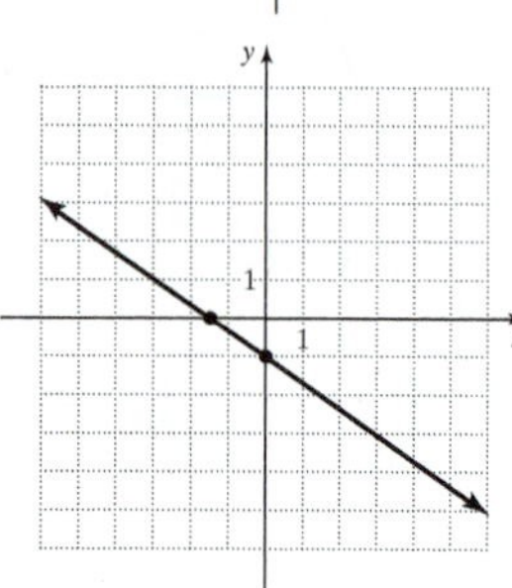

x-intercept: $(-\frac{3}{2}, 0)$
y-intercept: $(0, -1)$

39.

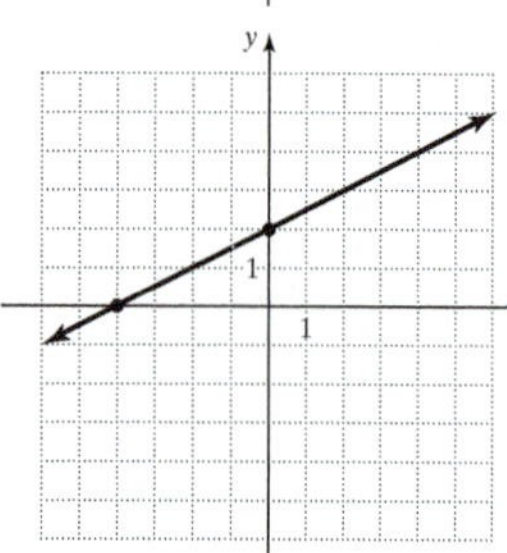

x-intercept: $(-4, 0)$
y-intercept: $(0, 2)$

41.

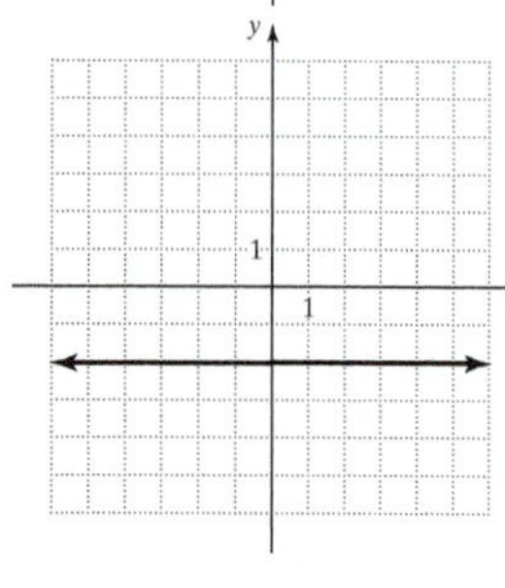

43.

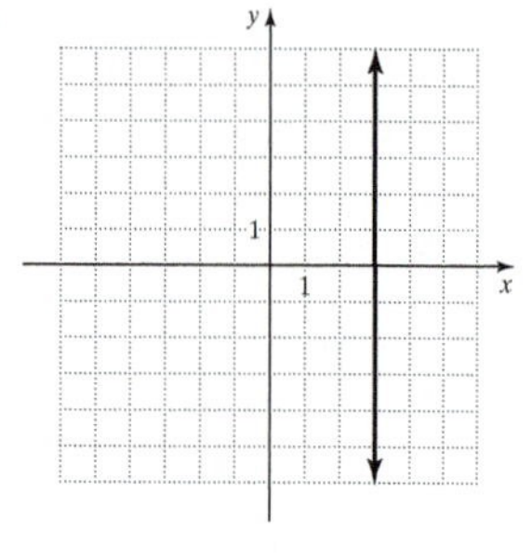

45.

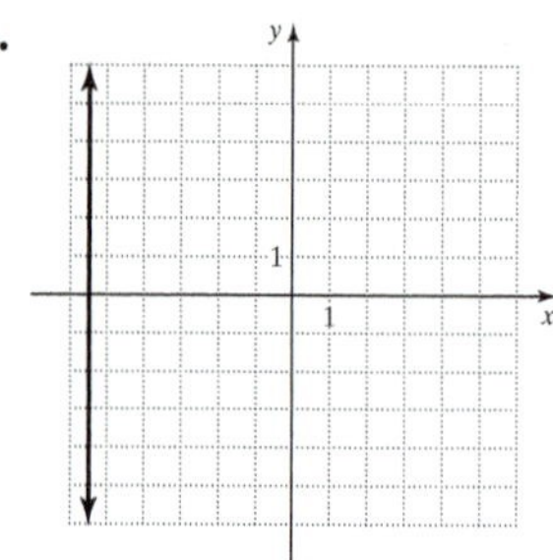

47.

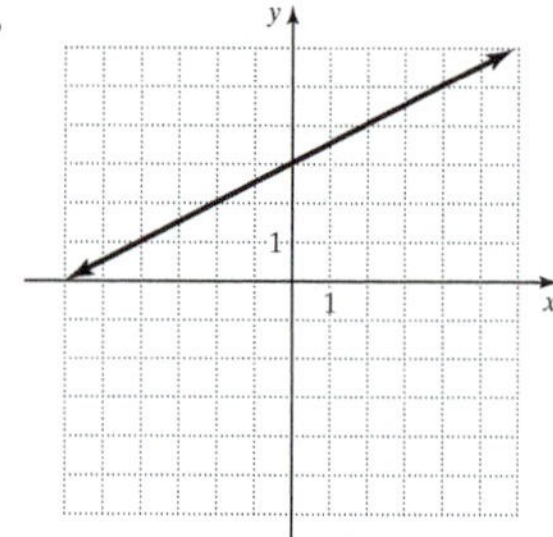

49.

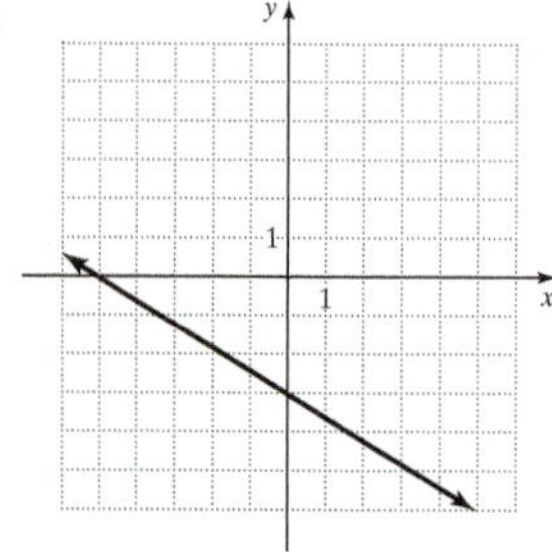

51.

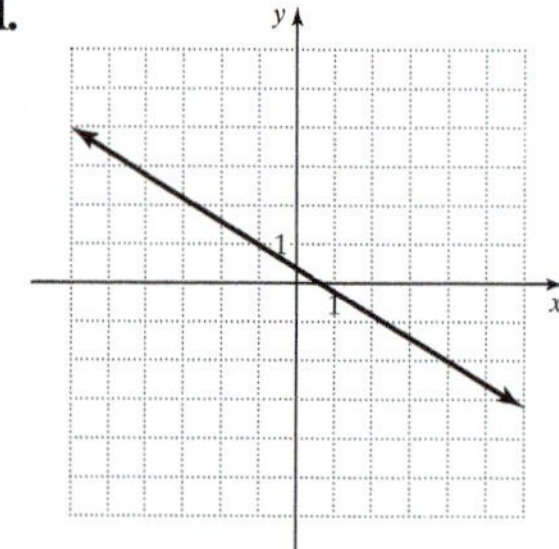

53.

x	-3	$\frac{5}{2}$	8	1
y	12	1	-10	4

55.

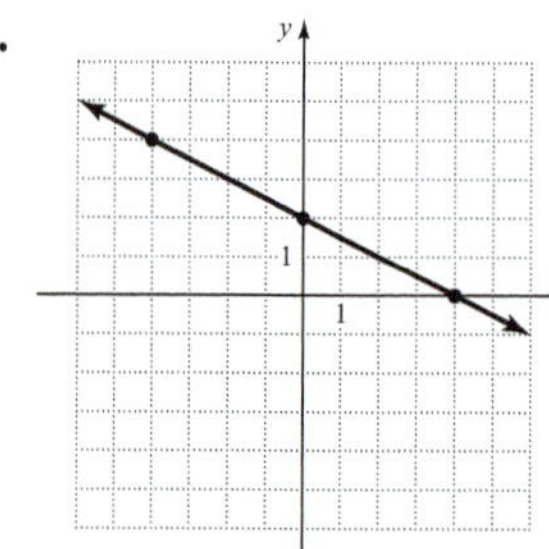

x-intercept: $(4, 0)$
y-intercept: $(0, 2)$
possible third point: $(-4, 4)$

57.

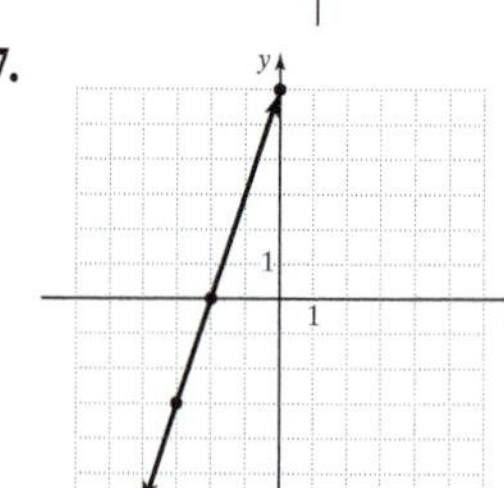

x-intercept: $(-2, 0)$
y-intercept: $(0, 6)$
possible third point: $(-3, -3)$

59.

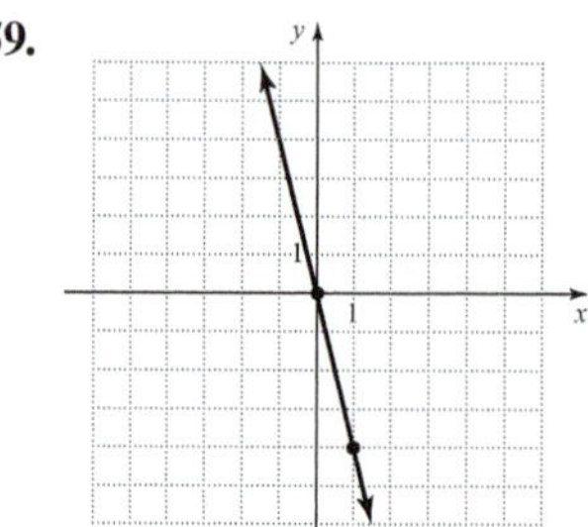

x-intercept: $(0, 0)$
y-intercept: $(0, 0)$
possible third point: $(1, -4)$

61. 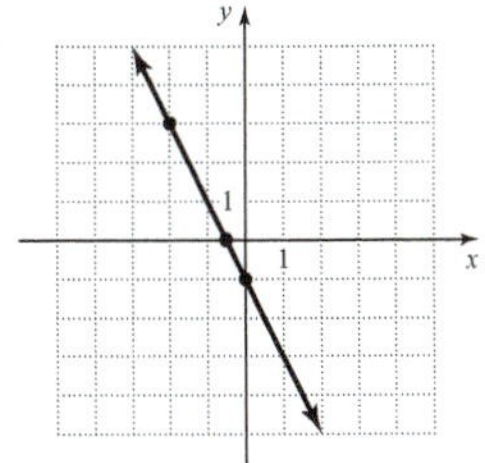

x-intercept: $(-\frac{1}{2}, 0)$
y-intercept: $(0, -1)$
possible third point: $(-2, 3)$

63. 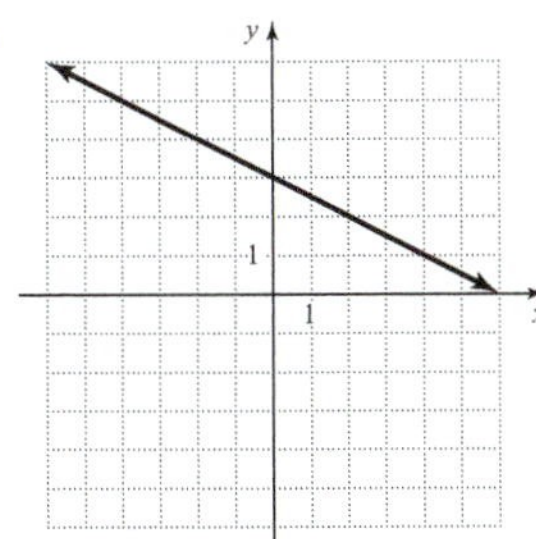

65. a.

t	0	0.5	1	1.5	2
v	10	−6	−22	−38	−54

A positive value of v means that the object is moving upward. A negative value of v means that the object is moving downward. **b.**

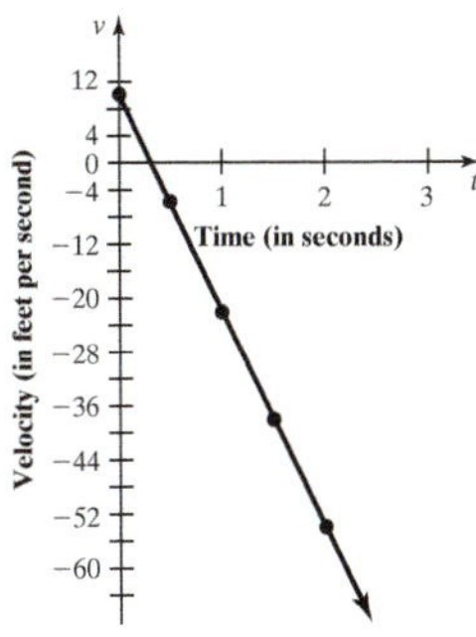

c. The v-intercept is the initial velocity of the object.
d. The t-intercept represents the time when the object changes from an upward motion to a downward motion.

67. a. $P = 100m + 500$ **b.**

m	1	2	3
P	600	700	800

c. 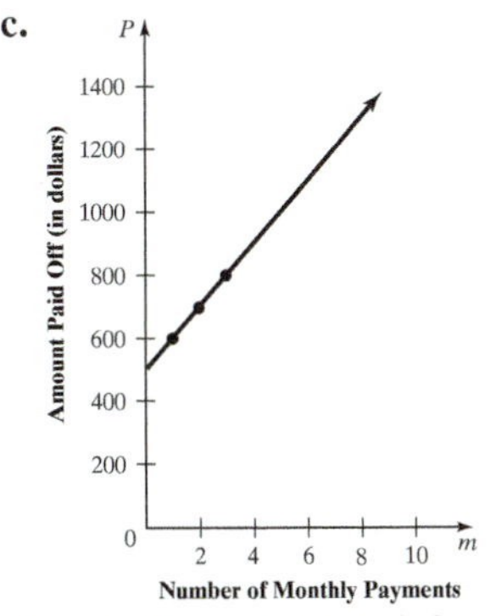

69. a. $0.05n + 0.1d = 2$ **b.**

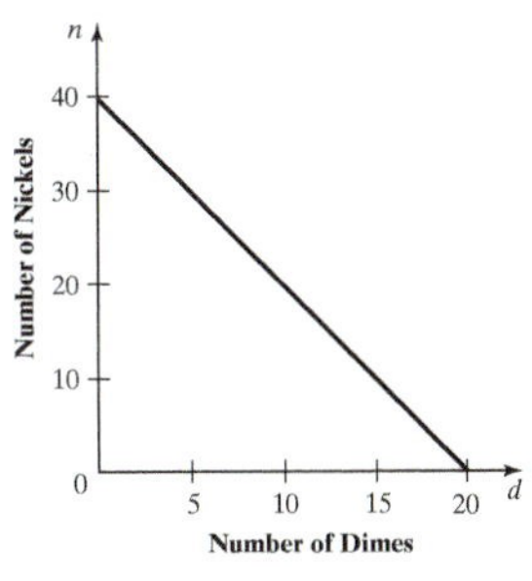

c. Only nonnegative integer values make sense, since you cannot have fractions of a nickel or a dime.

71. a. $F = 5d + 40$
b. 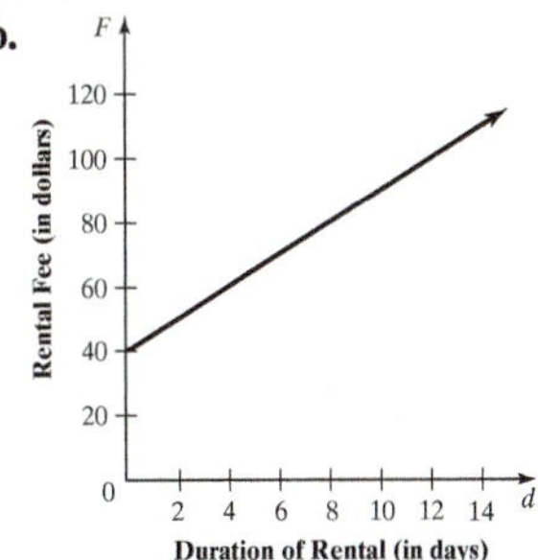

c. \$30

Section 13.4 Practices: pp. 703–709

1, *p. 703:* Slope is -2; y-intercept is $(0, 3)$. **2, *p. 704:*** Slope is -1; y-intercept is $(0, 0)$. **3, *p. 704:*** $y = \frac{3}{2}x - 2$ **4, *p. 704:*** $y = 4x + 6$ **5, *p. 705:*** $y = 1x + 2$, or $y = x + 2$ **6, *p. 705:*** $y = -2x - 1$ **7, *p. 705:*** $y = -\frac{1}{2}x - 2$ **8, *p. 705:*** $w = 45 - 3t$
9, *p. 706:*

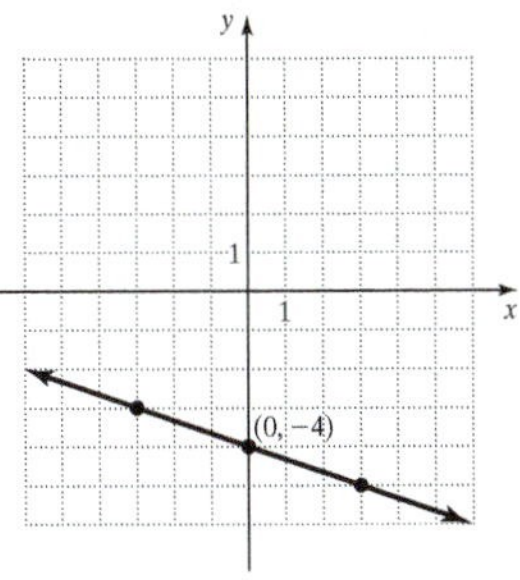

10, *p. 707:* $y - 0 = 2(x - 7)$ **11, *p. 708:*** $y - 7 = 1(x - 7)$ or $y - 0 = 1(x - 0)$ **12, *p. 708:*** Point-slope form: $w - 27 = 5(b - 4)$; slope- intercept form: $w = 5b + 7$ **13, *p. 709:***

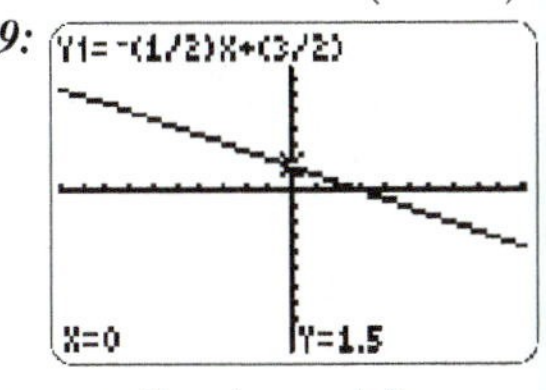

The displayed coordinates of the y-intercept are $x = 0$ and $y = 1.5$.

Exercises 13.4, pp. 710–717

1. Slope-intercept **3.** y-intercept **5.** For $y = 3x - 5$: 3, $(0, -5)$, ⁄, $(\frac{5}{3}, 0)$; for $y = -2x$: -2, $(0, 0)$, \, $(0, 0)$; for $y = 0.7x + 3.5$: 0.7, $(0, 3.5)$, ⁄, $(-5, 0)$; for $y = \frac{3}{4}x - \frac{1}{2}$: $\frac{3}{4}$, $(0, -\frac{1}{2})$⁄, $(\frac{2}{3}, 0)$; for $6x + 3y = 12$: -2, $(0, 4)$; \, $(2, 0)$; for $y = -5$: 0, $(0, -5)$, —, no x-intercept; for $x = -2$: undefined, no y-intercept, |, $(-2, 0)$
7. Slope: -1; y-intercept: $(0, 2)$ **9.** Slope: 3; y-intercept: $(0, -4)$
11. $y = x - 10$ **13.** $y = -\frac{1}{10}x + 1$ **15.** $y = -\frac{3}{2}x + \frac{1}{4}$ **17.** $y = \frac{2}{5}x - 2$
19. $y = 3x + 14$ **21.** b **23.** a

25. 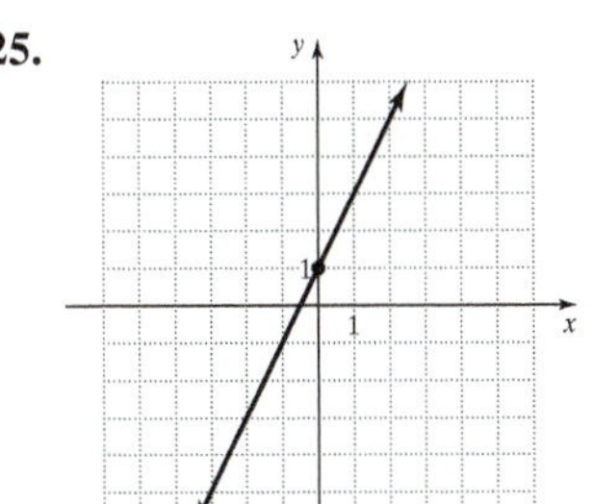

27. 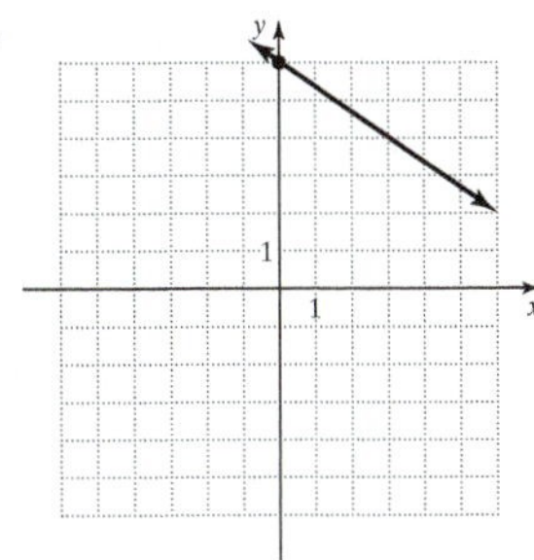

29.

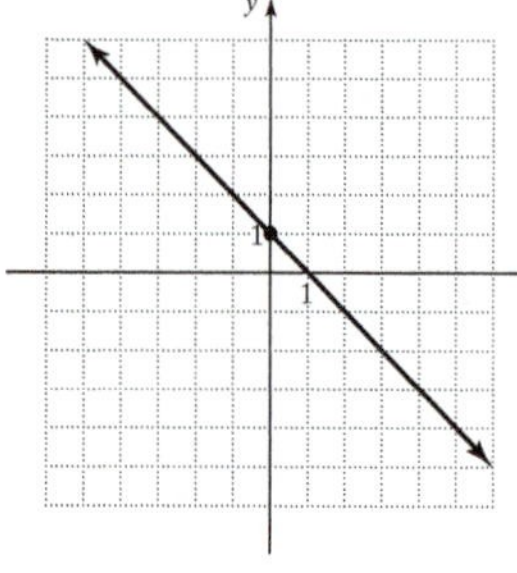

31.

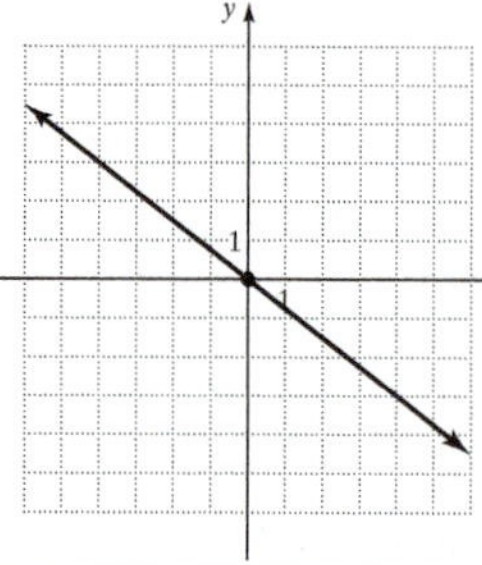

33.

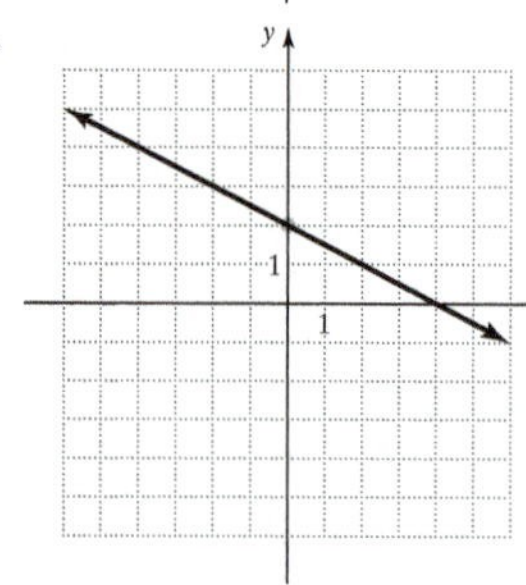

35.

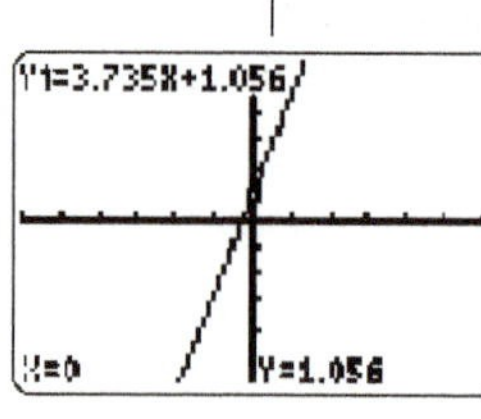

37. $y = 3x + 7$ **39.** $y = 5x - 20$ **41.** $y = -\frac{1}{2}x + 4$ **43.** $y = -x + 3$ **45.** $y = \frac{1}{6}x - \frac{29}{6}$ **47.** $x = -3$ **49.** $y = -6$ **51.** $y = \frac{3}{4}x + 3$ **53.** $y = -x - 2$ **55.** $y = 2$ **57.** $x = -2.5$ **59.** For $y = -7x + 2$: -7, $(0, 2)$, ╲, $(\frac{2}{7}, 0)$; for $y = 4x$: 4, $(0, 0)$, ╱, $(0, 0)$; for $y = 2.5x + 10$: 2.5, $(0, 10)$, ╱, $(-4, 0)$; for $y = \frac{2}{3}x - \frac{1}{4}$: $\frac{2}{3}$, $(0, -\frac{1}{4})$, ╱, $(\frac{3}{8}, 0)$; for $5x + 4y = 20$: $-\frac{5}{4}$, $(0, 5)$, ╲, $(4, 0)$; for $x = 9$: undefined, no y-intercept, |, $(9, 0)$; for $y = -3.2$: 0, $(0, -3.2)$, –, no x-intercept

61. $y = 4x - 5$ **63.** d **65.** $y = \frac{1}{2}x - 1$

67.

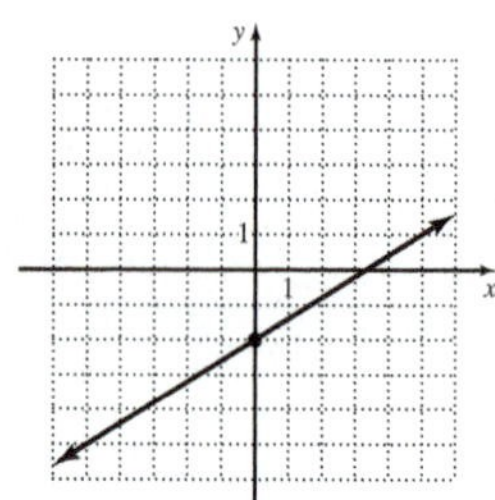

69. $y = -x + 5$

71. a. $\frac{9}{5}$ **b.** $F = \frac{9}{5}C + 32$ **c.** Water boils at 100°C.

73. a. $y - 4500 = 6(x - 500)$ **b.** $y = 6x + 1500$ **c.** The y-intercept represents the monthly flat fee the utility company charges its residential customers.

75. a. $I = 0.03S + 1500$ **b.**

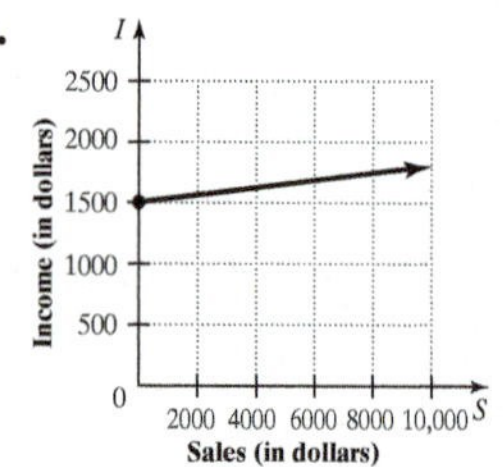

c. \$1686

77. $P = \frac{1}{33}d + 1$ **79.** $L = \frac{5}{6}F + 10$

Section 13.5 Practices: pp. 720–723

1, *p. 720:* No, $(1, 3)$ is not a solution to the inequality.

2, *p. 721:*

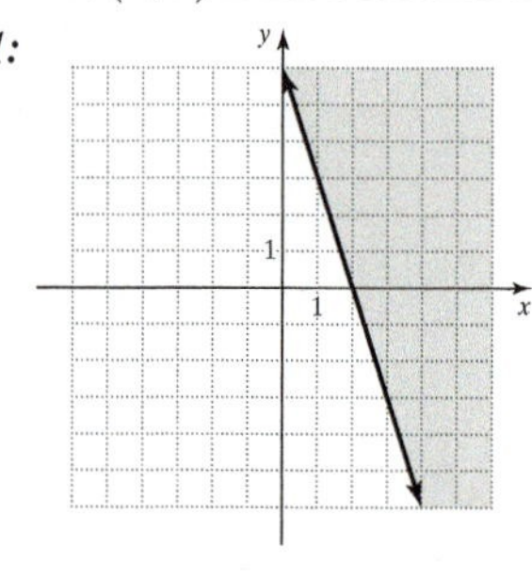

3, *p. 722:*

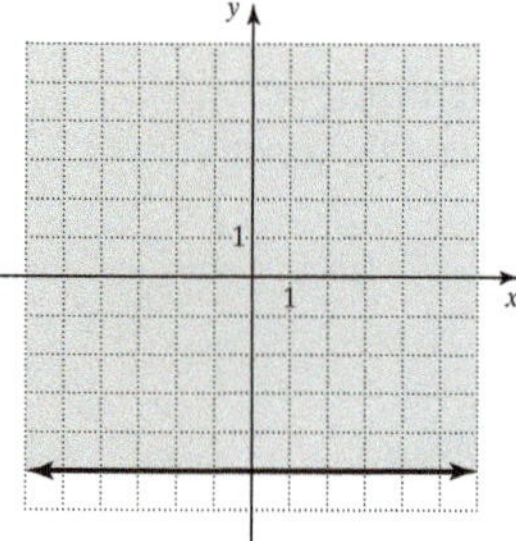

4, *p. 722:*

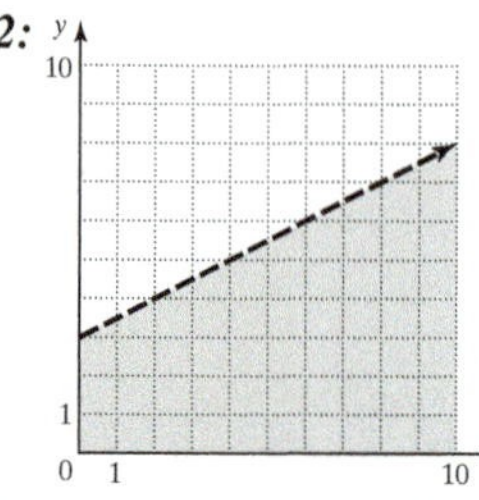

5, *p. 723:* **a.** $d + g \leq 3000$ **b.**

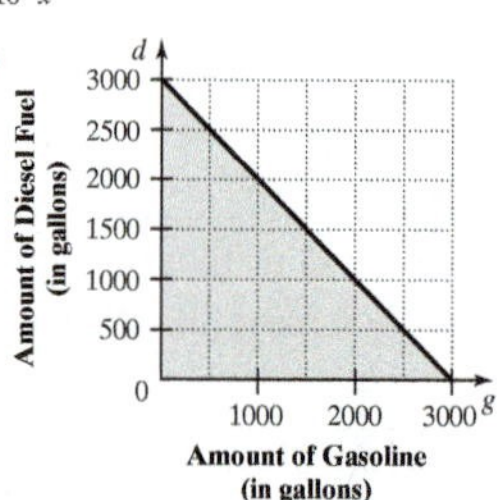

c. The d-intercept represents the maximum amount of diesel fuel the refinery can produce if no gasoline is produced. The g-intercept represents the maximum amount of gasoline the refinery can produce if no diesel fuel is produced.

Exercises 13.5, pp. 724–731

1. Half-plane **3.** Graph **5.** Broken **7.** No, not a solution **9.** Yes, a solution **11.** No, not a solution

13.

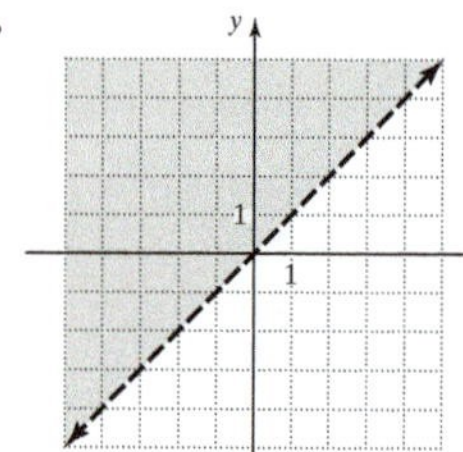

15.

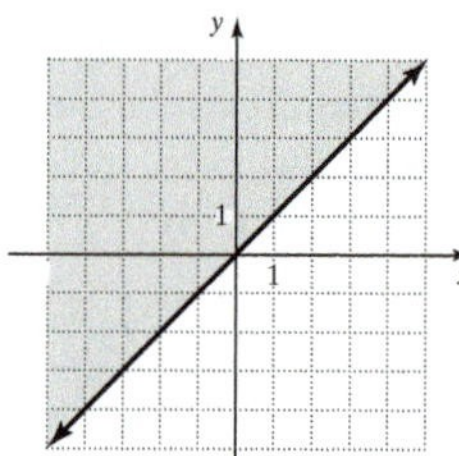

17.

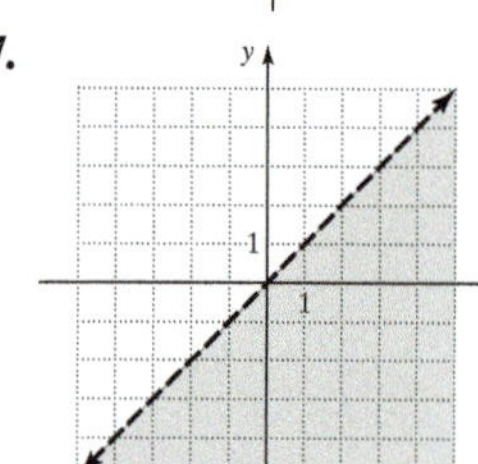

19. d **21.** b

23.

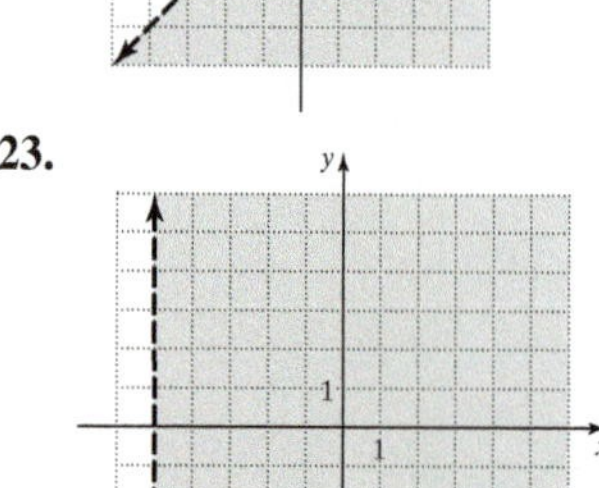

25.

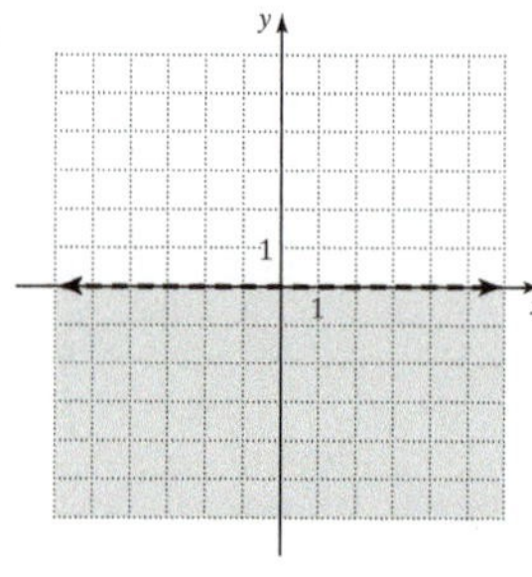

27.

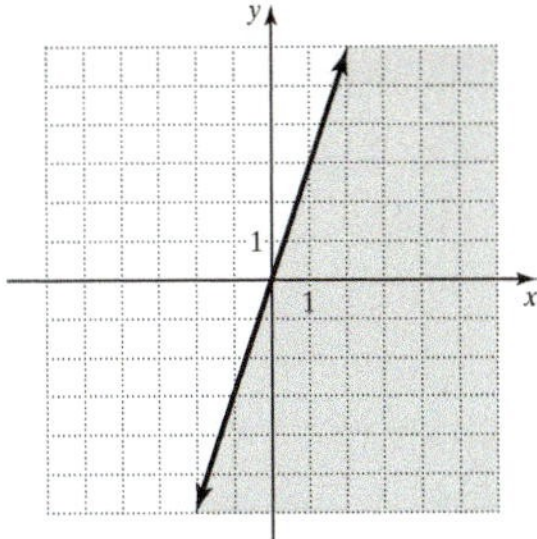

29.

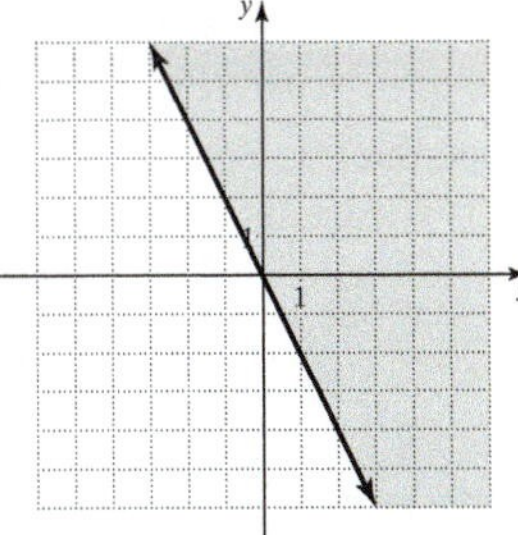

31.

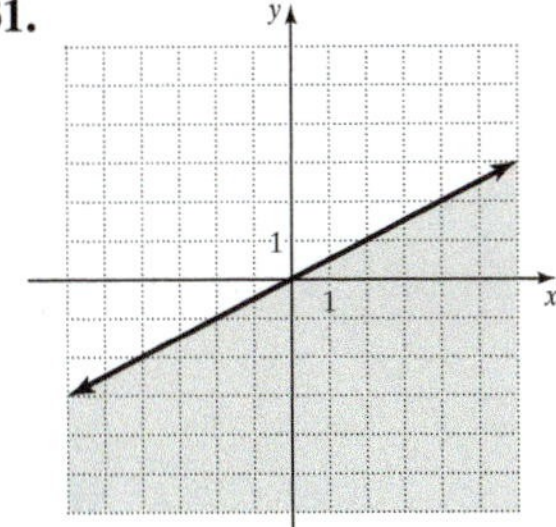

33.

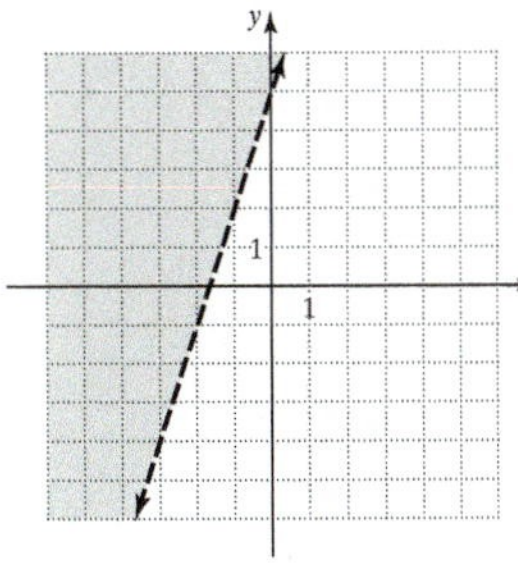

35.

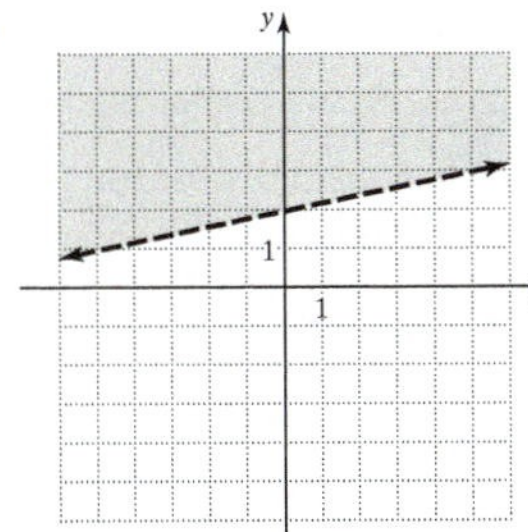

37.

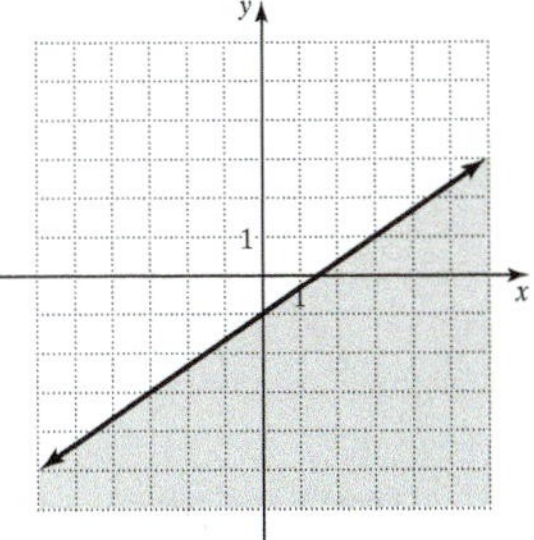

39.

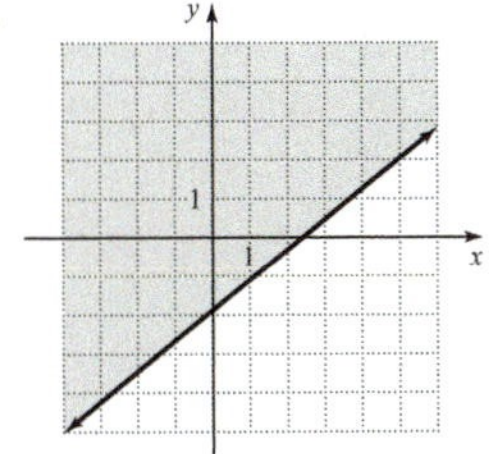

41. No, not a solution **43.** b

45.

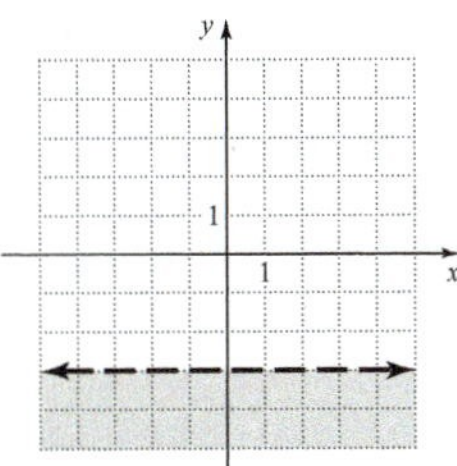

47.

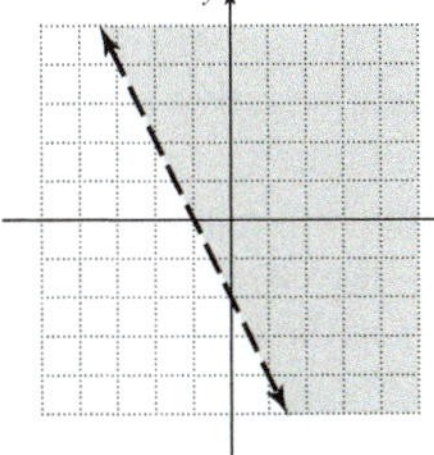

49. a. $h < \frac{1}{4}i$ **b.**

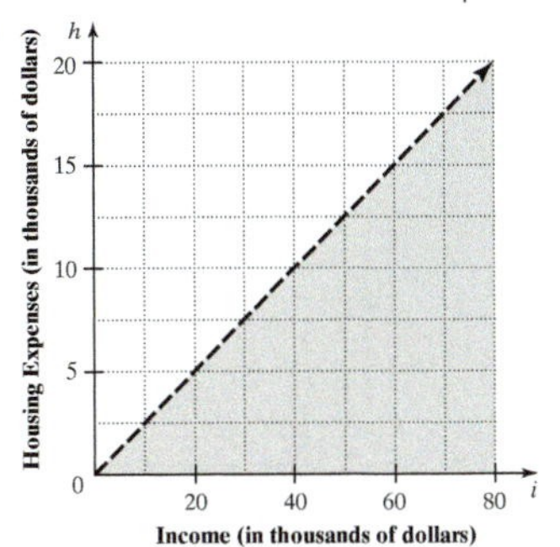

c. Choice of point may vary. Possible point: (20, 2.5). The guideline holds since the inequality is true when the values are substituted into the original inequality.

51. a. $x + y \geq 200$ **b.**

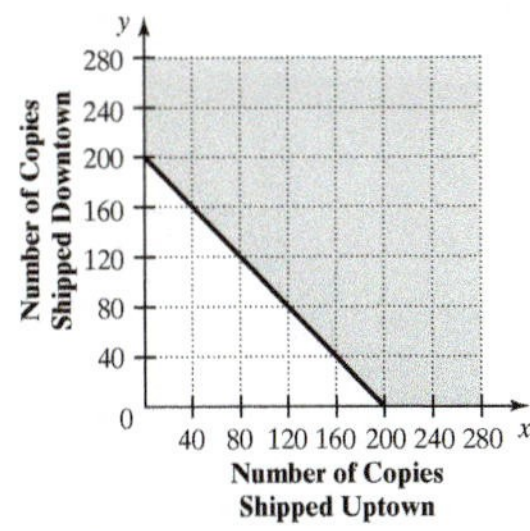

c. Choice of point may vary. Possible point: (200, 60). Check: $200 + 60 \overset{?}{\geq} 200$, $260 \geq 200$, True. At least 200 copies are shipped.

53. a. $30x + 75y \geq 1500$ or $2x + 5y \geq 100$

b.

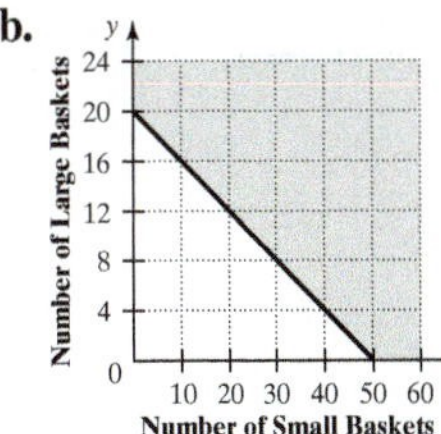

c. Since the point (20, 20) lies in the solution region, selling 20 small and 20 large gift baskets will generate the desired revenue.

55. a. $10w + 15m \leq 50{,}000$, or $2w + 3m \leq 10{,}000$

b.

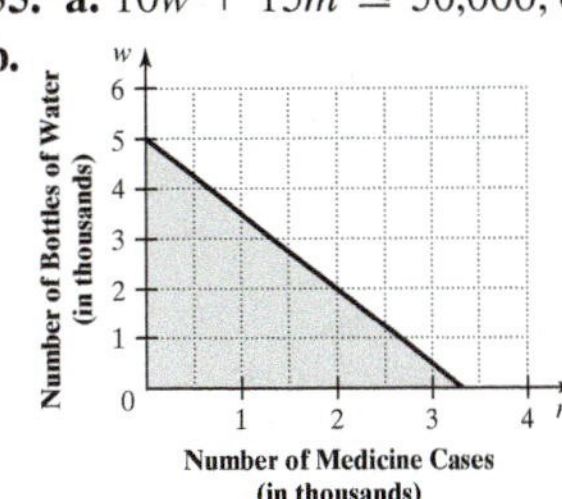

c. Answers may vary. Possible answers: 1500 cases of medicine and 1000 bottles of water: (1.5, 1); 300 cases of medicine and 500 bottles of water: (0.3, 0.5); 1200 cases of medicine and 2000 bottles of water: (1.2, 2)

57. a. $8x + 10y \geq 200$, or $4x + 5y \geq 100$

b.

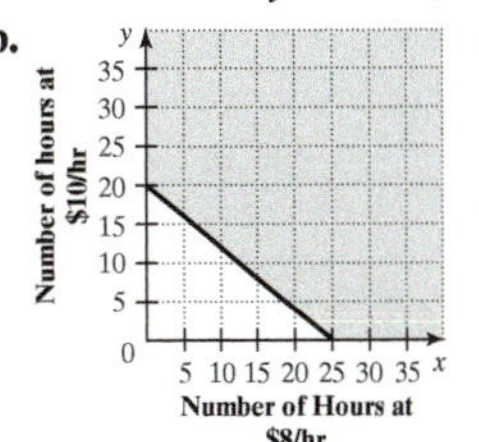

c. Answers may vary. Possible answers: 20 hr at the job paying \$8 per hour and 10 hr at the job paying \$10 per hour; 10 hr at the job paying \$8 per hour and 15 hr at the job paying \$10 per hour

59.

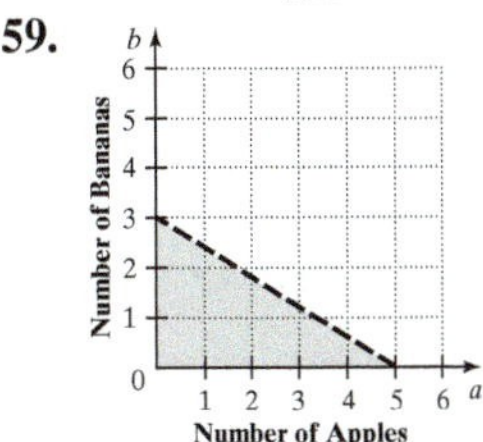

Chapter 13 Review Exercises, pp. 737–743

1. are not; possible answer: their y-coordinates have different signs **2.** does not; possible answer: the equation is false when 3 is substituted for x and -1 for y **3.** is not; possible answer: the coefficient of x is negative **4.** is; possible answer: the coefficient of x is 0 **5.** is; possible answer: $x = -3$ when $y = 0$ **6.** is; possible answer: its change in x-values is 0, and division by 0 is undefined **7.** is not; possible answer: the product of their slopes is 1, not -1 **8.** is; possible answer: substituting $x = 0$ and $y = 0$ into the inequality makes it true

ANSWERS

9.

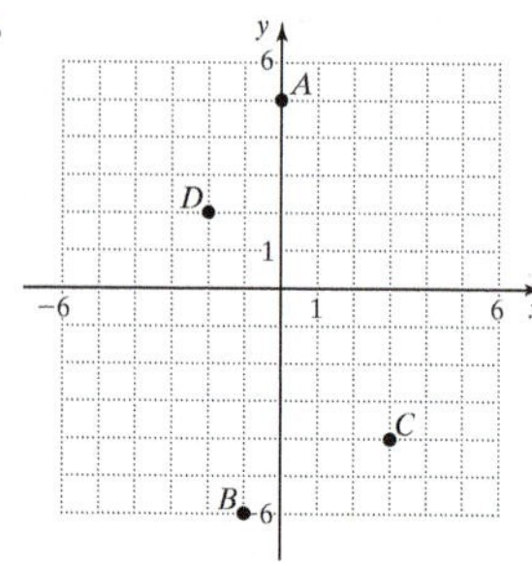

10.

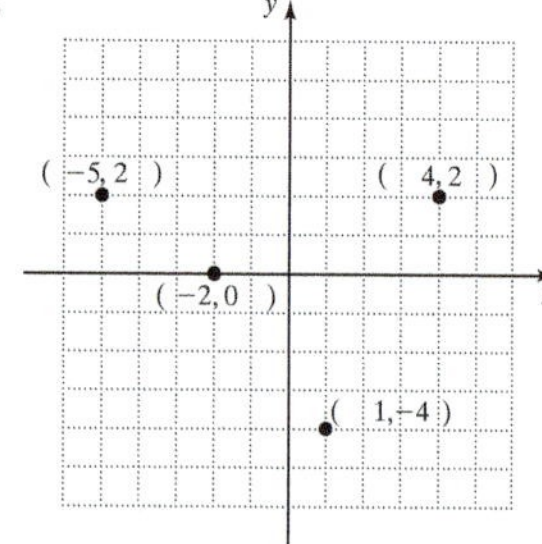

11. IV **12.** III

13. 5;

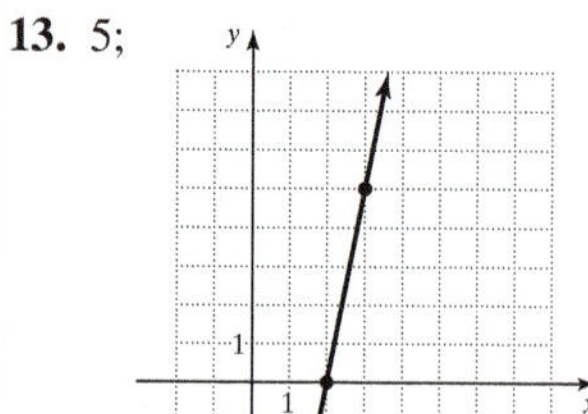

14. 0;

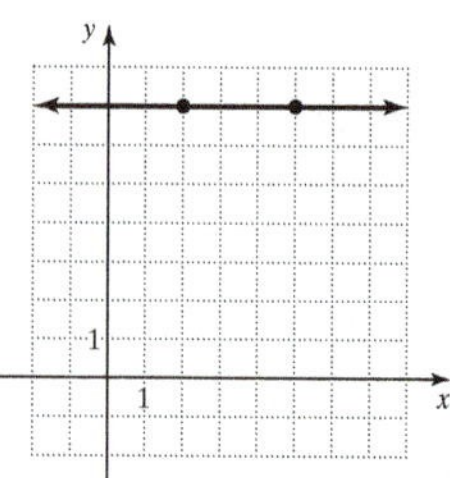

15.

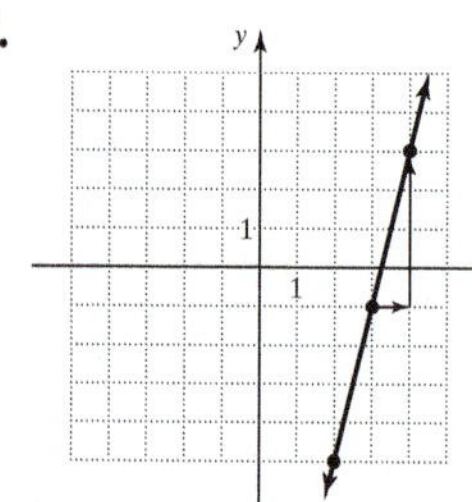

16.

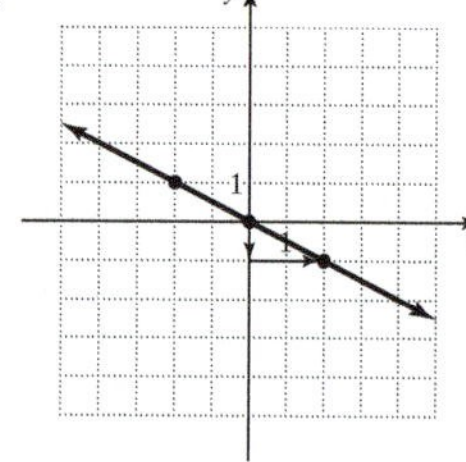

17. Positive slope **18.** Undefined slope **19.** Negative slope **20.** Zero slope **21.** Parallel **22.** Perpendicular **23.** $(30, 0)$ **24.** $(0, 50)$

25.

x	0	1	$\frac{5}{2}$	3
y	−5	−3	0	1

26.

x	2	5	−4	8
y	1	−2	7	−5

27.

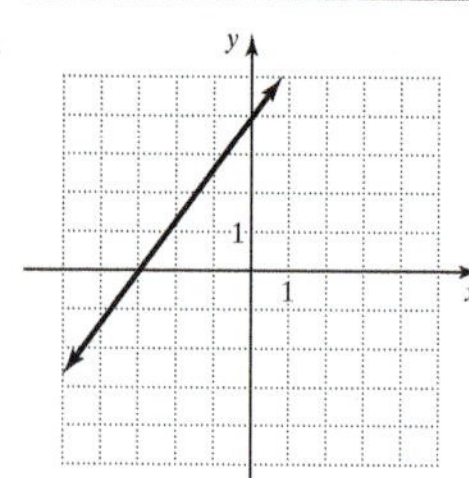

28.

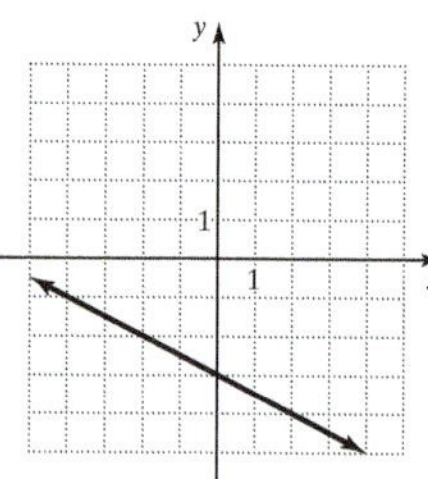

29.

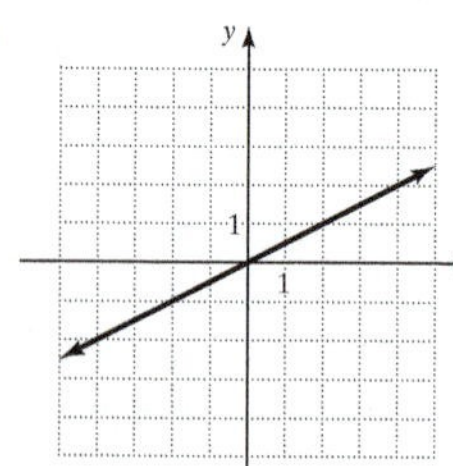

30.

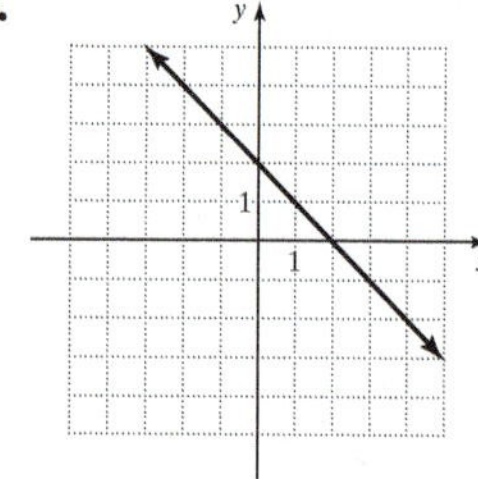

31. $y = x - 10$ **32.** $y = -\frac{1}{2}x - \frac{1}{2}$ **33.** $4, (0, -16),$ ╱, $(4, 0)$ **34.** $-\frac{1}{3}, (0, 0),$ ╲, $(0, 0)$ **35.** The slope of a line perpendicular to this line is -2. **36.** The slope of a line parallel to this line is 3. **37.** Point-slope form: $y - 5 = -(x - 3)$; slope-intercept form: $y = -x + 8$

38. $y = 0$ **39.** Point-slope form: $y - 5 = -5(x - 1)$; slope-intercept form: $y = -5x + 10$ **40.** Point-slope form: $y - 1 = \frac{1}{5}(x - 3)$; slope-intercept form: $y = \frac{1}{5}x + \frac{2}{5}$ **41.** $y = -\frac{3}{2}x + 3$ **42.** $y = \frac{3}{2}x - 3$ **43.** No, it is not a solution. **44.** Yes, it is a solution.

45.

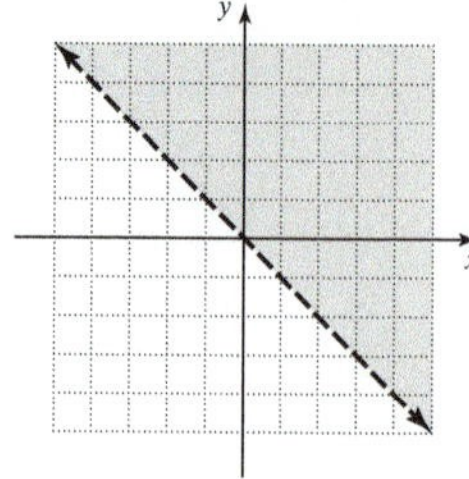

46.

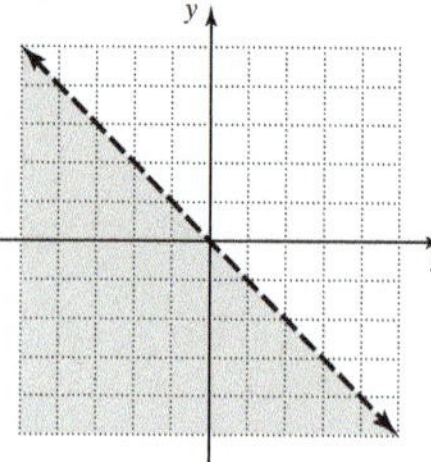

47.

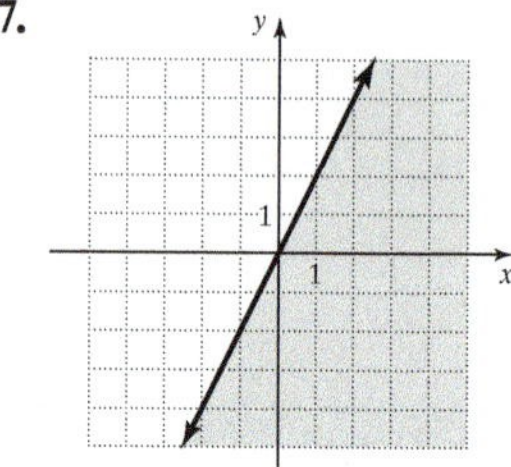

48.

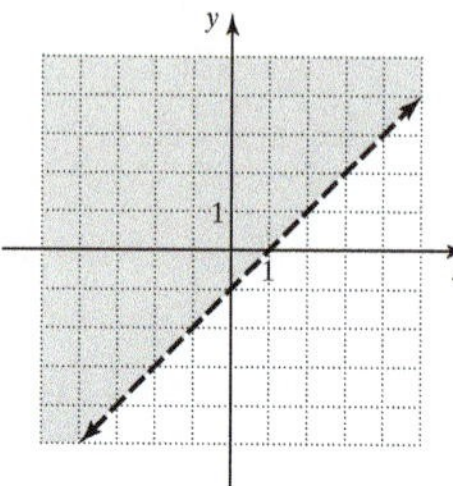

49. a.

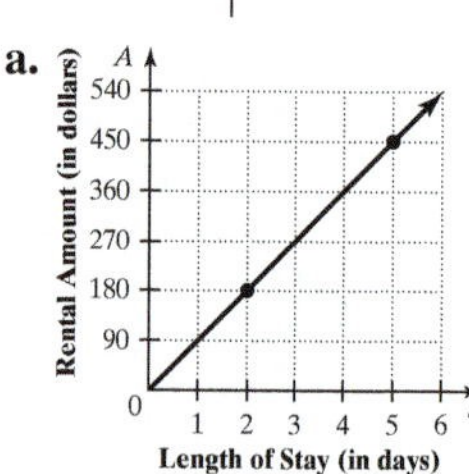

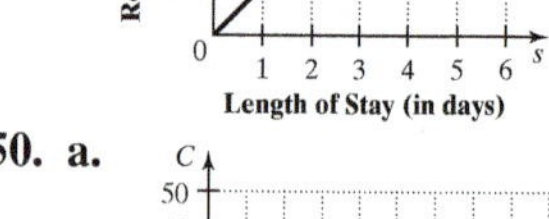

b. The A-intercept is $(0, 0)$. The A-intercept means that the cost for renting a room for 0 days is $0.

50. a.

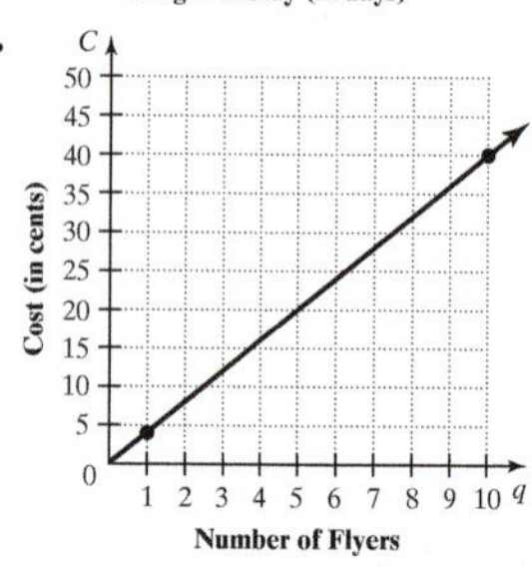

b. The slope of the line is 4. The slope represents the rate the print shop charges for each flyer, which is 4 cents.

51. The graph in part (a) could describe this motion. As the man drives toward the town, the distance between the man and the town decreases implying a negative slope. When the man stops, the distance between the man and the town remains the same, as indicated by the horizontal line segment. When the man drives toward the town again, the distance again decreases, implying a negative slope.

52. In the first part of the flight, the airplane takes off and ascends to a particular altitude (line segment slanting up to the right), then it flies at that same altitude during the second and longest part of the flight (horizontal line segment), and finally in the last part of the flight, it descends and lands (line segment slanting down to the right).

53. a. $i = 0.09s + 20{,}000$ **b.**

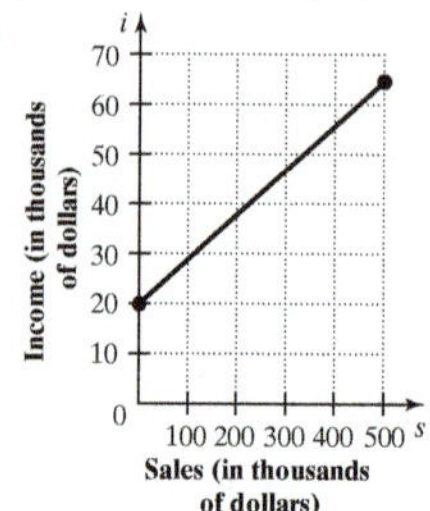

54. a.

b. The A-intercept of the graph is $(0, 100)$. The A-intercept represents the initial balance in the bank account.

55. a.

b. Answers may vary. Possible answer: 20 double jewel cases and 30 single jewel cases.

56. a.

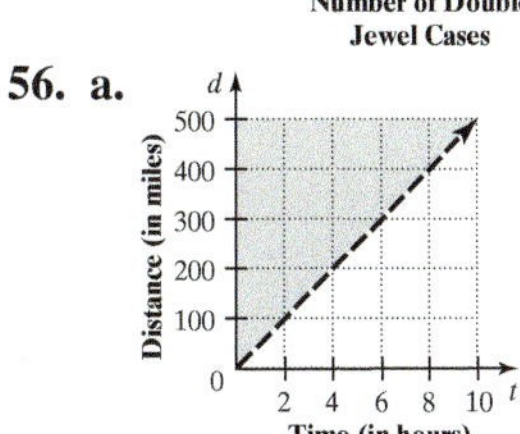

b. Choice of point may vary. Possible answer: $(2, 110)$; the coordinates mean that you caught up and passed your friend if you covered a distance of 110 mi in 2 hr.

Chapter 13 Posttest, pp. 744–746

1.

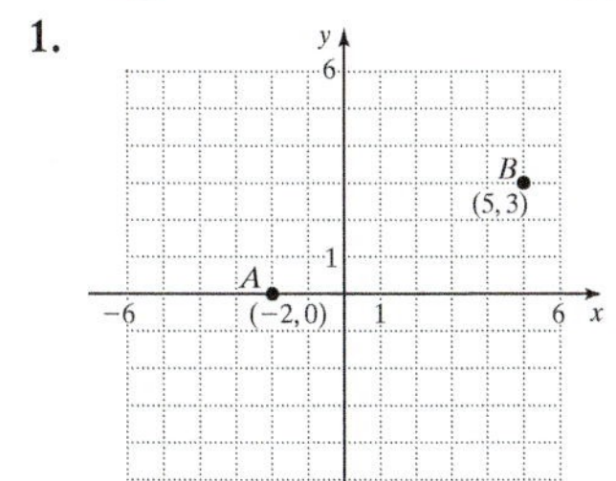

2. II **3.** $m = 1$

4. The graphs are parallel. The slope of $y = 3x + 1$ is 3 and the slope of $y = 3x - 2$ is 3. Since the slopes of the two lines are equal, their graphs are parallel. **5.** The slope of $\overleftrightarrow{AB}$ is $\frac{7}{2}$. The slope of $\overleftrightarrow{CD}$ is $-\frac{2}{7}$. $\overleftrightarrow{AB}$ is perpendicular to $\overleftrightarrow{CD}$, since the product of their slopes is -1. **6.** x-intercept: $(-5, 0)$, y-intercept: $(0, 2)$ **7.** The slope is positive. As the distance driven increases, the rental cost increases. **8.** Yes, the points do lie on the same line. The line containing $(0, 0)$ and $(-2, -4)$ is $y = 2x$. The line containing $(0, 0)$ and $(1, 2)$ is $y = 2x$.

9.

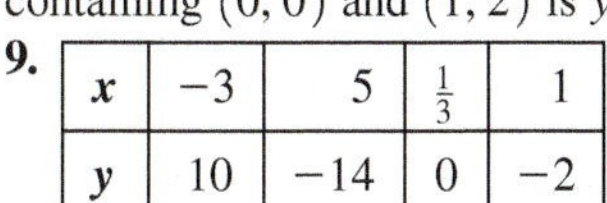

x	-3	5	$\frac{1}{3}$	1
y	10	-14	0	-2

10.

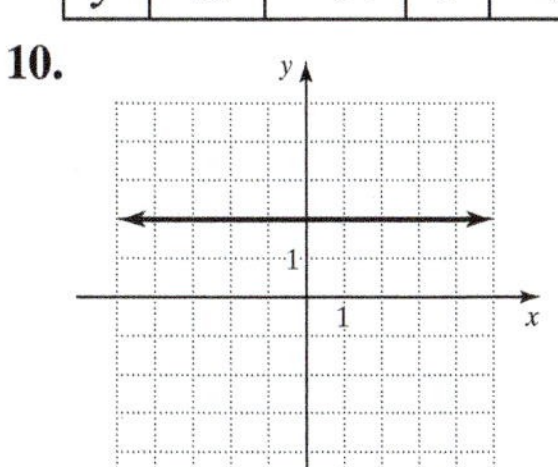

11.

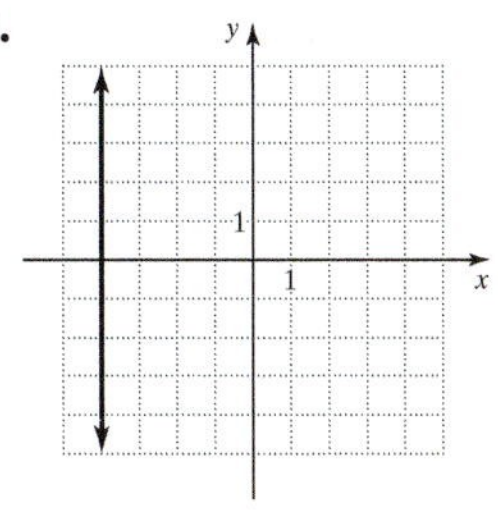

12.

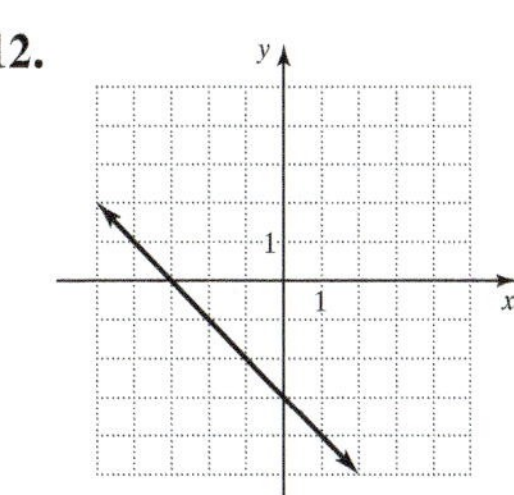

13.

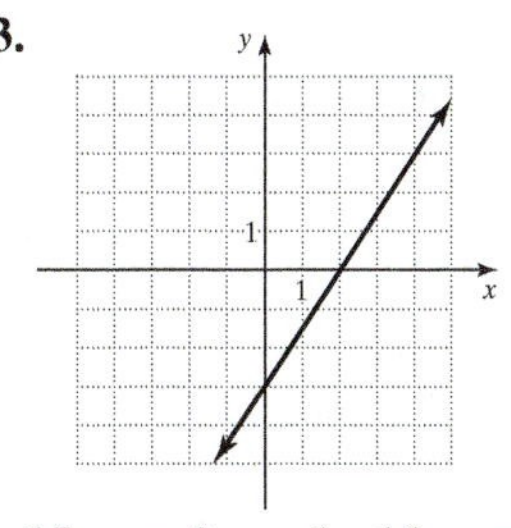

14. Slope: 3; y-intercept: $(0, 1)$ **15.** $y = 2x - 5$ **16.** $y = -x - 3$
17. Point-slope form: $y - 5 = \frac{3}{7}(x - 3)$; slope-intercept form: $y = \frac{3}{7}x + \frac{26}{7}$
18.

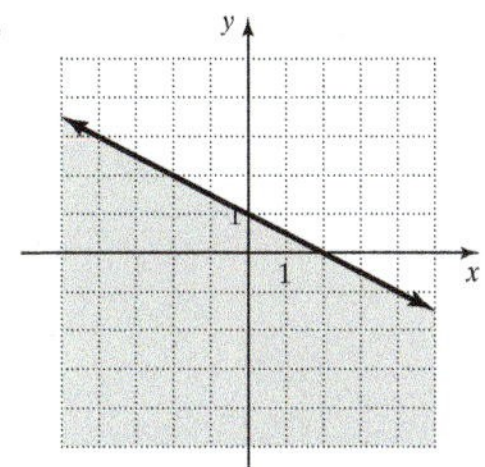

19.

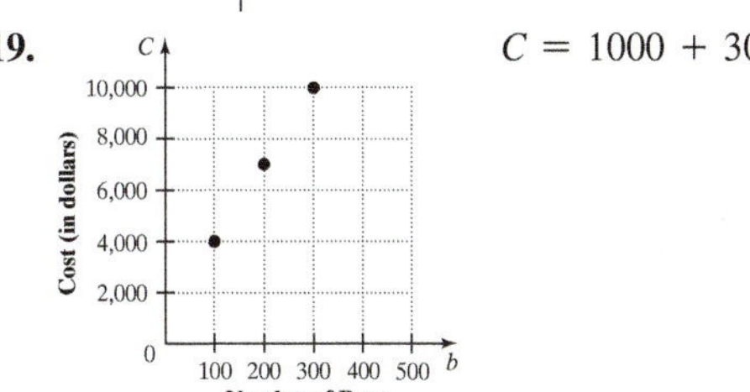

$C = 1000 + 30b$

20. $y > 124{,}000 + 8000x$

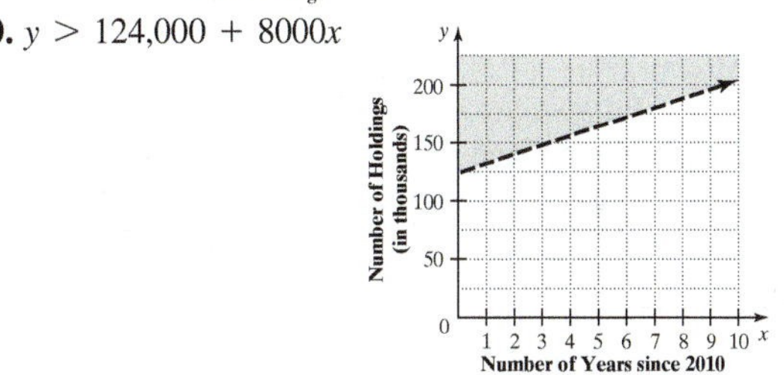

Chapter 13 Cumulative Review Exercises, pp. 747–748

1. $8\frac{3}{4}$ **2.** $4\frac{2}{3}$ **3.** $>$ **4.** -4.3 **5.** $2\frac{2}{5}$ **6.** $\frac{1}{4}$ **7.** 4 **8.** 13 **9.** 190 sec **10.** 5.11 **11.** 70° **12.** 125 m^3 **13.** 22 **14.** -4 **15.** Point-slope form: $y - 62 = -\frac{8}{5}(x - 10)$; slope-intercept form: $y = -\frac{8}{5}x + 78$ **16.** \$500 billion **17.** $T = \frac{S - 331}{0.6}$ **18.** On the average, males gain weight over time until age 40, when they begin to lose weight. **19.** In the summer

20. a. $y = 6x$

b.

c. x-intercept: $(0, 0)$, y-intercept: $(0, 0)$

ANSWERS

CHAPTER 14

Chapter 14 Pretest, pp. 750–751

1. a. Not a solution **b.** A solution **c.** Not a solution **2.** One solution **3.** $(-3, 1)$

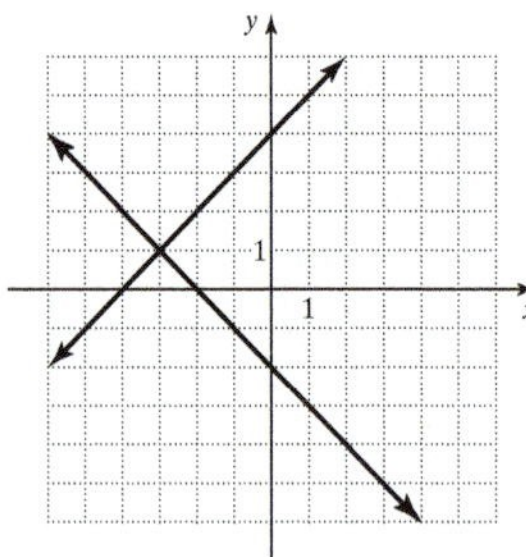

4. $(5, 2)$

5. Infinitely many solutions, namely all points on the line

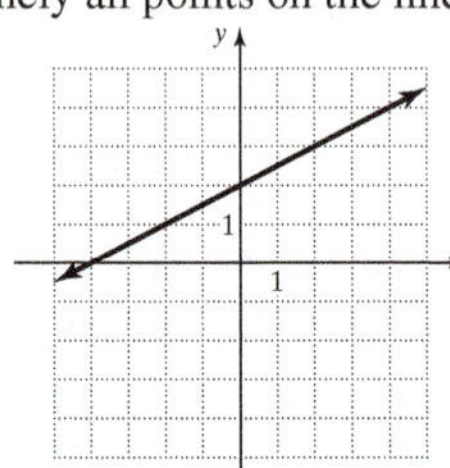

6. No solution

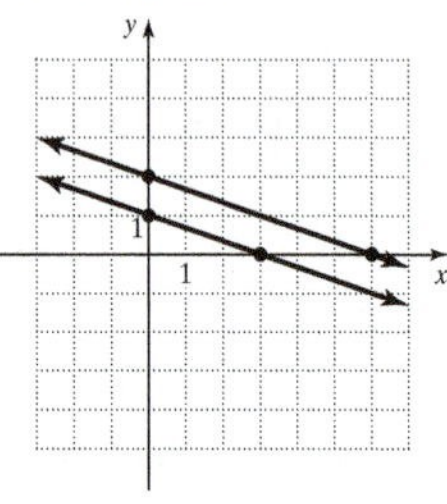

7. $(-5, -6)$ **8.** $(2, 1)$ **9.** $a = -5, b = 1$ **10.** $(1, -3)$ **11.** Infinitely many solutions, namely all ordered pairs that satisfy both equations **12.** $(-4, 1)$ **13.** $(-1, 4)$ **14.** $n = -15, m = -7$ **15.** No solution **16.** $(2, 0)$ **17.** The college awarded 2476 bachelor's degrees and 619 associate's degrees. **18.** Fifty \$5 tickets were printed. **19.** \$80,000 was invested in the fund at 5% interest, and \$120,000 was invested in the fund at 6%. **20.** The speed of the boat was 6 mph, and the speed of the current was 0.5 mph.

Section 14.1 Practices, pp. 753–761

1, *p. 753:* a. Yes, it is a solution of the system. **b.** No, it is not a solution of the system. **2, *p. 755:* a.** One solution **b.** Infinitely many solutions **c.** No solution **3, *p. 755:*** $(3, -1)$

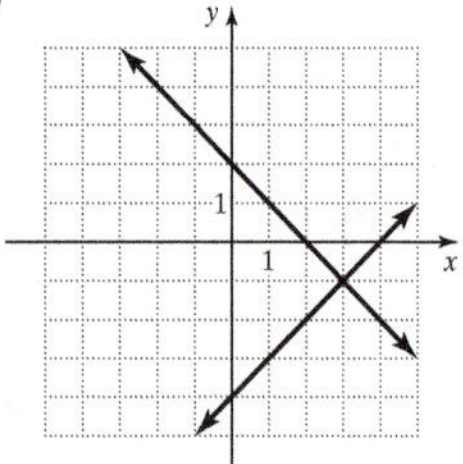

4, *p. 758:* No solution

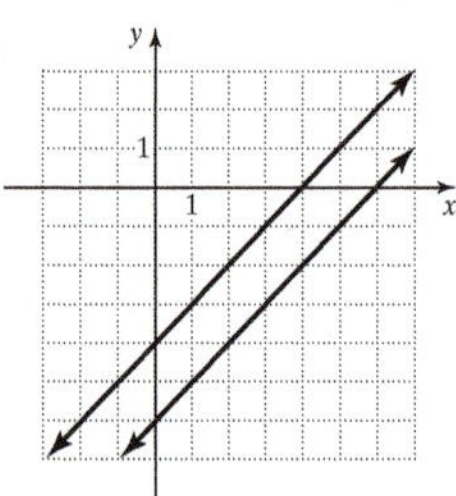

5, *p. 758:* Infinitely many solutions, namely all points on the line

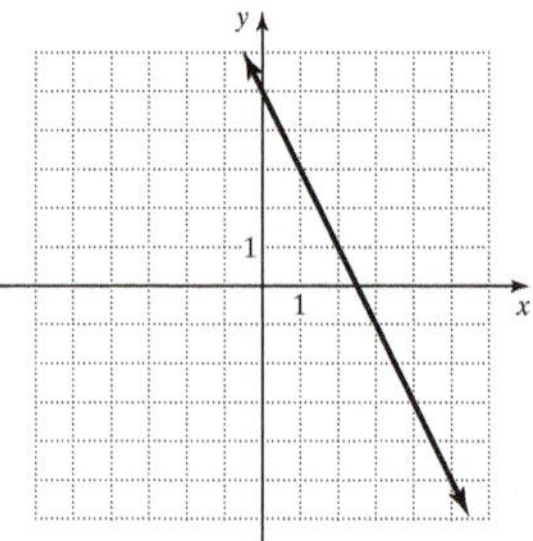

6, *p. 759:* a. $\begin{cases} m + v = 1150 \\ v = m - 100 \end{cases}$

b.

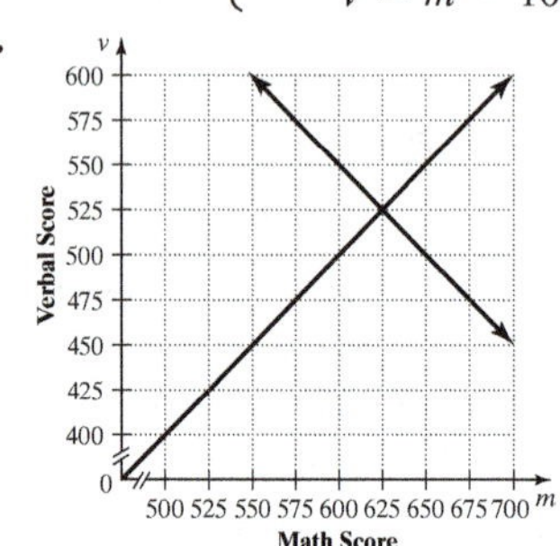

c. $(625, 525)$ **d.** The point of intersection indicates that she got a score of 625 on her test of math skills and a score of 525 on her test of verbal skills. **7, *p. 760:* a.** $y = 1.50x + 450$ **b.** $y = 3x$

c.

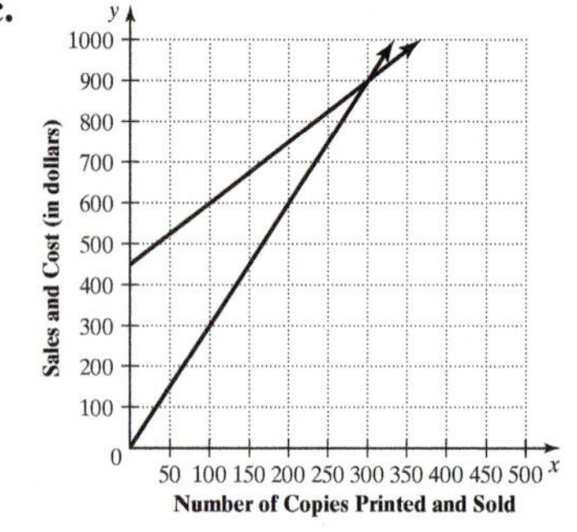

d. The break-even point is $(300, 900)$. So when 300 newsletters are printed and sold, the cost of printing the newsletter and the income from sales will be the same, \$900.

8, *p. 761:*

Intersection
X=.85714286 Y=5.8571429

The approximate solution is $(0.857, 5.857)$.

Exercises 14.1, pp. 762–773

1. system of equations **3.** are parallel **5. a.** Not a solution; **b.** Not a solution; **c.** A solution **7. a.** Not a solution; **b.** Not a solution; **c.** A solution **9. a.** III **b.** IV **c.** II **d.** I

11. $(3, 1)$

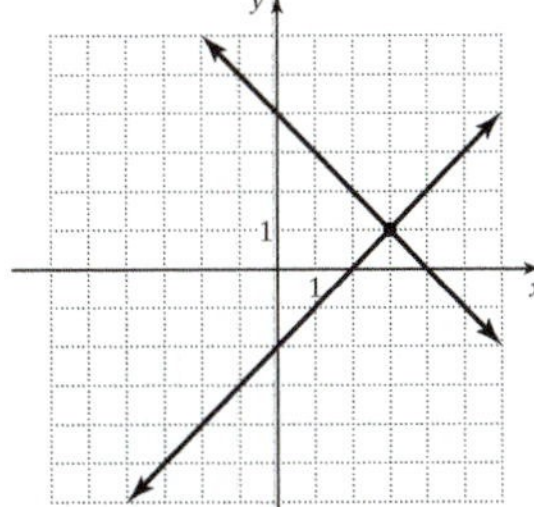

13. $(0, 4)$

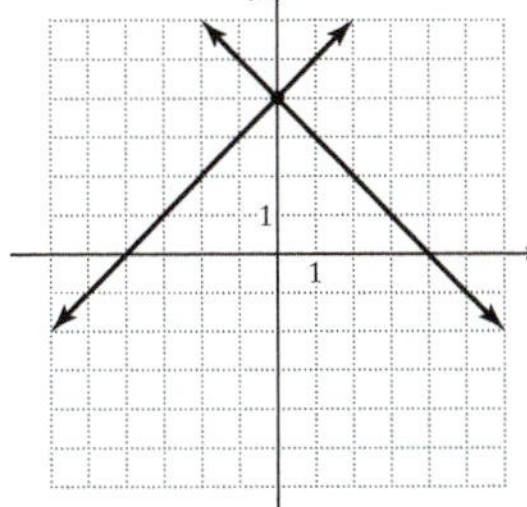

15. $(1, 5)$

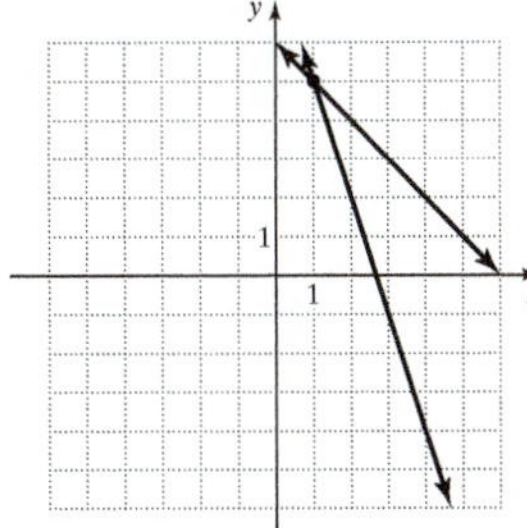

17. $(0, 1)$

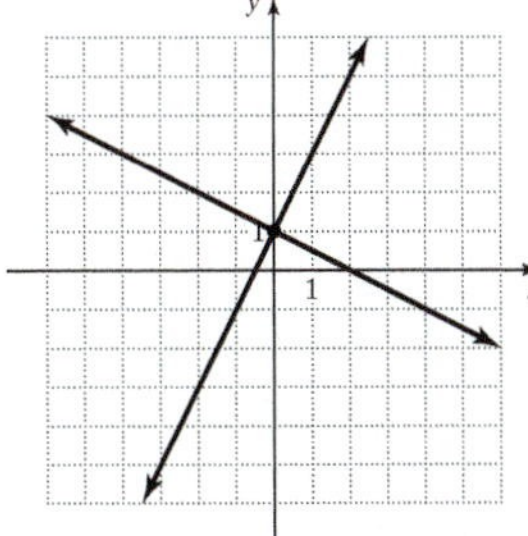

19. $(-2, -1)$

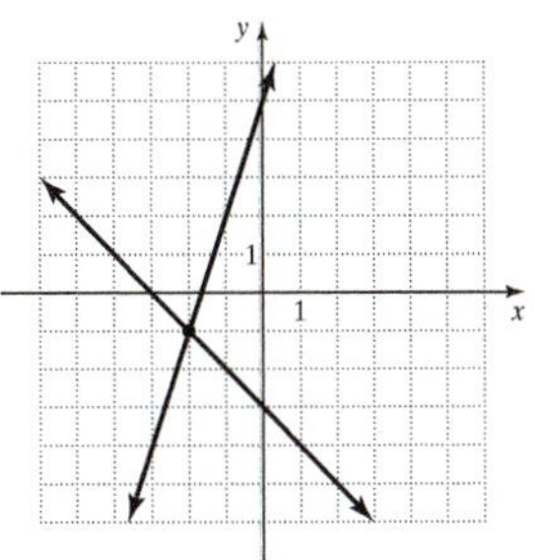

21. Infinitely many solutions, namely all points on the line

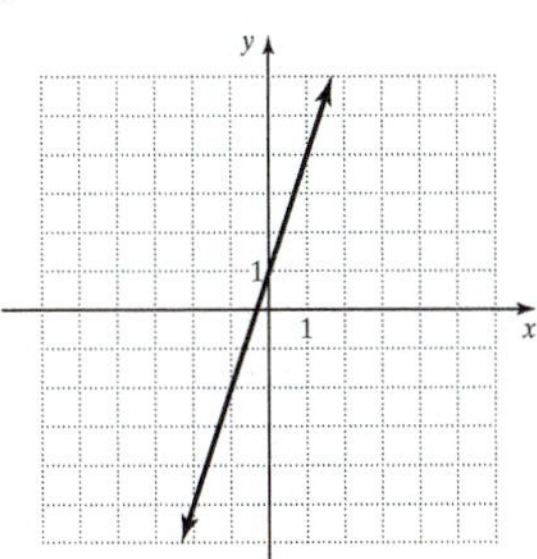

23. No solution

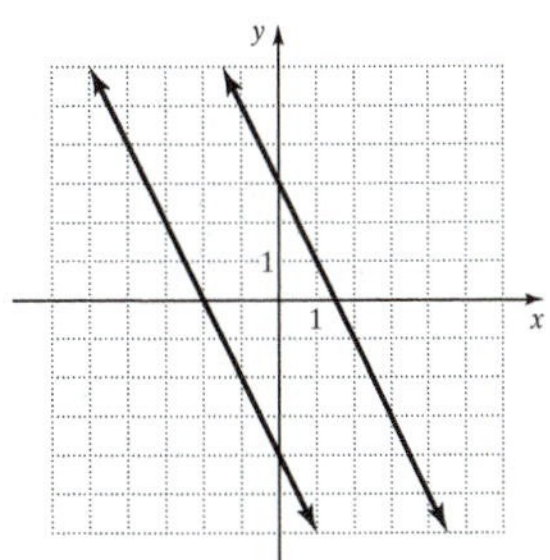

25. Infinitely many solutions, namely all points on the line

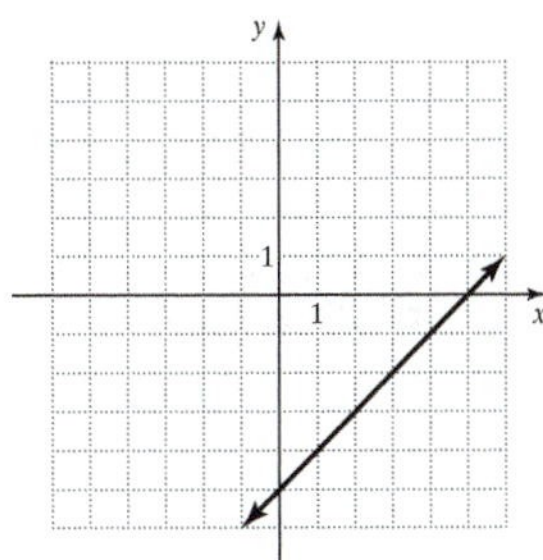

27. No solution

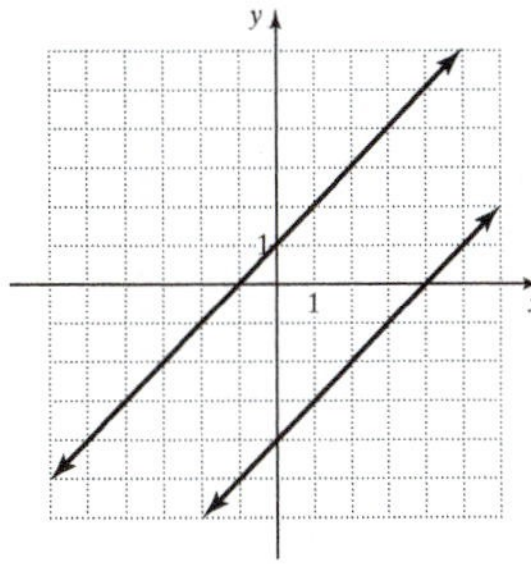

29. $(-2, -2)$

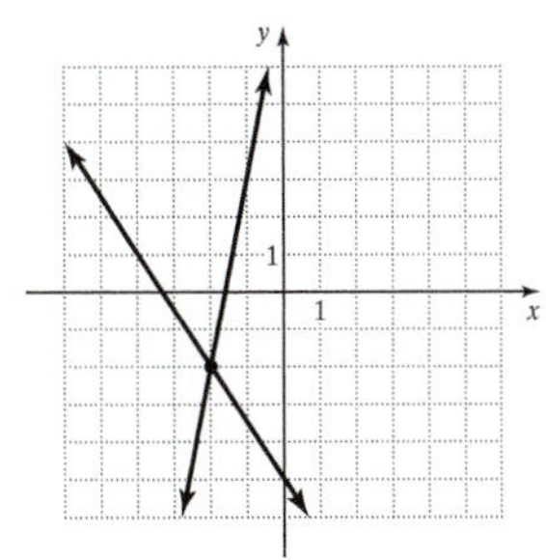

31. No solution

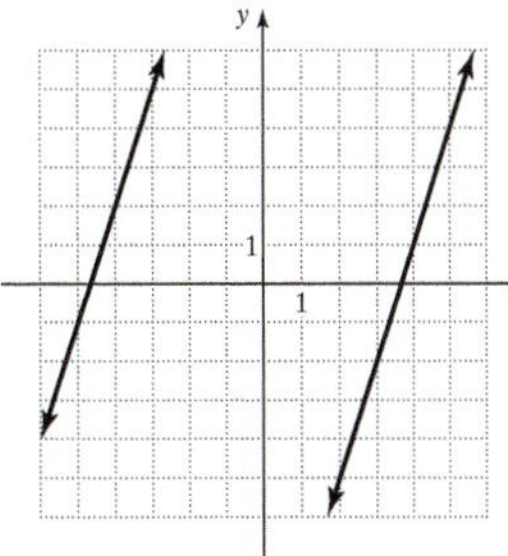

33. $(-1, 2)$

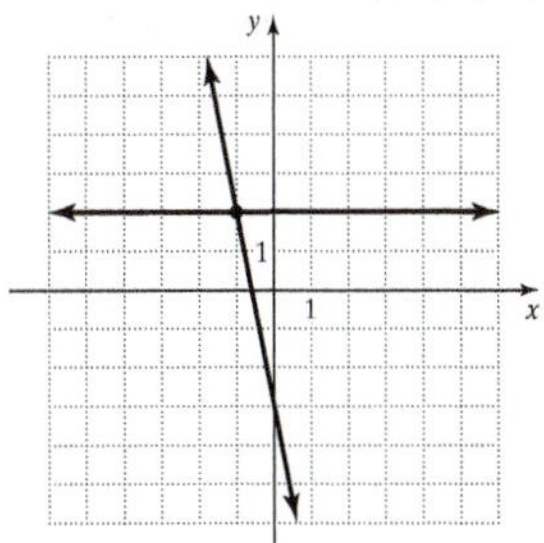

35. $(0, -2)$

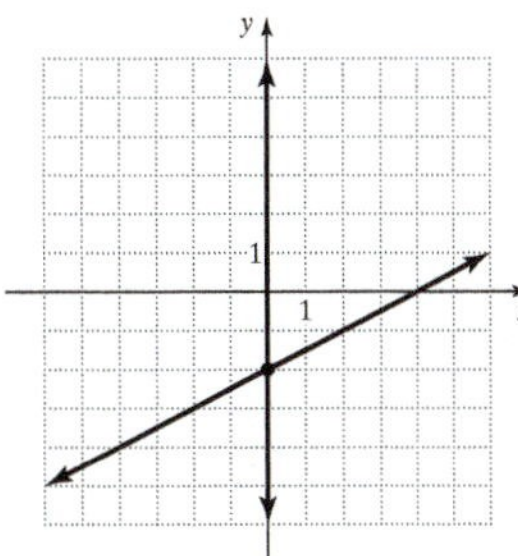

37. $(-6, -4)$

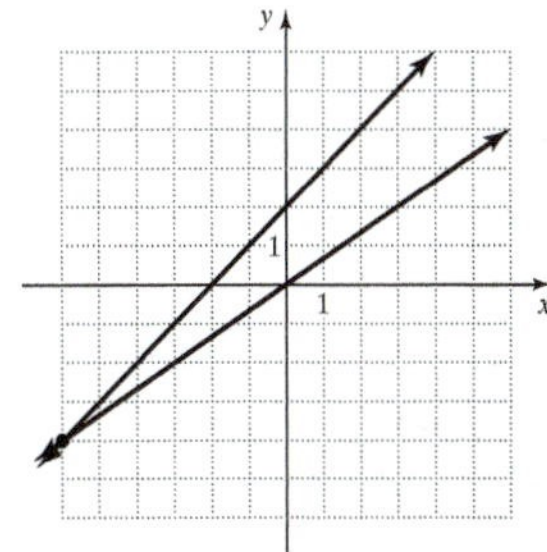

39. Infinitely many solutions, namely all points on the line

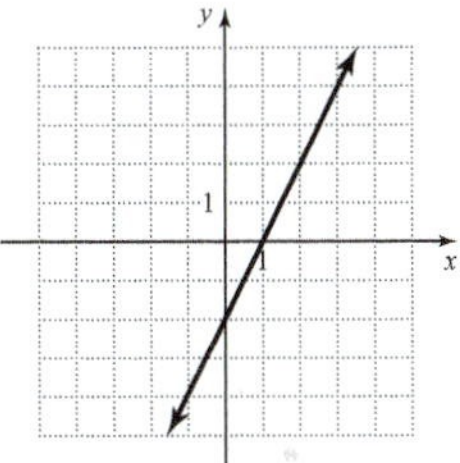

41. $(3, -2)$

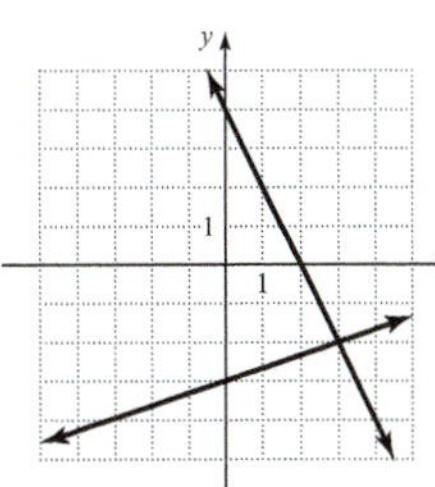

43. $(2, 2)$

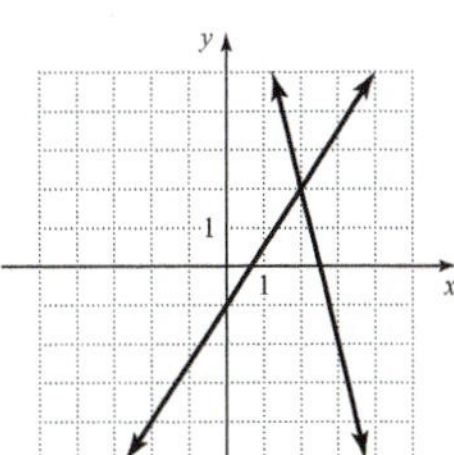

45. b **47. a.** $x + y = 57{,}000$
$y = x + 3000$

b.

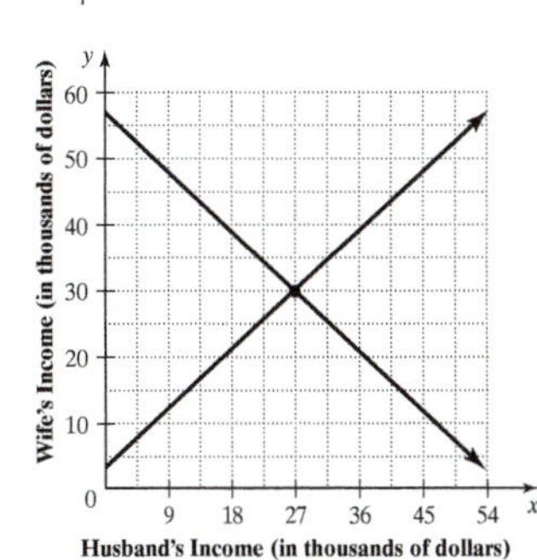

c. The husband made \$27,000 and the wife made \$30,000.

49. a. $y = 40x + 75$ (Mike)
$y = 30x + 100$ (Sally)

b.

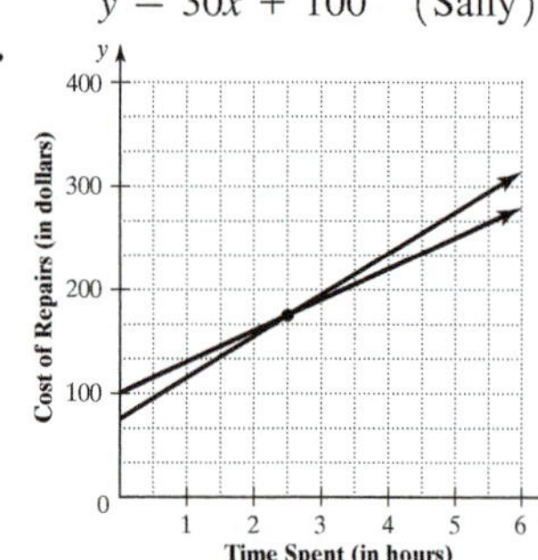

c. The plumbers would charge the same amount for 2.5 hr of work. **d.** Sally charges less.

51. The break-even point for duplicating DVDs is $(24, 36)$.

53.

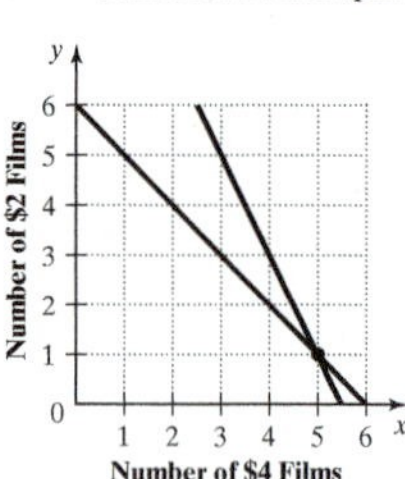

Five $4 films were rented.

55. The health clubs charge the same amount ($900) for 20 mo.

Section 14.2 Practices, pp. 774–780

1, ***p. 774:*** $(4, -3)$ **2,** ***p. 775:*** $m = \frac{32}{7}, n = -\frac{5}{7}$ **3,** ***p. 776:*** $(0, -1)$ **4,** ***p. 776:*** No solution **5,** ***p. 777:*** Infinitely many solutions, namely all ordered pairs that satisfy both equations **6,** ***p. 777:*** **a.** $c = 35n + 20$ (TV Deal), $c = 25n + 30$ (Movie Deal) **b.** $n = 1$ and $c = 55$ **c.** The cost is the same ($55) for both cable deals if you sign up for one month. **7,** ***p. 778:*** 12.5 oz of the 20% copper alloy and 2.5 oz of the 50% copper alloy are required. **8,** ***p. 780:*** $80,000 in McDonald's and $40,000 in Wendy's

Exercises 14.2, pp. 781–782

1. $(3, 7)$ **3.** $(-4, -3)$ **5.** $(-12, 4)$ **7.** $(2, 1)$ **9.** $(0, 0)$ **11.** $(-1, 2)$ **13.** $(6, -6)$ **15.** No solution **17.** $(\frac{7}{2}, -\frac{1}{2})$ **19.** No solution **21.** Infinitely many solutions, namely all ordered pairs that satisfy both equations **23.** Infinitely many solutions, namely all ordered pairs that satisfy both equations **25.** $p = 3, q = 5$ **27.** $s = -2, t = 1$ **29.** $(-3, 6)$ **31.** No solution **33.** $(-\frac{3}{2}, \frac{1}{2})$ **35. a.** $c = 1.25m + 3, c = 1.50m + 2$ **b.** $m = 4, c = 8$. The solution indicates that both companies charge the same amount ($8) for a 4-mi taxi ride. **37.** 80 full-price tickets were sold. **39.** She can combine 2.5 L of the antiseptic that is 30% alcohol with 7.5 L of the antiseptic that is 70% alcohol to get the desired concentration. **41.** There were 3 women in one department and 72 women in the other department. **43.** $4000 was loaned at 6%, and $1000 was loaned at 7%. **45.** $23,000 was invested at 7% and $17,000 was invested at 9%.

Section 14.3 Practices, pp. 784–789

1, ***p. 784:*** $(-2, 8)$ **2,** ***p. 785:*** $(2, -5)$ **3,** ***p. 785:*** $(0, 6)$ **4,** ***p. 786:*** $(2, -2)$ **5,** ***p. 787:*** $(10, 8)$ **6,** ***p. 787:*** Infinitely many solutions, namely all ordered pairs that satisfy both equations **7,** ***p. 788:*** The whale's speed in calm water is 30 mph and the speed of the current is 10 mph. **8,** ***p. 789:*** 100 adults and 75 students attended the game.

Exercises 14.3, pp. 791–793

1. $(5, -2)$ **3.** $(-\frac{3}{2}, -\frac{5}{2})$ **5.** $p = -3, q = -16$ **7.** $(-1, 0)$ **9.** No solution **11.** Infinitely many solutions, namely all ordered pairs that satisfy both equations **13.** $(-2, \frac{1}{2})$ **15.** $s = 0, d = -2$ **17.** $(7, 4)$ **19.** $(-1, 2)$ **21.** $p = \frac{9}{2}, q = -\frac{11}{2}$ **23.** $(-3.5, 2.5)$ **25.** $(2, -2)$ **27.** $(\frac{8}{3}, -\frac{4}{3})$ **29.** No solution **31.** $a = -3, b = -2$ **33.** Infinitely many solutions, namely all ordered pairs that satisfy both equations **35.** $(-\frac{3}{4}, \frac{1}{4})$ **37.** The speed of the pass if there were no wind would be 13 yd per sec. **39.** The zoo collected 83 full-price admissions and 140 half-price admissions. **41.** There were 738 Lords and 650 Members of Parliament. **43.** It takes the computer 3 nanoseconds to carry out one sum and 4 nanoseconds to carry out one product. **45.** The rate for full-page ads is $950 and for half-page ads is $645.

Chapter 14 Review Exercises, pp. 797–801

1. is not; Possible answer: the ordered pair does not satisfy both equations in the system **2.** does not have; Possible answer: there are no points of intersection **3.** has; Possible answer: all points on the line are solutions **4.** does; Possible answer: the coefficients of the x-terms are opposites **5.** No, it is not a solution of the system. **6. a.** No solution **b.** One solution **c.** Infinitely many solutions **7. a.** III **b.** IV **c.** II **d.** I **8.** $(1, 5)$

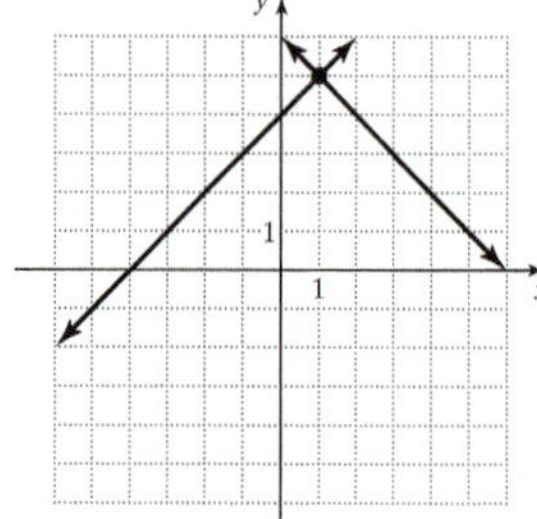

9. No solution

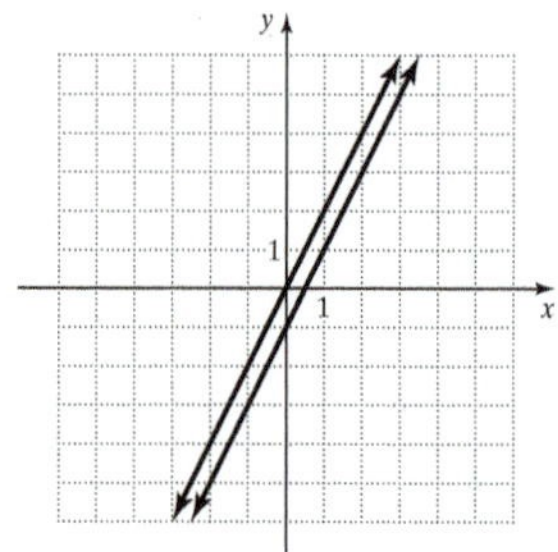

10. Infinitely many solutions, namely all ordered pairs that satisfy both equations

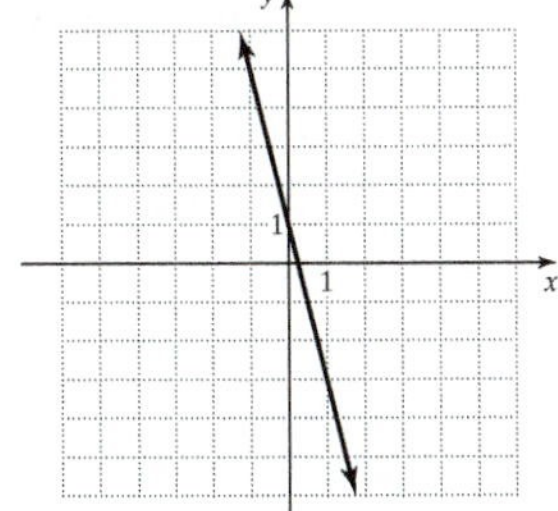

11. $(0, 5)$

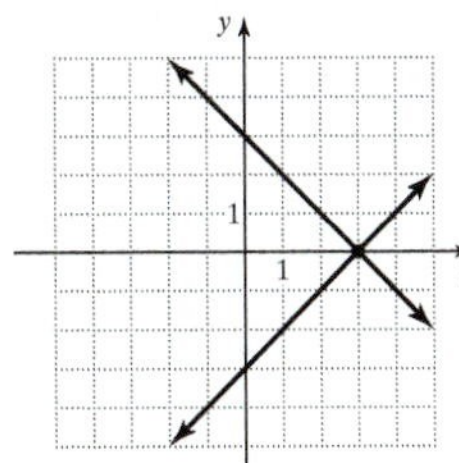

12. $(-1, 4)$ **13.** $a = 2, b = 2$ **14.** No solution **15.** Infinitely many solutions, namely all ordered pairs that satisfy both equations **16.** $(4, -3)$ **17.** Infinitely many solutions, namely all ordered pairs that satisfy both equations **18.** $(6, -3)$ **19.** $(2, -5)$ **20. a.** $y = 0.50x + 1750, y = 5.50x$ **b.** The student must type 350 pages in order to break even. **21. a.** $s = 10h, s = 8h + 50$ **b.** 25 hr **22.** The area of the screen is 6720 ft^2. **23.** The tennis court is 36 ft wide and 78 ft long. **24.** The coin box contained 200 nickels and 150 dimes. **25.** One train travels at a rate of 60 mph, and the other travels at a rate of 65 mph. **26.** The team made 818 two-point baskets and 267 three-point baskets. **27.** The pharmacist should mix 150 mL of the 30% solution and 50 mL of the 10% solution. **28.** 1400 L of the 50% solution and 600 L of water are needed to fill the tank. **29.** The client put \$40,000 in municipal bonds and \$10,000 in corporate stocks. **30.** \$7000 was invested in the high-risk fund, and \$3000 was invested in the low-risk fund. **31.** The speed of the slower plane was 400 mph. **32.** The bird flies at a speed of 21 mph in still air, and the speed of the wind was 5 mph. **33. a.** 57 senators **b.** 43 senators **34. a.** The two metals will be the same temperature after 12 min. **b.** The iron will be 14° colder than the copper after 14 min.

Chapter 14 Posttest, pp. 802–803

1. a. A solution **b.** Not a solution **c.** Not a solution **2.** The system has no solution. **3.** $(3, 0)$

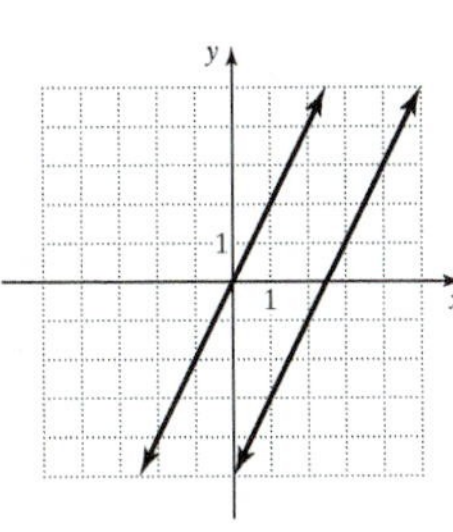

4. The system has no solution.

5. $(-1, 2)$

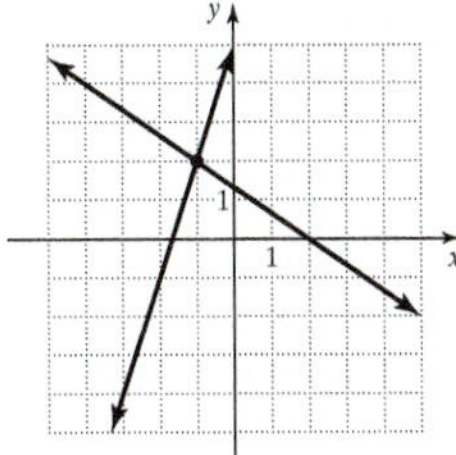

6. An infinite number of solutions, namely all points on the line

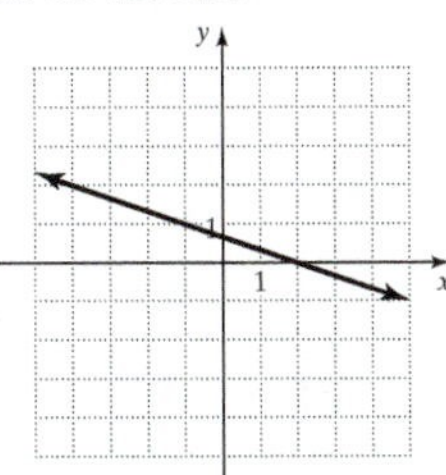

7. $(-4, 1)$ **8.** $(1, 3)$ **9.** $u = \frac{3}{4}, v = \frac{17}{4}$ **10.** $(2, -5)$ **11.** No solution **12.** $p = 0, q = 0.5$ **13.** $(-2, -7)$ **14.** Infinitely many solutions, namely all ordered pairs that satisfy both equations **15.** $(\frac{1}{7}, -\frac{6}{7})$ **16.** $(3, -2)$ **17.** The winning candidate got 4204 votes. **18.** One serving of turkey and two servings of salmon **19.** 1 gal of the 20% iodine solution and 3 gal of the 60% iodine solution **20.** The speed of the wind was 20 mph and the speed of the plane in still air is 150 mph.

Chapter 14 Cumulative Review Exercises, pp. 804–805

1. $\frac{5}{6}, \frac{2}{3}, \frac{7}{12}$ **2.** 10.603 **3.** $n = \frac{7}{20}$ **4.** No, 5 is not a solution. **5.** True **6.** True **7.** 2 yr 8 mo **8.** -26 **9.** $5x - 3y - 2$ **10.** $x = -3$ **11.** −2 −1 0 1 2 3 4 5 6 7 8 ; $x > 1$ **12.** Slope: $m = -\frac{1}{2}$; y-intercept: $(0, 2)$ **13.**

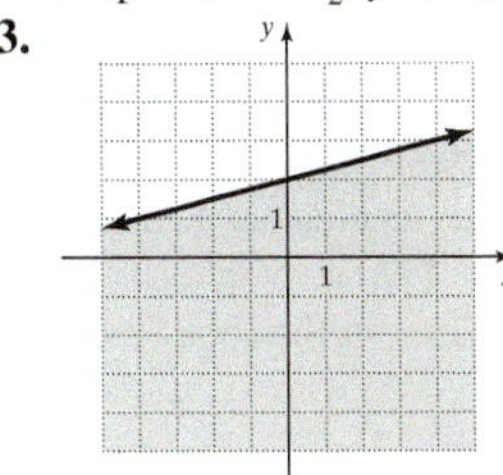

14. $(-4, 1)$ **15.** 0.0765 g **16.** The plane can fly 1554 mph ($S = 1554$). **17.** The two companies charge the same amount (\$95) for a one-day rental if the car is driven 300 mi. **18.** 3 L of plasma and 2 L of cells

19.

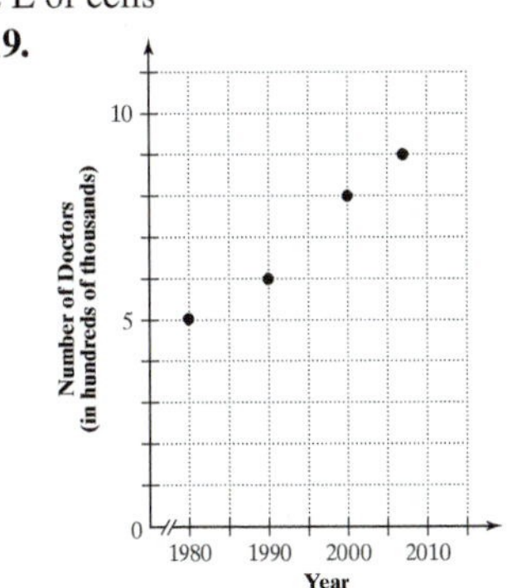

20. About 106,400 mi^2

CHAPTER 15

Chapter 15 Pretest, p. 807

1. x^9 **2.** y^4 **3.** -3 **4.** $16x^8y^6$ **5.** $\frac{a^3}{b^{15}}$ **6.** $\frac{x^2}{25y^8}$ **7. a.** $6x^4, 5x^3, x^2, -7x$, and 8 **b.** 6, 5, 1, -7, and 8 **c.** 4 **d.** 8 **8.** $3n^2 + n + 2$ **9.** $x^2 + x - 4$ **10.** $-6a^2 + 9a^2b + 3ab - 4b^2$ **11.** $3x^4 - 12x^3 + 27x^2$ **12.** $2n^3 + 7n^2 - 3n - 18$ **13.** $4x^2 - 3x - 27$

ANSWERS

14. $9y^2 - 49$ **15.** $25 - 20n + 4n^2$ **16.** $t^2 - 2t - 5$ **17.** $4x + 5$ **18.** 200 mol of hydrogen will contain 1.2×10^{26} molecules. **19.** $100 + 200r + 100r^2$ **20.** 8 billion people

Section 15.1 Practices, pp. 808–814

1, *p. 808:* **a.** 10,000 **b.** $-\frac{1}{32}$ **c.** y^6 **d.** $-y$ **2, *p. 809:*** **a.** 1 **b.** 1 **c.** 1 **d.** -1 **e.** 1 **3, *p. 810:*** **a.** 10^{12} **b.** $(-4)^6$ **c.** n^{10} **d.** y^5 **e.** We cannot apply the product rule because the bases are not the same. **4, *p. 811:*** **a.** 7^5 **b.** $(-9)^1$, or -9 **c.** s^0, or 1 **d.** r^7 **e.** We cannot apply the quotient rule because the bases are not the same. **5, *p. 811:*** **a.** y^9 **b.** x^5y^6 **c.** a^2 **6, *p. 812:*** **a.** $\frac{1}{9^2} = \frac{1}{81}$ **b.** $\frac{1}{n^5}$ **c.** $-\frac{1}{3y}$ **d.** $\frac{1}{5^3} = \frac{1}{125}$ **7, *p. 813:*** **a.** $\frac{s}{8}$ **b.** $\frac{3}{x}$ **c.** $\frac{1}{r^6}$ **d.** $\frac{3^2}{g^5} = \frac{9}{g^5}$ **e.** x^3 **8, *p. 813:*** **a.** a^3 **b.** $\frac{2x^2}{5}$ **c.** $\frac{r^3s}{2}$ **9, *p. 814:*** **a.** 2^{23} **b.** 2^{30}

Exercises 15.1, pp. 815–818

1. added **3.** 0 power **5.** 125 **7.** $\frac{9}{16}$ **9.** 0.16 **11.** -0.25 **13.** -8 **15.** 81 **17.** $-\frac{1}{8}$ **19.** x^4 **21.** pq **23.** 1 **25.** -1 **27.** 10^{11} **29.** 4^8 **31.** a^6 **33.** Cannot be simplified **35.** n^7 **37.** Cannot be simplified **39.** 8^2 **41.** 5^3 **43.** y^1, or y **45.** a^6 **47.** Cannot be simplified **49.** x^0, or 1 **51.** y^6 **53.** p^7q^5 **55.** $x^3y^2z^3$ **57.** a^1, or a **59.** x^2 **61.** $\frac{1}{5}$ **63.** $\frac{1}{x}$ **65.** $-\frac{1}{3a}$ **67.** $\frac{1}{2^4}$ **69.** $-\frac{1}{3^4}$ **71.** $\frac{8}{n^3}$ **73.** $\frac{1}{x^2}$ **75.** $-\frac{x}{3^2}$ **77.** $\frac{y^3}{x^2}$ **79.** $\frac{q}{r}$ **81.** $\frac{4y^2}{x}$ **83.** $\frac{1}{p^5}$ **85.** p^3 **87.** $\frac{1}{a}$ **89.** $2n^4$ **91.** p^4q **93.** $\frac{1}{t^5}$ **95.** x^7 **97.** a^1, or a **99.** $\frac{b^3}{a^3}$ **101.** s^3t^6 **103.** $-\frac{8}{27}$ **105.** Cannot be simplified **107.** $\frac{1}{y^6}$ **109.** $\frac{1}{q^5}$ **111.** x^4y^3 **113.** **a.** $35 \cdot 2^5 = 1120$ people were ill on the sixth day of the epidemic; $35 \cdot 2^9 = 17{,}920$ people were ill on the tenth day. **b.** The number of people ill on the tenth day was 2^4, or 16, times as great as the number ill on the sixth day. **115.** $60 \times (0.95)^{11}$ ppm **117.** **a.** Volume of small box: $8x^3$, Volume of large box: $125x^3$ **b.** The volume of the large box is $\frac{125}{8}$ times the volume of the small box.

Section 15.2 Practices, pp. 819–827

1, *p. 819:* **a.** $2^6 = 64$ **b.** $\frac{1}{7^3} = \frac{1}{343}$ **c.** q^8 **d.** $\frac{-1}{p^{15}} = -\frac{1}{p^{15}}$ **2, *p. 820:*** **a.** $49a^2$ **b.** $-64x^3$ **c.** $-64x^3$ **3, *p. 820:*** **a.** $36a^{18}$ **b.** $q^{16}r^{20}$ **c.** $-2a^3b^{21}$ **d.** $\frac{49}{a^2c^{10}}$ **4, *p. 821:*** **a.** $\frac{y^2}{9}$ **b.** $\frac{u^{10}}{v^{10}}$ **c.** $\frac{y^2}{9}$ **d.** $\frac{100a^{10}}{9b^4c^2}$ **e.** $125x^3y^6$ **5, *p. 822:*** **a.** $\frac{a^2}{25}$ **b.** $\frac{v}{4u}$ **c.** $\frac{b^6}{a^{10}}$ **6, *p. 823:*** 253.9 **7, *p. 823:*** 0.0000000043 **8, *p. 824:*** 8×10^{12} **9, *p. 824:*** 7.1×10^{-11} **10, *p. 824:*** **a.** 2.464×10^2 **b.** 4×10^{11} **11, *p. 825:*** 9×10^{14} **12, *p. 826:*** 1×10^{-3}, or 0.001 **13, *p. 826:*** 2.5E⁻17 (Answers may vary.) **14, *p. 827:*** 7.3E⁻10 (Answers may vary.) **15, *p. 827:*** 4.6×10^8, or 460,000,000

Exercises 15.2, pp. 828–832

1. factors **3.** raise each factor to that power **5.** raise the reciprocal of the quotient **7.** left **9.** 2^8, or 256 **11.** 5^4, or 625 **13.** 10^{10}, or 10,000,000,000 **15.** $\frac{1}{4^4}$, or $\frac{1}{256}$ **17.** x^{24} **19.** y^8 **21.** $\frac{1}{x^6}$ **23.** n^4 **25.** $64x^3$ **27.** $64y^2$ **29.** $-64n^{15}$ **31.** $64y^8$ **33.** $\frac{1}{9a^2}$ **35.** $\frac{1}{p^7q^7}$ **37.** $r^{12}t^6$ **39.** $4p^{10}q^2$ **41.** $-2m^{12}n^{24}$ **43.** $-\frac{64m^{15}}{n^{30}}$ **45.** $\frac{1}{a^{12}b^8}$ **47.** $\frac{16y^6}{x^4}$ **49.** $\frac{125}{b^3}$ **51.** $\frac{c^2}{b^2}$ **53.** $-\frac{a^7}{b^7}$ **55.** $\frac{a^6}{27}$ **57.** $-\frac{p^{15}}{q^{10}}$ **59.** $\frac{4}{a}$ **61.** $\frac{8x^{15}}{y^6}$ **63.** $\frac{1}{p^5q^5}$ **65.** $81x^4y^{12}$ **67.** $\frac{v^4}{16u^4}$ **69.** $\frac{y^4z^{16}}{16x^8}$ **71.** $\frac{t^{12}}{r^{10}}$ **73.** $\frac{b^6}{8a^{12}}$ **75.** 317,000,000 **77.** 0.000001 **79.** 6,200,000 **81.** 0.00004025 **83.** 4.2×10^8 **85.** 3.5×10^{-6} **87.** 2.17×10^{11} **89.** 7.31×10^{-9}

91.

Standard Notation	Scientific Notation (written)	Scientific Notation (displayed on a calculator)
975,000,000	9.75×10^8	9.75E8
487,000,000	4.87×10^8	4.87E8
0.0000000001652	1.652×10^{-10}	1.652E−10
0.000000067	6.7×10^{-8}	6.7E−8
0.0000000000001	1×10^{-13}	1E−13
3,281,000,000	3.281×10^9	3.281E9

93. 9×10^7 **95.** 2.075×10^{-4} **97.** 1.68×10^1 **99.** 1.25×10^{10} **101.** 3×10^2 **103.** 3×10^8 **105.** 0.0003067

107.

Standard Notation	Scientific Notation (written)	Scientific Notation (displayed on a calculator)
428,000,000,000	4.28×10^{11}	4.28E11
3,240,000	3.24×10^6	3.24E6
0.000005224	5.224×10^{-6}	5.224E−6
0.000000057	5.7×10^{-8}	5.7E−8
0.000000682	6.82×10^{-7}	6.82E−7
48,360,000	4.836×10^7	4.836E7

109. $\frac{9b^{10}}{a^4}$ **111.** $-\frac{1}{64y^3}$ **113.** $-\frac{16y^{12}}{x^8z^4}$ **115.** 7×10^{-8} **117.** The larger volume is 8 times the smaller volume. **119.** 4,000,000,000 bytes and 17,000,000,000 bytes **121.** 7×10^{-7} m **123.** 2×10^{11} cells **125.** 0.0000000000000000000000017 g **127.** 7,600,000,000 **129.** 1.6×10^{13} red blood cells **131.** **a.** 1.86×10^5 mi/sec **b.** About 8.495×10^8 sec, or about 27 yr

Section 15.3 Practices, pp. 834–838

1, *p. 834:* (a) Terms: $-10x^2$, $4x$, and 20 (b) Coefficients: -10, 4, and 20 **2, *p. 834:***

Polynomial	Monomial	Binomial	Trinomial
$5x^2 + 2x + 9$			✓
$-4x^2$	✓		
$12p - 1$		✓	

3, *p. 835:* **a.** Degree 1 **b.** Degree 2 **c.** Degree 3 **d.** Degree 0 **4, *p. 835:***

Polynomial	Constant Term	Leading Term	Leading Coefficient
$-3x^7 + 9$	9	$-3x^7$	-3
x^5	0	x^5	1
$x^4 - 7x - 1$	-1	x^4	1
$3x + 5x^3 + 20$	20	$5x^3$	5

5, *p. 836:* **a.** $9x^5 - 7x^4 + 9x^2 - 8x - 6$ **b.** $7x^5 + x^3 - 3x^2 + 8$ **6, *p. 837:*** $x^2 + 6x + 20$ **7, *p. 837:*** **a.** -1 **b.** 19 **8, *p. 837:*** 356 ft **9, *p. 838:*** 60°F

Exercises 15.3, pp. 839–842

1. monomial **3.** power **5.** leading term **7.** constant term **9.** Polynomial **11.** Not a polynomial **13.** Polynomial **15.** Not a polynomial **17.** (a) Terms: $5x^4$, $-2x^3$, and 1; (b) coefficients: 5, -2, and 1

19.

Polynomial	Monomial	Binomial	Trinomial
$5x - 1$		✓	
$-5a^2$	✓		
$-6a + 3$		✓	
$x^3 + 4x^2 + 2$			✓

21. $-4x^3 + 3x^2 - 2x + 8$; degree 3 **23.** $-3y + 2$; degree 1 **25.** $-p^4 + 3p^3 + 4p^2 - p + 10$; degree 4 **27.** $-4y^5 - y^3 - 2y + 2$; degree 5 **29.** $5a^2 - a$; degree 2 **31.** $3p^3$; degree 3

33.

Polynomial	Constant Term	Leading Term	Leading Coefficient
$-x^7 + 2$	2	$-x^7$	-1
$2x - 30$	-30	$2x$	2
$-5x + 1 + x^2$	1	x^2	1
$7x^3 - 2x - 3$	-3	$7x^3$	7

35. $10x^3 - 7x^2 + 10x + 6$ **37.** $r^3 + 3r^2 - 8r + 14$ **39.** $3x^2 - 5x - 6$ **41.** $2n^3 + 14n^2 + 20n + 10$ **43.** $0x^2$ **45.** $0x$ **47.** 11; -17 **49.** 37; 79 **51.** 13.73849; -4.74751 **53.** $5a^4 - a^3 + 2a - 4$; degree 4 **55.** For $4x - 20$: $-20, 4x, 4$; for $-7x^6 + 9$: $9, -7x^6, -7$; for $3x + 2 + x^2$: $2, x^2, 1$; for $6x - x^5 + 11$: $11, -x^5, -1$ **57.** Not a polynomial **59.** -1.89375; 83.02657 **61.** $0x^3$ and $0x^0$, or 0 **63.** A polynomial in x; degree 16 **65.** 120 ft **67.** \$50 **69.** 700 radio stations

Section 15.4 Practices, pp. 844–848

1, *p. 844:* $9x^2 + 3x - 43$ **2, *p. 844:*** $10p^2 + pq - 10q^2$ **3, *p. 845:*** $11n^2 + 2n - 3$ **4, *p. 845:*** $8p^3 + 2p^2q - 2pq^2 - 2q^3 + 25$ **5, *p. 846:*** **a.** $3r + 3s$ **b.** $-3q$ **6, *p. 847:*** $-3x^2 - 13x$ **7, *p. 847:*** $-5x^2 + 32x - 26$ **8, *p. 847:*** $-p^2 - 11pq + 17q^2$ **9, *p. 848:*** $(-0.04x + 1.22)$ million more attendees

Exercises 15.4, pp. 849–851

1. $2x^2 + 8x + 2$ **3.** $5n^3 + 9n$ **5.** $2p^2 + 3p - 1$ **7.** $11x^2 - 3xy + 2y^2$ **9.** $3p^3 + 2p^2q - 3pq^2 - 3q^3 + 5$ **11.** $10x^2 + 17x - 5$ **13.** $5x^3 + x^2 + 9x + 2$ **15.** $-3a^3 + 6a^2 + 7ab^2$ **17.** $-x^2 - 2x + 11$ **19.** $-2x^3 + 9x^2 - 13x + 11$ **21.** $x^2 - 2x - 9$ **23.** $5x^2 - 5y^2 + 4$ **25.** $-5p^2 + 7p + 6$ **27.** $t^3 - 12t^2 + 8$ **29.** $3r^3 - 17r^2s - 2$ **31.** $-x - r$ **33.** $2p - 3q - r$ **35.** $3y^2 + 3y + 4$ **37.** $m^3 - 12m + 16$ **39.** $2x^3 - 5x^2 - 10x + 9$ **41.** $9x^2 + 4x + 7$ **43.** $-2x + 12$ **45.** $8x^2y^2 - 12xy - 14$ **47.** $2m^2 - 5m - 9$ **49.** $5x^3 + 3x^2 + 2x - 12$ **51.** $5m - 8$ **53.** $-2x^2 + 6x + 3$ **55.** $5p^2 + 11p - 7$ **57.** **a.** $6.28r^2 + 6.28rh$ **b.** $12.56r^2$ or $12.56h^2$ **59.** $(-28x^3 - 61x^2 - 41x + 12{,}543)$ thousand more barrels per day **61.** **a.** $(0.8x^3 + 6.8x^2 - 20.6x + 21.9)$ millions of dollars more for the Giants **b.** \$43 million

Section 15.5 Practices, pp. 853–858

1, *p. 853:* $40x^5$ **2, *p. 853:*** $-350a^4b^5$ **3, *p. 853:*** $25x^2y^4$ **4, *p. 854:*** $70s^3 - 21s$ **5, *p. 854:*** $12m^6n^7 - 4m^4n^4 - 2m^3n^3$ **6, *p. 854:*** $-14s^5 + 34s^4 + 22s^3 + s^2$ **7, *p. 855:*** $2a^2 + a - 3$ **8, *p. 856:*** $16x^2 - 2x - 3$ **9, *p. 856:*** $14m^2 + 5mn - n^2$ **10, *p. 857:*** $n^3 - 3n^2 - 6n + 8$ **11, *p. 857:*** $8n^3 + 15n^2 + n + 6$ **12, *p. 858:*** $24x^4 + 56x^3 + 27x^2 + 60x - 7$ **13, *p. 858:*** $p^3 - 3p^2q + 3pq^2 - q^3$ **14, *p. 858:*** $P + 2Pr + Pr^2$

Exercises 15.5, pp. 859–861

1. $-24x^2$ **3.** $-9t^5$ **5.** $20x^6$ **7.** $-70x^8$ **9.** $16p^3q^3r^2$ **11.** $64x^2$ **13.** $\frac{1}{8}t^{12}$ **15.** $-350a^4$ **17.** $-24a^4b^3c$ **19.** $7x^2 - 5x$ **21.** $45t^2 + 5t^3$ **23.** $24a^5 - 42a^4$ **25.** $12x^3 - 8x^2$ **27.** $x^5 - 2x^4 + 4x^3$ **29.** $15x^3 + 25x^2 + 30x$ **31.** $-45x^3 + 27x^2 + 63x$ **33.** $6x^5 + 24x^4 - 6x^3 - 6x^2$ **35.** $28pq - 4p^3$ **37.** $-3p^4q^2 + 9pq^3$ **39.** $6a^6b^5 + 20a^3b^8$ **41.** $-6x^2 + 26x$ **43.** $8x^3 - 16x^2 + 7x$ **45.** $-23x^3 + 35x^2 - 45x$ **47.** $7x^4y - 8x^3y - 18x^2y^2$ **49.** $-45a^5b^4 + 51a^3b^6$ **51.** $y^2 + 5y + 6$ **53.** $x^2 - 8x + 15$ **55.** $a^2 - 4$ **57.** $2w^2 - w - 21$ **59.** $-10y^2 + 17y - 3$ **61.** $20p^2 - 18p + 4$ **63.** $u^2 - v^2$ **65.** $-2p^2 + 3pq - q^2$ **67.** $3a^2 - 7ab + 2b^2$ **69.** $4pq + 3p - 32q - 24$ **71.** $x^3 - 6x^2 + 10x - 3$ **73.** $2x^3 + 5x^2 - 13x + 5$ **75.** $a^3 - b^3$ **77.** $3x^3 - 6x^2 - 105x$ **79.** $12n^3 - 3n$ **81.** $8m^4n - 6m^3n + 8m^2n^2$ **83.** $40s^7$ **85.** $-36y^3 - 28y^2 + 32y$ **87.** $-24x^{11}$ **89.** $-4a^3b + 2a^2b^4$ **91.** $x^3 - 4x^2 + 5x - 2$ **93.** $(x^2 + 30x)$ mm^2 **95.** $(-3000x^2 + 55{,}000x + 1{,}500{,}000)$ dollars **97.** **a.** $(5000 + 15{,}000r + 15{,}000r^2 + 5000r^3)$ dollars **b.** $(5000r + 10{,}000r^2 + 5000r^3)$ dollars **c.** \$605

Section 15.6 Practices, pp. 863–866

1, *p. 863:* $p^2 + 20p + 100$ **2, *p. 864:*** $s^2 + 2st + t^2$ **3, *p. 864:*** $16p^2 + 40pq + 25q^2$ **4, *p. 865:*** $25x^2 - 20x + 4$ **5, *p. 865:*** $u^2 - 2uv + v^2$ **6, *p. 865:*** $4x^2 - 36xy + 81y^2$ **7, *p. 866:*** $t^2 - 100$ **8, *p. 866:*** **a.** $r^2 - s^2$ **b.** $64s^2 - 9t^2$ **9, *p. 866:*** $100 - 49k^4$ **10, *p. 866:*** $S^2 - s^2$

Exercises 15.6, pp. 867–869

1. plus **3.** negative **5.** $y^2 + 4y + 4$ **7.** $x^2 + 8x + 16$ **9.** $x^2 - 22x + 121$ **11.** $36 - 12n + n^2$ **13.** $x^2 + 2xy + y^2$ **15.** $9x^2 + 6x + 1$ **17.** $16n^2 - 40n + 25$ **19.** $81x^2 + 36x + 4$ **21.** $a^2 + a + \frac{1}{4}$ **23.** $64b^2 + 16bc + c^2$ **25.** $25x^2 - 20xy + 4y^2$ **27.** $x^2 - 6xy + 9y^2$ **29.** $16x^6 + 8x^3y^4 + y^8$ **31.** $a^2 - 1$ **33.** $16x^2 - 9$ **35.** $9y^2 - 100$ **37.** $m^2 - \frac{1}{4}$ **39.** $n^2 - 0.09$ **41.** $16a^2 - b^2$ **43.** $9x^2 - 4y^2$ **45.** $1 - 25n^2$ **47.** $x^3 - 25x$ **49.** $5n^4 + 70n^3 + 245n^2$ **51.** $n^4 - m^8$ **53.** $a^4 - b^4$ **55.** $36n^2 + 48n + 16$ **57.** $16p^2 - 81$ **59.** $64 - 16a + a^2$ **61.** $-32x^4 + 48x^3 - 18x^2$ **63.** $x^2 + y^2 - 10x - 2y + 26$ **65.** $A + \frac{AP}{50} + \frac{AP^2}{10{,}000}$ **67.** $\frac{3m^2 - 2am - 2bm - 2cm + a^2 + b^2 + c^2}{2}$ **69.** **a.** $4x$ **b.** Yes **c.** The area of the square is y^2 larger than the area of the rectangle.

Section 15.7 Practices, pp. 870–875

1, *p. 870:* $-4n^5$ **2, *p. 870:*** $-4pr^3$ **3, *p. 871:*** $3x^2 - 2x$ **4, *p. 871:*** $-7x^5 - 5x^2 + 4$ **5, *p. 871:*** $-a^6b^3 + \frac{ab}{5} - 3$ **6, *p. 873:*** $2x + 3$ **7, *p. 874:*** $x^2 + 2x + 3 + \frac{2}{3x + 1}$ **8, *p. 874:*** $3s - 5$ **9, *p. 875:*** $n^2 - 4n - 3 + \frac{-13}{4n - 3}$ **10, *p. 875:*** **a.** The future value of the investment after 1 yr is $10(1 + r)^1$, or $10(1 + r)$. The future value of the investment after 2 yr is $10(1 + r)^2$. **b.** $10r + 10$ and $10r^2 + 20r + 10$ **c.** The future value of the investment after 2 yr is $(r + 1)$ times as great as the future value of the investment after 1 yr.

Exercises 15.7, pp. 876–878

1. coefficients **3.** remainder **5.** $2x^2$ **7.** $-4a^7$ **9.** $\frac{4x}{3}$ **11.** $4q^2$ **13.** $3u^2v^2$ **15.** $-\frac{15ab^2}{7}$ **17.** $-\frac{3u^3v^2}{2}$ **19.** $3n + 5$

21. $2b^3 - 1$ **23.** $-6a - 4$ **25.** $3 - 2x^2$ **27.** $2a^2 - 3a + 5$ **29.** $-\frac{n^2}{5} + 2n + 1$ **31.** $\frac{5a}{2} + \frac{b^2}{2}$ **33.** $-4xy^2 + 3 + y$ **35.** $2q^2 - pq^2 + \frac{3p^2}{2}$ **37.** $x - 7$ **39.** $7x - 2$ **41.** $3x - 1$ **43.** $5x + 3$ **45.** $7x + 2$ **47.** $x + \frac{5}{x + 2}$ **49.** $2x + 1$ **51.** $2x - 3 + \frac{-2}{4x + 3}$ **53.** $x^2 - 6x + 5$ **55.** $2x^2 - x - 3 + \frac{7}{3x - 4}$ **57.** $5x + 20 + \frac{78}{x - 4}$ **59.** $2x^2 + 3x + 4 + \frac{15}{2x - 3}$ **61.** $x^2 - 3x + 9$ **63.** $-7r$ **65.** $-\frac{2}{3}ab$ **67.** $-3n^3 + 8n + \frac{1}{3}$ **69.** $2x + 1$ **71.** $x^2 - 1 + \frac{3}{4x + 1}$ **73.** $-4m^2 + 9$ **75.** **a.** 1100 words **b.** $\left(\frac{9x^2 - 340x + 3600}{x}\right)$, or $\left(9x - 340 + \frac{3600}{x}\right)$ words per point **c.** 110 words per point **77.** **a.** $\frac{d}{r} = t$, or $t = \frac{d}{r}$ **b.** It takes $(t^2 - 7t + 14)$ hr. **79.** $(x^2 + 16x + 150)$ dollars per person

Chapter 15 Review Exercises, pp. 884–887

1. is not; Possible answer: the product rule of exponents says to add the exponents when multiplying powers of the same base **2.** is; Possible answer: the quotient rule says to subtract the exponent in the denominator from the exponent in the numerator **3.** cannot; Possible answer: a simplified expression does not have any negative exponents **4.** is not; Possible answer: 0.24 is less than 1 **5.** is not; Possible answer: the power of the variable is negative **6.** does; Possible answer: it can be written as a coefficient of x^0. **7.** $-x^3$ **8.** -1 **9.** n^{11} **10.** x^7 **11.** n^3 **12.** p^3 **13.** y^7 **14.** a^3b^3 **15.** y **16.** n^2 **17.** $\frac{1}{5x}$ **18.** $-\frac{3}{n^2}$ **19.** $\frac{v^4}{64}$ **20.** y^4 **21.** $\frac{1}{x}$ **22.** $\frac{y^3}{5}$ **23.** a^{10} **24.** $\frac{1}{t^6}$ **25.** $\frac{1}{x^2y}$ **26.** x^2y **27.** $10^8 = 100{,}000{,}000$ **28.** $-x^9$ **29.** $4x^6$ **30.** $-64m^{15}n^3$ **31.** $\frac{3}{x^{12}}$ **32.** $\frac{b^8}{a^6}$ **33.** $\frac{x^4}{81}$ **34.** $\frac{a^2}{b^6}$ **35.** $\frac{y^6}{x^6}$ **36.** $x^{10}y^5$ **37.** $\frac{16a^6}{b^8c^2}$ **38.** $\frac{v^4}{49u^{10}w^2}$ **39.** 37,000,000,000 **40.** 1,630,000,000 **41.** 0.00005022 **42.** 0.00000000006 **43.** 1.2×10^{12} **44.** 4.27×10^8 **45.** 4×10^{-14} **46.** 5.6×10^{-7} **47.** 5.88×10^9 **48.** 6.3×10^3 **49.** 6×10^6 **50.** 6×10^{-10} **51.** Polynomial **52.** Not a polynomial **53.** Trinomial **54.** Binomial **55.** $-3y^3 + y^2 + 8y - 1$; degree 3, leading term: $-3y^3$, leading coefficient: -3 **56.** $n^4 - 7n^3 - 6n^2 + n$; degree 4, leading term: n^4, leading coefficient: 1 **57.** $-x^3 + x^2 + 2x + 13$ **58.** $3n^3 + 4n^2 - 6n + 4$ **59.** 12; 0 **60.** 0; -16 **61.** $x^2 - x + 13$ **62.** $-2y^3 - y^2 - y - 5$ **63.** $a^2 - 2ab - 3b^2$ **64.** $5s^3t + s^2t + 9s^2 - 6st - 3t^2$ **65.** $2x^2 - 8x - 8$ **66.** $-n^3 + 3n^2 + n$ **67.** $5y^4 - 5y^3 + 2y^2 - 6y - 3$ **68.** $2x^3 + 8x^2 - 12x + 3$ **69.** $4t^2 + 4t$ **70.** $-2x - y$ **71.** $2y^2 - 5y + 2$ **72.** $-5x^2 + 9$ **73.** $-6x^5$ **74.** $-144a^3b^5$ **75.** $8x^2y^2 - 10xy^3$ **76.** $-5x^4 + 15x^3 - 5x^2$ **77.** $n^2 + 10n + 21$ **78.** $3x^2 + 9x - 54$ **79.** $8x^2 - 6x + 1$ **80.** $9a^2 + 3ab - 2b^2$ **81.** $2x^4 + 6x^3 - 5x^2 - 13x + 6$ **82.** $y^3 - 9y^2 + 15y - 2$ **83.** $-6y^2 + 13y$ **84.** $-x^3 + 6x^2 - 6x$ **85.** $a^2 - 2a + 1$ **86.** $s^2 + 8s + 16$ **87.** $4x^2 + 20x + 25$ **88.** $9 - 24t + 16t^2$ **89.** $25a^2 - 20ab + 4b^2$ **90.** $u^4 + 2u^2v^2 + v^4$ **91.** $m^2 - 16$ **92.** $36 - n^2$ **93.** $49n^2 - 1$ **94.** $4x^2 - y^2$ **95.** $16a^2 - 9b^2$ **96.** $x^3 - 100x$ **97.** $-48t^4 + 120t^3 - 75t^2$ **98.** $p^4 - 2p^2q^2 + q^4$ **99.** $3x^2$ **100.** $-2a^2b^3c$ **101.** $6x^2 - 2$ **102.** $5x^3 + 3x^2 - 2x - 1$ **103.** $3x - 7$ **104.** $x^2 - 2x - 1 + \frac{12}{2x - 1}$ **105.** 1.39×10^{10} yr **106.** 6,240,000,000,000,000,000 eV **107.** 3×10^{-5} m **108.** 3,700,000 mi^2 **109.** There will be 36 handshakes. **110.** The object is 480.4 m above the ground. **111.** 77°F **112.** 1700 two-year colleges **113.** **a.** $(3b^2 - 16b)$ m^2 **b.** 3115 m^2 **114.** **a.** $(13x + 280)$ dollars per person **b.** \$300 **115.** **a.** lwh in^3 **b.** $12^3\,lwh$ in^3, or $1728\,lwh$ in^3 **116.** $V = \pi r^3$

Chapter 15 Posttest, p. 888

1. x^7 **2.** n^6 **3.** $\frac{7}{a}$ **4.** $-27x^6y^3$ **5.** $\frac{x^8}{y^{12}}$ **6.** $\frac{y^3}{27x^6}$ **7.** **a.** $-x^3, 2x^2, 9x, -1$ **b.** $-1, 2, 9$, and -1 **c.** 3 **d.** -1 **8.** $2y^2 - y + 5$ **9.** $-x^2 + x - 9$ **10.** $3x^2y^2 - 4x^2$ **11.** $10m^3n^3 - 20m^2n^3 + 2m^2n^4$ **12.** $y^4 - 3y^3 + 2y^2 + 4y - 4$ **13.** $6x^2 + 19x - 7$ **14.** $49 - 4n^2$ **15.** $4m^2 - 12m + 9$ **16.** $-4s^2 - 5s + 9$ **17.** $t^2 - t - 1 + \frac{4}{3t - 2}$ **18.** 1×10^{-7} m **19.** First house: $(1500x + 140{,}000)$ dollars; second house: $(800x + 90{,}000)$ dollars **20.** The account balance is \$1060.90.

Chapter 15 Cumulative Review Exercises, pp. 889–890

1. 210,000 **2.** $2^2 \cdot 3 \cdot 5$ **3.** $3\frac{1}{5}$ **4.** $x = 2.88$ **5.** 7.36 **6.** 61 **7.** $2y + 18$ **8.** $m = \frac{y - b}{x}$ **9.** $x \le -5$ **10.** Slope: $\frac{2}{3}$; y-intercept: $(0, -2)$

11. 

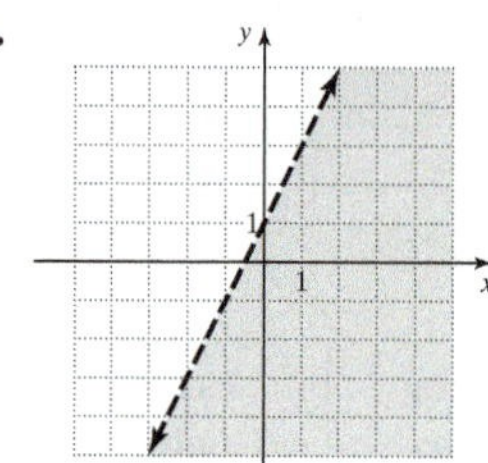

12. $(-1, -2)$

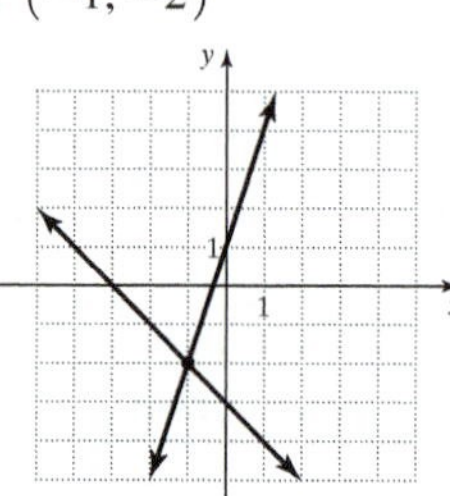

13. $(2, -2)$ **14.** $m^2 - 16m + 16$ **15.** **a.** $x + 4, x + 8$, and $x + 12$ **b.** Yes; $1980 - 1972 = 8$, which is a multiple of 4. **16.** **a.** $0.20b$ **b.** $c = 1.20b$ **17.** **a.** $C = 125d + 300$

b.

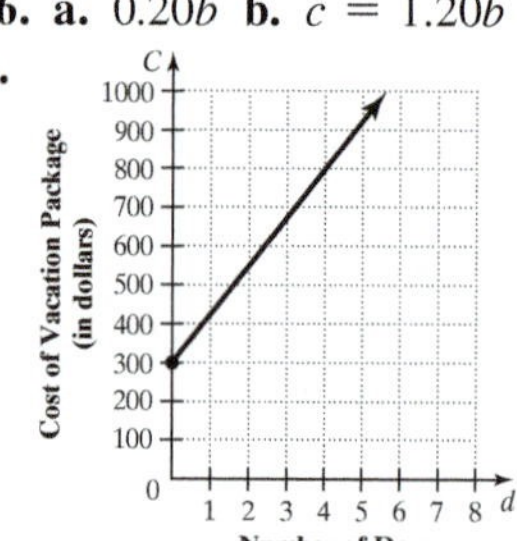

c. The C-intercept represents the cost of the vacation package with no hotel stay. **18.** **a.** $c = 1000m + 1500$ (small one-bedroom) $c = 900m + 1800$ (large studio) **b.** $m = 3$ and $c = 4500$ **c.** The cost (\$4500) is the same for renting the apartments for 3 mo. **19.** 9×10^{13} kg $\cdot$ m^2/sec^2 **20.** $(24\pi x^2 + 72\pi x + 54\pi)$ in^2

CHAPTER 16

Chapter 16 Pretest, p. 892

1. $18a$ **2.** $4p(q + 4)$ **3.** $5xy(2x - x^2y^2 + y)$ **4.** $(x + 2)(3x + 2)$ **5.** $(n - 3)(n - 8)$ **6.** $(a - 3)(a + 7)$ **7.** $3y(y - 1)(y - 3)$, or $3y(1 - y)(3 - y)$ **8.** $(5a - 4b)(a + 2b)$ **9.** $-2(3n + 1)(2n - 7)$ **10.** $(2x - 7)^2$ **11.** $(5n + 3)(5n - 3)$ **12.** $y(x + 2y)(x - 2y)$ **13.** $(y^3 - 4)(y^3 - 5)$ **14.** 0, 6 **15.** $\frac{2}{3}, -1$ **16.** $3, -5$ **17.** $h = \frac{A - 2lw}{2l + 2w}$ **18.** $-(16t + 1)(t - 4)$ ft **19.** $(S + 15)(S - 15)$ ft^2 **20.** The length of the screen is 32 in. and the height is 24 in.

Section 16.1 Practices, pp. 894–898

1, *p. 894:* 24 **2, *p. 894:*** a **3, *p. 895:*** $6xy^2$ **4, *p. 895:*** $2y^2(5 + 4y^3)$ **5, *p. 896:*** $7a(3ab - 2)$ **6, *p. 896:*** $2ab^2(4a - 3b)$ **7, *p. 896:*** $12(2a^2 - 4a + 1)$ **8, *p. 896:*** $a = \frac{s^2}{b + c}$ **9, *p. 897:*** $(y - 3)(4 + y)$, or $(y - 3)(y + 4)$ **10, *p. 897:*** $(x - 1)(3y - 2)$ **11, *p. 897:*** $(4 - 3x)(1 + 2x)$

12, ***p. 898:*** $(b - 5)(a + 4)$ **13,** ***p. 898:*** $(y - z)(5 - y)$
14, ***p. 898:*** $t(v_0 + \frac{1}{2}at)$

Exercises 16.1, pp. 899–901

1. factoring **3.** greatest common factor (GCF) **5.** 27 **7.** x^3 **9.** $4b$ **11.** $4y^3$ **13.** $3a^2b^2$ **15.** $3x - 1$ **17.** $x(x + 7)$ **19.** $3(x + 2)$ **21.** $8(3x^2 + 1)$ **23.** $9(3m - n)$ **25.** $x(7x - 2)$ **27.** $b^2(5 - 6b)$ **29.** $5x(2x^2 - 3)$ **31.** $ab(ab + 1)$ **33.** $xy(6y + 7x)$ **35.** $9pq(3q - 2p)$ **37.** $2x^3y(1 + 6y^3)$ **39.** $3(c^3 + 2c^2 + 4)$ **41.** $b^2(9b^2 - 3b + 1)$ **43.** $2m^2(m^2 + 5m - 3)$ **45.** $b^2(5b^3 - 3b + 2)$ **47.** $5x(3x^3 - 2x^2 - 5)$ **49.** $4ab(a + 2ab - 3)$ **51.** $3cd(3cd + 4c^2 + d^2)$ **53.** $(x - 1)(x + 3)$ **55.** $(a - 1)(5a - 3)$ **57.** $(s + 7)(r - 2)$ **59.** $(x - y)(a - b)$ **61.** $(y + 2)(3x - 1)$ **63.** $(b - 1)(b - 5)$ **65.** $(y - 1)(y + 5)$ **67.** $(t - 3)(1 + t)$ **69.** $(b - 7)(9a - 2)$ **71.** $(r + 3)(s + t)$ **73.** $(x + 6)(y - 4)$ **75.** $(5x - 3z)(3y + 4z)$ **77.** $(z + 4)(2x + 5y)$ **79.** $P = \frac{TM}{C + L}$ **81.** $l = \frac{S - 2wh}{2w + 2h}$ **83.** $8p(2p^2 + 3)$ **85.** $12rs(4s - 5r)$ **87.** $6(7j^2 - 1)$ **89.** $(s - 3)(t - 7)$ **91.** $(y - 4)(3x - 5)$ **93.** $4a^2b$ **95.** $m(v_2 - v_1)$ **97.** $0.5n(n - 1)$ **99.** $\frac{1}{2}n(n - 3)$ **101.** $n = \frac{P - D}{C + T}$ **103.** $50(14n + 18s - 5e)$ dollars

Section 16.2 Practices, pp. 904–910

1, ***p. 904:*** $(x + 1)(x + 4)$, or $(x + 4)(x + 1)$
2, ***p. 905:*** $(y - 4)(y - 5)$, or $(y - 5)(y - 4)$
3, ***p. 905:*** Prime polynomial **4,** ***p. 906:*** $(y - 4)(y - 8)$
5, ***p. 906:*** $(p - q)(p - 3q)$, or $(p - 3q)(p - q)$
6, ***p. 907:*** $(x - 2)(x + 3)$ **7,** ***p. 907:*** $(x + 2)(x - 23)$
8, ***p. 908:*** $(y - 4)(y + 6)$ **9,** ***p. 908:*** $(a + 3b)(a - 8b)$
10, ***p. 909:*** $y(y + 1)(y - 10)$ **11,** ***p. 909:*** $8x(x - 1)(x + 2)$
12, ***p. 909:*** $-(x - 1)(x + 11)$ **13,** ***p. 910:*** $-16(t + 1)(t - 3)$

Exercises 16.2, pp. 911–913

1. not factorable **3.** opposite signs **5.** f **7.** e **9.** b **11.** $x - 5$ **13.** $x + 4$ **15.** $x - 1$ **17.** $(x + 2)(x + 4)$ **19.** $(x - 1)(x + 6)$ **21.** Prime polynomial **23.** $(x - 2)(x + 4)$ **25.** $(x - 1)(x - 3)$ **27.** $(y - 4)(y - 8)$ **29.** $(t + 1)(t - 5)$ **31.** Prime polynomial **33.** $(y - 4)(y - 5)$ **35.** $(y + 3)(y - 16)$ **37.** $(b + 4)(b + 7)$ **39.** $(x - 7)(x - 7)$ **41.** $-(y + 5)(y - 10)$ **43.** $(x - 8)(x - 8)$ **45.** $(x - 2)(x - 8)$ **47.** $(w - 3)(w - 27)$ **49.** $(p - q)(p - 7q)$ **51.** $(p + q)(p - 5q)$ **53.** $(m - 5n)(m - 7n)$ **55.** $(x + y)(x + 8y)$ **57.** $5(x + 2)(x - 3)$ **59.** $2(x - 2)(x + 7)$ **61.** $6(t - 1)(t - 2)$ **63.** $3(x + 2)(x + 4)$ **65.** $y(y - 2)(y + 5)$ **67.** $a(a + 3)(a + 5)$ **69.** $t^2(t - 2)(t - 12)$ **71.** $4a(a - 1)(a - 2)$ **73.** $2x(x + 3)(x + 5)$ **75.** $4x(x - 3)(x - 4)$ **77.** $2s(s - 4)(s + 7)$ **79.** $2c^2(c - 5)(c + 7)$ **81.** $ax(x - 2)(x - 16)$ **83.** Prime polynomial **85.** $5x^2(x - 5)(x + 2)$ **87.** $-(w + 4)(w - 10)$ **89.** $6(m + 2)(m - 3)$ **91.** $(t - 5)(t - 12)$ **93.** b **95.** $(p + q)(p + q) = 1$, or $(p + q)^2 = 1$ **97.** $16(t - 2)(t + 5)$ **99. a.** $(6x^2 + 24x + 18)$ in^2 **b.** $6(x + 1)(x + 3)$ **101.** $\frac{57}{400{,}000}(x - 3000)(x + 1000)$

Section 16.3 Practices, pp. 916–920

1, ***p. 916:*** $(5x + 4)(x + 2)$ **2,** ***p. 916:*** $(6x - 7)(x - 3)$
3, ***p. 917:*** $(7y - 2)(y + 7)$ **4,** ***p. 918:*** $(2x - 5)(x + 2)$
5, ***p. 918:*** $3x(3x + 1)(2x - 3)$ **6,** ***p. 919:*** $3(6c - 5d)(2c + d)$
7, ***p. 919:*** $(2x + 1)(x - 4)$ **8,** ***p. 920:*** $x(2x - 5)(2x - 7)$

Exercises 16.3, pp. 921–923

1. e **3.** b **5.** c **7.** $3x + 1$ **9.** $x - 3$ **11.** $x - 3$ **13.** $(3x + 5)(x + 1)$ **15.** $(2y - 1)(y - 5)$ **17.** $(3x + 2)(x + 4)$ **19.** Prime polynomial **21.** $(6y + 5)(y - 1)$ **23.** $(2y - 7)(y - 2)$ **25.** $(3a + 2)(3a - 8)$ **27.** $(4x - 1)(x - 3)$ **29.** $(3y + 2)(4y + 3)$ **31.** $(2m - 3)(m - 7)$ **33.** $-(3a - 1)(2a + 3)$ **35.** $(8y - 11)(y + 2)$ **37.** Prime polynomial **39.** $(8a + 1)(a + 8)$ **41.** $(3x - 1)(2x + 9)$ **43.** $(4y - 3)(2y - 5)$ **45.** $2(7y - 5)(y - 2)$ **47.** $4(7a - 1)(a + 1)$ **49.** $-2(3b + 1)(b - 7)$ **51.** $2y(3y + 2)(2y + 7)$ **53.** $2a^2(7a - 5)(a - 2)$ **55.** $xy(2x + 3)(x + 5)$ **57.** $2ab(3b - 1)(b - 7)$ **59.** $(5c - d)(4c - d)$ **61.** $(2x + y)(x - 3y)$ **63.** $(4a - b)(2a - b)$ **65.** $3(3x + 2y)(2x - y)$ **67.** $2(4c - 5d)(2c - 3d)$ **69.** $3(3u + v)(3u + v)$ **71.** $3x(7x - 3y)(2x + 3y)$ **73.** $-5x^2y(3x + y)(2x - 3y)$ **75.** $a(5x + 2y)(x - 6y)$ **77.** $(m - 1)(3m - 2)$ **79.** $(4x - 3)(2x + 1)$ **81.** $2x(x + 3)(7x + 1)$ **83.** $(2m - 3n)(4m - 3n)$ **85.** $(r - 1)(7r - 2)$ **87.** d **89.** $-(5t - 4)(t + 5)$ **91. a.** $(2x^2 + 25x + 72)$ ft^2 **b.** $(2x + 9)(x + 8)$ ft^2 **93.** $(2n - 5)(2n - 1)$; since the difference of the factors is $(2n - 1) - (2n - 5) = 2n - 1 - 2n + 5 = 4$, the factors represent two integers that differ by 4 no matter what integer n represents.

Section 16.4 Practices, pp. 924–928

1, ***p. 924:*** **a.** The trinomial is a perfect square. **b.** The trinomial is not a perfect square. **c.** The trinomial is a perfect square. **d.** The trinomial is not a perfect square. **e.** The trinomial is a perfect square.
2, ***p. 925:*** $(n + 10)^2$ **3,** ***p. 926:*** $(t - 2)^2$ **4,** ***p. 926:*** $(5c - 4d)^2$
5, ***p. 926:*** $(x^2 + 4)^2$ **6,** ***p. 927:*** **a.** The binomial is a difference of squares. **b.** The binomial is not a difference of squares. **c.** The binomial is not a difference of squares. **d.** The binomial is a difference of squares.
7, ***p. 927:*** $(y + 11)(y - 11)$ **8,** ***p. 928:*** $(3x + 5y)(3x - 5y)$
9, ***p. 928:*** $(8x^4 + 9y)(8x^4 - 9y)$ **10,** ***p. 928:*** $16(4 + t)(4 - t)$

Exercises 16.4, pp. 929–931

1. perfect square trinomial **3.** difference of squares **5.** Perfect square trinomial **7.** Neither **9.** Perfect square trinomial **11.** Difference of squares **13.** Difference of squares **15.** Perfect square trinomial **17.** Neither **19.** Perfect square trinomial **21.** Neither **23.** $(x - 6)^2$ **25.** $(y + 10)^2$ **27.** $(a - 2)^2$ **29.** Prime polynomial **31.** $(2a - 9)^2$ **33.** $(7x + 2)^2$ **35.** $(6 - 5x)^2$ **37.** $(m + 13n)^2$ **39.** $(15a - b)^2$ **41.** $(y^2 + 1)^2$ **43.** $6(x + 1)^2$ **45.** $3m(3m - 2)^2$ **47.** $4s^2t(t + 10)^2$ **49.** $(m + 8)(m - 8)$ **51.** $(y + 9)(y - 9)$ **53.** $(12 + x)(12 - x)$ **55.** $(10m + 9)(10m - 9)$ **57.** Prime polynomial **59.** $(1 + 3x)(1 - 3x)$ **61.** $(x + 2y)(x - 2y)$ **63.** $(10x + 3y)(10x - 3y)$ **65.** $3k(k + 7)(k - 7)$ **67.** $4y^2(y + 3)(y - 3)$ **69.** $3x^2y(3 + y)(3 - y)$ **71.** $2(ab + 7)(ab - 7)$ **73.** $(16 + r^2)(4 + r)(4 - r)$ **75.** $5(x^2 + 4y^2)(x + 2y)(x - 2y)$ **77.** $(c - d)(x + 2)(x - 2)$ **79.** $(x - y)(4 + a)(4 - a)$ **81.** $(3c + 8d)^2$ **83.** $(a + 15b^2)(a - 15b^2)$ **85.** $6(3a^2b - 1)^2$ **87.** $(9 + w^2)(3 - w)(3 + w)$ **89.** $(6u + 5)^2$ **91.** Difference of squares **93.** $4\pi(r_1 + r_2)(r_1 - r_2)$ **95.** $4x(1 - x)^2$ **97.** $k(v_2 + v_1)(v_2 - v_1)$

Section 16.5 Practices, pp. 932–937

1, ***p. 932:*** **a.** Linear **b.** Quadratic **2,** ***p. 933:*** $\frac{1}{3}, -5$ **3,** ***p. 934:*** $0, -6$
4, ***p. 934:*** $-\frac{1}{4}, 3$ **5,** ***p. 935:*** $1, -5$ **6,** ***p. 936:*** The dimensions of the frame should be 8 in. by 10 in. **7,** ***p. 937:*** The scooter going north has traveled 12 mi.

Exercises 16.5, pp. 938–940

1. quadratic equation **3.** in standard form **5.** Quadratic **7.** Linear **9.** Quadratic **11.** $-3, 4$ **13.** 1 **15.** $0, -\frac{5}{3}$ **17.** $-\frac{1}{2}, 5$ **19.** $-\frac{3}{2}, \frac{3}{2}$ **21.** $0, \frac{2}{3}$ **23.** $0, 2$ **25.** $0, \frac{1}{5}$ **27.** $-2, -3$ **29.** $7, -8$ **31.** $-\frac{1}{2}, 3$ **33.** $\frac{2}{3}, -\frac{1}{2}$ **35.** $\frac{1}{6}$ **37.** $-11, 11$ **39.** $\frac{3}{2}$ **41.** $\frac{1}{2}$ **43.** $\frac{1}{3}, -2$ **45.** $-2, 3$ **47.** $3, 4$ **49.** $2, -4$ **51.** $-\frac{1}{3}, -1$ **53.** $-\frac{1}{2}, -1$ **55.** $0, -5$ **57.** $-\frac{1}{2}, \frac{1}{2}$ **59.** $-\frac{1}{2}, \frac{1}{2}$ **61.** -3 **63.** $-3, 4$ **65.** $-2, 3$ **67.** $4, -5$ **69.** $-\frac{1}{3}, -4$ **71.** $2, 4$ **73.** $1, 2$ **75.** $\frac{1}{4}, -3$ **77.** $\frac{1}{2}, -3$ **79.** $\frac{1}{5}, -4$ **81.** 8 **83.** $0, \frac{5}{3}$ **85.** $0, -\frac{1}{2}$ **87.** There were 15 teams in the league. **89.** One car traveled 6 mi and the other traveled 8 mi. **91.** 200 ft **93.** The average annual rate of return is 25%. **95. a.** $\frac{2}{105}(105 - x)(105 + x)$ **b.** It spans from -105 yd to 105 yd, and so is 210 yd wide.

Chapter 16 Review Exercises, pp. 944–945

1. is; Possible answer: it cannot be factored **2.** is not; Possible answer: $(2x^2 - x - 3)$ can be factored further (it is not a prime polynomial) **3.** is; Possible answer: it can be expressed as $(x - 2y)^2$ **4.** can; Possible answer: the binomial is a difference of squares **5.** 12 **6.** $3m^2$ **7.** $3(x - 2y)$ **8.** $2pq(8p^2q + 9p - 2q)$ **9.** $(n - 1)(1 + n)$ **10.** $(x - 5)(b - 2)$ **11.** $r = \frac{d}{t_1 + t_2}$ **12.** $x = \frac{c - y}{a - b}$ **13.** Prime polynomial **14.** Prime polynomial **15.** $(y + 6)(y + 7)$ **16.** $(m - 2n)(m - 5n)$ **17.** $-2(x - 2)(x + 6)$ **18.** $3x(x + y)(x - 5y)$ **19.** $(3x - 1)(x + 2)$ **20.** $(5n + 3)(n + 2)$ **21.** Prime polynomial **22.** $(3x + 4)(2x - 3)$ **23.** $(2a - 7b)(a + 5b)$ **24.** $-(2a - 3)(2a - 5)$ **25.** $3y(3y - 1)(y + 7)$ **26.** $q(2p + q)(p - 2q)$ **27.** $(b - 3)^2$ **28.** $(8 + x)(8 - x)$ **29.** $(5y - 2)^2$ **30.** $(3a + 4b)^2$ **31.** $(9p + 10q)(9p - 10q)$ **32.** $(2x^4 - 7)^2$ **33.** $3(4x^2 + y^2)(2x + y)(2x - y)$ **34.** $(x - 1)(x + 3)(x - 3)$ **35.** $-2, 1$ **36.** $0, 4$ **37.** $0, -6$ **38.** $-\frac{1}{2}$ **39.** $2, 8$ **40.** $-\frac{2}{3}, 1$ **41.** $-\frac{5}{2}, 1$ **42.** $3, -4$ **43.** $aL(t_2 - t_1)$ **44.** $(t_2 - t_1)[v - 16(t_2 + t_1)]$ **45.** The distance between the two intersections is 2500 ft. **46.** The length of the horizontal diagonal of the kite is 32 in. **47.** The rocket will reach a height of 18 ft above the launch in $\frac{1}{4}$ sec and $\frac{9}{2}$ sec, or in 0.25 sec and 4.5 sec. **48.** $h = \frac{2A}{b + B}$

Chapter 16 Posttest, p. 946

1. $3x^2$ **2.** $2y(x - 7)$ **3.** $2p(2p - 3q)(2p - q)$ **4.** $(a - b)(x - y)$ **5.** $(n + 3)(n - 16)$ **6.** $(x + 2)(x - 4)$ **7.** $-5x(x + 1)(x - 4)$ **8.** $(4x - 3y)(x + 4y)$ **9.** $-3(2x - 3)^2$ **10.** $(3x + 5y)^2$ **11.** $(11 + 2x)(11 - 2x)$ **12.** $(pq + 1)(pq - 1)$ **13.** $(y + 2)^2(y - 2)^2$ **14.** $-8, 1$ **15.** $\frac{1}{3}, -2$ **16.** $-\frac{3}{2}, 2$ **17.** $mg(y_2 - y_1)$ **18.** 8 m **19.** $2x(2x + 55)$ ft^2 **20.** The weight reaches the ground in $\frac{3}{4}$, or 0.75, sec.

Chapter 16 Cumulative Review Exercises, pp. 947–948

1. 320 **2.** $1\frac{2}{3}$ **3.** 0.064 **4.** \$0.08/min **5.** 12.5% **6.** 8

7.

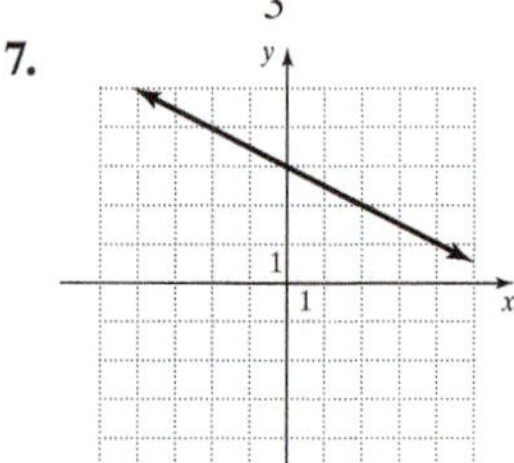

8.

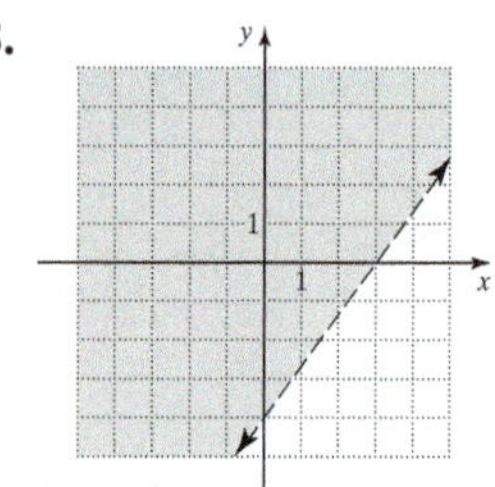

9. No solution **10.** $(-1, 7)$ **11.** n **12.** $21x^2 + x - 2$ **13.** $(4x - 3)^2$ **14.** $-3, \frac{1}{4}$ **15.** $-\$0.15$ **16.** The charge is $(f + 8c)$ dollars. **17.** The company must sell 50 shirts per day in order to break even. **18.** Positive **19.** 1.5×10^8 km **20.** $2\pi r(r + h)$

CHAPTER 17

Chapter 17 Pretest, p. 950

1. The expression is undefined when $x = -6$. **2.** $\frac{n-2}{3} = \frac{(n-2) \cdot n}{3 \cdot n} = \frac{n^2 - 2n}{3n}$ **3.** $\frac{4x}{y^2}$ **4.** $4a$ **5.** $-\frac{w}{w + 6}$ **6.** $\frac{10}{n^2}$ **7.** $\frac{6(a - 2)}{9a(a - 2)}$ and $\frac{a}{9a(a - 2)}$ **8.** $\frac{5y - 2}{y + 1}$ **9.** $\frac{15x + 2}{12x^2}$ **10.** $\frac{2c + 12}{(c - 3)(c + 3)}$ **11.** $\frac{3x - 1}{(x - 1)^2(x + 1)}$ **12.** $\frac{4a^2}{3n^2}$ **13.** $\frac{y - 4}{5y(y + 4)}$ **14.** $\frac{1}{x + 4}$ **15.** 5 **16.** $-\frac{2}{3}$ **17.** \$10/hr **18.** $\frac{2x + 1500}{x}$ dollars **19.** The average speed during the first part of the trip was 65 mph and the average speed during the second part was 55 mph. **20.** It takes the photocopier $2\frac{1}{2}$, or 2.5, min to make 30 copies.

Section 17.1 Practices, pp. 952–957

1, *p. 952:* a. The expression is undefined when n is equal to 4. **b.** The expression is undefined when n is equal to -3 or 3. **2, *p. 953:* a.** The expressions are equivalent. **b.** The expressions are equivalent. **3, *p. 954:* a.** $\frac{2n^2}{m}$ **b.** $-\frac{3x^2}{2}$ **4, *p. 955:* a.** $\frac{y - 4}{2y + 3}$ **b.** The expression cannot be simplified. **c.** $\frac{1}{3}$ **5, *p. 955:* a.** $\frac{t - x}{z - 3g}$ **b.** $\frac{4(n - 1)}{n + 2}$ **c.** $\frac{y - 1}{y + 3}$ **6, *p. 956:* a.** -1 **b.** -3 **c.** $-\frac{3}{n + 5}$ **d.** $-\frac{3s + 2}{s + 1}$ **7, *p. 957:* a.** $\frac{r}{2}$ **b.** The expression is undefined when $r = 0$.

Exercises 17.1, pp. 958–960

1. rational expression **3.** equivalent **5.** $x = 0$ **7.** $y = 2$ **9.** $x = -2$ **11.** $n = \frac{1}{2}$ **13.** $x = -1$ or $x = 1$ **15.** $x = -4$ or $x = 5$ **17.** The expressions are equivalent. **19.** The expressions are equivalent. **21.** The expressions are not equivalent. **23.** The expressions are equivalent. **25.** The expressions are not equivalent. **27.** $\frac{5a^3}{6}$ **29.** $\frac{1}{4x^3}$ **31.** $\frac{3t}{2s^2}$ **33.** $8a^3b^3$ **35.** $\frac{p}{5q}$ **37.** $\frac{5}{4}$ **39.** $-\frac{8x}{3}$ **41.** $x - 2$ **43.** $\frac{x + 1}{2x + 3}$ **45.** $\frac{a}{b}$ **47.** $\frac{2}{3}$ **49.** $t - 1$ **51.** -1 **53.** $-\frac{1}{2}$ **55.** $\frac{b - 4}{b + 4}$ **57.** $\frac{x + 3}{5}$ **59.** $\frac{a(a + 7)}{a - 12}$ **61.** $\frac{t + 1}{t + 2}$ **63.** $-\frac{4d + 3}{4d - 3}$ **65.** $-\frac{2s}{2s - 3}$ **67.** $\frac{2x + 1}{2x - 1}$ **69.** $\frac{2y - 1}{y(2y + 3)}$ **71.** $\frac{2ab}{2b + a}$ **73.** $\frac{m + 7n}{2(m + 6n)}$ **75.** $\frac{2y^2}{3x}$ **77.** $\frac{3(x - 1)}{x + 3}$ **79.** -1 **81.** $\frac{2x + 3}{3x}$ **83.** $p = \frac{3}{2}$ **85.** The expressions are equivalent. **87.** C in. **89. a.** $(x^2 + 7x + 10)$ ft^2 **b.** $(x^2 + 2x)$ ft^2 **c.** $\frac{x^2 + 7x + 10}{x^2 + 2x}$; $\frac{x + 5}{x}$ **d.** $\frac{13}{8}$ **91. a.** $\frac{\pi r_1^2}{\pi r_2^2 - \pi r_1^2}$; $\frac{r_1^2}{r_2^2 - r_1^2}$ **b.** $\frac{\pi r_2^2 - \pi r_1^2}{\pi r_3^2 - \pi r_2^2}$; $\frac{r_2^2 - r_1^2}{r_3^2 - r_2^2}$

Section 17.2 Practices, pp. 962–966

1, *p. 962:* a. $\frac{2n}{7m}$ **b.** $\frac{p}{15q}$ **2, *p. 963:* a.** $\frac{3}{2t}$ **b.** $\frac{2}{6g - 1}$ **3, *p. 963:* a.** $2(y + 7)(y - 2)$ **b.** $-\frac{x + 5}{x + 1}$ **4, *p. 965:* a.** $\frac{3a}{2b}$ **b.** $\frac{4p^2}{q}$ **5, *p. 965:* a.** $\frac{(x + 3)(x + 3)}{(x - 10)(x + 1)}$, or $\frac{(x + 3)^2}{(x - 10)(x + 1)}$ **b.** $\frac{1}{3}$ **c.** $\frac{y(y + 3)}{5(y + 2)(y - 1)}$ **6, *p. 966:*** $\frac{Wv^2}{gr}$

Exercises 17.2, pp. 967–969

1. $\frac{1}{4t}$ **3.** $\frac{6}{ab}$ **5.** $\frac{10}{3x^9}$ **7.** $-\frac{14}{x}$ **9.** $\frac{5}{x^3}$ **11.** $\frac{8n - 3}{n}$ **13.** $\frac{4}{5}$ **15.** $-\frac{1}{2}$ **17.** $\frac{3(x + 2y)}{4}$ **19.** $\frac{(p - 1)(p + 1)}{(p - 2)(p + 2)}$ **21.** $\frac{n + 3}{n + 1}$ **23.** $\frac{2y - 1}{2y - 5}$ **25.** $\frac{16}{x^4}$ **27.** $3x$ **29.** $\frac{1}{2}$ **31.** 2 **33.** $\frac{10}{x}$ **35.** $-\frac{3}{t^2}$ **37.** $\frac{6y^2}{x^4}$ **39.** $\frac{(c + 3)(c - 7)}{(c - 5)(c + 9)}$ **41.** $\frac{5}{12}$ **43.** $\frac{1}{2(1 - x)}$ **45.** $-p$ **47.** $\frac{x + y}{5x}$ **49.** $-\frac{1}{t + 2}$ **51.** $\frac{x + 5}{x + 6}$ **53.** $\frac{3(p + 2)(p - 4)}{(p + 5)^2}$ **55.** $-\frac{1}{r^2}$ **57.** $\frac{2}{3 - x}$ **59.** $-\frac{1}{6p^2}$ **61.** $\frac{3(y + 3)}{4}$ **63.** $\frac{3x(x - 1)}{3x + 1}$ **65.** $\frac{A - p}{pr}$ **67.** $\frac{pqB}{10{,}000}$ dollars **69.** $\frac{V^2r}{(R + r)^2}$ **71. a.** $\frac{4r}{3h}$ **b.** $\frac{2r}{h + r}$ **c.** $\frac{2(h + r)}{3h}$

Section 17.3 Practices, pp. 971–980

1, *p. 971:* a. $\frac{10}{y + 2}$ **b.** $\frac{5r}{s}$ **c.** 6 **d.** $3n - 1$ **2, *p. 972:* a.** $\frac{5}{3}$ **b.** $\frac{3t}{5}$ **c.** $-\frac{p}{3q}$ **3, *p. 972:* a.** 2 **b.** $2x$ **c.** $-\frac{3}{x - 5}$ **4, *p. 974:* a.** LCD $= 20y$ **b.** LCD $= 6t^2$ **c.** LCD $= 30x^2y^3$ **5, *p. 975:* a.** LCD $= 2n(n + 1)^2$ **b.** LCD $= (p + 2)(p - 1)(p + 5)$ **c.** LCD $= (s + 2t)^2(s - 2t)^2$ **6, *p. 976:* a.** $\frac{4}{14p^3}$ and $\frac{7p^2(p + 3)}{14p^3}$ **b.** $\frac{(3y - 2)(y - 3)}{(y + 3)(y - 3)^2}$ and $\frac{y(y + 3)}{(y + 3)(y - 3)^2}$ **7, *p. 977:* a.** $\frac{11}{12p}$ **b.** $\frac{3y - 2}{15y^2}$ **8, *p. 977:* a.** $\frac{9x + 6}{x(x + 3)} = \frac{3(3x + 2)}{x(x + 3)}$ **b.** $\frac{2x - 5}{x - 1}$ **9, *p. 978:* a.** $\frac{9x + 4}{4(x - 4)(x + 4)}$ **b.** $\frac{x - 9}{(x + 1)(x - 1)}$ **10, *p. 979:* a.** $\frac{-3y^2 - 12y - 3}{(y + 3)(y + 2)(y + 1)} = \frac{-3(y^2 + 4y + 1)}{(y + 3)(y + 2)(y + 1)}$ **b.** $\frac{8x^3 + 3x^2 - 20x + 5}{20x(x + 1)}$ **11, *p. 980:*** $\frac{100(C_1 - C_0)}{C_0}$

Exercises 17.3, pp. 981–984

1. $\frac{4a}{3}$ **3.** t **5.** $\frac{11}{x}$ **7.** $\frac{5}{7y}$ **9.** $\frac{3x}{y}$ **11.** $-\frac{p}{5q}$ **13.** $\frac{9}{x + 1}$ **15.** $-\frac{4}{x + 2}$ **17.** $\frac{a + 1}{a + 3}$ **19.** $\frac{5x + 1}{x - 8}$ **21.** $\frac{4x + 1}{5x + 2}$ **23.** 3

25. 4 **27.** $\frac{x-4}{x^2-4x-2}$ **29.** 1 **31.** $\frac{-x+7}{3x^2-x+2}$
33. LCD $= 15(x+2)$ **35.** LCD $= (p-3)(p+8)(p-8)$
37. LCD $= t(t+3)(t-3)$ **39.** LCD $= (t+2)(t+5)(t-5)$
41. LCD $= (3s-2)(s-3)(s+2)$ **43.** $\frac{4x}{12x^2}$ and $\frac{15}{12x^2}$
45. $\frac{35b}{14a^2b}$ and $\frac{2a(a-3)}{14a^2b}$ **47.** $\frac{8(n+1)}{n(n+1)^2}$ and $\frac{5n}{n(n+1)^2}$
49. $\frac{3n(n-1)}{4(n+1)(n-1)}$ and $\frac{8n}{4(n+1)(n-1)}$
51. $\frac{2n(n-3)}{(n+1)(n+5)(n-3)}$ and $\frac{3n(n+1)}{(n+1)(n+5)(n-3)}$
53. $\frac{13}{6x}$ **55.** $\frac{4-5x}{6x^2}$ **57.** $\frac{-2y+4x}{3x^2y^2}$, or $-\frac{2(y-2x)}{3x^2y^2}$ **59.** $\frac{2x}{(x+1)(x-1)}$
61. $\frac{p+39}{21}$ **63.** $3x-5$ **65.** $\frac{a+6}{6a^2}$ **67.** $\frac{a^2+1}{a-1}$ **69.** -1
71. $\frac{-8x-2}{x(x+1)}$, or $-\frac{2(4x+1)}{x(x+1)}$ **73.** $\frac{7x-6}{x-4}$ **75.** $\frac{11x-6}{(x-1)(x+2)}$
77. $\frac{n+20}{3(n-3)(n+5)}$ **79.** $\frac{3t-4}{(t+5)(t-5)}$ **81.** $\frac{2x^2+5x-5}{(x+1)^2(x+3)}$
83. $\frac{6t+5}{(2t+3)(t-1)}$ **85.** $\frac{12x^3+17x^2-2}{3x(x-1)(x+1)}$ **87.** $\frac{5y^2-28y+4}{(3y-1)(y-4)}$
89. $\frac{-9a^2-11a+32}{4(a+3)^2}$ **91.** $\frac{4q(p+3)}{24p^2q}$ and $\frac{15p}{24p^2q}$ **93.** $\frac{3(3x+2y)}{x^2y^2}$
95. $-\frac{3}{c(c+1)}$ **97.** $\frac{b^2-4b-13}{(b-3)^2(b+1)}$ **99.** $-\frac{5}{m+n}$ **101.** $\frac{r-1}{r+2}$
103. $\frac{2vt+gt^2}{2}$ **105.** $\frac{1000r}{(1+r)^2}$ dollars **107.** The trip took $\frac{30}{r}$ hr.
109. $\frac{3+0.1x}{x}$ dollars

Section 17.4 Practices, pp. 987–991

1, *p. 987:* **a.** $\frac{3}{5x^3}$ **b.** $\frac{8}{x+2}$ **2, *p. 988:*** $\frac{2n-1}{2n+1}$ **3, *p. 989:*** **a.** $2x^2$ **b.** $\frac{y+3}{2y^3}$ **4, *p. 990:*** **a.** $\frac{4y^2+y}{4y^2-1}$ **b.** $\frac{b-a}{10ab}$ **5, *p. 991:*** $\frac{3abc}{bc+ac+ab}$

Exercises 17.4, pp. 992–994

1. $\frac{2}{x}$ **3.** $\frac{a+1}{a-1}$ **5.** $\frac{x(3x+1)}{3x^2-1}$ **7.** $\frac{1}{3d(d+3)}$ **9.** $\frac{x-2y}{3x}$ **11.** $\frac{10-y}{5(5-y)}$
13. $\frac{x+2}{x+3}$ **15.** $\frac{3y+4}{5y+4}$ **17.** $\frac{xy^2}{16(y+1)}$ **19.** $\frac{3(x^2-2x-2)}{x^2(x+1)}$ **21.** $\frac{m+2}{3(m-1)}$
23. $\frac{4(u+2)}{u^2(u+1)}$ **25.** $\frac{4y+3}{3y(2y+1)}$ **27.** $\frac{2VR(R+1)}{2R+1}$ **29.** $\frac{9E}{I}$ **31.** $\frac{2ab}{a+b}$ mph
33. $\frac{W}{\left(1+\frac{h}{6400}\right)^2} = \frac{W}{\left(\frac{6400+h}{6400}\right)^2} = \frac{W}{\frac{(6400+h)^2}{6400^2}} = \frac{6400^2W}{(6400+h)^2}$

Section 17.5 Practices, pp. 995–1004

1, *p. 995:* $\frac{1}{2}$ **2, *p. 996:*** 8.5 **3, *p. 997:*** She would be paying \$225 less if she had a \$75,000 mortgage at the same interest rate. **4, *p. 998:*** 2, 5 **5, *p. 999:*** 2 **6, *p. 1000:*** 3 **7, *p. 1001:*** Working together, it will take both pumps $3\frac{3}{4}$ hr (or 3 hr 45 min) to fill the tank. **8, *p. 1002:*** The speed of the propeller plane was 150 mph. **9, *p. 1003;*** The speed of the plane in still air is 500 mph **10, *p. 1004:*** **a.** $x = \frac{500}{D-50}$ **b.** \$1.25 per unit

Exercises 17.5, pp. 1005–1007

1. -3 **3.** $\frac{1}{2}$ **5.** -2 **7.** 1 **9.** 4 **11.** 14 **13.** 11 **15.** $-\frac{2}{3}$ **17.** 3 **19.** 5 **21.** $-4, 4$ **23.** $8, -4$ **25.** $-\frac{6}{5}, 2$ **27.** No solution **29.** 3 **31.** 1, 2 **33.** $6, -4$ **35.** 3 **37.** 6 **39.** 2 **41.** -3 **43.** $-\frac{7}{2}$ **45.** $2, -8$ **47.** $4, -1$ **49.** $-\frac{5}{2}$ **51.** $-\frac{1}{3}, 3$ **53.** No solution **55.** 11 **57.** It would take the clerical worker 15 hr to finish the job working alone. **59.** The speed on the dry road was 60 mph. **61.** $D = \frac{P}{Lp}$ **63.** 3 fewer bags **65.** He was driving 55 mph. **67.** The boat travels 18 mph in still water. **69.** **a.** The evacuation would take $\frac{x^2+3x}{2x+3}$ hr. **b.** It would take 11 hr 12 min.

Chapter 17 Review Exercises, pp. 1013–1015

1. is not; Possible answer: the denominator is not 0 when $\frac{3}{2}$ is substituted for x **2.** is not; possible answer: the expression can be simplified by dividing out the common factor a **3.** is; possible answer: it is the product of the highest power of each factor in either denominator **4.** is; possible answer: it contains a rational expression in its denominator **5.** is not; possible answer: $a = 3$ is a solution of the original equation **6.** is; possible answer: They have the same value, for any value of the variable that makes neither denominator 0 **7.** **a.** $x = -1$ **b.** $x = 3$ and $x = -2$ **8.** **a.** Equivalent **b.** Equivalent **9.** $\frac{3}{5m}$ **10.** $\frac{5n-6}{3n+2}$ **11.** $-\frac{x+4}{x+2}$ **12.** $\frac{2x+5}{3x-1}$ **13.** $\frac{6n^2}{pm}$ **14.** $\frac{1}{2}$ **15.** $\frac{x-5}{2x+5}$ **16.** -1 **17.** $\frac{1}{2y}$ **18.** $\frac{5(7m-10)}{7(m-10)}$ **19.** $\frac{y}{5(x+6)}$ **20.** $\frac{x+7}{3x+2}$
21. $\frac{4x}{20x^2}$ and $\frac{3}{20x^2}$ **22.** $\frac{4(n+4)}{(n-1)(n+4)}$ and $\frac{n(n-1)}{(n-1)(n+4)}$
23. $\frac{x+1}{3(x+3)(x+1)}$ and $\frac{3x}{3(x+3)(x+1)}$
24. $\frac{2(x+2)}{(3x+1)(x-2)(x+2)}$ and $-\frac{3x+1}{(3x+1)(x-2)(x+2)}$
25. 2 **26.** 4 **27.** $\frac{y+4}{2y(2y-1)}$ **28.** $\frac{n^2+3n-6}{3n(n+5)}$ **29.** $\frac{8x+13}{(x-3)(x+3)}$
30. $\frac{2y+5}{(2+y)(2-y)}$ **31.** $\frac{8m-8}{(m+1)(m-3)}$, or $\frac{8(m-1)}{(m+1)(m-3)}$
32. $\frac{-x^2-3x-11}{(x+3)(x-4)}$ **33.** $\frac{x^2-9x+2}{(x+2)^2(x-4)}$ **34.** $\frac{2n^2+7n+11}{(2n-1)(n-1)(n+3)}$
35. $\frac{7}{6x}$ **36.** $\frac{y}{y+9}$ **37.** 2 **38.** $\frac{4x+1}{2x-3}$ **39.** 7 **40.** $\frac{1}{2}$ **41.** -3
42. $3, -1$ **43.** -1 **44.** 2 **45.** 33 **46.** $\frac{1}{8}$ **47.** $-2, 9$ **48.** $4, -9$
49. $\frac{0.72x+200}{x}$ dollars **50.** The total cost of the car rental is \$200.
51. $\frac{2rs}{s+r}$ **52.** It will take 30 min to fill the tub. **53.** The family drove 100 miles at 50 mph. **54.** $\frac{2x+1}{x(x+1)}$ of the job will be done in an hour.
55. $\frac{t_2}{t_1}$ **56.** $\frac{3n^2+6n+2}{n(n+1)(n+2)}$

Chapter 17 Posttest, pp. 1016–1017

1. The expression is undefined when $x = 8$.
2. $-\frac{3y-y^2}{y^2} = -\frac{y(3-y)}{y^2} = -\frac{\overset{1}{\cancel{y}}(3-y)}{\underset{y}{\cancel{y^2}}} = \frac{-(3-y)}{y} = \frac{y-3}{y}$ **3.** $\frac{5a^2}{4b}$
4. $\frac{x}{y}$ **5.** $\frac{3(b-3)}{b-7}$ **6.** $\frac{1}{x+3}$ **7.** $\frac{4(4n-1)}{4(n+8)(n-2)}$, $\frac{8(n-2)}{4(n+8)(n-2)}$, and $\frac{n(n+8)}{4(n+8)(n-2)}$ **8.** 2 **9.** $\frac{3y+1}{2y(y-4)}$ **10.** $\frac{4d+14}{(d-3)(d+2)} = \frac{2(2d+7)}{(d-3)(d+2)}$
11. $\frac{2x^2+6x+10}{(2x+1)(x-2)(x+2)}$, or $\frac{2(x^2+3x+5)}{(2x+1)(x-2)(x+2)}$ **12.** $\frac{1}{18n(n-1)}$
13. $\frac{a+5}{a+4}$ **14.** $\frac{2(x+4)(x+2)}{x-1}$ **15.** 6 **16.** -14 **17.** 4 ft
18. It would take $\frac{30m}{31}$ min to download the file. **19.** Working alone, the newer machine can process 1000 pieces of mail in 30 min and the older machine can process 1000 pieces of mail in 60 min. **20.** 10 mph

Chapter 17 Cumulative Review Exercises, pp. 1018–1019

1. 0.63 **2.** < **3.** 7% **4.** 12 m^2 **5.** 4 ft 3 in. **6.** $2y-3$
7.
8. (3, 5) **9.** False
10. $9k^2 - 36kl + 36l^2$ **11.** $-5, 2$
12. $(10y+9)(10y-9)$
13. $\frac{2x}{x+3y}$ **14.** $-3, 2$ **15.** 2^2p
16. The investments were \$15,000 in Stock A and \$7500 in Stock B.
17. 9 mi

18. **a.**

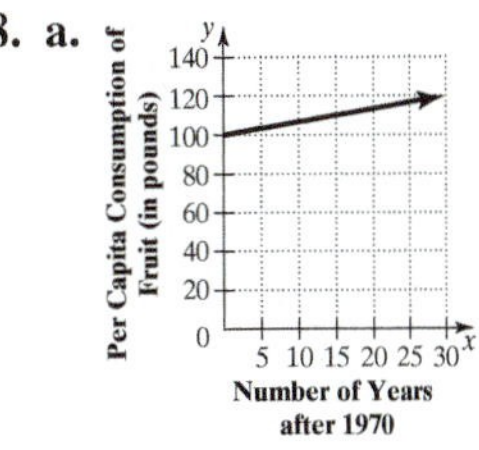

b. 0.6; the annual per capita consumption of fruit increased by 0.6 lb/yr **c.** The y-intercept is (0, 100); 100 represents the number of pounds of fruit consumed per capita in 1970. **19.** It will take the Reston bus 4 hr to overtake the Arlington bus. **20.** $(4.7x^3 - 50.1x^2 + 68.2x - 383.7)$ billion dollars

ANSWERS

CHAPTER 18

Chapter 18 Pretest, p. 1021

1. 9 2. $-3\sqrt{3}$ 3. $3a\sqrt{5}$ 4. $\frac{\sqrt{x}}{8}$ 5. $4\sqrt{2}$ 6. $12\sqrt{3}$ 7. $5x\sqrt{x}$ 8. $3\sqrt{2}$ 9. $2xy^2\sqrt{5}$ 10. $\sqrt{6}$ 11. $n + 2\sqrt{n}$ 12. $-1 + 3\sqrt{3}$ 13. $\frac{\sqrt{30x}}{6}$ 14. $2x\sqrt{5}$ 15. $\frac{8\sqrt{2} + \sqrt{14}}{2}$ 16. 36 17. 1, 3 18. The velocity of the car is 20 m/sec 19. The gymnast covers $12\sqrt{2}$, or approximately 17.0 m in the tumbling sequence. 20. $L = \frac{S^2}{30f}$

Section 18.1 Practices, pp. 1023–1028

1, *p. 1023:* a. 2 b. −35 2, *p. 1023:* 3.162 3, *p. 1024:* a. 6 b. 5 c. 7 d. 1 4, *p. 1024:* a. x^2 b. $8t^5$ c. $-11xy$ 5, *p. 1025:* a. $6\sqrt{2}$ b. $4\sqrt{10}$ c. $-\frac{\sqrt{3}}{3}$ 6, *p. 1026:* a. $x\sqrt{x}$ b. $3n^2\sqrt{2}$ c. $-5b\sqrt{2a}$ 7, *p. 1027:* a. $\frac{1}{4}$ b. $\frac{y}{2}$ c. $\frac{x^2}{y^3}$ 8, *p. 1027:* a. $\frac{\sqrt{3}}{4}$ b. $\frac{\sqrt{2y}}{7}$ c. $\frac{x^3y\sqrt{5x}}{2}$ 9, *p. 1028:* 1000 lb

Exercises 18.1, pp. 1029–1031

1. square root 3. irrational 5. take the square root 7. product 9. 6 11. 1 13. −10 15. 21 17. 2.236 19. 4.243 21. 16 23. 11 25. $5x$ 27. 2 29. 9 31. n^4 33. $7y$ 35. $3x^2$ 37. $5xy^5$ 39. $4\sqrt{2}$ 41. $-6\sqrt{3}$ 43. $18\sqrt{3}$ 45. $\frac{\sqrt{3}}{2}$ 47. $x\sqrt{11}$ 49. $n^2\sqrt{n}$ 51. $2x\sqrt{5x}$ 53. $-2p\sqrt{3q}$ 55. $9xy^2\sqrt{10x}$ 57. $\frac{2}{5}$ 59. $-\frac{1}{2}$ 61. $\frac{9}{n^3}$ 63. $\frac{x^2}{y}$ 65. $-\frac{\sqrt{3}}{2}$ 67. $\frac{\sqrt{5n}}{4}$ 69. $\frac{xy^3\sqrt{3}}{2}$ 71. $-\frac{3x^3\sqrt{3y}}{4}$ 73. $\frac{6}{y^3}$ 75. −12 77. $10\sqrt{3}$ 79. $-2a^2b\sqrt{10b}$ 81. r^3 83. a. $m = \sqrt{a \cdot b}$ b. $m = 4$ 85. a. It takes the object 2 sec to reach the ground. b. No; $\sqrt{\frac{40}{5}} = \sqrt{8} = 2\sqrt{2}$, which is not equal to $2 \cdot 2$. 87. The car was traveling at a speed of 60 mph at the time of the accident. 89. The distance between the towns is $\sqrt{146}$ mi, or about 12 mi.

Section 18.2 Practices, pp. 1033–1035

1, *p. 1033:* a. $6\sqrt{5}$ b. $11\sqrt{n}$ c. $11\sqrt{t^2 - 3}$ d. Cannot be simplified because the radicals are not like. 2, *p. 1034:* a. $12\sqrt{2}$ b. $-6\sqrt{3}$ c. $-9\sqrt{t}$ d. $12b^2\sqrt{a}$ 3, *p. 1035:* The length of the front yard is $20\sqrt{2}$ m, or approximately 28 m.

Exercises 18.2, pp. 1036–1037

1. like 3. unlike 5. $8\sqrt{7}$ 7. $-5\sqrt{2}$ 9. Cannot be combined 11. $-13\sqrt{11}$ 13. $9t\sqrt{3}$ 15. $23\sqrt{x}$ 17. Cannot be combined 19. $5\sqrt{x + 1}$ 21. $-2\sqrt{2}$ 23. $11\sqrt{2}$ 25. $3\sqrt{3}$ 27. $26\sqrt{3}$ 29. $11\sqrt{2} - 6\sqrt{3}$ 31. $14\sqrt{y}$ 33. $(3 - 4x)\sqrt{x}$ 35. $14\sqrt{p}$ 37. $\sqrt{x}$ 39. $(-5x^2y^2 + x^3y)\sqrt{2x}$, or $(x - 5y)x^2y\sqrt{2x}$ 41. $2\sqrt{3} + \sqrt{6}$ 43. $4\sqrt{7p}$ 45. $9\sqrt{3}$ 47. a. Each missing side measures $\sqrt{61}$ units. b. The perimeter of the triangle is $10 + 2\sqrt{61}$ units. 49. a. $3\sqrt{10}$ in. and $2\sqrt{10}$ in. b. The side of the larger tile is $\sqrt{10}$ in. longer. 51. The manufacturer will charge $90\sqrt{10}$ more dollars for 4000 machine parts than for 1000 machine parts.

Section 18.3 Practices, pp. 1040–1047

1, *p. 1040:* a. 5 b. $2y^3$ c. $9t + 9$ 2, *p. 1040:* a. $\sqrt{70}$ b. $-108\sqrt{2}$ c. $4y^3$ 3, *p. 1041:* a. $9\sqrt{2} - 4\sqrt{3}$ b. $\sqrt{ab} + 3\sqrt{a}$ 4, *p. 1041:* a. $2 + 2\sqrt{3}$ b. $3x + 4\sqrt{x} - 4$ 5, *p. 1042:* a. −2 b. $p - q$ 6, *p. 1042:* a. $2 + 2b\sqrt{2} + b^2$, or $b^2 + 2b\sqrt{2} + 2$ b. $x - 12\sqrt{x} + 36$ 7, *p. 1043:* It takes Mercury about 93 twenty-four-hour days to revolve around the Sun. 8, *p. 1043:* a. $\sqrt{7}$ b. $2x^2$ c. $\frac{1}{20y^2}$ 9, *p. 1044:* a. $\frac{m^2\sqrt{m}}{n^2}$ b. $\frac{y^3}{5x}$ c. $\frac{a\sqrt{5}}{3}$ 10, *p. 1044:* a. $\frac{\sqrt{2}}{2}$ b. $\frac{\sqrt{5s}}{s}$ c. $\frac{7r^2\sqrt{3}}{6}$ 11, *p. 1045:* a. $\frac{\sqrt{6}}{6}$ b. $\frac{\sqrt{5n}}{10}$ 12, *p. 1046:* a. $\frac{\sqrt{15} - \sqrt{3}}{3}$ b. $\frac{\sqrt{bc} + 2\sqrt{b}}{b}$ 13, *p. 1046:* a. $\frac{24 + 8\sqrt{2}}{7}$ b. $\frac{a\sqrt{b} - a\sqrt{5}}{b - 5}$ 14, *p. 1047:* $\frac{295\sqrt{2}}{6}$ beats per minute, or approximately 70 beats per minute

Exercises 18.3, pp. 1048–1051

1. are different from one another 3. bring the radicands under a single radical sign 5. 21 7. 15 9. $3n$ 11. $5y$ 13. $16x - 16$ 15. $4t + 20$ 17. $3\sqrt{6}$ 19. $-70\sqrt{2}$ 21. $4x^2$ 23. $r\sqrt{15}$ 25. $10\sqrt{xy}$ 27. $3 - \sqrt{3}$ 29. $x - 7\sqrt{x}$ 31. $4\sqrt{ab} + \sqrt{a}$ 33. $11 + 5\sqrt{5}$ 35. −2 37. $118 - 11\sqrt{3}$ 39. $3n + 14\sqrt{n} - 5$ 41. 33 43. 13 45. $x - 4$ 47. $a - b$ 49. $3x - y$ 51. $2 - 2x\sqrt{2} + x^2$ 53. $x - 2\sqrt{x} + 1$ 55. $\sqrt{5}$ 57. $\frac{1}{5}$ 59. $2a$ 61. $\frac{4}{3y^2}$ 63. $\frac{a^2}{b^3}$ 65. $\frac{4x^6}{y^4}$ 67. $\frac{x^5\sqrt{5}}{6}$ 69. $\frac{2\sqrt{3}}{3}$ 71. $\frac{\sqrt{5y}}{y}$ 73. $\frac{\sqrt{22}}{11}$ 75. $\frac{x\sqrt{5}}{5}$ 77. $\frac{\sqrt{2t}}{10}$ 79. $\frac{\sqrt{2a}}{2}$ 81. $\frac{\sqrt{15} + 2\sqrt{3}}{3}$ 83. $\frac{\sqrt{mn} - \sqrt{m}}{m}$ 85. $\frac{12 - 3\sqrt{6}}{2}$ 87. $4 + \sqrt{5}$ 89. $2\sqrt{5} + 2\sqrt{3}$ 91. $\frac{a\sqrt{b} + a\sqrt{3}}{b - 3}$ 93. $\frac{\sqrt{cd} - 6\sqrt{d}}{d}$ 95. $9a^2$ 97. $\frac{8s^4}{t^6}$ 99. $\frac{\sqrt{2c}}{12}$ 101. $\sqrt{r} + 3r$ 103. $7m - n$ 105. The standard deviation of the first data set is $\sqrt{2}$, or approximately 1.4, times that of the second. 107. The distance from (0, 0) to (3, 5) is $\sqrt{34}$, and the distance from (0, 0) to (6, 10) is $2\sqrt{34}$. 109. It will take the hailstone $\frac{5\sqrt{5}}{2}$, or about 5.6, sec to drop 500 ft. 111. $\frac{\sqrt{\pi hV}}{\pi h}$

Section 18.4 Practices, pp. 1052–1055

1, *p. 1052:* 16 2, *p. 1053:* No solution 3, *p. 1053:* 9 4, *p. 1054:* 2, 3 5, *p. 1054:* 4, 5 6, *p. 1055:* A radius of 1000 ft will permit a maximum safe speed of 50 mph. 7, *p. 1055:* $h = \frac{2}{3}d^2$

Exercises 18.4, pp. 1056–1057

1. radical equation 3. isolate 5. 9 7. 32 9. No solution 11. 64 13. 6 15. 3 17. 2 19. 3 21. No solution 23. 5 25. 0 27. −2 29. 8 31. No solution 33. −1, 2 35. 8, 17 37. No solution 39. 3 41. 4, −4 43. 64 45. 4 47. No solution 49. Its power is 2500 watts. 51. A mass of $\frac{8}{\pi^2}$ g will produce a period of 2 sec. 53. a. $A = \pi(gt + b)^2$ b. $A = 36\pi$, or approximately 113 in^2

Chapter 18 Review Exercises, pp. 1061–1063

1. is; Possible answer: it is the square root of a number that is not a perfect square 2. is; Possible answer: the square of the square root of a nonnegative number is that number 3. is not; Possible answer: the radicand 72 is divisible by the perfect square 36 4. is not; Possible answer: a radical is in the denominator 5. is not; Possible answer: the radicand does not contain a variable 6. is; Possible answer: squaring can eliminate the radical resulting in a simpler equation 7. −7 8. 6 9. $7x$ 10. $2\sqrt{7}$ 11. $-9\sqrt{2}$ 12. $4x\sqrt{2x}$ 13. $\frac{3}{5}$ 14. $-\frac{\sqrt{3t}}{4}$ 15. $\frac{12}{x^{50}}$ 16. $10a^2b\sqrt{ab}$ 17. $3\sqrt{5}$ 18. $8\sqrt{n}$ 19. $x\sqrt{3}$ 20. $-7\sqrt{3}$ 21. $5x\sqrt{x}$ 22. $7\sqrt{2a}$ 23. $\sqrt{15}$ 24. $4n$ 25. $5ab^3\sqrt{2a}$ 26. $x - 4\sqrt{x}$ 27. 6 28. $2y - \sqrt{y} - 3$ 29. $y + 10\sqrt{y} + 25$ 30. 3 31. $\frac{1}{2}$ 32. 4 33. $2a\sqrt{3a}$ 34. $\frac{2\sqrt{11}}{11}$ 35. $\frac{\sqrt{2x}}{2}$ 36. 9 37. $\frac{5\sqrt{7} + 5}{3}$ 38. 64 39. $\frac{15}{2}$ 40. 5 41. 8 42. The area of the city block is 5000 ft^2. 43. The astronaut can see approximately 2 mi. 44. No; an 8-inch screwdriver will not fit diagonally in the box. 45. $(4\sqrt{29} + 4\sqrt{74})$ in. 46. $r = \frac{\sqrt{S\pi}}{2\pi}$ 47. a. $D = \frac{2\pi r}{2\sqrt{\pi(\pi r^2)}} = \frac{2\pi r}{2\pi r} = 1$ b. $D = \frac{L\sqrt{\pi A}}{2\pi A}$ c. $A = \frac{L^2}{4\pi(1.58)^2}$ 48. $\frac{\sqrt{15V}}{15}$ in.

Chapter 18 Posttest, p. 1064

1. 12 2. $-9\sqrt{5}$ 3. $4x\sqrt{2xy}$ 4. $\frac{\sqrt{5n}}{4}$ 5. $3\sqrt{x}$ 6. $-15\sqrt{2}$ 7. $10t\sqrt{t}$ 8. $2\sqrt{6}$ 9. $5x^2y\sqrt{x}$ 10. 5 11. $5y - \sqrt{y}$ 12. $19 + 8\sqrt{3}$ 13. $\frac{\sqrt{10p}}{5}$ 14. $2x\sqrt{6x}$ 15. $1 - 2\sqrt{2}$ 16. 7 17. 2 18. The windchill temperature is 28°F. 19. The length of the ladder is $6\sqrt{17}$ ft. 20. A lookout must climb 25 ft to see a ship 6 mi away.

Chapter 18 Cumulative Review Exercises, pp. 1065–1066

1. 0.083, 0.83, 0.8625, and 0.875 **2.** 4 **3.** 2.5% **4.** $-\frac{2}{15}$ **5.** 43.5 **6.** $\frac{2}{3}$

7. Slope: -5

8. The system has one solution. **9.** $16x^3 - 4x^2 - 12x + 3$ **10.** $3x(x + 4)^2$ **11.** $\frac{10c + 6}{(c - 2)(c + 1)}$, or $\frac{2(5c + 3)}{(c - 2)(c + 1)}$ **12.** $-3, 6$ **13.** -2 **14.** $\frac{\sqrt{3}}{2}$ **15.** The speed is 343 m/sec when the temperature is 20°C. **16.** 4×10^1 **17.** $(4x^3 - 50x^2 + 68x - 384)$ billion dollars **18.** $\frac{7x}{6(x + 1)}$ **19. a.** Working together, the server and the computer would take 24 hr. **b.** It took 0.008 hr to process a sequence. **20.** The distance from the camp to the waterfall is $100\sqrt{514}$ m, or approximately 2300 m.

CHAPTER 19

Chapter 19 Pretest, pp. 1068–1069

1. $3\sqrt{2}, -3\sqrt{2}$ **2.** $r = \frac{1}{2}\sqrt{\frac{A}{\pi}} = \frac{\sqrt{A\pi}}{2\pi}$ **3.** 36 **4.** $2 + \sqrt{2}, 2 - \sqrt{2}$ **5.** $\frac{1 + \sqrt{11}}{2}, \frac{1 - \sqrt{11}}{2}$ **6.** 11, 1 **7.** $-3, -4$ **8.** $3 + \sqrt{5}, 3 - \sqrt{5}$ **9.** $\sqrt{3}, -\sqrt{3}$ **10.** $\frac{5 + 3\sqrt{5}}{5}, \frac{5 - 3\sqrt{5}}{5}$ **11.** $-2 + 3\sqrt{2}, -2 - 3\sqrt{2}$

12. (0, 0) (−1, −1) (1, −1) (−2, −4) (2, −4)

13. vertex: $(3, -2)$; axis of symmetry: $x = 3$

14.

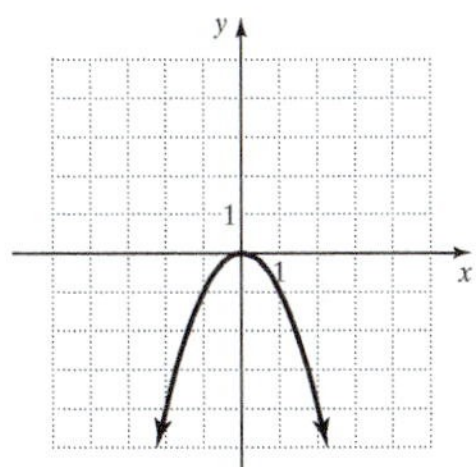

15.

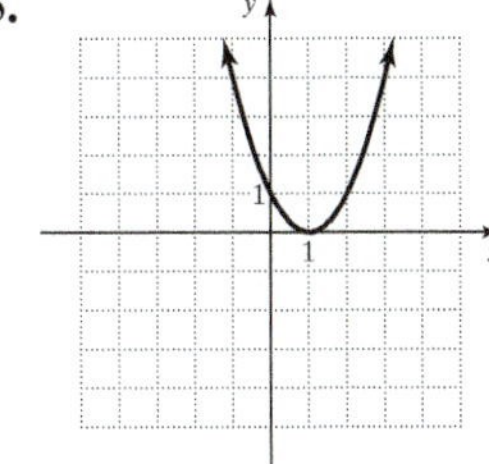

16.

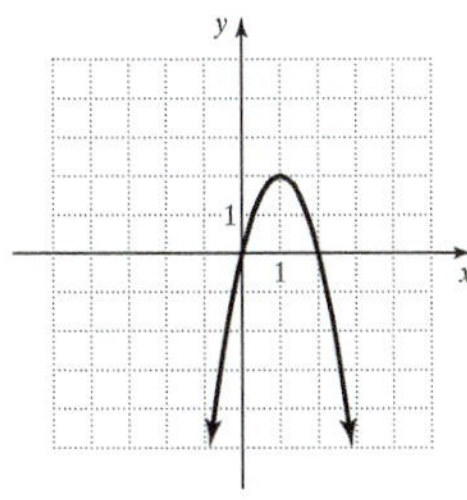

17. $v = \sqrt{2as}$ **18.** 3-in. by 3-in. squares should be cut from the corners to meet the specifications. **19.** The speed of the plane in still air is 150 mph. **20.** The ball reaches a maximum height of 41 ft.

Section 19.1 Practices, pp. 1070–1074

1, *p. 1070:* **a.** $10, -10$ **b.** $3\sqrt{2}, -3\sqrt{2}$ **c.** $\frac{5}{9}, -\frac{5}{9}$ **2, *p. 1071:*** **a.** $1, -5$ **b.** $4 + \sqrt{5}, 4 - \sqrt{5}$ **c.** $\frac{1 + 3\sqrt{6}}{4}, \frac{1 - 3\sqrt{6}}{4}$ **3, *p. 1072:*** $-4 + \sqrt{2}, -4 - \sqrt{2}$ **4, *p. 1072:*** **a.** $c = \sqrt{\frac{E}{m}} = \frac{\sqrt{Em}}{m}$ **b.** $a = \sqrt{c^2 - b^2}$ **5, *p. 1073:*** **a.** $r = \sqrt{\frac{2A}{\pi}} = \frac{\sqrt{2A\pi}}{\pi}$ **b.** The radius was about 12 mi. **6, *p. 1074:*** **a.** $d = 4 + \sqrt{B}$ **b.** 8 in.

Exercises 19.1, pp. 1075–1077

1. quadratic equation **3.** divide both sides by a **5.** $3, -3$ **7.** $\sqrt{2}, -\sqrt{2}$ **9.** $\frac{1}{2}, -\frac{1}{2}$ **11.** $2, -2$ **13.** $\sqrt{7}, -\sqrt{7}$ **15.** $2\sqrt{3}, -2\sqrt{3}$ **17.** $4\sqrt{5}, -4\sqrt{5}$ **19.** $\sqrt{6}, -\sqrt{6}$ **21.** $\sqrt{17}, -\sqrt{17}$ **23.** $\frac{3}{2}, -\frac{3}{2}$ **25.** No real solution **27.** $\frac{\sqrt{3}}{2}, -\frac{\sqrt{3}}{2}$ **29.** $1, -5$ **31.** 14, 0 **33.** $-6 + \sqrt{5}, -6 - \sqrt{5}$ **35.** $\frac{17}{4}, \frac{23}{4}$ **37.** 7, 1 **39.** $5 + \sqrt{3}, 5 - \sqrt{3}$ **41.** $-1 + 2\sqrt{2}, -1 - 2\sqrt{2}$ **43.** $\frac{1}{3}, -1$ **45.** $\frac{-5 + \sqrt{3}}{4}, \frac{-5 - \sqrt{3}}{4}$ **47.** $\frac{7 + 2\sqrt{5}}{2}, \frac{7 - 2\sqrt{5}}{2}$ **49.** $10 + 2\sqrt{10}, 10 - 2\sqrt{10}$ **51.** No real solution **53.** $-8 + \sqrt{5}, -8 - \sqrt{5}$ **55.** $3 + 2\sqrt{6}, 3 - 2\sqrt{6}$ **57.** $\sqrt{5}, -\sqrt{5}$ **59.** $x = \pm\sqrt{\frac{b}{a}}$ **61.** $v = \pm\frac{2\pi}{\sqrt{Kr}}$ **63.** $y = \pm\frac{5\sqrt{x^2 - 16}}{4}$ **65.** $\frac{6 + 2\sqrt{11}}{5}, \frac{6 - 2\sqrt{11}}{5}$ **67.** $\frac{\sqrt{3}}{5}, -\frac{\sqrt{3}}{5}$ **69.** $-2 + 2\sqrt{3}, -2 - 2\sqrt{3}$ **71.** $4 + 3\sqrt{7}, 4 - 3\sqrt{7}$ **73.** They can hear each other for 4 min. **75. a.** $H = 0.95h$ **b.** The new radius is approximately $0.97r$. **77.** $y = \pm\sqrt{r^2 - x^2}$ **79.** $d = 45\sqrt{\frac{2}{l}} = \frac{45\sqrt{2l}}{l}$

Section 19.2 Practices, pp. 1080–1084

1, *p. 1080:* **a.** 36 **b.** $\frac{25}{4}$ **2, *p. 1081:*** **a.** $3, -7$ **b.** $\frac{-7 + \sqrt{29}}{2}, \frac{-7 - \sqrt{29}}{2}$ **3, *p. 1082:*** **a.** $4 + 2\sqrt{5}, 4 - 2\sqrt{5}$ **b.** $\frac{1 + \sqrt{2}}{2}, \frac{1 - \sqrt{2}}{2}$ **4, *p. 1083:*** $\frac{4}{3}, -\frac{1}{3}$ **5, *p. 1084:*** **a.** The length of the table is $2w - 1$. **b.** The length is 8 ft and the width is 4.5 ft.

Exercises 19.2, pp. 1085–1086

1. 9 **3.** 1 **5.** $\frac{25}{4}$ **7.** $\frac{1}{4}$ **9.** $0, -4$ **11.** $5 + 2\sqrt{6}, 5 - 2\sqrt{6}$ **13.** $1, -15$ **15.** $3 + \sqrt{13}, 3 - \sqrt{13}$ **17.** No real solution **19.** $\frac{1 + \sqrt{13}}{2}, \frac{1 - \sqrt{13}}{2}$ **21.** $2, -6$ **23.** $-1 + \frac{\sqrt{66}}{6}, -1 - \frac{\sqrt{66}}{6}$ **25.** $\frac{5 + 3\sqrt{2}}{2}, \frac{5 - 3\sqrt{2}}{2}$ **27.** $\frac{3 + \sqrt{73}}{8}, \frac{3 - \sqrt{73}}{8}$ **29.** No real solution **31.** $1 + \sqrt{5}, 1 - \sqrt{5}$ **33.** $\frac{-1 + \sqrt{33}}{4}, \frac{-1 - \sqrt{33}}{4}$ **35.** No real solution **37.** $\frac{49}{4}$ **39.** There are 51 insects after 3 days. **41. a.** The fireworks will be 16 ft above the ground 1 sec after being shot into the air. **b.** They will be 8 ft above the ground in approximately 0.3 sec and again in 1.7 sec. **43.** One friend was 6 mi from the party and the other was 8 mi from the party. **45.** It would take the faster machine about 7 min to do the job alone.

Section 19.3 Practices, pp. 1089–1092

1, *p. 1089:* **a.** 1, 0 **b.** No real solution **2, *p. 1091:*** **a.** $\frac{4 + \sqrt{31}}{5}, \frac{4 - \sqrt{31}}{5}$ **b.** $\frac{-3 + \sqrt{93}}{6}, \frac{-3 - \sqrt{93}}{6}$ **3, *p. 1092:*** $2.5, -0.5$ **4, *p. 1092:*** Working alone, it takes one pump about 4.4 hr and the other about 9.4 hr to empty the tank.

Exercises 19.3, pp. 1094–1095

	Standard Form	$a =$	$b =$	$c =$
1.	$2x^2 + 9x - 1 = 0$	2	9	-1
3.	$-x^2 + 3x - 8 = 0$	-1	3	-8
5.	$x^2 - x + 8 = 0$	1	-1	8
7.	$\frac{1}{3}y^2 - \frac{1}{2}y + \frac{1}{4} = 0$	$\frac{1}{3}$	$-\frac{1}{2}$	$\frac{1}{4}$
9.	$6x^2 - 7x - 15 = 0$	6	-7	-15

11. $1, -3$ **13.** No real solution **15.** $-1, 7$ **17.** $2 + \sqrt{5}, 2 - \sqrt{5}$ **19.** $1, \frac{1}{3}$ **21.** $\frac{3 + \sqrt{5}}{4}, \frac{3 - \sqrt{5}}{4}$ **23.** $6, -\frac{5}{2}$ **25.** $-1, -2$ **27.** $-1, -3$ **29.** $0, \frac{2}{3}$ **31.** $-1, \frac{1}{4}$ **33.** $\frac{7 + \sqrt{29}}{2}, \frac{7 - \sqrt{29}}{2}$ **35.** $\frac{1 + 7\sqrt{5}}{2}, \frac{1 - 7\sqrt{5}}{2}$ **37.** No real solution **39.** $\frac{3 + \sqrt{3}}{2}, \frac{3 - \sqrt{3}}{2}$ **41.** No real solution **43.** $-2, 3$ **45.** $\frac{2}{5}d^2 - \frac{4}{5}d - \frac{1}{5} = 0$; $a = \frac{2}{5}, b = -\frac{4}{5}, c = -\frac{1}{5}$ **47.** n is 13. **49.** Working alone, it takes the lab coordinator about 91 min to set up the lab, and it takes the lab technician about 71 min to set up the lab. **51.** The speed of the car is approximately 27 mph. **53.** At about 11:40 A.M. and 5 P.M.

Section 19.4 Practices, pp. 1098–1102

1, *p. 1098:*

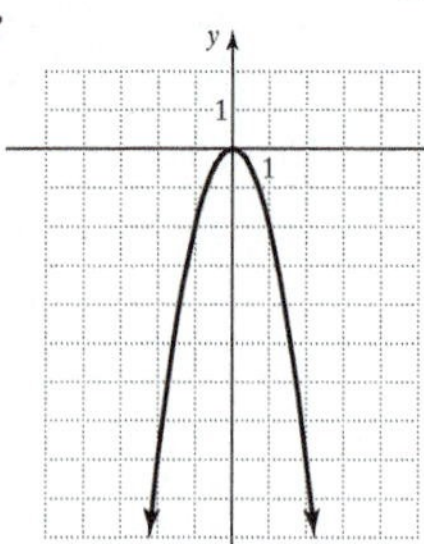

2, *p. 1099:*

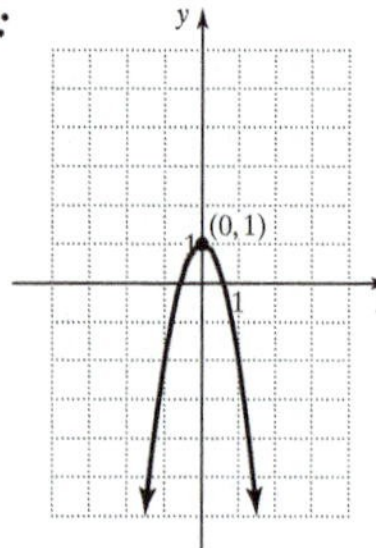

3, *p. 1100:* a.

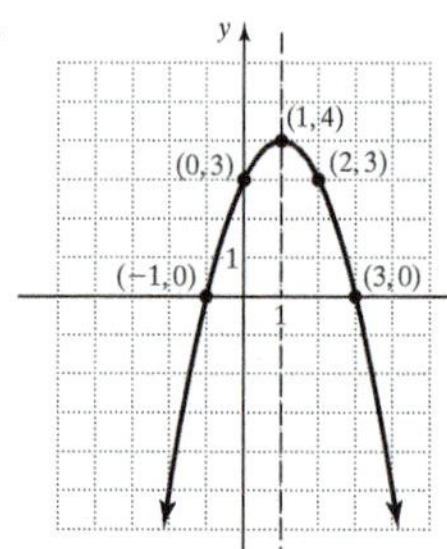

b. The graph of the parabola opens downward since $a = -1$ is negative. The curve turns at the vertex $(1, 4)$, the highest point of the graph. The equation of the axis of symmetry is $x = 1$.

4, *p. 1102:*

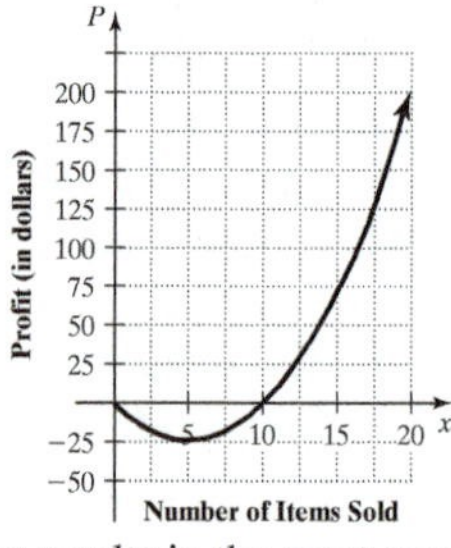

Selling 5 items results in the most money lost ($25).

Exercises 19.4, pp. 1103–1110

1. graph of an equation **3.** axis of symmetry **5.** positive

7.

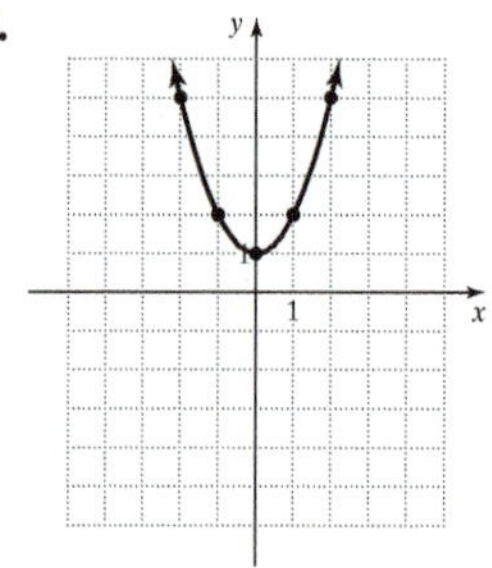

9.

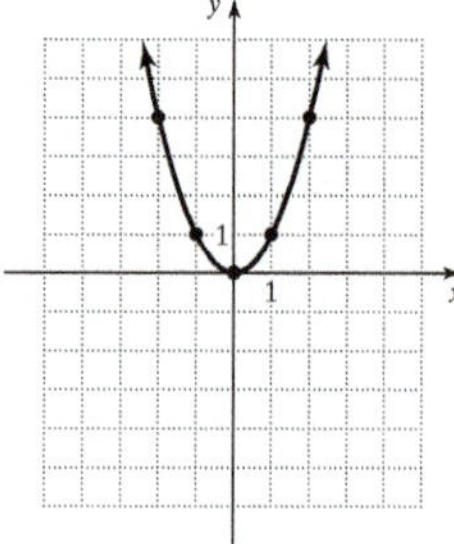

11.

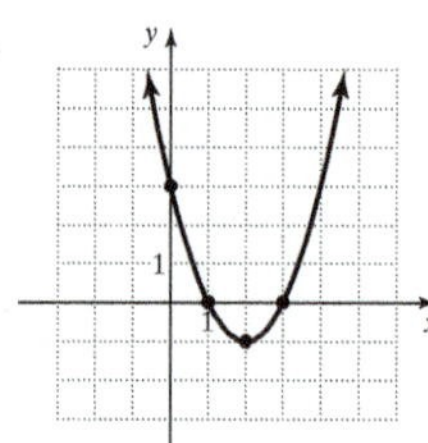

13. Axis of symmetry: $x = 0$; vertex: $(0, 0)$ **15.** Axis of symmetry: $x = 2$; vertex: $(2, 1)$

17.

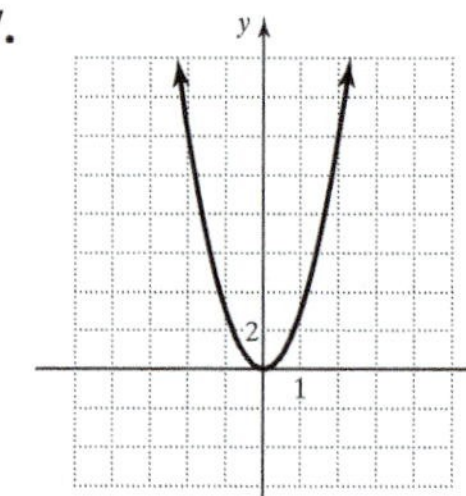

19.

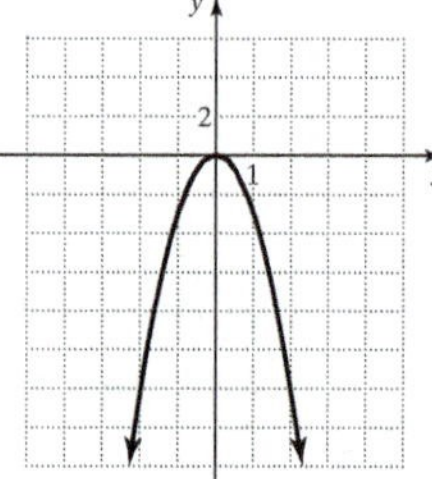

21.

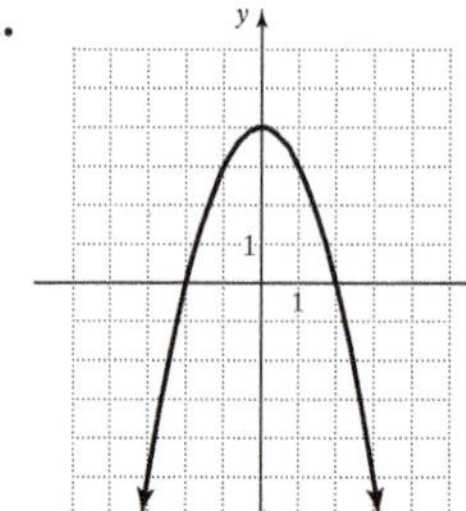

23.

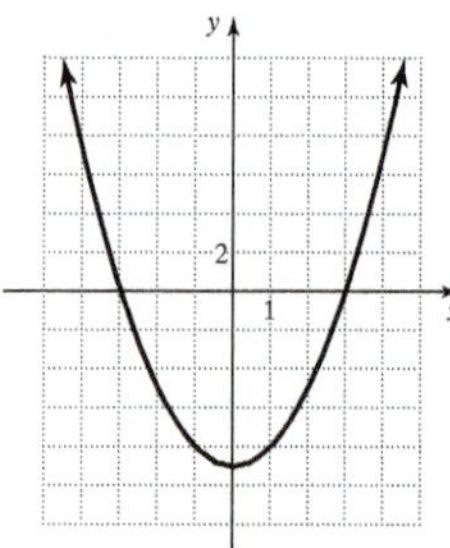

25.

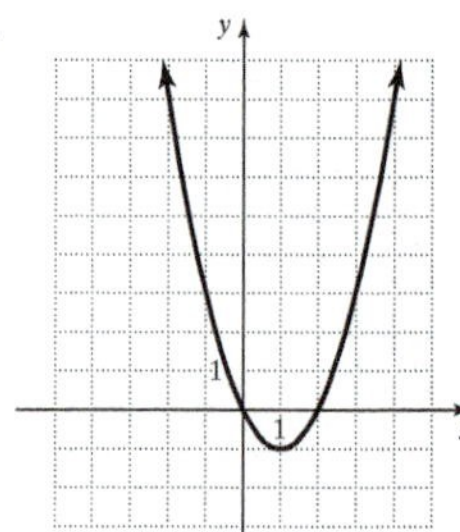

27.

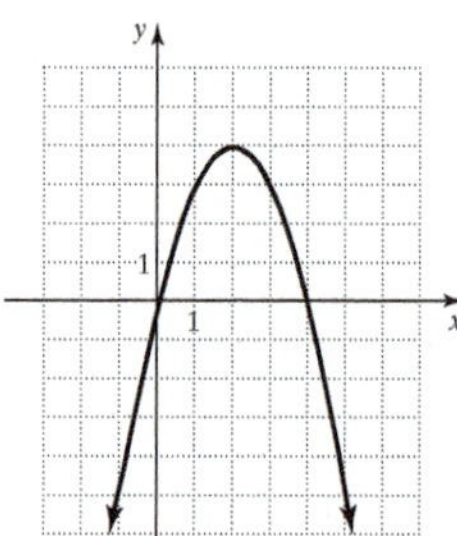

29.

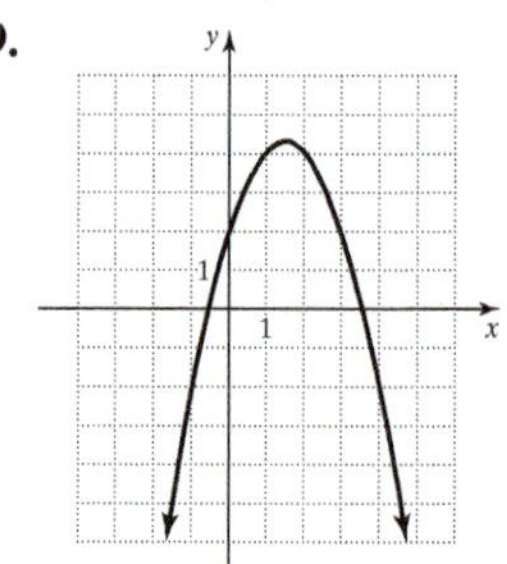

31.

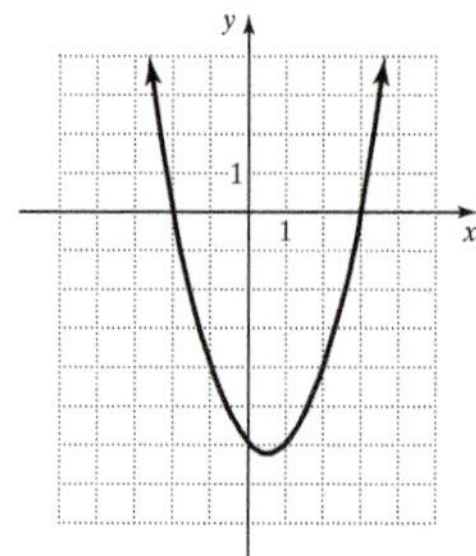

33.

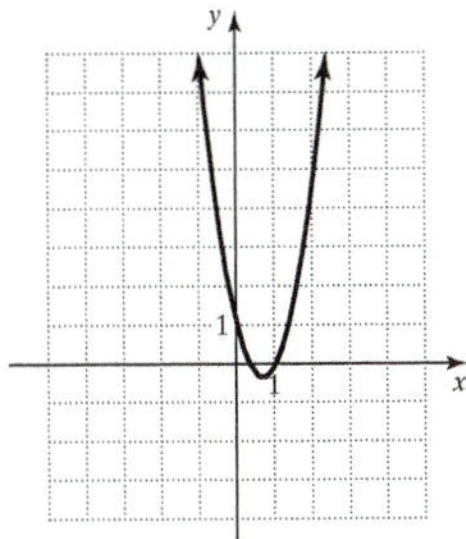

35.

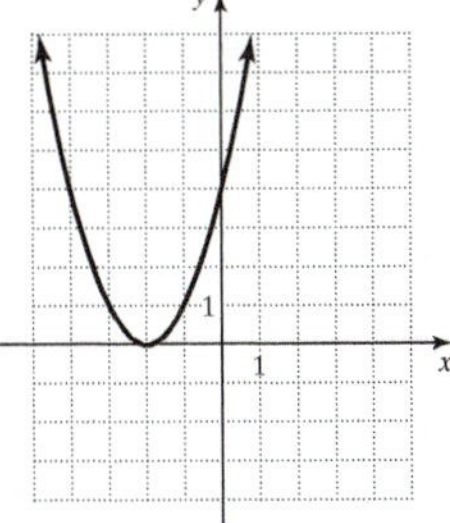

37.

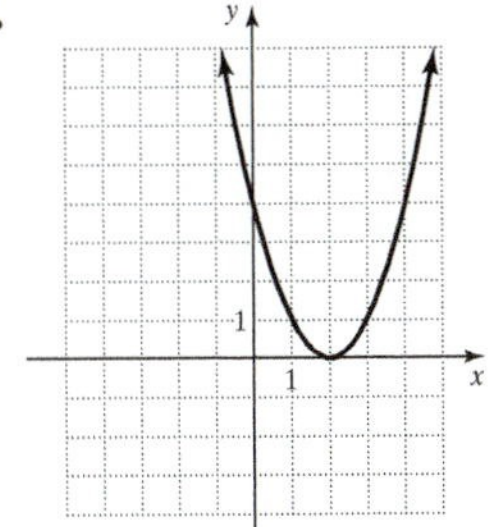

39.

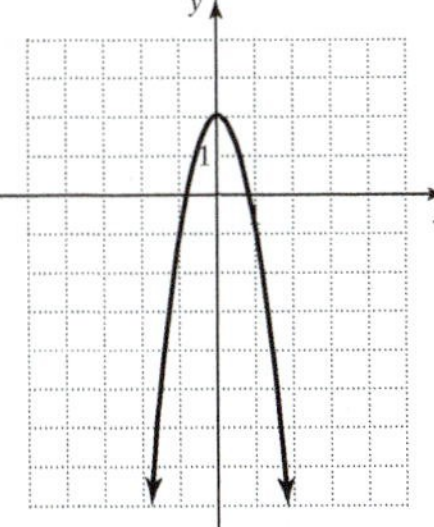

41.

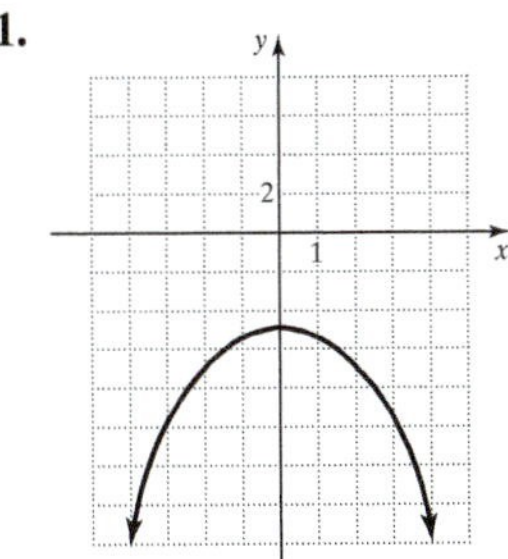

43.

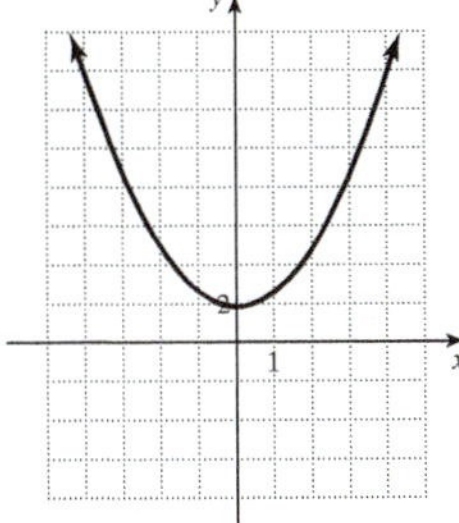

45.

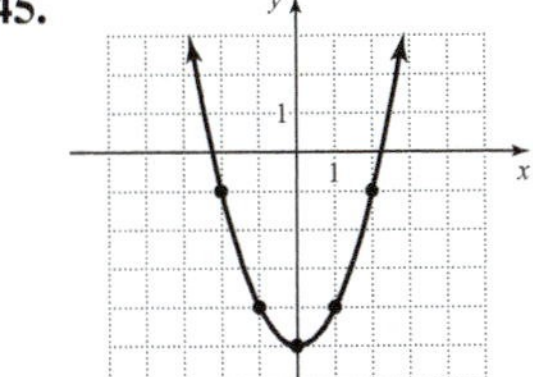

47.

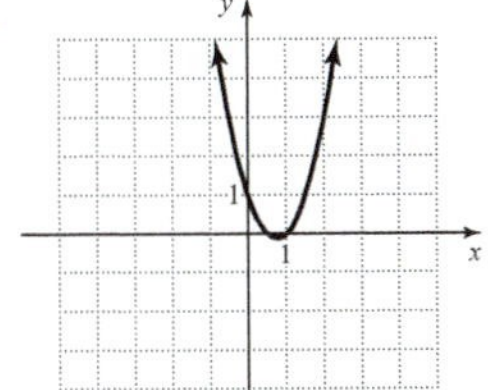

49.

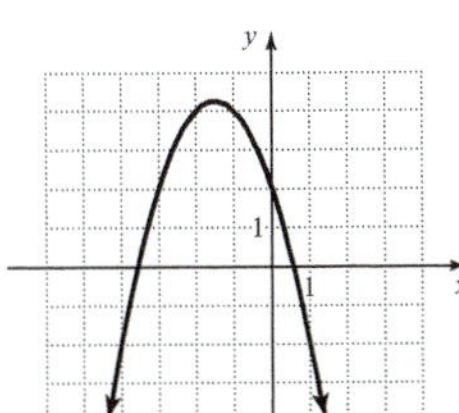

51. a.

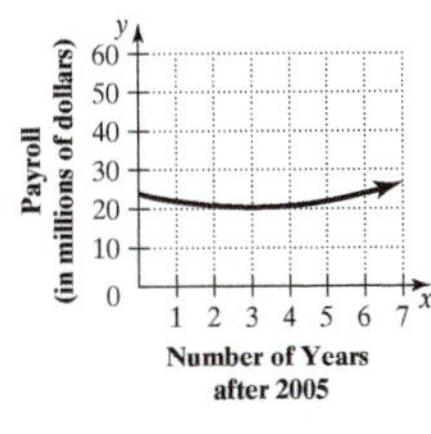

b. In the year 2008 **c.** The payroll decreases from 2005 to 2008 and increases thereafter.

53. a.

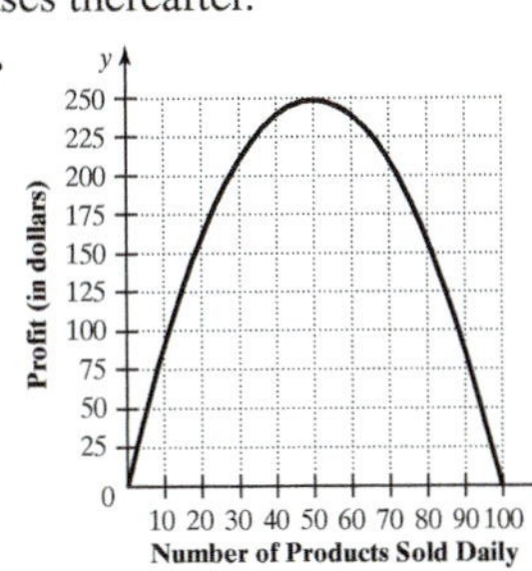

b. The company must sell 50 products each day to maximize the profit. **c.** The profit increases when 0 to 50 products are sold, but then decreases for sales of more than 50 products.

55. a. $R = x(100 - \frac{1}{4}x) = 100x - \frac{1}{4}x^2$

b.

x	0	100	200	300	400
R	0	7500	10,000	7500	0

c.

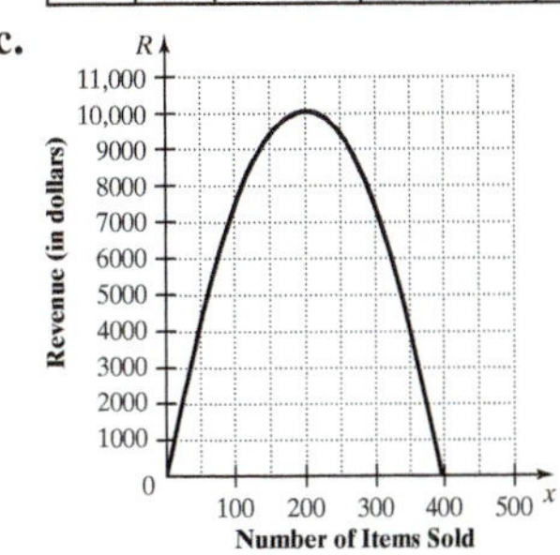

d. The store should sell 200 items to maximize the revenue. **e.** The maximum revenue is $10,000.

Chapter 19 Review Exercises, pp. 1115–1118

1. is; Possible answer: of the square root property of equality **2.** is; Possible answer: the constant term a^2 is the square of half the coefficient $2a$ of the linear term **3.** is; Possible answer: the equation has the form $y = ax^2 + bx + c$, where $a \neq 0$ **4.** is; Possible answer: its equation is $x = \frac{5}{8}$ **5.** downward; Possible answer: the coefficient of x^2 is negative **6.** minimum; Possible answer: the coefficient of the quadratic term is positive **7.** $2\sqrt{6}, -2\sqrt{6}$ **8.** $\sqrt{3}, -\sqrt{3}$ **9.** No real solution **10.** $\frac{5 + 3\sqrt{2}}{2}, \frac{5 - 3\sqrt{2}}{2}$ **11.** $r = \sqrt{\frac{A}{\pi}} = \frac{\sqrt{A\pi}}{\pi}$ **12.** $t = \sqrt{\frac{2d}{g}} = \frac{\sqrt{2dg}}{g}$ **13.** 25 **14.** $\frac{49}{4}$ **15.** 9, −3 **16.** 1, −4 **17.** $-2 + \sqrt{6}, -2 - \sqrt{6}$ **18.** $\frac{1 + \sqrt{2}}{2}, \frac{1 - \sqrt{2}}{2}$ **19.** $1 + \sqrt{2}, 1 - \sqrt{2}$ **20.** $-4 + \sqrt{15}, -4 - \sqrt{15}$ **21.** $\frac{-1 + \sqrt{41}}{4}, \frac{-1 - \sqrt{41}}{4}$ **22.** No real solution **23.** −5, 0 **24.** $\sqrt{6}, -\sqrt{6}$

25.

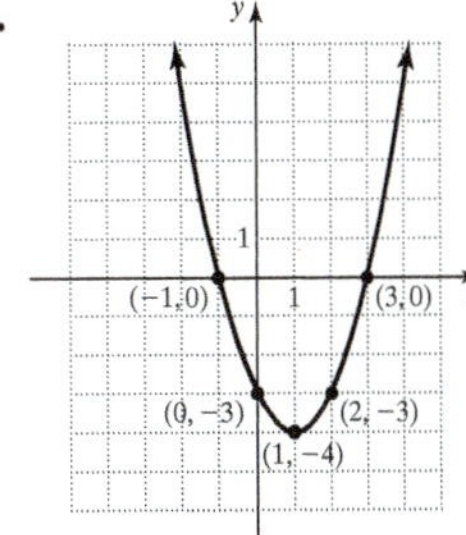

26.

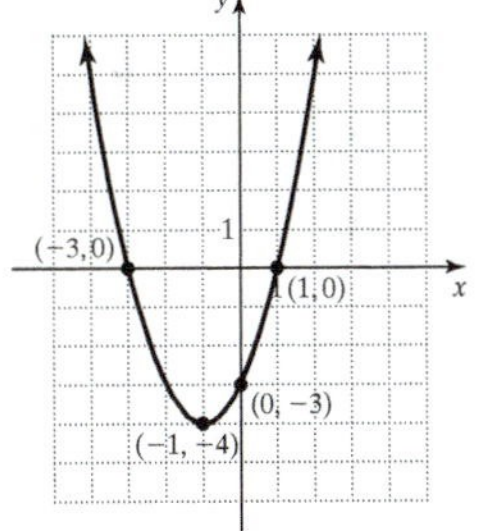

27. Axis of symmetry: $x = -2$; vertex: $(-2, -4)$
28. Axis of symmetry: $x = -1$; vertex: $(-1, 4)$

29.

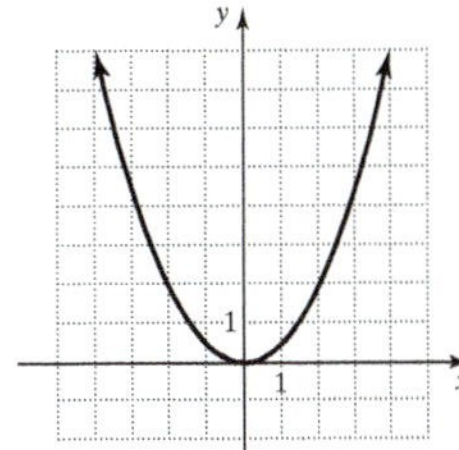

30.

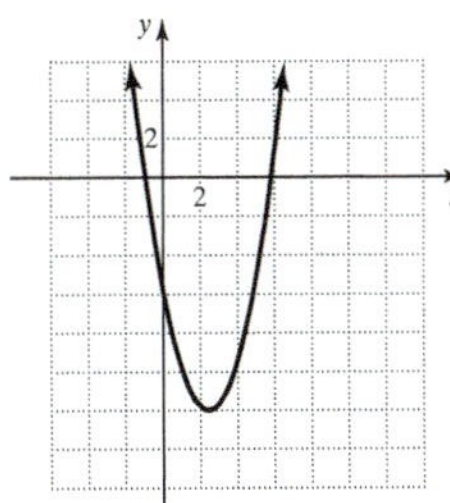

31.

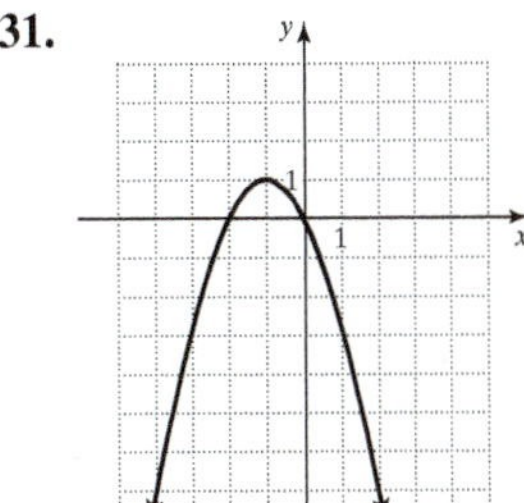

32.

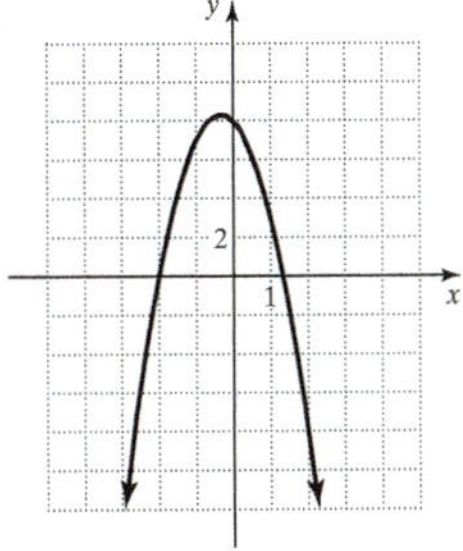

33. The radius of the sphere is approximately 7.0 in. **34.** The company had sold either 40 rugs or 50 rugs since the profit is \$2000 for both. **35.** The banner is approximately 2.2 ft by 9.3 ft. **36.** The building has 7 sides.

37. a.

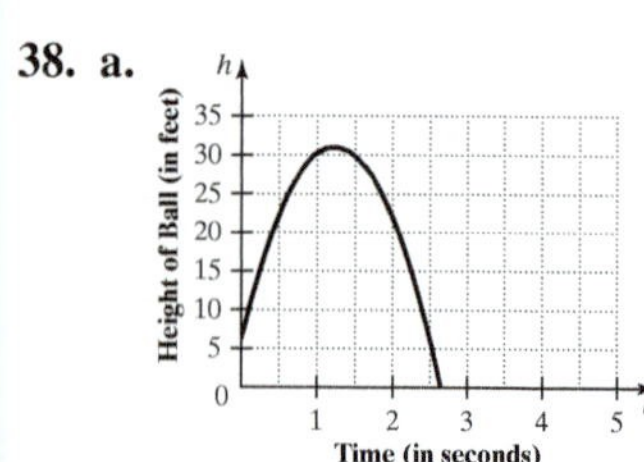

b. The profit per person if 16 people go is \$384. **c.** Either 13 people or 27 people went to the Bahamas since the profit per person is \$351 for both. **d.** The maximum profit per person that the agency can make is \$400 (when 20 people go to the Bahamas).

38. a.

b. The ball reaches its maximum height in $\frac{5}{4}$, or 1.25, sec. **c.** The maximum height of the ball is 31 ft. **39.** Working alone, it would take the experienced Web designer 4 days to develop the website and it would take the inexperienced designer 12 days to develop the website. **40.** Her average speed driving to work is approximately 36 mph.

Chapter 19 Posttest, pp. 1119–1120

1. $-1 + 2\sqrt{2}, -1 - 2\sqrt{2}$ **2.** $r = \sqrt{\frac{3V}{\pi h}} = \frac{\sqrt{3\pi hV}}{\pi h}$ **3.** $\frac{1}{4}$ **4.** $\frac{3 + \sqrt{13}}{2}, \frac{3 - \sqrt{13}}{2}$ **5.** $\frac{-3 + \sqrt{41}}{4}, \frac{-3 - \sqrt{41}}{4}$ **6.** $2\sqrt{3}, -2\sqrt{3}$ **7.** $4, -10$ **8.** $\frac{27}{5}, \frac{13}{5}$ **9.** $\frac{2 + \sqrt{7}}{2}, \frac{2 - \sqrt{7}}{2}$ **10.** $\frac{-4 + 2\sqrt{7}}{3}, \frac{-4 - 2\sqrt{7}}{3}$ **11.** $4 + \sqrt{15}, 4 - \sqrt{15}$ **12.**

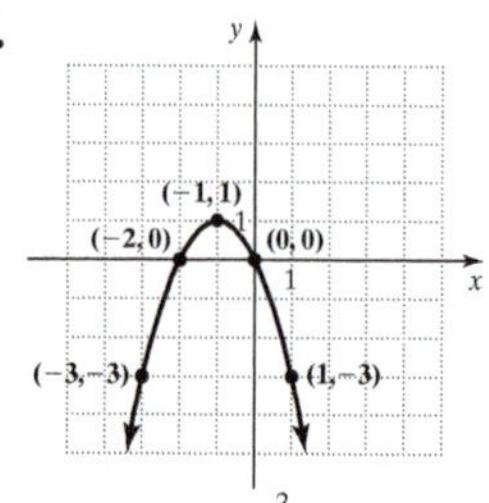

13. Vertex: $\left(-\frac{3}{2}, \frac{7}{4}\right)$; axis of symmetry $x = -\frac{3}{2}$

14.

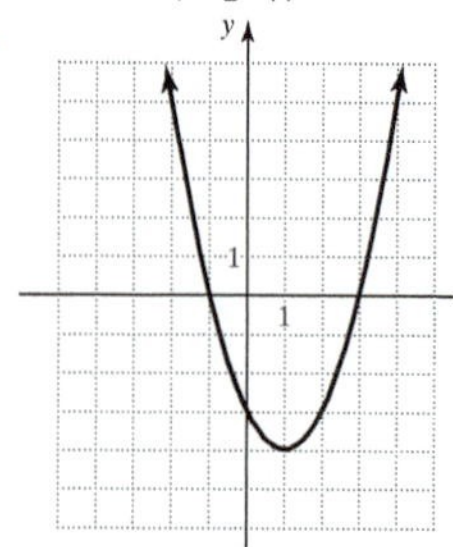

15.

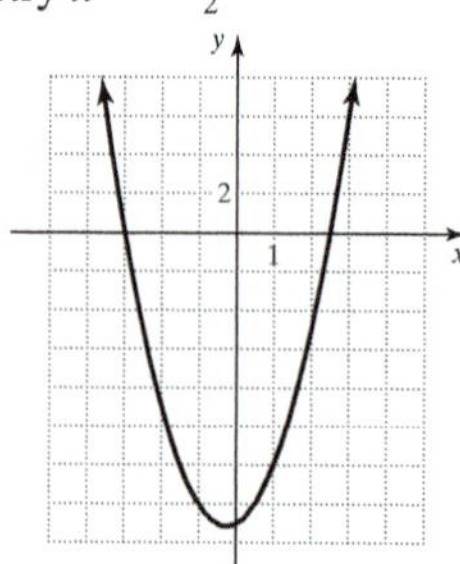

16.

17. $t = \frac{\sqrt{l}}{0.9}$ **18.** The maximum speed is 104 km/hr. **19.** The year 2032 **20.** The object reaches a maximum height of 100 ft (when $t = 2.5$ sec).

Chapter 19 Cumulative Review Exercises, pp. 1121–1122

1. -8 **2.** $-72x - 110$ **3.** $n = \frac{ax - c}{6}; -\frac{4}{5}$

4.

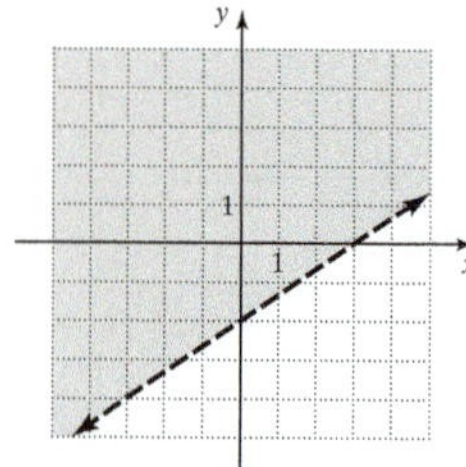

5. $(-1, 3)$ **6.** $18n^2 + 3n - 10$ **7.** $\frac{2x^2}{3(y - 2)}$ **8.** $3cd(2c + d)(c + d)$ **9.** $-\frac{2}{3}, 7$ **10.** $\frac{1}{4x(x + 4)}$ **11.** $-3, 2$ **12.** $xy\sqrt{y}$ **13.** $4 \pm 2\sqrt{3}$ **14.** $-2, \frac{1}{3}$ **15.** \$750

16. a. $n = \frac{5}{6}t + 2$

b.

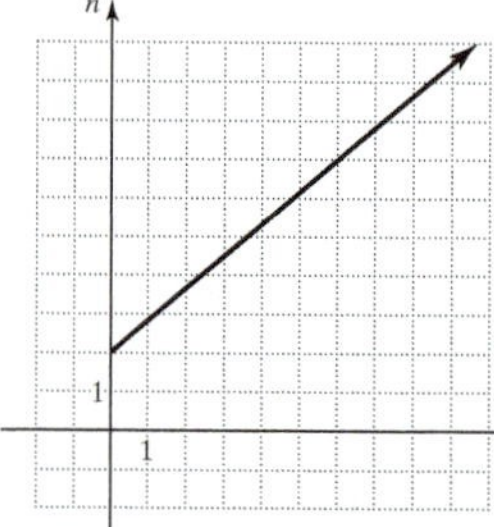

c. The slope of the graph, $\frac{5}{6}$, represents the rate of increase in the number of employees (5 employees every 6 months). In the equation it is the coefficient of t.

17. For $x = 0$, the number of bacteria is 400; for $x = 1$, the number of bacteria is 1200; for $x = 2$, the number of bacteria is 3600. **18.** $0.004a + 0.35b \le 80$ **19. a.** $300t$ meters away **b.** 2 sec

20. a.

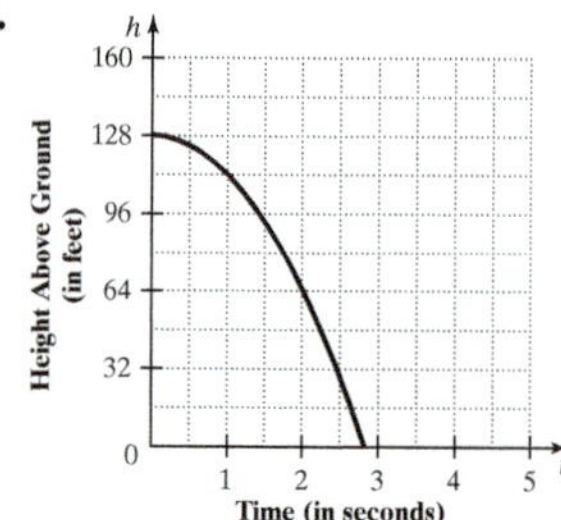

b. The sandbag has fallen 16 ft. At $t = 0$, the sandbag is at a height of 128 ft above the ground, and at 1 sec the sandbag has fallen to a height of 112 ft above the ground. So it has fallen a total of (128 – 112) ft, or 16 ft.

APPENDIX D

Appendix D Practices, pp. Ap-10–12

1, *Ap-10:* **a.** The binomial is a difference of cubes. **b.** The binomial is neither a sum of cubes nor a difference of cubes. **c.** The binomial is a sum of cubes. **d.** The binomial is a difference of cubes. **2, *Ap-11:*** **a.** $(r + s)(r^2 - rs + s^2)$ **b.** $(y - 4)(y^2 + 4y + 16)$ **3, *Ap-11:*** **a.** $(5 - x)(25 + 5x + x^2)$ **b.** $5(n - 2)(n^2 + 2n + 4)$ **c.** $2n(1 + 3n)(1 - 3n + 9n^2)$

Exercises D, pp. Ap-12–13

1. factorable **3.** difference of cubes **5.** $(p + q)(p^2 - pq + q^2)$ **7.** $(n + 2)(n^2 - 2n + 4)$ **9.** $(x + y^2)(x^2 - xy^2 + y^4)$ **11.** $(3n + 1)(9n^2 - 3n + 1)$ **13.** Prime polynomial **15.** $(2x^3)(4 + x)(16 - 4x + x^2)$ **17.** $2(p + 3)(p^2 - 3p + 9)$ **19.** $(a - b)(a^2 + ab + b^2)$ **21.** $(y - 1)(y^2 + y + 1)$ **23.** $(3 - n^2)(9 + 3n^2 + n^4)$ **25.** $(4 - 3y^2)(16 + 12y^2 + 9y^4)$ **27.** Prime polynomial **29.** $5(x - 1)(x^2 + x + 1)$ **31.** $2(y - 1)(y + 1)(y^2 - y + 1)(y^2 + y + 1)$ **33.** $(p + 1)(p^2 - p + 1)$ **35.** $(3x^2 - 2)(9y^4 + 6x^2 + 4)$ **37.** $(x + y)(x^2 - xy + y^2)$ in^3

39 a. $(x^3 + 3x^2 + 3x + 1) - x^3 = 3x^2 + 3x + 1$

b. $[(x + 1) - x)][(x + 1)^2 + x(x + 1) + x^2] = 1[(x^2 + 2x + 1) + (x^2 + x) + x^2] = 3x^2 + 3x + 1$

ANSWERS

Glossary

The numbers in brackets following each glossary term represent the section in which that term is discussed.

absolute value [7.1, ● App F] The absolute value of a number is its distance from 0 on the number line. The absolute value of a number n is written $|n|$.

absolute value function [● App G] A function in the form $f(x) = |x|$.

acute angle [10.1] An angle whose measure is less than 90°.

acute triangle [10.1] A triangle with three acute angles.

addends [1.2] Numbers being added.

addition property of equality [12.1] This property states that when adding any number to each side of an equation, the result is an equivalent equation. For any real numbers a, b, and c, $a = b$ and $a + c = b + c$ are equivalent.

additive identity property (or identity property of addition) [1.2, 11.2] This property states that the sum of a number and zero is the original number. For any real number a, $a + 0 = a$ and $0 + a = a$.

additive inverse property [11.2] This property states that for any real number a, there is exactly one real number $-a$ such that $a + (-a) = 0$ and $(-a) + a = 0$.

additive inverses [11.2] Two numbers that have a sum of 0.

algebraic expression [4.1, 11.4] An expression that combines variables, constants, and arithmetic operations.

altitude of a triangle [● App K] An altitude of a triangle is a line segment through a vertex that is perpendicular to the opposite side of the triangle.

angle [10.1] Two rays that have a common endpoint.

area [10.3] The number of square units that a figure contains.

ascending order of degree [15.3] The terms of a polynomial are written in ascending order of degree when the exponents get larger from left to right.

associative property of addition [1.2, 11.2] This property states that when adding three numbers, regrouping addends gives the same sum. For any three real numbers a, b, and c, $(a + b) + c = a + (b + c)$.

associative property of multiplication [1.3, 11.3] This property states that when multiplying three numbers, regrouping the factors gives the same product. For any three real numbers a, b, and c, $(a \cdot b)c = a(b \cdot c)$.

average (or mean) [1.5, 8.1] The sum of a set of numbers divided by however many numbers are in the set.

axis [13.1] Each number line in a coordinate plane.

bar graph [8.2] A graph in which quantities are represented by thin, parallel rectangles called bars. The length of each bar is proportional to the quantity that it represents.

base (exponent) [1.5, 11.4, 15.1] The number that is a repeated factor when written with an exponent.

binomial [15.3] A polynomial with two terms.

break-even point [14.1] The point at which the income for a business equals its expenses.

broken-line graph (or line graph) [8.2] A graph in which points are connected by straight-line segments and the position of any point on the line is read against the vertical axis and the horizontal axis.

center of a circle [10.1, ● App K] The given point that is a fixed distance from all the points on a circle.

circle [10.1, ● App K] A closed plane figure made up of points that are all the same distance from a fixed point called the center.

circle graph [8.2] A graph that resembles a pie (the whole amount) that has been cut into slices (the parts).

circumference [10.2] The distance around a circle.

coefficient [11.6] The numerical factor of a variable term.

commutative property of addition [1.2, 11.2] This property states that changing the order in which two numbers are added does not affect their sum. For any two real numbers a and b, $a + b = b + a$.

commutative property of multiplication [1.3, 11.3] This property states that changing the order in which two numbers are multiplied does not affect their product. For any two real numbers a and b, $a \cdot b = b \cdot a$.

complementary angles [10.1] Two angles the sum of whose measures is 90°.

composite figure [10.2] A figure combining two or more basic geometric figures.

composite number [2.1] A whole number that has more than two factors.

compound inequality [● App E] Two inequalities that are joined by the word *and* or the word *or* to form a compound inequality.

constant [4.1] A known number.

constant function [● App G] A linear function defined by $f(x) = k$, where k is a real number.

constant of variation (or constant of proportionality) [● App I] In the equations $y = kx$, $y = \frac{k}{x}$ and $y = kxz$, k is called the constant of variation or the constant of proportionality.

constant term [15.3] The term that is of degree 0 in a polynomial.

coordinate plane [13.1] The flat surface on which we draw graphs.

coordinates [13.1] A pair of numbers in a given order with a corresponding location.

●*Found in the online Appendix within MyMathLab.*

corresponding sides [10.5] In similar triangles, the sides opposite the equal angles.

cube [10.4] A solid in which all six faces are squares.

cube root [● App J] The number b is the cube root of a if $b^3 = a$, for any real numbers a and b.

cylinder [10.4] A solid in which the bases are circles and are perpendicular to the height.

decimal [3.1] A number that has a whole number part, which *precedes* the decimal point, and a fractional part, which *follows* the decimal point.

decimal places [3.1] The places to the right of the decimal point.

degree of a monomial [15.3] The power of the variable in the monomial.

degree of a polynomial [15.3] The highest degree of any of the terms of the polynomial.

denominator [2.2] The number below the fraction line in a fraction that stands for the number of parts into which the whole is divided.

descending order of degree [15.3] The terms in a polynomial are written in descending order of degree when the exponents get smaller from left to right.

diameter [10.1] A line segment that passes through the center of a circle and has both endpoints on the circle.

difference [1.2] The result in a subtraction problem.

difference of cubes [App D] An expression in the form $a^3 - b^3$, which can be factored as $(a - b)(a^2 + ab + b^2)$.

difference of squares [16.4] An expression in the form $a^2 - b^2$, which can be factored as $(a + b)(a - b)$.

digits [1.1] The numbers 0, 1, 2, 3, 4, 5, 6, 7, 8, and 9.

direct variation [● App I] Direct variation occurs if a relationship between two variables is described by an equation of the form $y = kx$, where k is a positive constant.

direct variation equation [● App I] The equation $y = kx$.

distance [● App F] For any real numbers a and b, the *distance* between them on a number line is $|a - b|$ or $|b - a|$.

distance formula [● App K] The formula $d = \sqrt{(x_2 - x_1)^2 + (y_2 - y_1)^2}$. In words, this formula states that the distance between two points (x_1, y_1) and (x_2, y_2) on the coordinate plane is the square root of the sum of the squares of the difference of the x-values and the difference of the y-values.

distributive property [1.3, 11.6] This property states that multiplying a factor by the sum of two numbers gives the same result as multiplying the factor by each of the two numbers and then adding. For all real numbers, a, b, and c, $a \cdot (b + c) = a \cdot b + a \cdot c$.

dividend [1.4] In a division problem, the number being divided.

divisor [1.4] In a division problem, the number that is being used to divide another number.

domain of a function [● App G] The set of all values of the independent variable.

element [● App E] An object in a set.

elimination (or addition) method [14.3] A method used to solve a system of equations that is based the addition property of equality states: The equations $a = b$ and $a + c = b + c$ are equivalent. That is, if we add the same number to both sides of an equation, we get an equivalent equation.

empty set [● App E] A set with no elements.

equation [4.2] A mathematical statement that two expressions are equal.

equilateral triangle [10.1] A triangle with three sides equal in length.

equivalent equations [12.1] Equations that have the same solution.

equivalent fractions [2.2] Fractions that represent the same value.

equivalent rational expressions [17.1] Rational expressions that have the same value, no matter what value replaces the variable.

exponent (or power) [1.5, 11.4, 15.1] A number that indicates how many times another number is used as a factor.

exponential notation [1.5, 11.4] A shorthand method for representing repeated multiplication of the same factor.

extraneous solutions [17.5, 18.4] Extraneous solutions are *not* solutions of the original equation.

factored completely [16.2] A polynomial is factored completely when it is expressed as the product of a monomial and one or more prime polynomials.

factoring by grouping [16.1] When trying to factor a polynomial that has four terms, it may be possible to group pairs of terms in such a way that a common binomial factor can be found. This method is called factoring by grouping.

factors [1.3, 2.1] In a multiplication problem, the whole numbers being multiplied.

finite set [● App E] A set whose elements we can count.

FOIL method [15.5] A method for multiplying two binomials: Multiply First terms, Outside terms, Inside terms, and Last terms. Then, combine like terms.

formula [11.5] A formula is an equation that indicates how variables are related to one another.

fraction [2.2] Any number that can be written in the form $\frac{a}{b}$, where a and b are whole numbers and b is nonzero.

fraction line [2.2] A line that separates the numerator from the denominator and stands for "out of" or "divided by."

function [● App G] A relation in which no two ordered pairs have the same first coordinates.

geometry [10.1] The branch of mathematics that deals with concepts such as point, line, angle, perimeter, area, and volume.

gram [9.2] A unit of mass.

graph [8.2, 13.3] A picture or a diagram of a set of data. The graph of an equation consists of all points whose coordinates make the equation true.

graphing method [14.1] A method for solving a linear system of equations in which we graph the equations that make up the system and any point of intersection is a solution of the system.

greatest common factor (GCF) of two of more integers [16.1] The greatest integer that is a factor of each integer.

greatest common factor (GCF) of two or more monomials [16.1] The product of the greatest common factor of the coefficients and for each variable, the variable to the lowest power to which it is raised in any of the monomials.

hexagon [10.1] A polygon with six sides and six angles.

histogram [8.2] The bar graph of a frequency table.

hypotenuse [10.6] In a right triangle, the side opposite the right angle.

index [● App J] In the radical $\sqrt[n]{a}$, n is called the index of the radical.

identity property of addition. See **additive identity property**.

identity property of multiplication. See **multiplicative identity property**.

improper fraction [2.2] A fraction whose numerator is greater than or equal to its denominator.

inequality [12.5] Any mathematical statement containing $<, \leq, \geq, \geq$, or $\neq$.

inequality symbols [11.1] The symbols $\neq, <, \leq, <$, and $\geq$, which are used to compare numbers.

infinite set [● App E] A set whose elements go on without end.

integers [7.1, 11.1] The set of numbers $\{\ldots -4, -3, -3, -1, 0, +1, +2, +3, +4, \ldots\}$ continuing indefinitely in both directions.

intersecting lines [10.1] Two lines that cross.

intersection of sets *A* and *B* [● App E] The set of all elements that are in *both* A and B, written $A \cap B$.

interval notation [● App E] A way of expressing inequalities in one variable using a pair of numbers inside parentheses or brackets to represent the set of all numbers between and sometimes including those two numbers.

inverse variation [● App I] Inverse variation occurs if a relationship between two variables is described by an equation of the form $y = \frac{k}{x}$, where k is a positive constant.

inverse variation equation [● App I] The equation $y = \frac{k}{x}$.

irrational numbers [11.1] Real numbers that cannot be written as the quotient of two integers.

isosceles triangle [10.1] A triangle with two or more sides equal in length.

joint variation [● App I] Joint variation occurs if a relationship among three variables is described by an equation of the form $y = kxz$, where k is a positive constant.

joint variation equation [● App I] The equation $y = kxz$.

leading coefficient [15.3] The coefficient of the leading term in a polynomial.

leading term [15.3] The term in a polynomial with the highest degree.

least common denominator (LCD) [2.2] For two or more fractions, the least common multiple of their denominators.

least common multiple (LCM) [2.1] The smallest nonzero whole number that is a multiple of two or more numbers.

legs of a right triangle [10.6] The two sides that form the right angle.

like fractions [2.2] Fractions with the same denominator.

like quantities [5.1] Quantities that have the same unit.

like radicals [18.2] Radical expressions that have the same radicand.

like terms [11.6] Terms that have the same variables with the same exponents.

line [10.1] A collection of points along a straight path that extends endlessly in both directions. A line has only one dimension.

linear equation in one variable [12.1] An equation that can be written in the form $ax + b = c$, where a, b, and c are real numbers and $a \neq 0$.

linear equation in two variables [13.3] An equation that can be written in the *general form* $Ax + By = C$, where A, B, and C are real numbers and A and B are not both 0.

linear function [● App G] The function $f(x) = mx + b$, where m and b are real numbers.

linear inequality in two variables [13.5] An inequality that can be written in the form $Ax + By < C$, where A, B, and C are real numbers and A and B are not both 0. The inequality symbol can be $<, >, \leq$, or $\geq$.

line graph (or broken-line graph) [8.2] A graph in which points are connected by straight-line segments and the position of any point on the line is read against the vertical axis and the horizontal axis.

line segment [10.1] A part of a line having two endpoints.

liter [9.2] A unit of capacity, that is, liquid volume.

literal equation [12.4] An equation involving two or more variables.

mean (or average) [1.5, 8.1] The sum of a set of numbers divided by however many numbers are in the set.

median [8.1] In a set of numbers arranged in numerical order, the number in the middle. If there are two numbers in the middle, the median is the mean of the two middle numbers.

GLOSSARY

member [● App E] An object in a set.

meter [9.2] A unit of length which gives the metric system its name.

midpoint formula [● App K] The line segment with endpoints (x_1, y_1) and (x_2, y_2) has as its midpoint $\left(\frac{x_1 + x_2}{2}, \frac{y_1 + y_2}{2}\right)$. This formula states that the coordinates of the midpoint are the average of the x-coordinates and the average of the y-coordinates of the endpoints.

minuend [1.2] In a subtraction problem, the number from which we subtract.

mixed number [2.2] A number with a whole number part and a proper fraction part.

mode [8.1] In a set of numbers, the number (or numbers) occurring most frequently.

monomial [15.3] An expression that is the product of a real number and variables raised to nonnegative integer powers.

multiples [2.1] The products of a number and the whole numbers.

multiplication property of equality [12.2] This property states that when multiplying each side of an equation by any nonzero number, the result is an equivalent equation. For any real numbers a, b, and c, $c \neq 0$, $a = b$ and $a \cdot c = b \cdot c$ are equivalent.

multiplication property of 0 [1.3, 11.3] This property states that the product of any number and 0 is 0. For any real number a, $a \cdot 0 = 0$ and $0 \cdot a = 0$.

multiplicative identity property (or **identity property of multiplication) [1.3, 11.3]** This property states that the product of any number and one is that number. For any real number a, $a \cdot 1 = a$ and $1 \cdot a = a$.

multiplicative inverse property [11.3] This property states that the product of a number and its multiplicative inverse is one. For any nonzero real number a, $a \cdot \frac{1}{a} = 1$ and $\frac{1}{a} \cdot a = 1$.

multiplicative inverses (or **reciprocals) [2.4, 11.3]** Two nonzero numbers that have a product of 1.

natural numbers [11.1] The set of numbers $\{1, 2, 3, 4, 5, 6, \ldots\}$.

negative number [7.1] A number less than 0.

negative slope [13.2] On a graph, the slope of a line that falls to the right.

***n*th root [● App J]** The number b is the nth root of a if $b^n = a$, for any real number a and for any positive integer n greater than 1.

numerator [2.2] The number above the fraction line in a fraction that tells us how many parts of the whole the fraction contains.

obtuse angle [10.1] An angle whose measure is more than 90° and less than 180°.

obtuse triangle [10.1] A triangle with one obtuse angle.

opposites [7.1] Two numbers that are the same distance from 0 on the number line but on opposite sides of 0.

ordered pair [13.1] A pair of numbers that represents a point in the coordinate plane.

order of operations [1.5] A rule we agree to follow when a mathematical expression involves more than one mathematical operation so that everyone gets the same value for an answer.

origin [11.1, 13.1] On the number line, the point at 0; in the coordinate plane, the point where the axes intersect, (0, 0).

parabola [19.4] A U-shaped graph of the equation of the form $y = ax^2 + bx + c$ that opens either upward or downward.

parallel lines [10.1, 13.2] Two lines in the same plane that do not intersect. Two nonvertical lines are parallel if and only if their slopes are equal. That is, if the slopes are m_1 and m_2, then $m_1 = m_2$.

parallelogram [10.1] A quadrilateral with both pairs of opposite sides parallel. Opposite sides are equal in length, and opposite angles have equal measures.

percent [6.1] A ratio or a fraction with denominator 100. A number written with the % sign means "divided by 100."

perfect cube (rational number) [● App J] A rational number that is the cube of another rational number.

perfect square (whole number) [10.6, 18.1] A whole number that is the square of a whole number.

perfect square trinomial [16.4] A trinomial that can be factored as the square of a binomial, for example, $(a + b)^2 = a^2 + 2ab + b^2$ and $(a - b)^2 = a^2 - 2ab + b^2$.

perimeter [1.2, 10.2] The distance around a polygon.

periods [1.1] Groups of three digits separated by commas.

perpendicular lines [10.1, 13.2] Two lines that intersect to form right angles. Two nonvertical lines are perpendicular if and only if the product of their slopes is -1. That is, if the slopes are m_1 and m_2, then $m_1 \cdot m_2 = -1$.

pictograph [8.2] A kind of graph in which images like those of people, books, or coins are used to represent and to compare quantities.

plane [10.1] A flat surface that extends endlessly in all directions.

plot [13.1] To plot a point in the coordinate plane, we find its location represented by its ordered pair.

point [10.1] An exact location in space, with no dimension.

point-slope form [13.4] The point-slope form of a linear equation is written as $y - y_1 = m(x - x_1)$, where x_1, y_1 and m are constants. In this form, m is the slope and (x_1, y_1) is a point that lies on the graph of the equation.

polygon [10.1] A closed plane figure made up of line segments.

polynomial [15.3] An algebraic expression with one or more monomials added or subtracted.

GLOSSARY

positive number [7.1] A number greater than 0.

positive slope [13.2] On a graph, the slope of a line that rises to the right.

power (or exponent) [1.5, 11.4. 15.1] A power (or exponent) is a number that indicates how many times another number is used as a factor.

prime factorization [2.1] The process of writing a whole number as a product of its prime factors.

prime number [2.1] A whole number that has exactly two factors: itself and 1.

prime polynomials [16.2] Polynomials that are not factorable.

principal square root [18.1] The square root of a number that is nonnegative.

product [1.1] The result in a multiplication problem.

proper fraction [2.2] A fraction whose numerator is smaller than its denominator.

proportion [5.2] A statement that two ratios are equal.

Pythagorean theorem [10.6, 16.5] This theorem states that for every right triangle, the sum of the squares of the lengths of the two legs equals the square of the length of the hypotenuse: $a^2 + b^2 = c^2$, where a and b are the lengths of the legs and c is the length of the hypotenuse.

quadrant [13.1] One of four regions of a coordinate plane separated by the x- and y-axes.

quadratic equation (second-degree equation) [16.5] An equation that can be written in the form $ax^2 + bx + c = 0$, where a, b, and c are real numbers and $a \neq 0$.

quadratic formula [19.3] This formula states that if $ax^2 + bx + c = 0$, where a, b, and c are real numbers and $a \neq 0$, then $x = \dfrac{-b \pm \sqrt{b^2 - 4ac}}{2a}$.

quadratic function [● App G] A function of the form $f(x) = ax^2 + bx + c$.

quadrilateral [10.1] A polygon with four sides.

quotient [1.4] The result in a division problem.

radical equation [18.4] An equation with a variable in one or more radicands.

radical expression [18.1] An algebraic expression that contains radicals.

radical sign [18.1] The symbol $\sqrt{\ }$.

radicand [18.1] The number under the radical sign.

radius of a circle [10.1, ● App K] A line segment with one endpoint on the circle and the other at the center.

range [8.1] In a set of numbers, the difference between the largest and smallest number.

range of a function [● App G] The set of all values of the dependent variable.

rate [5.1] A ratio of unlike quantities.

rate of change [13.2] An interpretation of **slope**. It indicates how fast a quantity is changing and if the quantity being graphed increases or decreases.

ratio [5.1] A comparison of two quantities expressed as a quotient.

rational equation (fractional equation) [17.5] An equation that contains one or more rational expressions.

rational exponents [● App J] Exponents that are rational numbers.

rational expression [17.1] An algebraic expression that can be written as the quotient of two polynomials, P and Q, where $Q \neq 0$.

rational function [● App G] A function of the form $f(x) = \dfrac{1}{x}$.

rational numbers [11.1] Numbers that can be written in the form $\dfrac{a}{b}$, where a and b are integers and $b \neq 0$.

rationalize the denominator [18.3] To rewrite an expression in an equivalent form that contains no radical in its denominator.

ray [10.1] A part of a line having only one endpoint.

real numbers [11.1] Numbers that can be represented as points on the number line.

reciprocal. See **multiplicative inverse**.

rectangle [10.1] A parallelogram with four right angles.

rectangular solid [10.4] A solid in which all six faces are rectangles.

regression line [13.1] A straight line that is closest to passing through the points on a graph.

relation [● App G] A set of ordered pairs.

right angle [10.1] An angle whose measure is 90°.

right triangle [10.1] A triangle with one right angle.

rounding [1.1] A technique to approximate the exact answer with a number that ends in a given number of zeros.

scalene triangle [10.1] A triangle with no sides equal in length.

scientific notation [15.2] A number is in scientific notation if it is written in the form $a \times 10^n$, where n is an integer and a is greater than or equal to 1 but less than 10 ($1 \leq a < 10$).

second-degree equation (quadratic equation) [16.5] An equation that can be written in the form $ax^2 + bx + c = 0$, where a, b, and c are real numbers and $a \neq 0$.

signed number [7.1] A number with a sign that is either positive or negative.

similar triangles [10.5] Triangles that have the same shape but not necessarily the same size.

simplified (or written in lowest terms) [2.2] A fraction is said to be simplified when the only common factor of its numerator and denominator is 1.

slope [13.2] The ratio of the change in y-values to the change in x-values along a line. The slope m of a line passing through the points (x_1, y_1) and (x_2, y_2) is defined to be $m = \frac{y_2 - y_1}{x_2 - x_1}$, where $x_1 \neq x_2$.

slope-intercept form [13.4] A linear equation is in slope-intercept form if it is written as $y = mx + b$, where m and b are constants. In this form, m is the slope and $(0, b)$ is the y-intercept of the graph of the equation.

solution of an equation [12.1] A value of the variable that makes the equation a true statement.

solution of an equation in two variables [13.3] An ordered pair of numbers that when substituted for the variables makes an equation true.

solution of an inequality in one variable [12.5] Any value of the variable that makes an inequality true.

solution of a system of inequalities in two variables [● App H] An ordered pair of numbers that makes both inequalities in the system true.

solution of a system of two linear equations in two variables [14.1] An ordered pair of numbers that makes both equations in the system true.

solution of an inequality in two variables [13.5] An ordered pair of numbers that when substituted for the variables makes the inequality a true statement.

solve an inequality [12.5] To find all of its solutions.

sphere [10.4] A three-dimensional figure made up of all points a given distance from the center.

square [10.1] A rectangle with four sides equal in length.

square root [10.6, 18.1] The (principle) square root of a number n, written $\sqrt{n}$ is the positive number whose square is n.

square root property of equality [19.1] This property states that if n is a nonnegative number and $x^2 = n$, then $x = \pm\sqrt{n}$, that is, $x = \sqrt{n}$ or $x = -\sqrt{n}$.

straight angle [10.1] An angle whose measure is 180°.

substitution method [14.2] A method for solving a system of equations in which one linear equation is solved for one of the variables and then the result is substituted into the other equation.

subtrahend [1.2] In a subtraction problem, the number being subtracted.

sum [1.2] The result in an addition problem.

sum of cubes [● App D] An expression in the form $a^3 + b^3$, which can be factored as $(a + b)(a^2 - ab + b^2)$.

supplementary angles [10.1] Two angles the sum of whose measures is 180°.

system of equations [14.1] Two or more equations considered simultaneously.

system of inequalities [● App H] Two or more inequalities considered simultaneously.

table [8.2] A rectangular display of data.

term [11.4] A number, a variable, or the product or quotient of numbers and variables.

trapezoid [10.1] A quadrilateral with only one pair of opposite sides parallel.

triangle [10.1] A polygon with three sides.

trinomial [15.3] A polynomial with three terms.

union of sets A and B [● App E] The set of all elements that are in either A or B or both, written $A \cup B$.

unit fraction [17.3] A fraction whose numerator is 1.

unit price [5.1] The price of one item or one unit.

unit rate [5.1] A rate in which the number in the denominator is 1.

unlike fractions [2.2] Fractions with different denominators.

unlike quantities [5.1] Quantities that have different units or are different kinds of measurement.

unlike radicals [18.2] Radical expressions with different radicands.

unlike terms [11.6] Terms that do not have the same variables with the same exponents.

variable [4.1] A letter that represents an unknown number.

varies directly (directly proportional) [● App I] In a relationship between two variables, y varies directly as x if there is a positive constant k such that $y = kx$.

varies inversely (inversely proportional) [● App I] In a relationship between two variables, y varies inversely as x if there is a positive constant k such that $y = \frac{k}{x}$.

varies jointly (jointly proportional) [● App I] In a relationship among three variables, y varies jointly as x and z, if there is a positive constant k such that $y = kxz$.

vertex [19.4] The highest or lowest point of a parabola.

vertex of an angle [10.1] The common endpoint of the two rays that form an angle.

vertical angles [10.1] Two opposite angles with equal measure formed by two intersecting lines.

vertical line test [● App G] A test that states that if any vertical line intersects a graph at more than one point, then the graph does not represent a function. If no such line exists, then the graph represents a function.

volume [10.4] The number of cubic units required to fill a three-dimensional figure.

whole numbers [11.1] The set of numbers $\{0, 1, 2, 3, 4, 5, \ldots\}$.

written in lowest terms (or simplified) [2.2] A fraction is said to be written in lowest terms when the only common factor of its numerator and denominator is 1.

x-axis [13.1] The horizontal number line in the coordinate plane.

GLOSSARY

x-coordinate [13.1] The first number in an ordered pair that represents a horizontal distance in the coordinate plane.

x-intercept [13.3] The x-intercept of a line is the point where the graph crosses the x-axis.

y-axis [13.1] The vertical number line in the coordinate plane.

y-coordinate [13.1] The second number in an ordered pair that represents a vertical distance in the coordinate plane.

y-intercept [13.3] The y-intercept of a line is the point where the graph crosses the y-axis.

zero-product property [16.5] This property states that if the product of two factors is zero, then either one or both of the factors must be zero. For any real numbers a and b, if $a \cdot b = 0$, then $a = 0$, or $b = 0$, or both a and $b = 0$.

Index

●Found in the online Appendices within MyMathLab.

D

E

F

G

Q

R

S

INDEX

T

U

V

W

X

Y

Z

INDEX

U.S. Customary Units

Length	Weight	Capacity	Time
12 in. = 1 ft	16 oz = 1 lb	16 fl oz = 1 pt	60 sec = 1 min
3 ft = 1 yd	2,000 lb = 1 ton	2 pt = 1 qt	60 min = 1 hr
5,280 ft = 1 mi		4 qt = 1 gal	24 hr = 1 day
			7 days = 1 wk
			52 wk = 1 yr
			12 mo = 1 yr
			365 days = 1 yr

Metric Units

Length	Weight	Capacity
1,000 mm = 1 m	1,000 mg = 1 g	1,000 mL = 1 L
100 cm = 1 m	1,000 g = 1 kg	1,000 L = 1 kL
1,000 m = 1 km		

Key U.S./Metric Conversions

Length	Weight	Capacity
1 in. ≈ 2.5 cm	1 oz ≈ 28 g	1 pt ≈ 470 mL
1 ft ≈ 30 cm	1 lb ≈ 450 g	1.1 qt ≈ 1 L
39 in. ≈ 1 m	2.2 lb ≈ 1 kg	1 gal ≈ 3.8 L
3.3 ft ≈ 1 m	1 ton ≈ 910 kg	
1 mi ≈ 1,600 m		
1 mi ≈ 1.6 km		